MASTER TRANSISTOR/IC SUBSTITUTION HANDBOOK

TAB EDITORIAL STAFF

TAB BOOKS
Blue Ridge Summit, Pa. 17214

FIRST EDITION

FIRST PRINTING—JANUARY 1977

Copyright © 1977 by TAB BOOKS

Printed in the United States
of America

Reproduction or publication of the content in any manner, without express permission of the publisher, is prohibited. No liability is assumed with respect to the use of the information herein.

Hardbound Edition: International Standard Book No. 0-8306-6970-1

Paperbound Edition: International Standard Book No. 0-8306-5970-6

Library of Congress Card Number: 76-45065

Library of Congress Cataloging in Publication Data

Tab Books.
Master transistor/IC substitution handbook.

1. Integrated circuits--Handbooks, manuals, etc. 2. Transistor circuits--Handbooks, manuals, etc. I. Title.
TK7874.T3 1976 621.381'73'0422 76-45065
ISBN 0-8306-6970-1
ISBN 0-8306-5970-6 pbk.

PREFACE

Over the years it has been a constant challenge to keep up with the never-ending flood of semiconductor devices on the consumer market. Simply knowing the equipment manufacturer's part number has not been enough to identify the device since a universal standardization of semiconductors has yet to be realized.

In the past few years, leading semiconductor suppliers—Sylvania, RCA, General Electric, International Rectifier, Workman, and Motorola—have individually introduced a "general replacement" line of semiconductors for wide ranges of applications that meet the needs of electronic service personnel, hobbyists, experimenters, hams, and engineers. It is to these wide-range semiconductor lines that we have cross-referenced thousands upon thousands of specialized in-house part numbers.

This cross-reference guide is composed of two parts:

- *Part I* is composed of six chapters which include transistor and IC base diagrams and outline drawings. Each chapter is devoted to a specific supplier.
- *Part II* is the actual cross-reference material. Indexed by the original part number, this list covers over 80,000 transistors and ICs, and will abbreviate your replacement problem in most cases as to the time it takes to find the appropriate entry.

Putting this Master Transistor/IC Cross-Reference guide together would have been impossible without the fine cooperation of the semiconductor suppliers. Our thanks goes to GTE Sylvania Electronic Component Group, RCA Distributor and Special Products Division, General Electric Tube Products Department, Workman Electronic Products, Inc. (International Rectifier devices are now supplied through Workman), and Motorola Semiconductor Products, Inc. In addition, special thanks must go to our two excellent typists—Ellen McCaffrey and Darlene Bellant—for the mammoth task of putting *Part II* on paper.

Mike Fair
Service Editor
TAB Editorial Staff

CONTENTS

PART 1

CHAPTER 1

SYLVANIA REPLACEMENT TRANSISTOR BASE DIAGRAMS

Outline diagrams for the Sylvania transistors that appear in the ECG column in *Part II* are shown on the following pages. To locate the proper diagram, refer to the listings which follow. ECG numbers are listed in the first column. Skim through this column until the appropriate number is found. Adjacent to the ECG number—in the next column—is the corresponding figure number. That is, for ECG100 the correct illustration is Fig. 1-1. Now turn to the page where this figure appears. Each of the pages containing illustrations is headed with the first and last figure number on that particular page.

ECG	Fig. 1-	APPLICATIONS
100	1	RF AMPL, OSC, MIX, IF AMPL
101	1	RF AMPL, OSC, MIX, IF AMPL
102	2	AF DRIVER, PREAMPL, POWER OUTPUT
102A	2	AF DRIVER, PREAMPL, POWER OUTPUT
103	1	AF DRIVER, PREAMPL, POWER OUTPUT
103A	2	AF DRIVER, PREAMPL, POWER OUTPUT
104	3	AF OUTPUT
105	4	AF OUTPUT
106	5	RF AMPL, OSC, MIX, IF AMPL
107	6	RF AMPL, OSC, MIX, IF AMPL FOR VHF
108	7	RF AMPL, OSC, MIX, IF AMPL, VIDEO FOR VHF/UHF
121	3	AF OUTPUT
123	8	AF PREAMPL, DRIVER, VIDEO AMPL
123A	5	AF/RF AMPL
124	9	HV AF OUTPUT
126	11	RF AMPL, OSC, MIX, IF AMPL
127	3	AF OUTPUT, HORZ/VERT DEFLECT AMPL
128	1	AF PREAMPL, DRIVER, VIDEO AMPL
129	1	AF PREAMPL, DRIVER, VIDEO AMPL
130	3	AF AMPL
131	10	AF OUTPUT
132	12	RF AMPL, MIX
133	12	GP AF AMPL, SW
152	13	AF OUTPUT
153	13	AF OUTPUT
154	14	TV VIDEO OUTPUT AMPL
155	10	AF AMPL
157	15	HV AF AMPL
158	16	AF AMPL
159	17	AF PREAMPL, DRIVER
160	18	RF AMPL, OSC, MIX, IF AMPL, FOR FM, TV
161	19	VIDEO IF AMPL
162	3	VERT DEFLECT
163	3	HORZ DEFLECT
164	3	VERT DEFLECT
165	3	HORZ DEFLECT
171	21	AF/VIDEO AMPL
172	20	DARLINGTON AMPL
175	9	AF AMPL
176	14	AF AMPL
179	3	AF AMPL
180	3	AF AMPL
181	3	AF AMPL
182	22	AF POWER AMPL
183	22	AF POWER AMPL
184	15	AF POWER AMPL
185	15	AF POWER AMPL
186	21	AF POWER AMPL
187	21	AF POWER AMPL
188	23	AF DRIVER, POWER AMPL

ECG	Fig. 1-	APPLICATIONS
189	23	AF DRIVER, POWER AMPL
190	23	AF POWER AMPL, HORZ DRIVER
191	23	HV AF AMPL, VIDEO AMPL
192	24	AF POWER OUTPUT
193	24	AF POWER OUTPUT
194	25	GP HV AMPL
195	14	CB RF POWER AMPL
195A	14	CB RF POWER AMPL, DRIVER
196	13	AF POWER OUTPUT
197	13	AF POWER OUTPUT
198	26	HV AF/SW
199	6	LOW-NOISE, HIGH-GAIN PREAMPL
210	21	AF OUTPUT, SW
211	21	AF OUTPUT, SW
213	4	HP HIGH-CURRENT OUTPUT
218	9	AF POWER OUTPUT
219	3	AF OUTPUT, SW
220	30	LOW-NOISE RF AMPL, MIX, IF
221	30	DUAL GATE RF, IF AMPL, MIX, TV/FM
222	30	DUAL GATE, LOW-NOISE RF, IF AMPL, MIX
223	27	AF OUTPUT, SW
224	28	CB RF POWER AMPL
225	28	AF, VIDEO, SW
226	29	AF POWER OUTPUT
228	35	AF/VIDEO OUTPUT, HIGH-SPEED SW
229	25	VHF OSC, MIX, TV IF AMPL
232	25	DARLINGTON AMPL
233	25	FINAL VIDEO IF
234	25	LOW-NOISE, HIGH-GAIN AF PREAMPL
235	37	HP HF OUTPUT
236	37	HP HF OUTPUT
237	36	CB RF POWER OUTPUT
238	3	HORZ OUTPUT
239	5A	VERT OSC, SW, TV
241	34	AF POWER, SW
242	34	AF POWER, SW
243	3	DARLINGTON AMPL
244	3	DARLINGTON AMPL
245	3	DARLINGTON AMPL
246	3	DARLINGTON AMPL
247	3	DARLINGTON AMPL
248	3	DARLINGTON AMPL
249	3	DARLINGTON AMPL
250	3	DARLINGTON AMPL
251	3	DARLINGTON AMPL
252	3	DARLINGTON AMPL
253	31	DARLINGTON AMPL
254	31	DARLINGTON AMPL

ECG	Fig. 1-	APPLICATIONS
257	32	DARLINGTON AMPL
258	32	DARLINGTON AMPL
259	32	DARLINGTON AMPL
260	32	DARLINGTON AMPL
261	13	DARLINGTON AMPL
262	13	DARLINGTON AMPL
263	13	DARLINGTON AMPL
264	13	DARLINGTON AMPL
265	21	DARLINGTON AMPL
266	21	DARLINGTON AMPL
267	21	DARLINGTON AMPL
268	21	DARLINGTON AMPL
269	21	DARLINGTON AMPL
270	33	DARLINGTON AMPL
271	33	DARLINGTON AMPL
272	23	DARLINGTON AMPL
273	23	DARLINGTON AMPL
274	9	DARLINGTON AMPL
275	9	DARLINGTON AMPL
276	46A	HORZ OUTPUT, TV
277	46	TV VERT OUTPUT
278	14	RF AMPL, AM/FM, CATV, MATV
279	8A	HORZ OUTPUT, TV
280	3	AF POWER AMPL
281	3	AF POWER AMPL
282	14	RF POWER AMPL, SW
283	3	HIGH-CURRENT SW, TV HORZ OUTPUT
284	3	AF POWER AMPL
285	3	AF POWER AMPL
286	9	POWER AMPL, SW, TV HORZ OUTPUT

ECG	Fig. 1-	APPLICATIONS
287	25	HV GP AMPL
288	25	HV GP AMPL
289	47	AF POWER AMPL
290	47	AF POWER AMPL
291	13	POWER AMPL, SW
292	13	POWER AMPL, SW
293	38	AF POWER AMPL
294	38	AF POWER AMPL
295	9	RF OUTPUT, DRIVER, CB
297	8	AF DRIVER, POWER AMPL
298	8	AF DRIVER, POWER AMPL
299	10	RF DRIVER, POWER AMPL
300	10	AF POWER OUTPUT
302	11	RF DRIVER, POWER AMPL, CB
306	11	RF DRIVER, POWER AMPL, CB
307	10	AF POWER OUTPUT
311	1	VHF/UHF OSC, AMPL, DRIVER
312	14	UHF/VHF AMPL, MIX, FM/TV
313	12	VHF TUNER, RF AMPL
314	3A	POWER REGULATOR, SW, TV
315	13	RF DRIVER, CB
6400	1A	
6401	5B	
6402	6A	
6409	5B	

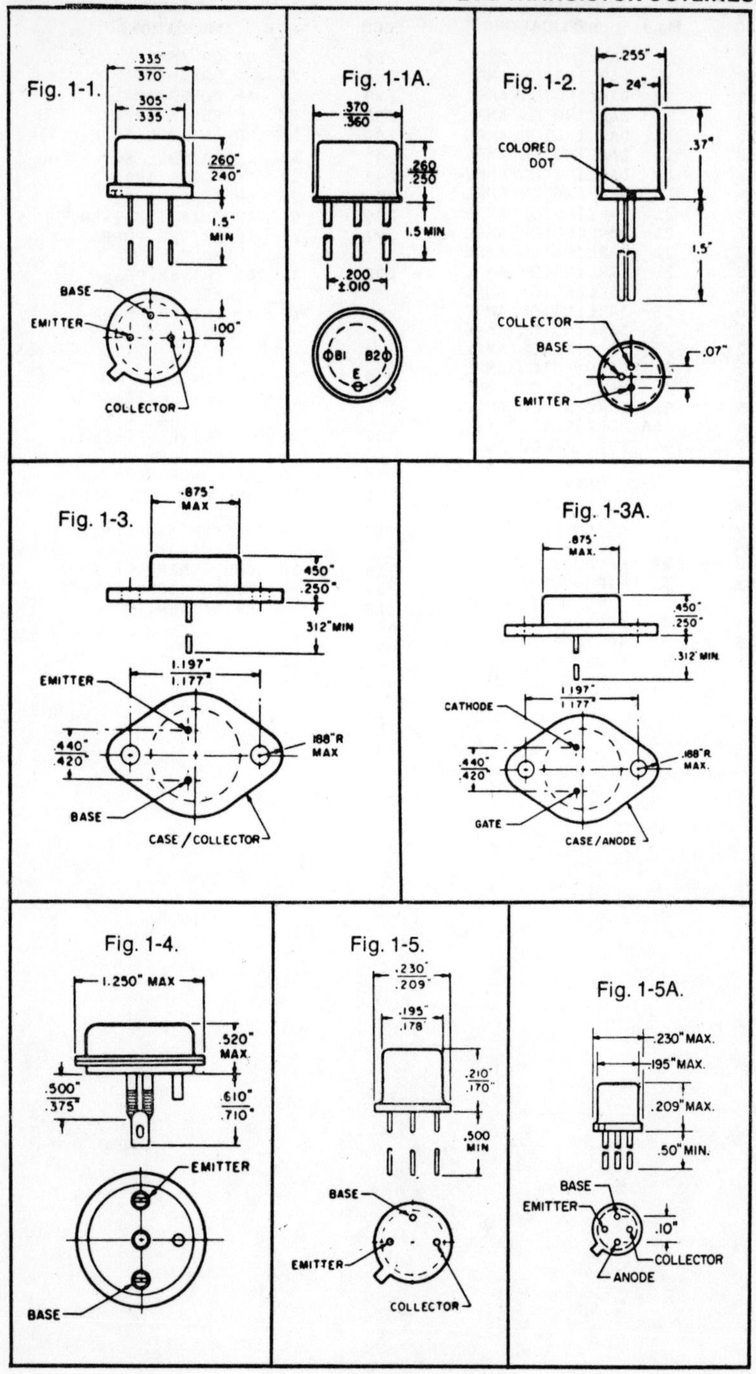
Fig. 1-1.
.335"
.370"
.305"
.335"
.260"
.240"
1.5"
MIN
BASE
EMITTER
100"
COLLECTOR
Fig. 1-1A.
.370
.360
.260
.250
1.5 MIN
.200
±.010
B1
B2
E
Fig. 1-2.
.255"
.24"
COLORED
DOT
.37"
1.5"
COLLECTOR
BASE
.07"
EMITTER
Fig. 1-3.
.875"
MAX
.450"
.250"
.312" MIN
1.197"
1.177"
EMITTER
.440"
.420
.188" R
MAX
BASE
CASE / COLLECTOR
Fig. 1-3A.
.875"
MAX.
.450"
.250"
.312" MIN.
1.197"
1.177"
CATHODE
.440"
.420"
.188" R
MAX.
GATE
CASE / ANODE
Fig. 1-4.
1.250" MAX
.520"
MAX.
.500"
.375"
.610"
.710"
EMITTER
BASE
Fig. 1-5.
.230"
.209"
.195"
.178"
.210"
.170"
.500
MIN
BASE
EMITTER
COLLECTOR
Fig. 1-5A.
.230" MAX.
.195" MAX.
.209" MAX.
.50" MIN.
BASE
EMITTER
.10"
COLLECTOR
ANODE

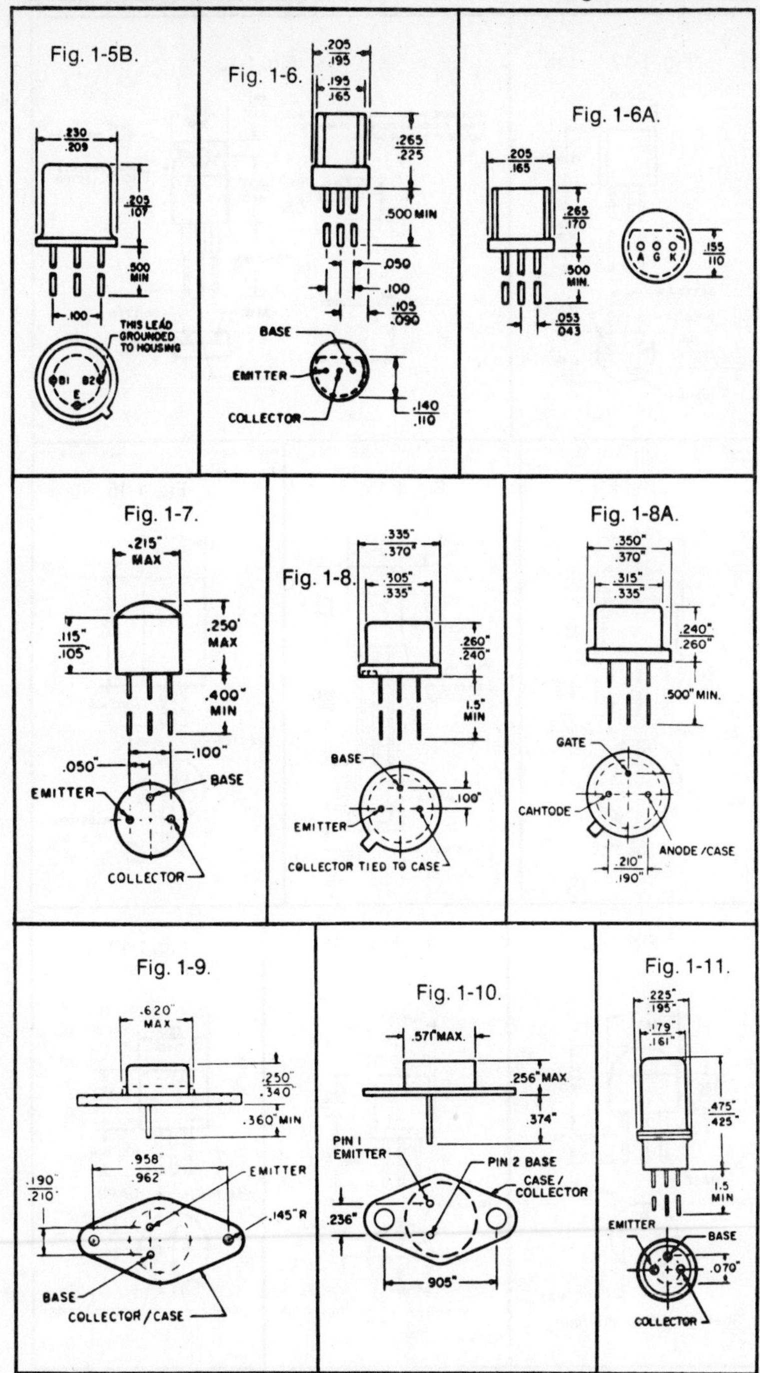
Fig. 1-5B.
.230
.209
.205
.107
.500
MIN
.100
THIS LEAD
GROUNDED
TO HOUSING
B1
B2
E
Fig. 1-6.
.205
.195
.195
.165
.265
.225
.500 MIN
.050
.100
.105
.090
BASE
EMITTER
COLLECTOR
.140
.110
Fig. 1-6A.
.205
.165
.265
.170
.500
MIN.
.053
.043
A G K
.155
.110
Fig. 1-7.
.215"
MAX
.115"
.105"
.250'
MAX
.400"
MIN
.050"
.100"
EMITTER
BASE
COLLECTOR
Fig. 1-8.
.335"
.370"
.305"
.335"
.260"
.240"
1.5"
MIN
BASE
.100"
EMITTER
COLLECTOR TIED TO CASE
Fig. 1-8A.
.350"
.370"
.315"
.335"
.240"
.260"
.500" MIN.
GATE
CAHTODE
ANODE /CASE
.210"
.190"
Fig. 1-9.
.620"
MAX
.250"
.340"
.360" MIN
.958"
.962"
EMITTER
.190"
.210"
.145" R
BASE
COLLECTOR / CASE
Fig. 1-10.
.571"MAX.
.256" MAX
.374"
PIN 1
EMITTER
PIN 2 BASE
CASE /
COLLECTOR
.236"
905"
Fig. 1-11.
.225"
.195"
.179"
.161"
.475"
.425"
1.5
MIN
EMITTER
BASE
.070"
COLLECTOR

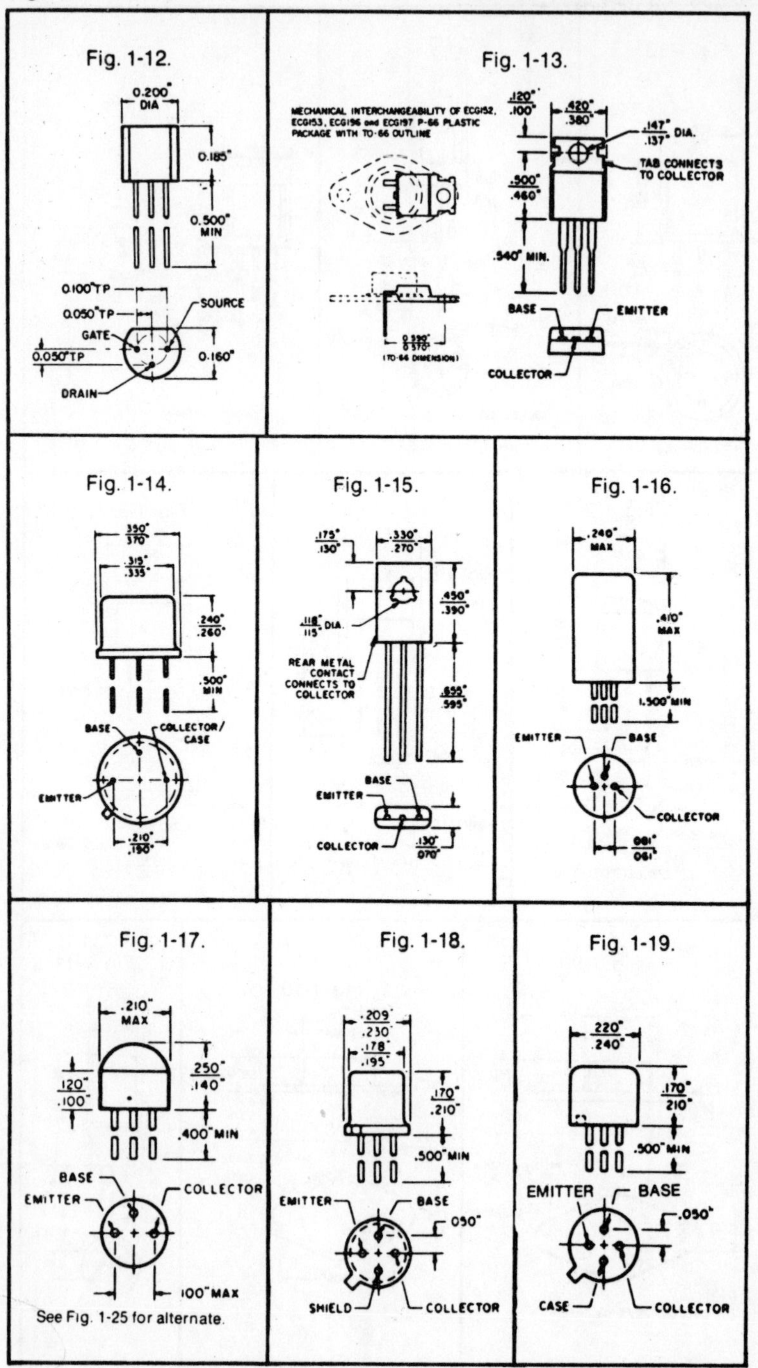

See Fig. 1-25 for alternate.

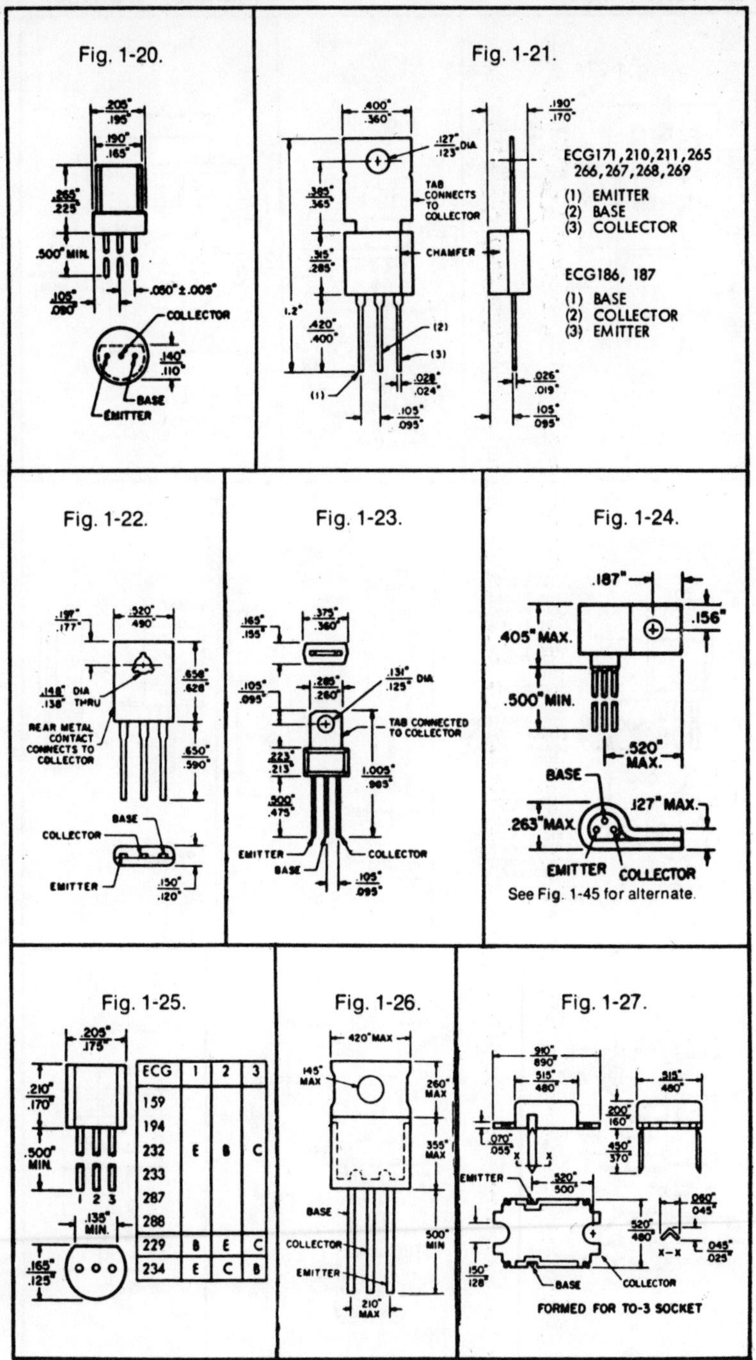

ECG	1	2	3
159			
194			
232	E	B	C
233			
287			
288			
229	B	E	C
234	E	C	B

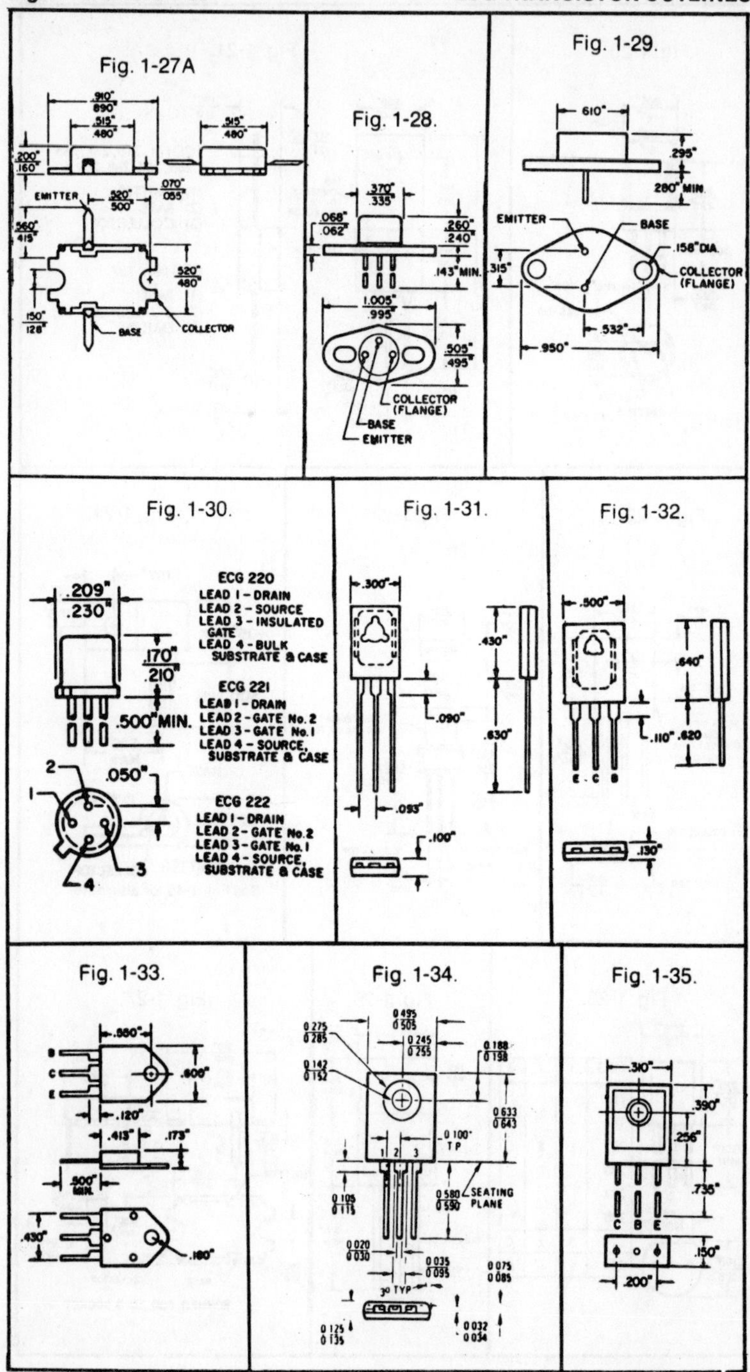
Fig. 1-27A.
EMITTER
BASE
COLLECTOR
Fig. 1-28.
COLLECTOR (FLANGE)
BASE
EMITTER
Fig. 1-29.
EMITTER
BASE
.158" DIA.
COLLECTOR (FLANGE)
Fig. 1-30.
ECG 220
LEAD 1 - DRAIN
LEAD 2 - SOURCE
LEAD 3 - INSULATED GATE
LEAD 4 - BULK SUBSTRATE & CASE
ECG 221
LEAD 1 - DRAIN
LEAD 2 - GATE No. 2
LEAD 3 - GATE No. 1
LEAD 4 - SOURCE, SUBSTRATE & CASE
ECG 222
LEAD 1 - DRAIN
LEAD 2 - GATE No. 2
LEAD 3 - GATE No. 1
LEAD 4 - SOURCE, SUBSTRATE & CASE
.500" MIN.
Fig. 1-31.
Fig. 1-32.
E C B
Fig. 1-33.
B
C
E
Fig. 1-34.
SEATING PLANE
3° TYP
Fig. 1-35.
C B E

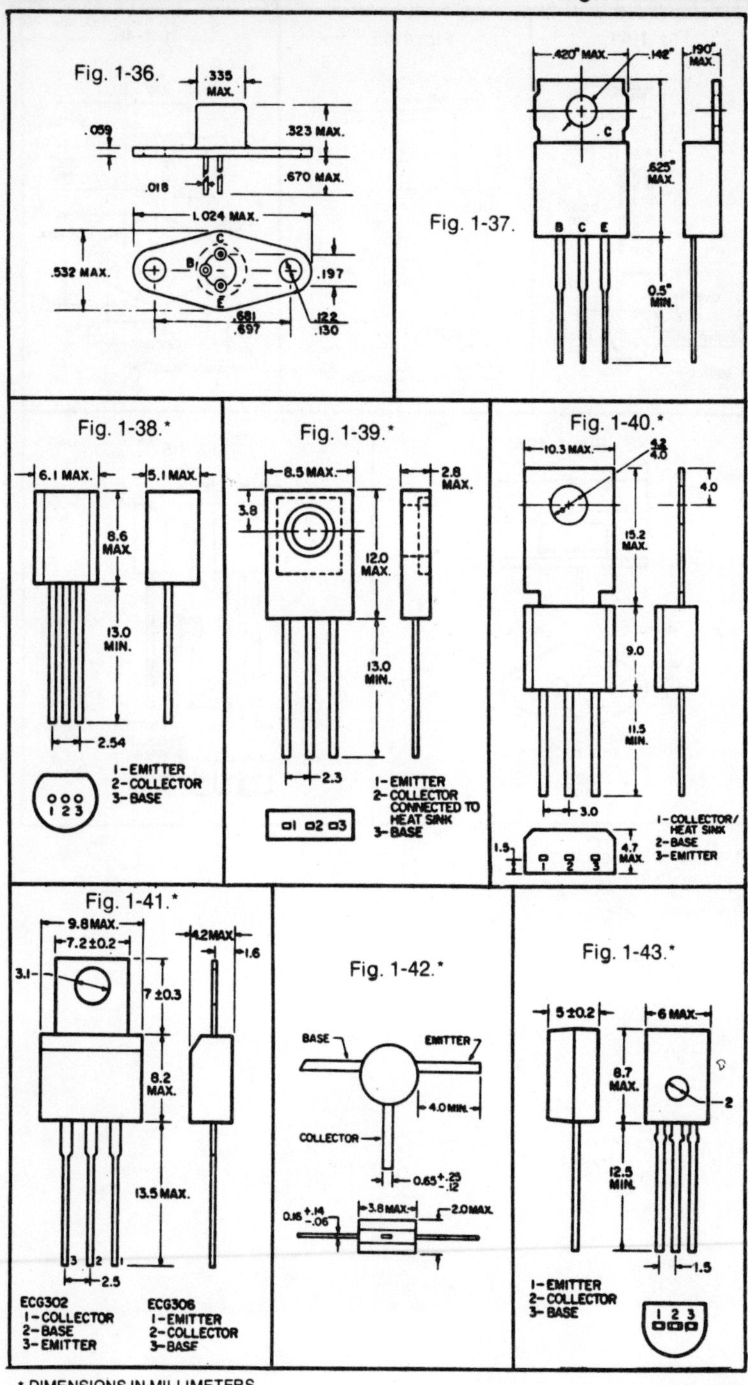

* DIMENSIONS IN MILLIMETERS

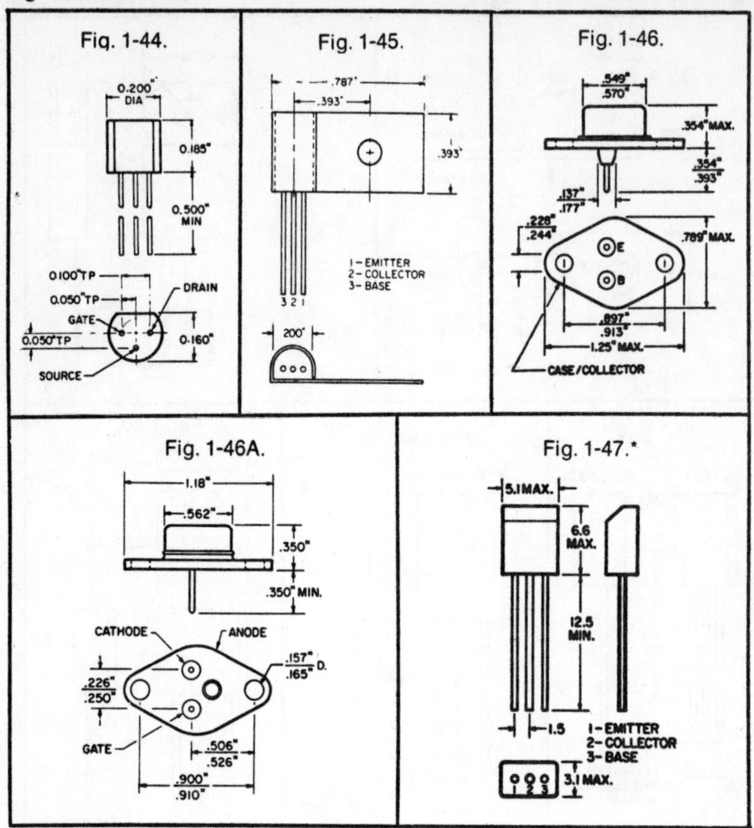

DIMENSIONS IN MILLIMETERS

CHAPTER 2
RCA REPLACEMENT TRANSISTOR BASE DIAGRAMS

Outline diagrams for the RCA transistors that appear in the SK column in *Part II* are shown on the following pages. To locate the proper diagram, refer to the listings which follow. SK numbers are listed in the first column. Skim through this column until the appropriate number is found. Just to the right of the SK number—in the next column—is the corresponding figure number. That is, for SK3003 the correct illustration is Fig. 2-1. Now turn to the page where this figure appears. Each page containing illustrations is headed at the top with the first and last figure number on that page.

SK	Fig. 2-	APPLICATIONS
3003	1	AF AMPL, DRIVER, LOW-LEVEL OUTPUT
3004	1	AF AMPL, DRIVER, LOW-LEVEL OUTPUT
3005	22	AM RF, IF AMPL
3006	24	FM RF, IF AMPL, TV SOUND IF AMPL
3007	22	RF, IF AMPL, AM/HF
3008	22	RF, IF AMPL, AM
3009	10	AF OUTPUT
3010	1	AF AMPL, DRIVER, LOW-LEVEL OUTPUT
3011	22	RF, IF AMPL, AM
3012	14	HF AF OUTPUT
3013	10	MATCHED AF OUTPUTS, SK3009
3014	10	HP AF OUTPUT
3015	10	MATCHED HP AF OUTPUTS, SK3014
3018	24	RF, IF AMPL, AM/FM/ TV, VIDEO IF
3019	24	TV UHF OSC
3020	2	MP AF DRIVER, OUTPUT
3021	31	AF OUTPUT @ 120V
3024	2	MP AF DRIVER, OUTPUT
3025	2	MP AF DRIVER, OUTPUT
3026	13	HP AF OUTPUT
3027	10	HP AF OUTPUT
3028	13	MATCHED HP AF OUTPUTS, SK3026
3029	10	MATCHED HP AF OUTPUTS, SK3027
3034	30	TV, VERT OUTPUT, HORZ DRIVER
3035	30	TV HORZ OUTPUT
3036	10	HP AF OUTPUT
3037	10	MATCHED HP AF OUTPUTS, SK3036
3038	2	AF AMPL, DRIVER LOW-LEVEL OUTPUT
3039	24	TV TUNER, UHF/VHF
3040	32	TV VIDEO AMPL
3041	9	HP AF OUTPUT
3044	32	HV REGULATOR, TV CHROMA, AGC AMPL
3045	3	HV AF OUTPUT, HV REGULATOR, VIDEO AMPL
3046	27	CB OSC, RF
3047	27	CB DRIVER, RF
3048	27	CB OUTPUT, RF
3049	26	CB OUTPUT, RF
3050	12	40 TV, VHF TUNER
3052	-	MP AF OUTPUT
3053	27	MP AF, DRIVER, OUTPUT
3054	9	AF OUTPUT, VERT DEFLECT
3065	13	40 RF, IF VHF RADIO
3079	30	VERT DEFLECT
3082	12	MP AF OUTPUT
3083	9	AF OUTPUT, VERT DEFLECT
3084	9	AF HP OUTPUT
3085	31	TV VERT OUTPUT DEFLECT
3086	12	MATCHED AF OUTPUTS SK3082
3103	29	AF OUTPUT @120V
3104	29	AF OUTPUT, HORZ DRIVER/OUTPUT, CHROMA AMPL
3111	30	TV HORZ OUTPUT, DEFLECT
3112	2	41 AF PREAMPL
3114	5	AF AMPL, DRIVER, LOW-LEVEL OUTPUT
3115	30	TV HORZ OUTPUT, DEFLECT
3116	7.5	39 FM RF, IF AMPL
3117	24	RF, VIDEO IF AMPL
3118	25	RF, IF AMPL, AM/FM /TV
3122	6	RF AMPL, AM/FM/HF

SK	Fig. 2-	APPLICATIONS
3123	8	AF OUTPUT
3124	5	AF AMPL, DRIVER, LOW-LEVEL OUTPUT
3131	31	AF OUTPUT, HV, VERT DEFLECT
3132	24	RF, VIDEO IF AMPL
3133	30	TV VERT OUTPUT, DEFLECT
3137	7	AF OUTPUT, DRIVER
3138	7	AF OUTPUT, MP DRIVER
3156	6	PREAMPL, LO-LEVEL DRIVER
3173	10	HP AF OUTPUT
3176	38	RF OUTPUT, TYP POWER 10W @ 156M
3177	38	RF OUTPUT, TYP POWER 25W @ 156M
3178	11	AF MP
3179	11	AF MP
3180	9	AF, HP/HV, DARL
3181	9	AF, HP, HV/POWER SW, DARLINGTON
3182	10	AF, HV/HP, DARL
3183	10	AF, HP, HV/POWER SW, DARLINGTON
3187	25	40 TV, VHF TUNER
3188	16	AF, HP OUTPUT
3189	16	AF, HP OUTPUT
3190	17	AF, HP OUTPUT
3191	17	AF, HP OUTPUT
3192	15	AF, HP OUTPUT
3193	15	AF, HP OUTPUT
3194	31	TV HORZ OUTPUT, DEFLECT
3195	32	RF OUTPUT, TYP POWER 1W @ 400M
3197	28	CB OUTPUT, RF
3198	10	AF OUTPUT, MP DRIVER
3199	19	AF OUTPUT, MP DRIVER
3200	19	AF OUTPUT, MP DRIVER
3201	20	AF OUTPUT, MP DRIVER
3202	20	AF OUTPUT, MP DRIVER
3203	20	AF OUTPUT, MP DRIVER
3510	30	AF POWER OUTPUT, HV/POWER SW
3511	30	AF POWER OUTPUT, HV/POWER SW
3512	2	AF MP OUTPUT, HV/ POWER SW
3513	2	AF MP OUTPUT, HV/ POWER SW
3528	27	AF POWER OUTPUT, HV/POWER SW
3529	32	HV, POWER SW
3530	33	AF HV OUTPUT, POWER SW
3531	6	39 CHOPPER MPX
3534	34	AF OUTPUT, POWER SW
3535	35	AF OUTPUT, POWER SW
3536	37	AF MP OUTPUT, HV/SW
3537	37	AF MP OUTPUT, HV/SW
3538	36	AF POWFR OUTPUT, HV/SW
3559	35	AF HV OUTPUT, HV POWER SW
3560	35	AF HV OUTPUT, HV POWER SW
3561	35	HV AF OUTPUT/ REGULATOR, SW
3562	36	HV AF OUTPUT/ REGULATOR, POWER SW
3563	35	AF HV OUTPUT, POWER SW

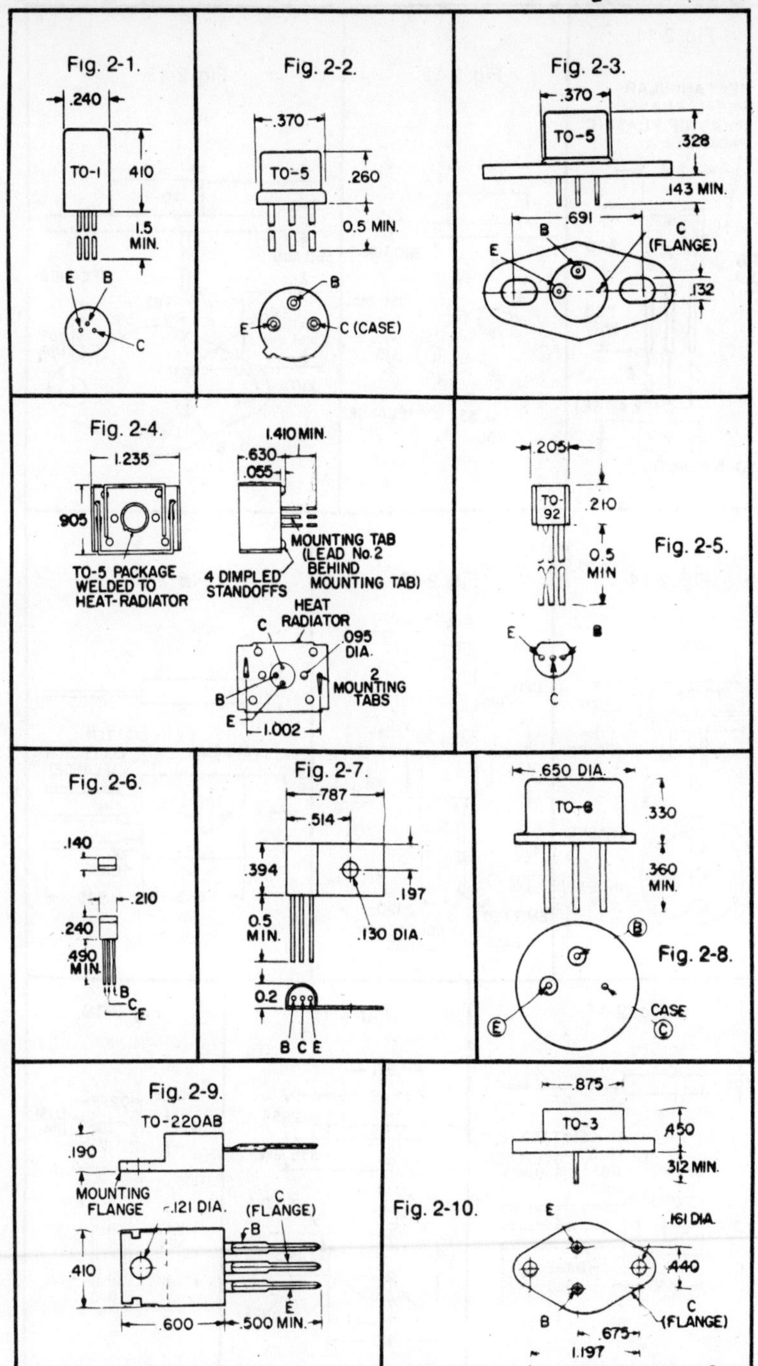
Fig. 2-1.
.240
TO-1
410
1.5
MIN.
E
B
C
Fig. 2-2.
.370
TO-5
.260
0.5 MIN.
B
E
C (CASE)
Fig. 2-3.
.370
TO-5
.328
.143 MIN.
.691
B
C
(FLANGE)
E
.132
Fig. 2-4.
1.235
905
TO-5 PACKAGE
WELDED TO
HEAT-RADIATOR
1.410 MIN.
.630
.055
MOUNTING TAB
(LEAD No. 2
BEHIND
MOUNTING TAB)
4 DIMPLED
STANDOFFS
HEAT
RADIATOR
C
.095
DIA.
2
MOUNTING
TABS
B
E
1.002
.205
TO-
92
.210
0.5
MIN
Fig. 2-5.
E
B
C
Fig. 2-6.
.140
.210
.240
.490
MIN
B
C
E
Fig. 2-7.
.787
.514
.394
.197
0.5
MIN.
.130 DIA.
0.2
B C E
.650 DIA.
TO-8
.330
.360
MIN.
B
Fig. 2-8.
E
CASE
C
Fig. 2-9.
TO-220AB
.190
MOUNTING
FLANGE
.121 DIA.
C
(FLANGE)
B
.410
E
.600
.500 MIN.
.875
TO-3
.450
.312 MIN.
Fig. 2-10.
E
.161 DIA.
.440
C
(FLANGE)
B
.675
1.197

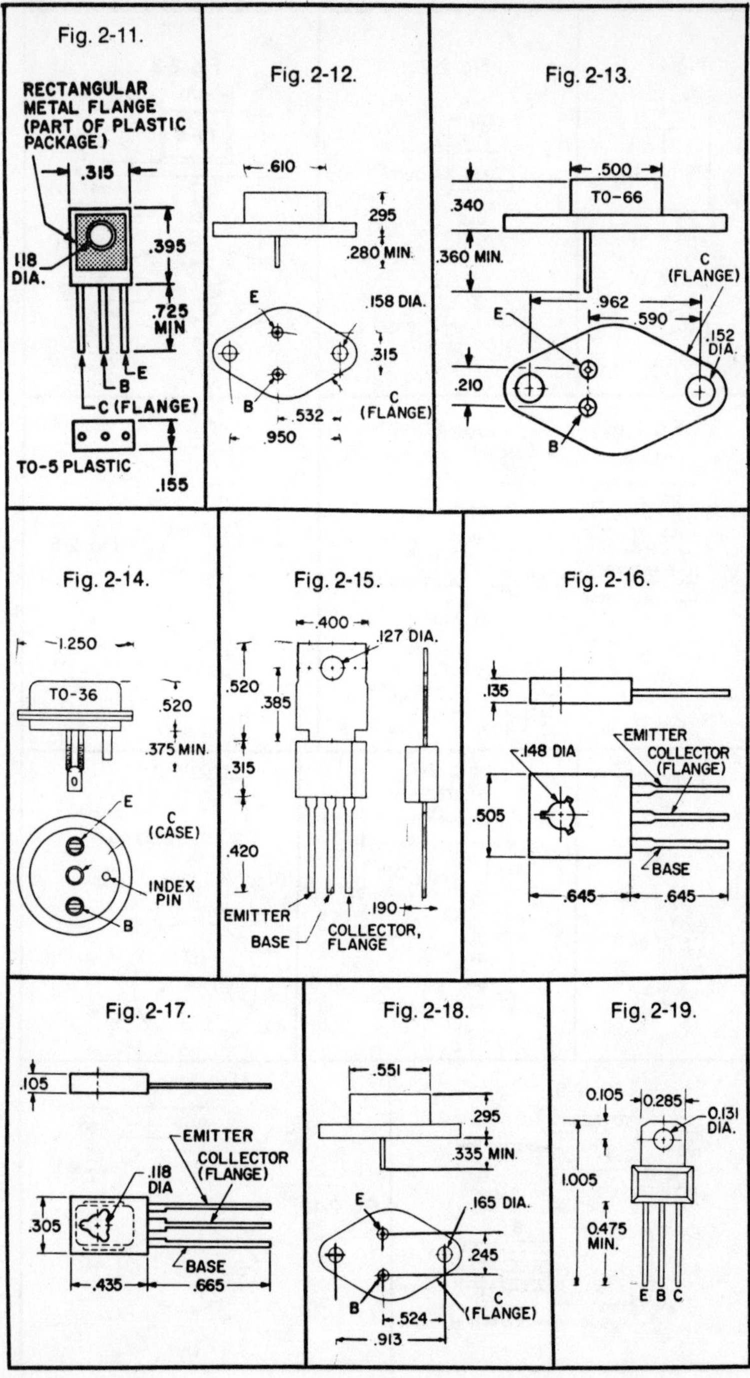

Fig. 2-11.

Fig. 2-12.

Fig. 2-13.

Fig. 2-14.

Fig. 2-15.

Fig. 2-16.

Fig. 2-17.

Fig. 2-18.

Fig. 2-19.

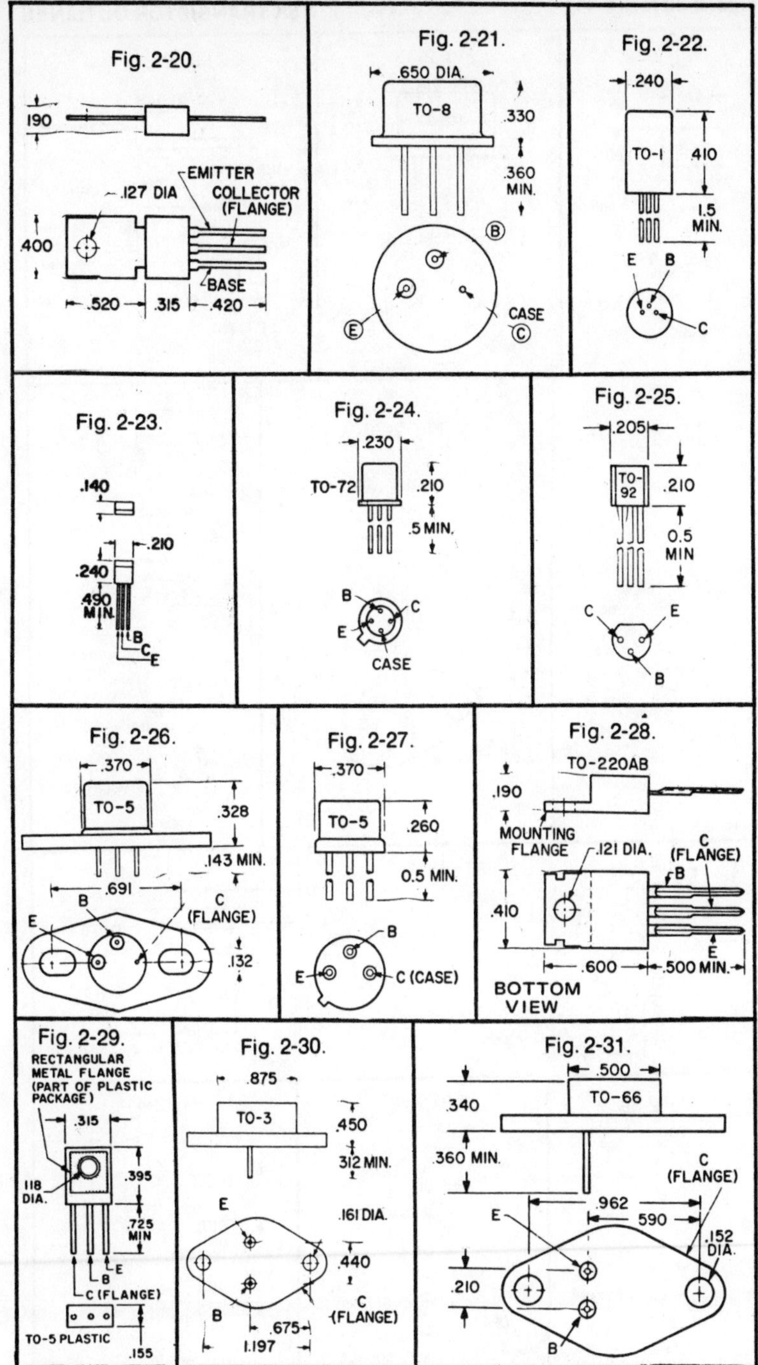
Fig. 2-20.
190
.127 DIA
EMITTER
COLLECTOR
(FLANGE)
.400
BASE
.520
.315
.420
Fig. 2-21.
.650 DIA.
TO-8
.330
.360
MIN.
B
E
CASE
C
Fig. 2-22.
.240
TO-1
410
1.5
MIN.
E
B
C
Fig. 2-23.
.140
.210
.240
.490
MIN.
B
C
E
Fig. 2-24.
.230
TO-72
.210
.5 MIN.
B
C
E
CASE
Fig. 2-25.
.205
TO-
92
.210
0.5
MIN
C
E
B
Fig. 2-26.
.370
TO-5
.328
.143 MIN.
.691
B
C
(FLANGE)
E
.132
Fig. 2-27.
.370
TO-5
.260
0.5 MIN.
B
E
C (CASE)
Fig. 2-28.
TO-220AB
.190
MOUNTING
FLANGE
.121 DIA.
C
(FLANGE)
B
.410
E
.600
.500 MIN.
BOTTOM
VIEW
Fig. 2-29.
RECTANGULAR
METAL FLANGE
(PART OF PLASTIC
PACKAGE)
.315
118
DIA.
.395
.725
MIN
E
B
C (FLANGE)
TO-5 PLASTIC
.155
Fig. 2-30.
.875
TO-3
.450
312 MIN.
E
.161 DIA.
.440
B
C
(FLANGE)
.675
1.197
Fig. 2-31.
.500
TO-66
.340
.360 MIN.
C
(FLANGE)
.962
E
590
.152
DIA.
.210
B

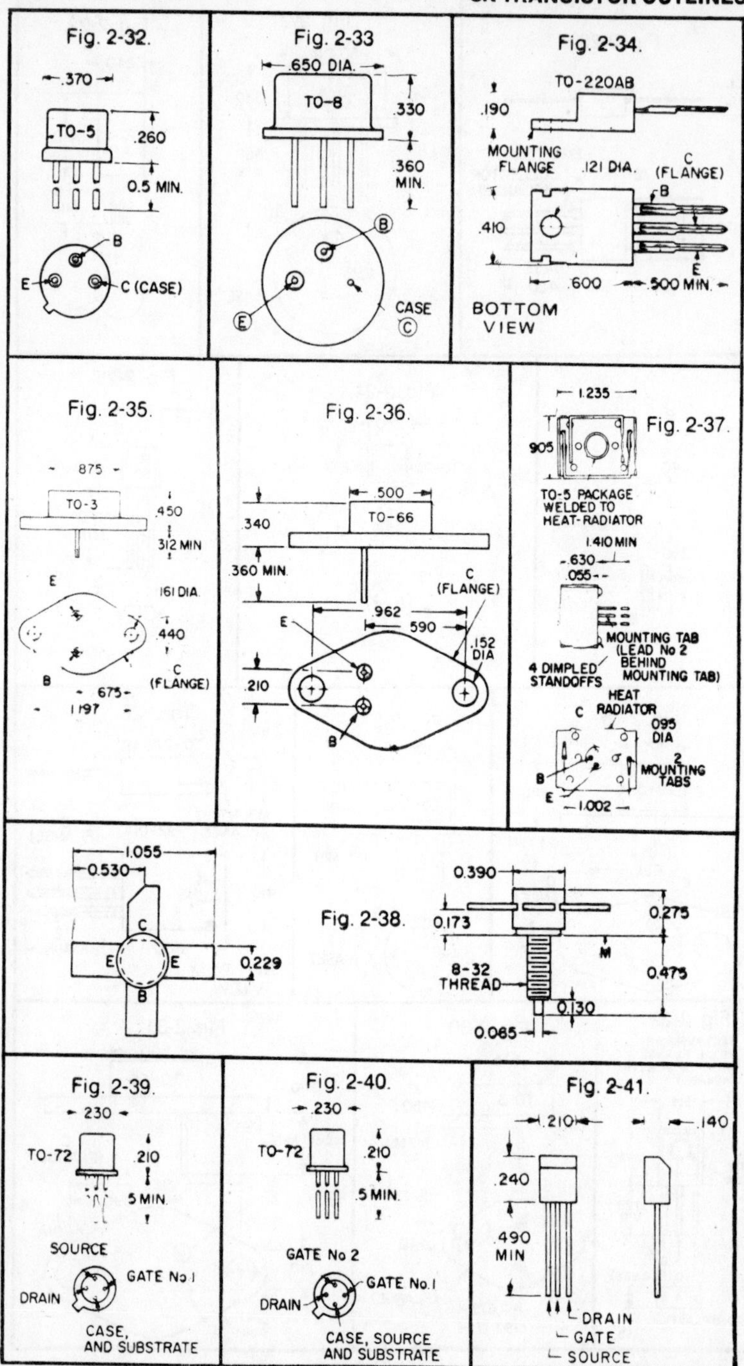
Fig. 2-32.
.370
TO-5
.260
0.5 MIN.
B
E
C (CASE)
Fig. 2-33
.650 DIA.
TO-8
.330
.360
MIN.
B
E
CASE
C
Fig. 2-34.
TO-220AB
.190
MOUNTING
FLANGE
.121 DIA.
C
(FLANGE)
B
.410
E
.600
.500 MIN.
BOTTOM
VIEW
Fig. 2-35.
875
TO-3
.450
312 MIN
E
161 DIA.
.440
C
(FLANGE)
B
675
1197
Fig. 2-36.
.500
TO-66
.340
.360 MIN.
C
(FLANGE)
.962
590
E
.152
DIA
.210
B
1.235
Fig. 2-37.
905
TO-5 PACKAGE
WELDED TO
HEAT-RADIATOR
1.410 MIN
.630
.055
MOUNTING TAB
(LEAD No 2
BEHIND
MOUNTING TAB)
4 DIMPLED
STANDOFFS
HEAT
RADIATOR
C
095
DIA
2
MOUNTING
TABS
B
E
1.002
1.055
0.530
C
E
E
B
0.229
Fig. 2-38.
0.390
0.173
0.275
M
8-32
THREAD
0.475
0.130
0.065
Fig. 2-39.
230
TO-72
.210
.5 MIN.
SOURCE
GATE No 1
DRAIN
CASE,
AND SUBSTRATE
Fig. 2-40.
.230
TO-72
.210
.5 MIN.
GATE No 2
GATE No. 1
DRAIN
CASE, SOURCE
AND SUBSTRATE
Fig. 2-41.
.210
.140
.240
.490
MIN
DRAIN
GATE
SOURCE

CHAPTER 3

GENERAL ELECTRIC REPLACEMENT TRANSISTOR BASE DIAGRAMS

Outline diagrams for the General Electric transistors that appear in the GE column in *Part II* are shown on the following pages. To locate the proper diagram, refer to the listings which follow. GE numbers are listed in the left column. In the adjacent column is the corresponding figure number. For example, the correct illustration for GE1 is Fig. 3-1. Now turn to the page where this figure appears. Each of the pages containing illustrations is headed at the top to the first and last figure number that is on that page.

GE	Fig. 3-	APPLICATIONS
1	6	RF, IF, OSC, MIX, AM RADIO
2	6	AF AMPL
3	5	AF OUTPUT
4	8	AF OUTPUT
5	6	RF, IF, OSC, MIX, AM RADIO
6	1	RF, OSC, MIX, AM RADIO
7	1	IF AMPL, AM RADIO
8		USE GE-51
10	12	RF, IF, AF, OSC, MIX, AM RADIO
11	12	RF, IF, OSC, MIX, FM, UHF/VHF TUNER
12	9	AF OUTPUT @ 120V
13MP	5	MATCHED AF OUTPUTS GE-3
14	5	HP AF OUTPUT
15MP	5	MATCHED AF OUTPUTS GE-14
16	5	HP AF OUTPUT, SW
17	2	FM RF, OSC, TV LOW-NOISE CIRCUITS
18	6	AF AMPL, OSC, OUTPUT
19	5	HP AF AMPL, OSC, OUTPUT
20	7	RF, IF, AF AMPL, OSC
21	6	RF, IF, AF AMPL, OSC
22	3	RF, IF, AF AMPL, OSC, AM/FM
23	9	CLASS A, B AF OUT PUT
24MP	9	MATCHED AF OUTPUTS GE-23
25	5	HORZ/VERT DEFLECT, HV AMPL
26	9	AF OUTPUT
27	15	TV VIDEO OUTPUT, HV
28	16	AF OUTPUT
29	16	AF OUTPUT
30	9	AF OUTPUT
31MP	9	MATCHED AF OUTPUTS GE-30
32	17	AC LINE AF AMPL
35	5	VERT DEFLECT, AF OUTPUT
36	5	HORZ DEFLECT
37	5	VERT DEFLECT
38	5	HORZ DEFLECT
39	10	VIDEO IF AMPL
40	21	VIDEO OUTPUT
43	22	AF OUTPUT
44	22	AF OUTPUT
45	21	POWER SW
46	35	CB RF OUTPUT, SW
47	32	AF DRIVER, OUTPUT
48	32	AF DRIVER, OUTPUT
49	33	AF POWER AMPL
50	20	FM RF AMPL, TV IF AMPL
51	10	RF, IF AMPL, AM/FM
52	4	LOW-NOISE AF AMPL
53	4	AF AMPL, OUTPUT
54	4	MATCHED COMPL AF DRIVER, OUTPUT, P, SIM TO GE-53 N, SIM TO GE-59

GE	Fig. 3-	APPLICATIONS
55	19	AF OUTPUT, SW
56	19	AF OUTPUT, SW
57	18	AF OUTPUT
58	18	AF OUTPUT
59	4	AF AMPL, OUTPUT
60	3	RF, IF AMPL, TV
61	11	3RD IF AMPL, TV
62	12	HIGH-GAIN, LOW-NOISE AMPL
63	13	AF AMPL, OUTPUT
64	12	DARLINGTON AMPL
65	12	HIGH-GAIN, LOW-NOISE AMPL
66	17	AF OUTPUT
67	13	AF OUTPUT
69	17	AF OUTPUT
72	5	AF AMPL
73	5	AF AMPL
74	5	AF AMPL
75	5	AF AMPL
76	5	AF AMPL, HV SW
80	6	AF OUTPUT
81	30	DRIVER, OUTPUT
82	30	DRIVER, OUTPUT
83	31	AF OUTPUT
84	31	AF OUTPUT
85	30	LOW-NOISE AMPL
86	26	RF, OSC, MIX, VHF
88	23	AF DRIVER, OUTPUT
89	23	AF DRIVER, OUTPUT
210	30	RF, MIX, OSC, VHF
211	30	RF, IF AMPL
212	12	VOLTAGE AMPL
213	12	3RD IF AMPL, TV
214	10	RF AMPL, UHF/VHF
215	24	RF DRIVER, OUTPUT, CB
216	24	RF DRIVER, OUTPUT, CB
217	31	HV AMPL, DRIVER
218	31	HV AMPL, DRIVER
219	21	RF OUTPUT, CB
220	30	GP HV AMPL
221	30	GP HV AMPL
222	30	GP HV AMPL
223	30	GP HV AMPL
224	31	HV VIDEO, LUM OUTPUT, TV
225	31	HV VIDEO, LUM OUTPUT, TV
226	31	HV AMPL, DRIVER
227	31	HV AMPL, DRIVER
228	21	HV AMPL TO VHF
229	21	VIDEO OUTPUT
230	21	GP HV AMPL
231	21	HV VIDEO
232	18	AF OUTPUT @ 120V
233	9	AMPL, DRIVER, SW
234	9	AMPL, DRIVER, SW
235	21	GP HV AMPL
FET1	11	GP AMPL, <100M
FET2	14	FM/TV MIX, VHF
FET3	28	AM/FM RF AMPL, MIX
FET4	27	FM/TV RF AMPL

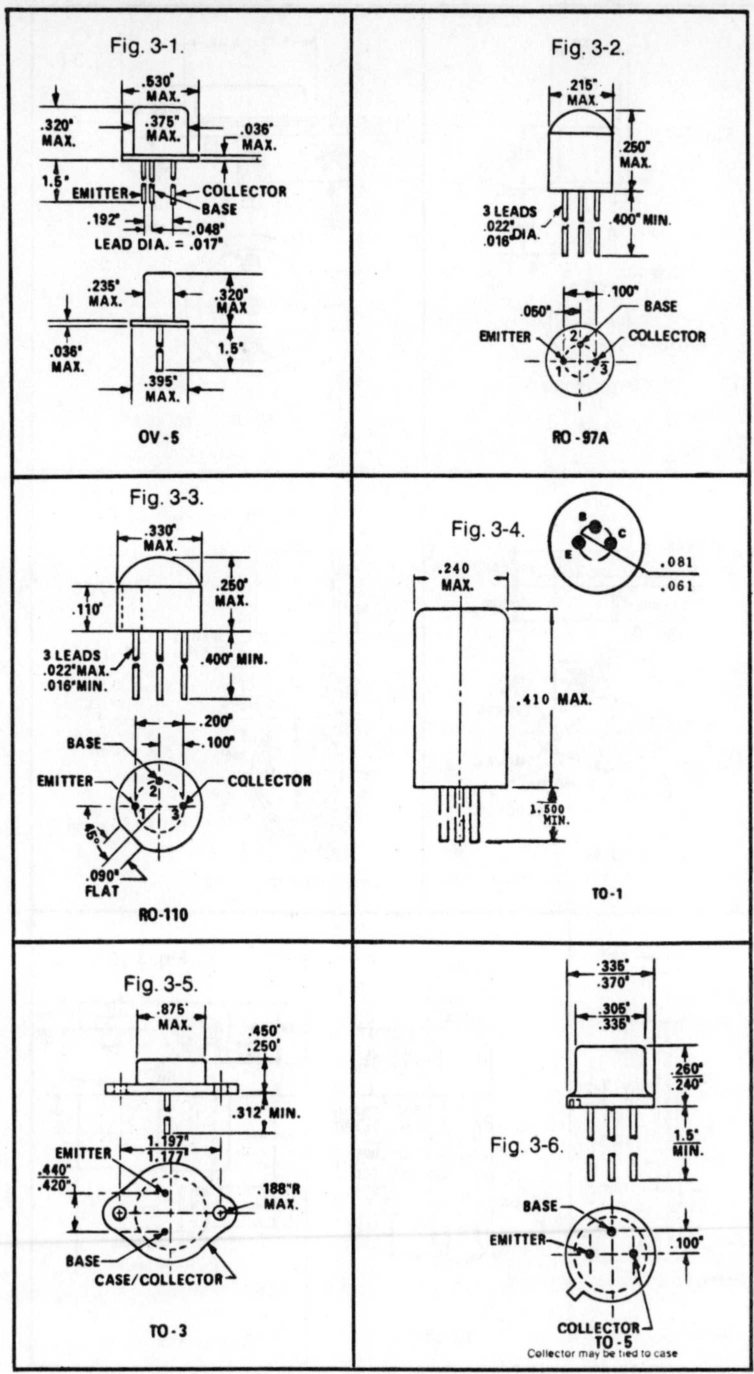
Fig. 3-1.
.530" MAX.
.375" MAX.
.320" MAX.
.036" MAX.
1.5"
EMITTER
COLLECTOR
BASE
.192"
.048"
LEAD DIA. = .017"
.235" MAX.
.320" MAX
.036" MAX.
1.5"
.395" MAX.
OV-5
Fig. 3-2.
.215" MAX.
.250" MAX.
3 LEADS
.022"
.016" DIA.
.400" MIN.
.100"
.050"
BASE
EMITTER
COLLECTOR
RO-97A
Fig. 3-3.
.330" MAX.
.250" MAX.
.110"
3 LEADS
.022" MAX.
.016" MIN.
.400" MIN.
.200"
.100"
BASE
EMITTER
COLLECTOR
45°
.090" FLAT
RO-110
Fig. 3-4.
B
C
E
.081
.061
.240 MAX.
.410 MAX.
1.500 MIN.
TO-1
Fig. 3-5.
.875" MAX.
.450"
.250"
.312" MIN.
1.197"
1.177"
EMITTER
.440"
.420"
.188"R MAX.
BASE
CASE/COLLECTOR
TO-3
.335"
.370"
.305"
.335"
.260"
.240"
1.5" MIN.
Fig. 3-6.
BASE
EMITTER
.100"
COLLECTOR
TO-5
Collector may be tied to case

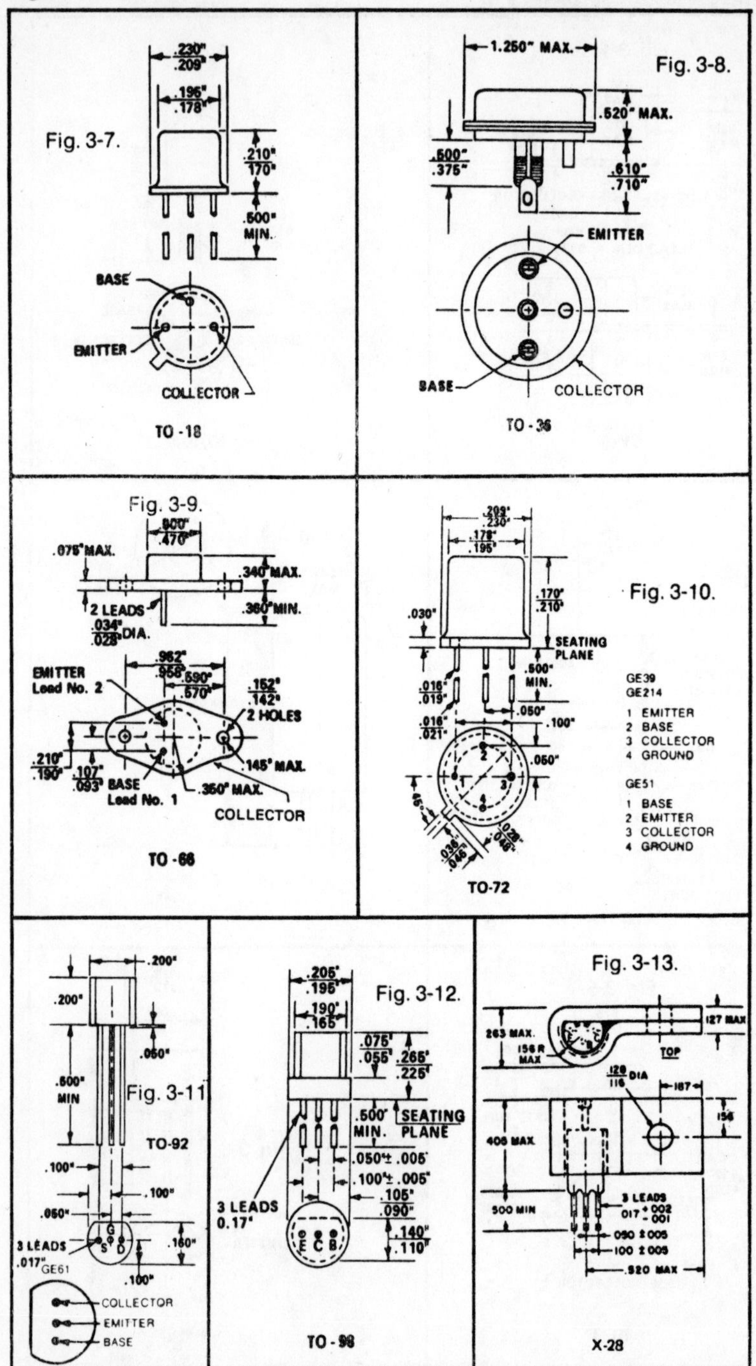
Fig. 3-7.
BASE
EMITTER
COLLECTOR
TO-18
Fig. 3-8.
1.250" MAX.
.520" MAX.
EMITTER
BASE
COLLECTOR
TO-36
Fig. 3-9.
2 HOLES
EMITTER Lead No. 2
BASE Lead No. 1
COLLECTOR
TO-66
Fig. 3-10.
SEATING PLANE
GE39
GE214
1 EMITTER
2 BASE
3 COLLECTOR
4 GROUND
GE51
1 BASE
2 EMITTER
3 COLLECTOR
4 GROUND
TO-72
Fig. 3-11
TO-92
3 LEADS
GE61
COLLECTOR
EMITTER
BASE
Fig. 3-12.
SEATING PLANE
3 LEADS 0.17"
TO-98
Fig. 3-13.
TOP
3 LEADS
X-28

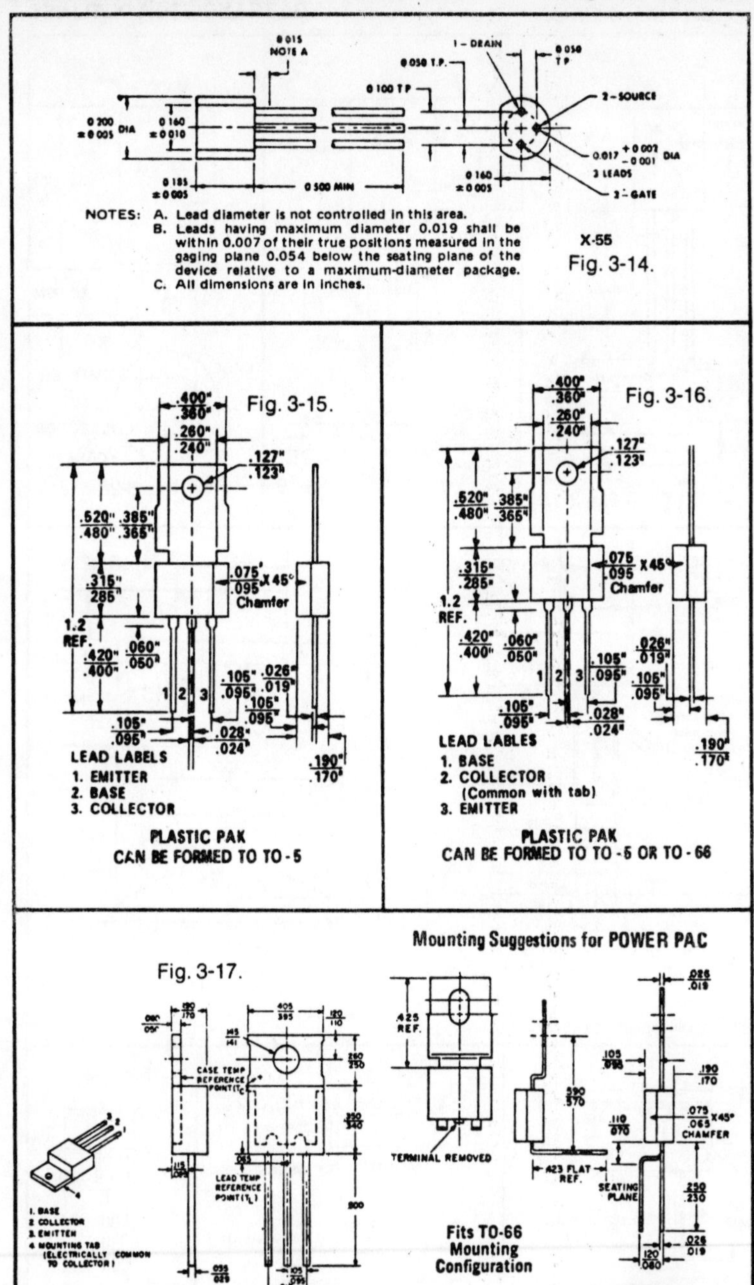

Fig. 3-14.

Fig. 3-15.

Fig. 3-16.

Fig. 3-17.

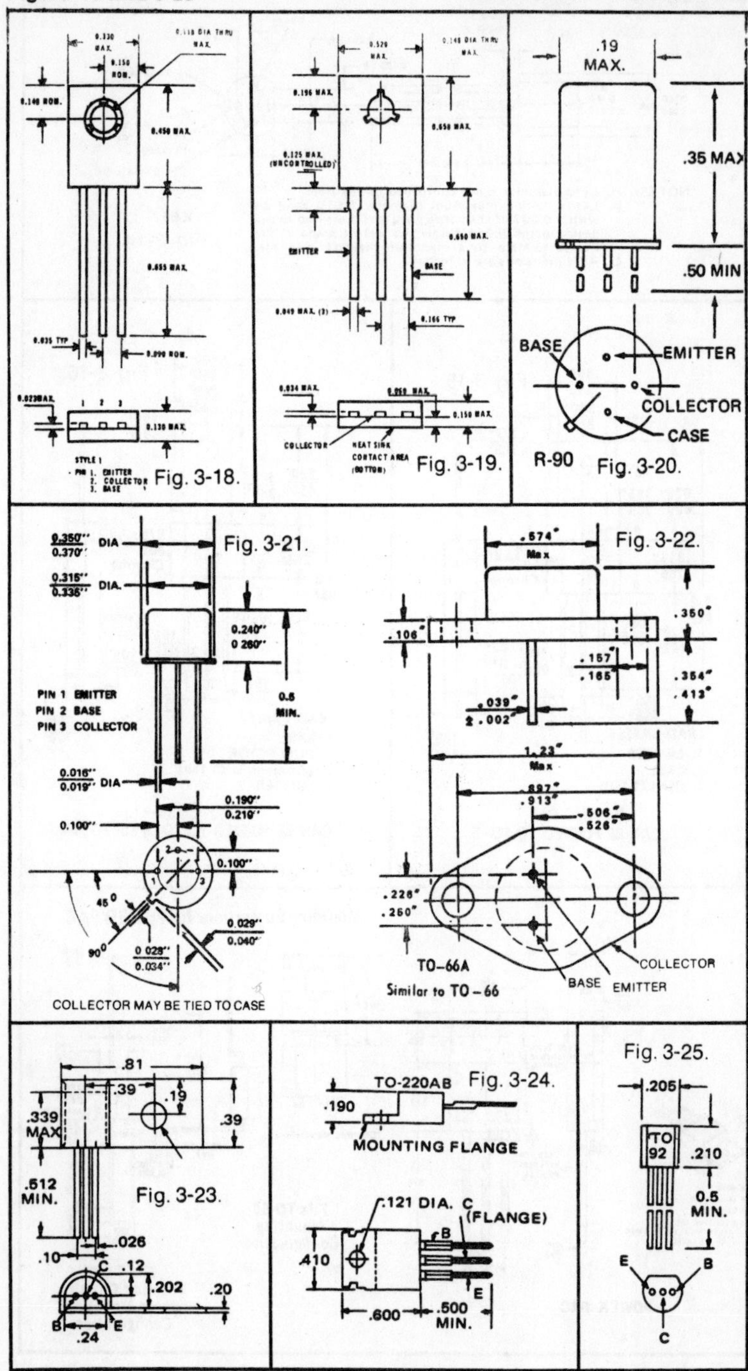

Fig. 3-18. Fig. 3-19. Fig. 3-20. Fig. 3-21. Fig. 3-22. Fig. 3-23. Fig. 3-24. Fig. 3-25.

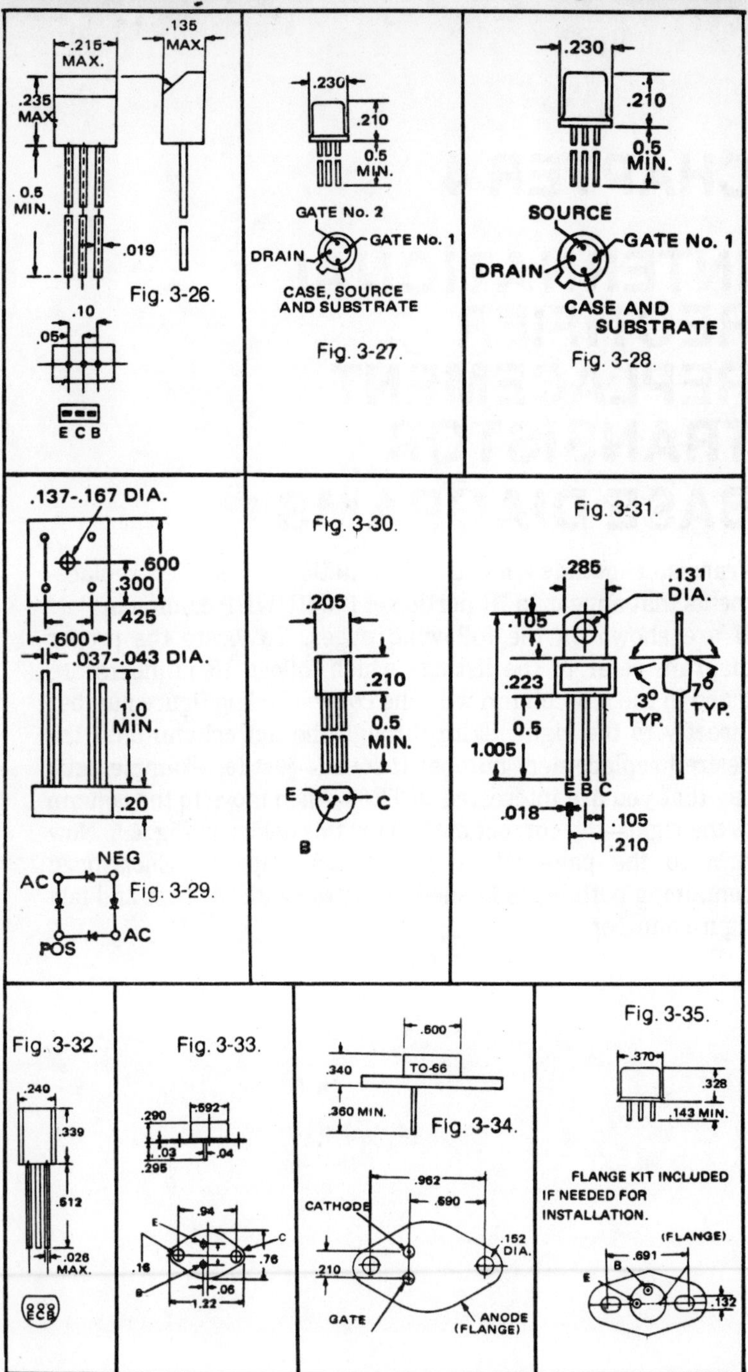

Fig. 3-26.

Fig. 3-27.

Fig. 3-28.

Fig. 3-29.

Fig. 3-30.

Fig. 3-31.

Fig. 3-32.

Fig. 3-33.

Fig. 3-34.

Fig. 3-35.

CHAPTER 4

INTERNATIONAL RECTIFIER REPLACEMENT TRANSISTOR BASE DIAGRAMS

Transistor outlines for the International Rectifier replacements that appear in IR portion of the IR/WEP column in *Part II* are shown on the following pages. To locate the proper diagram refer to the listings which follow. IR numbers are listed in the left column with the corresponding figure number directly to the right. Skim through the left column until the desired replacement number is found—just for example, let's say that you are interested in TR03—then move to the column to the right—the correct outline for this device is Fig. 4-5. Now turn to the page where this figure appears. Each page containing outlines is headed at the top with the first and last figure number.

IR	Fig. 4-	APPLICATIONS
TR01	2	HP AF AMPL, SW
TR01MP	2	HP AF AMPL, SW
TR02	2	MP AF AMPL, SW
TR03	5	HIGH-POWER, GAIN AMPL, SW
TR05	3	MP, HIGH-GAIN AF AMPL, SW
TR08	3	MP AMPL, RF/IF/OSC/MIX/DRIVER
TR12	1	AMPL, RF/IF/OSC/MIX
TR16	2	HP, LOW-LEAKAGE RADIO AMPL
TR17	10	HIGH-GAIN, LOW-NOISE AMPL, TV/FM/RF/IF
TR20	4	HIGH-SPEED SW, RF, IF, OSC, MIX, FM
TR21	4	HIGH-SPEED SW, RF, OSC, MIX, IF, FM
TR23	8	HV AF OUTPUT, HORZ/VERT OUTPUT
		GP LOW-NOISE RF, IF AMPL, AM/FM
TR24	12	GP LOW-NOISE AM/FM AMPL
TR25	3	AF OUTPUT HORZ/VERT DEFLECT
TR26	2	HP AF AMPL, SW
TR27	2	HV HORZ/VERT DEFLECT
TR28	3	AF, SW
TR29	2	HP AF AMPL, HF SW
TR30	12	GP LOW-NOISE AMPL
TR31	11	UNIVERSAL
TR32	11	HV AF VIDEO AMPL
TR33	13	OSC/MIX/RF/IF
TR34	2	AF OUTPUT AMPL, HORZ/VERT AMPL
TR35	2	AF POWER AMPL
TR36MP	2	MATCHED HP AF AMPL, TR36
TR36	2	HP AF AMPL
TR37	8	UNIVERSAL
TR50	8	HP AF AMPL, SW
TR51	15	GP AMPL
TR52	15	GP AMPL
TR53	15	GP AMPL, MED SPEED SW
TR54	12	GP SW, AMPL
TR55	7	MED POWER AF AMPL, SW
TR56	9	MED POWER AF AMPL, SW
TR57	8	MED POWER AF, SW
TR58	8	MED POWER AF, SW
TR59	2	HP AF AMPL, MED SPEED SW
TR60	7	MP HV AF OUTPUT
TR61	2	HP, HV HORZ/VERT DEFLECT
TR61	2	MATCHED PAIR OF HP, HV HORZ/VERT DEFLECT, TR61
TR62	3	CB RF BUFFER, OUTPUT--0.4W @ 50M
TR63	3	CB RF BUFFER, OUTPUT--0.4W @ 50M
TR64	6	CB RF BUFFER, OUTPUT-1.5W @ 50M
TR65	6	CB RF BUFFER, OUTPUT--3W @ 50M
TR66		CB RF BUFFER, OUTPUT--10W @ 50M
TR67	2	HV, HORZ DEFLECT
TR68	2	HV, TV VERT DEFLECT
TR69	13	DARLINGTON AMPL
TR70	10	RF AMPL
TR72	7	MP AF DRIVER, AMPL
TR73	7	MP AF DRIVER, AMPL
TR74	7	MP AF AMPL OUTPUT, TV HORZ DRIVER
TR75	7	HV MP AF, VIDEO AMPL
TR76	9	HP AF AMPL OUTPUT
TR77	9	HP AF AMPL
TR78	6	HV TV VIDEO AMPL
TR79	7	HV, VIDEO AF OUT
TR80	10	RF, OSC, MIX, UHF/VHF TUNER
TR81	8	AF POWER AMPL
TR81MP	8	MATCHED PAIR OF AF OUTPUTS, TR81
TR82	6	AF POWER AMPL
TR83	14	OSC UHF/TV TUNER
TR84	1	AF POWER AMPL
TR85	1	AF PREAMPL, DRIVER, VIDEO AMPL,
TR86	3	AF PREAMPL , DRIVER, VIDEO AMPL, SYNC SEPARATOR
TR87	3	AF PREAMPL, DRIVER OUTPUT VIDEO
TR88	3	AF PREAMPL, DRIVER OUTPUT VIDEO
TR91	8	HP AF AMPL
TR92	9	RF POWER OUTPUT
TR93	2	HORZ OUTPUT
TR94MP	8	AF AMPL
TR95	14	RF AMPL, MIX, OSC, IF AMPL
FE100	12	RF, OSC, MIX, AF AMPL, FM
2160	12	TIMING /TRIGGERING FOR SCR

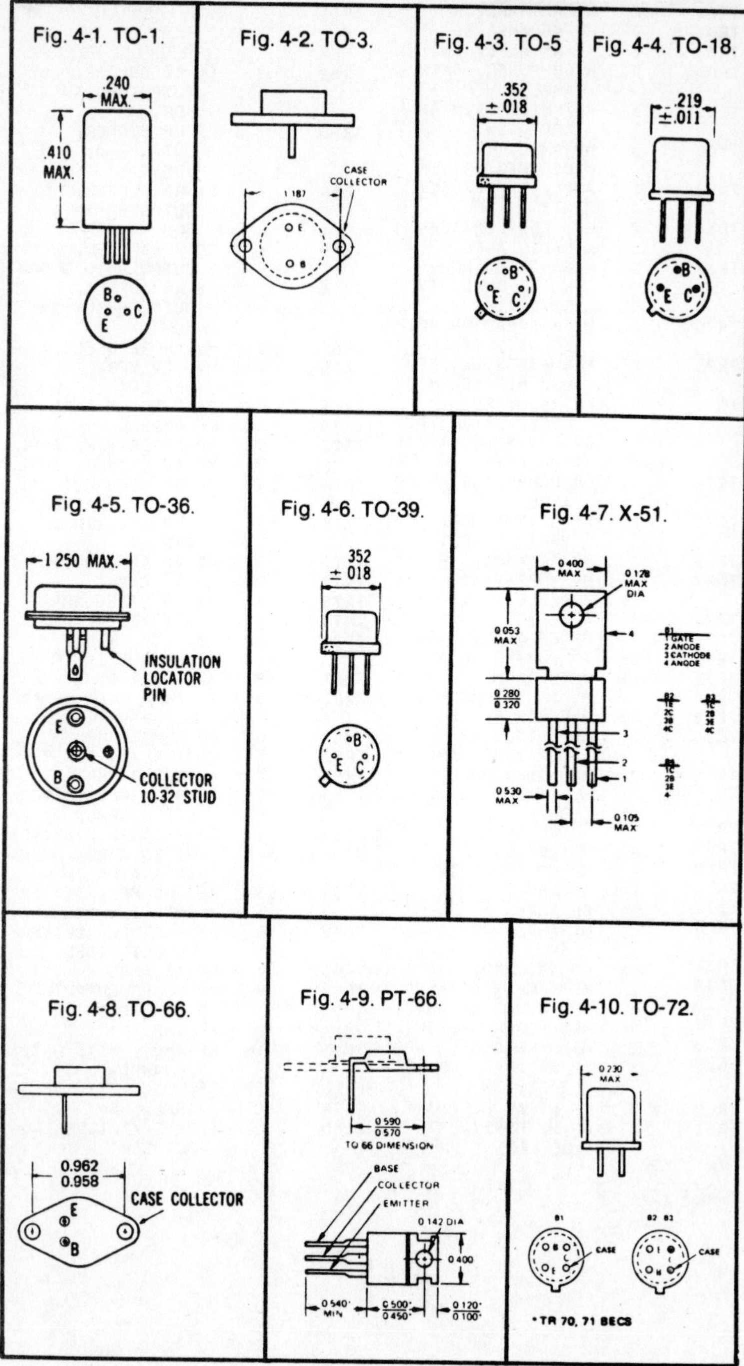

Fig. 4-1. TO-1.

Fig. 4-2. TO-3.

Fig. 4-3. TO-5

Fig. 4-4. TO-18.

Fig. 4-5. TO-36.

Fig. 4-6. TO-39.

Fig. 4-7. X-51.

Fig. 4-8. TO-66.

Fig. 4-9. PT-66.

Fig. 4-10. TO-72.

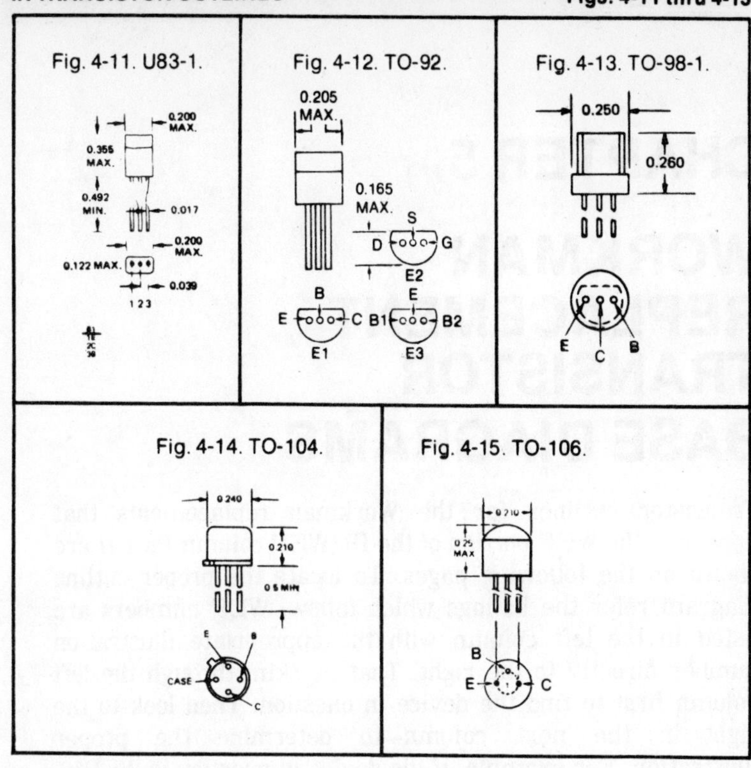

Fig. 4-11. U83-1.

Fig. 4-12. TO-92.

Fig. 4-13. TO-98-1.

Fig. 4-14. TO-104.

Fig. 4-15. TO-106.

CHAPTER 5

WORKMAN REPLACEMENT TRANSISTOR BASE DIAGRAMS

Transistor outlines for the Workman replacements that appear in the WEP portion of the IR/WEP column *Part II* are shown on the following pages. To locate the proper outline diagram refer the listings which follow. WEP numbers are listed in the left column with the appropriate illustration number directly to the right. That is, skim through the left column first to find the device in question. Then look to the right—in the next column—to determine the proper illustration. For example, if the device in question is WEP56, the corresponding outline number is Fig. 5-7. Now turn to the page where Fig. 5-7 appears. Each page containing outlines is headed at the top with the first and last figure number.

WEP	Fig. 5-	APPLICATIONS
52	11	RF, IF, OSC, MIX, AM/FM
53	4	AF AMPL, DRIVER, VIDEO, SYNC
56	7	RF, IF, OSC, MIX, VIDEO, VHF/UHF
230	2	AF OUTPUT
230MP	2	MATCHED AF OUTPUTS
231	1	AF OUTPUT
232	2	AF OUTPUT
233	1	AF OUTPUT
235	2	HORZ/VERT DEFLECT, AF OUTPUT
238	4	AF OUTPUT
240	8	HV AF OUTPUT @ 120V
241	8	AF OUTPUT
242	12	AF AMPL, DRIVER, OUTPUT, VIDEO AMPL
243	12	AF AMPL, DRIVER, OUTPUT, VIDEO AMPL
244	5	HV AF OUTPUT
247	2	AF OUTPUT
247MP	2	MATCHED AF OUTPUTS
250	4	AF AMPL, DRIVER, OUTPUT
254	4	RF, IF, OSC, MIX, AM RADIO
309	7	
310	7	
624	2	AF OUTPUT
628	2	AF OUTPUT
628MP	2	MATCHED AF OUTPUTS
630	4	AF OUTPUT
631	4	AF AMPL, DRIVER, OUTPUT
635	4	RF, IF, OSC, MIX, UHF
637	3	RF, IF, OSC, MIX, FM/TV
641	4	RF, IF, OSC, MIX, AM RADIO
641A	4	AF AMPL, DRIVER, OUTPUT
641B	4	AF AMPL, DRIVER, OUTPUT
642	8	AUTO AF OUTPUT
642MP	8	MATCHED AUTO OUTPUTS
700	10	AF OUTPUT
701	8	AF OUTPUT
707	2	VERT DEFLECT
712	4	TV VIDEO OUTPUT
717	7	AF AMPL, DRIVER
719	7	VIDEO IF AMPL
720	7	RF, IF, OSC, MIX, UHF/VHF
735	4	RF, AF AMPL
740	2	HORZ DEFLECT
740A	2	VERT DEFLECT
740B	2	HORZ DEFLECT
801	7	GP AF AMPL, SW
802	7	RF AMPL, MIX, VHF
S3002	4	POWER AMPL
S3003	4	POWER AMPL
S3020	6	AF DRIVER, OUTPUT
S3021	6	AF OUTPUT, HORZ DRIVER
S3023	9	AF OUTPUT
S3027	9	AF OUTPUT
S3031	6	AF OUTPUT, HORZ DRIVER
S5003	5	AF OUTPUT
S5004	5	AF OUTPUT
S5005	5	AF OUTPUT
S5007	5	AF OUTPUT
S7000	2	AF OUTPUT
S7001	2	AF OUTPUT
S9100	7	DARLINGTON AMPL
G6001	2	AF AMPL

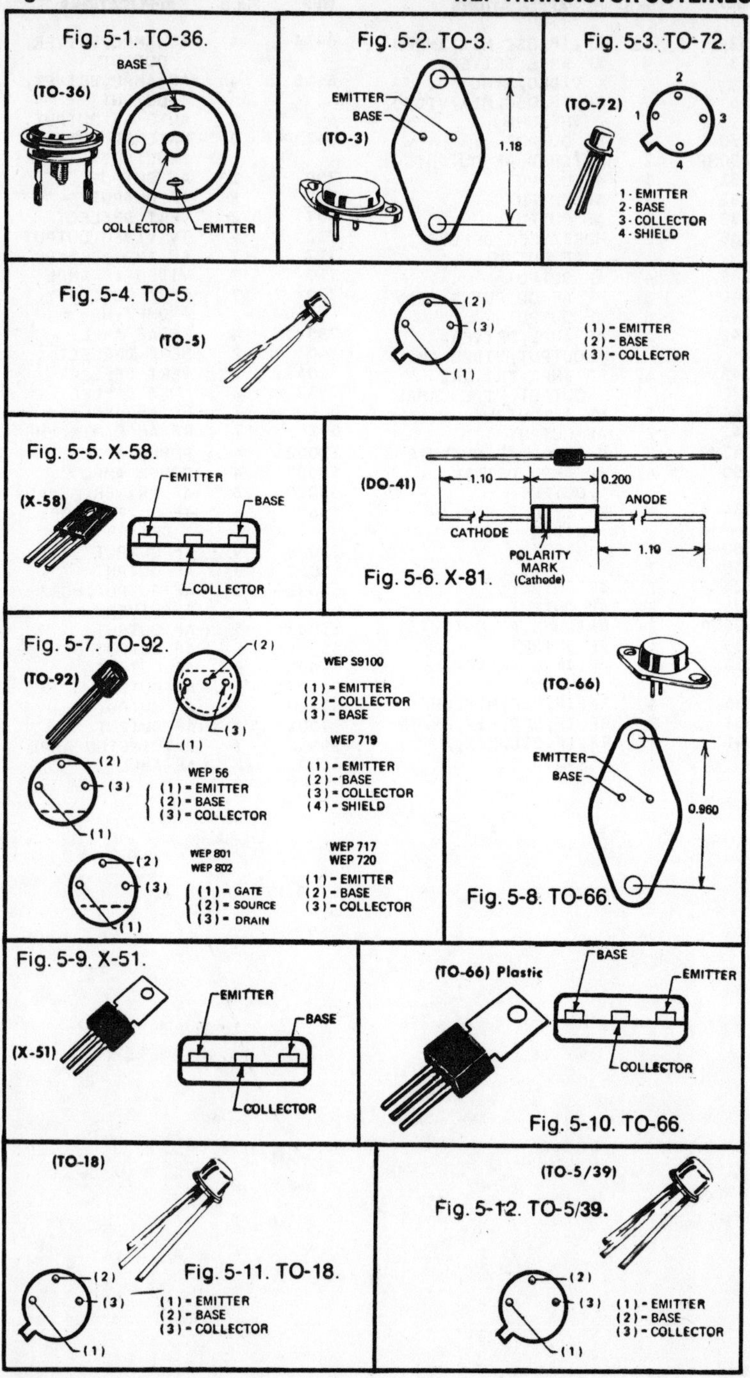

Fig. 5-1. TO-36.

Fig. 5-2. TO-3.

Fig. 5-3. TO-72.

Fig. 5-4. TO-5.

Fig. 5-5. X-58.

Fig. 5-6. X-81.

Fig. 5-7. TO-92.

Fig. 5-8. TO-66.

Fig. 5-9. X-51.

Fig. 5-10. TO-66.

Fig. 5-11. TO-18.

Fig. 5-12. TO-5/39.

CHAPTER 6

MOTOROLA REPLACEMENT TRANSISTOR BASE DIAGRAMS

Outline diagrams for the Motorola replacements that appear in the HEP column in *Part II* are shown on the following pages. To locate the proper illustration refer to the listings which follow. HEP numbers are shown in the left column with the appropriate figure number directly to the right. For example, if the device in question is HEPG0003, skim through the left column until this number is found. Adjacent to this number—in the next column—is the corresponding illustration number—Fig. 6-8. Now turn to the page containing this outline. Each page of outlines is headed at the top with the first and last illustration.

HEP	Fig. 6-	APPLICATIONS
G0001	2	RF AMPL
G0002	2	RF AMPL
G0003	8	RF AMPL
G0005	2	GP AMPL
G0006	2	GP AF AMPL
G0007	2	GP AF AMPL
G0008	3	GP RF,IF,OSC
G0009	3	RF AMPL
G0011	3	RF AMPL
G6000	1	HIGH-CURRENT POWER AMPL
G6002	4	HIGH-CURRENT POWER AMPL
G6003	1	HIGH-CURRENT POWER AMPL
G6004	4	HIGH-CURRENT POWER AMPL
G6005	1	HIGH-CURRENT POWER AMPL
G6006	4	HIGH-CURRENT POWER AMPL
G6007	1	HIGH-CURRENT POWER AMPL
G6009	6	HIGH-CURRENT POWER AMPL
G6010	4	HIGH-CURRENT POWER AMPL
G6011	2	HIGH-CURRENT POWER AMPL
G6012	2	HIGH-CURRENT POWER AMPL
G6013	1	GP AF POWER AMPL
G6015	1	HIGH-CURRENT POWER AMPL
G6016	7	HIGH-CURRENT POWER AMPL
G6018	1	HIGH-CURRENT POWER AMPL
S0005	11	GP HV AMPL
S0008	11	VHF AMPL, VIDEO IF
S0009	11	VHF AMPL MIX, 3RD VIDEO IF
S0010	11	VHF/UHF AMPL MIX, OSC
S0011	3	GP AMPL
S0012	2	RF AMPL
S0013	3	RF AMPL
S0014	2	RF AMPL
S0015	11	GP LP AMPL
S0016	11	VHF/UHF AMPL,FM, TV MIX,OSC
S0017	8	UHF OSC MIX, RF AMPL
S0019	11	GP MEDIUM-CURRENT AMPL,SW
S0020	11	VHF AMPL MIX
S0024	11	LOW-LEVEL,LOW-NOISE AMPL
S0025	11	GP AMPL,SW
S0026	11	GP AF DRIVER OUTPUT
S0027	11	HV AMPL
S0028	11	HV AMPL
S0029	11	GP HV AMPL, MIX DRIVER
S0030	11	GP RF AMPL
S0031	11	GP AMPL
S0032	11	MEDIUM SPEED SW
S0033	11	GP AMPL
S3001	5	RF OUTPUT,5W @ 175M
S3002	5	MP TRANSISTOR
S3003	5	MP TRANSISTOR
S3004	5	RF OUTPUT,3.5W @ 27M
S3005	14	RF OUTPUT,2W @ 175M
S3006	14	RF OUTPUT,10W @ 175M
S3007	14	RF OUTPUT,25W @ 175M

HEP	Fig. 6-	APPLICATIONS
S5005	10	GP POWER AMPL
S5006	9	GP POWER AMPL
S5011	7	POWER TRANSISTOR
S5012	7	HP OUTPUT
S5013	2	MP OUTPUT
S5014	2	MP OUTPUT
S5015	9	HV POWER OUTPUT
S5018	7	GP AF AMPL
S5019	7	GP AF AMPL
S5020	1	HV HP OUTPUT
S5021	1	HV HP OUTPUT
S5022	3	LP RF AMPL
S5023	2	LF RF/AF AMPL
S5024	5	GP AF AMPL
S5025	2	LF OSC,MIX,RF AMPL
S5026	2	LF OSC,MIX,RF AMPL
S7000	1	HP OUTPUT
S7001	1	HP OUTPUT
S7002	1	HP OUTPUT
S7003	1	HP OUTPUT
S7004	1	GP AF AMPL
S7005	1	HV HP OUTPUT
S9001	11	
S9002	11	
S9100	11	DARLINGTON AMPL
S9101	9	DARLINGTON AMPL
S9102	10	DARLINGTON AMPL
S9103	12	DARLINGTON AMPL
S9120	11	DARLINGTON AMPL
S9121	9	DARLINGTON AMPL
S9122	10	DARLINGTON AMPL
S9123	12	DARLINGTON AMPL
S9140	1	DARLINGTON AMPL
S9141	1	DARLINGTON AMPL
S9142	1	DARLINGTON AMPL
S9143	1	DARLINGTON AMPL
F0010	8	AF AMPL
F0015	11	RF AMPL
F0021	11	VHF/UHF AMPL
F1035	11	RF AMPL
F1036	8	AF OUTPUT
F2004	8	DUAL GATE
F2005	8	VHF/UHF AMPL
F2007A	15	DUAL GATE
S3008	5	RF OUTPUT,1.5W @ 150M
S3009	14	RF OUTPUT,40W @ 175M
S3010	5	MP OUTPUT
S3011	5	MP OUTPUT
S3012	5	MP OUTPUT
S3013	5	RF OUTPUT,1W @ 400M
S3014	5	RF OUTPUT,1W @ 400M
S3019	12	HV AF AMPL
S3020	12	MP OUTPUT
S3021	12	HV TRANSISTOR
S3024	12	GP AMPL
S3028	12	GP AMPL
S3032	12	MP OUTPUT
S3033	5	MP OUTPUT
S3034	5	MP OUTPUT
S3035	5	MP OUTPUT
S3036	13	RF OUTPUT,20W PEP @ 30M
S3037	13	RF OUTPUT,65W PEP @ 30M
S3038	5	RF OUTPUT,6W @ 150M
S3039	14	RF OUTPUT,15W @ 150M
S3040	14	RF OUTPUT,28W @ 150M
S3041	14	RF OUTPUT,40W @ 150M
S5000	9	GP POWER AMPL
S5004	10	GP POWER AMPL

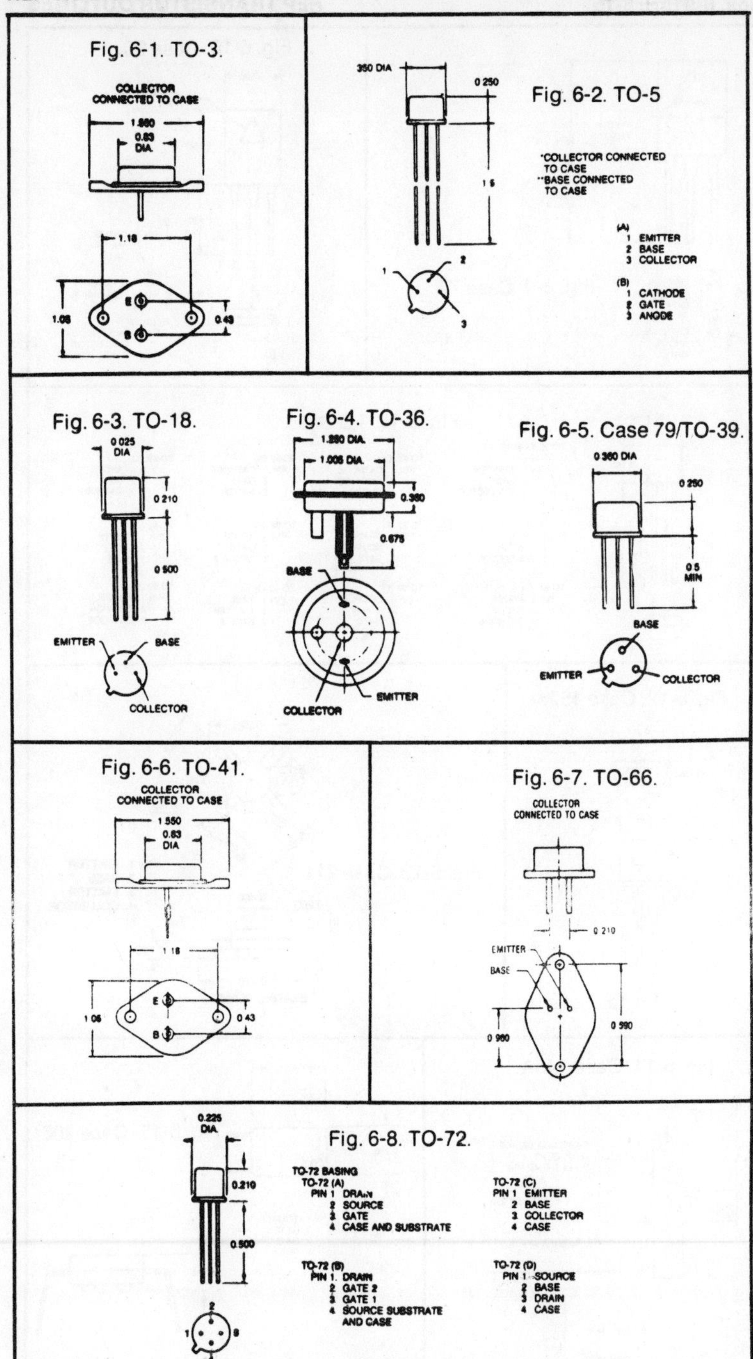

Fig. 6-1. TO-3.

Fig. 6-2. TO-5

Fig. 6-3. TO-18.

Fig. 6-4. TO-36.

Fig. 6-5. Case 79/TO-39.

Fig. 6-6. TO-41.

Fig. 6-7. TO-66.

Fig. 6-8. TO-72.

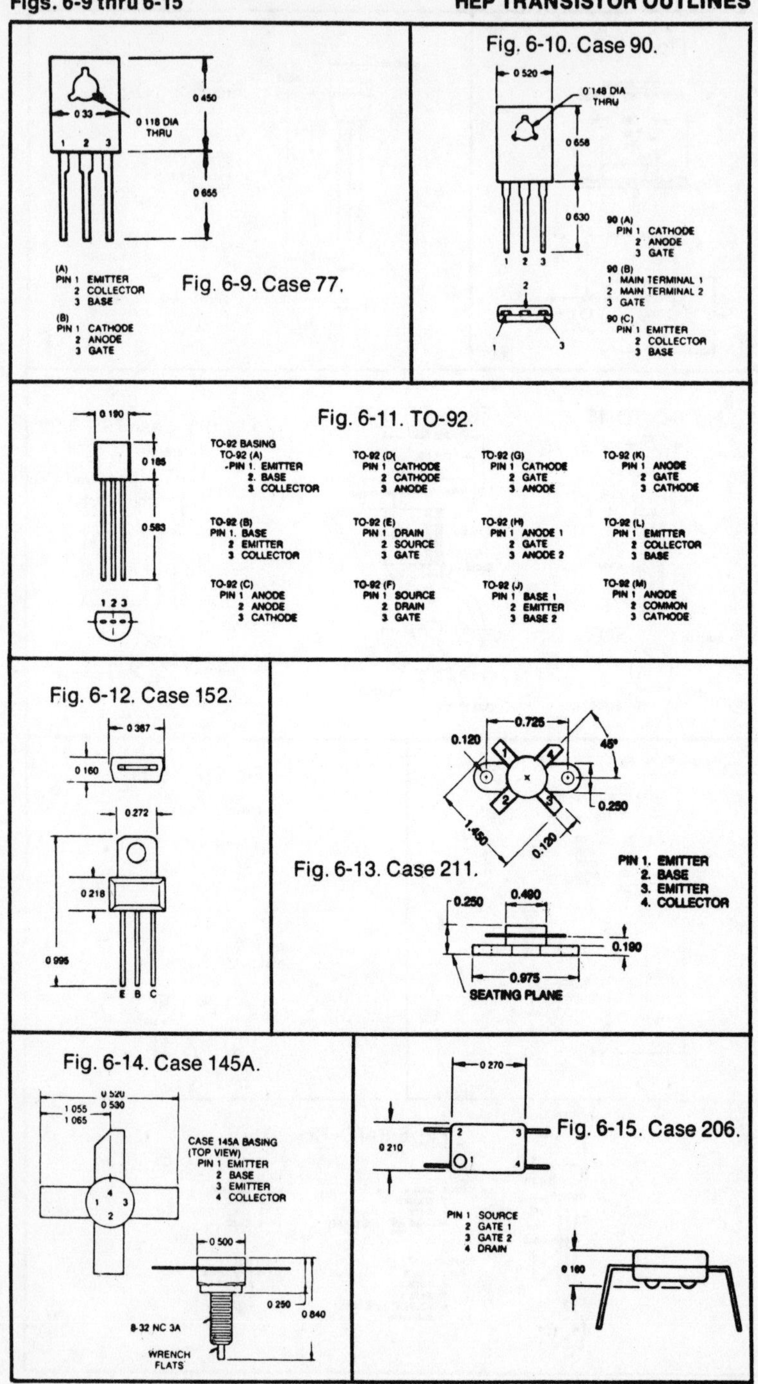

Fig. 6-9. Case 77.

Fig. 6-10. Case 90.

Fig. 6-11. TO-92.

Fig. 6-12. Case 152.

Fig. 6-13. Case 211.

Fig. 6-14. Case 145A.

Fig. 6-15. Case 206.

PART 2

MASTER CROSS REFERENCE OF OVER 80,000 TRANSISTORS & ICS

KEY TO MASTER CROSS-REFERENCE

Original Part Number	ECG	SK	GE	IR/WEP	HEP
Number shown in equipment parts list or schematic	Sylvania replacement; see Part I, Chapter 1 for transistor outlines	RCA replacement; see Part I, Chapter 2 for transistor outlines	General Electric; replacement; see Part I, Chapter 3 for transistor outlines	International Rectifier/Workman replacement; see Part I, Chapters 4 & 5 for transistor outlines	Motorola replacement; see Part I, Chapter 6 for transistor outlines

Original Part Number	ECG	SK	GE	IR/WEP	HEP
000000SV31	158	3004		TR34/630	
000002SB22	102A	3004	53	TR85/250	G0005
0000-00011-049	123A				
0000-00011-050	184	3190			
0000-00011-053	220				
0000-00011-054	132				
0000-00011-055	132				
00002S606	107	3039		TR70/720	
00002SA101	126	3007	1	TR17/635	G0009
00002SA202	102A	3004	2	TR85/	
00002SA550	151	3114	21	TR28/717	S0031
00002SA609	159				S0019
00002SB22	102A			TR85/	G0005
00002SB185	102A	3004	53	TR85/250	G0005
00002SB185-0	102A	3004		TR85/250	
00002SB186	102A	3004	53	TR85/250	G0005
00002SB186-0	102A	3004		TR85/250	
00002SB187	102A	3004	53	2SB187/250	G0005
00002SB303	102A	3004	52	TR85/250	G0005
00002SB303-3	102A	3004		TR85/250	
00002SB405	158	3004	53	2SB405/630	G6011
00002SB435	153	3083		TR77/700	S3028
00002SB460	107	3039		TR70/720	
00002SB474	226	3082	30		G6016
00002SB481	131MP	3052		TR94MP/642	G6016
00002SB492	176	3004		TR82	G6011
00002SC373	123A	3124	62	TR21/735	S0015
00002SC460	107	3018		TR70/720	S0014
00002SC460A	108	3018		TR95/	S0014
00002SC460B	108	3018		TR95/	S0014
00002SC460C	108	3018		TR95/	S0014
00002SC461	108		17		S0014
00002SC492	176	3123		TR82/238	S5004
00002SC535	107		11		S0016
00002SC536	199	3124	20	TR21/	S0016
00002SC537	123A	3018	20	TR21/735	S0016
00002SC606	108	3018		TR95/	S0016
00002SC609	195A	3114	22		S0019
00002SC644	199		62		S0015
00002SC668	108	3018		TR95/56	S0020
00002SC735	123A	3020		TR21/735	S0014
00002SC772	107	3018		TR70/720	S0016
00002SC828	123A		18		S0015
00002SC829	107	3018		TR70/720	S0011
00002SC838	123A	3122		2SC838/735	S0015
00002SC858	199	3124	62	TR21/720	S0015
00002SC870	123A	3020		TR21/735	S0015
00002SC870A	123A	3020		TR21/	S0015
00002SC870B	123A	3020		TR21/	S0015
00002SC870C	123A			TR21/	S0015
00002SC929	199	3018	20		S0016
00002SC930	199	3018	20		S0016
00002SC945	199	3124	20		S0015
00002SC968	123	3122		TR86/53	S5014
00002SD235	152	3021		TR76/701	
00002SC261	192		63		S3024
000062	129				
0000101	126		51	/635	
0000102	126		51	/635	
0000103	102A		2	/250	
0000104	102A		2	/250	
00007190	131	3052	30		G6016
000071120	187	3084	29		
000071130	187	3084	29		
000073100	123A	3020	10		S0014
000073110	128	3020	18		S5014
000073120	123A	3124			S0015
000073130	123A	3124			S0015
000073230	123A	3018	20		S0015
000073231	123A	3018	20		S0015
000073280	186	3122	18		S0003
000073290	123A	3124	18		S0015
000073303	192				
000073305	192				
000073320	186	3054	28		
000073350	199				
000073351	199				
000073360	199				
000073373	199				
000073374	199				
000073380	186	354	28		
0000HA1306	1029				
0000LA1201	1003				
0000LA1201B	1003				
0000LD3000	1015				
0002SB186-0	102A	3004		2SB186/	
0002SB187	102A	3004		2SB187/	G0005
0002SB187(RED)	102A				
0002SB303-0	102A	3004		TR85/	
0 02SB405R	158	3004	53		G6011
0002SC373	123A	3020		TR21/	S0015
0002SC373W	199	3020	62	TR95/56	S0015
0002SC458B	123A				S0014
0002SC458C	123A				S0014
0002SC531F	123A				
0002SC537F	123A	3124	10		S0014
0002SC644R	199				
0002SC644S	199				
0002SC668D	108				S0020
0002SC710B	123A	3018			S0016
0002SC710C	123A	3018			
0002SC772C	107		11		S0016
0002SC828	123A		20		S0015
0002SC828H	123A	3018	20		S0015
0002SC828Q	123A	3122			S0015
0002SC930D	199				S0016
0002SC930E	199				S0016
0002SC945Q	199				
0002SC968P	128		18		S5014
0002SC1023	107				S0017
0002SC1026	107	3018		TR70/720	S0025
0002SC1026A	107	3018		TR70/	S0025
0002SC1026B	107	3018		TR70/	S0025
0002SC1026C	107	3018		TR70/	S0025
0002SC1032	107	3039		TR70/720	S0015
0002SC1032A	107	3018		TR70/	S0015
0002SC1032B	107	3018		TR70/	S0015
0002SC1032C	107	3018		TR70/	S0015
0002SC1061	152	3054		TR76/701	S5000
0002SC1061A	152	3054		TR76/	S5000
0002SC1061B	152	3054		TR76/	S5000
0002SC1061C	152	3054		TR76/	S5000
0002SC1162	184	3190			S5000
0002SC1317	123A				S0003
0002SC1383	128				
000546-1	123A	3122	20	/735	
000653	123A				
0004201	102A	3004		/250	G0005
0004202	102A	3004		/250	G0007
0004203	131	3052	44	/642	G6016
000HA1306U	1029				
000LA1201	1003				
000LA1201B	1003		IC-2		C6060P
000LA3300	1005				
000LA3301	1006				
000LD1020A	1012				
00	199				
001	6400				
001-028-000	909				
001-01101-0	101	3011	6	TR08/	
001-01202-1	100	3005	1	TR05/	
001-01203-1	100	3005	1	TR05/	
001-01204-0	104	3009	3	TR01/	
001-01205-0	104	3009	3	TR01/	
001-01205-1	104		3	TR01/	
001-01206-0	102	3004	2	TR05/	
001-02010	158			TR84/	
001-02011	123	3122		TR86/	
001-02020	123A	3122		TR21/	
001-02101-0	123A	3122	20	TR21/	
001-02101-1	123A	3122	20	TR21/	
001-02102-0	123A	3122	20	TR21/	
001-02103-0	123A	3122	20	TR21/	
001-02104-0	123A	3122	20	TR21/	
001-02105-0	123A	3122	20	TR21/	
001-02106-0	123A	3122	20	TR21/	
001-02107-0	123A	3122	20	TR21/	
001-02108-0	123A	3122	20	TR21/	
001-02109-0	123A	3122	20	TR21/	
001-02110-0	123A	3122	20	TR21/	
001-02111-0				TR25/	
001-02111-1	123A	3122	20	TR21/	
001-02113-2	123A	3122	20	TR21/	
001-02113-3	123A	3122	20	TR21/	
001-02113-4	123A	3122	20	TR21/	
001-02113-5	123A	3122	20	TR21/	
001-02114-0	198		32		
001-02115-0	124	3021	12	TR81/	
001-02116-0	124	3021	12	TR81/	
001-02117-2	195	3138	67		

Original Part Number	ECG	SK	GE	IR/WEP	HEP
001-02119-0	128	3024	18	TR87/	
001-02121-0	123A	3122	20	TR21/	
001-02201-0	159	3025	21	TR20/	
001-011010	101	3011		TR08/	
001-012011	158	3004		TR84/	
001-012021	158	3123		TR84/	
001-012031	158	3123		TR84/	
001-012040	104	3009		TR01/	
001-012050	104	3009		TR01/	
001-012051	104	3009		TR01/	
001-012060	102A	3005		TR85/	
001-021010	123	3124		TR86/	
001-021020	123A	3124		TR21/	
001-021030	123A	3124		TR21/	
001-021040	123	3124		TR86/	
001-021050	123	3124		TR86/	
001-021060	123	3124		TR86/	
001-021070	123A	3124		TR21/	
001-021080	123A	3124		TR21/	
001-021090	123	3124		TR86/	
001-021100	154	3124		TR78/	
001-021110	123	3024		TR86/	
001-021111	123	3024		TR86/	
001-021132	123A	3124		TR21/	
001-021133	123	3124		TR86/	
001-021134	123A	3124		TR21/	
001-021135	123A	3124		TR21/	
001-021140	124	3044		TR81/	
001-021150	124	3021		TR81/	
001-021151	124	3021			
001-021160	124	3021		TR81/	
001-021162	124	3021			
001-021170	234	3118			
001-021171	234	3118			
001-021172	234	3118		TR20/	
001-021173	234	3118			
001-021190	128	3024		TR87/	
001-021210	123A	3124		TR21/	
001-021232	123A	3124		TR21/	
001-021270	130	3027			
001-022010	159	3114		TR20/	
001-022020	159	3025		TR20/	
001-027030	132	3116		FE100/	
001-533-00	159				
001-21011	123A	3122			
001B	726				
001	726				
002-005100	102A	3004	2	TR85/250	
002-006300	126	3008	52	TR17/635	
002-006500	123	3020	20	TR86/53	
002-006600	102A	3004	52	TR85/250	
002-006800	102A	3004	52	TR85/250	
002-006900	102A	3004	52	TR85/250	
002-007000	104	3009	3	TR01/	
002-007100	160	3007	1	TR17/637	
002-007200	160	3006	1	TR17/637	
002-007300	102A	3004	52	TR85/250	
002-007400	160	3007	1	TR17/637	
002-008100	104	3009	3	TR01/	
002-008400	102A	3004	52	TR85/250	
002-008800	121	3009	16	TR01/232	
002-009100	124	3021	12	TR81/240	
002-009500	123A	3020	20	TR21/735	
002-009501	123A	3020	20	TR21/735	S0024
002-009502	123A	3020	10	TR21/735	S0024
002-009600	108	3018	20	TR95/56	S0030
002-009601	108	3018		TR95/	S0016
002-009701	121	3009	16	TR01/232	
002-009800	159	3025	21	TR20/717	S5013
002-009900	123A	3020	20	TR21/735	S5014
002-0105-00	106				
002-010100	121	3009	16	TR01/232	G6005
002-010300	159	3118	21	TR20/717	S0012
002-010300-6	129	3025		TR88/	
002-010300A	159	3114		TR20/	
002-010400	123A	3020	20	TR21/735	S0015
002-010500	159	3114	21	TR21/717	S0019
002-010500A	159	3114		TR20/	
002-010600	128	3024	21	TR87/243	S0015
002-010700	129	3025	20	TR88/242	S0019
002-010800	123A	3024	20	TR21/735	S0015
002-010900	159	3025	21	TR20/717	S0019
002-010900A	159	3114		TR20/	
002-011000	102A	3004		TR85/	
002-011400	161	3018	20	TR83/719	
002-011500	161	3117	20	TR83/719	
002-011600	126	3008	51	TR17/635	
002-011700	103A	3010	8	TR08/724	
002-011800	176	3123	52	TR82/250	
002-011900	102A	3003	52	TR85/250	
002-012000	123	3124	20	TR86/53	
002-012100	185	3025	22	TR28/S5007	S5013
002-012200	184	3024	63	TR25/S5003	S5014
002-012300	153	3083	21	TR77/700	S5006
002-012400	152	3041	63	TR76/701	S5000
002-012500	128	3122		TR87/735	
002-012600	129	3025	21	TR88/242	S3028
002-012700	121	3009		TR01/232	G6003
002-012700-12	127	3024		TR27/	
002-012800	159			TR20/717	S0019
002-012800A	159	3114		TR20/	
002-1	196				
002-2	196				
002-3	241				
002-4	196				
002-104-000	199				
002-9501	199			TR21/	
002-9502	199			TR21/	
002-9502-12	199			TR21/	
002-9601	108	3039		TR95/	
002-9601-12	108	3039		TR95/	
002-9700	121	3009		TR01/	
002-9800-12	129	3025		TR88/	
002-9800-A	159	3025		TR20/	
002-11700	103A	3010		TR08/	
002-11800	102A	3004		TR85/	
002-11900	102A	3004		TR85/	
002-12000	123	3020		TR21/	
002D235RY	152				
002SA203AA	102A				G0005
002SB185AA	102A	3004			G0005
002SB435RY	153	3084	69		S3028
002SB18500	102A	3004		TR85/	
002SC1209C	192	3137	63		S5014
002SC7350Y	123A	3047	20		S0014
002SD235RY	152	3041	28		S5000
003	724				
003-H03	102	3004		TR05/631	
004	1098				
004-001	179			TR35/	
004-8000	121	3009	16	TR01/232	
005	724				
006-0000134	123A				
006-0000135	159				
006-0000146	7400				
006-0000147	7410				
006-0000151	7472				
006-0000162	74107				
006-0004270-001	909D				
006-0004443	133				
006-0004779	9109				
006-0004956	129				
006-0005191	129				
006-0005545	9112				
006-00055182	234				
006-640032	123A				
006B2M	722	3161	IC-9		
007-0030	175				
007-0040-00	130				
007-0051	128				
007-0074	175	3026	23	TR81/	
007-0112	152				
007-0112-03	152	3054			
007-0112-05	152	3054			
007-0214-00	133				
007-112-04	152				
007-74004-01	159				
007-74008-01	159				
007-74655-02	123A				
007-74655-06	123A				
007-74659-01	128				
007-74659-04	128				
007-74659-06	128				
007-74661-01	123A				
007-1668701	911				
007-1668601	910				
007-1668602	910D				
007-1669602	941M				
007-1669901	923D				

Original Part Number	ECG	SK	GE	IR/WEP	HEP
007-1681301	9112				
007-1695001	7400				
007-1695101	7420				
007-1695301	7404				
007-1695701	7440				
007-1695901	7410				
007-1696001	7485				
007-1696101	7414				
007-1696201	7402				
007-1696301	7437				
007-1696701	7411				
007-1696801	74145				
007-1696901	7406				
007-1697701	74180				
007-1697801	7441				
007-1698301	74192				
007-1698401	74193				
007-1698901	7450				
007-1699201	7411				
007-1699301	7408				
007-1699401	7403				
007-1699801	7474				
007-7450301	130	3027			
007-7466101	123A				
009-301003	102A	3004			
0019-003485	175	3026			
0036-001	159	3020		TR20/717	
0055	129	3025		TR88/242	
00234-B	175				
00415	159				
005575	181				
0022481	131	3082		TR94/	G6016
0023645	107	3039		TR70/	S0020
0023828	123A	3122		TR21/	S0015
0023829	107	3039		TR70/	S0011
00351980	128				
00352080	123A				
00444028-010	123A				
00444028-014	123A				
004440028-002	123A				
004440028-006	123A				
004440028-007	123A				
004440028-008	123A				
004440028-010	123A				
004440028-014	123A				
00HA1306P-U	1029				
0-008	128	3024		TR87/243	S5014
0-008(AMPEX)	129			TR88/242	
0-009	129	3025		TR88/242	S5013
01	159				
01-1201-0	102A	3004		TR85/	
01-1205-0P	179			TR35/	
01-2101	123A	3122		TR21/	
01-2101-0	123A	3122		TR21/	
01-2102	123A	3122		TR21/	
01-2104	123A	3122		TR21/	
01-2105	123A	3122		TR21/	
01-2106	123A	3122		TR21/	
01-2107	123A	3122		TR21/	
01-2108	123A	3122		TR21/	
01-2109	123A	3122		TR21/	
01-2110	128	3024		TR87/	
01-2111	128	3024		TR87/	
01-2114		3024		TR87/	
01-9011-52221-3	123A			TR21/	
01-9013-72221-3	123A			TR21/	
01-9014-22221-3	123A			TR21/	
01-9016-42221-3	108			TR95/	
01-9018-62221-3	108			TR95/	
01F	703A	3157			
01F-SL8020	703A	3157			
02-1078-01	123A				
02P1B	159	3114		TR20/717	
02P1BC	189	3114		TR73/	
02SC458LGC	123A				S0014
03-0018-6	124	3021		TR81/	
03-0020-0	102	3004	2	TR05/631	
03-0022	102	3004		TR05/631	
03-0023-0	102	3004	2	TR05/631	
03-57-001	126	3008		TR17/635	
03-57-002	126	3007		TR17/635	
03-57-003	160			TR17/637	G0002

Original Part Number	ECG	SK	GE	IR/WEP	HEP
03-460C	108	3122			
03-461B	108	3018			
03-535A	108	3018			
03-1585/G	123A	3124	10		
04-00072-01	103A	3010	54		
04-00156-03	158	3004	53		
04-00460-03	123A	3018	20		
04-00461-02	107	3018	17		
04-00535-02	107	3018	20		
04-00535-06	108		20		
04-01585-06	123A	3122	10		
04-01585-07	123A	3122	10		
04-01585-08	107	3018	20		
04-57-303	102	3004		TR05/631	
04-440032-002	159				
03-57-101	126	3008		TR17/635	
03-57-102	160			TR17/637	G0003
03-57-200	160	3006		TR17/	
03-57-201	126	3008		TR17/635	
03-57-202	160			TR17/637	G0003
03-57-301	102	3003		TR05/631	
03-57-302	102A	3004		TR85/	
03-57-304	102	3004		TR05/631	
03-57-501	121	3009		TR01/232	
03-156B	102A	3003			
04-440032-008	159				
04A1	121			TR01/	
04A1-12	121			TR01/	
04-1-12-7	121			TR01/	
05B1Z	712	3072			
05B2Z	712	3072			
06B1M	722	3161			
06B2Z	722	3161			
06B-3	121MP				
06B-3-12	121MP	3015			
06B-3-12-7	121MP	3015			
06P1C	121				
07-07113	159	3118			
07-07119	102A	3004			
07-07124	107	3018			
07-07125	123A	3124			
07-07129	107	3018			
07-07139	123A	3018			
07-07141	224	3049			
07-07156	123A	3018			
07-07158	132	3116			
07-07159	133	3112			
07-07161	724				
07-07163	123A	3038			
07-07164	224	3049			
07-07166	123A	3018			
07-07167	103A	3010			
07-1075-01	102A	3004		TR85/250	
07-1075-02	102A			TR85/250	
07-1156-03	102A			TR85/	
07-1485-85	123A	3124		TR21/735	
07-2012-04	102A	3005		TR85/250	
07-3012-04	102A	3006		TR85/250	
07-3015-05	102A	3008		TR85/250	
07-3080-06	160	3006		TR17/	
07-3350-57	126	3006		TR17/635	
07-4233-19	160	3006		TR17/	
07-4235-13	160	3006		TR17/	
07-4235-73	160	3006		TR17/635	
07-6015-16	102A	3004		TR85/	
07B1Z	708	3135			
07B2Z	788	3135			
07B3B	708	3135			
07B3C	708	3135			
07B3D	708	3135			
07B3M	708	3135			
07B3Z	708	3135			
08A8300-2	123A				
08B1M	725	3162			
08B2D	725				
08B2M	725	3162			
08P-2	102A	3004		TR85/	
08P-2-12-7	102A	3004		TR85/	
08P-12-12	102A			TR85/	
09-004	718	3159	IC-8		C6094P
09-005	703A				
09-006	703A				
09-007	750				
09-010	746				
09-011	720		IC-7		C6095P

Original Part Number	ECG	SK	GE	IR/WEP	HEP
09-017	722	3161			
09-018	788				
09-033006	103A	3010	8	TR08/724	G0011
09-30063	159				
09-30126	102A	3004		TR85/	
09-30313	102A	3004		TR85/	
09-30501	199				
09-32124	108				
09-32128				TR70/	
09-32129				TR21	
09-32130				TR78/	
09-32131				TR51/	
09-32146				TR23/	
09-300002	126	3008	1	TR17/635	G0008
09-300005	158	3005		TR84/	G0005
09-300006	126	3006	1	TR17/635	G0006
09-300007	126	3004		TR17/635	G0008
09-300011	126	3005	2	TR17/635	G0008
09-300012	126	3006	50	TR17/635	G0008
09-300015	126	3006	50	TR17/635	G0008
09-300016	126	3006	50	TR17/635	G0008
09-300017	102A	3005	1		G0005
09-300021	160	3006	50		G0003
09-300024	160	3006	51	TR17/637	G0008
09-300027	126	3006		TR17/635	G0009
09-300028	160	3006		TR17/	G0009
09-300029	126	3006		TR17/635	G0008
09-300037	159	3114		TR01/717	S0013
09-300037A	121	3114		TR01/	
09-300061	159	3025	21	TR20/	
09-300062	159	3025	21	TR20/	
09-300063	159	3114		TR20/	S0012
09-300064	159	3024	21	TR52/	S0012
09-300067	187	3083	29	TR56/	S3032
09-300068	187	3083	29	TR56/	S5006
09-300072	193		67		
09-300073	187	3084	29	TR56/	S5006
09-300074	159	3118	67	TR21/	
09-300076	159	3114	58	TR30/	S5013
09-300077	159	3114	21		S0013
09-300078	126	3083	69	TR17/	G0009
09-300079	126	3006		TR17/635	
09-300081	159	3114	21	TR52/	S0013
09-300307	159				S0013
09-301001	102A	3004	53	TR85/	G0005
09-301002	102A	3004		TR85/250	G0005
09-301002-6	102A	3004		TR85/	
09-301003	102A	3004	52	TR85/250	G0007
09-301004	102A	3123	52	TR85/250	G0005
09-301005	102A	3004	52	TR85/	G0005
09-301006	102A	3004		TR85/	G0005
09-301007	102A	3004		TR85/	G0005
09-301008	158	3004		TR84/630	G0005
09-301008-18	102A	3004		TR85/	
09-301009	102A	3004		TR85/250	G0007
09-301010	104	3009	16	TR01/230	G6003
09-301014	158	3004	2	TR84/630	G6011
09-301015	176	3123		TR84/238	
09-301016	102A	3004	2	TR85/250	
09-301019	158	3004		TR84/630	
09-301020	102A	3004	2	TR85/250	G0005
09-301022	158	3004	53	TR84/630	G0005
09-301023	102A	3004	52	TR85/	G0005
09-301024	131	3082		TR94/642	G6016
09-301025	102A	3004		TR85/250	G0003
09-301025-6	102A	3004		TR85/	
09-301026	102A	3004	52	TR85/	G0005
09-301027	158	3004	2	TR84/630	G6011
09-301030	226MP	3052	31MP	TR94/	G6016
09-301031	176	3004		TR82/	G0005
09-301032	102A	3004		TR85/250	G0007
09-301034	131	3082		TR94/642	G6016
09-301036	102A	3123	52	TR85/250	G0005
09-301048	102A	3004	52	TR85/	G0005
09-301052	121	3009	3	TR01/	G6005
09-301054	102A	3004	53	TR85/	G0005
09-301056	102A	3004	52	TR85/	
09-301066	158	3004			G6011
09-301071	127	3024		TR27/	G6018
09-301072	102A	3004	2		G0005
09-301073	127	3035	25	TR27/	G6008
09-301074	102A	3004			G0005
09-301075	226	3082	30		G6016
09-301077	153		29	TR56/	
09-301079	127	3035			G6007

Original Part Number	ECG	SK	GE	IR/WEP	HEP
09-302002	107	3122	20	TR21/	S0014
09-302003	108	3018	17	TR95/	S0014
09-302004	107	3018	20	TR70/720	S0014
09-302005	108	3039	20	TR95/	S0014
09-302006	108	3018	17	TR95/	S0016
09-302007	123A	3124	20	TR95/735	S0016
09-302009	108	3018	17	TR95/720	S0020
09-302010	108	3018	11	TR95/	S0016
09-302012	107	3018	11	TR21/735	S0016
09-302014	107	3124		TR70/720	S0011
09-302015	195	3024	18	TR65/243	S3001
09-302016	233	3117	20	TR95/720	S0016
09-302017	108	3018	20	TR95/	S0014
09-302020	123A	3124	20	TR21/735	S0014
09-302030	123A	3122	11	TR21/735	S5014
09-302032	107	3039		TR70/720	S0016
09-302033	107	3018	20	TR70/720	S0011
09-302034	123A	3124	10	TR21/735	S0015
09-302035	128				
09-302036	199	3039	60	TR33/	S0016
09-302037	107	3018	20	TR70/720	S0016
09-302037(IC)	1006				
09-302038	123A	3124	20	TR21/	S0015
09-302039	123A	3122	20	TR21/735	S0015
09-302040	123A	3122	63	TR21/735	S0016
09-302044	107	3124		TR70/720	S0015
09-302045	123A	3124	20	TR21/735	S5014
09-302045-12	123A	3124		TR21/	
09-302046	186	3083	28	TR55/S3023	S3020
09-302050	224	3049		TR87/243	S3002
09-302051	236	3049		TR87/243	S3002
09-302053	199	3124	20	TR95/735	S0016
09-302054	123A	3122	17	TR21/	S0011
09-302055	195	3048		TR65/	
09-302056	224	3049			
09-302058	123A		20		S0015
09-302060	108	3124	20	TR70/	S0016
09-302061	107	3019	11	TR70/	S0020
09-302062	123A	3024		TR21/735	S0015
09-302063	107	3018	11	TR70/	S0017
09-302068	195	3047			S3013
09-302072	107	3117	20	TR70/720	S0017
09-302073	107	3018		TR70/	S0017
09-302074	123A	3117	20	TR21/735	S0015
09-302075	192	3122	18	TR24/735	S0014
09-302078	123A	3117	11	TR83/719	S0015
09-302079	107	3018		TR70/720	S0025
09-302080	186	3046		TR55/S3023	S3020
09-302081	195	3047		TR65/	S3001
09-302082	195			TR65/	S3001
09-302083	152	3054	28	TR76/701	S5000
09-302085	199	3124	18		S0015
09-302086	199	3018	62	2SC536/	S0016
09-302090	128	3024			
09-302092	107	3122	17	TR70/	S0015
09-302093	199	3124	62	TR21/	S0015
09-302095	107	3018	17	TR21/	S0016
09-302099	154	3040	18		S5025
09-302101	123A	3038	20	TR21/	S0015
09-302102	210	3047	28	TR74/	S3020
09-302106	123A	3122	10	TR21/	S0014
09-302107	123A	3122	20	TR51/	S0020
09-302113	186	3054	28	TR55/	S5000
09-302114	107				
09-302115	108				
09-302116	192				
09-302117	237	3048			
09-302122	130	327	19	TR59/	S7002
09-302123	186	3054	28	TR55/	S5000
09-302124	123A	3122	20	TR21/	S0015
09-302124A		3018			
09-302125	123A	3122	20	TR21/	S0015
09-302126	186	3041	28	TR76/701	S5004
09-302127	123A	3020	20	TR21/	S0024
09-302128	161	3018	60	TR70/	S0017
09-302129	107	3132	61	TR21/	S0020
09-302130	194	3045		TR78/	S3021
09-302131	123A	3124	18	TR25/	S0015
09-302132	152	3054	28	TR55/	S5000
09-302135	190	3047			S3020
09-302136	184	3054	28		S5000
09-302138	107	3018	20		S0016
09-302139	123A				
09-302140	123A	3122	20	TR21/	S0015
09-302141	108	3019	11	TR83/	S0020

Original Part Number	ECG	SK	GE	IR/WEP	HEP
09-302142	107	3018	11	TR80/	S0016
09-302143	107	3018	11	TR80/	S0016
09-302146	124	3021	12		S5011
09-302148	123A	3018	10	TR51/	
09-302149	108	3018	17	TR95/	
09-302150	154		18	TR78/	S5016
09-302151	107	3018	11	TR80/	S0016
09-302152	107	3117	60	TR71/	S0020
09-302153	123A	3132	61	TR70/	S5026
09-302155	128	3024		TR25/	
09-302156	124	3021	12	TR23/	S5011
09-302157	164	3133	37	TR68/	S5020
09-302158	163	3111	36	TR93/	S5021
09-302159	162	3535	14	TR61/	S7004
09-302160	124	3021	12	TR23/	S5011
09-302161	193	3025	67		S5022
09-302162	108	3019	11	TR83/	
09-302164	152	3054			
09-302165	123A				S0015
09-302166	237		46		
09-302169	237	3049	46		
09-302170	187	3083	66	TR56/	
09-302171	128	3024	18	TR25/	
09-302172	123A	3124	17	TR25/	S0015
09-302173	128	3122	20	TR24/	S0014
09-302175	123A		67	TR24/	S5014
09-302177	163	3111	73		S5020
09-302185	124	3021	12	TR23/	S5011
09-302186	190		32	TR74/	S3021
09-302187	165	3115	38	TR93/	S5021
09-302188	124	3021	12	TR81/	S5011
09-302189	190		32	TR76/	S3021
09-302190	108	3018	20	TR24/	S0016
09-302192	235	3054			S3020
09-302193	236				
09-302194	199	3124			S0015
09-302200	233				
09-302201	161	3018			S0016
09-302202	195	3048			S5014
09-302203	107				
09-302204	107				
09-302206	229	3018			S0016
09-302207	107	3018			
09-302212	295	3054			S5000
09-302215	123A	3018			S0015
09-302216	107	3018			S0015
09-302218	175	3131			
09-302219	302				
09-303005	175		23	TR81/	S5019
09-303006	103A		59		
09-303012	103		6		G0011
09-303013	103A	3010	8	TR08/724	
09-303018	152	3054			
09-303019	192	3124	10	TR24/	S0014
09-303021	196	3054			S5000
09-303022	152	3054	28	TR55/	
09-303023	103A		54		G0011
09-303025	123A	3010	20	TR24/	S0015
09-303028	124	3021	12	TR23/	S5011
09-303029	198	3054			S5000
09-303058		3020			
09-304011	160	3008		TR17/637	G0008
09-304012	159	3114	21	TR20/	S0013
09-304017	132	3116			F0021
09-304042		3018			
09-304043		3018			
09-304044		3018			
09-304045		3018			
09-304046		3104			
09-304047		3114			
09-304048		3018			
09-304049		3114			
09-304050		3114			
09-304051		3114			
09-304052		3104			
09-304055		3104			
09-304056		3111			
09-304057		3115			
09-304058		3018			
09-304140	152				
09-305006	161	3117		TR83/	S5006
09-305011	161	3018		TR83/719	S0030
09-305014	132	3116	FET2	FE100/802	F0015
09-305021	133	3116	FET1	FE100/801	F0015
09-305023	133				F0010
09-305024	159	3114	21	TR23/	S0019
09-305031	133	3112	FET2	FE100/	F0010
09-305032	132	3116	FET2	FE100/	
09-305033	107	3018	17	TR24/	
09-305034	123A	3018	20	TR21/	
09-305041	107	3018			S0014
09-305048	199	3124			S0033
09-305049	300	3054			S5000
09-305050	229	3018			
09-305051	229	3018			
09-305052	199	3124			
09-305058	234	3025			S0032
09-305062	123A	3018			S0011
09-305063	123A	3018			S0015
09-305064	123A				
09-305065	123A				
09-305066	123A				
09-305067	123A				
09-305068	123A				
09-305069	108	3018	20	TR24/	S0015
09-305070	108	3018			
09-305071	108	3018			
09-305072	108	3018	20	TR24/	S0015
09-305073	159	3114			
09-305074	108	3018	20	TR21/	
09-305075	129	3025			
09-305076	128	3024			
09-305077	123A	3122			
09-305091		3049			
09-305092	154	3024			
09-305093	107	3018			
09-305094	107	3018			
09-305095	235	3197			
09-305096	229				
09-305123	199	3124			
09-305126	199	3122			S0016
09-305132	161	3039			S0020
09-305133	133	3112			F0010
09-305134	129	3083			S3014
09-305136	236				
09-305137	236				
09-305138	223	3027			
09-305139	123A	3018			
09-308002	724	3525			
09-308003	724	3525			
09-308004	703A	3157	IC12	IC500/	C6001
09-308008	1003		IC2	IC507/	C6060P
09-308009	1046				
09-308010	726	3129			
09-308011	1054		IC45		
09-308013	703A	3157	IC12		C6001
09-308017	704	3023			
09-308019	703A	3157			
09-308021	9946				
09-308022	7400				
09-308023	1004				
09-308024	1021				
09-308025	1076				
09-308026	1091*				
09-308027	1047				
09-308029	1086**				
09-308030	1089¤				
09-308031	1048				
09-308033	1045	3073	IC2		C6060P
09-308034	1006				
09-308036	1103				
09-308037	1029				
09-308038	1142				
09-308041	1100				
09-308043	1103				
09-308044	1120				
09-308045	715	3076	IC6	IC510/	C6071P
09-308046	715	3076	IC6	IC510/	C6071P
09-308047	713	3077	IC5	IC508/	C6057P
09-308048	713	3077	IC5	IC508/	C6057P
09-308049	1019				
09-308051	1113				
09-308052	1082				
09-308053	1075				
09-308058	1005				
09-308059	1100				
09-308061	1097				
09-308062	1135				
09-308063	1003	3102	IC2		C6060P
09-308064	1006				

Original Part Number	ECG	SK	GE	IR/WEP	HEP
09-308065	1016				
09-308066	1106				
09-308067	1107				
09-308069	1047				
09-308070	1108				
09-308071	720				
09-308076	1052				
09-308077	1012				
09-308078	732				
09-309006	123A	3018	17	TR22/	S0015
09-309007	108	3018	17	TR21/	S0011
09-309012	123A	3122	20	TR21/	S0015
09-309013	108	3122	20		S0015
09-309023	123A	3122	10	TR21/	
09-309024	108	3122	20		S0014
09-309027	108	3018	20	TR21/	S0014
09-309028	108	3122	20	TR21/	S0014
09-309029	172	3156			S9100
09-309030	193	3025	67	TR28/	S5013
09-309031	192	3024	63	TR25/	S5014
09-309032	108	3122	20	TR21/	S0015
09-309038	159	3025	21	TR20/	
09-309042	159	3114	21	TR30/	S0019
09-309049	123A	3124	62		
09-309050	123A	3124	20		
09-309059	199		212		
09-309061	289		88		
09-309062	195	3047			
09-309063	123A	3024			
09-309064	123A	3018			
09-309065	229				
09-309069	107	3018	20	TR21/	S0015
09-309070	199	3124	18	2SC945/	S0015
09-309071	186	3054	28	TR55/	S5000
09-309072	107	3122	17		S0015
09-309073	108	3018	20		
09-309074	132				
09-309075	103A	3010			
09-309076	123A				
09-309672	107				
09-310171			25		
09-3002006	229	3018			
09L-4	102A	3003		TR85/	
09L-4-12	102A	3003		TR85/	
09L-4-12-7	102A	3003		TR85/	
09N1	123A	3122		TR21/735	
010-001	912				
010-694(AMPEX)	154			TR78/706	
011-H01	100	3005		TR05/254	
012				TR24/	
012-1	123A	3020		TR21/	
012-1-12	123A	3020		TR21/	
012-1-12-7	123A	3020		TR21/	
012-103002	152				
012E	123A	3122		TR21/735	
012-H02	100	3005		TR05/254	
013-005002-06	754				C6007
013-005005-6	761				C6011
013-005007	761				
013-005007-6	761				C6011
013-005009-6	767				
014-556	105			TR03/	G6006
014-611	159	3025		TR20/717	S0019
014-652	159	3114		TR20/717	S0019
014-652C	159	3114		TR20/	
014-680	123A				
014-686	123A				
014-698	123A			TR21/735	S0015
014-754	234				
014-772	159				
014-784	123A			TR21/735	S0015
014-862	199	3122		TR21/735	S0024
014A	121	3009		TR01/	
014A-12	121	3009		TR01/	
014A-12-7	121	3009		TR01/	
015	108	3039		TR95/	
015(LAMBDA)	912				
016B12	123A	3122		TR21/735	S0015
016B810	123A	3122		TR21/735	S0015
016B812	123A	3122		TR21/735	S0015
017E824	123A	3122		TR21/735	
018-00001	159	3114	21	TR20/	S0019
018-00002	159	3025	21	TR20/	S0019
018-00003	123A	3124	20	TR21/	S0015
018-00004	187	3084	69	TR56/	S5006
018-00005	186	3054	20	TR55/	S5000
018-0003				TR21/	
018-0004				TR56/	
018-0005				TR55/	
018-0009				TR21/	
018-0010				TR21/	
018B	107	3039		TR70/720	S0020
019-00009	123A	3124	20	TR21/	S0015
019-00010	123A	3124	20	TR21/	S0015
019-003315	126	3007		TR17/635	
019-003317	103	3010		TR08/641A	
019-003318	103	3010		TR08/641A	
019-003319	103	3010		TR08/641A	
019-003324	100	3005		TR05/254	
019-003342	100	3005		TR05/254	
019-003343	100	3005		TR05/254	
019-003349	123A	3018	17	TR21/735	
019-003415	158	3004	53	TR84/630	
019-003416	102A	3004	2	TR85/250	
019-003485	175	3026		TR81/	
019-003637	128				
019-003675-196	123A				
019-003675-203	123A				
019-003675-205	199				
019-003675-207	123A				
019-003675-231	159				
019-003675-232	159				
019-003675-234	159				
019-003675-246	123A				
019-003675-257	159				
019-003691	195	3048		TR65/	
019-003692	195	3048	18	TR65/243	
019-003777	126	3007		TR17/635	
019-003778	126	3007		TR17/635	
019-003929	108	3018		TR95/	
019-003931	159				
019-003932	108	3018		TR95/	
019-003934	123A	3047		TR21/	S0015
019-003935	224	3049			S3019
019-004094	199				
019-004111	123A				
019-004428-002	123A				
019-004558	159				
019-005006	108	3018		TR95/	S0020
019-005010	123A	3020		TR21/	S0019
019-005021	123A				
019-005157	161				
019-005179	159				
020-00023	126	3008	51	TR17/	G0008
020-00024	107	3018	20	TR70/	S0014
020-00025	107	3018	20	TR70/	
020-00026	108	3018	17	TR95/	S0014
020-00027	108	3018	11	TR95/	S0014
020-00028	107	3018	17	TR70/	S0016
020-114-007	703A	3157			
020-114-008	703A	3157			
020-114-009	720				
020-1110-004	159	3114		TR30/717	S0019
020-1110-004 (SCOTT)	159			TR20/	
020-1110-005	234	3114	21	TR24/717	S0015
020-1110-005 (SCOTT)	159			TR20/	
020-1110-006 (SCOTT)	160			TR17/	
020-1110-007 (SCOTT)	129			TR88/	
020-1110-008	128	3024	18	TR87/243	S5014
020-1110-008 (SCOTT)	128			TR87/	
020-1110-009 (SCOTT)	128			TR88/	
020-1110-010 (SCOTT)	123A			TR21/	
020-1110-011				TR51/735	
020-1110-012	199	3122	20	TR51/735	S5013
020-1110-012 (SCOTT)	123A			TR21/	
020-1110-013	199	3122	20	TR51/735	
020-1110-014	129	3044	27	TR24/706	
020-1110-014 (SCOTT)	154			TR78/	
020-1110-015		3025		TR88/242	
020-1110-015 (SCOTT)	129			TR88/	

Original Part Number	ECG	SK	GE	IR/WEP	HEP
020-1110-025				TR85/250	
020-1110-025 (SCOTT)	102A			TR85/	
020-1111-002		3027	14	TR26/247	
020-1111-002 (SCOTT)	130			TR59/	
020-1110-018		3112		/801	
020-1110-018 (SCOTT)	133				
020-1110-021	133	3112	FET1	FE100/	F0010
020-1110-022	133				
020-1110-016	133	3112	FET1	/801	
020-1110-016 (SCOTT)	133				
020-1110-017		3112	20	TR51/735	
020-1110-017 (SCOTT)	123A			TR21/	
020-1111-003		3027	14	TR26/247	
020-1111-003 (SCOTT)	130			TR59/	
020-1111-004		3083	69	TR /S5007	
020-1111-004 (SCOTT)	185	3191			
020-1111-005		3054	66	TR55/S5003	
020-1111-005 (SCOTT)	184	3190			
020-1111-006 (SCOTT)	179			TR35/	
020-1111-007			14	TR26/247	
020-1111-007 (SCOTT)	130			TR59/	
020-1111-008	130	3027	35	TR26/247	
020-1111-008 (SCOTT)	130			TR59/	
020-1111-009	219			/S7001	
020-1111-009 (SCOTT)	180				
020-1111-010				/S5004	S5004
020-1111-010 (SCOTT)	182	3188			
020-1111-011				/S5005	S5005
020-1111-011 (SCOTT)	183	3189			
020-1111-017	129	3025		TR88/	S5013
020-1111-018	128	3024		TR87/	S5014
020-1111-019	130				
020-1111-038	128				
020-1111-080	219				
020-1112-001	123A	3024	18	TR87/243	
020-1112-001 (SCOTT)	128			TR87/	
020-1112-002		3116	FET2	/802	
020-1112-002 (SCOTT)	132			FE100/	
020-1112-003	161			TR83/	
020-1112-004	107	3122	20	TR51/735	
020-1112-004 (SCOTT)	123A			TR21/	
020-1112-005	133	3112	FET1	FE100/	F0010
020-1112-006	132	3116	FET2	FE100/	F0021
020-1112-007	133				
020-1112-008		3116	FET2	FE100/	F0021
020-1112-009	133				C6095P
020-1114-003	725	3162			
020-1114-005	725	3162			
020-1114-006	718	3159			
020-1114-007	703A		IC12	IC500/	
020-1114-007 (SCOTT)	703A				
020-1114-008	703A		IC12	IC500/	
020-1114-009	720		IC7	IC512/	
020-1114-013	725				
021	128			TR87/	
021-0121-00	123A				
021-0137-00	128				
021-0224-00	129				
022-006500	123	3020	20	TR86/53	
022-009502				TR25/	
022-1	703A	3157			
022-1110-005C	159			TR20/	
022-3504-040	102	3003		TR05/631	S0032
022-3504-060	102	3003		TR05/631	G0008
022-3505-910	102	3004	2	TR05/631	G0009
022-3511-770	180	3006		TR17/637	G0003
022-3511-780	160	3006		TR17/637	G0003
022-3511-790	160	3006		TR17/637	G0003
022-3516-360	126	3008			
022-3640-050	121	3009	16	TR01/232	
022-3640-080	108	3018	17	TR95/56	
022-3640-081	195	3047		TR65/	
022-3640-082	195	3048		TR65/	

Original Part Number	ECG	SK	GE	IR/WEP	HEP
022-3640-253	158	3004		TR84/	G0005
022-5311-770	126	3006		TR17/635	
022-5311-780	126	3006		TR17/635	
022-5511-790	126	3006		TR17/635	
022A	703A	3157			
023	175			TR81/241	S5012
024	218	3083	26	TR58/	
025	130	3027		TR59/247	S7004
025-1	730	3143			
025-10030	123A	3122		TR21/	
025-10031	102A	3004		TR85/	
025-100003	108	3018	17	TR95/56	
025-100004	108	3018	17	TR95/56	
025-100009	108	3039	17	TR95/56	
025-100012	161	3018	17	TR83/719	S0017
025-100013	108	3018	17	TR95/56	S0016
025-100014	108	3018	11	TR95/56	S0030
025-100015	128	3024	63	TR87/243	S0016
025-100017	199	3020	20	TR21/735	S0024
025-100018	123A	3020	20	TR21/735	S0015
025-100026	107	3016	11	TR70/720	S0016
025-100030	123A			TR21/56	S0011
025-100031	158	3004		TR84/630	G0005
025-100036	161	3018		TR83/56	S0017
025-100037	161	3018		TR83/56	S0017
025-100038	161	3018		TR83/56	S0017
025-100040	123A	3020	20	TR21/735	S0025
025B	123A	3122		TR21/735	S0015
025B1C	730				
025-YEL	123A	3122		TR21/735	
026-100003	121	3009	13	TR01/232	
026-100004	121MP	3013	13	/232MP	
026-100005	102	3004	2	TR05/631	
026-100012	102	3004	52	TR05/631	
026-100013	128	3024	20	TR87/243	
026-100017	123	3020	20	TR86/53	
026-100018	102	3004	52	TR05/631	
026-100019	129	3025		TR88/	
026-100020	121MP	3013	16	TR01/232MP	
026-100026	123A	3020		TR21/	
026-100028	121	3009		TR01/232	G6003
028	124			TR81/240	
028-300-226	110MP	3119			
031A	123A	3122		TR21/	
033-31(SYL)	123A			/735	
033A	199			TR21/	
034A	194	3104			
037	123A	3122		TR21/	
038	128	3122		TR21/243	
039	129	3025		TR20/242	
041	123A	3122		TR21/735	
041A	222	3050			
042	123A	3102		TR21/735	
042A	710	3102			
042(IC)	710				
044-9667-02	123A				
045	221	3050			
045-1(SYL)	108			TR21/56	
045-2	108	3039		TR21/56	
046-1(SYL)	186			TR55/S3023	
047	221	3050			
047-1(SYL)	187			TR56/S3037	S5006
048	220				
051-0010-00	1023				
051-0010-01	1023				
051-0010-02	1023				
051-0010-03	1023				
051-0010-04	1023				
051-0010-05	1023				
051-0011-00	1103				
051-0011-00-04	1103				
051-0011-00-05	1107				
051-0011-01	1103				
051-0011-02	1103				
051-0011-03	1103				
051-0011-04	1103				
051-0012-00	720		IC-7	IC512	
051-0017-00	722		IC-9		
051-0021-00	746		IC-26		
051-0022-00	709				
051-0038-00	1005				
051-0046	123A	3018		TR21/735	
051-0047	123A	3020		TR21/735	
051-0049	107	3020		TR70/720	
051-0062	160	3007		TR17/637	

Original Part Number	ECG	SK	GE	IR WEP	HEP
051-0063	160	3007		TR17/637	
051-0079	126	3008		TR17/635	
051-0107	159	3025		TR20/717	
051-0151	175	3026		TR81/	
051-0155	123A	3046		TR21	
051-0156	195	3047		TR65/	
051-0157	195	3048		TR65/	
052	172	3156		TR69/S9100	
052A	172	3156		TR69/S9100	
053	722				
053-1	722	3161			
054	123A	3122		TR21/735	
055	123A	3122		TR21/735	
056	159	3025		TR20/717	
057	123A	3132		TR21/735	
057B474H	161	3132		TR83/719	
059	124	3021		TR81/240	
062	123A	3122		TR21/735	
065	175			TR81/241	
065-1	160	3006		TR17/	
065-1-12	160	3006		TR17/	
065-1-12-7	160	3006		TR17/	
065-2	103A	3010		TR08/	
065-2-12	103A	3010		TR08/	
065A	160	3007		TR17	
065A-12	160	3007		TR17/	
065-12-7	160	3007		TR17/	
065B	126	3008		TR17/	
065B-12	126	3008		TR17/	
065B-12-7	126	3008		TR17/	
065C	121	3009		TR01/	
065C-12	121	3009		TR01/	
065C-12-7	121	3009		TR01/	
066	128			TR87/243	
066-1	102A	3003		TR85	
066-1-12	102A	3003		TR85	
066-1-12-7	102A	3003		TR85	
066-2	102A	3004		TR85/	
066-2-12	102A	3004		TR85/	
066-2-12-7	102A	3004		TR85/	
066-3	102A	3004		TR85/	
066-3-12	102A	3004		TR85/	
066-3-12-7	102A	3004		TR85/	
069	123A	3122		TR21/735	
070	130	3027		TR59/247	
070-001	160	3006		TR17/	
070-003	110MP				
070-020	100	3005		TR05/	
072	184	3102		/S5003	
072(IC)	710				
074-1(PHILCO)	186				
075-045037	7474				
075-046270	7472				
077A	105	3012		TR03/	
077A-12	105	3012		TR03/	
077A-12-7	105	3012		TR03/	
077B	121MP	3013			
077B-12	121MP	3013			
077B-12-7	121MP	3013			
077C	121	3014		TR01/	
077C-12	121	3014		TR01/	
077C-12-7	121	3014		TR01/	
079	185	3191		/S5007	
080	184	3190		/S5003	
081	184	3190		/S5003	
081-1	790				
081(IC)	790				
082	712	3072			
082-1	712	3072			
082A	712	3072			
082A(PB)	712				
082.115.015	159				
083A(PB)	196			TR92/	
084A	121	3009		TR01/	
084A-12	121	3009		TR01/	
084A-12-7	121	3009		TR01/	
085	130	3027		TR59/247	
086	104				
086-005132-02	123A				
087	104				
088B	160	3006		TR17/	
088B-12	160	3006		TR17/	
088B-12-7	160	3006		TR17/	
088C	100	3005		TR05/	
088C-12	100	3005		TR05/	

Original Part Number	ECG	SK	GE	IR WEP	HEP
088C-12-7	100	3005		TR05/	
089-2	228				
089-3	228				
089-4(SYL)	157			TR60/	
089-214	108	3018	17	IRTR95/	S0014
089-215	108	3018	17	IRTR95/	S0015
089-216	108	3122	20	TR21/	S0011
089-220	126	3006	50	TR17/	G0003
089-222	102A	3003	2	TR05/	G0005
089-223	123A	3122	62	IRTR51/	S0015
089-226	123A	3122	62	IRTR51/	S0015
089-231	158	3004	53	IRTR84/	G6011
089-233	103	3010	8	TR09/	G0011
092-1	105	3012		TR03/	
092-1-12	105	3012		TR03/	
094	714	3075			
094-1	714	3075			
094A	714	3075			C6070P
094A(PB)	714				
095	715	3076			
095-1	715	3076			
095A	715	3076			C6071P
095A(PB)	715				
096	713	3077			
096-1	713	3077			
096A	713	3077			C6057P
096A(PB)	713				
097	780				
097-1	783				C6100P
097A	780	3141			C6100P
Q97A(PB)	780				
099-1(PHILCO)	152				
0101-0034	126	3006	50	TR17/	G0009
0101-0060A	108				
0101-0222	126	3006	50		G0008
0101-0439	193	3025	67	TR28/	S301
0101-0448	187	3083	29	IRTR56/	S5006
0101-0448A	187	3083	29	IRTR56/	
0101-0466	187	3083	29	TR56/	
0101-0491	123A	3122		TR21/	
0101-0531	108	3039		TR95/	
0101-0540	123A	3122		TR21/	
0101-439	129	3025		TR88/	
0102-0371	102A	3004			
0103-0014	123A	3024	20	TR21/	S0015
0103-0014/4460	123A				
0103-0014R	128	3024		TR25 /	S0015
0103-0014S	128	3024	18	TR25/	S0015
0103-0051	128	3024	63	TR25/	S0015
0103-0060	108	3018	20	TR95/	S0011
0102-0060B	108	3018	20	TR95/	S0014
0103-0088	123A	3122	20	TR21/	S0015
0103-0088/4460	199				
0103-0088H	123A	3124	10	TR24/	
0103-0088R	123A	3122	62	IRTR51/	S0015
0103-0088S	123A	3122	62	IRTR51/	S0015
0103-0191	108	3018	17	TR95/	S0016
0103-0389	108			TR95/	S0016
0103-0419	186	3054	28	IRTR55/	S5000
0103-0419A	186	3054	28	IRTR55/	
0103-0473	123A	3124	20	TR24/	S0015
0103-0482	123A	3124			
0103-0491	123A	3122	17	TR21/	S0015
0103-0491/4460	123A				
0103-0492	199	3124			
0103-0503	128	3024	63	TR87/	S0015
0103-0503S	128	3024	18	TR25/	S5000
0103-0504	123A	3122		TR21/	
0103-0521	108			TR21/	S0016
0103-0521B	108	3018	17	IRTR95/	S0014
0103-0531	108	3039	20	TR95/	S0015
0103-0540	123A	3122	10	TR21/	S0015
0103-0607	128	3024	63	TR87/	S3001
0103-0616	128	3024	63	TR87/	S3001
0103-389	108	3039	20		S0016
0103-512M	186	3054	28	TR55/	S5000
0103-521	108			TR21/	S0016
0103-0531/4460	107				
0103-568	107				S0020
0103-568/4460	107				
0103-93	199				
0103-94	123A				
0103-95	158				
0103-9531/4460	108				
0104-0013	128	3024	63	TR87/	S5014

Original Part Number	ECG	SK	GE	IR/WEP	HEP
0123	185	3041		/S5007	S5006
0123(WARDS)	185			/S5007	S5006
0124	184	3024		IRTR55/717	
0124A	159	3114		TR20/	
0124(HOFFMAN)	184				S5000
0124(KNIGHT)	159			TR20/	S0019
0124(WARDS)	184				S5000
0125	123A	3122		TR21/717	S0015
0126	123A	3025		TR21/735	S0015
0126(WARDS)	189	3200		TR73/	S3028
0127	123A	3122		TR21/735	S0015
0128	123A	3122		TR21/735	S0014
0130	128	3024		TR87/243	
0131	123A	3122		TR21/735	S0015
0131-000100	102A	3004		TR85/	
0131-000101	102A	3004		TR85/	
0131-000102	102A	3004		TR85/	
0131-000192	121	3009		TR01/	
0131-000335	159	3098		TR20/	
0131-000336	121	3009		TR01/	
0131-000337	121	3009		TR01/	
0131-000418	160	3006		TR17/	
0131-000419	160	3006		TR17/	
0131-000473	123A	3122		TR21/	
0131-000498	160	3006		TR17/	
0131-000561	128	3024		TR87/	
0131-000562	121	3009		TR01/	
0131-000563	102A	3004		TR85/	
0131-000704	123A	3122		TR21/	
0131-000802	160	3006		TR17/	
0131-000859	160	3006		TR17/	
0131-000862	160	3006		TR17/	
0131-000863	160	3006		TR17/	
0131-001050	158			TR84/	
0131-001056	158	3004		TR84	
0131-001182	160	3006		TR17/	
0131-001314	160	3006		TR17/	
0131-001328	159	3025		TR20/	
0131-001329	159	3025		TR20/	
0131-001332	160	3006		TR17/	
0131-001417	123A	3122		TR21/	
0131-001418	123A	3122		TR21/	
0131-001419	102A	3004		TR85/	
0131-001420	159	3025		TR20/	
0131-001421	123A	3122		TR21/	
0131-001422	123A	3122		TR21/	
0131-001423	123A	3122		TR21/	
0131-001424	123A	3122		TR21/	
0131-001425	121	3009		TR01/	
0131-001426	102A	3004		TR85/	
0131-001427	129	3025		TR88/	
0131-001428	129	3025		TR88/	
0131-001429	128	3024		TR87/	
0131-001430	128	3024		TR87/	
0131-001433	160	3006		TR17/	
0131-001434	160	3006		TR17/	
0131-001435	160	3006		TR17/	
0131-001436	160	3006		TR17/	
0131-001438	106	3118		TR20/	
0131-001439	159	3025		TR20/	
0131-001464	123A	3122		TR21/	
0131-001597	130	3027		TR59/	
0131-001697	160	3006		TR17/	
0131-001864	123A	3122		TR21/	
0131-002049	153	3083		TR77/	
0131-002068	130	3027		TR59/	
0131-002656	102A	3004		TR85/	
0131-003029	160	3006		TR17/	
0131-004323	123A				
0131-004367	130				
0131-004746	159				
0131-004792	199				
0131-005347	199				
0131-005348	199				
0131-005349	199				
0131-005350	199				
0131-005351	159				
0131-005352	152				
0131-005353	153	3084			
0131-04560	175				

Original Part Number	ECG	SK	GE	IR/WEP	HEP
0105-0012	132	3116	FET2	FE100/	
0117-02	221				
0121	123A	3122		TR21/735	
0121(AIRLINE)	185			/S5007	
0122(AIRLINE)	184			/S5003	
0131-4328	159				
0132	130	3027		TR59/247	S7002
0133	159	3025		TR20/717	S0019
0133-008T	159				
0133-1003T	159				
0134	123A	3122		TR21/735	S0015
0137	219	3173			S7003
0140-5	199				
0140-6	123A				
0140-7	186				
0142	108	3039		TR95/56	S0020
0205	722	3161			C6056P
0205(DYNACO)	722				
0243-001	175				
0301-3055-00	130				
0418	121MP	3013			
0418(HK)	121MP				
0731	128				
0770	185	3083		TR77/S5006	S5006
0772	184	3054		/S5003	S5000
0773	185	3041		/S5007	S5006
0776-0160	123A				
0776-0195	159				
0831	234				S0019
01057-1	184	3054			S5000
02057-1	185	3083			S5006
02375-A	175			TR81/241	
03008-1	123A	3122		TR21/735	
04053	234				
05206-00	199				
06246-00	123A				
09004	718	3159			C6094P
09005	703A	3157			
09006	703A	3157			
09007	750				C6061P
09010	746				C6059P
09011	720				C6095P
09017	722	3161			
09018	788				
09500	199			TR21/	S0024
09501	199	3124	63	TR21/735	S0024
09502	199	3122		TR21/735	S0024
09502-8	123A			TR21/	
09800	159	3025		TR20/717	S0019
09800-12	129	3025		TR88/	
012013-1	123A				
012015	128				
012085	175				
012099-1	129				
014382	121	3009		TR01/	
014558	154	3044		TR78/	
018069	129	3025		TR88/	
018077	128	3024		TR87/	
020156	102A	3123		TR85/	
020425-3	196				
020426-3	197				
022939	157	3044		TR60/	
023606	132	3116		FE100/	
023754	175	3026		TR81/	
023762	130			TR59/247	
026237	123A	3019		TR21/	
030010	123A	3122	20	TR21/735	
030010-1	123A	3122	20	TR21/735	
030011	128	3024	18	TR87/243	
030011-1	128	3024	18	TR87/243	
030011-2	128	3024	18	TR87/243	
030254	128	3124		TR87/	
030512-1	123A	3122	20	TR21/735	
030512-2	123A	3122	20	TR21/735	
030515	123A	3122	20	TR21/735	
030515-4	123A	3122	20	TR21/735	
030527	123A	3122	20	TR21/735	
030531-1	105	3122	20	TR03/735	
030536	123A	3122	20	TR21/735	
030537	123A	3122	20	TR21/735	
030537-1	123A	3122	20	TR21/735	
030537-2	123A	3122	20	TR21/735	
030538	123A	3122	20	TR21/735	
030539-1	130		14	TR59/247	
030542	123A	3122	20	TR21/735	
030542-1	123A	3122	20	TR21/735	
030543	123A	3122	20	TR21/735	
030543-1	123A	3122	20	TR21/735	
030543-2	123A		20	TR21/735	
030812-1	103A	3010		TR08/724	

Original Part Number	ECG	SK	GE	IR/WEP	HEP
033563	161	3124		TR83/	
033571	129			TR88/	
033589	159	3025		TR20/	
036001	703A	3157	IC12	IC500/	
037077	175(2)				
037085	130	3036		TR59/	
040001	107	3020		TR70/	S0015
043001	108	3018		TR95/	S0016
061366	123A	3122		TR21/	
071818	129				
080001	126	3008		TR17/635	
080003	102A	3004	2	TR85/250	
080004	102A	3004	2	TR85/250	
080006	108	3018		TR95/56	
080021	108	3018	20	TR95/56	
080022	108	3018	20	TR95/56	
080023	108	3018	20	TR95/56	
080026	160	3006		TR17/637	
080027	160	3006		TR17/637	
080028	160	3006		TR17/637	
080041	108	3018		TR95/56	
080042	108	3018		TR95/56	
080043	102A	3004		TR85/250	
080047	102A	3004		TR85/	
080048	121	3009		TR01/	
080052	102A	3004		TR85/	
080059	108	3018		TR95/56	
080060	108	3018		TR95/56	
080061	160	3006		TR17/637	
080071	160	3006		TR17/637	
080072	126	3008		TR17/635	
080073	102A	3004	2	TR85/250	
080114	160	3005		TR17/637	
080206	126	3008		TR17/635	
080224	126	3008		TR17/	G0009
080225	126	3006		TR17/	
080228	126	3007		TR17/	G0009
080236	126	3008		TR17/635	
080244	160	3006		TR17/637	
080245	160	3005		TR17/637	
080253	126	3008		TR17/635	
080258	160	3007		TR17/	G0003
080266	160	3006		TR17/637	
080267	160	3006		TR17/637	
080269	160	3006		TR17/637	
080274	126	3005		TR17/635	
080275	126	3008		TR17/635	
080276	126	3008		TR17/635	
080277	126	3007		TR17/635	
081001	102A	3004	2	TR85/250	
081018	102A	3004		TR85/250	
081019	102A	3004		TR85/250	
081026	158	3004		TR84/	G0005
081027	158	3004		TR84/	G0005
081029	158	3004		TR84/	G0005
081038	102A	3004	2	TR85/250	
081042	131	3009		TR94/	G6016
081046	102A	3004		TR85/250	
081047	102A	3004	2	TR85/250	
081048	102A	3004		TR85/250	
081049	102A	3004		TR85/250	
081050	102A	3004		TR85/	
081056	102A	3004		TR85/250	
081059	158	3004		TR84/	G0005
082006	123A	3046		TR21/	S0014
082019	123A	3018		TR21/	S0014
082020	123A	3018		TR21/	S0014
082022	195	3048		TR65/	S0011
082028	123A	3124		TR21/	S0014
082029	195	3047		TR65/	
082033	195	3046		TR65/	S3001
084001	159	3025		TR20/717	S0019
084001C	121	3114		TR01/	
087003	102A	3004		/250	
01122945	186	3105			
0207046	1108				
0320031	123A	3122		TR21/	S0014
0320051	154	3044		TR78/	S5025
0404011-001	9937				
0404012-001	9949				
0440002-003	128				
0510079	160	3006		TR17/	
0510079H	160	3006		TR17/	
0563012H	102A	3004	52		G0005
0573001	102A		2	TR85/250	G0008

Original Part Number	ECG	SK	GE	IR/WEP	HEP
0573001-14	102A	3004		TR85/	
0573001H	102A	3004		TR85/	
0573003	102A	3004	52	TR85/631	G0005
0573003H	102A	3004	52	TR85/	G0005
0573004	102A	3004	53	TR85/	G0005
0573005	102A	3004	52	TR85/631	G0007
0573005-14	102A	3004		TR85/	
0573005H	102A	3004	52	TR85/	G0007
0573011	102A	3004	2	TR85/	G0005
0573012	102A	3004	2	TR85/250	G0005
0573012H	102A	3004		TR85/	G0005
0573018	102A	3004	2	TR85/250	G0008
0573018H	102A	3004	52	TR85/	G0003
0573022	102A	3004		TR85/250	G0006
0573022H	102A	3004		TR85/	
0573023	102A	3004	2	TR85/250	G0005
0573023A	102A			TR85/	
0573023H	102A	3004		TR85/	G0006
0573024	102A	3004	2	TR85/250	G0007
0573024-14	102A	3004		TR85/	
0573025	102A	3004		TR85/250	
0573030	131	3052	44	TR94/742	G6016
0573030-14	131	3052		TR94/	
0573031	131MP	3052	31MP	TR94MP/642	G6016
0573034	102A	3003		TR85/	G0006
0573036	102A	3004	52	TR85/250	G0007
0573036H	102A	3004		TR85/	G0007
0573037	013A	3010	8	TR08/734	G0011
0573037H	103A	3010		TR08/	G0011
0573040	121	3009	16	TR01/232	G6005
0573055	158	3004		TR84/631	G0007
0573056	102A	3004		TR85/	G0007
0573066	123A	3124	17	TR21/735	S0020
0573103	102	3004	52	TR05/631	G0005
0573114	102A	3004	2	TR85/250	G0007
0573114H	102A	3004	2	TR85/	G0005
0573117	102A	3004	2	TR85/250	G0005
0573117-14	102A	3004		TR85/	
0573119	102A	3004	2	TR85/250	G0007
0573125	102A	3004		TR85/250	G0005
0573131	102A	3004	2	TR85/	G0007
0573139	103A	3010		TR08/	G0011
0573142	102A	3004	52	TR85/250	G0005
0573142H	102A	3003	2	TR85/	G0005
0573152	102A	3004	2	TR85/250	G0005
0573153	102A	3004	2	TR85/250	G0005
0573153H	102A	3004		TR85/	G0005
0573166	104	3009	16	TR01/232	G6005
0573184	158	3004	53	IRTR84/	G6011
0573185	176	3123	52	TR82/250	G0005
0573187	102A	3004	52	TR85/250	G0006
0573199	127	3034	25	TR27/235	G6018
0573199H	127	3034	25	TR27/	G6018
0573200	102A		52	TR85/250	G0005
0573202	123A	3122		TR21/	S0011
0573204	158	3004		TR84/630	G0005
0573205	121	3009		TR01/	G6005
0573212	131	3035	25	TR94/235	G6008
0573212H	127	3035	25	TR27/	G6016
0573328	102A	3004	51	TR85/637	G0005
0573330	126	3006	50	TR17/	
0573335	160	3006	50	TR17/637	G0003
0573366	160	3006	50	TR17/637	G0003
0573398	126	3006	50	TR17/635	G0008
0573415	124	3021		TR81/240	S5011
0573418	123A	3122		TR21/735	
0573422	102A	3005	52	TR85/250	G0001
0573422H	102A	3004	1	TR85/	G0001
0573427	126	3006	50	TR17/	
0573428	160	3006	50	TR17/637	G0003
0573429	102A	3004		TR85/	
0573430	123A	3122		TR21/	S0030
0573460	123A				S0014
0573468	127	3018	25	TR27/235	S0020
0573468(HITACH)	107			TR70/	
0573469	123A	3024	20	TR21/735	S0020
0573469H	123A	3124	18	TR21/	S0015
0573471	126	3006	50	TR17/	G0003
0573474	161	3018	11	TR83/719	S0017
0573474H	161	3039		TR83	S0017
0573475	161	3018	20	TR83/719	S0020
0573479	123A	3124	20	TR21/735	S0014
0573479H	123A	3124	20	TR21/	S0030
0573480	123A	3044	20	TR21/735	S0030
0573480H	128	3124	18	TR87/	S0014

Original Part Number	ECG	SK	GE	IR/WEP	HEP
0573481	123A	3124	63	TR21/735	S0014
0573481H	123A	3018	20	TR21/	S0014
0573485	107	3018	20	TR70/720	S0014
0573486	107	3018	20	TR70/720	S0014
0573486H	108	3018	20	TR95/	S0014
0573487	107	3018	20	TR70/720	S0014
0573487H	107	3018		TR70/	S0014
0573490	123A	3122	20	TR21/735	S0014
0573491	107	3018	20	TR70/720	S0014
0573491H	123A	3018		TR21/	S0014
0573492	123A	3122	10	TR21/735	S0014
0573494	108	3039		TR95/56	S0016
0573495	108	3039		TR95/56	S0016
0573501	154	3045		TR78/712	S5024
0573506	108	3018	11	TR95/	S0016
0573506H	108	3039	11	TR95/	
0573507	107	3018	20	TR70/720	S0014
0573507H	108	3018		TR95	S0014
0573508	107	3018	20	TR24/	S0014
0573509	108		11	TR95/56	S0016
0573509H	107	3018		TR70/	S0016
0573510	107	3018	20	TR70/720	S0016
0573510H	107	3018	17	TR95/	S0016
0573511	108	3018	11	TR95/56	S0011
0573511H	107	3018		TR70/	S0016
0573515	124	3021	12	TR81/240	S5011
0573515H	124	3021		TR81/	S5011
0573517	128	3047	18	TR87/240	S3013
0573518	126	3007	50	TR17/635	G0008
0573519	154	3144		TR78/712	S5026
0573523	123A	3122		TR21/	S0030
0573525	175			TR81/	S5012
0573526	163	3111	36	TR67/	S5020
0573527	128	3024		TR87/	S5014
0573529	123A	3004	52	TR21/735	G0005
0573532	123A	3024	63	TR21/735	S0020
0573541	224	3049		TR25/243	S5014
0573542	159	3025		TR20/	S5022
0573556	123A	3024	17	TR21/	S5014
0573557	128		18	TR25/	
0573559	129	3025	67	TR88/	S5013
0573560	129		67	TR28/	
0573562	130	3027		TR59/	S7004
0573570	108	3018		TR95/56	S0016
0573607	107	3018		TR70/720	S0014
0573742	102A	3004		TR85/	G0005
0573981	123A	3020		TR21/	
0576001	102A	3004		TR85/250	
04040501-1	909D	3590			
04040503-1	910				
04040505	941D				
04040505-001	941D				
04040751-001	923-D				
04440028-001	123A				
04440028-003	123A				
04440028-006	123A				
04440028-008	123A				
04440028-013	123A				
04440032-003	159				
04440032-004	159				
04440032-005	159				
04440032-006	159				
04440032-007	159				
04440032-008	159				
04440032-009	159				
04440052-001	123A				
04440052-002	123A				
04450002-001	123A				
04450002-004	123A				
04450002-005	123A				
04450016-001	159				
04450016-002	159				
04450016-003	129				
04450023-001	130				
04450023-005	181				
04450023-006	130				
04450037-001	181				
04450037-002	182	3188			
05320074H	123A	3088			
09301020	102A	3004		TR85/250	
09308004	703A	3157		IC500/	
09308038	1142				
09391000A	925				
0A8-1	160	3007		TR17/	
0A8-1-12	160	3007		TR17/	

Original Part Number	ECG	SK	GE	IR/WEP	HEP
0A8-1-12-7	160	3007		TR17/	
0C238	123A				
0N020540	128				
0N047204-2	123A				
0N049874	9912				
0N47204-1	123A				
0N187840	941				
0S536G	123A	3122		TR24/	
1-000-099-00	7490				
1-0006-0021	123A				
1-0006-0022	123A				
1-0006-0023	159				
1.000.111-00	159				
1-001-003-15	123A	3122		TR21/	
1-002	130				
1-003	130				
1-034/2207	159	3114	67	TR28/	
1-035/2207	128	3124	18	TR25/	
1-041/2207	108	3018	17	IRTR95/	
1-042/2207	123A		20	TR21/	
1-043/2207	159	3114	21	TR30/	
1-044/2207	123A	3124	20	TR25/	
1-021151	124				
1-021162	124				
1-021221	128				
1-021270	130				
1-021280	181				
1-021290	128				
1-21-73	100	3005		TR05/	
1-21-74	100	3005		TR05/	
1-21-75	100	3005		TR05/	
1-21-76	100	3005		TR05/	
1-21-78	100	3005		TR05/	
1-21-83	100	3005		TR05/	
1-21-91	100	3005		TR05/	
1-21-92	100	3005		TR05/	
1-21-93	100	3005		TR05/	
1-21-95	100	3004		TR85/	
1-21-96	100	3004		TR85/	
1-21-100	100	3005		TR05/	
1-21-102	100	3005		TR05/	
1-21-103	100	3005		TR05/	
1-21-104	100	3005		TR05/	
1-21-105	100	3005		TR05/	
1-21-106	102A	3004		TR85/	
1-21-107	102A	3004		TR85/	
1-21-120	102A	3004		TR85/	
1-21-128	100	3005		TR05/	
1-21-135	160	3006		TR17/	
1-21-137	160	3006		TR17/	
1-21-138	126	3008		TR17/	
1-21-139	160	3006		TR17/	
1-21-148	102A	3004		TR85/	
1-21-150	160	3007		TR17/	
1-21-157	160	3006		TR17/	
1-21-161	100	3005		TR05/	
1-21-162	100	3005		TR05/	
1-21-164	102A	3004		TR85/	
1-21-179	100	3005		TR05/	
1-21-180	100	3005		TR05/	
1-21-184	102A	3004		TR85/	
1-21-186	100	3005		TR05/	
1-21-189	126	3008		TR17/	
1-21-190	160	3007		TR17/	
1-21-191	102A	3004		TR85/	
1-21-192	102A	3004		TR85/	
1-21-225	102A	3004		TR85/	
1-21-226	102A	3004		TR85/	
1-21-227	102A	3004		TR85/	
1-21-228	160	3006		TR17/	
1-21-229	160	3006		TR17/	
1-21-230	160	3006		TR17/	
1-21-231	160	3006		TR17/	
1-21-232	102A	3004		TR85/	
1-21-233	160	3006		TR17/	
1-21-234	100	3005		TR05/	
1-21-235	100	3005		TR05/	
1-21-236	100	3005		TR05/	
1-21-240	100	3005		TR05/	
1-21-241	100	3005		TR05/	
1-21-242	126	3008		TR17/	
1-21-243	126	3008		TR17/	
1-21-244	126	3008		TR85/	
1-21-246	102A	3004		TR85/	
1-21-254	100	3005		TR05/	

Original Part Number	ECG	SK	GE	IR/WEP	HEP
1-21-256	160	3007		TR17/	
1-21-257	126	3008		TR17/	
1-21-258	160	3007		TR17/	
1-21-259	160	3007		TR17/	
1-21-260	160	3007		TR17/	
1-21-266	102A	3004		TR85/	
1-21-267	102A	3004		TR85/	
1-21-270	121	3009		TR01/	
1-21-271	121	3009		TR01/	
1-21-272	102A	3004		TR85/	
1-21-273	100	3005		TR05/	
1-21-274	102A	3003		TR85/	
1-21-275	100	3005		TR05/	
1-21-276	123A	3124		TR21/	
1-21-277	123A	3124		TR21/	
1-21-278	123A	3124		TR21/	
1-21-279	123A	3124		TR21/	
1-21-289	100	3005		TR05/	
1 735 0060	159				
1 735 0061	159				
1 735 0066	159				
1 735 0068	129				
1-801-003	123A	3018	17	TR21/735	S0016
1-801-003-12	108	3122	20	TR21/	S0014
1-801-003-13	108	3018	17	TR95/	S0016
1-801-003-14	108	3018	20	TR95/	S0016
1-801-003-15	108	3018	20	TR21/735	S0016
1-801-004	123A	3124	62	TR21/735	S0015
1-801-004-17	123A	3124	10	TR24/	S0014
1-801-005	158	3004	53	TR84/630	G6011
1-801-005-23	102A	3004	2	TR24/	G0005
1-801-006	158	3004	53	TR86/	G6011
1-801-006-12	102A	3004	2		G0005
1-801-006-14	102A	3004	2	TR23/	G0005
1-801-301-13	124	3021	12	TR81/240	S5011
1-801-301-14	124	3021	12	TR23/	S5011
1-801-301-15	124	3021	12	TR23/	S5011
1-801-304-15	107	3039	20	TR21/	S0016
1-801-305-13	108	3018	17	TR95/	S0014
1-801-306	108	3018	17	TR95/	S0014
1-801-306-13	108	3018	17	TR95/	S0014
1-801-306-14	108	3018	17	TR95/	S0014
1-801-306-15	108	3018	17	TR95/	S0014
1-801-308	102A	3004		TR85/250	G0005
1-801-308-24	102A	3004	52		
1-801-309	103A	3018		TR08/724	G0011
1-801-310	158	3004		TR84/630	G6011
1-801-314	123A	3124		TR21/735	S0016
1-801-314-15	123A	3018	10	TR24/	S0014
1-801-314-16	123A	3018	10	TR24/	S0014
1-2114-0	128				
1-522221011	102A	3004		TR85/	
1-522210111	102A	3004		TR85/	
1-522210131	160	3005		TR17/	
1-522210300	160	3006		TR17/	
1-522210921	160	3007		TR17/	
1-522211021	160	3006		TR17	
1-522211200	102A	3004		TR85/	
1-522211328	102A	3004		TR85/	
1-522211921	160	3006		TR17/	
1-522214400	160	3006		TR17/	
1-522214411	160	3006		TR17/	
1-522214435	160	3006		TR17/	
1-522214821	160	3006		TR17/	
1-522214831	160	3006		TR17/	
1-522216500	102A	3004		TR85/	
1-522216600	160	3007		TR17/	
1-522217400	160	3006		TR17/	
1-522223720	123A	3124		TR21/	
1-6147191229	123A	3124		TR21/	
1-6171191368	123A	3124		TR21/	
1-6207190405	102A	3004		TR85/	
1A0013	199		10	TR21/	
1A0013(YAMAHA)	175			TR81/	
1A0020	123A	3124	20	TR21/	
1A0021	123A	3124	17	TR21/	
1A0022	123A	3124	17	TR21/	
1A0024	123A	3124		TR21/	
1A0025	123A	3124	17	TR21/	
1A0027	175	3024	28	TR81/	
1A0027(YAMAHA)	175			TR81/	
1A0029	123A	3124	20	TR21/	
1A0032	123A	3124	17	TR21/	
1A0033	123A	3124	17	TR21/	
1A0034	123A	3124	20	TR21/	

Original Part Number	ECG	SK	GE	IR/WEP	HEP
1A0035	123A	3124	20	TR21/	
1A0037	123A	3124	17	TR21/	
1A0038	124	3021	12	TR81/	
1A0043	123A	3039	17	TR21/	
1A0044	123A	3124	17	TR21/	
1A0046	152	3027	66	TR76/	
1A0048	175	3024	28	TR81/	
1A0048(YAMAHA)	175			TR81/	
1A0051	123A	3124	17	TR21/	
1A0055	102A	3003	2	TR85/	
1A0056	102A	3003	2	TR85/	
1A0058	152	3024	66	TR76/	
1A0059	152	3026	66	TR76/	
1A0063	123A	3124	20	TR21/	
1A0066	128	3024	18	TR87/	
1A0067	123A	3124	20	TR21/	
1A0070	123A	3124	17	TR21/	
1A0076	123A	3124	20	TR21/	
1A0077	123A	3124	20	TR21/	
1A0078	123A	3124	20	TR21/	
1A0079	123A	3124	17	TR21/	
1A0080	123A	3124	20	TR21/	
1A0081	123A	3124	20	TR21/	
1A0083	123A	3124	20	TR21/	
1A0084	123A	3124	20	TR21/	
1A348	128				S5014
1A348R	128	3024			S5014
1A475-1	123A				
1B3096-1	159				
1C3576	123A				
1C73045	235				
1E535A	229				
1E535A/T825B	229				
1E535B	107				
1E703E	703A	3157	IC12		
1J1	123A	3122	61	TR21/735	
2-36	103	3010		TR08/641A	
2-8454-031	123A				
2A	126	3006	51	TR17/635	G0005
2A12	129		67	TR88/	S5014
2AC187	176(2)			/238	G0011(2)
2AC188	158(2)			/630	G6011(2)
2AD140	121	3009	16	TR01/232	G6005(2)
2AD149	121MP	3014		/232MP	G6005
2AG	161	3018		TR83/719	
2AH	161	3018		TR83/719	
2B	100	3005		TR05/254	G0001
2C	100	3005		TR05/254	G0001
2C291	128				
2CD1985	130			TR59/	
2CS900	123A	3020		TR21/735	
2D	100	3005		TR05/254	G0001
2D001	121			TR01/	
2D002	123A	3124		TR21/	
2D002-41	123A	3124	20	TR21/	
2D002-168	123A	3124	20	TR21/	
2D002-169	123A	3124	20	TR21/	
2D002-170	123A	3124	20	TR21/	
2D002-171	123A	3124	20	TR21/	
2D002-175	123A	3124	20	TR21/	
2D004	121	3009		TR01/	
2D004-9		3009	16	TR01/	
2D010	130	3027	14	TR59/	
2D013	102A	3004		TR85/	
2D013-13	102A	3004	53	TR85/	
2D013-54	102A	3004	53	TR85/	
2D013-109	102A	3004	53	TR85/	
2D013-160	102A	3004	99	TR85/	
2D015	121	3009		TR01/	
2D016	102A	3004	53	TR85/	
2D016-45	102A	3009	53	TR85/	
2D016-54	102A	3009	53	TR85/	
2D017	123A	3025		TR21/	
2D017-165	159	3025	22	TR20/	
2D017-166	159	3025	22	TR20/	
2D017-167	159	3025	22	TR20/	
2D017-169	159	3025	22	TR20/	
2D020	194				
2D020-173	194		18	TR25/	
2D020-174	194		18	TR25/	
2D021	102A	3009		TR85/	
2D021-8	102A	3009	53	TR85/	
2D021-11	102A	3009	53	TR85/	
2D021-56	102A	3009		TR85/	
2D022	6400				

Original Part Number	ECG	SK	GE	IR/WEP	HEP
2D022-211	6400				
2D023	102A	3009		TR85/	
2D026	123A	3124	20	TR21/	
2D026-274	123A	3124	20	TR21/	
2D027	159	3114	21	TR20/	
2D030	6400				
2D031	133	3112	FET1		
2D033	123A	3019		TR21/	
2D036	158	3004	53	TR84/	
2D038	123A	3124	20	TR21/	
2D039	158	3004	53	TR84/	
2E	126	3006	51	TR17/635	G0001
2E1A20A22AAB	128				
2F	160			TR17/637	G0002
2G	126	3006	51	TR17/635	G0003
2G101	160		2	TR17/637	G0002
2G102	160		2	TR17/637	G0002
2G103	160			TR17/637	G0003
2G104	160			TR17/637	G0003
2G106	160			TR17/637	G0003
2G108	160	3123	2	TR17/637	G0002
2G109	160	3123	2	TR17/637	G0002
2G110	160			TR17/637	G0002
2G138	100	3005	2	TR05/254	
2G139	100	3005	2	TR05/254	
2G140	100	3005	2	TR05/254	
2G201	160	3123	52	TR17/637	G0002
2G202	160	3123	2	TR17/637	G0002
2G210	121	3014		TR01/232	G6005
2G220	121	3009		TR01/230	G6003
2G221		3014		/232	G6005
2G222	121	3014		TR01/232	G6005
2G223	104	3009		TR01/230	G6003
2G224	121	3014		TR01/232	G6005
2G225	121	3009		TR01/232	G6005
2G226	179			TR35/G6001	G6014
2G227	179			TR35/G6001	G6014
2G228	179			TR35/G6001	G6014
2G240	121	3014		TR01/232	G6005
2G270	102	3004	52	TR05/631	G0005
2G271	102	3004	2	TR05/631	G0005
2G301	160			TR17/637	G0003
2G302	102	3004		TR05/631	G0003
2G303	102A			TR85/250	G0003
2G304	102A			TR85/250	G0003
2G306	102	3004		TR05/631	G0003
2G308	102A	2004	2	TR85/250	G0003
2G309	102A			TR85/250	G0003
2G319	100	3005	2	TR05/254	G0005
2G320	100	3005	2	TR05/254	G0005
2G321	102	3004	2	TR05/631	G0005
2G322	102	3004	2	TR05/631	G0005
2G323	102	3004	2	TR05/631	G0005
2G324	102	3004	2	TR05/631	G0005
2G339	103	3010	8	TR08/641A	
2G339A	103	3010	8	TR08/641A	G0011
2G344	126	3006	2	TR17/635	G0003
2G345	160	3005	2	TR17/637	G0003
2G371	102A	3005	2	TR85/250	G0003
2G371A	102A	3005	1	TR85/250	
2G374	102A	3005	2	TR85/250	G0003
2G374A	102A	3004	2	TR85/250	G0003
2G376	102A			TR85/250	G0003
2G377	102A			TR85/250	G0003
2G381	102A	3004	52	TR85/250	G0003
2G381A	102A	3004	2	TR85/250	
2G382	160			TR17/637	G0003
2G383	100	3005		TR05/254	G0005
2G384	102	3004		TR05/631	G0005
2G385	102	3004		TR05/631	G0005
2G386	102	3004		TR05/631	G0005
2G387	102	3004		TR05/631	G0005
2G394	100	3005	2	TR05/254	G0005
2G395	158	3123	2	TR84/250	G0005
2G396	100	3005	2	TR05/254	G0005
2G397	100	3005	2	TR05/254	G0005
2G401	160		52	TR17/637	G0002
2G402	160		52	TR17/637	G0002
2G403	160		52	TR17/637	G0002
2G404	160		52	TR17/637	
2G413	160		52	TR17/637	G0003
2G414	160		52	TR17/637	G0003
2G415	160		52	TR17/637	G0003
2G416	160		52	TR17/637	G0003
2G417	160		52	TR17/637	G0003

Original Part Number	ECG	SK	GE	IR/WEP	HEP
2G508	102	3004	52	TR05/631	G0005
2G509	102	3004	52	TR05/631	
2G524	100	3005		TR05/254	G0005
2G525	100	3005		TR05/254	G0005
2G526	100	3005		TR05/254	G0005
2G527	100	3005		TR05/254	G0005
2G577	100	3005		TR05/254	G0005
2G601	126	3006	2	TR17/635	
2G602	126	3006	2	TR17/635	
2G603	100	3005	2	TR05/254	
2G604	100	3005	2	TR05/254	
2G605	100	3123	2	TR05/254	G0005
2G1024	100	3005		TR05/254	G0005
2G1025	100	3005		TR05/254	G0005
2G1026	100	3005		TR05/254	G0005
2G1027	102	3004		TR05/254	G0005
2G1055	130				
2H1254	106		65	TR20/52	S0013
2H1255	106		65	TR20/52	S0013
2H1256	106		65	TR20/52	S0013
2H1257	106		65	TR20/52	S0013
2H1258	106		65	TR20/52	S0013
2H1259	106		65	TR20/52	S0013
2J72	160			TR17/637	G0002
2J73	160			TR17/637	G0002
2K48	160		50	TR17/637	G0003
2M78	101			TR08/641	
2M1127	102			TR05/	
2MC	126	3006		TR17/635	
2N2J324	128	3024	63	TR87/243	
2N21			53		
2N23	102	3004	52	TR05/631	
2N24	102	3004	52	TR05/631	
2N25	102	3004	52	TR05/631	
2N26	102	3004	52	TR05/631	
2N27	100	3005	2	TR05/254	G0005
2N28	101	3011		TR08/641	G0011
2N29	101	11		TR08/641	G0011
2N30	100	3005	51	TR05/254	
2N31	100	3005	51	TR05/254	
2N32	102	3004	52	TR05/631	G0005
2N32A	102	3004	52	TR05/631	
2N33	126	3006	51	TR17/635	G0002
2N34	102A	3004	80	TR85/250	G0005
2N34A	102A		80	TR85/250	G0005
2N35	103A	3010	8	TR08/724	G0011
2N36	102A	3003	51	TR85/250	G0005
2N37	102A	3003	51	TR85/250	G0005
2N38	102A	3003	51	TR85/250	G0005
2N38A	102A	3003	51	TR85/250	G0005
2N39	102	3004	52	TR05/631	G0005
2N40	102	3004	52	TR05/631	G0005
2N41	102	3003	52	TR05/631	G0005
2N42	102	3004	52	TR05/631	G0005
2N43	102	3003	2	TR05/631	G0005
2N43A	102	3004	2	TR05/631	G0005
2N44	102	3004	2	TR05/631	G0005
2N44A	102	3004	2	TR05/631	G0005
2N45	102A		2	TR85/250	G0005
2N45A	102A			TR85/250	G0005
2N46	102	3004	52	TR05/631	G0005
2N47	102	3004	52	TR05/631	G0005
2N48	102	3004	52	TR05/631	G0005
2N48A	102	3004		TR05/631	G0005
2N49	102	3004	52	TR05/631	G0005
2N50	102	3004		TR05/631	G0005
2N51	102	3004	52	TR05/631	G0005
2N52	102	3004	53	TR05/631	G0005
2N53	102	3004	52	TR05/631	G0005
2N54	102	3004	2	TR05/631	G0005
2N55	102	3004	2	TR05/631	G0005
2N56	102	3004	52	TR05/631	G0005
2N57			16	TR27/	
2N58	102		53	TR05/	
2N59	102	3003	53	TR05/631	G0005
2N59A	102	3003	2	TR05/631	G0005
2N59B	102	3004	2	TR05/631	G0005
2N59C	102	3003	2	TR05/631	G0005
2N59D	102	3004	52	TR05/631	
2N60	102	3003	53	TR05/631	G0005
2N60A	102	3003	2	TR05/631	G0005
2N60B	102	3003	2	TR05/631	G0005
2N60C	102	3003	2	TR05/631	G0005
2N60R	102	3004		TR05/631	
2N61	102	3003	53	TR05/631	G0005

Original Part Number	ECG	SK	GE	IR/WEP	HEP
2N61A	102	3003	80	TR05/631	G0005
2N61B	102	3003	2	TR05/631	G0005
2N61C	102	3003	2	TR05/631	G0005
2N62	102A	3004	52	TR85/250	G0005
2N63	102A	3003	52	TR85/250	G0005
2N64	102A	3003	52	TR85/250	G0005
2N65	102	3003	52	TR05/631	G0005
2N66			16	TR27/	
2N67			16	TR27/	
2N68			16	TR27/	
2N71	102	3004	52	TR05/631	
2N72	126	3006	52	TR17/635	
2N73	102	3004	52	TR05/631	
2N74	102	3004	52	TR05/631	
2N75	126	3006	51	TR17/635	G0005
2N76	102	3004	50	TR05/631	G0005
2N77	102	3004	52	TR85/254	G0005
2N78	101	3011	5	TR08/641	G0011
2N78A	101	3011	5	TR08/641	G0011
2N79	102	3004	52	TR05/631	G0005
2N80	102	3004	51	TR05/631	G0005
2N81	102	3003	52	TR05/631	G0005
2N82	102	3003	52	TR05/631	G0005
2N83	100	3005	1	TR05/254	G6005
2N84			51	TR85/	
2N85	126	3006	53	TR17/635	G0002
2N86	126	3006	53	TR17/635	G0002
2N87	126	3006	53	TR17/635	G0002
2N88	126	3006	53	TR17/635	G0002
2N89	126	3006	53	TR17/635	G0002
2N90	126	3006	53	TR17/635	G0002
2N94	103A	3011	5	TR08/724	G0011
2N94A	103A	3011	5	TR08/724	G0011
2N95	176/411				
2N96	102	3004	52	TR05/631	G0005
2N97	101	3011	59	TR08/641	G0011
2N97A	101	3011	8	TR08/641	G0011
2N98	101	3011	7	TR08/641	G0011
2N98A	101	3011	8	TR08/641	G0011
2N99	101	3011	7	TR08/641	G0011
2N100	101	3011	8	TR08/641	G0011
2N101				TR84/	
2N102	101		54	TR08/	
2N103	101	3011	59	TR08/641	G0011
2N104	102A	3004	52	TR85/250	G0005
2N105	102A	3004	52	TR85/250	G0005
2N106	102A	3003	50	TR21/250	G0005
2N107	102A	3003	51	TR85/250	G0005
2N108	102	3003	50	TR05/631	G0005
2N109	102A	3004	53	TR85/250	G0005
2N109/5	102	3004	52	TR05/631	G0005
2N109BLU	102A	3123		TR85/250	G0005
2N109GRN	102A	3123		TR85/250	G0005
2N109M1	102A			TR85/250	G0005
2N109M2	102A			TR85/250	G0005
2N109WHT	102A			TR85/250	G0005
2N109YEL	102A			TR85/250	G0005
2N111	100	3005	52	TR05/254	G0005
2N111A	126	3005	1	TR05/254	G0005
2N111B	126	3006	51	TR17/635	G0005
2N111M1	126	3006		TR17/635	G0005
2N111M2	126			TR17/635	G0005
2N112	126	3005	1	TR17/635	G0009
2N112A	100	3005		TR05/254	G0005
2N112M1	226	3006		TR17/635	G0009
2N113		3004	1	TR05/254	G0009
2N114	100	3005	1	TR05/254	G0002
2N115			3	TR01/	
2N117	123	3124	60	TR86/53	S0014
2N118	123	3124	60	TR86/53	S0014
2N118A	123	3124	60	TR86/53	S0014
2N119	123	3124	60	TR86/53	S0014
2N120	123	3124	39	TR86/53	S0014
2N122			32	TR23/	
2N123	100	3005	1	TR05/254	G0002
2N123/5			51		G0002
2N123A	126	3006	51	TR17/635	G0002
2N123A5			51		
2N124	101	3011	5	TR08/641	G0011
2N125	101	3011	5	TR08/641	G0011
2N126	103A	3011	5	TR08/724	G0011
2N127	101	3004	8	TR08/641	G0011
2N128	126	3008	51	TR17/635	G0003
2N129	126	3008	51	TR17/635	G0003
2N130	102	3003	50	TR05/631	G0005
2N130A	102	3004	51	TR05/631	G0005
2N131	102	3003	50	TR05/250	G0005
2N131A	102	3003	52	TR05/250	G0005
2N132	102	3003	50	TR05/631	G0005
2N132A	102	3003	52	TR05/631	G00
2N133	102	3003	50	TR05/631	G0005
2N133A	102A	3003	52	TR05/250	G0005
2N134				TR85/	
2N135	100	3005	2	TR05/254	G0005
2N136	100	3005	1	TR05/254	G0005
2N137	100	3005	1	TR05/254	G0005
2N138	102A	3003	52	TR85/250	G0005
2N138A	102A	3003	52	TR85/250	G0005
2N138B	102A	3003	2	TR85/250	G0005
2N139	126	3005	50	TR17/635	G0003
2N140	126	3005	50	TR17/635	G0001
2N140M1	126	3123		TR17/635	G0001
2N140M2	126	3006		TR17/635	G0001
2N141			3	TR01/	
2N143			3	TR01/	
2N145	101	3011	7	TR08/641	G0011
2N146	101	3011	7	TR08/641	G0011
2N147	101	3011	10	TR08/641	G0011
2N148	123	3124	7	TR86/53	G0011
2N148A	103A	3124	7	TR08/724	G0011
2N148B	103A	3010	8	TR08/	G0011
2N148C/D	123A	3011	10	TR24/	
2N149	103A	3124	7	TR08/724	G0011
2N149A	103A	3124	7	TR08/724	G0011
2N149B	158	3124	1		G0005
2N150	101	3011	7	TR08/641	G0011
2N150A	101	3011	7	TR08/641	G0011
2N151	102	3004	52	TR05/631	G0005
2N155	121	3009	25	TR01/230	G6003
2N156	104	3009	3	TR01/230	G6003
2N157	121	3009	25	TR01/232	G6003
2N157A	121	3009	25	TR01/232	G6005
2N158	121	3009	3	TR01/232	G6005
2N158A		3009	3	TR01/232	G6012
2N159	102	3004	52	TR05/631	
2N160	123	3124	60	TR86/53	S0014
2N160A	123	3124	60	TR86/53	S0014
2N161	123	3124	60	TR86/53	S0014
2N161A	123	3124	60	TR86/53	S0014
2N162	107	3039	60	TR70/	S0014
2N162A	107	3039	60	TR70/720	S0014
2N163	123	3124	60	TR86/53	S0014
2N163A	123	3124	60	TR86/53	S0014
2N164	101	3011	5	TR08/641	G0011
2N164A	101	3011	5	TR08/641	G0011
2N165	101	3011	7	TR08/641	G0011
2N166	101	3011	5	TR08/641	G0011
2N167	101	3011	6	TR08/641	G0011
2N167A	101	3011	6	TR08/641	G0011
2N168	101	3011	5	TR08/641	G0011
2N168A	101	3011	5	TR08/641	G0011
2N169	101	3011	7	TR08/641	G0011
2N169A	101	3011	5	TR08/641	G0011
2N170	101	3011	7	TR08/641	G0011
2N171			8	TR08/	
2N172	101	3011	5	TR08/641	G0011
2N173	105	3012	4	TR03/233	G6006
2N174	105	3012	4	TR03/233	G6006
2N174A	105	3012	4	TR03/233	G6006
2N174RED	105	3123		TR03/	
2N175	102A	3003	51	TR85/250	G0005
2N176	104	3009	3	TR01/624	G6013
2N176-1	104	3009	16	TR01/	G6007
2N176-1BLU				TR01/624	G6013
2N176-1WHT	121			TR01/624	G6013
2N176-1YEL	121	3009		TR01/624	G6013
2N176-3PUR	121	3009		TR01/624	G6013
2N176-4PUR	121	3009		TR01/624	G6013
2N176-5WHT	121	3009		TR01/624	G6013
2N176-6WHT	121	3009		TR01/624	G6013
2N176A	104	3009	16	TR01/232	G6005
2N176BLK	121				G6013
2N176BLU	121			TR01/624	G6013
2N176G	104	3009	16	TR01/	G6007
2N176GRN	121			TR01/628	G6013
2N176M	121				
2N176PUR	121			TR01/624	G6013
2N176RED	121			TR01/624	G6013
2N176W	104	3009	16	TR01/	G6007
2N176WHT	121			TR01/624	G6013

Original Part Number	ECG	SK	GE	IR/WEP	HEP
2N176YEL	121			TR01/	G6013
2N178	104	3009	3	TR01/230	G6003
2N179	104	3009	16	TR01/230	G6003
2N180	102A	3004	53	TR85/250	G0005
2N181	102A	3004	53	TR85/250	G0005
2N181WHT				TR85/	
2N182	101	3011	5	TR08/641	G0011
2N183	101	3011	5	TR08/641	G0011
2N184	101	3011	6	TR08/641	G0011
2N185	102A	3003	52	TR85/250	G0005
2N185BLU	102A	3123		TR85/250	G0005
2N186	102A	3004	53	TR85/250	G0005
2N186A	102A	3004	53	TR85/250	G0005
2N187	102A	3004	53	TR85/250	G0005
2N187A	102	3004	53	TR05/631	G0005
2N188	102	3004	53	TR05/6314	G0005
2N188A	102	3004	53	TR05/631	G0005
2N189	102	3004	53	TR05/631	G0005
2N190	102	3004	53	TR05/631	G0005
2N191	102	3004	53	TR05/631	G0005
2N192	102	3004	53	TR05/631	G0005
2N193	103A	3011	5	TR08/724	G0011
2N194	103A	3011	5	TR08/724	G0011
2N194A	103A	3011	5	TR08/724	G0011
2N195	102	3004	52	TR05/631	G0005
2N196	102	3004	53	TR05/631	G0005
2N197	102	3004	53	TR05/631	G0005
2N198	102	3004	53	TR05/631	G0005
2N199	102	3004	53	TR05/631	G0005
2N200	102		2	TR05/631	G0005
2N204	102	3004	2	TR05/631	G0011
2N205	102	3004	2	TR05/631	G0011
2N206	102	3004	2	TR05/631	G0005
2N207	102	3003	52	TR05/631	G0005
2N207A	102	3003	52	TR05/631	G0005
2N207B	102	3003	52	TR05/631	G0005
2N207BLU	102	3004		TR05/631	G0005
2N211	103A	3011	5	TR08/724	G0011
2N211M2		3006			
2N212	103A	3011	5	TR08/724	G0011
2N213	103A	3010	8	TR08/724	G0011
2N213A	103A	3010	8	TR08/724	G0011
2N214	103A	3010	8	TR08/724	G0011
2N214A	103	3010		TR08/641A	G0011
2N215	102A	3004	52	TR85/250	G0005
2N216	103A	3011	5	TR08/724	G0011
2N217	102A	3004	53	TR85/250	G0008
2N217A	102A			TR85/250	
2N217RED	102A			TR85/250	G0008
2N217WHT	102A			TR85/250	G0008
2N217YEL	102A			TR85/250	G0008
2N218	160	3005	50	TR17/637	G0003
2N219	160	3005	50	TR17/637	G0005
2N220	1C2A	3004	51	TR85/250	G0005
2N222	102A	3004	51	TR85/250	G0005
2N222A	123A				
2N223	102A	3003	53	TR85/250	G0005
2N224	102A	3004	53	TR85/250	G0005
2N225	102A	3004	53	TR85/250	G0005
2N226	102A	3004	53	TR85/250	G0005
2N227	102A	3004	53	TR85/250	G0005
2N228	103A	3010	8	TR08/724	G0011
2N228A			53		
2N229	103A	3010	59	TR08/724	G0011
2N230	104	3009	16	TR01/232	G6005
2N231		3008	51	TR17/	G0008
2N231BLU	126	3006		TR17/635	G0008
2N231YEL	126	3006		TR17/635	G0008
2N231YEL-RED	126	3006		TR17/635	G0008
2N232	160	3008	51	TR17/637	G0003
2N233	103A	3011	5	TR08/724	G0011
2N233A	103A	3011	5	TR08/724	G0011
2N234	121	3009	16	TR01/232	G6003
2N234A	121	3009	3	TR01/232	G6003
2N235	121	3009	16	TR01/232	G6003
2N235A	121	3009	3	TR01/232	G6003
2N235B	121	3009	3	TR01/232	G6003
2N236	121	3009	3	TR01/232	G6005
2N236A	121	3009	3	TR01/232	G6003
2N236B	121	3009	3	TR01/232	G6003
2N237	102	3004	2	TR05/631	G0005
2N238	102A	3003	51	TR85/250	G0005
2N238D	102A	3123		TR85/250	G0005
2N238E	102A	3123		TR85/250	G0005
2N238F	102A	3123		TR85/250	G0005
2N238ORN	102A	3123		TR85/250	G0005
2N240	160	3008	50	TR17/637	G0003
2N241	102	3004	53	TR05/631	G0005
2N241A	102	3004	53	TR05/631	G0005
2N242	104	3009	76	TR01/230	G6003
2N243			47	TR63/	S5026
2N244			47	TR63/	S5026
2N245			63	TR63/	
2N246			63	TR63/	
2N247	126	3007	1	TR17/635	G0009
2N247/33	126	3006		TR17/635	G0003
2N248	126	3007	51	TR17/635	G0003
2N249	100	3005	53	TR05/254	G0005
2N250	104	3009	49	TR01/230	G6003
2N250A	104	3009	76	TR01/230	G6003
2N251	121	3009	25	TR01/232	G6005
2N251A	121	3009	76	TR01/232	G6005
2N252	126	3005	51	TR17/635	G0008
2N253	101	3011	5	TR08/641	G0011
2N254	101	3011	5	TR08/641	G0011
2N255	104	3009	3	TR01/230	G6003
2N255A	104	3009	25	TR01/230	G6003
2N256	104	3009	3	TR01/230	G6003
2N256A	104	3009	25	TR01/230	G6003
2N257	104	3009	3	TR01/230	G6003
2N257A	104	3009	16	TR01/230	G6003
2N257B	104	3009	25	TR01/230	G6003
2N257BLK			25		
2N257G	104	3009	25	TR01/230	G6003
2N257GRN			25	TR16/	
2N257W	104	3009	25	TR01/230	G6003
2N257WHT			76		
2N262	100	3005	2	TR05/254	G0005
2N263	108	3005	39	TR95/56	S0014
2N264	108	3039	39	TR95/56	S0014
2N265		3003	53	TR05/631	G0005
2N266	102	3004	2	TR05/631	G0005
2N267	160	3005	52	TR17/637	G0002
2N268	121	3009	25	TR01/232	G6005
2N268A	121	3009	25	TR01/232	G6005
2N269	160	3005	1	TR17/637	G0003
2N270	102	3004	53	TR05/631	G0005
2N270-5E	102			TR05/	
2N270-5M		3004		TR05/631	G0005
2N270A	102	3004	2	TR05/631	G0005
2N271	100	3005	1	TR05/254	G0005
2N271A	100	3005	1	TR05/254	G0005
2N272	102	3003	53	TR05/631	G0005
2N273	100	3004	53	TR05/254	G0005
2N274	160	3007	1	TR17/637	G0003
2N274BLU	160			TR17/637	G0003
2N274WHT	160			TR17/637	G0003
2N275	126		53	TR17/635	G0002
2N275W			16	TR16/	
2N276	160		53	TR17/637	G0003
2N277	105	3012	4	TR03/231	G6004
2N278	105	3012	4	TR03/231	G6004
2N79	102A	3004	52	TR85/250	G0005
2N280	102A	3004	52	TR85/250	G0005
2N281	102A	3003	52	TR85/250	G0005
2N282	102	3004	52	TR05/631	G0005
2N283	102A	3003	52	TR85/250	G0005
2N284	102A	3004	53	TR85/250	G0005
2N284A	102A	3004	2	TR85/250	G0005
2N285	104	3009	16	TR01/230	G6005
2N285A	104	3009	3	TR01/230	G6003
2N285B	104	3009	3	TR01/230	G6003
2N286	160		51	TR17/637	G0008
2N289			51	TR17/	
2N290	104	3012	4	TR01/233	G6006
2N291	102	3004	53	TR05/631	G0005
2N292	101	3011	7	TR08/641	G0011
2N292A	101		5	TR08/641	G0011
2N293	101	3011	7	TR08/641	G0011
2N296	104	3020	25	TR01/230	G6003
2N297	121	3009	25	TR01/232	G6005
2N297A	121	3009	25	TR01/232	G6005
2N299	160	3006	51	TR17/637	G0003
2N300	160	3006	51	TR17/637	G0003
2N301	121	3009	25	TR01/232	G6003
2N301A	121	3009	25	TR01/232	G6005
2N301B	121	3009	3	TR01/232	
2N301G		3009	52	/232	
2N301W	121	3009	52	TR01/232	
2N302	100	3003	1	TR05/254	G0005

Original Part Number	ECG	SK	GE	IR/WEP	HEP
2N303	102	3003	2	TR05/631	G0005
2N306	103A	3010	59	TR08/724	G0011
2N306A	103A	3010	8	TR08/724	
2N307		3009	76	TR01/104-1	G6003
2N307A	121	3009	76	TR01/104-1	G6003
2N307B	121	3009	3	TR01/104-1	G6003
2N308	100	3007	51	TR05/254	G0009
2N309	100	3007	51	TR05/254	G0009
2N310	102	3007	51	TR05/631	G0009
2N311	100	3005	50	TR05/254	G0002
2N312	101	3011	6	TR08/641	G0011
2N313	101	3011	8	TR08/641	G0011
2N314	101	3011	8	TR08/641	G0011
2N315	100	3005	1	TR05/254	G0005
2N315A	100	3004	2	TR05/254	G0005
2N315B	100	3005	53	TR05/254	G0005
2N316	160	3005	1	TR17/637	G0002
2N316A	102	3004	2	TR05/631	G0005
2N317	100	3005	1	TR05/254	G0002
2N317A	100	3005	51	TR05/254	G0002
2N318	126		52	TR17/635	
2N319	102	3003	2	TR05/631	G0005
2N320	102A	3003	2	TR05/631	G0005
2N321	102	3003	2	TR05/631	G0005
2N322	102	3003	2	TR05/631	G0005
2N323	102	3004	2	TR05/631	G0005
2N324	102	3003	53	TR05/631	G6003
2N325	104	3009	49	TR01/104-1	S0032
2N326				TR91/	
2N327	100	3005	65	TR05/254	S0032
2N327A	100	3114	65	TR05/254	S0032
2N327B	129	3025	82	TR88/242	S0032
2N328	129	3114	65	TR88/242	
2N328A	129	3114	65	TR88/242	S0032
2N328B	129	3025	82	TR88/242	S0032
2N329	129	3025	65	TR88/242	S0032
2N329A	129	3025	65	TR88/242	S0032
2N329B	129	3025	82	TR88/242	S0032
2N330	129	3025	65	TR88/242	
2N330A	129	3025	65	TR88/242	S0032
2N331	160	3008	53	TR17/637	G0005
2N331M		3123	2	TR06/	
2N332	123	3124	30	TR86/53	S0014
2N332A	123	3124	81	TR86/53	S0014
2N333	123	3124	39	TR86/53	S0014
2N333A	123	3124	81	TR86/53	S0014
2N334	123	3124	39	TR86/53	S0014
2N334A	123	3124	81	TR86/53	S0014
2N335	123	3124	39	TR86/53	S0014
2N335A	123	3124	81	TR86/53	S0014
2N335B	123	3124	81	TR86/53	S0014
2N336	123	3124	39	TR86/53	S0014
2N336A	123	3124	81	TR86/53	S0014
2N337	107	3039	39	TR70/720	S0014
2N337A	108		20	TR95/56	S0014
2N338		3124	39	TR21/53	S0014
2N338A	123	3124	20	TR86/53	S0014
2N339			63	TR25/	S5014
2N340			27	TR78/	S5026
2N340A			27	TR78/	S3010
2N341			27	TR78/	S5026
2N341A			27	TR78/	S3021
2N342			63	TR25/	S0005
2N342A			27	TR78/	S5026
2N342B			18		
2N343			63	TR25/	S0005
2N343A	128	3024	63	TR87/243	S0005
2N343B			63		S0005
2N344	126	3006	51	TR17/635	G0003
2N345	126	3006	51	TR17/635	G0003
2N346	126	3006	51	TR17/635	G0003
2N350	121	3009	3	TR01/232	G6003
2N350A	121	3009	25	TR01/232	G6003
2N351	121	3009	3	TR01/232	G6003
2N351A	121	3009	25	TR01/232	G6003
2N352	121	3009	16	TR01/232	G6005
2N353	121	3009	16	TR01/232	G6005
2N356	101	3011	7	TR08/641	G0011
2N356A	101	3011	5	TR08/641	G0011
2N357	101	3011	6	TR08/641	G0011
2N357A	101	3011	5	TR08/641	G0011
2N358	101	3011	6	TR08/641	G0011
2N358A	101	3011	5	TR08/641	G0011
2N359	102	3004	2	TR05/631	G0005
2N360	102	3004	2	TR05/631	G0005

Original Part Number	ECG	SK	GE	IR/WEP	HEP
2N361	102	3004	2	TR05/631	G0005
2N362	102	3004	2	TR05/631	G0005
2N362B	102	3004		TR05/631	G0007
2N363	102	3004	2	TR05/631	G0005
2N364	103	3010	59	TR08/641A	G0011
2N365	103	3010	59	TR08/641A	G0011
2N366	103	3010	59	TR08/641A	G0011
2N367	102	3004	53	TR05/631	G0005
2N368	102	3004	53	TR05/631	G0005
2N369	102	3004	52	TR05/631	G0005
2N370	126	3007	50	TR17/635	G0003
2N370/33	126	3007	52	TR17/635	G0003
2N370A	126	3007	52	TR17/635	G0003
2N371	126	3007	50	TR17/635	G0003
2N371/33	126	3007	52	TR17/635	G0003
2N372	126	3007	50	TR17/635	G0003
2N372/33	126	3007	52	TR17/635	G0003
2N373	126	3007	50	TR17/635	G0003
2N374	126	3007	50	TR17/635	G0003
2N375	104	3009	76	TR01/230	G6005
2N376	104	3009	76	TR01/230	G6003
2N376A	104	3009	25	TR01/230	G6003
2N377	101	3011	8	TR08/641	G0011
2N377A	101	3011	6	TR08/641	G0011
2N378	104	3009	76	TR01/230	G6003
2N379	121	3009	76	TR01/230	G6003
2N380	104	3009	76	TR01/230	G6003
2N381	102	3004	2	TR05/631	G0005
2N382	102	3004	2	TR05/631	G0005
2N383	102	3004	2	TR05/631	G0005
2N384	160	3007	51	TR17/637	G0008
2N384/33	160	3007		TR17/637	G0003
2N385	101	3011	8	TR08/641	G0011
2N385A	101	3011		TR08/641	G0011
2N386	104	3009	16	TR01/232	G6005
2N387	104	3009	16	TR01/232	G6005
2N388	101	3011	8	TR08/641	G0011
2N388A	101	3011	7	TR08/641	G0011
2N392	104	3009	16	TR01/232	G6005
2N393	126	3008	51	TR17/635	G0003
2N394	100	3005	1	TR05/254	G0005
2N394A	100	3005		TR05/254	G0005
2N395	100	3005	1	TR05/254	G0005
2N396	100	3005	1	TR05/254	G0005
2N396A	100	3005	1	TR05/254	G0005
2N397	100	3005	2	TR05/254	G0002
2N399	104	3009	3	TR01/230	G6003
2N400	104	3009	25	TR01/230	G6003
2N401	104	3009	3	TR01/230	G6003
2N402	102	3003	53	TR05/631	G0005
2N403	102	3003	53	TR05/631	G0005
2N404	102	3004	53	TR05/631	S0032
2N404A	102	3004	2	TR05/631	S0032
2N405	102	3003	52	TR05/631	G0005
2N406	102A	3004	52	TR85/250	G0005
2N406BLU	102A			TR85/250	G0005
2N406BRN	102A			TR85/250	G0005
2N406GRN	102A			TR85/250	G0005
2N406GRN-YEL	102A			TR85/250	G0005
2N406ORN	102A			TR85/250	G0005
2N406RED	102A			TR85/250	G0005
2N407	102A	3003	52	TR85/250	G0005
2N407BLK	102A			TR85/250	G0005
2N407GRN	102A			TR85/250	G0005
2N407J	102A			TR85/	
2N407RED	102A			TR85/250	G0005
2N407WHT	102A			TR85/250	G0005
2N407YEL	102A			TR85/250	G0005
2N408	102A	3003	52	TR85/250	G0005
2N408J	102A			TR85/	
2N408WHT	102A			TR85/250	G0005
2N409	126	3005	50	TR17/635	G0003
2N410	126	3005	50	TR17/635	G0003
2N411	126	3005	50	TR17/635	G0003
2N412	126	3005	50	TR17/635	G0003
2N413	100	3005	1	TR05/254	G0005
2N413A	100	3005	53	TR05/254	G0005
2N414	100	3005	1	TR05/254	G0001
2N414A	100	3005	1	TR05/254	G0005
2N414B	100	3005	1	TR05/254	G0002
2N414C	100	3005	1	TR05/254	G0002
2N415	100	3005	1	TR05/254	G0002
2N415A	100	3005	1	TR05/254	G0002
2N416	100	3005	1	TR05/254	G0002
2N417	100	3005	1	TR05/254	G0002

Original Part Number	ECG	SK	GE	IR/WEP	HEP
2N418	121	3009	25	TR01/232	G6005
2N419	104	3009	3	TR01/230	G6003
2N420	121	3009	25	TR01/232	G6005
2N420A	121	3009	25	TR01/232	G6005
2N422	102	3003	53	TR05/631	G0005
2N425	100	3005	1	TR05/254	G0005
2N425A		3118			
2N426	100	3005	1	TR05/254	G0005
2N427	100	3005	1	TR05/254	G0002
2N427A	102A	3123	2	TR05/	
2N428	100	3005	1	TR05/254	G0002
2N428A	100	3005	1	TR05/254	G0002
2N431	123	3124	11	TR86/53	S0011
2N432	123	3124	11	TR86/53	S0011
2N433	123	3124	11	TR86/53	S0011
2N435			53		G0002
2N438	101	3011	7	TR08/641	G0011
2N438A	100	3011	7	TR05/254	G0011
2N439	101	3011	7	TR08/641	G0011
2N439A	100	3011	7	TR05/254	G0011
2N439AA	100	3011	7	TR05/254	
2N440	101	3011	6	TR08/641	G0011
2N440A	100	3011	6	TR05/254	G0011
2N440BLU				TR03/	
2N441	105	3012	4	TR03/233	G6004
2N441BLU	105			TR03/233	G6004
2N442	105	3012	4	TR03/233	G6006
2N443	105	3012	4	TR03/233	G6006
2N444	101	3011	59	TR08/641	G0011
2N444A	103	3010	59	TR08/641A	G0011
2N445	101	3010	7	TR08/641	G0011
2N445A	101	3010	5	TR08/641	G0011
2N446	101	3011	54	TR08/641	G0011
2N446A	103	3010	5	TR08/641A	G0011
2N447	101	3011	5	TR08/641	G0011
2N447A	103	3010	8	TR08/641A	G0011
2N447B	103	3010		TR08/641A	G0011
2N448	101	3011	7	TR08/641	G0011
2N449	101	3011	7	TR08/641	G0011
2N450	100	3005	1	TR05/254	G0002
2N456	104	3009	25	TR01/230	G6003
2N456A	104	3009	16	TR01/230	G6003
2N456B	104	3009	3	TR01/230	G6003
2N457	104	3009	25	TR01/230	G6005
2N457A	104	3009	3	TR01/230	G6005
2N457B	104	3009		TR01/230	G6005
2N458	104	3009	25	TR01/230	G6005
2N458A	104	3009	3	TR01/230	G6005
2N458B	121	3009	3	TR01/230	G6005
2N459	121	3012	76	TR01/232	G6005
2N459A	121	3009		TR01/232	G6018
2N460	102	3004	2	TR05/631	G0005
2N461	102	3004	1	TR05/631	G0005
2N462	102	3004	80	TR05/631	G0005
2N464	102	3004	53	TR05/631	G0005
2N465	102	3004	2	TR05/631	G0005
2N466	102	3003	53	TR05/631	G0005
2N467	102	3003	53	TR05/631	G0005
2N468	102	3004	52	TR05/631	G0005
2N469			53		
2N470	123	3124	86	TR85/53	S0014
2N471			60		S0014
2N471A	123	3124	60	TR86/53	S0014
2N472	123A	3122	212	TR21/735	S0014
2N472A	123A	3124	212	TR21/735	S0014
2N473	123	3124	86	TR86/53	S0014
2N474	123A	3122	61	TR21/735	S0014
2N474A	123A	3124	60	TR21/735	S0014
2N475	123A	3122	212	TR21/735	S0014
2N475A	123A	3124	212	TR21/735	S0014
2N476	123	3124	86	TR86/53	S0014
2N476A			11		
2N477	123	3124	60	TR86/53	S0014
2N478	123	3124	86	TR86/53	S0014
2N479	123	3124	60	TR86/53	S0014
2N479A	123	3124	60	TR86/53	S0014
2N480	123	3124	212	TR86/53	S0014
2N480A	123A	3122	212	TR21/735	S0014
2N480B	123A	3122		TR21/735	
2N481	100	3005	52	TR05/254	G0005
2N482	100	3005	52	TR05/254	G0005
2N483	100	3005	1	TR05/254	G0005
2N483-6M	102	3123		TR05/631	G0005
2N483B	102	3004		TR05/631	G0005
2N484	100	3005	1	TR05/254	G0002

Original Part Number	ECG	SK	GE	IR/WEP	HEP
2N485	100	3005	1	TR05/254	G0005
2N486	100	3005	2	TR05/254	G0001
2N486B	100	3005		TR05/254	G0001
2N487	102	3004	1	TR05/631	G0001
2N495	106	3114	65	TR20/52	S0013
2N495/18	106			TR20/52	S0013
2N496	106	3114	65	TR20/52	S0013
2N497	128		18	TR87/243	S5014
2N497A	128		63	TR87/243	S5014
2N498	128		27	TR87/243	S0014
2N498A			32		S3019
2N499	126	3006	51	TR17/635	G0008
2N499A	126	3006		TR17/635	G0009
2N500	126	3008	51	TR17/635	G0008
2N500BLU	126	3006		TR17/635	G0008
2N500WHT	126	3006		TR17/635	G0008
2N501	126	3006	51	TR17/635	G0003
2N501/18	126	3006		TR17/635	G0003
2N501A	126	3006	51	TR17/635	G0003
2N502	126	3006	51	TR17/635	G0009
2N502A	126	3006	51	TR17/635	G0009
2N502B	126	3006		TR17/635	G0009
2N503	100	3005	51	TR05/254	G0002
2N504	126	3008	1	TR17/635	G0003
2N505	100	3005	2	TR05/254	G0002
2N506	102A	3004	80	TR85/250	G0005
2N507	103A	3010	8	TR08/724	G0011
2N508	102	3003	53	TR05/631	G0005
2N508A	102	3003	2	TR05/631	G0005
2N509	160		51	TR17/637	
2N511	104	3009	3	TR01/624	
2N511A	104	3009	3	TR01/624	
2N511B	104	3009	3	TR01/624	
2N512	104	3009	3	TR01/624	
2N512A	104	3009	3	TR01/624	
2N512B	104	3009	3	TR01/624	
2N513	104	3009	3	TR01/624	
2N513A	104	3009	3	TR01/624	
2N513B	104	3009	3	TR01/624	
2N514	121	3009	3	TR01/232	
2N514A	121	3009	3	TR01/232	
2N514B	121	3009	3	TR01/232	
2N515	103A	3011	5	TR08/724	G0011
2N516	103A	3010	5	TR08/724	G0011
2N517	103A	3011	5	TR08/724	G0011
2N518	100	3005	51	TR05/254	G0002
2N519	100	3005	2	TR05/254	G0005
2N519A	100	3005	50	TR05/254	G0005
2N520	100	3005	2	TR05/254	G0005
2N520A	100	3005	50	TR05/254	G0005
2N521	100	3005	1	TR05/254	G0005
2N521A	100	3005	50	TR05/254	G0002
2N522	100	3003	2	TR05/254	G0002
2N522A	100	3005	50	TR05/254	G0002
2N523	100	3003	2	TR05/254	G0002
2N523A	100	3005	50	TR05/254	G0002
2N524	102	3004	2	TR05/631	G0005
2N524A	102	3004	2	TR05/631	G0005
2N525	102	3004	2	TR05/631	G0005
2N525A	102	3004	2	TR05/631	G0005
2N526	102	3004	2	TR05/631	G0005
2N526A	102	3004	2	TR05/631	G0005
2N527	102	3004	2	TR05/631	G0005
2N527A	102	3004		TR05/631	G0005
2N529	100	3005	2	TR05/254	G0005
2N530	100	3005	1	TR05/254	G0005
2N531	100	3005	1	TR05/254	G0005
2N531/P	176	3123	50	TR82/	
2N532	100	3005	1	TR05/254	G0005
2N533	100	3005	1	TR05/254	G0005
2N534	102A	3123	51	TR85/250	G0003
2N535	102A	3003	52	TR85/250	G0005
2N535A	102A	3003	52	TR85/250	G0005
2N535B	102A	3003	52	TR85/250	G0005
2N536	102A	3003	52	TR85/250	G0005
2N537	160		2	TR17/637	G0003
2N538	121	3009	3	TR01/232	
2N538A	121	3009	3	TR01/232	
2N539	104	3009	3	TR01/624	G6012
2N539A	104	3009	3	TR01/624	G6006
2N540	104	3009	3	TR01/624	
2N540A	104	3009	3	TR01/624	
2N541	123A	3122	60	TR21/735	S0014
2N542	123A	3122	60	TR21/735	S0014
2N542A	123	3124	60	TR86/53	S0014

Original Part Number	ECG	SK	GE	IR/WEP	HEP
2N543	123A	3122	212	TR21/735	S0014
2N543A	123A	3122	212	TR21/735	S0014
2N544	126	3008	50	TR17/635	G0003
2N544/33	126	3008	50	TR17/635	G0003
2N546	123	3124	63	TR86/53	S0014
2N547	123	1324	63	TR86/53	S0014
2N548	123	3124	47	TR86/53	S0014
2N549	123	3124	63	TR86/53	S0014
2N550	123	3124	47	TR86/53	S0014
2N551	123	3124	81	TR86/53	S0014
2N552	123	3124	81	TR86/53	S0014
2N553	121	3009	3	TR01/232	G6005
2N554	121	3009	25	TR01/232	G6003
2N555	121	3009	76	TR01/232	G6003
2N556	3011	3011	8	TR08/641	G0011
2N557	101	3011	6	TR08/641	G0011
2N558	101	3011	6	TR08/641	G0011
2N559	160		51	TR17/637	G0003
2N560	123		81	TR86/53	S0014
2N561	121	3009	25	TR01/232	G6005
2N563	102	3004	53	TR05/631	G0005
2N564	102	3004	53	TR05/631	G0005
2N565	102	3004	53	TR05/631	G0005
2N566	102	3004	53	TR05/631	G0005
2N567	103A	3004	53	TR08/724	G0011
2N568	102	3004	53	TR05/631	G0005
2N569	102	3003	53	TR05/631	G0005
2N570	102	3003	53	TR05/631	G0005
2N571	100	3003	52	TR05/254	G0005
2N572	100	3003	53	TR05/254	G0005
2N573BRN	102	3004		TR05/631	G0005
2N573ORN	102	3004		TR05/631	G0005
2N573RED	102	3004		TR05/631	G0005
2N574	105	3004	4	TR03/233	G6006
2N575	105	3012	4	TR03/233	
2N575A	103	3012	4	TR03/233	
2N576	101	3011	5	TR08/641	G0011
2N576A	101	3011	5	TR08/641	G0011
2N578	100	3005	1	TR05/254	G0005
2N579	100	3005	1	TR05/254	G0005
2N580	100	3005	1	TR05/254	G0005
2N581	100	3005	1	TR05/254	G0005
2N582	100	3003	2	TR05/254	G0002
2N583	126	3005	1	TR17/635	G0005
2N584	126	3003	2	TR17/635	G0002
2N585	101	3011	8	TR08/641	G0011
2N586	100	3005	1	TR05/254	G0005
2N587	101	3011	5	TR08/641	G0011
2N588	126	3006	51	TR17/635	G0008
2N588A	126	3006		TR17/635	G0003
2N589	121	3009	3	TR01/232	G6005
2N591	102A	3004	53	TR85/250	G0009
2N591/5	102A		53	TR85/250	
2N591-6M	126	3006		TR17/635	G0008
2N591A	102A		50	TR85/250	G0005
2N592	100	3005	50	TR05/254	
2N593	100	3005	53	TR05/254	
2N594	101	3011	5	TR08/641	
2N595	101	3011	5	TR08/641	
2N596	101	3011	5	TR08/641	
2N597	100	3005	1	TR05/254	G0005
2N598	100	3005	1	TR05/254	G0005
2N599	100	3005	1	TR05/254	G0005
2N600	100	3005	51	TR05/254	
2N602	126	3008	50	TR17/635	G0002
2N602A	126	3008	2	TR17/635	G0002
2N603	126	3008	51	TR17/635	G0002
2N603A	126	3008	51	TR17/635	G0002
2N604	126	3008	51	TR17/635	G0002
2N604A	126	3006		TR17/635	G0002
2N605	126	3008	50	TR17/635	G0002
2N606	126	3008	50	TR17/635	G0002
2N607	126	3008	50	TR17/635	G0002
2N608	126	3008	50	TR17/635	G0002
2N609	102	3004	53	TR05/631	G0005
2N610	102	3004	53	TR05/631	G0005
2N611	102	3004	53	TR05/631	G0005
2N612	102	3004	53	TR05/631	G0005
2N613	102	3004	53	TR05/631	G0005
2N614	100	3005	52	TR05/254	G0005
2N615	100	3005	2	TR05/254	G0005
2N617	100	3005	1	TR05/254	G0005
2N619	108	3039	61	TR95/56	
2N620	108	3039	61	TR95/56	
2N621	108	3039	61	TR95/56	

Original Part Number	ECG	SK	GE	IR/WEP	HEP
2N622	123		61	TR86/53	S5025
2N623	126		20	TR17/635	G0009
2N624	126	3008	1	TR17/635	G0003
2N625	103		20	TR08/641A	
2N627	121	3009	76	TR01/628	G6013
2N628	121	3009	76	TR01/628	G6013
2N629	121	3009	76	TR01/628	G6013
2N630	179	3009	3	TR35/G6001	G6018
2N631	102	3003	53	TR05/631	G0005
2N632	102	3004	53	TR05/631	G0005
2N633	102	3004	53	TR05/631	G0005
2N633B	102	3014		TR05/631	G0005
2N634	101	3011	8	TR08/641	G0011
2N634A	101	3011	6	TR08/641	G0011
2N635	101	3011	6	TR08/641	G0011
2N635A	101	3011	6	TR08/641	G0011
2N636	101	3011	5	TR08/641	G0011
2N636A	101	3011	6	TR08/641	G0011
2N637	104	3009	76	TR01/624	G6005
2N637A	104	3009	76	TR01/624	G6005
2N637B	104	3009	3	TR01/624	G6005
2N638	104	3009	76	TR01/624	G6005
2N638A	104	3009	76	TR01/624	G6005
2N638B	104	3009	3	TR01/624	
2N639	104	3009	25	TR01/624	G6005
2N639A	104	3009	25	TR01/624	G6005
2N639B	104	3009	25	TR01/624	G6005
2N640	126	3008	51	TR17/635	G0009
2N641	126	3008	1	TR17/635	G0009
2N641REDM/F	126			TR17/635	G0008
2N642	126	3007	51	TR17/635	G0009
2N643	126	3007	53	TR17/635	G0002
2N644	126	3007	51	TR17/635	G0002
2N645	126	3007	51	TR17/635	G0002
2N646	103	3010	8	TR08/641A	G0011
2N647	103A	3010	5	TR08/724	G0011
2N647/22	103A		8	TR08/724	
2N648	103	3010	8	TR08/641A	
2N649	103A	3010	5	TR08/724	G0011
2N649/5	103A		52	TR08/724	
2N650	102	3004	2	TR05/631	G0005
2N650A	102	3004	2	TR05/631	G0005
2N651	102	3004	2	TR05/631	G0005
2N651A	102	3004	2	TR05/631	G0005
2N652	102	3004	2	TR05/631	G0007
2N652A	102	3004	2	TR05/631	G0005
2N654	102	3004	2	TR05/631	G0005
2N655	102	3004	2	TR05/631	G0007
2N655A	128			TR87/	
2N655GRN	102	3004		TR05/631	G0007
2N655RED	102	3004		TR05/631	G0007
2N656	128	3024	18	TR87/243	S5014
2N657	100	3024	32	TR05/254	S5026
2N658	176	48	48	TR82/238	G0005
2N659	176	3005	48	TR82/238	G0005
2N660	176	3005	1	TR82/238	G0005
2N661	176	3005	1	TR82/238	G0005
2N662	176	3005	1	TR82/	G0005
2N663	121	3009	25	TR01/232	G6003
2N665	121	3009	25	TR01/232	G6005
2N669	121	3009	25	TR01/232	G6003
2N672	176		53	TR82/238	G6011
2N674	100	3005	53	TR05/254	G6010
2N677	179	3009	76	TR35/G6001	G6005
2N677A	179	3009	76	TR35/G6001	G6005
2N677B	179	3009	76	TR35/G6001	G6005
2N677C	121	3009	16	TR01/232	G6005
2N678	179	3009	76	TR35/G6001	G6003
2N678A	179	3009	76	TR35/G6001	G6005
2N678B	121	3009	76	TR01/232	G6005
2N678C	121	3009	16	TR01/232	G6005
2N679	101	3011	8	TR08/641	G0011
2N680	126	3004	52	TR17/635	G0003
2N694	126	3008	52	TR17/635	G0003
2N695	126	3006	51	TR17/635	G0003
2N696	123	3124	81	TR86/53	S0014
2N697	123	3024	81	TR86/53	S0014
2N697A	123	3024	47	TR86/53	S3011
2N699	154	3024	18	TR78/712	S5026
2N699A	154	3024	18	TR78/712	S3019
2N699B	154	3024	18	TR78/712	S3019
2N700	160		51	TR17/637	G0003
2N700/18	160		51	TR17/637	G0003
2N700A	160		51	TR17/637	G0003
2N700A/18	160		51	TR17/637	G0003

Original Part Number	ECG	SK	GE	IR/WEP	HEP
2N701	123	3124	53	TR86/53	S0014
2N702	108	3039	61	TR95/56	S0011
2N703	108	3039	61	TR95/56	S0011
2N705	160		51	TR17/637	G0003
2N705A	160		51	TR17/637	G0003
2N706	108	3039	210	TR95/56	S0011
2N706A	108	3039	210	TR95/56	S0011
2N706B	108	3039	210	TR95/56	S0011
2N706C	108	3039	210	TR95/56	S0011
2N707	108	3039	212	TR95/56	S0011
2N707A	123	3124	20	TR86/53	S0011
2N708	123A		17	TR21/735	S0011
2N708A	123A	3122	17	TR21/735	S0011
2N709	108		17	TR95/56	S0011
2N709/52	108	3039	63	TR95/56	S0011
2N709A	108	3039	17	TR95/56	S0011
2N710	160		51	TR17/637	G0003
2N710A	160		51	TR17/637	G0003
2N711	160		51	TR17/637	G0003
2N711A	160		51	TR17/637	G0003
2N711B	160		51	TR17/637	G0003
2N715	123	3124	20	TR86/53	S0011
2N717	108	3039	210	TR95/56	S0011
2N717A	108	3039	17	TR95/56	S3011
2N718	108	3122	210	TR95/56	S0015
2N718A	108	3122	20	TR95/56	S3020
2N719	194		18		S0005
2N719A	194		18		S3019
2N720	128		18		S0005
2N720A	128		18		S3019
2N721	159	3114	82	TR20/717	S0013
2N721A	159	3114	82	TR20/717	S0013
2N722	159	3114	82	TR20/717	S0013
2N722A	159	3114	82	TR20/717	S0013
2N723	106		21	TR20/52	
2N725	160	3008	51	TR17/637	G0003
2N726	106		22	TR20/52	S0013
2N727	106		65	TR20/52	S0013
2N728	123	3124	81	TR86/53	S0011
2N729	123	3124	81	TR86/53	S0011
2N738	154	3045	18	TR78/712	S5025
2N739	154	3045	18	TR78/712	S5025
2N739A	154	3045	27	TR78/712	S5025
2N740	154	3045	18	TR78/712	S0005
2N740A	154	3045	27	TR78/712	S5025
2N741	160	3006	51	TR17/637	G0003
2N741A	160		51	TR17/637	G0003
2N742	123	3124	17	TR86/53	S0011
2N742A	123	3124	81	TR86/53	S0011
2N743	108	3039	20	TR95/56	S0011
2N743A	108	3039	17	TR95/56	S0025
2N744	108	3039	20	TR95/56	S0011
2N745	123A	3122	20	TR21/735	S0011
2N746	123A	3122	20	TR21/735	S0011
2N747	123A	3122	20	TR21/735	S0011
2N748	123A	3122	20	TR21/735	S0011
2N749	123A	3122	18	TR21/735	S0011
2N750	123A	3122	20	TR21/735	S0011
2N751	123A	3122	18	TR21/735	S0011
2N752	108	3039	17	TR95/56	S5026
2N753	123A	3122	61	TR21/735	S0011
2N754	123A	3122	81	TR21/735	S0011
2N756	108	3039	210	TR95/56	S0011
2N756A	123	3124	81	TR86/53	S0011
2N757	108	3039	210	TR95/56	S0011
2N757A	123	3124	81	TR86/53	S0011
2N758	108	3039	210	TR95/56	S0011
2N758A	123	3124	81	TR86/53	S0011
2N758B	123	3124	81	TR86/53	S0011
2N759	108	3039	81	TR95/56	S0011
2N759A	123	3124	81	TR86/53	S0011
2N759B	123	3124	81	TR86/53	S0011
2N760	108	3039	81	TR95/56	S0011
2N760A	123	3124	81	TR86/53	S0011
2N760B	123	3124	81	TR86/53	S0011
2N761	108	3039	17	TR95/56	S0011
2N762	108	3039	17	TR95/56	S0011
2N768	160		51	TR17/637	G0001
2N769	160		51	TR17/637	G0001
2N770	123	3124	20	TR86/53	S0011
2N771	123	3124	20	TR86/53	S0011
2N772	123	3124	20	TR86/53	S0011
2N773	123	3124	20	TR86/53	S0011
2N774	123	3124	20	TR86/53	S0011
2N775	123	3124	20	TR86/53	S0011
2N776	123	3124	20	TR86/53	S0011
2N777	123	3124	20	TR86/53	S0011
2N778	123	3124	20	TR86/53	S0011
2N779	160		51	TR17/637	G0003
2N779A	160		51	TR17/637	G0003
2N779B	160		51	TR17/637	G0003
2N780	107	3039	212	TR70/720	S0011
2N781	160	3004	1	TR17/637	G0003
2N782	160	3123	2	TR17/637	G0003
2N783	123A	3122	210	TR21/735	S0011
2N784	123A	3122	210	TR21/735	S0011
2N784A	123A	3122	20	TR21/735	S0011
2N789	123	3124	39	TR86/53	S0011
2N790	123	3124	20	TR86/53	S0011
2N791	123	3124	20	TR86/53	S0011
2N792	123	3124	20	TR86/53	S0011
2N793	123	3124	20	TR86/53	S0011
2N794	160		51	TR17/637	G0003
2N795	160		51	TR17/637	G0003
2N796	160		51	TR17/637	G0003
2N797	103	3010	7	TR08/641A	G0011
2N799	102	3004	1	TR05/631	G0005
2N800	102A	3123	1	TR85/250	G0005
2N801	100	3005	1	TR05/254	G0005
2N802	100	3005	1	TR05/254	G0005
2N803	158	3005	1	TR84/630	G0005
2N804	158	3005	1	TR84/630	G0005
2N805	158	3005	1	TR84/630	G0005
2N806	158	3005	1	TR84/630	G0006
2N807	102A	3005	1	TR85/250	G0005
2N808	102A	3005	1	TR85/250	G0005
2N809	100	3005	1	TR05/254	G0005
2N810	100	3005	1	TR05/254	G0005
2N811	100	3005	1	TR05/254	G0005
2N812	100	3005	1	TR05/254	G0005
2N813	102A	3005	1	TR85/250	G0005
2N814	102A	3005	1	TR85/250	G0007
2N815	102A	3011	8	TR85/250	G0005
2N816	102A	3011	8	TR85/250	G0005
2N817	158	3011	6	TR84/630	G0005
2N818	158	3011	6	TR84/630	G0005
2N819	102A	3011	6	TR85/250	G0005
2N820	102A	3011	6	TR85/250	G0005
2N821	101	3011	6	TR08/641	G0011
2N822	101	3011	6	TR08/641	G0011
2N823	101	3011	5	TR08/641	G0011
2N824	101	3011	5	TR08/631	G0011
2N825	102	3011	1	TR05/631	G0005
2N826	102	3004	1	TR05/637	G0005
2N827	160		53	TR17/637	G0003
2N828	160		51	TR17/637	G0003
2N828A	160			TR17/637	G0003
2N829	160		51	TR17/637	G0003
2N834	108		20	TR95/56	S0011
2N834A	123A	3122		TR21/735	S0011
2N835	123A	3122	20	TR21/735	S0011
2N837	160		52	TR17/637	G0003
2N838	160		53	TR17/637	G0003
2N839	123A	3122	212	TR21/735	S0011
2N840	123A	3122	212	TR21/735	S0011
2N841	108	3039	212	TR95/56	S0011
2N842	123A	3122	212	TR21/735	S0011
2N843	108	3039	212	TR95/735	S0011
2N844	123A	3122	18	TR21/735	S0011
2N846	160		52	TR17/637	G0003
2N846A	160		51	TR17/637	G0003
2N846B	160	3123	2	TR17/637	G0003
2N849	108	3039	61	TR95/56	S0016
2N850	108	3039	61	TR05/56	S0020
2N851	108	3039	20	TR95/56	S0016
2N852	108	3039	20	TR95/56	S0016
2N858	159	3114	65	TR20/717	S0013
2N859	159	3114	65	TR20/717	S0013
2N860	159	3114	65	TR20/717	S0013
2N861	159	3114	65	TR20/717	S0013
2N862	159	3114	65	TR20/717	S0013
2N863	159	3114	65	TR20/717	S0013
2N864	159	3114	65	TR20/717	S0013
2N864A	106	3114	65	TR20/52	S0013
2N865	106	3118	65	TR20/52	S0013
2N865A	106		65	TR20/52	S0013
2N866	123	3124	20	TR86/53	S0011
2N867	123	3124	20	TR86/53	S0011
2N869	106		65	TR20/52	S0013
2N869A	106			TR20/52	S0013

Original Part Number	ECG	SK	GE	IR/WEP	HEP
2N870	128		18		S5026
2N871	128		18		S5026
2N909	123		210	TR86/53	S0011
2N910	128		18		S0005
2N911	128		18		S5026
2N912	128		18		S5026
2N913	108	3039	17	TR95/56	S0011
2N914	108		17	TR95/56	S0011
2N914/51	108	3039	61	TR95/56	S0016
2N914A	108	3039		TR95/56	S0011
2N915	123A		20		S0025
2N916	108	3039	17	TR95/56	S0011
2N916A	108	3039	17	TR95/56	S0011
2N917	161	3019	60	TR83/719	S0016
2N917A	161	3117	86	TR83/719	S0016
2N918	108	3018	86	TR95/56	S0017
2N919	123A	3122	47	TR21/735	S0011
2N920	123A	3122	47	TR21/735	S0011
2N921	123A	3122	47	TR21/735	S0011
2N922	123A	3122	17	TR21/735	S0011
2N923	159	3114	65	TR20/717	S0013
2N924	159	3114	65	TR20/717	S0013
2N924A		3114			
2N925	159	3114	65	TR20/717	S0013
2N926	159	3114	65	TR20/717	S0013
2N926A		3114			
2N927	159	3114	82	TR20/717	S0013
2N928	159	3114	82	TR20/717	S0013
2N929	123A	3122	212	TR21/735	S0011
2N929A	123A	3122	81	TR21/735	S0011
2N930	123A	3122	212	TR21/735	S0015
2N930A	123A	3039	81	TR21/735	S0011
2N933	126		50	TR17/635	G0003
2N934	160		2	TR17/637	G0003
2N935	159	3114	65	TR20/717	S0013
2N936	159	3114	65	TR20/717	S0013
2N937	159	3114	65	TR20/717	S0013
2N938	159	3114	65	TR20/717	S0013
2N938A		3114			
2N939	159	3114	21	TR20/717	S0013
2N940	159	3114	65	TR20/717	S0013
2N940A		3114			
2N941	159	3114	65	TR20/717	S0013
2N941A		3114			
2N942	159	3114	65	TR20/717	S0013
2N942A		3114			
2N943	159	3114	65	TR20/717	S0013
2N943A		3114			
2N944	159	4114	65	TR20/717	S0013
2N944A		3114			
2N945	159	3114	65	TR20/717	S0013
2N945A		3114			
2N946	159	3114	82	TR20/717	S0013
2N946A		3114			
2N947	123A	3122	10	TR21/735	S0011
2N955	101	3011	11	TR08/641	G0011
2N955A	101	3011	11	TR08/641	G0011
2N957	123A	3122	61	TR21/735	S0015
2N960	160		51	TR17/637	G0003
2N960/46	160		51	TR17/637	G0003
2N961	160		51	TR17/637	G0003
2N962	160		51	TR17/637	G0003
2N963	160		51	TR17/637	G0003
2N964	160		51	TR17/637	G0003
2N964/46	160		51	TR17/637	G0003
2N964A	160		2	TR17/637	G0003
2N965	160		51	TR17/637	G0003
2N966	160		51	TR17/637	G0003
2N967	160		51	TR17/637	G0003
2N968	160		51	TR17/637	G0003
2N969	160		51	TR17/637	G0003
2N970	160		51	TR17/637	G0003
2N971	160		51	TR17/637	G0003
2N971	160		51	TR17/637	G0003
2N973	160		51	TR17/637	G0003
2N974	160		51	TR17/637	G0003
2N975	160		51	TR17/637	G0003
2N976	160		51	TR17/637	G0003
2N977	160	3123	2	TR17/637	G0003
2N978	106	3006	82	TR20/635	S0013
2N979	160		1	TR17/637	G0003
2N980	160		51	TR17/637	G0003
2N981	128		81	TR87/243	S0013
2N982	160	3006	51	TR17/637	G0003
2N983	160	3006	51	TR17/637	G0003

Original Part Number	ECG	SK	GE	IR/WEP	HEP
2N984	160	3006	51	TR17/637	G0003
2N985	160		2	TR17/637	G0003
2N986	126			TR17/635	G0003
2N987	160	3006	51	TR17/637	G0003
2N988	108	3039	20	TR95/56	S0011
2N989	108	3039	20	TR95/56	S0011
2N990	160	3006	51	TR17/637	G0008
2N992	160		50	TR17/637	G0001
2N993	160	3005	51	TR17/637	G0008
2N994	160	3008	2	TR17/637	G0003
2N995	106	3118	65	TR20/52	S0013
2N995A	106	3118	65	TR20/52	S0013
2N996	106		65	TR20/52	S0013
2N1000	101	3011	6	TR08/641	G0011
2N1003	126	3006	51	TR17/635	G0009
2N1004	126		51	TR17/635	G0009
2N1005	108	3039	214	TR95/56	S0016
2N1006	123A	3122	214	TR24/735	S0014
2N1007	104	3009	3	TR01/230	G6003
2N1008	102	3004	2	TR05/631	G0005
2N1008A	102	3004	2	TR05/631	G0005
2N1008B	102	3004	2	TR05/631	G0005
2N1009	102	3003	52	TR05/631	G0005
2N1010	103A	3011	7	TR08/724	G0011
2N1011	121	3009	25	TR01/ 232	G6005
2N1012	101	3011	6	TR08/641	G0011
2N1014	121	3009	25	TR01/232	G6005
2N1017	160	3011	51	TR17/637	G0002
2N1018	160	3011	51	TR17/637	G0002
2N1020	121	3009		TR01/232	
2N1021	121	3009	3	TR01/232	G6005
2N1021A	121	3009	3	TR01/232	G6005
2N1022	121	3009	3	TR01/232	G6005
2N1022A	121	3009	3	TR01/232	G6005
2N1023	160	3006	51	TR17/637	G0008
2N1024	106	3114	65	TR20/52	S0012
2N1025	159	3114	65	TR20/717	S0012
2N1026	159	3114	65	TR20/717	S0012
2N1026A	159	3114	65	TR20/717	S0012
2N1027	159	3114	65	TR20/717	S0012
2N1028	159	3114	65	TR20/717	S0012
2N1028A		3114			
2N1029	121	3009	16	TR01/232	G6005
2N1029A	121	3009	3	TR01/232	G6005
2N1029B	121	3009	3	TR01/232	G6005
2N1029C	121	3009	3	TR01/232	G6005
2N1030	121	3009	16	TR01/232	G6003
2N1030A	121	3009	16	TR01/232	G6005
2N1030B	121	3009	16	TR01/232	G6005
2N1030C	121	3009	16	TR01/232	G6005
2N1031	121	3009	76	TR01/232	G6003
2N1031A	121	3009	76	TR01/232	G6005
2N1031B	121	3009	76	TR01/232	G6005
2N1031C	121	3009	3	TR01/232	G6005
2N1032	121	3009	76	TR01/232	G6003
2N1032A	121	3009	76	TR01/232	G6005
2N1032B	121	3009	76	TR01/232	G6005
2N1032C	121	3009		TR01/232	G6005
2N1033	121	3009	16	TR01/232	G6005
2N1034	129	3114	65	TR88/232	S0012
2N1035	129	3114	65	TR88/242	S0012
2N1036	129	3114	65	TR88/242	S0012
2N1037	129	3114	65	TR88/242	S0012
2N1038	176	3123	2	TR82/624	G6011
2N1040	104	3009	3	TR01/624	G6012
2N1041		3009	3	TR01/	G6012
2N1042	126		2	TR17/635	
2N1043	126		2	TR17/635	
2N1044	102	3004	2	TR05/631	
2N1045	102	3004	2	TR05/631	
2N1046	121	3009	25	TR01/232	G6005
2N1046A		3009	25	TR01/232	G6005
2N1046B	121	3009	25	TR01/232	G6005
2N1051	123A	3122	20	TR21/735	S0014
2N1052	154	3045	32	TR78/712	S5025
2N1053	154	3045	32	TR78/712	S5025
2N1054	154	3045	32	TR78/712	S0014
2N1056	102	3004	2	TR05/631	G0005
2N1057	102	3004	2	TR05/631	G0005
2N1058	103A	3011	5	TR08/724	G0011
2N1059	103A	3010	59	TR08/724	G0011
2N1059-1			8	TR08/	
2N1060	108	3039	63	TR95/56	
2N1065	160	3008	51	TR17/637	G0002

Original Part Number	ECG	SK	GE	IR/WEP	HEP
2N1066	160	3006	51	TR17/637	G0003
2N1069	130	3027	77	TR59/247	S7002
2N1070	130	3027	77	TR59/247	S7002
2N1073	121	3034	76	TR01/232	G6003
2N1073A	121	3034	76	TR01/232	G6005
2N1073B	127	3034	25	TR27/235	G6008
2N1078	121	3009	54	TR01/232	
2N1081	123A	3122	63	TR21/735	S0014
2N1082	123A	3122	20	TR21/735	S0014
2N1084			29	TR77/	S5013
2N1085			46	TR76/	S5014
2N1086	101	3011	6	TR08/641	G0011
2N1086A	101	3011	6	TR08/641	G0011
2N1087	101	3011	6	TR08/641	G0011
2N1090	101	3011	6	TR08/641	G0011
2N1091	101	3011	6	TR08/641	G0011
2N1092	128	3024	47	TR87/243	S5014
2N1093	160		53	TR17/637	G0002
2N1094	160		52	TR17/637	G0002
2N1095	103	3010	17	TR08/641A	S3020
2N1096	103	3010	18	TR08/641A	S5026
2N1097	102	3003	2	TR05/631	G0005
2N1098	102	3003	2	TR05/631	G0005
2N1099	105	3012	4	TR03/233	G6006
2N1100	105	3012	4	TR08/233	G6006
2N1101	103A	3010	59	/724	G0011
2N1102	103A	3010	59	TR08/724	G0011
2N1102/5	101	3011		TR08/641	
2N1103	123A	3124	11	TR21/735	S0014
2N1104	123	3124	39	TR86/53	S0014
2N1107	126	3005	51	TR17/635	G0008
2N1108	126	3005	51	TR17/635	G0008
2N1108RED	126	3006		TR17/635	G0008
2N1109	126	3007	51	TR17/635	G0008
2N1110	126	3005	51	TR17/635	G0008
2N1111	126	3007	51	TR17/635	G0008
2N1111A	126	3006	51	TR17/635	G0008
2N1111B	126	3006	51	TR17/635	G0008
2N1111RED	126	3006		TR17/635	G0008
2N1112	101	3011	8	TR08/641	
2N1114	101	3011	8	TR08/641	G0011
2N1115	160	3005	1	TR17/637	G0002
2N1115A	160	3005	2	TR17/637	G0005
2N1116	123A	3122	63	TR21/735	S0014
2N1117	123A	3122	63	TR21/735	S0014
2N1118	159	3114	65	TR20/717	S0012
2N1118A	159	3114	65	TR20/717	S0012
2N1118AB		3114			
2N1119	159	3114	65	TR20/717	S0012
2N1119A		3114			
2N1120	121	3009	76	TR01/232	G6005
2N1121	101	3011	7	TR08/641	G0011
2N1122	160	3005	51	TR17/637	G0003
2N1122A	160	3005	51	TR17/637	G0003
2N1123			53		
2N1124	102	3004	80	TR05/631	G0005
2N1125	102	3004	80	TR05/631	G0005
2N1126	102	3004	53	TR05/631	G6011
2N1127		3004	53	/631	G6011
2N1128	102	3004	53	TR05/631	G0005
2N1129		3004	53	TR05/631	G0005
2N1130	102A	3004	80	TR85/250	G0005
2N1131	159	3114	21	TR20/717	S0012
2N1131A	159	3114	21	TR20/717	S0012
2N1131AD		3114			
2N1131/46			21		
2N1131/51			21		
2N1132	159	3114	21	TR20/717	S0012
2N1132/46	159	3114	21	TR20/717	S0013
2N1132A	159	3114	21	TR20/717	S0012
2N1132B	159	3114	67	TR20/717	S0012
2N1132AC		3114			
2N1132BC		3114			
2N1132/46L		3114			
2N1132/51			21		
2N1132A46			21		
2N1132B46			67		
2N1132B/51			67		
2N1135	159	3114		TR20/717	S0012
2N1135A	159	3114		TR20/717	S0012
2N1135AC		3114			
2N1136	121	3009	76	TR01/232	G6005
2N1136A	121	3009	76	TR01/232	G6005
2N1136B	127	3009	16	TR01/235	
2N1136C			16	TR01/	
2N1137	121	3009	16	TR01/232	G6005
2N1137A	121	3009	16	TR01/232	G6005
2N1137B	127	3009	16	TR01/235	G6018
2N1138	121	3009	16	TR01/232	G6005
2N1138A	121	3009	16	TR01/232	G6005
2N1138B		3009	16	TR92/	G6005
2N1139	123A	3122	16	TR21/735	S0014
2N1140			20	TR25/	S0014
2N1141	160			TR17/637	G0002
2N1141A	160			TR17/637	G0002
2N1142	160			TR17/637	G0009
2N1142A	160			TR17/637	G0002
2N1143	160			TR17/637	G0002
2N1143A	160			TR17/637	G0002
2N1144	160	3003	52	TR17/637	G0002
2N1145	160	3003	52	TR17/637	G0002
2N1146	104	3009	76	TR01/232	G6003
2N1146A	104	3009	76	TR01/232	G6005
2N1146B	104	3009	76	TR01/232	G6005
2N1146C	104	3009	3	TR01/232	G6005
2N1147	104	3009	76	TR01/232	G6003
2N1147A	104	3009	76	TR01/232	G6005
2N1147B	104	3009	76	TR01/232	G6005
2N1147C	104	3009	3	TR01/232	G6005
2N1149	123	3124	39	TR86/53	S0014
2N1150	123	3124	39	TR86/53	S0014
2N1150/904			17	TR24/	
2N1151	123	3124	39	TR86/53	S0014
2N1152	123	3124	39	TR86/53	S0014
2N1153	123	3124	39	TR86/53	S0014
2N1153/910			17	TR24/	
2N1154			47	TR63/	S0005
2N1155			18	TR25/	S0005
2N1155/952			47	TR25/	
2N1156			18		S5025
2N1157	105	3012	4	TR03/233	
2N1157A	105	3012	4	TR03/233	
2N1158	160			TR17/637	G0002
2N1158A	160			TR17/637	G0002
2N1159	121	3009	16	TR01/232	G6005
2N1160	121	3009	76	TR01/232	G6005
2N1162	104	3009	76	TR01/624	G6001
2N1162A	104	3009	76	TR01/624	G6001
2N1163	121	3009	76	TR01/232	G6014
2N1163A	121	3009	76	TR01/232	G6014
2N1164	104	3009	76	TR01/624	G6001
2N1164A	104	3009	76	TR01/624	G6001
2N1165	104	3009	76	TR01/624	G6009
2N1165A	121	3009	76	TR01/232	G6009
2N1166	104	3009	3	TR01/624	G6001
2N1166A	121	3009	16	TR01/232	G6001
2N1167		3009	25	TR01/	G6009
2N1167A		3009	25	TR01/	G6009
2N1168	121	3009	16	TR01/232	G6005
2N1169	103		5	TR08/641A	
2N1170	103		5	TR08/641A	
2N1171	100	3005	1	TR05/254	G0005
2N1172	104	3009	3	TR01/624	
2N1173	103A	3010	8	TR08/724	G0011
2N1173W	103A	3123	8	TR08/724	
2N1174	102	3010	2	TR05/631	G0005
2N1174W	176	3123	2	TR82/	
2N1175	102	3004	53	TR05/631	G0005
2N1175A	102	3004	53	TR05/631	G0005
2N1176	100	3005	53	TR05/254	G0005
2N1176A	100	3005	80	TR05/254	G0005
2N1176B	100	3005		TR05/254	G0005
2N1177	160	3006	51	TR17/637	G0009
2N1178	160	3006	51	TR17/637	G0009
2N1179	160	3006	51	TR17/637	G0009
2N1180	126	3006	51	TR17/635	G0009
2N1182	121	3014	76	TR01/232	G6005
2N1183	176	3123		TR82/232	G6011
2N1183A	176	3123		TR82/232	G6011
2N1183B	176			TR82/232	G6012
2N1184	176	3123		TR82/232	G6011
2N1184A				TR82/232	G6011
2N1184B	176			TR82/232	G6012
2N1185	100	3005	1	TR05/254	G0005
2N1186	100	3005	1	TR05/254	G0005
2N1187	100	3005	1	TR05/254	G0005
2N1188	100	3005	1	TR05/254	G0005
2N1189	102	3004	53	TR05/631	G0005
2N1190	102	3004	53	TR05/631	G0005
2N1191	102	3004	2	TR05/631	G0005

Original Part Number	ECG	SK	GE	IR/WEP	HEP
2N1192	102	3004	2	TR05/631	G0005
2N1193	102	3004	53	TR05/631	G0007
2N1194	102	3004	2	TR05/631	G0007
2N1195	160		51	TR17/637	G0002
2N1196	159		82	TR20/717	S5023
2N1197	129	3025	82	TR88/642	S5023
2N1198	103	3011	5	TR08/641A	G0011
2N1199	123A	3122	212	TR21/735	S0025
2N1199A	123A	3122		TR21/735	S0011
2N1200	123A	3124	11	TR21/735	S0014
2N1201	123	3124	11	TR86/53	S0014
2N1202	105	3012	16	TR03/233	
2N1203	105	3012	4	TR03/233	
2N1204	126	3006	51	TR17/635	G0005
2N1204A	126	3006		TR17/635	G0005
2N1205	123	3124	214	TR86/53	S0014
2N1206			63	TR25/	S5014
2N1207	154	3045	27	TR78/712	S5026
2N1213	126	3006	2	TR17/635	G0002
2N1214	126	3006	2	TR17/635	G0002
2N1215	126	3006	2	TR17/635	G0002
2N1216	126	3006	2	TR17/635	G0002
2N1217	103	3010	5	TR08/641A	G0011
2N1218	121			TR01/232	
2N1219	159	3114	65	TR20/717	S0012
2N1220	159	3114	65	TR20/717	S0012
2N1221	159	3114	65	TR20/717	S0012
2N1222	159	3114	65	TR20/717	S0012
2N1222A		3114			
2N1223	159	3114	65	TR20/717	S0012
2N1224	126	3007	51	TR17/635	G0009
2N1225	126	3006	51	TR17/635	G0008
2N1226	126	3008	51	TR17/635	G0008
2N1227	104	3009	25	TR01/232	G6003
2N1227-3	104	3009	16	TR01/232	G6005
2N1227-4	104	3014	16	TR01/232	G6005
2N1227-4R	104	3009	16	TR01/232	G6005
2N1227A	104	3009	16	TR01/232	G6005
2N1228	129	3025	22	TR88/242	S0012
2N1229	129	3025	22	TR88/242	S0012
2N1230	129	3025	82	TR88/242	S0012
2N1231	129	3025	82	TR88/242	S0012
2N1232	129	3025	82	TR88/242	S0012
2N1233	129	3025	82	TR88/242	S0012
2N1234	129	3025		TR88/242	
2N1238	106		67	TR20/52	S0032
2N1239	106		67	TR20/52	S0032
2N1240	106		67	TR20/52	S0032
2N1241	106		67	TR20/52	S0032
2N1242			67	TR28/	S5022
2N1243			67	TR28/	S5022
2N1245	104	3009	16	TR01/230	G6003
2N1246	104	3009	16	TR01/230	G6003
2N1247	123	3124	214	TR88/53	S0014
2N1248	123	3124	214	TR86/53	S0014
2N1249	123	3124	214	TR86/53	S0014
2N1251	103A	3010	5	TR08/724	G0011
2N1252	128	3024	81	TR87/243	S0014
2N1253	128	3024	81	TR87/243	S0014
2N1253A			63	TR25/	S3011
2N1254	129	3025	65	TR88/242	S0012
2N1255	129	3025	65	TR88/242	S0012
2N1256	129	3025	65	TR88/242	S0012
2N1257	129	3025	65	TR88/242	S0012
2N1258	129	3025	65	TR88/242	S0012
2N1259	129	3025	65	TR88/242	S0012
2N1261	105	3012	4	TR03/233	
2N1262	105	3012	4	TR03/233	
2N1263	105	3012	4	TR03/233	
2N1264	100	3006	51	TR05/254	G0005
2N1265	102	3003	52	TR05/631	G0005
2N1265A	102	3004		TR05/631	
2N1265/5			52	TR85/	
2N1266	102A	3005	50	TR85/250	G0005
2N1267	123	3124	212	TR86/53	S0011
2N1268			212	TR24/	S0011
2N1269			212	TR24/	S0011
2N1270			212	TR24/	S0011
2N1271			212	TR24/	S0011
2N1272			212	TR24/	S0011
2N1273	102	3004	52	TR05/631	G0005
2N1273BLU	102	3004		TR05/631	G0007
2N1273GRN	102	3004		TR05/631	G0005
2N1273ORN	102	3004		TR05/631	G0005
2N1273RED	102	3004		TR05/631	G0005
2N1273YEL	102	3004		TR05/631	G0005
2N1274	102	3004	53	TR05/631	G0005
2N1274BLU	102	3004		TR05/631	G0007
2N1274BRN	102	3004		TR05/631	G0005
2N1274GRN	102	3004		TR05/631	G0005
2N1274ORN	102	3004		TR05/631	G0005
2N1274PUR	102	3004		TR05/631	G0005
2N1274RED	102	3004		TR05/631	G0005
2N1274V10	176	3123			G0007
2N1275	129	3025	21	TR88/242	
2N1276	123	3124	60	TR86/53	S0014
2N1277	123	3124	60	TR86/53	S0014
2N1278	123	3124	60	TR86/53	S0014
2N1279	123	3124	39	TR86/53	S0014
2N1280	100	3005	1	TR05/254	G0005
2N1281	100	3005	1	TR05/254	G0005
2N1282	160	3005	1	TR17/637	G0002
2N1284	100	3005	1	TR05/254	G0005
2N1285	160	3006	51	TR17/637	G0003
2N1287	102	3004	52	TR05/631	G0005
2N1287A	102	3004	52	TR05/631	G0005
2N1288	101	3011	8	TR08/641	G0011
2N1289	101	3011	5	TR08/641	G0011
2N1291	104	3009	3	TR01/230	G6003
2N1292	121	3009	3	TR01/232	
2N1293	121	3009	25	TR01/232	G6005
2N1294	121	3014	16	TR01/232	
2N1295	121	3009	25	TR01/232	G6005
2N1296	121	3014	3	TR01/232	
2N1297	121	3009	25	TR01/232	G6005
2N1298			54		
2N1299	101	3011	6	TR08/641	G0011
2N1300	160	3005	1	TR17/637	G0002
2N1301	160	3005	1	TR17/637	G0002
2N1302	101	3011	8	TR08/641	G0011
2N1303	158	3004	1	TR88/630	G0005
2N1304	101	3011	8	TR08/641	G0011
2N1305	158	3123	1	TR84/630	G0005
2N1306	101	3011	5	TR08/641	G0011
2N1307	158	3123	1	TR84/630	G0005
2N1308	101	3011	5	TR08/641	G0011
2N1309	100	3123	1	TR05/637	G0005
2N1309A	160			TR17/	G0002
2N1310	101	3011	5	TR08/641	
2N1311	101	3011	5	TR08/641	
2N1312	103	3010	7	TR08/641A	
2N1313	126	3008	1	TR17/635	G0005
2N1314	104	3009	25	TR01/230	G6003
2N1314R	104	3009	16	TR01/230	
2N1315	176	3123	3	TR82/	G6003
2N1316	100	3005	1	TR05/254	G0002
2N1317	100	3005	1	TR05/254	G0002
2N1318	100	3005	1	TR05/254	G0002
2N1319	100	3005	1	TR05/254	G0005
2N1330		3124			
2N1335			18		
2N1336			18		
2N1338	123		18	TR86/53	
2N1339			18		
2N1340			18		
2N1341			18		
2N1343	100	3005	1	TR05/254	G0005
2N1344	100	3005	1	TR05/254	G0005
2N1345	100	3005	1	TR05/254	G0002
2N1346	100	3005	1	TR05/254	G0002
2N1347	100	3005	1	TR05/254	G0005
2N1348	102	3004	2	TR05/631	G0005
2N1349	100	3005	1	TR05/254	G0005
2N1350	100	3005	1	TR05/254	G0005
2N1351	100	3005	1	TR05/254	G0005
2N1352	102	3003	53	TR05/631	G0005
2N1353	102	3003	2	TR05/631	G0005
2N1354	100	3005	1	TR05/254	G0005
2N1355	100	3005	1	TR05/254	G0005
2N1356	100	3005	1	TR05/254	G0005
2N1357	100	3005	1	TR05/254	G0005
2N1358	105	3012	4	TR03/233	G6006
2N1358A	105	3012	4	TR03/233	
2N1358M	105	3012	4	TR03/233	G6006
2N1359	121	3009	76	TR01/232	G6005
2N1361	100	3009	2	TR05/254	G0005
2N1361A		3005	2	TR05/254	G0002
2N1362	121	3009	16	TR01/232	G6005
2N1363	121	3009	4	TR01/232	G6005
2N1364	121	3009	16	TR01/232	G6005

Original Part Number	ECG	SK	GE	IR/WEP	HEP
2N1365	121	3009	16	TR01/232	G6017
2N1366	101	3011	5	TR08/641	G0011
2N1367	101	3011	5	TR08/641	G0011
2N1370	102	3004	53	TR05/631	G0005
2N1371	102	3004	2	TR05/631	G0005
2N1372	102	3004	53	TR05/631	G0005
2N1373	102	3004	2	TR05/631	G0005
2N1374	102	3004	53	TR05/631	G0007
2N1375	102	3004	2	TR05/631	G0005
2N1376	102	3004	53	TR05/631	G0005
2N1377	102	3004	2	TR05/631	G0005
2N1378	102	3004	53	TR05/631	G0005
2N1379	102	3004	53	TR05/631	G0005
2N1380	102	3004	53	TR05/631	G0005
2N1381	102	3004	53	TR05/631	G0005
2N1382	102	3004	2	TR05/631	G0005
2N1383	102	3004	53	TR05/631	G0005
2N1384	126	3008	51	TR17/635	G0002
2N1385	160		51	TR17/637	G0002
2N1386	123	3124	61	TR86/53	S0014
2N1387	123	3124	61	TR86/53	S0014
2N1388	123	3124	212	TR86/53	S0014
2N1389	123	3124	212	TR86/53	S0014
2N1390			61		S0011
2N1391	103		214	TR08/641A	G0011
2N1392	102	3004	52	TR05/631	
2N1393	100	3005	1	TR05/254	
2N1394			51	TR12/	
2N1395	100	3006	51	TR05/254	G0003
2N1396	126	3006	51	TR17/635	G0003
2N1397	126	3006	51	TR17/635	G0003
2N1398	126	3006	51	TR17/635	
2N1399	126	3006	51	TR17/635	
2N1400	126	3006	51	TR17/635	
2N1401	126	3006	51	TR17/635	
2N1401A	126	3006	51	TR17/635	
2N1402	126	3006	51	TR17/635	G0009
2N1403	160	3123	2	TR17/637	G0002
2N1404	102		2	TR05/631	G0005
2N1404A	160			TR17/637	G0002
2N1405	160		50	TR17/637	G0003
2N1406	160		50	TR17/637	G0003
2N1407	160		50	TR17/637	G0003
2N1408	160		2	TR17/637	G0002
2N1409	160		81	TR17/637	G0002
2N1410	160		81	TR17/637	G0002
2N1410A			47	TR63/	
2N1411	160		51	TR17/637	G0003
2N1412	105		4	TR03/233	G6006
2N1413	102	3004	53	TR05/631	G0005
2N1414	102	3004	53	TR05/631	G0005
2N1415	102	3004	53	TR05/631	G0005
2N1416	102A		53	TR85/250	G0005
2N1417	123A	3124	39	TR21/735	S0014
2N1418	123A	3124	39	TR21/735	S0014
2N1419	104	3009	16	TR01/232	G6005
2N1420	123A		81	TR21/735	S0014
2N1420A	123A		47		S3011
2N1422	130			TR59/247	S5000
2N1423	130			TR59/247	S5000
2N1425	126	3008	50	TR17/635	G0003
2N1426	126	3008	50	TR17/635	G0003
2N1427	126	3003	2	TR17/635	G0003
2N1428	159	3114		TR20/717	S0012
2N1429	129	3025	210	TR88/242	S0012
2N1430	121	3014	25	TR05/232	G6005
2N1431	103A	3010	59	TR08/724	G0011
2N1432	102	3004	51	TR05/631	G0003
2N1433	105	3012	4	TR03/233	
2N1434	105	3012	4	TR05/233	
2N1435	105	3012	4	TR03/233	
2N1436	160		50	TR17/637	G0002
2N1437			3		
2N1438			3		
2N1439	159	3114	82	TR20/717	S0012
2N1439A		3114			
2N1440	159	3114	82	TR20/717	S0012
2N1440A		3114			
2N1441	159	3114	82	TR20/717	S0012
2N1442	159	3114	82	TR20/717	S0012
2N1443	159	3114	82	TR20/717	S0012
2N1443A		3114			
2N1444			63	TR25/	S3011
2N1445			32		S3019
2N1446	102	3004	2	TR05/631	G0005

Original Part Number	ECG	SK	GE	IR/WEP	HEP
2N1447	102	3004	2	TR05/631	G0005
2N1448	102	3004	2	TR05/631	G0005
2N1449	102	3004	2	TR05/631	G0005
2N1450	102	3004	52	TR05/631	G0005
2N1551	102	3004	2	TR05/631	G0005
2N1452	102	3004	2	TR05/631	G0005
2N1469	100	3005	65	TR05/254	G0005
2N1470	100	3005	72	TR05/254	S5000
2N1471	100	3005	1	TR05/254	G0005
2N1472	128	3024	212	TR87/243	S0011
2N1473	101	3011	7	TR08/641	G0011
2N1474	159	3114	82	TR20/717	S0012
2N1474A	159	3114	82	TR20/717	S0012
2N1474AC		3114			
2N1475	159	3114	82	TR20/717	S0012
2N1478	102		2	TR05/631	G0005
2N1479	128	3024	46	TR87/243	S5014
2N1480	128	3024	28	TR78/243	S5014
2N1481	128	3024	46	TR87/243	S5014
2N1482	128	3024		TR87/243	S5014
2N1483		3024	66		S5014
2N1484		3530	66		S5000
2N1485		3024	66		S5014
2N1486		3530	66		S5000
2N1487	130	3510	77	TR59/247	S7002
2N1488	130	3510	75	TR59/247	S7002
2N1489	130	3510	77	TR59/247	S7002
2N1490	130	3510	75	TR59/247	S7002
2N1491	195	3048	17	TR65/735	S0014
2N1492	195	3048	17	TR65/735	S0014
2N1493	154			TR78/712	S5026
2N1495	160		51	TR17/637	G0002
2N1499	126	3006	51	TR17/635	G0001
2N1499A	126	3006	51	TR17/635	G0001
2N1499B	126	3006	51	TR17/635	G0001
2N1500	126	3006	51	TR17/635	G0002
2N1500/18	160		52	TR17/637	G0003
2N1501	104	3012	4	TR01/624	
2N1502	104	3012	4	TR01/624	
2N1505	108	3039	47	TR95/56	S3001
2N1506	108	3039	47	TR95/56	S3001
2N1506A	108	3039	17	TR95/56	S3011
2N1507	108	3039	81	TR95/56	S3020
2N1510	101	3011	6	TR08/641	
2N1515	160	3006	52	TR17/637	G0003
2N1516	160	3006	2	TR17/637	G0003
2N1517	160	3006	50	TR17/637	G0003
2N1517A	160		52	TR17/637	
2N1518	213	3012	4	TR03/233	G6004
2N1519	213	3012	4	TR03/233	G6006
2N1520	213	3012	4	TR03/233	G6004
2N1521	213	3012	4	TR03/233	G6006
2N1522	213	3012	4	TR03/233	G6002
2N1523	105	3012	4	TR03/233	G6002
2N1524	160	3008	50	TR17/637	G0003
2N1524-1	160	3007	52	TR17/637	
2N1524-2	160	3007	52	TR17/637	
2N1524/33	160		52	TR17/637	G0002
2N1525	160	3008	50	TR17/637	G0003
2N1526	126	3007	50	TR17/635	G0009
2N1526/33	160		52	TR17/637	G0003
2N1527	160	3008	50	TR17/637	G0003
2N1528	123A	3124	63	TR21/735	S0014
2N1529	121	3009	76	TR01/232	G6003
2N1529A	121	3009	76	TR01/232	G6003
2N1530	121	3009	76	TR01/232	G6005
2N1530A	121	3009	76	TR01/232	G6005
2N1531	121	3009	76	TR01/232	G6005
2N1531A	121	3009	76	TR01/232	G6005
2N1532	121	3009	3	TR01/232	G6005
2N1532A	121	3009	3	TR01/232	G6005
2N1533	121	3009	3	TR01/232	G6018
2N1534	121	3009	76	TR01/232	G6003
2N1534A	121	3009	76	TR01/232	G6003
2N1535	121	3009	76	TR01/232	G6005
2N1535A	121	3009	76	TR01/232	G6005
2N1535B	121	3009		TR01/232	
2N1536	121	3009	76	TR01/232	G6005
2N1536A	121	3009	76	TR01/232	G6005
2N1537	121	3009	3	TR01/232	G6005
2N1537A	121	3009	3	TR01/232	G6005
2N1538	121	3009	3	TR01/232	G6018
2N1538A	121	3009	16	TR01/232	
2N1539	121	3009	76	TR01/232	G6003
2N1539A	121	3009	76	TR01/232	G6003

Original Part Number	ECG	SK	GE	IR/WEP	HEP
2N1540	121	3009	76	TR01/232	G6005
2N1540A	121	3009	76	TR01/232	G6005
2N1541	121	3009	76	TR01/232	G6005
2N1541A	121	3009	76	TR01/232	
2N1542	121	3009	3	TR01/232	G6005
2N1542A	121	3009	3	TR01/232	G6005
2N1543	121	3009	3	TR01/232	G6018
2N1543A	121	3009	16	TR01/232	
2N1544	121	3009	76	TR01/232	G6003
2N1544A	121	3009	76	TR01/232	G6003
2N1545	121	3009	76	TR01/232	G6013
2N1545A	121	3009	76	TR01/232	G6013
2N1546	121	3009	76	TR01/232	G6005
2N1546A	121	3009	76	TR01/232	G6005
2N1547	121	3009	3	TR01/232	G6005
2N1547A	121	3009	3	TR01/232	G6005
2N1548	121	3009	3	TR01/232	G6018
2N1548A	121	3009	16	TR01/232	
2N1549	121	3009	76	TR01/232	G6003
2N1549A	121	3009	76	TR01/232	G6003
2N1550	121	3009	76	TR01/232	G6005
2N1550A	121	3009	76	TR01/232	G6005
2N1551	121	3009	76	TR01/232	G6005
2N1551A	121	3009	76	TR01/232	G6005
2N1552	121	3009	3	TR01/232	G6005
2N1552A	121	3009	3	TR01/232	G6005
2N1553	121	3009	76	TR01/232	G6003
2N1553A	121	3009	76	TR01/232	G6003
2N1554	121	3009	16	TR01/232	G6005
2N1554A	121	3009	76	TR01/232	G6005
2N1555	121	3009	76	TR01/232	G6005
2N1555A	121	3009	76	TR01/232	G6005
2N1556	121	3009	3	TR01/232	G6005
2N1556A	121	3009	3	TR01/232	G6005
2N1557	121	3009	76	TR01/232	G6003
2N1557A	121	3009	76	TR01/232	G6003
2N1558	121	3009	76	TR01/232	G6005
2N1558A	121	3009	76	TR01/232	G6005
2N1559	121	3009	76	TR01/232	G6005
2N1559A	121	3009	76	TR01/232	G6005
2N1560	121	3009	3	TR01/232	G6005
2N1560A	121	3009	3	TR01/232	G6005
2N1561	160		51	TR17/637	G0003
2N1562	160		51	TR17/637	G0003
2N1564			81	TR25/	S5026
2N1565			81	TR25/	S5026
2N1566			81		S5026
2N1566A			81	TR78/	S5026
2N1567			4		
2N1568			4		
2N1569			4		
2N1570	100	3005	2	TR05/254	G0005
2N1572	154	3045	18	TR78/712	S5025
2N1573	154	3045	18	TR78/712	S5025
2N1574	154	3045	18	TR78/712	S5025
2N1581	100	3005	2	TR05/254	G0005
2N1583	100	3005	2	TR05/254	G0005
2N1584	100	3005	2	TR05/254	G0005
2N1585	101	3011	8	TR08/641	G0011
2N1586	107	3124	60	TR70/720	S0014
2N1587	107	3124	60	TR70/720	S0014
2N1588	123A	3122	212	TR21/735	S0005
2N1589	107	3124	60	TR70/720	S0014
2N1590	107	3124	60	TR70/720	S0014
2N1591	123A	3020	212	TR21/735	S0014
2N1592	107	3124	60	TR70/720	S0014
2N1593	107	3124	60	TR70/720	S0014
2N1594	123A	3124	212	TR21/735	S0014
2N1595		3577	5		
2N1596		3577	5		
2N1596A		3577			
2N1597		3577	5		
2N1597A		3577			
2N1598		3578			
2N1598A		3578			
2N1599		3578			
2N1599A		3578			
2N1605	101	3011	8	TR08/641	G0011
2N1605A	101	3011	5	TR08/641	G0011
2N1606	129	3114		TR88/717	S0012
2N1607	159	3114		TR20/717	S0012
2N1608	159	3114		TR20/717	S0012
2N1608A		3114			
2N1609			16		
2N1610			16		
2N1611			3		
2N1612			3		
2N1613	128	3024	216	TR87/243	S5026
2N1613/46			81		S3011
2N1613A			27	TR78/	S3011
2N1613B			27	TR78/	S3019
2N1614	158		2	TR84/630	G0005
2N1615			32		S5026
2N1622	101	3011	54	TR08/641	G0011
2N1623	129	3114	65	TR88/242	S0012
2N1624	101	3011	50	TR08/641	G0011
2N1625			50	TR17/	
2N1631	126	3008	51	TR17/635	G0001
2N1632	126	3008	51	TR17/635	G0001
2N1633	160	3008	51	TR17/637	G0002
2N1634	160	3008	51	TR17/637	G0002
2N1635	126	3008	51	TR17/635	G0001
2N1636	126	3008	51	TR17/635	G0001
2N1637	126	3008	51	TR17/635	G0001
2N1637/33	160	3123	52	TR17/	G0001
2N1638	160	3008	51	TR17/637	G0008
2N1638/33			52	TR85/	G0003
2N1639	160	3008	51	TR17/637	G0008
2N1638/33	160	3123	52	TR17/	G0001
2N1640	106	3118	65	TR20/52	
2N1641	106	3118	65	TR20/52	
2N1642	106	3118	65	TR20/52	
2N1643	159	3114	210	TR20/	S0012
2N1644	123A	3122	27	TR21/735	S0014
2N1644A	123A	3122	63	TR21/735	S0014
2N1646	160		51	TR17/637	G0003
2N1651	179	3014	76	TR35/G6001	G6005
2N1652	121	3014	3	TR01/232	G6005
2N1653	121	3009	3	TR01/232	G6009
2N1654			18		
2N1655			27	TR78/	
2N1656			27	TR78/	
2N1663	123A	3122	20	TR21/735	S0011
2N1664	100	3005	2	TR05/254	G0006
2N1665	160		52	TR17/637	G0002
2N1666	121	3009	3	TR01/232	G6005
2N1667	121	3009	25	TR01/232	G6003
2N1668	104	3009	25	TR01/230	G6003
2N1669	104	3009	25	TR01/230	G6003
2N1670	160	3005	1	TR17/637	G0002
2N1672	103A		6	TR08/724	G0011
2N1672A	103A			TR08/724	G0011
2N1673	102A	3005	1	TR85/250	G0005
2N1674	123A	3122	212	TR21/735	S0014
2N1676	159	3114	65	TR20/717	S0012
2N1677	159	3118	65	TR20/717	S0012
2N1677A		3114			
2N1678	160	3008	51	TR17/637	G0002
2N1681	102A		1	TR85/250	G0005
2N1682	123	3124	17	TR86/53	S0014
2N1683	100	3005	6	TR05/254	G0005
2N1684	100	3005	1	TR05/254	G0005
2N1685	101	3011	6	TR08/641	G0011
2N1691			32		S3019
2N1694	101	3011	32	TR08/641	G0011
2N1699	126		51	TR17/635	
2N1700	128	3512	46	TR87/243	S5014
2N1701		3530	66		S5014
2N1702	130	3510	77	TR59/247	S7002
2N1703	130		19	TR59/247	S5004
2N1704	123A	3122	81	TR21/735	S0014
2N1705	102	3004	2	TR05/631	G0005
2N1706	102	3004	2	TR05/631	G0005
2N1707	102	3004	2	TR05/631	G0005
2N1708	123A	3122	20	TR21/735	S0011
2N1708A	123A	3122	20	TR21/735	S0011
2N1709			66		S3010
2N1710			28	TR76/	S3010
2N1711	128	3024	18	TR87/	S0015
2N1711/46			81		S3020
2N1711A	128				S3020
2N1713	160				G0009
2N1714			63		S3020
2N1715			32		S3019
2N1716			63		S3020
2N1717			32		S3019
2N1718			66		S3020
2N1720			66		S3020
2N1726	160	3006	51	TR17/637	G0002
2N1727	160	3006	51	TR17/637	G0002

Original Part Number	ECG	SK	GE	IR/WEP	HEP
2N1728	160	3006	51	TR17/637	G0002
2N1729	100	3005	53	TR05/254	G0005
2N1730	101	3011	8	TR08/641	G0011
2N1731	100	3005	53	TR05/254	G0005
2N1732	101	3011	54	TR08/641	G0011
2N1742	160		51	TR17/637	G0003
2N1743	100	3005	51	TR05/254	G0005
2N1744	100	3005	51	TR05/254	G0005
2N1745	160	3006	51	TR17/637	G0003
2N1746	160	3006	51	TR17/637	G0003
2N1747	160	3011	51	TR17/637	G0003
2N1748	160	3006	51	TR17/637	G0002
2N1748A	160	3006		TR17/637	G0002
2N1749	160	3006	51	TR17/637	G0002
2N1750	160	3007	51	TR17/637	G0003
2N1751	179			TR35/G6001	
2N1752	160		51	TR17/637	G0002
2N1753	160		52	TR17/637	G0002
2N1754	160	3005	51	TR17/637	G0002
2N1755	104	3009	3	TR01/624	
2N1756	104	3009	3	TR01/624	
2N1757	104	3009	3	TR01/624	
2N1758	104	3009	3	TR01/624	
2N1759	104	3009	3	TR01/624	
2N1760	104	3009	3	TR01/624	
2N1761	104	3009	3	TR01/624	
2N1762	104	3009	3	TR01/624	
2N1763	123	3124	18	TR86/53	S0014
2N1764	128	3024	20	TR87/243	S0011
2N1768			66		S5000
2N1769			66		S5000
2N1779	101	3011	6	TR08/641	G0011
2N1780	101	3011	5	TR08/641	G0011
2N1781	101	3011	5	TR08/641	G0011
2N1782	160	3005	2	TR17/637	G0003
2N1783	101	3005	1	TR08/641	G0011
2N1784	160	3005	1	TR17/637	G0003
2N1785	160	3006	51	TR17/637	G0002
2N1786	160	3006	51	TR17/637	G0002
2N1787	160	3006	51	TR17/637	G0002
2N1788	160	3006	51	TR17/637	G0002
2N1789	160	3006	51	TR17/637	G0002
2N1790	160		51	TR17/637	G0002
2N1800			53		
2N1808	101	3011	8	TR08/641	G0011
2N1837			81		S3011
2N1838	123A	3122	81	TR21/735	S0014
2N1839	123A	3122	27	TR21/735	S0014
2N1840	123A	3122	27	TR21/735	S0014
2N1853	100	3005	2	TR05/254	G0005
2N1853/18	160	3123	2	TR17/	G0005
2N1854	102	3004	2	TR05/631	
2N1858	103	3010	8	TR08/631A	
2N1864	160	3008	51	TR17/637	
2N1865	160	3008	51	TR17637	
2N1866	160	3008	51	TR17/637	G0002
2N1867	160	3008	51	TR17/637	G0002
2N1868	160		51	TR17/637	G0002
2N1872			50	TR17/	
2N1873			50	TR17/	
2N1874			50	TR17/	
2N1875			50	TR17/	
2N1886			66		S5000
2N1889	128		18		S3019
2N1890	128		18		S3019
2N1891	101	3011	5	TR08/641	G0011
2N1892	101	3011	54	TR08/641	G0011
2N1893	128	3020	27	TR87/243	S3019
2N1905	121	3014	16	TR01/242	G6005
2N1906	179	3014	25	TR35/G6001	G6014
2N1907	179	3014	25	TR35/G6001	G6005
2N1907A	179			TR35/G6001	G6001
2N1908	179	3014	25	TR35/G6001	G6005
2N1908A	179			TR35/G6001	
2N1917	159	3114	65	TR20/717	S0012
2N1918	159	3114	65	TR20/717	S0012
2N1918A		3114			
2N1919	159	3114	65	TR20/717	S0012
2N1919A		3114			
2N1920	159	3114	65	TR20/717	S0012
2N1920A		3114			
2N1921	159	3114	65	TR20/717	S0012
2N1921A		3114			
2N1922	129		82	TR88/242	
2N1924	102A	3004	2	TR85/250	G0005

Original Part Number	ECG	SK	GE	IR/WEP	HEP
2N1925	102A	3004	2	TR85/250	G0005
2N1926	102A	3004	2	TR85/250	G0005
2N1940	100	3005	2	TR05/254	
2N1941			47		S3011
2N1943			18	TR25/	S3020
2N1944	123A	3122	47	TR21/735	S0014
2N1945	123A	3122	47	TR21/735	S0014
2N1946	123A	3122	47	TR21/735	S0014
2N1947	123A	3122	17	TR21/735	S0014
2N1948	123A	3122	17	TR21/735	S0014
2N1949	123A	3122	17	TR21/735	S0014
2N1950	123A	3122	17	TR21/735	S0014
2N1951	123A	3122	17	TR21/735	S0013
2N1952	123A	3122	17	TR21/735	S0014
2N1953	123A	3122	81	TR21/735	S0014
2N1954	102A	3005	2	TR85/250	G0005
2N1955	102A	3005	2	TR85/250	G0005
2N1956	102A	3005	2	TR85/250	G0005
2N1957	102A	3005	2	TR85/250	G0005
2N1958	123A		81	TR21/735	S0014
2N1958/18			210	TR25	S0011
2N1958A			47		S3001
2N1958A/51			47		
2N1959	123A	3122	81	TR21/735	S0014
2N1959/18			210		S0011
2N1959A			47		S3001
2N1959A/51			47		
2N1960	160		2	TR17/637	G0003
2N1960/46	160	3123	2	TR17/637	G0002
2N1961	160	3123	2	TR17/637	G0003
2N1961/46		3123	2	TR05/	G0002
2N1962	123A	3122	210	TR21/735	S0011
2N1962/46			63	TR25/	S0014
2N1963	123A	3122	210	TR21/735	S0011
2N1964	123A	3122	210	TR21/735	S0011
2N1964/46			63	TR25/	S0014
2N1965	123A	3122	210	TR21/735	S0011
2N1965/46			63	TR25/	S0014
2N1969	102A		1	TR85/250	G0005
2N1970	105	3012	4	TR03/233	G6006
2N1971	121	3009	25	TR01/232	G6005
2N1972	123A	3122	81	TR21/735	S0014
2N1973	123A	3020	18	TR21/735	S0014
2N1974	123A	3122	18	TR21/735	S0014
2N1975	123A	3122	18	TR21/735	S0014
2N1980	105	3012	4	TR03/233	G6004
2N1981	105	3012	4	TR03/233	G6004
2N1982	105	3012	4	TR03/233	G6006
2N1983	123A	3020	47	TR21/735	S0014
2N1984	123A	3122	47	TR21/735	S0014
2N1985	123A	3122	47	TR21/735	S0014
2N1986	123A	3122	81	TR21/735	S0014
2N1987	123A	3122	81	TR21/735	S0014
2N1988	123A	3122	32	TR21/735	S0014
2N1989	108	3039	32	TR95/56	S3019
2N1990	128		32	TR87/	S3019
2N1990/46			18		
2N1990R			18		
2N1990S			18		S3019
2N1991	159	3114	82	TR20/717	S0012
2N1991B		3114			
2N1992	123A	3122	63	TR21/735	S0011
2N1993	101	3011	5	TR08/641	G0011
2N1994	101	3011	5	TR08/641	G0011
2N1995	101	3011	8	TR08/641	G0011
2N1996	101	3011	5	TR08/641	G0011
2N1997	100	3005	2	TR05/254	G0005
2N1998	100	3005	2	TR05/254	G0005
2N1999	160	3005	1	TR17/637	G0002
2N2000	102A		1	TR85/250	G0005
2N2001	102A		1	TR85/250	G0005
2N2002	159	3114	65	TR20/717	S0012
2N2002B		3114			
2N2003	159	3114	65	TR20/717	S0012
2N2003A		3114			
2N2004	159	3114	65	TR20/717	S0012
2N2004A		3114			
2N2005	159	3114	65	TR20/717	S0012
2N2005A		3114			
2N2006	159	3114	65	TR20/717	S0012
2N2006A		3114			
2N2007	159	3114	65	TR20/717	S0012
2N2007A		3114			
2N2008	194		32		S3019
2N2009			51		

Original Part Number	ECG	SK	GE	IR/WEP	HEP
2N2017	128	3024	63	TR87/243	S5014
2N2022	160		52	TR17/637	G0002
2N2032	108	3039	14	TR95/56	S5004
2N2034			66		S5000
2N2035			66		S5000
2N2036			66		S5004
2N2038	123A	3122	81	TR21/735	S0014
2N2039	123A	3122	81	TR21/735	S0014
2N2040	123A	3122	81	TR21/735	S0014
2N2041	123A	3122	81	TR21/735	S0014
2N2048	160		51	TR17/637	G0002
2N2048A	160			TR17/637	G0002
2N2049	123A		18		S3011
2N2059	160		52	TR17/637	G0003
2N2060			18		
2N2061	104	3009	76	TR01/230	G6003
2N2061A	121	3009	76	TR01/232	G6003
2N2062	104	3009	76	TR01/230	G6003
2N2062A	104	3009	76	TR01/230	G6003
2N2063	104	3009	76	TR01/230	G6003
2N2063A	104	3009	76	TR01/230	G6003
2N2064	104	3009	76	TR01/230	G6003
2N2064A	104	3009	76	TR01/230	G6003
2N2065	104	3009	76	TR01/232	G6005
2N2065A	104	3009	76	TR01/232	G6005
2N2066	104	3009	76	TR01/232	G6005
2N2066A	104	3014	76	TR01/232	G6005
2N2067	104	3009	3	TR01/232	
2N2067B	104	3009	16	TR01/232	
2N2067C	104	3009	16	TR01/232	
2N2067W	104	3009	16	TR01/232	
2N2069	104	3009	16	TR01/232	
2N2070	104	3009	16	TR01/232	G6018
2N2071			3		G6018
2N2072			3		G6018
2N2075	105	3012	4	TR03/233	G6006
2N2075A	105	3012	4	TR03/233	G6006
2N2076	105	3012	4	TR03/233	G6006
2N2076A	105	3012	4	TR03/233	G6006
2N2077	105	3012	7	TR03/233	G6004
2N2077A	105	3012	4	TR03/233	G6004
2N2078	105	3012	4	TR03/233	G6004
2N2078A	105	3012	4	TR03/233	G6004
2N2079	105	3012	4	TR03/233	G6006
2N2079A	105	3012	4	TR03/233	G6006
2N2080	105	3012	4	TR03/233	G6006
2N2080A	105	3012	4	TR03/233	G6006
2N2081	105	3012	4	TR03/233	G6004
2N2081A	105	3012	4	TR03/233	G6004
2N2082	105	3012	4	TR03/233	G6004
2N2082A	105	3012	4	TR03/233	G6004
2N2083	160		52	TR17/637	G0009
2N2084	160	3006	51	TR17/637	G0003
2N2085	101	3011	59	TR08/641	G0011
2N2086			18		S3019
2N2087			81		S3019
2N2089	126	3006	50	TR17/635	G0001
2N2090	126	3006	50	TR17/635	G0001
2N2091	160	3006	50	TR17/637	G0001
2N2092	126	3007	50	TR17/635	G0001
2N2093	160	3008	50	TR17/637	G0003
2N2094	123A	3122	18	TR21/735	S0015
2N2094A	123A	3122		TR21/735	S0015
2N2095	123A	3122	51	TR21/735	
2N2095A	123A	3122		TR21/735	S0015
2N2096	123A	3122	20	TR21/735	
2N2096A	123A	3122		TR21/735	S0015
2N2097			51		
2N2097A	123A	3122		TR21/735	S0015
2N2098	160		51	TR17/637	G0002
2N2099	160		51	TR17/637	G0002
2N2100	160		51	TR17/637	G0002
2N2102	128	3024	66	TR87/243	S3019
2N2102A			27	TR78/	S3019
2N2104	159	3114		TR20/717	S0012
2N2104A		3114			
2N2105	159	3114		TR20/717	S0012
2N2105A		3114			
2N2106	128	3024	63	TR87/243	S5014
2N2107	128	3024	63	TR87/243	S5014
2N2108	128	3024	63	TR87/243	S5014
2N2137	104	3009	16	TR01/624	G6003
2N2137A	104	3009	16	TR01/624	G6003
2N2138	104	3009	3	TR01/624	G6003
2N2138A	104	3009	16	TR01/624	G6003

Original Part Number	ECG	SK	GE	IR/WEP	HEP
2N2139	104	3009	16	TR01/624	G6005
2N2139A	104	3009	16	TR01/624	G6005
2N2140	104	3009	16	TR01/624	G6005
2N2140A	104	3009	16	TR01/624	G6005
2N2141	104	3009	16	TR01/624	G6005
2N2141A	104	3009	16	TR01/624	G6005
2N2142	104	3009	16	TR01/624	G6003
2N2142A	104	3009	16	TR01/624	G6003
2N2143	104	3009	16	TR01/624	G6003
2N2143A	104	3009	16	TR01/624	G6003
2N2144	104	3009	16	TR01/624	G6005
2N2144A	104	3009	16	TR01/624	G6005
2N2145	104	3009	3	TR01/624	G6005
2N2145A	104	3009	3	TR01/624	G6005
2N2146	104	3009	3	TR01/624	G6005
2N2146A	104	3009	3	TR01/624	G6005
2N2147	104	3014	16	TR01/624	G6005
2N2148	121	3014	16	TR01/232	G6005
2N2152	213	3012	4	TR03/233	G6004
2N2152A	213	3012	4	TR03/233	G6004
2N2153	213	3012	4	TR03/233	G6010
2N2153A	213	3012	4	TR03/233	G6010
2N2154	213	3012	4	TR03/233	G6010
2N2154A	213	3012	4	TR03/233	G6010
2N2155	105	3012	4	TR03/233	G6010
2N2155A	105	3012	4	TR03/233	G6010
2N2156	213	3012	4	TR03/233	G6010
2N2156A	213	3012	4	TR03/233	G6010
2N2157	213	3012	4	TR03/233	G6010
2N2157A	213	3012	4	TR03/233	G6010
2N2158	213	3012	4	TR03/233	G6010
2N2158A	213	3012	4	TR03/233	G6010
2N2159	105	3012	4	TR03/233	G6006
2N2159A	105	3012	4	TR03/233	G6006
2N2160	6400			IR2160/	S9002
2N2161	123A	3122		TR21/735	S0014
2N2162	159	3114	65	TR20/717	S0012
2N2163	159	3114	65	TR20/717	S0012
2N2164	159	3118	65	TR20/717	S0012
2N2164A		3114			
2N2165	159	3114	65	TR20/717	S0012
2N2165A		3114			
2N2166	159	3114	65	TR20/717	S0012
2N2166A		3114			
2N2167	159	3114	65	TR20/717	S0012
2N2167A		3114			
2N2168	160		51	TR17/637	G0002
2N2170	160		51	TR17/637	G0002
2N2171	100	3005	1	TR05/254	G0005
2N2172	100	3005	1	TR05/254	G0005
2N2173	102A		2	TR85/250	G0005
2N2175	106	3118	65	TR20/52	G0006
2N2176	106	3118	65	TR20/52	G0006
2N2177	106	3118	65	TR20/52	G0006
2N2178	159	3114	65	TR20/717	S0013
2N2178A		3114			
2N2180	126		52	TR17/635	G0003
2N2181	159	3114	65	TR20/717	S0012
2N2181A		3114			
2N2182	159	3114	65	TR20/717	S0012
2N2182A		3114			
2N2183	159	3114	21	TR20/717	S0013
2N2183A		3114			
2N2184	159	3114	21	TR20/717	S0013
2N2184A		3114			
2N2185	159	3114	65	TR20/717	S0013
2N2185A		3114			
2N2186	159	3114	65	TR20/717	S0013
2N2186A		3114			
2N2187	159	3114	65	TR20/717	S0013
2N2187A		3114			
2N2188	160	3006	51	TR17/637	G0002
2N2189	160	3006	51	TR17/637	G0002
2N2190	160	3006	51	TR17/637	G0002
2N2191	160	3006	51	TR17/637	G0002
2N2192	128	3039	47	TR87/56	S3001
2N2192A	128	3039	47	TR95/56	S3001
2N2192B	128		47		S3001
2N2193	128	3039	32	TR87/56	S3011
2N2193A	108	3039	32	TR95/56	S3011
2N2193B	108	3039	32	TR95/56	S3011
2N2194	123A	3122	47	TR21/735	S0014
2N2194A	123A	3122	47	TR21/735	S0014
2N2194B	123A	3122	47	TR21/735	S0014
2N2195	108	3039	47	TR95/243	S3001

Original Part Number	ECG	SK	GE	IR/WEP	HEP
2N2195A	128	3039	47	TR87/243	S3001
2N2195B	108	3039	47	TR95/56	S3001
2N2196	108	3039	17	TR95/56	S3020
2N2197	108	3039	17	TR95/56	S3020
2N2198	123	3124	81	TR86/53	S5026
2N2199	160		51	TR17/637	G0002
2N2200	160		51	TR17/637	G0002
2N2204	124	3021	12	TR81/240	S3019
2N2205	123A	3124	210	TR21/735	S0011
2N2206	123A	3122	210	TR21/735	S0011
2N2207	126	3006	51	TR17/635	G0009
2N2208	126	3006	1	TR17/635	G0001
2N2209	100	3005	2	TR05/254	G0005
2N2210	105	3012	4	TR03/233	G6006
2N2212	121	3009	25	TR01/232	G6005
2N2216			27	TR01/	
2N2217	123A		47		S3001
2N2218	123	3124	47	TR86/53	S3011
2N2218A	123	3124		TR86/53	S3011
2N2218/51			20		
2N2219			47	TR21/	S3011
2N2219A	123A	3024	18	TR21/735	S3011
2N2219A/51			20		S0005
2N2220	123A	3122	20	TR21/735	S0015
2N2221	123A	3122	20	TR21/735	S0015
2N2221A	123A		20		
2N2222	123A	3122	20	TR21/735	S0015
2N2222A			20		S3001
2N2223			18		
2N2223A			18		
2N2224			47		S3011
2N2225	160		2	TR17/637	G0002
2N2234	123	3124	20	TR86/53	S3024
2N2235	123	3124	20	TR86/53	S3020
2N2236	123A	3122	81	TR21/735	S0014
2N2237	123A	3122	81	TR21/735	S0014
2N2238	160		52	TR17/637	G0002
2N2239			63		S3020
2N2240	123A	3024	81	TR21/735	S0014
2N2241	123A	3024	81	TR21/735	S0014
2N2242	123A	3122	210	TR21/735	S0011
2N2243			32		S3019
2N2243A			32		S3019
2N2244	123A	3122	17	TR21/735	S0011
2N2245	123A	3122	17	TR21/735	S0011
2N2246	123A	3122	47	TR21/735	S0011
2N2247	123A	3122	81	TR21/735	S0011
2N2248	123A	3122	81	TR21/735	S0011
2N2249	123A	3122	17	TR21/735	S0011
2N2250	123A	3122	17	TR21/735	S0011
2N2251	123A	3122	17	TR21/735	S0011
2N2252	123A	3122	47	TR21/735	S0011
2N2253	123A	3122	81	TR21/735	S0011
2N2254	123A	3122	81	TR21/735	S0011
2N2255	123A	3122	47	TR21/735	S0011
2N2256	123A	3122	10	TR21/735	S0011
2N2257	123A	3122	10	TR21/735	S0011
2N2258	160		51	TR17/637	G0003
2N2259	160		51	TR17/637	G0003
2N2266	105	3012	4	TR03/233	
2N2267	105	3012	4	TR03/233	
2N2268	105	3012	4	TR03/233	
2N2269	105	3012	4	TR03/233	
2N2270	128	3024	63	TR87/243	S3001
2N2271	102	3004	53	TR05/631	G0005
2N2272	123A	3122	18	TR21/735	S0014
2N2273	160		51	TR17/637	G0003
2N2274	159	3114	65	TR20/717	S0013
2N2275	159	3114	65	TR20/717	S0013
2N2276	159	3114	65	TR20/717	S0013
2N2277	159	3114	65	TR20/717	S0013
2N2278	159	3114	65	TR20/717	S0013
2N2279	159	3114	65	TR20/717	S0013
2N2280	159	3114	65	TR20/717	S0013
2N2281	159	3114	65	TR20/717	S0013
2N2282	104	3009	3	TR01/624	
2N2285	179	3014	76	TR35/G6001	G6005
2N2286	179	3014		TR35/G6001	G6005
2N2287	121	3014		TR01/232	G6005
2N2288	121	3009	16	TR01/232	G6003
2N2289	121	3009	3	TR01/232	G6005
2N2290	127	3009	3	TR27/235	G6005
2N2291	121	3009	25	TR01/232	G6003
2N2292	121	3009	25	TR01/232	G6005
2N2293	121	3009	25	TR01/232	G6005
2N2294	121	3009	25	TR01/232	G6003
2N2295	121	3009	25	TR01/232	G6005
2N2296	121	3009	25	TR01/232	G6005
2N2297	128		27	TR51/	S3019
2N2297/51			213	TR51/	S0005
2N2303	159	3114	82	TR20/717	S0012
2N2303A		3114			
2N2303/46			21		S0012
2N2305	130	3027	77	TR59/247	S7002
2N2309	123A	3122	81	TR21/735	S0014
2N2310	123A	3122	81	TR21/735	S0014
2N2311			27	TR78/	S5026
2N2312	123A	3122	81	TR21/735	S0014
2N2313			27	TR78/	S5026
2N2314	123A	3122	62	TR21/735	S0014
2N2315	123A	3122	62	TR21/735	S0014
2N2316			18		S5026
2N2317			81		S3020
2N2318	123A	3122	17	TR21/735	S0011
2N2319	123A	3122	61	TR21/735	S0014
2N2320	123A	3122	215	TR21/735	S0014
2N2330	123	3124	47	TR86/53	
2N2331	123A	3122	17	TR21/735	S0011
2N2332	159	3114	65	TR20/717	S0013
2N2332A		3114			
2N2333	159	3114	65	TR20/717	S0013
2N2333A		3114			
2N2334	159	3114	65	TR20/717	S0013
2N2334A		3114			
2N2335	159	3114	65	TR20/717	S0013
2N2335A		3114			
2N2336	159	3114	65	TR20/717	S0013
2N2336A		3114			
2N2337	159	3114	65	TR20/717	S0013
2N2337A		3114			
2N2339			66		S5014
2N2349	123A	3122	8	TR21/735	S0014
2N2350			47		S3020
2N2350A			47		S3020
2N2351			18		S3011
2N2352			47		S3020
2N2352A			47		S3020
2N2353	123A	3122	47	TR21/735	S0014
2N2353A	123A	3122	47	TR21/735	S0014
2N2354	103	3124	59	TR08/641A	G0011
2N2357	121	3009	3	TR01/232	
2N2358	121	3009	3	TR01/232	
2N2360	160		51	TR17/637	G0002
2N2361	160		51	TR17/637	G0002
2N2362	160		51	TR17/637	G0002
2N2363	126	3006	51	TR17/635	G0003
2N2364			18		S3019
2N2364A			18		S3019
2N2368	123A	3122	20	TR21/735	S0011
2N2368A	123A				
2N2368/51			63	TR25/	S0025
2N2369	123A	3122		TR21/735	S0011
2N2369A	123A	3122		TR21/735	S0011
2N2369/51			63	TR25/	S0025
2N2369/64				TR25/	
2N2370	159	3114	65	TR20/717	S0012
2N2371	159	3114	65	TR20/717	S0012
2N2372	159	3114	65	TR20/717	S0013
2N2373	159	3114	65	TR20/717	S0013
2N2373A		3114			
2N2374	102	3004	2	TR05/631	G0005
2N2375	102	3004	1	TR01/631	G0005
2N2376	102	3004	53	TR05/631	G0005
2N2377	159	3114	65	TR20/717	S0013
2N2377A		3114			
2N2378	159	3114	65	TR20/717	S0013
2N2378A		3114			
2N2379			4		G6006
2N2380			81		S0015
2N2380A			81	TR78/	S3011
2N2380ORN				TR85/	
2N2381	160		51	TR17/637	G0002
2N2382	160		51	TR17/637	G0002
2N2383			14		S5000
2N2384			14		S5004
2N2387	123A	3122	212	TR21/735	S0025
2N2388	123A	3122	212	TR21/735	S0015
2N2389	123	3124	81	TR86/53	S3020
2N2390	123	3124	81	TR86/53	S3020
2N2393	159	3114	82	TR20/717	S0019

Original Part Number	ECG	SK	GE	IR/WEP	HEP
2N2393A		3114			
2N2394	159	3114	82	TR20/717	S0019
2N2394A		3114			
2N2395	106		20	TR20/52	S0025
2N2396	123	3124	20	TR86/53	S0025
2N2397	123	3124	210	TR86/53	S0013
2N2398	160		51	TR17/637	G0002
2N2399	160		51	TR17/637	G0002
2N2400	160		1	TR17/637	G0003
2N2401	160		1	TR17/637	G0003
2N2402	160		51	TR17/637	G0003
2N2403			63	TR25/	S3011
2N2404			63	TR25/	S3011
2N2405	128	3024	27	TR87/243	S3019
2N2406	176	3123	52	TR82/	
2N2410			47	TR65/	S300.
2N2410/51			47	TR65/	
2N2411	159	3114	65	TR20/717	S0013
2N2411A		3114			
2N2412	159	3114	65	TR20/717	S0013
2N2412A		3114			
2N2413	123A	3122	20	TR21/735	S0011
2N2415	160		51	TR17/637	G0003
2N2416	160		51	TR17/637	G0003
2N2417	123A	3124		TR21/735	
2N2423	121	3009	16	TR01/232	G6005
2N2424	159	3114	65	TR20/717	S0012
2N2424A		3114			
2N2425	159	3114	65	TR20/717	S0012
2N2425A		3114			
2N2426	101	3011	8	TR08/641	G0011
2N2427	123A	3122	20	TR21/735	S0011
2N2428	102A	3004	52	TR85/250	G0006
2N2429	102A	3004	53	TR85/250	G0005
2N2430	103A	3010	54R	TR08/724	G0011
2N2431	158	3004	53	TR84/630	G6011
2N2432	123A	3122	212	TR21/735	S0011
2N2432A			212	TR5[illegible]/	S002[illegible]
2N2433			27	TR78/	S3020
2N2435			27	TR78/	S3019
2N2437			27	TR78/	S3019
2N2438			27	TR78/	
2N2439			27	TR78/	S3019
2N2443	128		27	TR78/	S5026
2N2446	121	3009	16	TR01/232	G6005
2N2447	102A	3004	2	TR85/250	G0005
2N2448	102A	3004	2	TR85/250	G0005
2N2449	102A	3004	53	TR85/250	G0007
2N2450	102A	3004	53	TR85/250	G0007
2N2451	126	3006	51	TR17/635	G0003
2N2455	160		51	TR17/637	G0002
2N2456	126		51	TR17/635	G0003
2N2463			27	TR78/	S0005
2N2466	123A	3122		TR21/735	G0005
2N2468	102	3004	51	TR05/631	G6012
2N2469	102	3004	2	TR05/631	
2N2472			20		S3019
2N2473			20		S3019
2N2474			65	TR30/	
2N2475			61		S0016
2N2475/46			61		S0014
2N2475/51			61		S0011
2N2476	123A	3122	47	TR21/735	S0014
2N2477	123A	3122	47	TR21/735	S0014
2N2478			18		S3019
2N2479			81		S3011
2N2481	123A	3122	20	TR21/735	S0011
2N2482	103	3010	54	TR08/641A	G0011
2N2483	123A	3122	81	TR21/735	S0025
2N2484	123A	3122	81	TR21/735	S0005
2N2484A			63		S0005
2N2487	126	3006	51	TR17/635	G0003
2N2488	126	3006	51	TR17/635	G0003
2N2489	126	3006	51	TR17/635	G0003
2N2489A			81		
2N2490	105	3012	4	TR03/233	G6006
2N2491	105	3012	4	TR03/233	G6006
2N2492	105	3012	4	TR03/233	G6006
2N2493	105	3012	4	TR03/233	G6006
2N2494	126	3006	51	TR17/635	G0003
2N2495	160	3006	51	TR17/637	G0002
2N2496	160	3006	51	TR17/637	G0002
2N250[illegible]	123A	3122	17	TR21/735	S0011
2N25[illegible]	154	3045	18	TR78/712	S0005
2N2510	154	3045	20	TR78/712	S0005

Original Part Number	ECG	SK	GE	IR/WEP	HEP
2N2511	154		20		S0005
2N2512	160			TR17/637	
2N2514			81	TR25/	S3011
2N2515			81		S3011
2N2517			18		S3019
2N2520	123A	3122	81	TR21/735	S0014
2N2521	123A	3122	81	TR21/735	S0014
2N2522	123A	3122	63	TR21/735	S0014
2N2523	123A	3122	210	TR21/735	S0014
2N2524	123A	3122	210	TR21/735	S0014
2N2526	127		76	TR27/235	G6015
2N2527	127		25	TR27/235	G6007
2N2528	127		25	TR27/235	G6017
2N2529	123A	3124	39	TR21/735	S0011
2N2530	123A	3124	39	TR21/735	S0011
2N2531	123A	3124	39	TR21/735	S0011
2N2532	123A	3124	39	TR21/735	S0011
2N2533	123A	3124	39	TR21/735	S0011
2N2534	123A	3124	39	TR21/735	S0011
2N2537			47		S3001
2N2538			18		S3001
2N2539	123A	3122	20	TR21/735	S0011
2N2540	123A	3122	20	TR21/735	S0011
2N2541	176		53	TR82/238	G6011
2N2551			27	TR78/	
2N2564	102	3004	80	TR05/631	G6011
2N2564/5			80		
2N2565	102	3004	3	TR05/631	G6011
2N2569	123A	3122	212	TR21/735	S0011
2N2570	123A	3122	212	TR21/735	S0011
2N2571	123A	3122	20	TR21/735	S0011
2N2572	123A	3122	20	TR21/735	S0011
2N2586	123A	3122	212	TR21/735	S0011
2N2587	160		51	TR17/637	G0003
2N2588	126		52	TR17/635	G0009
2N2590			21		
2N2591			21		S0005
2N2592			21		S0005
2N2593			21		S0005
2N2594	128	3024	32	TR87/	S3011
2N2595	159	3114	82	TR20/717	S0013
2N2595A		3114			
2N2596	159	3114	82	TR20/717	S0013
2N2596A		3114			
2N2597	159	3114	82	TR20/717	S0013
2N2597A		3114			
2N2598	159	3114	21	TR20/717	S0013
2N2598A		3114			
2N2599	159	3114	21	TR20/717	S0013
2N2599A		3114			
2N2600	159	3114	21	TR20/717	S0013
2N2600A		3114			
2N2601	159	3114	82	TR20/717	S0013
2N2601A		3114			
2N2602	159	3114	82	TR20/717	S0013
2N2602A		3114			
2N2603	159	3114	82	TR20/717	S0013
2N2603A		3114			
2N2604	159	3114	82	TR20/717	S0013
2N2604A		3114			
2N2605	159	3114	82	TR20/717	S0013
2N2605A	159	3114	48	TR20/	S0013
2N2605AC		3114			
2N2610	123A	3122	39	TR21/735	S0014
2N2611			27		S3019
2N2612	104	3009	3	TR01/624	G6018
2N2613	102A	3004	2	TR85/250	G0005
2N2614	102A	3004	2	TR85/250	G0005
2N2615		3018	61		S0020
2N2616	108	3019	61	TR95/56	S0016
2N2617	123A	3004	212	TR21/735	S0013
2N2618	154	3045	47	TR78/712	S5026
2N2618/46			20		S0011
2N2621	160	3123	2	TR17/637	G0002
2N2622	160	3123	2	TR17/637	G0002
2N2623	160	3123	2	TR17/637	G0002
2N2624	160	3123	2	TR17/637	G0002
2N2625	160	3123	2	TR17/637	G0002
2N2626	160	3123	2	TR17/637	G0002
2N2627	160	3123	2	TR17/637	G0002
2N2628	160	3123	2	TR17/637	G0002
2N2629	160	3123	2	TR17/637	G0002
2N2630	160	3123	22	TR17/637	G0003
2N2632			66	S5004	S5004
2N2635	160		53	TR17/637	G0003

Original Part Number	ECG	SK	GE	IR/WEP	HEP
2N2636	179	3009		TR35/G6001	G6009
2N2637	179	3009		TR35/G6001	G6009
2N2638	179			TR35/G6001	G6009
2N2639		3009	212	TR95/56	S0025(2)
2N2640		3039	212	TR95/56	S0025(2)
2N2641		3039	212	TR95/56	S0025(2)
2N2642		3039	212	TR95/56	S0015(2)
2N2643		3039	212	TR95/56	S0015(2)
2N2644		3039	212	TR95/56	S0015(2)
2N2645	123A		81		S0005
2N2646	6401			2160/	S9002
2N2647	6409			2160/	S9002
2N2648	176		2	TR82/238	G0002
2N2651	108	3039	20	TR95/56	S0016
2N2652			18		S0005(2)
2N2652A			18		S0005(2)
2N2654	160	3006	51	TR17/637	G0008
2N2655			32		S3019
2N2656	123A	3122	210	TR21/735	S0011
2N2657			66		S3002
2N2671	160	3006	51	TR17/637	G0003
2N2672	160	3006	51	TR17/637	G0001
2N2672A	160			TR17/637	G0002
2N2672BLK	160	3123		TR17/637	G0001
2N2672GRN	160	3123		TR17/637	G0001
2N2673	123A	3122	212	TR21/637	S0011
2N2674	123A	3122	212	TR21/735	S0013
2N2675	123A	3122	212	TR21/735	S0011
2N2676	123A	3122	212	TR21/735	S0011
2N2677	123A	3122	61	TR21/735	S0011
2N2678	123A	3122	61	TR21/735	S0011
2N2691	179			TR35/G6001	G6009
2N2691A	179			TR35/G6001	G6009
2N2692	123A	3122	212	TR21/735	S0011
2N2693	123A	3122	61	TR21/735	S0011
2N2694			61	TR51/	S0025
2N2695	159	3114	22	TR20/717	S0013
2N2696	159	3114	22	TR20/717	S0013
2N2696A		3114			
2N2699	101	3011	66	TR08/641	G00[illegible]
2N2706	158	3004	53	TR84/630	G0005
2N2707	158	3010	8	TR84/630	
2N2708	108	3019	17	TR95/56	S0016
2N2709	159	3114	21	TR20/717	S0012
2N2709A		3114			
2N2710	108	3039		TR95/56	S0016
2N2711	108	3124	10	TR95/56	S0015
2N2712	108	3124	17	TR95/56	S0015
2N2713	123A	3124	210	TR21/735	S0015
2N2714	123A	3124	210	TR21/735	S0015
2N2715	108	3124	86	TR95/56	S0015
2N2716	108	3124	211	TR95/56	S0016
2N2717	160		51	TR17/637	G0003
2N2719	123A	3122	20	TR21/735	S0011
2N2720			20		S0007(2)
2N2721			20		S0007(2)
2N2722			20		S0025(2)
2N2723		3156			
2N2724		3156			
2N2725		3156			
2N2726			27		S3021
2N2727			27		S3021
2N2728	105	3012	4	TR03/233	G6002
2N2729	108	3018	61	TR95/56	S0016
2N2730	105	3012	4	TR03/233	
2N2731	105	3012	4	TR03/233	G6002
2N2732	105	3012	4	TR03/233	G6002
2N2783	126			TR17/635	S3021
2N2784	108	3039	17	TR95/56	S0016
2N2784/52	108	3039	63	TR95/56	S0016
2N2784/TNT			20	TR21/	
2N2786	160			TR17/637	G0002
2N2786A	160			TR17/637	G0002
2N2787			20		S3011
2N2788			18		S3011
2N2789			18		S3011
2N2790			20		S3001
2N2791			20		S3001
2N2792			20		S3001
2N2793	105	3012	4	TR03/233	G6002
2N2795	160		51	TR17/637	G0003
2N2796	160		51	TR17/637	G0003
2N2797	160		51	TR17/637	G0002
2N2798	160		51	TR17/637	G0002
2N2799	160		51	TR17/637	G0002

Original Part Number	ECG	SK	GE	IR/WEP	HEP
2N2800	159	3114	48	TR20/717	S0012
2N2800/46	159	3114	48	TR20/717	S0013
2N2800/51			21	TR19/	S0025
2N2800A		3114			
2N2801	159	3114	48	TR20/717	S0012
2N2801/46		3114	48	/717	S0013
2N2801/46D		3114			
2N2801/51			21		
2N2802	106	3118	65	TR20/637	S0019(2)
2N2803	106	3118	65	TR20/637	S0019(2)
2N2804	106	3018	65	TR20/637	S0019(2)
2N2805	106	3018	65	TR20/637	S0019(2)
2N2806	106	3018	65	TR20/637	S0019(2)
2N2807	106	3018	65	TR20/637	S0019(2)
2N2828			66		S5000
2N2829			66		S5000
2N2831	123A	3024	17	TR21/724	S0011
2N2832	121			TR01/232	G6014
2N2835	131			TR95/642	G6016
2N2836	121	3009	25	TR01/232	G6005
2N2837	159	3114	48	TR20/717	S0013
2N2838	159	3114	48	TR20/717	S0013
2N2838A		3114			
2N2845	123A	3025	20	TR21/735	S0013
2N2846	128	3024	47	TR87/243	S3001
2N2847	123A	3025	210	TR21/735	S0013
2N2848	128	3024	47	TR87/243	S3001
2N2849	128	3024		TR87/243	S5000
2N2850	128	3024		TR87/243	S5000
2N2851	128	3024		TR87/243	S5000
2N2852	128	3024		TR87/243	S5000
2N2853	128	3024	4	TR87/243	S5014
2N2853-1	210		46		S5014
2N2854	128	3024		TR87/243	S5014
2N2854-1	210		46		S5014
2N2855	128	3024	4	TR87/243	S5014
2N2855-1	210		46		S5014
2N2856	128	3024	4	TR87/243	S5014
2N2856-1	210		46		S5014
2N2857	108		17	TR83/56	
2N2860	160		2	TR17/637	G0003
2N2861	159	3114	65	TR20/717	S0013
2N2861A		3114			
2N2862	159	3114	65	TR20/717	S0013
2N2862A		3114			
2N2863			47		S3011
2N2864			47		S3011
2N2865	161	3019	86	TR83/719	S0016
2N2868	123A		47		S3010
2N2869	121	3009	25	TR01/232	G6005
2N2869/2N301	121	3009	25	TR01/232	
2N2870	121	3009	25	TR01/232	G6005
2N2870/2N301A			25		
2N2873	160		52	TR17/56	G0003
2N2875			69		S5006
2N2876			66		
2N2877			66		S5004
2N2878			66		S5004
2N2883	108	3039	17	TR95/56	S3013
2N2884	108	3039	17	TR95/56	S3013
2N2890			18		S3019
2N2891			18		S3019
2N2894	159	3114	20	TR20/717	S0013
2N2894A	159	3114		TR20/717	S0019
2N2895	128	3024		TR87/	S0005
2N2897	128	3024	47	TR87/	S0025
2N2900			47		S3020
2N2901			11	TR21/	
2N2904	129	3025	21	TR88/242	S5022
2N2904A	129	3025	21	TR88/242	S5022
2N2905	129	3025	48	TR88/242	S5022
2N2905A	129	3025	67	TR88/242	S5022
2N2906	159	3114	21	TR20/717	S0013
2N2906A	159	3114	21	TR20/717	S0013
2N2906AC		3114			
2N2907	159	3114	48	TR20/717	S0013
2N2907A	159	3114	67	TR20/717	S0013
2N2907-07A				TR20/	
2N2907AD		3114			
2N2909			27	TR78/	S3010
2N2910			20		S0025
2N2913			20		S0025
2N2914			20		S0005
2N2917			20		S0005
2N2918			20		S0005

Original Part Number	ECG	SK	GE	IR/WEP	HEP
2N2921	108	3124	10	TR95/56	S0015
2N2922	108	3124	10	TR95/56	S0015
2N2923	123A	3124	212	TR21/735	S0015
2N2923/30227				TR21/	
2N2924	123A	3124	212	TR21/735	S0015
2N2925	123A	3039	212	TR21/735	S0015
2N2925/30229				TR21/	
2N2926	123A	3018	212	TR21/735	S0015
2N2926-6	123A	3122	17	TR21/735	S0015
2N2926GRN	123A		212	TR51/735	S0015
2N2926ORN	123A		212	TR21/735	S0015
2N2926G		3018	10		
2N2926BRN			212	TR51/	S0015
2N2926O			63		
2N2926R			63		
2N2926RED			212	TR51/	S0015
2N2926Y			10		
2N2926YEL			212	TR51/	S0015
2N2926,3111				TR21/	
2N2927	159	3114	48	TR20/717	S0012
2N2927/46	159	3114	22	TR20/717	S0013
2N2927/46D		3114			
2N2927/51			21		S0019
2N2928	160	3123	2	TR17/637	G0003
2N2929	160			TR17/637	G0002
2N2930	100	3010	2	TR05/254	G0005
2N2931	123A	3124	11	TR21/735	S0015
2N2932	123A	3124	11	TR21/	S0015
2N2933	123A	3124	11	TR21/	S0015
2N2934	123A	3124	11	TR21/	S0015
2N2935	123A	3124	11	TR21/	S0015
2N2936			20		S0025
2N2937			20		S0025
2N2938	123A	3122	20	TR21/	S0011
2N2942	160		51	TR17/637	G0002
2N2943	160		51	TR17/637	G0002
2N2944	159	3114	82	TR20/717	S0013
2N2944A	159	3114	82	TR20/717	
2N2944AB		3114			
2N2945	159	3113	22	TR20/717	S0013
2N2945A	159	3114	82	TR20/717	S0013
2N2945AC		3114			
2N2946	159	3114	82	TR20/717	S0013
2N2946A	159	3114	82	TR20/717	S0013
2N2946AC		3114			
2N2947			66		S3020
2N2948			19		S3020
2N2949		3024	63		S3001
2N2951	123A	3122	63	TR21/735	S0014
2N2952	123A	3122	63	TR21/735	S0011
2N2953	102	3004	51	TR05/631	G0005
2N2954			20		S0025
2N2955	160		51	TR17/637	G0003
2N2956	160		51	TR17/637	G0003
2N2957	160		51	TR17/637	G0003
2N2958	123A	3122	47	TR21/735	S0014
2N2959	123A	3122	47	TR21/735	S0014
2N2960	123A	3122	18	TR21/735	S0014
2N2961	123A	3122	18	TR21/735	S0014
2N2966	100		2	TR05/254	G0002
2N2968	106	3118	65	TR20/52	S0012
2N2968A	133				
2N2969	106	3118	65	TR20/52	S0013
2N2970	106	3118	65	TR20/52	S0012
2N2971	106	3118	65	TR20/52	S0013
2N2974			20		S0025
2N2987			32		S3019
2N2988			32		S3019
2N2989			32		S3019
2N2990			32		S3019
2N2996	160	3006	51	TR17/637	G0003
2N2997	160		51	TR17/637	G0003
2N2998	160		51	TR17/637	G0003
2N2999	160			TR17/637	
2N3009	123A	3122	20	TR21/735	S0011
2N3010	108	3039	61	TR95/56	S0016
2N3011	123A	3039	20	TR95/56	S0016
2N3012	159	3114	21	TR20/717	S0013
2N3012A		3114			
2N3013	123A	3122	20	TR21/735	S0016
2N3014	123A	3122	20	TR21/735	S0016
2N3015	123		47	TR86/53	S3001
2N3019	128		18		S3019
2N3020	128		27		S3019
2N3021			69		S5006

Original Part Number	ECG	SK	GE	IR/WEP	HEP
2N3022			69		S5006
2N3023			69		S5006
2N3024			69		S7003
2N3025			69		S7003
2N3026			69		S7003
2N3034			81	TR53/	
2N3035	108	3039	210	TR95/56	
2N3036			18		S3019
2N3037			18		S0005
2N3038			18		S0005
2N3039			82	TR28/	S0019
2N3040			82		S0019
2N3043			17	TR24/	S0015(2)
2N3044			17	TR24/	S0015(2)
2N3045			17	TR24/	S0015(2)
2N3046			17	TR24/	S0025(2)
2N3047			17	TR24/	S0025(2)
2N3048			17	TR24/	S0025(2)
2N3049			21		S0019(2)
2N3050			21		S0019(2)
2N3051			21		S0019(2)
2N3052			63	TR25/	S0025
2N3053	128	3024	63	TR87/243	S3011
2N3053/40053			63	TR25/	
2N3054	175	3026	66	TR81/701	S5019
2N3055	130	3027	14	TR59/247	S7004
2N3055-1			77		
2N3055-2			77		
2N3055-3			75		
2N3055-4			77		
2N3055-5			77		
2N3055-6			75		
2N3055-7			75		
2N3055-8			75		
2N3055-9			77		
2N3055-10			77		
2N3055S			77		
2N3056			18		S3019
2N3057			18		S3019
2N3058	159	3114	82	TR20/717	S0013
2N3058A		3114			
2N3059	159	3114	82	TR20/717	S0013
2N3059A		3114			
2N3060	159	3114	82	TR20/717	S0013
2N3060A		3114			
2N3061			82		S0032
2N3062	159	3114	21	TR20/717	S0013
2N3062A		3114			
2N3063			21		
2N3064			21		
2N3065			21		
2N3066	133	3112	FET1		
2N3067	133	3112	FET1		
2N3068	133	3112			
2N3069			FET1		F0015
2N3070	133	3112	FET1		F0015
2N3071	133	3112	FET1		F0015
2N3072	159	3114	67	TR20/717	S0013
2N3072A		3114			
2N3073	159	3114	82	TR20/717	S0013
2N3073A		3114			
2N3074	160	3006	51	TR17/637	G0003
2N3075	100	3006	51	TR05/254	G0005
2N3077	128	3124	81	TR87/243	S5026
2N3078	128	3124	81	TR87/717	S5026
2N3081	159	3114	67	TR20/717	S0012
2N3081/46			67		S0012
2N3081A		3114			
2N3082	107	3039	17	TR70/720	S0020
2N3083	107	3039	17	TR70/720	S0020
2N3084	133	3112	FET1		F0015
2N3085	133	3112	FET1		F0015
2N3086	133	3112	FET1		F0015
2N3087	133	3112	FET1		F0015
2N3088	133	3112	FET1		F0015
2N3088A	133	3112			F0015
2N3089	133	3112	FET1		F0015
2N3089A	133	3112			F0015
2N3107	128		18		S3002
2N3108	128		18		S3002
2N3109	128		18		S3001
2N3110	128		18	TR25/	S3001
2N3114	154	3104	32	TR78/712	S3019
2N3115	123A	3122	20	TR21/735	S0014
2N3116	123A	3122	20	TR21/735	S0014

Original Part Number	ECG	SK	GE	IR/WEP	HEP
2N3117	128	3124	18	TR87/243	S0015
2N3118			18		S3019
2N3119			27	TR78/	S3019
2N3120	129	3025	48	TR88/242	S0014
2N3121	159	3114	82	TR20/717	S0012
2N3121A		3114			
2N3122	123A	3122	47	TR21/735	S0014
2N3124			76		G6018
2N3125	121	3014	76	TR01/232	G6005
2N3126			3		G6018
2N3127	160			TR17/637	G0003
2N3128	123A	3124	17	TR21/735	S0015
2N3129	123A	3122	17	TR21/735	S0025
2N3130	123A	3122		TR21/735	A0007
2N3132	121	3009	16	TR01/232	G6005
2N3133	129	3025	21	TR88/242	S0012
2N3134	129	3025	48	TR88/242	S0012
2N3135	159	3114	21	TR20/717	S0012
2N3135A		3114			
2N3136	159	3114	48	TR20/717	S0012
2N3136A		3114			
2N3137	107	3039	11	TR70/720	S0016
2N3138			66		S5000
2N3140			66		S5000
2N3142			66		S5000
2N3144			66		S5000
2N3147			3		
2N3148	160		52	TR17/637	G0003
2N3153	160		212	TR17/637	G0003
2N3171			74	TR29/	S7003
2N3172			74	TR29/	S7003
2N3173			74		S5006
2N3174			74		S5005
2N3183			74	TR29/	S7003
2N3184			74	TR29/	S7003
2N3185			74		S5005
2N3186			74		S5005
2N3195			74	TR29/	S7003
2N3196			74	TR29/	S7003
2N3197			74		S5005
2N3198			74		S5005
2N3199			69		S5006
2N3200			69		S5006
2N3202	129	3025	84	TR88/242	S5013
2N3203	129	3025		TR88/242	S5013
2N3205			69		A5007
2N3206			69		S5006
2N3208	129	3025	84	TR88/242	S5013
2N3209	106	3118	21	TR20/52	S0013
2N3210	123A	3122	20	TR21/735	S0014
2N3211	123A	3122	20	TR21/735	S0014
2N3212	121	3009	25	TR01/232	
2N3213	121	3009	16	TR01/232	
2N3214	121	3009	16	TR01/232	
2N3215	121	3009	16	TR01/232	
2N3216	160		52	TR17/637	G0002
2N3217	159	3114	22	TR20/717	S0013
2N3217A		3114			
2N3218	159	3114	22	TR20/717	S0013
2N3218A		3114			
2N3219	159	3114	82	TR20/717	S0013
2N3219A		3114			
2N3224			21		S3032
2N3226	130		77	TR59/247	S7002
2N3227	108	3039		TR95/56	S0016
2N3229			66		
2N3232	130		77	TR59/247	S7002
2N3233	130	3027	75	TR59/247	S5004
2N3234	130			TR61/	S5020
2N3235			77		S7004
2N3236			75		S7004
2N3237			75		S7000
2N3238			75		S7004
2N3239			75		S7004
2N3241	123A	3124	17	TR21/735	S0014
2N3241A	123A	3124	17	TR21/735	S0014
2N3242	123A	3124	20	TR21/735	S0011
2N3242A	123A	3124	20	TR21/735	S0014
2N3244			67	TR77/	S3028
2N3245	106	3118	67	TR20/52	S3032
2N3246	123A	3124	62	TR21/735	S0011
2N3247	123A	3124	212	TR21/735	S0014
2N3248	159	3114	22	TR20/717	S0013
2N3248A		3114			
2N3249	159	3114	21	TR20/717	S0013
2N3249A		3114			
2N3250	159	3114	21	TR20/717	S0013
2N3250A	159	3114	21	TR20/717	S0013
2N3250AB		3114			
2N3251	159	3114	21	TR20/717	S0013
2N3251A	159	3114		TR20/717	S0013
2N3251AB		3114			
2N3252			63	TR65/	S3001
2N3253			216	TR76/	S3001
2N3261	123A	3122	20	TR21/735	S0014
2N3262	128	3024		TR87/	S3019
2N3267	160		51	TR17/637	G0003
2N3268	123	3124	27	TR86/53	S0014
2N3279	160		51	TR17/637	G0003
2N3280	160		51	TR17/637	G0003
2N3281	160		51	TR17/637	G0003
2N3282	160		51	TR17/637	G0003
2N3283	160		50	TR17/637	G0003
2N3284	160		50	TR17/637	G0003
2N3285	160		50	TR17/637	G0003
2N3286	160		50	TR17/637	G0003
2N3287	161	3018	61	TR83/719	S0016
2N3288	161	3018	61	TR83/719	S0016
2N3289	161	3019	86	TR83/719	S0016
2N3290	161	3019	86	TR83/719	S0016
2N3291	161	3019	61	TR83/719	S0016
2N3292	161	3019	61	TR83/719	S0016
2N3293	161	3019	86	TR83/719	S0016
2N3294	161	3019	86	TR83/719	S0016
2N3295			63		S3008
2N3296			28	TR76/	
2N3297			19		
2N3298			10	TR76/	S5026
2N3299			47	TR64/	S3008
2N3300	123A		47		S3008
2N3301	123A	3122	210	TR21/735	S0015
2N3302	123A	3122	210	TR21/735	S0015
2N3304	106		21	TR20/52	
2N3305	159	3114	82	TR20/717	S0012
2N3305A		3114			
2N3306	159	3114	82	TR20/717	S0012
2N3306A		3114			
2N3307	159	3118	21	TR20/717	S0013
2N3307B		3114			
2N3308	159	3118	21	TR20/717	S0013
2N3308B		3114			
2N3309			215		
2N3310	123		18	TR86/53	S0014
2N3311	105	3012	4	TR03/233	G6004
3N3312	105	3012	4	TR03/233	G6004
2N3313	105	3012	4	TR03/233	G6006
2N3314	105	3012	4	TR03/233	G6004
2N3315	105	3012	4	TR03/233	G6004
2N3317	159	3114	65	TR20/717	S0013
2N3317A		3114			
2N3318	159	3114	65	TR20/717	S0013
2N3319	159	3114	65	TR20/717	S0013
2N3319A		3114			
2N3320	160		51	TR17/637	G0003
2N3321	160		51	TR17/637	G0003
2N3322	160		51	TR17/637	G0003
2N3323	160		50	TR17/637	G0003
2N3324	160		51	TR17/637	G0008
2N3325	160		51	TR17/637	G0003
2N3326	123A				S3001
2N3328			50		F1036
2N3337	161		61		S0025
2N3338	161	3117	61	TR83/719	S0025
2N3339	161	3117	61	TR83/719	S0025
2N3340	123A	3122	210	TR21/735	S0011
2N3341	159	3114	82	TR20/717	S0011
2N3341A		3114			
2N3342	106		65	TR20/52	S0012
2N3343	129		65	TR88/242	S0012
2N3344	129	3025	65	TR88/242	S0012
2N3345	129	3025	65	TR88/242	S0012
2N3346	159	3114	65	TR20/717	S0012
2N3346A		3114			
2N3365			FET1		
2N3366		3112	FET1		
2N3367		3112	FET1		
2N3368		3112	FET1		
2N3369		3112	FET1		
2N3370		3112	FET1		
2N3371	160		51	TR17/637	G0003

Original Part Number	ECG	SK	GE	IR/WEP	HEP
2N3371A				TR25/	
2N3374			63	TR25/	S3011
2N3375			28		S3020
2N3388	154	3045	27	TR78/712	
2N3389	154	3045	27	TR78/712	
2N3390	123A	3124	212	TR21/735	S0024
2N3391	123A	3124	212	TR21/735	S0015
2N3391A	123A	3124	212	TR21/735	S0015
2N3391-U29			212		
2N3391A489751-0				TR21/	
2N3391A/3235				TR21/	
2N3392	123A	3124	212	TR21/735	S0015
2N3392B		3124			
2N3392-U29			212		
2N3393	123	3124	212	TR86/53	S0015
2N3393A		3124			
2N3393-U29			212		
2N3394	123A	3124	212	TR21/735	S0015
2N3394-U29			212		
2N3395	123A	3124	20	TR21/735	S0015
2N3395-WHT		3156	212	TR51/	
2N3395-YEL			212	TR51/	
2N3396	123A	3018	212	TR21/735	S0015
2N3396-ORG			212	TR51/	
2N3396-WHT		3156	212	TR51/	
2N3396-YEL			212	TR51/	
2N3397	123A	3124	212	TR21/735	S0015
2N3397-ORG			212	TR51/	
2N3397-RED			212	TR51/	
2N3397-WHT		3156	212	TR51/	
2N3397-YEL			212	TR51/	
2N3398	123A	3124	212	TR21/735	S0015
2N3398-BLU		3156	212		
2N3398-ORG			212	TR51/	
2N3398-RED			212	TR51/	
2N3398-WHT		3156	212	TR51/	
2N3399	160		50	TR17/637	G0003
2N3400	160	3123	2	TR17/637	G0003
2N3401	159	3114	65	TR20/717	S0012
2N3401A		3114			
2N3402		3124	81	TR25/735	S0015
2N3402, 3113				TR21/	
2N3403		3124	47	TR21/735	S0015
2N3404	192	3124	81	TR21/735	S0015
2N3405		3124	47	TR21/735	S0015
2N3407	108	3019	17	TR95/56	S0020
2N3409			20		S0025(2)
2N3410			20		S0025(2)
2N3411			20		S0025(2)
2N3412	160		51	TR17/637	S0024
2N3414	123A	3124	210	TR21/735	S0033
2N3415	123A	3124	47	TR21/735	S0024
2N3416	123A	3124	210	TR21/735	S0015
2N3417	123A	3124	47	TR21/735	S0015
2N3418			28	TR76/	S3002
2N3420			28	TR76/	S3002
2N3423			20		S0020(2)
2N3424			20		S0020(2)
2N3426			47	TR65/	S3001
2N3427	102	3004	53	TR05/631	G0006
2N3428	102	3004	53	TR05/631	G0006
2N3435			63	TR25/	S3011
2N3436	133	3112	FET1		
2N3437	133	3112	FET1		
2N3438	133	3112	FET1		
2N3439	128	3103	32		
2N3440	124	3044	32	TR81/240	S3021
2N3441	175	3131		TR81/241	S5012
2N3442		3079			S5020
2N3443	160		2	TR17/637	G0002
2N3445	130	3027	14	TR59/247	S7002
2N3446	130	3027	14	TR59/247	S5004
2N3447	130	3027	14	TR59/247	S7002
2N3448	130	3027	14	TR59/247	S7004
2N3449	160		51	TR17/637	G0003
2N3451	159	3114	22	TR20/717	S0013
2N3451A		3114			
2N3452	133	3112	FET1		
2N3453	133	3112	FET1		
2N3454	133	3112	FET1		
2N3455	133	3112	FET1		
2N3456	133	3112	FET1		
2N3457		3112	FET1		
2N3458			FET1		
2N3459			FET1		

Original Part Number	ECG	SK	GE	IR WEP	HEP
2N3460	133	3112	FET1		
2N3461				TR25/	G6011
2N3462	123A	3124	20	TR21/735	S0011
2N3463	123A	3124	20	TR21/735	S0011
2N3464	159	3114	20	TR20/717	S0012
2N3464A		3114			
2N3465	133	3112	FET1		
2N3466	133	3112	FET1		
2N3467	129		67	TR77/	S3028
2N3468	129		67		S3032
2N3469	128	3024	28	TR87/243	S5014
2N3478	161		86	TR83/719	S0016
2N3483	175			TR81/241	
2N3485	159	3114	21	TR20/717	S0012
2N3485A	159	3114	21	TR20/717	S0012
2N3485AB		3114			
2N3486	159	3114	48	TR20/717	S0012
2N3486A	159	3114	67	TR20/717	S0012
2N3486AB		3114			
2N3493	108	3039	60	TR95/56	S0016
2N3494	129	3025		TR88/242	S5023
2N3496	159	3114	21	TR20/717	S0013
2N3496A		3114			
2N3498	123		20	TR86/53	
2N3500		3104	32		S5026
2N3501		3104	18		S5026
2N3502	129	3025	48	TR88/242	S0012
2N3503	129	3025	67	TR88/242	S0012
2N3504	159	3114	48	TR20/717	S5022
2N3504A		3114			
2N3505	159	3114	67	TR20/717	S0012
2N3505A		3114			
2N3506	128	3024	46	TR87/243	S3002
2N3507	128	3024	28	TR87/243	S3002
2N3508	108	3039	210	TR95/56	S0016
2N3509	108	3124	210	TR95/56	S0016
2N3510	123A	3122	20	TR21/735	S0011
2N3511	108	3039	20	TR95/56	S0016
2N3512			47	TR64/	S0005(2)
2N3526	154		27		S3019
2N3527	159	3019	82	TR20/717	S0013
2N3527A		3114			
2N3542			20		
2N3544	108	3039		TR95/56	S0016
2N3545	159	3114	22	TR20/717	S0013
2N3545A		3114			
2N3546	159	3114		TR20/717	S0019
2N3546A		3114			
2N3547	159	3114	67	TR20/717	S0013
2N3547A		3114			
2N3548	159	3114	48	TR20/717	S0013
2N3548A		3114			
2N3549	159	3114	21	TR20/717	S0013
2N3549A		3114			
2N3550	159	3114	21	TR20/717	S0013
2N3550A		3114			
2N3553			28		S3001
2N3554		3197	215	TR65/	S3001
2N3562	108	3122	20	TR21/	
2N3563	107	3018	86	TR70/720	S0020
2N3563-1	107	3018	11	TR70/720	
2N3564	107	3018	20	TR70/720	S0020
2N3565	123A	3124	212	TR21/735	S0015
2N3566	123A	3124	64	TR21/735	S0015
2N3566/353				TR21/	
2N3567	128	3024	20	TR87/243	S0015
2N3568	128	3024	81	TR87/243	S0015
2N3569	128	3024	20	TR87/243	S0015
2N3569F		3124			
2N3570	108	3018		TR95/56	
2N3571	108	3018	86	TR95/56	S0017
2N3572	108	3018	86	TR95/56	S0017
2N3576	106		20	TR20/717	S0013
2N3579	159	3114	82	TR20/717	S0013
2N3579A		3114			
2N3580	159	3114	82	TR20/717	S0013
2N3580A		3114			
2N3581	159	3114	82	TR20/717	S0013
2N3581A		3114			
2N3582	159	3114	82	TR20/717	S0013
2N3582A		3114			
2N3583	124		12	TR81/240	S3021
2N3584	124		66	TR81/	
2N3585	124		66	TR81/	
2N3588	160	3006	50	TR17/	G0003

Original Part Number	ECG	SK	GE	IR/WFP	HEP
2N3589			32		S3021
2N3590			32		S3021
2N3591			32		A3021
2N3592			32		S3021
2N3593			32		
2N3594			32		S3021
2N2600	108		86	TR95/56	S0016
2N3605	123A	3019	20	TR21/735	S0015
2N3605A			20	TR64/	S0025
2N3606	123A	3019	20	TR21/735	S0015
2N3606A			20	TR64/	S0025
2N3607	123A	3019	20	TR21/735	S0011
2N3608			FET1		F1036
2N3611	121	3009	76	TR01/232	G6003
2N3612	121	3009	76	TR01/232	G6005
2N3613	121	3009	76	TR01/232	G6003
2N3614	121	3009	76	TR01/232	G6005
2N3615	121	3009	76	TR01/232	G6005
2N3616	121	3009	16	TR01/232	G6018
2N3617	121	3009	76	TR01/232	G6005
2N3618	121	3009	16	TR01/232	G6018
2N3619			28	TR76/	S3002
2N3620			28	TR76/	
2N3621			66		
2N3622			66		
2N3623			28	TR76/	S3002
2N3624			28	TR76/	
2N3625			66		
2N3626			66		
2N3627			66		S3002
2N3628			28	TR76/	
2N3629			66		
2N3630			66		
2N3632			66		S3007
2N3633	108	3039	66	TR95/56	S0016
2N3633/46			11	TR21/	S0024
2N3633/52			63	TR25/	S0024
2N3633/TNT			11	TR21/	
2N3638	159	3025	22	TR20/717	S0019
2N3638A	159	3025	82	TR20/717	S0019
2N3638AB		3114			
2N3639	159	3118	21	TR20/717	S0019
2N3639A		3114			
2N3640	159	3114	21	TR20/717	S0019
2N3640A		3114			
2N3641	123A	3018	210	TR21/735	S0015
2N3642	123A	3122	210	TR21/735	S0015
2N3643	123A	3124	210	TR21/735	S0015
2N3644	159	3114	48	TR20/717	S0019
2N3644A		3114			
2N3645	129		67	TR88/242	S5022
2N3646	123A	3019	20	TR21/735	S0011
2N3647	123		20	TR86/53	S0015
2N3648	108	3039	20	TR95/56	S0016
2N3659			32		S3021
2N3660			84		S5013
2N3662	107	3039	11	TR70/720	S0016
2N3663	107	3018	11	TR70/720	S0016
2N3665	128				S3019
2N3666	128				S3019
2N3667	130	3036	77	TR59/247	S7004
2N3671	129	3025	21	TR88/242	S0012
2N3672	159	3114	21	TR20/717	S0019
2N3672A		3114			
2N3673	159	3114	21	TR20/717	S0019
2N3673A		3114			
2N3675			66		S3002
2N3677	129	3025	82	TR88/242	S0013
2N3678	123A				S3001
2N3681			11		S0017
2N3682	108	3039	20	TR95/56	S0016
2N3683			86		S0017
2N3684	133	3112	FET1		
2N3684A	133	3112			
2N3685	133	3112	FET1		
2N3685A	133	3112			
2N3686	133	3112	FET1		
2N3686A	133	3112			
2N3687	133	3112	FET1		
2N3687A	133	3112			
2N3688	108	3039	60	TR95/56	S0015
2N3689	108	3039	60	TR95/56	S0025
2N3690	108	3018	60	TR95/56	S0025
2N3691	108	3018	60	TR95/56	S0016
2N3692	108	3124	210	TR95/56	S0025

Original Part Number	ECG	SK	GE	IR/WEP	HEP
2N3693	108	3122	210	TR95/56	S0025
2N3694	123A	3122	210	TR21/735	S0025
2N3700	154	3045		TR88/712	S3019
2N3701	154				S3019
2N3702	159	3114	82	TR20/717	S0019
2N3702A		3114			
2N3703	159	3114	82	TR20/717	S0019
2N3703B		3114			
2N3704	123A	3024	20	TR21/735	S0015
2N3705	123A	3124	20	TR21/735	S0030
2N3706	123A	3122	47	TR21/735	S0015
2N3707	123A	3124	212	TR21/735	S0015
2N3708	123A	3124	62	TR21/735	S0033
2N3708-V10		3156			
2N3708-BLU			212	TR51/	
2N3708-BRN			61	TR51/	
2N3708-GRN			212	TR51/	
2N3708-ORG			61	TR51/	
2N3708-RED			61	TR51/	
2N3708-VIO			212		
2N3708-YEL			61	TR51/	
2N3709	123A	3124	61	TR21/735	S0015
2N3710	123A	3124	61	TR21/735	S0015
2N3711	123A	3124	212	TR21/735	S0005
2N3712	154	3045	32	TR78/712	S3019
2N3713	130	3036	14	TR59/247	S7002
2N3714	130	3036	14	TR59/247	S7004
2N3715	130	3036	14	TR59/247	S7002
2N3716	130	3036	14	TR59/247	S7004
2N3719	129	3025	29	TR88/242	S5013
2N3720	129	3025		TR88/242	S5013
2N3721	107	3124	62	TR70/720	S0015
2N3723		3104	27	TR55/	S3019
2N3724			47		S3019
2N3724A			215		S3020
2N3725			18		S3019
2N3730	127	3034	25	TR27/235	G6007
2N3731	127	3035	25	TR27/235	G6008
2N3732	127	3034	25	TR27/235	S5005
2N3734			215		S3020
2N3736			20		S3020
2N3738	124	3021	32	TR81/240	S5011
2N3739	124	3021	32	TR81/240	S5011
2N3740			69	TR58/700	S5018
2N3740A			69	TR58/700	S5018
2N3741	218			TR58/	S5018
2N3741A	218			TR58/	S5018
2N3742	154		27		S5024
2N3743	154				
2N3744			66		
2N3745			66		
2N3747			66		
2N3748			66		
2N3762	129	3025	29	TR88/242	S5013
2N3763	129	3025		TR88/242	S5013
2N3764	129	3025		TR88/242	S5013
2N3765	129	3025		TR88/242	S5013
2N3766	175			TR81/241	S5019
2N3767	175	3538		TR81/341	S5012
2N3770	160		51	TR17/637	G0003
2N3771	181	3036	75	TR36/S7000	S7000
2N3772	181	3036	75	TR36/S7000	S7004
2N3773	181	3535			S7000
2N3774	129	3025	84	TR88/242	S5013
2N3775	129	3025		TR88/242	S5013
2N3776	129	3025		TR88/242	S3032
2N3778	129	3025	84	TR88/242	S5013
2N3779	175			TR81/241	S5013
2N3780	129	3025		TR88/242	S3032
2N3781	129	3025		TR88/242	S3032
2N3782	129	3025	83	TR88/242	S5013
2N3783	160		51	TR17/637	G0003
2N3784	160		51	TR17/637	G0003
2N3785	160		51	TR17/637	G0003
2N3788			36	TR67/	S5020
2N3789	219				S7003
2N3790	219				S7003
2N3791	219				S7003
2N3792	219				S7003
2N3793	123A	3124	210	TR21/735	S0016
2N3794	123A	3124	210	TR21/735	S0015
2N3798	159	3114	67	TR20/717	S0019
2N3798A		3114			
2N3799	159	3114	21	TR20/717	S0019
2N3799A		3114			

Original Part Number	ECG	SK	GE	IR/WEP	HEP
2N3812		3114			S0019(2)
2N3813		3114			S0019
2N3814		3114			S0019(2)
2N3815		3114			
2N3816		3114			S0019(2)
2N3818			66		
2N3819	132	3112	FET1	FE100/802	F0015
2N3820	132	3116		FE100/802	F1036
2N3821	133	3112			
2N3822	133	3112	FET1		
2N3823		3116	FET1		F0015
2N3825	107	3039	210	TR70/720	S0011
2N3826	107	3018	210	TR70/720	S0005
2N3827	107	3018	210	TR70/720	S0005
2N3828	123A	3112	20	TR21/735	S0015
2N3829	159	3114	21	TR20/717	S0013
2N3829A		3114			
2N3831			315		S3001
2N3832	108	3039	86	TR95/56	S0016
2N3839			86		
2N3840	159	3114	82	TR20/717	S0013
2N3843	123A	3122	212	TR21/735	S0011
2N3843A	123A	3122	212	TR21/735	S0011
2N3844	123A	3122	212	TR21/735	S0015
2N3844A	123A	3122	212	TR24/735	S0015
2N3845	107	3039	212	TR70/720	S0015
2N3845A	107	3039	212	TR70/720	S0015
2N3846	107	3039	17	TR70/720	
2N3852			28		
2N3853			28		
2N3854	108	3039	212	TR95/56	S0015
2N3854A	108	3039	212	TR95/56	S0015
2N3855	107	3039	212	TR70/720	S0015
2N3855A	107	3039	212	TR70/720	S0015
2N3856	107	3039	212	TR70/720	S0015
2N3856A	107	3039	212	TR70/720	S0015
2N3857	159	3114	82	TR20/717	S0012
2N3857A		3114			
2N3858	123A	3122	212	TR21/735	S0015
2N3858A	123A	3122	81	TR21/735	S0015
2N3859	123A	3124	212	TR21/735	S0015
2N3859A	123A	3122	81	TR21/735	S0015
2N3860	107	3039	212	TR70/720	S0011
2N3862	123A	3122	20	TR21/735	S0025
2N3863	130	3036	77	TR59/247	S7002
2N3864	130	3036	75	TR59/247	S7004
2N3866	128	3048	28	TR87/242	S3008
2N3867	129	3025	29	TR88/242	S5013
2N3868	129	3025		TR88/242	S5013
2N3869	123		28	TR86/53	S3001
2N3877	128	3024	81	TR87/243	S0005
2N3877A			81		S5025
2N3878		3021	66		S5012
2N3879		3021			S5012
2N3880	123A	3122	86	TR21/735	S0011
2N3881			47		S3001
2N3883	160			TR17/637	G0005
2N3885			20		
2N3900		3124	212	TR21/735	S0015
2N3900A	123A	3124	212	TR21/735	S0015
2N3900AF		3124			
2N3901	123A	3124	10	TR21/735	S0015
2N3902	163		73	TR67/	S5021
2N3903	123A		210	TR53/	S0015
2N3904	123A	3024	20	TR87/735	S0015
2N3905	159	3114	21	TR20/717	S0019
2N3905A		3114			
2N3906		3025	48	TR20/717	S0019
2N3906A		3114			
2N3910			82		
2N3911		3114	82	TR20/717	S0032
2N3911A		3114			
2N3912		3114	82	TR20/717	S0032
2N3912A		3114			
2N3913			82		
2N3914	159	3114	82	TR20/717	S0032
2N3914A		3114			
2N3915	106		82	TR20/52	S0032
2N3916			32		S3019
2N3917			19		S5000
2N3918			19		S5000
2N3919			28		S7004
2N3923	154	3045	27	TR78/712	S5024
2N3924	123		28	TR86/53	S3013
2N3925			28		

Original Part Number	ECG	SK	GE	IR/WEP	HEP
2N3926			28	TR66/	
2N3927			66	TR66/	
2N3930	159				
2N3931	159				
2N3932	108		86	TR95/56	S0016
2N3933	108			TR95/56	S0016
2N3945			215		S3010
2N3946			210		S5026
2N3947			20		S5026
2N3948	128	3024	28	TR87/243	S3013
2N3953			86		S0017
2N3961			28		S3001
2N3962	159	3114	67	TR20/717	S5022
2N3962A		3114			
2N3963	159	3114	21	TR20/717	
2N3963A		3114			
2N3964	159	3114	48	TR20/717	S0019
2N3964-2		3114			
2N3965	159	3114	21	TR20/717	S0019
2N3965A		3114			
2N3966			FET1		F0010
2N3967	133	3112	FET1		F0010
2N3967A	133	3112			F0010
2N3968	133	3112	FET1		F0010
2N3968A	133	3112			F0010
2N3969	133	3112	FET1		F0010
2N3969A	133	3112			F0010
2N3973		3122	210	TR21/735	S0025
2N3974		3122	210	TR21/735	S0025
2N3975	123A	3122	210	TR21/735	S0025
2N3976	123A	3024	210	TR21/735	S0015
2N3977	159	3114	22	TR20/717	S0032
2N3977A		3114			
2N3978	159	3114	22	TR20/717	S0032
2N3978A		3114			
2N3979	159	3114	82	TR20/717	S0032
2N3979A		3114			
2N3983	108	3039	86	TR95/56	S0016
2N3984	108	3018	86	TR95/56	S0016
2N3985	108	3018	86	TR95/56	S0016
2N3995	160			TR17/637	G0002
2N4000			32		S3019
2N4001			32		S3019
2N4006			22	TR20/	
2N4007			22	TR20/	
2N4008			82	TR28/	
2N4012			28		
2N4013	123A	3122	20	TR21/735	S0015
2N4014	123A	3122	20	TR21/735	S0005
2N4026	129	3025	67	TR88/242	S5022
2N4027	129	3025		TR88/242	S3003
2N4028	129	3025	67	TR88/242	S5022
2N4029	129	3025		TR88/242	S3032
2N4030	129	3025	67	TR88/242	S3001
2N4031	129	3025		TR88/242	S3003
2N4032	129	3025	67	TR88/242	S3032
2N4033	129	3025		TR88/242	S3032
2N4034	106		21	TR20/52	S0013
2N4035	106		21	TR20/52	S0019
2N4036	129	3025	29	TR88/	S3032
2N4037	129	3025	67	TR88/	S3012
2N4040			28		S3005
2N4041			28		S3005
2N4046			47		S3008
2N4054	157	3103	27	TR60/244	S5015
2N4055	157	3103	27	TR60/244	S5015
2N4056	157	3103	27	TR60/244	S5015
2N4057	157	3103	27	TR60/244	S5015
2N4058	106	3118	65	TR20/52	S0013
2N4059	106	3118	65	TR20/52	S0013
2N4060	106	3118	65	TR20/52	S0013
2N4061	106	3118	65	TR20/52	S0013
2N4062	106	3118	65	TR20/52	S0013
2N4063	123A	3045		TR21/	
2N4064	123A	3045	32	TR21/	
2N4068	154	3045	18	TR78/712	S0005
2N4070		3561			S5004
2N4072	108	3018	17	TR95/56	S0020
2N4073	108	3018	215	TR95/56	S3013
2N4074	123A	3122	210	TR21/735	S0015
2N4077	155				
2N4078	131	3052	30	TR94/642	G6014
2N4078B		3052			
2N4079	131, 155(CP)			/642	
2N4080		3118	21	/52	

Original Part Number	ECG	SK	GE	IR/WEP	HEP
2N4081			60		S0025
2N4086		3124	212	/735	S0015
2N4087		3124	212	/735	S0015
2N4087A		3124	212	/735	S0015
2N4100			210		S0007
2N4102		3503			
2N4103		3527			
2N4104		3124	20		
2N4105		3010	8	TR08/724	
2N4106		3004	2	TR05/630	G6011
2N4107			54		
2N4111			19		S7004
2N4112			19		
2N4113			19		S7004
2N4114			19		
2N4117	133	3112	FET1		
2N4117A	133	3112			
2N4118	133	3112	FET1		
2N4118A		3112			
2N4119	133	3112	FET1		
2N4119A	133	3112			
2N4121		3118		TR30/52	S0013
2N4122		3118	21	/52	S0019
2N4123	123A	3124	210	TR53/735	S0015
2N4124	123A	3124	20	TR53/735	S0014
2N4125	159	3114	21	TR54/735	S0013
2N4125A		3114			
2N4126	159	3114	48	/717	S0019
2N4126A		3114			
2N4127			66		
2N4128			66		
2N4130			14		S7002
2N4134	108	3039	60	TR95/56	S0017
2N4135	108		60	TR64/	S0017
2N4136			54		
2N4137	123A				S0025
2N4138			61	TR51/	S0020
2N4140		3122	210	TR53/735	S0011
2N4141	199	3122	210	TR53/735	S0015
2N4142		3114	21	/717	S0013
2N4142A		3114			
2N4143		3114	48	/717	S0019
2N4143A		3114			
2N4149	159			TR20/	
2N4207	159			/52	S0019
2N4208	159	3118		/52	S0019
2N4209	159	3118	21	/52	S0019
2N4220		3112	FET1	/801	F0010
2N4220A		3112		/801	F0010
2N4221	159	3112	FET1	/801	F0010
2N4221A		3112		/801	F0010
2N4222	133	3112	FET1	/801	F0010
2N4222A	133	3112		/801	F0010
2N4223	133	3112	FET1		F0010
2N4224		3116	FET1		F0010
2N4225			28		S3002
2N4226			28	TR76/	S3002
2N4227			210	TR53/	S0015
2N4228			21		S0019
2N4231	175	3131	66	TR81/241	S5012
2N4232	175	3131	66	TR81/241	S5012
2N4233	175	3131		/241	S5012
2N4234	129	3025	29	TR88/242	S5013
2N4235	129	3025		TR88/242	S5013
2N4236	129				S3032
2N4237	128	3024	47	TR87/243	S5014
2N4238			32		S5014
2N4239			32		S3019
2N4240	124	3021		TR81/	
2N4241	121	3014	16	TR01/232	G6003
2N4242			76		G6018
2N4243			76		G6018
2N4244			76		G6005
2N4245			76		G6018
2N4246			76		G6018
2N4247			76		G6005
2N4248	159	3114	65	TR20/717	S0019
2N4248A		3114			
2N4249	159	3118	67	TR20/717	S5022
2N4249A		3114			
2N4250	159	3118	65	TR20/717	S0019
2N4250A	159	3114	63		S0019
2N4251	108	3039		TR95/56	S0016
2N4252			86		S0017
2N4253			86		S0017

Original Part Number	ECG	SK	GE	IR/WEP	HEP
2N4265			20	TR53/	S0011
2N4269	154	3045	27	TR78/712	S5025
2N4270	154	3045	27	TR78/712	S5025
2N4271			32		S3019
2N4274	108	3039	20	TR95/56	S0011
2N4275	108	3039	20	TR95/56	S0011
2N4284			65	TR30/	S0032
2N4285			65	TR52/	S0032
2N4286	123A	3124	212	TR21/735	S0015
2N4287	123A	3124	212	TR21/735	S0015
2N4288	159	3118	65	TR20/717	S0019
2N4288G		3114			
2N4289	159	3114	65	TR20/717	S0019
2N4289Y		3114			
2N4290	159	3114	21	TR20/717	S0019
2N4290GN		3114			
2N4291	159	3114	48	TR20/717	S0025
2N4291A		3114			
2N4292			86		S0025
2N4293			86		S0025
2N4294			20	TR21/	S0025
2N4295			60	TR81/	S0025
2N4296	124	3021	32	TR81/240	S5011
2N4297	124	3021	32	TR01/240	S5011
2N4298	124	3021	32	TR81/240	S5011
2N4299	124	3021	32	TR81/240	S5011
2N4302	133	3112	FET1		F0010
2N4303	133	3112	FET1		F0010
2N4304	133	3112	FET1	FE100/	F0010
2N4307			66		S3002
2N4308			66		
2N4311			66		S3002
2N4312			66		
2N4313	106	3118		TR20/52	S0013
2N4314	129	3025		TR88/	S3032
2N4316		3572			
2N4317		3572			
2N4318		3573			
2N4338	133	3112	FET1		
2N4339	133	3112	FET1		
2N4340	133	3112	FET1		
2N4341			FET1		
2N4346	127	3035	25	TR27/235	G6008
2N4347	130	3079	73	TR61/	S5020
2N4348		3079		TR61/	S5020
2N4349		3197	215		S3020
2N4350			28		S3001
2N4354	159	3025	82	TR20/717	S5022
2N4354A		3114			
2N4355	159	3114	82	TR20/717	S5022
2N4355A		3114			
2N4356	159	3114	82	TR20/717	S5022
2N4356A		3114			
2N4359	159		65		S0019
2N4383			47		S3024
2N4384			47		
2N4385			47		S3024
2N4386			47		
2N4387	153	3083	69	TR77/700	S5018
2N4388	153	3083	69	TR77/700	S5018
2N4389	106	3118		TR20/52	S0019
2N4390			27		S5026
2N4395	130	3027	19	TR59/247	S7002
2N4396	130	3027	19	TR59/247	S7002
2N4397			60		
2N4254	107	3039	86	TR70/720	S0016
2N4255	107	3018	86	TR70/720	S0016
2N4256	123A	3122	212	TR21/735	S0015
2N4257	159	3118		TR20/717	S0013
2N4257A	159	3118			S0013
2N4258	159	3114		TR20/717	S0013
2N4258A	159	3118			S0013
2N4258R		3114			
2N4259	123A		39	TR21/735	S0011
2N4264			20	TR53/	S0011
2N4398				/S7001	S7001
2N4399				/S7001	S7001
2N4400	123A	3122	20	TR21/735	S0015
2N4401	123A	3122	20	TR21/735	S0015
2N4402	159	3114	21	TR20/717	S0019
2N4402A		3114			
2N4403	159	3025	21	TR20/717	S0019
2N4403A		3114			
2N4404	129	3025		TR88/242	S3032
2N4405	129	3025		TR88/242	S3032

Original Part Number	ECG	SK	GE	IR WEP	HEP
2N4406	129	3025		TR88/242	S3012
2N4407	129	3025		TR88/242	S3032
2N4409	194		81	TR53/	S0005
2N4410	154	3045	18	TR78/712	S0005
2N4411	159	3114		TR20/717	S0013
2N4411A		3114			
2N4412	129	3025	48	TR88/242	S0012
2N4412A	129	3025	67	TR88/242	S5013
2N4413	159	3114	48	TR20/717	S0019
2N4413A	159	3114	67	TR20/717	S5022
2N4414	129	3025	48	TR88/242	S0012
2N4414A	129	3025	67	TR88/242	S5013
2N4415	159	3114	48	TR20/717	S0019
2N4415A	159	3114	67	TR20/717	S5022
2N4415AB		3114			
2N4416	133	3112	FET2		F0015
2N4416A	133	3112			
2N4417	133	3112	FET2		
2N4417A		3116			
2N4418	108	3039	20	TR95/56	S0025
2N4419	108	3039	20	TR95/56	S0011
2N4420			20	TR64/	S0016
2N4421		3118	20	TR64/	S0011
2N4422			20	TR64/	S0016
2N4423			21		S0013
2N4424	123A	3124	210	TR21/735	S0015
2N4425	192	3124	47	TR21/735	S3024
2N4427	128	3024	18	TR87/243	S3008
2N4428			28		S3001
2N4429			28		
2N4430			28		
2N4432	123A	3122	47	TR21/735	S0015
2N4432A	123A	3122	47	TR21/735	S0015
2N4433	107	3018	210	TR70/720	S0016
2N4434	161	3117	39	TR83/719	S0017
2N4435	161	3117	39	TR83/719	S0017
2N4436	123A	3122	210	TR21/735	S0015
2N4437	123A	3122	210	TR21/735	S0015
2N4440			28		
2N4449	108	3039		TR95/56	S0011
2N4450	123A	3122	210	TR21/735	S0025
2N4451	106			TR20/52	S0019
2N4452	159	3114		TR20/717	S0019
2N4452A		3114			
2N4453	106	3118		TR20/52	S3014
2N4867	133	3112	FET1		
2N4867A	133	3112			
2N4868	133	3112	FET1		
2N4868A	133	3112	FET1		
2N4869	133	3112	FET1		
2N4869A	133	3112			
2N4871	6401			TR2160/	S9002
2N4872	106	3118		TR20/52	S0013
2N4873	108	3039		TR95/56	S0025
2N4877			66		S3010
2N4890	129		67	TR88/242	S5022
2N4898		3083	69	TR58/700	S5018
2N4899		3083	69	TR58/700	S5018
2N4900	218			TR58/	S5018
2N4901	218			TR29/	S7003
2N4902	218			TR29/	S7003
2N4903	218				A5005
2N4904	218			TR29/	S7003
2N4905	218			TR29/	S7003
2N4906	218				S7003
2N4907	218				S7003
2N4908	218				S7003
2N4909	218				S5005
2N4910	175	3131	216	TR81/241	S5019
2N4911	175	3131	216	TR81/241	S5019
2N4912	175	3131	32	TR81/241	S5019
2N4913	130	3027	73	TR59/247	S7002
2N4914	130	3027	73	TR59/247	S7002
2N4915	130	3124	73	TR59/247	S7002
2N4916	159	3124	22	TR20/717	S0013
2N4916A		3114			
2N4917	159	3114	21	TR20/717	S0013
2N4917A		3114			
2N4918	185	3191	29	/S5007	S5006
2N4919	185	3191	29	TR77/S5007	S5006
2N4920	185	3191	69	/S5007	S5006
2N4921	184	3190	28	/S5003	S5000
2N4922	184	3190	28	/S5003	S5000
2N4922A		3954			
2N4923	184	3190	28	/S5003	S5000

Original Part Number	ECG	SK	GE	IR WEP	HEP
2N4924	154	3045	27	TR78/712	S5026
2N4925	154	3045	32	TR78/712	S5025
2N4926	154	3045	40	TR78/712	S5025
2N4927	154	3045	40	TR/8/712	S5025
2N4928	129	3025		TR88/242	
2N4932				TR66/	
2N4934	108	3039		TR95/56	S0016
2N4935	108	3039		TR95/56	S0016
2N4936	108	3039		TR95/56	S0030
2N4937			21		S0019(2)
2N4038			21		S0019(2)
2N4940			21		S0019(2)
2N4941			21		S0019(2)
2N4943	128	3024		TR87/243	S3019
2N4944	128	3024	20	TR87/243	
2N4945	194	3024	81	TR87/243	
2N4946	128	3024	20	TR87/243	S3019
2N4950			20		
2N4951	123A	3124	210	TR21/735	S0025
2N4952	123A	3124	210	TR21/735	S0025
2N4953	123A	3124	210	TR21/735	S0015
2N4954	123A	3124	210	TR21/735	S0025
2N4960	128	3024		TR87/243	S3001
2N4961	128	3024		TR87/243	S3002
2N4962	128	3024		TR87/243	S3001
2N4963	128	3024		TR87/243	S3019
2N4964	128	3024	65	TR87/243	S0019
2N4965	159	3114	65	TR20/717	S0019
2N4965A		3114			
2N4966	123A	3124	61	TR21/735	S0015
2N4967	123A	3124	212	TR21/735	S0024
2N4968	123A	3124	60	TR21/735	S0024
2N4969	123A	3122	60	TR21/735	S0025
2N4970	123A	3122	210	TR21/735	S0015
2N4971	159	3114	21	TR20/717	S0019
2N4971A		3114	89		
2N4972	159	3114	48	TR20/717	
2N4972A		3114	89		
2N4976			28		
2N4980			82		
2N4981			82		
2N4982	159	3114	82	TR20/717	
2N4982A		3114	89		
2N4994	123A	3122	210	TR21/735	
2N4995	123A	3122	210	TR21/735	S0015
2N4996	108	3018	61	TR95/56	S0016
2N4997	108	3018	61	TR95/56	S0016
2N5010			32		
2N5022	129	3025	67	TR88/242	
2N5023	129	3025	67	TR88/242	S3014
2N5024		3039	86		S0017
2N5027			210	TR64/	S0014
2N5028			210		S0014
2N5030			20	TR21/	S0011
2N5031	107	3039	11	TR70/720	S0017
2N5032	107	3039	11	TR70/720	S0017
2N5034	223	3036	14	TR59/	S7002
2N5035	223			TR59/	S7002
2N5036	233	3027		TR59/	S7002
2N5037	223			TR59/	S7002
2N5038		3535			
2N5039		3511			S7000
2N5040	129		48	TR30/	S0012
2N5041	129		48	TR52/	S0019
2N5042	129		48		
2N5045	133	3112	FET1		
2N5046	133	3112	FET1		
2N5047	133	3112	FET1		
2N5050	175	3131		TR81/241	S5012
2N5051	175	3131		TR81/241	S5012
2N5053	108	3018	86	TR95/56	S0017
2N5054	108	3018	20	TR95/56	
2N5055		3118	89		
2N5058	154	3045	32	TR78/712	S5024
2N5059	154	3045	32	TR78/712	S5024
2N5066			210	TR22/	S0020
2N5067	130	3027	73	TR59/247	S7002
2N5068	130	3027	73	TR59/247	S7002
2N5069	130		73		S7004
2N5073			32		S3021
2N5078			FET2		
2N5078A		3116	FET2		
2N5079			28		S3001
2N5080			28		
2N5081	123A	3122	88	TR21/735	S0015

Original Part Number	ECG	SK	GE	IR/WEP	HEP
2N5082	123A	3122	88	TR21/735	S0015
2N5086	159	3114	65	TR20/717	S0019
2N5086A		3114	89		
2N5087	159	3114	65	TR20/727	S0019
2N5087A		3114	89		
2N5088	123A	3124	212	TR21/735	S0024
2N5089	123A	3124	85	TR21/735	S0024
2N5090				TR66/	
2N5092			32		
2N5095			32		
2N5103			17		
2N5103A		3116	FET2		
2N5104A		3116	FET2		
2N5105A		3116	FET2		
2N5106			47		S3001
2N5107	123A		210		S0015
2N5108	128	3024	18	TR87/243	S3001
2N5109	128			TR87/243	S3013
2N5110	129	3025	84	TR88/242	S5013
2N5111	129	3025		TR88/242	S3032
2N5112			69		
2N5120			18		
2N5126	161	3124	86	TR83/719	S0025
2N5127	123A	3124	212	TR21/735	S0025
2N5128	123A	3124	210	TR21/735	S0014
2N5129	128	3024	210	TR87/243	S0014
2N5130	161	3018	86	TR83/719	S0016
2N5131	108	3124	64	TR95/56	S0011
2N5132	107	3124	60	TR70/720	S0033
2N5133	161	3018	212	TR83/719	S0011
2N5134	123A	3124	20	TR21/735	S0015
2N5135	128	3124	64	TR87/243	S0014
2N5136	128	3124	210	TR87/243	S0014
2N5137	123A	3124	210	TR21/735	S0014
2N5138	159	3114	65	TR20/717	S0013
2N5138A		3114	89		
2N5139	159	3114	21	TR20/717	S0013
2N5139A		3114	89		
2N5140	106	3118	89	TR20/52	
2N5141	106	3118	22	TR20/52	
2N5142	159	3114	22	TR20/717	S0012
2N5142A		3114	89		
2N5143	159	3114	22	TR20/717	S0012
2N5143A		3114	89		
2N5144			20		S0015
2N5155			25		
2N5156			25	TR27/	G6018
2N5157				TR88/242	S5021
2N5160	129	3025	29	TR88/242	S3014
2N5161			69		
2N5163	132	3116	FET1	FE100/802	F0015
2N5163A		3116	FET2		
2N5172	123A	3124	212	TR70/720	S0016
2N5174	154	3045	32	TR78/712	S5026
2N5175	154	3045	27	TR78/712	S5025
2N5176	154	3045	32	TR78/712	S5025
2N5179	107	3039	86	TR70/720	S0017
2N5180	108	3018	60	TR95/56	S0016
2N5181	161		39	TR83/719	S0025
2N5182	161		39	TR83/719	S0016
2N5183	123	3124	47	TR86/53	S3001
2N5184	154	3040	27	TR78/712	S5025
2N5185	3040		27		S5025
2N5186	123A	3122	20	TR21/735	S0011
2N5187	123A	3122	20	TR21/735	S0014
2N5188	128	3024	18	TR87/243	S3001
2N5189		3529	28	TR64/	S3010
2N5190	184	3054	57	/S5003	S5000
2N5191	184	3054	57	/S5003	S5000
2N5192	184	3054	57	/S5003	S5000
2N5193	185	3083	58	/S5007	S5006
2N5194	185	3083	58	/S5007	S5006
2N5195	185	3083	56		S5006
2N5200	108	3124	47	TR95/56	S0020
2N5201	100	3020	210	TR05/254	S0020
2N5208	106	3118	21	TR20/52	S0013
2N5209	123A	3122	212	TR21/735	S0025
2N5210	123A	3122	212	TR21/735	S0003
2N5211	128	3024	18	TR87/243	
2N5219	123A	3122	212	TR21/735	S0015
2N5220	123A	210		TR21/735	S0015
2N5221	159	3114	82	TR20/717	S0012
2N5222	107	3039	11	TR70/720	S0020
2N5223	123A	3122	212	TR21/735	S0025
2N5224	123A	3122	10	TR21/735	S0025
2N5225	123A	3122	210	TR21/735	S0015
2N5226	159	3114	82	TR20/717	S0019
2N5226A		3114	89		
2N5227	159	3114	212	TR20/717	S0019
2N5227A		3114	89		
2N5228	106		65	TR20/52	
2N5229			22		
2N5230			82		
2N5231			82		
2N5232	123A	3124	212	TR21/735	S0015
2N5232A	123A	3122	212	/735	S0015
2N5233	128	3024	81	TR87/243	
2N5234	128	3024	18	TR87/243	S0005
2N5235	128	3024	18	TR87/243	
2N5239			73		S5020
2N5240	124	3021	73	TR81/	S5020
2N5241			73		S5020
2N5242	129		48	TR88/242	S0012
2N5243	129		48	TR88/242	S0019
2N5245	132	3116	FET2	FE100/	F0021
2N5245A		3116	FET2		
2N5246		3116	FET2		F0021
2N5246A		3116	FET2		
2N5247		3116	FET2		F0021
2N5247A		3116	FET2		
2N5248	132	3116	FET2	FE100/802	F0015
2N5248A		3116	FET2		
2N5249		3024	212	TR21/243	
2N5249A	199	3024	212	TR87/243	F0015
2N5252			32		S3021
2N5253			32		S3021
2N5258			FET2		
2N5262		3529	66		S3010
2N5264	162			TR67/707	S5020
2N5267	133	3112	FET1		
2N5268	133	3112	FET1		
2N5269	133	3112	FET1		
2N5270	133	3112	FET1		
2N5272	108			TR95/56	S0014
2N5277	133	3112	FET1		
2N5278A		3116	FET2		
2N5279			32		
2N5292	108			TR95/56	S0020
2N5293	152	3054	66	TR76/	S5000
2N5294	152	3054	66	TR76/	S5000
2N5295	152	3054	66	TR76/701	S5000
2N5296	152	3054	28	TR76/701	S5000
2N5297	152	3054	66	TR76/701	S5000
2N5298	152	3054	66	TR76/701	S5000
2N5301			75		S7000
2N5302			75		S7000
2N5303			75		S7000
2N5305	172		64	TR69/S9100	
2N5306	172		64	TR69/S9100	S9100
2N5307	172		64	TR69/S9100	
2N5308	172	3156	64	TR69/S9100	S9100
2N5309	128	3024	62	TR87/243	S5026
2N5310	128	3024	62	TR87/243	S5026
2N5311	128	3024	212	TR87/243	S0005
2N5320		3512		TR25/	S3002
2N5320HS	192	3024	18		
2N5321		3512	28		S3010
2N5322	129	3513	28	TR20/52	S3003
2N5323		3513			S3003
2N5324			25		G6007
2N5325			25		G6008
2N5334			66		S3010
2N5352	106			TR20/52	
2N5354	159	3114	22	TR20/717	S0012
2N5354A		3114	89		
2N5355	159	3114	48	TR20/717	S0019
2N5355A		3114	89		
2N5356	159	3114	89	TR20/717	S0019
2N5356A		3114	89		
2N5358		3112	FET1		
2N5359		3112	FET1		
2N5360			FET1		
2N5360A		3116	FET2		
2N5361		3116	FET2		
2N5362		3116	FET1		
2N5363		3116	FET1		
2N5364		3116	FET1		
2N5365	159	3114	21	TR20/717	S0019
2N5365A		3114	89		
2N5366	159	3114	48	TR20/717	S0019

Original Part Number	ECG	SK	GE	IR/WEP	HEP
2N5366A		3114	48		
2N5367	159	3114	89	TR20/717	S0019
2N5367A		3114	89		
2N5368	123A	3122	210	TR21/735	S0015
2N5369	123A	3122	210	TR21/735	S0015
2N5370	123A	3122	210	TR21/735	S0015
2N5371	123A	3122	210	TR21/717	S0014
2N5372	159	3114	82	TR20/717	S5022
2N5372A		3114	89		
2N5373	159	3114	82	TR20/717	S5022
2N5373A		3114	89		
2N5374	159	3114	48	TR20/717	S5022
2N5374A		3114	89		
2N5375	159	3114	82	TR20/717	S0019
2N5378	159	3114	48	TR20/717	S0019
2N5378A		3114	89		
2N5379	159	3114	48	TR20/717	S0019
2N5379A		3114	89		
2N5380	123A	3122	210	TR21/735	S0015
2N5381			20		S0015
2N5382	159	3114	21	TR20/717	S0019
2N5382A		3114	89		
2N5383	159	3114	48	TR20/717	S0019
2N5383A		3114	89		
2N5391		3112	FET1		
2N5392		3116	FET1		
2N5393		3116	FET1		
2N5394		3116	FET1		
2N5395		3116	FET1		
2N5396		3116	FET1		
2N5397		3116	FET2		
2N5398		3116	FET2		
2N5399	108	3039	86	TR95/56	S0020
2N5401		3114			
2N5415		3053			
2N5416		3528			
2N5417			20		S3008
2N5418	123A	3122	210	TR21/735	S0025
2N5419	123A	3122	210	TR21/735	S0025
2N5420	123A	3122	88	TR21/735	S0024
2N5421	128	3024	28	TR87/243	S3001
2N5422			28		S3001
2N5423			28		
2N5424			66		
2N5447	159	3114	67	TR20/717	S0019
2N5447A		3114	89		
2N5448	159	3114	67	TR20/717	S0019
2N5448A		3114	89		
2N5449	128	3024	20	TR87/243	S3001
2N5450	128	3024	20	TR87/243	S3001
2N5451	128	3024	20	TR87/243	S3001
2N5452		3112	FET1		
2N5453		3112	FET1		
2N5454		3112	FET1		
2N5457	133	3116	FET1		F0010
2N5458	133	3112	FET1	FE100/801	F0010
2N5459		3116	FET1		F0015
2N5466	163			TR67/740	S5021
2N5467	165			TR93/740B	S5021
2N5481			28		S3005
2N5482			28		S3005
2N5483			66		S3006
2N5484		3116	FET2		F0021
2N5485	132	3116	FET2	FE100/802	F0021
2N5486	132	3116	FET2	FE100/802	F0021
2N5489			28		
2N5490	196	3054	66	TR92/	S5004
2N5491	196	3054	66	TR92/	S5004
2N5492	152	3054	28	TR76/701	S5004
2N5493	196	3054	66	TR92/	S5004
2N5494	196	3054	66	TR92/	S5004
2N5495	196	3054	66	TR92/	S5004
2N5496	196	3054	28	TR92/	S5004
2N5497	196	3054	28	TR92/	S5004
2N5525	172	3156	64	TR69/S9100	S9100
2N5541			32		
2N5543		3112	FET1		
2N5544		3112	FET1		
2N5550	194	3045		TR78/	S0005
2N5555			FET1		
2N5556		3112	FET1		
2N5557		3112	FET1		
2N5558		3112	FET1		
2N5561		3112	FET1		
2N5562		3112	FET1		

Original Part Number	ECG	SK	GE	IR/WEP	HEP
2N5563		3112	FET1		
2N5581			20		S3001
2N5582			20		S3001
2N5589			28	TR76/	S3005
2N5590		3176	66		S3006
2N5591		3177			S3007
2N5592		3116	FET2		
2N5593		3116	FET2		
2N5594		3116	FET2		
2N5597			69		S5018
2N5598	175	3131	66	TR81/241	S5012
2N5600	175	3131		TR81/241	S5012
2N5602	175	3131		TR81/241	S5012
2N5603		3562			
2N5604	175	3131		TR81/241	S5012
2N5606			66		S5012
2N5614	130		66	TR59/247	S7002
2N5616		3561			S7004
2N5618		3561			S7004
2N5620		3561			
2N5621	180			/S7001	S7003
2N5622	130			TR59/247	S7002
2N5623	180			/S7001	
2N5624		3561			S7004
2N5625	180			/S7001	
2N5628		3561			
2N5629			75		
2N5632	162		75	TR67/707	S5020
2N5633	162			TR67/707	S5020
2N5634	162			TR67/707	S5020
2N5637			66		S3007
2N5642			66		
2N5644			28		S3005
2N5645			28		S3006
2N5646			66		S3006
2N5648		3112	FET1		
2N5649		3112	FET1		
2N5650	161	3117	211	TR83/719	S0017
2N5651	161	3117	211	TR83/719	S0017
2N5652	161	3039	86	TR83/719	S0017
2N5655	157	3103	32	TR60/244	S5015
2N5656	157	3103	32	TR60/244	S5015
2N5660			32		S3021
2N5661			32		
2N5662			32		S3021
2N5663			32		
2N5668	133	3116	FET1		F0015
2N5669		3116	FET2		F0015
2N5670		3116	FET2		F0015
2N5681			32		S3019
2N5682		3024	32		S3019
2N5687			28		S3001
2N5688			28		S3005
2N5689			66		
2N5690			66		
2N5697			28		S3013
2N5698			28		
2N5699			28		S3006
2N5700			66		S3007
2N5701			66		S3007
2N5702			18		S3013
2N5703			28		S3006
2N5704			66		S3006
2N5705			66		S3007
2N5710			28		S3013
2N5712			66		
2N5713			66		
2N5716		3112	FET1		
2N5717		3112	FET1		
2N5718		3112	FET1		
2N5735			210		
2N5736			210		
2N5738	180			/S7001	S5005
2N5741	180			/S7001	
2N5742	180			/S7001	
2N5745	180			/S7001	
2N5758	162		75	TR67/707	S5020
2N5759	162			TR67/707	S5020
2N5760	162			TR67/707	S5020
2N5763			21		S5022
2N5764			28		S3006
2N5765			66		S3006
2N5766			28		
2N5767			28		
2N5768			66		

Original Part Number	ECG	SK	GE	IR/WEP	HEP
2N5769	123A	3040		TR21/	S0033
2N5781		3025			S3003
2N5782		3025	29		S3003
2N5783		3025	29		S5013
2N5784		3024	18		S3002
2N5785		3024	216		S3010
2N5786		3024	46		S5013
2N5810			20		S0014
2N5811			48		S0012
2N5812			47		S0015
2N5813			48		S0019
2N5815			48	TR54/	S0019
2N5816			20		S0015
2N5817			48		S0019
2N5818			47		S0015
2N5819			48		S0019
2N5820	128	3024	63	TR87/243	S5014
2N5820HS	192			TR25/	
2N5821	129	3025	67	TR88/242	S5013
2N5821HS	193			TR28/	
2N5822			63		S5014
2N5823			67		S5013
2N5824			62		S0025
2N5825			62		S0025
2N5826			62		S0015
2N5827			62		S0015
2N5827A			62		S0015
2N5828			62		S0030
2N5828A			62		S0030
2N5830	123A		20	TR21/735	S0005
2N5830A		3018	61		
2N5838			73		S5020
2N5839			73		S5020
2N5840			73		S5020
2N5845			20		S0025
2N5845A			20		S0025
2N5855			67		S3032
2N5856			63		S3020
2N5874		3563			S5004
2N5878		3563			S7002
2N5881	128	3024	18	TR87/243	S7002
2N5882	128	3024	18	TR87/243	S7004
2N5883	180			/S7001	S7001
2N5884	180			/S7001	S7001
2N5893	131	3082	49	TR94/642	G6016
2N5897	131	3052	30	TR94/642	G6016
2N5913			45		S3008
2N5930		3563			
2N5933		3563			
2N5949		3116	FET2		
2N5950		3116	FET2		
2N5951		3116	FET2		
2N5952		3116	FET2		
2N5953		3116	FET2		
2N5954		3085	56		S5012
2N5955		3085	56		S5012
2N5956		3085	56		S5012
2N5961			81		S0005
2N5981		3189			S5005
2N5982		3189			S5005
2N5985		3188			S5004
2N5998			210		S0024
2N5999			82		S3028
2N6000			210		S0025
2N6001			48		S0012
2N6004			210		S0025
2N6005			48		S0019
2N6008			210	TR53/	S0025
2N6009			48		S3028
2N6011			48		S3028
2N6014			63		S3020
2N6015			67		S3032
2N6021	153			TR77/	S5006
2N6022	153			TR77/	S5006
2N6023	153		69	TR77/	S5006
2N6024	153		69	TR77/	S5006
2N6025	180		69		S5006
2N6026	153		69	TR77/	S5006
2N6027	6402		X17		S9001
2N6029			74		
2N6034	254	3181			
2N6035	254				
2N6036	254				
2N6037	253	3180			
2N6038	253				

Original Part Number	ECG	SK	GE	IR/WEP	HEP
2N6039	255	3180			
2N6040	262				S9122
2N6041	262				S9122
2N6043	261	3180			S9102
2N6044	261	3180			S9102
2N6050	248				
2N6051	248				
2N6052	248				
2N6053	244	3183			S9141
2N6054	244	3183			S9141
2N6055	243	3182			S9141
2N6056	243	3182			S9141
2N6057	247				
2N6058	247				
2N6059	247				
2N6067			82		S0013
2N6076	159		65	TR54/	S0015
2N6098		3534			
2N6099		3534			S5004
2N6100		3534			
2N6101		3534			S5004
2N6102		3534			
2N6103		3534			
2N6106	197	3083			S5005
2N6107	197	3083			S5005
2N6108	197	3083			S5005
2N6109	197	3083			S5005
2N6110	197	3084			S5005
2N6111	197	3084			S5005
2N6112			62		S0015
2N6121	152			TR76/701	S5000
2N6122	152			TR76/701	S5000
2N6124	153			TR77/700	S5006
2N6125	153			TR77/700	S5006
2N6126	153			TR77/	S5006
2N6132	197				S5005
2N6133	197				S5005
2N6134	197				S5005
2N6175	228	3104	32		S5015
2N6176		3103	32	TR60/	S5015
2N6177	228	3103	32		
2N6178		3024			S5000
2N6179		3024			S5000
2N6180		3025			S5006
2N6181		3025			S5006
2N6222			81		
2N6223			82		
2N6224			81		
2N6225			82		
2N6226			74		S5005
2N6229			74		S5005
2N6246	219	3167			S7001
2N6247	219				S7001
2N6249		3167			S5020
2N6250		3559			S5020
2N6251		3559			S5020
2N6253	130	3027	77	TR59/	S7004
2N6254		3511	75		S7004
2N6257	181	3036	75	TR36/	S7000
2N6258	181	3036	14	TR36/	S7000
2N6259		3535			
2N6260	175	3026	66	TR81/	S5019
2N6261	175	3026	66	TR81/	S5019
2N6263	175	3538	12	TR81/	S5012
2N6264		3538			S5012
2N6282	251				
2N6283	251				
2N6284	251				
2N6285	250				
2N6286	250				
2N6287	250				
2N6288		3054	28		S5004
2N6289		3054	28		S5004
2N6290		3054	28		S5004
2N6291		3054	28		S5004
2N6292		3054	28		S5004
2N6293		3054	28		S5004
2N6294	274	3182			S9101
2N6295	274	3182			S9101
2N6296	275	3183		TR58/	S9121
2N6297	275	3183		TR58/	S9121
2N6298	244				S9121
2N6299	144				S9121
2N6300	243				S9101
2N6301	243				

Original Part Number	ECG	SK	GE	IR/WEP	HEP
2N6305			86		S0017
2N6326			75		S7000
2N6327			75		S7000
2N6328			75		S7000
2N6329	180		74		S7001
2N6330			74		S7001
2N6331			74		S7001
2N6355	251				
2N6356	251				
2N6357	247				
2N6358	251				
2N6371		3511			S7004
2N6383	245	3182			
2N6384	245	3182			
2N6385		3182			
2N6386	263	3180			
2N6387	263	3180			
2N6388	263	3180			
2N6389			212		
2N6470		3563			
2N6471		3563			
2N6472		3563			
2N6492	243				
2N6493	243				
2N6494	243				
2N6510		3560			
2N6511		3560			
2N6512		3560			
2N6513		3560			
2N6514		3560			
2N17780				TR08/	
2N19616	160			TR17/	
2N29260RN				TR51/	
2N3392489751-02				TR21/	
2N-3391-A				TR24/	
2N-3394				TR24/	
2NJ5A	100	3005	2	TR05/254	
2NJ5D		3123	2	TR05/	
2NJ6	100	3005	51	TR05/254	G0005
2NJ8A	100	3005	2	TR05/254	G0005
2NJ9A	102	3004	52	TR05/631	G0005
2NJ9D	102	3004	52	TR05/631	G0005
2NJ50	126	3007	51	TR17/635	G0008
2NJ51	126	3007	52	TR17/635	G0005
2NJ52	126	3006	52	TR17/635	G0005
2NJ53	126	3006	52	TR17/635	G0005
2NJ58A				TR05/	
2NJ59D			52	TR85/	
2NJ559D				TR85/	
2NK11A				TR01/	
2NL48			63	TR25/	
2NS31	102			TR05/631	
2NS32	102	3004	53	TR05/631	
2NS121	102	3004	53	TR05/631	
2NSM-1	125	3030		8D4DR5A4D/	
2NU/9M1		3123	80		
2NU/9M2		3123	80		
2NU/9WHT		3123	80		
2NU/9YEL		3123	80		
2OC72	158				G0005
2PB187	102A	3004	53	TR85/	
2S001	123	3124	39	TR85/	S0012
2S002	123	3124	39	TR86/53	S0012
2S003	123	3124	39	TR86/53	S0014
2S004	123	3124	39	TR86/53	S0016
2S005	123A	3124	60	TR21/735	S0012
2S006	108	3039	17	TR95/56	
2S012			66		S5000
2S014	128	3024	60	TR87/243	S0012
2S017	128	3024	215	TR87/243	S5014
2S018	128	3024	32	TR87/243	S5026
2S019	128	3024	215	TR87/243	S5014
2S020	128	3024	32	TR87/243	S5026
2S021	129	3025		TR88/242	
2S022	129	3025		TR88/242	S0013
2S023	129	3025		TR88/242	S0013
2S024(E)				TR23/	
2S028				TR23/	
2S033	130	3027	75	TR59/247	S7004
2S034	130	3027	75	TR59/247	S7004
2S035	162			TR67/707	S5020
2S036	162			TR67/707	S5020
2S072				TR08/	
2S095A			20		S0011

Original Part Number	ECG	SK	GE	IR/WEP	HEP
2S0226Q				TR57/	
2S0261				TR25/	
2S0350				TR61/	
2S0371				TR95/	
2S0460				TR21/	
2S12	100	3005	2	TR05/254	
2S13	100	3005	2	TR05/254	G0005
2S14	102	3004	52	TR05/	G0005
2S15	102	3004	52	TR05/631	G0005
2S15A	102	3004	52	TR05/631	G0005
2S22	102	3004	52	TR05/631	G0005
2S24	102	3004	52	TR05/631	
2S25	100	3005	2	TR05/254	
2S26	121	3009	16	TR01/232	
2S26A	121	3009	16	TR01/232	
2S30	100	3005	52	TR05/254	G0008
2S31	100	3005	52	TR05/254	G0005
2S32	102	3004	52	TR05/631	G0005
2S33	102	3004	52	TR05/631	G0005
2S34	102	3004	2	TR05/631	G0005
2S35	126	3008	50	TR17/635	G0002
2S36	126	3008	50	TR17/635	G0002
2S37	102	3004	52	TR05/631	G0005
2S38	102	3004	2	TR05/631	G0005
2S39	102	3004	52	TR05/631	G0005
2S40	102	3004	2	TR05/631	G0005
2S41	104	3009	16	TR01/230	G6003
2S41A	104	3009	16	TR01/230	G6003
2S42	104	3009	3	TR01/230	G6003
2S43	102	3004	52	TR05/631	G0005
2S44	102	3004	52	TR05/631	G0005
2S45	100	3005	2	TR05/254	G0005
2S46	102	3003	52	TR05/631	G0005
2S47	102	3003	52	TR05/631	G0005
2S49	100	3005	2	TR05/254	G0005
2S51	100	3005	2	TR05/254	
2S52	100	3005	2	TR05/254	G0005
2S53	100	3005	2	TR05/254	G0005
2S54	102	3004	52	TR05/631	G0005
2S56	102	3004	52	TR05/631	G0005
2S57			51	TR17/	
2S58	126	3008	50	TR17/635	
2S60	100	3005	2	TR05/254	G0008
2S75B				TR85/	
2S91	100	3005	2	TR05/254	G0005
2S92	100	3008	50	TR05/254	G0005
2S92A	100	3008	50	TR05/254	G0005
2S93	100	3008	50	TR05/254	G0005
2S93A	100	3008	50	TR05/254	G0005
2S95A	123A	3122	20	TR21/735	S0011
2S96	126	3006	52	TR17/635	
2S97	126	3006	52	TR17/635	
2S98	185	3006	52	TR17/635	
2S101	123A	3122	61	TR86/735	S0011
2S102	123A	3122	210	TR86/735	S0011
2S103	128	3024	210	TR86/243	S0011
2S104	128	3024	210	TR86/243	S0011
2S109	126	3008	50	TR17/635	G0008
2S110	126	3007	50	TR17/635	G0008
2S111	100	3005	54	TR05/254	G0008
2S112	126	3008	50	TR17/635	G0008
2S131	123A	3122	210	TR21/735	S0013
2S134		3123	80		
2S141	126	3006	50	TR17/635	G0008
2S142	126	3006	51	TR17/635	G0008
2S143	126	3006	50	TR17/635	G0001
2S144			51		G0003
2S145	126	3006	50	TR17/635	G0001
2S146	126	3006	51	TR17/635	G0001
2S148			51		
2S155	100	3005	2	TR05/254	G0005
2S159	100	3005	2	TR05/254	G0005
2S160	100	3005	2	TR05/254	G0005
2S163	102	3004	52	TR05/631	G0005
2S167	100	3005	53	TR05/254	G0005
2S174	100	3005	53	TR05/254	G0005
2S175		3006	50	TR17/	
2S176	126	3006	2	TR17/635	
2S178	100	3005	2	TR05/254	G0005
2S179	102	3004	52	TR05/631	G0005
2S189	102A	3004	53	TR85/250	G0005
2S201			51	TR92/	
2S235D				TR92/	S5000
2S273	102A			TR85/250	
2S277C			54		

Original Part Number	ECG	SK	GE	IR/WEP	HEP
2S301	129	3025	82	TR88/242	
2S302	129	3025	82	TR88/242	S0012
2S302A	129	3025	22	TR88/242	S0013
2S303	129	3025	22	TR88/242	S0012
2S304	129	3025	22	TR88/242	S0012
2S305		3025			
2S306	159	3114	65	TR20/717	
2S307	159	3118	65	TR20/717	S0012
2S307A		3114	89		
2S321	159	3114	82	TR20/717	
2S321A		3114	89		
2S322	159	3114	65	TR20/717	S0013
2S322A	159	3114	65	TR20/717	S0013
2S322AB		3114	89		
2S323	159	3114	65	TR20/717	S0013
2S323A		3114	89		
2S324	159	3114	65	TR20/717	S0013
2S324A		3114	89		
2S326	159	3114	65	TR20/717	S0013
2S326A		3114	89		
2S327	159	3114	65	TR20/717	S0013
2S327A		3114	89		
2S351B				TR17/	
2S363			20	TR21/	
2S365		3004			
2S381BN				TR21/	
2S471-1			50	TR17/	
2S494-YE		3114	89		
2S501	123A	3122	61	TR21/735	S0011
2S502	123A		61	TR21/735	S0011
2S503	123A	3122	212	TR21/735	S0011
2S512	108	3039	210	TR95/56	S0013
2S608				TR21/	
2S644(S)			18	TR51/	
2S701	123	3124	60	TR86/53	S0012
2S702	123	3124	60	TR86/53	S0012
2S703	123	3124	60	TR86/53	S0012
2S711	123	3124	81	TR86/53	S0014
2S712	123	3124	81	TR86/53	S0014
2S731	123A	3122	210	TR21/53	S0013
2S732	123A	3122	210	TR21/53	S0013
2S733	123A	3122	210	TR21/53	S0013
2S741	128	3024	11	TR87/243	S0012
2S741A	123A	3122	61	TR21/735	S0020
2S742	128	3024	63	TR87/243	S0012
2S742A	128	3024	81	TR87/243	S0005
2S743	154	3045	18	TR78/712	
2S743A	154	3045	27	TR78/712	S0005
2S744	128	3024	11	TR87/243	S0012
2S744A	123A	3122	61	TR21/735	S0020
2S745	128	3024	63	TR87/243	S0012
2S745A	154	3045	81	TR78/712	S0005
2S746	154	3045	18	TR78/712	
2S746A	154	3045	27	TR78/712	S0005
2S828Q				TR21/	
2S838(E)				TR24/	
2S930C				TR51/	
2S930D				TR51/	
2S930E				TR24/	
2S945				TR21/	S0015
2S1014-D1				TR76/	
2S1018		3054			
2S1182D			20		
2S1760			2		
2S3010	129	3025	82	TR88/242	S0019
2S3020	159	3114	82	TR20/717	S0019
2S3020A		3114	89		
2S3021	159	3114	22	TR20/717	S0013
2S3030	159	3114	22	TR20/717	S0015
2S3030A		3114	89		
2S3040	159	3114	65	TR20/717	S0031
2S3040A		3114	89		
2S3210	159	3114	82	TR20/717	S0013
2S3210A		3114	89		
2S3220	159	3114	82	TR20/717	S0013
2S3220A		3114	89		S0013
2S3221	159	3114	22	TR20/717	S0013
2S3221A		3114	89		
2S3230	159	3114	22	TR20/717	S0013
2S3230A		3114	89		
2S3240	159	3114	22	TR20/717	S0013
2S3240A		3114	22	TR20/717	
2S3370			52	TR05/	
2S3734-Y				TR87/	
2S6344				TR21/	
2S6371				TR21/	
2S6387				TR83/	
2S6856		3040	27		
2SA01				TR11/	
2SA-2H		3051			
2SA12	102A	3005	50	TR85/250	G0005
2SA12A	102A	3005	1	TR85/250	G0005
2SA12B	102A	3005	50	TR85/250	G0005
2SA12C	102A	3005	1	TR85/250	G0005
2SA12D	102A	3005	50	TR85/250	G0005
2SA12E		3005			
2SA12F		3005			
2SA12G		3005			
2SA12H	102A	3005	50	TR85/250	G0005
2SA12J		3005			
2SA12K		3005			
2SA12L		3005			
2SA12M		3005			
2SA12O		3005			
2SA12R		3005			
2SA12R		3005			
2SA12V	102A	3005	51	TR85/250	G0005
2SA12X		3005			
2SA12Y		3005			
2SA12(D)				TR85/	
2SA12(V)				TR11/	
2SA12V/2320011				TR85/	
2SA13	160	3005	50	TR17/637	G0005
2SA13A		3005	51		
2SA13B		3005	51		
2SA13C		3005	51		
2SA13D		3005	51		
2SA13G		3005	51		
2SA13L		3005	51		
2SA13M		3005	51		
2SA13OR		3005	51		
2SA13R		3005	51		
2SA13X		3005	51		
2SA13Y		3005	51		
2SA14	160	3005	52	TR17/637	G0005
2SA14A		3005	51		
2SA14B		3005	51		
2SA14C		3005	51		
2SA14D		3005	51		
2SA14E		3005	51		
2SA14G		3005	51		
2SA14L		3005	51		
2SA14M		3005	51		
2SA14OR		3005	51		
2SA14R		3005	51		
2SA14X		3005	51		
2SA14Y		3005	51		
2SA15	102A	3006	50	TR85/250	G0005
2SA15BK	102A			TR85/	G0005
2SA15-6		3005	51		
2SA15A		3005	51		
2SA15B		3005	51		
2SA15C		3005	51		
2SA15D		3005	51		
2SA15E		3005	51		
2SA15F		3005	51		
2SA15G		3005	51		
2SA15H	102A	3005	50	TR85/250	G0005
2SA15L		3005	51		
2SA15M		3005	51		
2SA15OR		3005	51		
2SA15R		3005	51		
2SA15U		3005			
2SA15V		3005	52	TR85/254	G0001
2SA15VR	102A			TR05/	G0005
2SA15X		3005	51		
2SA15Y	102A	3005	2	TR85/250	G0005
2SA15(R)				TR17/	
2SA15(V)				TR17/	
2SA15(VR)				TR17/	
2SA15V/R				TR85/254	G0005
2SA15/2320514				TR85/	
2SA16	102A	3005	50	TR85/250	G0005
2SA16A		3005	51		
2SA16B		3005	51		
2SA16C		3005	51		
2SA16D		3005	51		
2SA16E		3005	51		
2SA16F		3005	51		
2SA16L		3005	51		

Original Part Number	ECG	SK	GE	IR/WEP	HEP
2SA16M		3005	51		
2SA16OR		3005	51		
2SA16R		3005	51		
2SA16S		3005			
2SA16X		3005	51		
2SA16Y		3005	51		
2SA17	102A	3005	50	TR85/250	G0006
2SA17A		3005	51		
2SA17B		3005	51		
2SA17C		3005	51		
2SA17D		3005	51		
2SA17E		3005	51		
2SA17F		3005	51		
2SA17G		3005	51		
2SA17H	102A	3005	50	TR85/250	G0006
2SA17L		3005	51		
2SA17OR		3005	51		
2SA17R		3005	51		
2SA17X		3005	51		
2SA17Y		3005	51		
2SA18	102A	3005	50	TR85/250	G0006
2SA18A		3005	51		
2SA18B		3005	51		
2SA18C		3005	51		
2SA18D		3005	51		
2SA18E		3005	51		
2SA18F		3005	51		
2SA18G		3005	51		
2SA18H	102A		50	TR85/250	G0006
2SA18L		3005	51		
2SA18M		3005	51		
2SA18OR		3005	51		
2SA18R		3005	51		
2SA18X		3005	51		
2SA18Y		3005	51		
2SA19	126	3008	51	TR17/635	G0008
2SA19A		3008	53		
2SA19B		3008	53		
2SA19C		3008	53		
2SA19D		3008	53		
2SA19E		3008	53		
2SA19F		3008	53		
2SA19G		3008	53		
2SA19L		3008	53		
2SA19M		3008	53		
2SA19OR		3008	53		
2SA19R		3008	53		
2SA19X		3008	53		
2SA19Y		3008	53		
2SA20	126	3008	51	TR17/635	G0008
2SA20A		3008	53		
2SA20B		3008	53		
2SA20C		3008	53		
2SA20D		3008	53		
2SA20E		3008	53		
2SA20F		3008	53		
2SA20G		3008	53		
2SA20L		3008	53		
2SA20M		3008	53		
2SA20OR		3008	53		
2SA20R		3008	53		
2SA20X		3008	53		
2SA20Y		3008	53		
2SA21	126	3008	51	TR17/635	G0008
2SA21A		3008	53		
2SA21B		3008	53		
2SA21C		3008	53		
2SA21D		3008	53		
2SA21E		3008	53		
2SA21F		3008	53		
2SA21G		3008	53		
2SA21L		3008	53		
2SA21M		3008	53		
2SA21OR		3008	53		
2SA21R		3008	53		
2SA21X		3008	53		
2SA21Y		3008	53		
2SA22			51		
2SA23			51		
2SA24			50	TR12/	G0002
2SA24(1)				TR12/	
2SA24(C)				TR23/	
2SA24(F)				TR23/	
2SA24(Y)				TR23/	
2SA25			50		G0002

Original Part Number	ECG	SK	GE	IR/WEP	HEP
2SA26	100	3005	52	TR05/254	G0005
2SA26A		3005	51		
2SA26B		3005	51		
2SA26C		3005	51		
2SA26D		3005	51		
2SA26E		3005	51		
2SA26F		3005	51		
2SA26G		3005	51		
2SA26L		3005	51		
2SA26M		3005	51		
2SA26OR		3005	51		
2SA26R		3005	51		
2SA26X		3005	51		
2SA26Y		3005	51		
2SA27			51		
2SA28	126	3005	51	TR17/635	G0008
2SA28A		3005	51		
2SA28B		3005	51		
2SA28C		3005	51		
2SA28D		3005	51		
2SA28E		3005	51		
2SA28F		3005	51		
2SA28G		3005	51		
2SA28L		3005	51		
2SA28M		3005	51		
2SA28OR		3005	51		
2SA28R		3005	51		
2SA28X		3005	51		
2SA28Y		3005	51		
2SA29	126	3008	51	TR17/635	G0008
2SA29-1		3008			
2SA29-3		3008			S0014
2SA29(2)				TR12/	
2SA29A		3008	53		
2SA29B		3008	53		
2SA29C		3008	53		
2SA29D		3008	53		
2SA29E		3008	53		
2SA29F		3008	53		
2SA29G		3008	53		
2SA29L		3008	53		
2SA29M		3008	53		
2SA29OR		3008			
2SA29R		3008	53		
2SA29X		3008	53		
2SA29Y		3008	53		
2SA30	126	3005	50	TR17/635	G0003
2SA30 OR		3005			
2SA30A		3005	51		
2SA30B		3005	51		
2SA30C		3005	51		
2SA30D		3005	51		
2SA30E		3005	51		
2SA30F		3005	51		
2SA30G		3005	51		
2SA30L		3005	51		
2SA30M		3005	51		
2SA30X		3005	51		
2SA30Y		3005	51		
2SA31	100	3005	50	TR05/254	G0005
2SA31A		3005	51		
2SA31B		3005	51		
2SA31C		3005	51		
2SA31D		3005	51		
2SA31E		3005	51		
2SA31F		3005	51		
2SA31G		3005	51		
2SA31L		3005	51		
2SA31M		3005	51		
2SA31OR		3005	51		
2SA31X		3005	51		
2SA31Y		3005	51		
2SA32	102A	3005	53	TR85/250	G0005
2SA32 OR		3005			
2SA32A		3005	51		
2SA32B		3005	51		
2SA32C		3005	51		
2SA32D		3005	51		
2SA32E		3005	51		
2SA32F		3005	51		
2SA32G		3005	51		
2SA32L		3005	51		
2SA32M		3005	51		
2SA32X		3005	51		
2SA32Y		3005	51		

Original Part Number	ECG	SK	GE	IR/WEP	HEP
2SA33	102A		1	TR85/250	G0005
2SA35	126	3005	50	TR17/635	G0003
2SA35 OR		3005			
2SA35A		3005	51		
2SA35B		3005	51		
2SA35C		3005	51		
2SA35D		3005	51		
2SA35E		3005	51		
2SA35F		3005	51		
2SA35G		3005	51		
2SA35L		3005	51		
2SA35M		3005	51		
2SA35X		3005	51		
2SA35Y		3005	51		
2SA36	100	3005	50	TR05/254	G0005
2SA36A		3005	51		
2SA36B		3005	51		
2SA36C		3005	51		
2SA36D		3005	51		
2SA36E		3005	51		
2SA36F		3005	51		
2SA36G		3005	51		
2SA36L		3005	51		
2SA36M		3005	51		
2SA36OR		3005	51		
2SA36R		3005	51		
2SA36X		3005	51		
2SA36Y		3005	51		
2SA37	126	3005	51	TR17/635	G0005
2SA37A		3005	51		
2SA37B		3005	51		
2SA37C		3005	51		
2SA37D		3005	51		
2SA37E		3005	51		
2SA37F		3005	51		
2SA37G		3005	51		
2SA37L		3005	51		
2SA37M		3005	51		
2SA37OR		3005	51		
2SA37R		3005	51		
2SA37X		3005	51		
2SA37Y		3005	51		
2SA38	126	3005	51	TR17/635	G0005
2SA38A		3005	51		
2SA38B		3005	51		
2SA38C		3005	51		
2SA38D		3005	51		
2SA38E		3005	51		
2SA38F		3005	51		
2SA38G		3005	51		
2SA38L		3005	51		
2SA38M		3005	51		
2SA38OR		3005	51		
2SA38R		3005	51		
2SA38X		3005	51		
2SA38Y		3005	51		
2SA39	126	3005	51	TR17/635	G0005
2SA39A		3005	51		
2SA39B		3005	51		
2SA39C		3005	51		
2SA39D		3005	51		
2SA39E		3005	51		
2SA39F		3005	51		
2SA39G		3005	51		
2SA39L		3005	51		
2SA39M		3005	51		
2SA39OR		3005	51		
2SA39R		3005	51		
2SA39X		3005	51		
2SA39Y		3005	51		
2SA40	100	3005	1	TR05/254	G0005
2SA40 OR		3005			
2SA40A		3005	51		
2SA40B		3005	51		
2SA40C		3005	51		
2SA40D		3005	51		
2SA40E		3005	51		
2SA40F		3005	51		
2SA40G		3005	51		
2SA40L		3005	51		
2SA40M		3005	51		
2SA40R		3005	51		
2SA40X		3005	51		
2SA40Y		3005	51		
2SA41	160	3005	51	TR17/637	G0005

Original Part Number	ECG	SK	GE	IR/WEP	HEP
2SA41A		3005	51		
2SA41B		3005	51		
2SA41C		3005	51		
2SA41D		3005	51		
2SA41E		3005	51		
2SA41F		3005	51		
2SA41G		3005	51		
2SA41L		3005	51		
2SA41M		3005	51		
2SA41OR		3005	51		
2SA41R		3005	51		
2SA41X		3005	51		
2SA41Y		3005	51		
2SA42	160	3123	51	TR17/637	G0005
2SA43	126	3007	51	TR17/635	G0009
2SA44	100	3005	2	TR05/254	G0005
2SA44A		3005	51		
2SA44B		3005	51		
2SA44C		3005	51		
2SA44D		3005	51		
2SA44E		3005	51		
2SA44F		3005	51		
2SA44G		3005	51		
2SA44L		3005	51		
2SA44M		3005	51		
2SA44R		3005	51		
2SA44X		3005	51		
2SA44Y		3005	51		
2SA45	126	3006	51	TR17/635	G0001
2SA45-1	126	3006	51	TR17/635	G0001
2SA45-2	126	3006	51	TR17/635	G0001
2SA45-3	126	3006	51	TR17/635	G0001
2SA48			51		
2SA49	126	3005	51	TR17/635	G0008
2SA49A		3005	51		
2SA49B		3005	51		
2SA49C		3005	51		
2SA49D		3005	51		
2SA49E		3005	51		
2SA49F		3005	51		
2SA49G		3005	51		
2SA49L		3005	51		
2SA49M		3005	51		
2SA49OR		3005	51		
2SA49R		3005	51		
2SA49X		3005	51		
2SA49Y		3005	51		
2SA50	102A	3005	53	TR85/250	G0005
2SA50A		3005	51		
2SA50B		3005	51		
2SA50C		3005	51		
2SA50D		3005	51		
2SA50E		3005	51		
2SA50F		3005	51		
2SA50G		3005	51		
2SA50L		3005	51		
2SA50M		3005	51		
2SA50R		3005	51		
2SA50X		3005	51		
2SA50Y		3005	51		
2SA51	160	3005	51	TR17/637	G0005
2SA51A		3005	51		
2SA51B		3005	51		
2SA51C		3005	51		
2SA51D		3005	51		
2SA51E		3005	51		
2SA51F		3005	51		
2SA51G		3005	51		
2SA51L		3005	51		
2SA51M		3005	51		
2SA51OR		3005	51		
2SA51R		3005	51		
2SA51X		3005	51		
2SA51Y		3005	51		
2SA52	126	3006	51	TR17/635	G0008
2SA52A		3005	51		
2SA52B		3005	51		
2SA52C		3005	51		
2SA52D		3005	51		
2SA52E		3005	51		
2SA52F		3005	51		
2SA52G		3005	51		
2SA52L		3005	51		
2SA52M		3005	51		
2SA52OR		3005	51		

Original Part Number	ECG	SK	GE	IR/WEP	HEP
2SA52R		3005	51		
2SA52X		3005	51		
2SA52Y		3005	51		
2SA53	126	3005	51	TR17/635	G0008
2SA53A		3005	51		
2SA53B		3005	51		
2SA53C		3005	51		
2SA53D		3005	51		
2SA53E		3005	51		
2SA53F		3005	51		
2SA53G		3005	51		
2SA53L		3005	51		
2SA53M		3005	51		
2SA53OR		3005	51		
2SA53R		3005	51		
2SA53X		3005	51		
2SA53Y		3005	51		
2SA54	160	3005	52	TR17/637	G0003
2SA54A		3005	51		
2SA54B		3005	51		
25A54C		3005	51		
2SA54D		3005	51		
2SA54E		3005	51		
2SA54F		3005	51		
2SA54G		3005	51		
2SA54L		3005	51		
2SA54M		3005	51		
2SA54OR		3005	51		
2SA54R		3005	51		
2SA54X		3005	51		
2SA54Y		3005	51		
2SA55	100	3005	2	TR05/254	G0005
2SA55A		3005	51		
2SA55B		3005	51		
2SA55C		3005	51		
2SA55D		3005	51		
2SA55E		3005	51		
2SA55F		3005	51		
2SA55G		3005	51		
2SA55L		3005	51		
2SA55M		3005	51		
2SA55OR		3005	51		
2SA55R		3005	51		
2SA55X		3005	51		
2SA55Y		3005	51		
2SA56	160		52	TR17/637	G0008
2SA56A			22		S0019
2SA57	126	3008	51	TR17/635	G0008
2SA57A		3008	53		
2SA57B		3008	53		
2SA57C		3008	53		
2SA57D		3008	53		
2SA57E		3008	53		
2SA57F		3008	53		
2SA57G		3008	53		
2SA57L		3008	53		
2SA57M		3008	53		
2SA57OR		3008	53		
2SA57R		3008	53		
2SA57X		3008	53		
2SA57Y		3008	53		
2SA58	126	3008	51	TR17/635	G0008
2SA58A		3006	51		
2SA58B		3006	51		
2SA58C		3006	51		
2SA58D		3006	51		
2SA58E		3006	51		
2SA58F		3006	51		
2SA58G		3006	51		G0008
2SA58H		3006	51		
2SA58J		3006	51		
2SA58K		3006	51		
2SA58L		3006	51		
2SA58M		3006	51		
2SA58OR		3006	51		
2SA58R		3006	51		
2SA58X		3006	51		
2SA58Y		3006	51		
2SA59	126	3007	51	TR17/635	G0008
2SA59A		3007	50		
2SA59B		3007	50		
2SA59C		3007	50		
2SA59D		3007	50		
2SA59E		3007	50		
2SA59F		3007	50		
2SA59G		3007	50		
2SA59L		3007	50		
2SA59M		3007	50		
2SA59OR		3007	50		
2SA59R		3007	50		
2SA59X		3007	50		
2SA59Y		3007	50		
2SA60	126	3006	51	TR17/635	G0008
2SA60A		3006	51		
2SA60B		3006	51		
2SA60C		3006	51		
2SA60D		3006	51		
2SA60E		3006	51		
2SA60F		3006	51		
2SA60G		3006	51		G0008
2SA60H		3006	51		
2SA60K		3006	51		
2SA60L		3006	51		
2SA60M		3006	51		
2SA60OR		3006	51		
2SA60R		3006	51		
2SA60X		3006	51		
2SA60Y		3006	51		
2SA61	160	3005	2	TR17/637	G0003
2SA61A		3005	51		
2SA61B		3005	51		
2SA61C		3005	51		
2SA61D		3005	51		
2SA61E		3005	51		
2SA61G		3005	51		
2SA61K		3005	51		
2SA61L		3005	51		
2SA61OR		3005	51		
2SA61R		3005	51		
2SA61X		3005	51		
2SA61Y		3005	51		
2SA64	102A	3008	51	TR85/250	G0005
2SA64A		3008	53		
2SA64AB				TR28/	
2SA64B		3008	53		S0019
2SA64C		3008	53		
2SA64D		3008	53		
2SA64E		3008	53		
2SA64F		3008	53		
2SA64G		3008	53		
2SA64GN		3008	53		
2SA64H		3008	53		
2SA64J		3008	53		
2SA64K		3008	53		
2SA64L		3008	53		
2SA64M		3008	53		
2SA64OR		3008	53		
2SA64R		3008	53		
2SA64X		3008	53		
2SA64Y		3008	53		
2SA65	102A	3005	2	TR85/250	G0005
2SA65A		3005	51		
2SA65B		3005	51		
2SA65C		3005	51		
2SA65D		3005	51		
2SA65E		3005	51		
2SA65F		3005	51		
2SA65G		3005	51		
2SA65K		3005	51		
2SA65L		3005	51		
2SA65M		3005	51		
2SA65OR		3005	51		
2SA65R		3005	51		
2SA65X		3005	51		
2SA65Y		3005	51		
2SA66	102A	3005	2	TR85/250	G0005
2SA66A		3005	51		
2SA66B		3005	51		
2SA66C		3005	51		
2SA66D		3005	51		
2SA66E		3005	51		
2SA66F		3005	51		
2SA66G		3005	51		
2SA66K		3005	51		
2SA66L		3005	51		
2SA66M		3005	51		
2SA66OR		3005	51		
2SA66R		3005	51		
2SA66X		3005	51		

Original Part Number	ECG	SK	GE	IR/WEP	HEP
2SA66Y		3005	51		
2SA67	126	3006	2	TR17/635	G0005
2SA67A		3006	51		
2SA67B		3006	51		
2SA67C		3006	51		
2SA67D	5	3006	51		
2SA67E		3006	51		
2SA67F		3006	51		
2SA67G		3006	51		
2SA67H		3006	51		
2SA67K		3006	51		
2SA67L		3006	51		
2SA67OR		3006	51		
2SA67R		3006	51		
2SA67X		3006	51		
2SA67Y		3006	51		
2SA69	126	3006	50	TR17/635	G0008
2SA69A		3006	51		
2SA69B		3006	51		
2SA69C		3006	51		
2SA69D		3006	51		
2SA69E		3006	51		
2SA69F		3006	51		
2SA69G		3006	51		
2SA69H		3006	51		
2SA69K		3006	51		
2SA69L		3006	51		
2SA69M		3006	51		
2SA69OR		3006	51		
2SA69X		3006	51		
2SA69Y		3006	51		
2SA70	160	3006	50	TR17/637	G0008
2SA70-OA		3006			
2SA70-OB			50		
2SA70-08			50	TR17/	
2SA70A		3006	51		
2SA70B		3006	51		
2SA70C		3006	51		
2SA70D		3006	51		
2SA70E		3006	51		
2SA70F	160	3006	50	TR17/637	G0008
2SA70G		3006	51		
2SA70H		3006	51		
2SA70K		3006	51		
2SA70L	160	3006	51	TR17/637	G0003
2SA70MA	160	3006	50	TR17/637	G0003
2SA70NB		3006			
2SA70OA			51		
2SA70OB		3006	51		
2SA70OR		3006	51		
2SA70R		3006	51		
2SA70X		3006	51		
2SA70Y		3006	51	TR17/	
2SA71	126	3006	50	TR17/635	G0008
2SA71A	126	3006	50	TR17/635	G0005
2SA71AB	126	3006	50	TR17/635	G0008
2SA71AC	126	3006	50	TR17/635	G0008
2SA71B	126	3006	50	TR17/635	G0001
2SA71BS	126	3006	50	TR17/635	G0001
2SA71C		3006	51		
2SA71D	126	3006	51	TR17/635	G0001
2SA71E		3006	51		
2SA71F		3006	51		
2SA71G		3006	51		
2SA71H		3006	51		
2SA71K		3006	51		
2SA71L		3006	51		
2SA71M		3006	51		
2SA71OR		3006	51		
2SA71R		3006	51		
2SA71X		3006	51		
2SA71Y	126	3006	50	TR17/635	G0001
2SA71TA		3006	51		
2SA72	126	3006	51	TR17/635	G0008
2SA72BLU	126	3006	51	TR17/635	G0008
2SA72BLU-BLU	126	3006	51	TR17/635	G0008
2SA72BRN	126	3006	51	TR17/635	G0008
2SA72ORN	126	3006	51	TR17/635	G0008
2SA72WHT	126	3006	51	TR17/635	G0008
2DA73	160	3007	51	TR17/637	G0008
2SA73A		3007	50		
2SA73B		3007	50		
2SA73C		3007	50		
2SA73D		3007	50		
2SA73E		3007	50		
2SA73F		3007	50		
2SA73G		3007	50		
2SA73H		3007	50		
2SA73K		3007	50		
2SA73L		3007	50		
2SA73M		3007	50		
2SA73OR		3007	50		
2SA73R		3007	50		
2SA73X		3007	50		
2SA73Y		3007	50		
2SA74	160	3006	2	TR17/637	G0009
2SA74A		3006	51		
2SA74B		3006	51		
2SA74C		3006	51		
2SA74D		3006	51		
2SA74E		3006	51		
2SA74F		3006	51		
2SA74G		3006	51		
2SA74H		3006	51		
2SA74K		3006	51		
2SA74L		3006	51		
2SA74M		3006	51		
2SA74OR		3006	51		
2SA74R		3006	51		
2SA74X		3006	51		
2SA74Y		3006	51		
2SA75	126	3007	2	TR17/635	G0009
2SA75A		3005	51		
2SA75B	102A	3005	51	TR85/250	G0005
2SA75C		3005	51		
2SA75D		3005	51		
2SA75E		3005	51		
2SA75F		3005	51		
2SA75G		3005	51		
2SA75H		3005	51		
2SA75K		3005	51		
2SA75L		3005	51		
2SA75OR		3005	51		
2SA75R		3005	51		
2SA75X		3005	51		
2SA75Y		3005	51		
2SA76	126	3006	50	TR17/635	G0008
2SA76A		3006	51		
2SA76B		3006	51		
2SA76C		3006	51		
2SA76D		3006	51		
2SA76E		3006	51		
2SA76F		3006	51		
2SA76G		3006	51		
2SA76H		3006	51		
2SA76K		3006	51		
2SA76L		3006	51		
2SA76OR		3006	51		
2SA76R		3006	51		
2SA76X		3006	51		
2SA76Y		3006	51		
2SA77	126	3006	51	TR17/635	G0008
2SA77A	126	3006	51	TR17/635	
2SA77B	126	3004	50	TR17/635	
2SA77C	126	3006	50	TR17/635	
2SA77D	126	3006	50	TR17/635	
2SA77E		3006	51		
2SA77F		3006	51		
2SA77G		3006	51		
2SA77H		3006	51		
2SA77K		3006	51		
2SA77L		3006	51		
2SA77M		3006	51		
2SA77OR		3006	51		
2SA77R		3006	51		
2SA77X		3006	51		
2SA77Y		3006	51		
2SA78	160	3006	51	TR17/635	G0005
2SA79	160	3123	2	TR17/637	G0005
2SA80	126	3006	50	TR17/635	G0003
2SA80A		3008	53		
2SA80B		3008	53		
2SA80C		3008	53		
2SA80D		3008	53		
2SA80E		3008	53		
2SA80F		3008	53		
2SA80G		3008	53		
2SA80H		3008	53		
2SA80K		3008	53		
2SA80L		3008	53		

Original Part Number	ECG	SK	GE	IR/WEP	HEP
2SA80M		3008	53		
2SA80OR		3008	53		
2SA80R		3008	53		
2SA80X		3008	53		
2SA80Y		3008	53		
2SA81	160	3008	50	TR17/637	G0009
2SA82	126	3008	50	TR17/635	G0009
2SA82A		3008	53		
2SA82B		3008	53		
2SA82C		3008	53		
2SA82D		3008	53		
2SA82E		3008	53		
2SA82F		3008	53		
2SA82G		3008	53		
2SA82H		3008	53		
2SA82K		3008	53		
2SA82M		3008	53		
2SA82OR		3008	53		
2SA82R		3008	53		
2SA82X		3008	53		
2SA82Y		3008	53		
2SA83	126	3007	50	TR17/635	G0009
2SA83A		3007	50		
2SA83B		3007	50		
2SA83C		3007	50		
2SA83D		3007	50		
2SA83E		3007	50		
2SA83F		3007	50		
2SA83G		3007	50		
2SA83H		3007	50		
2SA83K		3007	50		
2SA83L		3007	50		
2SA83M		3007	50		
2SA83OR		3007	50		
2SA83R		3007	50		
2SA83X		3007	50		
2SA83Y		3007	50		
2SA84	160	3007	50	TR17/637	G0009
2SA84A		3007	50		
2SA84B		3007	50		
2SA84C		3007	50		
2SA84D		3007	50		
2SA84E		3007	50		
2SA84F		3007	50		
2SA84G		3007	50		
2SA84H		3007	50		
2SA84K		3007	50		
2SA84L		3007	50		
2SA84M		3007	50		
2SA84OR		3007	50		
2SA84R		3007	50		
2SA84X		3007	50		
2SA84Y		3007	50		
2SA85	126		50	TR17/635	G0009
2SA85L		3008	53		
2SA86	160	3123	51	TR17/637	G0001
2SA87	126	3006	52	TR17/635	G0009
2SA87A		3006	51		
2SA87B		3006	51		
2SA87C		3006	51		
2SA87D		3006	51		
2SA87E		3006	51		
2SA87F		3006	51		
2SA87G		3006	51		
2SA87H		3006	51		
2SA87K		3006	51		
2SA87L		3006	51		
2SA87M		3006	51		
2SA87OR		3006	51		
2SA87R		3006	51		
2SA87X		3006	51		
2SA87Y		3006	51		
2SA88	160		52	TR17/637	G0009
2SA89	160		52	TR17/637	G0009
2SA90	126	3006	52	TR17/635	G0008
2SA92	126	3007	51	TR17/635	G0008
2SA92A		3007	50		
2SA92B		3007	50		
2SA92C		3007	50		
2SA92D		3007	50		
2SA92E		3007	50		
2SA92F		3007	50		
2SA92G		3007	50		G0008
2SA92H		3007	50		
2SA92K		3007	50		
2SA92L		3007	50		
2SA92M		3007	50		
2SA92OR		3007	50		
2SA92R		3007	50		
2SA92X		3007	50		
2SA92Y		3007	50		
2SA93	126	3007	51	TR17/635	G0008
2SA93A		3007	50		
2SA93B		3007	50		
2SA93C		3007	50		
2SA93D		3007	50		
2SA93E		3007	50		
2SA93F		3007	50		
2SA93G		3007	50		G0008
2SA93H		3007	50		
2SA93K		3007	50		
2SA93L		3007	50		
2SA93M		3007	50		
2SA93OR		3007	50		
2SA93R		3007	50		
2SA93X		3007	50		
2SA93Y		3007	50		
2SA94	126	3008	51	TR17/635	G0008
2SA94A		3008	53		
2SA94C		3008	53		
2SA94D		3008	53		
2SA94E		3008	53		
2SA94F		3008	53		
2SA94G		3008	53		
2SA94H		3008	53		
2SA94K		3008	53		
2SA94L		3008	53		
2SA94M		3008	53		
2SA94OR		3008	53		
2SA94R		3008	53		
2SA94X		3008	53		
2SA94Y		3008	53		
2SA95 0		3118			
2SA95A		3118			
2SA95A(0)				TR30/	
2SA100	126	3005	51	TR17/635	G0008
2SA100A	102A	3008	53	TR85/250	G0008
2SA100B	126	3004	52	TR17/635	G0008
2SA100C	126	3008	51	TR17/635	G0008
2SA100D		3005	51		
2SA100E		3005	51		
2SA100F		3005	51		
2SA100G		3005	51		
2SA100H		3005	51		
2SA100J		3005	51		
2SA100K		3005	51		
2SA100M		3005	51		
2SA100OR		3005	51		
2SA100R		3005	51		
2SA100X		3005	51		
2SA100Y		3005	51		
2SA101	126	3007	51	TR17/635	G0009
2SA101A	126	3007	50	TR17/635	G0009
2SA101AA	126	3008	50	TR17/635	G0009
2SA101AY	126	3008	1	TR17/	G0008
2SA101AT				/635	
2SA101B	126	3006	50	TR17/635	
2SA101BA	126	3008	50	TR17/635	G0001
2SA101BB	126	3008	50	TR17/635	G0001
2SA101BC	126	3007	50	TR17/635	G0005
2SA101BX	126	3007	51	TR17/635	G0005
2SA101BY		3008			
2SA101C	126	3006	51	TR17/635	G0001
2SA101CA	126	3007	50	TR17/635	G0005
2SA101CV	126	3006	51	TR17/635	G0001
2SA101CX	126	3006	51	TR17/635	G0005
2SA101D		3007	50		
2SA101E	126	3008	50	TR17/635	G0001
2SA101F		3007	50		
2SA101G		3007	50		
2SA101H		3007	50		
2SA101K		3007	50		
2SA101L		3007	50		
2SA101M		3007	50		
2SA101OR		3007	50		
2SA101QA	126			TR17/	
2SA101R		3007	50		
2SA101V	126				G0009
2SA101X	126	3006	51	TR17/635	G0009
2SA101-X		3008			

Original Part Number	ECG	SK	GE	IR/WEP	HEP
2SA101XBX		3006	51		
2SA101Y	126	3008	51	TR17/635	G0008
2SA101YA		3004	53	TR85/635	G0005
2SA101Z	100	3008	51	TR05/254	G0005
2SA101(4)				TR12/	
2SA101(AY)				TR12/	
2SA101(X)				TR17/	
2SA102	126	3007	51	TR17/635	G0009
2SA102A	126	3006	51	TR17/635	G0005
2SA102AA	126	3008	51	TR17/635	G0009
2SA102AB	126	3007	50	TR17/635	G0006
2SA102B	126	3006	51	TR17/	G0009
2SA102BA	126	3006	50	TR17/635	G0009
2SA102BA-2		3006	51		
2SA102BN	126	3008	53	TR17/635	G0005
2SA102C		3006	51		
2SA102CA		3006	50	TR17/635	G0009
2SA102CA-1		3006	51		
2SA102D		3006	51		
2SA102E		3006	51		
2SA102F		3006	51		
2SA102G		3006	51		
2SA102H		3006	1		
2SA102K		3006	51		
2SA102L		3006	51		
2SA102M		3006	51		
2SA102OR		3006	51		
2SA102TV	126			TR17/635	G0005
2SA102TV-2		3008	53		
2SA102X		3006	51		
2SA102Y		3006	51		
2SA102(AA)				TR85/	
2SA102(B,A)				TR85/	
2SA102(BA)				TR12/	
2SA102(BA)/21MO				TR85/	
2SA102(BN)				TR85/	
2SA102(CA)				TR12/	
2SA102BA/21M006				TR85/	
2SA103	126	3006	51	TR17/635	G0009
2SA103A	126	3007	50	TR17/635	G0009
2SA103B	126	3006	50	TR17/635	G0005
2SA103C	126	3007	50	TR17/635	G0009
2SA103CA	126	3006	50	TR17/242	G0009
2SA103CAK	126	3006	50	TR12/	G0009
2SA103CG	126	3006	51	TR17/635	G0009
2SA103D		3006	51		
2SA103DA	126	3006	50	TR17/635	G0005
2SA103E		3006	51		
2SA103F		3006	51		
2SA103G		3006	51		
2SA103GA			51		
2SA103K	126	3006	51		
2SA103L		3006	51		
2SA103M		3006	51		
2SA103OR		3006	51		
2SA103R		3006	51		
2SA103X		3007	50		
2SA103Y		3006	51		
2SA103(CA)				TR12/	
2SA103(CAK)				TR12/	
2SA104	126	3008	1	TR17/635	G0009
2SA104A		3007	50		
2SA104B		3007	50		
2SA104C		3007	50		
2SA104D	126	3006	51	TR17/635	G0005
2SA104E		3007	50		
2SA104F		3007	50		
2SA104G		3007	50		
2SA104H		3007	50		
2SA104K		3007	50		
2SA104L		3007	50		
2SA104M		3007	50		
2SA104OR		3007	50		
2SA104P	126	3006	50	TR17/635	G0005
2SA104R		3007	50		
2SA104X		3007	50		
2SA104Y		3007	50		
2SA105	160	3008	51	TR17/637	G0008
2SA105A		3008	53		
2SA105B		3008	53		
2SA105C		3008	53		
2SA105D		3008	53		
2SA105E		3008	53		
2SA105G		3008	53		
2SA105H		3008	53		
2SA105K		3008	53		
2SA105L		3008	53		
2SA105M		3008	53		
2SA105OR		3008	53		
2SA105R		3008	53		
2SA105X		3008	53		
2SA105Y		3008	53		
2SA106	160	3007	51	TR17/637	G0008
2SA106A		3008	53		
2SA106B		3007	50		
2SA106C		3007	50		
2SA106D		3007	50		
2SA106E		3007	50		
2SA106F		3007	50		
2SA106G		3007	50		
2SA106H		3007	50		
2SA106K		3007	50		
2SA106L		3007	50		
2SA106M		3007	50		
2SA106OR		3007	50		
2SA106R		3007	50		
2SA106X		3007	50		
2SA106Y		3007	50		
2SA107	160	3007	51	TR17/637	G0008
2SA107A		3007	50		
2SA107B		3007	50		
2SA107C		3007	50		
2SA107D		3007	50		
2SA107E		3007	50		
2SA107F		3007	50		
2SA107G		3007	50		
2SA107H		3007	50		
2SA107K		3007	50		
2SA107L		3007	50		
2SA107M		3007	50		
2SA107OR		3007	50		
2SA107R		3007	50		
2SA107X		3007	50		
2SA107Y		3007	50		
2SA108	126	3007	50	TR17/635	G0008
2SA108A		3007	50		
2SA108B		3007	50		
2SA108C		3007	50		
2SA108D		3007	50		
2SA108E		3007	50		
2SA108F		3007	50		
2SA108G		3007	50		
2SA108H		3007	50		
2SA108K		3007	50		
2SA108L		3007	50		
2SA108M		3007	50		
2SA108OR		3007	50		
2SA108R		3007	50		
2SA108X		3007	50		
2SA108Y		3007	50		
2SA109	126	3007	50	TR17/635	G0003
2SA109A		3007	50		
2SA109B		3007	50		
2SA109C		3007	50		
2SA109D		3007	50		
2SA109E		3007	50		
2SA109F		3007	50		
2SA109G		3007	50		
2SA109K		3007	50		
2SA109L		3007	50		
2SA109M		3009	50		
2SA109OR		3007	50		
2SA109R		3007	50		
2SA109X		3007	50		
2SA109Y		3007	50		
2SA109(BA)				TR85/	
2SA110	126	3007	50	TR17/635	G0003
2SA110A		3007	50		
2SA110B		3007	50		
2SA110C		3007	50		
2SA110D		3007	50		
2SA110E		3007	50		
2SA110F		3007	50		
2SA110G		3007	50		
2SA110K		3007	50		
2SA110L		3007	50		
2SA110M		3007	50		
2SA110OR		3007	50		
2SA110R		3007	50		
2SA110X		3007	50		

Original Part Number	ECG	SK	GE	IR/WEP	HEP
2SA110Y		3007	50		
2SA111	126	3007	50	TR17/635	G0003
2SA111A		3007	50		
2SA111B		3007	50		
2SA111C		3007	50		
2SA111D		3007	50		
2SA111E		3007	50		
2SA111F		3007	50		
2SA111G		3007	50		
2SA111K		3007	50		
2SA111L		3007	50		
2SA111M		3007	50		
2SA111OR		3007	50		
2SA111R		3007	50		
2SA111X		3007	50		
2SA111Y		3007	50		
2SA112	126	3007	50	TR17/635	G0003
2SA112A		3007	50		
2SA112B		3007	50		
2SA112C		3007	50		
2SA112D		3007	50		
2SA112E		3007	50		
2SA112F		3007	50		
2SA112G		3007	50		
2SA112GN		2007	50		
2SA112H		3007	50		
2SA112K		3007	50		
2SA112L		3007	50		
2SA112M		3007	50		
2SA112OR		3007	50		
2SA112R		3007	50		
2SA112X		3007	50		
2SA112Y		3007	50		
2SA113	126	3007	51	TR17/635	G0005
2SA113A		3007	50		
2SA113B		3007	50		
2SA113C		3007	50		
2SA113D		3007	50		
2SA113E		3007	50		
2SA113F		3007	50		
2SA113G		3007	50		
2SA113GN		3007	50		
2SA113H		3007	50		
2SA113J		3007	50		
2SA113L		3007	50		
2SA113M		3007	50		
2SA113R		3007	50		
2SA113X		3007	50		
2SA113Y		3007	50		
2SA114	102A	3007	51	TR85/250	G0005
2SA114A		3007	50		
2SA114B		3007	50		
2SA114C		3007	50		
2SA114D		3007	50		
2SA114E		3007	50		
2SA114F		3007	50		
2SA114G		3007	50		
2SA114H		3007	50		
2SA114K		3007	50		
2SA114L		3007	50		
2SA114M		3007	50		
2SA114OR		3007	50		
2SA114R		3007	50		
2SA114X		3007	50		
2SA114Y		3007	50		
2SA115	102A	3007	51	TR85/250	G0005
2SA115A		3007	50		
2SA115B		3007	50		
2SA115C		3007	50		
2SA115D		3007	50		
2SA115E		3007	50		
2SA115F		3007	50		
2SA115G		3007	50		
2SA115GN		3007	50		
2SA115H		3007	50		
2SA115J		3007	50		
2SA115K		3007	50		
2SA115L		3007	50		
2SA115N		3007	50		
2SA115OR		3007	50		
2SA115R		3007	50		
2SA115X		3007	50		
2SA115Y		3007	50		
2SA116	102A	3006	1	TR85/250	G0003
2SA116A		3006	51		
2SA116B		3006	51		
2SA116C		3006	51		
2SA116D		3006	51		
2SA116E		3006	51		
2SA116F		3006	51		
2SA116G		3006	51		
2SA116GN		3006	51		
2SA116H		3006	51		
2SA116J		3006	51		
2SA116K		3006	51		
2SA116L		3006	51		
2SA116M		3006	51		
2SA116OR		3006	51		
2SA116R		3006	51		
2SA116X		3006	51		
2SA116Y		3006	51		
2SA117	160	3006	1	TR17/637	G0003
2SA117A		3006	51		
2SA117B		3006	51		
2SA117C		3006	51		
2SA117D		3006	51		
2SA117E		3006	51		
2SA117F		3006	51		
2SA117G		3006	51		
2SA117GN		3006	51		
2SA117H		3006	51		
2SA117J		3006	51		
2SA117K		3006	51		
2SA117L		3006	51		
2SA117M		3006	51		
2SA117OR		3006	51		
2SA117R		3006	51		
2SA117X		3006	51		
2SA117Y		3006	51		
2SA118	160	3006	1	TR17/637	G0003
2SA118A		3006	51		
2SA118B		3006	51		
2SA118C		3006	51		
2SA118D		3006	51		
2SA118E		3006	51		
2SA118F		3006	51		
2SA118G		3006	51		
2SA118GN		3006	51		
2SA118H		3006	51		
2SA118J		3006	51		
2SA118K		3006	51		
2SA118L		3006	51		
2SA118M		3006	51		
2SA118OR		3006	51		
2SA118R		3006	51		
2SA118X		3006	51		
2SA118Y		3006	51		
2SA120				TR12/	
2SA121	126	3006	51	TR17/635	G0008
2SA121A		3006	51		
2SA121B		3006	51		
2SA121C		3006	51		
2SA121D		3006	51		
2SA121E		3006	51		
2SA121F		3006	51		
2SA121G		3006	51		
2SA121GN		3006	51		
2SA121H		3006	51		
2SA121J		3006	51		
2SA121K		3006	51		
2SA121L		3006	51		
2SA121M		3006	51		
2SA121OR		3006	51		
2SA121R		3006	51		
2SA121X		3006	51		
2SA121Y		3006	51		
2SA122	126	3006	51	TR17/635	G0008
2SA122A		3006	51		
2SA122B		3006	51		
2SA122C		3006	51		
2SA122D		3006	51		
2SA122E		3006	51		
2SA122F		3006	51		
2SA122G		3006	51		
2SA122GN		3006	51		
2SA122H		3006	51		
2SA122K		3006	51		
2SA122L		3006	51		
2SA122M		3006	51		
2SA122OR		3006	51		

Original Part Number	ECG	SK	GE	IR/WEP	HEP
2SA122R		3006	51		
2SA122X		3006	51		
2SA122Y		3006	51		
2SA123	126	3006	51	TR17/635	G0008
2SA123A		3006	51		
2SA123B		3006	51		
2SA123C		3006	51		
2SA123D		3006	51		
2SA123E		3006	51		
2SA123F		3006	51		
2SA123G		3006	51		
2SA123GN		3006	51		
2SA123H		3006	51		
2SA123J		3006	51		
2SA123K		3006	51		
2SA123L		3006	51		
2SA123M		3006	51		
2SA123OR		3006	51		
2SA123R		3006	51		
2SA123X		3006	51		
2SA123Y		3006	51		
2SA124	126	3006	51	TR17/635	G0008
2SA124A		3006	51		
2SA124B		3006	51		
2SA124C		3006	51		
2SA124D		3006	51		
2SA124E		3006	51		
2SA124F		3006	51		
2SA124G		3006	51		
2SA124GN		3006	51		
2SA124H		3006	51		
2SA124J		3006	51		
2SA124K		3006	51		
2SA124L		3006	51		
2SA124M		3006	51		
2SA124OR		3006	51		
2SA124R		3006	51		
2SA124X		3006	51		
2SA124Y		3006	51		
2SA125	126	3006	50	TR17/635	G0008
2SA125A		3006	51		
2SA125B		3006	50		
2SA125C		3006	50		
2SA125D		3006	50		
2SA125E		3006	50		
2SA125F		3006	50		
2SA125F		3006	50		
2SA125G		3006	50		
2SA125GN		3006	50		
2SA125H		3006	50		
2SA125J		3006	50		
2SA125K		3006	50		
2SA125L		3006	50		
2SA125M		3006	50		
2SA125OR		3006	50		
2SA125R		3006	50		
2SA125X		3006	50		
2SA125Y		3006	50		
2SA126	160		52	TR17/637	G0008
2SA127	160		51	TR17/637	
2SA128	158	3008	51	TR84/630	G0005
2SA128A		3008	53		
2SA128B		3008	53		
2SA128C		3008	53		
2SA128D		3008	53		
2SA128E		3008	53		
2SA128F		3008	53		
2SA128G		3008	53		
2SA128GN		3008	53		
2SA128H		3008	53		
2SA128J		3008	53		
2SA128K		3008	53		
2SA128L		3008	53		
2SA128M		3008	53		
2SA128OR		3008	53		
2SA128R		3008	53		
2SA128X		3008	53		
2SA128Y		3008	53		
2SA129	158	3008	51	TR84/630	G0005
2SA129A		3008	53		
2SA129B		3008	53		
2SA129C		3008	53		
2SA129D		3008	53		
2SA129E		3008	53		
2SA129F		3008	53		
2SA129G		3008	53		
2SA129GN		3008	53		
2SA129H		3008	53		
2SA129J		3008	53		
2SA129K		3008	53		
2SA129L		3008	53		
2SA129M		3008	53		
2SA129OR		3008	53		
2SA129R		3008	53		
2SA129X		3008	53		
2SA129Y		3008	53		
2SA130	160	3005	52	TR17/637	G0003
2SA130A		3005	51		
2SA130B		3005	51		
2SA130C		3005	51		
2SA130D		3005	51		
2SA130E		3005	51		
2SA130F		3005	51		
2SA130G		3005	51		
2SA130GN		3005	51		
2SA130H		3005	51		
2SA130J		3005	51		
2SA130K		3005	51		
2SA130L		3005	51		
2SA130M		3005	51		
2SA130R		3005	51		
2SA130X		3005	51		
2SA130Y		3005	51		
2SA131	126	3005	52	TR17/635	G0008
2SA131A		3005	51		
2SA131B		3005	51		
2SA131C		3005	51		
2SA131D		3005	51		
2SA131E		3005	51		
2SA131F		3005	51		
2SA131G		3005	51		
2SA131GN		3005	51		
2SA131H		3005	51		
2SA131J		3005	51		
2SA131K		3005	51		
2SA131L		3005	51		
2SA131M		3005	51		
2SA131OR		3005	51		
2SA131R		3005	51		
2SA131X		3005	51		
2SA131Y		3005	51		
2SA132	126	3005	52	TR17/635	G0003
2SA132A		3005	51		
2SA132B		3005	51		
2SA132C		3005	51		
2SA132D		3005	51		
2SA132E		3005	51		
2SA132F		3005	51		
2SA132G		3005	51		
2SA132H		3005	51		
2SA132J		3005	51		
2SA132K		3005	51		
2SA132L		3005	51		
2SA132M		3005	51		
2SA132OR		3005	51		
2SA132X		3005	51		
2SA132Y		3005	51		
2SA133	126	3006	51	TR17/635	G0003
2SA133A		3006	51		
2SA133B		3006	51		
2SA133C		3006	51		
2SA133D		3006	51		
2SA133E		3006	51		
2SA133F		3006	51		
2SA133G		3006	51		
2SA133GN		3006	51		
2SA133H		3006	51		
2SA133J		3006	51		
2SA133K		3006	51		
2SA133L		3006	51		
2SA133M		3006	51		
2SA133OR		3006	51		
2SA133R		3006	51		
2SA133X		3006	51		
2SA133Y		3006	51		
2SA134	126	3006	52	TR17/635	G0003
2SA134A		3006	51		
2SA134B		3006	51		
2SA134C		3006	51		
2SA134D		3006	51		
2SA134E		3006	51		

Original Part Number	ECG	SK	GE	IR/WEP	HEP
2SA134F		3006	51		
2SA134G		3006	51		
2SA134H		3006	51		
2SA134J		3006	51		
2SA134K		3006	51		
2SA134L		3006	51		
2SA134OR		3006	51		
2SA134R		3006	51		
2SA134X		3006	51		
2SA134Y		3006	51		
2SA135	160	3005	52	TR17/637	G0003
2SA135A		3005	51		
2SA135B		3005	51		
2SA135C		3005	51		
2SA135D		3005	51		
2SA135E		3005	51		
2SA135F		3005	51		
2SA135G		3005	51		
2SA135GN		3005	51		
2SA135H		3005	51		
2SA135J		3005	51		
2SA135K		3005	51		
2SA135L		3005	51		
2SA135M		3005	51		
2SA135OR		3005	51		
2SA135R		3005	51		
2SA135X		3005	51		
2SA135Y		3005	51		
2SA136	126	3006	50	TR17/635	G0008
2SA136A		3006	51		
2SA136B		3006	51		
2SA136C		3006	51		
2SA136D		3006	51		
2SA136E		3006	51		
2SA136F		3006	51		
2SA136G		3006	51		
2SA136GN		3006	51		
2SA136H		3006	51		
2SA136J		3006	51		
2SA136K		3006	51		
2SA136L		3006	51		
2SA136M		3006	51		
2SA136OR		3006	51		
2SA136R		3006	51		
2SA136X		3006	51		
2SA136Y		3006	51		
2SA137	126	3005	50	TR17/635	G0008
2SA137A		3005	51		
2SA137B		3005	51		
2SA137C		3005	51		
2SA137D		3005	51		
2SA137E		3005	51		
2SA137F		3005	51		
2SA137G		3005	51		
2SA137GN		3005	51		
2SA137H		3005	51		
2SA137J		3005	51		
2SA137K		3005	51		
2SA137L		3005	51		
2SA137N		3005	51		
2SA137OR		3005	51		
2SA137R		3005	51		
2SA137X		3005	51		
2SA137Y		3005	51		
2SA138	102A		51	TR85/250	G0005
2SA139	126	3006	51	TR17/635	G0005
2SA139A		3006	51		
2SA139B		3006	51		
2SA139C		3006	51		
2SA139D		3006	51		
2SA139E		3006	51		
2SA139F		3005	51		
2SA139G		3005	51		
2SA139GN		3005	51		
2SA139J		3005	51		
2SA139K		3005	51		
2SA139L		3005	51		
2SA139M		3005	51		
2SA139OR		3005	51		
2SA139R		3005	51		
2SA139X		3005	51		
2SA139Y		3005	51		
2SA140				/635	
2SA141	126	3006	52	TR17/635	G0005
2SA141A		3006	51		G0005
2SA141B	126	3006	50	TR17/635	G0005
2SA141C	126	3006	50	TR17/635	G0005
2SA141D		3006	51		G0001
2SA141E		3006	51		
2SA141F		3006	51		
2SA141G		3006	51		
2SA141GN		3006	51		
2SA141H		3006	51		
2SA141K		3006	51		
2SA141L		3006	51		
2SA141M		3006	51		
2SA141OR		3006	51		
2SA141R		3006	51		
2SA141X		3006	51		
2SA141Y		3006	51		
2SA142	126	3008	52	TR17/635	G0005
2SA142A	126	3006	51	TR17/635	G0005
2SA142B	126	3008	50	TR17/635	G0005
2SA142C	126	3008	50	TR17/635	G0005
2SA142D		3008	53		
2SA142E		3008	53		
2SA142F		3008	53		
2SA142G		3008	53		
2SA142GN		3008	53		
2SA142H		3008	53		
2SA142J		3008	53		
2SA142K		3008	53		
2SA142L		3008	53		
2SA142M		3008	53		
2SA142OR		3008	53		
2SA142R		3008	53		
2SA142X		3008	53		
2SA142Y		3008	53		
2SA143	126	3006	52	TR17/635	G0005
2SA143A		3006	51		
2SA143B		3006	51		
2SA143C		3006	51		
2SA143D		3006	51		
2SA143E		3006	51		
2SA143F		3006	51		
2SA143G		3006	51		
2SA143GN		3006	51		
2SA143H		3006	51		
2SA143J		3006	51		
2SA143K		3006	51		
2SA143L		3006	51		
2SA143M		3006	51		
2SA143OR		3006	51		
2SA143R		3006	51		
2SA143X		3006	51		
2SA143Y		3006	51		
2SA144	126	3005	50	TR17/635	G0009
2SA144A		3005	51		
2SA144B		3005	51		
2SA144C	126	3005	2	TR17/635	G0009
2SA144D		3005	51		
2SA144E		3005	51		
2SA144F		3005	51		
2SA144G		3005	51		
2SA144GN		3005	51		
2SA144H		3005	51		
2SA144J		3005	51		
2SA144K		3005	51		
2SA144L		3005	51		
2SA144M		3005	51		
2SA144OR		3005	51		
2SA144R		3005	51		
2SA144X		3005	51		
2SA144Y		3005	51		
2SA145	126	3005	50	TR17/635	G0009
2SA145A		3005	52	TR17/635	G0005
2SA145B		3005	51		
2SA145C		3005	2	TR17/635	G0009
2SA145D		3005	51		
2SA145E		3005	51		
2SA145G		3005	51		
2SA145GN		3005	51		
2SA145J		3005	51		
2SA145M		3005	51		
2SA145OR		3005	51		
2SA145R		3005	51		
2SA145X		3005	51		
2SA146	100	3005	50	TR05/254	G0008
2SA146A		3005	51		
2SA146B		3005	51		

Original Part Number	ECG	SK	GE	IR/WEP	HEP
2SA145C		3005	51		
2SA146D		3005	51		
2SA146E		3005	51		
2SA146F		3005	51		
2SA146G		3005	51		
2SA146GN		3005	51		
2SA146H		3005	51		
2SA146J		3005	51		
2SA146K		3005	51		
2SA146L		3005	51		
2SA146M		3005	51		
2SA146OR		3005	51		
2SA146R		3005	51		
2SA146X		3005	51		
2SA146Y		3005	51		
2SA147	100	3005	50	TR05/254	G0008
2SA147A		3005	51		
2SA147B		3005	51		
2SA147C		3005	51		
2SA147D		3005	51		
2SA147E		3005	51		
2SA147F		3005	51		
2SA147G		3005	51		
2SA147H		3005	51		
2SA147J		3005	51		
2SA147K		3005	51		
2SA147L		3005	51		
2SA147M		3005	51		
2SA147OR		3005	51		
2SA147R		3005	51		
2SA147X		3005	51		
2SA147Y		3005	51		
2SA148	100	3005	50	TR05/254	G0008
2SA148A		3005	51		
2SA148B		3005	51		
2SA148C		3005	51		
2SA148D		3005	51		
2SA148E		3005	51		
2SA148F		3005	51		
2SA148G		3005	51		
2SA148GN		3005	51		
2SA148H		3005	51		
2SA148J		3005	51		
2SA148K		3005	51		
2SA148L		3005	51		
2SA148M		3005	51		
2SA148OR		3005	51		
2SA148R		3005	51		
2SA148X		3005	51		
2SA148Y		3005	51		
2SA149	100	3005	50	TR05/254	G0008
2SA149A		3005	51		
2SA149B		3005	51		
2SA149C		3005	51		
2SA149D		3005	51		
2SA149E		3005	51		
2SA149F		3005	51		
2SA149G		3005	51		
2SA149GN		3005	51		
2SA149H		3005	51		
2SA149J		3005	51		
2SA149K		3005	51		
2SA149L		3005	51		
2SA149M		3005	51		
2SA149OR		3005	51		
2SA149R		3005	51		
2SA149X		3005	51		
2SA149Y		3005	51		
2SA150	126	3008	50	TR17/635	G0008
2SA150A		3008	53		
2SA150B		3008	53		
2SA150C		3008	53		
2SA150D		3008	53		
2SA150E		3008	53		
2SA150F		3008	53		
2SA150G		3008	53		
2SA150GN		3008	53		
2SA150H		3008	53		
2SA150J		3008	53		
2SA150K		3008	53		
2SA150L		3008	53		
2SA150M		3008	53		
2SA150OR		3008	53		
2SA150R		3008	53		
2SA150X		3008	53		
2SA150Y		3008	53		
2SA151	160	3005	52	TR17/637	G0005
2SA151A		3005	51		
2SA151B		3005	51		
2SA151C		3005	51		
2SA151D		3005	51		
2SA151E		3005	51		
2SA151F		3005	51		
2SA151G		3005	51		
2SA151GN		3005	51		
2SA151H		3005	51		
2SA151J		3005	51		
2SA151K		3005	51		
2SA151L		3005	51		
2SA151M		3005	51		
2SA151OR		3005	51		
2SA151R		3005	51		
2SA151X		3005	51		
2SA151Y		3005	51		
2SA152	160	3005	52	TR17/637	G0005
2SA152A		3005	51		
2SA152B		3005	51		
2SA152C		3005	51		
2SA152D		3005	51		
2SA152E		3005	51		
2SA152F		3005	51		
2SA152G		3005	51		
2SA152GN		3005	51		
2SA152H		3005	51		
2SA152J		3005	51		
2SA152K		3005	51		
2SA152L		3005	51		
2SA152M		3005	51		
2SA152OR		3005	51		
2SA152R		3005	51		
2SA152X		3005	51		
2SA152Y		3005	51		
2SA153	126	3006	51	TR17/635	G0008
2SA153A		3006	51		
2SA153B		3006	51		
2SA153C		3006	51		
2SA153D		3006	51		
2SA153E		3006	61		
2SA153F		3006	51		
2SA153G		3006	51		
2SA153GN		3006	51		
2SA153H		3006	51		
2SA153J		3006	51		
2SA153K		3006	51		
2SA153L		3006	51		
2SA153M		3006	51		
2SA153OR		3006	51		
2SA153R		3006	51		
2SA153X		3006	51		
2SA153Y		3006	51		
2SA154	126	3006	51	TR17/635	G0008
2SA154A		3006	51		
2SA154B		3006	51		
2SA154C		3006	51		
2SA154D		3006	51		
2SA154E		3006	51		
2SA154F		3006	51		
2SA154G		3006	51		
2SA154GN		3006	51		
2SA154H		3006	51		
2SA154J		3006	51		
2SA154K		3006	51		
2SA154L		3006	51		
2SA154M		3006	51		
2SA154OR		3006	51		
2SA154R		3006	51		
2SA154X		3006	51		
2SA154Y		3006	51		
2SA155	126	3006	51	TR17/635	G0008
2SA155A		3006	51		
2SA155B		3006	51		
2SA155C		3006	51		
2SA155D		3006	51		
2SA155E		3006	51		
2SA155F		3006	51		
2SA155G		3006	51		
2SA155GN		3006	51		
2SA155H		3006	51		
2SA155J		3006	51		
2SA155K		3006	51		

Original Part Number	ECG	SK	GE	IR/WEP	HEP
2SA155L		3006	51		
2SA155M		3006	51		
2SA155OR		3006	51		
2SA155R		3006	51		
2SA155X		3006	51		
2SA155Y		3006	51		
2SA156	126	3006	51	TR17/635	G0008
2SA156A		3006	51		
2SA156B		3006	51		
2SA156C		3006	51		
2SA156D		3006	51		
2SA156E		3006	51		
2SA156F		3006	51		
2SA156G		3006	51		
2SA156GN		3006	51		
2SA156H		3006	51		
2SA156J		3006	51		
2SA156K		3006	51		
2SA156L		3006	51		
2SA156M		3006	51		
2SA156OR		3006	51		
2SA156R		3006	51		
2SA156X		3006	51		
2SA156Y		3006	51		
2SA157	126	3006	51	TR17/635	G0008
2SA157A		3006	51		
2SA157B		3006	51		
2SA157C		3006	51		
2SA157D		3006	51		
2SA157E		3006	51		
2SA157F		3006	51		
2SA157G		3006	51		
2SA157GN		3006	51		
2SA157H		3006	51		
2SA157J		3006	51		
2SA157K		3006	51		
2SA157L		3006	51		
2SA157M		3006	51		
2SA157OR		3006	51		
2SA157R		3006	51		
2SA157X		3006	51		
2SA157Y		3006	51		
2SA159	126	3006	51	TR17/635	G0008
2SA159A		3006	51		
2SA159B		3006	51		
2SA159C		3006	51		
2SA159D		3006	51		
2SA159E		3006	51		
2SA159F		3006	51		
2SA159G		3006	51		
2SA159GN		3006	51		
2SA159H		3006	51		
2SA159J		3006	51		
2SA159K		3006	51		
2SA159L		3006	51		
2SA159M		3006	51		
2SA159OR		3006	51		
2SA159R		3006	51		
2SA159X		3006	51		
2SA159Y		3006	51		
2SA160	100	3005	51	TR05/254	G0008
2SA160A		3005	51		
2SA160B		3005	51		
2SA160C		3005	51		
2SA160D		3005	51		
2SA160E		3005	51		
2SA160F		3005	51		
2SA160G		3005	51		
2SA160GN		3005	51		
2SA160H		3005	51		
2SA160J		3005	51		
2SA160K		3005	51		
2SA160L		3005	51		
2SA160M		3005	51		
2SA160OR		3005	51		
2SA160R		3005	51		
2SA160X		3005	51		
2SA160Y		3005	51		
2SA161	126	3006	51	TR17/635	G0008
2SA161A		3006	51		
2SA161B		3006	51		
2SA161C		3006	51		
2SA161D		3006	51		
2SA161E		3006	51		
2SA161F		3006	51		
2SA161G		3006	51		
2SA161GN		3006	51		
2SA161H		3006	51		
2SA161J		3006	51		
2SA161K		3006	51		
2SA161L		3006	51		
2SA161M		3006	51		
2SA161OR		3006	51		
2SA161R		3006	51		
2SA161X		3006	51		
2SA161Y		3006	51		
2SA162	126	3006	51	TR17/635	G0008
2SA162A		3006	51		
2SA162B		3006	51		
2SA162C		3006	51		
2SA162D		3006	51		
2SA162E		3006	51		
2SA162F		3006	51		
2SA162G		3006	51		
2SA162GN		3006	51		
2SA162H		3006	51		
2SA162J		3006	51		
2SA162K		3006	51		
2SA162L		3006	51		
2SA162M		3006	51		
2SA162OR		3006	51		
2SA162R		3006	51		
2SA162X		3006	51		
2SA162Y		3006	51		
2SA163	126	3006	51	TR17/635	G0008
2SA163A		3006	51		
2SA163B		3006	51		
2SA163C		3006	51		
2SA163D		3006	51		
2SA163E		3006	51		
2SA163F		3006	51		
2SA163G		3006	51		
2SA163GN		3006	51		
2SA163H		3006	51		
2SA163J		3006	51		
2SA163K		3006	51		
2SA163L		3006	51		
2SA163M		3006	51		
2SA163OR		3006	51		
2SA163R		3006	51		
2SA163X		3006	51		
2SA163Y		3006	51		
2SA164	126	3006	51	TR17/635	G0008
2SA164A		3006	51		
2SA164B		3006	51		
2SA164C		3006	51		
2SA164D		3006	51		
2SA164E		3006	51		
2SA164F		3006	51		
2SA164G		3006	51		
2SA164GN		3006	51		
2SA164H		3006	51		
2SA164J		3006	51		
2SA164K		3006	51		
2SA164L		3006	51		
2SA164M		3006	51		
2SA164OR		3006	51		
2SA164R		3006	51		
2SA164X		3006	51		
2SA164Y		3006	51		
2SA165	126	3006	50	TR17/635	G0008
2SA165A		3006	51		
2SA165B		3006	51		
2SA165C		3006	51		
2SA165D		3006	51		
2SA165E		3006	51		
2SA165F		3006	51		
2SA165G		3006	51		
2SA165GN		3006	51		
2SA165H		3006	51		
2SA165J		3006	51		
2SA165K		3006	51		
2SA165L		3006	51		
2SA165M		3006	51		
2SA165OR		3006	51		
2SA165R		3006	51		
2SA165X		3006	51		
2SA165Y		3006	51		
2SA166	126	3006	51	TR17/635	G0008
2SA166A		3006	51		

Original Part Number	ECG	SK	GE	IR/WEP	HEP
2SA166B		3006	51		
2SA166C		3006	51		
2SA166D		3006	51		
2SA166E		3006	51		
2SA166F		3006	51		
2SA166G		3006	51		
2SA166GN		3006	51		
2SA166H		3006	51		
2SA166J		3006	51		
2SA166K		3006	51		
2SA166L		3006	51		
2SA166M		3006	51		
2SA166OR		3006	51		
2SA166R		3006	51		
2SA166X		3006	51		
2SA166Y		3006	51		
2SA167	100	3005	52	TR05/254	G0005
2SA167A		3005	51		
2SA167B		3005	51		
2SA167C		3005	51		
2SA167D		3005	51		
2SA167E		3005	51		
2SA167F		3005	51		
2SA167G		3005	51		
2SA167GN		3005	51		
2SA167H		3005	51		
2SA167J		3005	51		
2SA167K		3005	51		
2SA167L		3005	51		
2SA167M		3005	51		
2SA167OR		3005	51		
2SA167R		3005	51		
2SA167X		3005	51		
2SA167Y		3005	51		
2SA168	100	3005	52	TR05/254	G0001
2SA168A	100	3005	52	TR05/254	G0001
2SA168B		3005	51		
2SA168C		3005	51		
2SA168D		3005	51		
2SA168E		3005	51		
2SA168F		3005	51		
2SA168G		3005	51		
2SA168GN		3005	51		
2SA168H		3005	51		
2SA168J		3005	51		
2SA168K		3005	51		
2SA168L		3005	51		
2SA168M		3005	51		
2SA168OR		3005	51		
2SA168R		3005	51		
2SA168X		3005	51		
2SA168Y		3005	51		
2SA169	100	3005	52	TR05/254	G0005
2SA169A		3005	51		
2SA169B		3005	51		
2SA169C		3005	51		
2SA169D		3005	51		
2SA169E		3005	51		
2SA169F		3005	51		
2SA169G		3005	51		
2SA169GN		3005	51		
2SA169H		3005	51		
2SA169J		3005	51		
2SA169K		3005	51		
2SA169L		3005	51		
2SA169M		3005	51		
2SA169OR		3005	51		
2SA169R		3005	51		
2SA169X		3005	51		
2SA169Y		3005	51		
2SA170	100	3005	52	TR05/254	G0005
2SA170A		3005	51		
2SA170B		3005	51		
2SA170C		3005	51		
2SA170D		3005	51		
2SA170E		3005	51		
2SA170F		3005	51		
2SA170G		3005	51		
2SA170GN		3005	51		
2SA170H		3005	51		
2SA170J		3005	51		
2SA170K		3005	51		
2SA170L		3005	51		
2SA170M		3005	51		
2SA170OR		3005	51		
2SA170R		3005	51		
2SA170X		3005	51		
2SA170Y		3005	51		
2SA171	100	3005	52	TR05/254	G0005
2SA171A		3005	51		
2SA171B		3005	51		
2SA171C		3005	51		
2SA171D		3005	51		
2SA171E		3005	51		
2SA171F		3005	51		
2SA171G		3005	51		
2SA171GN		3005	51		
2SA171H		3005	51		
2SA171J		3005	51		
2SA171K		3005	51		
2SA171L		3005	51		
2SA171M		3005	51		
2SA171OR		3005	51		
2SA171R		3005	51		
2SA171X		3005	51		
2SA171Y		3005	51		
2SA172	100	3005	2	TR05/254	G0005
2SA172A	100	3005	2	TR05/254	G0005
2SA172B		3005	51		
2SA172C		3005	51		
2SA172D		3005	51		
2SA172E		3005	51		
2SA172F		3005	51		
2SA172G		3005	51		
2SA172GN		3005	51		
2SA172H		3005	51		
2SA172J		3005	51		
2SA172K		3005	51		
2SA172L		3005	51		
2SA172M		3005	51		
2SA172OR		3005	51		
2SA172R		3005	51		
2SA172X		3005	51		
2SA172Y		3005	51		
2SA173	100	3005	52	TR05/254	G0005
2SA173A		3005	51		
2SA173B	102A	3005	51	TR85/250	G0005
2SA173C		3005	51		
2SA173D		3005	51		
2SA173E		3005	51		
2SA173F		3005	51		
2SA173G		3005	51		
2SA173GN		3005	51		
2SA173H		3005	51		
2SA173J		3005	51		
2SA173K		3005	51		
2SA173L		3005	51		
2SA173M		3005	51		
2SA173OR		3005	51		
2SA173R		3005	51		
2SA173X		3005	51		
2SA173Y		3005	51		
2SA174	102	3005	52	TR05/631	G0005
2SA174A		3005	51		
2SA174B		3005	51		
2SA174C		3005	51		
2SA174D		3005	51		
2SA174E		3005	51		
2SA174F		3005	51		
2SA174G		3005	51		
2SA174GN		3005	51		
2SA174H		3005	51		
2SA174J		3005	51		
2SA174K		3005	51		
2SA174M		3005	51		
2SA174OR		3005	51		
2SA174R		3005	51		
2SA174X		3005	51		
2SA174Y		3005	51		
2SA175	126	3006	51	TR17/635	G0008
2SA175A		3006	51		
2SA175B		3006	51		
2SA175C		3006	51		
2SA175D		3006	51		
2SA175E		3006	51		
2SA175F		3006	51		
2SA175G		3006	51		
2SA175GN		3006	51		
2SA175J		3006	51		
2SA175K		3006	51		

Original Part Number	ECG	SK	GE	IR/WEP	HEP
2SA175M		3006	51		
2SA175OR		3006	51		
2SA175R		3006	51		
2SA175X		3006	51		
2SA175Y		3006	51		
2SA176	126	3008	52	TR17/635	G0008
2SA176A		3008	53		
2SA176B		3008	53		
2SA176C		3008	53		
2SA176D		3008	53		
2SA176E		3008	53		
2SA176F		3008	53		
2SA176G		3008	53		
2SA176GN		3008	53		
2SA176H		3008	53		
2SA176J		3008	53		
2SA176K		3008	53		
2SA176L		3008	53		
2SA176M		3008	53		
2SA176OR		3008	53		
2SA176R		3008	53		
2SA176X		3008	53		
2SA176Y		3008	53		
2SA178			51		
2SA179					G0005
2SA180	160	3005	52	TR17/637	G0005
2SA180A		3005	51		
2SA180B		3005	51		
2SA180C		3005	51		
2SA180D		3005	51		
2SA180E		3005	51		
2SA180F		3005	51		
2SA180G		3005	51		
2SA180GN		3005	51		
2SA180H		3005	51		
2SA180J		3005	51		
2SA180K		3005	51		
2SA180L		3005	51		
2SA180M		3005	51		
2SA180OR		3005	51		
2SA180R		3005	51		
2SA180X		3005	51		
2SA180Y		3005	51		
2SA181	102A	3005	52	TR85/250	G0005
2SA181A		3005	51		
2SA181B		3005	51		
2SA181C		3005	51		
2SA181D		3005	51		
2SA181E		3005	51		
2SA181F		3005	51		
2SA181G		3005	51		
2SA181GN		3005	51		
2SA181H		3005	51		
2SA181J		3005	51		
2SA181K		3005	51		
2SA181L		3005	51		
2SA181M		3005	51		
2SA181OR		3005	51		
2SA181R		3005	51		
2SA181X		3005	51		
2SA181Y		3005	51		
2SA182	100	3005	1	TR05/254	G0005
2SA182A		3005	51		
2SA182B		3005	51		
2SA182C		3005	51		
2SA182D		3005	51		
2SA182E		3005	51		
2SA182F		3005	51		
2SA182G		3005	51		
2SA182GN		3005	51		
2SA182H		3005	51		
2SA182J		3005	51		
2SA182K		3005	51		
2SA182L		3005	51		
2SA182M		3005	51		
2SA182OR		3005	51		
2SA182R		3005	51		
2SA182X		3005	51		
2SA182Y		3005	51		
2SA183	102A	3005	52	TR85/250	G0003
2SA183A		3005	51		
2SA183B		3005	51		
2SA183C		3005	51		
2SA183D		3005	51		
2SA183E		3005	51		
2SA183F		3005	51		
2SA183G		3005	51		
2SA183GN		3005	51		
2SA183H		3005	51		
2SA183J		3005	51		
2SA183K		3005	51		
2SA183L		3005	51		
2SA183M		3005	51		
2SA183OR		3005	51		
2SA183R		3005	51		
2SA183X		3005	51		
2SA183Y		3005	51		
2SA184			51		
2SA186			53		
2SA187TV	102A			TR85/	
2SA188	126	3005	50	TR17/635	G0009
2SA188A		3005	51		
2SA188B		3005	51		
2SA188C		3005	51		
2SA188D		3005	51		
2SA188E		3005	51		
2SA188F		3005	51		
2SA188G		3005	51		
2SA188GN		3005	51		
2SA188H		3005	51		
2SA188J		3005	51		
2SA188K		3005	51		
2SA188L		3005	51		
2SA188M		3005	51		
2SA188OR		3005	51		
2SA188R		3005	51		
2SA188X		3005	51		
2SA188Y		3005	51		
2SA189	126	3005	50	TR17/635	G0008
2SA189A		3005	51		
2SA189B		3005	51		
2SA189C		3005	51		
2SA189D		3005	51		
2SA189E		3005	51		
2SA189F		3005	51		
2SA189G		3005	51		
2SA189GN		3005	51		
2SA189H		3005	51		
2SA189J		3005	51		
2SA189K		3005	51		
2SA189L		3005	51		
2SA189M		3005	51		
2SA189OR		3005	51		
2SA189R		3005	51		
2SA189X		3005	51		
2SA189Y		3005	51		
2SA190			51		
2SA191			51		
2SA192			51		
2SA193			51		
2SA194			51		
2SA195			51		
2SA196			51		
2SA197	102	3006	52	TR05/631	G0008
2SA197A		3006	51		
2SA197B		3006	51		
2SA197C		3006	51		
2SA197D		3006	51		
2SA197E		3006	51		
2SA197F		3006	51		
2SA197G		3006	51		
2SA197GN		3006	51		
2SA197H		3006	51		
2SA197J		3006	51		
2SA197K		3006	51		
2SA197L		3006	51		
2SA197M		3006	51		
2SA197OR		3006	51		
2SA197R		3006	51		
2SA197X		3006	51		
2SA197Y		3006	51		
2SA198	100	3005	51	TR05/254	G0008
2SA198A		3005	51		
2SA198B		3005	51		
2SA198C		3005	51		
2SA198D		3005	51		
2SA198E		3005	51		
2SA198F		3005	51		
2SA198G		3005	51		
2SA198GN		3005	51		

Original Part Number	ECG	SK	GE	IR/WEP	HEP
2SA198H		3005	51		
2SA198J		3005	51		
2SA198K		3005	51		
2SA198L		3005	51		
2SA198		3005	51		
2SA198OR		3005	51		
2SA198R		3005	51		
2SA198X		3005	51		
2SA198Y		3005	51		
2SA199			51		
2SA200			53		
2SA201	126	3005	50	TR17/635	G0005
2SA201-O	126	3005	1	TR17/635	G0005
2SA201A	126	3008	51	TR17/635	G0008
2SA201B	126	3005	51	TR17/635	G0005
2SA201CL		3005	51		G0005
2SA201D		3005	51		
2SA201E	102A	3005	51	TR85/250	G0005
2SA201F		3005	51		
2SA201G		3005	51		
2SA201GN		3005	51		
2SA201H		3005	51		
2SA201J		3005	51		
2SA201K		3005	51		
2SA201L		3005	51		
2SA201M		3005	51		
2SA201-N			1		
2SA201O			1		
2SA201OR		3005	51		
2SA201R		3005	51		
2SA201TV	126		50	TR17/	G0005
2SA201X		3005	51		
2SA201Y		3005	51		
2SA202	126	3008	50	TR17/635	G0005
2SA202A	126	3005	51	TR17/635	G0005
2SA202AP		3005	51		G0005
2SA202B	126	3008	51	TR17/635	G0005
2SA202C	126	3008	51	TR17/635	G0005
2SA202D	126	3004	51	TR17/635	G0008
2SA202D-4		3004	53		
2SA202E		3008	53		
2SA202F		3008	53		
2SA202G		3008	53		
2SA202GN		3008	53		
2SA202H		3008	53		
2SA202J		3008	53		
2SA202K		3008	53		
2SA202L		3008	53		
2SA202M		3008	53		
2SA202OR		3008	53		
2SA202R		3008	53		
2SA202X		3008	53		
2SA202Y		3008	53		
2SA202(D)				TR11/	
2SA203	102A	3005	50	TR85/250	G0005
2SA203A	126			TR17/	G0005
2SA203AA	102A	3005	1	TR85/250	G0005
2SA203B	102A	3008	51	TR85/250	G0005
2SA203P	102A	3124		TR85/250	G0005
2SA204	100	3005	2	TR05/254	G0005
2SA204A		3005	51		
2SA204B		3005	51		
2SA204C		3005	51		
2SA204D		3005	51		
2SA204E		3005	51		
2SA204F		3005	51		
2SA204G		3005	51		
2SA204GN		3005	51		
2SA204H		3005	51		
2SA204J		3005	51		
2SA204K		3005	51		
2SA204L		3005	51		
2SA204M		3005	51		
2SA204OR		3005	51		
2SA204R		3005	51		
2SA204X		3005	51		
2SA204Y		3005	51		
2SA205	100	3005	2	TR05/254	G0005
2SA205A		3005	51		
2SA205B		3005	51		
2SA205C		3005	51		
2SA205D		3005	51		
2SA205E		3005	51		
2SA205F		3005	51		
2SA205G		3005	51		
2SA205GN		3005	51		
2SA205H		3005	51		
2SA205J		3005	51		
2SA205K		3005	51		
2SA205L		3005	51		
2SA205M		3005	51		
2SA205OR		3005	51		
2SA205R		3005	51		
2SA205X		3005	51		
2SA205Y		3005	51		
2SA206	100	3005	2	TR05/254	G0005
2SA206A		3005	51		
2SA206B		3005	51		
2SA206C		3005	51		
2SA206D		3005	51		
2SA206E		3005	51		
2SA206F		3005	51		
2SA206G		3005	51		
2SA206GN		3005	51		
2SA206H		3005	51		
2SA206J		3005	51		
2SA206K		3005	51		
2SA206L		3005	51		
2SA206M		3005	51		
2SA206OR		3005	51		
2SA206R		3005	51		
2SA206X		3005	51		
2SA206Y		3005	51		
2SA207	100	3005	2	TR05/254	G0005
2SA207A		3005	51		
2SA207B		3005	51		
2SA207C		3005	51		
2SA207D		3005	51		
2SA207E		3005	51		
2SA207F		3005	51		
2SA207G		3005	51		
2SA207GN		3005	51		
2SA207H		3005	51		
2SA207J		3005	51		
2SA207K		3005	51		
2SA207L		3005	51		
2SA207M		3005	51		
2SA207OR		3005	51		
2SA207R		3005	51		
2SA207X		3005	51		
2SA207Y		3005	51		
2SA208	100	3005	1	TR05/254	G0005
2SA208A		3005	51		
2SA208B		3005	51		
2SA208C		3005	51		
2SA208D		3005	51		
2SA208E		3005	51		
2SA208F		3005	51		
2SA208G		3005	51		
2SA208GN		3005	51		
2SA208H		3005	51		G0006
2SA208J		3005	51		
2SA208K		3005	51		
2SA208L		3005	51		
2SA208M		3005	51		
2SA208OR		3005	51		
2SA208R		3005	51		
2SA208X		3005	51		
2SA208Y		3005	51		
2SA209	100	3005	1	TR05/254	G0005
2SA209A		3005	51		
2SA209B		3005	51		
2SA209C		3005	51		
2SA209D		3005	51		
2SA209E		3005	51		
2SA209F		3005	51		
2SA209G		3005	51		
2SA209GN		3005	51		
2SA209H		3005	51		
2SA209J		3005	51		
2SA209K		3005	51		
2SA209L		3005	51		
2SA209M		3005	51		
2SA209OR		3005	51		
2SA209R		3005	51		
2SA209X		3005	51		
2SA209Y		3005	51		
2SA210	100	3005	1	TR05/254	G0005
2SA210A		3005	51		
2SA210B		3005	51		

Original Part Number	ECG	SK	GE	IR/WEP	HEP
2SA210C		3005	51		
2SA210D		3005	51		
2SA210E		3005	51		
2SA210F		3005	51		
2SA210G		3005	51		
2SA210GN		3005	51		
2SA210H		3005	51		G0005
2SA210J		3005	51		
2SA210K		3005	51		
2SA210L		3005	51		
2SA210M		3005	51		
2SA210OR		3005	51		
2SA210R		3005	51		
2SA210X		3005	51		
2SA210Y		3005	51		
2SA211	100	3005	1	TR05/254	G0005
2SA211A		3005	51		
2SA211B		3005	51		
2SA211C		3005	51		
2SA211D		3005	51		
2SA211E		3005	51		
2SA211F		3005	51		
2SA211G		3005	51		
2SA211GN		3005	51		
2SA211H		3005	51		
2SA211J		3005	51		
2SA211K		3005	51		
2SA211L		3005	51		
2SA211M		3005	51		
2SA211OR		3005	51		
2SA211R		3005	51		
2SA211X		3005	51		
2SA211Y		3005	51		
2SA212	100	3005	1	TR05/254	G0005
2SA212A		3005	51		
2SA212B		3005	51		
2SA212C		3005	51		
2SA212D		3005	51		
2SA212E		3005	51		
2SA212F		3005	51		
2SA212G		3005	51		
2SA212GN		3005	51		
2SA212H		3005	51		G0005
2SA212J		3005	51		
2SA212K		3005	51		
2SA212L		3005	51		
2SA212M		3005	51		
2SA212OR		3005	51		
2SA212R		3005	51		
2SA212X		3005	51		
2SA212Y		3005	51		
2SA213	126	3006	51	TR17/635	G0008
2SA213A		3006	51		
2SA213B		3006	51		
2SA213C		3006	51		
2SA213D		3006	51		
2SA213E		3006	51		
2SA213F		3006	51		
2SA213G		3006	51		
2SA213GN		3006	51		
2SA213H		3006	51		
2SA213J		3006	51		
2SA213K		3006	51		
2SA213L		3006	51		
2SA213M		3006	51		
2SA213OR		3006	51		
2SA213R		3006	51		
2SA213X		3006	51		
2SA213Y		3006	51		
2SA214	126	3006	51	TR17/635	G0008
2SA214A		3006	51		
2SA214B		3006	51		
2SA214C		3006	51		
2SA214D		3006	51		
2SA214E		3006	51		
2SA214F		3006	51		
2SA214G		3006	51		
2SA214GN		3006	51		
2SA214H		3006	51		
2SA214J		3006	51		
2SA214K		3006	51		
2SA214L		3006	51		
2SA214M		3006	51		
2SA214OR		3006	51		
2SA214R		3006	51		
2SA214X		3006	51		
2SA214Y		3006	51		
2SA215	160	3006	51	TR17/637	G0008
2SA215A		3006	51		
2SA215B		3006	51		
2SA215C		3006	51		
2SA215D		3006	51		
2SA215E		3006	51		
2SA215F		3006	51		
2SA215G		3006	51		
2SA215GN		3006	51		
2SA215H		3006	51		
2SA215J		3006	51		
2SA215K		3006	51		
2SA215L		3006	51		
2SA215M		3006	51		
2SA215OR		3006	51		
2SA215R		3006	51		
2SA215X		3006	51		
2SA215Y		3006	51		
2SA216	126	3006	51	TR17/635	G0008
2SA216A		3006	51		
2SA216B		3006	51		
2SA216C		3006	51		
2SA216D		3006	51		
2SA216E		3006	51		
2SA216F		3006	51		
2SA216G		3006	51		
2SA216GN		3006	51		
2SA216H		3006	51		
2SA216J		3006	51		
2SA216K		3006	51		
2SA216L		3006	51		
2SA216M		3006	51		
2SA216OR		3006	51		
2SA216R		3006	51		
2SA216X		3006	51		
2SA216Y		3006	51		
2SA217	100	3005	51	TR05/254	G0005
2SA217A		3005	51		
2SA217B		3005	51		
2SA217C		3005	51		
2SA217D		3005	51		
2SA217E		3005	51		
2SA217F		3005	51		
2SA217G		3005	51		
2SA217GN		3005	51		
2SA217H		3005	51		G0005
2SA217J		3005	51		
2SA217K		3005	51		
2SA217L		3005	51		
2SA217M		3005	51		
2SA217OR		3005	51		
2SA217R		3005	51		
2SA217X		3005	51		
2SA217Y		3005	51		
2SA218	160	3008	52	TR17/637	G0008
2SA218A		3008	53		
2SA218B		3008	53		
2SA218C		3008	53		
2SA218D		3008	53		
2SA218E		3008	53		
2SA218F		3008	53		
2SA218G		3008	53		
2SA218GN		3008	53		
2SA218H		3008	53		
2SA218J		3008	53		
2SA218K		3008	53		
2SA218L		3008	53		
2SA218M		3008	53		
2SA218OR		3008	53		
2SA218R		3008	53		
2SA218X		3008	53		
2SA218Y		3008	53		
2SA219	160	3006	50	TR17/637	G0003
2SA219A		3006	51		
2SA219B		3006	51		
2SA219C		3006	51		
2SA219D		3006	51		
2SA219E		3006	51		
2SA219F		3006	51		
2SA219G		3006	51		
2SA219GN		3006	51		
2SA219H		3006	51		
2SA219J		3006	51		

Original Part Number	ECG	SK	GE	IR/WEP	HEP
2SA219K		3006	51		
2SA219L		3006	51		
2SA219M		3006	51		
2SA219OR		3006	51		
2SA219R		3006	51		
2SA219X		3006	51		
2SA219Y		3006	51		
2SA220	160	3008	52	TR17/637	G0003
2SA220A		3008	53		
2SA220B		3008	53		
2SA220C		3008	53		
2SA220D		3008	53		
2SA220E		3008	53		
2SA220F		3008	53		
2SA220G		3008	53		
2SA220GN		3008	53		
2SA220H		3008	53		
2SA220J		3008	53		
2SA220K		3008	53		
2SA220L		3008	53		
2SA220M		3008	53		
2SA220OR		3008	53		
2SA220R		3008	53		
2SA220X		3008	53		
2SA220Y		3008	53		
2SA221	160	3008	50	TR17/637	G0003
2SA221A		3008	53		
2SA221B		3008	53		
2SA221C		3008	53		
2SA221D		3008	53		
2SA221E		3008	53		
2SA221F		3008	53		
2SA221G		3008	53		
2SA221GN		3008	53		
2SA221H		3008	53		
2SA221J		3008	53		
2SA221K		3008	53		
2SA221L		3008	53		
2SA221M		3008	53		
2SA221OR		3008	53		
2SA221R		3008	53		
2SA221X		3008	53		
2SA221Y		3008	53		
2SA222	160	3008	50	TR17/637	G0003
2SA222A		3008	53		
2SA222B		3008	53		
2SA222C		3008	53		
2SA222D		3008	53		
2SA222E		3008	53		
2SA222F		3008	53		
2SA222G		3008	53		
2SA222GN		3008	53		
2SA222H		3008	53		
2SA222J		3008	53		
2SA222K		3008	53		
2SA222L		3008	53		
2SA222M		3008	53		
2SA222OR		3008	53		
2SA222R		3008	53		
2SA222X		3008	53		
2SA222Y		3008	53		
2SA223	160	3006	50	TR17/637	G0003
2SA223A		3006	51		
2SA223B		3006	51		
2SA223C		3006	51		
2SA223D		3006	51		
2SA223E		3006	51		
2SA223F		3006	51		
2SA223G		3006	51		
2SA223GN		3006	51		
2SA223H		3006	51		
2SA223J		3006	51		
2SA223K		3006	51		
2SA223L		3006	51		
2SA223M		3006	51		
2SA223OR		3006	51		
2SA223R		3006	51		
2SA223X		3006	51		
2SA223Y		3006	51		
2SA224	160	3008	52	TR17/637	G0003
2SA224A		3008	53		
2SA224B		3008	53		
2SA224C		3008	53		
2SA224D		3008	53		
2SA224E		3008	53		
2SA224F		3008	53		
2SA224G		3008	53		
2SA224GN		3008	53		
2SA224H		3008	53		
2SA224J		3008	53		
2SA224K		3008	53		
2SA224L		3008	53		
2SA224M		3008	53		
2SA224OR		3008	53		
2SA224R		3008	53		
2SA224X		3008	53		
2SA224Y		3008	53		
2SA225	160	3006	52	TR17/637	G0003
2SA225A		3006	51		
2SA225B		3006	51		
2SA225C		3006	51		
2SA225D		3006	51		
2SA225E		3006	51		
2SA225F		3006	51		
2SA225G		3006	51		
2SA225GN		3006	51		
2SA225H		3006	51		
2SA225J		3006	51		
2SA225K		3006	51		
2SA225L		3006	51		
2SA225M		3006	51		
2SA225OR		3006	51		
2SA225R		3006	51		
2SA225X		3006	51		
2SA225Y		3006	51		
2SA226	160		52	TR17/637	G0003
2SA227	160	3006	52	TR17/637	G0003
2SA227A		3006	51		
2SA227B		3006	51		
2SA227C		3006	51		
2SA227D		3006	51		
2SA227E		3006	51		
2SA227F		3006	51		
2SA227G		3006	51		
2SA227GN		3006	51		
2SA227H		3006	51		
2SA227J		3006	51		
2SA227K		3006	51		
2SA227L		3006	51		
2SA227M		3006	51		
2SA227OR		3006	51		
2SA227R		3006	51		
2SA227X		3006	51		
2SA227Y		3006	51		
2SA228	160		51	TR17/637	
2SA229	160	3006	50	TR17/637	G0003
2SA229A		3006	51		
2SA229B		3006	51		
2SA229C		3006	51		
2SA229D		3006	51		
2SA229E		3006	51		
2SA229F		3006	51		
2SA229G		3006	51		
2SA229GN		3006	51		
2SA229H		3006	51		
2SA229J		3006	51		
2SA229K		3006	51		
2SA229L		3006	51		
2SA229M		3006	51		
2SA229OR		3006	51		
2SA229X		3006	51		
2SA229Y		3006	51		
2SA230	160	3006	50	TR17/637	G0005
2SA230A		3006	51		
2SA230B		3006	51		
2SA230C		3006	51		
2SA230D		3006	51		
2SA230E		3006	51		
2SA230F		3006	51		
2SA230G		3006	51		
2SA230GN		3006	51		
2SA230H		3006	51		
2SA230J		3006	51		
2SA230K		3006	51		
2SA230L		3006	51		
2SA230M		3006	51		
2SA230OR		3006	51		
2SA230R		3006	51		
2SA230X		3006	51		
2SA230Y		3006	51		

Original Part Number	ECG	SK	GE	IR/WEP	HEP
2SA231	158	3123	80	TR84/630	G0005
2SA232	158			TR84/630	G0006
2SA233	160	3007	50	TR17/637	G0003
2SA233A	160	3007	50	TR17/637	G0001
2SA233B	160	3007	50	TR17/637	G0008
2SA233C	160	3007	50	TR17/637	G0008
2SA233D		3007	50		
2SA233E		3007	50		
2SA233F		3007	50		
2SA233G		3007	50		
2SA233GN		3007	50		
2SA233H		3007	50		
2SA233J		3007	50		
2SA233K		3007	50		
2SA233L		3007	50		
2SA233M		3007	50		
2SA233OR		3007	50		
2SA233R		3007	50		
2SA233X		3007	50		
2SA233Y		3007	50		
2SA234	160	3008	50	TR17/637	G0003
2SA234A	160	3006	50	TR17/637	G0001
2SA234B	160	3006	50	TR17/637	G0003
2SA234C	160	3006	50	TR17/637	G0003
2SA234(B)				TR17/	
2SA234(C)				TR17/	
2SA234D		3006	51		
2SA234E		3006	51		
2SA234F		3006	51		
2SA234G		3006	51		
2SA234GN		3006	51		
2SA234H		3006	51		
2SA234J		3006	51		
2SA234K		3006	51		
2SA234L		3006	51		
2SA234M		3006	51		
2SA234OR		3006	51		
2SA234R		3006	51		
2SA234X		3006	51		
2SA234Y		3006	51		
2SA234C57B2-22				TR17/	
2SA235	160	3006	50	TR17/637	G0003
2SA235A	160	3006	50	TR17/637	G0003
2SA235B	160	3006	50	TR17/637	G0008
2SA235C	160	3006	50	TR17/637	G0003
2SA235D		3006	51		
2SA235E		3006	51		
2SA235F		3006	51		
2SA235G		3006	51		
2SA235GN		3006	51		
2SA235H		3006	50	TR17/637	G0003
2SA235J		3006			
2SA235K		3006	51		
2SA235M		3006	51		
2SA235OR		3006	51		
2SA235R		3006	51		
2SA235X		3006	51		
2SA235Y		3006	51		
2SA235A57B2-14				TR17/	
2SA235C57B2-13				TR17/	
2SA236	126	3007	51	TR17/635	G0008
2SA236A		3007	50		
2SA236B		3007	50		
2SA236C		3007	50		
2SA236D		3007	50		
2SA236E		3007	50		
2SA236F		3007	50		
2SA236G		3007	50		
2SA236GN		3007	50		
2SA236H		3007	50		
2SA236J		3007	50		
2SA236K		3007	50		
2SA236L		3007	50		
2SA236M		3007	50		
2SA236OR		3007	50		
2SA236R		3007	50		
2SA236X		3007	50		
2SA236Y		3007	50		
2SA237	126	3008	51	TR17/635	G0008
2SA237A		3008	53		
2SA237B		3008	53		
2SA237C		3008	53		
2SA237D		3008	53		
2SA237E		3008	53		
2SA237F		3008	53		

Original Part Number	ECG	SK	GE	IR/WEP	HEP
2SA237G		3008	53		
2SA237GN		3008	53		
2SA237H		3008	53		
2SA237J		3008	53		
2SA237K		3008	53		
2SA237L		3008	53		
2SA237M		3008	53		
2SA237OR		3008	53		
2SA237R		3008	53		
2SA237X		3008	53		
2SA237Y		3008	53		
2SA238	126	3006	52	TR17/635	G0003
2SA238A		3006	51		
2SA238B		3006	51		
2SA238C		3006	51		
2SA238D		3006	51		
2SA238E		3006	51		
2SA238F		3006	51		
2SA238G		3006	51		
2SA238GN		3006	51		
2SA238H		3006	51		
2SA238J		3006	51		
2SA238K		3006	51		
2SA238L		3006	51		
2SA238M		3006			
2SA238OR		3006			
2SA238R		3006			
2SA238X		3006			
2SA238Y		3006			
2SA239	160	3006	50	TR17/637	G0003
2SA239A		3006	51		
2SA239B		3006	51		
2SA239C		3006	51		
2SA239D		3006	51		
2SA239E		3006	51		
2SA239F		3006	51		
2SA239G		3006	51		
2SA239GN		3006	51		
2SA239H		3006	51		
2SA239J		3006	51		
2SA239K		3006	51		
2SA239L		3006	51		
2SA239M		3006	51		
2SA239OR		3006	51		
2SA239R		3006	51		
2SA239X		3006	51		
2SA239Y		3006	51		
2SA240	160	3006	50	TR17/637	G0003
2SA240A	160	3006	51	TR17/637	G0001
2SA240B	160	3006	51	TR17/637	G0008
2SA240B2	160	3006	50	TR17/637	G0008
2SA240BL	160			TR17/637	G0003
2SA240C		3006	51		
2SA240D		3006	51		
2SA240E		3006	51		
2SA240F		3006	51		
2SA240G		3006	51		
2SA240GN		3006	51		
2SA240H		3006	51		
2SA240J		3006	51		
2SA240K		3006	51		
2SA240L		3006	51		
2SA240M		3006	51		
2SA240ORG					G0003
2SA240R		3006	51		
2SA240X		3006	51		
2SA240Y		3006	51		
2SA241	160	3006	50	TR17/637	G0008
2SA241A		3006	51		
2SA241B		3006	51		
2SA241C		3006	51		
2SA241D		3006	51		
2SA241E		3006	51		
2SA241F		3006	51		
2SA241G		3006	51		
2SA241GN		3006	51		
2SA241H		3006	51		
2SA241J		3006	51		
2SA241K		3006	51		
2SA241L		3006	51		
2SA241M		3006	51		
2SA241OR		3006	51		
2SA241R		3006	51		
2SA241X		3006	51		
2SA241Y		3006	51		

Original Part Number	ECG	SK	GE	IR/WEP	HEP
2SA242	160		51	TR17/637	G0008
2SA243	160		51	TR17/637	G0008
2SA244	160		52	TR17/637	G0003
2SA245	160		52	TR17/637	G0003
2SA246	126	3006	51	TR17/635	G0003
2SA246A		3006	51		
2SA246B		3006	51		
2SA246C		3006	51		
2SA246D		3006	51		
2SA246E		3006	51		
2SA246F		3006	51		
2SA246G		3006	51		
2SA246GN		3006	51		
2SA246H		3006	51		
2SA246J		3006	51		
2SA246K		3006	51		
2SA246L		3006	51		
2SA246M		3006	51		
2SA246OR		3006	51		
2SA246R		3006	51		
2SA246V	126	3006	50	TR17/635	G0003
2SA246X		3006	51		
2SA246Y		3006	51		
2SA247	126	3006	51	TR17/635	G0008
2SA247A		3006	51		
2SA247B		3006	51		
2SA247C		3006	51		
2SA247D		3006	51		
2SA247E		3006	51		
2SA247F		3006	51		
2SA247G		3006	51		
2SA247GN		3006	51		
2SA247H		3006	51		
2SA247J		3006	51		
2SA247K		3006	51		
2SA247L		3006	51		
2SA247M		3006	51		
2SA247OR		3006	51		
2SA247R		3006	51		
2SA247X		3006	51		
2SA247Y		3006	51		
2SA248	100	3005	51	TR05/254	G0008
2SA248A		3005	51		
2SA248B		3005	51		
2SA248C		3005	51		
2SA248D		3005	51		
2SA248E		3005	51		
2SA248F		3005	51		
2SA248G		3005	51		
2SA248GN		3005	51		
2SA248H		3005	51		
2SA248J		3005	51		
2SA248K		3005	51		
2SA248L		3006	51		
2SA248M		3005	51		
2SA248OR		3005	51		
2SA248R		3005	51		
2SA248X		3005	51		
2SA248Y		3005	51		
2SA250	126	3008	2	TR17/635	G0003
2SA250A		3008	53		
2SA250B		3008	53		
2SA250C		3008	53		
2SA250D		3008	53		
2SA250E		3008	53		
2SA250F		3008	53		
2SA250G		3008	53		
2SA250GN		3008	53		
2SA250H		3008	53		
2SA250J		3008	53		
2SA250K		3008			
2SA250L		3008	53		
2SA250M		3008	53		
2SA250R		3008			
2SA250X		3008	53		
2SA250Y		3008	53		
2SA251	126	3006	51	TR17/635	G0001
2SA251A		3006	51		
2SA251B		3006	51		
2SA251C		3006	51		
2SA251D		3006	51		
2SA251E		3006	51		
2SA251F		3006	51		
2SA251G		3006	51		
2SA251H		3006	51		
2SA251J		3006	51		
2SA251K		3006	51		
2SA251L		3006	51		
2SA251M		3006	51		
2SA251OR		3006	51		
2SA251R		3006	51		
2SA251X		3006	51		
2SA251Y		3006	51		
2SA252	126	3006	51	TR17/635	G0001
2SA252A		3006	51		
2SA252B		3006	51		
2SA252C		3006	51		
2SA252D		3006	51		
2SA252E		3006	51		
2SA252F		3006	51		
2SA252G		3006	51		
2SA252H		3006	51		
2SA252J		3006	51		
2SA252K		3006	51		
2SA252L		3006	51		
2SA252M		3006	51		
2SA252OR		3006	51		
2SA252R		3006	51		
2SA252X		3006	51		
2SA252Y		3006	51		
2SA253	126	3006	52	TR17/635	G0002
2SA253A		3006	51		
2SA253B		3006	51		
2SA253C		3006	51		
2SA253D		3006	51		
2SA253E		3006	51		
2SA253F		3006	51		
2SA253G		3006	51		
2SA253GN		3006	51		
2SA253H		3006	51		
2SA253J		3006	51		
2SA253K		3006	51		
2SA253L		3006	51		
2SA253M		3006	51		
2SA253OR		3006	51		
2SA253R		3006	51		
2SA253X		3006	51		
2SA253Y		3006	51		
2SA254	126	3005	51	TR17/635	G0003
2SA254A		3005	51		
2SA254B		3005	51		
2SA254C		3005	51		
2SA254D		3005	51		
2SA254E		3005	51		
2SA254F		3005	51		
2SA254G		3005	51		
2SA254H		3005	51		
2SA254J		3005	51		
2SA254K		3005	51		
2SA254L		3005	51		
2SA254M		3005	51		
2SA254OR		3005	51		
2SA254R		3005	51		
2SA254X		3005	51		
2SA254Y		3005	51		
2SA255	126	3005	51	TR17/635	G0008
2SA255A		3005	51		
2SA255B		3005	51		
2SA255C		3005	51		
2SA255D		3005	51		
2SA255E		3005	51		
2SA255F		3005	51		
2SA255G		3005	51		
2SA255GN		3005	51		
2SA255H		3005	51		
2SA255J		3005	51		
2SA255K		3005	51		
2SA255L		3005	51		
2SA255M		3005	51		
2SA255OR		3005	51		
2SA255R		3005	51		
2SA255X		3005	51		
2SA255Y		3005	51		
2SA256	126	3008	51	TR17/635	G0008
2SA256A		3008	53		
2SA256B		3008	53		
2SA256C		3008	53		
2SA256D		3008	53		
2SA256E		3008	53		
2SA256F		3008	53		

Original Part Number	ECG	SK	GE	IR/WEP	HEP
2SA256G		3008	53		
2SA256GN		3008	53		
2SA256H		3008	53		
2SA256J		3008	53		
2SA256K		3008	53		
2SA256L		3008	53		
2SA256M		3008	53		
2SA256OR		3008	53		
2SA256R		3008	53		
2SA256X		3008	53		
2SA256Y		3008	53		
2SA257	126	3007	51	TR17/635	G0008
2SA257A		3007	50		
2SA257B		3007	50		
2SA257C		3007	50		
2SA257D		3007	50		
2SA257E		3007	50		
2SA257F		3007	50		
2SA257G		3007	50		
2SA257GN		3007	50		
2SA257H		3007	50		
2SA257J		3007	50		
2SA257K		3007	50		
2SA257L		3007	50		
2SA257M		3007	50		
2SA257OR		3007	50		
3DS257R		3007	50		
3DS257X		3007	50		
3SA257Y		3007	50		
2SA258	126	3007	51	TR17/635	G0008
2SA258A		3007	50		
2SA258B		3007	50		
2SA257C		3007	50		
2SA257D		3007	50		
2SA257E		3007	50		
2SA257F		3007	50		
2SA257G		3007	50		
2SA257GN		3007	50		
2SA257H		3007	50		
2SA257J		3007	50		
2SA257K		3007	50		
2SA257L		3007	50		
2SA257M		3007	50		
2SA257OR		3007	50		
2SA257R		3007	50		
2SA257X		3007	50		
2SA257Y		3007	50		
2SA258	126	3007	51	TR17/635	G0008
2SA258A		3007	50		
2SA258B		3007	50		
2SA258C		3007	50		
2SA258D		3007	50		
2SA258E		3007	50		
2SA258F		3007	50		
2SA258G		3007	50		
2SA258GN		3007	50		
2SA258H		3007	50		
2SA258J		3007	50		
2SA258K		3007	50		
2SA258L		3007	50		
2SA258M		3007	50		
2SA258OR		3007	50		
2SA258R		3007	50		
2SA258X		3007	50		
2SA258Y		3007	50		
2SA259	126	3008	51	TR17/635	G0003
2SA259A		3008			
2SA259B		3008	53		
2SA259C		3008	53		
2SA259D		3008	53		
2SA259E		3008	53		
2SA259F		3008	53		
2SA259G		3008	53		
2SA259GN		3008	53		
2SA259H		3008	53		
2SA259J		3008	53		
2SA259K		3008	53		
2SA259L		3008	53		
2SA259M		3008	53		
2SA259OR		3008	53		
2SA259R		3008	53		
2SA259X		3008	53		
2SA259Y		3008	53		
2SA260	126	3006	51	TR17/635	G0003
2SA260A		3006	51		
2SA260B		3006	51		
2SA260C		3006	51		
2SA260D		3006	51		
2SA260E		3006	51		
2SA260F		3006	51		
2SA260G		3006	51		
2SA260GN		3006	51		
2SA260H		3006	51		
2SA260J		3006	51		
2SA260K		3006	51		
2SA260L		3006	51		
2SA260M		3006	51		
2SA260R		3006	51		
2SA260X		3006	51		
2SA260Y		3006	51		
2SA261	126	3006	51	TR17/635	G0003
2SA261A		3006	51		
2SA261B		3006	51		
2SA261C		3006	51		
2SA261D		3006	51		
2SA261E		3006	51		
2SA261F		3006	51		
2SA261G		3006	51		
2SA261GN		3006	51		
2SA261H		3006	51		
2SA261J		3006	51		
2SA261K		3006	51		
2SA261L		3006	51		
2SA261M		3006	51		
2SA261OR		3006	51		
2SA261R		3006	51		
2SA261X		3006	51		
2SA261Y		3006	51		
2SA262	126	3006	51	TR17/635	G0003
2SA262A		3006	51		
2SA262B		3006	51		
2SA262C		3006	51		
2SA262D		3006	51		
2SA262E		3006	51		
2SA262F		3006	51		
2SA262G		3006	51		
2SA262GN		3006	51		
2SA262H		3006	51		
2SA262J		3006	51		
2SA262K		3006	51		
2SA262L		3006	51		
2SA262M		3006	51		
2SA262OR		3006	51		
2SA262R		3006	51		
2SA262X		3006	51		
2SA262Y		3006	51		
2SA263	126	3006	51	TR17/635	G0003
2SA263A		3006	51		
2SA263B		3006	51		
2SA263C		3006	51		
2SA263D		3006	51		
2SA263E		3006	51		
2SA263F		3006	51		
2SA263G		3006	51		
2SA263GN		3006	51		
2SA263H		3006	51		
2SA263J		3006	51		
2SA263K		3006	51		
2SA263L		3006	51		
2SA263M		3006	51		
2SA263OR		3006	51		
2SA263R		3006	51		
2SA263X		3006	51		
2SA263Y		3006	51		
2SA264	126	3006	51	TR17/635	G0003
2SA264(1)				TR12/	
2SA264A		3006	51		
2SA264B		3006	51		
2SA264C		3006	51		
2SA264D		3006	51		
2SA264E		3006	51		
2SA264F		3006	51		
2SA264G		3006	51		
2SA264GN		3006	51		
2SA264H		3006	51		
2SA264J		3006	51		
2SA264K		3006	51		
2SA264L		3006	51		
2SA264M		3006	51		
2SA264OR		3006	51		

Original Part Number	ECG	SK	GE	IR/WEP	HEP
2SA264R		3006	51		
2SA264X		3006	51		
2SA264Y		3006	51		
2SA265	126	3006	51	TR17/635	G0008
2SA265A		3006	51		
2SA265B		3006	51		
2SA265C		3006	51		
2SA265D		3006	51		
2SA265E		3006	51		
2SA265F		3006	51		
2SA265G		3006	51		
2SA265GN		3006	51		
2SA265H		3006	51		
2SA265J		3006	51		
2SA265K		3006	51		
2SA265L		3006	51		
2SA265M		3006	51		
2SA265OR		3006	51		
2SA265R		3006	51		
2SA265X		3006	51		
2SA265Y		3006	51		
2SA266	126	3008	50	TR17/635	G0008
2SA266A		3008	53		
2SA266B		3008	53		
2SA266C		3008	53		
2SA266D		3008	53		
2SA266E		3008	53		
2SA266F		3008	53		
2SA266G		3008	53		
2SA266GN		3008	53		
2SA266H		3008	53		
2SA266J		3008	53		
2SA266K		3008	53		
2SA266L		3008	53		
2SA266M		3008	53		
2SA266OR		3008	53		
2SA266R		3008	53		
2SA266X		3008	53		
2SA266Y		3008	53		
2SA267	126	3006	50	TR17/635	G0008
2SA267A		3006	51		
2SA267B		3006	51		
2SA267C		3006	51		
2SA267D		3006	51		
2SA267E		3006	51		
2SA267F		3006	51		
2SA267G		3006	51		
2SA267GN		3006	51		
2SA267H		3006	51		
2SA267J		3006	51		
2SA267K		3006	51		
2SA267L		3006	51		
2SA267M		3006	51		
2SA267OR		3006	51		
2SA267R		3006	51		
2SA267X		3006	51		
2SA267Y		3006	51		
2SA268	126	3006	50	TR17/635	G0008
2SA268A		3006	51		
2SA268B		3006	51		
2SA268C		3006	51		
2SA268D		3006	51		
2SA268E		3006	51		
2SA268F		3006	51		
2SA268G		3006	51		
2SA268GN		3006	51		
2SA268H		3006	51		
2SA268J		3006	51		
2SA268K		3006	51		
2SA268L		3006	51		
2SA268M		3006	51		
2SA268OR		3006	51		
2SA268R		3006	51		
2SA268X		3006	51		
2SA268Y		3006	51		
2SA269	126	3008	50	TR17/635	G0008
2SA269(2)				TR12/	
2SA269A		3008	53		
2SA269B		3008	53		
2SA269C		3008	53		
2SA269D		3008	53		
2SA269E		3008	53		
2SA269F		3008	53		
2SA269G		3008			
2SA269GN		3008	53		
2SA269H		3008	53		
2SA269J		3008	53		
2SA269K		3008	53		
2SA269L		3008	53		
2SA269M		3008	53		
2SA269OR		3008	53		
2SA269X		3008	53		
2SA269Y		3008	53		
2SA270	126	3007	50	TR17/635	G0008
2SA270A		3007	50		
2SA270B		3007	50		
2SA270C		3007	50		
2SA270D		3007	50		
2SA270E		3007	50		
2SA270F		3007	50		
2SA270G		3007	50		
2SA270GN		3007	50		
2SA270H		3007	50		
2SA270J		3007	50		
2SA270K		3007	50		
2SA270L		3007	50		
2SA270M		3007	50		
2SA270OR		3007			
2SA270R			50		
2SA270X		3007	50		
2SA270Y		3007	50		
2SA271	126	3008	50	TR17/635	G0008
2SA271(2)				TR12/	
2SA271A		3008	53		
2SA271B		3008	53		
2SA271C		3008	53		
2SA271D		3008	53		
2SA271E		3008	53		
2SA271F		3008	53		
2SA271G		3008	53		
2SA271GN		3008	53		
2SA271H		3008	53		
2SA271J		3008	53		
2SA271K		3008	53		
2SA271L		3008	53		
2SA271M		3008	53		
2SA271OR		3008	53		
2SA271R		3008	53		
2SA271X		3008	53		
2SA271Y		3008	53		
2SA272	126	3008	50	TR17/635	G0008
2SA272A		3008	53		
2SA272B		3008	53		
2SA272C		3008	53		
2SA272D		3008	53		
2SA272E		3008	53		
2SA272F		3008	53		
2SA272G		3008	53		
2SA272GN		3008	53		
2SA272H		3008	53		
2SA272J		3008	53		
2SA272K		3008	53		
2SA272L		3008	53		
2SA272M		3008	53		
2SA272OR		3008	53		
2SA272R		3008	53		
2SA272X		3008	53		
2SA272Y		3008	53		
2SA273	126	3008	51	TR17/635	G0009
2SA273A		3008	53		
2SA273B		3008	53		
2SA273C		3008	53		
2SA273D		3008	53		
2SA273E		3008	53		
2SA273F		3008	53		
2SA273G		3008	53		
2SA273GN		3008	53		
2SA273H		3008	53		
2SA273J		3008	53		
2SA273K		3008	53		
2SA273L		3008	53		
2SA273M		3008	53		
2SA273OR		3008	53		
2SA273X		3008	53		
2SA273Y		3008	53		
2SA274	126	3008	51	TR17/635	G0009
2SA274A		3008	53		
2SA274B		3008	53		
2SA274C		3008	53		
2SA274D		3008	53		

Original Part Number	ECG	SK	GE	IR/WEP	HEP
2SA274E		3008	53		
2SA274F		3008	53		
2SA274G		3008	53		
2SA274GN		3008	53		
2SA274H		3008	53		
2SA274J		3008	53		
2SA274K		3008	53		
2SA274L		3008	53		
2SA274M		3008	53		
2SA274OR		3008	53		
2SA274R		3008	53		
2SA274X		3008	53		
2SA274Y		3008	53		
2SA275	126	3008	51	TR17/635	G0009
2SA275A		3008	53		
2SA275B		3008	53		
2SA275C		3008	53		
2SA275D		3008	53		
2SA275E		3008	53		
2SA275F		3008	53		
2SA275G		3008	53		
2SA275GN		3008	53		
2SA275H		3008	53		
2SA275J		3008	53		
2SA275K		3008	53		
2SA275L		3008	53		
2SA275M		3008	53		
2SA275OR		3008	53		
2SA275R		3008	53		
2SA275X		3008	53		
2SA275Y		3008	53		
2SA276	160		51	TR17/637	G0003
2SA277	100	3005	1	TR05/254	G0003
2SA277A		3005	51		
2SA277B		3005	51		
2SA277C		3005	51		
2SA277D		3005	51		
2SA277E		3005	51		
2SA277F		3005	51		
2SA277G		3005	51		
2SA277GN		3005	51		
2SA277H		3005	51		
2SA277J		3005	51		
2SA277K		3005	51		
2SA277L		3005	51		
2SA277M		3005	51		
2SA277OR		3005	51		
2SA277R		3005	51		
2SA277X		3005	51		
2SA277Y		3005	51		
2SA278	100	3005	1	TR05/254	G0005
2SA278A		3005	51		
2SA278B		3005	51		
2SA278C		3005	51		
2SA278D		3005	51		
2SA278E		3005	51		
2SA278F		3005	51		
2SA278G		3005	51		
2SA278GN		3005	51		
2SA278H		3005	51		
2SA278J		3005	51		
2SA278K		3005	51		
2SA278L		3005	51		
2SA278M		3005	51		
2SA278OR		3005	51		
2SA278R		3005	51		
2SA278X		3005	51		
2SA278Y		3005	51		
2SA279	100	3005	50	TR05/254	G0009
2SA279A		3005	51		
2SA279B		3005	51		
2SA279C		3005	51		
2SA279D		3005	51		
2SA279E		3005	51		
2SA279F		3005	51		
2SA279G		3005	51		
2SA279GN		3005	51		
2SA279H		3005	51		
2SA279J		3005	51		
2SA279K		3005	51		
2SA279L		3005	51		
2SA279M		3005	51		
2SA279OR		3005	51		
2SA279R		3005	51		
2SA279X		3005	51		
2SA279Y		3005	51		
2SA280	126	3006	51	TR17/635	G0009
2SA280A		3006	51		
2SA280B		3006	51		
2SA280C		3006	51		
2SA280D		3006	51		
2SA280E		3006	51		
2SA280F		3006	51		
2SA280G		3006	51		
2SA280GN		3006	51		
2SA280H		3006	51		
2SA280J		3006	51		
2SA280K		3006	51		
2SA280L		3006	51		
2SA280M		3006	51		
2SA280OR		3006	51		
2SA280R		3006	51		
2SA280X		3006	51		
2SA280Y		3006	51		
2SA281	126	3006	51	TR17/635	G0009
2SA281A		3006	51		
2SA281B		3006	51		
2SA281C		3006	10		
2SA281D		3006	51		
2SA281E		3006	51		
2SA281F		3006	51		
2SA281G		3006	51		
2SA281GN		3006	51		
2SA281H		3006	51		
2SA281J		3006	51		
2SA281K		3006	51		
2SA281L		3006	51		
2SA281M		3006	51		
2SA281OR		3006	51		
2SA281R		3006	51		
2SA281X		3006	51		
2SA281Y		3006	51		
2SA282	100	3005	1	TR05/254	G0005
2SA282A		3005	51		
2SA282B		3005	51		
2SA282C		3005	51		
2SA282D		3005	51		
2SA282E		3005	51		
2SA282F		3005	51		
2SA282G		3005	51		
2SA282GN		3005	51		
2SA282H		3005	51		
2SA282J		3005	51		
2SA282K		3005	51		
2SA282L		3005	51		
2SA282M		3005	51		
2SA282OR		3005	51		
2SA282R		3005	51		
2SA282X		3005	51		
2SA282Y		3005	51		
2SA283	100	3005	1	TR05/254	G0005
2SA283A		3005	51		
2SA283B		3005	51		
2SA283C		3005	51		
2SA283D		3005	51		
2SA283E		3005	51		
2SA283F		3005	51		
2SA283G		3005	51		
2SA283GN		3005	51		
2SA283H		3005	51		
2SA283J		3005	51		
2SA283K		3005	51		
2SA283L		3005	51		
2SA283M		3005	51		
2SA283OR		3005	51		
2SA283R		3005	51		
2SA283X		3005	51		
2SA283Y		3005	51		
2SA284	100	3005	1	TR05/254	G0005
2SA284A		3005	51		
2SA284B		3005	51		
2SA284C		3005	51		
2SA284D		3005	51		
2SA284E		3005	51		
2SA284F		3005	51		
2SA284G		3005	51		
2SA284GN		3005	51		
2SA284H		3005	51		
2SA284J		3005	51		
2SA284K		3005	51		

Original Part Number	ECG	SK	GE	IR/WEP	HEP
2SA284M		3005	51		
2SA284OR		3005	51		
2SA284R		3005	51		
2SA284X		3005	51		
2SA284Y		3005	51		
2SA285	160	3008	51	TR17/637	G0008
2SA285A		3008			
2SA285B		3008	53		
2SA285C		3008	53		
2SA285D		3008			
2SA285E		3008	53		
2SA285F		3008	53		
2SA285G		3008	53		
2SA285GN		3008	53		
2SA285H		3008	53		
2SA285J		3008	53		
2SA285K		3008			
2SA285L		3008	53		
2SA285M		3008	53		
2SA285OR		3008	53		
2SA285R		3008	53		
2SA285X		3008	53		
2SA285Y		3008	53		
2SA286	160	3008	51	TR17/637	G0008
2SA286A		3008	53		
2SA286B		3008	53		
2SA286C		3008	53		
2SA286D		3008	53		
2SA286E		3008	53		
2SA286F		3008	53		
2SA286G		3008	53		
2SA286GN		3008	53		
2SA286H		3008	53		
2SA286J		3008			
2SA286K		3008	53		
2SA286L		3008			
2SA286M		3008	53		
2SA286OR		3008	53		
2SA286R		3008	53		
2SA286X		3008	53		
2SA286Y		3008	53		
2SA287	160	3008	51	TR17/637	G0008
2SA287A		3008	53		
2SA287B		3008	53		
2SA287C		3008	53		
2SA287D		3008	53		
2SA287E		3008	53		
2SA287F		3008	53		
2SA287G		3008	53		
2SA287GN		3008	53		
2SA287H		3008	53		
2SA287J		3008	53		
2SA287K		3008	53		
2SA287L		3008	53		
2SA287M		3008	53		
2SA287OR		3008	53		
2SA287X		3008	53		
2SA287Y		3008	53		
2SA288	126	3008	51	TR17/635	G0003
2SA288A	126	3006	17	TR17/635	G0005
2SA288B		3008	53		
2SA288C		3008	53		
2SA288D		3008	53		
2SA288E		3008	53		
2SA288F		3008	53		
2SA288G		3008	53		
2SA288GN		3008	53		
2SA288H		3008	53		
2SA288J		3008	53		
2SA288K		3008	53		
2SA288L		3008	53		
2SA288M		3008	53		
2SA288OR		3008	53		
2SA288R		3008	53		
2SA288X		3008	53		
2SA288Y		3008	53		
2SA289	126	3006	51	TR17/635	G0003
2SA289A		3006	51		
2SA289B		3006	51		
2SA289C		3006	51		
2SA289D		3006	51		
2SA289E		3006	51		
2SA289F		3006	51		
2SA289G		3006	51		
2SA289GN		3006	51		
2SA289H		3006	51		
2SA289J		3006	51		
2SA289K		3006	51		
2SA289L		3006	51		
2SA289M		3006	51		
2SA289OR		3006	51		
2SA289R		3006	51		
2SA289X		3006	51		
2SA289Y		3006	51		
2SA290	126	3006	51	TR17/635	G0003
2SA290A		3006	51		
2SA290B		3006	51		
2SA290C		3006	51		
2SA290D		3006	51		
2SA290E		3006	51		
2SA290F		3006	51		
2SA290G		3006	51		
2SA290GN		3006	51		
2SA290H		3006	51		
2SA290J		3006	51		
2SA290K		3006	51		
2SA290L		3006	51		
2SA290M		3006	51		
2SA290R		3006	51		
2SA290X		3006	51		
2SA290Y		3006	51		
2SA291	126	3006	53	TR17/635	G0009
2SA291A		3006	51		
2SA291B		3006	51		
2SA291C		3006	51		
2SA291D		3006	51		
2SA291E		3006	51		
2SA291F		3006	51		
2SA291G		3006	51		
2SA291GN		3006	51		
2SA291H		3006	51		
2SA291J		3006	51		
2SA291K		3006	51		
2SA291L		3006	51		
2SA291M		3006	51		
2SA291OR		3006	51		
2SA291R		3006	51		
2SA291X		3006	51		
2SA291Y		3006	51		
2SA292	126	3006	50	TR17/635	G0009
2SA292A		3006	51		
2SA292B		3006	51		
2SA292C		3006	51		
2SA292D		3006	51		
2SA292E		3006	51		
2SA292F		3006	51		
2SA292G		3006	51		
2SA292GN		3006	51		
2SA292H		3006	51		
2SA292J		3006	51		
2SA292K		3006	51		
2SA292L		3006	51		
2SA292M		3006	51		
2SA292OR		3006	51		
2SA292R		3006	51		
2SA292X		3006	51		
2SA292Y		3006	51		
2SA293	126	3006	50	TR17/635	G0009
2SA293A		3006	51		
2SA293B		3006	51		
2SA293C		3006	51		
2SA293D		3006	51		
2SA293E		3006	51		
2SA293F		3006	51		
2SA293G		3006	51		
2SA293GN		3006	51		
2SA293H		3006	51		
2SA293J		3006	51		
2SA293K		3006	51		
2SA293L		3006	51		
2SA293M		3006	51		
2SA293OR		3006	51		
2SA293R		3006	51		
2SA293X		3006	51		
2SA293Y		3006	51		
2SA294	126	3006	50	TR17/635	G0009
2SA294A		3006	51		
2SA294B		3006	51		
2SA294C		3006	51		

Original Part Number	ECG	SK	GE	IR/WEP	HEP
2SA294D		3006	51		
2SA294E		3006	51		
2SA294F		3006	51		
2SA294G		3006			
2SA294GN		3006	51		
2SA294H		3006	51		
2SA294J		3006	51		
2SA294K		3006	51		
2SA294L		3006	51		
2SA294M		3006	51		
2SA294OR		3006	51		
2SA294R		3006	51		
2SA294X		3006	51		
2SA294Y		3006	51		
2SA295	126	3006	52	TR17/635	G0009
2SA295A		3006	51		
2SA295B		3006	51		
2SA295C		3006	51		
2SA295D		3006	51		
2SA295E		3006	51		
2SA295F		3006	51		
2SA295G		3006	51		
2SA295GN		3006	51		
2SA295H		3006	51		
2SA295J		3006	51		
2SA295K		3006	51		
2SA295L		3006	51		
2SA295OR		3006	51		
2SA295R		3006	51		
2SA295X		3006			
2SA295Y		3006	51		
2SA296	160	3005	51	TR17/637	G0009
2SA296A		3005	51		
2SA296B		3005	51		
2SA296C		3005	51		
2SA296D		3005	51		
2SA296E		3005	51		
2SA296F		3005	51		
2SA296G		3005	51		
2SA296GN		3005	51		
2SA296H		3005	51		
2SA296J		3005	51		
2SA296K		3005	51		
2SA296L		3005	51		
2SA296M		3005	51		
2SA296OR		3005	51		
2SA296R		3005	51		
2SA296X		3005	51		
2SA296Y		3005	51		
2SA297	160	3005	51	TR17/637	G0009
2SA297A		3005	51		
2SA297B		3005	51		
2SA297C		3005	51		
2SA297D		3005	51		
2SA297E		3005	51		
2SA297F		3005	51		
2SA297G		3005	51		
2SA297GN		3005	51		
2SA297H		3005	51		
2SA297J		3005	51		
2SA297K		3005	51		
2SA297L		3005	51		
2SA297M		3005	51		
2SA297OR		3005	51		
2SA297R		3005	51		
2SA297X		3005	51		
2SA297Y		3005	51		
2SA298	160	3008	51	TR17/637	G0009
2SA298A		3008	53		
2SA298B		3008	53		
2SA298C		3008	53		
2SA298D		3009	25		
2SA298E		3009	25		
2SA298F		3009	25		
2SA298G		3008	53		
2SA298GN		3009	25		
2SA298H		3009	25		
2SA298J		3009	25		
2SA298K		3009	25		
2SA298L		3009	25		
2SA298M		3009	25		
2SA298OR		3008	53		
2SA298R		3009	25		
2SA298X		3008	53		
2SA298Y		3008	53		
2SA301	126	3006	53	TRL77635	G0009
2SA301A		3006	51		
2SA301B		3006	51		
2SA301C		3006	51		
2SA301D		3006	51		
2SA301E		3006	51		
2SA301F		3006	51		
2SA301G		3006	51		
2SA301GN		3006	51		
2SA301H		3006	51		
2SA301J		3006	51		
2SA301K		3006	51		
2SA301L		3006	51		
2SA301M		3006	51		
2SA301OR		3006	51		
2SA301R		3006	51		
2SA301X		3006	51		
2SA301Y		3006	51		
2SA302	102A			TR85/250	G0005
2SA303	102A	3004		TR85/250	G0005
2SA304	102A	3123	52	TR85/250	G0005
2SA304A		3123	80		
2SA304B		3123	80		
2SA304C		3123	80		
2SA304D		3123	80		
2SA304F		3123	80		
2SA304G		3123	80		
2SA304GN		3123	80		
2SA304U		3123	80		
2SA304K		3123	80		
2SA304L		3123	80		
2SA304M		3123	80		
2SA304OR		3123	80		
2SA304R		3123	80		
2SA304X		3123	80		
2SA304Y		3123	80		
2SA305	100	3005	1	TR05/254	G0005
2SA305A		3005	51		
2SA305B		3005	51		
2SA305C		3005	51		
2SA305D		3005	51		
2SA305E		3005	51		
2SA305F		3005	51		
2SA305G		3005	51		
2SA305GN		3005	51		
2SA305H		3005	51		
2SA305J		3005	51		
2SA305K		3005	51		
2SA305L		3005	51		
2SA305M		3005	51		
2SA305OR		3005	51		
2SA305R		3005	51		
2SA305X		3005	51		
2SA305Y		3005	51		
2SA306	160		53	TR17/637	G0009
2SA307	160	3008	51	TR17/637	G0009
2SA307A		3008	53		
2SA307B		3008	53		
2SA307C		3008	53		
2SA307D		3008	53		
2SA307E		3008	53		
2SA307F		3008	53		
2SA307G		3008	53		
2SA307GN		3008	53		
2SA307J		3008	53		
2SA307K		3008	53		
2SA307L		3008	53		
2SA307M		3008	53		
2SA307OR		3008	53		
2SA307R		3008	53		
2SA307X		3008	53		
2SA307Y		3008	53		
2SA308	126	3006	51	TR17/635	G0003
2SA308A		3006	51		
2SA308B		3006	51		
2SA308C		3006	51		
2SA308D		3006	51		
2SA308E		3006	51		
2SA308F		3006	51		
2SA308G		3006	51		
2SA308GN		3006	51		
2SA308H		3006	51		
2SA308J		3006	51		
2SA308K		3006	51		
2SA308L		3006	51		

Original Part Number	ECG	SK	GE	IR/WEP	HEP
2SA308M		3006	51		
2SA308OR		3006	51		
2SA308R		3006	51		
2SA308X		3006	51		
2SA308Y		3006	51		
2SA309	126	3006	51	TR17/635	G0008
2SA309A		3006	51		
2SA309B		3006	51		
2SA309C		3006	51		
2SA309D		3006	51		
2SA309E		3006	51		
2SA309F		3006	51		
2SA309G		3006	51		
2SA309GN		3006	51		
2SA309H		3006	51		
2SA309J		3006	51		
2SA309K		3006	51		
2SA309L		3006	51		
2SA309M		3006	51		
2SA309OR		3006	51		
2SA309R		3006	51		
2SA309X		3006	51		
2SA309Y		3006	51		
2SA310	126	3006	51	TR17/635	G0008
2SA310A		3006	51		
2SA310B		3006	51		
2SA310C		3006	51		
2SA310D		3006	51		
2SA310E		3006	51		
2SA310F		3006	51		
2SA310G		3006	51		
2SA310GN		3006	51		
2SA310H		3006	51		
2SA310J		3006	51		
2SA310K		3006	51		
2SA310L		3006	51		
2SA310M		3006	51		
2SA310OR		3006	51		
2SA310R		3006	51		
2SA310X		3006	51		
2SA310Y		3006	51		
2SA311	100	3008	51	TR05/254	G0005
2SA311A		3008	53		
2SA311B		3008	53		
2SA311C		3008	53		
2SA311D		3008	53		
2SA311E		3008	53		
2SA311F		3008	53		
2SA311G		3008	53		
2SA311GN		3008	53		
2SA311H		3008	53		
2SA311J		3008	53		
2SA311K		3008	53		
2SA311L		3008	53		
2SA311M		3008	53		
2SA311OR		3008	53		
2SA311R		3008	53		
2SA311X		3008	53		
2SA311Y		3008	53		
2SA312	100	3008	51	TR05/254	G0005
2SA312A		3008	53		
2SA312B		3008	53		
2SA312C		3008	53		
2SA312D		3008	53		
2SA312E		3008	53		
2SA312E		3008	53		
2SA312F		3008	53		
2SA312G		3008	53		
2SA312GN		3008	53		
2SA312H		3008	53		
2SA312J		3008	53		
2SA312K		3008	53		
2SA312L		3008	53		
2SA312M		3008	53		
2SA312OR		3008	53		
2SA312R		3008	53		
2SA312X		3008	53		
2SA312Y		3008	53		
2SA313	126	3008	51	TR17/635	G0009
2SA313A		3008	53		
2SA313B		3008	53		
2SA313C		3008	53		
2SA313D		3008	53		
2SA313E		3008	53		
2SA313F		3008	53		
2SA313G		3008	53		
2SA313GN		3008	53		
2SA313H		3008	53		
2SA313J		3008	53		
2SA313K		3008	53		
2SA313L		3008	53		
2SA313M		3008	53		
2SA313OR		3008	53		
2SA313R		3008	53		
2SA313X		3008	53		
2SA313Y		3008	53		
2SA314	126	3008	51	TR17/635	G0009
2SA314A		3008	53		
2SA314B		3008	53		
2SA314C		3008	53		
2SA314D		3008	53		
2SA314E		3008	53		
2SA314F		3008	53		
2SA314G		3008	53		
2SA314GN		3008	53		
2SA314H		3008	53		
2SA314J		3008	53		
2SA314K		3008	53		
2SA314L		3008	53		
2SA314M		3008	53		
2SA314OR		3008	53		
2SA314R		3008	53		
2SA314X		3008	53		
2SA314Y		3008	53		
2SA315	126	3008	51	TR17/635	G0009
2SA315A		3008	53		
2SA315B		3008	53		
2SA315C		3008	53		
2SA315D		3008	53		
2SA315E		3008	53		
2SA315F		3008	53		
2SA315G		3008	53		
2SA315GN		3008	53		
2SA315H		3008	53		
2SA315J		3008	53		
2SA315K		3008	53		
2SA315L		3008	53		
2SA315M		3008	53		
2SA315OR		3008	53		
2SA315R		3008	53		
2SA315X		3008	53		
2SA315Y		3008	53		
2SA316	126	3008	51	TR17/635	G0009
2SA316A		3008	53		
2SA316B		3008	53		
2SA316C		3008	53		
2SA316D		3008	53		
2SA316E		3008	53		
2SA316F		3008	53		
2SA316G		3008	53		
2SA316GN		3008	53		
2SA316H		3008	53		
2SA316J		3008	53		
2SA316K		3008	53		
2SA316L		3008	53		
2SA316M		3008	53		
2SA316OR		3008	53		
2SA316R		3008	53		
2SA316X		3008	53		
2SA316Y		3008	53		
2SA321	160	3006	50	TR17/637	G0008
2SA321-1		3006	51		
2SA321A		3006	51		
2SA321B		3006	51		
2SA321C		3006	51		
2SA321D		3006	51		
2SA321E		3006	51		
2SA321F		3006	51		
2SA321G		3006	51		
2SA321GN		3006	51		
2SA321H		3006	51		
2SA321J		3006	51		
2SA321K		3006	51		
2SA321L		3006	51		
2SA321M		3006	51		
2SA321OR		3006	51		
2SA321R		3006	51		
2SA321X		3006	51		
2SA321Y		3006	51		
2SA322	160	3008	50	TR17/637	G0008

Original Part Number	ECG	SK	GE	IR/WEP	HEP
2SA322/4-279				TR17/	
2SA322A		3008	53		
2SA322B		3008	53		
2SA322C		3008	53		
2SA322D		3008	53		
2SA322E		3008	53		
2SA322F		3008	53		
2SA322G		3008	53		
2SA322GN		3008	53		
2SA322H		3008	53		
2SA322L		3008	53		
2SA322M		3008	53		
2SA322OR		3008	53		
2SA322R		3008	53		
2SA322X		3008	53		
2SA322Y		3008	53		
2SA323	160	3008	50	TR17/637	G0003
2SA323A		3008	53		
2SA323B		3008	53		
2SA323C		3008	53		
2SA323D		3008	53		
2SA323E		3008	53		
2SA323F		3008	53		
2SA323G		3008	53		
2SA323GN		3008	53		
2SA323J		3008	53		
2SA323K		3008	53		
2SA323L		3008	53		
2SA323M		3008	53		
2SA323R		3008	53		
2SA323X		3008	53		
2SA323Y		3008	53		
2SA324	160	3006	50	TR17/637	G0003
2SA324A		3006	51		
2SA324B		3006	51		
2SA324C		3006	51		
2SA324D		3006	51		
2SA324E		3006	51		
2SA324F	158	3006	51		
2SA324G		3006	51		
2SA324GN		3006	51		
2SA324H		3006	51		
2SA324K		3006	51		
2SA324L		3006	51		
2SA324M		3006	51		
2SA324OR		3006	51		
2SA324R		3006	51		
2SA324X		3006	51		
2SA324Y		3006	51		
2SA325	126	3006	51	TR17/635	G0005
2SA325A		3006	51		
2SA325B		3006	51		
2SA325D		3006	51		
2SA325E		3006	51		
2SA325G		3006	51		
2SA325GN		3006	51		
2SA325H		3006	51		
2SA325J		3006	51		
2SA325K		3006	51		
2SA325L		3006	51		
2SA325M		3006	51		
2SA325OR		3006	51		
2SA325R		3006	51		
2SA325X		3006	51		
2SA325Y		3006	51		
2SA326	126	3006	53	TR17/635	G0005
2SA326A		3006	51		
2SA326B		3006	51		
2SA326C		3006	51		
2SA326D		3006	51		
2SA326E		3006	51		
2SA326F		3006	51		
2SA326G		3006	51		
2SA326GN		3006	51		
2SA326H		3006	51		
2SA326J		3006	51		
2SA326K		3006	51		
2SA326L		3006	51		
2SA326M		3006	51		
2SA326OR		3006	51		
2SA326R		3006	51		
2SA326X		3006	51		
2SA326Y		3006	51		
2SA327	160		50	TR17/637	G0009
2SA329	126	3006	2	TR17/635	G0008
2SA329A	126	3008	2	TR17/635	G0008
2SA329B	126	3008	2	TR17/635	G0008
2SA329C		3006	51		
2SA329D		3006	51		
2SA329E		3006	51		
2SA329F		3006	51		
2SA329G		3006	51		
2SA329H		3006	51		
2SA329J		3006	51		
2SA329K		3006	51		
2SA329L		3006	51		
2SA329M		3006	51		
2SA329OR		3006	51		
2SA329R		3006	51		
2SA329X		3006	51		G0008
2SA329Y		3006	51		G0008
2SA330	100	3005	50	TR05/254	G0001
2SA330A		3005	51		
2SA330B		3005	51		
2SA330C		3005	51		
2SA330D		3005	51		
2SA330E		3005	51		
2SA330F		3005	51		
2SA330G		3005	51		
2SA330GN		3005	51		
2SA330H		3005	51		
2SA330J		3005	51		
2SA330K		3005	51		
2SA330L		3005	51		
2SA330M		3005	51		
2SA330X		3005	51		
2SA330Y		3005	51		
2SA331	126	3006	50	TR17/635	G0003
2SA331A		3006	51		
2SA331B		3006	51		
2SA331C		3006	51		
2SA331D		3006	51		
2SA331E		3006	51		
2SA331F		3006	51		
2SA331G		3006	51		
2SA331GN		3006	51		
2SA331H		3006	51		
2SA331J		3006	51		
2SA331K		3006	51		
2SA331L		3006	51		
2SA331M		3006	51		
2SA331OR		3006	51		
2SA331R		3006	51		
2SA331X		3006	51		
2SA331Y		3006	51		
2SA332	100	3008	53	TR05/254	G0005
2SA332A		3008	53		
2SA332B		3008	53		
2SA332C		3008	53		
2SA332D		3008	53		
2SA332E		3008	53		
2SA332F		3008	53		
2SA332G		3008	53		
2SA332GN		3008	53		
2SA332H		3008	53		
2SA332J		3008	53		
2SA332K		3008	53		
2SA332L		3008	53		
2SA332M		3008	53		
2SA332OR		3008	53		
2SA332R		3008	53		
2SA332X		3008	53		
2SA332Y		3008	53		
2SA334M		3006			
2SA335	126	3006	50	TR17/635	G0001
2SA335A		3006	51		
2SA335B		3006	51		
2SA335C		3006	51		
2SA335D		3006	51		
2SA335E		3006	51		
2SA335F		3006	51		
2SA335G		3006	51		
2SA335GN		3006	51		
2SA335H		3006	51		
2SA335J		3006	51		
2SA335K		3006	51		
2SA335L		3006	51		
2SA335M		3006	51		
2SA335OR		3006	51		
2SA335R		3006	51		

Original Part Number	ECG	SK	GE	IR/WEP	HEP
2SA335X		3006	51		
2SA335Y		3006	51		
2SA337	126	3006	51	TR17/635	G0001
2SA337A		3006	51		
2SA337B		3006	51		
2SA337C		3006	51		
2SA337D		3006	51		
2SA337E		3006	51		
2SA337F		3006	51		
2SA337G		3006	51		
2SA337GN		3006	51		
2SA337H		3006	51		
2SA337J		3006	51		
2SA337K		3006	51		
2SA337L		3006	51		
2SA337M		3006	51		
2SA337OR		3006	51		
2SA337R		3006	51		
2SA337X		3006	51		
2SA337Y		3006	51		
2SA338	126	3008	51	TR17/635	G0008
2SA338A		3008	53		
2SA338B		3008	53		
2SA338C		3008	53		
2SA338D		3008	53		
2SA338E		3008	53		
2SA338F		3008	53		
2SA338G		3008	53		
2SA338GN		3008	53		
2SA338H		3008	53		
2SA338J		3008	53		
2SA338K		3008	53		
2SA338L		3008	53		
2SA338M		3008	53		
2SA338OR		3008	53		
2SA338R		3008	53		
2SA338X		3008	53		
2SA338Y		3008	53		
2SA339	126	3008	51	TR17/635	G0008
2SA339A		3008	53		
2SA339B		3008	53		
2SA339C		3008	53		
2SA339D		3008	53		
2SA339E		3008	53		
2SA339F		3008	53		
2SA339G		3008	53		
2SA339GN		3008	53		
2SA339H		3008	53		
2SA339J		3008	53		
2SA339K		3008	53		
2SA339L		3008	53		
2SA339M		3008	53		
2SA339OR		3008	53		
2SA339R		3008	53		
2SA339X		3008	53		
2SA339Y		3008	53		
2SA340	160		51	TR17/637	G0003
2SA341	160	3008	51	TR17/637	G0003
2SA341-OA	160			TR17/637	
2SA341-OB	160			TR17/637	
2SA241A		3008	53		
2SA341B		3008	53		
2SA341C		3008	53		
2SA341D		3008	53		
2SA341E		3008	53		
2SA341F		3008	53		
2SA341G		3008	53		
2SA341GN		3008	53		
2SA341H		3008	53		
2SA341J		3008	53		
2SA341K		3008	53		
2SA341L		3008	53		
2SA341M		3008	53		
2SA341OR		3008	53		
2SA341R		3008	53		
2SA341X		3008	53		
2SA341Y		3008	53		
2SA342	160	3008	50	TR17/637	G0003
2SA342A	160	3006	50	TR17/637	G0003
2SA342B		3006	51		
2SA342C		3006	51		
2SA342D		3006	51		
2SA342E		3006	51		
2SA342F		3006	51		
2SA342G		3006	51		
2SA342GN		3006	51		
2SA342H		3006	51		
2SA342J		3006	51		
2SA342K		3006	51		
2SA342L		3006	51		
2SA342M		3006	51		
2SA342OR		3006	51		
2SA342R		3006	51		
2SA342X		3006	51		
2SA342Y		3006	51		
2SA343	126	3006	50	TR17/635	G0008
2SA343A		3006	51		
2SA343B		3006	51		
2SA343C		3006	51		
2SA343D		3006	51		
2SA343E		3006	51		
2SA343F		3006	51		
2SA343G		3006	51		
2SA343H		3006	51		
2SA343J		3006	51		
2SA343K		3006	51		
2SA343L		3006	51		
2SA343M		3006	51		
2SA343OR		3006	51		
2SA343R		3006	51		
2SA343X		3006	51		
2SA343Y		3006	51		
2SA344	126	3006	51	TR17/635	G0009
2SA344A		3006	51		
2SA344B		3006	51		
2SA344C		3006	51		
2SA344D		3006	51		
2SA344E		3006	51		
2SA344F		3006	51		
2SA344G		3006	51		
2SA344GN		3006	51		
2SA344H		3006	51		
2SA344J		3006	51		
2SA344K		3006	51		
2SA344L		3006	51		
2SA344M		3006	51		
2SA344OR		3006	51		
2SA344R		3006	51		
2SA344X		3006	51		
2SA344Y		3006	51		
2SA345	160		51	TR17/637	G0003
2SA346	160		51	TR17/637	G0003
2SA347	160		51	TR17/637	G0003
2SA348	160	3006	51	TR17/637	G0003
2SA348A		3006	51		
2SA348B		3006	51		
2SA348C		3006	51		
2SA348D		3006	51		
2SA348E		3006	51		
2SA348F		3006	51		
2SA348G		3006	51		
2SA348GN		3006	51		
2SA348H		3006	51		
2SA348J		3006	51		
2SA348K		3006	51		
2SA348L		3006	51		
2SA348M		3006	51		
2SA348OR		3006	51		
2SA348R		3006	51		
2SA348X		3006	51		
2SA348Y		3006	51		
2SA349	160		51	TR17/637	G0003
2SA350	126	3007	50	TR17/635	G0008
2SA350(A)				TR17/	
2SA350A	126	3006	50	TR17/635	G0008
2SA350B		3006	51		
2SA350BK		3006	51		
2SA350C	126	3006	51	TR17/635	G0005
2SA350D		3006	51		
2SA350E		3006	51		
2SA350F		3006	51		
2SA350G		3006	51		
2SA350GN		3006	51		
2SA350H	126	3006	51	TR17/635	G0009
2SA350J		3006	51		
2SA350K		3006	51		
2SA350L		3006	51		
2SA350M		3006	51		
2SA350OR		3006	51		
2SA350R	126	3006	50	TR17/635	G0006

Original Part Number	ECG	SK	GE	IR/WEP	HEP
2SA350T	126	3006	50	TR17/635	G0005
2SA350TY	126	3006	50	TR17/635	G0008
2SA350X		3006	51		
2SA350Y	126	3006	50	TR17/635	G0001
2SA351	126	3007	50	TR17/635	G0008
2SA351A	126	3007	51	TR17/635	G0008
2SA351A-2		3007	50		
2SA351B	126	3007	50	TR17/635	G0008
2SA351C		3007	50		
2SA351D		3007	50		
2SA351E		3007	50		
2SA351F		3007	50		
2SA351G		3007	50		
2SA351GN		3007	50		
2SA351GR	126				
2SA351K		3007	50		
2SA351L		3007	50		
2SA351M		3007	50		
2SA351OR		3007	50		
2SA351R		3007	50		
2SA351X		3007	50		
2SA351Y		3007	50		
2SA351-A		3006	50	TR17/	
2SA351(B,GR)/05				TR17/	
2SA352	126	3006	50	TR17/635	G0008
2SA352A	126	3006	51	TR17/635	G0005
2SA352B	126	3006	51	TR17/635	G0008
2SA352C		3006	51		
2SA352D		3006	51		
2SA352E		3006	51		
2SA352F		3006	51		
2SA352G		3006	51		
2SA352GN		3006	51		
2SA352H		3006	51		
2SA352J		3006	51		
2SA352K		3006	51		
2SA352L		3006	51		
2SA352M		3006	51		
2SA352OR		3006	51		
2SA352R		3006	51		
2SA352X		3006	51		
2SA352Y		3006	51		
2SA353	125	3006	50	TR17/635	G0008
2SA353A	126	3006	51	TR17/635	G0005
2SA353-AC	126			TR85/	
2SA353AL		3006	51		
2SA353B		3006	51	TR17/	
2SA353C	126	3006	51	TR17/635	G0009
2SA353CL		3006	51		
2SA353D		3006	51		
2SA353E		3006	51		
2SA353F		3006	51		
2SA353G		3006	51		
2SA353GN		3006	51		
2SA353H		3006	51		
2SA353J		3006	51		
2SA353K		3006	51		
2SA353L		3006	51		
2SA353M		3006	51		
2SA353OR		3006	51		
2SA353R		3006	51		
2SA353X		3006	51		
2SA353Y		3006	51		
2SA353(A)				TR17/	
2SA353(A,C)				TR17/	
2SA353(C)				TR17/	
2SA354	126	3006	50	TR17/635	G0008
2SA354A	126	3006	51	TR17/635	G0008
2SA354B	126	3006	51	TR17/635	G0008
2SA354BK		3008	53		
2SA354C		3006	51		
2SA354D		3006	51		
2SA354E		3006	51		
2SA354F		3006	51		
2SA354G		3006	51		
2SA354GN		3006	51		
2SA354H		3006	51		
2SA354J		3006	51		
2SA354K		3006	51		
2SA354L		3006	51		
2SA354M		3006	51		
2SA354OR		3006	51		
2SA354R		3006	51		
2SA354X		3006	51		
2SA354Y		3006	51		
2SA355	126	3006	50	TR17/635	G0008
2SA355A	126	3006	51	TR17/635	G0009
2SA355B		3006	51		
2SA355C		3006	51		
2SA355D		3006	51		
2SA355E		3006	51		
2SA355F		3006	51		
2SA355G		3006	51		
2SA355H		3006	51		
2SA355J		3006	51		
2SA355K		3006	51		
2SA355L		3006	51		
2SA355M		3006	51		
2SA355OR		3006	51		
2SA355R		3006	51		
2SA355X		3006	51		
2SA355Y		3006	51		
2SA356	126	3008	50	TR17/635	G0008
2SA356A		3008	53		
2SA356B		3008	53		
2SA356C		3008	53		
2SA356D		3008	53		
2SA356E		3008	53		
2SA356F		3008	53		
2SA356G		3008	53		
2SA356GN		3008	53		
2SA356H		3008	53		
2SA356J		3008	53		
2SA356K		3008	53		
2SA356L		3008	53		
2SA356M		3008	53		
2SA356OR		3008	53		
2SA356R		3008	53		
2SA357	126	3008	50	TR17/635	G0008
2SA357A		3008	53		
2SA357B		3008	53		
2SA357C		3008	53		
2SA357D		3008	53		
2SA357E		3008	53		
2SA357F		3008	53		
2SA357G		3008	53		
2SA357GN		3008	53		
2SA357H		3008	53		
2SA357J		3008	53		
2SA357K		3008	53		
2SA357L		3008	53		
2SA357M		3008	53		
2SA357OR		3008	53		
2SA357R		3008	53		
2SA357X		3008	53		
2SA357Y		3008	53		
2SA358	126	3006	51	TR17/635	G0006
2SA358-3		3006	51		
2SA358A		3006	51		
2SA358B		3006	51		
2SA358C		3006	51		
2SA358D		3006	51		
2SA358E		3006	51		
2SA358F		3006	51		
2SA358G		3006	51		
2SA358GN		3006	51		
2SA358H		3006	51		
2SA358J		3006	51		
2SA358K		3006	51		
2SA358L		3006	51		
2SA358M		3006	51		
2SA358OR		3006	51		
2SA358R		3006	51		
2SA358X		3006	51		
2SA358Y		3006	51		
2SA359	126	3006	53	TR17/635	G0002
2SA359A		3006	51		
2SA359B		3006	51		
2SA359C		3006	51		
2SA359D		3006	51		
2SA359E		3006	51		
2SA359F		3006	51		
2SA359G		3006	51		
2SA359GN		3006	51		
2SA359H		3006	51		
2SA359J		3006	51		
2SA359K		3006	51		
2SA359L		3006	51		
2SA359M		3006	51		
2SA359OR		3006	51		

Original Part Number	ECG	SK	GE	IR/WEP	HEP
2SA359R		3006	51		
2SA359X		3006	51		
2SA359Y		3006	51		
2SA360	160	3006	50	TR17/637	G0003
2SA360A		3006	51		
2SA360B		3006	51		
2SA360C		3006	51		
2SA360E		3006	51		
2SA360F		3006	51		
2SA360G		3007	50		
2SA360GN		3006	50		
2SA360H		3007	50		
2SA360J		3006	51		
2SA360K		3007	50		
2SA360L		3006	51		
2SA360M		3006	51		
2SA360OR		3006	51		
2SA360R		3006	51		
2SA360X		3006	51		
2SA360Y		3006	51		
2SA361	160	3007	51	TR17/637	G0003
2SA361A		3007	50		
2SA361B		3007	50		
2SA361C		3007	50		
2SA361D		3007	50		
2SA361E		3007	50		
2SA361F		3007	50		
2SA361G		3007	50		
2SA361GN		3007	50		
2SA361H		3007	50		
2SA361J		3007	50		
2SA361K		3007	50		
2SA361L		3007	50		
2SA361M		3007	50		
2SA361OR		3007	50		
2SA361R		3007	50		
2SA361X		3007	50		
2SA361Y		3007	50		
2SA362	160		51	TR17/637	G0003
2SA363	160			TR17/637	G0003
2SA364	126	3006	51	TR17/635	G0003
2SA364A		3006	51		
2SA364B		3006	51		
2SA364C		3006	51		
2SA364D		3006	51		
2SA364E		3006	51		
2SA364F		3006	51		
2SA364G		3006	51		
2SA364GN		3006	51		
2SA364H		3006	51		
2SA364J		3006	51		
2SA364K		3006	51		
2SA364L		3006	51		
2SA364M		3006	51		
2SA364OR		3006	51		
2SA364R		3006	51		
2SA364X		3006	51		
2SA364Y		3006	51		
2SA365	126	3006	51	TR17/635	G0003
2SA365A		3006	51		
2SA365B		3006	51		
2SA365C		3006	51		
2SA365D		3006	51		
2SA365E		3006	51		
2SA365F		3006	51		
2SA365G		3006	51		
2SA365GN		3006	51		
2SA365H		3006	51		
2SA365J		3006	51		
2SA365K		3006	51		
2SA365L		3006	51		
2SA365M		3006	51		
2SA365OR		3006	51		
2SA365R		3006	51		
2SA365X		3006	51		
2SA365Y		3006	51		
2SA366	126	3006	52	TR17/635	
2SA366A		3006	51		
2SA366B		3006	51		
2SA366C		3006	51		
2SA366D		3006	51		
2SA366E		3006	51		
2SA366F		3006	51		
2SA366G		3006	51		
2SA366GN		3006	51		
2SA366H		3006	51		
2SA366J		3006	51		
2SA366K		3006	51		
2SA366L		3006	51		
2SA366M		3006	51		
2SA366OR		3006	51		
2SA366R		3006	51		
2SA366X		3006	51		
2SA366Y		3006	51		
2SA367	126	3006	50	TR17/635	G0008
2SA367A		3006	51		
2SA367B		3006	51		
2SA367C		3006	51		
2SA367D		3006	51		
2SA367E		3006	51		
2SA367F		3006	51		
2SA367G		3006	51		
2SA367GN		3006	51		
2SA367H		3006	51		
2SA367J		3006	51		
2SA367K		3006	51		
2SA367L		3006	51		
2SA367M		3006	51		
2SA367OR		3006	51		
2SA367R		3006	51		
2SA367X		3006	51		
2SA367Y		3006	51		
2SA368	126	3007	50	TR17/635	G0008
2SA368A		3007	50		
2SA368B		3007	50		
2SA368C		3007	50		
2SA368D		3007	50		
2SA368E		3007	50		
2SA368F		3007	50		
2SA368G		3007	50		
2SA368GN		3007	50		
2SA368H		3007	50		
2SA368J		3007	50		
2SA368K		3007	50		
2SA368L		3007	50		
2SA368M		3007	50		
2SA368OR		3007	50		
2SA368R		3007	50		
2SA368X		3007	50		
2SA368Y		3007	50		
2SA369	126	3008	51	TR17/635	G0008
2SA369A		3008	53		
2SA369B		3008	53		
2SA369C		3008	53		
2SA369D		3008	53		
2SA369E		3008	53		
2SA369F		3008	53		
2SA369G		3008	53		
2SA369GN		3008	53		
2SA369H		3008	53		
2SA369J		3008	53		
2SA369K		3008	53		
2SA369L		3008	53		
2SA369M		3008	53		
2SA369OR		3008	53		
2SA369R		3008	53		
2SA369X		3008	53		
2SA369Y		3008	53		
2SA370			53		
2SA371			53		G0005
2SA372	160		51	TR17/637	G0008
2SA373	160		53	TR17/637	G0002
2SA374			51		
2SA375	160	3123	80	TR17/637	G0001
2SA375A		3123	80		
2SA375A-2B					G0007
2SA375B		3123	80		
2SA375C		3123	80		
2SA375D		3123	80		
2SA375E		3123	80		
2SA375F		3123	80		
2SA375G		3123	80		
2SA375GN		3123	80		
2SA375H		3123	80		
2SA375J		3123	80		
2SA375K		3123	80		
2SA375L		3123	80		
2SA375M		3123	80		
2SA375OR		3123	80		
2SA375R		3123	80		

Original Part Number	ECG	SK	GE	IR/WEP	HEP
2SA375X		3123	80		
2SA375Y		3123	80		
2SA376	126	3006	50	TR17/635	G0008
2SA376A		3006	51		
2SA376B		3006	51		
2SA376C		3006	51		
2SA376D		3006	51		
2SA376E		3006	51		
2SA376F		3006	51		
2SA376GN		3006	51		
2SA376H		3006	51		
2SA376J		3006	51		
2SA376K		3006	51		
2SA376L		3006	51		
2SA376OR		3006	51		
2SA376R		3006	51		
2SA376X		3006	51		
2SA376Y		3006	51		
2SA377	160	3006	50	TR17/637	G0008
2SA377A		3006	51		
2SA377B		3006	51		
2SA377C		3006	51		
2SA377D		3006	51		
2SA377E		3006	51		
2SA377F		3006	51		
2SA377G		3006	51		
2SA377GN		3006	51		
2SA377H		3006	51		
2SA377J		3006	51		
2SA377[illegible]		3006	51		
2SA377L		3006	51		
2SA377M		3006	51		
2SA377OR		3006	51		
2SA377R		3006	51		
2SA377X		3006	51		
2SA377Y		3006	51		
2SA378	160	3006	50	TR17/637	G0008
2SA379	160	3006	50	TR17/637	
2SA380	126	3007	51	TR17/635	G0008
2SA380A		3007	50		
2SA380AO					S0016
2SA380AORN					S0016
2SA380B		3007	50		
2SA380C		3007	50		
2SA380D		3007	50		
2SA380E		3007	50		
2SA380F		3007	50		
2SA380G		3007	50		
2SA380GN		3007	50		
2SA380H		3007	50		
2SA380J		3007	50		
2SA380K		3007	50		
2SA380L		3007	50		
2SA380M		3007	50		
2SA380OR		3007	50		
2SA380R		3007	50		
2SA380X		3007	50		
2SA380Y		3007	50		
2SA381	126	3008	51	TR17/635	G0008
2SA381A		3008	53		
2SA381B		3008	53		
2SA381D		3008	53		
2SA381E		3008	53		
2SA381F		3008	53		
2SA381G		3008	53		
2SA381GN		3008	53		
2SA381H		3008	53		
2SA381K		3008	53		
2SA381L		3008	53		
2SA381M		3008	53		
2SA381OR		3008	53		
2SA381R		3008	53		
2SA381X		3008	53		
2SA381Y		3008	53		
2SA382	126	3008	51	TR17/635	G0008
2SA382A		3008	53		
2SA382B		3008	53		
2SA382C		3008	53		
2SA382D		3008	53		
2SA382E		3008	53		
2SA382F		3008	53		
2SA382G		3008	53		
2SA382GN		3008	53		
2SA382H		3008	53		
2SA382J		3008	53		
2SA382K		3008	53		
2SA382L		3008	53		
2SA382M		3008	53		
2SA382OR		3008	53		
2SA382R		3008	53		
2SA382X		3008	53		
2SA382Y		3008	53		
2SA383	126	3008	51	TR17/635	G0008
2SA383A		3008	53		
2SA383B		3008	53		
2SA383C		3008	53		
2SA383D		3008	53		
2SA383E		3008	53		
2SA383F		3008	53		
2SA383G		3008	53		
2SA383GN		3008	53		
2SA383H		3008	53		
2SA383J		3008	53		
2SA383K		3008	53		
2SA383L		3008	53		
2SA383M		3008	53		
2SA383OR		3008	53		
2SA383R		3008	53		
2SA383X		3008	53		
2SA383Y		3008	53		
2SA384	126	3008	51	TR17/635	G0008
2SA384A		3008	53		
2SA384B		3008	53		
2SA384C		3008	53		
2SA384D		3008	53		
2SA384E		3008	53		
2SA384F		3008	53		
2SA384G		3008	53		
2SA384GN		3008	53		
2SA384H		3008	53		
2SA384J		3008	53		
2SA384K		3008	53		
2SA384L		3008	53		
2SA384M		3008	53		
2SA384OR		3008	53		
2SA384R		3008	53		
2SA384X		3008	53		
2SA384Y		3008	53		
2SA385	126	3006	50	TR17/635	G0005
2SA385A	126	3006	50	TR17/635	G0006
2SA385B		3006	51		
2SA385C		3006	50	TR17/	
2SA385D	126	3006	51	TR17/	
2SA385E		3006	51		
2SA385F		3006	51		
2SA385G		3006	51		
2SA385GN		3006	51		
2SA385H		3006	51		
2SA385J		3006	51		
2SA385K		3006	51		
2SA385L		3006	51	/635	G0005
2SA385M		3006	51		
2SA385OR		3006	51		
2SA385R		3006	51		
2SA385X		3006	51		
2SA385Y		3006	51		
2SA391	126	3006	1	TR17/635	G0008
2SA391A		3006	51		
2SA391B		3006	51		
2SA391C		3006	51		
2SA391D		3006	51		
2SA391E		3006	51		
2SA391F		3006	51		
2SA391G		3006	51		
2SA391GN		3006	51		
2SA391H		3006	51		
2SA391J		3006	51		
2SA391K		3006	51		
2SA391L		3006	51		
2SA391M		3006	51		
2SA391OR		3006	51		
2SA391R		3006	51		
2SA391X		3006	51		
2SA391Y		3006	51		
2SA392	126	3006	1	TR17/635	G0008
2SA392A		3006	51		
2SA392B		3006	51		
2SA392C		3006	51		
2SA392D		3006	51		
2SA392E		3006	51		

Original Part Number	ECG	SK	GE	IR/WEP	HEP
2SA392F		3006	51		
2SA392G		3006	51		
2SA392GN		3006	51		
2SA392H		3006	51		
2SA392J		3006	51		
2SA392K		3006	51		
2SA392L		3006	51		
2SA392M		3006	51		
2SA392OR		3006	51		
2SA392R		3006	51		
2SA392X		3006	51		
2SA392Y		3006	51		
2SA393	126	3006	51	TR17/635	G0008
2SA393A	126	3006	51	TR17/635	G0008
2SA394	126	3006	51	TR17/635	G0008
2SA394A		3006	51		
2SA394B		3006	51		
2SA394C		3006	51		
2SA394D		3006	51		
2SA394E		3006	51		
2SA394F		3006	51		
2SA394G		3006	51		
2SA394GN		3006	51		
2SA394H		3006	51		
2SA394J		3006	51		
2SA394K		3006	51		
2SA394L		3006	51		
2SA394M		3006	51		
2SA394OR		3006	51		
2SA394R		3006	51		
2SA394X		3006	51		
2SA394Y		3006	51		
2SA395	126	3006	51	TR17/635	G0008
2SA395A		3006	51		
2SA395B		3006	51		
2SA395C		3006	51		
2SA395D		3006	51		
2SA395E		3006	51		
2SA395F		3006	51		
2SA395G		3006	51		
2SA395GN		3006	51		
2SA395H		3006	51		
2SA395J		3006	51		
2SA395K		3006	51		
2SA395L		3006	51		
2SA395M		3006	51		
2SA395OR		3006	51		
2SA395R		3006	51		
2SA395X		3006	51		
2SA395Y		3006	51		
2SA396	102		1	TR05/631	G0008
2SA397	102		1	TR05/631	G0008
2SA398	126	3006	51	TR17/635	G0008
2SA398A		3006	51		
2SA398B		3006	51		
2SA398C		3006	51		
2SA398D		3006	51		
2SA398E		3006	51		
2SA398F		3006	51		
2SA398G		3006	51		
2SA398GN		3006	51		
2SA398H		3006	51		
2SA398J		3006	51		
2SA398K		3006	51		
2SA398L		3006	51		
2SA398M		3006	51		
2SA398OR		3006	51		
2SA398R		3006	51		
2SA398X		3006	51		
2SA398Y		3006	51		
2SA399	126	3006	51	TR17/635	G0008
2SA399A		3006	51		
2SA399B		3006	51		
2SA399C		3006	51		
2SA399D		3006	51		
2SA399E		3006	51		
2SA399F		3006	51		
2SA399G		3006	51		
2SA399GN		3006	51		
2SA399H		3006	51		
2SA399J		3006	51		
2SA399K		3006	51		
2SA399L		3006	51		
2SA399M		3006	51		
2SA399OR		3006	51		
2SA399R		3006	51		
2SA399X		3006	51		
2SA399Y		3006	51		
2SA400	126	3008	50	TR17/635	G0008
2SA400A		3008	53		
2SA400B		3008	53		
2SA400C		3008	53		
2SA400D		3008	53		
2SA400E		3008	53		
2SA400F		3008	53		
2SA400G		3008	53		
2SA400GN		3008	53		
2SA400H		3008	53		
2SA400J		3008	53		
2SA400K		3008	53		
2SA400L		3008	53		
2SA400M		3008	53		
2SA400R		3008	53		
2SA400X		3008	53		
2SA400Y		3008	53		
2SA401	160		50	TR17/637	G0003
2SA402	159			TR20/717	S0013
2SA403	160	3006	51	TR17/637	G0003
2SA403A		3006	51		
2SA403B		3006	51		
2SA403C		3006	51		
2SA403D		3006	51		
2SA403E		3006	51		
2SA403F		3006	51		
2SA403G		3006	51		
2SA403GN		3006	51		
2SA403H		3006	51		
2SA403J		3006	51		
2SA403K		3006	51		
2SA403L		3006	51		
2SA403M		3006	51		
2SA403OR		3006	51		
2SA403R		3006	51		
2SA403X		3006	51		
2SA403Y		3006	51		
2SA404	160	3006	1	TR17/637	G0003
2SA404A		3006	51		
2SA404B		3006	51		
2SA404C		3006	51		
2SA404D		3006	51		
2SA404E		3006	51		
2SA404F		3006	51		
2SA404G		3006	51		
2SA404GN		3006	51		
2SA404H		3006	51		
2SA404K		3006	51		
2SA404L		3006	51		
2SA404M		3006	51		
2SA404OR		3006	51		
2SA404R		3006	51		
2SA404X		3006	51		
2SA404Y		3006	51		
2SA405	160		51	TR17/637	G0003
2SA406	102A	3005	1	TR85/250	G0005
2SA406A		3005	51		
2SA406B		3005	51		
2SA406C		3005	51		
2SA406D		3005	51		
2SA406E		3005	51		
2SA406F		3005	51		
2SA406G		3005	51		
2SA406GN		3005	51		
2SA406H		3005	51		
2SA406J		3005	51		
2SA406K		3005	51		
2SA406L		3005	51		
2SA406M		3005	51		
2SA406OR		3005	51		
2SA406R		3005	51		
2SA406X		3005	51		
2SA406Y		3005	51		
2SA407	102A	3005	1	TR85/250	G0005
2SA407A		3005	51		
2SA407B		3005	51		
2SA407C		3005	51		
2SA407D		3005	51		
2SA407E		3005	51		
2SA407F		3005	51		
2SA407G		3005	51		
2SA407GN		3005	51		

Original Part Number	ECG	SK	GE	IR/WEP	HEP
2SA407H		3005	51		
2SA407J		3005	51		
2SA407K		3005	51		
2SA407L		3005	51		
2SA407M		3005	51		
2SA407OR		3005	51		
2SA407R		3005	51		
2SA407X		3005	51		
2SA407Y		3005	51		
2SA408	160	3123	51	TR17/637	G0001
2SA408A		3123	80		
2SA408B		3123	80		
2SA408C		3123	80		
2SA408D		3123	80		
2SA408E		3123	80		
2SA408F		3123	80		
2SA408GN		3123	80		
2SA408H		3123	80		
2SA408J		3123	80		
2SA408K		3123	80		
2SA408L		3123	80		
2SA408M		3123	80		
2SA408OR		3123	80		
2SA408R		3123	80		
2SA408X		3123	80		
2SA408Y		3123	80		
2SA409	160	3123	51	TR17/637	G0001
2SA409A		3123	80		
2SA409B		3123	80		
2SA409C		3123	80		
2SA409D		3123	80		
2SA409E		3123	80		
2SA409F		3123	80		
2SA409G		3123	80		
2SA409GN		3123	80		
2SA409H		3123	80		
2SA409J		3123	80		
2SA409K		3123	80		
2SA409L		3123	80		
2SA409M		3123	80		
2SA409OR		3123	80		
2SA409R		3123	80		
2SA409X		3123	80		
2SA409Y		3123	80		
2SA410	160		51	TR17/637	G0003
2SA411	160		51	TR17/637	G0008
2SA412	126	3006	51	TR17/635	G0005
2SA412A		3006	51		
2SA412B		3006	51		
2SA412C		3006	51		
2SA412D		3006	51		
2SA412E		3006	51		
2SA412F		3006	51		
2SA412G		3006	51		
2SA412H		3006	51		
2SA412J		3006	51		
2SA412K		3006	51		
2SA412L		3006	51		
2SA412M		3006	51		
2SA412OR		3006	51		
2SA412R		3006	51		
2SA412X		3006	51		
2SA412Y		3006	51		
2SA413	160		51	TR17/637	G0003
2SA414	100	3005	1	TR05/254	G0005
2SA414A		3005	51		
2SA414B		3005	51		
2SA414C		3005	51		
2SA414D		3005	51		
2SA414E		3005	51		
2SA414F		3005	51		
2SA414G		3005	51		
2SA414GN		3005	51		
2SA414H		3005	51		
2SA414J		3005	51		
2SA414K		3005	51		
2SA414L		3005	51		
2SA414M		3005	51		
2SA414OR		3005	51		
2SA414R		3005	51		
2SA414X		3005	51		
2SA415	100	3005	1	TR05/254	G0005
2SA415A		3005	51		
2SA415B		3005	51		
2SA415C		3005	51		
2SA415D		3005	51		
2SA415E		3005	51		
2SA415F		3005	51		
2SA415G		3005	51		
2SA415H		3005	51		
2SA415J		3005	51		
2SA415K		3005	51		
2SA415L		3005	51		
2SA415M		3005	51		
2SA415OR		3005	51		
2SA415R		3005	51		
2SA415X		3005	51		
2SA415Y		3005	51		
2SA416	121	3014	16	TR01/232	G6005
2SA416A		3014	3		
2SA416B		3014	3		
2SA416C		3014	3		
2SA416D		3014	3		
2SA416E		3014	3		
2SA416F		3014	3		
2SA416G		3014	3		
2SA416GN		3014	3		
2SA416H		3014	3		
2SA416J		3014	3		
2SA416K		3014	3		
2SA416L		3014	3		
2SA416M		3014	3		
2SA416OR		3014	3		
2SA416R		3014	3		
2SA416X		3014	3		
2SA416Y		3014	3		
2SA417	160		51	TR17/637	G0003
2SA419	160		51	TR17/637	G0003
2SA420	160	3006	50	TR17/637	G0003
2SA420A		3006	51		
2SA420B		3006	51		
2SA420C		3006	51		
2SA420D		3006	51		
2SA420E		3006	51		
2SA420F		3006	51		
2SA420G		3006	51		
2SA420H		3006	51		
2SA420J		3006	51		
2SA420K		3006	51		
2SA420L		3006	51		
2SA420M		3006	51		
2SA420OR		3006	51		
2SA420R		3006	51		
2SA420X		3006	51		
2SA420Y		3006	51		
2SA421	160		51	TR17/637	G0008
2SA422	160		51	TR17/637	G0003
2SA425	160		53	TR17/637	G0002
2SA426	160		53	TR17/637	G0002
2SA426GN		3006	51		
2SA427	126	3006	50	TR17/635	G0009
2SA427A		3006	51		
2SA427B		3006	51		
2SA427C		3006	51		
2SA427D		3006	51		
2SA427E		3006	51		
2SA427F		3006	51		
2SA427G		3006	51		
2SA427GN		3006	51		
2SA427H		3006	51		
2SA427J		3006	51		
2SA427K		3006	51		
2SA427L		3006	51		
2SA427M		3006	51		
2SA427OR		3006	51		
2SA427R		3006	51		
2SA427X		3006	51		
2SA427Y		3006	51		
2SA428	126	3006	50	TR17/635	G0009
2SA428A		3006	51		
2SA428B		3006	51		
2SA428C		3006	51		
2SA428D		3006	51		
2SA428E		3006	51		
2SA428F		3006	51		
2SA428G		3006	51		
2SA428GN		3006	51		
2SA428H		3006	51		
2SA428J		3006	51		
2SA428K		3006	51		

Original Part Number	ECG	SK	GE	IR/WEP	HEP
2SA428L		3006	51		
2SA428M		3006	51		
2SA428OR		3006	51		
2SA428R		3006	51		
2SA428X		3006	51		
2SA428Y		3006	51		
2SA429		3118	89		
2SA429A		3118	89		
2SA429B		3118	89		
2SA429C		3118	89		
2SA429D		3118	89		
2SA429E		3118	89		
2SA429F		3118	89		
2SA429G		3118	89		
2SA429GN		3118	89		
2SA429H		3118	89		
2SA429J		3118	89		
2SA429K		3118	89		
2SA429L		3118	89		
2SA429M		3118	89		
2SA429OR		3118	89		
2SA429R		3118	89		
2SA429X		3118	89		
2SA429Y		3118	89		
2SA430	160		51	TR17/637	G000[illegible]
2SA431	160		50	TR17/637	G0003
2SA431A	160			TR17/637	G0003
2SA431-G					G0003
2SA431A-G					G0003
2SA432	160	3006	51	TR17/637	G0003
2SA432A	160	3006	51	TR17/637	G0003
2SA432B		3006	51		
2SA432C		3006	51		
2SA432D		3006	51		
2SA432E		3006	51		
2SA432F		3006	51		
2SA432G		3006	51		
2SA432GN		3006	51		
2SA432H		3006	51		
2SA432K		3006	51		
2SA432L		3006	51		
2SA432M		3006	51		
2SA432OR		3006	51		
2SA432R		3006	51		
2SA432X		3006	51		
2SA432Y		3006	51		
2SA433	160	3006	51	TR17/637	G0008
2SA433A		3006	51		G0008
2SA433B		3006	51		
2SA433C		3006	51		
2SA433D		3006	51		
2SA433E		3006	51		
2SA433F		3006	51		
2SA433G		3006	51		
2SA433GN		3006	51		
2SA433H		3006	51		
2SA433K		3006	51		
2SA433L		3006	51		
2SA433M		3006	51		
2SA433OR		3006	51		
2SA433R		3006	51		
2SA433X		3006	51		
2SA433Y		3006	51		
2SA434	160	3006	51	TR17/637	G0003
2SA434A		3006	51		
2SA434B		3006	51		
2SA434C		3006	51		
2SA434D		3006	51		
2SA434E		3006	51		
2SA434F		3006	51		
2SA434G		3006	51		
2SA434GN		3006	51		
2SA434H		3006	51		
2SA434J		3006	51		
2SA434K		3006	51		
2SA434L		3006	51		
2SA434M		3006	51		
2SA434OR		3006	51		
2SA434R		3006	51		
2SA434X		3006	51		
2SA434Y		3006	51		
2SA435	160	3008	50	TR17/637	G0008
2SA435A	160	3006	50	TR17/637	G0001
2SA435B	160	3006	50	TR17/637	G0008
2SA435C		3006	51		
2SA435D		3006	51		
2SA435E		3006	51		
2SA435F		3006	51		
2SA435G		3006	51		
2SA435GN		3006	51		
2SA435H		3006	51		
2SA435J		3006	51		
2SA435K		3006	51		
2SA435L		3006	51		
2SA435M		3006	51		
2SA435OR		3006	51		
2SA435R		3006	51		
2SA435X		3006	51		
2SA435Y		3006	51		
2SA436	126	3006	50	TR17/635	G0003
2SA436A		3006	51		
2SA436B		3006	51		
2SA436C		3006	51		
2SA436D		3006	51		
2SA436E		3006	51		
2SA436F		3006	51		
2SA436G		3006	51		
2SA436GN		3006	51		
2SA436H		3006	51		
2SA436J		3006	51		
2SA436K		3006	51		
2SA437	126	3006	50	TR17/635	G0003
2SA437A		3006	51		
2SA437B		3006	51		
2SA437C		3006	51		
2SA437D		3006	51		
2SA437E		3006	51		
2SA437F		3006	51		
2SA437G		3006	51		
2SA437GN		3006	51		
2SA437H		3006	51		
2SA437J		3006	51		
2SA437K		3006	51		
2SA437L		3006	51		
2SA437M		3006	51		
2SA437OR		3006	51		
2SA437R		3006	51		
2SA437X		3006	51		
2SA437Y		3006	51		
2SA438	126	3006	50	TR17/635	G0003
2SA438A		3006	51		
2SA438B		3006	51		
2SA438C		3006	51		
2SA438D		3006	51		
2SA438E		3006	51		
2SA438F		3006	51		
2SA438G		3006	51		
2SA438GN		3006	51		
2SA438H		3006	51		
2SA438J		3006	51		
2SA438K		3006	51		
2SA438L		3006	51		
2SA438M		3006	51		
2SA438OR		3006	51		
2SA438R		3006	51		
2SA438X		3006	51		
2SA438Y		3006	51		
2SA440	160	3006	51	TR17/637	G0003
2SA440A	160	3006	50	TR17/637	G0003
2SA440AL		3006	51		
2SA440B		3006	51		
2SA440C		3006	51		
2SA440D		3006	51		
2SA440E		3006	51		
2SA440F		3006	51		
2SA440G		3006	51		
2SA440GN		3006	51		
2SA440H		3006	51		
2SA440J		3006	51		
2SA440K		3006	51		
2SA440L		3006	51		
2SA440M		3006	51		
2SA440OR		3006	51		
2SA440R		3006	51		
2SA440X		3006	51		
2SA440Y		3006	51		
2SA440(A)				TR17/	
2SA446	126	3006	53	TR17/635	G0002
2SA446-1				TR12/	
2SA446-2				TR12/	

Original Part Number	ECG	SK	GE	IR/WEP	HEP
2SA446-3				TR12/	
2SA446A		3006	51		
2SA446B		3006	51		
2SA446C		3006	51		
2SA446D		3006	51		
2SA446E		3006	51		
2SA446F		3006	51		
2SA446G		3006	51		
2SA446GN		3006	51		
2SA446H		3006	51		
2SA446J		3006	51		
2SA446K		3006	51		
2SA446L		3006	51		
2SA446M		3006	51		
2SA446OR		3006	51		
2SA446R		3006	51		
2SA446X		3006	51		
2SA446Y		3006	51		
2SA447	126	3006	50	TR17/635	G0009
2SA447A		3006	51		
2SA447B		3006	51		
2SA447C		3006	51		
2SA447D		3006	51		
2SA447E		3006	51		
2SA447F		3006	51		
2SA447G		3006	51		
2SA447GN		3006	51		
2SA447H		3006	51		
2SA447J		3006	51		
2SA447K		3006	51		
2SA447L		3006	51		
2SA447M		3006	51		
2SA447OR		3006	51		
2SA447R		3006	51		
2SA447X		3006	51		
2SA447Y		3006	51		
2SA448	160		51	TR17/637	G0003
2SA450	160		51	TR17/637	G0008
2SA450H	160			TR17/637	G0008
2SA451	160		51	TR17/637	G0008
2SA451H	160			TR17/637	G0008
2SA452	160		51	TR17/637	G0008
2SA452H	160			TR17/637	G0008
2SA453	126	3006	51	TR17/635	G0003
2SA453A		3006	51		
2SA453B		3006	51		
2SA453C		3006	51		
2SA453D		3006	51		
2SA453E		3006	51		
2SA453F		3006	51		
2SA453G		3006	51		
2SA453G		3006	51		
2SA453H		3006	51		
2SA453J		3006	51		
2SA453K		3006	51		
2SA453L		3006	51		
2SA453M		3006	51		
2SA453OR		3006	51		
2SA453R		3006	51		
2SA453X		3006	51		
2SA453Y		3006	51		
2SA454	126	3006	51	TR17/635	G0003
2SA454A		3006	51		
2SA454B		3006	51		
2SA454C		3006	51		
2SA454D		3006	51		
2SA454E		3006	51		
2SA454F		3006	51		
2SA454G		3006	51		
2SA454GN		3006	51		
2SA454H		3006	51		
2SA454J		3006	51		
2SA454K		3006	51		
2SA454L		3006	51		
2SA454OR		3006	51		
2SA454R		3006	51		
2SA454X		3006	51		
2SA454Y		3006	51		
2SA455	126	3006	51	TR17/635	G0003
2SA455A		3006	51		
2SA455B		3006	51		
2SA455C		3006	51		
2SA455D		3006	51		
2SA455E		3006	51		
2SA455F		3006	51		
2SA455G		3006	51		
2SA455GN		3006	51		
2SA455H		3006	51		
2SA455J		3006	51		
2SA455K		3006	51		
2SA455L		3006	51		
2SA455M		3006	51		
2SA455OR		3006	51		
2SA455R		3006	51		
2SA455X		3006	51		
2SA455Y		3006	51		
2SA456	126	3006	51	TR17/635	G0003
2SA456A		3006	51		
2SA456B		3006	51		
2SA456C		3006	51		
2SA456D		3006	51		
2SA456E		3006	51		
2SA456F		3006	51		
2SA456G		3006	51		
2SA456GN		3006	51		
2SA456H		3006	51		
2SA456J		3006	51		
2SA456K		3006	51		
2SA456L		3006	51		
2SA456M		3006	51		
2SA456OR		3006	51		
2SA456R		3006	51		
2SA456X		3006	51		
2SA456Y		3006	51		
2SA457	126	3006	51	TR17/635	G0003
2SA457A		3006	51		
2SA457B		3006	51		
2SA457C		3006	51		
2SA457D		3006	51		
2SA457E		3006	51		
2SA457F		3006	51		
2SA457G		3006	51		
2SA457GN		3006	51		
2SA457H		3006	51		
2SA457J		3006	51		
2SA457K		3006	51		
2SA457L		3006	51		
2SA457M		3006	51		
2SA457OR		3006	51		
2SA457R		3006	51		
2SA457X		3006	51		
2SA457Y		3006	51		
2SA458					G0005
2SA459					G0005
2SA460	160			TR17/637	G0003
2SA460B					S0014
2SA461	160			TR17/637	S0003
2SA462	160			TR17/637	G0003
2SA463	160		61	TR17/637	G0003
2SA464	160			TR17/637	G0003
2SA465		3114	21		S0019
2SA466	126	3006	51	TR17/635	
2SA466-1	126	3006			
2SA466-2	126	3006	50	TR17/635	G0005
2SA466-3	126	3006	50	TR17/635	G0005
2SA466A		3006	51		
2SA466B		3006	51		
2SA466BLK	126	3006	51	TR17/635	G0008
2SA466BLU	126	3006	51	TR17/635	G0008
2SA466C		3006	51		
2SA466D		3006	51		
2SA466E		3006	51		
2SA466F		3006	51		
2SA466G		3006	51		
2SA466GN		3006	51		
2SA466H		3006	51		
2SA466J		3006	51		
2SA466K		3006	51		
2SA466L		3006	51		
2SA466M		3006	51		
2SA466OR		3006	51		
2SA466R		3006	51		
2SA466X		3006	51		
2SA466Y		3006	51		
2SA466YEL	126	3006	51	TR17/635	G0008
2SA467	159	3114	21	TR20/717	S0012
2SA467-0	159			TR20/	
2SA467A		3114	89		
2SA467B		3114	89		
2SA467C		3114	89		

Original Part Number	ECG	SK	GE	IR/WEP	HEP
2SA467D		3114	89		
2SA467E		3114	89		
2SA467F		3114	89		
2SA467G	159	3114	89	TR20/717	
2SA467G-O	159			TR20/	S5022
2SA467G-R	159			TR20/	
2SA467G-Y	159			TR20/	
2SA467H		3114	89		
2SA467J		3114	89		
2SA467K		3114	89		
2SA467L		3114	89		
2SA467M		3114	89		
2SA467OR		3114	89		
2SA467R		3114	89		
2SA467X		3114	89		
2SA467Y		3114	89		
2SA468	126	3007	51	TR17/635	G0008
2SA468A		3007	50		
2SA468B		3007	50		
2SA468C		3007	50		
2SA468D		3007	50		
2SA468E		3007	50		
2SA468F		3007	50		
2SA468G		3007	50		
2SA468GN		3007	50		
2SA468H		3007	50		
2SA468J		3007	50		
2SA468K		3007	50		
2SA468L		3007	50		
2SA468M		3007	50		
2SA468OR		3007	50		
2SA468R		3007	50		
2SA468X		3007	50		
2SA468Y		3007	50		
2SA469	126	3007	51	TR17/635	G0008
2SA469A		3007	50		
2SA469B		3007	50		
2SA469C		3007	50		
2SA469D		3007	50		
2SA469E		3007	50		
2SA469F		3007	50		
2SA469G		3007	50		
2SA469GN		3007	50		
2SA469H		3007	50		
2SA469J		3007	50		
2SA469K		3007	50		
2SA469L		3007	50		
2SA469M		3007	50		
2SA469OR		3007	50		
2SA469R		3007	50		
2SA469X		3007	50		
2SA469Y		3007	50		
2SA470	126	3007	51	TR17/635	G0008
2SA470A		3006	51		
2SA470B		3006	51		
2SA470C		3006	51		
2SA470D		3006	51		
2SA470E		3006	51		
2SA470F		3006	51		
2SA470G		3006	51		
2SA470GN		3006	51		
2SA470H		3006	51		
2SA470J		3006	51		
2SA470K		3006	51		
2SA470L		3006	51		
2SA470M		3006	51		
2SA470OR		3006	51		
2SA470R		3006	51		
2SA470X		3006	51		
2SA470Y		3006	51		
2SA471	126	3006	50	TR17/635	G0008
2SA471-1	126	3006	51	TR17/635	G0008
2SA471-2	126	3006	51	TR17/635	G0008
2SA471-3	126	3006	51	TR17/635	G0006
2SA471A		3006	51		
2SA471B		3006	51		
2SA471C		3006	51		
2SA471D		3006	51		
2SA471E		3006	51		
2SA471F		3006	51		
2SA471G		3006	51		
2SA471GN		3006	51		
2SA471H		3006	51		
2SA471J		3006	51		
2SA471K		3006	51		
2SA471L		3006	51		
2SA471M		3006	51		
2SA471OR		3006	51		
2SA471R		3006	51		
2SA471X		3006	51		
2SA471Y		3006	51		
2SA472	126	3007	51	TR17/635	G0008
2SA472-1			50	TR17/	G0009
2SA472-2			50		G0008
2SA472-3			50		
2SA472-4			50	TR17/	G0009
2SA472-5			50		G0008
2SA472-6			50		G0005
2SA472A	126	3006	51	TR17/635	G0008
2SA472B	126	3006	51	TR17/635	G0008
2SA472C	126	3006	51	TR17/635	G0009
2SA472D	126	3006	51	TR17/635	G0009
2SA472E	126	3006	51	TR17/635	G0009
2SA472F		3007	50		
2SA472G		3007	50		
2SA472GN		3007	50		
2SA472H		3007	50		
2SA472J		3007	50		
2SA472K		3007	50		
2SA472L		3007	50		
2SA472M		3007	50		
2SA472OR		3007	50		
2SA472R		3007	50		
2SA472X		3007	50		
2SA472Y		3007	50		
2SA473	153	3084	69	TR77/700	S5006
2SA473-GR	153			TR77/	
2SA473-O	153			TR77/	
2SA473-R	153			TR77/	
2SA473-Y	153			TR77/	
2SA473Y	153	3083		TR56/	S5006
2SA473(1)				TR56/	
2SA473(Y)				TR77(9)	
2SA474	126	3006	50	TR17/635	G0009
2SA474A		3006	51		
2SA474B		3006	51		
2SA474C		3006	51		
2SA474D		3006	51		
2SA474E		3006	51		
2SA474F		3006	51		
2SA474G		3006	51		
2SA474-G					G0009
2SA474GN		3006	51		
2SA474H		3006	51		
2SA474J		3006	51		
2SA474K		3006	51		
2SA474L		3006	51		
2SA474M		3006	51		
2SA474OR		3006	51		
2SA474R		3006	51		
2SA474X		3006	51		
2SA474Y		3006	51		
2SA475		3007	50		G0009
2SA475A		3007			
2SA475(O)				TR30/	
2SA476	126	3008	52	TR17/635	G0008
2SA476A		3008	53		
2SA476B		3008	53		
2SA476C		3008	53		
2SA476D		3008	53		
2SA476E		3008	53		
2SA476F		3008	53		
2SA476G		3008	53		
2SA476GN		3008	53		
2SA476H		3008	53		
2SA476J		3008	53		
2SA476K		3008	53		
2SA476L		3008	53		
2SA476M		3008	53		
2SA476OR		3008	53		
2SA476R		3008	53		
2SA476X		3008	53		
2SA476Y		3008	53		
2SA477	126	3006	1	TR17/635	G0008
2SA477A		3006	51		
2SA477B		3006	51		
2SA477C		3006	51		
2SA477D		3006	51		
2SA477E		3006	51		
2SA477F		3006	51		

Original Part Number	ECG	SK	GE	IR/WEP	HEP
2SA477G		3006	51		
2SA477GN		3006	51		
2SA477H		3006	51		
2SA477J		3006	51		
2SA477K		3006	51		
2SA477L		3006	51		
2SA477M		3006	51		
2SA477OR		3006	51		
2SA477R		3006	51		
2SA477X		3006	51		
2SA477Y		3006	51		
2SA478	126	3006	51	TR17/635	G0005
2SA478A		3006	51		
2SA478B		3006	51		
2SA478C		3006	51		
2SA478D		3006	51		
2SA478E		3006	51		
2SA478F		3006	51		
2SA478G		3006	51		G0005
2SA478GN		3006	51		
2SA478H		3006	51		
2SA478J		3006	51		
2SA478K		3006	51		
2SA478L		3006	51		
2SA478M		3006	51		
2SA478OR		3006	51		
2SA478R		3006	51		
2SA478X		3006	51		
2SA478Y		3006	51		
2SA479	126	3006	51	TR17/635	G0005
2SA479A		3006	51		
2SA479B		3006	51		
2SA479C		3006	51		
2SA479D		3006	51		
2SA479E		3006	51		
2SA479F		3006	51		
2SA479G		3006	51		G0005
2SA479GN		3006	51		
2SA479H		3006	51		
2SA479J		3006	51		
2SA479K		3006	51		
2SA479L		3006	51		
2SA479M		3006	51		
2SA479OR		3006	51		
2SA479R		3006	51		
2SA479X		3006	51		
2SA479Y		3006	51		
2SA480	159	3118	65	TR20/717	S0013
2SA480A		3118	89		
2SA480B		3118	89		
2SA480C		3118	89		
2SA480D		3118	89		
2SA480E		3118	89		
2SA480F		3118	89		
2SA480G		3118	89		
2SA480GN		3118	89		
2SA480H		3118	89		
2SA480J		3118	89		
2SA480K		3118	89		
2SA480L		3118	89		
2SA480M		3118	89		
2SA480OR		3118	89		
2SA480R		3118	89		
2SA480X		3118	89		
2SA480Y		3118	89		
2SA482	159	3114	89	TR20/717	S0012
2SA482A		3114	89		
2SA482B		3114	89		
2SA482C		3114	89		
2SA482D		3114	89		
2SA482E		3114	89		
2SA482F		3114	89		
2SA482G		3114	89		
2SA482GN		3114	89		
2SA482H		3114	89		
2SA482J		3114	89		
2SA482K		3114	89		
2SA482L		3114	89		
2SA482M		3114	89		
2SA482OR		3114	89		
2SA482R		3114	89		
2SA482X		3114	89		
2SA482Y		3114	89		
2SA483	218	3085		TR37/	
2SA484	189	3153		TR73/	G6011
2SA484A		3513			
2SA485	189	3513		TR73/242	
2SA485A		3513			
2SA485RED					S3032
2SA485YEL					S3032
2SA486	189		69	TR73/	
2SA486RED					S5013
2SA486YEL					S3032
2SA489	153	3083		TR77/	S5005
2SA489-O	153		69	TR77/	
2SA489-R	153		69	TR77/	
2SA489-Y				TR77/	
2SA490	159	3083	89	TR20/717	S5022
2SA490(POW)	153			TR77/	
2SA490A		3083	89		
2SA490B		3083	89		
2SA490C		3083	89		
2SA490D		3083	89		
2SA490E		3083	89		
2SA490F		3083	89		
2SA490G		3083	89		
2SA490GN		3083	89		
2SA490H		3083	89		
2SA490J		3083	89		
2SA490K		3083	89		
2SA490L		3083	89		
2SA490M		3083	89		
2SA490OR		3083	89		
2SA490R		3083	89		
2SA490X		3083	89		
2SA490Y	159	3083	89	TR20/717	S5022
2SA490YA		3083	89		
2SA493		3118	48	TR52/	S0019
2SA493A		3118	89		
2SA493B		3118	89		S0019
2SA493C		3118	89		
2SA493D		3118	89		
2SA493E		3118	89		
2SA493F		3118	89		
2SA493G		3118	89		
2SA493GRN					S0019
2SA493GN		3118	89		
2SA493H		3118	89		
2SA493J		3118	89		
2SA493K		3118	89		
2SA493L		3118	89		
2SA493M		3118	89		
2SA493OR		3118	89		
2SA493R		3118	89		
2SA493RED					S5022
2SA493X		3118	89		
2SA493Y		3118	89		
2SA493YEL					S5022
2SA494	159	3114	21	TR20/717	S0014
2SA494-O	159	3114	89	TR52/717	
2SA494A		3114	89		
2SA494B		3114	89		
2SA494C		3114	89		
2SA494D		3114	89		
2SA494E		3114	89		
2SA494F		3114	89		
2SA494G		3114	89		
2SA494GN		3114	89		
2SA494GR		3114	65		
2SA494-GR	159	3114	89	TR20/717	
2SA494GRN					S0013
2SA494-GR-1		3114	89		
2SA494H		3114	89		
2SA494J		3114	89		
2SA494K		3114	89		
2SA494L		3114	89		
2SA494M		3114	89		
2SA494-O	159	3114	89	TR20/717	
2SA494ORN					S0013
2SA494OR		3114	89		
2SA494R		3114	89		
2SA494X		3114	89		
2SA494Y		3114	89		
2SA494-Y	159	3114		/717	
2SA494(Y)			21	TR30/	
2SA494YEL					S0013
2SA495	159	3114	22	TR20/717	S0013
2SA495-1		3114	89		
2SA495A		3114	89		
2SA495B		3114	89		

Original Part Number	ECG	SK	GE	IR/WEP	HEP
2SA495C		3114	89		
2SA495D	159	3114	89	/717	
2SA495E		3114	89		
2SA495F		3114	89		
2SA495G		3118	22		
2SA495-G	159	3114		TR20/717	
2SA495GN		3118	89		
2SA495G-GR	159				
2SA495G-O	159				
2SA495G-R	159				
2SA495G-Y	159				
2SA495H		3114	89		
2SA495J		3114	89		
2SA495K		3114	89		
2SA495L		3114	89		
2SA495M		3114	89		
2SA495-O		3114	89	TR30/717	S0013
2SA495-OR		3114	89		
2SA495R		3114			
2SA495-R	159	3114	89		
2SA495W	159	3114	89	TR20/717	
2SA495-Y	159	3114		TR20/717	S5013
2SA495Y	159	3025	89	TR30/	S0013
2SA495ORG			65		S0113
2SA495RED			65		S0013
2SA495YEL			65		S5022
2SA495WHT					S0013
2SA495/G03007C				2SA495/	
2SA496	185	3083	29	TR77/S5007	S5006
2SA496A		3083			
2SA496-O	185	3191	29	TR77/S5007	
2SA496-R	185	3191		TR77/S5007	
2SA496-Y	185	3191		TR77/S5007	
2SA496Y	185	3083	29	TR77/	
2SA496-ORG			48		S5006
2SA496-RED			48		S5006
2SA496-YEL			48		S0013
2SA497	129	3025		TR88/242	S5013
2SA497R	129	3025		TR28/	
2SA497ORN					S3032
2SA497RED					S3032
2SA497YEL					S3032
2SA498		3025	48	TR88/242	S0019
2SA498A		3025			
2SA498ORN			48		S5013
2SA498RED			48		S0012
2SA498YEL			48		S5013
2SA498Y	129	3025			
2SA498(Y)			67	TR56/	
2SA499	159	3118	21	TR20/717	S0013
2SA499A		3118	89		S0019
2SA499-O	159			TR20/	
2SA499-OR	159			TR20/	
2SA499-Y	159			TR20/	
2SA499B		3118	89		
2SA499C		3118	89		
2SA499D		3118	89		
2SA499E		3118	89		
2SA499F		3118	89		
2SA499G		3118	89		
2SA499GN		3118	89		
2SA499H		3118	89		
2SA499J		3118	89		
2SA499K		3118	89		
2SA499L		3118	89		
2SA499M		3118	89		
2SA499OR		3118	89		
2SA499ORG			65		S0013
2SA499R		3118	89		
2SA499RED			65		S0012
2SA499X		3118	89		
2SA499Y		3118	89		
2SA499YEL			65		S0012
2SA500	159	3118	21	TR20/717	S0013
2SA500-O	159			TR20/	
2SA500A		3118	89		
2SA500B		3118	89		
2SA500C		3118	89		
2SA500D		3118	89		
2SA500E		3118	89		
2SA500F		3118	89		
2SA500G		3118	89		
2SA500GN		3118	89		
2SA500H		3118	89		
2SA500J		3118	89		
2SA500K		3118	89		
2SA500L		3118	89		
2SA500M		3118	89		
2SA500OR		3118	89		
2SA500-O	159			TR20/	
2SA500ORG			65		S0013
2SA500R		3118	89		
2SA500-R	159			TR20/	
2SA500RED			65		S0013
2SA500X		3118	89		
2SA500Y		3118	89		
2SA500-Y	159			TR20/	
2SA500YEL			65		S0013
2SA501	129	3025		TR88/242	S0012
2SA502	159	3025	82	TR20/717	S5023
2SA502A		3025	89		
2SA502B		3025	89		
2SA502C		3025	89		
2SA502D		3025	89		
2SA502E		3025	89		
2SA502F		3025	89		
2SA502G		3025	89		
2SA502-G					S0019
2SA502GN		3025	89		
2SA502H		3025	89		
2SA502J		3025	89		
2SA502K		3025	89		
2SA502L		3025	89		
2SA502M		3025	89		
2SA502OR		3025	89		
2SA502R		3025	89		
2SA502X		3025	89		
2SA502Y		3025	89		
2SA503	129	3025	48	TR88/242	S5022
2SA503GR	211	3203			
2SA503-O	129	3025		TR88/242	
2SA503-R	129	3025		TR88/242	
2SA503-Y	129	3025		TR88/242	
2SA503GRN			48		S3032
2SA503ORG			48		S3001
2SA503YEL			48		S3032
2SA504	129	3025	48	TR88/242	S5022
2SA504GR	211	3203			
2SA504-O	129	3025		TR88/242	
2SA504-R	129	3025		TR88/242	
2SA504-Y	129	3025		TR88/242	
2SA504GRN			48		S3032
2SA504ORG			48		S0012
2SA504YEL			48		S3032
2SA505	185	3191	29	TR77/S5007	S5006
2SA505-O	185	3191	69	TR77/S5007	
2SA505-R	185	3191	69	TR77/S5007	
2SA505-Y	185	3191	69	TR77/S5007	
2SA505ORG			48		S0012
2SA505RED			48		S5006
2SA505YEL			48		S0013
2SA506	160		51	TR17/637	G0003
2SA507	160	3006	51	TR17/637	G0003
2SA507A		3006	51		
2SA507B		3006	51		
2SA507C		3006	51		
2SA507D		3006	51		
2SA507E		3006	51		
2SA507F		3006	51		
2SA507G		3006	51		
2SA507GN		3006	51		
2SA507H		3006	51		
2SA507J		3006	51		
2SA507K		3006	51		
2SA507L		3006	51		
2SA507OR		3006	51		
2SA507R		3006	51		
2SA507X		3006	51		
2SA507Y		3006	51		
2SA508	160		51	TR17/637	G0003
2SA509	159	3114	82	TR30/717	S0013
2SA509A	159	3114	89		S0013
2SA509B		3114	89		
2SA509C		3114	89		
2SA509D		3114	89		
2SA509E		3114	89		
2SA509F		3114	89		
2SA509G		3114	89		
2SA509GN		3114	89		
2SA509GR	159	3114	89	TR20/717	

Original Part Number	ECG	SK	GE	IR/WEP	HEP
2SA509GR-1		3114	89		
2SA509H		3114	89		
2SA509J		3114	89		
2SA509K		3114	89		
2SA509L		3114	89		
2SA509M		3114	89		
2SA509-O	159	3114	89	TR20/717	
2SA509ORN					S0019
2SA509OR		3114	89		
2SA509R		3114	89		
2SA509-R	159	3114	89	TR20/717	
2SA509RED		3114	89		S0019
2SA509X		3114	89		
2SA509Y		3114	22	TR19/	S0013
2SA509-Y	159	3114	89	TR20/717	
2SA509YEL		3114	89		S0019
2SA510	159	3025	89	TR20/717	
2SA510A		3025	89		
2SA510B		3025	89		
2SA510C		3025	89		
2SA510D		3025	89		
2SA510E		3025	89		
2SA510F		3025	89		
2SA510G		3025	89		
2SA510GN		3025	89		
2SA510H		3025	89		
2SA510J		3025	89		
2SA510K		3025	89		
2SA510L		3025	89		
2SA510M		3025	89		
2SA510-O	159	3114	89	TR20/717	
2SA510OR		3025			
2SA510-OR		3114	89		
2SA510R		3025	89		
2SA510-R			89	TR20/717	
2SA510-RD		3025	89		
2SA510X		3025	89		
2SA510Y		3025	89		
2SA511	159	3114	89	TR20/717	S3032
2SA511A		3114	89		
2SA511B		3114	89		
2SA511C		3114	89		
2SA511D		3114	89		
2SA511E		3114	89		
2SA511F		3114	89		
2SA511G		3114	89		
2SA511-G		3114	89		
2SA511GN		3114	89		
2SA511H		3114	89		
2SA511J		3114	89		
2SA511K		3114	89		
2SA511L		3114	89		
2SA511M		3114	89		
2SA511-O	159	3114	89	TR20/717	
2SA511OR		3114	89		
2SA511ORN					S3032
2SA511R		3114	89		
2SA511-R	159	3114	89	TR20/717	
2SA511RED		3114	89		S3032
2SA511X		3114	89		
2SA511Y		3114	89		
2SA512	159	3025	67	TR20/717	S3032
2SA512A		3025	89		
2SA512B		3025	89		
2SA512C		3025	89		
2SA512D		3025	89		
2SA512E		3025	89		
2SA512F		3025	89		
2SA512G		3025	89		
2SA512GN		3025	89		
2SA512H		3025	89		
2SA512J		3025	89		
2SA512K		3025	89		
2SA512L		3025	89		
2SA512M		3025	89		
2SA512-O		3025	89	TR20/717	
2SA512OR			89		
2SA512ORN					S3032
2SA512-OR			89		
2SA512R		3025	89		
2SA512RED		3114	89		S3003
2SA512-R	159		89	TR20/717	
2SA512X		3025	89		
2SA512Y		3025	89		
2SA513	159	3114	48	TR20/717	S3032
2SA513A		3114	89		S3032
2SA513B		3114	89		
2SA513C		3114	89		
2SA513D		3114	89		
2SA513E		3114	89		
2SA513F		3114	89		
2SA513G		3114	89		
2SA513GN		3114	89		
2SA513H		3114	89		
2SA513J		3114	89		
2SA513K		3114	89		
2SA513L		3114	89		
2SA513M		3114	89		
2SA513-O	159	3114	89	TR20/717	
2SA513OR		3114	89		
2SA513ORN					S3032
2SA513R		3114	89		
2SA513RED		3114	89		S5013
2SA513-R	159	3114	89	TR20/717	
2SA513X		3114	89		
2SA513Y		3114	89		
2SA515	124				
2SA516	129	3025		TR88/242	S3032
2SA516A	129	3025		TR88/242	
2SA516B		3025			
2SA517	126	3008	50	TR17/635	G0008
2SA518	126	3007	51	TR17/635	G0008
2SA518A		3006	51		
2SA518B		3006	51		
2SA518C		3006	51		
2SA518D		3006	51		
2SA518E		3006	51		
2SA518F		3006	51		
2SA518G		3006	51		
2SA518-G					
2SA518GN		3006	51		
2SA518H		3006	51		
2SA518J		3006	51		
2SA518K		3006	51		
2SA518L		3006	51		
2SA518M		3006	51		
2SA518OR		3006	51		
2SA518R		3006	51		
2SA518X		3006	51		
2SA518Y		3006	51		
2SA522	159	3114	22	TR20/717	S0013
2SA522A	159	3114	21	TR20/717	S0013
2SA522AL		3114	89		
2SA522B		3114	89		
2SA522C		3114	89		
2SA522D		3114	89		
2SA522E		3114	89		
2SA522F		3114	89		
2SA522G		3114	89		
2SA522GN		3114	89		
2SA522H		3114	89		
2SA522J		3114	89		
2SA522K		3114	89		
2SA522L		3114	89		
2SA522M		3114	89		
2SA522OR		3114	89		
2SA522R		3114	89		
2SA522X		3114	89		
2SA522Y		3114	89		
2SA525	160	3118	21	TR17/	S0013
2SA525A	160	3006	51	TR17/637	S0013
2SA525B	160	3006	51	TR17/637	S0013
2SA525C		3006	51		
2SA525D		3006	51		
2SA525E		3006	51		
2SA525F		3006	51		
2SA525GN		3006	51		
2SA525H		3006	51		
2SA525J		3006	51		
2SA525K		3006	51		
2SA525L		3006	51		
2SA525M		3006	51		
2SA525OR		3006	51		
2SA525R		3118	51		
2SA525X		3006	51		
2SA525Y		3118	51		
2SA527	129	3025	69	TR88/242	S5013
2SA527A		3025			
2SA528	129	3025	69	TR88/242	S5013
2SA530	159	3114	21	TR20/717	S0019

Original Part Number	ECG	SK	GE	IR/WEP	HEP
2SA530A		3118	89		
2SA530B		3118	89		S0019
2SA530C		3118	89		
2SA530D		3118	89		
2SA530E		3118	89		
2SA530F		3118	89		
2SA530G		3118	89		
2SA530-G					S0019
2SA530GN		3118	89		
2SA530GR		3118	89		
2SA530GRN					S0019
2SA530H	159	3114	21	TR20/717	S0019
2SA530H1		3114	89		
2SA530J		3118	89		
2SA530K		3118	89		
2SA530L		3118	89		
2SA530M		3118	89		
2SA530OR		3118	89		
2SA530ORN					S0019
2SA530R		3118	89		
2SA530X		3118	89		
2SA530Y		3118	89		
2SA532	129	3025		TR88/242	S0012
2SA532A	129	3025		TR88/	
2SA532B	129	3025		TR88/	
2SA532C	129	3025		TR88/	
2SA532D	129	3025		TR88/	
2SA532E	129	3025		TR88/	
2SA532F	129	3025		TR88/	
2SA537	129	3025	48	TR88/242	S5022
2SA537A	129	3025		TR88/242	S3032
2SA537AA	129			TR88/	
2SA537AB	129			TR88/	
2SA537AC	129			TR88/	S5013
2SA537AH	129	3025		TR88/242	S3003
2SA537B	129	3118		TR88/	
2SA537C	129	3025		TR88/	
2SA537H	129	3025		TR88/242	S3003
2SA538	102A		53	TR85/250	G0005
2SA538-G					G0005
2SA539	159	3114	82	TR54/717	S0019
2SA539K	159	3114	21		S0019
2SA539L	159	3114		TR24/	S0031
2SA539M	159	3114	21	TR30/	S0019
2SA539RED					S0019
2SA539S	159			TR20/	
2SA539/2057A2-3				TR54/	
2SA542	159	3114	65	TR20/717	S0019
2SA542A		3114	89		
2SA542B		3114	89		
2SA542C		3114	89		
2SA542D		3114	89		
2SA542E		3114	89		
2SA542F		3114	89		
2SA542G		3114	89		
2SA542GN		3114	89		
2SA542H		3114	89		
2SA542J		3114	89		
2SA542K		3114	89		
2SA542L		3114	89		
2SA542M		3114	89		
2SA542OR		3114	89		
2SA542R		3114	89		
2SA542RED					S0019
2SA542X		3114	89		
2SA542Y		3114	89		
2SA543	192	3025	65	TR52/	S0019
2SA544	159	3114	21	TR20/717	S5022
2SA544A		3114	89		
2SA544B		3114	89		
2SA544C		3114	89		
2SA544D		3114	89		
2SA544E		3114	89		
2SA544F		3114	89		
2SA544G		3114	89		
2SA544GN		3114	89		
2SA544H		3114	89		
2SA544J		3114	89		
2SA544K		3114	89		
2SA544L		3114	89		
2SA544M		3114	89		
2SA544OR		3114	89		
2SA544R		3114	89		
2SA544RED					S0019
2SA544X		3114	89		

Original Part Number	ECG	SK	GE	IR/WEP	HEP
2SA544Y		3114	89		
2SA545	193	3114	82	TR52/717	S5022
2SA545GRN	193				S0019
2SA545KLM	193	3118	21	TR88/	S5022
2SA545L	193	3114	21	TR52/717	S502
2SA546	129	3025	67	TR88/242	S3032
2SA546A	129	3025		TR88/242	S3032
2SA546-A					S5013
2SA546B	129	3025		TR88/	
2SA546-B					S5013
2SA546-D					S5013
2SA546E	129	3025		TR88/	S3032
2SA546H	129	3025		TR88/	S3032
2SA547	193		69		S3032
2SA547A	193				S3032
2SA548	129	3118	21	TR88/242	S0019
2SA548G		3118	89		
2SA548GN		3118	89		
2SA548GRN					S0019
2SA548ORN					S0019
2SA548RED					S0019
2SA548YEL					S0019
2SA549AH			65		
2SA550	159	3025	21	TR20/717	S0031
2SA550A	159	3114	65	TR20/717	S0019
2SA550AQ	129	3025		TR88/242	S3032
2SA550A(R,Q,S)			22	TR88/	S0019
2SA550P	159			TR28/	S0013
2SA550Q	159	3114	22		S0032
2SA550R	129	3025	22	TR88/242	S0032
2SA550RED					S3032
2SA550S	159		22		S0019
2SA550YEL					S3032
2SA551	129	3025		TR88/242	S3032
2SA551C	129	3025		TR88/242	S3032
2SA551D	129	3025		TR88/242	S3032
2SA551E	129	3025		TR88/242	S3032
2SA552	129	3025	21	TR88/242	S5022
2SA560	129	3025		TR88/242	S3032
2SA561	159	3114	82	TR17/717	S0019
2SA561-O	159	3114	89	TR52/717	S0012
2SA561-R	159	3114	89	TR20/717	
2SA561GRN		3114	48	TR20/717	S0019
2SA561ORN			82		S0019
2SA561RED			82		S0013
2SA561-Y	159	3114	89	TR20/717	S0012
2SA562	159	3114	82	TR54/717	S0013
2SA562-O	159	3114	89	TR20/717	S0012
2SA562-R	159	3114	89	TR20/717	
2SA562-Y	159	3114		TR20/717	S0012
2SA562GR	159	3114	89	TR54/717	S0019
2SA562OR		3114	89		S0012
2SA562RED		3114	82		S0013
2SA562YEL		3114	82		S0013
2SA564		3114	65	TR20/717	S0019
2SA564-O	234	3114	89	TR20/717	S0019
2SA564A	234	3114	65	2SA564/717	S0019
2SA564ABQ	234	3114	89	TR20/717	S0019
2SA564ABQ-1		3114	89		
2SA564F	234	3114	89	TR20/717	S5022
2SA564FQ	234	3114	89	TR20/717	S0019
2SA564FR	159	3025		TR20/717	S0019
2SA564J	159	3114	89		
2SA564P	234			TR20/	S0019
2SA564Q	234	3025	21	TR30/	S0032
2SA564R	234	3025	21	TR30/	S0019
2SA564S	234	3114	21	TR52/	S0019
2SA564T		3114	21	TR20/	S0019
2SA564X		3114	89		
2SA564Y		3114	89		
2SA564(T)			21		
2SA564(O)			21	TR30/	
2SA564(P)			22	TR24/	
2SA564(Q)			21	TR51/	
2SA565	159	3114		TR20/717	S5022
2SA565A	159	3114	89	TR88/717	S5022
2SA565AB		3114	89		S5022
2SA565B	159	3007	21	TR20/717	S5022
2SA565BA		3114	89		S0019
2SA565C	159	3114	21	TR20/717	S5022
2SA565D	159	3114	89		S5022
2SA565E		3114	89		
2SA565F		3114	89		
2SA565G		3114	89		
2SA565GN		3114	89		

Original Part Number	ECG	SK	GE	IR/WEP	HEP
2SA565H		3114	89		
2SA565J		3114	89		
2SA565K		3114	65		
2SA565L		3114	89		
2SA565M		3114	89		
2SA565OR		3114	89		
2SA565R		3114	89		
2SA565X		3114	89		
2SA565Y		3114	89		
2SA566	218	3085		TR58/	
2SA566A	218	3085		TR58/	
2SA566B	218	3085		TR58/	
2SA566C	218	3085		TR58/	
2SA567	159	3114	82	TR20/717	S0019
2SA567A		3114	89		S0019
2SA567B		3118	89		S0019
2SA567C		3114	89		
2SA567D		3114	89		
2SA567E		3114	89		
2SA567F		3114	89		
2SA567G		3114	89		
2SA567GN		3114	89		
2SA567GRN		3118	89		S0019
2SA567H		3114	89		
2SA567J		3114	89		
2SA567K		3114	89		
2SA567L		3114	89		
2SA567M		3114	89		
2SA567OR		3114	89		
2SA567ORN					S0019
2SA567R		3114	89		
2SA567RED					S0019
2SA567X		3114	89		
2SA567Y		3114	89		
2SA567YEL					S0019
2SA568	159	3114	82	TR20/717	S0013
2SA568A		3118	89		S0013
2SA568B		3118	89		S0013
2SA568C		3118	89		
2SA568D		3118	89		
2SA568E		3118	89		
2SA568F		3118	89		
2SA568G		3118	89		
2SA568GN		3118	89		
2SA568GRN					S0019
2SA568H		3118	89		
2SA568J		3118	89		
2SA568K		3118	89		
2SA568L		3118	89		
2SA568M		3118	89		
2SA568OR		3118	89		
2SA568R		3118	89		
2SA568RED					S0019
2SA568X		3118	89		
2SA568Y		3118	89		
2SA568YEL					S0019
2SA569	159	3114	82	TR20/717	S0019
2SA569A		3114	89		S0019
2SA569B		3114	89		
2SA569C		3114	89		
2SA569D		3114	89		
2SA569E		3114	89		
2SA569F		3114	89		
2SA569G		3114	89		
2SA569GN		3114	89		
2SA569GRN					S0019
2SA569H		3114	89		
2SA569J	159	3114	89	TR20/717	
2SA569K		3114	89		
2SA569L		3114	89		
2SA569M		3114	89		
2SA569OR		3114	89		
2SA569R		3114	89		
2SA569RED					S0019
2SA569X		3114	89		
2SA569Y		3114	89		
2SA569YEL					S0019
2SA570	159	3114	82	TR20/717	S5022
2SA570A		3114	89		S5022
2SA570B		3114	89		
2SA570C		3114	89		
2SA570D		3114	89		
2SA570E		3114	89		
2SA570F		3114	89		
2SA570G		3114	89		
2SA570GN		3114	89		
2SA570GRN					S0019
2SA570H		3114	89		
2SA570J		3114	89		
2SA570K		3114	89		
2SA570L		3114	89		
2SA570M		3114	89		
2SA570OR		3114	89		
2SA570ORN					S0019
2SA570R		3114	89		
2SA570RED					S0019
2SA470X		3114	89		
2SA570Y		3114	89		
2SA570YEL					S0019
2SA571	129	3025		TR88/242	S3032
2SA572				TR31/	
2SA578			65		S0019
2SA579			65		S0019
2SA592Y	159	3114		TR20/	
2SA594	129	3025	21	TR88/242	S5022
2SA594-O	129	3025		TR88/242	
2SA594-R	129	3025		TR88/242	
2SA594-Y	129	3025		TR88/242	
2SA594A		3025			
2SA594ORN					S5022
2SA594RED					S5022
2SA594YEL					S5022
2SA595C	129			TR88/242	S5013
2SA597	129	3025	29	TR88/242	S5013
2SA603	159	3118	82	TR20/719	S5022
2SA603A		3118	89		
2SA603B		3118	89		
2SA603C		3118	89		
2SA603D		3118	89		
2SA603E		3118	89		
2SA603F		3118	89		
2SA603G		3118	89		
2SA603GN		3118	89		
2SA603H		3118	89		
2SA603J		3118	89		
2SA603K		3118	89		
2SA603L		3118	89		
2SA603M		3118	89		
2SA603OR		3118	89		
2SA603R		3118	89		
2SA603X		3118	89		
2SA603Y		3118	89		
2SA604	129	3025	27	TR88/242	
2SA605			27	TR78/	
2SA606	129	3025		TR88/242	
2SA606S	129			TR88/	
2SA607		3025	89		
2SA608	159	3114	65	TR20/717	S0013
2SA608A	159	3114	89	TR20/	S0013
2SA608B	159	3114	89	TR20/	
2SA608C	159	3114	89	TR20/	S0013
2SA608-C	159	3114	89	TR20/717	
2SA608D	159	3114		TR52/	S0013
2SA608-D	159			TR20/	
2SA608E	159	3114	21	TR20/	S0013
2SA608-E	159	3114	89	TR20/717	
2SA608F	159	3114	89	TR20/	S0013
2SA608-F	159			TR20/717	
2SA608G	159	3114	89	TR20/	
2SA609	159	3114	65	TR20/717	S0013
2SA609A	159	3114	89	TR20/	S0013
2SA609B	159	3114	89	TR20/	
2SA609C	159	3114	89	TR20/	
2SA609D	159	3114	89	TR20/	
2SA609E	159	3114	21	TR20/242	S0013
2SA609F	159	3025		TR20/242	S0013
2SA609G	159	3114	89	TR20/	
2SA610	159	3025	21	TR20/717	S0013
2SA610B	159	3025		TR20/717	S0013
2SA611	159	3025	21	TR20/717	S0013
2SA611-4E	159	3025		TR20/717	
2SA611-4E(2SC61)				TR30/	
2SA612	129	3025	22	TR54/	S0019
2SA613	218		26	TR58/700	S5018
2SA614	218		69	TR58/700	S5018
2SA616	218	3085		TR58/	S5018
2SA616(1,2)	218		26	TR58/	
2SA617K			65		
2SA618K			65		
2SA623	211	3083	29	TR56/S3027	S3028

Original Part Number	ECG	SK	GE	IR/WEP	HEP
2SA623-0	211	3083			
2SA623A	211	3083			
2SA623B	211	3083			
2SA623C	211	3083			
2SA623D	211	3083			
2SA623G	211	3083			
2SA623R	211	3083			
2SA623Y	211	3083			
2SA624		3083	29	TR56/S3027	S3032
2SA624A	211	3083			
2SA624B	211	3083			
2SA624C	211	3083			
2SA624D	211	3083			
2SA624E	211	3083			
2SA624GN	211	3083			
2SA624L	211	3083			
2SA624LG	211	3083			
2SA624R	211	3083			
2SA624Y	211	3083			
2SA626	219				S7003
2SA627	219				S5005
2SA628	159	3118	65	TR20/717	S0013
2SA628A	159	3114	82	TR20/717	S5022
2SA628AA		3114	89		S0019
2SA628B		3118	89		
2SA628C		3118	89		
2SA628D		3118	89		
2SA628E	159	3114	22	TR20/717	S0013
2SA628EF	159				S0013
2SA628F	159	3114	22	TR20/	S0019
2SA628G		3118	89		
2SA628GN		3118	89		
2SA628H		3118	89		
2SA628J		3118	89		
2SA628K		3118	89		
2SA628L		3118	89		
2SA628M		3118	89		
2SA628OR		3118	89		
2SA628R		3118	89		
2SA628X		3118	89		
2SA628Y		3118	89		
2SA629	159	3118	65	TR20/717	S0019
2SA629A		3114	89		
2SA629B		3114	89		S0019
2SA629C		3114	89		
2SA629D		3114	89		
2SA629E		3114	89		
2SA629F		3114	89		
2SA629G		3114	89		
2SA629GN		3114	89		
2SA629H		3114	89		
2SA629J		3114	89		
2SA629K		3114	89		
2SA629L		3114	89		
2SA629M		3114	89		
2SA629OR		3114	89		
2SA629R		3114	89		
2SA629X		3114	89		
2SA629Y		3114	89		
2SA634	187	3084		TR56/	S0015
2SA634A		3084		TR87/243	S0015
2SA634B	187	3084		TR56/	
2SA634C	187	3084		TR56/	
2SA634D	187	3084		TR56/	
2SA634K	187	3193			
2SA634L	187	3084		TR56/	S5006
2SA634M	187	3193			S3028
2SA634(M)			29	TR56/	
2SA634(L)			29	TR56/	
2SA636	187	3083	29	TR56/S3027	S5006
2SA636A	187	3193		TR56/	
2SA636B	187	3193		TR56/	
2SA636C	187	3193		TR56/	
2SA636D	187	3193		TR56/	
2SA636L	187	3193			S5006
2SA636M	187	3083		TR56/	S5006
2SA636(L)			29	TR56/	
2SA636(M)			29	TR56/	
2SA637	159			TR20/	
2SA638E,F			22		
2SA639				TR56/	
2SA639S		3114		TR30/	S0019
2SA640	159	3118	65	TR30/	S0019
2SA640A	149				
2SA640L	159	3114		TR30/	

Original Part Number	ECG	SK	GE	IR/WEP	HEP
2SA640M	159			TR30/	S0019
2SA641	234	3118	21	TR30/	S0019
2SA641A	234	3114	89	TR20/	
2SA641B	234	3114	89	TR20/	
2SA641BL		3114	89		
2SA641C	234	3114	89	TR20/	
2SA641D	234	3114	89	TR20/	
2SA641L	234	3114	89		S0019
2SA641M	234	3114	89	TR20/	S0019
2SA642	159	3083	21	TR30/	S3014
2SA642A	159	3118	89	TR20/	S0019
2SA642B	159	3118	89	TR20/	S0019
2SA642C	159	3118	89	TR20/	
2SA642D	159	3118	89	TR20/	
2SA642E	159	3118	89	TR20/	
2SA642F	159	3118	89	TR20/	
2SA642GRN					S0019
2SA642ORN					S0019
2SA642RED					S0019
2SA642S	123A	3114	17	TR21/	S0019
2SA642YEL					S0019
2SA643	193	3025	82	TR28/	S3028
2SA643A	193	3025			
2SA643B	193	3025			
2SA643C	193	3025			
2SA643D	193	3025			
2SA643E	193	3025			
2SA643F	193	3025			
2SA643R	193	3025	67	TR28/	S5013
2SA643V		3138	67	TR56/	S3028
2SA643W	129		67	TR88/	S0019
2SA645	211	3203		TR56/S3027	S3032
2SA646	187	3193		TR56/S3027	S3032
2SA648	219				S5005
2SA651GR					S0019
2SA653,L					S5019
2SA656				TR56/	S5005
2SA656L				TR56/	S5006
2SA656M				TR56/	S5006
2SA657		3173			S5005
2SA658	219	3173			S5005
2SA659	129	3025	82	TR88/242	S0019
2SA659A		3114	89		
2SA659B		3114	89		
2SA659C	129	3025		TR88/242	S0019
2SA659D	129	3025		TR88/242	S0019
2SA659E	129	3025		TR88/242	S0019
2SA659F	159	3114	89	TR20/	S0019
2SA659G		3114	89		
2SA661		3114		TR31/	S5022
2SA663	219	3173		TR27/	S5005
2SA666	234	3114	65	TR30/	S0019
2SA666A	234	3114	65	TR20/	S0019
2SA666B		3114	89		
2SA666BL		3114	89		
2SA666C		3114	89		
2SA666D		3114	89		
2SA666E		3114	89		
2SA666H	234	3114	89	TR30/	
2SA666Q	234				S0019
2SA666QRS	234		21	TR20/	
2SA666R	234		89		S0031
2SA666S	234		21	TR20/	S0031
2SA666Y		3114	89		
2SA670	153	3083		TR77/	S5006
2SA670A	153	3083		TR77/	
2SA670B	153	3083		TR77/	
2SA670C	153	3083		TR77/	
2SA671	153	3083		TR77/	S5006
2SA671A	153	3083		TR77/	
2SA671B	153	3083		TR77/	
2SA671C	153	3083		TR77/	
2SA672	159	3114	82	TR20/	S0019
2SA672A	159	3114		TR20/	S5022
2SA672B	159	3114		TR20/	S5022
2SA672C	159	3114		TR20/	S5022
2SA673	159	3114	48	TR20/242	S5013
2SA673A	159	3114	48	TR20/	S5022
2SA673AA	159			TR20/	
2SA673B	15		67	TR20/24.	S5013
2SA673C	159	3114	67	TR20/242	S5013
2SA673D	159	3114	89	TR20/	S5013
2SA675			82		S5023
2SA677	159	3118	48	TR30/242	S0019
2SA678	159	3118	48	TR20/717	S0019

Original Part Number	ECG	SK	GE	IR/WEP	HEP
2SA678A		3118	89		S0019
2SA678(C)	159		22	TR30/	
2SA678(SONY)	159			TR20/717	
2SA679					S7001
2SA680	180	3173			
2SA682		3025			
2SA683	129	3114	67	TR88/	S3028
2SA683P	129	3025	67	TR30/	S5013
2SA683Q	129	3513	67	TR30/	S5013
2SA683R	193	3513	67	TR30/	S5013
2SA683S	193	3025	29	TR56/	S5006
2SA684	129	3025	58	TR88/	S3032
2SA684Q	129	3025		TR28/	
2SA684R	129	3025	22	TR88/	
2SA685	159	3114	21		S5023
2SA695	129		48	TR28/	S5013
2SA695C	129	3025		TR28/	S5013
2SA696	159	3118	82		
2SA697	193		82		
2SA697C	193	3114	67	TR88/	S0012
2SA699	187	3084	69	TR56/	S5006
2SA699A	187		69	TR56/S3027	S5006
2SA699AP	187	3083	29	TR56/	S5006
2SA699P	187	3084	69	TR56/	S3028
2SA699Q	187	3193		TR56/	S5006
2SA699R	187	3084		TR56/S3027	S5006
2SA700	187	3083	29	TR56/637	S5006
2SA700B	160	3083	29	TR17/637	
2SA700Y	160			TR17/637	
2SA701	159	3118	65	TR21/	S0019
2SA701F	159	3114	21	TR30/	
2SA704		3118	48		S0019
2SA704A		3118	89		S0019
2SA705		3118	48		S0019
2SA706	187	3534	69		S5006
2SA707	193	3114	67		S3032
2SA707V	193				
2SA708	211	3203			
2SA708A	211	3203			
2SA708B	211	3203			
2SA708C	211	3202			
2SA715	185	3191	58		
2SA715A	185	3191			
2SA715B	185	3513		TR56/	S5000
2SA715C	185	3191		TR73/	S5000
2SA715D	185	3191			
2SA715WT	185	3513			S5000
2SA715WTB	185	3513	67	TR56/	S5000
2SA715WTC	185	3513	67	TR56/	S5000
2SA717	211	2303			
2SA718	159			TR20/	
2SA719	159	3114	21	TR30/	S0019
2SA719Q	159	3025	58	TR30/	
2SA719R	159	3114		TR30/	S0019
2SA719S	159	3114	21	TR20/	S0019
2SA720	159	3114	21	TR20/	S0019
2SA720R	159	3114	21		
2SA720S	159	3114	22		
2SA721T		3114			S0019
2SA723	159			TR20/	
2SA723A	159			TR20/	
2SA723B	159			TR20/	
2SA723C	159			TR20/	
2SA723D	159			TR20/	
2SA723E	159			TR20/	
2SA723F	159			TR20/	
2SA725			65		
2SA726			65		
2SA730	159		21	TR20/	S3028
2SA731	159	3138	21	TR20/	S3032
2SA733	159	3025	65	TR51/	S5022
2SA733A	159			TR20/	
2SA733B	159			TR20/	
2SA733C	159			TR20/	
2SA733D	159			TR20/	
2SA733E	159			TR20/	
2SA733F	159			TR20/	
2SA733H	159	3138		TR33/	
2SA733I	159			TR33/	
2SA733P	159	3138	22	TR88/	S0019
2SA733Q	159	3138		TR31/	
2SA734	129				
2SA735					S0019
2SA736	129	3118	22	TR30/	S0019
2SA738	185	3084			S5006

Original Part Number	ECG	SK	GE	IR/WEP	HEP
2SA738B	185	3191	58		
2SA738C		3083			S5006
2SA741H		3114	65		
2SA742	129	3025			
2SA742H		3025	21		
2SA743		3513			S5006
2SA743A		3513			S5006
2SA744	219				
2SA746	219				
2SA748Q		3083			S3028
2SA749					S3002
2SA751	193	3025			
2SA751P	193	3025	67		
2SA751Q	193	3025	67	TR56/	S5006
2SA751R	193	3025		TR56/	S5006
2SA752	193				
2SA754		3083			
2SA755B		3083			S3028
2SA756		3173			
2SA763	234				
2SA763-W	234				
2SA763-WL-3	234				
2SA763-WL-4	234				
2SA763-WL-5	234				
2SA763-WL-6	234				
2SA763-WN	234				
2SA763-Y	234				
2SA763-YN	234				
2SA764	218			TR58/	
2SA765	218			TR58/	
2SA772				TR30/	
2SA773	129	3114		TR30/	S0019
2SA775C		3083			
2SA778AK		3114	65		
2SA778K		3114			S0028
2SA779		3084			
2SA786			65		
2SA787			65		
2SA811			65		
2SA812			65		
2SA813			21		
2SA816		3083			
2SA828		3020			
2SA836		3114			
2SA841		3114			
2SA842		3114			
2SA909X		3114			
2SA945Y	159				
2SA976G		3006			
2SA1010V	126	3006	1	TR12/	
2SA1018			51		
2SA1024				TR12/	
2SA1028A	126				
2SA1226(R)				TR56/	
2SA2010				TR85/	
2SA3410A		3006	21		
2SA3410B		3006	51	TR17/	
2SA4551		3006	51		
2SA4561		3006	51		
2SA4728			51		
2SA4728B				TR12/	
2SA4730		3083			
2SA4930		3118	89		
2SA4940		3118	89		
2SA4950	159	3004	53	TR20/	
2SA4950R		3118	89		
2SA4964	185	3191			
2SA5090				TR19/	
2SA5620		3114	89	TR30/	
2SA5670		3118	89		
2SA5670R		3118	89		
2SA5680		3118	89		
2SA5680R		3118	89		
2SA5690		3118	89		
2SA5690R		3118	89		
2SA6086				TR30/	
2SA6111			21		
2SA6341			56		
2SA6661QRS				TR20/	
2SA7200	159				
2SA-NJ101	159	3114		TR52/	S0019
2SAJ15GN		3005	51		
2SAUUJ		3123	80		
2SAU03				TR17/	
2SAZ69				TR12/	

Original Part Number	ECG	SK	GE	IR/WEP	HEP
2SB-F1A			53		
2SB-2C	506			/186	
2SB13			7		
2SB15			53		
2SB16A	105	3012	4	TR03/238	G6011
2SB16B		3012	4		
2SB16C		3012	4		
2SB16D		3012	4		
2SB16E		3012	4		
2SB16F		3012	4		
2SB16G		3012	4		
2SB16GN		3012	4		
2SB16H		3012	4		
2SB16J		3012	4		
2SB16K		3012	4		
2SB16L		3012	4		
2SB16M		3012	4		
2SB16OR		3012	4		
2SB16R		3012	4		
2SB16X		3012	4		
2SB16Y		3012	4		
2SB17A	105	3012	4	TR03/238	G6011
2SB17B		3012	4		
2SB17C		3012	4		
2SB17D		3012	4		
2SB17E		3012	4		
2SB17ER					G0005
2SB17F		3012	4		
2SB17G		3012	4		
2SB17GN		3012	4		
2SB17H		3012	4		
2SB17J		3012	4		
2SB17K		3012	4		
2SB17L		3012	4		
2SB17M		3012	4		
2SB17OR		3012	4		
2SB17R		3012	4		
2SB17S				TR85/	
2SB17X		3012	4		
2SB17Y		3012	4		
2SB18A	105	3012	4	TR03/233	G6012
2SB18B		3012	4		
2SB18C		3012	4		
2SB18D		3012	4		
2SB18E		3012	4		
2SB18F		3012	4		
2SB18G		3012	4		
2SB18GN		3012	4		
2SB18H		3012	4		
2SB18J		3012	4		
2SB18L		3012	4		
2SB18LA			53		
2SB18M		3012	4		
2SB18OR		3012	4		
2SB18R		3012	4		
2SB18X		3012	4		
2SB18Y		3012	4		
2SB19	104	3009	16	TR01/624	G6011
2SB19A		3009	25		
2SB19C		3009	25		
2SB19D		3009	25		
2SB19E		3009	25		
2SB19F		3009	25		
2SB19G		3009	25		
2SB19GN		3009	25		
2SB19H		3009	25		
2SB19J		3009	25		
2SB19K		3009	25		
2SB19L		3009	25		
2SB19M		3009	25		
2SB19OR		3009	25		
2SB18R		3009	25		
2SB19X		3009	25		
2SB19Y		3009	25		
2SB20	104	3009	16	TR01/624	G6011
2SB20A		3009	25		
2SB20B		3009	25		
2SB20C		3009	25		
2SB20D		3009	25		
2SB20E		3009	25		
2SB20F		3009	25		
2SB20G		3009	25		
2SB20GN		3009	25		
2SB20H		3009	25		
2SB20J		3009	25		
2SB20K		3009	25		
2SB20L		3009	25		
2SB20M		3009	25		
2SB20OR		3009	25		
2SB20R		3009	25		
2SB20X		3009	25		
2SB20Y		3009	25		
2SB21	104	3009	2	TR01/624	G6011
2SB21A		3009	25		
2SB21B		3009	25		
2SB21C		3009	25		
2SB21D		3009	25		
2SB21E		3009	25		
2SB21F		3009	25		
2SB21G		3009	25		
2SB21GN		3009	25		
2SB21H		3009	25		
2SB21J		3009	25		
2SB21K		3009	25		
2SB21L		3009	25		
2SB21M		3009	25		
2SB21OR		3009	25		
2SB21R		3009	25		
2SB21X		3009	25		
2SB21Y		3009	25		
2SB22	102A	3004	53	TR84/250	G0005
2SB22/09-30100		3004	53		
2SB22A	102A	3004	52	TR85/250	G0005
2SB22B	102A	3004	52	TR85/250	G0007
2SB22C		3004	53		
2SB22D		3004	53		
2SB22E		3004	53		
2SB22F		3004	53		
2SB22G		3004	53		
2SB22GN		3004	53		
2SB22H		3004	53		
2SB22I	102A				G0005
2SB22J		3004	53		
2SB22K		3004	53		
2SB22L		3004	53		
2SB22M		3004	53		
2SB22OR		3004	53		
2SB22ORN					G0005
2SB22P		3004	53		G0005
2SB22R		3004	53		
2SB22X		3004	53		
2SB22Y		3004	53		
2SB22YEL					G0007
2SB22(I)					G0005
2SB23	102A	3004	2	TR85/250	G0008
2SB23A		3004	53		
2SB23B		3004	53		
2SB23C		3004	53		
2SB23D		3004	53		
2SB23E		3004	53		
2SB23F		3004	53		
2SB23G		3004	53		
2SB23GN		3004	53		
2SB23H		3004	53		
2SB23J		3004	53		
2SB23K		3004	53		
2SB23L		3004	53		
2SB23M		3004	53		
2SB23OR		3004	53		
2SB23R		3004	53		
2SB23X		3004	53		
2SB23Y		3004	53		
2SB24	102A	3004	2	TR85/250	G0008
2SB24A		3004	53		
2SB24B		3004	53		
2SB24C		3004	53		
2SB24D		3004	53		
2SB24E		3004	53		
2SB24F		3004	53		
2SB24G		3004	53		
2SB24GN		3004	53		
2SB24H		3004	53		
2SB24J		3004	53		
2SB24K		3004	53		
2SB24L		3004	53		
2SB24M		3004	53		
2SB24OR		3004	53		
2SB24R		3004	53		
2SB24X		3004	53		
2SB24Y		3004	53		

Original Part Number	ECG	SK	GE	IR/WEP	HEP
2SB25	121	3009	16	TR01/232	G6005
2SB25A		3009	25		
2SB25B	126	3006	51	TR17/635	G0005
2SB26	104	3009	16	TR01/624	G6003
2SB26A	104	3009	16	TR01/624	G6003
2SB27	104	3009	49	TR01/624	G6003
2SB28	104	3009	49	TR01/624	G6003
2SB29	104	3009	49	TR01/624	G6003
2SB30	104	3009	49	TR01/624	G6003
2SB31	104	3009	49	TR01/624	G6003
2SB32	102A	3004	2	TR85/250	G0005
2SB32-0	102A	3004	2	TR85/250	G0005
2SB32-1	102A	3004	2	TR85/250	G0005
2SB32-2	102A	3004	2	TR85/250	G0005
2SB32-4	102A	3004	2	TR85/250	G0005
2SB32N	102A			TR85/250	
2SB33	102A	3004	52	TR85/250	G0005
2SB33-4	102A	3004	2	TR85/250	G0005
2SB33A		3004	53		
2SB33B		3004	53		
2SB33BK		3004	53		G0005
2SB33C	102A	3004	52	TR85/250	G0006
2SB33D	102A	3004	52	TR85/250	G0006
2SB33E	102A	3004	52	TR85/250	G0005
2SB33F	102A	3004	52	TR85/250	G0005
2SB33G		3004	53		
2SB33GN		3004	53		
2SB33H		3004	53		
2SB33J		3004	53		
2SB33K		3004	53		
2SB33L		3004	53		
2SB33M		3004	53		
2SB33OR		3004	53		
2SB33R		3004	53		
2SB33X		3004	53		
2SB33Y		3004	53		
2SB34	102	3003	53	TR05/631	G0005
2SB34N	127			TR27/235	
2SB35		3123	52	TR05/	
2SB37	102A	3004	52	TR85/250	G0005
2SB37A	102A	3004	2	TR85/250	G0007
2SB37B	102A	3004	2	TR85/250	G0007
2SB37C	102A	3004	8	TR85/250	G0005
2SB37D		3004	53		
2SB37E	102A	3004	52	TR85/250	G0005
2SB37F	102A	3004	2	TR85/250	
2SB37G		3004	53		
2SB37GN		3004	53		
2SB37H		3004	53		
2SB37J		3004	53		
2SB37K		3004	53		
2SB37L		3004	53		
2SB37M		3004	53		
2SB37OR		3004	53		
2SB37R		3004	53		
2SB37X		3004	53		
2SB37Y		3004	53		
2SB38	102	3004	53	TR05/631	G0005
2SB38A		3004	53		
2SB38B		3004	53		
2SB38C		3004	53		
2SB38D		3004	53		
2SB38E		3004	53		
2SB38F		3004	53		
2SB38G		3004	53		
2SB38GN		3004	53		
2SB38H		3004	53		
2SB38J		3004	53		
2SB38K		3004	53		
2SB38L		3004	53		
2SB38M		3004	53		
2SB38OR		3004	53		
2SB38R		3004	53		
2SB38X		3004	53		
2SB38Y		3004	53		
2SB39	102A	3003	51	TR85/250	G0005
2SB40	102A	3004	2	TR85/250	G0005
2SB41	104	3009			
2SB42	104	3009	3	TR01/624	G6003
2SB43	102A	3003	52	TR85/250	G0005
2SB43A	102A	3004	2	TR85/250	G0005
2SB43B		3004	53		
2SB43C		3004	53	TR85/	
2SB43D		3004	53		
2SB43E		3004	53		
2SB43F		3004	53		
2SB43G		3004	53		
2SB43GN		3004	53		
2SB43H		3004	53		
2SB43J		3004	53		
2SB43K		3004	53		
2SB43L		3004	53		
2SB43M		3004	53		
2SB43OR		3004	53		
2SB43R		3004	53		
2SB43X		3004	53		
2SB43Y		3004	53		
2SB44	102A	3004	52	TR85/250	G0005
2SB46	102A	3003	52	TR85/250	G0006
2SB47	102A	3003	52	TR85/250	G0006
2SB48	102	3004	52	TR05/631	G0005
2SB49	102	3004	52	TR05/631	G0005
2SB50	102	3003	52	TR05/631	G0006
2SB51	102	3003	53	TR05/631	G0005
2SB52	102	3003	53	TR05/631	G0005
2SB53	102	3004	53	TR05/631	G0005
2SB54	102A	3004	52	TR85/250	G0007
2SB54B	102A	3004	50	TR85/250	G0007
2SB54F	102A	3004	53		
2SB54Y	102A	3004	53	TR85/250	
2SB55	102A	3004	2	TR85/250	G0006
2SB56	102A	3004	52	TR84/250	G0005
2SB56A	102	3004	2	TR05/631	G0005
2SB56C	102A	3004	52	TR85/250	G0005
2SB57	102A	3004	52	TR85/250	G0005
2SB58			53		G0005
2SB59	102A	3004	52	TR85/250	G0005
2SB60	102A	3004	52	TR85/250	G0005
2SB60A	102A	3004	52	TR85/250	G0005
2SB61	102A	3004	52	TR85/250	G0005
2SB62	104	3009	16	TR01/624	
2SB63	131	3082	30	TR94/642	G6016
2SB64	121	3009	25	TR01/232	G6017
2SB65	102A	3004	52	TR85/250	G0005
2SB66	102A	3004	52	TR85/250	G0005
2SB66H	102A	3123	53	TR85/250	G0005
2SB67	102A	3004	2	TR85/250	G0005
2SB67A	102A	3004	2	TR85/250	G0005
2SB67AH					G0005
2SB67B		3004	53		
2SB67C		3004	53		
2SB67D		3004	53		
2SB67E		3004	53		
2SB67F		3004	53		
2SB67G		3004	53		
2SB67GN		3004	53		
2SB67H		3004	53		G0005
2SB67J		3004	53		
2SB67K		3004	53		
2SB67L		3004	53		
2SB67M		3004	53		
2SB67OR		3004	53		
2SB67R		3004	53		
2SB67X		3004	53		
2SB67Y		3004	53		
2SB68		3004	2		
2SB69	121	3009	25	TR01/232	G6005
2SB70			53		G0005
2SB70A					G0006
2SB71	102	3004	52	TR05/631	G0005
2SB72	102	3004	2	TR05/631	
2SB73	102A	3004	52	TR85/250	G0008
2SB73A	102A	3004	2	TR85/250	G0008
2SB73-1		3004	53		
2SB73B	102A	3004	2	TR85/250	G0003
2SB73C	102A	3004	53	TR85/250	
2SB73D		3004	53		
2SB73E		3004	53		
2SB73F		3004	53		
2SB73G		3004	53		
2SB73GN		3004	53		
2SB73GR	102A			TR85/	
2SB73H		3004	53		
2SB73J		3004	53		
2SB73K		3004	53		
2SB73L		3004	53		
2SB73M		3004	53		
2SB73OR		3004	53		
2SB73R		3004	53		
2SB73S	102A				

Original Part Number	ECG	SK	GE	IR/WEP	HEP	Original Part Number	ECG	SK	GE	IR/WEP	HEP
2SB73X		3004	53			2SB95	102	3004	52	TR05/ 631	G0008
2SB73Y		3004	53			2SB96					G0005
2SB74	160	3004	52	TR17/637	G0009	2SB97	102A	3004	52	TR85/250	G0005
2SB75	102A	3004	52	TR85/250	G0005	2SB98	102	3004	52	TR05/631	G0005
2SB75A	102A	3004	2	TR85/250	G0005	2SB99		3004	52	TR05/631	G0006
2SB75AH	102A	3123	80	TR85/250	G0005	2SB100	102	3004	52	TR05/631	G0005
2SB75B	102A	3004	52	TR85/250	G0005	2SB101	102	3004	52	TR05/631	G0005
2SB75C	102A	3004	52	TR85/250	G0005	2SB102	102	3004	52	TR05/631	G0005
2SB75C-4		3004	53			2SB103	100	3004	2	TR05/254	G0005
2SB75D		3004	53			2SB104	100	3004	2	TR05/254	G0005
2SB75E		3004	53			2SB105	158	3123	53	TR84/630	G0005
2SB75F	102A	3004	2	TR85/250	G0005	2SB106	158		53	TR84/630	G6011
2SB75G		3004	53			2SB107	102	3009	52	TR05/631	G6011
2SB75GN		3004	53			2SB107A	121	3009	16	TR01/232	G6003
2SB75H	102A	3004	52	TR85/250	G0005	2SB108	158	3123	53	TR84/630	G0005
2SB75J		3004	53			2SB108A	158	3123	80	TR84/630	G0005
2SB75L		3004	53			2SB108B	158	3123	80	TR84/630	
2SB75LB	102A			TR85/250	G0005	2SB109	158		53	TR84/630	G6011
2SB75M		3004	53			2SB109A					G6011
2SB75OR		3004	53			2SB110	102A	3003	52	TR85/250	G0005
2SB75R		3004	53			2SB111	102A	3003	52	TR85/250	G0005
2SB75X		3004	53			2SB111A		3004	53		
2SB75Y		3004	53			2SB111B		3004	53		
2SB76	102A	3003	52	TR85/250	G0005	2SB111C		3004	53		
2SB77	102A	3004	52	TR85/250	G0005	2SB111D		3004	53		
2SB77A	102A	3004	2	TR85/250	G0005	2SB111E		3004	53		
2SB77AA	102A			TR85/250	G0005	2SB111F		3004	53		
2SB77AB	102A	3004	53	TR85/250	G0005	2SB111G		3004	53		
2SB77AC	102A			TR85/250	G0007	2SB111GN		3004	53		
2SB77AD	102A			TR85/		2SB111H		3004	53		
2SB77AH	102A	3123	80	TR85/250	G0005	2SB111J		3004	53		
2SB77AP	102A	3004	52	TR85/250	G0005	2SB111K	102A	3004	2	TR85/250	G0005
2SB77B	102A	3004	53	TR85/250	G0007	2SB112	102A	3004	52	TR85/250	G0005
2SB77B2				TR85/		2SB113	102A	3004	52	TR85/250	G0005
2SB77B-11	102A	3004	53	TR85/		2SB114	102A	3003	52	TR85/250	G0005
2SB77C	102A	3004	52	TR85/250	G0007	2SB115	102A	3003	52	TR85/250	G0005
2SB77(C)SHARP			52			2SB116	102A	3004	52	TR85/250	G0005
2SB77D	102A	3004	2	TR85/250	G0007	2SB117	102A	3004	52	TR85/250	G0005
2SB77E		3004	53			2SB117K	102A	3004	2	TR85/250	G0005
2SB77F		3004	53			2SB118			53		
2SB77G		3004	53			2SB119	121	3009	3	TR01/232	G6003
2SB77GN		3004	53			2SB119A	121	3009	25	TR01/232	G6005
2SB77H	102A	3123	53	TR85/250	G0005	2SB120	102A	3004	52	TR85/250	G0005
2SB77K		3004	53			2SB122	121	3009	2	TR01/232	G6005
2SB77L		3004	53			2SB123	104	3009	3	TR01/624	G6005
2SB77M		3004	53			2SB123A	104	3009	3	TR01/624	G6005
2SB77OR		3004	63			2SB124	121	3009	3	TR01/232	
2SB77PD			54			2SB126	121	3009	25	TR01/232	G6003
2SB77R		3004	53			2SB126A	121	3009	25	TR01/	
2SB77V	102A	3004	53	TR85/250	G0007	2SB126B		3009	25		
2SB77VRED	102A			TR85/		2SB126C		3009	25		
2SB77X		3004	53			2SB126D		3009	25		
2SB77Y		3004	53			2SB126E		3009	25		
2SB78	102A	3004	52	TR85/250	G0005	2SB126F	127	3035	16	TR27/235	G6008
2SB79	102A	3004	52	TR85/250	G0005	2SB126G		3009	25		
2SB80	121	3009	3	TR01/232	G6016	2SB126GN		3009	25		
2SB81	121	3009	25	TR01/232		2SB126H		3009	25		
2SB82	121	3009	25	TR01/232		2SB126J		3009	25		
2SB82A		3009	25			2SB126K		3009	25		
2SB83	104	3009	3	TR01/624	G6005	2SB126L		3009	25		
2SB84	121	3014	25	TR01/232	G6005	2SB126M		3009	25		
2SB85	102	3004	2	TR05/631		2SB126OR		3009	25		
2SB87	102	3004	2	TR05/631		2SB126R		3009	25		
2SB89	102A	3003	53	TR85/250	G0005	2SB126V	127	3009	25	TR27/235	G6008
2SB89A	102A	3004	2	TR85/250	G0005	2SB127	104	3009	25	TR01/624	G6003
2SB89AH	102A	3123	80	TR85/250	G0005	2SB127A	121	3009	25	TR01/	
2SB89C		3004	53			2SB128	127	3035	25	TR27/235	G6013
2SB89D		3004	53			2SB128A	121	3009	25	TR01/232	G6018
2SB89E		3004	53			2SB128B		3035	25		
2SB89F		3004	53			2SB128C		3035	25		
2SB89G		3004	53			2SB128D		3035	25		
2SB89GN		3004	53			2SB128E		3035	25		
2SB89H	102A	3123	53	TR85/250	G0005	2SB128F		3035	25		
2SB89J		3004	53			2SB128G		3035	25		
2SB89K		3004	53			2SB128GN		3035	25		
2SB89L		3004	53			2SB128H		3035	25		
2SB89M		3004	53			2SB128J		3035	25		
2SB89OR		3004	53			2SB128K		3035	25		
2SB89R		3004	53			2SB128L		3035	25		
2SB89X		3004	53			2SB128M		3035	25		
2SB89Y		3004	53			2SB128OR		3035	25		
2SB90	102A	3004	52	TR85/250	G0006	2SB128R		3035	25		
2SB91	102A	3004	52	TR85/250	G0005	2SB128V	127	3035	3	TR27/235	G6005
2SB92	102	3003	52	TR05/631	G0005	2SB128X		3035	25		
2SB94	102A	3004	52	TR85/250	G0006	2SB128Y		3035	25		

Original Part Number	ECG	SK	GE	IR/WEP	HEP
2SB129	121	3009	25	TR01/232	G6013
2SB129A		3009	25		G6018
2SB130	131	3035	30	TR94/642	G6016
2SB130A	131	3082	30	TR94/642	G6016
2SB131	121	3009	3	TR01/232	G6003
2SB131A	121	3009	25	TR01/232	G6005
2SB132	121	3009	3	TR01/232	G6005
2SB132A	121	3014	16	TR01/232	G6005
2SB134	102A	3004	53	TR85/250	G0005
2SB134-D	102A				
2SB134E	102A	3004	2		G0005
2SB134-E	102A				
2SB135	102A	3004	53	TR85/250	G0005
2SB135A		3004	52	TR05/	G0005
2SB135B	102A	3004	52	TR85/250	G0001
2SB135C	102A	3004	52	TR85/250	G0001
2SB135D		3004	53		G0007
2SB135E	102A	3004	2	TR85/	G0001
2SB136	102A	3004	53	TR84/250	G0005
2SB136-2	102A				
2SB136-3	102A				
2SB136A	102A	3123	80	TR85/250	G0006
2SB136B	102A	3004	52	TR85/250	G0005
2SB136C	102A	3004	52	TR85/250	G0005
2SB136D		3004	53	TR85/	G0005
2SB136E		3004	53		
2SB136F		3004	53		
2SB136G		3004	53		
2SB136GN		3004	53		
2SB136H		3004	53		
2SB136J		3004	53		
2SB136K		3004	53		
2SB136L		3004	53		
2SB136M		3004	53		
2SB136OR		3004	53		
2SB136R		3004	53		
2SB136U	102A	3004	2	TR85/250	G0005
2SB136(V)				TR85/	
2SB136X		3004	53		
2SB137	121	3009	16	TR01/232	G6003
2SB138	121	3009	16	TR01/232	G6005
2SB138A		3014	16	TR01/	G6005
2SB138B		3014	16	TR01/	G6005
2SB140	104	3009	3	TR01/232	G6003
2SB141	104	3009	25	TR01/232	G6005
2SB142	104	3009	49	TR01/232	G6003
2SB142A		3009	25		
2SB142B	100	3005	3	TR05/254	G0005
2SB142C	100	3005	3	TR05/254	G0005
2SB143	104	3009	49	TR01/230	G6003
2SB143P	104	3009	52	TR01/230	G6003
2SB144	104	3009	49	TR01/230	G6003
2SB144P	104	3009	52	TR01/230	G6003
2SB145	104	3009	49	TR01/230	G6003
2SB146	104	3009	49	TR01/230	G6003
2SB147	104	3009	52	TR01/230	G6005
2SB149	104	3009	25	TR01/230	G6003
2SB150		3004	2		
2SB151	104	3009	25	TR01/230	G6018
2SB152	104	3009	25	TR01/232	G6018
2SB153	102	3004	52	TR05/631	G0005
2SB154	102	3004	52	TR05/631	G0005
2SB155	102A	3003	53	TR85/250	G0005
2SB155A	102A	3004	52	TR85/250	G0005
2SB155B	102A	3004	53	TR85/	G0005
2SB156	102A	3004	53	TR84/250	G0005
2SB156A	102A	3004	53	TR84/250	G0005
2SB156AA	102A			TR85/250	G0005
2SB156AB	102A	3004	53	TR85/250	G0005
2SB156AC	102A			TR85/250	
2SB156B	102A	3004	52	TR85/250	G0005
2SB156C	102A	3004	52	TR85/250	G0005
2SB156D	102A	3004	2	TR85/630	G0005
2SB156E		3004	53		
2SB156F		3004	53		
2SB156G		3004	53		
2SB156GN		3004	53		
2SB156H		3004	53		
2SB156J		3004	53		
2SB156K		3004	53		
2SB156L		3004	53		
2SB156M		3004	53		
2SB156P	102A			TR85/	
2SB156OR		3004	53		
2SB156R		3004	53		

Original Part Number	ECG	SK	GE	IR/WEP	HEP
2SB156X		3004	53		
2SB156Y		3004	53		
2SB156/7825B		3004			
2SB157	102A	3004	51	TR85/250	G0008
2SB158	102A	3004	51	TR85/250	G0008
2SB159	102A	3004	51	TR85/250	G0008
2SB160	102A	3004	51	TR85/250	G0008
2SB161	102	3004	2	TR05/631	G0005
2SB162	102	3004	2	TR05/631	G0005
2SB163	102	3004	2	TR05/631	G0005
2SB164	102	3004	2	TR05/631	G0005
2SB165	102	3004	2	TR05/631	G0006
2SB166	102	3004	2	TR05/631	G0001
2SB167	158	3004	53	TR84/630	G0005
2SB168	102A	3003	52	TR85/250	G0005
2SB169	102A	3003	52	TR85/250	G0005
2SB170	102A	3004	2	TR84/250	G0005
2SB171	102A	3004	2	TR85/250	G0005
2SB171A	102A	3004	52	TR85/250	G0005
2SB171B	102A	3004	52	TR85/250	G0005
2SB171F		3004	53	TR85/	
2SB172	102A	3004	2	TR84/250	G0005
2SB172A	102A	3004	52	TR85/250	G0005
2SB172AF	102A	3004	53	TR85/250	
2SB172A-F		3123	52	TR05/	
2SB172B	102A	3004	52	TR85/250	
2SB172C	102A	3004	53	TR85/250	
2SB172D	102A	3004	52	TR85/250	G0005
2SB172E	102A	3004	52	TR85/250	
2SB172F	102A	3004	52	TR85/250	G0005
2SB172FN		3004	53		
2SB172G		3004	53		
2SB172GN		3004	53		
2SB172H	102A	3004	52	TR85/	
2SB172J		3004	53		
2SB172K		3004	53		
2SB172L		3004	53		
2SB172M		3004	53		
2SB172P	102A			TR85/250	
2SB172OR		3004	53		
2SB172R	102A	3004	2	TR85/250	
2SB172X		3004	53		
2SB172Y		3004	53		
2SB173	102A	3004	50	TR85/250	G0005
2SB173A	102A	3004	52	TR85/250	G0005
2SB173B	102A	3004	52	TR85/250	G0005
2SB173BL		3004	53		
2SB173C	102A	3004	53	TR85/250	G0005
2SB173CL		3004	53		
2SB173D		3004	53		
2SB173E		3004	53		
2SB173F		3004	53		
2SB173G		3004	53		
2SB173GN		3004	53		
2SB173H		3004	53		
2SB173J		3004	53		
2SB173K		3004	53		
2SB173L	102A	3004	2	TR85/250	
2SB173M		3004	53		
2SB173OR		3004	53		
2SB173R		3004	53		
2SB173X		3004	53		
2SB173Y		3004	53		
2SB174	102A	3004	52	TR85/250	G0005
2SB175	102A	3004	2	TR85/250	G0005
2SB175A	102A	3004	52	TR85/250	G0005
2SB175(A)			1	TR12/	
2SB175B	102A	3004	53	TR85/250	G0005
2SB175B-1		3004	53		
2SB175BL		3004	53		
2SB175C	102A	3004	2	TR85/250	G0005
2SB175CL		3004	53		
2SB175D		3004	53		
2SB175E	102A	3004	53	TR85/250	
2SB175F		3004	53		
2SB175G		3004	53		
2SB175GN		3004	53		
2SB175H		3004	53		
2SB175L		3004	53		
2SB175M		3004	53		
2SB175OR		3004	53		
2SB175R		3004	53		
2SB175X		3004	53	TR85/	
2SB175Y		3004	53		
2SB176	102A	3004	2	TR84/250	G0005

Original Part Number	ECG	SK	GE	IR/WEP	HEP
2SB176M	102A	3004	52	TR85/	G0005
2SB176(O)				TR85/	G0005
2SB176-O	102A			TR85/	
2SB176O		3004	52		
2SB176P	102A	3004	52	TR85/250	G0005
2SB176-P	102A			TR85/	
2SB176-PR	102A	3004	52	TR85/250	
2SB176PRC	102A	3004	52	TR85/250	
2SB176PRO				TR85/	
2SB176R	102A	3004	52	TR85/250	G0005
2SB176RED					G0005
2SB176RG		3004	53		
2SB176X		3004	53		
2SB176Y		3004	53		
2SB177	102A	3004	2	TR85/	G0005
2SB177A		3004	8	TR10/	
2SB177R		3004	53	TR85/	
2SB178	102A	3004	2	TR85/250	G0005
2SB178A	102A	3004	2	TR85/250	G0005
2SB178B		3004	53		
2SB178C	102A	3004	2	TR85/250	G0005
2SB178D	102A	3004	53	TR85/250	G0005
2SB178E		3004	53		
2SB178F		3004	53		
2SB178G		3004	53		
2SB178GN		3004	53		
2SB178H		3004	53		
2SB178J		3004	53		
2SB178K		3004	53		
2SB178L		3004	53		
2SB178M	102A	3004	2	TR85/250	G0005
2SB178N	102A	3004	2	TR85/250	G0005
2SB178(P)					G0005
2SB178-O	102A			TR85/	
2SB178P		3004			G0005
2SB178-S	102A				
2SB178S	102A	3004	2	TR51/250	G0005
2SB178T	102A	3004	52	TR85/250	G0005
2SB178TC		3004	53		
2SB178TS		3004	53		
2SB178U	102A	3004	2	TR85/250	
2SB178V	102A	3004	2	TR85/250	G0005
2SB178X	102A	3004	2	TR85/	G0005
2SB178Y	102A	3004	53	TR85/250	
2SB179			50	TR17/	G0005
2SB180	158		3	TR84/630	G6011
2SB180A	158			TR84/630	G6011
2SB181	158		3	TR84/630	G6011
2SB181A	158			TR84/630	G6011
2SB182			51		
2SB183	102A	3004	52	TR85/250	G0005
2SB184	102A	3004	52	TR85/250	G0006
2SB185	102A	3004	2	TR84/250	G0005
2SB185A		3004	53		
2SB185AA	102A	3004	1	TR85/	
2SB185B		3004	53		
2SB185C		3004	53		
2SB185D		3004	53		
2SB185E		3004	53		
2SB185F	102A	3004	2		
2SB185G		3004	53		
2SB185GN		3004	53		
2SB185H		3004	53		
2SB185J		3004	53		
2SB185L		3004	53		
2SB185M		3004	53		
2SB185OR		3004	53		
2SB185P	102A	3004	2	TR85/250	G0005
2SB185R		3004	53		
2SB185X		3004	53		
2SB185Y		3004	53		
2SB186	102A	3004	53	2SB186/250	G0007
2SB186(O)	102A		52		
2SB186-1	102A			TR85/250	
2SB186-7		3004	53		
2SB186A	102A	3004	52	TR85/250	G0007
2SB186AG	102A	3004	53	TR85/250	
2SB186B	102A	3004	52	TR85/250	G0007
2SB186BY	102A	3004	53	TR85/250	
2SB186C		3123	2		
2SB186D		3004	53		
2SB186E		3004	53		
2SB186F		3004	53		
2SB186G	102A	3004	52	TR85/250	
2SB186H	102A	3004	53		G0007
2SB186J		3004	53		
2SB186-K	102A			TR85/250	G0005
2SB186K		3004	53	TR85/	G0007
2SB186L	102A	3004	53		
2SB186M		3004	53		
2SB186OR		3004	53		
2SB186R		3004	53		
2SB186X		3004	53		
2SB186Y	102A	3004	53		
2SB187	102A	3004	53	2SB187/250	G0005
2SB187(1)	102A				
2SB187A		3004	53	TR05/	
2SB187AA	102A	3004	53	TR85/250	
2SB187B	102A	3004	2	TR85/250	
2SB187C	102A	3004	52	TR85/250	
2SB187CORD				TR17/	
2SB187D	102A	3004	53	TR85/735	G0005
2SB187E		3004	53		
2SB187F		3004	53		
2SB187G	102A	3004	2	TR85/250	
2SB187GN		3004	53		
2SB187H		3004	53		
2SB187J		3004	53		
2SB187K		3004	53		
2SB187L		3004	53		
2SB187M		3004	53		
2SB187OR		3004	53		
2SB187R	102A	3004	2	TR85/250	G0005
2SB187S	102A	3004	53	TR85/250	G0005
2SB187X		3004	53		
2SB187Y		3004	53		
2SB187YEL	102A				
2SB188	102A	3004	2	TR85/250	G0005
2SB189	102	3004	53	TR05/631	G0005
2SB190			53		G0005
2SB192			53		G0005
2SB193			53		G0005
2SB194			53		G0005
2SB195			53		G0005
2SB196			53		G0005
2SB197			53		G0005
2SB198			53		G0005
2SB199	102	3004	53	TR05/631	G0005
2SB200	102	3004	53	TR05/631	G0005
2SB200A	102	3004	2	TR05/631	G0005
2SB201	102A		53	TR85/250	G0005
2SB202	102	3004	53	TR05/631	G0005
2SB203A		3004	53		
2SB203AA	102	3004	53	TR05/631	
2SB204	179			TR35/G6001	
2SB205	179			TR35/G6001	
2SB206	179			TR35/G6001	
2SB215	121	3009	16	TR01/232	G6005
2SB216	121	3034	52	TR01/232	G6005
2SB216A	121	3034	3	TR01/232	
2SB217	121	3009	16	TR01/232	G6003
2SB217A	121	3009	52	TR01/232	
2SB217B		3009	25		
2SB217C		3009	25		
2SB217D		3009	25		
2SB217E		3009	25		
2SB217F		3009	25		
2SB217G	121	3009	52	TR01/232	
2SB217GN		3009	25		
2SB217H		3009	25		
2SB217K		3009	25		
2SB217L		3009	25		
2SB217M		3009	25		
2SB217OR		3009	25		
2SB217R		3009	25		
2SB217U	121	3009	16	TR01/232	
2SB217X		3009	25		
2SB217Y		3009	25		
2SB218	102	3004	52	TR05/631	G6012
2SB219	102	3004	2	TR05/631	G0005
2SB220	102	3004	2	TR05/631	G0005
2SB220A	102	3004	52	TR05/631	
2SB221	102	3004	52	TR05/631	G0005
2SB221A	102	3004	2	TR05/631	G0005
2SB222	102	3004	2	TR05/631	G0005
2SB223	102	3004	2	TR05/631	G0005
2SB224	102	3004	52	TR05/631	G0005
2SB225	102	3004	52	TR05/631	G0005
2SB226	102	3004	52	TR05/631	G0006
2SB227	102	3004	52	TR05/631	G0006

Original Part Number	ECG	SK	GE	IR/WEP	HEP
2SB228	104	3009	25	TR01/624	G6018
2SB229	121	3009	25	TR01/232	G6018
2SB230	104	3009	25	TR01/624	G6018
2SB231	127	3035	25	TR27/235	G6017
2SB232	127	3034	16	TR27/235	G6007
2SB233	104	3009	16	TR01/624	G6007
2SB234	127	3035	25	TR27/	G6007
2SB234B		3035	25	TR17/	
2SB234N	127	3035	25	TR27/235	
2SB235	105	3012	4	TR03/233	G6006
2SB235A	160	3012	4	TR17/233	
2SB236	105	3012	4	TR03/233	G6006
2SB237	105	3012	4	TR03/231	G6004
2SB237-12A	105	3012	4	TR03/231	G6004
2SB237-12B	105	3012	4	TR03/231	G6004
2SB238	158			TR84/630	G6011
2SB238-12A	105	3012	4	TR03/231	G6004
2SB238-12B	105	3012	4	TR03/231	G6004
2SB238-12C	105	3012	4	TR03/231	G6004
2SB239	121	3009	16	TR01/232	G6012
2SB239A	121	3009	16	TR01/232	G6012
2SB240	160	3009	52	TR17/637	G6011
2SB240A	160	3009	16	TR17/637	G6011
2SB241	102	3004	53	TR05/631	G6012
2SB241A		3004	53		G6012
2SB242		3009	52	TR01/	G6011
2SB242A		3009	16	TR01/	G6011
2SB243		3009	52	TR01/	G6011
2SB243A		3009	16	TR01/	G6011
2SB244		3009	16	TR01/	G6012
2SB245		3009	16	TR01/	G6012
2SB245RD					G0005
2SB246	104	3009	16	TR01/230	G6003
2SB247	121	3009	16	TR01/232	G6005
2SB248	121	3009	16	TR01/232	G6003
2SB248A	121	3009	16	TR01/232	G6005
2SB249	104	3009	16	TR01/232	G6005
2SB249A	104	3009	3	TR01/232	
2SB250	104	3009	16	TR01/230	G6003
2SB250A	104	3009	16	TR01/232	G6003
2SB251	127	3035	16	TR27/235	G6003
2SB251A	127	3035	16	TR27/235	G6003
2SB252	127	3009	16	TR27/235	G6005
2SB252A	127	3035	25	TR27/235	G6018
2SB253	127	3035	16	TR27/235	G6005
2SB253A	127	3009	16	TR27/235	G6018
2SB254	104	3009	16	TR01/624	G0342
2SB255	104	3009	16	TR01/624	G6016
2SB256	104	3052	16	TR01/624	G6016
2SB257	102	3004	1	TR05/631	G0008
2SB258	105	3012	4	TR03/233	
2SB259	105	3012	4	TR03/233	G6006
2SB260	105	3012	4	TR03/233	G6004
2SB261	102A	3004	52	TR85/250	G0005
2SB262	102A	3004	2	TR85/250	G0005
2SB263	102A	3004	2	TR85/250	G0005
2SB263(3)				TR05/	
2SB263/127297				TR85/	
2SB264	102A	3004	52	TR85/250	G0005
2SB265	102A	3004	80	TR85/250	G0005
2SB266	102A	3004	2	TR85/250	G0005
2SB266P	102A	3004	52	TR85/250	
2SB266Q	102A	3004	52	TR85/250	
2SB267	102A	3004	2	TR85/250	G0005
2SB268	102A	3004	2	TR85/250	G0005
2SB268H		3004	53	TR50/	
2SB269	102A	3004	2	TR85/250	G0005
2SB270	102A	3004	52	TR85/250	G0005
2SB270A	102A	3004	53	TR85/250	
2SB270B	102A	3004	2	TR85/250	
2SB270C	102A	3004	52	TR85/250	
2SB270D	102A	3004	52	TR85/250	
2SB270E	102A	3004	2	TR85/250	
2SB271	102A		51	TR85/250	G0005
2SB272	102A	3004	51	TR85/250	G0005
2SB273	102A	3004	52	TR85/250	G0005
2SB274	127	3035	25	TR27/235	G6005
2SB275	127	3035	25	TR27/235	G6017
2SB276	127	3035	25	TR27/235	G6017
2SB281C			18		
2SB282	121	3009	25	TR01/232	G6005
2SB283	121	3009	25	TR01/232	G6005
2SB284	121	3009	25	TR01/232	G6005
2SB285	121	3009	25	TR01/232	G6005
2SB287				TR33/	

Original Part Number	ECG	SK	GE	IR/WEP	HEP
2SB287A				TR33/	
2SB288				TR33/	
2SB289				TR24/	
2SB290	100	3004	1	TR05/254	G0005
2SB291	100	3004	52	TR05/254	G0006
2SB292	100	3004	52	TR05/254	G0005
2SB292A	100	3005	51	TR05/254	G0005
2SB293	102A	3004	2	TR85/250	G0005
2SB294	102A	3123	2	TR85/250	G0005
2SB295	121	3009	25	TR01/232	G6018
2SB296	127	3035	25	TR27/235	G6017
2SB299	102A	3004	2	TR85/250	G0005
2SB300	127	3035	25	TR27/235	G6015
2SB301	127	3035	16	TR27/235	G6005
2SB302	102A	3004	51	TR85/250	G0005
2SB303	102A	3004	52	TR84/250	G0005
2SB303-0	102A	3004	53	TR85/250	
2SB303A	102A	3004	52	TR85/250	G0005
2SB303B	102A	3004	53	TR85/250	
2SB303C	102A	3004	53	TR85/250	
2SB303H	102A	3004	52		G0005
2SB303K	102A	3004	53	TR85/250	
2SB304	158	3123	53	TR84/630	G0005
2SB304A	158	3123	80	TR84/630	G0005
2SB309	127	3035	25	TR27/235	G6005
2SB310	127	3035	25	TR27/235	G6017
2SB311	127	3035	25	TR27/235	G6017
2SB312	127	3035	25	TR27/235	G6007
2SB313	127	3035	25	TR27/235	G6017
2SB314	100	3004	53	TR05/254	
2SB315	102A	3004	52	TR85/250	G0005
2SB316	102A	3004	52	TR85/250	G0005
2SB317	102A		53	TR85/250	G0005
2SB318	127	3014	25	TR27/235	G6005
2SB319	127	3035	25	TR27/235	G6007
2SB320	127	3035	25	TR27/235	G6017
2SB321	102A		52	TR85/250	G0006
2SB322	102A		52	TR85/250	G0005
2SB323	102A		52	TR85/250	G0006
2SB324	158	3004	53	TR84/630	G0005
2SB324/4454C	158				
2SB324A	158	3004	52	TR84/630	
2SB324B	158	3004	52	TR84/630	G0005
2SB324D	158	3004	53	TR84/630	G0006
2SB324E	158	3004	2	TR84/630	G0005
2SB324E-1	102A			TR85/	
2SB324E-L			2		G0005
2SB324F	158	3004	53	TR84/	G0006
2SB324G	158	3004	53	TR84/630	G0005
2SB324H	158	3004	52	TR84/630	G0005
2SB324I	158	3004	53	TR84/630	G0005
2SB324J	158	3004	53	TR84/	
2SB324K	158	3004	53	TR84/630	G0005
2SB324L	158	3004	53	TR84/	G0005
2SB324M	158	3004	2	TR84/	G0005
2SB324P	158	3004	53	TR84/630	
2SB324S	158				
2SB324V	158	3004	2	TR84/630	G0005
2SB326	102	3004	53	TR05/631	G0005
2SB327	102	3004	52	TR05/631	G0005
2SB328	102	3004	52	TR05/631	G0005
2SB329	102A	3004	52	TR85/250	G0005
2SB329K	102A	3004	2	TR85/250	G0005
2SB331	105	3012	4	TR03/233	
2SB332	105	3012	4	TR03/233	G6006
2SB333	105	3012	4	TR03/233	G6006
2SB334	105	3012	4	TR03/233	G6006
2SB335			52	TR84/	G0005
2SB336	102A		52	TR84/250	G0005
2SB337	104	3009	25	TR02/232	G6005
2SB337A	104	3009	25	TR85/250	G6005
2SB337B	104	3009	16	TR01/232	G6005
2SB337BK	104	3009			
2SB338	121	3009	25	TR01/232	G6005
2SB338HA	102A			TR85/250	
2SB338HB	102A			TR85/250	
2SB339	179		25	TR35/G6001	G6018
2SB339H	127			TR27/	
2SB340	179		25	TR35/G6001	G6018
2SB340H	127			TR27/	
2SB341	179	3034	25	TR27/G6001	G6018
2SB341H	127	3034	25	TR27/	
2SB341S	127	3034			
2SB341V	127	3034	25	TR27/235	G6008
2SB342	127	3035	25	TR27/235	G6017

Original Part Number	ECG	SK	GE	IR/WEP	HEP
2SB343	127	3035	25	TR27/235	G6017
2SB345	102A	3004	52	TR85/250	G0005
2SB346	102A	3004		TR85/250	G0007
2SB346K	102A	3004	2	TR85/250	G0007
2SB346Q	102A	3004	53	TR85/	G0007
2SB347	102	3004	52	TR05/631	G0007
2SB348	102A	3004		TR85/250	G0006
2SB348Q	102A	3004	53		G0006
2SB348R	102A	3004	53	TR85/250	
2SB349	102A		51	TR85/250	G0006
2SB350	102	3004	52	TR05/631	G0005
2SB351	105	3012	4	TR03/233	G6006
2SB352	105	3012	4	TR03/233	G6006
2SB353	105	3012	4	TR03/233	G6006
2SB354	105	3012	4	TR03/233	G6006
2SB355	104	3009	16	TR01/642	G6016
2SB356	104	3009	16	TR01/624	
2SB357	127	3035	25	TR27/235	
2SB357B		3035	25		G6016
2SB358	127	3035	25	TR27/235	G6005
2SB359	127	3035	25	TR27/235	G6017
2SB360	127	3035	25	TR27/235	G6017
2SB361	127	3035	3	TR27/235	G6005
2SB362	127	3014	25	TR27/235	G6005
2SB364	158	3004	53	TR85/630	G0005
2SB364K		3004	53	TR85/	
2SB365	158	3004	53	TR84/630	G0005
2SB366	102A	3123	52	TR85/250	
2SB367	131	3052	30	TR94/642	G6016
2SB367A	131	3052	30	2SB367A/250	G6016
2SB367B	131	3052	30	TR94/242	G6016
2SB367BP		3052			G6016
2SB367C	131	3052	30	TR94/242	G6016
2SB367H	131	3052	30	TR94/242	G6016
2SB368	131	3052	30	TR50/242	G6016
2SB368A	131	3052	30	TR94/642	G6016
2SB368H	131	3052	30	TR94/242	G6016
2SB370	102A	3004	53	TR85/250	G0006
2SB370A	102A	3004	53	TR85/250	G0006
2SB370AA	102A			TR85/250	
2SB370AB	102A			TR85/250	
2SB370AC	102A			TR85/250	
2SB370AHA	102A			TR85/250	
2SB370AHB	102A			TR85/250	
2SB370B	102	3004	2	TR05/630	G0005
2SB370C	102A	3004	2	TR85/250	G0006
2SB370D	102A	3004	53	TR85/	
2SB370P	102A			TR85/	
2SB370V	102A		2	TR85/250	G0005
2SB371	102A	3004	1	TR85/250	G0005
2SB371D	102A	3004	53	TR85/250	
2SB372	158		53	TR84/630	G6011
2SB373	158	3009	2	TR84/630	G6011
2SB374					G6011
2SB375	127	3035	25	TR27/235	G6007
2SB375-2B	127			TR27/235	
2SB375-5B	127			TR27/235	
2SB375A	127	3035	25	TR27/235	G6007
2SB375A-2B	127	3035	25	TR27/235	G6007
2SB375A-5B	127	3035	25	TR27/235	G6007
2SB375A-NB	127	3035	25	TR27/235	G6007
2SB376	102A	3004	53	TR85/250	G0005
2SB376G	102A	3004	53	TR85/250	
2SB377	102A		53	TR85/250	G0005
2SB377B	102A			TR27/	
2SB378	100	3004	2	TR05/254	G0005
2SB378A	102	3004	53	TR05/631	
2SB379	102	3004	2	TR05/631	G0005
2SB379-2	102	3004	52	TR05/631	
2SB379A	102	3004	52	TR05/631	
2SB379B	102	3004	52	TR05/631	
2SB380	102	3004	53	TR05/631	G0006
2SB380A	102	3004	52	TR05/631	G0005
2SB381	102A	3004	53	TR85/250	G0005
2SB382	102A	3004	53	TR85/250	G0005
2SB383	102A	3004	53	TR85/250	G0005
2SB383-1	102A	3004	52	TR85/250	G0005
2SB383-2	102A	3004	52	TR85/250	
2SB384	126	3004	52	TR17/635	G0005
2SB385	126	3004	52	TR12/250	G0005
2SB386	102A	3123	80	TR85/250	G0005
2SB387	100	3004	51	TR05/254	G0008
2SB389	102A	3004	50	TR85/250	G0003
2SB390	127	3035	25	TR27/235	G6005
2SB391	121	3009	25	TR01/232	G6003
2SB392	100	3004	2	TR05/254	G0005
2SB393	100	3004	53	TR05/254	G0005
2SB394	100	3004	53	TR05/254	G0005
2SB395	100	3004	53	TR05/254	G0005
2SB396	100	3004	80	TR05/254	G0005
2SB398			53		
2SB400	102A	3004	52	TR85/250	G0005
2SB400A	102A	3004	53	TR85/250	G0005
2SB400B	102A		52	TR85/250	
2SB400K	102A	3004	53	TR85/250	
2SB401	100	3004	80	TR05/254	G0005
2SB402	100	3004	53	TR05/254	G0005
2SB403	100	3004	80	TR05/254	G0005
2SB405	158	3004	53	2SB405/630	G6011
2SB405-2C	158	3004	53	TR84/630	G6011
2SB405-3C	158	3004	53	TR84/630	G6011
2SB405-4C	158	3004	53	TR84/630	G6011
2SB405A	158	3004	52	TR84/630	G6011
2SB405B	158	3004	2	TR84/630	G6011
2SB405BR	158				
2SB405C	158	3004	2	TR84/630	G6011
2SB405D	158	3004	53	TR84/630	
2SB405E	158	3004	53	TR84/630	
2SB405G	158	3004	53		G6011
2SB405H	158	3004	53	TR84/	G6011
2SB405K	158	3004	53	TR84/	
2SB405P	158				
2SB405R	158	3004	53	TR84/630	
2SB405RED	158	3004	53	TR84/	G0005
2SB407	104	3009	25	TR02/230	G6005
2SB407-O	104	3009	16	TR01/230	G6003
2SB407TV	121	3009		TR01/232	
2SB408	100	3004	2	TR05/254	G0005
2SB410	127			TR27/235	G6017
2SB411	127		76	TR27/235	G6007
2SB411TV		3035	25		
2SB413	104	3014	16	TR01/624	
2SB414	104	3014	16	TR01/624	G6016
2SB415	158	3004	53	TR84/630	G6011
2SB415A	158	3004	53	TR84/630	G6011
2SB415B	158	3004	53		
2SB416	100	3004	53	TR05/254	G0005
2SB417	100	3004	2	TR05/254	G0005
2SB421	102A			TR85/250	G6012
2SB422	102	3004	52	TR05/631	
2SB423	102	3004	52	TR05/631	G0005
2SB424	121	3034	25	TR01/232	G6005
2SB425	121	3034	25	TR01/232	G6005
2SB425Y	127	3034	16	TR27/235	
2SB426	121	3009	25	TR01/232	G6003
2SB426BL	104			TR01/	
2SB426R	121	3009	16	TR01/232	
2SB426Y	121	3009	16	TR01/232	G6003
2SB427	102A	3123	80	TR85/250	G0005
2SB428	102A	3123	80	TR85/250	G0005
2SB430	105			TR03/233	G6010
2SB431	158		53	TR85/630	G0005
2SB432	127	3035	25	TR27/235	G6007
2SB433	105			TR03/233	G6006
2SB434	153	3083	69	TR77/700	S5013
2SB434-O	153			TR77/	
2SB434-R	153			TR77/	
2SB434-Y	153			TR77/	
2SB435	153	3084	69	TR77/700	S3028
2SB435-O	153			TR77/	
2SB435-R	153			TR77/	
2SB435-Y	153			TR77/	
2SB439	102A	3004	52	TR85/250	G0005
2SB439A	102	3004	52	TR05/	
2SB440	102A	3004	52	TR85/250	G0005
2SB443	102A	3004	50	TR85/250	G0008
2SB443A	102A	3004	50	TR85/250	G0006
2SB443B	102A	3004	50	TR85/250	G0006
2SB444	126	3004	50	TR17/635	G0008
2SB444A	126	3006	50	TR17/635	G0006
2SB444B	126	3006	50	TR17/635	G0006
2SB445	121	3014	16	TR01/232	G6016
2SB446	121	3014	16	TR01/232	
2SB447	127	3035	25	TR27/235	G6007
2SB448	131	3034	3	TR94/642	G6016
2SB449	121	3009	25	TR85/232	G6003
2SB449F	121	3009	16	TR01/232	G6003
2SB449P	121			TR01/232	G6003
2SB450	158		53	TR84/630	G0005
2SB450A	158			TR84/630	G0005

Original Part Number	ECG	SK	GE	IR/WEP	HEP
2SB451	158			TR84/630	G6011
2SB452	158		52	TR84/630	G6011
2SB452A	158			TR84/630	G6011
2SB453	158			TR84/630	G0005
2SB454	158			TR84/630	G0005
2SB455	158			TR84/630	G0005
2SB457	158	3004	53	TR85/630	G0005
2SB457A	158	3004	53	TR84/630	G0005
2SB457AC	158	3004	53		G0005
2SB457-C	158				
2SB458	131	3052	44	TR94/642	G6016
2SB458A	131	3082	16	TR94/642	G6016
2SB459	102A	3004	52	TR85/250	G0006
2SB459A	102A	3004	53	TR85/250	
2SB459B	102A	3004	53	TR85/250	G0006
2SB459C	102A	3004	2	TR85/250	G0006
2SB459D	102A	3004	53	TR85/250	G0006
2SB459-O	102A			TR85/250	
2SB460	102A	3123	80	TR85/250	G0001
2SB460A	102A	3123	80	TR85/250	
2SB460B	102A	3123	80	TR85/250	
2SB461	176	3123	53	TR82/238	G0005
2SB462	131	3052	30	TR94/642	G6012
2SB463	131	3082	30	TR50/642	G6016
2SB463BL	131	3198		TR50/642	G6016
2SB463E	131	3082	44	TR94/642	
2SB463R	131	3052	30	TR50/642	
2SB463RED			30		G6016
2SB463Y	131	3052	30	TR50/642	G6016
2SB464	127	3035	25	TR27/235	G6007
2SB465	127	3035	16	TR27/235	G6005
2SB466	131	3009	16	TR94/642	G6016
2SB467	131	3009	16	TR94/642	G6016
2SB468	127	3035	25	TR27/235	G6008
2SB468A	127	3035	25	TR27/235	G6008
2SB468B	127	3035	25	TR27/235	G6008
2SB468C	127	3035	25	TR27/235	G6008
2SB468D	127	3035	25	TR27/235	
2SB470	102	3004	52	TR05/631	G0006
2SB471	121	3009	25	TR01/232	G6005
2SB471-2	121	3009	25	TR01/232	
2SB471A	121	3009	25	TR01/232	G6005
2SB471B	121	3009	16	TR01/232	G6005
2SB472	121	3034	25	TR01/232	G6005
2SB472A	121	3009	25	TR01/232	G6005
2SB472B	121	3009	25	TR01/232	G6005
2SB473	131	3198	25	TR50/642	G6016
2SB473D	131	3052	44	TR94/642	
2SB473F	131	3052	44	TR94/642	
2SB473H	131	3052	30	TR50/	G6016
2SB474	226	3082	49	2SB474/642	G6016
2SB474-2	226	3082	49	TR94/642	G6016
2SB474-3	226	3082	49	TR94/642	G6016
2SB474-4	226	3082	49	TR94/642	G6016
2SB474-6D	226	3082	49	TR94/642	G6016
2SB474MP	226	3086	49	TR94/	
2SB474S	226				
2SB474V4	226	3082	49	TR94/	
2SB474V10	126MP	3082		TR94MP/	
2SB474Y	226	3082	13MP	TR94/	
2SB475	158	3004	53	TR84/630	G0005
2SB475A	158	3004	53	TR84/630	G0005
2SB475B	158	3004	2	TR84/630	G0005
2SB475C	158	3004	53	TR84/630	
2SB475D	158	3004	52	TR84/630	G0005
2SB475F	158	3004	52	TR84/630	
2SB475G	158	3004	53	TR84/630	
2SB475P	158	3004	53	TR84/630	G0005
2SB475Q	158			TR84/630	
2SB476	158			TR84/630	G6011
2SB477		3123		TR05/	G6010
2SB478					G6010
2SB479					G6010
2SB480					G6010
2SB481	131		25	2SB481/642	G6016
2SB481D	131	3082	44	TR94/	G6016
2SB481E	131	3082	44	TR94/642	
2SB482	102A	3004	2	TR85/250	G0005
2SB483	179		76	TR35/G6001	G6018
2SB484	179			TR35/G6001	G6018
2SB485	179			TR35/G6001	
2SB486	102	3004	53	TR05/631	G0007
2SB492	176	3004	80	TR82/238	G6011
2SB492B	176	3004	53	TR82/238	
2SB493					G6011

Original Part Number	ECG	SK	GE	IR/WEP	HEP
2SB494	158		53	TR84/630	G6011
2SB495	158	3004	53	TR85/630	G6011
2SB495A	158	3004	53	TR84/630	G6011
2SB495B		3004	53		
2SB495C	158	3004	53	TR84/630	G6011
2SB495D	158	3004	53	TR84/630	G6011
2SB495T	158	3004	53	TR84/630	G6011
2SB496	102A	3004	54	TR85/250	G0005
2SB497	102	3004	2	TR82/631	G0005
2SB498	102A	3004	52	TR85/250	G0005
2SB502	197				S5018
2SB503	197				S5018
2SB504					S3003
2SB507	153	3083		TR77/	S5006
2SB508	153	3083		TR77/	S5006
2SB509	153			TR77/	S5018
2SB510	129			TR88/	
2SB510S	211	3203			
2SB511	153	3084	69	TR77/	S3028
2SB511C	153	3084	29		S5006
2SB511D	153	3084	29	TR77/700	S3028
2SB512	153			TR77/	
2SB512A				TR77/	
2SB513	153			TR77/	
2SB513A	153			TR77/	
2SB513P	153				
2SB513Q	153				
2SB513R	153				
2SB514	153			TR77/	S3032
2SB515	153			TR77/	S3032
2SB516	102A				
2SB516C	102A	3004	52	TR85/	G0007
2SB516CD	102A	3004			G0005
2SB516CD(P)			2	TR85/	G0005
2SB516D	102A	3004	52	TR85/	G0005
2SB516P	102A	3004			
2SB518			74		
2SB519			74		
2SB524					S5006
2SB525			48		
2SB526	126				
2SB531		3173			
2SB534			53		
2SB535			53		
2SB539R	123A			TR21/735	
2SB540		3123			
2SB542			22		
2SB544			48		
2SB546				TR05/	
2SB558		3173			
2SB560	102A		2	TR85/250	
2SB561	102A				
2SB565		3004	53		
2SB638B				TR01/	
2SB639A				TR01/	
2SB643R					G6016
2SB669				TR01/	
2SB677A				TR01/	
2SB677C				TR01/	
2SB678A				TR01/	
2SB681A	163			TR67/	
2SB681B	163			TR67/	
2SB683			39		
2SB1060B		3041	55		
2SB1560				TR84/	
2SB1760				TR85/	
2SB1780		3004	53	TR85/	
2SB1785				TR85/	
2SB1860				TR85/	
2SB1871				TR85/	
2SB3030	102A			TR85/250	
2SB3224E				TR85/	
2SB3240		3004	53		
2SB3244				TR05/	
2SB3244H	158				
2SB3783		3004	53		
2SB3812		3004	53		
2SB3813		3004	53		
2SB3855				TR17/	
2SB4050				TR08/	
2SB4151				TR84/	
2SB4440D		3004	53		
2SB4631				TR50/	
2SB4744				TR01/	
2SB7513				TR05/	

Original Part Number	ECG	SK	GE	IR/WEP	HEP
2SB17313				TR85/	
2SBF1	102A	3004	2	TR85/250	G0005
2SBF1A	102A	3004	52	TR85/250	G0005
2SBF2	102A	3004	53	TR85/250	G0005
2SBF2A		3004	53		
2SBF5	121	3009	16	TR01/232	G6005
2SBM77		3004	53		
2SBU86				TR85/	
2SBZ		3035	25		
2SC-2C	506				
2SC3T2Y	123A				
2SC-F6			20		
2SC6	195				
2SC7	130				
2SC9	236				
2SC11	101	3011	8	TR08/641	G0011
2SC12	128	3024	63	TR87/243	S0014
2SC13	101	3011	5	TR08/641	G0011
2SC14	101	3011	6	TR08/641	G0011
2SC15	123A	3122	21	TR21/735	S0014
2SC15-1	123A	3122	47	TR21/735	S0005
2SC15-2	123A		27	TR21/735	S0005
2SC15-3	123A	3122	27	TR21/735	S0005
2SC16	123A	3122	11	TR21/735	S0011
2SC16A	123A	3122	11	TR21/735	S0011
2SC17	108	3018	20	TR95/56	S0011
2SC17A	108	3039	20	TR95/56	S0011
2SC18	123A	3122	8	TR21/735	S0011
2SC19	128	3024	63	TR87/243	S0014
2SC20	128	3024	18	TR87/243	S0014
2SC21	130	3027	72	TR59/247	S7002
2SC22	128	3024	66	TR87/243	S3020
2SC23	128	3024	66	TR87/243	S3020
2SC24	128	3024	66	TR87/243	S3021
2SC24C			18	TR23/	
2SC25G				TR83/	
2SC26	123A	3122	88	TR21/735	S0025
2SC27	128	3024	17	TR87/243	S0014
2SC28	123	3124	61	TR86/53	S0014
2SC29	123	3019	60	TR86/53	S0014
2SC30	128	3024	17	TR87/243	S0014
2SC31	128	3024	47	TR87/243	S0014
2SC32	128	3024	47	TR87/243	S0014
2SC32A	128	3024	18	TR87/243	S5026
2SC33	123A	3122	39	TR21/735	S0011
2SC34	103A	3011	8	TR08/724	G0011
2SC35	103A	3011	8	TR08/724	G0011
2SC36	101	3011	8	TR08/641	G0011
2SC37	108	3019	11	TR95/56	S0014
2SC38	108	3047	11	TR95/56	S0014
2SC39	108	3019	61	TR95/56	S0016
2SC39A	108	3019	61	TR95/56	S0016
2SC40	108	3019	17	TR95/56	S0016
2SC41	163	3111		TR67/740	S5020
2SC41TV	163	3111	36	TR67/740	
2SC42	162	3079	35	TR67/707	S5020
2SC42A	162	3079	35	TR67/707	S5020
2SC43	162		35	TR67/707	S5020
2SC44	162		35	TR67/707	S5020
2SC45	123A		11	TR21/735	S0014
2SC46	128	3024	81	TR87/243	S0015
2SC46F		3024	18	TR24/	
2SC47	123A	3122	47	TR21/735	S0014
2SC48	123	3124	20	TR86/53	S5026
2SC49	128	3024	20	TR87/243	S5026
2SC50	101	3124	5	TR08/641	G0011
2SC50A	101	3124	5	TR08/641	
2SC51	128	3024	63	TR87/243	S5014
2SC52	123A	3122	17	TR21/735	S0014
2SC53	123A	3122	215	TR21/735	S0014
2SC54	123A	3122	20	TR21/735	S0014
2SC55	123A	3018	17	TR21/735	S0011
2SC56	107	3124	11	TR78/720	S0011
2SC57			63	TR25/	S3001
2SC58	154	3040	11	TR78/712	
2SC58A	154	3040	27	TR78/712	S5025
2SC59	128	3024	18	TR87/243	S5026
2SC60	101	3011	8	TR08/641	G0011
2SC61	128	3024	20	TR87/243	S3013
2SC62	123A	3122	17	TR21/735	S0011
2SC63	108	3039	20	TR95/56	S0011
2SC64	154	3045	27	TR78/712	S5026
2SC65	154	3024	27	TR78/712	S5025
2SC65B	154	3044	32	TR78/	
2SC65N	154			TR78/712	
2SC65Y	154	3044	63	TR78/712	S5025
2SC65YA	154	3044	32	TR78/	
2SC65YB	154			TR78/	S5025
2SC65YTV	154	3045	32	TR78/712	
2SC66	154	3045	27	TR78/712	S5025
2SC67	123A	3122	20	TR21/735	S0014
2SC68	123A	3122	20	TR21/735	S0014
2SC69	128	3024	18	TR87/243	S5026
2SC70	154	3044	27	TR78/712	S5025
2SC71	101	3011	8	TR08/641	G0011
2SC72	101	3011	8	TR08/641	G0011
2SC73	101	3011	7	TR08/641	G0011
2SC74	108	3018	81	TR95/56	S0014
2SC75	101	3011	7	TR08/641	G0011
2SC76	101	3011	7	TR08/641	G0011
2SC77	101	3011	7	TR08/641	G0011
2SC77C	102	3004	53	TR05/631	
2SC78	101	3011	6	TR08/641	G0011
2SC79	108	3039	61	TR95/56	S0020
2SC80	108	3019	61	TR95/56	S0011
2SC81		3124	20	TR21/	S5004
2SC82					S5004
2SC83			11		
2SC84					G0011
2SC85					G0011
2SC86					G0011
2SC87	123A	3122	215	TR21/735	S0014
2SC88	154	3045	32	TR78/712	S5026
2SC88A	154	3045	17	TR78/712	
2SC89	101	3011	7	TR08/641	G0011
2SC89H		3011	59		G0011
2SC90	101	3011	54	TR08/641	G0011
2SC90H		3011	59		G0011
2SC91	101	3011	54	TR08/641	G0011
2SC91H		3011	59		G0011
2SC92			66		S3002
2SC93			66		S3010
2SC94			66		S3002
2SC95	154	3045	32	TR78/712	S0005
2SC97	128	3024	47	TR87/243	S3001
2SC97A	128	3024	18	TR87/243	S3001
2SC98	108	3039	17	TR95/56	S0011
2SC99	108	3039	17	TR95/56	S0011
2SC100	123A	3122	63	TR21/735	S0014
2SC100-OY	123A			TR21/735	
2SC101	124	3021	12	TR81/240	S5019
2SC101A	124	3021	12	TR81/240	S5019
2SC101X	124	3021	12	TR17/635	
2SC102	105			TR03/233	S5004
2SC103	123	3039	17	TR21/735	S0011
2SC103A	123A	3019	212	TR21/735	S0011
2SC104	123A	3122	17	TR21/735	S0011
2SC104A	123A	3122	61	TR21/735	S0011
2SC105	123A		61	TR21/735	S0020
2SC106		3048	11		S3001
2SC106C		3048	46		S5000
2SC107			66		S3001
2SC108	128	3024	20	TR87/243	S3019
2SC109	128	3024	47	TR87/243	S0014
2SC109A	128	3024	18	TR87/243	S3001
2SC110	123A	3122	18	TR21/735	S0014
2SC111	123A	3048	18	TR21/735	S0014
2SC112	128	3024	47	TR87/243	S3013
2SC113	128	3024	47	TR87/243	S3013
2SC114	128	3024	18	TR87/243	S3013
2SC115	128	3024	11	TR87/243	S0015
2SC116	128	3048	47	TR87/243	S3013
2SC116T	128	3048	18	TR87/243	S5026
2SC117	128	3024	18	TR87/243	S3001
2SC118	128	3024	18	TR87/243	S3001
2SC119	128	3024	18	TR87/243	S3001
2SC120	128	3024	61	TR87/243	S0014
2SC121	128	3024	61	TR87/243	S0014
2SC122	128	3024	61	TR87/243	S0014
2SC122D		3024	20		S5014
2SC123	128	3024	61	TR87/243	S0014
2SC124	128	3024	61	TR87/243	S0014
2SC125	160		27	TR17/637	G0003
2SC126			27	TR78/	
2SC127	108	3019	17	TR95/56	S0014
2SC128	101	3011	54	TR08/641	G0011
2SC129	101	3011	5	TR08/641	G0011
2SC130	128	3024	18	TR81/243	S3001
2SC131	123A	3122	20	TR21/735	S0014
2SC131T		3122			S0024

Original Part Number	ECG	SK	GE	IR/WEP	HEP
2SC132	123A	3122	20	TR21/735	S0011
2SC133	123A	3122	20	TR21/735	S0011
2SC134	123A	3018	20	TR21/735	S0014
2SC134B	123A	3122	88	TR21/735	
2SC135	123A	3018	20	TR21/735	S0011
2SC135B		3018	61		G0001
2SC136	123A		17	TR21/735	S5026
2SC137	123A	3018	20	TR21/735	S0011
2SC138	123A	3122	20	TR21/735	S3001
2SC138A	123A	3122	20	TR21/735	S3001
2SC138S	123A	3122	10	TR24/	
2SC139	123A	3122	47	TR21/735	S3001
2SC140	128	3024	63	TR87/243	S5014
2SC141			210	TR21/	
2SC142			210	TR21/	
2SC143			210	TR64/	
2SC144			210	TR64/	
2SC144A			210	TR64/	
2SC147	128	3024	63	TR87/243	S5014
2SC148	108	3018	61	TR95/56	
2SC149					S5026
2SC150	123A	3018	47	TR21/735	S0014
2SC150T	128	3047	47	TR87/243	S3013
2SC151	123A	3122	47	TR21/735	S3013
2SC151H		3122	88		S3013
2SC152	128	3024	47	TR87/243	S3008
2SC152H		3024	18		S3008
2SC153					S5026
2SC154	154	3040	27	TR78/712	S5026
2SC154A		3040	32	TR78/	
2SC154B	154	3040	27	TR78/712	
2SC154C	154	3040	20	TR78/712	S5024
2SC154H	154	3040	27	TR78/712	S5026
2SC155	108	3019	60	TR95/56	S0011
2SC156	108	3019	60	TR95/56	S0033
2SC157	123	3124	11	TR86/53	S0033
2SC158	123	3124	11	TR86/53	S0033
2SC159	123A	3018	11	TR21/735	S0033
2SC160	123A	3018	11	TR21/735	S0016
2SC162			81	TR63/	
2SC163	128	3024	47	TR87/243	S3021
2SC164		3047	47	TR64/	
2SC165			88	TR64/	
2SC166	123A	3122	60	TR21/735	S0011
2SC167	123A	3122	212	TR21/735	S0011
2SC170	108	3019	39	TR95/56	S0011
2SC171	107	3019	86	TR70/720	S0011
2SC172	108	3039	61	TR95/56	S0016
2SC172A	108	3039	17	TR95/56	S0016
2SC173	101	3011	8	TR08/641	G0011
2SC174	107	3019	11	TR70/720	S0017
2SC174A	107	3019	213	TR70/720	S0025
2SC175	101	3011	8	TR08/641	G0011
2SC175B	102A	3011	2	TR85/250	
2SC176	101	3011	8	TR08/641	G0011
2SC177	101	3011	8	TR08/641	G0011
2SC178	101	3011	8	TR08/641	G0011
2SC179	103A	3124	8	TR08/724	G0011
2SC180	103A	3011	8	TR08/724	G0011
2SC181	103A	3124	8	TR08/724	G0011
2SC182	123A	3020	212	TR21/735	S0016
2SC182Q	107	3124	47	TR70/	S0016
2SC182V					S0016
2SC183	108	3030	39	TR95/56	S0016
2SC183E	108	3018	20	TR95/56	
2SC183J	108	3018	11	TR95/56	
2SC183K	108	3018	11	TR95/56	
2SC183L	108	3018	11	TR95/56	
2SC183M	108	3039	11	TR95/	
2SC183N				/56	
2SC183O				/56	
2SC183P	108	3018	20	TR95/56	S0016
2SC183Q	108	3018	20	TR95/	S0016
2SC183R	108	3018	20	TR95/56	S0016
2SC183RED					S0016
2SC103W		3018			S0015
2SC184	108	3018	39	TR95/56	S0016
2SC184E		3018	20	TR86/	
2SC184H	108	3018	11	TR95/56	
2SC184J	108	3018	11	TR95/56	
2SC184L	108	3018	11	TR95/56	
2SC184R		3018	20	TR95/56	
2SC184Q		3018	61	TR21/	S0016
2SC184RED					S0016
2SC185	108	3019	39	TR95/56	S0016

Original Part Number	ECG	SK	GE	IR/WEP	HEP
2SC185A	107	3039	86	TR70/720	S0016
2SC185J	108	3039	11	TR95/56	
2SC185M	108	3018	11	TR95/	
2SC185N				/56	
2SC185Q	108	3018	20	TR95/56	S0016
2SC185R	108	3018	20	TR95/56	
2SC185V	108	3018	11	TR95/56	
2SC186	107	3019	60	TR70/720	S0033
2SC187	107	3019	60	TR70/720	S0033
2SC187I	102A	3004	52		
2SC188	128	3024	81	TR87/243	S0014
2SC188A	128	3024	17	TR87/243	
2SC188AB	128	3024	17	TR87/	
2SC188B		3024	18	/243	
2SC189	128	3024	81	TR87/243	S0014
2SC190	128	3024	81	TR87/243	S0014
2SC191	123A	3122	212	TR21/735	S0014
2SC192	123A	3122	212	TR21/735	S0014
2SC193	123A	3122	212	TR21/735	S0014
2SC194	123A	3122	212	TR21/735	S0014
2SC195	123A	3018	61	TR21/735	S0014
2SC196	123A	3018	61	TR21/735	S0014
2SC197	123A	3018	61	TR21/735	S0014
2SC198			14		
2SC199	108	3039	81	TR95/56	S5026
2SC200	123A	3122	215	TR21/735	S0014
2SC201	123A	3122	215	TR21/735	S0014
2SC202	123A	3122	20	TR21/735	S5026
2SC203	123A	3122	20	TR21/735	S0011
2SC204	123A	3122	20	TR21/735	S0011
2SC205	123A	3122	20	TR21/735	S5026
2SC206	108	3018	11	TR95/56	S0017
2SC208	128	3024	11	TR87/243	
2SC210	128	3024	81	TR87/243	S0014
2SC211	128	3024	81	TR87/243	S0014
2SC212	194	3024	81	TR87/243	S0005
2SC213	128	3024	28	TR87/243	S3001
2SC214	128	3024	20	TR87/243	S3013
2SC215	128	3024	28	TR87/243	S3001
2SC216	128	3024	81	TR87/243	S0014
2SC217	128	3024	81	TR87/243	S0014
2SC218	128	3024	81	TR87/243	S5026
2SC218A	123A	3024	18	TR21/735	S0020
2SC219					S5014
2SC220	128	3024	47	TR87/243	S0014
2SC221	128	3024	47	TR87/243	S0014
2SC222	128	3024	20	TR87/243	S3002
2SC223	128	3024	28	TR87/243	S3001
2SC224	128	3024	28	TR87/243	S3001
2SC225	128	3024	17	TR87/243	S3001
2SC226	128	3024	47	TR87/243	S0014
2SC227	128	3024	47	TR87/243	S0014
2SC228	128	3024	32	TR87/243	S3002
2SC229	128	3024	32	TR87/243	S3002
2SC230	123A	3122	20	TR21/735	S5026
2SC231	128	3024	47	TR87/243	S0014
2SC232	128	3024	47	TR87/243	S0014
2SC233	128	3024	32	TR87/243	S3002
2SC234	128	3024	18	TR87/243	S3002
2SC235	128	3024	18	TR87/243	S3002
2SC235-O	190			TR74/	
2SC235-Y				TR21/	
2SC236	128	3024	18	TR87/243	S3019
2SC237	123A	3122	20	TR21/735	S0011
2SC238	128	3024	17	TR87/243	S0014
2SC239	123A	3122	20	TR21/735	S0011
2SC240	162	3027	35	TR67/707	S7004
2SC241	162	3027	35	TR67/707	S7002
2SC242	162	3027	35	TR67/707	S7004
2SC243	162		35	TR67/707	S7004
2SC244	130	3027	14	TR59/247	S7002
2SC245	162	3024	18	TR67/707	S7004
2SC246	162			TR67/707	S5020
2SC247	128	3024	18	TR87/243	S5026
2SC248	128	3024	81	TR87/243	S5026
2SC249	128	3024	81	TR87/243	S5026
2SC250	161	3019	60	TR83/719	S0033
2SC251	161	3019	86	TR83/56	S0014
2SC251A	161	3039	86	TR83/56	S0011
2SC252	161	3019	86	TR83/56	S0016
2SC253	161	3019	86	TR83/56	S0016
2SC254	175			TR81/241	S5012
2SC260				TR65/	
2SC261	192				
2SC263	108	3039	212	TR95/56	

Original Part Number	ECG	SK	GE	IR/WEP	HEP
2SC264			210	TR53/	
2SC265			210	TR53/	
2SC266	107	3019	39	TR70/720	S0016
2SC267	123A	3122	64	TR21/735	S0016
2SC267A	123A	3122	88	TR21/735	S0014
2SC268	128	3019	81	TR87/243	S0005
2SC268A	128	3024	81	TR87/243	S0005
2SC269	108	3019	210	TR95/56	S0016
2SC270	162			TR67/707	S5020
2SC271	108	3019	39	TR95/56	S0016
2SC272	108	3039	214	TR95/56	S0016
2SC273	154			TR78/	S3019
2SC277C	103A			TR08/	
2SC280A-O(TV)				TR21/	
2SC281	123A	3124	212	TR51/735	S0020
2SC281A	123A	3124	86	TR51/735	S0011
2SC281B	123A	3124	20	2SC281B/735	S0020
2SC281C	123A	3124	20	TR21/735	S0015
2SC281C-EP	123A			TR21/	
2SC281D	123A	3124	86	TR21/735	
2SC281H		3124	212		S0011
2SC282	108	3039	62	TR95/56	S0016
2SC283	123A	3122	62	TR21/735	S0015
2SC284	123A	3122	62	TR21/735	S5026
2SC285	123A	3122	20	TR21/735	S0014
2SC285A	123A	3122	20	TR21/735	S0014
2SC286	108	3019	11	TR95/56	S0016
2SC287	107	3019	11	TR95/720	S0016
2SC287A	107	3018	60	TR95/720	S0016
2SC288	229	3019	11	TR70/720	S0016
2SC288A	107	3018	60	TR70/720	S0016
2SC289	108	3039	214	TR95/56	S0016
2SC290	195			TR65/243	
2SC291	128	3024	28	TR87/243	S5014
2SC292	128	3024	28	TR87/243	S3002
2SC293	128	3024	18	TR87/243	
2SC294				TR51/	
2SC296	108			TR95/	
2SC297			28	TR76/	S3010
2SC298			10		S3002
2SC298-4	128	3024	18	TR25/	
2SC299	225	3124	20	TR86/53	
2SC300	123A	3122	20	TR21/735	S0016
2SC301	123A	3122	20	TR21/735	S0016
2SC302	123A	3122	20	TR21/735	S0025
2SC303			63	TR25/	S3001
2SC304			63	TR25/	S3001
2SC305		3104	63	TR25/	S3019
2SC306	128	3024	47	TR87/243	S0014
2SC307	128	3024	18	TR87/243	S3001
2SC308	128	3024	63	TR87/243	S3019
2SC309	128	3024	18	TR87/243	S3019
2SC310	128	3024	32	TR87/243	S0005
2SC311		3047	45		
2SC312		3048	46		
2SC313	161	3019	11	TR83/719	S0016
2SC313C	161	3117	211	TR83/719	
2SC313H	161	3117	211	TR83/719	
2SC315	123A	3047	45	TR21/735	
2SC316	108	3039	212	TR95/56	S0015
2SC317	123A	3122	62	TR21/735	
2SC317C	123A	3122	20	TR21/735	
2SC317H		3122	62		S5026
2SC318	123A	3122	212	TR21/735	S0025
2SC318A	123A	3122	212	TR21/735	S0025
2SC319	123A	3122	20	TR21/735	S3013
2SC320	123A	3122	20	TR21/735	S3013
2SC321	123A	3122	20	TR21/735	
2SC321H		3122	20		S0011
2SC323	123A	3122	212	TR21/735	S0011
2SC324	123A	3004	20	TR21/735	S0015
2SC324A	123A	3018	20	TR21/735	
2SC324H	123A	3018	20	TR21/735	S0015
2SC324HA	123A	3018	20	TR21/735	
2SC325C	152		28		S5000
2SC325E	152				
2SC328A	123A				
2SC328H				TR21/	
2SC329	107				
2SC329B	107				
2SC329C	107				
2SC335	199				
2SC341					G6008
2SC341V	127			TR27/235	
2SC348	161	3117	211	TR83/719	
2SC348Q				TR85/	
2SC350	123A	3122	212	TR21/735	S0015
2SC350H	123A	3122	212	TR21/735	S0011
2SC351	107	3039	20	TR70/720	S0016
2SC351(FA)				TR70/	
2SC352	128	3024	47	TR87/243	S5026
2SC352A	128	3024	47	TR87/243	S0014
2SC353	128	3024	27	TR87/243	
2SC353A	128	3024	18	TR87/243	S5026
2SC354		3197	215	TR76/	S3001
2SC354A					S3001
2SC356	123A	3039	210	TR21/56	S3013
2SC359B				TR21/	S0025
2SC360	123A	3019	11	TR21/735	S0011
2SC361	107	3018	11	TR70/720	S0016
2SC362	107	3124	20	TR70/720	S0016
2SC363	123A	3019	11	TR21/735	S0015
2SC366	123A	3122	88	TR21/735	S0015
2SC367	123A	3124	210	TR21/735	S0014
2SC368	199	3122	212	TR21/735	S0015
2SC368BL	199			TR21/	
2SC368GR	199			TR21/	
2SC368V	199			TR21/	
2SC369	123A	3124	212	TR51/735	S0015
2SC369BL	123A	3038	212	TR21/	S0015
2SC369G	199	3124	47	TR21/	S0015
2SC369G-BL	199	3122	88	TR21/735	
2SC369G-GR	199	3122	88	TR21/735	S0020
2SC369V	199	3038		TR21/	
2SC370	123A	3018	212	TR21/735	S0011
2SC370F	108	3018	61	TR95/	
2SC370G	123A	3018	212	TR21/735	S0025
2SC370H	108	3018	61	TR95/	
2SC370J	108	3018	61	TR95/	
2SC370K	108	3018	61	TR95/	
2SC371	123A	3018	212	TR33/735	S0011
2SC371(O)	123A			TR21/	
2SC371-O		3018	20	TR21/735	S0011
2SC371B	123A	3018	20	TR21/735	
2SC371G	123A	3018	210	TR21/735	S0014
2SC371-O	123A	3018	20	TR21/735	S0011
2SC371-R	123			TR86/735	
2SC372	123A	3020	20	2SC372/735	S0015
2SC372-O	123A	3122	61	TR21/735	S0015
2SC372-1	123A			TR21/735	
2SC372-2	123A		10	TR21/735	
2SC372BL	123A	3018		TR24/	S0015
2SC372H	123A	3122	61	TR33/	S0015
2SC372-O	123A	3018	20	TR21/735	S0015
2SC372-R	123A	3122	88	TR21/735	
2SC372-Y	123A			TR21/735	S0015
2SC372-Z	123A			TR21/735	
2SC373	199	3020	212	2SC373/735	S0015
2SC373BL	199			TR21/735	
2SC373G	199	3124	212	TR21/735	S0014
2SC373GR	199			TR21/	S0015
2SC373W	199	3122	88	TR21/735	S0015
2SC374	199	3018	212	TR33/735	S0015
2SC374-BL	199		62	TR51/735	S0015
2SC374-V	199	3018	61	TR21/735	S0015
2SC375	107	3039	86	TR70/720	S0016
2SC375-O	107	3039	86	TR70/720	
2SC375-Y	107	3039	86	TR70/720	
2SC376	128	3024	212	TR87/243	S5026
2SC377	123A	3018	17	TR21/735	S0025
2SC378	123A	3018	212	TR21/735	S0025
2SC379	123A	3122	17	TR21/735	S0016
2SC380	107	3018	60	TR70/720	S0016
2SC380-O	107	3018	20	TR21/720	S0016
2SC380-O/4454C	107				
2SC380A	107	3018	20	TR70/720	S0016
2SC380A-O	107	3018	61	TR24/	S0016
2SC380A-O(TV)	107			TR70/	
2SC380A(O)	107		20	TR24/	
2SC380A-R	107	3039	20	TR70/720	S0016
2SC380A-R(TV)	123A			TR21/	
2SC380ATV	123A	3018	20	TR24/	S0016
2SC380AY		3018		TR24/	S0016
2SC380D	107	3018	61	TR24/	
2SC380R	107	3018	61	TR33/720	
2SC380RED			86		S0016
2SC380Y	107	3018	20	TR21/720	S0016
2SC380YEL			211		S0016
2SC381	107	3039	17	TR70/720	
2SC381BN	107	3018	17	TR70/720	S0011

Original Part Number	ECG	SK	GE	IR/WEP	HEP
2SC381-O	107	3018	20	TR70/720	
2SC381-R	107			TR70/720	
2SC381R	107	3018	61	TR33/	S0011
2SC382	107	3117	39	TR71/720	S0016
2SC382BK	107	3117	17	TR70/720	S0016
2SC382-BK(1)	107			TR70/	
2SC382-BK(2)	107			TR70/	
2SC382BL	107	3117	17	TR70/720	S0016
2SC382BN	107	3117	17	TR70/720	S0017
2SC382BR	107	3117	86	TR70/720	S0017
2SC382G	107	3117	39	TR70/720	S0016
2SC382-GR	107	3018	61	TR70/	S0016
2SC382-GY	107			TR70/	
2SC382R	107	3018	39	TR70/720	S0016
2SC382V	107	3132	17	TR70/720	S0020
2SC382W	107	3039	60	TR70/720	S0020
2SC383	123A	3132	61	TR24/735	S5026
2SC383(3RD IF)	233				
2SC383(FNL IF)	233				
2SC383G	123A	3122	61	TR21/	
2SC383T	123A			TR24/	S5026
2SC383T(1ST IF)	233				
2SC383W	123A	3132	61	TR71/	S5026
2SC383Y	123A	3018	61	TR24/735	
2SC384	108	3018	86	TR24/56	S0016
2SC384-O	108	3018	11	TR24/56	
2SC384Y	107	3018	61	TR70/720	S0016
2SC385	107	3039	86	TR70/720	S0016
2SC385A	107	3039	11	TR70/720	S0016
2SC386	108	3039	86	TR95/56	S0016
2SC386A		3039	86		S0033
2SC386A-O(TV)	123A			TR21/	
2SC387	108	3019	86	TR70/720	S0016
2SC387A	108	3019	11	TR70/720	S0016
2SC387A(FA-3)	108			TR95/	
2SC387G	108	3039	86	TR70/720	S0016
2SC388	107	3132	11	TR33/720	
2SC388A	107	3132	61	TR70/720	S0016
2SC388A(3RD IF)	233				
2SC389	161	3117	60	TR83/719	S0017
2SC390	161	3039	211	TR83/719	
2SC391		3039	214		S0017
2SC392	108	3039	39	TR95/56	S0017
2SC394	107	3018	210	TR33/720	S0014
2SC394-O	199	3018	17	TR21/720	S0014
2SC394GR	107		60		
2SC394R	107	3018	60	TR70/720	
2SC394RED			212		S0014
2SC394W	107	3018	61	TR70/720	
2SC394X	107	3018	61		S0014
2SC394Y	107	3018	17	TR70/	
2SC394YEL			212		S0014
2SC395	123A	3122	88	TR21/735	S0011
2SC395A	123A	3122	210	TR21/735	S0011
2SC395R	108	3122	88		S0016
2SC396	123A	3122	88	TR21/735	S0015
2SC397	108	3039	86	TR95/56	S0017
2SC398	161	3018	11	TR70/719	S0017
2SC398(FA-1)				TR95/	
2SC399	161	3018	11	TR83/719	S0017
2SC400	123A	3122	17	TR21/735	S0011
2SC400-O	123A			TR21/	
2SC400-GR				TR21/	
2SC400-R				TR21/	
2SC400-Y				TR21/	
2SC401	123A	3018	212	TR21/735	S0015
2SC401B		3018	61	TR21/	
2SC402	123A	3124	212	TR21/735	S0015
2SC402A	123A	3122	212	TR21/735	S0015
2SC402B		3124	212		S0025
2SC403	123A	3126	212	2SC403A/735	S0015
2SC403A	123A	3124	212	2SC403A/735	S0015
2SC403B(SONY)	128			TR87/243	
2SC403C	123A	3018	20	TR21/735	S0015
2SC403C(SONY)	128			TR87/243	
2SC404	123A	3018	61	TR21/735	S0015
2SC405	108	3039	54	TR95/56	G0011
2SC406	108	3039	54	TR95/56	G0011
2SC407	162			TR67/707	S7004
2SC408	163	3111		TR67/740	S5020
2SC409	163	3111		TR67/740	S5020
2SC410	163	3111		TR67/740	S5020
2SC411	163	3111		TR67/740	S5020
2SC412	163	3111		TR67/740	S5020
2SC423	128	3024	20	TR87/243	S0014

Original Part Number	ECG	SK	GE	IR/WEP	HEP
2SC423A		3024	18		
2SC423B	128	3024	18	TR87/243	
2SC423C	128	3024	18	TR87/243	
2SC423D	123A	3024	18	TR21/735	
2SC423E	123A	3024	18	TR21/735	S0014
2SC423F	123	3124	20	TR86/53	
2SC424	108		20	TR95/56	S0014
2SC424D	108			TR95/56	
2SC425	128	3024	20	TR87/243	S0011
2SC425A		3024	18		
2SC425B	128	3024	18	TR87/243	
2SC425C	128	3024	18	TR87/243	
2SC425D	123A	3122	88	TR21/735	
2SC425E	123A	3122	88	TR21/735	
2SC425F	123A	3122	88	TR21/735	
2SC426			17	TR64/	S0011
2SC427			17	TR64/	
2SC428			17	TR64/	
2SC429	107	3019	214	TR70/720	S0016
2SC429J	108	3018	11	TR95/56	S0016
2SC429X	108	3018	61	TR95/56	
2SC430	108	3018	214	TR95/56	S0016
2SC430H	108	3018	20	TR95/56	S0016
2SC430W	108	3018	61	TR95/56	
2SC433	123A		20	TR24/	
2SC437					S3010
2SC438					S3010
2SC440	123		11	TR86/53	S0014
2SC441	123	3124	11	TR86/53	S0014
2SC442	123	3124	11	TR86/53	S0014
2SC443	128	3024	63	TR87/243	S3008
2SC444			63	TR25/	S3001
2SC445		3104	63	TR25/	S3019
2SC446				TR21/	
2SC454	123A	3018	210	TR21/735	S0014
2SC454A	123A	3018	20	TR24/735	S0014
2SC454B	123A	3018	20	2SC454B/735	S0014
2SC454C	123A	3018	20	TR21/735	S0014
2SC454D	123A	3018	20	TR21/	S0015
2SC454L	123A	3018	210	TR21/735	S0014
2SC454LA	123A	3122	20	TR21/735	
2SC455	108	3018	210	TR95/56	S0025
2SC456	195	3048	47	TR65/	S3001
2SC456A	195	3048	46	TR65/	
2SC456D	195	3024	18	TR65/	
2SC458	123A	3020	210	TR21/735	S0015
2SC458(LG)		3020	20	TR21/	S0014
2SC458A	123A	3124	20	TR21/735	S0030
2SC458B	123A	3124	20	2SC458B/735	S0030
2SC458BC	123A	3124	47	TR21/735	S0030
2SC458BL	108	3124	47	TR95/56	
2SC458C	123A	3124	20	TR51/735	S0015
2SC458CLG	123A			TR21/735	
2SC458CM	123A	3124	47	TR21/735	
2SC458D	123A	3124	20	TR51/735	S0015
2SC458L	123A	3124	210	TR21/735	S0030
2SC458LB	123A	3124	20	TR21/735	
2SC458LG	123A	3020	210	TR51/735	S0015
2SC458LGA	123A			TR21/735	
2SC458LGB	123A	3124	20	TR21/735	S0030
2SC458LGBM	123A			TR21/735	
2SC458LGC	123A	3124	20	TR21/	S0014
2SC458LGD	123A		20	TR21/735	S0014
2SC458LGO	123A			TR21/735	
2SC458LGS	123A			TR21/736	
2SC458LGR		3124	47		
2SC458M	123A	3124	47		
2SC458P	123A				
2SC458RGS	123A			TR21/	
2SC458V	123A				
2SC459	128	3018	210	TR87/243	
2SC459B	102A			TR85/250	
2SC460	107	3018	210	TR70/720	S0014
2SC460A	107	3018	61	TR24/720	S0014
2SC460B	107	3018	61	2SC460B/720	S0014
2SC460C	107	3018	20	TR51/720	S0014
2SC460D	107	3039	86	TR70/720	
2SC460G	107	3018	61		S0020
2SC460GB	107			TR70/720	S0014
2SC460H	107	3018	61	TR70/720	S0014
2SC460K	107	3039	86	TR70/720	
2SC460L	107	3039	86	TR70/720	
2SC461	229	3018	210	TR24/56	S0014
2SC461A	229	3018	11	TR95/56	S0014
2SC461B	229	3018	11	TR95/56	S0014

Original Part Number	ECG	SK	GE	IR/WEP	HEP
2SC461C	229	3018	20	TR70/720	S0014
2SC461E	229			TR95/56	S0014
2SC461L	229	3039	17	TR70/720	
2SC462			20		
2SC463	107	3019	60	TR70/720	
2SC463H	161	3117	60	TR83/719	
2SC464	161	3018	86	TR83/719	S0016
2SC464C	161	3117	211	TR83/719	
2SC465	161	3018	86	TR83/719	S0016
2SC466	161	3018	86	TR83/719	S0017
2SC466H	161	3117	211	TR83/719	
2SC467			16		
2SC468	123A		20	TR21/735	
2SC468A	123A			TR21/735	
2SC468H			20		S0014
2SC469	108	3018	39	TR95/56	S0016
2SC469A	107	3018	61	TR70/720	
2SC469F	108	3018	20	TR95/56	
2SC469K	108	3018	20	TR24/	
2SC469Q	108	3018	61	TR95/56	S0016
2SC469R	108	3018	61	TR95/56	
2SC470	154	3045	18	TR78/712	S5025
2SC470R		3045	32	TR78/	
2SC470Y		3045	32	TR78/	
2SC470-3			18		
2SC470-4			27	TR78/	
2SC470-5			27	TR78/	
2SC470-6			27	TR78/	
2SC472		3018	61		S0015
2SC472Y	154	3018	61		S0015
2SC475	123A	3124	212	TR21/735	S0015
2SC475K	123A	3124	47	TR21/	S0015
2SC476	123A	3122	212	TR21/735	S0015
2SC477	108	3039	61	TR95/56	S0015
2SC478	195	3047	212	TR65/735	S0025
2SC478-4	195	3047	45	TR65/	
2SC478D	195	3047	45	TR65/735	S0025
2SC479	128	3024	47	TR87/243	
2SC479H		3024	18		S0015
2SC480		3124			
2SC481	195	3048	47	TR65/243	S3001
2SC482	195	3046	47	TR65/243	S3001
2SC482-O	195			TR65/	
2SC482-GR	195		47	TR65/	
2SC482-Y	195		47	TR65/	
2SC482Y	195	3024	18	TR65/243	S3013
2SC484		3104	27	TR78/	S3019
2SC485	128	3024	18	TR87/243	S5026
2SC485BC				TR25/	
2SC485C	128	3024	18	TR87/243	
2SC485Y	128	3024	18	TR87/243	
2SC486	128	3024	18	TR87/243	S5026
2SC486Y	128	3024	20	TR87/243	S5026
2SC487	175	3131		TR81/241	S5012
2SC488	175	3131		TR81/241	S5012
2SC489	175	3026	66	TR81/241	S5012
2SC489Y	175	3026	66		
2SC490	175	3021	23	TR81/241	S5019
2SC491	175	3026	23	TR81/241	S5019
2SC491BL	175	3131		TR81/241	
2SC491R	175	3131		TR81/241	
2SC491Y	175	3026	66	TR81/241	
2SC491YEL			23		S5011
2SC492	162	3124	10	TR67/707	S5004
2SC493	130	3027	14	TR59/247	S7002
2SC493-BL	223			TR59/	
2SC493-R	223			TR59/	
2SC493RED					S7002
2SC493-Y	223			TR59/	
2SC493YEL					S7002
2SC494	130	3029	14	TR26/247	S7002
2SC494-BL	223	3029		TR59/	S7002
2SC494-R	223			TR59/	
2SC494RED					S7002
2SC494-Y	223			TR59/	
2SC494YEL					S7002
2SC495	184	3104	57	TR76/S5003	S5000
2SC495-O	184	3054	57	TR76/S5003	
2SC495ORN					S5000
2SC495P					S5000
2SC495Q					S5000
2SC495-R	184	3104	57	TR76/S5003	
2SC495RED			47		S5000
2SC495T		3054			S5000
2SC495-Y	184	3054	57	TR76/S5003	
2SC496	184	3104	57	TR55/S5003	S5000
2SC496-O	184	3054	28	TR76/S5003	S5000
2SC496-R	184	3190	57		S5000
2SC496-Y	184	3104	57	TR76/S5003	
2SC496YEL					S5000
2SC497	128	3024	18	TR87/243	S5014
2SC497-O	128			TR87/	
2SC497-R	128			TR87/	S5014
2SC497-Y	128			TR87/	
2SC498	128	3024	66	TR87/243	S3001
2SC498-O	128			TR87/	
2SC498O				TR87/	
2SC498-R	128			TR87/	
2SC498-Y	128			TR87/	
2SC499	194	3040		TR80/712	S5026
2SC499R	194	3045	20	TR80/712	
2SC499-R		3122	88	TR78/	S5026
2SC499-R(FA-1)	154			TR78/	
2SC499-RY	154	3045	32	TR78/	
2SC488-Y	128		18	TR78/	S5026
2SC499-Y(FA-1)	154			TR78/	
2SC499Y	194	3040		TR78/712	
2SC499YEL					S5026
2SC500	154	3040	18	TR87/712	S5026
2SC500R	154	3040	27	TR78/712	
2SC500Y	154	3045	32	TR78/712	S5026
2SC501	128	3024	47	TR87/243	S3008
2SC502	128	3024	47	TR87/243	S3002
2SC503	128	3024	47	TR87/243	S0014
2SC504GR	128	3024	18	TR87/243	
2SC504-O	128	3024	18	TR87/243	
2SC504-Y	128	3024	18	TR87/243	
2SC505	154	3045	32	TR78/712	
2SC505-O	154			TR78/712	
2SC505-R	154			TR78/712	
2SC506	154	3045	32	TR78/712	
2SC506-O	154	3045	32	TR78/712	
2SC506-R	154	3045	32	TR78/712	
2SC507	154	3044	32	TR78/712	S5025
2SC507-O	154			TR78/	
2SC507-R	154			TR78/	
2SC507-Y	154			TR78/	
2SC508	175	3538	45	TR81/241	S5012
2SC509	123A	3122	81	TR21/735	S0015
2SC509(O)	128		20	TR24/	
2SC509O		3122		TR24/	S3002
2SC509Y	123A	3122		TR21/	S0015
2SC510		3104	32	TR78/	S3019
2SC510O					S3019
2SC510R		3104	32		S3019
2SC511		3104	32	TR78/	S3019
2SC511O					S3019
2SC511R		3104	32		S3019
2SC512	128	3024	18	TR87/243	S3019
2SC512-O	128	3024	28	TR21/243	
2SC512O		3104	32	TR87/	S3019
2SC512-R	128		28	TR87/243	S3019
2SC513	128	3024	63	TR87/243	
2SC514	124	3021	12	TR81/240	S5015
2SC515	124	3021	12	TR23/240	S5011
2SC515A	124	3021	32	TR81/240	S5011
2SC515AM	124	3021	12	TR23/	S5011
2SC515A(BK)	124	3021	12	TR23/	
2SC515BK	124	3021			
2SC516	128	3024	18	TR87/243	S3019
2SC516A		3104	32		S3019
2SC517	224	3049	45	TR65/243	S5014
2SC518					S5020
2SC518A					S5020
2SC519	162	3079		TR67/707	S5004
2SC519A	162	3079		TR67/707	S5020
2SC520	130	3510		TR59/247	S7002
2SC520A	130	3510		TR59/247	S7004
2SC521	130	3027		TR59/247	S7002
2SC521A	130	3027		TR59/247	S7002
2SC522	225			TR78/	S3019
2SC522-O	225			TR78/	S3019
2SC522-R	225			TR78/	S3019
2SC523	225			TR78/	S3019
2SC523-O	225			TR78/	S3019
2SC523-R	225			TR78/	S3019
2SC524	225			TR78/	S3019
2SC524-O	225			TR78/	S3019
2SC524-R	225			TR78/	S3019
2SC525	225		28	TR78/	

Original Part Number	ECG	SK	GE	IR/WEP	HEP
2SC525-0	225			TR78/	
2SC525-R	225			TR78/	
2SC526	154	3044	27	TR78/712	S5025
2SC527			60	TR24/	
2SC528			212	TR53/	
2SC529	123A			TR21/735	
2SC529A	123A	3018	10	TR21/735	S0015
2SC531			20		
2SC533				TR24/	
2SC535	229	3018	39	TR70/720	S0016
2SC535A	229	3018	11	TR70/720	S0016
2SC535B	229	3018	20	2SC535B/720	S0016
2SC535C	229	3018	11	TR70/720	S0016
2SC535G	229	3039	86	TR70/720	
2SC536	199	3124	212	2SC536/735	S0016
2SC536A	199	3124	20	TR21/735	
2SC536A(3RD IF)	233				
2SC536AG	199	3124	47	TR21/	
2SC536B	199	3124	20	TR21/735	S0016
2SC536C	199	3124	20	TR21/735	S0016
2SC536D	199	3124	20	TR21/735	S0016
2SC536DK	199	3124	47	TR21/735	
2SC536E	199	3124	20	TR21/735	S0016
2SC536ED	199			TR21/	
2SC536EH	199			TR21/	
2SC536EJ	199			TR21/	
2SC536EN	199	3124	47	TR21/	
2SC536EP	199	3122	47	TR21/735	S0016
2SC536ER	199			TR21/735	
2SC536ET	199			TR21/735	
2SC536EZ	199			TR21/735	
2SC536F	199	3124	47	TR21/735	S0016
2SC536F1	123A	3124	20	TR21/	S0016
2SC536F2	123A	3124	20	TR21/	S0016
2SC536FC	199	3018	20		
2SC536FS	199			TR21/735	
2SC536FS6	199	3124	47	TR21/	
2SC536G	199	3124	20	TR21/735	S0016
2SC536G2	123A		20	TR24/	
2SC536GF	199			TR21/	
2SC536GK	199	3124	47	TR21/735	S0016
2SC536GT	123A	3124	47	TR21/735	
2SC536GV	123A	3124	47	TR21/735	
2SC536GY	123A	3122	88	TR21/735	S0016
2SC537	123A	3020	212	TR21/735	S0016
2SC537-01	123A			TR24/	
2SC537C	123A	3010	20	TR21/735	
2SC537D	123A	3010	20	TR21/735	S0016
2SC537D2	123A			TR21/735	
2SC537E	123A	3124	20	TR21/735	S0015
2SC537EH	123A	3122	88	TR21/735	
2SC537EJ	123A	3018	20	TR21/735	
2SC537EK	123A			TR25/	
2SC537F	123A	3124	10	TR21/735	S0014
2SC537F1	123A	3124	20	TR21/	S0016
2SC537F2	123A	3124	20	TR21/	S0016
2SC537F-C7	123A	3124	62		S0016
2SC537FC	123A	3122	62	TR24/	S0016
2SC537FV	123A	3124	47	TR21/735	
2SC537G	123A	3124	20	TR21/735	S0016
2SC537G2	123A	3124	18	TR21/	S0015
2SC537GF	123A			TR21/735	
2SC537GI	123A	3124	47	TR21/735	S0015
2SC537H	123A		20	TR21/	S0014
2SC538	123A	3020	61	2SC538/735	S0015
2SC538A	123A	3020	61	2SC538A/735	S0030
2SC538AQ	123A	3122	88	TR21/735	S0030
2SC538P	123A	3122	61	TR21/	S0030
2SC538Q	123A	3018	11	TR21/735	S0030
2SC538R	123A	3024	61	TR21/735	S0015
2SC538S	123A	3124	47	TR21/735	S0015
2SC538T	123A	3124	18	TR21/	
2SC539	123A	3020	61	TR21/735	S0015
2SC539K	123A				
2SC539L	123A				
2SC539R	123A	3124	17	TR21/735	S0015
2SC539S	123A	3124	47	TR21/735	S0015
2SC539T		3124	47		
2SC540	123A	3122	212	TR21/735	S0016
2SC541			28	TR76/	S3001
2SC542			28	TR76/	
2SC543		3018	66	TR21/	
2SC544	108	3018	39	TR95/56	S0016
2SC544C	108	3039	20	TR95/56	
2SC544D	108	3018	20	TR95/56	S0016

Original Part Number	ECG	SK	GE	IR/WEP	HEP
2SC544E	108	3039	20	TR95/56	
2SC545	107	3018	39	TR70/720	S0016
2SC545A	107	3018	11	TR70/720	
2SC545B	107	3018	11	TR70/720	
2SC545C	107	3018	11	TR70/720	S0016
2SC545D	107	3018	20	TR70/720	S0016
2SC545E	107	3018	20	TR70/720	
2SC546			39		
2SC547			28	TR76/	S3001
2SC548			28	TR76/	S3008
2SC549			28	TR76/	
2SC550			28	TR76/	
2SC551			66		
2SC552			66		
2SC553			66		
2SC554			20	TR76/	S3013
2SC555				TR25/	S3008
2SC556					S3013
2SC557(F)				TR24/	
2SC558	163	3111	73	TR67/740	S3008
2SC559			47		S0014
2SC560	128	3104	32	TR87/243	S3019
2SC561	108	3018	11	TR95/56	S0011
2SC562	161	3018	60	TR70/719	S0017
2SC562Y	161	3018	211	TR83/719	
2SC563	161	3018	39	TR83/719	S0017
2SC563(3RD IF)	233				
2SC563A	161	3018	20	TR70/719	S0020
2SC563A(3RD)	233				
2SC564	159				
2SC564A	159	3024	21	TR24/	
2SC564P	159	3024	18	TR20/717	
2SC564Q	159	3024	18	TR20/717	
2SC564R	159	3024	18	TR24/	
2SC564S	159			TR24/	
2SC564T	159	3114		TR24/	
2SC565B					S5022
2SC565D					S5022
2SC566	123A	3122	88	TR21/735	S3001
2SC567	161	3039	86	TR83/719	S0017
2SC568	108	3039	86	TR95/56	S0017
2SC571	224		28	TR76/	S3001
2SC572			28	TR76/	
2SC573			66	TR21/	
2SC580	128	3104	63	TR87/243	S3019
2SC581			39	TR24/	
2SC582	124	3021	12	TR23/240	S5011
2SC582A	124	3021	12	TR81/240	S5011
2SC582B	124	3021	12	TR23/	S5011
2SC582BX	124	3021		TR23/	S5011
2SC582BY	124	3021		TR23/	
2SC582C	124	3021	12	TR81/240	S5011
2SC582EA	124		12		
2SC583			39		
2SC585			66		
2SC586	162		73	TR67/707	S5020
2SC587	123A	3122	212	TR21/735	S0015
2SC587A	123A	3122	212	TR21/735	S0015
2SC588	123A	3122	47	TR21/735	S0011
2SC589	154	3044	27	TR78/712	S5025
2SC590	128	3024	18	TR87/243	S3019
2SC591		3104	66		S3019
2SC592			28	TR76/	
2SC593			61	TR51/	S0025
2SC594	128	3024	47	TR87/243	S0014
2SC595	123A	3122	210	TR21/735	S0016
2SC596	123A	3122	47	TR21/735	S3001
2SC597			215	TR76/	S3001
2SC598			28	TR76/	
2SC599			66		S3006
2SC600			66		
2SC601			17		S0016
2SC602			86		S0016
2SC604			11		
2SC605	107	3018	60	TR95/720	S0016
2SC605L	161	3117	60	TR71/	S0020
2SC605M	161	3018	61	TR24/	S0020
2SC605Q	107	3018		TR70/	S0020
2SC606	107	3018	60	TR95/720	S0016
2SC608	224	3024	18	TR87/243	
2SC608E	123A	3024	20	TR21/735	
2SC608T	224	3049	18	TR87/243	S3001
2SC609	224	3024	45	TR87/243	
2SC609F	195	3024	22	TR52/	
2SC609T	224	3049	18	TR87/243	S3001

Original Part Number	ECG	SK	GE	IR/WEP	HEP
2SC610	159	3024	18	TR20/717	
2SC611	108	3039	11	TR95/56	S0016
2SC612	108	3039	39	TR95/56	S0017
2SC613	108	3039	63	TR95/56	S0016
2SC614	128	3024	20	TR87/243	S3001
2SC614C	195	3024	18	TR65/	
2SC614D	195	3024	20	TR65/735	S3001
2SC614E	195	3024	18	TR65/	
2SC614F	195	3024	18	TR65/	
2SC614G	195	3024	18	TR65/	
2SC615	195	3024	18	TR65/243	S3001
2SC615A	195	3024	18	TR65/	
2SC615B	195	3024	18	TR65/	
2SC615C	195	3024	18	TR65/	
2SC615D	195	3024	18	TR65/	
2SC615E	195	3024	18	TR65/	
2SC615F	195	3024	18	TR65/	
2SC615G	195	3024	18	TR65/	
2SC619	123A	3122	210	TR24/735	S0014
2SC619B	123A	3122	88	TR24/	S0015
2SC619C	123A	3122	88	TR20/	S0019
2SC619D	123A	3122	88	TR21/735	
2SC620	123A	3122	210	TR24/735	S0014
2SC620C	123A	3122	20	TR21/	S0014
2SC620CD	123A			TR21/735	
2SC620D	123A	3122	20	TR21/	S0014
2SC620DE	123A	3024	18	TR21/735	
2SC621	123A	3122	10	TR21/735	S0011
2SC622	123A	3122	10	TR21/735	S0011
2SC624				TR21/	
2SC626					S0014
2SC627	154		18	TR78/712	S3021
2SC628	128			TR87/	
2SC629	107	3018	17	TR70/720	S0016
2SC631	123A	3124	212	TR21/735	S0015
2SC631A	123A	3124	210	TR33/735	S0011
2SC632	123A	3124	212	TR51/735	S0015
2SC632A	123A	3124	210	TR21/735	S0015
2SC633	123A	3124	212	TR21/735	S0016
2SC633-7	123A	3124	20	TR21/735	S0016
2SC633A	123A	3124	210	2SC633A/735	S0011
2SC634	123A	3124	212	2SC634/735	S0014
2SC634A	123A	3124	210	2SC634A/735	S0011
2SC635			28		
2SC635A				TR23/240	
2SC636	187	3083	66	TR56/	
2SC637			28	TR76/	
2SC638			66	TR66/	
2SC638C		3018	61		
2SC639					S0014
2SC640	123A	3122	212	TR21/735	S0015
2SC640B	123A	3122	88	TR21/	S0015
2SC641	108	3039	210	TR95/56	
2SC641B	108	3018	20	TR95/56	
2SC641H		3039	210		S0014
2SC642	164	3133		TR93/	
2SC642A	164	3133	37	TR93/740A	
2SC643A	165	3115	38	TR93/740B	
2SC644	199	3124	212	2SC644/56	S0015
2SC644C	123A	3027	75		
2SC644F	199	3124	47	TR24/	S0015
2SC644FR	199	3018		TR21/735	S0015
2SC644FS	123A	3018	62	TR24/	S0015
2SC644P	199			TR24/	
2SC644PJ	199	3124	47	TR21/	
2SC644Q	199		17	TR21/	S0015
2SC644R	199	3124	18	TR21/735	S0015
2SC644S	199	3124	18	TR21/735	S0015
2SC644T	199	3024	18	TR21/243	S0015
2SC645	107	3018	39	TR21/720	S0016
2SC645A	107	3018	20	TR70/720	S0016
2SC645B	107	3018	20	TR70/720	S0016
2SC645C	107	3018	11	TR70/720	S0016
2SC645G	107	3018	61	TR70/	S0020
2SC645N	107	3018	61	TR70/720	
2SC646	223	3027	66	TR59/	S7002
2SC647	130	3111	73	TR67/247	S7002
2SC647R	130	3111		TR59/	
2SC647RED					S5021
2SC648	199	3124	86	TR21/720	
2SC648H	199	3124	39	TR21/	S0015
2SC649	108	3039	60	TR95/56	S0014
2SC650	199	3124	210	TR95/56	S0011
2SC650B	108	3122	20	TR95/56	
2SC651					S3001
2SC652					S3013
2SC653	161	3039	11	TR70/	
2SC654	123A	3122	18	TR21/735	S0014
2SC655	123A	3122	212	TR21/735	S0017
2SC656	107	3039	214	TR70/720	S0017
2SC657	107	3018	39	TR70/720	S0016
2SC658	108	3039	17	TR95/56	S0016
2SC658A	108	3039	86	TR95/56	
2SC659	108	3039	20	TR95/56	S0011
2SC660					S0020
2SC661					S0020
2SC662	108	3039	86	TR95/56	S0016
2SC663	161	3117	214	TR83/719	S0017
2SC664	162	3027	14	TR67/707	S7004
2SC664B	130	3027	75	TR59/247	
2SC664C	130	3027	75	TR59/	
2SC665	163	3111		TR67/740	S7004
2SC665H	163	3111		TR67/740	
2SC665HA	163	3111		TR67/740	
2SC665HB	163	3111		TR67/740	
2SC667			39		
2SC668	107	3018	39	TR95/56	S0016
2SC668-0	108				
2SC668A	107	3018	20	TR95/56	
2SC668B	107	3018	11	TR95/56	
2SC668BC2	108		17	TR21/	
2SC668C	107	3018	11	TR95/56	S0016
2SC668C1	107	3039	11	TR70/720	S0016
2SC668CD	107			TR95/56	
2SC668D	107	3018	11	TR95/56	S0020
2SC668DO	107	3039	11	TR33/	S0016
2SC668D1	107	3039	86	TR70/720	
2SC668DO	107	3039	11	TR33/	S0016
2SC668DV	107	3018	61	TR70/	
2SC668DX	107	3018	61	TR70/720	
2SC668E	107	3018	11	TR95/56	S0020
2SC668E1	107	3039	11	TR33/	S0016
2SC668E2	107	3039	11	TR33/	S0016
2SC668EP	107	3039	86	TR70/720	
2SC668EX	107			TR70/720	
2SC668F	107	3039	11	TR70/720	S0020
2SC670	124				
2SC673	159				S5013
2SC673B	129	3114	89		
2SC673C2	159				S5013
2SC673D	159			TR88/	S5013
2SC674	161	3018	20	TR83/719	S0016
2SC674B	161	3117	211	TR83/719	S0016
2SC674C	161	3117	211	TR83/719	S0016
2SC674CK	161	3117	20		
2SC674CL	161	3117	20		
2SC674D	161	3117	11	TR83/719	S0016
2SC674E	161	3132	11	TR83/719	S0016
2SC674F	161	3117	300	TR83/719	S0016
2SC674G	161			TR83/	S0016
2SC677			22	TR21/	
2SC678				TR26/	
2SC679H					S0011
2SC680	175	3131	12	TR81/241	S5012
2SC680A	175	3131	66	TR81/241	S5012
2SC680R	175	3131		TR81/	S5012
2SC681	163	3111		TR67/740	S5020
2SC681A	163	3111	36	TR67/740	S5020
2SC682	161	3018	60	2SC682A/719	S0017
2SC682A	161	3018	39	2SC682A/719	S0017
2SC682B	161	3018	61	TR83/719	S0020
2SC682E	161	3018	61		
2SC683	161	3018	86	TR83/719	S0017
2SC683A	161	3117	11	TR83/719	S0016
2SC683B	161	3118	39	TR83/719	S0016
2SC683V	108	3018	11	TR21/719	S0016
2SC684	107	3019	86	TR70/56	
2SC684A	107	3019	86	TR70/720(CP)	
2SC684B	108	3039	86	TR95/56	
2SC684BK	107	3019	86	TR70/720	S0016
2SC684F	107	3039	86	TR80/	
2SC685	124	3021	27	TR23/240	S5011
2SC685(O)	124	3021	12	TR23/	
2SC685A	124	3021	12	TR81/240	S5011
2SC685B	124	3021	12	TR81/	
2SC685BK	124	3021	12	TR23/	S5011
2SC685P	124	3021	12	TR81/240	S5011
2SC685S	124				
2SC685Y	124	3021	12	TR81/	
2SC685YEL					S5011

Original Part Number	ECG	SK	GE	IR/WEP	HEP
2SC686	154	3045	40	TR78/712	S5025
2SC687	162	3079	73	TR67/707	S5020
2SC688	108	3018	66	TR95/56	
2SC689	123A	3122	88	TR21/735	
2SC689H	123A	3122	61	TR21/735	S0011
2SC690			66	TR23/	
2SC691			28	TR76/	S3005
2SC692			28	TR76/	S3005
2SC693	199	3124	210	TR21/735	S0015
2SC693E	199	3122	20	TR21/735	S0015
2SC693EB	199	3124	47	TR21/735	
2SC693F	199	3124	20	TR21/	S0015
2SC693FU	199	3124	20	TR21/720	S0015
2SC693G	199	3124	20	TR21/735	S0015
2SC693GS	199	3124	47	TR21/735	
2SC693GU	199	3124	20	TR21/	S0015
2SC693GZ	199	3124	20	TR21/	
2SC694	199	3124	210	TR21/735	S0015
2SC694E	199	3124	20	TR21/735	S0015
2SC694F	199	3122	20	TR21/735	S0015
2SC694G	199	3122	88	TR21/735	
2SC694Z	199	3039	86	TR21/720	
2SC695	107		39	TR70/720	S0016
2SC696	128	3047	32	TR87/243	S3002
2SC696A	128	3124	32	TR78/	S5014
2SC696B	128	3024	18	TR25/	S3002
2SC696D	128	3047	18	TR87/243	S0025
2SC696E	128	3024	18	TR87/243	S3002
2SC696F	128	3024	18	TR87/243	S5014
2SC696G	128	3047	45		
2SC696H	128	3024	18	TR78/	S3002
2SC696I	128	3024	18	TR87/243	S5026
2SC697	225	3049	28	TR76/	
2SC697A	225	3049	46		
2SC697F	224	3049	46		
2SC699			63	TR25/	S3020
2SC700			63	TR76/	S3001
2SC701	123	3124	20	TR86/53	S3001
2SC702	123	3124	20	TR86/53	S3005
2SC703			66		S3006
2SC704			66		S3007
2SC705	107	3039	39	TR70/720	S0016
2SC705B	107	3039	86	TR70/	
2SC705C	107	3039	17	TR70/56	
2SC705D	107	3039	17	TR70/56	
2SC705E	107	3039	17	TR70/56	
2SC705F	107	3039	86	TR70/	
2SC705TV	107		11	TR71/	
2SC705TV(3RD IF)	233				
2SC705TW			11	TR21/720	
2SC707	107	3039	214	TR70/720	S0016
2SC707H	107	3039	214	TR70/720	S0016
2SC708	128	3024	47	TR87/243	S5014
2SC708A	128	3024	18	TR87/243	S5014
2SC708AA	128			TR87/243	
2SC708AB	128	3024	18	TR87/243	
2SC708AC	128	3024	18	TR87/243	
2SC708B	128	3048	18	TR87/243	
2SC708C	128	3024	18	TR87/243	
2SC708H		3024	18		S3001
2SC709B	123A	3122	88	TR21/	S0016
2SC709C	123A	3018	61	TR21/735	S0016
2SC709CD	123A	3018	61	TR21/735	
2SC709D	123A	3122	20	TR21/735	
2SC710	123A	3018	20	2SC710/735	S0016
2SC710-1	123A			2SC710/	
2SC710-2	123A			TR24/	
2SC710-4	123A			TR24/	
2SC710A					S0016
2SC710B	107	3122	88	TR21/735	S0016
2SC710BC	123A	3018	61	TR21/735	
2SC710C	107	3018	10	2SC710/	S0016
2SC710D	123A	3122	88	2SC710/	S0016
2SC710E	123A	3018	20	TR71/	S0016
2SC711	199	3018	212	TR21/735	S0020
2SC711A	199	3018	212	TR21/735	S0025
2SC711D	199	3122	21	TR21/	S0019
2SC711E	199	3018	18	TR21/735	S0020
2SC711F	199	3018	20	TR87/243	S0020
2SC711G	199	3018	62	TR21/735	S0024
2SC712	123A	3124	212	TR24/735	S0011
2SC712A	123A	3122	17	TR21/735	S0016
2SC712B	123A	3124	47		
2SC712C	123A	3124	47	TR21/735	S0011
2SC712CD	123A	3124	63	TR21/	

Original Part Number	ECG	SK	GE	IR/WEP	HEP
2SC712D	123A	3122	18	TR21/735	S0011
2SC712E	123A	3018	47	TR21/	S0016
2SC712W	123A	3122	88	TR21/	S0016
2SC713	123A	3122	212	TR21/735	S0014
2SC714	123A	3122	20	TR21/735	S5026
2SC715	199	3124	212	TR21/735	S0016
2SC715A	199	3124	47	TR21/	
2SC715B	123A	3122	88	TR21/735	
2SC715C	123A	3117	211	TR21/735	S0016
2SC715D	123A	3117	211	TR21/735	S0016
2SC715E	199	3117	20	TR21/735	S0016
2SC715EJ	199	3124	47	TR21/735	
2SC715EV	107	3046	45	TR70/720	
2SC715F	199	3124	20	TR21/735	S0016
2SC715XL	199	3124	47	TR21/	
2SC716	107	3039	212	TR20/720	S0016
2SC716B	123A	3039	86	TR21/	
2SC716C	123A	3039	86	TR21/	
2SC716D	123A	3039	86	TR21	
2SC716E	107	3039	17	TR70/720	
2SC716F	199	3039	86	TR21/	
2SC716G	199	3039	86	TR21/	
2SC717	233	3018	86	2SC717/56	S0020
2SC717(3RD IF)	233				
2SC717B	107	3039	86	TR70/720	S0020
2SC717BK	107			2SC717/	S0016
2SC717BLK	107	3132		2SC717/	S0016
2SC717E	161	3132		TR21/	S0020
2SC724				TR95/	
2SC727	154	3024	27	TR78/712	S5026
2SC728	154	3045	27	TR78/712	S5024
2SC730				TR76/	S3001
2SC731					S3001
2SC731R		3018			S3001
2SC732	199	3124	64	TR21/735	S0024
2SC732BL	199	3122	47	TR21/735	S0024
2SC732GR	199	3122	20	TR21/735	S0024
2SC732GR/4454C	199				
2SC732S	199	3124	47	TR21/735	
2SC732V	199			TR21/	S0024
2SC732Y	199	3124	20	TR21/	
2SC733	123A	3124	212	TR33/735	S0025
2SC733-O	123A			TR21/735	S0015
2SC733B	199	3122	88		
2SC733BL	199	3124	47	TR51/735	S0025
2SC733GR	199	3124	20	TR21/735	S0025
2SC733(GR)				TR51/	
2SC733Q	107				
2SC733Y	123A	3124	20	TR21/735	S0025
2SC733YEL			212		S0025
2SC734	128	3024	212	TR24/243	S5026
2SC734GR	128	3122	212	TR87/243	S5026
2SC734-O	128	3122	212	TR87/243	
2SC734-R	128	3122	212	TR87/243	S5026
2SC734-Y	128			TR87/243	S5026
2SC734YEL			212		S5026
2SC735	123A	3122	20	TR33/735	S0014
2SC735-O	123A	3124	47	TR21/	S0014
2SC735/4454C	123A				
2SC735F	123A	3047	45	TR25/	S0014
2SC735(FA-3)	108			TR95/	
2SC735H	123A	3047	45	TR25/	S0014
2SC735J	123A	3047	45	TR25/	S0014
2SC735K	123A	3047	45	TR25/	S0014
2SC735L	123A	3047	45	TR25/	S0014
2SC735-O	123A			TR21/	S0014
2SC735ORN	123A		210	TR21/735	S0014
2SC735-R	123A			TR21/735	
2SC735Y	123A	3124	10	TR21/735	S0014
2SC735Y/4454C	123A				
2SC736	130	3027	14	TR59/247	S7004
2SC737	123A		66		S3005
2SC737Y	123A				S0014
2SC738	107	3018	214	TR70/720	S0011
2SC738C	107	3039	86	TR70/720	
2SC738D		3018	61		S0011
2SC739	107	3018	214	TR70/720	S0011
2SC739C		3018	61		S0011
2SC740	108	3039	214	TR95/56	S0016
2SC741	123A	3122	88	TR21/735	S3013
2SC743A	154	3045	32	TR78/712	
2SC744			17		
2SC744A		3024	18		
2SC746A	154	3045	32	TR78/712	
2SC748	108	3019	86	TR95/56	

Original Part Number	ECG	SK	GE	IR/WEP	HEP
2SC752	123A	3122	17	TR21/735	S0011
2SC752G	123A	3122	20	TR21/735	S0014
2SC756		3024	28	TR87/243	S3021
2SC756G	123A	3024	18	TR25/	
2SC756-2-4	224	3049			
2SC756C	128	3024	20	TR25/	
2SC761	161	3018	39	TR83/719	S0017
2SC761Y	108	3039	39	TR95/	S0017
2SC761Z	161				
2SC762	161	3018	60	TR83/719	S0017
2SC763	107	3018	86	TR24/720	S0033
2SC763B	107	3018	61	TR70/	S0033
2SC763C	107	3018	61	TR70/	S0033
2SC763D	107	3018	61	TR70/	S0033
2SC765	130	3027	19	TR59/247	S7002
2SC768	130	3027	19	TR59/247	S7002
2SC769					S7004
2SC770					S5020
2SC771		3018	61		S5020
2SC772	107	3018	39	2SC772/720	S0016
2SC772B	107	3019	11	TR70/720	S0020
2SC772BY	107				S0016
2SC772C	107	3018	11	TR70/720	S0016
2SC772CX	107	3018	61	TR70/720	S0011
2SC772D	107	3018	11	TR70/720	S0016
2SC772E	107	3018	20	TR70/720	S0016
2SC772F	107	3018	20	TR70/720	S0011
2SC772K	107	3018	20	TR70/720	
2SC772R	107	3018	17	TR70/720	
2SC772RD	107	3039	20	TR70/720	S0011
2SC773	123A	3018	210	TR62/735	S0015
2SC774	128	3024	47	TR62/243	S3001
2SC775	195	3047	45	TR25/243	S3001
2SC776	128	3024	45	TR65/243	S3001
2SC776Y	128	3024	18		
2SC777	224	3024	46	TR65/243	S3002
2SC778	236	3048	46	TR65/243	S3002
2SC778B	236	3048	46		S3002
2SC780	154	3045	81	TR78/712	S5026
2SC780AG	154	3045	27	TR78/712	S5025
2SC780AG-O	194				
2SC780AG-R	194				
2SC780AG-Y	194				
2SC780G	154	3045	81	TR78/712	S0005
2SC781	195	3024	46	TR87/243	S3001
2SC782	124	3021	12	TR23/701	S3021
2SC783	198	3131	12	TR81/	S3021
2SC784	107	3018	60	TR70/720	S0016
2SC784-O	107	3018	61	/720	S0016
2SC784A	107	3018	61		
2SC784BN	107	3018	61	TR70/720	S0016
2SC784R	107	3018	61	TR70/720	S0016
2SC784R/4454C	107				
2SC784Y	107	3018	17	TR70/720	
2SC784YEL					S0016
2SC785	107	3018	60	TR71/720	S0016
2SC785(O)	107	3018	20	TR71/	S0016
2SC785BN	107	3018	60	TR70/720	S0016
2SC785D	107	3020	20	TR87/	
2SC785E	107	3020	210	TR71/	
2SC785O	107	3018	61	TR70/	S0016
2SC785R	107	3018	61	TR70/720	S0014
2SC786	161	3039	11	TR83/719	S0017
2SC787	161	3039	214	TR83/56	S0020
2SC788	154	3045	27	TR78/712	S5025
2SC789	107	3054	28	TR76/720	
2SC789-O	152	3054	66	TR76/	
2SC789-R	152		66	TR76/	
2SC789-Y	152			TR76/	
2SC790		3054			S5000
2SC790(O)			66	TR76/	
2SC791	175	3021	12	TR92/241	S5012
2SC792	162	3111	73	TR86/	S5020
2SC793	130	3036	14	TR59/247	S7004
2SC793BL	130			TR59/247	S7004
2SC793R	130	3036	14	TR59/247	S7004
2SC793Y		3036	14	TR59/247	S7004
2SC794		3027	75		
2SC794R	130	3027	19	TR59/247	
2SC795	124	3021		TR81/240	S5011
2SC795A	124	3021	12	TR81/240	
2SC796	123A	3122	20	TR21/735	S0014
2SC797	128	3024	20	TR87/243	S0015
2SC798	128	3024	46	TR87/243	S3020
2SC799	237	3049	46	TR65/	
2SC800	107	3039	60	TR70/720	S0016
2SC802			63		S3001
2SC803	128	3024	46	TR87/243	S3002
2SC804			214		
2SC805	154	3040	40	TR78/712	S5026
2SC806	162	3115	35	TR67/707	S5021
2SC806A	163	3115	36	TR67/740	S5021
2SC807	163	3111	36	TR67/240	S5021
2SC807A	163	3111	73	TR61/740	S5020
2SC812				TR53/	
2SC812A		3049	46		
2SC814	192	3124	210	TR65/735	S0014
2SC815	123A	3024	210	TR21/735	S0015
2SC815A	123A	3122	88	TR21/	
2SC815B	123A	3122	88	TR21/	
2SC815C	123A	3122	88	TR21/	
2SC815K	123A	3122	18	TR21/	S0015
2SC815L	123A	3024	18	TR25/	S0015
2SC815M	123A	3024	18	TR87/	S0015
2SC815S	123A			TR21/	
2SC815SA	123A			TR21/	
2SC815SC	123A			TR21/	
2SC816	128	3024	63	TR87/243	S3001
2SC816K	128	3024	18	TR87/243	
2SC817					S3019
2SC818	154	3045	32	TR78/712	S5025
2SC821			28	TR76/	
2SC822			28	TR76/	S3001
2SC823					S0020
2SC825		3021	12		
2SC826	128	3024	18	TR87/243	S5026
2SC827	128	3024	18	TR87/243	S3019
2SC828	123A	3124	39	2SC828A/735	S0015
2SC828A	123A	3024	39	2SC828A/735	S0015
2SC828C	123A	3124	47		
2SC828E	123A	3124	47	TR21/735	
2SC828F	123A	3124	18	TR21/	S0015
2SC828FR	123A			TR21/735	
2SC828H	123A	3018	20	TR24/	S0015
2SC828HR	123A	3122		TR24/	
2SC828N	123A				
2SC828-O	123A	3122	88	TR21/735	S0015
2SC828-P	123A			TR21/735	
2SC828P	123A	3122	88	TR24/	S0015
2SC828Q	123A	3020	20	TR21/	S0015
2SC828-R	123A			TR21/735	
2SC828R	123A	3122	18	TR24/	S0015
2SC828RST	123A	3124		TR24/	
2SC828-S	123A			TR21/735	
2SC828S	123A	3124	18	TR21/	S0015
2SC828-T	123A			TR21/735	
2SC828T	123A	3122	20	TR24/	S0015
2SC828W	123A				
2SC828Y	123A	3124	47	TR21/	S0015
2SC829	123A	3018	39	TR24/720	S0011
2SC829A	123A	3018	17	TR24/	S0033
2SC829-A				/720	S0015
2SC829B	123A	3018	20	2SC829B/	S0011
2SC829-B				/720	S0025
2SC829B/4454C	107				
2SC829B-Y	107			TR70/	
2SC829BC	107	3018			S0011
2SC829BY	123A				S0011
2SC829C	123A	3018	20	TR70/720	S0011
2SC829D	123A	3018	61	TR70/720	S0015
2SC829R	123A	3018	61	TR70/	
2SC829X	123A	3018	20	TR70/720	S0011
2SC829Y	123A	3018	20	TR70/720	S0011
2SC830	175	3026	216	TR81/241	S5019
2SC830A	175	3026	66	TR81/241	
2SC830B	175	3026	66	TR81/241	
2SC830BP		3026			S5000
2SC830C	175	3026	66	TR81/241	
2SC831			69		
2SC834L					S0015
2SC835	107	3122	88	TR21/	
2SC836M	123A			TR21/735	
2SC837	108	3018	61		S0020
2SC837F	107	3018	61	TR24/	
2SC837H	108	3018	61		
2SC837K	161	3018	61	TR83/	S0020
2SC837L	161	3018	61	TR83/	S0020
2SC837WF	123A				
2SC838	199	3122	20	2SC838/735	S0015
2SC838A	199	3018	61		S0020

Original Part Number	ECG	SK	GE	IR/WEP	HEP
2SC838B	199				
2SC838C	199			TR21/	S0015
2SC838D	199				
2SC838E	199	3122	88		S0015
2SC838F	123A	3122	88	TR24/	S0015
2SC838H	199	3122	20	TR21/	S0015
2SC838J	107				S0015
2SC838K	199	3124	47	TR21/735	S0014
2SC838L	199	47	17	TR33/720	S0015
2SC838M	199	3020	210	TR82/720	S0015
2SC838S	192				
2SC839	123A	3018	61	TR24/735	S0015
2SC839A	123A	3018	61	TR24/	
2SC839B	123A	3018	61		
2SC839C	123A	3018	61		
2SC839D	123A	3018	61		
2SC839E		3124	61		S0015
2SC839F	123A	3122	17	TR95/	S0020
2SC839H	123A	3122	88	TR95/	S0015
2SC839JI	123A	3018		TR24/	S0015
2SC839K	123A	3018	61		
2SC839L	123A	3018	61	TR70/720	S0015
2SC839M	123A	3018	61	TR70/720	S0015
2SC839Y		3018	61		
2SC840	175	3021	12	TR81/241	S5012
2SC840A	175	3021	12	TR23/241	S5011
2SC840AC	124			TR81/240	S5012
2SC844	123A			TR21/735	S3013
2SC845				TR23/	S3001
2SC847	123A	3122	20	TR21/735	S0011
2SC848	123A	3122	20	TR21/735	S0011
2SC849	123A	3122	20	TR21/735	S0011
2SC850	123A	3122	20	TR21/735	S0015
2SC851	130	3027	75	TR59/247	S7002
2SC852					S0020
2SC853	192	3122	210	TR21/735	S0015
2SC853A	192	3122	88		
2SC853B	192	3122	88		
2SC853KLM	123A	3020	63	TR21/	S0015
2SC853L	123A	3122	88	TR21/735	S0015
2SC854					S3008
2SC855					S3008
2SC856	194	3044	27	TR78/712	S5025
2SC856-02	194	3045	18	TR78/712	S5025
2SC856C	194	3045	63	TR25/712	
2SC857	154	3040		TR78/712	S0030
2SC857H	154	3040	47	TR78/712	
2SC858		3124	212	TR21/720	S0015
2SC858E	199	3124	47	TR21/	S0015
2SC858F	199	3124	17	TR21/735	S0015
2SC858FG	199	3124	62	TR24/	S0015
2SC858G	199	3124	62	TR21/735	S0016
2SC859	199	3124	212	TR53/720	S0015
2SC859E	199	3124	47	TR21/	
2SC859F	199	3124	47	TR21/	S0015
2SC859FG	199	3124	47	TR21/720	S0015
2SC859G	199	3124	47	TR21/	
2SC859GK	123A		47	TR21/735	
2SC860	161	3117	86	TR83/719	S0017
2SC860C	161	3117	211	TR83/719	
2SC860D	161	3117	211	TR83/719	
2SC860E	161	3117	211	TR83/719	
2SC862				TR61/	
2SC863	161	3117	211	TR83/719	
2SC864	161	3132	60	TR83/719	S0016
2SC866		3048	46	TR65/	S3001
2SC867	124	3021	32	TR23/240	S5012
2SC868	154	3045	32	TR78/712	S5025
2SC869	154	3045	32	TR78/712	S5025
2SC870	123A	3124	212	TR21/735	S0015
2SC870BL	123A			TR21/735	S0015
2SC870E	161	3124	62	TR83/719	S0015
2SC870F	123A	3124	47	TR21/735	S0015
2SC871	123A	3018	212	TR51/735	S0015
2SC871BL	123A			TR21/735	S0015
2SC871E	123A	3018	20	TR21/735	S0015
2SC871F	123A	3018	62	TR53/	S0015
2SC871G	199	3018	62		S0024
2SC875	128	3024	81	TR87/243	S3019
2SC875-1	191	3024	18	TR75/243	
2SC875-1C	191	3024	18	TR75/243	
2SC875-1D	128	3024	18	TR87/243	
2SC875-1E	191	3024	18	TR75/	
2SC875-1F	191			TR75/	
2SC875-2	191	3024	18	TR75/243	
2SC875-2C	191	3024	18	TR75/243	
2SC875-2D	191	3024	18	TR75/243	
2SC875-2E	191	3024	18	TR75/243	
2SC875-2F	191			TR87/	
2SC875-3	128	3024	18	TR87/243	
2SC875-3C	128	3024	18	TR87/243	
2SC875-3D	128	3024	18	TR87/243	
2SC875-3E	128	3024	18	TR87/243	
2SC875-3F	128			TR87/	
2SC875BR	107				
2SC875D	128	3024	18	TR87/243	
2SC875E	128	3024	18	TR87/243	S3019
2SC875F	128	3024	18	TR87/243	S3019
2SC876	128	3024	81	TR87/243	S0015
2SC876C	128	3024	18	TR87/243	
2SC876D	128	3024	18	TR87/243	S0015
2SC876E	128	3024	18	TR87/243	S0015
2SC876F	128	3024	18	TR25/	S0015
2SC876TV	128	3044	18	TR87/	S0015
2SC876TVD	128	3024	18	TR25/	S0015
2SC876TVE	128	3024	18	TR51/	S0015
2SC876TVEF	128			TR87/	
2SC879			11		
2SC881	192	3020	210	TR87/243	S5026
2SC881A	192	3124	47		
2SC881B	192	3124	47		
2SC881C	192	3124	47	TR21/	
2SC881D	192	3124	47		
2SC881K	192	3124	47	TR21/	S5026
2SC881L	192	3124	47	TR87/	S0015
2SC889	162		35	TR67/707	
2SC890			28	TR76/	S3013
2SC891			28	TR76/	S3005
2SC892			66		S3006
2SC893			28	TR76/	S3019
2SC894	123A	3122	10	TR21/735	S0011
2SC895	175	3111	12	TR81/241	
2SC896	123A	3122	210	TR21/735	S5026
2SC897	162	3535	35	TR67/707	S5020
2SC897A	162	3535	35	TR67/707	
2SC897B	162	3535	35	TR67/707	
2SC898	162	3535	35	TR67/707	S5020
2SC898A	162	3535	35	TR67/707	
2SC898B	162	3535	35	TR67/707	
2SC899	123A	3122	61	TR21/735	S0025
2SC899K	123A	3124	47	TR21/735	
2SC900	199	3124	62	TR21/735	S0015
2SC900A	199	3124	47	TR21/	
2SC900B	199	3124	47	TR21/	
2SC900C	199	3124	47	TR21/	
2SC900D	199	3124	47	TR21/	
2SC900E	199	3124	62	TR24/	S0015
2SC900F	199	3124		TR21/	S0015
2SC900L	199	3124	62	TR25/	S0015
2SC900M		3124	47		S0015
2SC900U	199		62	TR24/	S0015
2SC901	163	3111	35	TR67/740	S5020
2SC901A	163	3111	35	TR67/740	S5020
2SC903			210		S0014
2SC904			210	TR21/	S0015
2SC905			81	TR25/	S0015
2SC906	159				
2SC906F	159	3114	89		S0019
2SC907	199	3122	88	TR21/735	
2SC907A	199	3122	88	TR21/735	
2SC907AC	199			TR21/	
2SC907AD	199			TR21/	
2SC907AH	123A	3122	88	TR21/735	
2SC907C	199	3122	88	TR21/	
2SC907D	199	3122	88	TR21/	
2SC907H	123A	3122	88	TR21/735	
2SC907HA	123A	3122	88	TR21/735	
2SC908					S3001
2SC909			28	TR76/	
2SC910D					S0020
2SC911			28	TR76/	
2SC912	108		212	TR95/56	S0016
2SC913	123A	3122	64	TR21/735	
2SC914			64	TR53/	
2SC915			64	TR53/	
2SC916			66		S3002
2SC917	123A	3132	61	TR21/735	S0005
2SC917K	123A	3132	61	TR83/	S0017
2SC918	161	3018	20	TR83/719	S0020
2SC920	107	3018	61	TR24/720	S0016

Original Part Number	ECG	SK	GE	IR/WEP	HEP
2SC920Q	108	3122	20	TR21/	S0014
2SC920R	107	3122	61	TR21/	S0020
2SC921	107	3018	214	TR70/720	S0020
2SC921C1	107			TR70/720	
2SC921K	107	3018	61	TR21/	
2SC921L	107	3018	61	TR70/	S0020
2SC921M	107	3018	11	TR70/	S0020
2SC922	107	3122	17	TR70/	S0020
2SC922A	107			TR70/	
2SC922B	107			TR70/	
2SC922C	107			TR70/	
2SC923	199	3122	20	TR21/735	S0015
2SC923A	199	3124	47	TR21/	
2SC923B	199	3124	47	TR21/	
2SC923C	199	3124	47	TR21/	
2SC923D	199	3124	47	TR21/	
2SC923E	199	3124	47	TR21/	S0015
2SC923F		3124	47	TR21/	S0015
2SC924	108	3124	61	TR95/56	S0011
2SC924E	108	3124	47	TR95/56	
2SC924F	108	3124	47	TR95/56	S0011
2SC924M	123A	3124	47	TR21/	S0011
2SC925	123A	3124		TR21/735	S0030
2SC926	154	3045		TR70/712	S5026
2SC926A		3024	27	TR78/	S5026
2SC927	161	3117	60	TR70/719	S0016
2SC927C	107	3117		TR70/720	S0016
2SC927CU	161	3117	11	TR83/719	
2SC927CW	161			TR83/719	
2SC927D	108	3118	17	TR95/56	S0016
2SC927E	161	3117	39	TR83/719	S0016
2SC928	161	3117	60	TR83/719	S0016
2SC928B	161	3117	211	TR83/	
2SC928C	161	3117	211	TR83/	
2SC928D	161	3117	211	TR83/	
2SC928E	161	3117	211	TR83/	
2SC929	199	3039	39	TR24/720	S0016
2SC929-O	107			TR70/	
2SC929B	199	3039	86	TR21/	
2SC929C	199	3018	20	TR21/	S0016
2SC939C1	199	3018	61	TR21/720	
2SC929D	199	3018	20	TR21/719	S0016
2SC929D1	199	3039	20	TR33/	
2SC929DE	107	3018	61	TR70/720	
2SC929DP	199	3039	20	TR33/	S0016
2SC929DU	199	3018	61	TR21/720	S0011
2SC929DV	199	3018	61	TR21/	
2SC929E	199	3039	20	TR21/720	S0016
2SC929ED	199	3039	20		S0016
2SC929F	199	3039	20	TR21/	S0015
2SC929FC					S0033
2SC930	199	3018	39	TR24/720	S0016
2SC930B	199	3018	61	TR21/	S0016
2SC930BB	199				
2SC930BK	199	3018	61	TR21/	
2SC930BV	199				
2SC930C	199	3039	60	TR21/720	S0016
2SC930CK	199	3039	20	TR33/	S0016
2SC930CS	199	3018			
2SC930D	199	3018	20	TR21/720	S0016
2SC930DB		3018	60		S0016
2SC930DC		3018	60		S0016
2SC930DG					S0016
2SC930DS	199	3039	20	TR33/	S0016
2SC930DT	107	3039		TR70/720	
2SC930DT-2	199	3018	61	TR21/	
2SC930DX	199	3018	20	TR24/	S0016
2SC930E	199	3018	20	TR21/720	S0016
2SC930EP	199	3039	20	TR33/	S0016
2SC930ET		3018			S0016
2SC930EV	199	3019	20	TR21/720	S0016
2SC930EX	199	3018	20	TR51/	S0016
2SC930F	199	3018	60	TR21/	S0016
2SC931	152	3054	66	TR76/701	S5000
2SC931C	186	3192	28	TR55/	
2SC931D	186	3192		TR55/S3023	S5000
2SC931E	188	3054	28	TR55/	S3020
2SC932	152		28	TR76/701	S5000
2SC932E	152	3054	28	TR76/	S5000
2SC933	123A	3122	20	TR21/735	S3013
2SC933BB	123A	3018	61	TR21/	
2SC933C	123A	3122	20	TR21/	S3013
2SC933D	123A	3122		TR21/735	S3013
2SC933E	123A	3122	20	TR21/	S3013
2SC933F	123A	3122	88	TR21/735	S3013

Original Part Number	ECG	SK	GE	IR/WEP	HEP
2SC933G	123A	3122	88	TR21/	S3013
2SC934	128	3024	20	TR87/243	S0025
2SC934C	123A	3024	18	TR21/	
2SC934D	123A	3024	18	TR21/	
2SC934E	123A	3024	18	TR21/	
2SC934F	123A	3024	18	TR21/	S5014
2SC934G	123A	3024	18	TR21/	
2SC934P	128				
2SC935	162	3111	38	TR61/707	S5020
2SC936	164	3115	38	TR93/740A	S5021
2SC936BK	164	3133	37	TR68/	S5021
2SC937	165	3115		TR93/740B	S5021
2SC937-01	123A			TR21/735	
2SC937A	163	3115	38	TR67/	S5021
2SC937B	102A	3115	38	TR85/250	
2SC937(BK)	165				
2SC937YL	165	3115			
2SC938	123A	3124	81	TR21/735	
2SC938A	123A	3124	47	TR21/	
2SC938B	123A	3124	47	TR21/	
2SC938C	123A	3124	47	TR21/	
2SC939	163	3111	73	TR61/740	S5020
2SC939L	163	3111		TR61/	S5021
2SC940	165	3111	73	TR61/740B	S5020
2SC940L	165	3111			S5020
2SC940M	165	3111			S5020
2SC941	199	3122	60	TR24/735	S0016
2SC941-O	199	3122	88	TR21/735	S0016
2SC941-R	199	3122	88	TR21/735	
2SC941-Y	199			TR21/	
2SC941R		3122			S0016
2SC941Y		3122	60		S0016
2SC943	123A	3122	210	TR21/735	S0015
2SC943A	123A	3122	88	TR21/	
2SC943B	123A	3122	88	TR21/	
2SC943C	123A	3122	88	TR21/	
2SC944		3122	210		S0025
2SC945	199	3124	20	2SC945/735	S0015
2SC945A	199	3124	47	TR21/	
2SC945B	199	3124	47	TR21/	
2SC945C	199	3124	20	TR21/	S0015
2SC945CK	199				S0015
2SC945D	199	3124	47	TR21/	
2SC945E	199	3124	47	TR21/	
2SC945F	199	3124	47	TR21/	
2SC945G	199	3124		TR21/	S0015
2SC945K	199	3124	47	TR21/	S0015
2SC945M	199	3124	47	TR87/	S5014
2SC945P		3124	62	TR21/	S0015
2SC945Q	199	3020	20	TR21/735	S0015
2SC945QL	123A	3124			S0015
2SC945R	199	3124	47	TR21/735	S0014
2SC945RED					S0015
2SC945S		3124	20	TR24/	S0015
2SC945TK	199	3124			S0015
2SC945TP	199	3124			S0015
2SC945TQ	199	3124			S0015
2SC945TR	199	3124		TR21/	
2SC945X	199	3124	47		
2SC947	161	3039	214	TR83/719	S0017
2SC948	161	3019	214	TR83/719	S0017
2SC948E	161	3039	86	TR71/	S0017
2SC957	161	3039	20	TR83/719	S0017
2SC959	128	3104	32	TR87/	S3019
2SC959A	128	3104	32	TR87/	
2SC959D	128	3104	32	TR87/	
2SC959C	128	3104	32	TR87/	
2SC959D	128	3104	32	TR87/	
2SC959M	128	3104	32	TR87/243	
2SC959S	128			TR87/	
2SC959SA	128			TR87/	
2SC959SB	128			TR87/	
2SC959SC	128			TR87/	
2SC959SD	128			TR87/	
2SC960	225	3024	10	TR75/	S3021
2SC960A	225	3024			
2SC960B	225	3024			
2SC960C	225				
2SC960D	225	3024			
2SC961				TR25/	
2SC963F		3124			
2SC964				TR10/	
2SC966	123A	3122	88	TR21/735	S5014
2SC967	123A	3122	63	TR21/735	
2SC968	123	3124	63	TR86/53	S5014

Original Part Number	ECG	SK	GE	IR/WEP	HEP
2SC968P	128			TR87/243	
2SC968YEL					S5014
2SC968Y				TR25/	
2SC971	192	3018	8	TR08/735	S5014
2SC972	128	3024	18	TR87/243	
2SC972C	128	3024	18	TR87/243	
2SC972D	128	3024	18	TR87/243	
2SC972E	128	3024	18	TR87/243	
2SC979		3122			S5026
2SC980		3122			S5026
2SC980A/G					S5026
2SC982	172	3156	64	TR69/S9100	S9100
2SC983	171	3201		TR79/244A	
2SC983-O	171	3044	27	TR79/244A	S5025
2SC983R	171	3040			
2SC983-R	171	3044	27	TR32/	S5025
2SC983Y	171	3201			
2SC983-Y	171	3044	27	TR32/	S5025
2SC984	123A	3024	210	TR21/735	S0015
2SC984A	123A	3024	88	TR21/735	
2SC984B	123A	3024	18	TR21/735	S0015
2SC984C	123A	3024	18	TR21/735	S0015
2SC985			86	TR24/	
2SC987			39	TR24/	
2SC987A			86	TR24/	
2SC988			39	TR24/	
2SC988A			39	TR24/	
2SC988B			39	TR24/	
2SC990			66		S3006
2SC991	123A	3122	88	TR21/735	S0014
2SC992	123A	3122	88	TR21/735	S0014
2SC994					S3013
2SC994W				TR33/	
2SC995	154	3045	27	TR78/712	S3021
2SC996	225	3045	27	TR78/244	S5015
2SC997	161	3018	60	TR83/719	S0016
2SC998					S3008
2SC999	165	3115	38	TR93/740B	
2SC999A	165	3115	38	TR93/740B	
2SC1000	199	3122	212	TR53/	S0030
2SC1000-BL	199			TR21/	
2SC1000-GR	199	3124	62	TR21/	S0030
2SC1000Y	123A	3124	17	TR21/735	S0030
2SC1001					S3001
2SC1003					S5000
2SC1004	164	3133	37	TR93/740A	
2SC1004A	164	3133	37	TR93/740A	
2SC1005	165	3115	38	TR93/740B	
2SC1005A	165	3115	38	TR93/740B	
2SC1006	199		212	TR21/	S0030
2SC1006A	199			TR21/	
2SC1006B	199			TR21/	
2SC1006C	199			TR21/	
2SC1007	123A			TR21/	
2SC1008	128			TR87/	S3019
2SC1009			61		
2SC1010	199	3122	212	TR21/735	S0030
2SC1010A	199	3122	88	TR21/	
2SC1010B	199	3122	88	TR21/	
2SC1010C	199	3122	88	TR21/	
2SC1012	154	3044	40	TR78/712	S5024
2SC1012A	154	3045	40	TR78/712	S5024
2SC1012E	154	3045	32		S5024
2SC1013	210	3202	28	TR55/S3023	S3020
2SC1014	210	3041	28	TR55/S3023	S3020
2SC1014B	210	3054	28	TR55/	S3020
2SC1014C	210	3054	28	TR55/	S3020
2SC1014CD	210	3054	28	TR76/701	S3020
2SC1014D	210	3054	28		S3020
2SC1014D1	210	3202			
2SC1015					S3007
2SC1016					S5000
2SC1017	235	3047	45	TR74/S3021	S3020
2SC1018	235	3047	28	TR55/S3021	S3020
2SC1023	107	3018	20	TR70/720	S0017
2SC1023-O	107	3039	86	TR70/720	
2SC1023-Y	107	3039	86	TR70/720	
2SC1024	175	3026	216	TR81/241	S5019
2SC1024B	175	3026	66	TR81/241	
2SC1024C	175	3026	66	TR81/241	
2SC1024D	175	3026	66	TR81/241	
2SC1024-D2	175			TR81/241	S5019
2SC1024E	175	3026	66	TR81/241	S5019
2SC1025	175	3131		TR81/241	S5012
2SC1025D	175		12	TR81/	S5012

Original Part Number	ECG	SK	GE	IR/WEP	HEP
2SC1026	107	3018		TR70/720	S0025
2SC1026G	107	3018	61	TR70/720	
2SC1026Y	107	3039	20	TR70/720	S0015
2SC1027	163	3111		TR67/740	
2SC1030	130	3027	14	TR59/247	S5020
2SC1030A	130	3027	75	TR59/247	
2SC1030C	130	3027	14	TR59/247	S5020
2SC1030D	130	3027	14	TR59/247	
2SC1032	107	3018	20	TR70/720	S0015
2SC1032Y	107	3018	61	TR70/720	
2SC1033	123A	3044	18	TR21/	S5025
2SC1033A	123A	3044	32	TR21/	S5025
2SC1034		3021	12	TR93/740A	
2SC1035	161	3117	211	TR83/719	S0016
2SC1035C	161	3117	211	TR83/	
2SC1035D	161	3117	211	TR83/	
2SC1035E	161	3117	211	TR83/	
2SC1036	161	3117	211	TR83/719	S0016
2SC1038					S3005
2SC1039					S3005
2SC1040					S3005
2SC1041					S3005
2SC1042					S3005
2SC1044					S0017
2SC1045	164	3115	37	TR93/740A	S5021
2SC1045B	164	3115	37	TR93/740A	S5021
2SC1045C	164	3115	37	TR83/740A	
2SC1045D	164	3115	37	TR83/740A	
2SC1045E	164	3115	37	TR83/740A	
2SC1046	165	3115	38	TR93/740B	S5021
2SC1046K	165	3115	38	TR67/	S5021
2SC1046M	165		38	TR67/	
2SC1047	108	3018	60	TR21/56	S0016
2SC1047B	107	3018	61	TR70/720	S0016
2SC1047BCD	108	3018			
2SC1047C	108	3018	61	TR70/	S0016
2SC1047D	108	3018	61	TR95/	
2SC1048	154	3045	32	TR78/712	S5025
2SC1048B	154	3045	32	TR78/712	
2SC1048C	154	3045	32	TR78/712	S5025
2SC1048D	154	3045	32	TR78/712	S5025
2SC1048E	154	3045	32	TR78/712	
2SC1048F	154	3045	32	TR78/	
2SC1048N	154				
2SC1050	124				S5021
2SC1051					S5020
2SC1055		3538			
2SC1055H	218	3538	69	TR58/	
2SC1056	154	3044	40	TR78/712	S5024
2SC1059	124	3021	12	TR81/240	S5011
2SC1060	152	3041	28	TR76/701	S5000
2SC1060A	152	3054	28	TR76/701	
2SC1060B	152	3054	28	TR76/701	S5000
2SC1060BM	152	3054	28	TR76/701	
2SC1060C	152	3054	28	TR76/701	S5000
2SC1060D	152	3054	28	TR76/	S5000
2SC1061	152	3054		TR76/701	S5000
2SC1061A	152	3054	28	TR76/701	S5000
2SC1061B	152	3054	28	TR76/701	S5000
2SC1061BT	152			TR76/	
2SC1061C	152	3054	28	TR76/701	S5000
2SC1061D	152	3041	55	TR76/	
2SC1061T	152	3054			S5000
2SC1062					S5024
2SC1063			47		
2SC1070	161	3018		TR70/	S0017
2SC1071	123A		64	TR21/735	
2SC1072	128	3024	47	TR87/243	
2SC1072A	128	3024	63	TR87/243	
2SC1076				TR55/	S5000
2SC1078	124	3131		TR23/	
2SC1079	181	3079	35	TR36/S7000	S5020
2SC1080	181	3027		TR36/S7000	S7004
2SC1081					S3006
2SC1086	165	3115	38	TR93/740B	
2SC1088					S3021
2SC1089	171	3104	40		S3021
2SC1089C	191	3104		TR75/	
2SC1089D	171	3040	40	TR75/	S5024
2SC1090	154				
2SC1091A			X18		
2SC1095	186	3192			
2SC1096	186	3054	28	TR55/701	S5000
2SC1096A	186	3041	55	TR55/	
2SC1096B	186	3041	55	TR55/	

Original Part Number	ECG	SK	GE	IR/WEP	HEP
2SC1096C	186	3041	55	TR55/	
2SC1096D	186	3054	28	TR55/	
2SC1096K	186	3054	28	TR55/	
2SC1096L	186	3054	28	TR55/	S5000
2SC1096M	186	3054	28	TR55/	S5000
2SC1096N	186	3192		TR55/	
2SC1097				TR93/	
2SC1098	186	3192	28	TR55/S3023	S5000
2SC1098A	186	3192		TR55/	
2SC1098B	186	3192		TR55/	
2SC1098C	186	3192		TR55/	
2SC1098D	186	3192		TR55/	
2SC1098L	186	3192	28	TR55/	S5000
2SC1098M	186	3192	28		S5000
2SC1099K		3111	36	TR93/	S5021
2SC1100	165	3115	38	TR93/740B	
2SC1101	164	3115	37	TR68/740A	S5020
2SC1101A	164	3115	37	TR93/	
2SC1101B	164	3115	37	TR93/	
2SC1101C	164	3115	37	TR93/	
2SC1101D	164	3115	37	TR93/	
2SC1101E	164	3115	37	TR93/	
2SC1101F	164	3115	37	TR86/	
2SC1101L	164	3115	37	TR68/	
2SC1102	124			TR81/	S5011
2SC1102A	124			TR81/	
2SC1102B	124			TR81/	
2SC1102C	124			TR81/	
2SC1102K	124	3021	12	TR23/	S5011
2SC1102L	124	3021	12	TR23/	S5011
2SC1102M	124	3021	12	TR23/	S5011
2SC1102(M)	124		12	TR23/	
2SC1103	154			TR78/	
2SC1103A	154	3040		TR79/	
2SC1103(A)	154		40	TR78/	
2SC1103B	154			TR79/	
2SC1103C	154			TR79/	
2SC1103L	154			TR78/	S3021
2SC1104	124	3079	32	TR81/	S5011
2SC1104A	124			TR81/	
2SC1104B	124			TR81/	
2SC1104C	124			TR81/	
2SC1105	124	3021	12	TR81/240	S5011
2SC1105A	124	3021	12	TR81/	
2SC1105B	124	3021	12	TR81/	
2SC1105C	124	3021	12	TR81/	
2SC1105K	124	3021	12	TR23/	S5011
2SC1105L	124	3021	12	TR23/	S5011
2SC1105M	124	3021	12	TR23/	S5011
2SC1106	162	3535	14	TR61/707	S5020
2SC1106A	162			TR67/	
2SC1106B	162			TR67/	
2SC1106C	162			TR67/	
2SC1106K			14	TR61/	S7004
2SC1106L		3027	14	TR61/	S7004
2SC1106M		3027	14	TR61/	S7004
2SC1107Q	186	3192	28	TR55/	
2SC1111					S7004
2SC1112					S7004
2SC1114		3027		TR36/	S7004
2SC1115	162	3079	19	TR61/	S7004
2SC1116		3079	35		S7004
2SC1117	161	3039	214		S0017
2SC1118					S3006
2SC1120					S3006
2SC1121					S3007
2SC1122					S3007
2SC1123	107	3018	61	TR80/	S0020
2SC1124	198	3104	27	TR74/	S3019
2SC1126	107	3018	61	TR80/	S0016
2SC1127	190	3104	27	TR75/S3021	S3021
2SC1127A		3104	32		S3019
2SC1128		3018		TR24/719	S0016
2SC1128(3RD IF)	233				
2SC1128(FNL IF)	233				
2SC1128(M)	161		61	TR24/	
2SC1128S(3 IF)	233				
2SC1128(S)	161		61	TR70/	
2SC1129		3018	61	TR24/719	S0016
2SC1129(M)	161		60	TR24/	
2SC1129R	161	3018	61		
2SC1129(R)	161		60	TR24/	
2SC1140					S5021
2SC1151	164	3133	37	TR93/	S5021
2SC1151A	164	3133	37	TR68/	S5021

Original Part Number	ECG	SK	GE	IR/WEP	HEP
2SC1152			14	TR61/	S7004
2SC1152F	130	3027	14	TR59/	S7004
2SC1153	163	3115	38	TR93/	S5021
2SC1154	163	3111		TR67/	S7004
2SC1155	190			TR74/S3021	S3020
2SC1156	190	3104	32	TR74/S3021	S3019
2SC1157	190	3104	32	TR74/S3021	S3019
2SC1158	108	3019	39	TR95/	S0020
2SC1159	107		60	TR80/	S0020
2SC1160	175		32	TR23/	S3021
2SC1160K	124			TR81/	S5011
2SC1160L	175		12	TR23/	S3021
2SC1161	175	3021	32	TR81/	S3021
2SC1162	184	3190		TR55/	S5000
2SC1162A	184	3190			
2SC1162B	184	3041	28	TR55/	S3024
2SC1162C	184	3122	28	TR55/	S5000
2SC1162CP	184	3190	28	TR55/	
2SC1162D	184	3054	57	TR55/	S3024
2SC1165					S3008
2SC1166	194	3024	81	TR63/	S5026
2SC1166Y	154	3122	18	TR63/	S5026
2SC1167	165	3111	38	TR93/	S5021
2SC1168	124	3131	12	TR81/	S5011
2SC1170	165	3115	38	TR93/740B	
2SC1170A	155	3115	38	TR93/740B	
2SC1170B	165	3115	38	TR93/	S5021
2SC1171	165	3133	38	TR93/	
2SC1172	165	3115	38	TR93/740B	
2SC1172A	165	3115	38	TR93/740B	
2SC1172B	238	3115	38		
2SC1173	153	3054	28	TR55/701	S3010
2SC1173C	152	3054	28	TR76/	S5000
2SC1173-GR	152			TR76/	
2SC1173-O	152	3054		TR76/	S3024
2SC1173R	152	3054		TR76/	S5000
2SC1173-R	152		28	TR55/	
2SC1173X	152	3054	66	TR55/	S5000
2SC1173-Y	152			TR76/	
2SC1174	165	3115	38	TR93/740B	
2SC1175	128	3122	210	TR87/	S0025
2SC1175C	128			TR87/	
2SC1175D	128	3122	20	TR21/	S0025
2SC1175E	128	3122	20	TR87/	S0025
2SC1175F	128	3122	20	TR87/	S0025
2SC1178					S3010
2SC1180			39		
2SC1182					S0020
2SC1182B	161	3018	20	TR83/719	S0020
2SC1182C	161	3018	20	TR83/719	S0020
2SC1182D	161	3018	61	TR83/719	S0020
2SC1184	164		37	TR93/	
2SC1184A	164		37	TR93/	
2SC1184B	164		37	TR93/	
2SC1184C	164		37	TR93/	
2SC1184D	164		37	TR93/	
2SC1184E	164		37	TR93/	
2SC1185	162	3079	38	TR67/	S7004
2SC1185A	162		35	TR67/	
2SC1185B	162		35	TR67/	
2SC1185C	162		35	TR67/	
2SC1185K	162		14	TR61/	S5020
2SC1185L	162	3027	14	TR61/	S7004
2SC1185M	162		14	TR61/	S7004
2SC1187	161	3117	60	TR70/	S0016
2SC1190					S3007
2SC1195					S5020
2SC1204	199	3124	210	TR21/	S0015
2SC1204B	199	3124		TR21/	
2SC1204C	199	3124		TR21/	
2SC1204D	199	3124		TR21/	
2SC1205	108	3122	210	TR95/	
2SC1205A	108	3122		TR95/	
2SC1205B	108	3122		TR95/	
2SC1205C	108	3122		TR95/	
2SC1206		3024	18		S0015
2SC1209	128	3124	20	TR25/	S5014
2SC1209C	128	3024	63	TR87/	S5014
2SC1209D	192	3024		TR25/	S5014
2SC1210	123A	3122	20	TR51/	
2SC1211	123A	3020	81	TR21/735	S0015
2SC1212	184	3190			S5000
2SC1212A	184	3190			S5000
2SC1213	123A	3122	210	TR21/	S0025
2SC1213A	123A	3122	210	TR25/	S0025

Original Part Number	ECG	SK	GE	IR/WEP	HEP
2SC1213AA	123A			TR21/	
2SC1213AB	123A			TR21/	
2SC1213AC	123A			TR21/	S0015
2SC1213B	123A	3122	18	TR87/243	S5014
2SC1213C	123A	3122	63	TR87/243	S5014
2SC1213D	123A	3122	88	TR21/	S5014
2SC1214	128	3122	47	TR87/	S0025
2SC1214A	128	3122		TR87/	
2SC1214B	128	3122	20	TR24/	S0025
2SC1214C	128	3044		TR24/	S3002
2SC1214D	128	3122		TR21/	
2SC1215	107	3018	86	TR21/	S5000
2SC1216			61		
2SC1217	171	3201		TR79/	S3019
2SC1218	128			TR87/	S3011
2SC1220-003	128	3122			
2SC1220				TR21/	
2SC1220E	128	3024	18	TR87/	S5014
2SC1222	199	3122	212	TR21/	S0015
2SC1222A	199			TR21/	
2SC1222B	199			TR21/	
2SC1222C	199			TR21/	
2SC1222D	199			TR21/	
2SC1226	186	3192	28	TR55/S3023	S5000
2SC1226A	186	3054		TR55/S3023	S5000
2SC1226AP	186	3192	28	TR55/	S5000
2SC1226AQ					S5000
2SC1226C	186	3054	28		S5000
2SC1226C/F					S5000
2SC1226F	186	3192			S5000
2SC1226P	186	3054	20	TR55/	S5000
2SC1226Q	186	3192	28	TR55/	S5000
2SC1226R	186	3054	28		S5000
2SC1231C		3122			
2SC1235	124				S5011
2SC1235AL	124				
2SC1235AM	124	3021	12	TR23/	S5011
2SC1237	236	3054	28	TR55/	S3024
2SC1237E	152				
2SC1239		3049	45	TR65/	S3001
2SC1243		3041	28	TR55/	S5000
2SC1243C2		3054			S5000
2SC1244	123A		20	TR21/	
2SC1254			39		S0017
2SC1256					S3001
2SC1257					S3005
2SC1258					S3006
2SC1259					S3007
2SC1278	194				
2SC1278S	194				
2SC1279	194				
2SC1280	172			TR69/	
2SC1280A					S9100
2SC1280AS	172			TR69/	
2SC1280S	172			TR69/	
2SC1285					S0015
2SC1293	161	3132	61	TR70/	S0020
2SC1293(3RD IF)	233				
2SC1293A	161	3018			S0020
2SC1293(A)			61	TR71/	
2SC1293B		3018			S0020
2SC1293C	161		61	TR70/	S0020
2SC1293C(3RDIF)	233				
2SC1295		3111	38	TR61/	S5020
2SC1296	163	3115	38	TR61/	S5021
2SC1297					S3007
2SC1298					S3007
2SC1303			81		S3013
2SC1304	124	3021	12	TR81/	S5011
2SC1306	235	3197		TR55/	S3024
2SC1307	236	3197	216	TR55/	S5004
2SC1308	165		38		
2SC1308K	165	3115			S5021
2SC1308L				TR61/	S5021
2SC1309	163		38	TR61/	S5020
2SC1312	123A	3122	20	TR24/	S0015
2SC1312G					S0015
2SC1312(Y)(G)				TR24/	
2SC1313			212		
2SC1313B		3122	88		
2SC1316	175			TR23/	S5012
2SC1317	128	3122	210	TR24/	S0003
2SC1317(B)				TR21/	S0003
2SC1317B	123A	3122	88		S0003
2SC1317O					S0003
2SC1317P	123A				S0003
2SC1317Q	128	3122	18	TR21/	S0003
2SC1317R	128	3024		TR25/	S0003
2SC1317S	128	3024	17	TR25/	S5000
2SC1318	128	3122	210	TR25/	S0015
2SC1318P	128	3024			S0015
2SC1318Q	128	3122		TR21/	S0015
2SC1318R	128	3122	210		S0015
2SC1318S	123A	3122	210		S5014
2SC1319					S0015
2SC1319Q				TR25/	S0015
2SC1320K	107	3018	61	TR33/	S0020
2SC1321			39		
2SC1324(C)			20		
2SC1325	165	3115	38	TR93/	S5011
2SC1325A		3115			S5020
2SC1326					S3008
2SC1327	199		64	TR51/	S0024
2SC1327T	199	3122		TR51/	S0024
2SC1327U	199	3122	62	TR51/	S0024
2SC1328	199		47	TR51/	
2SC1328T	199			TR51/	S0015
2SC1328U	199	3124	62	TR24/	S0015
2SC1330	192				
2SC1330A	192				
2SC1333					S3005
2SC1335	199	3122	210	TR24/	S0025
2SC1335A	199	3122	210	TR21/	
2SC1335B	199	3122	210	TR21/	S0025
2SC1335C	199	3122	210	TR24/	S0025
2SC1335D	199	3122	210	TR51/	S0025
2SC1335E	199	3122	210	TR21/	S0025
2SC1335F	199	3122	210	TR21/	
2SC1342	107	3018	39	TR24/	S0016
2SC1342A	107	3018	61	TR70/	S0014
2SC1342B	107	3018	61	TR70/S0014	
2SC1342C	107	3018	61	TR70/	S0014
2SC1343	181	3535			S5020
2SC1344	199	3124	210	TR21/	S0015
2SC1344C	199	3124		TR21/	
2SC1344D	199	3124		TR21/	
2SC1344E	199	3124		TR21/	S0015
2SC1344F	199	3124		TR21/	
2SC1345	199	3122	210	TR21/	S0025
2SC1345C	199	3122		TR21/	
2SC1345D	199	3122		TR21/	
2SC1345E	199	3122		TR21/	S0015
2SC1345F	199	3122		TR21/	
2SC1346	192		47		S3013
2SC1346Q	192				
2SC1346R	192	3020		TR51/	S3013
2SC1346S	192			TR51/	S3013
2SC1347	192	3137	47	TR21/	S3001
2SC1347Q	192		63	TR21/	S3001
2SC1347R	192	3122	88	TR21/	S3001
2SC1348	163			TR93/	S5021
2SC1358	165	3115	36	TR93/	S5011
2SC1358A	165				
2SC1358K	165		36	TR93/	S5021
2SC1358L					S5021
2SC1359	107	3018	211	TR24/	S0015
2SC1359A	107	3018		TR95/	S0014
2SC1359B	107	3018	211	TR95/	S0014
2SC1359C	107	3018			S0025
2SC1360	107	3124	11	TR70/	S0025
2SC1362	199	3124	62	TR24/	S0015
2SC1363	123A	3124	20	TR21/	S0011
2SC1364	123A	3124	20	TR24/	S0015
2SC1364A	123A	3124	20	TR24/	
2SC1367		3115			S5021
2SC1367A		3131			S5021
2SC1368	184	3190			
2SC1368B	184	3190	57		
2SC1368C		3054			S5000
2SC1372					S0016
2SC1372Y	123			TR24/	S0015
2SC1376		3024			
2SC1377	236	3197	66		
2SC1378					S3006
2SC1380	3122		212		S0015
2SC1380A		3122	212		S0015
2SC1382		3104			S3011
2SC1383	128	3512	20	TR25/	S3024
2SC1383P	128	3024			S5014
2SC1383Q	128	3024			S3001

Original Part Number	ECG	SK	GE	IR/WEP	HEP
2SC1383R	128	3024	63	TR87/	S3001
2SC1383S		3512	20	TR25/	S3001
2SC1383X		3512		TR24/	S3001
2SC1384	128	3122		TR25/	S3001
2SC1384Q	128	3018		TR25/	S3001
2SC1384R	128		63	TR87/	S3001
2SC1384S	128				S3001
2SC1385		3024			
2SC1386H		3024	63		
2SC1390	123A	3122	20	TR24/	S0015
2SC1390J	123A	3018			
2SC1390K	123A				
2SC1390L	123A	3122		TR24/	S0015
2SC1390V	107	3018	20	TR24/	
2SC1390W	123A			TR24/	S0016
2SC1390WX	123A	3018	20	TR24/	S0016
2SC1390X	123A	3018		TR24/	S0016
2SC1391					S0015
2SC1393		3018			
2SC1394					S0025
2SC1395	107	3122	17	TR24/	S0011
2SC1398	152		66	TR76/	S5000
2SC1398Q	186	3054	66	TR76/	S5000
2SC1402					S5020
2SC1403					S5020
2SC1406	192				S3024
2SC1406P	192		63		
2SC1407					S3020
2SC1407Q	192		63		S3020
2SC1409	196	3021			S5000
2SC1409B	196	3021	66	TR75/	S3019
2SC1409C	196	3021	12		
2SC1413	163	3115	38	TR61/	S5020
2SC1416			212		S0030
2SC1416A			81		S0030
2SC1417	107	3018	61	TR21/	S0030
2SC1417C	123A				
2SC1417D	123A	3018			
2SC1417D(U)	123A		20	TR51/	
2SC1417G	123A	3018			
2SC1417VW	107	3018	20	TR24/	S0016
2SC1417W				TR24/	S0016
2SC1418		3054			S5000
2SC1419		3054			S5000
2SC1419B		3054			S3024
2SC1433		3111			S5021
2SC1444					S5019
2SC1445					S5012
2SC1446	190				S3019
2SC1446Q	190		32	TR74/	S3019
2SC1447	198	3104	32	TR74/	S3021
2SC1447R	198	3104		TR74/	S3021
2SC1448	190		32	TR76/	S3021
2SC1449	184	3054	57		
2SC1450	124				S5012
2SC1453			212		
2SC1454					S5020
2SC1456	124		32	TR81/	S5011
2SC1456L	124	3021	12	TR81/	S5011
2SC1456LM	124				
2SC1456M	124	3021	12	TR81/	S5011
2SC1471					S0015
2SC1474	128			TR87/	
2SC1475	128			TR71/	S5014
2SC1506		3103			
2SC1507		3104	32	TR21/	S3021
2SC1509					S3020
2SC1518			47		
2SC1520	154			TR78/	
2SC1521K		3103			S3021
2SC1537	123A				
2SC1537(S)	123A	3122	63	TR51/	S0015
2SC1538	123A	3122		TR24/	
2SC1538A	123A				
2SC1538SA	123A	3122	63	TR51/	S0015
2SC1542		3124			S0024
2SC1547			39		
2SC1550	154	3130		TR78/	
2SC1556		3024			
2SC1568	184	3190			S5000
2SC1568R	184	3054	55	TR95/	S5000
2SC1569		3104			
2SC1570			212		
2SC1571			212		
2SC1573Q		3045			

Original Part Number	ECG	SK	GE	IR/WEP	HEP
2SC1585F	107	3018	11	TR51/	
2SC1585H	107	3018	11	TR51/	
2SC1617		3115			
2SC1621			214		
2SC1622			39		
2SC1623			212		
2SC1626		3054			
2SC1675M		3018			
2SC1681		3122			
2SC1682		3124			
2SC1684LA		3104			
2SC1685Q		3137			
2SC1686		3018			
2SC1687		3018			
2SC1688		3137			
2SC1717		3049			
2SC1838S		3512			
2SC1919C			11		
2SC2123C		3024			
2SC2810				TR27/	
2SC2818			20		
2SC2879A				TR21/	
2SC3625				TR25/	
2SC3710			20	TR71/	S0015
2SC3720		3018	61		S0016
2SC3724	123A				
2SC3731				TR24/	
2SC3800		3018	61	TR24/	
2SC3808				TR21/	
2SC3826R				TR33/	
2SC3940		3018	61	TR21/	
2SC3948				TR21/	
2SC4012		3124	47		
2SC4033		3018	61		
2SC4116		3026	14		
2SC5100		3104	32		S3019
2SC5110		3104	32		S3019
2SC5220					S3019
2SC5359	108			TR95/	
2SC5370		3124	47		
2SC5376				TR21/	
2SC6440				TR24/	
2SC6680					S0020
2SC6845				TR21/	
2SC6849				TR21/	
2SC6961				TR87/	
2SC7090				TR25/	
2SC7330					S0025
2SC7335-BL	123A			TR21/	
2SC7350		3124		TR21/	S0014
2SC7354	123A	3124			S0003
2SC7382BN					S0016
2SC7720				TR21/	
2SC7728				TR21/	
2SC7840				TR33/	
2SC7850		3018			
2SC7890		3054			
2SC8146			47		
2SC8156				TR25/	
2SC8280				TR21/	
2SC8290			20	TR21/	
2SC8380	108				S0015
2SC9001					S0015
2SC9011E,F,H			20		
2SC9012HG			21		
2SC9012HH			21		
2SC9270				TR22/	
2SC9290				TR24/	
2SC9370-1				TR21/	
2SC9450L				TR24/	
2SC9451				TR21/	
2SC10004				TR24/	
2SC10471				TR70/	
2SC10898	171				
2SC10966				TR55/	
2SC11070			28		
2SC11730					S5000
2SC11734				TR76/	
2SC13838	128				
2SC13901	123A				
2SC13980				TR76/	S3024
2SC14070	192				
2SC15385(A)				TR51/	
2SC371008-02				TR53/	
2SC-NJ100	199				S0024

Original Part Number	ECG	SK	GE	IR/WEP	HEP
2SCNJ107	128			TR25/	S5014
2SCF1	108	3018	61	TR95/56	S0033
2SCF2	123A	3018	20	TR21/735	S0015
2SCF-2	123A		20	TR21/735	
2SCF3		3048	46		
2SCF5	123	3046	17	TR86/53	S0011
2SCF6	128	3024	18	TR87/243	S3001
2SCF8	224	3048	45	TR65/	S3001
2SCF11	108	3018	20	TR95/	S0016
2SCF12A		3049	46		
2SCJ02				TR21/	
2SCM39J		3018	61		
2SCNJ100		3122			S0024
2SCNJ107		3020		TR25/	S5014
2SCS183E	123	3124	20	TR86/53	
2SCS184E	123	3124	20	TR86/53	
2SCS184J	108	3018	11	TR95/56	
2SC429		3018	61		
2SCS429A		3018	61	/56	
2SCS429J	108	3018	11	TR95/	
2SCS430		3018	61		
2SCS430H	108	3018	11	TR95/56	
2SCS461		3018	61		
2SCS461F	108	3018	11	TR95/56	
2SCS469		3018	61		
2SCS469F	108	3018	11	TR95/56	
2SCU64M		3047	45		
2SD11	103	3010	8	TR08/641A	G0011
2SD12	130	3027	72	TR59/247	S5012
2SD13	105	3012	4	TR03/233	S5004
2SD14	105			TR03/233	S5004
2SD15	130	3027	75	TR59/247	S7002
2SD16	130	3027	75	TR59/247	S7004
2SD17	130	3027	75	TR59/247	S5020
2SD18	162		35	TR67/707	S5020
2SD19	103	3010	8	TR08/741A	S0011
2SD20	103	3010	8	TR08/641A	G0011
2SD21	103	3010	8	TR08/641A	S0011
2SD22	103	3010	8	TR08/641A	G0011
2SD23	103	3010	8	TR08/641A	G0011
2SD24	124	3021	12	TR23/240	S5011
2SD24B	124	3021	12	TR81/	
2SD24C	124	3021	12	TR81/240	S5011
2SD24CK	124	3021	12	TR81/	
2SD24D	124	3021	12	TR81/	S5011
2SD24E	124	3021	12	TR81/	S5011
2SD24F	124	3021	12	TR81/	S5011
2SD24K	124	3021	12	TR81/	S5011
2SD24KD	124	3021	12	TR81/	S5011
2SD24KE	124	3021	12	TR81/	S5011
2SD24Y	124	3021	12	TR23/	S5011
2SD24YK	124	3021			S5011
2SD24YM	124	3021			S5011
2SD25	103A	3010	8	TR08/724	G0011
2SD26		3027	19		S7002
2SD26A	130	3010	19	TR59/247	S7002
2SD26B	130	3027	75	TR59/247	S7004
2SD26C	130	3027	75	TR59/247	S5020
2SD27				TR21/	
2SD28	175	3026	23	TR81/241	S5019
2SD29	175	3026	23	TR81/241	S5019
2SD30	103A	3010	59	TR08/724	G0011
2SD30-O	103A	3010	59	TR08/724	G0011
2SD30N				/724	G0011
2SD30-N	103A	3010	59	TR08/	
2SD31	103A	3010	8	TR08/724	G0011
2SD31D	103A	3010	59	TR08/724	
2SD32	103A	3010	8	TR08/724	G0011
2SD33	103A	3010	5	TR85/724	G0011
2SD33C	103A	3010	8	TR08/724	
2SD34	103A	3124	59	TR08/724	G0011
2SD35	103A		5	TR08/724	G0011
2SD36	101	3011	5	TR08/641	G0011
2SD37	103A	3010	8	TR08/724	G0011
2SD37A	103A	3010	59	TR08/724	
2SD37B	103A	3010	59	TR08/724	
2SD37C	103A	3010	8	TR08/724	
2SD38	103A	3124	59	TR08/724	G0011
2SD41	130	3027	75	TR59/247	S7002
2SD43	101	3010	5	TR08/641	G0011
2SD43A	101	3010	8	TR08/641	G0011
2SD44	101	3011	59	TR08/641	G0011
2SD45	162		35	TR67/707	S5020
2SD46	162		35	TR67/707	S5020
2SD47	162		14	TR67/707	S7004

Original Part Number	ECG	SK	GE	IR/WEP	HEP
2SD48			66		S3002
2SD49	175	3026	28	TR81/241	S5000
2SD50	130	3027	14	TR59/247	S7004
2SD51	130	3027	14	TR59/247	S7004
2SD51A		3027	75		S7004
2SD53	130	3027	75	TR59/247	S7004
2SD54	105			TR03/233	S5004
2SD55	181		75	TR36/S7000	S7000
2SD55A	181			TR36/S7000	S7000
2SD56	175	3026	14	TR81/241	
2SD57	175	3131	66	TR81/241	S5000
2SD58	175	3131	66	TR81/241	S5000
2SD59	162		75	TR67/707	S7004
2SD60	162		35	TR67/707	S5020
2SD61	103A	3010	59	TR08/724	G0011
2SD62	103A	3010	59	TR08/724	G0011
2SD63	103A	3010	5	TR08/724	
2SD64	103A	3010	5	TR08/724	G0011
2SD65	103A	3010	5	TR08/724	G0011
2SD65-1	103A	3010	8	TR08/724	
2SD66	103A	3010	5	TR08/724	G0011
2SD67	162		35	TR67/707	S5020
2SD67B	162			TR67/	
2SD67C	162			TR67/	
2SD67D	162			TR67/	
2SD67E	162			TR67/	
2SD68	223	3027	75	TR59/247	S7002
2SD68B	223	3027	75	TR59/	
2SD68C	223	3027	75	TR59/	
2SD68D	223	3027	75	TR59/	
2SD68E	223	3027	75	TR59/	
2SD69	130	3111		TR59/247	S5020
2SD70	175	3027	23	TR81/247	S7002
2SD71	175			TR81/	S5012
2SD72	103A	3010	59	TR08/724	G0011
2SD72-2C	103A	3010	59	TR08/724	G0011
2SD72-3C	103A	3010	59	TR08/724	G0011
2SD72-4C	103A	3010	59	TR08/724	G0011
2SD72A	103A	3010	59	TR08/724	G0011
2SD72B	103A	3010	59	TR08/724	G0011
2SD72BR	103A				
2SD72C	103A	3124	47	TR08/724	G0011
2SD72K	103A	3010	59	TR08/	G0011
2SD72P	103A		59		
2SD72RE	103A			TR08/	
2SD72RED					G0011
2SD73	162		14	TR67/707	S7004
2SD74	162		73	TR67/707	S5020
2SD75	103A	3010	5	TR08/724	G0011
2SD75A	103A	3010	10	TR08/724	G0011
2SD75AH	103A			TR08/724	G0011
2SD75B	103A	3010	59	TR08/	G0011
2SD75H	103A	3010	59	TR08/724	G0011
2SD77	103A	3010	5	TR08/724	G0011
2SD77A	103A	3010	8	TR08/724	G0011
2SD77AH	103A			TR08/724	G0011
2SD77B	103A	3010	8	TR08/724	G0011
2SD77C	103A	3010	8	TR08/724	G0011
2SD77D	103A	3010	59	TR08/724	
2SD77H	103A	3010	59	TR08/724	G0011
2SD77P	103A		6		
2SD78			66		S3002
2SD79	225		66		
2SD79A	225				
2SD80	130	3027	75	TR59/247	S7002
2SD81	130	3027	75	TR59/247	S7002
2SD82	130	3027	75	TR59/247	S7004
2SD83	162		35	TR67/707	S5020
2SD84	162		35	TR67/707	S5020
2SD88	162		35	TR78/707	S5020
2SD88A	162		35	TR67/707	S5020
2SD90	152	3024	72	TR76/701	S5014
2SD91	152	3026	12	TR76/701	S5019
2SD91F	175	3026	66	TR81/241	
2SD92	175	3131	66	TR81/241	S5012
2SD92D	175	3131		TR81/	
2SD93	152		72	TR76/701	S5012
2SD94	124		72	TR81/	S5012
2SD96	103A	3010	59	TR08/254	G0011
2SD96A		3010	59	/724	
2SD100	103A	3010	54R	TR08/	G0011
2SD100A	103A	3010	59	TR08/	G0011
2SD101	101		54	TR08/641	
2SD102	175	3538	47	TR81/241	S5012
2SD102-0	175	3538		TR81/241	

Original Part Number	ECG	SK	GE	IR/WEP	HEP
2SD102-R	175	3538		TR81/241	
2SD102-Y	175	3538		TR81/241	
2SD103	175	3026		TR81/241	
2SD103-O	175	3026		TR81/241	
2SD103-R	175	3131		TR81/241	
2SD103-Y	175	3026		TR81/241	
2SD104	103A	3010	59	TR08/724	G0011
2SD105	103A	3010	8	TR08/724	G0011
2SD107					S9140
2SD108					S9140
2SD110	162	3535	35	TR67/707	S5020
2SD110-O	162		35	TR67/707	
2SD110-R	162		35	TR67/707	
2SD110-Y	162		35	TR67/707	
2SD111	162	3510	75	TR67/707	S7004
2SD111-O	162		35	TR67/707	
2SD111-R	162		35	TR67/707	
2SD111-Y	162		35	TR67/707	
2SD113	181	3535	75	TR36/S7000	S7000
2SD113-O	181			TR36/S7000	
2SD113-R	181			TR36/S7000	
2SD113-Y	181			TR36/S7000	
2SD114	181	3535	75	TR36/S7000	S7000
2SD114-O	181		75	TR36/S7000	
2SD114-R	181		75	TR36/S7000	
2SD114-Y	181		75	TR36/S7000	
2SD118	162	3036	35	TR67/707	S5020
2SD118BL	162	3036	35	TR67/707	S5020
2SD118R	162	3036	35	TR67/707	S5020
2SD118Y	162	3036	35	TR67/707	S5020
2SD119	162	3036	75	TR67/707	S7004
2SD119BL	162	3036	35	TR67/707	S7004
2SD119R	162	3036	35	TR67/707	S7004
2SD119Y	162	3036	35	TR67/707	S7004
2SD120	128	3024	46	TR87/243	S5014
2SD120H	210	3024	18		
2SD120HA	210	3202			
2SD120HB	210	3202			
2SD120HC	210	3202			
2SD121	128	3024	18	TR87/243	S3019
2SD121H	210	3024	18		
2SD121HA	210	3202			
2SD121HB	210	3202			
2SD122					S5014
2SD123					S3002
2SD124	130	3027	77	TR59/247	S7002
2SD124AH	130	3027	77	TR59/247	S7002
2SD125	163	3027	14	TR67/740	S7004
2SD125AH	163	3111	75	TR67/740	S7004
2SD126				TR61/	
2SD126H	163		36	TR67/	S5020
2SD126HA	163		36	TR67/	
2SD126HB	163		36	TR67/	
2SD127	103A	3010	54	TR21/724	G0011
2SD127A	103A	3010	54	TR08/724	G0011
2SD128	103A	3010	59	TR08/724	G0011
2SD128A	103A	3010	59	TR08/724	G0011
2SD129	175	3026		TR81/241	S5012
2SD129-BL	175			TR81/	
2SD129-R	175			TR81/	
2SD129-Y	175			TR81/	
2SD130	175	3026		TR81/241	S5019
2SD130BL	175	3026	23	TR81/241	
2SD130-R	175		216	TR81/	
2SD130-Y	175		216	TR81/	
2SD132	181			TR36/	S7004
2SD134	123	3124	20	TR86/53	
2SD136	124	3021	66	TR81/240	S5024
2SD137	124	3021	12	TR81/240	S5024
2SD141	152	3041	28	TR76/241	S5000
2SD141H01	175			TR57/241	
2SD141H9Z	175			TR81/241	
2SD142	175	3026	28	TR81/241	S5000
2SD142M	175	3024		TR23/	S5000
2SD143	175	3131	12	TR81/241	S5019
2SD144	175	3131		TR81/241	S5012
2SD145	175	3131		TR81/241	S5012
2SD146	130	3026	14	TR59/247	S5000
2SD146UK	175	3026	23	TR81/	
2SD147	130	3027	75	TR59/247	S5000
2SD150	175	3021	12	TR81/247	S5019
2SD150M		3021	12		S5019
2SD151	130	3026	75	TR59/247	S7004
2SD152	164	3026		TR93/740A	S5012
2SD152L		3026	23		S5012

Original Part Number	ECG	SK	GE	IR/WEP	HEP
2SD154	152	3021		TR76/247	S5019
2SD155	175	3131		TR81/241	S5019
2SD155H	175	3026	66	TR81/	S5019
2SD155K	175	3026	66	TR81/	S5019
2SD155L	175	3021	12	TR81/	S5019
2SD156	124	3021	12	TR81/240	S5011
2SD157	124	3021	12	TR81/240	S5011
2SD157B	124	3021	12	TR23/240	S5011
2SD158	124	3021	32	TR81/240	S5011
2SD159	124	3021	32	TR81/240	S5011
2SD161	101	3011	59	TR08/641	
2SD162	101	3010	5	TR08/641	G0011
2SD163	130	3027	77	TR59/247	S7002
2SD164	130	3027	14	TR59/247	S7004
2SD165	162		35	TR67/707	S5020
2SD166	162		35	TR67/707	S5020
2SD167	101	3010	59	TR08/641	G0011
2SD170	103A		54	TR08/	G0011
2SD170A	103A	3010	54	TR08/	G0011
2SD170AA	103A			TR08/	
2SD170APB		3011	59		G0011
2SD170B	103A	3010		TR08/	G0011
2SD170BC	103A	3010	54		G0011
2SD170C	103A		8	TR08/724	G0011
2SD171					S5020
2SD172		3027	75	TR61/247	S7002
2SD173	130	3027	75	TR59/247	S7004
2SD174	130	3027	73	TR59/247	S7002
2SD175	130	3027	73	TR59/247	S7004
2SD176	130	3027	75	TR59/247	S7004
2SD177	162			TR67/707	S7004
2SD178	103A	3124	59	TR08/724	G0011
2SD178A	103A	3124	59	TR08/724	G0011
2SD178Q	103A			TR08/724	
2SD178T	103A			TR08/724	
2SD180	223	3027	77	TR59/247	S7004
2SD180A	223	3027	75	TR59/	
2SD180B	223	3027	75	TR59/	
2SD180C	223	3027	75	TR59/	
2SD180D	223	3027	75	TR69/	
2SD180M	223	3027	75	TR26/	S7002
2SD181	162			TR67/	S5020
2SD182	128	3024	28	TR87/243	S5014
2SD183	128	3024	18	TR87/243	S3019
2SD184	152	3024	66	TR76/701	S5014
2SD185			66		S3019
2SD186	103A	3010	59	TR08/724	G0011
2SD186A	103A	3010	59	TR08/	
2SD186B	103A	3010	59	TR08/	
2SD187	103A	3010	59	TR08/641	G0011
2SD187A	103A	3010	59	TR08/641	
2SD188	130	3027	75	TR59/247	S7004
2SD188A	130	3027	75	TR59/	
2SD188B	130	3027	75	TR59/	
2SD188C	130	3027	75	TR59/	
2SD188M		3027	75		S5020
2SD189	130	3027	75	TR59/247	S7004
2SD189A	162	3027	75	TR67/707	S7004
2SD190	124	3021	12	TR81/240	
2SD191	103A		59	TR08/724	G0011
2SD192	103A		59	TR08/724	G0011
2SD193	103A		59	TR08/724	G0011
2SD194	103A		59	TR08/724	G0011
2SD195	103A	3010	5	TR08/724	G0011
2SD195A	101	3010	8	TR08/641	
2SD196					S5004
2SD196A					S5004
2SD198	162	3079	35	TR67/707	S5020
2SD198H	162	3111	36	TR67/740	S5020
2SD198HQ	162			TR61/	S5020
2SD198P	162	3115	37		S5020
2SD198Q	162	3133	37	TR61/	S5020
2SD198R	162	3079	35	TR61/	S5020
2SD198S	162	3036	14		S7002
2SD198V	162	3079	35	TR67/707	S5020
2SD199	162	3111	35	TR67/707	S5021
2SD200	165	3115	38	TR93/740B	
2SD200A	165	3115	36	TR93/740B	S5021
2SD201	130	3027	75	TR59/247	S7004
2SD201(O)	130	3027			
2SD201M	130	3027			S7004
2SD201O	130				
2SD201Y	130	3027	75		
2SD202	162		35	TR93/707	S7004
2SD203	162		35	TR67/707	S5020

Original Part Number	ECG	SK	GE	IR/WEP	HEP
2SD204	128	3024	47	TR87/243	S5014
2SD204L	128	3024	18	TR65/243	S5014
2SD205	225	3024	63	TR87/243	S5014
2SD205K					S5014
2SD211	130	3027	75	TR59/247	S7002
2SD212	130	3027	75	TR59/247	S7004
2SD213	162			TR59/	
2SD214					S5020
2SD215	103			TR08/247	
2SD217	162		14	TR67/707	S7004
2SD218	162		35	TR67/707	S5020
2SD219	128	3024	18	TR87/243	S5014
2SD220	128	3024	18	TR87/	S5014
2SD221	128			TR87/243	S3019
2SD222		3024	18		S5014
2SD223		3045	32		S5014
2SD224					S3019
2SD226	175	3026	66	TR92/241	S5019
2SD226-O	175			TR81/241	
2SD226A	175	3026	23	TR81/241	S5019
2SD226AP	175	3131		TR81/241	S5011
2SD226B	175	3026	66	TR81/	S5019
2SD226BP	175	3131		TR81/241	
2SD226O	175	3131			
2SD226P	175	3026	66	TR81/241	S5012
2SD226Q	175	3026	66	TR81/241	S5019
2SD227	123A	3054	64	TR24/735	S0015
2SD227(PAN)			63		
2SD227A	123A	3124	47	TR21/	
2SD227B	123A	3124	47	TR21/	
2SD227C	123A			TR21/	
2SD227D	123A	3124	47	TR21/	
2SD227E	123A	3124	47	TR21/	
2SD227F	123A	3124	47	TR21/	
2SD227L	123A	3124	47	TR21/735	
2SD227R	123A	3124	47	TR22/	S0011
2SD227V	123A				S0025
2SD228	192	3122	210	TR21/735	S0011
2SD234	152	3054	28	TR55/701	S5000
2SD234-O	152	3054	28	TR76/701	
2SD234-R	152		66	TR76/701	
2SD234-Y	152			TR76/701	
2SD235	152	3041		TR92/701	S5000
2SD235D	152	3041	55	TR92/	S5000
2SD235-O	152	3041	66	TR76/701	S5000
2SD235-R	152		66	TR76/701	
2SD235-Y	152			TR76/701	
2SD236	175	3026	23	TR94/241	S5019
2SD237					S5019
2SD238	175	3131	12	TR81/241	S5012
2SD238F	175	3131		TR81/	
2SD246	165	3115	38	TR93/740B	
2SD254	175	3026	23	TR57/	S5000
2SD255	175	3026	216	TR81/	S5019
2SD256			216		S5019
2SD257	175			TR81/	S5012
2SD258	124	3026	12		S5012
2SD259		3021	12		S5012
2SD261	192	3137	20	TR25/	S3024
2SD261A	192	3137	18		
2SD261B	192	3137	18		
2SD261C	192	3137	18		
2SD261D	192	3137	18		
2SD261E	192	3137	18		
2SD261F	192	3137	18		
2SD261L	128	3024	18		S0014
2SD261O	192	3024	18		S5014
2SD261Q	192				S5014
2SD261R	192	3137	18		S3014
2SD261(R)	192			TR25/	
2SD261(U)	192				
2SD261V	192	3137	63	TR25/	S5014
2SD261W	192	3137	18	TR87/	S0014
2SD283	162			TR67/	
2SD284	162		35	TR67/	
2SD285	162		35	TR67/	
2SD287					S5020
2SD288	152			TR76/	S5000
2SD288A	152			TR76/	
2SD288B	152			TR76/	
2SD288C	152			TR76/	
2SD288L		3054			
2SD289	152			TR76/	S5000
2SD289A	152			TR76/	
2SD289B	152			TR76/	

Original Part Number	ECG	SK	GE	IR/WEP	HEP
2SD289C	152			TR76/	
2SD290	175				S5019
2SD290L	175				
2SD291	175	3026		TR23/241	S5019
2SD291(R)	175		23	TR23/	
2SD292	175	3026	66	TR81/241	S5019
2SD297	175			TR81/	S5012
2SD299	165	3115	38	TR93/740B	S5020
2SD299V	165	3115	38		
2SD300	165	3115	38	TR93/	
2SD300B	165	3115	38	TR68/	
2SD312	165	3079	37	TR93/	S5021
2SD313	152	3054	28		S5000
2SD313D	196	3054	28		
2SD313(D,E)				TR76/	S5000
2SD314		3054	28		S5000
2SD314E	198	3054	28		
2SD315	175			TR81/241	S5019
2SD316					S5004
2SD317	152			TR76/	S5000
2SD317A	152	3054	28	TR76/	S5000
2SD317AF					S5000
2SD317AP					S5000
2SD317F	152	3054			
2SD317P	152	3054			
2SD318	152			TR76/	S5000
2SD318A	152			TR76/	S5000
2SD318B	152			TR76/	S5000
2SD318P	152				S5000
2SD318Q		3054	57		S5000
2SD319	181		75	TR36/	S7000
2SD320	162		73		S5021
2SD321		3115	38	TR67/	S5020
2SD322	162		35	TR67/	S7004
2SD322A	162		35	TR67/	
2SD322B	162		35	TR67/	
2SD322C	162		35	TR67/	
2SD323	162		36	TR67/	S5020
2SD323A	162		35	TR67/	
2SD323B	162		35	TR67/	
2SD323C	162		35	TR67/	
2SD324	124			TR81/	S5015
2SD325	152	3054	28	TR55/	S5000
2SD325C	152	3054	28	TR55/	S5000
2SD325D	152	3054	28	TR76/701	S5000
2SD325E	152	3054	28	TR55/	S5000
2SD326	124	3021	12	TR23/	
2SD327	123A	3024	20	TR21/	S0015
2SD327A	123A	3122	88	TR21/	
2SD327B	123A	3122	88	TR21/	
2SD327C	123A	3122	88	TR21/	
2SD327D	123A	3122	88	TR21/	
2SD327E	123A	3122	88	TR21/	
2SD327F	123A			TR21/	
2SD328					S3019
2SD328S	210				
2SD330					S3020
2SD330D	152				
2SD331					S3020
2SD334	162			TR67/	S7004
2SD334A	162			TR67/	S5020
2SD335			75		S5000
2SD336	192	3122	63		S3020
2SD336R	192				
2SD336Y	192				
2SD338			75		
2SD339			75		
2SD341H			14		
2SD343	152	3054	28	TR76/	
2SD348	165	3115	38	TR93/	S5021
2SD350	238	3111	36	TR61/	S5020
2SD352	103A	3010	59	TR08/	
2SD353	164	3133	37	TR68/	S5020
2SD356	190				
2SD359		3041			
2SD360		3041			
2SD362					S5012
2SD367		3124	47		
2SD370					S7004
2SD371		3027			S7004
2SD389	152	3054			
2SD389LP	152	3054			
2SD389(LP)	152	3054			
2SD389(L)	152	3054			
2SD389(O,LP,P)			66	TR76/	

Original Part Number	ECG	SK	GE	IR/WEP	HEP
2SD389P	152	3054			S3020
2SD389(P)	152	3054			
2SD390	152	3054	28	TR55/	S5000
2SD390(O)	152			TR55/	
2SD390P	152				
2SD392			210		
2SD400			47		
2SD400D					S3020
2SD425		3535			
2SD426		3535			
2SD427		3535			
2SD428		3027			
2SD435				TR56/	
2SD441				TR57/	
2SD461				TR25/	
2SD511					S3028
2SD562	108	3019	11	TR95/	
2SD720D	103A			TR08/724	
2SD720E	103A			TR08/724	
2SD720PJ		3010	59		
2SD735	103		8		
2SD735Y				TR92/	
2SD829					S0011
2SD1046N					S5021
2SD1283		3010	59		
2SD1318	128				
2SD1347		3124	47		
2SD2340			66		
2SD2350			66		
2SD3601	235	3197			
2SD3890		3054			
2SD-180M				TR26/	S7002
2SDF1	101	3010	5	TR08/641	G0011
2SDF2					S7002
2SD105K		3024			
2SDL2F		3010	59		
2SDU9OR		3010			
2SDU36C		3021	12		
2SDU37E		3021	12		
2SDU47X		3027	75		
2SDU57M		3021	12		
2SDU78OR		3124	47		
2SDU84GN		3024	18		
2SDU9OR			59		
2SE535				TR21/	
2SE629	108	3018	61	TR95/56	
2SE4002	199	3011	59	TR08/641	G0011
2SF248	152			TR76/	
2SF1188	230	3042	700		
2SF1188A	230	3042	700		
2SF1188B	230	3042	700		
2SF1188C	230	3042	700		
2SF1188D	230	3042	700		
2SF1188E	230	3042	700		
2SF1188F	230	3042	700		
2SF1188G	230	3042	700		
2SF1188H	230	3042	700		
2SF1188K	230	3042	700		
2SF1188L	230	3042	700		
2SF1188M	230	3042	700		
2SF1188N	230	3042	700		
2SF1188R	230	3042	700		
2SF1188Y	230	3042	700		
2SF1189	231	3042	700		
2SF1189A	231	3042	700		
2SF1189C	231	3042	700		
2SF1189D	231	3042	700		
2SF1189E	231	3042	700		
2SF1189F	231	3042	700		
2SF1189G	231	3042	700		
2SF1189H	231	3042	700		
2SF1189K	231	3042	700		
2SF1189L	231	3042	700		
2SF1189M	231	3042	700		
2SF1189N	231	3042	700		
2SF1189R	231	3042	700		
2SF1189Y	231	3042	700		
2SFT212	121	3009	25	TR01/232	
2SG536GN	123A	3124	47	TR21/	
2SH11	6401			TR2160/	
2SH12	6400				
2SH18	6401			TR2160/	
2SH18K	6401			TR2160/	
2SH18L	6401			TR2160/	
7SH18M	6401			TR2160/	

Original Part Number	ECG	SK	GE	IR/WEP	HEP
2SH18N	6401			TR2160/	
2SH19	6401			TR2160/	
2SH19K	6401			TR2160/	
2SH19L	6401			TR2160/	
2SH19M	6401			TR2160/	
2SH19N	6401			TR2160/	
2SH20	6401			TR2160/	S9002
2SH21	6400		160	TR2160/	
2SH22	6409			TR2160/	S9002
2SH203	100	3005		TR05/254	
2SH643	193		67		
2SJ11	133	3112	FET1		F1036
2SJ12	133	3112	FET1		F1036
2SJ15					F1035
2SJ16					F1035
2SK11	132	3116	FET2	FE100/802	
2SK12	132	3116	FET2	FE100/802	
2SK13	132	3116	FET2	FE100/802	
2SK15		3112			
2SK17	133	3112	FET2	FE100/802	F0010
2SK17-O	132	3112	FET1	FE100/802	
2SK17A	132	3112	FET1	FE100/	
2SK17B	132	3112	FET1	FE100/	
2SK17BL	132	3112	FET1	FE100/	
2SK17GR	132	3112	FET1	FE100/	
2SK17R	132	3112	FET1	FE100/	
2SK17Y	132	3112	FET1	FE100/	
2SK19	132	3116	FET1	FE100/802	F0015
2SK19BL	132	3116	FET2	FE100/802	F0015
2SK19GE	132			FE100/802	
2SK19GR	132	3116	FET2	FE100/802	F0021
2SK19Y	132	3116	FET2	FE100/802	F0021
2SK22					F0015
2SK22Y	132			FE100/802	
2SK23	132	3116	FET2	FE100/802	F0015
2SK23A	132	3116			
2SK24	133	3050	FET4	FE100/801	F0010
2SK24C	133	3116			F0021
2SK24D	133				F0021
2SK24IR	133	3112	FET2	FE100/801	
2SK24E	133	3112	FET1	FE100/802	F0021
2SK24F	133	3116			F0021
2SK24G	133	3116			F0021
2SK25	132	3116	FET2	FE100/802	F0015
2SK25C	132	3116		FE100/	
2SK25D	132	3116		FE100/	
2SK25E	132	3116		FE100/	
2SK25ET	132	3116	FET2	FE100/802	
2SK25F	132	3116		FE100/	
2SK25G	132	3116		FE100/	
2SK30	132	3112	FET1	FE100/801	F0010
2SK30D	133	3112	FET1	FE100/	F0010
2SK30-GR	133				
2SK30GR	133	3112	FET1	FE100/802	F0010
2SK30-O	133	3112	FET1	FE100/801	F0010
2SK30-R	133	3112		/801	
2SK30-Y	133				
2SK30Y	132	3112	FET2	FE100/802	F0010
2SK31	172				
2SK31C	172		FET2		
2SK32B	132	3112	FET1	FE100/	
2SK33	132	3050	FET1	FE100/802	F0015
2SK33E	132	3116		FE100/802	
2SK33F	132	3116	FET2	FE100/	
2SK34	132	3050	FET2	FE100/802	F0015
2SK34B	132	3112	FET1	FE100/	F0021
2SK34C	132	3112	FET2	FE100/	F0021
2SK34D	132	3116	FET2	FE100/	F0021
2SK34E	132	3112			F0021
2SK35	133	3112	FET1	/801	F0015
2SK35-O	133	3112	FET1		
2SK35-1	133	3112	FET1		
2SK35-2	133	3112	FET1		
2SK35A	133	3112	FET1		
2SK35BL	133	3112	FET1		
2SK3[illegible]	133	3112	FET1		
2SK35GN	133	3112	FET1		F0015
2SK35R	133	3112	FET1		
2SK35Y	133	3112	FET1		
2SK37	132	3116	FET2	FE100/	F0010
2SK37H	132	3116	FET2	FE100/	F0021
2SK39		3050	FET4		
2SK39B				FE100/	F2004
2SK40	133	3112	FET2	FE100/	F0010
2SK40A	133	3112			

Original Part Number	ECG	SK	GE	IR/WEP	HEP
2SK40B	133	3112			
2SK40C	133	3112		FE100/	F0010
2SK40D	133	3112			
2SK41	132	3116		FE100/802	F0015
2SK41E	132	3116	FET2	FE100/	F0021
2SK41F	132				F0021
2SK42	132				
2SK42-CM1	132	3116	FET2	FE100/	
2SK43		3112	FET1	FE100/	
2SK44	133				F0015
2SK44D	133	3112	FET1	FE100/	F0015
2SK47	132			FE100/	
2SK48		3112			
2SK49	132	3116	FET2		
2SK170	132	3112	FET1	FE100/	F0010
2SK170R	132	3116		FE100/	
2SK196R		3116			
2SK304	132	3116		FE100/	F0021
2SK339H				TR24/	S0015
2SK-19				FE100/	
2SL668(L)				TR21/	
2SM610B	159	3025		TR20/717	
2SO12			66		
2SO13					S5000
2SO33			19		
2SO34			19		
2SO35			19		
2SO36			19		
2SO43					S5012
2SO44					S5012
2SO45					S5012
2SO46					S5012
2SP405				TR05/	
2SQ371	108	3018	20	TR95/56	
2SR24			FET1		
2SR54				TR05/	
2SR173				TR05/	
2SS5184G		3124			
2SS772				TR21/	
2SV341V	127	3035	25	TR27/235	
2SX193Y		3122			
2T2P			2		
2T3	102A	3004	53	TR85/250	
2T11	158	3004	52	TR84/630	G0005
2T12	158	3004	52	TR84/630	G0005
2T13	158	3004	52	TR84/630	G0005
2T14	158	3004	4	TR84/630	G0005
2T14A	126		52	TR17/635	
2T15	158	3004	52	TR84/630	G0005
2T16	158	3004	52	TR84/630	G0005
2T17	158	3004	52	TR84/630	G0005
2T20	160	3004	2	TR17/637	G0002
2T21	158	3004	52	TR84/630	G0002
2T22	158	3004	52	TR84/630	G0005
2T23	158	3004	52	TR84/630	G0005
2T24	158	3004	52	TR84/630	G0005
2T25	158		52	TR84/630	G0005
2T26	158	3004	52	TR84/630	G0005
2T40	123A	3122	17	TR21/735	S0011
2T41	123A	3122	17	TR21/735	S0011
2T42	123A	3122	17	TR21/735	S0011
2T43	123A	3122	17	TR21/735	S0011
2T44	123A	3122	17	TR21/735	S0011
2T51	103	3010	8	TR08/641A	G0011
2T52	101	3011	8	TR08/641	G0011
2T53	101	3011	8	TR08/641	G0011
2T54	101	3011	54	TR08/641	G0011
2T55	101	3011	54	TR08/641	G0011
2T56	101	3011	54	TR08/641	G0011
2T57	101	3011	54	TR08/641	G0011
2T58	101	3011	54	TR08/641	G0011
2T61	103	3010	8	TR08/641A	
2T62	103	3010	8	TR08/641A	
2T63	103	3010	8	TR08/641A	G0011
2T64	103	3010	5	TR08/641A	G0011
2T64R	103	3010	8	TR08/641A	
2T65	103	3010	5	TR08/641A	G0011
2T65R	103	3010	8	TR08/641A	
2T66	103	3010	5	TR08/641A	G0011
2T66R	103	3010	8	TR08/641	
2T67	101	3011	8	TR08/641	G0011
2T69	103	3010	5	TR08/641A	G0011
2T71	101	3011	8	TR08/641	G0011
2T72	101	3011	8	TR08/641	G0011
2T73	101	3011	7	TR08/641	G0011

Original Part Number	ECG	SK	GE	IR/WEP	HEP
2T73R	101	3011	8	TR08/641	
2T74	101	3011	8	TR08/641	
2T75	101	3011	7	TR08/641	G0011
2T75R	101	3011	8	TR08/641	
2T76	101	3011	7	TR08/641	G0011
2T76R	101	3011	8	TR08/641	
2T77	101	3011	7	TR08/641	G0011
2T77R	101	3011	8	TR08/641	
2T78	101	3011	7	TR08/641	G0011
2T78R	101	3011	8	TR08/641	
2T82	101	3011	8	TR08/641	
2T83	101	3011	53	TR08/641	G0005
2T84	103	3010	8	TR08/641A	
2T85	103	3010	8	TR08/641A	G0011
2T85A	103	3010	8	TR08/641A	
2T86	103	3010	8	TR08/641A	
2T89	103	3010	8	TR08/641A	
2T172	123A	3122	8	TR21/735	
2T201	160	3006	51	TR17/637	G0003
2T202	123A	3122	20	TR21/735	S0014
2T203	160	3006	51	TR17/637	G0003
2T204	160	3007	51	TR17/637	G0003
2T204A	160	3007	50	TR17/637	
2T205	160	3007	51	TR17/637	G0008
2T205A	160	3007	51	TR17/637	
2T230	102	3004		TR05/631	G0005
2T231	102	3004	51	TR05/631	G0005
2T311	158	3004	52	TR84/630	G0005
2T312	158	3004	52	TR84/630	G0005
2T313	158	3004	52	TR84/630	G0005
2T314	158	3004	52	TR84/630	G0005
2T315	158	3004	52	TR84/630	G0005
2T321	158	3123	53	TR84/630	G0005
2T322	158	3004	53	TR84/630	G0005
2T323	158	3004	53	TR84/630	G0005
2T324	158	3004	53	TR84/630	G0005
2T383	158	3123	52	TR84/630	
2T402	123	3124	20	TR86/53	S0014
2T403	123A	3122	17	TR21/735	S0011
2T404	123A	3122	17	TR21/735	S0011
2T513	101	3011	8	TR08/641	
2T520	101	3011	8	TR08/641	
2T521	101	3011	8	TR08/641	
2T522	103	3010	8	TR08/641A	
2T523	103	3010	8	TR08/641A	
2T524	101	3011	8	TR08/641	
2T551	101	3011	8	TR08/641	
2T552	103	3010	8	TR08/641A	
2T650	101	3011	8	TR08/641	
2T681	103	3010	59	TR08/641A	G0011
2T682	103	3010	59	TR08/641A	G0011
2T918	123A	3122		TR21/735	S0015
2T919	107	3039	17	TR70/720	
2T2001	102	3004	52	TR05/631	
2T2102			63	TR25/	
2T2708	123A	3122	17	TR21/735	S0015
2T2785	123A	3122		TR21/735	
2T2857	123A	3122	17	TR21/735	S0015
2T3011	121	3009	76	TR01/232	G6003
2T3021	121	3009	76	TR01/232	G6005
2T3022	121	3009	16	TR01/232	G6005
2T3030	121	3009	25	TR01/232	G6003
2T3031	121	3009	49	TR01/232	G6005
2T3032	121	3009	49	TR01/232	G6005
2T3033	121	3009	49	TR01/232	G6005
2T3041	121	3009	49	TR01/232	G6005
2T3042	121	3009	49	TR01/232	G6005
2T3043	121	3009	49	TR01/232	G6005
2TN15	100	3005	1	TR05/254	
2TN32	100	3005	1	TR05/254	
2TN45A	102	3004	2	TR05/631	
2TN48	100	3005	1	TR05/254	
2TN49	100	3005	1	TR05/254	
2TN52	100	3005	1	TR05/254	
2TN53	100	3005	1	TR05/254	
2TN56	102	3004		TR05/631	
2TN95	102	3004	1	TR05/631	
2TN95A	102	3004	1	TR05/631	
2V362	102	3004		TR05/631	G0005
2V363	102	3004		TR05/631	G0005
2V464	100	3005	52	TR05/254	G0005
2V465	100	3005	52	TR05/254	G0005
2V466	100	3005	52	TR05/254	G0005
2V467	100	3005	52	TR05/254	G0005
2V482	100	3005	52	TR05/254	G0005

Original Part Number	ECG	SK	GE	IR/WEP	HEP
2V483	100	3005	52	TR05/254	G0005
2V484	100	3005	52	TR05/254	G0005
2V485	160		52	TR17/637	G0008
2V486	100	3005	52	TR05/254	G0005
2V559	160		52	TR17/637	G0008
2V560	160		52	TR17/637	G0008
2V561	160		52	TR17/637	G0008
2V563	160		52	TR17/637	G0008
2V631	100	3005	2	TR05/254	G0005
2V632	100	3005	2	TR05/254	G0005
2V633	100	3005	2	TR05/254	G0005
2.01.03.02	159				
3-0033	123A				
3-041	129				
3-7	129				
3-19	124	3021			
3-20	127				
3-215	129				
3-233	197				
3-30173	181				
3-12104733		3006			
3-121004900		3006			
3-121007744		3006			
3-121024033		3006			
3-121024044		3006			
3-121047111		3006			
3-121050611		3006			
3-121050622		3006			
3-122005400		3004			
3-539306001		3124			
3-539306002		3124			
3-539306003		3124			
3B15	102	3004	2	TR05/631	
3B15-1	102	3004	2	TR05/631	
3C38	106				
3G2	132	3112		FE100/	
3L4-6004	175			TR81/241	S5000
3L4-6001-01	130	3027		TR59/247	S7002
3L4-6005-1	152	3041	66	/S5003	S5000
3L4-6005-2	184	3190		/S5003	
3L4-6005-3	184	3190		TR76/S5003	S5000
3L4-6005-5	152	3041			
3L4-6005-55	152	3041			
3L4-6006-1	124	3021	12	TR23/	S5012
3L4-6007-02	123A				
3L4-6007-03	123A				
3L4-6007-04	123A				
3L4-6007-08	123A				
3L4-6007-09	123A				
3L4-6007-1	123A	3018	12	TR21/735	S0015
3L4-6007-2	123A	3018	20	TR21/735	S0015
3L4-6007-3	123A	3018	20	TR21/735	S0015
3L4-6007-4	128	3024	63	TR87/	S5014
3L4-6007-10	161	3018	60	TR83/	S0020
3L4-6007-11	161	3018	60	TR83/	S0020
3L4-6007-12	161	3018	60	TR83/	S0020
3L4-6007-13	161	3018	60	TR83/	S0020
3L4-6007-14	161	3018	62	TR83/	S0020
3L4-6007-15	199				
3L4-6007-16	199				
3L4-6007-19	161	3018	62	TR83/	S0020
3L4-6007-20	161	3018	60	TR71/	S0020
3L4-6007-21	107	3018	60	TR71/	S0020
3L4-6007-22	161	3018	60	TR71/	S0020
3L4-6007-23	107	3018	60	TR71/	S0020
3L4-6007-35	107	3018	60	TR71/	S0020
3L4-6007-37	123A				
3L4-6007-38	123A				
3L4-6007-51	108	3018			
3L4-6010-03	123A				
3L4-6010-4	129	3025	67	TR88/	S5013
3L4-6010-6	123A	3024	18	TR24/	S0014
3L4-6010-8	159	3025			
3L4-6011-02	129				
3L4-6011-1	196				S3028
3L4-6011-2	211	3084	29	TR56/	S3028
3L4-6011-3	211	3084	29	TR56/244A	S3028
3L4-6011-52	211	3084			
3L4-6011-53	211	3084			
3L4-6012-02	152				
3L4-6012-06	196				
3L4-6012-2	152		66	TR76/	S5000
3L4-6012-3	152	3041		TR76/701	S5000
3L4-6012-4	182	3041	66	TR55/	S5000
3L4-6012-5	196	3041	66	TR55/	S5000
3L4-6012-6	196	3041	66	TR76/	S5000
3L4-6012-8	196	3041	66	TR55/	S5000
3L4-6012-55	196	3041			
3L4-6012-56	196	3054			
3L4-6012-58	196	3041			
3L4-6013-02	153				
3L4-6013-2	153	3084	69	TR77/	S5006
3L4-6013-3	153	3084		TR77/700	S5006
3L4-6013-4	183	3084	69	TR56/	S5006
3L4-6013-5	153	3084	69	TR56/	S5006
3L4-6013-6	197	3084	69	TR77/	S5006
3L4-6013-8	197	3084	69	TR56/	S5006
3L4-6013-55	153	3084			
3L4-6013-56	153	3083			
3L4-6013-58	153	3084			
3L4-6015-01	123A				
3L4-6020-01	238				
3L4-6021-01	247				
3L4-9002-01	786	3140			
3L4-9002-1	786	3140			
3L4-9004-01	743				
3L4-9004-51	743				
3L4-9006-01	726				
3L4-9006-1	726				
3L4-9006-51	726				
3L4-9007-01	737	3144			
3L4-9007-1	737	3144	20		
3L4-9007-51	737	3144	16		
3L4-9008-01	789				
3L4-9008-51	789				
3L46007-1			20		
3L46007-2			20		
3L49007-51	737				
3MC	160	3007	2	TR17/	G0008
3N21		3123	52	TR05/	
3N22	101	3011	54	TR08/641	
3N23	101	3011	54	TR08/641	
3N23A	101	3011		TR08/641	
3N23B	101	3011		TR08/641	
3N23C	101	3011		TR08/641	
3N25/501			51	TR11/	
3N29	101	3011	54	TR08/641	
3N30	101	3011	54	TR08/641	
3N31	101	3011	54	TR08/641	
3N34	160		60	TR17/637	S0011
3N35	123A	3122	60	TR21/735	S0011
3N35A	160		11	TR17/637	S0011
3N36	101	3011	54	TR08/641	G0011
3N37	101	3011	54	TR08/641	
3N49	105	3012	4	TR03/233	
3N50	105	3012	4	TR03/233	
3N51	105	3012	4	TR03/233	
3N52	105	3012	4	TR03/233	
3N56					G0011
3N57					G0011
3N71	107	3039	214	TR70/720	S0011
3N72	107	3039	214	TR70/720	S0011
3N73	107	3039	214	TR70/720	S0011
3N74			61		
3N75			61		
3N76			61		
3N77			61		
3N78			61		
3N79			61		
3N87			11		S0015
3N88			11		S0015
3N90	159	3114	65	TR20/717	
3N91	159	3114	65	TR20/717	
3N92	159	3114	65	TR20/717	
3N93	159	3114	65	TR20/717	
3N94	159	3114	65	TR20/717	
3N95	159	3114	65	TR20/717	
3N112	106		21	TR20/52	
3N113	106		21	TR20/52	
3N114			65		
3N115			65		
3N116			65		
3N117			65		
3N118			65		
3N119			65		
3N120			11		
3N121			11		
3N123			65		
3N127			86		
3S004		3019	17	TR21/735	S0014

Original Part Number	ECG	SK	GE	IR/WEP	HEP
3SA324	160		51	TR17/	
3SB347	102A			TR85/	G0006
3SB-B732				/158	
3SK22	132	3116	FET2	FE100/802	F2005
3SK23	132	3116	FET2	FE100/802	F2005
3SK30	132	3050	FET3		F2005
3SK30A	132		FET2	FE100/	F2005
3SK30AB					F2004
3SK30B		3116			F0021
3SK34	133	3112			
3T201	101	3011		TR08/641	
3T202	101	3011		TR08/641	
3T203	101	3011		TR08/641	
3TE110					S5004
3TE120	130	3036		TR59/247	S7002
3TE140	130	3027	19	TR59/	
3TE150			63	TR25/	S3011
3TE160			63	TR25/	S3011
3TE230	130	3027		TR59/247	
3TE240	130	3027	19	TR59/247	
3TE250			63	TR25/	S3011
3TE260			63	TR25/	S3011
3TX002					S7004
3TX003	130	3027	14	TR59/247	S7004
3TX004	130	3027	19	TR59/247	S7002
4-00535-06		3018			S0015
4-005	725				
4-006	725				C6058P
4-007	703A	3157			
4-008	703A	3157			
4-009	720		IC-12		C6095P
4-003721-00	175				
4-082-664-0001	704	3023			
4-0294	130				
4-0485	108				
4-0498	128				
4-0563	130				
4-05737		3039			
4-08018-667	9910				
4-4A-1A7-1	121	3009		TR01/	
4-6B-3A7-1	121MP	3015			
4-8P-2A7-1	102A	3004		TR85/	
4-9L-4A7-1	102A	3003		TR85/	
4-12-1A7-1	123A	3124		TR21/	
4-14A17-1	121	3009		TR01/	
4-44-0012-PT2	160				
4-46	199			TR24/735	S0015
4-47(SEARS)	123A	3122	20	TR21/735	
4-47		3122	20	TR24/	S0016
4-48(SEARS)				TR21/735	
4-48	226MP	3052	31MP	TR50/	G6016
4-65-1A7-1	160	3006		TR17/	
4-65-2A7-1	103A	3010		TR08/	
4-65-4A7-1	101	3011		TR08/	
4-65A17-1	160	3007		TR01/	
4-65B17-1	160	3008			
4-65C17-1	121	3009			
4-66-1A7-1	102A	3003		TR85/	
4-66-2A7-1	102A	3004		TR85/	
4-66-3A7-1	102A	3004		TR05/	
4-74-3A7-1	100	3005		TR05/	
4-77A17-1	105	3012		TR03/	
4-77B17-1	121MP	3013			
4-77C17-1	121	3014		TR01/	
4-88A17-1	121	3009		TR01/	
4-88B17-1	160	3006		TR17/	
4-88C17-1	100	3005		TR05/	
4-92-1A7-1	105	3012		TR03/	
4-142-1(SEARS)	176			TR82/	
4-142-1				/238	
4-142	131	3052	44	TR94/	G6016
4-245	181				
4-265	175				
4-274	128				
4-279	126	3008	50	TR17/	G0008
4-280	126	3008	50	TR17/	G0009
4-283	185				S5000
4-288	128				
4-324	133				
4-397	128	3024		TR87/	S5014
4-398	128	3024		TR87/	S0015
4-399	108	3039		TR95/	S0016
4-400	108	3039		TR95/	S0016
4-432	160	3006		TR17/	G0003
4-433	108	3018	20	TR95/	S0016

Original Part Number	ECG	SK	GE	IR/WEP	HEP
4-434	108	3018	20	TR95/	S0016
4-435	104	3009	25	TR01/	G6003
4-438	124				
4-443	108	3018	20		S0016
4-458	181				
4-464	152				
4-490	181				
4-850	107	3018	60		S0016
4-851	123A	3124	62		S0016
4-1544	199	3124	18		S0015
4-1545	123A	3124	18		S0015
4-1546	176	3123	18		G6011
4-1790	123A	3124			S0015
4-1791	199		18		S0015
4-1792	192	3137			
4-1848	131MP	3086	31MP		G6016
4-2060-02300	1004	3102			
4-2060-02400	1004	3102			
4-2060-02600	1080				
4-2060-03900	1080				
4-2060-04000	712	3072			
4-2060-04200	1094				
4-2060-04300	1121				
4-2060-04600	712				
4-2060-04800	1122				
4-2060-04900	1122				
4-2060-07200	1094				
4-2060-07300	1121				
4-2060-2300	1004				
4-2060-2400	1004				
4-2073	160	3008	50		G0008
4-5145	123A				
4-68681-2	102A	3004		TR85/	
4-68681-3	102A	3004		TR21/	
4-68682-3	123A	3124		TR21/	
4-68695-3	108	3018		TR95/	
4-684120-3	108	3018		TR95/	
4-685285-3	108	3019		TR95/	
4-686107-3	108	3018		TR95/	
4-686108-3	108	3018		TR95/	
4-686112-3	108	3018		TR95/	
4-686114-3	108	3018		TR95/	
4-686118-3	108	3018		TR95/	
4-686119-3	108	3018		TR95/	
4-686120-3	108	3018		TR95/	
4-686126-3	108	3018		TR95/	
4-686127-3	108	3018		TR95/	
4-686130-3	175	3026		TR81/	
4-686131-3	108	3018		TR95/	
4-686132-3	123A	3020		TR21/	
4-686140-3	108	3018		TR95/	
4-686143-3	123A	3124		TR21/	
4-686144-3	123A	3124		TR21/	
4-686145-3	154	3040		TR78/	
4-686163-3	102A	3004		TR85/	
4-686169-3	108	3018		TR95/	
4-686170-3	129	3025		TR88/	
4-686171-3	108	3018		TR95/	
4-686172-3	108	3018		TR95/	
4-686173-3	123A	3124		TR21/	
4-686182-3	128	3024		TR87/	
4-686183-3	123A	3124		TR21/	
4-686195-3	102A	3004		TR85/	
4-686196-3	102A	3004		TR85/	
4-686207-3	108	3018		TR95/	
4-686208-3	108	3018		TR95/	
4-686209-3	108	3018		TR95/	
4-686213-3	121	3009		TR01/	
4-686224-3	108	3018		TR95/	
4-686226-3	175	3026		TR81/	
4-686228-3	108	3018		TR95/	
4-686229-3	129	3025		TR88/	
4-686230-3	129	3025		TR88/	
4-686231-3	123A	3124		TR21/	
4-686232-3	154	3040		TR78/	
4-686234-3	196	3041		TR92/	
4-686235-3	129	3025		TR88/	
4-686238-3	129	3025		TR88/	
4-686244-3	108	3018		TR95/	
4-686251-3	108	3018		TR95/	
4-686252-3	191	3044		TR75/	
4-686256-1	126	3008		TR17/	
4-686256-2	126	3008		TR17/	
4-686256-3	126	3008		TR17/	
4-686257-3	123A	3124		TR21/	

Original Part Number	ECG	SK	GE	IR/WEP	HEP
4-3022861	108	3018		TR95/	
4-3023190	181	3036		TR36/	
4-3023212	123A	3124		TR21/	
4-3023221	123A	3124		TR21/	
4-3023222	129	3025		TR88/	
4-3023843	130	3027		TR59/	
4-3023844	128	3024		TR87/	
4-3025763	108	3018		TR95/	
4-3025764	108	3018		TR95/	
4-3025765	108	3018		TR95/	
4-3025766	123A	3124		TR21/	
4-3025767	108	3018		TR95/	
4-8134842	191	3044		TR75/	
4-30203845	129	3025		TR88/	
4-202104770	158			TR84/	
4A-1	121	3009		TR01/	
4A-1-70	121	3009		TR01/	
4A-1-70-12	121	3009		TR01/	
4A-1-70-12-7	121	3009		TR01/	
4A-1A	121	3009		TR01/	
4A-1A0	121	3009		TR01/	
4A-1A0R	121	3009		TR01/	
4A-1A1	121	3009		TR01/	
4A-1A2	121	3009		TR01/	
4A-1A3	121	3009		TR01/	
4A-1A3P	121	3009		TR01/	
4A-1A4	121	3009		TR01/	
4A-1A4-7	121	3009		TR01/	
4A-1A5	121	3009		TR01/	
4A-1A5L	121	3009		TR01/	
4A-1A6	121	3009		TR01/	
4A-1A6-4	121	3009		TR01/	
4A-1A7	121	3009		TR01/	
4A-1A7-1	121	3009		TR01/	
4A-1-7B	121	3009		TR01/	
4A-1A8	121	3009		TR01/	
4A-1A9	121	3009		TR01/	
4A-1A9G	121	3009		TR01/	
4A-1A19	121	3009		TR01/	
4A-1A21	121	3009		TR01/	
4A-1A82	121	3009		TR01/	
4A8-1A5	160	3007			
4A8-1A7-1	160	3007			
4A-10	121	3009		TR01/	
4A-11	121	3009		TR01/	
4A-12	121	3009		TR01/	
4A-13	121	3009		TR01/	
4A-14	121	3009		TR01/	
4A-15	121	3009		TR01/	
4A-16	121	3009		TR01/	
4A-17	121	3009		TR01/	
4A-18	121	3009		TR01/	
4A-19	121	3009		TR01/	
4C28	123A	3124	60	TR21/735	S0014
4C29	123A	3124	20	TR21/735	S0014
4C30	123A	3124	60	TR21/735	S0014
4C31	123A	3124	60	TR21/735	S0014
4D20	123A	3122	60	TR21/735	S0014
4D21	123A	3122	60	TR21/735	S0014
4D22	123A	3122	46	TR21/735	S0014
4D24	123A	3124	60	TR21/735	S0014
4D25	123A	3124	60	TR21/735	S0014
4D26	123A	3124	81	TR21/735	S0014
4JD1A17	100	3005	2	TR05/254	G0005
4JD1A73	102	3004	52	TR05/631	G0002
4JD3B1	101	3011		TR08/631	G0011
4JD5E29	6400				
4JX1A520	102	3004	2	TR05/631	
4JX1A520B	102	3004	52	TR05/631	G0005
4JX1A520C	102	3004	52	TR05/631	G0005
4JX1A520D	100	3005	52	TR05/254	G0005
4JX1A520E	100	3005	52	TR05/254	G0005
4JX1A813	160	3011	2	TR17/637	G0002
4JX1C707	160			TR17/	
4JX1C850	102	3004	52	TR05/631	G0005
4JX1C850A	100	3005	52	TR05/254	G0005
4JX1C1224	102		2	TR05/	
4JX1D925	102A			TR85/	
4JX1E596	123A			TR21/	
4JX1E821	100	3004	52	TR05/254	G0005
4JX1E850	103	3010	52	TR08/641A	G0005
4JX2A60	100	3005	52	TR05/254	G0005
4JX2A601	103	3010	8	TR08/641A	
4JX2A616	103A			TR08/	
4JX2A801	101	3011	8	TR08/641	G0011

Original Part Number	ECG	SK	GE	IR/WEP	HEP
4JX2A816	103	3010	8	TR08/641A	
4JX2A822	103	3010	8	TR08/641A	
4JX5E670	6400				
4JX7A972	123A		20	TR21/	
4JX8D404	121	3009	16	TR01/232	G6005
4JX8P404	121	3009	16	TR01/232	
4JX8P409	121	3009		TR01/232	
4JX11C2848	128			TR87/	
4JX16A567	123A	3019	17	TR21/735	
4JX16A667	123A		11	TR21/735	
4JX16A667G	123A	3122	17	TR21/735	S0015
4JX16A667O	123A	3122	17	/735	S0011
4JX16A667R	123A	3122	17	TR21/735	S0011
4JX16A667Y	123A	3122	17	TR21/735	S0015
4JX16A668	123A		11	TR21/	
4JX16A668G	123A	3122	17	TR21/735	S0015
4JX16A668O	123A	3122	17	TR21/735	S0015
4JX16A668Y	123A	3122	17	TR21/735	S0015
4JX16A669	123A		11	TR21/	
4JX16A669G	123A	3122		TR21/735	S0015
4JX16A669Y	123A	3122		TR21/735	S0015
4JX16A670	123A		20	TR21/	
4JX16A670G	123A	3122		TR21/735	S0015
4JX16A670B	123A	3122	17	TR21/735	
4JX16A670R	123A	3122	17	TR21/735	S0011
4JX16A670Y	123A	3122	17	TR21/735	S0011
4JX16E3860	123A			TR21/	
4JX16E3890	123A			TR21/	
4JX16E3960	123A			TR21/	
4JX29A826	159			TR20/	
4JX29A829	159			TR20/	
4JX2816	103	3010		TR08/641A	
4JX2825	103	3010		TR08/641A	
5-8	123A	3122		TR21/735	
5-70004503	123A	3020		TR21/	
5-70005452	123A	3124		TR21/	
5-70005503	123A	3124		TR21/	
5-7000901504	123A			TR21/	
5AA5-1	105	3012		TR03/	
6-0000139	128				
6-0000155A	102				
6-0000158	121				
6-000105	102				
6-000140	175				
6-000555-2	175				
6-04	123A	3124	20	TR21/735	S0005
6-04GRN	123A			TR21/735	S0024
6-04ORN	123A			TR21/735	S0015
6-04S1	123A	3038		TR21/735	S0015
6-04S2	123A	3124		TR21/735	S0015
6-05	123A	3124	20	TR21/735	S0015
6-05F	123A	3122		TR21/735	
6-05YEL	123A			TR21/735	S0015
6-0451	123A				
6-04552	123A				
6-4	123A	3122			
6-11	123A	3122	18	TR21/735	S0015
6-13	102A	3004	53	TR85/250	G0005
6-19	123A	3122	20	TR21/735	S0024
6-24	175			TR81/	
6-30	123A	3122	20	TR21/735	S0024
6-31	159	3004	2	TR20/	S0019
6-31A	159	3114		TR20/	
6-32					S0014
6-33					S0019
6-38	159	3025		TR20/717	S0019
6-38A	159	3114		TR20/	
6-49	234	3118	21		S5022
6-50	126	3006	2	TR17/635	
6-53	102A	3004	52	TR85/250	G0005
6-53/63			2	TR85/250	G0005
6-53A	102A	3004	2	TR85/250	
6-53F	102A			TR85/	
6-60	102A	3008	50	TR85/250	G0001
6-60A	160	3008		TR17/	
6-60B	102A	3008	2	TR85/250	G0001
6-60C	160	3008		TR17/	
6-60D	102A	3008	2	TR85/250	G0001
6-60E	160	3008		TR17/	
6-60F	160	3008		TR17/	
6-60P	102A	3008	2	TR85/250	G0001
6-60T	160			TR17/637	
6-60X	160	3123	2	TR17/637	
6-61	102A	3008	50	TR85/250	G0001
6-61A	160	3008		TR17/	

Original Part Number	ECG	SK	GE	IR/WEP	HEP
6-61D	102A	3008	1	TR85/250	G0001
6-61E	160	3008		TR17/	
6-61F	160	3008	21	TR21/637	
6-61P	102A	3008	2	TR85/250	G0001
6-61T	160	3008		TR17/637	
6-61X	160	3123	2	TR17/637	
6-62	102A	3008	51	TR85/250	G0001
6-62A	102A	3008		TR85/250	
6-62B	102A	3008	2	TR85/250	
6-62C	160	3008		TR17/	
6-62D	102A	3008	2	TR85/250	G0003
6-62E	160	3008		TR17/	
6-62F	126	3008	21	TR17/635	S0012
6-62P	126	3008	2	TR17/635	G0008
6-62X	126	3006	2	TR17/635	
6-63	102A	3004	52	TR85/250	G0005
6-63A	102A	3004	2	TR85/250	
6-63T	102A		53	TR85/250	
6-65	102A		53	TR85/250	
6-65T	102A	53		TR85/250	
6-66	102A		53	TR85/250	
6-66T	102A		53	TR85/250	
6-67	102A		53	TR85/250	
6-67T	102A		53	TR85/637	
6-69	160	3006	50	TR17/637	G0002
6-69X	160	3006	50	TR17/637	G0002
6-70	126	3006	50	TR17/635	
6-71	126	3006	50	TR17/635	
6-72	126	3006	50	TR17/635	
6-84F	160		51	TR17/637	
6-85F	160		51	TR17/637	
6-87	102A		53	TR85/250	
6-88	102A	3009	52	TR85/250	G0005
6-88(AUTO)	131			TR94/642	
6-89	126	3006	51	TR17/635	
6-89X	101	3006		TR08/641	G0011
6-90	123A	3122	20	TR21/735	S0020
6-93	123A	3124	20	TR21/735	S0015
6-137	130	3510			
6-138		3511			
6-139	128				
6-140	175				
6-155	102				
6-158	121				
6-856	128				
6-5193	152				
6-5363	102A				
6-5552	175				
6-9029-15D	123A				
6-9029-15E	123A				
6-9029-20J	129				
6-490001	175				
6-1260039	160	3008		TR17/	
6-1260039A	160	3008		TR17/	
6-6450036	123A				
6-6490001	175				
6-6490004	130			TR59/	
6A8-1A5L	160	3007		TR17/	
6A10227	123A	3122		TR21/735	S0015
6A10228	128	3122		TR87/243	S0015
6A10229	131	3082	30	TR94/642	G6016
6A10422	123A	3124		TR21/735	S0015
6A10423	123A	3024	17	TR21/735	S0015
6A10520	123A			TR21/735	S0015
6A10622	158			TR84/630	G0005
6A10624	102A	3123		TR85/250	G0005
6A10851	123A	3124	63	TR21/735	S0015
6A10855	123A	3124	63	TR21/735	S0014
6A11180	123A	3122	17	TR21/735	S0014
6A11223	161	3132		TR83/719	S0020
6A11301	158	3008		TR84/	S0005
6A11665	102A	3123		TR85/250	
6A11668	102A	3123		TR85/250	G0005
6A12515	158			TR84/630	G0005
6A12516	102A	3123		TR85/250	G0005
6A12517	158			TR84/630	G0005
6A12677	108		20	TR95/	S0014
6A12678	100		51	TR05/	G0003
6A12679	108		20	TR95/	S0014
6A12680	126		50	TR17/	G0008
6A12681	123A		10	TR21/	S0014
6A12682	123A		17	TR21/	S0015
6A12683	123A		20	TR21/	S0030
6A12684	102A		52	TR85/	G0006
6A12685	102A		52	TR85/	G0005

Original Part Number	ECG	SK	GE	IR/WEP	HEP
6A12725	123A		20	TR21/735	S0015
6A12788	123A			TR21/735	S0015
6A12789	123A	3124	52	TR21/	G0003
6A12889	160	3124	52	TR17/	G0003
6A12988	154	3024	18	TR78/712	S5026
6A12989	102A	3004	52	TR85/250	G0005
6A12990	102A	3004		TR85/250	G0005
6A12992	103A	3010	8	TR08/724	S5014
6A12993	103			TR08/641A	
6A16399		3124			S0015
6B-3	121MP	3015			
6B-3-70	121MP	3015			
6B-3-70-12	121MP	3015			
6B-3A	121MP	3015			
6B-3AO	121MP	3015			
6B-3A1	121MP	3015			
6B-3A2	121MP	3015			
6B-3A3	121MP	3015			
6B-3A3P	121MP	3015			
6B-3A4	121MP	3015			
6B-3A4-7	121MP	3015			
6B-3A4-7B	121MP				
6B-3A5	121MP	3015			
6B-3A5L	121MP	3015			
6B-3A6	121MP	3015			
6B-3A7	121MP	3015			
6B-3A7-1	121MP	3015			
6B-3A8	121MP	3015			
6B-3A9	121MP	3015			
6B-3A9G	121MP	3015			
6B-3A19	121MP	3015			
6B-3A21	121MP	3015			
6B-3A82	121MP	3015			
6B-3AOR	121MP				
6B-30	121MP	3015			
6B-32	121MP	3015			
6B-33	121MP	3015			
6B-34	121MP	3015			
6B-35	121MP	3015			
6B-36	121MP	3015			
6B-37	121MP	3015			
6B-38	121MP	3015			
6B-39	121MP	3015			
6D0000105	102				
6D122	102A	3004	52	TR85/250	G0005
6D122R	102A	3004	52	TR85/250	G0005
6D122T	102A		53	TR85/250	
6D122TC	102A		53	TR85/250	
6D122TH	102A		53	TR85/250	
6D122U	102A		53	TR85/250	
6D122V	102A	3123	52	TR85/250	G0005
6D122W	102A		53	TR85/250	
6D122Y	102A		53	TR85/250	
6E4850/56-0001	123A				
6E4850/56-0002	128				
6I0148-1	128				
6L122	102A	3123	2	TR85/250	
6MC	160	3007	2	TR17/637	G0002
6X97047A01	123A	3100	300	TR21/53	
6X97047A02	102A			TR85/250	G0005
7-0002-00	128				
7-0011-00	128				
7-0012-00	129				
7-0030-00	175				
7-0051-00	128				
7-0112-00	152				
7-0112-03	152				
7-0112-04	152				
7-0112-05	152				
7-0115	223				
7-0115-000	223				
7-0197	223				
7-1(SARKES)	123A			TR21/	
7-1(STANDEL)	121			TR01/	
7-2(SARKES)	123A			TR21/	
7-2(STANDEL)	104MP			TR01MP/	
7-2		3013	3	/624MP	
7-3(SARKES)	123A			TR21/	
7-3(STANDEL)	129			TR88/	
7-3				/242	
7-4	108	3019	17	TR95/56	S0020
7-4(SARKES)				TR21/	
7-4(STANDEL)	129			TR88/	
7-5	123A	3020			
7-5(SARKES)	123A			TR21/	

Original Part Number	ECG	SK	GE	IR/WEP	HEP
7-6	108	3018	17	TR95/56	
7-6(SARKES)	123A			TR21/	
7-7	108	3018	17	TR95/56	
7-7(SARKES)	123A			TR21/	
7-8	108	3018	17	TR95/56	
7-8(SARKES)	123A			TR21/	
7-12(STANDEL)	130			TR59/	
7-13(STANDEL)	130			TR59/	
7-14A	129	3025		TR88/	
7-14(STANDEL)				TR59/	
7-15(SARKES)	123A			TR21/	
7-16	107				
7-16(SARKES)	123A			TR21/	
7-17	123A	3122		TR21/735	S0015
7-17(SARKES)	123A			TR21/	
7-18(SARKES)	123A			TR21/	
7-19			12		
7-19(SARKES)	123A			TR21/	
7-19(STANDEL)	124			TR81/	
7-20(SARKES)	123A			TR21/	
7-28		3039			S0017
7-28(STANDEL)	218			TR58/	
7-29(STANDEL)	175			TR81/	
7-39	107				
7-40	222	3050			F2004
7-44	108	3018	11	TR80/	S0016
7-59-001		3088			
7-59-001A		3004			
7-59-068	123A	3124	20		S0033
7-59-0103477	158		2	TR05/	G0005
7-59-0193477	108	3018	20	TR21/	S0015
7-59-0203477	108	3018	20	TR21/	S0015
7-59-0213477	108	3018	20	TR21/	S0014
7-59-0223477	108		17	TR21/	S0014
7-59-0233477	108	3018	17		S0014
7-59-0243477	123A	3124	10	TR24/	S0014
7-59-0293477	102A	3004	1	TR85/	
7-59-0603477	102A	3004	2	TR85/	
7-117-02	221				
7-73004-02	176			TR82/	
7-73004-03	176			TR82/	
7-73004-04	176			TR82/	
7-73004-1	176			TR82/	
7-7340102	126				
7-7466201	181				
7A30	107	3039	63	TR70/720	S5014
7A30(GE)	128			TR87/243	
7A30(SHERWOOD)	123A			TR21/	
7A31	107	3039	63	TR70/720	S3019
7A31(GE)	128			TR87/243	
7A31(SHERWOOD)	123A			TR21/	
7A32	107	3039	63	TR70/720	S0015
7A32(GE)	128			TR87/243	
7A32(SHERWOOD)	128			TR87/243	
7A35(GE)	128			TR87/243	
7A995(GE)	128			TR87/243	
7A995(SHERWOOD)	128			TR87/243	
7A1011(GE)	128			TR87/243	
7A1011(SHERWD)	128			TR87/243	
7B1	124	3021	12	TR81/240	S5011
7B2	124	3021	12	TR81/240	S5011
7B13	124	3021	63	TR81/240	S3019
7B33			12	TR23/	S3021
7B34			12	TR23/	S3021
7C1	124	3021	12	TR81/240	S5011
7C2	124	3021	12	TR81/240	S5011
7C3	124	3021	12	TR81/240	S5011
7C4					S3019
7C13			63	TR25/	S3019
7D1	124	3021	12	TR81/240	S5011
7D2	124	3021	12	TR81/240	S5011
7D3	124	3021	12	TR81/240	S5011
7D13			63	TR25/	S3019
7D33			12	TR23/	S3021
7D34			12	TR23/	S3021
7E1	124	3021	12	TR81/240	S5011
7E2	124	3021	12	TR81/240	S5011
7E3	124	3021	12	TR81/240	S5011
7E4					S3019
7E13			63	TR25/	S3019
7F1					S3020
7F2					S3020
7F3					S3019
7F4					S3019
7F13			63	TR25/	S3019

Original Part Number	ECG	SK	GE	IR/WEP	HEP
7G1	124	3021	12	TR81/240	S5011
7G2	124	3021	12	TR81/240	S5011
7G3	124	3021	12	TR81/240	S5011
7G4	124	3021	12	TR81/240	S5011
7G13			63	TR25/	S3019
7G33					S3021
7G34					S3021
7L6-0105	175			TR81/	S5012
8-0024-1	108	3018	17	/56	
8-0024-2	108	3018	17	/56	
8-0024-3	123	3124	20	/53	
8-0060	100	3006	50	/254	G0006
8-0062	102	3004	2	/631	
8-00243	123A		20		
8-005202	123A				
8-0050100	123A	3124			
8-0050300	103A	3010			
8-0050400	160	3006			
8-0050500	160	3008			
8-0050600	103A	3010			
8-0050700	121	3009			
8-0051500	123A	3124			
8-0051600	129	3025			
8-0052102	123A	3124			
8-0052302	123A	3124			
8-0052402	130	3027			
8-0052600	123A	3124			
8-0052700	129	3025			
8-0052800	103A	3010			
8-0053001	123A	3124			
8-0053300	128	3024			
8-0053400	123A	3124			
8-0053600	108	3018			
8-0053702	130MP	3029			
8-0104900	160	3008			
8-0105200	160	3008			
8-0105300	160	3008			
8-0205400	102A	3004			
8-0205600	102A	3004			
8-0222631U	102A	3004			
8-0236400	102A	3004			
8-0236430	102A	3004			
8-0243900	102A	3004			
8-0318250	123A	3124			
8-0337390	123A	3124			
8-0338030	108	3018			
8-0338040	108	3018			
8-0339430	108	3018			
8-0339440	108	3018			
8-0383840	108	3018			
8-0383930	108	3018			
8-0383940	123A	3124			
8-0389910	123A	3124			
8-0389930	123A	3124			
8-0414120	130	3027			
8-0414130	130	3027			
8-0421980	123A	3124			
8-1	175			TR81/	
8-1(BENDIX)	175			TR81/	
8-4(BENDIX)	123A			TR21/	
8-9B		3136			
8-605-606-817		3004			
8-619-030-007	102A	3004		TR85/	
8-619-030-008	102A	3004		TR85/	
8-619-030-009	102A	3004		TR85/	
8-169-030-014	102A	3004		TR85/	
8-169-030-015	131	3198		TR94/	
8-169-030-016	102A	3004		TR85/	
8-169-030-017	102A	3004		TR85/	
8-697-020-567	102A	3004		TR85/	
8-697-020-568	102A	3004		TR85/	
8-697-020-569	102A	3004		TR85/	
8-697-020-570	123A	3122		TR21/	
8-721-323-000	158	3004		TR84/	
8-722-923-000	107	3039		TR70/	
8-723-650	103	3010		TR08/641A	
8-724-034-000	107	3039		TR21/	
8-724-733-000	123A	3122		TR21/	
8-750-105-11	1096				
8-759-101-60	1045	3072			C6060P
8-759-424-00	712				C6063P
8-759-425-00	712				C6063P
8-759-651-34	1096				
8-902-0706-071	123A	3122		TR21/735	
8-905-014-017	123A	3122		TR21/	

Original Part Number	ECG	SK	GE	IR/WEP	HEP
8-905-605-016	102A	3004		TR85/	
8-905-605-030	102A	3004		TR85/	
8-905-605-032	102A	3004		TR85/	
8-905-605-050	102A	3004		TR85/	
8-905-605-051	102A	3004		TR85/	
8-905-605-075	102A	3004		TR85/	
8-905-605-090	102A	3004		TR85/250	
8-905-605-091	102A	3004		TR85/	
8-905-605-105	103A	3010		TR08/	
8-905-605-108	103A	3010		TR08/	
8-905-605-109	103A	3010		TR08/	
8-905-605-110	103A	3010		TR08/	
8-905-605-111	103A	3010		TR08/	
8-905-605-112	103A	3010		TR08/	
8-905-605-113	103A	3010		TR08/	
8-905-605-120	158	3004	53	TR84/630	
8-905-605-123	158	3004		TR84/	
8-905-605-124	158	3004		TR84/	
8-905-605-125	158	3004		TR84/	
8-905-605-126	158	3004		TR84/	
8-905-605-127	158	3004		TR84/	
8-905-605-128	158	3004		TR84/	
8-905-605-129	158	3004		TR84/	
8-905-605-230	102A	3004		TR85/	
8-905-605-232	102A	3004	53	TR85/250	
8-905-605-234	102A	3004	53	TR85/250	
8-905-605-250	158	3004		TR84/	
8-905-605-255	158	3004		TR84/	
8-905-605-260	158	3004		TR84/	
8-905-605-264	158	3004		TR84/	
8-905-605-266	158	3004		TR84/	
8-905-605-268	158	3004		TR84/	
8-905-605-269	158	3004	53	TR84/630	
8-905-605-292	102A	3004		TR85/	
8-905-605-305	102A	3004	53	TR85/250	
8-905-605-320	160	3006	51	TR17/637	
8-905-605-365	103A	3010		TR08/	
8-905-605-384	103A	3010		TR08/	
8-905-605-390	103A	3010		TR08/	
8-905-605-607	131	3082	44	TR94/642	
8-905-605-624	121	3009		TR01/	
8-905-605-635	121	3009		TR01/	
8-905-605-636	121	3009		TR01/	
8-905-605-637	121	3009		TR01/	
8-905-605-644	161	3132		TR83/	
8-905-605-650	131	3082	44	TR94/	
8-905-605-775	127	3035		TR27/	
8-905-605-908	127	3035		TR27/	
8-905-606-001	160	3006		TR17/637	
8-905-606-003	160	3006		TR17/	
8-905-606-105	160	3006	51	TR17/637	
8-905-606-007	160	3006		TR17/	
8-905-606-008	160	3006		TR17/	
8-905-606-010	160	3006		TR17/	
8-905-606-016	160	3006		TR17/	
8-905-606-051	160	3006		TR17/	
8-905-606-075	160	3006		TR17/	
8-905-606-077	160	3006		TR17/	
8-905-606-090	160	3006		TR17/	
8-905-606-106	160	3006		TR17/	
8-905-606-120	160	3006		TR17/	
8-905-606-142	160	3006		TR17/	
8-905-606-152	160	3006	51	TR17/637	
8-905-606-153	160			TR17/	
8-905-606-154	160	3006	51	TR17/637	
8-905-606-155	160	3006		TR17/	
8-905-606-158	160	3006	51	TR17/637	
8-905-606-165	160	3006		TR17/	
8-905-606-168	160	3006		TR17/	
8-905-606-180	160	3006		TR17/	
8-905-606-211	160	3006		TR17/	
8-905-606-225	160	3006		TR17/	
8-905-606-241	160	3006		TR17/	
8-905-606-255	160	3006		TR17/	
8-905-606-256	160	3006		TR17/	
8-905-606-349	160	3006		TR17/	
8-905-606-350	160			TR17/	
8-905-606-351	160	3006		TR17/	
8-905-606-352	160	3006		TR17/	
8-905-606-360	160	3006		TR17/	
8-905-606-375	160	3006		TR17/	
8-905-606-390	160	3006	51	TR17/637	
8-905-606-391	160	3006	51	TR17/637	
8-905-606-392	160	3006		TR17/	
8-905-606-405	160	3006		TR17/	

Original Part Number	ECG	SK	GE	IR/WEP	HEP
8-905-606-419	160	3006		TR17/	
8-905-606-420	160	3006		TR17/	
8-905-606-423	160	3006		TR17/	
8-905-606-720	121	3009		TR01/	
8-905-606-750	102A	3004		TR85/	
8-905-606-800	102A	3004		TR85/	
8-905-606-815	102A	3004		TR85/	
8-905-606-817	102A	3004		TR85/	
8-905-606-885	102A	3004		TR85/	
8-905-613-010	102A	3004		TR85/	
8-905-613-015	103A	3010		TR08/	
8-905-613-062	103A	3010		TR08/	
8-905-613-070	158			TR84/	
8-905-613-071	158	3004		TR84/	
8-905-613-131	158	3004		TR84/	
8-905-613-132	158	3004		TR84/	
8-905-613-133	158	3004		TR84/	
8-905-613-150	103A			/630(CP)	
8-905-613-150	158(CP)			/630(CP)	
8-905-613-160	158	3004		TR84/	
8-905-613-166		3082			
8-905-613-210	121			TR01/	
8-905-613-215	121	3009		TR01/	
8-905-613-232	127	3035		TR27/	
8-905-613-240	131	3082	44	TR94/	
8-905-613-241	131	3082	44	TR94/642	
8-905-613-242	131	3082	44	TR94/642	
8-905-613-245	131	3082	44	TR94/	
8-905-613-250	121	3009		TR01/	
8-905-613-254				/642	
8-905-613-265	131	3082	44	TR94/	
8-905-613-266	131	3198	44	TR94/	
8-905-613-277	131	3082	44	TR94/	
8-905-613-282	131	3082	44	TR94/	
8-905-613-283	131	3082	44	TR94/	
8-905-613-284	131	3082	44	TR94/	
8-905-613-295	127	3035		TR27/	
8-905-613-555	131	3082		TR94/	
8-905-613-640	102A	3004		TR85/	
8-905-613-710	102A	3004		TR85/	
8-905-613-724				/630(CP)	
8-905-613-955	102A	3004		TR85/	
8-905-615-156	102A	3004		TR85/	
8-905-705-112	123A	3122		TR21/	
8-905-705-403	123A	3122		TR21/	
8-905-705-405	123A	3122		TR21/	
8-905-705-410	128	3024		TR87/	
8-905-706-010	171	3103		TR79/	
8-905-706-044	161	3132		TR83/	
8-905-706-055	161	3132		TR83/	
8-905-706-060	161	3132		TR83/	
8-905-706-067	171	3103		TR79/	
8-905-706-068	171	3103		TR79/	
8-905-706-070	161	3132		TR83/	
8-905-706-071	161	3132		TR83/	
8-905-706-075	161	3132		TR83/	
8-905-706-080	161	3132		TR83/	
8-905-706-101	161	3132		TR83/	
8-905-706-104	123A	3122		TR21/	
8-905-706-110	161	3132		TR83/	
8-905-706-201	123A	3122		TR21/	
8-905-706-202	123A	3122		TR21/	
8-905-706-203	123A	3122		TR21/	
8-905-706-206	123A	3122		TR21/	
8-905-706-208	123A	3122	20	TR21/735	
8-905-706-211	123A	3122	20	TR21/735	
8-905-706-215	123A	3122		TR21/	
8-905-706-235	123A	3122		TR21/	
8-905-706-236	123A	3122		TR21/	
8-905-706-238	123A	3122		TR21/	
8-905-706-239	123A	3122		TR21/	
8-905-706-240	123A	3122		TR21/	
8-905-706-242	123A	3122	20	TR21/735	
8-905-706-244	123A	3122		TR21/	
8-905-706-245	123A	3122		TR21/	
8-905-706-246	123A	3122		TR21/	
8-905-706-247	159	3025		TR20/	
8-905-706-250	123A	3122		TR21/	
8-905-706-251	159	3025		TR20/	
8-905-706-253	159	3025		TR20/	
8-905-706-254	159	3025		TR20/	
8-905-706-255	159	3025		TR20/	
8-905-706-256	159	3025		TR20/	
8-905-706-257	123A	3122		TR21/	
8-905-706-260	123A	3122		TR21/	

Original Part Number	ECG	SK	GE	IR/WEP	HEP
8-905-706-263	123A	3122		TR21/	
8-905-706-280	159	3025		TR20/	
8-905-706-286	159	3025		TR20/	
8-905-706-287	159	3025		TR20/	
8-905-706-288	159	3025		TR20/	
8-905-706-289	159	3025		TR20/	
8-905-706-290	159	3025		TR20/	
8-905-706-336	123A	3122		TR21/	
8-905-706-545	129	3025		TR88/	
8-905-706-555	130	3027		TR59/	
8-905-706-556	130			TR59/	
8-905-706-557	130	3027			
8-905-706-606	123A	3122		TR21/	
8-905-706-730	161	3122		TR83/	
8-905-706-790	160	3006		TR17/	
8-905-706-801	184	3054			
8-905-706-901	133	3112			
8-905-707-254	123A	3122		TR21/	
8-905-707-265	123A	3122		TR21/	
8-905-707-313	123A	3122		TR21/	
8-905-713-058	159	3025		TR20/	
8-905-713-101	130	3027		TR59/	
8-905-713-110	184	3054			
8-905-713-556	130	3027		TR59/	
8-905-713-810	129	3025		TR88/	
8-906-706-112	161	3132		TR83/	
8-1074	152				
8-1075	223				
8-2409501	123A	3124		TR21/	
8-2410300	129	3025		TR88/	
8-81250108	199	3038		TR21/	
8-81250109	123A	3124		TR21/	
8-905706557		3027			
8A1002	172	3156		TR69/	
8A1003	172	3156		TR69/	
8A10521	131	3052	44	TR94/	G6016
8A10625	131	3198	30	TR94/	G6016
8A11083	131	3052	30	TR94/642	G6016
8A11721	131	3052	30	TR94/642	G6016
8A12359	131	3082	44	TR94/642	G6016
8A12789	123A			TR21/735	S0016
8A12991	104	3009	16	TR01/624	G6003
8A13164	131	3052	44	TR94/642	G6016
8A13718	158	3004		TR84/630	G0005
8C145-01	218				
8C200	159	3114		TR20/	
8C201	159	3114		TR20/	
8C202	159	3114		TR20/	
8C203	159	3114		TR20/	
8C204	159	3114		TR20/	
8C205	159	3114		TR20/	
8C206	159	3114		TR20/	
8C207	159	3114		TR20/	
8C430	159	3114		TR20/	
8C430K	159	3114		TR20/	
8C440	159	3114		TR20/	
8C440K	159	3114		TR20/	
8C443	159	3114		TR20/	
8C443K	159	3114		TR20/	
8C445	159	3114		TR20/	
8C445K	159	3114		TR20/	
8C449	159	3114		TR20/	
8C449K	159	3114		TR20/	
8C450	159	3114		TR20/	
8C460	159	3114		TR20/	
8C460K	159	3114		TR20/	
8C463	159	3114		TR20/	
8C463K	159	3114		TR20/	
8C465	159	3114		TR20/	
8C465K	159	3114		TR20/	
8C466	159	3114		TR20/	
8C466K	159	3114		TR20/	
8C467	159	3114		TR20/	
8C467K	159	3114		TR20/	
8C468	159	3114		TR20/	
8C468K	159	3114		TR20/	
8C469	159	3114		TR20/	
8C469K	159	3114		TR20/	
8C470	159	3114		TR20/	
8C470K	159	3114		TR20/	
8C700	159	3114		TR20/	
8C700A	159	3114		TR20/	
8C700B	159	3114		TR20/	
8C702	159	3114		TR20/	
8C702A	159	3114		TR20/	

Original Part Number	ECG	SK	GE	IR/WEP	HEP
8C702B	159	3114		TR20/	
8C704	159	3114		TR20/	
8C740	159	3114		TR20/	
8C740G	159	3114		TR20/	
8C740M	159	3114		TR20/	
8C742	159	3114		TR20/	
8C742G	159	3114		TR20/	
8C742M	159	3114		TR20/	
8C7400	159	3114		TR20/	
8C7420	159	3114		TR20/	
8D	160	3006	51	TR17/637	G0002
8E	160	3006	51	TR17/637	G0002
8E(AUTO)	126			TR17/635	
8F	160	3006	51	TR17/637	G0002
8F(AUTO)	126			TR17/635	
8H303	121	3014	16	TR01/232	G6003
8L	160		51	TR17/637	G0002
8L201	121	3009	16	TR01/232	G6005
8L201B	121	3009	16	TR01/232	G6005
8L201C	121	3009		TR01/232	
8L201R	121	3009	3	TR01/232	G6005
8L201V	121	3009	16	TR01/232	G6003
8L301V	104	3009	3	TR01/	
8L404	121	3009		TR01/232	
8P	160		51	TR17/637	G0002
8P-2	102A	3004		TR85/	
8P-2-70	102A	3004		TR85/	
8P-2-70-12	102A	3004		TR85/	
8P-2-70-12-7	102A	3004		TR85/	
8P-2A	102A	3004		TR85/	
8P-2A0	102A	3004		TR85/	
8P-2A0R	102A	3004		TR85/	
8P-2A1	102A	3004		TR85/	
8P-2A2	102A	3004		TR85/	
8P-2A3	102A	3004		TR85/	
8P-2A3P	102A	3004		TR85/	
8P-2A4	102A	3004		TR85/	
8P-2A4-7	102A	3004		TR85/	
8P-2A5	102A	3004		TR85/	
8P-2A5L	102A	3004		TR85/	
8P-2A6	102A	3004		TR85/	
8P-2A6-2	102A	3004		TR85/	
8P-2A7	102A	3004		TR85/	
8P-2A7-1	102A	3004		TR85/	
8P-2A8	102A	3004		TR85/	
8P-2A9	102A	3004		TR85/	
8P-2A9G	102A	3004		TR85/	
8P-2A19	102A	3004		TR85/	
8P-2A21	102A	3004		TR85/	
8P-2A82	102A	3004		TR85/	
8P-20	102A	3004		TR85/	
8P-21	102A	3004		TR85/	
8P-22	102A	3004		TR85/	
8P-23	102A	3004		TR85/	
8P-24	102A	3004		TR85/	
8P-25	102A	3004		TR85/	
8P-26	102A	3004		TR85/	
8P-27	102A	3004		TR85/	
8P-28	102A	3004		TR85/	
8P-29	102A	3004		TR85/	
8P40	104	3009	16	TR01/624	
8P50S	131	3082	30	TR94/642	
8P73BLU	152	3041		TR76/701	S5000
8P73GRN	152	3041		TR76/701	S5000
8P73YEL	152	3041		TR76/701	S5000
8P111	133				
8P202	131	3198	30	TR94/	G6016
8P345	152				
8P404	104	3009	16	TR01/232	G6005
8P404B	121	3009	16	TR01/	
8P404F	104	3009		TR01/232	
8P404M	121	3009	16	TR01/232	G6005
8P404M-1		3009	3	TR16/232	G6003
8P404N	121	3009	3	TR01/232	
8P404ORN	104	3009		TR01/230	G6003
8P404R	121	3009	16	TR01/232	G6005
8P404T	131	3009	16	TR01/232	G6005
8P404V	121	3009	16	TR01/232	G6005
8P415C	121	3009	3	TR01/232	
8P416C	121	3009	16	TR01/232	
8P445	153				
8P505	131	3052	30	TR94/642	G6016
8P880	105		4	TR03/	
8P880B	105	3012	4	TR03/233	G6006
8P1555	179			TR35/G6001	

Original Part Number	ECG	SK	GE	IR/WEP	HEP
8P9253	151	3084	69	TR56/	S5006
8PC60	121	3009	16	TR01/232	
8PF111		3112			
8PS60	121	3009	16	TR01/232	
8Q-000003-11	234				
8Q-3-01			62		
8Q-3-02			21		
8Q-3-04	103A		54	TR08/	
8Q-3-10	199		62	TR21/	
8Q-3-11		3025	67	TR19/	S0031
8Q-3-12			10		
8Q-3-13		3024	18	TR25/	G6007
8Q-3-14	159		21	TR20/	
8Q-3-23				TR91/	
8R-38				TR25/	
9-003	107				
9-006	123A				
9-5108	160	3006		TR17/	
9-5110	160	3006		TR17/	
9-5111	160	3006		TR17/	
9-5112	101	3011		TR08/	
9-5113	101	3011		TR08/	
9-5114	101	3011		TR08/	
9-5116	160	3006		TR17/	
9-5117	160	3006		TR17/	
9-5118	160	3006		TR17/	
9-5119	160	3006		TR17/	
9-5120	160	3006		TR17/	
9-5120A	100	3005		TR05/	
9-5121	160	3006		TR17/	
9-5122	160	3006		TR17/	
9-5123	160	3006		TR17/	
9-5124	160	3006		TR17/	
9-5125	108	3039		TR95/	
9-5126	108	3039		TR95/	
9-5127	108	3039		TR95/	
9-5128	108	3039		TR95/	
9-5129	108	3039		TR95/	
9-5130	108	3039		TR95/	
9-5131	108	3039		TR95/	
9-5201	102A	3004		TR85/	
9-5202	103A	3010		TR08/	
9-5203	102A	3004		TR85/	
9-5204	102A	3004		TR85/	
9-5208	102A	3004		TR85/	
9-5209	102A	3004		TR85/	
9-5212	102A	3004		TR85/	
9-5213	102A	3004		TR85/	
9-5214	102A	3004		TR85/	
9-5216	123A	3122		TR21/	
9-5217	102A	3004		TR85/	
9-5218	102A	3004		TR85/	
9-5220	128	3024		TR87/	
9-5221	123A			TR21/	
9-5222-1	102A	3004		TR85/	
9-5222-2	103A	3010		TR08/	
9-5223	108	3039		TR95/	
9-5224-1	102A	3004		TR85/	
9-5224-2	103A	3010		TR08/	
9-5225	123A	3122		TR21/	
9-5226-003	129	3025		TR88/	
9-5226-004	128	3024		TR87/	
9-5226-1	129	3025		TR88/	
9-5226-2	123A	3122		TR21/	
9-5226-3	129	3025		TR88/	
9-5226-4	128	3024		TR87/	
9-5227	123A	3122		TR21/	
9-5250	121	3009		TR01/	
9-5250-1	121MP	3013			
9-5251	121	3009		TR01/	
9-5252	124	3021		TR81/	
9-5252-1	124	3021		TR81/	
9-5252-2	124	3021		TR81/	
9-5252-3	124	3021		TR81/	
9-5252-4	124	3021		TR81/	
9-5257	121MP	3013			
9-5296	123A	3122		TR21/	
9-9101	126	3006		TR17/	
9-9102	126	3006		TR17/	
9-9103	126	3006		TR17/	
9-9104	102A	3004		TR85/	
9-9105	160	3006		TR17/	
9-9106	160	3006		TR17/	
9-9107	160	3006		TR17/	
9-9108	160	3006		TR17/	

Original Part Number	ECG	SK	GE	IR/WEP	HEP
9-9109-1	123A	3122		TR21/	
9-9109-2	123A	3122		TR21/	
9-9120	160	3006		TR17/	
9-9121	160	3006		TR17/	
9-9201	102A	3004		TR85/	
9-9202	102A	3004		TR85/	
9-9203	102A	3004		TR85/	
9-51141400	121	3009		TR01/	
9-511410100	102A	3004		TR85/	
9-511410200	102A	3004		TR85/	
9-511410900	102A	3004		TR85/	
9-511413500	102A	3004		TR85/	
9A8-1A64	160	3007		TR17/	
9D14	129			TR88/242	
9GR2	123A	3124		TR21/	S0015
9H00DC	74H00				
9H01DC	74H01				
9H04DC	74H04				
9H05DC	74H05				
9H08DC	74H08				
9H10DC	74H10				
9H11DC	74H11				
9H20DC	74H20				
9H21DC	74H21				
9H22DC	74H22				
9H30DC	74H30				
9H40DC	74H40				
9H51DC	74H51				
9H52DC	74H52				
9H53DC	74H53				
9H54DC	74H54				
9H55DC	74H55				
9H60DC	74H60				
9H61DC	74H61				
9H62DC	74H62				
9H71DC	74H71				
9H72DC	74H72				
9H73DC	74H73				
9H74DC	74H74				
9H76DC	74H76				
9H78DC	74H78				
9H101PC	74H101				
9H102PC	74H102				
9H103DC	74H103				
9H106DC	74H106				
9H108DC	74H108				
9L-4	102A	3003		TR85/	
9L-4-70	102A	3003		TR85/	
9L-4-70-12	102A	3003		TR85/	
9L-4-70-12-7	102A	3003		TR85/	
9L-4A	102A	3003		TR85/	
9L-4A0	102A	3003		TR85/	
9L-4A0R	102A	3003		TR85/	
9L-4A1	102A	3003		TR85/	
9L-4A2	102A	3003		TR85/	
9L-4A3	102A	3003		TR85/	
9L-4A3P	102A	3003		TR85/	
9L-4A4	102A	3003		TR85/	
9L-4A4-7	102A	3003		TR85/	
9L-4A4-7B	102A	3003		TR85/	
9L-4A5	102A	3003		TR85/	
9L-4A5L	102A	3003		TR85/	
9L-4A6	102A	3003		TR85/	
9L-4A6-1	102A	3003		TR85/	
9L-4A7	102A	3003		TR85/	
9L-4A7-1	102A	3003		TR85/	
9L-4A8	102A	3003		TR85/	
9L-4A9	102A	3003		TR85/	
9L-4A9G	102A	3003		TR85/	
9L-4A19	102A	3003		TR85/	
9L-4A21	102A	3003		TR85/	
9L-4A82	102A	3003		TR85/	
9L-40	102A	3003		TR85/	
9L-41	102A	3003		TR85/	
9L-42	102A	3003		TR85/	
9L-43	102A	3003		TR85/	
9L-44	102A	3003		TR85/	
9L-45	102A	3003		TR85/	
9L-46	102A	3003		TR85/	
9L-47	102A	3003		TR85/	
9L-48	102A	3003		TR85/	
9L-49	102A	3003		TR85/	
9N00DC	7400				
9N00PC	7400				
9N01DC	7401				

Original Part Number	ECG	SK	GE	IR/WEP	HEP
9N01PC	7401				
9N02DC	7402				
9N02PC	7402				
9N03DC	7403				
9N03PC	7403				
9N04DC	7404				
9N04PC	7404				
9N05DC	7405				
9N05PC	7405				
9N06DC	7406				
9N06PC	7406				
9N07DC	7407				
9N07PC	7407				
9N08DC	7408				
9N08PC	7408				
9N09DC	7409				
9N09PC	7409				
9N10DC	7410				
9N10PC	7410				
9N11DC	7411				
9N11PC	7411				
9N12DC	7412				
9N12PC	7412				
9N13DC	7413				
9N13PC	7413				
9N14DC	7414				
9N14PC	7414				
9N16DC	7416				
9N16PC	7416				
9N17DC	7417				
9N17PC	7417				
9N20DC	7420				
9N20PC	7420				
9N21DC	7421				
9N21PC	7421				
9N23DC	7423				
9N23PC	7423				
9N25DC	7425				
9N25PC	7425				
9N26DC	7426				
9N26PC	7426				
9N27DC	7427				
9N27PC	7427				
9N30DC	7430				
9N30PC	7430				
9N32DC	7432				
9N32PC	7432				
9N37DC	7437				
9N37PC	7437				
9N38DC	7438				
9N38PC	7438				
9N39DC	7439				
9N39PC	7439				
9N40DC	7440				
9N40PC	7440				
9N50DC	7450				
9N50PC	7450				
9N51DC	7451				
9N51PC	7451				
9N53DC	7453				
9N53PC	7453				
9N54DC	7454				
9N54PC	7454				
9N60DC	7460				
9N60PC	7460				
9N70DC	7470				
9N70PC	7470				
9N72DC	7472				
9N72PC	7472				
9N73DC	7473				
9N73PC	7473				
9N74DC	7474				
9N74PC	7474				
9N76DC	7476				
9N76PC	7476				
9N86DC	7486				
9N86PC	7486				
9N107DC	74107				
9N107PC	74107				
9N122DC	74122				
9N123DC	74123				
9N123PC	74123				
9N132DC	74132				
9N132PC	74132				
9RA2787H02		4001			

Original Part Number	ECG	SK	GE	IR/WEP	HEP
9S00DC	74S00				
9S00PC	74S00				
9S03DC	74S03				
9S03PC	74S03				
9S04DC	74S04				
9S04PC	74S04				
9S05DC	74S05				
9S05PC	74S05				
9S011	102A		2	TR85/	
9S037	123A		20	TR21/	
9S10DC	74S10				
9S10PC	74S10				
9S11DC	74S11				
9S11PC	74S11				
9S15DC	74S15				
9S15PC	74S15				
9S20DC	74S20				
9S20PC	74S20				
9S22DC	74S22				
9S22PC	74S22				
9S40DC	74S40				
9S40PC	74S40				
9S41DC	74S41				
9S41PC	74S41				
9S42DC	74S42				
9S42PC	74S42				
9S64DC	74S64				
9S64PC	74S64				
9S65DC	74S65				
9S65PC	74S65				
9S74DC	74S74				
9S74PC	74S74				
9S109DC	74S109				
9S109PC	74S109				
9S112DC	74S112				
9S112PC	74S112				
9S113DC	74S113				
9S113PC	74S113				
9S114DC	74S114				
9S114PC	74S114				
9S133DC	74S133				
9S133PC	74S133				
9S134DC	74S134				
9S134PC	74S134				
9S135DC	74S135				
9S135PC	74S135				
9S140DC	74S140				
9S140PC	74S140				
9TR1	107	3018	20	TR70/720	S0033
9TR2	123A	3124	20	TR21/735	S0015
9TR3	108	3018	11	TR95/56	S0016
9TR4	159	3038	21	TR20/717	S0019
9TR5	123A	3124	20	TR21/735	S0015
9TR6	195	3047	20	TR65/735	S3001
9TR7	195	3020	20	TR65/243	S3002
9TR8	195	3024	20	TR65/243	S3001
9TR9	175	3021	23	TR81/241	S5019
9TR10	123A		11	TR21/735	S0011
9TR11	195	3047	20	TR65/735	S3001
9TR11001-01	107			TR70/	
9TR21001-02	123A			TR21/735	
9TR31001-03	123A			TR21/735	
9TR61001-07	128			TR87/243	
9TR91001-09	130			TR59/247	
9TRZ1001-02	123A	3124		TR21/	
10-080009		3018			S0014
10-080010		3124			S0003
10-1	161	3117		TR83/719	S0017
10-2	161	3039		TR83/719	
10-2A49	126	3018		TR17/	G0008
10-2SB54	102A	3004		TR85/	G0007
10-2SB56	102A	3004		TR85/	G0005
10-2SC80	108			TR95/	
10-2SC94	108	3018		TR95/	
10-2SC380	107	3018		TR70/	S0016
10-2SC394					S0014
10-13-002-003	130	3027			
10-13-002-004	130	3027			
10-13-002-3	130				
10-13-002-4	130				

Original Part Number	ECG	SK	GE	IR/WEP	HEP
10-374101	130				
10A	160			TR17/637	G0002
10B	160			TR17/637	
10B551	108	3039	11	TR95/56	S0025
10B553	108	3039	11	TR95/56	S0025
10B555	108	3039	11	TR95/56	S0015
10B556	108	3039		TR95/56	S0015
10B556-2			214	TR51/	S0016
10B556-3			214	TR51/	S0016
10B1051	108	3018	11	TR95/56	S0025
10B1055	108	3018	11	TR95/56	
10C573	108	3039	17	TR95/56	S0016
10C574	108	3039	17	TR95/56	S0016
10D556-23					S0020
10D701					S0020
10D702					S0020
10E1051					S0015
10G1051	108	3018	11	TR95/56	S0033
10G1052	108	3018	11	TR95/56	S0033
10H551	108	3018	17	TR95/56	S0025
10H553	108	3018	17	TR95/56	S0015
10H1051	108	3039	11	TR95/56	S0020
10H1053	108	3039	11	TR95/56	S0015
10P1	159	3114	22	TR20/717	S0019
10P1A	159	3114		TR20/	
11-0399	121	3009		TR01/	
11-0400	121	3009		TR01/	
11-0422	123A	3018	17	TR21/	
11-0423	172	3124	64	TR69/	
11-0770	185	3191	69		
11-0772	184	3054	66		
11-0773	185	3083	29		
11-0774	123A	3122	20	TR21/	
11-0775	172	3134	64	TR69/	
11-0778	123A	3018	17	TR21/	
11-085010	123A	3122	18	TR21/735	S0014
11-11911-1	181				
11-27070	199				
11-691501	234				
11-691502	234				
11-691504	159				
11B551	108	3039	11	TR95/56	S0005
11B552	108	3039	11	TR95/56	S0005
11B554	108	3039	11	TR95/56	S0005
11B555	108	3039	11	TR95/56	S0005
11B1052	108	3039	17	TR95/56	S0005
11B1055	108	3039	17	TR95/56	
11B1257					S3011
11C1B1	124	3021	12	TR81/240	S3020
11C3B1	124	3021	12	TR81/240	S3020
11C3B3	124	3021	12	TR81/240	
11C5B1	124	3021	12	TR81/240	S3020
11C7B1	124	3021	12	TR81/240	
11C10B1	124	3021	12	TR81/240	S3019
11C10F1					S3019
11C11B1	124	3021	12	TR81/240	S3020
11C11F1					S3020
11C201B20					S3020
11C203B20					S3020
11C205B20					S3010
11C207B20				TR65/	
11C210B20					S3019
11C211B20			63	TR25/	S3020
11C551	108	3039	17	TR95/56	S0025
11C553	108	3039	17	TR95/56	S0025
11C557	108	3039	11	TR95/56	S0025
11C702			47		S3020
11C704					S3020
11C710					S3019
11C1051	108	3039	17	TR95/56	S0025
11C1053	108	3039	17	TR95/56	S0025
11C1057	108	3018	11	TR95/56	S0015
11C1536	128	3024	47	TR87/243	S3011
11CB1					S3020
11CB2					S3020
11CB3					S3020
11CB4					S3020
11CB5					S3020
11CB6					S3020
11CB7					S3024
11CB8					S3024
11CF1					S3020
11CF2					S3020
11CF3					S3020
11CF4					S3020

Original Part Number	ECG	SK	GE	IR/WEP	HEP
11CF5					S3020
11CF6					S3020
11CF7					S3024
11CF8					S3024
11G702					S0025
11G703					S0025
11G1052					S0025
11G1053					S0025
11RT1			21		
11T1					G6011
11T2					S0020
12-1	154	3045		TR78/712	S5025
12-1-70	123A	3124		TR21/	
12-1-70-12	123A	3124		TR21/	
12-1-70-12-7	123A	3124		TR21/	
12-1-73	100	3005		TR05/	
12-1-74	100	3005		TR05/	
12-1-75	100	3005		TR05/	
12-1-76	100	3005		TR05/	
12-1-78	100	3005		TR05/	
12-1-83	100	3005		TR05/	
12-1-91	100	3005		TR05/	
12-1-92	100	3005		TR05/	
12-1-93	100	3005		TR05/	
12-1-95	102A	3004		TR85/	
12-1-96	102A	3004		TR85/	
12-1-100	100	3005		TR05/	
12-1-102	100	3005		TR05/	
12-1-103	100	3005		TR05/	
12-1-104	100	3005		TR05/	
12-1-105	100	3005		TR05/	
12-1-106	102A	3004		TR85/	
12-1-107	102A	3004		TR85/	
12-1-120	102A	3004		TR85/	
12-1-128	100	3005		TR05/	
12-1-135	160	3006		TR17/	
12-1-137	160	3006		TR17/	
12-1-138	160	3008		TR17/	
12-1-139	160	3006		TR17/	
12-1-148	102A	3004		TR85/	
12-1-150	160	3007		TR17/	
12-1-157	160	3006		TR17/	
12-1-161	100	3005		TR05/	
12-1-162	100	3005		TR05/	
12-1-164	102A	3004		TR85/	
12-1-179	100	3005		TR05/	
12-1-180	100	3005		TR05/	
12-1-184	102A	3004		TR85/	
12-1-186	100	3005		TR05/	
12-1-189	126	3008		TR17/	
12-1-190	160	3007		TR17/	
12-1-191	102A	3004		TR85/	
12-1-226	102A	3004		TR85/	
12-1-227	102A	3004		TR85/	
12-1-228	160	3006		TR17/	
12-1-229	160	3006		TR17/	
12-1-230	160	3006		TR17/	
12-1-231	160	3006		TR17/	
12-1-232	102A	3004		TR85/	
12-1-233	160	3006		TR17/	
12-1-234	100	3005		TR05/	
12-1-235	100	3005		TR05/	
12-1-236	100	3005		TR05/	
12-1-240	100	3005		TR05/	
12-1-241	100	3005		TR05/	
12-1-242	126	3008		TR17/	
12-1-243	126	3008		TR17/	
12-1-244	126	3008		TR17/	
12-1-246	102A	3004		TR85/	
12-1-254	100	3005		TR05/	
12-1-256	160	3007		TR17/	
12-1-257	126	3008		TR17/	
12-1-258	160	3007		TR17/	
12-1-259	160	3007		TR17/	
12-1-260	160	3007		TR17/	
12-1-266	102A	3004		TR85/	
12-1-267	102A	3004		TR85/	
12-1-270	121	3009		TR01/	
12-1-271	121	3009		TR01/	
12-1-272	102A	3004		TR85/	
12-1-273	100	3005		TR05/	
12-1-274	102A	3003		TR85/	
12-1-275	100	3005		TR05/	
12-1-276	123A	3124		TR21/	
12-1-277	123A	3124		TR21/	

Original Part Number	ECG	SK	GE	IR/WEP	HEP
12-1-278	123A	3124		TR21/	
12-1-279	123A	3124		TR21/	
12-1-289	100	3005		TR05/	
12-1A	123A	3124		TR21/	
12-1AO	123A	3124		TR21/	
12-1AOR	123A	3124		TR21/	
12-1A1	123A	3124		TR21/	
12-1A2	123A	3124		TR21/	
12-1A3	123A	3124		TR21/	
12-1A3P	123A	3124		TR21/	
12-1A4	123A	3124		TR21/	
12-1A4-7	123A	3124		TR21/	
12-1A4-7B	123A	3124		TR21/	
12-1A5	123A	3124		TR21/	
12-1A5L	123A	3124		TR21/	
12-1A6	123A	3124		TR21/	
12-1A6A	123A	3124		TR21/	
12-1A7	123A	3124		TR21/	
12-1A7-1	123A	3124		TR21/	
12-1A8	123A	3124		TR21/	
12-1A9	123A	3124		TR21/	
12-1A9G	123A	3124		TR21/	
12-1A19	123A	3124		TR21/	
12-1A21	123A	3124		TR21/	
12-1A82	123A	3124		TR21/	
12-2	225	3045		/712	S5025
12-4	123A	3122		TR21/735	S0014
12-10	123A	3124		TR21/	
12-11	123A	3124		TR21/	
12-12	123A	3124		TR21/	
12-13	123A	3124		TR21/	
12-14	123A	3124		TR21/	
12-15	123A	3124		TR21/	
12-16	123A	3124		TR21/	
12-17	123A	3124		TR21/	
12-18	123A	3124		TR21/	
12-19	123A	3124		TR21/	
12-21-050	234				
12-28471-1(PL)	188				
12-100027	129				
12-100047	130				
12-101001	123A				
12A6240	100	3005		TR05/254	
12A7239P1	102				
12A9244-1	160				
12A9244-P2	160				
12A9275	102	3004	2	TR05/631	
12A9275-1	102	3004	2	TR05/631	
12A9275-4		3004			
12AA2	101	3011		TR08/	
12CLN	123A	3124	20	TR24/	S0015
12M2	121	3009	16	TR01/232	G6003
12MC	160	3007	50	TR17/637	G0002
12MZ	160			TR17/637	G0002
12X047	123A	3018	17	TR21/	S0014
13-0006	159	3025		TR20/717	S0019
13-0006A	159	3114		TR20/	
13-0009	123A	3018	20	TR21/735	S0030
13-0010	161	3018	20	TR83/719	S0025
13-0014	175	3026	23	TR81/241	S5012
13-0020	108	3018	17	TR95/56	S0020
13-0021	123A	3046	20	TR21/735	S0011
13-0022	123A	3018	20	TR21/735	S0015
13-0024	123A	3047		TR21/	
13-0032	130			TR59/247	S7004
13-0035	123A	3047		TR21/	S0030
13-0040	108	3018	20	TR95/56	S0020
13-0041	123A		20	TR21/	S0015
13-0043	159	3114	21	TR20/717	S0013
13-0043A	159	3114		TR20/	
13-0044	159	3114		TR20/717	S0019
13-0044A	159	3114		TR20/	
13-0048	123A			TR21/735	S0015
13-0049	184	3026	66	/S5003	S5000
13-0058	123A	3124	20	TR21/735	S0015
13-0061	159	3114	21	TR20/717	S0019
13-0061A	159	3114		TR20/	
13-0062	123A	3018	20	TR17/	S0030
13-0063	123A		20	TR21/	S0030
13-0064	108	3046	17	TR95/	S0014
13-0065	108	3018	20	TR95/	S0020
13-0079	195	3047	28	TR65/	S3001
13-0026666-001	128				
13-0321-5	123A	3122	20	TR21/735	
13-0321-6	123A	3122	20	TR21/735	
13-0321-7	123A	3122	20	TR21/735	
13-0321-8	123A	3122	20	TR21/735	
13-0321-9	123A	3122		TR21/735	
13-0321-10	123A		20	TR21/735	
13-0321-11	123A	3122	20	TR21/735	
13-0321-12	123A	3122	20	TR21/735	
13-0321-14	108	3039		TR95/56	
13-0321-15	108	3039		TR95/56	
13-0321-16	108	3039		TR95/56	
13-0331-17	108	3039		TR95/56	
13-0321-21	108	3039		TR95/56	
13-0321-81	123A	3122	20	TR21/735	
13-076001	724	3157			
13-2I606-1	127				
13-2P64	152	3054			
13-2P64-1	152				
13-3-6	726	3022			
13-4P63	152	3054			
13-4P63-I	152				
13-17-6	123A	3122		/735	
13-17-6(SEARS)	123A				
13-26-6	718	3159	8	IC511/	
13-27-6	711	3070		IC514/	
13-27-6(SEARS)	711				
13-28-6	706	3101		IC507/	
13-28-6(SEARS)	706				
13-29-6	712	3072	2	IC507/	
13-29-6(SEARS)	712				
13-30-6	780	3141	20		
13-38-6	906				
13-40-6	715	3076	6		
13-41-6	790	3077	18		
13-42-6	714	3075	4		
13-50-6	767				
13-52-6	722	3161	9	IC513/	
13-52-G	722				
13-53-6	909	3590			
13-56-6	738	3167	29		
13-57-6	739		30		
13-58-6	733				
13-59-6	804		27		
13-60-6	722	3161	9		
13-61-6	804				
13-1032-5	108	3019	17	TR95/56	
13-1836-1	123A	3122	20	TR21/735	
13-1847-1	196			TR92/	
13-2384-2	161			TR83/719	
13-5018-6	741M				
13-5020-6	923D				
13-10320-14	161			TR83/	S0017
13-10321-1	108	3039		TR95/56	
13-10321-2	108	3039		TR95/56	
13-10321-5	108	3019	17	TR95/56	S0016
13-10321-6	108	3018	17	TR95/56	S0020
13-10321-7	108	3018	17	TR95/56	S0020
13-10321-8	108	3018	17	TR95/56	S0020
13-10321-9	108	3039		TR95/56	
13-10321-10	108	3018	11	TR95/56	S0017
13-10321-11	108	3018	11	TR95/56	S0017
13-10321-12	108	3019	17	TR95/56	S0016
13-10321-14	108	3039	11	TR95/56	S0017
13-10321-15	108	3039		TR95/56	S0020
13-10321-16	108	3039	11	TR95/56	S0020
13-10321-17	108	3030	11	TR95/56	S0020
13-10321-20	108	3039		TR95/56	S0016
13-10321-21	108	3019	11	TR95/56	S0016
13-10321-26	108	3019	11	TR83/	
13-10321-29	107	3018	11	TR70/720	S0016
13-10321-30	108	3018		TR95/56	
13-10321-31	107	3018	11	TR70/720	S0016
13-10321-32	107	3018	11	TR70/720	S0016
13-10321-37	222	3065			
13-10321-41	108	3039	11	TR95/56	S0016
13-10321-43	108	3018	11	TR80/	S0016
13-10321-46	107	3018	11	TR80/	S0016
13-10321-47	161	3018	39	TR80/	S0017
13-10321-50	161				
13-10321-51	108	3039	11	TR83/	S0020
13-10321-62	229				
13-13021-15			11		
13-13203-1 0		3009			
13-13543-1			12		
13-13543-2			12		
13-14065-77			20		
13-14085-1	108	3018		TR95/	S0016

Original Part Number	ECG	SK	GE	IR/WEP	HEP
13-14085-2	108	3018		TR95/	S0016
13-14085-3	107	3018		TR70/720	S0016
13-14085-4	107	3018		TR70/720	S0016
13-14085-6	123A	3124		TR21/735	S0016
13-14085-7	123A	3124		TR21/735	S0016
13-14085-8		3004			G0005
13-14085-9	102A	3004		TR85/250	G0005
13-14085-10	102A	3004		TR85/250	G0005
13-14085-11	158	3004		TR84/630	G0005
13-14085-12	158	3004		TR84/630	G0005
13-14085-13	159	3025		TR20/717	S0012
13-14085-13A		3025			
13-14085-14	172	3124		TR69/S9100	S9100
13-14085-15	123A	3024		TR21/735	S0015
13-14085-15A	128	3024		TR87/	
13-14085-16	107	3018		TR70/720	S0016
13-14085-17	107	3018		TR70/720	S0016
13-14085-18	102A	3004		TR85/250	G0005
13-14085-23	102A	3004		TR85/	G0005
13-14085-24	108	3124		TR95/56	S0011
13-14085-25	158	3004	52	TR84/630	G0005
13-14085-26	107	3018	20	TR70/	S0016
13-14085-27	108	3018	20	TR95/	S0014
13-14085-28	160	3006	50	TR17/	G0003
13-14085-29	124	3021	12	TR81/	S5011
13-14085-30	160	3006	50	TR17/	G0003
13-14085-31	126	3006	50	TR17/	G0008
13-14085-32	126	3006	50	TR17/	G0008
13-14085-33	126	3006	50	TR17/	G0008
13-14085-34	123A	3020	63	TR21/	S0020
13-14085-35	102A	3004	52	TR05/	G0005
13-14085-41		3124			S0025
13-14085-49		3124			S0024
13-14085-50	123A		10	TR21/	
13-14085-54					S0015
13-14085-60	102A	3004	52	TR85/	G0005
13-14085-71	102A	3004	52	TR85/	G0005
13-14085-72	123A	3122	20	TR21/	S0015
13-14085-74	108		20	TR95/	
13-14085-75	108		20	TR95/	
13-14085-76	108		20	TR95/	
13-14085-77	108		20	TR95/	
13-14085-83		3124			
13-14085-84	123A		17	TR21/	
13-14085-85	123A		17	TR21/	
13-14085-86	187	3193	29	TR56/	
13-14085-87	159		21	TR20/	
13-14085-88	186		28	TR55/	
13-14085-89	123A		17	TR21/	
13-14085-91		3024			
13-14085-92		3025			
13-14085-93		3054			
13-14085-95	199	3122	20	TR21/	S0014
13-14085-96	199	3122	20	TR21/	S0014
13-14085-97	199	3122	20	TR21/	S0014
13-14085-121		3041			
13-14085-122		3124			S0025
13-14279-1	103A			TR08/724	
13-14604-1	121MP	3015	13MP	TR01/232MP	G6005
13-14604-1A	121MP	3013		/232MP	
13-14604-1B	121MP	3013		/232MP	
13-14604-1C	121MP	3013		/232MP	
13-14604-1D	121MP	3013		/232MP	
13-14604-1E	121MP	3013		/232MP	
13-14605-1	128	3024		TR87/243	S5014
13-14606-1	123A	3122		TR21/735	S0015
13-14735	121	3009	16	TR01/	
13-14735-1	121	3014	16	TR01/232	G6005
13-14735A	121	3009	16	TR01/232	
13-14777-1	121MP	3014		/232MP	G6005
13-14777-1A	121MP	3013		/232MP	
13-14777-1B	121MP	3013		/232MP	
13-14777-1C	121MP	3013		/232MP	
13-14777-1D	121MP	3013		/232MP	
13-14778-1	121MP	3015	13MP	TR27/232MP	G6005
13-14778-1A	121MP	3013		/232MP	
13-14778-1B	121MP	3013		/232MP	
13-14778-1C	121MP	3013		/232MP	
13-14778-1D	121MP	3013		/232MP	
13-14886-1	126	3006	51	TR17/	G0003
13-14887-1	126	3006	51	TR17/	G0003
13-14888-3	102	3004	2	TR05/631	G0001
13-14889-1	126	3006	51	TR17/635	G0003
13-15804-1	123A	3124	20	TR21/735	S0025
13-15805-1	102	3004	2	TR05/631	

Original Part Number	ECG	SK	GE	IR/WEP	HEP
13-15806-1	121	3034	25	TR01/232	
13-15808-1	161	3018	20	TR83/719	S0025
13-15808-2	161	3124	20	TR83/719	S0025
13-15809-1	154	3045		TR78/712	S5025
13-15810-1	108	3018	11	TR95/56	S0020
13-15833-1	128	3024	20	TR87/243	S0015
13-15835-1	161	3018	11	TR83/719	S0015
13-15836-1	102A	3004		TR05/631	
13-15840-1	123A	3124	20	TR21/735	S0015
13-15840-2	123A	3124	20	TR21/735	S0015
13-15841-1	108	3018	11	TR95/56	
13-15842-1	123A	3024	63	TR21/735	
13-15842-2		3122		TR51/735	
13-15865-1	123A	3124	63	TR21/	
13-16570-1	159	3114		TR20/717	S0019
13-16570-1A	159	3114		TR20/	
13-16570-2	159	3114		TR20/717	
13-16570-2A	159	3114		TR20/	
13-16592-1	127	3012		TR27/235	
13-16607-1	127	3035		TR87/235	G6008
13-16608-1	127	3034		TR27/235	G6007
13-16744-1	108	3039		TR95/56	S0016
13-16769-1	123A	3122		TR21/735	S0011
13-17607-1	127	3035	25	TR27/235	G6008
13-17607B	127	3035	25	TR27/235	
13-17608-1	127	3034	25	TR27/235	G6007
13-17608-2	127	3034	25	TR27/235	G6007
13-17608B	127	3034	25	TR27/235	
13-17608C	127	3034	25	TR27/235	
13-17609-1	127	3034		TR27/235	
13-17918-1	130	3027		TR59/247	
13-18032-1	102A			TR85/250	
13-18033-1	102A			TR85/250	
13-18034	121	3009	16	TR01/232	
13-18034-1	121	3009	16	TR01/232	G6005
13-18034A	121	3009	16	TR01/232	
13-18087-1	123A	3124	20	TR21/735	S0015
13-18087-2	123A	3124	20	TR21/735	S0014
13-18158-1	123A	3124	20	TR21/735	S0011
13-18198-1	121MP	3015	13MP	/232MP	G6005
13-18282	124	3021	12	TR81/240	
13-18282-1	124	3021	12	TR81/240	
13-18304-1	102A	3004	2	TR85/250	
13-18359	124	3021	12	TR81/240	
13-18359-1	124	3021		TR81/240	
13-18359-3	124	3021	12	TR81/240	S5011
13-18359A	124	3021	12	TR81/240	
13-18363	123	3124	20	TR86/53	
13-18363-1	123A	3124	20	TR21/735	S0025
13-18363-1A	123	3124	20	TR86/53	
13-18364-1	123A	3124	20	TR21/735	S0015
13-18365-1	199	3124	20	TR21/735	S0015
13-18642-1	179	3009	3	TR35/G6001	G6009
13-18642-2	179			TR35/G6001	G6009
13-18642-2D	179			TR35/	
13-18642-2E	179			TR35/	
13-18642-2F	179			TR35/	
13-18642-3	179			TR35/G6001	
13-18642-3A	179			TR35/	
13-18642-3B	179			TR35/	
13-18642-20	179(TWO)				
13-18642-21	179(TWO)				
13-18654-1	103A	3010	8	TR08/724	G0011
13-18671-1	102A			TR85/250	
13-18671-1A	102A			TR85/250	
13-18671-1B	102A			TR85/250	
13-18671-1C	102A			TR85/250	
13-18927-1	123A	3124	20	TR21/735	S0015
13-18927-1A	128	3124	20	TR87/243	
13-18944-1	100	3004	2	TR05/254	S0032
13-18944-2	100	3005		TR05/254	S0032
13-18946-1	160	3006	51	TR17/637	G0002
13-18946-2	160			TR17/637	
13-18947-1	160	3006	51	TR17/637	G0003
13-18948-1	160	3006	50	TR17/637	G0003
13-18948-2	160	3006	51	TR17/637	G0008
13-18949-1	108	3039		TR95/635	
13-18950-1	108	3039	11	TR95/635	S0017
13-18951-1	126	3006	52	TR17/635	G0008
13-18951-2	126	3006	51	TR17/635	G0008
13-19776-1	159				
13-21606-1	127	3034	25	TR27/235	
13-22581	123A	3122	20	TR21/53	
13-22581-1	123A	3124	63	TR21/53	S0015
13-22582-1	159	3025	22	TR20/717	S0019

Original Part Number	ECG	SK	GE	IR/WEP	HEP
13-22582-1A	159	3114		TR20/	
13-22690-1	132	3116	FET1	FE100/802	F0015
13-22692-1	132	3006	51	FE100/802	F0015
13-22692-2	132	3116		FE100/	F0015
13-22739-1	121MP	3009	16	TR01/232MP	G6005
13-22741	121	3009	3	TR01/232	
13-22741-1	121MP	3009	16	TR01/232MP	G6005
13-22741-2	121MP	3013		/232MP	
13-23543-1	124	3021	12	TR81/240	S5012
13-23543-2	124	3021	12	TR81/240	S5011
13-23594-1	130	3027		TR59/247	
13-23785-1	102A	3004	52	TR85/250	G0008
13-23822	108	3018	20	TR95/	
13-23822-1	161	3039	17	TR83/719	S0017
13-23824-1	233	3018	17	TR83/719	S0017
13-23824-1(3IF)	133				
13-23824-2	161	3039	20	TR83/	S0017
13-23825-1	154	3124	21	TR78/712	S5026
13-23826-1	159	3025	21	TR20/717	S0013
13-23826-1A	159	3114		TR20/	
13-23826-2	159	3025	21	TR20/717	S0013
13-23826-2A	159	3114		TR20/	
13-23826-3	159	3025	21	TR20/717	S0013
13-23826-3A	159	3114		TR20/	
13-23840-1	128	3024	20	TR87/243	S0015
13-23916-1	123A	3020	20	TR21/735	S0015
13-25343-1	124	3021	12	TR81/240	
13-26009-1	161	3039	11	TR83/719	S0017
13-26377-1	131	3082	44	TR94/642	G6016
13-26377-2	155				
13-26386-1	159	3114	22	TR20/717	S0019
13-26386-1A	159	3114		TR20/	
13-26386-2	159	3025		TR20/717	
13-26386-2A	159	3114		TR20/	
13-26386-3	159	3025	21	TR20/717	S5022
13-26576-1	161	3039	11	TR83/719	S0016
13-26576-2	161	3039	11	TR83/719	S0016
13-26577-1	161	3018	20	TR83/719	S0016
13-26577-2	161	3117		TR83/719	S0016
13-26577-3	161	3039	20	TR83/719	S0016
13-26666	128	3024		TR87/	
13-26666-1	128	3024	63	TR87/243	S5014
13-27050-1	103A	3010	8	TR08/724	G0011
13-27404-1	123A	3024	63	TR21/735	S0016
13-27404-2	123A	3018	20	TR21/735	S0025
13-27432-1	128	3024	63	TR87/243	S0005
13-27432-2	128	3024		TR87/	S0005
13-27433-1	123A	3018	11	TR21/	
13-27443-1	128	3024		TR87/	
13-27974-1	157	3103	27	TR60/244	S5015
13-27974-2	157	3103	27	TR60/244	S5015
13-27974-3	157	3103	27	TR60/244	S5015
13-27974-4	157	3044	27	TR60/244	S5015
13-28222-1	183	3189	30	TR77/S5005	S5004
13-28222-2	182	3188	55	TR76/S5004	S5004
13-28222-3	182	3054	55	/S5004	S5004
13-28222-4	182	3188	55	/S5004	
13-28336-1	185	3191	21	TR77/S5007	S5006
13-28336-2	184	3054	63	TR76/S5003	S5000
13-28336-20	185	3191			S5000
13-28386-1	129			TR88/242	
13-28391-1	159	3114		TR20/717	S0013
13-28391-1A	159	3114		TR20/	
13-28391-2	159	3114		TR20/717	S0013
13-28391-2A	159	3114		TR20/	
13-28392-1	128	3024		TR87/243	S5014
13-28392-1					S0015
13-28392-2	128	3024	18	TR87/243	S3021
13-28392-2(MTL)					S5014
13-28392-2(PLAS)					S0015
13-28392-3	128	3024		TR87/243	
13-28392-3(MTL)					S5014
13-28392-3(PLAS)					S0015
13-28393-1	129	3124	21	TR88/242	S5013
13-28393-1A	129	3025		TR88/	
13-28393-2	129	3025	21	TR88/242	S3032
13-28393-2A	129	3025		TR88/	
13-28393-3	193				
13-28394-1	128	3024		TR87/243	
13-28394-2	128	3024		TR87/243	
13-28394-3	128	3024		TR87/243	S5014
13-28394-4	129	3025		TR88/242	
13-28394-4A	129	3025		TR88/	
13-28394-5	129	3025		TR88/242	
13-28394-5A	129	3025		TR88/	

Original Part Number	ECG	SK	GE	IR/WEP	HEP
13-28394-6	129	3025		TR88/242	S3003
13-28396-1	223				
13-28432-1	194	3045		TR25/712	S0005
13-28432-2	154			TR78/712	
13-28469-2	152	3054		TR76/701	
13-28471-1	128	3024	63	TR87/243	S3024
13-28431-1(MTL)	128			TR87/3020	
13-28431-1(PL)				TR78/	
13-28532-1	182	3036	55	TR26/S5004	S5004
13-28583-1	222	3050			F2004
13-28584	108	3018		TR95/	
13-28584-1	108	3018	FET1	TR95/56	S0020
13-28654-1	132	3116	FET1	FE100/802	F0015
13-28654-2	132	3116	FET1	FE100/802	F0015
13-28654-3	132	3116	FET2	FE100/802	F0015
13-28654-4	133	3112	FET1	FE100/	F0010
13-29033-1	123A	3122	11	TR21/735	S0025
13-29033-2	199	3044	20	TR21/735	S0015
13-29033-3	199	3122	11	TR21/735	S0015
13-29033-4	199	3122	20	TR21/735	S0025
13-29033-5	199			TR21/	
13-29165-2	128	3051		TR87/186	
13-29432-1	123A	3124		TR21/735	S0015
13-29437-1	191	3044		TR75/	
13-29775-1	172		64	TR69/S9100	S9100
13-29775-2	172/410		64	TR69/	S9100
13-29775-3	172/410		64	TR69/	S9100
13-29776-1		3025	21	TR21/717	S0031
13-29776-1A	159	3114		TR20/	
13-29776-2	159	3114	21	TR30/	S0031
13-29776-3	159	3138			S0013
13-29947-1	161	3018	60	TR83/	S0025
13-29974-4	157	3104	27	TR60/	S5015
13-29907-1					S0025
13-31013-1	123	3018	22	TR86/53	S0015
13-31013-1/2	159	3114	22	TR53/	S0019
13-31013-1(SYL)	159				
13-31013-4	108	3019	11	TR95/	S0016
13-31013-5	161	3018	11	TR83/	S0016
13-32362-1	199	3018	20	TR21/56	S0016
13-32364-1	159	3118	21	TR20/717	S0013
13-32366-1	108	3018	20	TR95/56	S0020
13-32366-2	108	3018	20	TR95/56	S0016
13-32630-1	128			TR87/	
13-32630-2	192		63		S5014
13-32630-3	128	3024		TR87/243	S5014
13-32631		3025	67		S5013
13-32631-1	129			TR88/	
13-32631-2	193		67		S5013
13-32631-3	129	3025		TR88/242	S5013
13-32631-4	193				
13-32632-1	188/210	3054		TR56/3020	S3019
13-32634-1	186	3083	28	TR55/S3023	S5000
13-32635-1	187	3193		TR56/S3027	
13-32636-1	152	3054		TR76/701	
13-32638-1	184	3054	57	/701	
13-32640-1	152	3054	57	TR76/701	S5000
13-32642-1	184	3054	57	/701	
13-33174-1	171	3104	27	TR79/244A	S3021
13-33174-2		3103			S3021
13-33175-1	172	3124	64	TR69/S9100	S9100
13-33175-2	172		64	TR69/	
13-33176-1	171	3104	27	TR79/244A	S3021
13-33178-1	193	3025	67	TR73/243	S3032
13-33180-1	188	3104	28	TR72/3020	S3020
13-33181-1	165	3115	38	TR93/740B	S5021
13-33181-2	165	3115			S7005
13-33181-3	165				
13-33182-1	162	3079	35	TR67/707	S5020
13-33182-2	162	3079		TR67/	S3021
13-33185-1	210	3045	28	TR55/S3023	S3024
13-33188-1	182	3188		/S5004	
13-33188-2	130	3027	14	TR59/247	S5020
13-33350-1	123A		20	TR21/	S0015
13-33595-1	123A	3124	17	TR21/735	S0015
13-33595-2	123A	3020	17	TR21/735	S0025
13-33595-3	123A	3040			S0015
13-33742-1(MTL)	128/401	3054	18	TR56/3020	S5000
13-33742-1(PL)	188			TR72/	
13-34002-1	196	3054		TR92/701	S5005
13-34002-2	196			TR92/	
13-34002-3		3054			
13-34002-4	196			TR92/	
13-34003-1(PL)	184,188	3054	28	TR21/243	S3020
13-34004-1A	185	3025			

Original Part Number	ECG	SK	GE	IR/WEP	HEP
13-34004-1(MTL)	129	3083		TR88/S5007	
13-34004-1(PL)	189/185				
13-34045-1	108	3018	10	TR95/56	S0015
13-34045-2	108	3018	10	TR95/56	S0015
13-34046-1	186	3054	28	TR55/S3023	S5000
13-34046-2	186	3054	28	TR55/	S5000
13-34046-3	186	3054	28	TR55/	S3024
13-34046-4	186	3104	28	TR55/	S3024
13-34046-5	186	3054	28	TR76/	S3024
13-34047-1	187	3083	29	TR56/S3027	S5006
13-34047-2	187	3193		TR56/	
13-34047-3	187	3083	29	TR56/	S3028
13-34047-4	187	3193		TR56/	
13-34089-2	157	3103	27	TR60/	S5015
13-34089-4	157	3103	27		S5015
13-34369-1	106	3025	22	TR20/	S0013
13-34371-1	128	3024	18	TR87/	S5014
13-34372-1	188	3199		TR72/	
13-34372-2	186			TR55/	
13-34373-1	189	3200		TR73/	
13-34373-2	187	3193		TR56/	
13-34374-1	181			TR36/	
13-34375-1	132		FET2	FE100/	F0021
13-34378-1	132		FET2	FE100/	S0024
13-34381-1	199	3122	20	TR21/	
13-34616-1	196			TR92/	S5005
13-34617-1	197				S5015
13-34684-1	181			TR36/	
13-34706-1					S5000
13-34707-1					S5006
13-34838-1	196			TR92/	
13-34839-1	197				
13-35059-1	712	3072	2	IC507/	C6063P
13-35089-1	228	3104	27	TR79/	S3032
13-35089-2	228	3103	27	TR79/	S5015
13-35089-3	228	3103	27	TR79/	S5015
13-35089-4	228	3103	27	TR79/	S5015
13-35226-1	123A	3122	20	TR21/	S0025
13-35257-1	157			TR60/	
13-35257-2	157	3045	40	TR60/	S0005
13-35257-3	157			TR60/	
13-35257-4	157			TR60/	
13-35324-1	265		64	TR69/	
13-35550	108		20	TR21/	
13-35550-1	199	3122	20	TR21/	S0015
13-35792-1	102A	3004		TR85/	
13-35807-1	194	3044	20	TR21/	S0015
13-36386-1	159			TR20/	S0019
13-37526-1	188	3199		TR72/	
13-37527-1	189	3200		TR73/	
13-37708-1	165	3115	38	TR61/	S5021
13-37869-1	191	3045	40	TR75/	S5024
13-37870-1	165	3111	38	TR93/	S5021
13-37870-2	165	3111			
13-39004-1	196	3054	32	TR76/	S3020
13-39074-1	186	3054	28	TR55/	
13-39098-1	186	3054	27	TR55/	
13-39099-1	196	3054	66	TR77/	
13-39100-1	197	3083	69	TR76/	
13-39114-1	192	3024	63	TR24/	
13-39115-1	193	3025	67	TR30/	
13-44290	133				
13-44291	133				
13-50484-1	100	3005		TR05/254	G0005
13-50486-1	100	3005		TR05/254	G0001
13-50631-1	100	3005		TR05/254	G0005
13-50944-1	100	3005		TR05/254	G0005
13-55009-1	127	3035	25	TR27/235	G6007
13-55009-2	127	3035	25	TR27/235	G6008
13-55018-4	128	3024	63	TR87/	S5014
13-55020-1	108	3018	20	TR95/	S0020
13-55061-1	123A	3122	20	TR21/735	S0025
13-55061-2	123A	3122	20	TR21/735	
13-55062-1	154	3045		TR78/712	S3021
13-55063-1	108	3039	20	TR95/56	S0020
13-55064-1	128	3024	63	TR87/243	S5014
13-55065-1	108	3039	20	TR95/56	S0020
13-55066-1	123A	3122	63	TR21/735	S0025
13-55066-2	123A	3122	63	TR21/735	S0005
13-55067-1	123A	3122	20	TR21/735	S0015
13-55068-1	123A	3122	27	TR21/735	S0015
13-55069-1	159	3114	21	TR20/717	S0019
13-55069-1A	159	3114		TR20/	
13-67583-5	108	3018	17	TR95/	S0016
13-67583-6	123A	3018	20	TR21/	S0014

Original Part Number	ECG	SK	GE	IR/WEP	HEP
13-67583-6/3464	107				S0014
13-67585-4	123A	3018	20	TR21/	S0014
13-67585-4/3464	107				S0014
13-67585-5	123A	3018	20	TR21/	S0014
13-67585-5/2439-2	108		20	TR21/	
13-67585-5/3464	123A				S0014
13-67585-6/2439-2	108		17	TR95/	
13-67585-7/2439-2	123A		10	TR24/	
13-67586-3	123A	3124	20	TR21/	S0011
13-67586-3/3464	123A				S0030
13-67599-3	102A	3004	53		G0005
13-67599-3/2439-2	102A		2		
13-67599-3/3464	158				G0005
13-68617-1	123A	3018		TR21/	
13-86416-1	103	3010		TR08/641A	
13-86420-1	101	3011		TR08/641	G0011
13-87433-1	101	3011		TR08/641	G0011
13-94096-2	102A	3004	2	TR85/250	G0005
13-100000	732				
13-105698-1	128	3024		TR87/243	
13-283361-1	197	3083			
14-0086-1	129				
14-0104-1	152				
14-0104-2	153				
14-0104-3	129				
14-0104-4	128				
14-0104-5	172				
14-0104-7	123A				
14-1	123A	3122		TR21/735	S0015
14-2	123A	3122		TR21/735	S0015
14-3	123A	3122		TR21/735	S0015
14-2011-02	779				
14-557-10	102A	3004	2	TR85/	
14-569-09	160		51	TR17/637	
14-572-10			10		
14-573-10	121	3009	4	TR01/232	
14-574-10	121	3009	16	TR01/232	
14-575-10	123A	3122	8	TR21/735	
14-576-10			53		
14-577-10	102A	3004	53	TR85/250	G0005
14-578-10	121	3009	16	TR01/232	
14-579-10	121	3009	16	TR01/232	
14-580-01	160		50	TR17/637	
14-581-01	160		51	TR17/637	
14-582-01	126	3006	50	TR17/635	
14-583-01	123A	3122	8	TR21/735	
14-584-01	102A		53	TR85/250	
14-585-01	160		51	TR17/637	
14-586-01	104	3009	16	TR01/624	
14-587-01	160		50	TR17/637	
14-588-01	160			TR17/637	
14-589-01	121	3009	16	TR01/232	
14-590-01	121	3009	3	TR01/232	
14-591-01	160		51	TR17/637	
14-600-01	160	3006	51	TR17/637	G0002
14-600-02	160		51	TR17/637	G0002
14-600-04	160	3006	51	TR17/637	
14-600-07				TR17/637	
14-600-10		3006	51	TR17/637	
14-600-11	160		50	TR17/637	
14-600-13	160		50	TR17/637	
14-600-16	160	3006	51	TR17/637	
14-600-19	160	3008	51	TR17/637	G0002
14-600-20	160		51	TR17/637	
14-600-22	160		51	TR17/637	
14-601-01	121	3009	4	TR01/232	
14-601-02	179		16	TR35/G6001	
14-601-03	121	3009	16	TR01/232	
14-601-04	121	3009	16	TR01/232	
14-601-05	121	3009	16	TR01/232	
14-601-06	121	3009	16	TR01/232	
14-601-07	121	3009	16	TR01/232	
14-601-08	104	3009	13	TR01/624	
14-601-09	121	3009	16	TR01/232	
14-601-10	130	3027		TR59/247	
14-601-11	121	3009	16	TR01/232	
14-601-12	130	3027	14	TR59/247	
14-601-13	130	3027	14	TR59/247	
14-601-14	162		19	TR67/707	
14-601-15	130	3027	14	TR59/247	
14-601-15A	130	3027		TR59/	
14-601-16	130	3027		TR59/247	

Original Part Number	ECG	SK	GE	IR/WEP	HEP
14-601-16A	130	3027		TR59/	
14-601-17	184	3027	19	TR26/S5003	
14-601-18	223				
14-601-20	223		14		
14-601-23	223	3027			
14-601-24	223				
14-601-26	123A	3079			
14-601-27	165	3115			
14-601-28	123A				
14-601-29	123A				
14-601-30		3027			
14-602-01	123A	3124	18	TR21/735	
14-602-02	123A	3018	10	TR21/735	
14-602-03	123A	3124	20	TR21/735	S0015
14-602-04	102A		51	TR85/250	
14-602-05	102A	3004	2	TR85/250	
14-602-05A	102A	3004	2	TR85/250	
14-602-06	158		53	TR84/630	
14-602-07	158		53	/630	
14-602-08	158		53	TR84/630	
14-602-09	158		53	/630	
14-602-3	103A				
14-602-10	102A		52	TR85/250	
14-602-11	159	3114	22	TR20/717	
14-602-11A	159	3114		TR20/	
14-602-12	123A	3122	10	TR21/735	
14-602-13	123A	3020	20	TR21/735	S0015
14-602-14	123A	3122	62	TR21/735	
14-602-15	102A		54	/630	
14-602-16	123A	3122	60	TR21/735	
14-602-17	123A	3122	20	TR21/735	
14-602-18	128	3124	20	TR87/243	
14-602-19	154	3045	18	TR78/712	
14-602-20	159	3114	21	TR20/717	
14-602-20A	159	3114		TR20/	
14-602-21	103A		54	TR08/724	
14-602-22	123A	3122	20	TR21/735	
14-602-23	123A	3122	62	TR21/735	
14-602-24	128	3024	20	TR87/243	
14-602-25		3122	62	TR51/735	
14-602-26	123A	3122	20	TR21/735	
14-602-27	103A		54	/630	
14-602-28	129	3025	21	TR88/242	
14-602-28A	129	3025		TR88/	
14-602-29	128	3124	20	TR87/243	
14-602-30	128	3124	20	TR87/243	
14-602-31		3018	20	TR21/735	
14-602-32	159	3025	21	TR20/717	
14-602-32A	159	3114		TR20/	
14-602-33	103A		54	/630	
14-602-34		3018	20	TR21/735	
14-602-35	123A	3124	20	TR21/735	S0015
14-602-36	154	3045	27	TR78/712	
14-602-37	154	3045	27	TR78/712	
14-602-38	103A			TR08/724	
14-602-39	103A			TR84/630	
14-602-40	103A			/630	
14-602-41	108	3018	20	TR95/56	S0015
14-602-42	159	3114	21	TR20/717	
14-602-43	128	3024	63	TR87/243	
14-602-44	159	3114	21	TR20/717	
14-602-44A	159	3114		TR20/	
14-602-45		3122	60	TR51/735	
14-602-46	199	3018	20	TR21/735	S0015
14-602-46A	199	3124		TR21/	
14-602-47	159	3114	67	TR20/717	
14-602-47A	159	3114		TR20/	
14-602-48	123A	3122	62	TR21/735	S0016
14-602-49	128	3024	20	TR87/243	S0015
14-602-50	123A	3122	62	TR21/735	
14-602-51	158			TR84/	
14-602-52	103A			TR08/	
14-602-54	193	3114	21	/717	S0013
14-602-54A	159	3114		TR20/	
14-602-55	192	3024	63	TR22/243	S0014
14-602-55A	123A	3124		TR21/	
14-602-56	159	3114	21	TR20/717	S0019
14-602-56A	159	3114		TR20/	
14-602-57	128		67		
14-602-58	159	3025	21	TR20/717	S0031
14-602-58A	159	3025		TR20/	
14-602-59	128	3024	63	TR87/243	S0031
14-602-60		3025	21	TR20/717	S0019
14-602-61	123A	3124	20	TR21/735	S0015
14-602-62	123A	3122	10	TR21/735	S0015
14-602-63	199	3039	62	TR21/720	
14-602-65			63	TR87/	
14-602-66			67	TR88/	
14-602-67	175		27	TR81/241	
14-602-68	159		20	TR20/	
14-602-69	123A	3122	63	TR21/735	S0015
14-602-70	192		63		
14-602-71	193		67		
14-602-72	128		63	TR87/	
14-602-73	129		67	TR88/	
14-602-74	192	3024	63	TR25/	
14-602-75	193	3025	67	TR28/	S5014
14-602-76	198		27	TR78/	
14-602-77	107	3018	63	TR70/720	S0001
14-602-77B	107	3018		TR70/	
14-602-78	123A	3122	11	TR21/735	S0020
14-602-79	129	3025	21	TR88/242	S0012
14-602-79A	129	3025		TR88/	
14-602-80	123A			TR21/735	S0015
14-602-81	123A	3122	20	TR21/735	S0005
14-602-85				TR19/	S0031
14-602-87	123A				
14-602-88	159				
14-602-89	123A				
14-602-90	159				
14-602-580	159	3114		TR20/	
14-602-600	159	3114		TR20/	
14-603-01	128	3024	60	TR87/243	
14-603-02		3018	20	TR87/243	S0016
14-603-02A	128	3018		TR87/	
14-603-03	123A	3018	17	TR21/735	S0017
14-603-04	123A	3122	60	TR21/735	
14-603-05	107	3039	11	TR70/720	S0016
14-603-05-2	107	3018		TR70/	
14-603-06	107	3039	17	TR70/720	
14-603-07	128	3024	60	TR87/243	
14-603-08	161		60	TR83/	
14-603-09	161		60	TR83/	
14-603-10	123A	3122	17	TR21/	S0020
14-603-11	123A	3122	11	TR21/735	S0015
14-603-12		3122	20	TR21/	S0016
14-603-13	161				
14-604-01	155		43		
14-604-02	155	3009	16	TR01/232	
14-604-03	131				
14-604-07	121	3009	16	TR01/232	
14-604-08	121	3009		TR01/232	
14-606-04				TR51/	
14-606-06				TR21/	
14-606-07				TR21/	
14-607-29	129			TR88/242	S0012
14-607-29A	129	3025		TR88/	
14-608-01	184	3054	57	/S5003	
14-608-02	184	3054	57	/S5003	S5000
14-608-02A	196	3054		TR92/	
14-608-03	184	3190			
14-608-04	184	3190			
14-609-00	152				
14-609-01	191		27	TR75/	S3021
14-609-01A	191	3044		TR75/	
14-609-02	171	3044	57	TR79/244A	S3021
14-609-02A	191	3044		TR75/	
14-609-03	152	3054	57	TR76/701	S5000
14-609-03A	196	3054		TR92/	
14-609-04	152	3054	28	TR76/701	S5000
14-609-05	191	3044	27	TR75/	S3021
14-609-06	152				
14-609-08	152				
14-609-09	196				
14-609-49A	108	3018		TR95/	
14-651-12	123A				
14-652-12	161				
14-653-21		3017			
14-654-21		3032			
14-655-13	123A				
14-656-21	123A				
14-659-12	123A				
14-660-12		3122			
14-661-21		3122			
14-700-01	132		FET2		
14-700-02	133		FET2		
14-700-03	132		FET2	FE100/	
14-700-04	132	3116	FET2	FE100/802	
14-700-05	132	3116	FET2	FE100/802	
14-700-06	132				

Original Part Number	ECG	SK	GE	IR/WEP	HEP
14-710-21	132				
14-713-31	132				
14-713-32	132				
14-714-13	132				
14-800-32	123A			TR51/	
14-801-12	123A				
14-801-23	199	3122		TR08/	
14-802-12		3122		TR51/	
14-803-12	159	3114		TR52/	
14-803-32	234				
14-804-12	159			TR52/	
14-805-12	123A	3122		TR51/	
14-806-12	123A				
14-806-23	123A				
14-807-12	106	3004		TR54/	
14-807-23	123A				
14-808-12	159			TR52/	
14-809-12				TR08/	
14-809-22		3122			
14-809-23	123A				
14-809-32	123A				
14-811-12	194				
14-850-12	161			TR51/	
14-851-12	161			TR51/	
14-851-32	123A				
14-852-23				TR21/	
14-853-23	123A			TR08/	
14-854-12	123A			TR51/	
14-855-32	159				
14-856-23	159	3122		TR52/	
14-857-12	159	3118		TR52/	
14-857-32	129				
14-857-79	159				
14-858-12	123A	3122		TR51/	
14-860-23				TR08/	
14-861-12	106	3122		TR54/	
14-862-23	123A				
14-862-32	123A	3118		TR51/	
14-863-23	159			TR52/	
14-864-12	123A				
14-864-23	159				
14-865-12	123A			TR51/	
14-866-32	123A			TR51/	
14-867-32	159			TR52/	
14-900-12	191				
14-901-12	171	3104		TR60/	
14-902-23	152				
14-903-23	152	3054			
14-904-12	191	3104			
14-904-14				TR60/	
14-905-23		3054			
14-906-13	184	3190			
14-907-13	184	3190			
14-908-23	152	3054			
14-909-23	158	3054			
14-910-13	158	3104			
14-911-13	184	3190			
14-2000-01	172		64	TR69/	
14-2000-02	172		64	TR69/	
14-2000-03	172		64	TR69/	
14-2000-04	172		64	TR69/	
14-2000-05	172			TR69/	
14-2000-23	172				
14-2002-01	133	3112		/801	
14-2007-00	705A	3134	IC3	IC502/	C6089
14-2007-00B	707				
14-2007-01	707		IC3	IC502/	C6089
14-2007-02	705A	3134	IC3	IC502/	C6089
14-2007-03	707				
14-2007-04	705A	3134			
14-2008-01	708	3135	IC10		C6062P
14-2010-01	713	3077	IC5		C6027P
14-2010-02	713	3077			
14-2010-03			IC18		
14-2011-01	779				
14-2012-01	801	3160			
14-2014-01	814				
14-2052-23	172				
14-2053-23	172				
14-2054-01	731	3170			
14-32430	108				
14-40325A	130	3027		TR59/	
14-40363A	130	3027		TR59/	
14-40369A	130	3027		TR59/	
14-40421A	130	3027		TR59/	

Original Part Number	ECG	SK	GE	IR/WEP	HEP
14-40464A	130	3027		TR59/	
14-40465A	130	3027		TR59/	
14-40466A	130	3027		TR59/	
14-40471	130	3027		TR59/	
14-40934-1	130	3027		TR59/	
14-90412	191				
14A	121	3009		TR01/	
14A0	121	3009		TR01/	
14A1	121	3009		TR01/	
14A1-A82	121	3009		TR01/	
14A2	121	3009		TR01/	
14A3	121	3009		TR01/	
14A4	121	3009		TR01/	
14A5	121	3009		TR01/	
14A6	121	3009		TR01/	
14A7	121	3009		TR01/	
14A8	121	3009		TR01/	
14A8-1	160	3007		TR17/	
14A8-1-12	160	3007		TR17/	
14A8-1-12-8		3007			
14A9	121	3009		TR01/	
14A10	121	3009		TR01/	
14A10R	121	3009		TR01/	
14A11	121	3009		TR01/	
14A12	121	3009		TR01/	
14A13	121	3009		TR01/	
14A13P	121	3009		TR01/	
14A14	121	3009		TR01/	
14A14-7	121	3009		TR01/	
14A14-7B	121	3009		TR01/	
14A15	121	3009		TR01/	
14A15L	121	3009		TR01/	
14A16	121	3009		TR01/	
14A16-5	121	3009		TR01/	
14A17	121	3009		TR01/	
14A17-1	121	3009		TR01/	
14A18	121	3009		TR01/	
14A19	121	3009		TR01/	
14A19G	121	3009		TR01/	
14A-70		3009			
14A-70-12		3009			
14A-70-12-7		3009			
14A119		3009			
14A121		3009			
14A182		3009			
1414-180	175				
14MW69	102A	3004		TR85/250	
14T1					G6011
15-0009-05	196				
15-008-1	106				
15-01742	159				
15-01913-00	159				
15-01999	123A				
15-02155	128				
15-02757-00	199				
15-02762-00	159				
15-02762-1	159				
15-02762-2	159				
15-02979	159				
15-03014-00	123A				
15-03051-00	107				
15-03068	130				
15-03099	159				
15-03100	123A				
15-03409-0	159				
15-03409-02	159				
15-03409-1	159				
15-05302	123A				
15-05369	123A				
15-05415	218				
15-05650	123A				
15-08800U	108	3019			
15-09090-01	159				
15-09338	123A				
15-09587	199				
15-09650	128				
15-09980	123A				
15-082019	123A	3044		TR21/735	S0014
15-085042	174				
15-088002	159	3114		TR20/717	S0019
15-088002A	159	3114		TR20/	
15-088003	107	3019		TR70/720	S0016
15-088004	108	3019	11	TR95/	S0016
15-1	123A	3122		TR21/735	S0015
15-2	123A	3122		TR21/735	S0015

Original Part Number	ECG	SK	GE	IR/WEP	HEP
15-3	159	3114		TR20/717	S0019
15-4	159	3114		TR20/717	S0019
15-5	159	3114		TR20/717	S0031
15-30	159	3114		TR20/	
15-40	159	3114		TR20/	
15-50	159	3114		TR20/	
15-166N	107			TR70/	
15-875-075-001	159				
15-875-075-003	123A				
15-10062-0	128				
15-14471-1	1085				
15-26587	703A			IC500/	
15-26587-1	703A	3157	12	IC500/	C6001
15-31015-1	1022				
15-33201	712	3072			C6063P
15-33201-1	712	3072	2	IC507/	C6063P
15-33201-2	712	3072			C6063P
15-34005-1	727	3071			
15-34048-1	708	3135	10	IC504/	C6062P
15-34048-2	708	3135			
15-34049	722	3161			
15-34049-1	722	3135	9	IC513/	C6068P
15-34202-1	727	3071			
15-34379-1	720		7	IC512/	C6095P
15-34401-1	762				C6012
15-34408-2	708				
15-34452	709			IC505/	
15-34452-1	709		11		C6062P
15-34502-1	778				
15-34502-2	778				C6102P
15-34503	722	3161		IC513/	
15-34503-1	722	3161	9		C6056P
15-34906-1	799				
15-35059-1	712	3072	2	IC507/	C6063P
15-35059-2	712	3072			
15-36446-1	743		32		
15-36647-1	813				
15-37534-1	794				
15-37534-2	794				
15-37700-1	792				
15-37701-1	793				
15-37701-2	793				
15-37702-1	714	3075	4		C6070P
15-37703-1	715	3076	6		C6071P
15-37704-1	790	3077	18		C6057P
15-37833-1	909D	3590			
15-37833-2	909D	3590			
15-39060-1	747				C6079P
15-39061-1	795				C6099P
15-39075-1	738	3167	29		C6075P
15-39207-1	808				
15-39208-1	809				
15-39209-1	791	3149	33		
15-39600-1	794				
15-505753-2	911				
15-505753-3	911D				
15-2221011	102A	3004		TR85/	
15-22210111	102A	3004		TR85/	
15-22210131	160	3006		TR17/	
15-22210300	160	3006		TR17/	
15-22210921	160	3007		TR17/	
15-22211021	160	3006		TR17/	
15-22211200	102A	3004		TR85/	
15-22211328	102A	3004		TR85/	
15-22211921	160	3006		TR17/	
15-22214400	160	3006		TR17/	
15-22214411	160	3006		TR17/	
15-22214435	160	3006		TR17/	
15-22214821	160	3006		TR17/	
15-22214831	160	3006		TR17/	
15-22216500	102A	3004		TR85/	
15-22216600		3007		TR17/	
15-22217400	160	3006		TR17/	
15-22223720	123A	3020		TR21/	
1523	159	3114		TR20/118	S0019
16-17		3116	FET2	/802	
16-19(SYM)	129			TR88/242	
16-20		3024		TR21/	
16-20(SYM)	128			TR87/243	
16-736	108	3019	11	TR95/56	
16-21426	161	3018	11	TR83/719	
16-147191229	123A	3124		TR21/	
16-171191368	123A	3124		TR21/	
16-207190405	102A	3004		TR85/	
16A1	103	3122	10	TR08/641A	
16A1(FLTWD)	123A			TR21/735	
16A2	103	3010	17	TR08/641A	
16A2(FLTWD)				TR21/	
16A545-7	123A			TR21/	S0030
16A667-GRN			212	TR51/	
16A667-ORG			212	TR51/	
16A667-RED			212	TR51/	
16A667-YEL			212	TR51/	
16A668-GRN			212	TR51/	
16A668-ORG			212	TR51/	
16A668-RED			212	TR51/	
16A668-YEL			212	TR51/	
16A669-GRN			212	TR51/	
16A669-YEL			212	TR51/	
16A787	102		2	TR05/	
16A1938	123A	3122		TR21/735	S0015
16B-3A82	121MP	3015			
16B1			210	TR53/	
16B2			210	TR53/	
16B670-GRN			64	TR53/	
16B670-RED			64	TR53/	
16B670-YEL			64	TR53/	
16E1330(GE)	123A	3122	20	TR51/735	
16G2	123A	3122	11	TR21/735	S0015
16G27	133				
16GN	107	3019		TR70/720	S0020
16J1	108	3019	20	TR95/56	S0016
16J2	108	3017A	20	TR95/56	S0016
16K1	108	3018	60	TR95/	S0015
16K2	108	3018	60	TR95/56	S0015
16K3	108	3039	60	TR95/56	S0015
16L2	108	3039	212	TR95/56	S0016
16L3	108	3039	212	TR95/56	S0016
16L5	108	3039	20	TR95/56	S0016
16L22	108	3039	212	TR95/56	S0016
16L23	108	3039	212	TR95/56	S0016
16L42	123A	3122	212	TR21/735	S0011
16L43	123A	3019	212	TR21/735	S0011
16L44	123A	3019	212	TR21/735	S0011
16L45			212	TR51/	
16L62	123A	3122	212	TR21/735	S0011
16L63	123A	3122	212	TR21/735	S0011
16L64	107	3019	212	TR70/720	S0011
16L65			212	TR51/	
16P2881	172		64	TR69/S9100	S9100
16P3367	172		64	TR69/S9100	S9100
16U1(HEATH KIT)	123A	3122	20	TR21/	
16X1	123A	3122	81	TR21/735	S0011
16X2	123A	3122	81	TR21/735	S0011
17-50(FISHER)				TR59/	
17-443	128	3024	18	TR87/243	S3011
17-451	123A	3124	20	TR21/735	S0015
17-457	123A	3124	20	TR21/735	
17-458	129	3024		TR88/242	S5014
17-459	159	3114	21	TR20/717	S5022
17-459A	159	3114		TR20/	
17-12054-1	9930				
17-12056-1	9932				
17-12057-1	9933				
17-12058-1	9946				
17-12064-1	9962				
17-12065-1	9945				
17-12089-1	9951				
17-12096-1	909	3590			
17-12097-1	910				
17A4422-1	121	3009	16	TR01/232	
18-148A	123A				
18-161-2		3024			
18-161-3		3024			
18-161-4		3024			
18-177-1	175				
18-177-2		3026			
18-178-1		3027			
18-197-1		3044			
18-197-2		3044			
18-198-2	124				
18-605	912				
18-605-RD	912				
18-606	724				
18-3539-1	128				
18A8-1	160	3007		TR17/	
18A8-1-12	160	3007		TR17/	
18A8-1-127	160	3007		TR17/	
18A8-1L	160	3007		TR17/	
18A8-1L8	160	3007		TR17/	

Original Part Number	ECG	SK	GE	IR/WEP	HEP
18AA-1-82	121	3009		TR01/	
18P-2A82	102A	3004		TR85/	
19-00-3485	175				
19-020-001	126			TR17/	
190020-002	126			TR17/	
19-020-003	102A			TR85/	
19-020-005	126			TR17/	
19-020-007	102A			TR85/	
19-020-015	102			TR05/	
19-020-031	160	3006		TR17/637	
19-020-032	160	3006		TR17/637	
19-020-033	100	3005	1	TR05/254	
19-020-034	102A	3004	2	TR85/250	
19-020-035	102A	3004	2	TR85/250	
19-020-036	102A	3004	2	TR85/250	
19-020-037	108	3039	17	TR95/56	
19-020-038	128	3048		TR87/	
19-020-043	123A		10	TR24/	S0011
19-020-043A	123A		10	TR24/	S0015
19-020-044	108		11	TR22/	S0015
19-020-045	175	3021			
19-020-046	154				
19-020-048	108	3018	20	TR95/56	S0016
19-020-050	128				S3020
19-020-052	108		17	TR21/	S0015
19-020-056	152			TR65/	S5014
19-020-058	123A		20	TR21/	S0015
19-020-066	152		57	TR55/	S5014
19-020-070	107	3018		TR70/720	S0016
19-020-071	107	3018		TR70/720	S0015
19-020-072	107	3039		TR70/720	S0011
19-020-073	123A	3124		TR21/735	S0015
19-020-074	123A	3124		TR21/735	S0014
19-020-075	128	3024		TR87/243	S5026
19-020-076	123A	3018		TR21/735	S0025
19-020-077	195	3024		TR65/243	S3001
19-020-078	224	3049			S3001
19-020-079	703A	3157	IC12	IC500/	C6001
19-020-081					S9002
19-020-44	108	3018		TR95/56	
19-020-100	129		67	TR28/	S3012
19-020-101	186	3192	28	TR55/	S5014
19-020-102	187	3193	29	TR56/	
19-020-114	159		21	TR30/	S0012
19-020-115	132		FET2	FE100/	F0021
19-09234	941				
19-09973-0	7401				
19-09981-0	923				
19-076001	1003				
19-1	159	3114		TR20/717	S0019
19-2	159	3114		TR20/717	S0019
19-2-02616	123A				
19-3	159	3114		TR20/717	S5022
19-10	159	3114		TR20/	
19-20	159	3114		TR20/	
19-30	159	3114		TR20/	
19-130-001	781	3169			
19-130-004	74145				
19-130-005	7490				
19-3349	128	3047		TR87/243	S0014
19-3415	158	3004	52	TR84/630	G0005
19-3416	158	3004	52	TR84/630	G0005
19-3692	128	3048		TR87/243	
19-3934-643	128	3024		TR87/	S0015
19-3935-641	190	3104		TR74/	S3019
19-10298-00	941M				
19-10415-00	923D				
19-10476-0	74H01				
19-15840	128				
19-15850		3024			
19-19420	108				
19A11644P1	708	3135	10		C6062P
19A115056-P1	127	3035	25	TR27/	
19A115061-P1	123A	3132	17	TR21/	
19A115061-P2	123A	3004	17	TR21/	
19A115077-P1	107	3004	2	TR05/	
19A115077-P2	102	3004	2	TR05/	
19A115087-P1	126	3006	1	TR17/	
19A115094-P1	105	3012	4	TR03/	
19A115098	126	3006	1	TR17/	
19A115098-P1	126	3006	1	TR17/	
19A115099-P1	126	3006	1	TR17/	
19A115101	121	3009	3	TR01/	
19A115101-P1	121	3009	3	TR01/	
19A115102-P1	123A	3122	17	TR21/	
19A115103-P1	101	3011	5	TR08/	
19A115108-P1	123A	3122	20	TR21/	
19A115108-P2	123A	3122	20	TR21/	
19A115123-2	123A	3122	20	TR21/	
19A115123-P1	123A	3122	20	TR21/	
19A115123-P2	123A	3122	20	TR21/	
19A115129-2	103	3010	8	TR08/641A	
19A115129-P1	103	3010	8	TR08/	
19A115140-P1	160	3006	51	TR17/	
19A115140-P2	160	3006	51	TR17/	
19A115142-P1	123A	3122	20	TR21/	
19A115142-P2	123A	3122	20	TR21/	
19A115157-1	123A				
19A115167-2	123A				
19A115178-P1	159	3025	21	TR20/	
19A115178-P2	159	3025	21	TR20/	
19A115180-2	129	3025	22	TR88/242	
19A115184-P1	121		3	TR01/	
19A115192-P1	160	3006	51	TR17/	
19A115192-P2	160	3006	51	TR17/	
19A115200-P1	152	3054	66	TR76/	
19A115201-P1	101	3011	6	TR08/	
19A115201-P2	101	3011	6	TR08/	
19A115208	100		1	TR05/	
19A115208-P1	100		1	TR05/	
19A115208-P2	100		1	TR05/	
19A115238-2	128				
19A115238-P1	128				
19A115245-P1	123A				
19A115245-P2	123A				
19A115249-1	108	3019	11	TR95/56	
19A115253-P1	123A	3122	17	TR21/	
19A115253-P2	123A	3122	17	TR21/	
19A115267-P1	121	3009		TR01/	
19A115281-P1	102	3004	2	TR05/	
19A115300	128				
19A115300-1	128	3024	18	TR87/243	
19A115300-2	128	3024		TR87/243	
19A115300-P1	186	3054	28	TR55/	
19A115300-P2	186	3054	28	TR55/	
19A115300-93	186	3054		TR55/	
19A115301-P1	100	3005	1	TR05/	
19A115301-P2	100	3005	1	TR05/	
19A115304-2	128	3024		TR87/243	
19A115315-P1	123A	3122	17	TR21/	
19A115315-P2	123A	3122	17	TR21/	
19A115321-1	123A				S0011
19A115330	123A				
19A115341-P1	121	3009		TR01/	
19A115342-1	108	3018	20	TR95/56	
19A115342-2	161	3019			
19A115342-P1	123A	3122	20	TR21/	
19A115342-P2	123A	3122	20	TR21/	
19A115359-P1	123A	3122	20	TR21/	
19A115359-P2	123A	3122	20	TR21/	
19A115361-P1	121	3009	3	TR01/	
19A115362-P1	123A	3009	17	TR21/	
19A115362-P2	123A	3009	17	TR21/	
19A115385-P1	121	3009	3	TR01/	
19A115410-P1	123A	3122	17	TR21/	
19A115410-P2	123A	3122	17	TR21/	
19A115440-1	108	3019	11	TR95/56	
19A115440-2	108				
19A115441-1	108	3018	20	TR95/56	
19A115458-P1	159	3025	21	TR20/	
19A115458-P2	159	3025	21	TR20/	
19A115487-P1	105	3012	4	TR03/	
19A115527	175	3026	23	TR81/241	
19A115527-P1	175	3538	23	TR81/	
19A115531-P1	127	3025	25	TR27/	
19A115540-P1	105	3012	4	TR03/	
19A115546-P1	101	3011	5	TR08/	
19A115546-P2	101	3011	5	TR08/	
19A115552-P1	123A		17	TR25/	
19A115552-P2	123A		17	TR25/	
19A115553-P1			51	TR17/	
19A115554-P1			51		
19A115554-P2			51	TR17/	
19A115556-P1			1	TR85/	
19A115561	121	3009	16	TR01/232	
19A115561-1	121	3		TR24/	
19A115562-P2	159				
19A115567-P1			51	TR17/	
19A115567-P2			51	TR17/	
19A115591-P1	123A				

Original Part Number	ECG	SK	GE	IR/WEP	HEP
19A115591-P2	123A				
19A115623-P1	124	3021	12	TR81/	
19A115623-P2	124	3021	12	TR81/	
19A115628-P1	160	3006	51	TR17/	
19A115628-P2	160		51	TR17/	
19A115635-1	160	3006		TR17/	
19A115635-P1			51	TR17/	
19A115636-P1	160	3006	51	TR17/	
19A115653-P1	159	3025	21	TR17/	
19A115653-P2	159	3025	21		
19A115654-P1	159	3025	21	TR20/	
19A115654-P7	159	3025	21	TR20/	
19A115665-P1	160	3006	51	TR17/	
19A115665-P2	160	3006	51	TR17/	
19A115666-1	108	3019	11	TR95/56	
19A115673-P1	101	3011	7	TR08/	
19A115673-P2	101	3011	7	TR08/	
19A115674-P1	102A	3004	2	TR85/	
19A115674-P2	102A	3004	404	TR85/	
19A115688-P1	159	3025	21	TR20/	
19A115688-P2	159	3025	21	TR20/	
19A115706-1	159				
19A115706-2	159				
19A115706-P1	159	3025	22	TR20/	
19A115706-P2	159	3025	22	TR20/	
19A115720-1	123A	3124	20	TR21/735	
19A115720-2	123A				
19A115728-1	123A	3124	20	TR21/735	
19A115768-1	159				
19A115768-2	159				
19A115768-3	159				
19A115768-P1	159	3025	22	TR20/	
19A115768-P2	159	3025	22	TR20/	
19A115779-P1	159				
19A115783-1	124				
19A115786	123A	3124	20	TR21/735	
19A115786A	123A	3124	20	TR21/735	
19A115817-P1	124				
19A115818	181				
19A115852-P1	159				
19A115889-P1	128				
19A115889-P2	128				
19A115889-P3	128				
19A115910-P1	123A				
19A115913-1	9930				
19A115913-2	9951				
19A115913-3	9932				
19A115913-4	9933				
19A115913-10	9093				
19A115913-19	9946				
19A115913-20	9962				
19A115913-P2	9951				
19A115913-P14	9936				
19A115925-1	108				
19A115944-P1	123A				
19A115944-P2	123A				
19A115976-P1	129				
19A116118-2	152				
19A116118-3		3054			
19A116118-I	152				
19A116118-P1	152				
19A116118-P2	152				
19A116180-7	7440				
19A116180-8	7450				
19A116180-11	7454				
19A116180-18	7486				
19A116180-22	7405				
19A116180-24	7490				
19A116180-25	7491				
19A116180-27	7492				
19A116180-29	7495				
19A116180P1	7400				
19A116180P2	7401				
19A116180P3	7402				
19A116180P4	7410				
19A116180P5	7420				
19A116180P7	7440				
19A116180P8	7450				
19A116180P11	7454				
19A116180P13	7470				
19A116180P15	7473				
19A116180P16	7474				
19A116180P18	7486				
19A116180P20	7404				
19A116223P1	159				
19A116297P3	941				
19A116297P3-9	941				
19A116297P3-10	941				
19A116375	153				
19A116375P1	153				
19A116408-1	159				
19A116445	709				
19A116549P1	909	3590			
19A116623	912				
19A116631P1	123A				
19A116753P1	181				
19A116755P1	123A				
19A116761P1	130				
19A116774-P1	123A				
19A116796-1	704	3023			
19A116797-1	710	3102			
19A116841P1	923				
19A116865	123A				
19A123160-1	108	3019	11	TR95/56	
19A123160-2	108				
19A126265-1	160				
19A126265-2	160				
19A126813	130	3027		TR59/247	S7004
19A126813A	130	3027		TR59/	
19A126826-P2	181	3036			
19A129207P1	123A				
19AR6-1	103A	3010		TR08/	
19AR6-2	103A	3010		TR08/	
19AR6-3	103A	3010		TR08/	
19AR7-1	102A	3004		TR85/	
19AR7-2	102A	3004		TR85/	
19AR13-1	160	3006		TR17/	
19AR13-2	160	3006		TR17/	
19AR13-3	160	3006		TR17/	
19AR13-4	160	3006		TR17/	
19AR14-1	102A	3004		TR85/	
19AR14-2	102A	3004		TR85/	
19AR16-1	102A	3004		TR85/	
19AR16-2	102A	3004		TR85/	
19AR18	160	3006	51	TR17/637	
19AR19-1	102A	3004		TR85/	
19AR19-2	102A	3004		TR85/	
19AR20	123A	3122		TR21/	
19AR24	160	3006		TR17/	
19AR25	102A	3004		TR85/	
19AR26	102A	3004		TR85/	
19AR27	102A	3004		TR85/	
19AR31	121	3009		TR01/	
19AR32	102A	3004		TR85/	
19AR35	124	3021	12	TR81/240	
19AR36	123A	3122	20	TR21/735	
19B200054-P1	102A	3004	2	TR85/	
19B200061-P1	102A	3004	2	TR85/	
19B200061-P2	102A	3004	2	TR85/	
19B200061-P3	102A	3004	2	TR85/	
19B200061-P4	102A	3004	2	TR85/	
19B200063-P1	102A	3004	2	TR85/	
19B200065-P1	101	3011	5	TR08/	
19B200065-P2	101	3011	5	TR08/	
19B200129-P1			1	TR85/	
19B200130-P1			51	TR17/	
19B200130-P2			51	TR17/	
19B200132-P1			2	TR05/	
19B200132-P2			2	TR05/	
19B200132-P3			2	TR05/	
19B200132-P4			2	TR05/	
19B200210-P1	102A	3004	2	TR85/	
19B200210-P2	102A	3004	2	TR85/	
19B200210-P3	102A	3004	2	TR85/	
19B2000129-P1	100	3005		TR05/	
19B2000130-P1	160	3006		TR17/	
19B2000130-P2	160	3006		TR17/	
19B2000132-P1	102A	3004		TR85/	
19B2000132-P2	102A	3004		TR85/	
19B2000132-P3	102A	3004		TR85/	
19B2000132-P4	102A	3004		TR85/	
19C300115-1	128				
19C300073-P1	102	3004	2	TR05/	
19C300073-P2	102	3004	2	TR05/	
19C300073-P3	102	3004	2	TR05/	
19C300073-P4	102	3004	2	TR05/	
19C300073-P5	102	3004	2	TR05/	
19C300073-P6	102	3004	2	TR05/	
19C300074-P2	102	3004	2	TR05/	
19C300113-P1	121	3009	3	TR01/	

Original Part Number	ECG	SK	GE	IR/WEP	HEP
19C300114-P1	123A	3122	17	TR21/	
19C300114P1	123A				
19C300114-P2	123A	3122	17	TR21/	
19C300114P2	123A				
19C300114P3	123A				
19C300115-P1	128	3024	18	TR87/	
19C300128-P1	102	3004	2	TR05/	
19C300128-P2	102	3004	2	TR05/	
19C300128-P3	102	3004	2	TR05/	
19C300128-P4	102	3004	2	TR05/	
19C300128-P5	102	3004	2	TR05/	
19C300128-P6	102	3004	2	TR05/	
19C300128-P7	102	3004	2	TR05/	
19C300128-P8	102	3004	2	TR05/	
19C300138-P4	102	3004	2	TR05/	
19C300138-P8	102	3004	2	TR05/	
19C300216-P1	160	3006	51	TR17/	
19C300216-P2	160	3006	51	TR17/	
19E15-1	9905				
19E18-1	9904				
19E19-2	9903				
19L-4A82	102A	3003		TR85/	
19QC17	107				
19QC19	234				
20-00229-001	108				
20-00444-001	108				
20-0352	904				
20-1	108	3039		TR95/56	S0020
20-1110-014		3122			
20-1111-002	130				
20-1111-003	130				
20-1680-174	128	3024		TR87/	
20-1680-189	158	3004		TR84/	
20A0007	102A	3004	2	TR85/250	G0005
20A0009	102A	3004	2	TR85/250	G0005
20A0015	102A	3004	2	TR85/250	G0005
20A0017	121	3009	16	TR01/232	G6003
20A0041	121	3009		TR01/	
20A0042	121	3009		TR01/	
20A0053	123A	3122		TR21/	
20A0055	128				
20A0059	189	3083		TR73/	
20A0060	189	3083		TR73/	
20A0073	123A	3122		TR21/	
20A0074	121	3009		TR01/	
20A0075	129	3025		TR88/	
20A0076	128	3024		TR87/	
20A10849	123A	3122		TR21/735	S0014
20B1M	703A		12		
20B2M	703A	3157			
20B3AH	703A	3157			
20C71	102A	3123	52	TR85/250	G0005
20C72	102A	3004	52	TR85/250	G0005
20V-HG	102A	3123	52	TR85/250	G0005
21-0099		3124			
21-0101	192				
21-1	123A	3122		TR21/735	S0015
21-1AA(MGVX)	6402				
21-1L	123A	3124		TR21/	
21-28	121	3009	3	TR01/232	G6005
21-32	126	3008	51	TR17/635	
21-33	126	3008	51	TR17/635	
21-34	102	3004	2	TR05/631	
21-35	124	3021	12	TR81/240	
21-36	102	3003	2	TR05/631	G0005
21-37	102	3004	2	TR05/631	G0005
21A005-000	102A	3004	2	TR85/250	
21A015-001	102A	3004	2	TR85/250	S0019
21A015-002	155			TR50/	
21A015-003	131	3052		TR94/642	G6016
21A015-004	108	3018	11	TR95/56	
21A015-005	103A	3010	8	TR08/735	
21A015-006	102A	3004	2	TR85/250	
21A015-008	159	3025	22	TR20/717	S0013
21A015-008A	159	3114		TR20/	
21A015-009	159	3025		TR20/717	S5022
21A015-009A	159	3114		TR20/	
21A015-011	159	3025	21	TR20/717	S0019
21A015-011A	159	3114		TR20/	
21A015-012	159	3025		TR20/717	S0019
21A015-012A	159	3114		TR20/	
21A015-013	123A	3124	20	TR21/735	S0015
21A015-014	108	3018		TR95/56	S0030
21A015-016	108	3018		TR95/56	S0020
21A015-018	128	3024		TR87/243	S0024

Original Part Number	ECG	SK	GE	IR/WEP	HEP
21A015-019	128	3024		TR87/243	S0024
21A015-020	123A	3124		TR21/735	S0030
21A015-021	155				
21A015-022	131	3082		TR94/642	G6016
21A015-025	159	3025	22	TR20/717	S0013
21A015-026	128	3024	20	TR87/243	S0015
21A015-027	123A			TR21/	S0015
21A038-000	102A	3004	2	TR85/250	
21A039-000	102A	3004	2	TR85/250	
21A040-000	102A	3039	2	TR85/	
21A040-003	108	3039	90	TR95/56	S0016
21A040-004	108	3018	17	TR95/56	S0016
21A040-005	102A	3004		TR85/250	G0005
21A040-007	108	3018	17	TR95/56	S0016
21A040-010	108	3124		TR95/56	S0016
21A040-014	102A			TR85/250	G0005
21A040-015	132	3116		FE100/802	F0015
21A040-016	108	3018		TR95/56	S0016
21A040-017	108	3018		TR95/56	S0014
21A040-019	108	3018		TR95/56	S0014
21A040-020	123A	3124		TR21/735	
21A040-021	102A	3004		TR85/250	
21A040-022	102A	3004		TR85/250	G0006
21A040-023	107		11	TR70/720	S0016
21A040-024	107		11	TR70/720	S0011
21A040-025	107		11	TR70/720	S0011
21A040-031	126	3008		TR17/635	G0008
21A040-032	123A	3122	20	TR21/735	S0015
21A040-033	123A	3018		TR21/735	S0014
21A040-033A	123A	3122		TR21/	
21A040-034	123A		63	TR21/735	S0015
21A040-035	176	3123	52	TR82/238	G0005
21A040-036	158	3004		TR84/630	G0005
21A040-037	123A			TR21/735	S0015
21A040-045		3018			
21A040-046		3018			
21A040-047		3018			
21A040-049	129	3114		TR88/242	S5013
21A040-050	129	3114	67	TR88/242	S5013
21A040-051	128	3122	67	TR87/243	S5014
21A040-052	152	3054	28	TR76/701	S5000
21A040-053	107	3018	20	TR70/	S0011
21A040-054	107	3018	20	TR70/	S0011
21A040-055	107	3018	20	TR70/	S0015
21A040-056	123A	3018	17	TR21/	S0015
21A040-057	102A	3004	50	TR85/	G0007
21A040-058	102A	3004	50	TR85/	G0005
21A040-059	159	3114	21	TR20/	S0013
21A040-060	158	3004	54	TR84/	G6011
21A040-061	103A	3010	54	TR08/	G6011
21A040-063	107	3018	20		S0016
21A040-064	199		20		S0015
21A040-065	199	3018	20		S0015
21A040-066	199	3124	62		S0015
21A040-067	199	3124	18		S0015
21A040-068	1075				
21A040-077	123A	3018	18	TR21/	S0011
21A040-078	123A	3018	17	TR21/	S0011
21A040-079	102A	3004	1	TR85/	G0007
21A040-082		3020			
21A040-091		3018			
21A040-092		3124			
21A040-36	102A			TR85/	G0005
21A040-37	123A			TR21/	S0015
21A040-49				/242	
21A040-50				/242	
21A040-51				/243	
21A040-52				/701	
21A045-000	160		51	TR17/	
21A048-000	160	3006		TR17/637	
21A049-000	160	3006	51	TR17/637	
21A050-000	160	3006	51	TR17/637	
21A050-001	160	3006		TR17/637	
21A050-004	108	3018		TR95/56	
21A051-000	100		1	TR05/	
21A053-000	100	3004	2	TR85/250	
21A054-000	102A	3004	2	TR85/250	
21A055-000	102A	3004	2	TR85/250	
21A062-000	106	3006	51	TR20/637	
21A064-000	121	3014	3	TR01/232	
21A074-000	102A	3004	2	TR85/250	
21A097-000	121	3009	16	TR01/232	
21A101-001	706	3101	504A	IC507/	C6063P
21A101-002	711	3070	504A	IC514/	C6069G
21A101-004	1047				

Original Part Number	ECG	SK	GE	IR/WEP	HEP
21A101-005	1046				
21A101-006	1089				
21A101-007	1048				
21A101-008	1077				
21A101-009	1076				
21A101-010	1091				
21A101-011	1084				
21A101-012	1086				
21A101-1	706	3101			
21A101-2	711	3070			
21A105-001	108	3039		TR95/	S0020
21A105-004	128	3024	18	TR87/	S0005
21A105-006	128	3024	18	TR87/	S0005
21A112-001	159	3114	21	TR30/	S0031
21A112-004	193	3025	67	TR30/	S0019
21A112-006	161	3039	60	TR70/	S0017
21A112-007	107	3018	60	TR24/	S0020
21A112-015	123A	3046	18	TR24/	S0015
21A112-017		3122	20	TR24/	S0015
21A112-018		3018			
21A112-020	123A	3124	20	TR24/	S0015
21A112-023	164	3133	37	TR86/	
21A112-025	124	3021	12	TR23/	S5011
21A112-029	124	3021	12	TR81/	S5011
21A112-031		3027	14	TR61/	S7004
21A112-033	124	3079	14	TR61/	S5020
21A112-036		3111	36	TR93/	S5021
21A112-045		3124			
21A112-046		3124			
21A112-047		3138			
21A112-050		3018			
21A112-062		3122			S0015
21A112-063		3124			S0015
21A112-065		3114			
21A112-070		3103			S3021
21A112-074		3024			
21A112-075		3024			
21A112-084		3018			
21A112-085		3018			
21A112-086		3018			
21A112-087		3018			
21A112-088		3018			
21A112-089		3018			
21A112-090		3124			
21A112-091		3124			
21A112-092		3018			
21A112-093		3114			
21A112-094		3083			
21A112-095		3054			
21A112-098		3104			
21A112-099		3103			
21A112-100		3118			
21A112-101		3018			
21A112-102		3118			
21A112-103		3115			S5020
21A112-104		3020			S0016
21A113-002		3116			
21A404-066	199				
21-B1	718	3159		IC511	
21B1M	718	3159	8		
21B1Z	718	3159			
21B2Z	718	3159			
21M006	126	3006	50	TR52/	G0009
21M007	100	3006	50	TR85/	G0009
21M022	159	3025	21	TR52/	S0019
21M025	187	3083	69		S5006
21M026	187	3083	29	TR56/	S5006
21M028	187	3083	69	TR56/	S5006
21M084	123A	3018	20		S0015
21M085	123A	3018	20		
21M086	123A	3018	20	TR24/	S0015
21M087	199				
21M091	199	3018	61		S0016
21M093	107	3018	20	TR24/	S0016
21M094	107	3018	20	TR24/	S0016
21M095	199	3018	17		S0014
21M096	199				S0014
21M099	107	3018	17	TR21/	S0014
21M107		3018			
21M111		3018			
21M113		3018			S0030
21M122	123A	3124	10	TR21/	S0015
21M123	123A	3124	18	TR24/	S0015
21M124	199	3124	10	TR51/	S0015
21M125	123A	3124	10	TR21/	S0015

Original Part Number	ECG	SK	GE	IR/WEP	HEP
21M137	199	3124	20		S0024
21M138	199	3124	20	TR24/	S0025
21M139	123A	3018	20		S0015
21M140	107	3018	20		S0016
21M146		3122			S0015
21M149	123A	3124	20	TR24/	S0015
21M150	123A	3122	18	TR24/	S0016
21M151	107	3018	20	TR33/	S0033
21M152	107	3018	20	TR24/	S0011
21M153	107	3018	20	TR24/	S0011
21M154	107	3018	20	TR24/	S0011
21M160	192	3122	20	TR24/	S0015
21M161	192	3018	20	TR24/	S0015
21M170		3124			
21M174	199	3124	18	TR24/	S0015
21M178	107	3018	61	TR21/	S0030
21M179	107	3018	61	TR70/	S0016
21M180	186	3054	28	TR55/	S5000
21M181	186	3192	28	TR55/	S5000
21M182	107	3020	20	TR24/	S0014
21M183	186	3054	28	TR55/	S5000
21M184	186	3054	28	TR55/	S5000
21M185	128	3024	18	TR24/	S0015
21M186	123A	3024	20	TR21/	S0003
21M188	107	3018	17	TR21/	S0016
21M192	192	3137	63		S3024
21M193		3137			
21M196	132	3112	FET1	FE100/	F0021
21M200	123A	3018	10	TR21/	S0015
21M205	123A				
21M224	132	3124	20	TR21/	S0015
21M286	186	3192	28	TR55/	
21M345	187	3084	29	TR56/	S5006
21M355	159	3114	21		S0019
21M366	123A				
21M367	186	3192			
21M369	186	3054	28	TR55/	S5000
21M387	128	3018	18		S3001
21M395	187	3193			S3028
21M443	187	3083	29	TR56/	S5006
21M446	199	3124	20	TR21/	S0015
21M448	128	3024	18	TR25/	S3001
21M455	128	3122	18		S3001
21M459	193	3138	67		S3028
21M465	185	3084	29		S5006
21M466	184	28			
21M476	108				
21M481	108				
21M483					S3028
21M488	123A	3124	18		S3001
21M503		3018			S0030
21M506	799				
21M520	123A	3024	20		S0015
21M532	1006				
21M534	132	3116	FET2	FE100/	F0021
21M541	128	3122	18	TR25/	S0015
21M550	199	3124			S0015
21M556	186	3054	66		S5000
21M563	123A	3124	20		S0015
21M577	107	3018	61	TR21/	S0030
21M578	123A	3018	20	TR33/	S0030
21M579	123A	3018	20	TR33/	S0030
21M580		3114			S5013
21M581	159	3114	21	TR30/	S5013
21M599	722				
21M600	722				
21M603		3124			
21M605		3018			
21M628	187				
22-001001	123A	3122		TR21/	S0015
22-001002	108	3039		TR95/	S0016
22-001003	108	3039		TR95/	S0016
22-001004	108	3039		TR95/	S0016
22-001005	108	3039		TR95/	S0016
22-001006	123A	3122		TR21/	S0014
22-001007	123A	3122		TR21/	S0015
22-001008	175	3538		TR81/	S5019
22-001009	131	3538		TR81/	S5019
22-001010	159	3025		TR20/	S0019
22-002001	131	3052	44	TR94/	G6016
22-002006	102	3123		TR05/	G0007
22-002007	102A	3004		TR85/	G0005
22-002008	131	3052	44	TR94/	G6016
22-002009	131	3052	44	TR94/	G6016
22-1	107	3124		TR70/720	G6005

Original Part Number	ECG	SK	GE	IR/WEP	HEP	Original Part Number	ECG	SK	GE	IR/WEP	HEP
22B1B	739					24MW15	102	3004	2	TR05/631	
22E89	176	3004	52	TR82/238	G6011	24MW16	102	3004	2	TR05/	
23	123A	3122		TR21/735	S0015	24MW27	102A	3006	2	TR85/250	G0005
23-1	159	3114		TR20/717	S0019	24MW28	102A	3004	52	TR85/250	G0005
23-2	159	3114		TR20/717	S0019	24MW29	102	3004	52	TR05/631	G0007
23-3	159	3114		TR20/717	S0019	24MW34	102A	3008	2	TR85/250	G0005
23-10	159	3114		TR20/		24MW43	102	3004	2	TR05/631	G0005
23-20	159	3114		TR20/		24MW44	160	3006	51	TR17/637	G0003
23-30	159	3114		TR20/		24MW55	160	3006	1	TR17/637	G0003
23-79		3084				24MW59	160	3006	51	TR17/637	G0003
23-82		3054				24MW60	102	3004	2	TR05/631	
23-5006		3004				24MW61	126	3007	51	TR17/635	G0008
23-5007		3004				24MW69	102	3004	2	TR05/631	
23-5009	121	3009	3	TR01/		24MW70	102	3004	2	TR05/631	G0005
23-5014	158	3004	53	TR84/		24MW74	126	3006		TR17/635	
23-5017	102A	3004	2	TR85/		24MW77	100	3005	1	TR05/254	G0005
23-5020	123A	3124	17	TR21/		24MW78	102	3004	2	TR05/631	
23-5021	123A	3124	17	TR21/		24MW83	102	3004	2	TR05/631	
23-5022	123A	3124	17	TR21/		24MW84	102	3004	2	TR05/631	
23-5023	123A	3124	17	TR21/		24MW107	102	3004	2	TR05/631	
23-5024	123A	3124	17	TR21/		24MW111	102A	3008	51	TR85/250	G0005
23-5025	123A	3124	17	TR21/		24MW115	102	3004	1	TR05/631	G0005
23-5026	123A	3124	17	TR21/		24MW116	102	3004	52	TR05/631	G0005
23-5027	123A	3124	17	TR21/		24MW119		3018		TR24/	S0015
23-5029	123A	3124	17	TR21/		24MW130	103	3010	8	TR08/641A	G0011
23-5031	175	3026	66	TR81/		24MW132	102	3004	2	TR05/631	G0005
23-5032	131		30			24MW152	126			TR17/635	G0008
23-5033	123A	3122	20	TR21/		24MW157	160	3008		TR17/	
23-5034	176	3004	53	TR82/		24MW178	102A	3004		TR85/	G0005
23-5035	130	3027	14	TR59/		24MW179	102A	3004		TR85/250	G0005
23-5037	128/401	3024	18			24MW185	102A	3004		TR85/250	G0005
23-5038	130	3027	19	TR59/		24MW187	102A			TR85/250	G0005
23-5039	128	3024	18	TR87/		24MW205	126	3008	51	TR17/635	
23-5041	130	3027				24MW256	158	3004		TR84/630	G6011
23-5042	121	3014	16	TR01/		24MW263	102	3004	2	TR05/631	
23-5044	172			TR69/		24MW271	126	3006	51	TR17/635	
23-5045	159	3025	21	TR20/		24MW287	107	3018		TR70/720	S0016
23-5048	103A		8	TR08/		24MW303	126	3006	51	TR17/635	
23-5049				TR08/		24MW333	123A	3122		TR21/735	
23-5051	128/401	3045				24MW351	160		51	TR17/	
23-5052	123A	3124	20	TR21/		24MW352	126	3008	1	TR17/635	
23-6001-16			53			24MW353	126	3008	1	TR17/635	
23-6001-17			53			24MW361	107	3018		TR70/720	S0014
23-6001-20			53			24MW368	126	3007		TR17/635	G0008
23-6001-21			53			24MW370	102A	3004		TR85/	
23-6001-23			53			24MW372	123A	3124		TR21/735	S0015
23-6025				TR25/		24MW373				TR51/	
23B-210-025	121	3009	16	TR01/		24MW384	102A	3004		TR85/	
23B-210-230-2	127	3034	25	TR27/		24MW441	102A	3008		TR85/250	G0005
23B114044	123A	3122	17	TR21/		24MW454	123A	3122		TR21/735	S0014
23B114053	128	3024	63	TR87/		24MW458	123A	3124	20	TR21/735	S0030
23B114054	128	3024	63	TR87/		24MW460	123A	3122		TR21/735	S0014
23E001-1	123A	3122	10	TR21/		24MW461	123A	3122		TR21/735	S0014
23-PT274-120	161	3132		TR83/		24MW462				TR21/	
23-PT274-121	123A	3122		TR21/		24MW463				TR21/	
23-PT274-122	159	3025		TR20/		24MW525				TR17/	
23-PT274-123	161	3132		TR83/		24MW535	107	3018		TR70/56	S0016
23-PT275-121	161	3039		TR83/56	S0017	24MW593	107	3018	20		S0014
23-PT275-122	108	3018		TR95/56	S0020	24MW594	107	3039		TR70/720	S0016
23-PT275-123		3126				24MW595	107	3018	20	TR70/720	S0014
23-PT283-122	161	3132		TR83/		24MW596	107	3039		TR70/720	S0014
23-PT283-124	161	3132		TR83/		24MW597	107	3039	20	TR70/720	S0014
23-PT284-122	160	3006		TR17/		24MW598	102A	3004		TR85/250	G0005
23-PT284-123	160	3006		TR17/		24MW599	102A	3004	2	TR85/250	G0005
24-000451	123A					24MW600	102A			TR85/250	G0005
24-000452	128					24MW601	102A	3004		TR85/250	G0005
24-000457	123A					24MW608	102A			TR85/250	G0005
24-000653-1	123A					24MW609	123A	3122		TR21/735	S0014
24-0003714-1	123A					24MW613	102A			TR85/	G0005
24-002	123A	3122	17	TR21/735	S0015	24MW614	102A	3004		TR85/250	G0005
24-001326	199					24MW615	102A	3004		TR85/250	G0007
24-001327-1	123A					24MW618	131MP	3052		TR94MP/642MP	G6016
24-001354	199					24MW652		3116		FE100/	F0021
24-602-25	123A	3122		TR21/		24MW653	107	3018	20	TR70/720	S0016
24-3564	108	3039		TR95/	S0020	24MW654	108	3018	20	TR70/720	S0014
24A	123A	3122		TR21/735	S0015	24MW655	123A		18	TR51/	
24A1	108	3039		TR95/	S0015	24MW656	107	3018	61	TR21/720	
24B	123A	3018	20	TR21/735	S0015	24MW657	108	3018	20		
24B1	123A	3122	20	TR21/	S0015	24MW658	123A	3124	20	TR21/	
24B1AH	731	3170				24MW659	123A	3122	20	TR24/	
24B1B	731	3170				24MW660		3122			S5026
24B1Z	731	3170				24MW661		3114			S0019
24C385A				TR80/		24MW662		3026			S5019
24M125	199	3024				24MW663	128	3024	18		
24MW11	100	3005	1	TR05/		24MW673	107	3018	20	TR70/720	S0014

Original Part Number	ECG	SK	GE	IR/WEP	HEP
24MW674	128	3024	18	TR21/	
24MW675	108	3018	17	TR51/	
24MW676	123A	3122	20		
24MW677		3122			S5026
24MW700	107	3018	20	TR21/	
24MW714	128	3024	18	TR21/	
24MW723	132	3116	FET2	FE100/	
24MW724	108	3018	11	TR21/	
24MW725	108	3018	11	TR21/	
24MW726		3126	90		
24MW727	129	3025		TR21/	
24MW736		3116	FET2	FE100/	F0021
24MW737	107	3018	60	TR70/720	S0014
24MW738	107	3018	61	TR70/720	S0016
24MW739		3018	61	TR21/	S0016
24MW740		3124	63	TR21/	S0014
24MW741	158	3004	53	TR84/630	G6011
24MW760	123A	3004		TR21/735	G0005(TWO)
24MW773	123A	3122		TR21/	S0015
24MW774	123A	3122		TR21/	S0014
24MW775	123A	3122		TR21/	S0011
24MW776	123A	3122		TR21/	S0020
24MW777	102A	3004		TR85/	G0005
24MW778	152	3054		TR76/	S3020
24MW780	102A	3004		TR85/250	G0005
24MW781	102A	3004		TR85/250	G0005
24MW782	102A	3004		TR85/250	G0005
24MW783	158	3004		TR84/630	G0005
24MW789	102A	3004		TR85/250	G0005
24MW790	123A	3018		TR21/735	S0014
24MW793	107	3018	61	TR70/720	S0016
24MW795	123A	3122		TR21/735	S0014
24MW796	123A	3124		TR21/735	S0014
24MW797	123A	3124	20	TR21/	S0014
24MW799	102A	3123		TR85/250	G0006
24MW801	123A	3122		TR21/735	S0014
24MW805	107	3039		TR70/720	S0011
24MW807	123A	3124	20	TR21/	S0015
24MW808	123A	3122	20	TR21/735	S0025
24MW809	123A	3122		TR21/735	S0025
24MW812	107	3018		TR70/720	S0016
24MW813	107	3018	11	TR70/720	S0014
24MW814	107	3018	20	TR70/720	S0014
24MW815	107	3018	62	TR70/720	S0014
24MW816	126	3006		TR17/635	G0008
24MW817	123A	3122	18	TR21/735	S0025
24MW818	123A	3122	18	TR21/735	S0025
24MW819	158	3004		TR84/630	G0005
24MW823	123A	3124		TR21/735	S0016
24MW824	102A	3004		TR85/250	G0006
24MW826	199	3124		TR21/	S0024
24MW827	108	3124		TR95/	S0016
24MW828	176	3004		TR82/	G6011
24MW852	108	3018	20	TR95/56	S0014
24MW853	102A	3004	2	TR85/250	G0007(TWO)
24MW854	123A	3031	504	TR21/735	
24MW855	123A	3124	20	TR21/735	S0030
24MW856	102A	3004	52	TR85/250	G0005
24MW857	102A	3004	53	TR85/250	G0005
24MW863	107	3018	11	TR70/720	S0016
24MW865	108	3018		TR95/56	S0014
24MW892	102A	3004		TR85/250	G0005
24MW893	102A	3004		TR85/250	G0006
24MW899	123A	3124	18	TR24/	S0015
24MW953	107	3018		TR70/720	S0014
24MW954	123A	3124		TR21/735	S0015
24MW957	161	3018	60		S0016
24MW958	161	3039	60		S0016
24MW961	123A	3124	20		S0016
24MW964	199	3124	20		S0015
24MW965	199	3124	20		S0016
24MW973	158	3004	53		G6011
24MW976	159	3114	21		S5022
24MW977	152	3054	28		S5000
24MW978	184	3036	28		S5000
24MW988	123A	3124	20		S0016
24MW989	132	3116	FET2		F0021
24MW990	199	3018	20		S0016
24MW991	126	3008	9		G0009
24MW992	123A	3008	20		S0015
24MW994	131	3052	30		G6016
24MW1022	199	3018	60		S0016
24MW1023	123A	3124	20		S0015
24MW1024	123A	3124	20		S0015
24MW1025	199	3124	20		S0015

Original Part Number	ECG	SK	GE	IR/WEP	HEP
24MW1028	1003				
24MW1031	159				S5022
24MW1038		3018	20		S0015
24MW1040	158	3004	53	TR85/	
24MW1049	159	3025	21		S0032
24MW1059	123A	3122	10		S0014
24MW1060	199	3122	62		S0024
24MW1061	159	3114	21		S0019
24MW1068	123A	3124	20	TR51/	S0014
24MW1069	123A	3124	20	TR51/	S0014
24MW1081	107	3018	17	TR24/	S0025
24MW1082	108	3018	20	TR24/	S0015
24MW1083	102A	3004	52	TR85/	G0005
24MW1084	102A	3004	2	TR05/	G0005
24MW1089	123A	3124	20	TR21/	S0015
24MW1096	123A	3124	17	TR22/	S0011
24MW1106	107	3018	20	TR24/	S0014
24MW1110	1032				
24MW1115	102A	3004	52		G0007
24MW1116	102A	3010	54		
24MW1141	123A	3018	20		
24MW1143	184	3041	28	TR55/	
24MW1147	123A	3122	18	TR25/	S5014
24MW1152	192	3024	63	TR25/	
24MW1161	128	3122	20		S5014
24R			17		
24T-002	108	3019	17	/56	S0016
24T-009	199				
24T-011-001	161				
24T-011-003	107				
24T-011-008	108	3018		TR95/56	
24T-011-011	159	3114		/717	
24T-011-012	161	3018		TR83/56	
24T-013-003	161	3039		TR83/719	
24T-013-005	108	3039		TR95/56	S0017
24T016	107				
24T-016	107	3019	17	TR70/720	S0016
24T-016-001	108	3019	17	TR95/56	S0020
24T-016-005	108	3019	17	TR95/56	
24T-016-010	107	3019	11	TR70/720	S0016
24T-016-013	108	3039	20	TR95/56	S0016
24T-016-015	108	3018		TR95/56	S0020
24T-016-016	108	3039		TR95/56	S0016
24T021	161	3132		TR83/	
24T-026-001	133				
24T013003	107			TR70/	
24T013005	107			TR70/	
24T016001	107			TR70/	
24T016005	107			TR70/	
25-000453	159				
25-000456-1	234				
25-000462	159				
25-0060-4	123A				
25-001328	106				
25-37833-2	909D				
25-100015	108	3019	11	TR95/56	
25A	108	3018	20	TR95/56	S0015
25A1	108	3018	20	TR95/	S0015
25A2	123A	3018	20	TR21/	S0025
25A21				TR95/	
25A321				TR17/	
25A324				TR17/	
25A440A				TR17/	
25A473				TR56/	
25A561Y			22		
25A1262-005	108	3019		TR95/56	
25A1273-001	123A	3018	17	TR21/735	S0030
25A1281-001	108	3039		TR95/	S0016
25AM624	108	3018	20	TR95/56	
25B	108	3018	20	TR95/56	S0015
25B1	108	3018	20	TR95/	S0015
25B1C	730	3143			
25B1T	730	3143			
25B2	123A	3018	20	TR21/	S0025
25B2C	730				
25B2T	730	3143			
25B378	102A	3004	1	TR85/250	G0005
25C206	108	3018		TR95/	
25CS58LGBM	123A	3124		TR21/	
25R	108	3018	20	TR24/	
25T-002	107	3039	11	TR70/	
25T1	102	3004	52	TR05/631	G0008
25T2					S0020
26MW613	102A	3004		TR85/250	G0005
26T1	102	3004	52	TR05/631	G0008

Original Part Number	ECG	SK	GE	IR/WEP	HEP
26T2					S0025
26T2C					S0025
27A10446-101-119099					
27A10489-101-11199					
27A10533	159				
27D127	198	3103			
27T401	100	3005	1	TR05/254	
27T402	100	3005	1	TR05/254	
27T403	102	3004	2	TR05/631	
27T404	102	3004	2	TR05/631	
27T405	102	3004	2	TR05/631	
27T406	121	3009	16	TR01/232	
27T407	105	3012	4	TR03/624	
27T408	101	3011	5	TR08/641	
27T409	123	3124	20	TR86/53	
27T410	103	3010	8	TR08/641A	
27T411	123	3124	20	TR86/53	
27T412	126	3006	51	TR17/635	
28C828				TR21/	
28C839(F)				TR24/	
28C1317R				TR25/	
28C37200				TR24/	
29A4	129	3025		TR88/242	S0014
29A829					S0019
29B1B	790				C6057P
29B1Z	790				
29V008M01	100	3006	2	TR05/254	G0005
29V0038H03	102	3004	52	TR05/631	G0005
29V011H01	100	3006	2	TR05/254	G0005
29V012H01	100	3006	2	TR05/254	G0005
29V069C02	163				
29V12H01	102A	3004		TR85/	
30-004-001	179	3014	16	TR35/232	
30-005072	223				
30-090	184		28	TR76/S5003	G0005
30V-HG	102A	3123	52	TR85/250	G0005
31-0001	160	3006	50	TR17/	
31-0002	160	3006	50	TR17/	G0003
31-0003	160	3006	50	TR17/	G0008
31-0004	160	3008	50	TR17/	G0003
31-0005	126	3004	51	TR12/	G0005
31-0006	102A	3008	52	TR85/	
31-0007	123A	3124	20	TR21/	S0014
31-0008	102A	3004	52	TR85/	G0007
31-0009	123A	3020	17	TR51/	S3002
31-0010	175	3026	23	TR57/	S5019
31-0012	199	3124	20	TR21/	S0024
31-0013	199	3124	20	TR24/	S0025
31-0015	160	3008	51	TR17/637	
31-0016	160	3008	51	TR17/	
31-0017	158	3003	2	TR84/630	G0005
31-0018	102A	3004	2	TR85/250	G0005
31-0025	102A	3004	53	TR85/250	G0005
31-0026	102A	3004		TR85/250	G0005
31-0033	102A		53	TR85/250	
31-0035	158	3004	30	TR84/630	G0005
31-0041	126	3006		TR17/635	
31-0042	126	3006	51	TR17/635	
31-0048	107	3018	20	TR71/	S0016
31-0049	107	3018	17	TR71/	S0014
31-0050	107	3018	20	TR71/	S0011
31-0051		3122	20	TR33/	S0016
31-0052	199	3018	18	TR24/	S0020
31-0053	102A	3004	52	TR85/	G0001
31-0054	107	3018	61		S0016
31-0055	193	3114	67	TR88/	S0012
31-0060					S0015
31-0065	126	3008	51	TR17/635	G0008
31-0066	186	3041	28	TR55/	S5000
31-0068		3018	62	TR53/	S0015
31-0069		3018	62	TR21/	
31-0070	102A	3004	52	TR84/	G0001
31-0075	158		53	TR84/630	
31-0080	123A	3124	62	TR21/	S0015
31-0081	123A	3124	20	TR24/	S0015
31-0082	123A	3124	20	TR21/	S0015
31-0083	128	3024	63	TR25/	S0014
31-0084	123A	3124	20	TR21/	
31-0085	123A	3124	20	TR21/	
31-0097	108	3018	20		S0015
31-0098	107	3018	17		S0015
31-0099	199	3124	20		S0015
31-0100	199	3124	20		S0015
31-0101	192	3054	63		
31-0102	193	3083	67		
31-0103	108	3018	17		S0016
31-0104	123A	3020	20		S0024
31-0105	158	3004	2		G0005
31-0106	123A	3124	10		S0015
31-0107	102A	3004	63	TR85/250	G0005
31-0108	126	3006	51	TR17/635	
31-0115	123A	3122	10		S0005
31-0116	123A	3122	20		S0005
31-0123	126	3006	51	TR17/635	
31-0124	126	3006	51	TR17/635	
31-0132	126	3006	51	TR17/635	
31-0134	126	3008	51	TR17/635	G0009
31-0135	126	3008	51	TR17/635	
31-0139	126	3008	51	TR17/635	G0009
31-0141	160	3006	51	TR17/637	G0002
31-0148	102A		53	TR85/250	
31-0150	160	3006	51	TR17/637	G0008
31-0153	102A	3004	2	TR85/250	
31-0161	160	3004	51	TR17/637	G0002
31-0163	160		51	TR17/637	
31-0165	160		51	TR17/637	
31-0166	160		51	TR17/637	
31-0168	160	3006	51	TR17/637	
31-0170	160	3008	51	TR17/637	G0002
31-0171	160		51	TR17/637	G0002
31-0172	102A	3004	2	TR85/250	G0005
31-0175	127	3035	25	TR27/235	
31-0177	123A	3020	20	TR21/735	S0015
31-0178	126	3008	51	TR17/635	G0009
31-0180	160	3006	50	TR17/637	G0008
31-0181	160		51	TR17/637	
31-0182	158	3004	53	TR84/630	G0005
31-0183	158		53	TR84/630	
31-0184	126	3006	51	TR17/635	
31-0187	123	3124	20	TR86/53	
31-0188	102A	3004	2	TR85/250	
31-0189	102A		2	TR85/250	
31-0190	126	3008	51	TR17/635	
31-0191	126	3006	51	TR17/635	
31-0192	121	3009	16	TR01/232	
31-0196	121	3035	25	TR01/232	G6003
31-0205	102A		2	TR85/250	G0005
31-0206	107	3039	11	TR70/720	
31-0217	126	3008	51	TR17/635	
31-0228	126	3006		TR17/635	
31-0229	158	3004	2	TR84/630	
31-0230	123A		20	TR21/735	
31-0239	123A	3124	20	TR21/735	S0015
31-0240	121	3009	31MP	TR01/232	
31-0241	160	3006	16	TR17/637	G0008
31-0241-1	160	3006	16	TR17/637	
31-0242	108	3018		TR95/56	
31-0243	108	3018		TR95/56	
31-0246	123A		20	TR21/735	
31-0247	102	3004	52	TR05/631	
31-0253	100	3005	51	TR05/254	
31-025	102	3004			
31-058	123A				
31-1	123A	3124			
31-16	123A	3124			
31-1012	941M				
31-2104733	160	3006		TR17/	
31-21004900	160	3006		TR17/	
31-21007744	160	3006		TR17/	
31-21024033	160	3006		TR17/	
31-21024044	160	3006		TR17/	
31-21047111	160	3006		TR17/	
31-21050611	160	3006		TR17/	
31-21050622	160	3006		TR17/	
31-22005400	102A	3004		TR85/	
32-13843-2	123A	3124		TR21/	
32-16591	104MP		13MP	TR01MP/	
32-16599	121MP	3015	13MP	/232MP	
32-20738	123A	3124	18	TR21/735	S0015
32-20739	159	3024	21	TR20/717	
32-23555-1	704	3022		IC501/	
32-23555-2	704	3022		IC501/	
32-23555-3	704	3022		IC501/	
32-23555-4	704	3023		IC501/	
32-107261-1	910				
32-805483-1	910				
32-805483-2	910				
32-807071-1	911				
32-807072-1	909	3590			
32E64	129	3025		TR88/242	S0019

Original Part Number	ECG	SK	GE	IR/WEP	HEP
33-0039	105		4	TR03/	
33-00234-B	175				
33-00243	124				
33-00706A	123A				
33-00742	129				
33-016	159				
33-048	128				S5026
33-050	175				
33-052	130	3510			
33-070	123A	3122		/735	S0011
33-071	123A	3124		/735	
33-084	123A				
33-086	159				
33-090	184	3054		/S5003	S5000
33-096	185	3083		/S5007	S5006
33-0706	123A				
33-108	223				
33-1000-00	102	3004		TR05/631	G0005
33-1001-00	102	3004		TR05/631	G0005
33-1002-00	131MP	3013		TR94MP/232	G6016(TWO)
33-1004-00	121	3009		TR01/232	G6005
33-1009-01	102A	3004		TR85/250	G0005
33-1019-00	102A	3004		TR85/250	G0005
33-1020-00	102A	3004		TR85/250	G0005
33-1021-00	102A	3123		TR85/250	G0005
33H50	108	3018	17	TR95/56	S0015
34-6	160	3006		TR17/637	G0002
34-34-6015-43	123A			TR21/	
34-119	160	3006		TR17/637	G0002
34-143-12	159				
34-194	941M				
34-220	160	3006		TR17/637	G0002
34-221	160	3006		TR17/637	G0002
34-298	160	3006	51	TR17/637	G0002
34-1000	130	3027		TR59/247	S7002
34-1000A	130	3027		TR59/	
34-1001	219				S7003
34-1002	152			TR76/701	S5000
34-1003	153	3083	29	TR77/700	S5006
34-1004					S3002
34-1005					S3003
34-1006	157	3103		TR60/244	S5015
34-1007					S0019
34-1010					S0015
34-1011	106				
34-1013	159				S0019
34-1014					S0025
34-1017	194				
34-1019	199				
34-1020					S0030
34-1021					S0030
34-1022	159				S0019
34-1023					S5026
34-1024					S5026
34-1026	175			TR81/241	S5012
34-1027	218		26	TR58/	S5018
34-1028	130	3027		TR59/247	S7002
34-1028A	130	3027		TR59/	
34-1029	219	3173			S7003
34-1035					S5019
34-1036					S5018
34-1041	726	3129			
34-2001		3087			
34-3015-46	107				
34-3015-47	107				
34-3015-49	107				
34-6000-3	160	3005	2	TR17/637	G0002
34-6000-4	102A	3004	53	TR85/250	
34-6000-5	102A	3004	53	TR85/250	
34-6000-6	102A	3004	53	TR85/250	
34-6000-7	102A	3004	53	TR85/250	S0015
34-6000-8	102A	3004	53	TR85/250	
34-6000-9	160	3006	51	TR17/637	
34-6000-10	160	3006	2	TR17/637	G0002
34-6000-11	160	3006	51	TR17/637	
34-6000-12	160	3006	51	TR17/637	
34-6000-13	160	3006	51	TR17/637	
34-6000-14	160	3006	51	TR17/637	
34-6000-15	102A	3004	53	TR85/250	
34-6000-16	160	3004	51	TR17/637	G0005
34-6000-17	160	3006	51	TR17/637	
34-6000-18	160	3005	2	TR17/637	G0002
34-6000-19	160	3005	2	TR17/637	G0002
34-6000-20	160	3006	51	TR17/637	
34-6000-25	160	3006		TR17/	

Original Part Number	ECG	SK	GE	IR/WEP	HEP
34-6000-26	160	3006		TR17/	
34-6000-27	102A		53	TR85/250	
34-6000-28	102A	3004	53	TR85/250	
34-6000-29	102A	3004	53	TR85/250	
34-6000-30	102A		53	TR85/250	
34-6000-31	102A	3004	53	TR85/250	
34-6000-32	102A	3004	53	TR85/250	
34-6000-33	102A	3004	53	TR85/250	
34-6000-34	102A	3004	53	TR85/250	
34-6000-58	160	3006	51	TR17/637	
34-6000-59	160	3006	51	TR17/637	
34-6000-60	160	3006	51	TR17/637	
34-6000-61	160	3006		TR17/	
34-6000-62	160	3006	50	TR17/637	
34-6000-63	160	3006	51	TR17/637	
34-6000-64	123A	3018	17	TR21/735	
34-6000-65	160	3122	51	TR17/637	S0005
34-6000-66	160		51	TR17/637	
34-6000-68	160	3006	51	TR17/637	
34-6000-69	123A	3018	11	TR21/735	S0017
34-6000-70	123A	3018	11	TR21/735	S0020
34-6000-71	123A	3018	20	TR21/735	
34-6000-72	123A	3132	11	TR83/735	S0020
34-6000-76	160	3006	50	TR17/637	G0003
34-6000-77	160	3006	50	TR17/637	
34-6000-78	160	3006	50	TR17/637	
34-6000-79	160	3006	50	TR17/637	G0003
34-6000-80	160	3006	50	TR17/	G0003
34-6000-81	160	3006	50	TR17/637	G0003
34-6000-82	160	3006	50	TR17/250	
34-6000-83	102A	3008	51	TR85/250	G0005
34-6000-84	102A	3008	1	TR85/250	
34-6000-85	102A	3008	1	TR85/250	
34-6000L3		3006			
34-6001-1	104	3009	3	TR01/230	
34-6001-3	108	3018		TR95/56	
34-6001-5	123A	3122	20	TR21/735	
34-6001-6	108	3039	11	TR95/56	S0015
34-6001-7	102A	3004	53	TR85/250	
34-6001-8	102A	3004	53	TR85/250	
34-6001-9	102A	3004		TR85/250	
34-6001-10	102A	3004	53	TR85/250	
34-6001-11	102A	3004	53	TR85/250	
34-6001-12	102A	3018	20	TR85/250	
34-6001-13	102A	3004	53	TR85/250	S0015
34-6001-14	102A	3004	53	TR85/250	
34-6001-15	159	3025	53	TR20/717	
34-6001-16	102A	3004	51	TR85/250	G0005
34-6001-17	102A	3004		TR85/250	
34-6001-18	102A	3004	2	TR85/250	
34-6001-19	102A	3004	53	TR85/250	
34-6001-20	102A	3004		TR85/250	
34-6001-21	102A	3004		TR85/250	
34-6001-22	102A	3004	53	TR85/250	
34-6001-23	102A	3004		TR85/250	
34-6001-26	102A	3004	53	TR85/250	
34-6001-28	158	3004	2	/630	G0005
34-6001-29	102A	3004	53	TR85/250	
34-6001-30	102A	3004	53	TR85/250	
34-6001-31	102A	3004	53	TR85/250	
34-6001-33	102A	3004	51	TR85/250	G0005
34-6001-34	123A	3122	20	TR21/250	
34-6001-41	102A		53	TR85/250	
34-6001-42	102A	3004	53	TR85/250	
34-6001-43	102A	3005	2	TR85/250	G0005
34-6001-44	102A	3005	2	TR85/250	G0005
34-6001-47	102A	3004	53	TR85/250	
34-6001-48	123A	3122	8	TR21/735	
34-6001-49	123A	3122	8	TR21/735	
34-6001-50	128	3124	20	TR87/243	
34-6001-51	128	3024	18	TR87/243	G0011
34-6001-51MR-121	28				
34-6001-52	123A	3124	20	TR21/735	
34-6001-53	123A	3124	20	TR21/735	
34-6001-54	123A	3024	20	TR33/735	S0014
34-6001-55	123A	3124	20	TR33/735	S0014
34-6001-56	123A	3124	20	TR33/735	S0014
34-6001-57	123A	3124	20	TR25/735	S0014
34-6001-58	123A	3024	20	TR25/735	S0014
34-6001-60	123A	3124	20	TR25/735	S0014
34-6001-61	123A	3024	51	TR25/735	S5014
34-6001-62	123A	3024	17	TR25/735	S5014
34-6001-63	123A	3122	20	TR21/735	S0015
34-6001-64	191	3018	20	TR75/	S0015
34-6001-65	154	3040	20	TR78/712	S5026

Original Part Number	ECG	SK	GE	IR/WEP	HEP
34-6001-66	158	3004	52	TR82/630	G0005
34-6001-69	123A	3124	20	TR05/735	S0015
34-6001-70	123A	3124	20	TR21/735	S0015
34-6001-71	123		63	TR86/53	S5014
34-6001-72	102A	3004	52	TR85/250	
34-6001-73	123A	3124	17	TR24/735	S0014
34-6001-74	123A	3024	63	TR24/735	S0014
34-6001-76	102A	3004	52	TR85/250	G0005
34-6001-77	123A	3124	20	TR51/735	S0011
34-6001-78	128	3124	63	TR87/243	
34-6001-79	121	3018	63	TR01/232	S3021
34-6001-80	128	3124	63	TR87/243	S5014
34-6001-82	128	3045	63	TR87/243	
34-6001-83	128	3024	63	TR87/243	
34-6001-84	103A		8	TR08/724	
34-6001-85	128	3124	63	TR87/243	
34-6001-86	129	3025	63	TR88/242	S5013
34-6002-1	175			TR81/	
34-6002-2	121	3009	3	TR01/232	
34-6002-3	121	3009	3	TR01/232	
34-6002-4	121	3009	3	TR01/232	
34-6002-5	121	3009	3	TR01/232	
34-6002-6	121	3009	3	TR01/232	
34-6002-7	121	3009	3	TR01/232	
34-6002-8	121	3009	3	TR01/232	
34-6002-9	121	3009	3	TR01/232	
34-6002-10	121	3009	3	TR01/232	
34-6002-11	121	3009	3	TR01/232	
34-6002-13	121	3009	3	TR01/232	
34-6002-14	121	3009	3	TR01/232	
34-6002-17	104	3009	3	TR01/624	
34-6002-18	104	3009	16	TR01/624	
34-6002-18A	104	3009	16	TR01/624	
34-6002-19	121	3009	16	TR01/232	G6004
34-6002-20	104	3009	16	TR01/624	
34-6002-21	124	3021	12	TR57/240	S5011
34-6002-22	104	3009	16	TR01/230	
34-6002-22A	104	3009	16	TR01/230	
34-6002-24	127		25	TR27/	
34-6002-25	174				
34-6002-26	124	3021	12	TR57/240	S5011
34-6002-28	175	3131	23	TR81/241	S5000
34-6002-29	175	3131	14	TR57/241	S5012
34-6002-30	127	3035	25	TR27/235	G6008
34-6002-31	127	3035	25	TR27/235	G6008
34-6002-32	130	3027	14	TR59/247	
34-6002-32A	130	3027		TR59/	
34-6002-33	179	3035	25	TR27/G6001	G6007
34-6002-34	121	3009	16	TR01/232	
34-6002-35	181			TR36/S7000	
34-6002-36	180			/S7001	
34-6002-37	181			TR36/S7000	
34-6002-38	180			/S7001	
34-6002-41	186	3192	28	TR76/	S3024
34-6002-42	187	3193	29	TR77/	S3028
34-6002-43		3026	28	TR81/	S5000
34-6002-46	124	3021	12	TR81/	S5011
34-6002-49	127	3035	25	TR27/	G6008
34-6002-50	152	3054			
34-6002-52	152	3054			
34-6002-54			20		S5000
34-6002-56		3054			S3024
34-6002-57		3083			S3028
34-6002-58	163	3111	36	TR61/	S5021
34-6002-59		3111			
34-6002-61	162	3133	35	TR61/	S5021
34-6002-62	124	3021	12	TR23/	S5012
34-6002-63		3111		TR93/	S5020
34-6002-64		3111			S5020
34-6005-1	160	3124	66	TR17/637	
34-6005-2	175	3131	66	TR81/241	
34-6005-3	175	3131	66	TR81/241	
34-6007-1	123A	3122	60	TR21/735	
34-6007-2	123A	3122	60	TR21/735	
34-6007-3	123A	3122	60	TR21/735	
34-6007-4			63		
34-6007-5			10		
34-6007-6			20		
34-6007-7			20		
34-6007-8			63		
34-6007-14			62		
34-6008	102A	3004	52	TR85/250	G0005
34-6009	102A	3004	52	TR85/250	G0005
34-6010	128			TR21/243	
34-6015-1	123A	3018	11	TR33/735	S0011

Original Part Number	ECG	SK	GE	IR/WEP	HEP
34-6015-2	123A	3018	20	TR33/735	S0015
34-6015-3	123A	3018	20	TR33/735	S0015
34-6015-4	123A	3018	20	TR33/735	S0015
34-6015-5	123A	3018	11	TR33/735	S0011
34-6015-6	123A	3018	11	TR33/735	S0011
34-6015-7	123A	3018	20	TR33/735	S0015
34-6015-8	107	3018	11	TR51/735	S0015
34-6015-9	123A	3018	11	TR51/735	S0011
34-6015-10	123A	3018	20	TR51/735	S0011
34-6015-11	123A	3018	20	TR51/735	S0015
34-6015-12	123A	3018	20	TR21/735	S0015
34-6015-13	123A	3018	20	TR51/735	S0015
34-6015-14	123A	3018	20	TR51/735	
34-6015-15	161	3117	39	TR83/719	S0013
34-6015-16	161	3018	39	TR83/719	
34-6015-17	161	3018	20	TR83/719	S0017
34-6015-18	161	3018	20	TR83/719	S0015
34-6015-19	161	3018	20	TR83/719	S0011
34-6015-20	161	3018	20	TR83/719	S0011
34-6015-21	123A	3018	20	TR21/735	S0015
34-6015-22	161	3018	20	TR83/735	S0011
34-6015-23	128	3024	63	TR87/243	S5014
34-6015-24	128	3124	63	TR87/243	S5014
34-6015-25	123A	3018	20	TR21/735	S0011
34-6015-26	102A	3123	22	TR85/250	G0005
34-6015-27	161	3117	39	TR83/719	S0017
34-6015-28	191	3024	27	TR75/243	S3021
34-6015-29	161	3117	39	TR83/719	S0017
34-6015-30	128	3024	63	TR83/719	S5014
34-6015-31	161	3117	50	TR83/719	
34-6015-32	160		50	TR17/637	
34-6015-33	160		51	TR17/637	
34-6015-34	160	3005	51	TR80/637	G0003
34-6015-35	160		51	TR17/637	
34-6015-36	160		51	TR17/637	
34-6015-37	161	3018	39	TR83/735	S0017
34-6015-38	108	3039	61	TR95/56	
34-6015-39	160		50	TR17/637	
34-6015-40	160		51	TR17/637	
34-6015-41	123A	3018	61	TR21/735	S0011
34-6015-42	123A	3124	17	TR21/735	S0015
34-6015-42A	123A	3124		TR21/	
34-6015-43		3124	17	TR87/243	S0015
34-6015-43A	123A	3124		TR21/	
34-6015-44	123A	3024	63	TR21/243	S5014
34-6015-44A	102A	3004		TR85/	
34-6015-46		3018	20	TR33/	S0015
34-6015-47	107	3018	17	TR33/	S0025
34-6015-48	107	3018	17	TR33/	S0025
34-6015-49	108	3018	20	TR33/	S0015
34-6015-50	107	3018	20	TR33/	S0011
34-6015-51	128	3018	20	TR33/	S0011
34-6015-52	107	3018	17	TR33/	S0025
34-6015-54	123A	3122	20	TR24/	S0024
34-6015-60		3025			
34-6015-62	233	3117	60	TR71/	S0020
34-6015-63	123A				
34-6015-64	191	3054	27	TR74/	S3019
34-6015-80	123A	3018	20	TR51/	S0015
34-6016-2	123A	3122	62	TR21/735	S0014
34-6016-3	123A	3124	62	TR21/735	S0014
34-6016-4	123A	3122	20	TR21/735	
34-6016-7	123A	3122	62	TR21/735	
34-6016-8	123A		10		
34-6016-11	102A	3004	52	TR85/250	G0005
34-6016-12			28	TR76/	
34-6016-13			29	TR77/	
34-6016-14	123A	3020	20	TR21/735	S0015
34-6016-15	159	3114	21	TR20/717	S0014
34-6016-15A	159	3114		TR20/	
34-6016-16	123A	3024	63	TR21/735	S5014
34-6016-17	108	3018	20	TR95/56	S0011
34-6016-18	123A	3124	20	TR33/735	S5014
34-6016-19	123A	3124	20	TR87/735	
34-6016-22	128	3024	28	TR87/243	
34-6016-23	129	3025	29	TR88/242	
34-6016-23A	129	3025		TR88/	
34-6016-24	123A	3122	62	TR25/735	
34-6016-25	123A	3024	63	TR87/735	S5014
34-6016-26	123A	3124	63	TR21/735	S5014
34-6007-9			10		
34-6007-10			60		
34-6007-11			60		
34-6007-12			60		
34-6007-13			60		

Original Part Number	ECG	SK	GE	IR/WEP	HEP
34-6016-27	128	3024	63	TR87/243	S5014
34-6016-28	160	3008	51	TR85/637	G0003
34-6016-29	160	3008	51	TR85/637	G0003
34-6016-30	128	3024	63	TR25/243	S5014
34-6016-31	210	3054	66	TR55/S3023	S3024
34-6016-32	159	3008	51	TR85/242	G0003
34-6016-32A	129	3025		TR88/	
34-6016-33	128	3024	63	TR25/243	S5014
34-6016-41	194				
34-6016-44			63		
34-6016-45			66		
34-6016-46			69		
34-6016-47	159	3114	21	TR20/	S0019
34-6016-49	123A	3122	17	TR21/735	S0015
34-6016-49A	123A	3124		TR21/	
34-6016-50	102A	3003	2	TR85/	G0005
34-6016-51			39	TR88/	
34-6016-53	210	3054	28		S3024
34-6016-54	211	3083	29		
34-6016-56	194	3137	10	TR24/	S0015
34-6016-59			67	TR87/	
34-6016-60	159	3114	22	TR88/	S5022
34-6016-63	123A			TR21/	S0013
34-6016-64		3114	21	TR54/	S0019
34-6016-65	187	3083	29	TR56/	S3028
34-6017		3024			
34-6017-1				TR30/	
34-6017-2				TR30/	
34-6017-3	172	3024	64	TR28/	S9100
34-6017-4				TR28/	
34-6018-2		3116	FET2	FE100/	
34-8001-43	126	3006	2	TR17/635	
34E31	108	3018	17	TR95/56	S0015
34H31	108	3018		TR95/	
35-020-21	128				
35-39306001	123A	3124		TR21/	
35-39306002	123A	3124		TR21/	
35-39306003	123A	3124		TR21/	
35P1	102	3004	52	TR05/631	G0005
35P2	102	3004	2	TR05/631	
35P2C	102	3004	2	TR05/631	
35(RCA)	159				
35T1	102	3004	52	TR05/631	G0003
36-0041	1142				
36-1	107	3114			
36-6015-46	107				
36-6016-59	193				
36J003-1	102A	3004	52	TR85/	G0005
36P1	102	3004	52	TR05/631	G0005
36P1C	102	3004	2	TR05/631	G0005
36P1F	102	3004	52	TR05/631	G0005
36P2F	102	3004	52	TR05/631	G0005
36P3	102	3004	52	TR05/631	G0005
36P3A	102	3004	2	TR05/631	G0005
36P3C	102	3004	2	TR05/631	G0005
36P4	102	3004	2	TR05/631	G0005
36P4C	102	3004	52	TR05/631	G0005
36P5	102	3004	2	TR05/631	G0005
36P5C	102	3004	52	TR05/631	G0005
36P6C	102A				
36P7	102A	3004	2	TR85/250	
36P7C	102A	3004	52	TR85/250	G0005
36P7T	102A	3004	53	TR85/250	G0005
36P8	102A	3004	52	TR85/250	
36P8C	102A	3004	2	TR85/250	
36T1	102	3118	52	TR17/635	G0008
37-193MP	199				
37-19201	234				
37-21401	199				
37T1	126	3008	52	TR17/635	
38P1	121	3009	2	TR01/254	G0005
38P1C	121	3009		TR01/254	
38T1	102	3004	52	TR05/631	G0002
39A9	102	3004	52	TR05/631	G0005
39P1	121	3009	16	TR01/232	G6003
39P1C	121	3009	16	TR01/232	
39T1	102	3004	52	TR05/631	G0008
40-0068-2	128				
40-065-19-001	9961				
40-065-19-002	9963				
40-065-19-003	9949				
40-065-19-004	9937				
40-065-19-005	9944				
40-065-19-006	9932				
40-065-19-007	9948				

Original Part Number	ECG	SK	GE	IR/WEP	HEP
40-065-19-008	9933				
40-065-19-012	74H40				
40-065-19-013	9946				
40-065-19-014	9097				
40-065-19-016	7440				
40-065-19-027	7475				
40-065-19-029	7486				
40-065-19-030	7495				
40-09437	234				
40-09952	234				
40-601	102	3004		TR05/	G0007
40-11253	234				
40-13481		3122			
40-102427		3104			
40-102429		3053			
40-102430	124				
40C2PW8V1SP	106	3118		TR20/	
40D1547	160	3007	51	TR17/637	
40P1	102	3004	52	TR05/631	G0005
40P2	102	3004	52	TR05/631	G0005
41-0318	130	3027		TR59/247	
41-0318A	130	3027		TR59/	
41-0499	123A	3122	20	TR21/735	
41-0500	129	3025		TR88/242	
41-0500A	129	3025		TR88/	
41-0606	198			TR81/241	
41-0609	198			/241	
41-0909	175	3538		TR81/	
41B581014	128	3024	18	TR87/243	
41B581144-P001	175				
41N1	108	3019		TR95/56	
41N2	108	3018	20	TR95/	S0025
41N2A	108	3018		TR95/	
41N2AA	108	3018	20	TR95/56	S0025
41N2B	108	3018	20	TR95/56	
41N2M	161	3132		TR83/719	S0025
41N3	108	3018	20	TR95/56	S0025
42-1	1112				
42-2	1111				
42-9029-31B	199				
42-9029-31L	199				
42-9029-31M	123A				
42-9029-31P	199				
42-9029-31Q	199				
42-9029-31R	123A				
42-9029-31V	123A				
42-9029-40C	123A				
42-9029-40F	234				
42-9029-40L	123A				
42-9029-40T	128				
42-9029-40U	129				
42-9029-40X	159				
42-9029-40Y	123A				
42-9029-60A	123A				
42-9029-60B	234				
42-9029-60C	123A				
42-9029-60D	199				
42-9029-60Q	159				
42-9029-60W	129				
42-9029-70C	123A				
42-9029-70D	159				
42-9029-70E	159				
42-9029-70F	123A				
42-9029-70G	129				
42-9029-70J	128				
42-9029-70K	128				
42-9029-70P	123A				
42-9029-70Q	234				
42-16599	121	3009	16	TR01/232	
42-17143	102	3004	52	TR05/631	
42-17444	123	3124	20	TR86/53	S0015
42-18109	176	3004	52	TR82/238	G6011
42-18111	123	3124	20	TR86/53	S0011
42-18310	124	3021	12	TR81/240	
42-19642	128	3024	20	TR87/243	
42-19643	129	3025	21	TR88/242	
42-19644	123	3124	20	TR86/53	
42-19670	123	3124	20	TR86/53	S0015
42-19671	102A	3004	52	TR85/250	
42-19682	160	3005	52	TR17/637	
42-19683	108	3018	20	TR95/56	
42-19792	160	3006	50	TR17/637	
42-19840	123A	3124	20	TR21/735	
42-19862	103A	3010	8	TR08/724	G0011
42-19862A	158	3004	52	TR84/630	G0005

Original Part Number	ECG	SK	GE	IR/WEP	HEP
42-19863	102A	3004	52	TR85/250	G0005
42-19863A	102A	3004	2	TR85/250	
42-19864	102	3004	52	TR85/250	G0005
42-19864A	102A	3004	2	TR85/250	
42-20222	102A	3004		TR85/250	
42-20738	123	3124	20	TR86/53	
42-20739	129	3025	21	TR88/242	
42-20863		3024			
42-20960	175	3026	23	TR81/241	
42-20961	130	3027	14	TR59/247	
42-20961A	130	3027		TR59/	
42-20961BL		3510			
42-20961BLU		3510			
42-20961BR		3510			
42-20961GR		3510			
42-20961R		3510			
42-20961OR		3510			
42-20961Y		3510			
42-21232	129	3025		TR88/242	
42-21233	128	3024		TR87/243	
42-21234	123A	3124		TR21/735	
42-21401	161	3018	20	TR83/719	
42-21402	161	3018	20	TR83/719	
42-21403	126	3008	51	TR17/635	
42-21404	103A	3024		TR08/724	
42-21405	158	3025		TR84/630	
42-21406	102A	3004	52	TR85/250	
42-21407	123	3124	20	TR86/53	
42-21443	121MP	3015	13MP	/232MP	
42-22008	159	3025	21	TR20/717	S0012
42-22008A	159	3114		TR20/	
42-22009	172		64	TR69/S9100	S9100
42-22158	123A	3024	18	TR21/735	S0014
42-22532	107	3018	11	TR70/720	S0016
42-22533	123A	3124	20	TR21/735	S0015
42-22534	102A	3004	52	TR85/250	G0008
42-22535	158	3004	2	TR84/630	G0005
42-22539	106	3087		TR20/	
42-22778	160	3006		TR17/637	G0003
42-22779	160	3006		TR17/637	G0003
42-22780	160	3006		TR17/637	
42-22781	160	3006		TR17/637	G0003
42-22784	160	3006		TR17/637	G0003
42-22785	108	3018	20	TR95/56	S0015
42-22786	123A	3124	20	TR21/735	S0015
42-22787	123A	3124	20	TR21/735	S0015
42-22809	123A	3124	20	TR21/735	S0015
42-22810	159	3025	22	TR20/717	S0019
42-22810A	150	3114		TR20/	
42-22811	123A	3124	20	TR21/735	S0015
42-22812	123A	3124	20	TR21/735	S0015
42-22834	121MP	3009	13MP	TR01/232MP	G6013
42-22847	123A	3124	20	TR21/735	S0015
42-23348	123A	3124		TR21/735	S0024
42-23349	123A	3124		TR21/735	S0025
42-23459	152			TR76/701	S5000
42-23541	159	3114		TR20/717	S5013
42-23541A	159	3114		TR20/	
42-23542	123A			TR21/735	S5014
42-23622	102A		52	TR85/250	
42-23960	107			TR70/720	S0016
42-23960P	107	3018		TR70/	
42-23961	107			TR70/720	S0016
42-23961P	107	3018		TR70/	
42-23962	107			TR70/720	S0015
42-23962P	107	3018		TR70/	
42-23963	107			TR70/720	S0033
42-23963P	107	3018			
42-23964	123A	3124		TR21/735	S0015
42-23964P	123A	3124		TR21/	
42-23965	126			TR17/635	G0009
42-23965P	160	3006		TR17/	
42-23966	123A	3124		TR21/735	
42-23966P	123A	3124		TR21/	
42-23967	158			TR84/630	G0005
42-23967P	158	3004		TR84/	
42-23968	121		16	TR01/232	G6003
42-23968P	121	3009		TR01/	
42-24263	172			TR69/S9100	S9100
42-27277	199	3124	62		S0015
42-27372		3018	60		S0016
42-27373		3018	60		S0016
42-27374		3124	20		S0016
42-27375		3124	20		S0016
42-27376		3004	2		G0005
42-28056	123A	3124	10		S0015
42-28057	158	3004	2		G0005
42-28203	108	3018		TR95/56	S0016
42-28204	108	3018		TR95/56	S0020
42-28205	123A	3018		TR21/735	S0030
42-28206	108	3018		TR95/56	S0016
42-28207	123A	3124		TR21/717	S0003
42-28208	159	3114		TR20/	S0019
42-28209					S0024
42-28210	123A			TR21/735	S0003
42-28211	159	3114		TR20/717	S0019
42-28212	184	3054		/S5003	S5000
42-28213	185	3083		/S5007	S5006
42X210	103	3010	2	TR08/641A	
42X230	102A	3004	2	TR85/250	
42X233	102	3004	2	TR05/631	
42X309	102	3004	2	TR05/631	
42X310	103	3010	8	TR08/641A	
42X311	102	3004	2	TR05/631	
43-0203845	129	3025		TR88/	
43-022861	108	3018		TR95/	
43-023190	181	3036		TR36/	
43-023212	123A	3020		TR21/	
43-023221	123A	3124		TR21/	
43-023222	129	3025		TR88/	
43-023223	128	3024		TR87/	
43-023843	130	3027		TR59/	
43-023844	128	3024		TR87/	
43-025763	108	3018		TR95/	
43-025764	108	3018		TR95/	
43-025765	108	3018		TR95/	
43-025766	123A	3124		TR21/	
43-025767	108	3018		TR95/	
43-025834	131	3052	44	TR94/	
43A111449	160	3005			
43A111449G		3005			
43A126932	128				
43A126932-1		3024			
43A128340-1	123A				
43A128340-2	123A				
43A128340-3	123A	3122			
43A128342-1		3122			
43A128342-2		3122			
43A128342-4	123A				
43A128342-5	123A				
43A128342-6	123A				
43A128342-7	123A				
43A128625-1C		3008			
43A145291-1	159				
43A145291-2	159				
43A162445P1	199				
43A162455-1	128				
43A165137P1	130				
43A165137P3	130				
43A165137P4	130				
43A167207P1	159				
43A167207P2	159				
43A167851	123A				
43A167885-P1	181				
43A167885-P2	181	3036			
43A167885-P3	181				
43A167886P4	218				
43A168016P1	123A				
43A168064-1	159				
43A168064P1	159				
43A168135-1	9932				
43A168135-2	9930				
43A168135-3	9933				
43A168135-4	9946				
43A168135-6	9936				
43A168135-7	9944				
43A168135-8	9945				
43A168135-9	9093				
43A168135-10	9962				
43A168440-1	912				
43A168481	941D				
43A168481P1	941D				
43A176002	159				
43A180002-P1	123A				
43A180002-P2	123A				
43A212040-1	909	3590			
43A212040-2	909D	3590			
43A212042P1	910				
43A212042P2	910D				
43A212067	130				

Original Part Number	ECG	SK	GE	IR/WEP	HEP
43A212067-1		3510			
43A212090P1	199				
43A223006P1	74H00				
43A223007	7401				
43A223008	74H01				
43A223009	7402				
43A223012	74H10				
43A223015	7430				
43A223017	74H40				
43A223018	74H40				
43A223025	7473				
43A223026P1	7474				
43A223028	7476				
43A223029P1	7442				
43A223030	7495				
43A223031	7496				
43A223033P1	7483				
43A223034P1	7493				
43A223046-1	923				
43A223060-1	128				
43B140883-1	123A				
43B167250-12		3501			
43B167250-13		3501			
43B167250-14		3501			
43B167250-15		3501			
43B168450-1	159				
43B168495-1	159				
43B168566-P1	159				
43B168610	123A				
43B168613-1	199				
43C168567	123A				
43C216408P1	74H00				
43C216409P1	74H01				
43C216410P1	74H04				
43C216411P1	74H20				
43C216414P1	74H51				
43C216447	74193				
43C216447P1	74193				
43N3	123	3124	20	TR86/53	
43N6	123	3124	20	TR86/53	
43P1	100	3005	1	TR05/254	G0005
43P2	102	3004	2	TR05/631	G0005
43P3	100	3005	1	TR05/254	G0005
43P4	102	3004	2	TR05/631	G0005
43P4C	102	3004	51	TR05/631	G0005
43P6	102	3004	52	TR05/631	G0005
43P6A	102	3004	2	TR05/631	G0005
43P6C	100	3004	2	TR05/254	G0005
43P7	102	3004	52	TR05/631	
43P7A	102	3004	2	TR05/631	
43P7C	102	3004	2	TR05/631	G0005
43-VD-09	128				
43X16A567	123		20	TR86/53	
44-13	126	3006	51	TR17/635	
44A-1A5	121	3009		TR01/	
44A319819-1	128				
44A332168-001	941				
44A332168-002	941				
44A332169-002	941				
44A333413	923				
44A333463-001	123A				
44A333464	159				
44A333464-1	159				
44A335854		3543			
44A340309		3528			
44A353980-002	175	3131			
44A354637-001	175	3026			
44A354803		3024			
44A355565-001	130	3027			
44A358624-003	129				
44A359496		3036			
44A359497-001	128				
44A359497-002	128				
44A390202-001	909				
44A390202-003	909				
44A390205	716				
44A390243-001	175				
44A390244-001	130	3027			
44A390247	123A				
44A390248-001	159				
44A390249	123A				
44A390251	123A				
44A390251-001	123A				
44A390256-001	159				
44A390261	159				
44A390264-001	128				
44A391505	129				
44A391593-001	181	3036			
44A393605	941D				
44A393611	923D				
44A393611-001	923D				
44A395909-1	181				
44A395986-001	152				
44A395992-001	128				
44A395992-002	128				
44A395994-001	123A				
44A397905	159				
44A-417014-001	181				
44A417031-001	159				
44A417032-001	152				
44A417033-001	130				
44A417034-001		3512			
44A417034	192				
44A417063-001	128				
44A417390		3535			
44A417715-001		3079			
44A417716	130				
44A417716-001	130	3027			
44A417717-001		3535			
44A417756		3083			
44A417756-1	197				
44A417779-001	941M				
44A418041-001	159				
44A4417714	181				
44A4417714-001	181				
44A355565001	130				
44B1B	738	3167			C6075P
44B1Z	738	3167	29		
44B238203-1	159				
44B238208-001	175				
44B238208-002	175				
44B238246	159				
44B311097	123A				
44B238208002	175				
44P1	160	3008	51	TR17/637	G0002
45AN4AA	123A				C6090
45B1AH	732				C6090
45B1D	732				C6090
45B1Z	732				
45N1	103	3010	8	TR08/641A	
45N2	103	3010	8	TR08/641A	
45N2A	103A			TR08/	
45N2M	123A	3122	10	TR21/735	S0015
45N3	123A	3124	10	TR21/735	S0015
45N4	123A	3124	20	TR21/735	S0015
45N4M	123A	3018	20	TR21/735	S0015
45NP	108	3018	17	TR95/56	
45X1A502C	102	3004		TR05/631	
45X1A520C	102A	3123	52	TR85/	
45X2	100	3005	2	TR05/254	
46-06311-3	107				
46-119-3	161	3018		TR83/719	S0017
46-163-3	102A	3004	2		G0007
46-840-3	102A			TR85/250	G0005
46-862-3	160			TR17/637	G0001
46-864-3	108	3039		TR95/56	S0016
46-865-3	160			TR17/637	G0003
46-866-3	160			TR17/637	G0003
46-867-3	154	3045		TR78/712	S5025
46-868-3	160			TR17/637	G0003
46-869-3	102A			TR85/250	G0007
46-1340-3	711	3070			C6069G
46-1343-3		3102			C6060P
46-1346-3	711	3070			C6069G
46-1347-3	710	3102			
46-1348-3	710	3102			
46-1352-3	707				C60676
46-1356-3	710	3102	511		
46-1357-3	1004	3102	21		C6100P
46-1361-3	712	3072			
46-1364-3					C6072P
46-1365-3	749	3168			C6076P
46-1366-3	748				C6060P
46-1369-3	1004				
46-1370-3	1109				C6079P
46-1382-3	1108				
46-1384-3	712				
46-1393-3	1130				
46-1394-3	1131				
46-1395-3	1134				

Original Part Number	ECG	SK	GE	IR/WEP	HEP
46-1396-3	1128				
46-1397-3	1132				
46-1398-3	1133				
46-5002-1	703A		12	IC500/	C6001
46-5002-2	703A	3157			
46-5002-3	717		12	IC500/	C6001
46-5002-4	703A	3157	12	IC500/	C6001
46-5002-5	872				C6077P
46-5002-6	711	3070		IC514/	C6069G
46-5002-7	703A	3157	12	IC500/	C6001
46-5002-8	748				C6060P
46-5002-8B	748				
46-5002-9	718	3159			
46-5002-10	706	3101			
46-5002-11	790		3		
46-5002-12	714	3075	4	IC509/	C6070P
46-5002-13	715	3076	6	IC510/	C6071P
46-5002-14	739				C6072P
46-5002-15	712	3072	2	IC508/	C6063P
46-5002-16	780		20		C6069G
46-5002-17	746				C6059P
46-5002-18	747				C6079P
46-5002-19	779				
46-5002-20	923D	3165			
46-5002-21	782	3077			C6077P
46-5002-23	740		31		
46-5002-26	712				
46-5002-27	779		13		
46-5002-28	749	3168			
46-5002-31	1062		4		
46-5002-33	731	3170			
46-8257-3	123A	3122	20	TR25/	S3013
46-8610-3	102A	3004	52	TR85/250	G0005
46-8611	102A	3004		TR85/250	G0005
46-8611-3	102A	3004	52	TR85/250	G0005
46-8612-3	160			TR17/637	G0003
46-8613-3	126	3008	51	TR17/635	G0009
46-8614-3	101	3011		TR08/641	G0011
46-8615-3	121	3009		TR01/232	G6005
46-8617-3	121	3009		TR01/232	G6003
46-8625-1	126	3008		TR17/635	
46-8625-2	126	3008		TR17/635	
46-8625-3	126	3008		TR17/635	
46-8625-4	126	3008		TR17/635	
46-8625-5	126	3008		TR17/635	
46-8625-6	126	3008		TR17/635	
46-8629	107	3019	11	TR70/720	S0016
46-8631-3	102A			TR85/250	G0007
46-8634-3	121	3009		TR01/232	G6005
46-8636-3	126	3005	9	TR17/635	G0008
46-8638-3	121	3009		TR01/232	G6003
46-8646	222	3050			F2004
46-8647	107	3018	11	TR80/	S0020
46-8660-3	102A	3123		TR85/	
46-8664-3	102A	3004			G0005
46-8665-3	102A	3004	1	TR85/250	G0007
46-8666-3	102A	3004	2	TR85/250	G0005
46-8668-3	102A			TR85/250	G0005
46-8671-3	103A	3010		TR08/724	G0011
46-8672-3	161	3117		TR83/719	S0020
46-8677-2	108			TR95/	S0016
46-8677-3	107	3019	11	TR70/720	S0016
46-8679-1	102A	3004		TR85/250	
46-8679-2	102A	3004		TR85/	
46-8679-3	102A	3004	52	TR85/250	G0005
46-8680-1	102A	3004		TR85/250	
46-8680-2	102A	3004		TR85/250	
46-8680-3	102A	3004	52	TR85/250	G0007
46-8681-1	102A	3004		TR85/250	
46-8681-2	102A	3004		TR85/250	
46-8681-3	158	3004		TR84/630	G0005
46-8682-2	123A	3122	20	TR21/	
46-8682-3	199	3124	20	TR21/735	S0016
46-8695-3	123A	3018		TR21/735	S0015
46-13101-3	1080				
46-13103-3	1094				
46-13104-3	1121				
46-13105-3	1122				
46-81187-3	171	3201	300	TR79/	
46-84120-3	108	3018		TR95/56	S0016
46-85285-3	108	3018		TR95/56	S0020
46-86101-3	107	3039		TR70/720	S0016
46-86102-3	126			TR17/635	G0008
46-86103-3					S5020
46-86107-3	161	3018		TR83/719	S0017
46-86108-3	161	3018		TR83/719	S0017
46-86109-3	107	3039	61	TR70/720	S0017
46-86110-3	107	3039		TR70/720	S0016
46-86112-2	161	3117		TR83/719	
46-86112-3	107	3018	17	TR70/720	S0017
46-86113-3	161	3039	60	TR83/719	S0017
46-86114-3	161	3018		TR83/719	S0016
46-86117-3	161	3018		TR83/719	S0017
46-86118-3	161	3018		TR83/719	S0017
46-86119-3	161	3018		TR83/719	S0017
46-86120-3	107	3018		TR70/720	S0016
46-86121-3	123A	3122		TR21/735	S0015
46-86122-3	123A	3122		TR21/735	S0015
46-86123-3	160	3005	52	TR17/	G0008
46-86125-3	131		44	TR94/642	G6016
46-86126-3	107	3018		TR70/720	S0016
46-86127-3	107	3018		TR70/720	S0016
46-86130-3	175	3026	35	TR81/241	S5012
46-86131-3	233	3018	61	TR95/56	S0016
46-86132-3	123A	3124	20	TR21/735	S0014
46-86133-3	107	3039	11	TR70/720	S0016
46-86135-3	131	3082	44	TR94/642	G6016
46-86136-3	121	3009		TR01/232	G6003
46-86140-3	108	3018	20	TR95/56	S0020
46-86143-3	123A	3124	20	TR21/735	S0015
46-86144-3	123A	3018	11	TR21/735	S0016
46-86145-3	123A	3040	17	TR21/735	S0020
46-86152-3	123A	3122		TR21/735	S0015
46-86163-3	102A	3124		TR85/250	G0007
46-86169-3	123A	3018	40	TR78/712	S5026
46-86170-3	159	3114	21	TR20/717	S0013
46-86171-3	123A	3018	20	TR21/735	S0016
46-86172-3	107	3018		TR70/720	S0016
46-86173-3	154	3124	20	TR78/712	S5026
46-86182-3	154	3024		TR78/712	S5026
46-86183-3	154	3020		TR78/712	S5026
46-86189-3	175	3131		TR81/241	S5012
46-86192-3	123A	3124	8		S0014
46-86195-3	102A	3004		TR85/250	G0005
46-86196-3	158	3004	52	TR84/630	G6011
46-86198-3	176				G6011
46-86207-3	107	3018	11	TR70/720	S0016
46-86208-3	108	3018	11	TR95/56	S0016
46-86209-3	107	3018		TR70/720	S0016
46-86210-3	154	3045	63	TR78/712	S0014
46-86211-3	103A	3010	8	TR08/724	G0011
46-86213-3	121	3009	16	TR01/232	G6003
46-86224-3	161	3018	11	TR83/719	S0016
46-86226-3	175	3131	23	TR81/241	S5019
46-86228-3	123A	3018	20	TR21/735	S0016
46-86229-3	159	3025	21	TR20/717	S0013
46-86229-3A	159	3114		TR20/	
46-86230-3	159	3025	21	TR21/717	S0013
46-86231-3	123A	3122		TR21/735	S0014
46-86232-3	154	3040		TR78/712	S5024
46-86233-3	128	3024		TR87/243	S5026
46-86234-3	184	3054	57	TR76/S5003	S5000
46-86235-3	185	3083	29	TR77/S5007	S5006
46-86238-3	159	3114	17	TR20/717	S0012
46-86238-3A	159	3114		TR20/	
46-86240-3	107	3018	20	TR70/720	S0016
46-86244-3	108	3018	20	TR95/56	S0016
46-86247-2	123A	3124	20		S0016
46-86247-3	123A	3124	18	TR21/	S0016
46-86251-3	107	3018		TR70/720	S0016
46-86252-3	123A	3018	20	TR78/712	S0015
46-86256-1	126	3008		TR17/	
46-86256-2	126	3008		TR17/	
46-86256-3	102A	3008	51	TR85/250	G0005
46-86257-3	123A	3124	63	TR21/735	S3013
46-86262-3	108	3122	20		
46-86265-3	107	3018		TR70/	S0016
46-86268-3	123A	3118		TR21/735	S0013
46-86269-3	108	3018	11	TR95/	S0020
46-86274-3	123A	3124	20	TR21/735	S0015
46-86283-3	159	3114		TR20/	S0012
46-86285-3	161	3018	11	TR83/	S0020
46-86293-3	159	3004	52		S0012
46-86295-3	107	3018	20	TR33/	
46-86299-3	107	3018		TR70/720	S0016
46-86300-3	160	3006	50	TR17/637	G0009
46-86301-3	107	3018		TR70/720	S0016
46-86302-3	107	3018		TR70/720	S0016
46-86310-3	123A	3018	20	TR21/735	S0015
46-86314-3	108	3018		TR95/56	S0016

Original Part Number	ECG	SK	GE	IR/WEP	HEP
46-86314-3A	108	3018		TR95/	
46-86316-3	132	3116		FE100/802	F0015
46-86317-3	152	3054		TR76/701	S5000
46-86318-3	154			TR78/712	S3021
46-86319-3	157	3103		TR60/244	S5015
46-86335-3	152	3026		TR76/701	S3021
46-86344-3	171	3044	27		S5025
46-86345-3	171	3024	27	TR87/243	S5025
46-86346-3	171	3044	27		S5025
46-86347-3	196	3054	66	TR92/	S5000
46-86348-3	152	3054	66	TR92/	S5000
46-86349-3	175	3026	23	TR81/	S5019
46-86350-3	175	3026	23	TR57/	S5019
46-86352-3	107	3018	11		S0020
46-86353-3	107	3018	11		S0020
46-86354-3	107	3018	11		S0020
46-86357-3	108	3018	17		
46-86360-3	186	3054	28		
46-86371-3	128	3024	20		
46-86373-3	102A	3004	53	TR85/	G0005
46-86374-3	152	3054	28		S5000
46-86375-3	123A	3024	61		S0015
46-86376-3	108	3122	20	2SC829/	S0015
46-86377-3	159	3025	21	TR52/	S0019
46-86378-3	123A	3124	62		S0015
46-86381	128	3020	18	TR87/	S0005
46-86381-3	128	3024	18	TR87/	S0015
46-86384-3	124	3021	12	TR81/	S5011
46-86386-3	124	3021	12	TR81/	S5011
46-86387-3	124		12	TR81/	S5011
46-86388-3	130	3027	36	TR59/	S7004
46-86389-3	165	3115	38	TR93/	S5021
46-86390-3	164	3115	38	TR93/	S5021
46-86396-3	221	3050			F2004
46-86397-3	108	3018	11	TR95/	S0016
46-86399-3	159	3114	21		S0019
46-86400-3	152	3054	28	TR76/	S5000
46-86403-3	129	3025		TR28/	S5013
46-86404-3	123A		17	TR51/	S3002
46-86405-3	128	3020	18	TR25/	S5014
46-86406-3	129	3114	21	TR19/	S5013
46-86407-3	123A	3124	61	TR21/	S0030
46-86408-3	123A	3124	20	TR24/	S0015
46-86409-3	123A	3124	20	TR24/	S0015
46-86411-3	185	3083	58	TR77/	
46-86412-3	159	3114	21	TR30/	S0013
46-86415-3	165	3111	38		S5021
46-86419-3	123A		62		
46-86424-3	159	3114	22	TR52/	S0013
46-86425-3	129	3025	69	TR30/	S3003
46-86426-3	128	3024	20		S3002
46-86427-3	128	3024	17	TR87/	S3001
46-86429-3	159	3025	22	TR52/	S0019
46-86430-3	102A	3004	52		G0007
46-86434-3	199	3018	60	TR21/	G6005
46-86435-3	107	3018	20	TR21/	S0016
46-86439-3	124	3021	12		S5011
46-86446-3	124	3021	12	TR81/	S5011
46-86446-9	124				
46-86455-3	164	3133	37	TR68/	S5020
46-86466-3	191	3104	32	TR75/	S3021
46-96434-3	107				
46-136279P1	909	3590			
46-136284-P1	923				
46-136284-P2	923D				
46-156741P2	941D				
46-203200P39	949				
46B-3A5	121MP	3015			
47-2(BRADFORD)	108	3019	17	TR95/56	
47C23-3			21		
47P1	121	3009	14	TR01/232	
48-01-003	128				
48-01-004	108				
48-01-005	123A				
48-01-007		3024			
48-01-010	108				
48-01-017	107				
48-01-027	161				
48-01-031	107				
48-01-049	199				
48-03-002	159				
48-03-041840-2	130				
48-03-04046103	128				
48-03-04093403	130				
48-03-05013702	128				

Original Part Number	ECG	SK	GE	IR/WEP	HEP
48-03-10111102	129	3025			
48-03-10744702	192				
48-03-10744802	129				
48-15	130	3027			
48-35P1	102	3004		TR05/631	
48-36P1	102	3004		TR05/631	
48-36P3	102	3004		TR05/631	
48-39P1	121	3009		TR01/232	
48-39P3	102	3004		TR05/631	
48-43P3	102	3004		TR05/631	
48-43P4	102	3004		TR05/631	
48-45N2	199			TR21/	
48-56P1	100			TR05/254	
48-57B2	121	3009		TR01/232	
48-57B42	121	3009		TR01/232	
48-171-A06	102A			TR85/	
48-971A05	123A			TR21/	
48-971A13	124			TR81/	
48-971A95	123A				
48-971A203	160			TR17/	
48-1050-SL12318	9930				
48-1050-SL12319	9932				
48-1050-SL12320	9933				
48-1050-SL12321	9936				
48-1050-SL12322	9944				
48-1050-SL12323	9945				
48-1050-SL12324	9946				
48-1050-SL12325	9962				
48-1050-SL12326	9093				
48-1050-SL12327	9099				
48-1050-SL12328	9934				
48-4937	233				
48-8613-3	126			TR17/	G0009
48-10073A01	102A			TR85/	G0005
48-10073A02	102	3004		TR05/631	G0005
48-10074A01	102A			TR85/	G0005
48-10074A02	102	3004		TR05/631	G0005
48-10074A03	102A			TR85/	G0006
48-10075A01	121MP	3013	16	/232MP	
48-10075A02	121MP	3013	16	/232MP	
48-10075A03	121MP	3013	16	/232MP	
48-10075A04	121MP	3013	16	/232MP	
48-10075A05	121MP	3013	16	/232MP	
48-10075A06	121MP	3013	16	/232MP	
48-10075A07	121MP	3013	16	/232MP	
48-10075A08	121MP	3013	16	/232MP	
48-10079A01	160		50	TR17/637	G0003
48-10079A02	160		50	TR17/637	G0003
48-10103A01	121MP	3013	16	/232MP	G6005
48-10103A02	121MP	3013	16	/232MP	G6005
48-10103A03	121MP	3013	16	/232MP	G6005
48-10103A04	121MP	3013	16	/232MP	G6005
48-10103A05	121MP	3013	16	/232MP	G6005
48-10103A06	121MP	3013	16	/232MP	G6005
48-10103A07	121MP	3013	16	/232MP	G6005
48-10103A08	121MP	3013	16	/232MP	G6005
48-10103A09	121MP	3013	16	/232MP	G6005
48-10103A10	121MP	3013	16	/232MP	G6005
48-10103A11		3013	16	/232MP	G6005
48-12091A	181				
48-12443	102A	3123		TR85/	
48-13309X3	152				
48-13470	108		20	TR95/56	S0025
48-13474					G6013
48-13481	123A	3122		TR21/	
48-13494				TR51/	G6013
48-13707	220				
48-17162A06	102A			TR85/	
48-17162A10	158			TR84/	
48-17162A13	121			TR81/	
48-17162A17	102A			TR85/	
48-17162A22	102A			TR85/	
48-17271A03	102A			TR85/	
48-21598B01	102A	3024	53	TR85/250	
48-40118B01	159	3114		TR20/717	S0013
48-40118B01A	159	3114		TR20/	
48-40170-G01	199	3122	20	TR21/735	S0015
48-40171G01	123A	3122	20	TR21/735	S0014
48-40172G01	131	3082	30	TR94/242	G6016]
48-40212	128				
48-40246G01	199	3122	20	TR21/735	S0025
48-40246G02	123A			TR21/735	S0025
48-40247G01	107	3039	20	TR70/720	S0024
48-40247G02	199			TR21/735	S0025
48-40247G03					S0015

Original Part Number	ECG	SK	GE	IR/WEP	HEP
48-40382J01	186	3192			
48-40383J01	187	3193			
48-40606J01	123A				
48-40606J02	199				
48-40607J01	199				
48-40662G04					S5000
48-40938G01					S3020
48-40939G01					S3028
48-42098B01	159	3114		TR20/717	S5022
48-42098B01A	159	3114		TR20/	
48-43240G01	152				S3010
48-43241G01	153				S5006
48-43351A01	102A		53	TR85/250	G0005
48-43351A02	107	3039	11	TR70/720	S0016
48-43351A03	108	3039	20	TR95/56	S0016
48-43351A04	108	3039	11	TR95/56	S0016
48-43351A05	108	3039	11	TR95/56	S0016
48-43354A81	123A			TR21/	S0015
48-43354A82	123A			TR21/	S0015
48-43354A83	158			TR84/	G6011
48-43356A16					G6016
48-43357A88					S0016
48-43357A90					S0016
48-43357A91					S0016
48-44166G01					S0015
48-44883G01	152				
48-44884G01	153				
48-44885G01	123A	3122			S0015
48-44885G02	123A				
48-44886G01	128	3122			S0024
48-60022A13	123A	3122	20	TR21/735	S0015
48-60022A14	175	3131	12	TR81/241	S5012
48-63005A66	123A	3122	20	TR51/735	S0015
48-63005A72	123A	3122	20	TR21/735	S0015
48-63026A45	124	3021		TR81/240	S5011
48-63026A46	108	3039	20	TR95/56	S0015
48-63026A47	123A	3122	20	TR21/735	S0015
48-63026A48	123A	3122	20	TR21/735	S0015
48-63029A16	160		51	TR17/637	G0008
48-63029A17	158		51	TR84/630	G0005
48-63029A18	102	3004	53	TR05/631	G0005
48-63029A19	102	3004	53	TR05/631	G0005
48-63029A60	160		51	TR17/637	G0003
48-63029A90	126	3006	51	TR17/635	G0008
48-63019A91	102	3004	51	TR05/631	G0005
48-63029A92	100	3005	51	TR05/254	G0005
48-63029A93	102	3004	53	TR05/631	G0005
48-63029A94	102	3004	53	TR05/631	G0005
48-63044A05	102A		53	TR85/250	
48-63075A72	160		51	TR17/637	G0003
48-63075A73	160		51	TR17/637	G0003
48-63075A74	126	3006	51	TR17/635	G0008
48-63075A75	160		51	TR17/637	G0003
48-63075A76	100	3005	53	TR05/254	G0005
48-63076A52	123A	3122		TR21/735	S0014
48-63076A81	107	3039	20	TR70/720	S0016
48-63076A82	107	3039	20	TR70/720	S0016
48-63076A83	123A	3122	20	TR21/735	S0014
48-63077A03	100	3005	53	TR05/254	G0005
48-63077A10	123A	3122	20	TR21/735	S0014
48-63077A29	108	3039	20	TR95/56	S0016
48-63077A30	123A	3122	20	TR21/735	S0014
48-63077A31	123A	3122	20	TR21/735	S0014
48-63078A26					S0016
48-63078A52	107	3039	20	TR70/720	S0016
48-63078A54	107			TR70/720	S0016
48-63078A59	100	3005	53	TR05/254	G0005
48-63078A60	100	3005	53	TR05/254	G0005
48-63078A61	100	3005	53	TR05/254	G0005
48-63078A62	158	3123	53	TR84/630	G0007
48-63078A63	160		51	TR17/637	G0005
48-63078A64	100	3005		TR05/254	G0005
48-63078A65	126	3006	53	TR17/635	G0005
48-63078A66	158		53	TR84/630	G0005
48-63078A68	158	3123	53	TR84/630	G0007
48-63078A69	158		53	TR84/630	S0014
48-63078A70	123A	3122	20	TR21/735	S0014
48-63078A71	123A	3122	20	TR21/735	S0014
48-63078A86	102A			TR85/250	
48-63078A97		3122		TR51/735	
48-63079A97	123A	3122	20	TR21/735	S0014
48-63081A82	160		51	TR17/637	G0005
48-63082A15	100	3005	53	TR05/254	G0001
48-63082A16	158		53	TR84/630	G0005
48-63082A24	126	3006	53	TR17/635	G0001

Original Part Number	ECG	SK	GE	IR/WEP	HEP
48-63082A25	123A	3122	20	TR21/735	S0014
48-63082A26	123A	3122	20	TR21/735	S0014
48-63082A27	123A	3122	20	TR21/735	S0014
48-63082A45	123A	3122	20	TR21/735	S0014
48-63082A71	123A	3122	20	TR21/735	S0014
48-63084A03	102A	3123	53	TR85/250	G0005
48-63084A04	102A	3123	53	TR85/250	G0007
48-63084A05	102A	3123	53	TR85/250	G0005
48-63086A19	158			TR84/	G0005
48-64978A10	121MP	3013	16	/232MP	G6005
48-64978A11	121MP	3013	16	/232MP	
48-64978A24	121MP	3013	16	/232MP	
48-64978A27	160		50	TR17/637	G0003
48-64978A28	160		51	TR17/637	G0003
48-64978A29	160		51	TR17/637	G0003
48-64978A39	222	3050			F2004
48-64978A40	159	3114	21	TR20/717	S5022
48-64978A40A	159	3114		TR20/	
48-64978A41	159	3114	21	TR20/717	S0019
48-64978A41A	159	3114		TR20/	
48-65108A23	102A(2)		53	TR85/250	
48-65108A62	102A(2)		53	TR85/250	
48-65112A65	108	3039		TR95/56	S0016
48-65112A67	108	3039		TR95/56	S0016
48-65112A68	161			TR83/	
48-65113A88	108	3039		TR95/56	S0016
48-65118A64	108	3039		TR95/56	S0016
48-65123A67	108	3039		TR95/56	S0016
48-65123A94	123A	3122	20	TR21/735	
48-65123A95	108	3039		TR95/56	S0016
48-65132A79	160			TR17/637	G0002
48-65144A72	108	3039	18	TR95/56	S0016
48-65146A61	161	3117	20	TR83/719	S0017
48-65146A62	161	3117	20	/719	S0017
48-65146A63	161	3122	20	TR83/719	S0020
48-65147A72	123A		18	TR21/735	S0015
48-65173A78	161	3117		TR83/719	S0025
48-65174A24	161			TR83/	S0016
48-65177A77	129	3025		TR88/242	
48-65177A77A	129	3025		TR88/	
48-69394A01	123A			TR21/	
48-83750G01	123A				S0015
48-83750G02					G6005
48-83750G03					S3001
48-83750G04					S5022
48-86376-3	123A		20		
48-86904GF	160				
48-90165A01	159	3114	21	TR20/717	S0019
48-90165A01A	159	3114		TR20/	
48-90172A01	123A	3122	20	TR21/735	S0014
48-90232A01	161	3117		TR83/719	S0020
48-90232A03	108	3039		TR95/56	S0025
48-90232A04	108	3039		TR95/56	S0025
48-90232A05	123A	3122		TR21/735	S0015
48-90232A06	129	3025		TR88/242	S3032
48-90232A06A	129	3025		TR88/	
48-90232A07	124	3021		TR81/240	S5011
48-90232A08	128	3024		TR87/243	S3021
48-90232A09	129	3025		TR88/242	S3032
48-90232A09A	129	3025		TR88/	
48-90232A10	108	3039		TR95/56	S0025
48-90232A11	123A			TR21/735	S0015
48-90232A12	129	3122		TR88/242	S3032
48-90232A13	123A			TR21/735	S0015
48-90232A14	132	3116		FE100/802	F0015
48-90232A15	129	3025		TR88/242	S3032
48-90232A15A	129	3025		TR88/	
48-90232A16	124			TR81/240	S5011
48-90232A17	161	3117		TR83/719	S0017
48-90232A18	161	3117		TR83/719	S0017
48-90232A19	108	3039		TR95/56	S0016
48-90232A21					S0017
48-90232A22					S0020
48-90232A23					S0025
48-90232A24					S0016
48-97046A02	102A			TR85/250	G0005
48-97046A03	102A	3123		TR85/250	G0007
48-97046A04	108	3039		TR95/56	S0016
48-97046A05	107	3039		TR70/720	S0016
48-97046A06	107	3039		TR70/720	S0016
48-97046A07	107	3039		TR70/720	S0016
48-97046A08	126	3006		TR17/635	G0009
48-97046A09	126	3006		TR17/635	G0009
48-97046A10	102A	3123		TR85/250	G0005
48-97046A11					G0007

Original Part Number	ECG	SK	GE	IR/WEP	HEP
48-97046A12					G0005
48-97046A13					G0005
48-97046A15	131	3198		TR94/	G6016
48-97046A16	126			TR17/635	G0008
48-97046A17	107			TR70/720	S0016
48-97046A18	108			TR95/56	S0014
48-97046A19					S0014
48-97046A20	107			TR70/720	S0014
48-97046A21	107			TR70/720	S0014
48-97046A22	123A			TR21/735	S0015
48-97046A23	123A			TR21/735	S0015
48-97046A24	123A			TR21/735	S0015
48-97046A25	161			TR83/	S0015
48-97046A26	159			TR20/717	S0019
48-97046A27	159			TR20/717	S0012
48-97046A28	123A			TR21/735	S0014
48-97046A29	128			TR87/243	S5026
48-97046A30	152			TR76/701	S5000
48-97046A31	131	3052		TR94/	G6016
48-97046A32	102A			TR85/	G0006
48-97046A33	102A			TR85/	G0006
48-97046A34	102A	3123		TR85/	G0005
48-97046A36	129			TR88/	S0019
48-97046A37	158			TR84/	G0007
48-97046A38	152			TR76/	S5000
48-97046A39	129			TR88/	S5022
48-97046A40	128			TR87/	
48-97046A41					S0016
48-97046A42	123A			TR21/	S0014
48-97046A43	123A			TR21/	S0030
48-97046A44					G0005
48-97046A45	161			TR83/	S0011
48-97046A46	123A			TR21/	S0015
48-97046A47	132			FE100/	F0015
48-97046A48	132			FE100/	G0005
48-97046A50	123A			TR21/	S0014
48-97046A51	108			TR95/	S0016
48-97046A52	123A			TR21/	S0030
48-97046A53	102			TR05/	G0005
48-97046A54	102A			TR85/	G0005
48-97046A55	102A			TR85/	G0005
48-97046A56	102A			TR85/	G0007
48-97046A57	102A			TR85/	G0005
48-97046A58					S0015
48-97046A59					G0005
48-97046A60					S0015
48-97046A61					S0015
48-97046A62					S0015
48-97127A02	108	3039		TR95/56	S0015
48-97127A03	108	3039		TR95/56	S0015
48-97127A04	128	3024		TR87/243	S0024
48-97127A05	158			TR84/630	G0005
48-97127A06	107	3039		TR70/720	S0011
48-97127A08		3122		TR51/735	S0011
48-97127A09	158			TR84/	G0005
48-97127A12	123A	3039		TR21/720	S0014
48-97127A13	123A	3039		TR21/56	S0014
48-97127A14					S0015
48-97127A15	106			TR20/	S0019
48-97127A17					S5012
48-97127A18	123A	3039		TR21/720	S0016
48-97127A19	123A	3122		TR21/735	S0014
48-97127A20	102A			TR85/250	G0005
48-97127A21					G0005
48-97127A22	102A			TR85/250	G0007
48-97127A23	158			TR84/	G0005
48-97127A24	123A			TR21/	S0014
48-97127A25					G0005
48-97127A26					G0005
48-97127A27					G0005
48-97127A28					G0005
48-97127A29	123A			TR21/	S0011
48-97127A30	102A			TR85/	G0005
48-97127A31	102A			TR85/	G0005
48-97127A32	102			TR05/	G0005
48-97127A33	123A			TR21/	S0011
48-97162A01	108	3039		TR95/56	S0016
48-97162A02	108	3039		TR95/56	S0016
48-97162A03	126	3008		TR17/635	G0008
48-97162A04	123A	3122		TR21/735	S0014
48-97162A05	123A	3122		TR21/735	S0014
48-97162A06	158			TR84/250	G0005
48-97162A07	158			TR84/	G0005
48-97162A08	158			TR84/	S0014
48-97162A09	123A			TR21/	S0015
48-97162A10					S0014
48-97162A11	102A			TR85/250	G0005
48-97162A12	123A			TR21/735	S0030
48-97162A13					S5011
48-97162A15	123A			TR21/735	S0011
48-97162A16	102A			TR85/250	G0008
48-97162A17					G0005
48-97162A18	102A	3123		TR85/250	G0007
48-97162A19	102A	3123		TR85/250	G0006
48-97162A20	102A			TR85/250	G0008
48-97162A21	123A			TR21/250	S0014
48-97162A22				/250	G0006
48-97162A23	123A			TR21/	S0014
48-97162A24	102A			TR85/	G0005
48-97162A25	102A			TR85/	G0005
48-97162A26	107			TR70/720	S0016
48-97162A28	161			TR83/	S0015
48-97162A29					S5006
48-97162A30	161			TR83/	S0014
48-97162A31	161			TR83/	S0014
48-97162A32	161			TR83/	S0014
48-97162A33	123A			TR21/	S0014
48-97162A34	158			TR84/	G0007
48-97162A35	184				S5006
48-97162A36					G0005
48-97162A37					G0006
48-97162A38					S0015
48-97162A69	184				
48-97177A01	132	3116	FET2	FE100/802	
48-97177A02	161	3117	20	TR83/719	S0016
48-97177A03	161	3117		TR83/719	S0016
48-97177A04	108	3039		TR95/56	S0030
48-97177A06	132			FE100/	F0021
48-97177A07	108			TR95/	S0016
48-97177A08	108			TR95/	S0020
48-97177A09	123A			TR21/	S0030
48-97177A10	188	3199		TR72/3020	S3024
48-97177A11	189	3200		TR73/3021	S3028
48-97177A12	123A	3122		TR21/735	S0015
48-97177A13	123A	3122		TR21/735	S0015
48-97177A14	159	3114		TR20/717	S0019
48-97177A14A	159	3114		TR20/	
48-97177H01	133				
48-97221A01	102A			TR85/	G0006
48-97221A02	102A			TR85/250	G0009
48-97221A03	102A	3123		TR85/250	G0007
48-97221A04	102A			TR85/250	G0005
48-97221A05	102A			TR85/250	G0005
48-97238A01	126	3006		TR17/635	G0008
48-97238A02	126	3006		TR17/635	G0008
48-97238A03	126	3008		TR17/635	G0008
48-97238A04	123A	3122		TR21/735	S0014
48-97238A05	102A			TR85/250	G0005
48-97238A06	131	3082	44	TR94/642	G6016
48-97238A07	102A			TR85/250	G0006
48-97271A01	102A	3123		TR85/250	G0007
48-97271A02	102A	3123		TR85/250	G0007
48-97271A03	102A	3123		TR85/250	G0005
48-97271A04	102A	3006		TR85/635	G0008
48-97271A05	160	3008		TR17/635	G0008
48-97271A06	160	3006		TR17/635	G0008
48-97271A2	102A			TR85/	
48-97271A3	102A			TR85/	
48-97271A4	102A			TR85/	
48-97271A5	102A			TR85/	
48-97271A6	102A			TR85/	
48-97762A02	108			TR95/	
48-103083	105			TR03/	
48-123522	126	3008		TR17/	
48-123536	126	3008		TR17/	
48-123802	123A			TR21/	
48-123803	123A			TR21/	
48-124158	102A	3123		TR85/	
48-124159	102A	3123		TR85/	
48-124175	102A	3123		TR85/	
48-124204	121	3009	3	TR85/232	G6013
48-124216	101	3011	54	TR08/641	G0011
48-124217	101	3011	54	TR08/641	G0011
48-124218	101	3011	54	TR08/641	G0011
48-124219	102	3004	53	TR05/631	G0005
48-124220	101	3011	54	TR08/641	G0011
48-124221	101	3011	54	TR08/641	G0011
48-124246	121	3009	3	TR01/232	G6013
48-124247	121	3009	3	TR01/232	G6005
48-124255	126	3006	50	TR17/635	G0009

Original Part Number	ECG	SK	GE	IR/WEP	HEP
48-124256	126	3006	50	TR17/635	G0009
48-124258	158		53	TR84/630	G0005
48-124259	158		53	TR84/630	G0005
48-124275	158		53	TR84/630	G0005
48-124276	158	3123	53	TR84/630	G0005
48-124279	158	3123	53	TR84/630	G0005
48-124285	121	3009	3	TR01/232	G6013
48-124286	102	3004	53	TR05/631	G0005
48-124296	160		51	TR17/637	G0003
48-124297	102	3004	53	TR05/631	G0005
48-124300	158	3123	53	TR84/630	G0006
48-124302	121	3009	3	TR01/232	G6013
48-124303	102	3004	53	TR05/631	G0005
48-124304	102	3004	53	TR05/631	G0005
48-124305	126	3006	3	TR17/635	G0009
48-124306	102	3004	53	TR05/631	G0005
48-124307	100	3005		TR05/254	G0007
48-124308	100	3005	53	TR05/254	G0007
48-124309	102	3004	53	TR05/631	G0005
48-124310	126	3006	3	TR17/635	G0009
48-124311	126	3006	3	TR17/635	G0009
48-124312	126	3006	3	TR17/635	G0009
48-124314	100	3005	53	TR05/254	G0005
48-124315	100	3005	53	TR05/254	G0005
48-124316	126	3008	51	TR17/635	G0008
48-124318	102	3004	53	TR05/631	G0006
48-124319	102	3004	53	TR05/631	G0006
48-124322	158	3123		TR84/	G0005
48-124327	100	3005	53	TR05/254	G0005
48-124328	100	3005	53	TR05/254	G0005
48-124329	105			TR03/	
48-124332	121	3009	16	TR01/232	G6003
48-124343	100	3005	53	TR05/254	
48-124344	100	3005	53	TR05/254	G0006
48-124345	100	3005	53	TR05/254	G0006
48-124346	126	3006	50	TR17/635	G0008
48-124347	126	3008	50	TR17/635	G0008
48-124348	126	3008	50	TR17/635	G0008
48-124349	126	3006	50	TR17/635	G0008
48-124350	126	3006	50	TR17/635	G0008
48-124351	126	3008	50	TR17/635	G0008
48-124352	126	3008	50	TR17/635	G0008
48-124353	102	3004	53	TR05/631	GC005
48-124354	102	3004	53	TR05/631	G0005
48-124355	102	3004	53	TR05/631	G0005
48-124356	121	3009	3	TR01/232	G6018
48-124357	100	3005		TR05/254	G000[illegible]
48-124358	100	3005	53	TR05/254	G0005
48-124359	100	3005	53	TR05/254	G0005
48-124360	126	3006	50	TR17/635	G0009
48-124363	160			TR17/	
48-124364	126	3006	51	TR17/635	G0008
48-124365	126	3006	51	TR17/635	G0008
48-124366	126	3006	50	TR17/635	G0009
48-124367	126	3006	50	TR17/635	G0009
48-124368	160		51	TR17/637	G0003
48-124370	100	3005	53	TR05/254	G0005
48-124371	100	3005		TR05/254	
48-124373	102A	3123	53	TR85/250	G0005
48-124377	126	3008	51	TR17/635	G0008
48-124378	100	3005	53	TR05/254	G0005
48-124379	100	3005	53	TR05/254	G0005
48-124380	100	3005	53	TR05/254	G0005
48-124388	126			TR17/	
48-124389	100			TR05/	
48-124398	100	3005	53	TR05/254	G0005
48-124443	100	3005	53	TR05/254	G0006
48-124444	100	3005	53	TR05/254	G0006
48-124445	100	3005	53	TR05/254	G0006
48-124446	100	3005	53	TR05/254	G0006
48-124804	108			TR95/	
48-124805	108			TR95/	
48-124808	108			TR95/	
48-125204	121	3009	3	TR01/232	G6013
48-125208	121	3009	16	TR01/232	G6003
48-125228	126	3006	51	TR17/631	G0008
48-125229	100	3005	53	TR05/254	G0005
48-125230	100	3005	53	TR05/254	G0005
48-125231	100	3005	53	TR05/254	G0005
48-125232	100	3005	53	TR05/254	G0005
48-125233	101	3011	54	TR08/641	G0011
48-125234	101	3011	54	TR08/641	G0011
48-125235	101	3011	54	TR08/641	G0011
48-125236	101	3011		TR08/641	G0011
48-125237	100	3005	53	TR05/254	G0005

Original Part Number	ECG	SK	GE	IR/WEP	HEP
48-125238	100	3005	53	TR05/254	
48-125239	100	3005	53	TR05/254	G0005
48-125240	100	3005	53	TR05/254	G0005
48-125242	100	3005	53	TR05/254	G0005
48-125252	105			TR03/	
48-125267	121	3009		TR01/232	G6003
48-125271	158	3123	53	TR84/630	G0005
48-125276	158	3123	53	TR84/630	G0005
48-125278	126	3006	51	TR17/635	G0001
48-125282	102A			TR85/	G0005
48-125285	102A			TR85/	
48-125286	105			TR03/	
48-125288	121	3014		TR01/232	
48-125294	102A		53	TR85/250	G0005
48-125296	100	3005	53	TR05/254	G0005
48-125299	105			TR03/	
48-125332	121	3009	16	TR01/232	G6003
48-128093	126	3008	53	TR17/635	G0008
48-128094	158	3008	53	TR84/630	G0008
48-128095	126	3008	51	TR17/635	G0008
48-128096	126	3006	51	TR17/635	G0008
48-128219	160		51	TR17/637	
48-128239	101	3011		TR08/641	G0011
48-128303	100	3005	51	TR05/254	
48-129934	121	3009	3	TR01/232	G6013
48-129935	121	3009	3	TR01/232	G6013
48-129936	121	3009	3	TR01/232	G6013
48-129937	121	3009	3	TR01/232	G6013
48-134101	126	3006	51	TR17/635	G0008
48-134173	123A	3122		TR21/735	S0015
48-134190A1G	161	3018	60	TR71/	
48-134302	121	3009	3	TR01/232	G6013
48-134344					G0007
48-134372	126	3008	51	TR17/635	G0008
48-134373					S0015
48-134404	126	3006	51	TR17/635	G0008
48-134405	126	3006	51	TR17/635	G0008
48-134406	126	3006	51	TR17/635	G0008
48-134407	158	3123	53	TR84/630	G0007
48-134408	102A	3123	53	TR85/250	G0005
48-134411	160		53	TR17/637	G0002
48-134412	160			TR17/637	G0002
48-134413	160			TR17/	G0002
48-134414	126	3006	51	TR17/635	G0008
48-134415	100	3005	51	TR05/254	G0001
48-134416	100	3005	53	TR05/254	G0005
48-134417	100	3005	53	TR05/254	G0005
48-134418	100	3005	53	TR05/254	G0005
48-134419	100	3005	53	TR05/254	G0005
48-134420	100	3005	53	TR05/254	G0005
48-134421	100	3005	53	TR05/254	G0005
48-134422	100	3005	53	TR05/254	G0005
48-134423	100	3005	53	TR05/254	G0005
48-134424	100	3005	53	TR05/254	G0005
48-134425	100	3005	53	TR05/254	G0005
48-134426	100	3005	53	TR05/254	G0005
48-134427	100	3005	53	TR05/254	G0007
48-134428	100	3005	53	TR05/254	G0007
48-134429	100	3005	53	TR05/254	G0005
48-134430	105			TR03/	
48-134431	176			TR82/	G6011
48-134432	100	3005		TR05/254	G0007
48-134433	100	3005		TR05/254	G0007
48-134434	126	3006	51	TR17/635	
48-134439	160			TR17/	
48-134443	100	3006	53	TR05/254	G0006
48-134444	100	3005	53	TR05/254	G0006
48-134445	100	3005	53	TR05/254	G0006
48-134446	100	3005	53	TR05/254	G0006
48-134447	121	3009	3	TR01/232	G6013
48-134448	121	3009	3	TR01/232	G6013
48-134449	121	3009	3	TR01/232	G6013
48-134450	100	3005	53	TR05/254	G0006
48-134454	126	3006	50	TR17/635	G0009
48-134456	126	3006	50	TR17/635	G0009
48-134457	126	3006	50	TR17/635	G0009
48-134458	100	3005	53	TR05/254	G0005
48-134459	127			TR27/235	G6008
48-134462	100	3005	53	TR05/254	G0005
48-134463	121	3009	16	TR01/232	G6003
48-134464	123A	3122	18	TR21/735	S0014
48-134465	123A	3122	18	TR21/735	S0014
48-134466	100	3005	53	TR05/254	G0005
48-134468	100	3005	53	TR05/254	G0005
48-134469	100	3005	53	TR05/254	G0005

Original Part Number	ECG	SK	GE	IR/WEP	HEP
48-134470	100	3005	53	TR05/254	G0005
48-134471	100	3005	53	TR05/254	G0005
48-134472	100	3005	53	TR05/254	G0005
48-134473	100	3005	53	TR05/254	G0005
48-134474	100	3005	53	TR05/254	G0005
48-134475	100	3005	53	TR05/254	G0005
48-134476	100	3005	53	TR05/254	G0006
48-134477	100	3005	53	TR05/254	G0006
48-134478	129	3025		TR88/242	S5023
48-134478A	129	3025		TR88/	
48-134479	126	3006	51	TR17/635	G0009
48-134480	126	3006	51	TR17/635	G0009
48-134481	126	3006	51	TR17/635	G0009
48-134482	158	3123		TR84/	G0007
48-134483	158	3123		TR84/	G0007
48-134484	160		51	TR17/637	G0003
48-134485	160		51	TR17/637	G0003
48-134486	160		51	TR17/637	G0003
48-134487	121	3009		TR01/232	G6005
48-134488	121	3009	3	TR01/232	G6013
48-134493	121	3009	16	TR01/232	G6003
48-134494	100	3005	51	TR05/254	G0009
48-134495	100	3005	51	TR05/254	G0009
48-134496	100	3005	51	TR05/254	G0009
48-134499	100	3005	50	TR05/254	G0009
48-134500	100	3005		TR05/254	G0007
48-134501	100	3005	50	TR05/254	G0009
48-134504	160			TR17/	
48-134506	160			TR17/	
48-134507	160	3005	51	TR17/254	G0009
48-134508	126	3006	51	TR17/635	G0008
48-134509	100	3005	50	TR05/254	G0009
48-134510	100	3005	53	TR05/254	G0006
48-134512	100	3005	53	TR05/254	G0005
48-134514	126	3006	51	TR17/635	G0008
48-134519	121	3009	3	TR01/641	G6013
48-134520	101	3011	54	TR08/641	G0011
48-134521	126			TR17/	G0008
48-134522	126		51	TR17/635	G0008
48-134524	160		51	TR17/637	G0008
48-134525	159	3114		TR20/717	S0032
48-134525A	159	3114		TR20/	
48-134526	160			TR17/	
48-134535	100	3005	50	TR05/254	G0009
48-134536	126	3006	50	TR17/635	G0008
48-134538	100	3005	53	TR05/254	G0005
48-134539	100	3005	53	TR05/254	G0005
48-134540	100	3005	53	TR05/254	G0005
48-134541	100	3005	53	TR05/254	G0005
48-134542	100	3005	53	TR05/254	G0005
48-134543	100	3005		TR05/254	G0007
48-134544	100	3005		TR05/254	G0007
48-134545	126	3006	51	TR17/635	G0008
48-134547	126	3006	51	TR17/635	G0008
48-134548					G0008
48-134549					G0008
48-134553	100	3005	53	TR05/254	G0005
48-134554	100	3005	53	TR05/254	G0005
48-134555	100	3005	53	TR05/254	G0005
48-134556	100	3005	53	TR05/254	G0005
48-134557	100	3005	53	TR05/254	G0005
48-134558	100	3005	53	TR05/254	G0005
48-134559	100	3005		TR05/254	G0007
48-134560	121	3009	3	TR01/232	G6013
48-134561	126	3006	51	TR17/635	G0009
48-134562	100	3005	53	TR05/254	G0005
48-134563	100	3005	53	TR05/254	G0006
48-134564	100	3005	53	TR05/254	G0006
48-134565	100	3005	53	TR05/254	G0005
48-134567	100	3005	53	TR05/254	G0005
48-134570	121	3009	16	TR01/232	G6003
48-134572	102A		53	TR85/250	
48-134573	102	3004	53	TR05/631	
48-134574	121	3009	3	TR01/232	G6013
48-134575	121	3009	16	TR01/232	G6003
48-134576	126			TR17/	G0009
48-134577	126			TR17/	G0009
48-134578	126			TR17/	G0009
48-134579	160			TR17/637	G0003
48-134584	176			TR82/	
48-134585	176			TR82/	G6011
48-134591	126	3006	51	TR17/635	G0008
48-134592	121	3009	16	TR01/232	G6003
48-134600	126	3006	51	TR17/635	G0008
48-134601	126	3006	51	TR17/635	G0008

Original Part Number	ECG	SK	GE	IR/WEP	HEP
48-134602	126	3006	51	TR17/635	G0008
48-134603	100	3005	50	TR05/254	G0009
48-134604	100	3005	50	TR05/254	
48-134605	126	3008	51	TR17/635	G0008
48-134606					G0009
48-134610	100	3005	53	TR05/254	G0005
48-134611	121	3009	3	TR01/232	G6013
48-134612	121	3009	3	TR01/232	G6013
48-134613	121	3009	3	TR01/232	G6013
48-134621	102A	3004	53	TR05/631	
48-134622	105	3012	53	TR03/231	G6004
48-134623	127			TR27/235	G6007
48-134625	100	3005	53	TR05/254	G0007
48-134626	100	3005	53	TR05/254	G0007
48-134631	100	3005	53	TR05/254	G0007
48-134632	102A	3004	53	TR05/631	
48-134633	158	3123		TR84/	G0005
48-134634	121	3009	3	TR01/624	G6013
48-134635	126	3006	51	TR17/635	G0008
48-134636	100	3005	51	TR05/254	G0005
48-134637	100	3005	51	TR05/254	G0005
48-134638	121	3009	3	TR01/624	G6013
48-134639	121	3009	3	TR01/232	G6013
48-134640	179			TR35/G6001	G6009
48-134641	100	3005	53	TR05/254	G0007
48-134644	121	3009	3	TR01/628	G6013
48-134645	121	3009	3	TR01/628	G6013
48-134646	121	3009	3	TR01/232	G6013
48-134647	121	3009	3	TR01/624	G6013
48-134648	154	3045	27	TR78/712	S5025
48-134649	121	3009	16	TR01/230	G6003
48-134651	121	3014	16	TR01/232	G6005
48-134652	127			TR27/235	G6015
48-134654	123A	3122	20	TR21/735	S0015
48-134655	100	3005	50	TR05/254	G0009
48-134656	100	3005	50	TR05/254	G0009
48-134657	100	3005	50	TR05/254	G0009
48-134664	123A	3122	20	TR21/735	S0015
48-134665	123A	3122	20	TR21/735	S0015
48-134666	123A	3122	12	TR21/735	S0015
48-134667	123A	3122	12	TR21/735	S0015
48-134668	123A	3122	18	TR21/735	S0014
48-134669	123A	3122	18	TR21/735	S0014
48-134670	121	3009	3	TR01/232	G6013
48-134672	121	3009	16	TR01/232	G6003
48-134673	123A	3122	20	TR21/735	S0015
48-134674	123A		20	TR21/735	S0014
48-134675	123A	3122	20	TR21/735	S0015
48-134676	160		51	TR17/637	G0003
48-134677	160		51	TR17/637	G0003
48-134678	160		51	TR17/637	G0003
48-134679	160		51	TR17/637	G0003
48-134680	126	3008	51	TR17/635	G0008
48-134681	126	3006	51	TR17/635	G0008
48-134682	126	3006	51	TR17/635	G0008
48-134683	126	3006	51	TR17/635	G0008
48-134684	126	3008	51	TR17/635	G0008
48-134689	128	3024	18	TR87/243	
48-134690	123A	3122	18	TR21/735	S0014
48-134691	123A	3122	18	TR21/735	S0014
48-134692	179			TR35/G6001	G6009
48-134693	160		51	TR17/637	G0003
48-134694	160		51	TR17/637	G0003
48-134695	179		16	TR35/G6001	G6009
48-134696	121	3009	3	TR01/232	G6013
48-134697	126	3006	51	TR17/635	G0008
48-134700	101	3011	54	TR08/641	G0011
48-134701	130	3027	14	TR59/247	S7004
48-134701A	130	3027		TR59/	
48-134702	159	3114		TR20/717	G6009
48-134702A	159	3114		TR20/	
48-134703	123A	3122	18	TR21/735	S0014
48-134705	123A	3122	18	TR21/735	S0014
48-134706	108	3039	20	TR95/56	S0020
48-134709	108	3039		TR95/56	S0015
48-134710					S0025
48-134711	126	3006	51	TR17/635	G0008
48-134713	108	3039	20	TR95/56	S0015
48-134714	123A	3122	18	TR21/735	S0014
48-134715	130	3027	14	TR59/247	S7004
48-134715A	130	3027		TR59/	
48-134717	108	3039	20	TR95/56	S0015
48-134718	123A	3122	20	TR21/735	S0015
48-134719	108	3039	20	TR95/56	S0015
48-134720	123A	3122	20	TR21/735	S0015

Original Part Number	ECG	SK	GE	IR/WEP	HEP
48-134721	123A	3122	20	TR21/735	S0015
48-134722	121	3009	3	TR01/232	G6013
48-134723	121	3009	3	TR01/232	G6013
48-134724	108	3039	20	TR95/56	S0015
48-134725	108	3039	20	TR95/56	S0015
48-134726	123A	3122	20	TR21/735	S0015
48-134727	104	3009	16	TR01/230	G6003
48-134729	121	3		TR35/G6001	G6009
48-134730	121	3009	16	TR01/232	G6015
48-134731	104	3009	16	TR01/230	G6003
48-134732	123A	3122	20	TR21/735	S0024
48-134733	123A	3122	20	TR21/735	S0024
48-134733A	123A	3122		TR21/	
48-134734	123A	3122	20	TR21/735	S0015
48-134734A	123A	3122		TR21/	
48-134737	123A	3122	18	TR21/735	S0014
48-134738	121	3009	3	TR01/232	G6013
48-134739	123A	3122	12	TR21/735	S5024
48-134740	179		16	TR35/G6001	G6009
48-134741	179		3	TR35/G6001	G6013
48-134742	179		3	TR35/G6001	G6013
48-134743	179		3	TR35/G6001	G6013
48-134744	121		3	TR01/232	G6013
48-134745	159	3114		TR20/717	S5022
48-134745A	159	3114		TR20/	
48-134746	121	3009	3	TR01/232	G6013
48-134747	104	3009	16	TR01/232	G6013
48-134748	179		3	TR35/G6001	G6013
48-134749	179	3009	13MP	TR35/G6001	G6013
48-134750	121	3009	3	TR01/232	G6013
48-134751	121	3009	3	TR01/232	G6013
48-134752	179		3	TR35/G6001	G6013
48-134753	179		3	TR35/G6001	G6013
48-134756	161	3117	20	TR83/719	S0017
48-134757	121	3009	3	TR01/232	G6013
48-134758	121	3009	3	TR01/232	G6013
48-134759	121	3009	3	TR01/232	G6013
48-134760	179	3009	3	TR35/G6001	G6013
48-134761	179		3	TR35/G6001	G6013
48-134763	121	3009	3	TR01/232	G6013
48-134764	121	3009	3	TR01/232	G6013
48-134765	123A	3122	20	TR21/735	S0014
48-134766	121	3009	3	TR01/232	G6013
48-134767	121	3009	3	TR01/232	G6013
48-134768	123A	3122	18	TR21/735	S0015
48-134772	108	3039	20	TR95/56	S0015
48-134773	108	3039	20	TR95/56	S0015
48-134774	108	3039	20	TR95/56	S0015
48-134775	123A	3122	20	TR21/735	S0015
48-134776	123A	3122	20	TR21/735	S0015
48-134777	108	3039	53	TR95/56	S0015
48-134779	108	3039	53	TR95/56	S0015
48-134780	108	3039	53	TR95/56	S0015
48-134782	123A	3100		TR21/	S0015
48-134783	108	3039	20	TR95/56	S0015
48-134784	108	3039	20	TR95/56	S0015
48-134785	123A	3122	20	TR21/735	S0015
48-134786	108	3039	20	TR95/56	S0015
48-134787	108	3039	20	TR95/56	S0015
48-134788	179		3	TR35/G6001	G6018
48-134789	107	3039	20	TR70/720	S0033
48-134791	123A	3122	18	TR21/735	S0015
48-134792	6400				S9002
48-134795	126	3006	51	TR17/635	G0008
48-134796	126	3008	51	TR17/635	G0008
48-134797	126	3006	51	TR17/635	G0008
48-134798	126	3006	51	TR17/635	G0008
48-134800	108	3039	20	TR95/56	S0015
48-134801	123A	3122	12	TR21/735	S0015
48-134802		3122	12	TR51/735	S0015
48-134803		3122	20	TR51/735	S0015
48-134804	108	3018	20	TR95/56	S0015
48-134805	108	3018	20	TR95/56	S0015
48-134806	108	3039	20	TR95/56	S0015
48-134807	123A	3018	20	TR21/735	S0015
48-134808	123A	3039	20	TR95/56	S0015
48-134809	123A	3122	20	TR21/735	S0015
48-134810	123A	3124	20	TR21/735	S0015
48-134811	123A	3124	20	TR21/735	S0015
48-134813	199				
48-134814	108	3039	20	TR95/56	S0015
48-134815	159	3114		TR20/717	S0019
48-134815A	159	3114		TR20/	
48-134817	123A	3122	20	TR21/735	S0015
48-134818	108	3039	20	TR95/56	S0015
48-134819	154	3044	27	TR78/712	S5025
48-134820	108	3039	20	TR95/56	S0020
48-134821	108	3039	20	TR95/56	S0025
48-134822	123A		20	TR21/	S0024
48-134823	199	3124	20	TR21/735	S0024
48-134824	123A			TR21/	S0024
48-134825	107	3018	17	TR70/720	S0033
48-134826	108	3039	20	TR95/56	S0025
48-134827	108	3039	20	TR95/56	S0015
48-134828	108	3039	20	TR95/56	S0015
48-134829	159	3114		TR20/717	S0019
48-134829A	159	3114		TR20/	
48-134830	234	3114	21	TR20/717	S0019
48-134830A	234	3114		TR20/	
48-134831	234	3114	21	TR20/717	S0019
48-134831A	234	3114		TR20/	
48-134832	234	3114	21	TR20/717	S0019
48-134832A	234	3114		TR20/	
48-134833	159	3114		TR20/717	S0019
48-134833A	159	3114		TR20/	
48-134837	107	3039	20	TR70/720	S0020
48-134838	128	3024	20	TR87/243	S5026
48-134839	123A	3122	20	TR21/735	S0015
48-134840	123A	3122	20	TR21/735	S0015
48-134841	123A	3122	20	TR21/735	S0015
48-134842	123A	3044	20	TR21/735	S0015
48-134843	154	3044	27	TR78/712	S5025
48-134844	123A	3122	20	TR21/735	S0014
48-134845	107	3039	20	TR70/720	S0020
48-134846	123A	3122	20	TR21/735	S0015
48-134847	123A	3122	12	TR21/735	S0015
48-134848	123A		20	TR21/	
48-134852	123A	3122	20	TR21/735	S0015
48-134853	154	3045	27	TR78/712	S5025
48-134854	123A	3122	20	TR21/735	S0015
48-134855	108	3039	20	TR95/56	S0015
48-134856	176			TR82/238	G6011
48-134857	108	3018	17	TR95/56	S0020
48-134859	126	3008	51	TR17/635	G0008
48-134860	126	3006	51	TR17/635	G0008
48-134861	126	3008	51	TR17/635	G0008
48-134862	126	3008	51	TR17/635	G0008
48-134865	159	3114		TR20/717	S0019
48-134865A	159	3114		TR20/	
48-134866	159	3114		TR20/717	S0019
48-134866A	159	3114		TR20/	
48-134867	159	3114		TR20/717	S0019
48-134867A	159	3114		TR20/	
48-134868	159	3114		TR20/717	S0019
48-134868A	159	3114		TR20/	
48-134869	159	3114		TR20/717	S0019
48-134869A	159	3114		TR20/	
48-134870	159	3114		TR20/717	S0019
48-134870A	159	3114		TR20/	
48-134871	159	3114		TR20/717	S0019
48-134871A	159	3114		TR20/	
48-134872	124		12	TR81/240	S5011
48-134879	108	3039	20	TR95/56	
48-134880	126	3008	51	TR17/635	G0008
48-134882	130		14	TR59/247	S7004
48-134884	130		14	TR59/247	S7004
48-134885	124			TR81/240	S5011
48-134888	121	3014	16	TR01/232	G6005
48-134889	123A	3122	18	TR21/735	S0014
48-134891	108	3039	20	TR95/56	S0015
48-134892	108	3039	20	TR95/56	S0015
48-134893	108	3039	20	TR95/56	S0015
48-134894	123A	3122	20	TR21/735	S0015
48-134895	123A	3122	20	TR21/735	S0015
48-134896	123A	3122	20	TR21/735	S0015
48-134897	123A	3122	20	TR21/735	S0015
48-134898	154	3045	27	TR78/712	S5024
48-134899	123A	3122	20	TR21/735	S0014
48-134900	162			TR67/707	S5021
48-134901	163			TR67/740	S5021
48-134902	108	3039	20	TR70/720	S0020
48-134903	123A	3122	18	TR21/735	S0014
48-134904	107	3039	20	TR70/720	S0016
48-134904F	107			TR70/	
48-134905	123A	3122	20	TR21/735	S0015
48-134906	123A	3122	20	TR21/735	S0015
48-134907	104	3009	16	TR01/230	G6003
48-134908	108	3039	20	TR95/56	S0015
48-134909	159	3114	21	TR20/717	S0019
48-134909A	159	3114		TR20/	

Original Part Number	ECG	SK	GE	IR/WEP	HEP
48-134910	159	3114	21	TR20/717	S0019
48-134910A	159	3114		TR20/	
48-134910F	159			TR20/	
48-134911	159	3114		TR20/717	S0019
48-134913	159	3114	21	TR20/717	S0019
48-134913A	159	3114		TR20/	
48-134914	159	3114	21	TR20/717	S0019
48-134914A	159	3114		TR20/	
48-134915	159	3114	21	TR20/717	S0019
48-134915A	159	3114		TR20/	
48-134918	123A	3122	20	TR21/735	S0015
48-134919	154	3044	27	TR78/712	S5025
48-134920	124	3021	12	TR81/240	S5011
48-134922	107	3039	20	TR70/720	S0033
48-134923	107	3039	20	TR70/720	S0033
48-134924	107	3039	20	TR70/720	S0033
48-134925	107	3039	20	TR70/720	S0033
48-134926	107	3039	20	TR70/720	S0033
48-134927	154	3045	27	TR78/712	S5024
48-134928	123A	3122	20	TR21/735	S0015
48-134929	123A	3122	20	TR21/735	S0015
48-134930	121	3009	16	TR01/232	S7003
48-134931	101	3011	54	TR08/641	G0011
48-134932	108	3039		TR95/56	S0016
48-134933	123A	3122	18	TR21/735	S0015
48-134933E	123A			TR21/	
48-134934	127			TR27/235	G6008
48-134935	123A	3122	18	TR21/735	S0014
48-134936	175	3131	12	TR81/241	S5019
48-134937	108	3039	20	TR95/56	S0020
48-134938	121	3014		TR01/232	G6005
48-134940	159	3114	21	TR20/717	S0013
48-134940A	159	3114		TR20/	
48-134941	128	3024	20	TR87/243	S5026
48-134942	123A	3122	20	TR21/735	S0015
48-134943	159	3114	21	TR20/717	S0019
48-134943A	159	3114		TR20/	
48-134944	133	3112	FET1	/801	F0010
48-134945	107	3039	20	TR70/720	S0020
48-134946	108	3039		TR95/56	S0016
48-134947	121	3009	16	TR01/628	G6013
48-134948	108	3122	20	TR21/735	S0016
48-134949	108	3039	20	TR95/56	S0020
48-134950	108	3039	20	TR95/56	S0016
48-134951	129	3025		TR88/242	S5023
48-134952	123A	3122	20	TR21/735	S0015
48-134953	128	3024	27	TR87/243	S5026
48-134956	100	3005	50	TR05/254	G0006
48-134960	108	3039	20	TR95/56	S0025
48-134961	107	3039	20	TR70/720	S0016
48-134962	107	3039	20	TR70/720	S0016
48-134963	107	3039	20	TR70/720	S0016
48-134964	107	3039	20	TR70/720	S0016
48-134965	107	3039	20	TR70/720	S0016
48-134966	107	3039	20	TR70/720	S0016
48-134967	159	3114	21	TR20/717	S0019
48-134967A	159	3114		TR20/	
48-134969	130		14	TR59/247	S7004
48-134970	123A	3122	20	TR21/735	S0030
48-134972	124	3021	12	TR81/240	S5011
48-134973	159	3114	21	TR20/717	S0031
48-134974	121	3009	16	TR01/232	G6003
48-134975	159	3114	21	TR20/717	S0031
48-134975A	159	3114		TR20/	
48-134977	121	3009		TR01/232	G6018
48-134979	108	3039		TR95/56	S0016
48-134980	123A	3122	20	TR21/735	S0024
48-134981	107	3039	20	TR70/720	S0016
48-134982	128	3024	27	TR87/243	S5026
48-134983	108	3039		TR95/56	S0025
48-134985	108	3039		TR95/56	S0025
48-134987	185	3191	69	TR77/S5007	S5006
48-134988	123A	3122	16	TR21/735	S0015
48-134989	159	3114		TR20/717	S0019
48-134989A	159	3114		TR20/	
48-134992	123A	3122	20	TR21/735	S0015
48-134994	123A	3122	20	TR21/735	S0024
48-134995	163	3111		TR67/740	S5021
48-134996	123A			TR21/	S0015
48-134997	199	3122	20	TR21/735	S0024
48-134998	157	3103		TR60/244	S5015
48-136665	123A	3122		TR21/735	S0015
48-137001	127			TR27/235	G6008
48-137002	154	3045	27	TR78/712	S5024
48-137003	123A	3122	20	TR21/735	S0015

Original Part Number	ECG	SK	GE	IR/WEP	HEP
48-137004	108	3039		TR95/56	S0020
48-137005	128	3024		TR87/243	S0020
48-137006	108	3039	20	TR95/56	S5026
48-137007	123A	3122		TR21/735	S0015
48-137008	130	3027	14	TR59/247	S7004
48-137008A	130	3027		TR59/	
48-137010	123A	3122	20	TR21/735	S0030
48-137011	190			TR74/S3021	S3019
48-137013	123A	3122	18	TR21/735	S0015
48-137014	123A	3122	20	TR21/735	S0024
48-137015	199	3122	20	TR21/735	S0024
48-137019	123A			TR21/	S0015
48-137020	159	3114		TR20/717	S0019
48-137020A	159	3114		TR20/	
48-137021	159			TR20/	S0019
48-137022	123A	3122	18	TR21/735	S0024
48-137023	133				F0010
48-137025	104			TR01/	G0005
48-137026	104			TR01/	G6003
48-137027	130	3027	14	TR59/247	S7004
48-137027A	130	3027		TR59/	
48-137031	121	3009	16	TR01/232	G6013
48-137032	159	3114		TR20/717	S0019
48-137032A	159	3114		TR20/	
48-137033	108	3039	20	TR95/56	S0020
48-137035	154	3045	27	TR78/712	S5026
48-137036	130	3027		TR59/247	S7002
48-137036A	130	3027		TR59/	
48-137037	184	3190	66	/S5003	S5000
48-137038					G0011
48-137039	176			TR82/238	G6011
48-137040	107			TR70/	S0033
48-137041	128	3024	18	TR87/243	S0015
48-137043	123A			TR21/	S0025
48-137044	123A	3122		TR21/735	S0015
48-137045	159	3114	20	TR20/717	S0019
48-137045A	159	3114		TR20/	
48-137046	159			TR20/	S0019
48-137047	123A	3122		TR21/735	S0015
48-137049	180				S7001
48-137053	130	3027		TR59/247	S7004
48-137053A	130	3027		TR59/	
48-137054		3083			S5006
48-137055	108	3039		TR95/56	S0016
48-137056	123A	3122		TR21/735	S0016
48-137057	123A	3122		TR21/735	S0016
48-137058	6400				S9002
48-137059	107			TR70/	S0015
48-137061	159			TR20/	S0019
48-137066	106				
48-137067	159	3114	21	TR20/717	S0013
48-137067A	159	3114		TR20/	
48-137068	159	3114	21	TR20/717	S0013
48-137068A	159	3114		TR20/	
48-137069	159	3114	21	TR20/717	S0013
48-137069A	159	3114		TR20/	
48-137070	222	3116			F2005
48-137071	107	3039		TR70/720	S0020
48-137072	123A			TR21/	S0020
48-137073	123A			TR21/	S0015
48-137075	107	3039	20	TR70/720	S0015
48-137076	107	3039	20	TR70/720	S0015
48-137077	107	3039	20	TR70/720	S0015
48-137078		3009	16	TR01/232	G6005
48-137079	130	3027		TR59/247	S7002
48-137079A	130	3027		TR59/	
48-137080	196	3031	28	TR76/S5003	S5000
48-137083	123A	3122	20	TR21/735	
48-137088	128	3024		TR87/243	S5014
48-137089	123A	3122	18	TR21/735	S0015
48-137090	159	3114		TR20/717	S0019
48-137091	188			TR72/	S3020
48-137092	188			TR72/	S3020
48-137093	190			TR74/S3021	S3019
48-137095	184	3190			S5000
48-137096	123A			TR21/	S0014
48-137101	123A			TR21/	S0030
48-137102	104	3009		TR01/230	G6003
48-137104	108	3039	20	TR95/56	S0015
48-137105	108	3039	20	TR95/56	S0015
48-137106	123A	3122	20	TR21/735	S0015
48-137107	123A	3122		TR21/735	S0030
48-137108	123A	3122		TR21/735	S0015
48-137109	123A	3122		TR21/735	S0030
48-137110	123A	3122		TR21/735	S0015

Original Part Number	ECG	SK	GE	IR/WEP	HEP
48-137111	123A	3122	20	TR21/735	S0015
48-137113	191				S3021
48-137115	123A	3122	20	TR21/735	S0015
48-137116	181			TR36/	S7000
48-137118	121	3009		TR01/232	G6013
48-137119	121	3009		TR01/232	G6013
48-137120	121	3009		TR01/628	G6013
48-137121	179			TR35/G6001	G6013
48-137122	121	3009		TR01/232	G6013
48-137123	121	3009		TR01/121	G6013
48-137124		3009		TR01/121	G6013
48-137125	179			TR35/G6001	G6013
48-137126	108	3039		TR95/56	S0015
48-137127	159	3114		TR20/717	S0019
48-137127A	159	3114		TR20/	
48-137128	184				S5000
48-137134	163			TR67/	S5021
48-137136	108	3039	20	TR95/56	S0015
48-137137	123A	3122	12	TR21/735	S0015
48-137138	123A	3122	12	TR21/735	S0015
48-137139	123A	3122	20	TR21/735	S0015
48-137140	108	3039	20	TR95/56	S0015
48-137144	108	3039	20	TR95/56	S0015
48-137145	184	3054	57	TR76/S5003	S5000
48-137146	184	3054	57	/S5003	S5000
48-137147	184	3054	57	/S5003	S5000
48-137148	184	3054	57	/S5003	S5000
48-137149	188	3199		TR72/3020	S3020
48-137153	185	3083	29	TR77/S5007	S5006
48-137154	185	3083	58	/S5007	S5006
48-137155	185	3083	58	/S5007	S5006
48-137156	185	3083	58	/S5007	S5006
48-137157	185	3083	58	/S5007	S5006
48-137158	108	3039		TR95/56	S0020
48-137160	189	3084		TR73/3031	S3028
48-137165	6400				S9002
48-137166	108	3039		TR95/56	S0016
48-137168	189			TR73/3031	S3028
48-137169	188			TR72/3020	S3024
48-137171	123A	3122		TR21/735	S0015
48-137171D	123A			TR21/	
48-137172	123A	3122		TR21/735	S0030
48-137173	159	3114		TR20/717	S0019
48-137173A	159	3114		TR20/	
48-137174	123A	3122		TR21/735	S0030
48-137175	130	3027		TR59/247	S7004
48-137175A	130	3027		TR59/	
48-137176	159	3114		TR20/717	S0019
48-137176A	159	3114		TR20/	
48-137178	127			TR27/235	G6017
48-137179	163			TR67/740	S5021
48-137180	130	3027		TR59/247	S7002
48-137180A	130	3027		TR59/	
48-137185	197				S5000
48-137190	108	3039		TR95/56	S0015
48-137191	108	3039		TR95/56	S0015
48-137192	123A	3122		TR21/735	S0015
48-137193	158			TR84/630	G0005
48-137194	108	3039		TR95/56	S0020
48-137195	159	3114		TR20/717	S0019
48-137196	108	3039		TR95/56	S0020
48-137197	161	3039		TR83/720	S0020
48-137199	102	3005		TR05/254	G0005
48-137200	184	3190			S5000
48-137202	188			TR72/	S3020
48-137203	163			TR67/	S5021
48-137206	123A			TR21/735	S0015
48-137207	124	3021		TR81/240	S5011
48-137211	184			/S5007	S5000
48-137213	104	3009		TR01/230	G6003
48-137214	131	3082		TR94/230	G6003
48-137215	104	3009		TR01/230	G6003
48-137216	104	3009		TR01/230	G6003
48-137217	104	3009		TR01/230	G6003
48-137218	104	3009		TR01/230	G6003
48-137219	104	3009		TR01/230	G6003
48-137220	104	3009		TR01/230	G6003
48-137234	127			TR27/235	G6008
48-137235	127			TR27/235	G6007
48-137238	194				S0005
48-137239	191			TR75/	S3020
48-137240	189			TR73/	S3032
48-137251	130	3027		TR59/247	S7004
48-137252					S0015
48-137253					S0015

Original Part Number	ECG	SK	GE	IR/WEP	HEP
48-137254					S0015
48-137256	185	3041		/S5007	S5006
48-137257	123A	3122		TR21/735	S0015
48-137258	185			TR77/S5007	S5006
48-137259	242	3083	69	TR56/	S5006
48-137260	123A	3122		TR21/735	S0015
48-137261					S0016
48-137262					S0016
48-137263					S0016
48-137265	123A	3122		TR21/735	S0030
48-137267	131	3052		TR94/642	G6016
48-137268	131	3052		TR94/642	G6016
48-137269		3052		TR94/642	G6016
48-137270	131	3052		TR94/642	G6016
48-137271	131	3052		TR94/642	G6016
48-137277	184	3190			S5000
48-137278	194				S5000
48-137303	185			TR77/	S5006
48-137304	185			TR77/	S5006
48-137307	128	3024		TR87/	S5014
48-137308	131	3052		TR94/	G6016
48-137309	196				S5004
48-137310	197				S5005
48-137311	152	3054	57	TR55/	S5000
48-137312	153	3083	58	TR56/	S5006
48-137313	219				S7003
48-137314	189			TR73/	S3032
48-137315	128			TR87/	S0015
48-137318	159			TR20/	S0019
48-137319	188	3199		TR72/	S3024
48-137320	189	3200		TR73/	S5006
48-137321	193			TR20/	S0019
48-137323	184	3190			S5000
48-137324	159			TR20/	S0019
48-137325	199			TR21/	S0015
48-137326	162			TR67/	S5020
48-137329	104			TR01/	G6003
48-137331	185	3084	29	TR77/	S5005
48-137332	194				S0005
48-137333	130			TR59/	S7004
48-137336	123A			TR21/	S0030
48-137339	108			TR95/	S0016
48-137342	127			TR27/	G6008
48-137342-P47	127				
48-137343	132			FE100/	F0015
48-137344	130			TR59/	S7004
48-137350	123A	3018	20	TR21/	S0025
48-137351	123A	3018	20	TR21/	S0025
48-137352	108			TR95/	S0015
48-137353	123A	3124	20	TR21/	S0015
48-137354	123A	3018	20	TR21/	S0015
48-137355	108	3018		TR95/	S0015
48-137364	154	3044		TR78/	S5024
48-137366	159			TR20/	S0019
48-137367	127			TR27/	G6008
48-137368	130			TR59/	S7004
48-137369	152	3054			S5000
48-137370	197	3083	58		S5006
48-137371	108			TR95/	S0015
48-137372	108			TR95/	S0015
48-137373	123A			TR21/	S0015
48-137374	123A			TR21/	S0015
48-137375	108			TR95/	S0015
48-137376	108			TR95/	S0015
48-137377	123A			TR21/	S0015
48-137378	123A			TR21/	S0015
48-137379	159			TR20/	S0019
48-137380	159			TR20/	S0031
48-137381	159			TR20/	S0031
48-137382	159			TR20/	S0019
48-137383	159			TR20/	S0019
48-137384	123A			TR21/	S0014
48-137386	194				S0005
48-137388	108			TR95/	S0016
48-137390	199			TR21/	S0015
48-137391	159			TR20/	S0019
48-137392	172			TR69/	S9100
48-137396	152			TR95/	S0016
48-137399	123A			TR21/	S0024
48-137400	107			TR70/	S0025
48-137415	154	3044			S5024
48-137416					S7002
48-137417					S5020
48-137437	152	3054	66	TR76/	S5000
48-137439					S5006

Original Part Number	ECG	SK	GE	IR/WEP	HEP
48-137445					S0015
48-137447					S0015
48-137463					S5004
48-137464					S5005
48-137465					S5006
48-137467					S0015
48-137471					GJ011
48-137472				TR77/	S5006
48-137473					S5000
48-137474					S5006
48-137475					S5000
48-137476	191				S3021
48-137477					S0019
48-137483	108	3039		TR95/56	S0015
48-137485					S5000
48-137486					S3021
48-137488	222	3050	FET4		F2004
48-137491	161	3018	60		S0017
48-137498	123A				S0005
48-137500		3124			S5000
48-137501					S5006
48-137502		3114	67	TR54/	
48-137503					S0015
48-137504	159	3118	65	TR52/	
48-137505	188	3024		TR72/	
48-137506			66	TR76/	S5000
48-137507			69	TR77/	S5006
48-137509	123A	3124	20	TR51/	S0024
48-137511					S5021
48-137523					S9001
48-137524					S5021
48-137525					S0019
48-137526	152	3054	28	TR76/	S5000
48-137527	153	3052	30	TR50/	S5006
48-137528					S5021
48-137529					S5013
48-137530					S0024
48-137532					S3021
48-137535	124				S5015
48-137536					S9001
48-137538					S0005
48-137540				TR77/	
48-137542					S0025
48-137543					S0005
48-137544					S0015
48-137545					S0031
48-137547					S0015
48-137548	164				
48-137549					S5000
48-137550					S5006
48-137552					S5004
48-137553					S5005
48-137555					G6003
48-137558					G6005
48-137562		3083			
48-137610		3084			
48-137855	123A				
48-137978	121			TR01/	
48-137988	128	3024	18	TR87/243	S0015
48-137998	123A	3122		TR21/735	
48-155001	165				
48-217241	181				
48-232796	181				
48-644676	160	3008	51	TR17/637	G0008
48-644677	160	3008	51	TR17/637	G0008
48-644678	102A	3123	53	TR85/250	G0007
48-644679	158	3123	53	TR84/630	G0005
48-645867	160	3008	51	TR17/637	G0008
48-859248	128	3024		TR87/243	
48-859428			18		
48-869001	102A		51	TR85/250	
48-869087B	104			TR01/	
48-869090	127			TR27/	
48-869099B	104			TR01/	
48-869138	128	3024	18	TR87/243	
48-869148	102		53	TR05/631	
48-869170	128	3024	18	TR87/243	
48-869177	176				
48-869182	121			TR01/	
48-869184	128			TR87/	
48-869198	102A			TR85/250	
48-869205	179			TR35/	
48-869221	128				
48-869225	175			TR81/	
48-869226-0	123A				

Original Part Number	ECG	SK	GE	IR/WEP	HEP
48-869228	128	3024	20	TR87/243	
48-869237	121			TR01/	
48-869244	130			TR59/	
48-869248	123A	3124	18	TR86/53	
48-869249	102	3004	53	TR05/631	
48-869253	102	3004	53	TR05/631	
48-869254	103	3010		TR08/641A	
48-869259	130			TR59/	
48-869263	128			TR87/	
48-869265	198				
48-869266	161			TR83/	
48-869273C	163			TR67/	
48-869274	175			TR81/	
48-869278	130			TR59/	
48-869279C	162			TR67/	
48-869282	102	3004		TR05/631	
48-869283	103	3010		TR08/641A	
48-869286	198				
48-869287	124				
48-869301	175			TR81/	
48-869302	130			TR59/	
48-869308	129			TR88/	
48-869311					S0019
48-869316	175				
48-869320	198				
48-869321	130			TR59/	
48-869325	123A				
48-869337	162			TR67/	
48-869380	128			TR87/	
48-869393	175			TR81/	
48-869400	129			TR88/	
48-869408	162			TR67/	S5020
48-869413	159				
48-869426	129	3025	21	TR88/242	
48-869427	179			TR35/	
48-869435	129				
48-869437					G6015
48-869440					S0014
48-869443					S0019
48-869444	123A				
48-869450	161			TR83/	
48-869459					S0012
48-869464	128	3024	18	TR87/243	
48-869465	198				
48-869467	234				
48-869475	102	3004	53	TR05/631	
48-869475A	102	3004		TR05/	
48-869476	103	3010	54	TR08/641	
48-869476A	103	3010		TR08/	
48-869480	181				
48-869481	161			TR83/	
48-869491					S3001
48-869497	199				
48-869515	130				
48-869536	225				
48-869539					S0015
48-869540					S5011
48-869561	128/401				
48-869570					S0015
48-869571					S0019
48-869576	152				
48-869591	128				
48-869599	128				
48-869610	175				
48-869631	128				
48-869660	223				
48-869661	152				
48-869676	196				
48-869677	197				
48-869681	129				
48-869701	197				
48-869703-3	128				
48-869767	123A				
48-6497840					S0019
48-1243002				TR01/	
48-1370893				TR56/	
48-8690468				TR17/	
48A111626		3039			
48A124315		3004	2		
48A134404		3008			
48A134434		3008			
48A134923		3018			
48A134924		3018			
48B407624P1	123A				
48C90172A01		3018			

Original Part Number	ECG	SK	GE	IR/WEP	HEP
48C124246		3009		TR01/	
48C125233			8	TR08/	
48C125235			8	TR10/	
48C125236			8	TR10/	
48C125237		3004	52	TR05/	
48D67120A1A		3124			
48D67120A11		3124	20	TR21/	
48K35P1		3004	52	TR05/	
48K36P1		3004	52	TR05/	
48K36P3		3004	52	TR05/	
48K39P1		3009	16	TR01/	
48K43P3		3004	52	TR05/	
48K43P4		3004	52	TR05/	
48K45N2			8	TR08/	
48K56P1		3123	2	TR05/	
48K57B2		3009	16	TR01/	
48K57B42			16	TR01/	
48K86904GF		3006			
48K125230		3009	16	TR01/	
48K134450			2		
48K134458		3004	2		
48K134494		3008	51		
48K134495		3008	51		
48K134496		3008	51		
48K134601		3008	51		
48K134796		3008	51		
48K134798		3008	51		
48K869001		3005	1		
48K869001C		3005			
48K869228		3024			
48K869321		3027			
48K869435		3025			
48NSP1035	121			TR01/	
48P1	102A	3004	2	TR85/250	
48P-2A5	102A	3004			
48P63005A72		3124	20		
48P63029A16		3008			
48P63029A60		3008			
48P63029A91		3008			
48P63029A94		3004			
48P63044A05		3004			
48P63075A72		3006			
48P63075A74		3006			
48P63075A75		3006			
48P63075A76		3004			
48P63076A81		3018	17	TR21/	
48P63076A82		3018	17	TR21/	
48P63077A03			52	TR05/	
48P63077A31		3018	20	TR21/	
48P63078A52		3018	11		
48P63078A54		3018		TR21/	
48P63078A59		3008			
48P63078A61		3004			
48P63078A62			52	TR05/	
48P63078A63		3008			
48P63078A65		3004			
48P63078A66		3004			
48P63078A68		3004			
48P63078A69			20	TR21/	
48P63078A70		3018			
48P63078A71	123A		20	TR21/	
48P63078H52				TR24/	
48P63079A97		3018	20	TR21/	
48P63081A82		3008			
48P63082A15		3004			
48P63082A16		3004			
48P63082A24			20	TR21/	
48P63082A25		3018	10	TR51/	
48P63082A26			20	TR21/	
48P63082A27		3018	17	TR24/	
48P63082A45	108	3018	20	TR95/	
48P63082A71	108		20	TR21/	
48P63084A03		3004			
48P63084A04		3004			
48P63086A18			20	TR21/	
48P65113A88		3039	17		
48P65146A61	107	3018	20	TR21/	
48P65146A63	108	3018	20	TR21/	
48P65173A78	108	3018	11	TR21/	
48P65174A24	161	3039	11	TR83/	S0020
48P217241	181	3511			
48P232796	181	3036			
48R134407		3004	2		
48R134546	160				
48R134573		3004	2		

Original Part Number	ECG	SK	GE	IR/WEP	HEP
48R134621	102A	3004	2		
48R134632	102A	3004	2		
48R134665		3010	8		
48R134666			20		
48R134722		3009	16		
48R859428		3024	18		
48R869062J		3039			
48R869090	127			TR27/	
48R869138		3024	18		
48R869148		3004	2		
48R869170		3024	18		
48R869198		3004			
48R869248		3124	20		
48R869249		3004	2		
48R869253		3004	2		
48R869254		3010	8		
48R869282		3004	2		
48R869283		3010	8		
48R869312	123A				
48R869325	123A				
48R869389	123A				
48R869413	159				
48R869426		3025	21		
48R869444	123A				
48R869464		3024	18		
48R869467	234				
48R869475		3004	2		
48R869475A		3004	2		
48R869476		3010	8		
48R869476A		3010	8		
48R869481		3122			
48R869487		3014			
48R869677		3083			
48R869767	123A				
48R969681	129				
48S50	218			TR58/	S5018
48S13270	131	3052		TR94/	
48S40170G01	199	3124	63	TR25/	S0025
48S40246G01	123A	3124	20	TR21/	S0024
48S40382J01	186	3084	28		
48S40383J01	187	3084	29		
48S40607J01	199	3122	62		S0024
48S43240G01	152	3054	66		
48S43241G01	152	3083	69		
48S44883G01	152		66		
48S44884G01	153		69		
48S44885G01	123A	3124	20		S0016
48S44885G02	123A	3020	20		S0016
48S44886G01	128	3122	18		S0003
48S134405		3124	51	TR12/	
48S134406			51	TR12/	
48S134407		3004	51	TR12/	
48S134408		3004	2	TR85/	
48S134695		3009	16	TR01/	
48S134718		3124	20		
48S134719		3124	20		
48S134720		3124	20		
48S134732		3123	20	TR21/	
48S134733		3124	20	TR21/	
48S134733A		3124	20		
48S134734		3124	20		
48S134734A		3124	20		
48S134737		3124	20		
48S134739		3021	12	TR23/	
48S134739A		3021	12		
48S134740		3009		TR01/	
48S134746		3009			
48S134747		3009	16	TR01/	
48S134756		3019		TR83/	
48S134759		3009	16	TR01/	
48S134761		3009	16	TR01/	
48S134765		3024	20	TR25/	
48S134766		3009	16	TR01/	
48S134767		3009	16	TR01/	
48S134768		3124	20	TR21/	
48S134773		3018	20	TR21/	
48S134774		3018	20	TR21/	
48S134775		3018	20	TR21/	
48S134776		3018	20	TR21/	
48S134783		3018	17		
48S134785		3018	17		
48S134797		3008	17		
48S134804		3018	17		
48S134805		3018	17		
48S134807		3018	17		

Original Part Number	ECG	SK	GE	IR/WEP	HEP
48S134809		3124	20		
48S134810		3124	62		S0015
48S134811		3124	20		S0015
48S134815	159	3025	21		S0019
48S134823		3124	20	TR21/	S0024
48S134825		3018	11	TR21/	
48S134826		3018	11	TR21/	
48S134827		3018	17		
48S134831		3118	21		S0019
48S134832		3124	20	TR21/	
48S134837		3018	11	TR21/	S0020
48S134838		3124	20	TR21/	
48S134840		3018	20	TR21/	
48S134841		3020	20	TR21/	S0015
48S134842		3124	20	TR21/	S0015
48S134843		3124	27	TR78/	
48S134844		3018	20	TR	
48S134845		3018	20		
48S134846		3124	20	TR21/	S0015
48S134853		3124	20		
48S134854		3124	20		
48S134855		3018	20	TR21/	
48S134857		3018	11	TR21/	S0020
48S134860		3005	1		
48S134861		3005	1		
48S134862		3005	1		
48S134872		3021	12	TR23/	
48S134879		3018	17		
48S134888			16	TR01/	
48S134889			10		
48S134894		3018	20		
48S134898		3124	20		
48S134899		3024	20	TR21/	
48S134900		3079	36		
48S134901		3027	36	TR67/	
48S134902	108	3019	11	TR21/	
48S134903	123A	3024	20	TR21/	S0014
48S134904		3018	11	TR21/	
48S134905		3124	20	TR21/	
48S134906		3124	20	TR24/	
48S134908			10		
48S134909		3004	21	TR28/	
48S134910		3025	21		
48S134918		3020	20	TR21/	
48S134919	154	3124	63	TR25/	
48S134922		3018	11	TR21/	
48S134925		3018	20	TR21/	
48S134926		3018	20	TR21/	
48S134932	161	3018	17	TR21/	S0016
48S134933	123A	3124	20	TR25/	S0015
48S134935		3124	20	TR21/	
48S134936	175	3026	23	TR57/	S5019
48S134937		3018	11	TR21/	S0020
48S134938		3034	25	TR27/	
48S134941		3024	18	TR21/	
48S134942		3024	63	TR25/	
48S134943		3025	21		
48S134944	132	3116		FE100/	
48S134945		3018	20	TR21/	
48S134946	108	3018	20	TR21/	
48S134947		3009	16	TR01/	
48S134948		3018	11	TR21/	
48S134949		3018	11	TR21/	
48S134950		3018	11	TR21/	
48S134952		3124	20		
48S134953		3024	20	TR25/	
48S134956		3006	51		
48S134960	108	3018	20	TR21/	S0025
48S134961		3018	20		
48S134862		3018	20		
48S134963		3018	20		
48S134964		3018	20		
48S134970	108	3018	11	TR21/	
48S134974		3009	16	TR01/	
48S134979	108	3018	20	TR21/	S0016
48S134981	161	3018	60	TR21/	S0016
48S134983		3018		TR21/	
48S134985		3018		TR21/	
48S134988	192	3024	63	TR25/	S5014
48S134989	193	3025	21	TR28/	S5013
48S134992		3024	63	TR25/	
48S134995		3111	36	TR67/	
48S134997	123A	3024	20	TR21/	S0024
48S137003		3018	20	TR21/	
48S137004		3018		TR21/	

Original Part Number	ECG	SK	GE	IR/WEP	HEP
48S137005		3124		TR25/	
48S137006	108	3122	20	TR24/	
48S137007			63	TR25/	
48S137014		3124	20	TR21/	
48S137015	199	.3124	20	TR21/	S0024
48S137020		3025			
48S137031		3009	3	TR01/	
48S137032	159	3025	21		
48S137036		3027			
48S137041	128	3024	20	TR21/	
48S137045		3025	21		
48S137055			17	TR21/	
48S137056			17	TR21/	
48S137057			17	TR21/	
48S137070		3116	FET2	FE100/	F2005
48S137107	123A	3040	63	TR25/	S0015
48S137108		3122	20	TR51/	
48S137109		3122	20	TR51/	
48S137110		3018	17	TR21/	
48S137111		3124	20	TR21/	
48S137115	123A	3124	20	TR21/	S0015
48S137121		3009			
48S137122		3009			
48S137123		3009			
48S137124		3009			
48S137125		3009			
48S137127		3025	21		S0019
48S137158		3018	11	TR21/	
48S137168	189	3084	29	TR56/	S3028
48S137169	188	3054	28	TR76/	S3024
48S137171		3124	18	TR21/	S0015
48S137172	123A	3040	63	TR25/	S0030
48S137173	159	3025	21		S0019
48S137174		3122	18		
48S137175		3027			
48S137185		3084			
48S137190		3018	20	TR21/	
48S137191		3018	20	TR21/	
48S137192		3018	20	TR21/	
48S137203		3111	36	TR67/	
48S137206		3124	17	TR24/	
48S137207		3021	12	TR23/	
48S137234		3035			
48S137259		3084			
48S137260			20	TR51/	
48S137270		3052	30	TR50/	
48S137300	123A	3122	20	TR51/	
48S137309	152	3054	66	TR76/	
48S137310	153	3083	69	TR77/	
48S137311	152	3054	66	TR55/	
48S137312	153	3083	69	TR56/	
48S137314		3118		TR30/	
48S137315	123A	3122	18	TR21/	
48S137321		3118			S0019
48S137323	152	3054	66	TR76/	S5000
48S137341	165	3115		TR67/	
48S137342		3035	25	TR27/	
48S137343	132	3112	FET2	FE100/	
48S137344	130	3133	14	TR61/	
48S137369	152	3054	28		S5000
48S137370	153	3083	29		S5006
48S137386		3018			
48S137415		3103	40	TR60/	S3021
48S137415(I)			40	TR78/	
48S137437		3054			
48S137472					S5005
48S137473					S5000
48S137476	191	3103	27	TR24/	S3021
48S137498	123A	3018	20	TR24/	S0005
48S137524		3115			
48S137527A		3083			
48S137530		3018			
48S137535	124	3538		TR23/	S5012
48S137539	163	3111	36	TR61/	S5020
48S137548	162	3079	35	TR61/	S5020
48SP134804			10		
48SP134826			11		
48SP134837			11		
48SP134855			10		
48SP134857			11		
48SP134894			10		
48SP134897			10		
48SP134903			10		
48SP134904			11		
48SP134905			10		

Original Part Number	ECG	SK	GE	IR/WEP	HEP
48SP134906			10		
48SP134933			10		
48SP134937			11		
48X9271A03				TR85/	
48X971A04		3018			
48X971A05		3020			
48X971A06		3004			
48X971A13		3021			
48X971A203		3006			
48X90222A03		3087			
48X90232A03		3018	11	TR21/	
48X90232A04			20		
48X90232A05		3124	20	TR21/	
48X90232A06			21		
48X90232A07		3021	20	TR21/	
48X90232A08			11	TR21/	
48X90232A10		3018	63	TR21/	
48X90232A11			63	TR25/	
48X90232A12			21		
48X90232A13			20	TR21/	
48X90232A15			21		
48X90232A16			12	TR23/	
48X90232A17		3018	11	TR21/	
48X90232A18		3018	11	TR21/	
48X90232A19		3018	17	TR21/	
48X97046A15			30	TR50/	
48X97046A16		3006	51	TR17/	
48X97046A17		3018	17	TR21/	
48X97046A18		3018	20		
48X97046A19		3018	20		
48X97046A20		3018	20	TR21/	
48X97046A21		3018	20	TR21/	
48X97046A22		3124	20	TR21/	
48X97046A23		3124	20	TR21/	
48X97046A24			17	TR24/	
48X97046A25			17	TR24/	
48X97046A26		3025			
48X97046A27		3025			
48X97046A28		3124			
48X97046A29		3024			
48X97046A30		3054			
48X97046A31			30	TR50/	
48X97046A32			52	TR85/	
48X97046A34			52	TR05/	
48X97046A36			67	TR28/	
48X97046A48	158	3004	52	TR05/	
48X97046A50			20	TR21/	
48X97046A51	108	3018	17	TR95/	
48X97046A52	128	3124	18	TR51/	
48X97046A53	102A	3004		TR05/	
48X97046A54	102A	3004	2	TR05/	
48X97046A55	102A	3004	2	TR05/	
48X97046A60	123A	3124	62	TR51/	
48X97046A61	123A	3124	62	TR51/	
48X97046A62	123A	3018	60	TR24/	
48X97127A02		3018		TR21/	
48X97127A03		3018		TR21/	
48X97127A04		3124		TR25/	
48X97127A05		3004			
48X97127A06				TR21/	
48X97127A08				TR25/	
48X97127A12				TR21/	
48X97127A13				TR21/	
48X97127A18				TR21/	
48X97127A19		3124			
48X97127A22		3004			
48X97162A01	108	3018	17		
48X97162A02	108	3018	17	TR24/	
48X97162A03				TR17/	
48X97162A04	108	3018	20	TR21/	
48X97162A05	123A	3124	10	TR21/	
48X97162A06	102A	3004	52	TR05/	
48X97162A07	102A	3004	52		
48X97162A09	108	3122	20	TR21/	
48X97162A10	108	3018	20	TR21/	
48X97162A11		3006			
48X97162A12		3124			
48X97162A13		3021			
48X97162A15		3025			
48X97162A16		3004			
48X97162A17		3004			
48X97162A18		3004			
48X97162A19		3004			
48X97162A20		3004			
48X97162A21	123A	3124	10	TR24/	

Original Part Number	ECG	SK	GE	IR/WEP	HEP
48X97162A22		3004			
48X97162A26		3018			
48X97162A36	158	3004	53	TR84/	
48X°7162A37	158	3004	53	TR84/	
48X97177A02		3018		TR21/	
48X97177A03		3018	20	TR21/	
48X97177A04		3018		TR21/	
48X97177A12		3124	63	TR25/	
48X97177A13		3018	20	TR21/	
48X97177A14		3025	21		
48X97177H01	133		FET1	FE100/	
48X97221A01		3004			
48X97221A02		3004		TR17/	
48X97221A03		3004			
48X97221A04		3004			
48X97238A01	126	3006	50		G0008
48X97238A02	126	3006	50		G0008
48X97238A03	126	3006	50		G0008
48X97238A04	123A	3124	17		S0014
48X97238A05	102	3004	52		G0005
48X97238A06	131MP	3052	31MP		G6016
48X97271A04				TR11/	
48X97271A06				TR11/	
48X97762A02		3018			
48X134902		3019			
48X134970			20		
49-1	108	3039		TR95/56	S0016
49-62139	128				
49A0000	7475				
49A0002-000	7473				
49A0005-000	7410				
49A0006-000	7420				
49A0010	9944				
49A0012-000	7474				
49A0103-001	923				
49A0510	9944				
49L-4A5	102A	3003		TR85/	
49P1C	121	3009	16	TR01/232	G6003
49X90232A05			20	TR21/	
50A52	126	3008		TR17/	G0008
50A102	126	3008	51	TR17/635	G0008
50A103	129	3006		TR88/242	G0009
50A103K	126	3006	51	TR17/	G0003
50B54	102A	3004		TR85/250	G0005
50B173-C	102A	3004		TR85/250	G0005
50B173-S	102A	3004		TR85/	G0005
50B175A	102A	3004	2	TR85/250	G0005
50B175B	102A	3004		TR05/250	G0005
50B175C	102A			TR85/250	G0005
50B324	158	3004		TR84/630	G0005
50B364	158	3004		TR84/630	G0005
50B415	158	3004		TR84/	G6011
50B423	102A	3004	2	TR05/	G0005
50BU75-C	102A	3004		TR85/	
50C05	6020				
50C05R	6021				
50C10	6022				
50C10R	6023				
50C20	6026				
50C20R	6030				
50C30	6030				
50C30R	6031				
50C40	6034				
50C40R	6035				
50C50	6038				
50C60	6040				
50C70	6042				
50C80	6042				
50C90	6044				
50C100	6044				
50C371	123A	3124		TR21/	S0015
50C372	123A	3124		TR21/	S0015
50C373	123A	3124		TR21/	S0015
50C374	123A	3124		TR21/	S0015
50C380-O	107	3018		TR70/720	S0016
50C380-OR	107	3018		TR70/	
50C394-O	107	3018		TR70/720	S0016
50C394-R	107	3018		TR70/	
50C401	130	3027			
50C538	123A	3124	61	TR24/	
50C580-O					-
50C644	123A	3124		TR21/	S0015
50C784	108	3018	17	TR95/	
50C784-R	107	3018		TR70/720	S0016
50C828	123A	3124	20	TR21/735	S0015

Original Part Number	ECG	SK	GE	IR/WEP	HEP
50C829	108	3018	17	TR95/	
50C829B	108	3018	17	TR95/	S0014
50C829C	108	3018	17	TR95/	S0014
50C838	123A	3124		TR21/	S0015
50C1047	108	3018	17	TR95/	
50C3800R					S0016
50CJ139	123A	3124		TR21/	S0015
51	108	3039		TR95/56	S0025
51-0022-00	709				
51-04488D03	727	3071			
51-4	157			TR60/	
51-47-20	128				
51-47-21	159				
51-47-23	123A				
51-47-24	123A				
51-47-34	234				
51-1056FA01	722	3161			
51-1056GA01	722				
51-10276A01	704	3023			
51-10302A01	703A	3157		IC500/	C6001
51-10382A	718	3159			
51-10382A01	722	3161			C6056P
51-10393A01	748				C6060P
51-10408A01	710	3102			
51-10422A	718				
51-10422A01	718	3159		IC511/	C6094P
51-10422A02					C6094P
51-10425A01	798				
51-10432A01	717				
51-10437A01	718	3159			
51-10522A01	771				
51-10534A01	755				
51-10534A03	755				C6013
51-10534A04	755				C6013
51-10541A01	917				
51-10542A	771				
51-10542A01	762				C6004
51-10559A01	722	3161			C6056P
51-10566A01	722	3161		IC513/	C6056P
51-10566A02	720				
51-10592A01	722	3161			C6056P
51-10594A01	723	3144			C6101P
51-10600A	753				
51-10600A01	753				C6010
51-10611A09	760				C6001
51-10611A10	762				C6012
51-10611A11	7400				C3000P
51-10611A12	7404				C3004P
51-10611A15	7450				C3050P
51-10611A16	7475				C3075P
51-10617A01	718				
51-10631A01	708		IC10	IC504/	C6062P
51-10636A	796				
51-10636A01	796				C6016
51-10637A01	781				
51-10638A01	789				
51-10650A01	778	3018			
51-10655A01	798				
51-10655A05	749				
51-10658A01	788				
51-10672A	736				
51-10672A01	736				
51-10678A01	789				
51-10711A01	743				
51-10715A01	741M				
51-25789H	909	3590			
51-43639B02	755				
51-43684B	735				
51-43684B01	755				
51-43684B02	755				
51-43684B0P	755				
51-70177A01	798				
51-70177A02	748				
51-70177A03	738				
51-70177A05	749	3168			
51-70177A07	767				
51-70177A08	775				
51-70177B02	748				
51-84320A16	912				
51-84320A28	774				
51-84320A32	916				
51-110276A01	704	3023			
51-43639802	755				
51A180-4	161	3117			
51A524004-01	941D				

Original Part Number	ECG	SK	GE	IR/WEP	HEP
51C43684B	735				
51C43684B01	755				
51C43684B02	755				
51D0177A02	748				
51D170	102	3004	52	TR05/631	G0005
51D176	102A		2	TR85/	
51D188	100	3005	52	TR05/254	G0005
51D189	100	3005	52	TR05/254	G0005
51D70177A01	798				C6066P
51D70177A02	748	3101	2	IC507/	C6060P
51D70177A03	738	3167			
51D70177B02	748			IC507/	C6060P
51M70177A01	798				C6066P
51M70177A02	748				C6060P
51M70177A03	738	3167			C6075P
51M70177A05	749	3168			C6076P
51M70177A07	767				C6015
51M70177B01	798				C6066P
51M70177B02	748				C6060P
51N3M	180	3018	20	/S7001	S0015
51P2	126	3006	51	TR17/635	G0003
51P4	126	3006	51	TR17/635	
51R04488D03	727	3071			
51R8432A09	796				C6016
51R84320A02	724				
51R84320A03	912				
51R84320A05	912				
51R84320A16	912				
51R84320A28	774				C6002
51R84320A32	916				
51S1056FA01	722	3161			
51S1056GA01	722	3161			
51S10072A01	736				
51S10276A01	704	3023		IC501/	
51S10302A01	703A	3157	12	IC500/	C6001
51S10382A	718	3159	8		
51S10382A01	718	3159			C6056P
51S10393A01	748				C6060P
51S10408A01	710	3102	2	IC506/	
51S10422A	718	3159	8		
51S10422A01	718	3159	8	IC511/	C6094P
51S10425A01	798				
51S10432A01	717				
51S10437A01	718				C6094P
51S10534A01	955				C6013
51S10534A03	755				C6013
51S10536A01	912				
51S10541A01	917				
51S10542A01	771				C6004
51S10553A01	917				
51S10559A	722	3159			C6056P
51S10559A01	722	3161			
51S10566A01	722	3161		IC513/	C6068P
51S10566A02	722				C6068P
51S10592A01	722	3161	9		C6056P
51S10594A01	723	3144			C6101P
51S10600A	753				C6010
51S10600A01	753				C6010
51S10611A09	760				C6001
51S10611A10	762				C6012
51S10611A11	7400				C3000P
51S10611A12	7404				C3004P
51S10611A15	7450				C3050P
51S10611A16	7475				C3075P
51S10631A01	708				C6082P
51S10636A	796				C6016
51S10636A01	796				
51S10637A01	781	3169			
51S10638A01	789	3078			
51S10650A01	778				C6102P
51S10655A01	798				
51S10655A01-3	798				
51S10655A03	738				
51S10655A03A	738				
51S10655A05	749	3168			
51S10655B03	738				C6075P
51S10655B05					C6076P
51S10655B13	748				C6063P
51S10655C05	749				
51S10658A01	788				
51S10672A	736				
51S10672A01	736				
51S10711A01	743				
51S10715A01	741M				C6052P
51S70177A01	798				

Original Part Number	ECG	SK	GE	IR/WEP	HEP
51S110276A01	704	3023			
52-4	157	3103		TR60/244	S5015
52C04	102A	3004		TR85/250	
52D189	158	3123		TR84/	
53-1110	123A	3124	62		S0015
53-1173	172		64	TR69/	S9100
53-1362	186	3054	28	TR55/	S3020
53-1516	159	3114	22	TR20/	S0012
53-1967	186	3054	28	TR55/	S3020
53N49	105	3012		TR03/	
53P151	123	3124	20	TR86/53	
53P153	121	3014	16	TR01/232	
53P157	102	3004	2	TR05/631	
53P158	123	3124	20	TR86/53	
53P159	123	3124	20	TR86/53	
53P161	123	3124	20	TR86/53	
53P162	123	3124	20	TR86/53	
53P163	123	3124	20	TR86/53	
53P165	123	3124	10	TR86/53	
53P166	129	3124	22	TR88/242	
53P169	128	3024	18	TR87/243	
53P170	129	3025	21	TR88/242	
54	108			TR95/56	S0015
54-1	123A	3124			
54A	123A	3124	20	TR21/735	S0015
54B	123A	3124	20	TR21/735	S0015
54BLK	108	3039		TR95/56	S0015
54BLU	123A			TR21/735	S0015
54BRN	108	3039		TR95/56	S0015
54C	123A	3124	20	TR21/735	S0015
54D	123A	3124	20	TR21/735	S0015
54E	123A	3124	20	TR21/735	S0015
54F	123A	3124	20	TR21/735	S0015
54GRN	123A			TR21/735	S0015
54ORN	108	3039		TR95/56	S0015
54RED	108	3039		TR95/56	S0015
54WHT	123A			TR21/735	S0015
54YEL	123A			TR21/735	S0015
55-0014	9923				
55-641	172		64	TR59/S9100	S9100
55-642	128	3024		TR87/243	
55-643	129	3025		TR88/242	G6011
55-1016	102A	3004	2	TR85/250	G0005
55-1026	123A	3124	20	TR21/735	S0015
55-1027	103A	3124	20	TR08/724	G0011
55-1029	158	3004	2	TR84/630	G0005
55-1031	158	3004	2	TR84/630	
55-1032	103A	3020	20	TR08/724	
55-1034	123A	3124	20	TR21/735	
55-1082	123A	3124	10	TR21/735	S0015
55-1083	159	3025	22	TR20/717	S0019
55-1083A	159	3114		TR20/	
55-1084	128	3024	18	TR87/243	S0014
55-1085	159	3025	22	TR20/717	S0019
55-1085A	159	3114		TR20/	
55-152579	159				
55-509276-2	218				
55-509276-4	218				
55P2	126	3006	51	TR17/635	G0003
55P3	126	3006		TR17/635	
56-35	123A	3122	20	TR21/735	
56-234	108				
56-8086	107	3018	17	TR70/720	S0011
56-8086A	107	3018		TR70/	
56-8086B	107	3018		TR70/	
56-8086C	107	3018		TR70/	
56-8087	107	3018	20	TR70/720	S0011
56-8087B	107	3018		TR70/	
56-8087C	107	3018		TR70/	
56-8088	107	3018		TR70/720	S0016
56-8088A	107	3018		TR70/	
56-8088B		3018			
56-8088C	107	3018		TR70/	
56-8089	123A	3018	20	TR21/735	S0011
56-8089A	123A	3018		TR21/	
56-8089B		3018			
56-8089C	123A	3018		TR21/	
56-8090	123A	3018	20	TR21/735	S0015
56-8090A	123A	3018		TR21/	
56-8090B		3018			
56-8090C	123A	3018		TR21/	
56-8091	102A	3004	50	TR85/250	G0007
56-8091A	102A	3004		TR85/	
56-8091B	102A	3004		TR85/	
56-8091C	102A	3004		TR85/	
56-8091D	102A	3004		TR85/	
56-8091K		3004			
56-8092	102A	3004	1	TR85/250	G0005
56-8092A	102A	3004		TR85/	
56-8092B	102A	3004		TR85/	
56-8092C		3004			
56-8094	103A	3126	90		
56-8098	159	3114		TR20/717	S0013
56-8098A	159	3114		TR20/	
56-8098B	159	3114		TR20/	
56-8098C	159	3114		TR20/	
56-8098F		3114			
56-8099	158			TR84/630	G6011
56-8100	103A	3010		TR08/724	G0011
56-8100A	103A	3010		TR08/	
56-8100B	103A	3010		TR08/	
56-8100C	103A	3010		TR08/	
56-8100D	103A	3010		TR08/	
56-8101	158	3087		TR84/630	
56-8196	199	3124	20		S0024
56-8197	199	3124	10		S0025
56-86412-3	184				
56A1-1	703A	3157			C6001
56A3-1	712	3072			
56A4-1	714	3075			
56A5-1	715	3076			
56A6-1	790	3077			
56A9-1	748		IC26		
56C1-1	703A	3157		IC500/	C6001
56C3-1	712	3072			
56C4-1	714	3075			
56C5-1	790	3076			
56C6-1	713				
56C9-1	748				
56C-1				IC500/	C6001
56D-1	703A	3157			
56D3-1	712	3072	IC2	IC508	C6063P
56D4-1	714	3075	IC4	IC509	C6070P
56D5-1	715	3076	IC6		C6071P
56D6-1	790	3077	IC5	IC508	C6057P
56D6-Y				IC508	
56D9-1	748				C6060P
56D17-1	791	3149	IC33		
56L101	714	3075			
56L102	715	3076			
56L103	790				
56P1	160			TR17/637	
56P2	160	3008		TR17/637	G0002
56P3	160	3004	2	TR17/637	G0003
56P4	160	3008	51	TR17/637	
56P4P	160			TR17/	G0003
57-0004503	123A	3124			
57-0005452	123A	3124			
57-0005503	123A	3124			
57-0006		3016			
57-000901504	123A	3124			
57-01491-B	123A				
57-01491-C	128				
57-01494C	128				
57A1-3					G0011
57A1-4					G0011
57A1-5					G0011
57A1-6					G0011
57A1-7					G0005
57A1-8					G0008
57A1-9					G0003
57A1-10					G0005
57A1-11					G0005
57A1-12					G0008
57A1-13					G0009
57A1-14					G0005
57A1-15					G0001
57A1-16					G0003
57A1-17					G6011
57A1-18					G0005
57A1-19					G6011
57A1-20					G6011
57A1-21					G6011
57A1-22					G0003
57A1-23					G0003
57A1-24					G0003
57A1-25					G0003
57A1-26					G0005
57A1-27					G0005
57A1-28					G0005

Original Part Number	ECG	SK	GE	IR/WEP	HEP
57A1-30					G0003
57A1-31					G0003
57A1-32					G0003
57A1-33					G0008
57A1-34					G0003
57A1-35					G0008
57A1-36					G0008
57A1-37					G0008
57A1-38					G6011
57A1-39					G0008
57A1-40					G0003
57A1-41					G0008
57A1-42					G0005
57A1-43					G0008
57A1-44					G0005
57A1-45					G0003
57A1-46					G0005
57A1-47					G0008
57A1-48					G0008
57A1-49					G0008
57A1-50					G0008
57A1-51					S0015
57A1-52					S0019
57A1-53					G6011
57A1-55					S5022
57A1-56					G0005
57A1-57					G0001
57A1-58					G6011
57A1-59					G0005
57A1-60					G6011
57A1-61					G0001
57A1-66					G0005
57A1-67					G0008
57A1-68					G0005
57A1-69					G0001
57A1-70					G0005
57A1-71					G6011
57A1-72					G0008
57A1-73					G0008
57A1-74					G0008
57A1-75					S0015
57A1-76					S0012
57A1-77					G6011
57A1-78					G6011
57A1-79					G0003
57A1-80					G0003
57A1-81					G0003
57A1-82					G0005
57A1-83					G6011
57A1-84					G0001
57A1-85					G0001
57A1-86					G0002
57A1-87					G0001
57A1-88					G0001
57A1-89					G0001
57A1-90					G0005
57A1-91					G0005
57A1-92					G0005
57A1-93					G0005
57A1-94					G0009
57A1-95					G0009
57A1-96					G0001
57A1-97					G0009
57A1-98					G0009
57A1-99					G0009
57A1-100					G0008
57A1-101					G0008
57A1-102					G0008
57A1-103					G0008
57A1-104					G0005
57A1-105					G0005
57A1-106					G0005
57A1-107					G0009
57A1-108					G0008
57A1-109					G0008
57A1-110					G0008
57A1-111					G0005
57A1-112					G0003
57A1-113					G0003
57A1-114					G0009
57A1-115					G0008
57A1-116					G0005
57A1-117					G0005
57A1-118					G0005
57A1-119					G6003
57A1-120					G0005
57A1-121					G0005
57A1-122					S3021
57A1-123					S0015
57A1-124					S0015
57A2-1					G0009
57A2-2					G0008
57A2-3					S0019
57A2-4	102A	3004		TR85/250	G0005
57A2-5					G0011
57A2-6					G0005
57A2-7					G0005
57A2-8					G0011
57A2-9					G0009
57A2-10					G6011
57A2-11					G0001
57A2-12					G0005
57A2-13					G0003
57A2-14					G0003
57A2-15	102A	3004		TR85/	G0005
57A2-16					G6011
57A2-17					G0009
57A2-18					G0001
57A2-19	126	3008	51	TR17/	G0009
57A2-20					G0008
57A2-21					G0005
57A2-22					G0003
57A2-23					G0005
57A2-24	102A	3004	2		G0007
57A2-25					G0007
57A2-26					G0009
57A2-27					S0015
57A2-28					S0015
57A2-29					G0005
57A3-1					S0012
57A3-2					S0012
57A3-3					G0005(2)
57A3-4					G0005(2)
57A3-5					G0005(2)
57A3-6					G0005(2)
57A3-7					G6005(2)
57A3-8					G6005(2)
57A3-9					G6005(2)
57A3-10					G6005(2)
57A3-11					G6005(2)
57A3-12					G6003(2)
57A4-1					G6005
57A4-2					G6005
57A4-4					G6005
57A5-1					G0003
57A5-2					G0003
57A5-4					G0003
57A5-5					G0005
57A5-6					S0016
57A5-7					S0016
57A5-8					S0016
57A5-9					G0005
57A5-10					G0005
57A6-1					G0005
57A6-2					G6003
57A6-4					S0015
57A6-5					G0011
57A6-6					G0005
57A6-6A					G0005
57A6-7					S0024
57A6-8					G6005
57A6-9					S0015
57A6-10					S5011
57A6-11					S0015
57A6-12					G6005
57A6-14					S5011
57A6-15					S0019
57A6-16					G0005
57A6-18					S0012
57A6-19					S0015
57A6-20					S0015
57A6-21					G0011
57A6-22					G0005
57A6-23					G6003
57A6-24					S5011
57A6-25					G0005
57A6-26					S5013
57A6-27					S0015
57A6-28					S0012
57A6-29					S0015

Original Part Number	ECG	SK	GE	IR/WEP	HEP
57A6-30					S0015
57A6-31					S5013
57A6-32					S0014
57A6-33					S0019
57A7-1					S0016
57A7-2					S0016
57A7-3					S0025
57A7-4					S0025
57A7-5					S0025
57A7-6					S0025
57A7-7					S0025
57A7-8					S3021
57A7-9					S0015
57A7-10					S0015
57A7-15					S0019
57A7-17					S0019
57A7-18					S0019
57A7-20					S0015
57A9-1					G6003
57A9-2					G6003
57A10-1					S0017
57A10-2					S0017
57A11-1					S0014
57A12-1					S5025
57A12-2					S5024
57A12-3					S5012
57A12-4					S5014
57A12-5					S5012
57A14-1					S0015
57A14-2					S0015
57A14-3					S3021
57A15-1					S0015
57A15-2					S0015
57A15-3					S0019
57A15-4					S0019
57A15-5					S0031
57A16-1					S0015
57A19-1					S0019
57A19-2					S0019
57A20-1					S0020
57A21-1					S0016
57A21-2	108	3018	11	TR95/56	S0016
57A21-3					S0016
57A21-4	108	3019	17	TR95/	S0016
57A21-5	108	3019	17	TR95/56	S0016
57A21-6	108	3018		TR95/56	S0020
57A21-7	108	3018		TR95/56	S0020
57A21-8	123A	3018	20	TR21/	S0015
57A21-9	108	3018		TR95/56	S0016
57A21-10	108	3018	11	TR95/	S0016
57A21-12	107	3018	11	TR80/	S0020
57A21-13	107	3018	11	TR80/	S0020
57A21-14	107	3018	11	TR80/	S0020
57A21-16	161	3039	11	TR80/	S0020
57A21-17	161	3018	39	TR70/	S0017
57A21-18	161	3019	11	TR83/	S0020
57A21-45	107	3018	11	TR80/	S0020
57A22-1					G6001
57A22-2					G6001
57A23-1					S5013
57A23-2					S5013
57A23-3					S5013
57A24-1					S0020
57A24-2					S0015
57A24-3					S0015
57A24-4					S0015
57A27-1					S0015
57A27-2					S0025
57A29-2	706	3101	IC2	IC507/	C6060P
57A31-1	132		FET2	FE100/	
57A31-4	133	3112	FET1	FE100/	
57A32-1	724	3525			
57A32-2	1003	3101	IC2	IC507/	C6060P
57A32-3	1039				
57A32-10	722	3161		IC513/	
57A32-11	1075				
57A32-13	1103				
57A32-21	799		IC34		
57A32-22	720		IC7		
57A32-Z		3101			
57A33-1	123A				S0031
57A102-4					S0017
57A103-4					S0020
57A104-1	154	3040		TR78/	
57A104-2	154	3040		TR78/	
57A104-3	154	3040		TR78/	
57A104-4	154	3040		TR78/	
57A104-5	154	3040		TR78/	
57A104-6	154	3040		TR78/	
57A104-7	154	3040		TR78/	
57A104-8	154	3040		TR78/712	S5025
57A104-8-6	128	3024		TR87/	
57A105-12	123A	3122		TR21/735	S0015
57A106-12	159	3118		TR20/717	S0019
57A107-1	108	3018		TR95/	
57A107-2	108	3018		TR95/	
57A107-3	108	3018		TR95/	
57A107-4	108	3018		TR95/	
57A107-5	108	3018		TR95/	
57A107-6	108	3018		TR95/	
57A107-8	108	3018		TR95/56	S0015
57A108-1	106	3118		TR20/	
57A108-2	106	3118		TR20/	
57A108-3	106	3118		TR20/	
57A108-4	106	3118		TR20/	
57A108-5	106	3118		TR20/	
57A108-6	106	3118		TR20/52	S0019
57A108-6-8	129	3025		TR88/	
57A108-7	106	3118		TR20/	
57A108-8	106	3118		TR20/	
57A110-9					S0019
57A121-9					S0024
57A122-9					S0019
57A123-10					S5012
57A124-10					G6005
57A125-9					S0015
57A126-12	107			TR70/	S0015
57A128-9	128			TR87/	S5014
57A129-9	128			TR87/	S5014
57A130-9	129			TR88/	S5013
57A131-10	128			TR87/	S5014
57A132-9					G0008
57A132-10	129			TR88/	S5013
57A133-12	159	3025	21	RE20/717	S0019
57A134-12	108	3018	20	TR95/56	S0016
57A135-12	108	3018	20	TR95/56	S0015
57A136-12	123A	3018	20	TR21/735	S0015
57A137-12	159	3114	22	TR20/717	S0013
57A137-12A	159	3114		TR20/	
57A138-4	108	3132	20	TR95/56	S0024
57A138-6	108	3018		TR95/	
57A139-1	161	3117		TR83/	
57A139-2	161	3117		TR83/	
57A139-3	161	3117		TR83/	
57A139-4	161	3117	20	TR83/56	S0020
57A139-4-6	108	3018		TR95/	
57A140-12	123A	3018	20	TR21/56	S0015
57A141-4	161	3117	39	TR83/719	S0017
57A142-2	199	3018			
57A142-4	233	3132	39	TR83/719	S0015
57A143-12	199	3018	20	TR70/720	S0011
57A144-12	199	3018	20	TR70/720	S0011
57A145-12	159	3039	21	TR70/720	S0012
57A146-12	108	3018	11	TR95/56	S0011
57A147-12					S0019
57A148-12	159	3114	22	TR20/717	
57A149-12	132	3112	FET2	FE100/802	F0015
57A150-12	132	3116	FET2	FE100/	F0015
57A151-6	108	3018	11	TR95/56	S0011
57A152-12	108	3039	11	TR95/56	S0011
57A153-9	123A	3124	10	TR21/735	S0014
57A155-10					S5011
57A156	128		18	TR87/	
57A156-9	123A	3124		TR21/735	S0024
57A157	159		21	TR20/	
57A157-9	159	3025		TR20/717	S0019
57A158-10	124	3021	12	TR81/240	S5011
57A159-12	159	3114	21	TR20/717	S0019
57A160-8	154	3132	61	TR78/712	S0015
57A162-12					S0019
57A163-12					S5013
57A164-4	161	3018	39	TR83/	S0020
57A166-12	123A	3122	20	TR21/735	S0015
57A167-9	192	3024	63	TR25/243	S5014
57A168-9	193	3025	67	TR28/242	S5013
57A169-12					F0010
57A170-9					S0015
57A171-9					S0031
57A172-8	171	3103	27	TR79/	
57A174-8	159	3114	21	TR20/	S0019

Original Part Number	ECG	SK	GE	IR/WEP	HEP
57A175-12	159	3114	21	TR20/	S0019
57A177-12	161	3018	60	TR83/	S0020
57A178-12	159	3114	22	TR20/717	S0019
57A179-4	161	3117			S0020
57A180-4	199	3132		TR70/	S0017
57A181-12	123A	3018	20	TR24/	S0015
57A182-12	123A	3018	20	TR24/	S0015
57A184-12	199	3122	20	TR33/	
57A185-12	159	3513	21	TR19/	
57A186-12	165	3115	38	TR61/	
57A187-12	152	3054	66	TR76/	
57A188-12	153	3083	69	TR77/	
57A189-8	159	3114	22	TR52/	S0019
57A191-12	199	3122	20	TR33/	
57A194-11	123A	3040	20	TR24/	S5021
57A198-11	165	3111	38	TR61/	
57A199-4		3122			S0015
57A200-12	123A	3018	20	TR24/	S0015
57A201-14		3114	21	TR19/	
57A202-13		3122	20	TR24/	
57A203-14		3122	20	TR25/	
57A204-14		3122	20	TR25/	
57A205-14	175	3054	66	TR76/	
57A206-14	197	3083	69	TR77/	
57A207-8	154	3040	40	TR78/	
57A214-12		3054			
57A216-12	159	3114			S0019
57A219-14		3024			
57A220-14		3025			
57A235-12	159	3114	21		
57A236-11	171	3040	27	TR75/	
57A240-14		3025			
57A241-14		3024			
57A249-4		3018			S0017
57A250-14		3054			S3024
57A251-14		3082			S3028
57A252-1		3124			
57A253-14		3124			S0033
57A256-10		3027			
57B1-56					G0005
57B2-1	160		51	TR17/637	
57B2-2	160		51	TR17/637	
57B2-3	160	3004	2	/630	G0005
57B2-4	158		53	TR84/630	G0005
57B2-5	103A	3010	54	TR08/724	
57B2-6	102A	3008	53	TR85/250	
57B2-7	158	3004	2	TR84/630	
57B2-8	103A	3010	8	TR08/724	
57B2-9	160		51	TR17/637	
57B2-10	158		53	TR84/230	
57B2-11	160	3006	51	TR17/637	G0003
57B2-12	102A	3008	50	TR85/250	G0008
57B2-13	160	3006	50	TR17/637	G0003
57B2-14	160	3006	50	TR17/637	G0003
57B2-15	158	3004	1	TR84/630	G0005
57B2-16	158			TR84/630	
57B2-17	160	3008	51	TR17/637	
57B2-18	160	3008	51	TR17/637	
57B2-19	160	3008	51	TR17/637	G0009
57B2-20	160		50	TR17/637	
57B2-21	158		53	TR84/630	
57B2-22	160	3006	51	TR17/637	G0003
57B2-23	102A	3005	2	TR85/250	G0005
57B2-24	102A	3123	52	TR85/250	G0007
57B2-25	158	3004	2	TR84/630	G0007
57B2-26	160		51	TR17/637	
57B2-27	123A	3122	20	TR21/735	
57B2-28	123A	3122	20	TR21/735	
57B2-29	102A		53	TR85/250	G0005
57B2-30	160		51	TR17/637	G0008
57B2-31	160		51	TR17/637	G0008
57B2-32	102A		53	TR85/250	
57B2-33	102A(2)		53	TR85/250	
57B2-34	102A		53	TR85/250	
57B2-35	160		51	TR17/637	
57B2-36	102A		53	TR85/250	
57B2-37	160		51	TR17/637	
57B2-38	128		18	TR87/243	
57B2-39	102A		53	TR85/250	
57B2-40	160		51	TR17/637	
57B2-41	160		51	TR17/637	
57B2-42	160		51	TR17/637	
57B2-43	102A		53	TR85/250	
57B2-44	102A		53	TR85/250	
57B2-45	102A		53	TR85/250	

Original Part Number	ECG	SK	GE	IR/WEP	HEP
57B2-46	102A(2)		53	TR85/250	
57B2-47	175	3131	12	TR81/241	
57B2-48	160		51	TR17/637	
57B2-49	102A		53	TR85/250	
57B2-50	160		51	TR17/637	
57B2-51	160		51	TR17/637	
57B2-52	102A		53	TR85/250	
57B2-58	175	3131	12	TR81/241	
57B2-59	123A	3122	20	TR21/735	
57B2-60	102A		53	TR85/250	
57B2-61	158(2)		53	/630	
57B2-62	123A	3122	20	TR21/735	
57B2-63	123A	3122	20	TR21/735	
57B2-64	123A	3122	20	TR21/735	
57B2-65	160			TR17/637	
57B2-66	160			TR17/637	
57B2-67	160		51	TR17/637	
57B2-68	160		51	TR17/637	
57B2-70	159	3114	21	TR20/717	
57B2-70A	159	3114		TR20/	
57B2-71	159	3114	21	TR20/717	
57B2-71A	159	3114		TR20/	
57B2-72	102A		53	TR85/250	
57B2-73	123A	3122	20	TR21/735	
57B2-75	160		51	TR17/637	
57B2-77	160		51	TR17/637	
57B2-78	102A		53	TR85/250	
57B2-79		3122	20	TR51/735	
57B2-80	160		51	TR17/637	
57B2-83	102A		53	TR85/250	
57B2-84	175	3131	12	TR81/241	
57B2-85	123A	3122	20	TR21/735	
57B2-87	123A	3122	20	TR21/735	
57B2-88	102A		53	TR85/250	
57B2-89	160		50	TR17/637	
57B2-90	160		50	TR17/637	
57B2-93	160		50	TR17/637	
57B2-97	123A	3122	20	TR21/735	
57B2-101	123A	3122	20	TR21/735	
57B2-102	123A	3122	20	TR21/735	
57B2-103	123A	3122	20	TR21/735	
57B2-104	160		51	TR17/637	
57B2-105	160		51	TR17/637	
57B2-113	123A	3122	20	TR21/735	
57B2-116	123A	3122	20	TR21/735	
57B2-126	123A	3122	20	TR21/735	
57B2-149	160		50	TR17/637	
57B2-153	123A	3122	20	TR21/735	
57B2-157	160		51	TR17/637	
57B2-158	160		51	TR17/637	
57B2-159	160		51	TR17/637	
57B2-192	123A	3122	20	TR21/735	
57B3-1	123A	3122	20	/735	
57B3-2	123A	3122	20		S0012
57B3-3	102A			TR85/250	
57B3-4	102A		53	TR85/250	
57B3-5	102A		53	TR85/250	G6011
57B3-6	102A		53	TR85/250	G6011
57B3-7	104MP	3013	3	TR01/624MP	
57B3-8	104MP	3013	3	TR01/624MP	
57B3-9	104MP	3013	3	TR01/624MP	
57B3-10	104MP	3013	3	TR01/624MP	
57B3-11	104MP	3013	3	TR01/624MP	
57B3-12	121MP		16	/232MP	
57B3-13	123A	3122	20	/735	
57B4-1	121	3009		TR01/121	
57B4-2	121	3009	16	TR01/121	
57B4-4	121	3009	16	TR01/121	
57B6	102	3004	2	TR05/631	
57B21	108	3019	17	TR95/56	
57B21-1	108				
57B21-2	108	3019	11	TR95/56	S0016
57B21-3	108				
57B21-4	108	3019	17	TR95/56	S0016
57B21-5	108	3019	17	TR95/56	S0016
57B21-6	161	3018	11	TR83/719	S0017
57B21-7	161	3018	11	TR83/719	S0017
57B21-8		3018			S0015
57B21-9	107	3018	11	TR80/	S0016
57B21-10		3018			S0011
57B21-12	108	3018	11	TR80/	S0020
57B21-13	108	3018	11	TR80/	S0016
57B21-13A		3018			
57B21-14	108	3018	11	TR80/	S0016
57B21-15	108	3018	11	TR80/	S0016

Original Part Number	ECG	SK	GE	IR/WEP	HEP
57B21-16	108	3039	11	TR80/	S0020
57B21-17	161	3039	39	TR70/	S0017
57B21-18	108	3019	11	TR83/	S0020
57B23				TR84/	
57B100-3	102A		53	TR85/250	
57B100-5	102A		53	TR85/250	
57B100-7	102A		53	TR85/250	
57B100-11	131			TR94/642	
57B100-385	102A		54		
57B101-4	108	3018	11	TR95/56	
57B102-4	161	3018	11	TR83/719	S0017
57B103-4	161	3018	11	TR83/719	S0020
57B104-8	161	3045	27	TR83/719	S5025
57B105-12	123A	3122	18	TR21/735	S0015
57B106-12	159	3006	21	TR20/717	S0019
57B107-8	123A	3122	18	TR21/735	S0015
57B108-6	159	3006	21	TR20/717	S0019
57B108-6A	159	3114		TR20/	
57B109-9	128	3024	18	TR87/243	S0014
57B110-9	129	3025	21	TR88/242	S0019
57B111-9	128		18	/242	
57B112-9	129	3025		TR88/242	
57B112-9A	129	3025		TR88/	
57B113-9	128	3025	18	TR87/243	
57B114-9	129	3025		TR88/242	
57B114-9A	129	3025		TR88/	
57B115-9	129	3025		TR88/242	
57B115-9A	129	3025		TR88/	
57B116-9	129	3025		TR88/242	
57B116-9A	129	3025		TR88/	
57B117-9	123A	3122	62	TR21/735	
57B118-12	123A	3122	20	TR21/735	
57B119-2	161		20		
57B119-12	123A	3122		TR21/735	
57B120-12	123A	3122	20	TR21/735	
57B121-9	123A	3122	20	TR21/735	
57B122-9	159	3025	21	TR20/717	
57B122-9A	159	3114		TR20/	
57B123-10	175	3131	12	TR81/241	
57B124-10	104	3009	3	TR01/232	G6005
57B125-9	123A	3122	20	TR21/735	
57B126-12	123A	3122	18	TR21/735	S0015
57B128-9	128	3024		TR87/	S5014
57B129-9	123A	3024	18	TR21/735	S5014
57B130-9	159	3025	21	TR20/717	S5013
57B130-9A	159	3114		TR20/	
57B131-10	152	3024	18	TR76/701	S5014
57B132-10	153	3025	21	TR77/700	S5013
57B133-12	159	3118		TR20/	S0019
57B134-12	108			TR95/	S0016
57B135-12	123A	3118		TR21/	S0015
57B136-1	154	3044		TR78/712	
57B136-2	154	3044		TR78/712	
57B136-3	154	3044		TR78/712	
57B136-4	154	3044		TR78/712	
57B136-5	154	3044		TR78/712	
57B136-6	154	3044		TR78/712	
57B136-7	154	3044		TR78/712	
57B136-8	154	3044		TR78/712	
57B136-9	154	3044		TR78/712	
57B136-10	154	3044		TR78/712	
57B136-11	154	3044		TR78/712	
57B136-12	123A	3024	18	TR21/735	S0014
57B137-12	159	3114	21	TR20/717	S0012
57B137-12A	159	3114		TR20/	
57B138-4	123A	3122	61	TR21/735	S0024
57B139-4	108	3039	60	TR95/56	S0020
57B140-12	123A	3118		TR21/	S0015
57B141-1	108	3018		TR95/56	
57B141-2	108	3018		TR95/56	
57B141-3	108	3018		TR95/56	
57B141-4	161	3018	20	TR83/719	S0017
57B142-1	108	3018		TR95/56	
57B142-2	108	3018		TR95/56	
57B142-3	108	3018		TR95/56	
57B142-4	161	3018	11	TR83/719	S0020
57B143-1	108	3018		TR95/56	
57B143-2	108	3018		TR95/56	
57B143-3	108	3018		TR95/56	
57B143-4	108	3018		TR95/56	
57B143-5	108	3018		TR95/56	
57B143-6	108	3018		TR95/56	
57B143-7	108	3018		TR95/56	
57B143-8	108	3018		TR95/56	
57B143-9	108	3018		TR95/56	
57B143-10	108	3018		TR95/56	
57B143-11	108	3018		TR95/56	
57B143-12	123A	3018	20	TR21/735	S0011
57B144-12	123A	3122	17	TR21/735	S0011
57B145-12	159	3114	67	TR20/717	S0012
57B145-12A	159	3114		TR20/	
57B146-12	123A	3124	20	TR21/735	S0011
57B147-12	159	3114	21	TR20/717	
57B147-12A	159	3114		TR20/	
57B148-1	129	3025		TR88/242	
57B148-2	129	3025		TR88/242	
57B148-3	129	3025		TR88/242	
57B148-4	129	3025		TR88/242	
57B148-5	129	3025		TR88/242	
57B148-6	129	3025		TR88/242	
57B148-7	129	3025		TR88/242	
57B148-8	129	3025		TR88/242	
57B148-9	129	3025		TR88/242	
57B148-10	129	3025		TR88/242	
57B148-11	129	3025		TR88/242	
57B148-12	129	3025	21	TR88/242	S0012
57B148-12A	129	3025		TR88/	
57B149-12	132	3116	FET1	FE100/802	F0015
57B150-12	132	3116	FET1	FE100/802	F0015
57B151-6	108	3018	20	TR95/56	S0011
57B152-1	123A	3124		TR21/735	
57B152-2	123A	3124		TR21/735	
57B152-3	123A	3124		TR21/735	
57B152-4	123A	3124		TR21/735	
57B152-5	123A	3124		TR21/735	
57B152-6	123A	3124		TR21/735	
57B152-7	123A	3124		TR21/735	
57B152-8	123A	3124		TR21/735	
57B152-9	123A	3124		TR21/735	
57B152-10	123A	3124		TR21/735	
57B152-11	123A	3124		TR21/735	
57B152-12	108	3018	17	TR95/56	S0011
57B153-1	123A	3124		TR21/735	
57B153-2	123A	3124		TR21/735	
57B153-3	123A	3124		TR21/735	
57B153-4	123A	3124		TR21/735	
57B153-5	123A	3124		TR21/735	
57B153-6	123A	3124		TR21/735	
57B153-7	123A	3124		TR21/735	
57B153-8	123A	3124		TR21/735	
57B153-9	123A	3124	20	TR21/735	S0015
57B155-10	175	3131	12	TR81/241	
57B156-9	123A	3025	20	TR21/735	S0024
57B157-9	159	3025	21	TR20/717	S0019
57B157-9A	159	3114		TR20/	
57B158-1	124	3021		TR81/240	
57B158-2	124	3021		TR81/240	
57B158-3	124	3021		TR81/240	
57B158-4	124	3021		TR81/240	
57B158-5	124	3021		TR81/240	
57B158-6	124	3021		TR81/240	
57B158-7	124	3021		TR81/240	
57B158-8	124	3021		TR81/240	
57B158-9	124	3021		TR81/240	
57B158-10	175	3021	12	TR81/241	S5011
57B159-12	159	3114	21	TR20/717	S0012
57B159-12A	159	3114		TR20/	
57B160-1	108	3018		TR95/56	
57B160-2	108	3018		TR95/56	
57B160-3	108	3018		TR95/56	
57B160-4	108	3018		TR95/56	
57B160-5	108	3018		TR95/56	
57B160-6	108	3018		TR95/56	
57B160-7	108	3018		TR95/56	
57B160-8	108	3018	17	TR95/56	S0015
57B163-12	129	3025		TR88/242	
57B163-12A	129	3025		TR88/	
57B164-4		3018			
57B165-11	128				
57B166-12	108	3018		TR95/	S0016
57B167-9	128	3024	63	TR87/243	S5014
57B168	102A		53	TR85/250	
57B168-9	129	3025	67	TR88/242	S5013
57B169	102A		53	TR85/250	
57B169-12	132			FE100/802	F0010
57B170	102A		53	TR85/250	
57B170-9	128	3024		TR87/243	S0015
57B171-9	129	3025		TR88/242	S0031
57B175-9	130	3025	21	TR59/247	
57B175-9A	130	3027		TR59/	

Original Part Number	ECG	SK	GE	IR/WEP	HEP
57B175-12		3114			
57B178-12	159				S0013
57B180	160			TR17/	
57B181-12	194				
57B182-12	123A				
57B184	160			TR17/	
57B186			51	TR17/637	
57B186-11	165				
57B187	160			TR17/	
57B188	160			TR17/	
57B189-8	159				
57B194-11	123A				
57B198-11	165	3115			S7005
57B199-11	165				
57B201-14	159				
57B202-13	123A				
57B203-14	199				
57B205-14	152	3054			
57B206-14	153	3083			
57B213-11	165	3027	36		S5021
57B1127	102A			TR85/	
57B1130	160			TR17/	
57B1131	160			TR17/	
57B1143	102A			TR85/	
57B1186	160			TR17/	
57B2153-1				TR51/	
57B2153-3				TR51/	
57C5	103	3010		TR08/641A	
57C5-1	126	3006	51	TR17/635	G0003
57C5-2		3006	51	TR17/635	G0003
57C5-3	102A	3006	51	TR85/250	G0005
57C5-4	126	3006	51	TR17/635	G0003
57C5-5	102A	3008	1	TR85/250	G0005
57C5-6	108	3018	11	TR95/56	S0016
57C5-7	108	3018	11	TR95/56	S0016
57C5-8	108	3018	17	TR95/56	S0016
57C5-9	126	3007	51	TR17/635	
57C5-10	126	3007	51	TR17/635	
57C6	199				
57C6-1	102	3004	2	TR05/631	
57C6-2	121	3009	16	TR01/232	
57C6-3	121	3009	16	TR01/232	G6003
57C6-4	123A	3124	20	TR21/735	S0015
57C6-5	103A	3010	2	TR08/724	G0011
57C6-6	103	3004	17	TR08/641A	G0005
57C6-6A	100	3005		TR05/254	G0005
57C6-6B	100	3005		TR05/254	G0005
57C6-6C	100	3005		TR05/254	G0005
57C6-7	123	3124	17	TR86/53	
57C6-8	121	3009	3	TR01/232	G6003
57C6-9	123A	3124	20	TR21/735	S0015
57C6-10	124	3021	12	TR81/240	
57C6-11	123A	3124	20	TR21/735	S0015
57C6-12	121	3009	16	TR01/232	G6005
57C6-14	124	3021	32	TR81/240	
57C6-15	129	3025	21	TR88/242	S0019
57C6-16	126	3006		TR17/635	
57C6-17	123	3124	20	TR86/53	
57C6-18	123A	3122	20	/735	
57C6-19	123	3124	20	TR86/53	S0015
57C6-20	103A		54	TR08/724	
57C6-21	103A	3010	52	TR08/724	
57C6-22	102	3004	8	TR05/631	
57C6-23	121	3009	3	TR01/232	
57C6-24	124	3021	12	TR81/240	
57C6-25	102A	3004	53	TR85/250	
57C6-26	129	3025	22	TR88/242	
57C6-26A	129	3025		TR88/	
57C6-27	123	3124	20	TR86/53	
57C6-28	123A	3122	20	/735	
57C6-29	123	3124	20	TR86/53	
57C6-30	123	3124	20	TR86/53	
57C6-31	129	3025	21	TR88/242	
57C6-31A	129	3025		TR88/	
57C6-32	123	3124	20	TR86/53	S0014
57C6-33	129	3025	21	TR88/242	S0019
57C6-33A	129	3025		TR88/	
57C6-34	128		18	/243	
57C7-1	108	3039		TR95/56	
57C7-2	108	3039		TR95/56	
57C7-3	108	3039		TR95/56	
57C7-4	108	3039		TR95/56	
57C7-5	108	3039		TR95/56	
57C7-6	108	3039		TR95/56	
57C7-7	108	3039		TR95/56	

Original Part Number	ECG	SK	GE	IR/WEP	HEP
57C7-8	128	3024	18	TR87/243	
57C7-9	123A	3122	20	TR21/735	
57C7-10	123A	3122	20	TR21/735	
57C7-15	123A	3122	20	TR21/735	
57C7-17	123A	3122	20	TR21/735	
57C7-18	123A	3122	20	TR21/735	
57C7-20	123A	3122	20	TR21/735	
57C9-2	121	3009	16	TR01/232	
57C10-1	108	3018	20	TR95/56	S0017
57C10-2	108	3018	20	TR95/56	S0017
57C11-1	123A	3122	20	TR21/735	
57C12-1	154	3045	18	TR78/712	S5025
57C12-2	154	3044	27	TR78/712	S5025
57C12-3	175			TR81/241	
57C12-4	128	3024	20	TR87/243	S0015
57C12-5	175	3131	66	TR81/241	
57C14-1	128	3024	18	TR87/243	
57C14-2	128	3024	18	TR87/243	
57C14-3	128	3024	18	TR87/243	
57C15-1	123A	3122	20	TR21/735	S0015
57C15-2	123A	3122	20	TR21/735	S0015
57C15-3	123A	3006	21	TR21/735	S0019
57C15-4	123A	3122	21	TR21/735	S0019
57C15-5	159	3114	21	TR20/717	S0031
57C15-50	159	3114		TR20/	
57C16	126	3006		TR17/635	
57C16-1	123A	3122	20	TR21/735	
57C19-1	159	3114	21	TR20/717	
57C19-1A	159	3114		TR20/	
57C20-1	108	3018	20	TR95/56	S0020
57C21	107				
57C21-5	107				
57C22-1	179			TR35/G6001	
57C22-2	179			TR35/G6001	
57C23	129				
57C23-1	129	3025	21	TR88/242	
57C23-2	129	3025	21	TR88/242	
57C23-3	129	3025	22	TR88/242	
57C24-1	123A	3122	20	TR21/735	
57C24-2	123A	3018	20	TR21/735	S0011
57C24-3	123A	3122	20	TR21/735	
57C24-4	123A	3122	20	TR21/735	
57C27-1	123A	3122	20	TR21/735	
57C27-2	108	3018	11	TR95/56	
57C28	704	3023		IC501/	
57C29	706	3101			
57C29-1	706	3101		IC507/	
57C29-2	706	3101		IC507/	
57C68	102A		53	TR85/250	
57C109-9	128	3024		TR87/243	
57C110-9	129	3025		TR88/242	
57C121-9	123A	3124		TR21/735	
57C122-9	129	3025		TR88/242	
57C142-4	161	3117		TR83/719	S0011
57C148-12	129	3025		TR88/242	
57C148-12A	129	3025		TR88/	
57C156-9	123A	3124		TR21/735	
57C157-9	159	3114		TR20/717	
57C157-90	159	3114		TR20/	
57C164-4	161			TR83/	S0020
57D1-3	103A		7	TR08/724	
57D1-4	103A		7	TR08/724	
57D1-5	103A		7	TR08/724	
57D1-6	103A		7	TR08/724	
57D1-7	158		53	TR84/630	
57D1-8	102A		53	TR85/250	
57D1-9	160		1	TR17/637	
57D1-10	160		1	TR17/637	
57D1-11	102	3004	2	TR05/631	G0005
57D1-12	160		51	TR17/637	
57D1-13	160		51	TR17/637	
57D1-14	102A		53	TR85/250	
57D1-15	160		1	TR17/637	
57D1-16	160		1	TR17/637	
57D1-17	158		53	TR84/630	
57D1-18	158		53	TR84/630	
57D1-19	158		53	TR84/630	
57D1-20	158		53	TR84/630	
57D1-21	158		53	TR84/630	
57D1-22	158		1	TR17/637	G0008
57D1-23	102A		1	TR85/250	G0003
57D1-24	160		51	TR17/637	
57D1-25	160		51	TR17/637	
57D1-26	102A	3003	2	TR85/250	G0005
57D1-27	102A	3003	2	TR85/250	

Original Part Number	ECG	SK	GE	IR/WEP	HEP
57D1-28	102A		53	TR85/250	
57D1-30	160		50	TR17/637	
57D1-31	160		50	TR17/637	
57D1-32	160		50	TR17/637	
57D1-33	160		50	TR17/637	
57D1-34	102A		53	TR85/250	
57D1-35	160		51	TR17/637	
57D1-36	160		51	TR17/637	
57D1-37	160		51	TR17/637	
57D1-38	158		53	TR84/63	
57D1-39			51	TR17/637	
57D1-40	102A		1	TR85/250	
57D1-41	160		1	TR17/637	
57D1-42	158	3003	2	TR84/630	
57D1-43	102A	3004	2	TR85/250	
57D1-44	102A	3004	2	TR85/250	
57D1-45	160		51	TR17/637	
57D1-46	160		51	TR17/637	
57D1-47	160		51	TR17/637	
57D1-48	160		51	TR17/637	
57D1-49	160		51	TR17/637	
57D1-50	160		51	TR17/637	
57D1-51	123A		20	TR21/735	
57D1-52	106	3118	21	TR20/52	
57D1-53	102A		53	TR85/250	
57D1-55	123A	3122	20	TR21/735	
57D1-56	158	3004	2	TR84/630	G0005
57D1-57	160		51	TR17/637	
57D1-58	102A		53	TR85/250	
57D1-59	102A		1	TR85/250	
57D1-60	158		53	TR84/630	
57D1-61	160		50	TR17/637	
57D1-66	102A		53	TR85/250	
57D1-67	160		51	TR17/637	
57D1-68	102A		1	TR85/250	
57D1-69	160		1	TR17/637	
57D1-70	102A	3004	2	TR85/250	
57D1-71	102A		53	TR85/250	
57D1-72	160		51	TR17/637	
57D1-73	160		51	TR17/637	
57D1-74	160		51	TR17/637	
57D1-75	123A	3122	20	TR21/735	
57D1-76	159	3114	21	TR20/717	
57D1-76A	159	3114		TR20/	
57D1-77	158		53	TR84/630	
57D1-78	103A		54	TR08/724	
57D1-79	102A		51	TR85/250	
57D1-80	160	3005	1	TR17/637	G0005
57D1-81	160	3008	51	TR17/637	
57D1-82	102A		53	TR85/250	
57D1-83	102A		53	TR85/250	
57D1-84	160	3005	1	TR17/637	G0005
57D1-85	160	3008	51	TR17/637	
57D1-86	160	3008	1	TR17/637	G0005
57D1-87	160	3008	1	TR17/637	G0005
57D1-88		3007	51	TR17/637	
57D1-89	160	3007	1	TR17/637	
57D1-90	102A		53	TR85/250	
57D1-91	102A		53	TR85/250	
57D1-92	102A		53	TR85/250	
57D1-93	102A		53	TR85/250	
57D1-94	160		50	TR17/637	
57D1-95	102A		53	TR85/250	
57D1-96	160	3006	51	TR17/637	
57D1-97	102A	3008	51	TR85/250	
57D1-98	160		50	TR17/637	
57D1-99	160		50	TR17/637	
57D1-100	160		50	TR17/637	
57D1-101	160		50	TR17/637	
57D1-102	160		50	TR17/637	
57D1-103	160		51	TR17/637	
57D1-104	102A		53	TR85/250	
57D1-105	102A	3004	2	TR85/250	
57D1-106	102A		50	TR85/250	
57D1-107	160	3006	51	TR17/637	
57D1-108	160		51	TR17/637	
57D1-109	160		51	TR17/637	
57D1-110	160		51	TR17/637	
57D1-111	102A	3004	2	TR85/250	
57D1-112	160		51	TR17/637	
57D1-113	160		51	TR17/637	
57D1-114	160	3006	51	TR17/637	
57D1-115	160		51	TR17/637	
57D1-116	102A		1	TR85/250	
57D1-117	102A		53	TR85/250	
57D1-118	102A		53	TR85/250	
57D1-119	121	3009	16	TR01/232	
57D1-120	160	3004	2	TR17/637	G0005
57D1-121	102A	3004	2	TR85/250	G0005
57D1-122	154	3045	27	TR78/712	
57D1-123	123A	3124	20	TR21/735	S0015
57D1-124	123A	3124	20	TR21/735	S0015
57D3-6	102	3004	2	TR05/631	
57D4-1	121	3009	3	TR01/232	G0005
57D4-2	121	3009	3	TR01/232	
57D5-1	126	3006	51	TR17/635	
57D5-2	126	3006	51	TR17/635	
57D5-4	126	3006	51	TR17/635	
57D6-4	123A	3124		TR21/735	
57D6-10	124	3021	12	TR81/240	
57D6-12	121	3009	16	TR01/232	
57D6-19	123	3124	20	TR86/735	
57D9-1	160	3009	3	TR17/637	
57D9-2	121	3009	16	TR01/232	
57D14-1	123A	3122		TR21/735	S0015
57D14-2	123A	3122		TR21/735	S0015
57D14-3	123A	3122		TR21/735	S0015
57D19	159				
57D19-1	159	3006		TR20/717	S0019
57D19-2	159	3114		TR20/717	S0019
57D19-3	159	3114		TR20/717	S5022
57D19-10	159	3114		TR20/	
57D19-20	159	3114		TR20/	
57D19-30	159	3114		TR20/	
57D24	161				
57D24-1	108	3018	11	TR95/56	
57D24-2	108	3018	11	TR95/56	
57D24-3	108	3018	11	TR95/56	
57D29-2	706				
57D68	102	3004	52	TR05/631	G0005
57D107-8	108	3122	20	TR95/	S0014
57D126	102	3004	52	TR05/631	
57D127	157	3004	2	TR60/244	
57D130	126	3007	51	TR17/635	
57D131	126	3007	51	TR17/635	
57D132	126	3007	51	TR17/635	
57D132-9	160		51	TR17/637	G0008
57D136-12	123A	3122	20	TR21/735	
57D141-4					S0017
57D143	102	3004	2	TR05/631	
57D156	102	3004	52	TR05/631	G0005
57D164-4					S0020
57D168	126	3006		TR17/	G0002
57D169	102	3004	52	TR05/631	G0005
57D170	102	3004	52	TR05/631	G0005
57D180	100	3005	2	TR05/254	G0005
57D184	100	3005	2	TR05/254	G0005
57D186	126	3007	2	TR17/635	G0005
57D187	126	3006	2	TR17/635	G0005
57D188		3004	52	TR17/631	G0005
57D189	102	3004	52	TR05/631	G0005
57D1127	102	3004	52	TR05/631	G0005
57D1130	160		50	TR17/637	G0002
57D1131	160		50	TR17/637	G0002
57D1132	160		50	TR17/637	G0002
57D1143	102	3004	52	TR05/631	G0005
57D1186	160		50	TR17/637	G0002
57DG-23	121	3014		TR01/	G0005
57DG-32	121			TR01/	
57L1-1	160		51	TR17/637	
57L1-2	160		51	TR17/637	
57L1-3	160		51	TR17/637	
57L1-4	160		51	TR17/637	
57L1-5			53	TR85/250	
57L1-6	102A		53	TR85/250	
57L1-7	158(2)		53	/630	
57L1-8	102A		53	TR85/250	
57L1-9	160		51	TR17/637	
57L1-10	160		51	TR17/637	
57L1-11	160		51	TR17/637	
57L1-12	160		51	TR17/637	
57L2-1	128		66	TR87/243	
57L2-2	123A	3024	28	TR21/735	
57L3-1	123A	3122	20	TR21/735	
57L3-4	123A	3122	20	TR21/735	
57L5-1	104	3009	3	TR01/624	
57L103-13	717				
57L105-12	128				
57L106-9	133				
57M1-1	160		51	TR17/637	

Original Part Number	ECG	SK	GE	IR/WEP	HEP
57M1-2	160		51	TR17/637	
57M1-3	160		51	TR17/637	
57M1-4	160		51	TR17/637	
57M1-5	102A		53	TR85/250	
57M1-6	102A		53	TR85/250	
57M1-7	158(2)		53	/630	
57M1-8	158		53	TR84/630	
57M1-9	160		51	TR17/637	
57M1-10	160		51	TR17/637	
57M1-11	160		51	TR17/637	
57M1-12	160		51	TR17/637	
57M1-13	129	3025		TR88/242	
57M1-14	123A	3122	20	TR21/735	
57M1-15	123A	3122	20	TR21/735	
57M1-16	103A		54	TR08/724	
57M1-17	160		51	TR17/637	
57M1-18	103A		54	TR08/641A	
57M1-19	123A	3122	20	TR21/735	
57M1-20	123A	3122	20	TR21/735	
57M1-21	129	3025		TR88/242	
57M1-21A	129	3025		TR88/	
57M1-22	129	3025		TR88/242	
57M1-23	123A	3122	20	TR21/735	
57M1-24	123A	3122	20	TR21/735	
57M1-25	129	3025		TR88/242	
57M1-25A	129	3025		TR88/	
57M1-26	123A	3122	20	TR21/735	
57M1-27	123A	3122	20	TR21/735	
57M1-28	123A	3122	20	TR21/735	
57M1-29	123A	3122	20	TR21/735	
57M1-30	123A	3122	20	TR21/735	
57M1-31	123A	3122	20	TR21/735	
57M1-32	123A	3122	20	TR21/735	
57M1-33	128	3024	18	TR87/243	
57M1-34	128	3024	18	TR87/243	
57M1-35	158		53	TR84/630	
57M2-1	103A			TR08/724	
57M2-2	103A			TR08/724	
57M2-3	158			TR84/630	
57M2-4	158			TR84/630	
57M2-6	103A			TR08/724	
57M2-7	128	3122	18	TR87/243	
57M2-8	158			TR84/630	
57M2-9	103A			TR08/724	
57M2-10	129	3025		TR88/242	
57M2-10A	129	3025		TR88/	
57M2-11	128	3024	18	TR87/243	
57M2-14	128	3024	18	TR87/243	
57M2-15	129	3025		TR88/242	
57M2-15A	129	3025		TR88/	
57M2-16	154	3045	27	TR78/712	
57M2-17	154	3045	27	TR78/712	
57M2-18	128	3024	18	TR87/243	
57M2-506	103A			/724	
57M3-1	130	3027	14	TR59/247	
57M3-1A	130	3027		TR59/	
57M3-2	130		14	TR59/	
57M3-3	128	3024	18	TR87/243	
57M3-4	130	3027	14	TR59/247	
57M3-4A	130	3027		TR59/	
57M3-5	175			TR81/241	
57M3-6	130	3027	14	TR59/247	
57M3-6A	130	3027		TR59/	
57M3-7	104	3009	3	TR01/624	
57M3-8	104	3009	3	TR01/624	
57M3-9N	155				
57M3-9P	131		44	TR94/642	
57M3-10N	155				
57M3-10P	131	3082	44	TR94/642	
57M3-11	155				
57M3-12	131	3082	44	TR94/642	
57P1	102	3004	52	TR05/631	G0005
58-000516	724				
58-1(TRU)	123A	3124			
58B2-14	160	3006	51	TR17/637	
59B402781	9923				
59B402787	9914				
59B402788	9974				
59P2C	102A				
60-211040	181				
60P22204	191				
61-309-458	128				
61-607	102A		53	TR85/250	
61-608	102A		53	TR85/250	
61-654	102A		53	TR85/250	

Original Part Number	ECG	SK	GE	IR/WEP	HEP
61-655	102A		53	TR85/250	
61-656	102A		53	TR85/250	
61-746	123A	3122	20	TR21/735	S0015
61-747	128	3024	18	TR87/243	S5014
61-751	123A	3122	20	TR21/735	S0015
61-754	123A	3122	20	TR21/735	S0015
61-755	123A		20	TR21/735	
61-782	121	3009	16	TR01/232	G6003
61-813	128	3024	18	TR87/243	S5014
61-814	123A	3122	20	TR21/735	S0015
61-815	123A	3122	20	TR21/735	S0015
61-828	102A		53	TR85/250	
61-829	102A		53	TR85/250	
61-1053-1	196			TR92/	
61-1130	102A		53	TR85/250	
61-1131	102A		53	TR85/250	
61-1215	102A		53	TR85/250	
61-1400	123A	3122	20	TR21/735	S0015
61-1401	123A	3122	20	TR21/735	S0015
61-1402	123A	3122	20	TR21/735	S0015
61-1403	123A	3122	20	TR21/735	S0015
61-1404	123A	3122	20	TR21/735	S0015
61-1763	123A			TR21/	
61-1764	128			TR87/	
61-1906	131	3198		TR94/	
61-1907	102A			TR85/	
61-1934	102A			TR85/	
61-1935	102A			TR85/	
61-3096-90	197				
61-260039	126	3008		TR17/	G0008
61-260039A	126	3008		TR17/	
61-309449	130	3511			
61-309690	153	3083			
61-503512-2		3122			
61A023-2	1045				
61A030-6	712	3072	IC2	IC507/	C6063P
61A030-9	1083				
61B002-1	160		50	TR17/637	G0003
61B003-1	126	3006	50	TR17/635	G0008
61B004-1	102A	3004	52	TR85/250	G0005
61A005-1	102A	3004	52	TR85/250	G0007
61A006-1	102A	3004	52	TR85/250	G0005
61A007-1	108			TR95/56	
61A007-2	108	3039	11	TR95/56	S0016
61A009-1	102A	3004	2	TR85/250	G0005
61B0015-1	102A			TR85/	
61B015-1	102A	3004		TR85/250	G0005
61B016-1	102A	3123		TR85/250	G0007
61B017-1	102A	3004		TR85/250	G0005
61B018-1	102A	3123		TR85/250	G0007
61B019-1	102A	3004		TR85/250	G0005
61B020-1	102A	3004		TR85/250	G0007
61B021-1	102A	3004	11	TR85/250	G0005
61B022-2	102A	3004	2	TR85/250	G0007
61B022-3	102A	3004	2	TR85/250	G0005
61B023-1	102A	3004	2	TR85/250	G0005
61B026-1	102A	3005	51	TR85/250	G0005
61B027-1	158	3004	2	TR84/630	G0005
61B042-9	1003				
61B22	783	3141			
61B45-14	102A				
61C001-1	123A	3124	20	TR86/53	S0015
61C002-1	102	3004	2	TR05/631	G0005
61C003-1	102	3004	2	TR05/631	G0005
61C004-1	128	3024	18	TR87/243	S5014
61C005-1	121	3009	16	TR01/232	G6003
61D189-1		3111			S5020
61J001-1	123A	3124		TR21/735	S0015
61J002-1	123A	3124		TR21/735	S0015
61J003-1	123A	3124		TR21/735	S0015
61J004-1	102A	3004		TR85/250	G0005
61P1	126	3006	50	TR17/635	G0005
61P1D	160		51	TR17/	
61P10	126	3006		TR17/635	
62-3597-1			21		
62-3597-2			21		
62-7567	123A	3122	20	TR21/735	
62-8555			20		
62-8781	160			TR17/	
62-13258	102A			TR85/	
62-13259	103A			TR08/	
62-13494	102A			TR85/	
62-16905	123A	3122	20	TR21/735	
62-16918	102A		53	TR85/250	
62-16919	175	3131	12	TR81/241	

Original Part Number	ECG	SK	GE	IR/WEP	HEP
62-17390	160		51	TR17/637	
62-17391	160		51	TR17/637	
62-17550	123A	3122	20	TR21/735	
62-18415	100	3005	51	TR05/254	
62-18416	100	3005	51	TR05/254	
62-18417	100	3005	51	TR05/254	
62-18418	160		51	TR17/637	
62-18419	160		51	TR17/637	
62-18420	102A		53	TR85/250	
62-18421	102A		53	TR85/250	
62-18422	160		51	TR17/637	
62-18423	100	3005	51	TR05/254	
62-18424	100	3005	51	TR05/254	
62-18425	123A	3122	20	TR21/735	
62-18426	154		27	TR78/712	
62-18427	104	3009	3	TR01/624	
62-18428	121		3	TR01/232	
62-18429	179			TR35/G6001	
62-18430	102A		53	TR85/250	
62-18641	123A	3122	20	TR21/735	
62-18642	123A	3122	20	TR21/735	
62-18643	123A	3122	20	TR21/735	
62-18828	123A	3122	20	TR21/735	
62-19280	123A	3122	20	TR21/735	
62-19452	159		21	TR20/717	
62-19516	123A	3122	20	TR21/735	
62-19548	123A	3122	20	TR21/735	
62-19581	107	3039		TR70/720	
62-19837	123A	3122	20	TR21/735	
62-19838	123A	3122	20	TR21/735	
62-20154	159	3114	21	TR20/717	
62-20154A	159	3114		TR20/	
62-20155	123A	3122	20	TR21/735	
62-20240	123A	3122	20	TR21/735	
62-20241	123A	3122	20	TR21/735	
62-20242	123A	3122	20	TR21/735	
62-20243	123A	3122	20	TR21/735	
62-20244	159	3114	21	TR20/717	
62-20244A	159	3114		TR20/	
62-20360	123A	3122	20	TR21/250	
62-21496	123A	3122	20	TR21/735	
62-21586	711				
62-21587	712				
62-22038	123A			TR21/	
62-22039	123A		61	TR21/	
62-22250	123A		62	TR21/	
62-22251	123A		20	TR21/	
62-22529	106			TR20/	
62-26851	160		51	TR17/637	
62A11871	159				
62B046-1	152	3054			
62B046-2	152	3054			
62B046-3	152				
62B046-4	152	3054			
62-128343			1		
62-129604			11		
62-130139			21		
62-132497			21		
63-3954	160		51	TR17/637	
63-7246	158		53	TR84/630	
63-7247	102A		53	TR85/250	
63-7248	103A		54	TR08/641A	
63-7396	102A		53	TR85/250	
63-7397	102A		53	TR85/250	
63-7398	102A		53	TR85/250	
63-7399	102A		53	TR85/250	
63-7420	102A		53	TR85/250	
63-7421	123A	3122	20	TR21/735	
63-7538	160		51	TR17/637	
63-7541	160		51	TR17/637	
63-7547	100	3005	51	TR05/254	
63-7548	160		51	TR17/637	
63-7549	103A		54	TR08/724	
63-7564	102A		53	TR85/250	
63-7565	103A		54	TR08/724	
63-7567	123A	3122	20	TR21/735	
63-7579	160		51	TR17/637	
63-7580	160		51	TR17/637	
63-7581	160		51	TR17/637	
63-7582	160		51	TR17/637	
63-7596	102A		53	TR85/250	
63-7660	160		51	TR17/637	
63-7670	123A	3122	50	TR21/735	
63-7871	102A		53	TR85/250	
63-7872	102A		53	TR85/250	
63-7873	102A		53	TR85/250	
63-8119	160		51	TR17/637	
63-8120	158		53	TR84/630	
63-8376	160		51	TR17/637	
63-8377	160		51	TR17/637	
63-8378	160		51	TR17/637	
63-8379	160		51	TR17/637	
63-8380	102A		53	TR85/250	
63-8473	103A		54	TR08/724	
63-8512	175		12	TR81/241	

Original Part Number	ECG	SK	GE	IR/WEP	HEP
63-8555	123A	3122		TR21/735	
63-8590	121		3	TR01/232	
63-8699	160		51	TR17/637	
63-8700	160		51	TR17/637	
63-8701	123A	3122	20	TR21/735	
63-8702	123A		20	TR21/735	
63-8703	102A		53	TR85/250	
63-8704	102A		53	TR85/250	
63-8705	103A		54	TR08/724	
63-8706	121	3009	3	TR01/232	
63-8707	130	3027	14	TR59/247	
63-8707A	130	3027		TR59/	
63-8945	175		12	TR81/241	
63-8954	160		51	TR17/637	
63-9072	160		51	TR17/637	
63-9337	123A	3122	20	TR21/735	
63-9338	123A	3122	20	TR21/735	
63-9339	123A	3122	20	TR21/735	
63-9340	103A		54	TR08/724	
63-9341	123A	3122	20	TR21/735	
63-9516	123A	3122	20	TR21/735	
63-9517	160		51	TR17/637	
63-9518	123A	3122	20	TR21/735	
63-9519	102A		53	TR85/250	
63-9520	102A		53	TR85/250	
63-9521	102A		53	TR85/250	
63-9659	102A		53	TR85/250	
63-9664	160		51	TR17/637	
63-9665	160		51	TR17/637	
63-9829	123A	3122	20	TR21/735	
63-9830	123A	3122	20	TR21/735	
63-9831	123A	3122	20	TR21/735	
63-9832	123A	3122	20	TR21/735	
63-9833	123A	3122	20	TR21/735	
63-9834				TR51/	
63-9847	123A	3122	20	TR21/735	
63-9876	160		51	TR17/637	
63-9877	160		51	TR17/637	
63-9941	160		51	TR17/637	
63-10035	160		51	TR17/637	
63-10036	160		51	TR17/637	
63-10037	102A		53	TR85/250	
63-10038	102A		53	TR85/250	
63-10062	175		12	TR81/241	
63-10145	160		51	TR17/637	
63-10146	160		51	TR17/637	
63-10147	102A		53	TR85/250	
63-10148	160		51	TR17/637	
63-10149	160		51	TR17/637	
63-10150	160		51	TR17/637	
63-10151	102A		53	TR85/250	
63-10152	102A		53	TR85/250	
63-10153	102A		53	TR85/250	
63-10154	102A		53	TR85/250	
63-10156	102A		53	TR85/250	
63-10158	102A		53	TR85/250	
63-10159	102A		53	TR85/250	
63-10188	123A	3122	20	TR21/735	
63-10195	160		51	TR17/637	
63-10196	160		51	TR17/637	
63-10200	100	3005	51	TR05/254	
63-10375	160		51	TR17/637	
63-10376	160		51	TR17/637	
63-10377	123A	3122	20	TR21/735	
63-10378	121	3009	3	TR01/232	
63-10383	103A		54	TR08/724	
63-10384	158		53	TR84/630	
63-10408	102A		53	TR85/250	
63-10708	123A	3122	20	TR21/735	
63-10725	123A			TR21/	
63-10732	123A	3122	20	TR21/735	
63-10733	123A	3122	20	TR21/735	
63-10734	123A	3122	20	TR21/735	
63-10735	123A	3122	20	TR21/735	
63-10736	123A	3122	20	TR21/735	
63-10737	123A	3122	20	TR21/735	
63-10860	123A	3122	20	TR21/735	
63-11025	123A			TR21/	
63-11055	160		51	TR17/637	
63-11073	158		53	TR84/630	
63-11143	123A	3122	20	TR21/735	
63-11144	102A		53	TR85/250	
63-11289	123A	3122	20	TR21/735	
63-11290	175		12	TR81/241	
63-11468	123A	3122	20	TR21/735	
63-11469	123A	3122	20	TR21/735	
63-11470	123A	3122	20	TR21/735	
63-11471	123A	3122	20	TR21/735	
63-11472	123A	3122	20	TR21/735	
63-11496	160			TR17/637	
63-11497	158		53	TR84/630	
63-11582	160		51	TR17/637	
63-11584	160		51	TR17/637	
63-11585	100		51	TR05/254	
63-11586	102A		53	TR85/250	

Original Part Number	ECG	SK	GE	IR/WEP	HEP
63-11660	123A	3122	20	TR21/735	
63-11661	102A		53	TR85/250	
63-11757	123A	3122	20	TR21/735	
63-11758	123A	3122	20	TR21/735	
63-11759	123A	3122	20	TR21/735	
63-11825		3122	20	TR51/735	
63-11831	123A	3122	20	TR21/735	
63-11832	123A	3122	20	TR21/735	
63-11833	123A	3122	20	TR21/735	
63-11878	175	3131	12	TR81/241	
63-11916	123A			TR21/	
63-11934	123A	3122	20	TR21/735	
63-11935	123A	3122	20	TR21/735	
63-11936	128		18	TR87/243	
63-11937	123A	3122	20	TR21/735	
63-11938	175	3131	66	TR81/241	
63-11989	128	3024	18	TR87/243	
63-11990	185				
63-11991	130	3027	14	TR59/247	
63-11991A	130	3027		TR59/	
63-12003	123A	3122	20	TR21/735	
63-12004	123A	3122	20	TR21/735	
63-12062	123A	3122	20	TR21/735	
63-12154	159	3114	21	TR20/717	
63-12154A	159	3114		TR20/	
63-12156	159	3114	21	TR20/717	
63-12156A	159	3114		TR20/	
63-12157	159	3114	21	TR20/717	
63-12157A	159	3114		TR20/	
63-12272	123A	3122	20	TR21/735	
63-12273	162			TR67/707	
63-12316	102A		53	TR85/250	
63-12317	102A		53	TR85/250	
63-12605	123A	3122	20	TR21/735	
63-12608	123A	3122	20	TR21/735	
63-12609	123A	3122	20	TR21/735	
63-12610	160		51	TR17/637	
63-12641	123A	3122	20	TR21/735	
63-12642	123A	3122	20	TR21/735	
63-12669	158		53	TR84/630	
63-12670	158		53	TR84/630	
63-12696	123A	3122	20	TR21/735	
63-12697	123A	3122	20	TR21/735	
63-12698	158		53	TR84/630	
63-12706	123A	3122	20	TR21/735	
63-12707		3122	20	TR21/735	
63-12750	123A	3122	20	TR21/735	
63-12751	123A	3122	20	TR21/735	
63-12752	123A	3122	20	TR21/735	
63-12753	123A	3122	20	TR21/735	
63-12874	123A	3122	20	TR21/735	
63-12875	123A	3122	20	TR21/735	
63-12876	123A	3122	20	TR21/735	
63-12877	123A	3122	20	TR21/735	
63-12878	123A	3122	20	TR21/735	
63-12879	123A	3122	20	TR21/735	
63-12880	102A		53	TR85/250	
63-12881	102A		53	TR85/250	
63-12933	123A			TR21/	
63-12940	123A	3122	20	TR21/735	
63-12941	123A	3122	20	TR21/735	
63-12942	123A	3122	20	TR21/735	
63-12943	123A	3122	20	TR21/735	
63-12945	158		53	TR84/630	
63-12946	123A	3122	20	TR21/735	
63-12947	158		53	TR84/630	
63-12948	123A	3122	20	TR21/735	
63-12949	123A	3122	20	TR21/735	
63-12950	123A	3122	20	TR21/735	
63-12951	123A	3122	20	TR21/735	
63-12952	123A	3122	20	TR21/735	
63-12953	123A	3122	20	TR21/735	
63-12954	175		66	TR81/241	
63-12989	128	3024	18	TR87/243	
63-12990	128	3024	18	TR87/243	
63-13025	160			TR17/	
63-13214	128	3024	18	TR87/243	
63-13215	128	3024	18	TR87/243	
63-13216	162			TR67/707	
63-13322	159	3114	21	TR20/717	
63-13322A	159	3114		TR20/	
63-13323	158		53	TR84/	
63-13419	123A	3122	20	TR21/735	
63-13438	123A	3122	20	TR21/735	
63-13440	123A	3122	20	TR21/735	
63-13441	123A			TR21/	
63-13839	160			TR17/	
63-13840	102A			TR85/	
63-13864	123A			TR21/	
63-13899	160			TR17/	
63-13926	132			FE100/	
63-13927	123A			TR21/	
63-13954	162			TR67/	
63-14032	123A			TR21/	
63-14051	123A			TR21/	

Original Part Number	ECG	SK	GE	IR/WEP	HEP
63-14052	123A			TR21/	
63-14057	123A			TR21/	
63-14135	727				
63-15345	718				
63-16918	102A		53	TR85/250	
63-17390	160		51	TR17/637	
63-18416	100	3005	51	TR05/254	
63-18418	160		51	TR17/637	
63-18419	160		51	TR17/637	
63-18420	102A		53	TR85/250	
63-18421	102A		53	TR85/250	
63-18423	160			TR17/	
63-18424	160		51	TR17/637	
63-18426		3045	27	TR /712	
63-18427	104	3009	3	TR01/624	
63-18430	102A		53	TR85/250	
63-18643	123A	3122	20	TR21/735	
63-19280	123A			TR21/	
63-19282	123A			TR21/	
63-23041	102A		53	TR85/250	
63-25179	102A		53	TR85/250	
63-25180	102A		53	TR85/250	
63-25181	102A		53	TR85/250	
63-25182	102A		53	TR85/250	
63-25261	158		53	TR84/630	
63-25281	102A		53	TR85/250	
63-25282	102A		53	TR85/250	
63-25720	102A		53	TR85/250	
63-25726	160		51	TR17/637	
63-25727	102A		53	TR85/250	
63-25728	102A		53	TR85/250	
63-25729	102A		53	TR85/250	
63-25942	102A		53	TR85/250	
63-25944	102A		53	TR85/250	
63-25946	100	3005	51	TR05/254	
63-26849	100	3005	51	TR05/254	
63-26850	160		51	TR17/637	
63-26851	102A		53	TR85/250	
63-27278	100	3005	51	TR05/254	
63-27279	100	3005	51	TR05/254	
63-27280	100	3005	51	TR05/254	
63-27281	102A		53	TR85/250	
63-27366	160		51	TR17/637	
63-27367	100	3005	51	TR05/254	
63-27500	160		51	TR17/637	
63-28348	160		51	TR17/637	
63-28358	160		51	TR17/637	
63-28390	102A		53	TR85/250	
63-28399	102A		53	TR85/250	
63-28426	154			TR78/	
63-29451	121	3009	3	TR01/232	
63-29459	121	3009	16	TR01/232	
63-29461	123A		20	TR21/735	
63-29661	160		51	TR17/637	
63-29662	100	3005	51	TR05/254	
63-29663	100	3005	51	TR05/254	
63-29664	100	3005	51	TR05/254	
63-29665	102		53	TR85/250	
63-29666	102A		53	TR85/250	
63-29819	160		51	TR17/637	
63-29820	160		51	TR17/637	
63-29821	160		51	TR17/637	
63-29862	160		51	TR17/637	
63-29863	100	3005	51	TR05/254	
63N1	163	3115	36	TR67/740	S5021
63N50	105	3012		TR03/	
63P3	102	3004	52	TR05/631	
64A-1A5L		3009			
64EPA					S0030
64EPB					S0015
64N1	162		35	TR67/707	S5020
64T1	100	3005	2	TR05/	G0005
65-08001			31MP		
65-080001	226MP	3086	31MP		G6016
65-1	123A	3122		TR21/735	S0015
65-1A4-7B		3006			
65-1-70	160	3006		TR17/	
65-1-70-12	160	3006		TR17/	
65-1-70-12-7	160	3006		TR17/	
65-1A	160	3006		TR17/	
65-1A0	160	3006		TR17/	
65-1A0R	160	3006		TR17/	
65-1A1	160	3006		TR17/	
65-1A2	160	3006		TR17/	
65-1A3	160	3006		TR17/	
65-1A3P	160	3006		TR17/	
65-1A4	160	3006		TR17/	
65-1A4-7	160	3006		TR17/	
65-1A4-7B	160			TR17/	
65-1A5	160	3006		TR17/	
65-1A5L	160	3006		TR17/	
65-1A6	160	3006		TR17/	
65-1A6-5	160	3006		TR17/	
65-1A7	160	3006		TR17/	
65-1A7-1	160	3006		TR17/	

Original Part Number	ECG	SK	GE	IR/WEP	HEP
65-1A8	160	3006		TR17/	
65-1A9	160	3006		TR17/	
65-1A9G	160	3006		TR17/	
65-1A19	160	3006		TR17/	
65-1A21	160	3006		TR17/	
65-1A82	160	3006		TR17/	
65-2A4-7B		3010			
65-2	103A	3010		TR08/	
65-2-70	103A	3010		TR08/	
65-2-70-12	103A	3010		TR08/	
65-2-70-12-7	103A	3010		TR08/	
65-2A	103A	3010		TR08/	
65-2A0	103A	3010		TR08/	
65-2A0R	103A	3010		TR08/	
65-2A1	103A	3010		TR08/	
65-2A2	103A	3010		TR08/	
65-2A3	103A	3010		TR08/	
65-2A3P	103A	3010		TR08/	
65-2A4	103A	3010		TR08/	
65-2A4-7	103A	3010		TR08/	
65-2A4-7B	103A			TR08/	
65-2A5	103A	3010		TR08/	
65-2A5L	103A	3010		TR08/	
65-2A6	103A	3010		TR08/	
65-2A6-1	103A	3010		TR08/	
65-2A7	103A	3010		TR08/	
65-2A7-1	103A	3010		TR08/	
65-2A8	103A	3010		TR08/	
65-2A9	103A	3010		TR08/	
65-2A9G	103A	3010		TR08/	
65-2A19	103A	3010		TR08/	
65-2A21	103A	3010		TR08/	
65-2A82	103A	3010		TR08/	
65-4	101	3011		TR08/	
65-4A4-7B		3011			
65-4-70	101	3011		TR08/	
65-4-70-12	101	3011		TR08/	
65-4-70-12-7	101	3011		TR08/	
65-4A	101	3011		TR08/	
65-4A0	101	3011		TR08/	
65-4A0R	101	3011		TR08/	
65-4A1	101	3011		TR08/	
65-4A2	101	3011		TR08/	
65-4A3	101	3011		TR08/	
65-4A3P	101	3011		TR08/	
65-4A4	101	3011		TR08/	
65-4A4-7	101	3011		TR08/	
65-4A4-7B	101			TR08/	
65-4A5	101	3011		TR08/	
65-4A5L	101	3011		TR08/	
65-4A6	101	3011		TR08/	
65-4A6-2	101	3011		TR08/	
65-4A7	101	3011		TR08/	
65-4A7-1	101	3011		TR08/	
65-4A8	101	3011		TR08/	
65-4A9	101	3011		TR08/	
65-4A9G	101	3011		TR08/	
65-4A19	101	3011		TR08/	
65-4A21	101	3011		TR08/	
65-4A82	101	3011		TR08/	
65-10	160	3006		TR17/	
65-11	160	3006		TR17/	
65-12	160	3006		TR17/	
65-13	160	3006		TR17/	
65-14	160	3006		TR17/	
65-15	160	3006		TR17/	
65-16	160	3006		TR17/	
65-17	160	3006		TR17/	
65-18	160	3006		TR17/	
65-19	160	3006		TR17/	
65-20	103A	3010		TR08/	
65-21	103A	3010		TR08/	
65-22	103A	3010		TR08/	
65-23	103A	3010		TR08/	
65-24	103A	3010		TR08/	
65-25	103A	3010		TR08/	
65-26	103A	3010		TR08/	
65-27	103A	3010		TR08/	
65-28	103A	3010		TR08/	
65-29	103A	3010		TR08/	
65-40	101	3011		TR08/	
65-41	101	3011		TR08/	
65-42	101	3011		TR08/	
65-43	101	3011		TR08/	
65-44	101	3011		TR08/	
65-45	101	3011		TR08/	
65-46	101	3011		TR08/	
65-47	101	3011		TR08/	
65-48	101	3011		TR08/	
65-49	101	3011		TR08/	
65A14-7B		3007			
65A	159	3114		TR20/717	S0019
65A0	160	3007		TR17/	
65A1	160	3007		TR17/	
65A2	160	3007		TR17/	

Original Part Number	ECG	SK	GE	IR/WEP	HEP
65A3	160	3007		TR17/	
65A4	160	3007		TR17/	
65A5	160	3007		TR17/	
65A6	160	3007		TR17/	
65A7	160	3007		TR17/	
65A8	160	3007		TR17/	
65A9	160	3007		TR17/	
65A10	160	3007		TR17/	
65A10R	160	3007		TR17/	
65A12	160	3007		TR17/	
65A13	160	3007		TR17/	
65A13P	160	3007		TR17/	
65A14	160	3007		TR17/	
65A14-7	160	3007		TR17/	
65A14-7B	160			TR17/	
65A15	160	3007		TR17/	
65A15L	160	3007		TR17/	
65A16	160	3007		TR17/	
65A16-3	160	3007		TR17/	
65A17	160	3007		TR17/	
65A17-1	160	3007		TR17/	
65A18	160	3007		TR17/	
65A19	160	3007		TR17/	
65A19G	160	3007		TR17/	
65A-70	160	3007		TR17/	
65A-70-12	160	3007		TR17/	
65A-70-12-7	160	3007		TR17/	
65A119	160	3007		TR17/	
65A121	160	3007		TR17/	
65A182	160	3007		TR17/	
65A11573	199				
65B14-7B		3008			
65B	159	3008		TR20/717	S0019
65B0		3008		TR17/	
65B1	126	3114		TR17/	
65B2		3008			
65B3	126	3008		TR17/	
65B4	126	3008		TR17/	
65B5	126	3008		TR17/	
65B6	126	3008		TR17/	
65B7	126	3008		TR17/	
65B8	126	3008		TR17/	
65B9	126	3008		TR17/	
65B10	126	3008		TR17/	
65B10R	126	3008		TR17/	
65B11	126	3008		TR17/	
65B12	126	3008		TR17/	
65B13	126	3008		TR17/	
65B13P	126	3008		TR17/	
65B14	126	3008		TR17/	
65B14-7	126	3008		TR17/	
65B14-7B	126			TR17/	
65B15	126	3008		TR17/	
65B15L	126	3008		TR17/	
65B16	126	3008		TR17/	
65B16-2	126	3008		TR17/	
65B17	126	3008		TR17/	
65B17-1	126	3008		TR17/	
65B18	126	3008		TR17/	
65B19	126	3008		TR17/	
65B19G	126	3008		TR17/	
65B-70	126	3008		TR17/	
65B-70-12	126	3008		TR17/	
65B-70-12-7	126	3008		TR17/	
65B119	126	3008		TR17/	
65B121	126	3008		TR17/	
65B182	126	3008		TR17/	
65C14-7B		3009			
65C	159	3114		TR20/717	S0019
65C0	121	3009		TR01/	
65C1	159	3009		TR20/	
65C2	121	3009		TR01/	
65C3	121	3009		TR01/	
65C4	121	3009		TR01/	
65C5	121	3009		TR01/	
65C6	121	3009		TR01/	
65C7	121	3009		TR01/	
65C8	121	3009		TR01/	
65C9	121	3009		TR01/	
65C10	121	3009		TR01/	
65C10R	121	3009		TR01/	
65C11	121	3009		TR01/	
65C12	121	3009		TR01/	
65C13	121	3009		TR01/	
65C13P	121	3009		TR01/	
65C14	121	3009		TR01/	
65C14-7	121	3009		TR01/	
65C14-7B	121			TR01/	
65C15	121	3009		TR01/	
65C15L	121	3009		TR01/	
65C16	121	3009		TR01/	
65C16-4	121	3009		TR01/	
65C17	121	3009		TR01/	
65C17-1	121	3009		TR01/	
65C18	121	3009		TR01/	

Original Part Number	ECG	SK	GE	IR/WEP	HEP
65C19	121	3009		TR01/	
65C19G	121	3009		TR01/	
65C-70	121	3009		TR01/	
65C-70-12	121	3009		TR01/	
65C-70-12-7	121	3009		TR01/	
65C119	121	3009		TR01/	
65C121	121	3009		TR01/	
65C182	121	3009		TR01/	
65D	159	3114		TR20/717	S0019
65D1	159	3114		TR20/	
65E	159	3114		TR20/717	S0019
65E1	159	3114		TR20/	
65F	159	3114		TR20/717	S0019
65F1	159	3114		TR20/	
65T1	102A	3123	2	TR85/250	G0005
65(TRANS)	107				
66-A14-7B		3003			
66-1	102A	3003		TR85/	
66-1-70	102A	3003		TR85/	
66-1-70-12	102A	3003		TR85/	
66-1-70-12-7	102A	3003		TR85/	
66-1A	102A	3003		TR85/	
66-A10	102A	3003		TR85/	
66-1A0R	102A	3003		TR85/	
66-1A1	102A	3003		TR85/	
66-1A2	102A	3003		TR85/	
66-1A3	102A	3003		TR85/	
66-1A3P	102A	3003		TR85/	
66-1A4	102A	3003		TR85/	
66-1A4-7	102A	3003		TR85/	
66-1A4-7B	102A			TR85/	
66-1A5	102A	3003		TR85/	
66-1A5L	102A	3003		TR85/	
66-1A6	102A	3003		TR85/	
66-1A6-3	102A	3003		TR85/	
66-1A7	102A	3003		TR85/	
66-1A7-1	102A	3003		TR85/	
66-1A8	102A	3003		TR85/	
66-1A9	102A	3003		TR85/	
66-1A9G	102A	3003		TR85/	
66-1A19	102A	3003		TR85/	
66-1A21	102A	3003		TR85/	
66-1A82	102A	3003		TR85/	
66-2A4-7B		3004			
66-2	102A	3004		TR85/	
66-2-70	102A	3004		TR85/	
66-2-70-12	102A	3004		TR85/	
66-2-70-12-7	102A	3004		TR85/	
66-2A	102A	3004		TR85/	
66-2A0	102A	3004		TR85/	
66-2A0R	102A	3004		TR85/	
66-2A1	102A	3004		TR85/	
66-2A2	102A	3004		TR85/	
66-2A3	102A	3004		TR85/	
66-2A3P	102A	3004		TR85/	
66-2A4	102A	3004		TR85/	
66-2A47	102A	3004		TR85/	
66-2A4-7B	102A			TR85/	
66-2A5	102A	3004		TR85/	
66-2A5L	102A	3004		TR85/	
66-2A6	102A	3004		TR85/	
66-2A6-4	102A	3004		TR85/	
66-2A7	102A	3004		TR85/	
66-2A7-1	102A	3004		TR85/	
66-2A8	102A	3004		TR85/	
66-2A9	102A	3004		TR85/	
66-2A9G	102A	3004		TR85/	
66-2A19	102A	3004		TR85/	
66-2A21	102A	3004		TR85/	
66-2A82	102A	3004		TR85/	
66-3A4-7B		3004			
66-3	102A	3004		TR85/	
66-3-70	102A	3004		TR85/	
66-3-70-12	102A	3004		TR85/	
66-3-70-12-7	102A	3004		TR85/	
66-3A	102A	3004		TR85/	
66-3A0	102A	3004		TR85/	
66-3A0R	102A	3004		TR85/	
66-3A1	102A	3004		TR85/	
66-3A2	102A	3004		TR85/	
66-3A3	102A	3004		TR85/	
66-3A3P	102A	3004		TR85/	
66-3A4	102A	3004		TR85/	
66-3A4-7	102A	3004		TR85/	
66-3A4-7B	102A			TR85/	
66-3A5	102A	3004		TR85/	
66-3A5L	102A	3004		TR85/	
66-3A6	102A	3004		TR85/	
66-3A6C	102A	3004		TR85/	
66-3A7	102A	3004		TR85/	
66-3A7-1	102A	3004		TR85/	
66-3A8	102A	3004		TR85/	
66-3A9	102A	3004		TR85/	
66-3A9G	102A	3004		TR85/	
66-3A19	102A	3004		TR85/	

Original Part Number	ECG	SK	GE	IR/WEP	HEP
66-3A21	102A	3004		TR85/	
66-3A82	102A	3004		TR85/	
66-10	102A	3003		TR85/	
66-11	102A	3003		TR85/	
66-12	102A	3003		TR85/	
66-13	102A	3003		TR85/	
66-14	102A	3003		TR85/	
66-15	102A	3003		TR85/	
66-16	102A	3003		TR85/	
66-17	102A	3003		TR85/	
66-18	102A	3003		TR85/	
66-19	102A	3003		TR85/	
66-20	102A	3004		TR85/	
66-21	102A	3004		TR85/	
66-22	102A	3004		TR85/	
66-23	102A	3004		TR85/	
66-24	102A	3004		TR85/	
66-25	102A	3004		TR85/	
66-26	102A	3004		TR85/	
66-27	102A	3004		TR85/	
66-28	102A	3004		TR85/	
66-29	102A	3004		TR85/	
66-30	102A	3004		TR85/	
66-31	102A	3004		TR85/	
66-32	102A	3004		TR85/	
66-33	102A	3004		TR85/	
66-34	102A	3004		TR85/	
66-35	102A	3004		TR85/	
66-36	102A	3004		TR85/	
66-37	102A	3004		TR85/	
66-38	102A	3004		TR85/	
66-39		3004			
66-6023	102A	3004		TR85/	G0007
66-6033	102A	3004		TR85/250	G0007
66-6023-00	102A	3004	2	TR85/250	
66-6024-00	102A	3004	2	TR85/250	
66-6025-00	102A	3004	2	TR85/250	
66-6026-00	102A	3004	2	TR85/250	
66-6027-00	102A	3004	2	TR85/250	
66-6028-00	102A	3004	2	TR85/250	
66-127119	123A				
66A00008A	123A				
66A00010A	123A				
66A10298	128				
66A10310	234				
66B-3A5L	121MP	3015			
66F015	704	3023			
66F020-1	157			TR60/	
66F020-2	157			TR60/	
66F021-1	161			TR83/	
66F022-1	161			TR83/	
66F023-1	159			TR20/	
66F024-1	159			TR20/	
66F025-1	128			TR87/	
66F026-1	199			TR21/	
66F027-1	123A			TR21/	
66F028-1	123A			TR21/	
66F029-1	123A			TR21/	
66F039-1	172			TR69/	
66F041-1	159			TR20/	
66F042-1	161			TR83/	
66F057-1	123A			TR21/	
66F057-2	123A			TR21/	
66F058-2	188	3199		TR72/	
66F074-1	171	3201		TR79/	
66F074-2	171	3201		TR79/	
66F074-3	171	3201		TR79/	
66F074-4	171	3201		TR79/	
66F077-1	712	3072			
66F085				TR57/	
66F086				TR57/	
66F125-1	792				
66F136-1	793				
66F175-1	808				
66F176-1	809				
66F1751	808				
66F1761	809				
66-P11I20	159				
66-P1112-0001	199				
66-P11139	129				
66-P11141	129				
66S00000A	128				
67A8926	234				
67A9060	199				
67P1	121	3009	16	TR01/232	
67P1C	121				
67P2	121	3009	16	TR01/232	
67P2C	121				
67P3	121	3009	16	TR01/232	
67P3C	121				
68-110-02	159				
68-20102-2701	724				
68A7349-D32	9932				
68A7349-D46	9946				
68A7349-D62	9962				

Original Part Number	ECG	SK	GE	IR/WEP	HEP
68A7349PD30	9930				
68A7349-PD32	9932				
68A7349PD36	9936				
68A7349PD45	9945				
68A7349PD46	9946				
68A7349PD62	9962				
68A7355P1	234				
68A7358P1		3021			
68A7363P1		3044			
68A7366-1	199				
68A7368	128				
68A7370-1	159				
68A7370-P3	159				
68A7380-1	123A				
68A7380-2	128				
68A7382-1	159				
68A7652P1	9109				
68A7652P2	9110				
68A7657P1		3044			
68A7701P1		3025			
68A7707P1	194				
68A7711-1		3512			
68A7711-3		3512			
68A7711P2		3513			
68A7711P3		3512			
68A7715P1	123A				
68A7734P1	159				
68A7860-P1	923				
68A8318-P1	159				
68A8321	123A				
68A9025	7400				
68A9026	74H00				
68A9027	7402				
68A9028	7404				
68A9030	7410				
68A9031	74H10				
68A9032	7406				
68A9033	7420				
68A9034	7423				
68A9035	7430				
68A9036	7437				
68A9037	7438				
68A9038	7450				
68A9040	7460				
68A9041	7475				
68A9042	7476				
68A9047	74H106				
68A9048	74151				
68A9049	74153				
68A8319001	130				
68P1	121	3009	16	TR01/232	
68P1B	121	3009	16	TR01/232	G6005
68P-2A5L	102A	3004		TR85/	
69-1810		3018	17		S0016
69-1811		3018	17		S0020
69-1812		3018	17		S0030
69-1813		3018	17		S0016
69-1814		3020	18		S0003
69-1815		3114	21		S0019
69-1816		3122	62		S0024
69-1817		3114	21		S0019
69-1818		3054	57		S5000
69-1819		3083	58		S5006
69-2401	732				
69-2403	732				
69-3116	732				
69AJ110	218			TR58/	
69B1Z	715	3076			
69B2Z	715	3076			
69L-4A5L	102A	3003		TR85/	
69N1	108	3018	17	TR95/56	S0020
69SP112	129			TR88/	
70.00.730	126		50	TR17/	
70.01.704	123A		62	TR51/	
70-943-722-001	123A				
70-943-754-002	123A				
70-943-762-001	123A				
70-943-773-001	234				
70-943-780001		3535			
70-943-780002		3511			
70-270050		3031			
70B1Z	714	3075			
70N1	123	3124	20	TR86/53	S0015
70N2	123	3124	20	TR86/53	S0015
70N3	107			TR70/	S0015
70N4	123A	3124	10	TR21/735	S0015
71-126268	123A	3122	20	TR21/735	
71-70177A02	748				
71-70177A03	738				
71-70177A05	749				
71-70177A07	767				
71-70177B02	743				
71D70177A02	748				
71D70177A03	738	3167			
71M70177A02	748				

Original Part Number	ECG	SK	GE	IR/WEP	HEP
71M70177A03	738	3167			
71N1	124	3021	12	TR81/240	S5011
71N1B	123A			TR21/	
71N1T	124	3021	12	TR81/240	S5011
71N2	124	3021	12	TR81/240	
71N2T	124	3021	12	TR23/	S5011
72-27200-5	417				
72-27200-6	418				
72-27200-7	419				
72-27200-8	420				
72-34063-1	409				
72-34063-2	416				
72-34063-5	407				
72N1	108	3018	20	TR95/56	
72N1B	128	3024		TR87/243	S0033
72N2	108	3018	20	TR95/56	
72N2B	123A	3122		TR21/735	S0033
73A01	176	3123		TR82/	G0005
73A02	176	3123		TR82/	G0005
73B-140-003-5	123A				
73B140-004	127				
73B-140-005-1	159				
73B-140-005-4	159				
73B140385-001	128				
73B140585-21	194				
73B140585-22	194				
73B140585-23	194				
73B140585-24	194				
73B140585-25	194				
73B140585-26	194				
73B140585-27	194				
73C18028-11	152				
73C18038-1		3168			
73C18038-2		3168			
73C180475	712	3072	2	IC507/	C6063P
73C180475-1	712	3072			
73C180475-4	712	3072			
73C180475-7	712	3072			
73C180475-8	712	3072			
73C180475-9	712	3072			
73C180476-5	712	3072		IC507/	
73C180497-4	154				
73C180499-5	154				
73C180499-6	154				
73C180803-2		3168			
73C180828-12	152				
73C180829-11	152				
73C180829-12	152	3054			
73C180830-11	153				
73C180830-12	153	3083			
73C180831-1	159				
73C180831-2	159				
73C180831-3	234				
73C180837-1	713	3077	5	IC508/	C6057P
73C180837-2	713	3077	5	IC508/	C6057P
73C180837P2	713	3077			
73C180838-1	749	3168			
73C180838-2	749				
73C180838-3	749				
73C180843-1	783		21		C6100P
73C180843-2	783		21		C6100P
73C180843-3	783	3141	21		C6100P
73C180843-4	783		21		C6100P
73C182077	728	3073			
73C182080-33	154				
73C182081-31	123A				
73C182088-31	154				
73C182186	712		2		C6063P
73C182186-1	712	3072			
73C182186-2	712	3072			
73C182186-3	712	3072			
73C182186-4	712	3072			
73C182763-1			30		
73C182763-2			30		
73C182764-1			29		
73C182764-2			29		
73C182764-4			29		
73C180475004	712	3072			
73N1	161	3018	20	TR83/719	S0015
73N1B	123A	3122		TR21/735	S0025
73N51	105	3012		TR03/	
74	108	3039	11	TR95/56	
74-01-772	199				
74-3	100	3005		TR05/	
74-3-70	100	3005		TR05/	
74-3-70-12	100	3005		TR05/	
74-3-70-12-7	100	3005		TR05/	
74-3A	100	3005		TR05/	
74-3A0	100	3005		TR05/	
74-3A0R	100	3005		TR05/	
74-3A1	100	3005		TR05/	
74-3A2	100	3005		TR05/	
74-3A3	100	3005		TR05/	
74-3A3P	100	3005		TR05/	
74-3A4	100	3005		TR05/	

Original Part Number	ECG	SK	GE	IR/WEP	HEP
74-3A4-7	100	3005		TR05/	
74-3A4-7B	100			TR05/	
74-3A5	100	3005		TR05/	
74-3A5L	100	3005		TR05/	
74-3A6	100	3005		TR05/	
74-3A6-3	100	3005		TR05/	
74-3A7	100	3005		TR05/	
74-3A7-1	100	3005		TR05/	
74-3A8	100	3005		TR05/	
74-3A9	100	3005		TR05/	
74-3A9G	100	3005		TR05/	
74-3A19	100	3005		TR05/	
74-3A21	100	3005		TR05/	
74-3A82	100	3005		TR05/	
74-30	100	3005		TR05/	
74-31	100	3005		TR05/	
74-32	100	3005		TR05/	
74-33	100	3005		TR05/	
74-34	100	3005		TR05/	
74-35	100	3005		TR05/	
74-36	100	3005		TR05/	
74-37	100	3005		TR05/	
74-38	100	3005		TR05/	
74-39	100	3005		TR05/	
74A01	176	3123		TR82/	G0005
74A02	176	3123		TR82/	G0005
74A03	176	3123		TR82/	G0005
74H00PC	74H00				
74H01PC	74H01				
74H04PC	74H04				
74H05PC	74H05				
74H08PC	74H08				
74H10PC	74H10				
74H11PC	74H11				
74H20PC	74H20				
74H21PC	74H21				
74H22PC	74H22				
74H30PC	74H30				
74H40PC	74H40				
74H50PC	74H50				
74H51PC	74H51				
74H52PC	74H52				
74H53PC	74H53				
74H54PC	74H54				
74H55PC	74H55				
74H60PC	74H60				
74H61PC	74H61				
74H62PC	74H62				
74H71PC	74H71				
74H72PC	74H72				
74H73PC	74H73				
74H74PC	74H74				
74H76PC	74H76				
74H78PC	74H78				
74H87DC	74H87				
74H87PC	74H87				
74H101DC	74H101				
74H102DC	74H102				
74H103PC	74H103				
74H106PC	74H106				
74H108PC	74H108				
74H183DC	74H183				
74N1	123A	3122	20	TR21/735	
74P1	159	3025	21	TR20/717	S0019
74P1M	129		21	TR88/	S5013
74Q1262	160	3004	52	TR17/637	
74S00PC	74S00				
74S03PC	74S03				
74S04PC	74S04				
74S05PC	74S05				
74S10PC	74S10				
74S11PC	74S11				
74S15PC	74S15				
74S16DC	74S16				
74S16PC	74S16				
74S20PC	74S20				
74S22PC	74S22				
74S40PC	74S40				
74S64PC	74S64				
74S65PC	74S65				
74S74PC	74S74				
74S86PC	74S86				
74S112PC	74S112				
74S113PC	74S113				
74S114PC	74S114				
74S133PC	74S133				
74S134PC	74S134				
74S135PC	74S135				
74S138DC	74S138				
74S139DC	74S139				
74S140PC	74S140				
74S151DC	74S151				
74S153PC	74S153				
74S157PC	74S157				
74S158PC	74S158				
74S175PC	74S175				
74S194PC	74S194				
74S196A	74S196				
74S251DC	74S251				
74S258DC	74S258				
75-461	102A		53	TR85/250	
75N1	154	3124	20	TR78/712	S5025
75N5AA	123A	3122		TR21/735	
76	126	3006		TR17/635	G0003
76-0105	175		66	TR81/241	
76-1	159	3118		TR20/717	S3021
76-11770	104	3009	3	TR01/624	
76-13570-39	108	3019	17	TR95/56	S0016
76-13570-59	108	3019	17	TR95/56	S0016
76-13866-17	108	3018	17	TR95/56	
76-13866-18	108	3018	17	TR95/56	
76-13866-19	108	3018	17	TR95/56	S0020
76-13866-20	108	3039	20	TR95/56	
76-13866-59	108	3039	20	TR95/56	
76-13866-62	108	3018	17	TR95/56	
76-14090-1	128	3024	18	TR87/243	
76B1Z	742				
76N1	123A	3124	20	TR21/735	S0015
76N1B	128	3024		TR87/243	S0015
76N1M	123A	3124	17	TR21/735	S0015
76N2	123A	3124	20	TR21/735	S0015
76N2B	128	3024		TR87/243	S0015
76N3B	128	3024		TR87/243	S0015
76N2369-000	123A				
76N2369-001	123A				
76S1030-000	128				
77	100	3005		TR05/254	G0003
77-27166-2	160			TR17/	
77-27198-3	199				
77-270877-2	121		3	TR01/	
77-270878-2	121			TR01/	
77-271025-1	102A		53	TR85/	
77-271026-1	102A		52	TR85/	
77-271027-1	102A			TR85/	
77-271029-1	160		51	TR17/637	
77-271029-2	160		51	TR17/637	
77-271036-1	102A		52	TR85/	
77-271037-1	102A		52	TR85/	
77-271038-1	160		51	TR17/637	
77-271039-1	102A		53	TR85/250	
77-271166-3	160		51	TR17/637	
77-271453-1	123A		20	TR21/735	
77-271490	128	3024		TR87/	
77-271490-1	128	3024	18	TR87/243	
77-271491-1	121	3009	16	TR01/232	
77-271798-1	184		62		
77-271798-2	199				
77-271798-3	184		62		
77-271818-1	129		21	TR20/	
77-271819-1	123A				
77-271967-1	123A		60	TR21/	
77-272913-1	152		63	TR76/	
77-272914-1	153		21	TR77/	
77-272999-1	128		20	TR87/	
77-273001-2	123A		20	TR21/	
77-273001-3	160		1	TR17/	
77-273004-1	158		53	TR84/	
77-273715-1	152			TR76/	
77-273716-1	153			TR77/	
77-273738-1	152		28	TR76/	
77-273739-1	153		29	TR77/	
77A	105	3012		TR03/	
77A0	105	3012		TR03/	
77A1	105	3012		TR03/	
77A2	105	3012		TR03/	
77A3	105	3012		TR03/	
77A4	105	3012		TR03/	
77A5	105	3012		TR03/	
77A6	105	3012		TR03/	
77A7	105	3012		TR03/	
77A8	105	3012		TR03/	
77A9	105	3012		TR03/	
77A10	105	3012		TR03/	
77A10R	105	3012		TR03/	
77A11	105	3012		TR03/	
77A12	105	3012		TR03/	
77A13	105	3012		TR03/	
77A13P	105	3012		TR03/	
77A14	105	3012		TR03/	
77A14-7	105	3012		TR03/	
77A14-7B	105			TR03/	
77A15	105	3012		TR03/	
77A15L	105	3012		TR03/	
77A16	105	3012		TR03/	
77A16-1	105	3012		TR03/	
77A17	105	3012		TR03/	
77A17-1	105	3012		TR03/	
77A18	105	3012		TR03/	
77A19	105	3012		TR03/	
77A19G	105	3012		TR03/	

Original Part Number	ECG	SK	GE	IR/WEP	HEP
77A-70	105	3012		TR03/	
77A-70-12	105	3012		TR03/	
77A-70-12-7	105	3012		TR03/	
77A119	105	3012		TR03/	
77A121	105	3012		TR03/	
77A182	105	3012		TR03/	
77B	121MP	3013			
77B0	121MP	3013			
77B1	121MP	3013			
77B2	121MP	3013			
77B3	121MP	3013			
77B4	121MP	3013			
77B5	121MP	3013			
77B6	121MP	3013			
77B7	121MP	3013			
77B8	121MP	3013			
77B9	121MP	3013			
77B10	121MP	3013			
77B10R	121MP	3013			
77B11	121MP	3013			
77B12	121MP	3013			
77B12(IC)	788				
77B13	121MP	3013			
77B13P	121MP	3013			
77B14	121MP	3013			
77B14-7	121MP	3013			
77B14-7B	121MP				
77B15	121MP	3013			
77B15L	121MP	3013			
77B16	121MP	3013			
77B16-2	121MP	3013			
77B17	121MP	3013			
77B17-1	121MP	3013			
77B18	121MP	3013			
77B19	121MP	3013			
77B19G	121MP	3013			
77B-70	121MP	3013			
77B-70-12	121MP	3013			
77B-70-12-7	121MP	3013			
77B119	121MP	3013			
77B121	121MP	3013			
77B182	121MP	3013			
77C	121	3014		TR01/	
77C0	121	3014		TR01/	
77C1	121	3014		TR01/	
77C2	121	3014		TR01/	
77C3	121	3014		TR01/	
77C4	121	3014		TR01/	
77C5	121	3014		TR01/	
77C6	121	3014		TR01/	
77C7	121	3014		TR01/	
77C8	121	3014		TR01/	
77C9	121	3014		TR01/	
77C10	121	3014		TR01/	
77C10R	121	3014		TR01/	
77C11	121	3014		TR01/	
77C12	121	3014		TR01/	
77C13	121	3014		TR01/	
77C13P	121	3014		TR01/	
77C14	121	3014		TR01/	
77C14-7	121	3014		TR01/	
77C14-7B	121			TR01/	
77C15	121	3014		TR01/	
77C15L	121	3014		TR01/	
77C16	121	3014		TR01/	
77C17	121	3014		TR01/	
77C17-1	121	3014		TR01/	
77C18	121	3014		TR01/	
77C19	121	3014		TR01/	
77C19G	121	3014		TR01/	
77C-70	121	3014		TR01/	
77C-70-12	121	3014		TR01/	
77C-70-12-7	121	3014		TR01/	
77C119	121	3014		TR01/	
77C121	121	3014		TR01/	
77C182	121	3014		TR01/	
77C800-005	9930				
77C800-007	9932				
77C800-008	9946				
77C813-002	9945				
77C10891-2	941				
77N1	123A	3124	20	TR21/735	S0015
77N2	123A	3124	20	TR21/735	S0015
77N2B	128	3024		TR87/243	S0015
77N3	123	3124	18	TR86/53	
77N4	123A	3024	18	TR21/735	S0015
77N5	123A	3024		TR21/	S0015
77N5	123A	3122	20	TR21/	
78				TR05/	
78-001	231				
78-5009	121	3009	16	TR01/232	
78-272212-1	121			TR01/	
78A200010P4	7400				
78BLK	100	3005		TR05/254	G0003
78C01	108	3018		TR95/56	S0016

Original Part Number	ECG	SK	GE	IR/WEP	HEP
78C02	108	3018		TR95/56	S0016
78E67-2	199				
78E74-1		3024			
78E88-3		3024			
78GRN	100	3005		TR05/254	G0003
78N1	123A	3124	20	TR21/735	S0015
78N2B	123A	3122		TR21/735	S0015
78RED	100	3005		TR05/254	G0003
78YEL	100	3005		TR05/254	G0003
79F114-1	128	3024		TR87/243	
79F114-2A	129	3025		TR88/242	
79F114-3	128	3024		TR87/243	
79F114-4	129	3025		TR88/242	
79F114-4A	129	3025		TR88/242	
79P1	102A	3004	51	TR85/250	S0019
80-050100	123A	3124		TR21/	
80-050300	103A	3010		TR08/	
80-050400	160	3006		TR17/	
80-050500	160	3008		TR17/	
80-050600	103A	3010		TR08/	
80-050700	121	3009		TR01/	
80-051500	123A	3124		TR21/	
80-051600	129	3025		TR88/	
80-052102	123A	3124		TR21/	
80-052202	123A	3124		TR21/	
80-052302	123A	3124		TR21/	
80-052402	130	3027		TR59/	
80-052600	123A	3124		TR21/	
80-052700	129	3025		TR88/	
80-052800	103A	3010		TR08/	
80-053001	123A	3124		TR21/	
80-053300	128	3024		TR87/	
80-053400	123A	3124		TR21/	
80-053600	108	3018		TR95/	
80-053702	130MP	3029		TR59MP/	
80-308-2	123A	3124		TR21/	
80-104900	126	3008		TR17/	
80-105200	126	3008		TR17/	
80-105300	126	3008		TR17/	
80-205400	102A	3004		TR85/	
80-205600	102A	3004		TR85/	
80-236400	102A	3004		TR85/	
80-236430	102A	3004		TR85/	
80-243900	102A	3004		TR85/	
80-318250	123A	3124		TR21/	
80-337390	123A	3124		TR21/	
80-338030	108	3018		TR95/	
80-338040	108	3018		TR95/	
80-339430	108	3018		TR95/	
80-339440	108	3018		TR95/	
80-383840	108	3018		TR95/	
80-383930	108	3018		TR95/	
80-383940	123A	3124		TR21/	
80-389910	123A	3124		TR21/	
80-389930	123A	3124		TR21/	
80-414120	130	3027		TR59/	
80-414130	130	3027		TR59/	
80-421980	123A	3124		TR21/	
80-2226314	102A	3004		TR85/	
80P1	102A	3004	51	TR85/250	G0005
80P2	129			TR88/	
80P2B	129			TR88/	
80P3	129			TR88/	
80P3B	129			TR88/	
80T2				TR63/	G0005
81-23860400-3	188	3054		TR72/W3020	
81-23860400A	188	3054		TR72/	
81-23860400B	188	3054		TR72/	
81-27125140-7	123A	3124		TR21/735	S0015
81-27125140-7A	123A	3124		TR21/735	
81-27125140-7B	123A	3124		TR21/735	
81-27125160-5	123A	3124		TR21/735	S0015
81-27125160-5A	123A	3124		TR21/	
81-27125160-5B	123A	3124		TR21/	
81-27125270-2	123A	3124		TR21/735	S0015
81-27125270-2A	123A	3124		TR21/	
81-27125270-2B	123A	3124		TR21/	
81-27125300-7	123A	3124		TR21/735	S0015
81-27125530-9	188	3122		TR72/W3020	S0015
81-27125530-9A	188	3122		TR72/	
81-27125530-9B	188	3122		TR72/	
81-27126100-0	188	3199		TR72/W3020	
81-27126130-7	121	3014		TR01/232	G6005
81-27126130-7A	121	3014		TR01/	
81-27126130-7B	121	3014		TR01/	
81-46125001-1	126	3007		TR17/635	G0009
81-46125002-9	102A	3004		TR85/250	G0005
81-46125003-7	102A	3004		TR85/250	G0007
81-46125004-5	102A	3004		TR85/250	G0005
81-46125005-2	158	3004		TR84/630	G6011
81-46125006-0	108	3018		TR95/56	S0016
81-46125007-8	107	3018		TR70/720	S0016
81-46125009-4	102A	3004	2	TR85/250	G0005
81-46125010-2	102A	3004	2	TR85/250	G0005
81-46125011-0	102A	3004	53	TR85/250	G0005

Original Part Number	ECG	SK	GE	IR/WEP	HEP
81-46125012-8	107	3018	20	TR70/720	S0011
81-46125013-6	107	3018	17	TR70/720	S0016
81-46125016-9	123A	3020	20	TR24/	S0015
81-46125018-5	158	3004	53		G6011
81-46125019-3	123A	3018	62	TR21/	S0015
81-46125026-8	123A	3124	20	TR51/	S0014
81-46125027-6	123A	3124	20	TR51/	S0014
81-46125028-4	121	3004	53	TR84/	G0005
81-46125029-2	102A	3004	52	TR85/	G0005
81-46125030-0	107	3018	20	TR70/	S0016
81-46125032-6	107	3018	20	TR70/	
81-46125033-4	107	3018	20	TR70/	S0014
81-46128001-8	1097				
81-46128002-6	1110				
81P3	129		21	TR88/	
81T2	128	3024	63	TR87/243	S3001
81X0042-100	184(CP)				
82-409501	123A	3124		TR21/	
82-410300	129	3025		TR88/	
82M432B2	915				
82M667B2	925				
82T1					G6011
82T2					S3020
83	129				
83-1056	121	3009	3	TR01/121	G6003
83N52	105	3012		TR03/	
83P1	159	3114	21	TR20/717	S0019
83P1A	159	3114		TR20/	
83P1B	159	3114		TR20/717	S0019
83P1BC	159	3114		TR20/	
83P1M	159	3114		TR20/717	S0019
83P1MC	159	3114		TR20/	
83P2	159	3114	20	TR20/	S0019
83P2A	159	3114		TR20/	
83P2AA	159	3114	22	TR20/717	S0019
83P2AA1	159	3114		TR20/	
83P2B	129	3025		TR88/242	S0019
83P2M	159	3114	22	TR20/717	S0019
83P2M1	159	3114		TR20/	
83P2N	159	3025		TR20/717	S0019
83P3	159	3114		TR20/717	S0019
83P3A	159	3114		TR20/	
83P3AA	159	3114		TR20/717	
83P3AA1	159	3114		TR20/	
83P3B	159	3114		TR20/717	S0019
83P3B1	159	3114		TR20/	
83P3M	159	3025		TR20/717	S0031
83P3M1	159	3114		TR20/	
83P4		3114			S0019
83T2					S3020
84	121	3009		TR01/232	G6003
84A	121	3009		TR01/	
84A0	121	3009		TR01/	
84A1	121	3009		TR01/	
84A2	121	3009		TR01/	
84A3	121	3009		TR01/	
84A4	121	3009		TR01/	
84A5	121	3009		TR01/	
84A6	121	3009		TR01/	
84A7	121	3009		TR01/	
84A8	121	3009		TR01/	
84A9	121	3009		TR01/	
84A10	121	3009		TR01/	
84A10R	121	3009		TR01/	
84A12	121	3009		TR01/	
84A13	121	3009		TR01/	
84A13P	121	3009		TR01/	
84A14	121	3009		TR01/	
84A14-7	121	3009		TR01/	
84A14-7B	121			TR01/	
84A15	121	3009		TR01/	
84A15L	121	3009		TR01/	
84A16	121	3009		TR01/	
84A16B	121	3009		TR01/	
84A17	121	3009		TR01/	
84A17-1	121	3009		TR01/	
84A18	121	3009		TR01/	
84A19	121	3009		TR01/	
84A19G	121	3009		TR01/	
84A20	121	3009		TR01/	
84A-70	121	3009		TR01/	
84A-70-12	121	3009		TR01/	
84A-70-12-7	121	3009		TR01/	
84A121	121	3009		TR01/	
84A182	121	3009		TR01/	
84AA1	121	3009		TR01/	
84AA19	121	3009		TR01/	
84B	121	3009		TR01/232	G6018
84C01	157			TR60/	S5015
84G01	157			TR60/244	
84S157PC	74S157				
85-370-2BLU	121			TR01/	
85C04	123A			TR21/	S0015
86-0007-004	123A			/735	
86-0022-001				/735	

Original Part Number	ECG	SK	GE	IR/WEP	HEP
86-0029-001	123A			/735	
86-0031-001	123A			/735	
86-0033-007	131MP			/642MP	
86-0036-001	159			/717	
86-005135-2	199				
86-4-2	101	3011	8	TR08/641	G0011
86-5-2	103A			TR08/735	G0011
86-6-2	103A	3010	8	TR08/735	G0011
86-601-2	192				
86-8-2	121	3009	16	TR01/232	G6003
86-10-2	101	3088			
86-11-2	101	3010	8	TR08/641	G0011
86-12-2	101	3010	8	TR08/641	G0011
86-13-2	103A	3010	8	TR08/724	G0011
86-14-2	103A	3010	8	TR08/724	G0011
86-16-2	102A	3004	8	TR85/250	G0005
86-18-2	160	3007	50	TR17/637	G0002
86-19-2	121	3009	16	TR01/232	G6003
86-20-2	160	3008	50	TR17/637	G0003
86-21-2	102A	3005	2	TR85/250	G0005
86-22-2	102A	3005	2	TR85/250	G0005
86-23-2	102A	3003	52	TR85/250	G0005
86-24-2	103A	3010	8	TR08/724	G0011
86-25-2	103A	3010	8	TR08/724	G0011
86-26-2	101	3011	8	TR08/641	G0011
86-27-2	102A	3005	2	TR85/250	G0005
86-28-2	102A	3005	2	TR85/250	G0005
86-29-2	102A	3003	52	TR85/250	G0005
86-30-2	102A	3004	52	TR85/250	G0005
86-31-2	101	3011	8	TR08/641	G0011
86-32-2	102A	3005	2	TR85/250	G0005
86-33-2	102A	3005	2	TR85/250	G0005
86-35-2	103A	3010	8	TR08/724	G0011
86-36-2	160	3007	50	TR17/637	G0002
86-37-2	160	3007	50	TR17/637	G0002
86-38-2	160	3007	50	TR17/637	G0002
86-39-2	102A	3003	52	TR85/250	G0005
86-42-1	123A			TR21/735	
86-44-1	101	3011		TR08/641	G0011
86-45-2	102A	3003	52	TR85/250	G0005
86-46-2	102A	3005	2	TR85/250	G0005
86-47-2	102A	3005	2	TR85/250	G0005
86-48-2	102A	3005	2	TR85/250	G0005
86-49-2	102A	3003	52	TR85/250	G0005
86-50-2	102A	3004	52	TR85/250	G0005
86-54-2	102A			TR85/250	G0005
86-58-2	123A	3010	8	TR21/735	S0015
86-59-2	102A	3005	2	TR85/250	G0005
86-60-2	102A	3005	2	TR85/250	G0005
86-61-2	102A	3003	52	TR85/250	G0005
86-62-2	121	3009	16	TR01/232	G6003
86-63-2	121	3009	16	TR01/232	G6003
86-72-2	102A	3003	52	TR85/250	G0005
86-73-2	102A	3005	2	TR85/250	G0005
86-74-2	102A	3003	52	TR85/250	G0005
86-75-2	102A	3005	2	TR85/250	G0005
86-76-2	103A	3010	8	TR08/724	G0011
86-77-2	102A	3004	52	TR85/250	G0005
86-78-2	102A	3005	2	TR85/250	G0005
86-79-2	102A	3005	2	TR85/250	G0005
86-80-2	102A	3004	52	TR85/250	G0005
86-81-2	103A	3011	8	TR08/724	G0011
86-82-2	102A	3004	52	TR85/250	G0005
86-83-2	102A	3004	52	TR85/250	G0005
86-84-2	102A	3004	52	TR85/250	G0005
86-86-2	160	3008	50	TR17/637	G0003
86-87-2	160	3008	50	TR17/637	G0003
86-88-2	160	3006	50	TR17/637	G0002
86-89-2	160	3006	50	TR17/637	G0003
86-90-2	160	3006	50	TR17/637	G0002
86-92-2	160	3006	50	TR17/637	G0002
86-92-2	160	3006	50	TR17/637	G0002
86-93-2	102A	3004	52	TR85/250	G0005
86-95-2	102A	3123	52	TR85/250	G0005
86-98-2	102A	3004	52	TR85/250	
86-99-2	160	3008	50	TR17/637	
86-100-2	160	3006	50	TR17/637	G0003
86-101-2	160	3008	50	TR17/637	G0003
86-102-2	160	3006	50	TR17/637	G0002
86-103-2	102A			TR85/250	G0005
86-107-2	160			TR17/637	G0003
86-108-2	126	3008	50	TR17/635	G0003
86-109-2	126	3008	50	TR17/635	G0005
86-110-2	123A	3124	20	TR21/735	S0015
86-111-2	126	3008	50	TR17/635	G0005
86-112-2	160	3006	50	TR17/637	G0002
86-114-2	102A	3004	52	TR85/250	G0005
86-116-2	102A	3008	2	TR85/250	G0005
86-116-2	126	3008	2	TR85/250	G0005
86-117-2	160	3006	50	TR17/637	G0002
86-119-2	123A	3006	50	TR21/735	S0015
86-120-2	121	3009	16	TR01/232	G6003
86-121-2	128/411			/243	S3024
86-123-2	123A	3124	20	TR21/735	S0015
86-126-2	102A	3004	52	TR85/250	G0005

Original Part Number	ECG	SK	GE	IR/WEP	HEP
86-127-2	121	3009	16	TR01/232	G6005
86-128-2	102A	3004	52	TR85/250	G0005
86-129-2	102A	3004	52	TR85/250	G0005
86-130-2	102A	3004	52	TR85/250	G0005
86-131-2	102A	3008	51	TR85/250	G0005
86-132-2	102A	3006	2	TR85/250	G0005
86-133-2	102A	3005	2	TR85/250	G0005
86-135-2	160	3006	50	TR17/637	G0002
86-136-2	160	3006	50	TR17/637	G0003
86-138-2	108	3019	20	TR95/56	S0020
86-139-2	123A	3124	20	TR21/735	S0015
86-141-2	121	3009	16	TR01/232	G0005
86-142-2	121	3009	16	TR01/232	G6005
86-143-2	123A	3124	20	TR21/735	S0015
86-144-2	123A	3024		TR21/735	S0015
86-146-2	121	3009	16	TR01/232	G6005
86-147-2	121	3009	16	TR01/232	G6005
86-149-2	160	3006	50	TR01/637	G0003
86-150-2	160	3006	50	TR17/637	G0003
86-151-2	160	3007	50	TR17/637	G0003
86-152-2	102A			TR85/250	G0005
86-155-2	123A	3124	20	TR21/735	S0015
86-156-2	102A	3004	2	TR85/250	G0008
86-156-2A	102A	3004	2	TR05/631	
86-157-2		3124	20	TR21/735	S0015
86-158-2	123A	3124	20	TR21/735	S0015
86-159-2	102A	3004	52	TR85/250	G0005
86-160-2	129/411		53	/242	G0005
86-161-2	128	3124	20	TR21/735	S0005
86-162-2	160	3006	20	TR17/637	G0003
86-163-2	160	3006	50	TR17/637	G0003
86-164-2	160	3006	51	TR17/637	G0003
86-165-2	157			TR60/244	S3021
86-166-2	123A	3122		TR21/735	S0005
86-169-2	102A	3004	52	TR85/250	G0008
86-170-2	128	3124	20	TR87/243	
86-171-2	123A	3124	20	TR21/735	S0015
86-172-2	102A			TR85/250	G0005
86-173-2	121	3009	16	TR85/232	G6003(TWO)
86-173-9	121	3009	25		
86-175-2	123A	3124	20	TR21/735	S0015
86-176-2	102A			TR85/250	G0005
86-177-2	157	3021	12	TR60/244	S3021
86-178-2	159	3114	21	TR20/717	G0005
86-178-20	159	3114		TR20/	
86-179-2	160	3006	50	TR17/637	G0003
86-180-2	160	3006	50	TR17/637	G0003
86-181-2	160	3006	50	TR17/637	G0003
86-182-2	123A	3122		TR21/735	S0005
86-183-2	159	3114		TR20/717	S0019
86-183-20	159	3114		TR20/	
86-185-2	161	3018	11	TR83/719	S0015
86-186-2	161	3117		TR83/719	S0015
86-188-2	123A	3124	20	TR21/735	S0016
86-189-2	123A	3122		TR21/735	S0015
86-190-2	123A				
86-191-2	123A	3124	20	TR21/735	S0015
86-192-2	123A	3122		TR21/735	S0015
86-193-2	123A				
86-194-2	123A	3124	20	TR21/735	S0015
86-195-2	123A	3122		TR21/735	S0015
86-196-2	123A	3122		TR21/735	S0015
86-197-2	123A				
86-198-2	123A	3122		TR21/735	S0015
86-199-2	123A	3124	20	TR21/735	S0030
86-200-2					S0015
86-201-2	123A	3124	20	TR21/735	S0015
86-202-2	123A	3122		TR21/735	S0015
86-204-2	161	3018	11	TR83/719	S0015
86-205-2	161	3117		TR83/719	S0015
86-207-2	128	3024		TR87/243	S0016
86-208-2	128	3024	11	TR87/243	S0016
86-210-2	128	3024		TR87/243	S0019
86-211-2	128	3024		TR87/243	S5026
86-212-2	194				
86-213-2	154	3045		TR78/712	S5025
86-214-2	154	3045		TR78/712	S5025
86-215-2	154	3045		TR78/712	S5025
86-216-2	159				
86-217-2	159	3114		TR20/717	S0019
86-217-20	159	3114		TR20/	
86-218-2	159	3114		TR20/717	S0019
86-218-20	159	3114		TR20/	
86-219-2	159	3114	20	TR20/717	S0019
86-221-2	163	3111	19	TR67/740	S5020
86-222-2	163	3111		TR67/740	S5020
86-224-2	163	3111		TR67/740	S5020
86-225-2	163	3111		TR67/740	S5020
86-227-2	124	3021		TR81/240	S3021
86-228-2	157	3103	12	TR60/244	S3021
86-230-2	121	3009		TR01/232	G6003
86-231-2	121	3009		TR01/232	G6003
86-232-2	121	3009		TR01/232	G6003
86-233-2	159		21	TR30/	
86-234-2	128	3024	20	TR87/243	S5026

Original Part Number	ECG	SK	GE	IR/WEP	HEP
86-235-2	121	3009	16	TR01/232	G6003
86-236-2	124	3021		TR81/240	S3021
86-237-2	123A	3124		TR21/735	S0015
86-238-2	123A	3124	20	TR21/735	S3021
86-243-2	108	3018	11	TR95/56	S0016
86-244-2	108	3018	11	TR95/56	S0016
86-245-2	108	3018	11	TR95/56	S0016
86-246-2	159	3025	21	TR20/717	S0019
86-246-20	159	3114		TR20/	
86-247-2	123A	3024	20	TR21/735	S0015
86-248-2	121	3009		TR01/232	G6003
86-249-2	102A	3021	52	TR85/250	G0005
86-249-9					S5011
86-250-2	123A	3124	20	TR21/735	S0015
86-251-2	159	3114	21	TR20/717	S0012
86-251-20	159	3114		TR20/	
86-253-2	160	3006	50	TR17/637	G0003
86-254-2	160	3006	50	TR17/637	G0003
86-255-2	123A	3018	20	TR21/735	S0015
86-256-2	123A	3124		TR21/735	S0015
86-257-2	157	3021	12	TR60/244	S3021
86-259-2	124	3021		TR81/240	S3021
86-260-2	124	3021		TR81/240	S3021
86-261-2	124	3021		TR81/240	S3021
86-262-2	161	3117		TR83/719	S0015
86-263-2	161	3117		TR83/719	S0015
86-264-2	123A	3122		TR21/735	S0015
86-265-2	123A	3122		TR21/735	S0015
86-266-2	128	3024		TR87/243	S0016
86-267-2	128	3024		TR87/243	S0016
86-271-2	175			TR81/	
86-272-2	175			TR81/	
86-273-2	128	3024		TR87/243	
86-275-2	124	3021		TR81/240	S3021
86-276-2	159	3025	21	TR20/717	S5013
86-276-20	159	3114		TR20/	
86-277-2	123A	3124	20	TR21/735	S0015
86-278-2	160	3006	50	TR17/637	G0003
86-279-2	160	3006	50	TR17/637	G0003
86-280-2	126	3006	2	TR85/250	G0003
86-281-2	126	3008	2	TR17/635	G0005
86-282-2	126	3008	2	TR85/250	G0005
86-283-2	102A	3008	2	TR85/250	G0005
86-284-2	188	3024	20	TR87/243	S3021
86-286-2	159	3114		TR20/717	S0019
86-286-20	159	3114		TR20/	
86-287-2	157	3103	12	TR60/244	S3021
86-289-2	161	3117		TR83/719	S0015
86-290-2	161	3117		TR83/719	
86-291-2	128	3124	20	TR21/735	S0015
86-291-9	123A	3122	20	TR24/	S0015
86-292-2	127			TR27/235	G6007
86-293-2	123A	3122		TR21/735	S0015
86-294-2	159	3114		TR20/717	S0019
86-294-20	159	3114		TR20/	
86-295-2	102A	3008	2	TR85/250	G0005
86-296-2	160	3006	2	TR17/637	G0003
86-297-2	102A			TR85/250	G0005
86-298-2	159	3114		TR20/717	S0019
86-298-20	159	3114		TR20/	
86-300-2	102A			TR85/250	G0005
86-301-2	103A	3004	2	TR08/724	G0011
86-303-2	102A	3004	2	TR85/250	G0005
86-304-2	102A	3004	2	TR85/250	G0005
86-305-2	102A	3004	2	TR85/250	G0005
86-306-2				TR51/735	S0015
86-308-2	123A	3124	20	TR21/735	S0015
86-309-2	123A	3122		TR21/735	S0015
86-310-2	123A	3124	20	TR21/735	S0015
86-311-2	126	3008	51	TR17/635	G0003
86-312-2	160	3008	51	TR17/637	G0003
86-313-2	121	3009	16	TR01/232	G6003
86-316-2	154	3045		TR78/712	S5025
86-317-2	121	3009	16	TR01/232	G6003
86-319-2	121	3009	16	TR01/232	G6003
86-320-2	160			TR17/637	G0003
86-321-2	160			TR17/637	G0003
86-322-2	160			TR17/637	G0003
86-323-2	123A	3122		TR21/735	S0015
86-324-2	123A	3122		TR21/735	S0005
86-327-2	123A	3124	20	TR21/735	S5014
86-328-2	123A	3124	20	TR21/735	S0015
86-329-2	129	3025		TR88/242	S5013
86-330-2	128	3024		TR87/243	S5014
86-334-2	129	3025	21	TR88/242	S5013
86-336-2	128	3024	63	TR87/243	S5014
86-339-2	123A	3024	20	TR21/735	S5014
86-339-9	123A	3024	20	TR24/	S0014
86-340-2	159	3114	21	TR20/717	S5013
86-340-20	159	3114		TR20/	
86-342-2	123A	3024	63	TR21/735	S5014
86-344-2	184	3054	28	TR76/S5003	S5000
86-345-2	185	3025		/S5007	S5006
86-347-2	160			TR17/637	G0003
86-348-2	160	3008	52	TR17/637	G0009

Original Part Number	ECG	SK	GE	IR/WEP	HEP
86-353-2	121	3009		TR01/232	G6003
86-354-2	121	3009		TR01/232	G6003
86-359-2	123A	3122		TR21/735	S0015
86-362-2	123A	3122	20	TR21/735	S0015
86-363-2	160	3006	51	TR17/637	G0008
86-365-2	123A	3122		TR21/735	S0015
86-366-2	160			TR17/637	G0003
86-367-2	160	3006	50	TR17/637	G0003
86-368-2	160	3006	50	TR17/637	G0003
86-370-2	179	3009	25	TR35/G6001	G6008
86-370-2GRN	121	3009		TR01/232	G6013
86-370-2ORN	121	3009		TR01/232	G6013
86-370-2VIL	121	3009		TR01/232	G6013
86-370-2YEL	121	3009		TR01/232	G6013
86-373-2	160	3006	50	TR17/637	G0003
86-374-2	160	3006	50	TR17/637	G0003
86-376-2	160	3006		TR17/637	
86-379-2	123A	3122		TR21/735	
86-381-2	161		20	TR83/719	S0020
86-386-2	108	3018	20	TR95/56	S0011
86-389-2	123	3018	20	TR86/53	S0015
86-390-2	123A	3018	20	TR21/735	S0015
86-391-2	123A	3124	10	TR21/735	S0015
86-392-2	102A		53	TR85/250	G0005
86-393-2	128	3024		TR87/243	S5014
86-396-2	185	3025	29	TR77/S5007	S5006
86-399-1	123A				
86-399-2	123A	3024	63	TR21/735	S0015
86-399-9	123A	3020	20	TR24/	S0014
86-400-2	123A	3124	20	TR21/735	S5014
86-403-2	123A	3124	20	TR21/735	
86-406-2	159	3114	22	TR20/	S0015
86-407-2	159	3114	20	TR21/735	S0015
86-416-2	108	3018	20	TR95/56	S0016
86-417-2	108	3124	63	TR95/56	S0015
86-419-2	102A	3004	52	TR85/250	S0019
86-420-2	123A	3124	63	TR51/735	S0015
86-421-2	102A	3004	52	TR85/250	G6011
86-422-2	188	3024	28	TR72/W3020	S3024
86-423-2	189	3200	29	TR73/W3031	S3028
86-423-3	189	3200		TR73/W3031	
86-428-9	128	3024		TR87/243	S5014
86-431-2	129	3025		TR88/242	S5013
86-431-9	129	3025		TR88/242	S5013
86-440-2	128	3024	61	TR87/243	S0030
86-441-2	128	3024	63	TR87/243	S0030
86-442-2	161	3018	20	TR83/719	S0016
86-444-2		3020	20	TR24/	S0014
86-445-2		3018	20	TR24/	S0015
86-449-2	126	3006	50	TR17/635	G0003
86-449-9	160	3006		TR17/	
86-452-2	128	3024	63	TR87/243	S5014
86-457-2	123A	3124	20	TR21/735	S0014
86-458-2	123A	3124	20	TR21/735	S0014
86-459-2	159	3114	21	TR20/717	S0012
86-460-2	123A	3124	20	TR21/735	S0014
86-461-2	123A	3124	20	TR21/735	S0014
86-462-2	123A	3124	20	TR21/735	S0014
86-463-2	128	3024	63	TR87/243	S5014
86-463-2(SEARS)	128			TR87/	
86-464-2	233	3050	FET4	TR25/	F2007
86-465-2		3018	20	/56	S0011
86-467-2	108	3018	20	TR95/56	S0014
86-472-2	123A		20	TR21/735	S0014
86-475-2	159		21	TR20/717	S0012
86-476-2	102A	3004	2	TR85/	G0007
86-477-2	133	3112	FET1	FE100/	F0015
86-480-0				TR50/	G6016
86-481-1	123A	3018	20	TR21/735	S0016
86-481-2	123A	3018	17	TR21/735	S3019
86-482-2	159	3114	21	TR20/717	S0012
86-483-2	123A	3018	20	TR21/735	S0016
86-483-3	123A	3018	63	TR21/	S0016
86-484-2	123A	3044	20	TR21/735	S0015
86-485-2	123A	3122	20	TR21/735	S0016
86-486-2	123A	3018	63	TR21/735	S0016
86-487-2	124	3021	12	TR81/240	S5011
86-487-3	124	3021	12		S5011
86-488-2	108	3018	17	TR95/	S0016
86-490-2	108	3018	20	TR95/	S0016
86-491-2	108	3018	20	TR95/	S0016
86-493-2	123A	3124	20	TR21/	S0014
86-494-2	123A	3124	20	TR21/	S0014
86-495-2	123A	3020	20	TR21/	S0014
86-496-2	123A	3020	20	TR21/	
86-497-2	102A	3004	2	TR85/	G0006
86-500-2	133				
86-501-2	159	3118	21	TR30/	
86-502-2	123A	3124		TR24/	
86-506-2	184		55		S5000
86-507-2		3083	56		
86-509-2		3005	53	TR84/	G0008
86-510-2	128	3024	63		S3024
86-512-2	105	3012	4	TR03/233	
86-513-2	233	3065		FE100/	F2007

Original Part Number	ECG	SK	GE	IR/WEP	HEP
86-514-2	123A				
86-515-2	194	3018	20	TR25/735	S0016
86-515-2(SEARS)	123A			TR21/	
86-520-2	123A		10	TR21/	
86-525-2	107	3039	17	TR33/	S0011
86-526-2	199	3018	20	TR51/	S0003
86-527-2	159	3114	21	TR30/	S0019
86-528-2	159	3114	21	TR54/	S0019
86-529-2		3054	66	TR76/	S3024
86-530-2		3083	69	TR77/	S3028
86-533-2	159	3114		TR19/	S0012
86-534-2	123A	3018	20	TR24/	S0015
86-536-2			17		
86-537-2			17		
86-538-2			17		
86-539-2	123A				
86-540-2	221	3050			F2007
86-541-2	172	3156	64	TR69/	S9100
86-543-2	128				
86-544-2	152				
86-547-2	159	3114	22	TR30/	
86-548-2	123A				
86-549-2	172		64	TR69/	S9100
86-550-2	194	3122	20	TR51/	S0015
86-551-2	123A	3122	10		
86-552-2	159	3114	21	TR19/	S0012
86-554-2	123A	3018	20	TR21/	S0016
86-555-2	159	3025	21		S0012
86-556-2	171	3104	27	TR74/	S3021
86-557-2	194	3122	18	TR53/	S0015
86-559-2	123A				
86-560-2	123A				
86-561-2	123A	3124	20	TR24/	
86-563-2	165	3115	38	TR93/	S5021
86-553-9	165	3115	38	TR93/	S5021
86-564-2	123A	3115	38	TR93/	S0016
86-564-3	165	3115	38	TR93/	S5021
86-564-9		3115	38	TR93/	S5021
86-565-2	123A	3122	20	TR51/	S0015
86-566-2	171	3024	32	TR94/	S5026
86-567-2		3020	18	TR25/	S0014
86-568-2	196	3054	28	TR76/	S5004
86-570-2	159	3114	22	TR30/	S0019
86-572-2	6402			TR24/	
86-572-3	6402				
86-573-2		3124	20	TR24/	S0014
86-577-2		3122			
86-593-2	107	3018	17	TR70/	S0016
86-594-2	107	3018	11	TR70/	S0016
86-595-2	123A	3018	61	TR33/	S0015
86-596-2	107	3018	17	TR33/	S0015
86-597-2	107	3018	17	TR33/	S0015
86-598-2	123A	3124	10	TR24/	S0014
86-599-2		3124	10	TR24/	S0014
86-600-2	159	3114	21	TR24/	S0019
86-601-2	192	3054	63		S3020
86-602-2	193	3083	67		S3032
86-602-2/183035				TR56/	
86-605-2	233	3050	FET4		F2007
86-606-2	233	3050			F2004
86-607-2	199	3018	20	TR53/	S0020
86-608-2		3018	65	TR30/	
86-609-2	192	3122	63	TR24/	
86-610-2	129	3114		TR30/	
86-610-9	159	3114	21	TR19/	S0012
86-611-2	194	3124	20	TR24/	
86-612-2		3044		TR78/	S5024
86-613-2	152	3054	28	TR76/	
86-614-2	242	3083	29	TR77/	
86-615-2		3124	20	TR24/	S0014
86-616-2	129				
86-619-2		3018			S0015
86-620-2		3018			S0015
86-621-2		3018			S0015
86-622-2	159	3114			S0019
86-624-2	124	3104	32	TR74/	S3021
86-624-9	124	3021	12	TR23/	S5011
86-625-2		3050	FET4		F2007
86-626-2	165	3111	38	TR61/	
86-628-2		3104			S3021
86-629-2	154	3044	40	TR78/	S5024
86-630-2	194	3124			
86-631-2	171	3104	32	TR60/	
86-632-2	222	3050	FET4	TR70/	F2004
86-633-2	165	3111	38	TR61/	
86-646-2	123A	3018		TR24/	S0015
86-648-2		3054		TR74/	S3020
86-649-2		3124			
86-650-2		3122			
86-655-2		3018			
86-0022-001				TR21/	
86-0029-001				TR21/	
86-0031-001				TR21/	
86-0033-007				TR94MP/	
86-0036-001				TR20/	

Original Part Number	ECG	SK	GE	IR/WEP	HEP
86-1392	123A	3124	20	TR21/735	S0015
86-5000-2	102A	3004	1	TR85/	
86-5001-2	102A	3123	1	TR85/	
86-5003-2	103A	3011	5	TR08/	
86-5004-2	103A	3011	5	TR08/	
86-5005-2	103A	3123	5	TR08/	
86-5006-2	102A	3017		TR85/	
86-5007-2	103A	3011	5	TR08/724	G0011
86-5008-2	103A	3011	5	TR08/724	G0011
86-5009-2	767				C6015
86-5011-2	103A	3011	5	TR08/724	G0011
86-5012-2	103A	3011	5	TR08/	
86-5013-2	103A	3011	5	TR08/	
86-5015-2	103A	3011		TR08/	
86-5016-2	103A	3011	5	TR08/724	G0011
86-5017-2	103A	3011	5	TR08/	
86-5018-2	123A	3011	5	TR08/	
86-5026-2	103A	3011	5	TR08/724	G0011
86-5027-2	102A			TR85/250	G0011
86-5029-2	103A	3011	5	TR08/	
86-5034-2	103A		5	TR08/	
86-5039-2	121	3014	16	TR01/	
86-5040-2	123A	3124		TR21/	
86-5041-2	123A	3124	17	TR21/	
86-5042-2	102A	3123	52	TR85/250	G0005
86-5043-2	121	3009	16	TR01/250	G6005
86-5044-2	123A	3122	18	TR21/735	S0015
86-5045-2	123A	3122	17	TR21/735	S0015
86-5046-2	123A	3122		TR21/735	S0015
86-5047-2	103A	3011	5	TR08/724	G0011
86-5048-2	103A	3011	5	TR08/724	G00111
86-5049-2	123A	3122	17	TR21/735	S0015
86-5050-2	123A	3122	17	TR21/735	S0015
86-5051-2	123A	3122	18	TR21/735	S0015
86-5052-2	158			TR84/	S9002
86-5055-2	123A	3122	20	TR21/	
86-5056-2	123A	3124	17	TR21/	
86-5057-2	121		3	TR01/	
86-5058-2	121	3009	3	TR01/	
86-5060-2	103A	3011	5	TR08/	
86-5061-2	103A	3011	5	TR08/724	G0011
86-5062-2	103A	3011	5	TR08/	
86-5063-2	102A	3123		TR85/	
86-5064-2	129	3025	21	TR88/242	S5013
86-5064-2A	129	3025		TR88/	
86-5065-2	123	3124	17	TR86/	
86-5067-2	102A	3004	2	TR85/250	
86-5070-2	128		18	TR87/	
86-5073-2	128	3024	18	TR87/243	S5014
86-5074-2	128		18	TR87/	
86-5075-2	192		28	/243	S3024
86-5079-2	159(2)	3114	21	TR28/	
86-5080-2	103A	3011	5	TR08/	
86-5081-2	123A	3122	17	TR21/735	S0015
86-5082-2	129	3025	21	TR88/242	S5013
86-5082-2A	129	3025		TR88/	
86-5083-2	121		16	TR01/	
86-5084-2	130	3027	14	TR59/247	
86-5084-2A	130	3027		TR59/	
86-5085-2	175	3131	23	TR81/241	
86-5086-2	103A		5	TR08/724	G0011
86-5087-2	103A	3011	5	TR08/	
86-5088-2	121	3009		TR01/232	G6003
86-5089-2	121			TR01/	
86-5090-2	121			TR01/	
86-5091-2	102A	3123		TR85/250	G0005
86-5093-2	128	3024	18	TR87/243	S5014
86-5095-2	133	3112	FET1		
86-5096-2	133	3112	FET1		
86-5097-2	123A	3124	18	TR21/	
86-5099-2	123A(2)	3124	17	TR25/	
86-5100-2	184	3190	28		
86-5101-2	130	3027	14	TR59/247	
86-5101-2A	130	3027		TR59/	
86-5102-2	152			TR76/	
86-5103-2	123A	3124	20	TR21/	
86-5104-2	193	3114	67		
86-5105-2	172			TR69/	
86-5106-2	175			TR81/	
86-5107-2	152			TR76/	
86-5108-2	152			TR76/	
86-5109-2	128/411	3044	28	/243	S5024
86-5110-2	123A	3019	17	TR21/	
86-5111-2	123A	3124	20	TR21/	
86-5112-2	130		·14	TR59/	
86-5112-2(THMS)	130			TR59/	
86-5113-2	121			TR01/	
86-5114-2	123A	3124	17	TR21/	
86-5117-2	123A	3019			
86-5122-2	133	3112	FET1		
86-5125-2	121			TR01/	
86-5135-2		3019			
86-5943-2	121	3009		TR01/232	
86-100002		3018			
86-100003		3018			

Original Part Number	ECG	SK	GE	IR/WEP	HEP
86-100004		3018			
86-100005		3018			
86-100006		3018			
86-100007		3018			
86-100008		3018			
86-100009		3114			
86A318	102A	3004	2	TR85/250	
86A324					S0015
86A327	199			TR21/	S0024
86A332	130			TR59/	S7004
86A333					S3021
86A334	123A				
86A335	159				
86A336	123A				
86A338	175			TR81/	S5019
86A339	218			TR58/	S5018
86A349					S0015
86A350	123A				
86B332					S7000
86B3339					S5018
86C04	185			TR77/S5007	S5006
86P1AA	159				
86R164		3009			
86X0004-001		3004			
86X0005-001		3009			
86X0006-001	123A	3018		TR21/735	S0025
86X0007-001	123A	3124	20	TR21/735	
86X0007-004	108	3124	20	TR95/56	S0020
86X0007-204	108	3124	20	TR24/	
86X0008-001	123A	3124	20	TR21/735	S0015
86X0009-001	121	3009	16	TR01/232	
86X0011-001	126	3008	52	TR17/635	
86X0012-001	123	3124	20	TR86/53	
86X0013-001	126	3006	50	TR17/635	
86X0014-001	100	3006	50	TR05/254	
86X0015-001	121	3009	16	TR01/232	
86X0016-001	159	3008	50	TR20/717	
86X0016-001A	159	3114		TR20/	
86X0017-001	102A	3004	52	TR85/250	G0006
86X0018-001	102A	3004	52	TR85/250	
86X0019-001		3018		TR21/	
86X0022-001	123A	3124	20	TR21/735	S0015
86X0024-001	704	3023	20	IC501/	
86X0025-001	123	3124	20	TR86/53	
86X0027-001	704	3023		IC501/	
86X0028-001	124	3021	12	TR81/240	
86X0029-001	123A	3018	20	TR21/735	S0016
86X0030-001	121MP	3014	3	TR01/232MP	G6003
86X0030-100	121MP	3013		/232MP	G6003(2)
86X0031-001	123A	3124	20	TR21/735	S0024
86X0031-002	123A	3124	20	TR21/735	S0024
86X0031-003	123A	3124	20	TR21/735	S0024
86X0032-001	123A		17	TR21/	
86X0033-001	131	3052	30	TR94/642	G6016
86X0034-001		3024	20	TR83/735	S0020
86X0035-001	123A	3024	20	TR21/735	S0015
86X0036-001		3124	21	TR20/717	S0013
86X0036-001A	159	3114		TR20/	
86X0037-001	103A	3010	8	TR85/724	G0011
86X0037-002	102A	3004		TR85/250	G0005
86X0037-100	102A	3010			G0005
86X0038-001	108	3018	20	TR95/56	S0020
86X0040-001	123A	3020	10	TR21/717	S0015
86X0041-001	159	3114	22	TR20/717	S5013
86X0041-001A	159	3114		TR20/	
86X0042-001	242	3190	58	TR56/S5007	S5006
86X0042-002	241	3054	57	TR55/S5003	S5000
86X0042-100	184	3054			S5000
86X0043-001	108	3039	20	TR95/56	S0016
86X0044-001	159	3025	21	TR20/717	S0019
86X0044-001A	159	3114		TR20/	
86X0045-001	123A	3024	20	TR21/735	S0015
86X0047-001	159	3025		TR20/717	S0019
86X0048-001	123A	3124	20	TR51/	S0005
86X0049-001	194	3044		TR83/	S0005
86X0050-001	123A	3124	62	TR21/	S0024
86X0051-001	123A	3040	20	TR21/	S0020
86X0052-001	161	3117	60	TR21/	S0011
86X0053-001	712	3072	2	IC507/	C6063P
86X0054-001	123A	3124	62	TR51/	S0015
86X0055-001	728	3073	22		
86X0056-001	729	3074	23		
86X0058-001	123A	3124	20	TR51/	S0015
86X0058-002	123A	3040	20	TR51/	
86X0058-003	123A	3040	20	TR51/	
86X0059-001	184	3054	57	TR60/	S5000
86X0060-001	107	3018	17	TR33/	S0016
86X0061-001	107	3018	20	TR33/	S0016
86X0062-001	107	3039	17	TR33/	S0011
86X0063-001	123A	3018	20	TR51/	S0011
86X0064-001	799				
86X0065-001	171	3201	27	TR79/	
86X0066-001	159	3114	21	TR52/	S0019
86X0070-001	6402				S9001
86X0071-001	128	3024	18	TR87/	S5014

Original Part Number	ECG	SK	GE	IR/WEP	HEP
86X0072-001	159	3114	21	TR52/	S5022
86X0073-001	188	3054	18	TR72/	S3019
86X0074-001	184	3054	55		S5000
86X0075-001	185	3083	56		S5006
86X0076-001	228	3103	27	TR60/	S5015
86X0077-001	230				
86X0077-204					S0015
86X0078-001	231				
86X0079-001	123A				
86X0080-001	242	3054	58		
86X0080-002	241	3054	57		
86X0080-100		3054			
86X1035-001				TR21/	
86X6029-001	108	3039		TR95/56	S0016
86X2	121	3009	3	TR01/	
86X3	102A	3004	53	TR85/	
86X6	123A	3122		TR21/735	
86X6-1	123A	3122	20	TR21/735	S0015
86X6-4-518	123A	3122	20	TR21/735	S0015
86X7-2	123A	3124	20	TR21/735	S0015
86X7-3	123A	3122	20	TR21/735	
86X7-4	123A	3124		TR21/735	S0015
86X7-5					S0014
86X7-6	107	3024	20	TR70/	S0015
86X7-6013	108	3018		TR95/	S0015
86X8-1	123A	3124	20	TR21/735	S0015
86X8-2	123A	3122		TR21/735	S0015
86X8-3	123A	3124	20	TR21/735	S0015
86X8-4	123A	3122	17	TR21/735	S0011
86X34-1	123A	3040		TR21/735	
86X46	159	3025		TR20/717	
86X47	159	3025		TR20/717	
86X53-1	712	3072			
86X55-1	728	3073			
86X56-1	729	3074			
87-0002		3018			
87-0003		3039			
87-0005		3124			
87-0009		3124			
87-0013		3124			
87-0014		3124			
87-0015		3008	2		G0008
87-0016		3005	51		G0008
87-0017		3008	52		G0008
87-0018		3004	52		G0007
87-0019		3004	53		G0005
87-0020		3004	2		G0007
87-0021		3004	52		G0005
87-0027		3018			
87B02	188			TR72/W3020	S5000
87M42899B39	735				
88-9302	744				
88-9302R	744		24		
88-9302S	744		24		
88-9304	732				C6090
88-9574	737		IC16		
88-9575	734				
88-9779	736		IC17		
88-9779F	736				
88-9841R	789				
88-9842F	723				
88-9842R	723				
88-9842RS			IC15		
88-9842S	723		IC15		
88-1250108	199	3038		TR21/	
88-1250109	123A	3124		TR21/	
88A7522	949				
88A752271	949				
88B14-7B		3006			
88B	160	3006		TR17/	
88B0	160	3006		TR17/	
88B1	160	3006		TR17/	
88B2	160	3006		TR17/	
88B3	160	3006		TR17/	
88B4	160	3006		TR17/	
88B5	160	3006		TR17/	
88B6	160	3006		TR17/	
88B7	160	3006		TR17/	
8838	160	3006		TR17/	
88B9	160	3006		TR17/	
88B10	160	3006		TR17/	
88B10R	160	3006		TR17/	
88B11	160	3006		TR17/	
88B12	160	3006		TR17/	
88B13	160	3006		TR17/	
88B13P	160	3006		TR17/	
88B14	160	3006		TR17/	
88B14-7	160	3006		TR17/	
88B14-7B	160			TR17/	
88B15	160	3006		TR17/	
88B15L	160	3006		TR17/	
88B16	160	3006		TR17/	
88B16B	160	3006		TR17/	
88B17	160	3006		TR17/	
88B17-1	160	3006		TR17/	
88B18	160	3006		TR17/	
88B19	160	3006		TR17/	
88B19G	160	3006		TR17/	
88B-70	160	3006		TR17/	
88B-70-12	160	3006		TR17/	
88B-70-12-7	160	3006		TR17/	
88B119	160	3006		TR17/	
88B121	160	3006		TR17/	
88B182	160	3006		TR17/	
88B7258	709		IC11	IC505/	C6062P
88C14-7B		3005			
88C	100	3005		TR05/	
88C0	100	3005		TR05/	
88C1	100	3005		TR05/	
88C2	100	3005		TR05/	
88C3	100	3005		TR05/	
88C4	100	3005		TR05/	
88C5	100	3005		TR05/	
88C6	100	3005		TR05/	
88C7	100	3005		TR05/	
88C8	100	3005		TR08/	
88C9	100	3005		TR05/	
88C10	100	3005		TR05/	
88C10R	100	3005		TR05/	
88C11	100	3005		TR05/	
88C12	100	3005		TR05/	
88C13	100	3005		TR05/	
88C13P	100	3005		TR05/	
88C14	100	3005		TR05/	
88C14-7	100	3005		TR05/	
88C14-7B	100			TR05/	
88C15	100	3005		TR05/	
88C15L	100	3005		TR05/	
88C16	100	3005		TR05/	
88C16D	100	3005		TR05/	
88C17	100	3005		TR05/	
88C17-1	100	3005		TR05/	
88C18	100	3005		TR05/	
88C19	100	3005		TR05/	
88C19G	100	3005		TR05/	
88C-70	100	3005		TR05/	
88C-70-12	100	3005		TR05/	
88C-70-12-7	100	3005		TR05/	
88C119	100	3005		TR05/	
88C121	100	3005		TR05/	
88C182	100	3005		TR05/	
88CC-70-12	100	3005		TR05/	
88X0053-001			IC2		C6063P
90-30	123A	3124	20	TR21/	
90-31	128	3024	18	TR25/	
90-32	123A	3124	20	TR21/	
90-33	199	3124	62	TR21/	
90-37	160		51	TR17/637	
90-38	186	3054	28	TR55/	
90-45	161	3018	39	TR70/	
90-46	161	3019	11	TR83/	
90-47	123A	3124	10	TR24/	
90-48	123A	3122	20	TR24/	
90-49	108	3019	11	TR83/	
90-50	132	3116	FET2	FE100/	
90-54	160		51	TR17/637	
90-55	132	3116	FET2	FE100/	F0021
90-56	158	3124	10	TR84/630	S0015
90-57	123A	3122	20	TR24/	S0015
90-58	102A	3050	53	TR85/250	F2004
90-59	160		53	TR17/637	
90-60	102A	3018	53	TR85/250	
90-61	123A	3124	10	TR24/	
90-62	132	3116	FET2	FE100/	
90-65	123A	3124	20	TR21/	
90-66	128	3024	18	TR25/	
90-69	123A	3124	20	TR21/	
90-70	199	3124	62	TR21/	
90-71	123A	3124	10	TR24/	
90-75	186	3054	28	TR55/	
90-110		3049			
90-111		3054			
90-112		3045			
90-2213-00-18	123A				
90T2	107	3039	18	TR70/720	
90X9				TR84/	
91-4	160			TR17/637	
91A	108	3039		TR95/56	S0015
91AJ150	175			TR81/	
91B	108	3039		TR95/56	S0015
91BGRN	108	3039		TR95/56	S0015
91C	123A	3122		TR21/735	S0015
91D	123A	3122		TR21/735	S0015
91E	123A	3122		TR21/735	S0015
91F	108	3122		TR95/735	S0015
91N1B	161			TR83/	
91T6			212	TR51/	
92-1-70	105	3012		TR03/	
92-1-70-12	105	3012		TR03/	
92-1-70-12-7	105	3012		TR03/	

Original Part Number	ECG	SK	GE	IR/WEP	HEP
92-1A	105	3012		TR03/	
92-1A0	105	3012		TR03/	
92-1A0R	105	3012		TR03/	
92-1A1	105	3012		TR03/	
92-1A2	105	3012		TR03/	
92-1A3	105	3012		TR03/	
92-1A3P	105	3012		TR03/	
92-1A4	105	3012		TR03/	
92-1A4-7	105	3012		TR03/	
92-1A4-7B	105	3012		TR03/	
92-1A5	105	3012		TR03/	
92-1A5L	105	3012		TR03/	
92-1A6	105	3012		TR03/	
92-1A6-1	105	3012		TR03/	
92-1A7	105	3012		TR03/	
92-1A7-1	105	3012		TR03/	
92-1A8	105	3012		TR03/	
92-1A9	105	3012		TR03/	
92-1A9G	105	3012		TR03/	
92-1A19	105	3012		TR03/	
92-1A21	105	3012		TR03/	
92-1A82	105	3012		TR03/	
92-10	105	3012		TR03/	
92-11	105	3012		TR03/	
92-12	105	3012		TR03/	
92-13	105	3012		TR03/	
92-14	105	3012		TR03/	
92-15	105	3012		TR03/	
92-16	105	3012		TR03/	
92-17	105	3012		TR03/	
92-18	105	3012		TR03/	
92-19	105	3012		TR03/	
92-30942	123A				
92B1C	723				
92N1	108	3018	11	TR95/56	S0016
92N1B	108	3039		TR95/56	S0016
93A9	102	3004	2	TR05/631	
93A9-1	102A		53	TR85/250	
93A9-2	102A		53	TR05/250	
93A9-3	100	3005	1	TR05/254	
93A9-4	100	3005	1	TR05/254	
93A39-15	123A			TR21/	
93C39-11	123			TR86/53	
93H87DC	74H87				
93H87PC	74H87				
93H183DC	74H183				
93H193PC	74H183				
93P1AA	159	3114	21	TR20/	S0013
93S16DC	74S16				
93S16PC	74S16				
93S153DC	74S153				
93S157DC	74S157				
93S157PC	74S157				
93S158DC	74S158				
93S158PC	74S158				
93S174DC	74S174				
93S174PC	74S174				
93S175DC	74S175				
93S175PC	74S175				
93S194DC	74S194				
93S194PC	74S194				
93SC165133	130	3027		TR59/247	
93SC165133A	130	3027		TR59/	
93SE124	130				
93SE165	181	3036	14	TR36/	S7002
94A-1A6-4	121	3009		TR01/	
94B1C	724				
94N1	123A	3124	20	TR21/735	S0015
94N1B	123A	3124	20	TR21/735	S0015
94N1R	123A	3122		TR21/735	
94N1V	128	3024		TR87/243	S0015
94N2			20	TR24/	S0015
94N2P	155				
94T1	100	3005		TR05/254	
95-108	160	3006		TR17/	
95-110	160	3006		TR17/	
95-111	126	3006		TR17/	
95-112	101	3011		TR08/	
95-113	101	3011		TR08/	
95-114	101	3011		TR08/	
95-116	160	3006		TR17/	
95-117	160	3006		TR17/	
95-118	160	3006		TR17/	
95-119	160	3006		TR17/	
95-120	160	3006		TR17/	
95-120A	100	3005		TR05/	
95-121	160	3006		TR17/	
95-122	160	3006		TR17/	
95-123	160	3006		TR17/	
95-124	160	3006		TR17/	
95-125	108	3039		TR95/	
95-126	108	3039		TR95/	
95-127	108	3039		TR95/	
95-128	108	3039		TR95/	
95-129	108	3039		TR95/	

Original Part Number	ECG	SK	GE	IR/WEP	HEP
95-130	108	3039		TR95/	
95-131	108	3039		TR95/	
95-201	102A	3004		TR85/	
95-202	103A	3010		TR08/	
95-203	102A	3004		TR85/	
95-204	102A	3004		TR85/	
95-208	102A	3004		TR85/	
95-209	102A	3004		TR85/	
95-212	102A	3004		TR85/	
95-213	102A	3004		TR85/	
95-214	102A	3004		TR85/	
95-216	123A	3122		TR21/	
95-217	102A	3004		TR85/	
95-218	102A	3004		TR85/	
95-220	128	3024		TR87/	
95-221	123A	3122		TR21/	
95-222-1	102A	3004		TR85/	
95-222-2	103A	3010		TR08/	
95-223	108	30		TR95/	
95-224-1	102A	3004		TR85/	
95-224-2	103A	3010		TR08/	
95-225	123A	3122		TR21/	
95-226-003	129	3025		TR88/	
95-226-004	128	3024		TR87/	
95-226-1	129	3025		TR88/	
95-226-2	123A	3122		TR21/	
95-226-3	129	3025		TR88/	
95-226-4	128	3024		TR87/	
95-227	123A	3122		TR21/	
95-250	121	3009		TR01/	
95-251	121	3009		TR01/	
95-252	124	3021		TR81/	
95-252-1	124	3021		TR81/	
95-252-2	124	3021		TR81/	
95-252-3	124	3021		TR81/	
95-252-4	124	3021		TR81/	
95-257	121MP	3013			
95-296	123A	3122		TR21/	
95-11410100	102A	3004		TR85/	
95-11410200	102A	3004		TR85/	
95-11410900	102A	3004		TR85/	
95-11413500	102A	3004		TR85/	
95-11414000	102	3009		TR05/	
96-056-234	108				
96-138-2	108	3019	20	TR95/56	
96-5003-01		3004			
96-5005-01	171	3103		TR79/	
96-5026-01	121	3009		TR01/	
96-5032-01	102A	3004		TR85/	
96-5033-01	102A	3004		TR85/	
96-5033-02	102A	3004		TR85/	
96-5033-03	102A	3004		TR85/	
96-5033-04	102A	3004		TR85/	
96-5045-01	121	3009		TR01/	
96-5062-01	160	3006		TR17/	
96-5064-01	121	3009		TR01/	
96-5076-01	158	3004		TR84/	
96-5080-02	123A	3122		TR21/	
96-5081-01	121	3009		TR01/	
96-5085-01	102A	3004		TR85/	
96-5085-02	102A	3004		TR85/	
96-5086-02	121	3009		TR01/	
96-5095-01	160	3006		TR17/	
96-5098-01	102A	3004		TR85/	
96-5099-01	160	3004		TR17/	
96-5100-01	121	3009		TR01/	
96-5100-03	121	3009		TR01/	
96-5101-01	102A	3004		TR85/	
96-5102-01	102A	3004		TR85/	
96-5107-01	128	3024		TR87/	
96-5107-02	128	3024		TR87/	
96-5115-01	123A	3122		TR21/	
96-5115-02	123A	3122		TR21/	
96-5115-03	123A	3122		TR21/	
96-5115-04	123A	3122		TR21/	
96-5115-05	123A	3122		TR21/	
96-5117-01	130	3016		TR59/	
96-5117-91	130	3027		TR59/	
96-5125-01	121	3009		TR01/	
96-5131-01	161	3132		TR83/	
96-5132-01	124	3021		TR81/	
96-5135-01	124	3021		TR81/	
96-5138-01	160	3006		TR17/	
96-5139-01	160	3006		TR17/	
96-5140-01	160	3006		TR17/	
96-5141-01	160	3006		TR17/	
96-5143-01	121	3009	16	TR01/ 232	
96-5143-02	121	3009		TR01/ 232	
96-5148-01	121	3009		TR01/	
96-5152-01	123A	3122		TR21/	
96-5152-03	123A	3122		TR21/	
96-5153-01	123A	3122		TR21/	
96-5153-03	123A	3122		TR21/	
96-5155-01	121	3009		TR01/	
96-5161-01	123A	3122		TR21/	

Original Part Number	ECG	SK	GE	IR/WEP	HEP
96-5162-03	130	3027		TR59/	
96-5162-04	130	3027		TR59/	
96-5163-01	161	3132		TR83/	
96-5164-02	130	3027		TR59/	
96-5164-03	130	3027		TR59/	
96-5165-01		3024	18	TR87/53	
96-5170-01	128	3024		TR87/	
96-5174-01	161	3132		TR83/	
96-5175-01	161	3132		TR83/	
96-5176-01	129	3025		TR88/	
96-5177-01	123A	3122		TR21/	
96-5180-01	128	3024		TR87/	
96-5180-02	128	3024		TR87/	
96-5187-01	123A	3122		TR21/	
96-5190-01		3124	20	TR86/	
96-5191-01	175	3538		TR81/	
96-5192-01	121	3009		TR01/	G6005
96-5198-01	161	3132		TR83/	
96-5199-01	161	3132		TR83/	
96-5201-01	130	3027		TR59/	
96-5202-01	130	3027		TR59/	
96-5203-01	128	3024		TR87/	
96-5204-01	128	3024		TR87/	
96-5205-01	103A	3010		TR08/	
96-5207-01	130	3027		TR59/	
96-5208-01	128				
96-5209-01	129	3157		TR88/	
96-5213-01	123A	3122		TR21/	
96-5215-01	159	3025		TR20/	
96-5219-01	171	3103		TR79/	
96-5220-01(NPN)	123A			TR21/	
96-5220-01(PNP)	129			TR88/	
96-5225-01		3054			
96-5228-01	123A	3122		TR21/	
96-5229-01	123A	3122		TR21/	
96-5231-01	127	3035		TR27/	
96-5232-01	196	3054			
96-5232-02	152	3054			
96-5235-01	161	3132		TR83/	
96-5236-01	161	3132		TR83/	
96-5237-01	123A	3122		TR21/	
96-5238-01	703A	3157			
96-5238-02	703A	3157			
96-5244-01	128	3024		TR87/	
96-5252-01	128	3024		TR87/	
96-5255-01	123A	3122		TR21/	
96-5256-01	128	3024		TR87/	
96-5257-01	123A	3122		TR21/	
96-5258-01	159	3025		TR20/	
96-5259-01	161	3132		TR83/	
96-5260-01	161	3132		TR83/	
96-5267-01	196	3021			
96-5270-01		3054			
96-5284-01	162			TR67/	S7002
96-5315-01	181	3036			
96-51430-02	121	3009		TR01/	
96B-3A65	121MP	3015			
96EP					S3020
96N(WARDS)					S0030
96N1	124	3021		TR81/56	S5011
96N927	108	3039		TR95/56	
96N932	108	3039		TR95/56	
96N(AIRLINE)	108			TR95/	
96NPT	108	3018		TR95/	
96XZ801/06N	121	3009		TR01/	
96XZ801/10N	121	3009		TR01/	
96XZ801/14N	123A	3122		TR21/	
96XZ801/34X	121	3009		TR01/	
96XZ801/37N	160	3006		TR17/	
96XZ801/50N	102A	3004		TR85/	
96XZ6050/25N	161			TR83/	
96XZ6051-28N	126	3006		TR17/635	
96XZ6051-35N	126	3006		TR17/635	
96XZ6051-36N	126	3006		TR17/635	
96XZ6052/52N	123A	3122	20	TR21/735	
96XZ6053-09N	102A	3004	2	TR05/631	
96XZ6053-10N	126	3006	20	TR17/635	
96XZ6053/11N	123A	3122	20	TR21/735	
96XZ6053-24N	126	3006		TR17/635	
96XZ6053/24N	158			TR84/	
96XZ6053-27N	102	3004	2	TR05/631	
96XZ6053/27N	102A			TR85/	
96XZ6053/35N	123A	3122		TR21/	
96XZ6053/36N	123A	3122		TR21/	
96XZ6053/38N	128	3024		TR87/	
96XZ6053/51N	102A	3004	53	TR85/250	
96XZ6054/45X	131MP			TR94MP/247MP	
96X76065/25N		3004			
97A83	104	3009		TR01/230	G6003
97EP8					S0025
97EPA					S0025
97N2	101	3010	8	TR08/641	
97N2U	103A	3010		TR08/724	
97P1	102	3004	52	TR05/631	
97P1U	158	3004		TR84/630	

Original Part Number	ECG	SK	GE	IR/WEP	HEP
98-1	102A	3123		TR85/250	G0005
98-2	102A	3123		TR85/250	G0005
98-3	102A	3123		TR85/250	G0005
98-4	102A	3123		TR85/250	G0005
98-5	102A	3123		TR85/250	G0005
98-6	102A	3123		TR85/250	G0005
98-7	102A	3123		TR85/250	G0005
98-24320-2	175				
98P1	159	3114		TR20/717	G6016
98P1P	131	3198		TR94/	G6016
98P-2A6-2	102A	3004		TR85/	
98P10	159	3114		TR20/	
98T2			210	TR53/	S0015
99-101	126			TR17/	
99-102	126			TR17/	
99-103	126			TR17/	
99-104	102A			TR85/	
99-105	160			TR17/	
99-106	160			TR17/	
99-107	160			TR17/	
99-108	160			TR17/	
99-109-1	123A			TR21/	
99-109-2	123A			TR21/	
99-120	160			TR17/	
99-121	160			TR17/	
99-201	102A			TR85/	
99-202	102A			TR85/	
99-203	102A			TR85/	
99-P2				TR88/	
99-PWR				TR01/	
99AT6	102	3004	52	TR05/631	G0005
99B5	102	3004	52	TR05/631	G0005
99B16	100	3005	2	TR05/	G0005
99BE6	100	3005	2	TR05/254	G0005
99D074				TR24/	
99E14-1	9902				
99E16-1	9900				
99K7	101	3011		TR08/641	
99L-4A6-1	102A	3003		TR85/	
99L6	103	3010	8	TR08/641A	G0011
99L6(SHARP)	128			TR87/	
99P1	159	3025	21	TR20/717	S0012
99P1M	159	3025		TR20/717	S0012
99P1M1		3114			
99P2	129	3025	22	TR88/242	
99P2B		3025			
99P3	102A				
99P3AA	102A			TR85/	G0005
99P3C	129				
99P3M					S0025
99P10	159	3114		TR20/	
99P117	129			TR88/	
99PIM	129	3025		TR88/	
99-PWR			16		
99S001	121	3009	16	TR01/232	G6018
99S002	102A	3004	52	TR85/250	G0005
99S003	100	3004	52	TR05/254	G0005
99S004	158	3004	50	TR84/630	G6011
99S004A	102A	3004		TR85/	
99S005	102A	3004	2	TR85/250	G0005
99S006	160	3006		TR17/637	G0001
99S007	160	3006		TR17/637	G0001
99S07	101				G0011
99S010	102A	3004	53	TR85/250	G0005
99S010A	102A	3004		TR85/	
99S011	102A	3004	53	TR85/250	G0007
99S011A	102A	3004	52	TR85/250	G0007
99S012	123A	3124	20	TR21/735	S0015
99S012A	123A	3124	20	TR21/735	S0015
99S012E	123A	3122		TR21/	S0015
99S013	121MP	3013		/232MP	G6005
99S013A	121MP	3014		/232MP	G6005
99S014	121	3009	16	TR01/232	G6018
99S014A	121	3009		TR01/	
99S015	121	3009	16	TR01/232	G6005
99S016	108	3018	17	TR95/56	S0016
99S016-1	108	3018		TR95/	
99S017	108	3018	17	TR95/56	S0016
99S018	108	3018	17	TR95/56	S0020
99S018A	108	3018		TR95/	
99S019	108	3018	11	TR95/56	S0016
99S019A	108	3018		TR95/56	S0016
99S019B	108	3018		TR95/56	S0016
99S020	123	3124	20	TR86/53	S0014
99S022	703A	3157	IC12	IC500/	C6001
99S022-1	703A	3157			
99S025	123A	3124	20	TR21/735	S0015
99S025A	123A	3124		TR21/	
99S031	161	3018	39	TR83/719	S0015
99S032	161	3018	39	TR83/719	S0025
99S032(3RDIF)	233				
99S033	199	3040	20	TR21/735	S0015
99S033-40	123A+159				
99S034	194	3044	20	TR87/712	S0005
99S035	123A	3044	20	TR21/712	S0015

Original Part Number	ECG	SK	GE	IR/WEP	HEP
99S036	123A	3044	63	TR21/712	S0015
99S036(TELYNE)			18		
99S037	108	3018		TR95/56	S0015
99S038	123A	3122	20	TR21/735	S0030
99S039	159	3114	22	TR20/717	S0031
99S039A	159	3114		TR20/	
99S040	194	3044	20	TR25/	S0005
99S041	222	3050			F2004
99S042	710	3102		IC506/	
99S044	161	3132	39	TR83/719	S0017
99S045	221	3116			F2005
99S045A	222	3050			
99S046	222	3050			F2004
99S047	175	3131	12	TR81/241	S5011
99S049	726				
99S053	722	3161	IC9	IC513/	C6056P
99S053-1	722	3161			
99S055	161	3018	39	TR83/	S0017
99S056	161	3018	39	TR83/	
99S057	179			TR35/G6001	
99S060-1		3137	20	TR24/	S0015
99S061-1		3137	20	TR24/	S0015
99S062-1	159				
99S063-1	228		27	TR70/	S3021
99S067-1	161	3018	39	TR70/	S0017
99S070-1	194	3044	20	TR25/	S0015
99S072	710	3102			
99S073	129	3114	21	TR19/	S5013
99S074		3024	20	TR24/	S0015
99S075	152	3041	66	TR55/	S5000
99S077-1	194	3044	20	TR25/	S0005
99S079-1	130	3111	37	TR61/	S5021
99S081	790				
99S081-1	790		IC18		C6057P
99S082	712				
99S082-1	712	3072	IC2		C6063P
99S083-1	196	3054	66	TR92/	S5004
99S084-1	159	3025	21	TR28/	S0012
99S085	123A	3124	20	TR24/	S0015
99S087-1	154			TR78/	
99S090-1	108	3020	20	TR24/	S0016
99S091-1	188	3024	IC18	TR72/	S3024
99S092-1	189	3025	69	TR73/	S3028
99S094	714				
99S094-1	714	3075	IC4	IC509/	C6070P
99S095	715				
99S095-1	715	3076	IC6	IC510/	C6071P
99S096	713				
99S096-1	713	3077	IC5	IC508/	C6063P
99S097	783				
99S097-1	783	3141			C6100P
99S099	197	3083			
99S099-1	153		69	TR77/	S5006
99S100	196	3054			
99S100-1	152		66	TR76/	S5000
99S101-1	154	3045	18	TR25/	S5026
99S102-1	190	3103	27	TR74/	S3021
99S103-1	130	3111	37	TR61/	S5021
99S103-2	130	3111	37	TR61/	S5021
99S103-3	130	3111	37	TR61/	S5021
99S105-1	184	3104	57	TR60/	S5000
99SA7	103A	3011	8	TR08/	G0011
99SK5	101	3011		TR08/	G0011
99SK7	101	3011	8	TR08/	
99SQ7	101	3010	8	TR08/	G0011
100-1(PHILCO)	153				
100-198	106(4)				
100-4107	107				
100-4790	159				
100-4846-001	128				
100-5338	128				
100-5338-001	128				
100-5765-001	175				
100-6399-001		3545			
100-6399-002		3545			
100-6400-001		3548			
100-6400-002		3548			
100-6400-003		3548			
100B31	1004				
100B63	102	3004	2	TR05/631	
100N1	161	3018	20	TR83/719	S0017
100N1A5					S0014
100N1AS	161			TR83/719	S0017
100N1P	161	3018	20	TR83/	S0017
100N3P	161	3039	39	TR83/	S0020
100T2	130	3027	19	TR59/247	
100T2A	130	3027		TR59/	
100U		3112			
100W1	108	3018		TR95/56	
100X2	130	3027			
100X6	130	3027	14	TR59/247	
100X6A	130	3027		TR59/	
101-1				TR21/719	S0015
100-1(ADMIRAL)	161			TR83/	
101-2				TR21/56	
101-2(ADMIRAL)	108			TR95/	
101-3				TR21/56	
101-3(ADMIRAL)	108			TR95/	
101-4				TR21/56	
101-4(ADMIRAL)	108			TR95/	
101-12	100	3005	52	TR05/254	
101-15	102A	3123		TR85/	
101A	160	3123	2	TR17/637	G0003
101B	160	3123	2	TR17/637	G0003
101M	160	3123	2	TR17/637	G0003
101P1	159	3114		TR20/717	
101P10	159	3114		TR20/	
102-4	161	3117		TR83/719	S0017
102P1	159	3114		TR20/717	
102P10	159	3114		TR20/	
103-4	108	3039		TR95/56	S0020
103EP					S0015
103P935	159			TR20/717	
103P935A	159	3114		TR20/	
103P				TR30/	
103P(AIRLINE)	159			TR20/	
103PA	159	3114		TR20/	
103P(WARIS)					S0012
104	154	3045		TR78/712	
104(ADM)					S5025
104-5		3008			
104-8	154	3045		TR78/712	S5025
104-17	126	3006	51	TR17/635	
104-17(RCA)	159			TR20/	
104-19	126	3006	51	TR17/635	
104-21	126	3006	51	TR17/635	
104-170	159	3114		TR20/	
104H01	128	3024	18	TR87/243	
104T2	130	3027	72	TR59/247	S7002
104T2A	130	3027		TR59/	
105	159	3122		TR51/735	
105-001-04		3122	20		S0014
105-001-05		3018	10		S0015
105-001-07		3122	20		S0014
105-001-08		3018			
105-00106-00	108	3018	20	TR24/	S0015
105-00107-09	107	3018	20		S0015
105-00108-07	108	3018	20	TR24/	S0015
105-003-06	123A			TR51/	
105-003-09	123A			TR51/	
105-005-04/ 2228-3	108		17	TR21/	
105-006-08		3018		TR24/	S0015
105-008-04/ 2228-3	108		20	TR21/	
105-009-21/ 2228-3	108		20	TR21/	
105-085-54	123A	3122	10	TR51/	
105-02004-09	108	3018	20		S0015
105-02005-07	107	3018	17		S0025
105-02006-05	107	3018	17		S0025
105-02008-01	107	3018	20	TR21/	S0015
105-060-09		3122			
105-06004-00	108	3020	20		S0016
105-06007-05	123A	3122	20		S0015
105-08243-05	123A	3122	62		S0015
105-12	123A	3122		TR21/735	S0015
105-931-91/ 2228-3	123A		20	TR21/	
105-941-97/ 2228-3	108	20		TR21/	
105-24191-04	108	3018	20	TR24/	S0015
105-28196-07	123A	3122	10	TR24/	S0015
105(ADMIRAL)	123A			TR21/	S0015
106-12	159	3114		TR20/717	S0019
106-120	159	3114		TR20/	
106KBO	129	3025		TR88/242	
106KBA	129	3025		TR88/	
106M	132	3116	FET2	FE100/	
106P1	121	3009	16	TR01/232	G6005
106P1AG	121	3009	3	TR01/	
106P1T	121	3009		TR01/232	G6005
106RED	159	3114		TR20/717	S0019
107-8	123A	3122		TR21/	S0015
107A	160		52	TR17/637	G0002
107B	160		52	TR17/637	G0002
107BRN	123A	3122		TR21/735	
107BRN(ADM)					S0015
107M	160	3116	52	TR17/637	G0002
107N1	128	3024		TR87/243	
107N2	123A	3124	18	TR21/735	S5014
108-1	160	3006	50	TR17/637	G0002
108-2	160	3006	50	TR17/637	G0003
108-3	160	3006	51	TR17/637	G0002
108-4	160	3006	50	TR17/637	G0002
108-6	159	3114		TR20/717	S0019
108-60	159	3114		TR20/	
108GRN	159	3114		TR20/717	
108GRN(ADM)					S0019
109	123A			TR21/	
109-1		3122	20	TR51/735	

Original Part Number	ECG	SK	GE	IR/WEP	HEP
109-1(RCA)	123A			TR21/	
109(ADM)					S0014
110	159			TR20/	
110(ADM)					S0019
110-01563-00	102A	3004		TR05/	G0005
110P1	159	3025	21	TR20/717	S0012
110P1AA	159	3114		TR20/717	S0019
110P1M	159	3114		TR20/	S0012
110P1N				/717	
110U		3112			
111-1		3016			
111-6935	126	3006	51	TR17/635	G0003
111N2C	130				
111N4	130	3027		TR59/247	S7002
111N4A	130	3027		TR59/	
111N4B	130		75	TR26/	S7002
111N4C	130				
111N6	130				
111N6C	152				S7000
111N8C	196				
111P3					S7003
111P3B	218		74	TR29/	S7003
111P5	219				
111P5C	104				S7001
111T2	154	3045	27	TR78/712	
112-000088	123A				
112-000172	159	3008	51	TR20/635	
112-000185	159	3008	51	TR20/635	
112-000187	159	3114	16	TR20/717	
112-000267	160	3007	50	TR17/635	
112-001	100	3005	2	TR05/254	G0005
112-002	126	3006		TR17/635	G0002
112-003	102	3004	52	TR05/631	G0005
112-004	102	3004	52	TR05/631	G0005
112-011A	123A				
112-034923	102A	3004	53	TR85/	
112-1A82	123A	3124		TR21/	
112-2	129	3025	21	TR88/242	S0019
112-2A	129	3025		TR88/	
112-7	159	3025	67	TR20/717	S0019
112-8	159	3024	21	TR20/717	S0031
112-10	159	3025		TR20/717	S0031
112-361	128	3024		TR87/	
112-362	224				
112-363	130	3027		TR87/	
112-520	108	3039		TR95/	
112-521	108	3039		TR95/	
112-522	108	3039		TR95/	
112-523	123A	3122		TR21/	
112-524	121	3009		TR01/	
112-525	128	3024		TR87/	
112-526	224				
112-527	224				
112-200525	102A	3004	2	TR85/	
112-202147	121	3014	16	TR01/	
112-203055	130	3027	14	TR35/	
112-203391	199	3124	20	TR21/	
112-7292955	179	3009	3	TR35/	
113-2		3024			
113-4		3124			
113-118	123	3124	20	TR86/53	
113-398	108	3039		TR95/	
113-938	108	3039		TR95/	
113N1AG	172	3156		TR69/S9100	S9100
113N2	172	3156	20	TR69/S9100	S9100
114-1		3025			
114-1(PHILCO)	192				
114-118	108	3018	17	TR95/56	
114-267	108	3039		TR95/	
114A-1-82	121	3009		TR01/	
114N2P	184	3041		/S5003	
114N4P	103A/410			/250	
114N4U	103A				
114P1P	185	3084		/S5007	
114P3P	158/410			/630	
114P3U	158				
115-063	121	3009		TR01/	
115-1	123	3124	20	TR86/53	
115-2		3124			
115-3		3124			
115-4	123	3124	20	TR86/53	
115-5		3124			
115-6		3124			
115-13		3018			
115-14		3018			
115-225	123A	3122		TR21/	
115-227	160	3006		TR17/	
115-228	160	3006		TR17/	
115-229	160	3006		TR17/	
115-268	121	3009		TR01/	
115-269	121	3009		TR01/	
115-275	126	3006		TR17/	
115-281	121MP	3013			
115-282	121MP	3013			
115-283	121MP	3013			
115-284	121MP	3013			
115-875	123A	3122		TR21/	
115C5	784	3524			
115C7	786	3140			
115C-8	912				
116-068				TR27/	
116-072	126	3006		TR17/	
116-073	108	3039		TR95/	
116-074	123A	3122		TR21/	
116-075	124	3021		TR81/	
116-078	123A	3122		TR21/	
116-079	108	3039		TR95/	
116-080	108	3039		TR95/	
116-082	108	3039		TR95/	
116-083	108	3039		TR95/	
116-085	123A	3122		TR21/	
116-086	127	3035		TR27/	
116-087	127	3035		TR27/	
116-088	127	3035		TR27/	
116-089	127	3035		TR27/	
116-091	102A	3004		TR85/	
116-092	123A	3122		TR21/	
116-1	124	3021	12	TR81/240	
116-198	108	3039		TR95/	
116-199	108	3039		TR95/	
116-200	108	3039		TR95/	
116-201	102A	3004		TR85/	
116-202	160	3006		TR17/	
116-203	102A	3004		TR85/	
116-206	102A	3004		TR85/	
116-207	160	3007		TR17/	
116-208	160	3007		TR17/	
116-209	160	3007		TR17/	
116-588	123A	3122		TR21/	
116-683	160	3006		TR17/	
116-684	160	3006		TR17/	
116-685	102A	3004		TR85/	
116-686	102A	3004		TR85/	
116-687	103A	3010		TR08/	
116-756	126	3006		TR17/	
116-757	102A	3004		TR85/	
116-875	123A	3122		TR21/	
116-997	102A	3004		TR85/	
116C3475	181	3036			
117-1	124	3021	12	TR81/240	
117-2(HTHKT)	158(4)		53	/630	
117-5					S0014
117-6(HTHKT)	103A+158		54	/724(CP)	
117-9(HTHKT)	184+185		57/58	/S5007(CP)	
118-1	108	3018	17	TR95/56	
118-2	108	3018	17	TR95/56	
118-3	108	3018	17	TR95/56	
118-4	108	3018	17	TR95/56	
119		3122		TR51/735	S0024
119-0016	105	3012		TR03/	
119-0054	123A	3124		TR21/	
119-0055	159	3114		TR20/	
119-0056	123A	3124		TR21/	
119-0068	175	3026		TR81/	
119-0075	181	3036		TR36/	
119-0077	128	3024	18	TR87/243	
120-00-19	102	3004		TR05/631	G0005
120-00190	126	3006		TR17/635	
120-002	912				
120-00195	102A			TR85/250	
120-00213	126	3006		TR17/635	
120-001190	160	3006		TR17/637	G0008
120-001192	100	3005	1	TR05/254	G0005
120-001195	102	3004	2	TR05/631	G0005
120-001795	102			TR05/	G0005
120-001798	195			TR65/	S3013
120-002012	102A			TR85/250	G0005
120-002013	102A	3004	2	TR85/250	G0005
120-002014	102A	3004	51	TR85/250	G0005
120-002213	160	3007		TR17/637	G0002
120-002214	160	3006	51	TR85/637	
120-002216	160	3007	51	TR17/637	G0002
120-002513	126	3006	51	TR17/635	G0003
120-002515	126	3006	1	TR17/635	G0003
120-002518	126	3006	51	TR17/635	G0003
120-002520	158	3006	1	TR84/630	G0005
120-002521	102A	3004	2	TR85/250	G0005
120-002656	126	3006		TR17/635	G0003
120-002748	102A	3004	2	TR85/250	G0005
120-003150	131	3052		TR94/	G6016
120-003151	195			TR65/	S3013
120-004048	126	3006		TR17/635	G0008
120-004480	123A			TR21/	
120-004482	123A			TR21/	
120-004483	123A			TR21/	
120-004492	126	3006	50	TR17/635	G0009
120-004493	158	3004	52	TR84/630	G0005
120-004494	158	3004	52	TR84/630	G0005
120-004495	158	3004	52	TR84/630	G0005
120-004496	108	3018	11	TR95/56	S0016

Original Part Number	ECG	SK	GE	IR/WEP	HEP
120-004497	108	3018	11	TR95/56	S0016
120-004722	160			TR17/	G0003
120-004723	108			TR95/	S0016
120-004724	108	3122	20	TR21/	S0014
120-004725	108	3122	20	TR21/	S0014
120-004727	102A	3004	2	TR05/	G0005
120-004728	158	3123	2	TR84/	G0007
120-004729	102A	3004	2	TR05/	G0005
120-004880	123A	3018		TR21/735	S0014
120-004881	108	3018		TR95/56	S0025
120-004882	123A	3018		TR21/735	S0014
120-004883	123A	3124		TR21/735	S0014
120-004884	191	3047		TR75/	S3021
120-004885	224	3049			
120-004886	224	3049			
120-004887	131	3009		TR94/735	G6016
120-004888	101	3010		TR08/641	G0011
120-005291	108	3018	11	TR80/	S0016
120-005292	108	3018	11	TR80/	S0016
120-005293	108	3018	11	TR80/	S0016
120-005294	108	3018	17	TR95/	S0014
120-005295	108	3018		TR95/	S0014
120-005296	108	3122	20	TR21/	S0015
120-005297	108	3122	20	TR21/	S0014
120-005298	108	3018	17	TR95/	S0014
120-006604	159	3118	21		S0012
120-01193	102A				
120-02213	160				
120-1	123A	3124	20	TR21/735	S0015
120-2	123A	3122	20	TR21/735	S0015
120-3	123A	3122	20	TR21/735	S0015
120-7	123A	3024	63	TR21/	
120-8	123A	3024	18	TR21/	S0015
120-8A	123A	3124		TR21/	
120-190	102A		51	/250	G0005
120BLU	123A	3122		TR21/735	S0015
120P1	159		21	TR20/717	S0013
120P1M	159	3118	21	TR20/	S0013
120U		3112			
121		3018		/735	
121-1	159	3004	21	TR20/717	S0031
121-1RED	159	3114		TR20/717	S0031
121(SEARS)				TR21/	
121-6	103A	3011	5	TR08/724	G0011
121-7	103A	3011	7	TR08/724	G0011
121-8	103A		54	TR08/724	G0011
121-9	102A	3123	1	TR85/250	G0005
121-10	102A	3123	51	TR85/250	G0005
121-11	102A	3123	53	TR85/250	G0005
121-12	102A	3123	53	TR85/250	G0005
121-14		3123	1	TR85/243	G0005
121-14(COL)	128			TR87/	
121-14(ZEN)	102A			TR85/	
121-15	101	3011	54	TR08/641	G0011
121-16	101	3011	54	TR08/641	G0011
121-17	101	3011	54	TR08/641	G0011
121-18	124		12	TR81/240	
121-18(ZEN)	102A			TR85/	
121-19	102A	3004	52	TR85/250	G0005
121-21	101	3011	5	TR08/641	G0011
121-22	101	3011	5	TR08/641	G0011
121-24	101	3011	5	TR08/641	G0011
121-25	101	3011	5	TR08/641	G0011
121-26	101	3011	5	TR08/641	G0011
121-27	102	3004	53	TR05/631	G0005
121-33	102A		7	TR85/250	G0011
121-34	102A	3004	53	TR85/250	G0005
121-44	126	3007	50	TR17/635	G0003
121-45	126	3005	50	TR17/635	G0003
121-46	102A	3004	53	TR85/250	G0005
121-47	102A	3004	53	TR85/250	
121-47GRN					G0005
121-47RED					G0005(2)
121-47WHT					G0005(2)
121-47YEL					G0006(2)
121-48	160	3008	50	TR17/637	G0003
121-49	160	3007	50	TR17/637	G0003
121-50	101	3011	5	TR08/641	G0011
121-51	101	3005	5	TR08/641	G0011
121-52	102A	3004	52	TR85/250	G0005
121-52A		3018			
121-53	100	3005	2	TR05/254	G0005
121-54	126	3005	51	TR17/635	G0008
121-59	103A		54	TR08/724	G0011
121-60	103A	3011	6	TR08/724	G0011
121-61	102A	3003	52	TR85/250	G0005
121-62	126	3005	50	TR17/635	G0003
121-63	160	3005	1	TR17/637	G0009
121-64	102A	3003	52	TR85/250	G0005
121-65	126	3005	50	TR17/635	G0003
121-66	126	3005	50	TR17/635	G0003
121-67	126	3005	51	TR17/635	G0009
121-68	102A	3004	52	TR85/250	G0005
121-69	102A	3004	52	TR85/250	G0005
121-70	101	3011	5	TR08/641	G0011

Original Part Number	ECG	SK	GE	IR/WEP	HEP
121-71	101	3011	5	TR08/641	G0011
121-72	102A	3005	2	TR85/250	G0005
121-73	160	3005	50	TR17/637	G0003
121-74	160	3005	50	TR17/637	G0003
121-75	160	3005	50	TR17/637	G0003
121-76	160	3005	50	TR17/637	G0003
121-78	160	3005	50	TR17/637	G0003
121-79	160	3123	2	TR17/637	G0005
121-80	100	3005	2	TR05/254	G0005
121-81	100	3005	2	TR05/254	G0005
121-82	100	3005	2	TR05/254	G0005
121-83	100	3005	2	TR05/254	G0005
121-84	100	3005	2	TR05/254	G0005
121-85	100	3005	2	TR05/254	G0005
121-86	100	3005	51	TR05/254	G0005
121-87	100	3005	51	TR05/254	G0005
121-88	100	3005	51	TR05/254	G0005
121-89	100	3005	51	TR05/254	G0005
121-90	100	3005	51	TR05/254	G0005
121-91	100	3005	1	TR05/254	G0005
121-92	100	3005	1	TR05/254	G0005
121-93	100	3005	1	TR05/254	G0005
121-94	100	3005	52	TR05/254	
121-95	102	3004	2	TR05/631	G0005
121-96	102	3004	2	TR05/631	G0005
121-100	101		1	TR08/641	G0011
121-101	160		50	TR17/637	G0003
121-102	160	3005	50	TR17/637	G0003
121-103	160	3005	50	TR17/637	G0003
121-104	160	3005	50	TR17/637	G0003
121-105	160	3005	50	TR17/637	G0003
121-106	102	3004	52	TR05/631	G0005
121-107	102	3004	52	TR05/631	G0005
121-113	126	3006	51	TR17/635	G0009
121-119	160			TR17/637	
121-120	158	3004	52	TR84/630	G0005
121-128	100	3005	51	TR05/254	G0008
121-132	160	3006		TR17/	G0003
121-134	160		3	TR17/637	G0009
121-135	160	3006	51	TR17/637	G0009
121-136	160		3	TR17/637	G0009
121-137	160	3006	51	TR17/637	G0008
121-138	160	3008	51	TR17/637	G0009
121-139	160	3006	51	TR17/637	G0009
121-145	100	3008	51	TR05/254	G0008
121-146	100	3008	51	TR05/254	G0008
121-147	100	3005	51	TR05/254	G0008
121-148	102A	3003	52	TR85/250	G0001
121-150	126	3008	51	TR17/635	G0001
121-151	102A	3	53	TR85/250	G0005
121-152	102A		53	TR85/250	G0002
121-153	126	3008	51	TR17/635	G0008
121-154	126	3008	51	TR17/635	G0008
121-157	160	3006	50	TR17/637	
121-160	100	3005	2	TR05/250	G0005
121-161	126	3005	50	TR17/635	G0003
121-162	126	3005	50	TR17/635	G0003
121-163	102A	3004	52	TR85/250	G0005
121-164	102A	3004	52	TR85/250	G0005(2)
121-165					G0011
121-166					G0011
121-167					G0005
121-168		3004			
121-169		3010			
121-170		3008			S0019
121-171	121			TR01/	
121-172		3010			
121-178		3004			
121-179	160	3005	50	TR17/637	G0002
121-180	160	3005	50	TR17/637	G0003
121-181	160		50	TR17/637	G0002
121-184	102A	3004	2	TR85/250	G0007(2)
121-185	160		50	TR17/637	G0003
121-186	126	3005	2	TR17/635	G0008
121-182D		3004			
121-187	126	3006	2	TR17/635	G0009
121-189	126	3008	2	TR17/635	G0005
121-189B		3008			
121-190	102A	3004	52	TR85/250	G0005
121-190E		3004			
121-191	102A	3004	52	TR85/250	G0005
121-192	102A	3003	52	TR85/250	G0005
121-193	102A	3004	2	TR85/250	
121-193GRN					G0005(2)
121-193RED					G0005(2)
121-193WHT					G0005(2)
121-193YEL					G0005(2)
121-195B	123A			TR21/	
121-200					S0032
121-205	100	3005	53	TR05/254	G0007
121-206	100	3005	51	TR05/254	G0005
121-207	100	3005	51	TR05/254	G0005
121-208	100	3005	51	TR05/254	G0005
121-209	100	3005	51	TR05/254	
121-210	100	3005	53	TR05/254	

Original Part Number	ECG	SK	GE	IR/WEP	HEP
121-211	100	3005	53	TR05/254	
121-212	100	3005		TR05/254	
121-213	100	3005		TR05/254	
121-219	100	3005	51	TR05/254	G0006
121-220	100	3005	51	TR05/254	G0005
121-221	100	3005	53	TR05/254	G0005
121-222	100	3005	53	TR05/254	G0005
121-225	100	3004	52	TR05/254	G0005
121-226	102A	3004	52	TR85/250	G0005
121-227	102A	3004	52	TR85/250	
121-227GRN					G0005(2)
121-227ORN					G0005(2)
121-227RED					G0005(2)
121-227WHT					G0005(2)
121-227YEL					G0005(2)
121-228	126	3006	51	TR17/635	G0003
121-229	126	3006	51	TR17/635	G0003
121-230	126	3006	51	TR17/635	G0003
121-231	126	3006	51	TR 7/635	G0003
121-232	126	3004	2	TR17/635	G0003
121-233	126	3006	52	TR17/635	G0003
121-234	100	3005	2	TR05/254	G0005
121-235	100	3005	2	TR05/254	G0005
121-236	100	3005	2	TR05/254	G0005
121-237	103A		54	TR08/724	G0005
121-238	103A		54	TR08/724	G0011(2)
121-239	102A		51	TR85/250	G0005
121-240	126	3005	1	TR17/635	G0009
121-240X	102A			TR85/250	G0005
121-241	154	3005	47	TR78/712CP	G0009
121-241(ZEN)	126			TR17/	
121-242	160	3008	51	TR17/637	G0008
121-243	160	3008	51	TR17/637	G0008
121-244	160	3008	51	TR17/637	G0008
121-245	102A		50	TR85/250	G0005
121-246	102A	3004	52	TR85/250	G0005(2)
121-247	103A		16	TR08/724	
121-247BLK					G0011
121-247VIL					G0011
121-247WHT					G0011
121-248	103A		16	TR08/724	
121-248BLK					G0011(2)
121-248BLU					G0011(2)
121-248VIL					G0011(2)
121-254	100	3005	1	TR05/254	
121-156	126	3007	51	TR17/635	G0001
121-157	126	3008	50	TR17/635	G0009
121-258	126	3007	50	TR17/635	G0003
121-259	126	3007	50	TR17/635	G0003
121-260	126	3007	50	TR17/635	G0003
121-261	126	3006	53	TR17/635	G0002
121-262	126	3006	53	TR17/635	G0002
121-263	126	3006	53	TR17/635	G0002
121-266	102	3004	52	TR05/631	G0005
121-267	102	3004	52	TR05/631	G0005
121-268	160		51	TR17/637	G0003
121-269	160		51	TR17/637	G0003
121-270	121	3009	16	TR01/232	
121-270BLK					G6005
121-270BLU					G6005
121-270BRN					G6005
121-270GRN					G6005
121-270ORN					G6005
121-270RED					G6005
121-270VIL					G6005
121-270YEL					G6005
121-271	121	3009	16	TR01/232	G6005
121-272	102A	3004	52	TR85/250	G0005
121-273	102A	3005	2	TR85/250	G0006
121-274	102A	3003	52	TR85/250	G0006
121-275	102A	3005	2	TR85/250	G0006
121-276	123	3124	20	TR86/53	G0011
121-277	123	3124	20	TR86/53	G0011
121-278	123	3124	20	TR86/53	S0014
121-279	128/411	3124	20	TR21/243	S0014
121-283	161			TR83/	
121-284	160			TR17/	
121-286					S0025
121-287					S0032
121-288					S0014
121-289	126	3005	51	TR17/635	G0008
121-290	126	3007	51	TR17/635	G0005
121-291	102A	3004	2	TR85/250	G0005
121-292	126	3006		TR17/635	G0009
121-293	126	3004	47	TR17/635	G0009
121-294	126	3006	51	TR17/635	G0008
121-295	126	3006		TR17/635	G0008
121-296	126	3006	53	TR17/635	G0008
121-297	126	3006	51	TR17/635	G0003
121-298	126	3006	51	TR17/635	G0003
121-299	126	3006	51	TR17/635	G0003
121-300	102A	3123	53	TR85/250	G0006
121-301	102A	3123	52	TR85/250	G0006(2)
121-302	101	3124	8	TR08/641	G0011
121-303	108	3039	20	TR95/56	S0020
121-304	160	3006	17	TR17/635	G0003
121-305	102A	3004	52	TR85/250	G0007
121-306	102A	3004	52	TR85/250	G0007
121-307	102A	3004	52	TR85/250	G0007
121-308	121	3009	16	TR01/232	
121-308BLU					G6013
121-308GRN					G6013
121-308ORN					G6013
121-308RED					G6003
121-308YEL					G6013
121-309	102A	3004	53	TR85/250	G0006
121-310	102A	3005	52	TR85/250	G0006
121-311	102A	3004	53	TR85/250	
121-311A					G0007(2)
121-311B	102A		53	TR85/250	G0007(2)
121-311C					G0006(2)
121-311D					G0006(2)
121-311E					G0005(2)
121-311F					G0005(2)
121-312	160	3008	50	TR17/637	G0009
121-313	160	3008	50	TR17/637	G0009
121-314	102A	3004	2	TR85/250	G0007
121-315	124	3021	12	TR81/240	S5011
121-316	108	3039		TR95/56	
121-317	108	3039		TR95/56	
121-318	108	3039		TR95/56	
121-318L	108	3039		TR95/56	
121-319	102A	3004	52	TR85/250	G0005
121-320	102A	3004	52	TR85/250	G0005
121-321	108	3039	54	TR95/56	S0020
121-327	102A	3004	54	TR85/250	G0006
121-328	102A	3004	51	TR85/250	
121-328GRN					G0005(2)
121-328RED					G0005(2)
121-328WHT					G0005(2)
121-328YEL					G0005(2)
121-329	160		51	TR17/637	G0003
121-330	160		51	TR17/637	G0003
121-331	160		51	TR17/637	G0003
121-332	160		16	TR17/637	G0001
121-333	160	3005	51	TR17/637	G0008
121-334	160	3005	51	TR17/637	G0008
121-335	160	3005	51	TR17/637	G0008
121-336	160		50	TR17/637	G0003
121-345	108	3019	20	TR95/56	S0020
121-347	102A	3004	2	TR85/250	G0005
121-348	102A	3004	2	TR85/250	G0006
121-349	160	3006	50	TR17/637	G0001
121-350	160	3006	50	TR17/637	G0001
121-351	160	3006	50	TR17/637	G0001
121-352	160	3006	50	TR17/637	G0001
121-353	160	3006	50	TR17/637	G0001
121-354	100		51	TR05/254	G0005
121-356	160	3006	51	TR17/637	G0003
121-357	160		51	TR17/637	G0003
121-358	160	3006	51	TR17/637	G0003
121-359	160	3006	51	TR17/637	G0001
121-360	126	3006		TR17/635	G0007
121-361	154	3045		TR78/712	S5025
121-362	158			TR84/630	G0007
121-363	121	3014	76	TR01/232	G6009
121-364	123A	3124	47	TR21/735	S0024
121-365	123A	3124	20	TR21/735	G0011
121-366	123A	3124	47	TR21/735	S0024
121-367	123A	3124	17	TR21/735	S0011
121-368	158	3123		TR84/630	G0005
121-369	123A	3124	20	TR21/	S0015
121-370	127			TR27/235	G6008
121-371	121	3009	3	TR01/232	G6013
121-372	102	3004	2	TR05/631	G0006
121-373	158	3004	53	TR84/630	
121-373A					G6011(2)
121-373B					G6011(2)
121-373BLU					G6011(2)
121-373BRN					G6011(2)
121-373C					G6011(2)
121-373D,E,F					G6011(2)
121-373G					G6011(2)
121-373GRN					G6011(2)
121-373H,I,J					G6011(2)
121-373K,L,M					G6011(2)
121-373N,O					G6011(2)
121-373ORN					G6011(2)
121-373P,Q					G6011(2)
121-373RED					G6011(2)
121-373VIL					G6011(2)
121-373YEL					G6011(2)
121-374	102A	3004	53	TR85/250	G0006
121-375	102A	3004	52	TR85/250	G0006
121-377	161	3018	17	TR83/719	S0017
121-378	161	3018	17	TR83/719	S0017
121-379	161	3018	17	TR83/719	S0017
121-380	161	3018	20	TR83/719	S0017
121-381	126	3008	50	TR17/635	G0009
121-382	121	3009	3	TR01/232	

Original Part Number	ECG	SK	GE	IR/WEP	HEP
121-382BLU					G6005
121-382BRN					G6005
121-382GRN					G6005
121-382VIL					G6005
121-382YEL					G6005
121-383	161	3006	61	TR83/719	G0003
121-384	160	3006	51	TR17/637	G0002
121-385	160	3006	51	TR17/637	G0003
121-388	158	3004	2	TR84/630	G6011
121-389	121	3009	3	TR01/232	G6005
121-395	102A	3004	2	TR85/250	G0007
121-396	102A	3004	2	TR85/250	G0006
121-397	100	3005	1	TR05/254	G0001
121-398	121	3009	16	TR01/232	
121-398BL					G6013(2)
121-398BLU					G6013(2)
121-398GRN					G6013(2)
121-398ORN					G6013(2)
121-398RED					G6013(2)
121-398YEL					G6013(2)
121-399	102A	3004	52	TR85/250	G0006
121-400	158	3004	80	TR84/630	G6011
121-401	158	3004	80	TR84/630	
121-401BLK					G6011(2)
121-401BRN					G6011(2)
121-401ORN					G6011(2)
121-401RED					G6011(2)
121-401YEL					G6011(2)
121-403	158	3004	53	TR84/630	
121-403A					G6011(2)
121-403BLU					G6011(2)
121-403BRN					G6011(2)
121-403C,D,E,F,G					G6011(2)
121-403GRN					G6011(2)
121-403H,I,J,K					G6011(2)
121-403L,M,N,O					G6011(2)
121-403ORN					G6011(2)
121-403P,Q					G6011(2)
121-403RED					G6011(2)
121-403YEL					G6011(2)
121-404	123A	3124	20	TR21/735	S0015
121-406	179			TR35/	
121-408	102A	3004	52	TR85/250	G0006
121-409	102A	3004	52	TR85/250	G0006
121-410					G0005
121-411	160	3006	50	TR17/637	G0003
121-412	160	3006	50	TR17/637	G0003
121-413	160	3006	50	TR17/637	G0003
121-414	160	3006	50	TR17/637	G0003
121-415	160	3006	50	TR17/637	G0003
121-415B	160		50	TR17/	
121-417	159		22	TR20/717	S0019
121-418	179(2)	3009	16	TR01/G6001	
121-418BLU					G6013(2)
121-418GRN					G6013(2)
121-418ORN					G6005(2)
121-418RED					G6005(2)
121-418VIL					G6013(2)
121-418YEL					G6013(2)
121-419	179		22	TR35/	S0019
121-420	102A		52	TR85/250	G0006
121-421	158		53	TR84/630	
121-421A,B					G6011(2)
121-421BLU					G6011(2)
121-421BRN					G6011(2)
121-421C,D,E,F,G					G6011(2)
121-421GRN					G6011(2)
121-421H,I,J,K					G6011(2)
121-421L,M,N,O					G6011(2)
121-421ORN					G6011(2)
121-421P,Q					G6011(2)
121-421RED					G6011(2)
121-421VIL					G6011(2)
121-421YEL					G6011(2)
121-422	123	3124	20	TR86/53	
121-423	123A				
121-425	158	3004	53	TR84/630	
121-425A,B					G6011(2)
121-425BLU					G6011(2)
121-425BRN					G6011(2)
121-425C,D,E,F,G					G6011(2)
121-425GRN					G6011(2)
121-425H,I,J,L					G6011(2)
121-425M,N,O					G6011(2)
121-425ORN					G6011(2)
121-425P,Q					G6011(2)
121-425RED					G6011(2)
121-425VIL					G6011(2)
121-425YEL					G6011(2)
121-426	160	3006	50	TR17/637	
121-427	160	3006	51	TR17/637	
121-428	160	3006	50	TR17/637	
121-429	160	3006	50	TR17/637	
121-430	123A	3124	20	TR21/735	
121-430B	123A	3122		TR21/735	

Original Part Number	ECG	SK	GE	IR/WEP	HEP
121-430CL	123A	3124		TR21/	
121-431	128	3124	20	TR87/243	
121-432	160	3006	51	TR17/637	G0008
121-433	123A	3124	20	TR24/735	S0024
121-433CL	123A	3124		TR21/	
121-434	123A	3124	10	TR21/735	S0015
121-434H	123A	3018		TR25/	
121-435	123A	3122	10	TR21/735	S0024
121-436	124	3021	12	TR81/240	S5011
121-437	102A	3004	2	TR85/250	G0006
121-441	159	3114	22	TR20/717	S0019
121-442	123A	3020	82	TR21/735	S5022
121-444	159	3114		TR20/717	S0019
121-445	154	3040	27	TR78/712	S3019
121-446	159	3025	21	TR30/717	S0019
121-447	123A	3124	20	TR51/735	S0015
121-448	123A	3122	47	TR21/735	S0015
121-449	162		35	TR67/707	S5020
121-450	123A	3124	17	TR21/735	S0015
121-451	124	3021	12	TR81/240	S5011
121-452	163	3111	36	TR67/740	S5021
121-453	161	3018	20	TR51/719	S0016
121-460	161	3117		TR83/719	
121-461	161	3117		TR83/719	
121-462	161	3018	11	TR80/719	S0017
121-470	161	3018	11	TR70/719	S0016
121-471	161	3018		TR83/719	S0017
121-472	108	3018	11	TR95/56	S0020
121-473	154			TR78/712	
121-480	107	3039	11	TR70/720	S0020
121-481	108	3018	11	TR95/56	S0020
121-482	108	3039	11	TR95/56	S0020
121-483	108	3018	11	TR95/56	S0020
121-490	102A	3004	52	TR05/250	G0005
121-491	126	3004	52	TR85/635	G0008
121-492	126	3004	52	TR85/635	G0008
121-493	126	3008	52	TR85/635	G0005
121-494	126	3004	52	TR85/635	G0005
121-495	159	3114	21	TR20/717	S0019
121-496	159	3025	21	TR20/717	S0019
121-497	159	3025	22	TR20/717	S0019
121-497WHT	159	3114		TR20/717	S0019
121-498	108	3018	20	TR51/56	S0025
121-499	123A	3124	20	TR2 ./735	S0024
121-500	161	3018	39	TR24/719	S0017
121-501	161	3018	39	TR24/719	S0017
121-502	161	3018	17	TR24/719	
121-503	161	3117	20	TR24/630	S0017
121-504	108	3117	20	TR24/56	S0016
121-505	161	3018	20	TR24/	S0016
121-506	161	3018	11	TR24/719	S0016
121-507	161	3039	20	TR53/719	S0014
121-508	161	3039	39	TR53/719	S0014
121-509	161	3039	20	TR53/719	S0014
121-510	161	3018	11	TR83/719	S0016
121-520	108	3018	17	TR95/56	S0016
121-521	161	3018	17	TR21/56	
121-522	233	3018	61	TR21/720	S0015
121-523	161	3018	11	TR21/719	
121-524	233	3018	61	TR21/719	S0015
121-524A	233		39	TR83/	
121-526	233	3018	20	TR21/	S0015
121-527					S0016
121-538	160	3006	50	TR17/637	G0003
121-538B	160		50	TR17/637	
121-539	160	3006	50	TR17/637	G0008
121-540	160	3006	50	TR17/637	G0008
121-540B	160		50	TR17/637	
121-541	160		50	TR17/637	G0008
121-541B	160		50	TR17/637	G0008
121-542	160		50	TR17/637	G0008
121-542B	160		50	TR17/637	G0008
121-543	102A	3004	52	TR85/250	G0006
121-544	102A	3004	52	TR85/250	G0006
121-546	108	3018	20	TR95/56	S0020
121-546B	108	3018		TR95/	
121-547	108	3018	20	TR95/56	S0016
121-551	108	3019	20	TR83/56	S0020
121-552	160	3123	51	TR17/637	G0003
121-553	160	3123	51	TR17/637	G0008
121-554	126	3006		TR17/635	G0008
121-555	126	3006		TR17/635	G0008
121-556					G0009
121-557	103A			TR08/724	G0011
121-558	103A			TR08/724	G0011(2)
121-560	108			TR95/	S0020
121-580	161	3018	17	TR83/719	S0017
121-581	123A	3124	47	TR21/735	S0015
121-582	124	3021	12	TR81/240	
121-582GRN					S5011(2)
121-582RED					S5011(2)
121-583					S0020
121-584		3009			G6013
121-584BRN					G6003
121-584GRN					G6013

Original Part Number	ECG	SK	GE	IR/WEP	HEP
121-584ORN					G6003
121-584RED					G6003
121-584YEL					G6003
121-585	107	3039	39	TR70/720	S0016
121-858B	107	3039		TR70/720	
121-587	123A	3122	20	TR21/735	S0025
121-594					S0016
121-600	123A	3018	20	TR24/735	S0025
121-600(ZEN)	154			TR78/	
121-601	160		50	TR17/637	G0003
121-602	159	3025	67	TR30/717	S0019
121-603	159	3025	67	TR20/717	S0019
121-604					S0020
121-610	123A	3124	20	TR21/735	S0030
121-612	108	3018	11	TR95/56	S0020
121-612-16	108	3018		TR95	
121-613	108	3018	11	TR95/56	S0020
121-613-16	108	3018		TR95/	
121-614	108	3018	20	TR95/56	S0020
121-614-9	108	3018		TR95/	
121-615	106	3118		TR20/52	S0019
121-616	108	3039	60	TR95/56	S0020
121-617					S0015
121-619					S0015
121-620					S0015
121-622					S0015
121-623					S9002
121-624					S0019
121-625					S0020
121-626					S0020
121-627					S0020
121-629	123A	3124		TR21/735	
121-630	108	3019	17	TR83/56	S0020
121-632	102A	3004	2	TR85/250	G0007
121-633	102A	3004	52	TR85/250	G0007
121-634	102A		52	TR85/250	G0005
121-635	102A		52	TR85/250	G0007
121-636	102A	3123	53	TR85/250	
121-636A					G0005(2)
121-636B					G0005(2)
121-636C,D					G0005(2)
121-637	108	3018	210	TR95/56	S0015
121-638	108	3018	20	TR95/56	S0025
121-638B	108	3039	210	TR95/56	S0025
121-639	123A	3024	20	TR21/735	S0015
121-639CL	128	3024		TR87/	
121-640	102A	3004	2	TR85/250	
121-640BLU					G0005
121-640GRN					G0005
121-640ORN					G0005
121-640RED					G0005
121-640YEL					G0005
121-641	103A	3010	8	TR08/724	
121-641BLU					G0011
121-641GRN					G0011
121-641ORN					G0011
121-641RED					G0011
121-641YEL					G0011
121-642					S0016
121-643					S0016
121-644					S0016
121-645					S0025
121-646					S0015
121-647					S0015
121-648					S0025
121-649					S0015
121-650					G0011
121-651					S0005
121-652					S0015
121-653					S0020
121-654					S0019
121-655					S0030
121-656					S0030
121-657					S0015
121-658					S0016
121-659					S0030
121-660					S0015
121-661					S0019
121-662	123A	3122		TR21/735	S0015
121-663					S0032
121-664					S0005
121-665					S5004
121-666					F0015
121-667					S0019
121-668					S0015
121-670					S0024
121-671	123A	3124	20	TR51/735	S0015
121-672					S0015
121-674					G0005
121-675	123A	3018	20	TR53/735	S0015
121-676	128	3018	20	TR21/243	S0015
121-677	123A	3018	20	TR53/735	S0015
121-678	123A	3122		TR21/735	
121-678GRN					S0014
121-678RED					S0014

Original Part Number	ECG	SK	GE	IR/WEP	HEP
121-678YEL					S0014
121-679	159	3114		TR20/717	
121-679GRN					S0012
121-679RED					S0012
121-679YEL					S0012
121-680					S0019
121-681					S0019
121-682					S0019
121-683				TR52/	S0019
121-684					S0030
121-687	107	3039		TR70/720	S0016
121-692	161	3117		TR83/719	S0020
121-695	123A	3122	20	TR24/735	S0015
121-697	160	3006	51	TR17/637	G0002
121-698	160	3006		TR17/637	G0003
121-699	159	3114	65	TR20/717	S0019
121-701	123A	3122	60	TR21/735	S0025
121-702			17	TR21/	S0015
121-703		3137	47		S0015
121-704	161	3117	60	TR83/719	S0016
121-705			51		S0032
121-706	123A	3024	81	TR21/735	S3020
121-707	185	3025	58	TR77/S5007	S5006
121-708	184	3024	57	TR77/S5003	S5000
121-709	185	3025	58	TR77/S5007	S5006
121-710	184	3024	57	TR76/S5003	S5000
121-711	123A	3124	212	TR21/735	S0015
121-712	157	3103	12	TR23/244	S5015
121-713	124	3021	32	TR23/240	S5011
121-713(2)				TR23/	
121-714	160	3006	65	TR17/637	S0013
121-716			65	TR30/	S0032
121-719	152	3054	27	TR76/701	S5015
121-722	128	3024	81	TR25/243	S0005
121-723	107			TR21/	S0020
121-725					S0019
121-726	130	3027	72	TR59/247	S7002
121-726A	130	3027		TR59/	
121-730	123A	3122	39	TR21/735	S0015
121-731	132	3116	FET1	FE100/802	F0015
121-732	161	3018	20	TR83/719	S0020
121-733					S0025
121-734	102	3004	2	TR05/631	S0019
121-735	107	3018	210	TR70/720	S0025
121-737	123A	3124	20	TR21/250	S0005
121-737CL	128	3024		TR87/	
121-741					S0020
121-742	108	3019	11	TR95/	S0020
121-743	154	3040	40	TR83/712	S3021
121-744	154	3040	63	TR24/712	S0025
121-745	123A	3040	20	TR24/735	S0015
121-746	154	3114	21	TR88/717	S0012
121-748	123A	3124	20	TR25/735	S0020
121-750					F2004
121-751	123A	3124	63	TR21/735	S0015
121-752	172	3124	64	TR69/S9100	S9100
121-753	108	3018	20	TR95/56	S0014
121-754	229	3117	11	TR21/719	S0020
121-755	190	3104	27	TR78/637	S5015
121-756	132	3116		FE100/802	F0015
121-758	164	3133	37	TR68/740A	S5021
121-758X	164	3133	37	TR93/740A	
121-759	165	3115	38	TR61/740B	S5021
121-759X	165	3115	38	TR93/740B	
121-760	161	3117		TR83/719	
121-761	161	3117		TR83/719	G0009
121-762	101	3011	8	TR08/641	G0011
121-762CL	103A	3010		TR08/	
121-764	123A	3024		TR21/735	S0024
121-765	129	3025		TR88/242	S0019
121-766	128	3024	63	TR87/243	S0015
121-767	199	3124	20	TR21/735	S0015
121-767CL	123A	3124		TR21/	
121-768	199	3124	17	TR21/735	S0015
121-768CL	123A	3124		TR21/	
121-770	152		66	TR76/	S5003(2)
121-770CL	196	3054		TR92/	
121-770X	152(2)			/701	
121-772			66	/S5003	S5003(2)
121-772CL	196	3054		TR92/	
121-772X	184(2)	3190			
121-773	123A	3122	18	TR21/735	S0015
121-773CL	128	3024		TR87/	
121-774	159	3114	21	TR20/717	S0019
121-774CL	129	3025	21	TR88/	
121-775	161	3117		TR83/719	S0025
121-776	154	3103	40	TR78/712	S5024
121-777	159	3044	20	TR78/712	S5024
121-778					S0025
121-779		3018			S0020
121-782	222				F2004
121-783	222				F2004
121-784	222	3050			F2004
121-785	222	3050			F2004
121-786	222	3050			F2004

Original Part Number	ECG	SK	GE	IR/WEP	HEP
121-787	222	3050	39		F2004
121-792	154	3040		TR78/712	S3021
121-793	121	3009		TR01/232	G6005
121-800-401				TR60/	
121-801					S0019
121-802					S0017
121-803	153			TR77/700	S5006
121-804	152			TR76/701	S5000
121-805					S5000
121-806					S5000
121-807					S0025
121-808	184	3054	29	TR55/701	S5000
121-809					S5000
121-812					S0015
121-818					S0019
121-819	108	3022	20	TR21/	S0015
121-821	164	3133	37	TR68/740A	S5021
121-822	171	3054	27	TR60/244A	S3021
121-823	161			TR83/719	S0025
121-824	161	3132		TR83/719	S0020
121-825	108	3132	20	2SC535B/	S0030
121-826	222	3084			F2004
121-827	108	3018	20	2SC535B/56	S0020
121-829	162	3079	14	TR61/	S5020
121-830	100	3009	51	TR05/254	
121-830BLU					G6005
121-830GRN					G6005
121-830ORN					G6005
121-830RED					G6005
121-830YEL					G6005
121-831	165	3115	38	TR67/740B	S7005
121-834	108	3018	60	TR51/735	S0015
121-835					S0016
121-836	123A	3122	17	TR51/735	S0015
121-837	123A				
121-838	159				
121-840					S0016
121-841		3018	11	TR80/	S0025
121-843	190	3024	63	TR25/S3021	S0005
121-844	128			TR87/	S0015
121-845	129	3114		TR51/	S0019
121-846	108	3018	20	TR21/56	S0009
121-847					S0017
121-848	108	3018	11	TR95/56	S0016
121-849	107	3018		TR95/720	S0016
121-850	123A	3018	20	TR51/	S0015
121-851	108	3018	11	TR20/56	S0016
121-853	186	3054	28	TR55/	S5004
121-854					S5000
121-855	108	3018	11	TR80/	S0016
121-856	123A	3018	20	TR21/735	S0015
121-857	108	3018	20	2SC535B/56	S0010
121-858	132	3116	FET2	FE100/	F0015
121-861	159	3114	22	2SC535B/	S0019
121-862	123A	3020		TR24/	S0025
121-863	123A	3122	20	TR51/	S0015
121-865	159	3114	21	TR24/	S0019
121-868	171	3104	27	TR78/	S3021
121-869	108	3018	11	TR95/	
121-871					S0025
121-872					S0016
121-873	153		58		S5006
121-874	152				S5000
121-875	159				S0019
121-876					S0015
121-877		3124	20	TR51/	S0025
121-878		3124	20	TR53/	S0015
121-879		3114	21	TR28/	S0019
121-880	196	3054	28	TR76/	S5004
121-881	123A	3122	63	TR25/	S5014
121-883	161	3018	20	TR21/	S0015
121-884	108	3018	11	TR95/56	S0016
121-885	107	3018	11	TR80/	S0020
121-886	153	3083	69	TR77/	S5006
121-887	152	3054	66	TR76/	S5000
121-888	106	3122	20	TR24/	S0025
121-889	123A	3124	20	TR24/	S3010
121-895	108	3018	60	TR51/	S0015
121-895A	108		17		S0015
121-898	108	3018	11	TR80/	S0016
121-899	108	3018	11	TR80/	S0016
121-900	108	3018	20	TR95/	S0010
121-906					S0025
121-907					S0025
121-908					S0025
121-909					S0025
121-910					S0025
121-911	157	3103		TR	S5015
121-912					S5011
121-913					S0025
121-914					S0025
121-915					S0025
121-916					S0024
121-924	161	3117	20	TR83/	S0015
121-925	108	3018	20	2SC535B/	S0020

Original Part Number	ECG	SK	GE	IR/WEP	HEP
121-926	242	3084	58	TR77/	S5006
121-927	241	3054	57	TR76/	S5000
121-928					S0015
121-929					S0015
121-930					S0015
121-931		3122	20	2SC535B/	S0015
121-932	108	3018	11	TR80/	S0016
121-933	159	3118	21	TR20/717	S5022
121-943					S0015
121-944					S0020
121-945					S0015
121-946					S0015
121-950	123A	3018	20	TR24/	S0016
121-951	107	3018	20	TR21/	S0016
121-952	154	3114	22		S5022
121-953	222	3065			F2004
121-954	161	3018	20	TR70/	S0016
121-956					S5000
121-957					S5006
121-966	152	3054	57	TR55/	S5000
121-968	107	3018	11	TR80/	S3021
121-969	153		69	TR77/	S5006
121-970	152		66	TR76/	S5000
121-972		3122	20	TR24/	S0015
121-973	123A	3114	20	TR	S0015
121-974	107	3018	11	TR80/	S0020
121-975		3122	20	TR24/	S0025
121-976	152	3054	66	TR76/	S5000
121-977	153	3084	69	TR77/	S5006
121-978	159	3025	22	TR24/	S0019
121-980		3083	58	TR77/	S5015
121-982		3122	20	TR21/	S0025
121-985	165	3115	36	TR93/	S5021
121-986	159	3025	22	TR30/	S0019
121-987	152		66	TR76/	S5000
121-988	153		69	TR77/	S5005
121-989	171	3104	27	TR78/	
121-990	171	3104	27	TR78/	
121-992		3083			S3031
121-993		3131			S5011
121-994		3083			
121-996	152	3054	28	TR55/	
121-1003	238	3111			
121-1032	102	3004	50	TR05/631	G0005
121-1033	102	3004	52	TR05/631	G0005
121-1034	102	3004	52	TR05/631	G0005
121-1035	102	3004	52	TR05/631	G0005
121-1036	102	3004	52	TR05/631	G0005
121-1124	121	3009	16	TR01/232	G6005
121-1134	121	3014	16	TR01/232	G6005
121-1330	100	3005	2	TR05/254	G0005
121-1350	100	3005	2	TR05/254	G0005
121-1360	100	3005	2	TR05/254	G0005
121-1390	100	3005	2	TR05/254	G0005
121-1400	100	3005	2	TR05/254	G0005
121-1410	101	3011	8	TR08/641	
121-5065	123A			TR21/	
121-5461				TR21/	
121-5872				TR25/	
121-505501	123A			TR21/	
121G3019	123A	3122		TR21/735	S0011
121G3020	123A	3122		TR21/735	S0011
121G3115		3024			
121G3124		3024			
121G3383		3054			
121J688-1	107				
121J688-2	107				
121S2	6402				
121(SEARS)	123A				
121T2					S0025
122-1	123A	3122	20	TR21/735	S0025
122-2	123A	3122	20	TR21/735	S0015
122-6	123A	3018	20		
122-7		3018			
122-229	100	3005	2	TR05/254	G0005
122-1028	105	3012		TR03/233	
122-1028A	105	3012		TR03/	
122-1625	121	3009		TR01/232	
122-1648	126		17	TR17/635	
122-1962	103	3010		TR08/641A	
122-A484	108	3018		TR95/56	S0017
122GRN	159	3114		TR20/717	S0019
122T2					S0025
122YEL	159	3114		TR20/717	S0019
123-002		3116			F2005
123-003		3112			
123-004		3018			S0015
123-005		3124			S0015
123-006		3138			
123-007		3124			S0015
123-010		3018			S0016
123-011		3054			S5003
123-011A		3054			S5003
123-012		3047			S0014
123-013		3088			

Original Part Number	ECG	SK	GE	IR WEP	HEP
123-015		3088			
123-025		3088			
123B-001		3027			
123B-002		3054			S5003
123N1	164	3133	37	TR68/	
123S425	128				
123S437	128				
123T2					S0024
124-1	121	3009	16	TR01/232	G6005
124N1	101			TR08/641	G0011
124N16	123A	3018		TR08/641	S0016
124-N16	101				
125-4			25	TR27/235	G6007
125-4(RCA)	127			TR27/	
125-121	128				
125-138		3054			
125-402	121			TR01/	G6018
125-403	121			TR01/	G6013
125-406					G6001
125-410	175		8	TR81/	S5019
125A134	102A				
125A137	128	3024			
125A137A	128				
125AS251	175				
125B132	123A				
125B133	159				
125B139	128				
125B410	175	3026			
125B415	152	3054			
125C3	784				
125C3RB	784				
125C211	123A				
125P1	159	3114	22	TR17/637	G0003
125P116	159	3114	22	TR20/	S0019
125PL	159				
125P1M			21	TR30/	S0019
125T1					G0005
125U		3112			
126-12	123A	3122		TR21/735	S0015
126-40	909	3590			
126N1	103A			TR08/724	
126N2	103A	3024		TR08/724	G0011
126P1	158	3025		TR84/630	G0005
126T1					G0005
127	123A	3122		TR21/735	S0015
127-7	101	3011		TR08/641	
127-115	190				
127T1					G0005
128	108	3039		TR95/56	S0016
128-9050	130				
128C212H01	9951				
128C213H01	910				
128C830H05	7493				
128N1				/243	
128N2	128	3018	20	TR87/	S0014
128WHT	123A	3122		TR21/735	S0024
129					S0016
129-4	176	3004	2	TR82/631	G6011
129-5	121	3009	16	TR01/232	G6003
129-6	104	3009	2	TR82/230	G6003
129-7	104	3009	16	TR01/230	G6003
129-8	102A	3004	52	TR85/250	
129-8-1	102A	3004	2	TR85/250	
129-8-1A	102A	3004	2	TR85/250	
129-8-2	102A	3004	2	TR85/250	
129-9	121	3009	3	TR01/232	G6003
129-10	121	3014	16	TR01/232	
129-11	100	3006	51	TR05/254	
129-13	121	3009	3	TR01/232	
129-14	123	3124	63	TR86/53	S0015
129-13	123	3124	20	TR86/53	
129-16	123A	3024	20	TR21/735	S0015
129-17	102	3004	2	TR05/631	
129-18	102	3004	2	TR05/631	
129-20	159	3025	21	TR20/717	
129-21	123A	3124	63	TR21/735	
129-23	175	3026	23	TR81/241	
129-27	161	3018	20	TR83/719	
129-30	103A	3011		TR08/641	G0011
129-31	102A			TR85/250	G0005
129-32	102A	3004	52	TR85/250	
129-33	152	3122	63	TR76/701	
129-33(PILOT)	123A			TR21/735	
129-34	159	3025	21	TR20/717	G0005
129-34(PILOT)	159			TR20/717	
129BRN	123A	3122		TR21/735	S0015
129N1	161		61	TR83/719	S0020
129WHT	123A	3122		TR21/735	
130	107	3018		TR70/720	S0015
130(SEARS)					S0015
130-104	121	3014		TR01/	G6005
130-112	107				
130-138	108	3039	11	TR95/56	S0017
130-144	128	3024		TR87/243	S3008
130-149	159	3114		TR20/717	S0015
130-150	128	3024		TR87/243	S3013
130-150-1					S3008
130-152	161	3117		TR83/719	S0017
130-172	195	3024		TR65/243	S3013
130-174	128	3024		TR87/243	S3013
130-185	161	3132		TR83/	S0017
130-191-00	128				
130-220					S0017
130-240	161	3132		TR83/	S0017
130-245	195	3024		TR65/243	S3013
130-40089	101	3011		TR08/	G0011
130-40095	102A	3004	53	TR85/250	G0005
130-40096	103A	3010	59	TR08/724	G0011
130-40214	123A	3124		TR21/735	S0015
130-40215	123A	3124		TR21/735	S0015
130-40216	199			TR21/	S0024
130-40236	102A	3004		TR85/250	G0005
130-40294	123A	3024		TR21/735	S0014
130-40304	108	3018		TR95/56	
130-40311	123A	3124		TR21/735	S0015
130-40312	123A	3124		TR21/735	S0015
130-40313	123A	3124		TR21/735	S0015
130-40314	103A		59	TR08/	
130-40315	159	3025		TR20/717	S0019
130-40317	123A	3124		TR21/735	S0015
130-40318	123A	3124		TR21/735	S0015
130-40347	103A	3010		TR08/724	
130-40349	131+155	3052	30	/642	G6016
130-40352	102A	3004		TR85/250	
130-40357	123A	3122		TR21/	S0015
130-40362	108	3018		TR95/56	
130-40421	108	3018		TR95/56	S0016
130-40429	159	3025		TR20/717	
130-40439	131+155			/642	
130-40456	158	3123	53	TR82/	G6011
130-40459	108	3018		TR95/56	
130-40883	199	3122	62	TR51/	G0011
130-40896	123A	3122	62	TR53/	S0015
130-40901	199	3122	62	TR08/	S0015
130-40922	123A	3122	62	TR53/	S0024
130ORN	107	3039		TR82/720	S0015
130VIO	107	3039		TR21/720	S0019
131		3018		TR21/56	
131(RAVIN)	108			TR95/	
131(SEARS)	108			TR95/	S0016
131-000561	128				
131-000562	121				
131-001-007	130				
131-005-353-1	197				
131-005-807	128				
131-005-808	129	3025			
131-043-67	130				
131-045-60	128				
131-045-61	129				
131-04367	130				
131AS471	941				
131N2	191	3124	28	TR21/735	S3024
131N2G	152		62	TR76/701	S0005
131N2(MAGNAVOX)	188			TR72/	
132-001	102A	3004		TR85/	
132-002	123A	3122		TR21/	
132-003	171	3201		TR79/	
132-004	123A	3122	20	TR21/735	
132-005	123A	3122		TR21/	
132-007	129	3025		TR88/	
132-008	161	3132		TR83/	
132-009	161	3132		TR83/	
132-010	102A	3004		TR85/	
132-011	123A	3122	20	TR21/735	
132-014	171	3103		TR79/	
132-015	108	3039		TR95/56	
132-017	123A	3122	20	TR21/735	
132-018	123A	3122		TR21/	
132-019	160	3006		TR17/637	
132-020	160	3006		TR17/637	
132-021	123A	3122	20	TR21/735	
132-021B	128	3024		TR87/243	
132-022	128	3024		TR87/	
132-023	123A	3122	20	TR21/735	S5012
132-024	185	3083	26	/S5007	S5018
132-025	162			TR67/707	
132-026	123A	3122	20	TR21/735	
132-027	160	3006		TR17/637	
132-028	124	3021		TR81/240	
132-029	234	3025		TR20/	
132-030	123A	3122		TR21/	
132-031	234	3118		TR20/	
132-032	129	3025		TR88/242	
132-033	124	3021		TR81/	
132-038	128	3024		TR87/243	
132-039	129	3025		TR88/242	
132-041	123A	3122		TR21/735	
132-042	123A	3122		TR21/735	
132-045	221	3050			
132-046	184	3054			

Original Part Number	ECG	SK	GE	IR WEP	HEP
132-047	221	3050			
132-049	132	3116		FE100/	
132-050	123A	3122		TR21/	
132-051	123A	3122		TR21/	
132-052A	172	3156		TR69/S9100	
132-054	123A	3122		TR21/735	
132-055	123A	3122		TR21/735	
132-056	159	3024		TR20/717	
132-057	123A	3122		TR21/735	
132-059	124	3021		TR81/240	
132-062	123A	3122		TR21/735	
132-065	175	3538		TR81/241	
132-066	128	3024		TR87/243	
132-069	123A	3122		TR21/735	
132-070	130	3027		TR59/247	
132-072	184	3054		/S5003	
132-074	159	3025		TR20/	
132-075	123A	3122		TR21/	
132-076	161	3132		TR83/	
132-077	123A	3122		TR21/	
132-078	157	3103		TR60/	
132-079	185	3083		/S5007	
132-080	184	3054		/S5003	
132-081	184	3054		/S5003	
132-082	161	3132		TR83/	
132-085	130	3027		TR59/247	
132-087	161	3132			
132-090	102A	3004			
132-2	151				
132-3	152	3054		TR55/701	S5000
132-4	152				
132-5(RCA)		3054			S5000
132-90	102A	3004	53	TR85/	
132-185	161	3132		TR83/	S0017
132-501	123A	3122		TR21/	
132-502	123A	3122		TR21/	
132-503	123A	3122		TR21/	
132-504	123A	3122		TR21/	
132-515	124	3021		TR81/	
132-516	124	3021		TR81/	
132-521	124	3021		TR81/	
132-522	124	3021		TR81/	
132-523	124	3021		TR81/	
132-524	124	3021		TR81/	
132-525	124	3021		TR81/	
132-526	124	3021		TR81/	
132-539	123A	3122		TR21/	
132-540	123A	3122		TR21/	
132-541	130	3027		TR59/	
132-542	130MP	3029		TR59MP/	
132N1	128	3024		TR87/	S5014
133-001	726	3129			
133-001B	726	3129			
133-002	710	3102			
133-003	724	3525			
133-004	725	3162			
133-005	724	3525			
133-0021-0	904				
133-1	188	3054		TR72/3020	S3024
133-3	152	3054		TR76/	
133P100	9109				
133P80057	909	3590			
133P80067	941D				
133P80104	941M				
134-1	189	3083		TR73/3031	S3028
134B1037	717				
134B1038-4			62		
134B1038-8			62		
134B1038-13			62		
134B1038-21			10		
134B1038-22			62		
134B1040-7			63		
134P1	159	3025	21	TR20/242	S0019
134P1A	159	3118	22	TR20/	S0019
134P1AA	129	3118	21	TR88/	S0019
134P1M	159	3114		TR20/	S0019
134P4		3114	21	TR52/	S0019
134P4AA	159	3118	22	TR20/717	S0019
134P4M	159	3114	21	TR20/	S0019
135-1		3027			
135C44322-542	123A				
135N1	154	3044			S5025
135N1M	154				
135ORN	123A	3122		TR21/735	S0015
136P1	106	3118	21	TR20/52	S0012
136RED	123A	3122		TR21/735	S0030
137(ADMIRAL)	159	3114		TR20/717	S0013
137-12					S3032
137-607				TR59/	
138-4	123A	3122		TR21/735	S0024
139-4	108	3039		TR95/56	S0020
139N1	161	3018	20	TR83/719	S0025
139N2	161	3018			
140-0007	123A				
140N1C	130				

Original Part Number	ECG	SK	GE	IR WEP	HEP
140N2	130	3027	14	TR61/	S5012
140N2B					S5020
141-430		3124	20	TR24/	S0015
142-001		3018			
142-003		3122			
142-004		3054			
142-005		3124			
142-008		3054			
142-4	123A				
142N1	161	3018	20	TR83/719	S3019
142N1P	161	3018		TR83/	
142N3	123A	3122	20	TR21/735	S0015
142N3T	123A	3124			
142N4	123A	3018	10	TR21/735	S0015
142N5		3122			S0015
142N6	108	3020	20		S0016
143(KRACO)					G6016
143(MOTV)					S0025
143-1	912				
144-4	154	3044		TR78/	S3021
144-12					S0015
144A-1	121	3009		TR01/	
144A-1-12	121	3009		TR01/	
144A-1-12-8	121	3009		TR01/	
144N1	171/403	3013	27	TR60/244A	S3019
144N1G	171/403	3103	28	TR55/244A	S5015
144N2		3104			S3021
145-2(SYL)	171				
145-2		3201			
145AT1B				TR12/	
145N1		3018			S0017
145N1P	161	3018		TR83/719	S0017
145-T1B	126	3006	51	TR01/635	
146					S0015
146B-3	121MP	3015			
146B-3-12	121MP	3015			
146B-3-12-8	121MP	3015			
146N3	123A	3018	20	TR21/	S0015
146-T1	121	3009		TR01/232	
146T1		3014	16	TR02/	G6003
147-7009-01	159				
147-7016-01	199				
147N1	123A	3122	17	TR21/735	S0015
147P2	159	3114	21	TR20/717	S0012
147-T1	121	3009		TR01/232	
147T1		3014	16	TR02/	G6003
148-134622	105	3012		TR03/	
148-12					S3032
148N1	128	3124		TR87/243	S5014
148N2	123A	3124		TR21/735	S0015
148N3		3122			S3002
148N212	123A	3124		TR21/	
148P-2	102A	3004		TR85/	
148P-2-12	102A	3004		TR85/	
148P-2-12-8	102A	3004		TR85/	
149					F0015
149L-4	102A	3003		TR85/	
149L-4-12	102A	3003		TR85/	
149L-4-12-8	102A	3003		TR85/	
149N1	184	3054		TR55/S5003	S5000
149N2	184	3190	57		S5000
149N3		3083			
149N4	184	3054		/S5003	S5000
149N4B		3054			
149P1	185	3083	58	TR56/S5007	S5006
149P3	185	3191		/S5007	S5006
149P4B		3083			
150-1N	161				
150N1	161	3018	20	TR83/719	S0017
150N3	161	3018			S0015
151-0040	101	3011			
151-0040-00	101				
151-0087-00	159				
151-0096-00	128				
151-045	126				
151-096-1C	128				
151-0103-00	123A				
151-0124-00	159				
151-0127-00	123A				
151-0136	128				
151-0136-00	128				
151-0136-02	128				
151-0138	108				
151-0140	130				
151-0140-00	130				
151-0141	175				
151-0148	175				
151-0149-1A	175				
151-0188-00	159				
151-0190-00	123A				
151-0208	129				
151-0208-00-AA	129				
151-0210	124				
151-0210-00	124				
151-0211	128				

Original Part Number	ECG	SK	GE	IR/WEP	HEP
151-0211-00	128				
151-0211-01	128				
151-0217	175				
151-0217-1	175				
151-0219-00	234				
151-0221-00	159				
151-0221-02	159				
151-0223-00	123A				
151-0224-00	123A				
151-0238	101				
151-0241	124				
151-0241-00	124				
151-0250-00	194				
151-0251	124				
151-0259-00	107				
151-0274-00	194				
151-0275	181				
151-0275-00	181				
151-0302-00	123A				
151-0315	124				
151-0315-00	124				
151-0316	124				
151-0325-00	159				
151-0336-00	130				
151-0337-00	130				
151-0337-00-A	130				
151-0341-00	199				
151-0341-00-A	199				
151-0342-00	234				
151-0347-00	194				
151-0407-00	194				
151-0410-00	234				
151-0417-00	106				
151-0424-00	123A				
151-0427-00	107				
151-0453-00	234				
151-0456-00	199				
151-0458-00	159				
151-0459-00	159				
151-0471-00	107				
151-0I49	175				
151-148-00-BC	175				
151-150	128				
151-150-1B	128				
151-211	128				
151-1002	126				
151M11	108			TR95/56	
151N1	108	3018	20	TR95/56	S0016
151N2	123A	3122	17	TR21/735	S0015
151N4	123A	3122	20	TR21/	S0015
151N11	108	3018		TR95/	S0016
151N116	108	3018		TR95/	
152	132·5				S0025
152-221011	102A	3004		TR85/	
152N2	130		75	TR26/	S7002
152N2C	181	3036		TR36/S7000	S7002
152P1B	218		74	TR29/	S7003
152P1C	180			/S7001	S7003
153N1	196	3054		TR92/	S5000
153N1C	152				
153N2C	152				
153N4C	152				
153N5C	152	3041			S5000
154A3675-105	102	3004	2	TR05/631	
154A3676	126	3008	51	TR17/635	
154A3676-205	102	3004	2	TR05/631	
154A3677	126	3008	51	TR17/635	
154A3679	102	3004	2	TR05/631	
154A3679-5110	126	3008	51	TR17/635	
154A3680	121	3009	3	TR01/232	
154A5941	123A				
154A5943	124	3021	12	TR81/240	
154A5943-1	124	3021	12	TR81/240	
154A5946	123A				
154A5946-622			18		
154A5946-667	129				
154A5947-7732	106				
154A8681	102	3004	2	TR05/231	
154T1	160	3007	51	TR17/637	G0003
154T1A	160	3006	51	TR17/637	GJ002
154T1B	160	3006	51	TR17/637	
155N1	196	3041		TR92/S5003	S5000
155T1	160		51	TR17/637	G0003
155U		3112			
156	123A	3124	18	TR21/735	
156-0011-00	9914				
156-0012-00	9923				
156-0013-00-A	910				
156-0015-00	909	3590			
156-0017	903				
156-0017-00	903				
156-0048-00	912				
156-0049-00	941				
156-0053-00	923				
156-0065-00		3543			
156-0067-00	941M				
156-0067-06	941M				
156-0068-00		3543			
156-0071-00	923D				
156-04					S7002
156-06					S7004
156-032	130		14	TR59/	
156-043	130	3027	77	TR59/247	S7004
156-043A	130	3027		TR59/	
156-044			77		S7004
156-053	130		14	TR59/	
156-063	130		14	TR59/	
156-064					S7004
156-083			75		S7004
156-084			75		S7004
156-0148-00	7404				
156-0151-00	915				
156-0176-00	309K				
156-0197	912				
156-0197-00	912				
156-0197-3306	912				
156-10					S7004
156-83			75		
156-104			75		S7004
156T1	160	3006	51	TR17/637	G0003
156WHT	123A	3122		TR21/735	S0024
157	159	3025	18	TR20/717	
157N3	188	3054	28	TR72/	S3024
157P4	189	3083	29	TR73/	S3028
157T1	160	3006	51	TR17/637	G0003
157T1A			51		
157YEL	159			TR20/717	S0019
158-045-0027	128				
158-10	124	3021		TR81/240	S5011
158P1M	129	3025	22	TR88/	S5013
158P2M	159				
158(SEARS)	123A				
159T1	160		52	TR17/637	G0003
160P08908	165				
160T1	160		52	TR17/637	G0003
161-006-0001	9905				
161-006-0002	9905				
161-011-0001	9903				
161-011-0002	9903				
161-012-0002	9907				
161-118-0001	9910				
161-152-0101	9914				
161-152-0102	9914				
161-249-1001	949				
161-5			66	TR76/	S3020
161N1C	130				
161N2	130MP	3029	15MP	TR26/	S7002
161N4		3027			
161N4C	130	3036			
161T1	160	3019	52	TR17/637	G0003
161T2	108	3019	63	TR95/56	S0016
162N2	184		27		S5000
162P1	185	3191	58		S5006
162T1	160	3006	52	TR17/637	G0003
162T2	108	3019	63	TR95/56	S0016
165-1A82	160	3006		TR17/	
165-2A82	103A	3010		TR08/	
165-4A82	101	3011		TR08/	
165A-182	160	3007		TR17/	
165A4383	123A				
165B-182	160	3008		TR17/	
165C-182	121	3009		TR01/	
166-1A82	102A	3003		TR85/	
166-2A82	102A	3004		TR85/	
166-3A82	102A	3004		TR85/	
167-9	192	3024	63	TR25/243	S5014
167N1	123A	3122	10	TR21/	S0014
167N2	123A				
168-9	193	3025	67	TR28/242	S5013
168N1	123A	3122		TR21/735	S0025
168P1	121MP	3013		/232MP	
169-257	124	3021	12	TR81/240	
169-284	124	3021	12	TR81/240	
170(RCA)	161			TR83/719	
170-9	128			TR87/243	S0015
171					S3019
171-001-9-001	121	3009	3	TR01/232	
171-003-9-001	124	3021	12	TR81/240	
171-015-9-001	121	3009		TR01/	G6003
171-016-9-001	131	3052	30	TR94/642	G6016
171-9	129			TR88/242	S0031
172-001	224	3049			
172-001-9-001	224	3049	17		
172-003-9-001	130	3021	12	TR59/247	S7002
172-003-9-001A	130	3027		TR59/	
172-006-9-001	186	3021		TR55/S3023	S3070
172-007-9-001	195	3049		TR65/243	S3001
172-008-9-001	195	3049		TR65/243	S3001
172-009-9-001	224	3049	45	TR65/243	S3001
172-010-9-001	152	3054	66	TR76/	S5000

Original Part Number	ECG	SK	GE	IR/WEP	HEP
172-011-9-001	152	3054	66	TR76/	S5000
172-014-9-001	152	3054	66	FE100/	
172-014-9-002	195	3049	46		
172-014-9-003		3054	66	TR55/	
172-019-9-001		3054			S5000
172-024-9-005	130	3036			
173-1(SYL)	132	3112	FET1	FE100/802	F0010
173A04490-1	108				
173A04490-2	108				
173A419-2	121	3009		TR01/232	
173A3936	121	3009	16	TR01/232	
173A3963	121	3009	16	TR01/232	G6003
173A3970	102			TR05/	
173A4057	123A			TR21/	
173A4348	102A		2	TR85/	
173A4349	102A		2	TR85/	
173A4389-1	102A			TR85/	
173A4390	102A			TR85/	
173A4391	128			TR87/	
173A4399	123A	3122	18	TR21/735	
173A4416	123A			TR21/	
173A4419	121	3009	16	TR01/232	
173A4419-1	121	3009	16	TR01/232	
173A4419-2	121	3009	16	TR01/232	G6005
173A4419-3	121	3009	16	TR01/232	
173A4419-4	121	3009		TR01/	
173A4419-5	121			TR01/	
173A4419-6	121			TR01/	
173A4419-7	121			TR01/	
173A4419-8	121	3009	3	TR01/232	
173A4419-9	121	3009	3	TR01/232	
173A4419-10	121			TR01/	
173A4420	121	3009		TR01/232	
173A4420-1	121	3009	16	TR01/232	
173A4420-5	121	3009	16	TR01/232	G6005
173A4421-1	121	3009		TR01/232	
173A4422-1	121	3009	16	TR01/232	G6005
173A4436	121	3009	16	TR01/	
173A4469	121	3009		TR01/232	G6005
173A4470-11	123A	3122	20	TR21/735	
173A4470-13	123A	3122	20	TR21/735	
173A4470-32	123A			TR21/	
173A4483-1	159				
173A4483-2	159				
173A4489-2	128			TR87/	
173A4490-5	128	3024	18	TR87/243	
173A4490-7	152				
173A4491-2	130	3027	14	TR59/247	
173A4491-2A	130	3027		TR59/	
173A4491-4	130	3027			
173A4491-5	152	3036			
173A4491-7	130	3027			
173A4491-8	152	3027			
174-002-9-001	103A		6	TR08/724	G0011
174-1(PHILCO)	190				
174-2(SYL)	171				
174-3A82	100	3005		TR05/	
174-20989-22	130				
174-20989-23	124				
174-25566-01	129	3025		TR88/242	
174-25566-21	128				
174-25566-50	128				
174-25566-62	129				
174-25566-63	128				
174-25566-76	128				
175-006-9-001	102A	3004		TR85/	G0005
175-007-9-001	102A	3004		TR85/	G0005
175-008-9-001	102A	3004	2	TR85/	G0007
175-2(PHILCO)	172				
176-003	108	3018	17	TR95/56	S0011
176-003-9-001	108	3018	20	TR95/56	S0011
176-004	128	3024		TR87/243	
176-004-9-001	108	3047	17	TR95/	
176-005	108	3018	17	TR95/56	S0011
176-005-9-001	108	3018	20	TR95/56	S0011
176-006	108	3018	17	TR95/56	S0015
176-006-9-001	108	3018	20	TR95/56	S0011
176-007	108	3018	17	TR95/56	S0011
176-007-9-001	108	3018	20	TR95/56	S0011
176-008-9-001	123	3124	20	TR86/53	S0015
176-014-9-001	123A	3018	20	TR21/735	S0016
176-016-9-001	123A	3124	20	TR21/735	S0014
176-017-9-001	123A			TR21/735	S0014
176-018-9-001	195	3024		TR65/243	S3001
176-024-9-001	123A	3122	17	TR21/735	S0011
176-024-9-002	235	3197			
176-024-9-003	236	3197			
176-024-9-004	186	3192			
176-025-9-001	123A	3124	10	TR21/735	S0015
176-025-9-002	107	3039		TR70/720	
176-026-9-001	107	3018	11	TR70/	S0016
176-029-9-001	123A	3122	10	TR21/735	S0015
176-029-9-002	128	3048	20	TR87/243	S0014
176-029-9-003	195	3047	20	TR65/243	S3001
176-031-9-001	123A	3124	20	TR21/	S0014

Original Part Number	ECG	SK	GE	IR/WEP	HEP
176-031-9-002	123A	3124	20	TR21/	S0024
176-037-9-001	107	3018	20	TR71/	S0014
176-037-9-002	154	3024	18	TF63/	S5026
176-037-9-003	123A	3124	20	TR24/	S0025
176-042-9-001	123A	3018	17	TR33/	S0014
176-042-9-002	123A	3018	20	TR24/	S0014
176-042-9-003	199	3018	20	TR21/	S0015
176-042-9-005	186	3054	28	TR55/	S5000
176-042-9-006	123A	3124	17	TR25/	S0014
176-042-9-007		3054	66	TR92/	S5000
176-047-9-001	123A	3124		TR21/	
176-047-9-002	123A	3018			
176-054-9-001		3122			
177-001	129	3025	22	TR88/242	S0012
177-001-9-001	129	3025	21	TR88/242	S0017
177-006-9-001	159	3114	21	TR20/	S0013
177-006-9-002	129	3114	21	TR20/	S0012
177-007-9-001	159	3114	22	TR20/	S0013
177-012-9-001	159	3138		TR21/	
177A01	798				
177A01A	798				
177A02	748				
177A03	738				
177A05	749				
177A07	767				C6015
177A-1-82	105			TR03/	
177B01	798				
177B02	748				
177B-1-82	121MP	3013			
177C-1-82	121MP	3014			
178-1(PHILCO)	193				
179-46444-01	9910				
179-46444-02	9914				
179-46444-04	9923				
179-46444-05	9913				
179-46445-01	9932				
179-46445-02	9933				
179-46445-03	9937				
179-46445-05	9948				
179-46445-06	9949				
179-46445-08	9961				
179-46445-09	9963				
179-46445-10	9094				
179-46447-01	910				
179-46447-03	909D				
179-46447-07	923				
179-46447-08	941				
179-46447-16	911				
179-46447-17	940				
179-46447-21	915				
179A7453-5		3027			
180N1	108	3132			
180N1P		3132			
180T2	130	3027	19	TR59/247	S7002
180T2A	130	3027	14	TR59/247	S7002
180T2B	130	3027	19	TR59/247	S7002
180T2C	130	3027	19	TR59/247	S7002
181-000200	718			IC511/	C6094P
181N1		3117			
181T2					S7004
181T2A	130	3027	14	TR59/247	S7004
181T2B		3027	14		S7004
181T2C	130	3027	14	TR59/247	S7004
182-1					S0015
182-009-9-001	132	3116	FET2	FE100/802	F0015
182-014-9-002	133	3112			
182-014-9-003	133	3112	FET1	FE100/	F0010
182-015-9-001	132	3050	FET1	FE100/802	F0015
182B2003JDP1	128				
183P1	129	3025	22	TR88/242	
183T2					S5020
184A-1	121	3009		TR01/	
184A-1-12	121	3009		TR01/	
184A-1-12-7	121	3009		TR01/	
184A-1L	121	3009		TR01/	
184A-1L8	121	3009		TR01/	
184T2C	162			TR67/707	S5020
185-736	175		12	TR81/241	
185T2					S5021
185T2A					S5021
186-001		3018			S0014
186-002		3018			S0015
186-003		3122			S0015
186-004		3124			S0015
186-005		3122			S0025
186-006		3124			S0014
186-007		3018			S0015
186-008		3054			S5003
186-009		3054			S3024
186B-3	121MP	3015			
186B-3-12	121MP	3015			
186B-3-12-7	121MP	3015			
186B-3L	121MP	3015			
186B-3L8	121MP	3015			
186N1		3132			

Original Part Number	ECG	SK	GE	IR WEP	HEP
188-826	121			TR01/	
188B-1-82	160	3006		TR17/	
188C-1-82	100	3005		TR05/	
188P-2	102A	3004		TR85/	
188P-2-12	102A	3004		TR85/	
188P-2-12-7	102A	3004		TR85/	
188P-2L	102A	3004		TR85/	
188P-2L8	102A	3004		TR85/	
189	108	3039		TR95/56	S0020
189L-4	102A	3003		TR85/	
189L-4-12	102A	3003		TR85/	
189L-4-12-7	102A	3003		TR85/	
189L-4L	102A	3003		TR85/	
189L-4L8	102A	3003		TR85/	
189N1	163	3111	38	TR93/	S5021
189N1G		3111			S5021
190N1C	184	3190	57	TR76/	S5000
190N3	184	3190			
190N3C	184	3054	57	TR76/	S5000
192-1A82	105	3012		TR03/	
193		3122			
195	199				
195N1		3054			
195N1C		3054			
195N1D					S3024
195P2		3083			
195P2C		3083			S3028
197S		3112			
199-POWER	121	3009		TR01/232	
200-007	108	3018		TR95/56	
200-010	108	3018	17	TR95/56	
200-011	128	3047		TR87/243	S5014
200-015	108	3018	17	TR95/56	
200-016	123	3124	20	TR86/53	
200-018	175	3026		TR81/241	S5012
200-052	159	3025		TR20/717	S0019
200-053	132	3116		FE100/802	F0015
200-055	108	3018		TR95/56	S0016
200-056	108	3018		TR95/56	S0016
200-057	123A	3018		TR21/735	S0015
200-058	123A	3026		TR21/735	S0011
200-064	132	3116		FE100/802	F0015
200-076	152			TR76/	
200-12	128				
200-846	123A	3122		TR21/735	
200-856	103A			/724	
200-862	123A	3122		TR21/735	
200-863	123A	3122		TR21/735	
200A	103	3010	8	TR08/641	
201	159				
201-15	102A			TR85/	
201-25-4343-12	108	3019		TR95/56	S0016
201-254323-12	108			TR95/	
201-254323-13	108			TR95/	
201-254343-12	108			TR95/	
201-254343-13	107				
201-254343-22	161				
201-254343-26	161				
201-254343-28	107				
201-254343-30	107				
201-254343-33	107				
201-283818-1	222				
201-283818-2	222				
201-283818-3	222				
201A	160	3123	2	TR17/637	G0003
201B	160	3123	2	TR17/637	G0003
202-1	189	3083			
202A	103	3010	8	TR08/641A	
202N1		3083			S3031
203S		3112			
204S		3112			
205A2-210	102A	3004		TR85/	
206-709-001	909				
207A	160			TR17/637	G0002
207A1	160	3006	51	TR17/637	
207A3	102A			TR85/	
207A7	102A			TR85/	
207A8					G6013
207A9	108	3018	17	TR95/56	
207A10	123A	3018	17	TR95/56	
207A14	234				
207A15					S0019
207A15B					S0019
207A16	175			TR81/	
207A16A	175			TR81/	
207A17	199				
207A20	104			TR01/	
207A20A	104			TR01/	G6005
207A25	107				
207A27	161				
207A30	152				
207A31	123A				
207A33	152				
207A35	123A				
207B	160			TR17/637	G0002
207M	160			TR17/637	G0002
207V073C04	159		21	TR20/	
209-1		3114			S0019
209-30	152	3026	24MP	TR76/	S5012
209-846	123A		20	TR21/	
209-856	103A				
209-862	123A		20	TR21/	
209-863	123A		20	TR21/	
210-0398-010		3039			
210-0398-022		3039			
210ATTF3638	159				
210BWTF4121	106				
210U		3112			
211-40140-18	130	3027			
211-40140-41		3054			
211-4014018		3027			
211A6380-1	130				
211A6380-3	152	3054			
211A6381-2	152	3054			
211A6381-I	175				
211A6382-2	152				
211AESU3055	130				
211APPE3391	159				
211AVPF3415	123A				
211AVTE4275	123A				
211ON-132					S0014
211ON-133					S0015
211ON-134					S5000
212-695	123A	3018	20	TR24/	S0020
212-699	159		21	TR20/	
215-37567	199				
216-001-001	152				
217				TR21/719	
217-1	108	3039		TR95/56	S0015
217(RCA)	161			TR83/	
218-22	123A	3018		TR21/	S0014
218-23	123A	3124		TR21/	S0015
218-24	123A	3124		TR21/	S0015
218-25	123A	3124		TR21/	S0015
218-26	128	3028		TR87/	S5014
220	175				
220-001001	123A	3124		TR21/	S0015
220-001002	123A	3124		TR21/	S0016
220-001011	108	3018		TR95/	S0016
220-001012	108	3018		TR95/	S0014
220-002001	131	3052		TR94/	G6016
220-008001	132			FE100/	F0015
220AACC9960	9960				
221		3124		TR24/735	
221-30	909	3590			
221-31	703A	3157			C6001
221-32	703A	3157	IC12		C6001
221-34	708	3135	IC10		C6062P
221-35					C2009G
221-36	705A	3134	IC3		C6089
221-37	705A	3134	IC3		C6067G
221-37A		3134			
221-39	705A	3134	IC3		C6089
221-39A		3134			
221-40	707				C6067G
221-41	707				C6089
221-42	714	3075	IC4	IC509/	C6070P
221-42-1	714				
221-43	715	3076	IC6	IC510/	C6071P
221-43-1	715				
221-45	731	3170	13		C6085P
221-46	713	3077	IC5	IC508/	C6057P
221-46-1	713				
221-48	712	3072	IC2	IC507/	C6063P
221-49	739				C6072P
221-51	713	3077			C6057P
221-52	713	3077			C6057P
221-62	790	3077	IC5	IC508/	C6057P
221-62-1	790				
221-64	1112				
221-65	722	3161	IC9		C6056P
221-69	791	3149	IC33		
221-76	767				
221-77	810				
221-79	722	3161	IC9		C6056P
221-80	1112				
221-89			IC17		
221-90			IC15		
221G3114	903				
221G3114-1	903				
221(SEARS)	123A			TR21/	S0024
223	108	3018		TR95/56	S0015
223P1		3114			
223(SEARS)					S0015
224HACA0723	923D				
225A6946-P000	7400				
225A6946-P003	7403				
225A6946-P004	7404				
225A6946-P010	7410				
225A6946-P020	7420				

Original Part Number	ECG	SK	GE	IR/WEP	HEP
260-10-016	159	3114	21		
260-10-020	123A	3018	20		
260-10-021	123A	3124	17		
260-10-023	199				
260-10-024	152				
260-10-039	159	3114	21		
260-10-042	123A	3122	20		
260-10-051	107				
260-10-052	154	3024	18		
260-10-053	210	3054	28		
260-10-054	152	3054	66		
260-10-055	236		66		
260-10-056	108	3019	11		
260D00401	102A	3004	52		G0005
260D00402	102A		54		G0005
260D00403	102A	3004	52	TR14/	
260D00404	102A				
260D02501	102A	3004	2		G0005
260D02601	102A	3005	2	TR05/	
260D02701	158	3004	53		
260D04501		3004	52		G0005
260D04701		3004	53		G0005
260D05701		3018	17		S0016
260D05704	107	3039	20	TR21/	S0014
260D05707	108	3018	20		S0014
260D05709		3018	17		S0014
260D07201	154			TR78/	S5025
260D07412	123A	3018	62	TR21/	S0015
260D07901	128			TR87/	S0015
260D08013	107	3018	20	TR24/	
260D08214	210	3041	28	TR55/	
260D08514	102A	3004	53		G0005
260D08601	128			TR87/	S5014
260D08701	128			TR87/	S0015
260D08801	123A			TR21/	S0014
260D08912	158	3004	53	TR84/	G6011
260D09001	123A	3124	10	TR24/	
260D09301	130			TR59/	
260D09314	123A	3124	62	TR21/	S0011
260D09413	102A	3004	53	TR85/	G0005
260D09612	128	3024	18	TR21/	
260D106A1	123A	3039			
260D13701	199	3124	20	TR21/	
260D13702	123A	3018	20	2SC372/	
260D13704	102A	3004	52	TR84/	
260P01209	121	3009	16	TR01/232	G6005
260P02903	123A	3018		TR21/735	
260P02903A	123A	3018		TR21/	
260P02908	123A	3018	20	TR21/	S0016
260P03001	158	3004	53	TR84/	G6011
260P03201	161	3039	17	TR83/719	S5026
260P03201A	161	3039		TR83/	
260P04001	123A	3018	20	TR21/	S0014
260P04002	123A	3122	20	TR21/	S0014
260P04003	123A	3122	18	TR21/	S0014
260P04004	123A	3122	20	TR21/	S0014
260P04502	123A	3124	20	TR21/735	S0014
260P04503	123A	3020		TR21/	S0014
260P04505	123A	3124		TR21/	
260P05402	107	3018		TR70/720	S0011
260P05402A	107	3018		TR70/	
260P05801	108	3018	11	TR95/	S0017
260P05901	161	3039		TR83/719	S0017
260P05901A	161	3039		TR83/	
260P06901	108	3018		TR95/	
260P06902	108	3018	11	TR95/	S0014
260P06903	108	3018		TR95/	S0014
260P06904	123A	3018	20	TR21/735	S0014
260P07001	123A	3018	11	TR21/	S0014
260P07002	123A	3018		TR21/	S0014
260P07004	108	3018		TR95/	
260P07301	123A	3018	16	TR21/642	S0015
260P07502	131	3052	30	TR94/642	G6016
260P07601	158	3004	52	TR84/630	G6011
260P07701	123A	3124	20	TR21/	S0011
260P07702	123A	3124	20	TR21/735	S0011
260P07703	123A	3122	63	TR21/735	S0011
260P07704	123A	3124	63	TR21/735	S0016
260P07705	123A	3122	20	TR21/	S0016
260P07707	123A	3122	20	TR21/	S0016
260P07901	108	3039	11	TR95/56	S0020
260P08001	108	3018	17	TR95/56	S0030
260P08201	159	3025	21	TR20/717	S0019
260P08401	108	3018	11	TR95/	S0017
260P08601	175			TR81/	S5012
260P08801	123A	3124		TR21/	S0015
260P08801A	123A	3124		TR21/	
260P08901	165	3115	38	TR93/740B	S5020
260P08908	165	3111	35	TR61/	
260P09201	161	3019	11	TR83/	S0016
260P09402	124	3021	12	TR81/	S5011
260P09501	164	3133	37	TR93/	
260P09508	164			TR68/	
260P09701	165	3111	36	TR93/	S5020
260P09902	123A	3124		TR21/735	S0014

Original Part Number	ECG	SK	GE	IR/WEP	HEP
260P1060	108				
260P1300	126	3005			
260P1601	161	3018			
260P1760	107				
260P2100	102A	3005			
260P4002	123A	3122		/735	
260P10003	128	3024		TR87/243	S5026
260P10003A	128	3024		TR87/	
260P10005	128	3024	63	TR87/	S5026
260P10301	154	3040	20	TR78/712	S5024
260P10403	108	3018	20	TR95/	S0011
260P10501	108			TR95/	S0016
260P10502	108			TR95/	S0016
260P10601	161	3019	11	TR83/	S0016
260P10602	108	3019		TR95/56	S0016
260P10801	154	3045	63	TR78/	S5026
260P11101	108	3039	20	TR95/56	S5026
260P11101A	108	3039		TR95/	
260P11209	165	3111		TR93/	S5020
260P11302	123A	3018	20	TR21/	S0015
260P11303	123A	3044	20	TR21/	S0015
260P11304	123A	3020	20	TR21/	S0015
260P11305	123A	3018	20	TR21/	S0015
260P11403	159	3114	21	TR20/	S5013
260P11502	123A	3024	63	TR21/	S5014
260P11503	123A	3024	63	TR21/	S5014
260P11504	123A	3024	63	TR21/735	S5014
260P11505	123A	3124		TR21/735	S0015
260P12001	123A	3122	20	TR21/735	S0020
260P12002	123A	3124	10	TR21/735	S0020
260P12401	128	3024	18	TR25/	S5014
260P12481	128				
260P12701	152	3041	66	TR92/	
260P13001		3006			G0008
260P14101	123A	3005	20	TR21/735	S0030
260P14102	123A	3122	63	TR21/735	S0030
260P14103	123A	3122	63	TR21/	S0030
260P14105	123A				
260P15100	124	3021	12	TR81/	S5011
260P15108	124	3021	12	TR81/	S5011
260P15201	159	3114	21	TR20/	S5022
260P15202	159	3114	21	TR20/	S5022
260P15203	159	3114	21	TR20/	S5022
260P16009	175	3026	12	TR81/241	S5012
260P16101	161	3018	39	TR83/	S0017
260P16202	124	3021	12	TR81/	S5011
260P16208	124	3021	12	TR81/	S5011
260P16301	108	3018	11	TR95/	S0016
260P16302	108	3018	11	TR95/	S0016
260P16502	159	3114	22	TR52/	S0013
260P16503	159	3114	22	TR52/	S0013
260P17002	128	3024		TR87/	S5014
260P17101	123A	3122	60	TR21/	S0020
260P17102	123A	3018	17	TR21/735	S0016
260P17103	123A	3018	17	TR21/735	S0016
260P17104	123A	3018	20	TR21/	S0016
260P17105	123A	3018	20	TR21/	S0020
260P17106	161	3122	20	TR83/	S0020
260P17201	108	3018	11	TR95/	S0016
260P17501	123A	3024	61	TR21/735	S0025
260P17503	123A	3018	63	TR21/735	S0020
260P17601	107	3018	60	TR70/	S0033
260P17602	108	3018		TR95/	S0033
260P17603	108	3018		TR95/	S0033
260P17701	161	3132	62	TR83/	S0025
260P17702	161	3132	61	TR83/	S0020
260P19009	165	3079	38	TR61/	
260P19101	123A			TR21/735	
260P19103	123A				
260P19108	164	3133	37	TR68/	S0015
260P19208	163	3115	38	TR67/	S5021
260P19501	123A	3124	20	2SC945/	
260P19503	199	3124	18		
260P20101	175	3021	12		S5019
260P21001	100	3005	53	TR05/	G0001
260P21002	100	3005	52	TR05/254	G0001
260P21102	186	3054	28	TR55/3023	S3020
260P21106	210	3054	28	TR76/	S3020
260P21208	124	3021			S5011
260P21308	186	3041	28	TR55/3023	S3020
260P21608	165	3079	38	TR61/	S5020
260P21901	162	3027	14	TR67/	S5020
260P22001	132	3116	FET2	FE100/	F0021
260P22002	132	3116		FE100/	F0021
260P22003	132	3116	FET2	FE100/	F0021
260P22101	154	3040	40	TR78/	S5024
260P22204	191	3104	27	TR75/	
260P24008	130	3027	14	TR59/	S7004
260P24108	165				
260P24308	164	3133	37	TR68/	S5021
260P24408	163	3111	36	TR67/	S5021
260P24803	175	3021	12		S5019
260P24901	161	3117	60		S0016
260P33108	163	3115	38	TR61/	S5020
260P70403	108	3018		TR95/	

Original Part Number	ECG	SK	GE	IR WEP	HEP
225A6046-P050	7450				
225A6946-P093	7493				
225A6976-P095	7495				
226-1		3122			
226-1(SYL)	123A			TR21/	S0025
226-3	129	3025	67	TR88/242	S5013
226-4	128	3024	63	TR87/243	S5014
229-0-180-33			11		
229-0-180-34			11		
229-0026			51		
229-0027			53		
229-0028			53		
229-0029			53		
229-0030			53		
229-0038			51		
229-0041			53		
229-0050-13	123A	3124		TR21/735	S0015
229-0050-14	123A	3124		TR21/735	S0015
229-0050-15	123A	3124		TR21/735	S0015
229-0055			53		
229-0056			53		
229-0062			53		
229-0077			51		
229-0079			51		
229-0080			53		
229-0082			53		
229-0083			51		
229-0085			53		
229-0086			50		
229-0087			51		
229-0088			53		
229-0089			51		
229-0090			51		
229-0091			53		
229-0092			53		
229-0095			50		
229-0097			53		
229-0098			50		
229-0099			51		
229-0100			53		
229-0106			51		
229-0110			51		
229-0111			50		
229-0112			50		
229-0116			16		
229-0121			51		
229-0122	103A		54		
229-0123			51		
229-0124			53		
229-0125			53		
229-0129			50		
229-0130			53		
229-0131			51		
229-0132			51		
229-0133			53		
229-0134			54		
229-0136			51		
229-0137			53		
229-0138			53		
229-0139			53		
229-0140			53		
229-0142			53		
229-0143			53		
229-0144			20		
229-0145			51		
229-0146			53		
229-0149			20		
229-0150			20		
229-0151			20		
229-0151-3	108	3019	11	TR95/56	S0020
229-0152			20		
229-0154			20		
229-0180-32	161	3018	11	TR83/719	S0017
229-0180-33	161	3018	20	TR83/719	S0017
229-0180-34	108	3018	11	TR95/56	
229-0180-119	161	3018	11	TR83/719	S0020
229-0180-123	123A			TR21/	
229-0180-124	108	3019		TR95/	S0016
229-0180-149	108			TR95/	
229-0185-2	108		11	TR95/	S0020
229-0185-3	108		11	TR95/	S0020
229-0190-29	108			TR95/	
229-0190-30	161		39	TR83/	
229-0190-31	107	3018	39	TR70/	S0017
229-0190-90	123A	3122	20	TR21/735	
229-0191-29	107	3018	11	TR70/	S0020
229-0191-30	107	3018	39	TR70/	S0016
229-0192-18	222	3050			F2007
229-0192-19	108	3018	11	TR80/	S0016
229-0192-20	132	3116	FET2	FE100/	F0021
229-0204-4	108	3039		TR95/56	S0016
229-0204-6	161	3117	11	TR83/719	S0017
229-0204-23	108	3018	17	TR95/56	S0016
229-0210-14	108		11	TR95/56	S0016
229-0210-19	107		11	TR70/	
229-0214-40	108	3018	11	TR95/	S0016
229-0220-9	108	3039	11	TR95/56	S0016
229-0220-19	108	3039		TR95/56	
229-0240-25	161	3019	11	TR83/	S0020
229-0248-45	108	3039	20	TR24/	S0020
229-0250-10	108			TR95/56	S0016
229-0260-18	107	3018			
229-1200-36	123A	3122	20	TR21/735	
229-1301-22	161				
229-1301-23	199				
229-1301-24	123A				
229-1301-26	190				
229-1301-27	172				
229-1301-28	193				
229-1301-34	186				
229-1301-35	152				
229-1301-36	153				
229-1301-37	192				
229-1301-38	193				
229-1301-39	794				
229-1301-40	792				
229-1301-41	790				
229-1301-42	747				
229-1301-43	795				
229-1301-44	738				
229-1301-63	165				
229-1301-64	130				
229-1513-46			18		
229-5100-15U	108	3039	11	TR95/56	S0016
229-5100-15V	108	3019	11	TR95/	
229-5100-31V	161	3018		TR83/719	S0017
229-5100-32	161				
229-5100-32V	161	3018		TR83/719	S0017
229-5100-33V	108	3018	11	TR95/56	S0020
229-5100-224	108	3018	11	TR95/56	S0016
229-5100-225	108	3018	11	TR95/56	S0016
229-5100-227	126	3008		TR17/635	G0008
229-5100-228	108	3018	11	TR95/56	S0016
229-6011			18		
230-616					G6005
230-627					S0024
231-0000-01	158	3004	53	TR84/	
231-000-001	102A				
231-0004	130	3027	14	TR59/	
231-0004-01	123A	3124	20	TR21/	
231-0004-03	123A			TR21/	
231-0006-03	179	3009		TR35/	
231-0006B			3	TR16/	
231-0008	124	3026	66	TR81/	
231-0009	102A	3004	2	TR85/	
231-006B	121				
231-0011	121	3014	16	TR01/	
231-0013	128	3024	28	TR87/	
231-0015	121	3009	16	TR01/	
231-0025		3122			
231-0026		3114			
231-0028		3024			
231-0029		3025			
231-0032		3114			
233(SEARS)	128			TR87/243	S5014
234S		3112			
235	100	3005	2	TR05/254	G0005
235S		3112			
236-0002	9932				
236-0003	9946				
236-0005	7400				
236-0006	7403				
236-0007	7404				
236-0008	7405				
236-0009	7473				
236-0012	9807				
236-0017	9962				
239A7920	196				
239A7921-1	197				
240-170		3003			
241-15A	160		51	TR17/637	
241B	128				
241U		3112			
242-997	6401				
247-016-013	108	3019		TR95/56	
247-256	103A			TR08/	
247-257	123A	3122	20	TR21/735	
247-623	102A			TR85/	
247-624	121			TR01/	
247-625	128			TR87/	
247-626	128			TR87/	
247-629	123A	3122	20	TR21/735	
247AS-C0450-001	923				
247AS-C1249-001	123A				
248-38104-1	128	3024		TR87/243	
251M1	103	3010	8	TR08/641A	
252-113002-001	911				
258-1	188			TR72/3020	S5013
258-2	189			TR73/3031	S5014
260-10-006	132	3116	FET2		

Original Part Number	ECG	SK	GE	IR/WEP	HEP
260P70501	108	3018		TR95/	
260P70502	108	3018		TR95/	
260P107707				TR21/	
260P141103	123A	3124		TR21/	
260Z00109		3122	20	TR21/	S0015
260Z00209		3018	17	TR95/	S0014
260Z00309		3018	17	TR95/	S0014
260Z00401		3047	20	TR21/	S5014
260Z00402	123A	3124	10	TR24/	S0014
260Z00703	158	3004	53	TR85/	G0005
260Z01201	158	3004	53		G0005
264P06301	102A	3004		TR85/	G0005
266P001-01	704	3023			
266P001-02	704				
266P00101	704	3023	20		
266P00102	704	3023		IC501/	
266P00301	1096				
266P00602	1099				
266P00801	1096				
266P10101	749	3168			C6076P
266P10201	747		IC7		C6079P
266P30103	712		IC5		C6063P
266P30109	712	3072			C6063P
266P33108					S5020
266P30201	1075				
270-950-030	123A				
270-950-037-02	128	3024			
274	729	3074			
276A01	704	3023			
281	123A	3122		TR21/735	S0011
290V02H69	108				
294		3114			S5022
295V041H04	121	3009	3	TR01/232	
296	172	3124		TR69/	
296(SEARS)	172			TR69/	S9100
296-18-9	102A			TR85/	
296-19-9	102A			TR85/	
296-46-9	160			TR17/	
296-50-9	123A	3122		TR21/735	S0014
296-51-9	123A	3122		TR21/735	S0014
296-56-9	123A	3122	17	TR21/735	S0015
296-58-9	128	3024		TR87/243	S5014
296-59-9	123A	3122		TR21/735	S0014
296-60-9	102A			TR85/250	G0005
296-61-8	131	3082		TR94/642	G6016
296-62-9	102A	3123	52	TR85/250	G0005
296-77-9	123A	3024	18	TR25/	
296-81-9	128	3047			
296-86	107	3018	20	TR21/	S0016
296-98-9	123A	3122			S0015
296-779					S0015
297L001H01	103A		54	TR08/724	
297L001H02	103A		53	TR08/724	
297L001H03	102A(2)		53	TR85/250	
297L001M01	103A		54	TR08/724	
297L001M02	102A		53	TR85/250	
297L001M03	102A(2)		53	TR85/250	
297L002H01	102A		53	TR85/250	
297L005H01	102A		53	TR85/250	
297L006H01	123A	3122	20	TR21/735	
297L006H02	123A	3122	20	TR21/735	
297L007C	199				
297L007C02	123A	3122	20	TR21/735	
297L007C03	199				
297L007H01	123A	3122	20	TR21/735	
297L007H02	123A	3122	20	TR21/735	
297L007H03	123A(2)	3122	20	TR51/735	
297L008C02	128(2)		18	TR87/243	
297L008H01	128		18	TR87/243	
297L010C	234				
297L010C01	234		21	/717	
297L011C01	107				
297L012C01	159	3114	21	TR20/717	
297L012C-01	106				
297L013B01	123A	3122	20	TR21/735	
297L013B02	159		21		
297L015C01	199				
297V002H03	103A		54	TR08/724	
297V002H04	103A		54	TR08/724	
297V002H05	103A		54	TR08/724	
297V002M04	103A	3011	7	TR08/724	
297V002M05	103A	3011	54	TR08/724	
297V003H02	102A(2)		53	TR85/250	
297V003H03	102	3004	2	TR05/631	G0005
297V003H06	102A(2)	3004	2	TR85/250	
297V003H08	102A(2)		53	TR85/250	
297V003H09	102A	3004	52	TR85/250	G0005
297V003M01	102	3004	52	TR05/631	G0005
297V003M07	102	3004	53	TR05/631	
297V004H01	102A		53	TR85/250	
297V004H03	102A	3004	52	TR85/250	G0005
297V004H04	102A		53	TR85/250	
297V004H06	102A	3004	2	TR85/250	G0005
297V004H08	102A		53	TR85/250	
297V004H09	102A		53	TR85/250	

Original Part Number	ECG	SK	GE	IR/WEP	HEP
297V004H10	102	3004	52	TR05/631	G0005
297V004H11	102A		53	TR85/250	
297V004H14	102	3004	52	TR05/631	G0005
297V004H15	102	3004	2	TR05/631	
297V004H16	102	3004	2	TR05/631	
297V004M01	102	3004	52	TR05/631	G0005
297V005H01	102A		53	TR85/250	
297V008H01	100	3005	51	TR05/254	
297V008M01	160		51	TR17/637	
297V010H01	102A(2)		53	TR85/250	
297V010M01	102A(2)		53	TR85/250	
297V011			1		
297V011H01	126	3008	2	TR17/635	G0005
297V011H02	100	3005	51	TR05/254	
297V012			1		
297V012H01	126	3005	51	TR17/635	
297V012H02	100	3005	51	TR05/254	
297V012H03	100	3005	51	TR05/254	
297V012H04	160		51	TR17/637	
297V012H05	100	3005	51	TR05/254	
297V012H06	100	3005	1	TR05/254	
297V012H07	160		51	TR17/637	
297V012H08	100	3008	2	TR05/254	G0005
297V012H09	100	3005	2	TR05/254	G0005
297V012H10	126	3008	2	TR17/635	G0005
297V012H11	160	3008	51	TR17/637	
297V012H12	160		51	TR17/637	
297V012H13	160		51	TR17/637	
297V012H14	102	3008	51	TR05/631	
297V012H15	126	3008	51	TR17/635	
297V017H01	100	3005	51	TR05/254	
297V017H02	100	3005	51	TR05/254	
297V018H01	102A		53	TR85/250	
297V019B01	100	3005		TR05/254	
297V019H01	100	3005	51	TR05/254	
297V020H01	160		51	TR17/637	
297V020H02	100	3005	51	TR05/254	
297V020M01	100	3005	51	TR05/254	
297V021H01	100	3005	51	TR05/254	
297V021H02	100	3005	51	TR05/254	
297V021H03	100	3005	51	TR05/254	
297V022H01	100	3005	51	TR05/254	
297V024H01	160		51	TR17/637	
297V024H03	160	3006	50	TR17/637	
297V025H02	102	3004	2	TR05/631	G0005
297V025H03	102A(2)		53	TR85/250	
297V025H04	102	3004	52	TR05/631	G0005
297V025H05	102	3004	2	TR05/631	
297V025H15	102	3004	52	TR05/631	G0005
297V026H01	126	E008	51	TR17/635	
297V026H03	100	3005	51	TR05/254	G0005
297V027H01	102	3004	2	TR05/631	
297V027H01-A		3003			
297V027H01A		3004			
297V032H01	102	3004	2	TR05/631	G0005
297V033H01	102	3004	52	TR05/631	G0005
297V034H01	126	3008	2	TR17/635	G0005
297V035H01	126	3008	2	TR17/635	G0005
297V036H01	126	3008	2	TR17/635	G0005
297V036H02	126	3007	51	TR17/635	
297V037B02	102A	3004		TR85/250	
297V037H01	102	3004	52	TR05/631	G0005
297V037H02	102	3004		TR05/631	
297V038H01	100	3005	50	TR05/254	G0002
297V038H02	160	3005	51	TR17/637	G)002
297V038H03	160	3005	50	TR17/637	G0002
297V038H04	160	3004	50	TR17/637	G0002
297V038H05	100	3005	2	TR05/254	G0005
297V038H06	126	3008	2	TR17/635	G0)05
297V038H07	102	3005	2	TR05/631	G0005
297V038H09	100	3005	2	TR05/254	G0005
297V038H10	126	3008	2	TR17/635	
297V038H11	126	3008	2	TR17/635	
297V)38H12	126	3008	2	TR17/635	
297V040H01	102	3004	52	TR05/631	G0005
297V040H08	102	3004	52	TR05/631	G0005
297V040H10	102	3004	52	TR05/631	G0005
297V040H11	102	3004	52	TR05/631	G0005
297V040H12	102	3004	52	TR05/631	G0005
297V040H13	102	3004	52	TR05/631	G0005
297V040H15	121	3009	3	TR01/232	G6003
297V040H16	102	3004	52	TR05/631	G0005
297V041H01	121MP	3009	16	TR01/232MP	G6003
297V041H02	121	3009	16	TR01/232	
297V041H03	121	3009	16	TR01/232	
297V041H04	121	3009	16	TR01/232	
297V041H05	104	3014	16	TR01/624	
297V041H06	121	3014	16	TR01/232	G6005
297V041H07	104	3009	16	TR01/624	
297V041H15			16		
297V042C01	102	3004	52	TR05/631	
297V042C02	102	3004	52	TR05/631	
297V042C03	102	3004	52	TR05/631	
297V042C04	102	3004	52	TR05/631	
297V042H01	102	3004	52	TR05/631	

Original Part Number	ECG	SK	GE	IR WEP	HEP
297V042H02	102	3004	52	TR05/631	
297V042H03	126	3006	52	TR17/635	
297V042H04	126	3006	50	TR17/635	
297V043H01	102	3004	2	TR05/631	G0005
297V043H02	100	3005	51	TR05/254	
297V044H01	100	3005	51	TR05/254	
297V045H01	126	3006	50	TR17/635	
297V045H02	126	3006	50	TR17/635	
297V049H01	123	3124	8	TR86/53	S0015
297V049H03	123	3124	8	TR86/53	S0015
297V049H04	123	3124	8	TR86/53	S0015
297V049H05	123A	3122	20	TR21/735	
297V049H06	128	3024	18	TR87/243	
297V050C02	102A	3004	52	TR85/250	
297V050H03	102	3004	2	TR05/631	
297V051C03	102	3004	52	TR05/631	
297V051C04	102	3004	2	TR05/631	
297V051H01	102	3004	2	TR05/631	G0005
297V051H02	102	3004	52	TR05/631	G0005
297V051H03	102	3004	52	TR05/631	G0006
297V051H04	102A	3004	52	TR85/250	
297V052C01	102	3004		TR05/631	
297V052H01	102A	3004	52	TR85/250	G0005
297V052H02	102	3004	52	TR05/631	
297V052H04	102A	3004		TR85/631	G0005
297V053C01	102	3004	2	TR05/631	
297V053H01	102A	3004	52	TR85/250	G0005
297V053H02	102A		53	TR85/250	G0005
297V054C01	100	3005	1	TR05/254	
297V054C02	100	3005	1	TR05/254	
297V054H01	126	3005	50	TR17/635	G0002
297V054H02	126	3005	50	TR17/635	G0002
297V055C01	100	3005	1	TR05/254	
297V055H01	126	3005	50	TR17/635	G0002
297V057H01	158(2)	3123	52	TR06/630	G0005
297V057H02	158	3123	52	TR84/630	G0005
297V059H01	123	3020	20	TR86/53	S0015
297V059H02	123	3124	20	TR86/53	
297V059H03	123	3124	20	TR86/53	
297V060H01	124	3021	12	TR81/240	
297V060H02	124	3021	12	TR81/240	
297V060H03	124	3021	12	TR81/240	
297V061C01	130	3020	20	TR59/247	
297V061C01A	130	3027		TR59/	
297V061C02	130	3124	20	TR59/247	
297V061C02A	130	3027		TR59/	
297V061C03	123	3124	20	TR86/53	
297V061C04	123	3124	20	TR86/53	
297V061C05	128	3024	20	TR87/243	
297V061C06	123	3124	10	TR86/53	
297V061H01	123A	3122	20	TR21/735	
297V061H02	123A	3122	20	TR21/735	
297V061H03	123A	3122	20	TR21/735	
297V062C01	121	3009	16	TR01/232	
297V062C05	121	3009	3	TR01/232	
297V062C06	128	3024	20	TR87/243	
297V063C01	126	3006	50	TR17/635	
297V063H01			51	TR17/637	
297V064B01		3006	50	TR17/635	
297V064H01	160		51	TR17/637	
297V065C01	126	3008	51	TR17/635	
297V065C02	126	3008	51	TR17/635	
297V065C03	126	3008	51	TR17/635	
297V065H01	100	3005	51	TR05/254	
297V065H02	100	3005	51	TR05/254	
297V065H03	100	3005	51	TR05/254	
297V070C01	126	3006	51	TR17/635	
297V070H49	108	3039		TR95/56	
297V071C03	124	3021	12	TR81/240	S5011
297V071H03	124	3021	12	TR81/240	
297V072C01	108	3018	11	TR95/56	
297V072C03	108	3018	11	TR95/56	
297V072C04	108	3018	20	TR95/56	
297V072C05	128	3024	20	TR87/243	
297V072C06	123A	3124		TR21/735	S0030
297V073C01	129	3025	21	TR88/242	
297V073C02	129	3025	21	TR88/242	
297V073C03	159	3025	21	TR20/717	
297V073C04	159	3114	21	TR20/717	S0013
297V074C01	161	3117		TR83/719	
297V074C02	123	3124	20	TR86/53	S0015
297V074C03	123	3124	20	TR86/53	S0015
297V074C04	123	3124	20	TR86/53	
297V074C06	123A	3024	20	TR21/735	S0015
297V074C07	123A	3124	20	TR21/735	S0024
297V074C08	123A	3124	20	TR21/735	
297V074C09	108	3045	20	TR95/56	S0020
297V074C10	128	3124	20	TR87/243	
297V074C11	129	3025	21	TR88/242	
297V074C12	128	3024	20	TR87/243	
297V076B01	102	3004	52	TR05/631	
297V076C01	102	3004	52	TR05/631	
297V077C01	126	3006	50	TR17/635	
297V078C01	108	3018	11	TR95/56	S0016
297V078C02	107	3018	11	TR95/56	S0016
297V080C01	129	3025	21	TR88/242	
297V081C01	102A	3004	52	TR85/250	
297V082B01	128	3024	20	TR87/243	
297V082B02	129	3025	21	TR88/242	
297V082B03	129	3025	22	TR88/242	
297V083C01	159	3025	21	TR20/717	S0019
297V083C02	123A	3124	17	TR21/735	S0015
297V083C03	129	3025	21	TR88/242	S5013
297V083C04	128	3024	63	TR87/243	S5014
297V084C01	157		66	TR60/244	S5015
297V085C01	123A	3122		TR21/735	S0030
297V085C02	123A	3122		TR21/735	S0030
297V085C03	123A	3122		TR21/735	S0015
297V085C04	123A	3124		TR21/735	S0015
297V086C01	159	3025	21	TR20/717	S0031
297V086C02	123A	3124	10	TR21/735	S0015
297V086C03	123A	3024	18	TR21/735	S0015
297V086C04	185		29	TR77/S5007	S5006
297V087B02	188		28	TR72/3020	S5000
299 POWER	105	3012	4	TR03/	
300	102	3004	52	TR05/631	G0005
300C043	128				
301	102	3004	52	TR05/631	G0005
301-576-2	9158				
301-576-3	708		IC10		C6062F
301-576-4	7400				
301-576-14	740		IC31		
301-576-18					C6061F
301-679-1	773				
301-679-7					C6052F
301-680-1	753				C6010
301-695-1					S3006
301-696-3					S3013
301-733-2					S3007
302	102	3004	52	TR05/631	G0005
307-001-9-001	703A	3157			
307-005-9-001	1003		IC2		
307-007-9-001	1100				
307-007-9-002	1102				
307-008-9-001	724	3525			
307-009-9-002	1102				
307-029-1-001	1100				
307-047-9-002	724				
309-327-926	123A			TR21/	
309-327-931	102A			TR85/	
309(CATALINA)	159				
310-068	126	3006	51	TR17/635	G0008
310-68	126	3006	51	TR17/635	G0008
310-123	126	3006	51	TR17/635	G0003
310-124	126	3006	51	TR17/635	
310-139	126	3006	51	TR17/635	
310-187	123	3124	20	TR86/53	
310-188	102	3004	2	TR05/631	
310-189	100	3005	2	TR05/254	
310-190	126	3008	51	TR17/635	
310-191	126	3006	2	TR17/635	
310-192	121	3009	16	TR01/232	
311D589-P2	159				
311D916P01	199				
314-6005-1	184				S5000
314-6006	175	3026	23	TR81/241	
314-6007-1	123A			TR21/	S0015
314-6007-2	123A			TR21/	S0015
314-6007-3	123A			TR21/	S0015
316-0227-001	910				
317-0083-001	159				
317-0139-001	129				
317-2627-1	9946				
317-3556-001		3024			
317-4501-001A		3024			
317-8504-001	123A				
319C	123A	3122			
321-264	154				
322(CATALINA)	123A				
322T1	102A	3123	2	TR85/250	
323T1	102A	3123	2	TR85/250	
324	102A			TR85/250	G0005
324-0016	160			TR17/637	
324-0026	160			TR17/	
324-0027	126	3008		TR17/635	G0008
324-0028	126	3008		TR17/635	G0008
324-0029	102A	3004	2	TR85/250	G0007
324-0030	158			TR84/630	
324-0038	126	3008	51	TR17/635	
324-0041	102	3004	2	TR05/631	
324-0055	102	3004	2	TR05/631	
324-0056	158	3004	2	TR84/630	
324-0062	158			TR84/630	
324-0074	102A	3004		TR85/250	
324-0077	160			TR17/	
324-0079	126	3006		TR17/635	G0009
324-0080	158		53	TR84/630	
324-0082	160			TR17/637	
324-0083	160	3006		TR17/637	G0009
324-0085	158	3004	2	TR84/630	G0005

Original Part Number	ECG	SK	GE	IR/WEP	HEP
324-0086	160	3006	51	TR17/637	G0003
324-0087	160	3006		TR17/637	G0003
324-0088	102	3004	2	TR05/631	
324-0089	102A	3008	51	TR85/250	G0005
324-0090	100	3008	51	TR05/254	G0005
324-0091	102A	3004	2	TR85/250	G0005
324-0092	102A	3004	2	TR85/250	G0007
324-0093	102A	3004		TR85/250	
324-0095	160	3006	50	TR17/637	G0003
324-0097	158			TR84/630	
324-0098	126	3006	51	TR17/635	
324-0099	126	3006	51	TR17/635	
324-0100	102A			TR85/250	
324-0106	126	3006	51	TR17/635	
324-0110	160			TR17/637	
324-0111	160			TR17/637	
324-0112	160			TR17/637	
324-0115	179			TR35/G6001	
324-0116	127	3034	25	TR27/235	
324-0122	103A			TR08/724	
324-0123	160			TR17/637	
324-0124	158			TR84/630	
324-0125	158			TR84/630	
324-0126	131	3082		TR94/642	
324-0128	127	3035	25	TR27/235	
324-0130	126	3007	51	TR17/635	G0008
324-0131	160	3008	51	TR17/637	G0003
324-0132	160		51	TR17/637	G0002
324-0133	102	3004	2	TR05/631	G0005
324-0134	103	3010	8	TR08/641A	G0011
324-0136	126	3006	51	TR17/635	
324-0137	126	3007	51	TR17/635	
324-0138	126	3006	51	TR17/635	
324-0139	102		2	TR05/631	
324-0140	102		2	TR05/631	
324-0142	102	3004	2	TR05/631	
324-0143	102	3004	2	TR05/631	
324-0144	102	3004	2	TR05/631	G0005
324-0145	126	3007	51	TR17/635	
324-0146	102	3004	2	TR05/631	
324-0149	108	3018	20	TR95/56	
324-0150	108	3018	17	TR95/56	
324-0151	123	3124	20	TR86/53	
324-0152	123	3124	20	TR86/53	
324-0154	123	3124	20	TR86/53	
324-0187	160	3006		TR17/	G0003
324-019	128	3024		TR87/243	
324-1	108	3018			
324-132	160	3006			
324-144	102	3004		/631	
324-6005-5	123A	3124		TR21/	
324-6011	128	3024	18	TR87/243	
324-6013	128	3024		TR87/243	S5014
324T1	102A	3123	2	TR85/	G0005
324T2	102A			TR85/250	G0006
325-0025-329	102A	3004	52	TR85/250	G0005
325-0025-330	102A	3004		TR85/250	G0007
325-0025-331	102A	3004		TR85/250	G0005
325-0028-79	102A	3004	52	TR85/250	G0005
325-0028-80	102A	3004	52	TR85/250	G0005
325-0028-81	102A	3004	52	TR85/250	G0005
325-0028-83	102A	3004	52	TR85/250	G0005
325-0028-84	108	3018	11	TR95/56	S0016
325-0028-85	160	3006	51	TR17/637	G0009
325-0030-315	102A	3004		TR85/250	G0005
325-0030-317	102A	3004		TR85/250	G0005
325-0030-318	102A	3004		TR85/250	G0005
325-0030-319	102A	3004		TR85/250	G0005
325-0031-303	123A	3124	20	TR21/735	S0015
325-0031-304	123	3124	20	TR86/53	S0015
325-0031-305	123	3124	20	TR86/53	S0016
325-0031-306	158	3004	52	TR84/630	S5014
325-0031-310	123A	3124	20	TR21/735	S0011
325-0036-536	102A	3006		TR85/250	G0005
325-0036-564	126	3008		TR17/635	G0009
325-0036-565	126	3006		TR17/635	G0008
325-0042-351	123A	3124	20	TR21/735	S0014
325-0047-516	102A	3004		TR85/250	G0005
325-0054-310	102A	3004		TR85/250	G0006
325-0054-311	102A	3004		TR85/250	G0005
325-0076-306	123A	3124	20	TR21/735	S0015
325-0076-307	123A	3124	20	TR21/735	S0025
325-0076-308	123A	3124	17	TR21/735	S0011
325-0081-100	123A	3038	20	TR21/735	S0015
325-0081-101	123A	3124	20	TR21/735	S0016
325-0081-102	158	3004	53	TR84/630	G0005
325-0500-12	199	3124	20	TR21/	S0024
325-0500-13	199	3124	20	TR24/	S0025
325-0574-30	123A	3018	20	TR21/	S0011
325-0574-31	123A	3018	20	TR21/	S0011
325-0670	102A	3004		TR85/250	
325-0670-1	102A	3004		TR85/	G0005
325-0670-7	102A			TR85/250	
325-067A	102A			TR85/250	
325-1370-18	102A	3004		TR85/250	

Original Part Number	ECG	SK	GE	IR/WEP	HEP
325-1370-19	123A	3124		TR21/735	S0011
325-1370-20	123A	3124		TR21/735	S0030
325-1370-20	123A	3124	20	TR21/735	S0013
325-1375-10	126	3008		TR17/635	G0008
325-1375-11	126	3008		TR17/635	G0008
325-375-12	126	3008		TR17/635	G0008
325-1376-53	102A	3008		TR85/250	G0005
325-1376-54	126	3008		TR17/635	G0008
325-1376-55	126	3004		TR17/635	G0001
325-1376-56	102A	3004		TR85/250	G0005
325-1376-57	102A	3004		TR85/250	G0005
325-1376+58	102A	3004		TR85/250	G0005
325-1378-18	107	3018		TR70/720	S0015
325-1378-19	107	3018		TR70/720	S0015
325-1378-20	102A	3004		TR85/250	G0005
325-1378-21	158	3004		TR84/630	G6011
325-1378-22	102A	3004		TR85/250	G0007
325-1442-8	102	3004	2	TR05/631	
325-1442-9	124	3021	12	TR81/240	
325-1446-26	123	3124	20	TR86/53	
325-1446-27	123	3124	20	TR86/53	
325-1446-28	123	3124	20	TR86/53	
325-1513-29	123	3124	20	TR86/53	
325-1513-30	123	3124	20	TR86/53	
325-1513-46	128	3024	18	TR87/243	
325-1771-15	123A	3124	10	TR21/735	S0024
325-1771-16	107	3124	20	TR70/720	S0015
325T1	102A	3123	2	TR85/250	G0005
326T1	102A	3123	2	TR85/250	G0005
330-1304-8	108				
332-2816		3024			
332-2911	130				
332-2911C		3027			
332-2912	181				
332-2912D		3036			
332-3562	223				
332-4052A		3027			
334-377	162			TR67/	
335-1		3115			
339					S0015
344-6000-2	123A			TR21/735	
344-6000-3	108	3018		TR95/56	S0011
344-6000-3A	108	3018		TR95/	
344-6000-4	123A	3018		TR21/735	S0011
344-6000-5	123A	3124		TR21/735	S0015
344-6000-5A	123A	3124		TR21/	
344-6001-1	128	3124		TR87/243	
344-6001-2	128	3124		TR87/243	
344-6002-3	123A	3024		TR21/735	
344-6005-1	123A	3122		TR21/735	
344-6005-2	123A	3124		TR21/735	
344-6005-5	123A	3024		TR21/735	S5014
344-6011-1	128	3124		TR87/243	
344-6011-2	128	3024		TR87/243	S5014
344-6012-1	129	3024		TR88/242	S5013
344-6012-3	129				
344-6013-1B	128	3024		TR87/243	S5014
344-6013-4	128	3124		TR87/243	
344-6014-1B	129			TR88/242	S5013
344-6015-7	108			TR95/56	S0011
344-6015-7A	108	3018		TR95/	
344-6015-8		3018			S0025
344-6015-9		3018			S0025
344-6015-10		3018			S0014
344-6015-11		3018			S0014
344-6017-1	159	3025	22	TR20/717	S5013
344-6017-2	123A	3124	20	TR21/735	S5014
344-6017-3	123A		67	TR28/	S5013
344-6017-4	128		63	TR25/	S5014
344-6017-5		3124			S0016
344-6017-6		3020			S0016
345-002		3004			
345-111-001		3011			
349-113-011	9930				
349-113-012	9932				
349-113-013	9945				
349-113-014	9946				
349-113-015	9951				
349-113-022	9945				
349-113-024	9950				
349-113-025	9941				
349-212-003	909				
349-1		3024			
349-2		3024			
350	100	3005	2	TR05/254	G0005
351-1008-010	716				
351-1011-022	724	3525			
351-1011-032	724	3525			
351-1017-010	703A				
351-1027-010	716				
351-1029-020	941				
351-1035	923				
351-1035-020	923				
351-1041-010	912				
351-1042-020	910				

Original Part Number	ECG	SK	GE	IR/WEP	HEP
351-3006	9914				
351-7008-010	9910				
351-7008-020	9913				
351-7011-020	9900				
351-7011-030	9903				
351-7011-040	9903				
351-7011-050	9905				
351-7011-060	9905				
351-7011-070	9910				
351-7011-080	9911				
351-7011-090	9911				
351-7011-100	9914				
351-7011-110	9914				
351-7011-120	9914				
351-7011-150	9921				
351-7011-160	9921				
351-7011-170	9923				
351-7011-180	9923				
351-7015-010	9908				
351-7015-020	9909				
351-7015-030	9910				
351-7015-040	9911				
351-7015-050	9912				
351-7015-060	9913				
351-7015-070	9921				
351-7025-010	9907				
351-7025-020	9914				
351-7026-010	9908				
351-7026-030	9910				
351-7026-060	9913				
351-7026-070	9921				
351-7121-010	9923				
351-7121-020	9914				
351-7140-010	909				
351-7189-010	910				
351-7206-080	9908				
351-7206-090	9909				
351-7206-100	9910				
351-7206-110	9911				
351-7206-140	9921				
351-7206-150	9908				
351-7206-160	9909				
351-7206-170	9910				
351-7206-180	9911				
351-7206-190	9912				
351-7206-200	9913				
351-7206-210	9921				
351-7577-010	9109				
352	102	3004	52	TR05/631	G0005
352-0023-001	9932				
352-0092-020	128				
352-043-010	128	3024			
352-0195-000	123A				
352-0197-000	123A				
352-0206-001	123A				
352-0219-000	159				
352-0316-00	123A				
352-0318-00	123A				
352-0318-001	123A				
352-0319-000	123A				
352-0322-010	123A				
352-0349-000	123A				
352-0364-000	128				
352-0364-010	128				
352-0365-000	123A				
352-0400-00	123A				
352-0400-010	123A				
352-0400-030	123A				
352-0403-010	154				
352-0433-00	123A				
352-0477-00	123A				
352-0479-010	128				
352-0506-000	123A				
352-0519-00	123A				
352-0546-00	123A				
352-0549-000	199				
352-0551-010	159				
352-0551-021	159				
352-0569-00	123A				
352-0569-010	123A				
352-0569-020	123A				
352-0579-00	123A				
352-0579-010	123A	3024			
352-0579-020	123A				
352-0581-011	175				
352-0581-020	175	3026			
352-0581-021	175				
352-0581-030	175	3026			
352-0581-031	175				
352-0583-011	130				
352-0596-010	123A				
352-0596-020	123A				
352-0596-030	123A				
352-0606-011	175	3131			
352-0610-030	159				
352-0610-040	159				
352-0629-010	123A				
352-0630	107				
352-0630-010	107				
352-0636-010	159				
352-0636-020	150				
352-0638	199				
352-0653-010	107				
352-0653-020	107				
352-0658-010	161				
352-0658-020	161				
352-0658-030	161				
352-0658-040	161				
352-0658-050	161				
352-0661-010	123A				
352-0661-020	123A				
352-0667-010	123A				
352-0675-030	123A				
352-0675-040	123A				
352-0675-050	123A				
352-0677-010	130				
352-0677-011	130				
352-0677-020	130				
352-0677-021	130				
352-0677-030	130				
352-0677-031	130				
352-0677-041	130				
352-0677-051	130				
352-0677-40	130				
352-0680-010	123A				
352-0680-020	123A				
352-0711-021	124				
352-0713-030	123A				
352-0749-010	223				
352-0754-020	159				
352-0766-010	128				
352-0773-010	234				
352-0773-020	234				
352-0773-030	234				
352-0778-010	159				
352-0783-020	128				
352-0809	123A				
352-0816-010	128	3024			
352-0825-020		3103			
352-0848-020	159				
352-0950-010	106				
352-0950-020	106				
352-0959-010	159				
352-0959-020	159				
352-0959-030	159				
352-7500-010	123A				
352-7500-450	123A				
352-7500-861	124				
352-8000-010	123A				
352-8000-020	123A				
352-8000-030	123A				
352-8000-040	123A				
352-9014-00	128				
352-9036-00	123A				
352-9079-00	123A				
352-9094-00	194				
352-9103-000	123A				
353	102	3004	52	TR05/631	G0005
353-9001-001	126	3008	51	TR17/635	
353-9001-002	126	3007		TR17/635	
353-9001-003	126	3008	51	TR17/635	
353-9002-002	126	3008	51	TR17/635	
353-9008-001	129	3025		TR88/242	
353-9012-001	102A			TR85/250	G0005
353-9201-001	121MP	3009		TR01/232MP	
353-9203-001	152	3041		TR76/701	
353-9301-001	128	3024	3	TR87/243	
353-9301-002	160	3006	10	TR17/735	
353-9301-004	129	3025	21	TR88/242	
353-9306-001	123A	3006	20	TR21/735	S0015
353-9306-002	123A	3124	20	TR21/735	S0015
353-9306-003	123A	3124	20	TR21/735	S0015
353-9306-004	123A				
353-9306-007	199				
353-9310-001	123A	3124		TR21/735	S0015
353-9312-001	102A	3004		TR85/250	
353-9314-001	123A	3124		TR21/735	S0015
353-9315-001	123A	3124		TR21/735	S0015
353-9317-001	129	3114	21		S5013
353-9318-001	199	3122	62	TR51/	S0015
353-9318-002	199	3122	62	TR51/	S0015
353-9319-001	123A	3122	62	TR51/	S0015
353-9319-002	123A	3122	62	TR51/	S0015
353-9502-001		3054	28	TR55/	S3024
353(CHRYSLER)	734				
353(I.C.)	736				
354-3052	102			TR05/	
354-3127-1	123A				
354(CHRYSLER)	723				
355D6	161			TR83/	

Original Part Number	ECG	SK	GE	IR/WEP	HEP
355D7	222				
355D8	161			TR83/	
355D9	123A			TR21/	
360-1(RCA)	128	3024	18	TR51/	S0005
362A10	192	3512			
364-1(SYL)	159	3118	21	TR20/	S0013
364-6004	184	3041		/S5003	S5000
364-6005-1				TR55/	
364-6007-1				TR21/	
364-6011-3				TR56/	
364-6012-2				TR55/	
364-6013-2				TR56/	
364-10048	133	3112		/801	F0010
365-1	123A	3122	20	TR21/735	S0015
365T1	100	3005		TR05/254	G0005
366-1(SYL)	108	3018	20	TR95/	S0020
366-2(SYL)	108	3122		TR95/	S0016
367(SEARS)	160				
370-116764	923				
375-1005	123A				
376-0062	9800				
376-0099	7441				
378-44	130	3027	14	TR59/247	
378-44A	130	3027		TR59/	
380-0057-000	124				
380-0171-000	129	3025			
386-1(SYL)	159	3025	22	TR20/717	S0019
386-40	130				
386-1102-P1	123A				
386-1102-P2	123A				
386-1102-P3	123A				
386-7118P1	108	3039		TR95/56	
386-7178-P001	199				
386-7178P1	123	3124	17	TR86/53	
386-7181P2	128	3047		TR87/243	
386-7182-P001	195			TR65/	
386-7182P1	195	3048		TR65/	
386-7183P1	130	3027		TR59/247	
386-7183P1A	130	3027		TR59/	
386-7184P1	129	3025	21	TR88/242	
386-7185P1	123A	3046		TR21/735	
386-7188P1	108	3018	17	TR95/56	
386-7243-P001	161			TR83/	
386-7254-P202	129				
386-7270-P2	181				
386-7316-P1	128				
386-7316-P2	128				
3907/2N404					S0032
392-1		3024	63	TR25/	
392-1(SYL)	128			TR87/	S0015
393-1		3025	67	TR28/	S5013
393-1(SYL)	129			TR88/	
394-3003-1	123A				
394-3003-3	123A				
394-3003-7	123A				
394-3003-9	123A				
394-3005-2	128				
394-3074-2	102	3004	2	TR05/631	
394-3074-5	102	3004	2	TR05/631	
394-3097-1	102	3004	2	TR05/631	
394-3097-2	102	3004	2	TR05/631	
394-3097-2A		3004			
394-3102-1	101			TR08/	
394-3123-1		3039			
394-3127-1	128	3024	18	TR87/243	
394-3127-2	128	3024	18	TR87/243	
394-3127-3	128	3024	18	TR87/243	
394-3127-4	130				
394-3135A	130	3027		TR59/	
394-3137-1	198				
394-3141-1	128				
394-3145	159				
396-7178P1	108	3018	20	TR95/56	
398-8418-1	9904				
398-8972-1	7460				
398-13222-1	923D				
398-13223-1	7400				
398-13224-1	7404				
398-13225-1	7405				
398-13226-1	7401				
398-13227	941M				
398-13632-1	7404				
400-1362-101	128				
400-1362-102	128				
400-1362-201	128				
400-1371-101	123A				
400-1569-101	199				
400-1735	9915				
400-1736	9926				
400-2023-101	123A				
400-2023-201	123A				
401-465-4		3005			
403-009/07	130				
404-2		3122	20	TR25/735	
404-2(SYL)	123A			TR21/	S0025
404B(NCR)	129			TR88/	
410-012-0150	102A	3004	52	TR85/	G0005
410-013-0240	102A	3004	52	TR85/	G0005
411-237	128	3024		TR87/	
412	100	3005	1	TR05/254	
412-1A5	123A	3124		TR21/	
414A-15	121	3009		TR01/	
417-2	160		51	TR17/637	
417-5	102A		53	TR85/250	
417-6	158		53	TR84/630	
417-7	123A			TR21/	
417-11	160		51	TR17/637	
417-12	160		51	TR17/637	
417-13	160		51	TR17/637	
417-14	160		51	TR17/637	
417-17	102A			TR85/	
417-18	158		53	TR84/630	G0005
417-19	107	3039		TR70/720	
417-20	121	3009	3	TR01/232	
417-21	158		53	TR84/630	
417-22	160		51	TR17/637	
417-23	160		51	TR17/637	
417-25	160		51	TR17/637	
417-26	160			TR17/637	
417-27	160		51	TR17/637	
417-28	158		53	TR84/630	
417-29	179	3014		TR35/G6001	
417-29BLK	179			TR35/G6001	G6005
417-29GRN	179			TR35/G6001	G6005
417-29WHT	179			TR35/G6001	G6005
417-30	121		3	TR01/232	
417-31	160			TR17/	
417-32	121	3009	3	TR01/232	
417-33	160		51	TR17/637	
417-35	160		51	TR17/637	
417-36	160		51	TR17/637	
417-37	160			TR17/	
417-38	160		51	TR17/637	
417-39	160		51	TR17/637	
417-40	102A		53	TR85/250	
417-41	102A		53	TR85/250	
417-42	179			TR35/G6001	G6018
417-43	129	3025		TR88/242	
417-44	121	3009	3	TR01/232	
417-45	121	3009	3	TR01/232	
417-45L		3014			
417-46	121	3009	3	TR01/232	
417-47	102A		53	TR85/250	
417-48	102A	3004	53	TR85/250	
417-49	128	3024		TR87/	
417-50	160		51	TR17/637	
417-51	158		53	TR84/630	
417-52	102A		53	TR85/250	
417-53	160		51	TR17/637	
417-54	160		51	TR17/637	
417-56	160		51	TR17/637	
417-57	160		51	TR17/637	
417-58	160			TR17/	
417-59	128	3024	18	TR87/243	S0014
417-60	160	3014		TR17/637	G6005
417-62	104	3009	3	TR01/624	
417-66	160			TR17/	
417-67	123A	3122	20	TR21/735	S0015
417-68	160	3008		TR17/	
417-69	123A	3122	20	TR21/735	
417-70	160		50	TR17/637	
417-71	160		50	TR17/637	G0008
417-72	160		50	TR17/637	
417-73	102A		53	TR85/250	
417-74	102A		53	TR85/250	
417-75	102A			TR85/	
417-76	160		50	TR17/637	
417-77	123A	3122	20	TR21/735	
417-78	102A		53	TR85/250	
417-79	160	3123	51	TR17/637	G0001
417-81	6400				
417-87	128	3024	18	TR87/243	
417-88	128	3024	18	TR87/243	
417-89	128	3024	18	TR87/243	
417-90	121	3009	3	TR01/232	G6005
417-91	123A	3122	20	TR21/735	
417-92	123A	3122	20	TR21/735	
417-93	123A	3122	20	TR21/735	
417-93-12903	123A				
417-94	123A	3122	20	TR21/735	
417-99	121	3009	3	TR01/232	
417-100	128				
417-101	162			TR67/707	
417-101K		3027			
417-102	106		21	TR20/52	
417-103	102A		53	TR85/250	
417-104	175				S5019
417-105	123A	3122	20	TR21/735	
417-106	123A	3122	20	TR21/735	
417-107		3122	20	TR51/735	

Original Part Number	ECG	SK	GE	IR/WEP	HEP
417-108	123A	3122	20	TR21/735	S0015
417-108-13163	199				
417-109	128	3024	18	TR87/243	
417-109-13163	123A				
417-110	123A	3122	20	TR21/735	
417-110-13163	123A				
417-111	129			TR88/	
417-112	127			TR27/235	
417-113	179			TR35/G6001	G6005
417-114	128	3024	18	TR87/243	
417-114-13163	123A				
417-115	154	3045	27	TR78/712	
417-115-13173	128				
417-116	159	3114	21	TR20/717	
417-116-13165	159				
417-118	123A	3122	20	TR21/735	S0015
417-119	704	3023		IC501/	
417-120	179			TR35/G6001	
417-121	103A		54	TR08/724	
417-122	102A		53	TR85/250	
417-123	726				
417-124	108	3018	20	TR95/56	
417-125	108	3018	20	TR95/56	
417-125-12903	107				
417-126	123	3124	20	TR86/53	
417-126-12903	199				
417-127	123	3124	20	TR86/53	
417-128	128	3048	18	TR86/53	
417-129	108	3018	20	TR95/56	
417-132	159	3114	21	TR20/717	S0019
417-133	128	3024	18	TR87/243	
417-134	123A	3122	20	TR21/735	
417-134-13271	123A				
417-135	123A		20	TR21/735	
417-136	128	3024	18	TR87/243	
417-137	128	3024	18	TR87/243	
417-138	129	3025		TR88/242	
417-139	130	3036	14	TR59/247	S7000
417-139-13286	181				
417-139A	130	3027		TR59/	
417-140	133		FET1		
417-141	121	3009	3	TR01/232	
417-142	179			TR35/G6001	
417-143	160			TR17/637	
417-144	184		66	TR84/630	S5000
417-145	185	3191	69	TR77/G5007	S5006
417-146	158		53	TR84/630	
417-147	158		53	TR84/630	
417-148	158		53	TR84/630	
417-149	158		53	TR84/630	
417-150	158		53	TR84/630	G0005
417-151	158		53	TR84/630	
417-152	158		53	TR84/630	
417-153	159	3114	21	TR20/717	
417-153-13431	234				
417-154	107	3039		TR70/720	
417-155	128	3024	18	TR87/243	
417-155-13163	123A				
417-158	162	3021		TR67/707	
417-159	157	3103		TR60/244	
417-160	131	3082		TR94/642	
417-161	172]			TR69/	
417-162	130				
417-162L		3027			
417-168	159			TR20/	
417-169	133	3112	FET1	FE100/801	
417-170	129	3025		TR88/242	
417-171	123A	3122	20	TR21/735	
417-171-13163	123A				
417-172	123A	3122	20	TR21/735	
417-172-13271	123A				
417-175	196			TR55/	
417-175-12993	152				
417-176	159	3114	21	TR20/717	
417-177	105	3012	4	TR03/233	
417-178	128				
417-179		3021			
417-180	128				
417-181	129				
417-182	159	3114	21	TR20/717	
417-183	6400				
417-184	159	3114	21	TR20/717	
417-185	123A			TR21/	
417-187	6400				
417-190	107	3039		TR70/720	
417-192	123A	3122	20	TR21/735	
417-193	128				
417-194	133	3112	FET1	/801	
417-195	157	3103		TR60/244	
417-196	159	3114	21	TR20/717	
417-196-13262	234				
417-197	123A	3122	20	TR21/735	
417-199	175	3131		TR81/241	
417-200	159	3114	21	TR20/717	
417-201	159	3114	21	TR54/717	S0019

Original Part Number	ECG	SK	GE	IR/WEP	HEP
417-202	711	3070			
417-203	196			TR92/	
417-204	162			TR67/707	
417-205	107	3039		TR70/720	
417-206	220				
417-207	220				
417-211	132	3116	FET2	FE100/802	
417-212	130	3027	14	TR59/247	
417-212A	130	3027		TR59/	
417-213	123A	3122	20	TR21/735	
417-214-13286	181				
417-215	130		14	TR59/247	
417-215-13286	130				
417-215A	130	3027		TR59/	
417-215M		3027			
417-216	121	3009	3	TR01/232	
417-217	123A	3122	20	TR21/735	
417-221				TR73/	
417-222	172			TR69/	
417-224	128	3024		TR87/243	
417-225	185	3191		/S5007	
417-226-1	123A	3122	20	TR21/735	
417-226-13163	199				
417-227	157	3103		TR60/244	
417-228	123A	3122	20	TR21/735	
417-229	123A	3122	20	TR21/735	
417-231	133	3112	FET1	TR51/801	
417-232	198				
417-233	128	3024	18	TR87/243	
417-233-13163	123A				
417-234	129	3025		TR88/242	
417-234-13165	159				
417-235	159	3114	21	TR20/717	
417-235-13262	159				
417-237	128	3024	18	TR87/243	
417-237-13163	128				
417-238					S5011
417-239	163			TR67/	
417-242-8181	159				
417-243	161			TR83/	
417-244	123A			TR21/	
417-244-12903	199				
417-245	171	3201		TR79/	
417-246	133				
417-247	128			TR87/	
417-248	163			TR67/	
417-250	128				
417-252	133				
417-253	133				
417-254	181			TR36/	
417-255	129			TR88/	
417-257	128			TR87/	
417-258	161			TR83/	
417-260	129			TR88/	
417-260-50127	159				
417-262	161			TR83/	
417-263					S5006
417-273	130	3027			
417-273-13286	130				
417-283-13271	199				
417-286		3083			
417-289	197	3083			
417-296		3054			
417-801-12903	123A				
417-821-13163	128				
417-822-13262	129				
420T1	102A			TR85/250	G0005
421-6	160	3008	50	TR17/637	G0003
421-6B	160		50	TR17/637	G0003
421-7	160	3008	50	TR17/637	G0003
421-7B	160		50	TR17/637	G0003
421-8	160	3008	50	TR17/637	G0003
421-8B	160		50	TR17/637	G0003
421-9	102A			TR85/250	
421-10	102A			TR85/250	
421-11	102A	3004	52	TR85/250	G0005
421-11B	102A	3123	52	TR85/250	G0005
421-12	102A	3004	52	TR85/250	G0005
421-12B	102A	3123	52	TR85/250	G0005
421-13	102A	3004	52	TR85/250	G0005
421-13B	102A		52	TR85/250	G0005
421-13C	102		52	TR05/	
421-14	102A	3004	53	TR85/250	G0005
421-14B	102A	3123		TR85/250	G0005
421-15	102A	3004	52	TR85/250	G0005
421-15B	102A		52	TR85/250	G0005
421-16	126	3008	50	TR17/635	G0008
421-17	126	3008	50	TR17/635	G0008
421-18	124	3021	12	TR81/240	
421-19	102	3004	2	TR05/631	
421-20	126	3006	50	TR17/635	
421-20B	126	3006	50	TR17/635	G0003
421-21	126	3006		TR17/635	
421-21B	126	3006		TR17/635	
421-22	126	3006		TR17/635	

Original Part Number	ECG	SK	GE	IR/WEP	HEP
421-22B	126	3006		TR17/635	
421-24	121	3009		TR01/232	
421-25	121	3009		TR01/232	
421-26	126	3008	50	TR17/635	G0003
421-6599	121	3009		TR01/	
421-7143	102A	3004		TR85/	
421-7444	123A	3124		TR21/	
421-8109	102A	3004		TR85/	
421-8111	123A	3124		TR21/	
421-8310	124	3021		TR81/	
421-9644	123A	3124		TR21/	
421-9670	123A	3124		TR21/	
421-9671	102A	3004		TR85/	
421-9682	100	3005		TR05/	
421-9683	108	3018		TR95/	
421-9792	160	3006		TR17/	
421-9840	123A	3124		TR21/	
421-9862	103A	3010		TR08/	
421-9862A	102A	3004		TR85/	
421-9863	102A	3004		TR85/	
421-9863A	102A	3004		TR85/	
421-9864	102A	3004		TR85/	
421-9864A	102A	3004		TR85/	
421T1	102A			TR85/	G0005
422-0222	102A	3004		TR85/	
422-0738	123A	3124		TR21/	
422-0739	129	3025		TR88/	
422-0960	175	3026		TR81/	
422-0961	130	3027		TR59/	
422-1232	129	3025		TR88/	
422-1233	128	3024		TR87/	
422-1234	123A	3124		TR21/	
422-1401	108	3018		TR95/	
422-1402	108	3018		TR95/	
422-1403	126	3008		TR17/	
422-1404	128	3024		TR87/	
422-1405	129	3025		TR88/	
422-1406	102A	3004		TR85/	
422-1407	123A	3124		TR21/	
422-1443	121MP	3015			
422-2008	129	3025		TR88/	
422-2158	128	3024		TR87/	
422-2532	108	3018		TR95/	
422-2533	123A	3124		TR21/	
422-2534	123A	3004		TR21/	
422-2535	102A	3004		TR85/	
422-2778	160	3006		TR17/	
422-2779	160	3006		TR17/	
422-2780	160	3006		TR17/	
423-800-235	9933				
423-800175	9932				
423-800176	9962				
423-800177	9932				
423-800178	9946				
423-800194	9949				
423-800202	9936				
423-800223	9097				
423-800234	9930				
423-800236	9093				
424-9001	175				
429-0092-1	160	3006	50		G0008
429-0092-2	102A	3004	2		G0005
429-0092-3	102A	3004	2		G0005
429-0093-69	126	3006	51	TR17/635	G0008
429-0094-39	126	3008	51	TR17/635	
429-0910-50	160		51	TR17/637	G0003
429-0910-51	102A	3123	2	TR85/250	G0005
429-0910-52	102A	3123	2	TR85/250	G0005
429-0958-42	123A	3122	10	TR21/735	S0014
429-0981-12	107	3039		TR70/720	
429-0985-12	107	3018	20	TR24/	S0016
429-0986-12	108	3018		TR95/	S0016
430		3124	18	TR24/735	S0024
430-85	158		53		G0005
430-86	123A				S0015
430-87	123A	3018	18		S0015
430-1044-0A	123A				
430-10034	123A				
430-10034-06	123A				
430-10047-0C	128				
430-10053-0	123A				
430-10053-0A	123A				
430-20013-0B	159				
430-20018-0A	159				
430-20021	159				
430-20023-0A	159				
430-20026-0	159				
430-22861	108	3018		TR95/	
430-23190	181	3036		TR36/	
430-23212	123A	3124		TR21/	
430-23221	123A	3124		TR21/	
430-23222	129	3025			
430-23223	128	3024		TR87/	
430-23843	130	3027		TR59/	
430-23844	128	3024		TR87/	

Original Part Number	ECG	SK	GE	IR/WEP	HEP
430-25762	132			FE100/	
430-25763	108	3018		TR95/	
430-25764	108	3018		TR95/	
430-25765	108	3018		TR95/	
430-25766	123A	3124		TR21/	
430-25767	108	3018		TR95/	
430-25834	131	3052		TR94/	
430-203845	129	3025		TR88/	
430CL	123A	3124		TR21/	
430(ZENITH)	123A			TR21/	
431-1(SYL)	123A				
431-26551A	722	3161	IC9	IC513/	C6056P
432-1		3054	18	TR25/3020	S5014
433		3124	18	TR24/735	S0024
433-1		3122	20	TR21/735	S0015
433(ZENITH)	123A			TR21/	S0024
433CL		3124		TR21/	
433MB52	129	3025	21	TR88/242	
434	199				S0016
434-260-1		3025			
435-21026-0A	7400				
435-21027-0A	7402				
435-21028-0A	7404				
435-21029-0A	7408				
435-21030-0A	7410				
435-21033-0A	7420				
435-21034-0A	7451				
435-21035-0A	7486				
435-23006-0A	7473				
435-23007-0A	7474				
436-403-001	123A				
436-404-002	159				
436-10010-0A	7492				
436-10011-0A	7496				
442-4	785				
442-5	780	3141	IC20		
442-7	909D	3590			
442-8	703A	3157			
442-9	720		IC7	IC512/	C6095P
442-10	705A	3134	IC3	IC502/	C6089
442-11	748				
442-16	722	3161	IC9	IC513/	C6056P
442-20	703A	3157			
442-20-14299	703A				
442-21	778				C6102P
442-22					C6052P
442-28	708	3135	IC10	IC504/	C6062P
442-30	309K				
442-30-2897	309K				
442-33	713	3077	IC5	IC508/	C6057P
442-41	725	3162			
442-46	801				C6096P
442-52				IC507/	C6060P
442-55	795				C6099P
442-56	747				C6079P
442-57					C6070P
442-58					C6071P
442-59	783				C6100P
442-62					C6063P
443-1	7400				
443-2	7420				
443-4	7472				
443-5	7473				
443-6	7474				
443-7	7490				
443-7-16088	7490				
443-8					C2003P
443-9					C2004P
443-13	7475				
443-16	7476				
443-18	7404				
443-23	74122				
443-27	9093				
443-32	74121				
443-35	7415				
443-36	7447				
443-43	74H102				
443-44	7413				
443-44-2854	7413				
443-45	7408				
443-46	7402				
443-65	7427				
443-71	74H00				
443-77	7438				
443-87	74145				
443-92	9960				
443-162	74193				
443-623	7411				
443-623-51644	7411				
443-625	74132				
443-628	74196				
444-0012-PT2	160				
444-012-P1	160				
444(SEARS)	123A				
445-0023	130				

Original Part Number	ECG	SK	GE	IR/WEP	HEP
445-0023-P1	130				
445-0023-P3	130				
445-0023-P4	130				
445-0034-1	181				
447				TR21/735	
447(ZENITH)	123A			TR21/	
448A662	128	3024	18	TR87/243	S3019
449	184	3041	66	TR55/	
450-1167-1	159				
450-1167-2	123A				
450-1261	123A				
454A104	128	3024	18	TR87/243	
455-1	126	3006	51	TR17/635	
461-1002					S5018
461-1006	159				
461-1014	129				
461-1048	129				
461-1055-01	159				
461-2001	234				
462-0119	123A				
462-1000	123A				
462-1003					S5012
462-1009-01	123A				
462-1016	128				
462-1019	128				
462-1038-01	199				
462-1059	198				
462-1061	123A				
462-1063	123A				
462-1066-01	199				
462-2002	123A				
462-2004	199				
462-I007	128				
465-005-19	102A		53	TR85/250	
465-032-19	160		51	TR17/637	
465-036-19	102A		52	TR85/250	
465-042-19	160		51	TR17/637	
465-045-19	160		50	TR17/637	
465-049-19	160		51	TR17/637	
465-061-19	160		51	TR17/637	
465-067-19	158		53	TR84/630	
465-072-19	102A		53	TR85/250	
465-073-19	102A		53	TR85/250	
465-075-19	102A		53	TR85/250	
465-080-19	102A		53	TR85/250	
465-082-19	102A		53	TR85/250	
465-086-19	160		50	TR17/637	
465-1A5	160	3006		TR17/	
465-2A5	103A	3010		TR08/	
465-4A5	101	3011		TR08/	
465-106-19	123A	3122	20	TR21/735	
465-108-19	158		53	TR84/630	
465-115-19	102A		52	TR85/250	
465-132-19	102A		53	TR85/250	
465-137-19	131	3082		TR94/642	
465-146-19	160		51	TR17/637	
465-163-19	102A		53	TR85/250	
465-165-19	102A		53	TR85/250	
465-166-19	131MP	3086		TR94MP/642MP	
465-181-19	107	3039	20	TR70/720	
465-191-15	102A		53	TR85/250	
465-206-19	131	3082		TR94/642	
465-223-19	160		51	TR17/637	
465-260-15	131			/642	
465A-15	160	3007		TR17/	
465B-15	160	3008		TR17/	
465C-15	121	3009		TR01/	
466-1A5	102A	3003		TR85/	
466-2A5	102A	3004		TR85/	
466-3A5	102A	3004		TR85/	
469-646-3	128				
471-009	726				
471-1		3041			
471-1(METAL)	128/401				
471-1(PLASTIC)	188			TR72/	
471-1(SYL)					S5000
472-0009A		3039			
472-0309-001	128				
472-0445-001	128				
472-0491-001	123A				
472-0569-002		3039			
472-0766-001		3024			
472-0779-002		3024			
472-0779-003		3024			
472-0788-001		3027			
472-0788-002		3027			
472-0790-001		3026			
472-0842-001		3036			
472-0946-001	130	3027			
472-0946-001A		3027			
472-0946-002	130				
472-0954-002		3021			
472-1007-001		3024			
472-1049-A		3039			
472-1093-001		3045			
472-1132-001		3103			
472-1198-001	123A				
473A31	121	3009	16	TR01/232	G6005
473B5	102A			TR85/	
473B6-2	102	3123	52	TR05/631	G0005
473B6-2A	102	3004	52	TR05/	G0005
473B6-4	102	3004	52	TR05/631	G0005
473B6-5	102	3004	52	TR05/631	G0005
473B6-7	102	3004	2	TR05/631	G0005
474-3A5	100	3005		TR05/	
474A410AXP001	124				
474A410BEP2	152				
474A410BW-2	152				
477-0375-001	9932				
477-0376-001	909	3590			
477-0376-002	909D				
477-0377-001	9930				
477-0379-001	9962				
477-0380-001	9945				
477-0381-001	9946				
477-0404-002	910				
477-0412-004	7473				
477-0415-002	7450				
477-0417-002	7453				
477-0542-001	941				
477A15	105	3012		TR03/	
477B15	121MP	3013			
477C15	121MP	3014			
480-9(SEARS)	131			TR94/642	
481-201-A	128				
481-201-B	128				
481-34842	191	3044		TR75/	
484A15	121	3009		TR01/	
484B15		3006			
484C15		3005			
488-2(SEARS)	108			TR95/56	
488B15	160	3006		TR17/	
488C15	100	3005		TR05/	
490-2(SEARS)	108			TR95/56	
491-2(SEARS)	108			TR95/56	
491A948	128	3024	18	TR87/243	
492-1A5	105	3012		TR03/	
499-1	108			TR95/56	S0015
499(CHRYSLER)	152			TR55/	
499(ZENITH)	123A			TR21/735	S5000
501ES001M	108	3039	63	TR95/56	
501T1	160		52	TR17/637	G0009
503T1	160		52	TR17/637	G0003
504T1	160		52	TR17/637	G0003
505ES105	176	3123	2	TR82/	
505T1	160		52	TR17/637	G0003
506T1	160		52	TR17/637	G0003
507T1	160		52	TR17/637	G0003
508T1	160		52	TR17/637	G0003
509ES0258	176			TR82/	
509ES025P		3123	2	TR06/	
509(SEARS)	123A				
510ES030M			63	TR25/	
510ES031M			63	TR25/	
511-515	123	3124	20	TR86/53	
511-519	123	3124	20	TR86/53	
511ES035P			63	TR25/	
511ES036P			63	TR25/	
512ES040P	176	3123	2	TR82/	
512RED	123A	3122		TR21/735	S0015
513ES045P	176	3123	2	TR82/	
514-048	761				C6011
514-053	754				C6007
514-033338	123A				
514-044910	234				
514-047830	152				
514-054214	130				
515	123	3124	20	TR86/53	
515-521	108	3018	17	TR95/56	
515-10408A-01	710	3102			
515ES045M			21		
515ES046M			21		
515ORN	123A			TR21/735	S0015
516	123	3124	20	TR86/53	
516ES047M			21		
516ES048M			21		
517-518	123	3124	20	TR86/53	
519-1(RCA)	123A				
519ES067M			17	TR24/	
519ES068M			17	TR24/	
520-301	128				
520ES070M			21		
520T1	100	3005		TR05/254	G0005
521ES071M			21		
521T1	100	3005		TR05/254	G0005
522(ZENITH)	107			TR70/720	
522ES075M			63		
522ES076M			63		
523ES077M			63	TR25/	
523ES078M			63	TR25/	

Original Part Number	ECG	SK	GE	IR/WEP	HEP
524WHT	123A	3122		TR21/735	S0015
535A	107				
536-1(RCA)	161				
536-2(RCA)	123A	3122			
536D	123A	3018			
536D9	123A	3018			
536DP					S0016
536E					S0015
536F(JVC)	199	3124			
536FS	123A	3124		TR21/735	
536FU	123A	3122		TR21/735	
536G(WARDS)	123A	3124			
536GT	199		20	TR21/	
536GU(WARDS)	128	3124			
537D	123A	3018			
537E	123A	3124			S0015
537FS	108			TR95/56	
537FV	123A	3124			
537FY	123A	3122			
540				TR21/	
542-1033	128				
542-1034	130				
544-2002-004	726				
546	108	3018	20	TR95/56	S0020
546A		3018			
549-1		3118			
549-2		3118			
549X768		3027			
550-026-00	123A				
550-027-00	159				
551	108	3039		TR95/56	S0016
551-006	761				C6011
551-007	754				C6007
551-008-00	741M				
551-010-00	9946				
551-013-00	9945				
551A	761				C6011
552-1362		3021			
555-3(RCA)	128				
559-1492-001	9911				
559-1493-001	9921				
559-1494-001	9913				
559-1495-001	9909				
559-1496-001	9910				
559-1516-001	128	3026			
560-2		3122			
560-417		3009			
565-072	128		63	TR87/	
565-074	123A		17	TR21/	
570-004503	123A	3124		TR21/	
570-005503	123A	3124		TR21/	
570-1		3114			
571-844	130	3027		TR59/	
571-844A	130	3027		TR59/	
572-0040-051	121	3009		TR01/	
572-683	123A			TR21/	
573-469	123A	3124		TR21/	
573-472	108	3018		TR95/	
573-474	108	3018		TR95/	
573-474A	108	3018		TR95/	
573-475	108	3018		TR95/	
573-479	123A	3124		TR21/	
573-480	123A	3124		TR21/	
573-481	123A	3124		TR21/	
573-491	108	3018		TR95/	
573-494	108	3018		TR95/	
573-495	108	3018		TR95/	
573-507	108	3018		TR95/	
573-509	108	3018		TR95/	
573-515	124	3021		TR81/	
573-518	160	3006		TR17/	
573-529	102A	3004		TR85/	
573-532	128	3024		TR87/	
574		3114			
574-844	162			TR67/	
575C2	1140				
576-0001-002	159	3004	52	TR20/717	S0030
576-0001-003	102A	3004	2	TR85/250	G0008
576-0001-004	123	3124	20	TR86/53	
576-0001-005	123	3124	20	TR86/53	S0015
576-0001-008	123A	3024	20	TR21/735	S0014
576-0001-009	102A	3004	52	TR85/250	G0009
576-0001-012	123A			TR21/	S0015
576-0001-013	159	3114	21	TR20/717	S0032
576-0001-014	102A	3123	52	TR85/250	G0005
576-0001-018	123A				
576-0002-001	241	3041	28	TR76/701	S5000
576-0002-002	104	3009	3	TR01/232	G6005
576-0002-003	130			TR59/	S7004
576-0002-004	102A	3004	52	TR85/250	G0005
576-0002-005	121			TR01/	G6013
576-0002-006	123A	3122	63	TR21/735	S5014
576-0002-008	159			TR20/	S5022
576-0002-009	105			TR03/	G6006
576-0002-010	105			TR03/	G6006
576-0002-011	184	3190		TR76/	S5000
576-0002-012	102A			TR85/250	G6011
576-0002-013	103A			TR08/724	
576-0002-026	182	3534	55	TR66/	
576-0002-027					S3024
576-0003-001	108	3038		TR95/	S0020
576-0003-002	108			TR95/	S0020
576-0003-003	108			TR95/	S0015
576-0003-004	108	3039	20	TR95/	S0015
576-0003-005	108	3039	17	TR95/	S0016
576-0003-006	108	3122	17	TR95/	S0016
576-0003-007	108	3018	17	TR95/56	S0015
576-0003-008	126	3006	50	TR17/635	G0009
576-0003-009	126	3007	50	TR17/635	G0009
576-0003-010	160	3008	51	TR17/637	G0008
576-0003-011	108	3018	11	TR95/56	S0015
576-0003-012	106	3007	21	TR20/52	
576-0003-012NPN	123A			TR21/	S0016
576-0003-012PNP	159			TR20/	S0031
576-0003-013	126	3006	51	TR17/635	
576-0003-014	126	3006	51	TR17/635	
576-0003-015	126	3006	51	TR17/635	
576-0003-017	159	3114	21	TR20/	S0019
576-0003-018	108	3018	11	TR95/56	S0016
576-0003-019	159			TR20/	S0019
576-0003-020	108	3122		TR95/	S0016
576-0003-021	108			TR95/	S0016
576-0003-022	199			TR21/	
576-0003-023	161	3039		TR83/	S0017
576-0003-024	126	3006	51	TR17/635	G0003
576-0003-025					G0003
576-0003-026	161	3117		TR83/719	S0015
576-0003-027	108	3039	17	TR95/	S0017
576-0003-028	108		17	TR21/	S0015
576-0003-029	107	3018	11		S0020
576-0004-001	154				S5026
576-0004-002					S3013
576-0004-004	195	3024	17	TR64/243	S5026
576-0004-005	195	3048	63	TR65/243	S3001
576-0004-006	108	3046	18	TR95/56	S0014
576-0004-007	195	3048		TR65/	
576-0004-008	123A	3047		TR21/735	S3008
576-0004-009	195			TR65/	S3001
576-0004-010	123		20	TR86/53	S3005
576-0004-011	195	3024		TR65/243	S3001
576-0004-012	195	3024	18	TR65/243	S3001
576-0004-013	128	3024		TR87/243	
576-0004-015					S3007
576-0005-001	6400				S9002
576-0005-004	128			TR87/	
576-0006-003	133				F0010
576-0006-011	108		17	TR21/	F2004
576-0006-221	221	3050			F2004
576-0006-222	222	3050			F2004
576-0007-001	172	3122	63	TR69/	S9100
576-001	102A	3004		TR85/	
576-001-013			21	TR20/	
576-002-001	186		28	TR55/	
576-003-009	160	3006	50	TR17/	
576-003-010				TR11/	
576-005	102A			TR85/	
576-0036-212	123A	3047		TR21/735	S5026
576-0036-213	195	3048		TR65/	S3020
576-0036-847	123	3124	20	TR86/53	S3005
576-0036-913	195			TR65/	S3001
576-0036-916	123A			TR21/	S0014
576-0036-917	123A			TR21/	S0014
576-0036-918	108	3018	20	TR95/56	
576-0036-919		3018	20	/56	
576-0036-920	108	3018	20	TR95/56	
576-0036-921		3018	20	/56	
576-0040-051	121	3009	20	TR01/232	
576-0040-251	130			TR59/	S7004
576-0040-253	102A			TR85/	G0005
576-0040-254	121			TR01/	G6013
576-1	233	3132	61	TR83/	
576-1(RCA)	161				
576-3	723				
576-14	740				
576-2000-278	105			TR03/	G6006
576-2000-990	126	3006		TR17/635	G0008
576-2000-993	126	3006	51	TR17/635	G0008
576-2001-970	105			TR03/	G6006
577R819H01	128	3024		TR87/243	
579-0001-009				TR05/	
580R304H01	159				
586-024	909	3590			
586-2	123A	3038		TR21/735	S0015
586-151	9930				
586-152	9936				
586-153	9946				
586-154	9099				
586-155	9962				
586-187	9945				
586-303	9932				

Original Part Number	ECG	SK	GE	IR/WEP	HEP
586-308	9937				
586-331	9806				
586-412	9812				
586-415	9157				
586-416	9939				
586-425	9093				
586-442	9158				
586-517	9804				
586-528	9802				
586-546	9800				
586-547	9808				
586-551	9938				
586-780	9810				
586-847	74184				
587-033	74170				
588-40-202	1142				
588U	132	3116	FET2	FE100/	
590-591731	194				
590-591811	194				
590-593031	123A				
592-027	9910				
592-028	9911				
592-029-0	9914				
592-032-0	9923				
592-032-1	9923				
592-079-0	9923				
592-081-0	9914				
593D742-1	123A				
595-1	123A	3122	20	TR21/735	S0015
595-2	123A	3122	20	TR21/735	S0025
597-1(RCA)	159	3118			
600-188-1-20	123A				
600-188-1-21	128				
600-188-1-22	194				
600-188-1-23	123A				
600A39	724				
600X0091-086	123A	3045		TR21/	S5014
600X0092-086	108	3018		TR95/	S0015
600X0093-086	161	3117		TR83/719	S0017
600X0094-086	161	3117		TR83/719	S0017
600X0095-086	159	3025		TR20/717	S0014
600X0141-000					S0017
600X0143-000					S0017
600X0175					S0016
600X0195-000					S0020
601-040	160	3006		TR17/	
601-054	126	3008		TR17/	
601-065	103A	3010		TR08/	
601-0100792	128				
601-0100793	123A				
601-0100810	234				
601-0100865	7413				
601-1		3122			
601-2		3122			
601-113	108	3018		TR95/	
601X0149-0186	123A	3018	17	TR21/	S0025
601X0402-038	1046				
601X0417-086	159	3114	21	TR20/	S0031
602-032	121	3009		TR01/	
602-040	102A	3004		TR85/	
602-075	160	3006		TR17/	
602-56	159	3025		TR20/717	S0019
602-60	159	3025	21	TR20/717	S0019
602-61	108	3044		TR95/56	S0015
602-113	108	3018		TR95/	
602X0008-002	199	3124	18	TR21/	S0015
602X0018-000	123A	3124	62	TR21/	S0015
602X0018-002	199	3124	18	TR21/	S0015
603-020	160	3006		TR17/	
603-030	160	3006		TR17/	
603-040	160	3006		TR17/	
603-113	108	3018		TR17/	
604	123A	3124	10	TR24/735	S0015
604-030	160	3006		TR17/	
604-080	160	3006		TR17/	
604-113	108	3019		TR95/	
604(SEARS)	123A			TR21/	S0015
605	123	3124	20	TR86/53	S0015
605-030	102A	3004		TR95/	
605-113	108	3018		TR95/	
606-020	102A	3004		TR85/	
606-9601-101	123A				
606-9602-101	123A				
607-030	123A	3124		TR21/	
608-112	127	3034		TR27/	
609-020	160	3006		TR17/	
609-112	123A	3124		TR21/	
610-035	102A	3004		TR85/	
610-035-1	102A	3004		TR85/	
610-036	102A	3004		TR85/	
610-036-1	102A	3004		TR85/	
610-036-2	102A	3004		TR85/	
610-036-3	102A	3004		TR85/	
610-036-4	102A	3004		TR85/	
610-036-5	102A	3004		TR85/	

Original Part Number	ECG	SK	GE	IR/WEP	HEP
610-036-6	102A	3004		TR85/	
610-036-7	102A	3004		TR85/	
610-036-8	102A	3004		TR85/	
610-039	121	3009		TR01/	
610-039-1	121	3009		TR01/	
610-040	102A	3004		TR85/	
610-040-1	102A	3004		TR85/	
610-040-2,	102A	3004		TR85/	
610-041	108	3018		TR95/	
610-041-1	108	3018		TR95/	
610-041-2	108	3018		TR95/	
610-041-3	108	3018		TR95/	
610-042	108	3039		TR95/	
610-042-1	108	3019		TR95/	
610-043	102A	3004		TR85/	
610-043-1	102A	3004		TR85/	
610-043-2	102A	3004		TR85/	
610-043-3	102A	3004		TR85/	
610-043-4	102A	3004		TR85/	
610-043-6	102A	3004		TR85/	
610-043-7	102A	3004		TR85/	
610-045	108	3018		TR95/	
610-045-1	108	3018		TR95/	
610-045-2	108	3018		TR95/	
610-045-3	123A	3124		TR21/	
610-045-4	123A	3124		TR21/	
610-046-7	102A	3004		TR85/	
610-050	160	3006		TR17/	
610-050-1	160	3006		TR17/	
610-050-2	160	3006		TR17/	
610-050-3	160	3006		TR17/	
610-051	160	3006		TR17/	
610-051-1	160	3006		TR17/	
610-051-2	160	3006		TR17/	
610-051-4	160	3006		TR17/	
610-052	126	3008		TR17/	
610-052-1	126	3008		TR17/	
610-053	160	3006		TR17/	
610-053-1	160	3006		TR17/	
610-053-2	160	3006		TR17/	
610-055	160	3006		TR17/	
610-055-1	160	3006		TR17/	
610-055-2	160	3006		TR17/	
610-055-3	160	3006		TR17/	
610-056	126	3008		TR17/	
610-056-1	126	3008		TR17/	
610-056-2	126	3008		TR17/	
610-056-3	126	3008		TR17/	
610-056-4	126	3008		TR17/	
610-067	121	3009		TR01/	
610-067-1	121	3009		TR01/	
610-067-2	121	3009		TR01/	
610-067-3	121	3009		TR01/	
610-068	121	3009		TR01/	
610-068-1	121	3009		TR01/	
610-069	108	3018		TR95/	
610-069-1	108	3018		TR95/	
610-070	123A	3124		TR21/	
610-070-1	123A	3124		TR21/	
610-070-2	123A	3124		TR21/	
610-070-3	123A	3124		TR21/	
610-071	124	3021		TR81/	
610-071-1	124	3021		TR81/	
610-072	108	3018		TR95/	
610-072-1	108	3018		TR95/	
610-072-2	108	3018		TR95/	
610-073	108	3018		TR95/	
610-073-1	108	3018		TR95/	
610-074	126	3008		TR17/	
610-074-1	126	3008		TR17/	
610-076	123A	3124		TR21/	
610-076-1	123A	3124		TR21/	
610-076-2	123A	3124		TR21/	
610-077	123A	3124		TR21/	
610-077-1	123A	3124		TR21/	
610-077-2	123A	3124		TR21/	
610-077-3	123A	3124		TR21/	
610-077-4	123A	3124		TR21/	
610-077-5	123A	3124		TR21/	
610-078	123A	3124		TR21/	
610-078-1	123A	3124		TR21/	
610-079	102A	3004		TR85/	
610-079-1	102A	3004		TR85/	
610-080	102A	3004		TR85/	
610-080-1	102A	3004		TR85/	
610-083	129	3025		TR88/	
610-083-1	129	3025		TR88/	
610-083-2	129	3025		TR88/	
610-083-3	129	3025		TR88/	
612-1A5L	123A	3124		TR21/	
612-16A	108	3018		TR95/	
612-16(ZENITH)	108	3018	17	TR95/56	S0016
612-60039	126	3008		TR17/	
612-60039A	126	3008		TR17/	
613(ZENITH)	108	3018	17	TR95/56	S0016

Original Part Number	ECG	SK	GE	IR/WEP	HEP
613-72	108	3018		TR95/	
614-1	128			TR87/	
614(ZENITH)	108			TR95/	S0016
614-2		3122			
614-12	108	3018		TR95/	
614A-1-5L	121	3009		TR01/	
614X1	102A	3008	51	TR85/250	G0005
614X2	102A	3008	51	TR85/250	G0005
614X3	102A	3008	51	TR85/250	G0005
614X4	102A	3008	51	TR85/250	G0005
614X5	102A	3004	2	TR85/250	G0005
614X6	102A	3004	2	TR85/250	G0005
614X7	102A	3004	2	TR85/250	G0005
614X8	103A	3010	10	TR08/250	S0015
614X9	102A	3004	2	TR85/250	G0005
614X10	102A	3004	2	TR85/250	G0005
615-1		3024			
617-10	123A			TR21/735	S0015
617-29	123A			TR21/	S0030
617-50	102A	3004		TR85/250	G0005
617-52	102A	3004		TR85/250	G0005
617-54	160	3006		TR17/637	G0003
617-55	160			TR17/637	G0003
617-56	126	3006		TR17/637	G0008
617-57	160	3006		TR17/637	G0003
617-58	160			TR17/637	G0003
617-63	123A			TR21/	S0014
617-64	123A			TR21/	S0014
617-65	161			TR83/	S0020
617-67	123A			TR21/735	S0024
617-68	123A			TR21/735	S0025
617-69	158			TR84/630	G6011
617-70	102A	3123		TR85/250	G0005
617-71	123A	3122		TR21/735	S5014
617-87	123	3122		TR86/53	S5014
617-117	152	3054		TR76/701	S5000
617-161	123A	3122		TR21/735	S0020
622-1	128	3024		TR87/	
623(RCA)	152			TR76/	
623-1	186	3192		TR55/	S5000
624(RCA)	153	3031		TR77/	
624-1	185	3191			S5006
625(RCA)	128	3124		TR87/	
630-076	123A	3124	20	TR21/735	S0015
631-1	129	3025		TR88/	
631-3(SYL)	129	3025		TR88/242	S5013
632-1(SYL)	188	3054		TR72/3020	S3019
635-1(RCA)	221	3050			F2004
637-1(RCA)	162			TR67/	
637(RCA)	162				
638	123A	3018	20	TR21/735	S0025
638-1(RCA)	218				
638H	175			TR81/241	S5019
638HJ	175	3026		TR81/	
639(ZENITH)	123A	3024	20	TR21/735	S0015
639CL	128	3024		TR87/	
640-1(SYL)	152	3054		TR76/701	S5000
642-116	160		50	TR17/637	
642-117	102A		53	TR85/250	
642-147	160		51	TR17/637	
642-150	102A		53	TR85/250	
642-151	158		53	TR84/630	G0005
642-152	131	3082	16	TR94/642	
642-173	160		50	TR17/637	
642-174	123A	3122	20	TR21/735	
642-176	121	3009	3	TR01/232	
642-202	160		51	TR17/637	
642-206	104		16	TR01/624	G6013
642-207	160			TR17/	
642-217	131	3082		TR94/642	
642-229	107	3018	60	TR70/720	S0016
642-230	107		60	TR70/720	S0016
642-242	123A		20	TR21/735	S0011
642-246	123A			TR21/	
642-254	107	3039	20	TR70/720	
642-260	107	3039	20	TR70/720	
642-261	184	3190	29	TR77/G5003	S5006
642-264	104		3	TR01/232	G6005
642-266	185	3191	69	/G5007	
642-268	107	3124	20	TR70/720	S0015
642-269	107		20	TR70/720	S0015
642-270	107		20	TR70/720	S0024
642-271	121MP	3013	3	/232MP	G6003
642-272	121	3009	3	TR01/232	
642-274	107		20	TR70/720	S0015
642-277	103A		54	TR08/724	G0011
642-306	128			TR87/	S0015
642-316	104			TR01/	G6003
642-319	123A		20	TR21/735	S0016
642AS4076-101	107				
644-1	185	3191	29	TR56/S5007	S5006
650-105	102	3004	2	TR05/631	
650-106	102	3004	2	TR05/631	
650-107	102	3004	2	TR05/631	
650-108	102	3003		TR05/631	

Original Part Number	ECG	SK	GE	IR/WEP	HEP
650-109	103	3010	8	TR08/641A	
653-202	108				
654-1(SYL)	132			FE100/	F0015
656-136	102A		52	TR85/250	G0005
656-137	102A			TR85/250	G0005
656-138	158			TR84/630	G0005
656-139	102A			TR85/250	G0005
657-31	161				
660-077-0					G6003
660-125	128	3024		TR87/243	S5014
660-126	123A	3018	20	TR21/735	S0011
660-127	108	3018	20	TR95/56	S0011
660-128	128	3047		TR87/243	S3021
660-131	123A	3124	20	TR21/735	S0015
660-132					S7004
660-133					S0019
660-134	123	3124	20	TR86/53	
660-137					S7003
660-138					S5018
660-145	128	3024	20	TR87/243	S0015
660-220	123A	3124		TR21/735	S0015
660-221	123A	3124		TR21/735	S0015
660-222	123A	3124		TR21/735	S0020
660-224	158	3004		TR84/630	G0005
660-225	123A	3124		TR21/735	S0014
660-227	102A	3004		TR85/250	G0005
660-228	103A	3010		TR08/724	G0011
660B	102	3004		TR05/631	G0005
665-1A5L	160	3006		TR17/	
665-2A5L	103A	3010		TR08/	
665-4A5L	101	3011		TR08/	
665A-1-5L	160	3007		TR17/	
665B-1-5L	126	3008		TR17/	
665C-1-5L	121	3009		TR01/	
666-1		3114			
666-1-A-5L	102A	3003		TR85/	
666-1(RCA)	128				
666-2-A-5L	102A	3004		TR85/	
666-3-A-5L	102A	3004		TR85/	
668					S0016
668CS	108				
669	159	3114		TR20/717	S0019
669-1(RCA)	165		36	TR93/	S5021
669A459H01	910D				
669A464H01	9932				
669A469H01	911D				
669A471H01	7472				
669A492H01	9962				
671A290H01	9930				
671A291H01	9944				
671A292H01	9945				
671A293H01	9946				
674-3A5L	100	3005		TR05/	
675-153	158	3004		TR84/630	G0005
675-154	102A	3004		TR85/250	G0005
675-155	102A	3004		TR85/250	G0005
675-156	102A	3004		TR85/250	G0005
675-206	131MP			TR94MP/642	G6016
676-1		3118			
677A-1-5L	105	3012		TR03/	
677B-1-5L	121MP	3013			
677C-1-5L	121MP	3014			
679-1	773				
679-1493-219		3024			
680-1	753	3018			C6010
684A-1-5L	121MP	3009		TR01/	
686-0004-0		3004			
686-0012	128	3024	18	TR87/243	
868-0112	128	3024		TR87/243	
686-0130	128	3024	18	TR87/243	
686-0165	128				
686-0210	175	3054	23	TR81/701	
686-0210-0	175	3026			
686-0243	130	3027	14	TR59/247	
686-0243-0	130	3027			
686-0325-0	159				
686-143	130	3027	14	TR59/247	
686-143A	130	3027		TR59/	
686-229-0	128				
686-237		3122			
686-257-0	199				
686-2700	159				
688B-1-5L	160	3006		TR17/	
688C-1-5L	100	3005		TR05/	
690L-021H25	152				
690L-021H26	153				
690L270H02	121	3009	3	TR01/232	
690L297H01	160		51	TR17/637	
690L297H02	121	3009	3	TR01/232	
690V02H69	108	3019	11	TR95/56	
690V08H36			10		
690V09H19				TR85/	
690V010H40	161	3018	20	TR83/719	S0017
690V010H41	108	3018	20	TR95/56	S0016
690V010H42	126	3005	1	TR17/635	G0008

Original Part Number	ECG	SK	GE	IR/WEP	HEP
690V0103H27		3122			
690V028H28	108	3039	17	TR95/56	S0016
690V028H48	108	3039	17	TR95/56	S0016
690V028H69	108	3039	17	TR95/56	
690V028H89	108	3039	20	TR95/56	S0016
690V034H29	126	3008	51	TR17/635	G0009
690V034H30	102A	3004	53	TR85/250	G0005
690V034H31	102A	3004	53	TR85/250	G0005
690V034H39	102	3004	53	TR05/631	
690V040H57	126	3006	51	TR17/635	G0008
690V040H58	126	3006	51	TR17/635	G0009
690V040H59	126	3006	51	TR17/635	G0008
690V040H60	126	3006	51	TR17/635	G0008
690V040H61	102	3004	2	TR05/631	G0005
690V040H62	102	3004	2	TR05/631	G0005
690V043H62	158	3004	2	TR84/630	G0005
690V043H63	158	3004	2	TR84/630	G0005
690V047H54	126	3008	51	TR17/635	
690V047H55	126	3008	51	TR17/635	
690V047H56	100	3005	1	TR05/254	G0005
690V047H57	100	3005	1	TR05/254	G0005
690V047H58	102	3004	2	TR05/631	
690V047H59	102A	3004	2		G0007
690V047H60	102	3004	2	TR05/631	G0005
690V047H97	123A	3122	20	TR21/735	
690V049H81	108	3019	17	TR95/56	S0016
690V052H23	102	3003	53	TR05/631	G0005
690V052H24	102	3004	2	TR05/631	G0005
690V052H63	160		51	TR17/637	
690V054H20	102A			TR85/250	
690V054H21	102	3004		TR05/631	
690V056H27	160	3007		TR17/637	
690V056H29	160	3007		TR17/637	
690V056H30	160	3007		TR17/637	
690V056H31	160	3008	51	TR17/637	G0008
690V056H32	160	3008	51	TR17/637	G0008
690V056H33	102A	3004	53	TR85/250	G0007
690V056H34	102A	3004	53	TR85/250	G0005
690V056H89	126	3008	51	TR17/635	
690V056H90	102	3004	2	TR05/631	
690V057H25	160	3006	51	TR17/637	G0003
690V057H27	102	3004	2	TR05/631	G0005
690V057H28	102	3004	2	TR05/631	
690V057H59	126	3006	51	TR17/635	
690V057H62	126	3006	51	TR17/635	
690V059H20	102A	3004		TR85/250	G0005
690V059H21	158	3004	53	TR84/630	G0005
690V059H52	102A	3004		TR85/250	G0007
690V059H55	102A	3004		TR85/250	G0005
690V060H58	108	3018	20	TR95/56	
690V060H59	108	3018	20	TR95/56	
690V061H98	102	3004	2	TR05/631	G0005
690V061H99	102	3004	2	TR05/631	G0005
690V062H47	102A		53	TR85/735	
690V063H14	126	3008	51	TR17/635	
690V063H15	126	3008	51	TR17/635	
690V063H16	102	3004	2	TR05/631	
690V063H17	102	3004	2	TR05/631	
690V063H50	102A	3004	53		G0006
690V063H51	102	3004	2	TR05/631	
690V066H44	126	3006	51	TR17/635	
690V066H45	126	3006	51	TR17/635	
690V066H46	102	3004	2	TR05/631	
690V066H47	102	3004	2	TR05/631	
690V066H49	126	3006		TR17/635	
690V066H89	160	3007	51	TR17/637	G0009
690V067H35	103A		54	TR08/724	
690V068H29	126	3008	51	TR17/635	G0008
690V068H30	102	3004	2	TR05/631	G0005
690V068H31	102	3004	2	TR05/631	G0005
690V070H49	108	3019	17	TR95/56	S0016
690V070H98	108	3018	20	TR95/56	
690V073H59	126	3006	51	TR17/635	G0008
690V073H85	126	3008	51	TR17/635	G0008
690V075H62	128	3024	18	TR87/243	
690V075H68	108			TR95/56	S0016
690V077H34	126	3006	51	TR17/635	
690V077H35	126	3008	51	TR17/635	
690V077H36	126	3006	51	TR17/635	G0008
690V077H37	102	3004	2	TR05/631	
690V077H33	102A		53	TR85/250	G0005
690V080H36	161	3018	17	TR83/719	
690V080H37	126	3006	51	TR17/635	
690V080H38	154	3045	51	TR78/712	S5025
690V080H39	102	3006	51	TR05/631	G0005
690V080H40	126	3006	51	TR17/635	G0009
690V080H41	123	3124	18	TR86/53	S0015
690V080H42	127	3034	25	TR27/235	G6003
690V080H43	127	3034	25	TR27/235	
690V080H44	102	3004	2	TR05/631	G0005
690V080H45	163	3111	14	TR67/740	
690V081H07	108	3039	20	TR95/56	S0016
690V081H08	160		51	TR17/637	
690V081H09	121	3009	16	TR01/232	
690V081H96	103A		54	TR08/724	
690V081H97	104	3009	3	TR01/624	
690V082H47	158	3004	53	TR84/630	G0005
690V084H60	126	3004	53	TR17/635	G0008
690V084H61	102A	3004	53	TR85/250	G0005
690V084H62	123A	3020	20	TR21/735	S0015
690V084H63	102A	3004	53	TR85/250	G0005
690V084H94	108		20	TR95/56	
690V084H95	108		20	TR95/56	
690V084H96	108		20	TR95/56	
690V085	158	3004		TR84/630	
690V085H42	126	3006	50		G0003
690V085H44	158	3004	52	TR84/630	G6011
690V086H39	102	3004	53	TR05/631	
690V086H51	123A	3122		TR21/735	S0020
690V086H52	108	3039	17	TR95/56	S0016
690V086H86	159		21	TR20/717	
690V086H87	108	3018	20	TR95/56	S0016
690V086H88	123A	3122	18	TR21/735	
690V086H89	129	3025	21	TR88/242	
690V086H90	128	3024	20	TR87/243	
690V086H94	161	3039		TR83/719	S0016
690V086H95	161	3039		TR83/719	S0017
690V086H96	108	3039		TR95/56	S0016
690V088H44	108	3018		TR95/56	S0020
690V088H45	108	3018		TR95/56	S0020
690V088H46	107	3018		TR70/720	S0015
690V088H47	107	3018		TR70/720	S0016
690V088H48	108	3018		TR95/56	S0016
690V088H49	107	3018		TR70/720	S0011
690V088H50	123A	3124		TR21/735	S5014
690V088H51	123A	3024		TR21/735	S5014
690V088H52	102A	3004		TR85/250	G0005
690V089H46	107	3018		TR70/720	S0014
690V089H86	107	3018		TR70/720	S0016
690V089H89	123A	3122		TR21/735	S0030
690V089H90	102A	3004	2	TR85/250	G0005
690V092H52	123A	3122	20	TR21/735	
690V092H54	123A	3122	20	TR21/735	
690V092H81	123A	3122	20	TR21/735	
690V092H84	123A	3122	20	TR21/735	
690V092H96	123A	3122	20	TR21/735	
690V092H97	123A	3122	20	TR21/735	
690V094H17	130	3021		TR59/247	S5019
690V094H18	102A	3004		TR85/250	G0005
690V094H19	102A	3004		/250	G0005
690V094H20	102A	3004		TR85/250	G0005
690V094H21	123A	3024		TR21/735	S0015
690V097H59	102A	3004		TR85/	
690V097H62	123A	3124	20	TR21/	S0015
690V098H48	123A	3122		TR21/735	S0015
690V098H49	123A	3122	2	TR21/735	S0015
690V098H50	123A	3122		TR21/735	S0015
690V098H51	131	3082		TR94/642	G6016
690V099H59	158	3004		TR84/630	G6011
690V099H79	123A			TR21/735	S0024
690V099H80	103A			/724	G0011
690V0103H27	123A				
690V102H96	126	3010	5	TR08/724	G0011
690V102H71	123A		20	TR21/735	S0011
690V102H96	126	3005	51	TR17/	G0008
690V103H23	108		20	TR95/	S0016
690V103H24	108	3018	20	TR95/56	S0014
690V103H25	108		20	TR95/	S0014
690V103H26	108		20	TR95/	S0015
690V103H27	108	3018	20	TR95/735	S0011
690V103H28	107	3018	20	TR70/720	S0016
690V103H29		3020	20		S0015
690V103H30	128		63	TR87/	S0015
690V103H31	123A		23	TR21/	S5019
690V103H32		3020	20		S0015
690V103H33	123A		20	TR21/	S0030
690V104H53	102A	3004	52	TR85/	G0001
690V104H54	102A	3004	52	TR85/	G0005
690V105H19			20	TR21/	S0015
690V105H21			51	TR21/	G0008
690V105H22			20	TR21/	S5026
690V105H24			52	TR85/	G0005
690V105H25			20	TR21/	S0015
690V105H26			51	TR21/	G0008
690V105H27			17	TR21/	S0016
690V105H28			17	TR21/	S0016
690V105H29			20	TR21/	S0016
690V109H46	107	3018		TR70/720	S0014
690V110H30	108	3018		TR95/56	S0015
690V110H31	108	3018		TR95/56	S0015
690V110H32	108	3018		TR95/56	S0011
690V110H33	108	3018		TR95/56	S0011
690V110H34	128	3124		TR87/243	S5014
690V110H36	128	3124		TR87/	S5014
690V110H55	129	3025		TR88/243	S5013
690V110H89		3054	28		S5000
690V114H29	108	3122	20	TR21/	
690V114H30	123A	3018	10	TR51/	
690V114H31	108	3018	17	TR95/	
690V114H33	123A	3011	10	TR24/	

Original Part Number	ECG	SK	GE	IR/WEP	HEP
690V116H19	108	3122	11	TR95/	S0014
690V116H20	107	3018	11	TR70/720	S0014
690V116H21	123A		20	TR21/	S0014
690V116H22	133		FET2	FE100/	
690V116H23	159		21	TR20/	
690V116H24	128		63	TR87/	
690V116H25	185	3191	58		
690V116H26	184		57		
690V118H59	108		20	TR95/	
690V118H60	159		22	TR20/	
690V118H61	159		21	TR20/	
690V118H62	102A		2	TR85/	
690V119H54	158	3004	53	TR84/	
690V119H94	126	3008	50		G0005
690V119H95	160	3006	50		G0003
690V119H96	160	3006	50		G0008
690V119H97	160	3008	50		G0003
690V119H98					G0005
690V120H89		3018	20		S0014
690W02H96				TR12/	
691C844	128				
692-0014-00	9917				
692-0016-00	9924				
692-0020-00	9989				
692-1A5L		3012			
693		3024			
693EP	123A	3124			S0015
693FS	123A	3020		TR21/735	
693G	123A	3124	20	TR21/735	S0015
693GT	123A	3124		TR21/735	
693GV	199	3124			
693G4				TR21/	
69694D	123A	3122	18	TR21/735	
694E	123A		17	TR21/	S0015
695-1					S3006
698V102H39			8		
699	123A	3124	20	TR21/735	
699(GE)					S0015
700-04	133		FET1	/801	F0010
700-110	186	3024	28	TR55/	
700-113	130	3027	14	TR59/	
700-133	159		67	TR20/	
700-134	123A		18	TR21/	
700-135	128		63	TR87/	
700-136	129		21	TR88/	
700-154	123A	3018	17	TR21/	
700-155	199	3124	62	TR21/	
700-156	199		62	TR21/	
700-325	123A	3018	17	TR21/	
700-848A	908				
700A-858-318	199		20		
700A858-318	123A	3124	62	TR25/	S0015
700A858-319	199	3124	62	TR25/	S0015
700A-858-328	199	3124			
700A858-328	123A	3124	62	TR21/	S0024
702-0002	107				
702-70050		3031			
703-056(4)	123A				
703-1	128			TR87/243	S0015
703-2	128	3024			S5014
703B	128	3024	63	TR87/243	
703HC	703A				
703U		3112			
706BC	1114				
706BPC	1114				
709DC	909D	3590			
709HC	909	3551			
709PC	909D	3590			
710DC	910D	3554			
710HC	910	3553			
710PC	910D				
711-001	223				
711DC	911D	3556			
711HC	911	3555			
711PC	911D				
715EN	123A	3018			S0020
715FB	107	3039		TR70/720	
715HC	915				
716HC	716				
720-35019	102	3004	2	TR05/631	
720-35019A	102A				
720-19					S5012
720DC	744		IC24		
720PC	744				
720SDC	744				
720SPC	744				
723DC	923D	3165			
723HC	923	3164			
723PC	923D	3165			
725-008	726				
725EHC	925				
725HC	925				
729-3	128	3024			S5014
729DC	720				
729PC	720				

Original Part Number	ECG	SK	GE	IR/WEP	HEP
732DC	718	3159			
732PC	718	3159			
733-0100-699	187				
733-2					S3007
733W00021	941M				
733W00024	9930				
733W00025	9932				
733W00026	9933				
733W00027	9936				
733W00028	9946				
733W00029	9962				
733W00030	9093				
733W00035	941D				
733W00039	7490				
733W00046		3541			
733W00048	9937				
733W00049	9949				
733W00050	9963				
733W00133	7493				
734EU	132	3116	FET2	FE100/	
734U		3112			
735-40C		3054			
737(ZENITH)	123A	3024	18	TR21/735	S0015
737CL		3024			
739		3162			
739DC	725	3162	IC19		
739H01	123A	3122		TR21/735	S0011
739PC	725	3162			
740-2001-307	722	3161			C6056
740-2002-111	708	3135			C6062
740-5853-300	1005				
740-8120-160	1005				S0015
740-9000-554	720				
740-9007-046	1108				
740-9016-105	720	3160		IC512/	C6095
740HC	940				
741DC	941D				
741HC	941				
741PC	941D				
741TC	941M	3559			
742-1	188			TR72/	
742C2030-020	130				
744				TR25/	
746HC	707	3134			
746PC	790	3077			
747-107		3083			
747HC	947	3526			
748(ZENITH)	123A			TR21/735	
748				TR25/	
749	949	3166			
749(BENDIX)	949				
749D		3166			
749DHC	949	3166	IC25		
750-045	175		23	TR81/	
750-137	102	3004	2	TR05/631	
750-138	102	3004	2	TR05/631	
750-139	102	3004	2	TR05/631	
750-140	102	3004	2	TR05/631	
750-35019	102A		2	TR85/	
750A858-328	199	3124	20	TR21/	S0024
750A858-448	152	3041	66	TR92/	S5000
750C838-123	123A	3124		TR21/	S0011
750C838-124	123A	3124		TR21/	S0011
750C838-125	123A	3124		TR21/735	S0015
750D858-212	123A	3124		TR21/735	S0024
750D858-213	108	3124		TR95/56	S0015
750M63-104	160			TR17/	
750M63-105	158			TR84/	
750M63-115	102A			TR85/	
750M63-116	160			TR17/	
750M63-117	160			TR17/	
750M63-119	123A			TR21/	
750M63-120	123A			TR21/	
750M63-121	102A				
750M63-122	102A				
750M63-146	123A			TR21/	
750M63-147	158			TR84/	
750M63-148	158			TR84/	
752A					C6007
753-0100-699	187	3084	29		S5006
753-0101-047	108	3018	17		S0016
753-0101-226	186	3054	28		S5000
753-2000-003	123A	3124	20	TR21/	S0014
753-2000-004	123A	3124	10	TR21/	S0014
753-2000-006	175			TR81/	S5019
753-2000-007	107	3018	60	2SC829B/	S0011
753-2000-008	123A			TR21/	S0014
753-2000-009	123A			TR21/	S0025
753-2000-011	123A	3124	20	TR21/735	S0025
753-2000-100	199	3122		TR51/	S0024
753-2000-101	159	3114		TR52/	S0019
753-2000-107	128	3020		TR25/	S5014
753-2000-460		3018			S0014
753-2000-463	131MP	3086	31MP	TR50/	G6016
753-2000-535		3016			S0016

Original Part Number	ECG	SK	GE	IR/WEP	HEP
753-2000-710	108	3122	20	TR95/	S0014
753-2000-711	123A	3122	20	TR21/735	S0014
753-2000-735	123A	3124	20	TR25/	S0014
753-2000-870	123A	3124	62	TR21/	S0014
753-2000-871	123A	3124	62	TR21/	S0014
753-2001-173	153	3054		TR77/	S0019
753-2100-001	123A		44	TR94MP/	S0015
753-2100-002	131	3082	30	TR94/642	G6016
753-2100-008	123A			TR21/	S0015
753-3000-535	229			2SC535B/	
753-4000-010	199	3124		TR21/735	S0024
753-4000-011	123A		20	TR21/	
753-4000-024	133	3112		/801	F0010
753-4000-025	132	3116		FE100/802	F0015
753-4000-101	123A	3122		TR21/735	
753-4000-537	123A	3018	20	TR21/735	S0016
753-4000-668	107	3018		TR70/720	S0016
753-4000-929	107	3018	20	TR70/720	S0016
753-4001-474	131	3082		TR94/642	G6016
753-4001-931	184	3041			S5000
753-4001-932	152	3054		TR76/701	
753-4004-248	159	3025	21	TR20/717	S0019
753-5851-359	107			TR24/	
753-6000-002	199	3124	18		S0015
753-6000-019	132				
753-8400-230	199	3124	62		S0015
753-8500-380	199	3124	18	TR21/	S0015
753-8510-470	128	3024	63	TR25/	S0014
753-9000-019	132				
753-9000-096	186	3054			
753-9000-839	199	3018			
753-9000-922	107	3122			
753-9010-235	152	3041	28	TR92/	S5000
753C		3163			
753T	736				
753TC		3163			
754-5853-300	1005				
755-422494	159				
755-800398		3004			
758DC	743				
758PC	743				
759-019		3018			
759-020		3018			
759-023		3018			
759-024		3124			
759-029		3004			
759-060		3004			
762-105-00	129				
762-110	234				
762-120	129				
763-1	128				
766-100999	223				
767		3124	18	TR24/735	
767C		3161			
767CL	123A	3020		TR21/	
767DC	722	3161			
767PC	722	3161			
767(ZENITH)	123A			TR21/	S0024
770-045	152	3026	23	TR76/701	
772-101-00	199				
772-110	123A				
772-120-00	128				
772-121-00	128				
772A	107	3018		TR70/	
772B	107	3018		TR70/	S0016
772B1	107	3039		TR70/720	
772BJ	107	3018		TR70/	
772BL	107	3018		TR70/	
772BM	107	3018		TR70/	
772BN	107	3018		TR70/	
772BY	107	3018	20	TR70/720	S0016
772C	107	3018		TR70/	
772CC	107		20	TR70/720	S0016
772D	107	3018		TR70/	S0016
772D1	107	3039		TR70/720	
772DC	107		20	TR70/720	
772DG	107		20	TR70/720	
772DQ					S0016
772E	107	3018		TR70/	
772EH	107	3039		TR70/720	
772F	107	3018		TR70/	
772FE	107	3039		TR70/720	
772G	107	3018		TR70/	
773		3024	63	TR25/735	
773CL	128	3024		TR87/	
773RED	108	3039		TR95/56	S0015
773(ZENITH)	123A			TR21/	S5014
774		3025	67	TR28/717	
774CL	129	3025		TR88/	
774ORN	108	3039		/56	S0015
774(ZENITH)	159			TR20/	S5013
775-1(SYL)	172		64	TR69/S9100	S9100
775BRN	108	3039		TR95/56	S0015
776-1(SYL)	159	3025		TR20/717	S0031
776-2(PHILCO)	123A				
776-151	123A				
776-183	123A				
776-201		3027			
776GRN	123A	3122		TR21/735	S0015
776Y	224				
779BLU	108	3039		TR95/56	S0015
780DC	714	3075			
780PC	714	3075			
780WHT	123A	3122		TR21/735	S0015
781DC	715	3076			
781PC	715	3076			
783RED	108	3039		TR95/56	S0015
784ORN	108	3039		TR24/56	S0015
785YEL	123A	3122		TR21/735	S0015
786	108	3039		TR95/56	S0015
787BLU	108	3039		TR95/56	S0015
787PC	797				
791	123A	3122		TR21/735	S0015
792-286	102	3004	2	TR05/631	
792-287	102	3004	2	TR05/631	
792-288	102	3004	2	TR05/631	
792-289	102	3004	2	TR05/631	
792-290	102	3004	2	TR05/631	
800-0004-P090	724				
800-001-031-1	159				
800-001-034	123A				
800-001-106-1	128				
800-003-07	103A			/724	
800-101-101-1	123A				
800-101-101-2	128				
800-101-102-1	123A				
800-101-108-1	159				
800-101-114-1	129				
800-122	124	3021		TR81/240	
800-158	124	3021		TR81/240	
800-172	124	3021		TR81/240	
800-180	124			TR81/	
800-181	124			TR81/	
800-196	104MP	3013		TR01MP/624	G6005(2)
800-203	124	3021		TR81/240	
800-204	124	3021		TR81/240	
800-205	158		53	TR84/630	
800-208	102A+103A			TR85/724	
800-219	102A+103A			TR85/250	G0005
800-245	128+129			TR87/243	
800-250-102	123	3124	20	TR86/53	
800-253	121MP	3013	53		
800-256(METAL)	124	3103	12	TR81/240	S5011
800-256(PLAS)	157	3103		TR60/	
800-270	184+185			/S5003	
800-284	129	3025		TR88/242	
800-289	184+185	3054		TR28/701	S5000
800-294	128+129	3024		TR87/243	S5013
800-305	128+129			TR87/243	
800-310	103A+158		54	/719	G0005
800-315	128+129			TR87/243	
800-321	124	3021		TR81/240	
800-328	128+129			TR87/	
800-329	104			TR01/624	G6003(2)
800-362	152+153				
800-366	152+153				
800-401	124		12	TR81/	S5015
800-501-00	123A	3018	20	TR21/735	S0015
800-501-01	123A	3122	20	TR21/735	
800-501-02	123A	3122	20	TR21/735	
800-501-03	123A	3122	20	TR21/735	
800-501-04	123A	3122	20	TR21/735	
800-501-11	123A	3122	20	TR21/735	
800-501-22	123A		20	TR21/	
800-502-00	102A	3004	2	TR85/	G0005
800-503-00	103A	3010	8	/724	
800-504-00	160	3006	51	TR17/637	G0002
800-505-00	102A	3005	2	TR85/250	G0005
800-506-00	103A	3010	8	TR08/724	
800-507-00	121	3009	3	TR01/232	G6005
800-508-00	123A	3122	20	TR21/735	
800-509-00	123A	3122	20	TR21/735	
800-510-00	130	3027	14	TR59/247	
800-511-00	129	3025		TR88/242	
800-512-00	128	3024	18	TR87/243	
800-513-00	128	3024	18	TR87/243	
800-514-00	123A	3122	20	TR21/735	
800-515-00	128	3024	18	TR87/243	
800-516-00	129	3025	22	TR88/242	
800-518-00	121	3009	3	TR01/232	
800-521-00	123A	3124	20	TR21/735	
800-521-02	128	3124		TR21/735	S3019
800-521-02(PEN)	188			TR72/3020	
800-521-03	128	3024	18	TR87/243	
800-522-01	123A	3124	20	TR21/735	
800-522-02	123A	3124	20	TR72/3020	S5000
800-522-02(PEN)	188			TR72/	
800-522-03	128	3024	20	TR87/243	
800-522-04	123A	3122	20	TR21/735	S5000
800-523-01	159	3114	20	TR20/717	

Original Part Number	ECG	SK	GE	IR/WEP	HEP
800-523-02	159	3124	14	TR73/717	S5006
800-523-02(PEN)	189			TR73/3031	
800-524-02	130	3027	14	TR59/247	
800-524-02A	130	3027		TR59/	
800-524-03	130	3027	14	TR59/247	
800-524-03A	130	3027		TR59/	
800-524-04A	130			TR59/	
800-525-03	129	3124	14	TR88/242	
800-525-04	159	3114	14	TR20/242	S5022
800-526-00	123A	3124	20	TR21/735	S5014
800-527-00	159	3025	21	TR20/717	S0013
800-528	103A	3010		TR08/724	
800-528-00	103A	3010	54	TR08/724	G0011
800-529-00	123A	3124	20	TR21/735	
800-530-00	123A	3122	20	TR21/735	
800-530-01	123A	3124	20	TR21/735	S0025
800-533-00	152	3054		TR76/701	S5015
800-533-01	184	3190		/S5003	
800-534-00	123A	3122	10	TR21/735	S0015
800-534-01	123A	3122	20	TR21/735	
800-535-00	132	3116	FET2	FE100/802	F0010
800-535-01	132	3116	FET2	FE100/802	F0015
800-536-00	108	3018	61	TR95/56	S0020
800-537-01	103A	3027	14	TR08/724	G0011
800-537-02	130MP	3027		TR59MP/247	S7002
800-537-03	158	3027		TR84/630	S7002
800-538-00	123A	3122	20	TR21/735	
800-544-00	123A	3122	20	TR21/735	S0020
800-544-00(UJT)	6400				
800-544-10	123A	3122	20	TR21/735	S0019
800-544-20	123A	3122	28	TR21/735	S3024
800-544-30	123A	3122	20	TR21/735	S3028
800-546-00	152	3054	66	/S5003	S5000
800-548-00	123A	3122		TR21/735	
800-550-00	162			TR67/707	
800-550-10	163			TR67/	
800-552-00	191			TR75/	
800-557-00	161	3018	39		S0017
800-50100	123A	3124		TR21/	
800-50300	103A	3010		TR08/	
800-50400	160	3006		TR17/	
800-50500	126	3008		TR17/	
800-50600	103A	3010		TR08/	
800-50700	121	3009		TR01/	
800-51500	123A	3020		TR21/	
800-51600	129	3025		TR88/	
800-52102	123A	3124		TR21/	
800-52202	123A	3124		TR21/	
800-52302	123A	3124		TR21/	
800-52402	130	3027		TR59/	
800-52600	123A	3124		TR21/	
800-52700	129	3025		TR88/	
800-52800	103A	3010		TR08/	
800-53001	123	3124		TR86/	
800-53300	128	3024		TR87/	
800-53400	123A	3124		TR21/	
800-53600	108	3018		TR95/	
800-53702	130MP	3029		TR59MP/	
801-04900	126	3008		TR17/	
801-05200	126	3008		TR17/	
801-05300	126	3008		TR17/	
801B	123A	3122	17	TR21/735	S0014
802-05400	102A	3004		TR85/	
802-05600	102A	3004		TR85/	
802-22631U	102A	3004		TR85/	
802-36400	102A	3004		TR85/	
802-36430	102A	3004		TR85/	
802-43900	102A	3004		TR85/	
803-18250	123A	3124		TR21/	
803-37390	123A	3124		TR21/	
803-38030	108	3018		TR95/	
803-38040	108	3018		TR95/	
803-39430	108	3018		TR95/	
803-39440	108	3018		TR95/	
803-83840	108	3018		TR95/	
803-83930	108	3018		TR95/	
803-83940	123A	3124		TR21/	
803-89910	123A	3124		TR21/	
803-89930	123A	3124		TR21/	
804	123A			TR21/	
804-14120	130	3027		TR59/	
804-14130	130	3027		TR59/	
804-21980	123A	3124		TR21/	
807-1(PHILCO)	194				
808-304	102A	3004		TR85/250	G0005
808-305	102A	3004		TR85/250	G0005
808-306	102A	3004		TR85/250	G0005
808-307	102A	3004		TR85/250	G0005
808-308	102A	3004		TR85/250	G0005
808-309	103A	3010		TR08/724	S0014
808-310	102A	3004		TR85/250	G0008
808-311	102A	3004		TR85/250	G0005
810	788	3147			
815-181D	102	3004		TR05/631	
815-1810	102A		2	TR85/	

Original Part Number	ECG	SK	GE	IR/WEP	HEP
817-275	175				
818WHT	123A	3122		TR21/735	S0015
822	123A		20	TR21/735	
822-1(SYL)	161			TR83/	S0017
822A	123A	3124	20	TR21/735	S0024
822ABLU	123A			TR21/735	S0024
822B	123A	3124	20	TR21/735	S0024
823B	123A	3122		TR21/735	S0024
823WHT	123A	3122		TR21/735	S0024
824-09501	123A	3124		TR21/	
824-1(SYL)	161	3018	20	TR83/719	S0017
824-10300	129	3025		TR88/	
826-1	159			TR20/	
827BRN	123A	3122		TR21/735	S0015
828GRN	123A	3122		TR21/735	S0015
828S	123A				
829	159	3114		TR20/717	S0019
829A	159	3114	21	TR20/717	S0019
829B	159	3114	21	TR20/717	S0019
829C	159	3114	21	TR20/717	S0019
829D	159	3114	67	TR20/717	S0019
829E	159	3114	21	TR20/717	S0019
829F	159	3114	21	TR20/717	S0019
830	102A	3123	2	TR85/250	G0005
833	159	3114		TR20/717	S0019
834-250-011	152				
847BLK	123A	3122		TR21/735	S0015
853-0300-632	123A			TR51/	S0015
853-0300-634	187	3083	69	TR56/	S5006
853-0300-643	193	3025			
853-0300-644	199	3124	62		S0015
853-0300-900	123A	3124	20	TR25/	S0015
853-0300-923	123A	3124	62	TR21/	S0015
853-0301-096	186	3054	28	TR55/	S5000
853-0301-317	128	3122	18		S0003
853-0373-110	123A				S0015
853-0800-923					S0015
858	199				
858GS	123A			TR21/735	
859GK	199		20	TR21/	
860-022-01	181			TR36/S7000	S7002
866-6	130			TR59/	
866-6(BENDIX)	130			TR59/	
880-101-00	703A	3157	IC12	IC500/	C6001
880-102-00	726	3129			
880-103-00	9914				
880-250-102	123A	3122		TR21/735	
880-250-107	129	3025		TR88/242	
880-250-108	123A			TR21/735	
880-250-109	123A	123A		TR21/735	
881-250-102	123A		20	TR21/	
881-250-107	129	3025		TR88/	S0019
881-250-108	123A	3124		TR21/	S0015
881-250-109	123A	3124		TR21/	S0015
881-250108	199	3038		TR21/	
884-250-001	121	3009		TR01/	
884-250-010	121	3009		TR01/	
884-250-011	184	3024		TR76/	S5000
886-1A3002C		3007			
897		3027			
900HC	9900				
903			39	TR24/	
903-3	123A	3124		TR21/	
903-3G	123A	3122		TR21/	
903HC	9903				
903Y002149	123A	3024	61	TR24/	
903Y002150	199	3124	62	TR51/	
903Y002151	187	3084	29	TR56/	
903Y002152	186	3041	28	TR55/	
904			39	TR24/	
904-95	123A	3124		TR21/735	
904-95A	123A	3124		TR21/735	
904-95B	199	3124		TR21/	S0024
904-96B	123A	3122		TR21/735	S0015
904A			39	TR24/	
904HC	9904				
905			39	TR24/	
905-5	1024		720		
905-27			IC27		
905-28			IC27		
905-30B	1003				
905-38B	722	3161			
905-39B	1075				
905-46B		3161			
905HC	9905				
906HC	9906				
907HC	9907				
908HC	9908				
909-2715-160				TR21/	
909-27125-140	123A	3122	212	TR21/735	S0015
909-27125-160	107	3039	212	TR70/720	S0015
909HC	9909				
909(RCA)	107				
910			39	TR51/	
910X1	158			TR84/630	G0005

Original Part Number	ECG	SK	GE	IR/WEP	HEP
910X2	158			TR84/630	G0005
910X3	102A			TR85/250	G0005
910X4	102A			TR85/250	G0005
910X5	102A			TR85/250	G0005
910X6	102A			TR85/250	G0005
910X7	158			TR84/630	G0005
910X8	158			TR84/630	G0005
910X9	158			/630	G0005
910X10	158			TR84/630	G0005
911HC	9911				
912-1A6A	123A	3124		TR21/	
912HC	9912				
913HC	9913				
914A-1-6-5	121	3009		TR01/	
914F298-1	128	3024	18	TR87/243	
914F298-4		3124			
914F346-01		3024			
914F-346-1		3024			
914HC	9914				
915HC	9915				
916-31001-1	102A	3004	52	TR85/250	G0005
916-31001-1B	102A		52	TR85/250	
916-31001-7B	102A		52	TR85/250	G0007
916-31003-5B	102A		52	TR85/250	G0005
916-31007-5	102A			TR85/250	G0005
916-31007-5B	102A		53	TR85/250	G0005
916-31012-6	102A	3004	52	TR85/250	G0005
916-31012-6B	102A		52	TR85/250	G0005
916-31019-3	126	3008	51	TR17/635	G0008
916-31019-3B	126	3008	50	TR17/635	G0008
916-31024-3	123A	3018	20	TR21/735	S0014
916-31024-3B	107		210	TR70/	
916-31024-5	107	3018	11	TR70/720	S0014
916-31024-5B	108	3039	210	TR95/56	
916-31025-4	161	3018	11	TR83/719	S0017
916-31025-4B	161	3117	210	TR83/719	
916-31025-5	123A	3018	11	TR21/735	S0014
916-31025-5B	107	3039	210	TR70/720	
916-31026-8B	123A	3122		TR21/735	
916-31026-9B	102A		51	TR85/250	
921-1A	160	3006		TR17/637	
921-1B	160	3006		TR17/637	G0003
921-2A	160	3006		TR17/637	
921-2B	160	3006		TR17/637	G0003
921-3A	160	3006		TR17/637	
921-3B	126	3006		TR17/635	G0008
921-4A	160	3006		TR17/637	
921-4B	126	3006		TR17/635	G0008
921-5A	103A	3010		TR08/724	
921-5B	103A	3010		TR08/724	G0011
921-6A	103A	3010		TR08/724	
921-6B	103A	3010		TR08/724	G0011
921-7	123A	3124	20	TR21/735	S0024
921-7B	103A			TR09/724	
921-8	123A	3124	20	TR21/735	S0025
921-9B	160	3008		TR17/637	G0003
921-10B	160	3008		TR17/637	G0003
921-11B	160	3008		TR17/637	G0003
921-12B	160	3004		TR17/637	G0005
921-13B	160			TR17/637	G0003
921-14B	102A	3004		TR85/250	G0005
921-15B	160			TR17/637	G0003
921-16B	160			TR17/637	G0003
921-17B	160			TR17/637	G0003
921-19B					G0008
921-20	108	3018		TR95/56	
921-20A	108	3018		TR95/56	
921-20B	108	3018	20	TR95/56	S0015
921-20BK	123A	3124		TR21/	
921-21	108	3018		TR95/56	
921-21A	108	3018		TR95/56	
921-21B	108	3018	20	TR95/56	S0015
921-21BK	108	3018		TR95/	
921-22	108	3018		TR95/56	
9221-22A	108	3018		TR95/56	
921-22B	108	3018	20	TR95/56	S0015
921-22BG	123A	3124		TR21/	
921-23	108	3018		TR95/56	
921-23A	108	3018		TR95/56	
921-23B	108	3018	20	TR95/56	S0015
921-23BK	123A	3124		TR21/	
921-24B	102A		52	TR85/250	G0005
921-25B	160		50	TR17/637	G0005
921-26	123A	3124		TR21/735	
921-26A	123A	3124		TR21/735	
921-26B	160	3124	18	TR17/637	S0015
921-27	102A	3004		TR85/250	
921-27A	102A	3124		TR85/250	
921-27B	123A	3004	2	TR21/735	S0014
921-28	123A	3124		TR21/	
921-28A	123A	3124		TR21/	
921-28B	123A	3024		TR21/735	S0012
921-28BLU	123A	3124		TR21/	
921-29B	159	3114		TR20/717	S0012
921-30	108	3018		TR95/	

Original Part Number	ECG	SK	GE	IR/WEP	HEP
921-30A	108	3018		TR95/	
921-30B	108	3039		TR95/56	S0015
921-31	108	3018		TR95/	
921-31A	108	3018		TR95/	
921-31B	108	3018	20	TR95/56	S0015
921-32	108	3018		TR95/	
921-32A	108	3018		TR95/	
921-32B	108	3018	20	TR95/56	S0015
921-33	108	3018		TR95/	
921-33B	108	3039	20	TR95/56	S0011
921-34	108	3018		TR95/	
921-34A	108	3018		TR95/	
921-34B	108	3039	17	TR95/56	S0011
921-35	102A	3004		TR85/250	
921-35A	102A	3004		TR85/250	
921-35B	102A	3004		TR85/250	G0008
921-36B	102A	3004	53	TR85/250	G0005
921-37B	102A	3123		TR85/250	G0007
921-38	102A	3004		TR85/250	
921-38A	102A	3004		TR85/250	
921-38B	102A	3004		TR85/250	G0006
921-39	102A	3004		TR85/250	
921-39A	102A	3004		TR85/250	
921-39B	102A	3004		TR85/250	G0005
921-40	102A	3004		TR85/250	
921-40A	102A	3004		TR85/250	
921-40B	102A	3004		TR85/250	G0007
921-41B	102A			TR85/	G0005
921-43B	123A	3124	20	TR21/735	S0014
921-44B	158			TR84/	G0005
921-45	102A	3004		TR85/	
921-45A	102A	3004		TR85/	
921-45B	102A	3004		TR85/250	G0007
921-46	123A	3124		TR21/	
921-46A	123A	3124		TR21/	
921-46B	103A	3123	10	TR08/724	S5014
921-46BK	123A	3124		TR21/	
921-47	123A	3124		TR21/	
921-47A	123A	3124		TR21/	
921-47B	129	3025	67	TR88/242	S5013
921-47BL	123A	3124		TR21/	
921-48B	123A	3124		TR21/735	S0014
921-49B	123A	3124		TR21/735	S0014
921-50B	123A	3124		TR21/735	S0014
921-51B	102A	3004		TR85/250	G0005
921-53B	158			TR84/630	G0005
921-54B	103A			TR08/724	G0011
921-55B	161	3117		TR83/719	S0016
921-56B	161	3018		TR83/719	S0016
921-57B	161	3018		TR83/719	S0016
921-58B	161	3018		TR83/719	S0016
921-59B	107	3018		TR70/720	S0016
921-60B	128	3024		TR87/243	S0015
921-61B	124	3021		TR81/240	S5011
921-62	107	3018		TR70/	
921-62A	107	3018		TR70/	
921-62B	107	3018	20	TR70/720	S0015
921-63	107	3018		TR70/	
921-63A	107	3018		TR70/	
921-63B	107	3018		TR70/720	S0015
921-64	107	3018		TR70/	
921-64A	107	3018		TR70/	
921-64B	107	3018		TR70/720	S0015
921-64C	107	3018		TR70/	
921-65	126	3005		TR17/	
921-65A	126	3005		TR17/	
921-65B	126	3005		TR17/635	G0008
921-66	126	3005		TR17/	
921-66A	126	3005		TR17/	
921-66B	126	3005		TR17/635	G0008
921-67	102A	3004		TR85/	
921-67A	102A	3004		TR85/	
921-67B	102A	3004		TR85/250	G0005
921-68	102A	3004		TR85/	
921-68A	102A	3004		TR85/	
921-68B	102A	3004		TR85/250	G0005
921-69	158	3004		TR84/	
921-69A	158	3004		TR84/	
921-69B	158	3004		TR84/630	G6011
921-70	159	3004		TR20/	
921-70A	159	3004		TR20/	
921-70B	159	3004		TR20/717	S0013
921-71	103A	3124		TR08/	
921-71A	103A	3124		TR08/	
921-71B	103A	3124		TR08/724	G0011
921-72B	108			TR95/56	S0016
921-73B	123A	3018		TR21/735	S0014
921-84	107	3018		TR70/	
921-84A	107	3018		TR70/	
921-84B	107	3018	20	TR70/720	S0016
921-85	107	3018		TR70/	
921-85A	107	3018		TR70/	
921-85B	107	3018		TR70/720	S0014
921-86	107	3018		TR70/	
921-86A	107	3018		TR70/	

Original Part Number	ECG	SK	GE	IR/WEP	HEP
921-86B	107	3018		TR70/720	S0014
921-88	124	3021		TR81/	
921-88A	124	3021		TR81/	
921-88B	124	3021		TR81/240	S5011
921-92B	199	3018	20	TR70/	S0016
921-93B	123A	3122	20	TR21/	G0007
921-94B					G0005
921-95B					G0005
921-97B	108	3018	17		
921-98B	108	3122	20	TR21/	
921-99B	123A	3124	10	TR24/	
921-100B	102A	3004	2	TR85/	G0007
921-102	107	3018		TR70/	
921-102A	107	3018		TR70/	
921-102B	107	3018		TR70/720	S0016
921-103B	106	3118	22		
921-104B					G0007
921-105B	158	3004	53	TR84/	G0005
921-106B	108	3018	17	TR21/	S0016
921-109B	123A	3020	10	TR21/	S0016
921-110B	129		67	TR28/	S5013
921-111B	199	3122	20	TR33/	S0015
921-112B	106	3118	22		
921-113B					G0005
921-114B	199	3018	20	TR70/	S0016
921-115B	199	3018	60	TR70/	G0005
921-116B	199	3018	20	TR70/	S0016
921-117B	123A	3020	10	TR21/	S0014
921-118B	158	3004	52	TR84/	G6011
921-120B	123A		17	TR21/	S0016
921-121B					G0007
921-122B					G0005
921-123B	123A		17	TR21/	S0016
921-124B	123A		17	TR21/	S0016
921-125B	123A		17	TR21/	S5014
921-126B	132		FET2	FE100/	F0015
921-127B	123A	3039	20	TR21/	S0016
921-128B	123A		17	TR21/	
921-129B	108		20	TR95/	S0016
921-131B					S0016
921-132B					S0016
921-133B	199	3122			S0015
921-134B					S5011
921-135B					S0020
921-136B					S0015
921-137B					S0015
921-138B					G0005
921-139B					G0005
921-140B				2SB405/	G6011
921-141B	107	3018	17	TR70/	S0016
921-142B	107	3018	17	TR70/	S0016
921-143B	107	3018	17	TR70/	S0011
921-144B					S0015
921-145B	107	3018	20		S0011
921-147B	102A	3004	2		G0007
921-148B	102A		2		G0007
921-150B	102A	3004	54		G6011
921-152B			11	TR33/	S0016
921-153B	102A	3004	2	TR85/	G0005
921-154B	199	3122	20	TR24/	S0015
921-155B	123A	3018	20	TR21/	S0015
921-156B	124	3021	12	TR23/	S5011
921-157B		3116			
921-158B		3018			S0016
921-159B		3124			S0014
921-160B		3114			S5013
921-161B		3024			S0025
921-162B					S5013
921-163B		3054			S5000
921-164B					S0016
921-165B					S0014
921-168B					S0025
921-169B					G0005
921-170B	108		17	TR95/	
921-171B		3122	20	TR21/	
921-172B		3122	20	TR21/	
921-173B		3122	20	TR21/	
921-174B		3122	20	TR21/	
921-176B	107	3018	20		
921-177B	107	3018	20		S0011
921-180B					G0011
921-182B	129	3114	21		S5013
921-183B					G0006
921-188B	128	3024	18	TR25/	S5014
921-189B		3124	10		
921-191B	123A	3124	20		
921-193BX					G0011
921-194BX					G0011
921-195B	123A	3018	20	TR21/	S0011
921-196B	123A	3124	20	TR21/	
921-197		3114			
921-197B	159	3114	22	TR20/	S0012
921-198		3054			
921-198BX	184	3054	58	TR55/	S5000
921-199B					S0016
921-200B	123A	3124	20	TR24/	S0014
921-201B					S0014
921-202B	199	3039	60	TR51/	S0016
921-203B	132	3116	FET2	FE100/	F0021
921-204B	107	3039	11	TR33/	S0016
921-205B	199	3124	20	TR21/	S0016
921-206B	123A	3124	62	TR33/	S0025
921-207B	199	3124	62	TR24/	S0015
921-208B	199	3124	62	TR24/	S0015
921-209B	199	3124	20	TR21/	S0015
921-211B	107	3018	17	TR24/	
921-212B	108	3039	11		S0020
921-213B	107	3018	10		S0015
921-214B	123A		10		S0014
921-215B	123A	3122	62		S0015
921-216B	102A		53		G0005
921-217B	102A	3004	52		G0001
921-218B					G0005
921-222					G0005
921-222B	102A	3004	52	TR85/	G0005
921-223B	102A	3004	52	TR85/	G0005
921-224B	158	3004	53	TR84/	G0005
921-225B	123A	3124	20	TR24/	S0015
921-226B	107	3018	10		S0016
921-227B	158		53		G0005
921-228B	123A	3124	62		S0015
921-229B	123A	3018	62		S0020
921-230B	192	3024	63	TR25/	S5014
921-231B	132		FET2	FE100/	F0021
921-232B	107	3018	17		
921-233B	107	3018	11		
921-234B	123A	3018	20		
921-235B	107	3018	61		S0016
921-236B	128	3024	20		G0008
921-237B	123A	3018	20		S0011
921-238B	102A	3004	2		G0005
921-239B	199		18		S0015
921-240B	199		18		S0015
921-241B			18		S0003
921-242B	158		53	TR84/	G0005
921-243B	158	3004	53	TR84/	G0005
921-244B	158		53	TR84/	G0005
921-249				TR70/	
921-250				TR88/	
921-251B					S0015
921-252B		3025	20		S0033
921-253B					G0005
921-254B	159	3025	21	TR52/	S0019
921-255B	123A	3124	62	TR24/	S0016
921-256BX					G0005
921-256X	158	3004	2	TR05/	G0005
921-257B	107	3039	11	TR33/	S0016
921-258B	107	3039	11	TR33/	S0016
921-262B					S0012
921-263B					S0024
921-264B	107	3018	20	TR21/	S0015
921-265B	107	3018	17	TR24/	S0025
921-266B	108	3018	20	TR24/	S0015
921-267B	108	3018	20	TR24/	S0015
921-268B	123A	3018	20	TR21/	S0015
921-269B	123A	3018	20	TR21/	S0015
921-270B	129	3025		TR28/	S5013
921-271B					S0014
921-272B	123A	3122	10	TR24/	S0015
921-273B	102A	3004	52	TR14/	G0005
921-274B	102A	3004	2	TR85/	G0005
921-275B	123A	3024			
921-276B	123A	3024			
921-281B	199	3124	20	TR21/	S0016
921-282B	158	3004	53	2SB405/	
921-283B					S0024
921-296B	159	3114	22	TR88/	S0019
921-297B					G0009
921-301B		3018			S0015
921-303B		3122			
921-304B		3122			S0014
921-305B		3122			S0014
921-306B		3018			S0015
921-307B		3018			S0015
921-308B		3114			S0019
921-312B		3018	20		S0015
921-313B		3018	20		S0014
921-314B		3018	20		S0033
921-315B		3025			S5013
921-318B		3004	52		G0005
921-319B		3004	2		G0005
921-325B		3018			
921-326B		3018			
921-327B		3004			G6011
921-334B		3018			
921-342B		3100			
921-343B					S0020
921-344B					S0020
921-245B		3018			S0015
921-348B		3114			S0019

Original Part Number	ECG	SK	GE	IR/WEP	HEP
921-349B		3018			
921-350B		3018			S0015
921-351B		3018			
921-352B		3018			S0015
921-369		3122			
921-1528	107				
921-2100-001		3122			
921HC	9921				
923HC	9923				
924-8				TR85/	
924-244B		3004			
924-2209	172			TR69/S9100	
924-16598	102A			TR85/250	
924-17945	121	3009		TR01/232	
926C193-1	123A				
926C193-P1M4165	123A				
926HC	9926				
927HC	9927				
929C	199	3018			
929CA	199	3124			
929CU	199	3124			
929D(JVC)	199				
929DX	199	3039			
929(D)	199				
930B		3018	60		S0016
930C	199	3039			
930C(WARDS)	199				
930D	199	3018			
930D(WARDS)	199				
930DC	9930				
930DU	199				
930DX	199	3018			
930DZ	107		20	TR70/720	S0016
930E	199	3018			
930E(JVC)	199				
930E(WARDS)	199	3018			
930EX	199	3018			
930(OV)	199				
930PC	9930				
930X1	123A	3018		TR21/735	S0014
930X2	123A	3018		TR21/735	S0014
930X3	123A	3018		TR21/735	S0011
930X4	108	3018		TR95/56	S0016
930X5	108	3018		TR95/56	S0016
930X6	126	3005		TR17/635	G0008
930X7	102A	3004		TR85/250	G0007
930X8	102A	3004		TR85/250	G0007
930X9	102A	3004		TR85/250	G0005
930X10	102A	3004		TR85/250	G0005
931PC	9931				
932DC	9932				
932PC	9932				
933DC	9933				
933PC	9933				
935-1	123A	3122		TR21/735	S0014
935DC	9935				
935PC	9935				
936DC	9936				
936NPN	184	3190			S5000
936PC	9936				
936PNP	185				S5006
937DC	9937				
937PC	9937				
941DC	9941				
941PC	9941				
941T1	176	3123	2	TR82/	G0005
942-002	754				C6007
943-535-003		3005			
943-721-001	159				
943-728-001	128				
943-742-002	199				
944DC	9944				
944PC	9944				
945	6400				S9002
945DC	9945				
945PC	9945				
946DC	9946				
946PC	9946				
947-1	161	3018	20	TR83/	
947-1(SYL)	126			TR95/	S0025
948DC	9948				
948PC	9948				
949DC	9949				
949PC	9949				
950DC	9950				
950PC	9950				
951-1(SYL)	126	3004	52	TR85/637	G0003
951DC	9951				
951PC	9951				
952			18	TR25/	
955-1	160	3006	51	TR17/637	G0003
955-2	160	3006	51	TR17/637	G0003
955-3	160	3006	51	TR17/637	G0003
958-023	159				
961DC	9961				
961PC	9961				
962DC	9962				
962PC	9962				
963DC	9963				
963PC	9963				
964-2073B	123A	3122		TR21/	
964-2209	172		64	TR69/	S9100
964-16598	126	3006	53	TR17/635	G0009
964-16599	104	3013	25	TR01/230	G6003
964-17142	102	3004	53	TR05/631	G0005
964-14144	123A	3122	20	TR21/735	S0015
964-17887	104	3009	25	TR01/230	G6003
964-17945	121			TR01/	
964-19862	102A	3010	8	TR85/250	
964-19862A	102A			TR85/	
964-19863	102A	3004	52	TR85/250	
964-19864	102A	3004	52	TR85/250	
964-20738	123A	3124		TR21/735	
964-20739	129	3025		TR88/242	
964-22008	232	3025		TR88/	S3032
964-22009	172	3156	64	TR69/S9100	S9100
964-22158	192	3024		TR87/	S3021
964-24387	128			TR87/	S5014
964-24584	123A			TR21/735	S0014
964-25046	128			TR87/	S5014
964-27986	210	3122	20	TR21/735	S0014
965-1A6-5	160	3006		TR17/	
965-2A6-1	103A	3010		TR08/	
965-4A6-2	101	3011		TR08/	
965A16-3	160	3007		TR17/	
965B16-2	126	3008		TR17/	
965C16-4	121	3009		TR01/	
965T1	102	3004	52	TR05/631	G0006
966-1A6-3	102A	3003		TR85/	
966-2A6-4	102A	3004		TR85/	
966-3A6C	102A	3004		TR85/	
972X1	160	3006		TR17/	G0003
972X2	160	3006		TR17/	G0003
972X3	160	3006		TR17/	G0008
972X4	160	3006		TR17/	G0009
972X5	160	3006		TR17/	G0009
972X6	126	3008		TR17/	G0008
972X7	126	3008		TR17/	G0009
972X8	126	3008		TR17/	G0009
972X9	102A	3004		TR85/	G0005
972X10	102A	3004		TR85/	G0005
972X11	102A	3004		TR85/	G0005
972X12	102A	3004		TR85/	G0005
974-1				TR21/	
974-1(SYL)	161			TR83/	
974-2(SYL)	157	3103	27	TR60/244	S5015
974-3	157			TR60/	
974-3A6-3	100	3005		TR05/	
974-4(SYL)	157	3104	27	TR60/244	S5024
974HC	9974				
977-64197	197				
977A1-6-1	105	3012		TR03/	
977B1-6-2	121MP	3013			
977C1-6-3	121MP	3014			
978-1923	129			TR88/	
984A-1-6B	121	3009		TR01/	
987T1	102	3004	2	TR05/631	G0005
988T1	102	3004	2	TR05/631	G0005
989T1	102	3004	52	TR05/631	
990T1	102	3004	52	TR05/631	G0005
991-00-1219	123A	3122	20	TR21/735	
991-00-1221	102A		53	TR85/250	
991-00-1222	102A		53	TR85/250	
991-00-2232	123A	3122	20	TR21/735	
991-00-2248	123A			TR21/	
991-00-2298	123A	3122	20	TR21/735	
991-00-2356	123A	3122	20	TR21/735	
991-00-2356/K	123A	3122		TR21/	
991-00-2873	123A	3122	20	TR21/735	
991-00-2888	128	3024	18	TR87/243	
991-00-3144	123A	3122	20	TR21/735	
991-00-3304	123A	3122	20	TR21/735	
991-00-8393	123A	3122	20	TR21/735	
991-00-8393A	123A			TR21/	
991-00-8393M	123A	3122		TR21/735	
991-00-8394	123A	3122	20	TR21/735	
991-00-8394A	123A	3122		TR21/735	
991-00-8394AH	123A	3122		TR21/735	
991-00-8395	123A			TR21/	
991-01-0098	159	3114	16	TR20/717	
991-01-0099	121			TR01/	
991-01-0461	172			TR69/	
991-01-0462	159			TR20/	
991-01-1216	121	3009	16	TR01/232	
991-01-1217	158		53	TR84/630	
991-01-1219	123A	3122	20	TR21/735	
991-01-1220	123A	3122	20	TR21/735	
991-01-1221	102A		53	TR85/250	
991-01-1222	102A		53	TR85/250	
991-01-1223	102A		53	TR85/250	

Original Part Number	ECG	SK	GE	IR/WEP	HEP
991-01-1224	102A		53	TR85/250	
991-01-1225	159	3114	21	TR20/717	
991-01-1305	128	3024	18	TR87/243	
991-01-1306	123A	3122	20	TR21/735	
991-01-1312	123A	3122	20	TR21/735	
991-01-1314	128			TR87/	
991-01-1315	129			TR88/	
991-01-1316	107	3039	20	TR70/720	
991-01-1317	130			TR59/	
991-01-1318	123A	3122	20	TR21/735	
991-01-1319	159	3114	21	TR20/717	
991-01-1705	123A			TR21/	
991-01-1706	133	3112	FET1	/801	
991-01-2328	159			TR20/	
991-01-2686	129			TR88/	
991-01-3044	123A	3122	20	TR21/735	
991-01-3055	133	3112	FET1	/801	
991-01-3056	123A			TR21/	
991-01-3057	123A	3122	20	TR21/735	
991-01-3058	159	3114	21	TR20/717	
991-01-3063	130	3027	14	TR59/247	
991-01-3068	123A			TR21/	
991-01-3170	196			TR92/	
991-01-3543	172			TR69/	
991-01-3544	123A			TR21/	
991-01-3599	159			TR20/	
991-01-3683	123A			TR21/	
991-01-3740	123A			TR21/	
991-01-5000	197				
991-01-5001	196			TR92/	
991-01-5062	197				
991-01-5063	196			TR92/	
991-010462	123A	3019	17	TR21/	
991-011217	158	3004	53	TR84/	
991-011219	123A	3019	20	TR21/	
991-011220	123A	3019	20	TR21/735	
991-011221	160	3123	51	TR17/	
991-011222	102A	3004	2	TR85/	
991-011223	102A	3004	2	TR85/	
991-011225	159	3025	21	TR20/	
991-011305	128	3124	18	TR87/	
991-011306	123A	3124	20	TR21/	
991-011312	123A	3019	17	TR21/	
991-011313	192	3122			
991-011314	128	3122	18	TR87/	
991-011318	123A	3019	20	TR21/	
991-011319	193	3114	67		
991-011576	133	3118	FET1		
991-011706	133	3118	FET1		
991-012328	234	3114			
991-012686	159	3114	22	TR20/	
991-013056	123A	3019	17	TR21/	
991-013057	199		10	TR21/	
991-013068	123A	3019	17	TR21/	
991-015587	123A	3124	20	TR21/	
991-015614	159	3025	21	TR20/	
991-015615	123A	3024	20	TR21/	
991-015663	172	3124	64	TR69/	
991-015942	754				
991-015942-1	761				
991-016614		3025			
991-016274	123A		53	TR21/	
991-016727		3134			
991-016788		3124			
991-017456		3118			
991-018047		3124			
991-2P	129				
991-3N	128				
991T1	102	3004	52	TR05/631	G0005
992-008-890	121MP	3114	16	TR01/232	
992-00271	130				
992-00271A	130				
992-00-1192	104	3009	3	TR01/624	
992-00-2271	130		14	TR59/247	
992-00-2271A		3027			
992-00-2298	123A	3122	20	TR21/735	
992-00-3139	130	3027	14	TR59/247	
992-00-3139A	130	3027		TR59/	
992-00-3144	123A	3122	20	TR21/735	
992-00-3172	124	3021	12	TR81/240	
992-00-4091	130			TR59/	
992-00-4092	130			TR59/	
992-00-8870	121	3009	16	TR01/232	
992-00-8890	179			TR35/G6001	
992-00-8890L	179			TR35/	
992-01-1216	121	3009	16	TR01/232	
992-01-1218	121	3009	16	TR01/232	
992-01-1317	130			TR59/	
992-01-3684	154			TR78/	
992-01-3705	175			TR81/	
992-01-3738	123A			TR21/	
992-02271	130	3027	14	TR59/	
992-08890	121	3014	3		
992-017169	130	3027	14	TR59/	
992-1A6-1	105	3012		TR03/	

Original Part Number	ECG	SK	GE	IR/WEP	HEP
992T1	102A	3004	2	TR85/250	G0006
995-01-6130	153			TR77/	
995-01-6131	152			TR76/	
998-0061114	123A				
998-0200816	123A				
999-4601	128				
1000-25	703A	3157		IC500/	C6001
1000-135	107	3018		TR70/720	S0014
1000-136	123A	3124		TR21/735	S0030
1000-137	123A	3024		TR21/735	S0014
1000-138	152	3041		TR76/701	S3020
1000-139	195	3047		TR63/243	S3001
1000-140	190	3027		TR74/S3021	S3020
1000-141	195	3048		TR65/243	S3001
1000-142	152	3041		TR76/701	S5000
1001	102A	3004	51	TR85/250	
1001(JULIETTE)				TR85/250	G0005
1001(JULIETTE)*				TR17/635	
1001-0036/4460	720				
1001-0091/4460	1140				
1001-01	128	3018		TR87/243	S0033
1001-02	123A	3124	20	TR21/735	S0015
1001-02(COUR)	107				
1001-03	123A	3018		TR21/735	S0016
1001-04	123A	3024		TR21/735	S0020
1001-05	123A	3024		TR21/735	S0015
1001-06	123A	3047		TR21/735	S0015
1001-07	195	3048		TR65/243	S3001
1001-08	195	3048		TR87/243	S3002
1001-09	130			TR59/247	S5019
1001-25(LAF)	703A				
1001-7663	102				
1001(G.E.)	712				
1002	234	3018	52	TR85/250	
1002(JULIETTE)				TR85/250	S0011
1002(JOHNSON)	102A			TR85/250	S0032
1002-01	133	3050	FET1	FE100/801	F0010
1002-02	107	3018	11	TR70/720	S0016
1002-02A	108	3018		TR95/	
1002-03	123A	3124	20	TR70/720	S0016
1002-04	123A	3018	17	TR21/735	S0016
1002-04-1	108	3018		TR95/	
1002-05	102A	3004	52	TR85/S3023	G0007
1002-07	1012				
1002-25	726				
1002-68		3018		TR24/	S0016
1002A(JULIETTE)	107			TR70/	
1002A-1				TR30/	S3014
1002A-2				TR24/	S3013
1003	102A	3004	52	TR85/250	
1003(JOHNSON)	102A			TR85/250	G0008
1003(JULIETTE)	126			TR85/250	S0015
1003-01	107	3018		TR70/720	S0016
1003-02	1045				
1004	123	3004	20	TR86/53	
1004(G.E.)	161			TR83/56	S0020
1004(JULIETTE)	158			TR17/635	G0009
1004(JULIETTE)*	108			TR95/56	
1004-01	158	3004	52	TR84/630	G6011
1004-02	195	3047		TR65/243	S3001
1004-03	123A	3046	17	TR21/735	S0015
1004-17	108	3018	20	TR95/56	S0017
1004(2SC537)	123A			TR21/	
1004P	129				
1005		3124	20	TR21/735	S0015
1005(JULIETTE)	102A			TR85/250	S0015
1005-17		3004	2	TR85/631	G0005
1005M19	159				
1006		3124	52	TR85/250	
1006(G.E.)	161			TR83/719	S0020
1006(JULIETTE)	123A			TR21/735	S0016
1006(JULIETTE)*	102A			TR85/250	
1006-78	160			TR17/	
1006-93	102A			TR85/	
1007(JULIETTE)	126			TR17/635	S0011
1007(2SB405)	158			TR84/	
1007-17	102A	3088		TR85/250	G0005
1008	128	3052	63	TR87/243	
1008(JOHNSON)	102A			TR85/250	S0014
1008(JULIETTE)	131			TR94/	G6011
1008(POWER)	131			TR94/	
1008-02	108			TR95/	S0015
1008-17	104			TR01/	G6003
1008-25	726				
1009		3030	52	TR85/250	
1009(G.E.)	161			TR83/719	S0020
1009(JOHNSON)	102A			TR85/250	G0009
1009-01		3116	FET2	FE100/802	F0015
1009-02	123A	3024		TR21/735	S0015
1009-02-16	123A	3046		TR21/	
1009-03	195	3024	18	TR87/243	S3001
1009-03-17	128	3047		TR87/	
1009-04	234	3048		TR65/	S3002
1009-04-17	195	3048		TR65/	
1009-05	210	3041	28	TR55/S3023	S3020

Original Part Number	ECG	SK	GE	IR/WEP	HEP
1009-05A	196	3041		TR92/	
1009-11	1016				
1009-17	123A			TR21/	S0011
1010-14	195	3047		TR65/	
1010-17	191		20	TR75/	S3001
1010-17(R.F.)	195			TR65/	
1010-78	160			TR17/	
1010-87	160			TR17/	
1010-89	126			TR17/	
1011-01	195			TR65/	S3001
1011-11	191			TR75/	
1011-11(R.F.)	108			TR95/	
1011M57P01	129				
1011M62P01	129				
1012(G.E.)	159		21	TR20/	S0013
1012(IC)	1003				
1013		3114	21		S0019
1013-15		3018		TR24/	S0016
1014			52		
1014(JOHNSON)	102A			TR85/	G0005
1014-25	703A		IC12	IC500/	
1015(G.E.)					S0013
1016		3114	21	/717	
1016-17	161	3018		TR83/719	S0017
1016-81	152	3054		TR76/701	S5000
1016-83	108			TR95/56	S0016
1016-84	108			TR95/56	S0016
1016-85		3018			S0016
1016(G.E.)	159			TR10/	
1018-25	1056				
1019-25	1021				
1019-74	102A			TR85/	
1020-17	123A	3018	20	TR21/735	S0011
1020-25	1075				
1021-17	102	3004		TR05/631	G0005
1023-17	102	3004	2	TR05/631	G0005
1023G	123A	3124	20	TR51/	S0015
1023G(GE)	123A				S0015
1024-17	104	3009	3	TR01/624	
1024G		3018		TR51/	
1024G(GE)	123A				S0015
1025G		3018	20	TR21/	
1025G(GE)	123A				S0015
1026G		3044	20	TR22/	
1026G(GE)	123A				S0015
1027(GE)	128			TR95/56	S0015
1027G		3044	18	TR87/	
1027G(GE)	128				S0016
1028G		3018	20	TR21/	
1028(GE)	123A		20	TR21/	S0015
1028G(GE)	123A				S0015
1029G	123A	3018	20	TR21/	
1029(GE)	123A		20	TR21/	S0016
1029G(GE)	123A				S0015
1030	159	3114		TR20/717	S0031
1030(GE)	165			TR93/	S5021
1032	102	3004	52	TR05/631	G0005
1032(GE)					S5015
1032-1(FISHER)					S3003
1032-2(FISHER)					S3003
1032-3(FISHER)					S3003
1032-4(FISHER)					S3003
1032-6(FISHER)					S3003
1033-1	126	3008	51	TR17/635	S3019
1033-2	126	3008	51	TR17/635	S3019
1033-3	126	3008	51	TR17/635	S3019
1033-4	126	3008	51	TR17/635	S3019
1033-5	102	3004	2	TR05/631	S3019
1033-6	103	3010	8	TR08/641A	S3019
1033-7	102	3004	2	TR05/631	
1034	158	3031	52	TR84/630	G0005
1034-17	123A	3020	20	TR21/735	
1034-43	101			TR08/	
1035	158		52	TR84/630	G0005
1035(GE)			52		S5025
1036	158	3004	52	TR84/630	G0005
1036-1(FISHER)					S5006
1036-2(FISHER)					S5006
1036-3(FISHER)					S5006
1037-1(FISHER)					S5000
1037-2(FISHER)					S5000
1037-3(FISHER)					S5000
1038(GE)				TR87/243	S5000
1038-1		3122			
1038-1-10	123A			TR21/735	S0015
1038-4(FISHER)					S7003
1038-5(FISHER)					S5005
1038-6	123A			TR21/735	S5005
1038-6CL	123A	3124		TR21/	
1038-8	123A	3124		TR21/735	S0014
1038-10	123A			TR21/735	S0015
1038-15	123A			TR21/735	S0014
1038-15CL	123A	3124		TR21/	
1038-18	123A	3038		TR21/735	S0015
1038-18CL	123A	3124		TR21/	

Original Part Number	ECG	SK	GE	IR/WEP	HEP
1038-21	123A	3124		TR21/735	S0015
1038-23	123A			TR21/735	S0014
1038-23CL	123A	3124		TR21/	
1038-24	123A	3124		TR21/735	S0015
1039-01	107	3018	20	TR21/	S0015
1039-4(FISHER)					S7004
1039-5(FISHER)					G6017
1039-6(FISHER)					S7004
1040-01	123A	3124	62	TR25/	S0015
1040-02	175	3026	28	TR23/	S5019
1040-03	123A	3122	20	TR21/	S0015
1040-04		3024		TR65/	S3008
1040-05	198	3049		TR65/	S3001
1040-09	160			TR17/	
1040-2	123A			TR21/735	S0015
1040-7	128			TR87/243	S0015
1040-11	128			TR87/243	S0015
1040-59	160			TR17/	
1040-80	103A			TR08/	
1040-68	1013				
1040-69	1016				
1040-155		3024			S0015
1041-71	107	3018			S0014
1041-72	123A	3044		TR24/	S0015
1041-73	123A	3047			S0014
1041-74	152	3041			S5000
1041-75	123A	3029			S7002
1041-76	190	3047			S3020
1041-77	195	3049			S3001
1042-01		3116			F2004
1042-02	133	3116	FET2	FE100/	F0021
1042-03	123A	3038	20	TR24/	S0015
1042-04	107	3018	10	TR24/	S0015
1042-05	123A	3124			S0014
1042-06	159	3118	22		S0013
1042-07	123A	3018	10		S0015
1042-08	186	3192			
1042-09	236		66		
1042-10	226MP	3082			G6016
1042-11	724	3525			
1050-21		3124			
1050B		3114			
1052-17	102A	3004		TR85/250	G0007
1057-17	102A			TR85/250	G0005
1060-17	102A	3004	52	TR85/250	G0007
1064-4417	175				
1064-6032	130	3027			
1064(G.E.)	712	3072			
1065-4861	7400				
1065(G.E.)	748				
1067	128	3024	40	TR25/	
1067(GE)	154		40		S5026
1068-17	224	3048	45	TR65/243	S3001
1068-17A	224	3049			
1069(GE)	152	3054	66	S5000	
1071		3083	69		S5006
1071-3642	130				
1071(GE)	153		69		S5006
1072G		3138			S0019
1072K(GE)	193		67		
1073		3111			
1074		3115	36	TR93/	S5021
1074-03	199	3124			
1074-115	123A	3124			S0015
1074-116	186	3041			S5000
1074-117	130	3027			S7002
1075	233	3039	39	TR70/	
1075BLUE(GE)	749				
1075(GE)	749		39		S0017
1076(GE)	749				C6076P
1077		3077			
1077(GE)	713				
1077-07		3018			
1079		3141			C6100P
1079(GE)	783				
1079GREEN	783				C6100P
1080-01	107	3018			S0014
1080-03	123A				S0025
1080-05	224	3049			
1080-06	186	3054			
1080-20	107	3018			S0016
1080-21	108	3039		TR24/	S0015
1080-130	152	3054		TR55/	S3024
1080-5364	130	3027			
1080-6396	123A				
1080-7584	123				
1080G	129	3114	67	TR52/	
1080G(GE)	193				S0012
1081-4739	909	3540			
1081-7104	223	3027			
1081K94-6	941				
1081K94-7	909	3590			
1081K94-9	941				
1083-17		3004			
1084-9784	159				

Original Part Number	ECG	SK	GE	IR/WEP	HEP
1087-01		3054			S3020
1087-2380	128				
1089-6199	159				
1089-9307	912				
1092		3079	75	TR26/	S7004
1092(GE)	181		75		S7004
1097	783	3083	29	TR55/	
1097(GE)	186		29		S5000
1100		3083	29		
100(GE)	187		29		S5006
100-75	195			TR65/	
1101-8181	124				
1102-17	158	3004		TR84/630	G0005
1102-17A	102A	3004		TR85/	
1102-63	101			TR08/	
1104-94	102A			TR85/	
1104-95	103A			TR08/	
1105-15	121			TR01/	
1105(TOPAZ)					G6002
1106-97	108			TR95/	
1106-99	128			TR87/	
1109		3054	66	TR76/	S5000
1111(GE)	152	3054	66	TR76/	S5000
1111-17	160			TR17/	
1111-18	160			TR17/	
1112-8	159				
1112-78	128			TR87/	
1112-79	195			TR65/	
1113-03	123A			TR21/	
1113(GE)	153	3083	69	TR77/	S5006
1113-13	160			TR17/	
1113-2875	102				
1115(GE)	163	3063			S5021
1116		3054	28	TR55/	S5000
1116-6527	130				
1116-6535	128				
1116(GE)	186		28		S5000
1117MC		3018	20		
1117MD		3018			
1119		3104	27	TR60/	S3021
1119-54	160			TR17/	
1119-55	160			TR17/	
1119-56	160	3006		TR17/	
1119-57	102A			TR85/	
1119-58	103A			TR08/	
1119-59	102A			TR85/	
1119-4628	181				
1119-8132	123A				
1119(GE)	171				S3021
1122-96	126			TR17/	
1123-55	108			TR95/	
1123-56	108			TR95/	
1123-57	108			TR95/	
1123-58	108			TR95/	
1123-59	108			TR95/	
1123-60	123A			TR21/	
1123-3335	175	3026			
1124	121	3009	16	TR01/232	G6005
1124A	121	3009	16	TR01/232	G6005
1124B	121	3009	16	TR01/	G6005
1124C	104	3009	16	TR01/232	
1125(GE)	186	3054	28	TR55/	S5000
1127-2077	910				
1128-17	102A			TR85/250	G0005
1129(GE)	712	3072			
1132-2(RCA)	184	3190			
1133-14	785				
1136(GE)	783	3141			C6100P
1136GREEN	783				C6100P
1137(GE)	713	3077			C6057P
1138(GE)	728	3016			
1138RED(GE)	728	3073			
1140-17	102A	3123		TR85/250	G0005
1145	100	3005	1	TR05/254	
1145-17	102A			TR85/250	G0006
1146	100	3005	1	TR05/254	
1147(GE)	159	3114	21	TR20/	S0019
[illegible]8-17	158	3004		TR84/630	G0005
1151BLUE(GE)	749				
1151(GE)	749	3168			C6076P
1152	234				
1152BLUE(GE)	749				
1152(GE)	749	3168			
1153		3141			C6100P
1153(GE)	783				
1153GREEN	783				C6100P
1154		3141			C6100P
1154(GE)	783				
1154GREEN	783				C6100P
1157		3024	18	TR25	S5014
1164(GE)	712	3072			
1165	712	3072			
1166					S5011
1166-7812	102				
1169		3104			S5024

Original Part Number	ECG	SK	GE	IR/WEP	HEP
1173(GE)	712	3072			
1174(GE)	749				
1178-3453	102	3004			
1178(GE)	712	3072			
1179(GE)	712	3072			
1179-1084		3011			
1180-0182	102				
1183-17	195	3048		TR65/	
1184G		3025	21	TR54/	S0012
1184G(GE)	193				S0012
1186		3118			
1187(GE)	192	3044	18	TR25/	S5014
1190		3115			S5021
1192	102	3004	2	TR05/631	
1192(GE)	712	3072	2		
1193(GE)	712	3072			
1198(HAR KARI)	128			TR87/	
1202		3124			S0015
1203		3124			
1203-169	123A				
1204		3024			
1207		3018			S0015
1210-17	108	3018	11	TR95/56	S0016
1210+17B	123A	3124		TR21/	
1214	159	3114		TR20/717	S0031
1217SK333	128				
1227-17	123A			TR21/735	S0020
1228-17	123A	3038		TR21/735	S0011
1229H	108	3020	20	TR24/	S0016
1236-3750	159				
1236-3776	123A				
1239-5752	175				
1241-719	126	3008		TR17/635	
1241A	102	3004	2	TR05/631	G0005
1248	102A	3004	2	TR85/250	
1251-1-1	222	3050			F2004
1254(GE)	159	3114	21		S0019
1264(GE)	713	3077	IC18		C6057P
1269	184				
1272	123A	3122		TR21/735	S0011
1277-17	102A	3123		TR85/250	G0007
1278(GE)					C6076P
1278A		3112			
1279A		3112			
1280A		3112			
1282		3024			S5000
1284		3018			
1285		3025			
1285A		3112			
1294		3114			S5022
1285		3025			
1285A		3112			
1294		3114			
1300-1	128				
1300A	128				
1300RN				TR21/	
1301-1	160				
1301-2	160				
1307(PENNY)					G6016
1314		3114			
1315		3124			
1315(PENNY)	123A			TR21/	S0015
1316(PENNY)	123A	3124		TR21/	S0016
1316-17	102A	3004		TR85/250	G0005
1317-17	102A			TR85/250	G0005
1320	102	3004	52	TR05/631	G0005
1321-7724	102				
1321-7732	102				
1326PC	739				
1329	102	3004	8	TR05/631	G0011
1330	102	3004	52	TR05/631	G0005
1340	100	3005	52	TR05/254	G0005
1341		3118			
1344-3767	101				
1344-7321	102				
1347	739		IC30		
1347-17	102A	3004		TR85/250	G0005
1348	739		IC30		
1348A12H01	7417				
1348A14H01	7476				
1348A30H01	9930				
1348A32H01	9932				
1348A36H01	9936				
1348A44H01	9944				
1348A45H01	9945				
1348A46H01	9946				
1348A62H01	9962				
1349-17	103A	3010	8	TR08/641A	G0011
1350	100	3005	52	TR05/254	G0005
1351	738	3167	IC29		
1352	738	3167	IC29		C6075P
1360	102	3004	52	TR05/631	G0005
1362-17	102A	3004		TR85/250	G0005
1362-17A	102A	3004		TR85/	
1364-17	102A	3004		TR85/250	G0005

Original Part Number	ECG	SK	GE	IR/WEP	HEP
1368C/D	123A	3124	20	TR21/	S0033
1371-17	103A	3010		TR08/724	G0011
1373-17	107	3018	11	TR70/720	S0014
1373-17A	107	3018		TR70/	
1373-17AL	108	3018		TR95/	
1374-17	123A	3124	20	TR21/735	S0014
1374-17A	123A	3124		TR21/	
1374-17AC	123A	3124		TR21/	
1375-1		3039			
1390	100	3005	51	TR05/254	G0005
1396	101	3011		TR08/641	G0011
1400	100	3005	51	TR05/254	G0005
1410	100	3005	51	TR05/254	G0005
1412-1	123A	3124		TR21/	
1412-1-12	123A	3124		TR21/	
1412-1-12-8	123A	3124		TR21/	
1413-159	105	3012	4	TR03/	
1413-160	102A	3004	53	TR85/	
1413-168	121	3009	3	TR01/	
1413-172	121	3009	3	TR01/	
1413-175	158	3004	53	TR84/	
1413-178	121	3014	3	TR01/	
1414-157	128	3024	18	TR87/	
1414-158(RDG)	159	3114	32	TR20/	
1414-173	128	3025	18	TR87/	
1414-174	123A	3124	20	TR21/	
1414-176	159	3114	22	TR20/	
1414-179	130	3027	14	TR59/247	
1414-180	175	3026	23	TR81/241	
1414-183	123A	3124	20	TR21/	
1414-184	128	3044	18	TR87/	
1414-185	129	3025	29	TR88/	
1414-186	128	3024	18	TR87/	
1414-187	129	3025	21	TR88/	
1414-188	130	3036	14	TR81/	
1414-189	128	3024	18	TR87/	
1414A	121	3009		TR01/	
1414A-12	121	3009		TR01/	
1414A-12-8	121	3009		TR01/	
1415	123A				
1417-177	105	3012		TR03/	
1420-1		3018			
1420-1-1	107	3018	11	TR80/	S0020
1420-2-2	161	3019	11	TR83/	S0020
1424		3122			S0025
1428	129				
1431-7184	181	3036			
1431-8349	123A				
1436-17	102A			TR85/250	G0005
1449	160				
1455-7-4	107	3018	11	TR80/	S0020
1458CP1	778	3551			
1458P1	778	3551			
1459	102	3004	2	TR05/631	
1463		3018			
1465		3124			S0016
1465-1	160	3006		TR17/	
1465-1-12	160	3006		TR17/	
1465-1-12-8	160	3006		TR17/	
1465-2	103A	3010		TR08/	
1465-2-12	103A	3010		TR08/	
1465-2-12-8	103A	3010		TR08/	
1465-4	101	3011		TR08/	
1465-4-12	101	3011		TR08/	
1465-4-12-8	101	3011		TR08/	
1465A	160	3007		TR17/	
1465A-12	160	3007		TR17/	
1465A-12-8	160	3007		TR17/	
1465B	126	3008		TR17/	
1465B-12	126	3008		TR17/	
1465B-12-8	126	3008		TR17/	
1465C	121	3009		TR01/	
1465C-12	121	3009		TR01/	
1465C-12-8	121	3009		TR01/	
1466-1	102A	3003		TR85/	
1466-1-12	102A	3003		TR85/	
1466-1-12-8	102A	3003		TR85/	
1466-2	102A	3004		TR85/	
1466-2-12	102A	3004		TR85/	
1466-2-12-8	102A	3004		TR85/	
1466-3	102A	3004		TR85/	
1466-3-12	102A	3004		TR85/	
1466-3-12-8	102A	3004		TR85/	
1471-4356	9936				
1471-4364	9946				
1471-4372	9962				
1471-4380	9930				
1471-4398	9944				
1471-4406	910				
1471-4414	726	3022			
1471-4778	123A				
1471-4802	130				
1471-9860	9936				
1472-8349	199				
1473-4255	128				
1474-3	100	3005		TR05/	
1474-3-12	100	3005		TR05/	
1474-3-12-8	100	3005		TR05/	
1474-3736	911				
1476-17	102A	3004		TR85/250	G0005
1476-17-6	102A	3004		TR85/	
1476-1118	128				
1477-3352	906	3548			
1477-3436	904	3542			
1477A	105	3012		TR03/	
1477A-12	105	3012		TR03/	
1477A-12-8	105	3012		TR03/	
1477B	121MP	3013			
1477B-12	121MP	3013			
1477B-12-8	121MP	3013			
1477C	121MP	3014			
1477C-12	121MP	3014			
1477C-12-8	121MP	3014			
1478-6982	9112				
1479-0224	9937				
1479-0240	74H00				
1479-0257	74H21				
1479-0265	74H20				
1479-0273	941				
1479-7963	199				
1479-7971	74H04				
1479-7989	123A				
1479-8011	923				
1479-8029	159				
1481-B		3005			
1482		3124			S0015
1482-17	195	3047		TR65/243	S3001
1483		3122			
1484A	121	3009		TR01/	
1484A-12	121	3009		TR01/	
1484A-12-8	121	3009		TR01/	
1485-9896	9135				
1486-8780	74H10				
1487-17	195			TR65/	S3001
1488B	160	3006		TR17/	
1488B-12	160	3006		TR17/	
1488B-12-8	160	3006		TR17/	
1488C	100	3005		TR05/	
1488C-12	100	3005		TR05/	
1488C-12-8	100	3005		TR05/	
1492-1	105	3012		TR03/	
1492-1-12	105	3012		TR03/	
1492-1-12-8	105	3012		TR03/	
1493-17	123A	3124		TR21/735	S0015
1501		3018			
1505-9850	124				
1507-7183	175				
1510-2718	102				
1521B	218		75	TR26/	
1524	123	3124	8	TR86/53	
1526	160		51	TR17/637	G0002
1527-5282	9932				
1527-5308	9945				
1534-8931	175	3026			
1540	123A		20	TR24/	S0015
1548-17	175	3131	23	TR81/241	S5012
1553-17	159	3025	21	TR20/717	S5013
1558-17		3114			S5022
1559-17	121	3009	16	TR01/232	G6005
1559-17A	121	3009		TR01/	
1561-17	121		16	TR01/232	G6005
1561-0404		3027	77		
1561-0808		3036	75		
1561-0810		3036	75		
1561-1010		3036	75		
1561-A608		3036	75		
1561-A615		3036	75		
1567	123A	3124	20	TR21/735	S0015
1567-0	123A	3124	11	TR21/735	
1567-2	123A	3124	20	TR21/735	
1571-0401			66		
1571-0402			66		
1571-0420			66		
1571-0425			66		
1571-0601			66		
1571-0602			66		
1571-0620			66		
1571-0825-6		3026			
1571-32					S5019
1571-33					S5012
1571-34					S5012
1571-35					S5012
1571-36					S5012
1571-41					S5019
1571-42					S5019
1571-43					S5012
1571-44					S5012
1571-46					S5012
1571-1025		3538			
1573-00	128				

Original Part Number	ECG	SK	GE	IR/WEP	HEP
1573-01	128				
1574-01	234				
1579		3014			
1582		3114			S0019
1582-0403			77		
1582-0404			77		
1582-0405			77		
1582-0408			75		
1582-0410			75		
1582-0415			75		
1582-0508			75		
1582-0510			75		
1582-0603			77		
1582-0604			77		
1582-0605			77		
1582-0608			75		
1582-0610			75		
1582-0615			75		
1582-0803			75		
1582-0804			75		
1582-0805			75		
1582-0808			75		
1582-0810			75		
1582-0815			75		
1582-1003			75		
1582-1004			75		
1582-1005			75		
1582-1008			75		
1582-1010			75		
1582-1015			75		
1585H	107				
1607A80	7406				
1612SK24E	133	3112	FET1		
1627-1413	9110				
1627-2064	9093				
1634-17	108	3018	20	TR95/56	S0016
1634-17-14A	108	3018		TR95/	
1673-0475	128				
1676-1991	128				
1677-1149	923D				
1679-7391	159				
1687-17	107	3018		TR70/720	S0016
1700-5182	903				
1705-4834	123A				
1705-5351	128				
1711-17	123A	3122	17	TR21/735	S0030
1712-17	123A	3038	20	TR21/735	S0011
1714-0402	175	3026	66	TR81/241	
1714-0602	175	3026	66	TR81/241	
1714-0605	175	3026	66	TR81/241	
1714-0802	175	3026		TR81/241	
1714-0805	175	3026		TR81/241	
1714-1002	175	3026		TR81/241	
1714-1005	175	3026		TR81/241	
1716-0402					S5004
1716-0405					S5004
1716-0602					S5004
1716-0605					S5004
1716-0802					S5004
1716-0805					S5004
1716-1002					S5004
1716-1005					S5004
1717-0402					S5004
1717-0405					S5004
1717-0602					S5004
1717-0605					S5004
1717-0802					S5004
1717-0805					S5004
1721-0725	218				
1723-0405	181	3036		TR36/S7000	
1723-0410	181	3036		TR36/S7000	
1723-0605	181	3036		TR36/S7000	
1723-0610	181	3036		TR36/S7000	
1723-0805	181	3036		TR36/S7000	
1723-0810	181	3036		TR36/S7000	
1723-17	123A	3124		TR21/735	S0014
1723-1005	181	3036		TR36/S7000	
1723-1010	181	3036		TR36/S7000	
1723-1205	181	3036		TR36/S7000	
1723-1210	181	3036		TR36/S7000	
1723-1405	181	3036		TR36/S7000	
1723-1410	181	3036		TR36/S7000	
1723-1605	181	3036		TR36/S7000	
1723-1610	181	3036		TR36/S7000	
1723-1805	181	3036		TR36/S7000	
1723-1810	181	3036		TR36/S7000	
1726-0405					S5004
1726-0410					S5004
1726-0605					S5004
1726-0610					S5004
1726-0805					S5004
1726-0810					S5004
1729-7284	923				
1741-0051	7400				
1741-0069	74H00				
1741-0085	7401				
1741-0119	7402				
1741-0143	7404				
1741-0150	74H04				
1741-0176	7405				
1741-0184	74H05				
1741-0200	7408				
1741-0234	7410				
1741-0242	74H10				
1741-0275	74H11				
1741-0291	7417				
1741-0325	7420				
1741-0333	74H20				
1741-0366	74H22				
1741-0416	7430				
1741-0424	74H30				
1741-0440	7437				
1741-0473	7440				
1741-0481	74H40				
1741-0564	7451				
1741-0572	74H51				
1741-0598	7453				
1741-0606	74H53				
1741-0622	7454				
1741-0630	74H54				
1741-0663	74H55				
1741-0689	7460				
1741-0697	74H60				
1741-0721	74H74				
1741-0747	7475				
1741-0770	7483				
1741-0804	7486				
1741-0895	7493				
1741-0952	7495				
1741-1018	74107				
1741-1042	74150				
1741-1075-74151					
1741-1133	7411				
1741-1190	7441				
1741-1224	7442				
1741-1257	74121				
1741-1299	74H21				
1741-1323	74H183				
1741-1349	74180				
1741-1380	74H52				
1743-0620					S7000
1743-0810					S7000
1743-0830					S7000
1743-1020					S7000
1751-17	123A	3024	20	TR21/735	S0015
1751G036	123A		17	TR21/	
1761-17	108			TR95/56	S0016
1763-0415	181	3036		TR36/S7000	S7000
1763-0420	181	3036		TR36/S7000	S7000
1763-0425	181	3036		TR36/S7000	S7000
1763-0615	181	3036		TR36/S7000	S7000
1763-0620	181	3036		TR36/S7000	S7000
1763-0625	181	3036		TR36/S7000	S7000
1763-0815	181	3036		TR36/S7000	S7000
1763-0820	181	3036		TR36/S7000	S7000
1763-0825	181	3036		TR36/S7000	S7000
1763-1015	181	3036		TR36/S7000	S7000
1763-1020	181	3036		TR36/S7000	S7000
1763-1025	181	3036		TR36/S7000	S7000
1763-1215	181	3036		TR36/S7000	S7000
1763-1220	181	3036		TR36/S7000	S7000
1763-1225	181	3036		TR36/S7000	S7000
1763-1415	181	3036		TR36/S7000	S7000
1763-1420	181	3036		TR36/S7000	S7000
1763-1425	181	3036		TR36/S7000	S7000
1767-2					S5006
1768-0625					S7000
1768-0815					S7000
1768-0825					S7000
1768-1015					S7000
1768-1025					S7000
1777-17	102A	3004		TR85/250	G0005
1789-17	128			TR87/243	S5026
1792-17	108	3039		TR95/56	S0016
1799-17	123A	3024		TR21/735	S0014
1800	9924				
1800-17	123A	3024	20	TR21/735	S0014
1800DC	9800				
1800PC	9800				
1801DC	9801				
1801PC	9801				
1802DC	9802				
1802PC	9802				
1803DC	9803				
1803PC	9803				
1804	754				C6007
1804-17	154	3024	20	TR78/712	S5026
1804DC	9804				
1804PC	9804				
1805	7401				

Original Part Number	ECG	SK	GE	IR/WEP	HEP
1805DC	9805				
1805PC	9805				
1806	7404				
1806-17	130	3027		TR59/247	
1806-17A	130	3027		TR59/	
1806DC	9806				
1806PC	9806				
1807	7442				
1807DC	9807				
1807PC	9807				
1808	7490				
1808DC	9808				
1808PC	9808				
1809(SCRANTON)	754				C6007
1809DC	9809				
1809PC	9809				
1810DC	9810				
1811DC	9811				
1811PC	9811				
1812-1	123A	3124		TR21/	
1812-1-12	123A	3124		TR21/	
1812-1-127	123A	3124		TR21/	
1812-1L	123A	3124		TR21/	
1812-1L8	123A	3124		TR21/	
1812DC	9812				
1812PC	9812				
1813DC	9813				
1813PC	9813				
1814A	121	3009		TR01/	
1814A-12	121	3009		TR01/	
1814A-127	121	3009		TR01/	
1814AL	121	3009		TR01/	
1814AL-8	121	3009		TR01/	
1814DC	9814				
1814PC	9814				
1818(IC)	754				
1820-0058	909	3590			
1820-0087	9950				
1820-0095	9948				
1820-0122	9093				
1820-0125	911				
1820-0174	7404				
1820-0183		3525			
1820-0216-1	941M				
1820-0217-1	941M				
1820-0240	906				
1820-0248	909	3590			
1820-0306	724	3525			
1820-0341	9094				
1820-0352	904				
1820-0398	910D				
1820-0430	309K				
1820-0474	726				
1820-0476	915				
1820-0495	7411				
1820-0512	74H74				
1820-0861	9946				
1820-0862	9948				
1820-0863	9158				
1820-0864	7403				
1820-0865	9944				
1820-0869	9935				
1820-0870	7408				
1820-0894	7404				
1820-1064	74164				
1820-1068	74H30				
1820-1111	7440				
1820-1172	7416				
1821-0001	912				
1821-0002	912				
1826-0009	925				
1826-0010	923				
1826-0013		3514			
1826-0055	911D				
1826-0070	941D				
1826-0075	914				
1826-0166	910				
1827-17	102A			TR85/250	G0005
1834		3126			
1835-17	192	3018	10	TR51/	S0015
1840-17	131	3052		TR94/642	G6016
1841-17	123A			TR21/735	S0015
1843-17	161			TR83/719	S0030
1844-17	159	3114		TR20/717	S5013
1845-17	161	3132		TR83/719	S0017
1850-0040	102				
1850-0040-1	102				
1850-0060	102				
1850-0062	102				
1850-0062-1	102				
1850-0101	102				
1850-0184	102				
1850-0184-1	102				
1850-17	159	3114	21	TR30/	S0019
1852-17	108	3039	17	2SC717/	S0020

Original Part Number	ECG	SK	GE	IR/WEP	HEP
1853-0001-1	159				
1853-0041	129				
1853-0045	129			TR88/	
1853-0069	106				
1853-0081	159				
1853-0089	159				
1853-0215	129				
1853-0224		3053			
1853-0233-1		3083			
1854-0003	123A	3122		TR21/735	S0014
1854-0005	123A				
1854-0012-1	128				
1854-0033	123A				
1854-0060	199				
1854-0090	128				
1854-0090-1	128				
1854-0100		3024			
1854-0228		3027			
1854-0228A		3027			
1854-0231	161				
1854-0245	181	3036			
1854-0253		3021			
1854-0265	175				
1854-0274	128				
1854-0332	128				
1854-0353	123A				
1854-0387	199				
1854-0402-1	3054				
1854-0417	107				
1854-0432	123A				
1854-0438-1	124				
1854-0490-1	181				
1854-0492					S0025
1854-0498	128				
1854-0530-1		3027			
1854-0532		3529			
1854-0563	130	3027			
1854-SRK1-1	199				
1858	101	3011	8	TR08/641	
1858-0004	906				
1858-0019	917				
1858-0019-1		3544			
1858-0023	916				
1859-11				TR01/	
1859-14	121	3009		TR01/232	
1859-16	121	3009		TR01/232	
1859-17	132	3116	FET2	FE100/802	F0015
1860		3122			
1865-1	160	3006		TR17/	
1865-1-12	160	3006		TR17/	
1865-1-127	160	3006		TR17/	
1865-1L	160	3006		TR17/	
1865-2	103A	3010		TR08/	
1865-2-12	103A	3010		TR08/	
1865-2L	103A	3010		TR08/	
1865-2L8	103A	3010		TR08/	
1865-4	101	3011		TR08/	
1865-4-12	101	3011		TR08/	
1865-4-127	101	3011		TR08/	
1865-4L	101	3011		TR08/	
1865-4L8	101	3011		TR08/	
1865A	160	3007		TR17/	
1865A-12	160	3007		TR17/	
1865A-127	160	3007		TR17/	
1865AL		3007			
1865AL8	160	3007		TR17/	
1865B	126	3008		TR17/	
1865B-12	126	3008		TR17/	
1865B-127	126	3008		TR17/	
1865BL	126	3008		TR17/	
1865BL8	126	3008		TR17/	
1865C	121	3009		TR01/	
1865C-12	121	3009		TR01/	
1865C-127	121	3009		TR01/	
1865CL	121	3009		TR01/	
1865CL8	121	3009		TR01/	
1866-1	102A	3003		TR85/	
1866-1-12	102A	3003		TR85/	
1866-1-127	102A	3003		TR85/	
1866-1L	102A	3003		TR85/	
1866-1L8	102A	3003		TR85/	
1866-2	102A	3004		TR85/	
1866-2-12	102A	3004		TR85/	
1866-2-127	102A	3004		TR85/	
1866-2L	102A	3004		TR85/	
1866-2L8	102A	3004		TR85/	
1866-3	102A	3004		TR85/	
1866-3-12	102A	3004		TR85/	
1866-3-127	102A	3004		TR85/	
1866-3L	102A	3004		TR85/	
1866-3L8	102A	3004		TR85/	
1866-17	123A	3024	20	TR21/735	S0030
1867-17	159	3114		TR20/717	S0012
1874-3	100	3005		TR05/	
1874-3-12	100	3005		TR05/	

Original Part Number	ECG	SK	GE	IR/WEP	HEP
1874-3-127	100	3005		TR05/	
1874-3L	100	3005		TR05/	
1874-3L8	100	3005		TR05/	
1877A	105	3012		TR03/	
1877A-12	105	3012		TR03/	
1877A-127	105	3012		TR03/	
1877AL	105	3012		TR03/	
1877AL8	105	3012		TR03/	
1877B	121MP	3013			
1877B-12	121MP	3013			
1877B-127	121MP	3013			
1877BL	121MP	3013			
1877BL8	121MP	3013			
1877C	121MP	3014			
1877C-12	121MP	3014			
1877C-127	121MP	3014			
1877CL	121MP	3014			
1877CL8	121MP	3014			
1879-17	123A	3047	20	TR21/735	S3001
1879-17A	128	3047		TR87/	
1880-17	108	3124		TR95/56	S0016
1881-17	108	3018		TR95/56	S0016
1882-17		3122	62		S0015
1883-17		3124	61		S0015
1884-17		3020	20		S0025
1884A	121	3009		TR01/	
1884A-12	121	3009		TR01/	
1884A-127	121	3009		TR01/	
1884AL	121	3009		TR01/	
1884AL8	121	3009		TR01/	
1885-17			18		S3021
1888B	160	3006		TR17/	
1888B-12	160	3006		TR17/	
1888B-127	160	3006		TR17/	
1888BL	160	3006		TR17/	
1888BL8	160	3006		TR17/	
1888C	100	3005		TR05/	
1888C-12	100	3005		TR05/	
1888C-127	100	3005		TR05/	
1888CL	100	3005		TR05/	
1888CL8	100	3005		TR05/	
1889-17		3025			
1890-17	108			TR95/	S0020
1892-1	105	3012		TR03/	
1892-1-12	105	3012		TR03/	
1892-1-127	105	3012		TR03/	
1892-1L	105	3012		TR03/	
1892-1L8	105	3012		TR03/	
1893-17	123A			TR21/	S0030
1901-0557	905				
1901-1011	907				
1906-17	103A	3010	8	TR08/724	G0011
1907-6553	941D				
1915-17	123A	3046	20	TR21/735	S0015
1916-17	195	3047	45	TR65/	S3001
1917-17	102A		52	TR85/	
1919-17	102A	3004	2	TR85/250	G0005
1919-17A	102A	3004		TR05/	
1923-17	108	3018	20	TR70/720	S0016
1923-17-1	108	3018		TR95/	
1925-17	107	3018	20	2SC829B/	S0011
1929-17	123A			TR21/	S0015
1931-17	107	3018	20	TR70/720	S0011
1931-17A	108	3018		TR95/	
1932-17	123A	3024	20	TR21/735	S0015
1933-17	128	3024	18	TR87/243	S5014
1934-17	132	3112	FET1	FE100/802	F0010
1935-17	1599	3025		TR20/717	S0019
1936-17	175	3026	23	TR81/241	S5019
1937-17	103A	3031		TR08/724	
1940-17	159			TR20/717	S0019
1945-17	131	3052	30	TR94/	G6016
1946-17	102A			TR85/250	G0005
1951-17	126	3006		TR17/635	G0008
1952-17	161	3117		TR83/719	
1954-17	102A	3004		TR85/250	G0005
1955-17	131	3082		TR94/642	G6016
1958-17	103A	3010	8	TR08/724	G0011
1960-17	102A	3004		TR85/	G0005
1961-17	123A	3018	17	TR21/	S0015
1966-17	123A			TR21/735	S0014
1968-17	152	3054		TR76/701	S3020
1969-17	152	3054		TR76/701	S5014
1972-17	128			TR87/	S5014
1973-17	102A			TR85/	G0005
1974-17	102A			TR85/	G0005
1979-808-10	159				
1983-17	108			TR95/	S0016
1991-17		3004			
1998-17	107	3018	20	TR24/	
1999-17	107	3018	20	TR24/	
2000M		3112			
2001		3041	66	TR55/	
2001(JOHNSON)	152			TR76/	S5000
2001-17	123A	3124	20	TR21/	

Original Part Number	ECG	SK	GE	IR/WEP	HEP
2002		3009	16	TR16/232	
2002(JOHNSON)	121			TR01/	G6005
2002-2	717				
2003	130			TR59/	S7004
2003-17	123A	3020	20	TR24/	
2003JDP1	128				
2004				TR85/	
2004(JOHNSON)	102A		52	TR85/250	G0005
2005	121			TR01/	G6013
2005-1	712	3072			
2005-2	712	3072			
2006	123A	3534		TR21/735	S5014
2006-1	722	3161			
2006-2	722	3161			
2007-01	121	3009		TR01/121	
2007-1	708				
2007-2	708	3135			
2007-3	708	3135			
2008	175			TR81/	S5019
2008-1	725	3144			
2008-2	725	3144			
2008-17	123A	3124	62	TR21/	S0015
2009	105	3012		TR03/	G6006
2010	105	3012		TR03/	G6006
2010-5409	912	3543			
2010-17					F0010
2011	184	3054	28	TR76/S5003	S5000
2012		3004		TR85/250	G6011
2012-5472	923				
2012(JOHNSON)				TR05/	
2014-6684	941				
2015	160	3006		TR17/	
2015-1	130	3027	14	TR59/247	
2015-1A	130	3027		TR59/	
2015-2	130	3027	14	TR59/247	
2015-2A	130	3027		TR59/	
2015-3	130	3027	14	TR59/247	
2015-3A	130	3027		TR59/	
2015-4	130				
2015-5	130				
2015-6	130				
2017-107		3118			
2017-108		3054		TR55/	S5000
2017-109		3054		TR76/	
2017-110		3024			
2017-115		3018		2SC372/	S0015
2020	160	3006		TR17/637	
2020-00	126	3006	51	TR17/635	
2020-1	703A	3157			
2020-2	703A	3157			
2020-3	703A	3157			
2021	160	3006		TR17/635	
2021-00	126	3006	51	TR17/635	
2021-1	718	3159			
2021-2	718	3159			
2022-01		3114			
2022-03		3018			
2022-05		3122			
2022-06		3050			F2004
2022-244		3018			
2024-1	731	3170			
2025-1	730	3143			
2025-2	730	3143			
2025-3	730				
2026	123A	3047		TR21/735	
2026-00	123A	3047		TR21/735	
2026-5	182	3534	55	TR66/	
2027-00	195	3048		TR65/	
2027(RF)	195			TR65/	
2028	108	3018		TR95/56	
2028-00	108	3018	20	TR95/56	
2029-1	790				
2029-2	790				
2030-1	715	3076			
2031-1	714	3075			
2036-58	153				
2036-59	152				
2036-68	909	3590			
2036-72	909	3590			
2036-93	909	3590			
2039-2	101	3011	8	TR08/641	
2042-1	1112				
2042-2	1111				
2042-17	102A	3004	52	TR84/	
2043-17	159	3025	21	TR52/	
2044-1	738	3167			
2044-18	123	3124	20	TR24/	
2045-1	732				C6090
2047A2-288					G0007
2048-17	199				
2057A2					S0019
2057A2-27					S0014
2057A2-28					S0015
2057A2-29					G0005
2057A2-32					G0005

Original Part Number	ECG	SK	GE	IR/WEP	HEP
2057A2-34					G0005
2057A2-35					G0001
2057A2-37	126	3006		TR17/	G0008
2057A2-38					S0015
2057A2-39					G0005
2057A2-41					G0001
2057A2-42					G0001
2057A2-43					G0005
2057A2-44					G0005
2057A2-45					G0005
2057A2-46					G0011
2057A2-47					S5011
2057A2-48					G0005
2057A2-49					G0005
2057A2-50					G0001
2057A2-51					G0008
2057A2-52					G0005
2057A2-57					S0014
2057A2-58					S5011
2057A2-59					S0015
2057A2-60	126	3005	52	TR17/635	G0008
2057A2-61		3004			G0005
2057A2-62	123A	3018	20	TR21/735	S0015
2057A2-63					S0015
2057A2-64	123A	3018	11	TR21/735	S0016
2057A2-65	126	3006	52	TR17/635	G0008
2057A2-66	126	3006	52	TR17/635	G0008
2057A2-67					G0001
2057A2-68					G0001
2057A2-70					G0001
2057A2-71					G0001
2057A2-72					G0005
2057A2-73					S0015
2057A2-77					G0001
2057A2-78					G0005
2057A2-79					G0005
2057A2-80	126	3006	51	TR17/	G0005
2057A2-81	124	3021		TR81/240	S5011
2057A2-83					G0005
2057A2-84	124	3021		TR81/240	S5011
2057A2-85					S0025
2057A2-86					G0005
2057A2-87	123A	3124		TR21/735	S0014
2057A2-88					G0005
2057A2-89					G0001
2057A2-90					G0001
2057A2-92					S0015
2057A2-93					G0008
2057A2-94					G0005
2057A2-96					S0015
2057A2-97					S0015
2057A2-99					G0005
2057A2-101					S0025
2057A2-102					S0025
2057A2-103	123A	3124	17	TR21/735	S0025
2057A2-104					G0005
2057A2-105					G0005
2057A2-108					S0025
2057A2-109	107	3018	20	TR70/	S0025
2057A2-110	107		20	TR70/	S0025
2057A2-111					S0025
2057A2-112					S0025
2057A2-113	123A	3018		TR21/735	S0014
2057A2-114					S0025
2057A2-115					S0025
2057A2-116	107	3018	11	TR70/720	S0013
2057A2-117	107	3018	20	TR70/720	S0014
2057A2-118					G0006
2057A2-119	108	3018	20		S0025
2057A2-120	107	3018	20	TR70/720	S0016
2057A2-121	123A	3122		TR21/735	S0030
2057A2-122	123A	3124	20		S0015
2057A2-123					S0015
2057A2-124					G0008
2057A2-125					S0015
2057A2-127	108		20	TR95/	S0025
2057A2-128	107	3018	20	TR70/	S0025
2057A2-129					G0005
2057A2-130					S0015
2057A2-131		3122			S0015
2057A2-133					G6005
2057A2-134					G6008
2057A2-135					G0005
2057A2-136					G6008
2057A2-137					G0005
2057A2-138					S0015
2057A2-139					S0020
2057A2-140					S5024
2057A2-141					G0005
2057A2-142					G0005
2057A2-143					S0025
2057A2-145	123A	3024		TR21/735	S5014
2057A2-146	123A	3024		TR21/735	S5014
2057A2-147	102A	3123	2	TR85/250	G0007
2057A2-148	102A	3004	2	TR85/250	G0005

Original Part Number	ECG	SK	GE	IR/WEP	HEP
2057A2-149		3006			G0008
2057A2-150	129	3025	67	TR88/242	S5013
2057A2-151	128	3024	63	TR87/243	S5014
2057A2-152	123A	3122	20	TR21/735	S0015
2057A2-153	123A	3124	62	TR21/	S0015
2057A2-154	123A	3124	20	TR21/735	S0015
2057A2-155	123A	3124	62	TR21/735	S0015
2057A2-156	128	3124	63	TR87/243	S0015
2057A2-157	161	3018	39	TR70/	G0001
2057A2-158	161	3018	39	TR70/	G0001
2057A2-159	160	3018	39	TR17/637	G0008
2057A2-160					S0020
2057A2-161					S0016
2057A2-162					S0025
2057A2-163	108	3018		TR95/56	S0014
2057A2-165	102A	3004		TR85/	G0005
2057A2-166	160	3006		TR17/	G0001
2057A2-167	103A	3010		TR08/	G0011
2057A2-179	108		20	TR95/	S0020
2057A2-180	161		20	TR83/719	S0016
2057A2-181	161		20	TR83/719	S0020
2057A2-182	159		21	TR20/	
2057A2-183	159		21	TR20/	
2057A2-184	123A		17	TR21/	
2057A2-185	161			TR83/719	S0016
2057A2-187	161			TR83/719	S0011
2057A2-192	161	3124	20	TR21/	S0015
2057A2-193	161			TR83/719	S0011
2057A2-195	161	3117	60	TR83/719	S0015
2057A2-196	161			TR83/719	S0015
2057A2-197	161			TR83/719	S0015
2057A2-198	159	3114		TR20/717	S5013
2057A2-199	128			TR87/243	S5014
2057A2-200	106			TR20/52	S0012
2057A2-201	108			TR95/56	S0025
2057A2-202	161			TR83/719	S0020
2057A2-203	106			TR20/52	S0013
2057A2-204	161			TR83/719	S0011
2057A2-205	160			TR17/637	G0008
2057A2-206	102	3004		TR05/631	G0005
2057A2-207	161	3018		TR83/719	S0030
2057A2-208	123A	3038		TR21/735	S0014
2057A2-209	123A	3122		TR21/735	S0014
2057A2-210	102A			TR85/250	G0005
2057A2-211	131	3082		TR94/642	G6016
2057A2-212	199	3124	18	TR51/	S0015
2057A2-215	123A			TR21/735	S0015
2057A2-216	107	3018	20	TR70/720	S0016
2057A2-217	107	3018	20	TR70/720	S0014
2057A2-218	107	3018	20	TR70/720	S0016
2057A2-219	107	3018	20	TR70/720	S0011
2057A2-220	107	3018	20	TR70/720	S0016
2057A2-221	107	3018	20	TR70/720	S0014
2057A2-222	123A	3122	20	TR21/735	S0015
2057A2-223	124	3021	12	TR81/240	S5011
2057A2-224	108	3039	20	TR95/	S0014
2057A2-225	123A	3122	20	TR21/	
2057A2-226	123A	3122	20	TR21/	S0015
2057A2-231	160	3006	17	TR17/637	G0003
2057A2-232	160	3006	17	TR17/637	G0003
2057A2-237	107	3018	20	TR24/	S0011
2057A2-241	102A	3004		TR14/	G0007
2057A2-249	199	3122	17	TR21/735	S0015
2057A2-251	161	3124	20	TR21/	S0015
2057A2-252	160	3006	50	TR17/	G0003
2057A2-257	199	3018	20	TR51/	S0016
2057A2-258	161	3039		TR70/	S0017
2057A2-259	107	3039	60	TR71/	S0016
2057A2-260	199	3122	20	TR21/	S0016
2057A2-261	154	3044		TR78/	S5025
2057A2-262	199	3124	20	TR24/	S0016
2057A2-263	126	3005	53	TR05/	G0005
2057A2-264	123A	3122	18	TR25/	S3013
2057A2-265	127			TR27/	G6007
2057A2-272	199	3039	20	TR51/	S0016
2057A2-273	199	3018	20	TR24/	S0016
2057A2-274	199	3018	20	TR24/	S0016
2057A2-275	199	3018	20	TR24/	S0016
2057A2-276	123A	3124	18	TR21/	S0015
2057A2-277	159	3114	21	TR28/	S0013
2057A2-278	123A	3124	18	TR21/	S0015
2057A2-279	123A	3124	10	TR25/	S5014
2057A2-280	123A	3124	20	TR51/	S0015
2057A2-281	123A	3124	20	TR25/	S0015
2057A2-284	128	3124	10	TR87/	S5014
2057A2-285	123A	3124	20	TR51/	S0015
2057A2-288	102A	3004	52	TR85/250	G0005
2057A2-289	123A	3122	20	TR21/735	S0015
2057A2-290	199	3124	20		S0015
2057A2-294	123A	3122	62	TR21/	
2057A2-295	128	3024	18	TR87/	
2057A2-296	123A	3122	10	TR21/	
2057A2-297	123A	3122	10	TR21/	
2057A2-298					S5013
2057A2-300	123A	3122	20	TR21/	S5000

Original Part Number	ECG	SK	GE	IR/WEP	HEP
2057A2-301					S5000
2057A2-302	104	3009	16	TR02/	G6003
2057A2-303	123A	3124	20	TR21/	
2057A2-304	107	3018	17	TR24/	S0033
2057A2-305	107	3018	20	TR24/	S0011
2057A2-306	123A	3122	20	TR24/	S0015
2057A2-307	159	3118	21	TR30/	S0013
2057A2-309	108	3018	17	TR95/	
2057A2-310	108	3018	11	TR80/	
2057A2-311	108	3018	17	TR95/	
2057A2-313	108	3122	20	TR21/	
2057A2-314	108	3122	20	TR21/	
2057A2-316	123A	3124	10	TR24/	
2057A2-317	158	3004	53	TR84/	
2057A2-322	107	3018	17	TR70/	S0016
2057A2-323	107	3018	20	TR71/	S0016
2057A2-324	123A	3018	20	TR51/	S0011
2057A2-325	107	3018	17	TR33/	S0011
2057A2-326	107	3018	17	TR33/	S0011
2057A2-329	102A	3004	52	TR14/	G0005
2057A2-331	107	3018	20		S0016
2057A2-332	123A	3018	20		S0011
2057A2-333	199	3124	20		S0015
2057A2-334	199	3124	18		S0015
2057A2-352	199	3124	18	TR51/	S0015
2057A2-353	159	3114	22	TR19/	S0012
2057A2-356	107	3018	20	TR24/	S0016
2057A2-359	159	3114	21	TR54/	S0019
2057A2-370	199	3122	62		S0024
2057A2-385	199				
2057A2-387	159	3018			S0015
2057A2-390		3018			S0014
2057A2-391		3122			S0024
2057A2-392		3039			S0015
2057A2-393		3018			S0015
2057A2-394		3018			S0014
2057A2-395		3018			S0025
2057A2-396	123A	3122	18	2SC828/	
2057A2-397	159	3114	21	TR30/	
2057A2-398	123A	3124	18	2SC644/	
2057A2-399	123A	3122	18	TR24/	
2057A2-400	159	3114	21		
2057A2-401	123A	3018	17	TR95/	S0030
2057A2-402	108	3018	17	TR33/	S0016
2057A2-403	159	3114	21	TR33/	S0019
2057A2-404	199	3024	18	TR51/	S0015
2057A2-405	199	3122	62	TR51/	S0024
2057A2-406	159	3114	21	TR30/	S0019
2057A2-427	128				
2057A2-432	108	3018	17	TR24/	
2057A2-433	123A	3122	20	TR24/	
2057A2-434	123A	3122	20	TR24/	
2057A2-446		3114			
2057A2-448		3018			
2057A2-449		3122			
2057A2-452		3122			
2057A2-454		3018			
2057A2-457	159		22	TR30/	
2057A2-465		3018			S0020
2057A2-466		3018			S0016
2057A2-468		3044			S3002
2057A2-475		3018			
2057A2-477		3018			
2057A2-478		3018			
2057A2-479		3018			
2057A2-480		3114			
2057A2-501		3018	61		S0030
2057A2-502		3018	60		S0016
2057A2-503		3018	20		S0015
2057A2-504		3020	20		S0016
2057A2-505		3020	20		S0016
2057A2-518		3122			
2057A2-988		3004			G0005
2057A10-64		3122	10		S0015
2057A31-4		3112			
2057A100			14		
2057A100-1					G0005(2)
2057A100-3					G0005
2057A100-4	103A	3004	52	TR84/	G0011
2057A100-5					G0005
2057A100-7					G0005(2)
2057A100-8	102A	3004	2		G0005(2)
2057A100-9	102A		52	TR85/	G0005(2)
2057A100-11					G6016(2)
2057A100-12					G0006(2)
2057A100-13					G0005(2)
2057A100-14	103A	3010		TR85/724/630	G0005
2057A100-15					G6011
2057A100-16	130MP	3027	14	TR59/	S0025(2)
2057A100-17	128	3024	18	TR87/243	S5014(2)
2057A100-18	103A	3010			G6011
2057A100-21	102A			TR85/250	G0005
2057A100-23	158	3004		TR84/	G6011(2)
2057A100-24	102A	3004	52	TR85/	G6011
2057A100-26	130MP	3036		TR59MP/	S7002
2057A100-30	102A	3011	54		G0011
2057A100-34	158	3004	53		G0005
2057A100-35	103A	3010	54		G0011
2057A100-40	192				
2057A100-41	158	3004	53		G6011
2057A100-44		3004			G0005
2057A100-45		3083			
2057A100-47	184	3190	57		S5000
2057A100-48	158	3004	53	TR85/	G0005
2057A100-49	184	3054			
2057A100-50	186				
2057A100-53	107	3124	61	TR24/	
2057A100-66		3083			
2057A-120	108	3018		TR95/	
2057B0-3	102A				
2057B2-4	102	3004	2	TR05/631	G0005
2057B2-14	108			TR95/56	S0014
2057B2-23	102A	3003		TR85/250	G0005
2057B2-27			10		S0014
2057B2-28	102A	3124	10	TR85/250	S0015
2057B2-29	102A	3004	2	TR85/250	G0005
2057A2-32	102	3004	2	TR05/631	G0005
2057B2-34	102	3004	52	TR05/631	G0005
2057B2-35	126	3006	51	TR17/635	G0001
2057B2-37	126	3008		TR17/635	G0008
2057B2-38	123A	3124	20	TR21/735	S0011
2057B2-41	126	3008	51	TR17/635	G0001
2057B2-42	126	3008	51	TR17/635	G0001
2057B2-43	102	3004	2	TR05/631	G0005
2057B2-44	102	3004	2	TR05/631	G0005
2057B2-45	102	3004	2	TR05/631	G0005
2057B2-46	103	3010	8	TR08/641A	G0011
2057B2-47	124	3021	12	TR81/240	S5011
2057B2-48	126	3008	2	TR17/635	S0014
2057B2-49	102	3004	2	TR05/631	G0005
2057B2-50	126	3008	51	TR17/635	G0001
2057B2-51	126	3008	51	TR17/635	G0008
2057B2-52	102A	3004	2	TR85/	G0005
2057B2-57	102	3124	20	TR05/631	S0014
2057B2-58	124	3021	12	TR81/240	S5011
2057B2-59	123	3124		TR86/53	S0015
2057B2-60	126	3005	2	TR17/635	G0008
2057B2-61	158	3004	2	TR84/630	G6011
2057B2-62	123A	3018	10	TR21/735	S0015
2057B2-63	123	3124	10	TR86/53	S0015
2057B2-64	108	3018	11	TR95/56	S0016
2057B2-65	126	3006	51	TR17/635	G0008
2057B2-66	126	3006	51	TR17/635	G0008
2057B2-67	126	3006	51	TR17/635	G0003
2057B2-68	126	3006	51	TR17/635	G0003
2057B2-69	123A		63	TR21/	
2057B2-70	126	3006	51	TR17/635	G0003
2057B2-71	126	3006	51	TR17/635	G0003
2057B2-72	102	3004	2	TR05/631	G0007
2057B2-73	123	3124	20	TR86/53	S0016
2057B2-77	126	3006	51	TR17/635	S0014
2057B2-78	102	3004	2	TR05/631	G0005
2057B2-79	126	3008	51	TR17/635	G0005
2057B2-80	126	3008	51	TR17/635	G0009
2057B2-81	126	3006	51	TR17/635	G0008
2057B2-83	102	3004	2	TR05/631	G0005
2057B2-84	124	3021		TR81/240	S5011
2057B2-85	108	3018		TR95/56	S0025
2057B2-86	102	3004	2	TR05/631	G0005
2057B2-87	108	3018	20	TR95/56	S0011
2057B2-88	126	3006	51	TR17/635	G0008
2057B2-89	160	3006	51	TR17/637	G0003
2057B2-90	160	3006	51	TR17/637	G0003
2057B2-92					S0015
2057B2-93	160	3006	51	TR17/637	G0003
2057B2-94	102A		52	TR85/250	G0005
2057B2-96					S0015
2057B2-97	123A	3124	10	TR21/735	S0015
2057B2-99	102A			TR85/250	G0005
2057B2-101	107	3018		TR70/720	S0025
2057B2-102	107	3018		TR70/720	S0025
2057B2-103	108	3124	10	TR95/56	S0025
2057B2-104	126	3008	51	TR17/635	G0008
2057B2-105	126	3008	51	TR17/635	G0008
2057B2-107	102A		52	TR85/	
2057B2-108	108	3039		TR95/56	S0016
2057B2-109	108	3039		TR95/56	S0015
2057B2-110	108	3039		TR95/56	S0015
2057B2-111	108	3039		TR95/56	S0015
2057B2-112	108	3006		TR95/56	G0003
2057B2-113	123A	3122	20	TR21/735	S0011
2057B2-114	107			TR70/720	S0014
2057B2-115	107	3039		TR70/720	S0016
2057B2-116	107	3018	20	TR70/720	S0014
2057B2-117	107	3018		TR70/720	S0014
2057B2-118	126			TR17/635	G0008
2057B2-119	108	3018		TR95/56	S0016
2057B2-120	108	3018		TR95/56	S0016
2057B2-121	123A	3124		TR21/735	S0030
2057B2-122	123A	3020		TR21/735	S0014

Original Part Number	ECG	SK	GE	IR/WEP	HEP
2057B2-123	123A	3124		TR21/735	S0014
2057B2-124	102A	3005		TR85/250	G0005
2057B2-125	107	3018		TR70/720	S0014
2057B2-127	108	3039		TR95/56	S0025
2057B2-128	108	3039		TR95/56	S0025
2057B2-129	102A			TR85/250	G0007
2057B2-130	123A	3024		TR21/735	S0020
2057B2-133	121	3009		TR01/232	G6005
2057B2-134	127	3034		TR27/235	G6008
2057B2-135	102A	3004		TR85/250	G0005
2057B2-136	127	3035		TR27/235	G6008
2057B2-137	102A			TR85/250	G0001
2057B2-138	161	3018		TR83/719	S0016
2057B2-139	161	3018		TR83/719	S0016
2057B2-140	154	3044		TR78/712	S5025
2057B2-141	126	3006		TR17/635	G0005
2057B2-142	102A	3004		TR85/250	G0005
2057B2-143	107			TR70/720	S0016
2057B2-145					S5014
2057B2-146					S5014
2057B2-147					G0007
2057B2-148					G0005
2057B2-149	160	3006		TR17/637	G0008
2057B2-150	153	3025		TR77/700	S5013
2057B2-151	152	3024		TR76/701	S5014
2057B2-152	123A	3124		TR21/735	S0015
2057B2-153	123A	3024		TR21/735	S0015
2057B2-154	123A	3124		TR21/735	S0015
2057B2-155	123A	3122		TR21/735	S0015
2057B2-156					S0015
2057B2-157	160	3006		TR17/637	G0001
2057B2-158	160	3006		TR17/637	G0001
2057B2-159	160	3006		TR17/637	G0008
2057B2-160	108			TR95/	S0016
2057B2-161	108			TR95/	S0016
2057B2-162	108			TR95/56	S0016
2057B2-163					S0014
2057B2-165					G0005
2057B2-166					G0001
2057B2-167					G0011
2057B2-179					S0020
2057B2-180					S0016
2057B2-181					S0020
2057B2-185					S0016
2057B2-187					S0011
2057B2-192	161	3124		TR83/719	S0015
2057B2-193					S0011
2057B2-195					S0015
2057B2-196					S0015
2057B2-197					S0015
2057B2-198					S5013
2057B2-199					S5014
2057B2-200					S0012
2057B2-201					S0025
2057B2-202					S0020
2057B2-203					S0013
2057B2-205					G0008
2057B2-206	102A			TR85/	G0005
2057B2A2-118	160	3006		TR17/	
2057B45-14	102A			TR85/	
2057B-59	199		10		
2057B-84	124		12		
2057B-85	123A		20		
2057B100-1	102A			TR85/250	G0005
2057B100-3	102A			TR85/250	G0005
2057B100-4	103A			TR08/	G0011
2057B100-5	102A			TR85/250	G0005
2057B100-6	102A		52	TR85/	
2057B100-7		3004	52	TR85/250	G0005
2057B100-8	102A		2	TR85/250	G0005
2057B100-9	102A			TR85/250	G0005
2057B100-11	131	3082		TR94/642	G6016(2)
2057B100-12	108	3004		TR95/56	G0006
2057B100-13	102A			TR85/250	G0005
2057B100-14					G0005
2057B100-15					G6011(2)
2057B100-16	130MP	3027		TR59/247MP	S7004
2057B100-17	128	3024		TR87/243	S5014
2057B100-18					G6011
2057B100-21					G0005(2)
2057B101-4	123A	3122		TR21/735	
2057B102-4	123A	3122		TR21/735	
2057B103-4	123A	3122		TR21/735	
2057B104-8	154	3045		TR78/712	
2057B106-12	159	3114		TR20/717	
2057B107-8	128	3024		TR87/243	
2057B108-6	159	3114		TR20/717	
2057B109-9	128	3024		TR87/243	
2057B110-9	129	3025		TR88/242	
2057B111-9	128	3024		/243	
2057B112-9	129	3025		TR88/242	
2057B113	108	3018			
2057B113-9	128	3024		TR87/243	
2057B114-9	129	3025		TR88/242	
2057B115-9	129	3025		TR88/242	
2057B116-9	129	3025		TR88/242	
2057B117-9	123A	3122		TR21/735	
2057B118-12	123A	3122		TR21/735	
2057B119-2	123A	3122		TR21/735	
2057B120-12	123A	3122		TR21/735	
2057B121-9	129	3025		TR88/242	
2057B122-9	129	3025		TR88/242	
2057B123-10	124	3021		TR81/240	
2057B124-10	121	3009		TR01/232	
2057B125-9	123A	3122		TR21/735	
2057B126-12	128	3024		TR87/243	
2057B129-9	128	3024		TR87/243	
2057B141-4	123A	3122		TR21/735	
2057B142-4	123A	3122		TR21/735	
2057B143-12	123A	3122		TR21/735	
2057B144-12	128	3024		TR87/243	
2057B145-12	129	3025		TR88/242	
2057B146-12	123A	3122		TR21/735	
2057B147-12	159	3114		TR20/717	
2057B149-12	133	3112		/802	
2057B151-6	123A	3122		TR21/735	
2057B152-12	123A	3122		TR21/735	
2057B153-9	128	3024		TR87/243	
2057B155-10	175			TR81/241	
2057B156-9	128	3024		TR87/243	
2057B157-9	162			TR67/707	
2057B158-10	175			TR81/241	
2057B159-12	159	3114		TR20/717	
2057B163-12	129	3025		TR88/242	
2057B168	102A			TR85/250	
2057B169	102A			TR85/250	
2057B170				TR85/250	
2057B175-9	162			TR67/707	
2057B186	160			TR17/637	
2057B206	102A			TR85/	
2058	126		50	TR17/635	G0002
2058A2-331		3018			
2061-1	783				
2061A45-47	102			TR05/631	
2061B45-14	158			TR84/630	G0005
2062-17	224	3049	45	/243	S3002
2062-17A	224	3049			
2063-17	123A	3124	20	TR21/735	S0015
2063-17-12	123A	3124		TR21/	
2064(CROWN)	123A	3122	20	TR21/735	
2067-1	715				
2069-1	715				
2071	181	3036			
2074-17	132	3116	FET1		F0015
2076-1	742				
2077-1	788				
2081-17	186	3054	28	TR76/	S5000
2081A		3112			
2082-6		3054			
2085-17	184	3054	28	TR55/	S5000
2090(CROWN)	154	3045	27	TR78/712	
2093A2-289	108	3018		TR95/	
2093A9-3	100	3005		TR05/	
2093A9-4	100	3005		TR05/	
2093A38-23	102	3004		TR05/631	
2093A38-24	128	3024		TR87/243	
2093A41-40	102			TR05/631	G0005
2093A41-41	102			TR05/631	
2093A41-87	158				
2101	199		10	TR21/	
2106-119	102A	3004	53	TR85/250	
2106-120	160	3006	51	TR17/637	
2106-121	102A	3004	53	TR85/250	
2106-122	102A	3004	53	TR85/250	
2106-123	102A	3004	53	TR85/250	
2110N-132	161	3122	20	TR83/	
2110N-133	161	3122	20	TR83/	
2110N-134	175	3026	24MP	TR81/	
2112-17	102A	3004	52	TR85/	G0005
2114-0	154	3044		TR78/	S5025
2114-17	158	3004	52	TR84/	G0005
2121-17	123A	3124	20	TR21/	S0015
2132-17	123A	3124	20		S0015
2134-17	107	3018	10	TR70/	S0015
2135E	735				C6013
2136	737				
2136D	737				
2136P	737				
2136PC	737				
2158-1541	123A				
2158-1558	159				
2160	6400				S9002
2160-17	186	3041	28	TR55/	S5000
2163-17	175	3026	23	TR81/	S5019
2167-17	224	3049			S3001
2168-17	186	3054	28		S5000
2180-41	178	3087			
2180-151	108	3018		TR95/	S0014
2180-152	108	3018		TR95/	S0015
2180-153	123A	3124		TR21/	S0015

Original Part Number	ECG	SK	GE	IR/WEP	HEP
2180-154	123A	3124		TR21/	S0015
2180-155	175	3026		TR81/	S5019
2181-17	123A	3024	18	TR25/S5014	
2195-17	199	3124			
2197-17	107	3018	20	TR24/	
2202-17		3054			S5000
2204-17	123	3124		TR21/	S0015
2207-17	199	3124	20	TR21/	S0015
2208-17	128	3122	20	TR21/	S0025
2210-17	152		28	TR55/	S5000
2211-17	153		29	TR56/	S5006
2212-17	199	3018	20	TR24/	S0016
2213-17	107	3018	60	TR24/	S0016
2214-17	199	3039	26	TR33/	S0016
2215-17	160	3006	50	TR17/	G0003
2224-17		3018			
2225-17		3039			
2226-17		3018			
2243	121	3014		TR01/232	G6005
2244-1	187	3083	29	TR56/	
2245-17	186	3054	28	TR55/	
2246-17	192	3122	63	TR21/	
2263	123A	3122		TR21/735	S0014
2269	159	3025		TR20/	S0019
2270	123A	3122		TR21/	S0003
2270-5	128				
2271	199			TR21/	S0024
2272	159	3025		TR20/	S0019
2275-17	123A	3122	10	TR24/	S0016
2277P	732				
2284-17	107	3018	20	TR24/	
2290-17	123A	3122	10	TR24/	S0016
2291-17	107	3018	20	TR24/	
2295	175	3538		TR81/701	S5019
2300.036.096	159				
2300.037/096	123A				
2302-17	108	3019	17	TR24/	S0020
2312-17	184	3190			
2320-17	123A	3018	20	TR24/	S0015
2321-17	123A	3038	18	TR25/	S0015
2322-17	195	3048		TR64/	S5014
2334-17	186	3054			S5000
2335-17	132				
2336-17	132	3116	FET2	FE100/	
2337-17	123A	3018	20	TR24/	S0011
2338-17	123A	3124			S0015
2340-17	195	3048			S5014
2347-17	104	3009			
2359-17	222	3050			
2361-17	123A	3024	20	TR51/	
2362-1(SYL)	108	3122	20	TR95/56	S0015
2381-17		3118			
2382-17		3054			
2402-453	102A	3004		TR85/250	G0005
2402-454	102A	3004		TR85/250	G0005
2402-455	102A	3004		TR85/250	G0005
2402-456	131	3052		TR94/642	C6016
2402-457	102A	3004		TR85/250	G0005
2405-453	102A	3004		TR85/250	G0005
2405-454	102A	3123		TR85/250	G0005
2405-455	102A	3123		TR85/250	G0005
2405-456	102A	3123		TR85/250	G0007
2405-457	102A	3123		TR85/250	G0005
2408-326	102A	3004		TR85/250	G0005
2408-328	102A	3123		TR85/250	G0005
2408-329	158	3004		TR84/630	G6011
2417	124	3021		TR81/240	S5011
2418-17		3054			
2427(RCA)	108	3039	11	TR95/56	
2432-1(RCA)	710	3102			
2434-1(RCA)	710	3101			
2434(RCA)	710	3101		IC506/	
2443(RCA)	123A	3124	11	TR21/735	
2444(RCA)	124			TR81/240	
2445	108	3039	20	TR95/56	
2445-1(RCA)	711	3070			
2445(RCA)	711	3039			
2446				TR51/	
2446(RCA)	123A	3122	20	TR21/735	
2446-1(RCA)	121				
2447(RCA)	123A	3124	20	TR21/735	
2448(RCA)	159	3025	11	TR20/717	
2450(RCA)	108	3018	17	TR95/56	S0015
2470-1724	74H01				
2470-1732	74H05				
2472-5632	123A				
2473-2109	7408				
2473(RCA)	108	3018	17	TR95/56	
2475(RCA)	123A	3124	11	TR21/735	
2476(RCA)	108	3018	17	TR95/56	
2477(RCA)	108	3018	17	TR95/56	
2478	126	3008		TR17/635	
2478A	126	3008		TR17/635	
2478B	126	3008		TR17/635	
2482(RCA)	129	3025	52	TR88/242	S5013
2487B	160	3008	1	TR17/637	G0008
2488	126	3008		TR17/635	
2488A	126	3008	1	TR17/635	G0008
2489	126	3008		TR17/635	
2489A	160	3008	1	TR17/637	G0008
2490	102A	3004		TR85/250	
2490A	102A	3004	2	TR85/250	G0006
2491	124	3021		TR81/240	
2491A	124	3021		TR81/240	
2491B	124	3021	12	TR81/240	S5011
2494(RCA)	127	3025	25	TR27/235	
2495		3124	11	TR22/735	
2495-012	126	3008		TR17/	
2495-013	126	3008		TR17/	
2495-014	102A	3004		TR85/	
2495-078	160	3006		TR17/	
2495-079	160	3006		TR17/	
2495-080	102A	3004		TR85/	
2495-082	160	3006		TR17/	
2495-166-1	108	3018		TR95/	
2495-166-2	123A	3124		TR21/	
2495-166-4	108	3018		TR95/	
2495-166-8	108	3018		TR95/	
2495-166-9	108	3018		TR95/	
2495-200	126	3008		TR17/	
2495-376	160	3006		TR17/	
2495-377	160	3006		TR17/	
2495-378	160	3006		TR17/	
2495-388	102A	3004		TR85/	
2495-488-1	160			TR17/	
2495-488-2	160			TR17/	
2495-520	108	3018		TR95/	
2495-521	108	3018		TR95/	
2495-522-1	108	3018		TR95/	
2495-522-4	123A	3124		TR21/	
2495-523-1	108	3018		TR95/	
2495-529	123A	3124		TR21/	
2495-567-2	102A	3004		TR85/	
2495-567-3	102A	3004		TR85/	
2495-586-2	102A	3004		TR85/	
2496(RCA)	127	3034	25	TR27/235	
2496-125-2	123A	3124			
2497-473	102A	3004		TR85/	
2497-496	102A	3004		TR85/	
2498-163	128	3024		TR87/	
2498-507-2	108	3018		TR95/	
2498-507-3	108	3018		TR95/	
2498-508-2	108	3018		TR95/	
2498-508-3	108	3018		TR95/	
2498-903-2	108	3018		TR95/	
2498-903-3	108	3018		TR95/	
2500(RCA)	127	3035	25	TR27/235	
2501(RCA)		3034			G6008
2502	129		22	TR88/	
2502(RCA)	102A	3025		TR85/250	S5013
2510-101		3018			S0015
2510-102		3018			S0015
2510-103	199				S0015
2510-104	123A	3124			S0011
2510-105	186	3054			S5000
2516(RCA)	712	3072			C6083P
2516-1(RCA)	712	3072	IC2	IC507	
2546(RCA)	123A	3124	11	TR51/735	
2548A0157	124				
2549	219				S7003
2554-1(RCA)	730	3143			
2554-2(RCA)	730	3143			
2554-3(RCA)	730	3143			
2554-4(RCA)	730	3143			
2559	728	3073			
2559-1	728	3073			
2560(RCA)	729	3074			
2560-1(RCA)	729	3074			
2577	121	3014			G6007
2584	123	3124	11	TR86/53	
2603-180	102A	3123		TR85/250	G0005
2603-181	102A	3123		TR85/250	G0007
2603-182	102A	3123		TR85/250	G0005
2603-183	126	3123		TR17/635	G0005
2603-184	123A	3122		TR21/735	S0016
2603-286	102A	3004		TR85/	G0005
2603-287	102A	3004		TR85/	G0007
2603-288	176	3004		TR82/	G6011
2603-291	102A	3004		TR85/	G0005
2603-292	126	3008		TR17/	G0005
2603-294	108	3018		TR95/	S0016
2603-295	160	3006		TR17/	G0009
2612	102A	3004		TR85/250	
2620	154	3045	27	TR78/712	
2626-1	912				
2633(RCA)	108	3039		TR95/56	
2634(RCA)	108	3018	17	TR95/56	S0016
2634				TR80/	
2634-1	108	3018		TR95/56	S0015
2636	108	3039		TR95/56	

Original Part Number	ECG	SK	GE	IR/WEP	HEP
2664-1	726				
2665-1	723				
2665-2	723				
2665-3	723				
2670		3025			
2671-1		3009			
2700	160	3006	51	TR17/637	G0003
2703-384	102A			TR85/250	G0005
2703-385	102A	3123		TR85/250	G0005
2703-386	102A	3123		TR85/250	G0005
2703-387	176			TR82/238	G6011
2703-388	158			TR84/630	G6011
2704-384	102A	3004		TR85/250	G0005
2704-385	102A	3004		TR85/250	G0005
2704-386	102A	3004		TR85/250	G0005
2704-387	158	3004		TR84/630	G6011
2780	121	3009	16	TR01/232	
2780(SEARS)					G6007
2780-3	121MP	3009	13MP	TR01/23MP	
2780-4	121	3009	3	TR01/232	G6003
2780-5	121	3009	3	TR01/232	
2780(AIRLINE)	121MP				
2781	102	3004	2	TR05/631	G0005
2781(WARDS)					G0005
2781(SEARS)					G0005
2787	123A	3124	20	TR21/735	S0015
2787(WARDS)					S0015
2791	126	3006	51	TR17/635	
2797	160	3006	51	TR17/637	G0003
2798	159	3006	51	TR20/717	S0013
2799	126	3006	51	TR17/635	
2799-1	789				
2799-2	789				
2802-1	781	3169			
2802-1(BENDIX)	781				
2802-2	781	3169			
2842-056	175				
2842-875	181	3036			
2849-1					S3002
2849-2					S3002
2849-3					S3002
2850-1					S3002
2850-2					S3002
2850-3					S3002
2851-1					S3002
2851-2					S3002
2851-3					S3002
2852-1					S3002
2851-2					S3002
2851-3					S3002
2852-1					S3002
2852-2					S3002
2852-3					S3002
2853-1	186	3192	28	TR55/	S3002
2853-2	152		66	TR76/	S3002
2853-3	186	3192	28	TR55/	S3002
2854-1	186	3192	28	TR55/	S3002
2854-2	152		66	TR76/	S3002
2854-3	186	3192	28	TR55/	S3002
2855-1	186	3192	216	TR55/	S3002
2855-2	152		66	TR76/	S3002
2855-3	186		28	TR55/	S3002
2856-1	186	3192	216	TR55/	S3002
2856-2	152		66	TR76/	S3002
2856-3	186	3192	28	TR55/	
2871-1		3129			
2874(RCA)	726	3129			
2874(SEARS)	726				
2886-3		3027			
2900-007	108	3018	17	TR95/56	
2901-010	121	3009	3	TR01/232	G6005
2904-003			17		
2904-008	121	3009	3	TR01/232	G6005
2904-014	121	3009		TR01/	
2904-016	126	3006	51	TR17/635	
2904-029	126	3007	51	TR17/635	
2904-030	224	3049			
2904-032	128	3047		TR87/	
2904-033	108	3018	20	TR95/56	
2904-034	123	3124	20	TR86/53	
2904-035	123	3124	20	TR86/53	
2904-037	195	3048		TR65/	
2904-038	126	3008		TR17/	
2904-038H05	100	3005		TR05/	
2925(JOHNSON)	123A			TR21/735	
2957A100-25			23	TR57/	S5019
3000(SEARS)					S0020
3001	108	3018		TR95/	S0020
3002	108	3019	17	TR95/56	S0020
3003				TR21/	S0015
3003(SEARS)	107	3018		TR70/720	
3004		3018		TR21/	S0015
3004(SEARS)	107	3018		TR70/720	
3004-856	102A	3004		TR85/	
3005(SEARS)	123A	3124		TR21/735	S0016

Original Part Number	ECG	SK	GE	IR/WEP	HEP
3005-861	123A	3124		TR21/	
3006(SEARS)	123A	3124		TR21/735	S0016
3007(SEARS)	123A	3124		TR21/735	S0015
3008(JOHNSON)	160	3006	50	TR17/637	G0009
3009(JOHNSON)	160	3006	50	TR17/637	G0009
3010(JOHNSON)	126	3008	52	TR17/635	G0008
3010(SEARS)	102A			TR85/250	G0007
3011	123A		20	TR21/735	S0015
3011(JOHNSON)	108			TR95/	S0015
3012(JOHNSON)	106	3118	22	TR20/	
3012(NPN)	108			TR95/	S0016
3012(PNP)	159				
3013				TR85/	
3013(SEARS)	176			TR82/238	G6011
3014				TR85/	
3014(SEARS)	102A			TR85/250	G0007
3017(JOHNSON)	159	3114	21	TR20/	S0019
3018(JOHNSON)	108	3018	11	TR95/56	S0016
3019	159			TR20/635	S0019
3019(JOHNSON)	126			TR17/	
3020	108			TR95/	S0016
3021	108			TR95/	S0016
3022	128			TR87/	S0015
3023	161	3039		TR83/	S0017
3024(JOHNSON)					G0003
3025(JOHNSON)					G0003
3026	123A			TR21/	S0015
3027	161	3039		TR83/	S0017
3028	108			TR95/	S0015
3034(RCA)	128	3024	18	TR87/243	
3053		3024			
3054	175	3131		TR81/241	S5019
3055		3027			
3055-1	130				
3055-3	130				
3064T	780				
3064TC			IC20		
3065D	712				
3065E	712				
3066D	728				
3066DC			IC22		
3066E	728				
3067D	729				
3067DC			IC23		
3067E	729				
3075D	723				
3075DC			IC15		
3075E	723				
3075PC	723				
3076HC	781	3169			
3076PC	781				
3107-204-9000	123A			TR21/	S0011
3107-204-90010	123A			TR21/	S0015
3107-204-90020	123A			TR21/	S0015
3107-204-90070	121			TR01/	G6005
3107-204-90080	108			TR95/	S0016
3107-204-90100	107			TR70/	S0015
3107-204-90140	104			TR01/	G6003
3107-204-90150	199			TR21/	S0024
3107-204-90180	210				S3024
3107-204-90182	152				
3107-204-90190	104			TR01/	G6003
3111	123A	3018	20	TR21/735	S0015
3112	102A	3003	2	TR85/250	G0005
3113	123A	3124	20	TR21/735	S0015
3146-977	130	3027		TR59/	
3152-159	175	3026		TR81/	
3152-170	130	3027		TR59/	
3190(HAR KAR)	181			TR36/	
3202		3024			
3202-5H01	128	3024		TR87/243	
3202-51-01	128	3024		TR87/243	
3212(HAR KAR)	123A			TR21/	
3221(HAR KAR)	123A			TR21/	
3222	129				
3223(HAR KAR)	128			TR87/	
3227-E	108	3019	17	TR95/56	S0016
3270-10		3004			
3295-001	124				
3301(SEARS)	107			TR70/720	
3367	172			TR69/S9100	
3370	108			TR95/	S0016
3373	181			TR36/	
3391	199	3124	63	TR21/735	S0015
3391A	199	3124		TR21/735	S0015
3391(SEARS)	199			TR21/	S0015
3400-2412-3	903				
3418		3004			
3423		3004			
3425	100	3005	2	TR05/254	G0005
3426		3005			
3432		3005			
3432-1		3124			
3433		3005			
3434	100	3005	2	TR05/254	G0005

Original Part Number	ECG	SK	GE	IR/WEP	HEP
3435	100	3005	2	TR05/254	G0005
3442		3003			
3450		3004			
3451		3004			
3456	176				
3458	160	3005	2	TR05/254	G0005
3460		3008			
3461		3004			
3473		3003			
3475		3011			
3476(RCA)	161	3117		TR83/719	S0020
3477				TR21/	
3477A		3004			
3477G		3018			
3484		3003			
3499	130				
3500	100	3004	2	TR05/254	G0005
3502-1(RCA)	704	3023			
3502-2(RCA)	704	3023			
3502(RCA)	704	3023		IC501/	
3503(RCA)	124	3021	12	TR81/240	
3504(RCA)	160	3004	2	TR17/637	G0005
3505(RCA)	123A	3087	11	TR21/735	S0015
3506(RCA)	123A	3122	20	TR21/735	
3507(RCA)	102A	3124	63	TR85/250	
3507(SEARS)	107			TR70/	S0015
3507(WARDS)	123A			TR21/735	S0015
3508	128	3018	11	TR87/243	S0020
3508(RCA)	123A	3018		TR21/735	
3508(SEARS)	161			TR83/56	S0020
3508(WARDS)	108			TR95/56	
3509	108	3124	20	TR95/56	S0015
3509(SEARS)	123A			TR21/735	S0015
3509(WARDS)	123A			TR21/735	
3510(RCA)	123A	3018		TR21/735	
3510(SEARS)	108			TR95/	S0020
3510(WARDS)	108			TR95/	S0020
3511	108	3122		TR95/56	
3511(SEARS)	132		FET2	FE100/802	F0015
3511(WARDS)	132		FET2	FE100/802	F0015
3512	121	3009		TR01/	
3512(RCA)	132		FET2	FE100/802	F0015
3512(SEARS)	132		FET2	FE100/802	F0021
3512(WARDS)	132		FET2	FE100/802	F0015
3513	159	3025	21	TR20/717	S0019
3513(RCA)	123A			TR21/735	
3513(SEARS)	159			TR20/717	S0019
3513(WARDS)	159			TR20/717	S0019
3514	121	3124	20	TR01/235	
3514(SEARS)	107			TR70/	S0015
3514(WARDS)	123A			TR21/735	S0015
3515(RCA)	121	3009	16	TR01/232	
3516	121	3004	11	TR01/	
3516(RCA)	161	3004		TR83/719	
3516(WARDS)	108			TR95/56	
3517	102A	3004	2	TR85/	G0005
3518(RCA)	161	3004	11	TR83/719	
3519(RCA)	123A	3122	10	TR21/735	S0014
3519-1(RCA)	199	3122			
3520(RCA)	124	3021	12	TR81/240	S5011
3520-1	124	3021	12	TR23/	S5011
3521 (SEARS)	123A		17	TR21/735	S0024
3522(SEARS)	159		21	TR20/717	S0015
3523(RCA)	129	3025	52	TR88/242	
3523(SEARS)	123A			TR21/735	
3524-1	108	3018	11		S0016
3524-1(RCA)	108	3018	11	TR22/	S0016
3524-2(RCA)	108	3018		TR80/	S0016
3524(SEARS)	159			TR20/717	S5022
3525(RCA)	123A	3122	29	TR21/735	
3525(SEARS)	185				S5006
3526(RCA)	123A	3122	28	TR21/735	S0016
3526(SEARS)	184			/S5003	S5000
3527(RCA)	108	3018	20	TR95/56	
3528-1	785				
3528-1(RCA)	785				
3530(RCA)	108	3019	11	TR95/56	S0020
3530-2		3019			
3531-021-000	7413				
3531-022-000	74190				
3531-023-000	74141				
3531-030-000	788				
3531-031-000	720				
3531-033-000	1111				
3531-1		3039			
3532(RCA)	123A	3124	20	TR21/735	S0016
3532-1	128	3024	11	TR87/243	
3533	722	3025	IC14		
3533-1	129	3025	21	TR88/242	
3533(RCA)	129	3025		TR88/242	
3534(RCA)	160		51	TR17/637	
3535(RCA)	161	3117	17	TR83/719	S0017
3535-110-50008	732				
3535-110-50009	732				
3536(2CA)	123A		63	TR74/	S3019

Original Part Number	ECG	SK	GE	IR/WEP	HEP
3537(RCA)	108	3132	17	TR95/56	
3538(RCA)	123A	3122	20	TR21/735	
3539(RCA)	108	3039	17	TR95/56	S0015
3539-307-001	108	3018		TR95/	
3539-307-002	108	3018		TR95/	
3540(RCA)	159	3005		TR20/717	S0019
3540-1(GE)	159	3114	22	TR52/	S0019
3541(RCA)	123A	3122	18	TR51/735	
3543(RCA)	123A	20	20	TR21/735	
3544	102A	3005	2	TR85/	G0005
3544(RCA)	128	3005		TR87/243	
3544-1	123A	1	17	TR21/735	S0015
3545(RCA)	154	3045		TR78/	S5025
3546-1(RCA)	123A			TR21/	
3546-2(RCA)	123A			TR21/	
3546(RCA)	123A	3004	20	TR21/735	S0015
3547(RCA)	225	3045	27	TR78/712	
3548(RCA)	123A	3122	20	TR51/735	
3549(RCA)	159	3114	21	TR20/717	
3549-1(RCA)	159	3118		TR20/	S0013
3549-2(RCA)	159				
3550		3004			
3551	126	3122	20	TR17/635	G0008
3551A	126	3007		TR17/635	G0008
3551A BLU	126			TR17/635	G0008
3551A GRN	126			TR17/635	G0008
3551A(RCA)	123A	3007		TR21/735	
3552	225	3045	20		S5025
3552(RCA)	154	3045		TR78/712	
3552-1(RCA)	191			TR75/	
3553	225	3045	18		G0008
3553(RCA)	154	3045		TR78/712	
3554(RCA)	123A	3122	18	TR21/735	
3555-1(RCA)	128	3044			
3555(RCA)	123A	3124	18	TR25/240	S5014
3558(RCA)	123A	3122	20	TR21/735	
3559(RCA)	159	3114	21	TR20/717	S0019
3560		3122	20	TR21/735	
3560-1(RCA)	107			TR70/	S0015
3560-2	123A	3122		TR87/	S0015
3561(RCA)	123A	3007	20	TR51/735	S0015
3561-1(GE)	128		18		
3561-1(RCA)	123A	3122	18	TR21/	S0015
3562(RCA)	159	3114	21	TR20/717	
3563(RCA)	159	3114	21	TR20/717	S0019
3564(RCA)	162	3004		TR67/707	
3565	124	3007	63	TR81/240	
3565(RCA)	128	3122		TR87/243	S0015
3566	124	3021	63	TR81/240	
3566(RCA)	128	3011		TR87/243	
3567(RCA)	124	3054	12	TR81/240	
3567-2(RCA)	157	3041		TR60/	S5000
3568(RCA)	108	3005	18	TR95/56	S0017
3568(WARDS)	108			TR24/	
3569	161	3122		TR83/	
3569(RCA)	123A	3122		TR21/735	
3570P	159			TR20/717	S0019
3570(RCA)	159	3004	21	TR20/717	S0019
3571-1	161	3114	20	TR24/	S0030
3571R	123A			TR21/719	S0030
3571(RCA)	123A	3010		TR21/735	S0030
3572-3	108	3124	20	TR21/	S5014
3572(RCA)	123A	3124	20	TR21/735	S0020
3574	159	3114	20	TR20/717	S0019
3574-1	129	3114	22	TR88/	S0019
3574-1(RCA)	159	3114		TR20/	S0019
3575		3005			
3576	233	3132	61	TR83/	
3576(RCA)	108	3132		TR83/56	S0017
3577-1	123A	3018	20		S0030
3577(RCA)	123A	3009	16	TR21/735	S0030
3578-1	102A			TR85/250	G0005
3579	161	3117		TR83/719	
3579(RCA)	161	3117		TR83/	
3581(RCA)	159	3114	21	TR20/717	
3582	225	3045	27	TR78/712	
3582(RCA)	154	3045		TR78/712	
3583(RCA)	127			TR27/235	
3584(RCA)	124			TR81/240	
3586(RCA)	123A			TR21/	S0020
3588	123A	3050		TR21/735	
3588(RCA)	222	3050			S0016
3589	123A		20	TR21/735	S0014
3590	129	3025	53	TR88/242	S0019
3591	128	3024	53	TR87/243	S0012
3592(RCA)	129			TR88/	
3593		3004			
3595		3006			
3597	159	3118		TR20/717	S0013
3597-1	159	3118	21		S0013
3597-1(RCA)	159	3118		TR88/	S0013
3597-2	159	3025	21	TR88/	S0013
3598(RCA)	108			TR95/	
3598-2		3004			
3600	102	3004	52	TR05/631	G0005

Original Part Number	ECG	SK	GE	IR/WEP	HEP
3601	128	3122	18	TR21/735	S0015
3601(RCA)	107			TR70/	S0015
3601-1	123A		17	TR21/735	S0015
3601-1(RCA)	107			TR70/	S0015
3603(RCA)	108	3005			S5014
3604(RCA)	108	3007	64	TR95/56	S0011
3607	101	3011		TR08/641	G0011
3608-1(RCA)	190			TR74/	S5014
3608-2(RCA)	128	3024		TR74/	S5014
3609	101	3011		TR08/641	G0011
3610(RCA)	108		11	TR95/	S0016
3610				TR21/	
3612(RCA)	152			TR76/	
3613	191	3007		TR75/	
3613-2(GE)	191	3104	27	TR60/	
3613-3	191	3104		TR75/	S3019
3613-3(RCA)	191	3104		TR75/	S3019
3613(RCA)	171	3007		TR75/	
3614-1		3122			
3614A		3004			
3615(RCA)	128	3024		TR87/	S5014
3616		3025			
3616-1(RCA)	129	3025	22	TR88/242	S0019
3617		3019			
3618	221	3050		TR21/	
3618(RCA)	108			TR95/56	F2004
3618-1	121			TR01/	S5014
3620	159	3114		TR20/	
3620-1	159	3114	21	TR20/717	S0013
3621(RCA)	152			TR76/	
3622(RCA)	128	3024		TR87/	
3622-1	128		28	TR87/243	S5014
3622-2(RCA)	128	3122			
3623		3004			
3623-1		3054			
3624-1		3083			
3624-2		3054			
3625(RCA)	123A			TR21/	S0030
3627(RCA)	159	3114		TR20/	
3628		3011			
3628-2(RCA)	171	3201		TR79/	
3628-3(RCA)					S5006
3531	159		21	TR20/	
3631-1	123A	3044	27	TR21/	
3631-1(RCA)	128	3044		TR87/	S5014
3632-1	196	3104	27	TR60/	
3632-2(RCA)	186	3054		TR55/	
3633		3104			
3633-1	710				
3633-1(GE)	171	3104		TR60/	
3634		3111			
3634-1	163	3111	36	TR93/	S5021
3634.0011	123A				
3634.2011	159				
3635(RCA)	222	3050			
3635-1	222				
3636	124		12	TR81/	S5011
3637(RCA)	162	3004		TR67/	
3637-1		3079			
3640(RCA)	196	3006		TR92/	
3640-1		3054			
3645		3008			
3646-2(RCA)	108			TR95/	
3646-8		3009			
3647		3111			
3648(RCA)	127			TR27/	
3648-1		3035			
3652		3018			S0016
3657		3003			
3657-1	108	3018	20		
3657-2	108	3018	20		
3658		3005			
3662-2	165				
3663		3004			
3664		3008			
3665		3008			
3665-2		3027			
3666(RCA)	128	3008		TR87/	
3668		3005			
3669	165				
3671		3010			
3672		3008			
3673		3005			
3676	233	3004	60	TR71/	
3677		3076			C6061P
3677-1	780				
3677-1(RCA)	711	3141			
3677-2(RCA)	780	3141			
3679		3008			
3680		3117			
3681-1	904	3054			
3681-2		3054			
3682(RCA)	160	3122	51	TR17/637	
3683(RCA)		3114			
3683-1		3027			

Original Part Number	ECG	SK	GE	IR/WEP	HEP
3686(RCA)	102A	3072	53	TR85/	
3687		3044			S5025
3693(ARVINE)	108			TR95/56	
3695		3006			
3696A		3005			
3697A		3005			
3698		3007			
3700-300		3027			
3701(RCA)		3007			
3706		3008			
3707		3008			
3708		3008			
3709		3008			
3711		3006			
3714H1	130	3027		TR59/247	S7002
3714H1A	130	3027		TR59/	
3716		3004			
3718		3005			
3721	175				
3723A		3004			
3724		3007			
3726		3010			
3728		3024			
3730		3005			
3738		3008			
3739		3004			
3742	188			TR72/	
3744		3004			
3745		3008			
3746	100	3005	52	TR05/254	G0009
3746-00	129				
3746-01	129				
3747		3005			
3749		3004			
3750	160	3005			
3751		3007			
3752		3007			
3753		3007			
3757		3004			
3759		3003			
3760		3008			
3763		3007			
3765		3004			
3767		3003			
3768		3005			
3770		3039			
3771	181				
3771-1	181				
3771-2	181				
3774		3004			
3775		3010			
3776		3008			
3778		3004			
3782		3003			
3783		3003			
3786		3008			
3787		3010			
3791		3004			
3792		3004			
3799		3008			
3801		3007			
3804		3004			
3806		3005			
3813		3004			
3815		3009			
3816		3004			
3818		3007			
3819(RCA)	132	3116		FE100/802	
3820		3007			
3824		3008			
3825		3008			
3826		3008			
3830		3004			
3831		3005			
3835		3005			
3841 GX		3008			
3842		3011			
3843(HAR KAR)	130			TR59/	
3843(SEARS)	196			TR92/	
3844(HAR KAR)	128			TR87/	
3846		3009			
3847		3024			
3851	126	3008	2	TR17/635	G0005
3852	100	3004	2	TR05/254	G0005
3856		3004			
3857		3003			
3858		3004			
3859		3007			
3860		3005			
3864		3003			
3865		3003			
3867	123A			TR21/	S0014
3868	195			TR65/	S3001
3869		3004			
3872		3005			

Original Part Number	ECG	SK	GE	IR/WEP	HEP
3873		3005			
3875		3004			
3876		3004			
3877		3003			
3878	154			TR78/	S5026
3879		3004			
3881	161	3039		TR83/	S0017
3885		3004			
3887		3005			
3888		3004			
3889		3008			
3890		3008			
3891		3008			
3892		3003			
3895		3005			
3899		3004			
3907	102	3004		TR05/631	G0005
3907/2N404A	126	3114	51	TR17/	
3959		3114			
3961	126	3006	50	TR17/637	
3961(GE)	160	3006		TR17/637	G0003
3970	100	3004	52	TR05/254	
3970CL	102A	3004		TR85/	
3970(GE)	102A			TR85/250	G0005
3980-002	175				
3991-303-112	785				
3999	123A	3122		TR21/735	S0015
4001	154	3047		TR78/	S5026
4001-222	160			TR17/	G0003
4001-223	160			TR17/	G0003
4001-224	102A			TR85/	G0005
4001-225	102A			TR85/	G0005
4001-226	102A			TR85/	G0005
4001-228	190			TR74/	S3019
4001(MODULE)	1001				
4002	195			TR65/	
4002(PENN)	123A	3124		TR21/735	S0015
4004(JOHNSON)	128	3047	63	TR87/243	S5026
4004(PENN)	102A			TR85/250	G0005
4005(JOHNSON)	195	3048		TR65/	S3001
4005				TR87/243	
4006(JOHNSON)	128	3024	63	TR87/	S0014
4007	195	3048		TR65/	
4008(JOHNSON)	128	3024		TR87/243	S3008
4009	195	3004		TR65/	S3001
4009(PENN)	158	3004		TR84/630	G0005
4010(JOHNSON)	108	3088	20	TR95/56	S3005
4011	195			TR65/	S3001
4011(JOHNSON)	195			TR65/	
4012(JOHNSON)				TR65/	S3001
4013	159	3114		TR20/717	S0019
4014-000-10160	102A	3004		TR85/	
4015					S3007
4020	785				
4021	123A	3122		TR21/735	S0014
4022	123A	3122		TR21/735	S0014
4032P-8M4	1011				
4036	129/400				
4039-00	199				
4039-01	199				
4041-000-10160	102A		2	TR85/	G0005
4041-000-20120	102A	3004		TR85/	G0007
4041-000-30180	102A	3004	2	TR85/	G0005
4041-000-40270	107		17	TR70/	S0016
4041-000-40300	107	17	17	TR70/	S0016
4041-000-60170	107		20	TR70/	S0011
4041-000-60200	107		20	TR70/	S0011
4041-000-80100	126		51	TR17/	G0009
4046(SEARS)	123A			TR21/	
4050	755				C6013
4057	123A	3124	20	TR21/735	
4066	123A	3122		TR21/735	S0015
4075-1025-4		3005			
4080-187-0507	128	3024		TR87/	
4080-838-0001	175	3026		TR81/	
4080-838-2	175	3026		TR81/	
4080-838-3	175	3026		TR81/	
4080-866-0006	196	3041		TR92/	
4080-866-1	175	3026		TR81/	
4080-866-2	175	3026		TR81/	
4080-879-0001	196	3041		TR92/	
4080A		3004			
4082-501-0001	121	3009		TR01/	
4085	123A	3122		TR21/735	S0015
4086	159	3114		TR20/717	S0031
4087	159	3114		TR20/717	S0031
4096-204-1B		3005			
4096-3003		3005			
4096-3006		3005			
4096-3038		3005			
4099A	189			TR73/	S3028
4117-1	949				
4117-2	949				
4126(JOHNSON)	159	3114	21	TR20/717	
4150-01	123A				

Original Part Number	ECG	SK	GE	IR/WEP	HEP
4151-01	159				
4167(AIRLINE)	161	3018	20	TR83/719	
4167(PENN)	161			TR83/	S0016
4167(SEARS)	108			TR95/56	
4168(PENN)	161	3018	20	TR83/719	S0016
4168(SEARS)	108	3018		TR95/56	
4168(WARDS)	161			TR83/719	S0002
4168(WELLS)	161			TR83/	
4169(PENN)	161	3018		TR83/719	
4169(SEARS)	108			TR95/56	
4169(WARDS)	161			TR83/719	S0030
4169(WELLS)	161			TR83/	
4216	130				
4216(HAR KAR)	130			TR59/	
4218	192				
4218(HAR KAR)	184				
4219	129				
4219(HAR KAR)	185				
4247	104	3009	16	TR01/250	G6003
4295-1	756				C6014
4295-2	756				C6012
4295-2(RCA)	762				
4306(HAR KAR)	128			TR87/	
4309(AIRLINE)	123A			TR21/735	S0015
4310(AIRLINE)	159	3114		TR20/717	S0019
4310				TR30/	
4312(RCA)	124		12	TR81/240	
4313	102	3004		TR05/631	G0006
4315	102	3004	52	TR05/631	G0005
4322-542	123A				
4331	121	3009		TR01/232	G6018
4347	104	3009	3	TR01/624	G6003
4348	102	3004	52	TR05/631	G0005
4349	102	3004	2	TR05/631	G0005
4363	126	3008		TR17/635	G0008
4363BLU	126			TR17/635	G0008
4363GRN	126			TR17/635	G0008
4363ORN	126			TR17/635	G0008
4363WHT	126			TR17/635	G0008
4364	126	3006		TR17/635	G0008
4365	126	3008		TR17/635	G0008
4366	126	3005	2	TR17/635	G0009
4367	126	3005	2	TR17/635	G0009
4367-001	129	3025			
4368	160	3123		TR17/637	G0003
4381P1	9914				
4398	102	3123		TR05/631	G0005
4419-4	180			TR29/S7001	S7003
4437-1(RCA)	787	3146			
4437-2(RCA)	787	3146			
4437-3(RCA)	787	3146			
4438-1(RCA)	788	3147			
4438-2(RCA)	788	3147			
4438-3(RCA)	788	3147			
4439-1(RCA)	789	3078			
4439-2(RCA)	789	3078			
4442	159	3114		TR20/717	S5022
4442-3	188			TR72/W3020	
4442-3CL		3041			
4442-366	196			TR92/	
4450	102	3004	52	TR05/631	G0006
4451	100	3004	52	TR05/254	
4454	126	3006		TR17/635	G0009
4456	126	3006		TR17/635	G0009
4457	126	3006		TR17/635	G0009
4459	127		25	TR27/235	G6008
4460		3078			
4460-1		3078			
4460-2	789	3078			
4460-3		3078			
4462	102	3004	52	TR05/631	G0005
4463	121	3009	52	TR01/230	G6003
4464	123	3124	52	TR86/53	S0014
4465	123	3124	52	TR86/53	S0014
4466				TR05/	G0005
4466ORN	102	3004		TR05/631	G0005
4466YEL					G0005
4467	176	3123	52	TR82/	
4468BRN	102	3004		TR05/631	G0005
4469RED	102	3004		TR05/631	G0005
4470	123	3124	20	TR86/53	S0015
4470-31	123			TR86/53	S0015
4470-32	123	3124		TR86/53	S0015
4470-33				TR86/53	S0015
4470M-32	123	3124	20	TR86/53	
4471ORN	102	3004		TR05/631	G0005
4471YEL	102	3004		TR05/631	G0005
4472				TR05/	
4472GRN	102	3004		TR05/631	G0005
4473	102A	3123		TR85/250	G0005
4473-1	108	3039		TR95/56	S0015
4473-2	108	3039		TR95/56	S0015
4473-3	108	3039	20	TR95/56	S0015
4473-4	123A	3124	20	TR21/735	S0015
4473-5	123A	3124		TR21/735	S0015

Original Part Number	ECG	SK	GE	IR/WEP	HEP
4473-5X	123A	3124		TR21/735	S0015
4473-6	108	3039		TR95/56	S0015
4473-7	108	3039		TR95/56	S0015
4473-8	108	3039		TR95/56	S0015
4473-9	123A	3122		TR21/735	S0015
4473-11	108	3039		TR95/56	S0015
4473-12	123A	3124	20	TR21/735	S0015
4473-M3	128	3024		TR87/243	
4473-M-3			20	TR21/	
4473-M-12	128	3024		TR87/243	
4473-N	128	3024	20	TR87/243	
4474				TR05/	
4474YEK		3123			
4474YEL	102	3004		TR05/631	G0005
4475				TR05/	
4475GRN	102	3004		TR05/631	G0005
4476				TR05/	
4476BLU	102	3004		TR05/631	G0006
4477				TR05/	
4477PUR	102	3004		TR05/631	G0006
4477V10	102A	3123		TR85/	G0006
4478	129	3025		TR88/242	S5023
4483	128	3024		TR87/243	S0015
4484	100	3005		TR05/254	G0003
4484-1	106			TR20/52	S0013
4478-2	106			TR20/52	S0013
4485	100	3005		TR05/254	G0003
4485-1	106			TR20/52	S0013
4486	100	3005		TR05/254	G0003
4490	128				
4490-1	108				
4490-2					S5014
4490-7	152				
4491-4	130				
4491-5	152				
4491-6	184	3041	57	/S5003	S5000
4491-7	130				
4491-8	152				
4491-9	184	3041	57	/S5003	S5000
4501	126	3006		TR17/635	G0009
4509	126	3006		TR17/635	G0009
4510	102	3004		TR05/631	G0006
4542				TR05/	
4545	126	3006		TR17/631	G0008
4545BLU	126			TR17/631	G0008
4545WHT	126			TR17/631	G0008
4553				TR05/	
4553BLU	102	3004		TR05/631	G0005
4553BRN	102	3004		TR05/631	G0005
4553GRN	102	3004		TR05/631	G0005
4553ORN	102	3004		TR05/631	G0005
4553RED	102	3004		TR05/631	G0005
4553V10	102A	3123		TR85/	G0007
4553YEL	102	3004		TR05/631	G0005
4562	102	3004	2	TR05/631	G0005
4563	102	3004	2	TR05/631	G0006
4564	102	3004	2	TR05/631	G0006
4565	102	3004	2	TR05/631	G0005
4567	102	3004	2	TR05/631	G0005
4570	121			TR01/230	G6003
4573	105	3012		TR03/233	G6006
4582					
4582BRN	121	3009		TR01/628	G6013
4583				TR01/	
4583RED	121	3009		TR01/628	G6013
4584				TR01/	
4584GRN	121	3009		TR01/628	G6013
4586	126			TR17/631	G0008
4587	108	3018	17	TR95/56	
4589	126	3006		TR17/635	G0008
4590	106			TR20/52	S5022
4594	123			TR86/53	S0014
4595	126	3006		TR17/635	G0009
4596	102	3004		TR05/631	G0005
4597	105	3012		TR03/233	G6006
4597GRN	105	3012		TR03/233	G6006
4597RED	105	3012		TR03/233	G6006
4603	126	3006		TR17/635	G0009
4604	126	3006		TR17/635	G0009
4605	126	3006		TR17/635	G0008
4605RED	126			TR17/635	G0008
4607	102	3004		TR05/631	G0005
4608	121	3009		TR01/628	G6013
4619				TR01/	
4619RED	121	3009		TR01/628	G6013
4620				TR01/	
4620GRN	121	3009		TR01/628	G6013
4621	126	3006		TR17/635	G0008
4622	105	3012		TR03/231	G6004
4623	127			TR27/235	G6007
4624	123			TR86/53	S0014
4627	102	3004		TR05/631	G0005
4630	123			TR86/53	S0014
4632	126	3008		TR17/635	G0008
4640	179			TR35/G6001	G6009

Original Part Number	ECG	SK	GE	IR/WEP	HEP
4640P	179			TR35/G6001	G6009
4648	154	3045		TR78/712	S5025
4649	121	3009		TR01/230	G6003
4652	127			TR27/235	G6015
4677	126	3006		TR17/635	G0008
4684-120-3	108	3018		TR95/	
4685-285-3	108	3019		TR95/	
4686-81-2	102A	3004		TR85/	
4686-81-3	102A	3004		TR85/	
4686-82-3	123A	3124		TR21/	
4686-95-3	108	3018		TR95/	
4686-107-3	108	3018		TR95/	
4686-108-3	108	3018		TR95/	
4686-112-3	108	3018		TR95/	
4686-114-3	108	3018		TR95/	
4686-118-3	108	3018		TR95/	
4686-119-3	108	3018		TR95/	
4686-120-3	108	3018		TR95/	
4686-126-3	108	3018		TR95/	
4686-127-3	108	3018		TR95/	
4686-130-3	175	3026		TR81/	
4686-131-3	108	3018		TR95/	
4686-132-3	123A	3124		TR21/	
4686-140-3	108	3018		TR95/	
4686-143-3	123A	3124		TR21/	
4686-144-3	123A	3124		TR21/	
4686-145-3	194	3040			
4686-163-3	102A	3004		TR85/	
4686-169-3	108	3018		TR95/	
4686-170-3	129	3025		TR88/	
4686-171-3	108	3018		TR95/	
4686-172-3	108	3018		TR95/	
4686-173-3	123A	3124		TR21/	
4686-182-3	128	3024		TR87/	
4686-183-3	123A	3124		TR21/	
4686-195-3	102A	3004		TR85/	
4686-196-3	102A	3004		TR85/	
4686-207-3	108	3018		TR95/	
4686-208-3	108	3018		TR95/	
4686-209-3	108	3018		TR95/	
4686-210-3	225	3045			
4686-213-3	121	3009		TR01/	
4686-224-3	108	3018		TR95/	
4686-226-3	175			TR81/	
4686-228-3	108	3018		TR95/	
4686-229-3	129	3025		TR88/	
4686-230-3	129	3025		TR88/	
4686-231-3	123A	3124		TR21/	
4686-232-3	154	3040		TR78/	
4686-234-3	196	3041		TR92/	
4686-235-3	129	3025		TR88/	
4686-238-3	129	3025		TR88/	
4686-244-3	108	3018		TR95/	
4686-251-3	108	3018		TR95/	
4686-252-3	191	3044		TR75/	
4686-256-1	126	3008		TR17/	
4686-256-2	126	3008		TR17/	
4686-256-3	126	3008		TR17/	
4686-257-3	123A	3124		TR21/	
4689	128	3024		TR87/243	
4700	101	3011		TR08/641	G0011
4701	130	3027		TR59/247	S7004
4701A		3027			
4702	179			TR35/G6001	G6009
4705	123			TR86/53	S0014
4706	108	3039		TR95/56	S0020
4709	108	3039		TR95/56	S0025
4714	123	3124		TR86/53	S0014
4715	130	3027		TR59/247	S7004
4715A	130	3027		TR59/	
4722	121			TR01/	G6013
4722BLU	121	3009		TR01/624	G6013
4722GRN	121	3009		TR01/624	G6013
4722ORN	121	3009		TR01/232	G6013
4722PUR	121	3009		TR01/232	G6013
4722RED	121	3009		TR01/232	G6013
4722YEL	121	3009		TR01/624	G6013
4727	121	3009		TR01/230	G6003
4729	179			TR35/	G6009
4730	121	3009		TR01/232	G6015
4732	123A	3122		TR21/735	S0024
4733	123A	3122		TR21/735	S0024
4734	123A	3124	20	TR21/735	S0015
4737	123A	3122		TR21/735	S0014
4745	196	3114		TR20/717	S5022
4756	161	3039		TR83/719	S0017
4765	123A	3122		TR21/735	S0014
4768	123A	3122		TR21/735	S0015
4781	197				
4789	107	3039		TR70/720	S0033
4792	6400				S9002
4793(KLH)	722	3161			
4794(KLH)	781	3169			
4795	726				
4795(KLH)	726				

Original Part Number	ECG	SK	GE	IR/WEP	HEP
4800-200	102A			TR85/250	
4800-220	102A			TR85/	G0005
4800-221	102A			TR85/250	G0007
4800-222	102A			TR85/250	G0005
4800-223	131MP			TR94/642MP	G6016
4801-1100-011	121	3009	16	TR01/232	
4802-00002	108			TR95/56	
4802-00003	123A			TR21/735	
4802-00004	159	3114		TR20/717	
4802-00005	130	3027		TR59/247	S7002
4802-00005A	130	3027		TR59/	
4802-00006	123A			TR21/735	S0015
4802-00007	195	3024		TR65/243	
4802-00008					S3002
4802-00009	108			TR95/	S0016
4802-00010	132			FE100/	F0015
4802-00012	108			TR95/	S0016
4802-00014	123A			TR21/	S0014
4802-00015	123A			TR21/	S0014
4802-00016	188	3199		TR72/	
4802-00017	195			TR65/	
4802-00019					
4805-00329		3014			
4815	159	3114		TR20/717	S0019
4819	154	3044		TR78/712	S5025
4820	108	3039		TR95/56	S0020
4821	108	3039		TR95/56	S0025
4822-130-40095	158		53	TR84/630	
4822-130-40096	103A		54	TR08/724	
4822-130-40132	130			TR59/	
4822-130-40184	123A	3122	20	TR21/735	
4822-130-40213	131	3082		TR94/642	
4822-130-40214	107		20	TR70/720	
4822-130-40215	107		20	TR70/720	
4822-130-40216	107		20	TR70/720	
4822-130-40233	104		3	TR01/624	
4822-130-40235	102A		53	TR85/250	
4822-130-40236	102A			TR85/	
4822-130-40252	160		51	TR17/637	
4822-130-40255	160		51	TR17/637	
4822-130-40304	161			TR83/	
4822-130-40311	107		20	TR70/720	
4822-130-40312	107		20	TR70/720	
4822-130-40313	107		20	TR70/720	
4822-130-40314	103A			TR08/	
4822-130-40315	159	3114	21	TR20/717	
4822-130-40317	107		20	TR70/720	
4822-130-40318	107		20	TR70/720	
4822-130-40319	103A			/724	
4822-130-40329	103A			/630	
4822-130-40343	123A		20	TR21/	
4822-130-40354	123A			TR21/	
4822-130-40356	128	3024	18	TR87/243	
4822-130-40361	123A			TR21/	
4822-130-40369	106			TR20/	
4822-130-40382	102A				
4822-130-40385	131		51	TR17/637	
4822-130-40441	160		50	TR17/637	
4822-130-40454	123A			TR21/	
4822-130-40456	158		53	TR84/630	
4822-130-40477	106		21	TR20/52	
4822-130-40508	106			TR20/	
4822-130-40537	184	3190			
4822-130-40614	106			TR20/	
4823-0018	175				
4823-0031-01	234				
4824-0014	199				
4824-0014-02	199				
4824-28		3024			
4824-33	128				
4825	107	3018	17	TR70/720	S0033
4826	108	3039		TR95/56	S0025
4837	108	3039		TR95/56	S0020
4838	154	3045		TR78/712	S5026
4839	123A	3122		TR21/735	S0015
4840	123A	3122		TR21/735	S0015
4841	123A	3122		TR21/735	S0015
4842	123A	3122	17	TR21/735	S0015
4843	154	3045		TR78/712	S5025
4844	159	3114		TR20/717	S0014
4845	108	3039		TR95/56	S0020
4846-1	949				
4846-2	949				
4851	107		11	TR70/720	
4852	123A	3122		TR21/735	S0015
4853	154	3044		TR78/712	S5025
4854	123A	3122	20	TR21/735	S0015
4855	108	3039		TR95/56	S0015
4856-0101	123A				
4856-0106	159				
4856-0107	123A				
4856-0109	123A				
4856-0110	123A				
4857	108	3018	17	TR95/56	S0020
4860-1-3	128				

Original Part Number	ECG	SK	GE	IR/WEP	HEP
4872	124	3021		TR81/240	S5011
4882	130	3027		TR59/247	S7004
4888A	121	3027		TR01/232	G6005
4888B	121	3027		TR01/232	G6005
4907-976	102A	3004		TR85/	
5000	761				
5001	6400				S9002
5001-002		3018			S0016
5001-014		3122			S0015
5001-020		3018			S0015
5001-021		3018			S0014
5001-032		3018			
5001-038		3038			S0015
5001-044		3054			
5001-046		3116			F2005
5001-047		3116			F0021
5001-048		3114			S0013
5001-050		3054			S3024
5001-053		3054			
5001-064		3041			
5001-070				TR21/	
5001-072		3122			
5001-075		3054			
5002(THOMAS)	754				C6007
5002-001	724				
5005	761				
5005(THOMAS)	761				C6011
5006(THOMAS)					C6011
5007(THOMAS)	761				C6011
5009(THOMAS)	767				C6015
5036-1	223				
5036-2	223				
5052	102			TR05/	
5059-0236	159				
5085	160		51	TR17/637	
5093	108	3039		TR95/56	S0015
5096	132	3116		FE100/802	F0015
5113T	704	3023			
5118	219				S7003
5158	797				
5158-1	797				
51588(RCA)	797				
5226-1	159	3114		TR20/717	S0019
5226-2	123A	3122		TR21/735	S0015
5233-7100	7411				
5253	104	3009	3	TR01/624	
5259	196				
5313-461B	108				
5320-003	175	3026		TR81/	
5380-71	123A	3124		TR21/735	S0014
5380-72	123A			TR21/	S0015
5380-73	128	3027		TR87/241	S5014
5380-73(POWER)	175			TR81/	
5459	133				
5464	102A		2	TR85/	
5489(KLH)	718	3159			
5504	181				
5505	181				
5565-001	130				
5596		3004			
5608-1	814				
5611-628	159				S0019
5611-628F	159	3114	21	TR30/	S0019
5611-673	159	3114	21	TR20/	
5611-673D	159	3025	21	TR20/	S5013
5611-695	193	3025	67	TR28/	
5611-695C	193	3025	67	TR28/	S5013
5612-75C	102A	3004	1		G0005
5612-77C	102A	3004	54		
5612-370	102A	3004	52	TR85/	
5612-370C	102A			TR85/	G0006
5613-46B	108				
5613-458	123A	3122	17	TR21/	
5613-458B	123A	3122	17	TR24/	S0030
5613-458C	123A	3122		TR24/	S0014
5613-458D	123A	3122		TR21/	S0014
5613-458LGC	123A	3020		TR21/	S0014
5613-460	107	3018	20	TR70/	
5613-460A	107	3018		TR70/	S0014
5613-460B	107	3018		TR70/	S0014
5613-460C	107	3018	20	TR70/	S0014
5613-461	107	3018	20	TR70/	
5613-461(B)			11		S0014
5613-461(C)				TR21/	
5613-461A		3018			
5613-461B		3018			
5613-461C	107	3018	20	TR70/	S0014
5613-535	107	3018	20	TR70/	
5613-535(A)				TR21/	
5613-535(B)			20	TR21/	
5613-535(C)				TR21/	
5613-535A	107	3018		TR70/	S0016
5613-535B	107	3018		TR70/	S0016
5613-535C	107	3018	20	TR70/	S0016
5613-558C	123A				

Original Part Number	ECG	SK	GE	IR/WEP	HEP
5613-711	123A				S0014
5613-711E	123A	3122	10	TR24/	S0014
5613-870	123A				S0033
5613-870F	123A	3124	20	TR21/	S0033
5613-871	199				S0024
5613-871F	199	3122	62	TR51/	S0024
5613-1209	192	3024	63	TR25/	
5613-1209A		3024			
5613-1209B		3024			
5613-1209C	192	3024	63	TR25/	S5014
5613-1213		3024			
5613-1213A		3024			
5613-1213B		3024			
5613-1213C		3024			
5613-1213D	128	3024		TR25/	S5014
5613-1335	123A	3038	17	TR21/	
5613-1335D	123A			TR24/	S0025
5613-1335E		3038			S0025
5613-1342	107	3018	20	TR70/	
5613-1342A		3018			
5613-1342B		3018			
5613-1342C	107	3018	20	TR70/	S0014
5613-4581		3122			
5613-4581C	123A		17	TR21/	
5614-77C	103A	3010			
5680	129			TR88/	
5701-88		3008			
5701-89		3008			
5721	123A				
5723P	801				C6096P
5724P	801				C6096P
5757	781	3169			
5766-25	102	3004		TR05/631	
5815	723	3144			
5847(RCA)	184				
5847			66	/S5003	
5909-001	181				
5932	789				
6000-70				TR21/	
6000-72				TR21/	
6001-63				TR51/	
6001-64				TR21/	
6001-65				TR24/	
6001-71				TR25/	
6001-72				TR21/	
6001-78				TR25/	
6001-79				TR24/	
6001-80				TR25/	
6001-82				TR25/	
6001-83				TR25/	
6001-84				TR08/	
6001-85				TR25/	
6001-86				TR25/	
6002-22				TR02/	
6002-26				TR57/	
6002-27				TR70/	
6002-28				TR76/	
6002-30				TR27/	
6002-31				TR27/	
6002-43				TR76/	
6002-46				TR23/	
6002-49				TR27/	
6008		3006	50	TR17/637	
6008(JOHNSON)	160			TR17/	
6015-12				TR21/	
6015-15				TR70/	
6015-16				TR70/	
6015-17				TR70/	
6015-18				TR24/	
6015-19				TR53/	
6015-20				TR21/	
6015-21				TR51/	
6015-22				TR33/	
6015-23				TR51/	
6015-24				TR24/	
6015-25				TR53/	
6015-26				TR52/	
6015-27				TR70/	
6015-28				TR78/	
6015-29				TR70/	
6015-30				TR25/	
6015-37				TR70/	
6015-41				TR53/	
6015-42				TR24/	
6015-43				TR24/	
6015-54				TR33/	
6015-62				TR24/	
6016-11				TR82/	
6016-15				TR52/	
6016-16				TR51/	
6016-17				TR21/	
6016-26				TR24/	
6016-27				TR24/	
6016-47				TR52/	
6016-50				TR82/	

Original Part Number	ECG	SK	GE	IR/WEP	HEP
6016-64				TR54/	
6017-2				FE100/	
6017-3				TR69/	
6033-31					S7003
6033-61					S7003
6099-2	196	3054		TR92/	
6100	158			TR84/630	G0005
6100-35	158	3004	2	TR84/630	
6136	123A			TR21/	
6151(RCA)	128	3024		TR87/243	S0015
6154	126	3006	50	TR17/635	
6155	126	3006	50	TR17/635	
6158	108	3018	50	TR95/56	
6158-3(GE)	108			TR95/56	S0015
6162	126	3006	50	TR17/635	
6181-1	154	3045		TR78/712	S5025
6185-3	108			TR95/56	
6221	221				F2004
6284	128				
6285	129				
6313	160			TR17/637	G0003
6343-1	123A	3122		TR51/	S0005
6365(GE)					S0003
6367	123A				S0005
6367-1	123A	3122		TR51/	S0005
6377-1(SYL)	131				
6377-2(SYL)	155				
6380-1				TR36/S7000	S7002
6380-3	152	3054			
6381-1	175			TR81/	S5019
6381-2	152	3054			
6382-2	152				
6440	102	3004	2	TR05/631	G0005
6445	100	3004	2	TR05/254	G0005
6452	102	3004	2	TR05/631	G0005
6507				TR21/	
6507(AIRLINE)	108			TR95/56	
6531					S0015
6534-5T6		3024			
6551	722	3161			
6640-831016		3004			
6651-486	128	3024	18	TR87/243	
6818	105	3012		TR03/	
6854K90-062	159				
6855K90	106				
6900K91-002	9944				
6900K91-003	9946				
6900K91-004	9962				
6900K91-006	9936				
6900K91-007	9932				
6900K91-009	9157				
6900K93-001	9951				
6900K93-002	9941				
6900K95-001	9099				
6900K95-002	9093				
6954K90-074	123A				
6990	160	3006		TR17/637	G0003
7001(DARL)	172			TR21/	
7001(JOHNSON)	123A		63	TR21/	S9100
7005G				TR51/	
7005G(LOWREY)	123A	3122		TR21/735	
7106				TR21/	
7107				TR21/	
7112(RCA)		3018			
7113	108	3018		TR95/56	S0016
7115	108	3018		TR95/56	S0015
7115(IC)	708				
7115(STELMA)	708				
7116	108	3018		TR95/56	S0016
7117		3018			
7118	108	3018		TR95/56	S0015
7120A(GE)	712	3072			
7122	108	3039		TR95/56	S0015
7122-5	123A			TR21/	
7123	108	3039		TR95/56	S0015
7124	108	3039		TR95/56	S0015
7125	108	3039		TR95/56	S0015
7126	108	3039		TR95/56	S0015
7127	108	3039		TR95/56	S0015
7128	108	3039		TR95/56	S0015
7129	123A	3124	20	TR21/735	
7130		3072			
7130A		3072	IC2	IC507/	
7131	108	3039		TR95/56	S0015
7132	108	3039	18	TR95/56	S0015
7133	108	3039		TR95/56	S0015
7134	108	3039		TR95/56	S0015
7136				TR22/	
7149A	712				
7161	752				
7161(WURL)	752				C6003
7171		3018			
7172		3018			
7172-54	105	3012		TR03/233	
7173	108	3018		TR95/56	S0020

Original Part Number	ECG	SK	GE	IR/WEP	HEP
7174	108	3018		TR95/56	S0016
7175	108	3018		TR95/56	S0016
7176	108	3018		TR95/56	S0016
7177	108	3018		TR95/56	
7178	108	3018		TR95/56	S0011
7204A	732				
7204A(SEARS)	732				
7213	152	3031			
7214	108	3039	14	TR95/56	S0020
7214A	130	3027		TR59/	
7214(LOWREY)	130			TR59/	
7215	108	3039		TR95/56	S0020
7215-0	102A	3004		TR85/	
7216	108	3039		TR95/56	S0020
7217	108	3039		TR95/56	S0020
7218	108	3039		TR95/56	S0020
7219	108	3039		TR95/56	
7219-3	131	3052		TR94/	
7220	108	3039		TR95/56	S0020
7221	108	3039		TR95/56	S0020
7221A		3072			
7231		3018			
7232	108	3018		TR95/56	S0011
7233	108	3018		TR95/56	S0016
7234	108	3018		TR95/56	S0016
7235	108	3018		TR95/56	S0015
7236	108	3018		TR95/56	S0011
7237	108	3018		TR95/56	S0011
7239	102				S5004
7252	152	3054	66	TR92/	S5000
7253(LOWREY)	154	3045	27	TR78/712	
7261		3018			
7262		3018			
7264		3018			
7310(GE)	124		12	TR23/240	S5011
7311		3021			
7312	175	3026	23	TR81/	S5012
7313	159	3077			
7316	188				
7317	124	3021	12	TR23/	S5011
7318	123A				
7321	194				
7322	175				
7332		3018			
7364-6053P1	195			TR65/	S3001
7398-6117P1	108	3039		TR95/56	S0016
7398-6118P1	132	3116		FE100/802	F0015
7398-6119P1	123A	3122		TR21/735	S0014
7398-6120P1	188			TR72/	S3020
7400-6A	7400				
7400-9A	7400				
7400PC	7400				
7401-6A	7401				
7401-9A	7401				
7401PC	7401				
7402					C6062P
7402-6A	7402				
7402-9A	7402				
7402PC	7402				
7403-6A	7403				
7403-9A	7403				
7403PC	7403				
7404-6A	7404				
7404-9A	7404				
7404PC	7404				
7405-6A	7405				
7405-9A	7405				
7405PC	7405				
7406PC	7406				
7407A	7407				
7407PC	7407				
7408-6A	7408				
7408-9A	7408				
7408PC	7408				
7409PC	7409				
7410-6A	7410				
7410-9A	7410				
7410PC	7410				
7411-6A	7411				
7411-9A	7411				
7411PC	7411				
7412PC	7412				
7413		3054			
7413PC	7413				
7414		3054			
7414PC	7414				
7416PC	7416				
7417PC	7417				
7420		3083			
7420-6A	7420				
7420-9A	7420				
7420PC	7420				
7423		3054			
7423PC	7423				
7425	108	3039		TR95/56	S0015

Original Part Number	ECG	SK	GE	IR/WEP	HEP
7425PC	7425				
7426	108	3039		TR95/56	S0015
7426PC	7426				
7427	108	3039		TR95/56	S0015
7427PC	7427				
7428	108	3039		TR95/56	S0015
7429	123A	3122		TR21/735	S0015
7430	123A	3122		TR21/735	S0015
7430-6A	7430				
7430-9A	7430				
7430PC	7430				
7431	123A	3122		TR21/735	S0015
7432	123A	3122		TR21/735	S0015
7432PC	7432				
7433	123A	3122		TR21/735	S0015
7437		3054			
7437PC	7437				
7438PC	7438				
7439DC	7439				
7439PC	7439				
7440-6A	7440				
7440-9A	7440				
7440PC	7440				
7441-6A	7441				
7441-9A	7441				
7441DC	7441				
7442DC	7442				
7443DC	7443				
7444DC	7444				
7445DC	7445				
7445PC	7445				
7446ADC	7446				
7446APC	7446				
7446PC	7446				
7447BDC	7447				
7447BPC	7447				
74+7DC	7447				
7448DC	7448				
7450-6A	7450				
7450-9A	7450				
7450PC	7450				
7451-6A	7451				
7451-9A	7451				
7451PC	7451				
7453-6A	7453				
7453-9A	7453				
7453PC	7453				
7454-6A	7454				
7454-9A	7454				
7454PC	7454				
7460-6A	7460				
7460-9A	7460				
7460PC	7460				
7470-6A	7470				
7470-9A	7470				
7470PC	7470				
7472-6A	7472				
7472-9A	7472				
7472PC	7472				
7473-6A	7473				
7473-9A	7473				
7473PC	7473				
7474-6A	7474				
7474-9A	7474				
7474PC	7474				
7475-6A	7475				
7475-9A	7475				
7475DC	7475				
7475PC	7475				
7476-6A	7476				
7476PC	7476				
7480DC	7480				
7480PC	7480				
7482-6A	7482				
7482-9A	7482				
7482DC	7482				
7483DC	7483				
7483PC	7483				
7485DC	7485				
7486PC	7486				
7490-6A	7490				
7490-9A	7490				
7490DC	7490				
7491DC	7491				
7492-6A	7492				
7492-9A	7492				
7492DC	7492				
7493-6A	7493A				
7493-9A	7493A				
7493DC	7493				
7493PC	7493A				
7494DC	7494				
7495DC	7495				
7496DC	7496				
7501	123A	3018	17	TR21/	

Original Part Number	ECG	SK	GE	IR/WEP	HEP
7502	123A	3124	20	TR21/	
7503	159		67	TR20/	
7505	123A	3018	17	TR21/	
7506	123A		18	TR21/	
7507	128		63	TR87/	
7508	129		21	TR88/	
7509	123A	3124	62	TR21/	
7510	175			TR81/	S5019
7511		3019			
7513	186	3024	28	TR55/	
7514	130	3027	14	TR59/	
7515	123A	3124		TR21/735	
7515(GE)					S0015
7516	123A	3124	20	TR21/735	S0015
7517	123A	3124		TR21/735	S0015
7518	123A	3124		TR21/735	S0015
7519	123A	3019		TR21/735	S0015
7527				TR25/	
7553	188	3199		TR72/3020	
7585		3018			
7586		3124			
7587		3018			
7588		3018			
7589		3018			
7590		3018			
7603		3004			
7608		3004			
7611		3004			
7613		3004			
7623		3004			
7625				TR25/	
7635		3114			
7637	123A	3018		TR21/735	S0015
7638	123A	3018			
7639	123	3018		TR86/53	S0015
7641	123A	3018			
7642	108	3018		TR95/	S0015
7675		3122			
7720G				TR21/	
7740RN				TR95/	
7810	108	3039		TR95/56	S0015
7811	108	3039		TR95/56	S0015
7812	108	3039		TR95/56	S0015
7813	108	3039		TR95/56	S0015
7814	108	3039		TR95/56	S0015
7815	108	3039		TR95/56	S0015
7816	123A	3122		TR86/735	S0015
7817	123A	3122		TR21/735	S0015
7818	123A	3122		TR21/735	S0015
7839		3004			
7840RN				TR95/	
7874(WURL)	761				C6011
7885-1	181				
7885-2	181				
7885-3	181				
7909	128				
7920-1	196				
7921-1	197				
7991	199	3124	62	TR24/	S0015
7992	123A	3124	62	TR21/	S0016
7993	158	3004	53		G6011
8000-00001-068	102A	3004		TR85/	
8000-00003-033	107	3018	10		S0015
8000-00003-034	108	3039	17		S0020
8000-00003-035	107	3018	20		S0030
8000-00003-037	159	3114	21		S0012
8000-00003-038	102A	3004	52		G0005
8000-00003-039	102A	3004	52		G0005
8000-00003-040		3052	30		S5013
8000-00003-041	190	3047			S3020
8000-00003-042	190	3047			S3020
8000-00003-043	210	3048		TR55/	S3020
8000-00004-079	123A	3018	20	TR21/735	S0016
8000-00004-080	133	3112	FET2	FE100/801	F0010
8000-00004-081	132	3116	FET2	FE100/802	F0015
8000-00004-082	123A	3046		TR21/735	S0015
8000-00004-083	128	3047		TR87/243	S3002
8000-00004-084	195	3048		TR65/243	S3002
8000-00004-085	123A	3122	20	TR21/735	S0014
8000-00004-086	103A	3010	8	TR08/724	G0011
8000-00004-087	186	3054	66	TR55/S3023	S3020
8000-00004-088	102A	3004	2	TR85/250	G0007
8000-00004-089	159	3114	21	TR20/717	S0013
8000-00004-090	724				
8000-00004-85	123A	3122		TR21/735	
8000-00004-185	186	3041		TR55/S3023	
8000-00004-241	130	3027	14	TR59/247	S7002
8000-00004-242	123A	3122	63	TR21/735	S0014
8000-00004-243	123A	3122	20	TR21/735	S0011
8000-00004-305	703A				
8000-00004-306	1045				
8000-00004-307	1075				
8000-00004-308	1021				
8000-00004-P079	108	3018		TR21/735	S0016
8000-00004-P080	133	3018		/801	F0010
8000-00004-P081	132	3018		FE100/802	F0015
8000-00004-P083	195	3047		TR65/243	S3002
8000-00004-P084	195	3048		TR65/243	S3002
8000-00004-P085	123A	3124		TR21/735	S0014
8000-00004-P086	123A	3124		TR21/735	G0011
8000-00004-P088	102A	3004		TR85/250	G0007
8000-00004-P089	129	3025		TR88/717	S0014
8000-00004-P185	184	3041		/S3023	S5000
8000-00005-001	132	3116		FE100/	F0015
8000-00005-002	108	3018		TR95/	S0015
8000-00005-003	108	3018			S0015
8000-00005-004	123A	3122		TR21/	S0015
8000-00005-005	123A	3124		TR21/	S0015
8000-00005-007	123A	3124		TR21/	S0011
8000-00005-009	123A	3047		TR21/	S0025
8000-00005-010	123A	3047		TR21/	S3001
8000-00005-011	195	3048		TR65/	S3002
8000-00005-012					G0011
8000-00005-013					S3005
8000-00005-014	1104				
8000-00005-055	123A	3122		TR21/	S0015
8000-00006-001	128	3047	17	TR87/	S3001
8000-00006-002	128	3049	45	TR87/	S3002
8000-00006-003	123A	3018		TR21/	S0016
8000-00006-004	159	3025	21	TR20/	S5013
8000-00006-005	107	3046	20	TR70/	S0011
8000-00006-006	130	3027	14	TR59/	S7004
8000-00006-230	224	3049	46		
8000-0004-086	103A	3124		TR08/724	
8000-0004-P083	195	3048		TR87/243	S3002
8000-0004-P084	195	3048		TR87/243	S3002
8000-0004-P090	724	3525			
8000-0004-P185	186	3192		TR55/S3023	S5000
8000-0005-001	132			FE100/	
8000-0005-002	107			TR70/	
8000-0005-003	107			TR70/	
8000-0005-007	123A			TR21/	
8000-0005-008	130			TR59/	
8000-0005-009	123A			TR21/	
8000-0005-010	128			TR87/	
8000-0005-011	128			TR87/	
8000-00011-004	107	3018	10	TR24/	S0015
8000-00011-047	123A	3018	10	2SC710/	S0015
8000-00011-048	123A	3018	10	2SC710/	S0015
8000-00011-049	123A	3124	20		S0014
8000-00011-050	152	3041	66		S5000
8000-00011-051	224	3049	45	TR65/	S3001
8000-00011-052	224	3049	45	TR65/	S3001
8000-00011-053	220	3116			F2005
8000-00011-054	132		FET2		F0021
8000-00011-055	132		FET2		F0021
8000-00011-086	103A				
8000-00012-040	128	3024	63	TR25/	S5014
8000-00012-041	703A	3157			
8000-00029-006	108	3020		TR24/	
8000-00029-007	123A	3122			
8000-00030-007	123A			TR24/	
8000-00030-008	195			TR64/	
8000-00030-009	103A				
8000-00032-025	123A		20	TR24/	S0015
8000-00032-026	161		39	TR80/	S0017
8000-00032-027	103A		59	2SB187/	G0011
8000-00032-028	236		66		
8000-00032-030	1058				
8000-00032-031	724				
8002-1C					S5014
8003-114	123A	3124		TR21/735	S0015
8005-3	704	3023			
8005(PENN)	181			TR36/S7000	
8007-0	704	3022			
8007-1	704	3022			
8007-3	704	3023			
8007-4	704	3023			
8008-1	704	3022			
8008-3	704	3023			
8009-0	704	3023			
8009-4	704	3023			
8010-171	1045		IC2	IC507/	C6060P
8010-172	1120				
8010-173	132	3116	FET2	FE100/	F0021
8010-174	161	3018	20	TR83/	S0015
8010-175	161	3018	20	TR83/	S0015
8010-176	123A	3020	20	TR21/	G0005
8020-204	199			TR25/	S0015
8020-205	123A	3122		TR21/	S0014
8020-206	152	3054		TR55/	S5000
8070-4	102A	3004		TR85/	G0005
8071-4	102A	3004		TR85/	G0005
8072-4	102A	3004		TR85/	G0005
8073-4	102A	3004		TR85/	G0005
8074-4	123A	3124		TR21/	S0015
8075-4	123A	3124		TR21/	S0015
8098	749				
8102-207		3124	20	TR51/	S0015
8102-208		3124	62	TR21/	S0016

Original Part Number	ECG	SK	GE	IR/WEP	HEP
8102-209		3024	20	TR25/	S0025
8102-210		3086	31MP	TR50/	G6016
8200-202	199	3124	10	TR24/	S0015
8200-203	199	3124	62	TR21/	S0016
8200-204	226	3086	31MP	TR50/	G6016
8210-1203	123A				
8281-1		3122			
8301(PENNEY)	159	3025	21	TR20/717	S0019
8302(PENNEY)	123A	3124	17	TR21/735	S0015
8303(PENNEY)	129	3025	21	TR88/242	S5013
8304(PENNEY)	128	3024	63	TR87/243	S5014
8305P1					S0014
8310	762				
8319-001	130				
8394	199		10	TR21/	
8400-1	129	3025		TR88/	
8400-1A	129	3025		TR88/	
8400-1B	129	3025		TR88/	
8405	159	3114		TR20/717	S0019
8440P1	912				
8471(SYL)	188	3054	28	TR72/	S3024
8500-201	102A		52	TR85/250	G0005
8500-202	102A	3123	52	TR85/250	G0007
8500-203	102A	3123	52	TR85/250	G0005
8500-204	131	3082	30	TR94/642	G6016
8503	107			TR70/	S0015
8504	123A	3124		TR21/735	
8540	159	3114		TR20/717	S0019
8554-9	128	3024		TR87/	
8600	123A	3018			
8601	159	3025		TR20/717	
8602	123A	3124		TR21/735	
8603		3024		TR21/735	
8606		3018			
8607		3018			
8609		3018			
8611		3018			
8614-007-0	123A				
8624-003	129				
8710-162	108	3122			
8710-163	123A	3122			
8710-164	123A	3124			
8710-165	123A	3122			
8710-166	123A	3122			
8710-167	123A	3124			
8710-168	123A	3122			
8710-169	159	3118			
8710-170	186	3054			
8710-172	720				
8800-202	123A	3122		TR21/735	S0025
8800-203	123A	3122		TR21/735	S0015
8800-204	123A	3122		TR21/735	S0015
8800-205	152	3027		TR76/701	S5019
8840-161	132	3116	FET2	FE100/	F0021
8840-162	107	3018	17	TR24/	S0025
8840-163	123A	3122	20	TR24/	S0015
8840-164	123A	3124	20	2SC945/	S0015
8840-165	123A	3122	20	TR24/	S0015
8840-166	123A	3122	20	TR24/	S0015
8840-167	123A				
8840-168	123A				
8840-169	186	3192			
8840-171	720				
8867	159	3114		TR20/717	S0019
8868-6	128	3024		TR87/	
8868-7	199	3038		TR21/	
8868-8	123A	3124		TR21/	
8880-3	128	3024		TR87/	
8883-2	121	3009		TR01/	
8883-4	128	3024		TR87/	
8886-2	123A	3124		TR21/	
8999-115	104	3009	25	TR01/233	G6003
8999-201		3004	52	/250	G0005
8999-202	102A	3004	2	TR85/250	G0007
8999-203	158	3004	53	TR84/630	G6011
9000DC	74104				
9000PC	74104				
9001DC	74105				
9001PC	74105				
9005-0	121	3009		TR01/	
9011	123A				
9011E	108	3018	20	TR21/	S0014
9011F		3018	20	TR21/	S0014
9011G	108	3018	20	TR21/	S0014
9013H			10	TR24/	
9013HF	123A	3124	10	TR24/	S0015
9013HG	123A	3124		TR24/	S0015
9014	123A	3124			
9014B	123A	3124	10	TR24/	S0015
9014C	123A		10	TR21/	
9014D			10	TR24/	S0015
9016	108	3018			
9016D	108	3018	20	TR21/	S0014
9016E	108	3018	20	TR95/	S0014
9018	108	3018		TR95/	

Original Part Number	ECG	SK	GE	IR/WEP	HEP
9018D	108	3018	20	TR21/	S0014
9018E					S0014
9018F		3018			S0015
9018G	108		17	TR95/	
9033	123A	3124	10	TR24/735	S0015
9033-1	123A			TR21/	
9033-2	199			TR21/	
9033-3	199			TR21/	
9033-4	199			TR21/	
9033-5	199			TR21/	
9033BRN	123A			TR21/	
9033G(SYL)	123A	3018	20	TR21/735	S0015
9033GRN	199			TR21/	
9033ORN	199			TR21/	
9033Q	199		20	TR21/	
9033RED	199			TR21/	
9033(SYL)	123A			TR21/	S0015
9033WHT	199			TR21/	
9050-1	910D				
9093-1-6A	9093				
9093-9-6A	9093				
9093-9-7A	9093				
9093DC	9093				
9093PC	9093				
9094-1-6A	9094				
9094-9-6A	9094				
9094-9-7A	9094				
9094DC	9094				
9094PC	9094				
9097-1-6A	9097				
9097-9-6A	9097				
9097-9-7A	9097				
9097DC	9097				
9097PC	9097				
9099-1-6A	9099				
9099-9-6A	9099				
9099-9-7A	9099				
9099DC	9099				
9099PC	9099				
9109DC	9109				
9110DC	9110				
9111DC	9111				
9112DC	9112				
9135PC	9135				
9137-C-1004	9945				
9137-C-1020	9962				
9144-60	787				
9144-61	788				
9157PC	9157				
9158		3027			
9158PC	9158				
9279	128				
9300	108	3018		TR95/	
9300A	108	3018		TR95/	
9300B	108	3018		TR95/	
9300Z	108	3018		TR95/	
9314(GE)	108	3018		TR95/56	S0015
9315DC	7441				
9315PC	7441				
9330-011-70112	102A			TR85/	G0005
9330-229-60112	107			TR70/	S0016
9330-229-70112	107			TR70/	S0016
9330-688-30112	199			TR21/	S0024
9330-767-60112	159			TR20/	S0019
9330-908-10112	159			TR20/	S0019
9341DC	74181				
9341PC	74181				
9342DC	74182				
9345DC	7445				
9345PC	7445				
9352DC	7442				
9352PC	7442				
9353DC	7443				
9353PC	7443				
9354DC	7444				
9354PC	7444				
9357A	7446A				
9357ADC	7446				
9357APC	7446				
9357B	7447A				
9357BDC	7447				
9357BPC	7447				
9358	7448				
9358DC	7448				
9358PC	7448				
9359	7449				
9360DC	74192				
9360PC	74192				
9366DC	74193				
9366PC	74193				
9367-1	128	3024	18	TR87/243	
9375DC	7475				
9375PC	7475				
9377	7497				
9380DC	7480				

Original Part Number	ECG	SK	GE	IR/WEP	HEP
9380PC	7480				
9382DC	7482				
9382PC	7482				
9383DC	7483				
9383PC	7483				
9385DC	7485				
9385PC	7485				
9390	102	3004	2	TR05/631	
9390DC	7490				
9390PC	7490				
9391	102	3004	2	TR05/631	
9391DC	7491				
9391PC	7491				
9392DC	7492				
9392PC	7492				
9393DC	7493				
9393PC	7493				
9394DC	7494				
9394PC	7494				
9395DC	7495				
9395PC	7495				
9396DC	7496				
9396PC	7496				
9400-8	102A	3004		TR85/	
9400-9	102A	3004		TR85/	
9401-7	102A	3004		TR85/	
9403-2	121	3009		TR01/	
9403-3	160	3007		TR17/	
9403-6	160	3007		TR17/	
9403-7	102A	3004		TR85/	
9403-8	126	3008		TR17/	
9403-9	102A	3004		TR85/	
9404-0	121	3009		TR01/	
9404-2	128	3047		TR87/	
9404-3	195	3048		TR65/	
9404-9	175	3026		TR81/	
9405-0	128	3024		TR87/	
9405-1	128	3024		TR87/	
9405-2	129	3025		TR88/	
9409-4	175	3026		TR81/	
9501	199			TR21/	S0024
9502	199	3124	63	TR21/735	S0024
9510-1	160	3007		TR17/	
9510-2	160	3007		TR17/	
9510-3	126	3008		TR17/	
9510-7	160	3006		TR17/	
9513	108	3039		TR95/56	
9550-1	766				C6009
9564	102A		2	TR85/	
9600	123A	3122		TR21/735	S0025
9600-5	123A			TR21/	S0025
9600C	108			TR21/	S0015
9600F	108	3122	20	TR21/	S0015
9600G	108	3122	20	TR21/	S0015
9600H	108	3122	20	TR21/	S0015
9601	108	3031		TR95/	S0016
9601-12	108	3018		TR95/	
9604F		3018			
9617		3124			
9617K	128	3124	18	TR25/	S5014
9618		3024			
9623		3018			
9623F	108		20	TR21/	S0014
9623G	108	3018	20	TR21/	S0014
9623H	108	3122	20	TR21/	S0014
9625F	108	3018	17	TR21/	S0014
9625H	108	3018	17	TR21/	S0014
9630C	108	3122	20	TR21/	S0015
9654-030.31	152				
9654-031.31	152				
9692-1	723	3144			
9693-1	787	3146			
9694-1	724	3525			
9696H	128	3122	18	TR25/	S5014
9842FF	723				
9914-8-8A					C2010G
9914-9-5B					C2010G
9920-4	102A			TR85/	
9920-5	102A			TR85/	
9920-6-1	123A			TR21/	
9920-6-2	123A			TR21/	
9920-7-2	123A			TR21/	
9921-7	102A			TR85/	
9921-8	102A			TR85/	
9925-0	121			TR01/	
9925-2	124			TR81/	
9925-2-1	124			TR81/	
9925-2-2	124			TR81/	
9942PC	74182				
10003	181	3036		TR36/S7000	
10032	102A			TR85/250	
10036	102A	3004	52	TR85/250	G0005
10036-001	129	3025		TR88/	
10037	102A	3004	52	TR85/250	G0005
10038	102A	3123	2	TR85/250	G0005

Original Part Number	ECG	SK	GE	IR/WEP	HEP
10039	102A	3123	2	TR85/250	G0005
10096-001A		3024			
10140		3024			
10226/2	123A				
10300	159			TR20/	S0019
10300-12	129	3025		TR88/	
10302-01	7430				
10302-02	7420				
10302-03	7410				
10302-04	7400				
10302-05	7451				
10302-06	7405				
10416-009	123A				
10416-010	128				
10444					G0005
10508		3114			
10650A01	778				
10655A01	798				
10655A01-3	798				
10655A03	738		IC29		
10655A03A	738				
10655B02	748		IC2		
10655B03	738				
10658A01	788				
10795-5	9945				
10795-6	9946				
10795-8	9962				
10795-9	9093				
10795-10	9099				
11000	234				
11200-1	7410				
11202-1	7404				
11203-1	7493				
11204-1	74193				
11205-1	7420				
11206-1	7470				
11207-1	7402				
11208-1	7430				
11209-1	7483				
11211-1	7409				
11212-1	7472				
11213-1	7474				
11214-1	7440				
11216-1	7400				
11233-2	7454				
11236-1	128			TR87/	
11236-2	224				
11236-3	130			TR59/	
11252-0	108			TR95/	
11252-1	108			TR95/	
11252-2	108			TR95/	
11252-3	123A			TR21/	
11252-4	121			TR01/	
11252-5	128			TR87/	
11252-6	224				
11252-7	224				
11273-1	74121				
11274-1	7403				
11276-1	7493				
11327B		3021			
11339-8	108			TR95/	
11393-8	108			TR95/	
11426-7	108			TR95/	
11506-3	121			TR01/	
11522-5	123A			TR21/	
11522-7	160			TR17/	
11522-8	160			TR17/	
11522-9	160			TR17/	
11526-8	121			TR01/	
11526-9	121			TR01/	
11527-5	126			TR17/	
11528-1	121MP				
11528-2	121MP				
11528-3	121MP				
11528-4	121MP				
11587-5	123A			TR21/	
11606-8	127			TR27/	
11607-2	126			TR17/	
11607-3	108			TR95/	
11607-4	123A			TR21/	
11607-5	124			TR81/	
11607-8	123A			TR21/	
11607-9	108			TR95/	
11608-0	108			TR95/	
11608-2	108			TR95/	
11608-3	108			TR95/	
11608-5	123A			TR21/	
11608-6	127			TR27/	
11608-7	127			TR27/	
11608-8	127			TR27/	
11608-9	127			TR27/	
11609-1	102A			TR85/	
11609-2	123A			TR21/	
11619-8	108			TR95/	
11619-9	108			TR95/	

Original Part Number	ECG	SK	GE	IR/WEP	HEP
11620-0	108			TR95/	
11620-1	108			TR95/	
11620-2	160			TR17/	
11620-3	102A			TR85/	
11620-6	102A			TR85/	
11620-7	160			TR17/	
11620-8	160			TR17/	
11620-9	160			TR17/	
11658-8	123A			TR21/	
11668-3	160			TR17/	
11668-4	160			TR17/	
11668-5	102A			TR85/	
11668-6	102A			TR85/	
11668-7	103A			TR08/	
11675-6	126			TR17/	
11675-7	102A			TR85/	
11687-5	123A			TR21/	
11699-7	102A			TR85/	
11727		3004			
11854-A-B		3004			
11901-3		3036			
12000-1C					S0015
12020-02	152				
12044-0021	175	3026			
12047-0023	192				
12048-0011	129				
12074E					G0005
12077B					S0015
12110-0	100	3005		TR05/	
12110-3	100	3005		TR05/	
12110-4	100	3005		TR05/	
12110-5	100	3005		TR05/	
12110-6	102A	3004		TR85/	
12110-7	102A	3004		TR85/	
12112-0	102A	3004		TR85/	
12112-8	100	3005		TR05/	
12112-C	123A		20	TR21/735	S0015
12112-D	123A		20	TR21/735	S0015
12112-E	123A		20	TR21/735	S0015
12112-F	123A		20	TR21/735	S0015
12113-5	160	3006		TR17/	
12113-7	160	3006		TR17/	
12113-8	160	3008		TR17/	
12113-9	160	3006		TR17/	
12114-8	102A	3004		TR85/	
12115-0	160	3007		TR17/	
12115-7	160	3006		TR17/	
12116-1	100	3005		TR05/	
12116-2	100	3005		TR05/	
12116-4	102A	3004		TR85/	
12117-9	100	3005		TR05/	
12118-0	100	3005		TR05/	
12118-4	102A	3004		TR85/	
12118-6	100	3005		TR05/	
12118-9	126	3008		TR17/	
12119-0	160	3007		TR17/	
12119-1	102A	3004		TR85/	
12119-2	102A	3004		TR85/	
12122-5	102A	3004		TR85/	
12122-6	102A	3004		TR85/	
12122-7	102A	3004		TR85/	
12122-8	160	3006		TR17/	
12122-9	160	3006		TR17/	
12123-0	160	3006		TR17/	
12123-1	160	3006		TR17/	
12123-2	102A	3004		TR85/	
12123-3	160	3006		TR17/	
12123-4	100	3005		TR05/	
12123-5	100	3005		TR05/	
12123-6	100	3005		TR05/	
12124-0	100	3005		TR05/	
12124-1	100	3005		TR05/	
12124-2	126	3008		TR17/	
12124-3	126	3008		TR17/	
12124-4	126	3008		TR17/	
12124-6	102A	3004		TR85/	
12125-4	100	3005		TR05/	
12125-6	160	3007		TR17/	
12125-7	126	3008		TR17/	
12125-8	160	3007		TR17/	
12125-9	160	3007		TR17/	
12126-0	160	3007		TR17/	
12126-6	102A	3004		TR85/	
12126-7	102A	3004		TR85/	
12127-0	121	3009		TR01/	
12127-1	121	3009		TR01/	
12127-2	102A	3004		TR85/	
12127-3	100	3005		TR05/	
12127-4	102A	3003		TR85/	
12127-5	100	3005		TR05/	
12127-6	123A	3124		TR21/	
12127-7	123A	3124		TR21/	
12127-8	123A	3124		TR21/	
12127-9	123A	3020		TR21/	
12128-9	100	3005		TR05/	
12163	104	3009	16	TR01/624	
12173	100	3005		TR05/	
12174	100	3005		TR05/	
12175	100	3005		TR05/	
12176	100	3005		TR05/	
12178	104	3005	16	TR01/624	
12180	126		51	TR17/635	
12183	100	3005		TR05/	
12191	100	3005		TR05/	
12192	100	3005		TR05/	
12193	100	3005		TR05/	
12195	102A	3004		TR85/	
12196	102A	3004		TR85/	
12393				TR21/	
12536	130	3027		TR59/247	S7004
12536A	130	3027		TR59/	
12537	219				S7003
12538	175			TR81/241	S5012
12539	184	3190	26	/S5003	S5018
12546	154	3045		TR78/712	S5025
12593	123A	3122		TR21/735	S0015
12594	159	3114		TR20/717	S0019
12888	159	3114		TR20/717	S0013
12951-1	904				
13000-1C					S0024
13000-1D					S5014
13002-3	130				
13002-4	130				
13030-4	181				
13124Z		3025			
13159-2	175				
13162	159				
13217E1513	128				
13254		3024			
13286-139					S7000
13297					S0013
13298	130			TR59/	S7002
13299	211	5203			S7003
13314					S5026
14166		3024			
14173I8	194				
14207	947				
14303	123A	3122	20	TR21/735	
14573	105	3012		TR03/	
14692	128				
14867		3005			
14995	128				
14996-1	128				
14996-2	128				
15009	102A		53	TR85/250	
15024	121	3009	16	TR01/232	G6003
15027	121	3009	16	TR01/232	G6003
15039		3024			
15354-3	121	3009		TR01/	
15405-1	9946				
15405-2	9949				
15405-4	9932				
15486	128				
15809-1	123	3124	20	TR86/53	
15810-1		3124	20	/53	
15820-1	123			TR86/	
15835-1	123	3020	20	TR86/53	
15840-1	123A	3122	20	TR21/735	S0015
15841-1	123A	3122	20	TR21/735	
15927	104	3009	3	TR01/624	
15942	754				C6007
15942-1	761				C6011
16001	130				
16002-40				TR12/	
16002-41				TR12/	
16002-033				TR01/	
16002-040				TR12/	
16002-041				TR12/	
16002-042				TR12/	
16002-062				TR01/	
16002-062A				TR01/	
16002-090				TR24/	
16002-095				TR01/	
16029	196				
16039	196				
16065	128				
16082	128				
16083	130				
16088	7490				
16113	152				
16115	196				
16163	196				
16164	152				
16165	196				
16166	152				
16167	197				
16169	197				
16175	197				
16176	130				
16181	152				

Original Part Number	ECG	SK	GE	IR/WEP	HEP
16182	152				
16190	108				
16194	108				
16201	130				
16207	152				
16230	130				
16232	124				
16234	130				
16235	130				
16237	199				
16239	129				
16240	130				
16241	152				
16254	128				
16259	124				
16261	130				
16266	130				
16267	181				
16277	196				
16279	197				
16287	130				
16292	130				
16299	130				
16305	152				
16306	197				
16319	130				
16320	130				
16334	152				
16335	152				
16336	152				
15338	130				
16341	196				
16598	102	3004	2	TR05/631	
16599	121	3009	3	TR01/232	
16605	707				
16958	102	3004		TR05/631	
16959	121	3009		TR01/232	
17042				TR05/	
17043				TR05/	
17045	159				
17047-1	102	3004	2	TR05/631	
17142	102	3004	2	TR05/631	
17143	102	3004	2	TR05/631	G0005
17144	123	3124		TR86/53	
17412-5	123A		17	TR21/	
17444	123	3124	20	TR86/53	
17607-1	127				
17887	121	3009	16	TR01/232	
17945	121	3009	3	TR01/232	
18109	102	3004	2	TR05/631	
18310	124	3021	12	TR81/240	
18493	126	3006	51	TR17/635	
18509	123A	3124	20	TR21/735	S0015
18529	100	3004	2	TR05/254	G0005
18530	102	3004	2	TR05/631	G0005
18540	160	3005	51	TR17/637	G0002
18541	160	3006	51	TR17/637	G0002
18555	123A	3124	20	TR21/735	S0015
18600-151	107	3018	17	TR24/	S0015
18600-152	123A	3018	17	TR24/	S0015
18600-153	123A	3124	20	2SC945/	S0015
18600-155	720				
18600-156	1058				
18601	102	3004	2	TR05/631	G0005
18611	102	3004	2	TR05/631	
18731	102	3004	2	TR05/631	G0005
19278	124	3021		TR81/240	S5011
19420	108	3018	20	TR95/56	
19500-253	199				
19645	123A	3122		TR21/735	
19797A					S0014
20011	159				
20015		3019			
20103	199				
20503		3024			
20738	123	3020	20	TR86/53	
20739	129	3025	21	TR88/242	
20810-91		3124			
20810-92		3124			
20810-93		3024			
20810-94		3054			
20861-11		3004			
20989	124				
21221	128				
21280	181				
21290	128				
21606-1	127				
21676A	128				
22008	159	3025	21	TR20/717	
22009	172		64	TR69/	S9100
22158	123A	3024	18	TR21/735	S0015
22595-000	159				
22605-005	159				
22635-002	123A				
22635-003	123A				

Original Part Number	ECG	SK	GE	IR/WEP	HEP
22636-0001		3007			
22636-0002		3007			
22822		3036			
22823		3036			
22939	198		32		
23114-046	123A	3124		TR21/	
23114-050	129	3025		TR88/	
23114-051	129	3025		TR88/	
23114-052	154	3040		TR78/	
23114-053	123A	3124		TR21/	
23114-054	123A	3124		TR21/	
23114-056	108	3018		TR95/	
23114-057	108	3018		TR95/	
23114-060	108	3018		TR95/	
23114-061	102A	3004		TR85/	
23114-070	130	3027		TR59/	
23114-078	108	3018		TR95/	
23114-082	123A	3124		TR21/	
23114-095	123A	3124		TR21/	
23114-097	127	3034		TR27/	
23114-104	108	3018		TR95/	
23115-057	123A	3124		TR21/	
23115-058	123A	3124		TR21/	
23125-037	108			TR95/	
23201H		3018			
23311-006	121	3009		TR01/	
23316	123A				
23606	132		FET2	FE100/	
23612A		3036			
23648	128				
23754	152		66	TR76/	
23762	130		14	TR59/	
23785(SYL)	102A		2	TR85/250	G0008
23785-1(SYL)	102A			TR85/250	G0008
23826(SYL)	159			TR20/717	S0013
24198	126			TR17/	
24451	711	3070			
24560A	124				
24785				TR85/	
25011(HONWL)	123A				
25114-101	102A	3004		TR85/	
25114-102	102A	3004		TR85/	
25114-103	102A	3004		TR85/	
25114-104	102A	3004		TR85/	
25114-116	123A	3124		TR21/	
25114-121	108	3018		TR95/	
25114-130	123A	3124		TR21/	
25114-143	124	3021		TR81/	
25114-161	123A	3124		TR21/	
25566-01	126				
25566-21	128				
25566-50	128				
25566-62	129				
25566-63	128				
25566-76	128				
25584-00010		3005			
25642-020	126			TR17/	G0009
25642-030	126			TR17/	G0009
25642-031	126			TR17/	G0009
25642-040	126			TR17/	G0009
25642-041	126			TR17/	G0009
25642-110	126			TR17/	G0008
25642-115	126			TR17/	G0008
25642-120	126			TR17/	G0008
25651-020	102A			TR85/	G0005
25651-021	102A			TR85/	G0005
25651-033	102A			TR85/250	G0005
25655-055	102A			TR85/250	G0005
25655-056	102A			TR85/250	G0005
25657-050	102A			TR85/	G0005
25658-120	131	3198		TR94/	G6016
25658-121	131	3198		TR94/	G6016
25661-020	121	3009		TR01/232	G6003
25661-022	104			TR01/	G6003
25671-020	161	3039		TR83/	S0017
25671-021	161	3039		TR83/	S0017
25671-023	161	3039		TR83/	S0017
25672-016	154			TR78/	S5026
25762-010	225	3045		/712	S5026
25762-012	225	3045		/712	S5026
25810-107	1081				
25810-161	123A	3122			
25810-162	123A	3124			
25810-163	123A	3122			
25810-166	1087				
25810-167	1081		IC7		
25840-161	123A	3122	20		S0015
25840-162	199	3124	20		S0015
25840-163	192	3122			S0015
25840-167	1081				
26587	703A	3157			
26587-1	703A	3157			
27125-080	123A	3124		TR21/	
27125-090	123A	3124		TR21/	
27125-110	128			TR87/	

Original Part Number	ECG	SK	GE	IR/WEP	HEP
27125-120	102A			TR85/	
27125-140	123A	3124		TR21/	
27125-150	102A	3004		TR85/	
27125-160	123A	3124		TR21/	
27125-170	102A	3004		TR85/	
27125-270	123A	3124		TR21/	
27125-300	123A	3124		TR21/	
27125-310	103A	3010		TR08/	
27125-330	102A	3004		TR85/	
27125-340	102A	3004		TR85/	
27125-350	102A	3004		TR85/	
27125-360	100	3005		TR05/	
27125-370	123A	3124		TR21/	
27125-380	128	3024		TR87/	
27125-460	199	3038		TR21/	
27125-470	190	3038		TR21/	
27125-480	102A	3004		TR85/	
27125-490	103A	3010		TR08/	
27125-500	123A	3124		TR21/	
27125-530	123A	3124		TR21/	
27125-540	102A	3004		TR85/	
27125-550	102A	3004		TR85/	
27126-060	131	3052		TR94/	
27126-090	121			TR01/	
27126-100	130	3510		TR59/	
27127-550	123A	3124		TR21/	
27840-161	123A	3122	20	TR24/	S0015
27840-162	123A	3124	20	TR21/	S0015
27840-164	1081				
27910-12150	123				
27910-12153	102				
28134-B		3027			
28222-1					S5000
28222-2					S5004
28222-3(SYL)	182				S5004
28222-4					S5004
28336-1					S5006
28336-2					S5000
28396	223				
28474	130	3027			
28810-172		3018			
28810-173		3018			
28810-174		3124			
28977	128				
29076-005	161				
29076-006	161				
29076-023	159				
29109		3026			
30201	158	3004	52	TR84/630	G0005
30202	158	3123	52	TR84/630	G0005
30203	121	3009	16	TR01/232	G6003
30204	102	3004	2	TR05/631	G0005
30206	102	3004		TR05/631	G0005
30207	102	3004		TR05/631	G0005
30208	100	3004	52	TR05/254	G0005
30208-1	102A	3004		TR85/250	G0005
30208-2	102	3004	2	TR05/631	G0005
30210	123	3124	20	TR86/53	S0015
30211	121	3009	16	TR01/232	G6005
30213	126	3008	50	TR17/635	G0009
30214	126	3008	52	TR17/635	G0003
30215	126	3007	52	TR17/635	G0003
30215(RCA)	121			TR01/232	
30216	102	3007	50	TR05/631	G0001
30216(RCA)	121			TR01/	
30217	126	3008		TR17/635	G0008
30218	102	3008	52	TR05/631	G0003
30218(RCA)	160			TR17/637	
30219	123	3124	20	TR86/53	
30221	126	3006	50	TR17/635	G0003
30222	126	3006	50	TR17/635	G0008
30223	126	3006	50	TR17/635	G0002
30224	123	3124	20	TR86/53	
30226	123	3124	20	TR86/53	
30227	123	3124	20	TR21/735	S0015
30228	123A	3124	20	TR21/735	
30229	123A	3124	20	TR21/735	S0015
30230	126	3006	50	TR17/635	G0003
30231	100	3007	1	TR05/254	G0002
30234	124	3021	12	TR81/240	
30235	123A	3124	10	TR21/735	S0015
30236	175			TR81/	
30238	160	3006	51	TR17/637	G0003
30239		3006	51	TR17/637	G0003
30240	160	3006	51	TR17/637	G0003
30241	123A	3124	20	TR21/735	S0015
30242	123A	3124	20	TR21/735	S0015
30243	123A	3124	20	TR21/735	S0015
30244	102	3004	52	TR05/631	
30245	124	3021	12	TR81/240	
30246	121	3009	16	TR01/232	
30246A	121	3009	16	TR01/232	
30247	126		51	TR17/635	
30248	123	3124	20	TR86/53	
30253	123A	3124		TR21/	

Original Part Number	ECG	SK	GE	IR/WEP	HEP
30254	175	3026	23	TR81/	S5019
30256	175	3026	12	TR81/	S5019
30257	124	3021	14	TR81/240	S5011
30259	123	3124	20	TR86/53	
30263	102A			TR85/	
30267	102A	3004			
30268	123A	3122	20	TR21/735	S0015
30269	123A	3122	20	TR21/735	S0015
30270	159	3114		TR20/717	S0019
30271	185	3083		TR77/S5007	S5006
30272	184	3054	57	TR76/S5003	S5000
30273	160		51	TR17/637	G0003
30274	126		51	TR17/635	G0003
30276	130	3027		TR59/	S7002
30278	129	3025	21	TR88/242	
30289	123A	3024	63	TR21/735	S0014
30290	159	3025	21	TR20/717	S0012
30291	128		63	TR25/243	S5013
30292	108	3018	20	TR95/56	
30293	102A	3006	52	TR85/250	
30294	152	3041	28	TR76/701	S5000
30302	104	3009	52	TR01/230	G6003
30302(RCA)	102A			TR85/250	
30805		3026			
31001	123A	3124		TR21/	
31003	123A	3124		TR21/	
31004-1	153				
31005	159				
31006	129				
31015	234				
31032-0	129				
33037R2		3021			
33188-2	130				
33201-1	712	3072			
33201-2	712	3072			
33324		3126			
33509-1	159				
33563	123A		20	TR21/	
33989-2069	121				
34000		3006			
34001		3006			
34001PC	4001				
34002		3006			
34002PC	4002				
34003		3006			
34005-1	727	3071			
34006		3008			
34011PC	4011				
34012		3024			
34012PC	4012				
34013		3004			
34013PC	4013				
34014		3003			
34014PC	4014				
34015		3004			
34015PC	4015				
34016		3004			
34016PC	4016				
34017		3004			
34017PC	4017				
34018		3004			
34019		3008			
34019PC	4019				
34020		3004			
34020PC	4020				
34021PC	4021				
34022	121	3009	16	TR01/232	G6005
34023PC	4023				
34025		3008			
34025PC	4025				
34027		3010			
34027PC	4027				
34028		3004			
34028PC	4028				
34029		3003			
34029PC	4029				
34030		3003			
34030PC	4030				
34031		3007			
34034A		3122			
34035		3008			
34036G		3005			
34040PC	4040				
34042PC	4042				
34043		3005			
34044	128				
34047		3005			
34048	126				
34048-1	708				
34049		3008			
34049-1	722	3161			
34049PC	4049				
34050		3008			
34050PC	4050				
34051		3008			

Original Part Number	ECG	SK	GE	IR/WEP	HEP
34051PC	4051				
34052		3004			
34052PC	4052				
34053		3008			
34054		3039			
34055		3009			
34055PC	4055				
34057		3008			
34058		3005			
34059		3024			
34063		3003			
34064		3003			
34066PC	4016				
34067		3004			
34068PC	4068				
34069		3004			
34069PC	4069				
34070F		3005			
34070K		3005			
34070PC	4030				
34071PC	4071				
34072F		3005			
34073F		3005			
34074H		3004			
34075F		3004			
34077PC	4077				
34078PC	4078				
34081PC	4081				
34082		3004			
34085PC	4085				
34086PC	4086				
34088		3005			
34089		3004			
34090		3004			
34092		3005			
34093		3005			
34094		3005			
34095		3024			
34098		3004			
34099		3005			
34099PC	4099				
34100		3008			
34101		3004			
34102		3008			
34104A		3008			
34105		3005			
34106		3004			
34107		3011			
34109		3004			
34110		3011			
34111		3008			
34115		3007			
34118	160	3008	51	TR17/637	
34119	100	3005	51	TR05/254	
34123		3007			
34124		3007			
34125		3007			
34130		3011			
34133		3006			
34134		3006			
34135		3006			
34136		3004			
34137		3004			
34139		3024			
34142		3004			
34144		3004			
34147		3004			
34149		3024			
34153C		3007			
34154		3039			
34157		3003			
34158		3005			
34160		3011			
34162		3024			
34163		3024			
34169		3009			
34171		3005			
34172		3039			
34173		3010			
34177		3005			
34179		3007			
34180		3004			
34181		3003			
34182		3004			
34185		3004			
34188		3009			
34202-1	727	3071			
34208	130				
34211		3005			
34212		3005			
34213		3005			
34214		3004			
34215		3005			
34216F		3004			
34217F		3004			

Original Part Number	ECG	SK	GE	IR/WEP	HEP
34219	100	3005	1	TR05/254	
34220	100	3005	1	TR05/254	
34221	100	3005	1	TR05/	
34236		3039			
34239		3122			
34240		3004			
34244		3024			
34254		3024			
34262	102A				
34271		3007			
34275		3011			
34276		3012			
34279		3024			
34285		3011			
34288		3039			
34298	121	3009	3	TR01/232	
34299		3004			
34303		3024			
34304		3004			
34305		3007			
34306		3003			
34311		3007			
34312		3007			
34313		3009			
34314		3009			
34315	121	3009	16	TR01/232	
34321		3006			
34323		3006			
34324		3006			
34325		3006			
34326		3006			
34342	160	3008	51	TR17/637	
34343		3006			
34349		3024			
34353		3004			
34354		3005			
34355		3003			
34357		3004			
34358		3011			
34361		3008			
34363		3007			
34371		3122			
34373		3003			
34374		3003			
34375		3004			
34377		3011			
34379-1	720				
34380		3003			
34382		3008			
34383		3008			
34385		3003			
34388		3005			
34389	160	3008	51	TR17/637	
34390		3007			
34392		3024			
34397		3024			
34400		3005			
34401		3005			
34401-1	762				C6012
34402		3009			
34403		3010			
34407		3004			
34408		3024			
34409		3007			
34418		3005			
34422		3004			
34423	160				
34425	121	3009	3	TR01/232	
34427		3024			
34430		3005			
34433		3004			
34437		3024			
34438		3024			
34441		3024			
34442		3122			
34444		3024			
34446		3004			
34447		3007			
34449		3024			
34450		3024			
34452-1	709				
34454		3011			
34455		3004			
34457		3123			
34459		3014		TR01/	
34460		3004			
34461		3008			
34463		3004			
34464		3122			
34467		3004			
34468		3011			
34475		3122			
34476		3124			
34477		3124			
34478		3024			

Original Part Number	ECG	SK	GE	IR/WEP	HEP
34481		3009			
34482		3009			
34491		3004			
34493P	102	3004			
34494		3011			
34495		3024			
34496		3124			
34497		3009			
34502		3004			
34502-1	778				
34503		3011			
34503-1	722	3161			
34504		3122			
34505		3004			
34512PC	4512				
34513		3005			
34516		3009			
34518		3007			
34518PC	4518				
34520		3009			
34520PC	4520				
34524		3005			
34527PC	4527				
34529		3122			
34536		3005			
34539		3004			
34539PC	4539				
34544		3024			
34545		3004			
34546		3004			
34551		3009			
34552		3024			
34553	160	3004	51	TR17/637	
34554		3024			
34555PC	4555				
34556		3009			
34556PC	4556				
34557		3024			
34558		3004			
34563		3011			
34567		3009			
34569		3005			
34570		3005			
34572		3010			
34575		3124			
34577		3024			
34579		3004			
34580		3124			
34581		3124			
34582		3005			
34583		3005			
34586		3004			
34587		3004			
34588		3024			
34588A	128				
34592		3008			
34594		3006			
34595		3004			
34597		3004			
34598		3004			
34602		3004			
34603		3007			
34604		3004			
34607		3024			
34608		3004			
34611		3011			
34615		3122			
34619		3004			
34620		3011			
34623		3024			
34624		3024			
34626		3024			
34627		3122			
34628		3122			
34629		3039			
34630		3122			
34631		3004			
34633		3007			
34637		3009			
34638		3005			
34640		3039			
34641		3004			
34642		3009			
34643A		3004			
34646		3039			
34647		3004			
34654		3008			
34655		3003			
34656		3003			
34661		3009			
34665		3024			
34667		3004			
34669		3122			
34670		3005			
34671		3005			
34674		3009			
34675	160				
34679		3039			
34680		3004			
34682		3039			
34683		3024			
34685		3024			
34686		3007			
34688		3122			
34695		3024			
34701		3004			
34702		3011			
34706		3004			
34707		3024			
34714		3024			
34715	104	3009	3	TR01/624	
34721		3039			
34722		3124			
34724		3024			
34725		3005			
34729		3024			
34734		3008			
34736		3010			
34740		3122			
34742		3024			
34746		3009			
34751		3009			
34754		3014			
34758		3024			
34760		3024			
34761		3005			
34764		3024			
34766		3024			
34768		3024			
34771		3039			
34773		3004			
34774		3011			
34776		3024			
34777		3024			
34783		3004			
34786		3122			
34791		3003			
34793		3011			
34795		3024			
34796		3004			
34797		3011			
34798		3039			
34801		3039			
34802		3004			
34803		3008			
34808		3012			
34821		3009			
34813		3011			
34824		3006			
34825		3011			
34826		3039			
34835		3024			
34838		3006			
34842		3008			
34847		3004			
34851		3009			
34852		3004			
34857		3008			
34858		3004			
34860		3009			
34863		3004			
34869		3004			
34871	102				
34874B		3004			
34875		3024			
34876		3024			
34877		3005			
34879		3009			
34880		3004			
34881		3004			
34882		3009			
34884		3004			
34886A		3122			
34887A		3122			
34890A		3122			
34891A		3122			
34893		3009			
34897		3009			
34898		3005			
34901		3004			
34905		3008			
34908		3012			
34909		3009			
34914		3004			
34919		3008			
34920		3003			
34921		3004			
34923	158		53	TR84/	
34925		3012			
34926		3009			

Original Part Number	ECG	SK	GE	IR/WEP	HEP
34930		3024			
34937A		3024			
34938		3024			
34939		3024			
34942	160	3004		TR17/	
34945		3012			
34949		3003			
34954		3003			
34962		3005			
34963		3005			
34967		3007			
34971		3008			
34972		3008			
34973		3008			
34974		3003			
34975		3003			
34979		3024			
34983		3003			
34984		3003			
34988		3008			
34989		3008			
34990		3008			
34991		3122			
34992		3122			
34993		3122			
34995		3039			
34999		3122			
35001	130	3024			
35002	175	3007			
35003		3122			
35004	108	3039	11	TR95/56	
35009		3004			
35011		3011			
35014		3004			
35015		3024			
35019		3004			
35022		3004			
35028		3006			
35031		3122			
35032		3122			
35035		3004			
35039		3003			
35040		3010			
35041		3024			
35042		3007			
35044	121	3009	3	TR01/232	G6005
35045	102	3004	2	TR05/631	G0005
35047		3122			
35051		3004			
35052		3004			
35053		3009			
35059		3122			
35059-1	712	3072			
35065		3009			
35066		3004			
35067		3024			
35068		3009			
35069		3122			
35070	160			TR17/	
35073		3122			
35074		3122			
35075		3024			
35077		3024			
35079		3008			
35084	104	3009	16	TR01/624	
35085		3006			
35086	102				
35087		3122			
35090		3008			
35095		3011			
35096		3010			
35097		3008			
35104		3004			
35110		3024			
35117		3024			
35120		3011			
35121		3004			
35122		3011			
35123		3004			
35124X		3004			
35126		3008			
35127		3005			
35134		3024			
35138		3004			
35144	121	3014	3	TR01/232	G6005
35145		3004			
35146		3004			
35147		3004			
35148		3003			
35149		3003			
35150		3004			
35152		3008			
35153		3006			
35154		3008			
35155		3004			

Original Part Number	ECG	SK	GE	IR/WEP	HEP
35156		3009			
35157		3003			
35158A		3027			
35159		3004			
35160		3011			
35161		3004			
35164		3011			
35168	126	3008	51	TR17/635	G0008
35169	160	3008	51	TR17/637	G0002
35170	160	3008	51	TR17/637	G0002
35176		3004			
35178		3004			
35179		3004			
35182		3011			
35192		3004			
35199		3004			
35201	104	3009	3	TR01/624	
35203		3024			
35204		3024			
35209		3024			
35210	128				
35212	128	3122			
35216		3011			
35218	176	3123	52	TR82/	
35219X		3004			
35220		3003			
35227		3004			
35228		3024			
35230		3004			
35231	121	3009	16	TR01/232	
35232		3039			
35237		3004			
35238		3124			
35239		3122			
35240		3024			
35242			20	TR21/	
35243		3039			
35244		3122			
35245		3005			
35246		3008			
35247		3004			
35248		3004			
35249		3004			
35259			11	TR21/	
35260	104	3009	16	TR01/624	
35265		3122			
35266		3004			
35271		3122			
35272		3004			
35273A		3122			
35274		3004			
35276		3004			
35278		3004			
35279		3039			
35281		3004			
35282		3004			
35283		3006			
35284		3006			
35285		3004			
35286		3004			
35296		3011			
35298		3004			
35301		3008			
35302		3004			
35303	128				
35305		3003			
35308		3027			
35309A		3039			
35310		3123			
35311		3008			
35315		3004			
35316		3008			
35317		3008			
35336		3024			
35337		3011			
35338		3005			
35349	121	3009	3	TR01/232	
35354		3122			
35355		3004			
35356		3122			
35357A		3024			
35360		3024			
35364		3124			
35367		3011			
35369		3004			
35379		3009			
35383	128				
35388		3004			
35391		3011			
35392		3004			
35396		3024			
35400		3009			
35401		3004			
35402		3004			
35405	175	3004		TR81/	

Original Part Number	ECG	SK	GE	IR/WEP	HEP
35406		3004			
35407		3021			
35408		3009			
35409		3024			
35413		3027			
35414		3039			
35420		3009			
35437		3009			
35438		3024			
35449	108	3019	17	TR95/56	S0016
35452-2	158			TR84/630	G0005
35454	102	3004	2	TR05/631	
35454-1	158		2	TR84/630	G0005
35454-2	158		2	TR84/630	
35454-3	158		2	TR84/630	G0005
35463		3027			
35464		3027			
35465		3027			
35466B		3027			
35467		3027			
35470		3027			
35471		3024			
35473		3004			
35478		3004			
35479		3006			
35484		3021			
35487		3009			
35488		3008			
35489		3039			
35491		3122			
35496		3004			
35497		3004			
35498		3004			
35499A		3009			
35503		3027			
35504		3024			
35505		3027			
35507		3004			
35508		3009			
35511		3024			
35512		3024			
35513		3024			
35515		3008			
35518		3027			
35525		3024			
35527		3039			
35528		3039			
35529		3024			
35530		3024			
35540		3024			
35549		3039			
35550		3004			
35553		3024			
35555		3024			
35556		3003			
35557		3004			
35559		3004			
35560		3039			
35563		3007			
35565		3004			
35577		3122			
35578		3010			
35580		3024			
35582		3009			
35588		3011			
35589		3004			
35590	102A	3004	53	TR85/250	
35591		3004			
35594		3008			
35598		3007			
35600		3004			
35601		3024			
35603		3004			
35606		3004			
35607A		3021			
35619		3009			
35621		3039			
35628	102A	3004	2	TR85/250	G0005
35629		3009			
35634		3004			
35636		3024			
35640		3004			
35643		3027			
35645		3004			
35651		3021			
35652		3004			
35655		3008			
35657		3027			
35658		3027			
35659		3027			
35660		3027			
35661		3026			
35663		3026			
35666		3027			
35667		3027			
35674		3024			
35677	102A	3008	2	TR85/250	G0005
35678	102A	3008	2	TR85/250	G0005
35684		3009			
35686		3004			
35688		3004			
35691		3004			
35693		3009			
35694		3122			
35696		3009			
35699		3004			
35700		3004			
35702		3024			
35703		3024			
35704		3024			
35706		3012			
35708		3024			
35709		3024			
35710		3003			
35711		3008			
35712		3003			
35718		3004			
35720		3008			
35721		3008			
35722		3008			
35723		3008			
35725		3009			
35728	121	3008	16	TR01/254	G0005
35729		3004			
35734		3008			
35735		3024			
35737		3027			
35744		3004			
35745		3006			
35746		3024			
35747		3009			
35748		3011			
35750		3004			
35751		3004			
35756		3006			
35757		3009			
35760		3005			
35761A		3122			
35762		3005			
35764		3003			
35766		3009			
35767		3011			
35768		3004			
35770		3024			
35772		3039			
35774		3004			
35777		3024			
35782		3024			
35783		3026			
35785		3027			
35787		3024			
35790		3039			
35792	102A	3004	52	TR85/	G0008
35793		3024			
35800		3004			
35801		3027			
35810		3009			
35811		3024			
35812		3024			
35814		3009			
35815	160	3008	51	TR17/637	G0002
35816	102A	3008	51	TR85/250	G0005
35817	102A	3008	51	TR85/250	G0005
35818	126	3008	51	TR17/635	
35819	158	3004	2	TR84/630	G0005
35820	102	3004	2	TR05/631	
35820-1	158	3004	2	TR84/630	G0005
35820-2	158	3004	2	TR84/630	G0005
35820-3	158	3004	2	TR84/630	G0005
35824	102A	3008	1	TR85/250	G0005
35826		3008			
35827		3004			
35828		3024			
35831		3011			
35832		3122			
35833		3024			
35836		3009			
35840		3024			
35842		3024			
35851		3005			
35853		3004			
35856		3024			
35857		3024			
35858		3024			
35859		3024			
35866		3039			
35867		3039			
35868		3039			
35872		3122			
35873		3122			

Original Part Number	ECG	SK	GE	IR/WEP	HEP
35875		3004			
35876		3004			
35880		3009			
35881		3009			
35882		3123			
35883		3004			
35884		3004			
35885A	121	3014	3	TR01/232	G6003
35885B	121	3014	3	TR01/232	
35888	128				
35889		3004			
35893		3024			
35895		3004			
35900		3026			
35901		3027			
35906		3019			
35910		3004			
35912		3014			
35913		3003			
35914		3009			
35921		3122			
35924		3122			
35926		3024			
35927		3024			
35935		3027			
35938		3024			
35939		3039			
35947		3008			
35949		3004			
35950	102		2	TR05/631	
35951	121	3009	3	TR01/232	
35952	102	3004	2	TR05/631	
35953	102	3004	52	TR05/631	
35954	102	3004	52	TR05/631	
35955	102	3004	52	TR05/631	
35957		3004			
35959		3024			
35959		3024			
35960		3024			
35962		3009			
35964		3024			
36145	128				G0011
36148		3009			
36149		3039			
36151		3004			
36156		3024			
36158		3004			
36159		3024			
36164		3011			
36165		3011			
36175		3008			
36176		3003			
36177		3004			
36178		3004			
36179P		3011			
36182		3027			
36200		3122			
36203	121	3019		TR01/	
36208		3079			
36212	108	3039	20	TR95/56	S5026
36212V1	108	3039	20	TR95/56	S5026
36213	188	3048		TR72/	S3020
36216		3009			
36219		3011			
36220		3027			
36224		3122			
36226		3004			
36228		3024			
36229		3011			
36230		3024			
36240		3009			
36245		3009			
36252P		3010			
36254P		3009			
36258		3021			
36260P		3004			
36261P		3011			
36265P		3011			
36266		3004			
36274	175				
36284		3012			
36290		3024			
36292		3008			
36293		3021			
36294		3027			
36302		3024			
36303	121	3009	3	TR01/232	
36304	121	3009	3	TR01/232	G6005
36304-4	121	3009	3	TR01/232	
36305		3024			
36312	121	3009		TR01/	
36313		3004			
36315		3004			
36317		3004			
36320	175				

Original Part Number	ECG	SK	GE	IR/WEP	HEP
36323		3122			
36329		3024			
36334		3027			
36335		3027			
36336		3004			
36337		3122			
36338		3027			
36340	176				
36344		3021			
36350		3007			
36358		3011			
36359		3009			
36359-4	121MP				
36362		3004			
36363		3004			
36367		3122			
36370		3122			
36370-05490	175				
36371		3122			
36374		3003			
36375		3010			
36387	128		63	TR87/	
36388		3004			
36389		3024			
36390		3012			
36393		3024			
36395	121	3009	3	TR01/232	
36400		3027			
36403		3004			
36406		3004			
36409		3024			
36411		3024			
36412		3027			
36414		3008			
36415		3004			
36416		3009			
36420		3009			
36425		3117			
36426		3018			
36427		3117			
36432		3004			
36442		3004			
36443		3004			
36444		3004			
36446		3039			
36447		3039			
36449		3024			
36457		3024			
36459		3027			
36461		3124			
36462		3009			
36464		3004			
36466	128				
36473		3027			
36477	121	3009	16	TR01/232	
36479		3039			S0017
36481		3009			
36482		3009			
36495		3014			
36497		3024			
36498		3027			
36510		3036			
36513		3021			
36518		3039			
36519		3039			
36521		3039			
36534	102	3027	2	TR05/631	
36535		3027			
36540		3004			
36545	130				
36548		3024			
36552		3024			
36553		3024			
36555		3009			
36557	102	3004	2	TR05/631	
36558	102	3004	2	TR05/631	
36559	126	3006	51	TR17/635	
36560	126	3006	51	TR17/635	
36563	126	3006	51	TR17/635	
36567		3004			
36574		3009			
36577	129	3025		TR88/242	
36578	108	3039		TR95/56	
36579	128	3024		TR87/243	
36580	123A	3122		TR21/735	
36581	108			TR95/56	
36582	132	3116		FE100/802	
36585		3122			
36607		3009			
36613		3039			
36615		3008			
36616		3027			
36618		3027			
36620		3122			
36625		3024			

Original Part Number	ECG	SK	GE	IR/WEP	HEP
36627		3122			
36628		3122			
36629		3122			
36630		3122			
36634	124	3021	12	TR81/240	
36637		3004			
36638		3021			
36640		3005			
36642		3027			
36643		3024			
36646		3036			
36647		3024			
36649		3027			
36651		3122			
36655		3079			
36657		3027			
36662	128				
36665		3039			
36667		3009			
36670		3122			
36671		3122			
36672		3122			
36673	176				
36675		3014			
36678		3009			
36680	128				
36680A	128				
36682	128				
36685		3122			
36687	121				
36688		3009			
36689		3004			
36691		3025			
36693		3027			
36695	176				
36700		3025			
36701		3026			
36702		3036			
36703		3027			
36705		3024			
36706		3004			
36707		3021			
36709		3024			
36714		3027			
36715		3008			
36716		3008			
36717		3008			
36718		3009			
36720		3027			
36721		3021			
36722		3004			
36723		3004			
36726		3026			
36727		3004			
36729		3009			
36730		3004			
36731		3004			
36733		3026			
36734		3036			
36739		3024			
36741		3004			
36742		3011			
36745		3024			
36747		3027			
36748	128				
36752		3004			
36756		3124			
36762		3026			
36763		3024			
36764		3004			
36765		3021			
36768		3008			
36775		3027			
36777		3021			
36779		3026			
36781		3036			
36783		3004			
36784		3021			
36788		3081			
36790		3014			
36792		3009			
36796		3024			
36799		3021			
36800-2	121				
36800-3	121				
36800-4	121				
36800-5	121				
36800-6	121				
36800-7	121				
36801		3024			
36803		3024			
36808		3004			
36813		3014			
36815		3004			
36816	100	3005	52	TR05/254	
36818		3027			
36821		3024			
36827		3004			
36828		3021			
36831		3019			
36834		3122			
36838		3027			
36839		3009			
36841		3014			
36843		3122			
36845		3122			
36846	130				
36847	108	3039	20	TR95/56	S3005
36848		3024			
36849		3024			
36850		3024			
36851		3122			
36855	130				
36856		3036			
36860		3122			
36861		3004			
36865		3036			
36866		3036			
36868		3024			
36875		3027			
36876		3019			
36878		3006			
36880		3009			
36882		3024			
36886		3021			
36887		3004			
36891		3036			
36892	130				
36893		3027			
36896	104	3009	3	TR01/624	
36899		3014			
36904		3024			
36906		3027			
36910	121	3009	3	TR01/232	
36913	195			TR65/	S3001
36917	123A	3039		TR21/56	S0015
36918	108	3039	17	TR95/56	
36919	108	3039	17	TR95/56	
36920	123A	3122	17	TR21/735	
36921	123A	3122	17	TR21/735	
36924		3009			
36928		3009			
36940		3004			
36943		3009			
36944		3004			
36946	130				
36947		3027			
36953	130				
36961		3036			
36962		3044			
36963		3004			
36964		3004			
36966		3010			
36967		3024			
36968		3027			
36970		3036			
36971	121				
36972		3036			
36973		3036			
36986		3004			
36997		3024			
37077	175				S5012
37085(FENDER)		3036			
37260		3039			
37261		3004			
37262		3024			
37263		3025			
37264		3027			
37265		3024			
37267	130				
37269	129				
37272		3027			
37278	160	3004	51	TR17/637	
37279	103A	3010	54	TR08/724	
37280	128				
37282		3026			
37284		3045			
37287	128				
37288		3025			
37289		3024			
37290		3004			
37291		3021			
37292		3079			
37294		3027			
37301		3027			
37302		3036			
37306		3014			
37308		3034			
37312		3044			
37314		3027			

Original Part Number	ECG	SK	GE	IR/WEP	HEP
37326		3027			
37334	161			TR83/	
37339		3014			
37346		3122			
37347		3122			
37376		3027			
37377		3027			
37383	108	3039	20	TR95/56	S0015
37384	108	3039		TR95/56	
37389		3009			
37393	128				
37413		3024			
37420		3021			
37421		3009			
37422		3009			
37428		3024			
37429		3025			
37431		3024			
37432		3024			
37433		3024			
37437		3019			
37438		3004			
37440		3011			
37445	128				
37448		3007			
37449		3007			
37451		3034			
37452		3034			
37458		3024			
37460		3024			
37461		3025			
37464	128	3024	18	TR21/735	
37465		3009			
37466		3122			
37471		3009			
37475	130				
37476	181				
37478		3025			
37483		3009			
37484	175				
37486	159				
37487		3122			
37490		3113			
37494		3009			
37495		3117			
37496		3117			
37497		3117			
37498		3004			
37499		3124			
37500		3004			
37501		3021			
37502		3006			
37503		3006			
37504		3006			
37510-161	199	3122	20	TR21/	S0015
37510-162	123A	3123	20	2SC945/	S0015
37510-163	123A	3122	20	TR21/	S0015
37510-167	1081				
37511		3047			
37516		3004			
37517		3021			
37518		3124			
37519		3024			
37524		3005			
37526		3036			
37527		3124			
37533		3025			
37534		3036			
37536		3122			
37549	102A	3004		TR85/250	
37550	102A	3004		TR85/250	
37551	102A	3004		TR85/250	
37552	103A	3010		TR08/724	
37555		3122			
37556		3009			
37561		3027			
37563	130				
37567		3021			
37570		3004			
37572		3122			
37576		3021			
37579		3009			
37580		3036			
37584	124	3021	12	TR81/240	
37585	123	3124	10	TR86/53	
37589		3027			
37590		3009			
37591		3021			
37592		3021			
37597		3024			
37599	175				
37601		3124			
37614		3122			
37615		3011			
37633		3009			

Original Part Number	ECG	SK	GE	IR/WEP	HEP
37637		3021			
37639		3122			
37640		3008			
37641		3008			
37642		3008			
37643		3004			
37644		3003			
37645		3024			
37649	128	3024			
37651		3027			
37657		3024			
37663	130				
37664	129				
37670		3024			
37676		3004			
37677	102				
37684		3004			
37685		3004			
37686		3124			
37687		3124			
37693		3005			
37694	128	3024	20	TR87/243	S0015
37694A	108				
37694B	108				
37697		3039			
37702-1	714	3075			
37703-1	715	3076			
37704-1	790				
37712		3027			
37714		3079			
37724		3025			
37725	154				
37730	124	3021	12	TR81/240	
37732		3025			
37733		3024			
37734		3027			
37735		3024			
37740	129				
37741	128				
37747		3018			
37754		3009			
37756		3004			
37757		3004			
37758		3010			
37759		3003			
37763	175				
37764	129				
37767	128				
37769		3024			
37778		3004			
37779		3122			
37781		3122			
37782		3039			
37784		3009			
37792		3027			
37793	129				
37797		3122			
37798		3008			
37800	128	3024	20	TR87/243	S0011
37802		3024			
37803		3025			
37804		3012			
37805		3024			
37806	128				
37821		3079			
37823		3004			
37824		3004			
37827		3079			
37833	100				
37833-2	909D	3590			
37837		3024			
37838		3009			
37839		3021			
37840	128				
37844		3027			
37847	128		18	TR87/	
37851		3014			
37859		3019			
37860		3039			
37864		3010			
37876		3039			
37884	123A	3122	20	TR21/735	
37885		3024			
37888	130	3027			
37889		3021			
37890		3027			
37891		3018			
37894		3024			
37895		3024			
37896		3004			
37897		3014			
37899	128				
37900	175				
37906		3004			
37913	175				

Original Part Number	ECG	SK	GE	IR/WEP	HEP
37917		3024			
37918	129				
37919		3004			
37920		3004			
37921		3004			
37922		3004			
37923		3004			
37924		3008			
37925		3007			
37926		3018			
37932	128				
37936		3019			
37938		3036			
37940		3124			
37941		3124			
37949		3009			
37951		3021			
37952		3004			
37954		3004			
37959		3025			
37962		3011			
37966	129				
37967	181				
37974	181				
37975	128				
37980		3025			
37982	128				
37983		3024			
37986-3563	107				
37986-4040	107				
37986-4046	107				
37993		3007			
37995		3004			
37997		3044			
37998		3018			
38011		3003			
38014		3009			
38015		3079			
38018		3044			
38020		3009			
38023		3025			
38029		3027			
38031		3036			
38034		3021			
38045	128				
38048		3009			
38049	181				
38055		3004			
38057	102A				
38058	128				
38060		3003			
38061		3010			
38063		3024			
38066		3019			
38071		3009			
38091	158			TR84/630	G0005
38092		3124			
38093	126				
38094	121MP	3014			G6005
38095	159			TR20/	S0032
38096		3005			
38098		3018			
38105		3047			
38112	124				
38119A		3039			
38120	154			TR78/	S5026
38121	154			TR78/	S5026
38122	195			TR65/	S3001
38131		3024			
38134		3027			
38136		3009			
38137	130	3510			
38138	130	3510			
38142		3021			
38145		3021			
38149		3027			
38158		3021			
38159		3044			
38166	130	3510			
38171-61		3024			
38173		3510			
38175	103		8	TR08/641A	
38176	102A	3004	52	TR05/631	
38177	102	3004	52	TR05/631	G0005
38178	123	3124	8	TR86/53	S0015
38179		3021			
38180		3014			
38182	128				
38185-0001		3024			
38190		3024			
38199	101				
38200	101				
38203		3510			
38204		3021			
38207	108	3039	20	TR95/56	

Original Part Number	ECG	SK	GE	IR/WEP	HEP
38208	108	3039	20	TR95/56	
38209	100	3005	51	TR05/254	
38210		3009			
38214		3510			
38227		3112			
38229		3021			
38230		3122			
38231		3044			
38232		3021			
38233		3021			
38236		3036			
38246	108		17	TR21/	S0016
38246A	108				
38251		3018			
38252		3008			
38254		3021			
38255		3036			
38258		3004			
38264		3024			
38265		3044			
38265-00000	784	3524			
38265-00010	784	3524			
38265-00020	784	3524			
38265-00030	784	3524			
38266		3004			
38267	181				
38268	181				
38269	158			TR84/	
38270	128	3024		TR87/	
38271	128	3025		TR87/	
38272	130	3027		TR59/	
38276		3024			
38281		3024			
38283	123A				
38293		3122			
38294		3024			
38296		3021			
38300		3027			
38302		3024			
38303		3024			
38306		3510			
38308		3009			
38317		3018			
38321		3024			
38325		3027			
38330		3004			
38331		3004			
38333		3025			
38334	128				
38335	223				
38338		3024			
38343		3048			
38345		3021			
38354	128				
38357		3004			
38361	128				
38364		3027			
38367		3027			
38371		3021			
38374		3036			
38376		3510			
38378	220				
38386		3024			
38387		3024			
38388	129				
38394		3036			
38397	181				
38398	128				
38401		3004			
38402		3004			
38403		3004			
38404		3010			
38406		3004			
38409		3039			
38415		3024			
38418		3021			
38422		3004			
38424	128				
38428		3024			
38432	128				
38433		3018			
38435		3006			
38438		3027			
38443	175				
38446-00000	724	3525			
38446-00010	724	3525			
38446-00020	724	3525			
38446-00030	724	3525			
38448	186		28	TR55/	3020
38456		3036			
38457		3024			
38458	129				
38467		3011			
38468	128				
38473	181	3036			

Original Part Number	ECG	SK	GE	IR/WEP	HEP
38474	130	3027			
38475	128				
38476	128				
38478	199				
38482		3024			
38484		3018			
38487		3004			
38491	181	3036			
38492		3027			
38493		3036			
38494	130				
38495	128				
38496	129				
38497	128				
38504		3510			
38505		3027			
38511	108		17	TR21/	S0015
38513		3024			
38513-50360C	223				
38514		3024			
38517		3009			
38519		3014			
38522		3024			
38523		3024			
38533	126				
38534		3027			
38535		3027			
38536		3027			
38542		3011			
38551	128				
38555		3009			
38559		3036			
38563	222				
38564		3024			
38566		3011			
38570		3009			
38572		3014			
38576		3024			
38577		3025			
38581		3124			
38584		3027			
38595		3024			
38598		3025			
38599		3024			
38617		3122			
38618		3004			
38619		3004			
38620		3025			
38621		3024			
38626	130				
38632		3027			
38634		3025			
38635		3024			
38636		3004			
38645		3024			
38654	129				
38657		3019			
38659	128				
38662		3004			
38669		3036			
38674		3122			
38680	160				
38681	160				
38685	102A				
38686		3004			
38687		3124			
38688		3018			
38689		3117			
38690		3008			
38691		3010			
38692		3004			
38695		3025			
38705		3036			
38706		3024			
38707		3027			
38708		3003			
38709		3124			
38710		3018			
38711		3117			
38712		3008			
38713		3010			
38714		3004			
38716	128	3024		TR87/	S5014
38718		3122			
38720		3004			
38721		3004			
38722		3004			
38725	128				
38726		3025			
38727		3039			
38729		3014			
38730		3014			
38731(KALOF)	130			TR59/	
38733	152				
38734	129				

Original Part Number	ECG	SK	GE	IR/WEP	HEP
38735	128				
38736	128				
38737	129				
38741		3036			
38746		3027			
38747		3024			
38759		3009			
38774		3024			
38779		3024			
38783		3036			
38785	108	3018	17	TR95/56	S0016
38786	108	3018	17	TR95/56	S0016
38787	161	3018	20	TR83/719	S0015
38788	123A	3124	10	TR21/735	S0011
38789	128	3024		TR87/243	S5014
38790		3025			
38794		3021			
38795		3027			
38796		3024			
38799		3021			
38804	223				
38807		3005			
38808		3038			
38810		3036			
38821		3005			
388837	128				
38842		3036			
38844		3122			
38846		3011			
38863		3005			
38864		3005			
38865		3011			
38867		3021			
38869	128				
38870	129				
38871		3079			
38872		3036			
38875		3079			
38882		3024			
38884		3009			
38894		3122			
38897	130	3027			
38904		3027			
38906		3027			
38908		3027			
38909		3027			
38913		3079			
38914		3021			
38915		3024			
38916	128				
38919		3021			
38920	108	3039		TR95/56	
38921	108	3039		TR95/56	
38923		3024			
38927-00000	905				
38932		3004			
38935		3011			
38939		3024			
38941		3021			
38943		3027			
38950		3024			
38954		3036			
38959		3036			
38961		3124			
38962		3024			
38964		3024			
38965	181	3036			
38971	128				
38978		3025			
38981		3035			
38984		3035			
38994		3018			
38996	154				
39003		3006			
39004		3006			
39023		3021			
39024		3018			
39025		3018			
39034	123A				
39053	160				
39054		3021			
39055		3021			
39058		3039			
39060-1	747				
39074		3122			
39075-1	738	3167			
39076		3054			
39077		3036			
39081		3019			
39085		3027			
39096	123A		20	TR21/	S0015
39097	128				
39113		3021			
39114	129				
39123	124				

Original Part Number	ECG	SK	GE	IR/WEP	HEP
39127	130				
39136		3027			
39140	181				
39143		3024			
39146		3024			
39148	130				
39149	124				
39161		3021			
39173		3027			
39179		3021			
39190		3027			
39196	181				
39202		3009			
39204		3009			
39206		3009			
39207		3009			
39208		3009			
39223		3021			
39231	128				
39238	128				
39247		3124			
39248	128				
39250	129				
39251	130				
39252	128				
39255	128				
39265		3039			
39271		3025			
39272		3024			
39276		3122			
39277		3122			
39278		3021			
39280		3036			
39285	175				
39302	152				
39311	128				
39322		3004			
39329	128				
39331	108				
39334		3011			
39335		3027			
39343	223				
39352		3124			
39369	130				
39372		3036			
39376		3027			
39411		3036			
39412		3036			
39414	130				
39421		3027			
39443	128				
39447		3027			
39455	181				
39458	184	3054	28	TR76/S5003	S5000
39460		3021			
39462	128				
39465	181				
39466	181				
39477	124				
39485	128				
39486	128				
39492	130				
39510	157			TR60/	
39555		3024			
39561	128				
39581	223				
39582		3036			
39586		3027			
39587	128				
39603		3054			
39616	181				
39617	128				
39618	129				
39619	129				
39628		3036			
39635	181	3036			
39665		3027			
39696		3036			
39700		3027			
39705	128				
39713	128				
39715	124				
39730	108				
39731	108				
39741	192				
39742		3027			
39750	152				
39750(ORRTRON)	184	3190		TR76/S5003	S5000
39751	181	3036			
39765		3027			
39767	152		57	TR55/	S5014
39789	108				
39789(POWER)	152				
39800		3027			
39803A	223				

Original Part Number	ECG	SK	GE	IR/WEP	HEP
39819	223				
39824	152	3054			
39828		3036			
39835	128				
39837		3027			
39842	128				
39845		3027			
39846		3009			
39847		3009			
39848		3027			
39849		3021			
39850		3021			
39853	129				
39863	128				
39864	128				
39865	129				
39868	128				
39870		3036			
39876	192				
39885		3027			
39889		3036			
39893	121				
39901-0001	105	3102		TR03/	
39919	192				
39920	128				
39921		3027			
39922	223				
39923	194				
39940	128				
39948	152				
39954	130				
39970		3027			
39981	152				
39995		3036			
39996		3036			
40004	160	3008	52	TR17/637	G0008
40005	160		52	TR17/637	G0008
40006	160	3008	52	TR17/637	G0008
40015		3027			
40019		3036			
40020		3009			
40022	104	3009	25	TR01/628	G6013
40034	102A				G0005
40034-1	102A	3004		TR85/250	
40034-2	102A	3004		TR85/250	
40034-3	102A	3004		TR85/250	
40034VM	102A	3004		TR85/250	G0006
40035	102A			TR85/250	G0005
40035-1	102A	3004		TR85/250	
40035-2	102A	3004		TR85/250	
40035-3	102A	3004		TR85/250	
40036	102A			TR85/250	G0005
40036-1	102A	3004		TR85/250	
40036-2	102A	3004		TR85/250	
40036-3	102A	3004		TR85/250	
40037	103A			TR08/724	G0011
40037-1	103A	3010		TR08/724	
40037-2	103A	3010		TR08/724	
40037-3	103A	3010		TR08/724	
40037VM	103A	3010		TR08/724	G0011
40038	102A			TR85/250	G0005
40038-1	102A	3004		TR85/250	
40038-2	102A	3004		TR85/250	
40038-3	102A	3004		TR85/250	
40038VM	102A	3010		TR85/250	G6011
40050	104	3009	25	TR01/230	G6003
40051	121	3009	25	TR01/230	G6005
40051-2	121		3	TR01/232	G6005
40053	128		63	TR87/243	S5014
40080	128	3046	20	TR87/243	S0014
40081	128	3047	215	TR87/243	S5026
40082	195	3048	28	TR65/243	S3001
40084	123A	3122		TR21/735	S0011
40151	130				
40217	123A	3122	20	TR21/735	S0016
40218	123A	3122	20	TR21/735	S0011
40219	123A	3122	20	TR21/735	S0011
40220	123A	3122	20	TR21/735	S0011
40221	123A	3122	20	TR21/735	S0011
40222	123A	3122	20	TR21/735	S0011
40231	161	3124	17	TR83/719	S0011
40232	161	3124	20	TR83/719	S0011
40233	161	3124	20	TR83/719	S0011
40234	161	3124	210	TR83/719	S0011
40235	161	3018	17	TR83/719	S0016
40236	161	3018	17	TR83/719	S0016
40237	161	3018	17	TR83/719	S0016
40238	161	3117	39	TR83/719	S0017
40239		3117	39		S0016
40240	161	3117	39	TR83/719	S0017
40242	161	3122	39	TR83/719	S0015
40243	161	3122	39	TR83/719	S0016
40244	161	3122	39	TR83/719	S0016
40245	161	3122	39	TR83/719	S0015
40246	161	3122	39	TR83/719	S0015

Original Part Number	ECG	SK	GE	IR/WEP	HEP
40250	175	3026	66	TR81/241	S5012
40250V1		3026	28	TR76/	S5000
40251	175	3027	77	TR81/241	S7004
40251V1	175/404				
40253	102A	3004	53	TR85/250	G0005
40254	104	3009	25	TR01/628	G6013
40256	191			TR75/	S3021
40259	161		11	TR83/719	S0017
40260	161		11	TR83/719	S0016
40261	126		51	TR17/635	S0030
40262	126	3008	51	TR17/635	G0009
40263	126	3004	50	TR17/635	G0005
40263(RCA)	160				
40264	124	3021	27	TR81/240	S3021
40264V1	124	3021	12	TR81/240	
40268	160			TR17/637	G0003
40269	100	3005	1	TR05/254	G0005
40282	175		66	TR81/241	
40283	123A		210	TR21/735	S0014
40290	123		20	TR86/53	S3008
40292			66		
40294	108		86	TR95/56	S0016
40295	108	3039		TR95/56	S0016
40296	161		86	TR83/	S0017
40305				TR76/	S3001
40306				TR76/	
40307			66		
40309	128	3024	63	TR87/243	S0014
40309V1	128	3024		TR87/	
40309V2	128	3024		TR87/	
40310	175	3026	66	TR81/241	S5012
40310V1	175	3026		TR81/	
40311	188	3024	63	TR72/53	S0014
40311A	188	3199		TR72/	
40311B	188	3199		TR72/	
40312	175	3026	66	TR81/241	S5012
40312V1	175	3026		TR81/	
40313	124	3021	18	TR81/240	S5011
40314	128	3024	63	TR87/243	S0015
40314V1	188	3024		TR72/	
40314V2	188	3024		TR72/	
40315	128	3024	63	TR87/243	S0014
40315V1	188	3024		TR72/	
40315V2	188	3024		TR72/	
40316	175	3026	66	TR81/241	S5012
40317	128	3024	63	TR87/243	S0015
40317V1	188	3024		TR72/	
40317V2	188	3024		TR72/	
40318	124	3021		TR81/240	S5011
40319	129	3025	67	TR88/243	S5022
40319V1	188	3025		TR72/	
40319V2	188	3025		TR72/	
40319L			67		
40319S			67		
40320	128	3024	63	TR87/243	S0015
40320V1	188	3024		TR72/	
40320V2	188	3024		TR72/	
40320L			63		
40320S			63		
40321	198	3044	32	TR78/243	S3021
40321V1	188	3044		TR72/	
40321V2	188	3044		TR72/	
40321L			32		
40322	124	3021	12	TR81/240	S5011
40323	128	3024	63	TR87/243	S0014
40323V1	188	3024		TR72/	
40323V2	188	3024		TR72/	
40324	175	3026	66	TR81/241	S5012
40325	130	3027	77	TR59/247	S7004
40326	128	3024	63	TR87/243	S0014
40326V2	188	3024		TR72/	
40326L			63		
40326S			63		
40327	128	3044	32	TR87/243	
40327V2	188	3045		TR72/	
40327L			32		
40328	124	3021		TR81/240	S5011
40329	102A	3004	212	TR85/250	G0005
40338					S3004
40342					S5000
40343					S5000
40346	198	3044	32		S3021
40346V1	19I	3537		TR75/	S3021
40346V2	225	3045		TR78/	S3021
40346L			32		
40347	128		63	TR87/243	S5014
40347V1		3536	28	TR76/	S3020
40347V2	225		28	TR76/	S3020
40347L			46		
40348			32	TR87/243	S5014
40348V1		3536			S3019
40348V2	225				S3019
40349			32	TR21/	S3021
40349V1		3536			S3021
40349V2	225				S3021

Original Part Number	ECG	SK	GE	IR/WEP	HEP
40350	128		11	TR87/243	S0015
40351	108		11	TR95/56	S0015
40352	108		11	TR95/56	S0015
40354	154	3040	27	TR78/712	S5025
40355	154	3040	27	TR78/712	S5025
40359	102	3004	2	TR05/631	G0005
40360	128	3024	18	TR87/243	S3020
40360V1	188	3024		TR72/	
40360V2	188	3024		TR72/	
40361	128	3024	18	TR87/243	S5014
40361V1		3025			
40361V2	188	3024		TR72/	
40361V3	188	3199		TR72/	
40362	129	3025		TR88/242	S5022
40362V1	189	3025		TR73/	
40362V2	189	3025		TR73/	
40363	130	3027	75	TR59/247	S7004
40364	175	3026		TR81/	S5019
40366	128	3024	32	TR87/243	S3019
40367	128	3044		TR87/243	S3019
40369	130	3079	14	TR59/247	S7004
40372	175	3026	28	TR81/	S5019
40373	175			TR81/	S5012
40374	124	3021		TR81/	
40385	128		32	TR87/243	
40385V1	188	3199		TR72/	
40385V2	188	3199		TR72/	
40389		3045	28	TR76/	
40390	191	3044	32	TR75/	
40391	128/401		29	TR77/	
40392	225	3045	28	TR76/	
40395	102A	3004	53	TR85/250	G0005
40396N	103A		54	TR08/724	G0011
40396P	102A		53	TR85/250	
40397	123A	3124	210	TR21/735	S0015
40398	123A	3124	210	TR21/735	S0015
40399	123A	3124	210	TP21/735	S0015
40400	123A	3124	210	TR21/735	S0015
40403	100	3005		TR05/254	G0007
40404	123A	3122	20	TR21/735	S0015
40405	123A		20	TR21/735	S0014
40406	129	3025	67	TR88/242	S5013
40407	129	3024	63	TR88/242	S5014
40408	190	3024	20	TR74/3021	S3019
40409	128	3024	18	TR87/243	
40410	129	3025	21	TR88/242	
40411	181	3036	14	TR36/S7000	S7000
40412	198	3044	32	TR87/243	S3021
40412V1		3537			S3021
40412V2	225	3045		TR78/	S3021
40413	108	3039		TR95/	
40414	108	3039	86	TR95/	
40421	130	3014	16	TR59/247	G6005
40422	124	3021	12	TR81/240	S5011
40423	124	3021	12	TR81/240	S5011
40424	175	3021	12	TR81/241	S5012
40425	124	3021	12	TR81/240	S5011
40426	124	3021	12	TR81/240	S5011
40427	124	3021	12	TR81/240	S5011
40432	123A			TR21/735	S0015
40434		3024			
40437V1	188	3199		TR72/	
40437V2	188	3199		TR72/	
40438V1	190			TR74/	
40438V2	190			TR74/	
40439	191	3035	25	TR75/235	G6008
40439V1	191			TR75/	
40439V2	191			TR75/	
40440	127	3035	25	TR27/235	G6007
40444	181			TR36/S7000	S7000
40446	224		28	TR76/	
40450			63	TR76/	S0014
40451			63	TR76/	S0014
40452			63	TR76/	S5014
40453			63	TR76/	S0015
40454			63	TR76/	S0014
40455			63	TR76/	S0015
40456			63	TR76/	S0014
40457	128	3024	47	TR87/243	S3024
40458	195	3122	47	TR65/	S3001
40459	154	3040	63	TR78/712	S5014
40461	133		FET1	/801	F0010
40461-2	128		20	TR87/243	
40462	104	3009	25	TR01/232	G6005
40464	130	3027	73	TR59/247	S7002
40465	130	3027	75	TR59/247	S7002
40466	130		75	TR59/247	S7002
40468	133			/801	F0015
40469	108	3039	39	TR95/56	S0016
40470	108	3039	39	TR95/56	S0016
40471	130		39	TR59/247	S0016
40472	108		39	TR95/56	S0015
40473	123A		39	TR21/	S0030
40474	123A		39	TR21/	S0030
40475	108		39	TR95/56	S0015

Original Part Number	ECG	SK	GE	IR/WEP	HEP
40476	123		39	TR86/53	S0016
40477	123A	3122	39	TR21/735	S0015
40478	108	3018	39	TR95/56	S0016
40479	108	3018	39	TR95/	S0016
40480	108		39	TR95/56	S0016
40481	108		39	TR95/56	S0016
40482	108		39	TR95/56	S0016
40487	160		1	TR17/637	G0008
40488	126		50	TR17/635	G0008
40489	160		1	TR17/637	G0008
40490	102A	3004	2	TR85/250	G0006
40491	124	3021	12	TR81/240	S5011
40500	123A		20	TR21/	S0014
40501	128	3024	18	TR87/243	
40513	223			TR59/	S5004
40514	223	3036		TR59/	S7002
40517	123	3039	86	TR86/53	S0017
40518	123		86	TR86/53	S0017
40519	123		20	TR86/53	S0014
40537	129	3025	67	TR88/	S5022
40538	129	3025	67	TR88/	S5022
40539	128	3024	63	TR87/243	S0015
40542	223			TR59/	S7002
40543	223			TR59/	S7002
40544	225	3045			
40546	124	3021	12	TR81/240	S5011
40547	124	3021	12	TR81/240	S5011
40559A	220				F2005
40577	123A			TR21/	S0015
40578	128		28	TR87/243	S3008
40581	195	3048	28	TR65/	S3001
40582	224		28	TR76/	
40594	188	3512	66	TR72/	S3002
40595	189	3513		TR73/	S3032
40595VX	129				
40598A					P2002
40600	221	3050			F2004
40601	221	3050			F2004
40602	221	3050			F2004
40603	221	3050			F2004
40604	221	3050			F2004
40605			28	TR76/	S3008
40608			28	TR76/	S3008
40611	128	3024	63	TR87/	S0014
40612	121	3014	25	TR01/	G6003
40613	152	3054	66	TR76/701	S5000
40616	128	3024	63	TR87/	S0014
40618	152	3054	66	TR76/701	S5000
40621	152	3054	66	TR76/701	S5000
40622	152	3054	66	TR76/701	S5000
40623	121MP	3014	25	TR01/232	G6005
40624	152	3054	66	TR76/701	
40625	188	3024		TR72/	
40626	121	3014	25	TR01/	G6005
40627	152	3054	66	TR76/701	
40628	188	3024		TR72/	
40629	152	3054	66	TR76/701	S5000
40630	152	3054	66	TR76/701	S5000
40631	152	3054	66	TR76/701	S5000
40632	196	3054		TR92/	
40633	223			TR59/	S5004
40634	129	3025		TR88/	S5013
40635	128	3024		TR87/	S5014
40636	130	3027	75	TR59/	S7004
40637	123A		212	TR21/	S0015
40650	171		27	TR79/	
40664	175			TR81/	
40673	222	3065		TR85/	F2004
40763	102A			TR85/	
40782(IC)	732				
40816(RCA)	196				
40819	222	3065			
40820	222	3065	FET4		
40821	222	3065	FET4		
40822	222	3065			
40823	222	3065	FET4		
40828	102A			TR85/	
40841	221	3065			F2004
40850	124	3021		TR81/	
40852			36	/250	S5020
40852(VM)	102A			TR85/	
40853			36	/630	S5020
40853(VM)	158			TR84/	
40885	228	3104			S3021
40886	228	3103			
40887	228	3103			
40893					S3006
40894		3039	86		S0017
40895		3039	86		S0017
40896		3039	86		S0017
40897	161	3039	86	TR83/	S0017
40910	175	3026		TR81/	S5019
40911		3026			S5019
40912		3021			S5012
40913		3538			S5012
40934-1	130	3027	14	TR59/247	
40934-2		3027			
40953			45		S3013
40954					S3006
40955					S3007
40964					S3001
40965					S3001
40967					S3005
40968					S3005
40972			45		S3013
40973					S3006
40974					S3007
40975			45		
40976			45		
41051	123A	3124		TR21/735	S5000
41052	129	3025		TR88/242	S5006
41053	128	3024		TR87/243	S5000
41175	199				
41176	123A				
41177	159				
41178	184	3190	57	TR76/	S5000
41179	185		58		S5006
41180	909D				
41180(B+H)	909D				
41342	185	3041	58	/S5007	S5006
41344	184	3041	57	/S5003	S5000
41440	159	3114		TR20/717	S0019
41570	102A				
41689	108	3018		TR95/	
41694	108	3018		TR95/	
42065	175			TR81/	
42302	126			TR17/	G0008
42304	158			TR84/	G0005
42305	102A			TR85/	G0006
42311	126			TR17/	G0008
42321	176			TR82/	G6011
42322	102A			TR85/	G0005
42323	124			TR81/	S5011
42324	102			TR05/	G0007
42342	175	3026		TR81/241	
42396	133	3050		/802	F0015
42464	123A	3018	20	TR21/735	
42942	152	3054		TR76/	
43021-017	123A	3124		TR21/	
43021-198	128	3024		TR87/	
43021-860	108	3018		TR95/	
43044	123A			TR21/	
43045	123A				
43046	121			TR01/	
43054	123A	3124	10	TR21/	
43055	123A	3124		TR21/	
43060	130	3027	19	TR59/	
43074	121	3009	16	TR01/	
43082	128				
43088	128				
43089	129				
43090	706				
43095	130			TR59/	
43107	129	3025			
43114	130	3027			
43115	192	3024			
43116	129	3025			
43117	128	3024			
43118	182	3188			S5004
43119	182	3188			S5004
43120	182	3188			S0005
43121					S0005
43122	128				S3020
43123					S3020
43124					S3032
43125					S3032
43126					S3032
43127	159				
43128					S3032
43129					S0015
43131					S0015
43132					S3020
43133					S0030
43134					S0015
43135					S0015
43136					S0003
43138					S0003
43139	123A				S0015
43163	152	3054			
43165	128				
43168	130	3027			
43200	7400				
43201	7404				
43202	7408				
43205	7473				
43296	132			FE100/	
43992-2	101	3011	5	TR08/641	
44208	152				
44209	153				
44486		3004			

Original Part Number	ECG	SK	GE	IR/WEP	HEP
44616-1	102	3003		TR05/631	
44699	152	3054			
44763	188	3024		TR72/	S5014
44764	189			TR73/	S5013
44765	188+189				S5013+S5014
44766	133				F0010
44967-2	102	3004	2	TR05/631	
45122	159				
45184	123A				
45190		3054			
45191		3054			
45192		3054			
45193		3083			
45194		3083			
45195		3083			
45299	905				
45333		3054			
45337-A	199				
45337-C	159				
45354	128				
45380	718		IC8	IC511/	C6094P
45381	722	3161	IC9	IC513/	C6056P
45381(VOFM)	722				
45385	747				C6059P
45386	750				C6061P
45387	732				C6090
45390	743		IC32		
45391	788	3147			
45393			IC27		
45394	740				
45395	804		IC27		
45495-2	101	3011	5	TR08/641	
46490-2	101	3011	5	TR08/641	
46590-2	103	3010	8	TR08/641A	
46591-2	103	3010	8	TR08/641A	
46592-2	103	3010	8	TR08/641A	
46593-2	103	3010	8	TR08/641A	
46631-2	101	3011	5	TR08/641	
46774-1	101	3011	5	TR08/641	
46775-2	101	3011	5	TR08/641	
46776-2	102	3003		TR05/631	
47394-2	102	3004	2	TR05/631	
47645-2	103	3010	8	TR08/641A	
47737-2		3003			
48004-07	123A				
48004-08	123A				
48385-2	103	3010	8	TR08/641A	
48937-2	126	3008	51	TR17/635	
48939-2	126	3008		TR17/635	
49058-2	103	3010	8	TR08/641A	
49092	128				
49138-2	101	3011	5	TR08/641	
49139-2	102	3003	1	TR05/631	
49341	102A	3004		TR85/250	
49901		3122			
49902		3122			
49939-2	160		51	TR17/	
49974		3122			
49977		3122			
49980		3122			
49981		3039			
49982		3122			
49983		3039			
50009	732				
50137-2	128	3024	18	TR87/243	
50200-1B	152				
50200-6		3036			
50200-8	196	3534			
50200-9	152	3054			
50200-12	128				
50200-14		3534			
50200-24	152				
50201-4	153				
50202-1	123A				
50202-2	123A				
50202-9	194				
50202-12	128				
50202-13	123A				
50202-14	123A				
50202-23	123A				
50202-24	128				
50203-7		3053			
50203-8	159				
50203-12	159				
50208-0000		3054			
50210-2	9932				
50210-7	9094				
50210-8	9962				
50280-3	175				
50308-0100	152				
50447-4	121	3009	3	TR01/232	
50477-4	121	3014	3	TR01/	
50957-03	108	3124		TR95/56	S0015
51194	128	3024		TR87/243	S5014
51194-01	128	3024		TR87/243	
51194-02	128	3024		TR87/243	
51194-03	128	3024		TR87/243	
51213	123A	3124		TR21/735	S0015
51213-01	123A	3124		TR21/735	
51213-02	123A	3124		TR21/735	
51213-03	123A	3124		TR21/735	
51213-2	123A	3122		TR21/735	S0015
51300	175				
51428-01	123A	3124	10	TR21/735	S0015
51429	199				
51429-02	123A	3038	10	TR21/735	S0015
51429-03	123A	3038	10	TR21/735	S0015
51429-3	123A	3038	10	TR21/735	
51441	123A	3124		TR21/735	S0015
51441-01	123A	3124		TR21/735	
51441-02	123A	3124		TR21/735	
51441-03	123A	3124		TR21/735	
51442	123A	3124		TR21/735	S0015
51442-01	123A	3124		TR21/735	
51442-02	123A	3124		TR21/735	
51442-03	123A	3124		TR21/735	
51545	123A	3124		TR21/735	S0015
51547	123A	3124		TR21/735	S0015
51650	121	3014	16	TR01/121	G6005
52000-030	9930				
52000-032	9932				
52000-033	9933				
52000-036	9936				
52000-037	9937				
52000-044	9944				
52000-046	9946				
52000-048	9948				
52000-049	9949				
52000-052	9099				
52000-055	9097				
52000-057	9157				
52000-058	9158				
52000-061	9961				
52000-062	9962				
52000-063	9963				
52215-00	130				
52329	130	3027			
52360	130	3027			
52361		3079			
53200-22	123A			TR21/	
53200-23	123A			TR21/	
53200-51	123A			TR21/	
53200-74	199			TR21/	
53201-01	102A			TR85/	
53201-11	102A			TR85/	
53201-51	224				
53203-72	128			TR87/	
53400-01	123A			TR21/	
55001	7400				
55002	7402				
55003	7404				
55004	7405				
55005	7410				
55006	7420				
55007	7430				
55008	7442				
55009	7447				
55010	7450				
55011	7474				
55012	7476				
55017	74H183				
55021	7409				
55022	7412				
55023	7422				
55024	7416				
55027	7413				
55029	7454				
55032	7401				
55034	7440				
55035	7407				
55036	7406				
55606	123A				
55810-161	161	3124	20	TR21/	S0015
55810-162		3018		TR21/	S0015
55810-163	161	3124	20	TR25/	S0015
55810-164	161	3018	17	TR21/	S0014
55810-165		3018		TR21/	S0015
55810-166	175	3027	24	TR57/	S5014
55810-167	703A	3157	IC12	IC500/	C6001
55810-168	1045			IC504/	C6060P
55974-1	910				
55975-1	9945				
55976-1	9946				
55977	9930				
55978-1	9944				
55979-1	9951				
55980	9933				
55982-1	9932				
55986-1	909				
55987-1	9962				

Original Part Number	ECG	SK	GE	IR/WEP	HEP
56519	9946				
56552	9093				
56553	9930				
56557	9936				
56558	9945				
56571	9962				
57000-5452	108	3039		TR95/56	S0015
57000-5503	123			TR86/53	S0015
57001-01	128				
57004-503				TR51/	
57005-452				TR51/	
57005-503				TR51/	
57009	234				
58215-01	123A				
58810-160	132		FET2		F0021
58810-161	107		17		S0025
58810-162	123A		20		S0015
58810-163	123A		20		S0015
58810-164	199		20		S0015
58810-165	123A		20		S0015
58810-166	123A		20		
58810-167	199		20		
58810-168	123A		20		S0015
58810-169	186	3192	28		
58810-171	1142	3160			
58840-192	107	3018	17	TR24/	S0025
58840-193		3122	20	TR24/	S0015
58840-194		3018	20	TR24/	S0015
58840-195	123A	3124	20	2SC945/	S0015
58840-196		3018	20	TR24/	S0015
58840-197		3018	20	TR24/	S0015
58840-198		3124	20	2SC945/	S0015
58840-199	123A	3122	20	TR24/	S0015
58840-200	186	3054	28	TR55/	S5003
58840-202	720				
59557-48	102A			TR85/	
59625-1	159	3114	21	TR20/717	
59625-2	159	3114	21	TR20/717	
59625-3	159	3114	21	TR20/717	
59625-4	159	3114	21	TR20/717	
59625-5	159	3114	21	TR20/717	
59625-6	159	3114	21	TR20/717	
59625-7	159	3114	21	TR20/717	
59625-8	159	3114	21	TR20/717	
59625-9	159	3114	21	TR20/717	
59625-10	159	3114	21	TR20/717	
59625-11	159	3114	21	TR20/717	
50625-12	159	3114	21	TR20/717	
59987-1	102A			TR85/	
59988-1	128	3011		TR87/	
59989-1	129			TR88/	
59990-1	121			TR01/	
60006-416		3024			
60008	648				
60012		3036			
60013		3036			
60023		3003			
60024	223				
60031	128				
60038		3027			
60041	130				
60036	130				
60037	130				
60048	108				
60069		3036			
60076	181				
60083		3036			
60085	130				
60091	128				
60106	128				S3002
60117		3036			
60115	181				
60122Y				TR85/	
60126		3027			
60127	130				
60130	130	3027			
60133		3054	66		S5000
60142	181				S5020
60146		3027			
60152		3123			
60154	129	3027			
60172	192				
60175	152				
60187	181				
60194	128				
60201	175				
60205	130				
60208		3024			
60213	726				
60216	152				
60219	175				
60228	128				
60234	181				
60237	196				
60243	130				

Original Part Number	ECG	SK	GE	IR/WEP	HEP
60262		3027			
60263		3027			
60276		3036			
60294	128				
60308	124				
60314	108				
60327		3027			
60335	128				
60337	221				
60339	181				
60343		3018			
60346		3036			
60350	181				
60353	724				
60355		3027			
60380	181				
60395	123A				
60407		3054	28	TR55/	S3020
60408	152				
60413	186	3192	28	TR55/	
60416	196	3054		TR92/	
60417	128				
60423	128				
60428	128				
60457	188				
60458	189				
60465	130				
60469		3510			
60490		3124			
60559	124				
60597	128				
60632	175				
60659	128				
60677	128				
60678	196				
60679	152				
60680	128				
60682	128				
60684	154				
60685	124				
60697	128				
60700	128				
60701	129				
60703	128				
60710	130				
60719	152				
60719-1	159				
60720	128				
60761	124				
60764		3054			
60770	104				
60784	124				
60793	220				
60810	181				
60835	152				
60837	130				
60838	152				
60885	181				
60886	152				
60944	130				
60947	129				
60966	152				
60973	130				
60977	196				
60978	124				
60987	152		28	TR76/	S5000
60991	181				
60994	128				
61003-4	158	3004	2	TR84/630	G0005
61007	223				
61008-8	102A	3004		TR85/	
61008-8-1	102A	3004		TR85/	
61008-8-2	102A	3004		TR85/	
61009-1	108	3018		TR95/	
61009-1-1	108	3018		TR95/	
61009-1-2	108	3018		TR95/	
61009-2	108	3018		TR95/	
61009-2-1	108	3018		TR95/	
61009-4	123A			TR21/	
61009-4-1	123A	3124		TR21/	
61009-6	108	3018		TR95/	
61009-6-1	108	3018		TR95/	
61009-9	129	3025		TR88/	
61009-9-1	129	3025		TR88/	
61009-9-2	129	3025		TR88/	
61009-9-3	102A	3004		TR85/	
61010-0	108	3018		TR95/	
61010-0-1	108	3018		TR95/	
61010-2-1	126	3008		TR17/	
61010-6	121	3009		TR01/	
61010-6-1	121	3009		TR01/	
61010-7-1	108	3039		TR95/	
61010-7-2	123A	3124		TR21/	
61011-0	129	3025		TR88/	
61011-0-1	129	3025		TR88/	

Original Part Number	ECG	SK	GE	IR/WEP	HEP
61011-3-2	123A	3124		TR21/	
61012	130				
61012-4-1	101	3011		TR08/	
61012-5-1	160	3006		TR17/	
61013-2-1	123A	3124		TR21/	
61013-4-1	129	3025		TR88/	
61013-9-1	108	3018		TR95/	
61014-0-1	108	3018		TR95/	
61016B-2				TR24/	
61019	130				
61035	196				
61049	123A				
61086-1	233				
61102	152				
61102-0	126	3008		TR17/	
61133	108				
61173	181				
61193	152				
61200-5-2		3072			
61200S-1		3072			
61209	128				
61218	127				
61219	128				
61234	130				
61239	187	3193	29	TR56/	
61242			28	TR55/	S5014
61244	129		67	TR28/	S3012
61252	196				
61275	128				
61285	152				
61286	152				
61359	128				
61366				TR25/	
61367	130				
61369/4367	130				
61370/4560	128				
61371/4561	129				
61418	152				
61443	1003				
61451	181				
61456	181				
61534	196				
61558	161				
61562	128				
61566	722	3161	IC9		C6056P
61636	152				
61661	108				
61663	108				
61666	129				
61667	128				
61733	128				
61755	161				
61772	152				
61774	129				
61787-1-1		3009			
61813-1-9		3009			
61828	128				
61841	128				
61868	130				
61917	128				
61937	129				
61958	152				
61965	124				
61981	152				
61997	196				
62004	130				
62005-1	130				
62013	124				
62019	128				
62032	102A				
62119	124				
62142	175				
62143	130				
62144	196				
62177	121				
62185	128				
62192	128				
62203	192				
62204	129				
62229	124				
62243	128				
62277	197				
62279	197				
62287	181				
62382	130				
62398	128				
62404	128				
62446	128				
62449	108				
62452	128				
62455		3054			
62512	197				
62540	128				
62571	152				
62584	129				

Original Part Number	ECG	SK	GE	IR/WEP	HEP
62612	128				
62660	175				
62665	124				
62681	152				
62689	124				
62708	129				
62759	197				
62763	175				
62792	130				
62950	186	3054	28	TR55/	S3020
63234-1	909				
63282		3114			S0019
63900-229	128				
64071-1	102	3004	2	TR05/631	
64293		3079			
64385		3024			
64520		3024			
64597		3012			
64597GRN		3012			
64597RED		3012			
64622		3012			
65804-62	160			TR17/637	G0002
65804-63	102A		2	TR85/250	G0005
66005-9		3004			
66006-0		3004			
66007-0		3024			
66007-2		3004			
66007-4	128	3024		TR87/	
66008-2	102A	3004		TR85/	
66009-5	121	3014		TR01/	
66010-3	121	3009		TR01/	
66070G		3024			
67001	128				
67003	128				
67085-0	121MP	3013			
67085-0-1	121MP	3013			
67193-82	100			TR05/254	G0005
67193-85	102	3004	2	TR05/631	G0005
67586	123A	3124	20		S0033
67599	158	3004	53		G0005
67802	108	3039		TR95/56	S0025
68504-62	160	3006	51	TR17/637	
68504-63	102A	3004	2	TR85/250	
68617	123A	3018	20		S0015
68895-13	102	3004	2	TR05/631	
68994-1	941				
68995-1	9109				
69107-42	126	3006	50	TR17/635	G0005
69107-43	160	3006		TR17/637	G0008
69107-44	126	3006	50	TR17/635	
69107-45	102	3004	52	TR05/631	G0005
70008-0	130	3027		TR59/	
70008-3	130	3027		TR59/	
70019-1	175	3026		TR81/	
70019-5	175	3026		TR81/	
70023-0-00	123A	3124		TR21/	
70023-1-00	123A	3124		TR21/	
70087-31	128				
70158-9-00	129	3025		TR88/	
70167-8-00	108	3018		TR95/	
70177A01	798			TR57/	C6066P
70177A01	748			TR57/	C6060P
70177A03	738	3167			C6075P
70177A05	749	3168			C6076P
70177B02	748				C6060P
70231	108	3039	11	TR95/56	
70260-11	108			TR95/	
70260-12	108			TR95/	
70260-13	108			TR95/	
70260-14	123A			TR21/	
70260-15	123A			TR21/	
70260-16	123A	3124		TR21/	
70260-18	130			TR59/	
70260-19	129			TR88/	
70260-20	123A			TR21/	
70260-29	129			TR88/	
70398-1	123A	3122	20	TR21/735	
70399-1	6400				
70399-2	6400				
70434	121	3009	16	TR01/232	G6013
70435					G0005
70511	123A	3122		TR21/	S0015
71193-2	102A	3004	53	TR85/250	
71226-1	123A	3122	20	TR21/735	
71226-2	123A	3122	20	TR21/735	
71226-3	123A	3122	20	TR21/735	
71226-6	123A	3018	17	TR21/	
71226-10	123A		17	TR21/	
71226-10-1		3019			
71226-15	123A	3122	20	TR21/735	
71226-115		3019			
71266-4	123A			TR21/	
71412-4	123A	3018	17	TR21/	
71412-5	123A	3018		TR21/	
71447-1	128	3024	18	TR87/243	

Original Part Number	ECG	SK	GE	IR WEP	HEP
71447-2	123A	3124	10	TR21/	
71447-3	128	3024	18	TR87/243	
71448	121		16	TR01/	
71448-1	121	3009		TR01/232	
71448-2	121			TR01/	
71448-3	121			TR01/	
71448-4	121			TR01/	
71448-5	121			TR01/	
71448-6	121			TR01/	
71448-7	121			TR01/	
71488-4	121	3009	16	TR01/232	
71488-5	121	3009	16	TR01/232	
71488-6	121	3009	16	TR01/232	
71686-4	133	3112	FET1	/801	
71686-5	133	3116	FET1	/801	
71686-6	132	3116	FET1	FE100/	
71687	193		67		
71687-1	193	3114	21	/717	
71687-101	193		67		
71748-1	133		FET1		
71818-1	159	3025	21	TR20/	
71819-1	123A	3024		TR21/	
71963-1	123A	3124		TR21/	
72003	176	3123	2	TR82/	G0005
72004	176	3123	2	TR82/	G0007
72006	176	3123	2	TR82/	G0005
72054-1	172		64	TR69/	
72114	123A	3124		TR21/735	
72115	123A	3124		TR21/735	
72116	123A	3124		TR21/735	
72117	102A	3004		TR85/250	
72132-1	766				
72133-1	767				
72133-2	767				
72150	102A	3004		TR85/250	
-2151	123A	3124		TR21/735	
72181	74121				
72185	7476				
72190	158			TR84/	G0005
72191	103A			TR08/	G0011
72193	131	3082		TR94/642	G6016
72204	123A			TR21/	S0015
72206	123A			TR21/	S0015
72207	123A			TR21/	S0015
72503		3004			
72784-21	100	3005		TR05/	
72784-22	102A	3004		TR85/	
72784-23	102A	3004		TR85/	
72797-80	126			TR17/	
72797-81	160			TR17/	
72799-41	102A			TR85/	
72813-10	102A			TR85/	
72847-51	102A			TR85/	
72856-63	121	3009		TR01/	
72874-52	123A			TR21/	
72879-39	126			TR17/	
72879-40	126			TR17/	
72923-08	160			TR17/	
72941-33	102A			TR85/	
72949-10	108			TR95/	
72951-95	108			TR95/	
72951-96	108			TR95/	
72963-14	123A			TR21/	
72979-80	108			TR95/	
73100-9	102A	3004		TR85/	
74004-1	724	3525			
74004-2	724	3525			
74021-1	74H30				
74066		3039			
74107P	74107				
74116D	74116				
74121PC	74121				
74122PC	74122				
74123PC	74123				
74132PC	74132				
74141IC	74141				
74145I	74145				
74150C	74150				
74151I	74151				
74151PC	74151				
74152IC	74152				
74153IC	74153				
74154I	74154				
74155IC	74155				
74156I	74156				
74157D	74157				
74158IC	74158				
74164IC	74164				
74160I	74160				
74165P	74165				
74166IC	74166				
74170IC	74170				
74174IC	74174				
74175IC	74175				
-4175P	74175				

Original Part Number	ECG	SK	GE	IR WEP	HEP
74176DC	74176				
74177DC	74177				
74178DC	74178				
74179DC	74179				
74179PC	74179				
74180DC	74180				
74181DC	74181				
74182DC	74182				
74190DC	74190				
74190PC	74190				
74191DC	74191				
74192DC	74192				
74193DC	74193				
74194DC	74194				
74194PC	74194				
74196DC	74196				
74196PC	74196				
74197DC	74197				
74198DC	74198				
74198PC	74198				
74199DC	74199				
74651-02	123A				
74662	181				
75145-3	128				
75561-1	159				
75561-2	159				
75561-3	123A				
75561-16	123A				
75561-18	234				
75561-20	128				
75561-21	234				
75561-28	123A				
75561-31	159				
75561-32	128				
75561-33	123A				
75568-3	107				
75596-1	234				
75596-2	234				
75596-3	234				
75596-4	234				
75613-1	199				
75613-2	199				
75614-1	123A				
75616-6	108				
75617-1	159				
75617-2	159				
75700-03-01	121	3009		TR01/	
75700-04	123A			TR21/	
75700-04-01	123A	3124		TR21/	
75700-05	123A			TR21/	
75700-05-01	123A	3124		TR21/	
75700-05-02	123A	3124		TR21/	
75700-05-03	123A	3124		TR21/	
75700-08	123A			TR21/	
75700-08-02	123A	3124		TR21/	
75700-09-01	123A	3124		TR21/	
75700-09-21	123A	3124		TR21/	
75700-13-01	129	3025		TR88/	
75700-22-01	175	3026		TR81/	
75803-1	153				
75803-2	153				
75803-3	153				
75960CH	102	3004	2	TR05/631	
76236	123A				
76251	223				
76797-2	923				
77027-1	9093				
77027-2	9093				
77052-3	102A	3004		TR85/	
77052-4	102A	3004		TR85/	
77053-2	102A	3004		TR85/	
77068-3170756	124			TR81/240	S5011
77271-8	160	3006		TR17/	
77272-0	102A	3004		TR85/	
77272-1	102A	3004		TR85/	
77272-5	102A	3004		TR85/	
77272-7	102A	3004		TR85/	
77272-9	102A	3004		TR85/	
77273-2	102A	3004		TR85/	
77273-3	102A	3004		TR85/	
77273-6	102A	3004		TR85/	
77273-7	102A	3004		TR85/	
77561-27	159				
78008		3024			
78331	159	3114		TR20/717	S0019
78399	130				
78499					S0003
78527-75-01	102A	3004		TR85/	
78527-76-01	102A	3004		TR85/	
78527-78-01	102A	3004		TR85/	
78527-79-01	102A	3004		TR85/	
78684					S5004
79051		3008			
79689		3004			
79855	108	3039		TR95/56	S0025

Original Part Number	ECG	SK	GE	IR/WEP	HEP
79856	108	3039		TR95/56	S0015
79922	175	3131		TR81/	S5019
79992	152			TR76/701	
80000-00006-005				TR21/	
80004-K1		3027			
80050	121	3009		TR01/232	G6003
80053	704	3022		IC501/	
80054		3022			
80070	704	3022		IC501	
80071	704	3022		IC501/	
80073	704	3023		IC501/	
80074	704	3022		IC501/	
80081	726	3022		IC501/	
80083	704	3022		IC501/	
80089		3022			
80090	704	3022		IC501/	
80094	704	3022		IC501/	
80114	706	3101		IC507/	
80116		3022			
80131	241				
80249-910787	163	3111		TR67/740	S5020
80287	706			IC507/	
80416C	121	3014		TR01/232	G6003
80540	123A	3122		TR21/735	S0015
80544	123A	3122		TR21/735	S0015
80545	123A	3122		TR21/735	G0008
80557		3004			
80710	712	3072			
80755	781	3169			
80757	727				
80807	130				
80813VM	123A	3124	20	TR21/735	S0015
80814VM	123A	3124	20	TR21/735	S0015
80815VM	123A	3124	20	TR21/735	S0015
80816VM	123A	3124	20	TR21/735	S0015
80817VM	103A	3010	8	TR08/724	G0011
80818VM	102A	3004	52	TR85/250	G0005
80827	706				
80829	789				
80902-1	128				
80904-1	128				
81170-6	123A				
81177	731	3170			
81336-1	727	3071			
81336-2	727	3071			
81404-4A	102A	3004		TR85/	
81500-3	102A	3004		TR85/	
81501-5	102A	3004		TR85/	
81502-0	100	3005		TR05/	
81502-0A	100	3005		TR05/	
81502-0B	100	3005		TR05/	
81502-1	100	3005		TR05/	
81502-1A	100	3005		TR05/	
81502-1B	100	3005		TR05/	
81502-2	102A	3004		TR85/	
81502-2A	102A	3004		TR85/	
81502-2B	102A	3004		TR85/	
81502-3B	102A	3004		TR85/	
81502-4	102A	3004		TR85/	
81502-4A	102A	3004		TR85/	
81502-4B	102A	3004		TR85/	
81502-5	102A			TR85/	
81502-5A	100	3005		TR05/	
81502-5B	100	3005		TR05/	
81502-6	101	3011		TR08/	
81502-6A	101	3011		TR08/	
81502-6B	101	3011		TR08/	
81502-6C	101	3011		TR08/	
81502-6D	101	3011		TR08/	
81502-7	100	3005		TR05/	
81502-7A	100	3005		TR05/	
81502-7B	100	3005		TR05/	
81502-7C	100	3005		TR05/	
81502-8	100	3005		TR05/	
81502-8A	100	3005		TR05/	
81502-8B	100	3005		TR05/	
81502-8C	100	3005		TR05/	
81502-9	102A	3004		TR85/	
81502-9A	102A	3004		TR85/	
81502-9B	102A	3004		TR85/	
81502-9C	102A	3004		TR85/	
81503-0	102A	3004		TR85/	
81503-0A	102A	3004		TR85/	
81503-0B	102A	3004		TR85/	
81503-1	102A	3004		TR85/	
81503-1A	102A	3004		TR85/	
81503-1B	102A	3004		TR85/	
81503-3	102A			TR85/	
81503-3A	102A			TR85/	
81503-4	102A	3004		TR85/	
81503-4A	102A	3004		TR85/	
81503-4B	102A	3004		TR85/	
81503-4C	102A	3004		TR85/	
81503-6	100	3005		TR05/	
81503-6A	100	3005		TR05/	
81503-6B	100	3005		TR05/	
81503-6C	100	3005		TR05/	
81503-7	100	3005		TR05/	
81503-7A	100	3005		TR05/	
81503-7B	100	3005		TR05/	
81503-7C	100	3005		TR05/	
81503-8	100	3005		TR05/	
81503-8B	102A	3004		TR85/	
81503-8C	102A	3004		TR85/	
81504-1	100	3005		TR05/	
81504-1A	100	3005		TR05/	
81504-1B	100	3005		TR05/	
81504-1C	100	3005		TR05/	
81504-3	100	3005		TR05/	
81504-3A	100	3005		TR05/	
81504-3B	100	3005		TR05/	
81504-3C	100	3005		TR05/	
81505-8	102A	3004		TR85/	
81505-8A	102A	3004		TR85/	
81505-8B	102A	3004		TR85/	
81505-8C	102A	3004		TR85/	
81505-8X	102A	3004		TR85/	
81506-4	126	3008		TR17/	
81506-4A	126	3008		TR17/	
81506-4B	126	3008		TR17/	
81506-4C	126	3008		TR17/	
81506-5	100	3005		TR05/	
81506-5A	100	3005		TR05/	
81506-5B	100	3005		TR05/	
81506-5C	100	3005		TR05/	
81506-7	126	3009		TR17/	
81506-7A	126	3008		TR17/	
81506-7B	126	3008		TR17/	
81506-7C	126	3008		TR17/	
81506-8	126	3008		TR17/	
81506-8A	126	3008		TR17/	
81506-8B	126	3008		TR17/	
81506-8C	126	3008		TR17/	
81506-9	102A	3004		TR85/	
81506-9A	102A	3004		TR85/	
81506-9B	102A	3004		TR85/	
81506-9C	102A	3004		TR85/	
81507-0	102A	3004		TR85/	
81507-0A	102A	3004		TR85/	
81507-0B	102A	3004		TR85/	
81507-0C	102A	3004		TR85/	
81507-0D	102A	3004		TR85/	
81507-4	102A	3004		TR85/	
81507-5	103A	3010		TR08/	
81507-6	103A	3010		TR08/	
81510-3	100	3005		TR05/	
81510-4	102A	3004		TR85/	
81510-5	100	3005		TR05/	
81511-4	102A	3004		TR85/	
81511-5	100	3005		TR05/	
81511-6	100	3005		TR05/	
81511-7	100	3005		TR05/	
81511-8	102A	3004		TR85/	
81512-0	102A	3004		TR85/	
81512-0A	102A	3004		TR85/	
81512-0B	102A	3004		TR85/	
81512-0C	102A	3004		TR85/	
81512-0D	102A	3004		TR85/	
81512-0E	102A	3004		TR85/	
81513-3	123A	3020		TR21/	
81513-6	102A	3004		TR85/	
81513-7	121	3009		TR01/	
81513-9	102A	3003		TR85/	
81515-8	102A	3004		TR85/	
81516-0	102A	3004		TR85/	
81516-0A	102A	3004		TR85/	
81516-0B	102A	3004		TR85/	
81516-0C	102A	3004		TR85/	
81516-0D	102A	3004		TR85/	
81516-0E	102A	3004		TR85/	
81516-0F	102A	3004		TR85/	
81516-0G	102A	3004		TR85/	
81516-0H	102A	3004		TR85/	
81516-0I	102A	3004		TR85/	
81516-0J	102A	3004		TR85/	
81532				TR08/	
81759A		3008			
82716	108	3039		TR95/56	
83272	159				
84001	129	3025		TR88/242	
84001A	129	3025		TR88/242	
84001B	129	3025		TR88/242	
84323	9931				
84324	9962				
84353-2	9927				
84380-1	911				
84398-1		3021			
84626	9914				
84626-1	9914				
84626-2	9914				

Original Part Number	ECG	SK	GE	IR/WEP	HEP
84626-3	9914				
84626-4	9914				
84628	9921				
84628-1	9921				
84628-2	9921				
84628-3	9921				
84628-4	9921				
84630	9900				
84630-1	9900				
84631	9974				
84631-1	9974				
84631-3	9974				
84631-4	9974				
85549	123A	3024		TR21/735	
85601-01		3005			
86127-2		3009			
86257	129				
86287	123A				
86313	786	3140			
86452	102	3004	2	TR05/631	
86458	124				
86458-00	124				
86812	123	3124	20	TR86/53	
86822	123	3124	20	TR86/53	
86832	102	3004	2	TR05/631	
86842	102	3004	2	TR05/631	
86848				TR05/	
87532	128				
87757	123A	3122		TR65/	S0024
87758	159	3114		TR54/	S0019
87759	159	3114	17	TR52/	
87850D		3004			
88060-141		3116			F0021
88060-142		3018			S0020
88060-143		3018			S0015
88060-144		3124			S0015
88060-145		3124			S0011
88510-172		3018			
88510-173		3018			
88510-174		3124			
88510-175		3124			
88686	123	3024		TR86/53	
88687	123A	3038		TR21/735	S0015
88688	123A	3124		TR21/735	S0015
88700	176/411			TR25/238	
88801-3-1	175				
88803	128	3024		TR87/243	S0015
88803-2-1	128				
88803-3-1	128				
88832	121	3009		TR01/232	
88834	128	3024		TR87/243	S0015
88862	123A	3124		TR21/735	S0015
89028-2	9945				
89028-4	9946				
89028-4-1-4	9946				
89028-6	9933				
90050	121	3009		TR01/232	
90209-172	123A				
90209-182	123A				
90209-247	234				
90247A		3007			
90326-001	123A				
90330-001	159				
90429	123A				
90431	194				
90432	159				
90448	194				
90934-35	199				
91021	101			TR08/	
91271	128				
91272	128				
91273	128				
91274	130				
91371	198				
91411	128				
91600		3116			
92138-01	941				
92138-001		3514			
92140-001		3514			
92162-023		3011			
93116DC	74116				
93116PC	74116				
93141DC	74141				
93141PC	74141				
93145DC	74145				
93145PC	74145				
93150DC	74150				
93150PC	74150				
93151DC	74151				
93151PC	74151				
93152DC	74152				
93152PC	74152				
93153DC	74153				
93153PC	74153				
93154DC	74154				
93154PC	74154				
93155DC	74155				
93155PC	74155				
93156DC	74156				
93156PC	74156				
93157DC	74157				
93157PC	74157				
93164DC	74164				
93164PC	74164				
93165DC	74165				
93165PC	74165				
93166DC	74166				
93166PC	74166				
93170DC	74170				
93170PC	74170				
93174DC	74174				
93174PC	74174				
93175DC	74175				
93175PC	74175				
93176DC	74176				
93176PC	74176				
93177DC	74177				
93177PC	74177				
93178DC	74178				
93178PC	74178				
93179DC	74179				
93179PC	74179				
93180DC	74180				
93180PC	74180				
93190DC	74190				
93190PC	74190				
93191DC	74191				
93191PC	74191				
93194DC	74194				
93194PC	74194				
93196DC	74196				
93196PC	74196				
93197DC	74197				
93197PC	74197				
93198DC	74198				
93198PC	74198				
93199DC	74199				
93199PC	74199				
94000	102A		53	TR85/250	
94001	102A		53	TR85/250	
94002	102A		53	TR85/250	
94003	158		53	TR84/630	
94004	104	3009	3	TR01/624	
94005	158		53	TR84/630	
94006	102A		53	TR85/250	
94007	160		50	TR17/637	
94008	102A	3004	2	TR85/250	
94009	158	3004	2	TR84/630	
94010	179			TR35/G6001	
94013	160		51	TR17/637	
94014	102A		53	TR85/250	
94015	102A		53	TR85/250	
94016	102A		53	TR85/250	
94017	158	3004	2	TR84/630	G0005
94018	158		53	TR84/630	
94019	158		53	TR84/630	
94020	158		53	TR84/630	
94021	158		53	TR84/630	
94022	158		53	TR84/630	
94023	103A		54	TR08/624	
94024	121	3009	3	TR01/232	
94025	121	3009	3	TR01/232	
94026	121	3009	16	TR01/232	
94027	123A	3122	20	TR21/735	
94028	160		51	TR17/637	
94029	103A		54	TR08/724	
94030	102A		53	TR85/	
94032	179	3009	16	TR35/G6001	
94033	160	3007	50	TR17/637	G0003
94034	121	3009	3	TR01/232	
94035	160		50	TR17/637	
94036	160	3007	50	TR17/637	G0003
94037	159	3004	53	TR20/717	S0032
94037(EICO)	102A			TR85/250	
94038	126	3008	53	TR17/635	G0005
94038(EICO)	102A			TR85/250	
94039	102A	3004	53	TR85/250	G0005
94040	121	3009	3	TR01/232	G6005
94041	128	3024	18	TR87/243	
94042	128	3047		TR87/243	S5014
94043	195	3048		TR65/243	S5014
94044	108	3039	20	TR95/56	S0015
94047	123A	3122	20	TR21/735	
94048	123A	3122	20	TR21/735	
94049	175	3026	66	TR81/241	
94050	128	3024		TR87/243	
94051	128	3024	18	TR87/243	
94051(EICO)	154			TR78/712	
94052	128	3025	18	TR87/243	
94062	192	3512			

Original Part Number	ECG	SK	GE	IR/WEP	HEP
94063	129	3025		TR88/242	
94064	129	3025		TR88/242	
94065	130	3027	14	TR59/247	
94065A	130	3027		TR59/	
94066	128	3024	18	TR87/243	
94067	129			TR88/	
94068	129	3513		TR88/	
94070	128	3024	18	TR87/243	
94094	175	3026		TR81/241	
94094(EICO)	130			TR59/	
94152	7408				
94325	9937				
94327	9944				
94330	910				
94331	9933				
94333	9932				
94835-145-00	128				
95101	160	3007	50	TR17/637	G0002
95102	160	3008	50	TR17/637	G0002
95103	160	3005	50	TR17/637	G0002
95107	160	3006	50	TR17/637	G0002
95108	160	3006	51	TR17/637	G0003
95110	160	3006	51	TR17/637	G0003
95111	160	3008	51	TR17/637	G0002
95112	101	3011	10	TR08/641	G0011
95113	101	3011	17	TR08/641	G0011
95114	101	3011	5	TR08/641	G0011
95115	101	3011	5	TR08/641	G0011
95116	126	3006		TR17/635	
95117	101	3006	51	TR08/641	G0011
95118	126	3006	51	TR17/635	
95119	126	3006	51	TR17/635	
95120	126	3006	50	TR17/635	
95120A	100	3005	1	TR05/254	
95121	160	3006	50	TR17/637	G0002
95122	126	3006	51	TR17/635	
95123	126	3006	51	TR17/635	
95124		3006	51	TR12/635	
95125	108	3018	20	TR95/56	
95126	108	3018	20	TR95/56	
95127	108	3018	11	TR95/56	S0016
95128	108	3018	11	TR95/56	S0016
95129	108	3018	20	TR95/56	S0016
95130	108	3124	20	TR95/56	S0015
95131	108	3018	20	TR95/56	S0016
95131(95127)				TR21/	
95131(95129)				TR21/	
95151				TR17/	
95170-1	108	3018	20	TR95/	S0016
95170-2	108	3018	20	TR95/	S0016
95171-1	108	3122	17	TR95/	S0015
95171-2	123A	3020	17	TR21/	
95171-3	123A	3122	62		S0015
95171-4	107	3018	20	TR24/	S0015
95172-2	102A		52	TR85/	
95173-1	102A		2	TR85/	
95201	102	3004	52	TR05/631	G0005
95202	103	3010	5	TR08/641A	G0005
95203	102	3003	52	TR05/631	G0005
95204	102	3004	53	TR05/631	G0005
95208	102	3004	2	TR05/631	G0005
95209	102	3004	2	TR05/631	G0005
95211	101	3011		TR08/641	G0011
95212	102	3004	2	TR05/631	G0005
95213	102	3004	2	TR05/631	G0005
95214	102	3004	52	TR05/631	G0005
95216	123	3124	20	TR86/53	S0014
95216RED	123			TR86/53	S0014
95216YEL	123			TR86/53	S0014
95217	102	3004	2	TR05/631	
95218	102	3004	52	TR05/631	G0005
95219	102	3004	2	TR05/631	G0005
95220	128	3024	18	TR87/243	
95221	123	3018	20	TR86/53	S0024
95222-1	102	3004	52	TR05/631	
95222-2	103	3010	8	TR08/641A	
95223	123A	3018	20	TR21/735	S0015
95224-1	102A	3004	52	TR85/250	
95224-2	103A	3010	8	TR08/724	
95224-3	102A	3004		TR85/250	G0006
95224-4	103A	3010		TR08/724	G0011
95225	123	3124	20	TR86/53	
95226-003	193	3025	67	TR88/242	S5013
95226-004	192	3024	63	TR87/243	S5014
95226-1	128	3025		TR87/243	S3019
95226-2	123	3124		TR86/53	S0024
95226-3	129	3025		TR88/242	S5013
95226-4	128	3024		TR87/243	S5014
95227	159	3114	21	TR20/717	S0019
95228	133	3112		/801	
95229	106				
95231	133	3112		/801	F0010
95232	159	3114	21	TR20/717	S0012
95233	128			TR87/243	S5014
95239-1	129				

Original Part Number	ECG	SK	GE	IR/WEP	HEP
95250	121	3009	3	TR01/232	G6005
95250-1	121	3013	16	TR01/232	
95251	121	3009	16	TR01/232	G6005
95252	124	3021	12	TR81/	S5015
95252-1	124	3021	12	TR81/240	
95252-2	124	3021	12	TR81/240	S5011
95252-3	124	3021	12	TR81/240	S5011
95252-4	124	3021	12	TR81/240	
95253		3009			
95255-000	102				
95257	104MP			TR01MP/624	G6005
95258-1	152	3054	67	TR76/701	S5013
95258-2	153	3083	63	TR77/700	S5014
95263-1		3054			S5000
95263-2		3083			S5006
95293	732				
95294	722	3161	IC9		
95296	172		64	TR69/S9100	S9100
95298	704	3023			C6091
95299-1					C6093(
95980-2		3025			
96457-1	123A				
96458-1	159				
96481	128				
97680	106				
97759					S0019
98484-001	181	3036			
98576				TR51/	
98693				TR21/	
99101	126	3008	50	TR17/635	G0008
99102	126	3008	50	TR17/635	G0008
99103	160	3008	50	TR17/637	G0002
99104	160	3004	50	TR17/637	G0001
99105	160	3006	50	TR17/637	G0003
99106	160	3006	50	TR17/637	G0003
99107	160	3006	50	TR17/637	G0003
99108	160	3006	50	TR17/637	G0003
99109-1	123	3124	20	TR86/53	
99109-2	123	3124	20	TR86/53	
99120	126	3006	51	TR17/635	
99121	126	3007	51	TR17/635	
99201	102A	3004	52	TR85/250	G0005
99202	102A	3004	2	TR85/250	
99203	102A	3004	52	TR85/250	G0005
99204	102A	3004	53	TR85/250	G0005
99205	102A	3004	2	TR85/250	G0005
99206-1	123A	3124	20	TR21/735	
99206-2	123A	3124	20	TR21/735	
99207-2	123A	3124	20	TR21/735	S0015
99217	102	3004	52	TR05/631	
99218	102	3004	52	TR05/631	G0005
99240-132		3011			
99240-138		3007			
99240-164		3005			
99240-167		3024			
99240-174		3007			
99240-202A		3122			
99240-263		3019			
99240-269	199				
99240-292	219				S7003
99240-300		3039			
99250	121	3009	3	TR01/232	
99252	124	3021	12	TR81/	S5015
99252-002					S5011
99252-1	124	3021	12	TR81/240	
99252-2(METAL)	124	3021	12	TR81/240	S5011
99252-2(PLAS)	157			TR60/	
99252-3	124	3021		TR81/240	
99252-4	152			TR76/701	
99740-139		3004			
100092	123A				
100093	123A	3122			
100213		3004			
100292	199				
100678	126	3007	51	TR17/635	
100693	102A	3004	2	TR85/250	
101078	160	3007	51	TR17/637	
101087	160	3007		TR17/637	
101089	126	3008		TR17/635	
101119	123A				
101185	123A				
101306		3024			
101434	108	3018	20	TR95/56	
101435	128	3047		TR87/243	
101436	195	3048		TR65/	
101497	159				
101568	181	3036			
101677		3004			
101678	101	3011	5	TR08/641	
101679		3011			
101973	102A	3004		TR85/250	G0005
101974	102A	3004	2	TR85/250	G0005
102001	159				
102002	123A				
102005	7490				

Original Part Number	ECG	SK	GE	IR/WEP	HEP
102209	128				
102260	159				
102263	159				
102722-0		3004			
103443	101	3011	54	TR08/641	
103501		3011			
103508-2	9914				
103508-3	9915				
103521	123A	3124	20	TR25/	S0016
103562	102A	3004	2		
103705	9946				
103717	9962				
103729	9930				
103731	9932				
103743	9936				
103755	9945				
104009	160	3007	51	TR17/637	
104059	160	3007	51	TR17/637	
104080	103A	3010		TR08/724	G0011
104389	123A				
104444	102A	3004	2		
104719	128				
104830	9930				
104833	9933				
104836	9936				
104838		3004			
104844	9944				
104846	9946				
104862	9962				
105180	128				
105287A		3004			
105412(5)	9931				
105432	123A				
105468	106				
106719	1110				
107274	102	3004			
110075	195	3048		TR65/	
110240-005	9948				
110242-003	9962				
110242-004	9962				
110263	101	3011	5	TR08/641	
110472-003	9944				
110494	102A	3004	2	TR85/250	
110495	103A	3010	8	TR08/724	
110515	121	3009	3	TR01/232	
110669	128	3024	18	TR87/243	
110697	108	3018	20	TR95/56	
110699	128	3024	18	TR87/243	
110775					S9100
110957	102A		53	TR85/250	
110958	103A		54	TR08/724	
110959	102A		53	TR85/250	
111001	102				
111011	102				
111012	102				
111013	102				
111117	160	3007	51	TR17/637	G0001
111118	160	3007	51	TR17/637	G0003
111193-001		3024			
111278	128	3024		TR87/243	S0005
111279	195	3048		TR65/	
111303	123A	3046		TR21/735	S0011
111313	160	3007		TR17/637	G0003
111943	108				
111945	129				
111954	160	3006	51	TR17/637	G0008
111955	160	3006	51	TR17/637	G0008
111956	160	3006	51	TR17/637	G0003
111957	102A	3004	2	TR85/250	G0005
111958	103A	3010	8	TR08/724	
111959	102A	3004	2	TR85/250	
112001	160				
112002	160				
112011	160				
112041	126				
112071	102				
112296	160	3008	51	TR17/637	
112297	102A	3004		TR85/250	
112355	108	3018	20	TR95/56	S0016
112356	123A	3018	20	TR21/735	S0014
112357	123A	3018	20	TR21/735	S0014
112358	123A	3124	20	TR21/735	S0014
112359	123A	3124	20	TR21/735	S0014
112360	128	3024		TR87/243	S3013
112361	128	3024		TR87/243	
112362	224	3049			
112363		3027		TR59/S7000	S7002
112520	123A	3018	18	TR21/735	S0014
112521	123A	3018	18	TR21/735	S0014
112522	123A	3018	18	TR21/735	S0014
112523	123A	3124	18	TR21/735	S0014
112525	188	3024		TR72/	S3021
112527	224	3049			
113182	159	3118	21	TR19/	S0012
113348	123A	3122		TR21/735	
113398(TRAN)	123A	3019		TR21/	
113438	123A		20	TR21/	
113524	123A				
113875	130				
113876	181				
113938	108	3019	17	TR95/56	S0016
113942	128				
114143-1	108				
114267	222	3019	11	TR95/56	S0017
114504	176				
114504A03	176				
114525	108	3039	20	TR95/56	
115063	121MP	3009	16	TR01/232MP	G6005
115167	123A				
115225	123A	3124	10	TR21/735	G0005
115227	160	3006	50	TR17/637	G0003
115228	160	3006	50	TR17/637	G0003
115229	160	3006	50	TR17/637	G0002
115268	121	3009	3	TR01/628	G6013
115269	121	3009	16	TR01/232	
115270-101	106				
115275	126	3008	51	TR17/635	
115281	121	3014	16	TR01/232	
115282	121	3014	16	TR01/232	
115283	121	3014	16	TR01/232	
115284	121	3014	16	TR01/232	
115300-1	128				
115304	195			TR65/	S3001
115330-1					S0020
115342				TR21/	
115444	108				
115503		3004			
115504	160	3004			
115517-001	159				
115720	123A				
115728	123A				
115783	124				
115792	153		26	TR77/700	S5018
115810P2	128				
115817	124				
115875	123A	3124	20	TR21/735	S0015
115910	108	3039	20	TR95/56	S0015
115925	108				
116068	127	3035	25	TR27/235	
116072	126	3008	51	TR17/635	G0002
116073	108	3018	17	TR95/56	
116074	123A		20	TR21/735	S0015
116075	124	3021	12	TR81/240	S3021
116076	123A	3009	3	TR21/735	
116077	123A	3124	18	TR21/735	
116078	123A			TR21/735	
116078(RCA)	159	3025	21	TR20/717	
116079	108	3018	11	TR95/56	S0015
116080	108	3018	11	TR95/56	
116081	154	3040	20	TR78/712	S5025
116082	108	3018	20	TR95/56	
116083	108	3018	20	TR95/56	
116084	100			TR05/254	
116084(RCA)	129	3025		TR88/242	
116085	123A	3124	20	TR21/735	
116086	127	3035	25	TR27/235	
116087	127	3124	25	TR27/235	
116088	127	3034	25	TR27/235	
116089	127	3034	16	TR27/235	
116091	102A	3025	22	TR85/250	
116092	123A	3124	20	TR21/735	
116093	121	3009	3	TR01/232	G6003
116118	152				
116118-2	152				
116119	108		11	TR80/	S0020
116137		3031			
116148	123		17	TR86/53	
116198	108	3018	17	TR95/56	
116199	108	3018	61	TR95/56	S0016
116200	108	3018	17	TR95/56	
116201	102A	3004	2	TR85/250	G0005
116202	160	3006	51	TR17/637	G0008
116203	102A	3003	2	TR85/250	G0005
116204	103A	3124	8	TR08/724	G0011
116205	102A	3004	52	TR85/250	G0005
116206	102A	3003	2	TR85/250	G0005
116207	160	3006	51	TR17/637	G0002
116208	160	3006	51	TR17/637	G0003
116209	160	3006	51	TR17/637	G0003
116279	124		12	TR81/240	
116284	129	3025	22	TR88/242	
116286	102A	3004	52	TR85/	G0005
116588	123A	3124	20	TR21/735	S0015
116623	912				
116628	102A			TR85/250	
116683	160	3006	51	TR17/637	
116684	160	3006	51	TR17/637	
116685	102A	3004	2	TR85/250	G0007
116686	102A	3004	2	TR85/250	G0005
116687	103A	3010	8	TR08/724	G0005

Original Part Number	ECG	SK	GE	IR/WEP	HEP
116707		3009			
116707JF		3009			
116756	126	3006	51	TR17/635	
116757	102A	3004	2	TR85/250	G0005
116796P1	704				
116875	123A	3124	20	TR21/735	
116988	158	3004	53		G0005
116996	102A	3123		TR85/250	G0005
116997	102A	3004	2	TR85/250	
116998	102A	3004	2	TR85/250	G0005
117208	102A	3004	2	TR85/250	
117209	102A	3004	2	TR85/250	G0007
117210	102A	3004	2	TR85/250	
117616	100	3005	1	TR05/254	G0006
117617	100	3005	1	TR05/254	G0003
117618	160	3007	51	TR17/637	
117658	160	3007	51	TR17/637	
117724	126	3008	51	TR17/635	G0003
117725	160	3006	51	TR17/637	G0003
117726	160	3006	51	TR17/637	
117727	102A	3004	2	TR85/250	G0007
117728	102A	3004	2	TR85/250	G0005
117823	108	3018	20	TR95/56	
117824	126	3008		TR17/635	
117866	160	3006		TR17/	
117867	158	3004	53	TR84/630	G0005
118200	123A	3122	20	TR21/735	
118279	124	3021	12	TR81/240	
118284	159	3025		TR20/717	
118361	704	3023		IC501/	
118686	124	3021	12	TR81/240	
118713	123A	3122	20	TR21/735	
118822	108	3018	20	TR95/56	S0016
119013	160	3006	51	TR17/637	G0003
119228-001	159				
119232-001	123A				
119258-001	123A				
119290-001	129				
119414	108	3018	20	TR95/56	S0016
119526	126	3008	2	TR17/635	
119554	108	3018	20	TR95/56	S0015
119555	108	3018	20	TR95/56	S0015
119556	108	3018	11	TR95/56	G0009
119557	108	3018	11	TR95/56	S0011
119609	704	3023			
119635	123A	3122	20	TR21/735	S0015
119636	123A	3122	20	TR21/735	
119650	124	3021	12	TR81/240	S5011
119721	121	3014		TR01/232	G6003
119722	127	3035	25	TR27/235	
119723	127	3034	25	TR27/235	
119724	123A	3122		TR21/735	
119725	123A	3122		TR21/735	
119726	123A	3122		TR21/735	
119727	158			TR84/630	
119728	128			TR87/	S5014
119730	159	3114		TR20/717	
119822	128	3024		TR87/243	
119823	161	3117		TR83/719	
119824	161	3117		TR83/719	
119825	161	3117		TR83/719	
119982	123A	3024	20	TR21/735	S0015
119983	129	3025	22	TR88/242	
120073	123A	3124	10	TR21/735	S0015
120074	123A	3124	10	TR21/735	S0015
120075	102A	3004	2	TR85/250	
120085	123A				
120143	102A	3004		TR85/	
120144	102A	3004	2	TR85/250	G0005
120481	123A	3124	20	TR21/735	
120482	123A	3124	20	TR21/735	
120483	123A	3124	20	TR21/735	
120545	102A	3005		TR85/250	
120546	102A	3004		TR85/250	
120909-24.4	102A	3004		TR85/	
120956		3004			
121002		3054			
121151	102A		2	TR85/250	
121152	102A		2	TR85/250	
121153	100	3005	1	TR05/254	
121154	100	3005	1	TR05/254	
121243	121	3009	16	TR01/232	
121244	127	3034		TR27/235	G6007
121467	159	3114	21	TR20/717	S0031
121500				TR80/	
121501				TR71/	
121503				TR70/	
121507				TR71/	
121508				TR71/	
121509				TR71/	
121524				TR21/	
121526				TR21/	
121551				TR83/	
121587				TR24/	
121655	123A	3122	20	TR21/735	S0015

Original Part Number	ECG	SK	GE	IR/WEP	HEP
121658	123A	3122	21	TR21/735	S0015
121659	159	3114	21	TR20/717	
121660	123A	3124	20	TR21/735	
121661	128	3122	20	TR21/735	S0015
121651(RCA)	123A			TR21/735	
121662	123A	3124	20	TR21/735	S0015
121663	123A	3124	20	TR21/735	
121664	123A	3124	20	TR86/53	S0025
121695				TR21/	
121699				TR20/	
121713				TR23/	
121743				TR78/	
121744				TR25/	
121748				TR53/	
121754				TR80/	
121755				TR75/	
121779				TR80/	
121787				TR80/	
121808				TR55/	
121821				TR68/	
121831				TR68/	
121834				TR78/	
121841				TR80/	
121855				TR80/	
121857				TR21/	
121868				TR78/	
121883				TR24/	
121888				TR21/	
121895				TR78/	
121900				TR21/	
121924				TR24/	
121925				TR21/	
121931				TR21/	
121952				TR52/	
121966				TR55/	
121975				TR21/	
121978				TR20/	
122061	101				
122074	123A			TR21/	
122111	101				
122112	101				
122199	704	3023		IC501/	
122243	102A	3004	2	TR85/250	G0005
122244	102A	3004	2	TR85/250	G0007
122517	108	3018	20	TR95/56	
122518	108	3018	17	TR95/56	
122519	123A	3122	10	TR21/735	S0014
122664	123A	3124	20	TR21/735	S0015
122665	123A	3124	20	TR21/735	
122725	126	3006		TR17/635	
122792	121	3009	16	TR01/232	
122901	102A	3004	2	TR85/250	G0005
122902	107	3018	20	TR70/720	S0015
122904	107	3018	11	TR70/720	S0020
123139	128	3024	20	TR87/	S0005
123160	108				
123243	128				
123244	126	3004		TR17/635	G0007
123274	123A	3124	20	TR21/735	S0020
123275	124	3021	12	TR81/240	S5011
123275-14	124	3021		TR81/	
123375	124	3021		TR81/	S5011
123379	102A	3004		TR85/250	G0005
123429	108	3039		TR95/56	
123430	108	3039		TR95/56	
123431	108	3039		TR95/56	
123511	126	3006		TR17/635	G0003
123703	128			TR87/	
123791	102A	3004	2	TR85/250	G0005
123792	121	3009	16	TR01/232	G6005
123805	102A			TR85/250	G0005
123806	102A	3004	53	TR85/250	G0008
123807	123A	3122	20	TR21/735	S0011
123808	131	3052	30	TR94/735	G6016
123809	102A	3	53	TR85/250	G0005
123852	128				
123872		3004			
123877	102A	3123	30	TR85/250	G0006
123940	128	3024	21	TR87/243	S0015
123941	123A	3124	11	TR21/735	S0019
123944					S0012
123971	159	3114		TR20/	
123991	159		21	TR20/	S0016
124024	107	3018	20	TR70/	S0016
124047	159	3114		TR20/717	
124097	126	3004	51	TR17/635	G0003
124263	108	3019	11	TR95/56	S0016
124412	108	3039	20	TR95/56	S0016
124511	130				
124557	123A				
124616	129	3025		TR88/242	S5013
124623	107	3018	11	TR70/720	S0016
124624	107	3018	11	TR70/720	S0016
124625	160	3006	50	TR17/637	G0005
124626	102A	3004	52	TR85/250	

Original Part Number	ECG	SK	GE	IR/WEP	HEP
124634	106				
124753	123A	3122	20	TR83/56	S0019
124754	161	3132	20	TR83/719	S0020
124755	159	3114	21	TR20/717	S0019
124756	123A	3018	20	TR21/735	S0015
124757	161	3117	17	TR83/719	S0017
124759	123A		20	TR21/	
125135	123A	3122	20	TR21/735	
125137	108	3039	20	TR95/56	
125138	108	3039	20	TR95/56	S0020
125139	123A	3018	20	TR21/735	S0030
125140	123A	3018	20	TR21/735	S0015
125141	123A	3018	20	TR21/735	S0030
125142	159	3025	21	TR20/717	S0019
125143	123A	3122	20	TR21/735	S0020
125144	161	3018	17	TR83/719	S0020
125263	108	3039		TR95/56	
125264	108	3039		TR95/56	S0016
125329	107	3018	20	TR70/720	
125330	126	3007	51	TR17/635	G0009
125389	123A	3124	20	TR21/735	S0030
125389-14	123A	3124		TR21/	
125390	123A	3018	20	TR21/735	S0014
125390(RCA)	107	3018		TR70/720	
125392	108	3018		TR95/	
125394	123A	3018	20	TR21/	S0015
125474	123A	3018	20	TR21/	S0014
125475	123A	3018	20	TR21/735	S0011
125475-14	108	3018		TR95/	
125519	123A	3122	10	TR51/	S0015
125703	121	3014		TR01/	G6005
125761	121	3014	16	TR01/232	G6005
125790	160	3008	51	TR17/637	G0003
125944	108	3018	17	TR95/56	S0014
125972	160	3008	51	TR17/637	G0008
125994	108	3018	17	TR95/56	S0014
125994-14	108	3018		TR95/	
125994(RCA)			11		
125995	108	3018	11	TR95/56	S0016
126023	108	3018	17	TR95/56	S0020
126024	108	3018	17	TR95/56	S0020
126025	161	3018	11	TR83/719	S0017
126093	102A	3004	2	TR85/250	
126093-1	102A	3004	2	TR85/250	
126093-2	102A	3004	2	TR85/250	
126093-3	102A		53	TR85/250	
126093-4	102A	3004	2	TR85/250	
126138	124	3021	12	TR81/240	S5011
126150	123A	3124	62	TR21/735	S0015
126156	123A	3124	62	TR24/	S0016
126184	160		51	TR17/637	G0008
126185	126	3006	51	TR17/635	G0008
126186	160		51	TR17/637	G0008
126187	102A	3004	52	TR85/250	G0006
126188	124	63		TR81/240	S5011
126276	102A	3004		TR85/250	
126331	123A	3122		TR21/735	
126334	123A	3122	63	TR21/735	S0015
126524	159	3114	21	TR20/717	S0019
126525	123A	3124	20	TR21/735	S0015
126526	123A	3024	63	TR21/735	S0015
126604	711	3070		IC514/	C6069G
126609	711	3070			
126670	108	3039		TR95/56	
126697	102A		53	TR85/	
126698	108	3132	17	TR95/56	S0020
126699	123A	3122	20	TR21/735	S3019
126700	159	3114	22	TR20/717	S3019
126702	123A	3122	20	TR21/735	
126703	128	3024		TR87/243	S0016
126704	123A	3018	20	TR21/735	S0016
126705	154	3045	27	TR78/712	S0016
126706	123A	3122	20	TR21/735	
126707	159	3118	21	TR20/717	S0013
126708	123A	3122	20	TR21/735	G0008
126709	154	3044	20	TR78/712	G0008
126710	154	3044	18	TR78/712	G0008
126711	123A	3122	63	TR21/735	S0015
126712	123A	3044	17	TR81/240	S5014
126713	123A	3122	63	TR21/735	S0015
126714	123A	3050	20	TR21/735	S0015
126715	159	3114	21	TR20/717	S0015
126716	123A	3124	20	TR21/735	S0015
126717	123A	3122	20	TR21/735	S0015
126718	159	3114	21	TR20/717	S0019
126719	159	3114	21	TR20/717	S0019
126720	128	3122	63	TR87/243	S0015
126721	128	3024	63	TR87/243	S0015
126722	124	3045	12	TR81/240	S5011
126724	129	3025	21	TR88/242	S0019
126725	128	3024	27	TR87/243	S0015
126726	124	3045		TR81/240	S3021
126772	124			TR81/	S5011
126826(TRAN)	181	3125			
126863	123A	3061		TR21/735	

Original Part Number	ECG	SK	GE	IR/WEP	HEP
126871	710	3102		IC506/	C6060I
126900	162			TR67/707	S5020
126945	102A	3004	53	TR85/250	G0005
127112	102A		53	TR85/250	G0005
127114	102A	3004	53	TR85/250	G0005
127166	785				
127214	132	3050		FE100/	
127263	123A	3122	20	TR21/	
127297	102A		2	TR85/250	G0005
127303	102A			TR85/250	
127354	123A	3124	10	TR21/735	S0014
127355	123A	3124	63	TR21/735	S0014
127376	128	3024		TR87/243	S5014
127397	158	3004		TR84/630	G6011
127398		3124			
127399		3124			
127529	123A	3122	11	TR21/735	S0016
127589	102A	3004		TR85/250	G0005
127590	176			TR82/238	G6011
127693	108	3018		TR95/56	S0015
127712	154	3045		TR78/712	
127792	108	3039	20	TR95/56	
127793	108	3039	20	TR95/56	S0015
127794	108	3039	63	TR95/56	
127798	152			TR76/701	
127828	179			TR35/G6001	
127845	128	3024		TR87/243	
127899	123A				
127962	102A	3004	2	TR85/250	
127978	184	3054	27	/S5003	S5000
128056	152	3054	28	TR76/	
128057	153	3083	29	TR77/	
128343				TR05/631	G0005
128938	126	3006		TR17/	G0008
128939		3005			
128940		3024	53	TR87/243	G0005
128945		3018			
128972		3008		TR11/635	G0008
129029		3122		TR51/735	
129049		3122	20	TR21/	
129050	108	3018	20	TR95/	
129051	128	3024	63	TR25/	
129144	108	3039	17	/56	S0016
129145		3039	20	TR95/56	S0016
129146		3039	20	TR21/56	S0016
129147	123A	3039		/56	S0014
129199				TR21/	
129251				TR17/	
129284		3006		TR12/635	G0008
129286	158	3004	53	TR85/250	G0005
129289		3006		TR17/635	G0008
129291		3006		TR12/635	G0008
129314					S0014
129347		3006	50	TR17/637	G0003
129389	160	3008	62	TR17/	G0008
129392	107	3018	20	TR21/56	S0016
129392-14	108	3018		TR95/	
129393		3018	20	TR21/720	S0014
129393-14	108	3018		TR95/	
129394	108	3018	20	TR21/56	S0014
129394-14	108	3018		TR95/	
129398		3122		TR51/735	S0014
129424		3116	FET2	/802	F2004
129425		3122	20	TR51/735	S0014
129507			53	/630	G0005
129508			53	TR85/250	G0005
129509	107	3039		/720	S0015
129510		3039	17	/720	S0015
129511		3039	17	/720	S0015
129512		3039	17	/720	S0015
129513		3039	17	/720	S0015
129571	108	3039	20	/720	S0014
129572		3039	20	/720	
129573	107	3039	20	/720	
129574	108	3039	20	/720	S0016
129604		3019	11	/720	S0016
129618		3018		/56	
129698		3122	20	TR21/735	S0014
129699	159	3114	63	TR20/735	S0012
129802	158	3004		TR84/250	G0005
129821	904				
129839				TR12/635	
129871	710	3102		IC506/	
129897	108	3020	20	TR95/	S5014
129899	123A	3018	17	TR21/243	S0015
129949	123A				
129979	108	3018	11	TR95/56	S0016
129980	222	3050		FE100/	F2004
130007		3048			
130013	123A	3018		TR21/	S0015
130040	172			TR69/S9100	S9100
130048					G0011
130122	712	3072			
130130	780	3141			
130139	159	3118	21	TR20/717	S0013

Original Part Number	ECG	SK	GE	IR/WEP	HEP
130172	128				
130174	128				
130200-00	102A	3004		TR85/	
130200-02	102A	3004		TR85/	
130221	909	3590			
130253	234				
130278	108	3018		TR95/	S0016
130400-95	102A	3004		TR85/	
130400-96	103A	3010		TR08/	
130402-36	102A	3004		TR85/	
130403-04	108	3018		TR95/	
130403-13	123A	3124		TR21/	
130403-17	123A	3124		TR21/	
130403-18	123A	3124		TR21/	
130403-47	103A	3010		TR08/	
130403-52	102A	3004		TR85/	
130403-62	108	3018		TR95/	
130404-21	108	3018		TR95/	
130404-29	129	3025		TR88/	
130404-59	108	3018		TR95/	
130474	188			TR72/S3020	S5014
130536	159	3124	17	TR20/	
130537	123A		17	TR21/	
130537-1		3018			
130538					C2004P
130751	712	3144	IC2	IC507/	C6083P
131000-101		3004			
131075	152	3054	28	TR76/701	S5000
131095-2	175				
131139	191	3104	27	TR75/	S3019
131140	191	3104		TR75/	S3019
131161	184	3054	66	TR76/S5003	S5000
131221	108	3039	11	TR95/56	S0020
131239	196	3054	28	TR92/701	S5000
131240	123A	3024	63	TR21/735	S0015
131241	159	3025	67	TR20/717	S0019
131242	159	3025	21	TR20/717	S0031
131242-12	129	3025		TR88/	
131243	123A	3024	63	TR21/735	S0015
131257	181	3054		TR36/S7000	S7002
131262	106				
131300	9930				
131301	9932				
131303	9936				
131304	9944				
131306	9946				
131308	9962				
131311	123A			TR21/735	S0015
131312	9093				
131543	161	3132	11	TR83/719	S0017
131544	161	3132	11	TR83/719	S0020
131545	161	3018	11	TR83/719	S0020
131647	159	3118	67	TR20/	S0019
131648	108			TR95/	
131710	234				
131844	108	3018		TR95/56	
131848	108	3018	11	TR95/	S0016
131848(RCA)	152			TR76/701	
131849	152	3054	28	TR76/701	S5000
132175	128	3024	18	TR87/243	S0015
132176		3114	67	TR87/243	S0019
132285	159				
132313	730	3143			
132314	728	3073			
132315	729	3074			
132327	128	3024	63	TR87/243	S5014
132328	128	3024	63	TR87/243	S5014
132329	123A	3122	20	TR24/	S0015
132445	188	3054		TR72/3020	S3024
132446	188	3054		TR72/3020	S3024
132447	189	3083		TR73/3031	S3028
132448	197	3083		TR73/3031	S3028
132488	189	3200		TR73/3031	
132495	184	3041	28	TR55/S5003	S5000
132498	159	3114	21	TR20/717	S0013
132499	184	3054		TR55/	S5000
132500	128	3024	28	TR87/243	S5014
132558	213			TR27/	G6008
132571	185	3083	29	/S5007	
132573	152	3054	28	TR76/701	S5000
132574	197	3083	29	TR56/	S5006
132642	123A	3124	17	TR21/	
132643	123A	3124	17	TR21/	
132650	6400				
132776	130	3027	66	TR59/247	S5004
132823	128	3124	63	TR87/243	S0030
132830	159		21	TR20/	S0019
133171	161	3117		TR83/719	S0017
133176			63		
133177	128			TR87/	S5014
133178	123A	3044	63	TR21/735	S5014
133182	159	3025	21	TR20/717	S0013
133218	199	3038	20	TR21/735	S0015
133249	123A	3124	20	TR24/	S0015
133253	159	3114	21	TR20/	S0019

Original Part Number	ECG	SK	GE	IR/WEP	HEP
133265	171	3104	27	TR79/	S3021
133501		3124			
133550	181	3036			
133573	186	3054	28	TR55/	S5000
133576	128				
133600	730	3143			
133684	130				
133685	130				
133690	123A	3122	20	TR21/243	S0015
133743	123A	3044	18	TR25/	S5014
133823	175				
133823EL		3026			
133924		3036			
133925EL		3036			
134142	108	3018	11	TR95/	S0015
134143	123A	3018	11	TR95/	S0015
134144	108	3018		TR95/	S0016
134155	128			TR87/	
134195-001	9930				
134196-001	9932				
134197-001	9945				
134198-001	9946				
134254-001	9950				
134263	108	3019	11	TR95/56	S0020
134277		3031			
134279	185		58		S5006
134280	184		57		S5000
134281	180		74		S7001
134282	181		75		S7000
134340	762				C6012
134417	108	3019	11	TR95/56	S0016
134419	108			TR95/56	S0016
134442	108		64	TR95/56	S0011
134450	222	3050		/56	F2004
134771	171	3104	27	TR79/	
134772	171	3104	27	TR79/	
134773	163	3111	36	TR67/	S5021
134774	124	3054	12	TR81/	S5011
134989	128	3124	20	TR87/243	S0020
135324	221	3050			F2004
135347	106				
135351	218	3085	26	TR37/	S5018
135352	162	3079	35	TR67/	S5020
135716	171	3104	27	TR79/	S3021
135735	196	3054	28	TR92/	S3024
135739	186	3054	28	TR55/	S5000
135744	127	3035	25	TR27/	G6008
135963		3050			F2004
136066	154			TR78/	
136145	787	3146			
136146	788	3146			
136147	789	3146			
136165	108	3018	20	TR24/	S0015
136168		3018	20	TR24/	S0015
136239		3018	20	TR24/	S0015
136240	123A	3018	20	TR24/	S0015
136257		3024			
136281	234				
136282	199				
136324		3025			
136423	129	3025	67	TR30/	S3014
136424	128	3024	63	TR24/	S3013
136430	108	3018	20	TR87/	S0015
136648	152	3054	28	TR76/	S5000
136696	128	3024	27	TR87/	S3019
136766	106			TR20/	
137065	189	3054	69	TR73/	S5000
137066	188	3199	66	TR72/	S5006
137093	102A	3004		TR85/	S5013
137127	108	3018	11	TR95/	S0016
137155	129	3114	18	TR87/	S0015
137161	752				C6003
137241	128	3122	18	TR87/	S5014
137245	780	3070			C6069(
137338	233	3132	60	TR70/	S0020
137339	123A	3122	20	TR21/	S0015
137340		3118	21		S0012
137352		3111	36	TR61/	S5021
137369		3054	28		S5000
137383	161	3117	60	TR71/	S0016
137388	161	3132	60	TR24/	
137527		3083	29		S5006
137607	165	3130	35	TR93/	S5020
137614	123A	3024	10	TR24/	S0015
137648	128	3044	18	TR87/	S5026
137718	165	3115	36	TR93/	S5021
137875	123A	3122	10	TR24/	S0015
137875(IC)	941M				
138001	199				
138019-001	199				
138035-001	128				
138049-001	234				
138049-004	234				
138121	189			TR30/	S5013
138191	123A	3122	20	TR24/	S0015

Original Part Number	ECG	SK	GE	IR/WEP	HEP
138192	152	3054	66	TR55/	S5004
138193	185	3083	69	TR56/	S5005
138194	130	3027	19	TR26/	S7002
138195	130	3027	19	TR26/	S7002
138310	762				C6012
138311	7400				
138312	7401				
138313	7402				
138314	7404				
138315	7408				
138317	7411				
138318	7420				
138319	7421				
138320	74193				
138376	159				
138378	123A				
138379	152				
138381	7432				
138403	7473				
138681	949				
138699	797	3073			
138763	128	3024	18	TR51/	S0005
139017	191	3044	40	TR78/	S5024
139044	190	3104	27		S3021
139266	188	3199	28	TR72/	S3024
139267	189	3200	29	TR73/	S3028
139268	123A	3018	20	TR24/	S0015
139269	194	3022	20	TR24/	S0015
139270	175	2538	12	TR81/	S5012
139295	165	3535	36	TR61/	S5020
139618	172		64		
140259	154	3044	40	TR78/	S5025
140290	159				
140371	159				
140372	159				
140501	128	3020		TR25/	
140506	154			TR78/	
140612	181				
140622	123A				
140623	159				
140625	197	3083			
140626	196	3054			
140858-12	123A				
140869-1					F2004
140977		3115			
141008		3024			
141335		3024			S3019
142007-2			IC3		
142294	754				
142348	194				
142349	923D				
142648	778				
143041	915				
143062	9109				
144178	941M				
145134-526	121				
145140-250		3026			
145150-250		3026			
146144-2	123A				
146153-1	123A				
146286-01	124	3021			
146466-1	130				
146641	9946				
146642	9962				
146643	9930				
146644	9945				
147112-7		3025	67		S0019
147115	123A				
147115-5	123A				
147115-6	123A				
147115-7	199				
147115-8	199				
147115-9	161				
147122P7	199				
147245-0-1	108				
147256	9944				
147350-2-1	704				
147351-4-1	127			TR27/	
147351-5-1	121			TR01/	
147352-0-1	124			TR81/	
147353-0-1	161			TR83/	
147355	128				
147356-9-1	108			TR95/	
147357-0-1	129			TR88/	
147357-1-1	123A			TR21/	
147357-2-1	108			TR95/	
147357-4-1	129			TR88/	
147357-7-1	123A			TR21/	
147357-9-1	108			TR95/	
147359-0-1	129			TR88/	
147363-1	123A	3124		TR21/	
147368-1					S7002
147513	123A				
147549-1	159	3118	21		S0013
147549-2	159	3118	21	TR30/	S0013

Original Part Number	ECG	SK	GE	IR/WEP	HEP
147555-1	123A	3124		TR21/	
147624-1	197	3083			
147676-1	233				
148217		3004			
148651-147	108		20	TR95/	
148991	9112				
149100-02					S5006
150044		3026			
150045	128				
150046A	128				
150060	175				
150070	181				
150095	128				
150359		3025			
150439-1		3009			
150580	941M				
150580-2285	941M				
150622	912				
150714-1	199				
150730	128				
150741	123A				
150742	159				
150753	159				
150758	159				
150762	159				
150763	123A				
150768	123A				
150771	159				
150787	128				
150796	128				
151544	9932				
151548	9099				
152139B		3126			
153107	152				
153270	941				
155103	923				
156931	108			TR95/	
157004	129				
157008	123A				
157506	912				
157564	915				
157575	309K				
157644	923				
157681	912				
157800	923D				
157800-2760	923D				
158066	9093				
160196	106				
160201	909				
161705	102A	3004	2	TR85/250	
162002-033	121	3009	16	TR01/232	G6003
162002-040	126	3008	51	TR17/635	G0008
162002-041	126	3008	51	TR17/635	G0008
162002-042	126	3008	51	TR17/635	G0008
162002-062	121	3014	16	TR01/232	
162002-062A	121	3014	16	TR01/232	
162002-071	181				
162002-081	128				
162002-082	128				
162002-085	199				
162002-090	108	3124	20	TR95/56	S0015
162002-40	126	3008	51	TR17/635	
162002-41	126	3008	51	TR17/635	
162002-71	130				
162002-101	130				
162002-103				TR01/	
164981	9935				
164982	9936				
165392	107	3018	11		S0016
165618				TR51/	
165667	126	3006		TR17/635	G0005
165668	123A	3122		TR21/735	
165735	128	3024	63	TR87/243	S5014
165736	128	3024	18	TR87/243	S5026
165737	128	3024	18	TR87/243	S5026
165738	175	3026	23	TR81/241	S5019
165827	123A	3124		TR21/735	S0015
165828	123A	3124		TR21/735	S0016
165910		3018			
165931	107	3018	20	TR70/	S0014
165932	107	3018	11	TR70/720	S0016
165976	102A	3004		TR85/250	G0006
165995	108			TR95/	S0016
166272	108	3122	20	TR21/	
166400	158	3004	52	TR84/	G0005
166590		3124			
166753		3018			
166754		3025			
166755		3024			
166882	102A	3004		TR85/250	G0006
166883	102A	3004	2	TR85/250	G0006
166906	107	3018	17	TR70/720	S0011
166907	1002				
166908	126	3008	51	TR17/635	G0009
166909	126	3008	1	TR17/635	G0009

Original Part Number	ECG	SK	GE	IR/WEP	HEP
166917	128	3024	18	TR87/243	S5014
166918	184	3036	57	TR76/S5003	S5000
166919	185	3083	58	TR77/S5007	S5006
166997	158			TR84/	G0005
167263	123A	3122	20	TR21/735	S0014
167285	131	3052	30	TR94/642	G6016
167540	123A	3122	62		S0015
167541	123A	3020	20		S0025
167542	184	3036	28		S5000
167569	128	3024	18	TR25/	S5014
167679	102A		52	TR85/	G0005
167680	158	3004	53		G0005
167688	123A	3122	18	TR21/	S0015
167690	159		21	TR20/	S5022
167691	152	3026	23	TR76/	S5000
167956	123A	3124	10	TR24/	S0015
167957	123A	3124	20	TR21/	S0016
167958	175	3026	28	TR23/	
167998	102A	3004	54		G0005
167999	102A	3004	2		G0005
168165		3018			
168166		3018			
168212		3122			
168405	123A	3122	62	TR51/	S0015
168567	107	3018	11	TR80/	
168651		3124	20		S0015
168657	108	3018	20	TR80/	
168658	108	3018	20	TR95/	
168659	108	3018	20	TR95/	
168660	128	3122	18	TR87/	
168716	199				
168826		3005			
168906	158	3004	53	TR84/	G0005
168907	102A	3004	52		G0005
168953	103A	3010	59	TR08/	G0011
168954	102A	3004	53	TR85/	G0006
168983	102A		52	TR85/	G0005
168984	158	3004	52	TR84/	G0005
169175	158	3004		TR84/630	G0005
169194	107	3018		TR70/720	S0016
169195	108	3018	17	TR95/56	S0014
169196	107	3018	20	TR70/720	S0014
169197	123A	3124		TR21/735	S0030
169359	102A	3004	52	TR85/	G0005
169360	102A	3004	52	TR85/	G0005
169361	102A	3004	2	TR85/	G0005
169505	107	3124	17	TR24/	
169574	123A	3124	20	TR21/	S0014
169679	123A	3020	20		
169680	123A	3124	20		
169681		3024			
169771	123A	3122	20	TR21/	S0016
169773	102A	3004	[illegible]2	TR85/	G0005
170128	159	3114	22		S5013
170294	123A	3124	20	TR51/	S0014
170307-1	121	3009	16	TR01/232	
170308		3124			S0014
170376	121	3009	3	TR01/232	G6003
170376-1	121	3009	3	TR01/232	G6003
170388	107	3018	20	TR70/	S0016
170407-1	121	3014	3	TR01/232	G6003
170479-1	121	3009	3	TR01/232	G6003
170666-1	121MP	3009	3	TR01/232MP	G6005
170668-1	121MP	3013	13MP	/232MP	
170753-1	161	3018	11	TR83/719	S0020
170756-1	161	3018	11	TR83/719	S0017
170783-3	103A	3010	8	TR08/724	
170794	199	3018	20	TR24/	S0016
170827-1	131	3052			G6016
170850-1	121MP	3013	13MP	/232MP	
170890-1	131			TR01/642	S7001
170891-1	181			TR01/S7000	S7002
170906	161	3019		TR83/719	
170906-1	161	3019	20	TR83/719	S0016
170954	1042				
170956	720				
170964	1030				
170967		3124			
170967-1	123A	3122		TR21/735	
170968-1	123A		20	TR21/735	S0016
170994-1	184+185				
171003	127	3034	16	TR27/235	G6005
171003(SEARS)	108			TR01/232	
171003(TOSH)	123A			TR21/	
171004	127	3035	25	TR27/235	G6003
171004(SEARS)	121			TR01/232	
171005	126	3010	8	TR17/635	
171005(TOSH)	103A			TR08/724	
171009	107	3019	20	TR70/720	S0016
171009(SEARS)	108			TR95/56	
171009(TOSH)	123A			TR21/735	
171010-1	102A+103A				G0005
171015(TOSH)	124		12	TR81/240	
171016	160	3006	2	TR17/637	G0008
171016(SEARS)	102A			TR85/250	

Original Part Number	ECG	SK	GE	IR/WEP	HEP
171017	102A	3004	52	TR85/250	G0007
171018	102A		2	TR85/250	G0005
171026(SEARS)	102A			TR85/250	
171026(TOSHIBA)	123A	3124	10	TR21/735	S0015
171027	123A	3124	10	TR21/735	S0011
171028	108	3018	20	TR95/56	S0017
171029(SEARS)	108	3018	11	TR95/56	S0017
171029(TOSHIBA)	107			TR70/720	
171030	161	3018	20	TR83/719	S0017
171030(SEARS)	108	3018		TR95/56	
171030(TOSHIBA)	123A			TR21/735	
171031	161	3018	11	TR70/720	S0016
171031(SEARS)	108	3018		TR95/56	
171031(TOSHIBA)	107			TR21/735	
171032	108	3018	20	TR95/56	S0016
171032-1	102A		59		
171033	108	3018	11	TR95/56	S0020
171033-1	184			/S5003	S5000
171033-2	184+185		57	/S5003	S5000
171034	108	3122			S0014
171038	128	3018	17	TR87/243	S0016
171039(SEARS)	160	3006	50	TR17/637	G0008
171039(TOSHIBA)	160			TR17/637	
171040(SEARS)	123A	3124	20	TR21/735	S0015
171040(TOSHIBA)	123A			TR21/735	
171044		3018	20	TR87/243	S0016
171044(SEARS)	128			TR87/243	
171044(TOSHIBA)	123A			TR21/735	
171045	108	3039	20	TR95/56	S0016
171046	123A	3122	10	TR21/735	S0015
171047				TR61/	S5020
171047(TOSHIBA)	162			TR67/707	
171048	108	3039	11	TR95/56	
171049	102A	3004	2	TR85/250	
171052	108	3018	11	TR95/56	S0016
171053-1	180	3036		/56	S7002
171054	108	3039	11	TR95/	
171090-1	108	3019		TR95/56	S0016
171139-1	108	3018		TR95/56	S0017
171140-1	108	3018	11	TR95/56	S0016
171141-1	108	3018		TR95/56	S0016
171162-003		3124		TR21/	
171162-004	199	3038	17	TR21/735	S0015
171162-005	123A	3124	17	TR21/735	S0016
171162-006	123A	3124	17	TR21/735	S0016
171162-008	123A	3018	20	TR21/	S0015
171162-009	123A			TR21/	S0011
171162-013		3124		TR83/	
171162-026	192	3018	20	TR21/	S0015
171162-027	108	3018	20	TR21/	S0015
171162-072	158	3004		TR84/	G0005
171162-073	158	3004		TR84/	G0005
171162-074	102A	3004		TR85/	G0005
171162-075	102A	3004		TR85/250	G0007
171162-076	102A	3004		TR85/250	G0005
171162-080	102A	3004		TR85/250	G0005
171162-081	158	3004		TR84/	G0005
171162-082	121	3014		TR01/232	G6005
171162-083	131	3052		TR94/642	G6016
171162-086	121	3014		TR01/232	G6005
171162-089	131	3082		TR94/	G6016
171162-090	131MP	3052	30	TR94/642MP	G6016
171162-092					S0016
171162-095		3122			
171162-108	176	3004		TR82/238	G6011
171162-113		3024	20	TR95/	S0016
171162-118	161	3117		TR83/719	S0020
171162-119	123A	3122		TR21/735	S0014
171162-120	102A			TR85/250	G0005
171162-121	102A			TR85/250	G0006
171162-124	154			TR78/	S5024
171162-125	154			TR78/	S5024
171162-126	154			TR78/	S5024
171162-128	108			TR95/	S0016
171162-129	108			TR95/	S0016
171162-130	108			TR95/	S0016
171162-131	108			TR95/	S0016
171162-132	123A			TR21/	S0011
171162-143	123A			TR21/	S0015
171162-161	123A	3124		TR21/735	S5014
171162-162	123A	3124		TR21/735	S5014
171162-163	128	3124		TR87/243	S5014
171162-164	210	3041	28	TR55/S3023	S3020
171162-169	160	3008		TR17/	G0008
171162-180		3124			
171162-186	107	3018	20	TR21/	S0016
171162-187	123A		20	TR24/	S0016
171162-188		3024	18	TR25/	S0015
171162-190	123A	3124	20	TR21/	S0015
171162-191	123A	3124	20	TR21/	S0015
171162-193	129			TR28/	S5013
171162-195	211	3084	29	TR56/	S3032
171174-1	130MP	3029	15MP	TR26/	S0015
171174-3			66	TR76/	S3020
171175-1	184+185				S5000

Original Part Number	ECG	SK	GE	IR/WEP	HEP
171179-027	1045	3101	IC2	IC507/	C6060P
171179-028	1120				
171179-036	1003	3101	IC2	IC507/	C6060P
171179-038	1007				
171206-1	108	3018	11	TR95/	S0008
171206-2	108	3018	11	TR95/	S0009
171206-4	108	3018	11	TR95/	S0016
171206-5	108	3018	11	TR95/	S0016
171206-6	222	3050			F2007
171207-1	108	3018	11	TR95/	S0016
171207-2	108	3018	11	TR95/	S0016
171217-1	131	3198		TR94/	
171522	102A	3004	52		G0005
171553	199	3018	62	TR21/	S0033
171554	123A	3018	20	TR21/	S0033
171555	159	3118	21	TR52/	S0019
171556	129	3114	67	TR30/	S5013
171557	128	3124	63	TR24/	S5014
171558	123A	3124	20	TR21/	S0015
171559	123A	3018	62	TR21/	S0015
171676		3124	20	TR21/	S0024
171677		3122	20		S0015
171678	199	3124	20	TR24/	S0025
171915	108	3018	20	TR24/	S0015
171916	102A	3004	2	TR05/	G0005
171917	102A	3004			
171982	1072				
171983	107	3018	20	TR24/	S0016
172089		3124			
172090		3024			
172202		3136			
172252	1003		IC2		
172272	1006				
172336	129	3025			S0019
172463	152	3041	66	TR76/	S5000
172761		3018			S0015
172762					S0019
172763					S0015
172816		3004			G0005
175005.423		3004		TR85/	
175005.424		3009			
175005.546		3018		TR21/	
175005.547		3124		TR21/	
175006-181	160	3006	51	TR17/637	
175006-182	160	3006	51	TR17/637	
175006-183	160	3006	51	TR17/637	
175006-184	160	3006	51	TR17/637	
175006-185	160	3006	51	TR17/637	
175006-186	102A	3004	2	TR85/250	
175006-187	108	3018	20	TR95/56	
175007-275	123A	3124		TR21/	S0011
175007-276	123A	3124		TR21/735	S0011
175007-277	128	3026		TR87/243	S5014
175027-021	123A			TR21/	S0015
175027-022	104			TR01/	G6003
175043-023	121	3009		TR01/121	G6005
175043-058	123A	3124		TR21/735	S0024
175043-059	123A	3124		TR21/735	S0024
175043-060	123A	3020		TR21/735	S0024
175043-062	108	3039		TR95/56	S0016
175043-063	108	3039		TR95/56	S0015
175043-064	108			TR95/56	S0015
175043-065	131	3052		TR94/642	G6016
175043-081	131MP			TR94MP/	G6016(2)
175043-81	121	3009		TR01/232	
175043-100	108	3018		TR95/56	S0015
175043-107	108	3018		TR95/56	S0011
175846-01A		3004			
177105	102				
180010-001	912				
181003-7	108			TR95/	
181003-8	108			TR95/	
181003-9	108			TR95/	
181012	199				
181015	159				
181023	123A				
181034	159				
181038	219				
181149A-4		3025			
181214	123A				
181469					S0014
181470					S0012
181503-6	108			TR95/	
181503-7	108			TR95/	
181503-9	108			TR95/	
181504-1	108			TR95/	
181504-2	123A			TR21/	
181504-7	108			TR95/	
181506-7	108			TR95/	
181515-4	128			TR87/	
181515-6	128			TR87/	
181515-7	128			TR87/	
181515-9	128			TR87/	
181671		3004			
181675					G6013

Original Part Number	ECG	SK	GE	IR/WEP	HEP
182503	74H20				
182510	7438				
182840					S0012
182850					S0014
183013	718	3159	IC8	IC511/	C6094P
183014	722	3161	IC9	IC513/	C5066P
183015	107	3018	11	TR70/	S0016
183016	107	3018	11	TR71/	S0014
183017	123A	3018	20	TR33/	S0011
183018	107	3018	20	TR33/	S0011
183019	107	3018	20	TR33/	S0011
183030		3124	10	TR24/	S0014
183031		3124	10	TR24/	S0014
183032		3114	21	TR30/	S0016
183034	192	3054	63	TR30/	S3020
183035	193	3083	67	TR30/	S3028
183044	804				
183045	733				
183405			IC27		
183532	941				
185203XB		3009			
185236		3018		TR80/	S3024
185500		3039			
186342A	128	3024	18	TR87/243	
187217	159				
187218	123A				
188091D		3004			
188164		3026			
188165	130				
188180	159				
188226	129				
188537A		3004			
188660-01	941				
190425	121	3009	16	TR01/628	G6013
190425A	121	3009	16	TR01/628	
190426	123A	3122		TR21/735	S0015
190427	160			TR17/637	G0008
190428	123A	3122		TR21/735	S0015
190429	159	3114		TR20/717	S0019
190714	161				S0030
190715	123A	3039		TR95/56	S0025
193022	9158				
193207	941				
194086-3	176				
194243	128				
194474-8	121			TR01/	
195601-6	102A			TR85/	
196023-1	123A				
196023-2	123A				
196058-4	121			TR01/	
196064-3	102A			TR85/	
196148-0	121			TR01/	
196183-5	121			TR01/	
196183-7	102A			TR85/	
196501-7	121			TR01/	
196607-9	121			TR01/	
196779-9	128			TR87/	
196779-9-1	128			TR87/	
196780-1	128			TR87/	
197281		3036			
198002-1		3122			
198003-1	123A	3122			
198003-2	123A	3122			
198005-1	195				
198006-1		3011			
198007-3	123A	3122			
198010-1	126	3008			
198013-P1	123A				
198014-1	128				
198020-1	128				
198020-2	128				
198020-3	128				
198023-1	123A	3122			
198023-2	128			TR87/243	S5014
198023-3	123A				
198023-4	123A				
198023-5	123A				
198024	159				
198030	123A				
198030-2	123A				
198030-3	123A	3122			
198030-4	123A				
198030-6	123A	3122			
198030-7	123A	3122			
198031-1	123A				S0011
198031-2	123A				S0011
198034-1	130	3027			
198034-2	130	3510			
198034-3	130	3027			
198034-4	130	3027			
198034-5	130				
198035-1	128				
198035-3	128				
198036-1	159				
198038-1	175				

Original Part Number	ECG	SK	GE	IR/WEP	HEP
198038-3	175	3026			
198038-4	175	3026			
198039-0507	130	3027		TR59/247	S7002
198039-1	175				
198039-3	175				
198039-6	130	3027			
198039-7	130	3027			
198039-501	175				
198039-503	175				
198039-506	130				
198039-507	130				
198042-2	123A				
198042-3	123A				
198045-4	128				
198047-1	128				
198047-2	128				
198047-3	128				
198047-5	128				
198047-6	128				
198048-1	128				
198048-2	128				
198049-1	175				
198050	159				
198051-1	123A	3122			
198051-2	123A	3122			
198051-3	123A				
198051-4	123A				
198063-1	153		26	TR77/700	S5018
198064-1	181	3511			
198065-1	129				
198065-3	129				
198067-1	123A				
198072-1	128				
198074-1	129				
198075-1	198				
198077-1	128				
198078-1	129				
198079-1	181				
198079-2	130				
198350-2					S7003
198409-1	9949				
198409-5	9944				
198409-6	9932				
198409-11	9946				
198409-12	9962				
198409-13	9930				
198409-14	9936				
198410-1	909	3590			
198581-1	123A				
198581-2	123A				
198581-3	123A				
198794-1	101	3011		TR08/641	G0011
198801-1	905				
199807	194				
200028-7-28	102A			TR85/	
200062-5-31	102A			TR85/	
200062-5-32	102A			TR85/	
200062-5-33	102A			TR85/	
200062-5-34	102A			TR85/	
200064-6-103	108			TR95/	
200064-6-104	160			TR17/	
200064-6-105	108			TR95/	
200064-6-106	160			TR17/	
200064-6-107	123A			TR21/	
200064-6-108	102A			TR85/	
200064-6-109	127			TR27/	
200064-6-110	127			TR27/	
200064-6-111	102A			TR85/	
200067	159				
200076	123A				
200200	199				
200200-700	199				
200220	159				
200251-5377	123A	3124		TR21/	
200252	128				
200259-700	218				
200433	159				
200681-94				TR82/	
200781-702	124				
201034	124				
201059		3026			
202609-0713	123A	3124		TR21/735	
202862-947	123A				
202907-047P1	123A				
202909-577	159				
202909-587	159				
202909-827	130				
202911-737	159				
202913-057	128				
202914-010	9930				
202914-010/250-060	9945				
202914-010/250-070	9935				
202914-010/250-100	9936				
202914-020	9946				
202914-030	9962				
202914-040	9932				
202914-050	9944				
202914-417	123A				
202915-627	123A				
202917-137	129				
202920-150	7491				
202922-237	123A				
202922-280	716				
202925-047	219				
203364	159				
203718	129				
204201-001	128				
204210-002	123A				
204211-001	130				
204969	123A				
205032	159				
205048	159				
205049	159				
205367	159				
205702-149				TR17/	
205782-97			20	TR21/	
205782-103			20	TR21/	
207010		3024			
207119	124				
209185-962	121	3009		TR01/	
209417-0714	108	3018		TR95/56	
210074	123A				
210076	159				
211040-1	181				
214396	121	3009	16	TR01/232	
215002					G0005
215008					G0009
215018					G0009
215031					G0009
215038					G0009
215044					G0005
215053					G0005
215060					G0003
215061					G0006
215071	131MP	3009		TR94MP/232	G6016(2)
215072	123A		17	TR21/735	S0015
215074	123	3124		TR86/53	S0015
215075	128	3124		TR87/243	S0015
215079					S0016
215081	123A		20	TR21/735	S0015
215089	131	3082	30	TR94/642	G6016
216161-1		3039			
216161-2		3039			
216445-2	123A				
216986	121	3009	16	TR01/232	
217119	105	3012	4	TR03/233	
217124		3124			
217230	105	3012	4	TR03/233	
217892	121	3009	16	TR01/232	
218012DS	104	3009	3	TR01/233	
218502	102A	3003	2	TR85/250	
218503	102A	3004	2	TR85/250	
218511	128				
218537	129				
219016	101	3011	5	TR08/641	
219301	121	3009		TR01/232	
219361	121	3009	16	TR01/232	
219440	121	3009		TR01/232	
219940	121	3009	16	TR01/232	
221158	128	3024		TR87/243	S5014
221600	123A	3124	20	TR21/735	
221601	101	3011	5	TR08/641	
221602	121	3009	16	TR01/232	
221605	121	3009	16	TR01/232	
221856	126	3005	1	TR17/635	
221857	123A	3122		TR21/735	
221897	123A	3122	20	TR21/735	
221918	123A	3124	20	TR21/735	
221924	101	3011	5	TR08/641	
221940	121	3009	16	TR01/232	
221941	121	3009	16	TR01/232	
222131	123A				
222509	102A	3004		TR85/	
222915	121	3009	16	TR01/232	
223124	102A	3004	2	TR85/250	
223365	121	3009	16	TR01/232	
223366	102A	3004	2	TR85/250	
223367	101	3011	5	TR08/641	
223368	101	3011	5	TR08/641	
223369	160	3007		TR17/637	
223370	101	3011	5	TR08/641	
223371	102A	3003	2	TR85/250	
223372	100	3005	1	TR05/254	
223473	100	3005	1	TR05/254	
223474	160	3007	51	TR17/637	
223475	100	3005	1	TR05/254	
223482	103A	3010	8	TR08/724	
223483	102A	3004	2	TR85/250	

Original Part Number	ECG	SK	GE	IR/WEP	HEP
223484	102A	3003	2	TR85/250	G0008
223485	102A	3003	2	TR85/250	
223486	102A	3004	2	TR85/250	
223487	126	3008	51	TR17/635	
223490	121	3009	16	TR01/232	
223576	121	3009	16	TR01/232	
223684	101	3011	5	TR08/641	
223810	102A	3009	2	TR85/250	
224503	121	3009	16	TR01/232	
224506	123A	3124		TR21/	
224584	100	3005	1	TR05/254	
224586	160	3006	51	TR17/637	
224587	160	3006	51	TR17/637	
224696	102A	3004	2	TR85/250	
224820	103A	3011	8	TR08/724	
224857	102A	3004	2	TR85/250	
224873	121	3014	16	TR01/232	
225300	103A	3010	8	TR08/	G0011
225311	160	3007	51	TR17/637	
225593	102A	3003	2	TR85/250	
225594	160	3007	51	TR17/637	
225594A	126	3008	51	TR17/635	
225595	121	3009	16	TR01/232	
225596	121	3009	16	TR01/232	
225600	160	3006	51	TR17/637	
225925	121	3009	16	TR01/232	
225927	121	3009	16	TR01/232	
226181	100	3005	1	TR05/254	
226338	100	3005	1	TR05/254	
226441	101	3011	5	TR08/641	
226634	121	3009	16	TR01/232	
226789	105	3012	4	TR03/233	
226791	103A	3010	8	TR08/724	
226924	102A	3004	2	TR85/250	
226999	121	3009	16	TR01/232	
227000	108	3018		TR95/	
227566	121	3009	16	TR01/232	
227752	100	3005	1	TR05/254	
227804	121	3009	16	TR01/232	
228229	121	3009	16	TR01/232	
228230	121	3009	16	TR01/232	
228287	102A	3003	2	TR85/250	
228558	121	3009	16	TR01/232	
228559	121	3009	16	TR01/232	
229045	105	3012	4	TR03/233	
229133	160	3006	51	TR17/637	
229392	108			TR95/	S0016
230084	175	3026		TR81/241	
230208	121	3009	16	TR01/232	
230209	101	3011	5	TR08/641	
230214	128	3024	18	TR87/243	
230233	128				
230253	102A	3004	2	TR85/250	
230256	101	3011	5	TR08/641	
230259	102A	3004	2	TR85/250	
230523	121	3014	16	TR01/232	
230524	102A	3004	2	TR85/250	
230525	102A	3004	2	TR85/250	
231140-01	108	3018		TR95/	
231140-04	127	3034		TR27/	
231140-07	108	3018		TR95/	
231140-09	127	3034		TR27/	
231140-11	121	3009		TR01/	
231140-15	123A	3124		TR21/	
231140-21	102A	3004		TR85/	
231140-23	108	3018		TR95/	
231140-26	127	3034		TR27/	
231140-28	154	3040		TR78/	
231140-31	108	3018		TR95/	
231140-33	121	3009		TR01/	
231140-34	108	3018		TR95/	
231140-36	161	3019		TR83/	
231140-37	161	3019		TR83/	
231140-43	161	3019		TR83/	
231140-44	108	3018		TR95/	
231140-45	103A	3010		TR08/	
231374	123A	3124	20	TR21/735	
231375	128	3024		TR87/	
231378	130	3027		TR59/	
231588	102A	3004	2	TR85/250	G0005
231672	121	3009	16	TR01/232	
231706	102A	3008		TR85/250	G0005
231797	121	3009	16	TR01/232	
232194	121	3009	16	TR01/232	
232268	175			TR81/	
232359	130	3027	14	TR59/247	
232359A	130	3027		TR59/	
232631	159				
232674	121	3009	16	TR01/232	
232675	121	3009	16	TR01/232	
232676	160	3006	16	TR17/637	
232678	123A	3047		TR21/	
232680	160	3006	51	TR17/637	
232681	126	3008	51	TR17/635	
232840	108	3019	11	TR95/56	
232841	128	3024	18	TR87/243	
232949	101	3011	5	TR08/641	
233305	105	3012	4	TR03/233	
233307	105	3012	4	TR03/233	
233507	100	3005	1	TR05/254	
233508	105	3012	4	TR03/233	
233509	121	3014	16	TR01/232	
233735	128	3024		TR87/	
233944	128	3024		TR87/	
233945	100	3005	1	TR05/254	
233969	129	3025		TR88/	
234015	160	3006	51	TR17/637	
234024	128	3024		TR87/	
234076	102A	3004	2	TR85/250	
234077	121	3009	16	TR01/232	
234078	105	3012	4	TR03/233	
234566	121	3009	16	TR01/232	
234612	123A	3124	20	TR21/735	
234630	102A	3004	2	TR85/250	
234631	160	3006	51	TR17/637	
234758	123A	3124		TR21/	
234763	123A	3124		TR21/	
235192	123A	3124		TR21/	
235194	102A	3004	2	TR85/250	
235200	160	3006	51	TR17/637	
235205	123A	3124	20	TR21/735	
235206	123A	3124	20	TR21/735	
235312	121	3009	3	TR01/232	
235997	162	3079		TR67/	
236039	161	3039		TR83/	
236251	108	3019	11	TR95/56	
236265	101	3011	5	TR08/641	
236282	154	3040		TR78/	
236285	123A	3124	20	TR21/735	
236286	123A	3124	20	TR21/735	
236287	128	3024		TR87/	
236288	127	3035		TR27/	
236433	129	3025		TR88/	
236706	108	3019	11	TR95/56	
236709	102A	3004		TR85/	
236854	181	3036		TR36/	
236907	108	3018	20	TR95/56	
236935	121	3009	16	TR01/232	
237020	108	3018		TR95/	
237021	108	3018		TR95/	
237024	108	3018		TR95/	
237025	123A	3046		TR21/	
237026	108	3018		TR95/	
237028	195	3047		TR65/	
237075	191	3044		TR75/	
237223	123A	3124		TR21/	
237450	124	3021	12	TR81/240	
237452	121	3009	16	TR01/232	
237785	108	3018		TR95/	
237840	108	3019		TR95/	
238208-002	175				
238368	123A	3038		TR21/	
238417	102A	3004	2	TR85/250	
238418	102A	3004	2	TR85/250	
239103	225	3045			
239612	128	3024		TR87/	
239713	181	3036		TR36/	
239970	123A	3124		TR21/	
240003	102A			TR85/	G0005
240006	102A			TR85/	G0005
240388	124	3021		TR81/	
240401	123A	3124		TR21/	
240402	129	3025		TR88/	
240403	127	3034		TR27/	
240404	162	3079		TR67/	
241052	129	3025		TR88/	
241249	108	3018		TR95/	
241657	130	3027		TR59/	
241778	108	3018		TR95/	
241960	108	3018		TR95/	
242102	152				
242183	121	3009		TR01/	
242221	102A	3004		TR85/	
242422	129	3025		TR88/	
242460	129	3025		TR88/	
242590	108	3039		TR95/	
242758	123A	3020		TR21/	
242759	123A	3124		TR21/	
242838	121	3009		TR01/	
242958	129	3025		TR88/	
242960	108	3018		TR95/	
243028	724				
243168	127	3035		TR27/	
243215	105	3012		TR03/	
243318	108	3018		TR95/	
243645	108	3018		TR95/	
243815	105	3012		TR03/	
243837	103A	3010		TR08/	
243939	102A	3004		TR85/	
244007	100	3005		TR05/	

Original Part Number	ECG	SK	GE	IR/WEP	HEP
244357	124	3021		TR81/	
245010-02		3122			
245078-3	108	3018		TR95/	
245568-2	176	3123		TR82/	
247149					S0012
248020		3018			
249588	126	3006		TR17/635	G0008
250400	103A	3010	8	TR08/	G0011
250601					S0015
250602					S0019
250603					S5014
250604					S5013
252744	124				
253704	105	3012		TR03/	
255728	102A	3004		TR85/	
255821HS	193	3025			
256068	121	3009		TR01/	
256071	121	3009		TR01/	
256126	100	3005		TR05/	
256127	101	3011		TR08/	
256480	105	3012		TR03/	
257242	105	3012		TR03/	
257243	105	3012		TR03/	
257340	102A	3004		TR85/	
257341	121	3009		TR01/	
257385	101	3011		TR08/	
257403	121	3009		TR01/	
257470	100	3005		TR05/	
257473	102A	3004		TR85/	
257534	105	3012		TR03/	
257536	121	3009		TR01/	
257540	108	3019		TR95/	
258990	121	3009		TR01/	
258993	101	3011		TR08/	
260468	101	3011		TR08/	
260565	108	3039		TR95/	
261401	105	3012		TR03/	
261488	105	3012		TR03/	
261586	160	3008		TR17/	
261970	121	3009		TR01/	
262066	123A	3124		TR21/	
262113	102A	3004		TR85/	
262114	121	3009		TR01/	
262116	175	3026		TR81/	
262309	105	3012		TR03/	
262370	121	3009		TR01/	
262417	128				
262417-1	128				
262638	129	3025		TR88/	
263561	124	3021		TR81/	
263856	121	3009		TR01/	
263857	124	3021		TR81/	
265074	108	3019		TR95/	
265240	123A	3124		TR21/	
265241	108	3019		TR95/	
265771	160	3006		TR17/	
266685	123A	3124		TR21/	
266686	102A	3004		TR85/	
266702	100	3005		TR05/	
267704	128				
267791	130	3027		TR59/	
267797	108	3018		TR95/	
267838	159				
267878	181	3036		TR36/	
267898	123A	3124		TR21/	
267899	123A	3124		TR21/	
268003	199				
268044L	123A				
269367	103A	3010		TR08/	
269374	102A	3004		TR85/	
270744	121	3009		TR01/	
270745	121	3009		TR01/	
270746	121	3009		TR01/	
270780	121	3009		TR01/	
270781	101	3011		TR08/	
270785	121	3009		TR01/	
270819	123A				
272350C		3004			
275131	123A	3124		TR21/	
275612	121	3009		TR01/	
275845	127	3034		TR27/	
276160	128	3024		TR87/	
276331	123A	3124		TR21/	
276413	128	3024		TR87/	
276415	128	3024		TR87/	
281001-53	199				
281001-83	199				
286784				TR21/	
290458LGD	123A	3124	62	TR21/735	S0014
291509	181	3036		TR36/	
293118	74165				
297065C03			51		
297074C11	129	3025		TR88/242	
297240-1	102A	3004	2	TR85/250	
299371-1	123A				

Original Part Number	ECG	SK	GE	IR/WEP	HEP
299379	724				
300008	102				
300061-04	716				
300113	123A				
300486	101	3011	5	TR08/641	
300536	101	3011	5	TR08/641	
300538	102A	3004	2	TR85/250	
300540	102A	3004	2	TR85/250	
300541	102A	3004	2	TR85/250	
300542	101	3011	5	TR08/641	
300774	101	3011	5	TR08/641	
301591	123A	3124		TR21/	
301606	128	3024		TR87/	
301915-1	941				
302342	123A	3124		TR21/	
302865	129				
304581B	123A	3124		TR21/735	
304900	108				
307004					C6007
308055		3024			
308449	123A				
309004	754				
309412	121	3009		TR01/	
309418		3024			
309419		3114			
309421	102A	3004	2	TR85/250	
309442	123A				
309459	152	3054			
309683	234				
309684	159				
309685	234				
309689	197	3054			
309690	197	3083			
310017	102A	3004	2	TR85/250	
310030	160	3007	51	TR17/637	
310035	102A	3052	2	TR85/250	
310110	128				
310132	160	3006	51	TR17/637	
310157	126	3008	1	TR17/635	
310158	126	3008	1	TR17/635	
310159	102A	3004	2	TR85/250	
310160	102A		2	TR85/250	
310162	160	3006	51	TR17/637	
310201	102A	3004	2	TR85/250	
310204	160	3006	51	TR17/637	
310221	126	3006	51	TR17/635	
310223	100	3005	1	TR05/254	
310224	126	3006	51	TR17/635	
310225	102A		2	TR85/250	
310254	7408				
312194-1		3039			
312194-2		3039			
313309-1	123A				
313316-1		3039			
313319-1		3039			
313319-2		3039			
315930	128				
315932	128				
318835	130	3027			
319304	129				
320280	159				
320529	123A				
321006P1	703A				
321145	128				
321165	159				
321166	128				
321264-2	195				
321517	123A				
321573	123A				
322722		3054			
322968-17	102A			TR85/	
322968-140	121			TR01/	
322968-141	127			TR27/	
322968-167	102A			TR85/	
323934	199				
324144	159				
325077	129				
325079	123A				
325099	124				
325101	128				
326809	909				
326823	923				
326830	9930				
326832	9932				
326833	9933				
326836	9936				
326844	9944				
326845	9945				
326846	9946				
326852	9099				
326853	9093				
326862	9962				
328785	123				
329814	9945				
330278		3011			

Original Part Number	ECG	SK	GE	IR/WEP	HEP
330803	123A				
331378	9932				
331383	128				
332762	219				
333060-1029	159				
333241	123A				
334724-1	123A				
335288-4	123A				
335613	129				
335774	123A				
336637-20	9900				
336638-21	9923				
337342	159				
339002	74H00				
339003	74H10				
339009	74H40				
339300	7400				
339300-2	7400				
339486	7486				
340085PC	40085				
340097PC	40097				
340098PC	40098				
340160PC	40160				
340161PC	40161				
340162PC	40162				
340163PC	40163				
340174PC	40174				
340175PC	40175				
340192PC	40192				
340193PC	40193				
340194PC	40194				
340195PC	40195				
340866-2	128				
341569		3079			
346015-15	108	3018		TR95/	
346015-16	108	3018		TR95/	
346015-17	108	3018		TR95/	
346015-18	108	3018		TR95/	
346015-19	108	3018		TR95/	
346015-20	108	3018		TR95/	
346015-21	108	3018		TR95/	
346015-22	108	3018		TR95/	
346015-23	128	3024		TR87/	
346015-24	123A	3124		TR21/	
346015-25	108	3018		TR95/	
346015-30	128	3024		TR87/	
346015-37	108	3018		TR95/	
346016-1	102A	3004		TR85/	
346016-11	102A	3004		TR85/	
346016-14	123A	3020		TR21/	
346016-16	128	3124		TR87/	
346016-17	128	3024		TR87/	
346016-18	123A	3124		TR21/	
346016-19	123A	3124		TR21/	
346016-25	123A	3124		TR21/	
346016-26	123A	3124		TR21/	
346016-27	128	3024		TR87/	
346016-63	191	3044		TR75/	
346309-1		3124			
346607-4	160				
348846-1	128				
373003	102A	3004	2	TR85/250	
373117	102A	3004	2	TR85/250	
373119	102A	3004	2	TR85/250	
373199				TR85/	
373401-1	7400				
373404-1	7404				
373405-1	7410				
373406-1	7420				
373407-1	7430				
373408-1	7440				
373409-1	7474				
373410-1	7486				
373411-1	7470				
373412-1	7483				
373413-1	74165				
373414-1	7476				
373423-1	7409				
373424-1	7472				
373427-1	7490				
373428-1	7496				
373429-1	7406				
373708-1	74179				
373712-1	7492				
373713-1	7475				
373714-1	7453				
373714-2	7454				
373715-1	7451				
373716-1	7482				
373718-1	7493				
373721-1	7407				
374109-1	7408				
374110-1	7460				
379101K	123A				
379102	123				

Original Part Number	ECG	SK	GE	IR/WEP	HEP
380049	9158				
382237		3036			
386726-1	129				
388060	123A				
395253-1	130				
400107		3024			
400108	128				
400115		3027			
400119		3036			
400127	175				
400603		3005			
400909	128				
400931	9935				
400965		3005			
401003-0010	159				
401113-1	9946				
401113-2	9930				
401113-3	9932				
401113-4	9945				
401113-5	9962				
401182	9951				
401646	912				
405101-1	194				
405192	106				
405457	123A				
405965-8A	128				
405965-30A	179			TR35/	
405965-35A	129			TR88/	
417214	162				
425411		3044			
425411-01	128				
433836	123A	3124		TR21/735	
436119-002	159				
445023-P1	130				
445111	910				
446914		3004			
450041-1		3024			
450826-1	128				
452077	133	3112		/801	
454549	100			TR05/	
454760	160			TR17/	
463984-1	7440				
465002-30	799				
480040		3024			
480041		3025			
481335	128				
486271		3004			
489251-097					S0019
489571-131		3039			
489751-025	123A	3038	20	TR21/735	S0015
489751-026	123A	3020	20	TR21/735	S0015
489751-027	108	3018	20	TR95/56	S0015
489751-028	159	3114	21	TR20/717	S0019
489751-029	123A	3124	20	TR21/735	S0015
489751-030	123A	3024	20	TR21/735	S0015
489751-031	159	3114	21	TR20/717	S0019
489751-032	153	3025	21	TR77/700	S5006
489751-033	152	3024	63	TR76/701	S5000
489751-037	128	3018	63	TR87/	S5014
489751-038	128	3018	63	TR87/	S5014
489751-039	161	3018		TR83/	S0020
489751-040	123A		17	TR21/	S0015
489751-041	123A		17	TR21/	S0014
489751-042	159	3118		TR20/	S0013
489751-043	124	3021	12	TR81/	S5012
489751-044	152		23	TR76/	S5012
489751-045	102A	3004	52	TR85/	G0005
489751-047	161	3018	11	TR83/	S0016
489751-049		3019			S0016
489751-052	107	3018	20	TR70/	S0020
489751-058		3019	11	TR21/	S0017
489751-095				TR22/	
489751-097	159	3118	21	TR20/	
489751-107	123A	3040	17	TR21/	S0030
489751-108	102A	3008		TR85/	G0005
489751-109	102A	3004	52	TR85/	G0005
489751-113	102A	3005	52	TR85/	G0005
489751-114	102A	3004	52	TR85/	G0005
489751-115	127	3034	25	TR27/	G6008
489751-119	130	3111	14	TR59/	S5020
489751-120					S0017
489751-121	161	3039		TR83/	S0017
489751-122	123A	3018		TR21/	S0016
489751-123					S0020
489751-124	159	3118	21	TR20/	S0012
489751-125	123A	3020	10	TR21/	
489751-127	161	3018	60	TR83/	S0020
489751-128	161	3018	20	TR83/	S0015
489751-129	128	3024		TR87/	S5014
489751-130	159	3114	21	TR20/	S5013
489751-131	108		11	TR95/	S0020
489751-137	108	3019		TR95/56	S0016
489751-143	108	3018	20	TR95/	S0020
489751-144	128	3024	18	TR87/	S0005
489751-145	108	3018	20	TR95/	S0014

Original Part Number	ECG	SK	GE	IR/WEP	HEP
489751-146	159	3114	22	TR20/	S0019
489751-147	108	3018		TR95/	S0014
489751-148	108	3018	20	TR95/	S0014
489751-162	108	3019	11	TR95/	S0020
489751-163	131	3052	30	TR50/	
489751-164		3122	20	TR21/	
489751-165	108	3122	20	TR21/	
489751-166	123A	3124	10	TR24/	
489751-167	108	3122	20	TR21/	
489751-168	108	3018	17	TR21/	
489751-169	108	3018	20	TR21/	
489751-171	108	3018	11	TR95/	S0016
489751-172	123A	3122	17	TR21/	S0015
489751-173	107	3018	20	TR70/	S0030
489751-174	175	3026	23	TR81/	S5019
489751-175	1046	3026			
489751-206	108	3018	20	TR95/	S0014
489751-208	132	3116	FET2	FE100/	F0021
489752-095	161			TR83/	
500879	128				
502349	130				
505198	130				
505254	9941				
505256	175				
505257	175				
505287	128		18	TR87/	
505342	909	3590			
505434	175				
505469	175				
505568	181				
506902	199	3124			
508511	941				
508590	7404				
508762	123A				
510007	102A	3004	52	TR85/250	G0005
510584	199				
510901				TR60/	
510901-770				TR60/	
511806	128				
514023	123A	3124	20	TR21/735	S0015
514045	107	3018	11	TR70/720	S0011
514067S	129	3114	21	TR19/	
514068S	193	3114	67	TR88/	
514072S	129	3114	21	TR19/	
515039S	128	3122	18	TR25/	
515041S	107	3018	20	2SC710/	
515043S	123A	3020	20	TR08/	
515045S	123A	3124	20	2SC945/	
515268-1		3039			
516009S	132	3116	FET2	FE100/	
517999	909	3590			
518022S	720				
530130-1	199				
531298-001	106				
531841-002	199				
531972	128				
532003	175	3026		TR81/241	
532010					G0008
532775	199				
533802	123A				
537200	102A	3004		TR85/250	
537428			50	TR17/	
537790	102				
543995	159	3114		TR20/717	S5022
547684	106				
551015	102A			TR85/250	G0005
551051	102A	3003		TR85/250	
552308	123A	3122		TR21/735	S0011
552503	159			TR20/717	S0013
558875	7400				
558876	7402				
558877	7410				
558878	7420				
558879	7430				
558880	7440				
558881	7473				
558882	7474				
558883	7490				
558885	7493				
559507	7401				
559509	7450				
559510	7460				
559557	180				
559613	7453				
560004	121MP		13MP	/232MP	
564671	129	3025			
567312	123	3020	18	TR25/	S0014
570000-5452	123A		20	TR21/	
570000-5503	123A		20	TR21/	
570004-503	123A	3124	20	TR21/735	
570005-452	123A	3124	20	TR21/735	
570005-503	123A	3124	20	TR21/735	
570009-01-504	123A	3124	20	TR21/735	S0015
570029	130		14	TR59/	
570030	185	3083	58	TR77/	S5006

Original Part Number	ECG	SK	GE	IR/WEP	HEP
570031	184	3190		TR76/	S5000
571208-8					S3021
572001	185	3512			
572002		3512			
572670		3024			
572683	123A	3124			
573001	102A	3004	2	TR85/250	
573003	102A		2	TR05/	
573005	102A	3004	2	TR85/250	
573011	102A	3004	2	TR85/250	
573012	102A	3004	2	TR85/250	
573018	102A	3004	2	TR85/250	
573022	102A	3004	2	TR85/250	
573029	102A	3004	2	TR85/250	
573036	102A	3004		TR85/250	
573037	103A	3010		TR08/724	
573101	108	3018	20	TR95/56	
573103	102A	3004	2	TR85/250	
573110	102A	3		TR85/	
573114	102A	3	2	TR85/	
573117	102A	3004	2	TR85/250	
573118		3004	2	TR85/250	
573119	102A	3004	2	TR85/250	
573125	102A	3004	2	TR85/250	
573142	102A	3004	2	TR85/250	
573152	102A	3004	2	TR85/250	G0005
573153	102A	3004	2	TR85/250	G0005
573166	121	3009	3	TR01/232	G6005
573184	102A		52	TR85/	
573199	127	3034	25	TR27/235	
573200	102A		53	TR85/250	G0005
573212	127	3035	25	TR27/235	G6008
573303	126	3008	51	TR17/635	
573328	102A		52	TR85/	
573329	126	3008	51	TR17/635	
573330	126	3008	51	TR17/635	
573335	160	3006	51	TR17/637	
573336	160	3006	51	TR17/637	
573356	100		1	TR12/	
573366	126		50	TR17/	
573371	160	3006	51	TR17/637	
573398	160	3006	51	TR17/637	
573402	126	3008	51	TR17/635	
573405	160	3006	51	TR17/637	
573406	160	3006	51	TR17/637	
573422	102A	3004	2	TR85/250	G0001
573428	126		51	TR17/	
573432	102A			TR85/250	
573467	123A		17	TR21/	
573468	123A		10	TR21/	
573469	123A	3124	20	TR21/735	S0011
573471	126		50	TR17/	
573472	108	3018	20	TR95/56	
573474	108	3018	20	TR95/56	
573474A		3018		TR24/56	
573475		3018	20	/56	S0016
573479	123A	3124	20	TR21/735	
573480	123A	3124	20	TR21/735	S0016
573481	123A	3124		TR21/735	
573494	108	3018	20	TR95/56	S0016
573495	108	3018	20	TR95/56	S0016
573501	154	3045	27	TR78/712	
573515	124	3021		TR81/240	
573518	126	3006	51	TR17/635	G0008
573529	102A	3004	2	TR85/250	G0005
573532	128	3024		TR87/243	
574003	102A	3512	2	TR05/	
576001	102A	3004	2	TR85/250	
576005	102A	3004	2	TR85/250	
581005	102	3004			
581024	101				
581034A	123A	3122			
581042	102				
581054	123A				
581055	123A				
581070	130			TR59/	
585600-1		3024			
587012	726				
595819-1	704				
595819-2	726				
597568-1A		3024			
598274-2		3039			
599357		3024			
599391		3024			
600080-413-001	123A				
600080-413-002	123A				
600096-413	123				
600098-413-001	123A				
600115-413-001	130				
600343-08		3009			
601032	160		51	TR17/637	
601040	160	3006	50	TR17/637	G0003
601052	160		51	TR17/637	
601054	126	3008	51	TR17/635	
601054(SHARP)	105			TR03/233	

Original Part Number	ECG	SK	GE	IR/WEP	HEP
601065	103A	3010	8	TR08/724	
601113	108	3018	17	TR95/56	S0016
601122	123A	3122	20	TR21/735	
602032	121	3009	3	TR01/232	G6003
602040	158	3004	2	TR84/630	G0005
602051	158		53	TR84/630	
602075	160	3006	51	TR17/637	G0002
602113	108	3018	17	TR95/56	S0016
602113(SHARP)	123A			TR21/735	
602122	128	3024	18	TR87/243	
602909-2A	123A			TR21/	
602909-3A	129			TR88/	
602909-7A	128			TR87/	
603020	160	3006	51	TR17/637	G0003
603030	160	3006	51	TR17/637	G0002
603031	121	3009	3	TR01/232	G6003
603040	160	3006	1	TR17/637	G0005
603112	160		51	TR17/637	
603113	108	3018	17	TR95/56	S0016
603122	123A	3122	20	TR21/735	
603312	160			TR17/637	
604030	160	3006	51	TR17/637	G0003
604040	160		51	TR17/637	
604080	160	3006	51	TR17/637	
604112	126	3006	51	TR17/635	
604113	108	3019	17	TR95/56	S0016
604122	123A	3122	20	TR21/735	
605030	102A	3004	2	TR85/250	G0005
605112	160		51	TR17/637	
605113	128	3024	2	TR87/243	S0014
605122	131			TR94/642	
606020	158	3004	2	TR84/630	G0005
606112	102A		53	TR85/250	
607030		3124	20	TR08/724	G0011
607122	124		12	TR81/240	
608112	127	3034	16	TR27/235	G6008
608112(SHARP)	179			TR35/G6001	
608122	128	3024	18	TR87/243	
609020	160	3006	51	TR17/637	G0002
609112	123A	3124	20	TR21/735	
610030-5		3004			
610031-1		3019			
610034-1		3004			
610035	102A	3004		TR85/250	G0005
610035-1	102A	3004	2	TR85/250	G0005
610035-2	102A	3004	2	TR85/250	
610036	102A	3004		TR85/250	
610036-1	102A	3004	52	TR85/250	G0005
610036-2	102A	3004	52	TR85/250	G0005
610036-3	102A	3004	52	TR85/250	G0005
610036-4	102A	3004	2	TR85/250	G0005
610036-5	102A	3004	2	TR85/250	G0005
610036-6	102A	3004			
610036-7	102A	3004	52	TR85/250	G0005
610036-8	102A	3004	52	TR85/250	
610039	121	3009		TR01/232	
610039-1	121	3009	16	TR01/232	G6003
610040	102A	3004		TR85/250	
610040-1	102A	3004	2	TR85/250	G0005
610040-2	102A	3004	52	TR85/250	G0005
610041	108	3018		TR95/56	
610041-1	108	3018		TR95/56	
610041-2	161	3018	11	TR21/719	S0025
610041-3	108	3018	11	TR24/56	S0025
610042	161	3039		TR83/719	
610042-1	108	3019	17	TR95/56	S0016
610043	102A	3004		TR85/250	
610043-1	102A	3004		TR85/250	
610043-2	102A	3004		TR85/250	
610043-3	102A	3004	2	TR85/250	G0005
610043-4	102A	3004	2	TR85/250	G0005
610043-6	102A	3004	52	TR85/250	G0005
610043-7	102A	3004	52	TR85/250	G0005
610043-7H		3008			
610043-610P1				TR30/	
610043-783P2				TR12/	
610045	108	3018		TR95/56	
610045-1	108	3018	20	TR95/56	
610045-2	108	3018	17	TR95/56	
610045-3	123A	3124	20	TR21/735	S0015
610045-4	123A	3124	20	TR21/735	S0015
610045-5	123A	3122		TR21/735	S0015
610046-7	108	3004		TR95/56	
610047-1		3009			
610049-1	121	3009			
610050	160	3006		TR17/637	
610050-1	160	3006	51	TR17/637	G0003
610050-2	160	3006	51	TR17/637	
610050-3	160	3006	51	TR17/637	
610051	160	3006		TR17/637	
610051-1	160	3006	51	TR17/637	G0003
610051-2	160	3006	51	TR17/637	
610051-4	160	3006	51	TR17/637	
610052	126	3008		TR17/635	
610052-1	102A	3008	1	TR85/250	G0005
610053-2	160	3006	1	TR17/637	G0005
610055	160	3006	51	TR17/637	
610055-1	160	3006		TR17/637	G0003
610055-2	160	3006	51	TR17/637	
610055-3	160	3006	51	TR17/637	
610056	126	3008		TR17/635	
610056-1	160	3008	51	TR17/637	G0002
610056-2	160	3008	51	TR17/637	
610056-3	160	3008	52	TR17/637	G0003
610056-4	160	3008	1	TR17/637	G0003
610059-1		3004			
610059-2	102A	3004			
610061-1	160		51	TR17/637	
610063-1	163	3111	36	TR61/740	S5021
610064-1	162		35	TR67/707	S5020
610067	121	3009		TR01/232	
610067-1	121	3009		TR01/232	
610067-2	121	3009		TR01/232	
610067-3	121MP	3009	16	TR01/232MP	
610067-D	121	3009			
610068	121	3009		TR01/232	
610068-1	121MP	3009	16	TR01/232MP	G6005
610069	108	3018		TR95/56	
610069-1	108	3018	20	TR95/56	
610070	123A	3124		TR21/735	
610070-1	123A	3124	20	TR21/735	S0015
610070-2	123A	3124	20	TR21/735	S0015
610C70-3	123A	3124		TR21/735	S0015
610070-4	123A	3122	17	TR21/735	S0015
610071	124	3021		TR81/240	S5011
610071-1	124	3021	12	TR23/240	S5011
610071-2	124	3021	12	TR23/	
610072	108	3018		TR95/56	
610072-1	108	3018	20	TR95/56	S0016
610072-2	108	3018	20	TR95/56	
610073	108	3018		TR95/56	
610073-1	161	3018	11	TR83/719	S0020
610074	126	3008		TR17/635	
610074-1	159	3114	21	TR20/717	S0019
610074-2	100	3005		TR05/254	
610075-1	154	3044	20	TR78/712	S5025
610076	123A	3124		TR21/735	
610076-1	123A	3124	20	TR21/735	S0015
610076-2	123A	3124	20	TR21/735	S0015
610077	123A	3124		TR21/735	
610077-1	123A	3124	20	TR21/735	S0015
610077-2	123A	3124	20	TR21/735	S0015
610077-3	123A	3124	20	TR21/735	
610077-4	123A	3124	10	TR24/735	S0015
610077-5	123A	3124		TR21/735	S0015
610077-6	123A	3122	20	TR21/	S0015
610078	123A	3124		TR21/735	
610078-1	123A	3124	20	TR21/735	S0015
610078-2	123A				
610079	102A	3004		TR85/250	
610079-1	102A	3004	2	TR21/250	G0005
610079-2	199		10	TR21/	
610080	102A	3004		TR85/250	
610080-1	102A	3004	2	TR85/250	G0005
610083	159	3025		TR20/717	
610083-1	234	3114	21	TR20/717	S0019
610083-2	234	3025	21	TR54/717	S0019
610083-3	159	3025	21	TR52/717	S0031
610083-4		3114	21	TR30/	S0019
610083-183P1				TR30/	
610083-283P2				TR30/	
610088	102A	3004		TR85/250	
610088-1	102A	3004		TR85/250	
610088-2	102A	3004	2	TR85/250	
610091	108	3018		TR95/56	
610091-1	108	3018	17	TR95/56	
610091-2	108	3018	11	TR95/56	
610092	108	3018		TR95/56	
610092-1	108	3018	11	TR95/56	S0016
610092-2	108	3039		TR95/	
610093-1	159	3114	22	TR52/	S0019
610094	123A	3124		TR21/735	
610094-1	123A	3124	20	TR24/735	S0015
610094-2	123A	3018			S0015
610096	108	3018		TR95/56	
610096-1	108	3018	11	TR95/56	
610099	129	3025	21	TR88/242	
610099-1	129	3025	21	TR54/242	S0012
610099-2	129	3025	21	TR88/242	S0019
610099-3	102A	3004	52	TR85/250	G0005
610099-5		3114			S0019
610099-6		3114	21	TR	S0019
610100	108	3018		TR95/56	
610100-1	161	3018	39	TR70/719	S0017
610100-2	107				
610100-3	108	3117	39	TR95/	S0020
610102-1	106	3008	21	TR20/52	
610106	121	3009		TR01/232	
610106-1	121	3009	16	TR01/232	G6005
610106-1106P1T				TR01/	

Original Part Number	ECG	SK	GE	IR/WEP	HEP
610106-1106PT				TR01/	
610107-1	108	3039	17	TR95/56	
610107-2	128	3122	18	TR87/243	S5014
610110	129	3025		TR88/242	
610110-1	159	3025	21	TR20/717	S0012
610110-2	159	3118	21	TR30/	S0013
610111-1		3009			
610111-2	130	3027			
610111-4	130	3027			
610111-5	104	3009			
610111-6	152	3054			
610111-7	104	3009			
610111-8	196	3054			
610112-1	153			TR77/700	
610113-1	172	3124	20	TR69/S9100	S9100
610113-2	172		64	TR69/S9100	S9100
610120-1	159	3114	21	TR20/717	S0013
610121-1	6402				
610121-2	6402				
610123-1	164	3133	37	TR68/	
610124-1	101	3018		TR08/641	G0011
610125-1	159	3114	22	TR52/637	G0003
610126-1	102A			TR85/	
610126-2	103A			TR08/	
610128-1	199				
610128-2	199	3018	20	TR87/243	S0014
610128-4		3018	11	TR24/	S0016
610128-5	199				
610128-6	199				
610129-1	129	3025	21	TR88/719	S0019
610131-2	191	3054	28	TR55/735	S3020
610132	123A	3122	21	TR21/735	
610132-1	123A			TR21/735	S5014
610134-1	234	3025	21	TR20/717	S0019
610134-2	234				
610134-4	234	3118	21	TR52/717	S0012
610134-5	234				
610134-6	234	3114	21	TR30/	S0032
610135-1	154	3044	20	TR87/	S0015
610136-1	106	3118	21	TR20/52	S0012
610139-1	161	3018	17	TR83/719	S0025
610139-2	107				
610140-1	130	3027	14	TR61/	S7004
610142-1	161	3018	20	TR83/719	S3019
610142-2	123A				
610142-3	123A	3124	20	TR21/735	S0015
610142-4	123A	3122	20	TR21/735	S0025
610142-5	123A	3122	20	TR21/	S0015
610142-6	108	3020	20	TR	S0016
610142-7	123A	3122	20	TR21/	S0015
610144-1	171	3103	28	TR79/244A	S0015
610144-2	171	3104	27	TR79/	S3021
610144-3	154	3104		TR78/	S3021
610144-101	171/403	3104	27	TR60/	S3019
610145-1	161	3018	20	TR83/719	S0017
610146-3	123A	3018	20	TR21/	S0015
610146-5	123A	3124	20	TR24/	S0016
610147-1	123A	3124	63	TR21/	S5014
610147-2		3025	67	TR30/	S5013
610148-1	128	3124	18	TR24/243	S5014
610148-2	123A	3124	20	TR21/735	S0015
610148-3		3018	20	TR21/	S0015
610149-1	153				
610149-2	152				
610149-3	153			TR56/	
610149-4				TR55/	
610150	108				
610150-1	161	3018	17	TR83/719	S0015
610150-2	123A	3122	20	TR21/	S0024
610150-3	161	3018	20	TR21/	S0015
610151-1	199	3018	20	TR95/56	S0016
610151-2		3124	20	TR24/	S0015
610151-3	199	3018	20	TR24/	S0015
610151-4		3122	20	TR24/	S0015
610151-5	199	3010	8		
610152-1	104	3009			
610153-1	152	3054	28	TR76/701	S5000
610153-2	152	3054			
610153-3	152				
610153-4	152	3054			
610153-5	152	3054			
610153-6		3041			S5000
610155-1	196	3041		TR92/	S5000
610157-3		3054			S3024
610157-4		3083			S3028
610158-1	129				
610158-2	159	3114	21	TR28/	S0019
610161-1		3027			
610161-2	130	3027			
610161-3		3054			
610161-4	130	3027	14	TR59/	S7002
610161-5		3534			
610162-4	152				
610162-7		3083			
610165-1	123A	3124	20	TR21/	S0015

Original Part Number	ECG	SK	GE	IR/WEP	HEP
610165-2	123A				
610166-1	222	3050			F2007
610167-1	123A	3122	20	TR24/	S0015
610167-2	123A	3124	20	TR24/	S0015
610168-1	123A	3047	20	TR21/	S0015
610168-2	123A	3047	17	TR21/	S0014
610174-1	108	3018	20	TR21/	S0025
610180-1		3132	61	TR21/	S0015
610181-1	123A	3117	60	TR24/	S0016
610181-2	161	3117	60	TR71/	S0016
610186-1	161	3132	61	TR61/	S0020
610189-1	163	3111	38	TR61/	S5021
610189-2	165				
610190-1	184	3054	57	TR76/	S5000
610194-1	165	3111	36		S5020
610195-1		3054	66	TR76/	S3024
610195-2		3083	69	TR77/	S3028
610195-3		3054	66	TR76/	S3024
610195-4		3083	69	TR77/	S3028
610202-1	189	3083	29	TR73/	S3032
610202-2	189	3083	29	TR73/	S3028
610203-1		3050	FET4		F2004
610203-4		3050	FET4		F2004
610209-1		3114			S0019
610216-1	165				
610217-1		3027	36		S7004
610217-2		3027	36		S7004
610223-1		3114			S0019
611020	102A	3008	1	TR85/250	G0005
611064	7405				
611065	7475				
611066	7486				
611071	7442				
611233	234				
611428	123A				
611563	7400				
611564	7402				
611565	7404				
611566	7410				
611567	7420				
611568	7440				
611569	7450				
611570	7453				
611571	7474				
611572	7490				
611573	7492				
611730	74193				
611731	74192				
611844	7486				
611845	7405				
611870	7476				
611872	74107				
611878	74H55				
611900	74107				
611901	74180				
612002-2	717				
612005-1	712	3072		IC507/	C6063P
612005-2	712	3072			C6063P
612005-2A		3072			
612006-1	722	3161		IC513/	C6056P
612006-1H	722				C6056P
612006-1Z	722				C6056P
612006-2	722	3161			
612007-1	708	3135		IC504/	
612007-2	708	3135			C6062P
612007-3	708	3135		IC504/	C6062P
612008-1	725	3144			
612008-2	725	3144			
612020	179	3035		TR35/	
612020-1	703A	3157		IC511/	C6001
612020-2	703A	3157			
612020-3	703A	3157			
612021-1	718	3159		IC511/	C6074P
612024-1	731	3170			
612025-1	730	3143			
612025-2	730	3143			
612025-3	730				
612029-1	790				
612029-2	790				
612030-1	715	3076			
612031-1	714	3075			
612042-1	1112				
612042-2	1111				
612044-1	738	3167			
612045-1	732				C6090
612061-1	783	3141			
612067-1	715	3076	IC6		
612069-1	715	3076			
612070-1			IC4		
612072-1			IC30		
612075-1	801	3160			
612076-1	742				
612077-1	788	3147			
613112		3039	11	TR95/56	
615150-001		3024			

Original Part Number	ECG	SK	GE	IR/WEP	HEP
615179-1	123A				
615179-2	123A				
615180-1	159				
615180-2	159				
615180-3	159				
615180-4	159				
615246-1	941M				
615268-101	941M				
616010		3119			
617871-1	121	3009		TR01/232	
618072	123A				
618099-1		3024			
618099-3C		3024			
618126-1	123A				
618136-1	128				
618139-1	121	3009		TR01/232	G6003
618217-2	123A				
618241-2	910				
618312-1		3024			
618483-1	941				
618580	133	3112		/801	F0010
618810-2	123A				
618955-2	130				
618960-1	198				
618984-1	941				
618986-2	218				
619006	123A				
619006-1	128				
619006-7	123A				
619009-1	175			TR81/	
619010-1	198				
619050-1	910D				
619361-1	175	3131			
619550-1	766				C6009
619692-1	723	3144			
619693-1	787	3146			
619693-201		3146			
619694-1	724	3525			
619694-201		3525			
619695-201		3542			
619696-201		3545			
619883-901		3065			
619889-901		3548			
620782	176				
649002	130				
650060	159				
650175	129				
650196	102A	3004	1	TR85/	
650859-1	102A	3004	2	TR85/	
650859-2	102A	3004	2	TR85/	
650859-3	102A	3004	1	TR85/	
650860	103A	3011	5	TR08/	
650970	121			TR01/	
651012	158	3004	2	TR84/	
651202	121			TR01/	
651236	158	3004	2	TR84/	
651891	123A	3018	17	TR21/	
651955	123A				
651955-1		3124	17	TR25/	
651955-2		3124	17		
651955-3		3019	17	TR25/	
651956	186	3024	18	TR55/	
651995-1	123A			TR21/	
651995-2	123A			TR21/	
651995-3	123A			TR21/	
652072	123A	3018	17	TR21/	
652085	121	3009		TR01/232	G6005
652086	121			TR01/	
652091	123A	3018	17	TR21/	
652230	123A	3019	17	TR21/	
652231	128	3045	20	TR87/	
652321	128	3045	20	TR87/	
653406	128				
654000	123A	3018	17	TR21/	
654001	6400				
654003		3009			
654007		3004			
654008		3018			
654010		3019			
654011		3018			
654012		3018			
654013		3025			
654041		3004			
655319	121			TR01/	
656064	6400				
656204	123A	3124	20	TR21/	
656524	123A	3018	17	TR21/	
656719	123A	3018	17	TR21/	
656746	128		18	TR87/	
657161	752				
657179	185	3191	69		
657180	184	3054	66		
657181	185	3083	29		
657874	761				C6011
658577	123A	3124		TR21/	

Original Part Number	ECG	SK	GE	IR/WEP	HEP
658578	123A	3124		TR21	
658583	909D	3590			
658657	123A	3018		TR21/	
659140		3054			
659141	219				
659174	130				
660030	102A		53	TR85/250	
660031	121MP	3013	3	/232MP	
660059	102A	3004	2	TR85/250	
660060	102A	3004	2	TR85/250	
660064	160			TR17/	G0008
660070	128	3024	18	TR87/243	
660072	102A	3004	2	TR85/250	
660074	128	3024	18	TR87/243	
660077	121	3009		TR01/232	G6005
660082	102A	3004	2	TR85/250	
660084	160	3006		TR17/635	G0008
660085	160	3006		TR17/635	G0008
660094	121			TR01/	
660095	121	3014	16	TR01/232	
660097	121	3009		TR01/230	G6003
660100	128	3512			
660103	121	3009	3	TR01/232	
660138	218		26	TR58/700	S5018
660144	105	3012		TR03/233	G6006
660144A	105	3012		TR03/	
660327-1					C6052P
660388-2					C6052P
670850	121MP	3013		/232MP	
670850-1	121MP	3013	13MP	/232MP	
681266	129				
681266-1	129				
682085				TR01/	
696575-198	128				
697240-1-0		3004			
697240		3004			
698941-1	159				
699291	199				
699410-140	128				
699414-164	124	3045			
700047-47	123A				
700047-49	123A				
700080	130	3027	14	TR59/247	
700080A	130	3027		TR59/	
700083	130	3027	14	TR59/247	
700083A	130	3027		TR59/	
700181	123A				
700191	175	3026		TR81/241	
700195	175	3026		TR81/241	
700230-00	123A	3124		TR21/735	
700231-00	123A	3124		TR21/735	
700276-2A		3009			
701584-00	159	3114		TR20/	
701589-00	159	3025		TR20/717	
701678-00	161	3018		TR83/719	
702407-00	128	3024		TR87/243	
702415-00	210	3122		TR51/735	
702884	123A				
702885	121				
702885-00	121				
702886	152				
705784-1	175				
705784M1		3131			
710206	910				
710206-43	923				
710398-28	909	3590			
716128-1		3027			
716128M1		3027			
717101	152	3054			
717117		3084			
717126-505	74H04				
717136-1	74S00				
717136-15	74S20				
717191		3054			
717192		3054			
717399-3	923				
717399-4	941				
717399-22	9109				
717399-49	941				
717399-73	9112				
720240		3018			S0015
721272	159				
723000-18	101				
723001-19	101				
723005-8		3005			
723020-41	128				
723025-027		3025			
723043-1	129				
723060-29	175				
723423-7	175				
723423-9	175				
723423-16	175				
723423-20	175				
726359-1		3036			
730547	704	3023			

Original Part Number	ECG	SK	GE	IR/WEP	HEP
731009	102A	3004	2	TR85/250	
740247	131MP	3086	31MP	TR50/	G6016
740306		3124	62	TR25/	S0015
740417	102A	3004	53		G0006
740437		3124	20	TR21/	S0024
740438	199			TR25/	S0014
740439	199			TR25/	S0014
740440	199			TR25/	S0014
740441		3047	20	TR25/	S0014
740442	199			TR25/	S0014
740443	131MP	3086	31MP	TR50/	G6016
740461	123A	3124	10		S0030
740462	123A	3124	20		S0014
740463	123A	3124	20		S0014
740466	123A	3026	10		S0020
740470	123A	3124	20		S0015
740471	102A	3052	30		G6016
740502	1107				
740543	1107				
740583	1103				
740622	720		IC7	IC512/	C6095P
740622(RCA)	720				
740857		3047			
740885		3123			
740886		3122			
740887		3124			
740946		3006			G0008
740947		3006			G0008
740948		3006			G0008
740949		3018			S0014
740950		3018			S0014
740952		3087			
741050		3008			
741114		3124			
741115		3054			
744002		3119			
744006		3119			
749002	107				
749014	107				
750351		3027			
750858-328		3124			
752664-1		3012			
755722		3090			
757008-02	123A			TR21/735	
759019		3018			
759020		3018			
759021		3018			
759022		3018			
759023		3018			
759024		3124			
759029		3004			
759060		3004			
760005		3065			
760011					C3000P
760014					C3806P
760021					S7001
760236		3123			
760239		3018			
760249		3039			
760251		3124			S0033
760253		3018			S0020
760268		3116			
760269		3025			S0013
760275		3083			S5000
760276					S5006
760284		3048			
761113		3031			
762200-14	9093				
770339(IC)	760				
770523	102A	3004	1	TR85/250	G0005
770524	102A	3004	1	TR85/250	G0005
770525	158	3004	2	TR84/630	G0005
770730	102A			TR85/	G0005
770768-3170756	124	3021		TR81/	
772716	160			TR17/	G0009
772717	160			TR17/	G0001
772718	102A			TR85/250	G0009
772719	160			TR17/	G0009
772720	102A			TR85/250	G0005
772721	102A	3004	1	TR85/250	G0005
772722	102A	3004		TR85/250	G0005
772723	102A			TR85/250	G0005
772724	102A			TR85/250	G0005
772725	102A			TR85/250	G0005
772727	102A			TR85/250	G0005
772728	102A			TR85/250	G0005
772729	102A			TR85/250	G0005
772732	158	3004	53	TR84/630	G0005
772733	158			TR84/630	G0005
772736	102A			TR85/250	G0007
772737	102A			TR85/250	G0006
772738	107	3039		TR70/720	S0016
772739	107	3039		TR70/720	S0016
772768	126	3008	51	TR12/	G0005
779021	197				

Original Part Number	ECG	SK	GE	IR/WEP	HEP
785278-101	123A	3124		TR21/	
785897-01	160	3006		TR17/	
793356-1	130	3027		TR59/	
800019-001	128				
800020-001	7420				
800021-001	7430				
800022-001	7440				
800023-001	7410				
800024-001	7400				
800025-001	7453				
800026-001	7450				
800073-6	128	3024		TR87/	
800073-7	128	3024		TR87/	
800080-001	7402				
800382-001	7475				
800383-001	7483				
800385-001	7442				
800386-001	74193				
800387-001	7404				
800400-001	7474				
800491-001	74121				
800651-001	7406				
800747	102A				
800782		3044			
800806-001	7407				
800946-001	128				
801004A		3024			
801500	102A		2	TR85/	
801501	102A		2	TR85/	
801507	102A		53	TR85/	
801509	102A		2	TR85/	
801510	102A		2	TR85/	
801511	102A		2	TR85/	
801512	123A	3018	17	TR21/	
801513	123A	3019	17	TR21/	
801514	123A	3124	17	TR21/	
801515	123A	3124	17	TR21/	
801516	123A	3018	17	TR21/	
801517	123A	3018	17	TR21/	
801518	121	3009		TR01/	
801519	121	3009	3	TR01/	
801520	102A	3003	2	TR85/	
801522	121			TR01/	
801523	121			TR01/	
801524	123A	3019	17	TR21/	
801525	6400				
801527	199	3124	10	TR21/	
801529	123A	3019	17	TR21/	
801530	123A	3039	17	TR21/	
801531	6400				
801532	123A	3124		TR21/	
801533	172			TR69/	
801534	123A	3124		TR21/	
801535	172			TR69/	
801536	123A	3124	20	TR21/	
801537	130	3027	14	TR59/	
801538	121			TR01/	
801539		3009			
801540	159	3025	22	TR20/	
801541	171	3201		TR79/	
801543	123A	3019	17	TR21/	
801546		3122			
801547		3112			
801725		3018			
801729	123A	3018	17	TR21/	
801800	9924				
801804	754				
801805	7401				
801806	7404				
801807	7442				
801808	7490				
801809	754				
802032-2	102A	3004		TR85/	
802032-4	102A	3004		TR85/	
802033-3	102A	3004		TR85/	
802037-001	153				
802054-0	102A	3004		TR85/	
802056-0	102A	3004		TR85/	
802189-7	102A	3004		TR85/	
802189-8	102A	3004		TR85/	
802263-0	102A	3004		TR85/	
802263-1	102A	3004		TR85/	
802389-2	102A	3003		TR85/	
802415-2	102A	3004		TR85/	
802425-0	127	3034		TR27/	
802439-0	102A	3004		TR85/	
802560	102	3004	2		G0005
803182-5	123A	3124		TR21/	
803369-6	123A	3124		TR21/	
803372-0	123A	3124		TR21/	
803373-0	123A	3124		TR21/	
803696	199			TR21/	
803733-0	123A	3124		TR21/	
803733-3	123A	3124		TR21/	
803735-3	123A	3124		TR21/	

Original Part Number	ECG	SK	GE	IR/WEP	HEP
810000-373	123A				
810002-269	9930				
810002-733	129				
810002-736	128				
811790	9930				
811791	9946				
811793	9962				
811794	9932				
813362	727	3071			
814044A	102A	3004	2	TR85/250	G0005
815003	102A	3004	2	TR85/250	
815015	102A	3004	2	TR85/250	
815020	100	3005	1	TR05/254	G0005
815020A	100	3005	1	TR05/254	
815020B	100	3005	1	TR05/254	
815021	100	3005	1	TR05/254	G0005
815021A	100	3005	1	TR05/254	
815021B	100	3005	1	TR05/254	
815022	102A	3004	2	TR85/250	G0005
815022A	102A	3004	2	TR85/250	
815022B	102A	3004	2	TR85/250	
815023	102A	3004	2	TR85/250	G0005
815023A	102A	3004	2	TR85/250	
815023B	102A	3004	2	TR85/250	
815024	102A	3004	2	TR85/250	G0005
815024A	102A	3004	2	TR85/250	
815024B	102A	3004	2	TR85/250	
815025	100	3005	1	TR05/254	G0005
815025A	100	3005	1	TR05/254	G0005
815025B	100	3005	1	TR05/254	
815026	101	3011	7	TR08/641	
815026A	101	3011	7	TR08/641	
815026B	101	3011	7	TR08/641	
815026C	101	3011	7	TR08/641	
815026D	101	3011	1	TR08/641	
815027	100	3005	1	TR05/254	G0005
815027A	100	3005	1	TR05/254	
815027B	100	3005	1	TR05/254	
815027C	100	3005	1	TR05/254	
815028	100	3005	1	TR05/254	G0005
815028A	100	3005	1	TR05/254	
815028B	100	3005	1	TR05/254	
815028C	100	3005	1	TR05/254	
815029	102A	3004	2	TR85/250	G0005
815029A	102A	3004	2	TR85/250	
815029B	102A	3004	2	TR85/250	
815029C	102A	3004	2	TR85/250	
815030	102A	3004	2	TR85/250	G0005
815030A	102A	3004	2	TR85/250	
815030B	102A	3004	2	TR85/250	
815031	102A	3004	2	TR85/250	G0005
815031A	102A	3004	2	TR85/250	
815031B	102A	3004		TR85/250	
815033	102		2	TR05/	
815034		3004	2	TR85/250	G0005
815034A	102A	3004	2	TR85/250	
815034B	102A	3004	2	TR85/250	
815034C	102A	3004	2	TR85/250	
815036	100	3005	1	TR05/254	G0005
815036A	100	3005		TR05/254	
815036B	100	3005	1	TR05/254	
815036C	100	3005		TR05/254	
815037	100	3005	1	TR05/254	G0005
815037A	100	3005	1	TR05/254	
815037B	100	3005	1	TR05/254	
815037C	100	3005	1	TR05/254	
815038	102A	3005	1	TR85/250	G0005
815038A	102A		1	TR85/250	
815038B	102A	3004	2	TR85/250	
815038C	102A	3004	2	TR85/250	
815041	100	3005	1	TR05/254	G0005
815041A	100	3005	1	TR05/254	
815041B	100	3005	1	TR05/254	
815041C	100	3005	1	TR05/254	
815043	100	3005	2	TR05/254	G0005
815043A	100	3005	1	TR05/254	
815043B	100	3005	1	TR05/254	
815043C	100	3005	1	TR05/254	
815055	100	3005	2	TR05/254	
815056	100	3005	2	TR05/254	
815057	100	3005	2	TR05/254	
815058	102A	3004	2	TR85/250	
815058A	102A	3004	2	TR85/250	
815058B	102A	3004	2	TR85/250	
815058C	102A	3004	2	TR85/250	
815058X	102A	3004	2	TR85/250	
815064	126	3008	51	TR17/635	G0002
815064A	126	3008	51	TR17/635	
815064B	126	3008	51	TR17/635	
815064C	126	3008	51	TR17/635	
815065	100	3005	1	TR05/254	G0005
815065A	100	3005	1	TR05/254	
815065B	100	3005	1	TR05/254	
815065C	100	3005	1	TR05/254	
815066	100	3005	1	TR05/254	G0005

Original Part Number	ECG	SK	GE	IR/WEP	HEP
815066A	100	3005	1	TR05/254	
815066B	100	3005	1	TR05/254	
815066C	100	3005	1	TR05/254	
815067	160	3009	53	TR17/637	G0002
815067A	160	3008		TR17/637	G0002
815067B	160	3008		TR17/637	
815067C	160	3008		TR17/637	
815068	126	3008	2	TR17/635	G0002
815068A	126	3008		TR17/635	
815068B	126	3008		TR17/635	
815068C	126	3008		TR17/635	
815069	102A	3004	2	TR85/250	G0005
815069A	102A	3004	2	TR85/250	
815069B	102A	3004	2	TR85/250	
815069C	102A	3004	2	TR85/250	
815070	102A	3004	53	TR85/250	G0005
815070A	102A	3004	2	TR85/250	G0005
815070B	102A	3004	2	TR85/250	G0005
815070C	102A	3004	2	TR85/250	G0005
815070D	102A	3004	2	TR85/250	G0005
815074	102A	3004	2	TR85/250	G0005
815075	103A	3010	8	TR08/724	
815076	103A	3010	8	TR08/724	
815082	102A		53	TR85/250	
815083	102A			TR85/250	
815101	100	3005		TR05/254	
815103	100	3005	1	TR05/254	G0005
815104	102A	3004	2	TR85/250	G0005
815105	100	3005	1	TR05/254	G0005
815107	100	3005		TR05/254	
815108	100	3005		TR05/254	
815109	100	3005		TR05/254	
815114	102A	3004	2	TR85/250	G0005
815115	100	3005	1	TR05/254	G0005
815116	100	3005	1	TR05/254	G0005
815117	100	3005	1	TR05/254	G0005
815118	102A	3004	2	TR85/250	G0005
815120	102A	3004	53	TR85/250	G0005
815120A	102A	3004	2	TR85/250	G0005
815120B	102A	3004	2	TR85/250	G0005
815120C	102A	3004	2	TR85/250	G0005
815120D	102A	3004	2	TR85/250	G0005
815120E	102A	3004	2	TR85/250	G0005
815122	102A		53	TR85/250	
815133	123A	3124	20	TR21/735	S0015
815134	123A	3122	20	TR21/735	S0015
815136	102A	3004	2	TR85/250	G0005
815137	121MP	3009		TR01/232MP	
815139	102A	3003	2	TR85/250	G0005
815158	102A	3004	2	TR85/250	
815160	102A	3004	2	TR85/250	
815160A	102A	3004	2	TR85/250	
815160B	102A	3004	2	TR85/250	
815160C	102A	3004	2	TR85/250	
815160D	102A	3004	2	TR85/250	
815160E	102A	3004	2	TR85/250	
815160F	102A	3004	2	TR85/250	
815160H	102A	3004	2	TR85/250	
815160-I	102A	3004		TR85/250	
815160-J	102A	3004		TR85/250	
815160-K	102A	3004		TR85/250	
815160-L	102A	3004		TR85/250	
815160-O	102A	3004		TR85/250	
815160-P	102A	3004		TR85/250	
815160-Q	102A	3004		TR85/250	
815164	108	3018	20	TR95/56	
815165	108	3018	20	TR95/56	
815166	124	3021	12	TR81/240	
815166-4	124	3021	12	TR81/240	
815167-3	124	3021	12	TR81/240	
815170	108	3018	20	TR95/56	
815171	123A	3124	20	TR21/735	S0015
815171D	123A	3124	20	TR21/735	
815172	108	3018	17	TR95/56	
815172A	108	3018	17	TR95/56	
815173	199	3018	17	TR95/56	
815173A	108	3018	17	TR95/56	
815173C	108	3018	20	TR95/56	
815173F	108	3018	20	TR95/56	
815174	123A	3124	20	TR21/735	S0015
815174L	123A	3122	20	TR21/735	
815175	124	3021	12	TR81/240	
815175H	124	3021	12	TR81/240	
815177	102A	3004	2	TR85/250	
815178	102A	3004	2	TR85/250	
815179	102A	3004	2	TR85/250	
815180-3	124	3021	12	TR81/240	
815180-4	124	3021	12	TR81/240	
815180-7	124	3021	12	TR81/240	
815181	102A	3004	2	TR85/250	
815181A	102A	3004	2	TR85/250	
815181B	102A	3004	52	TR85/250	
815181C	102A	3004	2	TR85/250	
815181D	102A	3004	2	TR85/250	
815182	123A	3124	20	TR21/735	

Original Part Number	ECG	SK	GE	IR/WEP	HEP
815183	123A	3020	20	TR21/735	
815184	123A	3124	20	TR21/735	
815184E	123A	3124	20	TR21/735	
815185	129	3025	22	TR88/242	
815185E	129	3025	21	TR88/242	
815186	123A	3124	20	TR21/735	
815186C	123A	3122	20	TR21/735	
815186L	123A	3124	20	TR21/735	
815189	102A		2	TR85/250	G0008
815190	123A	3124	20	TR21/735	
815191	123A	3124	20	TR21/735	
815193	160	3006	50	TR17/637	
815195	102A	3004	52	TR85/250	
81519[illegible]	102A	3004	52	TR85/250	
8151[illegible]	160	3006	51	TR17/637	
8151[illegible]	123A	3124	20	TR21/735	
81519[illegible]	159	3025		TR20/717	
815199-6	159	3025	21	TR20/717	
815201	123A	3124	20	TR21/735	
815202	123A	3124	63	TR21/735	S0015
815203-3	121MP	3015	13MP	/232MP	G6005
815203-5	121MP	3013		/232MP	G6005
815206	161	3018	17	TR83/719	
815209	108	3039	17	TR95/56	S0025
815210	123A	3124	20	TR21/735	S0015
815211	159	3025	21	TR20/717	S0031
815212	123A	3024	63	TR21/735	S5014
815213	159	3114	21	TR20/717	S5013
815215	711				
815218					S5000
815218-3	102A			TR85/S5003	G0005
815218-4	103A			TR08/	G0011
815227	123A	3124	10	TR21/735	S0015
815228	188+189	3004		TR85/250	S5000
815228A	102A	3004		TR85/250	
815228A01	102		2	TR05/	
815228A1	102A		52	TR85/250	
815228B	102A				
815228B1	102A		52	TR85/250	
815228(EMERSON)	188+189				
815229	159	3025	21	TR20/717	S5013
815232	103A+158	3010	8	TR09/724	
815233	123A	3124		TR21/735	
815234	160	3006		TR17/637	
815236	159	3114	21	TR20/717	
815237	123A	3122	63	TR21/735	
815243	123A+159	3122	63	TR25/735	
815246-1	130	3027		TR26/	
815246-2	121				
815247	159	3114		TR20/717	
815259	711				
815260	724				
815308A	100	3005		TR05/254	
816135	124	3021	12	TR81/240	
824960-0	108	3018		TR95/	
825065	102A	3005	1	TR85/250	G0005
829704-6	74H21				
829704-7	74H21				
831013		3007			
838105	159				
845050	123A				
848082	123A				
851759-3	160				
851881	123A				
852440-3		3004			
853864-0	160	3.07		TR17/	
860001-8	128				
860003-45	911				
860003-99	9944				
860003-111	9931				
860003-121	9932				
860003-141	9945				
860003-161	9936				
862200-16	9093				
862209-16	9962				
865982				TR24/	
865992				TR24/	
866002				TR24/	
880092	131			TR94/	G6016
882028	130				
889132	1114				
889302	744				
889303	1114				
889779F	736				
891008	159				
891032	130				
900201-81	128				
900201-104	128				
900201-105	128				
900201-167	128				
900552-6	123A				
900552-8	123A				
900552-17	159				
900552-20	123A				
900552-30	123A				

Original Part Number	ECG	SK	GE	IR/WEP	HEP
900555-11		3024			
902521	160			TR17/	
907338		3007			
908098	909				
908106		3005			
908844-1	199				
908864-2	159				
910050-2	102A	3004		TR85/	
910062-1	102A	3004		TR85/	
910070-6	102A	3004		TR85/	
910088	128				
910094-4	102A	3004		TR85/	
910100A		3039			
910634	128				
910742		3004			
910799	108	3039		TR95/56	S0015
910807-11	175				
910952	185	3041		/S5007	S5006
911686B		3024			
911743-1	123A				
912002	754				
916005		3124			
916028	108	3122	20	TR95/	S0015
916029	108	3122	20	TR95/	S0015
916030	123A	3020	17	TR21/	S0015
916031	123A	3122	17	TR21/	S0015
916033	128	3024	63	TR87/	S0014
916034	128	3024	63	TR87/	S5014
916046	186	3054	28	TR76/	S5000
916049	107	3122	20	TR70/	S0016
916050	123A	3122	20	TR24/	
916051	159	3114	21	TR20/	S0019
916052	128	3024	63	TR25/	S5014
916055	199	3124	62	TR21/	S0015
916055(IC)	1010				
916060	108	3122	20	TR95/	S0020
916061	1029				
916063	703A	3157	IC12		
916064	1045		IC2		
916067	1052				
916068	107				S0016
916069	108				S0016
916070	1142	3160			
916071	1003				
916072	1006				
916081	1010				
916082	132	3116	FET2	FE100/	
916083	1081				
916091	123A	3122	17	TR24/	
916098	1081				
922114	128				
922125	128				
922376-1A		3024			
922896	100	3005		TR05/254	
928103-1	123A				
928291-101	128				
928291-102	128				
928335-1		3021			
928335-101		3021			
928335-103		3021			
928361-101		3024			
928408-101	159				
928506-101	910				
928510-1	9951				
928510-101	9951				
928514-1	9946				
928514-101	9946				
928515-101	9945				
928517-101	9933				
928533-101	9962				
928560-1	9944				
928560-101	9944				
928571-1	9944				
928608-101	703A				
930236	123A				
930347-1	7420				
930347-2	7430				
930347-3	7400				
930347-4	7472				
930347-5	7440				
930347-6	7454				
930347-7	7473				
930347-9	7491				
930347-10	7420				
930347-11	7402				
930347-12	7451				
930347-13	7404				
930347-15	7405				
930419	923D				
932017-0001	128				
932030-1	909	3590			
932040	159				
932055-1	128				
932081-1	175				
932107-1	159				

Original Part Number	ECG	SK	GE	IR/WEP	HEP
932292-IC	9930				
933044-2D	9932				
941298-2	123A				
941295-3	123A				
942002					C6007
943720-001	123A				
943722		3122			
950015		3027			
954330-2	123A				
959492-2	123A				
960106-3	159				
960201	107				
960202	107				
960494-1	128				
960494-2	128				
961544-1	123A				
964547-2	199				
964634	108	3019		TR95/56	
964688	103A+158			/724	G0011
964713	108	3019	20	TR95/56	
965000	123A		11	TR21/	
965074	108	3019	17	TR95/56	S0016
965632	161	3018	11	TR83/719	
965633	108	3018	11	TR95/56	
965634	108	3018	11	TR95/56	
970046	108	3019	11	TR95/641A	S0016
970046-1	108	3019		TR95/	
970046-2	108	3019		TR95/	
970046A	108	3019		TR95/	
970107	129				
970108	128				
970244	108	3018		TR21/	S0015
970245	108	3018	11	TR95/56	S0011
970246	159	3114	11	TR20/717	S0012
970247	123A	3044	20	TR21/735	S0015
970248	159	3114	21	TR20/717	S0012
970249	108	3018	20	TR95/56	S0011
970250	123A	3018	20	TR21/735	S0011
970251	159	3025	22	TR20/717	S0012
970252	123A	3018	11	TR21/735	S0011
970253	132	3116	FET2	FE100/802	F0015
970254	159	3025	21	TR20/717	S0012
970255	124	3021	12	TR81/	S5011
970309	108	3018		TR95/56	S0016
970309-1	108	3018		TR95/	
970309-2	108	3018		TR95/	
970309-3	108	3018		TR95/	
970309-4	108	3018		TR95/	
970309-5	108	3018		TR95/	
970309-12	108	3018		TR95/	
970310	108	3018		TR95/56	S0016
970310-1	108	3018		TR95/	
970310-2	108	3018		TR95/	
970310-3	108	3018		TR95/	
970310-4	108	3018		TR95/	
970310-5	108	3018		TR95/	
970310-12	108	3018		TR95/	
970332	108	3019	11	TR95/56	S0016
970332-12	108	3019		TR95/	
970659	123A	3118	20	TR53/	S0016
970660	123A	3018	20	TR51/	S0016
970661	123A	3018	20	TR51/	S0015
970662	123A	3018	20	TR51/	S0015
970662-6	129	3018			
970663	159	3025	21	TR52/	S0019
970762	159	3118		TR20/717	S0019
970762	129	3025		TR88/	
970911	108	3019		TR95/56	S0016
970916	123A	3124		TR21/735	S0015
970916-6	123A	3124		TR21/	
971059	159	3114		TR52/	S0019
971460	199	3018			S0016
971477					G0005
972155	123A	3018	20	TR21/	S0015
972156	123A	3018	20	TR21/	S0015
972214	123A	3020			
972215	123A	3124			
972305	161	3018	39		S0017
972306	107	3018	11		S0016
972307	107	3018	11		S0016
972417	107	3018	11	TR80/	S0020
972418	107	3018	11	TR80/	S0020
972419	107	3018	11	TR80/	S0020
972420	107	3018	11	TR80/	S0020
975171-1				TR21/	
980052	105	3012	4	TR03/233	
980052A	105	3012		TR03/	
980132	121	3009	16	TR01/232	
980134	121	3009	4	TR01/232	
980135	121	3009	4	TR01/232	
980136	100	3005	2	TR05/254	
980138	108	3018	17	TR95/56	
980139	108	3018	17	TR95/56	
980140	160	3006	51	TR17/637	
980142	160	3006	51	TR17/637	
980144	102A	3004	2	TR85/250	
980146	160	3006	51	TR17/637	
980147	123A	3124	20	TR21/735	
980148	102A	3004	2	TR85/250	
980149	102A	3004	2	TR85/250	
980150	127	3034	25	TR27/235	
980153	102A	3004	2	TR85/250	
980155	121	3009	3	TR01/232	
980316	100	3005		TR05/254	
980372	126	3008	51	TR17/635	
980373	126	3008	51	TR17/635	
980374	126	3008	51	TR17/635	
980375	102A	3003	2	TR85/250	
980376	102A	3003	2	TR85/250	
980426	100	3005	1	TR05/254	
980432	100	3005	1	TR05/254	
980434	100	3005	1	TR05/254	
980435	160	3006	51	TR17/637	
980437	121	3009	16	TR01/232	
980438	100	3005	1	TR05/254	
980439	100	3005	1	TR05/254	
980440	123A	3124	20	TR21/735	
980441	160	3006	51	TR17/637	
980462	105	3012	4	TR03/233	
980463	105	3012	4	TR03/233	
980463A	105	3012		TR03/	
980505	160	3006	51	TR17/637	G0003
980506	160	3006	51	TR17/637	G0003
980507	160	3006	51	TR17/637	G0008
980508	102A	3004	2	TR85/250	G0005
980509	160	3006	51	TR17/637	
980510	102A	3004	2	TR85/250	G0005
980511	102A	3004	1	TR85/250	G0007
980514A	160	3006	51	TR17/635	
980545A	160	3006		TR17/637	
980626	126	3006	50	TR17/635	
980636A	160	3006		TR17/637	
980833	126	3008	51	TR17/635	
980834	126	3008	51	TR17/635	
980835	126	3008	51	TR17/635	
980836	102A	3004	2	TR85/250	
980837	102A	3004	2	TR85/250	
980958	126	3008	51	TR17/635	
980959	126	3008		TR17/635	
980960	102A	3004	2	TR85/250	G0005
980961	102A	3004	2	TR85/250	
981143	160	3006	51	TR17/637	G0003
981144	160	3006	51	TR17/637	G0003
981145	160	3006	51	TR17/637	G0008
981146	160	3006	51	TR17/637	G0003
981147	102A			TR85/250	G0005
981148	102A	3004	2	TR85/250	G0005
981149	102A			TR85/250	G0005
981203	160		51	TR17/637	
981204	160		51	TR17/637	
981206	102A		2	TR85/250	
981672	102A	3004	2	TR85/250	
981673	102A	3004	2	TR85/250	
981674	102A	3004	2	TR85/250	
981675	102A	3004	2	TR85/250	G0005
981959	160	3032	509	TR17/637	
981969	105	3012	4	TR03/233	
981969A	105	3012		TR03/	
982150	126	3008	51	TR17/635	
982151	102A	3004	2	TR85/250	
982152	102A	3004	2	TR85/250	
982231	123A		17	TR21/	
982244	102A	3004		TR85/250	G0005
982267	126	3008		TR17/635	G0008
982268	108	3018	20	TR95/56	S0015
982269	108	3018	20	TR95/56	S0015
982279	722				
982283	102A	3004		TR85/250	G0007
982284	102A	3004		TR85/250	G0005
982285	102A	3004		TR85/250	G0005
982289	126	3008		TR17/635	G0005
982300	128	3004		TR87/243	S5014
982321	108	3018	11	TR95/56	S0016
982322	160	3006	50	TR17/637	G0003
982374	160	3006		TR17/637	G0008
982375	102A	3004	2	TR85/250	G0007
982376	124	3021		TR81/240	S5011
982497	160	3006		TR17/637	G0008
982510	123A	3124	10	TR21/735	
982511	123A	3124	20	TR21/735	S0015
982512	123A	3124	10	TR21/735	S0015
982523	175	3021	12	TR81/241	S5019
982528	124		12	TR81/	
982531	102A	3004	2	TR85/250	G0007
982532	102A	3004	2	TR85/250	G0007
982815	108			TR95/56	S0016
982816	108			TR95/56	S0016
982817	108			TR95/56	S0016
982818	108			TR95/56	S0016
982819	108	3039		TR95/56	S0016

Original Part Number	ECG	SK	GE	IR/WEP	HEP
982820	102			TR05/631	G0005
983012	176	3123		TR82/	G0007
983036	121	3009	16	TR01/232	
983095	108	3018	20	TR95/56	
983096	108	3018		TR95/56	
983097	123A	3124	18	TR21/735	
983233	107	3018	11	TR70/720	
983234	107	3018	11	TR70/720	
983235	107	3018	20	TR70/720	
983236	126	3008	51	TR17/635	G0008
983237	102A			TR85/250	G0005
983238	102A	3123		TR85/250	G0005
983271	126	3008	1	TR17/635	G0005
983272	126	3008	1	TR17/635	G0005
983405	102A	3004	2	TR85/250	G0005
983406	102A	3004	2	TR85/250	G0005
983407	102A	3004	2	TR85/250	G0007
983408	102A	3004	2	TR85/250	G0005
983409	102A	3004	2	TR85/250	G0005
983411	102A		2	TR85/250	G0005
983742	107	3018	10	TR70/720	S0015
983743	123A	3124		TR21/735	S0011
983795	121			TR01/	
983874	121	3009	16	TR01/232	
983945	105	3012	4	TR03/233	
983975	121	3009	16	TR01/232	
984156	108		20	TR95/	
984158	108		20	TR95/	
984159	108		20	TR95/	
984160	102A		52	TR85/	
984161	158		53	TR84/	
984191	161	3018	17	TR83/719	S0016
984192	161	3018	17	TR83/719	S0016
984193	159	3114	21	TR20/717	G0008
984194	108	3018	20	TR95/56	
984195	108	3018	20	TR95/56	
984196	128	3024	21	TR87/243	
984197	123A	3024	63	TR21/735	
984198	123A	3024	20	TR21/735	
984221	102A	3004	52	TR85/250	G0008
984222	123A	3018	20	TR21/735	
984224	123A	3124	20	TR21/735	
984227	154	3040		TR78/712	S5025
984228	102A	3004	52	TR85/250	G0005
984229	128	3024	63	TR87/243	S0014
984259	130	3027		TR59/247	
984259A	130	3027		TR59/	
984261	121	3009	16	TR01/232	G6005
984286	123A	3124	20	TR21/735	
984431	121	3009	16	TR01/232	
984521	131	3052	30	TR94/642	G6016
984577	108	3018	20	TR95/56	S0016
984590	123A	3124		TR21/735	S0015
984591	123A	3124		TR21/735	S0025
984593	123A	3122		TR21/735	S0014
984608	124	3021	12	TR81/240	S5011
984685	126			TR17/635	G0009
984686	123A			TR21/735	S0030
984687	123A			TR21/735	S0014
984688					G6011+G0011
984743	108		20	TR95/	
984744	108		17	TR95/56	
984745	123A			TR21/735	
984746	102			TR05/631	G0005
984851	108	3018		TR95/56	
984852	108	3018		TR95/56	
984853	108	3018		TR95/56	
984854	123A	3124		TR21/	
984875	108	3018		TR95/56	S0014
984876	107	3018	20	TR70/720	S0016
984877	107	3018	20	TR70/720	S0014
984878	107	3018	20	TR70/720	S0014
984879	123A	3124	17	TR21/735	S0030
984932	124	3021		TR81/240	S5011
985036	121	3009	16	TR01/232	
985087		3122			
985096	108	3018	20	TR95/	S0016
985097	108		20	TR95/	S0014
985098	123A	3018	17	TR21/	S0030
985099	123A	3020	17	TR21/	S0030
985100	123A		10	TR21/	S0014
985101	123A		17	TR21/	S0014
985102	123A		17	TR21/	
985103	131		30	TR94/	G6016(2)
985175	221				
985215	108		20	TR95/	S0016
985216	102A		2	TR85/	G0006
985217	102		51	TR05/	G0005
985431	121	3009	16	TR01/232	
985432	105	3012	4	TR03/233	
985442	107			TR70/720	S0011
985442A	108	3018		TR95/	
985443	121	3009	16	TR01/232	S0011
985443A	108	3018		TR95/	
985444	107			TR70/720	S0011

Original Part Number	ECG	SK	GE	IR/WEP	HEP
985444A	108	3018		TR95/	
985445	126	3007	51	TR17/635	G0009
985445A	160	3008		TR17/	
985446	126	3007	51	TR17/635	G0009
985446A	160	3006		TR17/	
985447	121	3009	16	TR01/232	
985449	121	3009	16	TR01/232	
985453	121	3009	16	TR01/232	
985455	121	3009	16	TR01/232	
985468	102A			TR85/250	G0005
985468A	102A	3004		TR85/	
985469	102A			TR85/250	G0005
985469A	102A	3004		TR85/	
985470	158			TR84/630	G6011
985470A	102A	3004		TR85/	
985471	158			TR84/630	
985543	123A			TR21/735	
985609	102A			TR85/250	G0005
985610	102A			TR85/250	G0005
985611	126	3006		TR17/635	
985619	107			TR70/720	S0016
985686	121	3009	16	TR01/232	
985688	103A+158				
985715	132	3116		FE100/802	F0015
985735	103A			TR08/724	G0011
985735A	103A	3010		TR08/	
986015	106				
986030	129				
986302	102A	3004		TR70/	G0005
986305	158	3004		TR84/	G6011
986542	123A	3114		TR21/	S0019
986543	158	3004			G0005
986576		3018	60	TR51/	S0016
986634	108		17	TR95/	
986635	108		20	TR95/	
986636	123A		10	TR21/	
986693	107	3018	60	TR21/	S0016
986694	107	3018	60	TR21/	S0016
986766	102A	3004	52		G0007
986767		3004			G0005
986779	158	3004	53		G6011
986930	222	3050			F2004
986931	129	3114	21		S0013
986932	186	3054	28		S5000
986933	187	3083	29		S5006
987010	123A				
987030	128				
988000	108	3018	17	TR95/	S0016
988001	108	3018	17	TR95/	S0014
988002	108	3122	20	TR21/	S0016
988003	123A	3124	10	TR24/	S0024
988004					G6011
988005	158	3004	53	TR84/	G6011
988006					S0014
988080	121	3009	16	TR01/232	
988336	121	3009	16	TR01/232	
988413	121	3009	16	TR01/232	
988414	105	3012	4	TR03/233	
988468	121	3009	16	TR01/232	
988977	105	3012	17	TR03/233	
988985	108	3122	20		
988986	108	3018	17		
988987	108	3018	17		
988988	108	3122	20		
988989	108	3018	20		
988990	159	3114	21		
988991	123A	3124	10		
988992	193	3025	67		
988993	128	3024	63		
989171	105	3012	4	TR03/233	
989387	121	3009	16	TR01/232	
989615	105	3012	4	TR03/233	
989692	105	3012	4	TR03/233	
989693	105	3012	4	TR03/233	
989709	909	3590			
991021					C2006G
991029					C2006G
991129					C2007G
991421					C2010G
991422					C2010G
991429					C2010G
992289	102		2	TR05/	
993570-4	129				
993624-2	128				
994634	108			TR95/56	
995001	121	3009	16	TR01/232	
995002	102A	3003	2	TR85/250	
995003	102A	3004	2	TR85/250	
995014	121	3009	16	TR01/232	
995015	121	3009		TR01/232	
995016	123A	3122		TR21/735	S0015
995017	123A			TR21/735	S0030
995022			IC12		
995025				TR53/	
995028				TR21/	

Original Part Number	ECG	SK	GE	IR/WEP	HEP
995029				TR53/	
995030	124	3122		TR81/240	S5011
995033				TR51/	
995035				TR51/	
995036				TR51/	
995037				TR51/	
995053-1	722	3161		IC513/	
995081					C6057P
995081-1	713			IC508/	
995870-1	123A				
995870-3	123A				
995928-1	128				
996746	123A				
996817	152				
998280-930	9930				
998280-945	9945				
998280-962	9962				
999106-2	911				
999106-3	911D				
1000100	7492				
1000100-000	7492				
1000101	7493				
1000101-000	7493				
1018715	124				
1018734-001	124	3021			
1022612	123A				
1061854-2	126	3007			
1122022/7611				TR85/	
1127859	123A				
1132724		3004			
1133725		3004			
1134552		3004			
1211016		3004			
1211017		3004			
1211018		3004			
1211243				TR01/	
1221028	105	3012	4	TR03/233	G6006
1221615	121		3	TR01/232	G6005
1221625	104		16	TR01/624	
1221648	100	3123	2	TR05/254	G0008
1221649	100	3005		TR05/254	
1221962	123A	3122	8	TR21/735	
1222123	123A	3122	20	TR21/735	
1222133	123A	3122	20	TR21/735	
1222136	160			TR17/637	
1222314	160			TR17/637	
1222371	160			TR17/637	
1222424	123A	3122	20	TR21/735	
1222463	108	3018		TR95/56	S0016
1232752SC685A				TR23/	
1253892SC458				TR25/	
1253902SC460				TR21/	
1254752SC454				TR21/	
1259952SC535				TR21/	
1261915-191	129				
1261915-383	123A				
1273542SC458				TR21/	
1273552SC458				TR25/	
1288055	128				
1289050	130				
1293922SC535				TR21/	
1293942SC454				TR21/	
1293942SC460				TR21/	
1320041				TR21/	
1320135	123A	3124		TR21/735	S0015
1320135A	123A	3124		TR21/	
1320135BC	123A	3124		TR21/	
1320135C	123A	3124		TR21/	
1330021-0	904				
1330021-1	904				
1355739				TR55/	
1407200		3004			
1407201-1		3009			
1407205-1	121			TR01/	
1407206-1	121				
1408615-1	108	3018	11	TR95/	S0016
1408640-1	108			TR95/	
1408694-1	222	3050			
1415742-5	797				
1417302-1	123A	3122	20	TR21/	S0015
1417303-1	159	3118	21		S0012
1417306-1		3024	18	TR72/	S0005
1417306-2	123A				
1417308-1	199				
1417308-2	199				
1417312-1	123A				
1417312-2	123A				
1417316-1	188	3199	28	TR72/	S3024
1417317-1	189	3200	29	TR73/	S3028
1417318-1	123A	3018	20	TR24/	
1417321-1	194	3022	12	TR81/	S0015
1417322-1	175	3538	12	TR81/	S5012
1417338-5		3024			S3019
1417349-2		3024			
1417366-1	165	3111			
1420427-1	102A			TR85/	
1420427-2	102A			TR85/	
1420427-3	102A			TR85/	
1443024-1	199				
1443200-3	102A			TR85/	
1445829-501	189	3083		TR73/3031	S3028
1445829-502	189	3083		TR73/3031	S3028
1445829-503	188	3054		TR72/3020	S3024
1445829-504	188	3054		TR72/3020	S3024
1452516		3072			
1462432-1	710	3102			
1462434-1	710	3102	IC2	IC507/	C6060P
1462445-1	711	3070		IC514/	C6069G
1462516	712	3072	IC2	IC507/	
1462516-1	712	3102	IC2	IC507/	C6083P
1462554	730				
1462554-1	730	3143			
1462554-2	730	3143			
1462554-3	730	3143			
1462554-4	730	3143			
1462559		3073			
1462559-1	728	3073		TR51/	
1462560		3067			
1462560-1	729	3074	IC23	TR51/	
1462577	780				
1462931-2		3072			
1463028-7		3039			
1463030-1		3024			
1463087-2B		3024			
1463633-1	710				
1463677-1	780	3141			C6060P
1463677-2	780	3141			C6069G
1463677-3	780	3070			C6069G
1463681-1	904	3542			
1463686-1	712				C6063P
1464295-1	756				C6014
1464295-2	762				C6012
1464299-1C		3501			
1464358-5B		3012			
1464437-1	787	3146			
1464437-2	787	3146			
1464437-3	787	3146			
1464438-1	788	3147			
1464438-2	788	3147			
1464438-3	788	3147			
1464439		3078			
1464439-2	789	3078			
1464439-3		3078			
1464460-2	789	3078			
1464460-3	789				
1464686-1	712	3072			
1464846-1	949				
1464846-2	949				
1465158-1	797	3073			
1465263		3525			
1471036-14	121				
1471036-20	121	3009			
1471036-502		3009			
1471100-1	102A			TR85/	
1471100-8	102A			TR85/	
1471100-9	102A			TR85/	
1471101-2	102A			TR85/	
1471101-3	102A			TR85/	
1471101-4	102A			TR85/	
1471101-8		3004			
1471101-11		3004			
1471101-12		3004			
1471101-13		3004			
1471101-15	102A	3004		TR85/	
1471102		3009			
1471102-25		3009			
1471102-41	121				
1471102-42		3009			
1471103-5		3014			
1471103-6		3014			
1471104-5	126	3018		TR17/	
1471104-6	126			TR17/	
1471104-7	126			TR17/	
1471104-8	126			TR17/	
1471104-9		3008			
1471104-10		3008			
1471104-11		3008			
1471104-12		3008			
1471104-13		3008			
1471104-17		3008			
1471104-20		3004			
1471104-21		3008			
1471112-3	129				
1471112-5		3114			
1471112-7	159	3025	67	TR20/717	S0019
1471112-8	159	3025	21	TR20/717	S0031
1471112-8-9	129	3025		TR88/	
1471112-9		3025	21		S0031
1471112-10	234	3114	21	TR20/	S0019
1471112-12	129	3025	67	TR30/	S5013

Original Part Number	ECG	SK	GE	IR/WEP	HEP
1471113-2	123A	3024		TR21/735	S0015
1471113-3	123A				
1471113-4		3124			
1471114-1	159	3114		TR20/717	
1471114-3		3024			
1471115-1	123A	3124			
1471115-2	199	3124			
1471115-3	199				
1471115-4	199				
1471115-6		3124			
1471115-10	161				
1471115-11	199				
1471115-12	123A				
1471115-13	108	3018	20	TR95/	S0015
1471115-14	108	3018	20		S0015
1471117-1	124	3021		TR81/	
1471117-1X		3021			
1471118-1		3018			
1471118-2		3018			
1471118-3		3018			
1471118-4		3018			
1471120-7	123A	3024	63	TR21/735	S0015
1471120-8	123A	3024	18	TR21/735	S0015
1471120-8-9	123A	3124		TR21/	
1471120-11	199				
1471120-14	128	3024	63	TR24/	S5014
1471120-15	123A	3024	20	TR24/	S0020
1471122	199				
1471122-6	199	3018	20	TR95/	S0015
1471122-7		3018			
1471123-3	128	3024		TR87/243	S5014
1471123-4	128	3024		TR87/243	
1471123-5	128	3024			
1471124-5	179			TR35/G6001	
1471125-3	127	3034		TR27/235	G6007
1471132-002	196	3041		TR92/	
1471132-2	196	3054	57	TR76/	S5000
1471132-3	196	3054	28	TR92/701	S5000
1471132-4	152	3054	28	TR87/243	S5000
1471132-5	184	3054	57	TR27/S5003	S5000
1471132-6	181	3054		TR36/S7000	S7002
1471133-1	188	3054	66	TR72/3020	S5014
1471134	153				
1471134-1	189	3083	69	TR30/	S5013
1471135-001	130				
1471135-1	130	3027	66	TR59/247	S5004
1471135-2	181		75		S7000
1471136-2		3124			
1471136-3	172		64		
1471139-1	180		74		S7001
1471140-1	184	3190	57		S5000
1471141-1	185		58		S5006
1472434-1	706	3101			
1472446	121				
1472446-1	121	3009			
1472450-1	108			TR95/56	
1472450-1(11607)				TR21/	
1472474-2	154			TR78/712	S5025
1472475-1	199				
1472476		3018			
1472477		3117			
1472482-1	102A			TR85/	
1472494-1	127				
1472495-1	199				
1472500-1	127				
1472501-1	159				
1472519-1		3122			
1472549-1		3118			
1472633	161				
1472634-1	108	3018	11	TR95/56	S0016
1472636-1	107				
1472671-1		3009			
1473432-1		3124			
1473500-1	123A				
1473501-1	159				
1473503	124				
1473503-1	124			TR81/	
1473505-1	123A				
1473506-1	199				
1473507-1		3004			
1473507-2		3009			
1473508-1	128			TR87/	S5014
1473510-1		3122			
1473512-1	121				
1473514-1	127	3035	25	TR27/235	G6008
1473515-1	121	3014		TR01/232	G6003
1473516		3117			
1473516-1	129	3025	67	TR88/	S5013
1473518-1		3117			
1473519-1	123A	3122	20	TR21/	S0015
1473520-1	124	3021	12	TR81/240	S5011
1473523-1	159				
1473524-1		3018			
1473524-2	108	3018	11	TR95/56	S0016
1473526-2		3124			

Original Part Number	ECG	SK	GE	IR/WEP	HEP
1473527-1	123A	3018			
1473528-1	785				
1473529-1	107	3019			
1473530-1	108	3019		TR95/56	S0016
1473530-2	108	3019	11	TR95/56	S0020
1473531-1		3039			
1473532-1	123A	3018	21	TR21/735	S0016
1473533-1	108			TR95/735	S0016
1473535-1	161	3117	20	TR83/719	S0017
1473535-1(RCA)			39		
1473536-1	123A	3124	63	TR74/735	S3019
1473536-2	161	3124	20	TR25/	S5014
1473537-1	108	3132		TR95/56	S0020
1473538-1	123A	3122		TR21/735	S3019
1473539-1	123A	3018	17	TR21/735	S0015
1473540-1	159	3114	22	TR78/712	S3019
1473541-1	154	3114	25	TR78/712	S3019
1473543-1	161				
1473544				TR53/	
1473544-1	107	3122		TR70/	S0015
1473545-1	128	3024		TR87/243	S0016
1473546-1	123A	3018		TR21/735	S0016
1473546-2	123A	3018		TR21/735	S0015
1473546-3	123A				
1473547-1	128	3045		TR87/243	S0016
1473548-1	123A				
1473549-1	159	3118	21	TR88/242	S0013
1473549-2	159	3118	67	TR20/717	S0019
1473550-1	123A				
1473551-1	123A	3122		TR21/735	S0015
1473552-1	154	3044	40	TR78/712	S5025
1473553-1	128	3044	18	TR21/735	S0014
1473554-1	123A	3122		TR21/735	S0015
1473555-1	123A	3044	63	TR25/735	S5014
1473555-2	123A	3104		TR21/	
1473555-3	128	3024	27	TR87/	S3019
1473555-I	128				
1473556				TR51/	
1473556-1	123A	3122		TR21/735	S0015
1473557-1	123A				
1473558-1	107	3050		TR70/	S0015
1473559-1	159	3114	21	TR20/717	S0015
1473560-1	128	3124	17	TR87/243	S0015
1473560-2	128	3122	20	TR87/243	S0015
1473561-1	128	3122	17	TR87/243	S0015
1473562-1	159	3114	21	TR20/717	S0019
1473563-1	159	3114	27	TR20/717	S0019
1473564-1	162			TR67/707	S5020
1473565-1	128	3122	63	TR87/243	S0015
1473566-1	128	3024		TR87/243	S0015
1473567-1	124	3045		TR81/240	S5011
1473567-2	184	3041	28	TR55/S5003	S5000
1473567-4	152	3054	28	TR76/	S5000
1473568-1	108	3018	60	TR95/56	S0020
1473569-1	123A	3122	20	TR83/719	S0020
1473570-1	159	3025	22	TR20/717	S0019
1473570-2	159				
1473571-1	161	3018	20	TR83/719	S0030
1473572-1	123A	3018	20	TR21/735	S0015
1473572-3	128	3124	20	TR87/243	S5014
1473573-1	103A			TR08/	
1473574				TR53/	
1473574-1	159	3114	21	TR20/717	S0019
1473576	161	3018	20	TR83/	
1473576-1	233	3132	20	TR83/719	S0020
1473577-1	161	3018	18	TR83/719	S0030
1473578-1	102A	3004		TR85/250	G0005
1473579-1	161	3018	20	TR83/719	
1473580-1	128				
1473581-1	159	3025		TR20/717	S0019
1473582-1	123A			TR21/735	S0015
1473583-7	230	3042			
1473583-8	230	3042			
1473584-1	124	3045		TR81/240	S3021
1473586	161	3117	60	TR71/	S0020
1473586-2	123A	3038	60	TR21/735	S0015
1473588-2		3050			
1473589-1	123A	3122		TR21/735	S0014
1473591-1	159	3114		TR20/717	S0012
1473592-1	129	3025		TR88/242	S5013
1473593-1	128				
1473594-1		3114			
1473595-1	123A	3018		TR21/735	S0016
1473597-1	159	3118	21	TR35/G6001	S0013
1473597-2	159	3025	21	TR20/717	S0013
1473598-1		3004			
1473598-2	102A	3004		TR85/	S5013
1473600				TR51/	
1473601	163	3020	10	TR24/	
1473601-1	123A	3122	17	TR21/735	S0015
1473601-2	123A	3122	20	TR24/	S0015
1473602				TR53/	
1473603				TR51/	
1473603-1	108	3020	20	TR95/	S5014
1473604-1	172			TR69/S9100	S9100

Original Part Number	ECG	SK	GE	IR/WEP	HEP
1473604-3	108		64	TR95/56	S0011
1473606-1	108			TR95/56	S0016
1473608-1	128	3024	63	TR87/243	S5014
1473608-2	128	3024	63	TR87/243	S5014
1473608-3	128	3020			
1473610-1		3018			
1473611	152				
1473611-1	152	3054	28	TR76/701	S5000
1473612	152				
1473612-1	196	3054	28	TR92/	S5000
1473612-11	152	3054		TR76/701	
1473613-2	191		27	TR75/	S3019
1473613-3	191			TR75/	
1473613-4	171	3104	27	TR79/	S3021
1473613-5	128				
1473614-1	123A	3122	20	TR24/	S0015
1473614-2	199				
1473614-3		3122	10		S0015
1473615-1	128	3024	18	TR87/243	S0015
1473616-1	129	3114	22	TR88/242	S0019
1473617-1	108	3018	11	TR95/56	S0016
1473618-1	222	3050		/56	F2004
1473620-1	159	3114	21	TR20/717	S0013
1473621-1	196	3054	28	TR92/	S5000
1473622	128				
1473622-1	123A	3024	28	TR21/735	S0015
1473622-2	128		18	TR87/	S5014
1473623-1	196	3054	28	TR92/701	S5000
1473624-1	197	3083	58	TR77/S5007	S5006
1473625-1	128	3024	63	TR87/243	S0030
1473626-1		3124	20	TR24/	S0015
1473627-1	159	3114	21	TR20/	S0019
1473628-1	171	3201	27	TR79/	
1473628-2	171	3104	27	TR79/	S3021
1473628-3	188	3083	58	TR78/	S5006
1473629-3	189	3054	57	TR77/	S5000
1473631-1	123A	3124	63	TR21/735	S5014
1473632-1	171	3104	27	TR79/	S5018
1473632-2	186	3054	28	TR55/	S5000
1473632-3	196	3054	66		
1473633-1	171	3104	27	TR79/	
1473634-1	165	3111	36	TR67/	S5021
1473635	222				
1473635-1	222	3050			
1473635-2		3050			F2004
1473636-1	124	3054	12	TR81/	S5011
1473637-1	162	3079	35	TR67/	S5020
1473638-1	218	3085	26	TR23/	S5018
1473640-1	196	3054	28	TR92/	S3024
1473647-1	165	3111	35	TR68/	S5020
1473648-1	127	3035	25	TR27/	G6008
1473649-1	165	3115	36	TR61/	S5021
1473651-1	106	3050		TR20/	
1473652-1	108	3018	11	TR95/	S0016
1473656-1	190	3104	27		S3021
1473656-2	182	3104	27		
1473656-4	154				
1473657-1	123A	3018	20	TR95/	S0015
1473657-2	123A	3018	20	TR95/	S0015
1473665-2	130	3027	19	TR26/	S7002
1473666-1	129	3114	18	TR87/	S0015
1473669-1	165	3111	36	TR93/	S0020
1473669-2	165	3535	36	TR61/	S5020
1473676-1	161	3132	60	TR83/	S0020
1473676-1(3 IF)	233				
1473679-1	128	3044	18	TR87/	S5026
1473680-1	161	3018	39		
1473681-1	152	3054	66	TR55/	S5004
1473681-I	152				
1473682-1	185		69	TR56/	S5005
1473683-1	130	3114	19	TR26/	
1473687-1	154	3044	40	TR78/	S5025
1476171-11	104	3043			
1476188-1	123A	3061		TR21/735	
1501883	123A				
1502039	128				
1503097-0	159				
1522237-20	123A	3124		TR21/	
1560048	912				
1563295-101	123A				
1582501	925				
1588031-12		3024			
1596408	123A				
1599504		3011			
1600242-1C		3024			
1601555-2		3005			
1604609-2	909				
1611708-2	123A				
1612738-1	199				
1614326-1B		3024			
1614326-3B		3024			
1616226-1	159				
1617032	159				
1617510-1	123A				
1618626-1		3024			
1640480		3004			
1656672MPS9600				TR11/	
1670001-1		3004			
1670010-1		3012			
1670017-1C		3009			
1692810		3011			
1701790-1	130				
1701790-1B		3024			
1701790-2B		3024			
1702601-1	130				
1717006		3024			
1721399		3012			
1724595-2B		3024			
1751192-1B		3024			
1780142	159				
1780145-1	123A				
1780145-2	123A				
1780145-2-001	123A				
1780507		3039			
1780507-2001		3039			
1780522-1	159				
1780522-2	159				
1780522-2-001	159				
1780536-2001		3024			
1780553		3039			
1780724-1	123A				
1780738-1	123A				
1802520-001	941				
1802677-1	923D				
1802765	940				
1810037	108	3018	10	TR95/56	
1810038	108	3018	10	TR95/56	
1810039	108	3018	10	TR95/56	
1810945		3018			
1815036	108	3018	20	TR95/56	
1815037	108	3018		TR95/56	
1815039	108	3018	20	TR95/56	
1815041	123A	3018	20	TR21/735	
1815042	123A	3124	20	TR21/735	
1815043	123A	3124	10	TR24/	
1815045	108	3018	20	TR24/	S0014
1815047	108	3018	11	TR95/56	
1815054	123A			TR21/735	
1815067	108	3018	17	TR95/56	
1815068	108	3122	20		S0014
1815139			8	TR09/	
1815153	128			TR87/	
1815154	123A	3024	63	TR21/735	
1815154-9	128	3024	63	TR25/	
1815156	128	3024		TR87/243	
1815157	128	3024		TR87/243	
1815157-9			18		
1815159	128	3024		TR87/243	
1817004	108	3018	17		
1817005	123A	3018	20	TR21/	S0015
1817005-3	108	3018	20	TR24/	S0015
1817006	128		18	TR87/	
1817006-3	108	3018	20	TR24/	S0015
1817007	123A	3122	20		
1817008	108	3018	17		
1817017	103A	3010	8	TR08/	G0011
1817045	108	3018	20	TR24/	S0014
1817108	123A	3122	62		
1819045	108		20	TR24/	S0014
1820829	128				
1827322	129	3025	67		
1833404	223				
1835667	197				
1840399-1	123A				
1846282-1	123A				
1851515	123A	3122			
1851516	175	3026			
1851517	181	3036			
1851518	175	3026			
1861223-1	159				
1872425-1	909	3590			
1895989-1	9094				
1895991-1	9946				
1895992-1	9962				
1895993-1	9930				
1895994-1	9932				
1895995-1	9933				
1914062-1	9936				
1944313A1	128				
1944748	179	3009	16	TR35/G6001	
1945030		3079			
1945294	159				
1950052	159				
1950056-1	159				
1950160	152				
1956016	102A	3004		TR85/250	
1960023	123A				
1960083-1	128	3024			
1960085-2	154				
1960177-2	123A				

Original Part Number	ECG	SK	GE	IR/WEP	HEP
1960584	121	3014	16	TR01/232	
1960632	179			TR35/G6001	
1960643	102A	3004		TR85/250	
1961479	121	3014	16	TR01/232	G6005
1961480	121	3009	16	TR01/232	G6003
1961835	121	3009		TR01/232	
1961837	102A	3004	2	TR85/	
1962323	127	3035			
1962326	179			TR35/G6001	
1965016	129			TR88/	
1965017	121	3009	16	TR01/232	
1966079	121	3014	16	TR01/232	
1967784	181	3511			
1967799	128	3024		TR87/243	
1967799-1	128	3024		TR87/243	
1967801	128	3024		TR87/243	
1968958	123A				
1968959	128	3044			
1969113	124	3021			
1969281	159				
1970372-1		3024			
1971281		3036			
1971296	175	3026			
1971481		3036			
1971487	233	3027			
1971488		3124			
1971489	128	3512			
1971503	223	3027			
2000287-28	102A	3004	2	TR85/250	G0005
2000625-31	102A	3004	2	TR85/250	G0005
2000625-32	102A	3004	2	TR85/250	G0007
2000625-33	102A	3004	2	TR85/250	
2000625-34	102A	3004	2	TR85/250	
2000646-103	123A	3018	20	TR21/735	
2000646-104	160	3006	51	TR17/637	
2000646-105	108	3018	20	TR95/56	
2000646-106	126	3008	2	TR17/635	
2000646-107	123A	3124	20	TR21/735	
2000646-108	102A	3004	2	TR85/250	
2000646-109	126	3034	25	TR17/235	
2000646-110	127	3034	25	TR27/235	
2000646-111	158	3004	2	TR84/630	G6011
2000646-113	121	3009	3	TR01/232	
2000648-21	160	3007		TR17/637	G0003
2000648-22	160	3007		TR17/637	G0005
2000648-23	100	3005		TR05/254	G0005
2000752-80	128		17	TR87/243	
2000757-80	108	3039	20	TR95/56	
2000804-7	108	3018	17	TR95/56	
2000804-8	108	3018	17	TR95/56	
2000804-9	102A	3004	2	TR85/250	
2001653	102A			TR85/631	G0005
2001653-20	126	3008	51	TR17/635	
2001653-21	126	3008	51	TR17/635	
2001653-22	100			TR05/254	G0006
2001653-23	102A	3004	2	TR85/250	
2001653-24	102A	3004	52	TR85/250	G0005
2001653-58	160		51	TR17/	
2001653-59	160		51	TR17/	
2001809-47	102A	3123		TR85/250	G0005
2001809-48	102A	3004		TR85/250	G0007
2001809-48A	102A	3004		TR85/250	
2001809-48B	102A	3004		TR85/250	
2001812-65	100	3005	1	TR05/254	G0003
2002151-18	160	3007		TR17/637	G0003
2002151-18A	160	3008		TR17/637	
2002151-19	102A	3004		TR85/250	G0005
2002152-14	102A			TR85/250	G0005
2002153-58	160	3006	51	TR17/637	
2002153-59	160	3006	51	TR17/637	
2002153-60	160	3006	51	TR17/637	G0008
2002153-71	102A	3004	53	TR85/250	
2002153-76	160	3006	51	TR17/637	
2002153-77	123A	3124	10	TR21/735	
2002153-78	102A	3004	2	TR85/250	G0005
2002153-83	103A	3124	20	TR08/724	G0011
2002210-110	102A	3004	2	TR85/250	
2002211-24	102A	3004	2	TR85/250	
2002211-25	102A	3004	2	TR85/250	
2002332-53	108	3018		TR95/56	
2002332-54	108	3018		TR95/56	
2002332-55	108	3018	20	TR95/56	
2002332-56	108	3018		TR95/56	
2002336-19	126	3004	2	TR17/635	G0005
2002403-19	126	3004	2	TR17/635	G0007
2002620-18	108	3018	20	TR95/56	
2002620-19	108	3018	20	TR95/56	
2002621-2	123A	3124	20	TR21/735	
2003073-0701	123A	3004		TR21/	
2003073-0702	159	3025		TR20/	
2003073-8	102A	3004	52	TR85/250	
2003073-9	123A	3124	20	TR21/735	S0016
2003073-10	123A	3124	20	TR21/735	
2003073-11	160	3006	51	TR17/637	
2003073-12	160	3006	50	TR17/637	
2003073-13	160	3006	51	TR17/637	
2003073-14	102A	3004	52	TR85/250	
2003073-15	102A	3004	52	TR85/250	
2003073-16	103A	3010	8	TR08/724	
2003073-91	130	3027	14	TR59/247	
2003073-91A	130	3027		TR59/	
2003168-135	123A	3124	20	TR21/735	
2003168-136	123A	3124	20	TR21/735	
2003229-25	123A	3122		TR51/735	
2003239-65	123A	3124		TR21/735	S0030
2003342-109	108	3039		TR95/56	S0016
2003342-244	107	3039		TR70/720	S0016
2003779-22	107	3018	20	TR70/720	S0016
2003779-23	107	3018	20	TR70/720	S0011
2003779-24	107	3018	20	TR70/720	S0011
2003779-25	107	3018	20	TR70/720	S0015
2004358-123	102A	3004	53	TR85/250	G0005
2004358-168	102A	3004	53	TR85/250	G0005
2004746-87	124	3021	12	TR81/240	
2004746-114	108	3018	20	TR95/56	
2004746-115	108	3018	20	TR95/56	
2004746-116	129	3025		TR88/242	
2004746-117	129	3025		TR88/242	
2006226-14	102A	3004	2	TR85/	G0005
2006227-51	123A	3004		TR21/	S0014
2006334-31	158	3123		TR84/630	G0006
2006334-115	123A			TR21/	S0014
2006334-155	128	3024		TR87/56	S0014
2006431-44	108	3039	20	TR95/	S0016
2006431-45	123A	3004	20	TR21/	S0020
2006431-46	123A	3004	20	TR21/	S0015
2006431-47					S5013
2006431-48					S5012
2006431-49	123A	3004	17	TR21/	S0015
2006436-35	186	3054	28	TR55/	
2006436-36	186	3054	28	TR55/	
2006436-37	129	3025	21	TR88/	
2006436-40	128		18	TR87/	
2006441-113	102A	3004	2	TR85/	G0005
2006513-19	108	3039		TR95/56	
2006513-39	107	3018		TR70/720	S0016
2006513-59	131	3052		TR94/	G6016
2006514-60	123A			TR21/	S0014
2006514-61	123A			TR21/	S0014
2006582-25	123A			TR21/	S0030
2006582-101	161	3018		TR83/719	
2006607-59	121	3009		TR01/230	G6003
2006607-60	123A	3046		TR21/735	S0015
2006607-61	195	3047		TR65/243	S3001
2006607-62	224	3049			
2006613-77	123A			TR21/	S0015
2006623-47	132	3116	FET1	FE100/802	F0015
2006623-88	184	3041	57	TR76/S5003	S5000
2006623-128	195	3048		TR65/	S3001
2006623-145	123A	3124	10	TR21/735	S0015
2006623-148	128	3024		TR87/243	
2006681-93	107	3018		TR70/720	S0016
2006681-94	107	3018		TR70/720	S0016
2006681-95	107	3039		TR70/720	S0014
2006681-96	123A	3018		TR21/720	S0016
2006681-120	158	3004		TR84/630	G0005
2008292-56	107			TR70/720	S0015
2008292-87	126			TR17/635	G0009
2008293-109		3004	2	TR85/	G0005
2008293-111	158	3004	53	TR84/	G0005
2008299-1	107			TR70/720	S0014
2008300-103		3018		TR70/720	S0016
2008300-104		3018		TR83/735	S0016
2008300-105	1003				
2010088-49	123A	3124	20	TR21/	S0015
2010494-4				TR21/	
2010499-52	123A	3124	20	TR21/	S0030
2010952-14	103A	3010	54		G0011
2016974-1		3122			
2016974-3		3122			
2016996-1		3122			
2016996-3		3122			
2026011				TR24/	
2026012				TR24/	
2026013				TR24/	
2026014				TR51/	
2026015				TR51/	
2026016				TR51/	
2026018				TR26/	
2026020				TR51/	
2041614	123A				
2047102	102	3004			
2047102-1		3004			
2047102-2		3004			
2047353		3004			
2056606-0701	159				
2057013-0004	159	3025	21	TR20/717	S0012
2057013-0007	159	3025	21	TR20/	S5013
2057013-0008	159	3114	21	TR20/	S5013
2057013-0012	159	3025	21	TR20/	S5013

Original Part Number	ECG	SK	GE	IR/WEP	HEP
2057013-0701	159	3025	21	TR20/717	S0012
2057013-0702	159	3025	21	TR20/717	S0012
2057013-0703	159	3025	21	TR20/717	S5013
2057199-070BA	130	3027		TR59/	
2057199-0700	130			TR59/	
2057199-0701	130	3027		TR59/247	
2057199-701	130	3027			
2057323-050BA		3027			
2057323-0500	130			TR59/	
2057323-0501	130	3027	28	TR59/247	
2068491-704	221	3065			
2068510-0701	9936				
2068510-0702	9946				
2068510-0703	9962				
2068510-0704	9093				
2068510-0705	9099				
2076393	126	3007	51	TR17/635	
2076403	160	3007	51	TR17/637	
2076403-0703	160	3007	51	TR17/637	
2076945-0701	102A	3004	2	TR85/250	
2077372-0701		3122			
2077372-0702		3122			
2081859-0713		3009			
2088576-4		3024			
2089962-0702		3024			
2090056-1	104	3009	16	TR01/230	G6003
2090056-5	104	3009	16	TR01/230	G6003
2090056-27	104	3009	16	TR01/230	G6003
2090070		3004			
2090241		3011			
2090924-0008	102A	3004	52	TR85/250	G0005
2090924-008	102A		2	TR85/250	
2090924-6	102A		53	TR85/250	
2090924-8	102A	3004	53	TR85/250	G0005
2090924-8A	102A	3004	2	TR85/250	G0005
2090924B	102A	3004		TR85/250	
2091211-0014	100	3005	1	TR05/254	
2091217-0014	160		51	TR17/637	
2091241-0005	160	3008	51	TR17/637	G0002
2091241-005	126			TR17/635	
2091241-0013	100	3005	2	TR05/254	G0005
2091241-0014	100	3008	2	TR05/254	G0005
2091241-0015	100	3008	2	TR05/254	G0005
2091241-0018	102A	3004	51	TR85/250	
2091241-0719	126	3008	51	TR17/635	G0008
2091241-1	100	3008	2	TR05/254	G0005
2091241-2	100	3008	2	TR05/254	G0005
2091241-3	100	3008	51	TR05/254	G0005
2091241-4	100			TR05/	
2091241-5	126			TR17/	
2091241-5A	126	3006	2	TR17/635	G0005
2091241-6	100			TR05/	
2091241-7	100	3005	2	TR05/254	G0005
2091241-8	100			TR05/	
2091241-9	102A		53	TR85/250	
2091241-10	100			TR05/	
2091241-11	100			TR05/	
2091241-12	100			TR05/	
2091241-13	100	3005	2	TR05/254	G0005
2091241-13A	100	3005	2	TR05/254	G0005
2091241-14	100	3005	2	TR05/254	G0005
2091241-15	100	3005	2	TR05/254	G0005
2091241-15A	100	3005	2	TR05/254	G0005
2091247-005	160	3006		TR17/	
2091260			8		
2091260-1(NPN)	103A	3010	8	TR08/250	G0005
2091260-1(PNP)	102A			TR85/	
2091260-2(NPN)	103A	3010	8	TR08/250	G0005
2091260-2(PNP)	102A			TR85/	
2091260-3(NPN)	103A	3010	8	TR08/250	G0011
2091260-3(PNP)	102A			TR85/	
2091578-0702	102A	3004	2	TR85/250	
2091578-1	102A	3004	52	TR85/250	G0005
2091858-0712	121	3009	3	TR01/232	
2091858-11	121	3009	16	TR01/232	G6003
2091859-0008	104	3009	16	TR01/230	G6003
2091859-0011	121	3009	16	TR01/232	G6005
2091859-0025	104	3009		TR01/230	G6005
2091859-0711	108	3039	11	TR95/56	
2091859-0712	121	3009	11	TR01/232	G6005
2091859-0713	121	3014	16	TR01/232	G6005
2091859-0714	121	3009	16	TR01/232	G6005
2091859-0715	121	3014	3	TR01/232	G6005
2091859-0716	121	3009	3	TR01/232	G6005
2091859-0717	121	3014	3	TR01/232	G6005
2091859-0718	121	3009	16	TR01/232	G6005
2091859-0720	104	3009	16	TR01/230	
2091859-0723	121	3009	16	TR01/232	
2091859-2	104	3014	16	TR01/230	G6003
2091859-4	104	3009	16	TR01/230	G6003
2091859-6	121	3009	16	TR01/232	G6005
2091859-8	121	3009	16	TR01/232	G6005
2091859-9	121	3009	16	TR01/232	G6005
2091859-10	104	3009	16	TR01/230	G6003
2091859-11	104	3009	16	TR01/230	G6003
2091859-16	121			TR01/	
2091859-25	121			TR01/	
2091959-16	121			TR01/	
2092417-005	160			TR17/	
2092417-0017	107	3018	11	TR70/720	S0015
2092417-0018	107	3018	17	TR70/720	S0015
2092417-0019	107	3018	11	TR70/720	S0015
2092417-0704	160		51	TR17/637	
2092417-0707	160	3006	51	TR17/637	
2092417-0708	160	3006	51	TR17/637	G0003
2092417-0709	160	3006	51	TR17/637	G0003
2092417-0710	160	3006	51	TR17/637	
2092417-0711	161	3018	11	TR83/719	S0015
2092417-0712	161	3018	11	TR83/719	S0015
2092417-0713	161	3018	11	TR83/719	S0015
2092417-0714	161	3117	17	TR83/719	S0015
2092417-0715	161	3018	11	TR83/719	S0015
2092417-0716	161	3018	11	TR83/719	S0015
2092417-0717	160		51	TR17/637	G0003
2092417-0719	123A	3122	20	TR21/735	
2092417-0720	123A		20	TR21/	
2092417-0721	123A	3122	20	TR21/735	
2092417-0724	123A	3122	20	TR21/735	
2092417-0725	123A	3122	20	TR21/735	
2092417-1	160	3006	50	TR17/637	G0003
2092417-2	160	3006	50	TR17/637	G0003
2092417-3	160	3006	51	TR17/637	G0003
2092417-4	160		51	TR17/637	
2092417-5	160		51	TR17/637	
2092417-6	160	3006	50	TR17/637	G0003
2092417-7	160		51	TR17/637	
2092417-8	160		51	TR17/637	
2092417-9	160		51	TR17/637	
2092417-17	123A			TR21/	
2092417-18	123A			TR21/	
2092417-19	123A			TR21/	
2092418-0022	161	3018	11	TR83/719	S0015
2092418-0023	161	3018	11	TR83/719	S0015
2092418-0024	161	3018	11	TR83/719	S0015
2092418-071	160		50	TR17/637	G0003
2092418-0710	160	3006	51	TR17/637	
2092418-0711	101	3011	50	TR08/641	G0011
2092418-0712	160		50	TR17/637	G0003
2092418-0715	108	3018	20	TR95/56	S0015
2092418-0716	161	3018	20	TR83/719	S0015
2092418-0717	161	3018	20	TR83/719	S0015
2092418-0718	161	3018	11	TR83/719	S0015
2092418-0719	161	3018	11	TR83/719	S0015
2092418-0720	161	3018	11	TR83/719	S0015
2092418-0721	161	3018	11	TR83/719	S0015
2092418-0724	108	3039		TR95/56	
2092418-1	160	3006	50	TR17/637	G0003
2092418-2	160	3006	50	TR17/637	G0003
2092418-5	160	3006	50	TR17/637	G0003
2092418-6	160	3006	50	TR17/637	G0003
2092418-7	160	3006	50	TR17/637	G0003
2092418-8	160		51	TR17/637	
2092418-10	160		51	TR17/637	
2092418-11	160		51	TR17/637	
2092605-0705	123A			TR21/	
2092608-22	123A	3122	20	TR21/735	
2092609	123A				
2092609-0001	123A		20	TR21/735	S0015
2092609-0002	123A	3122	20	TR21/735	S0015
2092609-001	123A	3124		TR21/735	
2092609-0022	123A	3124	20	TR21/735	S0015
2092609-0023	123A	3122	17	TR21/	S0014
2092609-0024	123A	3122	17	TR21/	S0014
2092609-0025	159	3122	17	TR20/	S0014
2092609-0026	123A	3122	17	TR21/	S0014
2092609-0027	123A	3122	17	TR21/	S0014
2092609-0028	123A	3122	20	TR21/	S0014
2092609-0705	123A	3124	20	TR21/735	S0015
2092609-0706	123A	3124	8	TR21/735	S0015
2092609-0707	123A	3124	20	TR21/735	S0015
2092609-0713	123A	3124	20	TR21/735	S0015
2092609-0715	123A	3124	20	TR21/735	S0015
2092609-0718	123A	3124	20	TR21/735	S0015
2092609-0720	123A	3124	20	TR21/735	S0015
2092609-0721	123A	3124	20	TR21/735	S0015
2092609-1	123A	3124	8	TR21/735	G0011
2092609-2	123A	3124	8	TR21/735	G0011
2092609-3	123A	3122	8	TR21/735	S0015
2092609-5	123A	3122	8	TR21/735	
2092693-0724	107	3039		TR70/720	
2092693-0725	107	3039		TR70/720	
2092693-0734	159				
2092693-1	102A	3004	2	TR85/250	
2092693-2	160	3007	51	TR17/637	
2092693-3	160	3007	51	TR17/637	
2092693-4	160	3007	51	TR17/637	
2092693-8	102A	3004	2	TR85/250	
2092693-9	160	3006	51	TR17/637	
2093308-070	108	3018	17	TR95/56	
2093308-0700	108	3039	20	TR95/56	

Original Part Number	ECG	SK	GE	IR/WEP	HEP
2093308-0701	123A	3018	8	TR21/735	G0011
2093308-0702	123A	3018	8	TR21/735	G0011
2093308-0703	123A	3018	8	TR21/735	G0011
2093308-0704	107	3018	20	TR70/720	S0015
2093308-0704A	108	3018		TR95/	
2093308-0705	107	3018	20	TR70/720	S0015
2093308-0705A	108	3018		TR95/	
2093308-0706	107	3018	20	TR70/720	S0015
2093308-0706A	108	3018		TR95/	
2093308-0708	123A	3122	20	TR21/735	
2093308-0725	107	3018	11	TR70/720	S0015
2093308-1	108	3018	17	TR95/56	
2093308-2	108	3018	17	TR95/56	
2093308-3	108	3018	17	TR95/56	
2096700	129				
2096700-TM18	129				
2097013-0702	128	3024		TR87/243	
2098574-9		3004			
2106974-5		3122			
2115237-1		3122			
2125310	199				
2132523-1	159				
2132763-1	912				
2160153	725				
2180803		3011			
2180810-1E		3039			
2180824-1		3004			
2181788		3014			
2182004		3024			
2182142		3005			
2182183C		3004			
2182184-1D		3024			
2182184-3E		3024			
2182184-4F		3024			
2182188-1D		3009			
2182196-1A		3006			
2182803-1H		3122			
2182827-1E		3011			
2182843-1C		3122			
2183219-1B		3005			
2183219-2A		3005			
2183279A		3122			
2184185-1C		3122			
2184185-2C		3122			
2184967-1		3008			
2184995-2		3005			
2184996-2		3003			
2184996-3B		3004			
2184998-1		3011			
2184999-1A		3009			
2185414-1		3005			
2185440-1		3008			
2185502-2		3011			
2185508-1		3004			
2185512-10		3005			
2185513-1A		3004			
2185514-10		3024			
2185515-10		3009			
2185516-10		3006			
2185518-10		3011			
2185520-10		3122			
2185591-2		3011			
2185593-20		3003			
2185593-30		3004			
2185594-1		3011			
2185599-10		3005			
2185600-1		3008			
2185602-2		3011			
2185617-10		3024			
2185796		3005			
2185796-1A		3005			
2185804-2		3014			
2185805-1		3027			
2187337-1		3027			
2227367	176				
2234494	9900				
2234497	9926				
2243255-1	102A	3123	53	TR85/250	G0005
2295361	9962				
2314009	127		25	TR27/	
2316183	161				
2320011	102A	3006	50	TR85/250	G0005
2320022	123A	3122	17	TR21/	S0015
2320031	107	3018	20	TR70/720	S0014
2320041	107	3124	63	TR70/720	S0014
2320041H	161	3018	62	TR83/720	S0014
2320042	107	3018	20	TR70/720	S0014
2320043	107	3018	20	TR21/	S0014
2320051	154	3040	18	TR78/712	S5025
2320051H	154	3040		TR78/712	S5025
2320062	108		20	TR95/	
2320063	108	3018	20	TR95/	S0014
2320073	108	3018	20	TR95/	S0020
2320083	152		23	TR76/	
2320084	175	3026		TR81/241	S5019

Original Part Number	ECG	SK	GE	IR/WEP	HEP
2320092	131	3052	30	TR24/642	G6016
2320111	123A	3124		TR21/735	S0014
2320123	102A	3004	50	TR85/	
2320123(3)				TR85/	
2320141	233	3039	20	TR70/720	S0020
2320141H	161	3039		TR83/56	S0017
2320154	100	3005		TR05/	
2320161	159	3007	21	TR20/717	S5022
2320161(RCA)	160			TR17/	
2320191	154	3040	20	TR78/712	S5024
2320201	127	3034	25	TR27/232	G6007
2320221	124	3021	12	TR81/240	S5011
2320222	124	3026		TR81/240	S5011
2320223		3021			S0020
2320228		3021			
2320233	128	3024	18	TR25/	S5014
2320242	129	3025	67	TR28/	S5013
2320243	129	3025	67	TR28/	S5013
2320261	102A	3004	52	TR85/250	G0007
2320271	162	3079	37	TR67/707	S5020
2320273	162	3079	38		S5020
2320281	164	3079		TR93/740A	S5021
2320291	165	3115		TR93/740B	S5021
2320299	165	3115		TR93/	S5021
2320302	102A	3004		TR85/250	G0005
2320302H	102A	3004	25	TR85/250	G0005
2320331	103A	3010	8	TR08/724	G0011
2320413	123A	3010	8	TR21/735	G0011
2320422	158	3004	67	TR84/250	G0007
2320422-1	102A	3004		TR85/	
2320423	102A	3004	2		G0007
2320432					S3024
2320441	123A			TR21/	S0030
2320471	107	3018	60	TR70/720	S0016
2320471-1	107	3018		TR70/	
2320471H	107	3124	18	TR70/	S0016
2320482	152	3041	28	TR76/701	S5000
2320482H	152	3054	28	TR76/	S5000
2320483	152	3054	28	TR76/	S5000
2320485	152	3054	28	TR76/	S5000
2320486	152	3054	28	TR76/	S5000
2320512	100	3005	50		G0005
2320513	102A	3005	50	TR85/	G0005
2320514	160	3005	1	TR17/254	G0001
2320514-1	100	3005		TR05/	
2320541	121	3009	16	TR01/	
2320591	123A	3124	10	TR21/735	S0014
2320591-1	123A	3124		TR21/	
2320595	123A	3124	20	TR51/	S0014
2320596	123A	3024	20	TR21/735	S5014
2320631		3025	67	TR88/242	S5013
2320632	193	3114	67	TR28/	S5013
2320637		3114			S5013
2320643	192	3122	63	TR25/	S5014
2320646	192	3024	63	TR24/243	S5014
2320646-1	128	3024		TR87/	
2320647		3024	18	TR24/243	S5014
2320647-1	128	3024		TR87/	
2320651	152	3041	28	TR76/701	S5000
2320652	152	3054	28	TR76/	S5000
2320664	192	3122	18	TR25/	S0025
2320671	159				
2320681		3115			S5013
2320696	123A	3124		TR21/	
2320696-1	123A	3124		TR21/	
2320843		3041			S5000
2320845	184	3054	57	TR55/	S5013
2320855	185	3083	50	TR56/	S5014
2320931		3021			S5011
2320946		3044			S3002
2320961		3115			S5020
2321001		3021			S5011
2321101		3045			S0028
2321111		3045			S0028
2321121		3115			S5020
2326953	123A				
2327022	128	3024	63	TR87/	S5014
2327052	130	3027	14	TR59/	S7002
2327053	130	3027	14		S7004
2327061	124		12	TR81/	
2327122	123A	3122	62	TR51/	S0015
2327132	132	3116	FET2	FE100/	F2004
2327142	132	3050		FE100/	F2004
2327152	186	3054	28	TR55/	S3024
2327153	186	3054	28	TR55/	S5000
2327172	130	3027	14	TR67/	S7004
2327182	124	3021	12	TR23/	S5011
2327203	186		28	TR55/	
2327206	152	3054	57	TR55/	
2327212	703A				
2327232		3050			F2004
2327262		3114	21	TR30/	S5013
2327282	129	3025	67	TR28/	
2327283		3025	67	TR28/	
2327292		3024		TR25/	

Original Part Number	ECG	SK	GE	IR/WEP	HEP
2327293	128	3024	18	TR25/	S0015
2327302(HITACH)	703A				
2327312	1039				
2327332	128	3024	18	TR25/	S5014
2327363	123A	3124	20		S0015
2327387	159	3114	21	TR30/	S0019
2327403	128	3024	18		S5014
2327411	1041				
2327422	720				
2327431	222	3050			F2004
2337017		3018			S0011
2360021	1061				
2360042	712	3072			C6063P
2360042(HITACH)	712				
2360092		3072			
2360201		3072			
2360322-1		3024			
2360796		3019			
2360924-5601	123A				
2391342	910				
2391345		3036			
2391346		3036			
2391346C		3036			
2391773	909	3590			
2392152	941M				
2396475	923				
2399179		3036			
2402277	732				
2412275	732				
2412949-0001	129				
2414811		3122			
2414892		3027			
2469749	123A				
2469755	123A				
2469936-1	128				
2479692	123A				
2479836	123A				
2485076-2	128	3124	10	TR87/243	S5014
2485076-3	128	3124	10	TR87/243	S5014
2485077-2	128	3024		TR87/243	S5014
2485077-3	128	3024		TR87/243	S5014
2485078-1	123A	3018		TR21/56	S0011
2485078-2	123A	3018	20	TR21/56	S0011
2485078-3	123A	3018	20	TR21/56	S0011
2485079-1	123A	3018		TR21/56	S0015
2485079-2	123A	3018	17	TR21/56	S0015
2485079-3	123A	3018		TR21/56	S0015
2487340		3114	22		S5022
2487341		3114	22		S5022
2487424		3122	63		S3020
2495012	126	3008	1	TR17/635	G0006
2495013	126	3008	51	TR17/635	G0006
2495014	102A	3004	2	TR85/250	G0005
2495078	160	3006	51	TR17/637	G0003
2495079	160	3006	51	TR17/637	G0003
2495080	102A	3004	53	TR85/250	G0008
2495082	160	3006	2	TR17/637	G0007
2495166-1	108	3018	20	TR95/56	
2495166-2	123A	3124	11	TR21/735	
2495166-4	123A	3018	20	TR21/735	
2495166-8	123A	3018	20	TR21/735	
2495166-9	123A	3018	20	TR21/735	
2495200	126	3008	1	TR17/635	G0006
2495376	160	3006	20	TR17/637	S0013
2495377	160	3006	51	TR17/637	G0003
2495378	126			TR17/635	G0008
2495379	126	3006		TR17/635	G0008
2495388	102A	3004	52	TR85/250	G0007
2495388-1	102A	3004		TR85/250	G0007
2495388-2	102A	3123		TR85/250	G0007
2495488-1	126	3006		TR17/635	G0008
2495488-2	126	3006		TR17/635	G0008
2495520	161	3132		TR83/719	S0011
2495521	161	3132		TR83/719	S0011
2495521-1	123A			TR21/	
2495522-1	161	3132		TR83/719	S0011
2495522-4	123A			TR21/735	S0011
2495523-1	161	3132		TR83/719	S0011
2495529(ARVIN)	128	3024		TR87/243	
2495567-2	102A	3123		TR85/250	G0005
2495567-3	102A	3123		TR85/250	G0005
2495568	158			TR84/	
2495568-2	102A	3004	2	TR85/250	G0007
2496125-2	123A			TR21/735	S0011
2497094-1	161	3018		TR83/	
2497094-2	161	3018		TR83/	
2497473	102A	3004	2	TR85/250	
2497473-1	102A	3004		TR85/250	G0005
2497496	102A	3123		TR85/250	G0005
2497888	158	3004	52	TR84/630	G0005
2498163	128			TR87/243	
2498456-2	107	3018	20	TR70/720	S0016
2498482-2	107	3018	17	TR70/	S0016
2498507-2	108	3018		TR95/56	S0014
2498507-3	108	3018	20	TR95/56	S0014
2498508-2	107			TR70/720	S0014
2498508-3	107			TR70/720	S0014
2498512	159	3114		TR20/717	S5014
2498665-1	199				
2498665-2	199				
2498665-3	199				
2498837	160	3008		TR17/	G0008
2498837-4	102A			TR85/	
2498902-1	107	3018	17	TR70/720	S0015
2498902-2	107	3018		TR70/720	S0014
2498903-1	107	3018	61	TR70/720	S0016
2498903-2		3018	61	TR70/720	S0016
2498903-3	107			TR70/720	S0016
2498904-3	123A	3018	20	TR21/735	S0015
2498904-4	123A	3124		TR21/243	S0014
2498904-6	123A	3124	10	TR21/735	S0014
2499950	199	3124	63	TR21/53	S0024
2500389-432	9932				
2500389-444	9944				
2500390-435	9935				
2500390-436	9936				
2500390-437	9937				
2500390-446	9946				
2500390-449	9949				
2500390-461	9961				
2500390-462	9962				
2500391-445	9945				
2500747	9932				
2501337-433	9933				
2501549-341	703A				
2501557-430	9930				
2501557-436	9936				
2501557-437	9937				
2501557-446	9946				
2501557-449	9949				
2501557-461	9961				
2501557-462	9962				
2503134-105	159				
2504640		3018			
2505207	199				
2505209	123A				
2509319	124				
2520063	123A		10	TR21/	
2521108-1	159				
2530733	123A				
2545989-2	128				
2546220-1	910				
2557736		3036			
2557737		3036			
2557738		3036			
2573480				TR21/	
2600109				TR21/	
2600209				TR21/	
2600309				TR21/	
2600401				TR62/	
2600402				TR24/	
2604688-3	909				
2605022	128				
2608169-1	128				
2610023-01	910				
2610023-02	910D				
2610043-03	941M				
2610153	725				
2610154-01	923				
2610783	7402				
2610784	7496				
2610786	7400				
2610788	74121				
2621567-1	123A				
2621570	159				
2621764	123A				
2621811	106				
2622284	123				
2640830-1	123				
2640843-1	123A				
2652613	9944				
2652614	9945				
2656211-1	9930				
2656212	9946				
2656212-1	9946				
2656213	9944				
2656213-1	9944				
2656214	9945				
2656214-1	9945				
2656747	9944				
2656748	9946				
2666293	7482				
2666294-1	7483				
2666307-1	128				
2709759	309K				
2710002	941				
2712080	123A	3122		TR21/735	S0015
2729789	176	3123		TR82/	
2777301	128				
2797658-616030A	941				

Original Part Number	ECG	SK	GE	IR/WEP	HEP
2808577	9932				
2855296-01	102A	3004		TR85/	
2865101	192				
2868536-1	7474				
2875409		3026			
2875493	152				
2898431-2	9946				
2899000-00	9946				
2899001-00	9962				
2899002-00	9930				
2899003-00	9936				
2899004-00	9945				
2899018	911				
2899414	923D				
2902798-2	915				
2903993-I	159				
2904014	121	3009		TR01/232	G6005
2904037					S3006
2928054-1	128				
2970038H05	100	3005		TR05/	
3002406-RT99C		3011			
3004005		3011			
3004800		3122			
3004856	102A	3004	2	TR85/250	
3005861	123A	3124	20	TR21/735	
3006206-00	784				
3006892-00	910				
3006892-01	910D				
3007359-00	9937				
3007359-01	9949				
3007359-02	9963				
3007359-03	9932				
3007472-00	9945				
3007473-00	9932				
3007473-01	9932				
3007474-00	9930				
3007474-01	9930				
3007475-00	9962				
3007475-01	9962				
3007476-00	9946				
3007476-01	9946				
3007477-00	9936				
3007477-01	9936				
3007572-00	9933				
3007573-00	9093				
3007579	912				
3007579-00	912				
3007604-00	9135				
3007680-00	941				
3007681-01	941D				
3008321-00	309K				
3008340	923D				
3010474B		3008			
3019410		3045			
3019438		3044			
3019438A		3044			
3034725-1	903				
3068305-2	123A				
3130006	121				
3130011	102				
3130025	102				
3130053	128				
3130057	175	3026		TR81/	
3130058	130	3510		TR59/	
3130060	102				
3130090	181				
3130091	130	3510			
3130092	128				
3130093	175				
3130104	181				
3130109	121				
3130672-2		3122			
3146977	130	3027	14	TR59/247	
3146977A	130	3027		TR59/	
3151180-2		3122			
3152159	175	3026		TR81/701	
3152170	130	3027	14	TR59/247	
3152170A	130	3027		TR59/	
3159036-1		3036			
3165092		3021			
3170717	154	3045		TR78/712	
3170757	154	3045		TR78/712	S5024
3172626	9933				
3172627-2	9949				
3172629-1	9930				
3172629-2	9961				
3172630	9944				
3172631	9948				
3172632-1	9962				
3172638	9950				
3176135-1	9936				
3176135-2	9937				
3177200	9932				
3181972	108				
3201104-10	123A				
3201110-10A		3122			
3267390-01	74H40				
3401921		3004			
3403787	123A			TR21/	S0025
3403866-3	128				
3404114-1	102				
3404114-2	102				
3404520-012		3024			
3404520-81	126				
3404520-211		3024			
3404520-301	128				
3404520-601	221				
3406519-31	726				
3406519-31(RCA)	726				
3410161-001		3021			
3412004-1912	133				
3412907-1	128				
3412907-2		3027			
3412951	904				
3412986-1	903				
3438095	152				
3438854	163				
3438867	152	3054			
3450842-10	154				
3450842-20	154				
3450842-30	154				
3453897-1		3027			
3457104		3122			
3457107-1	123A				
3457632-5	123A				
3457633-1	128				
3457633-2	128				
3457697-1	726				
3457936-1	234				
3458267-1	128				
3458573-1	192				
3459332-1	161			TR83/	
3460532-1		3004			
3460549-5		3124			
3460550-1	160				
3460550-3	160				
3460550-4	160				
3460552		3010			
3460553-2	121			TR01/	
3460553-4	121	3009		TR01/	
3460679-1	102A				
3460728		3004			
3462175-1		3005			
3462221-1	121			TR01/	
3462306-1	121			TR01/	
3463099-1	128				
3463100-1	175	3510			
3463101-1	130	3027			
3463604-1	175			TR81/	
3463604-2	175			TR81/	
3463609-2	161			TR83/	
3464343-1		3511			
3464482-1	102A				
3464648-1	223				
3464648-2	223				
3468068-1	161			TR83/	
3468068-2	161			TR83/	
3468068-3	161			TR83/	
3468068-4	161			TR83/	
3468071-1	175				
3468071-3		3038			
3468182-1	123A				
3468182-2	123A				
3468183-1	159				
3468242-1	159				
3468242-2	123A				
3468841-4	175	3026			
3520025-001	9936				
3520041-001	7400				
3520042-001	7410				
3520043-001	7473				
3520044-001	7440				
3520045-001	7451				
3520046-001	7474				
3520047-001	7430				
3520048-001	7404				
3520050-001	7472				
3539307-001	161	3018	20	TR83/719	
3539307-002	161	3018			
3596063	159	3114	22	TR28/	S0031
3596067		3117	17	TR24/	S0015
3596068		3117	17	TR24/	S0015
3596069		3117	17	TR24/	S0015
3596070	108	3018	17	TR95/	S0015
3596071	108	3018	17	TR95/	S0015
3596072	108	3018	17	TR95/	S0015
3596091	196	3041	66	TR92/	S5000
3596092	196	3041	66	TR76/	S5000
3596100	197	3084	69	TR56/	S5006
3596101	197	3084	69	TR77/	S5006

Original Part Number	ECG	SK	GE	IR/WEP	HEP
3596116	128	3122	20	TR87/	S0015
3596117		3122	20	TR53/	S0015
3596118	159	3114	22	TR20/	S0031
3596128					S5000
3596338	123A	3122	10	TR24/	S0015
3596339	123A	3122	10	TR24/	S0015
3596340	129	3114	20	TR24/	S0016
3596341	129	3114	20	TR24/	S0016
3596353	734				
3596354	723	3144			
3596401	222	3050	FET4		F2004
3596402	222	3050	FET4		F2004
3596440		3004	17	TR24/	G0007
3596446	152	3054		TR55/	
3596447	152	3041		TR55/	
3596448		3054	66	TR55/	S5000
3596449		3041	66	TR55/	S5000
3596451	153	3083		TR56/	
3596452	153	3083		TR56/	
3596453	153	3083	69	TR56/	S5006
3596454	153	3083	69	TR56/	S5006
3596570	123A				
3596808	726				
3596809	789				
3596810	788				
3597049	1115				
3598070			17		
3610001-001	9923				
3610003	9914				
3610005	9900				
3646007-1				TR21/	
3646007-2				TR51/	
3700072	123A				
3700085	121				
3700109	123A				
3700135	130	3027			
3700144	159				
3700150		3027			
3700162	128				
3700163	129				
3700164	130				
3700171	123A				
3700198		3025			
3700219	130	3027			
3700228	130	3027			
3700249	159				
3700258	129				
3700279	123A				
3720968-1	923				
3731132-1	123A				
3731133-1	159				
3731313-1	121	3009		TR01/	
3731418-1	129				
3731418-2	128				
3731418-3	129				
3731418-4	129				
3755171	152				
3755862	199				
3939307-002	161	3132		TR83/	
4000921	199				
4002862-0001	123A				
4013373-0701	9974				
4017621-0701	123A				
4028839	161				
4031986-0701	159				
4032122	716				
4036598-P1	121	3009	2	TR01/	
4036598-P2	102A	3004	3	TR85/	
4036612-P1	100	3005	1	TR05/	
4036612-P2	100	3005	1	TR05/	
4036707-P1	100	3005	1	TR05/	
4036707-P2	100	3005	1	TR05/	
4036715-P1	160	3006	51	TR17/	
4036715-P2	160	3006	51	TR17/	
4036733-P1	121	3009	3	TR01/	
4036749		3011			
4036749-P1	101		6	TR08/	
4036749-P2	101	3011	6	TR08/	
4036754-P1	103	3010	8	TR08/	
4036754-P2	103	3010	8	TR08/	
4036831-P1	105	3012	4	TR03/	
4036832-P1	105	3012	4	TR03/	
4036887-P8	123A	3122	20	TR21/	
4036923-P1	160	3006	51	TR17/	
4036923-P2	160	3006	51	TR17/	
4036924-P1	123A	3122	20	TR21/	
4036924-P2	123A	3122	20	TR21/	
4036937-P1	100	3005	1	TR05/	
4036937-P2	100	3005	1	TR05/	
4036962-P1	160	3006	51	TR17/	
4036962-P2	160	3006	51	TR17/	
4036963-P1	160	3006	51	TR17/	
4036963-P2	160	3006	51	TR17/	
4036965-P1	160	3006	51	TR17/	
4036965-P2	160	3006	51	TR17/	

Original Part Number	ECG	SK	GE	IR/WEP	HEP
4037145-P1	102A	3004	2	TR85/	
4037145-P2	102A	3004	2	TR85/	
4037289-P1	101	3011	5	TR08/	
4037289-P2	101	3011	5	TR08/	
4037289-P21				TR08/	
4037410-P1	160	3006	51	TR17/	
4037410-P2	160	3006	51	TR17/	
4037586-P1	123A	3122	20	TR21/	
4037586-P2	123A	3122	20	TR21/	
4037594-P1	121	3009	3	TR01/	
4037607-P1	121	3009	3	TR01/	
4037647-P1	186	3054	28	TR55/	
4037647-P2	186	3054	28	TR55/	
4037764-P1	160	3006	51	TR17/	
4037764-P2	160	3006	51	TR17/	
4037800-P1	123A	3122	20	TR21/	
4037800-P2	123A	3122	20	TR21/	
4037804-P1	102A	3004	2	TR85/	
4037804-P2	102A	3004	2	TR85/	
4037839-P1	101	3011	6	TR08/	
4037839-P2	101	3011	6	TR08/	
4037993-P1	100	3005	1	TR05/	
4037993-P2	100	3005	1	TR05/	
4038256-P1	103A	3010	5	TR08/	
4038256-P2	103A	3010	5	TR08/	
4038260-P1	100	3005	1	TR05/	
4038260-P2	100	3005	1	TR05/	
4038264-P1	101	3011	7	TR08/	
4038264-P2	101	3011	7	TR08/	
4038359-P1	160	3006	51	TR17/	
4038359-P2	160	3006	51	TR17/	
4038406-P1	160	3006	51	TR17/	
4038406-P2	160	3006	51	TR17/	
4080187-0502	152	3054		TR76/701	S7002(2)
4080187-0504	130			TR59/	
4080187-0506	196	3054		TR92/701	S5019
4080187-0507	130	3054		TR59/701	S7002
4080320-050B	130	3027		TR59/	
4080320-0501	130	3027		TR59/248	
4080320-0504	130			TR59/	
4080627-0501	175			TR81/	
4080835-0002	196	3054	23	TR92/701	S5019
4080838-0001	175	3026	12	TR81/241	S5019
4080838-0002	175	3026		TR81/	S5019
4080838-002	196			TR92/	S5019
4080838-2	175	3026		TR81/241	
4080838-3	175	3026		TR81/241	
4080866		3054			
4080866-0001		3054		TR26/701	
4080866-0001(M)	130			TR76/247	
4080866-0001(P)				TR75/	
4080866-0002(M)	130				
4080866-0002(P)				TR59/	
4080866-0004		3054			
4080866-0006		3054		TR76/701	S5000
4080866-0006(M)	130			TR59/247	
4080866-0006(P)	152			TR76/	
4080866-0007	175			TR81/	S5019
4080866-000A	130	3027		TR59/	
4080866-003	152				
4080866-006	152				
4080866-009	152				
4080866-0012	196	3054	28	TR92/	S5000
4080866-0013	152	3041	66	TR76/	S5000
4080866-1	175	3026		TR81/241	
4080866-2	175	3026		TR81/241	
4080866-4	152				
4080866-8006	130	3027		TR59/	
4080873-0001	184		29	TR76/	S5000
4080879-0001	152	3054	28	TR76/701	S5006
4080879-0006	196	3041	28	TR92/701	S5000
4080879-0011	196	3054			
4080879-0015	175	3054		TR81/	S5000
4082501-0001	104MP	3009	16	TR01MP/624	G6005
4082501-0001A	121	3009		TR01/	
4082501-001	104	3009		TR01/	
4082626-0001	912	3543			
4082664-0001	726				
4082665-0001	723	3144			
4082665-0002	723				
4082665-0003	723	3144			
4082665-1	723	3144			
4082665-2	723	3144			
4082665-3	723	3144			
4082671-0002	172			TR69/S9100	S9100
4082799-0001	789	3078			
4082799-0002	789	3078			
4082799-1	789	3078			
4082799-2	789	3078			
4082802-0001	781	3169			
4082802-0002	781	3169			
4082802-1	781	3169			
4082802-2	781	3169			
4082873-0001	197	3083	29	TR56/	S5006
4082873-0002		3083			

Original Part Number	ECG	SK	GE	IR/WEP	HEP
4082886-0002	184	3027			S5000
4082886-0003		3027			
4082886-001	130	3027			
4082886-002	130				
4082886-3	130				
4084114-0001	222	3050			F2004
4084114-0002	222	3050			F2004
4084117-0001	949				
4084117-0002	949				
4089549-0001	949				
4089549-001	949				
4090187-0502	130MP			TR59MP/	
4090605-1			IC27		
4222158				TR25/	
4223055		3027			
4302685		3025			
4350052-1	703A				
4361720	130				
4450023	130				
4450023-P3	130				
4450023-P4	130				
4450023-P5	130				
4450023-P6	130				
4450026-P5	128				
4450040P1	223				
4511424	7432				
4522322G		3005			
4526757C		3007			
4550106-001	152				
4550106-003	152				
4590076B		3005			
4663001A905	7474				
4663001A909	7404				
4663001A911	7440				
4663001A912	7410				
4663001A915	7430				
4663001D907	7400				
4686171-3				TR21/	
4720787-001		3024			
4813466	123A	3122	20	TR21/735	S0015
4822354	130				
4822354J	130				
4832800	181	3036			S7004
4906071	123A				
4906072	123A				
4906073	123A				
4906093	199				
4907975		3011			
4907975B		3011			
4907976	102A	3004	2	TR85/250	
4907977A		3011			
4907977C		3011			
4907978		3004			
4907981		3004			
4907982		3124			
4914118A		3004			
4914296	9906				
4915702	7416				
4915705	7411				
4999774	121			TR01/	G6005
4999775					G6005
4999885	213				G6010
4999887	102A			TR85/	G0006
5049911	128				
5073004	102A	3004	2		G0005
5076204	74S00				
5076205BZ	74S20				
5113642	909	3590			
5165440	9936				
5175460	7408				
5214096		3027			
5214534		3036			
5294477-1	123A				
5294477-2	123A				
5302347-1		3024			
5320003	175	3026		TR81/241	S5019
5320004					S0015
5320011	102A			TR85/	
5320022	123A	3124	20	TR21/735	S0030
5320023	123A	3124	20	TR21/735	S0014
5320023H	123A	3124	20	TR21/	S0014
5320024	123A	3124	20		S0014
5320026	123A	3124	20	TR24/	S0014
5320028					S0014
5320031	220	3116			
5320032					F0021
5320042H	159	3114	21	TR20/	S5022
5320043H	159	3114	21	TR20/	S5022
5320051	123A	3046	17	TR21/735	S0016
5320064	123A	3124	17	TR21/	S0030
5320064H	123A	3020	17	TR21/	S0014
5320067	123A	3124	20	TR21/	S0014
5320074	123A	3038		TR21/735	S0020
5320101	102A	3004	52	TR85/	G0005
5320111	159	3114		TR20/717	S0019

Original Part Number	ECG	SK	GE	IR/WEP	HEP
5320141	102A			TR85/	
5320151	128	3024	18	TR87/243	S3001
5320205					G0011
5320241	123A	3124		TR21/735	S0014
5320295	103A	3004	59	TR08/	G0011
5320295H	103A		54		
5320296	102A	3004	52	TR85/250	G0005
5320305	102A	3004	53	TR85/	G0007
5320305H	102A		54		
5320306	103A	3010	8	TR08/724	G0011
5320326	123A	3124	20		S0016
5320326H	107	3018		TR21/	S0016
5320328	108	3018	17		
5320361	103A	3010		TR08/724	G0011
5320372	123A	3122		TR21/735	S0015
5320372H	123A		18	TR21/	S0015
5320373	123A	3122	18	TR21/735	S0015
5320422H	152	3054	28	TR76/	S5000
5320432	152	3054		TR76/	S3001
5320433	152	3054	28	TR76/701	S5000
5320475	103A	3010	54		G0011
5320485	102A	3004	54		G0005
5320492H	152	3054	28	TR76/	S5000
5320501	128	3047	45	TR87/	S0015
5320511	195	3049	45	TR65/	S3001
5320583	133	3112	FET1	FE100/	F0010
5320592	159	3025	21	TR28/	S5013
5320612	128	3024	18	TR25/	S5014
5320622		3122	63	TR25/	S0025
5320632	187	3083	29	TR56/	
5320642	186	3054	28	TR55/	S3024
5320651	123A	3122	20	TR21/	S0015
5320671	152	3054	66	TR76/	S5000
5320702	132	3116	FET2	FE100/	F0021
5320791					S3024
5320851	107	3018	20	TR24/	S0016
5320861	107	3018	20	TR24/	S0016
5340001	123A	3124		TR21/	
5350121	1029				
5350132	1030				
5350135	1030				
5350136	1030				
5350141	1031				
5350151	1032				
5350152	1032				
5350161	1033				
5350182	1075				
5350211	1043				
5350231	1058				
5351041	722		IC9	IC513/	C6056P
5351042	722				
5351051	1041				
5351062	1072				
5406665-P2	102A		2	TR85/	
5406665-P5	102A		2	TR85/	
5490860P1	105			TR03/	
5492153-1	105			TR03/	
5492153-P1			4	TR03/	
5492639-P1	103		8	TR08/	
5492639-P2	103		8	TR08/	
5492653-P1	101		7	TR08/	
5492653-P2	101		7	TR08/	
5492655-P1	101		6	TR08/	
5492655-P2	101		6	TR08/	
5492655-P3	101		6	TR08/	
5492655-P4	101		6	TR08/	
5492655-P5	101		6	TR08/	
5492655-P6	101		6	TR08/	
5492659-P1	101		7	TR08/	
5492659-P2	101		7	TR08/	
5493158-1	121		3	TR01/	
5493957-P1	160		51	TR17/	
5493957-P2	160		51	TR17/	
5493957-P3	160		51	TR17/	
5493957-P5	160		51	TR17/	
5493957-P6	160		51	TR17/	
5496663-P1	121		3	TR01/	
5496665-P1	102		2	TR05/	
5496665-P2	102		2	TR05/	
5496665-P3	102		2	TR05/	
5496665-P4	102		2	TR05/	
5496665-P5	102		2	TR05/	
5496665-P6	102		2	TR05/	
5496666-P1	102		2	TR05/	
5496666-P2	102		2	TR05/	
5496666-P3	102		2	TR05/	
5496666-P4	102		2	TR05/	
5496666-P5	102		2	TR05/	
5496666-P6	102		2	TR05/	
5496666-P7	102		2	TR05/	
5496666-P8	102		2	TR05/	
5496667-P1	102		2	TR05/	
5496667-P2	102			TR05/	
5496688-P1	105		4	TR03/	
5496774-P1	102		2	TR05/	

Original Part Number	ECG	SK	GE	IR/WEP	HEP
5496774-P2	102		2	TR05/	
5496774-P3	102		2	TR05/	
5496774-P4	102		2	TR05/	
5496774-P5	102		2	TR05/	
5496774-P6	102		2	TR05/	
5496839-P1	102		2	TR05/	
5496939-P1	121		3	TR01/	
5496939-P2	121		3	TR01/	
5518017	220				
5724125		3024			
5729031	105			TR03/	
5760001-014		3004			
5760003-020		3122			
5955748	102A	3004	2	TR85/250	
5958539	176	3123		TR82/	G0007
5983945A	105	3012		TR03/	
5985432	105	3012		TR03/	
5988414	105	3012		TR03/	
5988977	105	3012		TR03/	
5989171	105	3012		TR03/	
5989615	105	3012		TR03/	
5989692	105	3012		TR03/	
5989693	105	3012		TR03/	
6088501-3	912				
6100724-2	100	3005		TR05/254	
6101161-1	130				
6212096	152	3054	28	TR76/	S5000
6212839	123A	3018	20	TR21/	S0015
6212922	108	3018	20	TR21/	S0020
6218945	123A	3124	20	TR21/	S0015
6460006	102				
6460037	102				
6480000	130				
6480001	121				
6480004	121				
6480006	128				
6490001	175				
6494030	223				
6494612		3054			
6860870		3007			
6902021H25	152				
6902021H26	153				
6933243-001	175				
6984590	123A				
6984600	123A				
6993400	159				
6993630	159				
6993650	123A				
7002453	123A				
7011200-02	9933				
7011201-02	9949				
7011203-02	7475				
7011203-03	7475				
7011507	123A				
7011507-00	123A				
7011515	129				
7012109-01	923D				
7012128	9936				
7012128-02	9936				
7012130	9109				
7012131	9112				
7012132	7482				
7012133-02	7483				
7012142-03	74193				
7012166	7402				
7012167-02	7490				
7012411	129				
7020202	159				
7020203	153				
7026011	108	3018	20	TR95/56	
7026012	108	3018	20	TR95/56	
7026013	108	3018	20	TR95/56	
7026014	123A	3124	20	TR21/735	
7026015	123A	3124	20	TR21/735	
7026016	123A	3124	20	TR21/735	
7026018	130	3027		TR59/247	
7026019	129	3025	22	TR88/242	
7026020	123A	3124	20	TR21/735	
7026024		3122			S5012
7026029		3025			
7030105	102A		53	TR85/250	
7071021	123A	3124	20	TR21/7	S0024
7071031	123A	3020	20	TR21/	S0025
7121105-01	123A				
7172.01		3004			
7172.13		3007			
7172.32		3011			
7172.54		3012			
7210027-003	102	3004			
7210036-001	102				
7269847	103A		54	TR08/724	
7274653	121	3009	16	TR01/232	
7276211	100	3005	2	TR05/254	
7276605	105	3012	4	TR03/233	
7277066	103A		54	TR08/724	

Original Part Number	ECG	SK	GE	IR/WEP	HEP
7278421	100	3005	1	TR05/254	
7278422	102A	3004	53	TR85/250	
7278423	102A	3004	2	TR85/250	
7279003	105			TR03/	
7279005	105	3012	4	TR03/233	
7279007	105	3012	4	TR03/233	
7279009	105	3012	4	TR03/233	
7279011	105	3012	4	TR03/233	
7279013	105			TR03/	
7279017	105	3012	4	TR03/233	
7279025	105	3012	4	TR03/233	
7279027	105	3012	4	TR03/233	
7279031		3012	4	TR03/233	
7279033	105	3012	4	TR03/233	
7279039	104	3009	3	TR01/624	
7279049	104	3009	3	TR01/624	
7279069	179			TR35/G6001	
7279073	105	3012	4	TR03/233	
7279076	105	3012	3	TR03/233	
7279281	176	3123		TR82/	
7279293	105	3012	4	TR03/233	
7279298	105	3012	4	TR03/233	
7279379	100	3005	2	TR05/254	
7279461		3008			
7279559		3005			
7279556	105	3012	4	TR03/233	
7279779	126	3006	50	TR17/635	
7279779(G.M.)	123A			TR21/735	
7279780	160	3008	51	TR17/637	
7279781	160	3007	2	TR17/637	
7279782	100	3005		TR05/254	
7279788	100	3005	2	TR05/254	
7279789	100	3005	2	TR05/254	
7279793	105	3012	4	TR03/233	
7279940	100	3005	2	TR05/254	
7279941	102A	3004	2	TR85/250	
7280281	105	3012	4	TR03/233	
7281307	100	3005	2	TR05/254	
7281308	100	3005	2	TR05/254	
7281309	100	3005	2	TR05/254	
7281310	102A	3004	2	TR85/250	
7281806	128				
7281891	100	3005	2	TR05/254	
7282315	105	3012	4	TR03/233	
7283801		3004			
7284137	123A	3122		TR21/735	
7284513	160	3006		TR17/637	
7284751	102A	3004	2	TR85/250	
7284941		3004			
7285663	121	3009	3	TR01/232	
7285774	104	3009	3	TR01/233	
7285776	121			TR01/	
7285778	104	3009	3	TR01/233	
7286858	123A	3122	20	TR21/735	
7287106				TR35/G6001	
7287107	179			TR35/G6001	
7287110	121	3009	16	TR01/232	
7287112	179			TR35/	
7287117	179			TR35/G6001	
7287452	123A		20	TR21/735	S0015
7287939		3008			
7287940	126	3008		TR17/635	
7288072	105	3012	4	TR03/233	
7288073	105	3012	4	TR03/233	
7288076	105	3012	4	TR03/233	
7288079	105	3012	4	TR03/233	
7289041	121	3009	16	TR01/232	
7289047	104	3009	3	TR01/824	
7289079			3	TR01/	
7289097	179			TR35/G6001	
7290593	121			TR01/	
7290594	104	3009	3	TR01/624	
7291252	104	3009	3	TR01/624	
7292308	160	3006	51	TR17/637	
7292683	179			TR35/G6001	
7292684	127			TR27/235	
7292689	179			TR35/G6001	
7292690	121			TR01/	
7292955	121		3	TR01/	
7293818	162			TR67/707	
7293819	162			TR67/707	
7294133	102A	3004	2	TR85/250	
7294796	121	3009	16	TR01/232	
7294910	161	3018	20	TR83/719	S0014
7295195	108	3018	20	TR95/56	
7295196	108	3018	20	TR95/56	
7295197	123A	3122	20	TR21/735	
7296314	123A	3124	20	TR21/735	
7296811	123A				
7297043	104	3009	3	TR01/624	
7297053	123A	3122	20	TR21/735	
7297054	123A	3122	20	TR21/735	
7297092	104	3009	3	TR01/624	
7297093	104	3009	3	TR01/624	
7297347	179			TR35/G6001	

Original Part Number	ECG	SK	GE	IR/WEP	HEP
7297980	161	3018	20	TR83/719	S0014
7298079	121	3009		TR01/	
7299720	105	3012	4	TR03/233	
7299771	121MP				
7299780	179			TR35/G6001	
7299803	121MP				
7301660	179			TR35/G6001	
7301661	179			TR35/	
7301664	162			TR67/707	
7301665	162			TR67/	
7301666	162			TR67/707	
7302024	159				
7302699	162			TR67/707	
7303105	102A		53	TR85/250	S0012
7303120	123A	3122	20	TR21/735	S0014
7303304	162			TR67/	
7304149	163			TR67/	
7304380	123A				
7305468	107				
7305476	726	3022			
7305783			3	TR16/	
7306982	130	3027		TR59/247	
7309160	130	3027			
7311074	128	3044			
7311325	709		IC11	IC505/	C6082l
7311350	128	3512			
7312294	718	3159			
7313063	162				
7313568	721				
7314584	123A				
7397027	703A				
7528046-P4	9932				
7528048-P4	9946				
7528153-P4	9948				
7528156P1	910				
7528156-P3	9910				
7528157P1	909	3590			
7528158-P4	9910				
7528159-P4	9944				
7528160-P4	9951				
7528374P3	9951				
7570003	121		3	TR01/	
7570003-01	121	3009	16	TR01/232	G6005
7570004	123A	3124	20	TR21/735	
7570004-01	123A	3124	20	TR21/735	S0014
7570005	123A	3124	20	TR21/735	
7570005-01	123A	3124	20	TR21/735	S0015
7570005-02	123A	3020	20	TR21/735	S0015
7570005-03	123A	3020	20	TR21/735	S0015
7570008	123A	3020	20	TR21/735	
7570008-01	123A				
7570008-02	123A	3020	20	TR21/735	S0015
7570009	128				
7570009-01	123A	3020	20	TR21/735	S0015
7570009-21	123A	3020	20	TR21/735	S0014
7570013	159				
7570013-01	159	3025	21	TR20/717	
7570022-01	175	3026	14	TR81/241	
7570030-01	185	3083		TR77/S5007	S5006
7570031-01	152	3041		TR76/701	S5000
7570032-01	129	3025		TR88/242	S5013
7576004-01	159	3114	21	TR20/717	
7576015-01	123A	3122	20	TR21/735	
7576015-02	199				
7576015-03	199				
7576015-04	199				
7576015-05	199				
7576016-01	128				
7601010	128				
7835297-01				TR05/	
7840540-1	123A				
7851316-01	102A			TR85/	
7851317-01	102A			TR85/	
7851318-01	103A			TR08/	
7851319-01	103A			TR08/	
7851320-01	158			TR84/	
7851321-01	103A			TR08/	
7851322-01	104			TR01/	
7851323	160			TR17/637	G0008
7851324-01	123A			TR21/	
7851325	123A	3122		TR21/735	S0015
7851326	123A	3122		TR21/735	S0011
7851327	123A	3122		TR21/735	S0015
7851379-01	123A			TR21/	
7851380-01	123A			TR21/	
7851467-01	103A			TR08/	
7851650-01	161			TR83/	
7851651-01	161			TR83/	
7851652-01	159			TR20/	
7851949-01	123A	3020		TR21/735	S0015
7851952-01	123A	3122		TR21/735	S0015
7851953-01	123A	3122		TR21/735	S0015
7851954-01	159			TR20/	
7851955-01	131	3052		TR50/642	G6016
7851956-01	128			TR87/	

Original Part Number	ECG	SK	GE	IR/WEP	HEP
7852452-01			63		S0015
7852454	123A			TR21/	
7852454-01	123A	3122		TR21/735	S0024
7852455-01	123A	3122		TR21/735	S0025
7852459-01	123A	3122		TR21/735	S0014
7852775-01	102A	3004		TR85/250	G0008
7852776-01	102A	3004		TR85/250	G0005
7852778-01	102A	3004	2	TR85/250	G0007
7852779-01	102A	3004	53	TR85/250	G0005
7852781-01	123A	3020	20	TR21/735	S0030
7852897-01	160			TR17/637	
7852899-01	126	3006		TR17/635	G0008
7852900-01	160	3006		TR17/637	G0003
7853092-01					S0015
7853094-01					S0015
7853351-01	102A	3004	2	TR85/	G0005
7853352-01	102A	3004	52	TR85/	G0007
7853354-01	102A	3004	53	TR85/	G0005
7853356-01	102A	3004	52	TR85/	G0005
7853463-01	123A	3020	20	TR21/735	S0014
7853464-01	123A			TR21/735	
7853465-01	123A	3122	18	TR21/	S0015
7855283-01(AMP)	102A			TR85/	
7855291-01	123A	3020		TR21/735	S0015
7855292-01	123A	3020		TR21/735	S0015
7855293-01	123A	3020		TR21/735	S0015
7855294-01	123A	3020		TR21/735	S0015
7855295-01	131	3052		TR94/642	G6016
7855296-01	102A	3004		TR85/250	G0005
7855297-01	102A	3004		TR85/250	G0005
7855298-01	152	3041		TR76/701	S5000
7902045-0		3027			
7902310	128	3024			
7910070-01	102A			TR85/	G0005
7910071-01	102A	3004	52	TR85/	G0005
7910072-01	121			TR01/	G6003
7910073-01	184	3190			S5000
7910108-01	161			TR83/	S0016
7910122-01	1003				
7910134-01	132			FE100/	F0015
7910267-01			23		S5012
7910268-01			FET2		F0015
7910270-01			20		S0011
7910271-01			17		S0015
7910272-01			17		S0015
7910273-01			63		S0015
7910274-01			18		S5014
7910584-01	123A	3020		TR21/735	S0014
7910585-01	123A	3020		TR21/735	S0014
7910586-01	123A	3020		TR21/735	S0014
7910587-01	123A	3020		TR21/735	S0014
7910588-01	102A	3004		TR85/250	G0007
7910589-01	102A	3004		TR85/250	G0006
7910780-01	108		20	TR95/	S0016
7910781-01	107		20	TR70/	S0016
7910783-01	126	3006	50	TR17/	G0009
7910801-01	102A	3004	52	TR85/	G0007
7910803-01	158		53	TR84/	G6011
7910804-01	123A	3004	52	TR21/	G0005
7910875-01			21		S5013
7913605	162			TR67/	
7914009-01	107	3018	20	TR70/	S0016
7914010-01	107	3018	20	TR70/	S0014
7930945		3054			
7932367	719				
7932515	159				
7932980	709				
7935181	744				
7936256	123A				
7936331	123A				
7936983		3025			
7937586	130	3027			
7937762	721				
7938318	133				
7939165	161				
7939185	734				
7939186	159				
8000736	128	3024	18	TR87/243	
8000737	128	3024	18	TR87/243	
8002866	175				
8004265	128				
8010490	126	3008	2		G0008
8010520	100	3008	51	TR12/	
8010530	126	3008	2		G0008
8014711	160	3008		TR17/	G0008
8014712	100	3008	51	TR12/	
8015613	129	3025		TR88/242	S0019
8020322	102A	3004	2	TR85/250	
8020324	102A	3004	2	TR85/250	
8020333	102A	3004	2	TR85/250	
8020334	102A	3004	2		G0005
8020540	102A	3004	2	TR85/250	G0007
8020560	102A	3004	2	TR85/250	G0005
8021897	102A	3004	2	TR85/250	
8021898	102A	3004	2	TR85/250	

Original Part Number	ECG	SK	GE	IR/WEP	HEP
8022630	102A	3004	2	TR85/250	
8022631	102A	3004		TR85/250	G0005
8023643	102A			TR85/	G0005
8023892	102A	3003		TR85/250	
8024152	158	3004		TR84/630	G6011
8024250	127	3034	25	TR27/235	
8024390	160	3004	2	TR17/637	G0005
8024390(AIWA)	102A			TR85/250	
8024400	102A			TR85/250	G0005
8031825	108	3020		TR95/56	S0016
8031836		3018			S0016
8031837		3018			S0015
8031839		3018			S0015
8033690	123A	3122		TR21/735	S0015
8033696	123A	3020	10	TR21/735	S0015
8033720	123A	3020	10	TR21/735	S0015
8033730	123A	3020	17	TR21/735	S0015
8033803	108	3018	61	TR95/	S0016
8033804	108	3039	20		S0016
8033943	108	3018	17	TR95/	
8033944	123A	3018	17	TR21/	S0014
8034903	152	3054		TR76/701	S5019
8035180					S5020
8036683	107	3018	17	TR24/	S0025
8037330	123A	3020	10	TR21/735	S0025
8037332	123A	3122		TR21/735	S0025
8037333	123A	3020	17	TR21/735	S0025
8037343	128			TR87/243	S5026
8037353	123A	3020	10	TR21/735	S0014
8037722		3018			S0016
8037723		3018			S0016
8111227	159				
8111230	159	3114	21		S0013
8112023	102A	3004	52		G0005
8112027	102A	3004	52		G0007
8112028	102A				
8112071	102A	3004	53		G0005
8112090	102A	3004	52		G0005
8112143	131	3052	30		G6016
8112146	102A	3004	53		G0006
8112161	102A	3004			
8112162	102A	3004	52		G0006
8113024	192				
8113034		3039	20		S0015
8113051	123A				
8113052	123A				
8113060	123A				
8113102	152				
8113134	123A				
8114024	192				
8114031	123A				
8115009	159	3114	22		S0013
8121897				TR85/	
8121898				TR85/	
8122630				TR85/	
8122631				TR85/	
8150485		3021			
8249600	161	3018		TR83/719	
8378759	128				
8398315	128				
8421133	128				
8469903		3005			
8505870-1	9910				
8507465-1	726				
8508309	129			TR88/	
8508331RB	903				
8508403-1	784				
8510671-1	160				
8510671-2	160				
8510671-3		3005			
8510671-4	160				
8510694-1	130				
8510744-1	101	3011			
8510744-2		3011			
8510747		3007			
8510747-4	126				
8511724-3	176				
8511759-1	160	3004			
8511759-2		3004			
8511759-3		3004			
8512001-2	101	3010			
8512049-5		3005			
8512049-6		3005			
8513410		3004			
8516861-1	102				
8516861-P3		3004			
8516986	102	3005			
8516986-1	102				
8516986-2	102				
8521502-1	101				
8521502-2	101				
8521502-3		3011			
8521502-4	101				
8522468-1	123A	3039			
8522468-2		3039			

Original Part Number	ECG	SK	GE	IR/WEP	HEP
8522468-3		3039			
8522468-5		3122			
8523435-3		3123			
8523438		3011			
8524402-1	101	3011			
8524402-2		3011			
8524402-3		3011			
8524402-4	101				
8524403-1		3005			
8524403-2		3005			
8524440-1		3004			
8524440-2	102				
8524457	128	3024			
8525521-1		3122			
8525521-3		3122			
8525521-4		3122			
8526849-1	100	3123			
8526849-2		3004			
8531261-2-0		3009			
8531261-4-0		3009			
8538640	160	3007		TR17/637	
8539622-1		3006			
8539650-1		3008			
8547892-1001		3039			
8555270-1001		3025			
8556188	108				
8590646		3036			
8722248-2	199				
8878302	744				
8935910-1F		3005			
8935910-1K		3005			
8935911-1D		3005			
8935913		3005			
8935914-1		3004			
8935915-1E		3004			
8935915-2		3004			
8935915-3		3004			
8935919-1DEP		3122			
8945077-12	102	3004			
8975102-1		3004			
8975102-2		3004			
8975102-3		3004			
8975103-2	101	3011			
8975106		3004			
8975106-1		3004			
8975106-2		3003			
8975106-3		3004			
8975106-5		3004			
8975106-6		3004			
8975106-P5K		3003			
8975158-1	102A			TR85/	
8975158-3		3005			
8978392		3003			
8989441-2	160	3005			
8989441-3		3005			
8989457-1	126				
9000630	181	3036			
9000940	159				
9001225-02	910D				
9001324	152				
9001345-02	7493				
9001346	7411				
9001349-02	7486				
9001349-03	7486				
9001549-02	7440				
9001551-02	7404				
9001551-03	7404				
9001567	9157				
9001567-02	9157				
9001570-02	7403				
9001630	123A				
9001638	159				
9001756	152				
9001757	153				
9002097-03	74107				
9002159	123A				
9003091-02	7410				
9003091-03	7410				
9003096-02	7450				
9003096-03	7450				
9003097-02	74107				
9003148-01	9930				
9003148-02	9930				
9003149-02	9946				
9003150	9962				
9003150-02	9962				
9003150-03	9962				
9003151	7400				
9003151-03	7400				
9003152	7474				
9003152-01	7474				
9003234-04	7438				
9003398-03	7408				
9003398-04	7408				
9003420-03	7496				

Original Part Number	ECG	SK	GE	IR/WEP	HEP
9003445-03	7492				
9003642-03	7430				
9003911	9932				
9003911-01	9932				
9003911-03	9932				
9004017-01	219				
9004075-03	7401				
9004076	7420				
9004076-03	7420				
9004076-04	7420				
9004093-03	7473				
9004300-03	7476				
9004360-03	7454				
9004508-01	159				
9004896-04	7453				
9004898-04	7411				
9004915-04	219				
9006527-01	219				
9007038	159				
9008964-01	123A				
9100502	102A	3004	2	TR85/250	
9100621	102A	3004	2	TR85/250	
9100706	102A	3004	2	TR85/250	
9100944	102A	3004	2	TR85/250	
9101600	915				
9112857		3039			
9176494	123A				
9216200-13J		3004			
9340311	130	3027			
9340311-D5514	130				
9340388	159				
9340924	743				
9341238	737				
9341510	152				
9341767	159				
9341834	129				
9341899	806				
9342291	159				
9671292		3054			
9837422SC372				TR21/	
9837432SC371				TR25/	
9845632SB56				TR05/	
9845632SC735				TR21/	
9845772SC380				TR21/	
9845902SC373				TR21/	
9845912SC733				TR21/	
9845932SB56				TR05/	
9845932SC735				TR21/	
9845982SC733				TR21/	
10000020	123A				
10015595	123A				
10017663	102				
10018807-002B		3024			
10022104-101	123A				
10027973-101	107				
10106058	123A				
10112562	159				
10112563	9900				
10176209	9914				
10180722	123A				
10182330	159				
10183001	199				
10525796		3004			
10541284	9910				
10545502	123A				
10545506	199				
10635241		3008			
10641140	159				
10644417	175	3026			
10644433	123A				
10646032	130				
10653086	175				
10658276	9930				
10658278	9936				
10658279	9944				
10658280	9945				
10658281	9946				
10658282	9951				
10669666	159				
10670399	159				
10713642	130	3027			
10728228		3026			
10814788	128				
10838753		3036			
10849792	123A				
10896074	123A				
10896173		3036			
11018181	124				
11040202	9902				
11041003-1	181				
11069934A	130			TR59/	
11076438		3039			
11132875	102				
11153162		3036			
11166527	130				

Original Part Number	ECG	SK	GE	IR/WEP	HEP
11166535	128	3024			
11194628	181	3036			
11198132	123A	3122			
11220009		3004			
11220009/7825	102A		52		
11220018		3004			
11220018/7611	102A	3004	2	TR85/	
11220022		3004			
11220022/7611	102A		2	TR85/	
11220046		3018			
11220046/7825	123A		17	TR22/	
11220061		3004			
11220061/7825	158		53		
11220076		3004			
11220076/7825	102A		52		
11220106		3004			
11220106/7611	158		53	TR84/	
11233335	175				
11242096	909				
11253441	128				
11253588	9912				
11292312	74H00				
11292313	74H40				
11292314	74H51				
11292315	74H54				
11369564	74H05				
11473717		3024			
11667821	102				
11706998-2	199				
11718319	128				
11728100	941				
11729469	716				
11783453	102				
11785417		3009			
11790516		3008			
11794864		3011			
11800182	102				
11802400	123A				
11802500	123A				
11858909	102				
12090924	102A	3004		TR85/250	
12363800	175	3026			
12873428		3044			
12900510B		3004			
12965471	123A				
12994851	123A				
13020000	102A	3004	2	TR85/250	
13020002	102A	3004	2	TR85/250	
13035807-1	123A				
13040089	103A	3011		TR08/724	G0011
13040095	102A	3004		TR85/250	G0005
13040096	103A	3010		TR08/724	G0011
13040216	123A			TR21/735	S0024
13040236	102A	3004		TR85/250	G0005
13040304	108	3018		TR95/56	S0016
13040313	123A	3020		TR21/735	S0015
13040317	123A	3020		TR21/735	S0015
13040318	123A	3020		TR21/735	S0015
13040347	102A+103	3010		TR85/250	G0011
13040349	131			TR94/642	G6016
13040352	102A	3004		TR85/250	G0005
13040357	123A			TR21/735	S0015
13040362	108	3018		TR95/56	S0016
13040421	108	3018		TR95/56	S0016
13040429	129	3025		TR88/242	S0013
13040456	158			TR84/630	G6011
13040459	108	3018		TR95/56	S0016
13104367	130				
13104560	128				
13104561	129				
13217724	102				
13217732	102				
13217765		3009			
13279744				TR78/	
13297761				TR60/	
13443767	102				
13447321	102				
14020416		3024			
14317184	181				
14500001-001	9946				
14500001-002	9949				
14500001-003	9936				
14500001-004	9961				
14500004-001D	9936				
14500004-001P	9936				
14500004-002	9937				
14500004-003	9935				
14500005-001	9944				
14500016-001	941D				
14500016-002	941M				
14500022-001	123				
14628601	124				
14711331-1				TR25/	
14714760	123A				
14714786	123A				

Original Part Number	ECG	SK	GE	IR/WEP	HEP
14724501				TR21/	
14734248	909				
14736221	159				
14798029	159				
15010121-1		3024			
15038443	123A				
15039456	123A				
15039464	123A				
15059850	124				
15077183	175				
15102718	102				
15105300	309K				
15109200	74S04				
15222101-11		3004			
15222101-31		3006			
15222103-00		3006			
15222109-21		3007			
15222110-21		3006			
15222112-00		3004			
15222113-28		3004			
15222119-21		3006			
15222144-00		3006			
15222144-11		3006			
15222144-35		3006			
15222148-21		3006			
15222148-31		3006			
15222165-00		3004			
15222166-00		3007			
15222174-00		3006			
15222237-20		3020			
15263817		3044			
15348931	175				
16147191-229		3020			
16171191-368		3020			
16207190-405		3004			
16270092	123A				
16271033	9936				
16271041	9936				
16353343	736				
16520001-1	123A				
16566722SA201A				TR11/	
16566722SA202B				TR11/	
16763333-001	909				
16797300	123A				
16797301	123A				
16906600-01	159A				
17184600	7405				
17350065		3024			
17771400	128				
17809000	130				
17809000B		3027			
17942800-01	159				
17942800-801	159				
18151417	128				
18179900	123A				
18458117	941				
18525200	126				
19020056		3024			
19020065	152				
19020095	130				
19166123	9961				
20025153-77	123A	3020		TR21/735	
20030703-0701	123A	3122		TR21/735	
20030703-0702	159	3114		TR20/717	
20052600	123A			TR21/	
20088308	199				
20278600	130	3027			
20912578-1	176	3123		TR82/	
20918596-2	121	3009	16	TR01/232	
21828001-1C		3005			
22114210	192	3024	63	TR87/	
22114225	186	3041	28	TR55/	
22114253	159	3114	65	TR51/	
22114254	187	3193	29	TR56/	
22130009	128	3024			
22901513	108	3019		TR95/	
22902046	108	3018		TR95/	
23111006	121		3	TR01/	
23114001	108	3018	11	TR95/56	S0016
23114004	127	3034	25	TR27/235	
23114007	154	3018	20	TR78/712	S5026
23114009	127	3034		TR27/235	
23114011	121	3009	16	TR01/232	G6003
23114015	123A	3020	20	TR21/735	S0014
23114017	123A	3018	10	TR21/735	S0014
23114021	102A	3004	52	TR85/250	
23114023	108	3018	20	TR95/56	S0016
23114026	127	3034	25	TR27/235	G6016
23114028	154	3040	63	TR78/712	
23114031	108	3018	17	TR95/56	S0017
23114033	121	3009	16	TR01/232	G6016
23114034	108	3018	17	TR95/56	S0017
23114036	108	3019		TR95/56	S0016
23114038	175	3026	45	TR81/241	S5012
23114043	108	3019	11	TR95/56	S0016
23114044	233	3018	61	TR95/56	S5026
23114045	103A	3010		TR08/724	
23114046	123A	3020	20	TR21/735	
23114050	159	3025	21	TR20/717	S0013
23114051	159	3025	21	TR20/717	S0013
23114052	154	3040	18	TR78/712	S5026
23114053	154	3020	18	TR78/712	S5026
23114054	154	3020	18	TR78/712	S5026
23114056	108	3018	20	TR95/56	S0016
23114057	108	3018	20	TR95/56	S0016
23114058	128	3024		TR87/243	
23114060	108	3018	17	TR95/56	
23114061	102A	3004	52	TR85/250	
23114070	130	3027	19	TR59/247	
23114070A	130	3027		TR59/	
23114078	108	3018	17	TR95/56	S0020
23114081	129	3025	69	TR30/	S3003
23114082	108	3020	20	TR95/56	S0016
23114084	175	3021	12	TR92/	S5012
23114086	175			TR81/241	S5019
23114088	175			TR81/241	S5012
23114095	154	3020	63	TR78/712	S5026
23114097	127	3034		TR27/235	
23114100	130	3027		TR59/247	
23114104	233	3004	61	TR21/735	S5026
23114108	130	3027		TR59/	
23114109		3018		TR22/	S0017
23114118		3018	17	TR21/	S0014
23114119		3018	17	TR21/	S0014
23114124	159	3025	21	TR20/717	S0013
23114125	154	3040		TR78/712	S5025
23114126	107	3018	20	TR70/720	S0016
23114127	108	3039	20	TR95/	S0016
23114131	185	3191		/S5007	S5006
23114132	184	3190		/S5003	S5000
23114133	185	3191			S5006
23114134	184				S5000
23114136	159	3025	21	TR20/717	S0013
23114137	129	3025	69	TR30/	S3003
23114138	159	3114			S0012
23114155	123A	3018		TR21/735	S0015
23114157	107	3018	61	TR33/	S0016
23114163	161	3117	17	TR83/	S0017
23114164		3132	60	TR21/	S0020
23114165	107	3132	60	TR33/	S0020
23114171	108	3019	11	TR95/	S0016
23114172	108	3018	11	TR80/	S0016
23114180	108	3018	11	TR22/	S0016
23114181	108	3018	11	TR22/	S0020
23114195	128	3024	18	TR24/	S5026
23114200		3021			
23114208	162	3133			S5020
23114211	102A	3004	2	TR85/250	G0005
23114212	123A	3020	18	TR21/735	S0015
23114214	123A	3018	18	TR21/735	S0015
23114215					S0015
23114216	199	3020	18	TR21/735	S0015
23114217	199				S0015
23114220	175	3026	23	TR57/	S5019
23114221	175	3026	23	TR57/	S5019
23114232	128	3024	18	TR24/	S5026
23114238	108	3019	11	TR80/	S0016
23114248					S5011
23114249	171	3044	18	TR32/	S5025
23114250	171	3044	18	TR32/	S5025
23114251	171	3044	18	TR32/	S5025
23114255	123A				
23114258	199				S0015
23114266		3021			S5011
23114268	124	3021			S5011
23114275	123A	3018	20	TR21/	S0015
23114276	233	3018	20	TR21/	S0015
23114277	124	3021	12	TR23/	S5011
23114282	107	3018	20	TR24/	S0016
23114293	159	3025			
23114296	123A	3124			S0014
23114301	159	3114	21	TR19/	S0013
23114302	159	3114	21	TR30/	S0013
23114313	193	3025	67	TR30/	S5013
23114314	192	3024	63	TR24/	S5014
23114315	124	3021	12	TR81/	S5011
23114321	165	3114	38	TR93/	S5021
23114325	159				S0013
23114344	171	3104			S3021
23114550	159	3114		TR20/717	S0013
23114983	190				
23114994	192	3024	63	TR24/	S5014
23114995	193	3025	67	TR30/	S5013
23114999					S3021
23115057	123A	3020		TR21/735	
23115058	123A	3020	20	TR21/735	
23119003	711	3070			
23119004	710	3102		IC506/	
23119005	711	3070			C6069G
23119007	710	3102		IC506/	

Original Part Number	ECG	SK	GE	IR WEP	HEP
23119011	1101				
23119012	1004				
23119013	1109				
23119014	747				C6079P
23119016	1105				
23119017	1062				
23119019	749	3168			C6076P
23119022	1101				
23119023	1109				
23119025	748				
23119028	1105				
23119032	1109				
23119033	1109				
23119039	1129				
23119981	1128				
23119988	1133				
23119989	1134				
23119990	1128				
23119993	1130				
23119994	1132				
23119995	1131				
23119999	748				C6060P
23124037	108	3039		TR95/56	S0016
23126177					S0020
23126183	161	3018	39	TR70/	S0020
23126184					S0014
23126289	107			TR70/720	S0016
23126290	107			TR70/720	S0016
23126291	107			TR70/720	S0016
23126620	108	3019	11	TR83/	S0020
23200596-1	128	3024		TR87/	
23311006	104	3009	3	TR01/624	
23311066	121		3	TR01/	
24501000	123A				
24539800	159				
24551302	128				
24553600	123A				
24562000	123A				
24562001	123A				
24562100	123				
24562101	123A				
24562200	123A				
24562300	159				
24733362	129				
25114101	102A	3004		TR85/250	G0005
25114102	102A	3004		TR85/250	G0007
25114103	102A	3004		TR85/250	G0005
25114104	102A	3004		TR85/250	G0005
25114105					G0005
25114116	123A	3020	10	TR21/735	S0015
25114121	123A	3018	20	TR21/735	S0015
25114130	154	3045	10	TR78/712	S5026
25114143	124	3021	23	TR81/240	S5012
25114161	108	3020	20	TR95/56	S0025
26004001	123A				
26004301	159				
26004961	159				
26005121	159				
26010006	132			FE100/802	F0015
26010010	195	3048		TR65/243	S3001
26010016	159	3025		TR20/717	S0013
26010020	123A	3018		TR21/735	S0015
26010021	123A	3047		TR21/735	S0014
26010022	195	3048		TR65/	S3002
26010023	123A	3020		TR21/735	S0024
26010024	152	3054		TR76/701	S5000
26010026	107	3018		TR70/720	S0016
26010027	159	3025		TR20/717	S0012
26010041		3024			
26010056		3039			
26011310	159				
26316032	159				
26316033					S7003
26501505	123				
27119004	710				
27125080	123A	3020	20	TR21/735	S0015
27125090	123A	3020	20	TR21/735	S0015
27125110	102A			TR85/	
27125120	102A			TR85/	
27125140	123A	3020	20	TR21/735	S0015
27125150	102A	3004	20	TR85/250	S0005
27125160	123A	3020	20	TR21/735	S0015
27125170	158	3004		TR84/630	G6011
27125180					G0005
27125190					G0005
27125210	108	3039	11	TR95/56	
27125230	126		51	TR17/635	
27125240	100	3005	1	TR05/254	
27125250	123A		10	TR21/735	
27125260	102A	2		TR85/250	
27125270	123A			TR21/735	S0015
27125300	123A			TR21/735	S0015
27125310	103A	3010	8	TR08/724	G0011
27125330	102A	3004		TR85/250	G0005
27125340	102A	3004		TR85/250	G0007

Original Part Number	ECG	SK	GE	IR WEP	HEP
27125350	102A	3004		TR85/250	G0005
27125360	102A	3005		TR85/250	G0005
27125370	123A			TR21/735	
27125380	123A			TR21/735	
27125460	123A	3038	10	TR21/735	S0011
27125470	123A	3038	17	TR21/735	S0015
27125480	158	3004	2	TR84/630	G0005
27125490	103A	3010	8	TR08/724	G0011
27125500	123A			TR21/735	S0015
27125540	176	3004		TR82/238	G6011
27125550	102A	3004		TR85/250	G0005
27126060	131+155			/642	
27126090	121			TR01/	
27126100	130	3027		TR59/	
27126130	121			TR01/	G6005
27126130-12	121	3014		TR01/	
27126130(TRAN)	121			TR01/	
27126220	108	3039	11	TR95/56	
28101149	234				
28102128-002	199				
28102128-004	199				
28105203-001	152				
28105203-002	152				
28105207-001	234				
28105249-001	153				
28105789-001	199				
29004423	947				
30677435-001	909	3590			
30682592-001	234				
31210049-00	160	3006		TR17/	
31210077-44	160	3006		TR17/	
31210240-33	160	3006		TR17/	
31210240-44	160	3006		TR17/	
31210471-11	160	3006		TR17/	
31210506-11	160	3006		TR17/	
31210506-22	160	3006		TR17/	
31220054-00	102A	3004		TR85/	
32600025-01-08A	123A				
34002412	903				
34002412-3	903				
35024712	197				
35050411	198				
35050412	198				
35062712	198				
35393060-01	123A	3020		TR21/	
35393060-02	123A	3020		TR21/	
35393060-03	123A	3020		TR21/	
36001200	128				
36171100	123A				
36186200	915T				
36188000	7475				
36188700	74H00				
37001003	1047				
37001009	712	3072			
37001011	712	3072			
37002001	1046				
37003001	1089 □				
37004001	1048				
37006001	1077				
37007001	1076				
37008002	1091*				
37009006	1086**				
38970300	123A				
39126500	941				
42126551					C6056P
43020418	121MP				
43020731	128				
43021017	123A	3020	20	TR21/735	S5014
43021067	107	3018	20		S0015
43021083	123A	3124	20		S0033
43021168	129	3114	21		S5013
43021198	128	3024	20	TR87/243	S5014
43021415	128	3020	18		S5014
43022055	160	3006	51		G0003
43022134	132	3116	FET2	FE100/	F0021
43022458	159				
43022577	121				
43022860	107	3039		TR70/720	S0015
43022861	123A	3018		TR21/735	S0015
43023190	181	3036	19	TR36/S7000	S7002
43023212	123A	3020	63	TR21/735	S0015
43023221	123A	3020	63	TR21/735	S0015
43023222	129	3025		TR88/242	S0015
43023223	128	3024		TR87/243	S0019
43023843	130	3027	14	TR59/247	S7002
43023844	123A	3024		TR21/	S0015
43023845	159	3114		TR20/717	S5013
43024216	130	3027		TR59/	S7004
43024218	184	3190			S3002
43024219	185	3191			S3003
43024225	133	3112	FET1	FE100/	F0010
43024306	190			TR74/	S3019
43024833		3005			
43024834		3008			
43024859		3124			

Original Part Number	ECG	SK	GE	IR/WEP	HEP
43024873		3124			
43024878		3124			
43024879		3020			
43024880		3025			
43025055		3124			S0030
43025056		3124			S0014
43025059		3124			S0014
43025538	199	3024	62		S0015
43025539	192	3024	63		S5014
43025620		3006			
43025762					F0015
43025763					S0016
43025764					S0016
43025765					S0016
43025766					S0016
43025767					S0016
43025832					F0010
43025834					G6016
43025972		3124			S0015
43026284	128	3020			S5014
43026285	129	3025			S5013
43026932					S5004
43027213	152	3054			S5000
43027214	152	3054			S5004
43027379		3124			S0014
43027571		3054			S5000
43027614		3018			S0015
43027615		3018			S0015
43027616		3018			S0011
43027617		3018			S0015
43027618		3018			S0015
43027619		3118			S0013
43027620		3122			S0015
43027987		3054			
43028064					S5000
43040313	128	3122		TR87/	
43122874	726	3022			
43126551	722	3161			
43126551-A	722	3161	IC9		
43623212				TR25/	
44005001		3124			
44007301		3018			S0015
44008401		3018			S0014
44011001		3018			S0014
44079004		3122			S0024
44084003		3018			
44089001		3018			S0025
44090004		3018			S0015
44101001		3020			
44211001		3004			G0005
46156741P1	941				
46156741P2	941D				
48134666	123A		20	TR21/735	
48134842	123A	3020		TR21/	
48134857					S0020
48134989P1W				TR30/	
48137005				TR87/	
48137041				TR87/	
48137195A	159	3114		TR20/	
48697195	181			TR36/	
50210104	123A				
50210300-00	123A				
50210300-00VF	123A				
50210300-01	123A				
50210300-01VF	123A				
50210300-11	123A				
50210310-10	123A				
50210310-10VF	123A				
50210310-11	123A				
50210400-00	159				
50210400-00VF	159				
50210400-01	159				
50210400-01VF	159				
50210510	123A				
50210600-00VF	159				
50210600-01	159				
50210600-01VF	159				
50210600-03	159				
50210610-10	159				
50210610-13	159				
50210700-00	199				
50210700-00VF	199				
50210700-01VF	199				
50210700-10	199				
50210710-10VF	199				
50210800-00	123A				
50210800-01	123A				
50210800-01VF	123A				
50210800-02	123A				
50210800-02VF	123A				
50210800-10	128				
50210800-11	123A				
50211210	159				
50211300-00	106				
50211300-00VF	106				
50211300-01	106				
50211300-10	106				
50211400-00	234				
50211400-01	234				
50211400-10	234				
50211410-11	234				
50211500-01	159				
50211500-01VF	159				
50211510-10	159				
50211510-10VF	159				
50211510-11	159				
50211600-00	106				
50211600-00VF	106				
50211600-01	106				
50211600-02	106				
50211600-02VF	106				
50211600-10	106				
50211600-12	106				
50211610-10	106				
50211610-12	106				
50211900	128				
50212100	159				
50220601	175				
50220602	175				
50221300-01	130				
50221400-01	218				
50221800	152				
50221800-01	152				
50251300	941				
50254200	7406				
50254400	915				
50254600	74S00				
50254700	74S10				
50254900	74S20				
50255000	74S113				
50613800	923				
51003059	123A				
51003092	123A				
51003108	159				
51122245	123A				
51122675		3024			
51122870	128				
51122880	128				
51126540	175				
51126850	130	3027			
51161325	159				
51310000	9932				
51310001	9933				
51310002	9936				
51310003	9944				
51310004	9946				
51310005	9948				
51310006	9962				
51310007	9094				
51320000	7400				
51320001	7402				
51320002	7404				
51320003	7410				
51320004	7420				
51320005	7440				
51320006	74170				
51320008	7472				
51320009	7403				
51320010	7426				
51320011	7430				
51320012	7400				
51320016	7451				
51320017	7405				
51320018	7486				
51330005	7408				
51565600VF	123A				
51577500	9930				
51577600	9932				
51577700	9933				
51577800	9944				
51577900	9946				
51578000	9948				
51578100	9962				
51581300	123A				
51624200	9944				
51715100	910D				
51717000	9094				
51717100	9097				
51753300	910				
52335500	7495				
52335600	74H04				
52335600HL	74H04				
52335700	74H20				
52335800	74H50				
52335800HL	74H50				
52335900	74H74				
52335900HL	74H74				
54967774-P5	102			TR05/	
55430001-001	159				
55430028-001	218				

Original Part Number	ECG	SK	GE	IR/WEP	HEP
55440007-001	123				
55440011-001	108				
55440023-001	108				
55440043-001	123				
55440048-001	123A				
55440063-001	128				
55440087-001	130				
55440091-001	130				
55440111-001	107				
55460014-001	132				
55517007	159				
56301500	123A				
56301600	159				
57000901-504	123A	3020		TR21/	
57276605	105	3012		TR03/	
57279005	105	3012		TR03/	
57279007	105	3012		TR03/	
57279009	105	3012			
57279011	105	3012		TR03/	
57279017	105	3012		TR03/	
57279025	105	3012		TR03/	
57279027	105	3012		TR03/	
57279031		3012			
57279033	105	3012		TR03/	
57279073	105	3012		TR03/	
57279076	105	3012		TR03/	
57279293	105	3012		TR03/	
57279298	105	3012		TR03/	
57279566	105	3012		TR03/	
57279793	105	3012		TR03/	
57280281		3012		TR03/	
57282315	105	3012		TR03/	
57288072	105	3012		TR03/	
57288073	105	3012		TR03/	
57288076	105	3012		TR03/	
57288079	105	3012		TR03/	
57299720	105	3012		TR03/	
59700278	123A				
61260039	126	3008	51	TR17/635	
61260039A	126	3008		TR17/635	
65372000	128				
65600500	9930				
65600600	9932				
65600700	9946				
65600800	9093				
65600900	9099				
65611000	9094				
65611100	9097				
67283700	128				
70260450	107	3018	20	TR24/	S0016
70270730	1054				
70270740	1095				
72035900	159				
72045600	9110				
72729031	105			TR03/	
74140226-001	130	3027			
74140226-002	130				
74140226-003	130	3027			
74210052-D		3007			
75700032-01					S5013
75857322	123A	3018			
80050100	123A	3020	20	TR21/735	
80050300	103A	3010		TR08/724	
80050400	160	3006		TR17/637	
80050500	126	3008		TR17/635	
80050600	129	3010		TR88/242	
80050700	121	3009		TR01/232	
80051001	130		14	TR59/	
80051500	123A	3020	20	TR21/735	
80051600	129	3025		TR88/242	
80052101	199		10	TR21/	
80052102	123A	3020	20	TR21/735	
80052201	128		18	TR87/	
80052202	123A	3020	20	TR21/735	
80052301	128		18	TR87/	
80052302	123A	3020	20	TR21/735	
80052402	130	3027		TR59/247	
80052504				TR19/	
80052600	123A	3020		TR21/735	
80052700	129	3025	21	TR88/242	
80052800	103A	3010	8	TR08/735	
80053001	123A	3020	10	TR21/735	
80053300	152	3024		TR76/701	
80053400	123A	3020	10	TR21/735	
80053501	132	3116		FE100/	
80053600	108	3018	17	TR95/56	S0020
80053701					S7002
80053702	130MP	3029	19	TR59MP/247	S7004
80053703					S7004
80104900	126	3008	2	TR17/635	G0008
80105200	126	3008	51	TR17/635	G0008
80105300	126	3008	2	TR17/635	G0008
80146620	160	3006	51	TR17/	G0003
80146630	160	3006	51	TR17/	G0003
80203350	102A	3004	2		G0005

Original Part Number	ECG	SK	GE	IR/WEP	HEP
80205400	102A	3004	52	TR85/250	G0007
80205600	102A	3004	2	TR85/250	G0005
80222631U	102A	3004		TR85/250	
80226310	102A			TR85/250	G0005
80236400	102A	3004		TR85/250	G0005
80236430	158	3004		TR84/630	G0005
80243900	102A	3004	2	TR85/250	G0005
80285400	176	3123		TR82/	
80318250		3020			S0016
80337390	128			TR87/243	S0015
80338030	108	3018	20	TR95/56	S0016
80338040	108	3018	20	TR95/56	S0016
80339430	108	3018	17	TR95/56	S0014
80339440	108	3018	17	TR95/56	S0014
80366830	107	3018	17		S0025
80366840	107	3018	17	TR24/	S0025
80383840	123A			TR21/	S0015
80383930	123A			TR21/735	S0015
80383940	123A			TR21/735	S0015
80389910	123A			TR21/735	S0025
80414120	130			TR59/247	S5014
80414130	130			TR59/247	S5019
80421980	128	3024		TR87/243	S5014
82409501	123A	3020		TR21/735	
82410300	129	3025		TR88/242	
83073205	123A	3122	20	TR21/	S0024
83073206	123A	3122	20	TR21/	S0024
83073304	123A	3122	20	TR21/	S0025
83073305	123A	3122	20	TR21/	S0025
83073504		3122	20	TR21/	S0014
83093005	107	3018		TR70/720	S0016
83094502	123A	3122		TR21/735	S0015
84026103	192	3024			S0014
84667800	909	3590			
86401000	123A	3122		TR21/735	S0015
88125010-9	123A	3020		TR21/	
91011500	909	3590			
91011900	923				
91013300	941D				
91013600	940				
91014400	949				
91015300	947				
91056140	159	3025	21	TR20/717	S0019
92005600	102A	3004	52	TR85/250	G0005
92162000J		3004			
93037230	123A	3018	20	TR24/	S0015
93037240	123A	3020	20	TR21/735	S0015
93038040		3018	20		S0015
93039440		3018	17	TR24/	S0025
93053650		3124			S0016
93063240	123A	3124	20	TR51/	S0015
93063270	123A	3124	20	TR51/	S0015
93063280	123A	3124	20	TR51/	S0015
93063470	123A	3124	20	TR24/	S0015
93064440	199	3124	10		S0015
93064450	123A	3124	18		S0015
93069360		3124			S0015
93073260	199	3020		TR21/	S0024
93073350	128	3020	63	TR87/243	S0014
93073440	154	3020	63	TR78/243	S5026
93073540	123A	3020	63	TR21/735	S0014
93078420	107	3124	20	TR70/	S0016
93082830	123A	3124	20		S0015
93082840	123A	3124	20		S0015
93082920	107	3018	20		S0011
93094502	123A			TR21/735	
93097120	128	3020	18	TR25/	S5014
93938040	123A			TR21/	
93939440	107				
94029220	124	3026	12	TR23/	S5019
94650700-00	159				
94650700-01	159				
94742308	128				
94809600		3036			
94813000	181	3036			
94818400	223				
94824100-00	123A				
94824100-01	123A				
94824101	123A				
94825000-05	9093				
94825000-11	9936				
94913000-01	941D				
95114101-00	102A	3004		TR85/	
95114102-00	102A	3004		TR85/	
95114109-00	102A	3004		TR85/	
95114135-00	102A	3004		TR85/	
95114140-00	121	3009		TR01/	
95349700	197				
95522800	128				
95524600	124				
115115108-P2		3122			
120001190	160	3006		TR17/	
120001192	100	3005		TR05/	
120001195	102A	3004		TR85/	
120002013	102A	3004		TR85/	

Original Part Number	ECG	SK	GE	IR/WEP	HEP
120002014	102A	3004		TR85/	
120002213	160	3007		TR17/	
120002214	160	3006		TR17/	
120002216	160	3007		TR17/	
120002513	160	3006		TR17/	
120002515		3006		TR17/	
120002518	160	3006		TR17/	
120002520	160	3006		TR17/	
120002521	102A	3004		TR85/	
120002656	160	3006		TR17/	
120002748	102A	3004		TR85/	
120004492	160	3006		TR17/	
120004493	102A	3004		TR85/	
120004494	102A	3004		TR85/	
120004495	102A	3004		TR85/	
120004496	108	3018		TR95/	
120004497	108	3018		TR95/	
120004880	108	3018		TR95/	
120004881	108	3018		TR95/	
120004882	108	3018		TR95/	
120004883	123A	3020		TR21/	
120004884	195	3047		TR65/	
120004885	224	3049			
120004886	224	3049			
120004887	121	3009		TR01/	
131000561	128				
131000562	121				
131001007	130				
131005353-1	197				
131005807	128				
131005808	129				
143117184		3036			
152221011	102A	3004		TR85/250	
152221483	160			TR17/637	
163547338	723				
181515417		3024	63		S5014
226021014	108	3018		TR95/	
229005013	123A	3020		TR21/	
229005014	123A	3020		TR21/	
229005015	123A	3020		TR21/	
229018032	108	3018		TR95/	
229018033	108	3018		TR95/	
229018034	108	3018		TR95/	
229020423	108	3018		TR95/	
229021014	108	3018		TR95/	
229025010	108	3019		TR95/	
229510015V	108	3019		TR95/	
229510031V	108	3018		TR95/	
229510032V	108	3018		TR95/	
229510033V	108	3018		TR95/	
238208001	175				
260917101		3018			
261002302	910D				
261004303	941M				
261015401	923				
312104732	160			TR17/	
312104733	160	3006		TR17/	
314743006	175				
346190001	128				
430203845	129	3025		TR88/242	
430233843	152				
436004601	199				
436005001	123A				
436005901	234				
436006101	128				
436006201	129				
436006401	199				
450010201	108			TR95/	
450010701	161			TR83/	
450013800					S0017
450014600					S0017
485134732				TR25/	
485134811				TR21/	
485134825				TR21/	
485134825				TR21/	
485134826				TR21/	
485134837				TR21/	
485134855				TR21/	
485134872				TR23/	
485134922	123A		17	TR21/	
485134923	123A		17	TR21/	
485134924	123A		20	TR21/	
485134925	123A		20	TR21/	
485134926	123A		20	TR21/	
485134932				TR21/	
485134946				TR21/	
485134956	160		51	TR17/	
485134970				TR21/	
485134997	199		10	TR21/	
485137002	128				
485137014				TR25/	
485137020				TR30/	
485137055				TR21/	
485137370	197				
570004503	123A	3020	20	TR21/735	
570005452	123A	3020	20	TR21/735	
570005503		3020	20	TR21/735	
70.00.730		3006			
70.01.704		3122			
791078201					S0014
802026310			2		
802226310				TR85/	
881250108	123A	3038		TR21/735	
881250109	123A	3020		TR21/735	
885540026-3	7420				
885540031-2	74H21				
910678870	159	3118	22	TR30/	S0019
11220009/7825					G0005
11220018/7611					G0005
11220022/7611					G0005
11220046/7825					S0016
11220061/7825					G0005
11220076/7825					G0005
11220106/6711					G0005
11220106/7611					G0005
1251393577				TR21/	
1251393578				TR21/	
1251403572				TR21/	
1251413571				TR21/	
1261844047				TR12/	
1296983589				TR21/	
1296993591				TR30/	
1522210111	102A	3004		TR85/	
1522210131	126	3006	1	TR17/635	
1522210300	102A	3006	1	TR85/250	G0005
1522210921	126	3007		TR17/635	
1522211021	126	3006	51	TR17/635	
1522211200	102A	3004	2	TR85/250	G0005
1522211221	102A	3004	52	TR14/	
1522211328	102A	3004	53	TR85/250	
1522211921	160	3006		TR17/637	
1522214400	160	3006		TR17/637	G0003
1522214411	126	3006	50	TR17/635	
1522214435	126	3006	50	TR17/635	
1522214821	126	3006	50	TR17/635	
1522214831	160	3006	50	TR17/	
1522216500	102A	3004	2	TR85/250	G0005
1522216600	126	3007	51	TR17/635	G0008
1522217400	160	3006	51	TR17/637	G0003
1522223720	123A	3020		TR21/735	
1611819064	123A	3124	20	TR24/	
1679003608	1003				
1720109001		3054			
1720139001		3054			
1760089001	123A				
1760319001		3124			
1760379003		3124			
1770069002		3114			
2290180119	108	3018		TR95/	
2295100224	108	3018		TR95/	
2295100225	108	3018		TR95/	
2295100226	108	3018		TR95/	
2295100227	160	3008		TR17/	
2295100228	108	3018		TR95/	
2791012150	123				
2791012153	102				
3121004900	160	3006	51	TR17/637	
3121007744	160	3006	51	TR17/637	
3121024033	160	3006	51	TR17/637	
3121024044	160	3006	51	TR17/637	
3121047111	160	3006	51	TR17/637	
3121050611	160	3006		TR17/637	
3121050622	160	3006		TR17/637	
3122005400	102A	3004	2	TR85/250	
3520743010	159				
3539306001	123A	3020		TR21/735	
3539306002	123A	3020	20	TR21/735	
3539306003	123A	3020	20	TR21/735	
4100104753		3004	53		G0005
4100206443		3124	62		S0015
4100208283		3124	18		S0015
4100208284					S0015
4100208291		3018	60		S0011
4100208292		3018			
4100208293		3018			
4100210472		3018	61		S0016
4100213591		3018	60		S0016
4100217172		3122			
4104007380		3084			S5006
4104204603		3018			S0014
4104204612		3018			S0014
4104204613		3018			S0020
4104205352		3018			S0016
4104206440		3124			S0015
4104206442		3124			
4104208282		3124			
4104208283		3124			
4104213421		3018			S0014
4109208280		3124			S0015
4109208284		3122			S0015

Original Part Number	ECG	SK	GE	IR/WEP	HEP
4109213174		3024			S0024
4109213354		3020			
4131101789A		3011			
4150002031	1054				
4202005000		3119			
4202104770	102A	3088		TR85/250	G0005
4206002300	1004	3102			
4206002400	1004				
4206002900	1080				
4206004600	712	3072			
4206004700	1122				
4206104970	712	3072			
4206105170					C6076P
4360021001	123A				
4803101109-02	121				
4851378555				TR21/	
5700045452	123	3020	20	TR86/53	
8001200001	123A				
8001200004	128				
8001200006		3122			
8905706557		3027			
9511410100	102A	3004		TR85/250	G0005
9511410200	102A	3004		TR85/250	G0007
9511410900	102A	3004		TR85/250	G0005
9511413500	102A	3004		TR85/250	
9511414000	131	3009		TR94/642	G6016
12613740490				TR85/	
12618440487				TR12/	
12618540488				TR12/	
12618640489				TR12/	
12618840491				TR25/	
16000190201	126	3006	51	TR17/	G0003
16009090545	193	3025	67		
16100190668	108		17	TR21/	
16102190693	123A	3122	62		S0015
16102190929	108		20	TR21/	
16102190930	123A		20	TR21/	
16102190931	196	3192	28	TR55/	
16103190668	108		17	TR21/	
16103190930	107	3018	60		
16104191168	123A	3122	20	TR21/	S0015
16104191225	108	3018	20	TR21/	S0020
16104191226	108	3122	20	TR21/	S0014
16105190536	123A	3124	20		
16105191229	128	3024	18	TR25/	S5014
16106190537	123A	3124	62		
16106190772		3018	17		S0014
16108190536	199	3124	20		S0016
16108290536	128		18	TR25/	
16109190536	199	3124	20	TR24/	S0016
16109209536	123A		20	TR24/	
16112190710	107		10	TR24/	
16112190772		3018	17		S0014
16114190772		3018	17		S0014
16116190634	123A	3124	20	TR24/	S0015
16147191229	123A	3020		TR21/735	
16156197229	128	3024	63	TR25/	S5014
16171190693	123A		62		S0015
16171190858	123A	3122	62		
16171191368	123A	3020	62	TR21/735	S0015
16172190693	123A	3122	62		S0015
16172190858	123A	3122	62		S0015
16201190022	102A	3004	53		
16201190187	158	3004	53	TR84/	G6011
16204190457	158	3004	53	TR85/	G0005
16207190405	158	3004		TR84/630	G6011
16208190187		3004	53		G0005
16211190022	158	3004	53		G0005
16212190022	158	3004	53		G0005
16256197228	129		67		
16307190632	123A	3124	20	TR51/	S0015
16343190142	186	3054	28	TR55/	
16356179229	128		63		
16377190632	123A	3124	20	TR51/	S0015
16400690060		3088			
16401190188		3088			
16740135809	1024				
20924120713				TR21/	
20924180716				TR21/	
20933080705				TR22/	
26341116199				TR21/	
57000901504	123A	3020	20	TR21/735	
310720440170	121MP	3013			G6003
310720490000	123A	3020		TR21/	S0011
310720490010	123A	3020		TR21/	S0015
310720490020	123A	3020		TR21/	S0015
310720490060	185	3191			S5006
310720490070	121	3009		TR01/	G6005
310720490080	108	3018		TR95/	S0016
310720490100	123A	3020		TR21/	S0015
310720490150	123A	3020		TR21/	
310720490190	121MP	3013			G6003(2)
353511050009	732				
400100010160	102A			TR85/250	G0005
400100020120	102A	3004		TR85/250	G0007

Original Part Number	ECG	SK	GE	IR/WEP	HEP
404100030180	158	3004		TR84/630	G0005
404100900160	123A	3124		TR21/	S0011
933001170112	102A	3018		TR85/	G0005
933022960012	108	3018		TR95/	S0016
933022970112	108	3018		TR95/	S5024
933076760112	191	3054			
3535110500009	732				
17511280369984		3124			
17511280369985		3124			
17511280369986		3122			
17511280369987		3122			
17511280369988		3122			
A1DJ	163			TR67/	
A1E	108	3019	11	TR95/56	S0020
A1F	123A	3124	20	TR21/735	S0014
A1G	107	3018	11	TR70/720	S0016
A1G-1	107	3039		TR70/720	
A1G-1A	108	3039		TR95/	
A1H	123A		17	TR21/735	S0015
A1H(MOTOROLA)	163			TR67/740	
A1J	123A	3124	10	TR21/735	S0015
A1K	108	3018	17	TR95/56	S0015
A1L	123A	3018	20	TR21/735	S0015
A1M	154	3044	63	TR78/712	S5025
A1M-1	107			TR70/720	S0016
A1N	124	3021	12	TR81/240	S5011
A1P	107	3039	11	TR70/720	S0033
A1P-1	107	3039	20	TR70/720	S0033
A1P-1A	107	3039		TR70/	
A1P-5	108	3039		TR95/	
A1P/4922	107	3018		TR70/720	
A1P-/4923	108	3018		TR95/720	
A1P/4923	107	3039		TR70/720	
A1P/4923-1	108	3039		TR95/	
A1R	107	3039	20	TR70/720	S0033
A1R-1	107	3039	20	TR70/720	S0033
A1R-1/4925	108	3018			
A1R-1A	108	3039		TR95/	
A1R-2	107	3039	20	TR70/720	S0033
A1R-2/4926	108	3018		TR95/	
A1R-2A	108	3039		TR95/	
A1R-5	108	3039		TR95/	
A1R/4924	107	3018		TR70/720	
A1R/4925	107	3039		TR70/720	
A1R/4925A	108	3039		TR95/	
A1R/4926	107	3039		TR70/720	
A1R/4926A	108	3039		TR95/	
A1R-24926	108				
A1S	154	3040	63	TR78/712	S5024
A1T	123A	3124	20	TR21/735	S0015
A1T-1	123A	3124	20	TR21/735	S0015
A1U	233	3018	17	TR95/56	S0016
A1V	123A	3124	20	TR21/735	S0015
A1VE	123A			TR21/	
A1W	123A	3124	20	TR21/735	S0014
A1Y	175	3026	23	TR81/241	S5019
A1Z	233	3018	11	TR95/56	S0020
A2A	154	3045	20	TR78/712	S5026
A2B	123A	3024	63	TR21/735	S0015
A2BRN	123A			TR21/	
A2C	108	3039	20	TR95/56	S0020
A2D	108	3018	20	TR95/56	S0016
A2E	130	3027	11	TR59/247	S7004
A2E-2	130	3027		TR59/	
A2EBLK	130			TR59/247	S7004
A2EBRN	130	3027		TR26/247	S7004
A2EBRN-1	130	3027		TR59/	
A2F	108	3018	20	TR21/735	S0016
A2FGRN	123A			TR21/735	
A2G	108	3018	20	TR95/56	S0020
A2H	108	3018	20	TR95/56	S0016
A2J	123A	3122	20	TR21/735	S0015
A2K	154	3045	27	TR78/712	S5026
A2L	108	3018	20	TR95/56	S0025
A2M	107	3018	20	TR70/720	S0016
A2M-1	107	3018	20	TR70/720	S0016
A2M9	129			TR88/242	
A061-108	107	3039		TR70/720	S0014
A061-109	107	3039		TR70/720	S0014
A061-110	126	3006		TR17/635	G0008
A061-111	126	3006		TR17/635	G0008
A061-112	107	3039		TR70/720	S0014
A061-114	102A	3004		TR85/250	G0006
A061-115	102A	3123		TR85/250	G0005
A061-116	131	3052		TR94/642	G6016
A065-102	123A	3020	20	TR21/735	
A065-103	123A	3020	20	TR21/735	
A065-104	123A	3020	20	TR21/735	
A065-105	158	3004	2	TR84/630	G0005
A065-106	102A	3004	2	TR85/250	G0005
A065-108	123A	3020	20	TR21/735	S0015
A065-109	123A	3020	20	TR21/735	S0025

Original Part Number	ECG	SK	GE	IR/WEP	HEP
A056-110	123A	3024	17	TR21/735	S0011
A065-111	131	3052		TR94/642	G6016
A065-112	158	3004	2	TR84/630	G0005
A065-113	123A	3020	20	TR21/735	S0015
A066-109	123A	3018		TR21/735	S0015
A066-110	132	3116		FE100/802	F0015
A066-111	107	3018		TR70/720	S0016
A066-112	123A	3020		TR21/	S0015
A066-113	159	3114		TR20/717	S0013
A066-113(2SC-)	123A			TR21/	
A066-113A	159	3114		TR20/717	
A066-113AB	159	3114		TR20/	
A066-114	152	3041		TR76/S5003	S5000
A066-115	107	3047		TR70/720	S0016
A066-116	128	3047		TR87/243	S0014
A066-117	224	3049		TR65/243	S5014
A066-118	159	3025		TR20/717	S0012
A066-118A	159	3114		TR20/	
A066-143	123A	3020		TR21/	S0024
A068-106	132	3116		FE100/802	F0015
A068-107	132	3116		FE100/802	F0015
A068-108	123A			TR21/735	S0015
A068-109	159	3025		TR20/717	S0019
A068-109A	159	3114		TR20/	
A068-111	108	3018		TR95/56	S0016
A068-112	108	3018		TR95/56	S0016
A068-113	123A	3124		TR21/735	S0011
A068-114	175	3026		TR81/241	S5012
A069-101	107	3039		TR70/	S0011
A069-102	107	3039		TR70/	
A069-102/103	123A	3122			
A069-103	107	3039		TR70/	
A069-104	123A	3122		TR21/	
A069-104/106	123A	3122			
A069-105	102A	3004		TR85/	G0009
A069-106	123A	3122		TR21/	
A069-107	158	3004		TR84/	G0005
A069-114	107	3039		TR70/720	S0016
A069-116	107	3039		TR70/720	S0011
A069-120	123A	3122		TR21/735	S0015
A069-121	126	3006		TR17/635	G0009
A069-122	123A	3122		TR21/735	S0016
A04201	133	3112		/801	
A072133-001	767				
A.184/5	123A				
A1A	154	3124	20	TR78/712	S5024
A1B	123A			TR21/735	S0014
A1C	162		35	TR67/707	S5021
A1D	163	3111	36	TR67/740	
A1DI	163			TR67/	
A00	7400				
A01	160			TR17/	
A02	7403				
A03	7404				
A04	7405				
A05	7410				
A06-1-12	123A			TR21/	
A08-105018	130				
A08-1050108		3027			
A08-1050115	181	3036			
A08	7430				
A09	7440				
A054-108	123A	3122	17	TR21/	S0015
A054-109	129	3025	18	TR88/	S5026
A054-114	123A	3020		TR21/735	S0015
A054-115	123A	3122	10	TR21/	S0015
A054-142	220	3116		/802	F2005
A054-148	107	3018		TR70/720	S0014
A054-154	130	3027	14	TR59/	S7004
A054-155	123A		17	TR21/	S0015
A054-156	128	3024	63	TR87/	S3021
A054-157	108	3039	20	TR95/	S0030
A054-158	108	3039	20	TR95/	S3032
A054-159	108	3039	20	TR95/	S0014
A054-160	128	3024	63	TR87/	S5026
A054-163	108	3039	20	TR95/	S0014
A054-164	108	3039	20	TR95/	S0014
A054-165	703A				
A054-170	107	3018		TR70/720	S0020
A054-173	123A	3020		TR21/735	S0015
A054-175	123A	3024		TR21/735	S0014
A054-186	128	3024		TR87/243	S5026
A054-195	123A	3020		TR21/735	S0015
A054-206	128	3024		TR87/243	S5026
A054-221	123A	3018		TR21/735	S0015
A054-222	123A	3018		TR21/735	S0015
A054-223	159	3025		TR20/717	S0019
A054-224	175	3026		TR81/241	S5019
A054-225	123A	3122		TR21/735	S5014
A054-233	123A	3020		TR21/735	S0014
A054-234	123A	3020		TR21/735	S0015
A054-470	108	3018		TR95/	
A059-100	160	3007		TR17/637	G0008
A059-101	160	3007		TR17/637	G0008
A059-102	160	3007		TR17/637	G0008

Original Part Number	ECG	SK	GE	IR/WEP	HEP
A059-103	160	3007		TR17/637	G0008
A059-104	126	3006		TR17/635	G0008
A059-105	126	3008		TR17/635	G0008
A059-106	102A	3004		TR85/250	G0008
A059-107	102A	3004		TR85/250	G0008
A059-108	121	3009		TR01/232	G6016
A059-109	123A	3046		TR21/735	S0014
A059-110	128	3024		TR87/243	S3002
A059-111	224	3049			S3002
A059-115	104	3009		TR01/624	G6003
A059-116	102A	3004		TR85/250	G0005
A060-100	107	3039		TR70/	S0011
A06 -105	107	3039		TR70/720	S0014
A061-106	107	3039		TR70/720	S0014
A061-107	126	3006		TR17/635	G0008
A2N	107	3018	20	TR70/720	S0016
A2N-1	107	3018	20	TR70/720	S0016
A2N-2	107	3039		TR70/720	S0016
A2N-2A	108	3039			
A2N2156	213				
A2P	107	3039	20	TR70/720	S0016
A2P-5	108	3039		TR95/	
A2S	130	3027	21	TR59/247	S7004
A2S-3	130	3027		TR59/	
A2SA550P	129		21	TR88/	S0013
A2SA564F	129		21	TR88/	S5022
A2SA564FR	159	3025		TR20/717	
A2SA666PQR	102A		20	TR85/	S0019
A2SB240A	104	3009	3	TR01/624	
A2SB242A	104	3009	3	TR01/624	
A2SB248A	104	3009	3	TR01/624	
A2SC538PQR	123A		20	TR21/	S0015
A2SC538R	199	3124		TR21/	
A2SD226PQ	175		23	TR81/	S5019
A2SD2260P	152		23	TR76/	S5019
A2T	108	3018	11	TR95/56	S0030
A2T682	103A	3010		TR08/	
A2T919	108	3039		TR95/	
A2U	124	3021	12	TR81/240	S5011
A2V	108	3018	20	TR95/56	S0016
A2W	108	3039	20	TR95/56	S0024
A2Y	107	3018	17	TR70/720	S0016
A2Z	154	3045		TR78/712	S5026
A3A	108	3039	20	TR95/56	S0025
A3C	108	3039		TR95/56	S0025
A3D	108	3018	20	TR24/	
A3E	123A	3024	63	TR21/735	S0015
A3F	123A	3024	63	TR21/735	S0015
A3G	123A	3124	20	TR21/735	S0024
A3H	108	3111	36	TR67/740	S5021
A3H-1	163			TR67/740	S5021
A3J	199			TR21/	S0015
A3K	199	3024	20	TR21/735	S0024
A3L	157			TR60/244	S5015
A3L4-6001	130			TR59/	
A3L4-6001-01	130	3027			
A3M	154	3045	40	TR78/712	S5024
A3MA	154			TR78/	
A3N	123A	3124	20	TR21/735	S0015
A3N71	108	3039		TR95/	
A3N72	108	3039		TR95/	
A3N73	108	3039		TR95/	
A3P	108	3039		TR95/56	S0020
A3R	108	3039		TR95/56	S0020
A3S	108	3039	20	TR95/56	S5026
A3SB	108			TR95/	
A3T	123A	3122	63	TR21/735	S0015
A3T201	101	3011		TR08/	
A3T202	101	3011			
A3T203	101	3011		TR08/	
A3T918	161		61	TR83/	S0020
A3T929	123A		212	TR21/	S0025
A3T930	123A		212	TR21/	S0015
A3T2221	123A	3122	210	TR21/	S0025
A3T2221A	123A	3122	20		S0005
A3T2222	123A	3122	210		S0015
A3T2222A	123A		20		S0005
A3T2484	192	3137	63	TR51/	S0005
A3T2894	159	3118		TR20/	S0013
A3T2906	159	3118	21	TR20/	S0019
A3T2907	159	3118	48	TR20/	S0019
A3T2907A	193	3138	67		S0019
A3T3011	123A	3122	20	TR64/	S0011
A3TE120	130	3027		TR59/	
A3TE230	130	3027		TR59/	
A3TE240	130	3027		TR59/	
A3TX003	130	3027		TR59/	
A3TX004	130	3027		TR59/	
A3U	130	3027		TR59/247	S7004
A3U-4	130	3027		TR59/	
A3W	123A	3122		TR21/735	S0030
A3Y	124	3104		TR81/240	S3019
A3Z	123A	3122		TR21/735	S0015
A4A	123A	3124	20	TR21/735	S0024
A4A-1-70	121	3009		TR01/	

Original Part Number	ECG	SK	GE	IR/WEP	HEP
A4A-1-705	121	3009		TR01/	
A4A-1A9G	121	3009		TR01/	
A4B	199	3038	20	TR21/735	S0024
A4E	107	3039		TR70/720	S0015
A4E-5	108	3039			
A4F	199	3122	18	TR21/735	S0024
A4G	108	3039	20	TR95/56	S0020
A4H	154		27	TR78/712	S5026
A4I7014	181				
A4J	130	3027		TR59/	S7002
A4J BLK	130			TR59/247	S7002
A4J RED	130			TR59/	
A4JBRN	130			TR59/	S7002
A4JD3B1	101				
A4JRED	130	3027		/247	S7004
A4JRED-1	130	3027		TR59/	
A4JX2A822	103A	3010		TR08/	
A4K	184		66	/S5003	S5000
A4L	103A			TR08/	G0011
A4M	123A	3024	20	TR21/	S0015
A4N	123A		18	TR21/	S0025
A4P	123A	3122		TR21/735	S0015
A4R	123A	3122		TR21/735	S0015
A4S	130	3027		TR59/247	S7004
A4S-1	130	3027		TR59/	
A4T	108	3039	17	TR95/56	S0016
A4U	123A	3122	17	TR21/735	S0016
A4V	123A	3122	17	TR21/735	S0016
A4Y-1	107	3039		TR70/720	S0020
A4Y-1A	108	3039		TR95/735	
A4Y-2	123A	3122		TR21/735	S0015
A4Z	130	3027		TR59/247	S7002
A5A	196	3041	28	TR76/S5003	S5000
A5A-1	184	3041		/S5003	S5000
A5A-1B	152				
A5A-2	184	3041	57	/S5003	S5000
A5A-3	184	3041	57	/S5003	S5000
A5A-4	184	3190	57	TR76/S5003	S5000
A5A-5	184	3190	57	TR76/S5003	S5000
A5B	128	3024		TR87/243	S5014
A5C	123A	3122		TR21/735	S0015
A5E	188			TR72/	S3020
A5F	190	3104		TR74/S3021	S3019
A5FF	190			TR74/	
A5G	184	3190		TR60/	S5000
A5H	123A			TR21/	S0014
A5J	108	3039	20	TR95/56	S0015
A5K	123A	3122	20	TR21/735	S0015
A5L	123A	3122	18	TR21/735	S0015
A5M	123A	3040	63	TR21/735	S0030
A5N	123A	3018		TR21/735	S0015
A5P	123A	3122	20	TR21/735	S0030
A5R	123A			TR21/735	S0015
A5S	123A	3122	20	TR21/735	S0015
A5T	191	3044	27	TR75/	S3021
A5T-1	191	3104		TR75/	S3021
A5T404			82		
A5T404A			82		
A5T2192			47		
A5T2222	123A	3122	20		
A5T2604			82		
A5T2605			82		
A5T2907		3025	21		S0019
A5T3391			47		
A5T3391A			47		
A5T3392			81		
A5T3504			48		
A5T3505			67		
A5T3565			81		
A5T3638			82		
A5T3638A			82		
A5T3644	159	3025	21		
A5T3645	159	3025	21		
A5T3707			62		
A5T3708			62		
A5T3709			62		
A5T3710			62		
A5T3711			62		
A5T3821	132	3116			F0015
A5T3903	123A	3122	210		
A5T3904	123A	3122	20		S0025
A5T3905	159	3025	21		
A5T3906	159	3025	21		
A5T4026			67		
A5T4028			67		
A5T4058			65		
A5T4059			65		
A5T4060			65		
A5T4061			65		
A5T4062			65		
A5T4123			210		
A5T4124			20		
A5T4125	159	3025	21		
A5T4126	159	3025	21		
A5T4248			82		
A5T4249			82		
A5T4250			48		
A5T4402			21		
A5T4403			21		
A5T4409			81		
A5T4410			18		
A5T5086			82		
A5T5087			48		
A5T5172			81		
A5T5209			81		
A5T5210			47		
A5T5219			81		
A5T5220			81		
A5T5221			82		
A5T5223			81		
A5T5225			81		
A5T5226			82		
A5T5227			82		
A5T6222			47		
A5TA	191			TR75/	
A5U	123A	3122	20	TR21/735	S0015
A5V	181			TR36/	S7000
A5W	123A	3122		TR21/735	S0015
A5Y	184	3190			S5000
A-6-67703	130				
A-6-67703-A-7	130				
A6A	163	3111		TR67/740	
A6B	108	3039		TR95/56	S0015
A6B-3-70	121MP	3015			
A6B-3-705	121MP	3015			
A6B-3A9G	121MP	3015			
A6C	184	3041	57	TR51/S5003	S5000
A6C-1	184	3041	57	/S5003	S5000
A6C-1 RED	184	3054		/S5003	S5000
A6C-2	184	3041	57	/S5003	S5000
A6C-2 BLK	184	3190		/S5003	S5000
A6C-2 BLACK		3054			
A6C-3	184	3041	57	/S5003	S5000
A6C-3 WHT	184	3190		/S5003	S5000
A6C-3 WHITE		3054			
A6C GRN	184	3190		/S5003	S5000
A6D	188	3041		TR72/3020	S3024
A6D-1	188	3054		TR72/	S3020
A6D-2	188	3054		TR72/	S3024
A6D-3	188	3054		TR72/	S3024
A6E	108	3018	11	TR95/56	S0020
A6F	108	3039		TR95/56	S0016
A6G	188	3054	28	TR72/3020	S3024
A6H	123A	3124	20	TR21/735	S0015
A6HD	123A	3122		TR21/	
A6J	123A	3040	20	TR21/735	S0030
A6K	123A	3018		TR21/735	S0030
A6L	130	3133	14	TR59/	S7004
A6LBLK	130	3027		TR59/247	S7004
A6LBLK-1	130	3027		TR59/	
A6LBRN	130	3027		TR59/247	S7004
A6LBRN-1	130	3027		TR59/	
A6LRED	130	3027		TR59/247	S7004
A6LRED-1	130	3027		TR59/	
A6M	163	3111		TR67/740	S5021
A6N	130	3027		TR59/247	S7002
A6N-6	130	3027		TR59/	
A6P	108	3039	20	TR95/56	S0015
A6R	123A	3018	20	TR21/735	S0015
A6S	123A	3018	20	TR21/735	S0015
A6T	108	3039		TR95/56	S0020
A6U	108	3039		TR95/56	S0020
A6V	107	3039		TR70/720	S0020
A6V-5	108	3039		TR95/	
A6W	184	3190			S5000
A6Y	190			TR74/S3021	S3020
A6Z	163	3111	36	TR67/	S5021
A6ZH	163			TR67/	
A7-12	130			TR59/	
A7-13	130			TR59/	
A7-67703		3027			
A7A	123A	3122	17	TR21/735	S0015
A7A30	108	3039		TR95/	
A7A31	108	3039		TR95/	
A7A32	108	3039		TR95/	
A7B	124	3021	12	TR81/240	S5011
A7B(MOTOROLA)	124			TR81/	
A7C(MOTOROLA)	184			/S5003	S5000
A7E	123A			TR21/	S0005
A7F	190			TR74/	S3020
A7M	130	3027		TR59/247	S7004
A7M-5	130	3027		TR59/	
A7M42394B39	755				
A7M42894B39	755				
A7N	108				S0015
A7P	108				S0015
A7R	123A				S0015
A7S	123A	3122		TR21/735	S0015
A7T	123A	3122	20	TR21/	S0015
A7T3391			47		

Original Part Number	ECG	SK	GE	IR/WEP	HEP
A7T3391A			47		
A7T3392			81		
A7T5172			81		
A7U	108				S0016
A7V	108				S0016
A7W	108				S0016
A7Y	123A	3122		TR21/735	S0030
A7Z	184				S5000
A8-1	160	3007		TR17/	
A8-1-70	160	3007		TR17/	
A8-1-70-1	160			TR17/	
A8-1-70-12	160	3007			
A8-1-70-12-7	160	3007		TR17/	
A8-1A	160	3007		TR17/	
A8-1A0	160	3007		TR17/	
A8-1A0R	160	3007		TR17/	
A8-1A1	160	3007		TR17/	
A8-1A2	160	3007		TR17/	
A8-1A3	160	3007		TR17/	
A8-1A3P	160	3007		TR17/	
A8-1A4	160	3007		TR17/	
A8-1A4-7	160	3007		TR17/	
A8-1A4-7B	160			TR17/	
A8-1-A-7B	160				
A8-1A5	160	3007		TR17/	
A8-1A5L	160	3007		TR17/	
A8-1A6	160	3007		TR17/	
A8-1A6-4	160	3007		TR17/	
A8-1A7	160	3007		TR17/	
A8-1A7-1	160	3007		TR17	
A8-1A8	160	3007		TR17/	
A8-1A9	160	3007		TR17/	
A8-1A9G	160	3007		TR17/	
A8-1A19	160	3007		TR17/	
A8-1A21	160	3007		TR17/	
A8-1A82	160	3007		TR17/	
A8-10	160	3007		TR17/	
A8-11	160	3007	TR1	TR17/	
A8-12	160	3007		TR17/	
A8-13	160	3007		TR17/	
A8-14	160	3007		TR17/	
A8-15	160	3007		TR17/	
A8-16	160	3007		TR17/	
A8-17	160	3007		TR17/	
A8-18	160	3007		TR17/	
A8-19	160	3007		TR17/	
A8A	194				S0005
A8B	123A	3122	20	TR21/	S0015
A8C	185	3191			S5006
A8D	128			TR87/	S5014
A8E	196	3054	66	TR76/	S5004
A8F	152	3054	57	TR55/	S5000
A8F1				TR55/	
A8G	123A	3122	18	TR21/	S0015
A8J	188	3199		TR72/	S3024
A8K	152	3054	66	TR76/	S5000
A8L	123A			TR21/	S0015
A8M	162			TR67/	S5020
A8N	194				S0005
A8P	130			TR59/	S7004
A8P-2-70	102A	3004		TR85/	
A8P-2-705	102A	3004		TR85/	
A8P-2A9G	102A	3004		TR85/	
A8P404F	121	3009		TR01/	
A8P404-ORN	121	3009		TR01/	
A8P-404-ORN	121				
A8R	199			TR21/	S0030
A8S	108			TR95/	S0016
A8T	163	3115	38	TR67/	S5021
A8T404			82		
A8T404A			82		
A8T3391			47		
A8T3391A			47		
A8T3392			81		
A8T3702			82		
A8T3703			82		
A8T3704			47		
A8T3705			47		
A8T3706			47		
A8T3707			81		
A8T3708			81		
A8T3709			81		
A8T3710			81		
A8T3711			81		
A8T4026			67		
A8T4028			67		
A8T4058			82		
A8T4059			82		
A8T4060			82		
A8T4061			82		
A8T4062			82		
A8T5172			81		
A8U	130	3133	14	TR59/	S7004
A8V	154	3044	40	TR78/712	S5024
A8VA	154			TR78/	

Original Part Number	ECG	SK	GE	IR/WEP	HEP
A8W	130			TR59/	S7004
A8Y	152	3054	28	TR76/	S5000
A8Z	194	3045	20	TR24/	S0005
A9-175	199				
A9-67703		3027			
A9A	108			TR95/	S0016
A9B	123A			TR21/	S0015
A9C	172			TR69/	S9100
A9D	108			TR95/	S0016
A9E	123A			TR21/	S0024
A9F	123A			TR21/	S0024
A9G	123A			TR21/	S0025
A9H	123A			TR21/	S0015
A9J	123A				S0015
A9K	162				S5020
A9L-4-70	102A	3003		TR85/	
A9L-4-705	102A	3003		TR85/	
A9L-4A9G	102A	3003		TR85/	
A9M		3054	27	TR60/	
A9N	181				S7002
A9P	162				S5020
A9R	163				S5021
A9S	123A				S0025
A9T	123A				S0016
A9U	123A				S0016
A9V	152	3054	66	TR76/	S5000
A9W	123A				S0015
A9Y	123A				S0015
A10	74H40				
A10-1050108		3079			
A10-1050115		3079			
A10G05-010-A	123A				S0015
A10G05-011-A	123A				S0015
A10G05-015-D	123A				S0015
A12	102A				
A12-1	225	3045			
A12-1-70	123A	3124		TR21/	
A12-1-705	123A	3124		TR21/	
A12-1A9G	123A	3124		TR21/	
A12-2	225	3045			
A12-1050108	3:	3535			
A12-1050115	3535	3535			
A12A	102A				
A12B	102A				
A12C	102A				
A12D	102A				
A12H	102A				
A12(IC)	7451				
A12V	102A				
A13	7454				
A13-0032	130	3027		TR59/	
A13-14604-1A	121MP	3013			
A13-14604-1B	121MP	3013			
A13-14604-1C	121MP	3013			
A13-14604-1D	121MP	3013			
A13-14604-1E	121MP	3013			
A13-14777-1	121MP	3013			
A13-14777-1A	121MP	3013			
A13-14777-1B	121MP	3013			
A13-14777-1C	121MP	3013			
A13-14777-1D	121MP	3013			
A13-14778-1A	121MP	3013			
A13-14778-1B	121MP	3013			
A13-14778-1C	121MP	3013			
A13-14778-1D	121MP	3013			
A13-15809-1	225	3045			
A13-17918-1	130	3027		TR59/	
A13-22741-2	121MP	3013			
A13-23594-1	130	3027		TR59/	
A13-28432-1	225	3045			
A13-33188-2	130	3027		TR59/	
A13-55062-1	225	3045			
A13-86416-1	103A	3010		TR08/	
A13-86420-1	101	3011		TR08/	
A13-87433-1	101	3011		TR08/	
A13N1	123A	3018		TR21/	
A14	160			TR57/	
A14(IC)	7472				
A14-586-01	121	3009		TR01/	
A14-601-10	130	3027		TR59	
A14-601-12	130	3027		TR59/	
A14-601-13	130	3027		TR59/	
A14-602-19	225	3045			
A14-602-36	225	3045			
A14-602-37	225	3045			
A14-602-63	108	3039		TR95/	
A14-503-05	108	3039		TR95/	
A14-603-06	108	3039		TR95/	
A14-1001	126	3006		TR17/635	G0008
A14-1002	126	3006		TR17/635	G0008
A14-1003	126	3006		TR17/635	G0008
A14-1004	102A	3123		TR85/250	G0007
A14-1005	102A	3123		TR85/250	G0007
A14-1006	102A	3123		TR85/250	G0007
A14-1007	102A	3123		TR85/250	G0005

Original Part Number	ECG	SK	GE	IR/WEP	HEP
A14-1008	102A	3123		TR85/	G0005
A14-1009	102A	3123		TR85/250	G0005
A14-1010	102A	3123		TR85/	G0005
A14-1050115		3535			
A14A8-1	160	3007		TR17/	
A14A8-19	160	3007		TR17/	
A14A8-19G	160	3009		TR17/	
A1410G	121				
A14A-70	121	3009		TR01/	
A14A-705	121	3009		TR01/	
A15	7474				
A15-1001	126	3006		TR17/635	G0008
A15-1002	126	3006		TR17/635	G0008
A15-1003	126	3006		TR17/635	G0008
A15-1004	102A	3123		TR85/250	G0007
A15-1005	102A	3123		TR52/250	G0005
A15BK	102A				
A15H	102A				
A15R	102A				
A15V	102A				
A15VR	102A				
A15Y	102A				
A16	102A				
A16A1	103A	3010		TR08/	
A16A2	103A	3010		TR08/	
A16-1050108		3535			
A16-1050115		3535			
A17	102A				
A17H	102A				
A17(IC)	7490				
A18		3027	14	TR59/247	
A18-4	130	3027		TR59/	
A18H	102A				
A18(IC)	7493				
A19	126				
A19-020-072	108	3039		TR95/	
A19(IC)	7495				
A20	126				
A20K	159	3114		TR20/717	
A20KA	159	3114		TR20/	
A21	126				
A24	108	3039		TR95/56	
A24MW594	108	3039		TR95/	
A24MW595	108	3039		TR95/	
A24MW596	108	3039		TR95/	
A24MW597	108	3039		TR95/	
A24T-016-016	108	3039		TR95/	
A25-1001	126	3008	51	TR17/635	G0008
A25-1002	126	3008	51	TR17/635	G0008
A25-1003	126	3008	51	TR17/635	G0008
A25-1004	102A	3004	52	TR85/250	G0008
A25-1005	102A	3004	52	TR85/250	G0007
A25-1006	102A	3004	52	TR85/250	G0005
A25A509-016-101	123A				
A25A305020101	128				
A26	100				
A27(RCA)	175				
A28	126				
A29	126				
A29V082B03	129			TR88/	
A30	126				
A31	100				
A31-0206	108	3039		TR95/	
A32	102A				
A32-2805-50-1	128				
A32-2809	108				
A33	102A				
A34-6001-1	121	3009		TR01/	
A34-6002-17	121	3009		TR01/	
A35	154	3040		TR78/712	
A36	100	3005			
A37	126	3006			
A38	126	3006			
A39	126	3006			
A40	100				
A41	160	3006			
A42	160	3006			
A42X00286-01	102				
A42X00290-01	156				
A42X00434-01	123A				
A42X210	103A	3010		TR08/	
A43	126	3006			
A44	100	3005			
A45	126	3006			
A45-1	126	3006			
A45-2	126	3006			
A45-3	126	3006			
A46-867-3	154	3045		TR78/	
A46-8614-3	101	3011		TR08/	
A46-86101-3	107	3039		TR70/	
A46-86109-3	107	3039		TR70/	
A46-86110-3	107	3039		TR70/	
A46-86133-3	107	3039		TR70/	
A46-86301-3	107	3039		TR70/	
A46-86302-3	107	3039		TR70/	

Original Part Number	ECG	SK	GE	IR/WEP	HEP
A48-10075A01	121MP	3013			
A48-10075A02	121MP	3013			
A48-10075A03	121MP	3013			
A48-10075A04	121MP	3013			
A48-10075A05	121MP	3013			
A48-10075A06	121MP	3013			
A48-10075A07	121MP	3013			
A48-10075A08	121MP	3013			
A48-10103A01	121MP	3013			
A48-10103A02	121MP	3013			
A48-10103A03	121MP	3013			
A48-10103A04	121MP	3013			
A48-10103A05	121MP	3013			
A48-10103A06	121MP	3013			
A48-10103A07	121MP	3013			
A48-10103A08	121MP	3013			
A48-10103A09	121MP	3013			
A48-10103A10	121MP	3013			
A48-10103A11	121MP	3013			
A48-40247G01	108	3039		TR95/	
A48-43351A02		3039		TR95/	
A48-63076A81	121	3039			
A48-63076A82	108	3039			
A48-64978A10	121MP	3013			
A48-64978A11	121MP	3013			
A48-64978A24	121MP	3013			
A48-97046A05	108	3039		TR95/	
A48-97046A06	108	3039		TR95/	
A48-97046A07	108	3039		TR95/	
A48-97127A06	108	3039		TR95/	
A48-97127A12	108	3039		TR95/	
A48-97127A18	108	3039		TR95/	
A48-124216	101	3011		TR08/	
A48-124217	101	3011		TR08/	
A48-124218	101	3011		TR08/	
A48-124220	101	3011		TR08/	
A48-124221	101	3011		TR08/	
A48-125233	101	3011		TR08/	
A48-125234	101	3011		TR08/	
A48-125235	101	3011		TR08/	
A48-125236	101	3011		TR08/	
A48-128239	101	3011		TR08/	
A48-134520	101	3011		TR08/	
A48-134648	225	3045			
A48-134700	101	3011		TR08/	
A48-134727	121	3009		TR01/	
A48-134731	121	3009		TR01/	
A48-134789	108	3039		TR95/	
A48-134818	154	3045		TR78/	
A48-134837	108	3039		TR95/	
A48-134843	154	3045		TR78/	
A48-134845	108	3039		TR95/	
A48-134853	154	3045		TR78/	
A48-134898	154	3045		TR78/	
A48-134902	108	3039		TR95/	
A48-134904	108	3039		TR95/	
A48-134907	121	3009		TR01/	
A48-134919	154	3045		TR78/	
A48-134922	108	3039		TR95/	
A48-134923	108	3039		TR95/	
A48-134924	108	3039		TR95/	
A48-134925	108	3039		TR95/	
A48-134926	108	3039		TR95/	
A48-134927	154	3045		TR78/	
A48-134931	101	3011		TR08/	
A48-134945	108	3039		TR95/	
A48-134961	108	3039		TR95/	
A48-134962	108	3039		TR95/	
A48-134963	108	3039		TR95/	
A48-134964	108	3039		TR95/	
A48-134965	108	3039		TR95/	
A48-134966	108	3039		TR95/	
A48-134981	108	3039		TR95/	
A48-137002	154	3045		TR78/	
A48-137035	154	3045		TR78/	
A48-137071	108	3039		TR95/	
A48-137075	108	3039		TR95/	
A48-137076	108	3039		TR95/	
A48-137077	108	3039		TR95/	
A48-137102	121	3009		TR01/	
A48-137197	108	3039		TR96/	
A48-137213	121	3009		TR01/	
A48-137214	121	3009		TR01/	
A48-137215	121	3009		TR01/	
A48-137216	121	3009		TR01/	
A48-137217	121	3009		TR01/	
A48-137218	121	3009		TR01/	
A48-137219	121	3009		TR01/	
A48-137220	121	3009		TR01/	
A48-869254	103A	3010		TR08/	
A48-869283	103A	3010		TR08/	
A48-869476	103A	3010		TR08/	
A48-869476A	103A	3010		TR08/	
A49	126				
A50	102A				G0005

Original Part Number	ECG	SK	GE	IR/WEP	HEP
A51	160	3006			
A52	126	3006			
A53	126	3006			
A54-3	152	3041		TR76/701	
A54-96-001	123A				
A54-96-002	123A				
A54-96-003	159				
A55	100	3005			
A56	160	3006			
A57	126	3006			
A57A144-12	108	3039		TR95/	
A57A145-12	108	3039		TR95/	
A57B124-10	121	3009		TR01/	
A57C5	103A	3010		TR08/	
A57C12-1	154	3045		TR78/	
A57C12-2	154	3045		TR78/	
A57D1-122	154	3045		TR78/	
A57L5-1	121	3009		TR01/	
A57M2-16	154	3045		TR78/	
A57M2-17	154	3045		TR78/	
A57M3-7	121	3009		TR01/	
A57M3-8	121	3009		TR01/	
A58	126	3006			
A59	126	3009			
A60	126	3006			
A61	160	3006			
A62-18427	121	3009		TR01/	
A62-19581	108	3039		TR95/	
A62(H)				TR67/	
A63-18426	154	3045		TR78/	
A63-18427	121	3009		TR01/	
A64	102A	3004			
A65	102A	3004			
A65-1-1A9G	160	3006		TR17/	
A65-1-70	160	3006		TR17/	
A65-1-705	160	3006	TR1	TR17/	
A65-1A9G	160	3006			
A65-2-70	103A	3010		TR85/	
A65-2-705	103A	3010		TR85/	
A65-2A9G	103A	3010		TR85/	
A65-4-70	101	3011		TR08/	
A65-4-705	101	3011		TR08/	
A65-4A9G	101	3011		TR08/	
A65A19G	160	3007		TR17/	
A65A-70	160	3007		TR17/	
A65A-705	160	3007		TR17/	
A65B19G	126	3008		TR17/	
A65B-70	126	3008		TR17/	
A65B-705	126	3008		TR17/	
A65C-19G	121	3009		TR01/	
A65C19G	121	3009			
A65C-70	121	3009		TR01/	
A65C-705	121	3009		TR01/	
A66	102A	3004			
A66-1-70	102A	3003		TR85/	
A66-1-705	102A	3003		TR85/	
A66-1A9G	102A	3003		TR85/	
A66-2-70	102A	3004		TR85/	
A66-2-705	102A	3004		TR85/	
A66-2A9G	102A	3004		TR85/	
A66-3-3A9G	102A			TR85/	
A66-3-70	102A	3004		TR85/	
A66-3-705	102A	3004		TR85/	
A66-3A9G	102A	3004			
A67	126	3006			
A69	126	3006			
A70	160	3006			
A70F	160	3006			
A70L	160	3006			
A70MA	160	3006			
A71	126	3006			
A71A	126	3006			
A71AB	126	3006			
A71AC	126	3006			
A71B	126	3006			
A71BS	126	3006			
A71D	126	3006			
A71Y	126	3006			
A72	126	3006			
A72BLU	126	3006			
A72BRN	126	3006			
A72ORN	126	3006			
A72WHT	126	3006			
A73	160	3006			
A74	160	3006			
A74-3-3A9G	100	3005		TR05/	
A74-3-70	100	3005		TR05/	
A74-3-705	100	3005		TR05/	
A74-3A9G	100	3005			
A75	126	3006			
A75B	102A	3004			
A76	126	3006			
A76-11770	121	3009		TR01/	
A77	126	3006			
A77A	126	3006			
A77A19G	105	3012.		TR01/	
A77A-70	105	3012		TR03/	
A77A-705	105	3012		TR03/	
A77B	126	3006			
A77B19G	121MP	3013			
A77B-70	121MP	3013			
A77B-705	121MP	3013			
A77C	126	3006			
A77C19G	121	3014		TR01/	
A77C-70	121	3014		TR01/	
A77C-705	121	3014		TR01/	
A77D	126	3006			
A78	126	3006			
A79	160	3006			
A80	126	3006			
A81	160	3006			
A82	126	3006			
A83	126	3006			
A82P2B	129	3025			
A84	160	3006		TR88/	
A84A19G	121	3009		TR01/	
A84A-70	121	3009		TR01/	
A84A-705	121	3009		TR01/	
A85	126	3006			
A86	160	3006			
A86-10-2	101	3011		TR08/	
A86-44-2	101	3011		TR08/	
A86-213-2	154	3045		TR78/	
A86-214-2	154	3045		TR78/	
A86-215-2	154	3045		TR78/	
A86-316-2	154	3045		TR78/	
A86-565-2	194				
A86X0030-100	121MP	3013			
A87	126	3006			
A88	123A	3006		TR21/	
A88-70	160			TR17/	
A88-705	160			TR17/	
A88B19G	160	3006		TR17/	
A88B-70	160	3006			
A88B-705	160	3006			
A88C19G	180	3005		TR05/	
A88C-70	100	3005		TR05/	
A88C-705	100	3005		TR05/	
A89	160	3006			
A90	126	3006			
A90T2	108	3039		TR95/	
A92	126	3006			
A92-1-70	105	3012		TR03/	
A92-1-705	105	3012		TR03/	
A92-1A9G	105	3012		TR03/	
A93	126	3006			
A94	126	3006			
A97A83	121	3009		TR01/	
A99S07	101	3011		TR08/	
A99SK5	101	3011		TR08/	
A99SK7	101	3011		TR08/	
A100A	102A	3004			
A100B	126	3006			
A100C	126	3006			
A101	126				
A101A	126				
A101AA	126	3006			
A101AY	126	3006			
A101B	126	3006			
A101BA	126	3006			
A101BB	126	3006			
A101BC	126	3006			
A101BX	126	3006			
A101C	126	3006			
A101CA	126	3006			
A101CV	126	3006			
A101CX	126	3006			
A101E	126	3006			
A101QA	126	3006			
A101V	126				
A101X	126				
A101Y	126	3006			
A101Z	100	3005			
A102	160	3006		TR17/637	
A-102	160	3006			
A102A	126	3006			
A102AA	126	3006			
A102AB	126	3006			
A102B	126	3006			
A102BA	126	3006			
A102BN	126	3006			
A102CA	126	3006			
A102TV	126	3006			
A103	126	3006			
A103A	126	3006			
A103B	126	3006			
A103C	126	3006			
A103CA	126	3006			
A103CAK	126				
A103CG	126	3006			

Original Part Number	ECG	SK	GE	IR/WEP	HEP
A103DA	126	3006			
A104	123A	3124	61	TR21/735	S0015
A104-8	225	3045			
A104A	123A	3124	20	TR21/735	
A104B	123A	3124		TR21/735	
A104B/353-9				TR21	
A104D	126	3006			
A104P	126	3006			
A104Y	123A	3124		TR21/735	
A104Y/353				TR21/	
A105	160	3006			
A106	123A	3124	61	TR21/735	S0033
A107	128	3024	20	TR87/243	
A108	123A	3124	61	TR21/735	S0015
A108A	123A	3124		TR21/735	
A108B	123A	3124		TR21/735	
A108(JAPAN)	160				
A109	126	3006			
A109R				TR17/	
A110	199	3156	212	TR21/	S0024
A111	123A	3122	61	TR21/735	S0015
A111(JAPAN)	126				
A111T2	154	3045		TR78/	
A112	126				
A112-000172	159	3114		TR20/	
A112-000185	159	3114		TR20/	
A112-000187	159	3114		TR20/	
A113	126	3006			
A114	102A				
A115	128	3124	61	TR87/243	S0015
A115(JAPAN)	102A				
A116	123A	3122	61	TR21/	S0015
A116(JAPAN)	102A				
A117	160	3006			
A118	160	3006			
A120P1	159	3114		TR20/	
A121	126				
A121-1	159	3114		TR20/	
A121-1RED	159	3114		TR20/	
A121-15	101	3011		TR08/	
A121-16	101	3011		TR08/	
A121-17	101	3011		TR08/	
A121-21	101	3011		TR08/	
A121-50	101	3011		TR08/	
A121-361	154	3045		TR78/	
A121-444	159	3114		TR20/	
A121-446	159	3114		TR20/	
A121-480	108	3039		TR95/	
A121-495	159	3114		TR20/	
A121-496	159	3114		TR20/	
A121-497	159	3114		TR20/	
A121-497WHT	159	3114		TR20/	
A121-585	107	3039		TR70/	
A121-585B	107	3039		TR70/	
A121-602	159	3114		TR20/	
A121-603	159	3114		TR20/	
A121-679	159	3114		TR20/	
A121-687	107	3039		TR70/	
A121-699	159	3114		TR20/	
A121-746	159	3114		TR20/	
A121-762	101	3011		TR08/	
A121-774	159	3114		TR20/	
A121-1410	101	3011		TR08/	
A122	160	3007		TR17/637	
A122-1962	103A	3010		TR08/	
A122GRN	159	3114		TR20/	
A122YEL	159	3114		TR20/	
A122(JAPAN)	126				
A123	126				
A124	126	3006			
A125	126	3006			
A126	160	3006			
A127	160	3006			
A127-7	101	3011			
A128	123A	3124	20	TR21/735	
A128A	123A	3124	20	TR21/735	
A128(JAPAN)	158				
A129-30	101	3011		TR08/	
A129-34	159	3114		TR20/	
A130	154	3039	18	TR78/	S5026
A130-149	159	3114		TR20/	
A130-40315	159	3114		TR20/	
A130-40429	159	3114		TR20/	
A130(JAPAN)	160				
A130-ORN	154			TR78/	
A130ORN	108	3039			
A130-V10	154			TR78/	
A130V100	108	3039			
A131	126	3006			
A132	154		18	TR78/	S5026
A132(JAPAN)	126				
A133	194	3045		/712	S0005
A133(JAPAN)	126				
A134	126	3006			
A135	160				
A136	126				
A137	123A	3114			S0015
A137(JAPAN)	126				
A137(NPN)	123A			TR21/	
A137(PNP)	159			TR20/	
A138	199			TR21/	S0024
A138(JAPAN)	102A				
A139	199			TR21/	S0024
A139(JAPAN)	126				
A141	123A	3122	39	TR51/735	S0015
A141B	126				
A141C	126				
A141(JAPAN)	126				
A142	123A	3122	39	TR21/735	S0015
A142A	126				
A142B	126				
A142C	126				
A142(JAPAN)	126				
A143	123A	3122	212	TR21/735	S0015
A143(JAPAN)	126				
A144	126				
A144A-1	121	3009		TR01/	
A144A-19	121	3009		TR01/	
A144C	126				
A145	126				
A145A	126				
A145C	126				
A146	100				
A146B-3	121MP	3015			
A146B-39	121MP	3015			
A147	100				
A148	100				
A148P2	102A			TR85/	
A148P-2	102A	3004			
A148P2-29	102A			TR85/	
A148P-29	102A	3004			
A149	100				
A149L-4	102A	3003		TR85/	
A149L-49	102A	3003		TR85/	
A150	126				
A151	123A		39	TR21/	S0015
A151(JAPAN)	160				
A152	123A		39	TR21/	S0015
A152(JAPAN)	160				
A153	123A		212	TR21/	S0015
A153(JAPAN)	126				
A154(NPN)	108		61		
A154(PNP)	126				
A155(NPN)	108		61		
A155(PNP)	126				
A156	123A	3122		TR21/735	
A-156	123A	3122	20	TR21/	
A156(JAPAN)	126				
A157	123A	3010		TR21/	S0015
A157A	123A	3122		TR21/735	S0015
A157B	123A	3122		TR21/735	S0015
A157C	123A			TR21/	S0015
A157(JAPAN)	126				
A158	123A	3124	20	TR21/735	S0015
A158A	123A	3038		TR21/735	
A158B	123A	3124	10	TR21/735	S0015
A-158B	123A	3124			
A158C	123A	3124	10	TR21/735	S0024
A-158C	123A	3124			
A159	123A	3124	18	TR21/735	S0024
A159A	123A	3124		TR21/735	
A159B	123A	3124	20	TR21/735	S0015
A159C	123A	3122		TR21/735	S0024
A160	159	3118	65	TR20/	S0013
A160(JAPAN)	100				
A161	159	3118	65	TR20/	S0013
A162	159	3118	65	TR20/	S0019
A163	160	3006	21	TR17/637	
A164(NPN)	108		61		
A164(PNP)	126				
A165(NPN)	108		61		
A165(PNP)	126				
A166	126				
A167	100				
A168	123A		20	TR21/735	S0015
A-168	123A	3124			
A168A	100				
A168(JAPAN)	100				
A168P1	121MP	3013			
A169	100	3005			
A170	129	3025		TR88/243	S0019
A170(JAPAN)	100				
A171	159	3025		TR20/	S0013
A171(JAPAN)	100				
A172	100	3005			
A172A	100	3005			
A173	100	3005			
A173B	102A	3005			
A174	102	3004			
A175	126	3024			

Original Part Number	ECG	SK	GE	IR/WEP	HEP
A176	126				
A176-025-9-002	108	3039		TR95/	
A177	159		82	TR20/	S0019
A177A	159	3114		TR20/717	
A177(A)	159		21	TR30/	S0031
A177AB	159	3114		TR20/	
A178A	159	3114	65	TR20/717	S0013
A178AB	159	3114		TR20/	
A178B	159	3114	65	TR20/717	S0031
A178BA	159	3114		TR20/	
A179A	159	3114	65	TR20/717	S0013
A179AC	159	3114		TR20/	
A179B	159	3114	65	TR20/717	S0031
A179BB	159	3114		TR20/	
A180	160	3006			
A181	102A	3004			
A182	100	3005			
A183	102A	3004			
A187TV	102A	3004			
A188	126	3006			
A189	126	3006			
A192	132	3116		FE100/	
A194	133	3112			F0010
A195	133	3112			F0010
A-195C	159	3114	22	TR20/	S0013
A196	133	3112			F0010
A197	102	3004			
A198	100				
A200-052	159	3114		TR20/	
A200(REACH)	724	3525			
A201			28	TR76/	S3020
A201-0	126				
A201A	126				
A201B	126				
A201E	102A				
A202	126		66	TR66/	S3006
A202A	126				
A202B	126				
A202C	126				
A202D	126				
A203			28	TR76/	S3001
A203A	126				
A203AA	102A				
A203B	102A				
A203(NPN)	186				
A203P	102A				
A203(PNP)	102A				
A204	100				
A205	100				
A206	100				
A207	100				
A208	186		28	TR76/	
A208(JAPAN)	100				
A209	100				
A210			28	TR76/	S3008
A210(JAPAN)	100				
A211			28	TR76/	S3008
A210(JAPAN)	100				
A212	100				
A213	126				
A214	126				S3008
A215	160				S3013
A216	126				
A217	100				
A218	160				
A219	160				
A220	160				
A221	160				
A222	160				
A223	160				
A224	160				
A225	160				
A226	160				
A227	160				
A228	160				
A229	160				
A230	160				S3008
A231	160				
A232	160				
A233	160				
A233A	160				
A233B	160				
A233C	160				
A234	160				
A234A	160				
A234B		3006	50	TR17/	
A234C		3006	50	TR17/	
A235	160				
A235A		3006	50	TR17/	
A235B	160				
A235C		3006	50	TR17/	
A236	126				
A237	126				
A238	126				
A239	160				

Original Part Number	ECG	SK	GE	IR/WEP	HEP
A240	160				
A240A	160				
A240B	160				
A240B2	160				
A240BL	160				
A241	160				
A242	160				
A243	160				
A244	161	3117	11	TR83/719	
A244(JAPAN)	160				
A245	108	3039	11	TR95/	S3006
A245(JAPAN)	160				
A246	123A	3122	20	TR21/735	S3006
A246(JAPAN)	126				
A246V	126				
A247	154	3044	28	TR83/719	S3006
A247(JAPAN)	126				
A248	161	3117	11	TR83/719	
A248(JAPAN)	100				
A249	123A	3122	17	TR21/735	
A249(AMC)	128			TR87/	S0015
A250	126				
A251	126				
A252	126				
A253			28	TR76/	S3006
A253(JAPAN)	126				
A254	126				
A255	126				
A256	126				
A257	126				
A258	126				
A259	126				
A260	126				
A261	126				
A262	126				
A263	126				
A264	126				
A265	126				
A266	126				
A267	126				
A268	126				
A269	126				
A270	195		28	TR76/	S3001
A270(JAPAN)	126				
A271			28		S3006
A271(JAPAN)	126				
A272	152		66		
A272(JAPAN)	126				
A273	152		66		
A273(JAPAN)	126				
A274(NPN)	195				
A274(PNP)	126				
A275			28	TR76/	S3006
A275(JAPAN)	126				
A276			66		S3006
A276(JAPAN)	160				
A277			66		S3007
A277(JAPAN)	100				
A278	171	3201	27	TR78/	
A278(JAPAN)	100				
A279	171	3201	27	TR78/	
A279(JAPAN)	100				
A280	126				
A281	126				
A282	100				
A283	100				
A284	100				
A285	160				
A286	160				
A287	160				
A288	126				
A288A	126				
A289	126				
A290	126				
A291	126				
A292	126				
A293	126				
A294	126				
A295	126				
A296	160				
A297	160				
A297L012C01	159	3114		TR20/	
A297V073C01	129	3025		TR88/	
A297V073C02	129	3025		TR88/	
A297V073C03	159	3114		TR20/	
A297V073C04	159	3114		TR20/	
A297V082B03	129	3025			
A298	160				
A301	123A	3122	212	TR21/735	S0011
A301(JAPAN)	126				
A302	102A				
A303	102A				
A304	102A				
A305	100				
A306	123A	3122	62	TR21/735	S0011

Original Part Number	ECG	SK	GE	IR WEP	HEP
A306(JAPAN)	160				
A307	123A	3122	62	TR21/735	S0011
A307(JAPAN)	160				
A308	126				
A309	126				
A310	154	3045	27	TR78/712	S5025
A310(JAPAN)	126				
A311	128	3024	213	TR87/243	S5026
A311(JAPAN)	100				
A312	100				
A313	126				
A314	126				
A315	126				
A316	126				
A321	108	3039	61	TR95/56	S0015
A321(JAPAN)	160				
A322	123A	3122	212	TR21/	S0015
A322(JAPAN)	160				
A323	123A	3122	212	TR21/	S0015
A323(JAPAN)	160				
A324	123A	3122	212	TR21/	S0015
A324(JAPAN)	160				
A325	126				
A326	126				
A327	160				
A329	126				
A329A	126				
A329B	126				
A330	100				
A331	126				
A332	100				
A335	126				
A337	126				
A338	126				
A339	126				
A340	160				
A341	160				
A341-0A	160				
A341-0B	160				
A342	160				
A342A	160				
A343	126				
A344	123A	3122	20	TR21/735	S0011
A344(JAPAN)	126				
A345	123A	3122	20	TR21/735	S0011
A345(JAPAN)	160				
A346	3122	20		TR21/735	S0011
A346(JAPAN)	160				
A347	160				
A348	160				
A349	160				
A350	126				
A350A		3008	1	TR12/	
A350C	126				
A350H	126				
A350R	126				
A350T	126				
A350TY	126				
A350Y	126				
A351	126				
A351A	126				
A351B	126				
A352	126				
A352A	126				
A352B	126				
A353	126				
A353-9008-001	129	3025		TR88/	
A353A	126				
A353C	126				
A354	126				
A354A	126				
A354B	126	3006	50	TR17/	
A355	126				
A355A	126				
A356	126				
A357	126				
A358	126				
A359	126				
A360	160				
A361	160				
A362	160				
A363	160				
A364	126				
A365	126				
A366	126				
A367	126				
A368	126				
A369	126				
A372	160				
A373	160				
A375	160				
A376	160	3006	50	TR17/637	
A376(JAPAN)	126				
A377	160				
A378	160				
A379	160				
A380	126				
A381	126				
A382	126				
A383	126				
A384	126				
A385	126				
A385A	126				
A385D	126				
A391	126				
A392	126				
A393	126				
A393A	126				
A394	126				
A395	126				
A396	102				
A397	126				
A398	126				
A399	126				
A400	126				
A401	160				
A402	159				
A403	160				
A404	160				
A405	160				
A406	102A				
A407	102A				
A408	160				
A409	160				
A410	160				
A411	160	3006		TR17/	G0009
A412	126				
A413	160				
A414	100				
A415	123A	3018	61	TR21/735	S0016
A415(JAPAN)	100				
A416	121				
A417	108	3039	39	TR95/56	S0016
A417-19	161	3039		TR83/	
A417-43	129	3025		TR88/	
A417-62	121	3009		TR01/	
A417-115	154	3045		TR78/	
A417-116	159	3114		TR20/	
A417-132	159	3114		TR20/	
A417-138	129	3025		TR88/	
A417-153	159	3114		TR20/	
A417-154	107	3039		TR70/	
A417-170	129	3025		TR88/	
A417-176	159	3114		TR20/	
A417-182	159	3114		TR20/	
A417-184	159	3114		TR20/	
A417-190	107	3039		TR70/	
A417-196	159	3114		TR20/	
A417-200	159	3114		TR20/	
A417-201	159	3114		TR20/	
A417-205	107	3039		TR70/	
A417-234	129	3025		TR88/	
A417-235	159	3114		TR20/	
A417(JAPAN)	160				
A418	108	3039	39	TR95/56	S0017
A419	108	3039	39	TR95/56	S0017
A419(JAPAN)	160				
A420	108	3039	39	TR95/56	S0017
A420(JAPAN)	160				
A421	160				
A422	160				
A425	160				
A426	160				
A427	108	3039	214	TR95/56	S0020
A428	126				
A429-0981-12	161	3039		TR83/	
A430	161	3039		TR83/	S0017
A430(JAPAN)	160				
A431	160				
A431A	160				
A432	160				
A432A	160				
A433	160				
A434	160				
A435	160	3006	51	TR17/637	
A435A	160				
A435B	160				
A436	126				
A437	126				
A438	126				
A440	160				
A440A	160				
A446	126				
A447	126				
A448	160				
A450	160				
A450H	160				
A451	161	3039	17	TR83/719	S0017
A451H	160				
A451(JAPAN)	160				

Original Part Number	ECG	SK	GE	IR/WEP	HEP
A452	160				
A452H	160				
A453	126				
A454	123A	3039	20	TR21/735	S0017
A454(JAPAN)	126				
A455	126	3039	17		S0017
A455(IC)	709				
A456	126				S0016
A457	126				
A460	160				
A461	160				
A462	160				
A463	160				
A464	160				
A465-181-19	161	3039		TR83/	
A466	128	3024	17	TR87/243	
A466-2	126				
A466-3	126				
A466BLK	126				
A466BLU	126				
A466YEL	126				
A466(JAPAN)	126				
A467	108	3039	39	TR95/56	S0016
A467G	159				
A467G-O	159				
A467G-R	159				
A467G-Y	159				
A467(JAPAN)	159				
A467-O	159				
A467-Y	159				
A468	126				
A469	126				
A470	126				
A471	126				
A471-1	126				
A471-2	126				
A471-3	126				
A472	123A				S0015
A472A	126				
A472B	126				
A472C	126				
A472D	126				
A472E	126				
A472(JAPAN)	126			TR21/	
A473	108	3018	61	TR95/56	S0016
A473-GR	153				
A473(JAPAN)	153				
A473-O	153				
A473-R	153				
A473-Y	153				
A473Y	153				
A474	126				
A476	126				
A477	126				
A478	126				
A479	126				
A480	108	3039	39	TR95/56	S0017
A480(JAPAN)	159				
A481	161	3039	60	TR83/	S0017
A481A0028	123A				
A481A0030	159				
A481A0031	123A				
A482	161	3117	214	TR83/719	S0017
A482(JAPAN)	159				
A483	161	3039	214	TR83/719	S0017
A484	161	3117	214	TR83/719	S0017
A484A		3039			
A484(JAPAN)	189				
A485	128	3024	86	TR87/243	S0017
A486	161	3039	39	TR83/	S0017
A486(JAPAN)	189				
A489	161	3039		TR83/	S0017
A489(JAPAN)	153				
A489-O	153				
A489-R	153				
A489-Y	153				
A490	128	3024	86	TR87/243	S0017
A490(JAPAN)	159				
A490(POWER)	153				
A490Y	159				
A492	161	3117	86	TR83/719	S0017
A494	123A	3018	39	TR21/735	S0016
A494-GR	159				
A494(JAPAN)	123A				
A494-O	159				
A494-Y	159				
A495	107	3018	39	TR88/720	S0016
A495D	159				
A495-G	159				
A495G-GR	159				
A495G-O	159				
A495G-R	159				
A495G-Y	159				
A495(JAPAN)	159				
A495-O	159				
A495-R	159				
A495W	159				
A495-Y	159				
A495Y	159				
A496	107	3191	61	TR70/	S0016
A496(JAPAN)	185				
A496-O	185	3191		TR77/	
A496-R	185	3191			
A496-Y	185	3191			
A496Y	185	3191			
A497	108		61	TR95/	S0016
A497(JAPAN)	129				
A498	222	3050			F2004
A498(JAPAN)	129				
A499	159				
A499-O	159				
A499-R	159				
A499-Y	159				
A500	159				
A500-O	159				
A500-R	159				
A500-Y	159				
A501	129				
A502	159				
A503	129				
A503GR	211				
A503-O	129				
A503-R	129				
A503-Y	129				
A504	129				
A504GR	211	3203			
A504-R	129				
A504-Y	129				
A505	185	3191			
A505-O	185	3191			
A505-R	185	3191			
A505-Y	185	3191			
A506	160				
A507	160				
A508	160				
A509	159				
A509GR	159				
A509-O	159				
A509R	159				
A509Y	159				
A509-Y	159				
A510	159				
A510-O	159				
A510-R	159				
A511	159				
A511-O	159				
A511-R	159				
A512	159				
A512-O	159				
A512-R	159				
A5113	159				
A513-O	159				
A513-R	159				
A514-023553	103	3010	8	TR08/	
A514-027662	102A	3004		TR85/250	
A514-033338	123A	3018	17	TR21/	
A514-033377		3009			
A514-040296	133	3118	FET1		
A514-044910	159	3114	21	TR20/	
A514-047830	186	3054	28	TR55/	
A515	130			TR59/	S5024
A516	129				
A516A	129				
A517	126				
A518	126				
A522	130	3027	19	TR59/247	S7002
A522-3	130	3027		TR59/	
A522A	159				
A522(JAPAN)	159				
A523	130			TR59/	S7002
A525	160	3006			
A525A	160	3006			
A525B	160	3006			
A527	129	3025			
A528	129	3025			
A530	159	3025			
A530H	159	3025			
A532	129	3025			
A532A	129	3025			
A532B	129	3025			
A532C	129	3025			
A532D	129	3025			
A532E	129	3025			
A532F	129	3025			
A537	129	3025			
A537A	129	3025			
A537AA	129	3025			
A537AB	129	3025			
A537AC	129	3025			
A537AH	129	3025			

Original Part Number	ECG	SK	GE	IR/WEP	HEP
A537B	129	3025			
A537C	129	3025			
A537H	129	3025			
A538	102A	3004			
A539	159	3025			
A539L	159	3025			
A539S	159	3025			
A542	159	3025			
A543	192	3137			
A544	159	3025			
A545	193	3138			
A545GRN	193	3138			
A545KLM	193	3138			
A545L	193	3138			
A546	129	3025			
A546A	129	3025			
A546B	129	3025			
A546E	129	3025			
A546H	129	3025			
A547	193	3138			
A547A	193	3138			
A548	129	3025			
A550	159	3025			
A550A	159	3025			
A550AQ	129	3025			
A550Q	159				
A550R	129	3025			
A550S	159				
A551	129	3025			
A551C	129	3025			
A551D	129	3025			
A551E	129	3025			
A552	129	3025			
A560	129	3025			
A561	159	3025			S0019
A561GR	159	3025			
A561-O	159	3025			
A561-R	159	3025			
A561-Y	159	3025			
A562	159				S0012
A562GR	159				
A562-O	159	3025			
A562-R	159	3025			
A562-Y	159	3025			
A564	234	3025			S0019
A564-O	234				
A564A	234	3025			
A564ABQ	234	3025			
A564F	234	3025			
A564FQ	234				
A564FR	159	3025			
A564-O	159	3025			
A564P	234	3025			
A564Q	234	3025			
A564R	234	3025			S0019
A564S	234				
A564T	234	3025			
A565	159	3025			
A565A	159	3025			
A565B	159	3025			
A565C	159	3025			
A565D	159				
A566	218	3083			
A566A	218	3083			
A566B	218	3083			
A566C	218	3083			
A567	123A	3124		TR21/735	S0015
A-567			8	TR09/	
A567A	123A		20	TR21/735	
A-567A	123A	3124			
A567(JAPAN)	159				
A568	159	3025			
A569	159	3025			
A569J	159	3025			
A570	159	3025			
A571	129	3025			
A572	130	3027	19	TR59/247	S7002
A572-1	130	3027		TR59/	
A576-0001-002	159	3114		TR20/	
A576-0001-013	159	3114		TR20/	
A580-040215	181	3535	75		
A580-040315	181	3535	75		
A580-080215	181				
A580-080315	181	3535			
A580-080515	181	3535	75		
A592Y	159	3025			
A593	123A			TR21/	S0015
A594	129	3025			
A594-O	129	3025			
A594-R	129	3025			
A594-Y	129	3025			
A595	102A			TR85/	G0005
A595C	129	3025			
A597	129	3025			
A603	159	3025			
A604	129	3025			
A606	129	3025			
A606S	129	3025			
A608A	159	3025			
A608B	159	3025			
A608C	159	3025			
A608-C	159	3025			
A608D	159	3025			
A608-D	159				
A608E	159	3025			
A608-E	159	3025			
A608F	159	3025			
A608-F	159	3025			
A608G	159	3025			
A609	159	3025			
A609A	159	3025			
A609B	159	3025			
A609C	159	3025			
A608D	159	3025			
A609E	159	3025			
A609F	159	3025			
A609G	159	3025			
A610	159	3025			
A610B	159	3025			
A611	159	3025			
A611-4E	159	3025			
A613	218	3083			
A614	218	3083			
A615-1008	160	3006		TR17/637	
A615-1009	160	3006		TR17/637	
A615-1010	102A	3004		TR85/250	
A615-1011	102A	3004		TR85/250	
A616	218	3083			
A623	211	3203			
A623-O	211	3203			
A623A	211	3203			
A623B	211	3203			
A623C	211	3203			
A623D	211	3203			
A623G	211	3203			
A623R	211	3203			
A623Y	211	3203			
A624	211	3203			
A624A	211	3203			
A624B	211	3203			
A624C	211	3203			
A624D	211	3203			
A624GN	211	3203			
A624L	211	3203			
A624LG	211	3203			
A624R	211	3203			
A624Y	211	3203			
A626	219	3173			
A627	219	3173			
A628	159	3025			
A628A	159	3025			
A628E	159	3025			
A628F	159	3025			
A629	159				
A634	187	3193			
A634A	187	3193			
A634B	187	3193			
A634C	187	3193			
A634D	187	3193			
A634L	187	3193			
A636	187	3193			
A636A	187	3193			
A636B	187	3193			
A636C	187	3193			
A636D	187	3193			
A637	159	3025			
A640	108	3039		TR95/	S0015
A640A	159	3025			
A640B	159	3025			
A640C	159	3025			
A640D	159	3025			
A640(JAPAN)	159				
A640L	123A	3039	60	TR95/	S0015
A640M	159	3025			
A640S	161	3132	60		
A641	234	3122		TR21/	S0015
A641A	234	3025			
A641B	234	3025			
A641C	234	3025			
A641D	234	3025			
A641(JAPAN)	159				
A641L	123A	3122	212	TR21/	S0015
A641M	234	3025			
A641(NPN)	123A				
A641(PNP)	234				
A641S	199		212		
A642	123A	3122		TR21/	S0015
A642-254	108	3039		TR95/	
A642-260	108	3039		TR95/	
A642-268	108	3039		TR95/	

Original Part Number	ECG	SK	GE	IR/WEP	HEP
A642-271	121MP	3013			
A642A	159	3025			
A642B	159	3025			
A642C	159	3025			
A642D	159	3025			
A642E	159	3025			
A642F	159	3025			
A642(JAPAN)	159				
A642L	199		212		S0015
A642S	199		212		
A643	193	3138			
A643A	193	3138			
A643B	193	3138			
A643C	193	3138			
A643D	193	3138			
A643E	193	3138			
A643F	193	3138			
A643L	108	3039	61		
A643R	193	3138			
A643S	108	3039	61		
A643W	129	3025			
A644L	199		212		
A644S	199		212		
A645	211	3203			
A645L	199		212		
A645S	199		212		
A646	187	3193			S0025(2)
A648	219	3173			
A648A	219	3173			
A648B	219	3173			
A648C	219	3173			
A649L	123A	3122	212	TR21/	S0015
A649S	123A	3122	212	TR21/	S0015
A658	219				
A659	129	3025			
A659C	129	3025			
A659D	129	3025			
A659E	129	3025			
A659F	159	3025			
A660			90		
A663	219	3173			
A666	234	3025			
A666A	234	3025			
A666H	234				
A666QRS	234	3025			
A666R	234				
A666S	234	3025			
A667-GRN	199		212	TR51/	
A667-ORN	199		212	TR51/	
A667RED	108	3039		TR95/56	
A667-RED	199		212	TR51/	
A667-YEL	199		212	TR51/	
A668-GRN	199		212	TR51/	
A668-ORG	199		212	TR51/	
A668-YEL	199		212	TR51	
A669	159	3114		TR20/	
A669-GRN	199		212	TR51/	
A669-YEL	199		212		
A670	153	3083			
A670A	153	3083			
A670B	153	3083			
A670C	153	3083			
A671	153	3083			
A671A	153	3083			
A671B	153	3083			
A671C	153	3083			
A672	159	3025			
A672A	159	3025			
A672B	159	3025			
A672C	159	3025			
A673	159	3025			
A673A	159	3025			
A673AA	159	3025			
A673AB	159	3025			
A673AC	159	3025			
A673AD	159	3025			
A673AE	159	3025			
A673B	159	3025			
A673C	159	3025			
A673D	159	3025			
A677	129	3025			
A678	159	3025			
A683	129	3025			
A683Q	193				
A683R	193				
A683S	193				
A684	129	3025			
A684Q	129				
A684R	129				
A690V081H97	121	3009		TR01/	
A695C	129	3025			
A699	187	3193		TR56/	
A699A	187	3193			
A699AP	187	3193			
A699P	187	3193			
A699R	187	3193			
A700B	160	3006			
A700Y	160	3006			
A701	159				
A701F	159				
A702	162			TR67/707	S5015
A705	163	3111	36	TR67/740	S5021
A707	193				
A707V	193				
A708	211	3203			
A708A	211	3203			
A708B	211	3203			
A708C	211	3203			
A715	185	3191			
A715A	185	3191			
A715B	185	3191			
A715C	185	3191			
A715D	185	3191			
A715FB	108	3039			
A717	211	3203			
A718	159	3025			
A719	159	3025			
A719S	159	3025			
A720	159	3025			
A723	159	3025			
A723A	159	3025			
A723B	159	3025			
A723C	159	3025			
A723D	159	3025			
A723E	159	3025			
A723F	159	3025			
A730	159	3025			
A731	159	3025			
A733	159	3025			
A733A	159	3025			
A733B	159	3025			
A733C	159	3025			
A733D	159	3025			
A733E	159	3025			
A733F	159	3025			
A733H	159				
A733I	159				
A733Q	159				
A736	129				
A742	129				
A742-100-44		3008			
A744	219				
A746	219				
A747	123A	3122	210	TR21/	S0015
A747A	123A	3122		TR21/	S0015
A747B	199			TR21/	S0015
A748	123A	3124	10	TR21/735	S0015
A748B	123A	3122		TR21/	S0015
A748C	199			TR21/	S0024
A749	199	3124	212	TR21/735	S0024
A749B	123A	3122		TR21/	S0015
A749C	199			TR21/	S0024
A751	193	3138			
A751R	193				
A752	193	3138			
A753-4004-248	159	3114		TR20/	
A757	123A	3122	82	TR21/	S0015
A758	193	3025	65	TR52/	
A758A	159			TR20/	S0019
A758B	159	3025		TR20/	S0031
A759			65	TR52/	
A759A	159	3025		TR20/	S0019
A759B	159	3025		TR20/	S0031
A764	218	3083			
A772B1	107	3039		TR70'	
A772D1	108	3039			
A772EH	107	3039		TR70/	
A772FE	107	3039		TR70/	
A772.01	107			TR70/	
A777	154	3044	81	TR78/	S5026
A778	154	3044	27	TR78/	S5025
A779	171	3201	27	TR78/	S5024
A800-511-00	129	3025			
A800-516-00	129	3025		TR88/	
A800-523-01	159	3114		TR20/	
A800-523-02	159	3114		TR20/	
A800-527-00	159	3114		TR20/	
A829	159	3114		TR20/	
A829A	159	3114		TR20/	
A829B	159	3114		TR20/	
A829C	159	3114		TR20/	
A829D	159	3114		TR20/	
A829E	159	3114		TR20/	
A829F	159	3114		TR20/	
A833	159	3114		TR20/	
A880-250-107		3025		TR88/	
A909-1008	160	3006		TR17/637	G0003
A909-1009	126	3006		TR17/635	G0008
A909-1010	126	3006		TR17/635	G0009
A909-1011	102A	3004		TR85/250	G0005

Original Part Number	ECG	SK	GE	IR/WEP	HEP
A909-1012	102A	3004		TR85/250	G0005
A909-1013	102A	3004		TR85/250	G0005
A909-27125-160	108	3039		TR95/	
A916-31025-5B	108	3039		TR95/	
A921-59B	108	3039		TR95/	
A921-62B	108	3039		TR95/	
A921-63B	108	3039		TR95/	
A921-64B	108	3039		TR95/	
A921-70B	159	3114		TR20/	
A937	123A	3124		TR21/	
A937-1	123A	3124	20	TR21/735	
A937-3	123A	3020	20	TR21/735	
A964-17887	121	3009		TR01/	
A991-01-0098	159	3114		TR20/	
A991-01-1225	159	3114		TR20/	
A991-01-1316	108	3039		TR95/	
A991-01-1319	159	3114		TR20/	
A991-01-3058	159	3114		TR20/	
A992-01-1192	121	3009		TR01/	
A-1005-725	159				
A1016	129	3025		TR88/242	
A1030	159	3114		TR20/	
A1109	108	3039	17	TR95/56	S0011
A1124C	121	3009		TR01/	
A-1141-5932	128				
A-1141-6062	123A				
A1170	108	3019	11	TR95/56	S0016
A1200	1003				
A1201	1003	3101			
A1201B	1003				
A1201C	1003				
A1201C-W	1003				
A1201T	1003				
A1214	159	3114		TR20/	
A1220	160	3006	52	TR17/637	G0008
A1238	199		62		
A1243	100	3005	50	TR05/254	G0003
A1314	128	3024	17	TR87/243	
A1341	128	3024	63	TR87/243	S0005
A1353	1080				
A1364	1004				
A1365	712				
A1373	1094				
A1374	1121				
A1376	1122				
A1377	160	3006	52	TR17/637	G0009
A1378	160	3123	52	TR17/637	G0001
A1379	123A		81	TR21/735	S0011
A-1379	123A	3124			
A1380	123A		81	TR21/735	S0015
A-1380	123A	3124			
A1383	126	3006	51	TR17/635	G0001
A1384	126	3123	51	TR17/635	G0001
A-1384	160	3007			
A1396	101	3011		TR08/	
A1409	154	3044		TR78/	S5025
A1412-1	123A	3124		TR21/	
A1412-19		3124			
A1414A	121	3009		TR01/	
A14A9	121	3009		TR01/	
A1460	199			TR21/	S0024
A1462	108	3039	63	TR95/56	S0011
A1465-1	160	3006		TR17/	
A1465-4	101	3011		TR08/	
A1465-19	160	3006		TR17/	
A1465-29	103A	3010		TR08/	
A1465-49	101	3011		TR08/	
A1465A	160	3007		TR17/	
A1465A9	160	3007		TR17/	
A1465B	160	3008		TR17/	
A1465B9	160	3008		TR17/	
A1465C	121	3009		TR01/	
A1465C9	121	3009		TR01/	
A1466-1	102A	3003		TR85/	
A1466-2	102A	3004		TR85/	
A1466-3	102A	3004		TR85/	
A1466-19	102A	3003		TR85/	
A1466-29	102A	3004		TR85/	
A1466-39	102A	3004		TR85/	
A1472-19	123			TR86/	
A1474-3	100	3005		TR05/	
A1474-39	100	3005		TR05/	
A1477A	105	3012		TR03/	
A1477A9	105	3012		TR03/	
A1477B	121MP	3013			
A1477B9	121MP	3013			
A1477C	121	3014		TR01/	
A1477C9		3014		TR01/	
A1484A	121	3009		TR01/	
A1484A9	121	3009		TR01/	
A1488B	160	3006		TR17/	
A1488B9	160	3006		TR17/	
A1488C	100	3005		TR05/	
A1488C9	100	3005		TR05/	
A1492-1	105	3012		TR03/	

Original Part Number	ECG	SK	GE	IR/WEP	HEP
A1492-19	105	3012		TR03/	
A1518	108	3039	11	TR95/56	S0015
A1519	108	3039	11	TR95/56	S0015
A1520	108	3039	11	TR95/56	S0015
A1521	108	3039	11	TR95/56	S0015
A1558-17	159	3114		TR20/	
A1567	123A		20	TR21/735	
A-1567	123A	3124			
A1567-1	123A	3124	20	TR21/735	
A1820-0203-1	941				
A1820-0219-1	941M				
A1821-0001-1	912				
A1826-0007-1	941				
A1844-17	159	3114		TR20/	
A-1853-0009-1	106				
A-1853-0010-1	106				
A-1853-0016-1	159				
A-1853-0020-1	159				
A-1853-0027-1	159				
A-1853-0034-1	106				
A-1853-0036-1	159				
A-1853-0039-1	159				
A-1853-0041-1	129				
A-1853-0049-1	159				
A-1853-0050-1	234				
A-1853-0058-1	159				
A-1853-0062-1	159				
A-1853-0063-1	218				
A-1853-0065-1	159				
A-1853-0066-1	234				
A-1853-0077-1	234				
A-1853-0092-1	159				
A-1853-0098-1	234				
A-1853-0099-1	159				
A-1853-0220-1	218				
A1853-0233-1	197				
A-1853-0233-1	153				
A-1853-0234-1	153				
A-1853-0254-1	153				
A-1853-0285-1	159				
A-1853-0300-1	234				
A-1853-0321-1	159				
A-1853-0404-1	199				
A-1854-0003-1	123A				
A-1854-0019-1	123A				
A-1854-0022-1	128				
A-1854-0023-1	199				
A-1854-0025-1	123A				
A-1854-0027-1	123A				
A-1854-0071-1	123A				
A-1854-0087-1	128				
A-1854-0088-1	199				
A-1854-0090-1	128				
A-1854-0092-1	107				
A-1854-0094-1	123A				
A-1854-0099-1	123A				
A-1854-0201-1	123A				
A-1854-0215-1	123A				
A-1854-0241-1	123A				
A-1854-0246-1	123A				
A-1854-0251-1	123A				
A-1854-0254-1	123A				
A-1854-0284-1	199				
A-1854-0291-1	130				
A-1854-0294-1	130				
A-1854-0354-1	123A				
A-1854-0358-1	194				
A-1854-0365-1	194				
A-1854-0408-1	123A				
A-1854-0420-1	152	3054			
A-1854-0434-1	123A				
A1854-0439-1		3511			
A-1854-0449-1	175				
A-1854-0458-1	181				
A-1854-0464-1	152				
A-1854-0471-1	123A				
A-1854-0474-1	194				
A-1854-0485	108				
A1854-0490-1		3036			
A-1854-0492-1	123A				S0025
A1854-0533	194				
A-1854-0533	194				
A-1854-0541-1	123A				
A-1854-0554-1	123A				
A1854-0569-1		3536			
A1854-0582-1		3103			
A-1854-JBD1	108				
A-1854-PJL3		3040			
A-1854-RLK1-1		3021			
A1858	101	3011		TR08/	
A1867-17	159	3114		TR20/	
A1901-5338	159				
A-2010-5409-B	912				
A-2010-5409-RB	912				
A20192C	123A				

Original Part Number	ECG	SK	GE	IR/WEP	HEP
A2039-2	101	3011		TR08/	
A2057A2-198	159	3114		TR20/	
A2057B2-115	107	3039		TR70/	
A2057B104-8	154	3045		TR78/	
A2057B106-12	159	3114		TR20/	
A2057B108-6	159	3114		TR20/	
A2057B110-9	129	3025		TR88/	
A2057B112-9	129	3025		TR88/	
A2057B114-9	129	3025		TR88/	
A2057B115-9	129	3025		TR88/	
A2057B116-9	129	3025		TR88/	
A2057B121-9	129	3025		TR88/	
A2057B122-9	129	3025		TR88/	
A2057B145-12	129	3025		TR88/	
A2057B163-12	129	3025		TR88/	
A2090	154	3045		TR78/	
A2311	234				
A2332	161			TR83/	S0017
A2410	123A	3124	11	TR21/735	S0015
A2411	123A		11	TR21/735	S0015
A2412	123A		11	TR21/735	S0015
A2413	123A		20	TR21/735	S0020
A2414	158		52	TR84/630	G6011
A2415	175	3131	23	TR81/241	S5012
A2416	195	3024	63	TR65/243	S3001
A2417	163	3111		TR67/740	
A2418	130	3027	14	TR59/247	S7004
A2428	159		21	TR20/	S0019
A2434	123A	3124		TR21/735	
A2448	159	3114		TR20/	
A2464	161	3117		TR83/719	S0017
A2465	161	3117		TR83/719	S0017
A2466	123A	3117		TR21/735	S5026
A2468	123A	3040		TR21/735	S0015
A2469	123A	3018		TR21/735	S0016
A2470	123A	3018		TR21/735	S0016
A2471	128	3024		TR87/243	S5026
A2479	161	3018		TR83/719	S0011
A2480	161	3124		TR83/719	S0015
A2482	129	3025			
A2498	108			TR95/	S0016
A2499	123A	3024		TR21/735	S025
A2620	154	3045		TR78/	
A2652-919	133				
A2746	161	3039		TR83/	S0017
A2798	159	3114		TR20/	
A3148	727	3071			
A3300	1005				
A3301	1006				
A3513	159	3114		TR20/	
A3523	129	3025		TR88/	
A3533	129	3025		TR88/	
A3533-1	129	3025		TR88/	
A3540	159	3114		TR20/	
A3545	225	3045			
A3547	225	3045			
A3549	159	3114		TR20/	
A3552	225	3045			
A3553	225	3045			
A3559	159	3114		TR20/	
A3562	159	3114		TR20/	
A3563	159	3114		TR20/	
A3574	159	3114		TR20/	
A3581	159	3114		TR20/	
A3582	225	3045			
A3607	101	3011		TR08/	
A3609	101	3011		TR08/	
A3616-1	129	3025		TR88/	
A4000	1007				
A4030	1008				
A4030P	1009				
A4031	1010				
A4031P	1010				
A4032P	1011				
A4050P	1150				
A4051P	1150				
A-4068		3131			
A4086	159	3114		TR20/	
A4087	159	3114		TR20/	
A4126	159	3114		TR20/	
A4131P	1010				
A4247	121	3009		TR01/	
A4310	159	3114		TR20/	
A4347	121	3009		TR01/	
A4442	159	3114		TR20/	
A4478	129	3025		TR88/	
A4648	154	3045		TR78/	
A4700	101	3011		TR08/	
A4745	159	3114		TR20/	
A4789	108	3039		TR95/	
A4802-00004	159	3114		TR20/	
A4815	159	3114		TR20/	
A4819	154	3045		TR78/	
A4822-130-40348	159	3114		TR20/	

Original Part Number	ECG	SK	GE	IR/WEP	HEP
A4838	154	3045		TR78/	
A4843	154	3045		TR78/	
A4844	159	3114		TR20/	
A4851	108	3039		TR95/	
A4853	154	3045		TR78/	
A4950	159				
A5226-1	159	3114		TR20/	
A5253	121	3009		TR01/	
A6181-1	154	3045		TR78/	
A6661QRS	159				
A7253	154	3045		TR78/	
A8405	159	3114		TR20/	
A8540	159	3114		TR20/	
A8867	159	3114		TR20/	
A11744	132	3116		FE100/802	F0015
A11745	132	3116		FE100/802	F0015
A12163	121	3009		TR01/	
A12178	121	3009		TR01/	
A12546	154	3045		TR78/	
A12594	159	3114		TR20/	
A12888	159	3114		TR20/	
A13658E	712				
A14665-2	103A	3010		TR08/	
A14743	175	3131		TR81/241	
A15927	121	3009		TR01/	
A20372	123A	3122		TR21/735	
A22008	159	3114		TR20/	
A24100	123A			TR21/	S0015
A25762-010	154	3045		TR78/	
A25762-012	154	3045		TR78/	
A30270	159	3114		TR20/	
A30278	129	3025		TR88/	
A30290	159	3114		TR20/	
A30302	121	3009		TR01/	
A34715	121	3009		TR01/	
A35084	121	3009		TR01/	
A35201	121	3009		TR01/	
A35260	121	3009		TR01/	
A36577	129	3025		TR88/	
A36896	121	3009		TR01/	
A40410	129	3025		TR88/	
A41440	159	3114		TR20/	
A46051-1	124				
A47392R-0	128				
A59625-1	159	3114		TR20/	
A59625-2	159	3114		TR20/	
A59625-3	159	3114		TR20/	
A59625-4	159	3114		TR20/	
A59625-5	159	3114		TR20/	
A59625-6	159	3114		TR20/	
A59625-7	159	3114		TR20/	
A59625-8	159	3114		TR20/	
A59625-9	159	3114		TR20/	
A59625-10	159	3114		TR20/	
A59625-11	159	3114		TR20/	
A59625-12	159	3114		TR20/	
A71687-1	159	3114		TR20/	
A76228	199				
A78331		3114		TR20/	
A94004		3009		TR01/	
A94037	159	3114		TR20/	
A94063	129	3025		TR88/	
A95115	101	3011		TR08/	
A95211	101	3011		TR08/	
A95227	159	3114		TR20/	
A95232	159	3114		TR20/	
A112363	130	3027		TR59/	
A-113110	159			TR20/	S0013
A116078	159	3114		TR20/	
A116081	154	3045		TR78/	
A116084	129	3025		TR88/	
A116284	129	3025		TR88/	
A118284	159	3114		TR20/	
A119730	159	3114		TR20/	
A119983	129	3025		TR88/	
A-120018	128			TR87/	S0015
A-120278	123A			TR21/	S0015
A-120304	175			TR81/	S5019
A-120327	130			TR59/	S7004
A-120417	159			TR20/	S0019
A-120526	159				
A121467	159	3114		TR20/	
A121659	159	3114		TR20/	
A124047	159	3114		TR20/	
A124623	108	3039		TR95/	
A124624	108	3039		TR95/	
A124755	159	3114		TR20/	
A125329	108	3039		TR95/	
A-125332	123A			TR21/	S0015
A126524	159	3114		TR20/	
A126700	159	3114		TR20/	
A126705	154	3045		TR78/	
A126707	159	3114		TR20/	
A126715	159	3114		TR20/	
A126718	159	3114		TR20/	

Original Part Number	ECG	SK	GE	IR/WEP	HEP
A126719	159	3114		TR20/	
A126724	129	3025		TR88/	
A127384	176				
A127712	154	3045		TR78/	
A129509	108	3039		TR95/	
A129510	108	3039		TR95/	
A129511	108	3039		TR95/	
A129512	108	3039		TR95/	
A129513	108	3039		TR95/	
A129571	108	3039		TR95/	
A129572	108	3039		TR95/	
A129573	108	3039		TR95/	
A129574	108	3039		TR95/	
A129697	159	3114		TR20/	
A129699	159	3114		TR20/	
A130139	159	3114		TR20/	
A-140605	130				
A188103	192				
A188538		3004			
A190429	159	3114		TR20/	
A198794-1	101	3011		TR08/	
A200611A		3122			
A218012DS	121	3009		TR01/	
A297074C11	129	3025		TR88/	
A300043-06	176				
A322805-50-1	128				
A335854	912				
A391593	181				
A417014	181				
A417032	152				
A417033	130				
A417034	192				
A417756	197				
A489751-028	159	3114		TR20/	
A489751-031	159	3114		TR20/	
A573501	154	3045		TR78/	
A581025		3004			
A610074-1	159	3114		TR20/	
A610075-1	154	3045		TR78/	
A610083	159	3114		TR20/	
A610083-1	159	3114		TR20/	
A610083-2	159	3114		TR20/	
A610083-3	159	3114		TR20/	
A610110-1	159	3114		TR20/	
A610120-1	159	3114		TR20/	
A660031	121MP	3013			
A660097	121	3009		TR01/	
A667703		3027			
A701584-00	159	3114		TR20/	
A701589-00	159	3114		TR20/	
A772738	108	3039		TR95/	
A772739	108	3039		TR95/	
A815185	129	3025		TR88/	
A815185E	129	3025		TR88/	
A815199		3114		TR20/	
A815199-6		3114		TR20/	
A815203-5	121MP	3013			
A815211	159	3114		TR20/	
A815213	159	3114		TR20/	
A815229	159	3114		TR20/	
A815247	159	3114		TR20/	
A970246	159	3114		TR20/	
A970248	159	3114		TR20/	
A970251	159	3114		TR20/	
A970254	159	3114		TR20/	
A984193	159	3114		TR20/	
A1471114-1	159	3114		TR20/	
A1473549-1	129	3025		TR88/	
A1473563-1	159	3114		TR20/	
A1473570-1	159	3114		TR20/	
A1473574-1	159	3114		TR20/	
A1473590-1	159	3114		TR20/	
A1473591-1	159	3114		TR20/	
A1473597-1	159	3114		TR20/	
A1473616-1	129	3025		TR88/	
A2006681-95	108	3039		TR95/	
A2057013-0004	159	3114		TR20/	
A2057013-0701	159	3114		TR20/	
A2057013-0702	159	3114		TR20/	
A2057013-0703	159	3114		TR20/	
A2090056-1	121	3009		TR01/	
A2090056-5	121	3009		TR01/	
A2090056-27	121	3009		TR01/	
A2091859-0025	121	3009		TR01/	
A2091859-0720	121	3009			
A2091859-10	121	3009			
A2091859-11	121	3009			
A2092418	101			TR08/	
A2092418-0711	101	3011		TR08/	
A2092693-0724	108	3039		TR95/	
A2092693-0725	108	3039		TR95/	
A2370773	199				
A2498512	159	3114		TR20/	
A3011112	128	3024	18	TR87/243	
A3011112A		3024			

Original Part Number	ECG	SK	GE	IR/WEP	HEP
A3170717	154	3045		TR78/	
A3170757	225	3045			
A3902441	130				
A4037764-2	129	3025	18	TR88/242	
A5320111	159	3114		TR20/	
A7279039	121	3009		TR01/	
A7279049	121	3009		TR01/	
A7285774	121	3009		TR01/	
A7285778	121	3009		TR01/	
A7289047	121	3009		TR01/	
A7290594	121	3009		TR01/	
A7291252	121	3009		TR01/	
A7297043	121	3009		TR01/	
A7297092	121	3009		TR01/	
A7297093	121	3009		TR01/	
A7570013-01	159	3114		TR20/	
A7576004-01	159	3114		TR20/	
A8015613	129	3025		TR88/	
A-11095924	123A				
A-11166527	130				
A11200482	159				
A-11237336	123A				
A11414257	123A				
A-18530053-1	218				
A20030703-0702	159	3114		TR20/	
A23114050	159	3114		TR20/	
A23114051	159	3114		TR20/	
A23114550	159	3114		TR20/	
A25114130	154	3045		TR78/	
A43021415	123A				
A43023843	130	3027		TR59/	
A43023845	159	3114		TR20/	
A80052402	130	3027		TR59/	
A80414120	130	3027		TR59/	
A80414130	130	3027		TR59/	
A101050108		3079			
A101050115		3079			
A514027662		3004			
AA1	100	3005	2	TR05/254	G0005
AA1M	225	3045			
AA2	101	3011	8	TR08/641	G0011
AA2A	225	3045			
AA2K	225	3045			
AA2SB240A	121	3009		TR01/	
AA2Z	225	3045			
AA3	160		50	TR17/637	G0003
AA3M	225	3045			
AA4	121		16	TR01/232	G6003
AA5	105	3012	4	TR03/233	G6004
AA8-1-70	160	3007		TR17/	
AA8-1-705	160	3007		TR17/	
AA133	225	3045			
AA310	225	3045			
AA113852-2	909				
AC100	283				
AC105	102A	3004	2	TR85/250	G6011
AC106	102A	3004	2	TR85/250	G6011
AC107	160		50	TR17/637	G0003
AC-107	102A	3004			
AC107M	160		51	TR17/637	G0008
AC108	102A		52	TR85/250	G0005
AC109	102A		52	TR85/250	G0005
AC110	102A		52	TR85/250	G0005
AC113	102A		52	TR85/250	G0002
AC-113	102A	3004			
AC113A				TR85/250	
AC-113A	102A	3004	2		
AC114	102A			TR85/250	G0005
AC-114	102A	3004	2		
AC115	102A		52	TR85/250	G0002
AC116	102A		53	TR85/250	G0005
AC-116	102A	3004			
AC117	102A		2	TR85/250	G0005
AC-117	102A	3004			
AC117A	102A			TR85/250	
AC-117A	102A	3004			
AC117B	102A			TR85/250	
AC-117B		3004			
AC117P	102A			TR85/250	
AC-117P	102A	3004			
AC118	102A		53	TR85/250	
AC119	102A		53	TR85/250	
AC120	102A		53	TR85/250	G0005
AC121	102A		53	TR85/250	G0005
AC121IV			53		
AC121-IV	102A			TR85/	
AC-121IV	158				
AC121V	158		53		
AC121-V	102A			TR85/	
AC121VI	158		53		
AC121-VI	102A			TR85/	
AC121VII	158		53		
AC121-VII	102A			TR85/	
AC122	102A		52	TR85/250	G0005
AC-122	102A	3004			

Original Part Number	ECG	SK	GE	IR/WEP	HEP
AC122-30	102A			TR85/	
AC122GRN	102A			TR85/250	
AC122-GRN		3123	53	TR05/	
AC122RED	102A			TR85/250	
AC122-RED		3123	53	TR05/	
AC122-VIO			53		
AC122YEL	102A			TR85/250	
AC122-YEL		3123	53	TR05/	
AC123	102A		2	TR85/250	G0005
AC-123	102A	3004			
AC124	102A		53	TR85/250	S0011
AC125	102A		53	TR85/250	G0005
AC-125	102A	3004			
AC126	102A		2	TR85/250	G0005
AC-126	102A	3004			
AC127	103A	3010	59	TR08/724	G0011
AC-127	103A	3010			
AC127-01	103A			TR08/	G0011
AC127-132	103A	3010		TR08/724	
AC128	158	3004	53	TR84/630	G0005
AC-128	102A	3004			
AC128-01	158		53		
AC128/01	158	3004		/630	G0005
AC128K	158		53	TR84/	G6011
AC129	160		51	TR17/637	G0003
AC130	101	3010	5	TR08/641	G0011
AC131	102A		2	TR85/250	G0005
AC131-30	158			TR84/	
AC132	102A		2	TR85/250	G0005
AC-132	102A	3004			
AC132-01	102A			TR85/	G0005
AC133				TR85/	
AC133A	102A			TR85/	
AC134	102A		52	TR85/250	G0005
AC135	102A	3123	2	TR85/250	G0005
AC136	102A		2	TR85/250	G0005
AC137	102A		52	TR85/250	G0005
AC138	158		1	TR84/630	G0005
AC139	158		53	TR84/	G0005
AC141	103A		54	TR08/724	G0011
AC141B	103A		54	TR08/724	G0011
AC-141B	103, 158		54		
AC141K	103A		54	TR08/724	G0011
AC-141K	103, 158		54		
AC142	158			TR84/	G6011
AC142K	121			TR84/	
AC148	121		3	TR01/232	
AC150	102A	3030	52	TR85/250	G0005
AC-150	102A	3004			
AC150GRN	102A			TR85/250	
AC150-GRN			51		
AC150YEL	102A			TR85/250	
AC150-YEL			51		
AC151	102A	3004	53	TR85/250	G0005
AC-151	102A	3004			
AC151IV	158		53		
AC151-IV	102A			TR85/	
AC151R	102A	3004	53	TR85/250	G0005
AC151RIV	158		53		
AC151-RIV	102A			TR85/	
AC151RV	158		53		
AC151-RV	102A			TR85/	
AC151RVI	158		53		
AC151-RVI	102A			TR85/	
AC151V	158		53		
AC151-V	102A			TR85/	
AC151VI	158		53		
AC151-VI	102A			TR85/	
AC151VII	158		53		
AC151-VII	102A			TR85/	
AC152	158		53	TR84/630	G0005
AC-152	102A	3004			
AC152IV			53		
AC152-IV	102A			TR85/	
AC152V	158		53		
AC152-V	102A			TR85/	
AC152VI	158		53		
AC152-VI	102A			TR85/	
AC153	158	3004	53	TR84/630	G6011
AC153K	158	3004	53	TR84/630	G6011
AC154	158		2	TR84/630	G0005
AC-154	102A	3004			
AC155	102A		52	TR85/250	G0005
AC-155	102A	3004			
AC156	102A		52	TR85/250	G0001
AC-156	102A	3004			
AC157	103		8	TR08/641A	G0011
AC-157	103A	3010			
AC160	102A		50	TR85/250	G0003
AC160A	102A		52	TR85/250	G0003
AC160B	102A		52	TR85/250	G0009
AC160GRN	102A			TR85/250	
AC160-GRN			51		
AC160RED	102A			TR85/250	

Original Part Number	ECG	SK	GE	IR/WEP	HEP
AC160-RED			51		
AC160-VIO			51		
AC160YEL	102A			TR85/250	
AC160-YEL			51		
AC161	102A	3123	2	TR85/250	G0005
AC-161	102A	3004			
AC162	102A		53	TR85/250	G0005
AC-162	102A	3004			
AC163	102A		53	TR85/250	G0005
AC164	126	3006	52	TR17/635	G0009
AC165	102A		52	TR85/250	G0009
AC-165	102A	3004			
AC166	158		2	TR84/630	G0005
AC-166	102A	3004			
AC167	158		2	TR84/630	G0005
AC-167	102A	3004			
AC168	102A		8	TR85/250	G0011
AC-168	102A	3004			
AC169	126		52	TR17/635	G0005
AC-169	102A	3004			
AC170	102A	3123	53	TR85/250	G0005
AC171	102A		53	TR85/250	G0005
AC172	103A		59	TR08/724	G0011
AC-172	103A	3010			
AC173	102A		53	TR85/250	G0005
AC175	103A	3124	20	TR08/724	G0011
AC175A	103A			TR08/	
AC-175A	123A	3124			
AC175B	103A		20	TR08/724	
AC-175B	123A	3124			
AC175P	103A			TR08/724	
AC-175P	123A	3124			
AC176	103A	3010		TR08/724	G0011
AC176K	158			TR84/630	G6011
AC178	158			TR84/630	
AC179	103A		54	TR08/724	G0011
AC180	158		53	TR84/630	G0005
AC180K	158			TR84/	G6011
AC181	103A	3011	59	TR08/724	
AC181K	103A			TR08/	
AC182	102A	3123	2	TR85/250	G0005
AC183	103A		54	TR08/724	G0011
AC-183	103A, 158		54		
AC184	102A		53	TR85/250	G0005
AC185	103A		59	TR08/724	G0011
AC186	103A		54	TR08/724	G0011
AC-186	103A, 158		54		
AC187	103A	3010	54R	TR08/724	G0011
AC187/01	101	3010	59	TR08/724	G0011
AC187K	103A				
AC187R	103A			TR08/724	
AC188	158	3004	53	TR84/630	G6011
AC188/01	158	3004	53	/630	G0005
AC188K	158			TR84/630	G6011
AC191	102A	3004	53	TR85/250	G0005
AC192	102A	3004		TR85/	G0005
AC193	158	3004	53	TR84/	G6011
AC193K	158	3004		TR84/	G6011
AC194	103A	3010		TR08/	
AC194K	103A	3010		TR08/	
AC9082	159	3114		TR20/	
AC9083	159	3114		TR20/	
AC9084	159	3114		TR20/	
AC9085	159	3114		TR20/	
AC12801	158			TR84/	
AC18701	103A			TR08/	
AC18801	158			TR84/	
AC-N7B	102A				
ACDC9000-1	159	3114		TR20/	
ACR83-1001	126	3008		TR17/635	G0008
ACR83-1002	126	3008		TR17/635	G0008
ACR83-1003	126	3008		TR17/635	G0008
ACR83-1004	100	3004		TR05/254	G0007
ACR83-1005	100	3004		TR05/254	G0005
ACR83-1006	100	3004		TR05/254	G0005
ACR810-101	126	3008		TR17/635	G0008
ACR810-102	126	3008		TR17/635	G0008
ACR810-103	126	3008		TR17/635	G0008
ACR810-104	100	3004		TR05/254	G0007
ACR810-105	100	3004		TR05/254	G0005
ACR810-106	100	3004		TR05/254	G0005
ACY16	121	3009	16	TR01/232	G0005
ACY17	102			TR05/631	G0005
ACY-17	102	3004	2		
ACY-17-1	102A	3004			
ACY18	102			TR05/631	G0005
ACY-18	102	3004	2		
ACY18-1	102A	3004			
ACY19	102			TR05/631	G0005
ACY-19	102	3004	2		
ACY19-1	102A	3004			
ACY20	102A		80	TR85/631	G0005
ACY-20	102	3004			

Original Part Number	ECG	SK	GE	IR/WEP	HEP
ACY20-1	102A	3004		TR85/	
ACY21	102A			TR85/631	G0005
ACY-21	102A	3004	2		
ACY21-1	102A	3004		TR85/	
ACY22	102A		53	TR85/631	G0005
ACY-22	102A	3003			
ACY22-1	102A	3004		TR85/	
ACY23	102A	3123	53	TR85/250	G0005
ACY-23	102A	3004			
ACY23V	158		53		
ACY23-V	102A			TR85/	
ACY23VI	158		53		
ACY23-VI	102A			TR85/	
ACY24	160		50		G6011
ACY27	102A	3004	51	TR85/250	G0005
ACY28	102A	3004	51	TR85/250	G0005
ACY29	102A	3004	51	TR85/250	G0005
ACY30	102A	3004	51	TR85/250	G0005
ACY31	102A	3004	52	TR85/250	G0005
ACY32	102A	3123	53	TR85/250	G0005
ACY-32	102A	3004			
ACY32V	158		53		
ACY32-V	102A			TR85/	
ACY32VI	158		53		
ACY32-VI	102A			TR85/	
ACY33	102A		53	TR85/250	G0005
ACY33VI	158		53		
ACY33-VI	158			TR84/	
ACY33VII	158		53		
ACY33-VII	158			TR84/	
ACY33-VIII	158			TR84/	
ACY34	102A	3004	53	TR85/250	G0005
ACY35	102A	3004	53	TR85/250	G0005
ACY36	102A	3004	53	TR85/250	G0005
ACY38	102A	3123	2	TR85/250	G0006
ACY39	176			TR82/	G6012
ACY40	100	3005	53	TR05/254	G0005
ACY41	102A	3004	53	TR85/631	G0005
ACY41-1	102A	3004		TR85/	
ACY44	102A	3004	53	TR85/631	G0005
ACY44-1	102A	3004		TR85/	
ACZ	160			TR17/637	
ACZ21			51		
AD29A-4	159	3114			
AD29A-5	159	3114			
AD29A-6	159	3114			
AD29A-9	159	3114			
AD29E-1	159	3114			
AD29E-2	159	3114			
AD29E4	159	3114			
AD29E5	159	3114			
AD29E6	159	3114			
AD29E7	159	3114			
AD29E8	159	3114			
AD29E9	159	3114			
AD29E10	159	3114			
AD30A1	159	3114			
AD30A2	159	3114			
AD30A3	159	3114			
AD30A4	159	3114			
AD30A5	159	3114			
AD103	121	3014	16	TR01/232	G6005
AD104	121	3009	16	TR01/232	G6005
AD105	121	3009	16	TR01/232	G6005
AD130	121	3009	25	TR01/232	G6003
AD131	121	3009	25	TR01/232	G6005
AD132	121	3014	25	TR01/232	G6005
AD133	179	3014	76	TR35/G6001	G6005
AD138	121	3009	76	TR01/232	G6003
AD138/50	121	3014	16	/232	G6005
AD139	104	3009		TR01/232	G6016
AD140	121	3014	3	TR01/232	G6005
AD-140	121	3009		TR01/	
AD142	179	3009	16	TR35/G6001	G6015
AD143	179	3014	16	TR35/G6001	G6005
AD143B	179			TR35/G6001	
AD143R	121			TR01/	G6005
AD145	121		16	TR01/	G6003
AD148	131		3	TR94/642	G6003
AD-148	121	3009		TR01/	
AD149	104	3013	25	TR01/230	G6003
AD-149	121	3009		TR01/	
AD149-01	121	3009		TR01/232	
AD149-02	121	3009		TR01/232	
AD149B	121	3009		TR01/232	
AD149C	121	3009		TR01/232	
AD150	121		25	TR01/232	G6003
AD-150	121	3009		TR01/	
AD152	131	3198	16	TR94/642	G6016
AD-152	121	3009		TR01/	
AD153	121		16	TR01/	G6005
AD155	131	3082		TR94/642	G6016
AD156	131	3198	16	TR94/642	G6011
AD-156	121	3009		TR01/	
AD157	155		44	TR50/642	G6011

Original Part Number	ECG	SK	GE	IR/WEP	HEP
AD-157	121	3009		TR01/	
AD159	121		16	TR01/232	
AD-159	121	3009		TR01/	
AD160	175		12	TR81/241	
AD161	155		43	TR76/	
AD162	131	3052	44	TR94/642	G6016
AD163	179		25	TR35/G6001	G6007
AD164	131		44	TR94/642	G6016
AD165	155				
AD166	127		16	TR27/235	
AD167	127		16	TR27/235	
AD169	131	3198	44	TR94/642	G6016
AD262	131	3082	44	TR94/642	G6016
AD263	131	3082	44	TR94/642	G6016
AD710CH	910				
AD710CN	910D				
AD711CH	911				
AD711CN	911D				
AD741CH	941	3514			
AD741CN	941D	3552			
AD13850	121			TR01/	
AD138150				TR01/	
ADY22	121	3009	16	TR01/232	G6003
ADY23	121	3014	16	TR01/232	G6005
ADY24	121	3009	16	TR01/232	G6005
ADY26	213	3012		TR05/233	G6010
ADY-26	105		4		
ADY27	121		25	TR01/232	G6003
ADY-27	102A	3004			
ADY28	121	3009		TR01/232	G6018
ADZ11	105	3012	4	TR03/233	G6004
ADZ12	105	3012	4	TR03/233	G6006
AE-50	102A	3004	2		
AE900	786				
AE902	786	3140			
AE904	743				
AE904-02	743				
AE904-51	743				
AE906	726				
AE-906	726				
AE906-51	726				
AE907		3144	IC16		
AE-907	737	3144			
AE907-51	737	3144	IC16		
AE907@	723				
AE907@	737				
AE908	789				
AE908-51	789				
AEX9846	130	3027			
AEX76164		3024			
AEX77230		3036			
AEX-78507 0		3036			
AEX-78507		3511			
AEX-78508		3026			
AEX79846	130			TR59/247	S7002
AEX-79922		3026			
AEX-82308	223	3027			
AEX-85715	129	3004			
AF1				TR51/	
AF2				TR51/	
AF3				TR51/	
AF4				TR51/	
AF5				TR51/	
AF6				TR51/	
AF7				TR51/	
AF8				TR51/	
AF101	160		50	TR17/637	G0005
AF-101	100	3005			
AF102	160	3006	51	TR17/637	G0003
AF105	126	3008	50	TR17/635	G0003
AF105A	160		51	TR17/637	
AF-105A	160	3005			
AF106	160	3006	50	TR17/637	G0003
AF-106		3006		TR17/	
AF106A	160		50	TR17/	
AF107	160		52	TR17/637	G0002
AF108	126	3006	52	TR17/635	G0002
AF109	160		50	TR17/637	G0003
AF-109	160	3006			
AF109R	160		50	TR17/637	G0003
AF110	160	3006	50	TR17/637	G0002
AF111	160	3006	52	TR17/637	G0009
AF112	160	3006	52	TR17/637	G0009
AF113	160	3006	52	TR17/637	G0009
AF114	160	3006	50	TR17/637	G0003
AF114N	160	3006	51	TR17/	G0008
AF115	160	3006	50	TR17/637	G0003
AF115N	160	3006	51	TR17/	G0008
AF116	160	3006	50	TR17/637	G0003
AF116N	160	3006	51	TR17/	G0008
AF117	160	3006	50	TR17/637	G0003
AF117C	160			TR17/637	
AF117N	160	3006	51	TR17/	G0008
AF118	160	3006	51	TR17/637	G0003
AF119	160	3006	51	TR17/637	

Original Part Number	ECG	SK	GE	IR/WEP	HEP
AF120	126	3006	51	TR17/635	G0008
AF121	160		50	TR17/637	G0003
AF-121	160	3006			
AF121S	160			TR17/	G0009
AF122	160		50	TR17/637	
AF124	160	3006	51	TR17/637	G0008
AF125	160	3006	51	TR17/637	G0008
AF126	160	3006	51	TR17/637	G0008
AF127	160	3005	51	TR17/637	G0008
AF127/01	160			TR17/637	
AF128	160	3005	50	TR17/637	G0005
AF129	126	3006	52	TR17/635	G0008
AF130	126	3006	52	TR17/635	G0008
AF131	126	3006	52	TR17/635	G0008
AF132	126	3006	52	TR17/635	G0008
AF133	126	3006	52	TR17/635	G0008
AF134	126	3008	50	TR17/635	G0008
AF135	126	3008	50	TR17/635	G0008
AF136	126	3008	50	TR17/635	G0008
AF137	126	3008	50	TR17/635	G0008
AF137A	160			TR17/637	
AF138	126	3006	50	TR17/635	G0008
AF138/20		3008	51	TR12/	
AF138/290	100				
AF139	160			TR17/637	G0003
AF142	160		52	TR17/637	G0008
AF143	160		52	TR17/637	G0003
AF144	126	3006	52	TR17/635	G0003
AF146	160	3006	52	TR17/637	G0003
AF147	160	3006	52	TR17/637	G0003
AF148	160	3006	52	TR17/637	G0003
AF149	160	3006	52	TR17/637	G0003
AF150	160	3006	52	TR17/637	G0003
AF164	160	3006	52	TR17/637	G0003
AF165	160	3006	52	TR17/637	G0003
AF166	160	3008	51	TR17/637	G0003
AF-166	160	3006			
AF167	160	3006		TR17/637	
AF168	160	3006	52	TR17/637	G0003
AF169	160	3006	52	TR17/637	G0003
AF170	160	3006	50	TR17/637	G0003
AF171	160	3006	52	TR17/637	G0003
AF172	160	3006	50	TR17/637	G0003
AF178	160	3006	50	TR17/637	G0002
AF179	160	3006	50	TR17/637	G0002
AF180	160	3006	53	TR17/637	G0002
AF181	160	3006	53	TR17/637	G0002
AF182	160	3006	52	TR17/637	G0002
AF-182	160	3006			
AF185	160	3006	51	TR17/637	G0002
AF186	160	3006	52	TR17/637	G0002
AF186G	160	3006	50	TR17/637	G0002
AF186W	160	3006	50	TR17/637	G0002
AF187	102A		52		G0005
AF188	102A	3123	2	TR85/250	G0005
AF192	101	3011	8	TR08/641	G0011
AF193	160	3006	52	TR17/637	G0003
AF200	160	3006	51	TR17/637	G0005
AF200U			53		
AF201	160	3006	51	TR17/637	G0005
AF201C	160	3006		TR17/637	
AF201U			53		
AF202	160	3006	51	TR17/637	G0005
AF202L	160	3006	53	TR17/637	G0002
AF202S	160	3006		TR17/637	G0005
AF239	160	3006		TR17/637	G0003
AF239S	160	3006		TR17/637	G0003
AF240	160	3006		TR17/637	G0003
AF251	160				G0002
AF252	160				G0002
AF253	160				G0002
AF256	160	3006	50	TR17/	G0002
AF267	160				G0002
AF279	160	3006	51	TR17/	G0003
AF280	121				G0003
AF306			50		
AF699	159	3114		TR20/	
AF3570	159	3114		TR20/	
AF3590	159	3114		TR20/	
AF21490	159	3114		TR20/	
AF20019180				FE100/	
AFCS1170F	159	3114		TR20/	
AFS24226	159	3114		TR20/	
AFT0019M	159	3114		TR20/	
AFT052	159	3114		TR20/	
AFT1341	159	3114		TR20/	
AFT1746	159	3114		TR20/	
AFY10	126	3006	52	TR17/635	G0002
AFY11	126	3006	50	TR17/635	G0002
AFY12	160	3006	50	TR17/637	G0003
AFY14	160	3006	50	TR17/637	G0008
AFY15	160	3006	50	TR17/637	G0008
AFY16	160	3006		TR17/637	G0003
AFY18	126	3006	53	TR17/635	G0002
AFY19			80		S0012

Original Part Number	ECG	SK	GE	IR/WEP	HEP
AFY34	160	3006	52	TR17/637	G0003
AFY37	160	3006	50	TR17/637	G0003
AFY39	160	3006	50	TR17/637	G0003
AFY40	160	3006	1	TR17/637	G0003
AFY40K	160	3006		TR17/637	G0003
AFY40R	160	3006		TR17/	G0003
AFY41	160	3006		TR17/637	G0003
AFY42	160	3006		TR17/	G0008
AFZ11	160	3006	50	TR17/637	G0003
AFZ12	160	3006	50	TR17/637	G0003
AFZ23	100	3005	2	TR05/254	
AG6	188	3054		TR72/3020	
AG134	126	3006	51	TR17/635	
AIE	161		39	TR70/	S0017
AIE/48X134902		3019			
AIT				TR21/	
AIT-1				TR21/	
AJ103	179			TR35/G6001	
AL100	127	3035	25	TR27/235	G6007
AL101	121		16		G6005
AL102	127	3035	25	TR27/235	G6007
AL103	127	3035	25	TR27/235	G6007
AL113	179				G6015
AL210	160	3035		TR17/637	G0008
AL40934-2		3027		TR51/	
ALB6494612	196				
ALS-8922	123A	3122		TR21/	
ALZ10	100	3005		TR05/254	
AM3235	130	3027		TR59/247	S7002
AMD746	713	3077		IC508/	C6057P
AMD780	714	3075		IC509/	C6070P
AMD781	715	3076		IC510/	C6071P
AMF104	130	3027	19	TR59/247	S7002
AMF105	130	3027	77	TR59/247	S7002
AMF106	163		73	TR61/	S5004
AMF115	130	3027	77	TR59/247	S7002
AMF116	130	3027	77	TR59/247	S7002
AMF117	130	3027	19	TR59/247	S7002
AMF117A	130	3027	19	TR59/247	S7002
AMF118	130	3027	19	TR59/247	S7002
AMF118A	130	3027	19	TR59/247	S7002
AMF119	130	3027	19	TR59/247	S7002
AMF119A	130	3027	19	TR59/247	S7002
AMF120	130	3027	19	TR59/247	S7002
AMF120A	130	3027	19	TR59/247	S7002
AMF-121	130		14		
AMF201	130		77		S5004
AMF201B	181		75		S5004
AMF201C	181		75		S5004
AMF210	130	3027	75	TR59/247	S5004
AMF210A	130	3027	75	TR59/247	S5004
AMF210B	161		75	TR61/	S5004
AMF210C				TR61/	
AMP2970-2	129				
AMP-2919-2	3027				
AMP-2971-4	128				
AMU5B7741393	941				
AMU5B7747393	947				
AMU5R7723393	923				
AMU6A7723393	923D				
AMU6A7741393	941D				
AN	123A				
AN136	1053		IC44		
AN203	1054				
AN203AA	1054				
AN203BA	1054				
AN203BB	1054		IC45		
AN203C	1054				
AN206	1057	3072		IC507/	
AN-206	1057		IC2		C6060P
AN206S	1057			IC507/	C6060P
AN208	1136				
AN208LL	1136				
AN210	1055				
AN210A	1055				
AN210B	1055				
AN210C	1055		IC47		
AN210D	1055				
AN211	1056		IC48		
AN211A	1056		IC48		
AN211B	1056				
AN213	1059				
AN213AB	1059				
AN214	1058				
AN214P	1058				
AN214Q	1058				
AN214R	1058				
AN216	1125				
AN217	1060		IC50		
AN217AA	1060				
AN217AB	1060				
AN217BA	1060				
AN217BB	1060		IC50		
AN217CA	1060				
AN217CB	1060				

Original Part Number	ECG	SK	GE	IR/WEP	HEP
AN220	1061	3070			
AN221	712	3072	IC2		
AN225	1062				C6072P
AN227	1063				
AN229	1064				
AN230	1065				
AN231	1066				
AN234	1067				
AN238	749	3168			
AN238S	749				
AN240	712	3072			C6063P
AN-240	712				
AN240D	712				
AN241	712	3072	IC2	IC507/	C6063P
AN241A		3072			
AN241B		3072			
AN241C		3072			
AN241D	712	3072		IC507/	C6063P
AN242	1068				
AN253	1072				
AN253AB	1072				
AN260	1074				
AN271B		3160			
AN271C		3160			
AN277	1073				
AN277B	1073				
AN288	1069				
AN328	1070				
AN342	1071				
AQ1	126	3008	1	/635	
AQ7	101	3011		/641	
APB-11A-1001				TR17/	
APB-11A-1004				TR17/	
APB-11A-1007				TR12/	
APB-11A-1008	102A	3004		TR85/	
APB-11H-1001	160	3006		TR17/637	G0003
APB-11H-1004	160	3006		TR17/637	G0003
APB-11H-1007	126	3006		TR17/637	G0008
APB-11H-1008	102	3004		TR05/631	G0005
APB-11H-1010	102A	3004		TR85/250	G0005
APE-CD2006-09		3009			
AQ2	128	3024	18	TR87/243	
AQ3	128	3024	18	TR87/243	
AQ4	123	3124	20	TR86/243	
AQ5	128	3024	18	TR87/243	
AQ6	123	3124	20	TR86/53	
AR4	104		16	TR01/230	G6003
AR-4	121	3009		TR01/	
AR5	104		16	TR01/230	G6003
AR-5	121	3009		TR01/	
AR6	104		16	TR01/230	G6003
AR-6	121	3009		TR01/	
AR7	104		16	TR01/230	G6003
AR-7	121	3009		TR01/	
AR8	121		16	TR01/232	G6003
AR-8	121	3009		TR01/	
AR8P404R	121	3009		TR01/232	
AR9	121		16	TR01/232	G6005
AR-9	121	3009		TR01/	
AR10	121		16	TR01/232	G6005
AR-10	121	3009		TR01/	
AR11	121		16	TR01/232	G6005
AR-11	121	3009		TR01/	
AR12	121		16	TR01/232	G6005
AR-12	121	3009		TR01/	
AR13	121		16	TR01/232	G6005
AR-13	121	3009		TR01/	
AR14	121		16	TR01/232	G6005
AR-14	121	3009		TR01/	
AR15	130			TR59/247	
AR-15	124	3021	12	TR23/	
AR15-L8-0026	130				
AR17	175	3041	66	TR81/241	
AR-17	152	3041	28	TR76/701	S5000
AR17A	175			TR81/241	
AR17B	175			TR81/241	
AR17(GREY)	152				
AR18	175	3026	12	TR81/241	S5012
AR-18	124	3021		TR81/240	
AR22	184				S5000
AR22(PHILCO)	182				
AR23	185				S5006
AR23(PHILCO)	183				
AR24	184	3041			S5000
AR24(PHILCO)	186			TR55/	
AR24(RED)	196				
AR25	187	3084		TR56/	S5006
AR-25	153			TR56/	
AR25(G)			69	TR56/	
AR25G	187	3084			S5006
AR25(GREEN)	153				
AR25(ORANGE)	108				
AR25(WHITE)	108				
AR26	211	3084	29	TR92/	S3028
AR27	197	3084	69	TR92/700	S5006
AR27(GREEN)	153				
AR28	196	3041	65	TR92/701	S5000
AR28(RED)	196				
AR29	211	3053	29	TR56/244A	S3028
AR30	196	3084		TR56/	S3028
AR35	196	3087	29	TR56/	S3028
AR37	197	3084	69	TR56/	S5006
AR37(GREEN)	153				
AR38	196	3041	66	TR55/	S5000
AR102	158	3123	52	TR84/630	G0005
AR-102	102A	3004			
AR103	160	3008		TR17/637	G0002
AR-103	102A	3123	2	TR06/	
AR104	160	3008		TR17/637	G0002
AR-104	102A	3123	2	TR06/	
AR105	160	3008		TR17/637	G0002
AR-105	102A	3123	2	TR06/	
AR107	123A		8	TR21/735	S0015
AR-107	123A	3124			
AR108	123A		8	TR21/735	S0015
AR-108	123A	3124			
AR111	161				
AR200	108	3018	20	TR95/56	S0014
AR-200	123A	3018		TR21/735	S0015
AR200(GREEN)	107				
AR200W	108	3018	20	TR21/	S0015
AR201	108	3018	20	TR95/56	S0014
AR-201	123A	3018		TR21/735	S0015
AR201Y	108	3018	20	TR51/	S0015
AR201(YELLOW)	108				
AR202	108	3018	20	TR95/56	S0014
AR-202	123A	3018		TR21/735	S0015
AR202G	108	3018		TR21/	S0015
AR202(GREEN)	108				
AR203	128	3024		TR87/	S5014
AR203R	128	3024	63	TR25/	S5014
AR203(RED)	128				
AR204	123A		10		
AR205	123A		20		
AR206	123A		20		
AR207	192		63		
AR208	123A		10		
AR209	161	3018	60	TR83/	S0020
AR210	161	3018	60	TR83/	S0020
AR211	161	3018	60	TR83/	S0020
AR212	108	3018	60	TR95/	S0020
AR213	161	3018		TR83/	S0020
AR213V	161	3018	62	TR71/	S0020
AR213(VIOLET)	161				
AR218	161	3018		TR83/	S0020
AR218(ORANGE)	161				
AR218(RED)	161				
AR218RO	161	3018	62	TR71/	S0020
AR219	161	3018		TR71/	S0020
AR219YY	107	3018	60	TR71/	S0020
AR220	161	3018		TR71/	S0020
AR220GY	107	3018	60	TR71/	S0020
AR220(YELLOW)	107				
AR221	108	3018	60	TR71/	S0020
AR222	161	3018		TR71/	S0020
AR222(BLUE)	107				
AR222BY	107	3018	60	TR71/	S0020
AR222(YELLOW)	107				
AR224	107	3018	60	TR71/	S0020
AR-224		3018			
AR224(WHITE)	107				
AR224(YELLOW)	107				
AR304	129	3025	67	TR88/	S5013
AR304(GREEN)	159				
AR304(RED)	129				
AR306	123A	3024	18	TR24/	S0014
AR306(BLUE)	123A				
AR306(ORANGE)	123A				
AR308		3025			
AR308(VIOLET)	159				
AR313	108	3018		TR95/	
AR632539		3007			
AS215	121	3009	16	TR01/232	
AS401		3024			
AS477	176	3123		TR82/	G0005
AS3428	101	3011		TR08/641	
AS3867		3004			
AS4280		3006			
AS33867	102A		2	TR85/250	G0005
AS33868	102A	3004	2	TR85/250	G0005
AS34280	160		8	TR17/637	G0008
ASL802001-1	912				
ASY12-1	158	3004	51	TR84/630	G0003
ASY12-2	158	3004	51	TR84/630	G0003
ASY13-1	158	3004	51	TR84/630	G0003
ASY13-2	158	3004	51	TR84/630	G0003
ASY14	158		53	TR84/630	G0005
ASY14-1	158	3123	2	TR84/630	G0005
ASY14-2	158	3123	2	TR84/630	G0005
ASY14-3	158	3123	2	TR84/630	G0005

Original Part Number	ECG	SK	GE	IR/WEP	HEP
ASY24	102A			TR85/250	G0008
ASY-24	160	3008	51		
ASY26	100	3123	51	TR05/254	G0005
ASY-26	100	3005			
ASY27	100		51	TR05/254	G0005
ASY-27	100	3005			
ASY28	101	3011	51	TR08/641	G0011
ASY29	101	3011	5	TR08/641	G0011
ASY30	126	3006		TR17/635	
ASY31	102A	3123	2	TR85/	G0005
ASY-31		3004			
ASY32	102A	3123	2	TR85/250	G0005
ASY48	102A	3004	53	TR85/250	G0005
ASY48-IV	102A	3004		TR85/	G0005
ASY48-V	102A	3004		TR85/	G0005
ASY48-VI	102A	3004		TR85/	G0005
ASY49	102A	3004	51	TR85/250	
ASY50	102A	3004	53	TR85/250	G0005
ASY51	102A	3123	52	TR85/250	G0005
ASY52	102A	3004	51	TR85/250	G0005
ASY53	101	3011	54	TR08/641	G0011
ASY54	102A	3004	51	TR85/631	G0005
ASY55	012A	3004	51	TR85/631	G0005
ASY56	102A	3004	51	TR85/631	G0005
ASY56N			1		
ASY57	126	3006	51	TR17/635	G0005
ASY57N			1		
ASY58	126	3006	51	TR17/635	G0005
ASY58N			1		
ASY59	126	3006	51	TR17/635	G0008
ASY61	103A	3010	51	TR08/724	G0011
ASY62	103A		8	TR08/724	G0011
ASY-62		3011			
ASY63	160	3006	51	TR17/637	G0008
ASY67	160	3123	52	TR17/637	G0001
ASY70	102A	3004	53	TR85/250	G0005
ASY70-IV	102A	3004		TR85/	
ASY70IV	158		53		
ASY70V	158		53		G0005
ASY70-V	120A				
ASY70VI	158		53		G0005
ASY70-VI	102A	3004		TR85/	
ASY71	158		53	TR17/	
ASY72	101		8	TR08/641	G0011
ASY-72	101	3011			
ASY73	101	3011	7	TR08/641	G0011
ASY74	101	3011	7	TR08/641	G0011
ASY75	101	3011	7	TR08/641	G0011
ASY76	100	3005	80	TR05/254	G0005
ASY77	100	3005	53	TR05/254	G0005
ASY80	100	3005	80	TR05/254	G0005
ASY80-RT			80		
ASY81	176	3123		TR82/	G0005
ASY86	101	3011	8	TR08/641	G0011
ASY87	101	3011	8	TR08/641	G0011
ASY88	101	3011	8	TR08/641	G0011
ASY89	101	3011	8	TR08/641	G0011
ASY90	102A	3004		TR85/	G0005
ASY91	126	3004		TR85/	G0005
ASZ10	126	3006	52	TR17/635	
ASZ11	126	3006	2	TR17/635	G0005
ASZ15	121	3009	25	TR01/232	G6005
ASZ16	121	3009	25	TR01/232	G6005
ASZ17	121	3009	25	TR01/232	G6005
ASZ18	121	3009	25	TR01/232	G6005
ASZ20	160	3006	1	TR17/637	G0003
ASZ20N	160	3006		TR17/637	G0001
ASZ21	160	3006	51	TR17/637	G0003
ASZ30	126	3006	2	TR17/635	
AT-1	160	3006	50	TR17/637	G0003
AT-2	160	3006	50	TR17/637	
AT-3	160	3006	51	TR17/637	
AT4	160		51		
AT-4	160	3005		TR80/637	G0003
AT5	160		51	TR85/	
AT-5	100	3008		TR05/254	G0003
AT-6	160			TR12/	G0003
AT6	102A	3123	51	TR85/250	G0005
AT6A	102A	3123	2	TR85/250	G0005
AT-6A	102A				
AT7	192		63	TR25/	S5014
AT-7	128	3024		TR87/243	S5014
AT-8	160	3006	51	TR17/637	
AT-9	160	3006	51	TR17/637	
AT-10	130	3027	14	TR59/247	
AT10H	102A	3004	52	TR85/250	G0006
AT10M	102A	3004	52	TR85/250	G0005
AT10N	102A	3004	52	TR85/250	G0005
AT11		3114	51	TR85/637	G0003
AT-11	159	3008		TR88/242	G0003
AT12	128		63	TR25/637	
AT-12	128	3024		TR25/243	S5014
AT-12(PHILCO)	128			TR87/	
AT13	160		51	TR17/637	G0003
AT-13		3006			

Original Part Number	ECG	SK	GE	IR/WEP	HEP
AT14	160		51	TR17/637	G0003
AT-14	160	3006			G0003
AT15	160		51	TR17/637	G0003
AT-15	100	3005			
AT16	160		212	TR17/637	G0003
AT-16		3006			
AT17	160	3006	212	TR17/637	G0003
AT-17		3006			
AT20H	102A	3004	52	TR85/250	G0007
AT20M	102A	3004	52	TR85/250	G0005
AT20N	102A	3004	52	TR85/250	G0005
AT30H	102A	3004	52	TR85/250	G0006
AT30M	102A	3004	52	TR85/250	G0005
AT30N	102A	3004	52	TR85/250	G0005
AT50	102A		2	TR85/250	
AT-50	102A	3004			
AT52	101	3011	8	TR08/641	
AT-52				TR08/	
AT53	101	3011	8	TR08/641	
AT-53				TR08/	
AT71	101	3011	8	TR08/641	
AT72	101	3011	8	TR88/641	
AT73R	101	3011	8	TR08/641	
AT74	102A		8	TR85/250	G0005
AT-74		3004			
AT74S	102A			TR85/250	G0005
AT-74S		3004			
AT75R	101	3011	8	TR08/641	
AT76R	101	3011	8	TR08/641	
AT77	101	3011	8	TR08/641	
AT83				TR87/	
AT84				TR88/	
AT100H	102A	3123	52	TR85/250	G6011
AT100M	102A	3123	52	TR85/250	G6011
AT100N	102A	3123	52	TR85/250	G6011
AT200	127	3035		TR27/235	G6008
AT310	107	3039		TR70/720	S0016
AT311	107	3039		TR70/720	S0016
AT312	107	3039	17	TR70/720	S0005
AT313	107	3039		TR70/720	S0020
AT314	107	3039		TR70/720	S0020
AT315	107	3039		TR70/720	S0020
AT316	107	3039	17	TR70/720	S0025
AT318	107	3039		TR70/720	S0016
AT319	107	3039		TR70/720	S0016
AT321	107	3039		TR70/720	S0025
AT322	107	3039		TR70/720	S0016
AT323	107	3039		TR70/720	S0016
AT324	107	3039		TR70/720	S0025
AT325	107	3039		TR70/720	S0025
AT326	107	3039		TR70/720	S0015
AT327	107	3039		TR70/720	S0015
AT328	107	3039		TR70/720	S0015
AT329	123A	3122		TR21/735	S0011
AT330	107	3039		TR70/720	S0015
AT331	159	3114		TR20/717	S0013
AT331A	159	3114		TR20/	
AT332	159	3114		TR20/717	S0019
AT332A	159	3114		TR20/	
AT333	159	3025		TR20/717	
AT335	123A	3122		TR21/735	S0016
AT335A	159	3114		TR20/	
AT336	123A	3122		TR21/735	S0011
AT337	123A	3122		TR21/735	S0015
AT338	107	3039		TR87/720	S0025
AT339	128	3024		TR87/243	S0005
AT340	161	3117		TR83/719	S0025
AT341	161	3117		TR83/719	S0005
AT342	161	3117		TR83/719	S0017
AT343	161	3117		TR83/719	S0017
AT344	161	3117		TR83/719	S0025
AT345	161	3117		TR83/719	S0025
AT346	161	3117		TR83/719	S0025
AT347	123A	3122		TR21/735	S0025
AT348	123A	3122		TR21/735	S0025
AT349	123A	3122		TR21/735	S0025
AT350	154	3044		TR78/712	S5024
AT351	154	3044		TR78/712	S5024
AT370	123A	3122		TR21/735	S0015
AT380	128	3024		TR87/243	S3011
AT381	128	3024		TR87/243	S3002
AT382	128	3024		TR87/243	S3002
AT383	128	3024		TR87/243	S3011
AT384	128	3024		TR87/243	S3011
AT385	128	3024		TR87/243	S3002
AT386	128	3024		TR87/243	S3011
AT387	128	3024		TR87/243	S3011
AT388	128	3024		TR87/243	S3002
AT391	129	3025		TR88/242	S3012
AT392	129	3025		TR88/242	S3012
AT393	129	3025		TR88/242	S3012
AT394	129	3025		TR88/242	S3012
AT395	129	3025		TR88/242	S3012
AT396	129	3025		TR88/242	S3012
AT397	129	3025		TR88/242	S3012

Original Part Number	ECG	SK	GE	IR/WEP	HEP
AT398	129	3025		TR88/242	S3012
AT400	123A	3122		TR21/735	S0025
AT401	123A	3122		TR21/735	S0025
AT402	123A	3122		TR21/735	S0025
AT403	123A	3122		TR21/735	S0025
AT404	123A	3122		TR21/735	S0025
AT405	123A	3122		TR21/735	S0025
AT406	123A	3122		TR21/735	S0025
AT407	123A	3122		TR21/735	S0025
AT410	159	3114		TR20/717	S0012
AT410-1	159	3114		TR20/	
AT412	159	3114		TR20/717	S0019
AT412-1	159	3114		TR20/	
AT413	159	3114		TR20/717	S0019
AT413-1	159	3114		TR20/	
AT414	159	3114		TR20/717	S0012
AT414-1	159	3114		TR20/	
AT415	159	3114		TR20/717	S0012
AT415-1	159	3114		TR20/	
AT416	159	3114		TR20/717	S0019
AT416-1	159	3114		TR20/	
AT417	159	3114		TR20/717	S0019
AT417-1	159	3114		TR20/	
AT418	159	3114		TR20/717	S0012
AT418-1	159	3114		TR20/	
AT419	159	3114		TR20/717	S0025
AT419-1	159	3114		TR20/	
AT420	123A	3122		TR21/735	S0025
AT421	123A	3122		TR21/735	S0025
AT422	123A	3122		TR21/735	S0025
AT423	123A	3122		TR21/735	S0025
AT424	123A	3122		TR21/735	S0025
AT425	123A	3122		TR21/735	S0025
AT426	123A	3122		TR21/735	S0025
AT427	123A	3122		TR21/735	S0025
AT430	159	3114		TR20/717	S0012
AT430-1	159	3114		TR20/	
AT431	159	3114		TR20/717	S0019
AT431-1	159	3114		TR20	
AT432	159	3114		TR20/717	S0012
AT432-1	159	3114		TR20/	
AT433	159	3114		TR20/717	S0019
AT433-1	159	3114		TR20/	
AT434	159	3114		TR20/717	S0012
AT434-1	159	3114		TR20/	
AT435	159	3114		TR20/717	S0019
AT435-1	159	3114		TR20/	
AT436	159	3114		TR20/717	S0012
AT436-1	159	3114		TR20/	
AT437	159	3114		TR20/717	S0012
AT437-1	159	3114		TR20/	
AT438	159	3114		TR20/717	S0019
AT438-1	159	3114		TR20/	
AT440	128	3024		TR87/243	S0014
AT441	128	3024		TR87/243	S3011
AT442	128	3024		TR87/243	S3011
AT443	128	3024		TR87/243	S3011
AT444	128	3024		TR87/243	S3011
AT445	128	3024		TR87/243	S3011
AT446	128	3024		TR87/243	S3011
AT451	159	3114		TR20/717	S0019
AT451-1	159	3114		TR20/	
AT452	159	3114		TR20/717	S0019
AT452-1	159	3114		TR20/	
AT453	159	3114		TR20/717	S0019
AT453-1	159	3114		TR20/	
AT454	159	3114		TR20/717	S0013
AT454-1	159	3114		TR20/	
AT455	159	3114		TR20/717	S0019
AT455-1	159	3114		TR20/	
AT460	129	3025		TR88/242	S3012
AT461	129	3025		TR88/242	S3012
AT462	129	3025		TR88/242	S3012
AT463	129	3025		TR88/242	S3012
AT464	129	3025		TR88/242	S3012
AT465	129	3025		TR88/242	S3012
AT466	129	3025		TR88/242	S3012
AT467	129	3025		TR88/242	S3012
AT468	129	3025		TR88/242	S3012
AT470	128	3024	X18	TR87/243	S3011
AT471	128	3024		TR87/243	S3011
AT472	128	3024		TR87/243	S3002
AT473	128	3024	X18	TR87/243	S3011
AT474	128	3024		TR87/243	S3011
AT475	128	3024		TR87/243	S3002
AT476	128	3024	X18	TR87/243	S3011
AT477	128	3024		TR87/243	S3011
AT478	128	3024		TR87/243	S3002
AT479	128	3024		TR87/243	S3011
AT480	129	3025		TR88/242	S5022
AT481	129	3025		TR88/242	S5022
AT482	129	3025		TR88/242	S5022
AT483	129	3025		TR88/242	S5022
AT484	129	3025		TR88/242	S5022
AT485	129	3025		TR88/242	S5022

Original Part Number	ECG	SK	GE	IR/WEP	HEP
AT490	123A	3122		TR21/735	S0011
AT491	123A	3122		TR21/735	S0025
AT492	123A	3122		TR21/735	S0025
AT493	123A	3122		TR21/735	S0025
AT494	123A	3122		TR21/735	S0020
AT495	123A	3122		TR21/735	S0025
AT520	108	3039		TR95/56	S0020
AT521	101	3011	8	TR08/641	
AT551	101	3011	8	TR10/641	
AT874	102A	3004		TR85/250	G0001
AT1138	179		16	TR35/G6001	G6005
AT1138A	179		16	TR35/G6001	G6005
AT1138B	179		25	TR35/G6001	G6005
AT1833	179		16	TR35/G6001	G6005
AT1834	179		16	TR35/G6001	G6005
AT1856	130	3027		TR59/247	
AT-1856	130	3027	14		
AT2848	129	3025		TR88/242	
AT3260	130	3027		TR59/	
AT3660				/247	
AT5156	129	3025		TR88/242	
AT-8946		3045			
AT/AF1				TR85/250	G0005
ATAF1	102A	3004		TR85/	
ATAF2	102A	3004		TR85/	
AT/AF2				TR85/250	G0005
AT/RF1	160	3006		TR12/635	G0005
AT/RF2	160	3006		TR12/635	G0006
AT/S13	160	3006		TR12/635	G0005
ATC-TR-4	128	3024		TR87/243	
ATC-TR-5	121	3009		TR01/232	
ATC-TR-6	121	3009		TR01/232	
ATC-TR-7	128	3024		TR87/243	
ATC-TR-13	128	3024		TR87/243	
ATC-TR-14	121	3009		TR01/232	
ATC-TR-15	130	3027		TR59/247	
ATC-TR-19	152	3041		TR76/701	
ATGP	102A	3004	52	TR85/250	
ATRF1	126	3006		TR17/	
ATRF2	126	3006		TR17/	
ATS13	126	3006		TR17/	
AU41				TR21/	
AU100N	158	3004		TR84/630	G0005
AU101	127	3014	25	TR27/235	G6005
AU102	121	3009	25	TR01/232	G6005
AU103	127	3035	25	TR27/235	G6008
AU104	127	3035	25	TR27/235	G6008
AU105	127	3035	25	TR27/235	G6008
AU106	127	3035	25	TR27/235	G6008
AU107	127	3035	25	TR27/235	G6008
AU108	127	3035	25	TR27/235	
AU110	127	3035	25	TR27/235	
AU111	127	3035	25	TR27/235	G6008
AU112	127	3035	25	TR27/235	G6008
AU113	127	3035	25	TR27/235	G6008
AU410				TR01/	
AUY10	121	3009	16	TR01/230	G6005
AUY19	121	3014	25	TR01/232	G6005
AUY20	121	3014	25	TR01/232	G6005
AUY21	104		16	TR01/230	G6005
AUY-21	121	3009		TR01/	
AUY21A	121	3009	25	TR01/	G6005
AUY22	104		16	TR01/230	G6005
AUY-22		3009		TR01/	
AUY22A	121	3009	25	TR01/	G6005
AUY28	127	3035	25	TR27/235	G6005
AUY29	179			TR35/	G6009
AUY31	121		16		G6005
AUY33	121		16		G6005
AUY37	179		25		G6018
AUY38	127	3035		TR27/	G6008
AV105	121	3009		TR01/232	
AV				TR95/	
AX91770	108	3018	20	TR95/56	
AZY	107	3039		TR70/	
B00	7442				
B01	7475				
B02	7476				
B03	74H76				
B0301-049	181	3036			
B1A	182	3188			S5004
B1C	184	3114	28		S5000
B1C-1	184				S5000
B1C-2	184				S5000
B1D	152	3114	66	TR76/	S5000
B1D-1	184				S5000
B1E	191	3103	27	TR74/	S3021
B1E-1	191				S3021
B1E1(QUASAR)	191		27		
B1F	184				S5000
B1G	191				S3021
B1H	161	3018	60		S0017
B1J	159	3114			S5000
B1K	123A	3018	20	TR24/	S0005
B1M	188			TR72/	

Original Part Number	ECG	SK	GE	IR/WEP	HEP
B1N	123A	3124	20	TR51/	S0024
B1N-1	159	3114			
B1N-2	159	3114			
B1N-3		3114			
B1N/48-137509				TR51/	
B1P	163	3114			S5021
B1P-1	163	3114			
B1P7201	123A	3124			
B1P/M7511				TR27/	
B1T	283	3115	36		S5021
B1U	152	3054	28	TR76/	S5000
B1U148	184	3054	28		S5000
B1U/48-137526				TR76/	
B1V	286	3131	12		S5011
B1W	159	3114	20		S0024
B1Z	175	3114	12	TR23/	S5015
B2			214		S0017
B2A	159	3114			S0005
B2B	175	3111	36	TR61/	S5020
B2D	123A	3018	20		S0015
B2E	159	3114			S0015
B2G	159	3114			S0015
B2H	162	3079	35	TR61/	S5020
B2M-1	159	3114			
B2M-2	159	3114			
B2M-3	159	3114			
B2P	159	3114			
B2S	159	3114			
B2SB241	104	3009	3	TR01/	
B2SB244	104	3009	3	TR01/	
B2W	121	3114			
B2Y	159	3114			
B3			214		
B4			214		
B4A-1-A-21	121	3009			
B5	102A	3004		TR85/250	
B-5	176	3123	2	TR05/	
B5A	102A	3004	2	TR85/250	
B5D	154	3045	40		
B6B-3A21	121MP				
B6B-3-A-21	121MP	3015			
B8P-2A21	102A			TR85/	
B8P-2-A-21	102A	3004			
B9L-4A21	102A			TR85/	
B9L-4-A-21	102A	3003			
B12-1A21	123A			TR21/	
B12-1-A-21	123A	3124			
B14A-1-21	121	3009		TR01/	
B16A	105				
B17A	105				
B18A	105				
B19	104				
B20	104				
B21	104				
B22	102A				
B22-3	102A			TR85/250	
B-22-3	102A	3004	2		
B22-4	102A		2	TR85/250	
B-22-4	102A	3004			G0005
B22A	102A				
B22B	102A				
B22I	102A				
B23	102A			TR85/250	
B-23	102A	3004	2		
B23-1	102A			TR85/250	
B-23-1	102A	3004	2		G0007
B23-2	102A		2	TR85/250	
B-23-2	102A	3004			
B23-79	197	3084			
B23-82	196	3054			
B24	102A				
B24-1	102A		2	TR85/250	
B-24-1	102A	3004			G0005
B25	121				
B25B	126				
B26	102A		2	TR85/250	
B-26	102A	3004			G0007
B-26-1	102A	3123		TR85/	G0005
B26A	104				
B26(JAPAN)	104				
B27	104				
B28	104				
B29	104				
B30	104				
B31	104				
B32	102A	3004			
B32-0	102A				
B32-1	102A	3004			
B32-2	102A	3004			
B32-4	102A	3004			
B32N	102A	3004			
B32-O		3004			
B33	102A				
B33-4	102A	3004			
B33C	102A	3004			

Original Part Number	ECG	SK	GE	IR/WEP	HEP
B33D	102A	3004			
B33E	102A	3004			
B33F	102A	3004			
B34	102	3004			
B34N	127	3035			
B37	102A	3004			
B37A	102A	3004			
B37B	102A	3004			
B37C	102A	3004			
B37D	102A	3004			
B37E	102A	3004			
B37F	102A	3004			
B38	102	3004			
B39	102A	3004			
B40	102A	3004			
B41	104	3009			
B42	104	3009			
B43	102A	3004			
B43A	102A	3004			
B44	102A	3004			
B46	102A	3004			
B47	102A	3004			
B48	102	3004			
B49	102A	3004	2	TR85/250	
B50	102	3004			
B51	160	3123	2	TR17/637	G0002
B51(JAPAN)	102				
B52	102	3004			
B53	102	3004			
B54	102A	3004			
B54B	102A	3004			
B54E	102A				
B54Y	102A	3004			
B55	102A	3004			
B56	102A	3004			
B56A	102	3004			
B56C	102A	3004			
B57	102A				
B57A		3004			
B59	102A	3004			
B60	102A	3004			
B60A	102A	3004			
B61	102A	3004			
B62	104	3009			
B63	131	3082			
B64	121	3009			
B65	102A	3004			
B65-1A21	160			TR17/	
B65-1-A-21	160	3006			
B65-2A21	103A			TR08/	
B65-2-A-21	103A	3010			
B65-4A21	101			TR08/	
B65-4-A-21	101	3011			
B65A-1-21	160	3007		TR17/	
B65B-1-21	160	3008		TR17/	
B65C-1-21	121	3009		TR01/	
B66	102A	3004			
B66-1A21	102A			TR85/	
B66-1-A-21	102A	3003			
B66-2A21	102A			TR85/	
B66-2-A-21	102A	3004			
B66-3A21	102A			TR85/	
B66-3-A-21	102A	3004			
B66H	102A	3004			
B66X0040-006	130				
B67	102A	3004			
B67A	102A	3004			
B69	121	3009			
B71	102	3004			
B72	102	3004			
B73	102A	3004			
B73A	102A				
B73B	102A	3004			
B73C	102A	3004			
B73GR	102A	3004			
B74	160	3006			
B74-3A21	100			TR05/	
B74-3-A-21	100	3005			
B75	102A	3004			
B75-100		3018			
B75A	100	3004	51	TR12/	
B75AH	102A	3004			
B75B	102A	3004	52		
B75C	102A	3004	53		
B75F	102A	3004			
B75H	102A	3004			
B75LB	102A	3004			
B76	102A	3004			
B77	102A	3004			
B77A	102A	3004			
B77A-1-21	105	3012		TR03/	
B77AA	102A	3004			
B77AB	102A	3004			
B77AC	102A	3004			
B77AD	102A	3004			

Original Part Number	ECG	SK	GE	IR/WEP	HEP
B77AH	102A	3004			
B77AP	102A	3004			
B77B	102A	3004			
B77B-1-21	121	3013		TR01/	
B77B-11	102A	3004			
B77C	158	3004	53	TR84/	
B77C-1-21	121MP	3014			
B77D	102A	3004			
B77H	102A	3004			
B77T0049	9914				
B77V	102A	3004			
B77VRED	102A	3004			
B78	102A	3004			
B79	102A	3004			
B80	121	3009			
B81	121	3009			
B82	121	3009			
B83	104	3009			
B84	121	3009			
B84A-1-21	121	3009		TR01/	
B85	102	3004			
B86-72-2		3004			
B86-74-2		3003			
B86-75-2		3008			
B87	102	3004			
B87J0007	128				
B88B-1-21	160	3006		TR17/	
B88C-1-21	100	3005		TR05/	
B89	102A	3004			
B89A	102A	3004			
B89AH	102A	3004			
B89H	102A	3004			
B90	102A	3004			
B91	102A	3004			
B92	102	3004			
B92-1A21	105			TR03/	
B92-1-A-21	101	3012			
B94	102A	3004			
B95	102	3004			
B97	102A	3004			
B98	102	3004			
B100	102				
B101	102	3004			
B102	102	3004			
B103	100	3005			
B104	100	3005			
B105	158	3004			
B106	158	3004			
B107	102	3004			
B107A	121	3009			
B108	158	3004			
B108A	158	3004			
B108B	158	3004			
B109	158	3004			
B110	102A	3004			
B111	102A	3004			
B111K	102A	3004			
B112	102A	3004			
B113	102A	3004			G6005
B114	102A	3004			G6003
B115	102A	3004			
B116	102A	3004			
B117	102A	3004			
B117K	102A	3004			
B119	121	3009			
B119A	121	3009			
B120	102A	3004			
B122	121	3009			
B123	104	3009			
B123A	104	3009			
B124	121	3009			
B126	121	3009			
B126A	121	3009			
B126F	127	3035			
B126V	127				
B127	104	3009			
B127A	121	3009			
B128	127	3035			
B128A	121	3009			
B128V	127	3035			
B129	121	3009			
B130	131	3082			
B130A	131	3082			
B131	152			TR76/701	S5000
B131A	121	3009			
B131(JAPAN)	121				
B132	153	3009		TR77/700	S5006
B132A	121				
B132(JAPAN)	121				
B134	121	3009	16	TR01/232	G6003
B134A	121	3014	16	TR01/232	G6003
B134C	121	3014	16	TR01/232	G6005
B134-D	102A				
B134-E	102A				
B134(JAPAN)	102A				
B135	102A	3004			
B135B	102A	3004			
B135C	102A	3004			
B135E	102A	3004			
B136	102A	3004			
B136-2	102A				
B136-3	102A				
B136A	102A	3004			
B136B	102A	3004			
B136C	102A	3004			
B136U	102A	3004			
B137	121	3009			
B138	121	3009			
B140	104	3009			
B141	104	3009			
B142	104	3009			
B142B	100	3005			
B142C	100	3005			
B143	104	3009			
B143P	104	3009			
B144	104	3009			
B144P	104	3009			
B145	104	3009			
B146	104	3009			
B147	104	3009			
B149	104	3009			G6003
B150-200		3124			
B151	104	3009			
B152	104	3009			
B153	102	3004			
B154	102	3004			
B155	102A	3004			
B155A	102A	3004			
B155B	102A	3004			
B156	102A	3004			
B156A	102A	3004			
B156AA	102A	3004			
B156AB	102A	3004			
B156AC	102A	3004			
B156B	102A	3004			
B156C	102A	3004			
B156D	102A	3004			
B156P	102A	3004			
B157	102A	3004			
B158	102A	3004			
B159	102A	3004			
B160	102A	3004			
B161	102	3004			
B162	102	3004			
B163	102	3004			
B164	102	3004			
B165	102	3004			
B166	102	3004			
B167	158	3004			
B168	102A	3004			
B169	123A		20	TR21/735	F0010
B-169	123A	3124			
B170	102A	3004			
B171	123A			TR21/	
B171A	102A	3004			
B171B	102A	3004			
B171(JAPAN)	102A				
B172	102A	3004			
B172A	102A	3004			
B172AF	102A	3004			
B172B	102A	3004			
B172C	102A	3004			
B172D	102A	3004			
B172E	102A	3004			
B172F	102A	3004			
B172G	102A	3004			
B172H	102A	3004			
B172P	102A	3004			
B172R	102A	3004			
B173	102A	3004			
B173A	102A	3004			
B173B	102A	3004			
B173C	102A	3004			
B173L	102A	3004			
B174	102A	3004			
B175	102A	3004			
B175A	102A	3004			
B175B	102A	3004			
B175C	102A	3004			
B175E	102A	3004			
B176	102A	3004			
B176M	102A	3004			
B176-D	102A	3004			
B176P	102A				
B176-P	102A	3004			
B176-PR	102A	3004			
B176PRC	102A	3004			
B176R	102A	3004			
B177	121	3009	16	TR01/232	G6005
B177(JAPAN)	102A				

Original Part Number	ECG	SK	GE	IR/WEP	HEP
B178	121	3009	16	TR01/232	G6005
B178A	102A	3004			
B178C	102A	3004			
B178D	102A	3004			
B178(JAPAN)	102A				
B178M	102A	3004			
B178N	102A	3004			
B178-O	102A	3004			
B178-S	102A	3004			
B178T	102A	3004			
B178U	102A	3004			
B178V	102A	3004			
B178X	102A	3004			
B178Y	102A	3004			
B179	121	3014	16	TR01/232	G6005
B180	158	3004			
B180A	158	3004			
B181	158	3004			
B181A	158	3004			
B183	102A	3004			
B184	102A	3004			
B185	102A	3004			
B185AA	102A				
B185F	102A				
B185(O)	102				
B185P	102A	3004			
B186	158	3004	53	TR85/250	
B186(O)	102A				
B186-1	102A	3004			
B186A	102A	3004			
B186AG	102A	3004			
B186B	102A	3004			
B186BY	102A	3004			
B186G	102A	3004			
B186H	102A				
B186-K	102A	3004			
B186L	102A	3004			
B186(SANYO)	102A			TR85/	
B186Y	102A	3004			
B187	158	3004	53	TR85/250	
B187(1)	102A				
B187AA	102A	3004			
B187B	102A	3004			
B187C	102A	3004			
B187D	102A	3004			
B187G	102A	3004			
B187K	102A				
B187R	102A	3004			
B187S	102A	3004			
B187(SANYO)	102A			TR85/	
B187YEL	102A				
B188	102A	3004			
B189	102	3004			
B199	102				
B200	102	3004			
B200A	102	3004			
B201	102A	3004			
B202	102	3004			
B203AA	102	3004			
B204	179				
B205	179				
B206	179				
B215	121	3009			
B216	121	3009			
B216A	121	3009			
B217	121	3009			
B217A	121	3009			
B217G	121	3009			
B217U	121	3009			
B218	102	3004			
B219	102	3004			
B220	102	3004			
B220A	102	3004			
B221	102	3004			
B221A	102	3004			
B222	102	3004			
B223	102	3004			
B224	121	3009	16	TR01/232	G6005
B224(JAPAN)	102				
B225	102	3004			
B226	102	3004			
B227	102	3004			
B228	104	3009			
B229	121	3009			
B230	104	3009			
B231	127	3035			
B232	127	3035			
B233	104	3009			
B234	127	3035			
B234N	127	3035			
B235	105	3012			
B235A	160	3006			
B236	105	3012			
B237	105	3012			
B237-12A	105	3012			

Original Part Number	ECG	SK	GE	IR/WEP	HEP
B237-12B	105	3012			
B238	158	3004			
B238-12A	105	3012			
B238-12B	105	3012			
B238-12C	105	3012			
B239	121	3009			
B239A	121	3009			
B240	160	3006			
B240A	160	3006			
B241	102	3004			
B246	104	3009			
B248	121	3009			
B248A	121	3009			
B249	104	3009			
B249A	104	3009			
B250	104				
B250A	104	3009			
B251	127	3035			
B251A	127	3035			
B252	127	3035			
B252A	127	3035			
B253	127	3035			
B253A	127	3035			
B254	104	3009			
B255	104	3009			
B256	104	3009			
B257	102	3004			
B258	105	3012			
B259	105	3012			
B260	105	3012			
B261	102A	3004			
B262	102A	3004			
B263	102A	3004			
B264	102A	3004			
B265	102A	3004			
B266	102A	3004			
B266A-1	159	3025			
B266A-2	159	3025			
B266P	102A	3004			
B266Q	102A	3004			
B267	102A	3004			
B268	102A	3004			
B269	102A	3004			
B270	102A	3004			
B270A	102A	3004			
B270B	102A	3004			
B270C	102A	3004			
B270D	102A	3004			
B270E	102A	3004			
B271	102A	3004			
B272	102A	3004			
B273	102A	3004			
B274(JAPAN)	127			TR25/	
B274(SYLVANIA)	128			TR87/	
B275	127	3035			
B276	127	3035			
B282	121	3009			
B283	121	3009			
B284	121	3009			
B285	121	3009			
B290	100	3005			
B291	100	3005			
B292	100	3005			
B292A	100	3005			
B293	102A	3004			
B294	102A				
B295	121	3009			
B296	127	3035			
B299	102A	3004			
B300	127	3035			
B301	127	3035			
B302	102A	3004			
B303	102A	3004			
B303-O	102A	3004			
B303A	102A	3004			
B303B	102A	3004			
B303C	102A	3004			
B303H	102A	3004			
B303K	102A				
B304	158	3004			
B304A	158	3004			
B309	127	3035			
B310	127	3035			
B311	127	3035			
B312	127	3035			
B313	127	3035			
B314	100	3005			
B315	102A	3004			
B-315-1	158	3004		TR84/	G0005
B316	102A	3004			
B317	102A	3004			
B318	127	3035			
B319	127	3035			
B320	127	3035			
B321	102A	3004			

Original Part Number	ECG	SK	GE	IR/WEP	HEP
B322	102A	3004			
B323	102A	3004			
B324	158	3004		TR84/630	G0005
B-324	158	3004			G0005
B324A	158	3004			
B324B	158	3004			
B324D	158	3004			
B324E	158	3004			
B324E-1	102A	3004			
B324F	158				
B324G	158	3004			
B324H	158	3004			
B324I	158	3004			
B324J	158	3004			
B324K	158	3004			
B324L	158	3004			
B324N	158	3004			
B324P	158	3004			
B324V	158	3004			
B326	102	3004			
B327	102	3004			
B328	102	3004			
B329	102A	3004			
B329K	102A	3004			
B331	105	3012			
B332	105	3012			
B333	105	3012			
B334	105	3012			
B336	102A	3004			
B337	104	3009			
B337A	104	3004			
B33B	104	3009			
B337BK	102A				
B337H	102A	3004			
B338	121	3009			
B338HA	102A	3004			
B338HB	102A	3004			
B339	179				
B339H	127	3035			
B340	179				
B340H	127	3035			
B341	179				
B341H	127	3035			
B341V	127	3035			
B342	127	3035			
B343	127	3035			
B345	012A	3004			
B346	102A	3004			
B346K	102A	3004			
B346Q	102A				
B347	102	3004			
B348	102A	3004			
B348Q	102A				
B348R	102A	3004			
B349	102A	3004			
B350	102	3004			
B351	105	3012			
B352	105	3012			
B353	105	3012			
B354	105	3012			
B355	104	3009			
B356	104	3009			
B357	12	3035			
B358	127	3035			
B359	127	3035			
B360	127	3035			
B361	127	3035			
B362	127	3035			
B364	158	3004			
B365	158	3004			
B366	102A	3004			
B367	131	3198			
B367A	131	3198			
B367B	131	3198			
B367C	131	3198			
B367H	131	3198			
B368	131	3198			
B368A	131	3198			
B368B	131	3198			
B368H	131	3198			
B370	102A	3004			
B370A	102A	3004			
B370AA	102A	3004			
B370AB	102A	3004			
B370AC	102A	3004			
B370AHA	102A	3004			
B370AHB	102A	3004			
B370B	102A	3004			
B370C	102A	3004			
B370D	102A	3004			
B370P	102A	3004			
B370PB	102A				
B370V	102A	3004			
B371	102A	3004			
B371D	102A	3004			
B372	158	3004			
B373	158	3004			
B375	127	3035			
B375-2B	127	3035			
B375-5B	127	3035			
B375A	127	3035			
B375A-2B	127	3035			
B375A-5B	127	3035			
B375A-NB	127	3035			
B376	102A	3004			
B376G	102A	3004			
B377	102A	3004			
B377B	102A				
B378	100	3005			
B378A	102	3004			
B379	102	3004			
B379-2	102	3004			
B379A	102	3004			
B379B	102	3004		TR85/	
B380	102	3004			
B380A	102	3004			
B381	102A	3004			
B382	102A	3004			
B383	102A	3004			
B383-1	102A	3004			
B383-2	102A	3004			
B384	126	3006			
B385	126	3006			
B386	102A	3004			
B387	100	3005			
B389	102A	3004			
B390	127	3035			
B391	121	3009			
B392	100	3005			
B393	100	3005			
B394	100	3005			
B395	100	3005			
B396	100	3005			
B400	102A	3004			
B400A	102A	3004			
B400B	102A	3004			
B400K	102A	3004			
B401	100	3005			
B401-423		3004			
B401-465-3		3005			
B402	100	3005			
B403	100	3005			
B405	158	3004			
B405-2C	158	3004			
B405-3C	158	3004			
B405-4C	158	3004			
B405A	158	3004			
B405B	158	3004			
B405C	158	3004			
B405D	158	3004			
B405E	158	3004			
B405G	158				
B405H	158	3004			
B405K	158				
B405R	158	3004			
B405RE	158	3004			
B407	104	3009			
B407-0	104	3009			
B407TV	121	3009			
B408	100	3005			
B410	127	3035			
B411	127	3035			
B413	104	3009			
B414	104	3009			
B415	158	3004			
B415A	158	3004			
B415B	158				
B416	100	3005			
B417	100	3005			
B421	102A	3004			
B422	102	3004			
B423	102	3004			
B424	121	3009			
B425	121	3009			
B425Y	127	3035			
B426	121	3009			
B426BL	104	3009			
B426R	121	3009			
B426Y	121	3009			
B427	102A	3004			
B428	102A	3004			
B430	105	3012			
B431	158	3004			
B432	127	3035			
B433	105	3012			
B434	153	3083			
B434-0	153	3083			
B434-R	153	3083			
B434-Y	153	3083			
B435	153	3083			

Original Part Number	ECG	SK	GE	IR/WEP	HEP
B435-0	153	3083			
B435-R	153	3083			
B435-Y	153	3083			
B439	102A	3004			
B439A	102	3004			
B440	102A	3004			
B443	102A	3004			
B443A	102A	3004			
B443B	102A	3004			
B444	126	3006			
B444A	126	3006			
B444B	126	3006			
B445	121	3009			
B446	121	3009			
B447	127	3035			
B448	131	3082			
B449	121	3009			
B449F	121	3009			
B449P	121	3009			
B450	158	3004			
B450A	158	3004			
B451	158	3004			
B452	158	3004			
B452A	158	3004			
B453	158	3004			
B454	158	3004			
B455	158	3004			
B457	158	3004			
B457A	158	3004			G0005
B457-C	158				
B458	131	3082			
B458A	131	3082			
B459	102A	3004			
B459A	102A	3004			
B459B	102A	3004			
B459C	102A	3004			
B459D	102A	3004			
B459-0	102A	3004			
B460	102A	3004			
B460A	102A	3004			
B460B	102A	3004			
B461	176	3123			
B462	131	3198			
B463	131	3198			
B463BL	131	3198			
B463E	131	3198			
B463R	131	3198			
B463Y	131	3198			
B464	127	3035			
B465	127	3035			
B466	131	3198			
B467	131	3198			
B468	127	3035			
B468A	127	3035			
B468B	127	3035			
B468C	127	3035			
B468D	127	3035			
B470	102	3004			
B471	121	3009			
B471-2	121	3009			
B471A	121	3009			
B471B	121	3009			
B472	121	3009			
B472A	121	3009			
B472B	121	3009			
B473	131	3198			
B473D	131	3198			
B473F	131	3198			
B473H	131				
B474	226	3082			
B474-2	226	3082			
B474-3	226	3082			
B474-4	226	3082			
B474-6D	226	3082			
B474MP	131	3198			
B474S	131	3198			
B474V4	131	3198			
B474V10	126MP	3198			
B474Y	131	3198			
B475	158	3004			
B475A	158	3004			
B475B	158	3004			
B475D	158	3004			
B475E	158	3004			
B475F	158	3004			
B475G	158	3004			
B475P	158	3004			
B475Q	158	3004			
B476	158				
B481	131	3198			
B481D	131	3198			
B481E	131	3198			
B482	102A	3004			
B483	179				
B484	179				
B485	179				
B486	102	3004			
B492	176	3123			
B492B	176	3123			
B494	158	3004			
B495	158	3004			
B495A	158	3004			
B495C	158	3004			
B495D	158	3004			
B495T	158	3004			
B496	102A	3004			
B497	102	3004			
B498	102A	3004			
B502	197	3085			
B503	197	3085			
B507	153	3083			
B508	153	3083			
B509	153	3083			
B510	129	3029			
B510S	211	3203			
B511	153	3083			
B511C	153				
B511D	153	3083			
B512	153	3083			
B512A	153	3083			
B513	153	3083			
B513A	153	3083			
B514	153	3083			
B515	153	3083			
B516C	102A				
B516CD	102A				
B516D	102A				
B516P	102A				
B539R	123A	3122			
B560	102A	3004			
B561	102A				
B601-1006	160	3006		TR17/637	G0003
B601-1007	160	3006		TR17/637	G0003
B601-1008	160	3006		TR17/637	G0003
B601-1009	102A	3004		TR85/735	G0005
B601-1010	158	3004		TR84/630	G0005
B681A	163				
B681B	163				
B1017	121	3009	16	TR01/232	G6005
B1022	102	3004	52	TR05/631	G0005
B1022-1	102A	3004		TR85/	
B1058	102		2	TR05/631	
B-1058	102A	3004			
B1058-1	102A	3004		TR85/	
B1085	121	3014	25	TR01/232	G6005
B1151	121	3009	16	TR01/232	G6005
B1151A	121	3009	16	TR01/232	G6005
B1151B	121	3009	16	TR01/232	G6005
B1152	121	3009	16	TR01/232	G6005
B1152A	121	3009	16	TR01/232	G6005
B1152B	121	3009	16	TR01/232	G6005
B1154	102	3004	52	TR05/631	G6005
B1154-1	102A	3004		TR85/	
B1178	127		25		
B1181	121	3009	76		
B1215			16	TR02/	
B1274	121	3009	25	TR01/232	G6005
B1274A	121	3009	25	TR01/232	G6005
B1274B	121	3009	25	TR01/232	G6005
B-1338	123A	3124			S0014
B1368	121	3014	16	TR01/232	G6005
B1368A	121	3009	16	TR01/232	G6005
B1368B	121	3009	16	TR01/232	G6005
B1368C	121	3009	16	TR01/232	G6005
B1368D	121	3009	16	TR01/232	G6005
B1368E	121	3014	16	TR01/232	G6005
B1368F	121	3014	16	TR01/232	G6005
B-1421	123A	3124	20	TR51/	S0014
B-1426	159	3114	67	TR19/	S0019
B-1433	123A	3124			S0014
B-1511	121	3014		TR01/232	G6005
B-1666	123A	3124	20	TR51/	S0014
B-1695	185	3083			S5006
B-1790	184	3054			S5000
B-1823	152	3054			S5000
B1823		3054			S5000
B-1842	123A	3124			S0015
B-1872	123A	3124			S0015
B1904	121	3009	16		
B-1910	199	3124			S0015
B-1914	121	3009		TR01/232	G6005
B3030	102A	3004			
B3465	195	3048			
B3468	195			TR65/	
B3531	186	3192	28		
B3533	186	3192	28	TR76/	
B3537	186	3192	216	TR76/	
B3538	186	3192	28	TR76/	
B3540	186	3192	28	TR76/	
B3541	186	3192	28	TR76/	

Original Part Number	ECG	SK	GE	IR/WEP	HEP
B143001	186	3192	28	TR76/	
B143003	186	3192	28	TR76/	
B143004	152	3054	66		
B143009	186	3192	28	TR76/	
B143010	186	3192	28		
B143011	152	3054	66		
B143012	152	3054	66		
B143015	186	3192	28	TR76/	
B143016	186	3192	28	TR76/	
B143018	152	3054	66		
B3542	186	3192	28	TR76/	
B3547	152	3054	66		
B3548	152	3054	66		
B3550	152	3054	66		
B3551	152	3054	66		
B3570	186	3192	28	TR76/	
B3576	186	3192	28	TR76/	
B3577	152	3054	66		
B3578	152	3054	66		
B3580	152	3054	66		
B3584	152	3054	66		
B3585	152	3054	66		
B3586	152	3054	66		
B3588	152	3054	66		
B3588	152	3054	66		
B3606	186	3192	216	TR76/	
B3607	186	3192	216	TR76/	
B3608	186	3192	28	TR76/	
B3609	186	3192	216	TR76/	
B3610	186	3192	216	TR76/	
B3611	186	3192	28	TR76/	
B3612	186	3192	28	TR76/	
B3613	186	3192	28	TR76/	
B3614	186	3192	28	TR76/	
B3746	128	3024	63	TR87/243	
B3747	186	3192	28	TR76/	
B3748	186	3192	28	TR76/	
B3750	186	3192	28	TR76/	
B5000	152		MR6	TR76/701	S5000
B5001	186	3192	MR6		
B5002	152	3054	66		
B5020	130	3027	14		
B5021	152	3054	66		
B5022	152	3054	66		
B5031	152	3054	66		
B5032	152	3054	66		
B-6001	133	3112	FET1	FE100/	F0010
B 6288	159				
B 6340	128				
B10064	104	3009		TR01/230	G6003
B10069	104	3009		TR01/230	G6003
B10142	127		25	TR27/235	G6008
B10142A	127		25	TR27/235	G6008
B10142B	127		25	TR27/235	G6008
B10143	127			TR27/235	G6008
B10143A	127			TR27/235	G6008
B10143B	127			TR27/235	G6008
B10162	121	3014	16	TR01/232	G6005
B10163	121	3014	16	TR01/232	G6005
B10474	121	3014	16	TR01/232	G6005
B10475	121	3014	25	TR01/232	G6005
B10912	121	3009	16	TR01/232	
B10913	121	3009	16	TR01/232	
B-12822-2	130				
B-12822-4	161				
B13517-2	912				
B17007					S0002
B17008					S5020
B17307	130			TR59/	S7002
B54731-30	159				
B-75561-31	159				
B-75568-2	107				
B-75583-1	123A				
B-75583-2	123A				
B-75583-102	123A				
B-75589-3	123A				
B-75589-13	123A				
B-75608-3	123A				
B102000	127	3035	25		
B102001	127	3035	25		
B102002	127	3035	25		
B102003	127	3035	25		
B103000	127	3035	25		
B103001	127	3035	25		
B103002	127	3035	25		
B103003	127	3035	25		
B103004	127	3035	25		
B113000	179			TR35/G6001	
B133550	130				
B133577	130				
B133578	123A				
B133684	130				
B133685	130				
B133823	175				
B143000	186	3192	28	TR76/	

Original Part Number	ECG	SK	GE	IR/WEP	HEP
B143019	152	3054	66		
B143024	186	3192	28	TR76/	
B143025	186	3192	28	TR76/	
B143026	152	3054	66		
B143027	152	3054	66		
B170000	130	3027	75	TR59/247	S7002
B170000BLK	130			TR59/247	
B170000-BLK			77	TR26/	
B170000BRN	130			TR59/247	
B170000-BRN			77	TR26/	
B170000-ORN	130		77	TR61/	
B170000-RED	130		77	TR61/	
B170000-YEL			75		
B170001	130	3027	75	TR59/247	S7002
B170001BLK	130			TR59/247	
B170001-BLK	130		77	TR26/	
B170001BRN	130			TR59/247	
B170001-BRN	130		77	TR26/	
B170001-ORG	130		77	TR51/	
B170001-RED	130		77	TR61/	
B170001-YEL			75		
B170002	130	3027	75	TR59/247	S7002
B170002-BLK			77		
B170002-BRN			77		
B170002-ORG	130		77		
B170002-RED	130		77		
B170002-YEL			75		
B170003	130	3027	75	TR59/247	
B170003-BLK	181		75	TR61/	
B170003-BRN	181		75	TR61/	
B170003-ORG			75	TR61/	
B170003-RED	181		75	TR61/	
B170003-YEL			75		
B170004	130	3027	75	TR59/247	S7002
B170004-BLK	181		75	TR61/	
B170004-BRN	181		75	TR61/	
B170004-ORG			75	TR61/	
B170004-RED	181		75	TR61	
B170004-YEL			75		
B170005	130	3027	75	TR59/247	S7002
B170005-BLK	181		75		
B170005-BRN	181		75		
B170005-ORG			75		
B170005-RED	181		75		
B170005-YEL			75		
B170006	130	3027	75	TR59/247	
B170006-BLK	181		75	TR61/	
B170006-BRN	181		75	TR61/	
B170006-ORG			75	TR61/	
B170006-RED	181		75	TR61/	
B170006-YEL			75		
B170007	130	3027	75	TR59/247	
B170007-BLK	181		75	TR61/	
B170007-BRN	181		75	TR61/	
B170007-ORG			75	TR61/	
B170007-RED	181		75	TR61/	
B170007-YEL			75		
B170008	162	3027	75	TR67/707	
B170008-BLK	181		75		
B170008-BRN	181		75		
B170008-ORG			75		
B170008-RED	181		75		
B170008-YEL			75		
B170009	130	3027	77	TR59/247	S7002
B170010	130	3027	77	TR59/247	S7002
B170011	130	3027	77	TR59/247	S7002
B170012	130	3027	75	TR59/247	
B170013	130	3027	75	TR59/247	
B170014	130	3027	75	TR59/247	
B170015	130	3027	75	TR59/247	
B170016	130	3027	75	TR59/247	
B170017	181	3027	75		
B170018	130	3027	77	TR59/247	S7002
B170019	130	3027	77	TR59/247	S7002
B170020	130	3027	77	TR59/247	S7002
B170021	130	3027	75	TR59/247	
B170022	130	3027	75	TR59/247	
B170023	181	3027	75		
B170024	130	3027	75	TR59/247	
B170025	130	3027	75	TR59/247	
B170026	181	3027	75		
B176000	163		73	TR61/	S5020
B176001	162		35	TR61/	S5020
B176002	162		35	TR61/	S5020
B176003	162		35	TR61/	S5020
B176004	163	73	73	TR61/	S5020
B176005	163		36	TR61/	S5020
B176006	163		36	TR61/	S5020
B176007	163		36	TR61/	S5020
B176008					S5021
B176009	163		36		S5021
B176010	163		36		S5021
B176011	163		36		S5021
B176012					S5021
B176013	163		36		S5021

Original Part Number	ECG	SK	GE	IR/WEP	HEP
B176014	163		36		S5021
B176015	163		36		S5021
B176024	163		36	TR61/	S5020
B176025	163		36	TR61/	S5020
B176026	163		36		S5021
B176027	163		36		S5021
B176028	163		36		S5021
B176029	163		36		S5021
B177000	181		75		
B232008	124		12	TR81/240	
B 722246-2	123A				
B5493957-4	129	3025		TR88/242	
B5493957-5	129	3025		TR88/242	
B5493957-6	129	3025		TR88/242	
B8780010	123				
BA6	158	3005	2	TR84/630	G0005
BA-6		3004			
BA6A	158	3005	2	TR84/630	G0005
BA8-1A-21	160	3007		TR17/	
BA67	123A	3122		TR21/735	S0015
BA71	123A	3122		TR21/735	S0025
BA301	1135				
BA1003	172		64		
BACSH2M1	123A				
BACSH2M2	123A				
BACSH2M3	123A				
BACT 1A1		3007			
BACT2F	123A				
BACT2M1		3039			
BB71	123A	3122		TR21/735	S0025
BC20C	199				
BC25BB	159	3114			
BC71	123A				S0025
BC-71	123A	3122		TR21/735	
BC100	154	3044		TR78/712	S5024
BC103	128	3024	18	TR87/243	
BC103C	128	3024		TR87/243	
BC107	123A	3122	20	TR21/735	S0015
BC-107	123A	3124			
BC107A	123A	3122	210	TR21/735	S0015
BC-107A	123A	3124			
BC107B	123A	3122	20	TR21/735	S0015
BC108	123A	3124	20	TR21/735	S0011
BC-108	123A	3124			
BC108A	123A	3122	210	TR21/735	S0015
BC108B	123A	3122		TR21/735	S0015
BC-108B	123A		20		
BC108C	123A	3122		TR21/735	S0015
BC109	123A	3124	62	TR21/735	S0024
BC109B	123A	3124		TR21/735	S0015
BC-109B	123A		10		
BC109BP	123A	3124		TR21/735	
BC109C	123A	3122	20	TR21/735	S0015
BC110	123A	3122	81	TR21/735	S0005
BC111	108	3039	11	TR95/56	S0015
BC112	108	3039	39	TR95/56	S0015
BC113	123A	3122	212	TR21/735	S0015
BC114	123A	3122	212	TR21/735	S0015
BC-114	123A				
BC114A					S0030
BC114TR	123A	3122	62	TR51/	
BC115	123A	3122	212	TR21/735	S0015
BC116	159	3025	21	TR20/717	S0013
BC116A	159	3025	48	TR20/717	S5022
BC117	154	3024	27	TR78/712	S0005
BC118	123A	3122	210	TR21/735	S0015
BC119	128	3024	47	TR87/243	S0011
BC-119	128		18		
BC120	128	3024	47	TR87/243	S0014
BC121	107		212	TR70/720	S0025
BC-121	123A	3124			
BC122	107		212	TR70/720	S0020
BC-122	123A	3124			
BC123	107		212	TR70/720	S0016
BC-123	123A	3124			
BC125	123A	3122	61	TR21/735	S0011
BC125A			210		S0025
BC125B	123A	3122	213		S0025
BC126	159	3114	21	TR20/717	S0013
BC126-1	159	3114			
BC126A			21		S0019
BC127	199	3039	11	TR95/56	S0033
BC128	199	3039	11	TR95/56	S0015
BC129	123A	3122	20	TR21/735	S0011
BC130	123A	3122	20	TR21/735	S0011
BC130A					S0015
BC130B					S0015
BC130C					S0024
BC131	123A	3122	20	TR21/735	S0011
BC131B					S0015
BC131C					S0024
BC132	123A	3122	212	TR21/735	S0011
BC132A					S0030
BC134	123A	3122	210	TR21/735	S0015
BC135	123A	3122	210	TR21/735	S0015
BC135A					S0025
BC136	123A	3122	20	TR21/735	S0025
BC137	159	3114	21	TR20/717	S0019
BC137-1	159	3114			
BC138	128	3024	18	TR87/243	S3011
BC139	129	3025	21	TR88/242	S0013
BC139A					S3014
BC140	128	3024	28	TR87/243	S3002
BC-140	128		18		
BC140-6	128	3024		TR87/243	S3001
BC140-10	128	3024		TR87/243	S3001
BC140-16	128	3024		TR87/243	S3001
BC-140A	128		18		
BC140A		3024		TR87/243	
BC140B		3024		TR87/243	
BC-140B	128		18		
BC140C	128	3024	18	TR87/243	S3011
BC140D	128	3024	18	TR87/243	S3011
BC141	128	3024	18	TR87/243	S3019
BC141-6	128	3024		TR87/243	S3019
BC141-10	128	3024		TR87/243	S3019
BC141-16	128	3024		TR87/243	S3019
BC142	128	3024	18	TR87/243	S3001
BC143	129	3025	21	TR88/242	S3012
BC144	128	3024	47	TR87/243	S3001
BC145	154	3044	27	TR78/712	S0005
BC146	123A	3122	39	TR21/735	S0015
BC147	123A	3124	210	TR21/735	S0015
BC147A	123A	3124		TR21/735	S0015
BC147B	123A	3124	10	TR21/735	S0015
BC148	123A	3124	210	TR21/735	S0015
BC148A	123A	3124	20	TR21/	S0015
BC148B	123A	3124	20	TR21/	S0015
BC148C	123A	3124	20	TR21/	S0024
BC149	123A	3124	20	TR21/	S0015
BC149A	123A	3124		TR21/735	
BC149B	123A	3124	20	TR21/735	S0015
BC149C	123A	3124	62	TR21/735	S0024
BC149G	123A			TR51/	
BC150	123A	3122		TR21/735	S0015
BC151	123A	3122		TR21/735	S0015
BC152	123A	3122		TR21/735	S0015
BC153	234	3114	65	TR20/	S0011
BC153-1	159	3114			
BC154	234	3114	65	TR20/717	S0019
BC154-1	159	3114			
BC155	107	3039	17	TR70/720	S0015
BC155A	108	3039	39		S0015
BC155B	123A	3122	39		
BC155C	199				S0024
BC156	107	3039	20	TR70/720	S0025
BC156A	108	3039	39		S0015
BC156B	123A	3122	39		S0015
BC156C	199		212		S0024
BC157	159	3114	65	TR20/717	S0019
BC157-1	159	3114			
BC157A	159		65	TR20/717	
BC157B	159	3025	65		
BC158	159	3118	65	TR20/717	S0013
BC158-1	159	3114			
BC158A	159	3118	65	TR20/717	S0013
BC158B	159	3114	65	TR20/717	S0031
BC159	159	3118	65	TR20/717	S0031
BC159-1	159	3114			
BC159A	159	3114	65	TR20/717	S0013
BC159B	159	3114	65	TR20/717	S0031
BC160	129	3025	29	TR88/242	S3028
BC160-6	129	3025	48	TR88/242	S3012
BC160-10	129	3025	48	TR88/242	S3012
BC160-16	129	3025	48	TR88/242	S3012
BC161	129	3025		TR88/242	S3012
BC161-6	129	3025	48	TR88/242	S3012
BC161-10	129	3025	67	TR88/242	S3012
BC161-16	129	3025	67	TR88/242	S3012
BC167	123A	3122	210	TR21/735	S0025
BC167B	123A	3122	210		
BC168	123A	3124	210	TR21/735	S0024
BC168A	123A	3124	10	TR21/735	
BC168B	123A	3124	10	TR21/735	
BC168C	123A			TR21/735	
BC169	123A	3124	62	TR21/735	S0015
BC169A	123A	3124	20	TR21/735	
BC169B	123A	3124	10	TR21/735	S0014
BC169C	123A	3018	62	TR21/735	S0015
BC169CL	123A	3124			
BC170	123A	3122		TR21/735	S0025
BC170A	123A	3122	212	TR21/735	S0025
BC170B	123A	3122	212	TR21/735	S0015
BC170C	123A	3122	212	TR21/735	S0015
BC171	123A	3122	212	TR21/735	S0015
BC171A	123A	3122	210	TR21/735	S0015
BC171B	123A	3122	210	TR21/735	S0030
BC172	123A	3122	212	TR21/735	S0015
BC172A	123A	3122	210	TR21/735	S0015
BC172B	123A	3122	210	TR21/735	S0015

Original Part Number	ECG	SK	GE	IR/WEP	HEP
BC172C	123A	3122		TR21/735	S0015
BC173	123A	3124	212	TR21/735	S0024
BC173A	123A	3124	212	TR21/735	
BC173B	123A	3124	17	TR21/735	S0015
BC173C	123A	3122	212	TR21/735	S0024
BC174	128	3024		TR87/243	S0015
BC174A	128	3024	63	TR87/243	S0025
BC174B	128	3024		TR87/243	S0015
BC175	123A	3122		TR21/735	S3024
BC177	234	3004	65	TR20/717	S0019
BC177-1	159	3114			
BC177A	234	3114	48	TR20/717	S0019
BC177B	234	3114	48	TR20/717	S0032
BC177V	159	3114	21	TR20/717	S0019
BC177VI	159	3114	21	TR20/717	S0019
BC178	234	3114	65	TR20/717	S0031
BC178-1	159	3114			
BC178A	234	3114	48	TR20/717	S0013
BC178B	234	3053	48	TR20/717	S0019
BC178C	234		65		
BC178D	159	3114		TR20/717	
BC178V	159	3114	21	TR20/717	S0013
BC178VI	159	3114	21	TR20/717	S0013
BC179	234	3114	65	TR20/717	S0019
BC179A	234	3114	65	TR20/717	S0013
BC179B	234	3114	48	TR20/717	S0031
BC179C	234		65		
BC180	123A	3122	20	TR21/735	S0015
BC181	159		82		S0019
BC181A	159	3114	21	TR20/717	
BC182	123A	3122		TR21/735	
BC182A	123A		210		S0015
BC182L	123A	3122	210	TR21/735	S0015
BC183	123A	3122		TR21/735	
BC183A	123A		210		S0015
BC183B	123A				S0015
BC183L	123A	3122	210	TR21/735	S0015
BC184	123A	3122	210	TR21/735	S0015
BC184B	123A		210		
BC184L	123A	3122		TR21/735	S0015
BC185	123A	3197	215		S0014
BC186	123A	3122	65	TR21/735	S0013
BC187	159	3114	65	TR20/717	S0013
BC187-1	159	3114			
BC188	107	3039		TR70/720	
BC189	107	3039		TR70/720	
BC190A			63		S0005
BC190B	192		63		
BC192	159	3114	82	TR20/717	S0019
BC192-1	159	3114			
BC194	108		20	TR95/56	S0025
BC194B	108			TR95/56	
BC195	108			TR95/56	
BC195CD	108			TR95/56	
BC196	159	3114		TR20/717	
BC196-1	159	3114			
BC196A	159	3114	65	TR20/717	
BC196B	159	3118	65	TR20/717	
BC196VI	159	3114	65	TR20/717	
BC197	123A	3122	69	TR21/735	S0025
BC197A	123A	3122		TR21/735	
BC197B	123A	3122		TR21/735	
BC198	123A	3122		TR21/735	S0015
BC198A			17		
BC199	123A	3122		TR21/735	S0015
BC200	159	3118	65	TR20/717	S0032
BC200-1	159	3114			
BC201	159	3114	65	TR20/717	
BC201-1	159	3114			
BC202	159	3114	65	TR20/717	
BC202-1	159	3114			
BC203	159	3114	65	TR20/717	
BC203-1	159	3114			
BC204	234	3114		TR20/717	
BC204-1	159	3114			
BC204A	234	3114	21	TR20/717	
BC204B	234	3114	48	TR20/717	
BC204V	159	3114	21	TR20/717	
BC204VI	159	3114	21	TR20/717	
BC205	159	3114	65	TR20/717	S0013
BC205A	159	3114	21	TR20/717	
BC205B	159	3114	48	TR20/717	
BC205C	106				
BC205L	106				
BC205V	159	3114	22	TR20/717	
BC205VI	159	3114	22	TR20/717	
BC206	234	3114	65	TR20/717	S0019
BC206-1	159	3114			
BC206A	234				
BC206B	234	3114	48	TR20/717	
BC206C	234				
BC207	123A	3122	210	TR21/735	
BC207A	123A	3122	210	TR21/735	
BC207B	123A	3122	210	TR21/735	
BC208		3122	20	TR21/735	S0005

Original Part Number	ECG	SK	GE	IR/WEP	HEP
BC208A		3122	10	TR21/735	
BC208AL	123A				
BC208B		3122	10	TR21/735	
BC208BL	123A				
BC208C		3122	10	TR21/735	
BC208CL	123A				
BC209		3122		TR21/735	S0015
BC209B		3122	10	TR21/735	
BC209BL	123A				
BC209C	123A	3122	10	TR21/735	S0005
BC209CL	123A				
BC210	123A	3122		TR21/735	S0015
BC211	128	3024	18	TR87/243	
BC212	159	3114	21	TR20/717	
BC212-1	159	3114			
BC212A	193		48		
BC212B	193		48		
BC212K	159	3114	21	TR20/717	
BC212KA	159	3114	21	TR20/717	
BC212KB	159	3114	48	TR20/717	
BC212L	159	3114	48	TR20/717	
BC212LA	159	3114	21	TR20/717	
BC212LB	159	3114	48	TR20/717	
BC213	159	3114	21	TR20/717	
BC213-1	159	3114			
BC213A	159		21		
BC213B	193		48		
BC213C			48		
BC213K	159	3114	21	TR20/717	
BC213KA	159	3114	21	TR20/717	
BC213KB	159	3114	48	TR20/717	
BC213KC	159	3114	48	TR20/717	
BC213L	159	3114	48	TR20/717	
BC213LA	159	3114	21	TR20/717	
BC213LB	159	3114	48	TR20/717	
BC213LC	159	3114	48	TR20/717	
BC214	159	3114	48	TR20/717	
BC214-1	159	3114			
BC214A	159		21		
BC214B	159		48		
BC214K	159	3114	21	TR20/717	
BC214KA	159	3114		TR20/717	
BC214KB	159	3114	48	TR20/717	
BC214KC	159	3114	48	/717	
BC214L	159	3114	48	TR20/717	
BC214LA	159	3114		TR20/717	
BC214LB	159	3114	48	TR20/717	
BC214LC	159	3114	48	TR20/717	
BC215A					S0015
BC215B					S0015
BC216	128	3024		TR87/243	
BC216A	128	3024		TR87/243	S0025
BC216B	128	3024		TR87/243	
BC220	123A	3122		TR21/735	S0015
BC221	159	3114	82	TR20/717	
BC221-1	159	3114			
BC222	123A	3122	210		S0025
BC223		3122			
BC223A	123A	3122	210		
BC223B	123A	3122	210		
BC224	193	3138	65		
BC225	234	3118	65	TR20/717	
BC225-1	159	3114			
BC226	192	3137	47		S3011
BC231A	159	3025	82		
BC231B	193	3138	82		
BC232A	128	3024	81		
BC232B	128	3024	81		
BC233A	123A	3122	20	TR21/735	
BC237	123A	3122		TR21/735	S0025
BC237A	123A		212	TR21/	S0030
BC237B	123A		212	TR21/	S0030
BC238	123A	3122	17	TR21/735	S0024
BC238A	123A	3122	212	TR21/	S0015
BC238B	123A	3122	212	TR21/	S0015
BC238C	199	3156	212	TR21/	S0024
BC239	123A	3122		TR21/735	S0015
BC239A			212	TR51/	
BC239B	199	3122	212	TR21/	S0015
BC239C	199			TR21/	S0024
BC250	159	3114		TR20/717	S0013
BC250-1	159	3114			
BC250A	159	3114	65	TR20/717	S0013
BC250B	159	3114	65	TR20/717	S0019
BC250C	159	3114	65	TR20/717	S0031
BC251	159	3114	21	TR20/717	S0019
BC251A	159	3114	48	TR20/717	
BC251B	159	3114		TR20/717	
BC251C	159	3114		TR20/717	
BC252	159	3114	21	TR20/717	S0019
BC252A	159	3114	48	TR20/717	
BC252B	159	3114		TR20/717	
BC252C	159	3114		TR20/717	
BC253	159	3114	21	TR20/717	S0019
BC253-1	159	3114			

Original Part Number	ECG	SK	GE	IR/WEP	HEP
BC253A	159	3114	22	TR20/717	
BC253B	159	3114	22	TR20/717	
BC253C	159	3114	48	TR20/717	
BC254	128	3024	18		
BC255	128	3024	18		
BC256	159	3114		TR20/717	
BC256-1	159	3114			
BC256A	159	3114	67	TR20/717	
BC256B	159	3114		TR20/717	
BC257	159	3118	65	TR20/717	S0031
BC257-1	159	3114			
BC257A	193	3138	65		
BC257B	193	3138	65		
BC257VI	159	3025	65		
BC258	159	3114	65	TR20/717	S0019
BC258-1	159	3114			
BC258A	193	3138	65		
BC258B	193	3138	65		
BC258VI	159	3025	65		
BC259	159	3114	65	TR20/717	S0019
BC259-1	159	3114			
BC259A	193	3138	65		
BC259B	193	3138	65		
BC260	159	3114		TR20/717	
BC260A	159	3114	65	TR20/717	S0013
BC260B	159	3114	65	TR20/717	S0013
BC260C	159	3114	65	TR20/717	S0019
BC261	159	3025	21	TR20/717	S0019
BC261A	159	3025	48	TR20/717	S0019
BC261B	159	3025		TR20/717	S0019
BC261C	159	3114		TR20/717	S0019
BC262	159	3114	82	TR20/717	S0019
BC262-1	159	3114			
BC262A	159	3114	48	TR20/717	
BC262B	159	3114		TR20/717	
BC262C	159	3114		TR20/717	
BC263	159	3114		TR20/717	S0019
BC263A	159	3114	22	TR20/717	
BC263B	159	3114	22	TR20/717	
BC263C	159	3114	48	TR20/717	
BC264C	133				
BC266	159	3114		TR20/717	
BC266-1	159	3114			
BC266A	159	3114	67	TR20/717	
BC266B	159	3114		TR20/717	
BC267	123A	3122	47	TR21/735	
BC268	123A	3122		TR21/735	
BC269	123A	3122		TR21/735	
BC270	123A	3122		TR21/735	
BC271					S0015
BC272					S0015
BC280	123A	3122		TR21/735	
BC280A	123A	3122	62	TR21/735	
BC280B	123A	3122	62	TR21/735	
BC280C	123A	3122	62	TR21/735	
BC281	159	3114		TR20/717	
BC281-1	159	3114			
BC281A	159	3114		TR20/717	S0019
BC281B	159	3114		TR20/717	S0019
BC281C	159	3114		TR20/717	S0019
BC282	123A	3122	20	TR21/735	
BC283	159	3114	21	TR20/717	
BC284	123A	3122	20	TR21/735	
BC284A	123A	3122	20	TR21/735	
BC284B	123A	3122	20	TR21/735	
BC285	123A	3122	27	TR21/735	
BC286	128	3024	63	TR87/243	S3011
BC287	129	3025	67	TR88/242	
BC288	128	3024		TR87/243	S3002
BC289	123A	3122		TR21/735	
BC289A	123A	3122		TR21/735	S0015
BC289B	123A	3122		TR21/735	S0015
BC290	123A	3122		TR21/735	
BC290B	123A	3122		TR21/735	S0015
BC290C	123A	3122		TR21/735	S0024
BC291	159	3114		TR20/717	
BC291-1	159	3114			
BC291A	159	3114		TR20/717	S0019
BC291D	159	3114		TR20/717	S0019
BC292	159	3114		TR20/717	
BC292A	159	3114		TR20/717	S0019
BC292D	159	3114		TR20/717	
BC295	108	3039	61		S0015
BC297	193	3138	48		
BC298	193	3138	48		
BC301	128	3024		TR87/243	
BC302-4			63		
BC302-5			63		
BC302-6			63		
BC303	129	3025		TR88/242	
BC304-4			67		
BC304-5			67		
BC304-6			67		
BC307	159		65	TR20/717	S0019
BC307A	159	3114	65	TR20/	S0019
BC307B	193	3138	65	TR52/	
BC307VI	159		65	TR20/	S0019
BC308	159	3114	65	TR20/717	S0013
BC308A	159	3114	65	TR20/	S0013
BC308B	159		65	TR20/	S0019
BC308VI	159		65	TR20/	S0013
BC309	159	3114	65	TR20/717	S0019
BC309A	159	3114	65	TR20/	S0013
BC309B	159	3114	65	TR20/	S0031
BC309C	193	3138	65	TR52/	
BC310	128				
BC311	129				
BC312	198		32		
BC313	128	3024	63	TR87/243	
BC315	159	3025	65		
BC317	123A				
BC317A	123A				
BC317B	123A				
BC318	123A		20		
BC318A	123A				
BC318B	123A		20		
BC318C	123A				
BC319	123A		20		
BC319B	123A				
BC319C	123A				
BC320	234				
BC320A	234				
BC320C	234				
BC321	234				
BC321A	234				
BC321B	234				
BC321C	234				
BC322	234				
BC322B	234				
BC322C	234				
BC325	159	3114	82	TR20/717	
BC325A	159	3114		TR20/	
BC326	159	3114	67	TR20/717	
BC326A	159	3114		TR20/	
BC327	189	3200	82	TR73/	S3032
BC327-16	193	3138	48		
BC328	189	3200	22	TR73/	S3028
BC328-16	193	3138	48		
BC337	188	3124	63	TR72/	S3020
BC338	188	3124	20	TR72/	S3024
BC338-16	192	3137	47		
BC340-6	128	3024	47	TR87/243	
BC340-10	128	3024	47	TR87/243	
BC340-16	128	3024	47	TR87/243	
BC341-6	128	3024	18	TR87/243	
BC341-10	128	3024	18	TR87/243	
BC360-6	129	3025	48	TR88/242	
BC360-10	129	3025	48	TR88/242	
BC360-16	129	3025	48	TR88/242	
BC361-6	129	3025	67	TR88/242	
BC361-10	129	3025	67	TR88/242	
BC370	193	3138	48		
BC377	192	3137	47		
BC378	192	3137	47		
BC381	159	3025	82		
BC382			210		
BC382B	199		212		
BC382C			212		
BC383			210		
BC383B	199		212		
BC383C	199		212		
BC384			210		
BC384B			212		
BC384C	199		212		
BC385A			212		
BC385B	199		212		
BC386A	199		212		
BC386B	199		212		
BC394			27		S0005
BC396	129		82	TR88/	S5023
BC400	159	3114		TR20/	S5023
BC404			82		S0019
BC404A	129	3025	82	TR88/	S5023
BC404VI	129	3025	82	TR88/	S5023
BC405	159	3025	82		S0019
BC405A	159	3025	82	TR20/	S5022
BC405B	159	3025	82	TR20/	S0019
BC406	193	3138	48		S0019
BC406B	159	3025	82	TR20/	S0019
BC408	123A	3122	17		S0015
BC409	123A	3122	20		S0015
BC411					S3002
BC412	195	3048		TR65/	S3001
BC413B	199		210	TR21/	S0015
BC413C	199			TR21/	S0024
BC414B	199		210	TR21/	S0015
BC414C	199			TR21/	S0015
BC415A	159	3025	48	TR20/	S0019
BC415B	159	3025	48	TR20/	S0019
BC416			21		

Original Part Number	ECG	SK	GE	IR/WEP	HEP
BC416A	159	3025	48	TR20/	S0019
BC416B	159	3025	48	TR20/	S0019
BC417	159	3025	65		S0019
BC418	159	3025	65		S0031
BC419	159	3025	65		S0031
BC429	128	3024	81	TR87/243	S3020
BC430	188	3054	63	TR72/	S3020
BC431			81		
BC432			82		
BC442	107	3039	11	TR22/	
BC460	189	3083		TR73/	
BC461	189	3083		TR73/	S3032
BC471-4			67		
BC461-5			67		
BC461-6			67		
BC477	129	3025		TR88/	S5023
BC477A	129	3025		TR88/	S5023
BC477VI	129	3025		TR88/	S5023
BC478	159	3025	65	TR20/	S0019
BC478A	159	3025	65	TR20/	S0019
BC478B	159		65	TR20/	S0019
BC479	159	3025	65	TR20/	S0032
BC479B	159	3025	65	TR20/	S0032
BC507					S0015
BC507A	123A	3122		TR21/	S0015
BC507B	123A	3122	210	TR21/	S0015
BC508			63		S0015
BC508A	123A	3122	210	TR21/	S0014
BC508B	123A	3122	210	TR21/	S0015
BC508C	123A	3122			S0024
BC509			63		S0015
BC509B	123A	3122	210		S0024
BC509C	123A	3122			S0025
BC510	192	3137	210		S0015
BC510B	123A	3122	210		S0024
BC510C	107	3039			S0015
BC512	159	3025	21		
BC512A	159	3025	21		
BC512B	193	3138	48		
BC513	159	3025	21		
BC513A	159	3025	21		
BC513B	193	3138	48		
BC514	193	3138	48		
BC514A	159	3025	21		
BC514B	193	3138	48		
BC520	199				
BC520B	199				
BC520C	199				
BC521	199				
BC521C	199				
BC521D	199				
BC522	199				
BC522C	199				
BC522D	199				
BC522E	199				
BC523	199				
BC523B	199				
BC523C	199				
BC526A	234				
BC526B	234				
BC526C	234				
BC527	129				
BC528	129				
BC532	194				
BC533	194				
BC534	129				
BC535	128				
BC537	128				
BC538	128				
BC548			17		
BC548VI			17		
BC556			82		
BC556A			82		
BC556VI			82		
BC557			82		
BC557A			82		
BC557B			48		
BC557VI			82		
BC558			82		
BC558A			48		
BC558B			48		
BC558VI			82		
BC559			82		
BC559A			48		
BC559B			48		
BC560A			48		
BC560B			48		
BC582	192	3137	210		
BC582A	192	3137	210		
BC582B	192	3137	210		
BC583	123A	3122	210		
BC583A	192	3137	210		
BC583B	192	3137	210		
BC584	192	3137	210		
BC682			81		

Original Part Number	ECG	SK	GE	IR/WEP	HEP
BC727	129				
BC728	129				
BC737	128				
BC738	128				
BC1029LB	917				
BC-1072	123A	3122	20		
BC1073	104	3009	16	TR01/232	G6003
BC1073A	104	3009	16	TR01/232	G6003
BC1076				TR21/	
BC-1082	123A	3122	20		
BC-1086	123A	3122	20		
BC1096	123A	3122	20	TR21/	
BC-1096			17		
BC1274	104	3009	25	TR01/232	G6003
BC1274A	104	3009	25	TR01/232	G6003
BC1274B	104	3009	25	TR01/232	G6003
BC1478	123A	3122	17		
BC1493				TR51/	
BC-1690	123A	3122	10		
BC3337			47		
BC6500	108	3039		TR95/56	S0020
BC14141				TR87/	
BC14141-6				TR87/	
BCM1002-1	160	3006	51	TR17/637	
BCM1002-2	107	3039	20	TR70/720	
BCM1002-3	158	3004	53	TR84/630	
BCM1002-4	102A	3004	53	TR85/250	
BCM1002-5	102A	3004	53	TR85/250	
BCM1002-6	131	3082	44	TR94/642	
BCM1002-18	158	3004	53	TR84/630	
BCW29	106	3118	65	TR20/52	
BCW29R	106	3118	65	TR20/52	
BCW30	106	3118	65	TR20/56	
BCW30R	106	3118	65	TR20/56	
BCW31	108	3039	39	TR95/56	
BCW31R	108	3039	39	TR95/56	
BCW32	108	3039	39	TR95/56	
BCW32R	108	3039	39	TR95/56	
BCW34	123A	3122	47	TR21/735	
BCW35	159	3114	48	TR20/717	
BCW36	123A	3122	47	TR21/735	
BCW37	159	3114	48	TR20/717	
BCW37A	159	3114			
BCW44	192	3137	63		
BCW45	193	3138	67		
BCW46	128	3024		TR87/243	
BCW47	128	3024		TR87/243	
BCW48	128	3024		TR87/243	
BCW48A			17		
BCW49	128	3024		TR87/243	
BCW50	171	3103	27		S0005
BCW51	192	3137	210		
BCW52	193	3138	48		
BCW56	159	3114	82	TR20/717	
BCW56A	159	3114	82		
BCW57	159	3118	65	TR20/717	
BCW57A	159	3114	65		
BCW58	159	3118	65	TR20/717	
BCW58A	159	3114	65		
BCW59	159	3118	65	TR20/717	
BCW60A	123A	3122	210		
BCW60AA	123A	3122	210		
BCW61	159	3118			
BCW61A	193	3118	65		
BCW61B	193	3118	65		
BCW61BA	193	3138	65		
BCW61BB	193	3138	65		
BCW61C	159	3118	65		
BCW61D	159	3118	65		
BCW62	159	3025	21		
BCW62A	159	3025	21		
BCW62B	193	3138	48		
BCW63	159	3025	21		
BCW63A	159	3025	21		
BCW63B	193	3138	48		
BCW64	193	3138	48		
BCW64A	159	3025	21		
BCW64B	193	3138	48		
BCW65EA			20		
BCW65EB			20		
BCW65EC			47		
BCW65EF			47		
BCW65EG			47		
BCW65EH			47		
BCW65EW			47		
BCW67DA			48		
BCW67DB			48		
BCW67DC			48		
BCW68DF			48		
BCW68DG			48		
BCW68DH			48		
BCW69	159	3025	65		
BCW69R	159	3025	65		
BCW70	193	3138	65		
BCW70R	193	3138	65		

Original Part Number	ECG	SK	GE	IR/WEP	HEP
BCW71	108	3039	61		
BCW71R	108	3039	61		
BCW72	108	3039	61		
BCW72R	108	3039	61		
BCW73-16	192	3137	47		
BCW74-16	192	3137	47		
BCW75-10	193	3138	48		
BCW75-16	193	3138	48		
BCW76-10	193	3138	48		
BCW76-16	193	3138	48		
BCW77-16	192	3137	63		
BCW78-16	192	3137	63		
BCW79-10	159	3025	67		S5022
BCW79-16	159	3025	67		S5022
BCW79-25	211	3203	67		S3028
BCW80-10	159	3025	67		S5022
BCW80-16	159	3025	67		S5022
BCW80-25	211	3203			S3028
BCW82	192	3137	210		
BCW82A	192	3137	210		
BCW82B	192	3137	210		
BCW83	123A	3122	210		
BCW83A	192	3137	210		
BCW83B	192	3137	210		
BCW84	192	3137	210		
BCW86	193	3138	67		
BCW87			212		
BCW88			65		
BCW90	192	3137	47		S0015
BCW90K	192	3137	47		S0015
BCW90KA			47		S0015
BCW90KB			47		S0015
BCW90KC			47		S0015
BCW91	192	3137	63		S0015
BCW91K	192	3137	63		S0015
BCW91KA			63		S0015
BCW91KB			63		S0015
BCW92	193	3138	48		S5022
BCW92K	193	3138	48		S5022
BCW92KA			48		S5022
BCW92KB			48		S5022
BCW93	193	3138	67		S5022
BCW93K	193	3138	67		S5022
BCW93KA			67		S5022
BCW93KB			67		S5022
BCW94	123A	3122	20		S3010
BCW94K	128	3024	47		S3010
BCW94KA			20		S0015
BCW94KB			20		S0015
BCW94KC			20		S0015
BCW95	128	3024	81	S3010	S3010
BCW95K	128	3024	18		S3010
BCW96	159	3025	82		S5022
BCW96K	159	3025	21		S5022
BCW97	159	3025	82		S5022
BCW97K	159	3025	67		S5022
BCW98A			212		
BCW99A			48		
BCX10	211	3203			S3028
BCX17			82		
BCX17R			82		
BCX18			82		
BCX18R			82		
BCX19			210		
BCX19R			210		
BCX20			210		
BCX20R			210		
BCX58IX			47		
BCX58VII			17		
BCX58VIII			17		
BCX58X			47		
BCX59IX			47		
BCX59VII			47		
BCX59VIII			47		
BCX58X			47		
BCX70AG			210		
BCX70AH			210		
BCX70AJ			210		
BCX71BG	193	3138	65		
BCX71BH	193	3138	65		
BCX73-16			47		S3011
BCX73-25			47		S3011
BCX73-40			47		S3011
BCX74-16			47		S3011
BCX74-25			47		S3011
BCX74-40			47		S3011
BCX75-16			48		S3012
BCX75-25			48		S3012
BCX75-40			48		S3012
BCX76-25			48		S3012
BCX76-40			48		S3012
BCX78IX			48		S0019
BCX78VII			21		S0019
BCX78VIII			48		S0019
BCX79IX			48		S0019

Original Part Number	ECG	SK	GE	IR/WEP	HEP
BCX79VII			21		S0019
BCX79VIII			48		S0019
BCY10	159	3203	82	TR20/717	S0012
BCY11	159	3114	82	TR20/717	S0012
BCY11A	159	3114			
BCY12	159	3114	82	TR20/717	S0013
BCY12A	159	3114			
BCY13	123A	3122	63	TR21/735	S0015
BCY15	123A	3122	63	TR21/735	S0015
BCY16	123A	3122	20	TR21/735	S0005
BCY17	129	3025	65	TR88/242	S0013
BCY18	129	3025	65	TR88/242	S0013
BCY19	129	3025	65	TR88/242	S0013
BCY21	129	3025	65	TR88/242	S0013
BCY22	129	3025	21	TR88/242	S0013
BCY23	129	3025	65	TR88/242	S0013
BCY24	129	3025	65	TR88/242	S0013
BCY25	129	3025	65	TR88/242	S0013
BCY26	129	3025	65	TR88/242	S0013
BCY27	129	3025	65	TR88/242	S0013
BCY28	129	3025	65	TR88/242	S0013
BCY29	129	3114	82	TR88/242	S5013
BCY30	129	3025	82	TR88/242	S0013
BCY31	129	3025	82	TR88/242	S0013
BCY32	129	3025	82	TR88/242	S0013
BCY33	129	3025	65	TR88/242	S0013
BCY34	129	3025	65	TR88/242	S0013
BCY35	159	3114		TR20/717	
BCY35A	159	3114			
BCY36	123A	3122		TR21/735	
BCY37	159	3114		TR20/717	
BCY37A	159	3114			
BCY38	159	3114	82	TR20/717	S0013
BCY38A	159	3114			
BCY39	159	3114	82	TR20/717	S0013
BCY39A	159	3114			
BCY40	159	3114	82	TR20/717	S0013
BCY40A	159	3114			
BCY42	123A	3122	212	TR21/735	S0011
BCY43	123A	3122	212	TR21/735	S0011
BCY46	128	3024		TR87/243	
BCY47	128	3024		TR87/243	
BCY48	128	3024		TR87/243	
BCY49	128	3024	21	TR87/243	S0013
BCY50	123A	3122	20	TR21/735	S0011
BCY50I	160	3122		/735	S0011
BCY51	123A	3122		TR21/735	S0011
BCY51I	160	3122		/735	S0011
BCY54	159	3114	210	TR20/717	S0013
BCY54A	159	3114			
BCY55			212		S0005
BCY56	123A	3122	210	TR21/735	S0015
BCY57	123A	3122		TR21/735	S0015
BCY58	123A	3124	20	TR21/735	S0014
BCY58A		3122		TR51/735	S0014
BCY58B	123A	3122		TR21/735	S0014
BCY58C	123A	3122		TR21/735	
BCY58D	123A	3122		TR21/735	
BCY58VII			88		S0015
BCY58VIII			88		S0015
BCY59	123A	3124	17	TR21/735	S0011
BCY59A	123A	3122		TR21/735	S0015
BCY59B	123A	3122		TR21/735	S0015
BCY59C	123A	3122		TR21/735	S0015
BCY59D	123A	3122		TR21/735	S0015
BCY59VII			88		S0015
BCY59X					S0015
BCY65	128	3024		TR87/243	S3020
BCY66	128	3024	88	TR87/243	
BCY67	129	3025	48	TR88/242	
BCY69	123A	3122	20	TR21/735	S0015
BCY70	159	3114	21	TR20/717	S0013
BCY71	159	3114	21	TR20/717	S5022
BCY71A	159	3114			S0019
BCY72	159	3114	22	TR20/717	S0013
BCY72A	159	3114			
BCY77VII	159	3025	67		S5022
BCY77VIII	159	3025	67		S5022
BCY78	234	3114	48	TR20/717	S0013
BCY78A	193	3114	82		
BCY78B	193		48		
BCY78VII	193	3138	67		
BCY78VIII	193	3138	67		
BCY79	234	3114	48	TR20/717	S5022
BCY79A	193	3114	82		
BCY79B	193		48		
BCY79C			48		
BCY79VII	193	3138	67		
BCY79VIII	193	3138	67		
BCY84A	123A			TR21/	
BCY85	194	3104			S0005
BCY86	194	3104			S0005
BCY87	108	3039	39		
BCY88	108	3039	39		
BCY89	108	3039	39		

Original Part Number	ECG	SK	GE	IR/WEP	HEP
BCY90	159	3114	65	TR20/717	S0013
BCY90A	159	3114			
BCY90B	159	3114	82	TR20/717	S0013
BCY90B-1	159	3114			
BCY91	159	3114	65	TR20/717	S0013
BCY91A	159	3114			
BCY91B	102A	3114	82	/717	S0013
BCY91B-1	159	3114			
BCY92	159	3114	65	TR20/717	S0013
BCY92A	159	3114			
BCY92B	159	3114	82	TR20/717	S0013
BCY92B-1	159	3114			
BCY93	159	3114	82	TR20/717	S5022
BCY93A	159	3114			
BCY93B	159	3114	82	TR20/717	S5023
BCY93B-1	159	3114			
BCY94	159	3114	82	TR20/717	S5022
BCY94A	159	3114			
BCY94B	159	3114	82	TR20/717	S5023
BCY94B-1	159	3114			
BCY95	159	3114	82	TR20/717	S5022
BCY95A	159	3114			
BCY95B	159	3114	82	TR20/717	S5023
BCY95B-1	159	3114			
BCY96	129	3025		TR88/242	S5023
BCY96B	129	3025		TR88/242	S5023
BCY97	129	3025		TR88/242	S5023
BCY97B	129	3025		TR88/242	S5023
BCY98	159	3114		TR20/717	
BCY98A	159	3114			
BCY98B	159	3114	82	TR20/717	S0032
BCY98B-1	159	3114			
BCY443	128	3024	17	TR87/243	
BCY501	123A			TR21/	
BCY511	123A			TR21/	
BCY581X					S0015
BCY581X					S0015
BCY771X	159		67		S5022
BCZ10	159	3114	65	TR20/717	S0013
BCZ10A	159	3114			
BCZ10B	159	3114			
BCZ10C	159	3114			
BCZ11	159	3114	65	TR20/717	S0032
BCZ11A	159	3114			
BCZ12	159	3114	82	TR20/717	S0013
BCZ12A	159	3114			
BCZ12B	159	3114			
BCZ12C	159	3114			
BCZ13	159	3114	65	TR20/717	S0013
BCZ13A	159	3114			
BCZ13B	159	3114			
BCZ14	159	3114	65	TR20/717	S0013
BCZ14A	159	3114			
BCZ14B	159	3114			
BCZ14F	159	3114			
BD71	108		20	TR95/56	S0025
BD106	152	3054	28	TR76/701	S5012
BD106A	152	3054	28	TR76/701	S5000
BD106B	152	3054	28	TR76/701	
BD107	152	3054	28	TR76/701	S5012
BD107A	152	3054	28	TR76/701	S5000
BD107B	152	3054	28	TR76/701	
BD109	152	3054	66	TR76/701	S5000
BD111	130	3027	19	TR59/247	S7002
BD111A	181				S7002
BD112	130	3027	19	TR59/247	S7002
BD113	130	3027	19	TR59/247	S7002
BD115	154	3045	27	TR78/712	S5024
BD116	130	3027	19	TR59/247	S7002
BD118	130	3027	19		S7002
BD119	124		12	TR81/240	S5011
BD120	124		12	TR81/240	S5011
BD121	130	3027	19	TR59/247	S5004
BD123	182	3188			S5004
BD127	124			TR81/240	S5011
BD128	124			TR81/240	S5011
BD129	124			TR81/240	S5011
BD130	130	3027	14	TR59/247	S7004
BD131	184	3054	57	/S5003	S5000
BD132	242	3083	58	/S5007	S5006
BD135	184	3054	57	TR76/S5003	S3024
BD136	185	3083	58	TR77/S5007	S3028
BD137	184	3054	57	/S5003	S5000
BD138	185	3083	69	/S5003	S5006
BD139	184	3054	57	/S5003	S5000
BD140	185	3083		/S5007	S5006
BD141	162				S5020
BD142	130	3027	75	TR59/247	S7004
BD144	162				S5020
BD145	130	3027	66	TR59/247	S7002
BD148	182	3188			S5000
BD149	184	3190			S5000
BD153	123A	3122	17	TR21/735	
BD155	184	3190			S5000
BD162	152	3054	66	TR76/701	S5000

Original Part Number	ECG	SK	GE	IR/WEP	HEP
BD163	152	3054	66	TR76/701	S5000
BD165			28		
BD181			77		S7002
BD182			77		S7002
BD183	130	3027	75		S7004
BD195					S5004
BD197					S5004
BD199					S5004
BD215	157	3103			S5015
BD216	157	3103			S5015
BD220	152				
BD221	152				
BD222	152				
BD223	153				
BD224	153				
BD225	153				
BD226					S5000
BD227					S5006
BD228					S5000
BD229					S5006
BD230					S5000
BD231					S5006
BD251					S7002
BDX10			75		
BDX12	162				S5020
BDX13	181	3535	75		
BDX18	219		74		
BDX20	219				
BDY10	181	3036	75	TR36/S7000	
BDY11	181	3036	75	TR36/S7000	
BDY12	152	3054	66	TR76/701	
BDY13	152	3054	66	TR76/701	
BDY15	155		66		S5012
BDY15A	155		66		
BDY15B	155		66		
BDY15C	155		66		
BDY16	175	3538			S5012
BDY16A	155		66		
BDY16B	155		66		
BDY17	130	3027	77	TR59/247	
BDY20	130	3027	75		S7004
BDY23	130	3027		TR59/247	
BDY34	152	3054	28	TR76/701	
BDY38	130	3027	77	TR59/247	S7002
BDY39	130	3027	75	TR59/247	
BDY53	130	3027			S7004
BDY54	162				S5020
BDY57	181	3561		TR36/S7000	
BDY58	181	3561		TR36/S7000	
BDY60	179		66	TR35/G6001	S7002
BDY61	179		66	TR35/G6001	S7002
BDY62	121	3009	66	TR01/121	S7002
BDY65	190	3104			S3019
BDY66	190	3104			S3019
BDY70	129	3025		TR88/242	
BDY71	175	3538			S5019
BDY72	175	3538			S5012
BDY73			75		
BDY78	175	3538			S5019
BDY79	175	3538			S5012
BDY82	153			TR77/	
BDY83	153			TR77/	
BE6	100	3005	2	TR05/254	S5012
BE6A	100	3005	2	TR05/254	G0005
BE71	128	3024		TR87/243	S0025
BE173	107			TR70/720	
BES23				TR76/	
BES50				TR76/	
BEX55				TR76/	
BEY70				TR76/	
BF1	102A	3004			
BF1A	102A	3004			
BF2	102A	3004			
BF5	121	3009			
BF65	234				
BF71	123A	3122		TR21/735	S0025
BF108	154	3045	27	TR78/712	S3019
BF109	154	3045	27	TR78/712	S0011
BF110	154	3045	27	TR78/712	S5025
BF111	154	3045	27	TR78/712	
BF114	154	3045	27	TR78/712	S5025
BF115	123A	3122	61	TR21/735	S0011
BF-115	108	3019			
BF117	154	3045	27	TR78/712	S5025
BF118	154	3045	27	TR78/712	S5025
BF119	154	3045	27	TR78/712	S5025
BF120			27	TR78/	S0005
BF121	107	3039	61	TR70/720	
BF123	107	3039	61	TR70/720	
BF125	107	3039	61	TR70/720	

Original Part Number	ECG	SK	GE	IR/WEP	HEP
BF127	107	3039	60	TR70/720	
BF137			27	TR78/	S5025
BF140	154	3045	27	TR78/712	S5025
BF140A	154	3045		TR78/712	S5025
BF140R	154	3045		TR78/712	S5025
BF140S	154	3045		TR78/712	S5025
BF152	108	3039	11	TR95/56	S0016
BF153	108	3039	11	TR95/56	S0011
BF154	108	3039	61	TR95/56	S0011
BF155	161	3117	60	TR83/719	S0016
BF155R	154	3045		TR78/712	
BF155S	154	3045		TR78/712	
BF156	154	3045	27	TR78/712	
BF157	154	3045	27	TR78/712	
BF157B	154	3045	27	TR78/712	S5025
BF158	108		11	TR95/56	S0016
BF159	108	3039		TR95/56	S0016
BF160	108	3039	11	TR95/56	S0016
BF161	161	3117	61	TR83/719	
BF162	108	3039	60	TR95/56	S0016
BF163	108	3039	60	TR95/56	S0016
BF164	108	3039	60	TR95/56	S0016
BF165	108	3039	61	TR95/56	S0011
BF166	161	3117	60	TR83/719	S0016
BF167	161	3117	60	TR83/719	S0011
BF168	161	3117	61	TR83/719	S0016
BF169	161	3117	11	TR83/719	S0011
BF173	161	3018	61	TR83/719	S0016
BF173A	108	3018			
BF174	154	3045	27	TR78/712	
BF175	161	3117	60	TR83/719	
BF176	108	3039	61	TR95/56	S0011
BF177	154	3045	27	TR78/712	S5026
BF178	154	3045	27	TR78/712	S5025
BF179	154	3045		TR78/712	S5025
BF179A	154	3045	27	TR78/712	S5025
BF179B	154	3045	27	TR78/712	S5024
BF179C	154	3045	27	TR78/712	S5024
BF180	161	3117	39	TR83/719	S0017
BF181	161	3117	60	TR83/719	S0017
BF182	161	3117	214	TR83/719	S0017
BF183	161	3117	214	TR83/719	S0017
BF183A	123A	3122	20	TR21/735	
BF184	161	3117	39	TR83/719	S0017
BF185	161	3117	39	TR83/719	S0017
BF186	154	3045		TR78/712	S0005
BF187	108	3039	20	TR95/56	S0016
BF188	108	3039	20	TR95/56	S0016
BF189	123A	3122	20	TR21/735	S0016
BF194	107	3018	60	TR70/720	S0016
BF194A	108	3018			
BF194B	107	3018		TR70/720	S0016
BF195	107	3018	86	TR70/720	S0016
BF195C	107	3039		TR70/720	S0016
BF195D	107	3018	TR	TR70/720	S0016
BF196	107	3039	60	TR70/720	S0016
BF197	107	3039	60	TR70/720	S0016
BF198	107	3039	20	TR70/720	
BF200	161	3018	60	TR83/719	S0017
BF206	161	3117		TR83/719	S0017
BF207	161	3117		TR83/719	S0017
BF208	161	3117		TR83/719	S0017
BF209	161	3117		TR83/719	S0017
BF212	161	3117		TR83/719	S0017
BF213	161	3117		TR83/719	S0017
BF214	161	3117		TR83/719	S0017
BF-214	123A		20		
BF215	161	3117		TR83/719	S0017
BF-215	123A		20		
BF216	107	3039	17	TR70/720	S0016
BF217	107	3039	17	TR70/720	S0005
BF218	107	3039	17	TR70/720	S0005
BF219	107	3039		TR70/720	S0005
BF220	107	3039		TR70/720	S0005
BF222	161	3117	61	TR83/719	
BF223	108	3039		TR95/56	
BF224	108	3039	17	TR95/56	S0025
BF224J	123A	3122	17		
BF225	108	3039	20	TR95/56	
BF225J	123A	3122	20		
BF226	161	3117		TR83/719	S0017
BF-226	123A		20		
BF227			60		
BF228			18		
BF229	107	3039	39	TR70/720	

Original Part Number	ECG	SK	GE	IR/WEP	HEP
BF230	107	3039	39	TR70/720	
BF232	161	3117	61	TR83/719	
BF233-2	108	3039	61		S0025
BF233-3	108	3039	61		S0025
BF233-4	108	3039	61		S0025
BF233-5	108	3039	61		S0017
BF233-6					S0017
BF234	108	3039	61	TR95/56	S0015
BF235	108	3039	61	TR95/56	S0015
BF237	108	3039	62	TR95/56	S0025
BF238	108	3039	62	TR95/56	S0025
BF240	107	3039	61	TR70/720	S0020
BF240B			61		
BF241	107	3039	61	TR70/720	
BF244	133	3112		/801	
BF244A	132	3116			
BF244B	132	3116			
BF244C	132	3116			
BF245	133	3112		/801	F0015
BF245A	133	3116	FET2	FE100/802	
BF246	133	3112		/801	
BF247	133	3112		/801	
BF248	123A	3122	20	TR21/735	S0014
BF249	123A	3122	21	TR21/735	S0014
BF250	123A	3122	20	TR21/735	S0014
BF251	161	3117		TR83/719	S0017
BF252	161	3117		TR83/719	S0025
BF254	107	3039	60	TR70/720	S0016
BF254B			39		
BF255	107	3039	86	TR70/720	S0016
BF256	132	3116		TR51/	
BF257	154	3045	27	TR78/712	S5025
BF258	154	3045	27	TR78/712	S3021
BF259	154	3045	27	TR78/712	S3012
BF260	161	3117	39	TR83/719	S0020
BF261	161	3117	39	TR83/719	S0020
BF262	108	3039	60	TR83/	S0020
BF262A		3039			
BF263	108	3039	60		S0020
BF264	108	3039	60		S0020
BF270	161	3117	60	TR83/719	S0020
BF271	161	3117		TR83/719	S0020
BF273	161	3117	39	TR83/719	
BF273C	161	3117		TR83/719	
BF273D	161	3117		TR83/719	
BF274	161	3117	39	TR83/719	
BF274B	161	3117		TR83/719	
BF274C	161	3117		TR83/719	
BF287	161	3117	39	TR83/719	
BF288	161	3117	60	TR83/719	
BF290	161	3117	39	TR83/719	
BF291		210			
BF291A	123A	3122	20	TR21/735	
BF291B	123A	3122	20	TR21/735	
BF292	171	3201	27		
BF292A	154	3045	27	TR78/712	
BF292B	154	3045	27	TR78/712	
BF292C	154	3045	27	TR78/712	
BF293A	123A	3122		TR21/735	
BF293D	123A	3122		TR21/	
BF294	154	3045	27	TR78/712	
BF297	171	3201	27		
BF298	171	3201	27		
BF299	171	3201	27		
BF302	161	3117	39	TR83/719	
BF303	161	3117	39	TR83/719	
BF304	161	3117	39	TR83/719	
BF305	154	3045	27	TR78/712	
BF306	161	3117	61	TR83/719	
BF310	107	3039	61	TR70/720	
BF311	107	3039		TR70/720	
BF314	108	3039	61		
BF316	106	3118		TR20/52	
BF317					S0019
BF322			47		
BF323			48		
BF332	107	3039	60	TR70/720	
BF333	107	3039	86	TR70/720	
BF334	107	3039	61	TR70/720	
BF335	107	3039	61	TR70/720	
BF336	154				
BF337	154				S5024
BF338	154				
BF340	199	3118	65	TR21/	
BF340A	159	3118			

Original Part Number	ECG	SK	GE	IR/WEP	HEP
BF340B	159	3118			
BF340C	159	3118			
BF340D	159	3118			
BF341	199	3118	65	TR21/	
BF341A	159	3118			
BF341B	159	3118			
BF341C	159	3118			
BF341D	159	3118			
BF342	199	3118	65	TR21/	
BF342A	159	3118			
BF342B	159	3118			
BF342C	159	3118			
BF342D	159	3118			
BF343	199		65	TR21/	
BF344	161		39	TR83/	
BF345	161		39	TR83/	
BF348	133	3112			
BF357	108	3039	86		
BF362			39		
BF363			39		
BF364			61		
BF365			61		
BF440	123A	3118	21	TR21/	
BF441	123A	3118	21	TR21/	
BF456	171	3201	27		
BF457	171	3201	27		
BF458	171	3201	27		
BF459	171	3201	27		
BF494			61		
BF495			61		
BF516	161	3118		TR83/	
BF540			65		
BF541			65		
BF542			65		
BF594	108	3039	61		
BF595	108	3039	61		
BF596			20		
BF597			62		
BFR11	123A	3122	17		
BFR16	192	3137	81		
BFR18	171	3201	61		
BFR19	171	3201	27		
BFR20	171	3201	27		
BFR22	190	3104			S3019
BFR25	199		62		
BFR26	123A	3122	20		
BFR36	128	3024	47		
BFR40			63		
BFR40T05	63				
BFR41			47		
BFR41T05			47		
BFR51			63		
BFR52			47		
BFR57	171	3201	27		
BFR58	171	3201	27		
BFR59	171	3201	27		
BFR61			67		
BFR62			48		
BFR80			67		
BFR80T05			67		
BFR81			48		
BFR81T05			48		
BFS10					S3001
BFS11			39		
BFS12					S3012
BFS13E	107	3039	60	TR70/720	
BFS13F	107	3039	60	TR70/720	
BFS13G	107	3039	60	TR70/720	
BFS14A	159	3118			
BFS14B	159	3118			
BFS14C	159	3118			
BFS14D	159	3118			
BFS14E	107	3039	65	TR70/720	
BFS14F	107	3039	65	TR70/720	
BFS14G	107	3039	65	TR70/720	
BFS15E	107	3039	60	TR70/720	
BFS15F	107	3039	60	TR70/720	
BFS15G	1C7	3039	60	TR70/720	
BFS16	159	3118			
BFS16A	159	3118			
BFS16B	159	3118			
BFS16C	159	3118			
BFS16D	159	3118			
BFS16E	107	3039	65	TR70/720	
BFS16F	107	3039	65	TR70/720	
BFS16G	107	3039	65	TR70/720	
BFS17	108	3039	86	TR95/56	S0020
BFS17R	108	3039	39	TR95/56	S0016
BFS18	107	3039	39	TR70/720	S0020
BFS18CA	161	3132	39		
BFS18R	107	3039	39	TR70/720	S0016
BFS19	107	3039	39	TR70/720	S0020
BFS19R	107	3039	39	TR70/720	S0016
BFS20	107	3039	60	TR70/720	S0020
BFS20R	107	3039	60	TR70/720	S0016

Original Part Number	ECG	SK	GE	IR/WEP	HEP
BFS21	133	3112			
BFS21A	133	3112			
BFS22	195	3048			S3001
BFS22R					S3001
BFS23	195	3048	28		S3001
BFS23R					S3001
BFS26	159	3118			
BFS26A	159	3118			
BFS26E	159	3118	65		
BFS26F	159	3118	65		
BFS26G	159	3118	65		
BFS27E	161	3132	214		
BFS27F	161	3132	214		
BFS27G	161	3132	214		
BFS28	221	3050			F2004
BFS28R	221	3050			F2004
BFS29	128	3024	210		
BFS29P	192	3137	210		
BFS30	192	3137	210		
BFS30P	192	3137	210		
BFS31	159	3118	31		
BFS31P	123A		210		
BFS32	159	3118	21		
BFS32P	159	3118	21		
BFS33	159		48		
BFS33P	159	3118	21		
BFS34	193		48		
BFS34P	159	3118	21		
BFS36	128	3122	210		
BFS36A	123A	3122	210		
BFS36B	123A	3122			
BFS36C	123A	3122			
BFS37	159	3025	82		
BFS37A	159	3025	82		
BFS38	123A	3122	210		
BFS38A	123A	3122	210		
BFS39	192	3137	210		
BFS40	159	3025	82		
BFS40A	159	3025	22		
BFS41	159	3025	82		
BFS42	123A	3122	47		
BFS42A	123A	3122			
BFS42B	123A	3122			
BFS42C	123A	3122			
BFS43	123A	3122	63		
BFS43A	123A	3122			
BFS43B	123A	3122			
BFS43C	123A	3122			
BFS44	193	3138	48		
BFS45	193	3138	67		
BFS50	186	3192	28		
BFS51	186	3192	28		
BFS55A			61		
BFS58			86		
BFS59	192	3137	47		
BFS60	192	3137	47		
BFS62	108	3039	61		
BFS68	132	3116			
BFS69	159	3025	65		
BFS96	193	3138	48		
BFS97	193	3138	48		
BFS99	171	3201	27		
BFT21			48		
BFT22			48		
BFT30			63		
BFT31			47		
BFT40			63		
BFT41			47		
BFT54			47		
BFT55			62		
BFT61			48		
BFT62			48		
BFT70			67		
BFT71			48		
BFT80			67		
BFT81			48		
BFV10			20		
BFV11			20		
BFV12			20		
BFV14			47		
BFV20	159	3118	21		
BFV21	159	3118	21		
BFV22	159	3118	21		
BFV25	159	3114	65		
BFV26	159	3114	65		
BFV29	159	3118			
BFV30	159	3118	65		
BFV33	159	3118	65		
BFV34			65		
BFV35			65		
BFV36			65		
BFV37			212		
BFV38			212		
BFV40			210		
BFV41			210		

Original Part Number	ECG	SK	GE	IR/WEP	HEP
BFV42			20		
BFV43			20		
BFV44			20		
BFV45			210		
BFV46			20		
BFV47			20		
BFV49			210		
BFV50			20		
BFV51			20		
BFV52			47		
BFV53			20		
BFV54			20		
BFV55			20		
BFV59			39		
BFV60			39		
BFV61			39		
BFV62			212		
BFV80			39		
BFV82	159	3025	65		
BFV82A	159	3025	65		
BFV82B	159	3025	65		
BFV82C	159	3025	65		
BFV83	108	3039	61		
BFV83A	108	3039	61		
BFV83B	123A	3122	20		
BFV83C	123A	3122	20		
BFV85	123A	3122	20		
BFV85A	123A	3122	20		
BFV85B	123A	3122	20		
BFV85C	123A	3122	20		
BFV85D	108		212		
BFV85E	108		212		
BFV85F	108	3039	81		
BFV85G	108	3039	81		
BFV86	159	3025	21		
BFV86A	159	3025	21		
BFV86B	159	3025	21		
BFV86C	159	3025	21		
BFV87	123A	3122	20		
BFV88	123A	3122	20		
BFV88A	123A	3122	20		
BFV88B	123A	3122	20		
BFV88C	123A	3122	210		
BFV89	199		212		
BFV98A	199		212		
BFW10	132	3116			F0021
BFW11	132	3116			
BFW12	132	3116			
BFW13	133	3112			
BFW16					S3013
BFW16A					S3013
BFW17					S3013
BGW17A					S3013
BFW20	193	3138	67		
BFW24					S3019
BFW25					S3011
BFW26					S3011
BFW29	192	3137	81		
BFW30	161	3132			S0017
BFW31	193		48		
BFW32	123A	3122	20		
BFW33	128	3024	18		
BFW37	154	3045		TR78/712	
BFW41			39		
BFW45	154	3045		TR78/712	S0005
BFW54	132	3116			
BFW55	132	3116			
BFW56	132	3116			
BFW57			81		S3010
BFW58			81		S3010
BFW59	123A	3122	210		S0014
BFW60	123A	3122	210		S0014
BFW61	132	3116			
BFW63	161	3132	60		
BFW64	108	3039	60		
BFW66					S3011
BFW68	123A	3122	20		
BFW69					S3010
BFW71					S0015
BFW73					S0025
BFW73A					S0025
BFW74					S0025
BFW75					S0025
BFW76					S0025
BFW76A					S0015
BFW77					S0017
BFW78					S0005
BFW79					S0005
BFW87	189	3083			S3032
BFW88	189	3083	67		S3032
BFW89	159	3059	48		S0019
BFW90	159	3025	48		S0019
BFW91	129	3025	48		S0012
BFX12	106		22	TR20/52	S0013
BFX13	106		22	TR20/56	S0013
BFX14					S0014
BFX17	128	3024	47	TR87/243	S3001
BFX18	108	3039	60	TR95/56	S0016
BFX19	108	3039	60	TR95/56	S0016
BFX20	108	3039	60	TR95/56	S0016
BFX21	108	3039	60	TR95/56	S0016
BFX29	159	3025	21		S5022
BFX30	159	3025	21		S5022
BFX31	161	3117	60	TR83/719	S0016
BFX32	107	3039		TR70/720	S0015
BFX33					S3008
BFX34	154	3045		TR78/712	
BFX35	159		48	TR20/717	
BFX37	234		82	TR20/717	S0013
BFX38	129	3025	48	TR88/242	
BFX39	129	3025	48	TR88/242	
BFX40	129	3025		TR88/242	
BFX41	129	3025		TR88/242	
BFX42					S0015
BFX43	108	3039		TR95/56	S0011
BFX44	108	3039		TR95/56	S0016
BFX45	108	3039	212	TR95/56	
BFX48	106		21	TR20/52	S0013
BFX50	128	3024		TR87/243	
BFX51	128	3024		TR87/243	
BFX52	128	3024	47	TR87/243	
BFX53	192	3137	47	TR22/	S0033
BFX55			28		S3008
BFX59F	192	3137	62		
BFX60	161	3117	61	TR83/719	
BFX61	128	3024	63	TR87/243	S3011
BFX62	161	3117	60	TR83/719	
BFX65	159	3025	65		
BFX68	128	3024	27	TR87/243	
BFX68A	128	3024	27	TR87/243	S3011
BFX69	128	3024	27	TR87/243	
BFX69A	128	3024	27	TR87/243	
BFX73	161	3117	86	TR83/719	
BFX74	128	3024	81	TR87/243	
BFX74A	129	3025	67	TR88/242	
BFX75					S0015
BFX76					S0025
BFX77	161	3117	61	TR83/719	S0020
BFX84	128	3024	32	TR87/243	S3019
BFX85	128	3024	32	TR87/243	S3019
BFX86	128	3024	47	TR87/243	S3001
BFX87	129	3025	21	TR88/242	S5022
BFX88	129	3025	21	TR88/242	S0012
BFX89			86		
BFX92	123A	3122	212	TR21/735	
BFX93	123A	3122	212	TR21/735	
BFX94	128	3024	20	TR87/243	
BFX95	128	3024	20	TR87/243	
BFX95A	123A	3122	210		
BFX96	128	3024	47	TR87/243	
BFX96A	186	3192	47		
BFX97	128	3024	47	TR87/243	
BFX98	154		27	TR78/712	
BFY	123A	3124			
BFY10	128	3018	212	TR87/243	S0014
BFY11	128	3018	212	TR87/243	S0014
BFY12	128	3024	47	TR87/243	S0014
BFY13	128	3024	63	TR87/243	S3011
BFY14	128	3024		TR87/243	S3019
BFY15	128	3024	63	TR87/243	S0014
BFY16	192	3024	63	TR87/243	S3013
BFY17	128	3024	47	TR87/243	S0014
BFY18	129	3025	210	TR88/242	S0011
BFY19	128	3024	20	TR87/243	S0011
BFY22	123A	3124	39	TR21/735	S0025
BFY23	123A	3124	39	TR21/735	S0025
BFY23A	123A	3124	20	TR21/735	S0015
BFY24	123A	3124	39	TR21/735	S0015
BFY25	123A	3122	47	TR21/735	S0014
BFY26	123A	3122	210	TR21/735	S0011
BFY27	128	3024		TR87/243	S0015
BFY28	123A	3122	11	TR21/735	S0011
BFY29	123A	3124	39	TR21/735	S0015
BFY30	123A	3124	39	TR21/735	S0015
BFY33	123A	3122	20	TR21/735	S0014
BFY34	128	3024	20	TR87/243	S3011
BFY37	123A		10	TR21/735	S0011
BFY-37	108	3019			
BFY37I	123A	3122			S0011
BFY39	123A	3124	212	TR21/735	S0011
BFY39-1	199	3122	212	TR65/	
BFY39/1	123A			TR21/735	
BFY39-2	199		212		
BFY39/2	123A	3122		TR21/735	
BFY39-3	199		212	TR51/	
BFY39/3	123A	3122		TR21/735	
BFY39I	123A	3122		TR51/	S0011
BFY40	128	3024	47	TR87/243	S3011
BFY41	198		32		S3019
BFY43	154	3045	27	TR78/712	S5024

Original Part Number	ECG	SK	GE	IR/WEP	HEP
BFY44	128	3024		TR87/243	S3002
BFY45	154	3045		TR78/712	S5025
BFY46	128	3024	18	TR87/243	S3011
BFY47	107	3024	11	TR70/720	S0015
BFY-47	108	3018			
BFY48	107		39	TR70/720	S0015
BFY-48	108	3018			
BFY49	107	3018	39	TR70/720	S0015
BFY50	128	3104	18	TR87/243	S3019
BFY51	128	3024	17	TR87/243	S3001
BFY52	128	3024	18	TR87/243	S3001
BFY53	128	3024	47	TR87/243	S3001
BFY55	128	3024	45	TR87/243	S3019
BFY56	128	3024	27	TR87/243	S3011
BFY56A	128	3024		TR87/243	S3011
BFY57	154	3045	27	TR78/712	S5026
BFY63	123A	3122	17	TR21/735	S0014
BFY64	129		17	TR88/242	S0012
BFY65	154	3045	27	TR78/712	S5026
BFY66	128	3024	86	TR87/243	
BFY67	128	3024	18	TR87/243	S3001
BFY67A	128	3024	47	TR87/243	S3011
BFY67C	128	3024	47	TR87/243	S3011
BFY68	128	3024	27	TR87/243	S0015
BFY68A	128	3024	47	TR87/243	
BFY69	107	3039	214	TR70/243	
BFY69A	107	3039	214	TR70/720	
BFY69B	107	3039	214	/720	
BFY70	128	3024	28	TR87/243	S3001
BFY72	128	3024	215	TR87/243	S3011
BFY73	123A	3122	20	TR21/735	
BFY74	123A	3122	17	TR21/735	S0011
BFY75	123A	3122	17	TR21/735	
BFY76	123A	3122	62	TR21/735	S0015
BFY77	123A	3122	62	TR21/735	S0015
BFY78	108	3039	17	TR95/56	S0016
BFY79	161	3117	61	TR83/719	S0016
BFY80	154	3045	27	TR78/712	S0005
BFY85					S0025(2)
BFY86					S0025(2)
BFY87	161				
BFY88					S0017
BFY90	108	3039	86	TR95/56	S0017
BFY90B	108	3039	86	TR95/56	S0017
BFY94	159				
BFY99	128	3024	63	TR87/243	S3001
BFY371	123A	3122	20	TR21/735	S0011
BFY391	123A	3122	20	TR21/735	S0025
BFY501					S0011
BFY511					S0011
BFZ10					S0013
BG71	123A	3122		TR21/735	S0025
BH71	123A	3122		TR21/735	S0025
BI71	123A	3122		TR51/735	S0025
BIP7201	123		17	TR86/53	
BJ1A	159	3114			
BJ1B	159	3114			
BJ1C	159	3114			
BJ2	159	3114			
BJ2A	159	3114			
BJ2B	159	3114			
BJ2C	159	3114			
BJ2D	159	3114			
BJ3	159	3114			
BJ3A	159	3114			
BJ3B	159	3114			
BJ4	159	3114			
BJ4A	159	3114			
BJ4B	159	3114			
BJ4C	159	3114			
BJ4D	159	3114			
BJ5	159	3114			
BJ5A	159	3114			
BJ5B	159	3114			
BJ6	159	3118			
BJ6A	159	3118			
BJ6B	159	3118			
BJ6C	159	3118			
BJ6D	159	3118			
BJ7	159	3118			
BJ7A	159	3118			
BJ7B	159	3118			
BJ7C	159	3118			
BJ7D	159	3118			
BJ8	159	3118			
BJ8A	159	3118			
BJ8B	159	3118			
BJ8C	159	3118			
BJ8D	159	3118			
BJ9	159	3118			
BJ9A	159	3118			
BJ9B	159	3118			
BJ9C	159	3118			
BJ9D	159	3118			
BJ10	159	3118			

Original Part Number	ECG	SK	GE	IR/WEP	HEP
BJ11	159	3118			
BJ11A	159	3118			
BJ11B	159	3118			
BJ12	159	3118			
BJ12B	159	3118			
BJ12C	159	3118			
BJ13	159	3118			
BJ13A	159	3118			
BJ13B	159	3118			
BJ14	159	3118			
BJ14A	159	3118			
BJ15	159	3118			
BJ56	159	3114			
BJ160	159	3118			
BJ161	159	3118			
BJ161A	159	3118			
BJ161B	159	3118			
BJ161C	159	3118			
BL410				TR26/	
BL411				TR26/	
BL412				TR26/	
BL415				TR26/	
BL415A				TR23/	
BL447				TR26/	
BL447A				TR23/	
BL448				TR26/	
BL448A				TR23/	
BL449				TR61/	
BL449A				TR23/	
BL450				TR61/	
BL461				TR21/	
BLX65					S3001
BLX88					S0025
BLY10	130	3027	63	TR59/247	
BLY11	130	3027	63	TR59/247	
BLY12	130	3027	19	TR59/247	S3020
BLY15	130	3027	19	TR59/247	S5000
BLY15A	175		66	TR81/241	S5012
BLY16					S5019
BLY17C					S5004
BLY20	186		28	TR75/	
BLY21	152		66		
BLY27	192		63	TR25/	
BLY28	192		63	TR25/	
BLY33			28	TR76/	S3008
BLY34			28	TR76/	S3013
BLY35	152		66		
BLY36	152		66		
BLY37	186	3192	28	TR76/	
BLY38	186	3192	28	TR76/	
BLY47	130	3027	19	TR59/247	S7002
BLY47A	175	3131		TR81/241	S5012
BLY48	130	3027	19	TR59/247	S7002
BLY48A	175	3131		TR81/241	S5012
BLY49	162		19	TR67/707	S5020
BLY49A	175	3131		TR81/241	S5012
BLY50	162		19	TR67/707	S5020
BLY53	186	3192	28	TR76/	
BLY61	128	3024	28	TR87/243	S5014
BLY62	186	3192	28	TR76/	
BLY63	152		66		
BLY72					S5004
BLY78	186	3192	28	TR76/	
BLY79					S3006
BLY83					S3007
BLY84					S3007
BLY86					S5011
BLY88	152		66		
BLY89	152				
BLY91	186	3192	28	TR76/	
BLY92			66		S3007
BLY93	192		66		
BMT1991	159	3114			
BMT2303	159	3114			
BMT2411	159	3114			
BMT2412	159	3114			
BN7133	130	3027	14	TR59/247	
BN7168	175	3131		TR81/241	
BN7214	130	3027	14	TR59/247	
BN7253	154	3045	27	TR78/712	
BN7517	123A	3122	20	TR21/735	
BN7518	123A	3122	20	TR21/735	
BO8875/2	159				
BP67	123A	3122		TR21/735	
BP1037		3004			
BP5263	160			TR17/637	G0003
BQ67	123A	3122		TR21/735	S0015
BR67	123A	3122		TR21/735	S0015
BR100B	186	3192	28	TR76/	
BR101B	152	3054			
BRC-116	130				
BRC5296	152				
BRC-5496	196				
BRC6109	197				
BRX65-1		3011			

Original Part Number	ECG	SK	GE	IR/WEP	HEP
BS67	123A	3122		TR21/735	S0015
BS475	123A	3122	20	TR21/735	
BSC1015					S5004
BSC1015A					S5004
BSC1015B					S5004
BSC1016					S5004
BSC1016A					S5004
BSC1016B					S5004
BSF17				TR24/	
BSF17R				TR24/	
BSF18				TR21/	
BSF18R				TR21/	
BSF19				TR21/	
BSF20				TR21/	
BSF20R				TR21/	
BSS10	123A	3122	20		S0025
BSS15					S3002
BSS16					S3002
BSS17					S3003
BSS18					S3003
BSS19	171	3201	27		S5025
BSS20	171	3201	27		
BSS21	123A	3122	20		S0011
BSS22					S0013
BSS23	192	3137	47		S3011
BSS26	108	3039	61		S3011
BSS27					S3002
BSS28					S3011
BSS29					S3002
BSS30					S3011
BSS31					S3011
BSS32					S0005
BSS33			32		S3021
BSS42					S3019
BSS43					S3019
BSV15	129	3025	28	TR88/242	S3012
BSV16	129	3025		TR88/242	S3012
BSV17					S3012
BSV21	159	3114		TR20/717	S0013
BSV21A	159	3114			
BSV35	108	3018			S0025
BSV35A	123A	3018	20		S0025
BSV36					S0014
BSV37					S0012
BSV40	123A	3122	17		S0011
BSV41	123A	3122	17		S0011
BSV42					S3012
BSV43A	159	3025	21		S3012
BSV43B	193	3138	67		S3012
BSV44A	159	3025	21		S3012
BSV44B	193	3138	48		S3012
BSV45A	159	3025	21		S0012
BSV45B	193	3138	48		
BSV47A	159	3025	21		
BSV47B	193	3138	67		
BSV48A	159	3025	21		S5022
BSV48B	193	3138	48		S5022
BSV49A	159	3025	21		S0012
BSV49B	193	3138	48		S0012
BSV51	128	3024	18		S5026
BSV52	108	3039	39		S0020
BSV52R	108	3039	39		S0016
BSV53	123A	3122	20		
BSV53P	107	3039	214		
BSV54	123A	3122	20		
BSV54P	107	3039	214		
BSV55A	159	3118			
BSV55AP	159	3118			
BSV55P	159	3118			
BSV59	123A	3122	17		
BSV60	186	3054	28		S3002
BSV65FA			20		S0011
BSV65FB			20		S0011
BSV69	192	3137	47		S3011
BSV83					S3012
BSV84					S3011
BSV85			47		S3011
BSV86	123A	3122	20		S3008
BSV87	123A	3122	20		S3008
BSV88	123A	3122	210		S3008
BSV89	123A	3122	20		
BSV90	123A	3122	20		
BSV91	123A	3122	20		
BSV96	159	3025	82		S0012
BSV97	159	3025	82		S0012
BSV98	159	3025	82		S0012
BSW10	128	3024		TR87/243	S3011
BSW11	123A	3122	20	TR21/735	
BSW12	123A	3122	210	TR21/735	
BSW13					S0011
BSW19	123A	3122	17	TR21/735	S0011
BSW19A			65		S0013
BSW19VI			65		S0013
BSW20					S0019
BSW20A			65		S0019

Original Part Number	ECG	SK	GE	IR/WEP	HEP
BSW20VI			65		S0013
BSW21	159	3114	21	TR20/717	S0019
BSW21A	159	3114	21	TR20/717	S0019
BSW22	159	3114		TR20/717	S0019
BSW22A	159	3114		TR20/717	S0019
BSW23	129	3025	21	TR88/242	
BSW24	159	3114	22	TR20/717	
BSW25					S0013
BSW26	128	3024	47	TR20/243	S3011
BSW27	128	3024	47	TR87/243	S3011
BSW28	128	3024	47	TR87/243	S3011
BSW29	128	3024	63	TR87/243	S3011
BSW32	154	3045	18	TR78/712	S5025
BSW33	123A	3122	17	TR21/735	S0025
BSW34	123A	3122		TR21/735	S0025
BSW35	128	3024		TR87/243	S0015
BSW39				TR87/	S3019
BSW40-6					S3032
BSW41	123A	3122	210	TR21/735	S0025
BSW42	123A	3122	20	TR21/735	S0015
BSW42A	123A	3122		TR21/735	S0025
BSW43	123A	3122		TR21/735	S0015
BSW43A	123A	3122		TR21/735	S0015
BSW44	159	3114	21	TR20/717	S0019
BSW44A	159	3114	21	TR20/717	S0019
BSW45	159	3114		TR20/717	S0019
BSW45A	159	3114		TR20/717	S0032
BSW49			47		S3011
BSW51					S3001
BSW52					S3001
BSW53					S3001
BSW54					S3001
BSW58	123A	3122	20		S0014
BSW59					S0014
BSW61					S0015
BSW62					S0015
BSW63					S0015
BSW64					S3001
BSW65	128	3024		TR87/243	S3032
BSW66	128	3024		TR87/243	S3019
BSW67	190	3104			S3019
BSW68	190	3104			S3019
BSW-68	198		32		
BSW69	154	3044			S5025
BSW70	154	3045	18	TR78/712	S5025
BSW72	159	3114	21	TR20/717	S0012
BSW73	159	3114	48	TR20/717	S0012
BSW74	159	3114	67	TR20/717	S3012
BSW75	159	3114	67	TR20/717	S3012
BSW78					S0011
BSW79					S0011
BSW80					S0011
BSW81					S0013
BSW82	123A	3122	20	TR21/735	
BSW83	123A	3122	20	TR21/735	
BSW84	123A	3122	20	TR21/735	
BSW85	123A	3122	20	TR21/735	
BSW88	123A	3122	210	TR21/735	S0025
BSW89	123A	3122	210	TR21/735	S0025
BSW92	123A	3122		TR21/735	S0011
BSX12	108	3039	63	TR95/56	
BSX12A	192	3137	47		
BSX19	123A	3122		TR21/735	S0025
BSX20	123A	3122		TR21/735	S0025
BSX21	154	3045	18	TR78/712	S5025
BSX22	128	3024	18	TR87/243	
BSX23	128	3024		TR87/243	
BSX24	123A	3122	210	TR21/735	S0011
BSX25	123A	3122	61	TR21/735	S0016
BSX26	108	3039		TR95/56	S0016
BSX27	108	3039	17	TR95/56	S0016
BSX28	108	3039	17	TR95/56	S0016
BSX29	106		21	TR20/52	S0016
BSX30	106	3122	215	TR21/735	
BSX32	195	3048	X18		S3001
BSX33	128	3024	20	TR87/243	
BSX35	108	3039	21	TR95/56	S0016
BSX36	159	3114	21	TR20/717	
BSX38	123A	3122	210	TR21/735	S0011
BSX39	123A	3122		TR21/735	
BSX40	129	3025	48	TR88/242	
BSX41	129	3025	48	TR88/242	S3028
BSX44	123A	3122			S0011
BSX45	128	3024	28	TR87/243	
BSX45-6	192	3137	63		
BSX45-10	192	3137	63		
BSX45-16	152	3137	63		
BSX46	128	3024	28	TR87/243	
BSX46-6	192	3137	63		
BSX46-10	192	3137	63		
BSX46-16	192	3137	63		
BSX48	123A	3122	28	TR21/735	
BSX49	123A	3122	28	TR21/735	
BSX51	123A	3122	20	TR21/735	
BSX51A	123A	3122		TR21/735	

Original Part Number	ECG	SK	GE	IR/WEP	HEP
BSX52	123A	3122		TR21/735	
BSX52A	123A	3122		TR21/735	
BSX53	123A	3122	214	TR21/735	
BSX54	123A	3122	214	TR21/735	
BSX59	195	3048	X18		S3001
BSX60	128	3024	18	TR87/243	S3001
BSX61	128	3024	18	TR87/243	S3001
BSX62	128	3024	28	TR87/243	
BSX62B	128	3024		TR87/243	S5014
BSX62C	128	3024		TR87/243	S5014
BSX63	128	3024		TR87/243	
BSX63B	128	3024		TR87/243	S3010
BSX63C	128	3024		TR87/243	S3010
BSX66	123A	3122	210	TR21/735	
BSX67	123A	3122	210	TR21/735	
BSX68	123A	3122	212	TR21/735	
BSX69	123A	3122	212	TR21/735	S0015
BSX70	128	3024	20	TR87/243	
BSX71	128	3024	20	TR87/243	
BSX72	128	3024	47	TR87/243	
BSX75	123A	3122	17	TR21/735	
BSX76	123A	3122	10	TR21/735	S0011
BSX77	123A	3122	210	TR21/735	S0025
BSX78	123A	3122	17	TR21/735	S0025
BSX79	123A	3122	10	TR21/735	
BSX80	123A	3122		TR21/735	S3019
BSX81	123A	3122	210	TR21/735	S0025
BSX87	123A	3122	17	TR21/735	
BSX87A	108	3039		TR95/56	
BSX88	108	3039	20	TR95/56	
BSX88A	108	3039	17	TR95/56	
BSX89	123A	3122	210	TR21/735	
BSX90	123A	3122	20	TR21/735	
BSX91	123A	3122	20	TR21/735	
BSX92	108	3039		TR95/56	
BSX93	108	3039		TR95/56	
BSX94					S0013
BSX94A	123A	3122	20		S0013
BSX95	128	3024	216	TR87/243	
BSX96	128	3024	216	TR87/243	
BSX97	123A	3122		TR21/735	S0011
BSY10	123	3124	81	TR86/53	S0014
BSY11	123	3124	81	TR86/53	S0014
BSY17	123A	3122	17	TR21/735	S0011
BSY18	123A	3122	17	TR21/735	S0011
BSY19	123A	3122	17	TR21/735	S0011
BSY20	123A	3122	63	TR21/735	S0011
BSY21	123A	3122	20	TR21/735	S0011
BSY22	108	3039	11	TR95/56	S0011
BSY23	108	3039	63	TR95/56	S0011
BSY24	123		81	TR86/53	S0011
BSY25	128	3024	81	TR87/243	S0014
BSY26	123A	3122	10	TR21/735	S0011
BSY27	123A	3122	10	TR21/735	S0011
BSY28	123A	3122	20	TR21/735	S0011
BSY29	123A	3122	20	TR21/735	S0011
BSY32					S0025
BSY33					S0025
BSY34	123A	3122	28	TR21/735	S0014
BSY36					S0025
BSY37					S0025
BSY38	123A	3122	20	TR21/735	S0011
BSY39	123A	3122	20	TR21/735	S0011
BSY40	159	3114	22	TR20/717	S0013
BSY41	159	3114	22	TR20/717	S0013
BSY44	128	3024	47	TR87/243	
BSY45	128	3024	47	TR87/243	
BSY46	128	3024	47	TR87/243	
BSY47					S0025
BSY48	123A	3122	20		
BSY49	123A	3122	20		
BSY50					S0025
BSY51	128	3024	47	TR87/243	
BSY52	128	3024	47	TR87/243	
BSY53	128	3024	20	TR87/243	
BSY54	128	3024	20	TR87/243	
BSY55	128	3024	20	TR87/243	
BSY56	128	3024		TR87/243	
BSY58	123A	3122	28	TR21/735	S0014
BSY59	123A	3122	48	TR21/735	
BSY61	123A	3122	210	TR21/735	S0025
BSY62	123A		210	TR21/735	S0011
BSY-62	108	3019			
BSY62A	123A		210		
BSY62B	192	3137	210		
BSY63	123A	3122	20	TR21/735	S0011
BSY68			18		S5026
BSY70	108	3039	61	TR95/56	S0016
BSY71	128	3024		TR87/243	
BSY72	108	3019	61	TR95/56	S0011
BSY73	123A		212	TR21/735	S0011
BSY-73	108	3019			
BSY74	123A		212	TR21/735	S0011
BSY-74	108	3019			
BSY75	123A	3122	210	TR21/735	S0011

Original Part Number	ECG	SK	GE	IR/WEP	HEP
BSY76	123A	3122	210	TR21/735	S0011
BSY77	128	3024	81	TR87/243	
BSY78	128	3024	81	TR87/243	
BSY79	128	3044	18	TR87/243	S5025
BSY80	123A		10	TR21/735	S0015
BSY-80	108	3019			
BSY81	128	3024	63	TR87/243	S3001
BSY82	128	3024	63	TR87/243	S3024
BSY83	128	3024	18	TR87/243	S3001
BSY84	128	3024		TR87/243	S3020
BSY85	128	3024		TR87/243	S3019
BSY86	128	3024		TR87/243	S3019
BSY87	128	3024	18	TR87/243	S5026
BSY88	128	3024	27	TR87/243	S5026
BSY89	123A	3122	212	TR21/735	
BSY90	128	3024	47	TR87/243	
BSY91	128	3024	47	TR87/243	S3013
BSY92	128	3024	47	TR87/243	S5014
BSY93	123A	3122	62	TR21/735	S0016
BSY95	123A		10	TR21/735	S0011
BSY-95	108	3019			
BSY95A	123A	3122	10	TR21/735	S0016
BSY112					S5021
BSY165	123A	3122	20	TR21/735	
BSY168	123A	3122	17	TR21/735	S0011
BT67	123A	3122		TR21/735	S0015
BT71	123A	3122		TR21/735	S0025
BT929			39		
BT930	108	3039	39		
BTX068	123A			TR21/735	S5014
BTX-68	199				
BTX070	102A			TR85/250	G0005
BTX-070	123A				
BTX071	103A			TR08/724	G0011
BTX-071	129				
BTX-084	234				
BTX-094	123A	3122			S0015
BTX-095					S0024
BTX-096	123A	3122			S0014
BTX-097	129	3025			S0012
BTX-2367B	123A		10	TR24/	S0015
BU67	123A	3122		TR21/735	S0015
BU71	123A	3122		TR21/735	S0025
BU105	165	3115	38	TR93/740B	
BU106	162			TR67/707	
BU107	162			TR67/707	
BU108	165	3115	38	TR93/740B	
BU109	163		36		
BU110	162				S5020
BU111	163				S5021
BU114		3115			
BU115	165	3115		TR93/740B	
BU120	162			TR67/707	
BU204		3115			
BU205		3115			
BU207		3115			
BU208		3115			
BUC 97704-2	123A				
BUY10	130	3027	19	TR59/247	
BUY11	130	3027	19	TR59/247	
BUY12					S5020
BUY13					S5020
BUY20	162				S5020
BUY21	162				S5020
BUY21A	162				S5020
BUY22	162				S5020
BUY24	152	3054	66		
BUY26					S5020
BUY27					S5020
BUY28					S5020
BUY35	162				S5020
BUY38	175	3538			S5019
BUY41					S3002
BUY43	181	3535	66		S7002
BUY44	162				S5020
BUY46	181	3535	66		S7002
BUY69B	163		36		
BUY69C	162		35		
BV67	123A	3122		TR21/735	S0015
BV71	123A	3122		TR21/735	S0025
BVF85D		3039			
BVF85E		3039			
BW67	123A	3122		TR21/735	S0015
BW71	123A	3122		TR21/735	S0025
BWS66	198		32		
BWS67	198		32		
BX65-312		3004			
BX65-327		3004			
BX67	123A	3122		TR21/735	S0015
BX71	123A	3122		TR21/735	S0025
BX-324	158	3004	53	TR84/	G0005
BX-495	158(2)	3004	52	TR85/	G6011(2)
BY67	123A	3122		TR21/735	S0015
BY71	123A	3122		TR21/735	S0025
BZ67	123A	3122		TR21/735	S0015

Original Part Number	ECG	SK	GE	IR/WEP	HEP
BZ71	123A	3122		TR21/735	S0025
C00 686 0241	159				
C00686-0258-0	159				
C00 68602300	123A				
C0068602720	159				
C0C13000-1C			63	TR25/	
C1-1002	724				
C1-1004	717				
C6			65		
C6-15					S0019
C7			65		
C8			65		
C9FF-10A580	128				
C11E		3122		/735	
C12	128				
C13	101				
C14	101	3011			
C15-1	123A	3122			
C15-2	123A	3122			
C15-3	123A	3122			
C16	123A	3122			
C16A	123A	3122			
C17	108	3039			
C17A	108	3039			
C18	123A	3122			
C19	128	3024			
C20	128	3024			
C21	130	3027			
C22	128	3024			
C23	128	3024			
C24	128	3024			
C26	123A	3122			
C27	128	3024			
C28	123	3122			
C29	123	3122			
C30	128	3024			
C31	128	3024			
C32	128	3024			
C32A	128	3024			
C33	123A	3122			
C34	103A	3010			
C35	103A	3010			
C36	101	3011			
C37	108	3039			
C38	108	3039			
C39	108	3039			
C39-207	123A	3122	20	TR21/735	
C39A	108	3039			
C40	108	3039			
C41	163				
C41TV	163				
C42	162				
C42A	162				
C43	162				
C44	162				
C45	123A	3122			
C46	128	3024			
C47	123A	3122			
C48	123	3122			
C50	101	3011			
C50A	101	3011			
C50BA042	121	3009		TR01/232	
C51	128	3024			
C52	123A	3122			
C53	123A	3122			
C54	123A	3122			
C55	123A	3122			
C56	107	3039			
C58	154	3044			
C58A	154	3044			
C59	128	3024			
C60	101	3011			
C61	128	3024			
C62	123A	3122			
C63	123A	3122		TR21/735	S0011
C63(JAPAN)	108				
C64	123A	3122	17	TR21/735	S0011
C64(JAPAN)	154				
C65	154	3044			
C65B	154	3044			
C65N	154	3044			
C65Y	154	3044			
C65YA	154	3044			
C65YB	154	3044			
C65YTV	154	3044			
C66	154	3044			
C66-P11111-0001	123A				
C66-P11150-0001	123A				
C67	123A	3122			
C68	123A	3122			
C69	128	3024			
C70	154	3044			
C71	101	3011			
C72	101	3011			
C73	100	3005	2	TR05/254	G0005

Original Part Number	ECG	SK	GE	IR/WEP	HEP
C73(JAPAN)	101				
C74	108	3039			
C75	100	3005	2	TR05/254	G0005
C75(JAPAN)	101				
C76	100	3011	2	TR05/254	G0005
C76(JAPAN)	101				
C77	101	3011			
C77C	102	3004			
C78	101	3011			
C79	108	3039			
C80	108	3039			
C82S	108	3018			
C86-156-2A		3004			
C87	123A	3122			
C87J0003		3024			
C87J0004	3024	3024			
C88	154	3044			
C88A	154	3044			
C89	101	3011			
C90	101	3011			
C91	101	3011			
C92-0025-R0	724	3524			
C94		3112			
C94A		3112			
C94E		3112			
C95	154	3112			
C95A		3112			
C95E		3112			
C96E		3112			
C97	128	3024			
C97A	128	3024			
C98	108	3039			
C99	108	3039			
C99CD	128				
C100	123A	3122			
C100-0Y	123A	3122			
C101	129	3025	21	TR88/242	S0012
C101A	124	3021			
C101(JAPAN)	124				
C101X	126	3006			
C102	129	3006	21	TR88/242	S0012
C102(JAPAN)	105				
C103	106	3122	21	TR20/52	
C103A	123A	3122			
C103(JAPAN)	123A				
C104	123A	3122			
C104A	123A	3122			
C105	123A	3122			
C106	129	3025	65	TR88/242	
C108	128	3024			
C109	128	3024			
C109A	128	3024			
C110	123A	3122			
C111	123A	3122			
C111B	108	3039	17	TR95/56	
C111E	123A	3122	61	TR21/735	S0011
C112	129	3025		TR88/242	S0012
C112(JAPAN)	128				
C113	128	3024			
C114	128	3024			
C115	128	3024			
C116	128	3024			
C116T	128	3024			
C117	128	3024			
C118	129	3025	21	TR88/242	S0012
C118(JAPAN)	128				
C119	129	3025	21	TR88/242	S0012
C119(JAPAN)	128				
C120	128	3024			
C121	128	3024			
C122	128	3018			S0015
C123	128	3024			
C124	128	3024			
C125	160	3006			
C127	108	3039			
C128	101	3011			
C129	101	3011			
C130	128	3024			
C131	123A	3122			
C132	123A	3122			
C133	123A	3122			
C134	123A	3122			
C134B	123A	3122			
C135	123A	3122			
C136	123A	3122			
C137	123A	3122			
C138	123A	3122			
C138A	123A	3122			
C139	123A	3122			
C140	128	3024			
C147	128	3024			
C148	108	3039			
C150	123A	3122			
C150T	128	3024			
C151	123A	3122			

Original Part Number	ECG	SK	GE	IR/WEP	HEP
C152	128	3024			
C154	154	3044			
C154B	154	3044			
C154C	154	3044			
C154H	154	3044			
C155	108	3039			
C156	108	3039			
C157	123	3122			
C158	123	3122			
C159	123A	3122			
C160	123A	3122			
C163	128	3024			
C166	123A	3122			
C167	123A	3122			
C170	108	3039			
C171	107	3039			
C172	108	3039			
C172A	108	3039			
C173	101	3011			
C174	107	3039			
C174A	107	3039			
C175	101	3011			
C175B	102A	3004			
C176	101	3011			
C177	101	3011			
C178	101	3011			
C179	103A	3010			
C180	103A	3010			
C181	103A	3010			
C182	123A	3122			
C182Q	107	3039			
C183	108	3039			
C183E	108	3039			
C183J	108	3039			
C183K	108	3039			
C183L	108	3039			
C183M	108	3039			
C183P	108	3039			
C183Q	108	3039			
C183R	108	3039			
C183W	123A				
C184	108	3039			
C184H	108	3039			
C184J	108	3039			
C184L	108	3039			
C185	108	3039			
C185A	107	3039			
C185J	108	3039			
C185M	108	3039			
C185Q	108	3039			
C185R	108	3039			
C185V	108	3039			
C186	107	3039			
C187	107	3039			
C188	128	3024			
C188A	128	3024			
C188AB	128	3024			
C189	128	3024			
C190	128	3024			
C191	123A	3122			
C192	123A	3122			
C193	123A	3122			
C194	123A	3122			
C195	123A	3122			
C196	123A	3122			
C197	123A	3122			
C199	108	3039			
C200	123A	3122			
C201	129	3025	21	TR88/242	
C201(JAPAN)	123A				
C202	129	3025	21	TR88/242	
C202(JAPAN)	123A				
C203	123A	3122			
C204	123A	3122			
C205	123A	3122			
C206	108	3039			
C208	128	3024			
C210	128	3024			
C211	128	3024			
C212	194	3104			
C213	128	3024			
C214	128	3024			
C215	128	3024			
C216	128	3024			
C217	128	3024			
C218	128	3024			
C218A	123A	3122			
C220	128	3024			
C221	128	3024			
C222	128	3024			
C223	128	3024			
C224	128	3024			
C225	128	3024			
C226	128	3024			
C227	128	3024			
C228	128	3024			
C229	128	3024			
C230	123A	3122			
C231	128	3024			
C232	128	3024			
C233	128	3024			
C234	128	3024			
C235	128	3024			
C235-O	190	3104			
C236	128	3024			
C237	123A	3122			
C238	128	3024			
C239	123A	3122			
C240	162				
C241	162				
C242	162				
C243	162				
C244	130	3027			
C245	162				
C246	162				
C247	128	3024			
C248	128	3024			
C249	128	3024			
C250	161	3132			
C251	161	3132			
C252	161	3132			
C253	161	3132			
C254	175	3538			
C261	192	3137			
C263	108	3039			
C266	107	3039			
C267	123A	3122			
C267A	123A	3122			
C268	128	3024			
C268A	128	3024			
C269	108	3039			
C270	162				
C271	108	3039			
C272	108	3039			
C273	154	3044			
C277C	103A	3010			
C281	123A	3122			
C281A	123A	3122			
C281B	123A	3122			
C281C	123A	3122			
C281C-EP	123A	3122			
C281D	123A	3122			
C282	108	3039			
C283	123A	3122			
C284	123A	3122			
C285	123A	3122			
C285A	123A	3122			
C286	108	3039			
C287	107	3039			
C287A	107	3039			
C288	107	3039			
C288A	107	3039			
C289	108	3039			
C290	195	3048			
C291	128	3024			
C292	128	3024			
C293	128	3048			
C293A		3048			
C296	108	3039			
C299	225	3122			
C300	123A	3122			
C301	123A	3122			
C301A	129	3025		TR88/242	
C302	129	3122	21	TR88/242	
C302(JAPAN)	123A				
C306	128	3024			
C307	128	3024			
C308	128	3024			
C309	128	3024			
C310	128	3024			
C313	108	3019	17	TR95/56	
C313C	161	3132			
C313H	161	3132			
C313(JAPAN)	161				
C315	123A	3122			
C316	108	3039			
C317	123A	3122			
C317C	123A	3122			
C318	108	3018		TR95/56	
C318A	108	3018		/56	
C318A(JAPAN)	123A				
C318(JAPAN)	123A				
C319	123A	3122			
C320	123A	3122			
C321	123A	3122			
C323	123A	3122			
C324	123A	3122			
C324A	123A	3122			
C324H	123A	3122			
C324HA	123A	3122			

Original Part Number	ECG	SK	GE	IR/WEP	HEP
C325E	152				
C328A	123A				
C329	107				
C329B	107				
C329C	107				
C335	199				
C341V	127	3035			
C348	161	3132			
C350	123A				
C350H	123A	3122			
C351	107	3039			
C351(FA)	107				
C352	123A	3047		TR21/735	
C352A	123A	3122		TR21/735	
C352A(JAPAN)	128				
C352(JAPAN)	123A				
C353	128	3024			
C353A	128	3024			
C356	123A	3122			
C360	123A	3122			
C361	107	3039			
C362	107	3039			
C363	123A	3122			
C366	123A	3122			
C367	123A	3122			
C368	199				
C368BL	199				
C368GR	199				
C368V	199				
C369	123A	3122			
C369BL	123A	3122			
C369G	199				
C369G-BL	199				
C369G-GR	199				
C369G-V	199				
C369GBL	123A	3122			
C369GGR	123A	3122			
C369GR	123A	3122			
C369V	123A	3122			
C370	123A	3122			
C370F	108	3039			
C370G	123A	3039			
C370H	108	3039			
C370J	108	3039			
C370K	108	3039			
C371	123A	3122			
C371(O)	123A				
C371B	123A				
C371G	123A	3122			
C371-O	123A	3122			
C371-R	123A	3122			
C372	123A	3122			S0015
C372-0	123A	3122			
C372-1	123A	3122			
C372-2	123A	3122			
C372GR	123A				
C372H	123A				
C372-O	123A	3122			
C372-R	123A	3122			
C372Y	123A				
C372-Y	123A	3122			
C372-Z	123A	3122			
C373	199				
C373BL	199				
C373G	199				
C373W	199				
C374	199				
C374-BL	199				
C374JA	199				
C374-V	199				
C375	107	3039			
C375-O	107	3039			
C375-Y	107	3039			
C376	128	3024			
C377	123A	3122			
C378	123A	3122			
C379	123A	3122			
C380	107	3039			S0016
C380A	107	3039			
C380AO	229				
C380A-O(TV)	107	3039			
C380A(O)	107	3039			
C380AY	229				
C380A-R	107	3039			
C380A-R(TV)	123A	3122			
C380ATV	123A				
C380D	107				
C380-O	107	3039			
C380R	107	3039			
C380-Y	107	3039			
C381	107	3039			
C381BN	107	3039			
C381-O	107	3039			
C381-R	107	3039			
C382	107	3039			

Original Part Number	ECG	SK	GE	IR/WEP	HEP
C382BK	107	3039			
C382-BK1	107	3039			
C382-BK2	107	3039			
C382BL	107	3039			
C382BN	107	3039			
C382BR	107	3039			
C382G	107	3039			
C382-GR	107	3039			
C382-GY	107	3039			
C382R	107	3039			
C382V	107	3039			
C383	123A	3122			
C383G	123A	3122			
C383T	123A				
C383Y	123A	3122			
C384	108	3039			
C384Y	107	3039			
C385	107	3039			
C385A	107	3039			
C386	108	3039			
C386A-OTV	123A	3122			
C387	108	3039			
C387A	108	3039			
C387A(FA-3)	108	3039			
C387G	108	3039			
C388	107	3039			
C388A	107	3039			
C389	161	3132			
C390	161	3132			
C392	108	3039			
C394	107	3039			
C394R	107	3039			
C394W	107	3039			
C394Y	107	3039			
C395	123A	3122			
C395A	123A	3122			
C396	123A	3122			
C397	108	3039			
C398	161	3132			
C398(FA-1)	108	3039			
C399	161	3132			
C400	123A	3122	210	TR21/735	
C400-O	123A	3122			
C400-GR	199				
C400-R	123A	3122			
C400-Y	123A	3122			
C401	106	3118	21	TR20/52	
C401(JAPAN)	123A				
C402	129	3025	21	TR88/242	
C402(JAPAN)	123A				
C402A	123A	3122			
C403	123A	3122			
C403A	123A	3122			
C403B	128	3024			
C403C	123A	3122			
C403C(SONY)	128				
C404	123A	3122			
C405	108	3039			
C406	108	3039			
C407	154	3044	27	TR78/712	
C407(JAPAN)	162				
C408	163				
C409	163				
C410	163				
C411	163				
C412	163				
C420	128	3024	47	TR87/243	
C423	128	3024			
C423B	128	3024			
C423C	128	3024			
C423D	123A	3122			
C423E	123A	3122			
C423F	123A	3122			
C424	123A	3122	61	TR21/735	
C424D	108	3039			
C425	128	3024	18	TR87/243	
C425B	128	3024			
C425C	128	3024			
C425D	123A	3122			
C425E	123A	3122			
C425F	123A	3122			
C426	128	3024	47	TR87/243	
C429	107	3039			
C429J	108	3039			
C429X	108	3039			
C430	108	3039			
C430H	108	3039			
C430W	108	3039			
C434					S7004
C440	123	3122			
C441	123	3122	210		
C442	123	3122	62		
C443	128	3024			
C444	123A	3122	61	TR21/735	S0011
C450	123A	3122	60	TR21/735	

Original Part Number	ECG	SK	GE	IR/WEP	HEP
C454		3122	20	TR21/	
C454A	123A	3122			
C454B	123A	3122			S0014
C454C	123A	3122			S0011
C454D	123A	3122			
C454L	123A	3122			
C454LA	123A	3122			
C455	108	3039			
C456	195	3048			
C456A	195	3048			
C456D	195	3048			
C458	123A	3122	20	TR21/735	S0015
C458A	123A	3122			
C458B	123A	3122			S0030
C458C	123A	3122			S0030
C458LG	123A	3122			S0014
C458LGB	123A	3122			S0014
C458LGC	123A	3122			S0014
C458LGD	123A	3122			S0014
C459	128	3024			
C459B	102A	3004			
C460		3122	20	TR21/	
C460A	107	3039			
C460B	107	3018	20		S0014
C460C	107	3039			S0014
C460D	107	3039			
C460G	107				
C460GB	107	3039			
C460H	107	3039			
C460K	107	3039			
C460L	107	3039			
C461	229	3028	17	TR95/	
C461A	229	3039			S0014
C461B	229	3039			S0014
C461C	229	3039			S0014
C461E	229	3039			
C461L	229	3039			
C463	107	3039			
C463H	161	3132			
C464	161	3132			
C464C	161	3132			
C465	161	3132			
C466	161	3132			
C466H	161	3132			
C468	123A	3122			
C468A	123A	3122			
C469	108	3039			
C469A	107	3039			
C469F	108	3039			
C469K	108	3039			
C469Q	108	3039			
C469R	108	3039			
C470	154	3044			
C472Y	154	3044			
C475	123A	3122			
C475K	123A	3122			
C476	123A	3122			
C477	108	3039			
C478	195	3048			
C478-4	195	3048			
C478D	195	3048			
C479	128	3024			
C481	195	3048			
C482	195	3048			
C482-GR	195	3048			
C482-O	195	3048			
C482-Y	195	3048			
C485	128	3024			
C485C	128	3024			
C485Y	128				
C486	128	3024			
C486Y	128	3024			
C487	175	3538			
C488	175	3538			
C489	175	3538			
C490	175	3538			
C491	175	3538			
C491BL	175	3538			
C491R	175	3538			
C491Y	175	3538			
C492	162				
C493	130	3027			
C493-BL	223				
C493-R	223				
C493-Y	223				
C494	130	3027			
C494-BL	223				
C494-R	223				
C494-Y	223				
C495	184	3190			
C495-O	184	3190			
C495-R	184	3190			
C495-Y	184	3190			
C496	184	3190			S5006
C496-O	184	3190		TR76/	
C496-R	184	3190			
C496-Y	184	3190			
C497	128	3024			
C497-O	128	3024			
C497-R	128	3024			
C497-Y	128	3024			
C498	128	3024			
C498-O	128	3024			
C498-R	128	3024			
C498-Y	128	3024			
C499	194	3104			
C499R	194	3104			
C499-R(FA-1)	154	3044			
C499Y	194				
C500	154	3044			
C500R	154	3044			
C500Y	154	3044			
C501	128	3024			
C502	129	3025		TR88/242	
C502(JAPAN)	128				
C503	128	3024			
C503GR	128				
C503-O	128	3024			
C503-Y	128	3024			
C504	128	3024			
C504GR	128	3024			
C504-O	128	3024			
C504-Y	128	3024			
C505	154	3044			
C505-O	154	3044			
C505-R	154	3044			
C506	154	3044			
C506-O	154	3044			
C506-R	154	3044			
C507	154	3044			
C507-O	154	3044			
C507-R	154	3044			
C507-Y	154	3044			
C508	175	3538			
C509	123A	3122			S0015
C512	128	3024			
C512-O	128	3024			
C512-R	128	3024			
C513	128	3024			
C513-O	128	3024			
C513R	128	3024			
C514	124	3021			
C515	124	3021			
C515A	124	3021			
C516	128	3024			
C517	224	3049			
C519	162				
C519A	162				
C520	130	3027			
C520A	130	3027			
C521	130	3027			
C521A	130	3027			
C522	225	3103			
C522-O	225	3103			
C522-R	225	3103			
C523	225	3103			
C523-O	225	3103			
C523-R	225	3103			
C524	225	3103			
C524-O	225	3103			
C524-R	225	3103			
C525	225	3103			
C525-O	225				
C525-R	225	3103			
C526	154	3044			
C529	123A	3122			
C529A	123A	3122			
C535	229	3039	20	TR70/720	S0016
C535A	229	3039			S0016
C535B	229	3039			S0016
C535C	229	3039			S0016
C535G	229	3039			
C536	123A	3122	20	TR21/735	
C536A	199				
C536B	199				
C536C	199				
C536D	199				
C536E	123A	3124	20	TR21/735	
C536E(JAPAN)	199				
C536F	123A	3124	20	TR21/735	
C536F(JAPAN)	199				
C536G	199				
C537	123A	3124	20	TR21/735	
C537-01	123A				
C537A	123A	3124		TR21/735	
C537B	123A	3124		TR21/735	
C537C	123A	3020		TR21/735	
C537F	123A	3020	20	TR21/735	
C537G	123A				
C537G	123A				

Original Part Number	ECG	SK	GE	IR/WEP	HEP
C538	123A	3024		TR24/	
C538A	123A	3024			
C538P	123A				
C538Q	123A				
C538R	123A	3122		TR21/735	
C538T	123A				
C539	123A	3122			
C539R	123A	3122			
C539S	123A	3122			
C540	123A	3122			
C544	108	3039			
C544C	108	3039			
C544D	108	3039			
C544E	108	3039			
C545	107	3039			
C545A	107	3039			
C545B	107	3039			
C545C	107	3039			
C545D	107	3039			
C545E	107	3039			
C555A	703A	3157			C6001
C558	163				
C560	128	3024			
C561	108	3039			
C562	161	3132			
C562Y	161	3132			
C563	161	3132			
C563A	161	3132			
C563A(3RD IF)	233				
C564	159				
C564P	159	3025			
C564Q	159	3025			
C566	123A	3122			
C567	161	3132			
C568	108	3039			
C580	128	3024			
C582	124	3021			
C582A	124	3021			
C582B	124				
C582C	124	3021			
C586	161				
C587	123A	3122			
C587A	123A	3122			
C588	123A	3122			
C589	154	3044			
C590	128	3024			
C594	128	3024			
C595	123A	3122			
C596	123A	3122			
C605	107	3039			
C605L	161				
C605M	161				
C605Q	107	3039			
C605TW	107	3039			
C606	107	3039			
C608		3024			
C608E	123A	3122			
C608T	224	3049			
C609	195	3024			
C609F	159				
C609T	224	3049			
C610	159	3025			
C611	108	3039			
C612	108	3039			
C613	108	3039			
C614	128	3024			
C614D	195	3048			
C614E	195	3048			
C614F	195	3048			
C614G	195	3048			
C615	195	3048			
C615A	195	3048			
C615B	195	3048			
C615C	195	3048			
C615D	195	3048			
C615E	195	3048			
C615F	195	3048			
C615G	195	3048			
C616		3048			
C616A	3048	3048			
C617C		3025			
C619	123A				
C619D	123A	3122			
C620	123A	3122			
C620C	123A	3122			
C620CD	123A	3122			
C620D	123A	3122			
C620DE	123A	3122			
C621	123A	3122			
C622	123A	3122			
C627	154	3044			
C628	128	3024			
C629	107	3039			
C631	123A	3122			
C631A	123A	3122			

Original Part Number	ECG	SK	GE	IR/WEP	HEP
C632	123A	3122			
C632A	123A	3122			
C633	123A	3122			
C633-7	123A	3122			
C633A	123A	3122			
C634	123A	3122			
C634A	123A	3122			
C635A	124	3021			
C636	187				
C640	123A	3122			
C640B	123A	3122			
C641	108	3039			
C641B	108	3039			
C642	164	3133			
C642A	164	3133			
C643A	165	3115			
C644	199	3020		TR95/56	S0015
C644C	123A	3020			
C644F	199				S0015
C644F/494		3124	62	TR24/	
C644FR	123A				
C644FS	123A				
C644R	199				S0015
C645	107	3039			
C645A	107	3039			
C645B	107	3039			
C645C	107	3039			
C646	223	3027			
C647	223	3027			
C647R	130	3027			
C648	199				
C648H	199				
C649	108	3039			
C650	108	3039			
C650B	108	3039			
C651	123A	3122		TR21/735	S3011
C652	108	3039		TR95/56	S0011
C654	123A	3122			
C655	123A	3122			
C656	107	3039			
C657	107	3039			
C658	108	3039			
C658A	108	3039			
C659	108	3039			
C660					F0015
C661					F0015
C662	108	3039			
C663	161	3132			
C664	162				F0015
C664B	130	3027			
C664C	130	3027			
C665	280				F0015
C665H	280				
C665HA	280				
C668		3039			
C668-O	108				
C668C1	107	3039			
C668DE	108				
C668EP	107	3039			
C668EX	107	3039			
C668F	107	3039			
C670	124				
C673	234	3122		/801	F0010
C673C2	159				
C673D	159				
C674	133	3112		/801	F0010
C674B	161	3132			
C674C	161	3132			
C674CV	161				
C674D	161	3132			
C674E	161	3132			
C674F	161	3132			
C674G	161	3132			
C674(JAPAN)	167				
C680	175	3538			
C680A	175	3538			
C680R	175	3538			
C681	163				
C681A	163				
C682	161	3132			
C682A	161	3132			
C682B	161	3132			
C683	161	3132			
C683A	161	3132			
C683B	161	3132			S0016
C683V	108				
C684	133	3112		/801	F0010
C684A	133	3112		/801	F0010
C684A(JAPAN)	107				
C684B	108				
C684BK	107	3039			
C684F	107				
C685	124	3112		/801	F0010
C685A	124	3112		/801	
C685B	124	3021			

Original Part Number	ECG	SK	GE	IR/WEP	HEP
C685P	124	3021			
C685S	124				
C685Y	124	3021			
C686	154	3044			
C686-248-0	159				
C687	162				
C688	108	3039			
C689	123A	3122			
C689H	123A	3122			
C693	123A	3020		TR21/735	
C693A	123A	3020		TR21/735	
C693B	123A	3020		TR21/735	
C693C	123A	3020		TR21/735	
C693D	123A	3020		TR21/735	
C693E	123A	3020	20	TR21/735	
C693E(JAPAN)	199				
C693EB	199				
C693F	199				
C693FU	199				
C693G	123A	3020	20	TR21/735	
C693G(JAPAN)	199				
C693GS	199				
C693GZ	199				
C694	199				
C694D	123A	3020		TR21/735	
C694E	199				
C694F	199				
C694G	199				
C694Z	199				
C695	107	3039			
C696	128	3024			
C696A	128				
C696B	128				
C696C	128				
C696D	128	3024			
C696E	128	3024			
C696F	128	3024			
C697	225	3103			
C697A	225	3103			
C697F	224	3049			
C701	123	3122			
C702	123	3122			
C705	107	3039			
C705B	107	3039			
C705C	107	3039			
C705D	107	3039			
C705E	107	3039			
C705F	107	3039			
C705TV	107				
C707	107	3039			
C707H	107	3039			
C708	128	3024			
C708A	128	3024			
C708AA	128	3024			
C708AB	128	3024			
C708AC	128	3024			
C708B	128	3024			
C708C	128	3024			
C709B	123A	3122			
C709C	123A	3122			
C709CD	123A	3122			
C709D	123A	3122			
C710	123A	3122			
C710B	123A	3122			
C710BC	123A	3122			
C710C	123A	3122			
C710D	123A	3122			
C710E	123A	3122			
C711	123A	3122			
C711A	123A	3122			
C711D	199				
C711F	128	3024			
C711FG	123A				
C711G	123A	3122			
C712	123A	3122			
C712A	123A	3122			
C712C	123A	3122			
C712D	123A	3122			
C712E	123A	3122			
C712W	123A	3122			
C713	123A	3122			
C714	123A	3122			
C715	123A	3122	20	TR21/735	
C715(JAPAN)	199				
C715A	199				
C715B	123A	3122			
C715C	123A	3122			
C715D	123A	3122			
C715E	199				
C715EJ	199				
C715EV	107	3039			
C715F	199				
C715XL	199				
C716	107	3039			
C716B	123A	3122			
C716B	123A	3122			
C716C	123A	3122			
C716D	123A	3122			
C716E	107	3039			
C716F	199				
C716G	199				
C717	233	3039			
C717B	107	3039			
C717BK	107	3039			
C717BLK	107	3039			
C717C	161				
C717E	161				
C717(FINAL IF)	233				
C720			20		S0011
C722			17		S0011
C727	154	3044			
C728	154	3044			
C731R	128				
C732	199				
C732BL	199				
C732GR	199				
C732S	199				
C732V	199				
C732Y	199				
C733	123A	3122			
C733-0	123A	3122			
C733BL	199				
C733GR	199				
C733Y	123A	3122			
C734	128	3024			S5026
C734GR	128	3024			
C734-O	128	3024			
C734-R	128	3024			
C734-Y	128	3024			
C735	123A	3122			S0014
C735F	123A				
C735(FA-3)	108				
C735H	123A				
C735J	123A				
C735K	123A				
C735L	123A				
C735-O	123	3122			
C735-R	123A	3122			
C735Y	123A	3122			
C736	130	3027			
C738	107	3039			
C738C	107	3039			
C739	107	3039			
C740	108	3039			S0015
C741	123A	3122			
C742			20		S3013
C743A	154	3044			
C744			63		
C746A	154	3044			
C748	108	3039			
C752	123A	3122			
C752G	123A	3122			
C756		3024		TR87/	
C756-2-4	282				
C760			17		
C761	161	3132			
C761Y	108	3039			
C762	161	3132	20		S0011
C763	107	3039			
C763B	107	3039			
C763C	107	3039			
C763D	107	3039			
C764	133	3112	27	/801	F0010
C765	130	3027			
C768	130	3027			
C772	107	3039			
C772B	107	3039			S0016
C772BG	107	3039			
C772BH	107	3039			
C772BV	107	3039			
C772BX	107	3039			
C772C	107	3039			S0016
C772C1	107	3039			
C772C2	107	3039			
C772CK	107	3039			
C772CL	107	3039			
C772CS	101	3039			
C772CU	107	3039			
C772CV	107	3039			
C772CX	107	3039			
C772D	107	3039			
C772DJ	108	3039			
C772DU	107	3039			
C772DV	107	3039			
C772DX	107	3039			
C772DY	107	3039			
C772E	107	3039			
C772F	107	3039			
C772K	107	3039			
C772R	123A	3122	20	TR21/735	

Original Part Number	ECG	SK	GE	IR/WEP	HEP
C772R(JAPAN)	107				
C773	123A	3122			
C774	128	3024			
C775	195	3024			
C776	128	3024			
C776Y	224				
C777	224				
C778	224	3049			
C778B	236				
C780	154				
C780AG	154	3044			
C780AG-O	194	3104			
C780AG-R	194	3104			
C780AG-Y	194	3104			
C780G	154	3049			
C781	195	3048			
C782	124				
C783	198				
C784	107	3039			
C784A	107				
C784BN	107	3039			
C784-O	107	3039			
C784R	107	3039			
C784Y	107	3039			
C785	107	3039			
C785BN	107	3039			
C785D	107	3039			
C785E	108	3039			
C785R	107	3039			
C786	161	3132			
C787	161	3132			
C788	154	3044			
C789	107	3039			
C789-O	152	3054			
C789-R	152	3054			
C789-Y	152	3054			
C791	175	3538			
C793	130	3027			
C793BL	130	3027			
C793R	130	3027			
C793Y	130	3027			
C794R	130	3027			
C795	124	3021			
C795A	124	3021			
C796	123A	3122			
C797	128	3024			
C798	128	3024			
C799	236	3048			
C800	107	3039			
C803	128	3024			
C805	154	3044			
C806	162				
C806A	163				
C807	163				
C807A	163				
C814	192	3137			
C815	123A	3137			
C815A	123A	3137			
C815B	123A	3137			
C815C	123A	3137			
C815K	123A				
C815L	123A				
C815M	123A				
C815S	123A	3137			
C815SA	123A	3137			
C815SC	123A	3137			
C816	128	3024			
C826K	128	3024			
C818	154	3044			
C825			20		
C826	128	3024			
C827	128	3024			
C828	123A	3020		TR21/735	S0015
C828A	123A	3020			
C828E	123A				
C828-P	123A				S0015
C828Q	123A	3122		TR21/735	S0015
C828R	123A			/735	S0015
C828R/494		3122	20	TR24/	
C829	123A	3018		TR70/720	S0011
C829B	123A	3039			S0011
C829C	123A	3039			S0011
C829Y	123A	3039			
C830	175	3538			
C830A	175	3538			
C830B	175	3538			
C830C	175	3538			
C835	107				
C836K		3132			
C836L		3132			
C836M	123A	3122			
C837F	107				
C837K	161	3132			
C837L	161	3132			
C837WF	123A				

Original Part Number	ECG	SK	GE	IR/WEP	HEP
C838	199	3137			S0015
C838A	199	3137			
C838B	199	3137			
C838C	199	3137			
C838F	192				
C838H	199	3122			
C838J	107				
C838K	199	3122			
C838L	199	3137			
C838M	199	3137			
C838R	123A			TR21/735	
C839	123A	3137		TR21/	
C839A	123A	3137			
C839M	123A	3039			
C840	175	3538			
C840A	175	3538			
C840AC	124	3021			
C844	123A	3122			
C847	123A	3122			
C848	123A	3122			
C849	123A	3122			
C850	123A	3122			
C851	130	3082			
C853	192	3137			
C853A	192	3137			
C853B	192	3137			
C853C	192	3137			
C853KLM	123A	3122			
C853L	123A	3122			
C856	194	3104			
C856-02	194	3044			
C856C	194	3104			
C857	154	3044			
C857H	154	3044			
C858	199				
C858G	199				
C859	199				
C859F	199				
C859FG	199				
C859G	199				
C859GK	123A	3122			
C860	161	3132			
C860C	161	3132			
C860D	161	3132			
C860E	161	3132			
C863	161	3132			
C864	161	3132			
C867	124	3035			
C868	154	3044			
C869	154	3044			
C870	123A	3122			
C870BL	123A	3122			
C870E	161	3132			
C870F	123A	3122			
C871	123A	3122			
C871BL	123A	3122			
C871E	123A	3122			
C871F	123A				
C875	128	3024			
C875-1	191	3024			
C875-1C	191	3024			
C875-1D	191	3024			
C875-1E	191	3024			
C875-1F	191	3024			
C875-2	191	3024			
C875-2C	191	3024			
C875-2D	191	3024			
C875-2E	191	3024			
C875-2F	191	3024			
C875-3	191	3024			
C875-3C	191	3024			
C875-3D	191	3024			
C875-3E	191	3024			
C875-3F	191	3024			
C875BR	107				
C875D	128	3024			
C875E	128	3024			
C875F	128	3024			
C876	128	3024			
C876C	128	3024			
C876D	128	3024			
C876E	128	3024			
C876F	128	3024			
C876TV	128	3024			
C876TVE	128	3024			
C876TVEF	128	3024			
C881	192	3137			
C881A	192	3137			
C881B	192	3137			
C881C	192	3137			
C881D	192	3137			
C881K	192	3137			
C881L	192				
C889	162				
C894	123A	3122			

Original Part Number	ECG	SK	GE	IR/WEP	HEP
C895	175	3538			
C896	123A	3122			
C897	162				
C897A	162				
C897B	162				
C898	162				
C899	123A	3122			S0025
C899K	123A	3122			
C900	199				
C900A	199				
C900E	199				
C900U	199				
C901	163				
C901A	163				
C907	199				
C907A	199				
C907AC	199				
C907AD	199				
C907AH	123A				
C907C	199				
C907D	199				
C907H	123A	3122			
C907HA	123A	3122			
C912	108	3039			
C913	123A	3122			
C917	123A	3122			
C917K	161	3132			
C918	161	3132			
C920	107	3039			
C920Q	108	3039			
C920R	107	3039			
C921	107	3039			
C921C1	107	3039			
C921L	107	3039			
C921M	107	3039			
C922	107	3039			
C922A	107	3039			
C922B	107	3039			
C922C	107	3039			
C923	199				
C923A	199				
C923B	199				
C923C	199				
C923D	199				
C923E	123A				
C923F	123A				
C924	108	3039			
C924E	108	3039			
C924F	108	3039			
C924M	123A	3122			
C925	123A	3122			
C926	154	3044			
C926A	194	3044			
C927	161	3132			
C927B	161	3132			
C927C	107	3132			
C927CU	161	3132			
C927CW	161	3132			
C927D	108	3039			
C927E	161	3132			
C928	161	3132			
C928B	161	3132			
C928C	161	3132			
C928D	161	3132			
C928E	161	3132			
C929	199				
C929-O	107	3039			
C929B	199				
C929C	199				
C929D	199				
C929DE	107	3039			
C929DP	199				
C929E	199				
C929F	199				
C930	199				
C930B	199				
C930BK	199				
C930C	107				
C930CK	199				
C930CS	199				
C930D	199				
C930DS	199				
C930DT	107	3039			
C930DT-2	199				
C930E	199				
C930F	199				
C931	152	3054			
C931D	186	3192			
C931E	188				
C932	152	3054			
C932E	152	3054			
C933	123A	3122			
C933C	123A	3122			
C933G	123A	3122			
C934C	123A	3122			

Original Part Number	ECG	SK	GE	IR/WEP	HEP
C934D	123A	3122			
C934E	123A	3122			
C934F	123A	3122			
C934G	123A	3122			
C935	162				
C936	164	3133			
C937	165	3115			
C937-01	123A	3122			
C937A	163				
C937B	102A	3004			
C938	123A	3122			
C938A	123A	3122			
C938B	123A	3122			
C938C	123A	3122			
C939	163				
C939L	163				
C940	165	3115			
C940L	165				
C940M	165				
C941	199				
C941-O	199				
C941-Q	123A				
C941-R	199				
C941-Y	199				
C943	123A	3122			
C943A	123A	3122			
C943B	123A	3122			
C943C	123A	3122			
C945	199				S0015
C935A	199				
C945B	199				
C945C	199				
C945QL	123A				
C945R	199				
C945S	123A				
C945X	199				
C947	161	3132			
C948	161	3132			
C957	161	3132			
C959	128	3024			
C959A	128	3024			
C959B	128	3024			
C959C	128	3024			
C959S	128	3024			
C959SD	128	3024			
C960	123A	3122	8	TR21/735	
C960A	225	3103			
C960B	225	3103			
C960S	225	3103			
C966	123A	3122			
C967	123A	3122			
C968	123	3122			
C968P	128	3024			
C971	192	3137			
C972	128	3024			
C972C	128	3024			
C972D	128	3024			
C972E	128	3024			
C982	172	3156			
C983	171	3201			
C983-O	171	3201			
C983-Q		3103			
C983R	171	3201			
C983Y	171	3201			
C984	123A	3122			
C991	123A	3122			
C992	123A	3122			
C995	154	3044			
C996	225	3103			
C997	161	3132			
C999	165	3115			
C999A	165	3115			
C1000	199				
C1000-BL	199				
C1000-GR	199				
C1000Y	123A				
C1001			20		S0011
C1002			212		S0024
C1003			81		S0015
C1004	164	3133	61		S0015
C1004A	164	3133			
C1005	165	3115			
C1005A	165	3115			
C1006	199				
C1006A	199				
C1007	123A	3122			
C1008	128	3024			
C1010	199				
C1012	154	3044			
C1012A	154	3044			
C1013	210	3202			
C1014	210	3202			
C1014B	210	3202			
C1014C	186	3192			
C1014CD	210	3202			

Original Part Number	ECG	SK	GE	IR/WEP	HEP
C1014D1	210	3202			
C1017	235	3197			
C1023	108	3039		TR95/56	
C1023-O	107	3039			
C1023-Y	107	3039			
C1024	175	3538			
C1024B	175	3538			
C1024C	175	3538			
C1024D	175	3538			
C1024E	175	3538			
C1024F	175	3538			
C1025	175	3538			
C1026	107	3039			
C1027	162				
C1030	130	3027			
C1030A	130	3027			
C1030B	130	3027			
C1030C	130	3027			
C1030D	130	3027			
C1030P	9930				
C1032	107	3039			
C1032P	9932				
C1032Y	107	3039			
C1033	123A	3122			
C1033A	123A	3122			
C1033P	9933				
C1034	164	3133			
C1035	161	3132			
C1035C	161	3132			
C1035D	161	3132			
C1035E	161	3132			
C1035P	9935				
C1036	161	3132			
C1036P	9936				
C1044P	9944				
C1045	164	3133			
C1045B	164	3133			
C1045C	164	3133			
C1045D	164	3133			
C1045E	164	3133			
C1045P	9945				
C1045R	164	3133			
C1046	165	3115			
C1046P	9946				
C1047	108	3039			
C1047B	107	3039			
C1047C	108	3039			
C1047D	108	3039			
C1048	154	3044			
C1048B	154	3044			
C1048C	154	3044			
C1048D	154	3044			
C1048E	154	3044			
C1048F	154	3044			
C1052P	9099				
C1053P	9093				
C1055H	218				
C1056	154	3044			
C1057P	9157				
C1058P	9158				
C1059	124	3021			
C1060	152	3054			
C1060A	152	3054			
C1060B	152	3054			
C106C	152	3054			
C1060D	152	3054			
C1061	152	3054			
C1061A	152	3054			S5000
C1061B	152	3054			S5000
C1061T	152				
C1062P	9962				
C1070	161				
C1071	123A	3122			
C1072	128	3024			
C1072A	128	3024			
C1079	181	3535			
C1080	181	3535			
C1086	165	3115			
C1096	186	3192		TR76/	
C1096M	186	3192			
C1098	186	3192			
C1098M	186	3192			
C1100	165	3115			
C1101	164	3133			
C1101A	164	3133			
C1101L	164				
C1102	124	3021			
C1102A	124	3021			
C1102B	124	3021			
C1102C	124	3021			
C1102M	124				
C1103	154	3103			
C1103A	154	3103			
C1103B	154	3103			
C1103L	154	3103			

Original Part Number	ECG	SK	GE	IR/WEP	HEP
C1104	124	3021			
C1104A	124	3021			
C1104B	124	3021			
C1104C	124	3021			
C1105	124	3021			
C1105A	124	3021			
C1105B	124	3021			
C1105M	124				
C1106	162				
C1106M	162				
C1123	107				
C1124	198	3103			
C1126	107				
C1127	190	3104			
C1128	161	3132			
C1128(3RD IF)	233				
C1128(S)	161				
C1129	161	3132			
C1129(R)	161				
C1151	164	3133			
C1152	162				
C1152F	130	3027			
C1154	163				
C1155	190	3104			
C1156	190	3104			
C1157	190	3104			
C1158	190	3039			
C1159	107				
C1160	175	3538			
C1160K	124	3021			
C1160L	175				
C1161	175	3538			
C1162	184	3190			
C1162A	184	3190			
C1162B	184	3190			
C1162C	184	3190			
C1162CP	184	3190			
C1162D	184	3190			
C1167	165				
C1168	124				
C1170	165	3115			
C1170A	165	3115			
C1171	165	3115			
C1172	165	3115			
C1172A	165	3115			
C1172B	238				
C1173	152	3054			
C1173C	152	3054			
C1173R	152	3054			
C1174	165	3115			
C1175	128	3024			
C1175C	128	3024			
C1175F	128	3024			
C1182B	161	3132			
C1182C	161	3132			
C1182D	161	3132			
C1184	164	3133			
C1184A	164	3133			
C1184B	164	3133			
C1184E	164	3133			
C1185	162				
C1185M	162				
C1187	161	3132			
C1204	199				
C1204D	199				
C1205	108	3039			
C1205A	108	3039			
C1205B	108	3039			
C1205C	108	3039			
C1209	192				
C1209C	128				
C1210	123A				
C1211	123A	3122			
C1212	184	3190			
C1212A	184	3190			
C1212AD	184	3190			
C1212B	184	3190			
C1212C	184	3190			
C1212D	184	3190			
C1213	123A	3122			
C1213A	123A	3122			
C1213AD	123A	3122			
C1213B	128	3024			
C1213C	128	3024			
C1213D	123A	3122			
C1214	128	3024			
C1214A	128	3024			
C1214B	128	3024			
C1214C	128	3024			
C1215	199				
C1217	171	3201			
C1218	128	3024			
C1220E	128	3024			
C1222	199	3117			S0015
C1226	186	3192			

Original Part Number	ECG	SK	GE	IR/WEP	HEP
C1226A	186	3192			
C1226AP	152				
C1226C	186	3192			
C1226F	186	3192			
C1226P	186	3192			
C1226Q	152				
C1235	124				
C1237	236				
C1239		3048			
C1244	123A	3122			
C1278	194	3104			
C1278S	194	3104			
C1279	194	3104			
C1279S	194	3104			
C1280	172	3156			
C1280AS	172	3156			
C1280S	172	3156			
C1304	124	3021			
C1306	235				
C1307	236				
C1308K	165				
C1317	128	3024			
C1317B	123A				
C1317P	123A				
C1317Q	128				
C1317R	128				
C1318		3024			
C1318R	123A				
C1318S	123A				
C1325	165	3115			
C1327	199				
C1327U	199				
C1328	199				
C1328U	199				
C1330	192	3137			
C1330A	192	3137			
C1330B	192	3137			
C1330C	192	3137			
C1303D	192	3137			
C1335	199				
C1335A	199				
C1335F	199				
C1342	107	3039			
C1342A	107	3039			
C1342B	107	3039			
C1342C	107	3039			
C1344	199				
C1344C	199				
C1344F	199				
C1345	199				
C1346	192	3137			
C1347	192	3137			
C1358	165	3115			
C1359	107	3039			
C1360	107	3039			
C1362	123A				
C1363	123A				
C1364	123A	3122			
C1368	184				
C1372Y	123				
C1383	188	3054		TR60/	
C1383Q	128				
C1383R	128	3024			
C1383R/494		3024	63	TR65/	
C1384	188	3054			
C1384R	128	3024			
C1390A		3018	20		S0015
C1390J	123A				
C1390X	123A				
C1398	152	3054			
C1398Q	186	3192			
C1417C		3018	61		S0030
C1417H	107				
C1417W	108				
C1437	100	3004	2	TR05/254	G0005
C1438	100	3004	2	TR05/254	G0005
C1446	190				
C1446Q	190				
C1447R	198				
C1475	195				
C1537S	123A				
C1538	123A				
C2538-11	123A				
C3000P	7400				
C3001P	7401				
C3002P	7402				
C3004P	7404				
C3010P	7410				
C3020P	7420				
C3030P	7430				
C3040P	7440				
C3041P	7441				
C3050P	7450				
C3071					C6071F
C3073P	7473				

Original Part Number	ECG	SK	GE	IR/WEP	HEP
C3075P	7475				
C3123	108				
C3800P	7490				
C3801P	7492				
C5359	108				
C6001	760	3157	IC12	IC500/	
C6002	774				
C6003	752				
C6004	771				
C6007	754				
C6008	769				
C6009	766				
C6010	753				
C6011	761				
C6012	762				
C6013	755				
C6014	756				
C6015	767				
C6017	773				
C6019	754				
C6052P	941M	3553			
C6055L	725	3162			
C6056P	722	3161		IC513/	
C6057P	790	3077	IC5	IC508/	
C6059P	746				
C6060P	748	3072		IC507/	
C6061P	750				
C6062P	708	3135		IC505/	
C6063P	712	3072		IC507/	
C6065P	719		IC28		
C6066P	798				
C6067G	707				
C6068P	720				
C6069G	780	3070			
C6070P	714	3075			
C6071P	715	3076			
C6072P	739				
C6074P	718	3159			
C6075P	738	3167			
C6076P	749	3168			
C6077P	782				
C6079P	747				
C6080P	721				
C6081P	779				
C6082P	708				
C6083P	712	3072			
C6085P	731	3170			
C6089	705A	3134			
C6090	732				
C6099P	795				
C6100P	783				
C6101P	723				
C6102P	778				
C6103P	909				
C6644H	199				
C6690		3112			
C6691		3112			
C6692		3112			
C7076			212		
C7335-BL	123A				
C7350RN	123A				
C9001					S0013
C9002					S0013
C9003					S0013
C9080	129		82	TR88/242	S0012
C9081	129		82	TR88/242	S0012
C9082	159	3114	82	TR20/717	S0013
C9083	159	3114	82	TR20/717	S0013
C9084	159	3114	82	TR20/717	
C9085	159	3114	82	TR20/717	S0013
C10215-2	160	3006		TR17/637	
C10227	102A	3004	2	TR85/250	
C10230-3	102A	3004	2	TR85/250	
C10258	160	3006		TR17/737	
C10260	160	3006		TR17/637	
C10261	160	3006		TR17/637	
C10262	160	3006		TR17/637	
C10279-1	123A	3020	20	TR21/735	
C10279-3	123A	3020	20	TR21/735	
C10291	126	3008		TR17/635	
C11021	102A	3004	2	TR85/250	
C13704		3024			
C13705		3024			
C13706		3024			
C13901		3124	20		S0033
C36566	223				
C36577	129	3025	21	TR88/242	S0019
C36578	161	3039	20	TR83/56	S0020
C36579	128	3024	63	TR87/243	
C36580	123A	3020	20	TR21/735	
C36583	152				
C44189	910				
C59012HE		3114			
C70501J		3074			
C610034 GRP1		3004			

Original Part Number	ECG	SK	GE	IR/WEP	HEP
C610034-1		3004			
C610035		3004			
C610036 GRP1		3003			
C610036-1		3004			
C610036-2		3004			
C610036 GRP2		3003			
C610040-1		3004			
C610040-2		3004			
C610040-2A		3004			
C610043-5		3004			
C-750517-72AB		3008			
C2475078-3	108			TR95/56	
C2485076-3	128	3024		TR87/243	S5014
C2485077-2	128	3024		TR87/243	S5014
C2485078		3018		TR21/	
C2485078-1	108	3018		TR95/56	S0011
C2485079-1	108	3018		TR95/56	S0015
C6862400	123A				
C24850772				TR25/	
C532000585	234				
CA2D2	102A	3004		TR85/250	G6011
CA344V1	711				
CA555CE	955M				
CA741CS	941M				
CA741CT	941				
CA747CE	947D				
CA747CT	947				
CA758E	743				
CA810M	1115				
CA810Q	1115	3184			
CA1310E	801				
CA1352E	749				
CA1398E	738				
CA1458S	778				
CA2111AE	708	3135			
CA2111AQ	708	3135			
CA3000	900	3547			
CA3001	901	3549			
CA3010	903	3540			
CA3011	726	3129			
CA3012	726				
CA3013	704	3023		IC501/	
CA3014	704	3022		IC501/	
CA3015	903	3540			
CA3018	904	3542			
CA3019	905	3546			
CA3020	784	3524			
CA3020B	784				
CA3026	906	3548			
CA3028	724	3525			
CA3028A	724	3525			
CA3028B	724	3525			
CA3029	908	3539			
CA3030	908	3539			
CA3035	785				
CA3035V	785				
CA3035VI	785				
CA3037	908	3539			
CA3037A	908	3539			
CA3038	908	3539			
CA3038A	908	3539			
CA3039	907	3545			
CA3041	706	3101		IC507/	
CA3042	710	3102		IC506/	
CA3043	786	3140			
CA3044	711	3070		IC514/	C6069G
CA3044VI	711	3070		IC514/	C6069G
CA3045	912	3543			
CA3045F	912				
CA3045L		3543			
CA3046	912	3543			
CA3047	913	3514			
CA3048	727	3071			
CA3049	906	3548			
CA3052	727	3071			
CA3053	724				C6001
CA3054	917	3544			
CA3056A	913	3514			
CA3058	914	3541			
CA3059	914	3541			
CA3064	780	3141	IC20		C6069G
CA3064/5A	780				C6069G
CA3065	712	3072	IC2	IC507/	C6083P
CA3065/7F	712				C6083P
CA3066	728	3073	IC22		
CA3067	729	3074	IC23		
CA3068	730	3143			
CA3070	714	3075	IC4	IC509/	C6070P
CA3071	715	3076	IC6	IC510/	C6071P
CA3072	713	3077	IC5	IC508/	C6057P
CA3075	723	3168	IC15		
CA3076	781	3169			
CA3078AS		3566			
CA3079	914	3541			
CA3081	916	3550			

Original Part Number	ECG	SK	GE	IR/WEP	HEP
CA3086	912	3543			
CA3088E	787	3146			
CA3089E	788	3147			
CA3090		3078			
CA3090E	789	3078			
CA3120E	731	3172			
CA3121		3149	IC33		
CA3121E	791	3149			
CA3123E	744				
CA3125E	798				
CA3126Q	797				
CA3130		3568			
CA3741	913	3514			
CA3147CT	947	3526			
CA5053					C6001
CA-9011H	123A				
CA30900	789				
CAC5028A			11		
CB1F4	105	3012	4	TR03/105	G6004
CB102					G0009
CB103	126		1		
CB104		3004		TR85/	G0005
CB156	160	3007		TR17/637	
CB157	126	3008		TR17/635	
CB158	126	3008		TR17/635	
CB161	102A	3004		TR85/250	
CB244	160	3007		TR17/637	
CB246	123A	3046		TR21/735	
CB248	102A	3004	TR85/250		
CB249	102A	3004		TR85/250	
CB254	160	3007		TR17/637	
CC1168F	123A	3124	10	TR25/	
CC59017G925				TR21/	
CCA2001H					S0015
CCA2006D					S0014
CCA2008F015					S0014
CCC2027		3024			
CCS1235G	123A	3122		TR21/735	
CCS2001H	123A	3020		TR21/735	
CCS2004B	199	3018	62	TR51/	S0024
CCS2004D303	199	3124			
CCS2005B		3114		TR30/	
CCS2006D	108	3018	17	TR95/	
CCS2008F015	108	3018	20	TR95/	
CCS4004		3018		TR24/	S0015
CCS6168	123A	3122		TR21/735	S0015
CCS6168F	123A	3122		TR21/735	S0015
CCS6168G	128	3024		TR87/243	S0015
CCS6225F	161	3117		TR83/719	S0015
CCS6226G	161	3117		TR83/719	S0015
CCS6227F	161	3117		TR83/719	S0015
CCS6228F	159				S5002
CCS6229H		3020			
CCS9015		3114			
CCS9016D	108	3018	20	TR24/	S0014
CCS9016E		3039			S0015
CCS9017	161	3117		TR83/719	S0015
CCS9017G925	161	3117		TR83/719	S0011
CCS9018E		3018			S0015
CSS9018H924	161	3117		TR83/719	S0016
CCS-200801		3018			
CCS-200802		3018			
CCS-200803		3018			
CCS-200804		3018			
CCS-200805		3018			
CD0000		3122		/735	
CD0014			10	/735	S0014
CD0014NA	123A	3020	20	TR21/735	
CD0014NG	123A	3122		TR21/735	
CD0015N	123A	3122		TR21/735	
CD0021	123A	3020	18	TR21/735	S0014
CD-0071		3126	90		
CD38	123A				
CD73/187/72	7493				
CD73/187/73	74H30				
CD91					S0019(2)
CD92					S0019(2)
CD93					S0019(2)
CD94					S0019(2)
CD95					S0019(2)
CD96					S0019(2)
CD97					S0019(2)
CD98					S0019(2)
CD103		3006			
CD437	159				
CD441	199				
CD445	159				
CD446	123A				
CD461	130				
CD461-014-614	130				
CD500	234				
CD522	218				
CD541	909	3590			
CD562	199				
CD912					S0019

Original Part Number	ECG	SK	GE	IR/WEP	HEP
CD922					S0019
CD932					S0019
CD942					S0019
CD952					S0019
CD962					S0019
CD972					S0019
CD982					S0019
CD2300E	9930				
CD2300E/830	9930				C1030P
CD2301E	9961				
CD2301E/861	9961				
CD2302E	9946				
CD2302E/846	9946				
CD2303E	9949				
CD2303E/849	9949				
CD2304E	9945				
CD2304E/845	9945				C1045P
CD2305E	9948				
CD2305E/848	9948				
CD2306E	9932				
CD2306E/832	9932				C1032P
CD2307E	9944				
CD2307E/844	9944				C1044P
CD2308E	9962				
CD2308E/862	9962				C1062P
CD2309E	9963				
CD2309E/863	9963				
CD2310E	9936				
CD2310E/836	9936				
CD2311E/837	9937				
CD2312E	9935				
CD2314E	9933				
CD2314E/833	9933				C1033P
CD2315E	9093				
CD2318E	9099				
CD4000AE	4000	4000			
CD4000AZ	4000	4000			
CD4001AE	4001	4001			
CD4002AE	4002	4002			
CD4007AE		4007			
CD4009AE	4049	4009			
CD4010AE		4010			
CD4011AE	4011	4011			
CD4012AE	4012	4012			
CD4013AE	4013	4013			
CD4014AE	4014				
CD4015AE	4015	4015			
CD4016AE	4016	4016			
CD4017AE	4017	4017			
CD4019AE	4019	4019			
CD4020AE	4020	4020			
CD4021AE	4021				
CD4023AE	4023	4023			
CD4024AE	4024	4024			
CD4025AE	4025	4025			
CD4027AE	4027	4027			
CD4028AE	4028				
CD4029AE	4029				
CD4030AE	4030	4030			
CD4040AE	4040				
CD4042AE	4042				
CD4049AE	4049				
CD4050AE	4050				
CD4051AE	4051				
CD4052AE	4052				
CD4055AE	4055				
CD4068BE	4068				
CD4069BE	4069				
CD4070BE	4030				
CD4071BE	4071				
CD4077BE	4077				
CD4078BE	4078				
CD4081BE	4081				
CD4085BE	4085				
CD4086BE	4086				
CD4099BE	4099				
CD4518B	4518				
CD4520B	4520				
CD4527BE	4527				
CD4555BE	4555				
CD4556BE	4556				
CD5328	9910				
CD6019	123A				
CD6039	9914				
CD6150	123A				
CD6153	128				
CD6153-2	128				
CD6157	123A				
CD6375	123A				
CD-6439	911				
CD8000	123A	3020	17	TR21/735	S0015
CD9525	123A				
CD10000-1E	159				
CD12000	123A	3020	17	TR21/735	
CD40192BE	40192				

Original Part Number	ECG	SK	GE	IR/WEP	HEP
CD40193BE	40193				
CD40194BE	40194				
CDC86X7-5	123A	3020		TR21/735	S0020
CDC-102		3007			
CDC-103		3007			
CDC-104		3007			
CDC430	123A	3124	20	TR24/	S0024
CDC496	159	3025	21	TR20/717	S0019
CDC587				TR25/	S0015
CDC731	132	FET	FET1	FE100/802	F0015
CDC744	154	3045		TR78/712	S3019
CDC745(ZENTH)	123A			TR21/735	S0020
CDC746	123A	3122		TR21/735	S0012
CDC746(ZENTH)	159			TR20/	
CDC1201BC	123A	3122		TR21/735	S5014
CDC1203B				TR21/	S0011
CDC1230B		3018			S0015
CDC2010	123A	3124	20	TR21/	S0015
CDC2010C	123A			TR21/	S0015
CDC2010D	123A	3124	20	TR21/	S0015
CDC2510C-G	123A	3124	20	TR21/	S0015
CDC4023A130	123A	3122		TR21/735	S0014
CDC5000	107	3018		TR70/720	S0015
CDC5000-18	108	3018		TR95/56	S0016
CDC5008	128	3024		TR87/243	
CDC5028A	128	3024		TR87/243	S0016
CDC5038A	108	3039		TR95/56	
CDC5071A	108			TR95/	
CDC5075B	108			TR95/	
CDC8000	123A	3020	17	TR21/735	S3002
CDC8000-1	128	3020	17	TR87/243	S5014
CDC8000-1B	123A	3020	17	TR21/	S0015
CDC8000-1C	128		63	TR87/243	S5014
CDC8000-CM	128	3020		TR87/243	S5014
CDC8001	123A	3124		TR21/735	
CDC8002	128	3020	18	TR25/	S5014
CDC8002-1	128	3020		TR87/243	S5014
CDC8011		3024			
CDC8011B	123A	3122	20	TR21/735	S5014
CDC8021	123A	3122	10	TR21/735	
CDC8054	123A	3122		TR21/735	S0025
CDC9000-1	159	3114		TR20/717	
CDC9000-1B	129	3025	22	TR88/242	S5013
CDC9000-1C			67	TR88/242	S5013
CDC9002	129	3025		TR28/	S5013
CDC9002-1C	128	3024		TR87/243	S3019
CDC9002-18	128	3020		TR87/243	S5014
CDC10000-1E		3114	22		S0013
CDC12000-1C	123A	3122		TR21/735	S0015
CDC12018C	123A	3122		TR21/735	S5014
CDC12030B	107	3018		TR70/720	S5014
CDC12077F	123A	3122	20	TR21/735	S0015
CDC12108		3020	20		
CDC12112C	108			TR95/	
CDC12112D	108			TR95/	
CDC12112E	108			TR95/	
CDC12112F	108			TR95/	
CDC13000		3124		TR25/	S0015
CDC13000-1	123A	3020	17	TR21/735	S5014
CDC13000-1B	123A	3059		TR21/735	S5014
CDC13000-1C	123A		18	TR21/735	S0015
CDC13000-1D	123A	3020		TR21/735	S0015
CDC13000-18	123A	3122	20	TR21/735	S0011
CDC13016A	123A	3122	20	TR21/735	
CDC13019B			20	TR24/	S0015
CDC13500-1	123A	3020		TR21/735	S5014
CDC25100-6	123A	3124	20	TR21/	S0015
CDC120700	128			TR87/243	
CDC120770	3020				S5014
CDC4306813	123A	3020		TR21/735	
CDQ1004	154			TR78/	
CDQ1018			20		
CDQ1021			20		
CDQ1024			20		
CDQ10001	123A	3020	17	TR21/735	S0014
CDQ10002	123A	3122	17	TR21/735	S0014
CDQ10003	123A	3020	17	TR21/735	S0014
CDQ10004	123A	3122	17	TR21/735	S0014
CDQ10005	123A	3020	17	TR21/735	S0014
CDQ10006	123A	3122	17	TR21/735	S0014
CDQ10007	123A	3122	17	TR21/735	S0014
CDQ10008	123A	3122	17	TR21/735	S0014
CDQ10009	123A	3020	17	TR21/735	S0014
CDQ10010	123A	3122	17	TR21/735	S0014
CDQ10011	128	3024	63	TR87/243	
CDQ10012	128	3024	63	TR87/243	
CDQ10013	154	3045		TR78/712	
CDQ10014	128	3024	63	TR87/243	
CDQ10015	154	3045		TR78/712	
CDQ10016	123A	3018	11	TR21/735	S0014
CDQ10017	123A	3018	11	TR21/735	S0014
CDQ10018	123A	3122	17	TR21/735	S0014
CDQ10019	123A	3018	11	TR21/735	S0014
CDQ10020	123A	3018	11	TR21/735	S0014
CDQ10021	123A	3122	17	TR21/735	S0014

Original Part Number	ECG	SK	GE	IR/WEP	HEP
CDQ10022	123A	3018	11	TR21/735	S0014
CDQ10023	123A	3018	11	TR21/735	S0014
CDQ10024	123A	3122	17	TR21/735	S0014
CDQ10025	123A	3122		TR21/735	S0014
CDQ10026	123A	3018	11	TR21/735	S0014
CDQ10027	123A	3122	17	TR21/735	S0014
CDQ10028	123A	3122		TR21/735	
CDQ10032	123A	3122	11	TR21/735	S0014
CDQ10033	128	3024	63	TR87/243	
CDQ10034	154	3045		TR78/712	
CDQ10035	123	3020	11	TR86/53	S0014
CDQ10036	123	3020	11	TR86/53	S0014
CDQ10037	154	3045	63	TR78/712	
CDQ10045	154	3024	63	TR78/712	
CDQ10046	154	3024	63	TR78/712	
CDQ10047	154	3045	63	TR78/712	
CDQ10048	128	3024		TR87/243	
CDQ10049	154	3045	12	TR78/712	
CDQ10051	128	3024		TR87/243	
CDQ10052	128	3024			
CDQ10053	128	3024		TR87/243	
CDQ10057	128	3024		TR87/243	
CDQ10058	128	3024		TR87/243	
CDR69/121/207	911				
CDT1309	121	3009	25	TR01/232	G6003
CDT1310	121	3009	25	TR01/232	G6003
CDT1311	121	3009	25	TR01/232	G6005
CDT1312			25	TR01/	
CDT1313			25		
CDT1315			25		
CDT1319	121	3009	25	TR01/232	G6003
CDT1320	121	3014	25	TR01/232	G6005
CDT1322	179		25	/G6001	G6018
CDT1349	121	3009	3	TR01/232	
CDT1349A	121	3009	3	TR01/232	G6005
CDT1350	121	3009	3	TR01/232	G6005
CDT1350A	121	3009	3	TR01/232	G6005
CDZ15000	128	3024		TR87/243	
CE0088					G0005
CE0089					G0005
CE0360/7839	102A	3004	2	TR85/	G0005
CE0361/7839	102A		2	TR85/	G0005
CE0362/7839	102A	3004	2	TR85/	G0005
CE0363/7839	158	3004	53	TR84/	G0005
CE4001B	123A	3018	20	TR21/	S0015
CE4001E	123A	3018			
CE4002D		3025			S5013
CE4003E	123A	3028	20		S0015
CE4004C	123A	3018	20	TR24/	S0015
CE4005C		3114			S0019
CE4008B	123A	3018	20	TR21/	S0015
CE4008C	123A	3018	20	TR21/	S0015
CE4010D		3018			S0014
CE4010E		3018			S0014
CE190555		3024			
CE213811	102A	3004		TR85/250	
CE216734		3004			
CE223408-0		3004			
CED99-13		3004			
CF1	108				
CF2	123A				
CF5	123				
CF6	128				
CF8	224				
CF11	108				
CF2386		3112			
CG1	123A	3122		TR21/735	
CG24015A	9944				
CGE-50	160			TR17/	
CGE-51	160			TR17/	
CGE-52	102A			TR85/	
CGE-53	158			TR84/	
CGE-54	103A/158				
CGE-60	161			TR83/	
CGE-61	161			TR83/	
CGE-62	199			TR21/	
CGE-63	192	3137			
CGE-64	172			TR69/	
CGE-66	152			TR76/	
CGE-67	193	3138			
CGE-68	124			TR81/	
CGE-69	153			TR77/	
CI2711	123A	3122		TR21/735	S0025
CI2712	123A	3122		TR21/735	S0015
CI2713	123A	3122		TR21/735	S0011
CI2714	123A	3122		TR21/735	S0011
CI2923	123A	3122		TR21/735	S0015
CI2924	123A	3122		TR21/735	S0015
CI2925	123A	3122		TR21/735	S0015
CI2926	123A	3122		TR21/735	S0025
CI3390	123A	3122		TR21/735	S0015
CI3391	123A	3122		TR21/735	S0015
CI3391A	123A	3122		TR21/735	S0015
CI3392	123A	3122		TR21/735	S0015
CI3393	123A	3122		TR21/735	S0015

Original Part Number	ECG	SK	GE	IR/WEP	HEP
CI3394	123A	3122		TR21/735	S0015
CI3395	123A	3122		TR21/735	S0015
CI3396	123A	3122		TR21/735	S0015
CI3397	123A	3122		TR21/735	S0015
CI3398	123A	3122		TR21/735	S0015
CI3402	123A	3122		TR21/735	S0024
CI3403	123A	3122		TR21/735	S0024
CI3403	123A	3122		TR21/735	S0024
CI3404	123A	3122		TR21/735	S0025
CI3405	123A	3122		TR21/735	S0015
CI3414	123A	3122		TR21/735	S0025
CI3415	123A	3122		TR21/735	S0024
CI3416	123A	3122		TR21/735	S0025
CI3417	123A	3122		TR21/735	S0015
CI3704	128	3024		TR87/243	S0025
CI3705	128	3024		TR87/243	S0015
CI3706	128	3024		TR87/243	
CI3900	123A	3122		TR21/735	S0015
CI3900A	123A	3122		TR21/735	S0015
CI3901	123A	3122		TR21/735	S0015
CI4256	123A	3122		TR21/735	S0024
CI4424	123A	3122		TR21/735	S0015
CI4425	123A	3122		TR21/735	S0015
CIC-300	124				
CII-225-Q	130				
CIL511	108	3018	17	TR95/56	
CIL512	108	3018	17	TR95/56	
CIL513	108	3018	17	TR95/56	
CIL521	108	3018	17	TR95/56	
CIL522	108	3018	17	TR95/56	
CIL523	108	3018	17	TR95/56	
CIL531	108	3018	20	TR95/56	
CIL532	108	3018	20	TR95/56	
CIL533	108	3018	17	TR95/56	
CJ5201	123A	3122	20	TR21/735	
CJ5202	123A	3122	20	TR21/735	
CJ5203	123A	3122	20	TR21/735	
CJ5204	102A		53	TR85/250	
CJ-5206	123A	3122	20	TR21/	S0015
CJ5206A	192			TR21/	S0015
CJ-5207	123A	3122	20	TR21/	S0015
CJ-5208	123A	3020	17	TR21/	S0014
CJ5209	129	3025	21	TR88/242	S0013
CJ5210	128	3024	63	TR87/243	S5026
CJ5211	123A	3122	20	TR21/735	S0014
CJ5212	123A	3122	20	TR21/735	S0015
CJ5213	128	3024	18	TR87/243	
CJ5214	128	3024	18	TR87/243	
CJ5215	128	3024	20	TR87/243	
CK4	126	3006		TR17/635	G0005
CK4A	126	3006		TR17/635	G0005
CK13	102A	3005	2	TR85/250	G0003
CK13A	102A	3005	1	TR85/250	G0005
CK14	100	3005	1	TR05/254	G0005
CK14A	100	3005	1	TR05/254	G0005
CK16	100	3005	1	TR05/254	G0005
CK16A	100	3005	1	TR05/254	G0005
CK17	100	3005	1	TR05/254	G0005
CK17A	100	3005	1	TR05/254	G0005
CK22	102A	3004	53	TR85/250	G0005
CK22A	102A	3004	53	TR85/250	G0005
CK22B	102A	3004	53	TR85/250	G0005
CK22C	102A	3004	53	TR85/250	G0005
CK25	100	3005	1	TR05/254	G0005
CK25A	100	3005	1	TR05/254	G0005
CK26	100	3005	1	TR05/254	G0005
CK26A	100	3005	2	TR05/254	G0005
CK27	100	3005	1	TR05/254	G0005
CK27A	100	3005	2	TR05/254	G0005
CK28	126	3008	51	TR17/635	G0005
CK28A	126	3008	1	TR17/635	G0005
CK64	102A	3004	2	TR85/250	G0005
CK64A	102A	3004	2	TR85/250	G0005
CK64B	102A	3004	2	TR85/250	G0005
CK64C	102A	3004	2	TR85/250	G0005
CK65	102A	3004	2	TR85/250	G0005
CK65A	102A	3004	2	TR85/250	G0005
CK65B	102A	3004	2	TR85/250	G0005
CK65C	102A	3004	2	TR85/250	G0005
CK66	102A	3004	53	TR85/250	G0005
CK66A	102A	3004	53	TR85/250	G0005
CK66B	102A	3004	2	TR85/250	G0005
CK66C	102A	3004	2	TR85/250	G0005
CK67	102A	3004	53	TR85/250	G0005
CK67A	102A	3004	2	TR85/250	G0005
CK67B	102A	3004	2	TR85/250	G0005
CK67C	102A	3004	2	TR85/250	G0005
CK83					G0005
CK86					G0003
CK256					G6011
CK258					G6011
CK261	101	3011	52	TR08/641	G0011
CK262	101	3011	2	TR08/641	G0011
CK311					G6012
CK312					G6012

Original Part Number	ECG	SK	GE	IR/WEP	HEP
CK398			27	TR78/	
CK411					G6012
CK412					G6012
CK419	128	3024	61	TR87/243	S0025
CK420			212	TR51/	S0014
CK421		3156	212	TR51/	S0014
CK422	128	3024	212	TR87/243	S0014
CK474	128	3024	61	TR87/243	S0014
CK475	128	3024	61	TR87/243	S0014
CK476	108	3039	61	TR95/56	S0014
CK477	108	3039	61	TR95/56	S0014
CK661	100	3005		TR05/254	
CK662	100	3005		TR05/254	
CK718				TR05/	
CK721	102	3004	2	TR05/631	G0005
CK722	102	3004	2	TR05/631	G0005
CK724					G0005
CK725	102	3004	53	TR05/631	G0005
CK727	102	3004	50	TR05/631	G0005
CK751	102	3004	53	TR05/631	G0005
CK754	102	3004	1	TR05/631	G0005
CK759	100	3005	52	TR05/254	G0005
CK759A	100	3004	52	TR05/254	G0005
CK760	100	3005	1	TR05/254	G0005
CK760A	100	3005	52	TR05/254	G0005
CK761	100	3005	1	TR05/254	G0005
CK762	126	3006	1	TR17/635	G0005
CK766	126	3005	52	TR17/635	G0005
CK766A	126	3006	52	TR17/635	G0005
CK768	100	3005	52	TR05/254	G0005
CK776	100	3005	2	TR05/254	
CK776A	100	3005		TR05/254	G0005
CK790	102	3004	65	TR05/631	G0005
CK791	102	3004	65	TR05/631	G0005
CK793	102	3004	65	TR05/631	G0005
CK794	102	3004	52	TR05/631	
CK870	102	3004	52	TR05/631	G0005
CK871	102	3004	52	TR05/631	G0003
CK872	102	3004	52	TR05/631	G0005
CK875	102	3004	52	TR05/631	G0005
CK878	102	3004	52	TR05/631	G0005
CK879	102	3004	52	TR05/631	G0005
CK882	102	3004	53	TR05/631	G0005
CK888	102	3004	52	TR05/631	G0005
CK891	102A		2	TR85/250	G0005
CK892	102A		2	TR85/250	G0005
CK942	159	3114	65	TR20/717	S0012
CM0770	128	3024	20	TR87/243	
CM-9	736				
CM-11		3034			
CM-13		3034			
CM2550	121	3014	16	TR01/232	
CM4518	4518				
CM4520	4520				
CM7163			10		
CM8640E			2		
CMC334-423	123A				
CN2484	128	3024	17	TR87/243	
COC13000-1C			63		
CP398					G6012
CP400	130		19	TR59/247	S5000
CP401	130		19	TR59/247	S5000
CP402					S5004
CP403					S0014
CP404	130		19	TR59/247	S5000
CP405	130		19	TR59/247	S5000
CP406	130		19	TR59/247	S7002
CP407	130		19	TR59/247	S7002
CP408	130	3024	19	TR59/247	S5000
CP409	128	3024	63	TR87/243	
CP701					S7002
CP702					S7002
CP703					S3011
CP704					S7002
CP800	158			TR84/630	G6011
CP801	158			TR84/630	G6011
CP802	158		TR84	TR84/630	G6011
CP803	158			TR84/630	G6011
CP2357	128				
CP2357P		3024			
CP3120E			IC13		
CP3611		3004			
CP5237A		3004			
CPS15553-101	9932				
CPS-16676-1	9951				
CQ1	102A			TR85/250	G0008
CQT940A	121	3009	16	TR01/232	G6005
CQT940B	121	3009	76	TR01/232	G6005
CQT940BA	121	3009	76	TR01/232	G6005
CQT1075	121	3009	16	TR01/232	
CQT1076	121	3009	16	TR01/232	
CQT1077	121	3009	16	TR01/232	
CQT1110	121	3014	76	TR01/232	G6003
CQT1110A	121	3009	76	TR01/232	G6003
CQT1111	121	3009	76	TR01/232	G6005
CQT1111A	121	3009	76	TR01/232	G6005
CQT1112	121	3014	76	TR01/232	G6005
CQT1129	121				
CRT1544	121	3009	76	TR01/232	
CRT1545	121	3009	76	TR01/232	
CRT1552	121	3009	76	TR01/232	
CRT1553	121	3009	16	TR01/232	
CRT1602	121	3014	16	TR01/232	G6005
CRT3602A	121	3009	16	TR01/232	G6005
CS1D14H				TR21/	
CS104A					S0016
CS104G					S0016
CS116D				TR21/	
CS183E	123				
CS184E	123				
CS184J	108				
CS360	123A	3122		TR21/735	
CS429J	108				
CS430H	108				
CS-460B	123A	3018	20		
CS-461B	108	3018	20		
CS461F	108				
CS469F	108				
CS696	123A	3122		TR21/735	S0025
CS697	123A	3122		TR21/735	
CS706	123A	3122		TR21/735	
CS718	123A	3122		TR21/735	S0025
CS718A	123A	3122		TR21/735	S0005
CS720A	123A	3122		TR21/735	S0005
CS9012HF	159		21	TR20/	
CS918	108	3039		TR95/56	S0011
CS925M	123A			TR21/	
CS929	161	3117		TR83/719	S0015
CS930	161	3117		TR83/719	S0015
CS956	128	3024		TR87/243	S0005
CS1014	108	3018	20	TR95/56	
CS1014D	108	3018	20	TR21/	
CS1014E	108	3039	20	TR95/	
CS1014F	108	3039	20	TR95/56	
CS1014G	161	3132		TR83/719	
CS1014H	161	3117		TR83/719	S0016
CS1018	108	3018	20	TR95/56	
CS1068			20		
CS1120C	161	3132	20	TR83/719	S0015
CS1120D	161	3132		TR83/719	S0015
CS1120E	161	3132		TR83/719	S0015
CS1120F	161	3117		TR83/719	
CS1120H	161	3132	20	TR83/719	S5014
CS1121G	159			TR20/717	S5013
CS1124G	106		21	TR20/52	
CS1129E	128	3024	63	TR87/243	
CS1166	123A	3122		TR21/735	
CS1166D	123A	3020	18	TR21/735	S0015
CS1166E	123A	3024	63	TR21/735	
CS1166F	123A	3024	18	TR21/	
CS1166G	123A	3024	18	TR21/735	
CS1166H	123A	3024	18	TR21/735	
CS1166H/F	123A	3122		TR21/735	
CS1168E	108	3028	20	TR95/56	S0015
CS1168F	123A	3018	20	TR21/735	S0015
CS1168G	123A	3018	20	TR21/735	S0015
CS1168H	123A		20	TR21/735	S0014
CS1170F	159	3025	21	TR20/717	
CS1221F	159		21	TR20/	
CS1225D	108	3018	17	TR95/56	S0011
CS1225E	108	3018	20	TR95/56	S0011
CS1225F	108	3018	11	TR95/56	S0011
CS1225H	128	3024		TR87/	S5014
CS1225HF	128	3024		TR87/243	
CS1226	108	3018	20	TR95/56	S0015
CS1226E	108	3018	20	TR95/56	
CS1226F	108	3018	20	TR95/56	
CS1226G	108	3018	20	TR95/56	
CS1226H	108	3039		TR95/56	S0014
CS1227	108	3039		TR95/56	
CS1227D	108	3039	20	TR95/56	S0011
CS1227E	108	3018	20	TR95/56	S0015
CS1227F	108	3018	20	TR95/56	S0015
CS1227G	108	3018	11	TR95/56	S0015
CS1228	159	3025	21	TR88/	
CS1228E	159	3114	21	TR20/717	S0012
CS1229	123A	3020	20	TR21/735	
CS1229A	123A	3124	20	TR21/735	
CS1229B	123A	3124	20	TR21/735	
CS1229C	123A	3124	20	TR21/735	
CS1229D	123A	3124	20	TR21/735	
CS1229E	123A	3024	18	TR21/735	S0015
CS1229F	123A	3024	18	TR21/735	S0015
CS1229G	123A	3024	63	TR21/735	
CS1229H	123A	3124	20	TR21/735	S0016
CS1229K	128	3024		TR87/243	
CS1235C	123A	3122	10		S0015
CS1235E	123A	3018	20		S0015
CS-1235F	123A	3122		TR21/735	
CS1235F				TR25/	

Original Part Number	ECG	SK	GE	IR/WEP	HEP
CS1235G	123A	3124	20	TR21/735	S0011
CS1236D	123A	3124	20	TR21/735	
CS1236H	123A			TR21/735	S0014
CS1237	129	3025	21	TR88/242	
CS1238	108	3018	11	TR95/56	
CS1238F	123A	3122	63	TR21/735	S0014
CS1238G	108	3039	20	TR95/56	S0011
CS1238H	108	3039	20	TR95/56	S0015
CS1238I	108	3039	20	TR95/56	S0011
CS1238P	123A		20	TR21/735	S0015
CS1243E	108	3028	18	TR95/56	
CS1243H	108	3018	20	TR95/56	
CS1244H	108	3018		TR95/56	
CS1244J	108	3018	18	TR95/56	S0014
CS1244X	108	3018	17	TR95/56	S0014
CS1245F	123A	3018	20	TR21/735	
CS1245G	123A	3018	20	TR21/735	
CS1245H	123A	3018	20	TR21/735	
CS1245I	123A	3124	10	TR21/735	
CS1245T	123A	3124	18	TR21/735	
CS1248	128	3024		TR87/243	
CS1248I	128	3024	18	TR87/243	
CS1248T	128	3024		TR87/243	
CS1250E	123A	3024	18	TR21/735	
CS1250F	128	3024	18	TR87/243	
CS1251E	159	3025	21	TR20/717	
CS1251F	159	3114	21	TR20/717	
CS1252B	108	3039	17	TR95/56	S0011
CS1252C	108	3039	17	TR95/56	S0011
CS1255H	128	3124	18	TR87/243	S5014
CS1255HF	193	3024	63	TR87/243	S5014
CS1256H	129	3025		TR88/242	
CS1256HG	192	3025	21	TR88/242	S5013
CS1257	123A			TR21/735	
CS1258	123A	3124	20	TR21/735	S0011
CS1259	123A	3124	20	TR21/735	S0011
CS1283A	123A	3124	18	TR21/735	
CS1284B	161			TR83/719	
CS1284F	161			TR83/719	S0011
CS1284G	161	3117		TR83/719	
CS1284H	161	3117		TR83/719	
CS1286			20		
CS1288	123A	3122	20	TR21/735	
CS1289	123A	3122	20	TR21/735	
CS1293	108	3018	11	TR95/56	
CS1294E	159	3114	21	TR20/717	S0012
CS1294H	159	3114	21	TR20/717	S0012
CS1295E	123A	3122	20	TR21/735	
CS1295G	123A	3122	20	TR21/735	
CS1295H	128	3024	63	TR87/243	S0014
CS1298	159				
CS1303	159	3114	21	TR20/717	S3032
CS1305	123A	3047	17	TR95/56	S0011
CS1308	159	3114		TR20/717	S5022
CS1312G	129	3114	21	TR19/	
CS1330	161	3018	20	TR83/719	S0014
CS1330A	108	3018		TR95/56	S0014
CS1330B	108	3018		TR95/56	
CS1330C	108	3018		TR95/56	
CS1340D	108	3018	20	TR95/56	
CS1340E	108	3018	20	TR95/56	S0015
CS1340F	108	3018	20	TR95/56	S0011
CS1340G	108	3018	20	TR95/56	S0015
CS1340H	108	3018	20	TR95/56	S0015
CS1340I	123A	3124	10	TR21/	
CS1344	123A		60	TR21/	S0020
CS1345	123A			TR21/	
CS1347	154			TR78/	S3021
CS1348	123A		18	TR21/	S0005
CS1349	123A	3122		TR21/735	S0014
CS1350	108		20	TR95/	S0020
CS1351	108		60	TR95/	S0020
CS1352	128		63	TR87/	S5014
CS1353	123A			TR21/	S5021
CS1354	159		22	TR20/	S0019
CS1359	108	3018	17	TR95/	S0014
CS1360	108	3018	17	TR95/	S0014
CS1361E	108	3018	20	TR95/	S0015
CS1361F	108	3018	20	TR95/	S0015
CS1361G	123A	3124	10	TR21/	S0014
CS1362	123A		18	TR21/	S0005
CS1363	123A		63	TR21/	S0025
CS1368	123A	3038		TR21/735	
CS1368A	123A	3038		TR21/735	
CS1368B	123A	3038		TR21/735	
CS1368C	123A	3124	10	TR21/735	
CS1368D	123A	3124		TR21/735	S0015
CS1369	129	3025	21	TR88/242	
CS1370	123A	3038	235	TR21/735	S0015
CS1371	123A	3124	17	TR21/735	S0016
CS1372	123A	3124	17	TR21/735	S0016
CS1383	123A		17	TR21/735	S0015
CS-1386E	123A	3018	20		
CS1386H	108	3018	20	TR95/	S0014
CS1420	123A	3122		TR21/735	

Original Part Number	ECG	SK	GE	IR/WEP	HEP
CS1453E	123A	3124	63	TR21/735	S5014
CS1453F	128	3024		TR87/243	
CS1453G	128	3024	63	TR87/243	S5014
CS1460E	161	3132		TR83/719	
CS1460H	161	3132		TR83/719	
CS1461J	161	3132		TR83/719	
CS1461X	161	3132		TR83/719	
CS1462F	161	3132		TR83/719	
CS1462I	128	3024		TR87/243	
CS1463A	123A			TR21/735	
CS1464H	128	3024		TR87/243	
CS1465H	129			TR88/242	
CS1508G	108			TR95/	
CS1509E	108			TR95/	
CS1509F	108			TR95/	
CS1518E	108			TR95/	
CS1555	108	3039	20	TR95/56	S0011
CS1585	123A		10	TR21/	
CS1585E/F	123A	3122	10		S0015
CS1585G	123A	3122	10		S0015
CS1585H	107	3018	20		S0015
CS1589E	161	3132		TR83/719	
CS1589F	161	3132		TR83/719	
CS1589S	161	3132		TR83/719	
CS1591LE	128	3024		TR87/243	
CS1594E	161	3132		TR83/719	
CS1596E	161	3132		TR83/719	
CS1609F	128	3024	63	TR87/243	
CS1613	128	3024		TR87/243	S0005
CS1625	123A				
CS1627	159				
CS1661	108		20	TR95/	S0020
CS1664	128		63	TR87/	S5014
CS1665	123A		18	TR21/	S0005
CS1711	128	3024		TR87/243	S0005
CS1834	108	3122	20		S0014
CS1893	128	3024		TR87/243	S0005
CS1909B	293		82		S0019
CS1910B	294		81		S0015
CS1990	128	3024		TR87/243	
CS2001H	123A	3124		TR21/	S0015
CS2004C	123A	3124		TR51/	S0015
CS-2004C	108	3020	20	TR24/	
CS2004D	123A	3124	20	TR51/	S0015
CS2005		3114	21	TR30/	
CS-2005B	159	3114	21	TR73/	
CS-2005C	159	3114	21	TR73/	
CS2006	123A	3018	62		S0015
CS2006F	229	3018			
CS2006G	108	3018		TR21/	S0020
CS-2007G	108	3122	20		S0014
CS-2007H	108	3122	20		S0020
CS-2008F	108	3018	20		
CS2008G	108	3018	20	TR95/	S0020
CS2008H	108	3018		TR95/	S0020
CS2008H052		3018		TR21/	S0015
CS2008H552	108			TR95/56	
CS2142	193	3025	67	TR28/	
CS2143	192	3024	63	TR25/	
CS2218	123A	3122		TR21/735	S0015
CS2219	123A			TR21/735	S0015
CS2221	123A	3122		TR21/735	S0015
CS2222	123A	3122		TR21/735	S0015
CS2369	123A	3122		TR21/735	
CS2481	123A	3122		TR21/735	
CS2484	128	3024		TR87/243	
CS2639					S0025(2)
CS2640					S0025(2)
CS2641					S0025(2)
CS2642					S0015(2)
CS2643					S0015(2)
CS2644					S0015(2)
CS2711	123A	3122	17	TR21/735	
CS2712	123A	3122	17	TR21/735	
CS2713	123A	3122	17	TR21/735	
CS2714	123A	3122	17	TR21/735	
CS2715	161	3117	17	TR83/719	
CS2716	161	3117	17	TR83/719	
CS2922	123A	3122		TR21/735	
CS2923	123A	3122	17	TR21/735	
CS2924	123A	3122	17	TR21/735	
CS2925	123A	3122		TR21/735	
CS2941				TR19/	
CS-3001B	108	3122	20	TR21/	
CS-3024	172				
CS3390	123A	3122		TR21/735	
CS3391	123A	3122		TR21/735	
CS3391A	123A	3122		TR21/735	
CS3392	123A	3122		TR21/735	
CS3393	123A	3122	17	TR21/735	
CS3394	123A	3122	17	TR21/735	
CS3395	123A	3122		TR21/735	
CS3396	123A	3122	17	TR21/735	
CS3397	123A	3122	17	TR21/735	
CS3398	123A	3122	17	TR21/735	

Original Part Number	ECG	SK	GE	IR/WEP	HEP
CS3402	123A	3122		TR21/735	S0024
CS3403	123A	3122		TR21/735	S0024
CS3404	123A	3122		TR21/735	S0015
CS3405	123A	3122		TR21/735	S0015
CS3414	123A	3122		TR21/735	
CS3415	123A	3122		TR21/735	
CS3416	123A	3122		TR21/735	
CS3417	123A	3122		TR21/735	
CS3605	123A	3122	17	TR21/735	
CS3606	123A	3122	17	TR21/735	
CS3607	123A	3122	17	TR21/735	
CS3662	161	3117		TR83/719	
CS3663	161	3117		TR83/719	
CS3702	159	3114		TR20/717	
CS3703	159	3114		TR20/717	
CS3704	128	3024		TR87/243	
CS3705	128	3024		TR87/243	
CS3706	128	3024		TR87/243	
CS3707	161	3117	17	TR83/719	
CS3708	161	3117	17	TR83/719	
CS3709	161	3117	17	TR83/719	
CS3710	161	3117	17	TR83/719	
CS3711	161	3117	17	TR83/719	
CS3843	123A	3122	17	TR21/735	
CS3844	123A	3122	17	TR21/735	
CS3845	123A	3122	17	TR21/735	
CS3854	123A	3122		TR21/735	
CS3854A	123A	3122		TR21/735	
CS3855	123A	3122		TR21/735	
CS3855A	123A	3122		TR21/735	
CS3859	123A	3122	17	TR21/735	
CS3859A	123A	3122		TR21/735	
CS3860	123A	3122	17	TR21/735	
CS3900	123A	3122		TR21/735	
CS3900A	123A	3122		TR21/735	
CS3901	123A	3122		TR21/735	
CS3903	123A	3122		TR21/735	
CS3904	123A	3122		TR21/735	
CS3906	159	3114		TR20/717	S0019
CS4001			11		
CS4003			61		
CS4005			213		
CS4006			81		
CS4007			20		
CS4012			22		
CS4013			82		
CS4021			60		
CS4060			60		
CS4061			212		
CS4193			11		
CS4194			11		
CS4424	123A	3122		TR21/735	
CS4425	123A	3122		TR21/735	
CS5088	123A	3122		TR21/735	
CS5369	123A	3122		TR21/735	
CS5447	159	3114		TR20/717	
CS5448	158	3114		TR20/717	
CS5449	128	3024		TR87/243	
CS5450	128	3024		TR87/243	
CS5451	128	3024		TR87/243	
CS5995	703A				
CS6168F	123A	3124	10	TR21/735	S5014
CS6168G	128	3124		TR87/243	S5014
CS-6168G	123A		10	TR24/	S0015
CS-6168H	123A	3122	10	TR24/	S0015
CS-6225E	107	3018	17	TR24/	S0025
CS6225F	108	3018	20	TR95/56	S0011
CS-6225G	123A	3018	20	TR21/	S0015
CS6226F	108	3039		TR95/56	S0011
CS6227E	108	3039	20	TR95/56	S0011
CS6227F	108	3039	20	TR95/56	S0011
CS-6227G	108	3018	20	TR24/	S0015
CS6228F	159				
CS-6228G	129	3025	67	TR28/	S5013
CS6229F	123A	3124		TR51/	S0015
CS6229G	123A	3124		TR51/	S0015
CS7228G		3114	67	TR28/	S0013
CS7229F	128		63	TR25/	
CS7229G	123A	3124	20	TR25/	S0015
CS9001	108	3122	20	TR21/	
CS9011	123A	3018	20	TR21/735	S0015
CS9011D	123A	3018	20	TR21/735	
CS9011E	123A	3018		TR21/735	S0015
CS9011F	123A	3018	20	TR21/735	S0015
CS9011G	123A	3124	20	TR21/735	S0015
CS9011H	123A	3018	20	TR21/735	S0015
CS9011L	123A			/735	
CS9012	159	3025	21	TR20/717	S5013
CS9012/3490	129				
CS9012E	159	3114			
CS9012E-F	159	3114	67	TR28/	
CS9012H	129	3114	21	TR30/	S0013
CS9012HG	159	3124	21	TR20/717	
CS9012HH	159	3114	21	TR20/717	S5013
CS9013	161	3132	20	TR83/719	S0015

Original Part Number	ECG	SK	GE	IR/WEP	HEP
CS9013A	123A	3124		TR21/735	
CS9013B	123A	3024	63	TR21/735	S5014
CS9013E	128	3024		TR87/243	
CS9013H	128	3124	63	TR87/243	S0015
CS9013HH	123A		20	TR21/735	S5014
CS9014	161	3124	20	TR83/719	S0015
CS9014/3490	123A				
CS9014A	123A	3124	20	TR21/735	S0030
CS9014B	123A	3024		TR21/735	S5014
CS9014C	123A	3124		TR21/	S0020
CS9014D	123A	3124	62	TR21/735	S0014
CS9014G	123A	3124		TR21/735	S5014
CS9015	123A	3124	21	TR21/	S0019
CS9015B	123A	3124		TR21/735	S0019
CS9015C	159	3114	21	TR20/	S0019
CS9015D	159	3114	21	TR20/	S0019
CS9016	161	3018	20	TR83/719	S0015
CS-9016	108	3018	11	TR22/	
CS9016/3490	108				
CS9016D	123A	3018		TR22/	S0015
CS-9016D	107	3018	17	TR24/	
CS9016E	108	3039	20	TR95/56	S0015
CS9016F	108	3018	11	TR95/56	S0011
CS9016G	108	3122	20	TR95/735	S0015
CS9016H	108		20	TR21/	
CS9017F	107	3122	20	TR70/	S0015
CS9017G	161	3122	20	TR83/719	S0011
CS9017H	161	3132	20	TR83/719	S0011
CS9018	108	3018	11	TR95/56	S0011
CS9018/3490	108				
CS9018D	108	3018	11	TR95/56	S0015
CS9018E	108	3122	17	TR95/56	S0015
CS9018F	108	3018	17	TR95/56	S0015
CS9018G	108	3018	20	TR95/56	S0011
CS9020E	159			TR20/	
CS9020F	159			TR20/	
CS9021G-1	108	3018	17	TR95/	
CS9022LE	128	3020		TR25/	S5014
CS9101B	123A	3018	20	TR24/	S0015
CS9102B	129	3025	67	TR30/	S5013
CS9103B	128	3024	63	TR24/	
CS9103C	128	3024	63	TR87/	S5014
CS-9104	123A		62		
CS9123C1			21		
CS9124B1	108		20	TR95/	S0016
CS9124-C2	108	3124	20	TR95/	S0016
CS9125B	123A	3124	17		S0014
CS9125-B1	108			TR95/	S0030
CS9126	123A	3124	20		
CS9128		3025		TR28/	S0019
CS9128-B2	159	3114		TR20/	S0019
CS9128C1	159		21	TR20/	S0019
CS9129	159	3025			S0019
CS9129B	159	3025	21	TR30/	S0019
CS9129B1	159	3114	21	TR20/	S0019
CS9129B2	159	3114	21	TR20/	S0019
CS9417	192	3122	63		
CS9600-4	123A	3124		TR21/735	G0011
CS9600-5	123A	3124		TR21/735	G0011
CS12250				TR95/	
CS12381				TR21/	
CS12481				TR87/	
CS13401	123A	3124	10		
CS14621				TR87/	
CS29008					S0005(2)
CS29009					S0005(2)
CS29010					S0005(2)
CS29011					S0005(2)
CS29012					S0005(2)
CS29013					S0005(2)
CS90111	123A		20	TR21/735	
CS90146		3122			
CS96180D					S0011
CST1739	104	3009	16	TR01/230	G6003
CST1740	104	3009	16	TR01/230	G6003
CST1741	104	3009	16	TR01/230	G6003
CST1742	104	3009	16	TR01/230	G6003
CST1743	121	3009	16	TR01/232	G6005
CST1744	121	3009	16	TR01/232	G6005
CST1745	121	3009	16	TR01/232	G6005
CST1746	121	3009	16	TR01/232	G6005
CST1773					G6011
CST1773A					G6011
CT-06C	230				
CT35				TR87/	
CT1009	102A	3004		TR85/250	G0005
CT1012	107	3018		TR70/720	S0011
CT1013	107	3018		TR70/720	S0011
CT1017	158	3004		TR84/630	G0005
CT1018		3018			S0016
CT1019		3018			S0016
CT1020		3018			S0016
CT1021		3018			S0016
CT1022		3004			G0005
CT1023		3004			G0005

Original Part Number	ECG	SK	GE	IR/WEP	HEP
CT1024		3004			G0005
CT1122	121	3009	16	TR01/232	G6005
CT1124	121	3009	16	TR01/232	G6005
CT1124A	121	3009	16	TR01/232	G6005
CT1124B	121	3009	16	TR01/232	G6003
CT1300	108	3018	17	TR95/56	
CT1500	108	3018	17	TR95/56	
CT6776	154				
CTD					G6005
CTP4					G0005
CTP1002					G6011
CTP1003					G6011
CTP1004					G6011
CTP1005					G6011
CTP1006					G6011
CTP1032	102A	3004		TR85/250	G0005
CTP1033	102A	3004		TR85/250	G0005
CTP1034	102A	3004		TR85/250	G0005
CTP1035	102A	3004		TR85/250	G0005
CTP1036	102A	3004		TR85/250	G0005
CTP1102					G6005
CTP1104	104	3009	16	TR01/230	G6003
CTP1106	104	3009		TR01/230	G6005
CTP1108	104	3009	16	TR01/230	G6005
CTP1109	104	3009	16	TR01/230	G6005
CTP1111	121	3009	25	TR01/232	G6005
CTP1112					G6012
CTP1117	104	3009	25	TR01/230	G6005
CTP1119	104	3009		TR01/230	
CTP1124	121	3009	16	TR01/232	G6005
CTP1133	121	3014	25	TR01/232	G6005
CTP1135	121	3014	25	TR01/232	G6005
CTP1136	121	3009	25	TR01/232	G6005
CTP1137	121	3009	16	TR01/232	G6005
CTP1226					G6005
CTP1265	121	3009		TR01/232	
CTP1266	121	3009		TR01/232	
CTP1296					G6005
CTP1297					G6005
CTP1306	121	3009		TR01/232	G6005
CTP1307	121	3009		TR01/232	G6005
CTP1314					G6018
CTP1320					G0005
CTP1330					G0005
CTP1340					G0005
CTP1350					G0005
CTP1360					G0005
CTP1410					G0005
CTP1500	121	3009	3	TR01/232	G6005
CTP1503	121	3009	76	TR01/232	G6005
CTP1504	121	3014	76	TR01/232	G6005
CTP1505					G6015
CTP1506					G6015
CTP1507					G6015
CTP1508	121	3009	76	TR01/232	G6005
CTP1509	105	3012	4	TR03/233	G6004
CTP1511	121	3009	16	TR01/232	G6005
CTP1512	105	3009	4	TR03/233	G6004
CTP1513	121	3009	16	TR01/232	G6005
CTP1514	121	3009	16	TR01/232	G6005
CTP1544	179	3009	76	TR35/G6001	G6009
CTP1545	179	3009	76	TR35/G6001	G6009
CTP1550	121	3009	16	TR01/232	G6005
CTP1551	121	3009	16	TR01/232	G6005
CTP1552	179	3009	76	TP35/G6001	G6009
CTP1553	179	3009	16	TR35/G6001	G6009
CTP1728	121	3009	16	TR01/232	G6005
CTP1729	121	3009	16	TR01/232	G6005
CTP1730	121	3009	16	TR01/232	G6005
CTP1731	121	3009	16	TR01/232	G6005
CTP1732	121	3009	16	TR01/232	G6005
CTP1733	121	3009	16	TR01/232	G6005
CTP1735	121	3009	16	TR01,232	G6005
CTP1736	121	3009	16	TR01/232	G6005
CTP1739	121	3009	16	TR01/232	G6005
CTP2001-1001	102A	3004		TR85/250	G0005
CTP2001-1002	102A	3004		TR85/250	G0005
CTP2001-1003	102A	3004		TR85/250	G0005
CTP2001-1004	102A	3004		TR85/250	G0005
CTP2001-1007	123A	3124		TR21/735	S0015
CTP2001-1008	123A	3124		TR21/735	S0015
CTP2001-1009	102A	3004		TR85/250	G0005
CTP2006-1001	102A	3004		TR85/250	G0005
CTP2006-1002	102A	3004		TR85/250	G0007
CTP2006-1003	102A	3004		TR85/250	GU005
CTP2010-1001		3123			G0005
CTP2010-1002		3123			G0005
CTP2010-1003					G0005
CTP2010-1004					G0005
CTP2010-1005					G0005
CTP201^-1006					G0005
CTP2010-1007					G0008
CTP2010-1008					G0008
CTP2010-1009					G0008
CTP2076-1001	102A	3004		TR85/250	

Original Part Number	ECG	SK	GE	IR/WEP	HEP
CTP2076-1002	102A	3004		TR85/250	
CTP2076-1003	102A	3004		TR85/250	
CTP2076-1004	102A	3004		TR85/250	
CTP2076-1005	102A	3004		TR85/250	
CTP2076-1006	102A	3004		TR85/250	
CTP2076-1007	102A	3004		TR85/250	
CTP2076-1008	102A	3004		TR85/250	
CTP2076-1009	102A	3004		TR85/250	
CTP2076-1010	102A	3004		TR85/250	
CTP2076-1011	102A	3004		TR85/250	
CTP2076-1012	102A	3004	T	TR85/250	
CTP3500	121	3009	16	TR01/232	G6005
CTP3503	121	3009	76	TR01/232	G6005
CTP3504	121	3009	16	TR01/232	G6005
CTP3508	121	3009	16	TR01/232	G6005
CTP3544	179	3009	16	TR35/G6001	G6009
CTP3545	179	3009	16	TR35/G6001	G6009
CTP3550					G6018
CTP3551					G6018
CTP3552	179	3009	16	TR35/G6001	G6009
CTP3553	179	3009	16	TR35/G6001	G6009
CX-093	712				
CX100B	1004				
CXU-2J		3054			
CYT1555					G6009
CYT1556					G6009
CYT1559					G6009
CYT1560					G6009
D006		3018	20		
D008	102A	3004	52	TR84/	
D009		3018	11		S0016
D018	102A	3004		TR85/635	G0005
D019	100	3005	2	TR05/254	
D020	126	3006		TR17/635	G0005
D021	102A	3004		TR85/250	G0005
D030	102A	3004		TR85/250	G0007
D031	102A	3004	20	TR85/250	S0011
D031(CHAN MAST)	123A			TR21/735	
D038	102A	3004	2	TR85/250	
D043	102A	3004	2	TR85/250	
D048	123A	3124	20	2SC945/	
D053	123A	3124	62	TR25/	
D057	199				
D058	108	3018		TR95/	S0014
D059	102A			TR84/	G0005
D063	160	3006	51	TR17/637	
D069	108	3039		TR95/56	S0014
D072		3018	11		S0016
D073	160	3006		TR17/637	G0003
D078	100	3005	2	TR05/254	
D079	160	3006	51	TR17/637	
D080	127	3024	16	TR27/235	
D081	127	3009	3	TR27/235	
D083	103A	3010	8	TR08/724	
D085	103A	3010	8	TR08/724	
D086	160	3006	51	TR17/637	S0024
D087	108	3018	11	TR95/56	
D088	108	3019	11	TR95/56	S0025
D0250	175				
D1A	123A		20	TR21/735	
D1R38	123A	3122		TR21/735	
D1U					S0019
D1Y	107	3039		TR70/720	S0033
D2R38	123A	3122		TR21/735	
D2U					S0030
D4BE102					G6003
D4BE103					G6003
D4BE104					G6005
D4C28	161	3117	20	TR83/719	S0016
D4C29	161	3117	20	TR83/719	S0016
D4C30	161	3117	20	TR83/719	S0016
D4C31	161	3117	20	TR83/719	S0020
D4D20	161	3117		TR83/719	S0014
D4D21	161	3117		TR83/719	S0014
D4D22	161	3117		TR83/719	S0014
D4D24		3122	17	TR21/56	S0033
D4D25	123A	3122	17	TR21/735	S0016
D4D26	123A	3122	17	TR21/735	S0033
D5E37	6401			2160/	
D5E43	6409			2160/	
D5E44	6401			2160/	
D5345	6409			2160/	
D6C					S0005
D7A30	128	3024	63	TR87/243	S5014
D7A31	128	3024	63	TR87/243	S5014
D7A32	128	3024	63	TR87/243	S5014
D7B1					S3020
D7B2					S3020
D7C1					S5014
D7C2					S5014
D7C3					S3019
D7D1					S3011
D7D2					S3011
D7D3					S3019
D7D13					S3019

Original Part Number	ECG	SK	GE	IR/WEP	HEP
D7E1					S3020
D7E2					S3020
D7F1					S3020
D7F2					S3020
D7F3					S3019
D7F4					S3019
D7G1					S3020
D7G2					S3020
D7G3					S3019
D10B55-2.3					S0025
D10B553-2.3					S0025
D10B555-2.3					S0015
D10B556-2.3					S0015
D10B1051	107	3039		TR70/720	S0025
D10B1055	107	3039		TR70/720	S0025
D10C573-2.3					S0025
D10G1051	107	3039	17	TR70/720	S0033
D10G1052	107	3039	17	TR70/720	S0033
D10H551-2.3					S0025
D10H553-2.3					S0015
D11	103				
D11B554-2.3					S0025
D11B555-2.3					S0025
D11B556-2.3					S0005
D11B560-2.3					S0005
D11B1052					S0005
D11B1055					S0005
D11C1B1	128	3024	17	TR87/243	S5014
D11C1F1			17	TR24/	
D11C2D1B20			17	TR24/	
D11C3B1	128	3024	63	TR87/243	
D11C3F1			63	TR25/	
D11C5B1	128	3024	17	TR87/243	
D11C5F1			17	TR24/	
D11C7B1	128	3024		TR87/243	
D11C10B1	128	3024	12	TR87/243	
D11C10F1			12	TR23/	
D11C11B1	128	3024	12	TR87/243	
D11C11F1			12	TR23/	
D11C201B20	124	3021		TR81/240	
D11C203B20	124	3021	63	TR81/240	
D11C205B20	124	3021	17	TR81/240	
D11C207B20	124	3021		TR81/240	
D11C210B20	124	3021	12	TR81/240	
D11C211B20	124	3021	12	TR81/240	
D12	130				
D13	105				
D13K1	6402				
D13T1	6402		X1		S9001
D14	106			TR20/52	
D15	130				
D16	130				
D16E7	123A	3122		TR21/735	S0015
D16E9	123A	3122		TR21/735	S0015
D16EC18	123A	3124	63	TR21/735	S0015
D16G6	107	3124	11	TR70/720	
D16K1	107	3039		TR70/720	S0016
D16K2	107	3039		TR70/720	S0016
D16K3	107	3039		TR70/720	S0016
D16K4	107	3039	60	TR70/720	
D16P1			64		
D16P2	172		64	TR69/S9100	S9100
D16P4			64		
D18A12	184	3054	57	/S5003	
D19	103				
D20	103				
D21	103				
D22	103				
D24	124				
D24A3391			212	TR51/	
D24A3391A			212	TR51/	
D24A3392			212	TR51/	
D24A3393			212	TR51	
D24A3394	108	3039	212	TR95/56	
D24A3900			212	TR51/	
D24A3900A			212	TR51/	
D24B	124				
D24C	124				
D24CK	124				
D24D	124				
D24E	124				
D24F	124				
D24Y	124				
D24YK	124				
D26A	130				
D26B	130				
D26B1	107	3039	60	TR70/720	
D26B2	107	3039	60	TR70/720	
D26C	130				
D26C1	107	3039	60	TR70/720	
D26C2	107	3039	60	TR70/720	
D26C3	107	3039	39	TR70/720	
D26E2	107	3039	214	TR70/720	
D26E-3			214	TR51/	
D26E-4			39	TR51/	

Original Part Number	ECG	SK	GE	IR/WEP	HEP
D26E-5			212	TR51/	
D26E-6			214	TR51/	
D26E-7			212		
D26G1	107	3039	60	TR70/720	
D27C1	152	3054	28	TR76/701	
D27C2	152	3041	28	TR76/701	
D27C3	152	3041	28	TR76/701	
D27C4	152	3041	28	TR76/701	
D27D1			29	TR76/	
D27D2			29	TR76	
D27D3			29	TR76/	
D27D4			29	TR76/	
D28A1	152	3054	28	TR76/701	
D28A2	152	3054	28	TR76/701	
D28A3	152	3041	28	TR76/701	
D28A4	152	3041	28	TR76/701	
D28A5	152	3041	63	TR76/701	S5015
D28A6	152	3041	88	TR76/701	S5015
D28A7	152	3041	28	TR76/701	
D28A9	152	3041	28	TR76/701	S5000
D28A10	152	3054	28	TR76/701	
D28A12	152	3054	63	TR76/701	S5015
D28A13	152	3054	88	TR76/701	S5015
D28B	171			TR79/244A	
D28D1	152	3054	63	TR76/701	S5015
D28D2	152	3054	63	TR76/701	S5015
D28D3	152	3054	28	TR76/701	S5015
D28D4	152	3054	63	TR76/701	S5015
D28D5	152	3054	63	TR76/701	S5015
D28D7	152	3054	28	TR76/701	
D28D10	152	3054		TR76/701	
D29	175				
D29A4	159	3025	210	TR20/717	S0012
D29A5	159	3114	210	TR20/717	S0013
D29A6	159	3118		TR20/717	S0019
D29A7		3118	210		
D29A8		3118	210		
D29A9	159	3114		TR20/717	
D29A10			81		
D29A11			81		
D29E1	159	3114	48	TR20/717	
D29E1J1	193		48		
D29E2	159	3114	48	TR20/717	
D29E2J1	193		48		
D29E4	159	3114	48	TR20/717	
D29E4J1	193		48		
D29E5	159	3114	48	TR20/717	
D29E5J1	193		48		
D29E6	159	3114	48	TR20/717	
D29E6J1	193		48		
D29E7	159	3114	48	TR20/717	
D29E7J1	193		48		
D29E8	159	3114	67	TR20/717	
D29E8J1			67		
D29E9	193	3114	67	/717	
D29E9J1	193		67		
D29E10	193	3114	67	/717	
D29E10J1	193		67		
D29F1			65		
D29F2			65		
D29F3			65		
D29F4			65		
D29F5			82		
D29F6			82		
D29F7			82		
D30-0	103A				
D30A1	159	3114	65	TR20/717	
D30A2	159	3118	65	TR20/717	
D30A3	159	3118	65	TR20/717	
D30A4	159	3114	65	TR20/717	
D30A5	159	3114	65	TR20/717	
D30-N	103A				
D31	103A				
D31A1		3118			
D31A2		3118			
D31A9		3118			
D31D	103A				
D32	103A				
D32K1			17		
D32K2			17		
D32P1			62		
D32P2			62		
D32P3			62		
D32P4			62		
D33	103A				
D33C	103A				
D33D21	123A	3122	20	TR21/735	
D33D21J1	192		47		
D33D22	123A	3122	47	TR21/735	
D33D22J1	192		47		
D33D24	123A	3122	20	TR21/735	
D33D24J1	192		47		
D33D25	123A	3122	20	TR21/735	
D33D25J1	192		47		
D33D26	123A	3122	47	TR21/735	

Original Part Number	ECG	SK	GE	IR/WEP	HEP
D33D26J1	192		47		
D33D27	123A	3122		TR21/735	
D33D27J1	192				
D33D28	123A	3122	63	TR21/735	
D33D29	192	3122	63	/735	
D33D29J1	192		63		
D33D30	123A	3122		/735	
D33D30J1	192		63		
D33E01J1			48		
D33E02J1			48		
D34	103A				
D35	103A				
D36	101				
D37	103A				
D37A	103A				
D38	103A				
D40C1	265			TR76/	
D40C2	266				
D40C3	267		23		
D40C4	265				
D40C5	266				
D40C7	265				
D40C8	266				
D40D1	210	3054	X18	TR76/701	
D40D2	210	3054	X18	TR76/701	
D40D3	210	3054	28	TR76/701	
D40D4	210	3054	X18	TR76/701	
D40D5	210	3054	X18	TR76/701	
D40D7	210	3054	28	TR76/701	
D40D8	210	3054	28	TR76/701	
D40D10	210	3202			
D40D11	210	3202			
D40K1	268				
D40K2	268				
D40N1	171	3201	27	TR79/244A	S3021
D40N2			27		S3021
D40N3	171	3201	27	TR79/244A	S3021
D40N4			27		S3021
D40N5	171	3201	27	TR79/	
D41	130				
D41D1	211	3203	29	TR77/S3027	S3028
D41D2	211	3083	29	TR77/700	S3028
D41D4	211	3203	29	TR77/S3027	S3028
D41D5	211	3203	29	TR77/S3027	S3028
D41D7	211	3203	29	TR77/S3027	S3032
D41D8	211	3203	29	TR77/S3027	S3032
D41D10	211	3203			
D41D11	311	3203			
D41K1	269				
D41K2	269				
D42C1	186	3083	28	TR55/S3023	S5000
D42C2	186	3083	28	TR55/S3023	S5000
D42C3	186	3083	28	TR55/S3023	S5000
D42C4	186	3083	28	TR55/S3023	S5000
D42C5	186	3083	28	TR55/S3023	S5000
D42C6	186	3083	28	TR55/S3023	S5000
D42C7	186	3083	28	TR55/S3023	S5000
D42C8	186	3083	28	TR55/S3023	S5000
D42C10					S5000
D42C11					S5000
D43	101				
D43A	101				
D43C1	187	3193	29	TR56/S3027	S5006
D43C2	187	3193	29	TR56/S3027	S5006
D43C3	187	3193	29	TR56/S3027	S5006
D43C4	187	3193	29	TR56/S3027	S5006
D43C5	187	3193	29	TR56/S3027	S5006
D43C6	187	3193	29	TR56/S3027	S5006
D43C7	153	3083	29	TR77/700	S5006
D43C8	187	3193	29	TR56/S3027	S5006
D43C9					S5006
D43C10					S5006
D43C11					S5006
D44	101				
D44BB					S3024
D44C1	152	3054	66	TR76/701	
D44C2	152	3054	66	TR76/701	
D44C3	152	3054	66	TR76/701	
D44C4	152	3054	66	TR76/701	
D44C5	152	3054	66	TR76/701	
D44C6	152	3054	66	TR76/701	
D44C7	152	3054	66	TR76/701	
D44C8	152	3054	66	TR76/701	
D44C9	152	3054	66	TR76/701	
D44C88	186	3054	28	TR55/	S3024
D44E1	263				
D44E2	263				
D44E3	263				
D44H1			55		
D44H2			55		S5004
D44H4			55		S5004
D44H5			55		S5004
D44H7			55		S5004
D44H8			55		S5004
D44H10			55		S5004

Original Part Number	ECG	SK	GE	IR/WEP	HEP
D44R1	198		32		
D44R2	198		32		
D44R3	198		32		
D44R4	198		32		
D44R5			32		
D44R6			32		
D45	162				
D45C1	153	3083	69	TR77/700	
D45C2	153	3083	69	TR77/700	
D45C3	153	3083	69	TR77/700	
D45C4	153	3083	69	TR77/700	
D45C5	153	3083	69	TR77/700	
D45C6	153	3083	69	TR77/700	
D45C7	153	3083	69	TR77/700	
D45C8	153	3083	69	TR77/700	
D45C9	153	3083	69	TR77/700	
D45E1	264				
D45E2	264				
D45E3	264				
D45H1			56		
D45H2			56		
D45H4			56		
D45H5			56		
D45H7			56		
D45H8			56		
D45H10			56		
D46	162				
D47	162				
D53	130				
D54	105				
D55	181				
D56	175				
D56H1	165				
D56H2	165				
D57	175				
D58	175				
D59	162				
D60	162				
D61	103A				
D62	103A				
D63	103A		8	TR08/	
D64	103A	3010		TR08/724	
D65	126		51	TR17/635	
D65-1	103A				
D66	126	3024	51	TR17/635	G0003
D67	162				
D68B	223				
D68C	223				
D68D	223				
D68E	223				
D69	130				
D70	175				
D71	175				
D72	103A				
D72RE	103A				
D73	162				
D74	162				
D75	103A				
D75H	103A				
D77A	103A				
D77P	103A				
D79	225				
D79D	225				
D80	130				
D81	130				
D82	130				
D83	162				
D84	162				
D88A	162				
D90	152				
D91	152				
D91F	175				
D92	175				
D92D	175				
D93	152				
D94	124				
D96	103A				
D100A	103A				
D101	101				
D101B	102A			TR85/250	G0005
D102	175				S0016
D103	175				
D104	103A				
D105	103A				
D105B	102A	3004	2	TR85/250	G0005
D110	162				
D111	162				
D113	181				
D114	181				
D116	226MP				
D117	158				G0005
D118	162				
D119	162				
D120	128				
D120H	210	3202			

Original Part Number	ECG	SK	GE	IR/WEP	HEP
D120HC	210	3202			
D121	128				
D121H	210	3202			
D121HB	210	3202			
D124AH	130				
D125	163				
D126	158	3004		TR84/	G6011
D126H	163				
D126HB	163				
D127	103A				
D128	103A				
D129	175				
D130	175				
D132	181				
D133(CHAN MAST)	108		17	TR95/	
D134	126	3004		TR17/635	G0005
D134(CHAN MAST)	102A				
D135	102A	3004	2	TR85/250	G0005
D136	124				
D137	124				
D141	152	3045			S0014
D141(CHAN MAST)	108			TR95/	
D141H01	175				
D141H02	175				
D142	175				
D143	175				
D144	175				
D145	175				
D146	130				
D146UK	175				
D147	130				
D149	160			TR17/	G0008
D150	175	3124			
D151	130				
D152	164	3048			
D152(CHAN MAST)	195			TR65/	
D154	152				
D155	175				
D156	158	3004		TR84/	G0005
D157	124				
D158	124				
D159	124	3018			
D160	107	3018			
D161	101				
D162	101				
D163	101				
D164	130				
D165	162				
D166	162				
D167	101				
D170	103A				
D170A	103A				
D170B	103A				
D170C	103A				
D171	102A			TR85/250	G0005
D172	160			TR17/637	G0009
D173	160			TR17/637	G0008
D174	160	3006		TR17/637	G0009
D175	130				
D177	162				
D178	103A				
D178T	103A				
D180	158	3004		TR84/	G0005
D180A	223				
D180B	223				
D180C	223				
D180D	223				
D180M	223				
D181	162				
D182	128				
D183	128				
D184	152				
D186	103A				
D187	103A				G0005
D188	130				S0016
D189	130				
D190	124				S0015
D191	103A				
D192	103A				
D193	102A	3004		TR85/	G0005
D194	103A				
D195	108	3018		TR95/	S0014
D198	162				
D198H	163				
D198HQ	162				
D198V	162				
D199	162				
D200A	165				
D201	130				
D201Y	130				
D202	162				
D203	162				
D204	128				
D205	225				
D211	130				

Original Part Number	ECG	SK	GE	IR/WEP	HEP
D212	130				
D213	162				
D215	103				
D217	162				
D218	162				
D219	128				
D221	128				
D226	175				
D226A	175				
D226B	175				
D226P	175				
D226Q	175				
D227	123A				S0011
D228	192				
D234	152				
D235	152				
D236	175				
D238	175				
D246	165				
D254	175				
D225	175				
D257	175				
D261	192				
D283	162				
D284	162				
D285	162				
D288	152				
D289	152				
D290	175				
D291	175				
D292	175				S0020
D292(CHAN MAST)	107			TR70/	
D294	123A				
D297	175				
D299	165				
D300B	165				
D304A				TR52/	
D308	199	3039			
D312	164				
D315	175				
D317	152				
D318	152				
D318A	152				
D318B				TR75/	
D319	181				
D320	162				
D322	162				
D323	162				
D324	124				
D325	152				
D327	123A				
D328S	210	3202			
D330D	152				
D334	162				
D336	192				
D342		3124			
D343	152				
D350	238				
D352		3004			G0005
D368	1047				
D369	1081[**]				
D372BL	123A				
D513	152			TR76/	
D562	108				
D720D	103A				
D720E	103A				
D912	123A				
D1101	133	3112	FET1	/801	F0010
D1102	133	3112	FET1	/801	F0010
D1180			FET1		
D1181			FET1		
D1301	133	3112		/801	F0010
D1302	133	3112		/801	F0010
D1303	133	3112		/801	F0010
D1420			FET1		
D1421			FET1		
D1666	108	3039		TR95/56	S0016
D-50492-01	159				
D911138-1	128				
D911138-2	128				
D911138-3	128				
D911138-4	128				
D911138-5	128				
D911138-6	128				
D911138-7	128				
D917254-2	123A				
D919695	9900				
D919696	9901				
D919697	9902				
D919698	9903				
D919699	9904				
D919700	9905				
D921881-1	123A				
D922376-2		3024			
D926640-1	123				

Original Part Number	ECG	SK	GE	IR/WEP	HEP
D928121	123A				
D4000411	909				
D34002410-001	9914				
D34013890-002	74H04				
D34014094	175				
D-TEX-8	904				
DAT1A	126	3006	52	TR17/635	S0013
DAT2	126	3006	52	TR17/635	S0013
DAT-PS-55		3004			
DC-9	102A			TR85/250	G0005
DC-10	102A	3123		TR85/250	G0005
DC-12	158			TR84/630	G6011
DC-13					
DD11C381				TR25/	
DD-79D107-1	130				
DE-3181	102				
DED4191	159				
DEP-17	3008	3008			
DF1	101				
DF-2	130	3029		/247	S7002
DM-1	726				
DM-9	734		IC17		
DM11	709	3135	IC11	IC505/	C6082P
DM11A	737		IC11		
DM14	718	3159	IC8	IC511/	C6074P
DM-19	805				
DM20	744		IC24		
DM24	719				C6065P
DM26	721				
DM30	721		IC14	IC513/	C6080P
DM31	709	3135	IC11	IC505/	C6062P
DM32	744		IC24		
DM35	806				
DM41	737		IC16		
DM44	743		IC32		
DM-44	743		IC35		
DM-51	788	3147			
DM54	719		IC28		C6056P
DM74H00J	74H00				
DM74H00N	74H00				
DM74H01J	74H01				
DM74H01N	74H01				
DM74H04J	74H04				
DM74H04N	74H04				
DM74H05J	74H05				
DM74H05N	74H05				
DM74H08J	74H08				
DM74H08N	74H08				
DM74H10J	74H10				
DM74H10N	74H10				
DM74H11J	74H11				
DM74H11N	74H11				
DM74H20J	74H20				
DM74H20N	74H20				
DM74H21J	74H21				
DM74H21N	74H21				
DM74H22J	74H22				
DM74H22N	74H22				
DM74H30J	74H30				
DM74H30N	74H30				
DM74H40J	74H40				
DM74H40N	74H40				
DM74H50J	74H50				
DM74H50N	74H50				
DM74H51J	74H51				
DM74H51N	74H51				
DM74H52J	74H52				
DM74H52N	74H52				
DM74H53J	74H53				
DM74H53N	74H53				
DM74H54J	74H54				
DM74H54N	74H54				
DM74H55J	74H55				
DM74H55N	74H55				
DM74H60J	74H60				
DM74H60N	74H60				
DM74H61J	74H61				
DM74H61N	74H61				
DM74H62J	74H62				
DM74H62N	74H62				
DM74H71J	74H71				
DM74H71N	74H71				
DM74H72J	74H72				
DM74H72N	74H72				
DM74H73J	74H73				
DM74H73N	74H73				
DM74H74J	74H74				
DM74H74N	74H74				
DM74H76J	74H76				
DM74H76N	74H76				
DM74H78J	74H78				
DM74H78N	74H78				
DM74S00J	74S00				
DM74S00N	74S00				
DM74S03J	74S03				

Original Part Number	ECG	SK	GE	IR/WEP	HEP
DM74S03N	74S03				
DM74S04J	74S04				
DM74S04N	74S04				
DM74S05J	74S05				
DM74S05N	74S05				
DM74S10J	74S10				
DM74S10N	74S10				
DM74S11J	74S11				
DM74S11N	74S11				
DM74S20J	74S20				
DM74S20N	74S20				
DM74S22J	74S22				
DM930N	9930				C1030P
DM932N	9932				
DM933N	9933				
DM935N	9935				
DM936N	9936				
DM937N	9937				
DM944N	9944				
DM945N	9945				
DM946N	9946				
DM948N	9948				
DM949N	9949				
DM961N	9961				
DM962N	9962				C1062P
DM963N	9963				
DM7400N	7400				
DM7401N	7401				
DM7402N	7402				
DM7403N	7403				
DM7404N	7404				
DM7405N	7405				
DM7406N	7406				
DM7407N	7407				
DM7408N	7408				
DM7409N	7409				
DM7410N	7410				
DM7411N	7411				
DM7413N	7413				
DM7414N	7414				
DM7416N	7416				
DM7417N	7417				
DM7420N	7420				
DM7423N	7423				
DM7425N	7425				
DM7427N	7427				
DM7430N	7430				
DM7432N	7432				
DM7437N	7437				
DM7438N	7438				
DM7440N	7440				
DM7441AN	7441A				
DM7442N	7442				
DM7445N	7445				
DM7450N	7450				
DM7451N	7451				
DM7453N	7453				
DM7454N	7454				
DM7460N	7460				
DM7470N	7470				
DM7472N	7472				
DM7473N	7473				
DM7474N	7474				
DM7475N	7475				
DM7476N	7476				
DM7483N	7483A				
DM7486N	7486				
DM7488N	7488				
DM7489N	7489				
DM7490N	7490				
DM7491AN	7491A				
DM7492N	7292A				
DM7493N	7493A				
DM7495N	7495				
DM7496N	7496				
DM8000N					C3000P
DM8001N					C3001P
DM8004N					C3004P
DM8010N					C3010P
DM8020N					C3020P
DM8030N					C3020P
DM8040N					C3040P
DM8050N					C3050P
DM8501N					C3073P
DM8550N					C3075P
DM8840N					C3041P
DM9093N	9093				
DM9094N	9094				
DM9097N	9097				
DM9099N	9099				
DM74107N	74107				
DM74121N	74121				
DM74145N	74145				
DM74150N	74150				
DM74151N	74151				

Original Part Number	ECG	SK	GE	IR/WEP	HEP
DM74153N	74153				
DM74154N	74154				
DM74155N	74155				
DM74156N	74156				
DM74174N	74174				
DM74175N	74175				
DM74180N	74180				
DM74181N	74181				
DM74182N	74182				
DM74190N	74190				
DM74191N	74191				
DM74196N	74196				
DM74197N	74197				
DM74198N	74198				
DM74199N	74199				
DN	123A				
DN20-00453	128				
DN40N3				TR78/	
DO18		3004			
DO20		3008			
DO21	3008	3008			
DO30		3004			
DO31			2		
DO43		3004			
DO57		3124			
DO69		3047			
DO73		3008			
DODONO64089C		3122			
DRC81252		3003			
DRC86486		3004			
DRC86487		3007			
DRC87040		3008			
DRC87540		3007			
DS-1	103A	3011	8	/724	
DS1B	123A			TR21/735	
DS2	101	3011	8	TR08/641	
DS3	102A	3004	52	TR85/250	G0005
DS4	101	3011	8	TR08/641	
DS5	101	3011	7	TR08/641	
DS6	101	3011	8	TR08/641	
DS7	101	3011	8	TR08/641	
DS8	101	3011	2	TR08/641	G0005
DS-8	102A			/250	
DS9	101	3011	8	TR08/641	
DS11	101	3011	8	TR08/641	
DS12	101	3011	8	TR08/641	
DS13	102A			TR85/250	
DS-14	102A	3004	52	TR05/250	G0005
DS16	102A	3123	52	TR85/250	G0005
DS17	102A		2	TR85/250	G0005
DS-19	102A		2	TR85/250	G0005
DS21	100	3008	50	TR05/254	G0005
DS22	100	3123	52	TR05/254	G0005
DS23	100	3005	2	TR05/254	G0005
DS24	126	3006	51	TR17/635	G0003
DS25	126	3008	52	TR17/635	G0008
DS26	102A	3004	2	TR85/250	G0005
DS-28		3005		/254	
DS-28(DELCO)	100			TR05/	
DS29	102A	3004	51	TR85/250	G0003
DS34	160	3006	50	TR17/637	G0003
DS35	160	3006	50	TR17/637	G0002
DS36	126	3006	50	TR17/635	G0008
DS37	160		50	TR17/637	G0003
DS38	126		50	TR17/635	G0003
DS38(DELCO)	160			TR17/	
DS41	160	3006	50	TR17/637	G0003
DS42	160	3006	50	TR17/637	G0003
DS44	123A	3010	8	TR21/735	G0011
DS45	123A	3122	8	TR21/735	S0015
DS46	123A	3124	8	TR21/735	S0015
DS47	123A	3124	8	TR21/735	S0015
DS51	126	3006	2	TR17/635	G0008
DS52	126	3006	2	TR17/635	G0008
DS53	100	3006	2	TR05/254	G0008
DS56	160	3006	52	TR17/637	G0003
DS62	160	3006	50	TR17/	G0003
DS63	160	3006	50	TR17/637	G0003
DS64	160	3006	50	TR17/637	G0003
DS65	160	3006	50	TR17/637	G0003
DS66	123A	3124	17	TR21/735	S0015
DS67	123A	3124	20	TR21/735	S0015
DS67H	123A	3020	20	TR21/735	S0015
DS68	106	3118	22	TR20/52	S0013
DS71	107	3018	11	TR70/720	S0014
DS72	107	3018	11	TR70/720	S0014
DS73	107	3018	17	TR70/720	S0014
DS74	107	3018	86	TR70/720	S0016
DS75	107	3124	20	TR70/720	S0016
DS76	123A	3124	210	TR21/735	S0014
DS77	123A	3124	20	TR21/735	S0015
DS78	161	3018	11	TR83/719	S0011
DS81	161	3018	39	TR83/719	S0011
DS82	159	3118	22	TR20/	S0019
DS82(CHEVY)					S0015

Original Part Number	ECG	SK	GE	IR/WEP	HEP
DS83	159	3025	82	TR20/717	S0012
DS85	161	3018	17	TR83/719	S0014
DS-86	129	3114	21	TR19/	S5013
DS88	132	3116	FET2	FE100/	F0021
DS96	106			TR20/52	
DS-410(MOTO)	123A			TR21/735	
DS501	105	3012	4	TR03/233	G6004
DS502	105	3012	4	TR03/233	G6004
DS503	104	3009	25	TR01/230	G6005
DS504	105	3012	4	TR03/233	G6006
DS505	105	3012	4	TR03/233	G6006
DS506	105	3012	4	TR03/233	G6006
DS509	130	3027	72	TR59/247	S7004
DS512	128	3048		TR87/	S5014
DS513	152	3041	66	TR76/	S5000
DS-514	130	3027	19		
DS515	121	3009	16	TR01/232	G6005
DS519	130	3036	14	TR59/247	
DS520	121	3009	25	TR01/232	G6005
DS525	105	3012	4	TR03/233	G6006
DS570	105	3012	16	TR03/233	
DSD80-03-003		3024			
DSD404-1		3004			
DT41	121	3009	16	TR01/232	G6005
DT80	105	3012	4	TR03/233	
DT100	105	3012	4	TR03/233	G6006
DT161	123A	3122	20	TR21/735	
DT401	104	3014	16	TR01/230	G6003
DT1003	154	3045	32	TR78/712	
DT1040	121	3014	16	TR01/230	G6003
DT1110	128	3024	83	TR87/243	S5014
DT1111	128	3024	215	TR87/243	S5014
DT1112	128	3024	32	TR87/243	
DT1120	128	3024	83	TR87/243	S5014
DT1121	128	3024	215	TR87/243	
DT1122	128	3024	32	TR87/243	S5014
DT1311	128	3024	215	TR87/243	S5014
DT1321	128	3024	215	TR87/243	S5014
DT1510	128	3024	47	TR87/243	
DT1511	128	3024	47	TR87/243	
DT1512	128	3024	18	TR87/243	
DT1520	128	3024	47	TR87/243	
DT1521	128	3024	47	TR87/243	
DT1522	128	3024	18	TR87/243	
DT1602	154	3045	81	TR78/712	
DT1603	154	3045	27	TR78/712	
DT1610	123	3027	81	TR86/53	S0016
DT1612	154	3045	81	TR78/712	
DT1613	154	3045	27	TR78/712	
DT1621	128	3024	47	TR87/243	
DT3301	175	3131	66	TR81/241	S5012
DT3302	175	3131		TR81/241	S5012
DT4011	130	3027	75	TR59/247	
DT4110	130	3027	19	TR59/247	
DT4111	130	3027	19	TR59/247	
DT4120	130	3027	19	TR59/247	
DT4121	130	3027	19	TR59/247	
DT4303	162			TR67/707	
DT4304	162			TR67/707	
DT4305	162		73	TR67/707	
DT4306	162			TR67/707	
DT6110	121	3009	16	TR01/232	G6005
DTG110	121	3009	16	TR01/232	G6005
DTG110A	179		76	TR35/G6001	
DTG110B	179	3009	76	TR35/G6001	G6009
DTG400M	179		76	TR35/G6001	
DTG600	179		76	TR35/G6001	G6009
DTG601	179		76	TR35/G6001	
DTG602	179		76	TR35/G6001	G6009
DTG603	179		76	TR35/G6001	G6009
DTG603M	179		76	TR35/G6001	
DTG1010	127		25	TR27/235	G6008
DTG1011	121	3009	16	TR01/232	
DTG1040	121	3009	16	TR01/232	
DTG1110	127		25	TR27/235	G6008
DTG1200	179		76	TR35/G6001	G6009
DTG2000	179		76	TR35/G6001	G6009
DTG2000A	179		76	TR35/G6001	
DTG2100	179		76	TR35/G6001	G6009
DTG2100A	179		76	TR35/G6001	
DTG2200	179		76	TR35/G6001	G6009
DTM-2					S0030
DTM-3					S0030
DTM-4					S0019
DTM-5					S5026
DTM-6					S0019
DTM-8					S5026
DTM-9					G6016
DTM-10					S0025
DTM-11					S5012
DTM-12					S5018
DTM-13					S7002
DTM-14					S7003
DTN206	128				
DTS013	164	3115	37	TR93/740A	

Original Part Number	ECG	SK	GE	IR/WEP	HEP
DTS-0709		3036			
DTS0710	162			TR67/	
DTS0713	164		37	TR93/	
DTS103	162		35	TR67/707	S7004
DTS104	162		35	TR67/707	S7004
DTS-105					S7004
DTS-106					S7004
DTS-107					S7004
DTS401	162		35	TR67/707	
DTS402	163	3111	73	TR67/740	S5021
DTS410	162	3560	72	TR67/707	S5020
DTS411	162	3560	73	TR67/707	S5020
DTS413	162	3560	73	TR67/707	S5020
DTS423	162	3560	35	TR67/707	S5020
DTS423M	162		73	TR67/707	S5020
DTS424	163	3111	36	TR67/740	
DTS425	163	3111	36	TR67/740	
DTS430	162		73	TR67/707	S5020
DTS431	162	3560	73	TR67/707	S5020
DTS431M	162		73	TR67/707	
DTS701	164	3115	37	TR93/740A	S5021
DTS702	165	3115	38	TR93/740B	
DTS704	165	3115	38	TR93/740B	
DTS801	164	3115	37	TR93/740A	
DTS802	165	3115	38	TR93/740B	
DTS804	165	3115	38	TR93/740B	
DTS3705	162		35	TR67/707	
DTS3705A	162		35	TR67/707	
DTS3705B	162		35	TR76/707	
DTμL9093	9093				
DTμL9099	9099				
DTμL9930	9930				
DTμL9932	9932				
DTμL9933	9933				
DTμL9935	9935				
DTμL9936	9936				
DTμL9944	9944				
DTμL9945	9945				
DTμL9946	9946				
DTμL9948	9948				
DTμL9949	9949				
DTμL9961	9961				
DTμL9962	9962				
DTμL9963	9963				
DU1	160			TR17/637	G0003
DU2	160			TR17/637	G0002
DU3	102A			TR85/250	G0005
DU4	158			TR84/630	G0005
DU5	102A			TR85/250	G0005
DU6	104	3009		TR01/230	G6003
DU7	105	3012		TR03/231	G6004
DU12	160			TR17/637	G0003
DU47	105	3012		TR03/233	
DU4340		3112			
DW6034/M	123A				
DW6195	128				
DW 6505	123A				
DW 6982	128				
DW 7375	123A				
DW 7655	234				
DX1018	108	3018		TR95/56	
DXX763-1000-2H		3005			
DXX763-1000-9H		3003			
E-044A	102A		2	TR85/250	
E-065	126	3008		TR17/635	
E-066	126	3008	2	TR17/635	
E-067	126	3008	2	TR17/635	
E-068	126	3008	2	TR17/635	
E-070	102A	3004	51	TR85/250	
E-01381	128				
E03090-002	167				
E1A	107	3039		TR70/720	S0020
E3			62		
E4	158		53	TR84/	G6011
E5			62		
E13-000-03	123A	3018	20		S0015
E13-000-04	123A	3018			
E13-001-02	159	3114	21		S0019
E13-001-03	159	3114	21		S0019
E13-001-04	159	3114	21		S0019
E13-002-03	123A	3018	20		S0015
E13-003-00	123A	3018	20		S0015
E13-003-01	123A	3018	20		S0015
E13-004-00	233	3018	20		S0015
E13-005-02	123A	3018	20		S0015
E13-006-02	159	3114	21		S0019
E13-007-00	171	3104	27		S5015
E13-008-00	162	3133	35		S5020
E13-009-00	165	3111	36		S5020
E13-010-00	191	3104	27		S5015
E13-011-00	191	3104	27		S5015
E13-012-00	182	3104	55		S5000
E102	126	3008		TR17/635	
E103	133	3112	FET1	/801	
E105	126	3008		TR17/635	

Original Part Number	ECG	SK	GE	IR/WEP	HEP
E132	102A	3004	51	TR85/250	G0005
E158	102A	3004	2	TR85/250	G0005
E-167-228	128				
E181	102A			TR85/250	
E181A	102A	3004		TR85/250	
E181D	102A	3004		TR85/250	
E185B121712	157			TR60/244	
E201		3112			
E202		3112			
E203		3112			
E210	123A	3124		TR21/735	S0015
E211	159	3025		TR20/717	S0031
E212	123A	3112		TR21/735	S5014
E213	159			TR20/717	S5013
E214B			2		
E230		3112			
E231		3112			
E232		3112			
E241	102A	3004	2	TR85/250	G0005
E241A	102A	3004	2	TR85/250	G0005
E241B	102A	3004		TR85/250	G0005
E300	312	3116	FET2		
E304		3116			
E305		3116			
E318-1	128				
E629	108				
E2412	100	3005		TR05/254	G0005
E2427	101	3018	8	TR08/641	
E2428	101	3018	8	TR08/641	
E2429	101	3018	8	TR08/641	
E2430	123A	3124	20	TR21/735	S0015
E2431	123A	3124	20	TR21/735	
E2434	108	3018	17	TR95/56	
E2435	108	3018	20	TR95/56	
E2436	123A	3124	20	TR21/735	
E2438	160	3006	51	TR17/637	
E2439	160	3006	50	TR17/637	
E2440	160	3006	50	TR17/637	
E2441	128	3024	18	TR87/243	
E2444	123A	3124	20	TR21/735	
E2445	102A	3004	52	TR85/250	
E2447	103A	3010	8	TR08/724	
E2448	102A		20	TR85/250	
E2449	129	3024	18	TR87/243	
E2450	160	3006	20	TR17/637	
E2451	126	3005	1	TR17/635	
E2452	123A	3122	10	TR21/635	
E2453	102A	3004	2	TR85/250	
E2454	123A	3124	20	TR21/735	
E2455	123A	3122	10	TR21/735	
E2459	123A	3124	10	TR21/735	
E2460	124	3021	12	TR81/240	
E2461	123A	3124	63	TR21/735	
E2462	160	3005	50	TR17/637	
E2465	102A	3004	52	TR85/250	G0005
E2466	103A	3010	8	TR08/724	G0011
E2467	102A	3004	52	TR85/250	G0011
E2474	160	3006	50	TR17/637	G0003
E2475	160	3006	50	TR17/637	G0003
E2476	102A	3004	52	TR85/250	G0005
E2477	160	3008	51	TR17/637	G0005
E2478	160	3006	50	TR17/637	G0003
E2479	160	3006	50	TR17/637	G0003
E2480	102A	3004	52	TR85/250	G0005
E2481	102A	3004	52	TR85/250	G0005
E2482	102A	3123	52	TR85/250	G0006
E2490A		3004			
E2491B	128				
E2495	727	3071			
E2496	152	3041		TR76/701	S5000
E2497	123A	3124		TR21/735	S5014
E2498	129	3025		TR88/242	S5013
E2499	123A	3124		TR21/735	S0030
E3530		3019			
E4002	101				
E24103	123A		17	TR21/	S0030
E24104	102A		52	TR85/250	G0007
E24105	103A		54	TR08/	G0011
E24106	102A	3004	54	TR85/250	G0005
E24107	131	3052		TR94/642	G6016
E570022-01	175	3131		TR81/241	S5012
EA0002	160	3006		TR17/637	
EA0007	160	3006		TR17/637	
EA0009	102A	3004		TR85/250	
EA0013	108	3018		TR95/56	
EA0053	160	3006		TR17/637	
EA0081	128	3024	18	TR87/243	
EA0086	129			TR88/242	
EA0087	129	3025	21	TR88/242	
EA0090	128	3024		TR87/243	
EA0091	108	3018	20	TR95/56	
EA0092	123A	3124	20	TR21/735	
EA0093	108	3018	20	TR95/56	
EA0094	108	3018		TR95/56	
EA0095	108	3018		TR95/56	

Original Part Number	ECG	SK	GE	IR/WEP	HEP	Original Part Number	ECG	SK	GE	IR/WEP	HEP
EA1-380			57	TR76/		EA15X140	126	3006		TR17/635	G0008
EA1SY11			50	TR17/		EA15X141	126	3006	52	TR17/635	G0008
EA15X1	123A	3124	8	TR21/735	S0015	EA15X142	123A	3124		TR21/735	S0030
EA15X2	102A	3004	8	TR85/250	G0005	EA15X143	123A	3122		TR21/735	S0015
EA15X3	102A	3004	52	TR85/250	G0005	EA15X144	128	3024		TR87/243	S0015
EA15X4	102A	3004	52	TR85/250	G0005	EA15X152	123A	3020		TR21/	S0015
EA15X5	160	3124	8	TR17/637	S0015	EA15X153	123A	3024		TR21/	S0015
EA15X6	105	3012	4	TR03/233	G6004	EA15X154	131	3052		TR94/	G6016
EA15X7	102A	3004	52	TR85/250	G0005	EA15X160	186	3054	28	TR55/	
EA15X8	102A	3004	52	TR85/250	G0005	EA15X185	159	3138		TR31/	S5022
EA15X9	123A	3124	8	TR21/735	S0015	EA15X203	102A	3004	53	TR85/	
EA15X10	121	3009	16	TR01/232	G6003	EA15X207	102A	3004	52	TR85/	
EA15X11	160	3006	50	TR17/637	G0003	EA15X212	102A	3004	52	TR85/	
EA15X12	121	3009	16	TR01/232	G6003	EA15X245	199	3020	210		
EA15X13	160	3006	50	TR17/637	G0003	EA15X246	123A		20		
EA15X15	121	3009	16	TR01/232	G6003	EA15X247	172		64		
EA15X18	123A	3124	20	TR21/735	S0015	EA15X248	186		28		
EA15X19	102A	3004	52	TR85/250	G0005	EA15X249	128				
EA15X20	123A	3124	20	TR21/735		EA15X256	289	3122	81		S0015
EA15X22	123A	3124	20	TR21/735	S0015	EA15X257	158	3004	53		G0005
EA15X23	102A	3004	52	TR85/250		EA15X325	123A	3122	88		
EA15X24	123A	3124	20	TR21/735	S0015	EA15X326	156	3004	53		
EA15X25	102A	3004	52	TR85/250		EA15X3118			69		
EA15X26	121	3009	16	TR01/232	G6005	EA15X6840	102A	3004	53	TR84/	G6011
EA15X27	160		50	TR17/637		EA15X7115	123A	3018	20	TR51/	S5015
EA15X28	102A	3004	51	TR85/250		EA15X7119	123A	3018	20	TR51/	S5015
EA15X29	160		51	TR17/637		EA15X7121		3054	66		
EA15X30	160		50	TR17/637		EA15X7125	107		20		
EA15X31	123A	3124	8	TR21/735	S0015	EA15X7141		3018	20		
EA15X33	121	3009	16	TR01/232	G6003	EA15X7175	123A	3018	20		
EA15X35	121	3009	16	TR01/232	G6003	EA15X7176	123A	3018	20		
EA15X36	102A	3004	52	TR85/250	G0005	EA15X7215	107	3018	20	TR24/	S0011
EA15X37	123A	3124	20	TR21/735	S0015	EA15X7228	108		20		
EA15X38	121	3009	16	TR01/232	G6003	EA15X7231	107	3018	17	TR24/	S0025
EA15X40	160	3006	51	TR17/637		EA15X7232	123A	3018	20	TR33/	S5015
EA15X41	160	3006	50	TR17/637	G0003	EA15X7233	107		17	TR24/	S0025
EA15X43	160	3006	50	TR17/637		EA15X7235	108	3018	20	TR33/	S0015
EA15X44	123A	3018	20	TR21/735	S0015	EA15X7262	108		20		
EA15X45	123A	3018	20	TR21/735	S0015	EA15X7264	107				
EA15X48	108	3018	11	TR95/56	S0016	EA15X7514	123A	3018	20	TR53/	S5015
EA15X49	108	3018	11	TR95/56	S0016	EA15X7517	123A	3122	10	TR24/	S0015
EA15X50	108	3018	11	TR95/56	S0016	EA15X7519	192	3137	63	TR21/	S5015
EA15X51	108	3124	20	TR95/56	S0015	EA15X7586	123A	3018	20		
EA15X52	123A	3124	20	TR21/735	S0015	EA15X7587	108	3018	20		
EA15X53	121	3009	16	TR01/232	G6003	EA15X7588	123A	3018	20		
EA15X54	108	3019	17	TR95/56	S0016	EA15X7589	123A	3018	20		
EA15X55	108	3019	17	TR95/56	S0016	EA15X7592		3024	210		
EA15X56	123A	3124	8	TR21/735	S0015	EA15X7638	123A	3018	20	TR21/	S5015
EA15X57	128		20	TR87/243		EA15X7639		3024	212		
EA15X58	123A	3124	20	TR21/735		EA15X7722	108	3020	20	TR24/	S0016
EA15X59	123A	3124	20	TR21/735		EA15X8118		3083			
EA15X63	123A	3124	8	TR21/735	S0015	EA15X8119		3054	66		
EA15X66	160		51	TR17/637		EA15X8443	103A		54		
EA15X67	102A	3004	52	TR85/250	G0005	EA15X8444	102A		54		
EA15X68	123A	3124	8	TR21/735	S0015	EA15X8511	123A	3018	20		
EA15X69	106	3118	22	TR20/52	S0013	EA15X8517	293	3020	18		
EA15X70	106	3118	21	TR20/52	S0013	EA15X8518		3122	10		
EA15X71	106	3118	21	TR20/52	S0013	EA15X8521		3137	82		
EA15X72	123A	3124	20	TR21/735		EA15X8522		3138	81		
EA15X73	123A	3124	20	TR21/735		EA15X8601	123A	3083	69	TR21/S0015	
EA15X75	123A	3124	10	TR21/735	S0014	EA15X8602		3054	66		
EA15X76	123A	3124	10	TR21/735	S0014	EA15X8605		3083	69		
EA15X77	123A	3124	10	TR21/735	S0014	EA15X8608		3018	20		
EA15X83	123A	3124	10	TR21/735	S0014	EA15X8610		3018			
EA15X84	123A	3124	10	TR21/735	S0014	EA33X8367	1003	3102			
EA15X85	123A	3124	10	TR21/735	S0014	EA33X8368	1036				
EA15X86	123A	3024	10	TR21/735	S0014	EA1080	123A	3122	20	TR21/735	
EA15X88	121	3009	3	TR01/232	G6005	EA1081	102A	3123	52	TR85/250	
EA15X89	123A	3122	18	TR21/735		EA1128	123A	3124	20	TR21/735	
EA15X90			20		S0015	EA1129	123A	3124	20	TR21/735	
EA15X91	123A	3124		TR21/	S0015	EA1135	123A	3124	20	TR21/735	
EA15X94	108	3018	20	TR95/56	S0015	EA1145	123A	3122	20	TR21/735	
EA15X95					S0015	EA1337	160	3006		TR17/637	G0003
EA15X96	123A	3124	18	TR21/735	S0014	EA1338	160	3006		TR17/637	G0003
EA15X98	123A	3024	18	TR21/735		EA1339	160	3006		TR17/637	G0003
EA15X99	152	3041	28	TR76/701		EA1340	160	3006		TR17/637	G0003
EA15X100	130	3027	14	TR26/247+701		EA1341	121	3009		TR01/232	G6003
EA15X101	123A	3124		TR21/735	S0015	EA1342	160	3006		TR17/637	G0003
EA15X102	128	3024		TR87/243	S0030	EA1343	108	3018	20	TR95/56	S0015
EA15X103	123A	3124		TR21/735	S0015	EA1344	123A	3018	20	TR21/735	S0015
EA15X111	123A	3124	20	TR21/735		EA1345	123A	3124	20	TR21/735	S0011
EA15X112	123A	3124	18	TR21/735		EA1346	102A	3004	52	TR85/250	G0005
EA15X113	108	3039	17	TR95/56		EA1406	123A	3124	20	TR21/735	
EA15X121	175	3025		TR81/241	S5019	EA1407	123A	3020	20	TR21/735	
EA15X123	181			TR36/S7000	S7002	EA1408	123A	3124	20	TR21/735	
EA15X124	180			/S7001	S7003	EA1451	123A	3124	20	TR21/735	
EA15X130	107	3018		TR70/720	S0016	EA1452	123A	3124	20	TR21/735	
EA15X131	108	3018		TR95/56	S0014	EA1499	123A	3124	20	TR21/735	
EA15X132	108	3018		TR95/56	S0014	EA1549	128	3124	20	TR87/243	
EA15X133	126	3006		TR17/635	G0008	EA1562	108	3018	11	TR95/56	S0016
EA15X134	107	3018		TR70/720	S0014	EA1563	108	3018	11	TR95/56	S0016
EA15X135	107	3018	20	TR70/720	S0014	EA1564	123A	3124	20	TR21/735	S0015
EA15X136	123A	3124	10	TR21/735	S0030	EA1578	123A	3124	20	TR21/735	
EA15X137	123A	3124		TR21/735	S0014	EA1581	123A	3124	20	TR21/735	
EA15X139	131	3052		TR94/642	G6016	EA1628	123A	3124	10	TR21/735	S0014

Original Part Number	ECG	SK	GE	IR/WEP	HEP
EA1629	123A	3018	10	TR21/735	S0014
EA1630	123A	3124	10	TR21/735	S0014
EA1638	123A			TR21/735	
EA1684	128	3024	20	TR87/243	S5014
EA1695	123A	3038		TR21/735	S0015
EA1696	123A	3124	20	TR21/735	S0014
EA1697	123A	3124	20	TR21/735	S0014
EA1698	128	3124	63	TR87/243	S0014
EA1700	121	3009	16	TR01/232	G6005
EA1703	123A	3124	20	TR21/735	S0014
EA1716	123A	3124	20	TR21/735	S5014
EA1718	123A	3122	20	TR21/735	S0015
EA1733	108	3039	20	TR95/56	S0015
EA1735	123A	3124	20	TR21/735	S0014
EA1740	130	3027	19	TR26/247+701	
EA1760	717				
EA1778			20	TR21/	
EA1793	108	3018		TR95/56	S0015
EA1872	123A	3122		TR21/735	S0015
EA1873	128	3020		TR87/243	S0030
EA2131	107	3018	17	TR70/720	S0020
EA2132	107	3018	17	TR70/720	S0016
EA2133	126	3006	1	TR17/635	G0009
EA2134	102A	3003	2	TR85/250	G0007
EA2135	102A	3004	53	TR85/250	G0005
EA2136	158	3004	53	TR84/	G0005
EA2176	102A	3004		TR85/250	G0005
EA2219		3018			S0015
EA2220		3018			S0015
EA2349		3018			S0011
EA2350		3018			S0011
EA2351		3018			S0011
EA2352		3018			S0015
EA2353		3018			S0015
EA2354		3010			G0011
EA2355		3004			G0008
EA2356		3004			G0008
EA2357		3018			S0015
EA2429	199	3018	20	TR24/	
EA2488	186	3054	28	TR55/	
EA2489	123A	3124	10	TR24/	
EA2490	123A	3020		TR21/	S0015
EA2491	126	3006		TR17/	G0008
EA2493	108	3018		TR95/	S0014
EA2494	107	3018	20	TR70/	S0014
EA2495	108	3018		TR95/	S0014
EA2496	107	3018		TR70/	S0014
EA2497	126	3006		TR17/	G0008
EA2498	126	3006	52	TR17/	G0008
EA2600	108	3018		TR95/	S0020
EA2601		3018		TR21/	S0020
EA2602	108	3018		TR95/	S0020
EA2603	108	3018		TR95/	S0020
EA2604	108	3018		TR95/	S0020
EA2605	108	3018		TR95/	S0020
EA2770	123A	3020	10	TR21/	S0014
EA2771	123A	3124	212	TR21/	S0015
EA2812	107	3018		TR70/	S0016
EA3149	123A	3122	10		
EA3211		3122	62		S0024
EA3674	186	3054	28		S5000
EA3713	107	3018	20		S0016
EA3714	159	3114	21		S0013
EA3715	153		69		
EA3716	152		66		
EA3763	123A	3018	20		S0015
EA3990	192	3122	63		
EA4112	123A	3122	10		
EA6801	1000				
EAI-380	184			/S5003	S5000
EB0001	102A	3004	2	TR85/250	
EB0003	102A	3004	2	TR85/250	
EB15X4		3114			
EB-134144	194				
ECC-01262	9932				
ECC-01263	9945				
ECC-01264	9946				
ECC-01265	9962				
ECC-01266	9951				
ECC-01267	9930				
ECG100		3005	1	TR05/254	G0005
ECG101		3011	8	TR08/641	G0011
ECG102		3004	2	TR05/631	G0005
ECG102A		3004	52	TR85/250	G0005
ECG103		3010	8	TR08/641A	G0011
ECG103A		3010	59	TR08/641B	G0011
ECG104		3009	76	TR01/624	G6003
ECG104MP		3013	13MP	TR01MP/230MP	
ECG105		3012	4	TR03/231	G6004
ECG106		3118	21	TR20/52	S0013
ECG107		3039	11	TR70/720	S0016
ECG108		3039	11	TR95/56	S0025
ECG121		3009	76	TR01/628	G6005
ECG121MP		3013		TR01MP/628MP	
ECG123		3122	20	TR86/53	S0024

Original Part Number	ECG	SK	GE	IR/WEP	HEP
ECG123A		3122	20	TR21/735	S0015
ECG124		3021	12	TR23/240	S5011
ECG126		3006	52	TR17/635	G0006
ECG127		3035	25	TR27/235	G6008
ECG128		3024	18	TR87/243	S3021
ECG129		3025		TR88/242	S3032
ECG130		3027	77	TR59/247	S7002
ECG130MP		3029	15MP	TR59MP/247MP	
ECG131		3082	78	TR94/642	G6016
ECG131MP		3086		TR94MP/642MP	
ECG132		3116	FET1	FE100/802	F0015
ECG133		3112	FET1	/801	F0010
ECG152		3054	28	TR76/701	S5000
ECG153		3083	29	TR77/700	S5006
ECG154		3044	40	TR78/712	S5024
ECG155			79	TR76/	
ECG157		3103	27	TR60/244	S5015
ECG158		3004	53	TR84/630	G6011
ECG159		3025	22	TR20/717	S5022
ECG160		3006	51	TR17/637	G0008
ECG161		3132	39	TR83/	S0020
ECG162			35	TR67/707	S5020
ECG163			36	TR67/740	S5021
ECG164		3133	37	TR93/740A	S5020
ECG165		3115	38	TR93/740B	S5021
ECG171		3103	27	TR79/	S3021
ECG172		3156	64	TR69/S9100	S9100
ECG175		3538		TR81/241	S5012
ECG176		3123	80	TR82/238	G6011
ECG179			76	TR35/G6001	
ECG180			74	/S7001	S7001
ECG181		3535	75	TR36/S7000	S7000
ECG182		3188	55	/S5004	S5004
ECG183		3189	56	/S5005	S5005
ECG184		3054	57	TR76/S5003	S5000
ECG185		3083	58	TR77/S5007	S5006
ECG186		3054	226	TR55/S3023	S5000
ECG187		3083	227	TR56/S3027	S5006
ECG188		3054	217	TR72/S3020	S3019
ECG189		3083	218	TR73/S3031	S3032
ECG190		3104		TR74/S3021	S3021
ECG191		3104		TR75/	S3021
ECG192		3137	63		S3020
ECG193		3138	67		S3032
ECG194		3104	220		S0005
ECG195		3048	219	TR65/	S3001
ECG196		3054		TR92/	S5004
ECG197		3085			S5005
ECG198					S5015
ECG199				TR21/	
ECG210		3202			S3020
ECG211		3203			S3032
ECG213					G6010
ECG218		3083		TR58/S50	S5018
ECG219		3173			S7003
ECG220					F2005
ECG221		3050			F2004
ECG222		3065			F2004
ECG223		3027		TR59/	S5004
ECG224		3049	46		
ECG225		3103			
ECG226		3082	49	TR94/	G6016
ECG226MP		3086		TR94MP/	
ECG235		3197			
ECG236		3197			
ECG246		3183			
ECG262		3181			
ECG263		3180			
ECG286		3194			
ECG288			221		
ECG297			222		
ECG298			223		
ECG308		3196			
ECG310		3196			
ECG311		3195			
ECG703		3157	IC12	IC500/	C6001
ECG703A		3157	IC12	IC500/	C6001
ECG704		3023		IC501/	
ECG705				IC502/	C6067G
ECG705A		3134	IC3	IC502/	C6089
ECG706		3101		IC507/	
ECG707				IC503/	C6067G
ECG708		3135	IC10	IC504/	C6062P
ECG709		3135	IC11	IC505/	C6062P
ECG710		3102		IC506/	
ECG711		3070			C6069G
ECG712		3072	IC2	IC507/	C6083P
ECG713		3077	IC5	IC508/	C6057P
ECG714		3075	IC4	IC509/	C6070P
ECG715		3076	IC6	IC510/	C6071P
ECG718		3159	IC8	IC511/	C6094P
ECG719			IC28		C6065P
ECG720		3160	IC7	IC512/	C6095P
ECG721			IC14		C6080P
ECG722		3161	IC9	IC513/	C6056P

Original Part Number	ECG	SK	GE	IR/WEP	HEP
ECG723	3075	3075	IC15		C6101P
ECG724		3525			
ECG725		3162	IC19		
ECG726		3129			
ECG727		3071			
ECG728		3073	IC22		
ECG729	3074	3074	IC23		
ECG730		3143			
ECG731		3172	IC13		
ECG732			IC27		C6090
ECG734		3163			
ECG736			IC17		
ECG737			IC16		
ECG738	3170	3170	IC29		C6075P
ECG739			IC30		C6072P
ECG740			IC31		
ECG744			IC24		
ECG746					C6059P
ECG747					C6079P
ECG748			IC26		C6060P
ECG749		3171			C6076P
ECG750					C6061P
ECG752					C6003
ECG753					C6010
ECG754					C6007
ECG755					C6013
ECG756					C6014
ECG757					C6014
ECG760					C6001
ECG761					C6011
ECG762					C6012
ECG766					C6009
ECG767					C6015
ECG769					C6008
ECG771					C6004
ECG774					C6002
ECG778		3557			C6102P
ECG780		3141	IC20		C6069G
ECG781		3169			
ECG782					C6077P
ECG783			IC21		C6100P
ECG784		3524			
ECG786		3140			
ECG787		3146			
ECG788		3147			
ECG789		3078			
ECG790			IC18		C6057P
ECG791		3149	IC33		
ECG799			IC34		
ECG801		3160			
ECG903		3540			
ECG909		3551			
ECG909D		3552			
ECG910		3553			
ECG910D		3554			
ECG911		3555			
ECG911D		3556			
ECG917		3544			
ECG923		3164			
ECG923D		3165			
ECG941		3514			
ECG941D		3558			
ECG941M		3559			C6052P
ECG947		3526			
ECG949		3166	IC25		
ECG955M		3564			
ECG1003			IC43		
ECG1004			IC36		
ECG1005			IC42		
ECG1006			IC38		
ECG1010			IC37		
ECG1024		3152	720		
ECG1025		3155	722		
ECG1027		3153	721		
ECG1053			IC44		
ECG1054			IC45		
ECG1056			IC48		
ECG1060			IC50		
ECG1080			IC41		
ECG1115		3184			
ECG1121			IC39		
ECG1122			IC40		
ECG6401				2160/	
ECG6402					S9001
ED44A				TR85/	
ED51	160		2	TR17/637	G0003
ED52	100	3005	2	TR05/254	G0005
ED53	100	3005	2	TR05/254	G0005
ED54B	100	3005	2	TR05/254	G0005
ED55	102A	3004	52	TR85/250	G0005
ED56	102A	3004	52	TR85/250	G0005
ED57	102A	3004	52	TR85/250	G0005
ED-1402	123A	3018	20	TR21/	
ED1402A	123A				
ED1402B	123A	3018	20		S0015

Original Part Number	ECG	SK	GE	IR/WEP	HEP
ED1402C	123A	3018	20	TR51/	
ED1502B	161	3018	20	TR22/	S0015
ED-1502C	107	3018	17	TR24/	
ED1502C	123A				
ED1502D	123A	3018	20		
ED1602C	159	3114	21		S0019
ED1702L	123A		20	TR24/	S0015
ED1702M	293		62	TR51/	
ED1802M	294		65	TR30/	
EDC-Q10-1	123A				
EDC TR11-1	199				
EDC TR11-4	128				
EDC TR11-5	128				
EDO 219	123A				
EDS-100	123A				
EE100	102				
EE301		3039			
EECG-1234		3124			
EF1	133	3112		FE100/	F0015
EF2	132	3116		FE100/	F0015
EF3	132			FE100/	F1036
EF4	222	3050			F2005
EF5	222	3065			G6005
EF6					F2007
EK136	102A		2	TR85/250	
EK159	100	3005	1	TR05/254	
EL214	128	3024	63		S0015
EL232	123A	3024	63		S0015
EL264	129	3025	67		S0019
EL401	172		64		S9100
EL611		3024			
EM9-8	914				
EM13531-1		3004			
EM-73531H		3024			
EM73531U		3024			
EMS73278	128				
EMS73279	128				
EMS-73500	123A				
EN10			17		S0015
EN30			10	TR24/	S0015
EN40			20	TR25/	S0015
EN697	123A	3122	212	TR21/735	S0015
EN706	123A	3122	86	TR21/735	S0033
EN708	123A	3122	60	TR21/735	S0025
EN718A	108		212	TR51/	
EN722	159	3118	65	TR52/	
EN744	108				
EN870			18	TR51/	
EN914	108		60	TR95/56	
EN916	108		61	TR51/	
EN918	108		86	TR95/56	
EN930	123A		212	TR51/	
EN956			212	TR51/	
EN1132	159		65	TR52/	
EN1613	128		212	TR53/	
EN1711	128		212	TR53/	
EN2219		3122	20		
EN2222	123A		20		
EN2484	123A		63	TR51/	
EN2894	159	3114		TR20/717	
EN2905	129		48		
EN2907	159	3118	48		
EN3009	123A	3122	60	TR21/735	
EN3011	108		11	TR64/	
EN3013	123A	3122	60	TR21/735	
EN3014	123A	3122	60	TR21/735	
EN3250	159		21		
EN3502	129		48		
EN3504	159	3118	48		
EN3904	128		20		
EN3905	159		21		
EN3906	159		21		
EN3962	159		67	TR52/	
EN5172	107				
EO44A	102A	3004	2	TR85/250	G0005
EO65	100	3005		TR05/254	G0005
EO66	100	3005		TR05/254	G0005
EO67	100	3005		TR05/254	G0005
EO68	100	3005		TR05/254	G0005
EO70	126			TR17/635	G0005
EO105	100	3005		TR05/254	G0005
EP15X1	123A	3117	63	TR21/735	S0015
EP15X2	123A	3124	20	TR21/735	S0015
EP15X3	199	3024	63	TR87/243	S3019
EP15X4	129	3025	21	TR20/717	S0012
EP15X5	128	3024	21	TR87/243	S3019
EP15X6	229	3018	20	TR21/735	S0015
EP15X7	123A	3124	17	TR21/735	S0015
EP15X8	123A	3124	17	TR21/735	S0014
EP15X9	123A	3124	17	TR21/735	S0015
EP15X10	191	3024	18	TR75/243	S3019
EP15X11	196	3054	66	TR76/	S5000
EP15X12	163	3115	36	TR67/	S5021
EP15X13	193	3138	67	TR20/	S3032
EP15X14	196	3054	66	TR76/	S5000

Original Part Number	ECG	SK	GE	IR/WEP	HEP
EP15X15	153	3083	69	TR77/	S5006
EP15X16	124	3026		TR23/	S5011
EP15X17	159	3114		TR19/	S0019
EP15X18	154	3024		TR32/	S5026
EP15X19	128	3044		TR87/	S5015
EP15X20	233	3039		TR70/	S0017
EP15X21	193	3114		TR52/	S0012
EP15X22	152	3054		TR55/	S5000
EP15X23	153	3083	69	TR77/	S5006
EP15X24	187	3083	29	TR56/	S5006
EP15X25	186	3054	28	TR55/	S5000
EP15X26	159	3114	22	TR52/	S0019
EP15X27	171	3104	27	TR60/	S3021
EP15X28	163	3111	36	TR93/	S5021
EP15X29	181	3079	75	TR26/	S7004
EP15X30	152	3054	66	TR76/	S5000
EP15X32	186	3054	28	TR55/	S5000
EP15X33	128	3024	18	TR25/	S5014
EP15X34	186	3054	28	TR55/	S5000
EP15X35	124	3021	12	TR23/	S5011
EP15X43	196				
EP15X44	197				
EP15X45	165				
EP15X47	123A				
EP15X48	123A	3114	21		S0019
EP15X49	123A				
EP15X50	198	3054	66		
EP15X51		3114	21		
EP15X53	159	3114	21		
EP15X90		3114			
EP16X7	128	3124	17	TR87/243	S0014
EP20					S0019
EP25			22	TR28/	S0019
EP35			67		S0019
EP51X9					S0015
EP84X1	749				
EP84X2	712	3072			
EP84X3	713	3077			
EP84X4	783		IC21		C6100P
EP84X5	783		IC21		C6100P
EP84X6	712				
EP84X7	728				
EP84X9	713				
EP84X10	749				
EP84X12	738				
EP100	186		66	TR76/	S3024
EP101A	187		69	TR77/	S3028
EPX15X17	159				
EQ5-60		3047			
EQF-3		3112	FET1		G0003
EQF-4	132	3116		FE100/	F0021
EQG-6	121	3009		TR01/	G6005
EQG-8	104	3009	16		G6005
EQG-9	102A	3004		TR85/	G0007
EQG-15		3004	53		G0005
EQG-20		3004		TR08/	G0005
EQH-1	103A	3010		TR08/	G0011
EQH-20		3010		TR08/	G0011
EQR-1		3114	21		S0032
EQS-5	107	3018	20		S0011
EQS-9	123A	3020	20		S0015
EQS-10		3124	62		S0015
EQS-11		3124	62		S0015
EQS-13	123A				
EQS-18	107	3018		TR70/720	S0014
EQS-19	108	3018		TR95/56	S0016
EQS-20	108	3124		TR95/56	S0016
EQS-21	108	3124		TR95/56	S0016
EQS-22	123A	3047		TR21/735	S0015
EQS-56	195	3049		TR65/	S3020
EQS-57	195	3048			S3001
EQS-60	224				S3013
EQS-61	123A	3124			S0015
EQS-62		3124	17		S0030
EQS-64		3020	20		S0015
EQS-66		3041	28		S5000
EQS-67		3041	28		S5000
EQS-78	199	3124		TR21/	S0015
EQS-86	237	3049			
EQS-89	186	3054	46		
EQS-100	123A	3018	20		S0015
EQS-131	128	3024	20		
EQS-139	199	3018		TR51/	S0016
EQS-140	152	3054			
EQS-141	236		66		
ER15X4	160	3008	51	TR17/637	
ER15X5	160	3008	51	TR17/637	
ER15X6	160	3008	51	TR17/637	
ER15X7	102A	3004	2	TR85/250	
ER15X9	102A	3004	2	TR85/250	
ER15X10	121	3009	16	TR01/232	G6005
ER15X11	160	3006	2	TR17/637	G0003
ER15X12	160	3006	51	TR17/637	G0003
ER15X13	160	3006	51	TR17/637	G0003
ER15X14	160	3008	2	TR17/637	G0003

Original Part Number	ECG	SK	GE	IR/WEP	HEP
ER15X15	160	3008	50	TR17/637	G0002
ER15X16	160	3008	50	TR17/637	G0002
ER15X17	102A	3004	52	TR85/250	G0005
ER15X18	102A	3004	52	TR85/250	G0005
ER15X19	160		50	TR17/637	G0003
ER15X20	160		50	TR17/637	G0003
ER15X21	160		50	TR17/637	G0003
ER15X22	102A	3123	52	TR85/250	G0005
ER15X23	102A	3123	52	TR85/250	G0005
ER15X24	160	3006	50	TR17/637	
ER15X25	160	3006	51	TR17/637	
ER15X26	160	3006	50	TR17/637	
ERS100			18		
ERS120			18		
ERS140			27		
ERS160			27		
ERS180			27		
ERS200			27		
ERS225			27		
ERS250			27		
ERS275			27		
ERS301			27		
ERS325			27		
ERS350			32		
ERS375			32		
ERS401			32		
ERS425			32		
ERS450			32		
ERS475			32		
ES1	160	3007		TR17/	G0008
ES2	102A	3004		TR85/	G0006
ES3	126	3008	51	TR17/635	G0008
ES3(ELCOM)	102A			TR85/	
ES4				TR01/	G6011
ES4(ELCOM)	158			TR84/	
ES5	101	3011	8	TR08/641	G0011
ES5(ELCOM)	103A			TR08/	
ES6	103A	3010		TR08/	G0011
ES7	104	3009	16	TR01/624	G6016
ES7(ELCOM)	131			TR94/	
ES8	160			TR17/	G0003
ES9	104	3009	16	TR01/624	G6003
ES10	105	3012	4	TR03/231	G6004
ES11	160	3006		TR17/	G0008
ES12		3006		TR52/	
ES13	104	3009	16	TR01/230	G6003
ES14	126	3006	51	TR17/635	G0008
ES14(ELCOM)	160			TR17/	
ES15	160			TR17/	G0003
ES15X1	123A	3018	52	TR21/735	S0015
ES15X2	108	3018		TR95/56	
ES15X3	108	3018		TR95/56	
ES15X4	102A	3004	52	TR85/250	G0005
ES15X6	108	3018		TR95/56	
ES15X7	123A	3124	8	TR21/735	S0015
ES15X8	102A	3004	52	TR85/250	G0005
ES15X9	159	3114			
ES15X10	108	3124	8	TR95/56	S0015
ES15X11	123A	3124	8	TR21/735	S0015
ES15X12	123A	3124	8	TR21/735	S0015
ES15X14	123A	3124	8	TR21/735	S0015
ES15X16	123A	3124	8	TR21/735	S0015
ES15X17	121	3009	16	TR01/232	G6003
ES15X18	108	3019	17	TR95/56	S0016
ES15X19	108	3019	17	TR95/56	S0016
ES15X20	123A	3124	8	TR21/735	S0015
ES15X22	233				
ES15X23	123A	3124	8	TR21/735	S0015
ES15X24	123A	3124	8	TR21/735	S0015
ES15X30	108	3019	17	TR95/56	S0016
ES15X31	102A	3004	52	TR85/250	G0005
ES15X32	102A	3004	52	TR85/250	G0005
ES15X37	123A	3124	8	TR21/735	S0015
ES15X42	123A	3124	8	TR21/735	S0015
ES15X43	121	3009	16	TR01/232	G6003
ES15X45	104	3009	3	TR01/230	
ES15X48	103A	3010	8	TR08/724	G0011
ES15X49	102A	3004	52	TR85/250	G0005
ES15X50	102A	3004	52	TR85/250	G0007
ES15X51	104	3009	16	TR01/232	G6005
ES15X52	127	3034	25	TR27/235	G6018
ES15X53	102A	3004	52	TR85/250	G0005
ES15X54	127	3035	25	TR27/235	G6008
ES15X55	102A	3004	52	TR85/250	G0001
ES15X56	161	3018		TR83/719	S0017
ES15X57	161	3018		TR83/719	S0016
ES15X58	123A	3044	20	TR21/735	S0030
ES15X59	154	3044		TR78/712	S5025
ES15X60	108	3018	20	TR95/56	S0014
ES15X61	126	3008	50	TR17/635	G0008
ES15X62	123A	3124	20	TR21/735	S0014
ES15X63	102A	3008	1	TR85/250	G0005
ES15X64	123A	3124	63	TR21/735	S0020
ES15X65	161	3117	20	TR83/719	S0016
ES15X66	107	3117		TR70/720	S0016

Original Part Number	ECG	SK	GE	IR/WEP	HEP
ES15X67	107	3117	17	TR70/720	S0016
ES15X68	123A	3040	18	TR21/735	S0014
ES15X69	108	3040	63	TR95/56	S0016
ES15X70	123A	3020	17	TR21/735	S0016
ES15X71	103A	3010	8	TR08/724	G0011
ES15X72	102A	3010	8	TR85/250	G0014
ES15X73	126	3005	51	TR17/635	G0005
ES15X74	103A	3010	59	TR08/724	G0011
ES15X75	158	3004	52	TR84/630	G6011
ES15X76	123A	3122		TR21/735	S3013
ES15X77	127	3035	25	TR27/235	G6007
ES15X78	104	3009	16	TR01/230	G6003
ES15X79	161	3117		TR83/719	S0020
ES15X80	107	3018		TR70/720	S0014
ES15X81	107	3018		TR70/720	S0014
ES15X82	107	3044		TR70/720	S0016
ES15X83	123A	3040		TR21/735	S0020
ES15X84	123A	3124		TR21/735	S0030
ES15X85	123A	3024		TR21/735	S5014
ES15X86	152	3041		TR76/701	S5000
ES15X87	161	3039	20	TR83/719	S0020
ES15X88	161	3039	20	TR83/719	S0020
ES15X89	154		18	TR78/712	S5025
ES15X90	159	3114	21	TR20/717	S0013
ES15X91		3104			S5015
ES15X92	132	3116	FET2	FE100/802	F0015
ES15X93	128	3024		TR87/243	S0005
ES15X94	165	3115		TR93/740B	S5021
ES15X95	124	3021	12	TR81/240	S5011
ES15X96	161	3018	20	TR83/	S0020
ES15X97	107	3018	20	TR70/	S0016
ES15X98	175	3026	66	TR81/	S5019
ES15X99	158	3082	30	TR84/	G6016
ES15X100	102A		53	TR85/	G0005
ES15X101	159		22	TR20/	S0013
ES15X102	107	3018	11		S0016
ES15X104	161	3117		TR95/	S0020
ES15X105	161	3117		TR95/	S0020
ES15X106	233	3117		TR95/	S0020
ES15X107	154	3044			S5025
ES15X122	161	3132	61		S0020
ES15X123	161	3117		TR71/	S0020
ES15X125	291	3054	27		S0015
ES15X126	163	3111	36		S5020
ES15X127	233	3018	20		S0015
ES15X128	159	3114	21		S0019
ES16	130			TR59/	
ES16X					S7002
ES17	102A			TR85/	G0005
ES18	121	3014	16	TR01/232	G6005
ES18(ELCOM)	104			TR01/	
ES19	102A	3123	52	TR85/250	G0001
ES19(ELCOM)	160			TR17/	
ES20		3018		TR51/	S0015
ES20(ELCOM)	123A			TR21/	
ES21	105	3012	4	TR03/233	G6005
ES21(ELCOM)	121			TR01/	
ES22		3024		TR55/	S3001
ES22(ELCOM)	128			TR87/	
ES23	102A	3123	52	TR85/250	G0003
ES23(ELCOM)	160			TR17/	
ES24		3021		TR60/	S5011
ES24(ELCOM)	124			TR81/	
ES25	100	3005	2	TR05/254	G0008
ES25(ELCOM)	160			TR17/	
ES26	100	3005	2	TR05/254	G0005
ES26(ELCOM)	102A			TR85/	
ES27				TR27/	G6009
ES27(ELCOM)	179			TR35/	
ES28	155			TR57/	
ES29		3082		TR50/	G6016
ES29(ELCOM)	131			TR94/	
ES30	158			TR50/	
ES31		3027		TR61/	S7004
ES31(ELCOM)	130			TR59/	
ES32		3044		TR74/	S5025
ES32(ELCOM)	154			TR78/	
ES33		3111		TR61/	S5020
ES33(ELCOM)	162			TR67/	
ES34		3118		TR52/	S0019
ES34(ELCOM)	159			TR20/	
ES35	130MP			TR59MP/	S7002(2)
ES36		3026		TR57/	S5019
ES36(ELCOM)	175			TR81/	
ES37				TR01/	G6011
ES37(ELCOM)	158			TR84/	
ES38				TR55/	
ES38(ELCOM)	103A			TR08/	
ES39	157	3103		TR60/	S5015
ES40	194			TR53/	S5026
ES41	126	3006	50	TR17/635	G0003
ES41(ELCOM)	160			TR17/	
ES42		3035		TR27/	G6008
ES42(ELCOM)	127			TR27/	
ES43	181			TR36/	S7000

Original Part Number	ECG	SK	GE	IR/WEP	HEP
ES44				TR23/	S5019
ES44(ELCOM)	175			TR81/	
ES45				TR59/	S5019
ES45(ELCOM)	175			TR81/	
ES46	123	3122	8	TR86/724	S0025
ES46(ELCOM)	123A			TR21/	
ES47	6400			2160/	S9002
ES-48	124			TR23/	
ES49	218	3083		TR58/	S5018
ES50		3052		TR50/	G6016
ES50(ELCOM)	131			TR94/	
ES51		3025		TR56/	S5013
ES51(ELCOM)	129			TR88/	
ES51X65	128			TR87/	
ES52		3039		TR54/	S0020
ES53		3018		TR51/	S0025
ES53(ELCOM)	123A			TR21/	
ES54		3039		TR51/	S0005
ES54(ELCOM)	161			TR83/	
ES55		3079		TR61/	S7004
ES55(ELCOM)	162			TR67/	
ES56		3104		TR57/	S3019
ES56(ELCOM)	128			TR87/	
ES57		3104		TR57/	S3019
ES57(ELCOM)	128			TR87/	
ES58		3054			S5000
ES58(ELCOM)	184		57		
ES59	179			TR35/	G6009
ES60	185	3083		TR56/	S5006
ES61		3052		TR01/	G6016
ES61(ELCOM)	226			TR94/	
ES62		3024		TR53/	S0005
ES62(ELCOM)	128			TR87/	
ES63		3123		TR50/	G6011
ES63(ELCOM)	176			TR82/	
ES64	219			TR29/	S7003
ES65		3118		TR52/	S0019
ES65(ELCOM)	159			TR20/	
ES66	182			TR59/	S5004
ES67	183	3189			S5005
ES68	185	3513		TR56/	S5013
ES69	223	3027		TR59/	S7002
ES70		3104		TR55/	S3019
ES71		3079		TR61/	S5015
ES71(ELCOM)	162			TR67/	
ES72				TR64/	S3008
ES73				TR64/	S0014
ES73(ELCOM)	161			TR83/	
ES74	180				S7001
ES75	172			TR69/	
ES75X1	712				
ES76		3044		TR63/	S5024
ES76(ELCOM)	171			TR79/	
ES77					S9001
ES80		3054		TR57/	S5000
ES80(ELCOM)	152			TR76/	
ES81	189	3054		TR57/	S5006
ES81(ELCOM)	153			TR77/	
ES82		3104		TR55/	S3020
ES82(ELCOM)	188			TR72/	
ES83		3053		TR56/	S3032
ES83(ELCOM)	189			TR73/	
ES84	163			TR80/	
ES84(ELCOM)	163			TR67/	
ES84X1	1008				
ES84X2	1018				
ES84X3	712				
ES85	123A	3039		TR21/	S0025
ES86		3039		TR21/	S0025
ES86(ELCOM)	161			TR83/	
ES87		3118		TR54/	S0019
ES88	195			TR55/	S3002
ES89	171	3201		TR60/	S3021
ES89(ELCOM)	171			TR79/	
ES90	219			TR67/	
ES90(ELCOM)	219				
ES91				TR01/	
ES91(ELCOM)	176			TR82/	
ES92	257				
ES93	258				
ES94	225			TR75/	
ES95	165			TR93/	
ES96	191			TR75/	
ES97	194			TR75/	
ES103	196				
ES104	197				
ES107	232				
ES112	234				
ES113	241				
ES114	242				
ES501	105	3012	4	TR03/233	G6006
ES503	121	3009	16	TR01/232	G6005
ES503(ELCOM)	104			TR01/	
ES544F544B				TR31/	
ES3110	126	3006	51	TR17/635	G0005

Original Part Number	ECG	SK	GE	IR/WEP	HEP
ES3111	126	3006	51	TR17/635	G0005
ES3112	126	3006	51	TR17/635	G0005
ES3113	126	3006	51	TR17/635	G0005
ES3114	126	3006	51	TR17/635	G0005
ES3115	126	3006	51	TR17/635	G0005
ES3116	126	3006	51	TR17/635	G0005
ES3120	102A	3004	52	TR85/250	G0005
ES3121	102A	3004	52	TR85/250	G0005
ES3122	102A	3004	52	TR85/250	G0005
ES3123	102A	3004	52	TR85/250	G0005
ES3124	102A	3004	52	TR85/250	G0005
ES3125	102A	3004	52	TR85/250	G0005
ES3126	102A	3004	52	TR85/250	G0005
ES3266	108	3019	11	TR95/	S0020
ES4053	722	3161			
ES4996				TR24/	
ES4997				TR24/	
ES10110	131	3052	44		G6016
ES10186	107	3018	20		S0016
ES10187	108	3018	11		S0014
ES10188	107	3018	20		S0014
ES10222	123A	3018			S0015
ES10223	123A	3018	18		S0015
ES10231	199	3124	62		S0015
ES10232	123A	3018	20		S0015
ES15046	107	3018	11	TR80/	S0020
ES15047	107	3018	20	TR24/	S0011
ES15048		3124	20	TR24/	S0015
ES15049	199	3124	62	TR51/	S0015
ES15050	123A	3122	20	TR21/	S0015
ES15051	187		29	TR56/	S5006
ES15052	199	3122	62	TR51/	S0024
ES15226	128	3122	18	TR21/	S0024
ES15227	186	3192	28	TR55/	S5000
ES618292		3018			
ES922376P		3024			
ESA213	126	3006		TR17/635	
ESA233	126	3006		TR17/635	
ET1	160	3005	51	TR17/637	G0003
ET2	160	3005	2	TR17/637	G0002
ET3	102	3004	52	TR05/631	G0005
ET4	102	3004	52	TR05/631	G0005
ET5	102	3004	52	TR05/631	G0005
ET6	104	3009	16	TR01/230	G6003
ET7	105	3012	4	TR03/231	G6004
ET8	101	3011	8	TR08/641	
ET9	101	3011	8	TR08/641	
ET10	103	3010	8	TR08/641A	
ET11	103	3010	8	TR08/641A	
ET12	126	3006	50	TR17/635	G0003
ET15X1	102A	3003	2	TR85/250	G0005
ET15X2	107	3019		TR70/720	S0016
ET15X3	107	3019	11	TR70/720	S0020
ET15X4	121	3009	3	TR01/232	S0020
ET15X5	121	3009	3	TR01/232	G6005
ET15X7	108	3018	10	TR95/56	S0015
ET15X8	128	3024	20	TR87/243	
ET15X9	108	3018	20	TR95/56	S0015
ET15X10	123A	3124	20	TR21/735	S0015
ET15X11	123A	3124	10	TR21/735	S0015
ET15X12	123A	3124	10	TR21/735	S0015
ET15X13	123A	3124	20	TR21/735	
ET15X14	123A	3122	10	TR21/735	S0015
ET15X15	123A	3124	20	TR21/735	
ET15X16	123A	3122	10	TR21/735	S0015
ET15X17	121	3009	3	TR01/232	G6005
ET15X18	108	3018	11	TR95/56	S0016
ET15X19	108	3018	20	TR95/56	S0016
ET15X20	123A	3122	10	TR21/735	S0015
ET15X21	108	3018	20	TR95/56	S0015
ET15X23	108	3018	20	TR95/56	S0015
ET15X24	123A	3124	10	TR21/735	S0015
ET15X25	102A		2	TR85/250	
ET15X26	127	3035		TR27/235	G6008
ET15X27	123A	3124	20	TR21/735	
ET15X29	160		51	TR17/637	G0003
ET15X30	108		11	TR95/56	S0016
ET15X31	102A		2	TR85/250	G0005
ET15X32	102A		2	TR85/250	G0005
ET15X33	129	3025	22	TR88/242	
ET15X34	154	3045		TR78/712	S5024
ET15X36	128	3024	21	TR87/243	
ET15X37	123A	3122	10	TR21/735	S0015
ET15X38	129		22	TR88/242	
ET15X39	129		22	TR88/242	
ET15X40	127	3035	16	TR27/235	G6005
ET15X41	123A	3122	20	TR21/735	S0015
ET15X42	123A	3122	10	TR21/735	S0015
ET15X43	121	3009	3	TR01/232	G6003
ET15X45	123A	3122	20	TR21/735	
ET300				TR23/	
ET1511	102A	3004	2	TR85/250	G0005
ET368021	123A		20		
ET379462	199		62		
ET398711	199		20		

Original Part Number	ECG	SK	GE	IR/WEP	HEP
ET495371	186		28		
ET517263	199		88		
ETP2008			27		
ETP3114			32		
ETP3923			27		
ETP5092			32		
ETP5095			32		
ETS-003	130	3027	14	TR59/247	
ETS-005	223				
ETS-017	175				
ETS-068	123A	3020		TR21/735	
ETS-069	129	3025		TR88/242	
ETS-070	128	3024		TR87/243	S5014
ETS-071	129	3025		TR88/242	S5013
ETT-CDC-12000	123A	3122		TR21/	
ETTB-2SB176	102A	3004		TR85/250	
ETTB-2SB176R	102A	3004		TR85/250	
ETTB-75LB	102A	3004		TR85/250	G0005
ETTB-367B	131	3198		TR94/642	G6016
ETTC-2SC490	175			TR81/241	
ETTC-458LG	123A	3124		TR21/735	S0014
ETTC-CD8000	123A	3124		TR21/735	
ETTC-CD12000	123A	3124		TR21/735	
ETTC-CD13000	123A	3124		TR21/	
ETTR-367		3052			
ETX18	123A	3122	20	TR21/735	S0015
EU15X1	108	3018	17	TR95/56	S0011
EU15X2	108	3039	11	TR95/56	S0016
EU15X3	108	3039		TR95/56	S0020
EU15X4		3089			
EU15X6	108			TR95/56	
EU15X27	128	3024	20	TR87/243	S0015
EU15X34	129	3024	20	TR87/243	S0015
EVS-828A					S0015
EW162	108	3018		TR95/56	S0030
EW163	107	3018		TR70/720	S0014
EW164	107	3018		TR70/720	S0014
EW165	107	3124		TR70/720	S0015
EW165V	123A	3018		TR21/735	S0015
EW181	123A	3124		TR21/735	S0015
EW182	123A	3124		TR21/735	S0015
EW183	153	3025		TR77/700	S5006
EW183B	175	3026		TR81/241	S5019
EW212	161			TR83/719	S0020
EWQ202	159	3114	21	TR20/717	
EWS-78	199	3124			
EX15X25	102A	3004	2	TR85/250	
EX39-X	705A				C6089
EX42-X	714				C6070P
EX46-X	713				C6057P
EX48-X	715				C6065P
EX62-X	790				C6057P
EX499-X					S0024
EX500-X					S0017
EX524-X					S0015
EX695-X					S0015
EX699-X					S0019
EX743-X					S0025
EX747-X					S0012
EX748-X					S0020
EX831-X					S5021
EX868-X					S3021
EX888-X					S0025
EX911-X					S5015
EX4035	722	3161	IC9		C6056P
EX4053	722	3161	IC9		C6056P
EX-141216	128				
EX-151513		3021			
EX-151514		3036			
EYZP-546	159				
EYZP-623	159				
EYZP-632	123A				
EYZP-791	123A				
EYZP-808	234				
EZT308	904				
F0021		3116			
F0082M	725				
F1			61		
F2			61		
F3			61		
F4			61		
F4-005	725				
F5			61		
F20			3		
F20-1001	108	3018	17	TR95/56	
F20-1002	108	3018	17	TR95/56	
F20-1003	108	3018	20	TR95/56	
F20-1004	108	3018	20	TR95/56	
F20-1005	108	3018	20	TR95/56	
F20-1006	102A	3004	2	TR85/250	
F20-1007	102A	3004	2	TR85/250	
F20-1008	102A	3004	2	TR85/250	
F20-1009	102A	3004	2	TR85/250	
F24A					S0015
F25A					S0015

Original Part Number	ECG	SK	GE	IR/WEP	HEP
F54A					S0019
F54B					S0031
F54C					S0031
F54D					S0019
F54E					S0019
F67E	121	3009	16	TR01/232	
F88-9779F	736				
F96N	108	3039	20	TR95/56	
F103P		3118	21		
F121-516		3018			
F121-546				TR21/	
F121-603					S5022
F121-60216					S5022
F121-433804	123A	3124		TR21/735	
F130A				IC507/	
F153-9				TR21/	
F161-249-1001	949				
F209	159	3119	22		S0023
F215-1001	108	3018	17	TR95/56	
F215-1002	108	3018	17	TR95/56	
F215-1003	108	3018	20	TR95/56	
F215-1004	108	3018	20	TR95/56	
F215-1005	108	3018	20	TR95/56	
F215-1006	102A	3004	2	TR85/250	
F215-1007	102A	3004	2	TR85/250	
F215-1008	102A	3004	2	TR85/250	
F215-1009	102A	3004	2	TR85/250	
F222	123A	3124	17	TR24/	S0014
F248	152				
F302-1	123A	3124	20	TR21/735	S0011
F302-2	123A	3124	20	TR21/735	S0011
F302-001				TR51/	
F302-1532	123A	3124	20	TR21/735	S0015
F302-2532	123A	3124	20	TR21/735	S0015
F306-001	123A	3122		TR21/735	
F306-022	123A	3124		TR21/735	
F318-1	128				
F366	123A	3024		TR21/735	
F385-2					S3019
F442-41	725				
F501	161			TR83/719	S0017
F501-16	161			TR83/719	
F502	161			TR83/719	S0020
F523	161	3117		TR83/719	S0020
F549-1	159		21	TR20/717	S0013
F572-1	123A	3122		TR21/735	S0015
F587	123A	3122		TR21/735	
F625-1	128		63	TR87/243	S0030
F699	159	3114		TR20/717	
F767PC	722		IC9		C6056P
F1462	132	3116	FET2	FE100/802	
F1463	132	3116	FET2	FE100/802	
F1826-0044	725				
F2004		3050			
F2007		3065			
F2041		3025	67		S5013
F2427	108	3018		TR95/56	
F2443	123A	3124	20	TR21/735	
F2448	123A	3124	20	TR21/735	
F2450	108	3018	20	TR95/56	
F2480			2		
F2560-1	729				
F2584	123A	3124	20	TR21/735	
F2633	108	3018	20	TR95/56	
F2634	108	3018	20	TR95/56	
F2636		3018	20	TR95/56	
F3519	123A			TR21/735	S0015
F3530	123A			TR95/56	S0016
F3532	123A	3018		TR21/735	S0016
F3535	161	3117		TR83/719	S0017
F3549	129			TR88/242	S0019
F3559	159		21	TR20/717	S0013
F3560	128	3024		TR87/243	S0015
F3561	128	3024	17	TR87/243	S0015
F3565	128		63	TR87/243	S0015
F3569	123A	3122		TR21/735	S0013
F3570	159	3114		TR20/717	
F3571	123A	3122		TR21/735	S0030
F3574	161	3025			S0020
F3589	128	3024		TR87/243	S0014
F3590	159	3114		TR20/717	S0019
F3597	159			TR20/717	S0013
F4117-1	949				
F4706	108	3039		TR95/56	S0020
F4709	128	3024		TR87/243	S0015
F4846-1	949				
F7316			17		
F9600	123A	3018	20	TR21/735	S0014
F9623	123A	3018	20	TR21/56	S0017
F9623(GE)	108			TR95/	
F9625	108	3018	20	TR95/	S0014
F15810	108	3018	20	TR95/56	
F15835	108	3018	20	TR95/56	
F15840	123A	3124	20	TR21/735	
F15841	108	3018	20	TR95/56	
F21490	159	3114		TR20/717	
F73216	160	3018		TR17/637	S0015
F75116	123A	3020		TR21/735	S0024
F136555	725				
FA-1	123A				
FB401	160	3006	51	TR17/637	
FB402	160	3006	51	TR17/637	
FB403	160	3006	51	TR17/637	
FB420	102A	3004	2	TR85/250	
FB421	102A	3004	2	TR85/250	
FB440	160	3006	51	TR17/637	
FB6853	123A	3122	20	TR21/735	
FBC237	123A				
FBN-2N1183	176	3123			
FBN-35469	175	3026			
FBN-35903	175	3079			
FBN-36220	130	3510			
FBN-36485	130	3510			
FBN-37486	175	3026			
FBN-36488	175				
FBN-36603	130	3510			
FBN-36972	181	3510			
FBN-36973	181	3510			
FBN-37605	124	3021			
FBN-38021	124				
FBN-38022	181	3511			
FBN-38982	124	3021			
FBN-CP2293	102				
FBN-CP2357P		3024			
FBN-CP2388		3011			
FBN-CP34634	126	3512			
FBN-CP34759	176	3123			
FBN-L108	124	3044			
FBN-L109	128	3512			
FBN-L113	128	3512			
FBN-L115	128				
FBN-L124		3021			
FBN-L148	128	3512			
FBNSP209					G6005
FBT-00-015	912				
FBT40				TR35/	
FBTX070	128			TR87/	
FCS006	108	3039		TR95/56	
FCS1168E	108	3018	20	TR95/56	
FCS1168E641	108	3018		TR95/56	
FCS1168F813	123A	3124		TR21/735	
FCS1168G	123A	3024	18	TR21/735	
FCS1168G704	123A	3124		TR21/735	
FCS1170F	159	3025	21	TR20/717	
FCS1225E	108	3018	17	TR95/56	
FCS1227E	108	3018	20	TR95/56	
FCS1227E814	108	3018	20	TR95/56	
FCS1227F	108	3018	20	TR95/56	
FCS1227F743	108	3018	20	TR95/56	
FCS1227G	108	3018	20	TR95/56	
FCS1227G810	108	3018	20	TR95/56	
FCS1229	128	3024	18	TR87/243	
FCS1229F	123A	3024	18	TR21/735	
FCS1229G	123A	3024	18	TR21/735	
FCS9011					S0015
FCS9011H	108	3018	20	TR21/	
FCS9012	129	3025		TR28/	
FCS9012HE	159	3114	21	TR90/	S0019
FCS9012HG	129	3025	58	TR28/	S5013
FCS9013HH	159	3024	57	TR25/	S5014
FCS9014	108	3124			
FCS9014B	108	3018	20	TR95/56	S0015
FCS9014D	108		20	TR21/	S0015
FCS9016D	108	3018	20	TR24/	S0014
FCS9018E	108	3018	11	TR80/	
FCS9066	108	3018	20	TR95/56	
FD-1029ET	121	3014			
FD-1029-EY		3011			
FD-1029-FJ		3500			
FD-1029-FY	128				
FD-1029-GE	128				
FD-1029-JA	123A				
FD-1029-JE		3500			
FD-1029-JN	128				
FD-1029-JP	159				
FD-1029-LL	123A				
FD-1029-ML	129				
FD-1029-MM	128	3024			
FD-1029-NG	123A				
FD-1029-NS	192	3512			
FD-1029-PA	128				
FD-1029-PD-2		3021			
FD-1029-PP	123A				
FD-1073-AP	716				
FD-1073-BF	7400				
FD-1073-BG	7402				
FD-1073-BH	7403				
FD-1073-BJ	7404				
FD-1073-BM	7408				
FD-1073-BN	7410				

Original Part Number	ECG	SK	GE	IR/WEP	HEP
FD-1073-BR	7420				
FD-1073-BS	7430				
FD-1073-BU	7440				
FD-1073-BW	7451				
FD-1073-CA	7486				
FD-4500-11	124				
FD4500AL	181				
FE0654B		3112			
FE0654C		3112			
FE100	133	3112	FET1	/801	F0010
FE100A	133	3112		/801	F0010
FE102	133	3112		/801	F0010
FE102A	133	3112		/801	F0010
FE104A	133	3112		/801	F0010
FE400	133	3112		/801	F0010
FE402	133	3112		/801	F0010
FE402A	133	3112		/801	F0010
FE404A	133	3112		/801	F0010
FE2201/512		3024			
FE2201/513		3024			
FE3819		3116			
FE4302		3116			
FE4303		3116			
FE4304		3116			
FE5243		3116			
FE5246		3116			
FE5247		3116			
FE5457		3116			
FE5458		3116			
FE5459		3116			
FE5484		3116			
FE5485		3116			
FE5486		3116			
FEL2113	709				
FF274	715	3076	IC6	IC510/	C6071P
FF400	133	3112		/801	F0010
FFC-4030	754				
FG79	199				
FH70401E	748				C6060P
FI-1007	159				
FI-1008	159				
FI-1019	106				
FI-1021	234				
FI 1023	107				
FJH101	7430				C3030P
FJH111	7420				C3020P
FJH121	7410				C3010P
FJH131	7400				C3000P
FJH141	7440				C3040P
FJH151	7450				C3050P
FJH161	7451				
FJH171	7453				
FJH181	7454				
FJH191	7480				
FJH201	7482				
FJH211	7483				
FJH221	7402				
FJH231	7401				C3001P
FJH241	7404				
FJH251	7405				
FJH261	7442				
FJH291	7403				
FJH301	7426				
FJH311	7401				
FJH321	7405				
FJJ101	7470				
FJJ111	7472				
FJJ121	7473				C3073P
FJJ131	7474				
FJJ141	7490				
FJJ152	7492				
FJJ181	7475				C3075F
FJJ191	7476				
FJJ211	7493A				
FJJ261	74107				
FJK101	74121				
FJL101	7441				
FJY101	7460				
FK914	108	3039	60	TR95/56	
FK918	108	3039	39	TR95/56	
FK2369A	108	3039	17	TR95/56	
FK2484	108	3039	63	TR95/56	
FK2894	108			TR20/52	S0013
FK3014	108	3039	60	TR95/56	
FK3299	108	3039	213	TR95/56	
FK3300	108	3039	210	TR95/56	
FK3484			17		
FK3494			17		
FK3502			82		S0019
FK3503			82		
FK3962					S0019
FL274	712		IC2	IC507/	C6063P
FLH101	7400				
FLH121	7420				
FLH131	7430				

Original Part Number	ECG	SK	GE	IR/WEP	HEP
FLH141	7440				
FLH151	7450				
FLH161	7451				
FLH171	7453				
FLH181	7454				
FLH191	7402				
FLH201	7401				
FLH211	7404				
FLH221	7480				
FLH231	7482				
FLH241	7483				
FLH271	7405				
FLH281	7442				
FLH291	7403				
FLH291U	7426				
FLH341	7486				
FLH351	7413				
FLH361	7443				
FLH371	7444				
FLH381	7408				
FLH391	7409				
FLH401	74181				
FLH411	74182				
FLH421	74180				
FLH431	7485				
FLH441	74H87				
FLH451	74H183				
FLH481	7406				
FLH481T	7416				
FLH491	7407				
FLH491T	7417				
FLH501	7412				
FLH511	7423				
FLH521	7425				
FLH531	7437				
FLH541	7436				
FLH551	7448				
FLH561	74184				
FLH581	7411				
FLH601	74132				
FLH611	7422				
FLH621	7427				
FLH631	7432				
FLJ101	7470				
FLJ111	7472				
FLJ121	7473				
FLJ131	7476				
FLJ141	7474				
FLJ151	7475				
FLJ161	7490				
FLJ171	7492				
FLJ181	7493A				
FLJ191	7495				
FLJ201	74190				
FLJ211	74191				
FLJ221	7491				
FLJ231	7494				
FLJ241	74192				
FLJ251	74193				
FLJ261	7496				
FLJ271	74107				
FLJ281	74104				
FLJ301	74100				
FLJ311	74198				
FLJ321	74199				
FLJ331	7497				
FLJ381	74196				
FLJ391	74197				
FLJ401	74160				
FLJ411	74161				
FLJ421	74162				
FLJ431	74163				
FLJ441	74164				
FLJ451	74165				
FLJ461	74166				
FLJ471	74167				
FLJ531	74174				
FLJ541	74175				
FLJ551	74194				
FLJ561	74195				
FLK101	74121				
FLK111	74122				
FLK121	74123				
FLL101	74141				
FLL111	7445				
FLL111T	74145				
FLL121	7446				
FLL121T	7447				
FLL121U	7446				
FLL121V	7447				
FLL151	74142				
FLQ101	7489				
FLQ111	7481				
FLQ121	7484				
FLQ131	74170				

Original Part Number	ECG	SK	GE	IR/WEP	HEP
FLQ141	74200				
FLY101	7460				
FLY111	74150				
FLY121	74151				
FLY131	74153				
FLY141	74154				
FLY151	74155				
FLY161	74156				
FLY171	74157				
FM708	108	3039	63	TR95/56	S0011
FM709	108	3039	63	TR95/56	S0016
FM220A	108	3039	63	TR95/56	
FM870	108	3039	18	TR95/56	
FM871	109	3039	18	TR95/56	
FM910	108	3039	63	TR95/56	
FM911	108	3039	63	TR95/56	
FM914	108	3039	63	TR95/56	S0016
FM1613	108	3039	63	TR95/56	
FM1711	108	3039	63	TR95/56	
FM1893	154		63	TR78/712	
FM2368	108	3039	63	TR95/56	S0016
FM2369	108	3039	63	TR95/56	S0016
FM2846	108	3039	63	TR95/56	S0011
FM2894	106		21	TR20/52	S0016
FM3014	108	3039	63	TR95/56	S0016
FMPSA20	123A	3124		TR51/	
FN-51-1A			20	TR25/	
FOS100			212		
FOS101			212		
FOS104			17		
FPR40-1001	123A	3122		TR21/735	S0014
FPR40-1003	108	3039		TR95/56	S0016
FPR40-1004	102A	3123		TR85/250	G0007
FPR40-1005	102A	3123		TR85/250	G0005
FPR50-1001	123A	3018		TR21/735	S0014
FPR50-1002	123A	3018		TR21/735	S0014
FPR50-1003	108	3018		TR95/56	S0016
FPR50-1004	108	3018		TR95/56	S0016
FPR50-1005	102A	3004		TR85/250	G0007
FPR50-1006	102A	3004		TR85/250	G0005
FS-1133			17	TR25/	
FS1168E641			20		
FS1168F813			20		
FS1221	123A	3124	20	TR21/735	
FS1308	107	3039	11	TR70/720	
FS1331	128	3024	20	TR87/243	
FS1682	181			TR83/	
FS1974	123A	3124	20	TR21/735	
FS1978	123A	3047		TR21/735	
FS2003-1	130	3027	14	TR59/247	S7002
FS2042		3024	63		S5014
FS2043		3122	20		S0015
FS2299	160	3007	21	TR17/637	
FS2523		3049			
FS3266	108	3019		TR95/56	S0016
FS3683	108	3018		TR95/	S0020
FS24226	159	3114	21	TR20/717	
FS24954	159	3118	21	TR20/	S0012
FS27233	128	3024	18	TR87/243	
FS27604		3124	20		S0015
FS35529	108	3018	11	TR95/	S0025
FS36383				TR80/	
FS36999	123A				
FS326690	108	3018		TR95/56	S0016
FSE1001	108	3018		TR95/56	
FSE3001	108	3018	20	TR95/56	
FSE4002	199				
FSE5002	108	3018	20	TR95/56	
FSE5020				TR21/	
FSP-1	108	3039	11	TR95/56	
FSP-42	108	3039	63	TR95/56	
FSP-42-1	108	3039	63	TR95/56	
FSP-164	108	3039	11	TR95/56	
FSP-165	108	3019	63	TR95/56	S0016
FSP-166	108	3039	63	TR95/56	
FSP-166-1	108	3039	63	TR95/56	
FSP-215	108	3019	63	TR95/56	
FSP-242-1	108	3124	63	TR95/56	S0016
FSP-270-1	108	3019	21	TR95/56	S0011
FSP-289-1	108	3039	63	TR95/56	
FT001	128	3024	20	TR87/243	
FT002	128	3024	20	TR87/243	S0014
FT003	128	3024	11	TR87/243	S0014
FT004	128	3024	11	TR87/243	S0014
FT004A	128	3024	11	TR87/243	S0014
FT005	123A	3122	11	TR21/735	S0012
FT006	123A	3122	11	RE21/735	S0012
FT008	123A	3122	11	TR21/735	S0012
FT008A	123A	3122	11	TR21/735	S0012
FT0019H	128	3024		/243	
FT0019M	159	3114		/717	S0019
FT023	123A	3122	210	TR21/735	S0011
FT024	123A	3122	210	TR21/735	S0011
FT025	123A	3122	212	TR21/735	S0011
FT026	123A	3122	212	TR21/735	S0011

Original Part Number	ECG	SK	GE	IR/WEP	HEP
FT027	128	3024	17	TR87/243	
FT052	159	3114	17	TR20/717	S0012
FT053	123A	3122	63	TR21/735	
FT0601	222				
FT0654A		3116			
FT0654D		3116			
FT19H			82		
FT19M			82		
FT34C	154	3045	17	TR78/712	
FT34D	154	3045	17	TR78/712	
FT34Y	132	3116	FET2		F0021
FT40			17	TR51/	
FT45	161		20	TR83/	
FT107B			85		
FT107C				TR51/	
FT118	161	3117	60	TR83/719	
FT300	124	3021	12	TR81/240	
FT431			36		
FT459					G6018
FT654E		3116			
FT709	108	3039	61	TR95/56	S0016
FT1315	108	3039	63	TR95/56	S0016
FT1324B	108	3039		TR95/56	S0016
FT1324C	108	3039		TR95/56	S0016
FT1341	159	3114	17	TR20/717	S0013
FT1702	106		21	TR20/52	
FT1746	159	3114	65	TR20/717	S0013
FT3567			20		
FT3568			81		
FT3569			20		
FT3638	159			TR20/	
FT3641	154		17	TR78/	
FT3642			47		
FT3643	123A		47	TR21/	
FT3644			67		
FT3645			48		
FT5040			22		S0019
FT5041			82		S0019
FU5B7709393	909	3590			
FU5B7710393	910				
FU5B7740393	940				
FU5B7741393	941	3514			
FU5B7749393	949				
FU5D770331X	703A	3157			
FU5D770339	703A	3157			
FU5E7064393	780	3141			
FU5F7711393	911				
FU5F7715393	915				
FU5F7747393	947				
FU5T7725393	925				
FU6A7709393	909D	3590			
FU6A7710393	910D				
FU6A7711393	911D				
FU6A7723393	923D				
FU6A7741393	941D	3552			
FU8B770339X	703A				
FU9A7723393	923D	3165			
FU9A7741393	941D	3552			
FU9T7741393	941M	3553			
FµL914-28	9914				
FµL923	9923				
FV914	108	3039	39	TR95/56	S0025
FV918	108	3039	17	TR95/56	S0015
FV2369A	108	3039	39	TR95/56	S0025
FV2484	108	3039	212	TR95/56	S0015
FV2747C	100	3005	52	TR05/254	
FV2894	106		212	TR20/52	S0015
FV3014	108	3039	60	TR95/56	S0016
FV3299	108	3039	213	TR95/56	S0016
FV3300	108	3039	210	TR95/56	S0030
FV3502			82		S0019
FV3503			82		S0019
FV3964					S0019
FX274	714				
FX709			11		
FX914			61	TR51/	
FX918			61		
FX2368			20	TR64/	
FX2483			63	TR51/	
FX2894				TR20/	
FX3013			61	TR51/	
FX3014			61	TR64/	
FX3299			213	TR64/	
FX3300			210		
FX3502			82		
FX3724			47		
FX3962			67	TR52/	
FX3963				TR52/	
FX3964			65	TR52/	
FX3965				TR52/	
FX4046			210		
FX4960			63	TR51/	
FZ101			20		
G0002	160				
G0005	158				

Original Part Number	ECG	SK	GE	IR/WEP	HEP
G0006	102				
G0007	102A				
G0008	126				
G004	102A				
G0010	126				
G03-007C	184	3054	28	TR50/	
G03-404-B	129	3513			S5013
G03-404-C	129	3513			S5013
G04-041B	107	3018	20	TR70/	
G04-701-A	121		16	TR01/232	G6003
G04-704-A	131	3198	30	TR94/642	G6016
G04-711-E, F, G	158	3004	53	TR84/	G0005
G05-003-A, B	107	3018	20	TR21/	S0014
G05-004A	108	3018	11	TR24/	
G05-010-A	123A		17	TR21/735	S0015
G05-011-A	123A	3038	17	TR21/735	S0015
G05-015C	123A	3047	20	TR21/	
G05-015-D	123A	3020	63	TR21/735	S0014
G05-034-D	123A	3124	18	TR51/	S0015
G05-035-D	123A	3124	62	2SC544/	S0015
G05-035-E	199	3124	62	2SC644/	S0015
G05-036-B	123A	3124	20	TR21/	S0015
G05-036-C	123A	3122	20	TR24/	S0015
G05-036-D	123A	3122	18	TR24/	S0015
G05-036-E	123A	3122	20	TR21/	S0015
G05-037-B	123A	3122	20	TR21/	
G05-050-C	107	3018	17	TR24/	S0016
G05-055-D	128	3024	18	TR21/	S0024
G05-413A	128	3024	18	TR87/	
G06-714C	186	3192	28	TR55/	
G06-717B	184	3041	28	TR55/	S3024
G06-717-C	184	3041			S5000
G06-717-D	184	3041			S5000
G08-005L	133	3112	FET1	FE100/	
G09-006-A	1029				
G09-006-B	1029				
G09-006-C	1029				
G09-009-A	1006				
G09-010-A	1029				
G09-011-A	1038				
G09-012-A	1038				
G09015		3102			
G09-036-G	1029				
G03007C	184				S0013
G03703C	187				
G04041B	107				
G05004A	106				
G05015C	123A				
G05035D	199				
G05035E	123A				
G05036B	123A				S0015
G05036C	123A				S0015
G05036D	123A				S0015
G05036E	123A				S0015
G05037B	123A				
G05413A	128				
G05413B	128				
G06714C	186				
G08005L	133				
G09007B	1103				
G09009A	1106				
G4N1	162			TR67/707	S5020
G5A7A66-2	128				
G11	102A		2	TR85/250	G0005
G12	102A		52	TR85/250	G0005
G13	160		2	TR17/637	G0003
G14	102A		52	TR85/250	G0005
G16			8	TR08	
G17			8	TR08/	
G18		3010	8	TR08/	
G19	121	3009	16	TR01/232	G6005
G23-45	130	3027			
G23-46	128				
G23-67	130	3027			
G23-76	223				
G23-112		3054			
G110	105	3012	4	TR03/233	G6004
G181-722-001		3024			
G181-725-001	181	3036			
G234		3525			
G1239				TR56/	
G1242				TR55/	
G4090	230				
G6 13	121				
G6 16	131				
G[illegible]23		3018			
G9600	123A	3018		TR51/735	S0030
G9623	123A	3018	20	TR21/	S0017
G9625		3018			
G9696	123A	3124	10	TR24/	S0014
G12711				TR51/	
G12 12				TR51/	
G12 3				TR51/	
G12715				TR51/	
G12716				TR51/	

Original Part Number	ECG	SK	GE	IR/WEP	HEP
G12721				TR51/	
G12722				TR51/	
G12723				TR51/	
G12724				TR51/	
G13638	159	3114		TR20/	
G13641				TR51/	
G13643				TR51/	
G13644		3114			
G13702		3114			
G13703		3114			
G13704				TR51/	
G13705				TR51/	
G13706				TR51/	
G13707				TR51/	
G13708				TR51/	
G13709				TR51/	
G13710				TR51/	
G13711				TR51/	
G16506	101	3011		TR08/641	G0011
G71071K		3072			
G101079	101	3011		TR08/641	G0011
G390076S1	909	3590			
G390077S1	910				
G390503S1	915				
G390506-S1	940				
G390507S2	923D				
G395967	123A				
G395967-2	123A				
G612994	9903				
G(C)5320023				TR21/	
GA-10		3004			
GA52829	100	3005	52	TR05/254	G0005
GA52830					G0006
GA53104					G0005
GA53149	100	3005	52	TR05/254	G0009
GA53194					G0009
GA53213					G0005
GA53233					G0008
GA53242	100	3005	52	TR05/254	G0006
GA53270	101	3011		TR08/641	G0011
GC31	100			TR05/	
GC32	100			TR05/	
GC33	100			TR05/	
GC34	100			TR05/	
GC35	100			TR05/	
GC60	100	3005	53	TR05/254	G0005
GC61	100	3005	53	TR05/254	G0005
GC148	103A		8	TR08/724	
GC181	100	3005	53	TR05/254	G0005
GC182	100	3005	53	TR05/254	G0005
GC250	102A	3004	2	TR85/250	
GC282	160	3008	2	TR17/637	G0002
GC283	160	3008	2	TR17/637	G0002
GC284	160	3008	2	TR17/637	G0002
GC285	103A	3010	8	TR08/724	G0011
GC286	103A	3010	8	TR08/724	G0011
GC343	102A	3004	2	TR85/250	
GC360	100	3005	52	TR05/254	
GC387	160		51	TR17/637	
GC388	160		51	TR17/637	
GC389	160		51	TR17/637	
GC408	102A	3004	52	TR85/250	G0005
GC452	101	3011	5	TR08/641	
GC453	101	3011	5	TR08/641	
GC454	101	3011	5	TR08/641	
GC460	160	3005	50	TR17/637	G0002
GC461	160	3005	50	TR17/637	G0002
GC462	160	3005	50	TR17/637	G0002
GC463	103A		8	TR08/724	
GC464	102A	3123	52	TR85/250	G0005
GC465	103A		8	TR08/724	
GC466	102A	3004	52	TR85/250	G0005
GC467	103A		8	TR08/724	
GC520	102A	3004	52	TR85/250	G0005
GC521	102A	3005	52	TR85/250	G0005
GC532	100	3005	52	TR05/254	
GC551	102A	3004	2	TR85/250	
GC552	102A	3004	2	TR85/250	
GC578	102A	3004	52	TR85/250	G0005
GC579	102A	3004	52	TR85/250	G0005
GC580	102A	3004	52	TR85/250	G0005
GC581	102A	3004	52	TR85/250	G0005
GC588	102A		53	TR85/250	
GC608	103A		8	TR08/724	G0011
GC609	103A		8	TR08/724	G0011
GC630	160		50	TR17/637	
GC630A	160	3123	52	TR17/637	
GC631	160			TR17/637	
GC639	102A	3004	52	TR85/250	G0005
GC640	102A	3123	52	TR85/250	G0005
GC641	121	3009	16	TR01/232	G6005
GC680	102A	3004	2	TR85/250	
GC681	102A			TR85/	
GC682	102A	3004	2	TR85/250	
GC691	121			TR01/	

Original Part Number	ECG	SK	GE	IR/WEP	HEP
GC692	121			TR01/	
GC733B			2	TR05/	
GC783	123			TR86/	
GC784	123			TR86/	
GC805					S5013
GC856	102A	3004	52	TR85/250	G0005
GC864	102A	3004	52	TR85/250	G0005
GC1003	160		50	TR17/637	
GC1004	160		50	TR17/637	
GC1005	160		51	TR17/637	
GC1006	160		51	TR17/637	
GC1007	160		51	TR17/637	
GC1034	101	3011	5	TR08/631	
GC1035	101	3011	5	TR08/641	
GC1036	101	3011	5	TR08/641	
GC1081	126			TR17/635	
GC1082A					G0005
GC1092	160	3006	51	TR17/637	
GC1093	160	3006	51	TR17/637	
GC1093X3	160	3006	51	TR17/637	
GC1097	102A		53	TR85/250	
GC1101	102A(2)		53	TR85/250	
GC1134	102A	3004	2	TR85/250	
GC1136	102A		53	TR85/250	
GC1137	103A		54	TR08/724	
GC1142	160	3008	51	TR17/637	
GC1143	102A	3004	2	TR85/250	
GC1144	123A	3124	20	TR21/735	
GC1145	102A	3025	53	TR85/250	
GC1146	160	3006	51	TR17/637	
GC1148	160	3008	51	TR17/637	
GC1149	160	3006	51	TR17/637	
GC1150	102A	3004	53	TR85/250	
GC1155	160			TR17/637	
GC1159	100	3005	51	TR05/254	
GC1182	160	3006	51	TR17/637	
GC1183	102A	3004	2	TR85/250	
GC1184	102A	3004	53	TR85/250	
GC1185	103A	3004	54	TR08/724	
GC1187	102A	3123	52	TR85/250	
GC1214		3006			
GC1215		3006			
GC1216		3008			
GC1217		3008			
GC1257	102A	3123	53	TR85/250	
GC1302	100	3005	52	TR05/254	
GC1422	102A		53	TR85/250	
GC1423	103A		54	TR08/724	
GC1573	160		50	TR17/637	
GC1615-1	128			TR87/243	S0015
GC4022	100	3004	16	TR05/254	G0005
GC4045	121	3009	16	TR01/232	G6005
GC4062	121	3009	16	TR01/232	G6005
GC4087	121	3009	3	TR01/232	
GC4094	131	3035	44	TR94/642	
GC4097	121	3009	16	TR01/232	
GC4111	121			TR01/	
GC4144	102A	3123	53	TR85/250	G0005
GC4156	121	3009	3	TR01/232	
GC4213	176/411			TR25/238	
GC4251	121	3009	3	TR01/232	
GC4267-2	121	3009	16	TR01/232	
GC5000	102A	3123	52	TR85/250	G0005
GE-1	100	3005		TR05/254	G0002
GE-2	102	3004		TR05/631	G0005
GE-3	121	3009		TR01/232	G6003
GE-4	105	3012		TR03/232	G6004
GE-5	101	3011		TR08/641	G0011
GE-6	101	3011		TR08/641	G0011
GE-7	101	3011		TR08/641	G0011
GE-8	103	3010		TR08/641A	G0011
GE-9	160	3006		TR17/637	G0003
GE-9A	160	3006		TR17/637	G0003
GE-10	123A	3122		TR21/735	S0015
GE-10A	199			TR21/	
GE-11	107	3018		TR70/720	S0016
GE-12	124	3021		TR81/240	S5011
GE-13MP	104MP	3013		TR01/642MP	G6003(2)
GE-14	130	3027		TR59/247	S7004
GE-15MP	130MP	3029		TR59/247MP	S0025(2)
GE-16	121			TR01/232	G6005
GE-17	123A	3122		TR21/735	S0025
GE-18	128	3024		TR87/243	S3021
GE-19	130	3027		TR59/247	S7002
GE-20	123A	3122		TR21/735	S0014
GE-20J		3124			
GE-21	159	3118		TR20/717	S0012
GE-21A	129			TR88/	
GE-22	159	3118		TR20/717	S0025
GE-23	175	3026		TR81/241	S5012
GE-24MP	175(2)	3028		/241	S5019(2)
GE-25	127	3035		TR27/235	G6008
GE-26	153	3085		TR77/700	S5006
GE-27	171	3104		TR78/244A	S3021
GE-28	186	3054		TR55/S3023	S5000

Original Part Number	ECG	SK	GE	IR/WEP	HEP
GE-29	187	3085		TR56/S3023	S3032
GE-30	131	3052		TR94/642	G6016
GE-31MP	131MP			TR94/642MP	G6016(2)
GE-32	198	3103			S3021
GE-33	182	3188			
GE-34	183	3189			
GE-35	162			TR67/	
GE-36	163			TR67/	
GE-37	164	3133		TR93/	S7005
GE-38	165	3115		TR93/	S7005
GE-39	161	3117			S0009
GE-40		3044			S5024
GE-43	131				
GE-44	131				G6016
GE-50	160	3006		TR17/637	G0009
GE-51	160	3006		TR17/637	G0008
GE-52	102A	3004		TR85/250	G0007
GE-53	158	3004		TR84/630	G6011
GE54	103A+158				
GE55	182	3188			S5004
GE56	183	3189			S5005
GE57	184	3054			S5000
GE58	185	3083			S5006
GE59	103A			TR08/	
GE-60	161	3018		TR83/719	S0025
GE-61	108	3132		TR95/56	S0030
GE-62	199	3122		TR21/	S0005
GE-63	192	3137			S3020
GE-64	172	3156		TR69/S9100	S9100
GE-65	234				S0025
GE-66	152	3054		TR76/701	S5004
GE-67	193	3114			S3032
GE-69	153	3083		TR77/700	S3032
GE-72	162			TR67/	S5020
GE-73	163			TR67/	S5020
GE-74	180	3035			S7001
GE-75	181	3036		TR36/	S7000
GE-76	179				G6009
GE-80					G6011
GE129	107	3039		TR70/720	
GE296X	172			TR69/S9100	S9100
GE-720	1024	3152			
GE-721	1027	3153/4			
GE-722	1025	3155			
GE-723	1090				
GE-724	1028				
GE3265	123A	3038	10	TR21/735	
GE-FET1	133	3112		FE100/801	F0010
GE-FET2	132	3116		FE100/802	F0021
GE-IC2	712	3072		IC507/	C6063P
GE-IC3	705A	3134		IC502/	C6089
GE-IC4	714	3075		IC509/	C6070P
GE-IC5	713	3077		IC508	C6057P
GE-IC6	715	3076		IC510/	C6056P
GE-IC8	712	3159			C6074P
GE-IC9	722	3161			C6068P
GE-IC10	708	3135		IC504/	C6062P
GE-IC11	709			IC505/	C6082P
GE-IC12	703A	3157			C6001
GE-IC14					C6080P
GE-IC15					C6101P
GE-IC18					C6057P
GE-IC20					C6069G
GE-IC21		3141			C6100P
GE-IC26					C6060P
GE-IC27					C6090
GE-M100	160			TR17/637	G0008
GE-MR6	186			TR55/S3023	
GE-X8	103		8	TR08/724	G0011
GE-X9	102A	3004		TR85/631	G0005
GE-X10	6401			2160/	S9002
GE-X16A1938	123A	3122		TR21/735	S0015
GE-X17	6402				S9001
GE-X18	186			TR55/	S3020
GEL234	717				
GEL234F1	717				
GEL234F2	717				
GEL239F2	721				
GEL277	732				
GEL2111	708	3135	IC10	IC504/	C6062P
GEL2111AL1	708	3135	IC10	IC504/	
GEL2111F1	708	3135	IC10	IC504/	
GEL2113	709		IC11	IC505/	C6082P
GEL2113AL1	709		IC11	IC505/	C6062P
GEL2113F				IC505/	
GEL2113F1	709	3135	IC11	IC505/	C6062P
GEL2114	713	3077		IC508/	
GEL3072F1	713	3077		IC508/	C6057P
GEMR-6	186	3192		TR55/S3023	S3020
GEN11760	174				
GER-A	102A	3004	2	TR85/250	
GER-A-D	160	3006		TR17/637	
GES2N2369					S0014
GET0-50P			53		
GET3					G0005

Original Part Number	ECG	SK	GE	IR/WEP	HEP
GET4					G0005
GET6					G0005
GET7					G6005
GET8					G6005
GET9					G6005
GET15					G0005
GET102					G6011
GET103	102A		2	TR85/250	G6011
GET104					G6011
GET105					G6011
GET106					G6011
GET110					G6011
GET111					G6011
GET113	102A	3004	2	TR85/250	G6011
GET113A	102A	3004	2	TR85/250	
GET114	102A	3004	2	TR85/250	G6011
GET115					G6011
GET116					G6011
GET119					G6011
GET120					G6011
GET535					G6011
GET536					G6011
GET538					G6011
GET571					G6015
GET572	121	3009	3	TR01/232	G6015
GET573					G6015
GET574					G6015
GET581					G6018
GET582					G6018
GET583					G6005
GET584					G6005
GET671	160	3007		TR17/637	
GET672	160	3006	51	TR17/637	
GET672A	160	3006	51	TR17/637	
GET673	160	3006	51	TR17/637	
GET691	160	3008	2	TR17/637	G0009
GET692	160	3008	51	TR17/637	G0009
GET693	160	3008	2	TR17/637	G0009
GET706	123A	3122	20	TR21/735	S0011
GET706A, B		3039			
GET708	123A	3122	20	TR21/735	S0025
GET718, A		3122			
GET753		3122			
GET870					G0005
GET871	126	3006	2	TR17/635	G0005
GET872	126	3006	2	TR17/635	G0005
GET873, A	126	3006	52	TR17/635	G0005
GET874	126	3005	52	TR17/635	G0009
GET875	126	3006	2	TR17/635	G0005
GET880	100	3005	52	TR05/254	G0006
GET881	100	3005		TR05/254	G0006
GET882	100	3005		TR05/254	G0006
GET883	126	3008	51	TR17/635	G0005
GET885	126	3006		TR17/635	G0006
GET887	100	3005	2	TR05/254	G0006
GET888	100	3005	2	TR05/254	G0007
GET889	100	3005	2	TR05/254	G0006
GET890	100	3005	2	TR05/254	G0007
GET891	100	3005		TR05/254	G0006
GET892	100	3005		TR05/254	G0006
GET895	100	3005	2	TR05/254	G0006
GET896	100	3005	52	TR05/254	G0005
GET897	100	3005	52	TR05/254	G0005
GET898	102	3004	52	TR05/631	G0006
GET914	123A	3122	20	TR21/735	S0025
GET929	128	3122	62	TR87/243	S0015
GET930	128	3122	62	TR87/243	S0015
GET931					G0001
GET995		3118			
GET1613		3024			
GET1708		3122			
GET1711		3024			
GET1893		3021			
GET2192		3039			
GET2218, A		3124			
GET2219		3024			
GET2221	123A	3122	210	TR21/735	S0015
GET2222	123A	3122	210	TR21/735	S0015
GET2369	123A	3122	20	TR21/735	S0011
GET2483		3122	62		
GET2484		3122			
GET2604		3114			
GET2605		3114			
GET2904		3025	21		S5022
GET2905		3025	21		S5022
GET2906		3114	48		S5022
GET2907		3114	48		S5022
GET3013	123A	3122	20	TR21/735	S0011
GET3014	123A	3122	20	TR21/735	S0011
GET3053		3024			
GET3153		3025			
GET3134		3025			
GET3135		3114			
GET3136		3114			
GET3210		3122			

Original Part Number	ECG	SK	GE	IR/WEP	HEP
GET3250, A		3114			
GET3251, A		3114			
GET3301		3122			
GET3302		3122			
GET3402		3124			
GET3403		3124			
GET3404		3124			
GET3405		3124			
GET3414		3124			
GET3415		3124			
GET3416		3124			
GET3417		3124			
GET3510		3122			
GET3511		3039			
GET3562			61		
GET3563			61		
GET3564		3018			
GET3565		3124			
GET3566		3124			
GET3567		3024			
GET3568		3124			
GET3569		3024			
GET3638, A	159	3114	22	TR20/717	S0012
GET3644		3114			
GET3646	123A	3122	20	TR21/735	S0011
GET3648		3039			
GET3691		3018			
GET3692		3124			
GET3693		3122			
GET3694		3122			
GET3903			210		
GET3904			20		
GET3905			21		
GET3906			21		
GET4123		3124			
GET4124		3124			
GET4125		3114			
GET4126		3114			
GET4248		3114			
GET4249		3118			
GET4400		3122			
GET4401		3122			
GET4402		3114			
GET4403		3025			
GET4410		3045			
GET4870	6401			2160/	
GET4871	6409			2160/	
GET4944		3024			
GET4945		3024			
GET4946		3024			
GET5086		3114			
GET5088		3124			
GET-5116		3008			
GET5116	160	3123	2	TR17/637	G0003
GET-5117		3008			
GET5117	160	3123	2	TR17/637	G0003
GET5127		3124			
GET5131		3124			
GET5133		3018			
GET5134		3124			
GET5136		3124			
GET5137		3124			
GET5138		3114			
GET5139		3114			
GET5172		3124			
GET5209		3122			
GET5210		3122			
GET5219		3122			
GET5220		3122			
GET5221		3114			
GET5223		3122			
GET5224		3122			
GET5225		3122			
GET5226		3114			
GET5227		3114			
GET5305	172			TR69/	
GET5306	172	3134		TR69/	
GET5307	172			TR69/	
GET5308	172	3134		TR69/	
GET5308A	172			TR69/	
GET5354		3114			
GET5355		3114			
GET5356		3114			
GET5365		3114			
GET5366		3114			
GET5367		3114			
GET5449		3024			
GET5450		3024			
GET5451		3024			
GETO-50P	158	3004			
GEX8			8	TR10/724	
GEX9				/631	
GEX16A1938				/735	
GF20	102	3123	52	TR05/631	G0005
GF21	102	3123	52	TR05/631	G0005

Original Part Number	ECG	SK	GE	IR/WEP	HEP
GF21A		3004			
GF32	102	3123	52	TR05/631	G0005
GF32A		3004			
GFT20R					G0005
GFT20/15					G0005
GFT20/30					G0005
GFT21					G0005
GFT21R					G0005
GFT21/15					G0005
GFT21/30					G0005
GFT22					G0007
GFT22R					G0007
GFT22/15					G0007
GFT22/30					G0007
GFT22/60					G0007
GFT25					G0005
GFT25R					G0005
GFT25/15					G0005
GFT25/30					G0005
GFT26					G6005
GFT30					G6003
GFT31					G0005
GFT31/15					G0005
GFT31/30					G0005
GFT31/60					G0005
GFT32					G0005
GFT32/15					G0005
GFT32/30					G0005
GFT34/30					G0005
GFT34/60					G0005
GFT41					G0002
GFT42A					G0002
GFT43					G0009
GFT43A,B					G0009
GFT44	160	3005	50	TR17/637	G0003
GFT44/30					G0005
GFT45	160	3005	50	TR17/637	G0003
GFT45/30					G0005
GFT2006					G6011
GFT2006/30					G6011
GFT2006/60					G6011
GFT2006/90					G6012
GFT3008/20					G6011
GFT3008/40	102	3009	52	TR05/631	G0005
GFT3008/60					G6011
GFT3008/80					G6012
GFT3408/20					G6011
GFT3408/40					G6011
GFT3408/60					G6011
GFT3408/80					G6012
GFT4012/30					G6003
GFT8024					G6005
GGE-18		3024			
GI1	100	3005	1	TR05/254	
GI2	102	3004	2	TR05/631	
GI2A		3004			
GI3	126		1	TR17/635	
GI4	102	3004		TR05/631	
GI4A		3004			
GI5	101	3011	5	TR08/641	
GI6	101	3011	7	TR08/641	
GI7		3011	7	TR08/641	
GI8	103		8	TR08/641A	
GI9	105		3	TR03/232	
GI10	123A	3012	4	TR21/233	
GI2711	123A	3122	10	TR21/735	
GI2712	123A	3122	17	TR21/735	
GI2713	123A	3122	210	TR21/735	
GI2714	123A	3122	210	TR21/735	
GI2715	123A	3122	86	TR21/735	
GI2716	123A	3122	60	TR21/735	
GI2921	123A	3122	60	TR21/735	
GI2922	123A	3122	60	TR21/735	
GI2923	123A	3122	60	TR21/735	
GI2924	123A	3122	212	TR21/735	
GI2925			212		
GI2926			212		
GI3391			212		
GI3391A			212		
GI3392			212		
GI3393			212		
GI3394			212		
GI3395			212		
GI3396			212		
GI3397			212		
GI3398			212		
GI3402			210		
GI3403			210		
GI3404			210		
GI3405			210		
GI3414			210		
GI3415			210		
GI3416			210		
GI3417			210		
GI3566			210		

Original Part Number	ECG	SK	GE	IR/WEP	HEP
GI3605			86		
GI3606			86		
GI3607			86		
GI3638	159	3114	65	TR20/717	
GI3638A	159	3114	65	TR20/717	
GI3641	123A	3122	210	TR21/735	
GI3643	123A	3122	210	TR21/735	
GI3644	159	3114	48	TR20/717	
GI3702	159	3114	82	TR20/717	
GI3703	159	3114	82	TR20/717	
GI3704	123A	3122	20	TR21/735	
GI3705	123A	3122	20	TR21/735	
GI3706	123A	3122	20	TR21/735	
GI3707	123A	3122	212	TR21/735	
GI3708	123A	3122	212	TR21/735	
GI3709	123A	3122	61	TR21/735	
GI3710	123A	3122	61	TR21/735	
GI3711	123A	3122	212	TR21/735	
GI3793	161		210	TR83/719	
GI3794			210		
GI3900			212		
GI3900A			212		
GI6506	101	3011		TR08/641	G0011
GI-2711				TR24/	
GI-2712				TR51/	
GI-2713				TR53/	
GI-2714				TR53/	
GI-2715				TR53/	
GI-2716				TR53/	
GI-2921				TR53/	
GI-2922				TR53/	
GI-2923				TR53/	
GI-2924				TR53/	
GI-2925				TR53/	
GI-2926				TR53/	
GI-3391				TR53/	
GI-3392				TR53/	
GI-3394				TR53/	
GI-3395				TR53/	
GI-3396				TR53/	
GI-3397				TR53/	
GI-3398				TR53/	
GI-3566				TR53/	
GI-3605				TR53/	
GI-3606				TR53/	
GI-3607				TR53/	
GI-3638				TR54/	
GI-3702				TR54/	
GI-3703				TR54/	
GI-3705				TR65/	
GI-3707				TR51/	
GI-3708				TR51/	
GI-3709				TR51/	
GI-3710				TR51/	
GI-3711				TR51/	
GI-3721				TR51/	
GI-3793				TR25/	
GI-3900				TR51/	
GM0290	160		52	TR17/637	G0003
GM0375	160		51	TR17/637	G0003
GM0376	160		51	TR17/637	G0003
GM0377	160		51	TR17/637	G0003
GM0378	168		52	TR17/	G0003
GM0380	108	3019	11	TR83/56	
GM0476B					G0003
GM290	160		52	TR17/637	G0003
GM290A	160		1	TR17/637	G0003
GM308	161	3019	17	TR83/719	
GM378	160		52	TR17/637	G0003
GM378A	160		1	TR17/637	G0003
GM378RED	160				
GM380	126		17	TR17/	
GM428	121	3009	16	TR01/232	
GM656A	160		1	TR17/637	G0003
GM760	159	3114	21	TR20/717	
GM770	107	3039	17	TR70/720	S0016
GM875	160			TR17/	
GM876	160			TR17/	
GM877	160			TR17/	
GM878	160			TR17/	
GM878A,B	160			TR17/	
GME040-1	159	3114		TR20/717	S0013
GME0404	106	3118	21	TR20/52	S0013
GME0404-1		3118	21	/52	S0019
GME0404-2	106	3118	21	TR20/52	S0013
GME404-1	106			TR20/	
GME1001	123A	3122		TR21/735	S0011
GME1002	123A	3122		TR21/735	S0015
GME2001	123A	3122		TR21/735	S0011
GME2002	123A	3122		TR21/735	S0015
GME3001	108	3039		TR95/56	S0016
GME3002	108	3039		TR95/56	S0016
GME4001	123A	3122	11	TR21/735	S0011
GME4002	123A	3122	11	TR21/735	S0015
GME4003	123A	3122	11	TR21/735	S0015

Original Part Number	ECG	SK	GE	IR/WEP	HEP
GME6003	123A	3122	63	TR21/735	S0011
GME9001	108	3039		TR95/56	S0016
GME9002	108	3039		TR95/56	S0016
GME9021	108	3039		TR95/56	S0016
GME9022	108	3039		TR95/56	S0016
G04-703-A	176			/238	
G05-010-A	123A			/735	
G05-011-A	123A			/735	
G05-015-D	128			/243	
GP139	102A	3004		TR85/250	
GP139A	102A	3004		TR85/250	
GP139B	102A	3004		TR85/250	
GP1432	121	3009	16	TR01/232	
GP1493	121	3009	3	TR01/232	
GP1494	121	3009	3	TR01/232	
GP1622	105	3009	3	TR03/231	G6005
GP1882	121	3009	3	TR01/232	
GPT-16	121	3009	16	TR01/232	
GRASS-R2982	130				
GRF-3	132				
GT1			50		G0005
GT2			50		G0005
GT3			50		G0005
GT4A					G0005
GT11	100	3123	50	TR05/254	G0005
GT12	100	3123	50	TR05/254	G0005
GT13	100	3005	50	TR05/254	G0005
GT14	102A	3004	52	TR85/250	G0005
GT14H	102A	3123	52	TR85/250	G0005
GT18	102A			TR85/250	
GT20	102A	3004	51	TR85/250	G0005
GT28	102A	3123	52	TR85/250	G0005
GT20R	102A	3123	52	TR85/250	G0005
GT24H			50		G0005
GT31	102A	3004	2	TR85/250	G0005
GT32	102A	3004	52	TR85/250	G0005
GT33	102A	3004	52	TR85/250	G0005
GT34	102A	3004	51	TR85/250	G0005
GT34HV	102A	3123	52	TR85/250	G0005
GT34S	100	3005	2	TR05/254	
GT35	103A		8	TR08/724	
GT38					G0005
GT40	160		51	TR17/637	G0003
GT41	102A	3005	51	TR85/250	G0005
GT42	102A	3005	51	TR85/250	G0005
GT43	126	3005	51	TR17/635	G0001
GT44	102A		51	TR85/250	G0005
GT45	102A		51	TR85/250	G0005
GT46	126	3006	51	TR17/635	G0001
GT47	126	3006	51	TR17/635	G0001
GT66	126	3006	52	TR17/635	
GT74	102A	3004	50	TR85/250	G0005
GT75	102A	3004	50	TR85/250	G0005
GT81	102A	3004	50	TR85/250	G0005
GT81H		3123	52		G0005
GT81HS	102A	3004	53	TR85/250	G0005
GT81R	102A	3004	52	TR85/250	G0005
GT82	102A	3004	50	TR85/250	G0005
GT83	100	3005	50	TR05/254	G0005
GT87	100	3005	50	TR05/254	G0005
GT88	100	3005	50	TR05/254	G0005
GT100			1		G0005
GT109	102A	3004	2	TR85/250	G0005
GT109R	102A	3123	52	TR85/250	G0005
GT122	102A	3004	50	TR85/250	G0005
GT123	102A	3005	50	TR85/250	G0005
GT132	102A	3123	52	TR85/250	G0005
GT153	100	3005	1	TR05/254	G0005
GT167	103A	3010	5	TR08/724	G0011
GT210H		3123	52		G0005
GT222	102A	3004	1	TR85/250	G0005
GT229	101	3011	8	TR08/641	G0011
GT269	100	3123	2	TR05/254	
GT310					G0005
GT336	103A	3010	8	TR08/724	
GT364	103A		59	TR08/724	G0011
GT365	103A		59	TR08/724	G0011
GT366	103A		59	TR08/724	G0011
GT751	102A	3004	52	TR85/250	G0005
GT758	102A	3004	1	TR85/250	G0005
GT759	102A	3005	1	TR85/250	G0005
GT759R	102A	3005	2	TR85/250	G0005
GT760	102A	3005	50	TR85/250	G0005
GT760R	102A	3005	2	TR85/250	G0005
GT761	100	3005	50	TR05/254	G0005
GT671R	100	3005	2	TR05/254	G0005
GT762	100	3005	50	TR05/254	G0005
GT762R	100	3005	2	TR05/254	G0005
GT763	102A	3004	50	TR85/250	G0005
GT764	100	3123	50	TR05/254	G0006
GT766	100	3123	2	TR05/254	G0005
GT766A	100	3004	52	TR05/254	G0005
GT792	101	3011	5	TR08/641	G0011
GT832	100	3005	2	TR05/254	G0003
GT903	103A		8	TR08/724	
GT904	101	3011	8	TR08/641	G0011
GT905	101	3011	8	TR08/641	
GT905R	101	3010	5	TR08/641	G0011
GT947	101	3011	8	TR08/641	
GT948	101	3011	8	TR08/641	G0011
GT948R		3011	8	TR08/	
GT949	101	3010	59	TR08/641	G0011
GT949R	101	3010	5	TR08/641	G0011
GT1079					G0011
GT1200	101	3011	5	TR08/641	
GT1202	101	3011	5	TR08/641	G0011
GT1223	102A	3123	52	TR85/250	G0005
GT1604	100	3005	50	TR05/254	G0005
GT1605	100	3005	50	TR05/254	G0005
GT1606	100	3005	50	TR05/254	G0001
GT1607	100	3005		TR05/254	G0001
GT1608	101	3011	5	TR08/641	G0011
GT1609	101	3011	5	TR08/641	G0011
GT1644	159	3114	65	TR20/717	S0012
GT1658	103A	3011	8	TR08/724	
GT1665	102A	3005	2	TR85/250	G0005
GT2693	100	3005	51	TR05/254	G0005
GT2694	100	3005	50	TR05/254	G0005
GT2695	100	3005	51	TR05/254	G0005
GT2696	102A		50	TR85/250	G0005
GT2765	101	3011	5	TR08/641	G0011
GT2766	101	3011		TR08/641	G0011
GT2767	101	3011		TR08/641	G0011
GT2768	103A	3010	8	TR08/724	G0011
GT2883	102A		50	TR85/250	G0005
GT2884	101	3011	5	TR08/641	G0011
GT2885	102A		50	TR85/250	G0005
GT2886	101	3011	5	TR08/641	G0011
GT2887	102A		50	TR85/250	G0005
GT2888	101	3011	5	TR08/641	G0011
GT2906	101	3011	5	TR08/641	G0011
GT3150	101	3011		TR08/641	G0011
GT5116	126	3008	50	TR17/635	G0001
GT6117	126	3006	50	TR17/635	G0001
GT5148	126	3006	51	TR17/635	G0001
GT5149	126	3006	50	TR17/635	G0001
GT5151	126	3006	51	TR17/635	G0005
GT6153	126	3006	51	TR17/635	G0001
GT-14				TR06/	
GT-14H				TR05/	
GT-20				TR08/	
GT-28				TR05/	
GT-34				TR08/	
GT-34HV				TR05/	
GT-34S				TR06/	
GT-35				TR08/	
GT-74				TR06/	
GT-75				TR85/	
GT-81H				TR05/	
GT-81HS				TR05/	
GT-82				TR05/	
GT-83				TR05/	
GT-87				TR05/	
GT-88				TR05/	
GT-109				TR08/	
GT-122				TR08/	
GT-123				TR08/	
GT-153				TR08/	
GT-167				TR08/	
GT-210H				TR05/	
GT-222				TR08/	
GT-229				TR08/	
GT-269				TR08/	
GT-758				TR85/	
GT-759				TR06/	
GT-759R				TR06/	
GT-760R				TR06/	
GT-761R				TR85/	
GT-762				TR85/	
GT-762R				TR85/	
GT-763				TR85/	
GT-764				TR05/	
GT-792				TR08/	
GT-903				TR08/	
GT-905				TR08/	
GT-947				TR08/	
GT-948				TR08/	
GT-949				TR08/	
GTA1					G0005
GTA2					G0006
GTA3					G0005
GTE1	100	3123	2	TR05/254	G0005
GTE2	100	3004	2	TR05/254	G0005
GTJ33141	126	3006		TR17/635	
GTJ33229	126	3006		TR17/635	
GTJ33230	126	3006		TR17/635	
GTJ33231	102A			TR85/250	
GTJ33232	102A			TR85/250	
GTL1					G6011
GTL3					G6011

Original Part Number	ECG	SK	GE	IR/WEP	HEP
GTV	100	3004	2	TR05/254	G0005
GTX2001	121	3009	3	TR01/232	
GV6063	123A	3124	10	TR21/735	S0015
GVL20002		3039			
GVL20077	123A	3024			
GVL20083		3039			
H1V	108	3039		TR95/56	
H2					G6003
H3					G6003
H3A					G6011
H4					G6003
H4A					G6011
H5					G6012
H5B2N3					G6012
H6					G6012
H6G				TR55/	
H7					G6011
H10	102A	3004	2	TR85/250	G0005
H12	105	3012	4	TR03/233	G6004
H12A	105	3012	4	TR03/233	G6004
H45					G6012
H57B2-6				TR85/	
H57B2-7				TR85/	
H57B2-15				TR85/	
H57B2-25				TR85/	
H102	123A	3124		TR21/735	
H103A	121			TR01/	
H200E					G6005
H221(GE)	783				C6100P
H442	108	3019		TR95/56	
H931	123A			TR21/735	S0016
H932	154			TR78/712	S5025
H933	123A			TR21/735	S5025
H934	123A			TR21/735	S3001
H1567	123A	3124	20	TR21/735	
H9423		3018			S0019
H9618		3018			
H9623	123A	3018	20	TR21/	S0014
H9625	108	3018	20	TR21/	S0014
H9696	123A	3124	18	TR24/	S0015
H43020109		3027			
H43020136		3027			
H43020136A		3027			
H43020158		3024			
H43020158A		3024			
H43020631		3024			
H43020631A		3024			
H43021579		3014			
HA00052	100	3005	51	TR05/254	
HA00053	100	3005	1	TR05/254	
HA-00102	100	3008	1	TR85/	
HA-00495	159	3114	21	TR30/	S0013
HA-00496	185	3083	58	TR56/	S5006
HA-00634	187		29		
HA-00643	193		67		
HA-00699	153	3083	29	TR56/	
HA1					G0005
HA3					G0005
HA12	160	3008	51	TR17/637	
HA-12	100			TR05/254	G0005
HA15	100	3005	1	TR05/254	
HA-15	126			TR17/635	G0008
HA30	160			TR17/637	
HA-30		3006			G0003
HA49	100		1	TR05/254	
HA-49	126	3008		TR17/635	G0008
HA52	126		1	TR17/635	
HA-52		3008		TR17/	G0008
HA53	126		1	TR17/635	
HA-53		3008			G0008
HA54	102A		2	TR85/250	
HA-54		3004			
HA56	102A		2	TR85/250	
HA-56		3004			
HA70	160		51	TR17/637	
HA-70		3006			
HA101				TR12/	
HA-101	126			TR17/635	G0009
HA102	126		1	TR17/635	
HA-102		3007		TR11/	G0008
HA103	126		51	TR17/635	
HA-103		3008			
HA104	126		51	TR17/635	
HA-104		3006			G0008
HA201	126	3008		TR17/635	
HA-201		3123	2	TR05/	G0005
HA202	100			TR05/254	
HA-202		3008			G0005
HA234	160	3006		TR17/	G0003
HA-234			51	TR17/637	
HA235	126		51	TR17/635	G0003
HA-235		3006			G0003
HA235A	160	3006	51	TR17/637	G0003
HA235C		3006	51	TR17/637	G0003
HA240	160	3006		TR17/637	

Original Part Number	ECG	SK	GE	IR/WEP	HEP
HA-240					G0003
HA266	160		51	TR17/637	
HA267	160		51	TR17/637	
HA-267		3006			G0008
HA-268	160	3004		TR17/637	G0008
HA-269	126	3006	51	TR17/635	G0008
HA330	126	3008		TR17/635	
HA-330		3123	2	TR05/	
HA-342		3006	51	TR17/635	
HA350			51	TR17/	
HA-350	126	3006		TR17/635	G0008
HA-350A	126	3006	51	TR17/635	G0008
HA-353	126	3006	51	TR17/635	G0008
HA-353C	126	3007	51	TR17/635	G0008
HA-354	126	3006	51	TR17/635	G0008
HA-354B	126	3008		TR17/635	G0008
HA460		3018			
HA471	126	3006		TR17/635	G0008
HA505	153		29	TR77/700	
HA525	160	3006		TR17/637	
HA1040	160	3006	51	TR17/637	
HA1101	726				
HA1102	726				
HA1103	704				
HA1104	704				
HA1108	1061	3070			
HA1110	901				
HA1115	720				
HA1115W	720				
HA1117	1122				
HA1118	1094				
HA1119	1121				
HA1124	712	3072			C6063P
HA1124S	712				
HA1125	712	3072			
HA1125A	712	3072			
HA1126	1004	3141			
HA1127	912				
HA1133	730				
HA1139	749				
HA1140	1080				
HA1154	712		IC2		
HA1156	801				
HA1157	1122				
HA1158	1094				
HA1159	1121				
HA1201	1039				
HA1202	1041				
HA1203	1042				
HA1301	903				
HA1302	784				
HA1306	1029				
HA1306P	1029				
HA1306P-U	1029				
HA1306U	1029				
HA1306W	1029				
HA1308	1030				
HA1309	1030				
HA1310	1031				
HA1311	1032				
HA1311W	1032				
HA1312	1033				
HA1312W	1033				
HA1313	1034				
HA1314	1035				
HA1316	1036				
HA1318	1038				
HA1318P	1038				
HA1318P-U	1038				
HA1319	1043				
HA1322	1037				
HA1350	102A	3004	2	TR85/250	
HA1360	102A	3004	2	TR85/250	
HA2001	133	3112		/801	F0010
HA2010	133	3112		/801	F0010
HA2190	160	3006	51	TR17/637	
HA2356	160			TR17/	
HA3210	160	3006	51	TR17/637	
HA3480	160	3006	51	TR17/637	
HA3670	160	3006	51	TR17/637	
HA4400	160	3006	51	TR17/637	
HA5001	101	3011		TR08/641	G0011
HA5002	101	3011	59	TR08/641	G0011
HA5003	101	3011	54R	TR08/641	G0011
HA5005	101	3011	59	TR08/641	G0011
HA5009	101	3011		TR08/641	G0011
HA5010	103A			TR08/724	G0011
HA5011	101	3011		TR08/641	G0011
HA5012	101	3011	59	TR08/641	G0011
HA5014	101	3011		TR08/641	G0011
HA5016	101	3011	59	TR08/641	G0011
HA5020	101	3011		TR08/641	G0011
HA5021	101	3011		TR08/641	G0011
HA5022	101	3011		TR08/641	G0011
HA5023	101	3011		TR08/641	G0011

Original Part Number	ECG	SK	GE	IR WEP	HEP
HA5024	101	3011		TR08/641	G0011
HA5025	101	3011		TR08/641	G0011
HA5026	101	3011		TR08/641	G0011
HA7206			65		
HA7207			65		
HA7501			82		
HA7506			82		S0032
HA7507			22		S0032
HA7510			82		S0032
HA7520			67	TR28/	S5022
HA7521			67	TR28/	S5022
HA7522			67	TR28/	S5022
HA5723			67	TR28/	S5022
HA7524			67	TR28/	S5022
HA7526			67	TR28/	S5022
HA7527			67	TR28/	S5022
HA7528			67	TR28/	S5022
HA7530	129	3025	82	TR88/242	S0032
HA7531	129	3025	82	TR88/242	
HA7532			48		S0032
HA7533			82	TR28/	S0032
HA7534			82	TR28/	
HA7536			22	TR20/	S0032
HA7537			82	TR28/	S0032
HA7538			82	TR28/	
HA7543			82		
HA7597			82	TR28/	S0032
HA7598			82	TR28/	S0032
HA7599			82	TR28/	S0032
HA7630	129	3025	82	TR88/242	S0032
HA7631	129	3025	82	TR88/242	
HA7632	129	3025	82	TR88/242	S0032
HA7723			67		S0032
HA7730			67		S0032
HA7732			67		S0032
HA7733					S5022
HA7734			67		S5022
HA7735			67		S5022
HA7736			67		S5022
HA7737			67		S5022
HA7804			22		S0032
HA7806			22	TR20/	S0032
HA7808			22	TR20/	S0032
HA7810			22		S0032
HA7815			82		S0032
HA9048			65		S0013
HA9049			65		S0013
HA9054			65		S0013
HA9055			65		S0013
HA9056			65		S0013
HA9057			65		S0013
HA9058			65		S0013
HA9059			65		S0013
HA9078			65		S0013
HA9079			65		S0013
HA9500			21		S0012
HA9501			21		S0012
HA9502			21		S0012
HA9531			65		S0013
HA9531A			65		S5022
HA9532			65		S5022
HA9532A			65		
HA9532B					S5022
HA11115W	720				
HA17711	911D				
HA17723	923D	3165			
HA17741M	941	3514			
HA177417		3514			
HAK026				TR08/	
HAM-1	102A			TR85/250	G0005
HAW1183B	176				
HAW1183C	176				
HB-00054	102A	3004	2	TR85/250	
HB-00056	102A	3004	2	TR85/250	
HB-00171	102A	3004		TR85/250	G0005
HB-00172	102A	3004	2	TR85/250	G0005
HB-00173	102A	3004			G0005
HB-00175	102A	3004	2	TR85/250	G0005
HB-00176	102A	3004	2	TR85/250	G0005
HB-00178			52	TR85/	
HB-00186	102A	3004	52	TR85/	G0007
HB-00187	102A	3004		TR17/	G0005
HB-00303			52	TR85/	
HB-00324	158	3004	2	TR84/630	G0005
HB-00370			52		G0006
HB-00405		3004	52	TR85/	
HB-00481					G6016
HB32	102A		2	TR85/250	
HB33	102A		2	TR85/250	
HB54	102A		2	TR85/250	
HB55	102A			TR85/250	
HB56	102A		53	TR85/250	
HB75	102A	3004	2	TR85/250	
HB75C	102A	3004	2	TR85/250	
HB77	102A	3004	2	TR85/250	C005

Original Part Number	ECG	SK	GE	IR WEP	HEP
HB77B	102A	3004		TR85/250	
HB77C	102A			TR85/250	
HB156	102A			TR85/250	
HB156C	102A			TR85/250	
HB171	102A		2	TR85/250	
HB172	102A		2	TR85/250	
HB173	102A			/250	
HB175	102A		2	TR85/250	
HB176	102A		2	TR85/250	
HB178	102A		2	TR85/250	
HB186	102A			TR85/250	
HB187	102A			TR85/250	
HB263	102A		2	TR85/250	
HB270	102A			TR85/250	
HB324	158			TR84/630	
HB365	158			TR84/630	
HB367	131			TR94/642	G6016
HB415	158			TR84/630	
HB422	126			TR17/635	
HB459	102A		2	TR85/250	G0006
HB461	176			TR82/238	
HB475	158			TR84/630	
HB-32		3004			G0005
HB-33		3004			G0005
HB-54		3004		TR85/	G0007
HB-542SB54				TR85/	
HB-56		3004		TR85/	G0005
HB-562SB56				TR85/	
HB-75				TR85/	G0005
HB-75C		3004			G0005
HB-77				TR85/	G0005
HB-77B		3123			G0007
HB-77C		3004	2	TR85/	G0007
HB-85		3004	2		G0007
HB-156		3004	2	TR85/	G0005
HB-156C					G0005
HB-171		3004			
HB-172		3004		TR85/	G0005
HB-172AP		3004			
HB-173		3004		TR85/	G0005
HB-175		3004		TR85/	G0005
HB-175G		3004			
HB-176		3004			G0005
HB-178		3004			G0005
HB-186		3123	52	TR85/	G0007
HB-187		3123	52	TR85/	G0005
HB-263		3004			G0005
HB-270					G0005
HB-324		3004			G0005
HB-365		3004		TR85/	G0005
HB-3652SB365				TR85/	
HB-367					G6016
HB-415		3004		TR85/	G6011
HB-422				TR21/	
HB-4222SB422				TR21/	
HB-459		3004		TR85/	G0006
HB-461		3123		TR85/	G0005
HB-475		3004	53	TR85/	G0005
HB-475-6		3004			
HBF4001A	4001				
HBF4002A	4002				
HBF4009A	4049				
HBF4011A	4011				
HBF4012A	4012				
HBF4013A	4013				
HBF4014A	4014				
HBF4015A	4015				
HBF4016A	4016				
HBF4017A	4017				
HBF4019A	4019				
HBF4020A	4020				
HBF4021A	4021				
HBF4023A	4023				
HBF4024A	4024				
HBF4025A	4025				
HBF4027A	4027				
HBF4028A	4028				
HBF4029A	4029				
HBF4030A	4030				
HBF4040A	4040				
HBF4042A	4042				
HBF4049A	4049				
HBF4050A	4050				
HBF4051A	4051				
HBF4052A	4052				
HC-0073J				TR08/	
HC-00183J				TR24/	
HC-00176	102A			TR85/250	
HC-00373	199	3122	62	2SC373/	S0015
HC-00380	107	3018	20	TR24/	S0016
HC-00458			17	TR24/	S0030
HC-00460			20	TR21/	S0015
HC-00461			20	TR53/	S0014
HC-00481		3014	31		G6016
HC-00496	184	3054	57	TR55/	S5000

Original Part Number	ECG	SK	GE	IR/WEP	HEP
HC-00535			20	TR21/	S0016
HC-00536	199	3124	62		S0016
HC-00537	123A	3122	62	TR51/	S0014
HC-00644		3020	20	TR21/	S0015
HC-00668		3018	17	TR95/	
HC-00693	123A	3124	20		S0015
HC-00711	199	3018	62	TR21/	S0020
HC-00730	101	3011	5	TR08/641	
HC-00732	199	3122	20	TR21/	S0024
HC-00735	123A	3124	10	TR21/	S0024
HC-00772		3018	17	TR95/	
HC-00784	107	3018	20	TR21/	S0016
HC-00828	123A	3122	17	TR21/735	S0015
HC-00829	107	3018	20	TR70/720	S0011
HC-00839	108		20	TR95/	
HC-00871	123A	3122	62	TR51/	S0015
HC-00900	199	3124	62		S0015
HC-00920	107				
HC-00921	123A		20	TR21/	
HC-00923	199	3124	62		S0015
HC-00924	123A		17	TR21/	
HC-00929	199	3018	20		S0016
HC-00930	199	3018	20	TR51/	S0016
HC-00945	123A	3124	20		S0015
HC-01047	107	3018	17	TR24/	S0025
HC-01060					S5000
HC-01096	186		28		
HC-01209	128	3024	63	TR25/	S5014
HC-01226	186	3054	28	TR55/	
HC-01317	128	3122	18	TR25/	
HC-01335	199	3122	62	TR51/	S0024
HC-01359	107	3018	17	TR21/	
HC-01390		3018			S0015
HC-01417		3018			S0015
HC-01820	123A	3124	20	TR21/735	
HC-01830	108	3018	17	TR95/56	
HC-56	123A	3124	20	TR21/735	S0011
HC206	108	3018	20	TR95/56	
HC371	123A			TR21/735	
HC-371		3018		TR21/	S0011
HC372	123A	3122		TR21/735	
HC-372					S0015
HC373	123A			TR21/735	
HC-373			63	TR25/	S0015
HC380	107			TR70/720	
HC-380		3018		TR21/	S0016
HC-3802SC350				TR21/	
HC-3802SC380				TR21/	
HC381	107	3018		TR70/720	
HC-381				TR21/	S0011
HC-3812SC381				TR21/	
HC394	107			TR70/720	
HC-394		3018		TR21/	S0017
HC-3942SC394				TR21/	
HC454	107		20	TR70/720	
HC-454					S0014
HC458	123A	3124	20	TR21/735	S0014
HC-458		3124		TR21/	S0014
HC460	107	3018	20	TR70/720	S0014
HC-460		3018		TR21/	S0014
HC461	107	3018	20	TR70/720	S0014
HC-461		3018		TR21/	S0014
HC495	152			TR76/701	
HC-495			28	TR76/	S5000
HC515	124	3021	12	TR81/240	
HC-515		3021			
HC535	107			TR70/720	
HC535A	107			TR70/720	
HC535B	107			TR70/720	
HC-535			17	TR21/	S0016
HC-535A		3018			S0016
HC-535B		3018			S0016
HC537	107	3039		TR70/720	
HC-537		3018	62		S0015
HC539	123A			TR21/735	
HC-539				TR24/	S0015
HC545	107	3039		TR70/720	
HC-545		3018			S0016
HC561	123A			TR21/735	
HC-561		3018	17		S0011
HC645	108		17	TR95/56	
HC-645		3018			S0016
HC668	108	3039		TR95/56	
HC-668			60	TR70/	S0016
HC-722					S0015
HC772	108	3039		TR95/56	
HC-772		3018	17	TR95/	
HC784	107	3018		TR70/	
HC-784				TR21/	S0016
HC-7842SC784				TR21/	
HC-789		3018			
HC829	108		20	TR95/56	
HC-829		3018			S0011
HC-838					S0015
HC-839					S0015

Original Part Number	ECG	SK	GE	IR/WEP	HEP
HC-924					S0011
HC-1359					S0014
HC100603	1019				
HC1000109	703A	3157			
HC1000109-0	703A	3157			
HC1000111-0	703A	3157			
HC1000114-0	726	3129			
HC1000117-0	718	3159	IC8		
HC1000217	941	3514			
HC1000403	1005				
HC1000503	1019				
HC1000505	1085				
HC1000703	1027				
HC10001090			IC12		
HD-00072	103A	3010	54		G0011
HD-00227	123A	3124	17	TR24/	S0011
HD-00261	192	3122	63		
HD74S00	74S00				
HD74S00P	74S00				
HD74S03	74S03				
HD74S03P	74S03				
HD74S04	74S04				
HD74S04P	74S04				
HD74S05	74S05				
HD74S05P	74S05				
HD74S10	74S10				
HD74S10P	74S10				
HD74S11	74S11				
HD74S11P	74S11				
HD74S15	74S15				
HD74S15P	74S15				
HD74S20	74S20				
HD74S20P	74S20				
HD74S22	74S22				
HD74S22P	74S22				
HD74S40	74S40				
HD74S40P	74S40				
HD74S151P	74S151				
HD74S181	74S181				
HD74S181P	74S181				
HD74S251	74S251				
HD74S251P	74S251				
HD-187	101	3011		TR08/641	G0011
HD197					G0005
HD2201	9932				
HD2201P	9932				
HD2202	9933				
HD2202P	9933				
HD2203	9946				
HD2203P	9946				
HD2204	9930				
HD2204P	9930				
HD2205	9945				
HD2205P	9945				
HD2206	9936				
HD2206P	9936				
HD2207	9962				
HD2207P	9962				
HD2208	9935				
HD2208P	9935				
HD2209	9944				
HD2209P	9944				
HD2210	9099				
HD2210P	9099				
HD2211	9093				
HD2211P	9093				
HD2213	9157				
HD2213P	9157				
HD2214	9158				
HD2214P	9158				
HD2215	9962				
HD2215P	9962				
HD2216	9937				
HD2216P	9937				
HD2501	7440				
HD2501P	7440				
HD2502	7460				
HD2502P	7460				
HD2503	7400				
HD2503P	7400				
HD2504	7420				
HD2504P	7420				
HD2505	7451				
HD2505P	7451				
HD2506	7450				
HD2506P	7450				
HD2507	7410				
HD2507P	7410				
HD2508	7430				
HD2508P	7430				
HD2509	7401				
HD2509P	7401				
HD2510	7474				
HD2510P	7474				
HD2511	7402				

Original Part Number	ECG	SK	GE	IR/WEP	HEP
HD2511P	7402				
HD2512	7453				
HD2512P	7453				
HD2513	7482				
HD2513P	7482				
HD2514	7454				
HD2514P	7454				
HD2515	7473				
HD2515P	7473				
HD2516	7476				
HD2516P	7476				
HD2517	7475				
HD2517P	7475				
HD2518	7441				
HD2518P	7441				
HD2519	7490				
HD2519P	7490				
HD2520	7493A				
HD2520P	7493A				
HD2521	7492				
HD2521P	7492				
HD2522	7404				
HD2522P	7404				
HD2523	7405				
HD2523P	7405				
HD2524	7491				
HD2524P	7491				
HD2526	7486				
HD2526P	7486				
HD2528	7403				
HD2528P	7403				
HD2529	7472				
HD2529P	7472				
HD2530	74107				
HD2530P	74107				
HD2531	7445				
HD2531P	7445				
HD2532	7447				
HD2532P	7447				
HD2533	7494				
HD2533P	7494				
HD2534	7495				
HD2534P	7495				
HD2535	7483				
HD2535P	7483				
HD2536	7442				
HD2536P	7442				
HD2537	7443				
HD2537P	7443				
HD2538	7444				
HD2538P	7444				
HD2539	7470				
HD2539P	7470				
HD2540	74170				
HD2540P	74170				
HD2541	74192				
HD2541P	74192				
HD2542	74193				
HD2542P	74193				
HD2543	74121				
HD2543P	74121				
HD2544	7438				
HD2544P	7438				
HD2545	7413				
HD2545P	7413				
HD2546	7496				
HD2546P	7496				
HD2547	74181				
HD2547P	74181				
HD2548	74150				
HD2548P	74150				
HD2549	74151				
HD2549P	74151				
HD2550	7408				
HD2550P	7408				
HD2551	7409				
HD2551P	7409				
HD2552	7437				
HD2552P	7437				
HD2555	74145				
HD2555P	74145				
HD2558	74141				
HD2558P	74141				
HD2560	7426				
HD2560P	7426				
HD2561	74123				
HD2561P	74123				
HD2562	74182				
HD2562P	74182				
HD2563	74H183				
HD2563P	74H183				
HD2564	74153				
HD2564P	74153				
HD2572	74196				
HD2572P	74196				

Original Part Number	ECG	SK	GE	IR/WEP	HEP
HD2573	74197				
HD2573P	74197				
HD2580	74154				
HD2580P	74154				
HD7400	7400				
HD7400P	7400				
HD7402	7402				
HD7402P	7402				
HD7403	7403				
HD7403P	7403				
HD7404	7404				
HD7404P	7404				
HD7405	7405				
HD7406	7406				
HD7406P	7406				
HD7407	7407				
HD7407P	7407				
HD7410	7410				
HD7410P	7410				
HD7412	7412				
HD7412P	7412				
HD7414	7414				
HD7414P	7414				
HD7416	7416				
HD7416P	7416				
HD7417	7417				
HD7420	7420				
HD7420P	7420				
HD7426	7426				
HD7426P	7426				
HD7427	7427				
HD7427P	7427				
HD7430	7430				
HD7430P	7430				
HD7432	7432				
HD7432P	7432				
HD7440	7440				
HD7440P	7440				
HD7442A	7442				
HD7442AP	7442				
HD7443A	7443				
HD7443AP	7443				
HD7444A	7444				
HD7444AP	7444				
HD7450	7450				
HD7450P	7450				
HD7451	7451				
HD7451P	7451				
HD7453	7453				
HD7453P	7453				
HD7454	7454				
HD7454P	7454				
HD7460	7460				
HD7460P	7460				
HD7471P	7471				
HD7472	7472				
HD7472P	7472				
HD7474	7474				
HD7474P	7474				
HD7475	7475				
HD7475P	7475				
HD7485	7485				
HD7485P	7485				
HD7486	7486				
HD7486P	7486				
HD7490A	7490				
HD7490AP	7490				
HD7492A	7492				
HD7492AP	7492				
HD7493A	7493A				
HD7493AP	7493A				
HD7496	7496				
HD7496P	7496				
HD74107	74107				
HD74107P	74107				
HD74121	74121				
HD74121P	74121				
HD74125	74125				
HD74125P	74125				
HD74126	74126				
HD74126P	74126				
HD74132	74132				
HD74132P	74132				
HD74136	74136				
HD74136P	74136				
HD74147	74147				
HD74147P	74147				
HD74148	74148				
HD74148P	74148				
HD74150	74150				
HD74150P	74150				
HD74151A	74151				
HD74151AP	74151				
HD74155	74155				
HD74155P	74155				

Original Part Number	ECG	SK	GE	IR/WEP	HEP
HD74156	74156				
HD74156P	74156				
HD74157	74157				
HD74157P	74157				
HD74160	74160				
HD74160P	74160				
HD74161	74161				
HD74161P	74161				
HD74162	74162				
HD74162P	74162				
HD74163	74163				
HD74164	74164				
HD74164P	74164				
HD74166	74166				
HD74166P	74166				
HD74174	74174				
HD74174P	74174				
HD74175	74175				
HD74175P	74175				
HD74180	74180				
HD74180P	74180				
HD74190	74190				
HD74190P	74190				
HD74191	74191				
HD74191P	74191				
HD74194	74194				
HD74194P	74194				
HD74198	74198				
HD74198P	74198				
HD74199	74199				
HD74199P	74199				
HE-00829	107				
HE-00930	199				
HEP1	126	3006	51	TR17/635	G0002
HEP-2	160		1	TR17/	G0002
HEP3	160	3006	51	TR17/637	G0003
HEP50	123A	3018	210	TR21/735	S0011
HEP51	129	3118	21	TR88/242	S0012
HEP52	159		21	TR20/717	S0013
HEP53	123A	3020	47	TR21/735	S0014
HEP54	123A	3018	20	TR21/735	S0015
HEP55	123A	3124	210	TR21/735	S0015
HEP56	108	3018	11	TR95/56	S0016
HEP57	159	3118	48	TR20/717	S0019
HEP75		3048	88	TR64/	S3013
HEP76			89		S3014
HEP200	104	3009	16	TR01/230	G6003
HEP201					G6004
HEP230	104		16	TR01/230	G6003
HEP231	105	3012	4	TR03/231	G6004
HEP232	121	3014	16	TR01/232	G6005
HEP233	105		4	TR03/233	G6006
HEP234	127	3034	25	TR27/235	G6007
HEP235	127	3035	25	TR27/235	G6008
HEP236	179			TR35/G6001	G6009
HEP237					G6010
HEP238	176	3123			G6011
HEP239					G6012
HEP240	124	3021	32	TR81/240	S5011
HEP241	175	3538		TR81/241	S5012
HEP242	129		29	TR88/242	S5013
HEP243	128		46	TR87/243	S5014
HEP244	157		27	TR60/244	S5015
HEP245	185	3054	57	TR76/701	S5000
HEP246	185	3083		TR77/S5007	S5006
HEP247	130	3027	19	TR59/	S7002
HEP250	102A		2	TR85/250	G0005
HEP251	102A		53	TR85/250	G0005
HEP252	102		2	TR05/	G0005
HEP253	158	3004	2	TR84/630	G0005
HEP254	158	3004	2	TR84/630	G0005
HEP280	102A			TR85/250	G0005
HEP281	102A			TR85/250	G0006
HEP310	6400				S9002
HEP570	9917				C2003P
HEP571					C2002P
HEP572	9976				C2004P
HEP573	9989				C2005P
HEP580					C2006G
HEP581					C2007G
HEP582					C2008G
HEP583					C2009G
HEP584					C2010G
HEP590					C6091
HEP591	704	3022		IC501/	
HEP592					C6092G
HEP593					C6093G
HEP594	718	3159	IC8		C6094P
HEP595	720	3160	IC7		C6095P
HEP623	121	3009	16	TR01/232	G6013
HEP624	121	3009	16	TR01/232	G6013
HEP625			25		G6018
HEP626			16		
HEP627	179		16	TR01/	G6015
HEP628	121	3009	16	TR01/232	G6013

Original Part Number	ECG	SK	GE	IR/WEP	HEP
HEP629	102		53	TR05/631	G0005
HEP630	102		2	TR05/631	G0006
HEP631	102	3004	2	TR05/631	G0005
HEP632	102	3004	2	TR05/631	G0005
HEP633	102	3004	2	TR05/631	G0007
HEP634	102A		2	TR85/250	G0007
HEP635	126	3006	51	TR17/635	G0001
HEP636	126	3006	51	TR17/635	G0008
HEP637	160		51	TR17/637	G0003
HEP638	126	3006	51	TR17/635	G0009
HEP639	126		51	TR17/635	G0008
HEP640	126		51	TR17/635	G0008
HEP641	101		5	TR08/641	G0011
HEP642	131			TR94/642	G6016
HEP643	131	3052	30	TR94/642	G6016
HEP643X2			31MP		
HEP700	185	3191	58	TR77/S5007	S5006
HEP701	184	3054	57	TR76/S5003	S5000
HEP702	218	3083	26	TR20/700	S5018
HEP703	175		66	TR81/241	S5019
HEP703X2			24MP		
HEP704	130		75	TR59/247	S7004
HEP704X2			15MP		
HEP705			19		S7003
HEP706	154	3103	32	TR78/712	S5024
HEP707	162	3111		TR67/707	S5020
HEP708	159			TR20/717	S5022
HEP709	161		61	TR83/719	S0017
HEP710	129	3025		TR88/242	S5023
HEP711			18		
HEP712	154		18	TR78/712	S5025
HEP713	154	3024	18	TR78/712	S5026
HEP714	190		18	TR74/S3021	S3019
HEP715	159	3118	48	TR20/717	S0019
HEP716	159	3118	21	TR20/717	S0019
HEP717	159		65	TR20/717	S0031
HEP718	108	3018	11	TR95/56	S0020
HEP719	161	3018	11	TR83/719	S0020
HEP720	108	3018	11	TR95/56	S0020
HEP721	107		210	TR70/720	S0015
HEP722	108		10	TR95/56	S0015
HEP723	108		10	TR95/56	S0015
HEP724	123A		10	TR21/735	S0015
HEP725	123A		10	TR21/735	S0015
HEP726	199		210	TR21/735	S0015
HEP727	108	3018	10	TR95/56	S0025
HEP728	123A	3124	210	TR21/735	S0015
HEP729	123A	3124	210	TR21/735	S0025
HEP730	199		212	TR21/	S0024
HEP731	107		61	TR70/720	S0033
HEP732	107	3124	61	TR70/720	S0033
HEP733	108		212	TR95/56	S0020
HEP734	107		61	TR70/720	S0016
HEP735	123A		47	TR21/735	S0015
HEP736	128		47	TR87/243	S0015
HEP737	199		212	TR21/	S0024
HEP738	123A	3124	212	TR21/735	S0030
HEP739	159	3114	212	TR20/717	S0032
HEP740	163	3115		TR67/740	S5021
HEP801	133	3112	FET1	FE100/801	F0010
HEP802	132	3116	FET2	FE100/802	F0015
HEP803					F1036
HEP6004				TR85/	
HEP-1		3006			
HEP-2	160			TR17/	
HEP-50		3018			
HEP-52		3118			
HEP-56		3019			
HEP-230		3009		TR01/	
HEP-231		3012			
HEP-237	213				
HEP-245				TR55/	
HEP-246			58		
HEP-248	219				
HEP-250		3004			
HEP-251		3004			
HEP-252		3004			
HEP-591		3022			
HEP-594	718				
HEP-595	720				
HEP-639		3008			
HEP-640		3008			
HEP-702				TR58/	
HEP-703		3026			
HEP-704		3027			
HEP-705	219				
HEP-707			35		
HEP-709		3018			
HEP-714		3103			
HEP-722		3124			
HEP-723		3124			
HEP-724		3124			
HEP-725		3124			
HEP-728		3124			
HEP-729		3122			

Original Part Number	ECG	SK	GE	IR/WEP	HEP
HEP-731		3124			
HEP-733		3124			
HEP-734		3124			
HEP-736		3124			
HEP-738		3122			
HEP-740			36		
HEPC6056P			IC9		
HEPC6057P			IC5		
HEPC6071P			IC6		
HEPC6089			IC3		
HEP-C1030P	9930				
HEP-C1032P	9932				
HEP-C1033P	9933				
HEP-C1035P	9935				
HEP-C1036P	9936				
HEP-C1044P	9944				
HEP-C1045P	9945				
HEP-C1046P	9946				
HEP-C1052P	9099				
HEP-C1053P	9093				
HEP-C1057P	9157				
HEP-C1058P	9158				
HEP-C1062P	9962				
HEP-C3000P	7400				
HEP-C3001P	7401				
HEP-C3002P	7402				
HEP-C3004P	7404				
HEP-C3010P	7410				
HEP-C3020P	7420				
HEP-C3030P	7430				
HEP-C3040P	7440				
HEP-C3041P	7441				
HEP-C3050P	7450				
HEP-C3073P	7473				
HEP-C3075P	7475				
HEP-C3800P	7490				
HEP-C3801P	7492				
HEP-C6001	760	3167		IC500/	
HEP-C6003	752				
HEP-C6004	771				
HEP-C6007	754				
HEP-C6008	769				
HEP-C6009	766				
HEP-C6010	753				
HEP-C6011	761				
HEP-C6012	762				
HEP-C6013	755				
HEP-C6014	756				
HEP-C6015	767				
HEP-C6017	773				
HEP-C6019	754				
HEP-C6052P	941M	3553			
HEP-C6055L	725	3162			
HEP-C6056P	722				
HEP-C6057P	790	3077			
HEP-C6059P	746				
HEP-C6060P	748				
HEP-C6061P	750				
HEP-C6062P	708	3135			
HEP-C6063P	712				
HEP-C6065P	719				
HEP-C6066P	798				
HEP-C6067P	707				
HEP-C6068P	720				
HEP-C6069P	780	3141			
HEP-C6070P	714	3075			
HEP-C6071P	715	3076			
HEP-C6077P	782				
HEP-C6083P		3072			
HEP-C6085	731				
HEP-C6089	705A	3134			
HEP-C6096P	801				
HEP-C6099P	795				
HEP-C6100P	783				
HEP-C6101P	723				
HEP-C6102P	778				
HEP-C6103P	909	3590			
HEP-F0021		3116	FET2		
HEP-F2004	221	3116			
HEP-F2005	220	3112			
HEP-F2007	222				
HEP-G0004					G0006
HEP-G0010					G0008
HEP-G6001				TR35/G6001	
HEP-S0001	194				S0005
HEP-S0002	192	3122	210		S0024
HEP-S0003	192				S0024
HEP-S0004			20		S0025
HEP-S0005	194				
HEP-S0006	234		65		S0019
HEP-S0007					S0005
HEP-S0021					S0015
HEP-S0022					S0015
HEP-S0023					S0015
HEP-S3001	195		45		
HEP-S3020	188	3024		TR72/	
HEP-S3021	191	3104		TR75/	
HEP-S3022	191			TR75/	S3021
HEP-S3023	210	3512			S3024
HEP-S3024	210	3202			
HEP-S3025	210	3202			S3024
HEP-S3026	188	3199			S3024
HEP-S3027	211	3513			S3028
HEP-S3028	211				
HEP-S3029	211				S3028
HEP-S3030	189	3200			S3028
HEP-S3031	189			TR73/	S3032
HEP-S5000	184	3054		/S5003	
HEP-S5001	182	3188	55	/S5004	
HEP-S5002	183	3189	56	/S5005	S5005
HEP-S5003	184	3054			S5000
HEP-S5004	182	3188		/S5004	S5004
HEP-S5005	183	3189		/S5005	
HEP-S5006	185	3083		/S5007	
HEP-S5007	185	3083	58		S5006
HEP-S5008	183	3189			S5005
HEP-S5009	183	3189			S5005
HEP-S5010	183	3189			S5005
HEP-S7000	181	3535	75	TR36/	
HEP-S7001	180		74		
HEP-S9001	6402				
HEP-S9100	172	3156	64	TR69/S9100	
HEPC0900P	9662				
HEPC0901P	9663				
HEPC0902P	9668				
HEPC0903P	9669				
HEPC0904P	9670				
HEPC0905P	9661				
HEPC0906P	9664				
HEPC0907P	9682				
HEPC0908P	9681				
HEPC0909P	9672				
HEPC0910P	9665				
HEPC0911P	9667				
HEPC0912P	9663				
HEPC6016C	9796				
HEPC6096P	801				
HF1	123A			TR21/735	
HF2	123A		20	TR21/735	
HF3	123A		20	TR21/735	
HF3H	160	3005	50	TR17/637	G0005
HF3M	160	3005	50	TR17/637	G0005
HF4	123A		61	TR21/735	
HF5	123A		20	TR21/735	
HF6	123A		20	TR21/735	
HF6H	160	3005	50	TR17/637	G0006
HF6M	160	3005	50	TR17/637	G0005
HF7	123A		20	TR21/735	
HF8	123A		20	TR51/735	
HF9	128		18	TR87/243	
HF10	129		21	TR88/242	
HF11			28		
HF12			29		
HF12H	160	3007	50	TR17/637	G0006
HF12M	160	3007	50	TR17/637	G0005
HF12N	160	3008	1	TR17/637	G0005
HF15			28	TR76/	
HF16			29		
HF17			62		
HF19	121	3009	16	TR01/232	
HF19D				TR27/G6001	G6008
HF20	121	3009	16	TR01/232	
HF20H	160	3007	50	TR17/637	G0006
HF20M	160	3007	50	TR17/637	G0005
HF22	130			TR59/247	
HF23	180			/S7001	
HF24	181			TR36/S7000	
HF25	180			/S7001	
HF35		3123			G0001
HF40	194	3137	10	TR24/	S0015
HF40/34-6016-56				TR24/	
HF43			27	TR78/	
HF47	106	3114	21	TR54/	S0019
HF50		3123			G0001
HF50H	160		50	TR17/637	G0002
HF50M	160	3006	50	TR17/637	G0002
HF51	187	3083	29	TR56/	S3028
HF57		3054			S3024
HF58		3083			S3028
HF75					G0008
HF100					G0009
HF200					G0009
HF200191A	132		FET2	FE100/	
HF200191A-0	132		FET2	FE100/	
HF200191A0	132	3116	FET2	FE100/	F0021
HF200191B-0	132		FET2	FE100/	
HF200191B0	132	3116	FET1	FE100/	F0021
HF200301B		3112			
HF200301B0	133	3112	FET1	FE100/	F0010
HF200301C-0	133		FET2	FE100/	

Original Part Number	ECG	SK	GE	IR/WEP	HEP
HF200301E		3112			F0010
HF200411B	132		G	FE100/	F0021
HF309301E	199	3018	20	TR24/	S0016
HF2000411B	132		FET2	FE100/	F0021
HF-1				TR33/	S0011
HF-2		3018		TR33/	S0015
HF-3		3018		TR33/	S0015
HF-4		3018		TR33/	S0015
HF-5		3018		TR33/	S0011
HF-6		3018		TR33/	S0011
HF-7		3018		TR33/	S0015
HF-8		3018			S0015
HF-9		3024			
HF-10		3025			
HF-19D	179	3035		TR27/	G6007
HF-35			51		
HF-40	123A				
HF-47	106	3114	21		
HJ15	102A	3123	52	TR85/250	G0005
HJ15D	160		50	TR17/637	G0003
HJ17	102A	3123	52	TR85/250	G0005
HJ17D	102A	3123	52	TR85/250	G0005
HJ17P					G0005
HJ22	102A	3123	2	TR85/250	G0005
HJ22D	102A		50	TR85/250	G0005
HJ23	102A	3123	2	TR85/250	G0005
HJ23D	102A		50	TR85/250	G0005
HJ32	160		50	TR17/637	G0003
HJ34	160		50	TR17/637	G0003
HJ34A	160		50	TR17/637	G0003
HJ35	104		16	TR01/230	G6003
HJ37	160	3123	2	TR17/637	G0003
HJ41	100	3005	1	TR05/254	
HJ43	102A	3004	2	TR85/250	
HJ50	102A		50	TR85/250	G0005
HJ51	102A		50	TR85/250	G0005
HJ54	102A	3123	2	TR85/250	G0005
HJ55	160		50	TR17/637	G0003
HJ56	160		50	TR17/637	G0003
HJ57	160		50	TR17/637	G0003
HJ60	102A		50	TR85/250	G0005
HJ60A	160	3123	2	TR17/637	G0003
HJ60C	160	3005	1	TR17/637	
HJ62	102A	3123	2	TR85/250	G0005
HJ70	160		50	TR17/637	G0002
HJ71	158		50	TR84/630	G0005
HJ72	158		50	TR84/630	G0005
HJ73	158		50	TR84/630	G0005
HJ74	158		50	TR84/630	G0005
HJ226	158	3123	2	TR84/630	G0005
HJ228	158	3005	2	TR84/630	G0005
HJ230	158	3123	2	TR84/630	G0005
HJ315	158	3123	52	TR84/630	G0005
HJ606	102A	3005	1	TR85/250	
HJ-15		3004			
HJ-17		3004			
HJ-17D		3004			
HJ-22		3005			
HJ-22D		3005			
HJ-23		3005			
HJ-23D		3005			
HJ-34A		3004			
HJ-50		3004			
HJ-51		3004			
HJ-54		3005			
HJ-56		3008			
HJ-60		3005			
HJ-62		3005			
HJ-70		3007			
HJ-71		3005			
HJ-73		3005			
HJ-74		3005			
HJ-226		3005			
HJ-230		3005			
HJ-315		3004			
HJX2	102A	3123	2	TR85/250	G0005
HJX-2		3005			
HK040519B				/724	G6011
HK015619AX2		3004			
HK-00330	133		FET2	FE100/	F0010
HK3405	407				
HK3406	408				
HK3407	409				
HK3408	416				
HK3409	416				
HKT-133E		3004			
HKT-158	123A	3122		TR21/735	S0011
HKT-161	123A	3122		TR21/735	S0011
HL18998	7400				
HL18999	7474				
HL19000	7404				
HL19001	7410				
HL19002	7473				
HL19003	7420				
HL19004	7402				
HL19005	7403				
HL19006	7493				
HL19008	74121				
HL19009	7442				
HL19010	7476				
HL19011	7440				
HL19012	7475				
HL19013	7430				
HL19014	7486				
HL19015	7490				
HL24510	941				
HL24592	925				
HL24593	641				
HL24630	909	3590			
HL53424	9930				
HL53426	9935				
HL53428	9944				
HL53429	9946				
HL55660	7411				
HL65661	74S00				
HL55663	74S138				
HL55664	74S199				
HL55723	74S153				
HL55763	7408				
HL55764	74153				
HL55861	74150				
HL55862	7404				
HL56320	74S10				
HL56420	7400				
HL56421	7404				
HL56422	7420				
HL56423	7430				
HL56424	7438				
HL56425	7474				
HL56426	7485				
HL56427	7495				
HL56429	74192				
HL56430	74193				
HL56431	74H106				
HL56842	7483				
HL56899	7410				
HM-00049	100	3005	2	TR05/254	
HM-08014	102A	3004	2	TR85/250	
HN100					S7002
HN150					S7004
HO300	127		25	TR27/235	G6008
HPF1025	723				
HPF1030					S0015
HR1			51	TR12/	G0005
HR2			51	TR12/	G0005
HR2A		3004	2		
HR3			2		
HR4			2	TR05/	G0005
HR4A		3005	1		
HR5			1		
HR6		3005	2	TR05/	G0005
HR7			51	TR12/	G0005
HR7A		3004	2		
HR8		3123	2	TR05/	G0005
HR8A		3005	1		
HR9		3123	2	TR05/	G0005
HR9A		3005	1		
HR11		3124	20		
HR11A		3124	20		
HR11B		3124	20		
HR13		3024	20		S0015
HR13A		3024	20		
HR14		3124	20	TR21/	
HR14A		3124	20		
HR15		3124	52	TR33/	
HR15A		3124	20		
HR16		3124	20	TR33/	
HR16A		3124	20		
HR17		3124	20	TR33/	S0015
HR17A		3124	20	TR25/	
HR18		3124	20	TR25/	
HR18A		3124	20		
HR19		3124	20	TR25/	
HR19A		3124	20		
HR20		3006	51		
HR20A		3006	51		
HR21		3006	51		
HR21A		3006	51		
HR22		3006	51		
HR22A		3006	51		
HR22B		3006	51		
HR24		3006	51		
HR24A		3006	51		
HR25		3006	51		
HR25A		3006	51		
HR26		3006	51		
HR26A		3006	51		
HR27		3006	51		
HR27A		3006	51		
HR28	128	3024	18	TR25/243	

Original Part Number	ECG	SK	GE	IR/WEP	HEP
HR29	128	3024	20	TR25/243	S5014
HR30	158	3004		TR82/630	
HR32			20		
HR36	123A	3124	20	TR05/	S0015
HR37		3124	20	TR21/	
HR39				TR85/	
HR40	160	3006	50	TR17/637	
HR41	160		51	TR17/637	
HR42	160	3006	50	TR17/637	
HR43	160	3006		TR17/637	
HR44	160	3006	50	TR17/637	
HR45	160	3006		TR17/637	
HR46	160	3006	50	TR17/637	
HR47	199	3124		TR54/	S0014
HR48		3020		TR24/	S0014
HR50	160	3008	51	TR85/637	
HR51	160	3008	51	TR17/637	
HR52	160	3008	51	TR17/637	
HR53	102A	[illegible]	[illegible]	[illegible]/250	
HR58	108	3018		TR51/56	
HR59	108	3018		TR51/56	
HR60	108	3018		TR51/56	
HR61	102A			TR85/250	
HR62	123A		20	TR21/735	
HR63	123A	3018	20	TR51/735	S0015
HR64	123A		20	TR51/735	
HR65	123A	3122	62	TR33/735	
HR66	123A	3122		TR87/735	
HR67		3024	63	TR87/	
HR68					S3024
HR69			66		
HR70			69		
HR71	159		21	TR52/	S0019
HR72	193		67	TR88/	
HR72/34-6016-51				TR56/	
HR73		3202			
HR74		3203			
HR76	107	3018	20	TR21/	S0015
HR77	107	3018	17	TR21/	S0025
HR78	107	3018	17	TR24/	S0025
HR79	107	3018	20	TR24/	S0015
HR80	107	3018	20	TR24/	S0011
HR81	128	3018	20	TR24/	S0011
HR82	192				
HR83	192		63	TR87/	
HR83/34-6016-59				TR55/	
HR84		3018	20	TR20/	S5022
HR84(NPN)	123A				
HR84(PNP)	159				
HR85	210	3202	28	TR76/	S3024
HR86	211	3203	29	TR77/	S3028
HR87	107	3018	17	TR24/	S0025
HR101	104		16	TR01/230	G6003
HR101A		3009	3	TR01/	
HR101C					G6005
HR101D					G6005
HR101F					G6005
HR102	121	3009	16	TR01/232	
HR102C	121	3009		TR01/232	
HR103	121	3009	3	TR01/232	
HR103A		3009	3	TR01/232	
HR103C				TR01/	
HR105	121		16	TR02/232	
HR105A	121			TR01/232	
HR105B	121		16	TR01/232	
HR106	124	3021	12	TR57/240	S5011
HR107	124	3021	66	TR81/240	
HR107H	175	3131		TR81/241	
HR40836	160			TR17/637	
HR40837	160			TR17/637	
HR43835	160			TR17/637	
HR43838				TR17/	
HR45838	160			TR17/637	
HR45910	160			TR17/637	
HR45913	160			TR17/637	
HR104961C	187			TR56/	
HR105611C	193				
HR448636	160			TR17/637	
HR-1	102A			TR85/250	
HR-2	102A	3004		TR85/250	
HR-2A	102A			TR85/250	
HR-3	102A	3004		TR85/250	
HR-4	102A	3005		TR85/250	
HR-4A	102A			TR85/250	
HR-5	102A	3005		TR85/250	
HR-6	102A			TR85/250	
HR-7	102A	3004		TR85/250	
HR-7A	102A			TR85/250	
HR-8	102A	3005		TR85/250	
HR-8A	102A			TR85/250	
HR-9	102A	3005		TR85/250	
HR-9A	102A			TR85/250	
HR-11	123A			TR21/735	
HR-11A	123A			TR21/735	
HR-11B	123A			TR21/735	

Original Part Number	ECG	SK	GE	IR/WEP	HEP
HR-13	123A			TR21/735	
HR-13A	123A			TR21/735	
HR-14	199			TR21/735	
HR-14A	123A			TR21/735	
HR-15	199	3124		TR21/735	S0014
HR-15A	123A			TR21/735	
HR-16	123A	3124		TR21/735	S0014
HR-16A	123A			TR21/735	
HR-17	123A	3124		TR21/735	S0014
HR-17A	123A	3018		TR21/735	S0014
HR-18	123A	3124		TR21/735	S0014
HR-18A	123A			TR21/735	
HR-19	123A			TR21/735	S0014
HR-19A	123A			TR21/735	
HR-19E	179		25	TR35/G6001	G6007
HR-20	160			TR17/637	
HR-20A	160			TR17/637	
HR-21	160			TR17/637	
HR-21A	160			TR17/637	
HR-22	160			TR17/637	
HR-22A	160			TR17/637	
HR-22B	160			TR17/637	
HR-24	160			TR17/637	
HR-24A	160			TR17/637	
HR-25	160			TR17/637	
HR-25A	160			TR17/637	
HR-26	160			TR17/637	
HR-26A	160			TR17/637	
HR-27	160			TR17/637	
HR-27A	160			TR17/637	
HR-28		3024		TR25/	S5014
HR-29		3024		TR25/	S5014
HR-30		3004	52		G0005
HR-32	123A			TR21/	
HR-36	123A	3124		TR21/735	S0015
HR-37	123A			TR21/735	S0015
HR-38	123A	3024	63	TR21/735	S5014
HR-39	102A	3004	52	TR85/250	
HR-40		3006		TR17/	G0003
HR-43		3006	50	TR17/	G0003
HR-44		3006		TR17/	G0003
HR-45		3006	50	TR17/	S0014
HR-47	199	3124	17	TR21/735	S0014
HR-48	123A	3020	20	TR21/735	S0014
HR-50					G0005
HR-58			11	TR22/	S0011
HR-59		3018	11	TR22/	S0011
HR-60		3018	20	TR21/	S0011
HR-61			52	TR05/	G0005
HR-61A		3004			
HR-62		3020			S0015
HR-63		3018			S0015
HR-64		3018			S0015
HR-65		3024			S50[illegible]
HR-67	192			TR25/735	S5014
HR-67F		3024			S5014
HR-68	196	3054	28	TR92/S3023	
HR-69	186	3192		TR55/	
HR-70	187	3193		TR56/	
HR-71	159	3114		TR20/	S0019
HR-72	193				
HR-73	210	3054	28		
HR-74	211	3083	29		
HR-76		3018		TR33/	
HR-77				TR33/	
HR-78				TR33/	
HR-79				TR33/	
HR-80				TR33/	
HR-81				TR33/	
HR-84					S0015
HR-101		3009		TR01/	
HR-101A	121			TR01/232	G6005
HR-102C				TR01/	G6004
HR-105		3009		TR01/	
HR-106				TR23/	S5011
HR-107				TR57/	S5012
HR-107H				TR57/	S5012
HR-107(PHILCO)	175			TR81/	
HR-40336					G0003
HR-40836				TR17/	G0003
HR-40837					G0003
HR-43835				TR17/	G0003
HR-45838				TR17/	G0003
HR-45910				TR17/	G0003
HR-45913					G0003
HR-448636				TR17/	G0003
HRKO				TR12/	
HS5	102A			TR85/250	G0005
HS-15	102	3004	2	TR05/250	
HS17D	102A	3123	52	TR05/	G0005
HS-17D		3004		/250	
HS22D		3123	52	TR05/	G0005
HS-22D	102	3004		TR05/250	
HS23D	102		2	TR05/	G0005
HS-23D		3004		TR05/250	

Original Part Number	ECG	SK	GE	IR/WEP	HEP
HS29D				TR05/	
HS102	102A	3004	2	TR85/	
HS-102				TR85/250	
HS170	102	3123	52	TR05/	
HS-170				TR85/250	
HS290	102			TR05/	
HS-290				TR85/250	
HS-1168	123A	3122	20	TR21/	
HS-1225	108	3018	17	TR21/	
HS-1226	108		20	TR21/	
HS-1227	108	3018	17	TR21/	
HS-1229	123A	3124	10	TR21/	
HS5810	192	3137	47		
HS5811	193	3138	48		
HS5812	192	3137	47		
HS5813	193	3138	48		
HS5814	192	3137	47		
HS5815	193	3138	48		
HS5816	192		47		
HS5817	193	3138	48		
HS5818	192	3137	47		
HS5819	193	3138	48		
HS5820	192	3137	63		
HS5821	193	3138	67		
HS5822	192	3137	63		
HS5823	193	3138	67		
HS6010	192				
HS6011	193	3138			
HS6012	193				
HS6013	193	3138			
HS6014	192	3137			
HS6015	193	3138			
HS6016	192	3137			
HS6017	193	3138			
HS-40014	108	3018	20	TR21/	
HS40021	161	3132		TR83/719	
HS-40021				TR21/	S0015
HS40022	161	3132		TR83/719	
HS-40022				TR21/	S0015
HS40023	161	3132		TR83/719	
HS-40023				TR21/	S0015
HS40024	161	3132		TR83/719	
HS-40024				TR21/	S0015
HS40025	161	3132		TR83/719	
HS-40025				TR21/	S0015
HS40026	128			TR87/243	
HS-40026				TR25/	S0015
HS-40027	159	3114	67	TR28/	
HS-40031	159	3125	22		S5022
HS40032	129	3025	67	TR28/	
HS-40037	123A	3018	20	TR24/	
HS-40039	123A	3124	20	TR25/	
HS-40045	108	3039	20	TR70/	
HS-40046		3018			
HS-40047	108	3018	17	TR95/	
HS40049	108		20		
HS-40049		3018		TR21/	
HS-40050	159	3025	21	TR28/	
HST1050			35	TR61/	
HST1051			36	TR61/	
HST1052			36	TR61/	
HST1053			36		
HST1054			36		
HST1055			35	TR61/	
HST1056			36	TR61/	
HST1057			36	TR61/	
HST1058			36		
HST1059			36		
HST1060			35	TR61/	
HST1061			36	TR61/	
HST1062			36	TR61/	
HST1063			36		
HST1064			36		
HST4451			27	TR78/	
HST4452			27	TR78/	
HST4453			27	TR78/	
HST4454			27	TR78/	
HST4483			27	TR78/	
HST5001			27	TR78/	
HST5002			27	TR78/	
HST5003			27	TR78/	
HST5004			27	TR78/	
HST5005			27	TR78/	
HST5006			27	TR78/	
HST5007			27	TR78/	
HST5008			27	TR78/	
HST5009			27	TR78/	
HST5010			27	TR78/	
HST5051			27	TR78/	
HST5052			27	TR78/	
HST5053			27	TR78/	
HST5054			27	TR78/	
HST5055			27	TR78/	
HST5056			27	TR78/	
HST5501			27	TR78/	
HST5502			27	TR78/	
HST5503			27	TR78/	
HST5504			27	TR78/	
HST5505			27	TR78/	
HST5506			27	TR78/	
HST5507			27	TR78/	
HST5508			27	TR78/	
HST5509			27	TR78/	
HST5510			27	TR78/	
HST5511			46		
HST5551			27	TR78/	
HST5552			27	TR78/	
HST5553			27	TR78/	
HST5554			27	TR78/	
HST5555			27	TR78/	
HST5556			27	TR78/	
HST5906			216		
HST7401			27	TR78/	
HST7402			27	TR78/	
HST7403			27	TR78/	
HST7411			27	TR78/	
HST7412			27	TR78/	
HST7413			27	TR78/	
HST7414			27	TR78/	
HST7415			27	TR78/	
HST7416			27	TR78/	
HST7417			46		
HST7418			32		
HST7419			32		
HST7901			32		
HST7902			32		
HST7903			32		
HST7904			32		
HST7905			32		
HST7907			32		
HST7908			32		
HST7909			32		
HST7910			32		
HST9001			27	TR78	
HST9002			27	TR78	
HST9003			27	TR78/	
HST9004			27	TR78/	
HST9005			27	TR78/	
HST9006			27	TR78/	
HST9007			27	TR78/	
HST9008			27	TR78/	
HST9009			27	TR78/	
HST9010			46		
HST-9201			14		
HST-9205			14		
HST-9206			14		
HST-9210			14		
HT0405190C	103A		8		G0011+G0005
HT30H861				TR21/	
HT100	159	3114	65	TR20/717	S0013
HT101	159	3114	65	TR20/717	S0013
HT102					S0014
HT103					S0014
HT400	123A	3122	61	TR21/735	S0011
HT401	123A	3122	61	TR21/735	S0011
HT20324A		3004			
HT20436A	131			TR94/	
HT20451A	158			TR84/	
HT30491C	128	3024	18		
HT30494	130	3027		TR59/	
HT30494X	130	3027		TR59/	
HT30821		3020			
HT30821-0		3020			
HT30821A,B,C,D		3020			
HT36441B	199	3124			
HT100160A		3005			
HT101011X	126	3006	1	TR17/635	G0009
HT101021A	126	3008	1	TR17/635	G0009
HT102341	160	3006		TR17/	
HT102341A	160	3006		TR17/	
HT102341B	160	3006	51	TR17/637	G0003
HT102341C	160	3006	51	TR17/637	G0003
HT102351	160	3006		TR17/	
HT102351A	160	3006	51	TR17/637	G0003
HT103501	160	3006		TR17/	
HT103501A	126	3006	51	TR17/635	G0008
HT103531A		3006			
HT103531C	126	3006		TR17/	
HT103541B	126	3006		TR17/	
HT104861	128	3024		TR87/	
HT104861A	128	3024		TR87/	
HT104861B	128	3024		TR87/	
HT104941B-0	159		21	TR20/	
HT104941C-0	159		21	TR20/	
HT104941C0	159	3114	21	TR30/	S0019
HT104942A	159		22	TR19/	S0013
HT104951A-0	159		21	TR20/	
HT104951B	159	3114		TR20/	S0013
HT104951B-0	159	3114		TR20/	S0013
HT104951C	159	3114	22	TR20/	S0013

Original Part Number	ECG	SK	GE	IR/WEP	HEP
HT104961B	185	3191			
HT104961C	184	3190			
HT104971A-O	129	3025		TR88/	
HT104971AO	129	3025			S5013
HT105611	159	3025		TR20/	
HT-105611		3114			
HT105611A	159	3025		TR20/	
HT-105611A		3114			
HT105611B	159	3025		TR20/	
HT-105611B		3114			
HT105611BO	159	3114	22		
HT105611C	159	3114	22		S0019
HT105612B	159		22	TR30/	S0019
HT105621B-O	159		21	TR20/	
HT105621BO	159	3114	22	TR19/	S0012
HT105641C		3025			S0019
HT105642B	234	3025			S0019
HT106731B	159	3025			S5013
HT107211T		3114			S0019
HT200540	102A	3004		TR85/	
HT200540A	102A	3004	2	TR85/250	G0007
HT200541	102A	3004		TR85/	
HT200541A	102A	3004		TR85/	
HT200541B	102A	3006	52	TR85/250	G0007
HT200541B-O	102A	3004		TR85/	G0007
HT303730A	123A	3020		TR21/735	
HT303801	107	3018		TR70/720	
HT303801A	107	3018	20	TR70/720	S0016
HT303801AO	107	3039	20	TR24/	S0016
HT303801B	107	3018	20	TR70/720	S0016
HT303801BO	107		20	TR24/	S0016
HT303801B-O	107	3018	20	TR70/720	S0016
HT303801C	107	3018	20	TR70/720	S0016
HT303801CO	107	3039	20	TR24/	S0016
HT303941	108	3018		TR95/56	
HT303941A	108	3018		TR95/56	
HT303941B	108	3018		TR95/56	
HT304531	123A	3020		TR21/735	
HT304531A	123A	3020		TR21/735	
HT304531B	123A	3020		TR21/735	
HT304531C	123A	3020	20	TR21/735	
HT304540AO	108		20	TR95/	
HT304540BO	123A		20	TR21/	
HT304540C		3018			
HT304580	123A	3020		TR21/735	
HT304580A	123A	3020		TR21/735	
HT304580B	123A	3020	20	TR21/735	S0014
HT304580CO	123A		10	TR21/	
HT304580K	123A	3020	63	TR21/735	S0030
HT304580YO	123A		10	TR21/	
HT304580Z	123A	3124	20		S0030
HT304581	123A	3020		TR21/735	
HT304581A	123A	3020		TR21/735	
HT304581B	123A	3020	20	TR21/735	S0014
HT304581B-O	123A		10	TR21/	
HT304581C	123A	3020	20	TR21/735	S0014
HT304601B		3018			
HT304601BO	108		20	TR95/	
HT304601C		3018			
HT304601CO	108		20	TR95/	
HT304611B	108	3018	17		S0014
HT304861	128	3024		TR87/243	
HT304861A	128	3024		TR87/243	
HT304861B	123A	3024	17	TR21/735	
HT304911	175	3026		TR81/241	
HT304911A	175	3026		TR81/241	
HT304911B	175	3026		TR81/241	
HT304941X	130		14	TR59/	
HT304961B	184	3190		/S5003	
HT304961C	184	3190		/S5003	
HT304961C-O	186	3192	28	TR55/	
HT304971A	128	3024	18	TR25/	S5014
HT304971AO	128	3024	18	TR25/	S5014
HT304971B	128	3020	18		S5014
HT305351B		3018			
HT305351CO	108		17	TR95/	
HT305361E	123A		20	TR24/	S0016
HT305371E	123A	3124			
HT305642B		3114			
HT306441	123A	3020		TR21/735	
HT200541C	160	3006		TR17/637	
HT200561	102A	3004		TR85/250	
HT200561A	102A	3004		TR85/250	
HT200561B	102A	3004	52	TR85/250	G0005
HT200561C	102A	3004	52	TR85/250	G0005
HT200561C-O	102A	3004		TR85/250	G0005
HT200751B	102A	3004		TR85/	G0005
HT200770B	102A	3004		TR85/250	G0005
HT200771	102A	3004		TR85/250	
HT200771A	102A	3004		TR85/250	
HT200771B	102A	3004	2	TR85/250	G0007
HT200771C	102A	3004		TR85/250	
HT201721		3004			
HT201721A	102A	3004	2	TR85/250	G0005
HT201721B		3004			

Original Part Number	ECG	SK	GE	IR/WEP	HEP
HT201721C		3004			
HT201721D	102A	3004		TR85/250	G0005
HT201725A		3004			G0007
HT201782A	102A	3004			G0005
HT201861A	102A			TR85/	
HT201871L	102A			TR85/	
HT203243A	158	3004			
HT203701	102A	3004		TR85/250	
HT203701A	102A	3004		TR85/250	
HT203701B	102A	3004	53	TR85/250	G0005
HT204051	158	3004		TR84/630	
HT204051B	158	3004		TR84/630	
HT204051C	158	3004		TR84/630	
HT204051D	158	3004		TR84/630	
HT204051E	158	3004	2	TR84/630	G6011
HT204053	158	3004		TR84/630	
HT204053A	158	3004	54	TR84/630	G6011
HT204053B	158	3004		TR84/630	G6011
HT2040710					G6003
HT204671		3052			
HT204671A		3052			
HT204671B		3052			
HT204671C		3052			
HT204671D		3052			
HT204671-O		3052			G6016
HT204736	131	3052		TR94/642	G6016
HT204736A					G6016
HT303620B	123A	3122		TR21/735	S0015
HT303711		3018			
HT303711A	107	3018	17	TR70/720	S0011
HT303711AO	123A	3018	20	TR24/	S0011
HT303711A-O	108		20	TR95/	
HT303711B	107	3018	20	TR70/720	S0011
HT303711BO	123A		20	TR24/	S0011
HT303711B-O	123A	3018	20	TR21/735	S0011
HT303711C	107	3018		TR70/720	
HT303720A	108	3018		TR95/56	
HT303721-O	123A	3018		TR21/735	S0015
HT303721D	123A	3122		TR21/735	
HT303730	123A	3020		TR21/735	
HT306441A	123A	3020		TR21/735	
HT306441AO	128		18	TR87/	
HT306441B	123A	3020	18	TR21/735	S0015
HT306441BO	123A	3124	18	TR21/	S0015
HT306441B-O	123A		17	TR21/	
HT306441C	123A	3020		TR21/735	
HT306441C-O	123A		17	TR21/	
HT306442A	199	3124			
HT306442B	199	3124			
HT306451	108	3018		TR95/56	
HT306451A	107	3018	11	TR70/720	S0016
HT306451B	108	3018	20	TR95/56	
HT306451H			17		
HT306830B		3018			
HT306962A-O	108		20	TR95/	
HT307321A	123A	3122			
HT307321B-O	123A		17	TR21/	
HT307322A	123A		20	TR21/	S0024
HT307331B	123A	3122			S0025
HT307331C	123A		17	TR21/	
HT307331CO	123A	3020	20	TR	
HT307341	128	3024		TR87/	
HT307341A	128	3024		TR87/	
HT307341B	123A	3024	17	TR21/	
HT307341C-O	123A		17	TR21/	
HT307342B	128		18	TR24/	S5026
HT307342C	128		18	TR24/	S5026
HT307720		3018			
HT307720A		3018			
HT307720B	107	3018	20	TR70/720	S0016
HT307720C, D		3018			
HT307721		3018			
HT307721A, B		3018			
HT307721C	107	3018	20	TR70/720	S0016
HT307721D	107	3018	20	TR70/720	S0016
HT307902B			66	TR76/	S5000
HT308281B	123A	3122	18	TR24/243	S0015
HT308281C	123A	3124			S0015
HT308281D	123A	3122	18	TR21/243	S0015
HT308281G	123A		20	TR21/	S0015
HT308281O	123A			/243	S0015
HT308282A	123A	3122			S0015
HT308282A-O	123A		17	TR21/	
HT308282B	123A	3122			
HT308291		3018			
HT308291A	107	3018	20	TR70/720	S0011
HT308291A-O	107	3018		TR70/720	S0011
HT308291B	107	3018	20	TR70/720	S0011
HT308291B-O	107	3018	20	TR70/720	S0011
HT308291BO	107	3018	20	TR24/	S0011
HT308291C	107	3018	20	TR70/720	S0011
HT308291D		3018			
HT308301BO	152		23	TR76/	
HT309291C	108		20	TR24/	S0015
HT309291E	107	3018	20	TR33/	S0016

Original Part Number	ECG	SK	GE	IR/WEP	HEP
HT309301		3018			
HT309301-0		3018			
HT309301A, B		3018			
HT309301C	199	3018	60	TR24/	S0016
HT309301D	107	3018	20	TR70/720	S0016
HT309301E	199	3018	20		S0016
HT309301F	199	3018	60		
HT309680-0		3024			
HT309680A		3024			
HT309680B	128	3024	18	TR87/243	S5014
HT309680C, D		3024			
HT309714A-0		3048	45		S5014
HT309841B0	128		18	TR87/	
HT309842A-0	123A		17	TR21/	
HT310002A	199		62	TR51/	S0024
HT311621B	152	3054	57		
HT313171R		3122			S0015
HT313172A		3122			
HT313271T	199	3122			S0024
HT313592B		3018			
HT313831X	128	3024			S3001
HT313832C	192	3512			
HT314071Q	192		63		S3020
HT340519C			8	TR08/	
HT400721	103A	3010		TR08/724	
HT400721A	103A	3010		TR08/724	
HT400721B	103A	3010		TR08/724	
HT400721C	103A	3010		TR08/724	
HT400721D	103A	3010	8	TR08/724	G0011
HT400721E	103A	3010	8	TR08/724	G0011
HT400723	103A	3010		TR08/724	
HT400723A	103A	3010	54	TR08/724	G0011
HT400723B	103A	3010		TR08/724	G0011
HT400770B	103A	3010		TR08/724	G0011
HT401191	130	3027		TR59/247	
HT401191A	130	3027		TR59/247	
HT401191B	130	3027		TR59/247	
HT401301B0	175	3026	23		
HT402352B	152	3054		TR76/	S5000
HT403152B	175				S5019
HT404001E		3137			S3020
HT600011		3025			
HT600011A, B		3025			
HT600011C, D, E		3025			
HT600011F	159	3025	21	TR20/717	S0013
HT600011H	159		21	TR20/717	S5013
HT600021		3025			
HT600021A, B		3025			
HT600021C, D		3025			
HT800011		3018			
HT800011A, B		3018			
HT800011C, D, E		3018			
HT800011F	123A	3018		TR21/735	S5025
HT800011G	123A	3018		TR21/735	S0015
HT800011H	123A	3018	20	TR21/735	S0015
HT800011K	123A	3018	20	TR21/735	S0015
HT800121		3024			
HT800121-0		3024			
HT800121A, B, C, D		3024			
HT800131		3024			
HT800131-0		3024			S5014
HT800131A, B, C, D		3024			
HT800161		3117			
HT800161-0		3117			S0017
HT800161A, B, C, D		3117			
HT800171		3117			
HT800171-0		3117			S0017
HT800171A, B, C, D		3117			
HT800181		3020			
HT800181-0		3020			S0015
HT800181A, B, C, D		3020			
HT800191		3020			
HT800191A, B, C, D		3020			
HT800191E, F, G		3020			
HT800191H		3020	18		S5014
HT1001510	102A	3005	50		G0005
HT2040710	121			TR01/232	
HT2046710	131	3198		TR94/	
HT3036201			20		
HT3036210	123A	3020	20	TR21/	S0016
HT3037010	108	3018	20	TR95/	S0011
HT3037201	108	3018		TR95/	S0015
HT3037201A	108	3018		TR95/	
HT3037201B	108	3018		TR95/	
HT3037210	123A	3018	20	TR21/735	S0015
HT3037210-0	123A	3018		TR95/	
HT3037310	107	3039	17	TR70/	S0015
HT3038013-B	108	3039		TR70/	
HT3073310		3124			S0025
HT3082810			18	TR21/	
HT3138331X					S5014
HT6000210	159	3025	21	TR20/717	
HT8000101	154	3040	18	TR78/712	
HT8000101A, B, C, D		3040			
HT8001210	128	3024		TR87/243	

Original Part Number	ECG	SK	GE	IR/WEP	HEP
HT8001310	128	3024		TR87/243	
HT8001610	161	3132	20	TR83/719	
HT8001710	161	3132	20	TR83/719	
HT8001810	123A	3122	18	TR21/735	
HT30373100	123A	3122	20	TR21/	S0015
HT30733100	199	3124	20	TR24/	S0025
HV0000102	102A		2	TR85/250	
HV0000202	158				
HV0000302	107	3039		TR70/720	
HV0000405	158	3100	52	TR84/630	
HV0000405-0			52	TR85/	G0005
HV12	102A	3123	52	TR85/250	G0005
HV15	102A	3004	52	TR85/250	G0005
HV16	102A	3123	52	TR85/	G0005
HV-16		3004		TR85/250	
HV-16/1238				TR85/	
HV-16/123805				TR85/	
HV17	102A	3123	52	TR85/250	G0005
HV17B	102A		2	TR85/250	
HV-17B		3004			
HV18					G0005
HV19	102A		2	TR85/735	G0005
HV23	123A	3122		TR21/735	
HV23A	123A	3122		TR21/735	
HV23F	123A	3122	20	TR21/735	
HV25	123A	3122		TR21/735	
HV46	123A	3030			
HVT200					S5026
HVY400					S5024
HX50001	108			TR95/56	
HX-50001		3018		TR21/	S0015
HX50002	123A			TR21/735	
HX-50002		3024		TR25/	S5014
HX50003	107			TR70/720	
HX-50003		3020		TR21/	S0011
HX-50004		3020			S5014
HX50112		3114			S0019
HX50113		3018			S0015
HX-SE40010					S0015
HY3045801C	123A	3122	20	TR21/735	
I9A115180-2	126				
I9115728-2	123A				
I51-0141-00	175				
I6114	175				
I6191	128				
I6342	197				
I9631	128	3018		TR87/243	S0015
I9680		3114			
I12032	160				
I33218				TR51/	
I50865	106				
I81030	159				
I472446-I	121				
I473608-2	128				
I473679-1	128				
IATR57		3026			
IC-06-1(MAGN)	722				
IC-06-2(MAGN)	722				
IC-1-3304	912				
IC-1(DYNALCO)	912				
IC1(RCA)	711				
IC-2-274(DYNAL)	912				
IC-2(DYNALCO)	912	3135			
IC2(ECS)	904				
IC2(ELCOM)	704				
IC2(RCA)	710				
IC3(PHILCO)	717		IC1		
IC-3(PHILCO)@	717				
IC-3(PHILCO)@	769				
IC-4(PHILCO)	703A	3157			
IC5		3070			
IC-5(ELCOM)	703A	3157	IC12		C6001
IC-5(PHILCO)	782				C6077P
IC-6	711	3070			
IC-6(ELCOM)	703A				
IC-6(PHILCO)	711				C6069G
IC-7		3157			
IC-7(PHILCO)	703A				
IC8					C6060P
IC-8	748	3101			C6060P
IC8B	748	3101	IC2		
IC-8B	704				
IC-8B(PHILCO)	748				
IC-8(ELCOM)	704				
IC-8(PHILCO)	748				
IC-9		3159			
IC-9(ELCOM)	711				
IC-9(PHILCO)	718				
IC10(ELCOM)	710				
IC-10(ELCOM)	710	3101			
IC-10(PHILCO)	706				
IC11(ELCOM)	726				
IC-11(ELCOM)	726	3077			
IC-11(PHILCO)	790				
IC12(ELCOM)	708				

Original Part Number	ECG	SK	GE	IR/WEP	HEP
IC-12(ELCOM)	708	3075	IC10		
IC-12(PHILCO)	714				C6070P
IC13A(ELCOM)	717				
IC13(ELCOM)	717				
IC-13(ELCOM)	717	3076			
IC-13(PHILCO)	715				C6071P
IC-14(PHILCO)	739	3077			C6072P
IC-15(PHILCO)	712	3072			C6063P
IC16(ELCOM)	725				
IC-16(ELCOM)	725				C6092G
IC-16(PHILCO)	780				C6069G
IC17(ELCOM)	718				
IC-17(ELCOM)	718	3159	IC8		C6094P
IC-17(PHILCO)	746				C6059P
IC18(ELCOM)	720				
IC-18(ELCOM)	720		IC7		C6095P
IC-18(PHILCO)	747				C6079P
IC19(ELCOM)	705A				
IC-19(ELCOM)	705A	3134	IC3		
IC-19(PHILCO)	779				
IC20(ELCOM)	784				
IC-20		3165			
IC-20(PHILCO)	923D				
IC21(ELCOM)	752				
IC-21		3077			
IC-21(ELCOM)	752				C6003
IC-21(PHILCO)	782				C6077P
IC23(ELCOM)	784				
IC-23			IC31		
IC-23(ELCOM)	784				
IC-23(PHILCO)	740				
IC24(ELCOM)	706				
IC-24(ELCOM)	706				
IC25(ELCOM)	707				
IC25A(ELCOM)	705A				
IC-25		3134	IC3		
IC-25(ELCOM)	705A				
IC26(ELCOM)	748				
IC-26(ELCOM)	748				C6060P
IC-26(PHILCO)	712				
IC27(ELCOM)	909D				
IC-27		3590			
IC-27(ELCOM)	909D				
IC-27(PHILCO)	779				
IC-28	749				
IC28(ELCOM)	785				
IC-28(ELCOM)	785				
IC29(ELCOM)	904				
IC-30	799				
IC30(ELCOM)	9930				C1030P
IC-31	1062	3075			
IC31(ELCOM)	9936				
IC32(ELCOM)	9946				
IC-33	731	3170			
IC33(ELCOM)	9093				
IC-33(PHILCO)	731				
IC34(ELCOM)	9962				C1062P
IC35(ELCOM)	915				
IC36(ELCOM)	925				
IC37(ELCOM)	940				
IC38(ELCOM)	949				
IC40	703A	3514			
IC40(ELCOM)	941				
IC-40(ELCOM)	941	3514			
IC41(ELCOM)	903				
IC42(ELCOM)	911				
IC-42(ELCOM)	911				
IC43(ELCOM)	923				
IC-43(ELCOM)	923	3164			
IC46		3526			
IC46(ELCOM)	947				
IC-46(ELCOM)	947				
IC48(ELCOM)	725				
IC-48(ELCOM)	725	3162			
IC49(ELCOM)	716				
IC-49(ELCOM)	716				
IC-51A(TEL-SIG)	909D				
IC53(ELCOM)	923D				
IC-53(ELCOM)	923D				
IC-55(ELCOM)	9924				C2003P
IC-56(ELCOM)	9900				C2002P
IC-58(ELCOM)	9910				C2006G
IC-59(ELCOM)	9911				C2007G
IC-61(ELCOM)	9914				C2009G
IC-62(ELCOM)	9914				C2010G
IC-63(ELCOM)	9923				
IC-64(ELCOM)	9925				
IC-65(ELCOM)	9989				C2005P
IC-71(ELCOM)	7430				C3030P
IC-74(ELCOM)	7401				C3001P
IC-76(ELCOM)	7450				
IC-79(ELCOM)	7495				
IC-80(ELCOM)	7400				C3000P
IC-81(ELCOM)	7493				
IC-82(ELCOM)	7402				
IC-83(ELCOM)	7403				
IC-84(ELCOM)	7404				C3004P
IC-85(ELCOM)	7405				
IC-86(ELCOM)	7410				C3010P
IC-87(ELCOM)	7420				C3020P
IC-88(ELCOM)	7440				C3040P
IC-89(ELCOM)	7441				
IC-90(ELCOM)	7442				
IC-91		3157	IC12		
IC-91(ELCOM)	7451				
IC-91(PHILCO)	703A				
IC-92(ELCOM)	7453				
IC-93(ELCOM)	7454				
IC-94(ELCOM)	7472				
IC-95(ELCOM)	7473				C3073P
IC-96(ELCOM)	7475				C3075P
IC-97(ELCOM)	7474				
IC-98(ELCOM)	7490				
IC099(ELCOM)	7476				
IC100		3072			
IC-100(ELCOM)	7492				
IC-101-004	1098				
IC101-109	703A	3157			
IC101(COLUMBIA)	172				
IC-101(ELCOM)	7447				
IC-102(ELCOM)	7408				
IC-103(ELCOM)	7413				
IC-104(ELCOM)	7406				
IC-105(ELCOM)	7416				
IC-106(ELCOM)	74121				
IC-107(ELCOM)	910D				
IC-108(ELCOM)	911D				
IC-109(ELCOM)	941D				
IC-112	1080				
IC-126	1105				
IC-132	1079				
IC-132(SHARP)	1079				
IC-136	1051				
IC-142	1050				
IC-146	1004				
IC-200(ELCOM)	745				
IC201(ELCOM)	722	3161			
IC-201(ELCOM)	722				C6056P
IC202(ELCOM)	747				
IC-202(ELCOM)	747				
IC203(ELCOM)	749				
IC-203(ELCOM)	749				
IC204(ELCOM)	750				
IC-204(ELCOM)	750				
IC205(ELCOM)	708				
IC-209(ELCOM)					C6050G
IC210(ELCOM)	909	3590			
IC-210		3590			
IC-210(ELCOM)	909				
IC-211(ELCOM)	903				
IC-212(ELCOM)	911				
IC213(ELCOM)	753				
IC-213(ELCOM)	753				C6010
IC217(ELCOM)	773				
IC-217(ELCOM)	773				
IC218(ELCOM)	713				
IC-218		3077			
IC-218(ELCOM)	713				
IC-220(ELCOM)	9662				
IC-221(ELCOM)	9665				
IC-222(ELCOM)	9666				
IC-223(ELCOM)	9667				
IC-224(ELCOM)	9672				
IC-225(ELCOM)	9663				
IC229(ELCOM)	9945				
IC-234(ELCOM)	9099				
IC-237(ELCOM)	9807				
IC-243(ELCOM)	766				
IC-244(ELCOM)	767				
IC-246(ELCOM)	946				
IC-276(ELCOM)	9944				
IC-279(ELCOM)	9948				
IC-280(ELCOM)	9932				
IC-281(ELCOM)	9933				
IC-282(ELCOM)	9917				
IC-283(ELCOM)	9922				
IC285(ELCOM)	754				
IC-285(ELCOM)	754				
IC286(ELCOM)	761				
IC-286(ELCOM)	761				
IC-287(ELCOM)	814				
IC288(ELCOM)	779				
IC-288(ELCOM)	779				
IC-289(ELCOM)	798				
IC290(ELCOM)	755				
IC-290(ELCOM)	755				
IC292(ELCOM)	771				
IC-292(ELCOM)	771				
IC293(ELCOM)	739				
IC-293(ELCOM)	739				

Original Part Number	ECG	SK	GE	IR/WEP	HEP
IC294(ELCOM)	746				
IC-294(ELCOM)	746				
IC-295(ELCOM)	941M				
IC-296(ELCOM)	778				
IC297(ELCOM)	738	3167			
IC-442(ELCOM)	755				
IC-443(ELCOM)	757				
IC-444(ELCOM)	758				
IC-445(ELCOM)	759				
IC-446(ELCOM)	763				
IC-447(ELCOM)	764				
IC-448(ELCOM)	765				
IC-449(ELCOM)	768				
IC-450(ELCOM)	770				
IC-451(ELCOM)	772				
IC-452(ELCOM)	775				
IC-453(ELCOM)	776				
IC-454(ELCOM)	777				
IC-455(ELCOM)	9093				
IC-456(ELCOM)	783				
IC500		3157			
IC500(IR)	703A	3157			C6001
IC-500(ELCOM)	7482				
IC501		3023			
IC501(ELCOM)	760				
IC501(IR)	704				
IC502	705A	3134	IC3		C6089
IC502(ELCOM)	719				
IC-502(ELCOM)	719				
IC502(IR)	705A				
IC503(IR)	707				C6067G
IC504	708		IC10		C6062P
IC504(IR)	708				
IC-504		3135			
IC505	709		IC11		
IC505(IR)	709				C6062P
IC-505(ELCOM)	1061				
IC506		3102			
IC506(IR)	710				
IC507	712	3072	IC2		
IC507(ELCOM)	774				
IC507(IR)	712				C6063P
IC-507(ELCOM)	774				
IC508	713	3077	IC5		
IC508(IR)	713				C6057P
IC-508(ELCOM)	1075				
IC509	714	3075	IC4		
IC509(IR)	714				C6070P
IC-509		3075			
IC-509(ELCOM)	1083				
IC510	715	3076	IC6		
IC510(IR)	715				C6071P
IC511	718	3159	IC8		C6094P
IC511(IR)	718				
IC512		3160			
IC512(IR)	720				C6095
IC-512(ELCOM)	1039				
IC513	722	3161	IC9		C6056P
IC513(IR)	722	3161			
IC514		3070			
IC-297(ELCOM)	738				
IC298(ELCOM)	760				
IC-298(ELCOM)	760				
IC301(ELCOM)	724				
IC-301(ELCOM)	724				
IC302(ELCOM)	711				
IC-304(ELCOM)	912				
IC305(ELCOM)	727				
IC-305(ELCOM)	727				
IC-306(ELCOM)	914				
IC307(ELCOM)	783				
IC-307(ELCOM)	780				
IC308(ELCOM)	712				
IC-308		3072			
IC309(ELCOM)	723				
IC310(ELCOM)	781	3169			
IC311(ELCOM)	730				
IC312(ELCOM)	714				
IC-312		3075			
IC313(ELCOM)	715				
IC314(ELCOM)	728				
IC315(ELCOM)	729				
IC317		3514			
IC317(ELCOM)	941	3514			
IC-317(ELCOM)	913				
IC-318(ELCOM)	913				
IC319(ELCOM)	786				
IC-319(ELCOM)	786				
IC320(ELCOM)	789				
IC-320(ELCOM)	729				
IC-322(ELCOM)	731				
IC-323(ELCOM)	787				
IC-324(ELCOM)	788				
IC-325(ELCOM)	791				
IC401	706				
IC-401	706				
IC-403(ELCOM)	1053				
IC405(ELCOM)	721				
IC-405(ELCOM)	721				
IC-408(ELCOM)	724				
IC-409(ELCOM)	9663				
IC-410(ELCOM)	9668				
IC411(ELCOM)	910				
IC-412(ELCOM)	9670				
IC-413(ELCOM)	9661				
IC-414(ELCOM)	9664				
IC-415(ELCOM)	9682				
IC-419(ELCOM)	9157				
IC-420(ELCOM)	9158				
IC-423(ELCOM)	9917				
IC-431(ELCOM)	769				
IC-432(ELCOM)	762				
IC-433(ELCOM)	756				
IC-435(ELCOM)	755				
IC-441(ELCOM)	9669				
IC-517(ELCOM)	1042				
IC-520(ELCOM)	955M				
IC-521(ELCOM)	799				
IC-523(ELCOM)	782				
IC-524(ELCOM)	740				
IC-526(ELCOM)	370				
IC-527(ELCOM)	371				
IC-528(ELCOM)	372				
IC-531(ELCOM)	923D				
IC-533(ELCOM)	1046				
IC-534(ELCOM)	1136				
IC-535(ELCOM)	1123				
IC-537(ELCOM)	1071				
IC-540(ELCOM)	1137				
IC-542(ELCOM)	1003				
IC-543(ELCOM)	1006				
IC-545(ELCOM)	1109				
IC-546(ELCOM)	712				
IC-547(ELCOM)	1013				
IC-548(ELCOM)	1015				
IC-549(ELCOM)	1012				
IC-550(ELCOM)	1152				
IC-551(ELCOM)	1018				
IC-552(ELCOM)	1056				
IC-553(ELCOM)	1073				
IC-554(ELCOM)	1054				
IC-600(ELCOM)	734				
IC-601(ELCOM)	735				
IC-602(ELCOM)	732				
IC-603(ELCOM)	733				
IC-604(ELCOM)	736				
IC-605(ELCOM)	740				
IC-606(ELCOM)	743				
IC-607(ELCOM)	744				
IC700	1024				
IC-724(IR)					C2003P
IC1303P	725	3162			
IC-1303P(IR)					C6092G
IC3014	704				
IC743040		3018			
IC743041		3018			
IC743042		3018			
IC743048		3030			
IC3012434		3102			
ICF-1	703A	3157		IC500/	C6001
ICF-1-6826	703A	3157			
IE703E	703A	3157			
IE850			20		
IE-850	103A	3010		TR08/724	
IF65			2		
IF-65	102A	3004		TR85/250	G0005
IG-100	105	3012		TR03/233	
ILD1020	1012				
IP20-0001		3018			
IP20-0002		3018			
IP20-0003		3018			
IP20-0004		3018			
IP20-0006		3018			
IP20-0007		3054			
IP20-0008		3010			
IP20-0009		3114			
IP20-0010		3112			
IP20-0011		3112			
IP20-0012		3116			
IP20-0022		3030			
IP20-0024		3030			
IR-11		3006			
IR-14		3004			
IR-28-R/508	910D				
IR2160	6400				S9002
IR-FE100/		3116		FE100/802	F0010
IR-RE50	131				
IRTR01			16		
IRTR02			3		
IRTR03			4		

Original Part Number	ECG	SK	GE	IR/WEP	HEP
IRTR04			2		
IRTR05			2		
IRTR06			51		
IRTR07			51		
IRTR08			6		
IRTR09			8		
IRTR10			7		
IRTR11			1		
IRTR12			51		
IRTR14			53		
IRTR16			3		
IRTR16X2			13MP		
IRTR17			51		
IRTR18			51		
IRTR19			21		
IRTR20			22		
IRTR21			20		
IRTR22			17		
IRTR23			12		
IRTR24			17		
IRTR25			18		
IRTR26			19		
IRTR27			25		
IRTR28			21		
IRTR30			21		
IRTR31			20		
IRTR50			30		G6016
IRTR51	199	3122		TR51/	S0005
IRTR52		3118	67		S0019
IRTR53			63		S0005
IRTR54					S0019
IRTR55		3054	28	TR55/	S5000
IRTR56		3083	29	TR56/	S5006
IRTR57	196	3026	23	TR57/	S5019
IRTR58	218		26	TR58/	S5018
IRTR59		3036		TR59/	S7002
IRTR60			27	TR60/	S5015
IRTR61			14	TR61/	S5020
IRTR62	123A	3046		TR62/	S3013
IRTR63	123A	3047		TR63/	S3013
IRTR64	195	3048		TR64/	S3008
IRTR65	195	3048		TR65/	S3001
IRTR66				TR66/	S3006
IRTR67	163			TR67/	S5021
IRTR68	164	3133		TR68/	
IRTR69	172	3156		TR69/	S9100
IRTR70	161	3132		TR70/	S0025
IRTR71	161	3018		TR71/	S0025
IRTR72	188	3199		TR72/	S3020
IRTR73	189	3200		TR73/	S3032
IRTR74	190	3104		TR74/	S3021
IRTR75	191	3103		TR75/	S3021
IRTR76	152	3054		TR76/	S5000
IRTR77	153	3083		TR77/	S5006
IRTR78	154	3045		TR78/	S5024
IRTR79	171	3104		TR79/	S3021
IRTR80		3039			
IRTR82	176	3123		TR82/	
IRTR83		3019			
IRTR84	158			TR84/	
IRTR85	102A	3004		TR85/	
IRTR86	123	3124		TR86/	
IRTR87	128	3024		TR87/	
IRTR88	129	3025		TR88/	
IRTR91	155				
IRTR92	196	3054		TR92/	
IRTR93	165	3115		TR93/	
IRTR94	226MP				
IRTR94MP	226MP			TR94MP/	
IRTR95	108			TR95/	
IR-TR50				TR50/642	
IR-TR51		3122		TR51/243	
IR-TR52		3114		TR52/717	
IR-TR53	128	3024		TR53/243	
IR-TR54		3114		TR54/717	
IR-TR55				TR55/S3023	
IR-TR56	187			TR56/S3027	
IR-TR57	175			TR57/241	
IR-TR59	130	3027		TR59/247	
IR-TR60	157	3103		TR60/244	
IR-TR61	162			TR67/707	
IRTR-51				TR51/	
IRTR-53				TR53/	
IRTR-55		3054			
IRTR-57		3026			
IRTR-59				TR60/	
IRTR-60		3104			
IRTR-61				TR61/	
IRTR-62			20		
IRTR-63		3047	215		
IRTR-65			45	TR65/	
IRTR-68				TR68/	
IRTR-70			61	TR70/	
IRTR-71			60	TR71/	
IRTR-78				TR78/	

Original Part Number	ECG	SK	GE	IR/WEP	HEP
IRTR-80				TR80/	
IRTR-83				TR83/	
IRTR-92					S5000
IRTR-93				TR93/	
IRTRFE100		3116	FET1		
IRTR-IC500		3157			
IRTR-IC501		3023			
IRTR-IC502		3134			
IRTR-IC506		3102			
IRTR-IC507		3101			
IRTR-IC508		3077			
IRTR-IC509		3075			
IRTR-IC510		3076			
IRTR-IC511		3159			
IRTR-IC512		3160			
IRTR-IC513		3161			
IRTR-IC514		3070			
IS188		3087			
ISB18A				TR03/	
ISB304				TR84/	
ISBF1	102A			TR85/250	G0005
ISC157H		3124			
ISC1004A		3133			
ISD162			8	TR08/	
ISD-162	103			TR08/641A	
IT20		3112			
IT108		3116	FET2		
IT120			210		
IT121			210		
IT122			61		
IT205A			50		
IT210		3112			
IT918			39		
IT918A			212		
IT929			39		
IT930			39		
IT2218			20		
IT2219			20		
IT2221			20		
IT2222			20		
IT2483			63		
IT2604		3118	65		
IT2605		3118	65		
IT2904			21		
IT2905			21		
IT2906			21		
IT2907			21		
ITC918A			212		
ITE3066		3112			
ITE3067		3112			
ITE3068		3112			
ITE4117		3112			
ITE4118		3112			
ITE4119		3112			
ITE4338		3112			
ITE4339		3112			
ITE4340		3112			
ITE4341		3112			
ITE4416		3116	FET2		
ITE4867		3112			
ITE4868		3112			
ITE4869		3112			
ITT41-5		3514			
ITT74H00N	74H00				
ITT74H01N	74H01				
ITT74H04N	74H04				
ITT74H05N	74H05				
ITT74H10N	74H10				
ITT74H11N	74H11				
ITT74H20N	74H20				
ITT74H21N	74H21				
ITT74H30N	74H30				
ITT74H40N	74H40				
ITT74H50N	74H50				
ITT74H51N	74H51				
ITT74H53N	74H53				
ITT74H54N	74H54				
ITT74H60N	74H60				
ITT74H72N	74H72				
ITT74H73N	74H73				
ITT74H74N	74H74				
ITT74H76N	74H76				
ITT709		3590			
ITT709-5		3590			
ITT709-5(D.I.P)	909D				
ITT709-5(MCAN)	909				
ITT709(D.I.P.)	909D				
ITT709(METCAN)	909				
ITT710-5(D.I.P)	909D				
ITT710-5(MCAN)	910				
ITT710(D.I.P)	910				
ITT710(METCAN)	910				
ITT711-5(D.I.P)	911D				
ITT711-5(MCAN)	911				
ITT711(D.I.P)	911D				

Original Part Number	ECG	SK	GE	IR/WEP	HEP
ITT711(METCAN)	911				
ITT723		3165			
ITT723-5		3165			
ITT723-K		3164			
ITT723-5(D.I.P)	923D				
ITT723-5(MCAN)	923				
ITT723(D.I.P)	923D				
ITT723(METCAN)	923				
ITT741		3552			
ITT741-5		3514			
ITT741-5(D.I.P)	941D				
ITT741-5(MCAN)	941				
ITT741(D.I.P.)	941D				
ITT741(METCAN)	941				
ITT930-5	9930				
ITT930N	9930				
ITT932-5	9932				
ITT932N	9932				
ITT933-5	9933				
ITT933N	9933				
ITT935-5	9935				
ITT935N	9935				
ITT936-5	9936				
ITT936N	9936				
ITT937-5	9937				
ITT937N	9937				
ITT938-5	9938				
ITT938N	9938				
ITT941-5	9941				
ITT941N	9941				
ITT944-5	9944				
ITT944N	9944				
ITT945-5	9945				
ITT945N	9945				
ITT946-5	9946				
ITT946N	9946				
ITT948-5	9948				
ITT948N	9948				
ITT949-5	9949				
ITT949N	9949				
ITT951-5	9951				
ITT951N	9951				
ITT961-5	9961				
ITT961N	9961				
ITT962-5	9962				
ITT962N	9962				
ITT963-5	9963				
ITT963N	9963				
ITT1800-5	9800				
ITT1800N	9800				
ITT1801N	9801				
ITT1806-5	9806				
ITT1806N	9806				
ITT1807-5	9806				
ITT1807N	9807				
ITT1808-5	9808				
ITT1808N	9808				
ITT1809-5	9809				
ITT1809N	9809				
ITT1810-5	9810				
ITT1810N	9810				
ITT1811-5	9811				
ITT1811N	9811				
ITT7400N	7400				
ITT7401N	7401				
ITT7402N	7402				
ITT7403N	7403				
ITT7404N	7404				
ITT7405N	7405				
ITT7406N	7406				
ITT7407N	7407				
ITT7408N	7408				
ITT7409N	7409				
ITT7410N	7410				
ITT7411N	7411				
ITT7412N	7412				
ITT7413N	7413				
ITT7416N	7416				
ITT7417N	7417				
ITT7420N	7420				
ITT7421N	7421				
ITT7425N	7425				
ITT7426N	7426				
ITT7428N	7428				
ITT7430N	7430				
ITT7432N	7432				
ITT7433N	7433				
ITT7437N	7437				
ITT7438N	7438				
ITT7440N	7440				
ITT7442N	7442				
ITT7443N	7443				
ITT7444N	7444				
ITT7445N	7445				
ITT7446AN	7446				
ITT7447AN	7447				
ITT7448N	7448				
ITT7450N	7450				
ITT7451N	7451				
ITT7453N	7453				
ITT7454N	7454				
ITT7460N	7460				
ITT7470N	7470				
ITT7472N	7472				
ITT7473N	7473				
ITT7475N	7475				
ITT7476N	7476				
ITT7480N	7480				
ITT7482N	7482				
ITT7483N	7483				
ITT7486N	7486				
ITT7490N	7490				
ITT7491AN	7491				
ITT7492N	7492				
ITT7493N	7493A				
ITT7494N	7494				
ITT7495AN	7495				
ITT7496N	7496				
ITT9093-5	9093				
ITT9093N	9093				
ITT9094-5	9094				
ITT9094N	9094				
ITT9097-5	9097				
ITT9097N	9097				
ITT9099-5	9099				
ITT9099N	9099				
ITT74104N	74104				
ITT74105N	74105				
ITT74107N	74107				
ITT74109N	74109				
ITT74121N	74121				
ITT74122N	74122				
ITT74123N	74123				
ITT74141N	74141				
ITT74145N	74145				
ITT74150N	74150				
ITT74151N	74151				
ITT74152N	74152				
ITT74153N	74153				
ITT74154N	74154				
ITT74155N	74155				
ITT74156N	74156				
ITT74157N	74157				
ITT74160N	74160				
ITT74161N	74161				
ITT74162N	74162				
ITT74163N	74163				
ITT74164N	74164				
ITT74165N	74165				
ITT74174N	74174				
ITT74175N	74175				
ITT74180N	74180				
ITT74181N	74181				
ITT74182N	74182				
ITT74190N	74190				
ITT74191N	74191				
ITT74192N	74192				
ITT74193N	74193				
ITT74194N	74194				
ITT74195N	74195				
J4-1000	7400				
J4-1002	7402				
J4-1004	7404				
J4-1010	7410				
J4-1047	7447				
J4-1075	7475				
J4-1076	7476				
J4-1090	7490				
J4-1092	7492				
J4-1121	74121				
J4-1555	955M				
J107	107	3018		TR70/720	S0016
J108	108			TR95/56	S0015
J139		3020			S0015
J139-1,2,4,5		3020			
J139A	123A	3020		TR21/735	S0015
J139B,C,D		3020			
J139E		3124			
J187	108			TR95/56	S0015
J193A		3124			
J460					S0033
J461					S0033
J462					S0033
J463					S0033
J464					S0033
J465					S0033
J466					S0033
J581					S0033
J582					S0016
J584					S0025

Original Part Number	ECG	SK	GE	IR/WEP	HEP
J585					S0020
J587					S0020
J588					S0005
J595					S0005
J623					S0011
J624					S0011
J625					S0005
J626					S0011
J627					S0011
J628					S0005
J629					S0011
J630					S0011
J631					S0005
J961B				TR24/735	
J961B(G.E.)	123A			TR21/	
J1000-7400	7400				
J1000-7402	7402				
J1000-7404	7404				
J1000-7410	7410				
J1000-7447	7447				
J1000-7476	7476				
J1000-7490	7490				
J1000-7492	7492				
J1000-74121	74121				
J1000-NE555	955M				
J5062	160			TR17/637	
J5063	102A			TR85/250	
J5064	102A			TR85/250	
J9618		3024			S0015
J9680	159	3114	22	TR20/	S0019
J9697	159	3114	21	TR20/	S0019
J24458	123A	3124	20	TR21/735	S0014
J24561	107	3018	11	TR70/720	S0016
J24562	107	3018	11	TR70/720	S0014
J24563	107	3018	20	TR70/720	S0014
J24564	123A	3124	20	TR21/735	S0011
J24565	123A	3124	20	TR21/735	S0014
J24566	124	3021	12	TR81/240	S5011
J24596	108	3018	20	TR24/	S0015
J24620	160		50	TR17/637	G0003
J24621	160		50	TR17/637	G0003
J24622	160		50	TR17/637	G0003
J24623	160		50	TR17/637	G0003
J24624	123A		17	TR21/735	S0020
J24625	123A		63	TR21/735	S0020
J24626	102A		52	TR85/250	G0006
J24635	108			TR95/56	S0011
J24636	108			TR95/56	S0015
J24637	108			TR95/56	S0015
J24639	102A			TR85/250	G0005
J24640	159	3114		TR20/717	S5013
J24641	123A			TR21/735	S0015
J24642	175			TR81/241	G6011
J24646					S0015
J24658		3020			
J24701		3018			
J24752	123A	3122		TR21/735	S0015
J24753	123A	3122		TR21/735	S0012
J24754	184+185			TR76/S5003	S5000
J24812	199	3018	20	TR24/	S0016
J24813		3018	20		S0020
J24814	108	3018	11		S0014
J24817	123A	3122	10	TR24/	
J24832			21		S0013
J24833			52		G0007
J24834			52		G0005
J24835					G0005
J24836					G0005
J24842			20	TR21/	S0033
J24843			20	TR22/	S0033
J24844			20	TR21/	S0016
J24845			20	TR51/	S0011
J24846			20	TR21/	S0015
J24852	108	3018	20	TR24/	S0015
J24855	123A	3124	20		S0030
J24863	108	3018	20	TR70/	S0015
J24868	103A	3010		TR08/724	G0011
J24869	102A	3004		TR85/250	G0007
J24870	102A	3004		TR85/250	G0005
J24874	123A	3124	20	TR51/	
J24875		3122			
J24878		3020			
J24903	107		17	TR24/	S0025
J24904	108		20	TR95/	S0015
J24905	108		20	TR24/	S0015
J24906	123A		10	TR25/	S5014
J24907	123A		10	TR24/	S0015
J24908	129		67	TR28/	S5013
J24909	123A		20	TR21/	S0015
J24915	108		20	TR24/	S0016
J24916	123A		10	TR24/	S0015
J24921	108		20	TR21/	S0011
J24923	108		20	TR24/	S0015
J24932	199	3122	62	TR51/	S0015
J24933	108	3018	20	TR24/	S0015
J24934	102A	3004	2	TR05/	G0005
J241015	129	3025			S5013
J241054	123A	3124	20	TR51/	
J241099	123A	3018	10	TR24/	
J241164	102A	3004	52		G0007
J241177	107	3018	20	TR24/	S0016
J241178	158	3004	53	TR84/	G0005
J241185	103A	3010			
J241188	107	3018	20	TR24/	S0014
J241189	107	3018	20	TR24/	S0016
J241190	102A	3004	54		
J241219	749				
J241220	1050				
J241221	1089¤				
J241222	712				
J241224	218				
J241225	159				
J241227	238				
J241228	124				
J241229	124				
J241230	123A				
J241231	175				
J241241	152	3041	66	TR76/	S5000
J241250	184	3104	57		
J241251	123A	3124	20		
J241252	196	3054	66		
J241253	159	3114	67		
J241255	128	3122	63		
J241256	128	3122	63		
J241258	218	3085			
J241259	159	3114	67		
J241270	748				
J310159	102A		53	TR85/250	
J310224	102A		53	TR85/250	
J310249	123A	3122	20	TR21/735	
J310250	123A	3122	20	TR21/735	
J310251	160		51	TR17/637	
J310252	102A		53	TR85/250	
JA1000		3114			
JA1000A		3114			S0019
JA1050	159	3114	22	TR20/	
JA1050GL		3114			
JA1100		3018			
JA1150		3018			S0016
JA1200	123A	3122	20	TR21/	S0030
JA1300		3024			
JA1350		3122			S0015
JA1350W		3122			
JA1400		3124			S0025
JA1500		3122			
JA1600		3124			
JA1650		3114			
JA5010		3041			
JA5010-1		3041			
JA5011		3041			
JA5040		3054			
JA5040-1		3054			
JA5041		3054			
JA7072		3104			
JA7102		3104			
JA7152		3104			
JA9301					S5005
JA9302		3104			S5004
JC100	105	3012		TR03/233	
JD19-3		3004			
JEM1	105	3012		TR03/233	G6006
JEM2	105	3012		TR03/233	G6006
JEM3	105	3012		TR03/233	G6006
JEM4	105	3012		TR03/233	G6006
JEM5	105	3012		TR03/233	G6006
JF-1033	132	3116	FET2	FE100/	F0021
JF1034		3112			
JN271	123A	3124		TR21/735	
JNJ61673	133	3112		/801	
JP-001		3122			
JP1					G0005
JP40	104	3009	16	TR01/624	
JP5062	160	3006	50	TR17/637	
JP5063	102A	3004	52	TR85/250	
JP5064	102A	3004	52	TR85/250	
JR05		3123	52	TR05/	
JR5	102A	3004	2	TR85/250	G0005
JR10	160	3007	50	TR17/637	G0002
JR15	102A	3005	52	TR85/250	G0007
JR30	160	3006	51	TR17/637	G0001
JR30X	160	3007	51	TR17/637	G0008
JR40	104	3009	3	TR01/230	G6003
JR100	160	3006	50	TR17/637	G0008
JR200	160	3006	51	TR17/637	G0008
JST1501-1	421				
JT-1601-40	123A	3122	20	TR21/735	
K04774	104	3009		/230	
K071687	106				
K071818-001	129				
K071961-001	128				

Original Part Number	ECG	SK	GE	IR/WEP	HEP
K071962-001	129				
K071964-001	130				
K4-500	158			TR84/630	G0005
K4-501	103A			TR08/724	G0011
K4-505	159	3114		TR20/717	S0013
K4-505A		3114			
K4-506	123A	3122		TR21/735	S0014
K4-507		3122		TR51/735	S0014
K4-510	108			TR95/56	S0016
K4-520	104	3009		TR01/230	G6003
K4-521	104	3009		TR01/230	G6003
K4-525	130	3027		TR59/247	S7002
K4-632				FE100/	
K11	132				
K12	132				
K13	132				
K14-0066-1			53		
K14-0066-4			1		
K14-0066-6			61		
K14-0066-12			10		
K14-0066-13			18		
K17	133				
K17-0	132				
K17A, B	132				
K17BL	132				
K17GR	132				
K17R, Y	132				
K19	132				F0015
K19BL	132				
K19GE	132				
K19GR	132				
K19Y	132				
K22Y	132				
K23	132				
K24	133				
K24C, D	133				
K24DR	133				
K24E, F, G	133				
K25	132				
K25C, D, E	132				
K25ET	132				
K25F, G	132				
K30	132				
K30-0	133				
K30-GR	133				
K30-R, Y	133				
K32B	132				
K33	132				
K33E, F	132				
K34	132				
K34B, C, D	132				
K35	133				
K35-0	133				
K35-1, 2	133				
K35A	133				
K35BL, C, GN, R, Y	133				
K37	132				
K37H	132				
K40	133				
K40A, B, C, D	133				
K41	132				
K45	222				
K118I	129			TR28/	S5013
K121J688-1	107				
K170	132				
K170R	132				
K417-68	126	3008		TR17/635	
K1181		3025			
K2001	107	3039		TR70/720	S0016
K2109			86		
K2110			86		
K2111			86		
K2112			86		
K2113			86		
K2114			86		
K2115			86		
K2116			86		
K2117			86		
K2118			86		
K2119			11		
K2120			11		
K2121			11		
K2122			11		
K2123			11		
K2124			11		
K2126			11		
K2127			11		
K2501	107	3039	17	TR70/720	S0016
K2502	107	3039	17	TR70/720	S0016
K2503	107	3039	17	TR70/720	
K2509	107	3039	17	TR70/720	S0016
K2601C			11		
K2602C			11		
K2603C			11		
K2604			11		
K2604C			11		
K2615			11		
K2616			11		
K2857C			39		
K2857P			39		
K3683C			39		
K3683P			39		
K3880C			39		
K3880P			86		
K4002	107	3039	11	TR70/720	S0016
K9682	159	3114	22	TR20/	S0019
K24154	1142				
K75508	160			TR17/	
K/417-66		3004			
KA4559	132		FET2	FE100/	F0015
KB8339	123A	3124		TR21/735	
KB8416			17	TR25/	
KD2101	102A	3004	2	TR85/250	
KD2102	123A	3124	20	TR21/735	
KD2114	904	3542			
KD2115	984	3524			
KD2117	784+904				
KD2118	128	3024		TR87/	
KD2119	108	3057		TR95/	
KD2119A		3057			
KD2120	129	3025		TR88/	
KD2121	176	3123		TR82/	
KD2122(IC)	912	3543			
KD2122(TRAN)	102			TR05/	
KD2124	103	3011		TR08/	
KD2130	222	3065			
KD6311	1046				
KE-1007	130				
KE1007-0004-00	129				
KE3684		3116			
KE4223		3116			
KE4224		3116			
KE4416		3116	FET2		
KE5103		3116			
KE5104		3116			
KE5105		3116	FET2		
KGE41054	128	3122	17	TR87/	S3001
KGE41055	123	3122	20	TR86/	S0015
KGE41061	186		28	TR55/	S3020
KGS1000	102A			TR85/250	G0005
KGS1001					G0005
KGS1002					G0005
KGS1003					G0005
KGS1004					G0005
KGS1005					G0005
KH3405	407				
KH3406	408				
KH3407	409				
KH3408	416				
KLH4745	196				
KLH4781	197				
KLH4793	722	3161	IC9		C6056I
KLH4794	781				
KLH4795	726	3129			
KLH5353	197				
KLH5489	718	3159	IC8		C6094
KLH5807	128				
KLH5808	129				
KO4774	104	3009	16	TR01/230	G6003
KPG6682	123A			TR21/	
KR8417				TR25/	
KR-Q0001	102A		53	TR85/250	
KR-Q0002	102A		53	TR85/250	
KR-Q0004	102A		53	TR85/250	
KR-Q1010	102A	3004	53	TR85/250	G0005
KR-Q1011	102A	3004	53	TR85/250	G0007
KR-Q1012	102A	3004	2	TR85/250	G0005
KR-Q1013	123A	3124	20	TR21/735	S0016
KS-172		3005			
KS10969-L5	7416				
KS-19938	130				
KS-20033L2	219				
KS20180-L1	128				
KS-20345-L1		3510			
KS-20345-L1ISS		3027			
KS20967-L1	7400				
KS20967-L2	7404				
KS20967-L3	7486				
KS20969-L2	7483				
KS20969-L3	7492				
KS20969-L4	7495				
KS-20971-L1	909D	3590			
KS-21177	941M				
KS21292-L1	7408				
KS21282-L2	7425				
KS21282-L3	7432				
KSD1051			77		
KSD1052			75		
KSD1055			77		
KSD1056			77		

Original Part Number	ECG	SK	GE	IR/WEP	HEP
KSD1057			75		
KSD2102			216	TR57/	
KSD2103			216	TR57/	
KSD2201				TR61/	
KSD2202				TR61/	
KSD2203			75	TR61/	
KSD3055			14		
KSD3771			75		
KSD3772			75		
KSD9701			75		
KSD9701A			75		
KSD9704			75		
KSD9707			77		
KT218			20	TR53/	
KT218F			64	TR53/	
KT600			47	TR65/	
KT600F			47		
KT600G			32		
KT600P			47		
KT1017	121		16	TR01/232	
KT-1017		3009		TR01/	
KV1	102A			TR85/250	G0005
KV-1	102A	3004	52		
KV2	102A		2	TR85/250	
KV4	102A	3123	52	TR85/250	G0005
KV-14				TR05/	
L4			212		
L4A-1AOR		3009			
L5			212		
L6			212		
L6B-3AOR		3015			
L7			212		
L8P-2AOR		3004			
L9L-4AOR		3003			
L12-1AOR		3020			
L14A1OR		3009			
L65-1AOR		3006			
L65-2AOR		3010			
L65-4AOR		3011			
L65A1OR		3007			
L65B1OR		3008			
L65C1OR		3009			
L66-1AOR		3003			
L66-2AOR		3004			
L66-3AOR		3004			
L74-3AOR		3005			
L77A1OR		3012			
L77B1OR		3013			
L77C1OR		3014			
L84A1OR		3009			
L88B1OR		3006			
L88C1OR		3005			
L92-1AOR		3012			
L103T1	703A	3157			
L110P1		3114			
L110P1M		3114			
L123-T1	923				
L-417-29		3014			
L-417-29BLK	121	3009		TR01/232	G6005
L-417-29GRN	121	3009		TR01/232	G6005
L-417-29WHT	121	3009		TR01/232	G6005
L-417-60	121	3009		TR01/232	G6005
L532-008-012	128				
L5021	102A	3003		TR85/250	G0005
L5022	102A	3012	2	TR85/250	G0005
L5022A	102A	3004	2	TR85/250	G0005
L5025	102A	3004	2	TR85/250	G0005
L5025A	102A	3004	2	TR85/250	G0005
L5108	160	3123	2	TR17/637	G0002
L-5108		3008			
L5121	160	3123	2	TR17/637	G0002
L-5121		3008			
L5122	160	3123	2	TR17/637	G0002
L-5122		3008			
L5181	160			TR17/637	
L5431					G0005
L-612099	7408				
L-612106	7425				
L-612107	7432				
L-612150	74153				
L-612158	7416				
L-612161	74164				
L10A					S7002
L10B					S7004
L10C					S5020
L10D					S5020
L20A					S7000
L2090924-4		3004			
L2091241-2	160	3007		TR17/637	
L2091241-3	160	3008		TR17/637	
L30A					S7000
L30B					S7000
L532000162	128				
LA8-1AOR		3007			
LA-120	1003				
LA703E	703A	3157			
LA1041	1013				
LA1110	1014				
LA1200	1003				
LA1201	1003	3101	IC2	IC507/	C6060P
LA1201B	1003		IC2	IC507/	C6060P
LA1201C	1003	3102		IC507/	
LA1201C-W	1003				
LA1201T	1003				
LA-1201	706	3101	IC2	IC507/	
LA1202	1003				
LA1342	710	3102	511		
LA1353	1080				
LA1364	1004	3102	511		
LA1364N		3102			
LA1365	712	3072			
LA1373	1094				
LA1374	1121				
LA1375	1122				
LA1376	1122				
LA3148	727	3071			
LA3300	1005				
LA3301	1006				C6060P
LA4000	1007				
LA4030	1008				
LA4030P	1009				
LA4031P	1010				
LA4032P	1011				
LA4050P	1150				
LA4051P	1150				
LAD010	1119				
LAD011	1120				
LAP-011	1000				
LAP011	1000				
LAP-012	1001				
LB2000	9946				
LB2001	9962				
LB2002	9930				
LB2003	9932				
LB2004	9944				
LB2005	9933				
LB2006	9936				
LB2007	9935				
LB2030	9945				
LB2031	9093				
LB2032	9099				
LB2100	9949				
LB2101	9963				
LB2102	9961				
LB2106	9937				
LB2130	9948				
LB2131	9094				
LB2132	9097				
LB3000	7400				
LB3001	7410				
LB3002	7420				
LB3003	7430				
LB3004	7450				
LB3005	7460				
LB3006	7404				
LB3008	7402				
LB3009	7440				
LB3060	7480				
LB3150	7490				
LB3175	7491				
LC711		3122			
LC871		3122			
LC1013		3041			
LD1011	1044				
LD1020	1012				
LD1020A	1012				
LD-1110	1014				
LD3000	1015				
LD3001	1016				
LD3001W	1016				
LD3040	1017				
LD3061	1026				
LD3080	1018				
LD3120	1019				
LD3141	1020				
LD3150	1021				
LD-3150	1021				
LD3150C	1021				
LDA400			210	TR64/	
LDA401			210		
LDA401MP			210		
LDA402			210		
LDA404	128	3024	20	TR87/243	
LDA405	128	3024	20	TR87/243	
LDA406	128	3024	61	TR87/243	
LDA408	128	3024	17	TR87/243	S0016
LDA410			62	TR63/	
LDA420			64	TR63/	
LDA450	129	3025	21	TR88/242	
LDA451			48		

Original Part Number	ECG	SK	GE	IR/WEP	HEP
LDA452			21	TR20/	
LDA453			48		
LDA454			65		
LDA455			65		
LDF603		3116			
LDF604		3116			
LDF605		3112			
LDS200	128	3024		TR87/243	
LDS201	128	3024		TR87/243	
LDS202			21		
LDS203			21		
LDS206			10		
LDS207			62		
LDS208			64	TR64/	
LDS210			210		
LDS257			65		
LED-15005	9914				
LED21085	9910				
LED21092	9951				
LG(C)921-43B				TR21/	
LH101H		3514			
LH201H		3514			
LH740A	940				
LH740AC	940				
LID929			212		
LID930			212		
LJ152		3114			S0019
LJ1520		3114			
LM65A,B,C		3114			
LM65D,E,F		3114			
LM101AH		3565			
LM101H		3565			
LM111H		3567			
LM117				TR24/	
LM201AH		3565			
LM201H		3565			
LM211H		3567			
LM301AH		3565			
LM301AN		3565			C6107P
LM307N		3076			
LM309K	309K				
LM309KC	309K				
LM311H		3567			
LM339AN		3569			
LM339D		3569			
LM370	370				
LM371	371				
LM372	372				
LM377	732				
LM378	733				
LM380	740				
LM446		3114			
LM644		3114			
LM703E			IC12		
LM703L	703A	3157			
LM703LH				IC500/	
LM703LN	703A	3157			
LM709C	909	3590			
LM709CH	909	3590			
LM709CN	909D	3552			
LM710C	910				
LM710CH	910	3553			
LM710CN	909D				
LM711C	911				
LM711CH	911	3555			
LM711CN	911D	3556			
LM716				IC503/	
LM723		3164			
LM723C		3164			
LM723CD	923D	3165			
LM723CH	923	3164			
LM741		3514			
LM741C	941	3514			
LM741CH	941	3514			
LM741CN	941D	3514			
LM741CN-14		3558			
LM741H		3514			
LM746				IC508/	
LM746N	790				C6057P
LM747C	947	3526			
LM747CH	947	3526			
LM829A,B,C		3114			
LM829D,E,F		3114			
LM833		3114			
LM1110A		3018			S0014
LM1110B	107	3018	17	TR24/	
LM1117	199		20		S0015
LM1117D		3018			
LM1120B		3018			S0025
LM1120C	107	3018	20	TR21/	
LM1123H	108	3018	20	TR24/	
LM-1129		3018			
LM-1130		3018			
LM-1132		3018			
LM-1133		3018			

Original Part Number	ECG	SK	GE	IR/WEP	HEP
LM-1147		3018			
LM-1148		3018			
LM-1149		3114			
LM-1150		3114			
LM-1151		3114			
LM-1153		3114			
LM-1154		3104			
LM-1155		3018			
LM-1156		3104			
LM-1157		3104			
LM-1158		3030			
LM-1160		3030			
LM1303	725	3162			
LM1303N	725	3162			
LM1303P		3162			
LM1304	718	3159	IC8		
LM1304N	718	3159	IC8		C6094P
LM1305	720		IC7		
LM1305N	720	3160	IC7		C6095P
LM1307	722	3161			
LM1307N	722	3161			C6056P
LM1310	801		IC35		
LM1351N					C6060P
LM1403	123A	3122	10	TR24/	
LM1404	159	3114	21	TR30/	
LM1415-6		3122			
LM1415-7		3122			
LM1458N	778	3557			
LM1496H					C6050G
LM1501H	128(2)	3024	63	TR25(4)/	S5014
LM1502H	129(2)	3025	67	TR28(4)/	S5013
LM1502GH					S5013
LM1596H					C6050G
LM1800	743		IC32		
LM1818					S0015
LM-1862		3030			
LM2111		3135			
LM2111N	708	3135			C6062P
LM3028	724	3525			
LM3028A	724	3525			
LM3028B	724	3525			
LM3053	724	3525			
LM3064	780	3141			
LM3064H	780				
LM3064M					C6069G
LM3065	712	3072	IC2	IC507/	
LM3065N	712	3072	IC2	IC507/	C6063P
LM3066	728				
LM3067	729				
LM3070	714				
LM3070N	714	3075			C6070P
LM3071	715				
LM3071N	715				C6071P
LM3075	723				
LM4442		3114			
LM4590		3114			
LM4745		3114			
LM4815		3114			
LM9514		3114			
LM9531		3114			
LM511160	181				
LM511160D		3036			
LME501		3114			
LMF3304		3114			
LN5497		3510			
LN5638		3510			
LN75116	181	3511			
LN75497	130				
LN76963	129				
LN78533	223				
LRL317A		3005			
LS-0031-AR-218	199				
LS0065		3083			
LS-0066	152				
LS-0067	153				
LS-0079-01	159				
LS-0079-02	159				
LS-0085-01	123A				
LS-0095-AR-213	199				
LS0099		3039			
LS52	104	3009	16	TR01/230	G6003
LS305					S0015
LS703L			IC12		
LS918		3018			
LS1011		3039			
LS2220		3122			
LS2222		3122			
LS3702		3004			
LS3704		3122			
LS3705	123A	3122	20	TR21/	
LS3706		3122			
LS4123		3122			
LS4124		3122			
LS5484		3116	FET2	FE100/	F0021
LS5485	132	3116	FET1	FE100/802	F0015

Original Part Number	ECG	SK	GE	IR/WEP	HEP
LS5485/21M224				FE100/	
LSD155				TR81/	
LSD204				TR25/	
LT51					G6005
LT55					G6005
LT1016	108	3018			
LT1016D		3018		TR24/	S0017
LT1016E	108	3018		TR21/	S0017
LT1016(E)			20		S0017
LT1016H		3018	20	TR21/	S0015
LT1016I	108	3018			S0017
LT1016I,H		3018	20	TR21/	
LT5012					G6005
LT5022					G6003
LT5025					G6003
LT5028					G6003
LT5031					G6005
LT5034					G6005
LT5039					G6007
LT5042					G6007
LT5045					G6007
LT5048					G6007
LT5051					G6007
LT5054					G6003
LT5057					G6003
LT5060					G6003
LT5063					G6005
LT5066					G6005
LT5069					G6005
LT5072					G6005
LT5075					G6005
LT5078					G6005
LT5081					G6007
LT5084					G6007
LT5087					G6018
LT5093					G6003
LT5096					G6003
LT5099					G6005
LT5102					G6005
LT5105					G6013
LT5108					G6005
LT5111					G6005
LT5114					G6013
LT5117					G6018
LT5120					G6018
LT5123					G6018
LT5157					G6018
LT5158					G6018
LT5159					G6018
LT5160					G6018
LT5161					G6018
LT5162					G6018
LT5201					G6011
LT5202					G6011
LT5209					G6011
LT5515					G6005
LTE1016	108	3018		TR21/56	S0017
LTE1016(G.E.)	108			TR95/	
LTH1016		3018			S0017
LTH1016(G.E.)	108			TR95/	
LU2N544	126			TR17/635	
M0A8-1		3007			
M012	108				
M012-1		3020			
M012-12		3018			
M014A		3009			
M024	108				
M024-12		3018			
M094-585-46	159				
M04A-1		3009			
M06B-3		3015			
M065-1		3006			
M065-2		3010			
M065-4		3011			
M065A		3007			
M065B		3008			
M065C		3009			
M066-1		3003			
M066-2		3004			
M066-3		3004			
M074-3		3005			
M077A		3012			
M077B		3013			
M077C		3014			
M08P-2		3004			
M084A		3009			
M088B		3006			
M088C		3005			
M09L-4		3003			
M092-1		3012			
MOTMJE371			58	TR77/	
MOTMJE521			57	TR76/	
M1					G0002
M1S-12795B	130				
M1X	128		63	TR87/243	

Original Part Number	ECG	SK	GE	IR/WEP	HEP
M2					G0002
M2N168A	101	3011	6	TR08/641	
M4			65		
M5			65		
M5A			73	TR59/	S5000
M5B			73		S5000
M5C			73		S5020
M5D			73		S5020
M6			65		
M7			65		
M8D03	727				
M8D03(MOTOROLA)	727				
M10A					S5004
M10B					S5004
M10C					S5020
M10D					S5020
M12					S0017
M24	123A	3018	20	TR21/735	S0017
M24A	123A	3018	17	TR21/735	S0015
M24B	123A	3018	17	TR21/735	S0015
M24ORN					S0015
M25	123A	3018	20	TR21/735	S0015
M25A	123A	3018	17	TR21/735	S0015
M25B	123A	3018	17	TR21/735	S0015
M25B2	123A	3018	20	TR95/	S0025
M25BLU					S0015
M26WHT					S0015
M48S134779					S0015
M54	123A	3124	20	TR21/735	S0015
M54A	123A	3124	20	TR21/735	S0015
M54B	123A	3124	20	TR21/735	S0015
M54BLK	123A			TR21/735	S0015
M54BLU	123A			TR21/735	S0015
M54BRN	123A			TR21/735	S0015
M54C	123A	3124	20	TR21/735	
M54D	123A		20	TR21/735	S0015
M54E	123A	3124	20	TR21/735	S0015
M54GRN	123A			TR21/735	S0015
M54ORN	123A			TR21/735	S0015
M54RED	123A			TR21/735	S0015
M54WHT	123A			TR21/735	S0015
M54YEL	123A			TR21/735	S0015
M65					S0019
M65A	159	3114		TR20/717	S0019
M65B	159	3114		TR20/717	S0019
M65C	159	3114		TR20/717	S0019
M65D	159	3114		TR20/717	S0019
M65E	159	3114		TR20/717	S0019
M65F	159	3114		TR20/717	S0019
M76	160	3006	51	TR17/637	G0003
M77	160	3006	51	TR17/637	G0003
M78	160	3006	50	TR17/637	G0003
M78A	160		50	TR17/637	G0003
M78B	160		50	TR17/637	G0003
M78BLK	160		50	TR17/637	G0003
M78C	160		50	TR17/637	G0003
M78D	160		50	TR17/637	G0003
M78GRN	160		50	TR17/637	G0003
M78RED	160		50	TR17/637	G0003
M78YEL	160		50	TR17/637	G0003
M84	104	3009		TR01/230	G6003
M84B	121	3009		TR01/232	G6018
M84C01					S5015
M91	107	3039		TR70/720	S0015
M91A	123A	3122	20	TR21/735	S0015
M91B	123A	3122	20	TR21/735	S0015
M91BGRN	123A		20	TR21/735	S0015
M91C	123A	3018	20	TR21/735	S0015
M91CM624	123A	3018	20	TR21/735	
M91D	123A	3122	20	TR21/735	S0015
M91E	123A	3122	20	TR21/735	S0015
M91F	123A	3124	20	TR21/735	S0015
M91FM624	123A	3124	20	TR21/735	
M100	133	3112	51	TR12/801	F0010
M101	133	3112		/801	F0010
M108	102A	3123	52	TR85/250	G0005
M-108		3004			
M-128J509-1	161				
M-128J510-1	107				
M-128J511-3	161				
M140-1	107	3018		TR70/720	
M140-3	123A	3018	20		S0015
M300-1300A	128				
M304/4050	755				
M351	160		51	TR17/637	
M-351		3007			
M400					S0015
M401	161	3018			
M430		3124		TR24/	S0015
M433	172	3124	18	TR69/S9100	S0024
M446	159	3114		TR20/717	
M447	123A	3122		TR21/735	
M501	121	3009	16	TR01/232	
M511-035-0001	923				
M530	128			TR87/243	

Original Part Number	ECG	SK	GE	IR/WEP	HEP
M546	161	3018		TR83/719	S0016
M612	161	3018		TR83/719	S0016
M613	161	3018		TR83/719	S0016
M614	161	3018		TR83/719	S0016
M644	159	3114	21	TR20/717	
M652P1C	129		21	TR88/242	
M671	123A	3122		TR21/735	
M773	123A	3122	20	TR21/735	S0015
M773RED	123A	3122		TR21/735	S0015
M774	123A	3122	20	TR21/735	S0015
M774ORN	123A	3122		TR21/735	S0015
M775	123A	3122	20	TR21/735	S0015
M775BRN	123A	3122		TR21/735	S0015
M776	123A	3122	20	TR21/735	S0015
M776GRN	123A	3122		TR21/735	S0015
M779	123A(2)	3122	20		S0015
M779BLU	123A	3122		TR21/735	S0015
M780	123A(2)	3122	20	/735	S0015
M780WHT	123A	3122		TR21/735	S0015
M783	123A	3122	20	TR21/735	S0015
M783RED	123A	3122		TR21/735	S0015
M784	123A	3122	20	TR21/735	S0015
M784ORN	123A	3122		TR21/735	S0015
M785	123A	3122	20	TR21/735	S0015
M785YEL	123A	3122		TR21/735	S0015
M786	123A(2)	3122	20	TR51/735	S0015
M787	123A(2)	3122	20	TR51/735	
M787BLU	123A	3122		TR21/735	S0015
M791	123A	3122		TR21/735	S0015
M818	123A	3122	20	TR21/735	
M818WHT	123A	3122		TR21/735	S0015
M819	154		27	TR78/712	
M822	123A	3124	20	TR21/735	S0015
M822A	123A	3124	20	TR21/735	S0024
M822/48-134823				TR51/	
M822A-BLU	123A			TR21/735	S0024
M822B	123A	3124	20	TR21/735	S0024
M823	123A	3122	20	TR21/735	
M823B	123A	3122		TR21/735	S0024
M823WHT	123A	3122		TR21/735	S0024
M824					S0025
M827	123A	3122	20	TR21/735	S0015
M827BRN	123A	3122		TR21/735	S0015
M828	129	3025		TR88/242	S0015
M828GRN	123A	3122		TR21/735	S0015
M829A	159	3114		TR20/717	S0019
M829B	159	3114		TR20/717	S0019
M829C-F	159	3114		TR20/717	S0019
M833	159	3114	21	TR20/717	S0019
M836					S0015
M844	184	3190	66	/S5003	S0014
M844-84C01					S5000
M844-8503					S0015
M847			20		
M847BLK	123A	3122		TR21/735	S0015
M852	184	3190		/S5007	
M882		3124			
M882BWHT					S0024
M912	157	3103	66	TR60/244	
M912(WEST.)					S5015
M924		3018			
M936					S5004
M-1002-2	123A				
M-1002-17NC	123A				
M1400-1	107	3018		TR70/720	S0015
M1400-I					S0011
M2002	724				
M2003	912				
M2009	796				C6016
M2016	912				
M2028	774				C6002
M2032	916				
M2032-330	916				
M2032-330-BA	916				
M3519	123A	3122	20	TR21/735	S0024
M3567-2	184		28	TR76/S5003	S5000
M4225					F0021
M4313	102A	3123	53	TR85/250	G0006
M4315	102A	3004	2	TR85/250	G0005
M4331	121	3009	3	TR01/232	G6018
M4363	126	3006	51	TR17/635	G0008
M4363BLU	126			TR17/635	G0008
M4363GRN	126			TR17/635	G0008
M4363ORN	126			TR17/635	G0008
M4363WHT	126			TR17/635	G0008
M4364	126	3005	1	TR17/635	G0008
M4365	126	3005	1	TR17/635	G0008
M4366	126	3006	50	TR17/635	G0009
M4367	126	3006	50	TR17/635	G0009
M4368	126	3006	51	TR17/635	G0005
M4388	126			TR17/	
M4389	100			TR05/	
M4398	102A	3123	53	TR85/250	
M4439	160			TR17/	
M4442	159	3114	20	TR20/717	S5022

Original Part Number	ECG	SK	GE	IR/WEP	HEP
M4450	158		53	TR84/630	G0006
M4454	126	3006	50	TR17/635	G0009
M4456	126	3006	50	TR17/635	G0009
M4457	126	3006	50	TR17/635	G0009
M4459	127			TR27/235	G6008
M4462	102A		53	TR85/250	G0005
M4463	104	3009	16	TR01/230	G6003
M4464	123A	3122	18	TR21/735	S0014
M4465	123A	3122	18	TR21/735	S0014
M4466	102A		53	TR85/250	
M4466ORN	158	3123		TR84/630	G0005
M4468	102A		53	TR85/250	G0005
M4468BRN	158	3123		TR84/630	G0005
M4469	102A		53	TR85/250	
M4469RED	158	3123		TR84/630	G0005
M4470	102A		53	TR85/250	
M4470ORN	158	3123		/630	G0005
M4471	102A		53	TR85/250	
M4471YEL	158	3123		TR84/630	G0005
M4472	102A		53	TR85/250	
M4472GRN	158	3123		TR84/630	G0005
M4473	158		53	TR84/630	G0005
M4474	102A		53	TR85/250	
M4474YEL	158	3123		TR84/630	G0005
M4475	102A		53	TR85/250	
M4475GRN	158	3123		TR84/630	G0005
M4476	102A		53	TR85/250	
M4476BLU	158	3123		TR84/630	G0006
M4477	102A		53	TR85/250	
M4477VIO	158	3123		TR84/630	G0006
M4478	129	3025	20	TR88/242	S5023
M4485	160		51	TR17/637	G0003
M4486	160		51	TR17/637	G0003
M4501	126	3006	50	TR17/635	G0009
M4504	160			TR17/	
M4506	160			TR17/	
M4507	160			TR17/	
M4509	126	3006	50	TR17/635	G0009
M4510	158	3123	53	TR84/630	G0006
M4521					G0008
M4524	160			TR17/	
M4525	159			TR20/	
M4526	160			TR17/	
M4545	126	3007	51	TR17/635	
M4545BLU	126			TR17/635	G0008
M4545WHT	126			TR17/635	G0008
M4553	102A		53	TR85/250	
M4553BLU	102A	3123		TR85/250	G0005
M4553BRN	102A	3123		TR85/250	G0005
M4553GRN	102A	3123		TR85/250	G0005
M4553ORN	102A	3123		TR85/250	G0005
M4553PUR	105	3012		TR03/233	G6006
M4553RED	102A	3123		TR85/250	G0005
M4553YEL	102A	3123		TR85/250	G0005
M4562	158		53	TR84/630	G0005
M4563	102A	3123	53	TR85/250	G0006
M4564	102A	3123	53	TR85/250	G0006
M4565	102A	3123	53	TR85/250	G0005
M4567	102	3004	53	TR05/631	G0005
M4570	104	3009	16	TR01/230	G6003
M4573	102A	3123	4	TR85/250	G0007
M4573BLU					G0005
M4576					G0009
M4577					G0009
M4578					G0009
M4582	121	3009	3	TR01/232	
M4582BRN	121	3009		TR01/232	G6013
M4583	121	3009	3	TR01/232	
M4583RED	121	3009		TR01/232	G6013
M4584	121	3009	3	TR01/232	
M4584GRN	121	3009		TR01/232	G6013
M4586	126	3006	51	TR17/635	G0008
M4589	126	3006	51	TR17/635	G0009
M4590	159	3114	20	TR20/717	S0011
M4594	123A	3122	18	TR21/735	S0014
M4595	102A		50	TR85/250	G0005
M4596	102	3004	53	TR05/631	G0005
M4596A		3004			
M4597	102	3004	4	TR05/631	G0007
M4597GRN	102	3004		TR05/631	G0007
M4597GRN-1		3004			
M4597RED	102	3004		TR05/631	G0007
M4597RED-1		3004			
M4603	126	3006	50	TR17/635	G0009
M4604	126	3006	50	TR17/635	G0009
M4605	126	3008	51	TR17/635	G0008
M4605RED	126			TR17/635	G0008
M4606	179			TR35/G6001	G6009
M4607	102	3004	53	TR05/631	G0005
M4608	121	3009	3	TR01/232	S0011
M4619	104	3009	3	TR01/230	
M4619RED	121	3009		TR01/232	G6013
M4620	104	3009	3	TR01/230	
M4620GRN	121			TR01/232	G6013
M4621	126	3008	51	TR17/635	G0008

Original Part Number	ECG	SK	GE	IR/WEP	HEP
M4622	105	3012	4	TR03/231	G6004
M4623	127			TR27/235	G6007
M4624	123A	3122	18	TR21/735	S0014
M4627	102	3004	53	TR05/631	G0005
M4630	123A	3122	18	TR21/735	S0014
M4632	126	3006	51	TR17/635	G0008
M4640	179			TR35/G6001	G6009
M4640P	179			TR35/G6001	G6009
M4648	154		27	TR78/712	S5025
M4649	104	3009	16	TR01/230	G6003
M4652	127			TR27/235	G6007
M4689	128	3024	18	TR87/243	
M4697	160		2	TR17/637	G0002
M4700	101	3011	54	TR08/641	G0011
M4701	179			TR35/G6001	S7004
M4702	179			TR35/G6001	G6009
M4705	123A	3122	18	TR21/735	S0014
M4709	108	3019	17	TR95/56	S0016
M4710					S0025
M4714	123A	3122	18	TR21/735	S0014
M4715	130	3027	14	TR59/247	S7004
M4722	121	3009	16	TR01/232	G6005
M4722BLU	121	3009		TR01/232	G6013
M4722GRN	121	3009		TR01/232	G6013
M4722ORN	121	3009		/232	G6013
N4722PUR	121	3009		TR01/232	G6013
M4722RED	121	3009		TR01/232	G6013
M4722YEL	121	3009		TR01/232	G6013
M4727	104		16	TR01/230	G6003
M4729					G6009
M4730	121	3009	16	TR01/232	G6015
M4732	123A	3124	20	TR21/735	S0024
M4733	108	3124	20	TR95/56	S0020
M4734	123A	3122	20	TR21/735	S0015
M4737	123A	3124	20	TR21/735	S0014
M4739	123A	3021	20	TR21/735	S0015
M4745	159	3114	20	TR20/717	S5022
M4756	161	3019	17	TR83/719	S0017
M4765	123A	3124	20	TR21/	S0014
M4766	121	3009	3	TR01/232	
M4767	121	3009	3	TR01/232	
M4768	123A	3124	20	TR21/735	S0015
M4789	107	3018	20	TR70/720	S0033
M4792					S9002
M4815	159	3025	21	TR20/717	S0019
M4815D	159			TR20/	
M4819	154	3044	27	TR78/712	S5025
M4820	108	3018	17	TR95/56	S0020
M4821	123A	3018	20	TR21/735	S0025
M4825	107	3018	11	TR70/720	S0033
M4826	108	3018	17	TR95/56	S0025
M4834	123A	3124	20	TR21/735	
M4837	108	3018	17	TR95/56	S0020
M4837/48-134837				TR21/	
M4838	154	3124	20	TR78/712	S5026
M4839	154	3045	20	TR78/712	S0015
M4840	123A	3018	20	TR21/735	S0015
M4840A	107	3018	17	TR70/720	
M4841	123A	3124	20	TR21/735	S0015
M4842	123A	3124	20	TR21/735	S0015
M4842A	123A	3124	20	TR21/735	
M4842C	123A			TR21/	
M4843	154	3044	27	TR78/712	S5025
M4844	123A	3018	20	TR21/735	S0011
M4845	108	3018	20	TR95/56	S0020
M4852	123A	3122	20	TR21/735	S0015
M4853	154	3020	20	TR78/712	S5025
M4854	123A	3122	20	TR21/735	S0015
M4855	108	3018	20	TR95/56	S0015
M4857	108	3018	20	TR95/56	S0020
M4857/48-134857				TR21/	
M4860	160		51	TR17/637	
M4872	124	3021	12	TR81/240	S5011
M4882	130	3027	14	TR59/247	S7002
M4885	124	3021	12	TR81/240	S5011
M4888	121	3009	16	TR01/232	G6005
M4888A	121	3014		TR01/232	G6005
M4888B	121	3009		TR01/232	G6005
M4898	123A	3124	20	TR21/735	
M4900	162		35	TR67/707	
M4901	163	3111	36	TR67/740	
M4904	107	3018			
M4906	123A	3124	20	TR21/735	
M4926	123A	3122	20	TR21/735	
M4927	154	3045		TR78/712	
M4933	123A	3122	20	TR21/735	
M4935	123A	3024			
M4937	123A	3122	20	TR21/735	
M4937(3RD IF)	233				
M4943	159	3025			
M4948					S0016
M4974	104				
M4974/P1R	104				
M4989	159	3025			
M4995	163	3111		TR67/740	

Original Part Number	ECG	SK	GE	IR/WEP	HEP
M4998	157			TR60/244	S5015
M5010					S0019
M5101	1095				
M5101P	1095				
M5106	1097				
M5106P	1097				
M5108	1099				
M5108-9170	1099				
M5108P	1099				
M5112	1098				
M5112Y	1098				
M5113	704	3023			
M5113T	704			IC501/	
M5115	1110				
M5115P	1110				
M5115P-9085	1110				
M5115PA	1110				
M5115PR	1110				
M5131P	1096				
M5133P	913				
M5134	1096				
M5134-8266	1096				
M5134P	1096				
M5135P	1004				
M5141T	941	3514			
M5143	712				
M5143P	712				
M5155P	1126				
M5169P	747				
M5183	749				C6076P
M5183-8098	749				
M5183P	749				
M5190	738				
M5190P	738				
M5191P	1050				
M5285	102A			TR85/	
M5286		3014			
M5304	7460				
M5304P	7460				
M5310	7430				
M5310P	7430				
M5352	7450				
M5352P	7450				
M5362	7442				
M5362P	7442				
M5372	7472				
M5372P	7472				
M5374	7474				
M5374P	7474				
M5375	7470				
M5375P	7470				
M5395	7495				
M5395P	7495				
M5930	9930				
M5930P	9930				
M5932	9932				
M5932P	9932				
M5933	9933				
M5933P	9933				
M5935	9935				
M5935P	9935				
M5936	9936				
M5936P	9936				
M5937	9937				
M5937P	9937				
M5944	9944				
M5944P	9944				
M5945	9945				
M5945P	9945				
M5946	9946				
M5946P	9946				
M5948	9948				
M5948P	9948				
M5949	9949				
M5949P	9949				
M5952	9909				
M5952P	9909				
M5953	9093				
M5953P	9093				
M5955P	9097				
M5956	9094				
M5956P	9094				
M5961	9961				
M5961P	9961				
M5962	9962				
M5962P	9962				
M5963	9963				
M5963P	9963				
M7002	154	3045		TR78/712	
M7003	123A	3122	20	TR21/735	
M7006		3122	20	TR21/735	
M7014	123	3122	20	TR86/735	
M7015	123A	3122	20	TR21/735	
M7031	121	3009			
M7033	123A	3122	20	TR21/735	

Original Part Number	ECG	SK	GE	IR/WEP	HEP
M7108	123A	3018			
M7108/A5N	123A				
M7109	123A	3122			
M7109/A5P	123A				
M7127	159	3114			
M7127/P2S	159				
M7171	123A	3124			
M7310	197	3083			
M7313					S7003
M7342	127				
M7342/P4F	127				
M7476	191	3044			
M7511	163	3111			
M8014		3123	2	TR05/254	G0005
M8014(TRAN)	100				
M8062A	102A		2	TR85/250	
M8062B	102A		2	TR85/250	
M8062C	102A		2	TR85/250	
M8073C		3123		TR05/	
M8105	123A	3122	10	TR21/735	
M8116	126		51	TR17/635	
M8120	101	3011	7	TR08/641	
M8124	160		51	TR17/637	G0003
M8221	123A	3124	20		S0015
M8604	102A		53	TR85/250	
M8604A	102A			TR85/250	
M8640	158	3004	2	TR84/630	G0005
M8640A	158	3123	52	TR84/630	G0005
M8640E	158		52	TR84/630	
M-8641	123A				
M-8641A	158				
M9002		3004			S0011
M9002RED					S0011
M9003					G0003
M9003BRN					G0003
M9003GRN					G0003
M9003ORN					G0003
M9003WHT					G0003
M9009					G0003
M9010	108		86	TR95/56	
M-9010		3019			
M9028					G0006
M9032	107	3018	20	TR70/720	S0014
M9038					G0003
M9052					S3001
M9058					G0003
M9059					S0011
M9080					G0003
M9081					G0003
M9083					G6005
M9085				FE100/	
M9090	127			TR27/	
M9091					G6005
M9092				TR08/	
M9095	123A	3122	60	TR21/735	
M9096					S0011
M9098					G0003
M9104					G6005
M9105					G0005
M9105BLU					G0005
M9125					S3001
M9134			28	TR76/	
M9135				TR35/	
M9138	128	3024	18	TR21/735	
M9141	121				
M9142					G6005
M9144					S0011
M9147					S3001
M9148	102	3004	2	TR05/735	
M9159	123A	3122	20	TR21/735	
M9170	128			TR87/	S0014
M9171					G0002
M9174					S3001
M9177	176				
M9181	179				
M9184	128			TR87/	
M9186					S3001
M9187					S3001
M9190					G0003
M9197					S0015
M9198	102A	3004		TR85/250	
M9199					S0011
M9209	123A			TR21/	
M9221	128				
M9225	175			TR81/	
M9226	123A				
M9228	128	3024	18	TR21/735	
M9237	121			TR01/	
M9244	130			TR59/	
M9245					S0015
M9248	123A				
M9249	102A			TR85/	
M9255					G6005
M9256	6401			IR2160/	
M9258					G0003

Original Part Number	ECG	SK	GE	IR/WEP	HEP
M9259	130			TR59/	
M9260					S3001
M9263	121			TR01/	
M9265	198				
M9266	161			TR83/	
M9271					S3001
M9274	175			TR81/	
M9278	130			TR59/	
M9282	123				
M9285					S3001
M9286	198				
M9287	124				
M9294					G6005
M9299					S0015
M9301	175			TR81/	
M9302	130			TR59/	
M9307					S0013
M9308	129			TR88/	
M9316	175				
M9320	198				
M9321	130			TR59/	
M9323					S0015
M9325					S0015
M9329	199			TR21/	S0015
M9334	159				
M9338					S0015
M9342					G6005
M9344	180			/S9100	S7003
M9348	153	3083	26	TR77/700	S5018
M9355					S3001
M9357					S0015
M9359	180			/S7001	S7003
M9366					S0011
M9377					S0015
M9378					S0015
M9380	128			TR87/	
M9389	123A				S0015
M9390					S0015
M9393	175			TR81/	
M9400	129			TR88/	
M9408	162		35	TR67/707	S5020
M9418					S0015
M9420					G6005
M9423(DONM)					S0015
M9423(GOLD)					S0015
M9423(MOTC)					G6008
M9426	129				
M9430					S3001
M9435	129				
M9436	121			TR01/	
M9437					G6015
M9440					S0014
M9443					S0019
M9450	161			TR83/	
M9453					S0015
M9458					G6005
M9459					S0012
M9465	198				
M9475	123				
M9480	181				
M9481	161	3039		TR83/56	S0016
M9482	108	3039		TR95/56	S0020
M9484					S7003
M9491					S0014
M9514	159	3114		TR20/717	S0031
M9515	130				
M9517					G6005
M9518					G0005
M9519	128			TR87/	S3001
M9525	123A			TR21/	
M9526	159			TR20/	
M9527	158			TR84/	
M9531	159	3114		TR20/717	S0019
M9532	123A	3122		TR21/735	S0015
M9536	225				
M9539					S0015
M9540					S5011
M9556	184	3041	57	/S5003	S5000
M9561	128/401				
M9570					S0015
M9571					S0019
M9575	108				
M9576	152				
M9582	184	3041	57	/S5003	S5000
M9591	128				
M9599	128				
M9610	175				
M9618	184				
M9628	181				
M9631	128				
M9661	152				
M9666	223				
M9676	196				
M9676/NPN	196				
M9677	197				

Original Part Number	ECG	SK	GE	IR/WEP	HEP
M9677/PNP	197				
M9701	197				
M9703	128				
M9715	181			TR36/	
M13298					S7004
M31001	123A	3124		TR21/735	
M181675					G6013
M51709T	909				
M53200	7400				
M53200P	7400				
M53201	7401				
M53201P	7401				
M53202	7402				
M53202P	7402				
M53203	7403				
M53203P	7403				
M53204	7404				
M53204P	7404				
M53205	7405				
M53205P	7405				
M53210	7410				
M53210P	7410				
M53220	7420				
M53220P	7420				
M53230	7430				
M53230P	7430				
M53240	7440				
M53240P	7440				
M53241	7441				
M53241P	7441				
M53242	7442				
M53242P	7442				
M53243	7443				
M53243P	7443				
M53244	7444				
M53244P	7444				
M53247	7447				
M53247P	7447				
M53248	7448				
M53248A	7448				
M53248P	7448				
M53250	7450				
M53250P	7450				
M53253	7453				
M53253P	7453				
M53260	7460				
M53260P	7460				
M53270	7470				
M53270P	7470				
M53272	7472				
M53272P					
M53273	7473				
M53273P	7473				
M53274	7474				
M53274P	7474				
M53275	7475				
M53275P	7475				
M53276	7476				
M53276P	7476				
M53280	7480				
M53280P	7480				
M53283	7483				
M53283P	7483				
M53284	7484				
M53284P	7484				
M53286	7486				
M53286P	7486				
M53290	7490				
M53290P	7490				
M53291	7491				
M53291P	7491				
M53292	7492				
M53292P	7492				
M53293	7493A				
M53293P	7493A				
M53295	7495				
M53295P	7495				
M53296	7496				
M53296P	7496				
M53307	74107				
M53307P	74107				
M53321	74121				
M53321P	74121				
M53351	74151				
M53351P	74151				
M53380	74180				
M53380P	74180				
M53392	74192				
M53392P	74192				
M53393	74193				
M53393P	74193				
M75516-2	160	3007		TR17/637	
M75516-2P	160	3007		TR17/637	
M75516-2R	160	3007		TR17/637	
M75517-1	102A	3004	2	TR85/250	

Original Part Number	ECG	SK	GE	IR/WEP	HEP
M75517-2	102A			TR85/	
M75535-1A		3004			
M-75536-1	128				
M-75536-2	128				
M75537-1A		3009			
M75537-2	176				
M75543-1	152	3026	23	TR76/701	
M75545-1	161			TR83/	
M75547-1	108	3019	11	TR95/56	
M75547-2	108	3019	11	TR95/56	
M75549-1		3026			
M75549-2	130				
M-75557-1-6	123A				
M75561-7	102A			TR85/	
M75561-8	129			TR88/	
M75561-17	129			TR88/	
M75561-23	221				
M75561-23RN	221				
M75565-1	123				
M755162-P	160	3007		TR17/637	
M755162-R	160	3007		TR17/637	
M-322					S5013
M-502				TR01/	
M-4722				TR01/	
M-P3D	102				
MA1	160		50	TR17/637	G0003
MA2					G0008
MA10	172			TR69/S9100	
MA-10			6,		
MA23	102A	3123	2	TR85/250	G0011
MA-23		3004			
MA23A	160	3123	51	TR17/637	
MA23B	102A		52	TR85/250	G0005
MA-23B		3004			
MA25	102A	3004	53	TR85/250	
MA25A			2	TR85/250	G0005
MA-25A		3004			
MA100	100	3051		TR05/254	
MA112	158	3123	2	TR84/630	G0005
MA113	158	3025	2	TR84/630	G0005
MA114	158		53	TR84/630	G0005
MA115	158	3123	2	TR84/630	G0005
MA116	158	3123	2	TR84/630	G0005
MA117	158	3123	2	TR84/630	G0005
MA206	102A			TR85/250	
MA240	102A		52	TR85/250	
MA286	100	3005	2	TR05/254	G0005
MA287	100	3005	2	TR05/254	G0005
MA288	100	3005		TR05/254	G0005
MA393	102A	3008	52	TR85/250	G0008
MA393A, B, C	102A	3008	52	TR85/250	G0008
MA393E	102A		52	TR85/250	G0008
MA393G, R	102A	3008	52	TR85/250	G0008
MA703				IC500/	
MA703E				IC500/	
MA709		3590			
MA709A		3551			
MA709AHM		3591			
MA709C		3551			
MA709HC		3591			
MA709HM		3591			
MA729				IC512/	
MA732				IC511/	
MA737				IC502/	
MA747				IC502/	
MA767				IC513/	
MA780				IC509/	
MA781				IC510/	
MA815	102A	3004	2	TR85/250	G0005
MA881	102A			TR85/250	G0005
MA882	102A			TR85/250	G0005
MA883	102A			TR85/250	G0006
MA884	102A			TR85/250	G0006
MA885	102A			TR85/250	G0005
MA886	102A			TR85/250	G0005
MA887	102A			TR85/250	G0005
MA888	102A			TR85/250	G0006
MA889	102A			TR85/250	
MA890	102A	3123	2	TR85/250	G0005
MA891	102A	3123	2	TR85/250	G0005
MA892	102A	3123	2	TR85/250	G0005
MA893	102A	3123	2	TR85/250	G0005
MA894	102A	3123	2	TR85/250	G0005
MA895	102A	3123	2	TR85/250	G0005
MA896	102A	3123	2	TR85/250	G0005
MA897	102A	3123	2	TR85/250	G0005
MA898	102A	3123	2	TR85/250	G0005
MA899	102A	3123	2	TR85/250	G0005
MA900			2		G0005
MA901	102A	3123	2	TR85/250	G0005
MA902	102A	3123	2	TR85/250	G0005
MA903	102A	3123	2	TR85/250	G0005
MA904	102A	3123	2	TR85/250	G0005
MA909	102A			TR85/250	G0005
MA910	102A			TR85/250	[illegible]

Original Part Number	ECG	SK	GE	IR/WEP	HEP
MA0401			21		
MA0402			21		
MA0404			22		
MA0404-1			21		
MA0404-2			21		
MA0411			21		
MA0412			48		
MA0413			21		
MA0414			22		
MA1318	102A	3004	2	TR85/250	
MA1700	102	3004		TR05/631	G0006
MA1700A		3004			
MA1702	158	3005	1	TR84/630	G0007
MA1703	158		2	TR84/630	G0007
MA-1703		3004			
MA1704	158		1	TR84/630	G0007
MA-1704		3005			
MA1705	126		1	TR17/630	G0007
MA-1705		3005			
MA1706	158		2	TR84/630	G0007
MA-1706		3004			
MA1707	158		1	TR84/630	G0007
MA-1707		3005			
MA1708	158			TR84/630	G0007
MA3065		3072		IC507/	
MA4101			61		
MA4103			61		
MA4104			61		
MA4670	121	3009	16	TR01/232	G6005
MA6001			210		
MA6002			210		
MA6003			210		
MA6101		3197	215		
MA6102		3197	215		
MA7805		3591			S0012
MA7811					S0012
MA7812		3592			
MA7815		3593			
MA7816					S0032
MA7817					S0032
MA8001			47		
MA9426		3018			
MA0065				IC507/	
MAQ7786	123A			TR21/	
MAS20	126	3006	52	TR17/635	G0008
MAS21	126	3006	52	TR17/635	G0008
MAS22	126	3008	52	TR17/635	G0008
MAS23	126	3006	52	TR17/635	G0008
MB400	7400				
MB401	7410				
MB402	7420				
MB403	7430				
MB404	7440				
MB405	7450				
MB406	7460				
MB407	7472				
MB408	7480				
MB410	74107				
MB411	7453				
MB416	7401				
MB417	7402				
MB418	7404				
MB420	7474				
MB433	7438				
MB435	7437				
MB442	7442				
MB443	74145				
MB445	74151				
MB447	74180				
MB448	7485				
MB449	7486				
MB452	7496				
MB453	7495				
MB454	7491				
MB455	74198				
MB456	74191				
MB457	74190				
MB458	74181				
MB459	74182				
MB601	7400				
MB602	7410				
MB603	7420				
MB604	7430				
MB605	7440				
MB606	7450				
MB607	7460				
MB609	7472				
MB613	74H53				
MB614	74H21				
MB618	74H78				
MC101	160	3007	2	TR17/637	G0002
MC103	160	3008	51	TR17/637	G0002
MC660P	9660				
MC661P	9661				C0905P
MC662P	9662				C0900P

Original Part Number	ECG	SK	GE	IR/WEP	HEP
MC663P	9663				C0901P
MC664P	9664				C0906P
MC665P	9665				C0910P
MC666P	9666				
MC667P	9667				C0911P
MC668P	9668				C0902P
MC669P	9669				C0903P
MC670P	9670				C0904P
MC671P	9671				
MC672P	9672				C0909P
MC673P	9673				C0912P
MC674P	9674				
MC675P	9675				
MC676P	9676				
MC677P	9677				
MC678P	9678				
MC679P	9679				
MC680P	9680				
MC681P	9681				C0908P
MC682P	9682				C0907P
MC683P	9683				
MC684P	9684				
MC685P	9685				
MC686P	9686				
MC688P	9688				
MC689P	9689				
MC690P	9690				
MC691P	9691				
MC693P					C0912P
MC696P	9696				
MC700G	9900				
MC703G	9903				
MC704G	9904				
MC705G	9905				
MC706G	9906				
MC707G	9907				
MC708G	9908				
MC709G	9909				
MC710G	9910				C2006G
MC711G	9911				C2007G
MC712G	9912				
MC713G	9913				
MC714G	9914				C2010G
MC715G	9915				
MC717P	9917				C2502P
MC721G	9921				
MC723G	9923				
MC724P	9924				C2003P
MC726G	9926				
MC727G	9927				
MC770P					C2501P
MC774G	9974				
MC776P	9976				
MC781G					C2008G
MC782G					C2009G
MC785P					C2001P
MC786P					C2500P
MC787P					C2503P
MC789P	9989				C2005P
MC790P					C2004P
MC791P					C2004P
MC799P					C2002P
MC800G	9900				
MC803G	9903				
MC804G	9904				
MC805G	9905				
MC806G	9906				
MC807G	9907				
MC808G	9908				
MC809G	9909				
MC810G	9910				C2007G
MC811G	9911				C2007G
MC812G	9912				
MC813G	9913				
MC814G	9914				C2010G
MC815G	9915				
MC817P	9917				C2003P
MC821G	9921				
MC823G	9923				
MC824P	9924				C2003P
MC826G	9926				
MC827G	9927				
MC830L	9930				C1030P
MC830P	9930				C1030P
MC831L	9931				
MC831P	9931				
MC832L	9932				C1032F
MC832P	9932				C1032P
MC833L	9933				C1033P
MC833P	9933				C1033P
MC835L	9935				C1035P
MC835P	9935				C1035P
MC836L	9936				
MC836P	9936				
MC837L	9937				

Original Part Number	ECG	SK	GE	IR/WEP	HEP
MC837P	9937				
MC838L					C1038P
MC838P					C1038P
MC839L					C1039P
MC839P					C1039P
MC840L	9935				
MC840P	9935				
MC844L	9944				C1044P
MC844P	9944				C1044P
MC845L	9945				C1045P
MC845P	9945				C1045P
MC846L	9946				
MC846P	9946				
MC848L	9948				
MC848P	9948				
MC849L	9949				
MC849P	9949				
MC850L	9950				
MC850P	9950				
MC851L	9951				
MC851P	9951				
MC852L	9099				C1052P
MC852P	9099				C1052P
MC853L	9093				
MC853P	9093				
MC855L	9097				
MC855P	9097				
MC856L	9094				
MC856P	9094				
MC857L	9157				
MC857P	9157				
MC858L	9158				C1058L
MC858P	9158				C1058P
MC861L	9961				
MC861P	9961				
MC862L	9962				C1062L
MC862P	9962				C1062P
MC863L	9963				
MC863P	9963				
MC870P					C2501P
MC874G	9974				
MC876P	9976				C2004P
MC881G					C2008G
MC885P					C2001P
MC886P					C2500P
MC887P					C2503P
MC889P	9989				C2005P
MC890P					C2004P
MC891P					C2004P
MC899P					C2002P
MC910G					C2006G
MC911G					C2007G
MC914G					C2010G
MC917P					C2502P
MC930L					C1030P
MC932L					C1032P
MC932P					C1032P
MC933L,P					C1033P
MC935L,P					C1035P
MC938L,P					C1038P
MC939L,P					C1039P
MC944L,P					C1044P
MC945L,P					C1045P
MC952L,P					C1052P
MC953L					C1053P
MC958L,P					C1058P
MC962L,P					C1062P
MC981G					C2008G
MC982G					C2009G
MC985P					C2001P
MC986P					C2500P
MC999P					C2002P
MC1302G					C6092G
MC1303	725	3162	IC19		
MC1303L	725	3162	IC19		
MC1303P	725		IC19		
MC1304			IC8		
MC1304P	718	3159	IC8	IC511/	C6094P
MC1304PQ	718	3159	IC8		C6074P
MC1305			IC7		
MC1305P	720	3160	IC7	IC512/	C6095P
MC1305P-C	720			IC512/	
MC1305PQ	720	3160	IC7		
MC1306P	745				
MC1307	722		IC9	IC513/	
MC1307P	722	3161	IC9	IC513/	C6056P
MC1307PQ	722	3161	IC9		C6068P
MC1310P	801	3160	IC35	IC513/	C6096P
MC1311P	743				
MC1312P	799		IC34		
MC-1312P	799				
MC1314G	704			IC501/	
MC1314P	802				
MC1315P	803				
MC1316	814				

Original Part Number	ECG	SK	GE	IR/WEP	HEP
MC1316P	814				C6073P
MC1316PC	814				
MC1324			IC30		
MC1324P	739				C6072P
MC1324PQ	739				
MC1326P	739				C6072P
MC1326PQ	739				
MC1328		3077			
MC1328G	707				C6067G
MC1328P	790	3077	IC18	IC508/	C6057P
MC1328P2		3077			
MC1328PQ	790	3077	IC18		
MC1329P	713	3077			
MC1329PQ	713				
MC1330P	747		IC7		C6079P
MC1339P	721				C6080P
MC1339PQ	721				
MC1344P					C6081P
MC1345P	779				
MC1345PQ	779				
MC1349P	795				C6099P
MC1350P	746				C6059P
MC1350PQ	746				
MC1351P	748		IC26	IC507/	C6060P
MC1351PQ	748		IC26		
MC1352P	749	3171			C6076P
MC1352PQ	749	3171			
MC1352PQ-1		3171			
MC1355P	750				C6061P
MC1355PQ	750				
MC1357	708	3135	IC11	IC504/	
MC1357A		3135			
MC1357P	708	3135	IC10		C6062P
MC1357PQ	708	3135			C6082P
MC1358		3072	IC2	IC507/	
MC1358P	712	3072	IC2		C6063P
MC1358PQ	712	3072			C6083P
MC1364		3141			
MC1364G	780	3141			C6069G
MC1364P	783		IC21		C6100P
MC1364PQ	783		IC21		
MC1370P	714	3075			C6070P
MC1370PQ	714	3075			
MC1371P	715	3076			C6071P
MC1371PQ	715	3076			
MC1375P	723				C6101P
MC1375PQ	723				C6101P
MC1389P	788				
MC1398P	738	3167			C6075P
MC1398PQ	738	3167			
MC1398PQ-1		3170			
MC1437L					C6051P
MC1438R					C6078R
MC1439G					C6053G
MC1439L					C6053L
MC1454G					C6093G
MC1455		3564			
MC1455P1	955M				
MC1458CP1	778	3551			C6102P
MC1458P1	778	3551			
MC1461R	946				
MC1463G					C6054G
MC1463R					C6054R
MC1469G					C6049G
MC1469R					C6049R
MC1496G					C6050G
MC1538R					C6078R
MC1539G					C6053G
MC1550G					C6078R
MC1554G					C6093G
MC1555		3564			
MC1563G					C6054G
MC1563R					C6054R
MC1569G					C6049G
MC1569R					C6049R
MC1596G					C6050G
MC1709CG	909	3551			
MC1709CP1	909	3590			C6103P
MC1709CP2	909D	3552			
MC1709P2	909D	3552			
MC1710CG	910	3553			
MC1710CL	910D	3554			
MC1710CP2	910D				
MC1710P2	910D				
MC1711CG	911	3555			
MC1711CL	911D	3556			
MC1723		3164			
MC1723C		3164			
MC1723CG	923	3164			
MC1723CL	923D	3165			C6014L
MC1723G	923	3164			
MC1741		3514			
MC1741CG	941	3514			
MC1741GP1	941M	4553			
MC1741CP2	941D	3514			

Original Part Number	ECG	SK	GE	IR/WEP	HEP
MC1741G		3514			C6052G
MC1800P	9800				
MC1801P	9801				
MC1802P	9802				
MC1803P	9803				
MC1804P	9804				
MC1805P	9805				
MC1806P	9806				
MC1807P	9807				
MC1808P	9808				
MC1809P	9809				
MC1810P	9810				
MC1811P	9811				
MC1812P	9812				
MC1813P	9813				
MC1814P	9814				
MC1816L	9111				
MC1816P	9111				
MC3000L	74H00				
MC3000P	74H00				
MC3001L	74H08				
MC3001P	74H08				
MC3004L	74H01				
MC3004P	74H01				
MC3005L	74H10				
MC3005P	74H10				
MC3006L	74H11				
MC3006P	74H11				
MC3008L	74H04				
MC3008P	74H04				
MC3009L	74H05				
MC3009P	74H05				
MC3010L	74H20				
MC3010P	74H20				
MC3011L	74H21				
MC3011P	74H21				
MC3012L,P	74H22				
MC3016L,P	74H30				
MC3018L,P	74H62				
MC3019L,P	74H61				
MC3020L,P	74H50				
MC3021L,P	74H86				
MC3023L,P	74H51				
MC3024L,P	74H40				
MC3030L,P	74H60				
MC3031L,P	74H52				
MC3032L,P	74H53				
MC3033L,P	74H54				
MC3034L,P	74H55				
MC3054L,P	74H71				
MC3055L,P	74H72				
MC3061P	74S114				
MC3062P	74S113				
MC3063L,P	74H73				
MC3302P		3569			
MC3346P	912				
MC3403P		3594			
MC4016L,P					C3804P
MC4024L,P					C3805P
MC4039P					C3802P
MC4044L,P					C3806P
MC4080	1083				
MC4080-1,2,3	1083				
MC4316L					C3804P
MC4324L					C3805P
MC4344L					C3806P
MC5400L,P					C3000P
MC5401L,P					C3001P
MC5404L,P					C3004P
MC5410L,P					C3010P
MC5420L,P					C3020P
MC5430L,P					C3030P
MC5440L,P					C3040P
MC5441L,P					C3041P
MC5450L,P					C3050P
MC5473P					C3073P
MC5475L,P					C3075P
MC5490L,P					C3800P
MC5492L					C3801P
MC7400L,P	7400				C3000P
MC7401L,P	7401				C3001P
MC7402L,P	7402				
MC7403L,P	7403				
MC7404L,P	7404				C3004P
MC7405L,P	7405				
MC7406L,P	7406				
MC7407L,P	7407				
MC7408L,P	7408				
MC7409L,P	7409				
MC7410L,P	7410				C3010P
MC7416L,P	7416				
MC7417L,P	7417				
MC7420L,P	7420				C3020P
MC7426L,P	7426				
MC7430L,P	7430				C3030P

Original Part Number	ECG	SK	GE	IR/WEP	HEP
MC7437L,P	7437				
MC7438L,P	7438				
MC7440L,P	7440				C3040P
MC7441AL	7441				
MC7441AP	7441A				C3041P
MC7441L,P					C3041P
MC7442L,P	7442				
MC7443L,P	7443				
MC7444L,P	7444				
MC7445L,P	7445				
MC7446L,P	7446				
MC7447L,P	7447				
MC7448L,P	7448				
MC7450L,P	7450				C3050P
MC7451L,P	7451				
MC7453L,P	7453				
MC7454L,P	7454				
MC7460L,P	7460				
MC7470L,P	7470				
MC7472L,P	7472				
MC7473L,P	7473				C3073P
MC7475L,P	7475				C3075P
MC7476L,P	7476				
MC7479L,P	7479				
MC7480L,P	7480				
MC7483L,P	7483				
MC7486L,P	7486				
MC7490L,P	7490				C3800P
MC7491AL,AP	7491				
MC7492L,P	7492				C3801P
MC7493L,P	7493				
MC7494L,P	7494				
MC7495L,P	7495				
MC7496L,P	7496				
MC7724CP					C6105P
MC7805CP					C6110P
MC7806CP					C6111P
MC7808CP					C6112P
MC7812CP					C6113P
MC7815CP					C6114P
MC7818CP					C6115P
MC7824CP					C6116P
MC7902CP					C6117P
MC7905CP					C6118P
MC7906CP					C6120P
MC7908CP					C6121P
MC7912CP					C6122P
MC7915CP					C6123P
MC7918CP					C6124P
MC7924CP					C6125P
MC8601L,P					C3803P
MC9427		3018			
MC9600					S0015
MC9601L					C3803P
MC14001CP	4001				
MC14002CP	4002				
MC14011CP	4011				
MC14012CP	4012				
MC14013CP	4013				
MC14014CP	4014				
MC14015CP	4015				
MC14016CP	4016				
MC14017CP	4017				
MC14020CP	4020				
MC14021CP	4021				
MC14023CP	4023				
MC14024CP	4024				
MC14025CP	4025				
MC14027CP	4027				
MC14028CP	4028				
MC14029CP	4029				
MC14040CP	4040				
MC14042CP	4042				
MC14049CP	4049				
MC14050CP	4050				
MC14071CP	4071				
MC14512CP	4512				
MC14518CP	4518				
MC14520CP	4520				
MC14527CP	4527				
MC14539CP	4539				
MC14555CP	4555				
MC14556CP	4556				
MC74107P	74107				
MC74121P	74121				
MC74145P	74145				
MC74150P	74150				
MC74151P	74151				
MC74152P	74152				
MC74153P	74153				
MC74155P	74155				
MC74156P	74156				
MC74164AP	74164				
MC74165P	74165				
MC74176P	74176				

Original Part Number	ECG	SK	GE	IR/WEP	HEP
MC74177P	74177				
MC74180P	74180				
MC74181P	74181				
MC74182P	74182				
MC74192P	74192				
MC74193P	74193				
MC74195P	74195				
MCS2135			81	TR51/	S0005
MCS2136				TR51/	S0005
MCS2137			82	TR52/	S0005
MCS2138				TR52/	S0030
MCT492P	7492				C3801P
MD420	160			TR17/637	G0003
MD501	102A		52	TR85/250	G0003
MD501B	102A		52	TR85/250	G0003
MD3133F			21		
MD3134F			21		
MD8001					S0025(2)
MD8002					S0025(2)
MD8003					S0005(2)
MDS31	160		52	TR17/637	G0003
MDS32	160		52	TR17/637	G0003
MDS33	160		52	TR17/637	G0003
MDS33A,C,D	160		52	TR17/637	G0003
MDS34	160		52	TR17/637	G0003
MDS35	126	3008	51	TR17/635	G0008
MDS36	160	3123	2	TR17/637	G0003
MDS37	160		51	TR17/637	G0003
MDS38	160		52	TR17/637	G0003
MDS39	160		52	TR17/637	G0003
MDS40	160		52	TR17/637	G0003
ME0401			21		
ME0402			21		
ME0404	159	3025	65	TR20/717	S0019
ME0404-1	159	3114	21	TR20/717	S0013
ME0404-1A		3114			
ME0404-2	159	3114	21	TR20/717	S0019
ME0404-2A		3114			
ME0404A		3114			
ME0411		3114	65		
ME0412		3114	65		
ME0413		3114	65		
ME0414		3118	65		
ME0463			21		
ME0475		3118	82		
ME-1	123A	3018	62	TR21/	S0015
ME-2	123A	3018	62	TR25/	S0020
ME-3	123A	3124	62	TR25/	S0011
ME-5	210	3041	28	TR76/	S3020
ME213	123A	3122	61	TR21/735	S0019
ME213A	123A	3122	61	TR21/735	S0013
ME216	123A	3122	64	TR21/735	S0013
ME217	123A	3122	64	TR21/735	S0013
ME353					S0011
ME495					S0011
ME501	159	3114	22	TR20/717	
ME502			22		
ME503			21		
ME511			21		
ME512			21		
ME513			67		
ME900	123A	3122	62	TR21/735	S0013
ME900A	123A	3122	62	TR21/735	S0019
ME901	123A	3122	62	TR21/735	S0019
ME901A	123A	3122	62	TR21/735	S0019
ME1001	123A	3122	210	TR21/735	S0011
ME1002	123A	3018	210	TR21/735	S0011
ME1075	128	3024	81	TR87/243	
ME1110	154	3045		TR78/712	
ME1120	154	3045	17	TR78/712	
ME2001	123A	3122	61	TR21/735	S0011
ME2002	123A		61	TR21/735	S0011
ME-2002		3018			
ME3001	108		17	TR95/56	S0016
ME-3001		3018			
ME3002	108	3039		TR95/56	S0016
ME3011	108	3039		TR95/56	
ME3440			32		
ME4001	123A	3124	61	TR21/735	S0011
ME4002	123A	3122	212	TR21/735	S0015
ME4003	123A	3122	212	TR21/735	S0015
ME4003C	123A	3122	62	TR21/	S0015
ME4101	123A	3122	212	TR21/735	S0011
ME4102	123A	3122	212	TR21/735	S0011
ME4103	123A	3122	212	TR21/735	
ME4104	123A	3122	61	TR21/735	
ME5001	108	3039		TR95/56	S0016
ME6001	123A	3122	62	TR21/735	S0011
ME6002	123A	3122	62	TR21/735	S0011
ME6003	123A	3122	62	TR21/735	S0011
ME8001	128	3024	210	TR87/243	
ME8002	128	3024	27	TR87/243	
ME8003	128	3024	81	TR87/243	
ME8101					S0011
ME8201	108	3039	21	TR95/56	S0016

Original Part Number	ECG	SK	GE	IR/WEP	HEP
ME9001	108	3039	61	TR95/56	S0016
ME9002	108	3018	61	TR95/56	S0016
ME9021	108	3039	61	TR95/56	S0011
ME9022	108	3039	211	TR95/56	S0016
ME32865-00001		3524			
MEM564C	222				F2004
MF-55-62	121				
MF521		3112			
MF1161	107	3039	11	TR70/720	
MF1162	107	3039	11	TR70/720	
MF1163	107	3039	11	TR70/720	
MF1164	107	3039	11	TR70/720	
MF3304	159	3114	21	TR20/717	S0013
MFC4000	752				C6003
MFC4000A,B	752				C6003
MFC4010	753				C6010
MFC4010A	753				C6010
MFC4030					C6007
MFC4040	754				C6007
MFC4050	755				C6013
MFC4052	755				C6013
MFC4060	756				C6014
MFC4060A	756				C6014
MFC4062	757				C6014
MFC4062A	757				C6014
MFC4063A	758				
MFC4064A	759				
MFC6010	760		IC12	IC500/	C6001
MFC6020	761				C6011
MFC6030	762				C6012
MFC6030A	762				C6012
MFC6032	763				
MFC6032A	763				
MFC6033	764				
MFC6033A	764				
MFC6034	765				
MFC6034A	765				
MFC6040	766				C6009
MFC6050	767				C6015
MFC6060	768				
MFC6070	769				C6008
MFC6080	770				
MFC8010	771				C6004
MFC8020	772				
MFC8021	796				C6016
MFC8021A	796				C6016
MFC8030	773				
MFC8040	774				C6002
MFC8050	775				
MFC8070	776				
MFC9000					C6006
MFC9010					C6005
MFC9020	777				
MFE2000		3116			
MFE2001		3116			
MFE2093		3112			
MFE2094		3112			
MFE2095		3112			
MFE3001		3112			
MFE3002					F2005
MFE3004	221				F2005
MFE3005	221				F2005
MFE3006	222	3065			F2004
MFE3007	222				F2004
MFE3008	222				F2007
MFE4001					F1036
MFE4007		3112			
MFE4008		3112			F1036
MFE4009		3116			F1036
MFE4010		3116			F1036
MFE4011		3116			F1036
MFE4012		3116			
MH1501	192	3024	63	TR25(4)/	S5014
MH1502	193	3025	67	TR28/	S5013
MH9410A		3024			
MH9460A		3025			
MH9600					S0015
MH9630		3124			
MHM1001	108		17	TR95/56	
MHM1101	108		17	TR95/56	
MHT180	105	3012	4	TR03/233	G6004
MHT181	105	3012	4	TR03/233	G6004
MHT230	105	3012	4	TR03/233	G6004
MHT1802	105	3012	4	TR03/233	G6006
MHT1803	105	3012	4	TR03/233	G6006
MHT1804	105	3012	4	TR03/233	G6006
MHT1807	105	3012	4	TR03/233	G6006
MHT1808	105	3012	4	TR03/233	G6002
MHT1809	105	3012	4	TR03/233	G6002
MHT1810					G6002
MHT2002	101	3011		TR08/641	G0011
MHT2003	101	3011		TR08/641	G0011
MHT2004	101	3011		TR08/641	G0011
MHT2008	101	3011		TR08/641	G0011
MHT2009	101	3011		TR08/641	G0011

Original Part Number	ECG	SK	GE	IR/WEP	HEP
MHT2010	101	3011		TR08/641	G0011
MHT2305	105	3012	4	TR03/233	G6002
MHT2414	128	3024		TR87/243	
MHT2418	128	3024		TR87/243	
MHT4401	128	3024	18	TR87/243	S3011
MHT4402			27	TR78/	S3019
MHT4411	128	3024	47	TR87/243	S0015
MHT4412	128	3024	47	TR87/243	S0015
MHT4413	128	3024	47	TR87/243	S0015
MHT4414			32		S3011
MHT4415			32		S3011
MHT4416			32		S3011
MHT4417			32		S3019
MHT4418			32		S3019
MHT4419			32		S3019
MHT4451	128	3024	27	TR87/243	S3002
MHT4452			27	TR78/	S3002
MHT4453			27	TR78/	S3002
MHT4454			27	TR78/	S3002
MHT4455					S3002
MHT4456					S3002
MHT4483	128	3024	27	TR87/243	S5014
MHT4501					S3019
MHT4502					S3019
MHT4511	128	3024	53	TR87/243	S3019
MHT4512	128	3024	63	TR87/243	S3019
MHT4513	128	3024	63	TR87/243	S3019
MHT4514					S3019
MHT4515					S3019
MHT4515					S3019
MHT4517					S3019
MHT4518					S3019
MHT4519					S3019
MHT4551					S5000
MHT4552					S5000
MHT4553					S5000
MHT4554					S5000
MHT4555					S5000
MHT4556					S5000
MHT4583					S5000
MHT4611					S3002
MHT4612					S3002
MHT4613					S3002
MHT4614					S3002
MHT4615					S3002
MHT4616					S3019
MHT4617					S3002
MHT4618					S3002
MHT4619					S3019
MHT5001			27	TR78/	S3020
MHT5002			27	TR78/	S3020
MHT5003			27	TR78/	S3019
MHT5004			27	TR78/	S3019
MHT5005			27	TR78/	S3021
MHT5006					S3020
MHT5007					S3020
MHT5008					S3019
MHT5009					S3019
MHT5010					S3021
MHT5011					S3020
MHT5012					S3020
MHT5013					S3019
MHT5014					S3019
MHT5015					S3021
MHT5051					S3021
MHT5052					S3021
MHT5053					S3021
MHT5054					S3021
MHT5055					S3021
MHT5056					S3021
MHT5501			46		S3001
MHT5502					S3002
MHT5503					S3019
MHT5504					S3019
MHT5506			46		S3001
MHT5507					S3002
MHT5508					S3019
MHT5509					S3019
MHT5510					S3021
MHT5511			46		S3001
MHT5512					S3002
MHT5513					S3019
MHT5514					S3019
MHT5515					S3021
MHT5551					S3021
MHT5552					S3021
MHT5553					S3021
MHT5554					S3021
MHT5555					S3021
MHT5556					S3021
MHT5901			23	TR78/	S5019
MHT5902					S5019
MHT5903					S5012
MHT5904					S5012
MHT5905					S5012

Original Part Number	ECG	SK	GE	IR/WEP	HEP
MHT5906	152		23	TR76/701	S5019
MHT5907					S5019
MHT5908					S5012
MHT5909					S5012
MHT5910					S5012
MHT5911			23		S5019
MHT5912					S5019
MHT5913					S5012
MHT5914					S5012
MHT5915					S5012
MHT6001					S5000
MHT6011					S5000
MHT6012					S5000
MHT6013					S5000
MHT6014					S5000
MHT6015					S5000
MHT6016					S5000
MHT6031					S5000
MHT6308					S5000
MHT6309					S5000
MHT6310					S5000
MHT6311					S5000
MHT6312					S5000
MHT6313					S5000
MHT6314					S5000
MHT6315					S5000
MHT6316					S5000
MHT6408					S5000
MHT6409					S5000
MHT6410					S5000
MHT6411					S5000
MHT6412					S5000
MHT6413					S5000
MHT6414					S5000
MHT6415					S5000
MHT6416					S5000
MHT6901					S5012
MHT6902					S5012
MHT6905					S5012
MHT6906					S5012
MHT7011					S5004
MHT7012					S5004
MHT7013					S5004
MHT7014					S5004
MHT7015					S5004
MHT7016					S5004
MHT7017					S5004
MHT7018					S5004
MHT7019					S5004
MHT7201					S5020
MHT7202					S5020
MHT7203					S5020
MHT7204					S5020
MHT7205					S5020
MHT7206			35	TR61/	
MHT7207			35	TR61/	
MHT7208			35	TR61/	
MHT7209			36	TR61/	
MHT7401	128	3024		TR87/243	S5014
MHT7402					S3002
MHT7403					S3002
MHT7411	128	3024	27	TR87/243	S5014
MHT7412	128	3024	27	TR87/243	S3010
MHT7413			27	TR78/	S3002
MHT7414	128	3024	27	TR87/243	S5014
MHT7415			27	TR78/	S3010
MHT7416			27	TR78/	S3002
MHT7417	128	3024	46	TR87/243	S5014
MHT7418					S3010
MHT7419					S3002
MHT7511					S5004
MHT7512					S5004
MHT7513					S5004
MHT7514					S5004
MHT7515					S5004
MHT7516					S5004
MHT7517					S5004
MHT7518					S5004
MHT7519					S5004
MHT7601	130	3027	14	TR59/247	S7002
MHT7602	130	3027	14	TR59/247	S7004
MHT7603	130	3027	14	TR59/247	S7004
MHT7604					S5020
MHT7605					S5020
MHT7606					S5020
MHT7607	130	3027	14	TR59/247	S7002
MHT7608	130	3027	14	TR59/247	S7004
MHT7609	130	3027	14	TR59/247	S7004
MHT7610					S5020
MHT7611					S5020
MHT7612					S5020
MHT9001	128	3024	27	TR87/243	S5014
MHT9002	128	3024	27	TR87/243	S3010
MHT9003			27	TR78/	S3002
MHT9004	128	3024	27	TR87/243	S5014

Original Part Number	ECG	SK	GE	IR/WEP	HEP
MHT9005	128	3024	27	TR87/243	S3010
MHT9006			27	TR78/	S3002
MHT9007			27	TR78/	S5014
MHT9008			27	TR78/	S3010
MHT9009			27	TR78/	S3002
MHT9010			46		S5014
MHT9011					S3010
MHT9012					S3002
MHT18010	105	3012	4	TR03/233	
MI404		3004			
MI9623		3122			
MIC709-5(DIP)	909D				
MIC709-5(MET C)	909				
MIC710-5(DIP)	910D				
MIC710-5(MET C)	910				
MIC711-5(DIP)	911D				
MIC711-5(MET C)	911				
MIC723-5		3164			
MIC723-5(DIP)	923D				
MIC723-5(MET C)	923				
MIC741		3514			
MIC741-5		3514			
MIC741-5(DIP)	941D				
MIC741-5(MET C)	941				
MIC930-5D,5P	9930				
MIC931-5D,5P	9931				
MIC932-5D,5P	9932				
MIC933-5D,5P	9933				
MIC935-5D,5P	9935				
MIC936-5D,5P	9936				
MIC937-5D,5P	9937				
MIC941-5D,5P	9941				
MIC944-5D,5P	9944				
MIC945-5D,5P	9945				
MIC946-5D,5P	9946				
MIC948-5D,5P	9948				
MIC949-5D,5P	9949				
MIC950-5D,5P	9950				
MIC951-5D,5P	9951				
MIC961-5D,5P	9961				
MIC962-5D,5P	9962				
MIC963-5D,5P	9963				
MIC7400J,N	7400				
MIC7401J,N	7401				
MIC7402J,N	7402				
MIC7403J,N	7403				
MIC7404J,N	7404				
MIC7405J,N	7405				
MIC7410J,N	7410				
MIC7413J,N	7413				
MIC7420J,N	7420				
MIC7426J,N	7426				
MIC7428J,N	7428				
MIC7430J,N	7430				
MIC7440J,N	7440				
MIC7441AJ	7441				
MIC7441AN	7441				
MIC7442J,N	7442				
MIC7443J,N	7443				
MIC7444J,N	7444				
MIC7445J,N	7445				
MIC7450J,N	7450				
MIC7451J,N	7451				
MIC7453J,N	7453				
MIC7454J,N	7454				
MIC7460J,N	7460				
MIC7470J,N	7470				
MIC7472J,N	7472				
MIC7473J,N	7473				
MIC7474J,N	7474				
MIC7475J,N	7475				
MIC7476J,N	7476				
MIC7481J,N	7481				
MIC7482J,N	7482				
MIC7483J,N	7483				
MIC7484J,N	7484				
MIC7486J,N	7486				
MIC7490J,N	7490				
MIC7491AJ	7491				
MIC7491AN	7491				
MIC7492J,N	7492				
MIC7493J,N	7493A				
MIC7494J,N	7494				
MIC7495J,N	7495				
MIC7496J	7496				
MIC9093-5D,5P	9093				
MIC9094-5D,5P	9094				
MIC9097-5D,5P	9097				
MIC9099-5D,5P	9099				
MIC74107J,N	74107				
MIC74121J,N	74121				
MIC74145J,N	74145				
MIC74150J,N	74150				
MIC74151J,N	74151				
MIC74154J,N	74154				

Original Part Number	ECG	SK	GE	IR/WEP	HEP
MIC74155J,N	74155				
MIC74156J,N	74156				
MIC74180J,N	74180				
MIL PROP		3039			
MIS13674/47	234				
MIS14150-18	101				
MIS-14150-18A	101				
MIS14150-37	154	3027			
MIS-18101-3	9930				
MIS-18101-4	9932				
MIS-18101-5	9945				
MJ105	165			TR93/	
MJ139A		3124			
MJ400	124	3021	32	TR81/240	S5011
MJ410					S5020
MJ411					S5020
MJ413					S5020
MJ420	154	3045	27	TR78/712	S5024
MJ421	154	3045	32	TR78/712	S5024
MJ423					S5020
MJ424					S5021
MJ425					S5021
MJ430					S5013
MJ431					S5020
MJ432			73	TR61/	
MJ440			83		S5014
MJ450	180		74	/S7001	S7001
MJ480	130	3027	19	TR59/247	S7002
MJ481	130	3027	19	TR59/247	S7002
MJ490	180			/S7001	S7003
MJ491	180			/S7001	S7003
MJ703N	703A				
MJ802	181		75	TR59/S7000	S7000
MJ900	244				S9141
MJ901	244				S9141
MJ1000	243	3182			S9140
MJ1001	243	3182			S9140
MJ1800	162		35	TR67/707	S5021
MJ2249	175	3131	66	TR81/241	S5012
MJ2250	175	3131	23	TR81/241	S5012
MJ2251	124	3024	32	TR81/240	S5011
MJ2252	124	3021	32	TR81/240	S5011
MJ2253	218		69	TR58/	S5018
MJ2254	218			TR58/	S5018
MJ2267	180		74	/S7001	S7003
MJ2268	180		74	/S7001	S7003
MJ2500	246	3183			
MJ2501	246	3183			
MJ2801	130	3027	77	TR59/247	S7002
MJ2802	130	3027		TR59/247	S7002
MJ2840	130	3027	75	TR59/247	S7002
MJ2841	130	3027	75	TR59/247	S7002
MJ2901	180		74	/S7001	S7003
MJ2940			74		S7003
MJ2941			74		S5005
MJ2955	219				S7003
MJ3000	245	3182			
MJ3001	245	3182			
MJ3010	162		35	TR67/740	S5020
MJ3011	163		36	TR67/	
MJ3026	162		35		S5021
MJ3027	162		36		S5021
MJ3028			36		S5021
MJ3029	162		73	TR67/707	S5020
MJ3030	163	3111	73	TR67/740	S5021
MJ3055					S7004
MJ3101	175	3131	66	TR81/241	S5012
MJ3201	124	3021	32	TR81/240	S5011
MJ3202	124	3021	32	TR81/240	S5011
MJ3430			73		S5020
MJ3520	245				
MJ3521	245				
MJ3701	218	3083	69	TR58/	S5018
MJ3714					S7004
MJ3716					S7004
MJ3771			75		S7000
MJ3772			75		S7000
MJ3789					S7003
MJ4000		3182			S9140
MJ4001	243	3182			
MJ4010	244	3183			S9141
MJ4011	244	3183			
MJ4030	250				
MJ4031	250				
MJ4032	250				
MJ4033	249				
MJ4034	249				
MJ4035	249				
MJ4101	175	3131	32	TR81/241	S5019
MJ4102	175	3131		TR81/241	S5019
MJ4502	180		74	/S7001	S7001
MJ5202	175	3131	12	TR81/241	S5011
MJ5203	175	3131	12	TR81/241	S5011
MJ5204	175	3131	12	TR81/241	S5011
MJ8100			29		S3003

Original Part Number	ECG	SK	GE	IR/WEP	HEP
MJ8101					S3003
MJ8400	165	3115	38	TR93/740B	
MJ9600					S5012
MJ34000	243				
MJE101			69		S5005
MJE102			69		S5005
MJE103			69		S5005
MJE104	183		56	/S5005	S5005
MJE105	183	3189	56	/S5005	S5005
MJE105K	242				
MJE170					S5006
MJE171					S5006
MJE172					S5006
MJE181	182	3054	57	TR60/	S5000
MJE182					S5000
MJE200					S5000
MJE201			66		S5004
MJE202			66		S5004
MJE203			66		S5004
MJE204	182		55	/S5004	S5004
MJE205	182		55	/S5004	S5004
MJE205K	241				
MJE210					S5006
MJE240					S5000
MJE241					S5000
MJE242					S5000
MJE250					S5006
MJE251					S5006
MJE252					S5006
MJE340	157	3103	27	TR60/244	S5015
MJE-340		3021			
MJE341	157		32	TR60/244	S5015
MJE344	157		66	TR60/244	S5015
MJE344D			32		
MJE345			27	TR78/	
MJE370	185	3083	58	TR77/S5007	S5006
MJE370K	242				
MJE371	185	3083	58	TR76/S5007	S5006
MJE371K		3083	58		
MJE423	162		35	TR67/707	S5020
MJE482	184	3190			
MJE483	184	3190			
MJE488	184	3190	66		
MJE492	185	3191			
MJE493	185	3131			
MJE520	184	3054	57	TR76/S5003	S5000
MJE-520		3041			
MJE520K	241				
MJE521	184	3054	57	TR76/S5003	S5000
MJE-521		3041			
MJE521K	241				
MJE700	254				S9121
MJE701	254				S9121
MJE702	254				S9121
MJE703	254				S9121
MJE710					S5006
MJE711			58		S5006
MJE712			56		S5006
MJE720			57		S5000
MJE721			57		S5000
MJE722			55		S5000
MJE800	253	3180			S9101
MJE801	253	3180			S9101
MJE802	253	3180			S9101
MJE803	253	3180			S9101
MJE1090	258	3181			S9122
MJE1091	258				S9122
MJE1092	258				S9122
MJE1093	258				S9122
MJE1100	257	3180			S9102
MJE1101	257	3180			S9102
MJE1102	257	3180			S9102
MJE1103	257	3180			S9102
MJE1290					S5005
MJE1291					S5005
MJE1291A					S5005
MJE1660					S5004
MJE1661					S5004
MJE2010					S5005
MJE2011					S5005
MJE2020					S5004
MJE2021					S5004
MJE2055					S5004
MJE2090		3181			
MJE2100		3180			
MJE2101		3180			
MJE2102		3180			
MJE2103		3180			
MJE2370	242				S5005
MJE2371	242				S5005
MJE2480	241		69		S5004
MJE2481	241		66		S5004
MJE2482	241		66		S5004
MJE2483	241		66		S5004
MJE2490	242				S5005

Original Part Number	ECG	SK	GE	IR/WEP	HEP
MJE2491	242				S5005
MJE2520	241	3054	66		S5004
MJE2521	241	3054	66		S5004
MJE2522	241	3054	66		S5004
MJE2523	241	3054	66		S5004
MJE2801	182		55	/S5004	S5004
MJE2901	183		56	/S5005	S5005
MJE2940	130			TR59/247	
MJE2955	183		56	/S5005	S5005
MJE3054	241	3054	66		S5004
MJE3055	182		55	/S5004	S5004
MJE3370			58		
MJE3371			58		
MJE3439			80		
MJE3440	157		80	TR78/	
MJE3520			57		
MJE3521			57		
MJE3740	242				S5005
MJE3741	242				S5005
MJE4918					S5006
MJE4919					S5006
MJE4920					S5006
MJE4921					S5000
MJE4922					S5000
MJE4923					S5000
MJE5190					S5000
MJE5191					S5000
MJE5192					S5000
MJE5193					S5006
MJE5194					S5006
MJE5195					S5006
MJE5655					S5015
MJE5656					S5015
MJE5974					S5005
MJE5975					S5005
MJE5976					S5005
MJR5977					S5004
MJE5978					S5004
MJE5979					S5004
MJE5980					S5005
MJE5981					S5005
MJE5982					S5005
MJE5983					S5004
MJE5984					S5004
MJE5985					S5004
MJE6040	260				
MJE6041	260				
MJE6042	260				
MJE6043	259				
MJE6044	259				
MJE6045	259				
MJES191	241				
MJF10335	132	3116	FET2	FE100/	F0021
MJG5194	242				
MK10	132	3116	FET2	FE100/802	F0015
MK-10			FET1		
MK10-2	132	3116		FE100/802	
MK-10-2			FET1	FE100/	F0015
MK-10-E	132	3116	FET2	FE100/	
MK102		3116	FET2	FE100/	
MK3800	1136				
MK5485	132	3112			
MLM101AG		3565			
MLM111G		3567			
MLM201AG		3565			
MLM211G		3567			
MLM301AP1					C6107P
MLM309K	309K				
MM380	160			TR17/637	
MM486	128	3024	63	TR87/243	
MM487	128	3024	63	TR87/243	
MM488	128	3024	63	TR87/243	
MM511	128	3024	63	TR87/243	
MM512	128	3024	63	TR87/243	
MM513	128	3024	63	TR87/243	
MM709	108	3039	17	TR95/56	S0016
MM999	159			TR20/	
MM1105					G6002
MM1139	160		51	TR17/637	G0003
MM1151	102A		52	TR85/250	
MM1152	102A		52	TR85/250	
MM1153	102A		52	TR85/250	
MM1154	102A		52	TR85/250	
MM1199	160			TR17/	
MM1367/2SC684	108	3019		TR95/56	
MM1382	108	3019	11	TR95/56	
MM1387	108		17	TR95/56	S0020
MM-1387		3019			
MM1601					S3005
MM1602					S3006
MM1603					S3007
MM1612					S3001
MM1619			66		
MM1742	102A			TR85/250	G0005
MM1755	123A	3122	213	TR21/735	S0011

Original Part Number	ECG	SK	GE	IR/WEP	HEP
MM1756	123A	3122	213	TR21/735	
MM1757	123A	3122	210	TR21/735	S0011
MM1758	123A	3122	20	TR21/735	
MM1803	108	3039	17	TR95/56	
MM1809	128	3047		TR87/243	S0014
MM1809A	128	3047		TR87/243	
MM1810	128	3048		TR87/243	
MM1810A	128	3024		TR87/243	
MM1941	108	3039	20	TR95/56	S0016
MM1943	128	3024		TR87/243	S0025
MM1945	108	3039	17	TR95/56	S0025
MM2193A			63		
MM2258			32		S3019
MM2259			32		S3019
MM2260	154		32	TR78/	S3019
MM2261			63		
MM2263			32		
MM2264			46		S3001
MM2266	128	3024	18	TR87/243	
MM2503	160		52	TR17/637	G0003
MM2550	160	3123	2	TR17/637	G0003
MM2552	160	3123	2	TR17/637	G0003
MM2554	160	3123	2	TR17/637	G0003
MM2894	160		21	TR17/637	S0013
MM3000	154	3045		TR78/712	S5026
MM3001			32		S5026
MM3002	154	3045	27	TR78/712	S5025
MM3003			27	TR78/	
MM3004			28	TR76/	S5014
MM3005	154	3045		TR78/712	S5014
MM3006					S5014
MM3007					S5014
MM3008			32		
MM3009	154	3045	32	TR78/712	
MM3014	190			TR74/S3021	S3019
MM3053					S3011
MM3100	154	3045		TR78/712	S5026
MM3101	154	3045		TR78/712	S5025
MM3724			215	TR65/	
MM3725			215		
MM3726	159	3114	21	TR20/717	
MM3726A		3114			
MM3903			210		S0025
MM3904			20		S0025
MM3905	159	3114	21	TR20/717	S0013
MM3905A		3114			
MM3906	159	3114	48	TR20/717	S0013
MM3906A		3114			
MM4005	129	3025	67	TR88/242	S3001
MM4006	129	3025		TR88/242	S3019
MM4007					S3019
MM4008			67		S5022
MM4019	129	3025	29	TR88/242	
MM4020			69		
MM4048	159	3114	65	TR20/717	
MM4048A		3114			
MM4052			82		
MM4429			28	TR76/	
MM4430			28	TR76/	
MM5000	160			TR17/637	G0003
MM5001	160			TR17/637	G0003
MM5002	160			TR17/637	G0003
MM5005					S5013
MM5006					S5013
MM5007					S5013
MM7087	154	3045		TR78/712	S5026
MM7088	154	3045		TR78/712	S5026
MM8003					S3005
MM8004	195	3024		TR87/243	S3001
MM8006			11		
MM8007			11		
MMCM918			86		
MMCM930			212		
MMCM2222			213		
MMCM2484			81		
MMCM2907		3118	21		
MMT70			61		
MMT71		3118	65		
MMT72			20		
MMT73		3118		TR20/	
MMT75		3118	82		
MMT76			64		
MMT918			61		
MMT2222			210	TR51/	
MMT2907			48		
MMT3014			20	TR64/	
MMT3546		3118			
MMT3798		3114	67	TR52/	
MMT3799		3114			
MMT3903			210	TR53/	
MMT3904			20	TR53/	
MMT3905		3118	21	TR54/	
MMT3906		3118	48		
MMT3907		3118			
MMT8015			11		

Original Part Number	ECG	SK	GE	IR/WEP	HEP
MN13-B					G0005
MN21			76		S5019
MN22	121	3009	16	TR01/232	G6003
MN23	121	3009	16	TR01/232	G6003
MN24	121		3	TR01/232	G6005
MN-24		3009		TR01/	
MN25	121		3	TR01/232	G6005
MN-25		3009		TR01/	
MN26	121	3009	76	TR01/232	G6005
MN28					G6003
MN29	121	3009	76	TR01/232	G6013
MN29BLK	121	3009	16	TR01/232	G6013
MN29GREY					G6013
MN29GRN	121	3009	16	TR01/232	G6013
MN29PUR	121	3009	16	TR01/232	G6013
MN29WHT	121	3009	16	TR01/232	G6013
MN32	121	3009	16	TR01/232	G6005
MN46	121	3009	16	TR01/232	G6003
MN48	121	3009	25	TR01/232	G6005
MN49	121	3009	76	TR01/232	
MN52	102A		53	TR85/250	G0006
MN53	102A	3004	2	TR85/250	G0005
MN53BLU	102A	3123		TR85/250	G0007
MN53GRN	102A	3123		TR85/250	G0006
MN53RED	102A	3123		TR85/250	G0006
MN60	102A			TR85/250	G0005
MN61	127		3	TR27/235	G6018
MN63	127		3	TR27/235	G6018
MN63A			76		
MN64	127		3	TR27/235	G6018
MN73	121	3009	16	TR01/232	G6005
MN73BLK	121	3009		TR01/232	G6003
MN73WHT	121	3009		TR01/232	G6003
MN76	104	3009		TR01/230	G6003
MN86					G6005
MN194	121	3009	3	TR01/232	G6013
MO4A-16		3009			
MO6B-36		3015			
MO8P-26		3004			
MO9L-46		3003			
MO12-16		3124			
MO14A6		3009			
MO65-16		3006			
MO65-26		3010			
MO65-46		3011			
MO65A6		3007			
MO65B6		3008			
MO65C6		3009			
MO66-16		3003			
MO66-26		3004			
MO66-36		3004			
MO74-36		3005			
MO77A6		3012			
MO77B6		3013			
MO77C6		3014			
MO84A6		3009			
MO88B6		3006			
MO88C6		3005			
MO92-16		3012			
MOA8-16		3007			
MOS3635	221	3050			
MOS3635/135324		3050			
MOS6571		3124			
MOTMJE371	185				
MOTMJE371(9)				TR77(9)/	
MOTMJE521	184	3190			
MP110B	179			TR35/G6001	
MP110B-GRN			76		
MP110B-RED			76		
MP500	213	3012	4	TR03/233	G6002
MP500A	105	3012	4	TR03/233	
MP501	105	3012	4	TR03/233	
MP501A	105	3012	4	TR03/233	
MP502	105	3012	4	TR03/233	
MP502A	105	3012	4	TR03/233	
MP503	105	3012	4	TR03/233	
MP503A	105	3012	4	TR03/233	
MP504	105	3012	4	TR03/233	G6002
MP504A	105	3012	4	TR03/233	
MP505	105	3012	4	TR03/233	
MP505A	105	3012	4	TR03/233	
MP506	105	3012	4	TR03/233	G6010
MP506A	105	3012	4	TR03/233	
MP507	105	3012	4	TR03/233	
MP507A	105	3012	4	TR03/233	
MP525	105	3012	16	TR03/233	
MP525-1-6			76		
MP549		3051	510		
MP600	179			TR35/G6001	
MP601	179			TR35/G6001	
MP602	179			TR35/G6001	
MP603	179			TR35/G6001	
MP1014	121	3009	3	TR01/232	
MP1014-1	102A		53	TR85/250	
MP1014-2	123A	3122	20	TR21/735	

Original Part Number	ECG	SK	GE	IR/WEP	HEP
MP1014-4	102A		53	TR85/250	
MP1014-5	102A		53	TR85/250	
MP1014-6	102A		53	TR85/250	
MP1509-1	121	3014		TR01/232	G6005
MP1509-2	121	3009		TR01/232	G6005
MP1509-3	121	3009		TR01/232	G6005
MP1529			76		G6003
MP1529A			76		G6003
MP1530			76		G6005
MP1530A			76		G6005
MP1531			76		
MP1531A			76		G6005
MP1532					G6018
MP1532A					G6018
MP1533					G6018
MP1534			76		G6003
MP1534A			76		G6003
MP1535			76		G6005
MP1535A			76		G6005
MP1536,A			76		G6005
MP1537,A					G6018
MP1538					G6018
MP1539A			76		
MP1540A			76		
MP1541A			76		
MP1544A			76		
MP1545A			76		
MP1546A			76		
MP1549			76		
MP1550			76		
MP1551			76		
MP1552					G6018
MP1553,A			76		
MP1554,A			76		
MP1555,A			76		
MP1557,A			76		
MP1558,A			76		
MP1559,A			76		
MP1612	127			TR27/235	G6009
MP1612A,B	127			TR27/235	
MP1613	127			TR27/235	G6018
MP2000A			76		
MP2060	121	3009	76	TR01/232	G6005
MP2060-1	121	3009		TR01/232	
MP2061	121	3014	76	TR01/232	G6005
MP2061-3,4					G6013
MP2062	121	3009	76	TR01/232	G6005
MP2062-9					G6013
MP2063			76		G6005
MP2076		3014			G6005
MP2137					G6003
MP2137A	121	3009	16	TR01/232	G6003
MP2138					G6005
MP2138A	121	3009	16	TR01/232	G6005
MP2139					G6005
MP2139A	121	3009	16	TR01/232	G6005
MP2140					G6005
MP2140A					G6005
MP2141,A					G6018
MP2142					G6003
MP2142A	121	3009		TR01/232	G6003
MP2143					G6005
MP2143A	121	3009	16	TR01/232	G6005
MP2144					G6005
MP2144A	121	3009	16	TR01/232	G6005
MP2145,A					G6005
MP2146,A					G6018
MP2200A	179		76	TR35/G6001	G6009
MP2300A	179			TR35/G6001	G6009
MP2526			76		
MP3611	121	3009	76	TR01/232	
MP3612	121	3009	76	TR01/232	
MP3613	121	3009	76	TR01/232	
MP3614	121	3009	76	TR01/232	
MP3615	121	3009	76	TR01/232	
MP3617	121	3009	76	TR01/232	
MP3730	127		25	TR27/235	G6007
MP3731	127		25	TR27/235	G6008
MP5695					G6009
MP8111			66		
MP8112			66		
MP8211			215		
MP8212			215		
MP8213		3197	215		
MP8221		3197	215		
MP8222		3197	215		
MP8223		3197	215		
MP8231		3197	215		
MP8232		3197	215		
MP9423H				TR24/	
MP4906063	128				
MPC17C	1047				
MPC20C	1075				
MPC23C	1046				
MPC29C	1089□				

Original Part Number	ECG	SK	GE	IR/WEP	HEP
MPC29C2	1089□				
MPC31C	1048				
MPC32C	1050				
MPC41C	1093				
MPC46C	1077				
MPC47C	1076				
MPC48C	1091*				
MPC48C1	1091*				
MPC48C-1	1091*				
MPC554	1142				
MPC554C	1142				
MPC558	1051				
MPC561	1079				
MPC561C	1079				
MPC562C	1050				
MPC570C	1086**				
MPC571C	1078				
MPC575C2	1140				
MPF101	133	3112	FET1	/801	
MPF102	132	3116	FET1	FE100/802	F0015
MPF103	133	3112	FET2	/801	F0010
MPF104	133	3112	FET2	/801	F0010
MPF105	133	3112	FET1	FE100/801	F0010
MPF106	132	3116	FET2	FE100/802	F0015
MPF107	132	3116	FET2	FE100/802	F0015
MPF108		3116	FET1		F0015
MPF108BLU					F0015
MPF108GRN					F0015
MPF108ORN					F0015
MPF108YEL					F0015
MPF109	133	3112	FET1	/801	F0021
MPF111	132	3819	FET1	FE100/	
MPF112		3819	FET1		F0015
MPF120					F2007
MPF121	222				F2007
MPF122					F2007
MPF151					F1036
MPF161					F1035
MPM5006			20		S0033
MPS25	121	3009	20	TR01/232	
MPS404	159	3114	82	TR20/717	S0032
MPS404A	159	3114	82	TR20/717	S0032
MPS404AB		3114			
MPS706	108	3018	61	TR95/56	S0025
MPS706A	108	3039	61	TR95/56	S0025
MPS834	107	3039	20	TR70/720	S0025
MPS918	107	3039	17	TR70/720	S0020
MPS1097	160	3006	21	TR17/637	
MPS1572	159	3114	62		S0015
MPS2369	107	3039	18	TR70/720	S0025
MPS2711	123A	3122	212	TR21/735	S0015
MPS2712	123A	3122	212	TR21/735	S0015
MPS2713	123A	3122	210	TR21/735	S0025
MPS2714	123A	3122	210	TR21/735	S0015
MPS2715	123A	3122	61	TR21/735	S0015
MPS2716	123A	3124	61	TR21/735	S0015
MPS2823	107	3039	20	TR70/720	
MPS2894	107	3039	21	TR70/720	
MPS2923	123A	3122	212	TR21/735	S0015
MPS2924	123A	3122	212	TR21/735	S0015
MPS2925	123A	3122	212	TR21/735	S0015
MPS2926	123A	3122	20	TR21/735	S0015
MPS2926BRN	123A	3122		TR21/735	S0015
MPS2926GRN	123A	3122		TR21/735	S0015
MPS2926ORN	123A	3122		TR21/735	S0015
MPS2926RED	123A	3122		TR21/735	S0015
MPS2926YEL	123A	3122		TR21/735	S0015
MPS2926-BRN			17	TR53/	
MPS2926-GRN				TR53/	
MPS2926-ORN			17	TR53/	
MPS2926-RED			17	TR53/	
MPS2926-YEL			17	TR53/	
MPS3392	123A	3124	212	TR21/735	S0015
MPS3393	123A	3122	212	TR21/735	S0015
MPS3394	123A	3122	212	TR21/735	S0015
MPS3395	123A	3122	212	TR21/735	S0015
MPS3396	123A	3122	61	TR21/735	S0015
MPS3397	123A	3122	212	TR21/735	S0015
MPS3398	123A	3122	85	TR21/735	S0015
MPS3536	108	3039		TR95/56	
MPS3563	107	3039	61	TR70/720	S0020
MPS3638	159	3114	22	TR20/717	S0019
MPS3638A	159	3114	82	TR20/717	S0019
MPS3638AC		3114			
MPS3638L		3114			
MPS-3638A		3025			
MPS3639	159	3114	21	TR20/717	
MPS3639A		3114			
MPS3640	159	3114	21	TR20/717	S0015
MPS3640A		3114			
MPS3642	128	3024	210	TR87/243	S0015
MPS3646	108	3046	20	TR95/56	S0014
MPS3693	123A		210	TR21/735	S0030
MPS-3693		3018			
MPS3694	123A		210	TR21/735	S0030

Original Part Number	ECG	SK	GE	IR/WEP	HEP
MPS-3694		3018			
MPS3702	159	3114	82	TR20/717	S0019
MPS3702A		3114			
MPS3703	159	3114	82	TR20/717	S0019
MPS3703A		3114			
MPS3704	123A	3122	20	TR21/735	S0015
MPS3705	123A	3122	20	TR21/735	S0015
MPS3706	123A	3122	20	TR21/735	S0015
MPS3707	123A	3122	212	TR21/735	S0015
MPS3708	123A	3122	212	TR21/735	S0025
MPS3709	123A	3122	61	TR21/735	S0025
MPS3710	123A	3122	61	TR21/735	S0025
MPS3711	123A	3122	212	TR21/735	S0015
MPS3721	123A	3122	212	TR21/735	S0015
MPS3731			25		
MPS3826	123A	3122	210	TR21/735	S0011
MPS3827	123A	3122	210	TR21/735	S0011
MPS3992	123A	3122	20	TR21/735	
MPS5086	159	3114	21	TR20/717	
MPS5086A		3114			
MPS5172	123A	3114	212	TR51/	S0015
MPS5668	133	3126	FET1		F0015
MPS6351	123A	3122		TR21/735	
MPS6413	123A	3122	20	TR21/735	
MPS6507	108		20	TR95/56	S0020
MPS-6507		3018			
MPS6511	108	3039	210	TR95/56	S0020
MPS6512	123A	3122	210	TR21/735	S0025
MPS6513	123A	3122	210	TR21/735	S0015
MPS6514	123A	3018	17	TR21/735	S0015
MPS-6514		3018			
MPS6515	123A	3	62	TR21/735	S0015
MPS6516	159	3025	21	TR20/717	S0019
MPS6516A		3114			
MPS6517	159	3114	48	TR20/717	S0019
MPS6518	159	3114	67	TR20/717	S0019
MPS6518A		3114			
MPS6519	159	3114	67	TR20/717	S0019
MPS6519A		3114			
MPS6520	123A	3122	62	TR21/735	S0015
MPS6521	123A	3122	62	TR21/735	S0015
MPS6522	159	3114	67	TR20/717	S0019
MPS6522A		3114			
MPS6523	159	3114	67	TR20/717	S0019
MPS6523A		3114			
MPS6528	107	3039	20	TR70/720	S0016
MPS6529	107	3039		TR70/720	S0016
MPS6530	123A	3024	20	TR21/735	S0015
MPS6531	107	3124	20	TR70/720	S0015
MPS6532	107	3039	20	TR70/720	S0025
MPS6533	159	3114	21	TR20/717	S0019
MPS6533A		3114			
MPS6533M			21		
MPS6534	159	3114	21	TR20/717	S0019
MPS6534A		3114			
MPS6534M			21		
MPS6535	159	3114	21	TR20/717	S0019
MPS6535A		3114			
MPS6535M			21		
MPS6539	108	3039	61	TR95/56	S0016
MPS6540	107	3039	61	TR70/720	S0016
MPS6541	108	3039	61	TR95/56	S0020
MPS6542	108	3039		TR95/56	S0016
MPS6543	108	3039	11	TR95/56	S0016
MPS6544	123A	3018	212	TR21/735	S0016
MPS6545	123A	3122	212	TR21/735	S0016
MPS6546	108	3039	61	TR95/56	S0016
MPS6547	108	3039	61	TR95/56	S0016
MPS6548	108	3039	61	TR95/56	S0016
MPS6552	123A	3122	20	TR21/735	S0015
MPS6553	123A	3124	20	TR21/735	S0015
MPS6554	123A	3122	20	TR21/735	S0015
MPS6555	123A	3122	20	TR21/735	S0015
MPS6560	123A	3122	20	TR21/735	S0024
MPS6561	123A	3124	20	TR21/735	S0024
MPS6562	159	3114	21	TR20/717	S0019
MPS6562A		3114			
MPS6563	159	3114	21	TR20/717	S0019
MPS6563A		3114			
MPS6564			212		
MPS6565	123A	3122	210	TR21/735	S0015
MPS6566	123A	3124	210	TR21/735	S0015
MPS6566A		3018			
MPS6567	123A	3122	61	TR21/735	S0025
MPS6568	123A	3122	61	TR21/735	S0016
MPS6568A			61	TR21/	S0016
MPS6569	107	3039	61	TR70/720	S0015
MPS6570	107	3039	61	TR70/720	S0024
MPS6571	123A		212	TR21/735	S0024
MPS6572		3124	62		
MPS6573	123A	3124	20	TR21/735	S0015
MPS6574	123A	3122	210	TR21/735	S0015
MPS6575	123A	3122	210	TR21/735	S0015
MPS6576	123A	3122	210	TR21/735	S0015
MPS6580	106	3118	22	TR20/52	S0019

Original Part Number	ECG	SK	GE	IR/WEP	HEP
MPS6590	123A		18	TR21/	
MPS6591			210	TR53/	S0025
MPS8000			81		
MPS8001			62		
MPS8097			210		
MPS8098			81		
MPS8099			81		
MPS8598			82		
MPS8599			82		
MPS9185	123A	3122	20	TR21/735	
MPS9410A		3024		TR24/	S0015
MPS9410AJ		3024			S0024
MPS9416		3122			
MPS9416A		3122	63		S3020
MPS9417		3122			
MPS9417A		3138			
MPS9417A,T		3137			
MPS9418S,T		3137			
MPS9418T					S0015
MPS9423	123A	3018	20	TR33/	S0030
MPS9423F	108				
MPS9423G	108	3018			
MPS9423H	108	3018	20		S0015
MPS9423I	108	3018	20	TR24/	S0015
MPS9423/21M579				TR33/	
MPS9426A.B		3018		TR24/	S0015
MPS9426B		3018			S0015
MPS9426C		3018			
MPS9427B.C		3018		TR24/	S0015
MPS9460A		3025		TR30/	S5013
MPS9466		3114			
MPS9467		3114			
MPS9467A		3138			
MPS9467A,T		3138			
MPS-9467T		3114			
MPS9467T					S0019
MPS9468S,T		3138			
MPS9468T					S0019
MPS9600	123A			TR95/	S0014
MPS9600F	123A	3018	20	TR21/	S0014
MPS9600G	123A	3018	20	TR21/	S0014
MPS9600-5	160	3018		TR17/637	
MPS9601	108	3030		TR95/	S0015
MPS9604I	123A	3018	20	TR21/	S0015
MPS9604R		3018		TR24/	S0015
MPS9611-5	123A	3122		TR21/735	
MPS9616	123A	3122	10	TR24/	
MPS9616A		3137			
MPS9616J	128	3124		TR87/243	S5014
MPS9617		3122			
MPS9618	123A	3122			S0015
MPS9618H	123A	3122		TR24/	S0015
MPS9618I	123A	3122		TR24/	S0015
MPS9618J	123A	3122	20	TR51/	S0015
MPS9619		3122			
MPS9623	123A				
MPS9623C	107	3018	20	TR21/	S0015
MPS9623F	123A	3122	20	TR21/	S0014
MPS9623G	123A	3018	20	TR21/	S0014
MPS9623H	123A	3018	20	TR21/	S0014
MPS9623I		3018			S0015
MPS9623(F,G)				TR21/	
MPS9623L		3122			
MPS9623OH.1		3124			
MPS9625	108	3018	20	TR95/	S0030
MPS9625D		3018			S0015
MPS9625E	107	3018	20	TR21/	S0015
MPS9625F	107	3018	20	TR95/	S0014
MPS9625H	108	3122	20	TR95/	S0014
MPS9630		3124			
MPS9630I	123A	3018	20	TR21/	S0015
MPS9630H.I		3124		TR21/	S0015
MPS9630J		3122			
MPS9630K		3122			
MPS9630T		3124			S0015
MPS9631	123A	3124			S0025
MPS9631I	123A	3122	20	TR21/	S0015
MPS9631(I)		3124	20	TR51/	
MPS9631J		3018			
MPS9632	123A	3124	20	TR33/	S0030
MPS9632I	123A	3124			
MPS9632(I)			20	TR24/	S0015
MPS9632K	123A	3124			
MPS9632(K)				TR24/	S0015
MPS9632/21M503				TR33/	
MPS9633		3124			
MPS9633G		3018			
MPS9634		3124			S0030
MPS9634B		3124			S0015
MPS9634/21M503				TR33/	
MPS9666	159	3114	21	TR30/	
MPS9667		3114		TR30/	
MPS9668		3114			
MPS9668T		3114			S0019
MPS9669		3114			

Original Part Number	ECG	SK	GE	IR/WEP	HEP
MPS9680	159	3114	21	TR20/	
MPS9680I		3114			S0019
MPS9680J	159	3114	22	TR20/	S0019
MPS9680T		3114			S0019
MPS9681	159	3114	21	TR30/	S5013
MPS9681J		3114			
MPS9681K		3114			
MPS9681/21M581				TR30/	
MPS9681/2057A2				TR30/	
MPS9682	159	3114			
MPS9682I	159	3114	22	TR30/	S0019
MPS9682K	159	3114	22	TR20/	S0019
MPS9696H	123A	3124	18	TR21/	S0015
MPS9700					S0015
MPS96800			22		
MPS96801I		3114			
MPS6513489751-0				TR21/	
MPS6530489751-0				TR21/	
MPS6573489751-0				TR21/	
MPS-2716				TR21/	
MPS-6531				TR25/	
MPS-9600-5				TR21/	
MPS-A05		3122	63	TR25/	S0015
MPS-A06			63		S0005
MPS-A09			62		S0015
MPS-A10			20		S0015
MPS-A12			64		S9100
MPS-A13					S9100
MPS-A14					S9100
MPS-A16					S0030
MPS-A17					S0030
MPS-A18					S0024
MPS-A20		3124			S0015
MPS-A42					S0027
MPS-A43					S0005
MPS-A55		3114	67		S5022
MPS-A56			67		S9100
MPS-A66					S9120
MPS-A70					S0019
MPS-A92					S0028
MPS-H02					S0016
MPS-H07					S0025
MPS-H08					S0025
MPS-H10		3122		TR21/	S0016
MPS-H11					S0016
MPS-H17	229				
MPS-H19					S0016
MPS-H20					S0016
MPS-H24					S0016
MPS-H30					S0016
MPS-H31					S0016
MPS-H32					S0016
MPS-H37					S0016
MPS-H54					S0005
MPS-H55					S0005
MPS-H83					S0020
MPS-L01					S0005
MPS-U01		3024			S3024
MPS-U01A					S3024
MPS-U02					S3020
MPS-U03					S3019
MPS-U04					S3019
MPS-U05					S3020
MPS-U06					S3019
MPS-U07					S3019
MPS-U10					S3021
MPS-U45	272				
MPS-U51		3025			S3028
MPS-U51A					S3028
MPS-U52					S3032
MPS-U55					S3032
MPS-U56					S3032
MPS-U57					S3032
MPSA05	123A	3122	81	TR21/735	
MPSA06	128	3024	81	TR87/	S5014
MPSA09			212	TR51/	
MPSA10			212	TR51/	
MPSA10-BLU			212	TR51/	
MPSA10-GRN			212	TR51/	
MPSA10-RED			212	TR51/	
MPSA10-WHT			212	TR51/	
MPSA10-YEL			212	TR51/	
MPSA12	172			TR69/S9100	
MPSA13	172		64	TR69/S9100	S9100
MPSA13(11)			64	TR69/	
MPSA14	172			TR69/S9100	
MPSA16		3122	62		
MPSA16-1, 2, 5		3122			
MPSA16-18, 26, 28		3122			
MPSA17		3122	62		
MPSA17-1, 2, 5		3122			
MPSA17-18, 26, 28		3122			
MPSA18			85		
MPSA20	123A	3124	20	TR51/	S0015
MPSA20-BLU			212	TR51/	
MPSA20-GRN			212	TR51/	
MPSA20-RED			212	TR51/	
MPSA20-WHT			212	TR51/	
MPSA20-YEL			212	TR51/	
MPSA55	159	3114	82	TR20/717	
MPSA55A		3114			
MPSA56	159	3114	82	TR20/717	S5022
MPSA56A		3114			
MPSA70	159	3114	65	TR20/717	S0019
MPSA70-BLU			65	TR52/	
MPSA70-GRN			65	TR52/	
MPSA70-RED			65	TR52/	
MPSA70-WHT			65	TR52/	
MPSA70-YEL			65	TR52/	
MPSA-56					S5022
MPSD05			210		
MPSD06			62		
MPSD55			22		
MPSD56			65		
MPSH02			17	TR22/	
MPSH04			81	TR78/	
MPSH05			81	TR78/	
MPSH07			20		
MPSH10			61		
MPSH11			61		
MPSH17			81		
MPSH19			61		
MPSH20			20	TR64/	
MPSH30			61	TR64/	
MPSH31			61	TR64/	
MPSH32			17	TR64/	
MPSH37			61		
MPSH54			82		
MPSH55			82		
MPSK20			212		
MPSK21			212		
MPSK22			212		
MPSK70			65		
MPSK71			65		
MPSK72			65		
MPSL01	194			/3020	
MPSU01	188	3199	83	TR72/	
MPSU01A	188	3024	83	TR72/3020	
MPSU02	188	3199	63	TR72/3020	
MPSU03	190		32	TR74/S3021	
MPSU04	190		32	TR74/S3021	
MPSU05	188	3199	18	TR72/3020	
MPSU06	188	3054		TR72/3020	S3019
MPSU-06		3199			
MPSU10	191		27	TR75/	
MPSU11	191		27	TR75/	
MPSU11A		3044			
MPSU51	189	3200	84	TR73/3031	
MPSU51A	189	3200	84	TR73/3021	
MPSU52	189	3025	29	TR73/3031	
MPSU55	189	3025		TR73/3031	
MPSU56	189	3025		TR73/3031	
MPSU-56		3200			
MPSU95					S9123
MPSU6515		3020			
MPT4003-10	417				
MPT6003-7	419				
MPT6004-2	420				
MPT6004-4	418				
MPU131	6402				S9001
MPX3693					S0030
MQ1	123A	3124	20	TR21/735	
MQ2	123A		20	TR21/735	
MQ3799-1		3118			
MQ3799-2		3118			
MQ3799-3		3118			
MQ3799-4		3118			
MQ3799A		3117			
MQ3799A-1, 2, 3, 4		3118			
MQ3799B		3118			
MQ3799B		3118			
MQ3799B-1, 2, 3, 4		3118			
M13799C		3118			
MQ3799C-1, 2, 3, 4		3118			
MR3932	123A	3124	17	TR21/735	
MR3933	128	3024	18	TR87/243	
MR3934	129	3025	21	TR88/242	
MR9604		3018			
MRD100			213		
MRD150			60		
MRD450			60		
MRD-1263-100		3019			
MRD-1263-104		3019			
MRF209					S3006
MRF501			86		
MRF502		3039			
MRF8004	195	3047	28	TR65/	S3001
MRF8004A		3048			
MRF8004(B)				TR25/	
MRSA70					S0019

Original Part Number	ECG	SK	GE	IR/WEP	HEP
MS22B	123A	3124		TR21/735	
MS510			63		
MS701T	108			TR95/56	
MS1010			63		
MS2991	128	3024		TR87/243	
MS3694		3018	20		
MS7501S	108	3018		TR95/56	S0015
MS7501T	108	3018		TR95/56	S0015
MS7502R	123A	3018		TR21/735	S0011
MS7502S	108	3018		TR95/	S0011
MS7502T	108	3018		TR95/56	S0011
MS7503R	123A	3018		TR21/735	S0015
MS7504	172	3124		TR69/S9100	S5014
MS7505		3114		TR21/	
MS7505J					S5013
MS7506G	129			TR88/242	S5014
MS7506H	128	3124		TR87/243	S5014
MS7506J	128			TR87/243	
MS7507T		3024			S5014
MS7508T		3025			S5013
MS7991		3047			
MS9667		3114			
MS9681		3114			
MSA7505			66		
MSA8505			66		
MSA8506			28	TR76/	
MSA8508			28	TR76/	
MSG7506	128	3124		TR87/243	S5014
MSJ7505	129	3025		TR88/242	S5013
MSK5405	128	3124		TR87/243	S5014
MSP10A	225				
MSP15			32		
MSP15A	225				
MSP20			32		
MSP20A	225				
MSP25			32		
MSP25A	225				
MSP30			32		
MSP999058-1	128				
MSR7502	108	3018		TR95/56	S0011
MSR7503	123A	3018		TR21/735	S0015
MSS7501	108	3018		TR95/56	S0015
MSS7502	108	3018		TR95/56	S0011
MST10			32	TR21/	
MST10S			32		
MST-10	128	3024		TR87/243	
MST15			32		
MST20			32		
MST20B			32		
MST20S			32		
MST25			32		
MST30			32		
MST30B			32		
MST30S			32		
MST35			32		
MST40			32		
MST40B,S			32		
MST45			32		
MST50			32		
MST50B,S			32		
MST55			32		
MST105			32		
MST7501	108	3018		TR95/56	S0015
MST7502		3018			S0011
MT0404	159	3114	65	TR20/717	
MT0404-1	159	3114	21	TR20/717	
MT0404-2	159	3114	21	TR20/717	
MT0404-2A		3114			
MT0404A		3114			
MT0411	159	3114	65	TR20/717	
MT0412	159	3118	65	TR20/717	
MT0413	159	3114	65	TR20/717	
MT0414		3118	65		
MT0461		3118			
MT0462		3118			
MT0463		3118	21		
MT100	108		63	TR95/56	
MT-100		3019			
MT101	108	3039	11	TR95/56	S0016
MT102	108		63	TR95/56	S0016
MT-102		3019			
MT104	123A	3122	63	TR21/735	S3020
MT106	108		11	TR95/56	S0025
MT-106		3019			
MT107	108	3039	17	TR95/56	S0025
MT181					S0025
MT420				TR78/	
MT421				TR78/	
MT696	123A	3122		TR21/735	S0025
MT697	123A	3122		TR21/735	S0025
MT698	154	3045		TR78/712	S0005
MT699	154	3045		TR78/712	S0005
MT706	123A	3122		TR21/735	S0015
MT706A	123A			TR21/735	S0015
MT706B	123A	3122		TR21/735	S0015

Original Part Number	ECG	SK	GE	IR/WEP	HEP
MT707	123A	3122		TR21/735	S0025
MT708	123A	3122		TR21/735	S0025
MT726	159	3114		TR20/717	S0025
MT726A		3114			
MT743	108	3039		TR95/56	S0025
MT744	108	3039		TR95/56	S0020
MT753	108	3039		TR95/56	S0020
MT869	159	3114		TR20/717	S0025
MT869A		3114			
MT870	154	3045		TR78/712	S0005
MT871	154	3045		TR78/712	S0005
MT910	154	3045		TR78/712	S0005
MT911	154	3045		TR78/712	
MT912	154	3045		TR78/712	
MT914					S0025
MT995					S0013
MT1060					S0016
MT1070			28	TR76/	
MT1075			81		
MT1100			27		
MT1131	159	3114		TR20/717	S0019
MT1131A	159	3114		TR20/717	S0019
MT1131A-1		3114			
MT1132	159	3114		TR20/717	S0019
MT1132A	159	3114		TR20/717	S0019
MT1132A-1		3114			
MT1132B	159	3114		TR20/717	S0019
MT1132B-1		3114			
MT1254	159	3114		TR20/717	S0013
MT1254A		3114			
MT1255	159	3114		TR20/717	S0013
MT1255A		3114			
MT1256	159	3114		TR20/717	S0019
MT1256A		3114			
MT1257	159	3114		TR20/717	S0019
MT1257A		3114			
MT1258	159	3114		TR20/717	S0013
MT1258A		3114			
MT1259	159	3114		TR20/717	S0019
MT1259A		3114			
MT1420	159	3114		TR20/717	S0019
MT1420A		3114			
MT1613	128	3024		TR87/243	S0005
MT1711	128	3024		TR87/243	S0005
MT1893	154	3045		TR78/712	S0005
MT1991	159	3114		TR20/717	S0013
MT2303	159	3114		TR20/717	S0019
MT2411	159	3114		TR20/717	S0013
MT2412	159	3114		TR20/717	S0013
MT2736		3011			
MT3001			60		
MT3002			39		
MT3011			214		
MT3202	124	3021	12	TR81/240	
MT4101	123A	3122	212	TR21/735	
MT4102	123A	3122	212	TR21/735	
MT4102A	123A	3122		TR21/735	S0030
MT4103	123A	3122	212	TR21/735	
MT6001	123A	3122	210	TR21/735	
MT6002	123A	3122	210	TR21/735	
MT6003	123A	3122	60	TR21/735	
MT9001	123A	3122		TR21/735	
MT9002	123A	3122		TR21/735	
MT9003			214		
MT102351A	160			TR17/637	G0003
MTJ00109		3024			
MU20					S9002
MU739					S5024
MU9610	188				
MU9610F	188				
MU9610T	188				S3024
MU9611		3054			
MU9611T	188	3054	66	TR72/	S3024
MU9660	189				
MU9660S	189				
MU9660T	189	3083			S3028
MXC-1312	799				
MXC1312A	799				
MXC1312P	799				
N-020	160			TR17/637	G0008
N020	126				G0008
N0282CT	181				
N0400	162				
N1X	154	3045	81	TR78/712	
N2XA	154	3045	27	TR78/712	
N4T		3018			
N13T1	6402				
N13T2	6402				
N57B2-3	102A	3004		TR85/250	
N57B2-6	102A	3004		TR85/250	
N57B2-7	102A	3004		TR85/250	
N57B2-8	103A	3010		TR08/724	
N57B2-11	160	3006		TR17/637	
N57B2-13	160	3006		TR17/637	
N57B2-14	160	3006		TR17/637	

Original Part Number	ECG	SK	GE	IR/WEP	HEP
N57B2-15	102A	3004		TR85/250	
N57B2-17	126	3008		TR17/635	
N57B2-18	126	3008		TR17/635	
N57B2-19	126	3008		TR17/635	
N57B2-22	160	3006		TR17/637	
N57B2-23	126	3008		TR17/635	
N57B2-25	102A	3004		TR85/250	
N57B4-2	121	3009		TR01/232	
N57B4-4	121	3009		TR01/232	
N74H00A	74H00				
N74H00F	74H00				
N74H01A,F	74H01				
N74H04A,F	74H04				
N74H05A,F	74H05				
N74H08A,F	74H08				
N74H10A,F	74H10				
N74H11A,F	74H11				
N74H20A,F	74H20				
N74H21A,F	74H21				
N74H22A,F	74H22				
N74H30A,F	74H30				
N74H40A,F	74H40				
N74H50A,F	74H50				
N74H51A,F	74H51				
N74H52A,F	74H52				
N74H53A,F	74H53				
N74H54A,F	74H54				
N74H55A,F	74H55				
N74H60A,F	74H60				
N74H61A,F	74H61				
N74H62A,F	74H62				
N74H71A,F	74H71				
N74H72A,F	74H72				
N74H73A,F	74H73				
N74H74A,F	74H74				
N74H76B	74H76				
N74H101A,F	74H101				
N74H102A,F	74H102				
N74H103A,F	74H103				
N74H106A,F	74H106				
N74H108A,F	74H108				
N74S00A	74S00				
N74S00F	74S00				
N74S03A,F	74S03				
N74S04A,F	74S04				
N74S05A,F	74S05				
N74S10A,F	74S10				
N74S11A,F	74S11				
N74S15A,F	74S15				
N74S20A,F	74S20				
N74S22A,F	74S22				
N74S40A,F	74S40				
N74S64A,F	74S64				
N74S65A,F	74S65				
N74S74A,F	74S74				
N74S86A,F	74S86				
N74S112B	74S112				
N74S112F	74S112				
N74S113A,F	74S113				
N74S114A,F	74S114				
N74S133B,F	74S133				
N74S134B,F	74S134				
N74S135A,B	74S135				
N74S138B,F	74S138				
N74S139B,F	74S139				
N74S140A,F	74S140				
N74S151B,F	74S151				
N74S153J,N	74S153				
N74S157B,F	74S157				
N74S158B,F	74S158				
N74S174B,F	74S174				
N74S175B,F	74S175				
N74S181J,N	74S181				
N74S194B,J	74S194				
N74S195B,J	74S195				
N74S196F	74S196				
N74S251B,F	74S251				
N74S258B,J	74S258				
N3563	123A	3124	10	TR21/735	
N4967	176	3123			
N5070	714				
N5070B	714				C6070P
N5071	715				
N5071A	715			IC509/	C6071P
N5072	713				
N5072A	713			IC508/	C6057P
N5111	708		IC10	IC504/	C6062P
N5111A	708		IC10	IC504/	
N5558V	778				
N5596K					C6050G
N5709A	909D	3590			
N5709T	909	3590			
N5710A	910D				
N5710T	910				
N5711A	911D				

Original Part Number	ECG	SK	GE	IR/WEP	HEP
N5711K	911				
N5723A	923D	3165			
N5723L	923	3164			
N5740T	940				
N5741		3514			
N5741A	941D				
N5741T	941	3514			
N5741V	941M	3514			C6052P
N5747K	947				
N7400A	7400				C3000P
N7400F	7400				
N7401A	7400				C3001P
N7401F	7400				
N7402A	7402				
N7403A	7403				
N7403F	7403				
N7404A,F	7404				
N7405A,F	7405				
N7406A,F	7406				
N7407A,F	7407				
N7408A,F	7408				
N7409A,F	7409				
N7410A	7410				C3010P
N7410F	7410				
N7411A,F	7411				
N7413A,F	7413				
N7414B,F	7414				
N7416A,F	7416				
N7417A,F	7417				
N7420A,F	7420				
N7421A,F	7421				
N7426A,F	7426				
N7430A	7430				C3030P
N7430F	7430				
N7432A,F	7432				
N7437A,F	7437				
N7438A,F	7438				
N7439A,F	7439				
N7440A	7440				C3040P
N7440F	7440				
N7441A					C3041P
N7441B,F	7441				
N7442A	7442				
N7442BA	7442				
N7442BF	7442				
N7442F	7442				
N7443A,F	7443				
N7444B,F	7444				
N7445B	7445				
N7446B	7446				
N7447B	7447				
N7448B	7448				
N7450A	7450				C3050P
N7450F	7450				
N7451A,F	7451				
N7453A,F	7453				
N7454A,F	7454				
N7460A,F	7460				
N7470A,F	7470				
N7472A,F	7472				
N7473A	7473				C3073P
N7473F	7473				
N7474A,F	7474				
N7475B	7475				C3075P
N7476B,F	7476				
N7480A,F	7480				
N7483B,F	7483				
N7485A,F	7485				
N7486A,F	7486				
N7490A,F	7490				
N7491A,F	7491				
N7492A,F	7492				
N7493A,F	7493A				
N7494B,F	7494				
N7495A,F	7495				
N7496B,F	7496				
N8808A					C3030P
N8840A					C3050P
N8881A					C3001P
N-52329	130				
N74100F	74100				
N74100N	74100				
N74107A,F	74107				
N74109B,F	74109				
N74121A,F	74121				
N74122A,F	74122				
N74123B,F	74123				
N74132B,F	74132				
N74141B	74141				
N74145B	74145				
N74147	74147				
N74148	74148				
N74150F,N	74150				
N74151B,F	74151				
N74153B,F	74153				

Original Part Number	ECG	SK	GE	IR/WEP	HEP
N74154F,N	74154				
N74155B,F	74155				
N74156B,F	74156				
N74157B,F	74157				
N74158F	74158				
N74160B,F	74160				
N74161B,F	74161				
N74162B,F	74162				
N74163B,F	74163				
N74164A,F	74164				
N74165B,F	74165				
N74166B,F	74166				
N74170B,F	74170				
N74172F,N	74172				
N74174	74174				
N74175B,F	74175				
N74180A,F	74180				
N74181F,N	74181				
N74182B,F	74182				
N74190B,F	74190				
N74191B,F	74191				
N74192B,F	74192				
N74193B,F	74193				
N74194B,F	74194				
N74195B,F	74195				
N74198F,N	74198				
N74199F,H	74199				
N121122-	130				
N-121122-	130				
N910016-A		3005			
N2088576-3		3024			
N-EA15X130	107	3018		TR70/720	
N-EA15X131	108	3018		TR95/56	
N-EA15X132	108	3018		TR95/56	
N-EA15X133	126	3006		TR17/	
N-EA15X134	107	3018		TR70/720	
N-EA15X135	107	3018		TR70/720	
N-EA15X136	123A	3124		TR21/735	
N-EA15X137	123A	3124		TR21/735	
N-EA15X138	123A			TR21/735	
N-EA15X139	131MP	3052		TR94MP/247MP	
NA20	101	3011	8	TR08/641	G0011
NA30	103A	3010	8	TR08/724	G0011
NA1022-1001	160	3008		TR17/637	G0008
NA1022-1007	102A			TR85/250	G0005
NA-1114-1001	160			TR17/637	G0003
NA-1114-1002	160	3008		TR17/637	G0008
NA-1114-1004	102A			TR85/250	G0005
NA-1114-1005	102A			TR85/250	G0005
NA-1114-1006	102A			TR85/250	G0005
NA-1114-1007	102A			TR85/250	G0005
NA-1114-1008	102A			TR85/250	G0005
NA-1114-1009	102A			TR85/250	G0005
NA-1114-1010	102A			TR85/250	G0005
NA-1114-1011	102A			TR85/250	G0005
NA5018-1001	160			TR17/637	G0003
NA5018-1002	126			TR12/635	G0008
NA5018-1003	126			TR17/635	G0008
NA5018-1004	126			TR17/635	G0008
NA5018-1005	126			TR17/635	G0008
NA5018-1006	126			TR17/635	G0008
NA5018-1007	126			TR17/635	G0008
NA5018-1008	126			TR17/635	G0008
NA5018-1009	126			TR17/635	G0009
NA5018-1010	126			TR17/635	G0009
NA5018-1011	126			TR17/635	G0009
NA5018-1012	126			TR17/635	G0009
NA5018-1013	102A			TR85/250	G0005
NA5018-1014	102A			TR85/250	G0005
NA5018-1015	102A			TR85/250	G0005
NA5018-1016	102A			TR85/250	G0005
NA5018-1022	126			TR17/	
NA5018-1219	126			TR17/635	G0009
NA5018-1220	126			TR17/635	G0009
NAP-T-Z-10	102A			TR85/	
NAP-TZ-8	121	3009		TR01/232	
NAP-TZ-10	102A	3004		TR85/250	
NAS29			32		
NAS29A,B,C			32		
NC30	102A			TR85/250	G0005
NC32	158			TR84/630	G6011
NC33	103A			TR08/724	G0011
NC34	176			TR82/238	G6011
NC207AL	123A				
NCR046	128				
NCR047	128				
NCS9018D		3018			S0025
NE555		3564			
NE555V	955M	3564			
NE565A		3595			
NE4304	133	3112			
NEA15X133				TR12/	
NF500		3116			
NF501		3116			
NF506		3116			

Original Part Number	ECG	SK	GE	IR/WEP	HEP
NF520	133	3112			
NF522	133	3112			
NF523	133	3112			
NF531	133	3112			
NF533	133	3112			
NF550		3116			
NF4302	133	3112			
NF4303	133	3112			
NF5485	133	3112			
NF5486	133	3112			
NJ100A	107	3018	20	TR51/	
NJ100AB				TR51/	S0014
NJ100B	123A	3122	10	TR51/	S0015
NJ101B	159	3114	21	TR30/	S0019
NJ102C		3124			
NJ107	128		18	TR25/	S5014
NJ181B	102A		2	TR85/	G0007
NJ202B	107	3124	20	TR24/	
NJ703N	703A	3157		IC500C/	
NJM-703N	703A				
NK4A-1A19	121	3009			
NK6B-3A19	121MP	3015			
NK8P-2A19	102A	3004			
NK9L-4A19	102A	3003			
NK12-1A19	123A	3124			
NK14A119	121	3009			
NK65-1A19	160	3006			
NK65-2A19	103A	3010			
NK65-4A19	101	3011			
NK65A119	160	3007			
NK65B119	126	3008			
NK65C119	121	3009			
NK66-1A19	102A	3003			
NK66-2A19	102A	3004			
NK66-3A19	102A	3004			
NK74-3A19	100	3005			
NK77A119	105	3012			
NK77B119	121MP	3013			
NK77C119	121	3014			
NK84AA19	121	3009			
NK88B119	160	3006			
NK88C119	100	3005			
NK92-1A19	105	3012			
NK111				IC504/	
NK1302	160			TR17/	
NK1404	160			TR17/	
NKA8-1A19	160	3007			
NKT4	102A	3123		TR85/250	G0001
NKT5	102A	3123		TR85/250	G0001
NKT11	102A		52		
NKT12	100		1		
NKT24	102A	3123		TR85/250	G0001
NKT25	102A			TR85/250	G0001
NKT32	102A	3123	52	TR85/250	G0001
NKT33	102A		52	TR85/250	G0005
NKT42	100	3005	52	TR05/254	G0001
NKT43	100	3005	52	TR05/254	G0005
NKT52	102A		52	TR85/250	G0005
NKT53	102A		52	TR85/250	G0005
NKT54	102A		52	TR85/250	G0005
NKT62	100	3005	52	TR05/254	G0005
NKT63	100	3005	52	TR05/254	G0005
NKT64	100	3005	52	TR05/254	G0005
NKT72	100	3005	50	TR05/254	G0005
NKT73	100	3005	50	TR05/254	G0005
NKT74	100	3005	52	TR05/254	G0005
NKT101					G6011
NKT102	158			TR84/630	G0001
NKT103	126	3006		TR17/635	G0005
NKT104	158			TR84/630	G0005
NKT105	158	3005	1	TR84/630	G0005
NKT106	158	3005	1	TR84/630	G0005
NKT107	158	3004		TR84/630	G0005
NKT108	158	3004		TR84/630	G0005
NKT109	158	3004		TR84/630	G0005
NKT121	160	3006	51	TR17/637	G0002
NKT122	160	3006	51	TR17/637	G0002
NKT123	102A	3004	51	TR85/250	G0005
NKT124	160	3006	51	TR17/637	G0002
NKT125	160	3006	51	TR17/637	G0002
NKT126	102A	3004	51	TR85/250	G0005
NKT127	126	3006	51	TR17/635	G0002
NKT128	100	3005	51	TR05/254	G0002
NKT129	100	3005	51	TR05/254	G6011
NKT131	126	3006	52	TR17/635	G0002
NKT132	126	3005	52	TR17/635	G0005
NKT133	102A	3004	1	TR85/250	G0005
NKT141	100	3005	52	TR05/254	G0005
NKT142	100	3005	52	TR05/254	G0005
NKT143	100	3005	52	TR05/254	G0005
NKT144	100	3005	52	TR05/254	G0005
NKT151	126	3006	52	TR17/635	G0005
NKT152	126	3006	52	TR17/635	G0005
NKT153/25	158	3006	52	TR85/635	G0005
NKT154/25	158	3006	52	TR85/635	G0005

Original Part Number	ECG	SK	GE	IR/WEP	HEP
NKT162	100	3005	52	TR05/254	G0005
NKT163	100	3005	52	TR05/254	G0005
NKT163/25	158	3005	52	TR85/254	G0005
NKT164	100	3005	52	TR05/254	G0005
NKT164/25	158	3005	52	TR85/254	G0005
NKT201					G6011
NKT202	100	3005	2	TR05/254	G0005
NKT203	100	3004	2	TR05/254	G0005
NKT204	100	3004	2	TR05/254	G0005
NKT205	100	3004	2	TR05/254	G0005
NKT206	100	3005	2	TR05/254	G0005
NKT207	100	3005	52	TR05/254	G0005
NKT208	102A			TR85/250	G0005
NKT211	158	3004	53	TR84/630	G0005
NKT212	158	3123	2	TR84/630	G0005
NKT213	158	3004	53	TR84/630	G0005
NKT214	158	3004	53	TR84/630	G0005
NKT215	158	3004	53	TR84/630	G0005
NKT216	158	3004	53	TR84/630	G0005
NKT217	158	3004	2	TR84/630	G0005
NKT218	158	3004	53	TR84/630	G0005
NKT219	158		53		
NKT221	100	3005	53	TR05/254	G0005
NKT222	102A	3004	53	TR85/250	G0005
NKT222S1	100	3005		TR05/254	G0005
NKT222S2	100	3005		TR05/254	G0005
NKT223	102	3004	53	TR05/631	G0005
NKT223A	102A	3004			
NKT224	102	3004	53	TR05/631	G0005
NKT224A	102A	3004			
NKT224J	176	3123			
NKT225	102	3004	53	TR05/631	G0005
NKT225A	102A	3004			
NKT225J	176	3123			
NKT226	102	3004	53	TR05/631	G0005
NKT226A	102A	3004			
NKT226J	176	3123			
NKT227	102	3004	2	TR05/631	G0005
NKT227A	102A	3004			
NKT228	102	3004	53	TR05/631	G0005
NKT228A	102A	3004			
NKT229			53		
NKT231	102	3004	53	TR05/631	G0005
NKT231A	102A	3004			
NKT232	102	3004	53	TR05/631	G0005
NKT232A	102A	3004			
NKT238					G6011
NKT239					G6011
NKT240			80		G6011
NKT241			80		G6011
NKT242	126	3123	53	TR17/635	G0005
NKT243	100	3005	2	TR05/254	G0005
NKT244	102A	3004	53	TR85/250	G0005
NKT245			53		G6011
NKT246	102A	3004	52	TR85/250	G0005
NKT247	102	3004	52	TR05/631	G0005
NKT247A	102A	3004			
NKT247J	176	3123			
NKT249	126	3006	52	TR17/635	G0005
NKT251	160	3006		TR17/637	G0005
NKT251A					G0005
NKT252	126	3006	52	TR17/635	G0005
NKT253	126	3006		TR17/635	G0005
NKT254	126	3006	52	TR17/635	G0005
NKT255	126	3006	52	TR17/635	G0005
NKT261	100	3005	53	TR05/254	G0005
NKT262	100	3005	52	TR05/254	G0005
NKT263	100	3005		TR05/254	G0005
NKT264	100	3005	52	TR05/254	G0005
NKT265	126	3006	52	TR17/	G0005
NKT270	126	3006	53	TR17/635	G0005
NKT271	158	3004	53	TR84/630	G0005
NKT272	158		53	TR84/630	G0005
NKT-272		3004			
NKT273	102A	3004		TR85/250	G0005
NKT274	158	3123	53	TR84/630	G0005
NKT275	158	3004	53	TR84/630	G0005
NKT275A	158	3004	52	TR84/630	G0005
NKT275E	158	3004	52	TR84/630	G0005
NKT275J	158	3004	52	TR84/630	G0005
NKT278	102A	3004	52	TR85/250	G0005
NKT281			53		
NKT301					G6011
NKT301A					G6011
NKT302,A					G6011
NKT303	102A	3004	2	TR85/250	G6011
NKT304					G6011
NKT308	102A	3004		TR85/250	
NKT351					G6011
NKT352					G6011
NKT361					G6011
NKT401	121		25	TR01/232	G6005
NKT-401	121	3009			
NKT402	121	3009	25	TR01/232	G6005
NKT403	121	3009	25	TR01/232	G6005
NKT404	121	3009	25	TR01/232	G6005
NKT405	121	3009	25	TR01/232	G6005
NKT406	121		16		
NKT415	121	3009	16	TR01/232	G6003
NKT416	121	3009	16	TR01/232	G6005
NKT450	121	3009		TR01/232	G6003
NKT451	121	3009	49	TR94/642	G6003
NKT452	121	3009	49	TR94/642	G6003
NKT452-S1	131	3014		TR94/642	G6005
NKT453	131	3009	49	TR94/642	G6003
NKT454	121	3009	16	TR01/232	G6005
NKT501	121	3009	16	TR01/232	G6001
NKT502					G6001
NKT503	121	3009	16	TR01/232	
NKT504	121	3009	16	TR01/232	G6001
NKT618	126	3008	52	TR17/635	G0003
NKT618J	176	3123			
NKT674F	160		50		
NKT675	126	3006	52	TR17/635	G0008
NKT676	126	3006	52	TR17/635	G0008
NKT677	126	3006	52	TR17/635	G0008
NKT677F	160		50		
NKT701	103A	3010	8	TR08/724	G0011
NKT703	103A	3010	8	TR08/724	G0011
NKT713	103A	3010	8	TR08/724	G0011
NKT717	103A	3010		TR08/724	G0011
NKT734	101	3011	8	TR08/641	G0011
NKT736	101	3011		TR08/641	G0011
NKT751	103A	3010	8	TR08/724	G0011
NKT752	103A	3010	8	TR08/724	G0011
NKT753	101	3011		TR08/641	G0011
NKT773	103A		8	TR08/724	G0011
NKT774					G0011
NKT781	103A	3010	59	TR08/724	G0011
NKT4055	219	3173	74		
NKT10339	123A	3122	20	TR21/735	
NKT10419	123A	3122	212	TR21/735	
NKT10439	123A	3122	20	TR21/735	
NKT10519	123A	3122	212	TR21/735	
NKT12329	123A	3122	20	TR21/735	
NKT12429	123A	3122	20	TR21/735	
NKT13329	123A	3122	20	TR21/	
NKT13429	123A	3122	20	TR21/	
NKT15325	126	3006		TR17/	
NKT15425	126			TR17/	
NKT16229	161	3132	86	TR83/719	
NKT16325	100	3005		TR05/	
NKT16425	100	3005		TR05/	
NKT20329	159	3114	65	TR20/717	
NKT20329A	159	3114			
NKT20339	159	3114	82	TR20/717	
NKT20339A		3114			
NKT35219	161	3117	86	TR83/719	S0017
NKT80111	133	3112			
NKT80211	133	3112			
NKT80212	133	3112			
NKT80213	133	3112			
NKT80214	133	3112			
NKT80215	133	3112			
NKT80216	133	3112			
NKT800112	133	3112			
NKT800113	133	3112			
NKT-401				TR01/	
NKT-402				TR01/	
NKT-403				TR01/	
NKT-404				TR01/	
NKT-405				TR01/	
NKT-415				TR01/	
NKT-416				TR01/	
NKT-451				TR01/	
NKT-452				TR01/	
NKT-453				TR01/	
NKT-454				TR01/	
NKT-501				TR01/	
NKT-503				TR01/	
NKT-504				TR01/	
NKTU32				TR85/	
NL100B	107	3018	20	TR51/	
NL-102	123A				
NN7000	192		47		
NN7001	192		47	TR55/	S3024
NN7001/183034				TR55/	
NN7002	192		47		
NN7003	192		47		
NN7004	192		47		
NN7005	192		47		
NN7500	193		48		
NN7501	193		48		
NN7502	193		48		
NN7503	193		48		
NN7504	193		48		
NN7505	193		48		
NN7511	193		67	TR56/	S3028
NN7511/183035				TR56/	
NO400					S5020

Original Part Number	ECG	SK	GE	IR/WEP	HEP
NPC069			212		
NPC069-98			212		
NPC079			65		
NPC079-98			65		
NPC115	128	3024	61	TR87/243	
NPC167	161		60		
NPC173	108		61		
NPC187	128	3024	60	TR87/243	
NPC188	108		61		
NPC189	128	3024	61	TR87/243	
NPC211N	133	3112			
NPC212N	133	3112			
NPC214N	133	3112			
NPC215N	133	3112			
NPC216N	133	3112			
NPC312N	133	3112			
NPC737	123A	3124	20	TR53/	S0005
NPC1075	233	3039	39	TR70/	S0017
NPS404			22		
NPS404A			82		
NPS6512			210		
NPS6513			210		
NPS6514			20		
NPS6516			21		
NPS6517			21		
NPS6518	153	3083		TR77/700	S5006
NPS6520			20		
NPSA20			62		
NR071AU		3018			
NR-071AU		3124	20	TR24/	S0015
NR091ET		3018			S0015
NR05	101		8	TR10/	
NR5	101	3011	7	TR01/641	G0011
NR10	101		8	TR08/641	G0011
NR-10	101	3011			
NR20	103	3010	5	TR08/641A	
NR30	101	3010	5	TR08/641	
NR-141ES		3137			
NR-141ET	192				
NR201AY		3124		TR24/	S0015
NR261AS					S0015
NR-261AS			20	TR24/	
NR271AY		3124		TR24/	S0015
NR-431AG		3018			
NR-431AS		3018	20	TR24/	S0015
NR461AF		3018		TR24/	S0011
NR-461AS			20	TR24/	S0015
NR-601AT	159				
NR601BT		3114	22		S5022
NR621AT		3114	22		S5022
NR-621AU		3114			
NR621EU		3114			S0019
NR621U		3114			
NR631AY		3114		TR30/	S0019
NR-671ET	193				
NR700	101				G0011
NR7916	132			FE100/802	F0015
NS060					S0016
NS061					S0016
NS063					S0016
NS064					S0016
NS066					S0016
NS067					S0016
NS069					S0016
NS072					S0015
NS073					S0016
NS075					S0016
NS078					S0005
NS32	102		2		
NS100					G0003
NS121	102		2		
NS200					S0011
NS316					F0021
NS345	161				S0015
NS381	108	3039	61	TR95/56	S0011
NS382	108	3039	63	TR95/56	S0011
NS383					S0011
NS384					S0011
NS404	159	3114		TR20/717	S0012
NS404A		3114			
NS406	161				
NS430					S0011
NS431					S0011
NS432					S0011
NS433					S0011
NS434					S0011
NS435					S0011
NS436					S0011
NS437					S0011
NS438					S0011
NS475	123A	3122	210	TR21/735	S0011
NS476	123A	3122	210	TR21/735	S0011
NS477	123A	3122	210	TR21/735	S0015
NS478	123A	3122	81	TR21/735	S0011
NS479	123A	3122	81	TR21/735	S0011

Original Part Number	ECG	SK	GE	IR/WEP	HEP
NS480	123A	3122	81	TR21/735	S0015
NS661	159	3114	21	TR20/717	S0012
NS661A		3114			
NS662	159	3114	21	TR20/717	S0012
NS662A		3114			
NS663	159	3114	21	TR20/717	S0012
NS663A		3114			
NS664	159	3114		TR20/717	S0012
NS664A		3114			
NS665	159	3114	21	TR20/717	S0013
NS665A		3114			
NS666	159	3114	21	TR20/717	S0013
NS666A		3114			
NS667	159	3114	21	TR20/717	S0013
NS667A		3114			
NS668	159	3114		TR20/717	S0013
NS668A		3114			
NS731	123A	3122	210	TR21/735	S0011
NS731A	123A	3122	11	TR21/735	S0011
NS732	159	3114	210	TR20/717	S0013
NS732A	159	3114	11	TR20/717	S0013
NS733	123A	3122	210	TR21/735	S0011
NS733A	123A	3122	11	TR21/735	S0011
NS792			215		
NS793			215		
NS949	123A	3122	63	TR21/735	S3020
NS950	192		63	TR25/	S3020
NS1000	159	3114	17	TR20/717	S0012
NS1000A	159	3114			
NS1001	159	3114	17	TR20/717	S0012
NS1001A	159	3114			
NS1110	124		12	TR23/	
NS1355	192		63	TR25/	
NS1356	108	3039	17	TR95/56	S3001
NS1500	123A	3122	11	TR21/735	S0011
NS1510	107	3039	11	TR70/720	S0011
NS1535					S3001
NS1672	159	3114		TR20/717	S0013
NS1672A	159	3114			
NS1673	159	3114		TR20/717	S0013
NS1673A		3114			
NS1674	159	3114		TR20/717	S0013
NS1674A	159	3114			
NS1675	159	3114		TR20/717	S0013
NS1675A	159	3114			
NS1861	159	3114		TR20/717	S0012
NS1861A	159	3114			
NS1862	159	3114		TR20/717	S0012
NS1862A		3114			
NS1863	159	3114		TR20/717	S0013
NS1863A	159	3114			
NS1864	159	3114		TR20/717	S0013
NS1864A	159	3114			
NS1900	192		63	TR25/	S0005
NS1960	192		63	TR25/	
NS1972	123A	3122	11	TR21/735	S0011
NS1973	123A	3122	11	TR21/735	S0011
NS1974	123A	3122	11	TR21/735	S0011
NS1975	123A	3122	11	TR21/735	S0011
NS2100	128	3024	63	TR87/243	
NS2101	128	3024	63	TR87/243	S3011
NS2525					S0025
NS3000					S0033
NS3001					S0033
NS3039	107	3039	11	TR70/720	
NS3040	107	3039	11	TR70/720	
NS3041	107	3039	11	TR70/720	
NS3050					S0033
NS3051					S0033
NS3052					S0033
NS3053					S0033
NS-3065	712	3072			
NS3300	108	3039	17	TR95/56	S0005
NS3903			210		
NS3904			20		
NS3905			21		
NS3906			21		
NS6001		3114	22		S5022
NS6062	159	3114	21	TR20/717	S0013
NS6062A	159	3114			
NS6063	159	3114	21	TR20/717	S0013
NS6063A	159	3114			
NS6064	159	3114	21	TR20/717	S0013
NS6064A	159	3114			
NS6065	159	3114	21	TR20/717	S0013
NS6065A	159	3114			
NS6112	108	3039		TR95/56	S0015
NS6113	108	3039		TR95/56	S0015
NS6114	123A	3122		TR21/735	S0015
NS6115	123A	3122		TR21/735	S0015
NS6201					S3028
NS6203					S3024
NS6205					S3032
NS6207	123A	3122		TR21/735	
NS6210	123A	3122		TR21/735	S0015

Original Part Number	ECG	SK	GE	IR/WEP	HEP
NS6211	159	3114		TR20/717	S0032
NS6211A	159	3114			
NS6212	154	3045		TR78/712	S0005
NS6213					S0025
NS6214					S0025
NS6241		3114	22		S5022
NS7261	108	3018	11		S0015
NS7262	123A	3018	11		S0015
NS7267	108	3018	11		S0025
NS9001					S3002
NS9004					S3001
NS9400	128	3024	63	TR87/243	
NS9420	128	3024	63	TR87/243	
NS9500	128	3024	63	TR87/243	
NS9540	128	3024	63	TR87/243	
NS9608					S0025
NS9609					S0025
NS9710	108	3039	11	TR95/56	S0016
NS9726					S0015
NS9728	128	3024	17	TR87/243	S0025
NS9729	128	3024	17	TR87/243	S0015
NS9730	128	3024	17	TR87/243	S0015
NS9731	128	3024	17	TR87/243	S0015
NSQ8050L2	716				
NSQ8050L2-W2	716				
00415	159				
01F-SL8020	703A				
0-I5551	194				
OC3H,K,L					G0005
OC3LP,LR,N					G0005
OC4-0					G0005
OC4H,K,L					G0005
OC4LP,LR					G0006
OC4N					G0005
OC5-0					G0006
OC5K,L,LP,LR					G0006
OC5N					G0006
OC13					G0005
OC16	104		3	TR01/230	G6003
OC-16	121	3009		TR01/	
OC19	121	3009	16	TR01/232	
OC20	121	3009	25	TR01/232	G6017
OC22	104		16	TR01/	G6003
OC-22	121	3009		TR01/	
OC23	104		16	TR01/230	G6003
OC-23	121	3009		TR01/	
OC24	104		16	TR01/230	G6003
OC-24	121	3009		TR01/	
OC25	104		25	TR01/230	G6003
OC-25	121	3009		TR01/	
OC26	104		25	TR01/230	G6003
OC-26	104	3009		TR01/	
OC27	104	3009	25	TR01/230	G6003
OC28	104		25	TR01/230	G6003
OC-28	121	3009		TR01/	
OC29	104		25	TR01/230	G6003
OC-29	121	3009		TR01/	
OC30	131	3009	30	TR94/642	G6016
OC30A	131	3198	30	TR94/642	G6016
OC-30A	121	3009		TR01/	
OC30B	131	3082	30	TR94/642	
OC32	102				G0005
OC33	102A		53	TR85/250	G0009
OC34	102A		52	TR85/250	G0005
OC-34	102A	3004			
OC35	104		25	TR01/230	G6003
OC-35	121	3009		TR01/	
OC36	104		25	TR01/230	G6003
OC-36	121	3009		TR01/	
OC38	102A		52	TR85/250	G0005
OC-38	102A	3004			
OC40	126				G0001
OC41	102	3004	2	TR05/631	G0005
OC41A	102A	3004			
OC41N	126		2		G0009
OC42	126	3006	2	TR17/635	G0008
OC42N	126				G0009
OC43	126	3006	2	TR17/635	G0008
OC43N	126				G0009
OC44	126		50	TR17/635	G0008
OC-44	100	3005			
OC44N	126		50		G0008
OC45	102A	3123	50	TR85/250	G0008
OC-45	100	3005			
OC45N	126		50		G0008
OC46	102A	3123	2	TR85/250	G0005
OC-46	100	3005			
OC46N	126				G0001
OC47	102A		51	TR85/250	G0005
OC-47	100	3005			
OC47N	126				G0001
OC50	126				G0008
OC53	160	3123	52	TR17/637	G0008
OC54	160	3123	52	TR17/637	G0008
OC55	160	3123	52	TR17/637	G0008
OC56	102A	3123	52	TR85/250	G0008
OC57	102A		51	TR85/250	G0005
OC58	102A	3003	51	TR85/250	G0005
OC59	102A	3123	51	TR85/250	G0005
OC60	102A	3003	51	TR85/250	G0005
OC65	102A	3123	51	TR85/250	G0005
OC-65	102A	3004			
OC66	102A	3123	51	TR85/250	G0005
OC-66	102A				
OC70	102A	3123	52	TR85/250	G0005
OC-70	102A	3003			
OC70N	102A		53	TR85/250	
OC71	102A	3123	53	TR85/250	G0005
OC-71	102A	3004			
OC71A	102A		2	TR85/250	G0005
OC-71A	102A	3004			
OC71N	102A	3123	53	TR85/250	G0005
OC-71N	102A	3004			
OC72	102A	3003	53	TR85/250	G0005
OC-72	102A	3003			
OC73	102A		53	TR85/250	G0005
OC-73	102A	3003			
OC74	102A	3123	53	TR85/250	G0005
OC-74	102A	3003			
OC74N	102A		53	TR85/250	G0005
OC75	102A		53	TR85/250	G0005
OC-75	102A	3003			
OC75N	102A		53	TR85/250	G0005
OC-75N	102A	3004			
OC76	102A	3004	53	TR85/250	G0005
OC-76		3004			
OC77	102A		2	TR85/250	G0005
OC-77	102A	3004			
OC77M	102A	3123	52	TR85/250	G0005
OC79	102A		53	TR85/250	G0005
OC-79	102A	3004			
OC80	102A			TR85/250	G0005
OC81	102A	3004	52	TR85/250	G0005
OC81D	102A	3004	52	TR85/250	G0005
OC81DD	102A	3004	2	TR85/250	
OC-81DD	102A	3004			
OC81DM					G0005
OC81DN			53		
OC81M					G0005
OC81N			53		
OC82					G0005
OC83	158		53	TR84/630	G0005
OC83N			53		
OC84	158		53	TR84/630	G0005
OC84N			53		
OC85					G0005
OC110	102A		53	TR85/250	
OC120	102A		53	TR85/250	
OC122	102A		53	TR85/250	G0005
OC123	102A	3123	2	TR85/250	G0005
OC130	126	3006	2	TR17/635	G0008
OC-130	100	3005			
OC139	101	3011	8	TR08/254	G0011
OC140	101		7	TR08/254	G0011
OC-140	100	3005			
OC141	101	3011	54	TR08/254	G0011
OC169	160		51	TR17/637	G0003
OC-169	160	3006	51		
OC169R	160	3006	51	TR17/637	
OC170	160		50	TR17/637	G0003
OC-170	160	3006			
OC170N	160		51	TR17/637	
OC170R	160	3006	51	TR17/637	
OC170V	160	3006	51	TR17/637	
OC171	160		50	TR17/637	G0003
OC-171	160	3006	51		
OC171N	160			TR17/637	
OC171R	160	3006	51	TR17/637	
OC171V	160	3006	51	TR17/637	
OC200	159	3114	65	TR20/717	S0013
OC201	159	3114	65	TR20/717	S0013
OC202	159	3114	65	TR20/717	S0013
OC203	159	3114	65	TR20/717	S0013
OC204	159	3114	82	TR20/717	S0013
OC205	159	3114	82	TR20/717	S0013
OC206	159	3114	82	TR20/717	S0013
OC207	159	3114	82	TR20/717	S0013
OC302	102A		53	TR85/250	G0005
OC303	102A		52	TR85/250	G0005
OC304	102A	3123	52	TR85/250	G0005
OC-304	176	3004			
OC304-1	102A	3004	52	TR85/250	
OC-304/1	158				G0005
OC304-2	102A		52	TR85/250	
OC-304/2	158				G0005
OC304-3	102A		52	TR85/250	
OC-304/3	158				G0005
OC304N	102A		53	TR85/250	
OC305	102A	3004	52	TR85/250	G0005
OC305-1	102A	3004	52	TR85/250	

Original Part Number	ECG	SK	GE	IR/WEP	HEP
OC-305/1	158				G0005
OC305-2	102A	3004	52	TR85/250	
OC-305/2	158				G0005
OC306	102A		52	TR85/250	G0002
OC306-1	102A		52	TR85/250	
OC-306/1	158				G0005
OC306-2	102A		52	TR85/250	
OC-306/2	158				G0005
OC306-3	102A		52	TR85/250	
OC-306/3	158				G0005
OC307	102A	3123	2	TR85/250	G0005
OC-307	102A	3004			
OC307-1	102A	3123	2	TR85/250	G0005
OC307-2	102A	3123	2	TR85/250	G0005
OC307-3	102A	3123	2	TR85/250	G0005
OC308	102A	3123	2	TR85/250	G0005
OC-308	102A	3004			
OC309	102A	3123	52	TR85/250	G0005
OC309-1	102A	3123	2	TR85/250	G0005
OC309-2	102A	3123	2	TR85/250	G0005
OC309-3	102A	3123	2	TR85/250	G0005
OC318	102A		52	TR85/250	G0005
OC-318	102A	3003			
OC320	160		51	TR17/637	G0009
OC330	102A	3123	52	TR85/250	G0005
OC-330	102A	3004			
OC331	126	3006	50	TR17/635	G0003
OC340	102A	3123	52	TR85/250	G0005
OC-340	102A	3004			
OC341	126		50	TR17/635	G0003
OC-341	102A	3004			
OC342	126		50	TR17/635	G0003
OC-342	102A	3004			
OC343	126		50	TR17/635	G0003
OC-343	102A	3004			
OC350	102A	3123	52	TR85/250	G0005
OC-350	102A	3004			
OC351	126		50	TR17/635	G0003
OC-351	102A	3004			
OC360	102A		52	TR85/250	G0005
OC-360	102A	3004			
OC361	126		50	TR17/635	G0003
OC-361		3004			
OC362	126		50	TR17/635	G0003
OC-362	102A	3004			
OC363	126		50	TR17/635	G0003
OC-363	102A	3004			
OC364	102A		51	TR85/250	G0005
OC-364	102A	3004			
OC390	126		50	TR17/635	G0003
OC-390	126	3008			
OC400	160		51	TR17/637	G0001
OC410	160		50	TR17/637	G0002
OC-410	100	3005			
OC430	159	3114	21	TR20/717	S0013
OC430K	159	3114	21	TR20/717	S0013
OC440	159	3114	21	TR20/717	S0013
OC440K	159	3114	21	TR20/717	S0013
OC443,K	159	3114	21	TR20/717	S0013
OC445	159	3114	21	TR20/717	S0013
OC445K	159	3114		TR20/717	S0013
OC449	159	3114	21	TR20/717	S0013
OC449K	159	3114		TR20/717	S0013
OC450	159	3114	21	TR20/717	S0013
OC460,K	159	3114	21	TR20/717	S0013
OC463,K	159	3114	21	TR20/717	S0013
OC465,K	159	3114	21	TR20/717	S0013
OC466,K	159	3114	21	TR20/717	S0013
OC467,K	159	3114	21	TR20/717	S0013
OC468,K	159	3114	21	TR20/717	S0013
OC469,K	159	3114	21	TR20/717	S0013
OC470,K	159	3114	21	TR20/717	S0013
OC601	102A		53	TR85/250	G0005
OC602	102A		52	TR85/250	G0005
OC-602	102A	3004			
OC602SP	102A			TR85/250	G0005
OC602SQ	102A			TR85/250	
OC603	102A		50	TR85/250	G0005
OC-603		3004			
OC604	102A		2	TR85/250	G0005
OC-604	102A	3004			
OC604SP	102A			TR85/250	G0005
OC612	126		50	TR17/635	G0002
OC-612		3008			
OC613	126		50	TR17/635	G0008
OC-613	126	3008			
OC614	126		51	TR17/635	G0008
OC-614	126	3008			
OC615	126		51	TR17/635	G0008
OC-615	160	3006			
OC615N	160			TR17/637	
OC615V					G0008
OC700	159	3114	21	TR20/717	S0032
OC700A,B	159	3114	21	TR20/717	S0032
OC702	159	3114	21	TR20/717	S0032

Original Part Number	ECG	SK	GE	IR/WEP	HEP
OC702A,B	159	3114	21	TR20/717	S0013
OC704	159	3114	21	TR20/717	S0013
OC711	102A	3004	2	TR85/250	
OC740	159	3114	21	TR20/717	S0032
OC740G,M	159	3114	21	TR20/717	
OC7400	159	3114	21	/717	
OC742	159	3114	21	TR20/717	S0032
OC742G,M	159	3114	21	TR20/717	
OC7420	159	3114	21	/717	
OC810	102A		53	TR85/250	G0005
OC811					G0005
OC975					G0009
OC-975	126				
OC7400				TR20/	
OC7420				TR20/	
OC6015					G0002
OD603					G6003
OD603/50					G6003
OD604					G6003
OD605					G6003
OF129	102A			TR85/	
OF-129	102A	3004	52	TR05/	
ON0161646		3004			
ON0161705		3004			
ON047204-2	102A				
ON097159		3024			
ON174	160	3006	51	TR17/637	G0008
ON271	123A	3124	20	TR21/735	S0015
ON274	123A	3124		TR21/735	S0015
ON161705		3004			
ON47204-1	123A				
ONO-056748-1		3024			
OOV60529	128	3024			
ORF-2	128	3024			
OS536G			10	TR24/	
OX3003					G0005
OX3004					G0005
P00347100	159				
P00347101	159				
P04-44-0012		3005			
P04-44-0013		3007			
P04-44-0028	123A				
P04-44-0028-3		3122			
P04-44-0028-7		3122			
P04-45-0014-P2	123A				
P04-45-0014-P5	123A				
P04-45-0014PT		3122			
P04-45-0014PT2		3122			
P04-45-0015-P1	159				
P04-45-0016-P1	159				
P04-45-0023		3027			
P04-45-0023-P6		3027			
P04-45-0026		3024			
P04-45-0026-P5	128				
P04-45-0028-P2		3505			
P04-45-0034-1		3036			
P04-45-0034-2		3036			
P04-45-0037-0		3036			
P04-450016-002	159				
P0445-0034-1	181				
P0445-0034-2	181				
P04440028-001	123A				
P04440028-009	123A				
P04440028-014	123A				
P04440028-8	123A				
P04440032-001	123A				
P04450016-004	159				
P04450026P5	128				
P04450032-1	218				
P04450034-1	181				
P04450034-2	181				
P04450037	181				
P04450040-002	223	3510			
P1A	121	3009		TR01/232	G6003
P1B	159	3004	21	TR20/717	S0019
P1C	159	3114	21	TR20/717	
P1CG	159			TR20/	
P1D	159	3114		TR20/717	S0019
P1E	219		16	TR29/	S7003
P1E-1	121	3009	16	TR01/232	
P1E-1BLK	219				S7003
P1E-1BLU	219				S7003
P1E-1GRN	219				S7003
P1E-1RED	219				S7003
P1E-1VIO	219				S7003
P1E-2BLK	219				S7003
P1E-2BLU	219				S7003
P1E-2GRN	219				S7003
P1E-2RED	219				S7003
P1E-2VIO	219				S7003
P1E-3BLK	219				S7003
P1E-3BLU	219				S7003
P1E-3GRN	219				S7003
P1E-3RED	219				S7003
P1E-3VIO	219				S7003

Original Part Number	ECG	SK	GE	IR/WEP	HEP
P1F	127	3035	25	TR27/235	G6008
P1G	121	3034	25	TR01/232	G6005
P1H	159	3025	21	TR20/717	S0013
P1J	159	3025	21	TR20/71	S0019
P1J/M4943				TR19/	
P1K	121	3009	16	TR01/232	G6013
P1K48S134947				TR01/	
P1KBLK	121	3009		TR01/232	G6013
P1KBLU	121	3009		TR01/232	G6013
P1KBRN	121	3009		TR01/232	G6013
P1KGRN	121	3009		TR01/232	G6013
P1KORN	121	3009		TR01/232	G6013
P1KRED	121	3009		TR01/232	G6013
P1KYEL	121	3009		TR01/232	G6013
P1L	102A	3006	53	TR85/250	G0006
P1L4956	102A			TR85/	
P1L/4956		3006		TR85/250	
P1M	129			TR88/242	S5023
P1N	159	3114		TR20/717	S0019
P1N-1	159	3114		TR20/717	S0019
P1N-2	159	3114		TR20/717	S0019
P1N-3	159	3114		TR20/717	S0019
P1P	159	3114		TR20/717	S0031
P1P-1	159	3114	21	TR20/717	S0031
P1R	104	3009	16	TR01/230	G6003
P1T	121	3009		TR01/232	G6018
P1V	185	3041	69	/S5007	S5006
P1V-1	185			/S5007	
P1V-2	185	3041		/S5007	S5006
P1V-3	185	3041		/S5007	S5006
P1V-4	185	3041		TR77/700	S5006
P1W	159	3025	21	TR20/717	S0019
P1Y	127	3035	25	TR27/235	G6008
P2A	159	3114		TR20/717	S0019
P2B	153				G0005
P2C	104				G6003
P2D	121	3009	16	TR01/232	G6013
P2DBLU	121	3009		TR01/232	G6013
P2DBRN	121	3009		TR01/232	G6013
P2DGRN	121	3009		TR01/232	G6013
P2DORN	121	3009		TR01/232	G6013
P2DRED	121	3009		TR01/232	G6013
P2DYEL	121	3009		TR01/232	G6013
P2E	159	3025	21	TR20/717	S0019
P2F	176			TR82/238	G6011
P2G	159	3114		TR20/717	S0019
P2GE	159			TR20/	
P2H	159				S0019
P2J	180				S7001
P2K	153	3083		TR77/700	S5006
P2L	159				S0019
P2M-1-2-3	159	3114		TR20/717	S0013
P2N					S3003
P2P	159	3114		TR20/717	S0019
P2R	104	3009		TR01/230	G6003
P2S	159	3114	21	TR20/717	S0019
P2S/M7127				TR19/	
P2T	185	3041	29	TR77/S5007	S5006
P2T-1	185	3041		/S5007	S5006
P2T-2-3-4	185	3083		/S5007	S5006
P2U	189	3084		TR73/3031	S3028
P2U-1	211	3203			S3028
P2U-2	211	3084			S3028
P2U-3					S3028
P2U/48-137160				TR73/	
P2V	189	3084	29	TR73/3031	S3028
P2W	159	3025	21	TR20/717	S0019
P2Y	159	3114		TR20/717	S0019
P2Z	179			TR35/G6001	G6017
P3A	197				S5005
P3B	158			TR84/630	G0005
P3C	159	3114		TR20/717	S0019
P3CA	159	3114			
P3D	102A			TR85/250	G0005
P3E	131			TR94/642	
P3EBLK	104	3009		TR01/230	G6003
P3EBLU	104			TR01/230	G6003
P3EGRN	104			TR01/230	G6003
P3ERED	104			TR01/230	G6003
P3H	127			TR27/235	G6008
P3J	127			TR27/235	G6007
P3K	189				S3032
P3M	185	3083		/S5007	S5006
P3N	185	3083		/S5007	
P3N-1-2-3-4	185	3083		TR77/S5007	S5006
P3N-5	183	3189			S5005
P3P		3083	69	TR56/	S5006
P3P-1-2-3-4-5	185	3083			S5006
P3R	131	3082		TR94/642	G6016
P3R-1	131	3082		TR94/642	G6016
P3R-2	131	3082	30MP	TR94/642	G6016
P3R-3	131	3052	30	TR94/642	G6016
P3R-4	131	3082		TR94/642	G6016
P3S	185	3191			S5006
P3T	131	3052		TR50/	G6016
P3T-1-2	131	3198			G6016
P3U	197	3083	69	TR77/	S5005
P3UA	183	3189			
P3V	185	3083	58	TR77/	S5006
P3W	219				S7003
P3Y	189	3118		TR30/	S3032
P3Z	159				S0019
P4A					S5006
P4B	193	3118		TR30/	S0019
P4C	159				S0019
P4D	104				G6003
P4E	185	3083	29	TR77/	S5005
P4E-1-2-3-4	183	3189			S5005
P4E/48-137331				TR77/	
P4F	127	3035	25	TR27/	G600[illegible]
P4G	159				S001
P4H	127				G6008
P4J		3083	29	TR77/	S5006
P4J1+8		3083	29		S5006
P4J/48-137370				TR77/	
P4K	159				S0019
P4L	104				G6003
P4M	121				G6005
P4N	104				G6003
P4P	159				S0019
P4R	159				S0019
P4S	185	3191			S5006
P4T	183	3189			S5005
P4U	185	3191			S5006
P4V	183	3189	29		S5006
P4V-1	183	3189		TR77/	S5006
P4V-2	183	3189			S5006
P4W			69	TR77/	S5006
P4W-1-2	185	3191			S5006
P4W/48-137507				TR77/	
P4Y	159				S0019
P4Z	128				S0015
P5B		3025			S0019
P5C		3114	67	TR54/	
P5D	159	3118	65	TR52/	
P5D/48-137504				TR52/	
P5E					S0019
P5F		3052	30	TR50/	S5006
P5F/48-137527				TR50/	
P5G					S5013
P5H				TR77/	S5006
P5J					S0031
P5K					S5006
P5K-1-2					S5006
P5L					S5006
P5L-1-2					S5006
P5M					S5005
P5N					G6003
P5P					G6005
P13T1	6402				
P218-1	188	3199		TR72/3020	S5000
P218-2	189	3200		TR73/3031	S5006
P247(185936)				TR30/	
P346	108	3039		TR95/56	S0016
P480A0018	128				
P480A0022	159				
P480A0023	159				
P480A0027	159				
P480A0028	123A				
P480A0029	123A				
P1087		3112	FET1		F1035
P1901-48	199				
P1901-50	123A				
P1901-70	159				
P2015	727				
P2271	130	3027		TR59/247	
P3139	130	3027	14	TR59/247	S7002
P-3139B,OR,V,W,Y		3510			
P-3139D		3027			
P-3139BL,GR		3510			
P3172	175	3131	12	TR81/241	
P4069	128				
P5034	130	3510			
P5148	175			TR81/	
P5149	130	3510		TR59/	
P-5149B		3027			
P5152	123A	3124	20	TR21/735	
P5153	123A	3124	20	TR21/735	
P6022A	175				
P6128	175				
P6500A	181	3036			
P6786	128				
P6804	175				
P6940		3024			
P8393	123A	3122	20	TR21/735	
P-8393	123A				
P8394	123A	3122	20	TR21/735	
P8870	121	3009	16	TR01/232	
P8890	121	3014		TR01/232	
P8890A	121	3014		TR01/232	

Original Part Number	ECG	SK	GE	IR/WEP	HEP
P8890L				TR35/G6001	
P9623		3122			
P9962-1-2-4-5	128				
P10619-1	130				
P-10954-1	130				
P-10954-2	130				
P-11748-1	128				
P-11810-1	181				
P-11810-1B		3036			
P-11901-1	130				
P-11901-3	181				
P-11903-1	128				
P-11907-1C		3027			
P12407-1	234				
P15153	123A				
P31898	121			TR01/232	
P-31898	121	3009		TR01/	
P50200-11	130				
P50913/007		3024			
P64447	123A				
P69941	199				
P75534	121	3009	16	TR01/232	
P75534-1	121	3009		TR01/	
P75534-1D		3009			
P75534-2				TR01/	
P75534-3				TR01/	
P75534-4	121			TR01/	
P75534-5	121			TR01/	
P633024	128	3024			
P633024C(SYL)	128			TR87/243	
P633024G	128	3024		TR25/	S5014
P633567	123A				
P6450025		3011			
P6450026	128				
P6460006	102				
P6460037	102				
P6480001	121				
P/FTV/117	152				
P/N14-603-02	161				
P/N297L010C01	234				
P/N10000020	123A				
P-T-30	121	3009		TR01/	
PA234	717				
PA239	721				
PA243	717				
PA277	732				
PA306	733				
PA-400	1090		723		
PA713					C6091
PA1000	159	3114	21	TR20/717	S0013
PA1001	159	3114	21	TR20/717	S0013
PA1001A	159	3114			
PA7000/591	909D	3590			
PA7001/0001	123A				
PA7001/591	909	3590			
PA7001/502	909D	3590			
PA7001/503	941				
PA7001/505	923				
PA7001/518	7430				
PA7001/519	7420				
PA7001/520	7410				
PA7001/521	7400				
PA7001/522	7440				
PA7001/523	7451				
PA7001/524	7453				
PA7001/525	7402				
PA7001/526	7401				
PA7001/527	7404				
PA7001/528	7405				
PA7001/529	7474				
PA7001/530	7470				
PA7001/531	7473				
PA7001/532	7472				
PA7001/533	7460				
PA7001/539	7454				
PA7001/591	909D				
PA7001/593	74145				
PA7601					C6091
PA7703		3157			C6001
PA7703E	703A	3157			
PA7703X	703A	3157			C6001
PA7709	909	3590			
PA7709C	909	3590			
PA7710-31	910				
PA7710-39	910				
PA7711-31	911				
PA7741	941	3514			
PA7741A,B		3514			
PA7741C	941	3514			
PA8260	199		62		
PA8543	199		62		
PA8900	186		28		
PA9004	199		62		
PA9005	199		62		
PA9006	123A		10		
PA9154	160		50		
PA9155	160		50		
PA9156	102A		52		
PA9157	102A		52		
PA9158	158		53		
PA9483	192		63		
PA-10556	108				
PA10880	160		51		
PA10889-1	158	3123	52	TR84/630	G0005
PA-10889-1	121	3009		TR01/	
PA10889-2	158		2	TR84/630	G0005
PA-10889-2	121	3009		TR01/	
PA10890	104		16	TR01/230	G6003
PA-10890	121	3009		TR01/	
PA10890-1	121		16	TR01/232	
PA-10890-1	121	3009		TR01/	
PADT20	160	3006	52	TR17/637	G0003
PADT21	160	3006	52	TR17/637	G0003
PADT22	160	3006	52	TR17/637	G0002
PADT23	160	3005	2	TR17/637	G0003
PADT24	160	3006	2	TR17/637	G0003
PADT25	160	3006	2	TR17/637	G0003
PADT26	160	3006	2	TR17/637	
PADT27	160	3006	2	TR17/637	G0002
PADT28	160	3006	52	TR17/637	G0003
PADT30	160	3006	52	TR17/637	G0003
PADT31	160	3006	52	TR17/637	G0003
PADT35	160	3123	52	TR17/637	
PADT40	160		52	TR17/637	G0008
PADT50	104	3009	16	TR01/230	G6005
PADT51	160	3006	52	TR17/637	
PAR12	121			TR01/232	G6005
PAR-12	121	3009	16	TR01/	
PB110	121	3009	3	TR01/232	G6005
PBC107	199		212		
PBC107A,B	199		212	TR51/	
PBC108	199		212		
PBC108A,B	199		212	TR51/	
PBC108C	199		212		
PBC109	199		212		
PBC109B	199		212	TR51/	
PBC109C	199		212		
PBC182	108		210		
PBC183	123A		210		
PBC184	192		210		
PBE3014-1-2	158	3004	2	TR84/630	G0005
PBE3020-1	158	3004	2	TR84/630	G0005
PBE3020-2	103A	3124	20	TR08/724	G0011
PBE3162	158	3004		TR84/630	G0005
PBE3162-1	102A	3004		TR85/250	
PBE3162-2	102A	3004	2	TR85/250	
PBX103	102				
PBX113	102				
PC554	1142				
PC1066T	158		53		
PC1067T	158		53		
PC1068T	158		53		
PC3002	158		53		
PC3003	158		53		
PC3004	121		3		
PC3005	158		53		
PC3006	158		53		
PC3007	158		53		
PC3009	158		53		
PC3010	131		30		
PC4900-1	128				
PC7327C	9914				
PC-20003	1003				
PC-20004	1021				
PC-20005	1027	3153	721		
PC-20006	1024	3152	720		
PC-20008	720				
PC-20012	1030				
PC-20018	720		IC7		
PC20018	720				
PC-20023	1010				
PD9093-59	9093				
PD9094-59	9094				
PD9097-59	9097				
PD9099-59	9099				
PD9930-59	9930				
PD9932-59	9932				
PD9933-59	9933				
PD9935-59	9935				
PD9936-59	9936				
PD9937-59	9937				
PD9944-59	9944				
PD9945-59	9945				
PD9946-59	9946				
PD9948-59	9948				
PD9949-59	9949				
PD9950-59	9950				
PD9951-59	9951				
PD9961-59	9961				
PD9962-59	9962				

Original Part Number	ECG	SK	GE	IR/WEP	HEP
PD9963-59	9963				
PE3001	123A				S0015
PE3015			17		
PE5010			17		
PE5013			17		
PE5025	233		17		
PE9093-59	9093				
PE9094-59	9094				
PE9097-59	9097				
PE9099					C1052P
PE9099-59	9099				
PE9930					C1030P
PE9930-59	9930				
PE9932					C1032P
PE9932-59	9932				
PE9933					C1033P
PE9933-59	9933				
PE9935-59	9935				
PE9936-59	9936				
PE9937-59	9937				
PE9944					C1044P
PE9944-59	9944				
PE9945					C1045
PE9945-59	9945				
PE9946-59	9946				
PE9948-59	9948				
PE9949-59	9949				
PE9950-59	9950				
PE9951-59	9951				
PE9961-59	9961				
PE9962					C1062P
PE9962-59	9962				
PE9963-59	9963				
PEI6					S0011
PEP2	123A	3122		TR21/735	S0011
PEP5	123A	3122	63	TR21/735	S0011
PEP6	123A	3122	20	TR21/735	S0011
PEP7	123A	3122	20	TR21/735	S0011
PEP8	123A	3122	20	TR21/735	S0011
PEP9	123A	3122	210	TR21/735	S0011
PEP95	172		64	TR53/	
PEP1001	108	3039	17	TR95/56	
PEP2001	128	3024	17	TR87/243	
PET0404-1					S0019
PET0404-2					S0019
PET0414					S0012
PET-101-1	108	3018	20		
PET1001	192	3137	210		
PET1002	123A	3122	210		S0025
PET1075	108	3039	18	TR95/56	
PET1075A	154			TR78/712	
PET2001	123A	3122	210	TR63/	
PET2002	123A	3122	210		
PET3001	108	3018	61	TR95/56	S0020
PET3703					S0012
PET3704	123A	3122	20	TR	
PET3705	123A	3122	20	TR65/	
PET3706	123A	3122	20		
PET3903					S0025
PET3904					S0015
PET3905					S0019
PET3906					S0019
PET4001	128	3024	62	TR87/243	S0015
PET4002	123A	3122	62		S0015
PET4003	199	3156	62		S0015
PET4059					S0013
PET4060					S0013
PET4061					S0019
PET4123					S0025
PET4124					S0024
PET4125					S0019
PET4126					S0019
PET6001	123A	3122	20	TR65/	S0025
PET6002	123A	3122	20		S0025
PET6003					S0024
PET8000	123A	3122	20		S0015
PET8001	123A	3122	47		S0030
PET8002	123A	3122			S0015
PET8003	123A	3122	47		S0030
PET8004	123A	3122			S0030
PET8005					S0025
PET8005A					S0025
PET8006,A					S0015
PET8007,A					S0025
PET8101				TR24/	S0025
PET8200					S0020
PET8201	108	3039	20	TR95/56	S0020
PET8202					S0015
PET8203					S0015
PET8204					S0030
PET8250	108	3039	61		S0025
PET8251	108	3039	61		S0015
PET8300	108	3039	21		S0019
PET8301					S0019
PET8302					S0019
PET8303					S0013
PET8304					S0013
PET8350					G6011
PET8351					G6011
PET8352					G6011
PET8353					G6011
PET9001					S0025
PET9002	123A	3046		TR21/735	S0011
PET9002A					S0011
PET9021	128	3024	17	TR87/243	S0025
PET9022	128	3024	17	TR87/243	S0025
PF-AR15	130				
PF-AR18	175				
PF143/60A		3007			
PG33024G	128	3024			
PI-10,131	159				
PIC	159	3004	22		
PIK	121		16		
PIL/4956	160		50	TR17/	
PIT-37	199				
PIT-50	159				
PIT-74	128				
PIT-79	159				
PIT-81	159				
PIV	197	3083			
PIV-1-2-3	197	3083			
PL-172-010-9-001		3054			S5000
PL-172-013-9-001		3054			
PL-172-014-9-001	152	3054	28		S3024
PL-172-014-9-002	224				
PL-172-024-9-001	186			TR55/	
PL-172-024-9-002	235				
PL-172-024-9-003	236				
PL-176-025-9-001	199	3122	20		S0015
PL-176-026-9-001		3018			
PL-176-029-9-002		3018			S3001
PL-176-031-9-001		3122			
PL-176-031-9-002		3020			
PL-176-037-9-001		3018			
PL-176-037-9-002		3024			
PL-176-042-9-001	107	3018	17		S0014
PL-176-042-9-002	123A	3018	20	TR21/	S0015
PL-176-042-9-003	107	3018	17		S0014
PL-176-042-9-004	123A	3122	20		S0025
PL-176-042-9-005	186	3054	28		S5000
PL-176-042-9-006	123A	3124	17		S0014
PL-176-047-9-001	123A			TR21/	
PL-176-049-9-002	107			TR24/	
PL-177-006-9-001		3114			
PL-177-006-9-002		3114			S0012
PL-182-009-9-001		3116			
PL-182-014-9-002	133	3112	FET1	FE100/	
PL-307-047-9-002	724				
PL1021	161	3132	60		
PL1022	161	3132	60		
PL1023	161	3132	60		
PL1024	107	3039	214		
PL1025	107	3039	214		
PL1026	107	3039	60		
PL1031	159	3025	65		
PL1032			82		
PL1033	159	3025	65		
PL1034	159	3025	82		
PL1051	108	3039	212		
PL1052	123A	3122	212		
PL1053	108	3039	212		
PL1054	123A	3122	212		
PL1055	108	3039	212		
PL1061	108	3039	39		
PL1062	108	3039	39		
PL1063	108	3039	81		
PL1064	108	3039	39		
PL1065	108	3039	81		

Original Part Number	ECG	SK	GE	IR/WEP	HEP
PL1066	161	3132	60		
PL1067	161	3132	60		
PL1081	108	3039	212		
PL1082	108	3039	212		
PL1083	128	3024	212		
PL1084	128	3024	212		
PL1085	171	3103	27		
PL1091	132	3116			
PL1092	132	3116			
PL1093	132	3116			
PL1094	132	3116			
PL1101	159	3025	65		
PL1102	159	3025	65		
PL1103	159	3025	65		
PL1104	159	3025	65		
PL1111	161	3132	60		
PL1112	161	3132	60		
PL1113	107	3039	60		
PL4021			20		
PL4031			21		
PL4032			21		
PL4033			21		
PL4034			21		
PL4051			20		
PL4052			20		
PL4053			20		
PL4054			20		
PL4055			20		
PL4061			212		
PL4062			212		
PL991021					C2006G
PL991023					C2006G
PL991029					C2006G
PL991121					C2007G
PL991123					C2007G
PL991129					C2007G
PL991429					C2010G
PLE37	185			/S5007	
PLE-48	152	3054	66		
PLE52	184			/S5003	
PL252	184			/S5003	
PM194	107	3039	17	TR70/720	S0016
PM195	107	3018	20	TR70/720	S0016
PM195A	108	3018			
PM1120	234				
PM1121	123A				
PM12066		3005			
PMC-QP0010	130				
PMC-QP0012	130				
PMC-QP0040	181				
PMC-QS-0280	129				
PMC-QS-0320	128				
PMC-QS-0400	128				
PMT020					S0016
PMT023					S0025
PMT025					S0005
PMT115					S0005
PMT116					S0025
PMT123					S0015
PMT124					S0015
PMT125					S0005
PMT211					S0014
PMT212					S5014
PMT213					S5014
PMT214					S5014
PMT215					S5014
PMT216					S0025
PMT218					S0005
PMT219					S0005
PMT220					S0016
PMT221					S0025
PMT222					S0025
PMT223					S0025
PMT224					S0025
PMT225					S0005
PMT1767	108	3039		TR95/56	S0016
PMT1767M,P,T					S0011
PMT1787M,P,T					S0011
PN	159				
PN26	175	3538	66		S5019
PN66	152	3054	66		
PN70			21		
PN71			21		
PN72			22		
PN107			210		
PN108			210		
PN109			210		
PN350	130	3027			S7004
PN929			212		
PN930			212		
PN2221			47		
PN2222			47		
PN2369,A			81		
PN2484			81		
PN2904			21		

Original Part Number	ECG	SK	GE	IR/WEP	HEP
PN2905			21		
PN2906,A			21		
PN2907,A			21		
POWER12	121	3009	3	TR01/232	G6003
POWER25	121	3009	3	TR01/232	G6003
POWER40	105	3012	4	TR03/233	G6004
POWER60	105	3012	4	TR03/233	G6006
POWER80	105	3012	4	TR03/233	G6006
POWER99	121	3009	3	TR01/232	G6005
POWER299	121	3014	3	TR01/232	G6005
POWER500	105	3012		TR03/233	G6004
PP3000	130	3027	77		
PP3001	181	3535	75		
PP3003	130	3027	77		
PP3004	181	3535	75		
PP3006	130		77		
PP3007	181	3535	75		
PP3250	152	3054	66		
PP3310	152	3054	66		
PP3312	152	3054	66		
PPR1006	186	3054	28		
PPR1008	186	3054	28		
PPT720					S0014
PQ27	160	3006		TR17/637	G0003
PQ28	158	3004		TR84/630	G0005
PQ29	158	3004		TR84/630	G0005
PQ30	160	3006		TR17/637	G0003
PQ31	104	3009		TR01/230	G6003
PRT-101	123A	3124		TR21/735	
PRT101	154				S5024
PRT-104	123A	3124		TR21/735	
PRT-104-1	123A	3124		TR21/735	
PRT-104-2	123A	3124		TR21/735	
PRT-104-3	123A	3124		TR21/735	S5024
PRT-104-4	128	3024		TR87/243	
PS1	121		16	TR01/	G6005
PS-1	121	3009		TR01/	
PS-801	923D				
PS6010-1	199				
PS11327	124				
PS209800	123A				
PT06	121		16	TR01/	
PT0-139	102A				
PT0139	158				G0005
PT2A	160	3006	51	TR17/637	G0003
PT2S	160	3006	51	TR17/637	G0003
PT3	121	3009	16	TR01/232	
PT3A	121		16	TR01/232	
PT-3A	121	3009			
PT4	155				
PT4-7007		3124			
PT4-7059-001		3124			
PT4-7158	123A				
PT4-7158-01A	123A				
PT4-7158-02A	123A				
PT4-7158-012	123A				
PT4-7158-013	123A				
PT4-7158-021	123A				
PT4-7158-022	123A				
PT4-7158-023	123A				
PT6	121		16	TR01/232	G6003
PT-6	121	3009		TR01/	
PT12	121		16	TR01/232	G6003
PT-12	121	3009		TR01/	
PT25	121		16	TR01/232	G6003
PT-25	121	3009		TR01/	
PT30	121	3009	16	TR01/232	G6005
PT32	131				G6016
PT40	121		16	TR01/232	G6005
PT-40	121	3009		TR01/	
PT50	121		16	TR01/232	G6018
PT-50	121	3009		TR01/	
PT150	104		4	TR01/624	G6018
PT-150	121	3009	4	TR01/	
PT155	121		16	TR01/232	G6005
PT-155	121	3009		TR01/	
PT176	121	3009	16	TR01/232	G6005
PT201	105				G6004
PT234	121	3009	16	TR01/232	G6005
PT235	121	3009	16	TR01/232	G6005
PT235A	104		16	TR01/230	G6003
PT-235A	121	3009		TR01/	
PT236	104		16	TR01/230	G6003
PT-236	121	3009		TR01/	
PT236A	104		16	TR01/230	G6003
PT-236A	121	3009		TR01/	
PT236B	121	3009	16	TR01/232	G6005
PT236C	121	3009		TR01/232	G6005
PT242	121	3009	16	TR01/232	G6005
PT250	105		4	TR03/233	G6009
PT255	121	3009	16	TR01/232	G6005
PT256	121	3009	16	TR01/232	G6005
PT285	121	3009	16	TR01/232	G6005
PT285A	121	3009	16	TR01/232	G6005
PT301	121	3009	16	TR01/232	G6005

Original Part Number	ECG	SK	GE	IR/WEP	HEP
PT301A	121	3009	16	TR01/232	G6005
PT307	121	3009	16	TR01/232	G6005
PT307A	121	3009	16	TR01/232	G6005
PT366B	121		16		
PT501	121		4	TR01/232	G6006
PT-501	105	3012			
PT515	105		4	TR03/233	
PT515(SEMITRON)	213				G6010
PT530	102A	3005	1	TR85/250	
PT530A	102A	3005	1	TR85/250	G0005
PT554	104		16	TR01/230	G6003
PT-554	121	3009			
PT555	121	3009	16	TR01/232	G6005
PT600	186	3192	28	TR76/	
PT601	186	3192	28	TR76/	
PT612	128	3024	66	TR87/243	
PT627	123A	3124	20	TR21/735	
PT665	152		66		
PT703	123A	3122	62	TR21/735	S0011
PT720	123A	3122	210	TR21/735	
PT850	128	3024	63	TR87/243	S0005
PT850A	128	3024	63	TR87/243	S0005
PT851	123A	3122	20	TR21/735	
PT855	160	3046		TR17/637	S3013
PT856	160	3047	81	TR17/637	S3013
PT857	195	3048			
PT886	123A	3124	81	TR21/735	S0014
PT887	123A	3024	81	TR21/735	S0014
PT888	123A	3024	81	TR21/735	S0014
PT896	128	3024	27	TR87/243	S5014
PT897	123A	3122	81	TR21/735	S0014
PT898	123A	3122	27	TR21/735	S0014
PT900-1					S5004
PT1515			32		
PT1537	195	3048			
PT1544	128		215	TR25/	
PT1545	128		215	TR25/	
PT1558	123A	3122		TR21/735	
PT1559	123A	3122		TR21/735	
PT1610	123A	3122	63	TR21/735	
PT1835	123A	3122	63	TR21/735	S0011
PT1836	123A	3122	63	TR21/735	S0011
PT1837	123A	3122	63	TR21/735	S0005
PT1941	130		19	TR59/247	
PT2040A	123A	3122	20	TR21/735	
PT2523	171	3201	27	TR78/	S3021
PT2524	171	3201	27	TR78/	S3021
PT2525	171	3201	27	TR78/	S3021
PT2525A	198		32		S3021
PT2540	192	3192	63	TR63/	S3001
PT2620	186		28	TR76/	
PT2630			32		
PT2635	152		66		
PT2640	186	3192	32	TR76/	
PT2660	186	3192	215	TR76/	
PT2677C		3048			S3006
PT2760	123A	3122	210	TR21/735	S0011
PT2896	123A	3046		TR21/735	
PT3141	123A	3124	20	TR21/735	
PT3141A	123A	3046	20	TR21/735	
PT3141B	123A	3047		TR21/735	
PT3141C	195	3048			
PT3151A	123A	3046	20	TR21/735	S0015
PT3151B	123A	3047	20	TR21/735	S0015
PT3151C	123A	3048		TR21/735	S3001
PT3473					S3019
PT3500	123A	3122		TR21/735	
PT3501	123A	3122	20	TR21/735	S3011
PT3502	128	3024	28	TR87/243	
PT3503	186	3192	28	TR76/	
PT3540					S3005
PT3691					S3020
PT3760					S0025
PT4690	186	3192	28	TR76/	
PT4800	123A	3024	83	TR21/735	S0025
PT4816	107	3039	27	TR70/720	
PT4830	107	3039	27	TR70/720	S5014
PT5690					S5000
PT5693	152		66		
PT6618	186	3192	28	TR76/	
PT6669	186	3192	28	TR76/	
PT6696	186	3192	X18		
PT7930			75		
PT7931			75		
PT31961	195				S3001
PT31962					S3005
PT31963					S3006
PT-2677C					S3006
PT-31961					S3013
PT-31962					S3005
PT-31963					S3006
PTC101	128	3122	20	TR22/	S0015
PTC102	100	3005	1	TR05/	G0005
PTC103	159	3118	21	TR52/	S3012
PTC104	124	3021	12	TR23/	

Original Part Number	ECG	SK	GE	IR/WEP	HEP
PTC105	121	3009	16	TR01/	G6005
PTC106	105	3012	4	TR03/	G6004
PTC107	160	3006	51	TR17/	G0003
PTC108	101	3011	5	TR08/	G0011
PTC109	102		52		G0005
PTC110	184	3054	28	TR55/	S5000
PTC111	185	3083	29	TR56/	S5006
PTC112	175	3026	66	TR57/	S5019
PTC113	218		69	TR58/	S5018
PTC114	104	3014	16	TR16/	G6003
PTC115	123A	3018		TR21/	S0025
PTC116	181	3036		TR59/	S7002
PTC117	154	3044	27		S5024
PTC118	162		73	TR61/	S5020
PTC119	130	3027	77		S7004
PTC120	131	3052	30	TR50/	G6016
PTC121	123A	3122	61	TR51/	S0030
PTC122	127		25	TR27/	S5020
PTC123	194			TR53/	S0005
PTC124	228	3103	32	TR60/	S5015
PTC125	194	3024	32	TR62/	S0005
PTC126	161	3018	60		S0017
PTC127	194	3053	18		
PTC128				TR66/	S3006
PTC129	163		36	TR67/	S5021
PTC130	164	3133	37	TR68/	
PTC131	159	3118	67		S0019
PTC132	161	3132	61	TR21/	S0020
PTC133		3019			S0020
PTC134		3010			
PTC135		3004			
PTC136					S0025
PTC137		3054			S5004
PTC138		3014			G6013
PTC139		3124			S0015
PTC140		3027			S7004
PTC141		3025			S3032
PTC142					S5013
PTC143					S5014
PTC144		3024			S3019
PTC151	132	3112	FET1	FE100/	F0015
PTC152	132	3116	FET2		F0021
PTC153	172	3156	64	TR69/	S9100
PTC701	708	3135	IC10		C6062P
PTC703	709		IC11		C6062P
PTC-703	709				
PTC705	713	3077	IC5		C6057P
PTC707	707				C6067G
PTC708	705A	3134	IC3		C6089
PTC709	718	3159	IC8		C6094P
PTC711	719		IC28		C6065P
PTC713	720		IC7		C6095P
PTC715	714	3075	IC4		C6070P
PTC-715	714				
PTC717			IC14		C6080P
PTC719	715	3076	IC6		C6071P
PTC721	722	3161	IC9		C6056P
PTC723		3144			
PTC726	712	3072	IC2		C6063P
PTG132	161	3132			
PTO-6	121			TR01/232	
PTO139	102A	3123	52	TR05/250	
PTO-139		3004			
PXB103	102		2		
PXB-103	102A	3004		TR85/250	
PXB113	102		2		
PXB-113	102A	3004		TR85/250	
PXC101	102		2		
PXC-101	102A	3004		TR85/250	
PXC101A	102		2		
PXC-101A		3004		TR85/250	
PXC101AB	102		2		
PXC-101AB	102A	3004		TR85/250	
Q-00169	132	3116			
Q-00169A	132	3116			
Q-00169B	132	3116			
Q-00169C	132	3116		FE100/802	F0015
Q-00184R	132			FE100/802	F0015
Q-00269	108	3018			
Q-00269A, B	108	3018			
Q-00269C	107	3018		TR70/720	S0015
Q-00284R	108			TR95/56	S0016
Q-00284R-3	108	3018			
Q-00369	108	3018			
Q-00369A, B	108	3018			
Q-00369C	107	3018		TR70/720	S0015
Q-00384R	107			TR70/720	S0014
Q-00384R-3	108	3018			
Q-00469	123A	3020			
Q-00469A, B	123A	3124			
Q-00469C	123A	3124		TR21/735	S0015
Q-00484R	123A			TR21/735	S0014
Q-00484R-1	108	3018			
Q-00569	123A	3124			
Q-00569A, B	123A	3124			

Original Part Number	ECG	SK	GE	IR/WEP	HEP
Q-00569C	123A	3124		TR21/735	S0015
Q-00584R	107			TR70/720	S0014
Q-00584R-3	108	3018			
Q-00669	123A	3124			
Q-00669A,B	123A	3124			
Q-00669C	123A	3124		TR21/735	S0011
Q-00684R	123A	3124			S0015
Q-00769	175	3026			
Q-00769A,B	175	3026			
Q-00769C	175	3026		TR81/241	S5019
Q-00784R	123A	3124			S0014
Q-00869	128	3047			
Q-00869A,B	128	3047			
Q-00869C	123A	3047		TR21/735	S0025
Q-00969	128	3047			
Q-00969A,B	128	3047			
Q-00969C	128	3047		TR87/243	S3001
Q-00984R	159				S0013
Q-0172	128				
Q-0-172	128				
Q0415	159				
Q0-419	159				
Q-01069	195	3048			
Q-01069A,B	195	3048			
Q-01069C	128	3048		TR87/243	S3002
Q-01084R	131				G6016
Q-01115C	107	3018	10	TR24/	S0015
Q-01169	128	3024			
Q-01169A,B	128	3024			
Q-01169C	128			TR87/243	S5014
Q-01184R	123A				S0025
Q-01269	175	3026			
Q-01269B	175	3026			
Q-01269C	175	3026		TR81/241	S5019
Q-01284R					S3002
Q-01369C	1104				
Q-01384R					S3001
Q-01484R	724				
Q-02115C	107	3018	10	TR24/	S0015
Q-03115C	107	3018	10	TR24/	S0015
Q-04115C	107	3018	10	TR24/	S0015
Q-05115C	107	3018	10	TR24/	S0015
Q-06115C	107	3018	10	TR24/	S0015
Q-07115C	107	3018	10	TR24/	S0015
Q-08115C	199	3122	62	TR51/	S0015
Q-09115C	199	3122	62	TR51/	S0015
Q-10115C	159	3020	20	TR86/	S0015
Q-10484R	724				
Q-11115C	152	3054	66	TR76/	S5000
Q-12115C	152	3054	66	TR76/	S5000
Q-13115C	199	3122	62	TR51/	S0015
Q-14115C	107	3018	10	TR24/	S0015
Q-15115C	107	3018	10	TR24/	S0015
Q-16115C	123A	3047	20	TR62/	S0025
Q-17115C	190	3047	28	TR64/	S3020
Q-18115C	195	3048	45	TR65/	S3001
Q0V60526	102A	3004		TR85/	
Q0V60527	103A	3010			
Q0V60528	102A	3004		TR85/	
Q0V60529	123A	3020			
Q0V60530	123A	3020			
Q0V60537	103A	3010			
Q0V60538	123A	3004		TR85/	
Q1	102A	3124		TR85/250	G0005
Q-1	102	3005	2		
Q1-7C	102A	3004	2	TR85/250	
Q-1A	100	3005		TR05/254	G0006
Q2		3124	39		
Q-2	101	3011		TR08/	
Q2-7C	102A	3004	2	TR85/250	
Q2N406	102A	3004	2	TR85/250	
Q2N1526	126	3008	51	TR17/635	
Q2N2428	102A	3004		TR85/	
Q2N2613	102A	3004		TR85/250	
Q2N4105	103A	3010		TR08/724	
Q2N4106	158	3004		TR84/630	
Q2N4107	103A+158			/724,630	
Q3			39		
Q-3	101	3011		TR08/	
Q3-Q4		3024			
Q4	102A		39	TR85/250	G0005
Q-4	102	3005		TR05/	
Q5			39		
Q-5	101	3011		TR08/641	G0011
Q6	158			TR84/630	G0005
Q-6	102A	3004	52	TR05/	
Q7	158	3045		TR84/630	G0005
Q-7	102A	3004	52	TR05/	
Q8	158			TR84/630	G0005
Q-8	102A	3004	52	TR05/	
Q-9	101				G0011
Q16	158			TR84/630	G0005
Q-16	102A	3004	52	TR05/	
Q-35	123A	3122		TR21/735	S0015
Q-36	159	3114		TR20/717	S5022

Original Part Number	ECG	SK	GE	IR/WEP	HEP
Q-36A	159	3114			
Q205			IC9		
Q205(DYNACO)	722				
Q301	107			TR70/720	
Q514Z		3115			
Q591-8003		3024			
Q591-8080		3024			
Q591-8081		3024			
Q591-8082		3024			
Q5030	127	3035		/235	
Q5039	103A	3010		/724	
Q5044	160	3006		TR17/637	
Q5050	103A	3010		/724	
Q5053	123A	3124		/735	
Q5053D		3124			
Q5053E		3124	20	TR21/	S0015
Q5053F		3124	20	TR21/	S0015
Q5053G		3124			
Q5075ZMY		3021			S5011
Q5078Z		3132			
Q5083C		3079	37	TR61/	S5020
Q5087Z	159	3114			
Q5100A	129	3114	67	TR28/	S0019
Q5102P		3025			
Q5102Q		3025			S0032
Q5102R		3025			
Q5102Z					S5011
Q5104Z		3021	12	TR23/	
Q5104-Z		3021			
Q5110ZK					S5021
Q5111ZK	165	3115	38	TR67/	
Q5113ZMM		3021			
Q5116CA		3114			
Q5120P		3115			
Q5120Q		3133			
Q5120R		3111			
Q51210		3124			
Q5121Q		3025			
Q5121R		3124			
Q5134Z		3131			
Q5137BA				TR75/	S3019
Q5138L					S5015
Q5140ZXP		3115			
Q5140ZXQ		3111			
Q5140ZXR		3111			
Q5161Z		3115			
Q6010H		3520			
Q6015H		3520			
Q6025H		3521			
Q34450	124	3021	12	/240	
Q34451		3004			
Q35218	102A	3004		TR85/250	
Q35242	123A	3124		TR21/735	
Q35259	161	3018		TR83/719	
Q35524,B		3014			
Q36642		3027			
Q36643		3024			
Q40263	102A	3004		TR85/250	
Q40359	160	3008	51	TR17/637	G0002
Q51200Q					S5020
QA-1	160		51	TR17/637	
QA-3		3039			
QA-8	128	3024	18	TR87/243	S0015
QA-9	129	3025		TR88/242	S5013
QA-10	128	3024	18	TR87/243	S5014
QA11	124		21		
QA-11	129	3025		TR88/242	S5013
QA-12	123A	3122	20	TR21/735	
QA-13	123A	3122	20	TR21/735	
QA-14	123A	3122	20	TR21/735	S0015
QA-15	123A	3122	20	TR21/735	S0015
QA-16	123A	3122	27	TR21/735	S5026
QA-17	129	3025		TR88/242	S5013
QA-18	133	3116	FET1	/801	F0015
QA-19	123A	3122	20	TR21/735	
QA-20	133	3112	FET1	/801	F0010
QA-21	159	3114	21	TR20/717	
QA-21A	159	3114			
QA22				TR28/	
QA703E	703A	3157	IC12	IC500/	
QB4		3045			
QB400428	9914				
QG0074		3009	3	TR01/230	
QG0076	102A		2	TR85/250	
QG0254	123A	3122		TR21/735	
QN2613	102		2		
QOC61209				TR84/	
QOC61210				TR85/	
QOC61689				TR20/	
QOV60526	102A		2	TR85/250	
QOV60527	103A		8	TR08/724	
QOV60528	102A		2	TR85/250	
QOV60529	123A	3124	20	TR87/735	
QOV60530	123A	3124	20	TR21/735	
QOV60537	103A		8	TR08/724	

Original Part Number	ECG	SK	GE	IR/WEP	HEP
QOV60538	102A		2	TR85/250	
QOV61772				TR08/	
QP001200A	130				
QP0010		3027			
QP0020		3079			
QP0040		3036			
QP-1	179	3014		TR35/G6001	
QP1	121		16		
QP-1A	179	3014		TR35/G6001	
QP1A	121		16		
QP-2	179	3014		TR35/G6001	
QP2	121		16		
QP-3	179	3014		TR35/G6001	
QP-4	179			TR35/G6001	
QP-5	179			TR35/G6001	
QP-6	179	3009		TR35/G6001	
QP6	121		16		
QP-7	179	3009		TR35/G6001	
QP7	121		16		
QP-8	130	3027		TR59/247	S7004
QP8	130		14		
QP-8-1		3027		TR59/247	S7004
QP-8-P	233	3027		TR59/247	S7004
QP8-6623N	105	3012	4	TR03/230	
QP-8B,BL,BR,G		3027			
QP-8GR,OR,V,W,Y		3027			
QP-10	179			TR35/G6001	
QP-11	130	3027	14	TR59/247	
QP-12	130	3027	14	TR59/247	S7004
QP-13	185	3041	58	/700	S5006
QP-14	184	3041	57	/S5003	S5000
QP-31	197	3083	58	/701	
QQC60528				TR85/	
QQC60539				TR85/	
QQC61209	158			/630	
QQC61210	102A	3124		TR85/250	
QQC61689	159	3114		/717	
QQC61689A	159	3114			
QQV60526	158			/630	
QQV60527	103A			/724	
QQV60528	102A			TR85/250	
QQV60529	128	3024		TR87/243	
QQV60537	103A			/724	
QQV60538	158			/630	
QQV60539	102A			TR85/250	
QQV61772	103A			/724,630	
QR2378	102A	3004	2	TR85/250	
QRF-2	128		18	TR87/243	
QRF3	132				F0015
QRF-3	132	3116	FET2	FE100/802	
QRG-3	132			FE100/	
QS0254			10		
QS0280		3025			
QS0320		3024			
QS0400		3024			
QS054	123A	3122	10	TR21/735	
QS313					S0031
QS316	159				
QSE1001	123A	3018	20	TR21/735	S0030
QSE3001	108	3018	20	TR95/56	S0020
QSE5020	161	3018	20	TR83/719	S0017
R06-1001	108	3018		TR95/56	
R06-1002	108	3018		TR95/56	
R06-1003	108	3018		TR95/56	
R06-1004	108	3018		TR95/56	
R06-1005	108	3018		TR95/56	
R06-1006	108	3018		TR95/56	
R06-1007	102A	3004		TR85/250	
R06-1008	102A	3004		TR85/250	
R06-1009	102A	3004		TR85/250	
R06-1010	102A	3004		TR85/250	
ROG01					S0015
ROG02					G6005
ROG03					S3001
ROG04					S0011
ROG05					G0005
ROG06					G0003
R1N1102	722				
R-2SA222	160	3008		TR85/250	
R-2SB186	102A	3004		TR85/250	
R-2SB187	102A	3004		TR85/250	
R-2SB303	102A	3004		TR85/250	
R-2SB405	158	3004		TR84/630	
R-2SB474	226MP	3082		TR94MP/	G6016(2)
R-2SB492	176	3004	2	TR82/238	
R2SB492	102				
R-2SC535	107	3018		TR70/720	
R-2SC537	123A	3018		TR21/735	G6016
R-2SC545	108	3018		TR95/56	
R-2SC668	108	3018		TR95/56	
R-2SC772	107	3018		TR70/	S0016
R-2SC858	107	3124		TR70/	S0015
R-2SD187	103A	3010		TR08/724	
R6-0A	176	3123			
R12	101			TR08/	
R14	101			TR08/	
R15	124			TR81/	
R16	102A		52	TR85/250	G0005
R-16	102A	3004			
R23-1003	102A	3004		TR85/250	
R23-1004	102A	3004		TR85/250	
R24-1001	102A	3004	52	TR85/250	G0001
R24-1002	102A	3004	52	TR85/250	
R24-1003	102A	3004	52	TR85/250	G0005
R24-1004	102A	3004	52	TR85/250	G0005
R33	103A		8	TR08/724	
R-33	101	3011			
R34	103A		8	TR08/724	
R-34	101	3011			
R34-6016-58	123A				
R35	102A	3123	52	TR85/250	G0011
R36	176	3123			
R41	101			TR08/	
R46	102			TR05/	
R52	102A	3123		TR85/250	G0005
R56	102A		52	TR85/250	G0005
R-56	158	3004			
R60-1001	126	3008			G0008
R60-1002	160	3008		TR17/637	G0008
R60-1003	160	3008		TR17/637	G0008
R60-1004	102A	3004		TR85/250	G0005
R60-1005	102A	3004		TR85/250	G0005
R60-1006	102A	3004		TR85/250	G0005
R61	103A		8	TR08/724	
R62	103A		8	TR08/724	
R-62	101	3011			
R63	103A		8	TR08/724	
R-63	101	3011			
R64	102A	3123	52	TR85/250	G0005
R65	102A	3004	52	TR85/250	G0005
R66	102A	3123	52	TR85/250	G0005
R-66	102A	3004			
R67	102A	3123	52	TR85/250	G0005
R-67	102A	3004			
R79	103A			TR08/724	G0011
R80	103A			TR08/724	G0011
R83	102A	3123	52	TR85/250	
R-83	102A	3004			
R87	102A			TR85/250	G0005
R98	102A	3123		TR85/250	G0005
R100-1	102A			TR85/	
R100-8	102A			TR85/	
R100-9	102A			TR85/	
R101-2	102A			TR85/	
R101-3	102A			TR85/	
R101-4	102A			TR85/	
R102-41	121				
R104-5	126			TR17/	
R104-6	126			TR17/	
R104-7	126			TR17/	
R104-8	126			TR17/	
R105				TR12/	
R117	101	3011		TR08/641	
R118	107	3039	20	TR70/720	
R119	100	3005	2	TR05/254	G0005
R120	102A	3123	52	TR85/250	G0005
R-120	102A	3004			
R123	128		63	TR87/243	S5014
R123-2	128				
R123-3,5	128				
R125	101		8	TR08/641	
R-125	101	3011			
R135	101		8	TR08/641	
R-135	101	3011			
R135-1	130				
R136	101		8	TR08/641	
R-136	101	3011			
R137	101		8	TR01/641	
R-137	101	3011			
R152	102A	3123	52	TR85/250	G0005
R-152	102A	3004			
R163	100		2	TR05/254	G0005
R-163	100	3005			
R164	102A	3123	52	TR85/250	G0005
R-164	102A	3004			
R177	103A			TR08/724	G0011
R186	100		2	TR05/254	G0005
R-186	100	3005			
R202	101		8	TR08/641	
R-202	101	3011			
R203	101		8	TR08/641	
R-203	101	3011			
R212					G0005
R-213		3124			
R227	100		2	TR05/254	G0005
R-227	100	3005			
R242	102A	3123	52	TR85/250	G0005
R-242	102A	3004			
R244	100		2	TR05/254	G0005
R-244	100	3005			

Original Part Number	ECG	SK	GE	IR/WEP	HEP
R245	102A		2	TR85/250	
R-245	102A	3004			
R255	102A	3123	52	TR85/250	G0005
R258	100	3005	8	TR05/254	
R264-1	197				
R265A	121		3	TR01/232	G6005
R289	102A		2	TR85/250	
R290	102A	3123	52	TR85/250	G0005
R291	102A	3123	52	TR85/250	G0005
R-291	102A	3004			
R324	102A	3123	2	TR85/250	G0005
R336	126				G0008
R337	126				G0008
R338	160		2	TR17/637	G0008
R339	160		2	TR17/637	G0008
R340	123A	3122	8	TR21/735	
R341	160		2	TR17/637	G0002
R364	102A		53	TR85/250	
R393					G6011
R424	160		2	TR17/637	G0002
R-424	100	3005			
R424-1	160	3123	2	TR17/637	G0002
R425	160	3123	2	TR17/637	G0002
R-425	100	3005			
R428	102A	3123	52	TR85/250	G0005
R440					G6011
R488	100		52	TR05/254	G0005
R-488	100	3005			
R497	160	3123	2	TR17/637	G0002
R506	100	3123	2	TR05/254	G0005
R-506	100	3005			
R515	160	3123	2	TR17/637	G0008
R515A	160			TR17/637	
R516	121	3009	16	TR01/232	G0008
R516A	160			TR17/637	
R516(T.I.)	160			TR17/637	
R530	102A		51	TR85/250	G0008
R-530	102A	3004			
R537	102A		53	TR85/250	
R539	160		50	TR17/637	G0002
R-539	126	3008			
R558	160	3004	53	TR17/637	G0005
R558(T.I.)	102A			TR85/250	
R563	160		53	TR17/637	G0002
R653(T.I.)	102A			TR85/250	
R564	160				G0002
R565	160				G0002
R579	160		1	TR17/637	G0002
R579(T.I.)	102A			TR85/250	
R581	126	3006		TR17/637	G0008
R582	123A	3122	20	TR21/735	
R592	101	3010	8	TR08/641	S0015
R593	160	3123	2	TR17/637	G0002
R-593	102A	3004			
R593A	160			TR17/637	
R608	102A		53	TR85/250	
R608A	102A		52	TR85/250	G0005
R-608A	102A	3004			
R612-1	152				
R621-1	152		28	TR76/701	S5000
R623-1	196		28	TR76/701	S5000
R632-1	152				
R632-2	152				
R640-1	196				
R684	126			TR17/635	
R714	126	3008	53	TR17/635	G0005
R715	126	3008	53	TR17/635	G0005
R868	102A		53	TR85/250	
R1117-1	124			TR81/	
R1273	102A		2	TR85/250	
R1274	102A		2	TR85/250	
R1530	103A		54	TR08/724	
R1531	103A		54	TR08/724	
R1532	103A		54	TR08/724	
R1533	101		5	TR08/641	
R-1533	101	3011			
R1534	103A		54	TR08/724	
R1537	103A		54	TR08/724	
R1538	103A		54	TR08/724	
R1539	160		51	TR17/637	
R1540	102A		53	TR85/250	
R1541	102A		53	TR85/250	
R1542	102A		53	TR85/250	
R1543	102A		53	TR85/250	
R1544	102A		53	TR85/250	
R1545	103A		54	TR08/724	
R1546	102A		53	TR85/250	
R1547	103A		54	TR08/724	
R1548	102A		53	TR85/250	
R1549	103A		54	TR08/724	
R1550	160		51	TR17/637	
R1553	103A		54	TR08/724	
R1554	160		51	TR17/637	
R1555	102A		53	TR85/250	
R2001	127			TR27/235	G6008
R-2001	127				
R2003	127			TR27/235	G6005
R2004				TR01/	
R2096	175			TR81/241	S5012
R2270-60106	128				
R2270-75116	181				
R2270-75497	130				
R2270-75638		3027			
R2270-76963	129	3025			
R2270-77873D	128				
R2270-78399	130				
R2350	102A		53	TR85/250	
R2351	102A		53	TR85/250	
R2352	102A		53	TR85/250	
R2353	102A		53	TR85/250	
R2355	102A		53	TR85/250	
R2356	103A		54	TR08/724	
R2359	103A		54	TR08/724	
R2360	103A		54	TR08/724	
R2364	103A		54	TR08/724	
R2365	103A		54	TR08/724	
R2366	102A		53	TR85/250	
R2367	102A		53	TR85/250	
R2373	102A		53	TR85/250	
R2374	103A		54	TR08/724	
R2375	103A		54	TR08/724	
R2432-1	710	3102			
R2434		3102			
R2432-1	706	3101			
R2434-1(RCA)	706				
R2444	124	3021	12	TR81/240	
R2445		3070			
R2445-1		3070			C6069G
R2445-1(RCA)	711				
R2446	121				
R2446-1	121				
R2460-9	127				
R2473	108	3018	20	TR95/56	
R2474-2	154				
R2476	108	3018	20	TR95/56	
R2477	108	3018	20	TR95/56	
R2482-1	102A			TR85/	
R2494-1	127				
R2500-1	127				
R2516		3144			
R2516-1		3072			S0015
R2516-1(RCA)	712				
R2559-1	728				
R2560-1	729				
R2675	102A		53	TR85/250	
R2677	102A		53	TR85/250	
R2683	160		51	TR17/637	
R2684	160		51	TR17/637	
R2685	160		51	TR17/637	
R2686	160		51	TR17/637	
R2687	160		51	TR17/637	
R2688	160		51	TR17/637	
R2689	102A		53	TR85/250	
R2694	160		51	TR17/637	
R2695	160		51	TR17/637	
R2696	160		51	TR17/637	
R2697	160		53	TR17/637	
R2749	102A		2	TR85/250	
R2749M	102A		2	TR85/250	
R2964	127		25	TR27/235	
R2982	130	3027	14	TR59/247	
R3273-P1	123A				
R3273-P2	123A				
R3275	102A		53	TR85/250	
R3276	102A		53	TR85/250	
R3277	160		51	TR17/637	
R3278	160		51	TR17/637	
R3279	160		51	TR17/637	
R3280(RCA)	102A		53	TR85/250	
R3282	102A		53	TR85/250	
R3283	123A	3122	20	TR21/735	
R3284	102A		53	TR85/250	
R3286	102A		53	TR85/250	
R3287	160		51	TR17/637	
R3288	160		51	TR17/637	
R3293	103A	3122	8	TR08/724	
R3293(GE)	123A			TR21/735	
R3299	102A		53	TR85/250	
R3301	102A		53	TR85/250	
R3309	160		51	TR17/637	
R3502	704				
R3502-1-2	704	3023			
R3502(RCA)	704	3023			
R3503	124			TR81/	
R3508	128		63	TR87/243	S5014
R3512-1	121				
R3514-1	127			TR27/235	G6008
R3515	104	3009	16	TR16/232	G6003
R3515(RCA)	121			TR01/	
R3520				TR23/	

Original Part Number	ECG	SK	GE	IR/WEP	HEP
R3520-1	124			TR81/	
R3528-1	785				
R3528-1(RCA)	785				
R-3530-1	108				
R-3552-1	128				
R-3553-1	128				
R-3555	128				
R3555-3	128				
R3573-1	103A			TR08/	
R3576-1	233				
R3578-1	102A			TR85/	
R-3580-1	128				
R3593	128				
R3598-2	102A			TR85/	
R3608	128		18		
R3608-1		3024			S5014
R3608-2	128	3024	63	TR25/	S5014
R3611-1	152				
R3613-3(RCA)	191			TR75/	
R3647	124				
R3651-1	222				
R3676-1	233				
R3677-1	780	3141			
R3677-1(RCA)	780				
R3677-2	780	3141			
R3677-2(RCA)	780				
R3677-3	780	3141			
R3677-3(RCA)	780				
R3679	128				
R3681-1	152				
R4057	123A	3122	20	TR21/735	
R4192	124		12	TR81/240	
R4193	124		12	TR81/240	
R4194	124		12	TR81/240	
R4195	124		12	TR81/240	
R4196	124		12	TR81/240	
R4295-1	756				
R4295-1(RCA)	756				
R4295-2	756				
R4295-2(RCA)	762				
R4348	102A		53	TR85/250	
R4349	102A		53	TR85/250	
R4369	130	3027	14	TR59/247	
R4437-1	787	3146			
R4437-1(RCA)	787				
R4437-2(RCA)	797	3146			
R4438-1(RCA)	788	3147			
R4438-2(RCA)	788	3147			
R4439-1(RCA)	789				
R4439-2(RCA)	789				
R5048	128	3024	18	TR87/243	
R5050	103A		54	TR08/724	
R5051	102A		53	TR85/250	
R5052	102A		53	TR85/250	
R5053	102A		53	TR85/250	
R5054	103A		54	TR08/724	
R5055	102A		53	TR85/250	
R5056	103A		54	TR08/724	
R5097	102A		53	TR85/250	
R5098	102A		53	TR85/250	
R5099	102A		53	TR85/250	
R5100	102A		53	TR85/250	
R5101	158		53	TR84/630	
R5102	160		51	TR17/637	
R5103	160		51	TR17/637	
R5158-1	797				
R5179	103A		54	TR08/724	
R5180	103A		54	TR08/724	
R5181	102A		53	TR85/250	
R5182	158		53	TR84/630	
R5523	102A		53	TR85/250	
R5524	102A		53	TR85/250	
R5525	102A		53	TR85/250	
R5708	102A		53	TR85/250	
R6553	102A		53	TR85/250	
R6922	102A		53	TR85/250	
R7048	102A		53	TR85/250	
R7124	102A		53	TR85/250	
R7127	102A		53	TR85/250	
R7163	123A	3122	10	TR21/735	
R7164	102A		53	TR85/250	
R7165	123A	3122	10	TR21/735	
R7166	102A		53	TR85/250	
R7167	104	3009	16	TR01/232	
R7249	123A	3122	20	TR21/735	
R7253	121	3009	3	TR01/232	
R7343	123A	3122	20	TR21/735	
R7359	123A	3122	20	TR21/735	
R7360	123A	3122	20	TR21/735	
R7361	123A	3122	20	TR21/735	
R7362	103A		54	TR08/724	
R7363	102A		53	TR85/250	
R7489	102A		53	TR85/250	
R7490	102A		53	TR85/250	
R7491	102A		53	TR85/250	
R7582	123A	3122	10	TR21/735	
R7612	102A		62	TR85/250	
R7613	128	3024	18	TR87/243	
R7620	104	3009	16	TR01/230	
R7885	160		51	TR17/637	
R7886	160		51	TR17/637	
R7887	123A	3122	20	TR21/735	
R7888	102A		53	TR85/250	
R7889	102A		53	TR85/250	
R7890	105		4	TR03/233	
R7891	160		51	TR17/637	
R7953	123A	3122	20	TR21/735	
R7962	160		50	TR17/637	
R8066	123A	3122	20	TR21/735	
R8067	123A	3122	20	TR21/735	
R8068	123A	3122	20	TR21/735	
R8069	123A	3122	20	TR21/735	
R8070	123A	3122	20	TR21/735	
R8115	123A	3122	20	TR21/735	
R8116	123A	3122	20	TR21/735	
R8117	123A	3122	20	TR21/735	
R8118	123A	3122	20	TR21/735	
R8119	123A	3122	20	TR21/735	
R8120	123A	3122	20	TR21/735	
R8121	102A		53	TR85/250	
R8158	124		12	TR81/240	
R8223	123A	3122	20	TR21/735	
R8224	123A	3122	20	TR21/735	
R8225	123A	3122	20	TR21/735	
R8240	160		50	TR17/637	
R8241	160		50	TR17/637	
R8242	160		51	TR17/637	
R8243	123A	3122	20	TR21/735	
R8244	123A	3122	20	TR21/735	
R8259	123A	3122	62	TR21/735	
R8260	123A	3122	62	TR21/735	
R8261	123A	3122	63	TR21/735	
R8305	123A	3122	20	TR21/735	
R8210	102A		53	TR85/250	
R8311	102A		53	TR85/250	
R8312	123A	3122	20	TR21/735	
R8313	104	3009	3	TR01/230	
R8528	123A	3122	20	TR21/735	
R8529	107	3039	20	TR70/720	
R8530	123A	3122	20	TR21/735	
R8543	123A	3122	62	TR21/735	
R8551	123A	3122	20	TR21/735	
R8552	123A	3122	20	TR21/735	
R8553	123A	3122	20	TR21/735	
R8554	123A	3122	20	TR21/735	
R8555	123A	3122	20	TR21/735	
R8556	123A	3122	20	TR21/735	
R8557	123A	3122	20	TR21/735	
R8559	160		50	TR17/637	
R8620	123A	3122	20	TR21/735	
R8646	123A	3122	62	TR21/735	
R8647	123A	3122	62	TR21/735	
R8648	123A	3122	10	TR21/735	
R8649	128(2)		18	/243	
R8658	123A	3122	20	TR21/735	
R8659	121	3009	3	TR01/232	
R8685	160		51	TR17/637	
R8686	160		51	TR17/637	
R8687	102A		53	TR85/250	
R8688	102A		53	TR85/250	
R8692	160		51	TR17/637	
R8693	160		51	TR17/637	
R8694	160		51	TR17/637	
R8695	102A		53	TR85/250	
R8697	102A		53	TR85/250	
R8703	160		51	TR17/637	
R8704	160		51	TR17/637	
R8705	160		51	TR17/637	
R8706	102A		53	TR85/250	
R8707	102A		53	TR85/250	
R8881	160		51	TR17/637	
R8882	160		51	TR17/637	
R8883	102A		53	TR85/250	
R8884	102A		53	TR85/250	
R8885	102A		53	TR85/250	
R8886	102A		53	TR85/250	
R8889	123A	3122	20	TR21/735	
R8900	123A(2)	3122	28	TR76/735	
R8914	123A	3122	20	TR21/735	
R8915	128	3024	18	TR87/243	
R8916	123A	3122	20	TR21/735	
R8963	123A	3122	20	TR21/735	
R8964	123A	3122	20	TR21/735	
R8965	123A	3122	20	TR21/735	
R8966	123A	3122	20	TR21/735	
R8967	159	3114	21	TR20/717	
R8968	123A	3122	20	TR21/735	
R8969	159	3114	21	TR20/717	
R8969A	159	3114			
R8971	102A		53	TR85/250	

Original Part Number	ECG	SK	GE	IR/WEP	HEP
R8989	124		12	TR81/	
R9004	123A	3122	62	TR21/735	
R9005	123A	3122	62	TR21/735	
R9025	123A	3122	20	TR21/735	
R9071	123A	3122	20	TR21/735	
R9382	171	3103	27	TR78/	
R9383	171	3103	27	TR78/	
R9384	123A	3122	20		
R9385	123A	3122	20		
R9483	123A	3122	20		
R9531	160	3006	51		
R9532	160	3006	51		
R9533	158	3004	53		
R9534	158	3004	53		
R9600	160	3006	51		
R9601	160	3006	51		
R9602	160	3006	51		
R9603	158	3004	53		
R9604	158	3004	53		
R10254P936	9936				
R10254P945	9945				
R10254P945B	9945				
R10254P946	9946				
R10255P946B	9946				
R10256P962	9962				
R10256P962B	9962				
R15003	128				
R15003P1	128				
R22707-8399	130				
R24451	711	3070			
R25161	712	3072			
R64194	108				
R332389P		3004			
R430550B		3006			
R227060106					S3001
R227075116B		3036			
R227075497	130	3027			
R227077873D	128				
R227078533	223				
R227078684					S5005
RB6-PS-801	923D				
RC679		3036			
RC709T	909	3590			
RC710T	910				
RC711T	911				
RC741T	941				
RC741TE		3514			
RC1700	130	3027	14	TR59/247	
RC1750		3027			
RC2270	128	3024		TR87/243	
RC6979		3511			
RC6980		3079			
RCA1B06		3560			
RCA1B07		3182			
RCA1B09		3560			
RCA29		3054			
RCA29A, B		3054			
RCA30		3083			
RCA30A, B		3083			
RCA31		3054			
RCA31A, B		3054			
RCA32		3083			
RCA32A, B		3083			
RCA106A, B		3570			
RCA106C. D. E		3571			
RCA106F		3570			
RCA106M		3571			
RCA106Q, Y		3570			
RCA120	261	3180			
RCA125	262				
RCA126	262				
RCA370		3083			
RCA371		3083			
RCA410		3560			
RCA411		3560			
RCA413		3560			
RCA423		3560			
RCA431		3560			
RCA520		3054			
RCA521		3054			
RCA1000	243	3182			
RCA1001	243	3182			
RCA3517	102A			TR85/	
RCA3858	102A			TR85/	
RCA8203	262	3181			
RCA8203A, B	264				
RCA8350	246	3183			
RCA8350A, B	246				
RCA34098	126			TR17/	
RCA34099	126			TR17/	
RCA34100	126			TR17/	
RCA34101	102A			TR85/	
RCA34106	102A			TR85/	
RCA35953	102A		2	TR85/250	G0005
RCA35954	102A		2	TR85/250	

Original Part Number	ECG	SK	GE	IR/WEP	HEP
RCA36789				TR05/	
RCA40231	103A		8	TR08/724	
RCA40245	161			TR83/	
RCA40246	161			TR83/	
RCA40395	102A			TR85/	
RCA40396N	103A			TR08/	
RCA40396P	102A			TR85/	
RCA44098	126			TR17/	
RCP111A, B, C, D	171	3201			
RCP113A, B, C, D	171	3201			
RCP115	171	3201			
RCP115A, B	171	3201			
RCP117, B	171	3201			
RCS242	130				
RE1					G0005
RE2			8		G0011
RE3					G0005
RE4		3004			G0006
RE5					G0011
RE6					G0011
RE7			76		G6013
RE8					G6004
RE9					S0020
RE10					S0020
RE11			76		G6005
RE12			20		S0014
RE13			20		S0015
RE14			12		S5011
RE15					G0001
RE16			25		G6008
RE17					S3019
RE18					S3032
RE19					S7004
RE20			78		G6016
RE21					S5000
RE22					S5006
RE23		3045	40		S5024
RE24		3103			S5015
RE25					G6011
RE26		3025			S5023
RE27					G0008
RE28			39		S0020
RE29			36		S5020
RE30					S5021
RE31			37		
RE33		3156			S9100
RE34		3538			S5012
RE35					G6011
RE36			76		G6001
RE37		3535	75		S7000
RE40					S5000
RE41					S5006
RE42					S5000
RE43					S5006
RE44					S3021
RE45					F0015
RE46					F0010
RE58					S3032
RE59					S3019
RE60					S3021
RE61					S5005
RE62			65		S0019
RE63			82		S0019
RE64			212		S0030
RE1001	108		61		
RE1002	108		61		
RE1102		3039			
RE2001	161		60		
RE2002	161		60		
RE3001	107		11		
RE3002	107		11		
RE4001	199		212		
RE4002	199		212		
RE4010	199		212		
RE5001	161		60		
RE5002	161		60		
RF69Z					S3007
RF200	161			TR83/719	
RF7313				TR23/	
RH-1X0004CEZZ	749				
RH-1X0023CEZZ	1079				
RH-1X0032CEZZ	1089¤				
RH-1X0038CEZZ	712				
RH120	123A	3122	63	TR21/	S0033
RH-1X0020CEZZ	1004	3102			
RH-1X0021CEZZ	1080				
RH-1X0022CEZZ	1051				
RH-1X0023CEZZ	1074				
RH-1X0024CEZZ	1105				
RH-1X0025CEZZ	1050				
RH-1X0043CEZZ		3072			
RIN1102	722				
RL709T	909				
RL711T	911				
RLH111	7410				

Original Part Number	ECG	SK	GE	IR WEP	HEP
RR7504	123A	3122	20	TR21/735	
RR8068	123A		20		
RR8070	108		61		
RR8116	108		61		
RR8118	108		61		
RR8119	108		61		
RR8914	123A		20		
RR8989	108		61		
RR12766	717				
RS-101	160	3006			G0003
RS-102	158	3004			G0005
RS-103	160	3006			G0002
RS-104	101	3011			G0011
RS-105	104	3009			G6003
RS-106	105	3012			G6004
RS-107	123A	3122			S0014
RS-108	123A	3122			S0015
RS-109	108	3039			S0016
RS-110	159	3025			S0013
RS113				TR21/	
RS115				TR21/	
RS128	123A	3122			S0015
RS132	128	3035			S5014
RS136	123A	3122			S0015
RS174				TR21/	
RS175				TR21/	
RS176				TR21/	
RS177				TR21/	
RS231(A)				TR21/	
RS238				TR21/	
RS310(C)				TR23/	
RS406	102A	3123	2	TR85/250	G0005
RS-406	102A	3004			
RS593	160	3123	2	TR17/637	G0002
RS684	160	3123	2	TR17/637	G0002
RS-684	126	3008			
RS685	160	3123	2	TR17/637	G0002
RS-685	126	3008			
RS686	160	3123	52	TR17/637	G0002
RS-686	102A	3004			
RS687	160	3123	52	TR17/637	G0002
RS-687	102A	3004			
RS1049	123A	3122	8	TR21/735	
RS1059	123A	3122	8	TR21/735	
RS1192	102A		2	TR85/250	G0005
RS-1192	102A	3004			
RS1513	101	3011	8	TR08/254	
RS1524	103A		8	TR08/724	
RS-1524	103A	3010			
RS1530	101	3011	51	TR08/254	
RS1531	101	3011	8	TR08/254	
RS1532	101	3011	8	TR08/254	
RS1533	103A		8	TR08/724	
RS1534	101	3011	8	TR08/641	
RS1536	101		8	TR08/641	
RS-1536	101	3011			
RS1537	101	3011	8	TR08/641	
RS1538	101	3011	8	TR08/641	
RS1539	160	3123	2	TR17/637	G0002
RS-1539	100	3005			
RS1540	102A	3123	52	TR85/250	G0005
RS-1540	102A	3004			
RS1541	102A		52	TR85/250	G0005
RS-1541	102A	3004			
RS1542	102A	3123	52	TR85/250	G0005
RS-1542	102A	3004			
RS1543	102A	3123	52	TR85/250	G0005
RS-1543	102A	3004			
RS1544	102A	3123	52	TR85/250	G0005
RS-1544	102A	3004			
RS1545	102A		8	TR85/250	
RS-1545	101	3011			
RS1546	102A	3123	52	TR85/250	G0005
RS-1546	102A	3004			
RS1547	101		8	TR08/641	
RS1548	102A	3123	52	TR85/250	G0005
RS-1548	102A	3004			
RS1549	103A		8	TR08/724	
RS1550	160	3123	2	TR17/637	G0002
RS-1550	100	3005			
RS1553	101		8	TR08/641	
RS-1553	101	3011			
RS1554	160	3008	50	TR17/637	G0002
RS-1554	126	3008			
RS1555	102A	3123	52	TR85/250	G0005
RS-1555	102A	3004			
RS1726	108	3018			
RS2001	101	3011		TR08/	G0011
RS2002	160	3006		TR08/	G0003
RS2003	126	3006		TR12/	G0008
RS2004	102	3004		TR85/	G0005
RS2005	102	3004		TR85/	G0007
RS2006	104	3009		TR02/	G6003
RS2007	158	3004		TR82/	G0005
RS2008	154	3044		TR78/	S5024
RS2009	123A	3122		TR21/	S0025
RS2010				TR21/	S0005
RS2011	161	3132		TR71/	S0017
RS2013	123A	3122			S0015
RS2014	128	3024		TR70/	S0015
RS2015	108	3039		TR21/	S0015
RS2016	123A	3122		TR24/	S0025
RS2017	184	3054		TR76/	S5000
RS2018	210	3202		TR76/	S3024
RS2019	182			TR59/	S5004
RS2020	184	3054		TR36/	S5004
RS2021	159	3025		TR20/	S0013
RS2022	159	3025			S0013
RS2023	159	3025		TR20/	S0019
RS2024	159	3025		TR20/	S0019
RS2025	185	3083		TR77/	S5006
RS2026	211	3203		TR58/	S3028
RS2027	183			TR29/	S5005
RS2028	132	3116		FE100/	F0015
RS2029	6409			IR2160/	S9002
RS2350	102A	3123	52	TR85/250	G0005
RS-2350	102A	3004			
RS2351	102A	3123	52	TR85/250	G0005
RS-2351	102A	3004			
RS2352	102A	3123	52	TR85/250	G0005
RS-2352	102A	3004			
RS2353	102A	3123	52	TR85/250	G0005
RS-2353	102A	3004			
RS2354	102A	3123	52	TR85/250	G0005
RS-2354	102A	3004			
RS2355	102A	3123	52	TR85/250	G0005
RS-2355	102A	3004			
RS2356	101	3011	52	TR08/641	
RS2359	101		8	TR08/641	
RS-2359	101	3011			
RS2360	101		8	TR08/641	
RS-2360	101	3011			
RS2364	101		8	TR08/641	
RS-2364	101	3011			
RS2365	101		8	TR08/641	
RS-2365	101	3011			
RS2366	101		8	TR08/641	
RS-2366	101	3011			
RS2367	102A	3123	52	TR85/250	G0005
RS-2367	102A	3004			
RS2373	102A	3123	52	TR85/250	G0005
RS-2373	101	3011			
RS2374	102A		8	TR85/250	
RS-2374	101	3011			
RS2375	101		8	TR08/641	
RS-2375	101	3011			
RS2675	102A	3123	52	TR85/250	G0005
RS-2675	102A	3004			
RS2677	102A	3123	52	TR85/250	G0005
RS-2677	102A	3004			
RS2679	160	3123	2	TR17/637	G0002
RS2680	160	3123	2	TR17/637	G0002
RS2683	160	3123	2	TR17/637	G0002
RS-2683	100	3005			
RS2684	160	3123	2	TR17/637	G0002
RS-2684	100	3005			
RS2685	160	3123	2	TR17/637	G0002
RS-2685	100	3005			
RS2686	160	3123	2	TR17/637	G0002
RS-2686	100	3005			
RS2687	160	3123	2	TR17/637	G0002
RS-2687	100	3005			
RS2688	160	3123	2	TR17/637	G0002
RS-2688	100	3005			
RS2689	102A	3123	2	TR85/250	G0005
RS-2689	102A	3004			
RS2690	100		1	TR05/254	
RS-2690	100	3005			
RS2691	100		1	TR05/254	
RS-2691	100	3005			
RS2692	100		1	TR05/254	
RS-2692	100	3005			
RS2694	160	3123	2	TR17/637	G0002
RS-2694	100	3005			
RS2695	160	3123	2	TR17/637	G0002
RS-2695	100	3005			
RS2696	100	3123	2	TR05/254	G0002
RS-2696	100	3005			
RS2697	102A	3123	2	TR85/250	G0005
RS-2697	102A	3004			
RS2867	102A			TR85/	
RS2914	123A		20	TR21/	
RS3211	102A	3123	2	TR85/250	G0005
RS3275	102A	3123	52	TR85/250	G0005
RS-3275	102A	3004			
RS3276	102	3123	52	TR05/250	G0005
RS-3276	102A	3004			
RS3277	160	3123	2	TR17/637	G0002
RS-3277	100	3005			
RS3278	160	3123	2	TR17/637	G0002

Original Part Number	ECG	SK	GE	IR/WEP	HEP
RS-3278	100	3005			
RS3279	160	3123	2	TR17/637	G0002
RS-3279	100	3005			
RS3280	102A	3123	52	TR85/250	G0005
RS3281	100	3005		TR05/254	G0005
RS3282	102A	3123	52	TR85/250	G0005
RS-3282	102A	3004			
RS3283	102A	3123	52	TR85/250	G0005
RS-3283	102A	3004			
RS3284	102A		2	TR85/250	
RS-3284	102A	3004			
RS3285	102A	3123	52	TR85/250	G0005
RS-3285	102A	3284			
RS3286	102A	3123	52	TR85/250	G0005
RS-3286	102A	3004			
RS3287	100	3005	1	TR05/254	
RS3288	160	3123	2	TR17/637	G0002
RS-3288	100	3005			
RS3289	102A		2	TR85/250	G0005
RS-3289	102A	3004			
RS3293	102A			TR85/250	G0005
RS3299	102A	3123	52	TR85/250	G0005
RS-3299	102A	3004			
RS3301	102A	3123	52	TR85/250	G0005
RS-3301	102A	3004			
RS3306	101	3011	8	TR08/641	
RS3308	102A		2	TR85/250	
RS-3308	102A	3004			
RS3309	160	3123	2	TR17/637	G0002
RS-3309	100	3005			
RS3310	102A	3003	2	TR85/250	
RS-3310	102A	3004			
RS3316	102A		2	TR85/250	G0005
RS-3316	102A	3004			
RS3316-1	102A		2	TR85/250	
RS-3316-1	102A	3004			
RS3316-2	102A		2	TR85/250	
RS-3316-2	102A	3004			
RS3318	102A		2	TR85/250	
RS-3318	102A	3004			
RS3322	160			TR17/637	G0002
RS-3322	126	3008			
RS3323	160		51	TR17/637	G0002
RS-3323	126	3008			
RS3324	160		51	TR17/637	G0002
RS-3324	126	3008			
RS3358-1	121	3009		TR01/232	
RS3359-1	121	3009	16	TR01/232	
RS3668	160			TR17/	
RS3717	102A	3004	2	TR85/250	
RS3726	102A			TR85/250	G0005
RS3857	102A			TR85/250	G0005
RS3858	121	3009		TR01/232	
RS3858-1	121	3009	3	TR01/232	G6005
RS-3858-1	121	3009			
RS3862	160		50	TR17/637	G0003
RS-3862	126	3008			
RS3863	160		50	TR17/637	G0003
RS-3863	126	3008			
RS3864	160			TR17/637	G0003
RS3866	102A		53	TR85/250	G0005
RS-3866	126	3008			
RS3867	158		1	TR85/250	G0005
RS-3867	100	3005			
RS3868	160		50	TR17/637	G0003
RS-3868	100	3005			
RS3880	102A	3004	52	TR85/250	G0005
RS3892	100		51	TR05/254	
RS-3892	160	3006			
RS3897	102A	3123	52	TR85/250	G0005
RS3898	160	3123	2	TR17/637	G0003
RS-3898	160	3006			
RS3900	160		50	TR17/637	G0003
RS-3900	160	3006			
RS3901	160		50	TR17/637	
RS3902	160		50	TR17/637	G0003
RS-3902	160	3006			
RS3903	160		50	TR17/637	G0003
RS-3903	160	3006			
RS3904	102A		52	TR85/250	G0005
RS-3904	102A	3004			
RS3905	160	3006	50	TR17/637	G0002
RS3906	160	3123	2	TR17/637	G0002
RS3907	160		50	TR17/637	G0003
RS-3907	100	3005			
RS3911	160		50	TR17/637	G0003
RS-3911	160	3006			
RS3912	160	3006	50	TR17/637	G0003
RS3913	102A		50	TR85/250	G0005
RS-3913	100	3005			
RS3914	100	3005	1	TR05/254	G0005
RS3915	100	3005	1	TR05/254	G0005
RS3925	102A		2	TR85/250	G0005
RS-3925	102A	3004			
RS3926	102A		50	TR85/250	

Original Part Number	ECG	SK	GE	IR/WEP	HEP
RS3929	160		50	TR17/637	G0003
RS-3929	100	3005			
RS3931	103A		8	TR08/724	
RS3959	121	3009	16	TR01/	
RS3959-1	121	3014		TR01/232	G6003
RS3986	160	3006	51	TR17/637	
RS3995	160	3006	51	TR17/637	
RS5008	102A		2	TR85/250	G0005
RS-5008	102A	3004			
RS5101	160		50	TR17/637	G0003
RS5102	102A		53	TR85/250	G0005
RS5103	102A		53	TR85/250	G0005
RS5104	100	3123	2	TR05/254	G0005
RS-5104	100	3005			
RS5105	100		50	TR05/254	G0002
RS-5105	100	3005			
RS5106	160	3123	2	TR17/637	G0002
RS-5106	100	3005			
RS5107	160		51	TR17/637	
RS-5107	126	3008			
RS5108	160	3123	2	TR17/637	
RS-5108	126	3008			
RS5109	160	3008	51	TR17/637	G0003
RS-5109	126	3008			
RS5110		3008			
RS5201	160		50	TR17/637	G0003
RS-5201	126	3008			
RS5202	102A		53	TR85/250	G0005
RS5203	102A		53	TR85/250	G0005
RS5204	160		50	TR17/637	G0003
RS5205	160		50	TR17/637	G0003
RS-5205	126	3008			
RS5206	160	3004	50	TR17/637	G0003
RS-5206	102A	3004			
RS5207	160	3008		TR17/637	G0003
RS-5207	126	3008			
RS5208	160	3123		TR17/637	
RS-5208	160	3006			
RS5209	160			TR17/637	G0005
RS-5209	126	3008			
RS5243-2	102A			TR85/250	
RS5301	160		50	TR17/637	G0003
RS-5301	126	3008			
RS5302	100	3005	53	TR05/254	G0005
RS5303	100	3005	53	TR05/254	G0005
RS5305	160		50	TR17/637	G0003
RS-5305	126	3008			
RS5306	160		50	TR17/637	G0003
RS-5306	126	3008			
RS5307		3004			
RS5311	160	3004	50	TR17/637	G0003
RS-5311	102A	3004			
RS5312	160	3123	52	TR17/637	G0003
RS-5312	126	3008			
RS5313	160	3123	2	TR17/637	G0003
RS-5313	126	3008			
RS5314	160	3008	50	TR17/637	
RS5317			51		
RS5401	102A		52	TR85/250	G0003
RS-5401	100	3005			
RS5402	100	3005	53	TR05/254	G0005
RS5403	100	3005	52	TR05/254	G0005
RS5406	102A	3003	52	TR85/250	G0005
RS-5406	102A	3004			
RS5502	102A	3123	52	TR85/250	G0005
RS-5502	102A	3004			
RS5503	102A	3123	52	TR85/250	G0005
RS5504	100	3123	52	TR05/254	G0005
RS5505	102A	3123	52	TR85/250	G0005
RS-5505	102A	3004			
RS5506	102A	3123	52	TR85/250	G0005
RS-5506	102A	3004			
RS5507	102A	3123	52	TR85/250	G0005
RS5511	100		1	TR05/254	G0005
RS-5511	100	3005			
RS5530	102A	3123	52	TR85/250	G0005
RS-5530	102A	3004			
RS5531	102A		2	TR85/250	
RS-5531	102A	3004			
RS5532	102A	3123	52	TR85/250	G0005
RS-5532	102A	3004			
RS5533	102A	3004	52	TR85/250	G0005
RS-5533	102A	3004			
RS5534	102A	3004	52	TR85/250	G0005
RS5535	102A	3123	52	TR85/250	G0005
RS-5535	102A	3004			
RS5536	102A	3004	52	TR85/250	G0005
RS-5536	102A	3004			
RS5540	100		1	TR05/254	
RS-5540	100	3005			
RS5541	102A	3004	52	TR85/250	
RS5542	102A	3123	52	TR85/250	G0005
RS-5542	102A	3004			
RS5543	102A	3004	52	TR85/250	G0005
RS5544	102A	3123	52	TR85/250	G0005

Original Part Number	ECG	SK	GE	IR/WEP	HEP
RS-5544	102A	3004			
RS5545	102A	3004	52	TR85/250	G0005
RS5551	102A		2	TR85/250	
RS-5551	102A	3004			
RS5552	102A		2	TR85/250	
RS-5552	102A	3004			
RS5553	102A		2	TR85/250	
RS-5553	102A	3004			
RS5554	102A		2	TR85/250	
RS-5554	102A	3004			
RS5555	102A		2	TR85/250	
RS-5555	102A	3004			
RS5556	102A		2	TR85/250	
RS-5556	102A	3004			
RS5557	102A		52	TR85/250	
RS-5557	102A	3004			
RS5558	102A		2	TR85/250	
RS-5558	102A	3004			
RS5563	102A	3123	52	TR85/250	G0005
RS5564	102A	3123	52	TR85/250	G0005
RS5565	102A	3123	52	TR85/250	G0005
RS5566	102A	3123	52	TR85/250	G0005
RS5567	102A	3123	52	TR85/250	G0005
RS5568	102A	3123	52	TR85/250	G0005
RS5602	102A		2	TR85/250	G0005
RS-5602	102A	3004			
RS5603	102A		53	TR85/250	G0005
RS5605	102A		53	TR85/250	G0005
RS5607	102A	3123	52	TR85/250	G0005
RS5608	102A	3123	52	TR85/250	G0005
RS5610	102A	3123	52	TR85/250	G0005
RS5612	121	3009	16	TR01/232	G6005
RS5613	121	3009	16	TR01/232	G6005
RS5614	121	3009	16	TR01/232	G6005
RS5616	121	3009	16	TR01/232	
RS5704	102A	3123	52	TR85/250	G0005
RS-5704	102A	3004			
RS5704-2	102A			TR85/250	G0005
RS5708	102A		2	TR85/250	G0005
RS-5708	102A	3004			
RS5708-2	102A		2	TR85/250	G0005
RS-5708-2	102A	3004			
RS5709	102A		2	TR85/250	G0005
RS-5709	102A	3004			
RS5711	102A		2	TR85/250	
RS-5711	102A	3004			
RS5715	176	3123			
RS5715-1	176	3123			
RS5717	102A	3004	52	TR85/250	G0005
RS-5717	102A	3004			
RS5717-1	102A		52	TR85/250	G0005
RS5717-3	102A		2	TR85/250	G0005
RS5717-6	102A		2	TR85/250	G0005
RS-5717-3	102A	3004			
RS-5717-6	102A	3004			
RS5720	102A		2	TR85/250	G0005
RS-5720	102A	3004			
RS5731	102A	3004	2	TR85/250	
RS5732	102A	3004	2	TR85/250	
RS5733	102A	3123	52	TR85/250	G0005
RS-5733	102A	3004			
RS5734	102A	3123	52	TR85/250	G0005
RS-5734	102A	3004			
RS5735	102A	3123	52	TR85/250	G0005
RS-5735	176	3004			
RS5737	102A	3004	2	TR85/250	G0005
RS5738	102A			TR85/250	
RS5740	102A		53	TR85/250	
RS5740-1	102A	3123	52	TR85/250	G0005
RS5742	102A		2	TR85/250	G0005
RS-5742	102A	3004			
RS5743	102A	3123	52	TR85/250	G0005
RS-5743	102A	3004			
RS5743-1	102A	3004	2	TR85/250	G0005
RS-5743-1	102A	3004			
RS5743-2	102A	3004	2	TR85/250	G0005
RS5743-3	102A	3004		TR85/250	G0005
RS5743. 3	100		52		
RS5744	102A		52	TR85/250	
RS-5744	102A	3004			
RS5744-3	102A	3004	2	TR85/250	
RS5745	102A	3004	2	TR85/250	
RS5746	102A	3004	52	TR85/250	G0005
RS5747	102A	3004	52	TR85/250	G0005
RS5748	102A	3004	52	TR85/250	G0005
RS5749	102A	3123	52	TR85/250	G0005
RS-5749	102A	3004			
RS5750	102A	3004	2	TR85/250	
RS5751	102A	3004	2	TR85/250	
RS5752	102A		2	TR85/250	G0005
RS-5752	126	3008			
RS5753	126		50	TR17/635	G0005
RS-5753	126	3008			
RS5753-2	126	3006	2	TR17/635	G0005
RS5754	126		2	TR17/635	G0005

Original Part Number	ECG	SK	GE	IR/WEP	HEP
RS-5754	126	3008			
RS5755	126		2	TR17/635	G0005
RS-5755	126	3008			
RS5756	126		2	TR17/635	G0005
RS-5756	126	3008			
RS5757	126		51	TR17/635	
RS-5757	126	3008			
RS5758	126		51	TR17/635	
RS-5758	126	3008			
RS5759	126		51	TR17/635	
RS-5759	126	3008			
RS5760	126		51	TR17/635	
RS-5760	126	3008			
RS5761	126		51	TR17/635	G0005
RS-5761	126	3008			
RS5762	126		51	TR17/635	
RS-5762	126	3008			
RS5765	102A	3123	52	TR85/250	G0005
RS5766	102A	3004	52	TR85/250	G0005
RS5767	102A	3004	52	TR85/250	G0005
RS5768	102A	3123	52	TR85/250	G0005
RS5788	176	3004		TR82/238	
RS5802	126		50	TR17/635	G0003
RS-5802	126	3008			
RS5818	160		50	TR17/635	G0003
RS-5818	126	3008			
RS5825	102		2		
RS5835	121		16	TR01/232	
RS-5835	121	3009			
RS5851	123A		20	TR21/735	
RS-5851	123A	3124			
RS5852	102A	3123	52	TR85/250	G0005
RS-5852	102A	3004			
RS5853	123A		20	TR21/735	S0015
RS-5853	123A	324			
RS5854	102A	3123	52	TR85/250	G0005
RS-5864	102A	3004			
RS5855	121MP		16	TR01/232MP	G6005
RS-5855	121	3009			
RS5856	123A		20	TR21/735	
RS-5856	123A	3124			
RS5857	123A		20	TR21/735	S0015
RS-5857	123A	3124			
RS6523	108			TR95/	
RS6821	126	3006	51	TR17/635	
RS6822	126	3006	51	TR17/635	
RS6824	100	3005	51	TR05/254	
RS6840	102A	3004	52	TR85/250	
RS6843	102A	3004		TR05/631	
RS6843A	102A	3004			
RS6846	102	3004		TR05/631	
RS6846A	102A	3004			
RS7101	108	3018	17	TR95/56	S0016
RS7102	108	3018	11	TR95/56	S0016
RS-7102	108	3018			
RS7103	123A		20	TR21/	
RS-7103	123A	3124			
RS7104	108		17	TR95/	S0016
RS-7104	108	3018			
RS7105	123A		20	TR21/	S0016
RS-7105	123A	3124			
RS7106	108		11	TR95/	S0016
RS-7106	108	3018			
RS7107	108		20	TR95/56	S0016
RS-7107	108	3018			
RS7108	123A		11	TR21/735	S0016
RS-7108	108	3018			
RS7109	108		17	TR95/56	
RS-7109	108	3018			
RS7110	108		17	TR95/56	
RS-7110	108	3018			
RS7111	123A	3124	20	TR21/735	S0011
RS7112	108	3018	17	TR95/56	S0015
RS7113	108		20	TR95/56	S0011
RS-7113	108	3018			
RS7114	108	3018	11	TR95/56	S0015
RS-7114	108	3018			
RS7115	108		11	TR95/56	S0015
RS-7115	108	3018			
RS7116	108	3018	20	TR95/56	S0015
RS7117	108	3018	20	TR95/56	S0015
RS7118	108	3018	20	TR95/56	S0016
RS7119	108	3018	17	TR95/56	
RS7120	108	3018	20	TR95/56	
RS7121	123A	312	20	TR21/735	
RS7122	108	3018	20	TR95/56	
RS7123	108	3018	20	TR95/56	S0015
RS7124	108	3018	20	TR95/56	
RS7125	108	3018	20	TR95/56	S0011
RS7126	108	3018	11	TR95/56	S0015
RS7127	123A	3018	11	TR21/56	S0011
RS7128	108	3124	11	TR95/56	
RS7129	123A	3124	20	TR21/735	
RS7132	123A	3124	11	TR21/735	
RS7133	123A	3124	20	TR21/735	

Original Part Number	ECG	SK	GE	IR/WEP	HEP
RS7135	108	3124	20	TR95/56	S0011
RS7136	123A	3124	20	TR21/735	
RS7138	108	3018	11	TR95/56	S0011
RS7139	108	3018		TR95/56	
RS7140	108	3018		TR95/56	S0015
RS7141	108	3018		TR95/56	
RS7142	108	3018	20	TR95/56	
RS7143	107	3018		TR70/720	S0015
RS7144	108	3018	20	TR95/56	
RS7145	108	3018		TR95/56	
RS7160	123A	3018	20	TR21/735	
RS7161	108	3018	20	TR95/56	
RS7162	108	3018	20	TR95/56	
RS7163	108	3018	20	TR95/56	
RS7164	108	3018	20	TR95/56	
RS7165	108	3018	20	TR95/56	
RS7166	108	3018	20	TR95/56	
RS7167	108	3018	20	TR95/56	
RS7168	108	3018	20	TR95/56	
RS7169	108	3018	20	TR95/56	
RS7170	108	3018	20	TR95/56	
RS7174	108	3018	20	TR95/	S0016
RS7175	108	3018	20	TR95/56	S0016
RS7176		3018	20	TR21/	S0016
RS7177	108	3018	20	TR21/	S0015
RS7201	108	3018	17	TR95/56	S0011
RS7202	108		17	TR95/56	S0016
RS-7202	108	3018			
RS7209	108	3018	20	TR95/56	
RS7210	108	3018	20	TR95/56	S0015
RS7211	108	3018	20	TR95/56	
RS7212	108	3018	17	TR95/56	S0016
RS7214	108	3018	20	TR95/56	
RS7215	108	3018	11	TR95/56	
RS7216	108	3018	20	TR95/56	
RS7217	108	3018	20	TR95/56	S0015
RS7218	108	3018	11	TR95/56	S0015
RS7219	108	3018	11	TR95/56	S0011
RS7220	108	3018	20	TR95/56	S0016
RS7221	108	3018	11	TR95/56	
RS7222	107	3018		TR70/720	
RS7223	123A	3122		TR21/735	S0015
RS7224	123A	3018		TR21/735	S0015
RS7225	108	3018	11	TR95/56	S0015
RS7226	123A	3018	11	TR21/735	S0015
RS7227	108	3018	11	TR95/56	
RS7228	108	3018	11	TR95/56	S0005
RS7229	108	3018		TR95/56	
RS7230	108	3018		TR95/56	
RS7231	108	3018	20	TR21/	S0015
RS7232	123A				S0015
RS7233	123A	3018			S0015
RS7234		3040	17	TR21/	S0016
RS7235	123A	3040	17	TR21/	S0015
RS7236	123A				S0015
RS7237	108	3018	11		S0025
RS7238		3018	20	TR21/	S0015
RS7241	123A	3020	20	TR21/	S0015
RS7242	123A	3020	20	TR21/	S0015
RS7310	124	3021	12	TR81/240	S5011
RS7311	124	3021	12	TR81/240	S5011
RS7312	124	3021	12	TR81/240	S5011
RS7313	124	3021	12	TR81/240	S5011
RS7314		3021			
RS7315	124	3021	12	TR81/240	
RS-7315	124	3021			
RS7316	124	3021	12	TR81/240	
RS-7316	124	3021			
RS7317	124	3021	12	TR81/240	
RS-7317	124	3021			
RS7318	124	3021	12	TR81/240	
RS-7318	124	3021			
RS7320	124	3021	12	TR81/240	
RS7321	124	3021	12	TR81/240	
RS7327	124	3021	12	TR81/240	S5011
RS7328	124	3021	12	TR81/240	S5011
RS7329	124	3021	12	TR81/240	S5011
RS7330	124	3021	12	TR81/240	S5011
RS7333	108			TR95/56	S0015
RS7334	108			TR95/56	S0015
RS7365	124	3021		TR81/240	
RS7366	124	3021		TR81/240	
RS7367	124	3021		TR81/240	
RS7368	124	3021		TR81/240	
RS7405	123A	3124		TR21/735	
RS7406	123A	3124		TR21/735	
RS7407	123A	3124	20	TR21/735	
RS7408	123A	3124	20	TR21/735	
RS7409	123A		20	TR21/735	
RS-7409	123A	3020			
RS7410	123A	3020	20	TR21/735	
RS7411	123A		20	TR21/735	
RS-7411	123A	3020			
RS7412	123A		20	TR21/735	
RS-7412	123A	3020			

Original Part Number	ECG	SK	GE	IR/WEP	HEP
RS7413	123A			TR21/735	
RS-7413	123A	3020			
RS7415	123A	3020		TR21/735	
RS7421	123A	3020		TR21/735	
RS7504	123A		20	TR21/735	
RS-7504	123A	3020			
RS7510	123A	3020	20	TR21/735	
RS7511	108		20	TR95/56	S0015
RS-7511	123A	3020			
RS7512	108		17	TR95/56	S0015
RS-7512	108	3018			
RS7513	123A	3020	20	TR21/735	
RS7513-15	123A	3020		TR21/735	
RS7514	123A	3018	20	TR21/735	S5014
RS7515	123A	3020	20	TR21/735	S0015
RS7516	123A	3020	20	TR21/735	S0015
RS7517	123A	3018	20	TR21/735	S0015
RS7517-19	123A	3020		TR21/735	
RS7518	123A	3020	20	TR21/735	S5014
RS7519	123A	3020	20	TR21/735	S0015
RS7520	108	3018	11	TR95/56	S0015
RS7521	123A	3020	20	TR21/735	
RS7522	108	3018	11	TR95/56	
RS7523	107	3018	11	TR /56	
RS7524	108	3018		TR95/56	
RS7525	123A	3020	20	TR21/735	
RS7526	123A	3020	20	TR21/735	
RS7527	123A	3020	20	TR21/735	S0011
RS7528	123A	3020	20	TR21/735	
RS7529	123A	3020	20	TR21/735	
RS7530	123A	3020	20	TR21/735	
RS7532	108	3018	20	TR95/56	
RS7533	108	3018	20	TR95/56	
RS7542	123A	3020	20	TR21/735	
RS7543	123A	3020		TR21/735	
RS7544	123A	3020		TR21/735	
RS7555	123A	3020	63	TR21/735	
RS7568	102A	3004	2	TR85/250	
RS7606	123A		20	TR21/735	
RS-7606	123A	3020			
RS7607	123A		20	TR21/735	
RS-7607	123A	3020			
RS7609	123A		20	TR21/735	S0015
RS-7609	123A	3020			
RS7610	123A		20	TR21/735	S0015
RS-7610	123A	3020			
RS7611	123A		20	TR21/735	S0015
RS-7611	123A	3020			
RS7612	123A		20	TR21/735	S0015
RS-7612	123A	3020			
RS7613	123A		20	TR21/735	S0015
RS-7613	123A	3020			
RS7614	123A		20	TR21/735	S0015
RS-7614	123A	3020			
RS7620	123A	3020	20	TR21/735	
RS7621	123A	3020		TR21/735	
RS7622	123A		20	TR21/735	
RS-7622	123A	3020			
RS7623	123A		20	TR21/735	S0015
RS-7623	123A	3020			
RS7624	123A	3020	20	TR21/735	S0015
RS7625	123A	3020	20	TR21/735	S0015
RS7626	123A	3020	20	TR21/735	S0015
RS7627	123A	3020	20	TR21/735	S0015
RS7628	123A	3020	20	TR21/735	S0015
RS7634	123A	3020		TR21/735	
RS7635	123A	3020		TR21/735	
RS7636	123A	3020		TR21/735	
RS7637	123A	3020		TR21/735	S0015
RS7638	123A	3020	20	TR21/735	
RS7639	123A	3020	20	TR21/735	S0015
RS7640	123A	3020		TR21/735	
RS7641	123A	3020		TR21/735	
RS7642	123A	3020		TR21/735	
RS7643	123A	3124		TR21/735	
RS7665	159	3114		TR20/717	
RS7665A	159	3114			
RS7672	128				
RS7678	128	3024	63	TR87/243	
RS7814	123A	3122	10	TR21/735	
RS7916	132			FE100/802	F0015
RS8100	129	3025		TR88/242	
RS8101	128	3024		TR87/243	
RS8102	129	3025		TR88/242	
RS8103	128	3024		TR87/243	
RS8104	129	3025		TR88/242	
RS8105	128	3024		TR87/243	
RS8106	129	3025		TR88/242	
RS8107	128	3024		TR87/243	
RS8108	129	3025		TR88/242	
RS8109	128	3024		TR87/243	
RS8110	129	3025		TR88/242	
RS8111	128	3024		TR87/243	
RS8112	129	3025		TR88/242	
RS8113	128	3024		TR87/243	

Original Part Number	ECG	SK	GE	IR/WEP	HEP
RS8406	158	3004	52	TR84/630	
RS8407	103A	3010	8	TR08/724	
RS8420	103A			TR08/724	
RS8421	102A			TR85/250	
RS8424	158				G0005
RS8441	103A	3010		TR08/724	G0011
RS8442	123A	3004	53	TR21/735	G0005
RS8443	103A			TR08/724	G0011
RS8444	102A			TR85/250	G0005
RS8445	103A	3010			
RS8446	158	3004			G0005
RS8503	123A	3020	20	TR21/	S0015
RS9510	107			TR70/720	S0015
RS9511	107			TR70/720	
RS9512	107			TR70/720	S0015
RS15048	123A				
RS57042	102A		2	TR85/250	G0005
RS-57042	102A	3004			
RS57062	102A		2	TR85/250	G0005
RS-57062	102A	3004			
RS57433	102A			TR85/250	
RS86057332	123A	3018			
RS 7113				TR21/	
RS 7115				TR21/	
RS 7117				TR21/	
RS 7174				TR21/	
RS 7175				TR21/	
RS 7176				TR21/	
RS 7176(B)				TR21/	
RS 7177				TR21/	
RS 7231(A)				TR21/	
RS 7238				TR21/	
RS 7310(C)				TR23/	
RS 7512				TR24/	
RSKK36	176	3123			
RT36		3007			
RT-44		3008			
RT52		3007			
RT100	123A	3038		TR21/735	
RT-100	123A	3122	47		S0014
RT-100SPRAGUE		3020			
RT-101			48		S0012
RT-101SPRAGUE		3025			
RT102				TR25/	
RT-102			20		S0011
RT-102SPRAGUE		3122			
RT-103			82		S0013
RT-103SPRAGUE		3114			
RT104				TR51/	
RT-104			212		S0015
RT-104SPRAGUE		3124			
RT105				TR24/	
RT-105			212		S0015
RT-105SPRAGUE		3124			
RT-106			65		S0019
RT-106SPRAGUE		3114			
RT107				TR21/	
RT-107			212		S0016
RT-107SPRAGUE		3018			
RT108				TR21/	
RT-108			212		S0020
RT-108SPRAGUE		3039			
RT-109			81		S0025
RT-109SPRAGUE		3122			
RT-110					S5024
RT-110SPRAGUE		3045			
RT-111			27		S3021
RT-111SPRAGUE		3104			
RT112	107			TR70/720	
RT-112			212		S0020
RT-112SPRAGUE		3132			
RT-113			39		S0017
RT-113SPRAGUE		3117			
RT114	123A	3018	18	TR21/735	
RT-114					S3019
RT115			21		
RT-115SPRAGUE		3513			
RT-118			65		S0013
RT-118SPRAGUE		3118			
RT-119		3008	8		G0011
RT-119SPRAGUE		3011			
RT-120			65		S0024
RT-120SPRAGUE		3114			
RT121	158	3004		TR84/630	
RT-121					S0032
RT-121SPRAGUE		3025			
RT-122					G0011
RT-122SPRAGUE		3010			
RT124				TR01/	
RT-124			76		G6011
RT-126			21		S0015
RT-126SPRAGUE		3118			
RT-127			76		G6013
RT-128			12		S5011
RT-128SPRAGUE		3021			
RT-131			77		S7004
RT-131SPRAGUE		3027			
RT-133					S5000
RT-133SPRAGUE		3054			
RT-135			32		S5015
RT-135SPRAGUE		3103			
RT141	128	3024	18	TR87/243	
RT-147			76		G6001
RT-148			74		S7001
RT-149			75		S7000
RT-149SPRAGUE		3036			
RT150			66		
RT-150					S5004
RT151			69		S5005
RT152			28	TR76/	
RT-152					S5000
RT-152SPRAGUE		3054			
RT-153					S5006
RT-153SPRAGUE		3083			
RT154	128	3024	28	TR87/243	
RT-154	130	3027			S5000
RT-154SPRAGUE		3054			
RT155		3011	69		
RT-155					S5006
RT-155SPRAGUE		3083			
RT156		3005			
RT-156			32		S3020
RT-156SPRAGUE		3054			
RT-157					S3032
RT-157SPRAGUE		3083			
RT159			27		
RT-159					S3021
RT-159SPRAGUE		3104			
RT-164					S5004
RT-165					S5005
RT175	132		FET2	FE100/	
RT-175SPRAGUE		3116			
RT176			FET1		
RT-176SPRAGUE		3112			
RT-177SPRAGUE		3112			
RT180	222				
RT-180SPRAGUE		3065			
RT-181SPRAGUE		3065			
RT182		3007			
RT185	102A		2	TR85/250	
RT-185	102A	3004			
RT188	128	3024	18	TR87/243	
RT-188	128	3024			
RT194		3510			
RT233		3004			
RT409E					S5014
RT482	128	3024	81	TR87/243	S0014
RT483	128	3024	81	TR87/243	S0014
RT484	128	3024	81	TR87/243	S0014
RT697M	123A		210		
RT697AM					S5014
RT-698					S3019
RT698M					S3019
RT699M	128		18		
RT719M					S3019
RT730M					S5014
RT731M					S5014
RT910M					S0005
RT929H	123A	3124	20	TR21/735	S0015
RT930H	108		17	TR95/56	S0015
RT-930H	123A	3124			
RT1115					S5026
RT1116	171	3201	27		
RT1210					S3021
RT1252M					S0014
RT1253M					S0014
RT1409M					S0014
RT1410M					S5014
RT1420M					S5014
RT1890M					S3019
RT1893	171	3201	27	TR78/	
RT2016	123A	3122	20	TR21/735	
RT2230	102A				G0007
RT2329	102A	3004	2	TR85/250	G0005
RT2330	102A	3004	51	TR85/250	G0007
RT2331	102A	3004	53	TR85/250	G0005
RT2332	123A	3124	20	TR21/735	
RT2459					S5022
RT2460					S0013
RT2461					S0012
RT2462					S0013
RT2463					S0012
RT2709	102A		52	TR85/250	
RT2914	123A	3124	10	TR21/735	
RT2915	108	3018	20	TR95/56	
RT3063	123A	3024	11	TR21/735	
RT3064	123A	3024	11	TR21/735	
RT3065	159	3114	21	TR20/717	
RT3065A	159	3114			
RT3069	108	3018	20	TR95/56	

Original Part Number	ECG	SK	GE	IR/WEP	HEP
RT3070	108	3018	20	TR95/56	
RT3071	159	3114	21	TR20/717	
RT3071A	159	3114			
RT3095	108	3018	10	TR95/56	
RT3096	103A	3010	8	TR08/724	
RT3097	102A	3004	52	TR85/250	
RT3098	102A	3004	52	TR85/250	
RT3225	108	3018	17	TR95/	S0014
RT3226	108	3018	20	TR21/	S0020
RT3227	108	3122	20	TR21/	S0014
RT3228	123A	3122	62	TR51/	S0015
RT3229	102A	3003	2	TR05/	G0005
RT3230	102A	3004	2	TR85/250	G0007
RT3231	102A	3004	2		G0005
RT3232	108	3122	20	TR21/	S0014
RT3361	126	3008	51	TR17/635	
RT3362	126	3008	51	TR17/635	
RT3363	102A	3004	2	TR85/250	
RT3364	102A	3004	2	TR85/250	
RT3365	102A	3004	2	TR85/250	
RT3449	102A	3004	2	TR85/250	
RT3466	160	3008		TR17/637	G0003
RT3467	102A	3008	51	TR85/250	G0005
RT3468	102A	3004	52	TR85/250	G0005
RT3500					S0012
RT3564	102A			TR85/250	G0005
RT3565	123A	3122	20	TR21/735	S0024
RT3566	102A	3004	52	TR85/250	G0005
RT3567	123A	3122	20	TR21/735	S0024
RT3568	102A	3004		TR85/250	G0005
RT4525	160	3004		TR17/637	G0008
RT4624	102A	3123	2	TR85/250	G0007
RT-4624		3004			
RT4625	158		2	TR84/630	G6011
RT-4625	102A	3004			
RT4760	123A	3124	20	TR21/735	S0015
RT4761	123A	3124	20	TR21/735	S0015
RT4762	131	3052	16	TR94/642	G6016
RT5001					S3020
RT5002					S5020
RT5003					S3019
RT5004					S3019
RT5061	108	3018		TR95/56	S0016
RT5063	126	3006		TR17/635	G0009
RT5151	128	3024	81	TR87/243	S3024
RT5152	128	3024	81	TR87/243	S3024
RT5200	108		20	TR95/	
RT5201	108		20	TR95/	
RT5202	123A		10	TR21/	
RT5203	128		81	TR63/	S3024
RT5204	128	3039	81	TR63/	S0016
RT5205	108		20	TR95/	
RT5206	123A		10	TR21/	
RT5207	123A		10	TR21/	
RT5208	123A	3020	10	TR21/	
RT5230	129	3025	82	TR88/242	
RT5401	128	3024	47	TR87/243	S3024
RT5402	128	3024	47	TR87/243	S3024
RT5403	128	3024	47	TR87/243	S3021
RT5404	128	3024	47	TR87/243	S3021
RT5411					S5024
RT5412					S5024
RT5413					S5024
RT5434					S0015
RT5435	199	3124	20	TR21/	S0015
RT5464	107	3018		TR70/720	S0011
RT5465	107	3018		TR70/720	S0011
RT5466	126	3006		TR17/635	G0009
RT5467	126	3008		TR17/635	G0009
RT5468	102A	3004		TR85/250	G0005
RT5520	126			TR17/635	G0008
RT5521	102A			TR85/250	G0005
RT5522	102A			TR85/250	G0005
RT5551	123A	3124	20		S0015
RT5637	102A	3004	2	TR85/250	G0005
RT5804					S0014
RT5900	108	3018		TR95/56	S0015
RT5901	108	3018		TR95/56	S0015
RT5902	108	3018		TR95/56	S0015
RT5903	108	3018		TR95/56	S0015
RT5904	108	3018		TR95/56	S0015
RT5905	128	3124		TR87/243	S5014
RT5906	128	3124		TR87/243	S5014
RT5907	128	3124	63	TR87/243	S5014
RT6157	108				S0016
RT6158	108				S0016
RT6159	108				S0016
RT6160	108				S0016
RT6201	108	3018	20	TR95/56	S0016
RT6202	108	3018	17	TR95/56	S0014
RT6203	108	3018	20	TR95/56	S0014
RT6205	102A	3004	2	TR85/250	G0005
RT6600	108	3018	20	TR95/	S0020
RT6601		3018	20		S0025
RT6602		3018	20		S0016

Original Part Number	ECG	SK	GE	IR/WEP	HEP
RT6604	102A	3004	52		G0005
RT6732	107	3018		TR70/720	S0011
RT6733	107	3018		TR70/720	S0011
RT6734	102A	3004		TR85/250	G0005
RT6735	102A	3004		TR85/250	G0005
RT6736	102A	3004		TR85/250	G0005
RT6787	107	3018	20	TR70/720	S0016
RT6921		3124	20		S0014
RT6988	160	3006		TR17/637	G0009
RT6989	123A	3122		TR21/735	S0016
RT6990	102A	3004		TR85/250	G0005
RT6991	108	3018		TR95/56	S0016
RT7311	124	3021			
RT7320	108		20	TR95/	
RT7321	108		20	TR95/	
RT7322	123A		20	TR21/	
RT7323	108		20	TR95/	
RT7324	108		20	TR95/	
RT7325	123A		17	TR21/	
RT7326	123A	3024	17	TR21/	S5014
RT7327	123A		17	TR21/	
RT7399	1003	3101	IC2		C6060P
RT7400	103A	3010		TR08/724	G0011
RT7401	102A	3004		TR85/250	G0005
RT7511	123A		20	TR21/735	
RT7514	123A	3122	20	TR21/735	
RT7515	123A	3122	20	TR21/735	
RT7517	123A	3122	20	TR21/735	
RT7518	123A	3122	20	TR21/735	
RT7528	123A	3122	20	TR21/735	
RT7557	123A	3124	63	TR21/735	S5014
RT7558	158	3004	52	TR84/630	G0005
RT7559		3124	20	TR21/	S0015
RT7638	123A	3122	20	TR21/735	
RT7703	107	3018	20	TR70/	S0016
RT7704	107	3018	20	TR70/	S0016
RT7845	123A	3124			S0015
RT7846	158	3004			G0005
RT7943	123A	3124	20	TR21/	S0014
RT7944	103A	3010	8	TR08/	G0011
RT7945	128	3024	18		S5014
RT8047		3124	18		S0015
RT-8047		3122			
RT8193	108		20	TR21/	
RT8195	123A		10	TR24/	
RT8197		3124	63		S0014
RT8198		3122	20		S0015
RT8201	123A		20	TR21/	
RT8330		3018	17		S0014
RT8331		3116	FET2		F0021
RT8332		3122	20		S0014
RT8333	107	3018	17	TR70/720	S0016
RT8335					S5000
RT8336					S5006
RT8337		3122	62		S0015
RT8338					S5006
RT8442	102A			TR85/250	
RT8527	108	3018	20		S0015
RT8602	102A	3004	2	TR85/	G0007
RT8666		3122			
RT8667		3115			
RT8668		3018			
RT8669		3018			
RT8670		3114			
RT8842		3004			
RT8863		3122			
RT8895	159	3114		TR20/717	S0013
RT-61012	190				
RT61014	126			TR17/635	
RT-61014	126	3008			G0009
RT61015	102A		2	TR85/250	
RT-61015	158	3004			G0005
RT61016	102A		2	TR85/250	
RT-61016	158	3004			G0005
RT69221	123A				
RTN1001	804				
RTN1102	722				
RV1467	108	3018		TR95/56	S0017
RV1468	108	3018		TR95/56	S0017
RV1469	108	3018		TR95/56	S0017
RV1470	108	3018		TR95/56	S0017
RV1471	123A	3018		TR21/735	S0030
RV1472	129	3025		TR88/242	S5013
RV1473	128	3018		TR87/243	S0015
RV1474	123A	3024		TR21/735	S0015
RV1475	102A	3004		TR85/250	G0005
RV2068	128	3024			S5014
RV2069	129	3025			S5013
RV2070	199	3124	62		S0015
RV2248	199	3124	62	TR25/	S0015
RV2249	123A	3124	20	TR21/	S0015
RV2351	159	3114			S0019
RV2353	159	3025	21		S0032
RV2354	123A	3018	60		S0015
RV2355	159	3114	21		S0019

Original Part Number	ECG	SK	GE	IR/WEP	HEP
RV2356	187	3084	69		S5006
RVILA3301	1006				
RVIMC4080	1083				
RVIUPC20C2	1075				
RVIUPC22C	1078				
RVS2SC645	108				S0016
RVTCS1381	123A		20	TR24/	S0015
RVTCS1382	159		21	TR30/	S0019
RVTCS1383	123A		10	TR24/	S0014
RVTCS1384		3020	20		S0016
RVTCS1473	199		62	TR51/	S0024
RVTMK10-2	132		FET2	FE100/	F0021
RVTMK10-E		3116	FET2		F0021
RVTS22410		3124	20		S0015
RVTS22411		3114	21		S0019
RYN12104	128				
RYN121105	123A				
RYN121105-3	123A				
RYN121J05-4	123A				
S0002		3122			
S0015	123A				
S0016	108				
S0020	108				
S0021	108				
S0022	123A				
S0023	199				
S0024	199				
S0025	123A				
S0026	159				
S001465	199				
S001466	123A				
S001683	128				
S006793	199				
S006927	199				
S007220	128				
S007764	199				
S01	160	3006		TR17/637	G0003
S02	160	3006		TR17/637	G0003
S03	160	3006		TR17/637	
S025	102A	3123		TR85/250	G0005
S028	101			TR08/	
S031A	123A				
S037	123A				
S063-1					S3021
S065	158			TR05/250	
S065A	102A			TR85/250	G0003
S073					S5022
S074					S0015
S083-1					S5004
S085					S0015
S088	102A			TR85/250	G0005
S096-1					C6063P
S0704	123A				
S017446	159				
S019843	159				
S022010	123A				
S022011	123A				
S022012	159				
S023735	159				
S024428	123A				
S024987	123A				
S025232	123A				
S025289	123A				
S026094	234				
S1B02C			53		
S2			21		
S3			21		
S4			21		
S4A-1-A-3P	121	3009			
S4E				TR51/	
S6B-3-A-3P	121MP	3015			
S6S87231				TR70/	
S8P-2-A-3P	102A	3004			
S9L-4-A-3P	102A	3003			
S12-1-A-3P	123A	3124			
S14A-1-3P	121	3009			
S21		3124			
S-39T	121	3009	16	/232	
S-40T	121	3009	16	/232	
S40T				TR01/	
S40TB				TR01/	
S-40TB	121	3009	16	/232	
S41T				TR01/	
S-41T	121	3009	16	/232	
S-42T	121	3009	16	TR01/232	
S-43T	121	3009	16	/232	
S-46T	121	3009	16	/232	
S-48T	121	3009	16	/232	
S-49T	121	3009	16	/232	
S-55TB	126	3006		/635	
S-58TB	121	3009	16	/232	
S65-1-A-3T	160	3006			
S65-2-A-3P	103A	3010			
S65-4-A-3P	101	3011			
S65A-1-3P	160	3007			

Original Part Number	ECG	SK	GE	IR/WEP	HEP
S65B-1-3P	126	3008			
S65C-1-3P	121	3009			
S66-1-A-3P	102A	3003			
S66-2-A-3P	102A	3004			
S66-3-A-3P	102A	3004			
S-70T	126	3008		/635	
S70T	160		51		
S74-3-A-3P	100	3005			
S77A-1-3P	105	3012			
S77B-1-3P	121	3013			
S77C-1-3P	121	3014			
S-80T	126	3007		/635	
S84A-1-3P	121	3009			
S-85	184	3190			
S-86	184	3190			
S87TB	160		51		
S-87TB	126	3008		/635	
S88B-1-CP	160	3006			
S88C-1-3P	100	3005			
S88TB	160		51	TR12/	
S-88TB	126	3008		M /635	
S92-1-A-3P	105	3012			
S93SE133	130	3027	14	TR59/247	
S93SE140	130MP			TR59MP/	
S93SE165	130	3027	14	TR59/247	
S103-2					S5021
S103-3					S5021
S130-138	161				S0017
S130-152					S0017
S130-191		3510			
S130-251	161				S0017
S133-1	123A	3122		TR21/735	
S-141		3004			
S-142		3004			
S-143		3004			
S169N	123A	3122	62		
S305	130		14	TR59/247	
S-305	130	3027			
S305A	130	3027	14	TR59/247	
S-305-ABL		3510			
S-305-AO		3027			
S-305-AG		3510			
S305D	130				
S305-DW		3027			
S-305-PD	130				
S306A	175			TR81/	
S-310E	152				
S320C		3024			
S320D		3025			
S320E		3024			
S-320F	129				
S-350-AR		3510			
S353	130	3510			
S-353-P		3027			
S-353-PQ		3027			
S354	175			TR81/	
S355		3027			
S355B		3027			
S-355-4		3027			
S356	130	3027		TR59/	
S-356-7OR,GR,BL		3510			
S-356-7R,BR,Y		3510			
S-371	160				
S409		3044			
S409F	128	3024	18	TR87/	
S437	129	3045			
S437F	129				
S500	106			TR20/52	
S501	106			TR20/52	
S504-O	129				
S520	159			TR20/717	
S-522	192	3513			
S608	159				
S684	160		51	TR17/637	
S685	102A		53	TR85/250	
S686	102A		53	TR85/250	
S687	102A		53	TR85/250	
S715	186		28	TR76/	
S801	186		28	TR76/	
S-873TB	160			TR17/637	
S-874TB	160	3007		/637	
S939					S3001
S1000			28	TR76/	
S1009	108	3019	17	TR95/56	
S1016	123A	3122	20	TR21/735	
S1019		3019	17	TR95/56	S0016
S1019(UHF)			11		
S1037	108	3019	17	TR95/56	S0016
S1041	107		17	TR70/720	S0020
S-1041	108	3019			
S1041-16GN	107	3019		TR70/720	
S1044	108	3019	17	TR95/56	S0016
S1047	159				
S1058	108	3018	17	TR95/56	S0015
S1059	108	3018	17	TR95/56	S0016

Original Part Number	ECG	SK	GE	IR/WEP	HEP
S1060	108	3018	17	/56	S0016
S1061	123A	3124	10	TR21/735	S0015
S1062	108	3018	17	TR95/56	S0015
S1065	123A	3124	20	TR21/735	S0015
S1066	123A	3124	20	TR21/735	
S1068	123A	3124	20	TR21/735	
S1069	123A	3122	20	TR21/735	S0015
S1074(R)			20	TR24/	S0015
S1076	108	3124	20	TR95/56	
S1078	108	3018	20	TR95/56	S0016
S1079	108	3018	20	TR95/56	S0011
S1080		3124	20	/56	S0011
S1090		3122	20	TR51/735	
S1122	107	3039		TR70/720	
S1126	107	3039		TR70/720	
S1128	123A	3124	20	TR21/735	
S1142	108	3039	12	TR95/56	
S1143	123A	3124	20	TR21/735	S0015
S1153	108	3018	20	TR95/56	S0015
S1202	157				
S1211N	133	3112			
S1212N	133	3112			
S1213N	133	3112			
S1214N	133	3112			
S1215N	133	3112			
S1216N	133	3112			
S1221	123A	3124	20	TR21/735	
S1221A	123A	3020	20	TR21/735	
S1221N	133	3112			
S1222N	133	3112			
S1223N	133	3112			
S1224N	133	3112			
S1225N	133	3112			
S1226	123A	3122	20	TR21/735	
S1227	108	3018	17	TR95/56	S0015
S1231N	133	3112			
S1232N	133	3112			
S1233N	133	3112			
S1234N	133	3112			
S1235N	133	3112			
S1236N	133	3112			
S1241	123A	3122	20	TR21/735	
S1241N	132	3116			
S1242	123A	3122	20	TR21/735	
S1242N	132	3116			
S1243	123A	3122		TR21/735	
S1245	123A	3122	20	TR21/735	
S1272	123A	3122	20	TR21/735	
S1276	108	3018	17	TR95/56	
S1286	161	3018	17	TR83/719	S0017
S1296	108	3018	12	TR95/56	
S1307	123A	3122	20	TR21/735	
S1308	107	3039		TR70/720	
S1309	123A	3122	20	TR21/735	
S1310					S0019
S1313	108	3039	17	TR95/56	S0016
S1316	108	3018	17	TR95/56	
S1317	108	3018	17	TR95/56	
S1318	108	3018	17	TR95/56	S0020
S1331	123A	3124	20	TR21/735	S0014
S1331N	123A	3122		TR21/735	
S1331W	123A	3020	20	TR21/735	
S1332	126			TR17/	
S1348	102A	3004	2	TR85/250	
S1349	102A	3004	2	TR85/250	
S1350	159	3114	21	TR20/717	
S1350A	159	3114			
S1360	108	3018	17	TR95/56	
S1361	108	3018	17	TR95/56	S0015
S1362	108	3018	17	TR95/56	S0020
S1363	123A	3124	20	TR21/735	S0025
S1364	123A	3124	20	TR21/735	S0015
S1366	154	3044		TR78/712	S5025
S1367	159	3114		TR20/717	S0019
S1367A	159	3114			
S1368	128	3024		TR87/243	S0015
S1369	123A	3122	20	TR21/735	S0025
S1373	123A	3122	20	TR21/735	
S1374	123A	3122	20	TR21/735	
S1403	123A	3124	17	TR21/735	
S1405	123A	3122	20	TR21/735	
S1407	154			TR78/712	
S1408	108	3018	17	TR95/56	
S1409	108	3018	17	TR95/56	
S1419	123A	3122	20	TR21/735	
S1420	123A	3122	20	TR21/735	
S1421		3024	18	TR87/243	
S1429-3	123A	3038		TR21/735	
S1430	129	3025		TR88/242	
S1431	129	3025		TR88/242	
S1432	123A	3122	20	TR21/735	
S1443	123A	3122	20	TR21/735	
S1453	123A			TR21/	
S1475	123A	3122	10	TR21/735	
S1476	123A	3122	20	TR21/735	

Original Part Number	ECG	SK	GE	IR/WEP	HEP
S1477		3025			
S1482					S0025
S1487	123A	3122	20	TR21/735	
S1502	123A	3122	20	TR21/735	
S1510	123A	3122	20	TR21/735	
S1512	123A	3122	20	TR51/735	
S1514	128	3024	18	TR87/243	
S1516	128	3024	18	TR87/243	
S1517	128	3024	18	TR87/243	
S1520	129	3025	21	TR88/242	
S1523	128	3024	18	TR87/243	
S1525	128	3024	18	TR87/243	
S1526	123A	3122	20	TR21/735	
S1527	123A	3122	20	TR21/735	
S1529	123A	3122	20	TR21/735	
S1530	123A	3122	20	TR21/735	
S1533	123A	3122	20	TR51/735	
S1556-2	121		16	TR01/	
S1559	123A	3124	20	TR21/735	
S1568	123A	3122	10	TR21/735	
S1570	123A	3124	20	TR21/735	
S1619	123A	3122	20	TR21/735	
S1620	123A	3122	20	TR21/735	
S1629	123A	3124	10	TR21/735	
S1636	108			TR95/56	
S1639	102A	3004	2	TR85/250	
S1640	160	3006	51	TR17/637	
S1642	128	3024		TR87/243	S0015
S1644	128			TR87/243	
S1671	128			TR87/630	
S1672	158			TR84/630	
S1674	108	3024	20	TR95/56	
S1674A	108	3018	20	TR95/56	
S1682	108	3018	17	TR95/56	S0025
S1689	128	3024	18	TR87/243	
S1691	130	3027	14	TR59/247	
S1692	130	3027	14	TR59/247	
S1697	123A	3122	20	TR21/735	
S1698	129	3025		TR88/242	
S1761	123A	3122		TR21/735	
S1761A, B, C	123A	3122	20	TR21/735	
S1762	128	3024	18	TR87/243	
S1764	123A	3122	20	TR21/735	
S1765	123A	3122	20	TR21/735	
S1766	123A	3122	20	TR21/735	
S1768	123A	3122	20	TR21/735	
S1769	154		27	TR78/712	
S1770	123A	3122	20	TR21/735	
S1772	123A	3122	20	TR21/735	
S1773	128	3024	18	TR87/243	
S1777	128	3024	18	TR87/243	
S1784	123A	3122	20	TR21/735	
S1785	123A	3122	20	TR21/735	
S1835	123A	3122	20	TR21/735	S0015
S1863	129	3025		TR88/242	
S1864	128	3024	18	TR87/243	
S1865	130	3027	14	TR59/247	
S1871	123A	3122	20	TR21/735	
S1874	128	3024	10	TR87/243	
S1889	159		67	TR20/717	
S1891	123A	3124	20	TR21/735	
S1891A	123A	3124	20	TR21/735	
S1891B	123A	3124	20	TR21/735	
S1897	107	3039		TR70/720	S0015
S1905	130	3027	14	TR59/247	
S1905A	130	3027		TR59/247	
S1907	130	3027	14	TR59/247	
S1955	123A	3122	20	TR21/735	
S1978					S3001
S1983	129	3025		TR88/242	
S1993	123A	3122	20	TR21/735	
S2002	161	3124	20	TR21/	
S2003-1	130	3027	14	TR59/247	
S2020		3019	17	/56	S0016
S-2020		3019			
S2034	123A	3122	20	TR21/735	
S2038		3019			S0016
S2041	153	3083	67	TR77/700	S5013
S-2041-8661	9944				
S2042	152	3024	63	TR76/701	S5014
S2043	123A	3122	20	TR21/735	S0015
S2044	123A	3124	62	TR21/	S0015
S2045		3122	20	TR21/735	S0015
S2059	124	3021	12	TR81/240	S5011
S2090		3122	20	TR51/735	
S2091	159	3114	21	TR20/717	
S2091A		3114			
S2104		3024	63		
S2117	129	3025		TR88/242	
S2118	128	3024	18	TR87/243	
S2121	123A	3122	20	TR21/735	
S2122	123A	3122	20	TR21/735	
S2123	123A	3122	20	TR21/735	
S2124	123A	3122	20	TR21/735	
S2125		3122		TR51/735	

Original Part Number	ECG	SK	GE	IR/WEP	HEP
S2128	159	3114	21	TR20/717	
S2128A		3114			
S2129	159	3114	21	TR20/717	
S2129A		3114			
S2130		3114		/717	
S2130A		3114			
S2131	107	3039		TR70/720	
S2132	107	3039		TR70/720	
S2133	107	3039		TR70/720	
S2134	107	3039		TR70/720	
S2135		3039		TR21/720	
S2140				TR51/735	
S2159	108			TR95/56	
S2171	123A	3122	20	TR21/735	
S2172	123A		17	TR21/735	
S-2172		3018			
S-2200-1135	74H04				
S2209	128	3024	18	TR87/243	
S2224	108		20	TR95/	S0016
S2225	123A		10	TR21/	S0015
S2241	130	3027	14	TR59/247	
S2274	129		21	TR88/	S0031
S2321	175			TR81/241	
S2368	129	3025		TR88/242	
S2369	128	3024	18	TR87/243	
S2370	129	3025		TR88/242	
S2371	128	3024	18	TR87/243	
S2392	130	3027		TR59/247	
S2397	123A	3122		TR21/735	
S2398C	129	3025		TR88/242	
S2400	128	3024		TR87/243	
S2400A	128	3024	18	TR87/243	
S2400B	128	3024		TR87/243	
S2401	128	3024		TR87/243	
S2401A	128	3024	18	TR87/243	
S2401B	128	3024		TR87/243	
S2401C	128	3024		TR87/243	
S2402	128	3024		TR87/243	
S2402A	128	3024	18	TR87/243	
S2402B,C	128	3024		TR87/243	
S2403B	130	3027	14	TR59/247	
S2403C	130	3027		TR59/247	
S2427	128	3024	18	TR87/243	
S2486	175			TR81/241	
S2487	128	3024	18	TR87/243	
S2525	159	3114	21	TR20/717	
S2525A		3114			
S2526	128	3024	18	TR87/243	
S2527	175		66	TR81/241	
S2581	123A	3122	20	TR21/735	
S2582	123A	3122	20	TR21/735	
S2590	123A	3122		TR21/735	
S2593	123A	3124	20	TR21/735	
S2617	108	3019	17	TR95/56	S0016
S2617(UHF)			11		
S2635	123A	3122	20	TR21/735	
S2636	123A	3122	20	TR21/735	
S2645	159	3114	21	TR20/717	
S2645A	159	3114			
S2648	128	3024	18	TR87/243	
S2741	130	3027	14	TR59/247	
S2771	129	3025		TR88/242	
S2794	128	3024	63	TR87/243	
S2935	123A	3122	20	TR21/735	
S2944	123A	3122	20	TR21/735	
S2984	123A	3122	20	TR21/735	
S2985	123A	3122	20	TR21/735	
S2986	154	3045	27	TR78/712	
S2988	159	3114	21	TR20/717	
S2988A	159	3114			
S2989	123A	3122	20	TR21/735	
S2991	129	3025		TR88/242	
S2992	128	3024	18	TR87/243	
S2993	129	3025		TR88/242	
S2994	129	3025		TR88/242	
S2995	129	3025		TR88/242	
S2996	123A	3122	20	TR21/735	
S2997	123A	3122	20	TR21/735	
S2998	123A	3122	20	TR21/735	
S2999	123A	3122	20	TR21/735	
S3001	195	3048			
S3002	154	3040		TR78/712	
S3004	159			TR20/717	
S3008		3048			
S3011		3048			
S3012	129				
S3019	108	3019	17	TR95/56	
S3020	108	3039		TR95/	
S3033	154				
S3034	154				
S3035	154				
S3098A		3024			
S3386	129			TR88/242	
S3639	159	3114		TR20/717	
S3639A		3114			

Original Part Number	ECG	SK	GE	IR/WEP	HEP
S3640		3114		/717	
S3640A	159	3114			
S3655	159	3114	2	TR20/717	
S3655A	159	3114			
S3771			77		
S4002	161	3124	20	TR21/	S0015
S4248	102A	3004	2		G0007
S4249	159		21		
S5000		3054			
S5003		3054			
S5006		3083			
S5007		3083			
S5021	161	3018	20	TR83/719	S0017
S5327E	108	3039	17	TR95/56	S0016
S5328E	108		17	TR95/56	S0016
S-5328E	108	3019			
S5596K					C6050G
S5679E	108		17	TR95/56	S0016
S-5670-E	108	3019			
S6801	123A	3122	17	TR21/735	S0011
S6804	128	3024	18	TR87/243	S0014
S8660	184		57		
S9100	172	3156		TR69/S9100	
S9102	257				
S9121	254				
S9122	258				
S9631		3018			
S10153	184				S5000
S12020-04	152	3054			
S15649	123A	3122	212	TR21/735	
S15650	161	3117	212	TR83/719	
S15657	161	3117	60	TR83/719	S0016
S15658	161	3117	60	TR83/719	S0016
S15659	161	3117	17	TR83/719	
S15660	128	3024		TR87/243	
S17862	154	3045	27	TR78/719	
S17900	128	3024		TR87/243	
S18000	128	3024	21	TR87/243	
S18100	159		82		S5022
S18200	128		18		
S19386	128	3024		TR87/243	
S20445	128	3024		TR87/243	S3001
S20446	195	3049			S3001
S-20446	195				
S21520	175	3026	23	TR81/	S5012
S21549	128	3024	18	TR87/243	
S21648	107	3039		TR70/720	
S22543	123A	3122	20	TR21/735	
S24226	159		21	TR20/717	
S24591	123A	3122	20	TR21/735	
S24592	199		62		
S24594	129		21	TR88/242	
S24596	123A	3122	20	TR21/735	
S24597	129	3025	67	TR88/242	
S24598	128	3024	20	TR87/243	
S24612	129	3025	21	TR88/242	S0031
S24612A	129	3025			
S24614	128	3024	63	TR87/243	
S24615	129	3025	21	TR88/242	
S24616	128	3024	21	TR87/243	
S25805	108	3018	11	TR95/56	S0016
S27233	128	3024	18	TR87/243	
S27604	161	3124	20	TR21/	S0015
S29445	123A				
S32669	108	3019	11	TR83/	S0020
S33291	124			TR81/240	
S33886	129		21	TR88/242	S0012
S33886A	129	3025			
S34540	123A			TR21/735	S0015
S35486	180			/S7001	
S35487	181			TR36/S7000	
S36999	123A				
S67794	124			TR81/	
S67809	121			TR01/	
S95101	160		50	TR17/637	G0003
S-95101	126	3008			
S95102	160		50	TR17/637	G0003
S-95102	126	3008			
S95103	160		50	TR17/637	G0003
S-95103	126	3008			
S95104	160		51	TR17/637	
S95106	160		51	TR17/637	
S95125	108		20	TR95/56	
S-95125	108	3018			
S95125A	108		20	TR95/56	
S-95125A	108	3018			
S95126	108		20	TR95/56	
S-96126	108	3018			
S95126A	108		20	TR95/56	
S-95126A	108	3018			
S95201	102A		2	TR85/250	G0005
S-95201	102A	3004			
S95202	123A		20	TR21/735	S0015
S-95202	103A	3010			
S95203	102A	3123	52	TR85/250	G0005

Original Part Number	ECG	SK	GE	IR/WEP	HEP
S95204	102A	3004	2	TR85/250	G0005
S95206	102A		2	TR85/250	
S95207	123A		2	TR85/250	
S95214	102A	3123	52	TR85/250	G0005
S95218	102A			TR85/250	
S95252	157	3103		TR60/244	
S95253	121	3009	16	TR01/232	
S95253-1	121	3009	16	TR01/232	
S99101	160	3008	50	TR17/637	G0003
S99102	160	3008	50	TR17/637	G0003
S99103	160	3008	50	TR17/637	G0003
S99104	160	3008	51	TR17/637	
S99201	102A	3004	52	TR85/250	G0005
S99203	102A	3004	52	TR85/250	G0005
S99218	102A			TR85/250	
S99252	157	3103		TR60/244	
S326690	108	3018	11	TR80/	S0016
S413796				TR05/	
S1157449				TR51/	
S1158504				TR51/	
S1977634	130	3027	14	TR59/247	
S2041635	102A		53	TR85/250	
S2042634	103A		54	TR08/724	
S3867171-P2		3047			
S8660121-808	184				
SA7	101	3011	8	TR08/641	G0011
SA8-1-A-3P	160	3007			
SA29	158			TR84/630	G0006
SA33	102A			TR85/250	G0005
SA33BRN	102A			TR85/250	G0005
SA33RED	102A			TR85/250	G0005
SA50	159	3114	21	TR20/717	S0013
SA50A	159	3114		TR20/717	
SA51	159	3114	21	TR20/717	S0013
SA52	159	3114	21	TR20/717	S0013
SA52A, AC, B, BC	159	3114	21	TR20/717	S0013
SA53	159	3114	21	TR20/717	S0013
SA53A	159	3114			
SA54	159	3114	21	TR20/717	S0013
SA55	159	3114	21	TR20/717	S0013
SA55A	159	3114			
SA56	159	3114	21	TR20/717	S0013
SA56A	159	3114			
SA70	159	3114	21	TR20/717	S0013
SA70A	159	3114			
SA102	126			TR17/635	
SA128	102A			TR85/250	
SA128-1	102A	3123		TR85/250	G0005
SA197	102A	3123		TR85/250	G0005
SA197-1, 2, 3	102A	3123		TR85/250	G0005
SA204	102A	3123		TR85/250	G0005
SA205BLU	102A	3123		TR85/250	G0005
SA205BRN	102A	3123		TR85/250	G0005
SA205GRN	102A	3123		TR85/250	G0005
SA205ORN	102A	3123		TR85/250	G0006
SA205RED	102A	3123		TR85/250	G0006
SA205VIO	102A	3123		TR85/250	G0005
SA205WHT	102A	3123		TR85/250	G0007
SA205YEL	102A	3123		TR85/250	G0005
SA240	159	3123		TR85/250	G0005
SA310	159	3114	65	TR20/717	S0012
SA310A	159	3114			
SA311	159	3114	65	TR20/717	S0012
SA311A	159	3114			
SA312	159	3114	65	TR20/717	S0012
SA312A	159	3114			
SA313	159	3114	65	TR20/717	S0012
SA313A		3114			
SA314	159	3114	65	TR20/717	S0012
SA314A	159	3114			
SA315	159	3114	65	TR20/717	S0012
SA315A	159	3114			
SA316	159	3114	65	TR20/717	S0012
SA316A	159	3114			
SA318-2	158				G0005
SA318-3	158				G0005
SA354B	103				
SA410	159	3114	65	TR20/717	S0013
SA410A	159	3114			
SA411	159	3114	65	TR20/717	S0013
SA411A	159	3114			
SA412	159	3114	65	TR20/717	S0013
SA412A	159	3114			
SA413	159	3114	65	TR20/717	S0013
SA413A	159	3114			
SA414	159	3114	65	TR20/717	S0013
SA414A	159	3114			
SA415	159	3114	65	TR20/717	S0013
SA415A	159	3114			
SA416	159	3114	65	TR20/717	S0013
SA416A	159	3114			
SA495	106		21	TR20/52	S0013
SA495A	106		21	TR20/52	S0013
SA496	106		21	TR20/52	S0013
SA496A, B	106		21	TR20/52	S0013

Original Part Number	ECG	SK	GE	IR/WEP	HEP
SA529	102A				G0006
SA537	106		65	TR20/52	S0013
SA538	106		65	TR20/52	S0013
SA539	106		65	TR20/52	S0013
SA540	106		65	TR20/52	S0013
SA565	102A			TR85/250	G0005
SA646	102A			TR85/250	G0005
SA681	158				G0005
SA853	102A			TR85/250	
SA2644					S0033(2)
SA2648					S0033(2)
SA2711					S0015(2)
SA2712					S0015(2)
SA2713					S0015(2)
SA2714					S0015(2)
SA2715					S0025(2)
SA2716					S0015(2)
SA2717					S0015(2)
SA2718					S0015(2)
SA2719					S0025(2)
SA2720					S0025(2)
SA2721					S0025(2)
SA2722					S0025(2)
SA2723					S0025(2)
SA2724					S0025(2)
SA2725					S0030(2)
SA2726					S0015(2)
SA2738					S0025(2)
SA2739					S0025(2)
SAB1044	161				S0017
SAB3469	161				S0017
SAC40	106		21	TR20/52	S0032
SAC40A, B	106		21	TR20/52	S0032
SAC42	106		21	TR20/52	S0032
SAC42A	106		21	TR20/52	S0032
SAC42B	106		21	TR20/52	
SAC44	106		21	TR20/52	
SAJ72155	1076				
SAJ72156	1091				
SAJ72157	1046				
SAJ72158	1047				
SAJ72159	1077				
SAJ72160	1086**				
SAJ72161	1089¤				
SAJ72162	1048				
SATCHZ339	108	3039			
SB0319	152		66		
SB0419	218			TR58/	
SB5	130				
SB6	130				
SB7	130				
SB100	126	3008	51	TR17/635	G0002
SB101	160		51	TR12/	
SB102	160		51	TR17/	
SB103	160		51	TR17/	
SB168	102A			TR85/250	G0005
SB169	102A			TR85/250	G0005
SB200	126		51	TR17/635	G0002
SB364				TR85/	
SB5122	126	3008	51	TR17/635	G0002
SC12	102A	3004	52	TR85/250	G0005
SC43	100	3005	1	TR05/254	
SC44	100	3005	1	TR05/254	
SC45	102A		2	TR85/250	
SC46	100	3005	1	TR05/254	
SC56	103A	3010	8	TR08/724	
SC63	102A	3004	2	TR85/250	
SC65	123A	3124	20	TR21/735	
SC66	102A	3004	2	TR85/250	G0005
SC68	102A	3004	2	TR85/250	
SC69	102A	3004	2	TR85/250	
SC70	121	3009	3	TR01/232	
SC71	160	3006	51	TR17/637	
SC72	160	3006	51	TR17/637	
SC73	102A	3004	2	TR85/250	
SC74	160	3006	51	TR17/637	
SC78	160	3006	51	TR17/637	
SC79	160	3006	51	TR17/637	
SC80	160	3006	51	TR17/637	
SC108			17		
SC108A			17		
SC108B			20		
SC109A			20		
SC147A, B			212		
SC148			212		
SC148A, B, C			212		
SC149, B			212		
SC157			65		
SC157A			65		
SC157V1			65		
SC158			65		
SC158A, B, V1			65		
SC159			65		
SC159A, B			65		
SC174			81		

Original Part Number	ECG	SK	GE	IR/WEP	HEP
SC174A,B			81		
SC256			82		
SC256A			82		
SC256B			67		
SC257			65		
SC257A,V1			65		
SC258			65		
SC258A,B,V1			65		
SC259			65		
SC259A,B			65		
SC350	123A			TR21/	
SC365	129			TR88/	
SC400Y	124		12		
SC441	198		32		
SC5150P	755				
SC700		3124			
SC-701		3122			
SC-702		3025			
SC-703		3025			
SC-704		3124			
SC-705		3039			
SC706		3024			
SC-707		3124			
SC727	124	3021	12	TR81/240	
SC777	124	3021	12	TR81/240	
SC785	123			TR86/	
SC786	128/411				
SC832	123A	3124	20	TR21/735	S0015
SC842	123A	3122	20	TR21/735	
SC843	129	3045	27	TR88/242	
SC843(T.I.)	154			TR78/712	
SC1001	123A			TR21/735	
SC1007	160		51	TR17/637	
SC1010	123A			TR21/735	
SC1168G	123A	3124		TR21/735	
SC1168H	123A	3124		TR21/735	
SC1182	914				
SC1227F	108	3018		TR95/56	
SC1227G	108	3018		TR95/56	
SC2929E	128	3024		TR87/243	
SC1229G	123A				S0014
SC1294H	129				S0012
SC1940G	704				
SC2906					C6091
SC2914	725				
SC2914P	725	3162			
SC3288PK					C2004P
SC3289PK					C2003P
SC4004	124	3021	12	TR81/240	S5011
SC-4004	124	3021			
SC4010	123A			TR21/735	
SC4044	123A		20	TR21/735	S0015
SC-4044	123A	3124			
SC4073	103A/410				
SC4131	128/411		12	/243	
SC-4131	124	3021			
SC4131-1	128/411		20	TR21/243	
SC-4131-1	124	3021			
SC4133	184		66	/S5007	
SC4167	128/411			/243	
SC-4167	124	3021			
SC4244	128/411		12	/243	
SC-4244	123A	3021			
SC4274	131			TR94/642	
SC4303	152	3041		TR76/701	
SC4303-1,2	152	3041		TR76/701	
SC4308	152			TR76/	
SC5116L	725	3162			
SC5117P	718		IC8		
SC5118P	720		IC7	IC512/	C6095P
SC5150P	755				C6013
SC5150R	735				
SC5172P	719				C6065P
SC5175G	941	3514			
SC5177P	718	3159			
SC5182P	720				
SC5199P	718	3159			C6094P
SC5204P	798				C6066
SC5740PQ	722	3161			
SC5741P	720				
SC5743P	722	3161			
SC5747	799				
SC8705P	722			IC513/	C6056P
SC8705P/99S053				IC513/	
SC8707	782				
SC8707P	782				
SC8718	748				
SC9314P	718				C6094P
SC9426P	722	3161			
SC7430P	747		IC7		C6079P
SC9431P	749				C6076P
SC9436P	712		IC5		C6063P
SC9961P	9917				
SC9963P	9924				
SC9964P	9976				

Original Part Number	ECG	SK	GE	IR/WEP	HEP
SC9965P	9989				
SC17063L	725				
SC0321			75		
SC0328			75		
SC0421			74		
SC0428			74		
SCA3021			86		
SCA3244			212		
SCC321			75		
SCC421			74		
SCD-1062013	9158				
SCD321			75		
SCD421			74		
SCD2011					C6062P
SCD-T320	130				
SCD-T322	123A				
SCD-T326	159				
SCD-T330	152				
SCD-T334	153				
SCDT323	123A				
SCE321			75		
SCE421			74		
SCF40501	755				
SCI44191005	128				
SCI444103053	128				
SCI444204037	129				
SCI444291004	129				
SCS1173(O)				TR55/	
SD109	123A	3124		TR21/735	
SD-109	123A				
SD345	152	3054		TR76/701	
SD445	153			TR77/700	
SD1023	186		28		
SD1345	152		66		
SD1445	153		69		S5006
SD3720					C2010G
SD3730					C2006G
SD3745					C2502P
SDA345	152	3054		TR76/701	
SDA445	153	3083		TR77/700	S5006
SDB345	152	3054		TR76/701	
SDB445	153	3083		TR77/700	
SDD320					S5000
SDD420	128	3024	20	TR87/243	S0014
SDD421	123A	3122		TR21/735	S0011
SDD821	123A	3122	11	TR21/735	S0011
SDD1220	128	3024	63	TR87/243	S0014
SDD3000	123A	3122	11	TR21/735	S0012
SDI345			66		
SDI445	153		69		
SDJ345	152	3054	66	TR76/701	
SDJ445	153	3083	69	TR77/700	
SDK345	152	3054	66	TR76/701	
SDK445	153	3083	69	TR77/701	
SDL345	152	3054	66	TR76/701	
SDL445	153	3083	69	TR77/700	
SDM345	152	3054	66	TR76/701	
SDM445	153	3083	69	TR77/700	
SDN345	152	3054	66	TR76/701	
SDN445	153	3083	69	TR77/700	
SDS-200		3039			
SDS202		3122			
SDS240	161	3019		TR83/719	
SDS420	161	3132			
SDT402			73		
SDT-410	162	3560	72		
SDT-411	163	3560	73		
SDT-413	163	3560	73		
SDT-423	163	3560	73		
SDT-430	163		73		
SDT431		3560	73		
SDT-445	153		69		
SDT1000					S5020
SDT1001			73		S5020
SDT1002			73		S5020
SDT1003			73		S5020
SDT1004			73		S5021
SDT1005			73		S5021
SDT1006					S5021
SDT1007					S5021
SDT1011			73		S5020
SDT1012			73		S5020
SDT1013			73		S5020
SDT1014			73		S5021
SDT1015			73		S5021
SDT1016					S5021
SDT1017					S5021
SDT1611	163		73		
SDT1612	163		73		
SDT1613	163		73		
SDT1614	163		73		
SDT1615	163		73		
SDT1616	163		73		
SDT1617	163		73		
SDT1618	163		73		

Original Part Number	ECG	SK	GE	IR/WEP	HEP
SDT1621			75		
SDT1622			75		
SDT1623			75		
SDT1631			75		
SDT1632			75		
SDT1633			75		
SDT-3048	121				
SDT3321	129	3025	28	TR88/242	
SDT3322	129	3025	29	TR88/242	
SDT3325	187	3193	29		
SDT3326	186	3192	28		
SDT3421	186	3192	216		
SDT3422	152		216		
SDT3425	186	3192	216		
SDT3426	186	3192	216		
SDT3501	129	3025	84	TR88/242	
SDT3502	129	3025	29	TR88/242	
SDT3503	129	3025		TR88/242	
SDT3505	187	3513	84		
SDT3506	187	3513	29		
SDT3509	153	3085	69		
SDT3510	153		69		
SDT3513	153	3085	69		
SDT3514	153		69		
SDT3550	187	3193	29		
SDT3552	187	3193	84		
SDT3553	187	3193	29		
SDT3575	218		26	TR58/	
SDT3576	218		69	TR58/	
SDT3577	218			TR58/	
SDT3578	218		69	TR58/	
SDT3579	218		69	TR58/	
SDT3701	153	3085	69		
SDT3702	153		69		
SDT3703	153	3085	69		
SDT3704	153		69		
SDT3706	153	3085	69		
SDT3707	153	3085	69		
SDT3708	153		69		
SDT3709	153	3085	69		
SDT3710	153	3085	69		
SDT3711	153		69		
SDT3712	153	3085	69		
SDT3713	153		69		
SDT3715	153	3085	69		
SDT3716	153	3085	69		
SDT3717	153		69		
SDT3720	153	3085	69		
SDT3721	153	3085	69		
SDT3722	153		69		
SDT3725	153	3085	69		
SDT3726	153	3085	69		
SDT3727	153		69		
SDT3729	153	3085	69		
SDT3730	153		69		
SDT3733	153	3085	69		
SDT3760	180		74		
SDT3764	180		74		
SDT3765	180		74		
SDT3766	180		74		
SDT3775	187	3193	29		
SDT3776	187	3193	29		
SDT3778	187	3193	29		
SDT3826	180		74		
SDT3827	180		74		
SDT3875	180		74		
SDT3876	180		74		
SDT3877	180		74		
SDT4301	186	3192	83		
SDT4302	186	3192	215		
SDT4303	186	3192	83		
SDT4305	186	3192	215		
SDT4307	186	3192	83		
SDT4308	186	3192	215		
SDT4310	186	3192	83		
SDT4311	186	3192	215		
SDT4455	186	3192	28		
SDT4483	186	3192	214		
SDT4551	186	3192	28		
SDT4553	186	3192	28		
SDT4583	186	3192	28		
SDT4611	186	3192	28		
SDT4612	186	3192	28		
SDT4614	186	3192	28		
SDT4615	186	3192	28		
SDT5001	186	3512	28		
SDT5002		3512			
SDT5003		3512			
SDT5006	186	3512	28		
SDT5007		3512			
SDT5008		3512			
SDT5011	186	3512	28		
SDT5012		3512			
SDT5013		3512			
SDT5501	186	3512	46		
SDT5502		3512			
SDT5503		3512			
SDT5506	186	3512	46		
SDT5507		3512			
SDT5508		3512			
SDT5511	186	3512	46		
SDT5512		3512			
SDT5513		3512			
SDT5901	186	3026	28		
SDT5902	186	3026	28		
SDT5906	186	3192	28		
SDT5907	152	3026	66		
SDT6001	152		66		
SDT6011	152		66		
SDT6013	152		66		
SDT6031	152		66		
SDT6101	186	3192	28		
SDT6102	186	3192	28		
SDT6103	152		66		
SDT5104	186	3192	28		
SDT6105	186	3192	28		
SDT6106	186	3192	28		
SDT6901	175	3131		TR81/241	S5012
SDT6905	175	3131		TR81/241	S5012
SDT7401	186	3192	28		
SDT7402	152		66		
SDT7411	186	3192	28		
SDT7412	152		66		
SDT7414	186		28		
SDT7415	152		66		
SDT7511	152		66		
SDT7512	152		66		
SDT7514	152		66		
SDT7515	152		66		
SDT7601		3027			
SDT7602		3027			
SDT7603		3027			
SDT7604		3079			
SDT7607		3027			
SDT7608		3027			
SDT7609		3027			
SDT7610		3079			
SDT7731		3027			
SDT7732		3027			
SDT7733		3027			
SDT7734		3079			
SDT7735		3079			
SDT9001	186		216		
SDT9002	186	3192	216		
SDT9003	186	3192	28		
SDT9004	187	3193	216		
SDT9005	186	3192	216		
SDT9006	186	3192	28		
SDT9007	186	3192	28		
SDT9008	186	3192	28		
SDT9009	152		66		
SDT9201	130	3027	77	TR59/247	
SDT9202	180	3027	75		
SDT9203		3079			
SDT9204		3079			
SDT9205	130	3027	77		
SDT9206	130	3027	77		
SDT9207	180	3027	75		
SDT9208		3079			
SDT9209		3079			
SDT9210	130	3027	77		
SDT9261	130			TR59/247	S7002
SDT9301			77		
SDT9301-09		3511			
SDT9302			77		
SDT9303			75		
SDT9304			77		
SDT9305			77		
SDT9306			75		
SDT9307			77		
SDT9308			77		
SDT9309			75		
SDT9701	180		75		
SDT9704	180		75		
SDT9707	180		75		
SDT9801					S7004
SDT9802					S7004
SDT9803					S7004
SDT9901	162		35		
SDT9902	162		35		
SDT9903	162		35		
SDT9904	162		35		
SE002(1)			11		S0020
SE-0566	123A				
SE5-0127	199				
SE5-0128	123A				
SE5-0249	161				
SE5-0250	161				
SE5-0253	123A				
SE5-0274	123A				

Original Part Number	ECG	SK	GE	IR/WEP	HEP
SE5-0366	194				
SE5-0367	123A				
SE5-0370	159				
SE-5-0399	127			TR27/235	
SE5-0399	127	3035			
SE5-0452	128				
SE5-0545	199				
SE5-0567	123A				
SE5-0569	199				
SE5-0608	123A				
SE5-0745	194				
SE5-0798	159				
SE-5-0819	102	3004		TR05/631	G0005
SE-5-0819A		3004			
SE5-0831	159				
SE5-0854	123A				
SE5-0887	123A				
SE5-0888	123A				
SE5-0930		3075			
SE5-0931		3076			
SE5-0938	199				
SE5-0938-54	123A				
SE5-0938-55	199				
SE5-0938-56	199				
SE5-0938-57	199				
SE5-0944		3141			
SE5-0949	159				
SE5-0958	128				
SE5-0963	152				
SE5-0964	153				
SE5-0996	133				
SE5-1057	159				
SE5-1120		3054			
SE5-1223	159				
SE500G	108		17	TR21/	
SE504	108		12	TR95/56	
SE521	108			TR95/56	
SE1000				TR24/	
SE1001	161	3122	210	TR83/719	S0030
SE-1001	123A	3122			
SE1001-1	161	3122		TR83/719	S0025
SE1001-2	161	3122		TR83/719	S0025
SE1002	161	3124	210	TR83/719	S0030
SE-1002	123A	3124		TR24/	
SE1002-1	161	3117		TR83/719	S0015
SE1002-2	161	3117		TR83/719	
SE1010	192	3019	86	TR83/719	S0020
SE1019	108	3019	17	TR95/56	S0016
SE1044	108	3019	17	TR95/56	S0016
SE1331	123A	3124	20	TR21/735	
SE1419	108	3018	17	TR95/56	
SE2001	161	3132	60	TR83/719	S0025
SE-2001	123A	3124			
SE2002	161	3039	210	TR83/719	S0015
SE2020	107		20	TR70/720	
SE2397	107	3039	20	TR70/720	
SE2400	160			TR17/637	
SE2401	123A	3122	20	TR21/735	
SE2402	123A	3122	20	TR21/735	
SE3001	161	3018	11	TR83/719	S0020
SE-3001	108	3018		TR22/	
SE3001R	108			TR80/	
SE3001Y	108	3019			
SE3002	108	3018	11	TR95/56	S0020
SE-3002	108	3019			
SE3003	108	3039			S0017
SE3005	161	3117	86	TR83/719	S0020
SE-3005	108	3018			
SE3019	161	3132	39	TR83/719	
SE-3019	108	3019			
SE3030	162		19	TR67/707	S5020
SE3031	162		19	TR67/707	S5020
SE3032	130	3027	19	TR59/247	S7002
SE3033	130		19	TR59/247	S7002
SE-3033	130	3027			
SE3034	130		18	TR59/247	S7002
SE-3034	224	3049			
SE3035	130	3027	19	TR59/247	S7002
SE3036	130	3027	19	TR59/247	S7002
SE3040	175			TR81/241	S5019
SE3041	175			TR81/241	S5012
SE3646			19		S0025
SE-3646	123A				
SE3819	133	3112			
SE4001	123A		60	TR21/735	
SE-4001	123A	3124			
SE4002	199	3124	212	TR21/735	S0015
SE-4002	123A	3124			
SE4010	123A	3124	212	TR21/735	S0015
SE-4010	123A	3124		TR21/	S0015
SE4020	161	3117	39	TR83/719	S0005
SE4021	161	3117	85	TR83/719	
SE4022	161	3117	39	TR83/719	
SE4172	123A		61		S0011
SE5001	161	3132	60	TR83/719	S0020

Original Part Number	ECG	SK	GE	IR/WEP	HEP
SE-5001	108	3019			
SE5002	161	3132	60	TR83/719	S0020
SE-5002	108	3018			
SE5003	161	3132	60	TR83/719	S0020
SE-5003	108	3018			
SE5004	161	3018	39	TR83/719	
SE5006	108	3018	60	TR95/56	S0016
SE-5006	123A	3018	17	TR21/	
SE-5006-14	108	3018			
SE5010	108		61		S0016
SE5015	108		61		S0016
SE5020	108	3018	60	TR95/56	S0017
SE-5020	108	3018			
SE5021	161	3018	60	TR83/719	S0017
SE-5021	108	3018			
SE5022	161	3117	60	TR83/719	S0017
SE5023	161	3018	60	TR83/719	S0017
SE-5023	161	3039		TR21/	S0017
SE5024	161	3018	60	TR83/719	S0017
SE5025	161	3018	61	TR83/719	S0020
SE-5025	161	3018	61		S0020
SE5029	108		61		S0016
SE5030	108	3018	61	TR95/	S0025
SE5030A, B	123A		61		S0025
SE5031	108		61		S0016
SE5032	161		60		
SE5035	108	3018	39	TR95/56	S0017
SE5036	108				S0016
SE5040	108		60		S0016
SE5050	161	3018	60	TR83/719	S0017
SE-5050	108	3018			
SE5051	161	3117	60	TR83/719	S0017
SE5052	161	3117	60	TR83/719	S0017
SE5055	123A	3117	60	TR21/735	S0011
SE5056	108	3018	61		S0016
SE5151	123A	3122	20	TR21/735	
SE6001	161	3124	47	TR83/719	S0015
SE-6001	123A	3124		TR25/	
SE6002	161	3124	47	TR83/719	S0015
SE-6002	123A	3020		TR21/	
SE6006	128	3018	20	TR87/243	
SE6010	123A	3122	20	TR21/735	S0011
SE6020	128	3024	18	TR87/243	S3001
SE6020A	128	3024		TR87/243	S3011
SE6021	128	3024	18	TR87/243	S3019
SE6021A	128	3024	18	TR87/243	S3011
SE6022	128	3024	18	TR87/243	S3001
SE6023	128	3024	18	TR87/243	S3019
SE7001	154		40	TR78/712	S5026
SE-7001	103A	3010			
SE7002	154	3045	40	TR78/712	S5026
SE7005	128	3024	18	TR87/243	S5025
SE7006	124	3021	32	TR81/240	S5011
SE7010	154	3045	20	TR78/712	S5025
SE7015	128	3024	27	TR87/243	S0005
SE7016	154	3045	27	TR78/712	S0005
SE7017	154	3045	27	TR78/712	S0005
SE7020	124	3021	12	TR81/240	S5011
SE7030	124	3021	12	TR81/240	S3021
SE7055	154	3045	40	TR78/712	
SE7056	154	3045	27	TR78/712	
SE8001	128	3024	18	TR87/243	S0015
SE-8001	128	3024			
SE8002	128	3024	18	TR87/243	S0015
SE8010	154		18	TR78/712	S0005
SE-8010	128	3047			
SE8012	128	3024	18	TR87/243	S0005
SE8040	123A	3122	20	TR21/735	S3001
SE8041	128	3024	47	TR87/243	S3001
SE8042	128	3024	63	TR87/243	S3001
SE8510	128	3024	18	TR87/243	
SE8520	128	3024	18	TR87/243	
SE8521	128	3024	18	TR87/243	
SE8540	129	3025	48	TR88/242	S0019
SE8541	129	3025	48	TR88/242	
SE8542	129	3025	67	TR88/242	
SE9002	130	3027	19	TR59/247	
SE9020	162			TR67/707	S5020
SE9060	175				S5012
SE9061	175				S5012
SE9062	175				S5012
SE9063	175				S5012
SE9080	130	3027	14	TR59/247	
SE9300		3182			
SE9301		3182			
SE9302		3182			
SE9303		3183			
SE9560	218				S5018
SE9561	218				S5018
SE9562	218				S5018
SE9563	218				S5018
SE9570	185	3191			
SE9571	185	3191			
SE9572	185	3191			S5006
SE9573	185	3191			S5006

Original Part Number	ECG	SK	GE	IR/WEP	HEP
SE40022	121		16	TR01/232	
SE-40022	121	3009		TR01/	
SE50399	127			TR27/235	
SEC1078	192		63	TR25/	
SEC1079	192		63	TR25/	
SEC1477	192		63	TR25/	
SEC1479	192		63	TR25/	
SES632	130				S7004
SES881	130		14		
SES3819	132	3116		FE100/802	
SF.T124	158		53		
SF.T130	158		53		
SF.T131P	158		53		
SF.T163	160		50		
SF.T171	100		1	TR85/	
SF.T172	100		1	TR85/	
SF.T173	100		1	TR85/	
SF.T174	100		1	TR85/	
SF.T184	101	3123	5	TR85/	
SF.T187	171		27		
SF.T191			16	TR02/	
SF.T212	127		25	TR16/	
SF.T213	127		25	TR16/	
SF.T214	127		25	TR16/	
SF.T221	158		53		
SF.T222	102		2		
SF.T223	158		53		
SF.T227	102		2		
SF.T237	102A	3123	1	TR05/	
SF.T238	127		25	TR01/	
SF.T239	127		25	TR01/	
SF.T240	127		25	TR01/	
SF.T250	127		25		
SF.T251	102A	3123	52	TR05/	
SF.T252	102A	3123	52	TR05/	
SF.T253	102A	3123	52	TR05/	
SF.T264	105		4	TR03/	
SF.T265				TR03/	
SF.T266				TR03/	
SF.T306	102A	3123	2	TR05/	
SF.T316	160		50		
SF.T317	160		50		
SF.T318	102	3123	2	TR06/	
SF.T319	160		50		
SF.T320	160		50		
SF.T321	158	3123	53	TR05/	
SF.T322	158		53		
SF.T337	102A	3123	2	TR05/	
SF.T351	158	3123	53	TR05/	
SF.T352	158	3123	53	TR05/	
SF.T353	176	3123		TR05/	
SF.T354	160		50		
SF.T357	160		50		
SF.T358	160		50		
SF.T377	103A		59		
SF.T440	192		63	TR25/	
SF.T443	192		63	TR25/	
SF.T443A	192		63	TR25/	
SF.T445	192		63	TR25/	
SF.T714	192		63	TR25/	
SF100P					C3073P
SF101P					C3073P
SF102P					C3073P
SF103P					C3073P
SF115			61		
SF115A,B,C,D,E			61		
SF167			60		
SF173			60		
SF194,B			39		
SF195,C,D			39		
SF196			60		
SF197			60		
SF294,B			39		
SF295,C,D			39		
SF310			60		
SF314			60		
SF334,B			60		
SF335,C,D			60		
SF1001	123A	3018		TR21/735	
SF1713	123A	3122	8	TR21/735	S0014
SF1714	123A	3122	8	TR21/735	S0014
SF1726	123A	3122	20	TR21/735	S0014
SF1730	123A	3122	18	TR21/735	S0014
SF4010				TR21/	
SF8014	159	3114			
SFB6183		3116			
SFB8970	222				
SFC3012	771				C6004
SFC4050	755				
SFC4052	755				C6013
SFC6032	723	3144			
SFC6050	767				C6015
SFC8012A	771				
SFC8021A	771				
SFC8999		3050			F2004

Original Part Number	ECG	SK	GE	IR/WEP	HEP
SFD-23	129	3025		TR88/242	
SFE145	133	3112	FET1		F1035
SFE303	221	3050	FET4		F2004
SFE425	222				
SFE106					G0005
SFE107					G0005
SFE108					G0005
SFE113					G6011
SFE114					G6011
SFE115					G0009
SFE120	160		51	TR17/637	
SFE121	158		53	TR84/630	G0005
SFE122	158		53	TR84/630	G0005
SFE123	158		53	TR84/630	G0005
SFE124	158		53	TR84/630	G0005
SFE125	158		53	TR84/630	G0005
SFE125P	158			TR84/630	G0005
SFE126					G0005
SFE127					G0005
SFE128					G0005
SFE130	158		53	TR84/630	G0005
SFE131	158		53	TR84/630	
SFE131P	158			TR84/630	
SFE135					G0005
SFE136					G0005
SFE141					G0005
SFE142					G0005
SFE143	158		53	TR84/630	G0005
SFE144	158		53	TR84/630	G0005
SFE145	158			TR84/630	
SFE146	158			TR84/630	
SFE150					G6012
SFE151	102A		53	TR85/250	G0005
SFE152	102A		53	TR85/250	G0005
SFE153					G0005
SFE155					G0009
SFE162	160			TR17/637	G0003
SFE163	160		51	TR17/637	G0003
SFE-163	160	3006			
SFE168					G6006
SFE171	160		50	TR17/637	G0003
SFE172	160		50	TR17/637	G0003
SFE173	160		50	TR17/637	G0003
SFE174	160		50	TR17/637	G0003
SFE184	101		8	TR08/641	G0011
SFE-184	103A	3010			
SFE186	154	3045	27	TR78/712	
SFE187	154	3045	32	TR78/712	
SFE190	121	3009		TR01/232	G0005
SFE191	121	3009		TR01/232	
SFE192	121	3009	16	TR01/232	G6005
SFE211					G6005
SFE212	121	3009	3	TR01/232	G6003
SFE213	121	3014	16	TR01/232	G6003
SFE214	121	3009	3	TR01/232	G6005
SFE221	102	3004	51	TR05/631	G0005
SFE221A	102A	3004			
SFE222	102	3004	53	TR05/631	G0005
SFE222A	102A	3004			
SFE223	100	3005	51	TR05/254	G0005
SFE226	100	3005	51	TR05/254	G0005
SFE227	100	3005	51	TR05/254	G0001
SFE228	100	3005	51	TR05/254	G0001
SFE229	100	3005	51	TR05/254	G0001
SFE232	102A			TR85/250	
SFE235					G6012
SFE237	100		51	TR05/254	G0005
SFE238	121	3009	16	TR01/232	G6003
SFE239	121	3014	16	TR01/232	G6005
SFE240	121	3009	25	TR01/232	G6005
SFE241	158		53	TR84/630	G0005
SFE242	158			TR84/630	
SFE243	158		53	TR84/630	G0005
SFE244					G6012
SFE245					G6012
SFE250	121	3009	25	TR01/232	G6005
SFE251	100	3005	53	TR05/254	G0005
SFE252	100	3005	53	TR05/254	G0005
SFE253	100	3005	53	TR05/254	G0005
SFE259	101	3011	54	TR08/641	G0011
SFT260	101	3011	54	TR08/641	G0011
SFT261	101	3011	54	TR08/641	G0011
SFT264	105	3012	4	TR03/233	G6004
SFT265	105		4	TR03/233	G6004
SFT-265	105	3012			
SFT266	105		4	TR03/233	G6006
SFT-266	105	3012			
SFT267	105		4	TR03/233	G6006
SFT-267	105	3012			
SFT268	160		4	TR17/637	G6006
SFT288	100		51	TR05/254	G0001
SFT298	101		5	TR08/641	G0011
SFT-298	101	3011			
SFT306	102A		51	TR85/250	G0005
SFT-306	102A	3004			

Original Part Number	ECG	SK	GE	IR/WEP	HEP
SFT3007	126		53	TR17/635	G0005
SFT-307	100	3005			
SFT308	126	3008	51	TR17/635	G0008
SFT315	160		51	TR17/637	G0003
SFT-315	126	3008			
SFT316	160		51	TR17/637	G0003
SFT-316	126	3008			
SFT317	126		51	TR17/635	G0008
SFT-317	126	3008			
SFT318	126		53	TR17/635	G0002
SFT319	126		51	TR17/635	G0008
SFT-319	100	3005			
SFT320	126		51	TR17/635	G0008
SFT-320	126	3008			
SFT321	102A		53	TR85/250	G0005
SFT322	102A		53	TR85/250	G0005
SFT-322	102A	3004			
SFT323	102A		53	TR85/250	G0005
SFT-323	102A	3004			
SFT325	158/410		53	/630	G0005
SFT327	102A		53	TR85/250	
SFT-327	102A	3004			
SFT337	102A	3123	53	TR85/250	G0005
SFT-337	102A	3004			
SFT337B	102A			TR85/250	
SFT337V	102A			TR85/250	
SFT351	102A	3004	53	TR85/250	G0005
SFT352	102A		53	TR85/250	G0005
SFT-352	102A	3004			
SFT353	102A	3003	2	TR85/250	G0005
SFT-353	102A	3004			
SFT354	126		51	TR17/635	G0003
SFT-354		3008			
SFT357	126		51	TR17/635	G0003
SFT-357	126	3008			
SFT357P	126	3006		TR17/637	G0003
SFT358	160		51	TR17/637	G0003
SFT-358	160	3006			
SFT367	158		53	TR84/630	
SFT377	103A	3024	54	TR08/724	G0011
SFT440	128/411			/243	
SFT443	128	3024		TR87/243	S3011
SFT443A	128	3024		TR87/243	
SFT445	128	3024		TR87/243	
SFT523	158		53	TR84/630	
SFT526	102A		53	TR85/250	G0005
SFT713	123A	3122	20	TR21/735	S0011
SFT714	123A	3122	20	TR21/735	S0011
SG40P					C3020P
SG41P					C3020P
SG42P					C3020P
SG43P					C3020P
SG60P					C3030P
SG61P					C3030P
SG62P					C3030P
SG101AT,M,T		3565			
SG111T		3567			
SG130P					C3040P
SG131P					C3040P
SG132P					C3040P
SG133P					C3040P
SG140P					C3000P
SG141P					C3000P
SG142P					C3000P
SG143P					C3000P
SG190P					C3010P
SG191P					C3010P
SG192P					C3010P
SG193P					C3010P
SG201AT,T		3565			
SG211T		3567			
SG301AM,AT		3565			
SG311T		3567			
SG333					C3001P
SG339AN,N		3569			
SG383					C3004P
SG709CN	909D				
SG709CT	909	3590			
SG710CD,CN	910D				
SG710CT	910				
SG711CD,CN	911D				
SG711CT	911				
SG723CD	923D	3165			
SG723CN	923D				
SG723CT	923	3164			
SG741		3514			
SG741CD	941D	3552			
SG741CM	941M	3514			
SG741CN	941D				
SG741CT	941	3514			
SG741M,T		3514			
SG747CT	947	3526			
SG2182	102A+103A				
SG2183	102A+103A				
SG5013	128	3024		TR87/	

Original Part Number	ECG	SK	GE	IR/WEP	HEP
SGS17139XRH	1113				
SGS87231	161	3039	39		S0017
SH17910L					C2006G
SHA7520	193		67	TR28/	
SHA7521	193		67	TR28/	
SHA7522	193		67	TR28/	
SHA7523	193		67	TR28/	
SHA7524	193		67	TR28/	
SHA7526	193		67	TR28/	
SHA7527	193		67	TR28/	
SHA7528	193		67	TR28/	
SHA7530	159	3114	82	TR28/	
SHA7531	159	3114	82	TR28/	
SHA7532	159	3114	22	TR28/	
SHA7533	159	3114	82	TR28/	
SHA7534	159	3114	82	TR28/	
SHA7536	159	3114	22	TR20/	
SHA7537	159	3114	82	TR28/	
SHA7538	159	3114	82	TR28/	
SHA7597	193		67	TR28/	
SHA7598	193		67	TR28/	
SHA7599	193		67	TR28/	
SI341P			21		S0012
SI342P			21		S0012
SI343P			21		S0012
SI351P			21		S0013
SI352P			21		S0013
SI353P			21		S0013
SJ130	124			TR81/240	S5011
SJ285	185	3191		/S5007	
SJ570	123A	3122	20	TR21/735	
SJ619	130			TR59/247	S7002
SJ619-1	130			TR59/247	S7002
SJ652	218				S5018
SJ805	124				S5011
SJ806	124	3021	12	TR81/240	S5011
SJ811	175			TR81/241	S5019
SJ820	130			TR59/247	S7002
SJ821	219			/S5005	S7003
SJ822	218			TR58/700	S5018
SJ1106	130			TR59/247	S7004
SJ1152	153		26	TR77/700	S5018
SJ1165	124	3021	12	TR81/240	S5011
SJ1171	153		26	TR77/700	S5018
SJ1172	175	3131		TR81/241	S5012
SJ1201	124	3021	12	TR81/240	S5011
SJ1272	180				S7001
SJ1284	153		26	TR77/700	S5018
SJ1286	124	3021	12	TR81/240	S5011
SJ1470	130			TR59/247	S7004
SJ1902	247				
SJ1903	248				
SJ2000	130		19	TR26/	
SJ2001	219			TR29/	S7003
SJ2008	130				
SJ2009	218				
SJ2023	180				S7001
SJ2024	180				S7000
SJ2031	129				S5013
SJ2032	128				S5014
SJ2047	181				
SJ2064	181				
SJ2095	152	3026	66	TR76/	S5019
SJ2095(A)				TR23/	
SJ2283		3191			
SJ2515					S5012
SJ2516					S5018
SJ2519					S5020
SJ3408	175	3131		TR81/241	S5019
SJ3423	162		35	TR67/707	S5020
SJ3447	175	3131		TR81/241	S5012
SJ3464	130				S7004
SJ3477	162		35	TR67/707	S5020
SJ3478	162			TR67/707	S5020
SJ3507	180			/S7001	S7003
SJ3519	181				S7002
SJ3520	219				S7003
SJ3604	130			TR59/247	S7004
SJ3607					S7003
SJ3608					S7002
SJ3636	180				S7000
SJ3637	180				S7001
SJ3648	175	3131		TR81/241	
SJ3678	130			TR59/247	S7002
SJ3679	219				S7003
SJ3680	175				
SJ5106					S7004
SJ5196	162				S5020
SJ5231					S5018
SJ5525		3115			
SJ5526		3111			
SJ8701	130			TR59/247	S7002
SJ9110	130			TR59/247	S7004
SJE42	152	3054		TR76/701	S5000
SJE100	184	3054	57	TR76/S5003	S5000

Original Part Number	ECG	SK	GE	IR/WEP	HEP
SJE103	190	3104	27		S3021
SJE106	184	3054	57	TR76/S5003	S5000
SJE108	185	3083	58	TR77/S5007	S5006
SJE111	185	3083	58	TR77/S5007	S5006
SJE112	185	3083	58	TR77/S5007	S5006
SJE113	184	3054	57	TR76/S5003	S5000
SJE114	185	3083	58	TR77/S5007	S5006
SJE133	184	3190		TR76/	S5000
SJE202	185	3083	58	TR77/S5007	S5006
SJE203	184	3054	57	TR76/S5003	S5000
SJE205	157	3103		TR60/244	S5015
SJE210	185	3083	58	TR77/S5007	S5006
SJE211	184	3054	57	TR76/S5003	S5000
SJE218	157	3103		TR60/244	S5015
SJE220	182			/S5004	S5000
SJE221	185	3083	58	TR77/S5007	S5006
SJE-221		3083			
SJE222	184	3054	57	TR76/S5003	S5000
SJE-222		3010			
SJE227	185	3083	58	TR77/S5007	S5006
SJE228	184	3054	57	TR76/S5003	S5000
SJE229	184	3054	57	TR76/S5003	S5000
SJE231	185	3083	58	/S5007	S5006
SJE232	157	3103		TR60/244	S5015
SJE237	184	3054	57	TR76/S5003	S5000
SJE241	185	3083	58	TR77/S5007	S5006
SJE242	184	3054	57	TR76/S5003	S5000
SJE243	185	3083	58	TR77/S5007	S5006
SJE244	184	3054	57	TR76/S5003	S5000
SJE245	185	3083	58	TR77/S5007	S5006
SJE246	184	3054	57	TR76/S5003	S5000
SJE247					S5006
SJE248	184	3054	57	TR76/S5003	S5000
SJE248MK					S5004
SJE252MK					S5004
SJE253	184	3054	57	TR76/S5003	S5000
SJE254	184	3054	57	/S5003	S5000
SJE255	184	3041	57	/S5003	S5000
SJE256	185	3041	58	/S5007	S5006
SJE257	185	3083	58	TR77/S5007	S5006
SJE261	184	3054	57	TR76/S5003	S5000
SJE262	184	3054	57	TR76/S5003	S5000
SJE264	180			/S7001	
SJE265	185	3083	58	TR77/S5007	S5006
SJE267	185	3083	58	TR77/S5007	S5006
SJE271	184	3054	57	TR76/S5003	S5000
SJE272	184	3054	57	TR76/S5003	S5000
SJE273	185	3083	58	TR77/S5007	S5006
SJE274	184	3054	57	TR76/S5003	S5000
SJE275	185	3083	58	TR77/S5007	S5006
SJE276	185	3083	58	TR77/S5007	S5006
SJE277	185	3083	58	TR77/S5007	S5006
SJE278	184	3054	57	TR76/S5003	S5000
SJE279	185	3083	58	TR77/S5007	S5006
SJE280	184	3054	57	TR76/S5003	S5000
SJE283	185	3083	58	/S5007	S5006
SJE284	184	3041	57	/S5003	S5000
SJE288	185	3041	58	/S5007	S5006
SJE289	184	3041	57	/S5003	S5000
SJE290	157				S5015
SJE305	184	3054	57	TR76/S5003	S5000
SJE320	184	3041	57	/S5003	S5000
SJE340	184	3054	57	TR76/S5003	S5000
SJE400	157	3103		TR60/244	S5015
SJE401	184	3041	57	/S5003	S5000
SJE402	184	3041	57	/S5003	S5000
SJE403	185	3083	58	/S5007	S5006
SJE404	184	3054	57	/S5003	S5000
SJE405	184	3054	57	/S5003	S5000
SJE407	184	3054	57	/S5003	S5000
SJE408	185	3083	58	TR77/S5007	S5006
SJE513	152	3054		TR76/701	S5000
SJE-513	152	3041			
SJE514	153	3025		TR77/700	S5006
SJE515	152	3054		TR76/701	S5000
SJE-515	152	3041			
SJE516	182	3188			
SJE517	183				
SJE527	184	3190		TR76/	S5000
SJE583	184	3054	57	TR76/	S5000
SJE584	185	3083	58	TR77/	S5006
SJE587					S5004
SJE607					S5006
SJE633	185	3191	58		S5006
SJE634	184	3190	57		S5000
SJE669	184	3104	57	TR60/	S5000
SJE669/86X0059				TR60/	
SJE721	184	3054	57	TR76/S5003	S5000
SJE723	185	3083	58	/S5007	S5006
SJE724	184	3041	57	/S5003	S5000
SJE736	185	3083	58	TR77/S5007	S5006
SJE737	184	3054	57	TR76/S5003	S5000
SJE743	185	3041	58	/S5007	S5000
SJE747MK					S5005
SJE764	180			/S7001	S7003
SJE768	185	3083	58	/S5007	S5006
SJE769	184	3054	57	TR76/S5003	S5000
SJE781	184	3041	57	/S5003	S5000
SJE783	152	3041		TR76/701	S5000
SJE784	184	3054	57	/S5003	S5000
SJE785	184	3041	57	/S5003	S5000
SJE797	185	3083	58	TR77/S5007	S5006
SJE799	185	3083	58	TR77/S5007	S5006
SJE1032	190	3104		TR74/	
SJE1518	185	3083	58	TR77/S5007	S5006
SJE1519	184	3054	57	TR76/S5003	S5000
SJE1520	184	3054	57	TR76/S5003	S5000
SJE1767-2					S5006
SJE3702					S5004
SJE3702-1					S5001
SJE3702-2					S5004
SJE3754	157				S5015
SJE5018	182	3188			S5004
SJE5019	182	3188			S5004
SJE5020	182	3188			S5004
SJE-5038	186		28		
SJE5402	184	3190	57	/S5003	
SJE-5402	186				
SJE5439	241				
SK7	101			TR08/641	G0011
SK-7	101	3011			
SK19	132	3116		FE100/802	
SK342	904				
SK1320	108		11	TR95/56	S0016
SK1639	159	3114	21	TR20/717	
SK1639D	159	3114			
SK1640	159	3114	21	TR20/717	
SK1640A	123A	3114			
SK1641	123A	3122	20	TR21/735	
SK1700	716				
SK1856	159	3114		TR20/717	S0019
SK1856A	159	3114			
SK2604	159	3114		TR20/717	S0019
SK2604A	159	3114			
SK3003	102A		52	TR85/250	G0005
SK3004	102A		53	TR85/250	G0005
SK3005	100		1	TR85/250	G0007
SK3006	160		51	TR17/637	G0001
SK3007	160		50	TR17/637	G0001
SK3008	126		53	TR85/635	G0008
SK3009	121		25	TR01/232	G6003
SK3010	103A		5	TR08/641B	G0011
SK3011	101		8	TR08/641	G0011
SK3012	105		4	TR03/231	G6004
SK3013	121MP		13MP	TR01/232	G6003(2)
SK3014	121		3	TR01/232	G6005
SK3015	121MP		13MP	TR01MP/232	G6005(2)
SK3018	108		39	TR95/56	S0015
SK3019	108		11	TR95/56	S0017
SK3020	123A		20	TR83/735	S0015
SK3021	124		12	TR87/240	S5011
SK3022	704			IC501/	
SK3023	704			IC501/	
SK3024	128	3024	18	TR87/243	S5014
SK3025	129		22	TR88/242	S5013
SK3026	175		66	TR81/241	S5012
SK3027	130		75	TR81/247	S7002
SK3028	175(2)		24MP	TR57MP/241	S5012(2)
SK3029	130MP		75	TR81MP/247	S7002(2)
SK3034	127		25	TR36MP/235	G6007
SK3035	127		25	TR27/235	G6008
SK3036	181		75	TR27/S7000	S7004
SK3037	181(2)		75	TR36MP/S7000	S0025(2)
SK3038	123A		210	TR21/735	S0011
SK3039	108		11	TR80/56	S0017
SK3040	154		27	TR78/712	S5026
SK3041	152		55	TR92/701	S5004
SK3044	191		32	TR75/712	S3021
SK3045	225		32	TR78/	S5015
SK3046	123A		20	TR64/735	S3013
SK3047	128		215	TR63/243	S3013
SK3048	195		215	TR65/	S5014
SK3049	224		215	TR76/	S5000
SK3050	222		FET4		F2004
SK3052	131		30	TR94/642	G6016
SK3053	191			TR75/	
SK3054	196		55	TR92/S5004	
SK3065	222				F2004
SK3070	711				C6069G
SK3071	727				
SK3072	712		IC2	IC508/	C6063P
SK3073	728				
SK3074	729				
SK3075	714		IC4	IC509/	C6070P
SK3076	715		IC6	IC510/	C6071P
SK3077	713		IC5		C6057P
SK3078	789				
SK3079	162		35	TR61/707	S5020
SK3082	226		49	TR94/	G6016
SK3083	197			/701	S5006

Original Part Number	ECG	SK	GE	IR/WEP	HEP
SK3084	197			TR76/701	S5005
SK3085			56		S5006
SK3086	226MP			TR01MP/	G6016(2)
SK3101	706				
SK3102	710				
SK3103			32	TR75/	
SK3104	228		32	TR75/	S3021
SK3104-SK3003				TR75/	
SK3111	165			TR93/740A	S7005
SK3112	133		FET1		F0010
SK3114	159		48	TR20/	S0019
SK3115	165		38	TR93/740B	S7005
SK3116	132		FET2	FE100/	F0015
SK3117	108		60	TR70/56	S0020
SK3118	159		48	TR20/	S0019
SK3122	123A		63	TR21/	S0015
SK3123	176			TR82/	G6011
SK3124	123A		210	TR21/	S0015
SK3129	726				
SK3133	164		37	TR68/	S7005
SK3134	705A				C6089
SK3135					C6062P
SK3137	192				S5014
SK3138	193				S5013
SK3140	786				
SK3141	780		IC20		C6069G
SK3143	730				
SK3149	791				
SK3434A	123A	3122	20	TR21/735	
SK3510	130		75	TR59/	S7004
SK3511	181		75	TR36/	S7004
SK3512					S3002
SK3513					S3032
SK3524	784				
SK3525	724				
SK3526	947				
SK3529					S3011
SK3534					S5004
SK3538					S5012
SK3539	908				
SK3540	903				
SK3541	914				
SK3542	904				
SK3543	912				
SK3544	917				
SK3545	907				
SK3546	905				
SK3547	900				
SK3548	906				
SK3549	901				
SK3550	916				
SK3551	123A	3122	20	TR21/735	
SK3554	917				
SK3730	160	3006	51	TR17/637	G0003
SK3770	160		50	TR17/	
SK3803	159			TR20/	
SK3960	123A		20	TR21/735	
SK-3960	123A	3020			
SK4000	4000				
SK4001	4001				
SK4002	4002				
SK4007	4007				
SK4009	4049				
SK4010	4010				
SK4011	4011				
SK4012	4012				
SK4013	4013				
SK4015	4015				
SK4017	4017				
SK4019	4019				
SK4023	4023				
SK4024	4024				
SK4025	4025				
SK4027	4027				
SK4030	4030				
SK5184A	124	3021		TR81/240	
SK5797	159			TR20/	
SK5798	159			TR20/	
SK5801	123A			TR21/	
SK5814A			12		
SK5915	123A	3020	20	TR21/735	
SK6345	159	3114	21	TR20/717	
SK6345A		3114			
SK6346	159	3114	21	TR20/717	
SK6346A		3114			
SK6347	159	3114	21	TR20/717	
SK6347A	159	3114			
SK7181	108	3039	20	TR95/56	S0016
SK7664	159			TR20/	
SK8215	123A	3122			
SK8251	123A		20	TR21/735	
SK8261	154			TR78/	
SK-31024-3	108	3018		TR95/56	
SK32543-001		3024			
SK32543-002		3024			

Original Part Number	ECG	SK	GE	IR/WEP	HEP
SK-56072-427-1		3004			
SK16510006-2	129				
SK16510006-4	129				
SKA1079	129			TR88/	
SKA1080	123A	3020	20	TR21/735	S0024
SKA1117	123A		20	TR21/	
SKA1279	159		22		
SKA1395	123A			TR21/735	
SKA1416	107			TR70/	
SKA4074	108	3018	11	TR95/56	S0016
SKA4075	108		11	TR95/	
SKA4076	108			TR95/	
SKA4129	159	3114		TR20/717	
SKA4129A		3114			
SKA4141	123A			TR21/	
SKA4410	192		63		
SKA4525	108			TR95/	
SKA4616	128			TR87/	
SKA4621	129			TR88/	
SKA4768	161			TR83/	
SKA9013	108	3018	20		S0015
SKA-4075				TR21/	
SKB8339	123A			TR21/735	
SKWHG7006	159				
SL01640	9914				
SL02518	9900				
SL02519	9914				
SL02734	9914				
SL02779	9914				
SL02781	9923				
SL03019	9914				
SL03021	9923				
SL03667	9913				
SL03911	9949				
SL03912	9961				
SL03913	9963				
SL03914	9932				
SL03915	9944				
SL03916	9937				
SL03917	9951				
SL04194	9910				
SL04217	9900				
SL04218	9914				
SL04563	9930				
SL04567	9936				
SL04568	9932				
SL04570	9951				
SL04732	9914				
SL07040	909	3590			
SL07055	909	3590			
SL07059	703A				
SL07682	716				
SL07684	910				
SL08058	703A				
SL08066	909	3590			
SL08797	941D				
SL56					S3032
SL100	108		20	TR95/56	
SL-100	108	3019			
SL200	106		21	TR20/52	
SL201	106		21	TR20/52	
SL300	123A	3122	20	TR21/735	
SL301C	128		81		
SL301CE	128		81		
SL404					S0013
SL3101	129	3025		TR88/242	
SL3111	129			TR88/	
SL7059	703A	3157	IC12	IC500/	
SL7283	703A	3157	IC12		
SL7308	703A	3157	IC12		
SL7531	703A	3157	UC12		
SL7593	703A	3157	IC12		
SL8020	703A	3157	IC12	IC500/	
SL11877	9930				
SL11878	9936				
SL11879	9946				
SL11880	9948				
SL11881	9962				
SL14959	74H10				
SL14960	74H11				
SL14971	7408				
SL14972	7438				
SL16121	9923				
SL16122	9910				
SL16201	9930				
SL16203	9951				
SL16204	9946				
SL16206	9932				
SL16208	9937				
SL16209	9099				
SL16210	9936				
SL16211	9945				
SL16212	9962				
SL16215	9944				
SL16216	9949				

Original Part Number	ECG	SK	GE	IR/WEP	HEP
SL16218	9963				
SL16516	9930				
SL16517	9933				
SL16518	9936				
SL16519	9946				
SL16520	9962				
SL16521	9945				
SL16522	9099				
SL16584	9930				
SL16585	9931				
SL16586	9932				
SL16587	9936				
SL16588	9945				
SL16589	9946				
SL16590	9948				
SL16591	9962				
SL16793	7400				
SL16794	7401				
SL16795	7402				
SL16796	7404				
SL16797	7405				
SL16798	7408				
SL16799	7411				
SL16800	7420				
SL16801	7410				
SL16802	7430				
SL16803	7440				
SL16804	7450				
SL16805	7453				
SL16806	7473				
SL16807	7474				
SL16808	7476				
SL16809	7492				
SL16810	7496				
SL16811	9944				
SL17242	7473				
SL17284	9932				
SL17289	9936				
SL17869	7408				
SL17887	7411				
SL18386	7483				
SL18387	74107				
SL18699	9911				
SL20575	703A				
SL20721	705A	3134	IC3		
SL20755	713		IC5		
SL20783	923				
SL20927	909	3590			
SL20929	941				
SL21017	715		IC6		
SL21122	714	3075	IC4		
SL21384	915				
SL21385	923D				
SL21436	910				
SL21441	707	3134	IC3		
SL21577	915				
SL21584	949				
SL21619	923				
SL21654	712	3072	IC2		
SL21673	941				
SL21823	910				
SL21829	911				
SL21864	722	3161	IC9		
SL21885	923				
SL21895	703A				
SL21923	909	3590			
SL22044	911				
SL22108	725	3162			
SL22211	725				
SL22273	912				
SL22310	923D				
SL22348	925				
SL22623	912				
SL22745	941M				
SL22756	718				
SL22757	722				
SL22819	703A				
SL22935	923D				
SL23059	941M				
SL23145	722				
SL23147	712				
SL23252	941M				
SL23256	912				
SL23296	923				
SL23297	911				
SL23299	736				
SL23324	909	3590			
SL23325	923D				
SL23326	941D				
SL23418	705A				
SL23421	703A				
SL23422	910				
SL23423	911				
SL23424	725				
SL23425	949				
SL23426	722				
SL23482	909	3590			
SL23485	923				
SL23486	941				
SL23496	941M				
SL23546	923				
SL23560	705A				
SL23648	923				
SL23649	781				
SL23829	912				
SL23903	713				
SL23904	714				
SL23905	715				
SL23985	725				
SL53424	9930				
SL53425	9932				
SL53426	9935				
SL53427	9936				
SL53428	9944				
SL53429	9946				
SL53431	7401				
SL-174024		3039			
SL-175798		3039			
SM0843	102A			TR85/250	G0005
SM63	7475				C3075P
SM73	7475				C3075P
SM140				TR59/	
SM150				TR59/	
SM217	160	3123	2	TR17/637	
SM-217	100	3005			
SM576-1	123A				S0014
SM576-2	123A				S0014
SM716	123A	3122		TR21/735	
SM-716	123A				
SM843	102A			TR85/250	G0005
SM862	126				G0008
SM1297	160			TR17/637	G0008
SM1507	159				S0013
SM1600	160			TR17/637	G0002
SM2491	160		50	TR17/637	
SM2492	160			TR17/637	G0003
SM2700	123A	3122	20	TR21/735	S0014
SM2701	123A	3122	20	TR21/735	S0014
SM2716					S3001
SM3014	160			TR17/637	G0003
SM3104	123A	3122		TR21/735	S0014
SM3117A	123A	3122	20	TR21/735	
SM3505	123A				S0014
SM3788					S0014
SM3915					S0014
SM3916					S0012
SM3917					S0012
SM3978	128	3024	18	TR87/243	
SM3986	123A	3122		TR21/735	S0014
SM3987	129	3025		TR88/242	S0012
SM4304-S		3018			
SM4508-B	123A	3020		TR21/735	
SM4547	159	3114		TR20/717	S0013
SM4547A	159	3114			
SM4719	159	3114		TR20/717	S0013
SM4719A		3114			
SM5379	123A	3122		TR21/735	S0011
SM5380					S0005
SM5564	123A		20	TR21/735	G0011
SM-5564	123A	3020			
SM5643	123A		20	TR21/735	S0015
SM-5643	123A	3020			
SM5796	108	3039	20	TR95/56	
SM6251	128	3024	20	TR87/243	
SM6727	154			TR78/	
SM6728	129			TR88/	
SM6773	123A	3122	62	TR21/735	
SM6814	198		32		
SM7545	123A	3122		TR21/735	S0014
SM7815	123A		20	TR21/735	
SM-7815	123A	3020			
SM7836	123A		20	TR21/735	
SM-7836	123A	3020			
SM7989	224	3049			
SM7991	128	3047	20	TR87/243	
SM-7991	195				
SM8112	123A	3122	20	TR21/735	
SM8113	123A	3122	20	TR21/735	
SM8341	102A		53	TR85/250	
SM8978	123A	3122	20	TR21/735	
SM9008	123A	3122	20	TR21/735	
SM9135	123A	3020	20	TR21/735	
SM9253	123A	3122		TR21/735	S0015
SM9906					S5026
SM-716					S0014
SM-7991					S3001
SM-4304-S				TR21/	
SM-A-595819-1	704	3022			
SM-A-595830-12	107				
SM-A-618687-1	175				

Original Part Number	ECG	SK	GE	IR/WEP	HEP
SM-A-726655	123A				
SM-A-726658	159				
SM-A-726664	123A				
SM-B-523974	159				
SMA748700		3541			
SMB416401		3039			
SMB416405		3122			
SMB447610	102	3004			
SMB447883		3123			
SMB448902		3019			
SMB454549	102	3004		TR05/	
SMB454760	160			TR17/	
SM-B-610342	123A				
SM-B-686767	123A				
SMB603953		3542			
SMB620782-1	176				
SMB621960	102				
SMB653078		3542			
SMB-706009D	128				
SM-C-583256	123A				
SMC449077	129				
SMC502575		3036			
SMC-503225		3036			
SMC-583259	128	3044			
SMC-620774-1	128				
SMC750123-1	941				
SMD-413796		3004			
SN60	101		20	TR08/641	S0011
SN74H00N	74H00				
SN74H01N	74H01				
SN74H04N	74H04				
SN74H05N	74H05				
SN74H10N	74H10				
SN74H11N	74H11				
SN74H20N	74H20				
SN74H21N	74H21				
SN74H22N	74H22				
SN74H30N	74H30				
SN74H40N	74H40				
SN74H50N	74H50				
SN74H51N	74H51				
SN74H52N	74H52				
SN74H53N	74H53				
SN74H54N	74H54				
SN74H55N	74H55				
SN74H60N	74H60				
SN74H61N	74H61				
SN74H62N	74H62				
SN74H71N	74H71				
SN74H72N	74H72				
SN74H73N	74H73				
SN74H74N	74H74				
SN74H76N	74H76				
SN74H78N	74H78				
SN74H87N	74H87				
SN74H101N	74H101				
SN74102N	74H102				
SN74H103N	74H103				
SN74H106N	74H106				
SN74H108N	74H108				
SN74H183N	74H183				
SN74L73N					C3073P
SN74S00N	74S00				
SN74S03N	74S03				
SN74S04N	74S04				
SN74S05N	74S05				
SN74S10N	74S10				
SN74S11N	74S11				
SN74S15N	74S15				
SN74S20N	74S20				
SN74S22N	74S22				
SN74S30N	74S30				
SN74S40N	74S40				
SN74S51N	74S51				
SN74S64N	74S64				
SN74S65N	74S65				
SN74S74N	74S74				
SN74S86N	74S86				
SN74S112N	74S112				
SN74S113N	74S113				
SN74S114N	74S114				
SN74S133N	74S133				
SN74S134N	74S134				
SN74S138J	74S138				
SN74S138N	74S138				
SN74S140N	74S140				
SN74S151J,N	74S151				
SN74S153J,N	74D153				
SN74S157J,N	74S157				
SN74S158J,N	74S158				
SN74S174N	74S174				
SN74S175J,N	74S175				
SN74S194J,N	74S194				
SN74S251J,N	74S251				
SN74S258J,N	74S258				
SN80	101		20	TR08/641	S0015
SN166	128	3024	63	TR87/243	
SN167	128	3024	63	TR87/243	
SN238S	749				
SN-400-319-P1	159				
SN962-2M	9962				
SN7400N	7400N				C3000P
SN7401N	7401				C3001P
SN7402N	7402				
SN7403N	7403				
SN7404N	7404				C3004P
SN7405N	7405				
SN7406N	7406				
SN7407N	7407				
SN7408N	7408				
SN7409N	7409				
SN7410N	7410				C3010P
SN7412N	7412				
SN7413N	7413				
SN7414N	7414				
SN7416N	7416				
SN7417N	7417				
SN7420N	7420				C3020P
SN7422N	7422				
SN7423N	7423				
SN7425N	7425				
SN7426N	7426				
SN7427N	7427				
SN7428N	7428				
SN7430N	7430				C3030P
SN7432N	7432				
SN7433N	7433				
SN7437N	7437				
SN7438N	7438				
SN7440N	7440				C3040P
SN7441N					C3041P
SN7442N	7442				
SN7443N	7443				
SN7444N	7444				
SN7445N	7445				
SN7446AN	7446				
SN7447AN	7447				
SN7448N	7448				
SN7450N	7450				C3050P
SN7451N	7451				
SN7453N	7453				
SN7454N	7454				
SN7460N	7460				
SN7470N	7470				
SN7472N	7472				
SN7473N	7473				C3073P
SN7474N	7474				
SN7475N	7475				C3075P
SN7476N	7476				
SN7480N	7480				
SN7481A	7481				
SN7482N	7482				
SN7483AN	7483				
SN7484AN	7484				
SN7485N	7485				
SN7486N	7486				
SN7488AN	7488				
SN7489N	7489				
SN7490AN	7490				
SN7491AN	7491				
SN7492AN	7492				
SN7492N					C3801P
SN7493AN	7493A				
SN7494N	7494				
SN7495AN	7495				
SN7496N	7496				
SN7497N	7497				
SN7541P					C6052P
SN7610N-07	722	3161			
SN15830J	9930				
SN15830N	9930				C1030P
SN15831J,N	9931				
SN15832J	9932				
SN15832N	9932				C1032P
SN15833J	9932				
SN15833N	9933				C1033P
SN15835J,N	9935				
SN15836J,N	9936				
SN15837N	9937				
SN15838J,N	9135				
SN15844J	9944				
SN15844N	9944				C1044P
SN15845J	9945				
SN15845N	9945				C1045P
SN15846J,N	9946				
SN15848J,N	9948				
SN15849J,N	9949				
SN15850J,N	9950				
SN15851J,N	9951				
SN15857J,N	9157				

Original Part Number	ECG	SK	GE	IR/WEP	HEP
SN15858J,N	9158				
SN15861J,N	9961				
SN15862J	9962				
SN15862N	9962				C1062P
SN15863J,N	9963				
SN17810L					C2006G
SN17811L					C2007G
SN17910L					C2006G
SN52101AL,L		3565			
SN52741,L,P		3514			
SN72301AL,AN,AP		3565			
SN72471L		3514			
SN72471P	941M				
SN72558	778	3557			
SN72558P	778	3551			
SN72709L	909	3551			
SN72709N	909D	3552			
SN72710L	910	3553			
SN72710N	910D	3554			
SN72711L	911	3555			
SN72711N	911D	3556			
SN72741L	941	3514			
SN72741N	941D	3558			
SN72741P	941M	3514			C6052P
SN74090N					C3800P
SN74107N	74107				
SN74121N	74121				
SN74122N	74122				
SN74123N	74123				
SN74132N	74132				
SN74141N	74141				
SN74145N	74145				
SN74150N	74150				
SN74151N	74151				
SN74152N	74152				
SN74153N	74153				
SN74154N	74154				
SN74155N	74155				
SN74156N	74156				
SN74160N	74160				
SN74161N	74161				
SN74162N	74162				
SN74163N	74163				
SN74164N	74164				
SN74165N	74165				
SN74166N	74166				
SN74170N	74170				
SN74174N	74174				
SN74175N	74175				
SN74176N	74176				
SN74177N	74177				
SN74178N	74178				
SN74179N	74179				
SN74180N	74180				
SN74181N	74181				
SN74182N	74182				
SN74190N	74190				
SN74191N	74191				
SN74192N	74192				
SN74193N	74193				
SN74196J,N	74196				
SN74197N	74197				
SN74198N	74198				
SN74199N	74199				
SN75110N	722	3161			
SN76001	1113				
SN76001A,AN	1113				
SN76007N	812				
SN76021N,ND	810				
SN76104	718	3159			
SN76104N	718	3159		IC511/	C6094P
SN76105	720				C6095P
SN76105N	720	3160		IC512/	C6095P
SN76107					C6056P
SN76107N	722				
SN76110	722	3161	IC9	IC513/	
SN76110N	722	3161	IC9		
SN76111	719		IC28		
SN76115			IC35		
SN76115N	801	3160			
SN76131	725	3162			
SN76131N	719	3162			
SN76177	732				
SN76177ND	732				
SN76242	714	3075			
SN76242N	714	3075	IC4	IC509/	C6070P
SN76242N-07	714				
SN76243	715	3076			
SN76243N	715		IC6	IC510/	C6071P
SN76246	713	3077			
SN76246N	713	3077	IC5	IC508/	C6057P
SN76266	728	3073			
SN76266N	728				
SN76267	729	3074			
SN76267N	729				

Original Part Number	ECG	SK	GE	IR/WEP	HEP
SN76530P	747				C6079P
SN76564		3141	IC21		
SN76564N	783				
SN76600	746				C6059P
SN76600N	746				
SN76600P	746		IC26		
SN76642	709				
SN76642N	709	3135		IC505/	C6059P
SN76643		3135			
SN76643N	708	3135	IC10	IC504/	
SN76650	749				C6076P
SN76650N	749	3171			C6076P
SN76651N	748				C6060P
SN76653N	708		IC10	IC504/	
SN76665	712	3072		IC507/	
SN76665N	712	3072	IC2	IC507/	C6063P
SN76666	712				
SN76666N	712	3072			
SN76675N	723				
SN76676L	781	3169			
SN151800J	9800				
SN151800N	9800				
SN151801J,N	9801				
SN151802J,N	9802				
SN151803J,N	9803				
SN151804J,N	9804				
SN151805J,N	9805				
SN151806J,N	9806				
SN151807J,N	9807				
SN151808J,N	9808				
SN151809J,N	9809				
SN151810J,N	9810				
SN151811J,N	9811				
SN151812J,N	9812				
SN158093J,N	9093				
SN158094J,N	9094				
SN158097J,N	9097				
SN158099J	9099				
SN158099N	9099				C1052P
SN764246N		3077			
SNT204	159	3114	21	TR20/717	
SNT204A	159	3114			
S01	160		51	TR17/635	
S02	160		51	TR17/635	G0008
S03	160		51	TR17/635	G0008
S025	102A	3004	52	TR85/250	
S065	102A	3005	2	TR85/254	
S065A	126			TR17/635	
S0-65A	160	3007			
S088	102A	3004	2	TR85/250	
S0-1				TR12/	
S0-2				TR12/	
S0-3				TR12/	
S0-4295-10		3027			
S019806	199				
S025094	199				
S0S0121	129	3025			
SP26	104				G6003
SP39	127	3035		TR27/235	G6018
SP47	104	3009		TR01/230	G6003
SP62	127	3035		TR27/235	G6018
SP70	159		21	TR28/	S0019
SP-70	198		32		
SP90	159				S0013
SP115					G6005
SP148-3	121	3009		TR01/232	
SP176	104	3009	3	TR01/230	
SP180C1		3542			
SP209					C6018
SP230	104	3009		TR01/230	G6003
SP300,A					G6005
SP301,A					G6005
SP302,A					G6005
SP305,A					G6003
SP306,A					G6004
SP314,A					G6006
SP317,A					G6006
SP321,A					G6018
SP333,A					G6005
SP334	121	3009	16	TR01/232	G6001
SP334A					G6001
SP337,A					G6006
SP356,A					G6006
SP362,A					G6013
SP365,A					G6005
SP367					G6006
SP370,A					G6013
SP374,A					G6006
SP375,A					G6006
SP377,A					G6006
SP380,A					G6018
SP384,A					G6010
SP387,A					G6013
SP404	121		16		G6013
SP404T	121	3009	16	TR01/232	G6005

Original Part Number	ECG	SK	GE	IR/WEP	HEP
SP441	121	3009	16	TR01/232	G6003
SP441D	104	3009		TR01/230	G6003
SP441G	121	3014		TR01/232	G6005
SP441S	104	3009		TR01/230	G6003
SP485	121	3009	16	TR01/232	G6003
SP485B	121	3009		TR01/232	G6013
SP485BLK,BL,BRN	121	3009		TR01/232	G6013
SP485W	121	3009	16	TR01/232	G6013
SP485WHT	121	3009		TR01/232	G6013
SP486	104		16	TR01/230	G6003
SP-486	121	3009			
SP486W	121		16	TR01/232	G6013
SP-486W	121	3009			
SP486WHT	121	3009		TR01/232	G6013
SP541					G6005
SP634	121	3009	16	TR01/232	G6005
SP649	121	3009	16	TR01/232	
SP649-1	121	3009	16	TR01/232	G6005
SP744	121	3014		TR01/232	G6005
SP777					G6003
SP819R	121	3009		TR01/232	G6013
SP834	121	3009	16	TR01/232	G6005
SP838	179	3009		TR01/232	G6009
SP838-1	179			TR17/637	
SP875	121	3009		TR01/232	G6003
SP880	104		16	TR01/230	G6003
SP-880	121	3009			
SP880-1	121	3009	16	TR01/232	G6005
SP880-2					G6005
SP880-3	121	3009	16	TR01/232	G6005
SP880-4,5,6,7,8					G6005
SP880-9,10,11					G6005
SP891	104	3014	16	TR01/230	G6003
SP-891	121	3009			
SP891B	121	3009	16	TR01/232	G6013
SP891BLU	121	3009		TR01/232	G6013
SP891G	121	3009		TR01/232	G6013
SP891GRN	121	3009		TR01/232	G6013
SP891R	121	3009		TR01/232	G6013
SP891W	121	3009	16	TR01/232	G6013
SP891WHT	121	3009		TR01/232	G6013
SP1013B	104	3009		TR01/230	G6003
SP1029	179				
SP1081	213				G6010
SP1105	213				G6010
SP1108	104		16	TR01/634	G6013
SP-1108	121	3009			
SP1118	121				G6005
SP1137	104	3009	16	TR01/624	
SP1271	121	3014		TR01/232	G6005
SP1323	104	3009		TR01/230	G6003
SP1378					G6018
SP1403	104	3009		TR01/230	G6003
SP1481	121	3009	3	TR01/232	G6013
SP1481-1	121	3009		TR01/232	
SP1481-2,3,4,5	121	3009		TR01/232	G6013
SP1482	121	3009	3	TR01/232	G6013
SP1482-2	121	3009		TR01/232	G6013
SP1482-3	121	3009	16	TR01/232	G6013
SP1482-4	121	3009		TR01/232	G6013
SP1482-5	121	3009	16	TR01/232	G6013
SP1482-6,7	121	3009		TR01/232	G6013
SP1483	121	3009	3	TR01/232	G6013
SP14831,2,3	121	3009		TR01/232	G6013
SP1484	179	3009	3	/G6001	G6009
SP1484(1-5)	179			TR35/G6001	G6009
SP1550-3	121	3009	16	TR01/232	
SP1556	121	3009		TR01/232	G6013
SP1556(1-4)	121	3009	16	TR01/232	G6013
SP1563-2	121	3009		TR01/232	G6013
SP1595BLK	121	3009		TR01/232	G6013
SP1595BLU	121	3009		TR01/232	G6013
SP1595GRN	121	3009		TR01/232	G6013
SP1595RED	121	3009		TR01/232	G6013
SP1596BLK	121	3009		TR01/232	G6013
SP1596BLU	121	3009		TR01/232	G6013
SP1596GRN	121	3009		TR01/232	G6013
SP1596RED	121	3009		TR01/232	G6013
SP1600	104				G6003
SP1603	121	3009	16	TR01/232	G6013
SP1603-1,2,3	121	3009	16	TR01/232	G6013
SP1619	104				G6003
SP1650	179			TR35/G6001	G6009
SP1651	104	3009		TR01/230	G6003
SP1657	121				G6013
SP1742	127		3	TR27/235	G6018
SP1775					G6018
SP1801	121				G6005
SP1817	179			TR35/G6001	G6009
SP1844	121				G6005
SP1927	104				G6003
SP1938	121	3009		TR01/232	G6005
SP1950	121	3009		TR01/232	G6005
SP1994	213				G6010
SP2045	121	3009	3	TR01/232	

Original Part Number	ECG	SK	GE	IR/WEP	HEP
SP2046	121	3009	3	TR01/232	
SP2048	121	3009	3	TR01/232	
SP2072	121	3009		TR01/232	G6005
SP2076	121	3009		TR01/232	G6005
SP2077	179			TR35/G6001	
SP2094	121				G6013
SP2155	121	3009		TR01/232	G6005
SP2158	103A		12	TR08/724	
SP-2158	124	3021			
SP2233					G6013
SP2234	104	3009		TR01/230	G6003
SP2247	104				G6003
SP2341	121				G6005
SP2358	104	3009	12	TR01/624	G6005
SP2361	121	3009		TR01/232	G6013
SP2361BLU,BRN	121	3009		TR01/232	G6013
SP2361GRN,ORN	121	3009		TR01/232	G6013
SP2361RED,YEL	121	3009		TR01/232	G6013
SP2395	121				G6005
SP2426	105				G6004
SP2431	104				G6003
SP2431P	105		4	TR03/	
SP2493	121	3014		TR01/232	G6005
SP2541	121	3014			G6005
SP2551	131	3052		TR94/642	G6016
SP2597				TR27/	
SP4168	108	3018			
SP4231	130				
SP4436	161			TR83/	
SP7056			27		
SP8400	154	3045	63	TR78/712	S5026
SP8401	154	3045	63	TR78/712	S5026
SP8402	192		63	TR25/	S5026
SP8416	152			TR76/	
SP8660	152	3054	66	TR76/	S5000
SP8660/121-808		3054			
SP8918	184				S5000
SP8919					S5006
SP12271	106				
SP15363				TR01/	
SP273184		3036			
SPC40	123A	3122	11	TR21/735	S0011
SPC42	123A	3122	11	TR21/735	S0011
SPC50	123A	3122	17	TR21/735	S0011
SPC51	123A	3122	17	TR21/735	S0011
SPC52	123A	3122	17	TR21/735	S0011
SPC410		3560			
SPC411		3560			
SPC413		3560			
SPC423		3560			
SPC430			73		
SPC431		3560	73		
SPC40411			75		
SPD-80059	181	3511			
SPD-80060	121	3511			
SPD-80061	181	3511			
SPD-80062	130	3027			
SPD-80123	128				
SPF024	133	3112		/801	F0010
SPF215	222				F1036
SPF274	222	3050			F2007
SPF512	222	3050			F2007
SPS0121	129	3025			S3028
SPS-0121	129			TR88/242	
SPS0122	128	3024			S3024
SPS-0122	128	3024		TR87/243	
SPS020		3589			
SPS08		3589			
SPS4	108	3039		TR95/56	S0015
SPS12	159				S0013
SPS20	108	3039	20	TR95/56	S0025
SPS22	159				S0013
SPS-29	129	3025		TR88/242	
SPS38	108	3039	20	TR95/56	S0025
SPS40		3018	20		S0033
SPS-41	123A	3122	20	TR21/735	S0015
SPS42	159	3114	21	TR20/717	S0019
SPS42A	159	3114			
SPS43-1	107	3039		TR70/720	S0033
SPS47	159			TR20/	S0019
SPS49	128			TR87/	
SPS91	121				G6013
SPS120		3589			
SPS220	108	3018	11	TR80/	S0016
SPS401K	159				S0019
SPS428	108	3019	11	TR95/56	S0016
SPS514	159	3114		TR20/717	S0019
SPS514A	159	3114			
SPS817		3122	62		S0024
SPS817N	123A	3122	62		
SPS837	189				S3032
SPS856	107	3039		TR70/720	
SPS-856	107		11		
SPS860	107	3039	11	TR70/720	
SPS868	123A				S0025

Original Part Number	ECG	SK	GE	IR WEP	HEP
SPS-952		3114	62		
SPS1000					S0015
SPS1001					S0015
SPS1004					S0015
SPS1005					S0015
SPS1006					S0015
SPS1011					S0015
SPS1035					S0015
SPS1041					S0019
SPS1042					S0019
SPS1043					S0019
SPS1045	123A	3122	20	TR21/735	
SPS1095					S0020
SPS1097	159	3114	21	TR20/717	
SPS1097A	159	3114			
SPS1351	108	3018	11	TR95/56	
SPS1352	108	3018	11	TR95/56	
SPS1353	108	3018	11	TR95/56	
SPS1436		3054	66	TR76/	S3024
SPS1437		3083	69	TR77/	S3028
SPS-1473		3018	17		
SPS-1475	123A				
SPS-1475YT	123A	3122	10		
SPS1523	159	3114		TR20/717	S0019
SPS-1539WT		3114	62		
SPS1750					S0025
SPS2110	108	3018	11	TR95/	S0008
SPS2111	108	3018	11	TR95/	S0016
SPS2224	108	3018		TR95/56	S0016
SPS2225	123A	3020	20	TR21/735	S0015
SPS2226	129	3114	22	TR88/	S5013
SPS2265	108	3018	11	TR95/	S0016
SPS2265-1		3018			
SPS2265-2	108	3018	11	TR95/	S0016
SPS2266	108		11	TR95/	S0016
SPS2269	159	3114	21	TR20/717	S0019
SPS2270	123A	3020	18	TR21/735	S0003
SPS2271	199	3122	62	TR51/	S0024
SPS2272	159	3114	21	TR20/717	S0019
SPS2274	159	3025	21	TR20/717	S0031
SPS2279	159			TR20/717	
SPS2320	108	3018	11	TR95/	S0009
SPS3003	108	3039		TR95/56	S0020
SPS3015	123A	3122		TR21/735	S0015
SPS3329	159		63	TR25/	S0031
SPS3370	108	3018	17	TR95/56	S0016
SPS3724	159	3114		TR20/717	S0019
SPS3725					S0020
SPS3735	123A	3122		TR21/735	S0015
SPS3763					S0015
SPS3764					S0033
SPS3767					S0033
SPS3777					S0020
SPS3779					S0015
SPS3783					S0025
SPS3786	159	3114		TR20/717	S0019
SPS3787	107	3039		TR70/720	S0016
SPS3788					S0016
SPS3792					S0020
SPS3798					S0031
SPS3900	123A	3122		TR21/735	S0015
SPS3907	123A	3122		TR21/735	S0015
SPS3908	123A	3122		TR21/735	S0015
SPS3909	123A	3122		TR21/735	S0015
SPS3912	128				S0015
SPS3914	128				S0015
SPS3915	123A				S0014
SPS3923	123A	3122		TR21/735	S0015
SPS3924	159	3114		TR20/717	S0019
SPS3925	123A	3122		TR21/735	S0015
SPS3926	123A	3122		TR21/735	S0015
SPS3927	159	3114		TR20/717	S0019
SPS3929	108	3039		TR95/56	S0020
SPS3930	123A	3122		TR21/735	S0015
SPS3931	159	3114		TR20/717	S0019
SPS3936	123A			TR21/735	S0015
SPS3937	108	3122		TR95/56	S0015
SPS3938	123A	3122		TR21/735	S0015
SPS3940	123A	3122		TR21/735	S0015
SPS3948	108	3039		TR95/56	S0015
SPS3951	123A	3122		TR21/735	S0015
SPS3952	108	3039		TR95/56	S0015
SPS3957C	123A	3122		TR21/735	S0015
SPS3967	123A	3122		TR21/735	S0015
SPS3968	108	3039		TR95/56	S0015
SPS3971	108	3039		TR95/56	S0015
SPS3972	123A	3122		TR21/735	S0015
SPS3973	123A	3122		TR21/735	S0015
SPS3987	159	3114		TR20/717	S0019
SPS3988	159	3114		TR20/717	S0019
SPS3988A	159	3114			
SPS3990	159	3114		TR20/717	S0019
SPS3990A	159	3114			
SPS3999	123A	3122		TR21/735	S0015
SPS4002	108	3039		TR95/56	S0020
SPS4003	123A	3122		TR21/735	S0015
SPS4004	123A	3122		TR21/735	S0015
SPS4005	108	3039		TR95/56	S0015
SPS4006	123A	3122		TR21/735	S0015
SPS4007	159	3114		TR20/717	S0019
SPS4008	108	3039		TR95/56	S0015
SPS4009	123A	3122		TR21/735	S0015
SPS4010	129	3025		TR88/242	
SPS4013	159	3114		TR20/717	S0019
SPS4014	159	3114		TR20/717	S0019
SPS4016	108	3039		TR95/56	S0015
SPS4017	123A	3122	20	TR21/735	S0024
SPS4018	159	3114		TR20/717	S0019
SPS4019	159	3114		TR20/717	S0019
SPS4020	123A	3122	20	TR21/735	S0015
SPS4025	159	3114	21	TR20/717	S0019
SPS4026	159	3114	21	TR20/717	S0019
SPS4027	159	3114	21	TR20/717	S0019
SPS4028	159	3114		TR20/717	S0032
SPS4029	123A	3122	20	TR21/735	S0015
SPS4030	108	3039		TR95/56	S0016
SPS4031	159	3114		TR20/717	S0019
SPS4032	123A	3122		TR21/735	S0015
SPS4034	123A	3122		TR21/735	S0015
SPS4037	123A	3122	20	TR21/735	
SPS4038	128	3024		TR87/243	S3014
SPS4039	123A	3122		TR21/735	S0015
SPS4040	123A	3122		TR21/735	S0015
SPS4041	123A	3122		TR21/735	S0025
SPS4042	123A	3122		TR21/735	S0015
SPS4043	108	3039		TR95/56	S0016
SPS4044	123A	3122		TR21/735	S0024
SPS4045	123A	3122		TR21/735	S0015
SPS4049	123A	3122		TR21/735	S0015
SPS4050	108	3039		TR95/56	S0016
SPS4051	108	3039		TR95/56	S0016
SPS4052	123A	3122		TR21/735	S0015
SPS4053	123A	3122		TR21/735	S0015
SPS4054	159	3114		TR20/717	S0019
SPS4055	123A	3122		TR21/735	S0015
SPS4056	159	3114		TR20/717	S0019
SPS4059	123A	3122		TR21/735	S0024
SPS4060	123A	3122		TR21/735	S0015
SPS4061	123A	3122		TR21/735	S0015
SPS4062	123A	3122		TR21/735	S0015
SPS4063	123A	3122		TR21/735	S0015
SPS4064	159	3114		TR20/717	S0031
SPS4066	123A	3122		TR21/735	S0015
SPS4067	123A	3122		TR21/735	S0015
SPS4068	108	3039		TR95/56	S0020
SPS4069	123A	3122		TR21/735	S0015
SPS4072	159	3114		TR20/717	S0019
SPS4073	159	3114		TR20/717	S0019
SPS4074	123A	3122		TR21/735	S0015
SPS4075	123A		20	TR21/735	S0024
SPS-4075	123A	3020			
SPS4076	159	3114	21	TR20/717	S0019
SPS-4076	129	3025			
SPS4077	123A		18	TR21/735	S0015
SPS-4077	128	3024			
SPS4078	159		21	TR20/717	S0031
SPS-4078	129	3025			
SPS4079	108			TR95/56	S0015
SPS4080	108	3039		TR95/56	S0020
SPS4081	123A	3122		TR21/735	S0015
SPS4082	159	3114		TR20/717	S0019
SPS4083	123A	3122		TR21/735	S0024
SPS4084	123A	3122		TR21/735	S0024
SPS4085	123A	3122		TR21/735	S0015
SPS4086	159	3114		TR20/717	S0031
SPS4087	159	3114		TR20/717	S0031
SPS4088	123A	3122		TR21/735	S0015
SPS4089	123A	3122		TR21/735	S0015
SPS4090	159	3114		TR20/717	S0019
SPS4091	108	3039		TR95/56	S0015
SPS4092	211	3203			S3028
SPS4095	123A	3122		TR21/735	S0030
SPS4099					S3028
SPS4143	107	3018		TR80/	
SPS4145	108	3039	11	TR95/56	S0016
SPS4167	108	3018	17	TR95/56	S0016
SPS4168	107	3018	20	TR70/720	S0020
SPS4169	123A	3018	20	TR21/735	S0030
SPS4199	123A	3038	20	TR21/735	S0005
SPS4237	185				
SPS4300	128	3024	63	TR87/243	S0015
SPS4301	129	3025	21	TR88/242	S0019
SPS4302	159	3114	21	TR20/717	S0019
SPS4303	123A	3020	63	TR21/735	S0015
SPS4309	128	3024		TR87/243	S0015
SPS4310	129	3025		TR88/242	S0019
SPS4311	128		18	TR87/243	S0014
SPS4312	129	3025		TR88/242	S0019
SPS4313	123A	3122	20	TR21/735	S0024
SPS4314	159	3114	21	TR20/717	S0019

Original Part Number	ECG	SK	GE	IR/WEP	HEP
SPS4343	161			TR83/	
SPS4345	123A				S0025
SPS4347	123A	3020	20	TR21/735	
SPS4348	159	3114	21	TR20/717	
SPS4354	159	3114		TR20/717	S0019
SPS4355	159	3114	21	TR20/717	S0031
SPS4356	123A	3122	20	TR21/735	S0024
SPS4359	123A	3122	18	TR21/735	S0015
SPS4360	123A	3122	20	TR21/735	
SPS4361	128		18	TR87/243	
SPS4363	123A	3122	20	TR21/735	S0024
SPS4365	159	3114	21	TR20/717	S0019
SPS4367	123A	3122	18	TR21/735	S0030
SPS4368	123A	3122	18	TR21/735	S0015
SPS4382	123A	3122	20	TR21/735	S0024
SPS4399	108			TR95/	
SPS4401			21	TR30/	
SPS4423			11	TR95/	
SPS4446	123A	3122	20	TR21/735	S0015
SPS4450	123A	3122	20	TR21/735	S0024
SPS4451	123A	3122	20	TR21/735	S0015
SPS4452	159	3114	21	TR20/717	S0019
SPS4453	123A	3122	20	TR21/735	S0025
SPS4455	123A	3122	20	TR21/735	S0015
SPS4456	123A	3122		TR21/735	S0025
SPS4457	123A	3122	18	TR21/735	S0015
SPS4458	159	3114	21	TR20/717	S0019
SPS4459	123A	3122	20	TR21/735	S0015
SPS4460	159	3114	21	TR20/717	S0019
SPS4461	128		18	TR87/243	
SPS4462	129	3025		TR88/242	
SPS4472	123A	3122	18	TR21/735	S0024
SPS4473	159	3114	21	TR20/717	S0019
SPS4476	123A	3122	20	TR21/735	S0030
SPS4478	123A	3122	20	TR21/735	
SPS4480	159	3114	21	TR20/717	S0019
SPS4489	159	3114	21	TR20/717	
SPS4490	128		62	TR87/243	
SPS4491	123A	3122	20	TR21/735	
SPS4492	129	3025	21	TR88/242	
SPS4493	123A	3122	20	TR21/735	
SPS4494	123A	3122	20	TR21/735	
SPS4495	128		20	TR87/243	
SPS4497	129	3025	67	TR88/242	
SPS4610	108	3039	20	TR95/56	S0015
SPS4813	159				S0019
SPS4814	199				S0015
SPS4815	159				S0019
SPS4920	123A	3122		TR21/735	S0025
SPS4925					S0030
SPS4927					S0015
SPS4929					S0024
SPS5000	123A	3122		TR21/735	
SPS5006	123A	3122		TR21/735	S0024
SPS5006-1,2	123A	3122		TR21/735	S0024
SPS5007	159	3114		TR20/717	S0031
SPS5007-1,1A	159	3114		TR20/717	S0031
SPS5007-2,2A,A	159	3114		TR20/717	S0031
SPS5008	159	3114	21	TR20/717	S0019
SPS5380					S0020
SPS5450		3024			S0005
SPS5451		3114			S0019
SPS5809	128				S0015
SPS6109	159	3114		TR20/717	S0019
SPS6111	123A	3122		TR21/735	S0030
SPS6112	123A	3122		TR21/735	S0030
SPS6113	123A	3122		TR21/735	S0015
SPS6124	128	3024		TR87/243	
SPS6125	129			TR88/242	
SPS6562					S0015
SPS6571	123A	3122	20	TR21/735	
SPS7652	123A	3122	20	TR21/735	S0015
SPT3440	198		32		
SPT3713	130		77		
SQ7	101			TR08/641	G0011
SQ-7	103A	3011			
SQD-2170	123				
SR1	102A			TR85/250	
SR3003				TR85/	
SR3004				TR85/	
SR75844	123A	3122		TR21/735	S0015
SRS1004			32		
SRS2004			32		
SRS2504			32		
SRS2804S			32		
SRS3014			32		
SRS3204S			32		
SRS3504			32		
SRS3604S			32		
SRS4014			32		
SRS4014S			32		
SRS4404S			32		
SRS4504			32		
SRS5014			32		
SRS5504			32		

Original Part Number	ECG	SK	GE	IR/WEP	HEP
SS0001	102A	3004		TR85/250	G0005
SS0002	102A	3004		TR85/250	G0005
SS0003	158	3004		TR84/630	G0005
SS0004	102A	3004		TR85/250	G0005
SS0005	102A	3004		TR85/250	G0005
SS1-145128	123A				
SS29A4			65		
SS29A5			65		
SS155	160			TR17/637	
SS524	154	3045		TR78/712	S5025
SS1122				TR72/	S3032
SS1123	191				S3020
SS1241					S3011
SS1606	159	3114		TR20/717	S5022
SS1606A	121	3114			
SS1805					S3002
SS1806					S3003
SS1821					S5014
SS1822					S5013
SS1831					S5023
SS1906	159	3114		TR20/717	S5022
SS1912	154		27	TR78/712	S5025
SS2308	123A	3122		TR21/735	S0011
SS2409	213				G6010
SS2503	159	3114		TR20/717	S0013
SS2504	123A	3122		TR21/735	S0011
SS3519					S3013
SS3526					S3013
SS3534-4	132	3116		FE100/802	F0015
SS3540					F0010
SS3586	133	3112		/801	F0010
SS3588					S3013
SS3672	133				F0010
SS3694	123A	3122	20	TR21/735	
SS3704	132	3116	FET2	FE100/802	
SS-3704	133		FET1		
SS3905					S3008
SS3917					S3008
SS3935	195				S3001
SS4042	161	3039		TR83/719	S0017
SS4165				TR87/	S0017
SS6111	154				S5026
SS9327	190				S3019
SSA43,A	159	3114	21	TR20/717	S0013
SSA43A-1	159	3114			
SSA46	159	3114	21	TR20/717	S0013
SSA48	159	3114	21	TR20/717	S0013
ST01	123A	3122	63	TR21/735	S0011
ST02	123A	3122	63	TR21/735	S0011
ST03	123A	3122	63	TR21/735	S0011
ST04	123A	3122	63	TR21/735	S0011
ST05	123A	3122	63	TR21/735	S0011
ST06	123A	3122	63	TR21/735	S0011
ST-021660	129				
ST.082.112.005	123A				
ST.082.114.015	123A				
ST.082.114.016	199				
ST.082.115.015	159				
ST2T003	199				
ST9	108	3039	11	TR95/56	
ST10	108	3039	11	TR95/56	
ST11	107		11	TR21/	
ST12	108		11	TR95/56	
ST13	108	3039	11	TR95/56	
ST14	108	3031	11	TR95/56	
ST15	108	3039	11	TR95/56	S0011
ST25A,B,C	123A	3122	20	TR21/735	S0011
ST28A	102A	3123	2	TR85/250	G0005
ST-28A	176	3005			
ST28B	102A	3123	2	TR85/250	G0005
ST-28B	100	3005			
ST28C	102A	3123	2	TR85/250	G0005
ST-28C	100	3005			
ST29	108	3039	11	TR95/56	
ST30	108	3039	11	TR95/56	
ST31	108	3039	11	TR95/56	
ST32	108	3039	11	TR95/56	
ST33	108	3039	11	TR95/56	
ST34	108	3039	11	TR95/56	
ST35	108	3039	11	TR95/56	S0011
ST37C	102A	3123	2	TR85/250	G0005
ST-37C	100	3005			
ST37D	102A	3123	2	TR85/250	G0005
ST-37D	100	3005			
ST37E	102A			TR85/250	G0005
ST370	176	3123			
ST40	108	3039	17	TR95/56	
ST41	108	3038	17	TR95/56	
ST42	108	3039	17	TR95/56	
ST43	108	3039	63	TR95/56	S0005
ST44	108	3039	17	TR95/56	
ST45	108	3039	17	TR95/56	S0011
ST50	123A	3122	63	TR21/735	S0011
ST51	123A	3122	63	TR21/735	S0011
ST53	123A	3122	63	TR21/735	S0011

Original Part Number	ECG	SK	GE	IR/WEP	HEP
ST54	123A	3122		TR21/735	S0011
ST55	123A	3122	63	TR21/735	S0011
ST56	123A	3122	63	TR21/735	S0011
ST57	123A	3122	63	TR21/735	S0011
ST58	123A	3122	63	TR21/735	S0011
ST59	123A	3122	63	TR21/735	S0011
ST60	108	3039	63	TR95/56	S0016
ST61	108	3039	63	TR95/56	S0016
ST62	108	3039	63	TR95/56	S0016
ST63	123A	3122	63	TR21/735	S0011
ST64	123A	3122	63	TR21/735	S0011
ST66					S7002
ST70	108	3039	11	TR95/56	S0016
ST71	108	3039	11	TR95/56	S0016
ST72	108	3039	11	TR95/56	S0016
ST80	108	3039	63	TR95/56	S0016
ST82	108	3039	63	TR95/56	S0016
ST101	181				
ST-103	126				
ST106	105	3012	4	TR03/233	G6004
ST107	105	3012	4	TR03/233	G6004
ST108	105	3012	4	TR03/233	G6004
ST109	105	3012	4	TR03/233	G6006
ST110	105	3012	4	TR03/233	G6006
ST111	105	3012	4	TR03/233	G6006
ST112	105	3012	4	TR03/233	G6004
ST122	102A	3123	52	TR85/250	G0005
ST-122	102A	3004			
ST123	102A	3123	52	TR85/250	G0005
ST-123	102A	3004			
ST-125	160				
ST129-1	159				
ST150	123A	3122	63	TR21/735	S0014
ST151	123A	3122	11	TR21/735	S0014
ST152	123A	3122	11	TR21/735	S0014
ST153	123A	3122	11	TR21/735	S0014
ST154	123A	3122	11	TR21/735	S0014
ST155	123A	3122	11	TR21/735	S0014
ST156	123A	3122	11	TR21/735	S0014
ST157	123A	3122	11	TR21/735	S0014
ST160	123A	3122	63	TR21/735	S0014
ST161	123A	3122	63	TR21/735	S0014
ST162	123A		63	TR21/735	S0014
ST-162		3011			
ST163	123A	3122	63	TR21/735	S0014
ST172	101		5	TR08/641	
ST-172	101	3011			
ST175	123A	3122	63	TR21/735	S3020
ST176	123A	3122	63	TR21/735	S3020
ST177	123A	3122	63	TR21/735	S3020
ST178	123A	3122	63	TR21/735	S3020
ST180	123A	3122	63	TR21/735	S3019
ST181	123A	3122	63	TR21/735	S3019
ST182	123A	3122	63	TR21/735	S3019
ST185					S3019
ST186					S3019
ST187					S3019
ST-201	128				
ST-205		3011			
ST213	128				
ST/217/Q	123A				
ST235	121			TR01/232	
ST-235	121	3009			
ST250	123A	3122	63	TR21/735	S0025
ST251	123A	3122	63	TR21/735	S0025
ST 254 Q	128				
ST290				TR21/	
ST301	102A	3123	52	TR85/250	G0005
ST-301	102A	3004			
ST302	102A	3123	52	TR85/250	G0005
ST-302	102A	3004			
ST303	102A	3123	52	TR85/250	G0005
ST-303		3004			
ST304	102A	3123	52	TR85/250	G0005
ST-304	102A	3004			
ST332	102A	3123	52	TR85/250	G0005
ST-332	102A	3004			
ST370	102A		2	TR85/250	G0005
ST-370	100	3005			
ST382	102A			TR85/250	G0005
ST402	192		63	TR25/	S5000
ST403	123A		17	TR24/	S5000
ST415	108	3039	63	TR95/56	S5004
ST440					S5000
ST450					S5000
ST501	123A	3122	63	TR21/735	S0011
ST502	123A	3122	63	TR21/735	S0011
ST503	123A	3122		TR21/735	S0011
ST504	123A	3122		TR21/735	S0011
ST610					S7004
ST615					S5020
ST721					S0016
ST722					S0016
ST723					S0016
ST903	108	3039	11	TR95/56	

Original Part Number	ECG	SK	GE	IR/WEP	HEP
ST904	108	3039	11	TR95/56	
ST904A	108	3039	11	TR95/56	
ST905	108	3039	11	TR95/56	
ST910	108	3039	11	TR95/56	
ST1026	108	3039	11	TR95/56	
ST1036A					S7002
ST1050	108	3039	11	TR95/56	
ST1051	108	3039	11	TR95/56	
ST1151				TR21/	
ST1152				TR21/	
ST1153				TR21/	
ST1154				TR21/	
ST1155				TR21/	
ST1156				TR21/	
ST1157				TR21/	
ST1242	123A	3020	17	TR21/735	S0012
ST1243	123A	3020	17	TR21/735	S0012
ST1244	123A	3020	17	TR21/735	S0012
ST1290	123A	3020	11	TR21/735	S0012
ST1336	108	3039	17	TR95/56	
ST1504	192		63	TR25/	
ST1505	192		63	TR25/	
ST1506	123A	3122		TR21/735	S0011
ST1607	123A	3122		TR21/735	S0011
ST1633					S0005
ST1694	108	3039	11	TR95/56	S0014
ST1700					S0025
ST2110					S0011
ST2120					S0017
ST2130					S0017
ST3030	108	3039	63	TR95/56	S0014
ST3031	108	3039	11	TR95/56	S0014
ST4150	128		81	TR25/	
ST4201	128		81	TR25/	
ST4202	128		81		
ST4203	128		81	TR25/	
ST4204	128		81		
ST4341	192		63	TR25/	S3011
ST4402					S3019
ST5060	123A		17	TR24/	S3011
ST5061	192		63	TR25/	S3011
ST5641	161		60	TR24/	
ST6110	159		63	TR20/717	S0013
ST6120	108	3039	63	TR95/56	S0011
ST6125					S0011
ST6130					S0011
ST6510	107		11	TR22/	S0014
ST6511	123A		17	TR24/	S0014
ST6512	123A		17	TR24/	S0014
ST6573	128	3024	11	TR87/243	S3011
ST6574	128	3024	11	TR87/243	S3011
ST6593					S0025
ST6594					S0025
ST6600					S0025
ST6601					S5014
ST7100	199		62		S0015
ST7120					S5000
ST7130					S5000
ST7200					S5004
ST7430					S5000
ST8014	159	3025	21	TR20/717	S0012
ST8033	159	3114	21	TR20/717	S0012
ST8034	159	3114	21	TR20/717	S0012
ST8035	159	3114	21	TR20/717	S0013
ST8036	159	3114	21	TR20/717	S0013
ST8065	159	3114	21	TR20/717	S0012
ST8181					S5022
ST8182					S5022
ST8183					S3012
ST8184					S3012
ST8190	159		21		S3012
ST8191					S3012
ST8229					S5022
ST8230					S3012
ST8500	159	3114	21	TR20/717	S0012
ST8509	159	3114	21	TR20/717	S0012
ST8700					S0019
ST8704					S0019
ST8705					S0019
ST8709					S0019
ST16573				TR21/	
ST27020	153		26		
ST29045	180			/S7001	
ST29046	180			/S7001	
ST29047	180			/S7001	
ST30100					S3019
ST61000	193		65	TR52/	S0032
ST72039	129	3025		TR88/242	
ST72040	129	3025		TR88/242	
ST74000			32		
ST84000			32		
ST84027	198		32		
ST84028	198		32		
ST84029	198		32		
ST984539-006	9948				

Original Part Number	ECG	SK	GE	IR/WEP	HEP
STC1035,A	130	3027		TR59/247	S7002
STC1036,A	130	3027		TR59/247	S7002
STC1080	130	3027	72	TR59/247	S7002
STC1081	130	3027	72	TR59/247	S7002
STC1082	130	3027	72	TR59/247	
STC1083	130	3027	35	TR59/247	S7002
STC1084	130	3027	35	TR59/247	S7002
STC1085	130	3027	35	TR61/247	
STC1094	181		75		
STC1102					S5004
STC1103					S5004
STC1104					S5004
STC1105,A					S5004
STC1106,A					S5004
STC1300	152		66		
STC1336	192		63	TR25/	
STC1800	186		28		
STC1850	152		66		
STC1862	186	3192	28	TR76/	
STC2220	181		75		
STC2221	181		75		
STC2224	181		75		
STC2225	181		75		
STC2228	181		75		
STC2229	181		75		
STC4252	130	3036	77	TR59/247	S7002
STC4253	130	3027	77	TR59/247	S7002
STC4254	130	3036	75	TR59/247	
STC4255	130	3036	75	TR59/247	
STC4401	175	3131	216	TR57/241	S5012
STC5202	153		69		
STC5203	153		69		
STC5205	153		69		
STC5206	153		69		
STC5303	153		69		
STC5610	129	3025	69	TR88/242	S5013
STC5611	129	3025	21	TR88/242	S5013
STC5612	129	3025		TR88/242	S5013
STC5802			69		
STC5803			69		
STC5805	153		69		
STC5806	153		69		
STD9007	186	3192	28	TR76/	
STE400	108		61		
STI10			32		
STI20			32		
STI30			32		
STI40			32		
STI50			32		
STK-011	1024		720		
STK-011A,B,C,D		3152	720		
STK-011DA,E,F,K		3152	720		
STK-011L,Y		3152	720		
STK-015	1027		721		
STK-015A,B,C,D		3153	721		
STK-015DA,E,F		3153	721		
STK-015K,L,Y		3153	721		
STK-016A,B,C,D		3154	721		
STK-016DA,E,F		3154	721		
STK-016K,L,Y		3154	721		
STK-018	1139				
STK-020	1025		722		
STK-020A		3155	722		
STK-020B	1025	3155	722		
STK-020C		3155	722		
STK-020D	1025	3155	722		
STK-020DA		3155	722		
STK-020E,F,K,L	1025	3155	722		
STK-020Y		3155	722		
STK-032	1090		723		
STK-036	1148				
STK-056,A	1028		724		
STK-401	1151				
STM730	159				
STS402			73		
STS403			73		
STS409			73		
STS430			73		
STS431			73		
STS1121				TR61/	S5020
STS1152					S5020
STT2300			75	TR61/	
STT2405			216		
STT2406			216		
STT3500			75	TR61/	
STT4451	186		28	TR76/	
STT4483			216		
STT9001			216		
STT9002			216		
STT9004			216		
ST19005			216		
STX0010	175	3131		TR81/241	S5012
STX0011	129	3025		TR88/242	S5013
STX0013	184	3054	57	/S5003	S5000
STX0014	130	3027		TR59/247	S7002
STX0015	124	3021	12	TR81/240	S5011
STX0016	175	3131		TR81/241	S5012
STX0020	185	3083	58	TR77/S5007	S5006
STX0022					S7003
STX0026	184	3041	57	/S5003	S5000
STX0027	130	3027		TR59/247	S7004
STX0028	154			TR78/712	S5024
STX0029	153		26	TR77/700	S5018
STX0030	153		26	TR77/700	S5018
STX0032	130	3027		TR59/247	S7004
STX0033	126			TR17/635	G0008
STX0034	126			TR17/635	G0008
STX0036	160			TR17/637	G0001
STX0085	160			TR17/637	G0008
STX0087	160			TR17/637	G0001
STX0089	160			TR17/637	G0001
STX0090	160			TR17/637	G0008
STX0096	102A			TR85/250	G0007
STX0099	102A			TR85/250	G0007
STX0104	102A			TR85/250	G0007
STX0105	102A			TR85/250	G0007
STX0110	102A			TR85/250	G0007
STX0114	102A			TR85/250	G0005
STX0121	158			TR84/630	G0006
STX0123	102A			TR85/250	G0005
STX0224	158			TR84/630	G0006
STX0260	102A			TR85/250	G0005
STX0263	102A			TR85/250	G0005
STX0264	102A			TR85/250	G0005
STX0265	102A			TR85/250	G0007
STX0268	102A			TR85/250	G0007
STX0269	102A			TR85/250	G0007
STX3326	196	3054			
STX49007	941M				
STX5/3010					S5000
STX5/3025					S5000
STX5/5010					S5004
STX5/5025					S5004
STX5/6010					S5004
STX5/6025					S5004
STX5/7010					S5004
STX5/7025					S5000
SU44	6409				S9002
SU110	6401				
SU2076		3112			
SU2077		3112			
SU2081		3112			
SU9808					S9002
SV31	158	3123	52	TR84/630	G0005
SVI-LA3300	1005				
SVI-LD3120	1019				
SVI-ST020F	1025		722		
SVI-STK018	1139				
SVI-STK020F	1025				
SW705-2M	9093				
SW705-2P	9093				
SW706-2M,2P	9099				
SW708-2M,2P	9094				
SW709-2M,2P	9097				
SW930-2M,2P	9930				
SW932-2M,2P	9932				
SW933-2M,2P	9933				
SW935-2M,2P	9935				
SW936-2M,2P	9936				
SW937-2M,2P	9937				
SW941-2M,2P	9941				
SW944-2M,2P	9944				
SW945-2M,2P	9945				
SW946-2M,2P	9946				
SW948-2M,2P	9948				
SW949-2M,2P	9949				
SW950-2M,2P	9950				
SW951-2M,2P	9951				
SW957-2M,2P	9157				
SW958-2M,2P	9158				
SW961-2M,2P	9961				
SW962-2P	9962				
SW963-2M,2P	9963				
SW-10548	9937				
SW-10549	9158				
SWT1728			50		
SWT3588			50		
SX55	123A	3122		TR21/735	S0011
SX60M	154	3045	27	TR78/712	
SX61	129	3025		TR88/242	
SX61M	159	3114	21	TR20/717	
SX61MA	159	3114			
SX3001	108	3039		TR95/56	S0016
SX3702	159	3114		TR20/717	S0019
SX3709	108	3039	20	TR95/56	S0025
SX3711	123A	3020	20	TR21/735	S0015
SX3819	132			FE100/	
SX3825	108	3018		TR95/56	S0011
SX3826	108	3018		TR95/56	S0030
SY101			6		

Original Part Number	ECG	SK	GE	IR WEP	HEP
SY61583	176	3123			
SYL101	101	3011	8	TR08/641	
SYL102	101	3011	8	TR08/641	
SYL103	103		8	TR08/641A	
SYL-103	103A	3010			
SYL104	103		8	TR08/641A	
SYL-104	103A	3010			
SYL105	100	3123	2	TR05/254	G0005
SYL-105	100	3005			
SYL106	100	3123	2	TR05/254	G0005
SYL-106	100	3005			
SYL107	102	3123	52	TR05/631	G0005
SYL-107	102A	3004			
SYL107A	102A	3004			
SYL108	102	3123	52	TR05/631	G0005
SYL-108	102A	3004			
SYL108A	102A	3004			
SYL109	121	3009	16	TR01/232	G6005
SYL152	123		20	TR86/53	S0015
SYL-152	103A	3010			
SYL160	100	3123	2	TR05/254	G0005
SYL-160	100	3005			
SYL792	101	3005	50	TR08/254	
SYL1182	123A	3020		TR21/735	
SYL1279	101	3011	8	TR08/641	
SYL1297	103		8		
SYL-1297	101	3011			
SYL1310	101		8	TR08/641	
SYL-1310	101	3011			
SYL1311	101		8	TR08/641	
SYL-1311	101	3011			
SYL1312	101	3011	8	TR08/641	
SYL1313	101	3011	8	TR08/641	
SYL1326	101	3011		TR08/641	G0011
SYL1327	101	3011		TR08/641	G0011
SYL1329	103		8	TR08/641A	
SYL-1329	103A	3010			
SYL1380	101	3011	8	TR08/641	G0011
SYL1396	103		8	TR08/641A	
SYL-1396	103A	3010			
SYL1408	101	3011		TR08/641	G0011
SYL1430	102			TR05/	
SYL1454	101	3011		TR08/641	G0011
SYL1468	103A			TR08/724	G0011
SYL1524	103		8	TR08/641A	
SYL-1524	103A	3010			
SYL1535	102				
SYL1536	103	3010	8	TR08/641A	
SYL1537	101	3011	8	TR08/641	
SYL1538	103	3010	8	TR08/641A	
SYL1539	103	3010	8	TR08/641A	
SYL1547	103	3010	8	TR08/641A	
SYL1583	102		52	TR05/631	G0005
SYL-1583	102A	3004			
SYL1583A	102A	3004			
SYL1588	100	3005	52	TR05/254	
SYL1591	101	3011		TR08/641	G0011
SYL1592					G0005
SYL1608	100	3123	2	TR05/254	G0005
SYL-1608	100	3005			
SYL1617	101	3011		TR08/641	G0011
SYL1655	102	3004		TR05/631	G0001
SYL1655	102A	3004			
SYL1665	102	3004		TR05/631	G0005
SYL1665A	102A	3004			
SYL1668	102	3123	52	TR05/631	G0005
SYL-1668	102A	3004			
SYL1668A	102A	3004			
SYL1690	100	3005	2	TR05/254	G0005
SYL1697	100	3005		TR05/254	G0005
SYL1717	100	3005		TR05/254	G0005
SYL1750	101	3011	8	TR08/641	G0011
SYL1987	101	3011		TR08/641	
SYL2120	100	3005	52	TR05/254	G0008
SYL2130	101	3011		TR08/641	
SYL2131	101	3011		TR08/641	
SYL2132	101	3011		TR08/641	
SYL2134	103	3010		TR08/641A	
SYL2135	103	3010		TR08/641A	
SYL2136	103	3010		TR08/641A	
SYL2189	160		52	TR17/637	G0003
SYL2245	101	3011		TR08/641	
SYL2246	101	3011		TR08/641	
SYL2247	100	3005		TR05/254	
SYL2248	102			TR05/631	G0005
SYL-2248	100	3005			
SYL2248A	102A	3004			
SYL2249	102			TR05/631	G0005
SYL-2249	100	3005			
SYL2249A	102A	3004			
SYL2250	100	3005		TR05/254	G0005
SYL2300	102			TR05/631	G0005
SYL-2300	108	3019			
SYL2300A	102A	3004			
SYL2650	103	3010	8	TR08/641A	

Original Part Number	ECG	SK	GE	IR WEP	HEP
SYL3460	123A				
SYL3613	102	3004		TR05/631	G0003
SYL3613A	102A	3004			
SYL4131	108		17	TR95/56	S0016
SYL-4131	108	3019			
SYL4275	159			TR20/	
SYL4280	128	3024		TR87 243	S5014
SYL4315	103		8	TR08/641A	G0011
SYL-4315	103A	3010			
SYL4339	101			TR08/	G0011
T0003	102A	3003	52	TR85/250	
T0004	102A	3003	52	TR85/250	
T0005	102A	3003	52	TR85/250	
T00014	102A	3003		TR85/250	
T0012	102A	3003		TR85/250	
T0014	102A		52		
T0015	102A	3003	52	TR85/250	
T0031	102A	3123		TR85/250	
T0033	102A	3003	52	TR85/250	
T0038	102A	3004	52	TR85/250	
T0039	102A	3004	52	TR85 250	
T0040	102A	3004	52	TR85/250	
T0041	102A	3004	2	TR85/250	
T0051	102A	3123	52	TR85/250	
T01-013	123A				
T01-014	123A	3024			
T01-022	128				
T01-023	159				
T01-030	152				
T01-044	133				
T01-047	199				
T01-101	123A	3122		TR51/735	S0015
T01-104	123A	3122		TR51/735	S0015
T01-105	123A	3122		TR21/735	S0015
T0-003				TR17/	
T0-004				TR17/	
T0-005				TR85/	
T0-012				TR01/	
T0-014				TR85/	
T0-015				TR01/	
T0-033				TR51/	
T0-038				TR51/	
T0-039				TR51/	
T0-040				TR51/	
T0-041				TR85/	
T03323	100		52	TR14/	
T0-101	100	3005		TR05/254	
T0101	100		1	TR85/	
T0-102	100	3005		TR05/254	
T0102	100		1	TR85/	
T0-103				TR85/250	
T0-104				TR85/250	
T-04689	128				
T1-1A6	123A	3124	10	TR24/	S0015
T1-503	159	3114			
T1-503A	159	3114			
T1-741	132	3116			
T1-743	159	3114			
T1-743A	159	3114			
T1-744	159	3114			
T1-752	159	3114			
T1-906	159	3114			
T1A	704	3023		IC501/	
T1B	703A	3157	IC12	IC500/	C6001
T1B151-10302A01				IC500/	
T1(DM)					G0008
T1E	722	3161			C6056P
T1F	748				C6060P
T1H	710	3102		IC506/	
T1J	718	3159	IC8	IC511/	C6094P
T1J/51-10422A01				IC511/	
T1J/51-10422A02				IC511/	
T1K	798				
T1M	717				
T1N	718	3159	IC8	IC511/	C6094P
T1P29	152	3041		TR55/	
T1P29X	152			TR76/	
T1P30C					S5005
T1P31	184				S5000
T1P31A	184				S5000
T1P33	182				S5000
T1P33A	182			TR59/	
T1P34				TR29/	
T1P36A	183				
T1P-41B				TR92/	
T1P140		3180			
T1P145		3181			
T1P645		3183			
T1P2955	219				
T1P3055	130				
T1S03	159	3114			
T1S04	159	3114			
T1S5A	159	3114			
T1S14	133	3112			
T1S18	108	3019		TR83/	S0016

Original Part Number	ECG	SK	GE	IR/WEP	HEP
T1S34	132	3116			
T1S37A	159	3114			
T1S38	159	3114			
T1S53	159	3114			
T1S54	159	3114			
T1S58	132	3116			
T1S59	132	3116			
T1S60A,B,C,D,E				TR87/	
T1S60M				TR87/	
T1S61A	159	3114			
T1S61B,C,D,E,M				TR88/	
T1S71				TR51/	
T1S72				TR51/	
T1S83				TR51/	
T1S86				TR82/	
T1S87				TR82/	
T1S88	132	3116		FE100/	F0021
T1S90				TR51/	
T1S91,A	159	3114			
T1S92				TR51/	
T1S92M				TR87/	
T1S93	159	3114		TR51/	
T1S93A	159	3114			
T1S94	108	3018			
T1S95			10	TR24/	G0009
T1S96				TR51/	
T1S97				TR82/	
T1S98	108	3018	63		
T1S99				TR82/	
T1S100				TR78/	
T1S101				TR78/	
T1S402				TR22/	
T1S412				TR82/	
T1T	755				
T1V	917				
T1W	771				C6004
T1X-M14	161	3018			S0017
T1X-M15	161	3018			S0017
T1X-M16	161				S0017
T1XM15	108				S0016
T1XM17	108				S0016
T1XS12				TR51/	
T1XS13				TR51/	
T1Y	917				
T1Z	722	3161			C6056P
T2A	172			TR69/	
T2A(I.C.)	798		IC9		C6068P
T2A(MOTOROLA)	722				
T2C	722	3161	IC9		C6056P
T2D	723	3144	IC15		C6101P
T2(DM)					G0002
T2E	753				C6010
T2F	718				
T2G	723				
T2J	708	3159	IC10	IC504/	C6082P
T2J/51-10631A01				IC504/	
T2K					C6087P
T2L	796				C6016
T2M	781	3169			
T2N	789				
T2R	778				C6102P
T2T	788				
T2V	736				
T3B	941M				C6052P
T3(DM)					G0003
T-23	102A	3004	2	TR85/250	G0005
T-23-71	223				
T23-93	196	3054			
T23-94	128	3024			
T33Z1	102A	3004	52	TR85/250	
T33Z3	102A	3004		TR85/250	
T35A-5	123A	3020			
T39	102A	3123	2	TR85/250	G0005
T-39	102A	3004			
T40	106				
T45	102A	3123	52	TR85/250	G0005
T46	102A	3123	2	TR85/250	G0005
T-46	100	3005			
T46D				TR21/	
T47	102A	3123	52	TR85/250	G0005
T-47	100	3005			
T48	102A	3123	2	TR85/250	G0005
T-48	100	3005			
T50	102A		2	TR85/250	
T50(DM)					S0011
T51(DM)					S0012
T52(DM)					S0013
T53(DM)					S0014
T54(DM)					S0015
T55				TR51/	
T55(DM)					S0015
T56(DM)					S0016
T57(DM)					S0019
T59	102A		1	TR85/250	
T60	102A		1	TR85/250	

Original Part Number	ECG	SK	GE	IR/WEP	HEP
T61	102A	3004	2	TR85/250	G0005
T72	102A	3004	2	TR85/250	G0005
T74	102A	3004	2	TR85/250	G0005
T75(DM)					S3013
T76	123A	3122	20	TR21/735	
T76(DM)					S3014
T77	102A		2	TR85/250	
T78	102A		1	TR85/250	G0005
T-78	100	3005			
T81	103A	3010	8	TR08/724	
T82	102A	3004	52	TR85/250	G0005
T83	102A		2	TR85/250	G0005
T84	102A	3004	2	TR85/250	G0005
T87	102A			TR85/250	G0005
T95	102A	3004	2	TR85/250	G0005
T99	102A		51	TR85/250	G0005
T-99	160	3006			
T100	102A			TR85/250	G0005
T101	102A		16	TR85/250	G0005
T-101	121	3009		TR01/	
T102A	102A			TR85/250	G0005
T108	102A	3005	1	TR85/250	G0005
T109	102A		1	TR85/250	G0005
T-109	100	3005			
T112	159				
T116	102A	3008	51	TR85/250	G0005
T-116	100	3005			
T126	102A	3004	2	TR85/250	G0005
T127	102A		3	TR85/250	G0005
T-127	121	3009		TR01/	
T129	102A	3004	52	TR85/250	G0005
T130	102A	3004	52	TR85/250	G0005
T131	100		51	TR05/254	G0005
T-131	158				G0005
T139	104	3020	20	TR01/230	G6005
T142	121	3009	16	TR01/232	
T143	123A	3020	20	TR21/735	
T154A,B,C,D,E					S0015
T157	123A	3020	20	TR21/735	
T158	123A	3020	20	TR21/735	
T158/86-3999-9				TR21/	
T160	102A			TR85/250	
T163	160		50	TR17/637	G0003
T170	123A	3020	20	TR21/735	
T171	123A	3020	20	TR21/735	
T185	123A	3020	63	TR21/735	
T200(DM)					G6000
T-203	107				S0016
T222	162			TR67/707	S5020
T225	162			TR67/707	S5020
T230(DM)					G6003
T232(DM)					G6005
T233(DM)					G6006
T234(DM)					G6007
T-235	121				G6013
T235A013-2	123A				
T235(DM)					G6008
T236(DM)					G6009
T237	123A	3020		TR21/735	
T237(DM)					G6010
T238(DM)					G6011
T239(DM)					G6012
T240(DM)					S5011
T241(DM)					S5012
T242(DM)					S5013
T243(DM)					S5014
T244(DM)					S5015
T245(DM)					S5000
T246	129	3025		TR88/242	
T-246	159				S0019
T246(DM)					S5006
T247	128	3024		TR87/243	
T247(DM)					S7002
T248(DM)					S7003
T250(DM)					G0005
T-251	159				S0019
T251(DM)					G0006
T252(DM)					G0005
T253	160	3006	50	TR17/637	
T253(SEARS)	160			TR17/	
T253(DM)					G0005
T254(DM)					G0005
T255	123A	3020	20	TR21/735	S0015
T256	123A	3020		TR21/735	
T257	124		12	TR81/240	S5011
T261	124		12	TR81/240	S5011
T271	175			TR81/241	S5012
T276	159	3025	21	TR20/717	
T277	123A	3020	17	TR21/735	S0015
T278	160		50	TR17/637	G0003
T279	160		50	TR17/637	G0003
T280	160	3006	50	TR17/635	
T280(SEARS)	126			TR17/	
T281	160	3008	51	TR11/635	
T281(SEARS)	126			TR17/	

Original Part Number	ECG	SK	GE	IR/WEP	HEP
T282	160	3008	51	TR11/250	
T282(SEARS)	102A			TR85/	
T287	157			TR60/244	S5015
T291	123A	3122	20	TR21/735	
T-291	128			TR21/	
T308	108	3039		TR95/56	
T309	234				
T310(DM)					S9002
T327	123A	3020	63	TR21/735	
T327-2	123A	3020		TR21/735	
T328	123A	3020		TR21/735	
T334-2	129			TR88/242	
T336-2	128		63	TR87/243	
T339	123A		63	TR21/735	
T-339	128	3024		TR25/	S5014
T340	159	3114	21	TR20/717	
T-340	129			TR30/	S5013
T342	123A		63	TR21/735	
T-342	184	3024		TR25/	S5014
T344	184	3054	28	TR76/S5003	S5000
T-344	184	3190		TR76/	S5000
T345	185	3083		TR77/S5007	
T348	160	3008	50	TR17/637	G0009
T-348	126			TR11/	G0009
T367	160	3006	51	TR17/637	
T368	160	3006	51	TR17/637	
T370	179	3034	25	TR35/G6001	G6008
T373	160	3006	51	TR17/637	
T374	160	3006	51	TR17/637	
T381	161		20	TR21/719	S0020
T381(SEARS)	161			TR83/	
T386	123A		20	TR21/56	S0011
T386(SEARS)	108			TR95/	
T396	185	3083	29	TR77/S5007	
T-396	185	3191		TR77/	S5006
T399	123A		63	TR21/735	S0015
T416-16(SEARS)	123A			TR21/735	
T417	123A	3020	20	TR21/735	S0015
T422	188		28	TR72/3020	S3024
T423	189		29	TR73/3031	S3028
T449	160	3006	50	TR17/635	
T449(SEARS)	126			TR17/	
T452	128	3024	63	TR87/243	
T457-16(SEARS)	123A	3020		TR21/735	S0014
T458-16	123A	3020		TR21/735	S0014
T459	129	3025		/717	S0012
T459(SEARS)	123A			TR21/	
T460	123A	3020		TR21/735	S0014
T460(SEARS)	159			TR20/	
T461-16(SEARS)	123A	3020		TR21/	S0014
T462(SEARS)	123A	3020		TR21/	S0014
T464	222				F2007
T472(SEARS)	123A		20	TR21/735	S0014
T475	129		21	/717	S0012
T475(SEARS)	159			TR20/	
T481(SEARS)	154			TR78/712	
T-481	154	3040		TR25/	
T482(SEARS)	159			TR20/717	
T-482	129	3025			
T483(SEARS)	123A			TR21/735	
T-483	108	3018		TR21/	
T484(SEARS)	123A			TR21/735	
T-484	108	3018			
T485(SEARS)	123A			TR21/735	
T486(SEARS)	123A			TR21/735	
T-486	108	3018		TR25/	
T-510					S3024
T576-1	161	3117		TR83/719	S0020
T611-1	152	3054		TR76/701	
T611-1(RCA)	184				S5000
T612-1	152	3054		TR76/701	
T612-1(RCA)	184				S5000
T615A002	123A				
T615A006-1	123A				
T623(DM)					G6013
T624(DM)					G6013
T625(DM)					G6018
T626(DM)					G6014
T627(DM)					G6015
T628(DM)					G6013
T629(DM)					G0005
T630(DM)					G0006
T631(DM)					G0005
T632(DM)					G0005
T633(DM)					G0007
T634(DM)					G0007
T635(DM)					G0001
T636(DM)					G0008
T637(DM)					G0003
T638(DM)					G0009
T639(DM)					G0008
T640(DM)					G0008
T641(DM)					G0011
T642(DM)					G6016
T643(DM)					G6016

Original Part Number	ECG	SK	GE	IR/WEP	HEP
T644(DM)					G6017
T700(DM)					S5006
T701(DM)					S5000
T702(DM)					S5018
T703(DM)					S5019
T704(DM)					S7004
T705(DM)					S7003
T706(DM)					S5024
T707(DM)					S5020
T708(DM)					S5022
T709(DM)					S0017
T710(DM)					S5023
T712(DM)					S5025
T713(DM)					S5026
T714(DM)					S3019
T715(DM)					S0019
T716(DM)					S0019
T717(DM)					S0031
T718(DM)					S0020
T719(DM)					S0020
T720(DM)					S0020
T721(DM)					S0015
T722(DM)					S0015
T723(DM)					S0015
T724(DM)					S0015
T725(DM)					S0015
T726(DM)					S0015
T727(DM)					S0025
T728(DM)					S0015
T729(DM)					S0025
T730(DM)					S0024
T731(DM)					S0033
T732(DM)					S0033
T733(DM)					S0020
T734(DM)					S0016
T735(DM)					S0015
T736(DM)					S0015
T737(DM)					S0024
T738(DM)					S0030
T739(DM)					S0032
T740(DM)					S5021
T801(DM)					F0010
T802(DM)					F0015
T803(DM)					F1036
T811	160	3008	50	TR17/637	G0003
T814	158	3004	52	TR84/630	G0005
T815	158	3004	52	TR84/630	G0005
T841	181	3036	14	TR36/S7000	
T842	181	3036	14	TR36/S7000	
T843	181	3036	14	TR36/S7000	
T844	181	3036	14	TR36/S7000	
T 900	159				
T1000	158	3033	52	TR84/630	G0005
T1001	102A	3004	52	TR85/250	G0005
T1002	102A		53	TR85/250	
T1002A	102A			TR85/250	
T1003	102A	3005	52	TR85/250	G0005
T1003-521	108	3039		TR95/56	
T1004	102A		53	TR85/250	G0005
T1005	102A	3004	52	TR85/250	G0005
T1006	102A	3004	52	TR85/250	G0005
T1007	102A	3024	52	TR85/250	G0005
T1007(ZENITH)	199				S0024
T1008	102A	3004	52	TR85/250	G0005
T1008-834	123A				S0015
T1009	102A		53	TR85/250	
T1010	102A	3123	52	TR85/250	G0005
T1011	160	3005	2	TR17/637	G0002
T1012	160	3005	2	TR17/637	G0002
T1013	102A	3003	53	TR85/250	G0005
T1023	102A	3123	52	TR85/250	G0005
T1028	160	3008	51	TR17/637	G0002
T1032	160	3006	51	TR17/637	
T1033	160	3007	52	TR17/637	G0002
T1034	160	3008	51	TR17/637	G0002
T1036	102A		53	TR85/250	
T1037	102A		53	TR85/250	
T1038	160	3008	50	TR17/637	G0002
T1040	104	3009	16	TR01/230	G6003
T1041	104	3009	16	TR01/230	G6003
T1042	102A	3004	53	TR85/250	G0005
T1043	102A	3004	2	TR85/250	G0005
T1046	102A	3004	53	TR85/250	G0005
T1047	102A	3004	2	TR85/250	G0005
T1076	102A		53	TR85/250	
T1166	160	3008	51	TR17/637	G0002
T1167	121	3009	16	TR01/232	G6005
T1168	121	3009	16	TR01/232	G6005
T1202	102A	3039	39	TR85/250	G0005
T1202(GE)	161				S0017
T1203	102A			TR85/250	G0005
T1208	132	3116	FET2	FE100/	F0021
T1224	160	3006	47	TR17/637	G0003
T1225	160	3006	51	TR17/637	G0003
T1232	160	3005	2	TR17/637	G0002

Original Part Number	ECG	SK	GE	IR/WEP	HEP
T1233	160	3005	2	TR17/637	G0002
T1250	160	3006	51	TR17/637	G0003
T1251	100			TR05/254	
T1275	129	3025	65	TR88/242	
T1276	129	3025	65	TR88/242	
T1289	100	3005	51	TR05/254	
T1291	100	3005	51	TR05/254	
T1298	160	3005	2	TR17/637	G0002
T1299	160	3005	2	TR17/637	G0002
T1300	102A	3123	52	TR85/250	G0005
T1305	160	3005	2	TR17/637	G0002
T1306	160	3005	2	TR17/637	G0002
T1310	102A	3004	52	TR85/250	G0005
T1312	100	3005		TR05/254	
T1314	160	3008	51	TR17/637	G0002
T1322	100	3005		TR05/254	
T1326	100	3005	1	TR05/254	G0005
T1327	102A	3005	1	TR85/250	G0005
T1328	102A	3005	1	TR85/250	G0005
T1334	102A	3005	1	TR85/250	G000[illegible]
T1340A3H	128	3024	18	TR25/	S5014
T1340A3I	123A		20	TR21/	S0015
T1340A3J	123A		20	TR21/	S0015
T1340A3K	123A		20	TR21/	S0015
T1340A3L	123A	3124	10	TR24/	
T1341A3K		3122			S0024
T1342	102A			TR85/250	
T1346	102A	3005	1	TR85/250	G0005
T1352	102A	3004	52	TR85/250	G0005
T1363	102A		51	TR85/250	G0005
T-1363	126	3008			
T1364	102A		51	TR85/250	G0005
T-1364	126	3008			
T1366,A	104	3009		TR01/624	
T1367,A	104	3009		TR01/624	
T1368	104	3009	16	TR01/624	
T1368A	104	3009		TR01/624	
T1369,A	104	3009		TR01/624	
T1370,A	104	3009		TR01/624	
T1387	160	3006	50	TR17/637	G0003
T1388	160	3006	51	TR17/637	G0003
T1389	160	3006	51	TR17/637	G0003
T1390	160	3006	51	TR17/637	G0003
T1391	160	3006	51	TR17/637	G0003
T1396			80		
T1397			80		
T1400	160	3006	52	TR17/637	G0003
T1401	160	3006	51	TR17/637	G0003
T1402	160			TR17/637	G0003
T1403	160	3006	51	TR17/637	G0003
T1413	123A			TR21/	
T1414	123A		18	TR21/	
T1415	123A	3020	20	TR21/735	
T1416	123A	3020	20	TR21/735	S0015
T1417	123A	3020	20	TR21/735	
T1454	160	3005	2	TR17/637	G0002
T1459	160	3005	2	TR17/637	G0002
T1460	126	3008	51	TR17/635	
T1461	160		51	TR17/637	
T1474	100	3005		TR05/254	
T1486	152		66		
T1487	152		66		
T1495	123A	3122	20	TR21/735	S0015
T1510	100	3005		TR05/254	
T1524	126	3008			
T1524BRN	126	3006		TR17/635	G0008
T1524BRN/RED	126			TR17/635	G0008
T1546	102A	3004	53	TR85/250	G0005
T1548	160	3008	51	TR17/637	G0002
T1559	102A		53	TR85/250	
T1573	102A	3004	80	TR85/250	G0005
T1574	102A	3004	53	TR85/250	G0005
T1577	102A	3004	52	TR85/250	G0005
T1583	102A	3004	52	TR85/250	G0005
T1593	102A		53	TR85/250	
T1594	102A		53	TR85/250	
T1595	102A		53	TR85/250	G0005
T1596	102A		53	TR85/250	
T1597	102A		53	TR85/250	
T1598	102A		53	TR85/250	G0005
T1599	102A		53	TR85/250	
T1601	121	3009	3	TR01/232	
T1602	121	3009	3	TR01/232	
T1618	160	3008	2	TR17/637	G0002
T1642B	123A	3122	10	TR21/735	
T1654	126	3006		TR17/635	G0003
T1654BLU	126			TR17/635	G0003
T1657	160	3006	51	TR17/637	G0003
T1679					S0015
T1690	160	3006	51	TR17/637	G0003
T1691	160	3006	51	TR17/637	G0003
T1692	160	3006	51	TR17/637	G0003
T1706	128	3024		TR87/243	
T1706A,B,C	128	3024		TR87/243	
T1737	160	3006	51	TR17/637	G0003
T1738	160	3006	51	TR17/637	G0003
T1740	102A	3003	53	TR85/250	G0005
T1746	123A	3020		TR21/735	
T1746A,B,C	123A	3020		TR21/735	
T1748	123A	3020		TR21/735	
T1748A,B,C	123A	3020		TR21/735	
T1788	160	3008	51	TR17/637	G0002
T1796					G0005
T1802	123A	3020		TR21/735	
T1802A	123A	3020		TR21/735	
T1802B	123A	3122		TR21/735	
T1804	123A	3020		TR21/735	
T1805	123A	3020		TR21/735	
T1808	129	3025		TR88/242	
T1808A,B,C,D,E	129	3025		TR88/242	
T1810	123A		20	TR	
T1810B	123A	3020		TR21/735	
T1811	128		18		
T1811E	128	3024		TR87/243	
T1811G	128	3024		TR87/	
T1814	160	3006	51	TR17/637	G0003
T1831	160	3007	51	TR17/637	G0003
T1877	100	3005	52	TR05/254	
T1902	100	3005	1	TR05/254	
T1903	102A	3004	2	TR85/250	
T1904	102A	3005	2	TR85/250	G0005
T1909	123A	3018	20	TR21/735	S0015
T1961	102A	3005	2	TR85/250	G0005
T2015	160	3006	51	TR17/637	G0003
T2016	160	3006	51	TR17/637	G0003
T2017	160	3006	51	TR17/637	G0003
T2019	160	3006	51	TR17/637	G0003
T2020	160	3008	51	TR17/637	G0002
T2021	160	3006	51	TR17/637	G0003
T2022	160	3006	51	TR17/637	G0003
T2024	160	3008	51	TR17/637	G0002
T2025	160	3008	51	TR17/637	G0002
T2026	160	3008	51	TR17/637	G0002
T2028	160		51	TR17/637	
T2029	160		51	TR17/637	
T2030	160		51	TR17/637	
T2038	100	3005	52	TR05/254	
T2039	100	3005	52	TR05/254	
T2040	100	3005	52	TR05/254	
T2062	783	3102	511		
T2091	100	3005	52	TR05/254	
T2122	100	3005	2	TR05/254	
T2159	102A	3005	1	TR85/250	G0005
T2172	100	3005	1	TR05/254	G0005
T2173	100	3005	1	TR05/254	G0005
T2191	160		51	TR17/637	
T2256	100	3005	2	TR05/254	
T2257	100	3005	2	TR05/254	
T2258	100	3005	2	TR05/254	
T2259	100	3005	2	TR05/254	
T2260	100	3005	2	TR05/254	
T2261	100	3005	2	TR05/254	
T2322	126	3008	2	TR17/635	G0002
T2323	126	3008	2	TR17/635	G0002
T2324	126	3008	2	TR17/635	G0002
T2351					G0009
T2357	106		21	TR20/52	
T2364	160		51	TR17/637	G0005
T2379	160	3006	51	TR17/637	G0003
T2384	160	3006	51	TR17/637	G0003
T2439	102A		51	TR85/250	G0005
T-2439	100	3005			
T2440	102A		51	TR85/250	G0005
T-2440	100	3005			
T2441	102A		51	TR85/250	G0005
T-2441	100	3005			
T2446	123A	3020	18	TR21/735	S0015
T2515	102A			TR85/	
T2517	102A			TR85/	
T2634	108	3018	11	TR80/	
T2788	160		52	TR17/637	G0001
T2857					S0014
T2878	160		52	TR17/637	G0005
T2896	160		51	TR17/637	G0003
T2945	160		52	TR17/637	G0003
T2946	160		52	TR17/637	G0003
T3005	102A	3005	51	TR85/250	G0005
T3321	102A	3004	2	TR85/250	
T3322	102A	3004	51	TR85/250	
T3530	108	3019		TR95/56	
T3535	108	3117			
T3536	108	3018		TR95/56	S3019
T3539	108	3018		TR95/56	S0015
T3568	107	3039	60	TR70/720	
T3568(RCA)	108				S0020
T3570	159	3025		TR20/717	
T3570A	159	3114			
T3601	123A	3122		TR21/735	
T3601(RCA)	107				S0015
T 4205 L1	128				

Original Part Number	ECG	SK	GE	IR/WEP	HEP
T6028	126	3008	51	TR17/635	G0005
T6029	126	3008	51	TR17/635	G0005
T6030	126	3008	51	TR17/635	G0005
T6031	126	3008	51	TR17/635	G0005
T6032	126	3008	51	TR17/635	G0005
T6058	160		51	TR17/637	
T7400B1	7400				
T7401B1	7401				
T7402B1	7402				
T7403B1	7403				
T7404B1	7404				
T7405B1	7405				
T7406B1	7406				
T7407B1	7407				
T7410B1	7410				
T7416B1	7416				
T7420B1	7420				
T7430B1	7430				
T7441AB1	7441				
T7450B1	7450				
T7451B1	7451				
T7453B1	7453				
T7454B1	7454				
T7460B1	7460				
T7472B1	7472				
T7473B1	7473				
T7474B1	7474				
T7475B1	7475				
T7476B1	7476				
T7486B1	7486				
T7490B1	7490				
T7493B1	7493				
T9011A1C		3018			S0014
T9011A1G	123A	3018	20	TR21/	S0015
T9011AZ	123A	3018	10	TR51/	
T9011CD	108				
T9011EF					S0015
T9011G	108				S0015
T9011G(CD)		3122		TR21/	
T9011G(EF)	108	3018	20	TR24/	
T9011H	108				S0015
T9011H(EF)	108	3018	20	TR24/	
T9011I(EF)		3018			S0015
T9011J(GH)		3018			S0014
T9016F	108		20	TR24/	S0014
T9016H	108	3122			
T9418		3137			
T9423		3018			
T9468		3138			
T9631		3018			
T9681		3114			
T10010	102A	3004	2	TR85/250	G0005
T10085	102A		53	TR85/250	
T10618				TR85/	
T11618	102A	3005	1	TR85/250	G0005
T13000	102A	3004	2	TR85/250	G0005
T13015	128	3024		TR87/	
T13029	121			TR01/	
T38763				TR51/	
T50339A	102A	3004	2	TR85/250	G0005
T50631	102A		2	TR85/250	G0005
T50818	126	3008		TR17/635	G0008
T50931B	102				G0005
T50933B	158				G0005
T50944	100	3005	1	TR05/254	
T51304					G6011
T51573A	102				G0005
T51648		3005			
T52054	126				G0001
T52147,Z	100	3005		TR05/254	
T52148Z	100	3005		TR05/254	
T52149	100	3005	1	TR05/254	
T52149Z	100	3005		TR05/254	
T52150,Z	102A	3004	2	TR85/250	
T52151,Z	102A	3004	2	TR85/250	
T52159	102A	3004	2	TR85/250	G0005
T52247				TR85/	
T52249				TR85/	
T59235A	123A	3020	10	TR21/735	S0015
T59247	102A	3004	52	TR85/250	G0005
T59249	102A	3004	52	TR85/250	G0005
T59276	101	3020	20	TR08/641	G0011
T59277	101	3020	20	TR08/641	G0011
T74107B1	74107				
T74180B1	74180				
T74193B1	74193				
T89412	911				
T90116F		3018			
T152148	100	3005		TR05/254	
T164213	128	3024		TR87/	
T1003521	108				S0020
T1004671	123A	3122		TR21/735	S0015
T1008834	123A	3122		TR21/735	S0015
T-H2SC213		3039			
T-H2SC313	108	3019		TR95/	S0016

Original Part Number	ECG	SK	GE	IR/WEP	HEP
T-H2SC387	108	3019		TR95/	
T-H2SC536	123A	3122		TR21/	
T-H2SC693	123A	3020		TR21/	
T-H2SC715	123A	3122		TR21/	
T-HU60U		3004			
T-Q5019	163	3115	38	TR61/	S5021
T-Q5020		3005	1		
T-Q5021		3006			
T-Q5022		3006			
T-Q5025		3004			
T-Q5026		3004			
T-Q5027		3004	53		
T-Q5028		3034	16		
T-Q5030		3035	25		
T-Q5031		3011	8		
T-Q5032		3011	18		
T-Q5034		3006	51		
T-Q5035		3006			
T-Q5036		3009	16	TR01/	
T-Q5038		3006			
T-Q5039		3010			
T-Q5044		3006			
T-Q5049		3018			
T-Q5050		3010			
T-Q5053	108	3020	20	TR21/	S0016
T-Q5053C	123A		20	TR21/	S0016
T-Q5055	108	3117	20		S0016
T-Q5057	124	3021	12	TR81/	S5011
T-Q5063		3020	63	TR51/	S5025
T-Q5071	161	3018		TR83/	S0017
T-Q5073	123A	3122	20	TR21/735	S3013
T-Q5075	124	3021	12	TR81/	S5011
T-Q5077	159	3114	21	TR20/717	S0013
T-Q5078	161	3018	60	TR83/	S5026
T-Q5079	107	3018	20	TR70/	S0016
T-Q5080	175	3021	12	TR81/	S5013
T-Q5081	128	3024	63	TR87/243	S3019
T-Q5082	154	3045		TR78/	S5025
T-Q5083	164	3079	38	TR93/	S5021
T-Q5084	165	3111	38	TR93/	S0030
T-Q5086	161	3018	20	TR83/	S0030
T-Q5087	159	3114	21	TR20/	S0019
T-Q5093	199	3124	61	TR21/	S0030
T-Q5099	128	3024	18	TR87/	S0015
T-Q5104	124	3021	12	TR23/	S5011
T-Q5105	130	3027	14	TR61/	S7002
T-Q5106	107	3018	20	TR24/	S0011
T-Q7037		3102			
T-SK967		3024			
T+CA05		3123			
TA-1	105	3012		TR03/233	
TA-2	121	3009		TR01/232	
TA-3	105	3012		TR03/233	
TA-4	102A			TR85/250	
TA-5	126			TR17/635	
TA-6	123A	3122	20	TR21/735	
TA7	123A		17		
TA-7	108	3039		TR95/56	
TA117	704	3023			
TA1575	102A	3004	2	TR85/250	G0005
TA1575B	102A		2	TR85/250	G0005
TA-1575B	100	3005			
TA1614	121	3009	16	TR01/232	G6005
TA1620A,B	103	3010	8	TR08/641A	
TA1628	160	3007		TR17/637	G0003
TA1650A	160			TR17/637	G0002
TA-1650A	126	3008			
TA1655B	102A		1	TR85/250	G0005
TA-1655B	100	3005			
TA1658	160	3007		TR17/637	G0003
TA1659	160	3007		TR17/637	G0003
TA1660	160	3007		TR17/637	G0003
TA1662	160	3007		TR17/637	G0003
TA1682,A	121		16	TR01/232	G6005
TA-1682,A	121	3009		TR01/	
TA1697	102A	3003		TR85/250	G0005
TA1704	100	3005	1	TR05/254	G0005
TA1705	121	3009	16	TR01/232	G6005
TA1706	102A	3003		TR85/250	G0005
TA1730	102A	3004	2	TR85/250	G0005
TA1731	160	3007		TR17/637	G0003
TA1755	160		51	TR17/637	G0002
TA-1755	126	3008			
TA1756	160		51	TR17/637	G0002
TA-1756	126	3008			
TA1757	160	3007		TR17/637	G0003
TA1759	101	3011	5	TR08/641	
TA1763,A	100		1	TR05/254	G0005
TA-1763,A	100	3005			
TA1765	121	3009	16	TR01/232	G6005
TA1766	121	3009	16	TR01/232	G6005
TA1767	101	3011	5	TR08/641	
TA1771	101	3011	5	TR08/641	
TA1772	101	3011	5	TR08/641	
TA1773	121	3009	16	TR01/232	G6005

Original Part Number	ECG	SK	GE	IR/WEP	HEP
TA1778	100	3005	1	TR05/254	G0005
TA1782	100	3005	1	TR05/254	G0005
TA1783	100	3005	1	TR05/254	G0005
TA1794	121	3009	16	TR01/232	G6005
TA1796	160	3007		TR17/637	G0003
TA1797	160	3007		TR17/637	G0003
TA1798	160	3007		TR17/637	G0003
TA1828	160	3007		TR17/637	G0003
TA1828A		3008			
TA1828B		3006			
TA1829	160	3006		TR17/637	
TA-1829	160	3007			
TA1829A		3006			
TA1830	160		51	TR17/637	G0002
TA-1830	126	3008			
TA1846	160	3006	51	TR17/637	G0003
TA1847	160	3006	51	TR17/637	G0003
TA1860	160	3006	51	TR17/637	G0003
TA1860A		3006			
TA1861	160	3006	51	TR17/637	G0003
TA1881	121	3009	16	TR01/232	G6005
TA1890	121	3009	16	TR01/232	G6005
TA1891	121	3009	16	TR01/232	G6005
TA1928A	127	3035	3	TR27/235	G6008
TA1932,A,B		3009			
TA1934		3006			
TA1936,A		3006			
TA1945,A		3024			
TA1946,A		3024			
TA1947		3024			
TA1948		3024			
TA1965		3008			
TA1968		3008			
TA1969		3008			
TA1970		3008			
TA1971		3008			
TA1972		3008			
TA1990		3006			
TA1996		3006			
TA2045		3014			
TA2046		3014			
TA2047		3014			
TA2048		3014			
TA2053,A,B		3024			
TA2083	127	3034	3	TR27/235	G6007
TA2188	127	3034	25	TR27/235	
TA2192A		3024			
TA2235		3124			
TA2235A		3024			
TA2277		3122			
TA2301	121	3009	12	TR01/232	
TA2322	160			TR17/637	G0003
TA2333		3039			
TA2358A		3018			
TA2401	108			TR95/	S0017
TA2402,A	175	3026			
TA2403A		3027			
TA2412,A		3009			
TA2458		3103			
TA2468A		3079			
TA2470		3044			
TA2501		3024			
TA2503	108	3039		TR95/56	S0016
TA2509,A	124	3021			
TA2510		3021			
TA2511		3021			
TA2512		3021			
TA2544		3036			
TA2554	161			TR83/	
TA2555	161			TR83/	
TA2577A	130	3027	14	TR59/247	
TA2606		3019			
TA2626					S5014
TA2644	222	3065			
TA2650		3036			
TA2651		3025			
TA2658		3024			
TA2669,A		3561			
TA2670,A		3025			
TA2672	121	3014		TR01/	
TA2700	124			TR81/	
TA2703A		3044			
TA2710		3024			
TA2729		3122			
TA2733		3024			
TA2733A		3025			
TA2765A		3021			
TA2800		3024			
TA2808		3079			
TA2809		3079			
TA2819	191	3053			
TA2840	220				
TA2862		3014			
TA2871		3021			
TA2911	152	3054	57	TR55/	S5014
TA2920		3035			
TA2921		3035			
TA2928		3035			
TA5220		3524			
TA5234		3022			
TA5235		3022			
TA5262		3525			
TA5274(RCA)	711				
TA5345,A		3525			
TA5360		3070			
TA5361,A,BX		4000			
TA5388,B,BX		4007			
TA5455,BX		4001			
TA5456,BX		4002			
TA5460,A,AV,AW,AX		4016			
TA5517C		3141			
TA5551		4000			
TA5553		4007			
TA5555		4002			
TA5579,V,X		4015			
TA5625,A		3073			
TA5628,C		3147			
TA5649	714	3075	IC4	IC509/	
TA5649A		3075			
TA5649(RCA)	714				
TA5652,V,X		4019			
TA5660,V,X		4009			
TA5668,V,X		4010			
TA5672		3071			
TA5675,V,W,X		4013			
TA5681,V,W,X		4011			
TA5682,V,W,X		4012			
TA5684,V,X		4017			
TA5702	715		IC6	IC510/	
TA5702(RCA)	715				
TA5720		3072			
TA5752		3074			
TA5797		3514			
TA5807,X		3566			
TA5814	712	3072	IC2	IC507/	C6063P
TA5814(RCA)	712				
TA5842		3146			
TA5866		3144			
TA5912	713	3077	IC5	IC508/	
TA5912A,B		3077			
TA5912(RCA)	713				
TA5914,C		3143			
TA5932,A,B,E,Q		3078			
TA6157		3526			
TA6200	128	3024	63	TR87/243	S3020
TA6220	708	3135			
TA6243	731				
TA6243(RCA)	731				
TA6319(RCA)	797				
TA6404	798				
TA6405	738	3167			
TA6465A		3568			
TA6523	749				
TA6548		3565			
TA6613		3569			
TA6615		3567			
TA6710A		3564			
TA6746		3184			
TA6804		3184			
TA7005		3559			
TA7006		3559			
TA7006P	1149				
TA7007		3559			
TA7027M	726				
TA7031M	901				
TA7037M	726				
TA7038M	704				
TA7045M	724	3525			S0016
TA7046P	1108				
TA7047M	904				
TA7050M	711	3070			C6069G
TA7050P	1061				
TA7051P	710	3102		IC506/	
TA7051/01			511		
TA7053M	707				C6067G
TA7054P	1106				
TA7057M	905				
TA7068	130	3027			
TA7069	130	3027			
TA7070	1004				
TA7070P	1004	3102	511		
TA7071P	712				
TA7072P	748	3102			
TA7073AP	748				C6060P
TA7073P	748				C6060P
TA7074P	749	3168			C6076P
TA7075,P	1080				
TA7076P	1109				C6079P
TA7076P(FA-1)	1109				
TA7076P(FA-6)	1109				

Original Part Number	ECG	SK	GE	IR/WEP	HEP
TA7076P(FA-7)	1109				
TA7080		3024			
TA7092AP	1023				
TA7092AP-B-C	1023				
TA7092P	1107				
TA7092P-A-D	1107				
TA7092P-H	1107				
TA7102P	1105				
TA7102P(FA-1)	1105				
TA7102P(FA-2)	1105				
TA7103P	1062				C6072P
TA7106P	1149				
TA7117P	1122				
TA7120B,P	1087				
TA7120P-E	1087				
TA7121		3112			
TA7122		3512			
TA7122AP	1085				
TA7122AP-D	1085				
TA7122P-B	1085				
TA7124P	1128				
TA7125		3513			
TA7134	228	3103			
TA7137	152	3054		TR76/701	
TA7145P	1134				
TA7146P	1133				
TA7148P	1131				
TA7149	221	3050			
TA7149P	1132				
TA7150	221	3050			
TA7150P	1130				
TA7151	221	3050			
TA7153P	1129				
TA7155	196	3054			
TA7156	152	3054		TR76/701	
TA7189	222	3050			
TA7199	223	3027		TR59/	
TA7200	223			TR59/	
TA7201	223	3036		TR59/	
TA7202	223			TR59/	
TA7262	152	3050		TR76/701	
TA7262(RCA)	221				
TA7264		3085			
TA7265		3085			
TA7266		3085			
TA7274	222	3065			
TA7280		3167			
TA7281		3167			
TA7285		3021			
TA7292	154	3040			
TA7293	154	3040			
TA7303	108	3039			
TA7311	182	3054			S5004
TA7312	182	3054			S5004
TA7313	182	3054			S5004
TA7314	182	3054			S5004
TA7315	182	3054			S5004
TA7316	182	3054			S5004
TA7317		3054			
TA7318	182	3054			S5004
TA7319	108	3039			
TA7362	196	3054			
TA7363	152	3054		TR76/701	
TA7374	222	3065			
TA7399	222				
TA7502M	909				
TA7520	185				S5006
TA7554	152				
TA7555	152		66		
TA7556	153	3179	69		
TA7557	153	3179	69		
TA7669	222	3065			
TA7684	222				
TA7739	228	3104	27	TR78/	
TA7740	228	3103	27	TR78/	
TA7741		3083	29	TR56/	
TA7742	197	3083			
TA7743	197	3084			
TA7782	196	3054			
TA7783	196	3054			
TA7784	196	3054			
TA7936		3065			
TA7999		3065			
TA8000		3065			
TA8001		3065			
TA8002		3065			
TA8201		3180			
TA8202		3180			
TA8204		3181			
TA8210	197	3083			
TA8211	197	3084			
TA8212	197	3083			
TA8231	196	3054			
TA8232	196	3054			
TA8233	196	3054			

Original Part Number	ECG	SK	GE	IR/WEP	HEP
TA8242	222	3065			
TA8247		3103			
TA8248		3104			
TA8249		3103			
TA8344		3039			
TA8345		3039			
TA8346		3039			
TA8247		3039			
TA8348		3182			
TA8349		3182			
TA8351		3183			
TA8428		3511			
TA8429		3027			
TA8430		3036			
TA8431		3036			
TA8434		3026			
TA8442		3563			
TA8443		3563			
TA8485		3180			
TA8486		3182			
TA8559		3176			
TA8561		3177			
TA8724		3167			
TA8726		3563			
TA8847A,B,C,D,E		3560			
TA17044				TR05/	
TA198030-4	123A				
TA198035-1	128				
TA198036-2	159				
TAA320	133	3112		/801	
TAA521	909	3590			
TAA521/709	909	3590			
TAA521A	909D				
TAA521A/709C	909D				
TAA522	909				
TAA522/709	909				
TAA611A12	1113				
TAA611B12	1113				
TAA611C11	1118				
TAA621A11	1111				
TAA621A12	1112				
TAC047	123A	3020	20	TR21/735	
TBA221	941	3514			
TBA221/741C	941				
TBA221A	941D				
TBA221A/741C	941D	3552			
TBA222	941	3514			
TBA222/741	941	3514			
TBA231	725	3162			
TBA281	923	3164			
TBA281/723	923	3164			
TBA641B11	1114				
TBA741,C		3514			
TBA800	1116				
TBA810AS,S	1115	3184			
TBA820	1117				
TBRC147B			10		
TC0914	108	3039	11	TR95/56	S0025
TC0918	108	3019	63	TR95/56	S0011
TC2369A	108	3039	63	TR95/56	S0025
TC2483	108	3039	63	TR95/56	S0025
TC2484	108	3039	63	TR95/56	S0005
TC4001P	4001				
TC4002P	4002				
TC4011P	4011				
TC4013,P	4013				
TC4015P	4015				
TC4017P	4017				
TC4019P	4019				
TC4021P	4021				
TC4023P	4023				
TC4025P	4025				
TC4027P	4027				
TC4028P	4028				
TC4030P	4030				
TC4040P	4040				
TC4042P	4042				
TC4049P	4049				
TC4050P	4050				
TC4512P	4512				
TC4539P	4539				
TC3123036722	123A	3020	10	TR21/735	S0014
TC3123036900	123A	3020	10	TR21/735	S0015
TC3123037111	123A	3020	17	TR21/735	S0011
TC3123037222	123A	3020		TR21/735	S0015
TC3123037412	123A	3020	10	TR21/735	S0015
TC3123041557	102A	3004		TR85/250	G0005
TCH98	199		65		
TCH99			82		
TCH99B			82		
TCS100			63		
TCS101			67		
TCS102			63		
TCS103			67		
TCS614				TR53/	

Original Part Number	ECG	SK	GE	IR/WEP	HEP
TD100			210		
TD101	123A		210		
TD102	123A		210		
TD200			210		
TD201	123A		210		
TD202	123A		210		
TD250			210		
TD400	159	3114	82		
TD401	159	3114	82		
TD402	159	3114	82		
TD500	159	3114	82		
TD501	159	3114	82		
TD502	159	3114			
TD550	159	3114	82		
TD1060,P	9930				
TD1061,P	9935				
TD1062,P	9932				
TD1063,P	9933				
TD1064,P	9944				
TD1065,P	9946				
TD1066,P	9962				
TD1067,P	9948				
TD1070,P	9945				
TD1072	9936				
TD1073,P	9093				
TD1074,P	9099				
TD1080,P	9961				
TD1085,P	9949				
TD1086,P	9963				
TD1401,P	7400				
TD1402,P	7410				
TD1403,P	7420				
TD1404,P	7430				
TD1405,P	7440				
TD1406,P	7450				
TD1407,P	7460				
TD1408,P	7472				
TD1409,P	7473				
TD1419,P	7451				
TD2001P	9661				
TD2002P	9670				
TD2003P	9668				
TD2004P	9664				
TD2005P	9663				
TD2006P	9669				
TD2008P	9660				
TD2009P	9671				
TD2010P	9672				
TD2011P	9662				
TD2012P	9678				
TD2013P	9677				
TD2015P	9667				
TD2219	123A		210		
TD2905	159	3114	21		
TD3400A	7400				
TD3400AP	7400				
TD3400P	7400				
TD3401A,AP	7401				
TD3402A,AP	7402				
TD3404A,AP	7404				
TD3405A,AP	7405				
TD3409A,AP	7409				
TD3410A,AP,P	7410				
TD3420A,AP,P	7420				
TD3430A,AP,P	7430				
TD7440A,AP,P	7440				
TD3441A,AP	7441				
TD3442A,AP	7442				
TD3447A,AP	7447				
TD3450A,AP,P	7450				
TD3451A,AP,P	7451				
TD3460A,AP,P	7460				
TD3472A,AP	7472				
TD3473A,AP	7473				
TD3474A,AP,P	7474				
TD3475A,AP	7475				
TD3480A,AP	7480				
TD3482P	7482				
TD3483P	7483				
TD3490A,AP,P	7490				
TD3491A,AP	7491				
TD3492A,AP,P	7492				
TD3493A,AP	7493A				
TD3493P	7493				
TD3495A,AP	7495				
TD3503A,AP	74164				
TD34121A,AP	74121				
TD34192A,AP	74192				
TDA1200	788				
TE-500-E		3116			
TE697	123A		20		
TE706	108		211		
TE1420	123A		17		
TE1990	128		18		
TE2369	123A		20		

Original Part Number	ECG	SK	GE	IR/WEP	HEP
TE2484	108		81		
TE2711	199		212		
TE2712	199		212		
TE2713	172		64		
TE2714	172		64		
TE2715	108		61		
TE2716	108		61		
TE2921	199		212		
TE2922	199		212		
TE2923	199		212		
TE2924	199		212		
TE2925	199		212		
TE2926	199		212		
TE3390	199		212		
TE3391	199		212		
TE3391A	199		212		
TE3392	199		212		
TE3393	199		212		
TE3394	199		212		
TE3395	199		212		
TE3396	199		212		
TE3397	199		212		
TE3398	199		212		
TE3414	123A		210		
TE3415	123A		210		
TE3416	128		210		
TE3417	192		210		
TE3605	123A	3122	20		
TE3605A	123A	3122	20		
TE3606		3122	20		
TE3606A		3122	20		
TE3607	123A		20		
TE3702	193		48		
TE3703	193		82		
TE3704	123A		20		
TE3705	123A		20		
TE3707	108		61		
TE3708	108		61		
TE3709	108		61		
TE3710	108		61		
TE3711	108		61		
TE3843	199		212		
TE3844	199		212		
TE3845	199		212		
TE3854	199		212		
TE3854A	199		212		
TE3855	199		212		
TE3859	199		212		
TE3860	199		212		
TE3900	199		212		
TE3901	199		212		
TE3903	123A		210		
TE3904	123A		20		
TE3905	159		21		
TE3906	123A		210		
TE4123	123A		210		
TE4124	123A		20		
TE4125	159		21		
TE4126	159		21		
TE4256	199		212		
TE4424	123A		210		
TE4425	192		81		
TE4951	123A		210		
TE4952	123A		210		
TE4953	123A		210		
TE4954	123A		210		
TE5086	193		65		
TE5087	193		65		
TE5088	199		212		
TE5089	199		212		
TE5249	199		212		
TE5309	199	3122	212		
TE5309A	123A	3122			
TE5310	199	3122	212		
TE5311	199	3122	212		
TE5311A	123A	3122			
TE5365	159		82		
TE5366	159		82		
TE5367	193		82		
TE5368	123A		210		
TE5369	123A		210		
TE5370	123A		210		
TE5371	123A		210		
TE5376	123A		20		
TE5377	123A		20		
TE5378	159		21		
TE5379	159		21		
TE5447	159		82		
TE5448	159		82		
TE5449	123A		20		
TE5450	123A		20		
TE5451	123A		20		
TE6606					S0011
TEH0147					S0005
TF0021(DM)					F0021

Original Part Number	ECG	SK	GE	IR/WEP	HEP
TF7	124			TR81/	
TF30	102A	3004	2	TR85/250	G0005
TF49	102A			TR85/250	G0005
TF65	102A	3004	52	TR85/250	G0005
TF65/30	102A		52	TR85/250	G0005
TF65/M	102A			TR85/250	
TF65M	158	3004	2		
TF65/S/30	102A		53	TR85/250	
TF66	102A	3123	52	TR85/250	G0005
TF-66	102A	3004			
TF66/30	102A			TR85/250	G0005
TF66/60	102A			TR85/250	G0005
TF70	101	3011		TR08/641	G0011
TF71	101	3011		TR08/641	G0011
TF72	101	3011		TR08/641	G0011
TF75	102A	3004	52	TR85/250	G0005
TF77	102A	3004	52	TR85/250	G0005
TF77/30					G6011
TF78	104		20	TR01/230	G6003
TF-78	123A	3020		TR21/	
TF78/30	104	3009		TR01/230	G6003
TF78/30Z	121	3009		TR01/232	
TF78/60	121	3009		TR01/232	
TF80					G6011
TF80/30	121	3009		TR01/232	G6011
TF80/30Z	121		16	TR01/	
TF80/302	102A	3004		TR85/250	G6005
TF101-A,B,D	128				
TF251					F0033
TF252					S0033
TF260					S5026
TF1035(DM)					F1035
TF2004(DM)					F2004
TF2005(DM)					F2005
TF7473E					C3073P
TF66160				TR85/	
TF-70				TR01/	
TF-71				TR01/	
TF-72				TR01/	
TG2SA201		3005			
TG2SA201A		3005			
TG2SA201C		3005		TR17/	G0005
TG2SA608		3114			
TG2SA608A		3114			
TG2SA608C		3114		TR52/	S0013
TG2SA2010		3005			
TG2SC65A		3044			
TG2SC65N		3044			
TG2SC65Y		3044		TR78/	S5025
TG2SC536		3124			
TG2SC536B		3124			
TG2SC536C		3124		TR24/	S0016
TG2SC536E		3124		TR24/	S0016
TG2SC927		3117			
TG2SC927A		3117			
TG2SC927C		3117		TR70/	S0016
TG2SC1025A		3131			
TG2SC1025D		3131		TR81/	S5012
TG2SC1175		3122			
TG2SC1175A		3122			
TG2SC1175C		3122		TR21/	S0025
TG2SC1175E		3122			
TG2SC1293		3018			
TG2SC1293A		3018		TR71/	S0020
TG2SC1295,A		3111			
TG2SC1295O		3111		TR61/	S5020
TGD24A,B		3021			
TSD24Y		3021		TR23/	S5011
TG48	158			/250	G0005
TG-48		3004			
TG6001(DM)					G6001
TG7400E					C3000P
TG7401E					C3001P
TG7410E					C3010P
TG7420E					C3020P
TG7430E					C3030P
TG7440E					C3040P
TG7450E					C3050P
TH0600				TR85/	
T-H2SC313				/56	
T-H2SC387				/56	
TH2SC536	123A		20	/735	
TH2SC693	123A		20	/735	
TH2SC715	123A		20	/735	
TH-H2SC313	161	3117		TR83/719	
THP36					S0033
THP61					S0033
THP62					S0033
THP106					S0025
THP501					G0003
THP502					G0003
THU60U	102A		2	TR85/250	
TI1A6	123A		10		
TI-1A6	131		10	TR94/642	
TI-7A	131			TR94/642	

Original Part Number	ECG	SK	GE	IR/WEP	HEP.
TI24A,B	123A				S0015
TI25A,B	108				S0015
TI54A,B,C	123A				S0015
TI54D	199				S0015
TI54E	123A				S0015
TI-92	199		212		
TI266A	121		16		
TI-266A	121	3009		TR01/232	
TI269	121	3009	16	TR01/232	
TI320					G0005
TI321					G0001
TI338	160	3006	51	TR17/637	
TI363	160			TR17/637	G0003
TI-363	100	3005	1	TR85/	
TI364	160			TR17/637	G0003
TI-364	100		1	TR85/	
TI365	160		51	TR17/637	G0003
TI366	121	3009	16	TR01/232	G6005
TI366A	121	3009	16	TR01/232	G6005
TI367,A	121	3009	16	TR01/232	G6005
TI368,A	121	3009	16	TR01/232	G6005
TI369,A	121	3009	16	TR01/232	G6005
TI370,A	104		16	TR01/230	G6003
TI-370,A	121	3009		TR01/	
TI-3776					G0006
TI377					G0005
TI378					G0008
TI383					G0008
TI384					G0008
TI385					G0008
TI386					G0008
TI387	160	3006	51	TR17/637	G0003
TI388	160	3006	51	TR17/637	G0003
TI389	160		51	TR17/637	G0003
TI390	160		51	TR17/637	G0003
TI391	160		51	TR17/637	G0003
TI393	160			TR17/637	
TI395	160		51	TR17/637	G0003
TI396	160			TR17/637	
TI397	160		51	TR17/637	G0003
TI398	160		51	TR17/637	G0003
TI399	160		51	TR17/637	G0003
TI400	160	3006	51	TR17/637	G0003
TI401	160	3006	51	TR17/637	G0003
TI402	160	3006	51	TR17/637	G0003
TI403	160	3006	51	TR17/637	G0003
TI407	108			TR95/56	S0016
TI-407	161	3019	86		
TI408	108			TR95/56	S0016
TI-408	161	3019	86		
TI409	108			TR95/56	S0016
TI-409	161	3019	86		
TI410	108	3019	17	TR95/56	S0016
TI411	123A	3122	20	TR21/735	S0015
TI-412	123A	3122	20	TR21/735	S0015
TI-414	123A	3122	47	TR21/735	
TI415	123A			TR21/735	S0015
TI-415	199	3020	212	TR51/	
TI416	123A			TR21/735	S0015
TI-416	199	3020	62	TR51/	
TI417	123A			TR21/735	S0015
TI-417	108	3019	61		
TI418	123A			TR21/735	S0015
TI-418	199	3019	212		
TI419	123A			TR21/	S0015
TI-419	199	3019	212		
TI420	123A	3122		TR21/735	
TI-420	108		212	TR51/	
TI-421	199		212	TR51/	
TI422	123A	3122		TR21/735	
TI-422	123A		20	TR53/	
TI-423	123A		20	TR53/	
TI424	123A	3122		TR21/735	
TI-424	128		81	TR25/	
TI-425	128		81		
TI-428	159		82	TR28/	
TI-429	159		82	TR28/	
TI430	123A	3122		TR21/735	
TI-430	108		61	TR65/	
TI431	108	3039		TR95/56	
TI-431	108		61	TR65/	
TI432	123A	3122		TR21/735	
TI-432	123A		20		
TI-433	123A		20		
TI-474	108		212	TR51/	
TI-475	128		212	TR51/	
TI480	123A	3122		TR21/735	S0014
TI-480	128		81	TR25/	
TI481	123A	3122		TR21/735	
TI-481			81		
TI482	123A	3024		TR21/735	S0014
TI-482	128		81	TR25/	
TI483	123A	3024		TR21/735	S0014
TI-483	128		81	TR25/	
TI484	123A	3024		TR21/735	S0014

Original Part Number	ECG	SK	GE	IR/WEP	HEP
TI-484	128		81	TR65/	
TI485	123A	3122		TR21/735	S0011
TI-485	192		47	TR65/	
TI486	152				
TI-486			32		
TI487	152		66		
TI-490	108		212	TR51/	
TI492	123A			TR21/735	S0014
TI-492	161	3020	60	TR51/	
TI493	123A			TR21/735	S0014
TI-493	161	3020	60	TR51/	
TI494	123A			TR21/735	S0014
TI-494	161	3020	60	TR51/	
TI495	123A			TR21/735	S0014
TI-495	108	3020	39	TR51/	
TI496	123A	3122		TR21/735	S0014
TI-496	128		81	TR25/	
TI503	159		21		
TI-503	159	3114	21	TR20/717	
TI714	123A	3122	20	TR21/735	
TI-714A	123A	3122		TR21/735	
TI722	171		27		
TI-722	154			TR78/712	
TI741	132		FET2	FE100/802	
TI743	159	3114	21	TR20/717	
TI744	159	3114	21	TR20/717	
TI751	123A	3122	20	TR21/735	
TI752	159	3114	21	TR20/717	
TI802B	123A	3020		TR21/735	
TI803B	123A	3122		TR21/735	
TI806G				TR51/	
TI-806G	123A	3122		TR21/735	
TI808E	129			TR88/242	
TI810B	123A	3122		TR21/735	
TI811G	128	3024		TR87/243	
TI-890	159		65	TR52/	
TI-891	193		65	TR52/	
TI904	123A			TR21/	S0014
TI-905	159			TR20/	
TI906				TR30/	
TI-906	159	3114		/717	
TI907	123A		20		
TI-907	123A	3122		TR21/735	
TI908	123A	3122	20	TR21/735	
TI3010	160		53	TR17/637	G0002
TI3011	160		53	TR17/637	G0002
TI3012	121	3009	16	TR01/232	G6005
TI3015	128	3024		TR87/243	
TI3016	108	3039		TR95/56	
TI-3016	107		11		
TI3027	121	3009	16	TR01/232	G6003
TI3028	121	3009	3	TR01/232	G6005
TI3029	121	3014	16	TR01/232	G6005
TI-3030	179			TR35/G6001	
TI-3031	179			TR35/G6001	
TI64213	128	3024	20	TR87/243	
TIA01	102A	3004		TR85/250	
TIA02	160	3123	50	TR17/637	
TIA03	100	3005	53	TR05/254	G0005
TIA04	102	3004	53	TR05/631	G0005
TIA04A		3004			
TIA05	100	3005	53	TR05/254	G0005
TIA05A	100				
TIA06	123A	3122	20	TR21/735	
TIA35	176	3123			
TIA102	123A			TR21/735	
TIE	718		IC8		
TIJ	718		IC8		
TIM-01	160		51	TR17/637	
TIM-10	160		51	TR17/637	
TIM-11	160		51	TR17/637	
TIN	718		IC8		
TIP04			73	TR61/	
TIP3A	197	3083			
TIP14	184		28		S5000
TIP-14	152	3054		TR76/701	
TIP24	152		66	TR76/701	S5000
TIP-24	196	3054			
TIP27	157	3103		TR60/244	S5015
TIP29	152	3041		TR76/701	S5000
TIP29A	152	3054	28	TR76/701	S5000
TIP29B	198		32		S5000
TIP29C	198		32		
TIP29L					S5000
TIPXA	196	3054			
TIP30,A	153		69	TR77/700	S3032
TIP30B	189	3200			S3032
TIP30C					S3032
TIP31	152	3054		TR76/701	S5000
TIP-31	196	3054			
TIP31A	152	3054	66	TR76/701	S5000
TIP-31A	196	3054			
TIP31B	184				S5000
TIP-31B	196	3054			
TIP31C					S5004
TIP32	153		69	TR77/700	S5006
TIP32A	153		69	TR77/700	S5006
TIP32B	197				S5006
TIP32C					S5005
TIP34	183		56	/S5005	S5005
TIP34A	183	3189	56	/S5005	S5005
TIP34B	183	3189	56	/S5005	S5005
TIP34C					S5005
TIP41	196		55	TR92/	S5004
TIP41A	196		55	TR92/	S5004
TIP41B	196			TR92/	S5004
TIP41C					S5004
TIP42	197		56	/S5005	S5005
TIP42A	197		56	/S5005	S5005
TIP42B	197		56	/S5005	S5005
TIP42C					S5005
TIP47	198				
TIP48	198				
TIP49	198				
TIP120	261				
TIP121	261				
TIP125	262				
TIP126	262				
TIP140	270				
TIP141	270				
TIP142	270				
TIP145	271				
TIP146	271				
TIP147	271				
TIP503	175			TR81/	
TIP544			74		
TIP550		3111			
TIP551		3115			
TIP552		3115			
TIP553		3115			
TIP640	245				
TIP641	245				
TIP645	246				
TIP646	246				
TIP2955	219				
TIP3055		3027			S5004
TIS					S0005
TIS03	159	3114	82	TR20/717	S0013
TIS04	159	3114	82	TR20/717	S0013
TIS14	133	3112	FET1	/801	F0010
TIS18	161	3039	86	TR83/719	S0016
TIS22	123A	3122	39	TR21/735	S0011
TIS23	123A	3122	39	TR21/735	S0011
TIS24	108	3039	63	TR95/56	S0005
TIS34	132	3116	FET2	FE100/802	F0015
TIS37	159	3025	65	TR20/717	S0013
TIS38	159	3114	65	TR20/717	S0013
TIS39					S3001
TIS42	132				F0015
TIS44	123A	3122	61	TR21/735	S0025
TIS45	123A	3122	20	TR21/735	S0011
TIS46	123A	3122	20	TR21/735	S0011
TIS47	123A	3122	20	TR21/735	S0025
TIS48	123A	3122		TR21/735	S0011
TIS49	123A	3122		TR21/735	S0011
TIS50	159	3122			S0013
TIS51	123A	3122	20	TR21/735	S0011
TIS52	123A	3122	20	TR21/735	S0025
TIS53	159	3114		TR20/717	S0013
TIS54	159	3114	22	TR20/717	S0013
TIS55	123A	3046	20	TR21/735	S0025
TIS56	123A	3122	86	TR21/735	S0017
TIS57	123A	3122	86	TR21/735	S0017
TIS58	132	3116	FET2	FE100/802	F0015
TIS59	132	3116	FET2	FE100/802	F0015
TIS60	123A	3024	210	TR21/735	S0015
TIS60A,B,C,D,E	128	3024		TR87/243	
TIS60M	128	3024	210	TR87/243	
TIS61	159	3025	82	TR20/717	S0015
TIS61A,B,C,D,E	129	3025		TR88/242	
TIS61M	129	3025	82	TR88/242	
TIS62	108	3039	86	TR95/56	S0016
TIS63	108	3039	86	TR95/56	S0016
TIS64	108	3039	86	TR95/56	S0016
TIS71	123A	3122		TR21/735	
TIS72	123A	3122		TR21/735	
TIS78	133	3112			
TIS79	133	3112			
TIS82	186		215		
TIS83	123A	3122	61	TR21/735	S0016
TIS84	108	3039	61	TR95/56	S0016
TIS85	108	3039	61	TR95/56	
TIS86	107	3039		TR70/720	S0016
TIS87	107	3039		TR70/720	S0016
TIS88	132	3116	FET2	FE100/802	F0015
TIS89					S0015
TIS90	123A	3122	47	TR21/735	S0014
TIS90-2	123A	3020			
TIS91	159	3114	48	TR20/717	S0012
TIS92	123A	3122	81	TR21/735	S0014

Original Part Number	ECG	SK	GE	IR/WEP	HEP
TIS92BLU	123A		81		
TIS92GRN	123A		81		
TIS92GRY	123A		47		
TIS92M	128	3024	63	TR87/243	
TIS92VIO, YEL	123A		81		
TIS93	159	3114	82	TR20/717	S0012
TIS93BL, GRN	159		82		
TIS93GRY	159		48		
TIS93M	129	3025	21	TR88/242	
TIS93VIO, YEL	159		82		
TIS94	123A	3122	210	TR21/735	S0015
TIS-94	199		62		
TIS95	123A	3122	10	TR21/735	
TIS96	123A	3122	20	TR21/735	S0005
TIS97	107	3039	20	TR70/720	S0015
TIS-97	199		210		
TIS98	107	3122	63	TR70/720	S0005
TIS98A	108	3018			
TIS99	107	3039	20	TR70/720	S0005
TIS100	154	3045	27	TR78/712	S0005
TIS101	154	3045	27	TR78/712	S0005
TIS102	154	3045		TR78/712	S5025
TIS103	154	3045		TR78/712	S5025
TIS104	159		67		S0019
TIS105	108				S0016
TIS106	194				S0005
TIS107	128		210		S0015
TIS108	107	3039	61	TR70/720	S0016
TIS109	195		20		S3001
TIS110	195		20		S3001
TIS111	195		20		S3001
TIS112	159		21		S0019
TIS113	123A		47		S0025
TIS114	123A		47		S0025
TIS125			61		
TIS128			21		
TIS133			47		
TIS134			47		
TIS138			21		
TIS412	107	3039	20	TR70/720	
TIU3304	912				
TIUG1888				TR22/	
TIX90	102A			TR85/250	
TIX91	160			TR17/637	G0003
TIX92	160			TR17/637	G0003
TIX155					S5004
TIX316	160			TR17/637	G0003
TIX528				TR24/	
TIX529				TR24/	
TIX530				TR24/	
TIX531				TR24/	
TIX613					S0025(2)
TIX614					S0025(2)
TIX615					S0025(2)
TIX616					S0015(2)
TIX617					S0015(2)
TIX618					S0015(2)
TIX619					S0019(2)
TIX620					S0019(2)
TIX621					S0019(2)
TIX622					S0019(2)
TIX623					S0019(2)
TIX624					S0019(2)
TIX712	123A				S0011
TIX804	159				S0019
TIX805	159				S0019
TIX806					S0015(2)
TIX807					S0015(2)
TIX808					S0015(2)
TIX809					S0025(2)
TIX810					S0025(2)
TIX811					S0025(2)
TIX812					S0019(2)
TIX813					S0019(2)
TIX814					S0019(2)
TIX876	108				S0016
TIX880	108				S0016
TIX888	195		215	TR65/	S3001
TIX890	159				S0013
TIX891	159				S0013
TIX895	100		52		
TIX896	101				G0011
TIX1392	195				S3001
TIX1393	195				S3001
TIX2000					G0003
TIX3016	160			TR17/637	
TIX3016A	160		86	TR17/637	
TIX3022				TR17/	
TIX3032	160			TR17/637	G0003
TIX3033					S3002
TIX3035					S3002
TIX-M01	160		50		
TIX-M02	160		50		
TIX-M03	160		50		
TIX-M04	160		50		

Original Part Number	ECG	SK	GE	IR/WEP	HEP
TIX-M05	160		50		
TIX-M06	160		50		
TIX-M07	160		50		
TIX-M08	160		51		
TIX-M11	160		50		
TIX-M14		3006	39	TR70/	
TIX-M15		3006	39	TR70/	
TIX-M16		3006	39	TR70/	
TIX-M17	160		51		
TIX-M101	160		50		
TIX-M201	160		50		
TIX-M202	160		50		
TIX-M203	160		50		
TIX-M204	160		50		
TIX-M205	160		50		
TIX-M206	160		50		
TIX-M207	160		50		
TIXA01	100	3005	2	TR05/254	G0005
TIXA02	100	3005	2	TR05/254	G0005
TIXA03	100	3123	2	TR05/254	G0001
TIXA04	100	3123	2	TR05/254	G0001
TIXA05	100	3123	2	TR05/254	G0001
TIMM01	160		51	TR17/637	G0003
TIXM02	160			TR17/637	G0003
TIXM03	160			TR17/637	G0003
TIXM04	160			TR17/637	G0003
TIXM05	160		51	TR17/637	G0003
TIXM06	160			TR17/637	G0003
TIXM07	160			TR17/637	G0003
TIXM08	160			TR17/637	G0008
TIXM10	160			TR17/637	G0001
TIXM11	160		51	TR17/637	G0003
TIXM13	160			TR17/637	G0003
TIXM14	160			TR17/637	G0002
TIXM15	160			TR17/637	G0003
TIXM16	160			TR17/637	G0002
TIXM17	160			TR17/637	G0003
TIXM18	160			TR17/637	G0003
TIXM19	160			TR17/637	G0003
TIXM101	160			TR17/637	G0003
TIXM103	160			TR17/637	
TIXM104	160			TR17/637	
TIXM105	160			TR17/637	G0003
TIXM106	160			TR17/637	G0003
TIXM107	160			TR17/637	G0003
TIXM108	160			TR17/637	G0003
TIXM201	160	3006	51	TR17/637	G0003
TIXM202	160	3006	51	TR17/637	G0003
TIXM203	160	3006	51	TR17/637	G0003
TIXM204	160	3006	51	TR17/637	G0003
TIXM205	160	3006	51	TR17/637	G0003
TIXM206	160	3006	51	TR17/637	G0003
TIXM207	160		51	TR17/637	G0009
TIXP07					S5005
TIXS09	108	3039	86	TR95/56	G0005
TIXS10	108	3039	86	TR95/56	
TIXS12	123A	3122	20	TR21/735	
TIXS13	123A	3122	20	TR21/735	
TIXS28	108	3039	61	TR95/56	S0016
TIXS29	108	3039	60	TR95/56	S0016
TIXS30	108	3039	60	TR95/56	S0016
TIXS31	108	3039	60	TR95/56	S0016
TJ					G0005
TJ1					G0005
TJ2					G0005
TK20A, B					G0005
TK21A, B					G0005
TK23, A					G0005
TK23C	102A	3004	2	TR85/250	G0003
TK24A, B					G0005
TK25A, B					G0005
TK26, A, B					G0005
TK27, A, B					G0005
TK28C					G0005
TK30C					G0005
TK31, C					G0005
TK33C	101			TR08/641	G0011
TK34C					G0005
TK35C					G0005
TK36C					G0005
TK37C					G0005
TK38C					G0005
TK40	102	3004	2	TR05/631	
TK40A	102A	3004			G0005
TK40C	102	3004	2	TR05/631	G0005
TK-40C	102A	3004			
TK40CA	102A	3004			
TK41	102	3004	2	TR05/631	
TK41A	102A	3004			
TK41C	160		2	TR17/612	G0003
TK-41C	102A	3004			
TK42	102	3004	2	TR05/631	G0005
TK42A	102A	3004			
TK42C	160		2	TR17/637	G0003
TK-42C	102A	3004			

Original Part Number	ECG	SK	GE	IR/WEP	HEP
TK44C					G0005
TK45C	102	3004	2	TR05/631	G0003
TK-45C	102A	3004			
TK45CA		3004			
TK46C					G0005
TK47C					G0005
TK48C					G0005
TK49C	102A	3010	8	TR85/250	G0005
TK70					S0014
TK71					S0014
TK72					S0011
TK82			210		
TK200A					S5000
TK201A					S3024
TK251A					S0012
TK254A					S0014
TK255A					S0011
TK256A					S0011
TK257A					S0011
TK258A					S0011
TK259A					S0011
TK264A					S0011
TK400A					G6005
TK401A					G6005
TK402A					G6005
TK403A					G6005
TK582			48		
TK1228-001	160	3006			
TK1228-1001	126	3008		TR17/635	G0008
TK1228-1002	102A			TR85/250	G0005
TK1228-1003	102A			TR85/250	G0005
TK1228-1004	102A			TR85/250	G0005
TK1228-1005	102A			TR85/250	G0005
TK1228-1006	102A			TR85/250	G0005
TK1228-1007	102A			TR85/250	G0005
TK1228-1008	123	3010		TR86/53	S0011
TK1228-1009	123	3020		TR86/53	S0011
TK1228-1010	123A	3122		TR21/735	S0015
TK1228-1011	123A	3122		TR21/735	S0015
TK1228-1012	123A	3122		TR21/735	S0015
TK3055					S7004
TK9201	181		75		S7002
TK30551			75		S7004
TK30552			75		S7004
TK30553					S7004
TK30555			75		S7004
TK30556			75		S7004
TK30557			75		S7004
TK30558					S7004
TK30560			75		S7004
TK5800400-3062		3007			
TL74H87N	74H87				
TL74H183N	74H183				
TL7400N	7400				
TL7401N	7401				
TL7402N	7402				
TL7403N	7403				
TL7404N	7404				
TL7405N	7405				
TL7406N	7406				
TL7407N	7407				
TL7408N	7408				
TL7409N	7409				
TL7410N	7410				
TL7412N	7412				
TL7413N	7413				
TL7416N	7416				
TL7420N	7420				
TL7423N	7423				
TL7426N	7426				
TL7430N	7430				
TL7437N	7437				
TL7438N	7438				
TL7440N	7440				
TL7442N	7442				
TL7443N	7443				
TL7444N	7444				
TL7445N	7445				
TL7446AN	7446				
TL7447AN	7447				
TL7448N	7448				
TL7450N	7450				
TL7451N	7451				
TL7453N	7453				
TL7454N	7454				
TL7460N	7460				
TL7470N	7470				
TL7472N	7472				
TL7473N	7473				
TL7474N	7474				
TL7475N	7475				
TL7476N	7476				
TL7480N	7480				
TL7481N	7481				
TL7482N	7482				
TL7483N	7483				
TL7484N	7484				
TL7485N	7485				
TL7486N	7486				
TL7489N	7489				
TL7490N	7490				
TL7491N	7491				
TL7492N	7492				
TL7493N	7493A				
TL7494N	7494				
TL7495AN	7495				
TL7496N	7496				
TL7497N	7497				
TL74107N	74107				
TL74121N	74121				
TL74122N	74122				
TL74123N	74123				
TL74141N	74141				
TL74145N	74145				
TL74150N	74150				
TL74151N	74151				
TL74153N	74153				
TL74154N	74154				
TL74155N	74155				
TL74156N	74156				
TL74162N	74162				
TL74163N	74163				
TL74164N	74164				
TL74165N	74165				
TL74166N	74166				
TL74180N	74180				
TL74181N	74181				
TL74182N	74182				
TL74190N	74190				
TL74191N	74191				
TL74192N	74192				
TL74193N	74193				
TL74196N	74196				
TL74197N	74197				
TL74198N	74198				
TL74199N	74199				
TM13				TR53/	
TM-22	106				
TM300		3004			
TM308		3005			
TM1614	159		21		
TM1711			47		
TM1712	159		21		
TM2613	123A		20		
TM2614	159		21		
TM2711	123A		20		
TM2712	159		21		
TM12025		3011			
TM12050		3005			
TM12051		3007			
TM-12067		3005			
TMT696	108	3039	17	TR95/56	S0025
TMT697	108	3039	17	TR95/56	S0025
TMT839	108	3039	17	TR95/56	S0025
TMT840	108	3039	17	TR95/56	S0025
TMT841	108	3039	17	TR95/56	S0015
TMT842	108	3039	63	TR95/56	S0025
TMT843	108	3039	63	TR95/56	S0025
TMT1131					S0019
TMT1132					S0019
TMT1543	123A		11	TR21/735	
TMT-1543	123A	3020			
TMT2427	108	3019	63	TR95/56	S0016
TN061690	159	3114			
TN061703	159				
TN51					S5000
TN52					S5000
TN53	123A	3024	18	TR21/735	
TN55	123A	3024	18	TR21/735	
TN56	123A	3122		TR21/735	S0011
TN59	123A	3024	47	TR21/735	
TN60	123A		20	TR21/735	S0011
TN61	123A	3024	47	TR21/735	
TN62	123A	3122	20	TR21/735	S0011
TN63	123A	3024	47	TR21/735	
TN64	123A	3122	20	TR21/735	S0011
TN71					S5014
TN72					S5014
TN79	192		47		
TN80	123A		20		
TN81	192		47	TR63/	
TN237	123A	3024	47	TR21/735	S3024
TN238					S0025
TN301					S5000
TN302					S5000
TN303					S5000
TN591	102A		53	TR85/250	
TN3200			60		
TN83487D		3004			
TN83488D		3004			

Original Part Number	ECG	SK	GE	IR/WEP	HEP
TNC61688	172			TR69/	S9100
TNC61689	123A			TR21/735	S0014
TNC61690	159			TR20/717	S0019
TNC61702	123A			TR21/735	S0024
TNC61703	159			TR20/717	S0019
TNJ1034	199				
TNJ1173	161				
TNJ6167(2SB186)				TR85/	
TNJ6172				TR21/	
TNJ6172(2SC722)	107			TR70/	
TNJ60063	160		51	TR17/637	
TNJ60064	160		51	TR17/637	
TNJ60065	160		51	TR17/637	
TNJ60066	108	3019	17	TR95/56	G0003
TNJ60067	160	3006	51	TR17/637	
J60068	160	3006	51	TR17/637	
TNJ60069	160	3006	51	TR17/637	
TNJ60069(2SC74)	107			TR70/720	
TNJ60070	102A	3004	2	TR85/250	S0015
TNJ60071	160	3006	51	TR17/637	S0015
TNJ60072	154			TR78/712	
TNJ60073	160	3006	51	TR17/637	S0015
TNJ60074	102A	3004	2	TR85/250	S0015
TNJ60075	127	3034	25	TR27/235	G0005
TNJ60076	123A	3020	20	TR21/735	
TNJ60077	160	3006	51	TR17/637	
TNJ60078	162			TR67/707	
TNJ60079	102A	3004	2	TR85/250	G0005
TNJ60080	127	3034	3	TR27/235	G0005
TNJ60279	160	3006		TR17/637	
TNJ60280	160	3006	51	TR17/637	
TNJ60281	160	3006	51	TR17/637	
TNJ60282	102A	3004	52	TR85/250	
TNJ60283	102A	3004	2	TR85/250	
TNJ60362	126	3008	51	TR17/635	
TNJ60363	126			TR17/635	
TNJ-60363	160	3008	51		
TNJ60364	126			TR17/635	
TNJ-60364	160	3008	51		
TNJ60365	102A			TR85/250	
TNJ-60365	160	3004	51		
TNJ60447	107	3039		TR70/720	
TNJ60448	107	3039		TR70/720	
TNJ60449	107	3039		TR70/720	
TNJ60450	160			TR17/637	
TNJ60451	175	3131		TR81/	
TNJ60453	175	3131		TR81/	
TNJ60454	121	3009		TR01/232	
TNJ60455	162			TR67/707	
TNJ60456	160			TR17/637	
TNJ60457	174				
TNJ60604	107	3018	11	TR70/720	S0020
TNJ60605	107	3018	11	TR70/720	
TNJ60606	107			TR70/720	
TNJ-60606	123A	3018	20		
TNJ60607	107			TR70/720	
TNJ-60607	123A	3018	20		
TNJ60608	100			TR05/254	
TNJ-60608	160	3008	51		
TNJ60610	102A			TR85/250	
TNJ-60610	100	3004	52		
TNJ60611	102A	3123		TR85/250	G0007
TNJ-60611	100	3004	52		
TNJ60612	158			TR84/630	G6011
TNJ-60612	100	3004	52		
TNJ60728	102A	3004	2	TR85/250	
TNJ61217	107	3018	11	TR70/720	S0016
TNJ61218	107	3018	20	TR70/720	S0016
TNJ61219	123A	3020	20	TR21/735	
TNJ61220	123A	3020	20	TR21/735	S0011
TNJ61221		3004	52	TR85/250	G0005
TNJ61222	102A	3004	52	TR85/250	G0005
TNJ61223	176	3004	52	TR82/	
TNJ61282	102A	3004	52	TR85/250	G0005
TNJ61671	101	3010		TR08/720	G0011
TNJ61671 (2SC688)	107			TR70/	
TNJ61671(2SD72)	103A			TR08/724	
TNJ61672		3116		FE100/	
TNJ61672(2SK25)	132			FE100/802	
TNJ61673	133	3112		FE100/801	F0010
TNJ61673 (2SB186)	102A			TR85/	
TNJ61673(2SK24)	133				
TNJ61674	158	3004		TR84/630	G6011
TNJ61679	107			TR70/720	S0016
TNJ61729	108	3039		/720	
TNJ61730	107	3039		TR70/720	
TNJ61731	107	3039		TR70/720	
TNJ61734	103A			TR08/724	
TNJ70450	186			TR55/	
TNJ70478	108	3018	17	TR95/	S0015
TNJ70478-1		3018	17	TR95/	S0011
TNJ70479	108	3122	20	TR21/	S0025
TNJ70479-1		3122	20	TR21/	S0015

Original Part Number	ECG	SK	GE	IR/WEP	HEP
TNJ70480		3018	17	TR24/	S0015
TNJ70481	193	3114	21	TR30/	S0019
TNJ70482	128	3024	18	TR25/	S5014
TNJ70483	131	3052	30	TR50/	G6016
TNJ70484	108	3018	17	TR21/	S0015
TNJ70537	123A	3020		TR21/735	S0016
TNJ70539	123A	3122		TR21/735	S5014
TNJ70540	184	3054		/S3023	S5000
TNJ70541	131	3082		TR94/642	G6016
TNJ70634	102		52		
TNJ70635	102		52		
TNJ70637	123A	3124	17	2SC828/	S0015
TNJ70638	123A		20		
TNJ70639	123A		20		
TNJ70640	123A		20		
TNJ70641	160		51		
TNJ70688	158		52		
TNJ70691	199	3124	17	2SC644/	S0015
TNJ71034	199	3124	20	TR24/	S0016
TNJ71035	128	3024	18	TR78/	S0005
TNJ71036	123A	3122	20	TR21/	S0015
TNJ71037	159	3122	20	TR30/	S0015
TNJ71173	108	3019	11	TR83/	S0020
TNJ71234	128	3024	18		S3019
TNJ71248	126	3006	50	TR85/	G0009
TNJ71252	192	3054	63	TR55/	S3023
TNJ71271	199	3124	62	TR24/	S0015
TNJ71277	199	3124	20	TR24/	S0030
TNJ71498	108	3018	11	TR80/	S0016
TNJ71629	107			TR70/	
TNJ71773	159	3114	21	TR30/	S0019
TNJ71774	159	3114	21	TR52/	S0019
TNJ71937	107	3018	11	TR80/	S0020
TNJ71963	107	3018	11	TR80/	S0020
TNJ71964	161	3018	39	TR70/	S0017
TNJ71965	199	3018	60	TR33/	S0016
TNJ72146	164	3133	37	TR68/	
TNJ72147	124	3021	12	TR23/	S5011
TNJ72148	130	3027	14	TR61/	S7004
TNJ72149	165	3111	36	TR93/	S5021
TNJ72150	161	3039		TR70/	S7004
TNJ72151		3117	60	TR71/	S0020
TNJ72153	124	3021	12	TR23/	S5011
TNJ72154	159	3114	21	TR30/	S5013
TNJ72275	161	3039		TR70/	S0017
TNJ72276	233	3132	39	TR80/	S0020
TNJ72277	107	3018	20	TR21/	S0014
TNJ72278	102A	3005	1	TR14/	G0001
TNJ72279	107	3018	20	TR70/	S0016
TNJ72280	123A	3124	20	TR21/	S0030
TNJ72281	123A	3124	20	TR21/	S0020
TNJ72282	154		40	TR78/	S5024
TNJ72283	102A	3004	1	TR14/	
TNJ72284	103A	3010	8	TR09/	G0011
TNJ72285	102A		2	TR14/	G0007
TNJ72286	175		12	TR23/	S5012
TNJ72287	102A	3006	50	TR23/	G0005
TNJ72288	128	3024	18	TR22/	S5014
TNJ72289	102A	3004	2	TR14/	G0007
TNJ72318	121	3009	16	TR01/	G6005
TNJ72319	163	3111	36	TR93/	S5020
TNJ72320	162	3027	14	TR59/	S7004
TNJ72368	161	3039	39	TR70/	
TNJ72701	161	3039	39	TR70/	S0017
TNJ72773		3114			S0019
TNJ72774	187	3083	29	TR56/	S5006
TNJ72775	186	3054	28	TR55/	S5000
TNJ72783	123A	3122	20	TR24/	S0015
TNJ72784	123A	3122	20	TR24/	S0015
TNJ77279				TR70/	
TNT839	108	3018	17	TR95/56	S0020
TNT840	108	3018	17	TR95/56	S0020
TNT841	108	3018	17	TR95/56	S0020
TNT842	123A		17	TR24/	S0020
TNT843	108		17	TR95/56	S0020
TNT-843	123A	3018			
TNT1121					S0019
TNT1131					S0019
TNTN304					S5000
TO-003	160			TR17/637	G0003
TO-004	160			TR17/637	G0003
TO-005	102A			TR85/250	G0005
TO-012	121	3009		TR01/232	G6005
TO-014	102A			TR85/250	G0005
TO-015	121	3009		TR01/232	G6005
TO-033	123A	3122		TR21/735	S0015
TO-038	123A	3122		TR21/735	S0015
TO-039	123A	3122		TR21/735	S0015
TO-040	123A	3122		TR21/735	S0015
TO-041	102A			TR85/250	G0005
TO1-101	123A	3020		TR21/735	S0015
TO1-104	123A	3020		TR21/735	S0015
TO1-105	123A	3020		TR21/735	S0015
TO101	102A	3005		TR85/250	G0005
TO102	102A	3005		TR85/250	G0005

Original Part Number	ECG	SK	GE	IR/WEP	HEP
TO103	102A	3003		TR85/250	G0005
TO104	102A	3003		TR85/250	G0005
TP1					G0003
TP2					G0003
TP3638	159		65	TR20/	
TP3638A	193		65		
TP4067-410			62	TR51/	S0024
TP4067-411	199		20	TR33/	S0025
TP4123	123A		210		S0015
TP4124	123A		20		S0014
TP4125	159		21		S0013
TP4126	159		21		S0019
TP4257	159	3118			
TP4258	159	3118			
TP4274	101		61	TR21/	
TP4275	108		61	TR21/	
TP4512	4512				
TPS6512	123A		62		S0025
TPS6513	123A		62		S0015
TPS6514	123A		62		S0015
TPS6515	123A		62		S0015
TPS6516	159		65		S0019
TPS6517	159		65		S0019
TPS6518	159		65		S0019
TPS6519	159		65		S0019
TPS6520	123A				S0015
TPS6521	123A				S0015
TPS6522	159		65		S0019
TPS6523	159		65		S0019
TQ1	108	3018	20	TR95/56	
TQ2	108	3018	20	TR95/56	
TQ3	108	3018	20	TR95/56	
TQ4	123A	3020	20	TR21/735	
TQ5	108	3018	20	TR95/56	
TQ6	108	3018	20	TR95/56	
TQ7	108	3018	20	TR95/56	
TQ8	108	3018	20	TR95/56	
TQ9	108	3018	20	TR95/56	
TQ53	193		67		
TQ54	193		67		
TQ55	193		48		
TQ56			48		
TQ57	193		48		
TQ58	193		48		
TQ59	193		48		
TQ59A	193		67		
TQ60	193		48		
TQ61	159		21		
TQ62	159		21		
TQ63	129		21	TR88/242	
TQ-63	159		21		
TQ63A	129		21	TR88/242	
TQ64	129		21	TR88/242	
TQ-64	159		21		
TQ64A	129		21	TR88/242	
TQ5020	100	3119	2	TR05/254	G0005
TQ5021	160		51	TR17/637	G0003
TQ5022	160		51	TR17/637	G0003
TQ5023	102A		2	TR85/250	G0003
TQ5025	102A		2	TR85/250	G0005
TQ5026	102A		2	TR85/250	
TQ5027			2	TR85/250	G0005
TQ5028	127		16	TR27/235	G6005
TQ5030	179			TR35/	G6007
TQ5031	101	3011	8	TR08/641	G0011
TQ5032	101	3011		TR08/641	S5025
TQ-5032	103A+158		54		
TQ5034	160		51	TR17/637	
TQ5035	160		51	TR17/637	G0003
TQ5036	121	3009	3	TR01/232	G6003
TQ5038	160		51	TR17/637	G0003
TQ5039	101	3011	8	TR08/641	G0011
TQ5044	103A			TR08/724	G0003
TQ-5044	103A+158		54		
TQ5049	108	3039	17	TR95/56	S0020
TQ5050	103A			TR08/724	
TQ-5050	103A+158		54		
TQ5051	102A			TR85/250	
TQ-5051	158		53		
TQ5052	123A	3122	20	TR21/735	
TQ5053	123A	3122	20	TR21/735	
TQ5054	123A	3122	20	TR21/735	
TQ5055	128	3024	18	TR87/243	
TQ5060	123A	3122	20	TR21/735	
TQ5061	102A			TR85/250	
TQ-5061	158		53		
TQ5062	103A			TR08/724	
TQ-5062	103A+158		54		
TQ5063	154			TR78/712	
TQ-5063	171	3201	27	TR78/	
TQ5064	121	3009		TR01/232	
TQ-5064	121		3		
TQ7037	1004				
TQ7038	1004				
TQPD3055	130	3027			

Original Part Number	ECG	SK	GE	IR/WEP	HEP
TQPD3053	128				
TQSAO-222	102				
TR0055	159	3114		TR20/	
TR-01	121	3009	16	TR01/232	G6003
TR-01(PENCRST)	128				S5014
TR-01B	108	3009		TR01/56	
TR-01B(PENCRST)	123A				S0015
TR-01C	121	3009		TR01/232	
TR-01C(PENCRST)	123A				S0015
TR-01E	128	3009		TR87/243	
TR-01E(PENCRST)	128				S5014
TR-02	104		16	TR01/230	G6003
TR02	121	3009		TR01/	
TR02C	121	3009		TR01/232	
R-03	105	3012	4	TR03/231	G6004
R-03(PENCRST)	175				S5012
R03	105	3012	8		G0011
R03C	105	3012		TR03/233	
R-04	102A			TR85/250	G0005
R04	102	3003	8		G0011
TR04C	102A	3003		TR85/250	
TR-04C(PENCRST)	129				S5013
TR-05	100			TR05/254	G0006
TR05	102	3004	8		G0011
TR05C	100	3004		TR05/254	
TR-06	100	3004		TR05/254	G0005
TR-06(PENCRST)	123A				S0015
TR06C	160	3004		TR17/637	
TR-07	160			TR17/637	G0002
TR07	160	3007	8		G0011
TR07C	100			TR05/254	
TR-07(PENCRST)	128				S5014
TR-08	101			TR08/641	G0011
TR08	103		8		G0011
TR-08C	101			TR08/641	
TR-08(PENCRST)	129				S5013
TR-09	101			TR08/641	G0011
TR09	103	3011	59		G0011
TR-09C	101			TR08/641	
TR09C	101	3011			
TR-016	108				
TR01015		3122			
TR01026	108	3018	20	TR24/	
TR01037	123A	3122	62		S0015
TR01040		3122			
TR01042	161	3039			S0017
TR01053-1	159	3114	21		S0019
TR01054-7	128				S5014
TR01060-1		3020			
TR01060-7	181				S7000
TR01064-5	247				
TR02012	126	3006			G0008
TR02020-2	234	3118			
TR02050-5		3200			
TR02051-1	159	3114	21		S0031
TR02053-5	189				S3032
TR02053-7	189	3200			S3032
TR02054-1		3118			
TR02054-7	129				S5013
TR02060-7	180				S7001
TR02062-1	159		21		S0019
TR02064-5	248				
TR02064-6	248				
TR02064-7,8	248				
TR06011	132		FET2		F0021
TR09004	718				C6094P
TR09005	703A				
TR09006	703A				
TR09007	750				C6061P
TR09008	760				C6001
TR09009	760				C6001
TR09010	746				C6059P
TR09011	720				C6095P
TR09012	703A				
TR09017	722				
TR09018	788	3147			
TR09019	801				
TR09022	1148				
TR09027	801	3160			
TR09029	708				
TR09032	799				
TR09033	802				
TR09034	803				
TR010602-1	108		20		S0016
TR-IR26	160	3006	51	TR17/637	
TR-IR31	108			TR95/56	S0015
TR-IR33	123A			TR21/735	S0011
TR-IR35	108	3018		TR95/56	
TR2N2614C	102A				
TR-2R26	160	3006	51	TR17/637	
TR-2R31	108			TR95/56	S0015
TR-2R33	108			TR95/56	S0015
TR-2R35	108	3018		TR95/56	
TR-2SC367	123A	3046		TR21/735	
TR-2SC371	108	3018		TR95/56	

Original Part Number	ECG	SK	GE	IR/WEP	HEP
TR-2SC372	108	3018		2SC372/56	
TR-2SC373	123A	3020		TR21/735	
TR-2SC384	108	3018		TR95/56	
TR-2SC482	128	3024		TR87/243	
TR-2SC671	195	3048			
TR-2SC735	123A	3122		TR21/735	
TR2SC735	123A	3020			
TR2SC3677	123A				
TR-3R26	160	3006	51	TR17/637	
TR-3R31	108			TR95/56	S0015
TR-3R33	108			TR95/56	S0015
TR-3R35	108	3018		TR95/56	
TR-3R38	123A	3018		TR21/735	
TR-4R26	160	3006	51	TR17/637	
TR-4R31	128	3024		TR87/243	S0014
TR-4R33	123A			TR21/735	S0015
TR-4R35	108	3018		TR95/56	
TR-4R38	159	3024		TR20/717	
TR-5	121				
TR-5R26	102A	3004	2	TR85/250	
TR-5R31	128	3024		TR87/243	S5014
TR-5R33	123A			TR21/735	S0015
TR-5R35	123A	3018		TR21/735	
TR-5R38	123A	3024		TR21/735	
TR-6R26	102A			TR85/250	
TR6R26	102	3004	2		
TR-6R33	123A			TR21/735	S0015
TR-6R35	159	3114		TR20/717	
TR-6R35A	159	3114			
TR-TR31	128	3024		TR87/243	S5014
TR-TR35	123A	3020		TR21/735	
TR-8R35	123A	3024		TR21/735	
TR10	101	3011	7		G0011
TR-10	101			TR08/641	G0011
TR-10C	101	3011		TR08/641	
TR-11	126		51	TR17/635	G0008
TR11	100				
TR-11C	160			TR17/637	
TR11C	160	3006			
TR-12	160		51	TR17/637	G0003
TR12	160	3007	50		G0005
TR-12C	160			TR17/637	
TR-13	160			TR17/637	G0003
TR13	160	3007	50		G0005
TR-13C	160	3007		TR17/637	
TR14	102A	3004	50		G0005
TR-14	102A		52	TR85/250	G0006
TR-14C	102A			TR85/250	
TR14C	102A	3004			
TR-15	102A			TR85/250	G0005
TR15	102A	3004	52		G0005
TR-15C	102A	3004		TR85/250	
TR-16	104		76	TR01/230	G6003
TR16	127	3034	52	TR01/	G0007
TR-16C	104			TR01/230	
TR16C	121	3009		TR01/	
TR16X2	104M		13MP		
TR-17	160			/637	G0003
TR17	160	3006	52		G0007
TR-17A	160				G0003
TR-17C	160			TR17/637	
TR17C		3006			
TR-18	160	3006	50	TR17/637	G0002
TR18	160	3004	50		G0005
TR-18C	160			TR17/637	
TR18C	160	3006			
TR-19	152	3025		TR20/717	S0012
TR19	159		50		
TR-19A	159	3114			
TR19A	199				
TR-20	159	3114		/717	
TR20	159	3025	1	TR30/	G0005
TR-20A	159	3114			
TR-21	123A	3018	20	TR21/735	S0014
TR21	123A	3020	52	TR25/	S0014
TR-21-6	123A	3020			
TR-21C	123A			TR21/735	
TR21C	108	3018			
TR-22	123A	3019	20	TR21/735	S0011
TR22A	129	3025			
TR-22C	123A			TR21/735	
TR22C	108	3018			
TR22(FORD)					S5013
TR-23	124	3021		/240	S5011
TR23		3021	12	TR25/	
TR23A	129	3024		TR88/242	
TR-23C	124	3021		/240	
TR23C	124	3021			
TR23(FORD)					S5014
TR-24	108	3018	17	/56	S0016
TR24	108	3039		TR21/	
TR-24(PHILCO)	123A			TR21/735	
TR-25	128	3024	215	/243	S5014
TR25	128	3024		TR81/	
TR-26	130			/247	S7002

Original Part Number	ECG	SK	GE	IR/WEP	HEP
TR26	130	3027	19		
TR26-1	175	3026			
TR26C	130	3027		/247	
TR-27			25	TR27/235	G6008
TR-28	129	3025		TR88/242	S5012
TR28	159	3025	21		
TR29	219	3173			
TR-29	219				S7003
TR-30	159	3114	21	TR20/717	S0019
TR30	159	3114			
TR-30A	159	3114			
TR-31			82		
TR31		3044			
TR-31B	128				
TR32	194	3047			
TR33		3122	20		
TR34	127	3123		TR85/250	G0005
TR-34			25		
TR35	179				G0005
TR-35			76		
TR36	181	3018		TR95/56	
TR-36			75	TR24/	
TR37	218				
TR38	108	3018		TR95/56	
TR-38				TR24/	
TR40				TR01/	
TR43	102A		2	TR85/250	G0005
TR43A					G0005
TR-43B	121				G6005
TR44	102A			TR85/250	G0005
TR45	102A	3123	2	TR85/250	G0005
TR50	131	3052		TR94/642	G6016
TR51	160			TR17/637	G0003
TR52	160			TR17/637	G0002
TR53	100	3122		TR05/254	G0005
TR54	102A	3118		TR85/250	G0005
TR55	100	3005		TR05/254	G0005
TR56	104	3009		TR01/233	G6003
TR57	105	3054		TR03/231	G6004
TR58		3083			
TR58(IR)					S5018
TR59(IR)					S5013
TR60		3103			
TR60(IR)					S5015
TR61(IR)					S5020
TR62	160			TR17/637	G0003
TR63			52		G0005
TR64			52		G0005
TR65			52		G0005
TR-65	195	3048			
TR66(IR)					S3006
TR67	165	3111			
TR67(IR)					S5021
TR68	162	3079			
TR69(IR)					S9100
TR70(IR)					S0025
TR71	102A	3004		TR85/250	G0005
TR72	102A	3004	52	TR85/250	G0005
TR73(IR)					S3032
TR74(IR)					S3021
TR76(IR)					S5000
TR77					G0005
TR78(IR)					S5024
TR79(IR)					S3021
TR80(IR)					S0020
TR81	102A	3004	50	TR85/250	G0005
TR82(IR)					G6011
TR83(IR)					S0020
TR84(IR)					G6011
TR85(IR)					G6011
TR86(IR)					S0014
TR87			50		G0005
TR88			90		G0005
TR91(IR)					G6016
TR92(IR)					S5000
TR95(IR)					S0017
TR104			52		G0005
TR105			50		G0005
TR109			52		G0005
TR112	108	3018			
TR123					G0005
TR139			90		G0005
TR-147		3082			
TR-157(OLSON)	158				G0005
TR-158(OLSON)	158				G0005
TR-159(OLSON)	101				G0011
TR-160(OLSON)	101				G0011
TR-161(OLSON)	160				G0003
TR-162(OLSON)	123A				S0011
TR-163(OLSON)	108				S0020
TR-164(OLSON)	128				S5014
TR-165(OLSON)	129				S0012
TR-166(OLSON)	160				G0003
TR167	101	3011		TR08/641	G0011
TR-167(OLSON)	159				S0013

Original Part Number	ECG	SK	GE	IR/WEP	HEP
TR-168(OLSON)	126				G0008
TR-169(OLSON)	102A				G0005
TR-170(OLSON)	158				G0005
TR-171	161				S0017
TR-172(OLSON)	104				G6003
TR-173(OLSON)					G6003(2)
TR-174(OLSON)	105				G6004
TR-175(OLSON)	124				S5011
TR-176(OLSON)	181				S7002
TR-177(OLSON)					S7002(2)
TR-178(OLSON)	121				G6005
TR-179(OLSON)					G6005(2)
TR-180(OLSON)	175				S5012
TR-181(OLSON)					S5012(2)
TR182	101	3011	5	TR08/641	G0011
TR-182(OLSON)	185				S5006
TR183	101	3011	5	TR08/641	G0011
TR-183(OLSON)	127				G6008
TR184	101	3011		TR08/641	G0011
TR-184(OLSON)	131				G6016
TR-185(OLSON)	150				S5024
TR-186(OLSON)	162				S5020
TR-187(OLSON)	163				S5021
TR-188(OLSON)	175				S5012
TR-189(OLSON)					G6011
TR193	101	3011	5	TR08/641	G0011
TR194	101	3011	5	TR08/641	G0011
TR-196(OLSON)					G0002
TR-197(OLSON)					G0003
TR-198(OLSON)					S0011
TR-199(OLSON)					S0012
TR-200(OLSON)					S0013
TR-201(OLSON)					S0014
TR-202(OLSON)					S0015
TR-203(OLSON)					S0015
TR-204(OLSON)					S0016
TR211	101	3011	5	TR08/641	G0011
TR212	101	3011	5	TR08/641	G0011
TR213	101	3011	7	TR08/641	G0011
TR214	101	3011	5	TR08/641	G0011
TR215			52		G0005
TR216	101	3011	5	TR08/641	G0011
TR217			52		G0005
TR218			50		G0005
TR-224(OLSON)					G6003
TR-225(OLSON)					G6004
TR-226(OLSON)					G6005
TR-227(OLSON)					G6006
TR-228(OLSON)					G6011
TR-229(OLSON)					S5011
TR-230(OLSON)					S5013
TR-231(OLSON)					S5014
TR-232(OLSON)					S5000
TR-233(OLSON)					S5006
TR-234(OLSON)					S7002
TR-235(OLSON)					G0005
TR-236(OLSON)					G0006
TR-237(OLSON)					G0005
TR-238(OLSON)					G0005
TR-240(OLSON)					S9002
TR-246(OLSON)					C2003P
TR-248(O_SON)					C2004P
TR-249(OLSON)					C2006G
TR-251(OLSON)					C2009G
TR-258(OLSON)					G0011
TR262-2	124	3021	12	TR81/240	
TR-263(OLSON)					S3019
TR-264(OLSON)					S0019
TR266-2	124	3021	12	TR81/240	
TR-266(OLSON)					S0015
TR-267(OLSON)					S0015
TR-288(OLSON)					S0015
TR269					G0005
TR-270(OLSON)					F0010
TR271TR26	130	3027		TR59/247	
TR-271(OLSON)					F0015
TR-272(OLSON)					C6002
TR-273(OLSON)					S6003
TR-274(OLSON)					C6004
TR-275(OLSON)					C6008
TR-276(OLSON)					C6010
TR-277(OLSON)					C6049G
TR-278(OLSON)					C6049R
TR-279(OLSON)					C6051P
TR-183(OLSON)					F2007
TR-292(OLSON)					S3001
TR-300					S3013
TR301	154	3045		TR78/735	
TR•301(OLSON)					S5012
TR302	123A			TR21/	
TR-302(OLSON)					S5015
TR-303(OLSON)					C2007G
TR-304(OLSON)					C6091
TR-305(OLSON)					C6092G
TR-306(OLSON)					C6093G

Original Part Number	ECG	SK	GE	IR/WEP	HEP
TR-308(OLSON)					G0008
TR-309(OLSON)					G6016
TR-310(OLSON)					G6016
TR-311(OLSON)					S7004
TR-312(OLSON)					S0019
TR-313(OLSON)					S0015
TR-314(OLSON)					S0015
TR-315(OLSON)					F1036
TR-316(OLSON)					C3000P
TR-318(OLSON)					C3004P
TR-319(OLSON)					C3041P
TR320	102	3123	1	TR05/631	G0005
TR-320	102A	3004			
TR-320(OLSON)					C3073P
TR320A	102	3004	2	TR05/631	
TR-320A	102A	3004			
TR320AN	102A	3004			
TR321	102	3123	1	TR05/631	G0005
TR-321	102A	3004			
TR-321(OLSON)					C3800P
TR-321A	102A	3004			
TR321A		3004	2	TR05/250	
TR321(HFGH1)	100			TR05/	
TR-322(OLSON)					C3801P
TR323	102	3004	1	TR05/631	G0005
TR323A	102	3004	2	TR05/	
TR-323(OLSON)					C3802P
TR-323A	102A	3004			
TR323AN	102A	3004			
TR-324(OLSON)					C6001
TR-325(OLSON)					C6007
TR-326(OLSON)					C6009
TR-327(OLSON)					C6011
TR-328(OLSON)					C6012
TR-329(OLSON)					C6014
TR-330(OLSON)					C6015
TR331	160		2	TR17/637	G0002
TR-331(OLSON)					C6054G
TR332	102	3123	52	TR05/631	G0005
TR-332(OLSON)					C6054R
TR333	121	3009	16	TR01/232	G6005
TR-333(OLSON)					C6073P
TR334	105	3012	4	TR03/233	G6006
TR-334(OLSON)					F0021
TR335	101	3011	5	TR08/641	
TR-335(OLSON)					F2004
TR336	101	3011	5	TR08/641	
TR337	101	3011	8	TR08/641	
TR338	103	3010	8	TR08/641A	
TR346					G6005
TR-371(OLSON)					S0024
TR-372(OLSON)					S0025
TR-373(OLSON)					S0019
TR-374(OLSON)					S0005
TR375					G6005
TR-375(OLSON)					S3011
TR-376(OLSON)					S3020
TR-377(OLSON)					S3024
TR-378(OLSON)					S9001
TR-379(OLSON)					S9100
TR-380(OLSON)					S9120
TR381			53		G0005
TR382			53		G0005
TR383	158	3123	53	TR05/631	G0005
TR383(HGFH-2)	102			TR05/	
TR460					G0005
TR461					G0005
TR482	100	3004	1	TR05/254	G0005
TR482R	100		2	TR05/254	
TR-482A	102A	3004			
TR508	102	3004	1	TR05/631	G0005
TR508A	102	3004	2	TR05/631	
TR-508A	102A	3004			
TR508AN	102A	3004			
TR526					G0005
TR527					G0006
TR601	123A	3501	20	TR21/735	
TR650	102	3004	2	TR05/631	G0005
TR650A	102A	3004			
TR653	102	3004	1	TR05/631	G0005
TR653A	102A	3004			
TR721	102	3004	1	TR05/631	G0005
TR721A	102A	3004			
TR722	102	3004	1	TR05/631	G0005
TR722A	102A	3004			
TR758A			1		G0005
TR759			1		G0005
TR760	126	3007	51	TR17/635	G0003
TR761	126	3007	1	TR17/635	G0003
TR762	126	3007	2	TR17/	G0003
TR763	102	3004	52	TR05/	G0005
TR763A	102A	3004			
TR764			1		G0007
TR792			52		G0005
TR801			52		G0005

Original Part Number	ECG	SK	GE	IR/WEP	HEP
TR802			1		G0005
TR803					G0005
TR804					G0005
TR1000	129	3114	21	TR20/	S30 2
TR-1000-2	159		21		
TR-1000-3	128		18		
TR-1000-7	130		14		
TR1000A	159	3114			
TR1001	128	3114	20	TR87/	S3020
TR1001A		3114			
TR1002	129	3114	21	TR88/	
TR1002A		3114			
TR1003	128	3122	20	TR21/	
TR1004	129	3025		TR88/	
TR1005	128	3024	18	TR87/	
TR1007	130	3027		TR59/	
TR1007BL,GR		3036			
TR1007BR,OR,R,Y		3036			
TR1009	162	3027	14	TR67/	
TR1009A	181			TR36/	
TR1009BL,BR,GR, OR,R,Y		3036			
TR1011	123A	3122	20	TR21/	
TR1012	129			TR88/	
TR1025	223		66	TR59/	
TR1030	159	3114	21	TR20/	
TR-1030-1	159	3114	21		
TR1030-1	159				S0031
TR-1030-2	159	3114	21		
TR1030-2	159				S0031
TR1030A	159	3114			
TR1031	123A	3122	20	TR21/	
TR1031-1-2	194				S0005
TR1032	159	3114	21	TR20/	
TR1032-1	159				S3003
TR-1032-2	159	3114	21		
TR1032A	159	3114			
TR1032-2,3,4,6					S3003
TR1033	123A	3122	20	TR21/	
TR1033-1	190				S3019
TR1033-2					S3019
TR1033-3,4,5,6	190				S3019
TR-1033-1	123A	3122	20		
TR-1033-2	123A	1322	20		
TR-1033-3			18	TR56/	
TR1034	159	3114	21	TR20/	
TR1034A	159	3114			
TR1035-3					S0030(?
TR1036	179			TR35/	
TR1036-1				TR77/	S5005
TR1036-2	183	3189		TR77/	S5005
TR1036-3	183	3189		TR77/	S5005
TR-1036-1			56		
TR-1036-2	183	3189	56		
TR-1036-3	187	3193	29	TR56/	
TR1037-1	183	3189		TR76/	S5004
TR1037-2	182	3188		TR76/	S5004
TR1037-3	182	3188		TR76/	S5004
TR-1037-1-2	182	3188	55		
TR-1037-3	186	3192	28	TR55/	
TR1038	179			TR35/	
TR1038-4	219				S7003
TR1038-5-6	183	3189			S5005
TR-1038-4				TR29/	
TR1039-4	130				S7004
TR1039-5	179			TR35/	G6017
TR1039-6	130				S7004
TR-1039-4	130		19	TR26/	
TR1077	130	3027		TR59/	
TR-1330		3114			
TR-1347	123A				
TR1490	130	3027	14	TR59/	
TR1491	130	3027	14	TR59/	
TR1492	130	3027	14	TR59/	
TR1493	130	3027	14	TR59/	
TR1494		3027			
TR1496		3079			
TR1512-80	108	3019		TR95/	S0016
TR-1512-80				TR24/	
TR1590		3026			
TR1591	175	3026	23	TR81/	
TR1592		3026			
TR1593	175	3026	23	TR81/	
TR1605LP	124	3021	12	TR81/	
TR-1993	199				
TR1993-2	123A	3122	20	TR21/	
TR2083-70		3116	FET2		F0021
TR2083-71		3018	17		S0014
TR2083-72		3122	20		S0014
TR2083-73		3122	20		S0014
TR2083-74		3124	60		S0016
TR2083-75		3004	2		G0005
TR4010	199				
TR4010-2	123A	3122	20	TR21/	
TR4083-2327311	1039				
TR4104-2327411	1041				
TR4104-2327421	720				
TR5528	132	3116	FET2	FE100/	
TR8001	160	3006	51	TR17/	
TR8002	160	3006	51	TR17/	
TR8003	160	3006	51	TR17/	
TR8004	108			TR95/	
TR-8004	123A	3018	20		
TR-8004-4	108			TR95/	S0015
TR8004-4	123A	3039	20		
TR-8004-5	108	3018		TR95/	S0015
TR8005	124	3021	12	TR81/	
TR8006	121	3009	3	TR01/	
TR8007	102		2	TR05/	
TR-8007	159	3004			S0019
TR8007A	159	3114			
TR-8007(FISHER)	159			TR20/	
TR8010	123A		20		
TR-8010	107				S0015
TR8014	123A	3124	63	TR21/	
TR-8014	123A	3020		TR25/	
TR8015		3018			
TR8016		3117			
TR8017		3008			
TR8018	181		19	TR36/	
TR-8018	181	3036		TR26/	
TR8019	153		26		
TR-8019	153			TR77/	S5018
TR8020	159		21		
TR-8020	159	3025		TR88/	S5013
TR8021	123A		20		
TR-8021	128	3024		TR87/	S5014
TR8022	190		20	TR74/	
TR-8022	190	3024		TR21/	S3019
TR8023	190		63	TR74/	
TR-8023	128	3024		TR25/	
TR8024	190		21	TR74/	
TR-8024	128	3024		TR30/	
TR8025	123A		20		
TR-8025	123A	3020		TR21/	S0015
TR8026A	159	3114			
TR-8026	159	3114		TR20/	S0019
TR-8027	132	3116		FE100/	F0015
TR8028	123A		20		
TR-8028	108	3018		TR95/	S0016
TR8029	123A		17		
TR-8029	108	3018		TR95/	S0016
TR8030	123A		17		
TR-8030	108	3018		TR95/	S0016
TR8031	123A		20		
TR-8031	108	3018		TR95/	S0016
TR-8032	108	3018		TR95/	S0016
TR8034	123A		62	TR21/	
TR-8034	199	3018			S0024
TR8035	123A		20	TR21/	
TR8036	128	3122	62	TR87/	
TR8037	129		20	TR88/	
TR-8037	159				S5022
TR8038			20		
TR-8038	108				S0020
TR8039	123A		20	TR21/	
TR8040	123A		17	TR21/	
TR-8040	199				S0015
TR8042	108			TR95/	
TR-8042	123A	3122			S0030
TR8043	123A	3018		TR21/	
TR-8043	107				S0016
TR8055	159	3114			
TR8330			17		
TR9100	123A	3122	20	TR21/	
TR9100-18					S0015
TR-9100-18	123A				
TR35144	121	3014			G6005
TR35144A	121	3014			
TR35524	121				G6005
TR36643	128		18	TR87/	
TR38117	102A	3004	53	TR85/	G0005
TR39453	221				
TR40245		3018			
TR40603	221				
TR310011	102A	3004	2	TR85/	
TR310012	102A	3004	2	TR85/	
TR310015	100	3005	1	TR05/	
TR310017	102A	3004	2	TR85/	
TR310018	102A	3004	2	TR85/	
TR310019	160	3008	51	TR17/	
TR310025	160	3004	2	TR17/	G0008
TR310026	102A	3004	52	TR85/	G0005
TR310065	160	3008	51	TR17/	G0008
TR310068	160	3006	51	TR17/	
TR310069	160	3007		TR17/	
TR310075	102A	3004	2	TR85/	G0005
TR310107	102A	3004	2	TR85/	
TR310123	160	3006	51	TR17/	

Original Part Number	ECG	SK	GE	IR/WEP	HEP
TR310124	160	3006		TR17/	
TR310125	102A	3004	2	TR85/	
TR310136	102A	3004	2	TR85/	
TR310139	160	3006	51	TR17/	
TR310147	160	3006	51	TR17/	
TR310149	102A	3004		TR85/	
TR310150	160	3006	51	TR17/	
TR310153	102A	3004	2	TR85/	
TR310155	160	3008	51	TR17/	
TR310156	160	3008	51	TR17/	
TR310157	160	3008	51	TR17/	
TR310158	160	3008	51	TR17/	
TR310159	102A	3004	2	TR85/	G0007
TR310160	103A	3004	2	TR08/	
TR310161	100	3005	1	TR05/	
TR310164	102A	3004	2	TR85/	
TR310193	160	3007		TR17/	
TR310224	160			TR17/	G0003
TR310225	102	3004			G0005
TR310227	102A			TR85/	
TR310230	108	3018	20	TR95/	
TR310231	123A	3018	20	TR21/	S0011
TR310232	160	3007	51	TR17/	
TR310235	102A	3004	52	TR85/	
TR310236	103A	3010		TR08/	
TR310243	123A	3018	20		S0014
TR310244	108	3018		TR95/56	S0016
TR310245	123A	3018	20	TR21/735	S0014
TR310249	108	3039		TR95/56	S0016
TR310250	108	3018		TR95/56	S0016
TR310251	102A	3008		TR85/250	G0005
TR310252	102A	3004		TR85/250	G0005
TR310255	102A	3008		TR85/	G0005
TR3100227		3004	2		
TR22873700104		3160			
TR228735045311		3018			
TR228735046011		3018			
TR228735048617		3018			
TR228735048618		3018			
TR228735120325		3116			
TR228736003026		3061			
TR-BRC149C	123A	3122			
TR-C44	100	3005	50	TR05/254	G0008
TR-C44A	100	3005		TR05/254	
TR-C45	102A	3004	50	TR85/250	G0005
TR-C45A	100	3005		TR05/254	
TR-C70	102A		50	TR85/250	G0005
TR-C71	102A		50	TR85/250	G0005
TR-C72	102A		50	TR85/250	G0005
TR-FET-1	132	3116			
TR-FE100			FET2		
TR-RR38	123A	3018		TR21/735	
TR-TR38	123A	3018		TR21/	
TR-U1650E	132	3116	FET2		F0021
TR-U1650E-1			FET2	FE100/	
TR-U1835E	132	3116	FET2	FE100/	F0021
TRA-2	126	3008		TR17/635	
TRA-4	123A	3020	20	TR21/735	
TRA-4A	123A	3020	20	TR21/735	
TRA-4B	123A	3020	20	TR21/735	
TRA-7R	121	3009	3	TR01/232	
TRA-7RM	121	3009		TR01/232	
TRA-8R	121	3009	3	TR01/232	
TRA-9R	123A	3020	20	TR21/735	
TRA-10R	126	3008		TR17/635	
TRA10R	160		51		
TRA-11R	126	3008		TR17/635	
TRA11R	160		51		
TRA-12R	126	3008		TR17/635	
TRA12R	160		51		
TRA-22	126	3006		TR17/635	
TRA-22A	126	3008		TR17/	
TRA22A	160		51		
TRA-22B	126	3008		TR17/635	
TRA22B	160		51		
TRA-23	126	3008		TR17/635	
TRA23	160		51		
TRA-23A	126	3008		TR17/635	
TRA23A	160		51		
TRA-23B	126	3008		TR17/635	
TRA23B	160		51		
TRA-24	126	3008		TR17/635	
TRA24			51		
TRA-24A	126	3008	51	TR17/635	
TRA-24B	126	3008		TR17/635	
TRA24C	160				
TRA32	102A	3004	2	TR85/250	
TRA33	102A	3004	2	TR85/250	
TRA34	123A	3020	20	TR21/735	
TRA36	123A	3020	20	TR21/735	
TRA-3437-1		3012			
TRAPLC711	123A	3122		TR21/735	S0020
TRAPLC871	123A	3122		TR21/735	S0015
TRAPLC871A	123A	3122			
TRAPLC1013	186	3041		TR55/S3020	S3020

Original Part Number	ECG	SK	GE	IR/WEP	HEP
TRBC147B	123A	3124			S0014
TRC44	100		1		
TRC44A	100		1		
TRC45,A	100		1		
TRC70	102		2		
TRC71	102		2		
TRC72	102		2		
TRFE100		3116			
TR031	176	3123			
TR01054-7	128				
TR0-9004	718	3159	IC8		
TR0-9005	703A	3157	IC12	IC500/	
TR0-9006	703A	3157	IC12	IC500/	
TR0-9011	720		IC7		
TR0-9017		3161			
TR0-9018		3147			
TR09019		3160			
TR09029		3135			
TRM13			50		G0005
TRM14			50		G0005
TRM15			52		G0005
TRM16			52		G0007
TRM17			52		G0007
TRM21			52		G0005
TRM34					G0005
TRM81			50		G0005
TRPLC711		3018			S0015
TRS25X			32		
TRS30X			32		
TRS35X			32		
TRS100	154	3045	27	TR78/712	
TRS100A			27		G6012
TRS101	154	3045	27	TR78/712	
TRS120	154	3045	27	TR78/712	
TRS140	154	3045	27	TR78/712	
TRS140HP	171		27	TR78/	
TRS140MP	198		32		
TRS160	154	3045	27	TR78/712	
TR160HP	171		27	TR78/	
TRS160MP	198		32		
TRS180	154	3045	27	TR78/712	
TRS180HP	171		27	TR78/	
TRS180MP	198		32		
TRS200	154	3045	27	TR78/712	
TRS200HP	171		27	TR78/	
TRS200MP	198		32		
TRS225	154	3045	27	TR78/712	
TRS225HP	171		27	TR78/	
TRS225MP	198		32		
TRS250	154	3045	27	TR78/712	
TRS250HP	171		27	TR78/	
TRS250MP	198		32		
TRS275	154	3045	27	TR78/712	
TRS275HP	171		27	TR78/	
TRS275MP	198		32		
TRS301	154	3045	27	TR78/712	
TRS301HF	171				
TRS301HP			27	TR78/	
TRS301LC	198		32		
TRS301MP	198		32		
TRS325HP			27		
TRS325MP	198		32		
TRS350			32		
TRS350HP,MP			32		
TRS375			32		
TRS375HP,MP			32		
TRS401			32		
TRS401HP,LC,MP			32		
TRS425			32		
TRS425HP,MP			32		
TRS3254LP			32		
TRS3255	198		32		
TRS3504,LP			32		
TRS3604S			32		
TRS3742	198		32		
TRS3754,LP			32		
TRS4004			32		
TRS4014,LP			32		
TRS4014S			32		
TRS4016LC	124	3021	32	TR81/240	S5011
TRS4254,LP			32		
TRS4404S			32		
TRS4504,LP			32		
TRS4754,LP			32		
TRS4926	198		32		
TRS4927	198		32		
TRS5014			32		
TRS5016LC	124	3021	32	TR81/240	S5011
TRS5254			32		
TRS5504			32		
TRS6016LC	124	3021	12	TR81/240	S5011
TS0001(DM)					S0005
TS0002(DM)					S0024
TS0003(DM)					S0003
TS0004(DM)					S0025

Original Part Number	ECG	SK	GE	IR/WEP	HEP
TS0005(DM)					S0005
TS0006(DM)					S0019
TS0007(DM)					S0005
TS-1	102A	3004		TR85/250	
TS-2	102A	3004		TR85/250	
TS-3	102A	3004		TR85/250	
TS3					G0005
TS7					G0005
TS9					G0005
TS13	102		2		G0005
TS-13	102A	3004		TR85/250	
TS14	102		2		G0005
TR-14	102A	3004		TR85/250	
TS15	102		2	TR05/	G0005
TS-15	102A	3004		TR85/250	
TS67BB	176	3123			
TS162	158	3123	52	TR05/	G0005
TS-162	102A			TR85/250	
TS163	158	3123	52	TR05/	G0005
TS-163	102A			TR85/250	
TS164	158	3004	52	TR05/	G0005
TS-164	102			TR05/250	
TS165	158	3004	52	TR05/	G0005
TS-165	102A			TR85/250	
TS166	158	3004	52	TR05/	G0005
TS-166	102A			TR85/250	
TS173	121	3014	3	TR16/	G6005
TS-173	121	3009		TR01/232	
TS176	121	3009	16	TR01/	G6005
TS-176	121			TR01/232	
TS-203				TR85/	
TS-204				TR85/	
TS301				TR76/	
TS517					G0005
TS601			53		G0002
TS-601	100	3005		TR05/254	
TS602			53		G0002
TS-602	100	3005		TR05/254	
TS603	158		53		G0005
TS-603	102A			TR85/250	
TS604	158		53		G0005
TS-604	102A			TR85/250	
TS605					G0005
TS606					G0005
TS609	105		4	TR03/	G6006
TS-609	105	3012		TR03/233	
TS610	121	3014	16	TR02/	G6003
TS-610	104	3009		TR01/624	
TS612	121	3009	3	TR01/	G6003
TS-612	104	3009		TR01/624	
TS613	121	3014	25	TR01/	G6003
TS-613	104	3009		TR01/624	
TS614	121	3009	16	TR01/	G6003
TS-614	104	3009		TR01/624	
TS615			1		G0002
TS-615	160			TR17/637	
TS616	158	3004	53		G0005
TS-616	102A			TR85/250	
TS617	158	3004	53		G0005
TS-617	102A			TR85/250	
TS618	158	3004	53		G0005
TS-618	102A			TR85/250	
TS619	158	3004	53	TR05/	G0005
TS-619				TR85/250	
TS620		3008	53	TR05/	G0002
TS-620	160			TR17/637	
TS621		3008	53	TR05/2	G0002
TS-621	160			TR17/637	
TS627	158	3004	52	TR05/	G0005
TS627A,B		3123	2	TR05/	G0002
TS-627	102A			TR85/250	
TS-627A,B	160			TR17/637	
TS629	158	3004	52	TR05/	G0005
TS-629	102A			TR85/250	
TS630	160	3123	1	TR05/	G0002
TS-630	160			TR17/637	
TS669A,B,D,E,F	100			TR05/	
TS669C	160			TR17/	
TS672A		3005	2	TR05/	G0002
TS672B	160	3005	2	TR05/	G0002
TS-672A,B	160			TR17/637	
TS673A,B	160	3005	2	TR05/	G0002
TS-673A,B	160			TR17/637	
TS739	158	3004	52	TR05/	G0005
TS739B	158				G0005
TRS450			32		
TRS451,MP			32		
TRS475,MP			32		
TRS501,LC,MP			32		
TRS525,MP			32		
TRS560,MP			32		
TRS1004			32		
TRS1004LP			32		S0005
TRS1005	124	3021	12	TR81/240	
TRS1005LP	124	3021	12	TR81/240	S3021
TRS1203LP					S0005
TRS1204	198		32		
TRS1205	124	3021	12	TR81/240	
TRS1205LP	124	3021			S3021
TRS1404	198		32		
TRS1404LP			32		S0005
TRS1405	124	3021	12	TR81/240	
TRS1405LP	124	3021	12	TR81/240	S3021
TRS1604	198		32		
TRS1604LP			32		S3021
TRS1605	124	3021	12	TR81/240	
TRS1605LP					S3021
TRS1804	198		32		
TRS1804LP			32		S0005
TRS1805	124	3021	12	TR81/240	
TRS1805LP	124	3021	12	TR81/240	S3021
TRS2004	198		32		
TRS2004LP			32		S3021
TRS2005	124	3021	12	TR81/240	
TRS2005LP	124	3021	12	TR81/240	S3021
TRS2006	198		32		
TRS2254	198		32		
TRS2254LP			32		S3021
TRS2255	124	3021	12	TR81/240	
TRS2255LP	124	3021	12	TR81/240	S3021
TRS2504	198		32		
TRS2504LP			32		S3021
TRS2505	124	3021	12	TR81/240	
TRS2505LP	124	3021	12	TR81/240	S3021
TRS2754	198		32		
TRS2754LP			32		S3021
TRS2755	1234	3021	12	TR81/240	
TRS2755LP	124	3021	12	TR81/240	S3021
TRS2804S	198		32		
TRS2805S	198		32		
TRS2006	198		32		
TRS3011	154	3045	27	TR78/712	
TRS3012	154	3045	27	TR78/712	
TRS3014	171	3201	27	TR78/	
TRS3015	198		32		
TRS3015LP	124	3021	214	TR81/240	
TRS3016LC	124	3021	12	TR81/240	S5011
TRS3204S	198		32		
TRS3205S	198		32		
TRS3254	198		32		
TS-739,B	102A			TR85/250	
TS740	158	3004	2		G0005
TS-740	102A			TR85/250	
TS748					G0005
TS765	158	3123	52	TR05/	G0005
TS-765	158			TR85/250	
TS1000					G0005
TS1007	158	3004	52	TR05/	G0005
TS-1007	102A			TR85/250	
TS-1193-736	130				
TS1266	158	3123	52	TR05/	G0005
TS-1266	102A			TR85/250	
TS1541	176	3123			
TS1657	121	3009	3	TR01/	G6005
TS-1657	121			TR01/232	
TS1727	158	3004	2		G0005
TS-1727	102A			TR85/250	
TS1728	158	3004	2		G0005
TS-1728	102A			TR85/250	
TS1792	102A				G0005
TS2218	192		47		
TS2219	192		47		
TS2221	123A		20		
TS2222	123A		20		
TS2904	159		21		
TS2905	159		21		
TS2906	159		21		
TS2907	159		21		
TS3001(DM)					S3001
TS3002(DM)					S3002
TS3003(DM)					S3003
TS3005(DM)					S3005
TS3006(DM)					S3006
TS3007(DM)					S3007
TS3008(DM)					S3008
TS3010(DM)					S3010
TS3020(DM)					S3020
TS3021(DM)					S3021
TS3022(DM)					S3021
TS3023(DM)					S3024
TS3024(DM)					S3024
TS3025(DM)					S3024
TS3026(DM)					S3024
TS3027(DM)					S3028
TS3028(DM)					S3028
TS3029(DM)					S3028
TS3030(DM)					S3028
TS3031(DM)					S3032
TS5000(DM)					S5000
TS5002(DM)					S5005

Original Part Number	ECG	SK	GE	IR/WEP	HEP
TS5003(DM)					S5000
TS5004(DM)					S5004
TS5005(DM)					S5005
TS5006(DM)					S5006
TS5007(DM)					S5006
TS5008(DM)					S5005
TS5009(DM)					S5005
TS5010(DM)					S5005
TS7000(DM)					S7000
TS7001(DM)					S7001
TS9001(DM)					S9001
TS9013		3122	20		S0015
TS9100(DM)					S9100
TS9101(DM)					S9120
TS9140(DM)					S9140
TS9141(DM)					S9141
TS-10144NC		3027			
TS-10144		3510			
TS-21756640	159				
TSC499	123A			TR21/735	
TSC-499	199		20	TR21/	
TSC614	108	3018	20	TR53/	S0020
TSC695	123A	3020	17	TR21/	S0015
TSC722	128			TR87/243	
TSC-722	128		18	TR25/	S0015
TSC767	199	3124	20	TR24/	S0015
TSG-499				TR21/	
TSS-616-1	358		42		
TST705899A	123A				
TT204	108	3018	17	TR95/56	
TT204A	108	3018	17	TR95/56	
TT204AB	108	3018	17	TR95/56	
TT204B	108	3018	17	TR95/56	
TT204C	108	3018	17	TR95/56	
TT1083	121	3009	3	TR01/232	
TT1097	123A	3020	20	TR21/735	
TT500					S5000
TT501					S5000
TT502					S5000
TUS34	132	3116			
TV2+TV2S				TR69(PAIR)	
TV2SB126	127			TR27/235	
TV2SB126F	121	3009		TR01/232	
TV2SB126V	127			TR27/235	
TV2SB448	127			TR27/235	
TV2SC208	123A		17		
TV-6		3124	20	TR21/	S0015
TV-7		3124	20	TR21/	S0015
TV10				TR27/	
TV15	161	3132	60	TR83/719	
TV-15	108	3018			
TV15A	161	3018	60	TR21/719	S0017
TV15B	161	3018	60	TR21/719	S0017
TV-15B	108				S0020
TV15C	161	3018		TR83/719	
TV16	108	3018	17	TR95/56	S0015
TV17	123A	3018	20	TR51/735	S0015
TV-17		3045		TR21/	S0015
TV17A	108	3018			
TV17/34-6001-63				TR51/	
TV18	108	3018	20	TR21/56	S0015
TV-18	123A	3018		TR21/	S0015
TV19	194	3040	40	TR24/712	S5026
TV-19	154	3040		TR21/	S5026
TV19/34-6001-65				TR53/	
TV20	161		20	TR83/719	S0020
TV-20	233	3018		TR21/	S0020
TV20-10K80	108	3018			
TV21	192	3020	63	TR25/	
TV-21	123A			TR21/735	
TV22	191	3018	63	TR24/	S3021
TV-22	108			TR95/56	
TV23	128	3024	63	TR25/	S5014
TV-23	123A			TR21/735	
TV-25	191	3045	63	TR75/	
TV25		3045	63	TR25/	
TV26	192	3024	63	TR25/	
TV-26	128	3024		TR87/243	
TV27	103		8	TR08/	
TV-27	103A			TR08/724	
TV28	192	3020	63	TR25/	
TV-28	128	3024		TR87/243	
TV29	129	3025	63	TR25/	S5013
TV-29	129			TR88/242	
TV29/34-6001-86				TR88/	
TV32		3018	20	TR21/	S0015
TV-32	108	3018		TR95/56	
TV33	161	3018	39	TR71/	S0017
TV-33	161	3117		TR83/719	
TV-34	161	3117		TR83/719	
TV34			39		S0017
TV35	161	3018	20	TR70/	S0017
TV-35	161	3132		TR83/719	
TV36		3018	20	TR21/	S0015
TV-36	161	3132		TR83/719	

Original Part Number	ECG	SK	GE	IR/WEP	HEP
TV36/34-6015-1		3044			
TV36/34-6015-18				TR24/	
TV37	123A	3018	20	TR21/	S0011
TV-37	161	3132		TR83/719	
TV38	123A	3018	20	TR25/	S0011
TV-38	161	3132		TR83/719	
TV39	123A	3018	20	TR21/	S0015
TV-39	161	3132		TR83/719	
TV40	161	3018	20	TR83/719	S0011
TV-40				TR21/735	
TV40/34-6015-22				TR33/	
TV41	128	3024		TR21/	S5014
TV-41	128	3024		TR87/243	
TV41/34-6015-23				TR51/	
TV42	128	3020	63	TR87/243	S5014
TV-42				TR21/735	
TV43	199	3018	20	TR21/	S0011
TV-43	123A			TR21/735	
TV44	159	3018	21	TR52/	S5013
TV-44	159	3025		TR20/717	
TV45	128	3024	63	TR51/	S5014
TV45A				TR71/	
TV-45	128	3024		TR87/243	
TV46	123A	3018	20	TR21/	S0011
TV-46	123A	3018		TR21/735	S0011
TV47	158		52	TR85/	G0005
TV-47	159	3114	22	TR20/717	
TV-47A	159	3114			
TV48	123A		39	TR70/	S0017
TV-48	161	3117		TR83/719	
TV49	191		63	TR78/	
TV-49	128	3024		TR87/243	
TV50	161	3039	39	TR21/	S0017
TV-50	161	3117		TR83/719	
TV50/34-6015-29				TR70/	
TV51	128	3024	63	TR25/	S5014
TV-51	123A	3122		TR21/735	
TV52	128	3024	63	TR24/	S5014
TV-52	123A			TR21/735	
TV53	128	3024	63	TR24/	S5014
TV-53	123A	3024		TR21/735	S5014
TV54	161	3039	39	TR70/	S0017
TV-54	161	3117		TR83/719	
TV54/34-6015-15				TR70/	
TV54/34-6015-37				TR24/	
TV55	108		61		
TV-55	107	3039		TR70/720	
TV56	123A	3018	20	TR53/	S0011
TV-56	123A	3122		TR21/735	
TV57	107	3020	17	TR24/	S0015
TV-57	123A	3122		TR21/735	
TV57A	123A	3020			
TV58	107	3020	17	TR24/	S0015
TV-58	128	3024		TR87/243	
TV58A	123A	3020			
TV59	128	3020	63		S5014
TV-59	128	3024	18	TR87/243	
TV59A	123A	3020			
TV60	107	3020	17		S0015
TV-60	123A	3122		TR21/735	
TV60A	123A	3020			
TV61	158	3004	52	TR82/	G0005
TV-61	102	3004		TR05/631	
TV61A	102A	3004			
TV62	172	3024	64	TR69/	
TV-62			20	TR21/	
TV62(PHILCO)			64		
TV62/34-60173		3122			
TV65	123A		20	TR33/	S0024
TV-65	123A	3122		TR21/735	S0015
TV-66		3124	20	TR24/	S0014
TV-67		3024	18	TR25/	S5014
TV-68		3018			S0015
TV-70		3044	40		S5025
TV-70B		3044			
TV-73		3054	28	TR55/	S5000
TV-74		3083			S3028
TV-75		3054	28	TR55/	S5000
TV80	132	3116	FET2	FE100/	F0021
TV81	233	3117	60	TR24/	S0020
TV81/34-6015-62				TR95/	
TV83				FE100/	
TV-83	132	3116	FET2	FE100/	F0021
TV-84		3018			S0015
TV85	191	3054	27	TR74/	S3019
TV-87		3114			S0019
TV-92		3018			S0015
TV-93		3114			S0019
TV106	127		25		
TV108	163	3111	19	TR67/740	
TV109	175	3026	28	TR76/241	S5000
TV110	127	3035	25	TR27/235	G6008
TV111	127	3035	25	TR27/235	G6008
TV112	196	3026	28	TR76/241	S5000
TV-112	196	3054		TR76/	

Original Part Number	ECG	SK	GE	IR WEP	HEP
TV113	124	3021	12	TR23/240	S5011
TV113/34-6002-2				TR57/	
TV114	127	3035	25	TR27/	G6008
TV-114	127				
TV115	108	3018		TR95/56	
TV-115			66	TR76/	
TV116	108	3018		TR95/56	
TV-116			69	TR77/	
TV117	196			TR76/	
TV-117		3054	66	TR76/	
TV118	163	3111	36	TR61/	S5021
TV-118	165				
TV-119	165	3111			
TV121	162	3133	35	TR61/	S5021
TV-121	164				
TV122	124	3021	12	TR23/	S5012
TV-124	165	3111		TR93/	S5020
TV-125	165	3111			S5020
TV132				TR21/	
TV133				TR70/	
TV134				TR70/	
TV135				TR70/	
TV136				TR24/	
TV137				TR53/	
TV138				TR21/	
TV139				TR51/	
TV140				TR33/	
TV141				TR51/	
TV142				TR24/	
TV143				TR53/	
TV147				TR52/	
TV148				TR70/	
TV149				TR78/	
TV150				TR70/	
TV151				TR25/	
TV1000	108	3039	17	TR95/56	G0008
TV2403	108	3039		TR95/56	
TV2404	108	3039		TR95/56	
TV2428	102A	3004		TR85/250	
TV2429	102A	3004	2	TR85/250	60007
TV2434	102A	3005	2	TR85/250	G0001
TV2455	160	3006	51	TR17/637	G0003
TV2479	160	3006		TR17/637	
TV3843				TR21/	
TV4152	100	3005			
TV24102	108	3018	11	TR95/56	S0020
TV24115	102A	3004		TR85/250	
TV24137	160		51	TR17/637	G0003
TV24142	127	3035	25	TR27/235	G6005
TV24143	103A	3010	8	TR08/724	
TV24148	108	3039	11	TR95/56	
TV24152	100	3005	51	TR05/254	
TV24154	102A	3004	2	TR85/250	G0005
TV24156	102A	3004	2	TR85/250	G0005
TV24158	160	3006	51	TR17/637	G0002
TV24160	161	3018	11	TR83/719	S0017
TV24161	161	3018	11	TR83/719	S0017
TV-24161				TR22/	
TV24162	127	3034	16	TR27/235	
TV24163	127	3034	25	TR27/235	
TV24164	154	3045		TR78/712	S5024
TV24166	160		51	TR17/637	
TV24172	160	3006	51	TR17/637	G0003
TV24189	102A	3004	2	TR85/250	G0005
TV24194	102A	3004	2	TR85/250	G0005
TV24203	108	3018	20	TR95/56	
TV24204	108	3018	20	TR95/56	
TV24209	161	3018	17	TR83/719	S0016
TV24210	107	3018	17	TR70/720	S0016
TV24211	175	3021	12	TR81/241	S5012
TV24214	159	3009	21	TR20/717	S0019
TV24214A	159	3114			
TV24215	123A	3020	20	TR21/735	S0014
TV24216	123A	3020	20	TR21/735	S0015
TV24229	160		51	TR17/637	
TV24230	160		51	TR17/637	
TV24239	160		51	TR17/637	
TV24281	123A		20	TR21/735	S0011
TV24313	108	3039		TR95/56	S0016
TV24337	121	3009		TR01/	G6005
TV24341	127	3034	25	TR27/235	G6018
TV24351	160	3006		TR17/637	
TV24363	159	3114	21	TR20/717	S0019
TV24363A	159	3114			
TV24370	102A	3004	52	TR85/250	G0006
TV24372	123A	3018	20	TR21/735	S0015
TV24380	107	3018	20	TR70/720	S0016
TV24382	107	3117	60	TR70/720	S0017
TV24383	107	3018	20	TR70/720	S5026
TV24385	107			TR70/720	S0016
TV24387	108	3019	17	TR95/56	S0020
TV24399	161			TR83/719	S0017
TV24435	154	3040	27	TR78/712	S5025
TV24436	161	3132		TR83/719	S0016
TV24437	161	3117		TR83/719	S0016

Original Part Number	ECG	SK	GE	IR WEP	HEP
TV24438	107			TR70/720	S0016
TV24453	123A	3122		TR21/735	S0014
TV24454	123A	3018	20	TR21/735	S0014
TV24458	123A		20	TR21/735	S0014
TV24468	127	3035	25	TR27/235	G6008
TV24487	175			TR81/241	S5012
TV24495	159		22	TR20/717	S0013
TV24495A	159	3114			
TV24499	154			TR78/712	S5026
TV24568	127	3035		TR27/235	
TV24571	161	3117	17	TR83/719	S0016
TV24573	108	3039	17	TR95/56	S0016
TV24574	108	3039	20	TR95/56	S0011
TV24576	123A	3122	20	TR21/735	S0011
TV24589	108	3039		TR95/56	S0016
TV24599	102A			TR85/	G0005
TV24655	123A			TR21/	S0020
TV24678	121			TR01/232	G6005
TV24684	108			TR95/56	S0016
TV24806	107	3019	11	TR70/720	S0016
TV24859				TR21/	
TV24945	102A			TR85/	S5012
TV24983	103A			TR08/	G0011
TV24984	102A			TR85/	G0005
TV34458				TR21/	
TV241072	1004				
TV241077	123A	3122	20	TR21/	
TV241078	161	3132		TR83/	
TVC-MK3800	1136				
TVCM-1	705A	3134	IC3	IC503/	C6089
TVCM-2	713	3077	IC5	IC508/	C6057P
TVCM-3	707	3134		IC503/	C6067G
TVCM-4	708	3135	IC10	IC504/	C6062P
TVCM-5	709		IC11	IC505/	C6062P
TVCM-6	718	3159	IC8	IC511/	C6094P
TVCM-7	720	3160	IC7	IC512/	C6095P
TVCM-8	714	3075	IC4	IC509/	C6070P
TVCM-9	715	3076	IC6	IC510/	C6071P
TVCM-10	722	3161	IC9	IC513/	C6056P
TVCM-11	712	3072	IC2	IC507/	C6063P
TVCM-12	719		IC28		
TVCM-13	721		IC14		
TVCM-14	734	3163			
TVCM-15			IC13		
TVCM-16		3144	IC15		
TVCM-18			IC16		
TVCM-19			IC24		
TVCM-20			IC17		
TVCM-21			IC30		
TVCM-22			IC18		
TVCM-26			IC27		
TVCM-27			IC29		
TVCM-30		3141			
TVCM-33		3074			
TVCM-34		3149			
TVCM-35			IC31		
TVCM-39		3147			
TVCM-551		3512			
TVCM-S	709				
TVS-1S18	108				S0016
TVS2SC125HG				TR28/	
TVS-2SC645A	123A		20		
TVS2CS1256HG			67		
TVS-2S288A	107	3039			
TVS-2SA71B	102A	3004	2	TR85/250	
TVS-2SA103	160	3006		TR17/637	
TVS-2SA171	100	3005	2	TR05/254	
TVS-2SA385	102A	3006	51	TR85/250	G0005
TVS-2SA385A	102A	3006		TR85/250	
TVS-2SA385L	102A		51	TR85/250	G0005
TVS2SA543	192	3025	67	TR28/	S5013
TVS2SA546	129	3025		TR88/	S3032
TVS-2SA564	159	3114	21	TR20/717	S0019
TVS2SA564A	159	3114	21	TR20/	S0019
TVS2SA564C	159				S0019
TVS2SA564-O	159	3114			
TVS-2SA564P	159	3114		TR20/717	S0019
TVS2SA564PY	159	3114			
TVS2SA564Q	159	3114		TR20/717	S0019
TVS2SA607	159	3118			
TVS2SA609	159	3118			
TVS2SB126	121		16		
TVS-2SB126	121	3009		TR01/232	
TVS2SB126B				TR21/	
TVS-2SB126F	121	3009	16	TR01/232	G6005
TVS2SB126V			25	TR27/	
TVS-2SB126V	127	3034		TR27/235	G6005
TVS2SB171	158		52	TR85/	G0005
TVS-2SB171	102A	3004		TR85/250	G0005
TVS2SB171A	102A	3004		TR85/250	
TVS-2SB171A	158				G0005
TVS2SB171B	102A	3004		TR85/250	G0005
TVS-2SB171F				TR85/	
TVS2SB172				TR85/	
TVS2SB172A			52	TR85/	

Original Part Number	ECG	SK	GE	IR/WEP	HEP
TVS-2SB172	102A	3004		TR85/250	
TVS-2SB172A	100				
TVS-2SB172F	102A	3004	2	TR85/250	G0005
TVS2SB176					G0005
TVS-2SB176	102A	3004	2	TR85/250	
TVS-2SB234	102A			TR85/250	
TVS-2SB324	158	3004		TR84/630	
TVS2SB324	102A				
TVS-2SB448	127	3035		TR27/235	
TVS2SB449	121	3009	16	TR01/	G6003
TVS2SB449F		3009	16	TR27/	G6003
TVS-2SB449F	121		25	TR01/232	
TVS2SC58		3044			
TVS2SC58A			11	TR21/	S5025
TVS-2SC58	154	3040		TR17/712	
TVS-2SC58A	154	3040		TR17/712	S5025
TVS-2SC185A	108	3039		TR95/56	
TVS-2SC208	108	3018		TR95/56	
TVS-2SC287A	108				S0016
TVS2SC288A	108		17	/720	S0016
TVS-2SC313	161	3039		TR83/719	S0016
TVS-2SC429A	107	3039		TR70/720	
TVS-2SC446	108	3018		TR83/56	
TVS2SC466	108				S0016
TVS-2SC466	161	3039		TR83/719	S0016
TVS-2SC469A	107			TR70/720	
TVS-2SC526	154	3045	18	TR78/712	
TVS2SC526				TR25/	
TVS2SC538	108	3018			
TVS2SC538A	123A		63	TR25/	S0030
TVS-2SC538	123A	3020	18	TR21/735	S0015
TVS-2SC538A	123A	3024		TR21/735	S0030
TVS2SC562	161	3039	20	TR21/	S0017
TVS-2SC562	161	3018		TR83/719	S0011
TVS2SC563	161		20	TR21/	
TVS2SC563A				TR21/	S0017
TVS-2SC563	161	3018		TR83/719	S0017
TVS-2SC563A	161	3018		TR83/719	S0017
TVS2SC564	159			TR20/	
TVS-2SC564	159		21	TR28/717	
TVS2SC564-0	159			TR20/	
TVS2SC5640	159	3025		/717	S0019
TVS2SC564R	159	3025		TR20/717	S0019
TVS-2SC582	128	3024		TR87/243	S5011
TVS-2SC582A	128	3024		TR87/243	
TVS-2SC605	108				S0016
TVS-2SC606	108				S0016
TVS2SC644	123A		20	TR21/	
TVS-2SC644	107			2SC644/720	
TVS2SC645	108		20	TR21/	S0016
TVS2SC645A	108				S0016
TVS2SC645B	123A			TR21/	S0016
TVS2SC645C	108				S0016
TVS-2SC645	107	3018		TR70/720	S0014
TVS-2SC645A	107	3018		TR70/720	S0016
TVS-2SC645B	107	3018		TR70/720	S0014
TVS-2SC645C	107	3044		TR70/720	S0014
TVS-2SC646	181				S7002
TVS2SC647	130	3111		TR59/247	S5021
TVS2SC647A	101	3111			
TVS2SC647B-D	165	3111			
TVS2SC647-O,P	165	3111			
TVS2SC647Q,R	165	3111			
TVS2SC683	108		11	TR21/	S0016
TVS-2SC683	161	3018		TR83/719	S0016
TVS-2SC683V	161	3018		TR83/719	
TVS2SC684	123A		17	TR24/	
TVS-2SC684	161	3039		TR83/719	
TVS-2SC687	162		25	TR67/707	S5020
TVS2SC696	128	3024		TR87/	S3002
TVS2SC762	108		11	TR22/	S0016
TVS-2SC762	161	3018		TR83/719	
TVS2SC828	123A		20	2SC828/	
TVS-2SC828	123A	3018		2SC828A/735	
TVS2SC828A	123A			2SC828A/735	S0015
TVS-2SC828A	123A	3020		2SC828A/735	
TVS2SC828P	123A	3020		TR21/735	S0016
TVS2SC828Q	108	3122	20	TR21/735	S0016
TVS-2SC828Q	123A	3122		TR21/735	
TVS2SC828R	123A	3020		TR21/735	S0016
TVS2SC829B	107		20	2SC829B/	S0011
TVS2SC840	124		12	TR23/	S5011
TVS-2SC840	175	3021		TR81/241	S5012
TVS2SC840A	124		12	TR23/	S5011
TVS-2SC840A	175	3021		TR81/241	S5012
TVS2SC840/840A				TR23/	
TVS2SC840C	175				S5012
TVS2SC901	162				S5020
TVS-2SC901	163	3111		TR67/740	
TVS2SC920-0Q	108	3018			
TVS-2SC948	161	3132		TR83/719	
TVS2SC968	123	3020		TR86/53	S5014
TVS2SC1255	128	3024			
TVS2SC1255HF	128	3024			
TVS2SC1256	129	3025			
TVS2SC1256HG	129	3025			
TVS2SC1520		3103			
TVS2SC5640	159				S0019
TVS2SD226	175	3026			
TVS2SD226A	175	3026	23	TR81/241	S5019
TVS2SD226B	175	3026			
TVS2SD226C	175	3026			
TVS2SD226D	175	3026			
TVS2SD2260	175	3026			S5019
TVS2SD226P	175	3026			S5019
TVS2SD968	128	3024			
TVS25A103	160		51		
TVS25B172	102		2		
TVS25B172F				TR85/	
TVS25B448	127		25		
TVS25C208	123A		20		
TVS25C582	123A		20		
TVS25C645	123A		20		
TVS828A	123A	3122	20	TR21/735	S0015
TVS-1303	159		21		
TVS25126F	121		16		
TVS-AN227	1063				
TVS-AN241			IC2		
TVS-AN241D	712				
TVS-CS1255H	123A	3020	18	TR21/735	
TVS-CS1255HF	123A	3122	63	TR21/735	S0014
TVS-CS1256HG	192	3025	67	TR28/242	S0012
TVS-CS1303	159	3114		TR20/717	S3032
TVS-CS1303A	159	3114			
TVS-SN76665N	712		IC2		
TVS-TIS18	108	3039		TR83/719	
TVS-UPC23C	1046				
TVSAN220	711	3070			
TVSAN225	739	3072			C6072P
TVSAN241	712	3072		IC507/	C6063P
TVSAN241D	712	3072			C6063P
TVSCS1255H	128			TR25/	S5014
TVSCS1255HF	192			TR25/	S5013
TVSCS1256HG	193			TR28/	S5013
TVSJA1200			20	TR21/	S0030
TVSMPC-23C	1046				
TVSSA71B	102				
TVSSJE5472-1		3054			
TVSSN76665	712	3072			C6063P
TW135	106		82	TR20/52	S0019
TX100-1	123A	3020	20	TR21/735	
TX100-2	123A	3020	20	TR21/735	
TX100-3	123A		12	TR21/735	
TX-100-3	124	3021		TR21/	
TX100-4	128	3024	18	TR87/243	
TX100-5	124	3021	12	TR81/240	
TX101-8	124	3021	12	TR81/240	
TX101-9	128	3024	18	TR87/243	
TX101-11	124	3021	12	TR81/240	
TX101-12	123A	3020	20	TR21/735	
TX102-1-2	123A	3020	20	TR21/735	S0015
TX102-4	124	3021	12	TR81/240	
TX103-1	104	3009	3	TR01/624	
TX104-3	158	3004	2	TR84/630	G0005
TX105-4	128	3020		TR21/	
TX106-1	102A		2	TR85/250	G0005
TX-106-1	158				
TX107-1	123A	3020	20	TR21/735	
TX107-3	123A			TR21/735	S0015
TX-107-3	107	3020	20	TR21/	
TX107-4	123A	3020	20	TR21/735	S0014
TX107-5,6,10	123A	3020	20	TR21/735	
TX107-12	123A			TR21/735	S0015
TX-107-12	107	3020	20	TR21/	
TX107-13	124	3021		TR81/241	S5011
TX-107-13	124	3020	12	TR81/	
TX107-16	123A	3020		TR21/735	
TX108-1	123A	3020	20	TR21/735	
TX111-1	124	3021	12	TR81/240	S5011
TX112-1	123A	3020	20	TR21/735	
TX116-1,2,3					S0019
TX-119-1	123A	3020	20	TR21/735	S0024
TX120	107				S0015
TX121	159				S0031
TX122	159				S0019
TX122-1	159	3025	21	TR20/717	S0019
TX122-1A	159	3114			
TX123				TR69/	S9100
TX123-1	172		64	TR69/S9100	S9100
TX124	128	3024	18	TR22/	S0014
TX-124	128	3024		TR87/243	
TX125	159		21		
TX125-1	129				S0012
TX-125	129	3025		TR88/242	
TX-126-1	103A			TR08/724	
TX128-1	128	3020			S5014
TX-128-1	123A		63	TR21/735	
TX-131-1		3114		TR20/	
TX134-1	129	3025			S5013
TX-134-1	159	3114	21	TR20/717	

Original Part Number	ECG	SK	GE	IR/WEP	HEP
TX-134-1A	159	3114			
TX135	128				S5014
TX-135	128	3024	63	TR87/243	
TX136	129		21		S5013
TX-136	129	3025		TR88/242	
TX138	128	3024		TR24/	S5014
TX-138	123A			TR21/735	
TX139	129	3025		TR28/	S5013
TX-139	129	3025		TR88/242	
TX140	128	3024		TR25/	S5014
TX-140	128	3024		TR87/243	
TX141	107	3024		TR25/	S0015
TX-141		3024		TR21/735	
TX-183S	162	3079	19		
TX1005	124		12	TR20/	
TX-1005	175	3021		TR81/241	
TY-107-4	123A		20	TR21/	
TY-107-12	123A		20	TR21/	
TZ-1		3003			
TZ6	121	3009		TR01/	
TZ7	121	3009		TR01/	
TZ8	121	3009	16	TR01/232	
TZ10	102A	3004	2	TR85/250	
TZ81	199		62		
TZ115I	199				
TZ543	159			TR20/	
TZ551	159	3114	65	TR20/	S0019
TZ552	193		65		
TZ553	193		65		
TZ554	193		65		
TZ581	211	3203	65		S3028
TZ582	211	3203	65		S3028
TZ1151	199	3124	20	TR21/	S0024
TZ1151/183030				TR21/	
TZ1152	199	3124	20	TR21/	S0024
TZ1180	123A			TR21/	
TZ1182	123A			TR21/	
TZ7001					S0025
TZ7002					S0025
TZ7003					S0015
TZ7500					S0019
TZ7501					S0019
TZ7502					S0019
TZ7503					S0019
U2N34	102	3004		TR05/631	
U2N34A	102A	3004			
U2N96	102	3004		TR05/631	
U2N96A	102A	3004			
U2N474A	123	3122		TR86/53	
U2S93	126	3006		TR17/635	
U2SB267	102A	3004		TR85/250	
U2T85	103A	3010		TR08/724	
U5A7064354	780				
U5B770939X	909	3590			
U5B771039X	910				
U5B771639X	716				
U5B7741312		3514			
U5B7741393		3514			
U5B990029	9900				
U5B990329,X	9903				
U5B990429,X	9904				
U5B990529	9903				
U5B990529X	9905				
U5B990629,X	9906				
U5B990729,X	9907				
U5B990829,X	9908				
U5B990929,X	9909				
U5B991021X					C2006G
U5B991029	9910				
U5B991029X	9910				C2006G
U5B991121X					C2007G
U5B991129	9911				
U5B991129X	9911				C2007G
U5B991229,X	9912				
U5B991329,X	9913				
U5B991421X					C2010G
U5B991422X					C2010G
U5B991429	9914				
U5B991429X	9914				C2010G
U5B992129	9921				
U5B992159X	9921				
U5B992329	9923				
U5B992359X	9923				
U5B997429,X	9974				
U5B7709393	909				
U5B7710393	910				
U5B7716393	716				
U5B7716394	716				
U5B7725391	925				
U5B7725393	925				
U5B7740393	940				
U5B7741393	941				
U5B7749394	949				
U5D770331X	703A		IC12		
U5D770339X	703A	3157		IC500/	C6091

Original Part Number	ECG	SK	GE	IR/WEP	HEP
U5D990021	9900				
U5D990029X	9900				
U5D7703312	703A	3157			C6001
U5D7703393	703A	3157			C6001
U5D7703394	703A	3157		IC500/	C6001
U5E7064394	780	3141			
U5E7746394	707		IC3		C6050P
U5E7796312					C6050G
U5F771139X	911				
U5F991529,X	9915				
U5F992629	9926				
U5F992659X	9926				
U5F992729	9927				
U5F992759X	9927				
U5F7711393	911				
U5F7715393	915				
U5F7723393	923				
U5F7746394	707				
U5F7747312		3514			
U5F7747393	947	3514			
U5M7749394	949				
U5R7723393	923	3164			
U5R7723394		3164			
U5T7725393	925				
U5Z7703394	703A				
U6A180659X	9806				
U6A740059X					C3000P
U6A740159X					C3001P
U6A740459X					C3004P
U6A741059X					C3010P
U6A742059X					C3020P
U6A743059X					C3030P
U6A744059X					C3040P
U6A745059X					C3050P
U6A747359X					C3073P
U6A900059X	74104				
U6A900159X	74105				
U6A909359X	9093				
U6A909459X	9094				
U6A909759X	9097				
U6A909959X	9099				C1052P
U6A910959X	9109				
U6A911059X	9110				
U6A911159X	9111				
U6A911259X	9112				
U6A913559X	9135				
U6A913759X	9137				
U6A915759X	9157				
U6A915859X	9158				
U6A993059X	9930				C1030P
U6A993159X	9931				
U6A993259X	9932				
U6A993359X	9933				C1033P
U6A993559X	9935				
U6A993659X	9936				
U6A993759X	9937				
U6A994159X	9941				
U6A994459X	9944				C1044P
U6A994559X	9945				
U6A994659X	9946				
U6A994859X	9948				
U6A994959X	9949				
U6A995059X	9950				
U6A995159X	9951				
U6A996159X	9961				
U6A996259X	9962				C1062P
U6A7045312	912				
U6A7046312	912				
U6A7075354	723				
U6A7136354	737				
U6A7709393	909D				
U6A7710393	910D				
U6A7711393	911D				
U6A7720354	744				
U6A7720395	744				
U6A7723393	723D	3165			
U6A7729394	720	3160	IC7		
U6A7730393	725				
U6A7732394	718	3159	IC8		C6094P
U6A7739312	725				
U6A7739393	725				
U6A7739394	725	3162			
U6A7741393	941D				
U6A7746393	713				
U6A7746394	713	3077	IC5		C6057P
U6A7767394	722	3161	IC9		
U6A7780394	714				C6070P
U6A7781394	715	3076	IC6		C6071P
U6AA7780394	714				
U6AH00059X	74H00				
U6AH00159X	74H01				
U6AH00459X	74H04				
U6AH00559X	74H05				
U6AH00859X	74H08				
U6AH01059X	74H10				

Original Part Number	ECG	SK	GE	IR/WEP	HEP
U6AH01159X	74H11				
U6AH02159X	74H21				
U6AH03059X	74H30				
U6AH04059X	74H40				
U6AH05059X	74H50				
U6AH05159X	74H51				
U6AH05259X	74H52				
U6AH05359X	74H53				
U6AH05459X	74H54				
U6AH05559X	74H55				
U6AH06059X	74H60				
U6AH06159X	74H61				
U6AH06259X	74H62				
U6AH07159X	74H71				
U6AH07159X	74H72				
U6AH07359X	74H73				
U6AH07459X	74H74				
U6AH07859X	74H77				
U6AH10159X	74H101				
U6AH10259X	74H102				
U6AH10359X	74H103				
U6AH10859X	74H108				
U6AH18359X	74H183				
U6AN00059X	7400				
U6AN00159X	7401				
U6AN00259X	7402				
U6AN00359X	7403				
U6AN00459X	7404				
U6AN00559X	7405				
U6AN00659X	7406				
U6AN00759X	7407				
U6AN00859X	7408				
U6AN00959X	7409				
U6AN01059X	7410				
U6AN01159X	7411				
U6AN01259X	7412				
U6AN01359X	7413				
U6AN01459X	7414				
U6AN01659X	7416				
U6AN01759X	7417				
U6AN02059X	7420				
U6AN02159X	7421				
U6AN02559X	7425				
U6AN02659X	7426				
U6AN02759X	7427				
U6AN03059X	7430				
U6AN03259X	7432				
U6AN03759X	7437				
U6AN03859X	7438				
U6AN03959X	7439				
U6AN04059X	7440				
U6AN05059X	7450				
U6AN05159X	7451				
U6AN05359X	7453				
U6AN05459X	7454				
U6AN06059X	7460				
U6AN07059X	7470				
U6AN07259X	7472				
U6AN07359X	7473				
U6AN07459X	7474				
U6AN08069X	7480				
U6AN08259X	7482				
U6AN08659X	7486				
U6AN09059X	7490				
U6AN09159X	7491				
U6AN09359X	7493				
U6AN09559X	7495				
U6AN10759X	74107				
U6AN12259X	74122				
U6AN13259X	74132				
U6AN17659X	74176				
U6AN17759X	74177				
U6AN18059X	74180				
U6AN19659X	74196				
U6AN19759X	74197				
U6AN993259X	9932				
U6AN993359X	9933				
U6AS00059X	74S00				
U6AS00359X	74S03				
U6AS00459X	74S04				
U6AS00559X	74S05				
U6AS01059X	74S10				
U6AS01159X	74S11				
U6AS01559X	74S15				
U6AS02059X	74S20				
U6AS02259	74S22				
U6AS04059X	74S40				
U6AS04159X	74S41				
U6AS04259X	74S42				
U6AS06459X	74S64				
U6AS06559X	74S65				
U6AS07459X	74S74				
U6AS10959X	74S109				
U6AS11359X	74S113				
U6AS11459X	74S114				
U6AS14059X	74S140				
U6B744159X					C3041P
U6B7066354	728				
U6B7780394	714	3075	IC4		
U6BA01559X	7441				
U6BA04259X	74182				
U6BH02259X	74H22				
U6BH07659X	74H76				
U6BH10659X	74H106				
U6BN02359X	7423				
U6BN04259X	7442				
U6BN04359X	7443				
U6BN04459X	7444				
U6BN07559X	7475				
U6BN07659X	7476				
U6BN08359X	7483				
U6BN08559X	7485				
U6BN09459X	7494				
U6BN15359X	74153				
U6BN15559X	74155				
U6BN15659X	74156				
U6BN17559X	74175				
U6BS11259X	74S112				
U6BS13359X	74S133				
U6BS13459X	74S134				
U6BS13559X	74S135				
U6BS15359X	74S153				
U6BS15759X	74S157				
U6BS15859X	74S158				
U6BS17559X	74S175				
U6BS19559X	74S194				
U6E7729394	720		IC7		
U6E7739393	725	3162			
U6NA00859X	74116				
U6NA01159X	74154				
U6NA04159X	74181				
U6NN15059X	74150				
U6NN19859X	74198				
U6NN19959X	74199				
U7AH08759X	74H87				
U7AN15259X	74152				
U7AN16459X	74164				
U7AN17859X	74178				
U7B7758354	743				
U7BA02259X	74157				
U7BA04559X	7445				
U7BN04859X	7448				
U7BN09659X	7496				
U7BN046591	7446				
U7BN12359X	74123				
U7BN14159X	74141				
U7BN14559X	74145				
U7BN15159X	74151				
U7BN16559X	74165				
U7BN16659X	74166				
U7BN17059X	74170				
U7BN17459X	74174				
U7BN17959X	74179				
U7BN19059X	74190				
U7BN19159X	74191				
U7BN19259X	74192				
U7BN19359X	74193				
U7BN19459X	74194				
U7F7065354	712	3072	IC2		
U7F7065394	712	3072			
U7F7729394	720	3160	IC7		
U7F7732394	718	3159	IC8		
U7F7746394	713	3077	IC5		
U7F7767394	722	3161	IC9		
U7F7780394	714	3075	IC4		
U7F7781394	715	3076	IC6		
U7F77463394	713	3077			
U7F77803394	714	3075			
U8A990028	9900				
U8A991428	9914				
U8A992328	9923				
U8B770339X	703A	3157	IC12	IC500/	
U8B7703394	703A	3157	IC12	IC500/	
U8F7737394	705A	3134		IC502/	
U8F7746394	705A	3134	IC3		
U9A180059X	9800				
U9A180159X	9801				
U9A180259X	9802				
U9A180359X	9803				
U9A180459X	9804				
U9A180559X	9805				
U9A180659X	9806				
U9A180759X	9807				
U9A180859X	9808				
U9A180959X	9809				
U9A181059A	9810				
U9A181159X	9811				
U9A181259X	9812				
U9A181359X	9813				

Original Part Number	ECG	SK	GE	IR/WEP	HEP
U9A181459X	9814				
U9A900059X	74104				
U9A900159X	74105				
U9A909359X	9093				
U9A909459X	9094				
U9A909759X	9097				
U9A909959X	9099				
U9A913559X	9135				
U9A913759X	9137				
U9A915759X	9157				
U9A915859X	9158				
U9A993059X	9930				
U9A993159X	9931				
U9A993259X	9932				
U9A993359X	9933				
U9A993559X	9935				
U9A993659X	9936				
U9A993759X	9937				
U9A994159X	9941				
U9A994459X	9944				
U9A994559X	9945				
U9A994659X	9946				
U9A994859X	9948				
U9A994959X	9949				
U9A995059X	9950				
U9A995159X	9951				
U9A996159X	9961				
U9A996259X	9962				
U9A996359X	9963				
U9A7064354	783				
U9A7075394	723				
U9A7136354	737				
U9A7709393	909D				
U9A7711393	911D				
U9A7720354	744				
U9A7720395	744				
U9A7723393	923D	3165			
U9A7729394	720				
U9A7741393	941D				
U9A7746394	713	3077	IC5		
U9A7767394	722				
U9A7781394	715		IC6		
U9AC00159X	4001				
U9AC00259X	4002				
U9AC01159X	4011				
U9AC01259X	4012				
U9AC01359X	4013				
U9AC02359X	4023				
U9AC02559X	4025				
U9AC02759X	4027				
U9AC03059X	4030				
U9AC04959X	4049				
U9AC05059X	4050				
U9AC81159X	4811				
U9AH00059X	74H00				
U9AH00159X	74H01				
U9AH00459X	74H04				
U9AH00559X	74H05				
U9AH00859X	74H08				
U9AH01059X	74H10				
U9AH01159X	74H11				
U9AH02059X	74H20				
U9AH02159X	74H21				
U9AH02259X	74H22				
U9AH03059X	74H30				
U9AH04059X	74H40				
U9AH05059X	74H50				
U9AH05159X	74H51				
U9AH05259X	74H52				
U9AH05359X	74H53				
U9AH05459X	74H54				
U9AH05559X	74H55				
U9AH06059X	74H60				
U9AH06159X	74H61				
U9AH06259X	74H62				
U9AH07159X	74H71				
U9AH07259X	74H72				
U9AH07359X	74H73				
U9AH07459X	74H74				
U9AH07659X	74H76				
U9AH07859X	74H78				
U9AH08759X	74H87				
U9AH10159X	74H101				
U9AH10259X	74H102				
U9AH10359X	74H103				
U9AH10859X	74H108				
U9AH18359X	74H183				
U9AN00059X	7400				
U9AN00159X	7401				
U9AN00259X	7402				
U9AN00359X	7403				
U9AN00459X	7404				
U9AN00559X	7405				
U9AN00659X	7406				
U9AN00759X	7407				
U9AN00859X	7408				
U9AB00959X	7409				
U9AN01059X	7410				
U9AN01159X	7411				
U9AN01259X	7412				
U9AN01359X	7413				
U9AN01459X	7414				
U9AN01659X	7416				
U9AN01759X	7417				
U9AN02059X	7420				
U9AN02159X	7421				
U9AN02559X	7429				
U9AN02659X	7426				
U9AN02759X	7427				
U9AN03059X	7430				
U9AN03259X	7432				
U9AN03759X	7437				
U9AN03859X	7438				
U9AN03959X	7439				
U9AN04059X	7440				
U9AN05059X	7450				
U9AN05159X	7451				
U9AN05359X	7453				
U9AN05459X	7454				
U9AN06059X	7460				
U9AN07059X	7470				
U9AN07259X	7472				
U9AN07359X	7473				
U9AN07459X	7474				
U9AN08059X	7480				
U9AN08259X	7482				
U9AN08659X	7486				
U9AN09059X	7490				
U9AN09159X	7491				
U9AN09359X	7493A				
U9AN09559X	7495				
U9AN10759X	74107				
U9AN12259X	74122				
U9AN13259X	74132				
U9AN015259X	74152				
U9AN16459X	74164				
U9AN17659X	74176				
U9AN17759X	74177				
U9AN17859X	74178				
U9AN18059X	74180				
U9AN19659X	74196				
U9AN19759X	74197				
U9AN993259X	9932				
U9AN993359X	9933				
U9AS00059X	74S00				
U9AS00359X	74S03				
U9AS00459X	74S04				
U9AS00559X	74S05				
U9AS01059X	74S10				
U9AS01159X	74S11				
U9AS01559X	75S15				
U9AS02059X	74S20				
U9AS02259X	74S22				
U9AS04059X	74S40				
U9AS04159X	74S41				
U9AS04259X	74S42				
U9AS06459X	74S64				
U9AS06559X	74S65				
U9AS07459X	74S74				
U9AS10959X	74S109				
U9AS11359X	74S113				
U9AS11459X	74S114				
U9AS14059X	74S140				
U9B7780394	714	3075	IC4		
U9B7787354	797				
U9BA01559X	7441				
U9BA02259X	74157				
U9BA04259X	74182				
U9BA04559X	7445				
U9BC01959X	4019				
U9BC02859X	4028				
U9BC51259X	4512				
U9BH10659X	74H106				
U9BN02359X	7423				
U9AN04259X	7442				
U9AN04359X	7443				
U9AN04459X	7444				
U9AN04859X	7448				
U9AN07559X	7475				
U9AN07659X	7476				
U9AN08359X	7483				
U9AN08559X	7485				
U9AN09459X	7494				
U9AN09659X	7496				
U9AN046591	7446				
U9AN12359X	74123				
U9AN14159X	74141				
U9AN14559X	74145				
U9AN15159X	74151				
U9AN15359X	74153				

Original Part Number	ECG	SK	GE	IR/WEP	HEP
U9AN15559X	74155				
U9AN15659X	74156				
U9AN16559X	74165				
U9AN16659X	74166				
U9AN17059X	74170				
U9AN17459X	74174				
U9AN17559X	74175				
U9AN17959X	74179				
U9AN19059X	74190				
U9AN19159X	74191				
U9AN19259X	74192				
U9AN19359X	74193				
U9AN19459X	74194				
U9AS11259X	74S112				
U9AS13359X	74S133				
U9AS13459X	74S123				
U9AS13559X	74S135				
U9AN15359X	74S153				
U9AN15759X	74S157				
U9AN15859X	74S158				
U9BS17459X	74S154				
U9BS17559X	74S175				
U9BS19459X	74S194				
U9BS31659X	74S16				
U9C7065354	712	3072	IC2		
U9D7067354	729				
U9D7758354	743				
U9J7706354	1114				
U9NA00859X	74116				
U9NA01159X	74154				
U9NA04159X	74181				
U9NN15059X	74150				
U9NN19859X	74198				
U9NN19959X	74199				
U9T7741393	941M				C6052P
U9T7753354	736				
U9T7753395	734				
U10	191				
U51	189	3025			
U110					F1036
U112					F1036
U114					F1036
U128K		3112			
U146					F1036
U148					F1036
U149					F1036
U197	133	3112			
U235	133	3112			
U460A,B	107	3018	11	TR51/	
U535A	107	3018	11	TR70/	
U535B	229	3018	20	TR70/	S0015
U535B/7825B	229	3018			
U737	705A				
U1177	133	3112		/801	F0010
U1178	133	3112		/801	F0010
U1180	133	3112		/801	F0010
U1181	133	3112		/801	F0010
U1277	133	3112			
U1278	133	3112			
U1279	133	3112			
U1280	133	3112			
U1282	132	3116			
U1283	133	3112			
U1284	133	3112			
U1285	133	3112		/801	F0010
U1286	133	3112			
U1322	133	3112		/801	F0010
U1323	133	3112		/801	F0010
U1324	133	3112		/801	F0010
U1325	133	3112			
U1420	133		FET1		
U1421	133		FET1		
U1422	133		FET1		
U1585E	123A	3018	20		S0015
U1585F,H	107	3018	11	TR51/	
U1650E	132				F0021
U1714	133	3112			
U1715	133	3112			
U1835E	132				F0021
U1837E	132	3116	FET2		
U1916	133		FET1		
U1994E	132	3116	FET2		
U2047E	132	3116			
U2848-1	123A	3020	20	TR21/735	
U3012	133	3112			
U7003					S0020
U-45054-1	941				
U2091858-11	121	3009		TR01/232	
U513991121X					C2007C
UA703	703A	3157	IC12	IC500/	C6001
UA703C	703A	3157	IC12		C6001
UA703E	703A	3157	IC12	IC500/	C6001
UA706	1114				
UA706BPC	1114				
UA709A		3552			

Original Part Number	ECG	SK	GE	IR/WEP	HEP
UA709ADM		3552			
UA709C		3552			
UA709C(DIP)	909D				
UA709C(METCAN)	909				
UA709CA	909D				
UA709CT	909	3590			
UA709DC,DM		3552			
UA710		3553			
UA710C		3553			
UA710C(DIP)	910D				
UA710C(METCAN)	910				
UA710CA	910D				
UA710CT	910				
UA710DC,DM		3554			
UA710HC,HM		3553			
UA711		3555			
UA711C		3555			
UA711C(DIP)	911D				
UA711C(METCAN)	911				
UA711CA	911D				
UA711CK,CT	911				
UA711DM,DC		3556			
UA711HC		3555			
UA715(METCAN)	915				
UA716	716				
UA720	744				
UA720DC	744		IC24		
UA720PC	744				
UA723		3164			
UA723C		3164			
UA723C(DIP)	923D				
UA723C(METCAN)	923				
UA723CA	923D	3165			
UA723CL	923	3164			
UA723DC,DM		3165			
UA723HC,HM		3164			
UA725(METCAN)	925				
UA729	720	3160	IC7		C6095P
UA729CA		3160			
UA732	718	3159	IC8	IC511/	C6094P
UA737E	705A	3134		IC502/	C6057P
UA739	725	3162			
UA739C	725	3162			C6055L
UA739DC		3162	IC19		
UA739PC		3162			
UA740(METCAN)	940				
UA740CT	940				
UA741		3514			
UA741(8-PIN)					C6052P
UA741A		3558			
UA741ADM		3558			
UA741C		3514			
UA741C(MINIDIP)	941M				
UA741CA	941D	3552			
UA741CT	941	3514			
UA741CV	941M	3553			
UA741DC,EC		3558			
UA741TC		3559			
UA746	707	3077	IC3	IC503/	C6057P
UA746(DIP)	713	3077		IC508/	
UA746C		3134			
UA746DC		3077			
UA746E	705A			IC502/	
UA746(METCAN)	707				
UA746HC		3134			
UA746PC		3077			
UA747		3526			
UA747C		3526			
UA747CK	947	3526			
UA747T		3526			
UA747(METCAN)	947				
UA749		3166	IC25		
UA749(METCAN)	949				
UA749C(METCAN)	949				
UA749D,DHC		3166			
UA753	736	3163			
UA753C	734	3163			
UA753TC		3163			
UA758	743		IC32		
UA767	722	3161	IC9		C6056P
UA767D,DC		3161			
UA767PC	722	3161			
UA780	714	3075	IC4	IC509/	C6070P
UA781	715	3076	IC6	IC510/	C6071P
UA796					C6050G
UA910					C2006G
UA911					C2007G
UA914					C2010G
UA923					C2004P
UA3064TC			IC20		
UA3065	712	3072	IC2		
UA3066	728				
UA3066DC			IC22		
UA3067	729				
UA3067DC			IC23		

Original Part Number	ECG	SK	GE	IR/WEP	HEP
UA3075	723				
UB747559X					C3075P
UBB770339X	703			IC500/	
UBB7703393X	703A	3157			
UBFY11	108	3039		TR95/56	
UC20	133	3112		/801	F0010
UC100	133	3112		/801	F0010
UC105	133	3112		/801	F0010
UC110	133	3112		/801	F0010
UC115	133	3112		/801	F0010
UC120	133	3112		/801	F0010
UC125	133	3112		/801	F0010
UC155	132	3116			
UC201	132	3116			
UC210	132	3116			
UC220	133	3112			
UC241	133	3112			
UC410					F1036
UC588	132	3116			
UC701	133		FET1		
UC703	133		FET1		
UC704	133		FET1		
UC705	133		FET1		
UC714	133	3116	FET1		
UC734	132	3116	FET1	FE100/802	
UC734E	132	3116	FET2		F0015
UC750	133		FET1		
UC751	133		FET1		
UC752	133		FET1		
UC753	133		FET1		
UC756	133	3112			
UC900					C2002P
UC1100	159	3114		TR20/717	S0019
UC1100A	159	3114			
UC2136	133	3112			
UC2138	133	3112			
UC2139	133	3112			
UC2147	133	3112			
UC2148	133	3112			
UC2149	133	3112			
UC4741		3514			
UC4741C		3514			
UL910					C2006G
UL914	9914				
UL923	9923				
ULC3037		3135			
ULM2111A		3135			
ULM2114A		3077			
ULN228P	741				
ULN2111	708	3135	IC10	IC504/	
ULN2111A	708	3135	IC10	IC504/	C6062P
ULN2111N	708	3135	IC10	IC504/	
ULN2113	709		IC11	IC505/	
ULN2113A	709		IC11	IC505/	C6062P
ULN2113N	709		IC11	IC505/	
ULN2114	713	3077			C6057P
ULN2114A	713	3077	IC5	IC508/	C6057P
ULN2114K	707			IC505/	C6067G
ULN2114N	713	3077	IC5		
ULN2114W	705A	3134	IC3	IC502/	C6089
ULN2118A					C6056P
ULN2120A	718	3159	IC8	IC511/	C6094P
ULN2120N	718	3159			
ULN2121A,N	719		IC28		
ULN2122	720		IC7	IC512	C6095P
ULN2122N	720		IC7		
ULN2124	714	3075		IC509/	
ULN2124A	714	3075	IC4		C6070P
ULN2124N	714	3075	IC4		
ULN2125A	731	3170	IC13		
ULN2125N	731	3170			
ULN2126A	721		IC14		C6080P
ULN2126N	721				
ULN2127	715	3076		IC510/	
ULN2157A	715	3076	IC6		C6071P
ULN2127N	715	3076	IC6		
ULN2128	722				C6056P
ULN2128/183014				IC513/	
ULN2128A	722	3161	IC9	IC513/	
ULN2128N	722	3161	IC9		
ULN2129A	723	3140	IC15		C6101P
ULN2129N	723	3140			
ULN2131M	734				
ULN2134A	806				
ULN2135E	735				
ULN2136A	737		IC16		
ULN2136N	737				
ULN2137A	744		IC24		
ULN2137N	744				
ULN2165	712	3072			
ULN2165A	712	3072	IC2	IC507/	C6063P
ULN2165N	712	3072	IC2	IC507/	
ULN2208M	805				
ULN2209M	736		IC17		
ULN2210			IC35		
ULN2210A,N	801				
ULN2211P	742				
ULN2212A	807				
ULN2224	1062				
ULN2224A	1062		IC30		
ULN2224N	1062				
ULN2226A,N	739				
ULN2228A	790		IC18		C6057P
ULN2228N	790				
ULN2244A,N	743		IC32		
ULN2262A,N	797				
ULN2264A	783	3141	IC21		C6100P
ULN2264N	783				
ULN2267A		3074	IC23		
ULN2269S	791	3149	IC33		
ULN2269N	791				
ULN2275P	732				
ULN2275Q	732	3161			
ULN2276P	733				
ULN2277P	732				C6090
ULN2277Q	732				
ULN2278P	804		IC27		
ULN2280A	740		IC31		
ULN2280N	740				
ULN2289A,N	788				
ULN2298A	738	3167	IC29		
ULN2298N	738	3167			
ULN2741D		3514			
ULN21111A	708				
ULS2741D		3514			
ULX2228A,N	790				
ULX2298N	790				
UMP3202	124	3021		TR81/240	
UP16C				IC504/	
UP12217	186				
UP12218	186				
UP14046	186				
UP14047	186				
UPB2S00D	74S00				
UPB2S10D	74S10				
UPB2S20D	74S20				
UPB2S112D	74S112				
UPB120D	9669				
UPB121D	9668				
UPB123D	9661				
UPB124D	9678				
UPB125D	9662				
UPB201C,D	7400				
UPB202C,D	7410				
UPB203C,D	7420				
UPB204C,D	7430				
UPB205C,D	7440				
UPB206C,D	7450				
UPB207C,D	7451				
UPB208D	7453				
UPB209D	7454				
UPB210C,D	7460				
UPB211C	7472				
UPB211D	7470				
UPB212C	7473A				
UPB212D	7472				
UPB213C	7474				
UPB214C,D	7474				
UPB215C,D	7401				
UPB217C,D	7475				
UP218C,D	7441				
UPB219C,D	7490				
UPB222C,D	7492				
UPB223C	7493				
UPB223D	7493A				
UPB224C,D	7476				
UPB225C,D	7473				
UPB226C,D	7495				
UPB227D	7442				
UPB230D	7483				
UPB232C,D	7402				
UPB234C,D	7408				
UPB235D	7404				
UPB236D	7405				
UPB237D	7437				
UPB238D	7438				
UPB2047D	7447				
UPB2080D	7480				
UPB2084D	7484				
UPB2085D	7485				
UPB2086D	7486				
UPB2091D	7491				
UPB2150D	74150				
UPB2151D	74151				
UPB2154D	74154				
UPB2161D	74161				
UPB2170D	74170				
UPB2180D	74180				
UPB2181D	74181				
UPB2182D	74182				

Original Part Number	ECG	SK	GE	IR/WEP	HEP
UPB2192D	74192				
UPB2193D	74193				
UPB2195D	74195				
UPB2198D	74198				
UPC16A	1045				
UPC16C	1045	3101	IC2	IC507/	C6060P
UPC-16C	748	3072	IC2		
UPC17C	1047				
UPC17C/D368	1047				
UPC20C	1075				
UPC22C	1078				
UPC23C	1046				
UPC29-2	1089¤				
UPC29C	1089¤				
UPC29C3	1089¤				
UPC30C	1049				
UPC31A,C	1048				
UPC31C1	1048				
UPC32C	1050				
UPC41C	1093				
UPC46C	1077				
UPC47C	1076				
UPC48C	1091				
UPC48C-1	1091				
UPC49C	1086				
UPC50C	1084				
UPC55A				IC507/	C6001
UPC81C	1075				
UPC206	1075				
UPC554	1142				
UPC554C	1142	3160	IC7		
UPC555	703A	3157			
UPC555A	703A	3157	IC12	IC500/	C6061
UPC-555A	703A				C6001
UPC555C	703A	3157			
UPC555H	1092				
UPC558	1051				
UPC558C	1051				
UPC561,C	1079				
UPC562C	1050				
UPC562C1	1050				
UPC563C	1081¤¤				
UPC563H	1081		IC7		
UPC563H2	1081				
UPC563H1D369	1081¤¤				
UPC564C	1088				
UPC566H	1052				
UPC570C	1086**				
UPC571C	1078				
UPC575C	1141				
UPC575C2	1140				
UPC577H	1082				
UPC1001H	1127				
UPC1001H2	1127				
UPI706,A	108		61		
UP1706B			61		
UPI718A			20		
UPI956			81		
UPI1303	158		53		
UPI1305	158		53		
UPI1307	158		53		
UPI1309	158		53		
UPI1345	158		53		
UPI1347	100		1		
UPI1353	102		2		
UPI1613			18		
UPI2217			47		
UPI2218			47		
UPI2222	123A		20		
UPI2222B,P			20		
UPI4046-46	123A		210		
UPI4047-46	123		20		
UPT011			32		
UPT012			32		
UPT013			32		
UPT014			32		
UPT015			32		
UPT021			32		
UPT022			32		
UPT023			32		
UPT024			32		
UPT025			32		
UPT111			215		
UPT112			32		
UPT113			32		
UPT114			32		
UPT115			32		
UPT121			216		
UPT122			32		
UPT123			32		
UPT124			32		
UPT125			32		
UPT211			46		
UPT221			23		
UPT611			216		

Original Part Number	ECG	SK	GE	IR/WEP	HEP
UPT621			216		
US74H00A,J	74H00				
US74H01A,J	74H01				
US74H04A,J	74H04				
US74H05A,J	74H05				
US74H08A,J	74H08				
US74H10A,J	74H10				
US74H11A,J	74H11				
US74H20A,J	74H20				
US74H21A,J	74H21				
US74H22A,J	74H22				
US74H30A,J	74H30				
US74H40A,J	74H40				
US74H50A,J	74H50				
US74H51A,J	74H51				
US74H52A,J	74H52				
US74H53A,J	74H53				
US74H54A,J	74H54				
US74H55A,J	74H55				
US74H60A,J	74H60				
US74H61A,J	74H61				
US74H62A,J	74H62				
US74H71A,J	74H71				
US74H72A,J	74H72				
US74H73A,J	74H73				
US74H74A,J	74H74				
US74H76A	74H76				
US74H78A,J	74H78				
US7400A,J	7400				
US7401A,J	7401				
US7402A,J	7402				
US7403A,J	7403				
US7404A,J	7404				
US7405A,J	7405				
US7408A,J	7408				
US7409A,J	7409				
US7410A,J	7410				
US7411A,J	7411				
US7418A,J	7418				
US7420A,J	7420				
US7426A	7426				
US7427A	7427				
US7430A,J	7430				
US7432A,J	7432				
US7438A,J	7438				
US7440A,J	7440				
US7441A	7441				
US7442A	7442				
US7443A	7443				
US7444A	7444				
US7445A	7445				
US7446A	7446				
US7447A	7447				
US7448A	7448				
US7450A,J	7450				
US7451A,J	7451				
US7453A,J	7453				
US7454A,J	7454				
US7460A,J	7460				
US7470A,J	7470				
US7472A,J	7472				
US7473A,J	7473				
US7474A,J	7474				
US7475A,J	7475				
US7476A	7476				
US7480A	7480				
US7482A	7482				
US7483A	7483				
US7486A,J	7486				
US7490A,J	7490				
US7492A,J	7492				
US7493A,J	7493A				
US74107A	74107				
US74121A,J	74121				
US74145A	74145				
US74153A	74153				
US74154A	74154				
US74180A,J	74180				
USA1012					G6004
USAF12ES040P					G0008
USAF501ES001M					S3013
USAF505ES105					G0008
USAF506ES0M					G6006
USAF508ES020P					S3012
USAF508ES021P					S3012
USAF509ES025P					G0008
USAF510ES030M					S5025
USAF510ES031M					S5025
USAF511ES035P					S3011
USAF511ES036P					S3011
USAF513ES045P					G0008
USD7703394	703A		IC12		
USN7400A					C3000P
USN7401A					C3001P
USN7410A					C3010P

Original Part Number	ECG	SK	GE	IR/WEP	HEP
USN7420A					C3020P
USN7430A					C3030P
USN7440A					C3040P
USN7441A					C3041P
USN7450A					C3050P
USN7473A					C3073P
UST10					G0005
UST19					G0005
UST81					G0005
UST87					G0005
UST88					G0005
UST722					G0005
UST760					G0005
UST761					G0005
UST762					G0005
UST763					G0007
UST764					G0007
UTRA-7RM	121	3009		TR01/232	
VOC236-001	221				
VOC236-00I	221				
V5B7710393	710				
V6/2R	126	3005	52	TR17/635	G0005
V6/2RC	102A	3005	1	TR85/250	G0005
V6/2RJ	102A		52	TR85/250	G0005
V6/4R	126	3005	52	TR17/635	G0005
V6/4RC	102A	3005		TR85/250	G0005
V6/4RJ	102A		52	TR85/250	G0005
V6/8R	126	3005	52	TR17/635	G0005
V6/8RJ	102A			TR85/250	G0005
V6-RC	102A			TR85/250	
V10/1S	126	3005	1	TR17/635	G0005
V10/1SJ	126	3006		TR17/635	G0001
V10/2S	100	3005	1	TR05/254	G0005
V10/2SJ	100	3005		TR05/254	G0005
V10/15					G0005
V10/15A	102	3004	2	TR05/631	G0005
V10/15A18	102A	3004			
V10/30					G0005
V10/30A	102	3004	2	TR05/631	G0005
V10/30A18	102A	3004			
V10/50					G0005
V10/50A	102	3004	2	TR05/631	G0005
V10/50A18	102A	3004			
V10/50B					G0005
V13/11	100	3004	2	TR05/254	
V15/10DP	121	3009		TR01/230	G6003
V15/10P					G6011
V15/20DP	121	3009		TR01/230	G6003
V15/20P					G6011
V15/20R	126	3006	51	TR17/635	G0003
V15/30DP	121	3009		TR01/230	G6003
V15/30P					G6011
V30/10DP	121	3009		TR01/230	G6003
V30/10P					G6011
V30/20DP	121	3009		TR01/230	G6003
V30/20P					G6011
V30/30DP	104	3009		TR01/230	G6003
V30/30P					G6011
V51	102A		53	TR85/250	
V58	160			TR17/637	
V60/10DP	121	3009		TR01/232	G6005
V60/10P	121	3009	16	TR01/232	G6011
V60/20DP	121	3014		TR01/232	G6005
V60/20P	121	3009	16	TR01/232	G6011
V60/30DP	121	3009		TR01/232	G6005
V60/30P	121	3009	16	TR01/232	G6011
V75	160		50	TR17/637	
V118	107	3039		TR70/720	
V119	123A	3122	20	TR21/735	
V120	160		51	TR17/637	G0003
V120PH	108	3039		TR95/56	
V120RH	128	3024	11	TR87/243	
V129	107	3039		TR70/720	
V139	128	3024	18	TR87/243	
V143	107	3039		TR70/720	
V144	124	3021	12	TR81/240	
V145	121	3009	3	TR01/232	
V146	128	3024	18	TR87/243	
V151	9914				
V152	159	3114	21	TR20/717	
V152A	121	3114			
V154	128	3024	18	TR87/243	
V159	133	3112	FET1	/801	
V162	159	3114	21	TR20/717	
V162A	121	3114			
V166	128	3024	18	TR87/243	
V169	123A	3122	20	TR21/735	
V172	128	3024	18	TR87/243	
V176	184	3054	66	/S5003	
V177	128	3024	18	TR87/243	
V178	703A	3157			
V180	129	3025		TR88/242	
V181(BENCO)	9989				
V182	725	3162			
V183	132	3116	FET2	FE100/802	

Original Part Number	ECG	SK	GE	IR/WEP	HEP
V184	703	3157			
V205	160	3006	65	TR17/637	G0003
V208					G6003
V220	108	3039	20	TR95/56	S0016
V221	108	3039		TR95/56	S0016
V222	108	3039	20	TR95/56	S0016
V297	123A	3020	18	TR21/735	
V405	108	3039		TR95/56	S0016
V409	152		66		
V410	159	3114	21	TR20/717	S0012
V410A	129	3025	21	TR88/242	
V415	108		61	TR24/	
V417	108		211	TR24/	
V435	108	3039	63	TR95/56	
V435A	159		65		
V600					S5014
V601					S5014
V602					S5014
V653					S5022
V654	193		48		S3012
V655	193	3118	55		S0019
V741	193		65		
V743					S0013
V745			48		
V761	159		22		
V763	159		22		
V1650E-1	133		FET1		
V1650E-4	133	3112		/801	
V1833E	133	3112	FET1	/801	
V904218		3004			
V907511		3008			
V908015H		3003			
V908980G		3008			
V910533		3007			
VB11	102A		52	TR85/250	G0005
VB701					G0005
VB704					G0005
VB709					G0006
VD13		3004	52		
VE79092	9949				
VE79093	9944				
VE79141	9937				
VFG2745B	104	3014	3	TR01/230	G6003
VFG-274513	104	3009		TR01/624	
VFL2744K	160		51		
VFL-2744K	126	3006		TR17/635	
VFP2746C	121	3014	16		G6003
VFP-2746C	104	3009		TR01/624	
VFP6537C	121	3014	3		G6005
VFP-6537C	121	3009		TR01/232	
VFQ2745F	102A		1	TR85/250	G0005
VFQ-2745F	100	3005			
VFS5K	126	3008	51	TR17/635	
VFS2745, J	158		1		G0005
VFS-2745, J	102A	3005		TR85/250	
VFSGK				TR12/	
VFT-2745H	102A	3004	2	TR85/250	G0005
VFU2746B	121	3014	3		G6005
VFU-2746B	121	3009		TR01/232	
VFU65326B	121				
VFW2745D	160		51		
VFW-2745D	126	3006		TR17/635	
VFY-2745E	100	3005	1	TR05/254	
VL/8RJ	100		52		
VL18RJ	126			TR17/635	
VM20233	124	3021	12		
VM30203	121	3009	16	TR01/232	
VM30209	123A	3020	20	TR21/735	
VM30233	124	3021	12	TR81/240	
VM30234	124	3021	12	TR81/240	
VM30241	123A	3020	20	TR21/735	
VM30242	123A	3020	20	TR21/735	
VM30244	102A	3004	2	TR85/250	
VM30245	124	3021	12	TR81/240	
VS-2SA71	160	3006	51	TR17/637	
VS-2SA71B, BS	160	3006	51	TR17/637	
VS-2SA103	160	3007	51	TR17/637	
VS-2SA358	126	3006		TR17/635	
VS-2SA378	126	3006		TR17/635	
VS2SA378	160		51		
VS-2SA379	126	3006		TR17/635	
VS2SA379	160		51		
VS-2SA385	126	3008		TR17/635	
VS-2SA385L	160	3006			
VS2SA385	102		2		
VS2SA448	127		25		
VS-2SB126	121		16	TR01/232	
VS-2SB126F		3009	16	TR27/235	
VS-2SB126V	127	3034	16	TR27/235	
VS-2SB128V	127	3034		TR27/235	
VS-2SB171	102A	3004		TR85/250	
VS2SB171	102		2		
VS-2SB172	102A			TR85/250	
VS-2SB172FN	102A	3004		TR85/250	
VS2SB172	102		2		

Original Part Number	ECG	SK	GE	IR/WEP	HEP
VS-2SB176	102A	3004		TR85/250	
VS2SB176	102		2		
VS-2SB178,A	102A	3004		TR85/250	
VS2SB178,A	102		2		
VS-2SB324	102A	3004		TR85/250	
VS2SB324	123A		20		
VS-2SB448	127	3035		TR27/235	
VS-2SC58,A,B,C	154	3041		TR17/712	
VS-2SC206	108	3019		TR95/56	
VS2SC206	123A		17		
VS-2SC208	108	3018		TR95/56	
VS2SC208	123A		20		
VS-2SC288A	108	3019		TR95/56	
VS2SC288A	123A		17		
VS-2SC324H	123A	3122	20	TR21/735	
VS2SC374-B-1	123A				
VS-2SC385L	160	3006	51	TR17/637	
VS2SC394-O-1	107				
VS2SC394-Y-1	107				
VS-2SC446	161	3039		TR83/719	
VS-2SC458	123A	3020	20	TR21/735	
VS2SC481-1	195				
VS-2SC538	123A	3020	20	TR21/735	
VS-2SC563	108	3018	17	TR95/56	
VS-2SC645,A	108	3019	17	TR95/56	
VS2SC645A	123A		20		
VS-2SC683	161	3039		TR83/719	
VS-2SC684	108	3019		TR95/56	
VS2SC684	107		11		
VS2SC732-V1F	123A				
VS2SC735-Y-1	123A				
VS-2SC762	108	3018	17	TR95/56	
VS2SC784-R1F	107				
VS2SC1166-Y-1	154				
VS2SC1173-Y-3	152				
VS2SC1237-1	152				
VS9-0003-913	199				
VS9-0004-923	234				
VS9-0005-913	123A				
VS9-0006-913	123A				
VS9-0008-923	234				
VS-CS1255H	128	3024	18	TR87/243	
VS-CS1255HF	128	3024		TR87/243	
VS-CS1256HG	129	3025	21	TR88/242	
VSCS1256HG	159				
VSF2745	100	3005	2	TR05/254	
VTV-BMC				TR85/	
VTVAT6				TR85/	
VTVV5A				TR85/	
VX3375	186		28		
VX3733	152		66		
W1	100	3005		TR05/	
W1A	133	3112	FET1	FE100/801	F0010
W1B					F0010
W1C					S9002
W1E	132	3116	FET3	FE100/	F2005
W1H					S9002
W1M					S9002
W1P	132	3112	FET2	FE100/	
W1S					S9001
W1U	222	3050			F2004
W1V					S9001
W1Y					S9001
W2	101	3011		TR08/	
W3	102A	3004		TR85/	
W4	103A	3010		TR08/	
W5	104	3009		TR01/	
W6	105	3012		TR03/	
W7	106	3118		TR20/	
W8	108	3019		TR95/	
W9	121	3009		TR01/	
W9MP	121MP	3013			
W10	123	3122		TR86/	
W11	124	3021		TR81/	
W12	160	3006		TR17/	
W13	127	3035		TR27/	
W14	128	3104	FET4	TR87/	
W15	129	3025		TR88/	
W16	130	3027		TR59/	
W16MP	130MP	3029		TR59MP/	
W17	131	3082		TR94/	
W17MP	131MP	3086		TR94MP/	
W18	152	3041		TR76/	
W19	153	3083		TR77/	
W20	123A			TR21/	
W21	159	3114		TR20/	
W22	160			TR17/	
W23	161	3018		TR83/	
W24	123A	3122		TR21/	
W25	158	3004		TR84/	
W26	157	3103		TR60/	
W27	155				
W28	154	3044		TR78/	
W29	107	3039		TR70/	
WC19862	103A	3010		TR08/724	G0011

Original Part Number	ECG	SK	GE	IR/WEP	HEP
WC19862A	158	3004		TR84/630	G0005
WC19863	102A	3004		TR85/250	G0005
WC19864	102A	3004		TR85/250	G0005
WE4110	752				C6003
WED8748-2	909				
WEP3		3006			
WEP50		3122			
WEP51		3114			
WEP52		3118			S0013
WEP53		3020			S0014
WEP56		3018			S0016
WEP230					G6003
WEP231		3012			G6004
WEP232					G6005
WEP233					G6006
WEP235					G6008
WEP238					G6011
WEP240		3021			S5011
WEP241					S5012
WEP242		3025			S5013
WEP243		3024			S5014
WEP244		3103			S5015
WEP246		3083			
WEP247		3036			S7002
WEP247MP		3037			
WEP250		3003			G0005
WEP253		3004			
WEP254		3004			G0005
WEP3010					S9002
WEP624					G6013
WEP628					G6013
WEP630					G0006
WEP631		3004			G0005
WEP632		3004			
WEP635		3006			G0001
WEP637		3007			G0003
WEP641		3011			G0011
WEP641A,B		3010			G0011
WEP642		3052			G6016
WEP643		3052			
WEP700		3084			S5006
WEP701					S5000
WEP703		3026			
WEP704		3027			
WEP707					S5020
WEP709		3132			
WEP712		3044			S5025
WEP715		3114			
WEP716		3114			
WEP717		3114			S0031
WEP719		3132			S0020
WEP720		3018			S0020
WEP723		3122			
WEP724		3122			
WEP728		3122			
WEP729		3122			
WEP735		3122			S0015
WEP736		3122			
WEP740					S5021
WEP740A		3133			S5021
WEP740B		3115			S5021
WEP801		3112			F0010
WEP802		3116			F0015
WEPG6001					G6001
WEPS3002					S3002
WEPS3003					S3003
WEPS3020					S3020
WEPS3021		3104			S3021
WEPS3023					S3024
WEPS3027					S3028
WEPS3031		3083			S3032
WEPS5003					S5000
WEPS5004					S5004
WEPS5005					S5005
WEPS5007					S5006
WEPS7000					S7000
WEPS7001					S7001
WEPS9100					S9100
WF1	133	3112			
WF2	132	3116		FE100/	
WIE	220				
WIE/48-13707				FE100/	
WIE/48-137070				FE100/	
WK-5457	105	3112			
WK5458	105	3112			
WK5459	105	3112			
WM1146T					C6091
WRR1952	123A				S0015
WRR1953	123A				S0015
WRR1954	123A				S0015
WRT1114	102				G0005
WS6945-2	9945				
WTV1PWR				TR01/	
WTV3MC	160	3123	2	TR17/637	G0002
WTV6MC	160	3005	2	TR17/637	G0002

Original Part Number	ECG	SK	GE	IR/WEP	HEP
WTV6PWR	121	3014	16	TR01/232	G6005
WTV12MC	160	3005	2	TR17/637	G0002
WTV12PWR	121	3014	16	TR01/232	G6005
WTV15MG	102A	3004		TR85/250	
WTV15VMG	102A	3123	52	TR85/250	G0005
WTV20MC	160	3005	2	TR17/637	G0002
WTV20MG	102A			TR85/250	
WTV20VH6	102A	3123	52	TR85/250	G0005
WTV20VHG	102A			TR85/250	
WTV20VMG	102A	3004	52	TR85/250	G0005
WTV25PWR	121	3009	16	TR01/232	G6005
WTV30LVG	176	3123			
WTV30VH6	158				G0005
WTV30VHG	102A	3004	52	TR85/250	
WTV30VLG	102A	3004	52	TR85/250	
WTV30VMG	102A	3004	52	TR85/250	G0005
WTV40PWR	121	3014	16	TR01/232	G6005
WTV99PWR	121	3014	16	TR01/232	G6005
WTV129PWR	105	3012	4	TR03/233	
WTV199PWR	121	3009	16	TR01/232	G6005
WTV299PWR	105	3012	4	TR03/233	G6006
WTV-BMC	102A			TR85/250	G0005
WTV-L6	101	3011	8	TR08/641	
WTVAT6	102A	3004	52	TR85/250	G0005
WTVAT6A	160	3005	2	TR17/637	G0002
WTVB5	160	3005	2	TR17/637	G0002
WTVB5A	102A	3004	52	TR85/250	G0005
WTVB6	100	3123	1		
WTVB6A	102A		2		
WTVBA6	160	3005	2	TR17/637	G0002
WTVBA6A	102A	3004	52	TR85/250	G0005
WTVBE6	160	3005	2	TR17/637	G0002
WTVBE6A	102A	3123	52	TR85/250	G0005
WTVBMC	158	3005	1		G0005
WTVSA7	101	3011	8	TR08/641	
WTVSK7	101	3011	8	TR08/641	
WTVSQ7	101	3011	8	TR08/641	
WU2N1307		3004			
X1C1644	102A	3004	52	TR85/250	
X16A545-7	123A	3122		TR21/735	S0030
X16A1938	123A	3122		TR21/735	S0015
X16E3860	123A		20		
X16E3890	192		63		
X16E3960	123A		20		
X16N1485	123A	3020	20	TR21/735	
X29A829					S0019
X32C5099	129			TR87/	S0005
X32C5111	128			TR87/	S0005
X42	160	3123	2	TR17/637	G0002
X45C-H06		3004			
X78	100	3005		TR05/254	
X-78	102A	3123	52		
X137	105	3012	4	TR03/233	G6014
X300-1300A	128				
X733-8934		3026			
X733-8935		3044			
X735-40C		3054			
X735-41		3124			
X1000	213				G6010
X1005	121	3009	16	TR01/232	G6005
X2919				TR36/	
X6584-C	123A				
X6661A	726				
X19001-A	123A				
X19001-B	128				
X19001-C	128				
X19001-D	199				
X19002	941D				
X253223		3021			
X330302		3062			
X-439623	912	3543			
X908922-001		3525			
X-2408473	9914				
X-2408784	9951				
XA49ZD					S5013
XA101	160	3005	50	TR17/637	G0003
XA102	160	3005	50	TR17/637	G0003
XA103	160	3005	1	TR17/637	
XA104	160	3005	1	TR17/637	
XA111	160	3005	50	TR17/637	G0003
XA112	160	3005	50	TR17/637	G0003
XA122	102A	3007		TR85/250	G0005
XA123	160	3007	50	TR17/637	G0003
XA124	160	3007	50	TR17/637	G0003
XA126	160	3007	50	TR17/637	G0003
XA131	160	3006	51	TR17/637	G0003
XA141	160			TR17/637	G0001
XA142	160			TR17/637	G0008
XA143	160			TR17/637	G0008
XA151					G0005
XA152					G0005
XA161					G0002
XA494	159	3025		TR20/717	S0013
XA495	159	3025	52	2SA495/	S5022
XA495AC	159		21	2SC492/	S0013

Original Part Number	ECG	SK	GE	IR/WEP	HEP
XA495C	159	3114		TR52/717	S5022
XA-495C	129		30		
XA495D	129	3114	21	TR52/	
XA701	101	3011		TR08/641	G0011
XA702	101	3011		TR08/641	G0011
XA703	101	3011		TR08/641	G0011
XA1018	128				
XA-1071	123A				
XA-1072	159				
XA-1078	130				
XA-1093	124				
XA-1095	128				
XA-1139	123A				
XA-1140	159				
XA-1160	152				
XA-1161	130	3036			
XA-1164	234				
XA1198		3054			
XA1199		3083			
XAA104	7402				
XAA105	7430				
XAA106	7440				
XAA107	7473				
XAA108	7476				
XAA109	7490				
XB1	102	3004	52	TR05/631	G0005
XB1A	102A	3004			
XB2	102	3004	52	TR05/631	G0005
XB2A	102A	3004			
XB2/XB4	103A+158			/641A	
XB2.XB4	103A+158		54		
XB3	102	3004	52	TR05/631	G0005
XB3A	102A	3004			
XB3B	102	3004	2	TR05/631	
XB3BN	102A	3004			
XB3C	102	3004	2	TR05/631	
XB3C-1		3004			
XB4	103	3010	52	TR08/641A	G0011
XB4-1	102A	3004			
XB5	104	3009	16	TR01/230	G6003
XB-5	121				
XB7	104	3009	16	TR01/230	G6003
XB-7	121				
XB8	160	3006	50	TR17/637	G0003
XB9	160	3005	2	TR17/637	G0002
XB10	160	3005	2	TR17/637	G0002
XB12	128		18	TR87/243	
XB13	102A		53	TR85/250	
XB14	121	3009	3	TR01/232	
XB102	102A	3004	53	TR85/250	G0005
XB103	102A	3004	53	TR85/250	G0005
XB104	102A	3004	50	TR85/250	G0003
XB112	102A	3004	53	TR85/250	G0005
XB113	102A	3004	53	TR85/250	G0005
XB114	102A	3004	2	TR85/250	
XB401	186		28	TR76/	
XB404	152		66		
XB408	152		66		
XB413					S3006
XB476	152		66		
XC101	102A	3004	53	TR85/250	G0005
XC121	102A	3004	53	TR85/250	G0005
XC131	102A		53	TR85/250	G0005
XC141	121	3009	16	TR01/232	G6003
XC142	121	3009	16	TR01/232	G6005
XC155	104	3009	25	TR01/230	G6003
XC156	104	3014	25	TR01/230	G6003
XC171	102A	3004	53	TR85/250	G0005
XC371	102	3039	20	TR70/720	S0014
XC372	123A	3018	20	2SC372/735	S0015
XC373	123A	3122		2SC373/735	S0015
XC374	123A	3122		TR21/735	S0015
XC703					S5014
XC713					S5000
XC723					S7002
XC1312A,P	799				
XG1	160	3006		TR17/	
XG2	160	3006		TR17/	
XG3	160	3006		TR17/	
XG5	160	3006		TR17/	
XG8	102A	3004		TR85/	
XG10	160	3006		TR17/	
XG11	160	3006		TR17/	
XG12	160	3006		TR17/	
XG24	160	3006		TR17/	
XG28	103A	3004		TR08/	
XG29	103A	3004		TR08/	
XG30	123A	3122		TR21/	
XG32	158	3004		TR84/	
XG33	103A	3010		TR08/	
XI-548		3027			
XI-549		3026			
XJ13	102	3004	52	TR05/631	G0005
XJ13-1	102A	3004			
XJ71	160	3008	51	TR17/637	G0002

Original Part Number	ECG	SK	GE	IR WEP	HEP
XJ72	160	3006		TR17/637	
XJ73	160	3006		TR17/637	
XN12A, B, C, E	104	3009		TR01/230	G6003
XN12F					G6003
XN-400-318-P1	123A				
XN-400-319-P2	159				
XNC101	103		8	TR09/	
XPS30800		3021			
XS1	108	3039		TR95/	
XS2	108	3039		TR95/	
XS3	108	3039		TR95/	
XS4	108	3039		TR95/	
XS6	108	3039		TR95/	
XS14	108	3039		TR95/	
XS15	108	3039		TR95/	
XS19	159	3025		TR20/	
XS21	123A	3122		TR21/	
XS22	123A			TR21/	
XS26	103A	3010		TR08/	
XS30	128	3024		TR87/	
XS36	161	3132		TR83/	
XS37	161	3132		TR83/	
XS38	161	3132		TR83/	
XS39	161	3132		TR83/	
XS40	107	3039		TR70/	
XS101					G0005
XS104					G0005
XS121					G0005
XT1A					S3021
XT1B					S3021
XT15X3	108	3019	11	TR95/56	
XT100					G0005
XT200					G0005
XT300					G0008
XT400					G0008
XT515					S3019
XT516					S3019
XT517					S3019
XT518					S3019
XT519					S3019
XT520					S3019
XT194-3005829A		3027			
XT-548A		3027			
XT-549A		3026			
Y363	102A	3004	53	TR85/250	G0005
Y410	121	3114			
Y633					G0005
Y49001-21	123A				
Y56001-86	123A				
Y56601-08	123A				
Y56601-44	234				
Y56601-45	123A				
Y56601-46	129				
Y56601-47	128				
Y56601-48	234				
Y56601-49	123A				
Y56601-50	159				
Y56601-51	123A				
Y56601-63	159				
Y56601-73	123A				
Y56601-74	159				
Y56601-75	123A				
Y56601-76	129				
Y56601-79	159				
Y56601-80	123A				
Y56601-82	159				
Y56601-84	159				
Y56601-86-AD	123A				
Y56601-86-AH (GRN, ORN, YEL)	123A				
Y56601-93	123A				
YAAM003	1006				
YAAN2SC1096	186	3192			
YAAN2SC1096K, L	186	3192		TR55/	
YAAN2SC1096M	186	3192			
YAAN2SC1096N	186			TR55/	
YAAN2SD141	175	3538		TR81/	S5014
YAANL2SC1096	186	3054			
YAANZ	186	3054			
YAANZ2S1096	186	3192		TR55/	
YAANZ2SC1096	186	3054	28		S5000
YAANZ2SC1096 (BLU, GRN)	186	3192			
YAANZ2SC1096 (K, L)	186	3192		TR55/	
YAANZ2SC1096 (M, RED)	186	3192			
YEAM004	1098				
YV1	102A	3004	52	TR85/250	G0005
YV1A	102A		2	TR85/250	
YV2	102A	3004		TR85/250	G0005
YVIA		3004			
Z4MW333	123A		10		
Z20	104				
Z10003	923D				

Original Part Number	ECG	SK	GE	IR/WEP	HEP
Z-28058-1	123A				
ZA100962E		3039			
ZA105604		3003			
ZAG-9673	159				
ZC2SA70	160		51		
ZC2SA71, A	160		51		
ZC2SA101, BA	160		51		
ZC2SA102, CA	160		51		
ZC2SA103, CA	160		51		
ZC2SA377	160		51		
ZC2SA700A, B	160		51		
ZC2SB172, A	158		53		
ZDP-D22-69:54	129				
ZDT					S0011
ZDT10	108		61	TR95/56	
ZDT11	108		61	TR95/56	
ZDT20	108		61	TR95/56	
ZDT21	108		61	TR95/56	
ZDT30	108		39	TR95/56	
ZDT-30	108	3019			
ZDT31	108	3019	39	TR95/56	
ZEN100	123A				S0011
ZEN101	129				S0012
ZEN102	123A				S0014
ZEN103	123A				S0015
ZEN104	108				S0016
ZEN105	161				S0017
ZEN106	159				S0019
ZEN107	159				S0019
ZEN108	108				S0020
ZEN109	108				S0020
ZEN110	123A				S0015
ZEN111	123A				S0015
ZEN112	123A				S0015
ZEN113	123A				S0015
ZEN114	123A				S0015
ZEN115	123A				S0025
ZEN116	199				S0024
ZEN117	229				S0033
ZEN118	108				S0016
ZEN119	123A				S0015
ZEN120	123A				S0015
ZEN121	229				S0030
ZEN122	159				S0032
ZEN123	132				F0015
ZEN124	221				F2004
ZEN125	194				
ZEN127	123A				S0025
ZEN128	172				S9100
ZEN129	6409				S9002
ZEN200	124				S5011
ZEN201	225				S5015
ZEN202	185				S5000
ZEN203	185				S5006
ZEN204	162				S5020
ZEN205	154				S5025
ZEN206	165				S5021
ZEN207	282				S3010
ZEN208	191				S3021
ZEN209	182				S5004
ZEN210	184				S5000
ZEN211	185				S5006
ZEN300	160				G0002
ZEN301	160				G0003
ZEN302	158				G0005
ZEN303	102				G0006
ZEN304	158				G0005
ZEN305	158				G0005
ZEN306	102				G0006
ZEN307	158				G0005
ZEN308	158				G0005
ZEN309	102A				G0007
ZEN310	102A				G0007
ZEN311	126				G0001
ZEN312	126				G0008
ZEN313	126				G0009
ZEN314	126				G0008
ZEN315	101				G0011
ZEN325	104				G6003
ZEN326	121				G6005
ZEN327	213				G6006
ZEN328	127				G6008
ZEN329					G6011
ZEN330	121				G6013
ZEN331	121				G6013
ZEN600					C2009G
ZEN601	760				C6001
ZEN602	722				C6056P
ZEN603	790				C6057P
ZEN604	708				C6026P
ZEN605	712				C6063P
ZEN606	707				C6067P
ZEN607	714				C6070P
ZEN608	715				C6071P
ZEN609	739				C6072P

Original Part Number	ECG	SK	GE	IR WEP	HEP
ZEN900	413				
ZEN902	414				
ZEN905	405				
ZEN907	401				
ZJ13					G0005
ZJ40	108	3018	51	TR95/56	
ZJ72	160	3008	51	TR17/637	
ZJ73	160	3008	51	TR17/637	
ZN3416				TR51/	
ZN7400E	7400				
ZN7401E	7401				
ZN7402E	7402				
ZN7404E	7404				
ZN7405E	7405				
ZN7410E	7410				
ZN7420E	7420				
ZN7430E	7430				
ZN7440E	7440				
ZN7441AE	7441				
ZN7450E	7450				
ZN7451E	7451				
ZN7453E	7453				
ZN7454E	7454				
ZN7460E	7460				
ZN7470E	7470				
ZN7472E	7472				
ZN7473E	7473				
ZN7474E	7474				
ZN7475E	7475				
ZN7476E	7476				
ZN74107E	74107				
ZN35024712	197				
ZN35050411	198				
ZN35050412	198				
ZN35050511	198				
ZN35062712	198				
ZN37001003	1047				
ZN37002001	1046				
ZN37003008	738				
ZN37004002	1050				
ZN37007003	1109				
ZN37011002	749				
ZS34					G0005
ZS38					G0005
ZS56					G0005
ZSA49				TR17/	
ZSA706				TR73/	
ZSC380				2SC380/	
ZSC535B		3018			
ZSC619C				TR86/	
ZSC620C				TR62/	
ZSC645(C)				TR21/	
ZSC710C				TR24/	
ZSC710D				TR24/	
ZSC711D,E				TR51/	
ZSC756				TR65/	
ZSC839				TR21/	
ZSC921				TR21/	
ZSC924				TR51/	
ZSC1018				TR64/	
ZSC11730				TR76/	
ZSD235Y		3054			
ZT20	123A	3122	62	TR21/735	S0014
ZT20-1	123A	3122			
ZT20-12	123A	3122			
ZT20-55	123A	3122			
ZT20A,B,C	123A	3122			
ZT21	123A	3122	62	TR21/735	S0014
ZT21-1,12,55	123A	3122			
ZT21A,B,C	123A	3122			
ZT22	123A	3122	62	TR21/735	S0014
ZT22-1,12,55	123A	3122			
ZT22A,B,C	123A	3122			
ZT23	123A	3122	62	TR21/735	S0014
ZT23-1,12	123A	3122			
ZT23-55		3122			
ZT23A,B,C	123A	3122			
ZT24	123A	3122	62	TR21/735	S0014
ZT24-1,12,55	123A	3122			
ZT24A,B,C	123A	3122			
ZT40	123A	3018	61	TR21/735	S0011
ZT41	123A	3018	61	TR21/735	S0011
ZT42	123A	3018	212	TR21/735	S0011
ZT43	123A	3018	212	TR21/735	S0011
ZT44	123A	3018	212	TR21/735	S0011
ZT50	123A	3122	11	TR21/735	
ZT60	123A	3122	210	TR21/735	S0014
ZT60-1,12,55	123A	3122			
ZT60A,B,C	123A	3122			
ZT61	123A	3122	210	TR21/735	S0014
ZT61-1,12,55	123A	3122			
ZT61A,B,C	123A	3122			
ZT62	123A	3122	210	TR25/611	
ZT62-1,12,15	123A	3122			
ZT62C	123A	3122			
ZT63	123A	3122	210	TR21/735	S0014
ZT63-1,12,55	123A	3122			
ZT63A,B,C	123A	3122			
ZT64	123A	3122	210	TR21/735	S0014
ZT64-1,5,12,55	123A	3122			
ZT64A,B,C	123A	3122			
ZT66	123A	3122	17	TR21/735	
ZT66A,B,C	128	3024			
ZT68	123A	3122	210	TR21/735	
ZT80	123A	3122	210	TR21/735	S0011
ZT81	123A	3122	210	TR21/735	S0011
ZT82	123A	3122	210	TR25/735	S0025
ZT83	123A	3122	210	TR21/735	S0011
ZT84	123A	3122	210	TR21/735	S0011
ZT86	123A	3122	17	TR21/735	
ZT87	123A	3122	210	TR21/735	S0011
ZT88	123A	3122	17	TR21/735	
ZT89	123A	3122	17	TR21/735	
ZT90	128	3024		TR87/243	
ZT93	128	3024		TR87/243	
ZT94	128	3024	215	TR87/243	
ZT110	123A	3122	210	TR25/735	
ZT111	123A	3122	210	TR21/735	S0011
ZT112	123A	3122	210	TR21/735	S0011
ZT113	123A	3122	210	TR21/735	S0011
ZT114	123A	3122	210	TR21/735	S0011
ZT116	123A		17	TR21/735	
ZT117	123A	3122	210	TR21/735	S0011
ZT118	123A	3122	17	TR21/735	
ZT119	123A	3122	17	TR21/735	
ZT131	159	3114		TR20/717	S0013
ZT131A	159	3114			
ZT152	159	3114	22	TR20/717	S0013
ZT153	159	3114	21	TR20/717	S0013
ZT154	159	3114	21	TR20/717	S0013
ZT180	159	3114	22	TR20/717	S0013
ZT181	159	3114	82	TR20/717	S0019
ZT182	159	3114	82	TR20/717	S0013
ZT183	159	3114	82	TR20/717	S0013
ZT184	159	3114	82	TR20/717	S0013
ZT187	159	3114	22	TR20/717	S0013
ZT190	128	3024	63	TR87/243	S0011
ZT191	128	3024	63	TR87/243	S0011
ZT192	128	3024	63	TR87/243	S0025
ZT193	128	3024	63	TR87/243	S0025
ZT202	123A	3018	61	TR21/735	S0014
ZT203	123A	3018	61	TR21/735	S0014
ZT204	123A	3018	61	TR21/735	S0014
ZT210	128	3024	67	TR87/243	S5014
ZT211	128	3024	63	TR87/243	
ZT280	159	3114	22	TR20/717	S0013
ZT280A	159	3114			
ZT281	159	3114	82	TR20/717	S0013
ZT282	159	3114	82	TR20/717	S0013
ZT283	159	3114	82	TR20/717	S0013
ZT284	159	3114	82	TR20/717	S0013
ZT287	159	3114	22	TR20/717	S0013
ZT287A	159	3114			
ZT402	123A	3018	61	TR21/735	S0014
ZT403	123A	3018	61	TR21/735	S0011
ZT404	123A	3018	61	TR21/735	S0011
ZT406	123A	3122	17	TR21/735	
ZT600	186		47		
ZT696	123A	3024	81	TR21/735	S0014
ZT697	123A	3024	81	TR21/735	S0014
ZT706	123A	3122	61	TR21/735	S0011
ZT706A	123A	3122	61	TR21/735	S0011
ZT708	123A	3122	17	TR21/735	S0011
ZT709	108	3039	61	TR95/56	S0016
ZT917	108	3039	11	TR95/56	S0016
ZT918	108	3018	17	TR95/56	S0016
ZT930					S0024
ZT1420					S0015
ZT1479	128	3024	18	TR87/243	
ZT1481	128	3024	18	TR87/243	S5014
ZT1483	152		66		
ZT1484	152		66		
ZT1485	152		66		
ZT1486	152		66		
ZT1487	130	3027	77	TR59/242	S7002
ZT1488	130	3027	75	TR59/242	
ZT1489	130	3027	77	TR59/242	S7002
ZT1490	130	3027	75	TR59/242	
ZT1511					S5004
ZT1512					S5004
ZT1513					S5004
ZT1613	128	3024	216	TR87/243	
ZT1700	128	3024	46	TR87/243	S5014
ZT1701	152		66		
ZT1702	130	3027	77	TR59/247	S7002
ZT1708	123A	3122	210	TR21/735	S0011
ZT1711	123A	3122	18	TR21/735	
ZT2102	102		28		
ZT2205	123A	3122	210	TR21/735	S0011
ZT2206	123A	3122	210	TR21/735	S0011

Original Part Number	ECG	SK	GE	IR/WEP	HEP
ZT2270	192		63	TR76/	
ZT2368	108	3039	20	TR95/56	S0016
ZT2369	108	3039	20	TR95/56	S0016
ZT2475	108	3039	61	TR95/56	S0016
ZT2476	123A	3122	47	TR21/735	S0014
ZT2477	123A	3122	47	TR21/735	S0014
ZT2631					S3011
ZT2708	108	3019	17	TR95/56	S0016
ZT2857	108	3019	86	TR95/56	
ZT2876	152		66		
ZT2938	108	3039	20	TR95/56	S0016
ZT3053	192		63	TR25/	
ZT3269A	108	3039	17	TR95/56	
ZT3375	186		28	TR76/	
ZT3440	198		32		
ZT3512	128	3024	63	TR87/243	
ZT3600	108	3039	11	TR95/56	S0016
ZT3866	128	3024	20	TR87/243	
ZT-62				TR21/	
ZT-82				TR21/	
ZT-110				TR21/	
ZTR-B54	10 A	3004	52		
ZTR-B56	102A	3004	52		
ZTX107	199		212		
ZTX108	199		212		
ZTX109	199		212		
ZTX300	123A		210		S0014
ZTX301	123A		210		
ZTX302	128		210		S0015
ZTX303	192		210		

Original Part Number	ECG	SK	GE	IR/WEP	HEP
ZTX304			81		
ZTX310	123A		210		
ZTX311	123A		210		
ZTX312	123A		20		
ZTX320	108		61		
ZTX321	108		61		
ZTX325			86		
ZTX326			86		
ZTX330	123A		210		
ZTX331	128		210		
ZTX341	171		27		
ZTX342	171		27		
ZTX350					F1036
ZTX360			47		S3020
ZTX500	129		22		S0012
ZTX501	159		82		S0019
ZTX502	159		82		S0019
ZTX503	159		82		S0019
ZTX504			82		S0019
ZTX510					S0013
ZTX530	159	3114	82		
ZTX530A, B, D	159	3114			S0019
ZTX531	159	3114			
ZTX531A, B	159	3114			S0019

SPECIAL NOTES FOR ECG DEVICES

*—ECG1091. When using ECG1091 to replace MPc46C, ground pin 15 through a 5.6K resistor (1/2) watt).

**—ECG1086. When using ECG1086 to replace MPc49C, ground pin 7 to the nearest point through a 3.3K resistor.

[ ]—ECG1089. Devices marked ECG1089-3 when used to replace a MPc29C may require the addition of a 0.01 μF capacitor between pins 13 and 17.

[ ] [ ]—ECG1081. This device is an improved replacement for MPc563H. The following modifications are required in order to assure proper operation:

1. Remove the 680Ω resistor between pins 3 and 4, the 1S953 or SD46 diode, and the 10K resistor between pins 9 and 10.
2. Change the 220 pF capacitor to 33 pF to obtain the same frequency response.
3. No modification is required when used to replace MPc563Hz.